Mechanical Costs with RSMeans data

2019
42nd annual edition

Joseph Kelble, Senior Editor

Chief Data Officer
Noam Reininger

Engineering Director
Bob Mewis *(1, 3, 4, 5, 11, 12)*

Contributing Editors
Brian Adams *(21, 22)*
Christopher Babbitt
Sam Babbitt
Michelle Curran
Matthew Doheny *(8, 9, 10)*
John Gomes *(13, 41)*
Derrick Hale, PE *(2, 31, 32, 33, 34, 35, 44, 46)*
Michael Henry
Joseph Kelble *(14, 23, 25,)*
Charles Kibbee
Gerard Lafond, PE
Thomas Lane *(6, 7)*
Jake MacDonald
Elisa Mello
Michael Ouillette *(26, 27, 28, 48)*
Gabe Sirota
Matthew Sorrentino
Kevin Souza
David Yazbek

Product Manager
Andrea Sillah

Production Manager
Debbie Panarelli

Production
Jonathan Forgit
Mary Lou Geary
Sharon Larsen
Sheryl Rose

Data Quality Manager
Joseph Ingargiola

Technical Support
Kedar Gaikwad
Todd Klapprodt
John Liu

Cover Design
Blaire Collins

Data Analytics
Tim Duggan
Todd Glowac
Matthew Kelliher-Gibson

**Numbers in italics are the divisional responsibilities for each editor. Please contact the designated editor directly with any questions.*

RSMeans data from Gordian
Construction Publishers & Consultants
1099 Hingham Street, Suite 201
Rockland, MA 02370
United States of America
1.800.448.8182
RSMeans.com

Printed in the United States of America
ISSN 1524-3702
ISBN 978-1-946872-63-0

0029 $293.99 per copy (in United States)
Price is subject to change without prior notice.

Related Data and Services

Our engineers recommend the following products and services to complement *Mechanical Costs with RSMeans data:*

Annual Cost Data Books

2019 Assemblies Costs with RSMeans data
2019 Square Foot Costs with RSMeans data

Reference Books

Estimating Building Costs
RSMeans Estimating Handbook
Green Building: Project Planning & Estimating
How to Estimate with RSMeans data
Plan Reading & Material Takeoff
Project Scheduling & Management for Construction
Universal Design Ideas for Style, Comfort & Safety

Seminars and In-House Training

Unit Price Estimating
Training for our online estimating solution
Practical Project Management for Construction Professionals
Scheduling with MSProject for Construction Professionals
Mechanical & Electrical Estimating

RSMeans data Online

For access to the latest cost data, an intuitive search, and an easy-to-use estimate builder, take advantage of the time savings available from our online application. To learn more visit: RSMeans.com/2019online.

Enterprise Solutions

Building owners, facility managers, building product manufacturers, and attorneys across the public and private sectors engage with RSMeans data Enterprise to solve unique challenges where trusted construction cost data is critical. To learn more visit: RSMeans.com/Enterprise.

Custom Built Data Sets

Building and Space Models: Quickly plan construction costs across multiple locations based on geography, project size, building system component, product options, and other variables for precise budgeting and cost control.

Predictive Analytics: Accurately plan future builds with custom graphical interactive dashboards, negotiate future costs of tenant build-outs, and identify and compare national account pricing.

Consulting

Building Product Manufacturing Analytics: Validate your claims and assist with new product launches.

Third-Party Legal Resources: Used in cases of construction cost or estimate disputes, construction product failure vs. installation failure, eminent domain, class action construction product liability, and more.

API

For resellers or internal application integration, RSMeans data is offered via API. Deliver Unit, Assembly, and Square Foot Model data within your interface. To learn more about how you can provide your customers with the latest in localized construction cost data visit: RSMeans.com/API.

Table of Contents

Foreword

The Value of RSMeans data from Gordian

Since 1942, RSMeans data has been the industry-standard materials, labor, and equipment cost information database for contractors, facility owners and managers, architects, engineers, and anyone else that requires the latest localized construction cost information. More than 75 years later, the objective remains the same: to provide facility and construction professionals with the most current and comprehensive construction cost database possible.

With the constant influx of new construction methods and materials, in addition to ever-changing labor and material costs, last year's cost data is not reliable for today's designs, estimates, or budgets. Gordian's cost engineers apply real-world construction experience to identify and quantify new building products and methodologies, adjust productivity rates, and adjust costs to local market conditions across the nation. This adds up to more than 22,000 hours in cost research annually. This unparalleled construction cost expertise is why so many facility and construction professionals rely on RSMeans data year over year.

About Gordian

Gordian originated in the spirit of innovation and a strong commitment to helping clients reach and exceed their construction goals. In 1982, Gordian's chairman and founder, Harry H. Mellon, created Job Order Contracting while serving as chief engineer at the Supreme Headquarters Allied Powers Europe. Job Order Contracting is a unique indefinite delivery/indefinite quantity (IDIQ) process, which enables facility owners to complete a substantial number of repair, maintenance, and construction projects with a single, competitively awarded contract. Realizing facility and infrastructure owners across various industries could greatly benefit from the time and cost saving advantages of this innovative construction procurement solution, he established Gordian in 1990.

Continuing the commitment to providing the most relevant and accurate facility and construction data, software, and expertise in the industry, Gordian enhanced the fortitude of its data with the acquisition of RSMeans in 2014. And in an effort to expand its facility management capabilities, Gordian acquired Sightlines, the leading provider of facilities benchmarking data and analysis, in 2015.

Our Offerings

Gordian is the leader in facility and construction cost data, software, and expertise for all phases of the building life cycle. From planning to design, procurement, construction, and operations, Gordian's solutions help clients maximize efficiency, optimize cost savings, and increase building quality with its highly specialized data engineers, software, and unique proprietary data sets.

Our Commitment

At Gordian, we do more than talk about the quality of our data and the usefulness of its application. We stand behind all of our RSMeans data—from historical cost indexes to construction materials and techniques—to craft current costs and predict future trends. If you have any questions about our products or services, please call us toll-free at 800.448.8182 or visit our website at gordian.com.

Estimating with RSMeans data: Unit Prices

Following these steps will allow you to complete an accurate estimate using RSMeans data Unit Prices.

1. Scope Out the Project

- Think through the project and identify the CSI divisions needed in your estimate.
- Identify the individual work tasks that will need to be covered in your estimate.
- The Unit Price data have been divided into 50 divisions according to CSI MasterFormat® 2016.
- In printed versions, the Unit Price Section Table of Contents on page 1 may also be helpful when scoping out your project.
- Experienced estimators find it helpful to begin with Division 2 and continue through completion. Division 1 can be estimated after the full project scope is known.

2. Quantify

- Determine the number of units required for each work task that you identified.
- Experienced estimators include an allowance for waste in their quantities. (Waste is not included in our Unit Price line items unless otherwise stated.)

3. Price the Quantities

- Use the search tools available to locate individual Unit Price line items for your estimate.
- Reference Numbers indicated within a Unit Price section refer to additional information that you may find useful.
- The crew indicates who is performing the work for that task. Crew codes are expanded in the Crew Listings in the Reference Section to include all trades and equipment that comprise the crew.
- The Daily Output is the amount of work the crew is expected to complete in one day.
- The Labor-Hours value is the amount of time it will take for the crew to install one unit of work.
- The abbreviated Unit designation indicates the unit of measure upon which the crew, productivity, and prices are based.
- Bare Costs are shown for materials, labor, and equipment needed to complete the Unit Price line item. Bare costs do not include waste, project overhead, payroll insurance, payroll taxes, main office overhead, or profit.
- The Total Incl O&P cost is the billing rate or invoice amount of the installing contractor or subcontractor who performs the work for the Unit Price line item.

4. Multiply

- Multiply the total number of units needed for your project by the Total Incl O&P cost for each Unit Price line item.
- Be careful that your take off unit of measure matches the unit of measure in the Unit column.
- The price you calculate is an estimate for a completed item of work.
- Keep scoping individual tasks, determining the number of units required for those tasks, matching each task with individual Unit Price line items, and multiplying quantities by Total Incl O&P costs.
- An estimate completed in this manner is priced as if a subcontractor, or set of subcontractors, is performing the work. The estimate does not yet include Project Overhead or Estimate Summary components such as general contractor markups on subcontracted work, general contractor office overhead and profit, contingency, and location factors.

5. Project Overhead

- Include project overhead items from Division 1-General Requirements.
- These items are needed to make the job run. They are typically, but not always, provided by the general contractor. Items include, but are not limited to, field personnel, insurance, performance bond, permits, testing, temporary utilities, field office and storage facilities, temporary scaffolding and platforms, equipment mobilization and demobilization, temporary roads and sidewalks, winter protection, temporary barricades and fencing, temporary security, temporary signs, field engineering and layout, final cleaning, and commissioning.
- Each item should be quantified and matched to individual Unit Price line items in Division 1, then priced and added to your estimate.
- An alternate method of estimating project overhead costs is to apply a percentage of the total project cost—usually 5% to 15% with an average of 10% (see General Conditions).
- Include other project related expenses in your estimate such as:
  - Rented equipment not itemized in the Crew Listings
  - Rubbish handling throughout the project (see section 02 41 19.19)

6. Estimate Summary

- Include sales tax as required by laws of your state or county.
- Include the general contractor's markup on self-performed work, usually 5% to 15% with an average of 10%.
- Include the general contractor's markup on subcontracted work, usually 5% to 15% with an average of 10%.
- Include the general contractor's main office overhead and profit:
  - RSMeans data provides general guidelines on the general contractor's main office overhead (see section 01 31 13.60 and Reference Number R013113-50).
  - Markups will depend on the size of the general contractor's operations, projected annual revenue, the level of risk, and the level of competition in the local area and for this project in particular.
- Include a contingency, usually 3% to 5%, if appropriate.
- Adjust your estimate to the project's location by using the City Cost Indexes or the Location Factors in the Reference Section:
  - Look at the rules in "How to Use the City Cost Indexes" to see how to apply the Indexes for your location.
  - When the proper Index or Factor has been identified for the project's location, convert it to a multiplier by dividing it by 100, then multiply that multiplier by your estimated total cost. The original estimated total cost will now be adjusted up or down from the national average to a total that is appropriate for your location.

Editors' Note:
We urge you to spend time reading and understanding the supporting material. An accurate estimate requires experience, knowledge, and careful calculation. The more you know about how we at RSMeans developed the data, the more accurate your estimate will be. In addition, it is important to take into consideration the reference material such as Equipment Listings, Crew Listings, City Cost Indexes, Location Factors, and Reference Tables.

How to Use the Cost Data: The Details

What's Behind the Numbers? The Development of Cost Data

RSMeans data engineers continually monitor developments in the construction industry in order to ensure reliable, thorough, and up-to-date cost information. While overall construction costs may vary relative to general economic conditions, price fluctuations within the industry are dependent upon many factors. Individual price variations may, in fact, be opposite to overall economic trends. Therefore, costs are constantly tracked and complete updates are performed yearly. Also, new items are frequently added in response to changes in materials and methods.

Costs in U.S. Dollars

All costs represent U.S. national averages and are given in U.S. dollars. The City Cost Index (CCI) with RSMeans data can be used to adjust costs to a particular location. The CCI for Canada can be used to adjust U.S. national averages to local costs in Canadian dollars. No exchange rate conversion is necessary because it has already been factored in.

G The processes or products identified by the green symbol in our publications have been determined to be environmentally responsible and/or resource-efficient solely by RSMeans data engineering staff. The inclusion of the green symbol does not represent compliance with any specific industry association or standard.

Material Costs

RSMeans data engineers contact manufacturers, dealers, distributors, and contractors all across the U.S. and Canada to determine national average material costs. If you have access to current material costs for your specific location, you may wish to make adjustments to reflect differences from the national average. Included within material costs are fasteners for a normal installation. RSMeans data engineers use manufacturers' recommendations, written specifications, and/or standard construction practices for the sizing and spacing of fasteners. Adjustments to material costs may be required for your specific application or location. The manufacturer's warranty is assumed. Extended warranties are not included in the material costs. **Material costs do not include sales tax.**

Labor Costs

Labor costs are based upon a mathematical average of trade-specific wages in 30 major U.S. cities. The type of wage (union, open shop, or residential) is identified on the inside back cover of printed publications or selected by the estimator when using the electronic products. Markups for the wages can also be found on the inside back cover of printed publications and/or under the labor references found in the electronic products.

- If wage rates in your area vary from those used, or if rate increases are expected within a given year, labor costs should be adjusted accordingly.

Labor costs reflect productivity based on actual working conditions. In addition to actual installation, these figures include time spent during a normal weekday on tasks, such as material receiving and handling, mobilization at the site, site movement, breaks, and cleanup.

Productivity data is developed over an extended period so as not to be influenced by abnormal variations and reflects a typical average.

Equipment Costs

Equipment costs include not only rental but also operating costs for equipment under normal use. The operating costs include parts and labor for routine servicing, such as the repair and replacement of pumps, filters, and worn lines. Normal operating expendables, such as fuel, lubricants, tires, and electricity (where applicable), are also included. Extraordinary operating expendables with highly variable wear patterns, such as diamond bits and blades, are excluded. These costs are included under materials. Equipment rental rates are obtained from industry sources throughout North America—contractors, suppliers, dealers, manufacturers, and distributors.

Rental rates can also be treated as reimbursement costs for contractor-owned equipment. Owned equipment costs include depreciation, loan payments, interest, taxes, insurance, storage, and major repairs.

Equipment costs do not include operators' wages.

Equipment Cost/Day—The cost of equipment required for each crew is included in the Crew Listings in the Reference Section (small tools that are considered essential everyday tools are not listed out separately). The Crew Listings itemize specialized tools and heavy equipment along with labor trades. The daily cost of itemized equipment included in a crew is based on dividing the weekly bare rental rate by 5 (number of working days per week), then adding the hourly operating cost times 8 (the number of hours per day). This Equipment Cost/Day is shown in the last column of the Equipment Rental Costs in the Reference Section.

Mobilization, Demobilization—The cost to move construction equipment from an equipment yard or rental company to the job site and back again is not included in equipment costs. Mobilization (to the site) and demobilization (from the site) costs can be found in the Unit Price Section. If a piece of equipment is already at the job site, it is not appropriate to utilize mobilization or demobilization costs again in an estimate.

Overhead and Profit

Total Cost including O&P for the installing contractor is shown in the last column of the Unit Price and/or Assemblies. This figure is the sum of the bare material cost plus 10% for profit, the bare labor cost plus total overhead and profit, and the bare equipment cost plus 10% for profit. Details for the calculation of overhead and profit on labor are shown on the inside back cover of the printed product and in the Reference Section of the electronic product.

General Conditions

Cost data in this data set are presented in two ways: Bare Costs and Total Cost including O&P (Overhead and Profit). General Conditions, or General Requirements, of the contract should also be added to the Total Cost including O&P when applicable. Costs for General Conditions are listed in Division 1 of the Unit Price Section and in the Reference Section.

General Conditions for the installing contractor may range from 0% to 10% of the Total Cost including O&P. For the general or prime contractor, costs for General Conditions may range from 5% to 15% of the Total Cost including O&P, with a figure of 10% as the most typical allowance. If applicable, the Assemblies and Models sections use costs that include the installing contractor's overhead and profit (O&P).

Factors Affecting Costs

Costs can vary depending upon a number of variables. Here's a listing of some factors that affect costs and points to consider.

Quality—The prices for materials and the workmanship upon which productivity is based represent sound construction work. They are also in line with industry standard and manufacturer specifications and are frequently used by federal, state, and local governments.

Overtime—We have made no allowance for overtime. If you anticipate premium time or work beyond normal working hours, be sure to make an appropriate adjustment to your labor costs.

Productivity—The productivity, daily output, and labor-hour figures for each line item are based on an eight-hour work day in daylight hours in moderate temperatures and up to a 14' working height unless otherwise indicated. For work that extends beyond normal work hours or is performed under adverse conditions, productivity may decrease.

Size of Project—The size, scope of work, and type of construction project will have a significant impact on cost. Economies of scale can reduce costs for large projects. Unit costs can often run higher for small projects.

Location—Material prices are for metropolitan areas. However, in dense urban areas, traffic and site storage limitations may increase costs. Beyond a 20-mile radius of metropolitan areas, extra trucking or transportation charges may also increase the material costs slightly. On the other hand, lower wage rates may be in effect. Be sure to consider both of these factors when preparing an estimate, particularly if the job site is located in a central city or remote rural location. In addition, highly specialized subcontract items may require travel and per-diem expenses for mechanics.

Other Factors—

- season of year
- contractor management
- weather conditions
- local union restrictions
- building code requirements
- availability of:
  - adequate energy
  - skilled labor
  - building materials
- owner's special requirements/restrictions
- safety requirements
- environmental considerations
- access

Unpredictable Factors—General business conditions influence "in-place" costs of all items. Substitute materials and construction methods may have to be employed. These may affect the installed cost and/or life cycle costs. Such factors may be difficult to evaluate and cannot necessarily be predicted on the basis of the job's location in a particular section of the country. Thus, where these factors apply, you may find significant but unavoidable cost variations for which you will have to apply a measure of judgment to your estimate.

Rounding of Costs

In printed publications only, all unit prices in excess of $5.00 have been rounded to make them easier to use and still maintain adequate precision of the results.

How Subcontracted Items Affect Costs

A considerable portion of all large construction jobs is usually subcontracted. In fact, the percentage done by subcontractors is constantly increasing and may run over 90%. Since the workers employed by these companies do nothing else but install their particular products, they soon become experts in that line. As a result, installation by these firms is accomplished so efficiently that the total in-place cost, even with the general contractor's overhead and profit, is no more, and often less, than if the principal contractor had handled the installation. Companies that deal with construction specialties are anxious to have their products perform well and, consequently, the installation will be the best possible.

Contingencies

The allowance for contingencies generally provides for unforeseen construction difficulties. On alterations or repair jobs, 20% is not too much. If drawings are final and only field contingencies are being considered, 2% or 3% is probably sufficient and often nothing needs to be added. Contractually, changes in plans will be covered by extras. The contractor should consider inflationary price trends and possible material shortages during the course of the job. These escalation factors are dependent upon both economic conditions and the anticipated time between the estimate and actual construction. If drawings are not complete or approved, or a budget cost is wanted, it is wise to add 5% to 10%. Contingencies, then, are a matter of judgment.

Important Estimating Considerations

The productivity, or daily output, of each craftsman or crew assumes a well-managed job where tradesmen with the proper tools and equipment, along with the appropriate construction materials, are present. Included are daily set-up and cleanup time, break time, and plan layout time. Unless otherwise indicated, time for material movement on site (for items that can be transported by hand) of up to 200' into the building and to the first or second floor is also included. If material has to be transported by other means, over greater distances, or to higher floors, an additional allowance should be considered by the estimator.

While horizontal movement is typically a sole function of distances, vertical transport introduces other variables that can significantly impact productivity. In an occupied building, the use of elevators (assuming access, size, and required protective measures are acceptable) must be understood at the time of the estimate. For new construction, hoist wait and cycle times can easily be 15 minutes and may result in scheduled access extending beyond the normal work day. Finally, all vertical transport will impose strict weight limits likely to preclude the use of any motorized material handling.

The productivity, or daily output, also assumes installation that meets manufacturer/designer/standard specifications. A time allowance for quality control checks, minor adjustments, and any task required to ensure proper function or operation is also included. For items that require connections to services, time is included for positioning, leveling, securing the unit, and making all the necessary connections (and start up where applicable) to ensure a complete installation. Estimating of the services themselves (electrical, plumbing, water, steam, hydraulics, dust collection, etc.) is separate.

In some cases, the estimator must consider the use of a crane and an appropriate crew for the installation of large or heavy items. For those situations where a crane is not included in the assigned crew and as part of the line item cost, then equipment rental costs, mobilization and demobilization costs, and operator and support personnel costs must be considered.

Labor-Hours

The labor-hours expressed in this publication are derived by dividing the total daily labor-hours for the crew by the daily output. Based on average installation time and the assumptions listed above, the labor-hours include: direct labor, indirect labor, and nonproductive time. A typical day for a craftsman might include but is not limited to:

- Direct Work
  - ☐ Measuring and layout
  - ☐ Preparing materials
  - ☐ Actual installation
  - ☐ Quality assurance/quality control
- Indirect Work
  - ☐ Reading plans or specifications
  - ☐ Preparing space
  - ☐ Receiving materials
  - ☐ Material movement
  - ☐ Giving or receiving instruction
  - ☐ Miscellaneous
- Non-Work
  - ☐ Chatting
  - ☐ Personal issues
  - ☐ Breaks
  - ☐ Interruptions (i.e., sickness, weather, material or equipment shortages, etc.)

If any of the items for a typical day do not apply to the particular work or project situation, the estimator should make any necessary adjustments.

Final Checklist

Estimating can be a straightforward process provided you remember the basics. Here's a checklist of some of the steps you should remember to complete before finalizing your estimate.

Did you remember to:

- factor in the City Cost Index for your locale?
- take into consideration which items have been marked up and by how much?
- mark up the entire estimate sufficiently for your purposes?
- read the background information on techniques and technical matters that could impact your project time span and cost?
- include all components of your project in the final estimate?
- double check your figures for accuracy?
- call RSMeans data engineers if you have any questions about your estimate or the data you've used? Remember, Gordian stands behind all of our products, including our extensive RSMeans data solutions. If you have any questions about your estimate, about the costs you've used from our data, or even about the technical aspects of the job that may affect your estimate, feel free to call the Gordian RSMeans editors at 1.800.448.8182.

Unit Price Section

Table of Contents

RSMeans data: Unit Prices—How They Work

All RSMeans data: Unit Prices are organized in the same way.

03 30 Cast-In-Place Concrete

03 30 53 – Miscellaneous Cast-In-Place Concrete

03 30 53.40 Concrete In Place		Crew	Daily Output	Labor-Hours	Unit	Material	2019 Bare Costs Labor	Equipment	Total	Total Incl O&P
0010	CONCRETE IN PLACE									
0020	Including forms (4 uses), Grade 60 rebar, concrete (Portland cement									
0050	Type I), placement and finishing unless otherwise indicated									
0500	Chimney foundations (5000 psi), over 5 C.Y.	C-14C	32.22	3.476	C.Y.	178	170	.83	348.83	455
0510	(3500 psi), under 5 C.Y.	"	23.71	4.724	"	206	232	1.13	439.13	580
3540	Equipment pad (3000 psi), 3' x 3' x 6" thick	C-14H	45	1.067	Ea.	46.50	53.50	.59	100.59	133
3550	4' x 4' x 6" thick		30	1.600		72.50	80.50	.88	153.88	201
3560	5' x 5' x 8" thick		18	2.667		132	134	1.47	267.47	350
3570	6' x 6' x 8" thick		14	3.429		181	172	1.89	354.89	460
3580	8' x 8' x 10" thick		8	6		385	300	3.30	688.30	880
3590	10' x 10' x 12" thick		5	9.600		665	485	5.30	1,155.30	1,475
3800	Footings (3000 psi), spread under 1 C.Y.	C-14C	28	4	C.Y.	193	196	.96	389.96	510
3825	1 C.Y. to 5 C.Y.		43	2.605		227	128	.63	355.63	445
3850	Over 5 C.Y.		75	1.493		212	73	.36	285.36	345

It is important to understand the structure of RSMeans data: Unit Prices so that you can find information easily and use it correctly.

1 Line Numbers

Line Numbers consist of 12 characters, which identify a unique location in the database for each task. The first 6 or 8 digits conform to the Construction Specifications Institute MasterFormat® 2016. The remainder of the digits are a further breakdown in order to arrange items in understandable groups of similar tasks. Line numbers are consistent across all of our publications, so a line number in any of our products will always refer to the same item of work.

2 Descriptions

Descriptions are shown in a hierarchical structure to make them readable. In order to read a complete description, read up through the indents to the top of the section. Include everything that is above and to the left that is not contradicted by information below. For instance, the complete description for line 03 30 53.40 3550 is "Concrete in place, including forms (4 uses), Grade 60 rebar, concrete (Portland cement Type 1), placement and finishing unless otherwise indicated; Equipment pad (3000 psi), 4' × 4' × 6" thick."

3 RSMeans data

When using **RSMeans data**, it is important to read through an entire section to ensure that you use the data that most closely matches your work. Note that sometimes there is additional information shown in the section that may improve your price. There are frequently lines that further describe, add to, or adjust data for specific situations.

4 Reference Information

Gordian's RSMeans engineers have created **reference** information to assist you in your estimate. **If** there is information that applies to a section, it will be indicated at the start of the section. The Reference Section is located in the back of the data set.

5 Crews

Crews include labor and/or equipment necessary to accomplish each task. In this case, Crew C-14H is used. Gordian's RSMeans staff selects a crew to represent the workers and equipment that are typically used for that task. In this case, Crew C-14H consists of one carpenter foreman (outside), two carpenters, one rodman, one laborer, one cement finisher, and one gas engine vibrator. Details of all crews can be found in the Reference Section.

Crews - Standard

Crew No.	Bare Costs		Incl. Subs O&P		Cost Per Labor-Hour	
Crew C-14H	**Hr.**	**Daily**	**Hr.**	**Daily**	**Bare Costs**	**Incl. O&P**
1 Carpenter Foreman (outside)	$53.65	$429.20	$81.15	$649.20	$50.29	$75.79
2 Carpenters	51.65	826.40	78.15	1250.40		
1 Rodman (reinf.)	54.85	438.80	83.00	664.00		
1 Laborer	41.05	328.40	62.10	496.80		
1 Cement Finisher	48.90	391.20	72.20	577.60		
1 Gas Engine Vibrator		26.55		29.20	.55	.61
48 L.H., Daily Totals		$2440.55		$3667.20	$50.84	$76.40

6 Daily Output

The **Daily Output** is the amount of work that the crew can do in a normal 8-hour workday, including mobilization, layout, movement of materials, and cleanup. In this case, crew C-14H can install thirty 4' × 4' × 6" thick concrete pads in a day. Daily output is variable and based on many factors, including the size of the job, location, and environmental conditions. RSMeans data represents work done in daylight (or adequate lighting) and temperate conditions.

7 Labor-Hours

The figure in the **Labor-Hours** column is the amount of labor required to perform one unit of work–in this case the amount of labor required to construct one 4' × 4' equipment pad. This figure is calculated by dividing the number of hours of labor in the crew by the daily output (48 labor-hours divided by 30 pads = 1.6 hours of labor per pad). Multiply 1.6 times 60 to see the value in minutes: 60 × 1.6 = 96 minutes. Note: the labor-hour figure is not dependent on the crew size. A change in crew size will result in a corresponding change in daily output, but the labor-hours per unit of work will not change.

8 Unit of Measure

All RSMeans data: Unit Prices include the typical **Unit of Measure** used for estimating that item. For concrete-in-place the typical unit is cubic yards (C.Y.) or each (Ea.). For installing broadloom carpet it is square yard and for gypsum board it is square foot. The estimator needs to take special care that the unit in the data matches the unit in the take-off. Unit conversions may be found in the Reference Section.

9 Bare Costs

Bare Costs are the costs of materials, labor, and equipment that the installing contractor pays. They represent the cost, in U.S. dollars, for one unit of work. They do not include any markups for profit or labor burden.

10 Bare Total

The **Total column** represents the total bare cost for the installing contractor in U.S. dollars. In this case, the sum of $72.50 for material + $80.50 for labor + $.88 for equipment is $153.88.

11 Total Incl O&P

The **Total Incl O&P column** is the total cost, including overhead and profit, that the installing contractor will charge the customer. This represents the cost of materials plus 10% profit, the cost of labor plus labor burden and 10% profit, and the cost of equipment plus 10% profit. It does not include the general contractor's overhead and profit. Note: See the inside back cover of the printed product or the Reference Section of the electronic product for details on how the labor burden is calculated.

National Average

The RSMeans data in our print publications represent a "national average" cost. This data should be modified to the project location using the ***City Cost Indexes*** *or* ***Location Factors*** *tables found in the Reference Section. Use the Location Factors to adjust estimate totals if the project covers multiple trades. Use the City Cost Indexes (CCI) for single trade projects or projects where a more detailed analysis is required. All figures in the two tables are derived from the same research. The last row of data in the CCI—the weighted average—is the same as the numbers reported for each location in the location factor table.*

RSMeans data: Unit Prices— How They Work (Continued)

Project Name: Pre-Engineered Steel Building		Architect: As Shown						
Location:	Anywhere, USA						01/01/19	STD
1 Line Number	Description	Qty	Unit	Material	Labor	Equipment	SubContract	Estimate Total
03 30 53.40 3940	Strip footing, 12" x 24", reinforced	15	C.Y.	$2,460.00	$1,710.00	$8.40	$0.00	
03 30 53.40 3950	Strip footing, 12" x 36", reinforced	34	C.Y.	$5,372.00	$3,111.00	$15.30	$0.00	
03 11 13.65 3000	Concrete slab edge forms	500	L.F.	$155.00	$1,305.00	$0.00	$0.00	
03 22 11.10 0200	Welded wire fabric reinforcing	150	C.S.F.	$3,000.00	$4,275.00	$0.00	$0.00	
03 31 13.35 0300	Ready mix concrete, 4000 psi for slab on grade	278	C.Y.	$35,584.00	$0.00	$0.00	$0.00	
03 31 13.70 4300	Place, strike off & consolidate concrete slab	278	C.Y.	$0.00	$5,184.70	$133.44	$0.00	
03 35 13.30 0250	Machine float & trowel concrete slab	15,000	S.F.	$0.00	$9,750.00	$300.00	$0.00	
03 15 16.20 0140	Cut control joints in concrete slab	950	L.F.	$47.50	$408.50	$57.00	$0.00	
03 39 23.13 0300	Sprayed concrete curing membrane	150	C.S.F.	$1,890.00	$1,035.00	$0.00	$0.00	
Division 03	Subtotal			$48,508.50	$26,779.20	$514.14	$0.00	$75,801.84
08 36 13.10 2650	Manual 10' x 10' steel sectional overhead door	8	Ea.	$10,600.00	$3,680.00	$0.00	$0.00	
08 36 13.10 2860	Insulation and steel back panel for OH door	800	S.F.	$4,120.00	$0.00	$0.00	$0.00	
Division 08	Subtotal			$14,720.00	$3,680.00	$0.00	$0.00	$18,400.00
13 34 19.50 1100	Pre-Engineered Steel Building, 100' x 150' x 24'	15,000	SF Flr.	$0.00	$0.00	$0.00	$367,500.00	
13 34 19.50 6050	Framing for PESB door opening, 3' x 7'	4	Opng.	$0.00	$0.00	$0.00	$2,240.00	
13 34 19.50 6100	Framing for PESB door opening, 10' x 10'	8	Opng.	$0.00	$0.00	$0.00	$9,200.00	
13 34 19.50 6200	Framing for PESB window opening, 4' x 3'	6	Opng.	$0.00	$0.00	$0.00	$3,300.00	
13 34 19.50 5750	PESB door, 3' x 7', single leaf	4	Opng.	$2,700.00	$716.00	$0.00	$0.00	
13 34 19.50 7750	PESB sliding window, 4' x 3' with screen	6	Opng.	$2,910.00	$612.00	$44.10	$0.00	
13 34 19.50 6550	PESB gutter, eave type, 26 ga., painted	300	L.F.	$2,160.00	$840.00	$0.00	$0.00	
13 34 19.50 8650	PESB roof vent, 12" wide x 10' long	15	Ea.	$562.50	$3,360.00	$0.00	$0.00	
13 34 19.50 6900	PESB insulation, vinyl faced, 4" thick	27,400	S.F.	$12,330.00	$9,864.00	$0.00	$0.00	
Division 13	Subtotal			$20,662.50	$15,392.00	$44.10	$382,240.00	$418,338.60
		Subtotal		$83,891.00	$45,851.20	$558.24	$382,240.00	$512,540.44
Division 01	General Requirements @ 7%			5,872.37	3,209.58	39.08	26,756.80	
2		Estimate Subtotal		$89,763.37	$49,060.78	$597.32	$408,996.80	$512,540.44
		3 Sales Tax @ 5%		4,488.17		29.87	10,224.92	
		Subtotal A		94,251.54	49,060.78	627.18	419,221.72	
		4 GC O & P		9,425.15	25,560.67	62.72	41,922.17	
		Subtotal B		103,676.69	74,621.45	689.90	461,143.89	$640,131.94
		5 Contingency @ 5%						32,006.60
		Subtotal C						$672,138.53
		6 Bond @ $12/1000 +10% O&P						8,872.23
		Subtotal D						$681,010.76
		7 Location Adjustment Factor			113.90			94,660.50
		Grand Total						$775,671.26

This estimate is based on an interactive spreadsheet. You are free to download it and adjust it to your methodology.
A copy of this spreadsheet is available at **RSMeans.com/2019books.**

Sample Estimate

This sample demonstrates the elements of an estimate, including a tally of the RSMeans data lines and a summary of the markups on a contractor's work to arrive at a total cost to the owner. The Location Factor with RSMeans data is added at the bottom of the estimate to adjust the cost of the work to a specific location.

1 Work Performed

The body of the estimate shows the RSMeans data selected, including the line number, a brief description of each item, its take-off unit and quantity, and the bare costs of materials, labor, and equipment. This estimate also includes a column titled "SubContract." This data is taken from the column "Total Incl O&P" and represents the total that a subcontractor would charge a general contractor for the work, including the sub's markup for overhead and profit.

2 Division 1, General Requirements

This is the first division numerically but the last division estimated. Division 1 includes project-wide needs provided by the general contractor. These requirements vary by project but may include temporary facilities and utilities, security, testing, project cleanup, etc. For small projects a percentage can be used—typically between 5% and 15% of project cost. For large projects the costs may be itemized and priced individually.

3 Sales Tax

If the work is subject to state or local sales taxes, the amount must be added to the estimate. Sales tax may be added to material costs, equipment costs, and subcontracted work. In this case, sales tax was added in all three categories. It was assumed that approximately half the subcontracted work would be material cost, so the tax was applied to 50% of the subcontract total.

4 GC O&P

This entry represents the general contractor's markup on material, labor, equipment, and subcontractor costs. Our standard markup on materials, equipment, and subcontracted work is 10%. In this estimate, the markup on the labor performed by the GC's workers uses "Skilled Workers Average" shown in Column F on the table "Installing Contractor's Overhead & Profit," which can be found on the inside back cover of the printed product or in the Reference Section of the electronic product.

5 Contingency

A factor for contingency may be added to any estimate to represent the cost of unknowns that may occur between the time that the estimate is performed and the time the project is constructed. The amount of the allowance will depend on the stage of design at which the estimate is done and the contractor's assessment of the risk involved. Refer to section 01 21 16.50 for contingency allowances.

6 Bonds

Bond costs should be added to the estimate. The figures here represent a typical performance bond, ensuring the owner that if the general contractor does not complete the obligations in the construction contract the bonding company will pay the cost for completion of the work.

7 Location Adjustment

Published prices are based on national average costs. If necessary, adjust the total cost of the project using a location factor from the "Location Factor" table or the "City Cost Index" table. Use location factors if the work is general, covering multiple trades. If the work is by a single trade (e.g., masonry) use the more specific data found in the "City Cost Indexes."

Estimating Tips

01 20 00 Price and Payment Procedures

- Allowances that should be added to estimates to cover contingencies and job conditions that are not included in the national average material and labor costs are shown in Section 01 21.
- When estimating historic preservation projects (depending on the condition of the existing structure and the owner's requirements), a 15–20% contingency or allowance is recommended, regardless of the stage of the drawings.

01 30 00 Administrative Requirements

- Before determining a final cost estimate, it is good practice to review all the items listed in Subdivisions 01 31 and 01 32 to make final adjustments for items that may need customizing to specific job conditions.
- Requirements for initial and periodic submittals can represent a significant cost to the General Requirements of a job. Thoroughly check the submittal specifications when estimating a project to determine any costs that should be included.

01 40 00 Quality Requirements

- All projects will require some degree of quality control. This cost is not included in the unit cost of construction listed in each division. Depending upon the terms of the contract, the various costs of inspection and testing can be the responsibility of either the owner or the contractor. Be sure to include the required costs in your estimate.

01 50 00 Temporary Facilities and Controls

- Barricades, access roads, safety nets, scaffolding, security, and many more requirements for the execution of a safe project are elements of direct cost. These costs can easily be overlooked when preparing an estimate. When looking through the major classifications of this subdivision, determine which items apply to each division in your estimate.
- Construction equipment rental costs can be found in the Reference Section in Section 01 54 33. Operators' wages are not included in equipment rental costs.
- Equipment mobilization and demobilization costs are not included in equipment rental costs and must be considered separately.
- The cost of small tools provided by the installing contractor for his workers is covered in the "Overhead" column on the "Installing Contractor's Overhead and Profit" table that lists labor trades, base rates, and markups. Therefore, it is included in the "Total Incl. O&P" cost of any unit price line item.

01 70 00 Execution and Closeout Requirements

- When preparing an estimate, thoroughly read the specifications to determine the requirements for Contract Closeout. Final cleaning, record documentation, operation and maintenance data, warranties and bonds, and spare parts and maintenance materials can all be elements of cost for the completion of a contract. Do not overlook these in your estimate.

Reference Numbers

Reference numbers are shown at the beginning of some major classifications. These numbers refer to related items in the Reference Section. The reference information may be an estimating procedure, an alternate pricing method, or technical information.

Note: Not all subdivisions listed here necessarily appear. ■

01 11 Summary of Work

01 11 31 – Professional Consultants

01 11 31.10 Architectural Fees

	01 11 31.10 Architectural Fees		Crew	Daily Output	Labor-Hours	Unit	Material	Labor (2019 Bare Costs)	Equipment	Total	Total Incl O&P
0010	**ARCHITECTURAL FEES**	R011110-10									
2000	For "Greening" of building	G				Project				3%	3%

01 11 31.30 Engineering Fees

	01 11 31.30 Engineering Fees		Crew	Daily Output	Labor-Hours	Unit	Material	Labor	Equipment	Total	Total Incl O&P
0010	**ENGINEERING FEES**	R011110-30									
0020	Educational planning consultant, minimum					Project				.50%	.50%
0100	Maximum					"				2.50%	2.50%
0200	Electrical, minimum					Contrct				4.10%	4.10%
0300	Maximum									10.10%	10.10%
0400	Elevator & conveying systems, minimum									2.50%	2.50%
0500	Maximum									5%	5%
0600	Food service & kitchen equipment, minimum									8%	8%
0700	Maximum									12%	12%
1000	Mechanical (plumbing & HVAC), minimum									4.10%	4.10%
1100	Maximum									10.10%	10.10%

01 21 Allowances

01 21 16 – Contingency Allowances

01 21 16.50 Contingencies

	01 21 16.50 Contingencies		Crew	Daily Output	Labor-Hours	Unit	Material	Labor	Equipment	Total	Total Incl O&P
0010	**CONTINGENCIES**, Add to estimate										
0020	Conceptual stage					Project				20%	20%
0050	Schematic stage									15%	15%
0100	Preliminary working drawing stage (Design Dev.)									10%	10%
0150	Final working drawing stage									3%	3%

01 21 53 – Factors Allowance

01 21 53.50 Factors

	01 21 53.50 Factors		Crew	Daily Output	Labor-Hours	Unit	Material	Labor	Equipment	Total	Total Incl O&P
0010	**FACTORS** Cost adjustments	R012153-10									
0100	Add to construction costs for particular job requirements										
0500	Cut & patch to match existing construction, add, minimum					Costs	2%	3%			
0550	Maximum						5%	9%			
0800	Dust protection, add, minimum						1%	2%			
0850	Maximum						4%	11%			
1100	Equipment usage curtailment, add, minimum						1%	1%			
1150	Maximum						3%	10%			
1400	Material handling & storage limitation, add, minimum						1%	1%			
1450	Maximum						6%	7%			
1700	Protection of existing work, add, minimum						2%	2%			
1750	Maximum						5%	7%			
2000	Shift work requirements, add, minimum							5%			
2050	Maximum							30%			
2300	Temporary shoring and bracing, add, minimum						2%	5%			
2350	Maximum						5%	12%			

01 21 53.60 Security Factors

	01 21 53.60 Security Factors		Crew	Daily Output	Labor-Hours	Unit	Material	Labor	Equipment	Total	Total Incl O&P
0010	**SECURITY FACTORS**	R012153-60									
0100	Additional costs due to security requirements										
0110	Daily search of personnel, supplies, equipment and vehicles										
0120	Physical search, inventory and doc of assets, at entry					Costs		30%			
0130	At entry and exit							50%			
0140	Physical search, at entry							6.25%			
0150	At entry and exit							12.50%			
0160	Electronic scan search, at entry							2%			

01 21 Allowances

01 21 53 – Factors Allowance

01 21 53.60 Security Factors

		Crew	Daily Output	Labor-Hours	Unit	2019 Bare Costs Material	Labor	Equipment	Total	Total Incl O&P
0170	At entry and exit				Costs		4%			
0180	Visual inspection only, at entry						.25%			
0190	At entry and exit						.50%			
0200	ID card or display sticker only, at entry						.12%			
0210	At entry and exit						.25%			
0220	Day 1 as described below, then visual only for up to 5 day job duration									
0230	Physical search, inventory and doc of assets, at entry				Costs		5%			
0240	At entry and exit						10%			
0250	Physical search, at entry						1.25%			
0260	At entry and exit						2.50%			
0270	Electronic scan search, at entry						.42%			
0280	At entry and exit						.83%			
0290	Day 1 as described below, then visual only for 6-10 day job duration									
0300	Physical search, inventory and doc of assets, at entry				Costs		2.50%			
0310	At entry and exit						5%			
0320	Physical search, at entry						.63%			
0330	At entry and exit						1.25%			
0340	Electronic scan search, at entry						.21%			
0350	At entry and exit						.42%			
0360	Day 1 as described below, then visual only for 11-20 day job duration									
0370	Physical search, inventory and doc of assets, at entry				Costs		1.25%			
0380	At entry and exit						2.50%			
0390	Physical search, at entry						.31%			
0400	At entry and exit						.63%			
0410	Electronic scan search, at entry						.10%			
0420	At entry and exit						.21%			
0430	Beyond 20 days, costs are negligible									
0440	Escort required to be with tradesperson during work effort				Costs		6.25%			

01 21 55 – Job Conditions Allowance

01 21 55.50 Job Conditions

		Crew	Daily Output	Labor-Hours	Unit	Material	Labor	Equipment	Total	Total Incl O&P
0010	**JOB CONDITIONS** Modifications to applicable									
0020	cost summaries									
0100	Economic conditions, favorable, deduct				Project				2%	2%
0200	Unfavorable, add								5%	5%
0300	Hoisting conditions, favorable, deduct								2%	2%
0400	Unfavorable, add								5%	5%
0700	Labor availability, surplus, deduct								1%	1%
0800	Shortage, add								10%	10%
0900	Material storage area, available, deduct								1%	1%
1000	Not available, add								2%	2%
1100	Subcontractor availability, surplus, deduct								5%	5%
1200	Shortage, add								12%	12%
1300	Work space, available, deduct								2%	2%
1400	Not available, add								5%	5%

01 21 57 – Overtime Allowance

01 21 57.50 Overtime

		Crew	Daily Output	Labor-Hours	Unit	Material	Labor	Equipment	Total	Total Incl O&P
0010	**OVERTIME** for early completion of projects or where R012909-90									
0020	labor shortages exist, add to usual labor, up to				Costs		100%			

01 21 Allowances

01 21 63 – Taxes

01 21 63.10 Taxes		Crew	Daily Output	Labor-Hours	Unit	Material	2019 Bare Costs Labor	Equipment	Total	Total Incl O&P
0010	**TAXES** R012909-80									
0020	Sales tax, State, average				%	5.08%				
0050	Maximum R012909-85					7.50%				
0200	Social Security, on first $118,500 of wages						7.65%			
0300	Unemployment, combined Federal and State, minimum R012909-86						.60%			
0350	Average						9.60%			
0400	Maximum						12%			

01 31 Project Management and Coordination

01 31 13 – Project Coordination

01 31 13.20 Field Personnel

		Crew	Daily Output	Labor-Hours	Unit	Material	Labor	Equipment	Total	Total Incl O&P
0010	**FIELD PERSONNEL**									
0020	Clerk, average				Week		495		495	750
0100	Field engineer, junior engineer						1,175		1,175	1,775
0120	Engineer						1,525		1,525	2,325
0140	Senior engineer						1,725		1,725	2,650
0160	General purpose laborer, average						1,650		1,650	2,475
0180	Project manager, minimum						2,175		2,175	3,300
0200	Average						2,500		2,500	3,800
0220	Maximum						2,850		2,850	4,325
0240	Superintendent, minimum						2,125		2,125	3,225
0260	Average						2,325		2,325	3,525
0280	Maximum						2,650		2,650	4,025
0290	Timekeeper, average						1,350		1,350	2,050

01 31 13.30 Insurance

		Crew	Daily Output	Labor-Hours	Unit	Material	Labor	Equipment	Total	Total Incl O&P
0010	**INSURANCE** R013113-40									
0020	Builders risk, standard, minimum				Job				.24%	.24%
0050	Maximum R013113-50								.64%	.64%
0200	All-risk type, minimum								.25%	.25%
0250	Maximum R013113-60								.62%	.62%
0400	Contractor's equipment floater, minimum				Value				.50%	.50%
0450	Maximum				"				1.50%	1.50%
0600	Public liability, average				Job				2.02%	2.02%
0800	Workers' compensation & employer's liability, average									
0850	by trade, carpentry, general				Payroll		11.97%			
1000	Electrical						4.91%			
1150	Insulation						10.07%			
1450	Plumbing						5.77%			
1550	Sheet metal work (HVAC)						7.56%			

01 31 13.50 General Contractor's Mark-Up

		Crew	Daily Output	Labor-Hours	Unit	Material	Labor	Equipment	Total	Total Incl O&P
0010	**GENERAL CONTRACTOR'S MARK-UP** on Change Orders									
0200	Extra work, by subcontractors, add				%				10%	10%
0250	By General Contractor, add								15%	15%
0400	Omitted work, by subcontractors, deduct all but								5%	5%
0450	By General Contractor, deduct all but								7.50%	7.50%
0600	Overtime work, by subcontractors, add								15%	15%
0650	By General Contractor, add								10%	10%

01 31 Project Management and Coordination

01 31 13 – Project Coordination

01 31 13.80 Overhead and Profit		Crew	Daily Output	Labor-Hours	Unit	2019 Bare Costs Material	Labor	Equipment	Total	Total Incl O&P
0010	**OVERHEAD & PROFIT** Allowance to add to items in this									
0020	book that do not include Subs O&P, average				%				25%	
0100	Allowance to add to items in this book that R013113-55									
0110	do include Subs O&P, minimum				%				5%	5%
0150	Average								10%	10%
0200	Maximum								15%	15%
0300	Typical, by size of project, under $100,000								30%	
0350	$500,000 project								25%	
0400	$2,000,000 project								20%	
0450	Over $10,000,000 project								15%	

01 31 13.90 Performance Bond		Crew	Daily Output	Labor-Hours	Unit	Material	Labor	Equipment	Total	Total Incl O&P
0010	**PERFORMANCE BOND** R013113-80									
0020	For buildings, minimum				Job				.60%	.60%
0100	Maximum				"				2.50%	2.50%

01 32 Construction Progress Documentation

01 32 33 – Photographic Documentation

01 32 33.50 Photographs		Crew	Daily Output	Labor-Hours	Unit	Material	Labor	Equipment	Total	Total Incl O&P
0010	**PHOTOGRAPHS**									
0020	8" x 10", 4 shots, 2 prints ea., std. mounting				Set	545			545	600
0100	Hinged linen mounts					550			550	605
0200	8" x 10", 4 shots, 2 prints each, in color					510			510	560
0300	For I.D. slugs, add to all above					5.10			5.10	5.60
1500	Time lapse equipment, camera and projector, buy				Ea.	2,775			2,775	3,050
1550	Rent per month				"	1,300			1,300	1,425
1700	Cameraman and processing, black & white				Day	1,275			1,275	1,400
1720	Color				"	1,450			1,450	1,600

01 41 Regulatory Requirements

01 41 26 – Permit Requirements

01 41 26.50 Permits		Crew	Daily Output	Labor-Hours	Unit	Material	Labor	Equipment	Total	Total Incl O&P
0010	**PERMITS**									
0020	Rule of thumb, most cities, minimum				Job				.50%	.50%
0100	Maximum				"				2%	2%

01 45 Quality Control

01 45 23 – Testing and Inspecting Services

01 45 23.50 Testing		Crew	Daily Output	Labor-Hours	Unit	Material	Labor	Equipment	Total	Total Incl O&P
0010	**TESTING** and Inspecting Services									
5820	Non-destructive metal testing, dye penetrant				Day				310	340
5840	Magnetic particle								310	340
5860	Radiography								450	495
5880	Ultrasonic								310	340
9000	Thermographic testing, for bldg envelope heat loss, average 2,000 S.F. G				Ea.				500	500

01 51 Temporary Utilities

01 51 13 – Temporary Electricity

01 51 13.80 Temporary Utilities		Crew	Daily Output	Labor-Hours	Unit	Material	2019 Bare Costs Labor	Equipment	Total	Total Incl O&P
0010	**TEMPORARY UTILITIES**									
0350	Lighting, lamps, wiring, outlets, 40,000 S.F. building, 8 strings	1 Elec	34	.235	CSF Flr	5.70	14.15		19.85	27.50
0360	16 strings	"	17	.471		11.40	28.50		39.90	54.50
0400	Power for temp lighting only, 6.6 KWH, per month								.92	1.01
0430	11.8 KWH, per month								1.65	1.82
0450	23.6 KWH, per month								3.30	3.63
0600	Power for job duration incl. elevator, etc., minimum								47	51.50
0650	Maximum								110	121
0675	Temporary cooling				Ea.	1,000			1,000	1,100

01 52 Construction Facilities

01 52 13 – Field Offices and Sheds

01 52 13.20 Office and Storage Space

		Crew	Daily Output	Labor-Hours	Unit	Material	Labor	Equipment	Total	Total Incl O&P
0010	**OFFICE AND STORAGE SPACE**									
0020	Office trailer, furnished, no hookups, 20′ x 8′, buy	2 Skwk	1	16	Ea.	9,050	855		9,905	11,300
0250	Rent per month					201			201	222
0300	32′ x 8′, buy	2 Skwk	.70	22.857		14,500	1,225		15,725	17,800
0350	Rent per month					254			254	280
0400	50′ x 10′, buy	2 Skwk	.60	26.667		29,900	1,425		31,325	35,100
0450	Rent per month					365			365	400
0500	50′ x 12′, buy	2 Skwk	.50	32		25,500	1,700		27,200	30,600
0550	Rent per month					455			455	500
0700	For air conditioning, rent per month, add					51			51	56
0800	For delivery, add per mile				Mile	12.15			12.15	13.35
0890	Delivery each way				Ea.	2,800			2,800	3,075
0900	Bunk house trailer, 8′ x 40′ duplex dorm with kitchen, no hookups, buy	2 Carp	1	16		88,000	825		88,825	98,000
0910	9 man with kitchen and bath, no hookups, buy		1	16		90,000	825		90,825	100,500
0920	18 man sleeper with bath, no hookups, buy		1	16		97,000	825		97,825	108,000
1000	Portable buildings, prefab, on skids, economy, 8′ x 8′		265	.060	S.F.	25.50	3.12		28.62	32.50
1100	Deluxe, 8′ x 12′		150	.107	"	28	5.50		33.50	39.50
1200	Storage boxes, 20′ x 8′, buy	2 Skwk	1.80	8.889	Ea.	3,250	475		3,725	4,300
1250	Rent per month					85			85	93.50
1300	40′ x 8′, buy	2 Skwk	1.40	11.429		3,800	610		4,410	5,100
1350	Rent per month					111			111	122

01 52 13.40 Field Office Expense

		Crew	Daily Output	Labor-Hours	Unit	Material	Labor	Equipment	Total	Total Incl O&P
0010	**FIELD OFFICE EXPENSE**									
0100	Office equipment rental average				Month	207			207	227
0120	Office supplies, average				"	85			85	93.50
0125	Office trailer rental, see Section 01 52 13.20									
0140	Telephone bill; avg. bill/month incl. long dist.				Month	87			87	95.50
0160	Lights & HVAC				"	162			162	179

01 54 Construction Aids

01 54 16 – Temporary Hoists

01 54 16.50 Weekly Forklift Crew		Crew	Daily Output	Labor-Hours	Unit	Material	2019 Bare Costs Labor	Equipment	Total	Total Incl O&P
0010	**WEEKLY FORKLIFT CREW**									
0100	All-terrain forklift, 45′ lift, 35′ reach, 9000 lb. capacity	A-3P	.20	40	Week		2,075	2,325	4,400	5,650

01 54 19 – Temporary Cranes

01 54 19.50 Daily Crane Crews

		Crew	Daily Output	Labor-Hours	Unit	Material	Labor	Equipment	Total	Total Incl O&P
0010	**DAILY CRANE CREWS** for small jobs, portal to portal R015433-15									
0100	12-ton truck-mounted hydraulic crane	A-3H	1	8	Day		460	710	1,170	1,475
0200	25-ton	A-3I	1	8			460	785	1,245	1,550
0300	40-ton	A-3J	1	8			460	1,275	1,735	2,100
0400	55-ton	A-3K	1	16			855	1,450	2,305	2,850
0500	80-ton	A-3L	1	16			855	2,200	3,055	3,700
0900	If crane is needed on a Saturday, Sunday or Holiday									
0910	At time-and-a-half, add				Day		50%			
0920	At double time, add				"		100%			

01 54 19.60 Monthly Tower Crane Crew

		Crew	Daily Output	Labor-Hours	Unit	Material	Labor	Equipment	Total	Total Incl O&P
0010	**MONTHLY TOWER CRANE CREW**, excludes concrete footing									
0100	Static tower crane, 130′ high, 106′ jib, 6200 lb. capacity	A-3N	.05	176	Month		10,100	26,000	36,100	43,800

01 54 23 – Temporary Scaffolding and Platforms

01 54 23.70 Scaffolding

		Crew	Daily Output	Labor-Hours	Unit	Material	Labor	Equipment	Total	Total Incl O&P
0010	**SCAFFOLDING** R015423-10									
0015	Steel tube, regular, no plank, labor only to erect & dismantle									
0090	Building exterior, wall face, 1 to 5 stories, 6′-4″ x 5′ frames	3 Carp	8	3	C.S.F.		155		155	234
0200	6 to 12 stories	4 Carp	8	4			207		207	315
0301	13 to 20 stories	5 Clab	8	5			205		205	310
0460	Building interior, wall face area, up to 16′ high	3 Carp	12	2			103		103	156
0560	16′ to 40′ high		10	2.400			124		124	188
0800	Building interior floor area, up to 30′ high		150	.160	C.C.F.		8.25		8.25	12.50
0900	Over 30′ high	4 Carp	160	.200	"		10.35		10.35	15.65
0906	Complete system for face of walls, no plank, material only rent/mo				C.S.F.	33			33	36.50
0908	Interior spaces, no plank, material only rent/mo				C.C.F.	3.80			3.80	4.18
0910	Steel tubular, heavy duty shoring, buy									
0920	Frames 5′ high 2′ wide				Ea.	92			92	101
0925	5′ high 4′ wide					97.50			97.50	107
0930	6′ high 2′ wide					99			99	109
0935	6′ high 4′ wide					116			116	128
0940	Accessories									
0945	Cross braces				Ea.	17.10			17.10	18.80
0950	U-head, 8″ x 8″					18.90			18.90	21
0955	J-head, 4″ x 8″					13.75			13.75	15.15
0960	Base plate, 8″ x 8″					15.15			15.15	16.65
0965	Leveling jack					36			36	39.50
1000	Steel tubular, regular, buy									
1100	Frames 3′ high 5′ wide				Ea.	89.50			89.50	98.50
1150	5′ high 5′ wide					108			108	119
1200	6′-4″ high 5′ wide					106			106	117
1350	7′-6″ high 6′ wide					157			157	173
1500	Accessories, cross braces					18.05			18.05	19.85
1550	Guardrail post					19.90			19.90	22
1600	Guardrail 7′ section					8.15			8.15	8.95
1650	Screw jacks & plates					25.50			25.50	28.50
1700	Sidearm brackets					22.50			22.50	25
1750	8″ casters					37.50			37.50	41

01 54 Construction Aids

01 54 23 – Temporary Scaffolding and Platforms

01 54 23.70 Scaffolding		Crew	Daily Output	Labor-Hours	Unit	2019 Bare Costs Material	Labor	Equipment	Total	Total Incl O&P
1800	Plank 2" x 10" x 16'-0"				Ea.	67			67	74
1900	Stairway section					284			284	310
1910	Stairway starter bar					32.50			32.50	35.50
1920	Stairway inside handrail					53			53	58.50
1930	Stairway outside handrail					84.50			84.50	93
1940	Walk-thru frame guardrail					42			42	46
2000	Steel tubular, regular, rent/mo.									
2100	Frames 3' high 5' wide				Ea.	4.45			4.45	4.90
2150	5' high 5' wide					4.45			4.45	4.90
2200	6'-4" high 5' wide					5.40			5.40	5.95
2250	7'-6" high 6' wide					9.85			9.85	10.85
2500	Accessories, cross braces					.89			.89	.98
2550	Guardrail post					.89			.89	.98
2600	Guardrail 7' section					.89			.89	.98
2650	Screw jacks & plates					1.78			1.78	1.96
2700	Sidearm brackets					1.78			1.78	1.96
2750	8" casters					7.10			7.10	7.80
2800	Outrigger for rolling tower					2.67			2.67	2.94
2850	Plank 2" x 10" x 16'-0"					9.85			9.85	10.85
2900	Stairway section					31.50			31.50	34.50
2940	Walk-thru frame guardrail					2.22			2.22	2.44
3000	Steel tubular, heavy duty shoring, rent/mo.									
3250	5' high 2' & 4' wide				Ea.	8.35			8.35	9.20
3300	6' high 2' & 4' wide					8.35			8.35	9.20
3500	Accessories, cross braces					.89			.89	.98
3600	U-head, 8" x 8"					2.46			2.46	2.71
3650	J-head, 4" x 8"					2.46			2.46	2.71
3700	Base plate, 8" x 8"					.89			.89	.98
3750	Leveling jack					2.46			2.46	2.71
5700	Planks, 2" x 10" x 16'-0", labor only to erect & remove to 50' H	3 Carp	72	.333			17.20		17.20	26
5800	Over 50' high	4 Carp	80	.400			20.50		20.50	31.50
6000	Heavy duty shoring for elevated slab forms to 8'-2" high, floor area									
6100	Labor only to erect & dismantle	4 Carp	16	2	C.S.F.		103		103	156
6110	Materials only, rent/mo.				"	43			43	47.50
6500	To 14'-8" high									
6600	Labor only to erect & dismantle	4 Carp	10	3.200	C.S.F.		165		165	250
6610	Materials only, rent/mo				"	63			63	69.50

01 54 23.75 Scaffolding Specialties		Crew	Daily Output	Labor-Hours	Unit	Material	Labor	Equipment	Total	Total Incl O&P
0010	**SCAFFOLDING SPECIALTIES**									
1200	Sidewalk bridge, heavy duty steel posts & beams, including									
1210	parapet protection & waterproofing (material cost is rent/month)									
1220	8' to 10' wide, 2 posts	3 Carp	15	1.600	L.F.	45.50	82.50		128	175
1230	3 posts	"	10	2.400	"	70	124		194	265
1500	Sidewalk bridge using tubular steel scaffold frames including									
1510	planking (material cost is rent/month)	3 Carp	45	.533	L.F.	8.35	27.50		35.85	50.50
1600	For 2 uses per month, deduct from all above					50%				
1700	For 1 use every 2 months, add to all above					100%				
1900	Catwalks, 20" wide, no guardrails, 7' span, buy				Ea.	151			151	166
2000	10' span, buy					212			212	233
3720	Putlog, standard, 8' span, with hangers, buy					81.50			81.50	90
3730	Rent per month					16.15			16.15	17.75
3750	12' span, buy					101			101	111

01 54 Construction Aids

01 54 23 – Temporary Scaffolding and Platforms

01 54 23.75 Scaffolding Specialties		Crew	Daily Output	Labor-Hours	Unit	Material	2019 Bare Costs Labor	Equipment	Total	Total Incl O&P
3755	Rent per month				Ea.	20			20	22
3760	Trussed type, 16′ span, buy					266			266	293
3770	Rent per month					24			24	26.50
3790	22′ span, buy					279			279	305
3795	Rent per month					32.50			32.50	35.50
3800	Rolling ladders with handrails, 30″ wide, buy, 2 step					294			294	325
4000	7 step					965			965	1,050
4050	10 step					1,375			1,375	1,525
4100	Rolling towers, buy, 5′ wide, 7′ long, 10′ high					1,325			1,325	1,450
4200	For additional 5′ high sections, to buy					253			253	278
4300	Complete incl. wheels, railings, outriggers,									
4350	21′ high, to buy				Ea.	2,275			2,275	2,500
4400	Rent/month = 5% of purchase cost				″	199			199	219

01 54 36 – Equipment Mobilization

01 54 36.50 Mobilization		Crew	Daily Output	Labor-Hours	Unit	Material	Labor	Equipment	Total	Total Incl O&P
0010	**MOBILIZATION** (Use line item again for demobilization) R015436-50									
0015	Up to 25 mi. haul dist. (50 mi. RT for mob/demob crew)									
1200	Small equipment, placed in rear of, or towed by pickup truck	A-3A	4	2	Ea.		103	30.50	133.50	189
1300	Equipment hauled on 3-ton capacity towed trailer	A-3Q	2.67	3			155	55.50	210.50	294
1400	20-ton capacity	B-34U	2	8			395	218	613	835
1500	40-ton capacity	B-34N	2	8			410	335	745	985
1600	50-ton capacity	B-34V	1	24			1,250	920	2,170	2,875
1700	Crane, truck-mounted, up to 75 ton (driver only)	1 Eqhv	4	2			115		115	173
1800	Over 75 ton (with chase vehicle)	A-3E	2.50	6.400			335	48.50	383.50	555
2400	Crane, large lattice boom, requiring assembly	B-34W	.50	144			7,175	6,850	14,025	18,400
2500	For each additional 5 miles haul distance, add						10%	10%		
3000	For large pieces of equipment, allow for assembly/knockdown									
3100	For mob/demob of micro-tunneling equip, see Section 33 05 23.19									

01 55 Vehicular Access and Parking

01 55 23 – Temporary Roads

01 55 23.50 Roads and Sidewalks		Crew	Daily Output	Labor-Hours	Unit	Material	Labor	Equipment	Total	Total Incl O&P
0010	**ROADS AND SIDEWALKS** Temporary									
0050	Roads, gravel fill, no surfacing, 4″ gravel depth	B-14	715	.067	S.Y.	3.39	2.90	.45	6.74	8.60
0100	8″ gravel depth	″	615	.078	″	6.80	3.37	.52	10.69	13.10
1000	Ramp, 3/4″ plywood on 2″ x 6″ joists, 16″ OC	2 Carp	300	.053	S.F.	1.65	2.75		4.40	6
1100	On 2″ x 10″ joists, 16″ OC	″	275	.058	″	2.36	3.01		5.37	7.15

01 56 Temporary Barriers and Enclosures

01 56 13 – Temporary Air Barriers

01 56 13.60 Tarpaulins		Crew	Daily Output	Labor-Hours	Unit	Material	Labor	Equipment	Total	Total Incl O&P
0010	**TARPAULINS**									
0020	Cotton duck, 10-13.13 oz./S.Y., 6′ x 8′				S.F.	.86			.86	.95
0050	30′ x 30′					.60			.60	.66
0100	Polyvinyl coated nylon, 14-18 oz., minimum					1.47			1.47	1.62
0150	Maximum					1.47			1.47	1.62
0200	Reinforced polyethylene 3 mils thick, white					.04			.04	.04
0300	4 mils thick, white, clear or black					.13			.13	.14
0400	5.5 mils thick, clear					.22			.22	.24

01 56 Temporary Barriers and Enclosures

01 56 13 – Temporary Air Barriers

01 56 13.60 Tarpaulins

	01 56 13.60 Tarpaulins	Crew	Daily Output	Labor-Hours	Unit	2019 Bare Costs Material	Labor	Equipment	Total	Total Incl O&P
0500	White, fire retardant				S.F.	.63			.63	.69
0600	12 mils, oil resistant, fire retardant					.51			.51	.56
0700	8.5 mils, black					.20			.20	.22
0710	Woven polyethylene, 6 mils thick					.06			.06	.07
0730	Polyester reinforced w/integral fastening system, 11 mils thick					.19			.19	.21
0740	Polyethylene, reflective, 23 mils thick					1.37			1.37	1.51

01 56 13.90 Winter Protection

	01 56 13.90 Winter Protection	Crew	Daily Output	Labor-Hours	Unit	Material	Labor	Equipment	Total	Total Incl O&P
0010	**WINTER PROTECTION**									
0100	Framing to close openings	2 Clab	500	.032	S.F.	.46	1.31		1.77	2.49
0200	Tarpaulins hung over scaffolding, 8 uses, not incl. scaffolding		1500	.011		.25	.44		.69	.94
0250	Tarpaulin polyester reinf. w/integral fastening system, 11 mils thick		1600	.010		.21	.41		.62	.85
0300	Prefab fiberglass panels, steel frame, 8 uses		1200	.013		2.58	.55		3.13	3.67

01 56 16 – Temporary Dust Barriers

01 56 16.10 Dust Barriers, Temporary

	01 56 16.10 Dust Barriers, Temporary	Crew	Daily Output	Labor-Hours	Unit	Material	Labor	Equipment	Total	Total Incl O&P
0010	**DUST BARRIERS, TEMPORARY**									
0020	Spring loaded telescoping pole & head, to 12′, erect and dismantle	1 Clab	240	.033	Ea.		1.37		1.37	2.07
0025	Cost per day (based upon 250 days)				Day	.22			.22	.24
0030	To 21′, erect and dismantle	1 Clab	240	.033	Ea.		1.37		1.37	2.07
0035	Cost per day (based upon 250 days)				Day	.68			.68	.74
0040	Accessories, caution tape reel, erect and dismantle	1 Clab	480	.017	Ea.		.68		.68	1.04
0045	Cost per day (based upon 250 days)				Day	.32			.32	.35
0060	Foam rail and connector, erect and dismantle	1 Clab	240	.033	Ea.		1.37		1.37	2.07
0065	Cost per day (based upon 250 days)				Day	.10			.10	.11
0070	Caution tape	1 Clab	384	.021	C.L.F.	2.69	.86		3.55	4.25
0080	Zipper, standard duty		60	.133	Ea.	7.40	5.45		12.85	16.45
0090	Heavy duty		48	.167	″	9.95	6.85		16.80	21.50
0100	Polyethylene sheet, 4 mil		37	.216	Sq.	2.66	8.90		11.56	16.40
0110	6 mil		37	.216	″	3.87	8.90		12.77	17.70
1000	Dust partition, 6 mil polyethylene, 1″ x 3″ frame	2 Carp	2000	.008	S.F.	.33	.41		.74	.99
1080	2″ x 4″ frame	″	2000	.008	″	.35	.41		.76	1.02
1085	Negative air machine, 1800 CFM				Ea.	910			910	1,000
1090	Adhesive strip application, 2″ width	1 Clab	192	.042	C.L.F.	6.90	1.71		8.61	10.15

01 56 23 – Temporary Barricades

01 56 23.10 Barricades

	01 56 23.10 Barricades	Crew	Daily Output	Labor-Hours	Unit	Material	Labor	Equipment	Total	Total Incl O&P
0010	**BARRICADES**									
0020	5′ high, 3 rail @ 2″ x 8″, fixed	2 Carp	20	.800	L.F.	6.40	41.50		47.90	69.50
0150	Movable		30	.533		5.35	27.50		32.85	47.50
1000	Guardrail, wooden, 3′ high, 1″ x 6″ on 2″ x 4″ posts		200	.080		1.34	4.13		5.47	7.70
1100	2″ x 6″ on 4″ x 4″ posts		165	.097		2.66	5		7.66	10.50
1200	Portable metal with base pads, buy					14.50			14.50	15.95
1250	Typical installation, assume 10 reuses	2 Carp	600	.027		2.37	1.38		3.75	4.69
1300	Barricade tape, polyethylene, 7 mil, 3″ wide x 300′ long roll				Ea.	8.05			8.05	8.90
3000	Detour signs, set up and remove									
3010	Reflective aluminum, MUTCD, 24″ x 24″, post mounted	1 Clab	20	.400	Ea.	2.66	16.40		19.06	28
4000	Roof edge portable barrier stands and warning flags, 50 uses	1 Rohe	9100	.001	L.F.	.06	.03		.09	.12
4010	100 uses	″	9100	.001	″	.03	.03		.06	.09

01 56 26 – Temporary Fencing

01 56 26.50 Temporary Fencing

	01 56 26.50 Temporary Fencing	Crew	Daily Output	Labor-Hours	Unit	Material	Labor	Equipment	Total	Total Incl O&P
0010	**TEMPORARY FENCING**									
0020	Chain link, 11 ga., 4′ high	2 Clab	400	.040	L.F.	1.79	1.64		3.43	4.45
0100	6′ high		300	.053		4.68	2.19		6.87	8.45

01 56 Temporary Barriers and Enclosures

01 56 26 – Temporary Fencing

01 56 26.50 Temporary Fencing		Crew	Daily Output	Labor-Hours	Unit	Material	Labor	Equipment	Total	Total Incl O&P
						2019 Bare Costs				
0200	Rented chain link, 6′ high, to 1000′ (up to 12 mo.)	2 Clab	400	.040	L.F.	3.04	1.64		4.68	5.80
0250	Over 1000′ (up to 12 mo.)		300	.053		3.82	2.19		6.01	7.50
0350	Plywood, painted, 2″ x 4″ frame, 4′ high	A-4	135	.178		6.45	8.70		15.15	20
0400	4″ x 4″ frame, 8′ high	″	110	.218		12.35	10.65		23	29.50
0500	Wire mesh on 4″ x 4″ posts, 4′ high	2 Carp	100	.160		10.50	8.25		18.75	24
0550	8′ high	″	80	.200		15.80	10.35		26.15	33

01 56 29 – Temporary Protective Walkways

01 56 29.50 Protection		Crew	Daily Output	Labor-Hours	Unit	Material	Labor	Equipment	Total	Total Incl O&P
0010	**PROTECTION**									
0020	Stair tread, 2″ x 12″ planks, 1 use	1 Carp	75	.107	Tread	5.45	5.50		10.95	14.35
0100	Exterior plywood, 1/2″ thick, 1 use		65	.123		1.90	6.35		8.25	11.70
0200	3/4″ thick, 1 use		60	.133		2.81	6.90		9.71	13.50
2200	Sidewalks, 2″ x 12″ planks, 2 uses		350	.023	S.F.	.91	1.18		2.09	2.79
2300	Exterior plywood, 2 uses, 1/2″ thick		750	.011		.32	.55		.87	1.18
2400	5/8″ thick		650	.012		.39	.64		1.03	1.39
2500	3/4″ thick		600	.013		.47	.69		1.16	1.56

01 58 Project Identification

01 58 13 – Temporary Project Signage

01 58 13.50 Signs		Crew	Daily Output	Labor-Hours	Unit	Material	Labor	Equipment	Total	Total Incl O&P
0010	**SIGNS**									
0020	High intensity reflectorized, no posts, buy				Ea.	25.50			25.50	28

01 66 Product Storage and Handling Requirements

01 66 19 – Material Handling

01 66 19.10 Material Handling		Crew	Daily Output	Labor-Hours	Unit	Material	Labor	Equipment	Total	Total Incl O&P
0010	**MATERIAL HANDLING**									
0020	Above 2nd story, via stairs, per C.Y. of material per floor	2 Clab	145	.110	C.Y.		4.53		4.53	6.85
0030	Via elevator, per C.Y. of material		240	.067			2.74		2.74	4.14
0050	Distances greater than 200′, per C.Y. of material per each addl 200′		300	.053			2.19		2.19	3.31

01 74 Cleaning and Waste Management

01 74 13 – Progress Cleaning

01 74 13.20 Cleaning Up		Crew	Daily Output	Labor-Hours	Unit	Material	Labor	Equipment	Total	Total Incl O&P
0010	**CLEANING UP**									
0020	After job completion, allow, minimum				Job				.30%	.30%
0040	Maximum				″				1%	1%
0050	Cleanup of floor area, continuous, per day, during const.	A-5	24	.750	M.S.F.	2.34	31	2.03	35.37	52
0100	Final by GC at end of job	″	11.50	1.565	″	2.48	65	4.24	71.72	106

01 91 Commissioning

01 91 13 – General Commissioning Requirements

01 91 13.50 Building Commissioning		Crew	Daily Output	Labor-Hours	Unit	Material	2019 Bare Costs Labor	Equipment	Total	Total Incl O&P
0010	**BUILDING COMMISSIONING**									
0100	Systems operation and verification during turnover				%				.25%	.25%
0150	Including all systems subcontractors								.50%	.50%
0200	Systems design assistance, operation, verification and training								.50%	.50%
0250	Including all systems subcontractors								1%	1%

01 93 Facility Maintenance

01 93 13 – Facility Maintenance Procedures

01 93 13.15 Mechanical Facilities Maintenance		Crew	Daily Output	Labor-Hours	Unit	Material	2019 Bare Costs Labor	Equipment	Total	Total Incl O&P
0010	**MECHANICAL FACILITIES MAINTENANCE**									
0100	Air conditioning system maintenance									
0130	Belt, replace	1 Stpi	15	.533	Ea.		34		34	51
0170	Fan, clean		16	.500			32		32	48
0180	Filter, remove, clean, replace		12	.667			42.50		42.50	64
0190	Flexible coupling alignment, inspect		40	.200			12.80		12.80	19.20
0200	Gas leak, locate and repair		4	2			128		128	192
0250	Pump packing gland, remove and replace		11	.727			46.50		46.50	70
0270	Tighten		32	.250			16		16	24
0290	Pump, disassemble and assemble		4	2			128		128	192
0300	Air pressure regulator, disassemble, clean, assemble	1 Skwk	4	2			107		107	163
0310	Repair or replace part	"	6	1.333			71		71	108
0320	Purging system	1 Stpi	16	.500			32		32	48
0400	Compressor, air, remove or install fan wheel	1 Skwk	20	.400			21.50		21.50	32.50
0410	Disassemble or assemble 2 cylinder, 2 stage		4	2			107		107	163
0420	4 cylinder, 4 stage		1	8			425		425	650
0430	Repair or replace part		2	4			214		214	325
0700	Demolition, for mech. demolition see Section 23 05 05.10 or 22 05 05.10									
0800	Ductwork, clean									
0810	Rectangular									
0820	6″ G	1 Shee	187.50	.043	L.F.		2.60		2.60	3.95
0830	8″ G		140.63	.057			3.47		3.47	5.25
0840	10″ G		112.50	.071			4.33		4.33	6.60
0850	12″ G		93.75	.085			5.20		5.20	7.90
0860	14″ G		80.36	.100			6.05		6.05	9.20
0870	16″ G		70.31	.114			6.95		6.95	10.55
0900	Round									
0910	4″ G	1 Shee	358.10	.022	L.F.		1.36		1.36	2.07
0920	6″ G		238.73	.034			2.04		2.04	3.10
0930	8″ G		179.05	.045			2.72		2.72	4.14
0940	10″ G		143.24	.056			3.40		3.40	5.15
0950	12″ G		119.37	.067			4.08		4.08	6.20
0960	16″ G		89.52	.089			5.45		5.45	8.25
1000	Expansion joint, not screwed, install or remove	1 Stpi	3	2.667	Ea.		171		171	256
1010	Repack	"	6	1.333	"		85.50		85.50	128
1200	Fire protection equipment									
1220	Fire hydrant, replace	Q-1	3	5.333	Ea.		305		305	455
1230	Service, lubricate, inspect, flush, clean	1 Plum	7	1.143			72		72	108
1240	Test	"	11	.727			46		46	69
1310	Inspect valves, pressure, nozzle	1 Spri	4	2	System		123		123	185
1800	Plumbing fixtures, for installation see Section 22 41 00									
1801	Plumbing fixtures									

01 93 13.15 Mechanical Facilities Maintenance		Crew	Daily Output	Labor-Hours	Unit	Material	2019 Bare Costs Labor	Equipment	Total	Total Incl O&P
1850	Open drain with toilet auger	1 Plum	16	.500	Ea.		31.50		31.50	47.50
1870	Plaster trap, clean	"	6	1.333			84		84	126
1881	Clean commode	1 Clab	20	.400		.14	16.40		16.54	25
1882	Clean commode seat		48	.167		.07	6.85		6.92	10.40
1884	Clean double sink		20	.400		.14	16.40		16.54	25
1886	Clean bathtub		12	.667		.27	27.50		27.77	42
1888	Clean fiberglass tub/shower		8	1		.34	41		41.34	62.50
1889	Clean faucet set		32	.250		.05	10.25		10.30	15.60
1891	Clean shower head		80	.100		.01	4.11		4.12	6.20
1893	Clean water heater		16	.500		.48	20.50		20.98	31.50
1900	Relief valve, test and adjust	1 Stpi	20	.400			25.50		25.50	38.50
1910	Clean pump, heater, or motor for whirlpool	1 Clab	16	.500		.07	20.50		20.57	31
1920	Clean thermal cover for whirlpool	"	16	.500		.68	20.50		21.18	32
2000	Repair or replace, steam trap	1 Stpi	8	1			64		64	96
2020	Y-type or bell strainer		6	1.333			85.50		85.50	128
2040	Water trap or vacuum breaker, screwed joints		13	.615			39.50		39.50	59
2100	Steam specialties, clean									
2120	Air separator with automatic trap, 1" fittings	1 Stpi	12	.667	Ea.		42.50		42.50	64
2130	Bucket trap, 2" pipe		7	1.143			73		73	110
2150	Drip leg, 2" fitting		45	.178			11.35		11.35	17.05
2200	Thermodynamic trap, 1" fittings		50	.160			10.25		10.25	15.35
2210	Thermostatic		65	.123			7.85		7.85	11.80
2240	Screen and seat in Y-type strainer, plug type		25	.320			20.50		20.50	30.50
2242	Screen and seat in Y-type strainer, flange type		12	.667			42.50		42.50	64
2500	Valve, replace broken handwheel		24	.333			21.50		21.50	32
3000	Valve, overhaul, regulator, relief, flushometer, mixing									
3040	Cold water, gas	1 Stpi	5	1.600	Ea.		102		102	154
3050	Hot water, steam		3	2.667			171		171	256
3080	Globe, gate, check up to 4" cold water, gas		10	.800			51		51	77
3090	Hot water, steam		5	1.600			102		102	154
3100	Over 4" ID hot or cold line		1.40	5.714			365		365	550
3120	Remove and replace, gate, globe or check up to 4"	Q-5	6	2.667			153		153	230
3130	Over 4"	"	2	8			460		460	690
3150	Repack up to 4"	1 Stpi	13	.615			39.50		39.50	59
3160	Over 4"	"	4	2			128		128	192

Division Notes

		CREW	DAILY OUTPUT	LABOR-HOURS	UNIT	BARE COSTS MAT.	BARE COSTS LABOR	BARE COSTS EQUIP.	BARE COSTS TOTAL	TOTAL INCL O&P

Estimating Tips

02 30 00 Subsurface Investigation

In preparing estimates on structures involving earthwork or foundations, all information concerning soil characteristics should be obtained. Look particularly for hazardous waste, evidence of prior dumping of debris, and previous stream beds.

02 40 00 Demolition and Structure Moving

The costs shown for selective demolition do not include rubbish handling or disposal. These items should be estimated separately using RSMeans data or other sources.

- Historic preservation often requires that the contractor remove materials from the existing structure, rehab them, and replace them. The estimator must be aware of any related measures and precautions that must be taken when doing selective demolition and cutting and patching. Requirements may include special handling and storage, as well as security.
- In addition to Subdivision 02 41 00, you can find selective demolition items in each division. Example: Roofing demolition is in Division 7.
- Absent of any other specific reference, an approximate demolish-in-place cost can be obtained by halving the new-install labor cost. To remove for reuse, allow the entire new-install labor figure.

02 40 00 Building Deconstruction

This section provides costs for the careful dismantling and recycling of most low-rise building materials.

02 50 00 Containment of Hazardous Waste

This section addresses on-site hazardous waste disposal costs.

02 80 00 Hazardous Material Disposal/Remediation

This subdivision includes information on hazardous waste handling, asbestos remediation, lead remediation, and mold remediation. See reference numbers R028213-20 and R028319-60 for further guidance in using these unit price lines.

02 90 00 Monitoring Chemical Sampling, Testing Analysis

This section provides costs for on-site sampling and testing hazardous waste.

Reference Numbers

Reference numbers are shown at the beginning of some major classifications. These numbers refer to related items in the Reference Section. The reference information may be an estimating procedure, an alternate pricing method, or technical information.

Note: Not all subdivisions listed here necessarily appear. ■

02 41 Demolition

02 41 13 – Selective Site Demolition

02 41 13.17 Demolish, Remove Pavement and Curb		Crew	Daily Output	Labor-Hours	Unit	Material	2019 Bare Costs Labor	Equipment	Total	Total Incl O&P
0010	DEMOLISH, REMOVE PAVEMENT AND CURB R024119-10									
5010	Pavement removal, bituminous roads, up to 3" thick	B-38	690	.058	S.Y.		2.69	1.62	4.31	5.85
5050	4"-6" thick		420	.095			4.42	2.67	7.09	9.60
5100	Bituminous driveways		640	.063			2.90	1.75	4.65	6.30
5200	Concrete to 6" thick, hydraulic hammer, mesh reinforced		255	.157			7.30	4.39	11.69	15.80
5300	Rod reinforced		200	.200			9.30	5.60	14.90	20
5400	Concrete, 7"-24" thick, plain		33	1.212	C.Y.		56	34	90	122
5500	Reinforced		24	1.667	"		77.50	46.50	124	169
5600	With hand held air equipment, bituminous, to 6" thick	B-39	1900	.025	S.F.		1.09	.12	1.21	1.79
5700	Concrete to 6" thick, no reinforcing		1600	.030			1.29	.15	1.44	2.12
5800	Mesh reinforced		1400	.034			1.48	.17	1.65	2.42
5900	Rod reinforced		765	.063			2.71	.31	3.02	4.43

02 41 13.23 Utility Line Removal		Crew	Daily Output	Labor-Hours	Unit	Material	Labor	Equipment	Total	Total Incl O&P
0010	UTILITY LINE REMOVAL									
0015	No hauling, abandon catch basin or manhole	B-6	7	3.429	Ea.		153	45.50	198.50	282
0020	Remove existing catch basin or manhole, masonry		4	6			267	80	347	495
0030	Catch basin or manhole frames and covers, stored		13	1.846			82.50	24.50	107	151
0040	Remove and reset		7	3.429			153	45.50	198.50	282
0900	Hydrants, fire, remove only	B-21A	5	8			410	74.50	484.50	695
0950	Remove and reset	"	2	20			1,025	186	1,211	1,725

02 41 13.30 Minor Site Demolition		Crew	Daily Output	Labor-Hours	Unit	Material	Labor	Equipment	Total	Total Incl O&P
0010	MINOR SITE DEMOLITION R024119-10									
4000	Sidewalk removal, bituminous, 2" thick	B-6	350	.069	S.Y.		3.06	.91	3.97	5.60
4010	2-1/2" thick		325	.074			3.29	.99	4.28	6.05
4050	Brick, set in mortar		185	.130			5.80	1.73	7.53	10.60
4100	Concrete, plain, 4"		160	.150			6.70	2	8.70	12.30
4110	Plain, 5"		140	.171			7.65	2.29	9.94	14.05
4120	Plain, 6"		120	.200			8.90	2.67	11.57	16.40
4200	Mesh reinforced, concrete, 4"		150	.160			7.15	2.13	9.28	13.10
4210	5" thick		131	.183			8.15	2.44	10.59	15
4220	6" thick		112	.214			9.55	2.86	12.41	17.55

02 41 19 – Selective Demolition

02 41 19.13 Selective Building Demolition		Crew	Daily Output	Labor-Hours	Unit	Material	Labor	Equipment	Total	Total Incl O&P
0010	SELECTIVE BUILDING DEMOLITION									
0020	Costs related to selective demolition of specific building components									
0025	are included under Common Work Results (XX 05)									
0030	in the component's appropriate division.									

02 41 19.19 Selective Demolition		Crew	Daily Output	Labor-Hours	Unit	Material	Labor	Equipment	Total	Total Incl O&P
0010	**SELECTIVE DEMOLITION**, Rubbish Handling R024119-10									
0020	The following are to be added to the demolition prices									
0600	Dumpster, weekly rental, 1 dump/week, 6 C.Y. capacity (2 tons)				Week	415			415	455
0700	10 C.Y. capacity (3 tons)					480			480	530
0725	20 C.Y. capacity (5 tons) R024119-20					565			565	625
0800	30 C.Y. capacity (7 tons)					730			730	800
0840	40 C.Y. capacity (10 tons)					775			775	850
2000	Load, haul, dump and return, 0'-50' haul, hand carried	2 Clab	24	.667	C.Y.		27.50		27.50	41.50
2005	Wheeled		37	.432			17.75		17.75	27
2040	0'-100' haul, hand carried		16.50	.970			40		40	60
2045	Wheeled		25	.640			26.50		26.50	39.50
2050	Forklift	A-3R	25	.320			16.55	5.75	22.30	31.50
2080	Haul and return, add per each extra 100' haul, hand carried	2 Clab	35.50	.451			18.50		18.50	28

02 41 19 – Selective Demolition

02 41 19.19 Selective Demolition		Crew	Daily Output	Labor-Hours	Unit	2019 Bare Costs Material	Labor	Equipment	Total	Total Incl O&P
2085	Wheeled	2 Clab	54	.296	C.Y.		12.15		12.15	18.40
2120	For travel in elevators, up to 10 floors, add		140	.114			4.69		4.69	7.10
2130	0′-50′ haul, incl. up to 5 riser stairs, hand carried		23	.696			28.50		28.50	43
2135	Wheeled		35	.457			18.75		18.75	28.50
2140	6-10 riser stairs, hand carried		22	.727			30		30	45
2145	Wheeled		34	.471			19.30		19.30	29
2150	11-20 riser stairs, hand carried		20	.800			33		33	49.50
2155	Wheeled		31	.516			21		21	32
2160	21-40 riser stairs, hand carried		16	1			41		41	62
2165	Wheeled		24	.667			27.50		27.50	41.50
2170	0′-100′ haul, incl. 5 riser stairs, hand carried		15	1.067			44		44	66
2175	Wheeled		23	.696			28.50		28.50	43
2180	6-10 riser stairs, hand carried		14	1.143			47		47	71
2185	Wheeled		21	.762			31.50		31.50	47.50
2190	11-20 riser stairs, hand carried		12	1.333			54.50		54.50	83
2195	Wheeled		18	.889			36.50		36.50	55
2200	21-40 riser stairs, hand carried		8	2			82		82	124
2205	Wheeled		12	1.333			54.50		54.50	83
2210	Haul and return, add per each extra 100′ haul, hand carried		35.50	.451			18.50		18.50	28
2215	Wheeled		54	.296			12.15		12.15	18.40
2220	For each additional flight of stairs, up to 5 risers, add		550	.029	Flight		1.19		1.19	1.81
2225	6-10 risers, add		275	.058			2.39		2.39	3.61
2230	11-20 risers, add		138	.116			4.76		4.76	7.20
2235	21-40 risers, add		69	.232			9.50		9.50	14.40
3000	Loading & trucking, including 2 mile haul, chute loaded	B-16	45	.711	C.Y.		30.50	12.60	43.10	60
3040	Hand loading truck, 50′ haul	″	48	.667			28.50	11.80	40.30	56.50
3080	Machine loading truck	B-17	120	.267			12.05	5.50	17.55	24
5000	Haul, per mile, up to 8 C.Y. truck	B-34B	1165	.007			.32	.49	.81	1.03
5100	Over 8 C.Y. truck	″	1550	.005			.24	.37	.61	.77

02 41 19.20 Selective Demolition, Dump Charges		Crew	Daily Output	Labor-Hours	Unit	Material	Labor	Equipment	Total	Total Incl O&P
0010	**SELECTIVE DEMOLITION, DUMP CHARGES** R024119-10									
0020	Dump charges, typical urban city, tipping fees only									
0100	Building construction materials				Ton	74			74	81
0200	Trees, brush, lumber					63			63	69.50
0300	Rubbish only					63			63	69.50
0500	Reclamation station, usual charge					74			74	81

02 41 19.27 Selective Demolition, Torch Cutting		Crew	Daily Output	Labor-Hours	Unit	Material	Labor	Equipment	Total	Total Incl O&P
0010	**SELECTIVE DEMOLITION, TORCH CUTTING** R024119-10									
0020	Steel, 1″ thick plate	E-25	333	.024	L.F.	.88	1.39	.04	2.31	3.23
0040	1″ diameter bar	″	600	.013	Ea.	.15	.77	.02	.94	1.41
1000	Oxygen lance cutting, reinforced concrete walls									
1040	12″-16″ thick walls	1 Clab	10	.800	L.F.		33		33	49.50
1080	24″ thick walls	″	6	1.333	″		54.50		54.50	83

02 65 Underground Storage Tank Removal

02 65 10 – Underground Tank and Contaminated Soil Removal

02 65 10.30 Removal of Underground Storage Tanks			Crew	Daily Output	Labor-Hours	Unit	2019 Bare Costs Material	Labor	Equipment	Total	Total Incl O&P
0010	REMOVAL OF UNDERGROUND STORAGE TANKS R026510-20										
0011	Petroleum storage tanks, non-leaking										
0100	Excavate & load onto trailer										
0110	3,000 gal. to 5,000 gal. tank	G	B-14	4	12	Ea.		520	80	600	870
0120	6,000 gal. to 8,000 gal. tank	G	B-3A	3	13.333			585	305	890	1,225
0130	9,000 gal. to 12,000 gal. tank	G	"	2	20			875	455	1,330	1,825
0190	Known leaking tank, add					%				100%	100%
0200	Remove sludge, water and remaining product from bottom										
0201	of tank with vacuum truck										
0300	3,000 gal. to 5,000 gal. tank	G	A-13	5	1.600	Ea.		82.50	145	227.50	283
0310	6,000 gal. to 8,000 gal. tank	G		4	2			103	181	284	355
0320	9,000 gal. to 12,000 gal. tank	G		3	2.667			138	241	379	470
0390	Dispose of sludge off-site, average					Gal.				6.25	6.80
0400	Insert inert solid CO_2 "dry ice" into tank										
0401	For cleaning/transporting tanks (1.5 lb./100 gal. cap)	G	1 Clab	500	.016	Lb.	1.29	.66		1.95	2.41
1020	Haul tank to certified salvage dump, 100 miles round trip										
1023	3,000 gal. to 5,000 gal. tank					Ea.				760	830
1026	6,000 gal. to 8,000 gal. tank									880	960
1029	9,000 gal. to 12,000 gal. tank									1,050	1,150
1100	Disposal of contaminated soil to landfill										
1110	Minimum					C.Y.				145	160
1111	Maximum					"				400	440
1120	Disposal of contaminated soil to										
1121	bituminous concrete batch plant										
1130	Minimum					C.Y.				80	88
1131	Maximum					"				115	125
2010	Decontamination of soil on site incl poly tarp on top/bottom										
2011	Soil containment berm and chemical treatment										
2020	Minimum	G	B-11C	100	.160	C.Y.	8.05	7.70	3.20	18.95	24
2021	Maximum	G	"	100	.160		10.45	7.70	3.20	21.35	26.50
2050	Disposal of decontaminated soil, minimum									135	150
2055	Maximum									400	440

02 81 Transportation and Disposal of Hazardous Materials

02 81 20 – Hazardous Waste Handling

02 81 20.10 Hazardous Waste Cleanup/Pickup/Disposal		Crew	Daily Output	Labor-Hours	Unit	Material	Labor	Equipment	Total	Total Incl O&P
0010	HAZARDOUS WASTE CLEANUP/PICKUP/DISPOSAL									
0100	For contractor rental equipment, i.e., dozer,									
0110	Front end loader, dump truck, etc., see 01 54 33 Reference Section									
1000	Solid pickup									
1100	55 gal. drums				Ea.				240	265
1120	Bulk material, minimum				Ton				190	210
1130	Maximum				"				595	655
1200	Transportation to disposal site									
1220	Truckload = 80 drums or 25 C.Y. or 18 tons									
1260	Minimum				Mile				3.95	4.45
1270	Maximum				"				7.25	7.98
3000	Liquid pickup, vacuum truck, stainless steel tank									
3100	Minimum charge, 4 hours									
3110	1 compartment, 2200 gallon				Hr.				140	155
3120	2 compartment, 5000 gallon				"				200	225

02 81 Transportation and Disposal of Hazardous Materials

02 81 20 – Hazardous Waste Handling

02 81 20.10 Hazardous Waste Cleanup/Pickup/Disposal		Crew	Daily Output	Labor-Hours	Unit	2019 Bare Costs Material	Labor	Equipment	Total	Total Incl O&P
3400	Transportation in 6900 gallon bulk truck				Mile				7.95	8.75
3410	In teflon lined truck				"				10.20	11.25
5000	Heavy sludge or dry vacuumable material				Hr.				140	155
6000	Dumpsite disposal charge, minimum				Ton				140	155
6020	Maximum				"				415	455

02 82 Asbestos Remediation

02 82 13 – Asbestos Abatement

02 82 13.39 Asbestos Remediation Plans and Methods		Crew	Daily Output	Labor-Hours	Unit	Material	Labor	Equipment	Total	Total Incl O&P
0010	**ASBESTOS REMEDIATION PLANS AND METHODS**									
0100	Building Survey-Commercial Building				Ea.				2,200	2,400
0200	Asbestos Abatement Remediation Plan				"				1,350	1,475

02 82 13.41 Asbestos Abatement Equipment		Crew	Daily Output	Labor-Hours	Unit	Material	Labor	Equipment	Total	Total Incl O&P
0010	**ASBESTOS ABATEMENT EQUIPMENT** R028213-20									
0011	Equipment and supplies, buy									
0200	Air filtration device, 2000 CFM				Ea.	800			800	880
0250	Large volume air sampling pump, minimum					325			325	355
0260	Maximum					345			345	380
0300	Airless sprayer unit, 2 gun					2,500			2,500	2,750
0350	Light stand, 500 watt					38.50			38.50	42.50
0400	Personal respirators									
0410	Negative pressure, 1/2 face, dual operation, minimum				Ea.	27			27	29.50
0420	Maximum					30.50			30.50	33.50
0450	P.A.P.R., full face, minimum					680			680	745
0460	Maximum					1,425			1,425	1,575
0470	Supplied air, full face, including air line, minimum					375			375	410
0480	Maximum					530			530	580
0500	Personnel sampling pump					243			243	268
1500	Power panel, 20 unit, including GFI					445			445	490
1600	Shower unit, including pump and filters					1,075			1,075	1,200
1700	Supplied air system (type C)					3,500			3,500	3,825
1750	Vacuum cleaner, HEPA, 16 gal., stainless steel, wet/dry					1,125			1,125	1,250
1760	55 gallon					1,425			1,425	1,575
1800	Vacuum loader, 9-18 ton/hr.					98,500			98,500	108,000
1900	Water atomizer unit, including 55 gal. drum					290			290	320
2000	Worker protection, whole body, foot, head cover & gloves, plastic					8.25			8.25	9.10
2500	Respirator, single use					29.50			29.50	32.50
2550	Cartridge for respirator					4.30			4.30	4.73
2570	Glove bag, 7 mil, 50" x 64"					8			8	8.80
2580	10 mil, 44" x 60"					8.90			8.90	9.80
2590	6 mil, 44" x 60"					5.45			5.45	6
6000	Disposable polyethylene bags, 6 mil, 3 C.F.					.90			.90	.99
6300	Disposable fiber drums, 3 C.F.					18.50			18.50	20.50
6400	Pressure sensitive caution labels, 3" x 5"					2.98			2.98	3.28
6450	11" x 17"					7.70			7.70	8.45
6500	Negative air machine, 1800 CFM					910			910	1,000

02 82 13.42 Preparation of Asbestos Containment Area		Crew	Daily Output	Labor-Hours	Unit	Material	Labor	Equipment	Total	Total Incl O&P
0010	**PREPARATION OF ASBESTOS CONTAINMENT AREA** R028213-20									
0100	Pre-cleaning, HEPA vacuum and wet wipe, flat surfaces	A-9	12000	.005	S.F.	.01	.31		.32	.49
0200	Protect carpeted area, 2 layers 6 mil poly on 3/4" plywood	"	1000	.064		2.06	3.67		5.73	7.90

02 82 Asbestos Remediation

02 82 13 – Asbestos Abatement

02 82 13.42 Preparation of Asbestos Containment Area		Crew	Daily Output	Labor-Hours	Unit	Material	Labor	Equipment	Total	Total Incl O&P
						2019 Bare Costs				
0300	Separation barrier, 2" x 4" @ 16", 1/2" plywood ea. side, 8' high	2 Carp	400	.040	S.F.	2.87	2.07		4.94	6.30
0310	12' high		320	.050		2.82	2.58		5.40	7
0320	16' high		200	.080		2.79	4.13		6.92	9.30
0400	Personnel decontam. chamber, 2" x 4" @ 16", 3/4" ply ea. side		280	.057		3.43	2.95		6.38	8.25
0450	Waste decontam. chamber, 2" x 4" studs @ 16", 3/4" ply ea. side		360	.044		3.43	2.30		5.73	7.25
0500	Cover surfaces with polyethylene sheeting									
0501	Including glue and tape									
0550	Floors, each layer, 6 mil	A-9	8000	.008	S.F.	.04	.46		.50	.75
0551	4 mil		9000	.007		.03	.41		.44	.66
0560	Walls, each layer, 6 mil		6000	.011		.04	.61		.65	.99
0561	4 mil		7000	.009		.03	.52		.55	.84
0570	For heights above 14', add						20%			
0575	For heights above 20', add						30%			
0580	For fire retardant poly, add					100%				
0590	For large open areas, deduct					10%	20%			
0600	Seal floor penetrations with foam firestop to 36 sq. in.	2 Carp	200	.080	Ea.	13.20	4.13		17.33	21
0610	36 sq. in. to 72 sq. in.		125	.128		26.50	6.60		33.10	39
0615	72 sq. in. to 144 sq. in.		80	.200		53	10.35		63.35	73.50
0620	Wall penetrations, to 36 sq. in.		180	.089		13.20	4.59		17.79	21.50
0630	36 sq. in. to 72 sq. in.		100	.160		26.50	8.25		34.75	41.50
0640	72 sq. in. to 144 sq. in.		60	.267		53	13.75		66.75	79
0800	Caulk seams with latex	1 Carp	230	.035	L.F.	.17	1.80		1.97	2.91
0900	Set up neg. air machine, 1-2k CFM/25 M.C.F. volume	1 Asbe	4.30	1.860	Ea.		107		107	165
0950	Set up and remove portable shower unit	2 Asbe	4	4	"		229		229	355

02 82 13.43 Bulk Asbestos Removal		Crew	Daily Output	Labor-Hours	Unit	Material	Labor	Equipment	Total	Total Incl O&P
0010	**BULK ASBESTOS REMOVAL**									
0020	Includes disposable tools and 2 suits and 1 respirator filter/day/worker									
0200	Boiler insulation	A-9	480	.133	S.F.	.42	7.65		8.07	12.25
0210	With metal lath, add				%				50%	50%
0300	Boiler breeching or flue insulation	A-9	520	.123	S.F.	.32	7.05		7.37	11.25
0310	For active boiler, add				%				100%	100%
0400	Duct or AHU insulation	A-10B	440	.073	S.F.	.19	4.18		4.37	6.65
0500	Duct vibration isolation joints, up to 24 sq. in. duct	A-9	56	1.143	Ea.	2.98	65.50		68.48	104
0520	25 sq. in. to 48 sq. in. duct		48	1.333		3.47	76.50		79.97	122
0530	49 sq. in. to 76 sq. in. duct		40	1.600		4.17	92		96.17	147
0600	Pipe insulation, air cell type, up to 4" diameter pipe		900	.071	L.F.	.19	4.08		4.27	6.50
0610	4" to 8" diameter pipe		800	.080		.21	4.59		4.80	7.35
0620	10" to 12" diameter pipe		700	.091		.24	5.25		5.49	8.35
0630	14" to 16" diameter pipe		550	.116		.30	6.70		7	10.65
0650	Over 16" diameter pipe		650	.098	S.F.	.26	5.65		5.91	9.05
0700	With glove bag up to 3" diameter pipe		200	.320	L.F.	9.50	18.35		27.85	39
1000	Pipe fitting insulation up to 4" diameter pipe		320	.200	Ea.	.52	11.50		12.02	18.30
1100	6" to 8" diameter pipe		304	.211		.55	12.10		12.65	19.25
1110	10" to 12" diameter pipe		192	.333		.87	19.15		20.02	30.50
1120	14" to 16" diameter pipe		128	.500		1.30	28.50		29.80	46
1130	Over 16" diameter pipe		176	.364	S.F.	.95	21		21.95	33
1200	With glove bag, up to 8" diameter pipe		75	.853	L.F.	6.50	49		55.50	82.50
2000	Scrape foam fireproofing from flat surface		2400	.027	S.F.	.07	1.53		1.60	2.44
2100	Irregular surfaces		1200	.053		.14	3.06		3.20	4.88
3000	Remove cementitious material from flat surface		1800	.036		.09	2.04		2.13	3.25
3100	Irregular surface		1000	.064		.12	3.67		3.79	5.80
6000	Remove contaminated soil from crawl space by hand		400	.160	C.F.	.42	9.20		9.62	14.65

02 82 13 – Asbestos Abatement

	02 82 13.43 Bulk Asbestos Removal	Crew	Daily Output	Labor-Hours	Unit	Material	2019 Bare Costs Labor	Equipment	Total	Total Incl O&P
6100	With large production vacuum loader	A-12	700	.091	C.F.	.24	5.25	1.03	6.52	9.50
7000	Radiator backing, not including radiator removal	A-9	1200	.053	S.F.	.14	3.06		3.20	4.88
9000	For type B (supplied air) respirator equipment, add				%				10%	10%

02 82 13.44 Demolition In Asbestos Contaminated Area

		Crew	Daily Output	Labor-Hours	Unit	Material	Labor	Equipment	Total	Total Incl O&P
0010	**DEMOLITION IN ASBESTOS CONTAMINATED AREA**									
0200	Ceiling, including suspension system, plaster and lath	A-9	2100	.030	S.F.	.08	1.75		1.83	2.79
0210	Finished plaster, leaving wire lath		585	.109		.28	6.30		6.58	10
0220	Suspended acoustical tile		3500	.018		.05	1.05		1.10	1.67
0230	Concealed tile grid system		3000	.021		.06	1.22		1.28	1.95
0240	Metal pan grid system		1500	.043		.11	2.45		2.56	3.90
0250	Gypsum board		2500	.026		.07	1.47		1.54	2.34
0260	Lighting fixtures up to 2′ x 4′		72	.889	Ea.	2.32	51		53.32	81.50
0400	Partitions, non load bearing									
0410	Plaster, lath, and studs	A-9	690	.093	S.F.	.88	5.30		6.18	9.15
0450	Gypsum board and studs	"	1390	.046	"	.12	2.64		2.76	4.21
9000	For type B (supplied air) respirator equipment, add				%				10%	10%

02 82 13.45 OSHA Testing

		Crew	Daily Output	Labor-Hours	Unit	Material	Labor	Equipment	Total	Total Incl O&P
0010	**OSHA TESTING**									
0100	Certified technician, minimum				Day				200	220
0110	Maximum								300	330
0121	Industrial hygienist, minimum	1 Asbe	1.75	4.571			262		262	405
0130	Maximum								400	440
0200	Asbestos sampling and PCM analysis, NIOSH 7400, minimum	1 Asbe	8	1	Ea.	16.20	57.50		73.70	106
0210	Maximum		4	2		28	115		143	208
1000	Cleaned area samples		8	1		203	57.50		260.50	310
1100	PCM air sample analysis, NIOSH 7400, minimum		8	1		15.80	57.50		73.30	106
1110	Maximum		4	2		2.34	115		117.34	180
1200	TEM air sample analysis, NIOSH 7402, minimum								80	106
1210	Maximum								360	450

02 82 13.46 Decontamination of Asbestos Containment Area

		Crew	Daily Output	Labor-Hours	Unit	Material	Labor	Equipment	Total	Total Incl O&P
0010	**DECONTAMINATION OF ASBESTOS CONTAINMENT AREA**									
0100	Spray exposed substrate with surfactant (bridging)									
0200	Flat surfaces	A-9	6000	.011	S.F.	.41	.61		1.02	1.40
0250	Irregular surfaces		4000	.016	"	.46	.92		1.38	1.93
0300	Pipes, beams, and columns		2000	.032	L.F.	.60	1.84		2.44	3.50
1000	Spray encapsulate polyethylene sheeting		8000	.008	S.F.	.41	.46		.87	1.16
1100	Roll down polyethylene sheeting		8000	.008	"		.46		.46	.71
1500	Bag polyethylene sheeting		400	.160	Ea.	1.01	9.20		10.21	15.30
2000	Fine clean exposed substrate, with nylon brush		2400	.027	S.F.		1.53		1.53	2.36
2500	Wet wipe substrate		4800	.013			.77		.77	1.18
2600	Vacuum surfaces, fine brush		6400	.010			.57		.57	.89
3000	Structural demolition									
3100	Wood stud walls	A-9	2800	.023	S.F.		1.31		1.31	2.03
3500	Window manifolds, not incl. window replacement		4200	.015			.87		.87	1.35
3600	Plywood carpet protection		2000	.032			1.84		1.84	2.84
4000	Remove custom decontamination facility	A-10A	8	3	Ea.	15.15	173		188.15	283
4100	Remove portable decontamination facility	3 Asbe	12	2	"	14.30	115		129.30	193
5000	HEPA vacuum, shampoo carpeting	A-9	4800	.013	S.F.	.11	.77		.88	1.30
9000	Final cleaning of protected surfaces	A-10A	8000	.003	"		.17		.17	.27

02 82 Asbestos Remediation

02 82 13 – Asbestos Abatement

02 82 13.47 Asbestos Waste Pkg., Handling, and Disp.		Crew	Daily Output	Labor-Hours	Unit	Material	Labor	Equipment	Total	Total Incl O&P
						2019 Bare Costs				
0010	**ASBESTOS WASTE PACKAGING, HANDLING, AND DISPOSAL**									
0100	Collect and bag bulk material, 3 C.F. bags, by hand	A-9	400	.160	Ea.	.90	9.20		10.10	15.20
0200	Large production vacuum loader	A-12	880	.073		1.01	4.18	.82	6.01	8.45
1000	Double bag and decontaminate	A-9	960	.067		.90	3.83		4.73	6.90
2000	Containerize bagged material in drums, per 3 C.F. drum	"	800	.080		18.50	4.59		23.09	27.50
3000	Cart bags 50′ to dumpster	2 Asbe	400	.040			2.29		2.29	3.54
5000	Disposal charges, not including haul, minimum				C.Y.				61	67
5020	Maximum				"				355	395
9000	For type B (supplied air) respirator equipment, add				%				10%	10%

02 82 13.48 Asbestos Encapsulation With Sealants		Crew	Daily Output	Labor-Hours	Unit	Material	Labor	Equipment	Total	Total Incl O&P
0010	**ASBESTOS ENCAPSULATION WITH SEALANTS**									
0100	Ceilings and walls, minimum	A-9	21000	.003	S.F.	.42	.18		.60	.73
0110	Maximum		10600	.006	"	.52	.35		.87	1.11
0300	Pipes to 12″ diameter including minor repairs, minimum		800	.080	L.F.	.51	4.59		5.10	7.65
0310	Maximum		400	.160	"	1.21	9.20		10.41	15.55

02 87 Biohazard Remediation

02 87 13 – Mold Remediation

02 87 13.16 Mold Remediation Preparation and Containment		Crew	Daily Output	Labor-Hours	Unit	Material	Labor	Equipment	Total	Total Incl O&P
0010	**MOLD REMEDIATION PREPARATION AND CONTAINMENT**									
6010	Preparation of mold containment area R028213-20									
6100	Pre-cleaning, HEPA vacuum and wet wipe, flat surfaces	A-9	12000	.005	S.F.	.01	.31		.32	.49
6300	Separation barrier, 2″ x 4″ @ 16″, 1/2″ plywood ea. side, 8′ high	2 Carp	400	.040		3.47	2.07		5.54	6.95
6310	12′ high		320	.050		3.47	2.58		6.05	7.75
6320	16′ high		200	.080		2.38	4.13		6.51	8.85
6400	Personnel decontam. chamber, 2″ x 4″ @ 16″, 3/4″ ply ea. side		280	.057		4.50	2.95		7.45	9.40
6450	Waste decontam. chamber, 2″ x 4″ studs @ 16″, 3/4″ ply each side		360	.044		4.51	2.30		6.81	8.45
6500	Cover surfaces with polyethylene sheeting									
6501	Including glue and tape									
6550	Floors, each layer, 6 mil	A-9	8000	.008	S.F.	.04	.46		.50	.75
6551	4 mil		9000	.007		.03	.41		.44	.66
6560	Walls, each layer, 6 mil		6000	.011		.04	.61		.65	.99
6561	4 mil		7000	.009		.03	.52		.55	.84
6570	For heights above 14′, add						20%			
6575	For heights above 20′, add						30%			
6580	For fire retardant poly, add					100%				
6590	For large open areas, deduct					10%	20%			
6600	Seal floor penetrations with foam firestop to 36 sq. in.	2 Carp	200	.080	Ea.	13.20	4.13		17.33	21
6610	36 sq. in. to 72 sq. in.		125	.128		26.50	6.60		33.10	39
6615	72 sq. in. to 144 sq. in.		80	.200		53	10.35		63.35	73.50
6620	Wall penetrations, to 36 sq. in.		180	.089		13.20	4.59		17.79	21.50
6630	36 sq. in. to 72 sq. in.		100	.160		26.50	8.25		34.75	41.50
6640	72 sq. in. to 144 sq. in.		60	.267		53	13.75		66.75	79
6800	Caulk seams with latex caulk	1 Carp	230	.035	L.F.	.17	1.80		1.97	2.91
6900	Set up neg. air machine, 1-2k CFM/25 M.C.F. volume	1 Asbe	4.30	1.860	Ea.		107		107	165

02 87 13.33 Removal and Disposal of Materials With Mold		Crew	Daily Output	Labor-Hours	Unit	Material	Labor	Equipment	Total	Total Incl O&P
0010	**REMOVAL AND DISPOSAL OF MATERIALS WITH MOLD**									
0015	Demolition in mold contaminated area									
0200	Ceiling, including suspension system, plaster and lath	A-9	2100	.030	S.F.	.08	1.75		1.83	2.79
0210	Finished plaster, leaving wire lath		585	.109		.28	6.30		6.58	10

02 87 13 – Mold Remediation

02 87 13.33 Removal and Disposal of Materials With Mold		Crew	Daily Output	Labor-Hours	Unit	Material	2019 Bare Costs Labor	Equipment	Total	Total Incl O&P
0220	Suspended acoustical tile	A-9	3500	.018	S.F.	.05	1.05		1.10	1.67
0230	Concealed tile grid system		3000	.021		.06	1.22		1.28	1.95
0240	Metal pan grid system		1500	.043		.11	2.45		2.56	3.90
0250	Gypsum board		2500	.026		.07	1.47		1.54	2.34
0255	Plywood		2500	.026		.07	1.47		1.54	2.34
0260	Lighting fixtures up to 2′ x 4′		72	.889	Ea.	2.32	51		53.32	81.50
0400	Partitions, non load bearing									
0410	Plaster, lath, and studs	A-9	690	.093	S.F.	.88	5.30		6.18	9.15
0450	Gypsum board and studs		1390	.046		.12	2.64		2.76	4.21
0465	Carpet & pad		1390	.046		.12	2.64		2.76	4.21
0600	Pipe insulation, air cell type, up to 4″ diameter pipe		900	.071	L.F.	.19	4.08		4.27	6.50
0610	4″ to 8″ diameter pipe		800	.080		.21	4.59		4.80	7.35
0620	10″ to 12″ diameter pipe		700	.091		.24	5.25		5.49	8.35
0630	14″ to 16″ diameter pipe		550	.116		.30	6.70		7	10.65
0650	Over 16″ diameter pipe		650	.098	S.F.	.26	5.65		5.91	9.05
9000	For type B (supplied air) respirator equipment, add				%				10%	10%

Division Notes

	CREW	DAILY OUTPUT	LABOR-HOURS	UNIT	BARE COSTS MAT.	LABOR	EQUIP.	TOTAL	TOTAL INCL O&P

Estimating Tips

General

- Carefully check all the plans and specifications. Concrete often appears on drawings other than structural drawings, including mechanical and electrical drawings for equipment pads. The cost of cutting and patching is often difficult to estimate. See Subdivision 03 81 for Concrete Cutting, Subdivision 02 41 19.16 for Cutout Demolition, Subdivision 03 05 05.10 for Concrete Demolition, and Subdivision 02 41 19.19 for Rubbish Handling (handling, loading, and hauling of debris).
- Always obtain concrete prices from suppliers near the job site. A volume discount can often be negotiated, depending upon competition in the area. Remember to add for waste, particularly for slabs and footings on grade.

03 10 00 Concrete Forming and Accessories

- A primary cost for concrete construction is forming. Most jobs today are constructed with prefabricated forms. The selection of the forms best suited for the job and the total square feet of forms required for efficient concrete forming and placing are key elements in estimating concrete construction. Enough forms must be available for erection to make efficient use of the concrete placing equipment and crew.
- Concrete accessories for forming and placing depend upon the systems used. Study the plans and specifications to ensure that all special accessory requirements have been included in the cost estimate, such as anchor bolts, inserts, and hangers.
- Included within costs for forms-in-place are all necessary bracing and shoring.

03 20 00 Concrete Reinforcing

- Ascertain that the reinforcing steel supplier has included all accessories, cutting, bending, and an allowance for lapping, splicing, and waste. A good rule of thumb is 10% for lapping, splicing, and waste. Also, 10% waste should be allowed for welded wire fabric.
- The unit price items in the subdivisions for Reinforcing In Place, Glass Fiber Reinforcing, and Welded Wire Fabric include the labor to install accessories such as beam and slab bolsters, high chairs, and bar ties and tie wire. The material cost for these accessories is not included; they may be obtained from the Accessories Subdivisions.

03 30 00 Cast-In-Place Concrete

- When estimating structural concrete, pay particular attention to requirements for concrete additives, curing methods, and surface treatments. Special consideration for climate, hot or cold, must be included in your estimate. Be sure to include requirements for concrete placing equipment and concrete finishing.
- For accurate concrete estimating, the estimator must consider each of the following major components individually: forms, reinforcing steel, ready-mix concrete, placement of the concrete, and finishing of the top surface. For faster estimating, Subdivision 03 30 53.40 for Concrete-In-Place can be used; here, various items of concrete work are presented that include the costs of all five major components (unless specifically stated otherwise).

03 40 00 Precast Concrete
03 50 00 Cast Decks and Underlayment

- The cost of hauling precast concrete structural members is often an important factor. For this reason, it is important to get a quote from the nearest supplier. It may become economically feasible to set up precasting beds on the site if the hauling costs are prohibitive.

Reference Numbers

Reference numbers are shown at the beginning of some major classifications. These numbers refer to related items in the Reference Section. The reference information may be an estimating procedure, an alternate pricing method, or technical information.

Note: Not all subdivisions listed here necessarily appear. ■

03 01 Maintenance of Concrete

03 01 30 – Maintenance of Cast-In-Place Concrete

03 01 30.62 Concrete Patching		Crew	Daily Output	Labor-Hours	Unit	Material	Labor	Equipment	Total	Total Incl O&P
						2019 Bare Costs				
0010	CONCRETE PATCHING									
0100	Floors, 1/4" thick, small areas, regular grout	1 Cefi	170	.047	S.F.	1.54	2.30		3.84	5.10
0150	Epoxy grout	"	100	.080	"	8.75	3.91		12.66	15.40
2000	Walls, including chipping, cleaning and epoxy grout									
2100	1/4" deep	1 Cefi	65	.123	S.F.	8.60	6		14.60	18.35
2150	1/2" deep		50	.160		17.20	7.80		25	30.50
2200	3/4" deep		40	.200		26	9.80		35.80	43

03 11 Concrete Forming

03 11 13 – Structural Cast-In-Place Concrete Forming

03 11 13.40 Forms In Place, Equipment Foundations

		Crew	Daily Output	Labor-Hours	Unit	Material	Labor	Equipment	Total	Total Incl O&P
0010	FORMS IN PLACE, EQUIPMENT FOUNDATIONS									
0020	1 use	C-2	160	.300	SFCA	2.99	15.05		18.04	26.50
0050	2 use		190	.253		1.64	12.70		14.34	21
0100	3 use		200	.240		1.20	12.05		13.25	19.55
0150	4 use		205	.234		.98	11.75		12.73	18.85

03 11 13.45 Forms In Place, Footings

		Crew	Daily Output	Labor-Hours	Unit	Material	Labor	Equipment	Total	Total Incl O&P
0010	FORMS IN PLACE, FOOTINGS									
0020	Continuous wall, plywood, 1 use	C-1	375	.085	SFCA	7	4.18		11.18	14.05
0050	2 use		440	.073		3.85	3.56		7.41	9.65
0100	3 use		470	.068		2.80	3.34		6.14	8.15
0150	4 use		485	.066		2.28	3.23		5.51	7.40
5000	Spread footings, job-built lumber, 1 use		305	.105		2.31	5.15		7.46	10.35
5050	2 use		371	.086		1.29	4.23		5.52	7.80
5100	3 use		401	.080		.93	3.91		4.84	6.90
5150	4 use		414	.077		.75	3.79		4.54	6.60

03 11 13.65 Forms In Place, Slab On Grade

		Crew	Daily Output	Labor-Hours	Unit	Material	Labor	Equipment	Total	Total Incl O&P
0010	FORMS IN PLACE, SLAB ON GRADE									
3000	Edge forms, wood, 4 use, on grade, to 6" high	C-1	600	.053	L.F.	.31	2.61		2.92	4.29
6000	Trench forms in floor, wood, 1 use		160	.200	SFCA	2.05	9.80		11.85	17.10
6050	2 use		175	.183		1.13	8.95		10.08	14.80
6100	3 use		180	.178		.82	8.70		9.52	14.10
6150	4 use		185	.173		.67	8.50		9.17	13.55

03 15 Concrete Accessories

03 15 05 – Concrete Forming Accessories

03 15 05.75 Sleeves and Chases

		Crew	Daily Output	Labor-Hours	Unit	Material	Labor	Equipment	Total	Total Incl O&P
0010	SLEEVES AND CHASES									
0100	Plastic, 1 use, 12" long, 2" diameter	1 Carp	100	.080	Ea.	1.89	4.13		6.02	8.35
0150	4" diameter		90	.089		5.30	4.59		9.89	12.80
0200	6" diameter		75	.107		9.30	5.50		14.80	18.60
0250	12" diameter		60	.133		27.50	6.90		34.40	41
5000	Sheet metal, 2" diameter G		100	.080		1.51	4.13		5.64	7.90
5100	4" diameter G		90	.089		1.89	4.59		6.48	9.05
5150	6" diameter G		75	.107		2.03	5.50		7.53	10.60
5200	12" diameter G		60	.133		3.97	6.90		10.87	14.75
6000	Steel pipe, 2" diameter G		100	.080		3.12	4.13		7.25	9.70
6100	4" diameter G		90	.089		10.85	4.59		15.44	18.90
6150	6" diameter G		75	.107		35	5.50		40.50	47

03 15 Concrete Accessories

03 15 05 – Concrete Forming Accessories

03 15 05.75 Sleeves and Chases			Crew	Daily Output	Labor-Hours	Unit	Material	Labor	Equipment	Total	Total Incl O&P
							2019 Bare Costs				
6200	12″ diameter	G	1 Carp	60	.133	Ea.	83.50	6.90		90.40	102

03 15 16 – Concrete Construction Joints

03 15 16.20 Control Joints, Saw Cut

		Crew	Daily Output	Labor-Hours	Unit	Material	Labor	Equipment	Total	Total Incl O&P
0010	**CONTROL JOINTS, SAW CUT**									
0100	Sawcut control joints in green concrete									
0120	1″ depth	C-27	2000	.008	L.F.	.04	.39	.06	.49	.68
0140	1-1/2″ depth		1800	.009		.05	.43	.06	.54	.77
0160	2″ depth		1600	.010		.07	.49	.07	.63	.88
0180	Sawcut joint reservoir in cured concrete									
0182	3/8″ wide x 3/4″ deep, with single saw blade	C-27	1000	.016	L.F.	.05	.78	.11	.94	1.34
0184	1/2″ wide x 1″ deep, with double saw blades		900	.018		.10	.87	.12	1.09	1.52
0186	3/4″ wide x 1-1/2″ deep, with double saw blades		800	.020		.21	.98	.14	1.33	1.82
0190	Water blast joint to wash away laitance, 2 passes	C-29	2500	.003			.13	.03	.16	.23
0200	Air blast joint to blow out debris and air dry, 2 passes	C-28	2000	.004			.20	.01	.21	.30
0300	For backer rod, see Section 07 91 23.10									
0342	For joint sealant, see Section 07 92 13.20									

03 15 19 – Cast-In Concrete Anchors

03 15 19.10 Anchor Bolts

			Crew	Daily Output	Labor-Hours	Unit	Material	Labor	Equipment	Total	Total Incl O&P
0010	**ANCHOR BOLTS**										
0015	Made from recycled materials										
0025	Single bolts installed in fresh concrete, no templates										
0030	Hooked w/nut and washer, 1/2″ diameter, 8″ long	G	1 Carp	132	.061	Ea.	1.48	3.13		4.61	6.35
0040	12″ long	G		131	.061		1.64	3.15		4.79	6.55
0050	5/8″ diameter, 8″ long	G		129	.062		4.10	3.20		7.30	9.35
0060	12″ long	G		127	.063		5.05	3.25		8.30	10.45
0070	3/4″ diameter, 8″ long	G		127	.063		5.05	3.25		8.30	10.45
0080	12″ long	G		125	.064		6.30	3.31		9.61	11.95
0090	2-bolt pattern, including job-built 2-hole template, per set										
0100	J-type, incl. hex nut & washer, 1/2″ diameter x 6″ long	G	1 Carp	21	.381	Set	5.30	19.70		25	36
0110	12″ long	G		21	.381		5.95	19.70		25.65	36.50
0120	18″ long	G		21	.381		6.95	19.70		26.65	37.50
0130	3/4″ diameter x 8″ long	G		20	.400		12.75	20.50		33.25	45.50
0140	12″ long	G		20	.400		15.30	20.50		35.80	48.50
0150	18″ long	G		20	.400		19.05	20.50		39.55	52.50
0160	1″ diameter x 12″ long	G		19	.421		24.50	22		46.50	60
0170	18″ long	G		19	.421		29	22		51	65
0180	24″ long	G		19	.421		35	22		57	71.50
0190	36″ long	G		18	.444		47	23		70	86
0200	1-1/2″ diameter x 18″ long	G		17	.471		43	24.50		67.50	84.50
0210	24″ long	G		16	.500		51	26		77	95
0300	L-type, incl. hex nut & washer, 3/4″ diameter x 12″ long	G		20	.400		15.90	20.50		36.40	49
0310	18″ long	G		20	.400		19.60	20.50		40.10	53
0320	24″ long	G		20	.400		23.50	20.50		44	57
0330	30″ long	G		20	.400		29	20.50		49.50	63.50
0340	36″ long	G		20	.400		32.50	20.50		53	67.50
0350	1″ diameter x 12″ long	G		19	.421		24	22		46	59.50
0360	18″ long	G		19	.421		29	22		51	65
0370	24″ long	G		19	.421		35.50	22		57.50	72
0380	30″ long	G		19	.421		41.50	22		63.50	78.50
0390	36″ long	G		18	.444		47	23		70	86
0400	42″ long	G		18	.444		56.50	23		79.50	96.50
0410	48″ long	G		18	.444		63	23		86	104

03 15 19.10 Anchor Bolts			Crew	Daily Output	Labor-Hours	Unit	Material	Labor	Equipment	Total	Total Incl O&P
							2019 Bare Costs				
0420	1-1/4" diameter x 18" long	G	1 Carp	18	.444	Set	37	23		60	75
0430	24" long	G		18	.444		43.50	23		66.50	82
0440	30" long	G		17	.471		49.50	24.50		74	91.50
0450	36" long	G		17	.471		56	24.50		80.50	98.50
0460	42" long	G	2 Carp	32	.500		63	26		89	109
0470	48" long	G		32	.500		71.50	26		97.50	118
0480	54" long	G		31	.516		84	26.50		110.50	133
0490	60" long	G		31	.516		92	26.50		118.50	142
0500	1-1/2" diameter x 18" long	G		33	.485		44.50	25		69.50	87
0510	24" long	G		32	.500		51.50	26		77.50	95.50
0520	30" long	G		31	.516		58	26.50		84.50	104
0530	36" long	G		30	.533		66	27.50		93.50	115
0540	42" long	G		30	.533		75	27.50		102.50	124
0550	48" long	G		29	.552		84	28.50		112.50	136
0560	54" long	G		28	.571		102	29.50		131.50	157
0570	60" long	G		28	.571		111	29.50		140.50	167
0580	1-3/4" diameter x 18" long	G		31	.516		67	26.50		93.50	115
0590	24" long	G		30	.533		78.50	27.50		106	128
0600	30" long	G		29	.552		90.50	28.50		119	143
0610	36" long	G		28	.571		103	29.50		132.50	158
0620	42" long	G		27	.593		115	30.50		145.50	174
0630	48" long	G		26	.615		127	32		159	187
0640	54" long	G		26	.615		156	32		188	220
0650	60" long	G		25	.640		169	33		202	236
0660	2" diameter x 24" long	G		27	.593		131	30.50		161.50	192
0670	30" long	G		27	.593		148	30.50		178.50	210
0680	36" long	G		26	.615		162	32		194	226
0690	42" long	G		25	.640		180	33		213	249
0700	48" long	G		24	.667		207	34.50		241.50	280
0710	54" long	G		23	.696		246	36		282	325
0720	60" long	G		23	.696		264	36		300	345
0730	66" long	G		22	.727		283	37.50		320.50	365
0740	72" long	G		21	.762		310	39.50		349.50	400
1000	4-bolt pattern, including job-built 4-hole template, per set										
1100	J-type, incl. hex nut & washer, 1/2" diameter x 6" long	G	1 Carp	19	.421	Set	7.95	22		29.95	41.50
1110	12" long	G		19	.421		9.25	22		31.25	43
1120	18" long	G		18	.444		11.20	23		34.20	47
1130	3/4" diameter x 8" long	G		17	.471		23	24.50		47.50	62
1140	12" long	G		17	.471		28	24.50		52.50	67.50
1150	18" long	G		17	.471		35.50	24.50		60	76
1160	1" diameter x 12" long	G		16	.500		46.50	26		72.50	90.50
1170	18" long	G		15	.533		56	27.50		83.50	103
1180	24" long	G		15	.533		67.50	27.50		95	116
1190	36" long	G		15	.533		91.50	27.50		119	142
1200	1-1/2" diameter x 18" long	G		13	.615		83.50	32		115.50	140
1210	24" long	G		12	.667		99	34.50		133.50	161
1300	L-type, incl. hex nut & washer, 3/4" diameter x 12" long	G		17	.471		29	24.50		53.50	69
1310	18" long	G		17	.471		36.50	24.50		61	77
1320	24" long	G		17	.471		44	24.50		68.50	85.50
1330	30" long	G		16	.500		55	26		81	99.50
1340	36" long	G		16	.500		62.50	26		88.50	108
1350	1" diameter x 12" long	G		16	.500		45.50	26		71.50	89
1360	18" long	G		15	.533		56	27.50		83.50	103

03 15 19.10 Anchor Bolts			Crew	Daily Output	Labor-Hours	Unit	Material	2019 Bare Costs Labor	Equipment	Total	Total Incl O&P
1370	24" long	G	1 Carp	15	.533	Set	68	27.50		95.50	117
1380	30" long	G		15	.533		80	27.50		107.50	130
1390	36" long	G		15	.533		91	27.50		118.50	142
1400	42" long	G		14	.571		110	29.50		139.50	166
1410	48" long	G		14	.571		123	29.50		152.50	181
1420	1-1/4" diameter x 18" long	G		14	.571		71	29.50		100.50	123
1430	24" long	G		14	.571		84	29.50		113.50	137
1440	30" long	G		13	.615		96.50	32		128.50	154
1450	36" long	G		13	.615		109	32		141	168
1460	42" long	G	2 Carp	25	.640		123	33		156	186
1470	48" long	G		24	.667		140	34.50		174.50	207
1480	54" long	G		23	.696		165	36		201	237
1490	60" long	G		23	.696		181	36		217	254
1500	1-1/2" diameter x 18" long	G		25	.640		86.50	33		119.50	145
1510	24" long	G		24	.667		100	34.50		134.50	162
1520	30" long	G		23	.696		113	36		149	179
1530	36" long	G		22	.727		130	37.50		167.50	200
1540	42" long	G		22	.727		147	37.50		184.50	219
1550	48" long	G		21	.762		165	39.50		204.50	242
1560	54" long	G		20	.800		201	41.50		242.50	284
1570	60" long	G		20	.800		220	41.50		261.50	305
1580	1-3/4" diameter x 18" long	G		22	.727		132	37.50		169.50	202
1590	24" long	G		21	.762		154	39.50		193.50	229
1600	30" long	G		21	.762		179	39.50		218.50	257
1610	36" long	G		20	.800		203	41.50		244.50	287
1620	42" long	G		19	.842		228	43.50		271.50	315
1630	48" long	G		18	.889		251	46		297	345
1640	54" long	G		18	.889		310	46		356	410
1650	60" long	G		17	.941		335	48.50		383.50	445
1660	2" diameter x 24" long	G		19	.842		260	43.50		303.50	350
1670	30" long	G		18	.889		293	46		339	390
1680	36" long	G		18	.889		320	46		366	425
1690	42" long	G		17	.941		360	48.50		408.50	470
1700	48" long	G		16	1		410	51.50		461.50	535
1710	54" long	G		15	1.067		490	55		545	625
1720	60" long	G		15	1.067		525	55		580	665
1730	66" long	G		14	1.143		565	59		624	710
1740	72" long	G		14	1.143		615	59		674	765
1990	For galvanized, add					Ea.	75%				

03 15 19.45 Machinery Anchors			Crew	Daily Output	Labor-Hours	Unit	Material	Labor	Equipment	Total	Total Incl O&P
0010	**MACHINERY ANCHORS**, heavy duty, incl. sleeve, floating base nut,										
0020	lower stud & coupling nut, fiber plug, connecting stud, washer & nut.										
0030	For flush mounted embedment in poured concrete heavy equip. pads.										
0200	Stud & bolt, 1/2" diameter	G	E-16	40	.400	Ea.	57.50	23	2.39	82.89	103
0300	5/8" diameter	G		35	.457		67	26	2.73	95.73	118
0500	3/4" diameter	G		30	.533		79.50	30.50	3.19	113.19	140
0600	7/8" diameter	G		25	.640		89.50	36.50	3.83	129.83	161
0800	1" diameter	G		20	.800		89.50	45.50	4.78	139.78	176
0900	1-1/4" diameter	G		15	1.067		128	61	6.40	195.40	245

03 21 Reinforcement Bars

03 21 11 – Plain Steel Reinforcement Bars

03 21 11.60 Reinforcing In Place		Crew	Daily Output	Labor-Hours	Unit	Material	Labor	Equipment	Total	Total Incl O&P
						2019 Bare Costs				
0010	**REINFORCING IN PLACE**, 50-60 ton lots, A615 Grade 60									
0020	Includes labor, but not material cost, to install accessories									
0030	Made from recycled materials									
0502	Footings, #4 to #7 G	4 Rodm	4200	.008	Lb.	.51	.42		.93	1.20
0552	#8 to #18 G		7200	.004		.51	.24		.75	.94
0602	Slab on grade, #3 to #7 G		4200	.008		.51	.42		.93	1.20
0900	For other than 50-60 ton lots									
1000	Under 10 ton job, #3 to #7, add					25%	10%			
1010	#8 to #18, add					20%	10%			
1050	10-50 ton job, #3 to #7, add					10%				
1060	#8 to #18, add					5%				
1100	60-100 ton job, #3 to #7, deduct					5%				
1110	#8 to #18, deduct					10%				
1150	Over 100 ton job, #3 to #7, deduct					10%				
1160	#8 to #18, deduct					15%				

03 22 Fabric and Grid Reinforcing

03 22 11 – Plain Welded Wire Fabric Reinforcing

03 22 11.10 Plain Welded Wire Fabric		Crew	Daily Output	Labor-Hours	Unit	Material	Labor	Equipment	Total	Total Incl O&P
0010	**PLAIN WELDED WIRE FABRIC** ASTM A185									
0020	Includes labor, but not material cost, to install accessories									
0030	Made from recycled materials									
0050	Sheets									
0100	6 x 6 - W1.4 x W1.4 (10 x 10) 21 lb./C.S.F. G	2 Rodm	35	.457	C.S.F.	15.90	25		40.90	55.50

03 22 13 – Galvanized Welded Wire Fabric Reinforcing

03 22 13.10 Galvanized Welded Wire Fabric		Crew	Daily Output	Labor-Hours	Unit	Material	Labor	Equipment	Total	Total Incl O&P
0010	**GALVANIZED WELDED WIRE FABRIC**									
0100	Add to plain welded wire pricing for galvanized welded wire				Lb.	.24			.24	.27

03 22 16 – Epoxy-Coated Welded Wire Fabric Reinforcing

03 22 16.10 Epoxy-Coated Welded Wire Fabric		Crew	Daily Output	Labor-Hours	Unit	Material	Labor	Equipment	Total	Total Incl O&P
0010	**EPOXY-COATED WELDED WIRE FABRIC**									
0100	Add to plain welded wire pricing for epoxy-coated welded wire				Lb.	.42			.42	.46

03 30 Cast-In-Place Concrete

03 30 53 – Miscellaneous Cast-In-Place Concrete

03 30 53.40 Concrete In Place		Crew	Daily Output	Labor-Hours	Unit	Material	Labor	Equipment	Total	Total Incl O&P
0010	**CONCRETE IN PLACE**									
0020	Including forms (4 uses), Grade 60 rebar, concrete (Portland cement									
0050	Type I), placement and finishing unless otherwise indicated									
0500	Chimney foundations (5000 psi), over 5 C.Y.	C-14C	32.22	3.476	C.Y.	178	170	.83	348.83	455
0510	(3500 psi), under 5 C.Y.	"	23.71	4.724	"	206	232	1.13	439.13	580
3540	Equipment pad (3000 psi), 3′ x 3′ x 6″ thick	C-14H	45	1.067	Ea.	46.50	53.50	.59	100.59	133
3550	4′ x 4′ x 6″ thick		30	1.600		72.50	80.50	.88	153.88	201
3560	5′ x 5′ x 8″ thick		18	2.667		132	134	1.47	267.47	350
3570	6′ x 6′ x 8″ thick		14	3.429		181	172	1.89	354.89	460
3580	8′ x 8′ x 10″ thick		8	6		385	300	3.30	688.30	880
3590	10′ x 10′ x 12″ thick		5	9.600		665	485	5.30	1,155.30	1,475
3800	Footings (3000 psi), spread under 1 C.Y.	C-14C	28	4	C.Y.	193	196	.96	389.96	510

03 30 Cast-In-Place Concrete

03 30 53 – Miscellaneous Cast-In-Place Concrete

03 30 53.40 Concrete In Place		Crew	Daily Output	Labor-Hours	Unit	2019 Bare Costs Material	Labor	Equipment	Total	Total Incl O&P
3825	1 C.Y. to 5 C.Y.	C-14C	43	2.605	C.Y.	227	128	.63	355.63	445
3850	Over 5 C.Y.	↓	75	1.493		212	73	.36	285.36	345
3900	Footings, strip (3000 psi), 18″ x 9″, unreinforced	C-14L	40	2.400		149	115	.67	264.67	340
3920	18″ x 9″, reinforced	C-14C	35	3.200		174	157	.77	331.77	430
3925	20″ x 10″, unreinforced	C-14L	45	2.133		146	103	.60	249.60	315
3930	20″ x 10″, reinforced	C-14C	40	2.800		166	137	.67	303.67	390
3935	24″ x 12″, unreinforced	C-14L	55	1.745		143	84	.49	227.49	286
3940	24″ x 12″, reinforced	C-14C	48	2.333		164	114	.56	278.56	355
3945	36″ x 12″, unreinforced	C-14L	70	1.371		139	66	.38	205.38	253
3950	36″ x 12″, reinforced	C-14C	60	1.867		158	91.50	.45	249.95	310
4000	Foundation mat (3000 psi), under 10 C.Y.		38.67	2.896		231	142	.70	373.70	470
4050	Over 20 C.Y.	↓	56.40	1.986		205	97.50	.48	302.98	375
4650	Slab on grade (3500 psi), not including finish, 4″ thick	C-14E	60.75	1.449		147	72	.43	219.43	270
4700	6″ thick	″	92	.957	↓	141	47.50	.29	188.79	228
4701	Thickened slab edge (3500 psi), for slab on grade poured									
4702	monolithically with slab; depth is in addition to slab thickness;									
4703	formed vertical outside edge, earthen bottom and inside slope									
4705	8″ deep x 8″ wide bottom, unreinforced	C-14L	2190	.044	L.F.	3.99	2.11	.01	6.11	7.60
4710	8″ x 8″, reinforced	C-14C	1670	.067		6.40	3.29	.02	9.71	12.05
4715	12″ deep x 12″ wide bottom, unreinforced	C-14L	1800	.053		8.20	2.56	.01	10.77	12.90
4720	12″ x 12″, reinforced	C-14C	1310	.086		12.60	4.19	.02	16.81	20
4725	16″ deep x 16″ wide bottom, unreinforced	C-14L	1440	.067		13.90	3.20	.02	17.12	20
4730	16″ x 16″, reinforced	C-14C	1120	.100		19.15	4.90	.02	24.07	28.50
4735	20″ deep x 20″ wide bottom, unreinforced	C-14L	1150	.083		21	4.01	.02	25.03	29
4740	20″ x 20″, reinforced	C-14C	920	.122		28	5.95	.03	33.98	39.50
4745	24″ deep x 24″ wide bottom, unreinforced	C-14L	930	.103		30	4.96	.03	34.99	40.50
4750	24″ x 24″, reinforced	C-14C	740	.151	↓	38.50	7.40	.04	45.94	53.50
4751	Slab on grade (3500 psi), incl. troweled finish, not incl. forms									
4760	or reinforcing, over 10,000 S.F., 4″ thick	C-14F	3425	.021	S.F.	1.62	.98	.01	2.61	3.24
4820	6″ thick		3350	.021		2.37	1	.01	3.38	4.11
4840	8″ thick		3184	.023		3.24	1.05	.01	4.30	5.15
4900	12″ thick		2734	.026		4.87	1.23	.01	6.11	7.20
4950	15″ thick	↓	2505	.029	↓	6.10	1.34	.01	7.45	8.70

03 31 Structural Concrete

03 31 13 – Heavyweight Structural Concrete

03 31 13.35 Heavyweight Concrete, Ready Mix		Crew	Daily Output	Labor-Hours	Unit	Material	Labor	Equipment	Total	Total Incl O&P
0010	**HEAVYWEIGHT CONCRETE, READY MIX**, delivered									
0012	Includes local aggregate, sand, Portland cement (Type I) and water									
0015	Excludes all additives and treatments									
0020	2000 psi				C.Y.	110			110	121
0100	2500 psi					113			113	125
0150	3000 psi					124			124	136
0200	3500 psi					125			125	137
0300	4000 psi					128			128	141
1000	For high early strength (Portland cement Type III), add					10%				
1300	For winter concrete (hot water), add					5.35			5.35	5.90
1410	For mid-range water reducer, add					4.01			4.01	4.41
1420	For high-range water reducer/superplasticizer, add					6.25			6.25	6.90
1430	For retarder, add					3.25			3.25	3.58
1440	For non-Chloride accelerator, add				↓	6.40			6.40	7.05

03 31 Structural Concrete

03 31 13 – Heavyweight Structural Concrete

03 31 13.35 Heavyweight Concrete, Ready Mix		Crew	Daily Output	Labor-Hours	Unit	2019 Bare Costs Material	Labor	Equipment	Total	Total Incl O&P
1450	For Chloride accelerator, per 1%, add				C.Y.	3.91			3.91	4.30
1460	For fiber reinforcing, synthetic (1 lb./C.Y.), add					8.05			8.05	8.85
1500	For Saturday delivery, add					8.25			8.25	9.10
1510	For truck holding/waiting time past 1st hour per load, add				Hr.	105			105	115
1520	For short load (less than 4 C.Y.), add per load				Ea.	92			92	101
2000	For all lightweight aggregate, add				C.Y.	45%				

03 31 13.70 Placing Concrete

		Crew	Daily Output	Labor-Hours	Unit	Material	Labor	Equipment	Total	Total Incl O&P
0010	**PLACING CONCRETE**									
0020	Includes labor and equipment to place, level (strike off) and consolidate									
1900	Footings, continuous, shallow, direct chute	C-6	120	.400	C.Y.		17.10	.44	17.54	26
1950	Pumped	C-20	150	.427			18.80	6	24.80	35
2000	With crane and bucket	C-7	90	.800			35.50	11.80	47.30	66.50
2100	Footings, continuous, deep, direct chute	C-6	140	.343			14.65	.38	15.03	22.50
2150	Pumped	C-20	160	.400			17.60	5.60	23.20	32.50
2200	With crane and bucket	C-7	110	.655			29	9.65	38.65	54.50
2400	Footings, spread, under 1 C.Y., direct chute	C-6	55	.873			37.50	.97	38.47	57
2450	Pumped	C-20	65	.985			43.50	13.80	57.30	80.50
2500	With crane and bucket	C-7	45	1.600			71.50	23.50	95	133
2600	Over 5 C.Y., direct chute	C-6	120	.400			17.10	.44	17.54	26
2650	Pumped	C-20	150	.427			18.80	6	24.80	35
2700	With crane and bucket	C-7	100	.720			32	10.60	42.60	60
2900	Foundation mats, over 20 C.Y., direct chute	C-6	350	.137			5.85	.15	6	8.95
2950	Pumped	C-20	400	.160			7.05	2.24	9.29	13.05
3000	With crane and bucket	C-7	300	.240			10.70	3.53	14.23	20

03 35 Concrete Finishing

03 35 13 – High-Tolerance Concrete Floor Finishing

03 35 13.30 Finishing Floors, High Tolerance

		Crew	Daily Output	Labor-Hours	Unit	Material	Labor	Equipment	Total	Total Incl O&P
0010	**FINISHING FLOORS, HIGH TOLERANCE**									
0012	Finishing of fresh concrete flatwork requires that concrete									
0013	first be placed, struck off & consolidated									
0015	Basic finishing for various unspecified flatwork									
0100	Bull float only	C-10	4000	.006	S.F.		.28		.28	.41
0125	Bull float & manual float		2000	.012			.56		.56	.83
0150	Bull float, manual float & broom finish, w/edging & joints		1850	.013			.60		.60	.89
0200	Bull float, manual float & manual steel trowel		1265	.019			.88		.88	1.31
0210	For specified Random Access Floors in ACI Classes 1, 2, 3 and 4 to achieve									
0215	Composite Overall Floor Flatness and Levelness values up to FF35/FL25									
0250	Bull float, machine float & machine trowel (walk-behind)	C-10C	1715	.014	S.F.		.65	.02	.67	.99
0300	Power screed, bull float, machine float & trowel (walk-behind)	C-10D	2400	.010			.46	.05	.51	.75
0350	Power screed, bull float, machine float & trowel (ride-on)	C-10E	4000	.006			.28	.06	.34	.48
0352	For specified Random Access Floors in ACI Classes 5, 6, 7 and 8 to achieve									
0354	Composite Overall Floor Flatness and Levelness values up to FF50/FL50									
0356	Add for two-dimensional restraightening after power float	C-10	6000	.004	S.F.		.19		.19	.28
0358	For specified Random or Defined Access Floors in ACI Class 9 to achieve									
0360	Composite Overall Floor Flatness and Levelness values up to FF100/FL100									
0362	Add for two-dimensional restraightening after bull float & power float	C-10	3000	.008	S.F.		.37		.37	.55
0364	For specified Superflat Defined Access Floors in ACI Class 9 to achieve									
0366	Minimum Floor Flatness and Levelness values of FF100/FL100									
0368	Add for 2-dim'l restraightening after bull float, power float, power trowel	C-10	2000	.012	S.F.		.56		.56	.83

03 54 Cast Underlayment

03 54 16 – Hydraulic Cement Underlayment

03 54 16.50 Cement Underlayment		Crew	Daily Output	Labor-Hours	Unit	Material	Labor	Equipment	Total	Total Incl O&P
						2019 Bare Costs				
0010	CEMENT UNDERLAYMENT									
2510	Underlayment, P.C. based, self-leveling, 4100 psi, pumped, 1/4" thick	C-8	20000	.003	S.F.	1.67	.13	.04	1.84	2.07
2520	1/2" thick		19000	.003		3.34	.13	.04	3.51	3.92
2530	3/4" thick		18000	.003		5	.14	.05	5.19	5.75
2540	1" thick		17000	.003		6.65	.15	.05	6.85	7.60
2550	1-1/2" thick		15000	.004		10	.17	.06	10.23	11.30
2560	Hand placed, 1/2" thick	C-18	450	.020		3.34	.83	.13	4.30	5.05
2610	Topping, P.C. based, self-leveling, 6100 psi, pumped, 1/4" thick	C-8	20000	.003		2.12	.13	.04	2.29	2.58
2620	1/2" thick		19000	.003		4.25	.13	.04	4.42	4.92
2630	3/4" thick		18000	.003		6.35	.14	.05	6.54	7.25
2660	1" thick		17000	.003		8.50	.15	.05	8.70	9.60
2670	1-1/2" thick		15000	.004		12.75	.17	.06	12.98	14.30
2680	Hand placed, 1/2" thick	C-18	450	.020		4.25	.83	.13	5.21	6.05

03 63 Epoxy Grouting

03 63 05 – Grouting of Dowels and Fasteners

03 63 05.10 Epoxy Only

		Crew	Daily Output	Labor-Hours	Unit	Material	Labor	Equipment	Total	Total Incl O&P
0010	EPOXY ONLY									
1500	Chemical anchoring, epoxy cartridge, excludes layout, drilling, fastener									
1530	For fastener 3/4" diam. x 6" embedment	2 Skwk	72	.222	Ea.	5.20	11.85		17.05	24
1535	1" diam. x 8" embedment		66	.242		7.75	12.95		20.70	28.50
1540	1-1/4" diam. x 10" embedment		60	.267		15.55	14.25		29.80	38.50
1545	1-3/4" diam. x 12" embedment		54	.296		26	15.80		41.80	52.50
1550	14" embedment		48	.333		31	17.80		48.80	61
1555	2" diam. x 12" embedment		42	.381		41.50	20.50		62	76.50
1560	18" embedment		32	.500		52	26.50		78.50	97.50

03 82 Concrete Boring

03 82 13 – Concrete Core Drilling

03 82 13.10 Core Drilling

		Crew	Daily Output	Labor-Hours	Unit	Material	Labor	Equipment	Total	Total Incl O&P
0010	CORE DRILLING									
0015	Includes bit cost, layout and set-up time									
0020	Reinforced concrete slab, up to 6" thick									
0100	1" diameter core	B-89A	17	.941	Ea.	.18	44.50	6.60	51.28	75
0150	For each additional inch of slab thickness in same hole, add		1440	.011		.03	.52	.08	.63	.92
0200	2" diameter core		16.50	.970		.29	46	6.80	53.09	77.50
0250	For each additional inch of slab thickness in same hole, add		1080	.015		.05	.70	.10	.85	1.22
0300	3" diameter core		16	1		.39	47	7	54.39	79.50
0350	For each additional inch of slab thickness in same hole, add		720	.022		.07	1.05	.16	1.28	1.83
0500	4" diameter core		15	1.067		.51	50.50	7.50	58.51	85.50
0550	For each additional inch of slab thickness in same hole, add		480	.033		.08	1.57	.23	1.88	2.74
0700	6" diameter core		14	1.143		.78	54	8	62.78	91.50
0750	For each additional inch of slab thickness in same hole, add		360	.044		.13	2.10	.31	2.54	3.67
0900	8" diameter core		13	1.231		1.09	58	8.65	67.74	98.50
0950	For each additional inch of slab thickness in same hole, add		288	.056		.18	2.62	.39	3.19	4.61
1100	10" diameter core		12	1.333		1.34	63	9.35	73.69	107
1150	For each additional inch of slab thickness in same hole, add		240	.067		.22	3.15	.47	3.84	5.55
1300	12" diameter core		11	1.455		1.87	68.50	10.20	80.57	117
1350	For each additional inch of slab thickness in same hole, add		206	.078		.31	3.67	.54	4.52	6.50

03 82 Concrete Boring

03 82 13 – Concrete Core Drilling

03 82 13.10 Core Drilling		Crew	Daily Output	Labor-Hours	Unit	Material	2019 Bare Costs Labor	Equipment	Total	Total Incl O&P
1500	14" diameter core	B-89A	10	1.600	Ea.	2.14	75.50	11.20	88.84	130
1550	For each additional inch of slab thickness in same hole, add		180	.089		.36	4.20	.62	5.18	7.45
1700	18" diameter core		9	1.778		2.99	84	12.45	99.44	144
1750	For each additional inch of slab thickness in same hole, add		144	.111		.50	5.25	.78	6.53	9.35
1754	24" diameter core		8	2		4.18	94.50	14	112.68	163
1756	For each additional inch of slab thickness in same hole, add		120	.133		.70	6.30	.93	7.93	11.35
1760	For horizontal holes, add to above						20%	20%		
1770	Prestressed hollow core plank, 8" thick									
1780	1" diameter core	B-89A	17.50	.914	Ea.	.24	43	6.40	49.64	73
1790	For each additional inch of plank thickness in same hole, add		3840	.004		.03	.20	.03	.26	.36
1794	2" diameter core		17.25	.928		.39	44	6.50	50.89	74
1796	For each additional inch of plank thickness in same hole, add		2880	.006		.05	.26	.04	.35	.49
1800	3" diameter core		17	.941		.52	44.50	6.60	51.62	75.50
1810	For each additional inch of plank thickness in same hole, add		1920	.008		.07	.39	.06	.52	.73
1820	4" diameter core		16.50	.970		.68	46	6.80	53.48	78
1830	For each additional inch of plank thickness in same hole, add		1280	.013		.08	.59	.09	.76	1.09
1840	6" diameter core		15.50	1.032		1.04	49	7.25	57.29	83
1850	For each additional inch of plank thickness in same hole, add		960	.017		.13	.79	.12	1.04	1.46
1860	8" diameter core		15	1.067		1.45	50.50	7.50	59.45	86.50
1870	For each additional inch of plank thickness in same hole, add		768	.021		.18	.98	.15	1.31	1.85
1880	10" diameter core		14	1.143		1.79	54	8	63.79	93
1890	For each additional inch of plank thickness in same hole, add		640	.025		.22	1.18	.18	1.58	2.23
1900	12" diameter core		13.50	1.185		2.49	56	8.30	66.79	97
1910	For each additional inch of plank thickness in same hole, add		548	.029		.31	1.38	.20	1.89	2.66

03 82 16 – Concrete Drilling

03 82 16.10 Concrete Impact Drilling		Crew	Daily Output	Labor-Hours	Unit	Material	Labor	Equipment	Total	Total Incl O&P
0010	**CONCRETE IMPACT DRILLING**									
0020	Includes bit cost, layout and set-up time, no anchors									
0050	Up to 4" deep in concrete/brick floors/walls									
0100	Holes, 1/4" diameter	1 Carp	75	.107	Ea.	.07	5.50		5.57	8.45
0150	For each additional inch of depth in same hole, add		430	.019		.02	.96		.98	1.47
0200	3/8" diameter		63	.127		.05	6.55		6.60	9.95
0250	For each additional inch of depth in same hole, add		340	.024		.01	1.22		1.23	1.86
0300	1/2" diameter		50	.160		.06	8.25		8.31	12.55
0350	For each additional inch of depth in same hole, add		250	.032		.01	1.65		1.66	2.52
0400	5/8" diameter		48	.167		.10	8.60		8.70	13.15
0450	For each additional inch of depth in same hole, add		240	.033		.03	1.72		1.75	2.63
0500	3/4" diameter		45	.178		.13	9.20		9.33	14.05
0550	For each additional inch of depth in same hole, add		220	.036		.03	1.88		1.91	2.88
0600	7/8" diameter		43	.186		.18	9.60		9.78	14.75
0650	For each additional inch of depth in same hole, add		210	.038		.05	1.97		2.02	3.03
0700	1" diameter		40	.200		.17	10.35		10.52	15.85
0750	For each additional inch of depth in same hole, add		190	.042		.04	2.18		2.22	3.34
0800	1-1/4" diameter		38	.211		.28	10.85		11.13	16.75
0850	For each additional inch of depth in same hole, add		180	.044		.07	2.30		2.37	3.55
0900	1-1/2" diameter		35	.229		.51	11.80		12.31	18.40
0950	For each additional inch of depth in same hole, add		165	.048		.13	2.50		2.63	3.93
1000	For ceiling installations, add						40%			

Estimating Tips

04 05 00 Common Work Results for Masonry

- The terms mortar and grout are often used interchangeably—and incorrectly. Mortar is used to bed masonry units, seal the entry of air and moisture, provide architectural appearance, and allow for size variations in the units. Grout is used primarily in reinforced masonry construction and to bond the masonry to the reinforcing steel. Common mortar types are M (2500 psi), S (1800 psi), N (750 psi), and O (350 psi), and they conform to ASTM C270. Grout is either fine or coarse and conforms to ASTM C476, and in-place strengths generally exceed 2500 psi. Mortar and grout are different components of masonry construction and are placed by entirely different methods. An estimator should be aware of their unique uses and costs.
- Mortar is included in all assembled masonry line items. The mortar cost, part of the assembled masonry material cost, includes all ingredients, all labor, and all equipment required. Please see reference number R040513-10.
- Waste, specifically the loss/droppings of mortar and the breakage of brick and block, is included in all unit cost lines that include mortar and masonry units in this division. A factor of 25% is added for mortar and 3% for brick and concrete masonry units.
- Scaffolding or staging is not included in any of the Division 4 costs. Refer to Subdivision 01 54 23 for scaffolding and staging costs.

04 20 00 Unit Masonry

- The most common types of unit masonry are brick and concrete masonry. The major classifications of brick are building brick (ASTM C62), facing brick (ASTM C216), glazed brick, fire brick, and pavers. Many varieties of texture and appearance can exist within these classifications, and the estimator would be wise to check local custom and availability within the project area. For repair and remodeling jobs, matching the existing brick may be the most important criteria.
- Brick and concrete block are priced by the piece and then converted into a price per square foot of wall. Openings less than two square feet are generally ignored by the estimator because any savings in units used are offset by the cutting and trimming required.
- It is often difficult and expensive to find and purchase small lots of historic brick. Costs can vary widely. Many design issues affect costs, selection of mortar mix, and repairs or replacement of masonry materials. Cleaning techniques must be reflected in the estimate.
- All masonry walls, whether interior or exterior, require bracing. The cost of bracing walls during construction should be included by the estimator, and this bracing must remain in place until permanent bracing is complete. Permanent bracing of masonry walls is accomplished by masonry itself, in the form of pilasters or abutting wall corners, or by anchoring the walls to the structural frame. Accessories in the form of anchors, anchor slots, and ties are used, but their supply and installation can be by different trades. For instance, anchor slots on spandrel beams and columns are supplied and welded in place by the steel fabricator, but the ties from the slots into the masonry are installed by the bricklayer. Regardless of the installation method, the estimator must be certain that these accessories are accounted for in pricing.

Reference Numbers

Reference numbers are shown at the beginning of some major classifications. These numbers refer to related items in the Reference Section. The reference information may be an estimating procedure, an alternate pricing method, or technical information.

Note: Not all subdivisions listed here necessarily appear. ■

04 05 Common Work Results for Masonry

04 05 23 – Masonry Accessories

04 05 23.19 Masonry Cavity Drainage, Weepholes, and Vents		Crew	Daily Output	Labor-Hours	Unit	2019 Bare Costs Material	Labor	Equipment	Total	Total Incl O&P
0010	**MASONRY CAVITY DRAINAGE, WEEPHOLES, AND VENTS**									
0020	Extruded aluminum, 4" deep, 2-3/8" x 8-1/8"	1 Bric	30	.267	Ea.	39	13.60		52.60	64
0050	5" x 8-1/8"		25	.320		51	16.35		67.35	81
0100	2-1/4" x 25"		25	.320		88.50	16.35		104.85	123
0150	5" x 16-1/2"		22	.364		61	18.55		79.55	96
0175	5" x 24"		22	.364		92.50	18.55		111.05	131
0200	6" x 16-1/2"		22	.364		99.50	18.55		118.05	138
0250	7-3/4" x 16-1/2"	▼	20	.400		89	20.50		109.50	129
0400	For baked enamel finish, add					35%				
0500	For cast aluminum, painted, add					60%				
1000	Stainless steel ventilators, 6" x 6"	1 Bric	25	.320		228	16.35		244.35	276
1050	8" x 8"		24	.333		255	17		272	305
1100	12" x 12"		23	.348		292	17.75		309.75	345
1150	12" x 6"		24	.333		277	17		294	330
1200	Foundation block vent, galv., 1-1/4" thk, 8" high, 16" long, no damper	▼	30	.267		16.30	13.60		29.90	39
1250	For damper, add				▼	3.29			3.29	3.62

04 21 Clay Unit Masonry

04 21 13 – Brick Masonry

04 21 13.15 Chimney		Crew	Daily Output	Labor-Hours	Unit	Material	Labor	Equipment	Total	Total Incl O&P
0010	**CHIMNEY**, excludes foundation, scaffolding, grout and reinforcing									
0100	Brick, 16" x 16", 8" flue	D-1	18.20	.879	V.L.F.	25.50	40		65.50	89.50
0150	16" x 20" with one 8" x 12" flue		16	1		40	45.50		85.50	114
0200	16" x 24" with two 8" x 8" flues		14	1.143		58.50	52.50		111	145
0250	20" x 20" with one 12" x 12" flue		13.70	1.168		49	53.50		102.50	136
0300	20" x 24" with two 8" x 12" flues		12	1.333		66.50	61		127.50	166
0350	20" x 32" with two 12" x 12" flues	▼	10	1.600	▼	86.50	73		159.50	208

Estimating Tips

05 05 00 Common Work Results for Metals

- Nuts, bolts, washers, connection angles, and plates can add a significant amount to both the tonnage of a structural steel job and the estimated cost. As a rule of thumb, add 10% to the total weight to account for these accessories.
- Type 2 steel construction, commonly referred to as "simple construction," consists generally of field-bolted connections with lateral bracing supplied by other elements of the building, such as masonry walls or x-bracing. The estimator should be aware, however, that shop connections may be accomplished by welding or bolting. The method may be particular to the fabrication shop and may have an impact on the estimated cost.

05 10 00 Structural Steel

- Steel items can be obtained from two sources: a fabrication shop or a metals service center. Fabrication shops can fabricate items under more controlled conditions than crews in the field can. They are also more efficient and can produce items more economically. Metal service centers serve as a source of long mill shapes to both fabrication shops and contractors.
- Most line items in this structural steel subdivision, and most items in 05 50 00 Metal Fabrications, are indicated as being shop fabricated. The bare material cost for these shop fabricated items is the "Invoice Cost" from the shop and includes the mill base price of steel plus mill extras, transportation to the shop, shop drawings and detailing where warranted, shop fabrication and handling, sandblasting and a shop coat of primer paint, all necessary structural bolts, and delivery to the job site. The bare labor cost and bare equipment cost for these shop fabricated items are for field installation or erection.
- Line items in Subdivision 05 12 23.40 Lightweight Framing, and other items scattered in Division 5, are indicated as being field fabricated. The bare material cost for these field fabricated items is the "Invoice Cost" from the metals service center and includes the mill base price of steel plus mill extras, transportation to the metals service center, material handling, and delivery of long lengths of mill shapes to the job site. Material costs for structural bolts and welding rods should be added to the estimate. The bare labor cost and bare equipment cost for these items are for both field fabrication and field installation or erection, and include time for cutting, welding, and drilling in the fabricated metal items. Drilling into concrete and fasteners to fasten field fabricated items to other work is not included and should be added to the estimate.

05 20 00 Steel Joist Framing

- In any given project the total weight of open web steel joists is determined by the loads to be supported and the design. However, economies can be realized in minimizing the amount of labor used to place the joists. This is done by maximizing the joist spacing and therefore minimizing the number of joists required to be installed on the job. Certain spacings and locations may be required by the design, but in other cases maximizing the spacing and keeping it as uniform as possible will keep the costs down.

05 30 00 Steel Decking

- The takeoff and estimating of a metal deck involve more than the area of the floor or roof and the type of deck specified or shown on the drawings. Many different sizes and types of openings may exist. Small openings for individual pipes or conduits may be drilled after the floor/roof is installed, but larger openings may require special deck lengths as well as reinforcing or structural support. The estimator should determine who will be supplying this reinforcing. Additionally, some deck terminations are part of the deck package, such as screed angles and pour stops, and others will be part of the steel contract, such as angles attached to structural members and cast-in-place angles and plates. The estimator must ensure that all pieces are accounted for in the complete estimate.

05 50 00 Metal Fabrications

- The most economical steel stairs are those that use common materials, standard details, and most importantly, a uniform and relatively simple method of field assembly. Commonly available A36/A992 channels and plates are very good choices for the main stringers of the stairs, as are angles and tees for the carrier members. Risers and treads are usually made by specialty shops, and it is most economical to use a typical detail in as many places as possible. The stairs should be pre-assembled and shipped directly to the site. The field connections should be simple and straightforward enough to be accomplished efficiently, and with minimum equipment and labor.

Reference Numbers

Reference numbers are shown at the beginning of some major classifications. These numbers refer to related items in the Reference Section. The reference information may be an estimating procedure, an alternate pricing method, or technical information.

Note: Not all subdivisions listed here necessarily appear. ■

05 05 19 – Post-Installed Concrete Anchors

05 05 19.10 Chemical Anchors		Crew	Daily Output	Labor-Hours	Unit	Material	2019 Bare Costs Labor	Equipment	Total	Total Incl O&P
0010	CHEMICAL ANCHORS									
0020	Includes layout & drilling									
1430	Chemical anchor, w/rod & epoxy cartridge, 3/4" diameter x 9-1/2" long	B-89A	27	.593	Ea.	10.90	28	4.15	43.05	59
1435	1" diameter x 11-3/4" long		24	.667		18.80	31.50	4.67	54.97	73.50
1440	1-1/4" diameter x 14" long		21	.762		37.50	36	5.35	78.85	102
1445	1-3/4" diameter x 15" long		20	.800		72	38	5.60	115.60	143
1450	18" long		17	.941		86	44.50	6.60	137.10	169
1455	2" diameter x 18" long		16	1		114	47	7	168	204
1460	24" long		15	1.067		148	50.50	7.50	206	248

05 05 19.20 Expansion Anchors			Crew	Daily Output	Labor-Hours	Unit	Material	Labor	Equipment	Total	Total Incl O&P
0010	EXPANSION ANCHORS										
0100	Anchors for concrete, brick or stone, no layout and drilling										
0200	Expansion shields, zinc, 1/4" diameter, 1-5/16" long, single	G	1 Carp	90	.089	Ea.	.45	4.59		5.04	7.45
0300	1-3/8" long, double	G		85	.094		.59	4.86		5.45	8
0400	3/8" diameter, 1-1/2" long, single	G		85	.094		.70	4.86		5.56	8.10
0500	2" long, double	G		80	.100		1.18	5.15		6.33	9.10
0600	1/2" diameter, 2-1/16" long, single	G		80	.100		1.37	5.15		6.52	9.30
0700	2-1/2" long, double	G		75	.107		2.17	5.50		7.67	10.75
0800	5/8" diameter, 2-5/8" long, single	G		75	.107		2.19	5.50		7.69	10.75
0900	2-3/4" long, double	G		70	.114		2.97	5.90		8.87	12.20
1000	3/4" diameter, 2-3/4" long, single	G		70	.114		3.78	5.90		9.68	13.10
1100	3-15/16" long, double	G		65	.123		5.55	6.35		11.90	15.70
2100	Hollow wall anchors for gypsum wall board, plaster or tile										
2300	1/8" diameter, short	G	1 Carp	160	.050	Ea.	.37	2.58		2.95	4.32
2400	Long	G		150	.053		.32	2.75		3.07	4.52
2500	3/16" diameter, short	G		150	.053		.52	2.75		3.27	4.74
2600	Long	G		140	.057		.68	2.95		3.63	5.20
2700	1/4" diameter, short	G		140	.057		.76	2.95		3.71	5.30
2800	Long	G		130	.062		.86	3.18		4.04	5.75
3000	Toggle bolts, bright steel, 1/8" diameter, 2" long	G		85	.094		.26	4.86		5.12	7.65
3100	4" long	G		80	.100		.29	5.15		5.44	8.10
3200	3/16" diameter, 3" long	G		80	.100		.34	5.15		5.49	8.15
3300	6" long	G		75	.107		.45	5.50		5.95	8.85
3400	1/4" diameter, 3" long	G		75	.107		.46	5.50		5.96	8.85
3500	6" long	G		70	.114		.59	5.90		6.49	9.60
3600	3/8" diameter, 3" long	G		70	.114		.86	5.90		6.76	9.90
3700	6" long	G		60	.133		1.71	6.90		8.61	12.30
3800	1/2" diameter, 4" long	G		60	.133		1.68	6.90		8.58	12.25
3900	6" long	G		50	.160		2.25	8.25		10.50	15
5000	Screw anchors for concrete, masonry,										
5100	stone & tile, no layout or drilling included										
5700	Lag screw shields, 1/4" diameter, short	G	1 Carp	90	.089	Ea.	.48	4.59		5.07	7.50
5800	Long	G		85	.094		.55	4.86		5.41	7.95
5900	3/8" diameter, short	G		85	.094		.73	4.86		5.59	8.15
6000	Long	G		80	.100		.93	5.15		6.08	8.80
6100	1/2" diameter, short	G		80	.100		1.04	5.15		6.19	8.95
6200	Long	G		75	.107		1.38	5.50		6.88	9.85
6300	5/8" diameter, short	G		70	.114		1.56	5.90		7.46	10.65
6400	Long	G		65	.123		2.08	6.35		8.43	11.90
6600	Lead, #6 & #8, 3/4" long	G		260	.031		.19	1.59		1.78	2.61
6700	#10 - #14, 1-1/2" long	G		200	.040		.40	2.07		2.47	3.57
6800	#16 & #18, 1-1/2" long	G		160	.050		.42	2.58		3	4.37

05 05 19 – Post-Installed Concrete Anchors

05 05 19.20 Expansion Anchors		Crew	Daily Output	Labor-Hours	Unit	Material	Labor	Equipment	Total	Total Incl O&P
						2019 Bare Costs				
6900	Plastic, #6 & #8, 3/4" long	1 Carp	260	.031	Ea.	.05	1.59		1.64	2.46
7000	#8 & #10, 7/8" long		240	.033		.06	1.72		1.78	2.67
7100	#10 & #12, 1" long		220	.036		.08	1.88		1.96	2.93
7200	#14 & #16, 1-1/2" long		160	.050		.07	2.58		2.65	3.99
8000	Wedge anchors, not including layout or drilling									
8050	Carbon steel, 1/4" diameter, 1-3/4" long G	1 Carp	150	.053	Ea.	.54	2.75		3.29	4.76
8100	3-1/4" long G		140	.057		.71	2.95		3.66	5.25
8150	3/8" diameter, 2-1/4" long G		145	.055		.54	2.85		3.39	4.91
8200	5" long G		140	.057		.95	2.95		3.90	5.50
8250	1/2" diameter, 2-3/4" long G		140	.057		1.05	2.95		4	5.65
8300	7" long G		125	.064		1.80	3.31		5.11	7
8350	5/8" diameter, 3-1/2" long G		130	.062		1.79	3.18		4.97	6.80
8400	8-1/2" long G		115	.070		3.81	3.59		7.40	9.65
8450	3/4" diameter, 4-1/4" long G		115	.070		2.79	3.59		6.38	8.50
8500	10" long G		95	.084		6.35	4.35		10.70	13.60
8550	1" diameter, 6" long G		100	.080		9	4.13		13.13	16.15
8575	9" long G		85	.094		11.70	4.86		16.56	20
8600	12" long G		75	.107		12.65	5.50		18.15	22.50
8650	1-1/4" diameter, 9" long G		70	.114		22	5.90		27.90	33
8700	12" long G		60	.133		28	6.90		34.90	41.50
8750	For type 303 stainless steel, add					350%				
8800	For type 316 stainless steel, add					450%				
8950	Self-drilling concrete screw, hex washer head, 3/16" diam. x 1-3/4" long G	1 Carp	300	.027	Ea.	.19	1.38		1.57	2.29
8960	2-1/4" long G		250	.032		.21	1.65		1.86	2.73
8970	Phillips flat head, 3/16" diam. x 1-3/4" long G		300	.027		.19	1.38		1.57	2.29
8980	2-1/4" long G		250	.032		.21	1.65		1.86	2.73

05 05 21 – Fastening Methods for Metal

05 05 21.15 Drilling Steel		Crew	Daily Output	Labor-Hours	Unit	Material	Labor	Equipment	Total	Total Incl O&P
0010	**DRILLING STEEL**									
1910	Drilling & layout for steel, up to 1/4" deep, no anchor									
1920	Holes, 1/4" diameter	1 Sswk	112	.071	Ea.	.09	4		4.09	6.50
1925	For each additional 1/4" depth, add		336	.024		.09	1.33		1.42	2.23
1930	3/8" diameter		104	.077		.09	4.31		4.40	6.95
1935	For each additional 1/4" depth, add		312	.026		.09	1.44		1.53	2.39
1940	1/2" diameter		96	.083		.10	4.67		4.77	7.55
1945	For each additional 1/4" depth, add		288	.028		.10	1.56		1.66	2.59
1950	5/8" diameter		88	.091		.17	5.10		5.27	8.30
1955	For each additional 1/4" depth, add		264	.030		.17	1.70		1.87	2.89
1960	3/4" diameter		80	.100		.19	5.60		5.79	9.15
1965	For each additional 1/4" depth, add		240	.033		.19	1.87		2.06	3.19
1970	7/8" diameter		72	.111		.27	6.20		6.47	10.25
1975	For each additional 1/4" depth, add		216	.037		.27	2.07		2.34	3.61
1980	1" diameter		64	.125		.25	7		7.25	11.40
1985	For each additional 1/4" depth, add		192	.042		.25	2.33		2.58	3.99
1990	For drilling up, add						40%			

05 05 21.90 Welding Steel		Crew	Daily Output	Labor-Hours	Unit	Material	Labor	Equipment	Total	Total Incl O&P
0010	**WELDING STEEL**, Structural									
0020	Field welding, 1/8" E6011, cost per welder, no operating engineer	E-14	8	1	Hr.	4.68	58	11.95	74.63	111
0200	With 1/2 operating engineer	E-13	8	1.500		4.68	84	11.95	100.63	149
0300	With 1 operating engineer	E-12	8	2		4.68	110	11.95	126.63	188
0500	With no operating engineer, 2# weld rod per ton	E-14	8	1	Ton	4.68	58	11.95	74.63	111
0600	8# E6011 per ton	"	2	4		18.75	232	48	298.75	445

05 05 21 – Fastening Methods for Metal

05 05 21.90 Welding Steel		Crew	Daily Output	Labor-Hours	Unit	Material	2019 Bare Costs Labor	Equipment	Total	Total Incl O&P
0800	With one operating engineer per welder, 2# E6011 per ton	E-12	8	2	Ton	4.68	110	11.95	126.63	188
0900	8# E6011 per ton	"	2	8		18.75	440	48	506.75	755
1200	Continuous fillet, down welding									
1300	Single pass, 1/8" thick, 0.1#/L.F.	E-14	150	.053	L.F.	.23	3.09	.64	3.96	5.90
1400	3/16" thick, 0.2#/L.F.		75	.107		.47	6.20	1.28	7.95	11.75
1500	1/4" thick, 0.3#/L.F.		50	.160		.70	9.30	1.91	11.91	17.65
1610	5/16" thick, 0.4#/L.F.		38	.211		.94	12.20	2.52	15.66	23.50
1800	3 passes, 3/8" thick, 0.5#/L.F.		30	.267		1.17	15.45	3.19	19.81	29.50
2010	4 passes, 1/2" thick, 0.7#/L.F.		22	.364		1.64	21	4.35	26.99	40
2200	5 to 6 passes, 3/4" thick, 1.3#/L.F.		12	.667		3.04	38.50	7.95	49.49	73.50
2400	8 to 11 passes, 1" thick, 2.4#/L.F.		6	1.333		5.60	77.50	15.95	99.05	147
2600	For vertical joint welding, add						20%			
2700	Overhead joint welding, add						300%			
2900	For semi-automatic welding, obstructed joints, deduct						5%			
3000	Exposed joints, deduct						15%			
4000	Cleaning and welding plates, bars, or rods									
4010	to existing beams, columns, or trusses	E-14	12	.667	L.F.	1.17	38.50	7.95	47.62	71.50

05 05 23 – Metal Fastenings

05 05 23.30 Lag Screws

			Crew	Daily Output	Labor-Hours	Unit	Material	Labor	Equipment	Total	Total Incl O&P
0010	**LAG SCREWS**										
0020	Steel, 1/4" diameter, 2" long	G	1 Carp	200	.040	Ea.	.10	2.07		2.17	3.24
0100	3/8" diameter, 3" long	G		150	.053		.28	2.75		3.03	4.48
0200	1/2" diameter, 3" long	G		130	.062		.69	3.18		3.87	5.55
0300	5/8" diameter, 3" long	G		120	.067		1.31	3.44		4.75	6.65

05 05 23.35 Machine Screws

			Crew	Daily Output	Labor-Hours	Unit	Material	Labor	Equipment	Total	Total Incl O&P
0010	**MACHINE SCREWS**										
0020	Steel, round head, #8 x 1" long	G	1 Carp	4.80	1.667	C	4.13	86		90.13	135
0110	#8 x 2" long	G		2.40	3.333		8.40	172		180.40	270
0200	#10 x 1" long	G		4	2		5.05	103		108.05	162
0300	#10 x 2" long	G		2	4		9	207		216	325

05 05 23.50 Powder Actuated Tools and Fasteners

			Crew	Daily Output	Labor-Hours	Unit	Material	Labor	Equipment	Total	Total Incl O&P
0010	**POWDER ACTUATED TOOLS & FASTENERS**										
0020	Stud driver, .22 caliber, single shot					Ea.	162			162	178
0100	.27 caliber, semi automatic, strip					"	460			460	505
0300	Powder load, single shot, .22 cal, power level 2, brown					C	6.50			6.50	7.15
0400	Strip, .27 cal, power level 4, red						9.30			9.30	10.20
0600	Drive pin, .300 x 3/4" long	G	1 Carp	4.80	1.667		4.43	86		90.43	135
0700	.300 x 3" long with washer	G	"	4	2		13.50	103		116.50	171

05 05 23.55 Rivets

			Crew	Daily Output	Labor-Hours	Unit	Material	Labor	Equipment	Total	Total Incl O&P
0010	**RIVETS**										
0100	Aluminum rivet & mandrel, 1/2" grip length x 1/8" diameter	G	1 Carp	4.80	1.667	C	8.05	86		94.05	139
0200	3/16" diameter	G		4	2		11	103		114	168
0300	Aluminum rivet, steel mandrel, 1/8" diameter	G		4.80	1.667		10.35	86		96.35	141
0400	3/16" diameter	G		4	2		16.80	103		119.80	175
0500	Copper rivet, steel mandrel, 1/8" diameter	G		4.80	1.667		10.20	86		96.20	141
1200	Steel rivet and mandrel, 1/8" diameter	G		4.80	1.667		8.15	86		94.15	139
1300	3/16" diameter	G		4	2		11.50	103		114.50	169
1400	Hand riveting tool, standard					Ea.	75.50			75.50	83
1500	Deluxe						425			425	470
1600	Power riveting tool, standard						540			540	595
1700	Deluxe						1,550			1,550	1,700

05 05 Common Work Results for Metals

05 05 23 – Metal Fastenings

05 05 23.70 Structural Blind Bolts			Crew	Daily Output	Labor-Hours	Unit	Material	2019 Bare Costs Labor	Equipment	Total	Total Incl O&P
0010	**STRUCTURAL BLIND BOLTS**										
0100	1/4″ diameter x 1/4″ grip	G	1 Sswk	240	.033	Ea.	1.76	1.87		3.63	4.92
0150	1/2″ grip	G		216	.037		1.87	2.07		3.94	5.35
0200	3/8″ diameter x 1/2″ grip	G		232	.034		3.46	1.93		5.39	6.90
0250	3/4″ grip	G		208	.038		3.33	2.15		5.48	7.10
0300	1/2″ diameter x 1/2″ grip	G		224	.036		6.15	2		8.15	10
0350	3/4″ grip	G		200	.040		8.35	2.24		10.59	12.75
0400	5/8″ diameter x 3/4″ grip	G		216	.037		11.10	2.07		13.17	15.55
0450	1″ grip	G		192	.042		15.50	2.33		17.83	21

05 12 Structural Steel Framing

05 12 23 – Structural Steel for Buildings

05 12 23.40 Lightweight Framing

			Crew	Daily Output	Labor-Hours	Unit	Material	Labor	Equipment	Total	Total Incl O&P
0010	**LIGHTWEIGHT FRAMING**										
0015	Made from recycled materials										
0400	Angle framing, field fabricated, 4″ and larger	G	E-3	440	.055	Lb.	.77	3.09	.22	4.08	6
0450	Less than 4″ angles	G		265	.091		.79	5.15	.36	6.30	9.45
0600	Channel framing, field fabricated, 8″ and larger	G		500	.048		.79	2.72	.19	3.70	5.40
0650	Less than 8″ channels	G		335	.072		.79	4.06	.29	5.14	7.70
1000	Continuous slotted channel framing system, shop fab, simple framing	G	2 Sswk	2400	.007		4.10	.37		4.47	5.10
1200	Complex framing	G	″	1600	.010		4.63	.56		5.19	6
1250	Plate & bar stock for reinforcing beams and trusses	G					1.45			1.45	1.60
1300	Cross bracing, rods, shop fabricated, 3/4″ diameter	G	E-3	700	.034		1.59	1.94	.14	3.67	4.99
1310	7/8″ diameter	G		850	.028		1.59	1.60	.11	3.30	4.41
1320	1″ diameter	G		1000	.024		1.59	1.36	.10	3.05	4.02
1330	Angle, 5″ x 5″ x 3/8″	G		2800	.009		1.59	.49	.03	2.11	2.55
1350	Hanging lintels, shop fabricated	G		850	.028		1.59	1.60	.11	3.30	4.41
1380	Roof frames, shop fabricated, 3′-0″ square, 5′ span	G	E-2	4200	.013		1.59	.74	.40	2.73	3.34
1400	Tie rod, not upset, 1-1/2″ to 4″ diameter, with turnbuckle	G	2 Sswk	800	.020		1.72	1.12		2.84	3.68
1420	No turnbuckle	G		700	.023		1.65	1.28		2.93	3.86
1500	Upset, 1-3/4″ to 4″ diameter, with turnbuckle	G		800	.020		1.72	1.12		2.84	3.68
1520	No turnbuckle	G		700	.023		1.65	1.28		2.93	3.86

05 12 23.60 Pipe Support Framing

			Crew	Daily Output	Labor-Hours	Unit	Material	Labor	Equipment	Total	Total Incl O&P
0010	**PIPE SUPPORT FRAMING**										
0020	Under 10#/L.F., shop fabricated	G	E-4	3900	.008	Lb.	1.77	.46	.02	2.25	2.72
0200	10.1 to 15#/L.F.	G		4300	.007		1.74	.42	.02	2.18	2.61
0400	15.1 to 20#/L.F.	G		4800	.007		1.72	.38	.02	2.12	2.51
0600	Over 20#/L.F.	G		5400	.006		1.69	.34	.02	2.05	2.41

Division Notes

		CREW	DAILY OUTPUT	LABOR-HOURS	UNIT	BARE COSTS MAT.	LABOR	EQUIP.	TOTAL	TOTAL INCL O&P

Estimating Tips

06 05 00 Common Work Results for Wood, Plastics, and Composites

- Common to any wood-framed structure are the accessory connector items such as screws, nails, adhesives, hangers, connector plates, straps, angles, and hold-downs. For typical wood-framed buildings, such as residential projects, the aggregate total for these items can be significant, especially in areas where seismic loading is a concern. For floor and wall framing, the material cost is based on 10 to 25 lbs. of accessory connectors per MBF. Hold-downs, hangers, and other connectors should be taken off by the piece.

  Included with material costs are fasteners for a normal installation. Gordian's RSMeans engineers use manufacturers' recommendations, written specifications, and/or standard construction practice for the sizing and spacing of fasteners. Prices for various fasteners are shown for informational purposes only. Adjustments should be made if unusual fastening conditions exist.

06 10 00 Carpentry

- Lumber is a traded commodity and therefore sensitive to supply and demand in the marketplace. Even with "budgetary" estimating of wood-framed projects, it is advisable to call local suppliers for the latest market pricing.
- The common quantity unit for wood-framed projects is "thousand board feet" (MBF). A board foot is a volume of wood—1" x 1' x 1' or 144 cubic inches. Board-foot quantities are generally calculated using nominal material dimensions—dressed sizes are ignored. Board foot per lineal foot of any stick of lumber can be calculated by dividing the nominal cross-sectional area by 12. As an example, 2,000 lineal feet of 2 x 12 equates to 4 MBF by dividing the nominal area, 2 x 12, by 12, which equals 2, and multiplying that by 2,000 to give 4,000 board feet. This simple rule applies to all nominal dimensioned lumber.
- Waste is an issue of concern at the quantity takeoff for any area of construction. Framing lumber is sold in even foot lengths, i.e., 8', 10', 12', 14', 16', and depending on spans, wall heights, and the grade of lumber, waste is inevitable. A rule of thumb for lumber waste is 5–10% depending on material quality and the complexity of the framing.
- Wood in various forms and shapes is used in many projects, even where the main structural framing is steel, concrete, or masonry. Plywood as a back-up partition material and 2x boards used as blocking and cant strips around roof edges are two common examples. The estimator should ensure that the costs of all wood materials are included in the final estimate.

06 20 00 Finish Carpentry

- It is necessary to consider the grade of workmanship when estimating labor costs for erecting millwork and an interior finish. In practice, there are three grades: premium, custom, and economy. The RSMeans daily output for base and case moldings is in the range of 200 to 250 L.F. per carpenter per day. This is appropriate for most average custom-grade projects. For premium projects, an adjustment to productivity of 25–50% should be made, depending on the complexity of the job.

Reference Numbers

Reference numbers are shown at the beginning of some major classifications. These numbers refer to related items in the Reference Section. The reference information may be an estimating procedure, an alternate pricing method, or technical information.

Note: Not all subdivisions listed here necessarily appear. ■

06 16 36 – Wood Panel Product Sheathing

06 16 36.10 Sheathing		Crew	Daily Output	Labor-Hours	Unit	2019 Bare Costs Material	Labor	Equipment	Total	Total Incl O&P
0010	**SHEATHING**									
0012	Plywood on roofs, CDX									
0030	5/16" thick	2 Carp	1600	.010	S.F.	.59	.52		1.11	1.43
0035	Pneumatic nailed		1952	.008		.59	.42		1.01	1.29
0050	3/8" thick		1525	.010		.61	.54		1.15	1.50
0055	Pneumatic nailed		1860	.009		.61	.44		1.05	1.35
0100	1/2" thick		1400	.011		.63	.59		1.22	1.59
0105	Pneumatic nailed		1708	.009		.63	.48		1.11	1.43
0200	5/8" thick		1300	.012		.78	.64		1.42	1.82
0205	Pneumatic nailed		1586	.010		.78	.52		1.30	1.65
0300	3/4" thick		1200	.013		.94	.69		1.63	2.07
0305	Pneumatic nailed		1464	.011		.94	.56		1.50	1.88
0500	Plywood on walls, with exterior CDX, 3/8" thick		1200	.013		.61	.69		1.30	1.72
0505	Pneumatic nailed		1488	.011		.61	.56		1.17	1.52
0600	1/2" thick		1125	.014		.63	.73		1.36	1.81
0605	Pneumatic nailed		1395	.011		.63	.59		1.22	1.60
0700	5/8" thick		1050	.015		.78	.79		1.57	2.05
0705	Pneumatic nailed		1302	.012		.78	.63		1.41	1.82
0800	3/4" thick		975	.016		.94	.85		1.79	2.31
0805	Pneumatic nailed	▼	1209	.013	▼	.94	.68		1.62	2.06

Estimating Tips

07 10 00 Dampproofing and Waterproofing

- Be sure of the job specifications before pricing this subdivision. The difference in cost between waterproofing and dampproofing can be great. Waterproofing will hold back standing water. Dampproofing prevents the transmission of water vapor. Also included in this section are vapor retarding membranes.

07 20 00 Thermal Protection

- Insulation and fireproofing products are measured by area, thickness, volume, or R-value. Specifications may give only what the specific R-value should be in a certain situation. The estimator may need to choose the type of insulation to meet that R-value.

07 30 00 Steep Slope Roofing 07 40 00 Roofing and Siding Panels

- Many roofing and siding products are bought and sold by the square. One square is equal to an area that measures 100 square feet.

  This simple change in unit of measure could create a large error if the estimator is not observant. Accessories necessary for a complete installation must be figured into any calculations for both material and labor.

07 50 00 Membrane Roofing 07 60 00 Flashing and Sheet Metal 07 70 00 Roofing and Wall Specialties and Accessories

- The items in these subdivisions compose a roofing system. No one component completes the installation, and all must be estimated. Built-up or single-ply membrane roofing systems are made up of many products and installation trades. Wood blocking at roof perimeters or penetrations, parapet coverings, reglets, roof drains, gutters, downspouts, sheet metal flashing, skylights, smoke vents, and roof hatches all need to be considered along with the roofing material. Several different installation trades will need to work together on the roofing system. Inherent difficulties in the scheduling and coordination of various trades must be accounted for when estimating labor costs.

07 90 00 Joint Protection

- To complete the weather-tight shell, the sealants and caulkings must be estimated. Where different materials meet—at expansion joints, at flashing penetrations, and at hundreds of other locations throughout a construction project—caulking and sealants provide another line of defense against water penetration. Often, an entire system is based on the proper location and placement of caulking or sealants. The detailed drawings that are included as part of a set of architectural plans show typical locations for these materials. When caulking or sealants are shown at typical locations, this means the estimator must include them for all the locations where this detail is applicable. Be careful to keep different types of sealants separate, and remember to consider backer rods and primers if necessary.

Reference Numbers

Reference numbers are shown at the beginning of some major classifications. These numbers refer to related items in the Reference Section. The reference information may be an estimating procedure, an alternate pricing method, or technical information.

Note: Not all subdivisions listed here necessarily appear. ■

07 65 Flexible Flashing

07 65 10 – Sheet Metal Flashing

07 65 10.10 Sheet Metal Flashing and Counter Flashing		Crew	Daily Output	Labor-Hours	Unit	2019 Bare Costs Material	Labor	Equipment	Total	Total Incl O&P
0010	**SHEET METAL FLASHING AND COUNTER FLASHING**									
0011	Including up to 4 bends									
0020	Aluminum, mill finish, .013" thick	1 Rofc	145	.055	S.F.	.86	2.49		3.35	5.10
0030	.016" thick		145	.055		1.01	2.49		3.50	5.25
0060	.019" thick		145	.055		1.42	2.49		3.91	5.70
0100	.032" thick		145	.055		1.36	2.49		3.85	5.65
0200	.040" thick		145	.055		2.30	2.49		4.79	6.65
0300	.050" thick		145	.055		2.75	2.49		5.24	7.15
0400	Painted finish, add					.33			.33	.36
1600	Copper, 16 oz. sheets, under 1000 lb.	1 Rofc	115	.070		8.10	3.13		11.23	14.10
1700	Over 4000 lb.		155	.052		8.10	2.33		10.43	12.75
1900	20 oz. sheets, under 1000 lb.		110	.073		10.75	3.28		14.03	17.30
2000	Over 4000 lb.		145	.055		10.20	2.49		12.69	15.40
2200	24 oz. sheets, under 1000 lb.		105	.076		14.75	3.43		18.18	22
2300	Over 4000 lb.		135	.059		14	2.67		16.67	19.85
2500	32 oz. sheets, under 1000 lb.		100	.080		19.60	3.60		23.20	27.50
2600	Over 4000 lb.		130	.062		18.65	2.77		21.42	25
5800	Lead, 2.5 lb./S.F., up to 12" wide		135	.059		6.20	2.67		8.87	11.25
5900	Over 12" wide		135	.059		3.99	2.67		6.66	8.85
8900	Stainless steel sheets, 32 ga.		155	.052		3.41	2.33		5.74	7.60
9000	28 ga.		155	.052		4.75	2.33		7.08	9.10
9100	26 ga.		155	.052		4.70	2.33		7.03	9
9200	24 ga.		155	.052		5.25	2.33		7.58	9.60
9400	Terne coated stainless steel, .015" thick, 28 ga.		155	.052		8.20	2.33		10.53	12.85
9500	.018" thick, 26 ga.		155	.052		9.15	2.33		11.48	13.90
9600	Zinc and copper alloy (brass), .020" thick		155	.052		10.10	2.33		12.43	15
9700	.027" thick		155	.052		12.50	2.33		14.83	17.60
9800	.032" thick		155	.052		15.80	2.33		18.13	21.50
9900	.040" thick		155	.052		20.50	2.33		22.83	26.50

07 65 13 – Laminated Sheet Flashing

07 65 13.10 Laminated Sheet Flashing		Crew	Daily Output	Labor-Hours	Unit	Material	Labor	Equipment	Total	Total Incl O&P
0010	**LAMINATED SHEET FLASHING**, Including up to 4 bends									
4300	Copper-clad stainless steel, .015" thick, under 500 lb.	1 Rofc	115	.070	S.F.	6.70	3.13		9.83	12.55
4400	Over 2000 lb.		155	.052		6.80	2.33		9.13	11.35
4600	.018" thick, under 500 lb.		100	.080		8.05	3.60		11.65	14.85
4700	Over 2000 lb.		145	.055		7.85	2.49		10.34	12.75
8550	Shower pan, 3 ply copper and fabric, 3 oz.		155	.052		4.05	2.33		6.38	8.35
8600	7 oz.		155	.052		4.87	2.33		7.20	9.20
9300	Stainless steel, paperbacked 2 sides, .005" thick		330	.024		3.96	1.09		5.05	6.20

07 65 19 – Plastic Sheet Flashing

07 65 19.10 Plastic Sheet Flashing and Counter Flashing		Crew	Daily Output	Labor-Hours	Unit	Material	Labor	Equipment	Total	Total Incl O&P
0010	**PLASTIC SHEET FLASHING AND COUNTER FLASHING**									
7300	Polyvinyl chloride, black, 10 mil	1 Rofc	285	.028	S.F.	.27	1.26		1.53	2.41
7400	20 mil		285	.028		.26	1.26		1.52	2.40
7600	30 mil		285	.028		.33	1.26		1.59	2.47
7700	60 mil		285	.028		.86	1.26		2.12	3.06
7900	Black or white for exposed roofs, 60 mil		285	.028		1.25	1.26		2.51	3.49
8060	PVC tape, 5" x 45 mils, for joint covers, 100 L.F./roll				Ea.	180			180	198
8850	Polyvinyl chloride, 30 mil	1 Rofc	160	.050	S.F.	1.53	2.25		3.78	5.45

07 71 Roof Specialties

07 71 16 – Manufactured Counterflashing Systems

07 71 16.20 Pitch Pockets, Variable Sizes		Crew	Daily Output	Labor-Hours	Unit	Material	2019 Bare Costs Labor	Equipment	Total	Total Incl O&P
0010	PITCH POCKETS, VARIABLE SIZES									
0100	Adjustable, 4" to 7", welded corners, 4" deep	1 Rofc	48	.167	Ea.	19.50	7.50		27	34
0200	Side extenders, 6"	"	240	.033	"	2.87	1.50		4.37	5.65

07 71 23 – Manufactured Gutters and Downspouts

07 71 23.10 Downspouts		Crew	Daily Output	Labor-Hours	Unit	Material	Labor	Equipment	Total	Total Incl O&P
0010	DOWNSPOUTS									
0020	Aluminum, embossed, .020" thick, 2" x 3"	1 Shee	190	.042	L.F.	.93	2.57		3.50	4.92
0100	Enameled		190	.042		1.37	2.57		3.94	5.40
4800	Steel, galvanized, round, corrugated, 2" or 3" diameter, 28 ga.		190	.042		2.11	2.57		4.68	6.20
4900	4" diameter, 28 ga.		145	.055		2.12	3.36		5.48	7.45
5100	5" diameter, 26 ga.		130	.062		3.65	3.75		7.40	9.70
5400	6" diameter, 28 ga.		105	.076		3.57	4.64		8.21	11
5500	26 ga.	↓	105	.076	↓	3.79	4.64		8.43	11.20

07 72 Roof Accessories

07 72 26 – Ridge Vents

07 72 26.10 Ridge Vents and Accessories		Crew	Daily Output	Labor-Hours	Unit	Material	Labor	Equipment	Total	Total Incl O&P
0010	RIDGE VENTS AND ACCESSORIES									
2300	Ridge vent strip, mill finish	1 Shee	155	.052	L.F.	3.94	3.15		7.09	9.10

07 72 53 – Snow Guards

07 72 53.10 Snow Guard Options		Crew	Daily Output	Labor-Hours	Unit	Material	Labor	Equipment	Total	Total Incl O&P
0010	SNOW GUARD OPTIONS									
0100	Slate & asphalt shingle roofs, fastened with nails	1 Rofc	160	.050	Ea.	12.40	2.25		14.65	17.40
0200	Standing seam metal roofs, fastened with set screws		48	.167		17.70	7.50		25.20	32
0300	Surface mount for metal roofs, fastened with solder		48	.167	↓	7.50	7.50		15	21
0400	Double rail pipe type, including pipe	↓	130	.062	L.F.	34	2.77		36.77	42

07 76 Roof Pavers

07 76 16 – Roof Decking Pavers

07 76 16.10 Roof Pavers and Supports		Crew	Daily Output	Labor-Hours	Unit	Material	Labor	Equipment	Total	Total Incl O&P
0010	ROOF PAVERS AND SUPPORTS									
1000	Roof decking pavers, concrete blocks, 2" thick, natural	1 Clab	115	.070	S.F.	3.56	2.86		6.42	8.25
1100	Colors		115	.070	"	3.82	2.86		6.68	8.50
1200	Support pedestal, bottom cap		960	.008	Ea.	2.64	.34		2.98	3.42
1300	Top cap		960	.008		4.84	.34		5.18	5.80
1400	Leveling shims, 1/16"		1920	.004		1.21	.17		1.38	1.59
1500	1/8"		1920	.004		1.21	.17		1.38	1.59
1600	Buffer pad		960	.008	↓	2.52	.34		2.86	3.29
1700	PVC legs (4" SDR 35)		2880	.003	Inch	.14	.11		.25	.32
2000	Alternate pricing method, system in place	↓	101	.079	S.F.	7.15	3.25		10.40	12.75

07 84 Firestopping

07 84 13 – Penetration Firestopping

07 84 13.10 Firestopping		Crew	Daily Output	Labor-Hours	Unit	Material	2019 Bare Costs Labor	Equipment	Total	Total Incl O&P
0010	**FIRESTOPPING** R078413-30									
0100	Metallic piping, non insulated									
0110	Through walls, 2″ diameter	1 Carp	16	.500	Ea.	3.95	26		29.95	43.50
0120	4″ diameter		14	.571		6.90	29.50		36.40	52
0130	6″ diameter		12	.667		10.15	34.50		44.65	63
0140	12″ diameter		10	.800		20	41.50		61.50	85
0150	Through floors, 2″ diameter		32	.250		2.01	12.90		14.91	22
0160	4″ diameter		28	.286		3.53	14.75		18.28	26.50
0170	6″ diameter		24	.333		5.15	17.20		22.35	31.50
0180	12″ diameter		20	.400		10.35	20.50		30.85	43
0190	Metallic piping, insulated									
0200	Through walls, 2″ diameter	1 Carp	16	.500	Ea.	6.90	26		32.90	46.50
0210	4″ diameter		14	.571		10.15	29.50		39.65	55.50
0220	6″ diameter		12	.667		13.60	34.50		48.10	67
0230	12″ diameter		10	.800		23	41.50		64.50	88
0240	Through floors, 2″ diameter		32	.250		3.53	12.90		16.43	23.50
0250	4″ diameter		28	.286		5.15	14.75		19.90	28
0260	6″ diameter		24	.333		6.95	17.20		24.15	33.50
0270	12″ diameter		20	.400		10.60	20.50		31.10	43
0280	Non metallic piping, non insulated									
0290	Through walls, 2″ diameter	1 Carp	12	.667	Ea.	49	34.50		83.50	106
0300	4″ diameter		10	.800		97	41.50		138.50	170
0310	6″ diameter		8	1		183	51.50		234.50	280
0330	Through floors, 2″ diameter		16	.500		32.50	26		58.50	75
0340	4″ diameter		6	1.333		61	69		130	171
0350	6″ diameter		6	1.333		118	69		187	234
0370	Ductwork, insulated & non insulated, round									
0380	Through walls, 6″ diameter	1 Carp	12	.667	Ea.	13.60	34.50		48.10	67
0390	12″ diameter		10	.800		23	41.50		64.50	88
0400	18″ diameter		8	1		28	51.50		79.50	109
0410	Through floors, 6″ diameter		16	.500		6.95	26		32.95	46.50
0420	12″ diameter		14	.571		11.85	29.50		41.35	57.50
0430	18″ diameter		12	.667		14.40	34.50		48.90	68
0440	Ductwork, insulated & non insulated, rectangular									
0450	With stiffener/closure angle, through walls, 6″ x 12″	1 Carp	8	1	Ea.	18.15	51.50		69.65	98
0460	12″ x 24″		6	1.333		35.50	69		104.50	143
0470	24″ x 48″		4	2		76	103		179	240
0480	With stiffener/closure angle, through floors, 6″ x 12″		10	.800		9.05	41.50		50.55	72.50
0490	12″ x 24″		8	1		18.15	51.50		69.65	98
0500	24″ x 48″		6	1.333		38	69		107	146
0510	Multi trade openings									
0520	Through walls, 6″ x 12″	1 Carp	2	4	Ea.	49.50	207		256.50	370
0530	12″ x 24″	″	1	8		172	415		587	815
0540	24″ x 48″	2 Carp	1	16		590	825		1,415	1,900
0550	48″ x 96″	″	.75	21.333		2,200	1,100		3,300	4,100
0560	Through floors, 6″ x 12″	1 Carp	2	4		46.50	207		253.50	365
0570	12″ x 24″	″	1	8		63	415		478	695
0580	24″ x 48″	2 Carp	.75	21.333		151	1,100		1,251	1,850
0590	48″ x 96″	″	.50	32		360	1,650		2,010	2,900
0600	Structural penetrations, through walls									
0610	Steel beams, W8 x 10	1 Carp	8	1	Ea.	75	51.50		126.50	161
0620	W12 x 14		6	1.333		109	69		178	224
0630	W21 x 44		5	1.600		147	82.50		229.50	287

07 84 Firestopping

07 84 13 – Penetration Firestopping

07 84 13.10 Firestopping		Crew	Daily Output	Labor-Hours	Unit	Material	Labor	Equipment	Total	Total Incl O&P
						2019 Bare Costs				
0640	W36 x 135	1 Carp	3	2.667	Ea.	241	138		379	475
0650	Bar joists, 18" deep		6	1.333		44.50	69		113.50	153
0660	24" deep		6	1.333		59	69		128	169
0670	36" deep		5	1.600		89	82.50		171.50	223
0680	48" deep		4	2		119	103		222	286
0690	Construction joints, floor slab at exterior wall									
0700	Precast, brick, block or drywall exterior									
0710	2" wide joint	1 Carp	125	.064	L.F.	9.85	3.31		13.16	15.85
0720	4" wide joint	"	75	.107	"	16.35	5.50		21.85	26.50
0730	Metal panel, glass or curtain wall exterior									
0740	2" wide joint	1 Carp	40	.200	L.F.	10.65	10.35		21	27.50
0750	4" wide joint	"	25	.320	"	16.65	16.55		33.20	43.50
0760	Floor slab to drywall partition									
0770	Flat joint	1 Carp	100	.080	L.F.	14.55	4.13		18.68	22.50
0780	Fluted joint		50	.160		16.15	8.25		24.40	30.50
0790	Etched fluted joint		75	.107		20.50	5.50		26	31
0800	Floor slab to concrete/masonry partition									
0810	Flat joint	1 Carp	75	.107	L.F.	14.50	5.50		20	24.50
0820	Fluted joint	"	50	.160	"	16.55	8.25		24.80	30.50
0830	Concrete/CMU wall joints									
0840	1" wide	1 Carp	100	.080	L.F.	20.50	4.13		24.63	29.50
0850	2" wide		75	.107		25.50	5.50		31	36.50
0860	4" wide		50	.160		35.50	8.25		43.75	51.50
0870	Concrete/CMU floor joints									
0880	1" wide	1 Carp	200	.040	L.F.	20.50	2.07		22.57	25.50
0890	2" wide		150	.053		25.50	2.75		28.25	32
0900	4" wide		100	.080		36.50	4.13		40.63	46.50

07 91 Preformed Joint Seals

07 91 13 – Compression Seals

07 91 13.10 Compression Seals		Crew	Daily Output	Labor-Hours	Unit	Material	Labor	Equipment	Total	Total Incl O&P
0010	**COMPRESSION SEALS**									
4900	O-ring type cord, 1/4"	1 Bric	472	.017	L.F.	.41	.87		1.28	1.77
4910	1/2"		440	.018		1.06	.93		1.99	2.59
4920	3/4"		424	.019		2.05	.96		3.01	3.73
4930	1"		408	.020		4.15	1		5.15	6.10
4940	1-1/4"		384	.021		7.90	1.06		8.96	10.30
4950	1-1/2"		368	.022		9.70	1.11		10.81	12.35
4960	1-3/4"		352	.023		13.60	1.16		14.76	16.70
4970	2"		344	.023		22.50	1.19		23.69	26.50

07 91 16 – Joint Gaskets

07 91 16.10 Joint Gaskets		Crew	Daily Output	Labor-Hours	Unit	Material	Labor	Equipment	Total	Total Incl O&P
0010	**JOINT GASKETS**									
4400	Joint gaskets, neoprene, closed cell w/adh, 1/8" x 3/8"	1 Bric	240	.033	L.F.	.33	1.70		2.03	2.96
4500	1/4" x 3/4"		215	.037		.64	1.90		2.54	3.60
4700	1/2" x 1"		200	.040		1.51	2.04		3.55	4.78
4800	3/4" x 1-1/2"		165	.048		1.65	2.47		4.12	5.60

07 91 Preformed Joint Seals

07 91 23 – Backer Rods

07 91 23.10 Backer Rods		Crew	Daily Output	Labor-Hours	Unit	Material	Labor	Equipment	Total	Total Incl O&P
						2019 Bare Costs				
0010	BACKER RODS									
0030	Backer rod, polyethylene, 1/4" diameter	1 Bric	4.60	1.739	C.L.F.	2.42	89		91.42	139
0050	1/2" diameter		4.60	1.739		4.09	89		93.09	141
0070	3/4" diameter		4.60	1.739		6.95	89		95.95	144
0090	1" diameter		4.60	1.739		12.45	89		101.45	150

07 91 26 – Joint Fillers

07 91 26.10 Joint Fillers		Crew	Daily Output	Labor-Hours	Unit	Material	Labor	Equipment	Total	Total Incl O&P
0010	JOINT FILLERS									
4360	Butyl rubber filler, 1/4" x 1/4"	1 Bric	290	.028	L.F.	.23	1.41		1.64	2.40
4365	1/2" x 1/2"		250	.032		.91	1.63		2.54	3.49
4370	1/2" x 3/4"		210	.038		1.37	1.95		3.32	4.47
4375	3/4" x 3/4"		230	.035		2.05	1.78		3.83	4.96
4380	1" x 1"		180	.044		2.73	2.27		5	6.45
4390	For coloring, add					12%				
4980	Polyethylene joint backing, 1/4" x 2"	1 Bric	2.08	3.846	C.L.F.	13.85	196		209.85	315
4990	1/4" x 6"		1.28	6.250	"	29.50	320		349.50	520
5600	Silicone, room temp vulcanizing foam seal, 1/4" x 1/2"		1312	.006	L.F.	.46	.31		.77	.99
5610	1/2" x 1/2"		656	.012		.92	.62		1.54	1.97
5620	1/2" x 3/4"		442	.018		1.38	.92		2.30	2.93
5630	3/4" x 3/4"		328	.024		2.08	1.25		3.33	4.19
5640	1/8" x 1"		1312	.006		.46	.31		.77	.99
5650	1/8" x 3"		442	.018		1.38	.92		2.30	2.93
5670	1/4" x 3"		295	.027		2.77	1.38		4.15	5.15
5680	1/4" x 6"		148	.054		5.55	2.76		8.31	10.30
5690	1/2" x 6"		82	.098		11.10	4.98		16.08	19.80
5700	1/2" x 9"		52.50	.152		16.60	7.80		24.40	30
5710	1/2" x 12"		33	.242		22	12.40		34.40	43.50

07 92 Joint Sealants

07 92 13 – Elastomeric Joint Sealants

07 92 13.20 Caulking and Sealant Options		Crew	Daily Output	Labor-Hours	Unit	Material	Labor	Equipment	Total	Total Incl O&P
0010	CAULKING AND SEALANT OPTIONS									
0050	Latex acrylic based, bulk				Gal.	30			30	33
0055	Bulk in place 1/4" x 1/4" bead	1 Bric	300	.027	L.F.	.09	1.36		1.45	2.18
0060	1/4" x 3/8"		294	.027		.16	1.39		1.55	2.29
0065	1/4" x 1/2"		288	.028		.21	1.42		1.63	2.40
0075	3/8" x 3/8"		284	.028		.24	1.44		1.68	2.46
0080	3/8" x 1/2"		280	.029		.32	1.46		1.78	2.58
0085	3/8" x 5/8"		276	.029		.39	1.48		1.87	2.69
0095	3/8" x 3/4"		272	.029		.47	1.50		1.97	2.81
0100	1/2" x 1/2"		275	.029		.42	1.49		1.91	2.73
0105	1/2" x 5/8"		269	.030		.53	1.52		2.05	2.90
0110	1/2" x 3/4"		263	.030		.63	1.55		2.18	3.06
0115	1/2" x 7/8"		256	.031		.74	1.60		2.34	3.25
0120	1/2" x 1"		250	.032		.84	1.63		2.47	3.42
0125	3/4" x 3/4"		244	.033		.95	1.67		2.62	3.60
0130	3/4" x 1"		225	.036		1.26	1.82		3.08	4.16
0135	1" x 1"		200	.040		1.68	2.04		3.72	4.97
0190	Cartridges				Gal.	33			33	36
0200	11 fl. oz. cartridge				Ea.	2.82			2.82	3.10

07 92 Joint Sealants

07 92 13 – Elastomeric Joint Sealants

07 92 13.20 Caulking and Sealant Options		Crew	Daily Output	Labor-Hours	Unit	Material	2019 Bare Costs Labor	Equipment	Total	Total Incl O&P
0500	1/4" x 1/2"	1 Bric	288	.028	L.F.	.23	1.42		1.65	2.42
0600	1/2" x 1/2"		275	.029		.46	1.49		1.95	2.78
0800	3/4" x 3/4"		244	.033		1.04	1.67		2.71	3.70
0900	3/4" x 1"		225	.036		1.38	1.82		3.20	4.29
1000	1" x 1"	↓	200	.040	↓	1.73	2.04		3.77	5
1400	Butyl based, bulk				Gal.	37.50			37.50	41
1500	Cartridges				"	37.50			37.50	41
1700	1/4" x 1/2", 154 L.F./gal.	1 Bric	288	.028	L.F.	.24	1.42		1.66	2.44
1800	1/2" x 1/2", 77 L.F./gal.	"	275	.029	"	.49	1.49		1.98	2.80
2300	Polysulfide compounds, 1 component, bulk				Gal.	84.50			84.50	93
2600	1 or 2 component, in place, 1/4" x 1/4", 308 L.F./gal.	1 Bric	300	.027	L.F.	.27	1.36		1.63	2.38
2700	1/2" x 1/4", 154 L.F./gal.		288	.028		.55	1.42		1.97	2.77
2900	3/4" x 3/8", 68 L.F./gal.		272	.029		1.24	1.50		2.74	3.65
3000	1" x 1/2", 38 L.F./gal.	↓	250	.032	↓	2.22	1.63		3.85	4.93
3200	Polyurethane, 1 or 2 component				Gal.	55.50			55.50	61
3300	Cartridges				"	68			68	75
3500	Bulk, in place, 1/4" x 1/4"	1 Bric	300	.027	L.F.	.18	1.36		1.54	2.28
3655	1/2" x 1/4"		288	.028		.36	1.42		1.78	2.57
3800	3/4" x 3/8"		272	.029		.82	1.50		2.32	3.19
3900	1" x 1/2"	↓	250	.032	↓	1.44	1.63		3.07	4.08
4100	Silicone rubber, bulk				Gal.	58.50			58.50	64
4200	Cartridges				"	52.50			52.50	58

07 92 16 – Rigid Joint Sealants

07 92 16.10 Rigid Joint Sealants		Crew	Daily Output	Labor-Hours	Unit	Material	Labor	Equipment	Total	Total Incl O&P
0010	**RIGID JOINT SEALANTS**									
5800	Tapes, sealant, PVC foam adhesive, 1/16" x 1/4"				C.L.F.	9.20			9.20	10.10
5900	1/16" x 1/2"					9.20			9.20	10.10
5950	1/16" x 1"					16.30			16.30	17.95
6000	1/8" x 1/2"				↓	9.20			9.20	10.10

07 92 19 – Acoustical Joint Sealants

07 92 19.10 Acoustical Sealant		Crew	Daily Output	Labor-Hours	Unit	Material	Labor	Equipment	Total	Total Incl O&P
0010	**ACOUSTICAL SEALANT**									
0020	Acoustical sealant, elastomeric, cartridges				Ea.	8.60			8.60	9.45
0025	In place, 1/4" x 1/4"	1 Bric	300	.027	L.F.	.35	1.36		1.71	2.47
0030	1/4" x 1/2"		288	.028		.70	1.42		2.12	2.94
0035	1/2" x 1/2"		275	.029		1.40	1.49		2.89	3.81
0040	1/2" x 3/4"		263	.030		2.11	1.55		3.66	4.69
0045	3/4" x 3/4"		244	.033		3.16	1.67		4.83	6.05
0050	1" x 1"	↓	200	.040	↓	5.60	2.04		7.64	9.30

Division Notes

		CREW	DAILY OUTPUT	LABOR-HOURS	UNIT	BARE COSTS MAT.	LABOR	EQUIP.	TOTAL	TOTAL INCL O&P

Estimating Tips

08 10 00 Doors and Frames

All exterior doors should be addressed for their energy conservation (insulation and seals).

- Most metal doors and frames look alike, but there may be significant differences among them. When estimating these items, be sure to choose the line item that most closely compares to the specification or door schedule requirements regarding:
  - ☐ type of metal
  - ☐ metal gauge
  - ☐ door core material
  - ☐ fire rating
  - ☐ finish
- Wood and plastic doors vary considerably in price. The primary determinant is the veneer material. Lauan, birch, and oak are the most common veneers. Other variables include the following:
  - ☐ hollow or solid core
  - ☐ fire rating
  - ☐ flush or raised panel
  - ☐ finish
- Door pricing includes bore for cylindrical locksets and mortise for hinges.

08 30 00 Specialty Doors and Frames

- There are many varieties of special doors, and they are usually priced per each. Add frames, hardware, or operators required for a complete installation.

08 40 00 Entrances, Storefronts, and Curtain Walls

- Glazed curtain walls consist of the metal tube framing and the glazing material. The cost data in this subdivision is presented for the metal tube framing alone or the composite wall. If your estimate requires a detailed takeoff of the framing, be sure to add the glazing cost and any tints.

08 50 00 Windows

- Steel windows are unglazed and aluminum can be glazed or unglazed. Some metal windows are priced without glass. Refer to 08 80 00 Glazing for glass pricing. The grade C indicates commercial grade windows, usually ASTM C-35.
- All wood windows and vinyl are priced preglazed. The glazing is insulating glass. Add the cost of screens and grills if required and not already included.

08 70 00 Hardware

- Hardware costs add considerably to the cost of a door. The most efficient method to determine the hardware requirements for a project is to review the door and hardware schedule together. One type of door may have different hardware, depending on the door usage.
- Door hinges are priced by the pair, with most doors requiring 1-1/2 pairs per door. The hinge prices do not include installation labor because it is included in door installation. Hinges are classified according to the frequency of use, base material, and finish.

08 80 00 Glazing

- Different openings require different types of glass. The most common types are:
  - ☐ float
  - ☐ tempered
  - ☐ insulating
  - ☐ impact-resistant
  - ☐ ballistic-resistant
- Most exterior windows are glazed with insulating glass. Entrance doors and window walls, where the glass is less than 18" from the floor, are generally glazed with tempered glass. Interior windows and some residential windows are glazed with float glass.
- Coastal communities require the use of impact-resistant glass, dependent on wind speed.
- The insulation or 'u' value is a strong consideration, along with solar heat gain, to determine total energy efficiency.

Reference Numbers

Reference numbers are shown at the beginning of some major classifications. These numbers refer to related items in the Reference Section. The reference information may be an estimating procedure, an alternate pricing method, or technical information.

Note: Not all subdivisions listed here necessarily appear. ■

08 34 Special Function Doors

08 34 13 – Cold Storage Doors

08 34 13.10 Doors for Cold Area Storage		Crew	Daily Output	Labor-Hours	Unit	Material	2019 Bare Costs Labor	Equipment	Total	Total Incl O&P
0010	**DOORS FOR COLD AREA STORAGE**									
0020	Single, 20 ga. galvanized steel									
0300	Horizontal sliding, 5' x 7', manual operation, 3.5" thick	2 Carp	2	8	Ea.	3,125	415		3,540	4,050
0400	4" thick		2	8		3,375	415		3,790	4,350
0500	6" thick		2	8		3,425	415		3,840	4,375
0800	5' x 7', power operation, 2" thick		1.90	8.421		5,600	435		6,035	6,825
0900	4" thick		1.90	8.421		5,700	435		6,135	6,925
1000	6" thick		1.90	8.421		6,475	435		6,910	7,775
1300	9' x 10', manual operation, 2" insulation		1.70	9.412		4,450	485		4,935	5,600
1400	4" insulation		1.70	9.412		4,725	485		5,210	5,925
1500	6" insulation		1.70	9.412		5,600	485		6,085	6,900
1800	Power operation, 2" insulation		1.60	10		7,725	515		8,240	9,250
1900	4" insulation		1.60	10		7,875	515		8,390	9,425
2000	6" insulation	↓	1.70	9.412	↓	8,125	485		8,610	9,650
2300	For stainless steel face, add					25%				
3000	Hinged, lightweight, 3' x 7'-0", 2" thick	2 Carp	2	8	Ea.	1,300	415		1,715	2,050
3050	4" thick		1.90	8.421		1,625	435		2,060	2,450
3300	Polymer doors, 3' x 7'-0"		1.90	8.421		1,350	435		1,785	2,125
3350	6" thick		1.40	11.429		2,350	590		2,940	3,475
3600	Stainless steel, 3' x 7'-0", 4" thick		1.90	8.421		1,725	435		2,160	2,550
3650	6" thick		1.40	11.429		2,775	590		3,365	3,950
3900	Painted, 3' x 7'-0", 4" thick		1.90	8.421		1,325	435		1,760	2,100
3950	6" thick	↓	1.40	11.429	↓	2,350	590		2,940	3,475
5000	Bi-parting, electric operated									
5010	6' x 8' opening, galv. faces, 4" thick for cooler	2 Carp	.80	20	Opng.	7,250	1,025		8,275	9,550
5050	For freezer, 4" thick		.80	20		7,975	1,025		9,000	10,400
5300	For door buck framing and door protection, add		2.50	6.400		670	330		1,000	1,250
6000	Galvanized batten door, galvanized hinges, 4' x 7'		2	8		1,825	415		2,240	2,650
6050	6' x 8'		1.80	8.889		2,475	460		2,935	3,425
6500	Fire door, 3 hr., 6' x 8', single slide		.80	20		8,575	1,025		9,600	11,000
6550	Double, bi-parting	↓	.70	22.857	↓	11,400	1,175		12,575	14,400

08 91 Louvers

08 91 16 – Operable Wall Louvers

08 91 16.10 Movable Blade Louvers		Crew	Daily Output	Labor-Hours	Unit	Material	Labor	Equipment	Total	Total Incl O&P
0010	**MOVABLE BLADE LOUVERS**									
0100	PVC, commercial grade, 12" x 12"	1 Shee	20	.400	Ea.	82	24.50		106.50	127
0110	16" x 16"		14	.571		95.50	35		130.50	158
0120	18" x 18"		14	.571		98	35		133	161
0130	24" x 24"		14	.571		140	35		175	207
0140	30" x 30"		12	.667		169	40.50		209.50	248
0150	36" x 36"		12	.667		199	40.50		239.50	280
0300	Stainless steel, commercial grade, 12" x 12"		20	.400		225	24.50		249.50	284
0310	16" x 16"		14	.571		272	35		307	350
0320	18" x 18"		14	.571		288	35		323	370
0330	20" x 20"		14	.571		330	35		365	420
0340	24" x 24"		14	.571		395	35		430	490
0350	30" x 30"		12	.667		440	40.50		480.50	545
0360	36" x 36"	↓	12	.667	↓	535	40.50		575.50	650

08 91 19 – Fixed Louvers

08 91 19.10 Aluminum Louvers		Crew	Daily Output	Labor-Hours	Unit	Material	Labor	Equipment	Total	Total Incl O&P
						2019 Bare Costs				
0010	**ALUMINUM LOUVERS**									
0020	Aluminum with screen, residential, 8" x 8"	1 Carp	38	.211	Ea.	21	10.85		31.85	39.50
0100	12" x 12"		38	.211		16.50	10.85		27.35	34.50
0200	12" x 18"		35	.229		22	11.80		33.80	42
0250	14" x 24"		30	.267		30	13.75		43.75	54
0300	18" x 24"		27	.296		33	15.30		48.30	59.50
0500	24" x 30"		24	.333		63.50	17.20		80.70	96
0700	Triangle, adjustable, small		20	.400		60.50	20.50		81	98
0800	Large	↓	15	.533	↓	85	27.50		112.50	135
1200	Extruded aluminum, see Section 23 37 15.40									
2100	Midget, aluminum, 3/4" deep, 1" diameter	1 Carp	85	.094	Ea.	.94	4.86		5.80	8.40
2150	3" diameter		60	.133		3.07	6.90		9.97	13.80
2200	4" diameter		50	.160		5.75	8.25		14	18.85
2250	6" diameter	↓	30	.267		3.80	13.75		17.55	25
3000	PVC, commercial grade, 12" x 12"	1 Shee	20	.400		96.50	24.50		121	143
3010	12" x 18"		20	.400		109	24.50		133.50	157
3020	12" x 24"		20	.400		146	24.50		170.50	198
3030	14" x 24"		20	.400		153	24.50		177.50	205
3100	Aluminum, commercial grade, 12" x 12"		20	.400		166	24.50		190.50	219
3110	24" x 24"		16	.500		228	30.50		258.50	298
3120	24" x 36"		16	.500		325	30.50		355.50	405
3130	36" x 24"		16	.500		340	30.50		370.50	415
3140	36" x 36"		14	.571		420	35		455	520
3150	36" x 48"		10	.800		510	49		559	635
3160	48" x 36"		10	.800		485	49		534	610
3170	48" x 48"		10	.800		445	49		494	565
3180	60" x 48"		8	1		505	61		566	650
3190	60" x 60"	↓	8	1	↓	730	61		791	900

08 91 19.20 Steel Louvers		Crew	Daily Output	Labor-Hours	Unit	Material	Labor	Equipment	Total	Total Incl O&P
0010	**STEEL LOUVERS**									
3300	Galvanized steel, fixed blades, commercial grade, 18" x 18"	1 Shee	20	.400	Ea.	159	24.50		183.50	212
3310	24" x 24"		20	.400		213	24.50		237.50	271
3320	24" x 36"		16	.500		277	30.50		307.50	350
3330	36" x 24"		16	.500		283	30.50		313.50	355
3340	36" x 36"		14	.571		350	35		385	440
3350	36" x 48"		10	.800		410	49		459	525
3360	48" x 36"		10	.800		400	49		449	515
3370	48" x 48"		10	.800		495	49		544	620
3380	60" x 48"		10	.800		540	49		589	665
3390	60" x 60"	↓	10	.800	↓	645	49		694	785

08 95 Vents

08 95 13 – Soffit Vents

08 95 13.10 Wall Louvers		Crew	Daily Output	Labor-Hours	Unit	2019 Bare Costs Material	Labor	Equipment	Total	Total Incl O&P
0010	**WALL LOUVERS**									
2340	Baked enamel finish	1 Carp	200	.040	L.F.	5.40	2.07		7.47	9.05
2400	Under eaves vent, aluminum, mill finish, 16" x 4"		48	.167	Ea.	1.95	8.60		10.55	15.20
2500	16" x 8"	↓	48	.167	"	2.92	8.60		11.52	16.25

Estimating Tips

General

- Room Finish Schedule: A complete set of plans should contain a room finish schedule. If one is not available, it would be well worth the time and effort to obtain one.

09 20 00 Plaster and Gypsum Board

- Lath is estimated by the square yard plus a 5% allowance for waste. Furring, channels, and accessories are measured by the linear foot. An extra foot should be allowed for each accessory miter or stop.
- Plaster is also estimated by the square yard. Deductions for openings vary by preference, from zero deduction to 50% of all openings over 2 feet in width. The estimator should allow one extra square foot for each linear foot of horizontal interior or exterior angle located below the ceiling level. Also, double the areas of small radius work.
- Drywall accessories, studs, track, and acoustical caulking are all measured by the linear foot. Drywall taping is figured by the square foot. Gypsum wallboard is estimated by the square foot. No material deductions should be made for door or window openings under 32 S.F.

09 60 00 Flooring

- Tile and terrazzo areas are taken off on a square foot basis. Trim and base materials are measured by the linear foot. Accent tiles are listed per each. Two basic methods of installation are used. Mud set is approximately 30% more expensive than thin set. The cost of grout is included with tile unit price lines unless otherwise noted. In terrazzo work, be sure to include the linear footage of embedded decorative strips, grounds, machine rubbing, and power cleanup.
- Wood flooring is available in strip, parquet, or block configuration. The latter two types are set in adhesives with quantities estimated by the square foot. The laying pattern will influence labor costs and material waste. In addition to the material and labor for laying wood floors, the estimator must make allowances for sanding and finishing these areas, unless the flooring is prefinished.
- Sheet flooring is measured by the square yard. Roll widths vary, so consideration should be given to use the most economical width, as waste must be figured into the total quantity. Consider also the installation methods available—direct glue down or stretched. Direct glue-down installation is assumed with sheet carpet unit price lines unless otherwise noted.

09 70 00 Wall Finishes

- Wall coverings are estimated by the square foot. The area to be covered is measured—length by height of the wall above the baseboards—to calculate the square footage of each wall. This figure is divided by the number of square feet in the single roll which is being used. Deduct, in full, the areas of openings such as doors and windows. Where a pattern match is required allow 25–30% waste.

09 80 00 Acoustic Treatment

- Acoustical systems fall into several categories. The takeoff of these materials should be by the square foot of area with a 5% allowance for waste. Do not forget about scaffolding, if applicable, when estimating these systems.

09 90 00 Painting and Coating

- A major portion of the work in painting involves surface preparation. Be sure to include cleaning, sanding, filling, and masking costs in the estimate.
- Protection of adjacent surfaces is not included in painting costs. When considering the method of paint application, an important factor is the amount of protection and masking required. These must be estimated separately and may be the determining factor in choosing the method of application.

Reference Numbers

Reference numbers are shown at the beginning of some major classifications. These numbers refer to related items in the Reference Section. The reference information may be an estimating procedure, an alternate pricing method, or technical information.

Note: Not all subdivisions listed here necessarily appear. ■

09 22 Supports for Plaster and Gypsum Board

09 22 03 – Fastening Methods for Finishes

09 22 03.20 Drilling Plaster/Drywall		Crew	Daily Output	Labor-Hours	Unit	Material	Labor	Equipment	Total	Total Incl O&P
						2019 Bare Costs				
0010	DRILLING PLASTER/DRYWALL									
1100	Drilling & layout for drywall/plaster walls, up to 1" deep, no anchor									
1200	Holes, 1/4" diameter	1 Carp	150	.053	Ea.	.01	2.75		2.76	4.18
1300	3/8" diameter		140	.057		.01	2.95		2.96	4.48
1400	1/2" diameter		130	.062		.01	3.18		3.19	4.82
1500	3/4" diameter		120	.067		.02	3.44		3.46	5.20
1600	1" diameter		110	.073		.02	3.76		3.78	5.70
1700	1-1/4" diameter		100	.080		.04	4.13		4.17	6.30
1800	1-1/2" diameter		90	.089		.06	4.59		4.65	7
1900	For ceiling installations, add						40%			

09 69 Access Flooring

09 69 13 – Rigid-Grid Access Flooring

09 69 13.10 Access Floors

		Crew	Daily Output	Labor-Hours	Unit	Material	Labor	Equipment	Total	Total Incl O&P
0010	ACCESS FLOORS									
0015	Access floor pkg. including panel, pedestal, & stringers 1500 lbs. load									
0100	Package pricing, conc. fill panels, no fin., 6" ht.	4 Carp	750	.043	S.F.	17.20	2.20		19.40	22
0105	12" ht.		750	.043		17.40	2.20		19.60	22.50
0110	18" ht.		750	.043		17.60	2.20		19.80	22.50
0115	24" ht.		750	.043		17.85	2.20		20.05	23
0120	Package pricing, steel panels, no fin., 6" ht.		750	.043		13.45	2.20		15.65	18.15
0125	12" ht.		750	.043		18.30	2.20		20.50	23.50
0130	18" ht.		750	.043		18.45	2.20		20.65	24
0135	24" ht.		750	.043		18.55	2.20		20.75	24
0140	Package pricing, wood core panels, no fin., 6" ht.		750	.043		13.55	2.20		15.75	18.30
0145	12" ht.		750	.043		13.80	2.20		16	18.55
0150	18" ht.		750	.043		13.80	2.20		16	18.55
0155	24" ht.		750	.043		14.05	2.20		16.25	18.85
0160	Pkg. pricing, conc. fill pnls, no stringers, no fin., 6" ht, 1250 lbs. load		700	.046		10.90	2.36		13.26	15.55
0165	12" ht.		700	.046		10.10	2.36		12.46	14.65
0170	18" ht.		700	.046		11.05	2.36		13.41	15.70
0175	24" ht.		700	.046		12	2.36		14.36	16.75
0250	Panels, 2' x 2' conc. fill, no fin.	2 Carp	500	.032		23	1.65		24.65	28
0255	With 1/8" high pressure laminate		500	.032		56.50	1.65		58.15	65
0260	Metal panels with 1/8" high pressure laminate		500	.032		67	1.65		68.65	76.50
0265	Wood core panels with 1/8" high pressure laminate		500	.032		44	1.65		45.65	51
0400	Aluminum panels, no fin.		500	.032		34	1.65		35.65	39.50
0600	For carpet covering, add					9.20			9.20	10.10
0700	For vinyl floor covering, add					9.45			9.45	10.40
0900	For high pressure laminate covering, add					7.80			7.80	8.55
0910	For snap on stringer system, add	2 Carp	1000	.016		1.61	.83		2.44	3.02
1000	Machine cutouts after initial installation	1 Carp	50	.160	Ea.	20.50	8.25		28.75	35
1050	Pedestals, 6" to 12"	2 Carp	85	.188		8.80	9.70		18.50	24.50
1100	Air conditioning grilles, 4" x 12"	1 Carp	17	.471		71.50	24.50		96	116
1150	4" x 18"	"	14	.571		98	29.50		127.50	153
1200	Approach ramps, steel	2 Carp	60	.267	S.F.	27	13.75		40.75	51
1300	Aluminum	"	40	.400	"	34.50	20.50		55	69
1500	Handrail, 2 rail, aluminum	1 Carp	15	.533	L.F.	120	27.50		147.50	174

09 91 23.52 Miscellaneous, Interior		Crew	Daily Output	Labor-Hours	Unit	Material	2019 Bare Costs Labor	Equipment	Total	Total Incl O&P
0010	**MISCELLANEOUS, INTERIOR** R099100-10									
3800	Grilles, per side, oil base, primer coat, brushwork	1 Pord	520	.015	S.F.	.13	.67		.80	1.14
3850	Spray		1140	.007		.14	.30		.44	.60
3880	Paint 1 coat, brushwork		520	.015		.22	.67		.89	1.25
3900	Spray		1140	.007		.25	.30		.55	.72
3920	Paint 2 coats, brushwork		325	.025		.43	1.06		1.49	2.07
3940	Spray		650	.012		.50	.53		1.03	1.35
3950	Prime & paint 1 coat		325	.025		.35	1.06		1.41	1.98
3960	Prime & paint 2 coats		270	.030		.35	1.28		1.63	2.30
4500	Louvers, 1 side, primer, brushwork		524	.015		.09	.66		.75	1.09
4520	Paint 1 coat, brushwork		520	.015		.10	.67		.77	1.11
4530	Spray		1140	.007		.11	.30		.41	.57
4540	Paint 2 coats, brushwork		325	.025		.20	1.06		1.26	1.81
4550	Spray		650	.012		.22	.53		.75	1.04
4560	Paint 3 coats, brushwork		270	.030		.29	1.28		1.57	2.24
4570	Spray	↓	500	.016	↓	.33	.69		1.02	1.40
5000	Pipe, 1"-4" diameter, primer or sealer coat, oil base, brushwork	2 Pord	1250	.013	L.F.	.09	.55		.64	.93
5100	Spray		2165	.007		.09	.32		.41	.58
5200	Paint 1 coat, brushwork		1250	.013		.12	.55		.67	.96
5300	Spray		2165	.007		.10	.32		.42	.59
5350	Paint 2 coats, brushwork		775	.021		.21	.89		1.10	1.57
5400	Spray		1240	.013		.23	.56		.79	1.09
5420	Paint 3 coats, brushwork		775	.021		.31	.89		1.20	1.68
5450	5"-8" diameter, primer or sealer coat, brushwork		620	.026		.19	1.12		1.31	1.88
5500	Spray		1085	.015		.31	.64		.95	1.30
5550	Paint 1 coat, brushwork		620	.026		.32	1.12		1.44	2.02
5600	Spray		1085	.015		.35	.64		.99	1.35
5650	Paint 2 coats, brushwork		385	.042		.42	1.80		2.22	3.15
5700	Spray		620	.026		.46	1.12		1.58	2.18
5720	Paint 3 coats, brushwork		385	.042		.62	1.80		2.42	3.37
5750	9"-12" diameter, primer or sealer coat, brushwork		415	.039		.28	1.67		1.95	2.81
5800	Spray		725	.022		.38	.95		1.33	1.85
5850	Paint 1 coat, brushwork		415	.039		.32	1.67		1.99	2.85
6000	Spray		725	.022		.36	.95		1.31	1.82
6200	Paint 2 coats, brushwork		260	.062		.62	2.66		3.28	4.67
6250	Spray		415	.039		.69	1.67		2.36	3.26
6270	Paint 3 coats, brushwork		260	.062		.92	2.66		3.58	5
6300	13"-16" diameter, primer or sealer coat, brushwork		310	.052		.38	2.23		2.61	3.76
6350	Spray		540	.030		.42	1.28		1.70	2.38
6400	Paint 1 coat, brushwork		310	.052		.43	2.23		2.66	3.81
6450	Spray		540	.030		.48	1.28		1.76	2.44
6500	Paint 2 coats, brushwork		195	.082		.83	3.55		4.38	6.20
6550	Spray	↓	310	.052	↓	.93	2.23		3.16	4.36
6600	Radiators, per side, primer, brushwork	1 Pord	520	.015	S.F.	.09	.67		.76	1.10
6620	Paint, 1 coat		520	.015		.08	.67		.75	1.09
6640	2 coats		340	.024		.20	1.02		1.22	1.74
6660	3 coats	↓	283	.028	↓	.29	1.22		1.51	2.15

Division Notes

I	CREW	DAILY OUTPUT	LABOR-HOURS	UNIT	BARE COSTS MAT.	LABOR	EQUIP.	TOTAL	TOTAL INCL O&P

Estimating Tips

General

- The items in this division are usually priced per square foot or each.
- Many items in Division 10 require some type of support system or special anchors that are not usually furnished with the item. The required anchors must be added to the estimate in the appropriate division.
- Some items in Division 10, such as lockers, may require assembly before installation. Verify the amount of assembly required. Assembly can often exceed installation time.

10 20 00 Interior Specialties

- Support angles and blocking are not included in the installation of toilet compartments, shower/dressing compartments, or cubicles. Appropriate line items from Division 5 or 6 may need to be added to support the installations.
- Toilet partitions are priced by the stall. A stall consists of a side wall, pilaster, and door with hardware. Toilet tissue holders and grab bars are extra.
- The required acoustical rating of a folding partition can have a significant impact on costs. Verify the sound transmission coefficient rating of the panel priced against the specification requirements.
- Grab bar installation does not include supplemental blocking or backing to support the required load. When grab bars are installed at an existing facility, provisions must be made to attach the grab bars to a solid structure.

Reference Numbers

Reference numbers are shown at the beginning of some major classifications. These numbers refer to related items in the Reference Section. The reference information may be an estimating procedure, an alternate pricing method, or technical information.

Note: Not all subdivisions listed here necessarily appear. ■

10 31 Manufactured Fireplaces

10 31 13 – Manufactured Fireplace Chimneys

10 31 13.10 Fireplace Chimneys		Crew	Daily Output	Labor-Hours	Unit	2019 Bare Costs Material	Labor	Equipment	Total	Total Incl O&P
0010	**FIREPLACE CHIMNEYS**									
0500	Chimney dbl. wall, all stainless, over 8'-6", 7" diam., add to fireplace	1 Carp	33	.242	V.L.F.	88	12.50		100.50	116
0600	10" diameter, add to fireplace		32	.250		111	12.90		123.90	142
0700	12" diameter, add to fireplace		31	.258		173	13.35		186.35	210
0800	14" diameter, add to fireplace		30	.267		223	13.75		236.75	266
1000	Simulated brick chimney top, 4' high, 16" x 16"		10	.800	Ea.	455	41.50		496.50	565
1100	24" x 24"		7	1.143	"	560	59		619	705

10 31 23 – Prefabricated Fireplaces

10 31 23.10 Fireplace, Prefabricated		Crew	Daily Output	Labor-Hours	Unit	Material	Labor	Equipment	Total	Total Incl O&P
0010	**FIREPLACE, PREFABRICATED**, free standing or wall hung									
0100	With hood & screen, painted	1 Carp	1.30	6.154	Ea.	1,625	320		1,945	2,250
0150	Average		1	8		1,800	415		2,215	2,625
0200	Stainless steel		.90	8.889		3,225	460		3,685	4,250
1500	Simulated logs, gas fired, 40,000 BTU, 2' long, manual safety pilot		7	1.143	Set	540	59		599	685
1600	Adjustable flame remote pilot		6	1.333		1,275	69		1,344	1,500
1700	Electric, 1,500 BTU, 1'-6" long, incandescent flame		7	1.143		262	59		321	380
1800	1,500 BTU, LED flame		6	1.333		345	69		414	485

10 35 Stoves

10 35 13 – Heating Stoves

10 35 13.10 Wood Burning Stoves		Crew	Daily Output	Labor-Hours	Unit	Material	Labor	Equipment	Total	Total Incl O&P
0010	**WOOD BURNING STOVES**									
0015	Cast iron, less than 1,500 S.F.	2 Carp	1.30	12.308	Ea.	1,375	635		2,010	2,475
0020	1,500 to 2,000 S.F.		1	16		2,350	825		3,175	3,850
0030	greater than 2,000 S.F.		.80	20		2,800	1,025		3,825	4,650
0050	For gas log lighter, add					48			48	53

Estimating Tips

General

- The items in this division are usually priced per square foot or each. Many of these items are purchased by the owner for installation by the contractor. Check the specifications for responsibilities and include time for receiving, storage, installation, and mechanical and electrical hookups in the appropriate divisions.
- Many items in Division 11 require some type of support system that is not usually furnished with the item. Examples of these systems include blocking for the attachment of casework and support angles for ceiling-hung projection screens. The required blocking or supports must be added to the estimate in the appropriate division.
- Some items in Division 11 may require assembly or electrical hookups. Verify the amount of assembly required or the need for a hard electrical connection and add the appropriate costs.

Reference Numbers

Reference numbers are shown at the beginning of some major classifications. These numbers refer to related items in the Reference Section. The reference information may be an estimating procedure, an alternate pricing method, or technical information.

11 11 Vehicle Service Equipment

11 11 13 – Compressed-Air Vehicle Service Equipment

11 11 13.10 Compressed Air Equipment		Crew	Daily Output	Labor-Hours	Unit	2019 Bare Costs Material	Labor	Equipment	Total	Total Incl O&P
0010	**COMPRESSED AIR EQUIPMENT**									
0030	Compressors, electric, 1-1/2 HP, standard controls	L-4	1.50	16	Ea.	495	775		1,270	1,725
0550	Dual controls		1.50	16		1,025	775		1,800	2,325
0600	5 HP, 115/230 volt, standard controls		1	24		2,075	1,175		3,250	4,075
0650	Dual controls		1	24		3,175	1,175		4,350	5,275

11 11 19 – Vehicle Lubrication Equipment

11 11 19.10 Lubrication Equipment		Crew	Daily Output	Labor-Hours	Unit	Material	Labor	Equipment	Total	Total Incl O&P
0010	**LUBRICATION EQUIPMENT**									
3000	Lube equipment, 3 reel type, with pumps, not including piping	L-4	.50	48	Set	10,400	2,325		12,725	15,000
3700	Pump lubrication, pneumatic, not incl. air compressor									
3710	Oil/gear lube	Q-1	9.60	1.667	Ea.	880	94.50		974.50	1,100
3720	Grease	"	9.60	1.667	"	1,150	94.50		1,244.50	1,425

11 11 33 – Vehicle Spray Painting Equipment

11 11 33.10 Spray Painting Equipment		Crew	Daily Output	Labor-Hours	Unit	Material	Labor	Equipment	Total	Total Incl O&P
0010	**SPRAY PAINTING EQUIPMENT**									
4000	Spray painting booth, 26' long, complete	L-4	.40	60	Ea.	10,500	2,925		13,425	16,100

11 21 Retail and Service Equipment

11 21 73 – Commercial Laundry and Dry Cleaning Equipment

11 21 73.16 Drying and Conditioning Equipment		Crew	Daily Output	Labor-Hours	Unit	Material	Labor	Equipment	Total	Total Incl O&P
0010	**DRYING AND CONDITIONING EQUIPMENT**									
0100	Dryers, not including rough-in									
1500	Industrial, 30 lb. capacity	1 Plum	2	4	Ea.	3,325	253		3,578	4,050
1600	50 lb. capacity	"	1.70	4.706	"	4,500	297		4,797	5,400

11 21 73.26 Commercial Washers and Extractors		Crew	Daily Output	Labor-Hours	Unit	Material	Labor	Equipment	Total	Total Incl O&P
0010	**COMMERCIAL WASHERS AND EXTRACTORS**, not including rough-in									
6000	Combination washer/extractor, 20 lb. capacity	L-6	1.50	8	Ea.	6,625	495		7,120	8,050
6100	30 lb. capacity		.80	15		10,100	930		11,030	12,500
6200	50 lb. capacity		.68	17.647		12,800	1,100		13,900	15,700
6300	75 lb. capacity		.30	40		18,900	2,475		21,375	24,500
6350	125 lb. capacity		.16	75		29,200	4,650		33,850	39,200

11 21 73.33 Coin-Operated Laundry Equipment		Crew	Daily Output	Labor-Hours	Unit	Material	Labor	Equipment	Total	Total Incl O&P
0010	**COIN-OPERATED LAUNDRY EQUIPMENT**									
0990	Dryer, gas fired									
1000	Commercial, 30 lb. capacity, coin operated, single	1 Plum	3	2.667	Ea.	3,450	168		3,618	4,050
1100	Double stacked	"	2	4	"	8,200	253		8,453	9,375
5290	Clothes washer									
5300	Commercial, coin operated, average	1 Plum	3	2.667	Ea.	1,325	168		1,493	1,700

11 21 83 – Photo Processing Equipment

11 21 83.13 Darkroom Equipment		Crew	Daily Output	Labor-Hours	Unit	Material	Labor	Equipment	Total	Total Incl O&P
0010	**DARKROOM EQUIPMENT**									
0020	Developing sink, 5" deep, 24" x 48"	Q-1	2	8	Ea.	490	455		945	1,225
0050	48" x 52"		1.70	9.412		1,225	535		1,760	2,150
0200	10" deep, 24" x 48"		1.70	9.412		1,575	535		2,110	2,550
0250	24" x 108"		1.50	10.667		3,350	605		3,955	4,600

11 30 13 – Residential Appliances

11 30 13.15 Cooking Equipment

	11 30 13.15 Cooking Equipment	Crew	Daily Output	Labor-Hours	Unit	Material	2019 Bare Costs Labor	Equipment	Total	Total Incl O&P
0010	**COOKING EQUIPMENT**									
0020	Cooking range, 30" free standing, 1 oven, minimum	2 Clab	10	1.600	Ea.	470	65.50		535.50	620
0050	Maximum		4	4		2,325	164		2,489	2,825
0150	2 oven, minimum		10	1.600		1,025	65.50		1,090.50	1,225
0200	Maximum		10	1.600		3,400	65.50		3,465.50	3,825

11 30 13.16 Refrigeration Equipment

	11 30 13.16 Refrigeration Equipment	Crew	Daily Output	Labor-Hours	Unit	Material	Labor	Equipment	Total	Total Incl O&P
0010	**REFRIGERATION EQUIPMENT**									
5200	Icemaker, automatic, 20 lbs./day	1 Plum	7	1.143	Ea.	1,400	72		1,472	1,625
5350	51 lbs./day	"	2	4	"	1,400	253		1,653	1,925

11 30 13.17 Kitchen Cleaning Equipment

	11 30 13.17 Kitchen Cleaning Equipment	Crew	Daily Output	Labor-Hours	Unit	Material	Labor	Equipment	Total	Total Incl O&P
0010	**KITCHEN CLEANING EQUIPMENT**									
2750	Dishwasher, built-in, 2 cycles, minimum	L-1	4	4	Ea.	305	246		551	705
2800	Maximum		2	8		455	495		950	1,225
2950	4 or more cycles, minimum		4	4		420	246		666	835
2960	Average		4	4		560	246		806	985
3000	Maximum		2	8		1,900	495		2,395	2,825

11 30 13.18 Waste Disposal Equipment

	11 30 13.18 Waste Disposal Equipment	Crew	Daily Output	Labor-Hours	Unit	Material	Labor	Equipment	Total	Total Incl O&P
0010	**WASTE DISPOSAL EQUIPMENT**									
3300	Garbage disposal, sink type, minimum	L-1	10	1.600	Ea.	109	98.50		207.50	266
3350	Maximum	"	10	1.600	"	208	98.50		306.50	375

11 30 13.19 Kitchen Ventilation Equipment

	11 30 13.19 Kitchen Ventilation Equipment	Crew	Daily Output	Labor-Hours	Unit	Material	Labor	Equipment	Total	Total Incl O&P
0010	**KITCHEN VENTILATION EQUIPMENT**									
4150	Hood for range, 2 speed, vented, 30" wide, minimum	L-3	5	3.200	Ea.	96	179		275	375
4200	Maximum		3	5.333		965	299		1,264	1,500
4300	42" wide, minimum		5	3.200		155	179		334	440
4330	Custom		5	3.200		1,625	179		1,804	2,075
4350	Maximum		3	5.333		1,975	299		2,274	2,625
4500	For ventless hood, 2 speed, add					17.80			17.80	19.60
4650	For vented 1 speed, deduct from maximum					64			64	70.50

11 30 13.24 Washers

	11 30 13.24 Washers	Crew	Daily Output	Labor-Hours	Unit	Material	Labor	Equipment	Total	Total Incl O&P
0010	**WASHERS**									
5000	Residential, 4 cycle, average	1 Plum	3	2.667	Ea.	980	168		1,148	1,325
6650	Washing machine, automatic, minimum		3	2.667		590	168		758	905
6700	Maximum		1	8		1,350	505		1,855	2,250

11 30 13.25 Dryers

	11 30 13.25 Dryers	Crew	Daily Output	Labor-Hours	Unit	Material	Labor	Equipment	Total	Total Incl O&P
0010	**DRYERS**									
0500	Gas fired residential, 16 lb. capacity, average	1 Plum	3	2.667	Ea.	715	168		883	1,050
7450	Vent kits for dryers	1 Carp	10	.800	"	46.50	41.50		88	114

11 30 15 – Miscellaneous Residential Appliances

11 30 15.13 Sump Pumps

	11 30 15.13 Sump Pumps	Crew	Daily Output	Labor-Hours	Unit	Material	Labor	Equipment	Total	Total Incl O&P
0010	**SUMP PUMPS**									
6400	Cellar drainer, pedestal, 1/3 HP, molded PVC base	1 Plum	3	2.667	Ea.	141	168		309	410
6450	Solid brass	"	2	4	"	239	253		492	645
6460	Sump pump, see also Section 22 14 29.16									

11 30 15.23 Water Heaters

	11 30 15.23 Water Heaters	Crew	Daily Output	Labor-Hours	Unit	Material	Labor	Equipment	Total	Total Incl O&P
0010	**WATER HEATERS**									
6900	Electric, glass lined, 30 gallon, minimum	L-1	5	3.200	Ea.	895	197		1,092	1,275
6950	Maximum		3	5.333		1,250	330		1,580	1,875
7100	80 gallon, minimum		2	8		1,650	495		2,145	2,550
7150	Maximum		1	16		2,300	985		3,285	4,025

11 30 Residential Equipment

11 30 15 – Miscellaneous Residential Appliances

11 30 15.23 Water Heaters		Crew	Daily Output	Labor-Hours	Unit	Material	Labor	Equipment	Total	Total Incl O&P
						2019 Bare Costs				
7180	Gas, glass lined, 30 gallon, minimum	2 Plum	5	3.200	Ea.	1,600	202		1,802	2,050
7220	Maximum		3	5.333		2,225	335		2,560	2,950
7260	50 gallon, minimum		2.50	6.400		1,725	405		2,130	2,500
7300	Maximum		1.50	10.667		2,400	675		3,075	3,625

11 30 15.43 Air Quality		Crew	Daily Output	Labor-Hours	Unit	Material	Labor	Equipment	Total	Total Incl O&P
0010	**AIR QUALITY**									
2450	Dehumidifier, portable, automatic, 15 pint	1 Elec	4	2	Ea.	208	120		328	410
2550	40 pint		3.75	2.133		240	128		368	455
3550	Heater, electric, built-in, 1250 watt, ceiling type, minimum		4	2		119	120		239	310
3600	Maximum		3	2.667		185	160		345	440
3700	Wall type, minimum		4	2		199	120		319	400
3750	Maximum		3	2.667		194	160		354	455
3900	1500 watt wall type, with blower		4	2		189	120		309	385
3950	3000 watt		3	2.667		490	160		650	780
4850	Humidifier, portable, 8 gallons/day					164			164	180
5000	15 gallons/day					198			198	218

11 32 Unit Kitchens

11 32 13 – Metal Unit Kitchens

11 32 13.10 Commercial Unit Kitchens		Crew	Daily Output	Labor-Hours	Unit	Material	Labor	Equipment	Total	Total Incl O&P
0010	**COMMERCIAL UNIT KITCHENS**									
1500	Combination range, refrigerator and sink, 30" wide, minimum	L-1	2	8	Ea.	1,400	495		1,895	2,275
1550	Maximum		1	16		1,150	985		2,135	2,750
1570	60" wide, average		1.40	11.429		1,275	705		1,980	2,450
1590	72" wide, average		1.20	13.333		1,800	820		2,620	3,225
1600	Office model, 48" wide		2	8		1,400	495		1,895	2,250
1620	Refrigerator and sink only		2.40	6.667		2,650	410		3,060	3,550
1640	Combination range, refrigerator, sink, microwave									
1660	Oven and ice maker	L-1	.80	20	Ea.	4,900	1,225		6,125	7,250

11 41 Foodservice Storage Equipment

11 41 13 – Refrigerated Food Storage Cases

11 41 13.10 Refrigerated Food Cases		Crew	Daily Output	Labor-Hours	Unit	Material	Labor	Equipment	Total	Total Incl O&P
0010	**REFRIGERATED FOOD CASES**									
0030	Dairy, multi-deck, 12' long	Q-5	3	5.333	Ea.	11,700	305		12,005	13,300
0100	For rear sliding doors, add					1,875			1,875	2,075
0200	Delicatessen case, service deli, 12' long, single deck	Q-5	3.90	4.103		8,675	236		8,911	9,900
0300	Multi-deck, 18 S.F. shelf display		3	5.333		8,025	305		8,330	9,275
0400	Freezer, self-contained, chest-type, 30 C.F.		3.90	4.103		4,425	236		4,661	5,200
0500	Glass door, upright, 78 C.F.		3.30	4.848		9,325	279		9,604	10,700
0600	Frozen food, chest type, 12' long		3.30	4.848		7,775	279		8,054	8,975
0700	Glass door, reach-in, 5 door		3	5.333		10,200	305		10,505	11,700
0800	Island case, 12' long, single deck		3.30	4.848		7,600	279		7,879	8,800
0900	Multi-deck		3	5.333		9,125	305		9,430	10,500
1000	Meat case, 12' long, single deck		3.30	4.848		7,825	279		8,104	9,025
1050	Multi-deck		3.10	5.161		10,000	297		10,297	11,400
1100	Produce, 12' long, single deck		3.30	4.848		6,625	279		6,904	7,725
1200	Multi-deck		3.10	5.161		8,425	297		8,722	9,725

11 41 Foodservice Storage Equipment

11 41 13 – Refrigerated Food Storage Cases

11 41 13.20 Refrigerated Food Storage Equipment		Crew	Daily Output	Labor-Hours	Unit	Material	Labor	Equipment	Total	Total Incl O&P
						2019 Bare Costs				
0010	**REFRIGERATED FOOD STORAGE EQUIPMENT**									
2350	Cooler, reach-in, beverage, 6' long	Q-1	6	2.667	Ea.	3,225	152		3,377	3,775
4300	Freezers, reach-in, 44 C.F.		4	4		4,650	227		4,877	5,475
4500	68 C.F.	↓	3	5.333	↓	5,950	305		6,255	7,000

11 44 Food Cooking Equipment

11 44 13 – Commercial Ranges

11 44 13.10 Cooking Equipment

		Crew	Daily Output	Labor-Hours	Unit	Material	Labor	Equipment	Total	Total Incl O&P
0010	**COOKING EQUIPMENT**									
0020	Bake oven, gas, one section	Q-1	8	2	Ea.	5,850	114		5,964	6,600
0300	Two sections		7	2.286		9,550	130		9,680	10,700
0600	Three sections	↓	6	2.667		12,200	152		12,352	13,700
0900	Electric convection, single deck	L-7	4	7		5,175	350		5,525	6,225
6350	Kettle, w/steam jacket, tilting, w/positive lock, SS, 20 gallons		7	4		8,500	199		8,699	9,650
6600	60 gallons	↓	6	4.667	↓	20,300	232		20,532	22,800

11 46 Food Dispensing Equipment

11 46 83 – Ice Machines

11 46 83.10 Commercial Ice Equipment

		Crew	Daily Output	Labor-Hours	Unit	Material	Labor	Equipment	Total	Total Incl O&P
0010	**COMMERCIAL ICE EQUIPMENT**									
5800	Ice cube maker, 50 lbs./day	Q-1	6	2.667	Ea.	1,700	152		1,852	2,100
6050	500 lbs./day	"	4	4	"	2,675	227		2,902	3,275

11 48 Foodservice Cleaning and Disposal Equipment

11 48 13 – Commercial Dishwashers

11 48 13.10 Dishwashers

		Crew	Daily Output	Labor-Hours	Unit	Material	Labor	Equipment	Total	Total Incl O&P
0010	**DISHWASHERS**									
2700	Dishwasher, commercial, rack type									
2720	10 to 12 racks/hour	Q-1	3.20	5	Ea.	3,325	284		3,609	4,075
2730	Energy star rated, 35 to 40 racks/hour G		1.30	12.308		5,000	700		5,700	6,550
2740	50 to 60 racks/hour G	↓	1.30	12.308		11,100	700		11,800	13,300
2800	Automatic, 190 to 230 racks/hour	L-6	.35	34.286		15,200	2,125		17,325	20,000
2820	235 to 275 racks/hour		.25	48		28,900	2,975		31,875	36,300
2840	8,750 to 12,500 dishes/hour	↓	.10	120	↓	48,100	7,450		55,550	64,000

11 53 Laboratory Equipment

11 53 13 – Laboratory Fume Hoods

11 53 13.13 Recirculating Laboratory Fume Hoods

		Crew	Daily Output	Labor-Hours	Unit	Material	Labor	Equipment	Total	Total Incl O&P
0010	**RECIRCULATING LABORATORY FUME HOODS**									
0600	Fume hood, with countertop & base, not including HVAC									
0610	Simple, minimum	2 Carp	5.40	2.963	L.F.	565	153		718	855
0620	Complex, including fixtures		2.40	6.667		1,075	345		1,420	1,725
0630	Special, maximum	↓	1.70	9.412	↓	1,175	485		1,660	2,025
0670	Service fixtures, average				Ea.	345			345	380
0680	For sink assembly with hot and cold water, add	1 Plum	1.40	5.714	"	775	360		1,135	1,400

11 53 Laboratory Equipment

11 53 19 – Laboratory Sterilizers

11 53 19.13 Sterilizers		Crew	Daily Output	Labor-Hours	Unit	Material	Labor	Equipment	Total	Total Incl O&P
						2019 Bare Costs				
0010	**STERILIZERS**									
0700	Glassware washer, undercounter, minimum	L-1	1.80	8.889	Ea.	6,875	550		7,425	8,400
0710	Maximum	"	1	16		14,900	985		15,885	17,900
1850	Utensil washer-sanitizer	1 Plum	2	4	↓	8,775	253		9,028	10,000

11 53 33 – Emergency Safety Appliances

11 53 33.13 Emergency Equipment		Crew	Daily Output	Labor-Hours	Unit	Material	Labor	Equipment	Total	Total Incl O&P
0010	**EMERGENCY EQUIPMENT**									
1400	Safety equipment, eye wash, hand held				Ea.	425			425	470
1450	Deluge shower				"	840			840	920

11 53 43 – Service Fittings and Accessories

11 53 43.13 Fittings		Crew	Daily Output	Labor-Hours	Unit	Material	Labor	Equipment	Total	Total Incl O&P
0010	**FITTINGS**									
1600	Sink, one piece plastic, flask wash, hose, free standing	1 Plum	1.60	5	Ea.	2,075	315		2,390	2,750
1610	Epoxy resin sink, 25" x 16" x 10"	"	2	4	"	234	253		487	635
8000	Alternate pricing method: as percent of lab furniture									
8050	Installation, not incl. plumbing & duct work				% Furn.				22%	22%
8100	Plumbing, final connections, simple system								10%	10%
8110	Moderately complex system								15%	15%
8120	Complex system								20%	20%
8150	Electrical, simple system								10%	10%
8160	Moderately complex system								20%	20%
8170	Complex system				↓				35%	35%

11 71 Medical Sterilizing Equipment

11 71 10 – Medical Sterilizers & Distillers

11 71 10.10 Sterilizers and Distillers		Crew	Daily Output	Labor-Hours	Unit	Material	Labor	Equipment	Total	Total Incl O&P
0010	**STERILIZERS AND DISTILLERS**									
0700	Distiller, water, steam heated, 50 gal. capacity	1 Plum	1.40	5.714	Ea.	24,900	360		25,260	27,900
5600	Sterilizers, floor loading, 26" x 62" x 42", single door, steam					137,500			137,500	151,500
5650	Double door, steam					230,000			230,000	253,000
5800	General purpose, 20" x 20" x 38", single door					12,700			12,700	14,000
6000	Portable, counter top, steam, minimum					2,825			2,825	3,100
6020	Maximum					5,050			5,050	5,575
6050	Portable, counter top, gas, 17" x 15" x 32-1/2"					43,800			43,800	48,200
6150	Manual washer/sterilizer, 16" x 16" x 26"	1 Plum	2	4	↓	60,000	253		60,253	66,500
6200	Steam generators, electric 10 kW to 180 kW, freestanding									
6250	Minimum	1 Elec	3	2.667	Ea.	11,400	160		11,560	12,800
6300	Maximum	"	.70	11.429		22,900	685		23,585	26,100
8200	Bed pan washer-sanitizer	1 Plum	2	4	↓	9,800	253		10,053	11,200

11 73 Patient Care Equipment

11 73 10 – Patient Treatment Equipment

11 73 10.10 Treatment Equipment		Crew	Daily Output	Labor-Hours	Unit	Material	Labor	Equipment	Total	Total Incl O&P
						2019 Bare Costs				
0010	**TREATMENT EQUIPMENT**									
1800	Heat therapy unit, humidified, 26" x 78" x 28"				Ea.	4,325			4,325	4,750
8400	Whirlpool bath, mobile, sst, 18" x 24" x 60"					6,100			6,100	6,700
8450	Fixed, incl. mixing valves	1 Plum	2	4		4,925	253		5,178	5,775

11 74 Dental Equipment

11 74 10 – Dental Office Equipment

11 74 10.10 Diagnostic and Treatment Equipment

		Crew	Daily Output	Labor-Hours	Unit	Material	Labor	Equipment	Total	Total Incl O&P
0010	**DIAGNOSTIC AND TREATMENT EQUIPMENT**									
0020	Central suction system, minimum	1 Plum	1.20	6.667	Ea.	1,075	420		1,495	1,800
0100	Maximum	"	.90	8.889		5,675	560		6,235	7,100
0600	Chair, electric or hydraulic, minimum	1 Skwk	.50	16		2,425	855		3,280	3,950
0700	Maximum		.25	32		4,275	1,700		5,975	7,300
2000	Light, ceiling mounted, minimum		8	1		910	53.50		963.50	1,075
2100	Maximum		8	1		2,075	53.50		2,128.50	2,375
2200	Unit light, minimum	2 Skwk	5.33	3.002		775	160		935	1,100
2210	Maximum		5.33	3.002		1,525	160		1,685	1,925
2220	Track light, minimum		3.20	5		1,650	267		1,917	2,225
2230	Maximum		3.20	5		2,100	267		2,367	2,700
2300	Sterilizers, steam portable, minimum					2,000			2,000	2,200
2350	Maximum					6,275			6,275	6,900
2600	Steam, institutional					2,750			2,750	3,025
2650	Dry heat, electric, portable, 3 trays					1,375			1,375	1,525

11 78 Mortuary Equipment

11 78 13 – Mortuary Refrigerators

11 78 13.10 Mortuary and Autopsy Equipment

		Crew	Daily Output	Labor-Hours	Unit	Material	Labor	Equipment	Total	Total Incl O&P
0010	**MORTUARY AND AUTOPSY EQUIPMENT**									
0015	Autopsy table, standard	1 Plum	1	8	Ea.	10,200	505		10,705	12,000
0020	Deluxe	"	.60	13.333		16,800	840		17,640	19,800
3200	Mortuary refrigerator, end operated, 2 capacity					9,075			9,075	10,000
3300	6 capacity					16,900			16,900	18,600

11 78 16 – Crematorium Equipment

11 78 16.10 Crematory

		Crew	Daily Output	Labor-Hours	Unit	Material	Labor	Equipment	Total	Total Incl O&P
0010	**CREMATORY**									
1500	Crematory, not including building, 1 place	Q-3	.20	160	Ea.	78,500	9,625		88,125	101,000
1750	2 place	"	.10	320	"	112,500	19,200		131,700	152,500

11 81 Facility Maintenance Equipment

11 81 19 – Vacuum Cleaning Systems

11 81 19.10 Vacuum Cleaning		Crew	Daily Output	Labor-Hours	Unit	2019 Bare Costs Material	Labor	Equipment	Total	Total Incl O&P
0010	VACUUM CLEANING									
0020	Central, 3 inlet, residential	1 Skwk	.90	8.889	Total	1,200	475		1,675	2,050
0200	Commercial		.70	11.429		1,375	610		1,985	2,450
0400	5 inlet system, residential		.50	16		1,725	855		2,580	3,200
0600	7 inlet system, commercial		.40	20		2,200	1,075		3,275	4,050
0800	9 inlet system, residential		.30	26.667		4,175	1,425		5,600	6,775
4010	Rule of thumb: First 1200 S.F., installed								1,425	1,575
4020	For each additional S.F., add				S.F.				.26	.26

11 82 Facility Solid Waste Handling Equipment

11 82 19 – Packaged Incinerators

11 82 19.10 Packaged Gas Fired Incinerators		Crew	Daily Output	Labor-Hours	Unit	Material	Labor	Equipment	Total	Total Incl O&P
0010	PACKAGED GAS FIRED INCINERATORS									
4400	Incinerator, gas, not incl. chimney, elec. or pipe, 50 lbs./hr., minimum	Q-3	.80	40	Ea.	41,700	2,400		44,100	49,500
4420	Maximum		.70	45.714		43,300	2,750		46,050	52,000
4440	200 lbs./hr., minimum (batch type)		.60	53.333		71,500	3,200		74,700	84,000
4460	Maximum (with feeder)		.50	64		80,500	3,850		84,350	94,500
4480	400 lbs./hr., minimum (batch type)		.30	107		77,500	6,400		83,900	94,500
4500	Maximum (with feeder)		.25	128		95,500	7,700		103,200	116,500
4520	800 lbs./hr., with feeder, minimum		.20	160		122,000	9,625		131,625	148,500
4540	Maximum		.17	188		205,500	11,300		216,800	243,000
4560	1,200 lbs./hr., with feeder, minimum		.15	213		146,000	12,800		158,800	179,500
4580	Maximum		.11	291		195,000	17,500		212,500	240,500
4600	2,000 lbs./hr., with feeder, minimum		.10	320		409,500	19,200		428,700	479,000
4620	Maximum		.05	640		654,000	38,500		692,500	777,000
4700	For heat recovery system, add, minimum		.25	128		86,000	7,700		93,700	106,000
4710	Add, maximum		.11	291		269,000	17,500		286,500	322,000
4720	For automatic ash conveyer, add		.50	64		35,800	3,850		39,650	45,200
4750	Large municipal incinerators, incl. stack, minimum		.25	128	Ton/day	21,700	7,700		29,400	35,400
4850	Maximum		.10	320	"	58,000	19,200		77,200	92,500

11 82 26 – Facility Waste Compactors

11 82 26.10 Compactors		Crew	Daily Output	Labor-Hours	Unit	Material	Labor	Equipment	Total	Total Incl O&P
0010	COMPACTORS									
0020	Compactors, 115 volt, 250 lbs./hr., chute fed	L-4	1	24	Ea.	12,400	1,175		13,575	15,500
0100	Hand fed		2.40	10		16,400	485		16,885	18,800
0300	Multi-bag, 230 volt, 600 lbs./hr., chute fed		1	24		16,400	1,175		17,575	19,800
0400	Hand fed		1	24		16,500	1,175		17,675	19,900
0500	Containerized, hand fed, 2 to 6 C.Y. containers, 250 lbs./hr.		1	24		16,400	1,175		17,575	19,900
0550	For chute fed, add per floor		1	24		1,500	1,175		2,675	3,425
1000	Heavy duty industrial compactor, 0.5 C.Y. capacity		1	24		11,000	1,175		12,175	13,900
1050	1.0 C.Y. capacity		1	24		16,800	1,175		17,975	20,300
1100	3.0 C.Y. capacity		.50	48		28,100	2,325		30,425	34,500
1150	5.0 C.Y. capacity		.50	48		35,400	2,325		37,725	42,600
1200	Combination shredder/compactor (5,000 lbs./hr.)		.50	48		69,000	2,325		71,325	79,000
1400	For handling hazardous waste materials, 55 gallon drum packer, std.					21,400			21,400	23,500
1410	55 gallon drum packer w/HEPA filter					26,400			26,400	29,000
1420	55 gallon drum packer w/charcoal & HEPA filter					35,200			35,200	38,700
1430	All of the above made explosion proof, add					1,550			1,550	1,700
5800	Shredder, industrial, minimum					25,500			25,500	28,100
5850	Maximum					136,500			136,500	150,000

11 82 Facility Solid Waste Handling Equipment

11 82 26 – Facility Waste Compactors

11 82 26.10 Compactors		Crew	Daily Output	Labor-Hours	Unit	Material	Labor	Equipment	Total	Total Incl O&P
						2019 Bare Costs				
5900	Baler, industrial, minimum				Ea.	10,200			10,200	11,200
5950	Maximum				↓	595,500			595,500	655,000

Division Notes

		CREW	DAILY OUTPUT	LABOR-HOURS	UNIT	BARE COSTS MAT.	LABOR	EQUIP.	TOTAL	TOTAL INCL O&P

Estimating Tips

General

- The items and systems in this division are usually estimated, purchased, supplied, and installed as a unit by one or more subcontractors. The estimator must ensure that all parties are operating from the same set of specifications and assumptions, and that all necessary items are estimated and will be provided. Many times the complex items and systems are covered, but the more common ones, such as excavation or a crane, are overlooked for the very reason that everyone assumes nobody could miss them. The estimator should be the central focus and be able to ensure that all systems are complete.
- It is important to consider factors such as site conditions, weather, shape and size of building, as well as labor availability as they may impact the overall cost of erecting special structures and systems included in this division.
- Another area where problems can develop in this division is at the interface between systems. The estimator must ensure, for instance, that anchor bolts, nuts, and washers are estimated and included for the air-supported structures and pre-engineered buildings to be bolted to their foundations. Utility supply is a common area where essential items or pieces of equipment can be missed or overlooked because each subcontractor may feel it is another's responsibility. The estimator should also be aware of certain items which may be supplied as part of a package but installed by others, and ensure that the installing contractor's estimate includes the cost of installation. Conversely, the estimator must also ensure that items are not costed by two different subcontractors, resulting in an inflated overall estimate.

13 30 00 Special Structures

- The foundations and floor slab, as well as rough mechanical and electrical, should be estimated, as this work is required for the assembly and erection of the structure. Generally, as noted in the data set, the pre-engineered building comes as a shell. Pricing is based on the size and structural design parameters stated in the reference section. Additional features, such as windows and doors with their related structural framing, must also be included by the estimator. Here again, the estimator must have a clear understanding of the scope of each portion of the work and all the necessary interfaces.

Reference Numbers

Reference numbers are shown at the beginning of some major classifications. These numbers refer to related items in the Reference Section. The reference information may be an estimating procedure, an alternate pricing method, or technical information.

Note: Not all subdivisions listed here necessarily appear. ■

13 11 Swimming Pools

13 11 13 – Below-Grade Swimming Pools

13 11 13.50 Swimming Pools		Crew	Daily Output	Labor-Hours	Unit	2019 Bare Costs Material	Labor	Equipment	Total	Total Incl O&P
0010	**SWIMMING POOLS** Residential in-ground, vinyl lined									
0020	Concrete sides, w/equip, sand bottom	B-52	300	.187	SF Surf	27	8.85	2	37.85	45.50
0100	Metal or polystyrene sides	B-14	410	.117		22.50	5.05	.78	28.33	33.50
0200	Add for vermiculite bottom					1.73			1.73	1.90
0500	Gunite bottom and sides, white plaster finish									
0600	12' x 30' pool	B-52	145	.386	SF Surf	50.50	18.35	4.14	72.99	87.50
0720	16' x 32' pool		155	.361		45.50	17.15	3.88	66.53	80.50
0750	20' x 40' pool		250	.224		40.50	10.65	2.40	53.55	63
0810	Concrete bottom and sides, tile finish									
0820	12' x 30' pool	B-52	80	.700	SF Surf	51	33	7.50	91.50	114
0830	16' x 32' pool		95	.589		42	28	6.35	76.35	95.50
0840	20' x 40' pool		130	.431		33.50	20.50	4.62	58.62	73
1100	Motel, gunite with plaster finish, incl. medium									
1150	capacity filtration & chlorination	B-52	115	.487	SF Surf	62	23	5.25	90.25	109
1200	Municipal, gunite with plaster finish, incl. high									
1250	capacity filtration & chlorination	B-52	100	.560	SF Surf	80.50	26.50	6	113	135
1350	Add for formed gutters				L.F.	118			118	130
1360	Add for stainless steel gutters				"	350			350	385
1700	Filtration and deck equipment only, as % of total				Total				20%	20%
1800	Deck equipment, rule of thumb, 20' x 40' pool				SF Pool				1.18	1.30
1900	5,000 S.F. pool				"				1.73	1.90
3000	Painting pools, preparation + 3 coats, 20' x 40' pool, epoxy	2 Pord	.33	48.485	Total	1,750	2,100		3,850	5,075
3100	Rubber base paint, 18 gallons	"	.33	48.485		1,350	2,100		3,450	4,650
3500	42' x 82' pool, 75 gallons, epoxy paint	3 Pord	.14	171		7,425	7,425		14,850	19,300
3600	Rubber base paint	"	.14	171		5,600	7,425		13,025	17,300

13 11 23 – On-Grade Swimming Pools

13 11 23.50 Swimming Pools		Crew	Daily Output	Labor-Hours	Unit	Material	Labor	Equipment	Total	Total Incl O&P
0010	**SWIMMING POOLS** Residential above ground, steel construction									
0100	Round, 15' diam.	B-80A	3	8	Ea.	810	330	82	1,222	1,475
0120	18' diam.		2.50	9.600		910	395	98.50	1,403.50	1,700
0140	21' diam.		2	12		1,025	495	123	1,643	2,000
0160	24' diam.		1.80	13.333		1,125	545	137	1,807	2,225
0180	27' diam.		1.50	16		1,325	655	164	2,144	2,625
0200	30' diam.		1	24		1,450	985	246	2,681	3,350
0220	Oval, 12' x 24'		2.30	10.435		1,500	430	107	2,037	2,425
0240	15' x 30'		1.80	13.333		2,075	545	137	2,757	3,250
0260	18' x 33'		1	24		2,350	985	246	3,581	4,375

13 11 46 – Swimming Pool Accessories

13 11 46.50 Swimming Pool Equipment		Crew	Daily Output	Labor-Hours	Unit	Material	Labor	Equipment	Total	Total Incl O&P
0010	**SWIMMING POOL EQUIPMENT**									
0020	Diving stand, stainless steel, 3 meter	2 Carp	.40	40	Ea.	17,300	2,075		19,375	22,200
0300	1 meter	"	2.70	5.926		10,500	305		10,805	12,000
2100	Lights, underwater, 12 volt, with transformer, 300 watt	1 Elec	1	8		365	480		845	1,125
2200	110 volt, 500 watt, standard		1	8		320	480		800	1,075
2400	Low water cutoff type		1	8		296	480		776	1,050
2800	Heaters, see Section 23 52 28.10									

13 18 Ice Rinks

13 18 13 – Ice Rink Floor Systems

13 18 13.50 Ice Skating		Crew	Daily Output	Labor-Hours	Unit	Material	Labor	Equipment	Total	Total Incl O&P
						2019 Bare Costs				
0010	**ICE SKATING** Equipment incl. refrigeration, plumbing & cooling									
0020	coils & concrete slab, 85′ x 200′ rink									
0300	55° system, 5 mos., 100 ton				Total	593,000			593,000	652,000
0700	90° system, 12 mos., 135 ton				"	670,500			670,500	737,500
1200	Subsoil heating system (recycled from compressor), 85′ x 200′	Q-7	.27	119	Ea.	41,200	7,225		48,425	56,000
1300	Subsoil insulation, 2 lb. polystyrene with vapor barrier, 85′ x 200′	2 Carp	.14	114	"	30,900	5,900		36,800	42,900

13 18 16 – Ice Rink Dasher Boards

13 18 16.50 Ice Rink Dasher Boards		Crew	Daily Output	Labor-Hours	Unit	Material	Labor	Equipment	Total	Total Incl O&P
0010	**ICE RINK DASHER BOARDS**									
1000	Dasher boards, 1/2″ H.D. polyethylene faced steel frame, 3′ acrylic									
1020	screen at sides, 5′ acrylic ends, 85′ x 200′	F-5	.06	533	Ea.	144,500	27,800		172,300	201,000
1100	Fiberglass & aluminum construction, same sides and ends	"	.06	533	"	160,000	27,800		187,800	218,000

13 21 Controlled Environment Rooms

13 21 13 – Clean Rooms

13 21 13.50 Clean Room Components		Crew	Daily Output	Labor-Hours	Unit	Material	Labor	Equipment	Total	Total Incl O&P
0010	**CLEAN ROOM COMPONENTS**									
1100	Clean room, soft wall, 12′ x 12′, Class 100	1 Carp	.18	44.444	Ea.	16,400	2,300		18,700	21,600
1110	Class 1,000		.18	44.444		15,000	2,300		17,300	20,000
1120	Class 10,000		.21	38.095		13,600	1,975		15,575	18,000
1130	Class 100,000	↓	.21	38.095	↓	13,000	1,975		14,975	17,300
2800	Ceiling grid support, slotted channel struts 4′-0″ OC, ea. way				S.F.				5.90	6.50
3000	Ceiling panel, vinyl coated foil on mineral substrate									
3020	Sealed, non-perforated				S.F.				1.27	1.40
4000	Ceiling panel seal, silicone sealant, 150 L.F./gal.	1 Carp	150	.053	L.F.	.29	2.75		3.04	4.49
4100	Two sided adhesive tape	"	240	.033	"	.13	1.72		1.85	2.74
4200	Clips, one per panel				Ea.	1.01			1.01	1.11
6000	HEPA filter, 2′ x 4′, 99.97% eff., 3″ dp beveled frame (silicone seal)					365			365	400
6040	6″ deep skirted frame (channel seal)					480			480	525
6100	99.99% efficient, 3″ deep beveled frame (silicone seal)					450			450	495
6140	6″ deep skirted frame (channel seal)					480			480	530
6200	99.999% efficient, 3″ deep beveled frame (silicone seal)					470			470	520
6240	6″ deep skirted frame (channel seal)				↓	515			515	570
7000	Wall panel systems, including channel strut framing									
7020	Polyester coated aluminum, particle board				S.F.				18.20	18.20
7100	Porcelain coated aluminum, particle board								32	32
7400	Wall panel support, slotted channel struts, to 12′ high				↓				16.35	16.35

13 21 26 – Cold Storage Rooms

13 21 26.50 Refrigeration		Crew	Daily Output	Labor-Hours	Unit	Material	Labor	Equipment	Total	Total Incl O&P
0010	**REFRIGERATION**									
0020	Curbs, 12″ high, 4″ thick, concrete	2 Carp	58	.276	L.F.	3.35	14.25		17.60	25
1000	Doors, see Section 08 34 13.10									
2400	Finishes, 2 coat Portland cement plaster, 1/2″ thick	1 Plas	48	.167	S.F.	.94	7.95		8.89	12.95
2500	For galvanized reinforcing mesh, add	1 Lath	335	.024		.62	1.20		1.82	2.44
2700	3/16″ thick latex cement	1 Plas	88	.091		1.56	4.34		5.90	8.20
2900	For glass cloth reinforced ceilings, add	"	450	.018		.63	.85		1.48	1.96
3100	Fiberglass panels, 1/8″ thick	1 Carp	149.45	.054		2.12	2.76		4.88	6.50
3200	Polystyrene, plastic finish ceiling, 1″ thick		274	.029		2.85	1.51		4.36	5.40
3400	2″ thick		274	.029		3.21	1.51		4.72	5.80
3500	4″ thick	↓	219	.037	↓	4.07	1.89		5.96	7.35

13 21 Controlled Environment Rooms

13 21 26 – Cold Storage Rooms

13 21 26.50 Refrigeration		Crew	Daily Output	Labor-Hours	Unit	Material	Labor	Equipment	Total	Total Incl O&P
						2019 Bare Costs				
3800	Floors, concrete, 4″ thick	1 Cefi	93	.086	S.F.	1.60	4.21		5.81	7.95
3900	6″ thick	″	85	.094		2.51	4.60		7.11	9.55
4000	Insulation, 1″ to 6″ thick, cork				B.F.	1.48			1.48	1.63
4100	Urethane					.61			.61	.67
4300	Polystyrene, regular					.56			.56	.62
4400	Bead board					.27			.27	.30
4600	Installation of above, add per layer	2 Carp	657.60	.024	S.F.	.44	1.26		1.70	2.38
4700	Wall and ceiling juncture		298.90	.054	L.F.	1.43	2.76		4.19	5.75
4900	Partitions, galvanized sandwich panels, 4″ thick, stock		219.20	.073	S.F.	10.15	3.77		13.92	16.85
5000	Aluminum or fiberglass		219.20	.073	″	6.55	3.77		10.32	12.90
5200	Prefab walk-in, 7′-6″ high, aluminum, incl. refrigeration, door & floor									
5210	not incl. partitions, 6′ x 6′	2 Carp	54.80	.292	SF Flr.	107	15.10		122.10	141
5500	10′ x 10′		82.20	.195		86	10.05		96.05	110
5700	12′ x 14′		109.60	.146		77	7.55		84.55	96
5800	12′ x 20′		109.60	.146		114	7.55		121.55	136
6100	For 8′-6″ high, add					5%				
6300	Rule of thumb for complete units, w/o doors & refrigeration, cooler	2 Carp	146	.110		164	5.65		169.65	190
6400	Freezer		109.60	.146		114	7.55		121.55	137
6600	Shelving, plated or galvanized, steel wire type		360	.044	SF Hor.	8.50	2.30		10.80	12.80
6700	Slat shelf type		375	.043		17.75	2.20		19.95	23
6900	For stainless steel shelving, add					300%				
7000	Vapor barrier, on wood walls	2 Carp	1644	.010	S.F.	.19	.50		.69	.97
7200	On masonry walls	″	1315	.012	″	.33	.63		.96	1.31
7500	For air curtain doors, see Section 23 34 33.10									

13 24 Special Activity Rooms

13 24 16 – Saunas

13 24 16.50 Saunas and Heaters		Crew	Daily Output	Labor-Hours	Unit	Material	Labor	Equipment	Total	Total Incl O&P
0010	**SAUNAS AND HEATERS**									
0020	Prefabricated, incl. heater & controls, 7′ high, 6′ x 4′, C/C	L-7	2.20	12.727	Ea.	5,250	635		5,885	6,725
0050	6′ x 4′, C/P		2	14		4,350	695		5,045	5,825
0400	6′ x 5′, C/C		2	14		5,425	695		6,120	7,000
0450	6′ x 5′, C/P		2	14		5,150	695		5,845	6,700
0600	6′ x 6′, C/C		1.80	15.556		7,100	775		7,875	8,975
0650	6′ x 6′, C/P		1.80	15.556		5,450	775		6,225	7,175
0800	6′ x 9′, C/C		1.60	17.500		7,850	870		8,720	9,975
0850	6′ x 9′, C/P		1.60	17.500		6,450	870		7,320	8,425
1000	8′ x 12′, C/C		1.10	25.455		10,700	1,275		11,975	13,700
1050	8′ x 12′, C/P		1.10	25.455		8,850	1,275		10,125	11,700
1200	8′ x 8′, C/C		1.40	20		8,600	995		9,595	11,000
1250	8′ x 8′, C/P		1.40	20		7,675	995		8,670	9,925
1400	8′ x 10′, C/C		1.20	23.333		9,400	1,150		10,550	12,100
1450	8′ x 10′, C/P		1.20	23.333		8,325	1,150		9,475	10,900
1600	10′ x 12′, C/C		1	28		12,600	1,400		14,000	16,000
1650	10′ x 12′, C/P		1	28		13,600	1,400		15,000	17,000
2500	Heaters only (incl. above), wall mounted, to 200 C.F.					960			960	1,050
2750	To 300 C.F.					1,050			1,050	1,150
3000	Floor standing, to 720 C.F., 10,000 watts, w/controls	1 Elec	3	2.667		2,475	160		2,635	2,975
3250	To 1,000 C.F., 16,000 watts	″	3	2.667		3,600	160		3,760	4,200

13 24 Special Activity Rooms

13 24 26 – Steam Baths

13 24 26.50 Steam Baths and Components		Crew	Daily Output	Labor-Hours	Unit	Material	Labor	Equipment	Total	Total Incl O&P
						2019 Bare Costs				
0010	**STEAM BATHS AND COMPONENTS**									
0020	Heater, timer & head, single, to 140 C.F.	1 Plum	1.20	6.667	Ea.	2,300	420		2,720	3,150
0500	To 300 C.F.		1.10	7.273		2,625	460		3,085	3,600
1000	Commercial size, with blow-down assembly, to 800 C.F.		.90	8.889		5,750	560		6,310	7,175
1500	To 2,500 C.F.		.80	10		8,225	630		8,855	9,975
2000	Multiple, motels, apts., 2 baths, w/blow-down assm., 500 C.F.	Q-1	1.30	12.308		6,675	700		7,375	8,400
2500	4 baths	"	.70	22.857		10,700	1,300		12,000	13,800
2700	Conversion unit for residential tub, including door					3,900			3,900	4,275

13 34 Fabricated Engineered Structures

13 34 23 – Fabricated Structures

13 34 23.10 Comfort Stations		Crew	Daily Output	Labor-Hours	Unit	Material	Labor	Equipment	Total	Total Incl O&P
0010	**COMFORT STATIONS** Prefab., stock, w/doors, windows & fixt.									
0100	Not incl. interior finish or electrical									
0300	Mobile, on steel frame, 2 unit				S.F.	190			190	209
0350	7 unit					325			325	355
0400	Permanent, including concrete slab, 2 unit	B-12J	50	.320		251	15.75	17.25	284	320
0500	6 unit	"	43	.372		189	18.35	20	227.35	258
0600	Alternate pricing method, mobile, 2 fixture				Fixture	6,675			6,675	7,350
0650	7 fixture					11,600			11,600	12,800
0700	Permanent, 2 unit	B-12J	.70	22.857		20,700	1,125	1,225	23,050	25,900
0750	6 unit	"	.50	32		17,800	1,575	1,725	21,100	23,900

13 34 23.16 Fabricated Control Booths		Crew	Daily Output	Labor-Hours	Unit	Material	Labor	Equipment	Total	Total Incl O&P
0010	**FABRICATED CONTROL BOOTHS**									
0100	Guard House, prefab conc. w/bullet resistant doors & windows, roof & wiring									
0110	8′ x 8′, Level III	L-10	1	24	Ea.	53,500	1,375	470	55,345	61,500
0120	8′ x 8′, Level IV	"	1	24	"	61,000	1,375	470	62,845	70,000

13 47 Facility Protection

13 47 13 – Cathodic Protection

13 47 13.16 Cathodic Prot. for Underground Storage Tanks		Crew	Daily Output	Labor-Hours	Unit	Material	Labor	Equipment	Total	Total Incl O&P
0010	**CATHODIC PROTECTION FOR UNDERGROUND STORAGE TANKS**									
1000	Anodes, magnesium type, 9 #	R-15	18.50	2.595	Ea.	36	152	15.30	203.30	284
1010	17 #		13	3.692		76	217	22	315	435
1020	32 #		10	4.800		119	282	28.50	429.50	580
1030	48 #		7.20	6.667		164	390	39.50	593.50	810
1100	Graphite type w/epoxy cap, 3″ x 60″ (32 #)	R-22	8.40	4.438		124	244		368	500
1110	4″ x 80″ (68 #)		6	6.213		240	340		580	775
1120	6″ x 72″ (80 #)		5.20	7.169		1,525	395		1,920	2,275
1130	6″ x 36″ (45 #)		9.60	3.883		770	213		983	1,175
2000	Rectifiers, silicon type, air cooled, 28 V/10 A	R-19	3.50	5.714		2,150	345		2,495	2,875
2010	20 V/20 A		3.50	5.714		2,200	345		2,545	2,950
2100	Oil immersed, 28 V/10 A		3	6.667		2,850	400		3,250	3,725
2110	20 V/20 A		3	6.667		3,175	400		3,575	4,075
3000	Anode backfill, coke breeze	R-22	3850	.010	Lb.	.29	.53		.82	1.11
4000	Cable, HMWPE, No. 8		2.40	15.533	M.L.F.	590	855		1,445	1,925
4010	No. 6		2.40	15.533		845	855		1,700	2,200
4020	No. 4		2.40	15.533		1,275	855		2,130	2,675

13 47 13 – Cathodic Protection

13 47 13.16 Cathodic Prot. for Underground Storage Tanks		Crew	Daily Output	Labor-Hours	Unit	2019 Bare Costs Material	Labor	Equipment	Total	Total Incl O&P
4030	No. 2	R-22	2.40	15.533	M.L.F.	1,975	855		2,830	3,450
4040	No. 1		2.20	16.945		2,700	930		3,630	4,350
4050	No. 1/0		2.20	16.945		3,500	930		4,430	5,250
4060	No. 2/0		2.20	16.945		5,025	930		5,955	6,950
4070	No. 4/0		2	18.640		6,250	1,025		7,275	8,400
5000	Test station, 7 terminal box, flush curb type w/lockable cover	R-19	12	1.667	Ea.	79	100		179	237
5010	Reference cell, 2" diam. PVC conduit, cplg., plug, set flush	"	4.80	4.167	"	156	251		407	545

Estimating Tips

General

- Many products in Division 14 will require some type of support or blocking for installation not included with the item itself. Examples are supports for conveyors or tube systems, attachment points for lifts, and footings for hoists or cranes. Add these supports in the appropriate division.

14 10 00 Dumbwaiters
14 20 00 Elevators

- Dumbwaiters and elevators are estimated and purchased in a method similar to buying a car. The manufacturer has a base unit with standard features. Added to this base unit price will be whatever options the owner or specifications require. Increased load capacity, additional vertical travel, additional stops, higher speed, and cab finish options are items to be considered. When developing an estimate for dumbwaiters and elevators, remember that some items needed by the installers may have to be included as part of the general contract.

  Examples are:
  - ☐ shaftway
  - ☐ rail support brackets
  - ☐ machine room
  - ☐ electrical supply
  - ☐ sill angles
  - ☐ electrical connections
  - ☐ pits
  - ☐ roof penthouses
  - ☐ pit ladders

  Check the job specifications and drawings before pricing.
- Installation of elevators and handicapped lifts in historic structures can require significant additional costs. The associated structural requirements may involve cutting into and repairing finishes, moldings, flooring, etc. The estimator must account for these special conditions.

14 30 00 Escalators and Moving Walks

- Escalators and moving walks are specialty items installed by specialty contractors. There are numerous options associated with these items. For specific options, contact a manufacturer or contractor. In a method similar to estimating dumbwaiters and elevators, you should verify the extent of general contract work and add items as necessary.

14 40 00 Lifts
14 90 00 Other Conveying Equipment

- Products such as correspondence lifts, chutes, and pneumatic tube systems, as well as other items specified in this subdivision, may require trained installers. The general contractor might not have any choice as to who will perform the installation or when it will be performed. Long lead times are often required for these products, making early decisions in scheduling necessary.

Reference Numbers

Reference numbers are shown at the beginning of some major classifications. These numbers refer to related items in the Reference Section. The reference information may be an estimating procedure, an alternate pricing method, or technical information.

Note: Not all subdivisions listed here necessarily appear. ■

14 91 Facility Chutes

14 91 33 – Laundry and Linen Chutes

14 91 33.10 Chutes		Crew	Daily Output	Labor-Hours	Unit	2019 Bare Costs Material	Labor	Equipment	Total	Total Incl O&P
0011	**CHUTES**, linen, trash or refuse									
0050	Aluminized steel, 16 ga., 18" diameter	2 Shee	3.50	4.571	Floor	1,800	279		2,079	2,425
0100	24" diameter		3.20	5		2,000	305		2,305	2,675
0200	30" diameter		3	5.333		2,250	325		2,575	3,000
0300	36" diameter		2.80	5.714		2,800	350		3,150	3,600
0400	Galvanized steel, 16 ga., 18" diameter		3.50	4.571		1,075	279		1,354	1,625
0500	24" diameter		3.20	5		1,225	305		1,530	1,825
0600	30" diameter		3	5.333		1,375	325		1,700	2,000
0700	36" diameter		2.80	5.714		1,625	350		1,975	2,300
0800	Stainless steel, 18" diameter		3.50	4.571		3,175	279		3,454	3,925
0900	24" diameter		3.20	5		3,350	305		3,655	4,150
1000	30" diameter		3	5.333		3,975	325		4,300	4,875
1005	36" diameter		2.80	5.714		4,200	350		4,550	5,150
1200	Linen chute bottom collector, aluminized steel		4	4	Ea.	1,525	244		1,769	2,050
1300	Stainless steel		4	4		1,925	244		2,169	2,500
1500	Refuse, bottom hopper, aluminized steel, 18" diameter		3	5.333		1,100	325		1,425	1,700
1600	24" diameter		3	5.333		1,325	325		1,650	1,975
1800	36" diameter		3	5.333		2,625	325		2,950	3,400

14 91 82 – Trash Chutes

14 91 82.10 Trash Chutes and Accessories		Crew	Daily Output	Labor-Hours	Unit	Material	Labor	Equipment	Total	Total Incl O&P
0010	**TRASH CHUTES AND ACCESSORIES**									
2900	Package chutes, spiral type, minimum	2 Shee	4.50	3.556	Floor	2,575	217		2,792	3,175
3000	Maximum	"	1.50	10.667	"	6,750	650		7,400	8,400

14 92 Pneumatic Tube Systems

14 92 10 – Conventional, Automatic and Computer Controlled Pneumatic Tube Systems

14 92 10.10 Pneumatic Tube Systems		Crew	Daily Output	Labor-Hours	Unit	Material	Labor	Equipment	Total	Total Incl O&P
0010	**PNEUMATIC TUBE SYSTEMS**									
0020	100' long, single tube, 2 stations, stock									
0100	3" diameter	2 Stpi	.12	133	Total	3,375	8,525		11,900	16,500
0300	4" diameter	"	.09	178	"	4,275	11,400		15,675	21,800
0400	Twin tube, two stations or more, conventional system									
0600	2-1/2" round	2 Stpi	62.50	.256	L.F.	42.50	16.35		58.85	71.50
0700	3" round		46	.348		38	22		60	75.50
0900	4" round		49.60	.323		48	20.50		68.50	84
1000	4" x 7" oval		37.60	.426		92	27		119	142
1050	Add for blower		2	8	System	5,200	510		5,710	6,500
1110	Plus for each round station, add		7.50	2.133	Ea.	1,350	136		1,486	1,700
1150	Plus for each oval station, add		7.50	2.133	"	1,350	136		1,486	1,700
1200	Alternate pricing method: base cost, economy model		.75	21.333	Total	5,750	1,375		7,125	8,375
1300	Custom model		.25	64	"	11,500	4,100		15,600	18,800
1500	Plus total system length, add, for economy model		93.40	.171	L.F.	8.25	10.95		19.20	25.50
1600	For custom model		37.60	.426	"	25	27		52	68.50
1800	Completely automatic system, 4" round, 15 to 50 stations		.29	55.172	Station	22,500	3,525		26,025	30,100
2200	51 to 144 stations		.32	50		15,600	3,200		18,800	22,000
2400	6" round or 4" x 7" oval, 15 to 50 stations		.24	66.667		27,600	4,275		31,875	36,700
2800	51 to 144 stations		.23	69.565		21,100	4,450		25,550	29,900

Estimating Tips

22 10 00 Plumbing Piping and Pumps

This subdivision is primarily basic pipe and related materials. The pipe may be used by any of the mechanical disciplines, i.e., plumbing, fire protection, heating, and air conditioning.

Note: CPVC plastic piping approved for fire protection is located in 21 11 13.

- The labor adjustment factors listed in Subdivision 22 01 02.20 apply throughout Divisions 21, 22, and 23. CAUTION: the correct percentage may vary for the same items. For example, the percentage add for the basic pipe installation should be based on the maximum height that the installer must install for that particular section. If the pipe is to be located 14' above the floor but it is suspended on threaded rod from beams, the bottom flange of which is 18' high (4' rods), then the height is actually 18' and the add is 20%. The pipe cover, however, does not have to go above the 14' and so the add should be 10%.
- Most pipe is priced first as straight pipe with a joint (coupling, weld, etc.) every 10' and a hanger usually every 10'. There are exceptions with hanger spacing such as for cast iron pipe (5') and plastic pipe (3 per 10'). Following each type of pipe there are several lines listing sizes and the amount to be subtracted to delete couplings and hangers. This is for pipe that is to be buried or supported together on trapeze hangers. The reason that the couplings are deleted is that these runs are usually long, and frequently longer lengths of pipe are used. By deleting the couplings, the estimator is expected to look up and add back the correct reduced number of couplings.
- When preparing an estimate, it may be necessary to approximate the fittings. Fittings usually run between 25% and 50% of the cost of the pipe. The lower percentage is for simpler runs, and the higher number is for complex areas, such as mechanical rooms.
- For historic restoration projects, the systems must be as invisible as possible, and pathways must be sought for pipes, conduit, and ductwork. While installations in accessible spaces (such as basements and attics) are relatively straightforward to estimate, labor costs may be more difficult to determine when delivery systems must be concealed.

22 40 00 Plumbing Fixtures

- Plumbing fixture costs usually require two lines: the fixture itself and its "rough-in, supply, and waste."
- In the Assemblies Section (Plumbing D2010) for the desired fixture, the System Components Group at the center of the page shows the fixture on the first line. The rest of the list (fittings, pipe, tubing, etc.) will total up to what we refer to in the Unit Price section as "Rough-in, supply, waste, and vent." Note that for most fixtures we allow a nominal 5' of tubing to reach from the fixture to a main or riser.
- Remember that gas- and oil-fired units need venting.

Reference Numbers

Reference numbers are shown at the beginning of some major classifications. These numbers refer to related items in the Reference Section. The reference information may be an estimating procedure, an alternate pricing method, or technical information.

Note: Not all subdivisions listed here necessarily appear. ■

Note: "Powered in part by CINX™, based on liscensed proprietary information of Harrison Publishing House, Inc."

Note: Trade Service, in part, has been used as a reference source for some of the material prices used in Division 22.

22 01 02 – Labor Adjustments

22 01 02.10 Boilers, General		Crew	Daily Output	Labor-Hours	Unit	Material	2019 Bare Costs Labor	Equipment	Total	Total Incl O&P
0010	**BOILERS, GENERAL**, Prices do not include flue piping, elec. wiring,									
0020	gas or oil piping, boiler base, pad, or tankless unless noted									
0100	Boiler H.P.: 10 KW = 34 lb./steam/hr. = 33,475 BTU/hr.									
0150	To convert SFR to BTU rating: Hot water, 150 x SFR;									
0160	Forced hot water, 180 x SFR; steam, 240 x SFR									

22 01 02.20 Labor Adjustment Factors		Crew	Daily Output	Labor-Hours	Unit	Material	2019 Bare Costs Labor	Equipment	Total	Total Incl O&P
0010	**LABOR ADJUSTMENT FACTORS** (For Div. 21, 22 and 23) R220102-20									
0100	Labor factors: The below are reasonable suggestions, but									
0110	each project must be evaluated for its own peculiarities, and									
0120	the adjustments be increased or decreased depending on the									
0130	severity of the special conditions.									
1000	Add to labor for elevated installation (Above floor level)									
1080	10′ to 14.5′ high R221113-70						10%			
1100	15′ to 19.5′ high						20%			
1120	20′ to 24.5′ high						25%			
1140	25′ to 29.5′ high						35%			
1160	30′ to 34.5′ high						40%			
1180	35′ to 39.5′ high						50%			
1200	40′ and higher						55%			
2000	Add to labor for crawl space									
2100	3′ high						40%			
2140	4′ high						30%			
3000	Add to labor for multi-story building									
3010	For new construction (No elevator available)									
3100	Add for floors 3 thru 10						5%			
3110	Add for floors 11 thru 15						10%			
3120	Add for floors 16 thru 20						15%			
3130	Add for floors 21 thru 30						20%			
3140	Add for floors 31 and up						30%			
3170	For existing structure (Elevator available)									
3180	Add for work on floor 3 and above						2%			
4000	Add to labor for working in existing occupied buildings									
4100	Hospital						35%			
4140	Office building						25%			
4180	School						20%			
4220	Factory or warehouse						15%			
4260	Multi dwelling						15%			
5000	Add to labor, miscellaneous									
5100	Cramped shaft						35%			
5140	Congested area						15%			
5180	Excessive heat or cold						30%			
9000	Labor factors: The above are reasonable suggestions, but									
9010	each project should be evaluated for its own peculiarities.									
9100	Other factors to be considered are:									
9140	Movement of material and equipment through finished areas									
9180	Equipment room									
9220	Attic space									
9260	No service road									
9300	Poor unloading/storage area									
9340	Congested site area/heavy traffic									

22 05 05 – Selective Demolition for Plumbing

22 05 05.10 Plumbing Demolition		Crew	Daily Output	Labor-Hours	Unit	Material	2019 Bare Costs Labor	Equipment	Total	Total Incl O&P
0010	**PLUMBING DEMOLITION** R220105-10									
0400	Air compressor, up thru 2 HP	Q-1	10	1.600	Ea.		91		91	136
0410	3 HP thru 7-1/2 HP R024119-10		5.60	2.857			162		162	244
0420	10 HP thru 15 HP		1.40	11.429			650		650	975
0430	20 HP thru 30 HP	Q-2	1.30	18.462			1,100		1,100	1,625
0500	Backflow preventer, up thru 2" diameter	1 Plum	17	.471			29.50		29.50	44.50
0510	2-1/2" thru 3" diameter	Q-1	10	1.600			91		91	136
0520	4" thru 6" diameter	"	5	3.200			182		182	273
0530	8" thru 10" diameter	Q-2	3	8			470		470	705
1900	Piping fittings, single connection, up thru 1-1/2" diameter	1 Plum	30	.267			16.85		16.85	25.50
1910	2" thru 4" diameter		14	.571			36		36	54
1980	Pipe hanger/support removal		80	.100			6.30		6.30	9.50
2000	Piping, metal, up thru 1-1/2" diameter		200	.040	L.F.		2.53		2.53	3.79
2050	2" thru 3-1/2" diameter		150	.053			3.37		3.37	5.05
2100	4" thru 6" diameter	2 Plum	100	.160			10.10		10.10	15.15
2150	8" thru 14" diameter	"	60	.267			16.85		16.85	25.50
2153	16" thru 20" diameter	Q-18	70	.343			20.50	.80	21.30	31.50
2155	24" thru 26" diameter		55	.436			26	1.02	27.02	40
2156	30" thru 36" diameter		40	.600			36	1.40	37.40	55
2160	Plastic pipe with fittings, up thru 1-1/2" diameter	1 Plum	250	.032			2.02		2.02	3.03
2162	2" thru 3" diameter	"	200	.040			2.53		2.53	3.79
2164	4" thru 6" diameter	Q-1	200	.080			4.55		4.55	6.80
2166	8" thru 14" diameter		150	.107			6.05		6.05	9.10
2168	16" diameter		100	.160			9.10		9.10	13.65
2180	Pumps, all fractional horse-power		12	1.333	Ea.		76		76	114
2184	1 HP thru 5 HP		6	2.667			152		152	227
2186	7-1/2 HP thru 15 HP		2.50	6.400			365		365	545
2188	20 HP thru 25 HP	Q-2	4	6			355		355	530
2190	30 HP thru 60 HP		.80	30			1,775		1,775	2,650
2192	75 HP thru 100 HP		.60	40			2,350		2,350	3,525
2194	150 HP		.50	48			2,825		2,825	4,250
2230	Temperature maintenance cable	1 Plum	1200	.007	L.F.		.42		.42	.63
3100	Tanks, water heaters and liquid containers									
3110	Up thru 45 gallons	Q-1	22	.727	Ea.		41.50		41.50	62
3120	50 thru 120 gallons		14	1.143			65		65	97.50
3130	130 thru 240 gallons		7.60	2.105			120		120	180
3140	250 thru 500 gallons		5.40	2.963			168		168	253
3150	600 thru 1,000 gallons	Q-2	1.60	15			885		885	1,325
3160	1,100 thru 2,000 gallons		.70	34.286			2,025		2,025	3,025
3170	2,100 thru 4,000 gallons		.50	48			2,825		2,825	4,250
9100	Valve, metal valves or strainers and similar, up thru 1-1/2" diameter	1 Stpi	28	.286			18.25		18.25	27.50
9110	2" thru 3" diameter	Q-1	11	1.455			82.50		82.50	124
9120	4" thru 6" diameter	"	8	2			114		114	171
9130	8" thru 14" diameter	Q-2	8	3			177		177	265
9140	16" thru 20" diameter		2	12			705		705	1,050
9150	24" diameter		1.20	20			1,175		1,175	1,775
9200	Valve, plastic, up thru 1-1/2" diameter	1 Plum	42	.190			12.05		12.05	18.05
9210	2" thru 3" diameter		15	.533			33.50		33.50	50.50
9220	4" thru 6" diameter		12	.667			42		42	63
9300	Vent flashing and caps		55	.145			9.20		9.20	13.80
9350	Water filter, commercial, 1" thru 1-1/2"	Q-1	2	8			455		455	680
9360	2" thru 2-1/2"	"	1.60	10			570		570	855
9400	Water heaters									

22 05 05 – Selective Demolition for Plumbing

22 05 05.10 Plumbing Demolition		Crew	Daily Output	Labor-Hours	Unit	Material	2019 Bare Costs Labor	Equipment	Total	Total Incl O&P
9410	Up thru 245 GPH	Q-1	2.40	6.667	Ea.		380		380	570
9420	250 thru 756 GPH		1.60	10			570		570	855
9430	775 thru 1,640 GPH		.80	20			1,125		1,125	1,700
9440	1,650 thru 4,000 GPH	Q-2	.50	48			2,825		2,825	4,250
9470	Water softener	Q-1	2	8			455		455	680

22 05 23 – General-Duty Valves for Plumbing Piping

22 05 23.10 Valves, Brass

		Crew	Daily Output	Labor-Hours	Unit	Material	Labor	Equipment	Total	Total Incl O&P
0010	**VALVES, BRASS**									
0032	For motorized valves, see Section 23 09 53.10									
0500	Gas cocks, threaded									
0510	1/4"	1 Plum	26	.308	Ea.	15.60	19.45		35.05	46
0520	3/8"		24	.333		15.60	21		36.60	48.50
0530	1/2"		24	.333		13.20	21		34.20	46
0540	3/4"		22	.364		17	23		40	53
0550	1"		19	.421		33.50	26.50		60	76.50
0560	1-1/4"		15	.533		51.50	33.50		85	107
0570	1-1/2"		13	.615		73	39		112	139
0580	2"		11	.727		111	46		157	191
0672	For larger sizes use lubricated plug valve, Section 23 05 23.70									

22 05 23.20 Valves, Bronze

		Crew	Daily Output	Labor-Hours	Unit	Material	Labor	Equipment	Total	Total Incl O&P
0010	**VALVES, BRONZE** R220523-90									
1020	Angle, 150 lb., rising stem, threaded									
1030	1/8"	1 Plum	24	.333	Ea.	138	21		159	184
1040	1/4"		24	.333		138	21		159	183
1050	3/8"		24	.333		151	21		172	198
1060	1/2"		22	.364		159	23		182	210
1070	3/4"		20	.400		216	25.50		241.50	276
1080	1"		19	.421		310	26.50		336.50	380
1090	1-1/4"		15	.533		305	33.50		338.50	385
1100	1-1/2"		13	.615		450	39		489	555
1102	Soldered same price as threaded									
1110	2"	1 Plum	11	.727	Ea.	845	46		891	1,000
1300	Ball									
1304	Soldered									
1312	3/8"	1 Plum	21	.381	Ea.	17.25	24		41.25	55
1316	1/2"		18	.444		17.25	28		45.25	61
1320	3/4"		17	.471		28.50	29.50		58	76
1324	1"		15	.533		36	33.50		69.50	90
1328	1-1/4"		13	.615		50	39		89	114
1332	1-1/2"		11	.727		76.50	46		122.50	154
1336	2"		9	.889		96	56		152	190
1340	2-1/2"		7	1.143		430	72		502	585
1344	3"		5	1.600		500	101		601	700
1350	Single union end									
1358	3/8"	1 Plum	21	.381	Ea.	24.50	24		48.50	63
1362	1/2"		18	.444		25.50	28		53.50	70
1366	3/4"		17	.471		45	29.50		74.50	94
1370	1"		15	.533		60	33.50		93.50	117
1374	1-1/4"		13	.615		92	39		131	160
1378	1-1/2"		11	.727		116	46		162	197
1382	2"		9	.889		199	56		255	300
1398	Threaded, 150 psi									

22 05 23 – General-Duty Valves for Plumbing Piping

22 05 23.20 Valves, Bronze		Crew	Daily Output	Labor-Hours	Unit	Material	Labor (2019 Bare Costs)	Equipment	Total	Total Incl O&P
1400	1/4"	1 Plum	24	.333	Ea.	17.40	21		38.40	50.50
1430	3/8"		24	.333		17.40	21		38.40	50.50
1450	1/2"		22	.364		15.15	23		38.15	51
1460	3/4"		20	.400		28.50	25.50		54	69.50
1470	1"		19	.421		31	26.50		57.50	74
1480	1-1/4"		15	.533		52.50	33.50		86	108
1490	1-1/2"		13	.615		70.50	39		109.50	136
1500	2"		11	.727		82.50	46		128.50	160
1510	2-1/2"		9	.889		277	56		333	390
1520	3"		8	1		420	63		483	555
1522	Solder the same price as threaded									
1600	Butterfly, 175 psi, full port, solder or threaded ends									
1610	Stainless steel disc and stem									
1620	1/4"	1 Plum	24	.333	Ea.	20	21		41	54
1630	3/8"		24	.333		16.20	21		37.20	49.50
1640	1/2"		22	.364		17.30	23		40.30	53.50
1650	3/4"		20	.400		28	25.50		53.50	68.50
1660	1"		19	.421		34	26.50		60.50	77.50
1670	1-1/4"		15	.533		55	33.50		88.50	111
1680	1-1/2"		13	.615		71	39		110	137
1690	2"		11	.727		90	46		136	168
1750	Check, swing, class 150, regrinding disc, threaded									
1800	1/8"	1 Plum	24	.333	Ea.	74.50	21		95.50	114
1830	1/4"		24	.333		66	21		87	104
1840	3/8"		24	.333		70	21		91	109
1850	1/2"		24	.333		69.50	21		90.50	108
1860	3/4"		20	.400		91	25.50		116.50	138
1870	1"		19	.421		143	26.50		169.50	197
1880	1-1/4"		15	.533		206	33.50		239.50	277
1890	1-1/2"		13	.615		277	39		316	365
1900	2"		11	.727		315	46		361	420
1910	2-1/2"	Q-1	15	1.067		785	60.50		845.50	955
1920	3"	"	13	1.231		1,225	70		1,295	1,450
2000	For 200 lb., add					5%	10%			
2040	For 300 lb., add					15%	15%			
2060	Check swing, 300 lb., lead free unless noted, sweat, 3/8" size	1 Plum	24	.333	Ea.	105	21		126	148
2070	1/2"		24	.333		105	21		126	148
2080	3/4"		20	.400		140	25.50		165.50	192
2090	1"		19	.421		208	26.50		234.50	269
2100	1-1/4"		15	.533		294	33.50		327.50	375
2110	1-1/2"		13	.615		345	39		384	440
2120	2"		11	.727		510	46		556	630
2130	2-1/2", not lead free	Q-1	15	1.067		775	60.50		835.50	945
2140	3", not lead free	"	13	1.231		995	70		1,065	1,200
2350	Check, lift, class 150, horizontal composition disc, threaded									
2430	1/4"	1 Plum	24	.333	Ea.	142	21		163	188
2440	3/8"		24	.333		180	21		201	230
2450	1/2"		24	.333		156	21		177	204
2460	3/4"		20	.400		191	25.50		216.50	248
2470	1"		19	.421		290	26.50		316.50	360
2480	1-1/4"		15	.533		365	33.50		398.50	450
2490	1-1/2"		13	.615		435	39		474	540
2500	2"		11	.727		770	46		816	915

22 05 23.20 Valves, Bronze		Crew	Daily Output	Labor-Hours	Unit	2019 Bare Costs Material	Labor	Equipment	Total	Total Incl O&P
2850	Gate, N.R.S., soldered, 125 psi									
2900	3/8"	1 Plum	24	.333	Ea.	66	21		87	104
2920	1/2"		24	.333		59.50	21		80.50	97
2940	3/4"		20	.400		68	25.50		93.50	113
2950	1"		19	.421		76.50	26.50		103	125
2960	1-1/4"		15	.533		142	33.50		175.50	207
2970	1-1/2"		13	.615		163	39		202	238
2980	2"		11	.727		186	46		232	273
2990	2-1/2"	Q-1	15	1.067		500	60.50		560.50	640
3000	3"	"	13	1.231		565	70		635	725
3350	Threaded, class 150									
3410	1/4"	1 Plum	24	.333	Ea.	82	21		103	122
3420	3/8"		24	.333		82	21		103	122
3430	1/2"		24	.333		77	21		98	116
3440	3/4"		20	.400		89	25.50		114.50	136
3450	1"		19	.421		111	26.50		137.50	163
3460	1-1/4"		15	.533		151	33.50		184.50	217
3470	1-1/2"		13	.615		216	39		255	297
3480	2"		11	.727		258	46		304	350
3490	2-1/2"	Q-1	15	1.067		595	60.50		655.50	745
3500	3"	"	13	1.231		845	70		915	1,025
3600	Gate, flanged, 150 lb.									
3610	1"	1 Plum	7	1.143	Ea.	1,175	72		1,247	1,375
3620	1-1/2"		6	1.333		1,450	84		1,534	1,725
3630	2"		5	1.600		2,500	101		2,601	2,900
3634	2-1/2"	Q-1	5	3.200		2,100	182		2,282	2,575
3640	3"	"	4.50	3.556		3,850	202		4,052	4,525
3850	Rising stem, soldered, 300 psi									
3900	3/8"	1 Plum	24	.333	Ea.	160	21		181	207
3920	1/2"		24	.333		127	21		148	172
3940	3/4"		20	.400		145	25.50		170.50	198
3950	1"		19	.421		194	26.50		220.50	253
3960	1-1/4"		15	.533		272	33.50		305.50	350
3970	1-1/2"		13	.615		330	39		369	425
3980	2"		11	.727		530	46		576	650
3990	2-1/2"	Q-1	15	1.067		1,125	60.50		1,185.50	1,350
4000	3"	"	13	1.231		1,750	70		1,820	2,025
4250	Threaded, class 150									
4310	1/4"	1 Plum	24	.333	Ea.	74.50	21		95.50	114
4320	3/8"		24	.333		74.50	21		95.50	114
4330	1/2"		24	.333		67.50	21		88.50	106
4340	3/4"		20	.400		79	25.50		104.50	125
4350	1"		19	.421		106	26.50		132.50	157
4360	1-1/4"		15	.533		144	33.50		177.50	209
4370	1-1/2"		13	.615		182	39		221	259
4380	2"		11	.727		244	46		290	340
4390	2-1/2"	Q-1	15	1.067		570	60.50		630.50	720
4400	3"	"	13	1.231		795	70		865	980
4500	For 300 psi, threaded, add					100%	15%			
4540	For chain operated type, add					15%				
4850	Globe, class 150, rising stem, threaded									
4920	1/4"	1 Plum	24	.333	Ea.	105	21		126	147
4940	3/8"		24	.333		103	21		124	145

22 05 23 – General-Duty Valves for Plumbing Piping

22 05 23.20 Valves, Bronze		Crew	Daily Output	Labor-Hours	Unit	Material	Labor	Equipment	Total	Total Incl O&P
						2019 Bare Costs				
4950	1/2"	1 Plum	24	.333	Ea.	103	21		124	145
4960	3/4"		20	.400		140	25.50		165.50	192
4970	1"		19	.421		200	26.50		226.50	260
4980	1-1/4"		15	.533		291	33.50		324.50	370
4990	1-1/2"		13	.615		490	39		529	600
5000	2"		11	.727		660	46		706	795
5010	2-1/2"	Q-1	15	1.067		1,250	60.50		1,310.50	1,475
5020	3"	"	13	1.231		1,825	70		1,895	2,100
5120	For 300 lb. threaded, add					50%	15%			
5130	Globe, 300 lb., sweat, 3/8" size	1 Plum	24	.333	Ea.	131	21		152	176
5140	1/2"		24	.333		134	21		155	179
5150	3/4"		20	.400		182	25.50		207.50	238
5160	1"		19	.421		252	26.50		278.50	315
5170	1-1/4"		15	.533		400	33.50		433.50	490
5180	1-1/2"		13	.615		495	39		534	600
5190	2"		11	.727		835	46		881	985
5200	2-1/2"	Q-1	15	1.067		1,375	60.50		1,435.50	1,625
5210	3"	"	13	1.231		1,775	70		1,845	2,075
5600	Relief, pressure & temperature, self-closing, ASME, threaded									
5640	3/4"	1 Plum	28	.286	Ea.	210	18.05		228.05	258
5650	1"		24	.333		335	21		356	400
5660	1-1/4"		20	.400		680	25.50		705.50	790
5670	1-1/2"		18	.444		1,300	28		1,328	1,500
5680	2"		16	.500		1,500	31.50		1,531.50	1,700
5950	Pressure, poppet type, threaded									
6000	1/2"	1 Plum	30	.267	Ea.	77	16.85		93.85	111
6040	3/4"	"	28	.286	"	88	18.05		106.05	124
6400	Pressure, water, ASME, threaded									
6440	3/4"	1 Plum	28	.286	Ea.	144	18.05		162.05	185
6450	1"		24	.333		305	21		326	365
6460	1-1/4"		20	.400		470	25.50		495.50	560
6470	1-1/2"		18	.444		660	28		688	770
6480	2"		16	.500		955	31.50		986.50	1,100
6490	2-1/2"		15	.533		4,250	33.50		4,283.50	4,725
6900	Reducing, water pressure									
6920	300 psi to 25-75 psi, threaded or sweat									
6940	1/2"	1 Plum	24	.333	Ea.	460	21		481	535
6950	3/4"		20	.400		460	25.50		485.50	550
6960	1"		19	.421		715	26.50		741.50	825
6970	1-1/4"		15	.533		1,275	33.50		1,308.50	1,450
6980	1-1/2"		13	.615		1,925	39		1,964	2,150
6990	2"		11	.727		2,825	46		2,871	3,200
7100	For built-in by-pass or 10-35 psi, add					46			46	50.50
7700	High capacity, 250 psi to 25-75 psi, threaded									
7740	1/2"	1 Plum	24	.333	Ea.	790	21		811	900
7780	3/4"		20	.400		800	25.50		825.50	920
7790	1"		19	.421		945	26.50		971.50	1,100
7800	1-1/4"		15	.533		1,625	33.50		1,658.50	1,850
7810	1-1/2"		13	.615		2,375	39		2,414	2,675
7820	2"		11	.727		3,475	46		3,521	3,900
7830	2-1/2"		9	.889		5,050	56		5,106	5,625
7840	3"		8	1		5,775	63		5,838	6,450
7850	3" flanged (iron body)	Q-1	10	1.600		5,850	91		5,941	6,550

22 05 23.20 Valves, Bronze		Crew	Daily Output	Labor-Hours	Unit	Material	Labor	Equipment	Total	Total Incl O&P
						2019 Bare Costs				
7860	4" flanged (iron body)	Q-1	8	2	Ea.	7,450	114		7,564	8,375
7920	For higher pressure, add					25%				
8000	Silent check, bronze trim									
8010	Compact wafer type, for 125 or 150 lb. flanges									
8020	1-1/2"	1 Plum	11	.727	Ea.	435	46		481	550
8021	2"	"	9	.889		460	56		516	590
8022	2-1/2"	Q-1	9	1.778		485	101		586	685
8023	3"		8	2		540	114		654	765
8024	4"		5	3.200		925	182		1,107	1,300
8025	5"	Q-2	6	4		875	236		1,111	1,325
8026	6"		5	4.800		1,200	283		1,483	1,725
8027	8"		4.50	5.333		2,100	315		2,415	2,800
8028	10"		4	6		5,650	355		6,005	6,750
8029	12"		3	8		12,800	470		13,270	14,800
8050	For 250 or 300 lb. flanges, thru 6" no change									
8051	For 8" and 10", add				Ea.	40%	10%			
8060	Full flange wafer type, 150 lb.									
8061	1"	1 Plum	14	.571	Ea.	325	36		361	415
8062	1-1/4"		12	.667		340	42		382	440
8063	1-1/2"		11	.727		395	46		441	505
8064	2"		9	.889		530	56		586	670
8065	2-1/2"	Q-1	9	1.778		600	101		701	810
8066	3"		8	2		665	114		779	900
8067	4"		5	3.200		980	182		1,162	1,350
8068	5"	Q-2	6	4		1,275	236		1,511	1,750
8069	6"	"	5	4.800		1,675	283		1,958	2,250
8080	For 300 lb., add					40%	10%			
8100	Globe type, 150 lb.									
8110	2"	1 Plum	9	.889	Ea.	605	56		661	750
8111	2-1/2"	Q-1	9	1.778		750	101		851	970
8112	3"		8	2		895	114		1,009	1,150
8113	4"		5	3.200		1,225	182		1,407	1,625
8114	5"	Q-2	6	4		1,525	236		1,761	2,025
8115	6"	"	5	4.800		2,100	283		2,383	2,750
8130	For 300 lb., add					20%	10%			
8140	Screwed end type, 250 lb.									
8141	1/2"	1 Plum	24	.333	Ea.	54.50	21		75.50	91.50
8142	3/4"		20	.400		54.50	25.50		80	98
8143	1"		19	.421		62	26.50		88.50	109
8144	1-1/4"		15	.533		86	33.50		119.50	145
8145	1-1/2"		13	.615		96.50	39		135.50	165
8146	2"		11	.727		130	46		176	212
8350	Tempering, water, sweat connections									
8400	1/2"	1 Plum	24	.333	Ea.	105	21		126	148
8440	3/4"	"	20	.400	"	144	25.50		169.50	197
8650	Threaded connections									
8700	1/2"	1 Plum	24	.333	Ea.	149	21		170	196
8740	3/4"		20	.400		1,025	25.50		1,050.50	1,175
8750	1"		19	.421		1,150	26.50		1,176.50	1,300
8760	1-1/4"		15	.533		1,800	33.50		1,833.50	2,025
8770	1-1/2"		13	.615		1,950	39		1,989	2,200
8780	2"		11	.727		2,925	46		2,971	3,300

22 05 Common Work Results for Plumbing

22 05 23 – General-Duty Valves for Plumbing Piping

22 05 23.60 Valves, Plastic		Crew	Daily Output	Labor-Hours	Unit	Material	Labor	Equipment	Total	Total Incl O&P
						2019 Bare Costs				
0010	**VALVES, PLASTIC** R220523-90									
1100	Angle, PVC, threaded									
1110	1/4"	1 Plum	26	.308	Ea.	58.50	19.45		77.95	93.50
1120	1/2"		26	.308		80	19.45		99.45	117
1130	3/4"		25	.320		94.50	20		114.50	135
1140	1"		23	.348		115	22		137	159
1150	Ball, PVC, socket or threaded, true union									
1230	1/2"	1 Plum	26	.308	Ea.	43.50	19.45		62.95	76.50
1240	3/4"		25	.320		45	20		65	80
1250	1"		23	.348		53.50	22		75.50	92
1260	1-1/4"		21	.381		82	24		106	126
1270	1-1/2"		20	.400		84.50	25.50		110	131
1280	2"		17	.471		133	29.50		162.50	191
1290	2-1/2"	Q-1	26	.615		228	35		263	305
1300	3"		24	.667		275	38		313	355
1310	4"		20	.800		450	45.50		495.50	565
1360	For PVC, flanged, add					100%	15%			
1450	Double union 1/2"	1 Plum	26	.308		39	19.45		58.45	72
1460	3/4"		25	.320		47.50	20		67.50	83
1470	1"		23	.348		56	22		78	94.50
1480	1-1/4"		21	.381		83.50	24		107.50	128
1490	1-1/2"		20	.400		95	25.50		120.50	143
1500	2"		17	.471		95.50	29.50		125	150
1650	CPVC, socket or threaded, single union									
1700	1/2"	1 Plum	26	.308	Ea.	59	19.45		78.45	94
1720	3/4"		25	.320		75	20		95	113
1730	1"		23	.348		89	22		111	131
1750	1-1/4"		21	.381		149	24		173	200
1760	1-1/2"		20	.400		142	25.50		167.50	194
1770	2"		17	.471		196	29.50		225.50	260
1780	3"	Q-1	24	.667		810	38		848	945
1840	For CPVC, flanged, add					65%	15%			
1880	For true union, socket or threaded, add					50%	5%			
2050	Polypropylene, threaded									
2100	1/4"	1 Plum	26	.308	Ea.	39.50	19.45		58.95	72.50
2120	3/8"		26	.308		39.50	19.45		58.95	72.50
2130	1/2"		26	.308		39	19.45		58.45	72
2140	3/4"		25	.320		47	20		67	82
2150	1"		23	.348		55	22		77	93.50
2160	1-1/4"		21	.381		74	24		98	117
2170	1-1/2"		20	.400		90.50	25.50		116	138
2180	2"		17	.471		122	29.50		151.50	179
2190	3"	Q-1	24	.667		273	38		311	355
2200	4"	"	20	.800		495	45.50		540.50	615
2550	PVC, three way, socket or threaded									
2600	1/2"	1 Plum	26	.308	Ea.	54.50	19.45		73.95	89
2640	3/4"		25	.320		64.50	20		84.50	102
2650	1"		23	.348		77	22		99	118
2660	1-1/2"		20	.400		115	25.50		140.50	164
2670	2"		17	.471		156	29.50		185.50	217
2680	3"	Q-1	24	.667		440	38		478	540
2740	For flanged, add					60%	15%			

22 05 23.60 Valves, Plastic		Crew	Daily Output	Labor-Hours	Unit	Material	2019 Bare Costs Labor	Equipment	Total	Total Incl O&P
3150	Ball check, PVC, socket or threaded									
3200	1/4"	1 Plum	26	.308	Ea.	49	19.45		68.45	83
3220	3/8"		26	.308		49	19.45		68.45	83
3240	1/2"		26	.308		44.50	19.45		63.95	78
3250	3/4"		25	.320		50.50	20		70.50	86
3260	1"		23	.348		63	22		85	103
3270	1-1/4"		21	.381		106	24		130	152
3280	1-1/2"		20	.400		106	25.50		131.50	154
3290	2"		17	.471		144	29.50		173.50	203
3310	3"	Q-1	24	.667		400	38		438	495
3320	4"	"	20	.800		565	45.50		610.50	690
3360	For PVC, flanged, add					50%	15%			
3750	CPVC, socket or threaded									
3800	1/2"	1 Plum	26	.308	Ea.	62.50	19.45		81.95	97.50
3840	3/4"		25	.320		85	20		105	124
3850	1"		23	.348		101	22		123	144
3860	1-1/2"		20	.400		173	25.50		198.50	229
3870	2"		17	.471		213	29.50		242.50	279
3880	3"	Q-1	24	.667		630	38		668	745
3920	4"	"	20	.800		825	45.50		870.50	980
3930	For CPVC, flanged, add					40%	15%			
4340	Polypropylene, threaded									
4360	1/2"	1 Plum	26	.308	Ea.	49	19.45		68.45	83
4400	3/4"		25	.320		67.50	20		87.50	105
4440	1"		23	.348		71.50	22		93.50	112
4450	1-1/2"		20	.400		139	25.50		164.50	190
4460	2"		17	.471		176	29.50		205.50	238
4500	For polypropylene flanged, add					200%	15%			
4850	Foot valve, PVC, socket or threaded									
4900	1/2"	1 Plum	34	.235	Ea.	61.50	14.85		76.35	90
4930	3/4"		32	.250		69.50	15.80		85.30	100
4940	1"		28	.286		90.50	18.05		108.55	127
4950	1-1/4"		27	.296		171	18.70		189.70	216
4960	1-1/2"		26	.308		173	19.45		192.45	220
4970	2"		24	.333		201	21		222	253
4980	3"		20	.400		480	25.50		505.50	565
4990	4"		18	.444		835	28		863	955
5000	For flanged, add					25%	10%			
5050	CPVC, socket or threaded									
5060	1/2"	1 Plum	34	.235	Ea.	76.50	14.85		91.35	107
5070	3/4"		32	.250		97	15.80		112.80	131
5080	1"		28	.286		119	18.05		137.05	158
5090	1-1/4"		27	.296		189	18.70		207.70	236
5100	1-1/2"		26	.308		189	19.45		208.45	237
5110	2"		24	.333		237	21		258	293
5120	3"		20	.400		480	25.50		505.50	570
5130	4"		18	.444		875	28		903	1,000
5140	For flanged, add					25%	10%			
5280	Needle valve, PVC, threaded									
5300	1/4"	1 Plum	26	.308	Ea.	64.50	19.45		83.95	100
5340	3/8"		26	.308		74.50	19.45		93.95	111
5360	1/2"		26	.308		74.50	19.45		93.95	111
5380	For polypropylene, add					10%				

22 05 23 – General-Duty Valves for Plumbing Piping

22 05 23.60 Valves, Plastic		Crew	Daily Output	Labor-Hours	Unit	Material	2019 Bare Costs Labor	Equipment	Total	Total Incl O&P
5800	Y check, PVC, socket or threaded									
5820	1/2"	1 Plum	26	.308	Ea.	65.50	19.45		84.95	101
5840	3/4"		25	.320		70.50	20		90.50	108
5850	1"		23	.348		82	22		104	123
5860	1-1/4"		21	.381		129	24		153	178
5870	1-1/2"		20	.400		140	25.50		165.50	192
5880	2"		17	.471		174	29.50		203.50	237
5890	2-1/2"		15	.533		370	33.50		403.50	460
5900	3"	Q-1	24	.667		350	38		388	440
5910	4"	"	20	.800		605	45.50		650.50	735
5960	For PVC flanged, add					45%	15%			
6350	Y sediment strainer, PVC, socket or threaded									
6400	1/2"	1 Plum	26	.308	Ea.	57	19.45		76.45	92
6440	3/4"		24	.333		60.50	21		81.50	98
6450	1"		23	.348		65.50	22		87.50	105
6460	1-1/4"		21	.381		110	24		134	157
6470	1-1/2"		20	.400		110	25.50		135.50	159
6480	2"		17	.471		136	29.50		165.50	195
6490	2-1/2"		15	.533		335	33.50		368.50	420
6500	3"	Q-1	24	.667		335	38		373	420
6510	4"	"	20	.800		570	45.50		615.50	695
6560	For PVC, flanged, add					55%	15%			

22 05 29 – Hangers and Supports for Plumbing Piping and Equipment

22 05 29.10 Hangers & Supp. for Plumb'g/HVAC Pipe/Equip.		Crew	Daily Output	Labor-Hours	Unit	Material	Labor	Equipment	Total	Total Incl O&P
0010	**HANGERS AND SUPPORTS FOR PLUMB'G/HVAC PIPE/EQUIP.**									
0011	TYPE numbers per MSS-SP58									
0050	Brackets									
0060	Beam side or wall, malleable iron, TYPE 34									
0070	3/8" threaded rod size	1 Plum	48	.167	Ea.	4.24	10.55		14.79	20.50
0080	1/2" threaded rod size		48	.167		3.88	10.55		14.43	20
0090	5/8" threaded rod size		48	.167		11.45	10.55		22	28.50
0100	3/4" threaded rod size		48	.167		20.50	10.55		31.05	39
0110	7/8" threaded rod size		48	.167		12.70	10.55		23.25	30
0120	For concrete installation, add						30%			
0150	Wall, welded steel, medium, TYPE 32									
0160	0 size, 12" wide, 18" deep	1 Plum	34	.235	Ea.	180	14.85		194.85	221
0170	1 size, 18" wide, 24" deep		34	.235		216	14.85		230.85	261
0180	2 size, 24" wide, 30" deep		34	.235		285	14.85		299.85	340
0200	Beam attachment, welded, TYPE 22									
0202	3/8"	Q-15	80	.200	Ea.	9.85	11.35	.70	21.90	28.50
0203	1/2"		76	.211		12.75	11.95	.74	25.44	33
0204	5/8"		72	.222		10.85	12.65	.78	24.28	31.50
0205	3/4"		68	.235		13.70	13.35	.83	27.88	36
0206	7/8"		64	.250		18.20	14.20	.88	33.28	42.50
0207	1"		56	.286		39.50	16.25	1	56.75	69
0300	Clamps									
0310	C-clamp, for mounting on steel beam flange, w/locknut, TYPE 23									
0320	3/8" threaded rod size	1 Plum	160	.050	Ea.	4.04	3.16		7.20	9.20
0330	1/2" threaded rod size		160	.050		4.88	3.16		8.04	10.10
0340	5/8" threaded rod size		160	.050		5.35	3.16		8.51	10.65
0350	3/4" threaded rod size		160	.050		6.85	3.16		10.01	12.30
0352	7/8" threaded rod size		140	.057		16.65	3.61		20.26	24

22 05 29.10	Hangers & Supp. for Plumb'g/HVAC Pipe/Equip.	Crew	Daily Output	Labor-Hours	Unit	Material	2019 Bare Costs Labor	Equipment	Total	Total Incl O&P
0400	High temperature to 1050°F, alloy steel									
0410	4" pipe size	Q-1	106	.151	Ea.	28.50	8.60		37.10	44
0420	6" pipe size		106	.151		48.50	8.60		57.10	66.50
0430	8" pipe size		97	.165		53.50	9.35		62.85	73
0440	10" pipe size		84	.190		82	10.85		92.85	106
0450	12" pipe size		72	.222		95	12.65		107.65	123
0460	14" pipe size		64	.250		248	14.20		262.20	294
0470	16" pipe size		56	.286		264	16.25		280.25	315
0480	Beam clamp, flange type, TYPE 25									
0482	For 3/8" bolt	1 Plum	48	.167	Ea.	4	10.55		14.55	20
0483	For 1/2" bolt		44	.182		5.15	11.50		16.65	23
0484	For 5/8" bolt		40	.200		5.15	12.65		17.80	24.50
0485	For 3/4" bolt		36	.222		5.15	14.05		19.20	26.50
0486	For 1" bolt		32	.250		5.15	15.80		20.95	29
0500	I-beam, for mounting on bottom flange, strap iron, TYPE 21									
0510	2" flange size	1 Plum	96	.083	Ea.	20.50	5.25		25.75	30.50
0520	3" flange size		95	.084		24.50	5.30		29.80	35
0530	4" flange size		93	.086		6.60	5.45		12.05	15.40
0540	5" flange size		92	.087		7.25	5.50		12.75	16.25
0550	6" flange size		90	.089		8.25	5.60		13.85	17.45
0560	7" flange size		88	.091		9.20	5.75		14.95	18.70
0570	8" flange size		86	.093		9.85	5.85		15.70	19.65
0600	One hole, vertical mounting, malleable iron									
0610	1/2" pipe size	1 Plum	160	.050	Ea.	1.23	3.16		4.39	6.10
0620	3/4" pipe size		145	.055		1.32	3.48		4.80	6.70
0630	1" pipe size		136	.059		1.42	3.71		5.13	7.10
0640	1-1/4" pipe size		128	.063		2.49	3.95		6.44	8.65
0650	1-1/2" pipe size		120	.067		2.87	4.21		7.08	9.45
0660	2" pipe size		112	.071		4.14	4.51		8.65	11.30
0670	2-1/2" pipe size		104	.077		8.15	4.86		13.01	16.30
0680	3" pipe size		96	.083		9.65	5.25		14.90	18.50
0690	3-1/2" pipe size		90	.089		9.35	5.60		14.95	18.70
0700	4" pipe size		84	.095		13.55	6		19.55	24
0750	Riser or extension pipe, carbon steel, TYPE 8									
0756	1/2" pipe size	1 Plum	52	.154	Ea.	2.29	9.70		11.99	17.10
0760	3/4" pipe size		48	.167		3.38	10.55		13.93	19.50
0770	1" pipe size		47	.170		3.51	10.75		14.26	20
0780	1-1/4" pipe size		46	.174		5.50	11		16.50	22.50
0790	1-1/2" pipe size		45	.178		10.15	11.25		21.40	28
0800	2" pipe size		43	.186		5.60	11.75		17.35	24
0810	2-1/2" pipe size		41	.195		5.50	12.30		17.80	24.50
0820	3" pipe size		40	.200		5.30	12.65		17.95	25
0830	3-1/2" pipe size		39	.205		5.50	12.95		18.45	25.50
0840	4" pipe size		38	.211		9	13.30		22.30	30
0850	5" pipe size		37	.216		11.45	13.65		25.10	33
0860	6" pipe size		36	.222		15.70	14.05		29.75	38.50
0870	8" pipe size		34	.235		32	14.85		46.85	57.50
0880	10" pipe size		32	.250		30	15.80		45.80	56.50
0890	12" pipe size		28	.286		43.50	18.05		61.55	75
0900	For plastic coating 3/4" to 4", add					190%				
0910	For copper plating 3/4" to 4", add					58%				
0950	Two piece, complete, carbon steel, medium weight, TYPE 4									
0960	1/2" pipe size	Q-1	137	.117	Ea.	2.62	6.65		9.27	12.85

22 05 29 – Hangers and Supports for Plumbing Piping and Equipment

22 05 29.10 Hangers & Supp. for Plumb'g/HVAC Pipe/Equip.		Crew	Daily Output	Labor-Hours	Unit	2019 Bare Costs Material	Labor	Equipment	Total	Total Incl O&P
0970	3/4" pipe size	Q-1	134	.119	Ea.	2.69	6.80		9.49	13.15
0980	1" pipe size		132	.121		2.47	6.90		9.37	13.05
0990	1-1/4" pipe size		130	.123		3.22	7		10.22	14.05
1000	1-1/2" pipe size		126	.127		4.10	7.20		11.30	15.35
1010	2" pipe size		124	.129		5.05	7.35		12.40	16.60
1020	2-1/2" pipe size		120	.133		5.50	7.60		13.10	17.40
1030	3" pipe size		117	.137		5.30	7.75		13.05	17.50
1040	3-1/2" pipe size		114	.140		12.15	8		20.15	25.50
1050	4" pipe size		110	.145		8	8.25		16.25	21
1060	5" pipe size		106	.151		20.50	8.60		29.10	35.50
1070	6" pipe size		104	.154		22	8.75		30.75	37
1080	8" pipe size		100	.160		26.50	9.10		35.60	42.50
1090	10" pipe size		96	.167		52.50	9.45		61.95	72
1100	12" pipe size		89	.180		64	10.20		74.20	86
1110	14" pipe size		82	.195		85	11.10		96.10	110
1120	16" pipe size	↓	68	.235		110	13.35		123.35	141
1130	For galvanized, add				↓	45%				
1150	Insert, concrete									
1160	Wedge type, carbon steel body, malleable iron nut, galvanized									
1170	1/4" threaded rod size	1 Plum	96	.083	Ea.	8.15	5.25		13.40	16.90
1180	3/8" threaded rod size		96	.083		19.90	5.25		25.15	30
1190	1/2" threaded rod size		96	.083		10.65	5.25		15.90	19.65
1200	5/8" threaded rod size		96	.083		55	5.25		60.25	68.50
1210	3/4" threaded rod size		96	.083		12.10	5.25		17.35	21
1220	7/8" threaded rod size	↓	96	.083	↓	12.45	5.25		17.70	21.50
1250	Pipe guide sized for insulation									
1260	No. 1, 1" pipe size, 1" thick insulation	1 Stpi	26	.308	Ea.	140	19.70		159.70	184
1270	No. 2, 1-1/4"-2" pipe size, 1" thick insulation		23	.348		171	22		193	222
1280	No. 3, 1-1/4"-2" pipe size, 1-1/2" thick insulation		21	.381		171	24.50		195.50	225
1290	No. 4, 2-1/2"-3-1/2" pipe size, 1-1/2" thick insulation		18	.444		171	28.50		199.50	232
1300	No. 5, 4"-5" pipe size, 1-1/2" thick insulation	↓	16	.500		194	32		226	261
1310	No. 6, 5"-6" pipe size, 2" thick insulation	Q-5	21	.762		221	44		265	310
1320	No. 7, 8" pipe size, 2" thick insulation		16	1		286	57.50		343.50	400
1330	No. 8, 10" pipe size, 2" thick insulation	↓	12	1.333		425	76.50		501.50	580
1340	No. 9, 12" pipe size, 2" thick insulation	Q-6	17	1.412		425	84.50		509.50	590
1350	No. 10, 12"-14" pipe size, 2-1/2" thick insulation		16	1.500		500	89.50		589.50	685
1360	No. 11, 16" pipe size, 2-1/2" thick insulation		10.50	2.286		500	136		636	755
1370	No. 12, 16"-18" pipe size, 3" thick insulation		9	2.667		730	159		889	1,050
1380	No. 13, 20" pipe size, 3" thick insulation		7.50	3.200		730	191		921	1,100
1390	No. 14, 24" pipe size, 3" thick insulation	↓	7	3.429	↓	990	205		1,195	1,375
1400	Bands									
1410	Adjustable band, carbon steel, for non-insulated pipe, TYPE 7									
1420	1/2" pipe size	Q-1	142	.113	Ea.	.36	6.40		6.76	10
1430	3/4" pipe size		140	.114		.36	6.50		6.86	10.15
1440	1" pipe size		137	.117		.36	6.65		7.01	10.35
1450	1-1/4" pipe size		134	.119		.39	6.80		7.19	10.65
1460	1-1/2" pipe size		131	.122		.39	6.95		7.34	10.85
1470	2" pipe size		129	.124		.39	7.05		7.44	11.05
1480	2-1/2" pipe size		125	.128		.70	7.25		7.95	11.65
1490	3" pipe size		122	.131		.75	7.45		8.20	12.05
1500	3-1/2" pipe size		119	.134		.82	7.65		8.47	12.35
1510	4" pipe size		114	.140		1.22	8		9.22	13.30
1520	5" pipe size	↓	110	.145	↓	1.91	8.25		10.16	14.50

22 05 29.10 Hangers & Supp. for Plumb'g/HVAC Pipe/Equip.		Crew	Daily Output	Labor-Hours	Unit	Material	Labor	Equipment	Total	Total Incl O&P
						2019 Bare Costs				
1530	6" pipe size	Q-1	108	.148	Ea.	2.29	8.40		10.69	15.15
1540	8" pipe size		104	.154		3.48	8.75		12.23	16.95
1550	For copper plated, add					50%				
1560	For galvanized, add					30%				
1570	For plastic coating, add					30%				
1600	Adjusting nut malleable iron, steel band, TYPE 9									
1610	1/2" pipe size, galvanized band	Q-1	137	.117	Ea.	5.15	6.65		11.80	15.65
1620	3/4" pipe size, galvanized band		135	.119		5.15	6.75		11.90	15.75
1630	1" pipe size, galvanized band		132	.121		5.30	6.90		12.20	16.20
1640	1-1/4" pipe size, galvanized band		129	.124		5.30	7.05		12.35	16.40
1650	1-1/2" pipe size, galvanized band		126	.127		5.50	7.20		12.70	16.90
1660	2" pipe size, galvanized band		124	.129		5.70	7.35		13.05	17.30
1670	2-1/2" pipe size, galvanized band		120	.133		9.05	7.60		16.65	21.50
1680	3" pipe size, galvanized band		117	.137		9.45	7.75		17.20	22
1690	3-1/2" pipe size, galvanized band		114	.140		27.50	8		35.50	42
1700	4" pipe size, cadmium plated band		110	.145		24	8.25		32.25	39
1740	For plastic coated band, add					35%				
1750	For completely copper coated, add					45%				
1800	Clevis, adjustable, carbon steel, for non-insulated pipe, TYPE 1									
1810	1/2" pipe size	Q-1	137	.117	Ea.	1.09	6.65		7.74	11.15
1820	3/4" pipe size		135	.119		1.15	6.75		7.90	11.35
1830	1" pipe size		132	.121		1.20	6.90		8.10	11.65
1840	1-1/4" pipe size		129	.124		1.38	7.05		8.43	12.10
1850	1-1/2" pipe size		126	.127		1.01	7.20		8.21	11.95
1860	2" pipe size		124	.129		1.92	7.35		9.27	13.10
1870	2-1/2" pipe size		120	.133		3.06	7.60		10.66	14.70
1880	3" pipe size		117	.137		3.50	7.75		11.25	15.50
1890	3-1/2" pipe size		114	.140		3.97	8		11.97	16.30
1900	4" pipe size		110	.145		2.75	8.25		11	15.45
1910	5" pipe size		106	.151		3.76	8.60		12.36	17
1920	6" pipe size		104	.154		4.66	8.75		13.41	18.25
1930	8" pipe size		100	.160		10.30	9.10		19.40	25
1940	10" pipe size		96	.167		16.30	9.45		25.75	32
1950	12" pipe size		89	.180		21	10.20		31.20	38.50
1960	14" pipe size		82	.195		35.50	11.10		46.60	55.50
1970	16" pipe size		68	.235		56	13.35		69.35	82
1971	18" pipe size		54	.296		85	16.85		101.85	119
1972	20" pipe size		38	.421		137	24		161	186
1980	For galvanized, add					66%				
1990	For copper plated 1/2" to 4", add					77%				
2000	For light weight 1/2" to 4", deduct					13%				
2010	Insulated pipe type, 3/4" to 12" pipe, add					180%				
2020	Insulated pipe type, chrome-moly U-strap, add					530%				
2250	Split ring, malleable iron, for non-insulated pipe, TYPE 11									
2260	1/2" pipe size	Q-1	137	.117	Ea.	4.98	6.65		11.63	15.45
2270	3/4" pipe size		135	.119		5.05	6.75		11.80	15.65
2280	1" pipe size		132	.121		5.35	6.90		12.25	16.25
2290	1-1/4" pipe size		129	.124		6.65	7.05		13.70	17.90
2300	1-1/2" pipe size		126	.127		8.10	7.20		15.30	19.75
2310	2" pipe size		124	.129		9.15	7.35		16.50	21
2320	2-1/2" pipe size		120	.133		13.20	7.60		20.80	26
2330	3" pipe size		117	.137		16.50	7.75		24.25	30
2340	3-1/2" pipe size		114	.140		16.90	8		24.90	30.50

22 05 29.10 Hangers & Supp. for Plumb'g/HVAC Pipe/Equip.		Crew	Daily Output	Labor-Hours	Unit	2019 Bare Costs Material	Labor	Equipment	Total	Total Incl O&P
2350	4″ pipe size	Q-1	110	.145	Ea.	17.20	8.25		25.45	31.50
2360	5″ pipe size		106	.151		23	8.60		31.60	38
2370	6″ pipe size		104	.154		51	8.75		59.75	69
2380	8″ pipe size		100	.160		76	9.10		85.10	97.50
2390	For copper plated, add					8%				
2400	Channels, steel, 3/4″ x 1-1/2″	1 Plum	80	.100	L.F.	2.63	6.30		8.93	12.40
2404	1-1/2″ x 1-1/2″		70	.114		3.43	7.20		10.63	14.60
2408	1-7/8″ x 1-1/2″		60	.133		17.60	8.40		26	32
2412	3″ x 1-1/2″		50	.160		17.95	10.10		28.05	35
2416	Hangers, trapeze channel support, 12 ga. 1-1/2″ x 1-1/2″, 12″ wide, steel		8.80	.909	Ea.	23.50	57.50		81	112
2418	18″ wide, steel		8.30	.964		25.50	61		86.50	120
2430	Spring nuts, long, 1/4″		120	.067		.63	4.21		4.84	7
2432	3/8″		100	.080		.71	5.05		5.76	8.40
2434	1/2″		80	.100		.79	6.30		7.09	10.35
2436	5/8″		80	.100		4.36	6.30		10.66	14.30
2438	3/4″		75	.107		7.50	6.75		14.25	18.35
2440	Spring nuts, short, 1/4″		120	.067		1.28	4.21		5.49	7.70
2442	3/8″		100	.080		1.58	5.05		6.63	9.35
2444	1/2″		80	.100		1.66	6.30		7.96	11.35
2500	Washer, flat steel									
2502	3/8″	1 Plum	240	.033	Ea.	.08	2.10		2.18	3.25
2503	1/2″		220	.036		.18	2.30		2.48	3.65
2504	5/8″		200	.040		.34	2.53		2.87	4.16
2505	3/4″		180	.044		.56	2.81		3.37	4.83
2506	7/8″		160	.050		.42	3.16		3.58	5.20
2507	1″		140	.057		1.41	3.61		5.02	6.95
2508	1-1/4″		120	.067		1.27	4.21		5.48	7.70
2520	Nut, steel, hex									
2522	3/8″	1 Plum	200	.040	Ea.	.21	2.53		2.74	4.02
2523	1/2″		180	.044		.48	2.81		3.29	4.74
2524	5/8″		160	.050		.83	3.16		3.99	5.65
2525	3/4″		140	.057		1.29	3.61		4.90	6.80
2526	7/8″		120	.067		1.86	4.21		6.07	8.35
2527	1″		100	.080		2.94	5.05		7.99	10.85
2528	1-1/4″		80	.100		7.50	6.30		13.80	17.75
2532	Turnbuckle, TYPE 13									
2534	3/8″	1 Plum	80	.100	Ea.	4.97	6.30		11.27	14.95
2535	1/2″		72	.111		5.55	7		12.55	16.65
2536	5/8″		64	.125		9.85	7.90		17.75	22.50
2537	3/4″		56	.143		12.60	9		21.60	27.50
2538	7/8″		48	.167		30.50	10.55		41.05	49.50
2539	1″		40	.200		43.50	12.65		56.15	67
2540	1-1/4″		32	.250		80.50	15.80		96.30	112
2650	Rods, carbon steel									
2660	Continuous thread									
2670	1/4″ thread size	1 Plum	144	.056	L.F.	2.30	3.51		5.81	7.80
2680	3/8″ thread size		144	.056		2.46	3.51		5.97	7.95
2690	1/2″ thread size		144	.056		3.87	3.51		7.38	9.50
2700	5/8″ thread size		144	.056		5.50	3.51		9.01	11.30
2710	3/4″ thread size		144	.056		9.65	3.51		13.16	15.90
2720	7/8″ thread size		144	.056		12.15	3.51		15.66	18.60
2721	1″ thread size	Q-1	160	.100		20.50	5.70		26.20	31.50
2722	1-1/8″ thread size	″	120	.133		23.50	7.60		31.10	37.50

22 05 29.10 Hangers & Supp. for Plumb'g/HVAC Pipe/Equip.		Crew	Daily Output	Labor-Hours	Unit	Material	2019 Bare Costs Labor	Equipment	Total	Total Incl O&P
2725	1/4" thread size, bright finish	1 Plum	144	.056	L.F.	1.33	3.51		4.84	6.70
2726	1/2" thread size, bright finish	"	144	.056	↓	4.35	3.51		7.86	10.05
2730	For galvanized, add					40%				
2820	Rod couplings									
2821	3/8"	1 Plum	60	.133	Ea.	1.70	8.40		10.10	14.50
2822	1/2"		54	.148		2.38	9.35		11.73	16.65
2823	5/8"		48	.167		3.01	10.55		13.56	19.10
2824	3/4"		44	.182		4.23	11.50		15.73	22
2825	7/8"		40	.200		7.15	12.65		19.80	27
2826	1"		34	.235		9.45	14.85		24.30	33
2827	1-1/8"	↓	30	.267	↓	16.10	16.85		32.95	43
2860	Pipe hanger assy, adj. clevis, saddle, rod, clamp, insul. allowance									
2864	1/2" pipe size	Q-5	35	.457	Ea.	38	26.50		64.50	81.50
2866	3/4" pipe size		34.80	.460		39.50	26.50		66	83
2868	1" pipe size		34.60	.462		24	26.50		50.50	66.50
2869	1-1/4" pipe size		34.30	.466		24.50	27		51.50	67.50
2870	1-1/2" pipe size		33.90	.472		25	27		52	68.50
2872	2" pipe size		33.30	.480		26	27.50		53.50	70.50
2874	2-1/2" pipe size		32.30	.495		29.50	28.50		58	75.50
2876	3" pipe size		31.20	.513		31.50	29.50		61	79
2880	4" pipe size		30.70	.521		34	30		64	82.50
2884	6" pipe size		29.80	.537		47	31		78	98
2888	8" pipe size		28	.571		61.50	33		94.50	117
2892	10" pipe size		25.20	.635		70	36.50		106.50	132
2896	12" pipe size	↓	23.20	.690	↓	103	39.50		142.50	173
2900	Rolls									
2910	Adjustable yoke, carbon steel with CI roll, TYPE 43									
2918	2" pipe size	Q-1	140	.114	Ea.	12.25	6.50		18.75	23.50
2920	2-1/2" pipe size		137	.117		12.25	6.65		18.90	23.50
2930	3" pipe size		131	.122		13.05	6.95		20	25
2940	3-1/2" pipe size		124	.129		18.25	7.35		25.60	31
2950	4" pipe size		117	.137		17.90	7.75		25.65	31.50
2960	5" pipe size		110	.145		25.50	8.25		33.75	41
2970	6" pipe size		104	.154		34.50	8.75		43.25	51
2980	8" pipe size		96	.167		49.50	9.45		58.95	68
2990	10" pipe size		80	.200		60.50	11.35		71.85	83.50
3000	12" pipe size		68	.235		97	13.35		110.35	127
3010	14" pipe size		56	.286		183	16.25		199.25	226
3020	16" pipe size	↓	48	.333	↓	220	18.95		238.95	271
3050	Chair, carbon steel with CI roll									
3060	2" pipe size	1 Plum	68	.118	Ea.	14.20	7.45		21.65	27
3070	2-1/2" pipe size		65	.123		15.20	7.75		22.95	28.50
3080	3" pipe size		62	.129		16.30	8.15		24.45	30
3090	3-1/2" pipe size		60	.133		21	8.40		29.40	35.50
3100	4" pipe size		58	.138		19	8.70		27.70	34
3110	5" pipe size		56	.143		22	9		31	38
3120	6" pipe size		53	.151		29.50	9.55		39.05	47
3130	8" pipe size		50	.160		40.50	10.10		50.60	59.50
3140	10" pipe size		48	.167		54.50	10.55		65.05	76
3150	12" pipe size	↓	46	.174		83.50	11		94.50	108
3170	Single pipe roll (see line 2650 for rods), TYPE 41, 1" pipe size	Q-1	137	.117		9.50	6.65		16.15	20.50
3180	1-1/4" pipe size		131	.122		9.85	6.95		16.80	21
3190	1-1/2" pipe size	↓	129	.124	↓	10.05	7.05		17.10	21.50

22 05 29.10 Hangers & Supp. for Plumb'g/HVAC Pipe/Equip.		Crew	Daily Output	Labor-Hours	Unit	Material	2019 Bare Costs Labor	Equipment	Total	Total Incl O&P
3200	2" pipe size	Q-1	124	.129	Ea.	10.05	7.35		17.40	22
3210	2-1/2" pipe size		118	.136		10.70	7.70		18.40	23.50
3220	3" pipe size		115	.139		11.25	7.90		19.15	24.50
3230	3-1/2" pipe size		113	.142		12.30	8.05		20.35	25.50
3240	4" pipe size		112	.143		12.40	8.10		20.50	26
3250	5" pipe size		110	.145		13.85	8.25		22.10	27.50
3260	6" pipe size		101	.158		31	9		40	47.50
3270	8" pipe size		90	.178		31	10.10		41.10	49
3280	10" pipe size		80	.200		40.50	11.35		51.85	61.50
3290	12" pipe size		68	.235		66.50	13.35		79.85	93
3291	14" pipe size		56	.286		101	16.25		117.25	136
3292	16" pipe size		48	.333		130	18.95		148.95	172
3293	18" pipe size		40	.400		185	22.50		207.50	237
3294	20" pipe size		35	.457		182	26		208	239
3296	24" pipe size		30	.533		310	30.50		340.50	385
3297	30" pipe size		25	.640		560	36.50		596.50	670
3298	36" pipe size	▼	20	.800	▼	575	45.50		620.50	705
3300	Saddles (add vertical pipe riser, usually 3" diameter)									
3310	Pipe support, complete, adjust., CI saddle, TYPE 36									
3320	2-1/2" pipe size	1 Plum	96	.083	Ea.	108	5.25		113.25	126
3330	3" pipe size		88	.091		109	5.75		114.75	129
3340	3-1/2" pipe size		79	.101		109	6.40		115.40	130
3350	4" pipe size		68	.118		158	7.45		165.45	185
3360	5" pipe size		64	.125		161	7.90		168.90	189
3370	6" pipe size		59	.136		162	8.55		170.55	191
3380	8" pipe size		53	.151		167	9.55		176.55	198
3390	10" pipe size		50	.160		195	10.10		205.10	230
3400	12" pipe size	▼	48	.167	▼	205	10.55		215.55	242
3450	For standard pipe support, one piece, CI, deduct					34%				
3460	For stanchion support, CI with steel yoke, deduct					60%				
3550	Insulation shield 1" thick, 1/2" pipe size, TYPE 40	1 Asbe	100	.080	Ea.	3.21	4.59		7.80	10.65
3560	3/4" pipe size		100	.080		3.57	4.59		8.16	11.05
3570	1" pipe size		98	.082		3.87	4.68		8.55	11.50
3580	1-1/4" pipe size		98	.082		4.13	4.68		8.81	11.80
3590	1-1/2" pipe size		96	.083		4.14	4.78		8.92	11.95
3600	2" pipe size		96	.083		4.30	4.78		9.08	12.15
3610	2-1/2" pipe size		94	.085		4.44	4.88		9.32	12.45
3620	3" pipe size		94	.085		5.30	4.88		10.18	13.35
3630	2" thick, 3-1/2" pipe size		92	.087		5.65	4.99		10.64	13.95
3640	4" pipe size		92	.087		8.05	4.99		13.04	16.55
3650	5" pipe size		90	.089		10.35	5.10		15.45	19.25
3660	6" pipe size		90	.089		12.20	5.10		17.30	21.50
3670	8" pipe size		88	.091		19.50	5.20		24.70	29.50
3680	10" pipe size		88	.091		24.50	5.20		29.70	35
3690	12" pipe size		86	.093		28	5.35		33.35	39.50
3700	14" pipe size		86	.093		47	5.35		52.35	60.50
3710	16" pipe size		84	.095		50	5.45		55.45	63.50
3720	18" pipe size		84	.095		55.50	5.45		60.95	70
3730	20" pipe size		82	.098		61.50	5.60		67.10	76
3732	24" pipe size	▼	80	.100	▼	72	5.75		77.75	88.50
3750	Covering protection saddle, TYPE 39									
3760	1" covering size									
3770	3/4" pipe size	1 Plum	68	.118	Ea.	5.35	7.45		12.80	17.05

22 05 29.10 Hangers & Supp. for Plumb'g/HVAC Pipe/Equip.		Crew	Daily Output	Labor-Hours	Unit	Material	2019 Bare Costs Labor	Equipment	Total	Total Incl O&P
3780	1" pipe size	1 Plum	68	.118	Ea.	5.35	7.45		12.80	17.05
3790	1-1/4" pipe size		68	.118		5.35	7.45		12.80	17.05
3800	1-1/2" pipe size		66	.121		5.75	7.65		13.40	17.80
3810	2" pipe size		66	.121		5.75	7.65		13.40	17.80
3820	2-1/2" pipe size		64	.125		5.75	7.90		13.65	18.20
3830	3" pipe size		64	.125		7.65	7.90		15.55	20.50
3840	3-1/2" pipe size		62	.129		8.50	8.15		16.65	21.50
3850	4" pipe size		62	.129		8.50	8.15		16.65	21.50
3860	5" pipe size		60	.133		8.65	8.40		17.05	22
3870	6" pipe size		60	.133		10.35	8.40		18.75	24
3900	1-1/2" covering size									
3910	3/4" pipe size	1 Plum	68	.118	Ea.	6.40	7.45		13.85	18.20
3920	1" pipe size		68	.118		6.35	7.45		13.80	18.15
3930	1-1/4" pipe size		68	.118		6.40	7.45		13.85	18.20
3940	1-1/2" pipe size		66	.121		6.40	7.65		14.05	18.55
3950	2" pipe size		66	.121		6.95	7.65		14.60	19.15
3960	2-1/2" pipe size		64	.125		8.35	7.90		16.25	21
3970	3" pipe size		64	.125		8.35	7.90		16.25	21
3980	3-1/2" pipe size		62	.129		7.65	8.15		15.80	20.50
3990	4" pipe size		62	.129		8.80	8.15		16.95	22
4000	5" pipe size		60	.133		8.80	8.40		17.20	22.50
4010	6" pipe size		60	.133		11.95	8.40		20.35	26
4020	8" pipe size		58	.138		14.95	8.70		23.65	29.50
4022	10" pipe size		56	.143		14.95	9		23.95	30
4024	12" pipe size		54	.148		36.50	9.35		45.85	54.50
4028	2" covering size									
4029	2-1/2" pipe size	1 Plum	62	.129	Ea.	8.80	8.15		16.95	22
4032	3" pipe size		60	.133		9.90	8.40		18.30	23.50
4033	4" pipe size		58	.138		9.90	8.70		18.60	24
4034	6" pipe size		56	.143		14.15	9		23.15	29
4035	8" pipe size		54	.148		16.75	9.35		26.10	32.50
4080	10" pipe size		58	.138		17.95	8.70		26.65	33
4090	12" pipe size		56	.143		41	9		50	59
4100	14" pipe size		56	.143		41.50	9		50.50	59.50
4110	16" pipe size		54	.148		54.50	9.35		63.85	74
4120	18" pipe size		54	.148		54.50	9.35		63.85	74
4130	20" pipe size		52	.154		61	9.70		70.70	81.50
4150	24" pipe size		50	.160		71	10.10		81.10	93
4160	30" pipe size		48	.167		78	10.55		88.55	102
4180	36" pipe size		45	.178		90	11.25		101.25	116
4186	2-1/2" covering size									
4187	3" pipe size	1 Plum	58	.138	Ea.	10.10	8.70		18.80	24
4188	4" pipe size		56	.143		11	9		20	25.50
4189	6" pipe size		52	.154		16.85	9.70		26.55	33
4190	8" pipe size		48	.167		18.70	10.55		29.25	36.50
4191	10" pipe size		44	.182		20.50	11.50		32	40
4192	12" pipe size		40	.200		44.50	12.65		57.15	68
4193	14" pipe size		36	.222		44.50	14.05		58.55	70
4194	16" pipe size		32	.250		57.50	15.80		73.30	86.50
4195	18" pipe size		28	.286		64	18.05		82.05	97.50
4200	Sockets									
4210	Rod end, malleable iron, TYPE 16									
4220	1/4" thread size	1 Plum	240	.033	Ea.	1.63	2.10		3.73	4.95

22 05 29.10 Hangers & Supp. for Plumb'g/HVAC Pipe/Equip.		Crew	Daily Output	Labor-Hours	Unit	2019 Bare Costs Material	Labor	Equipment	Total	Total Incl O&P
4230	3/8" thread size	1 Plum	240	.033	Ea.	1.64	2.10		3.74	4.96
4240	1/2" thread size		230	.035		2.16	2.20		4.36	5.70
4250	5/8" thread size		225	.036		4.42	2.25		6.67	8.25
4260	3/4" thread size		220	.036		5.75	2.30		8.05	9.80
4270	7/8" thread size		210	.038		8.30	2.41		10.71	12.70
4290	Strap, 1/2" pipe size, TYPE 26	Q-1	142	.113		1.85	6.40		8.25	11.65
4300	3/4" pipe size		140	.114		1.89	6.50		8.39	11.85
4310	1" pipe size		137	.117		2.71	6.65		9.36	12.95
4320	1-1/4" pipe size		134	.119		2.76	6.80		9.56	13.25
4330	1-1/2" pipe size		131	.122		2.97	6.95		9.92	13.65
4340	2" pipe size		129	.124		3.01	7.05		10.06	13.90
4350	2-1/2" pipe size		125	.128		4.67	7.25		11.92	16.05
4360	3" pipe size		122	.131		6	7.45		13.45	17.80
4370	3-1/2" pipe size		119	.134		6.55	7.65		14.20	18.65
4380	4" pipe size		114	.140		6.85	8		14.85	19.45
4400	U-bolt, carbon steel									
4410	Standard, with nuts, TYPE 42									
4420	1/2" pipe size	1 Plum	160	.050	Ea.	1.17	3.16		4.33	6.05
4430	3/4" pipe size		158	.051		1.17	3.20		4.37	6.10
4450	1" pipe size		152	.053		1.21	3.32		4.53	6.30
4460	1-1/4" pipe size		148	.054		1.52	3.41		4.93	6.75
4470	1-1/2" pipe size		143	.056		1.62	3.53		5.15	7.10
4480	2" pipe size		139	.058		1.75	3.63		5.38	7.40
4490	2-1/2" pipe size		134	.060		2.86	3.77		6.63	8.80
4500	3" pipe size		128	.063		3.21	3.95		7.16	9.45
4510	3-1/2" pipe size		122	.066		3.46	4.14		7.60	10
4520	4" pipe size		117	.068		3.52	4.32		7.84	10.35
4530	5" pipe size		114	.070		4.08	4.43		8.51	11.15
4540	6" pipe size		111	.072		7.80	4.55		12.35	15.45
4550	8" pipe size		109	.073		8.60	4.63		13.23	16.40
4560	10" pipe size		107	.075		14.95	4.72		19.67	23.50
4570	12" pipe size		104	.077		21.50	4.86		26.36	31
4580	For plastic coating on 1/2" thru 6" size, add					150%				
4700	U-hook, carbon steel, requires mounting screws or bolts									
4710	3/4" thru 2" pipe size									
4720	6" long	1 Plum	96	.083	Ea.	.98	5.25		6.23	9
4730	8" long		96	.083		1.26	5.25		6.51	9.30
4740	10" long		96	.083		1.36	5.25		6.61	9.40
4750	12" long		96	.083		1.49	5.25		6.74	9.55
4760	For copper plated, add					50%				
7000	Roof supports									
7006	Duct									
7010	Rectangular, open, 12" off roof									
7020	To 18" wide	Q-9	26	.615	Ea.	160	34		194	228
7030	To 24" wide	"	22	.727		186	40		226	266
7040	To 36" wide	Q-10	30	.800		213	45.50		258.50	305
7050	To 48" wide		28	.857		239	49		288	335
7060	To 60" wide		24	1		266	57		323	380
7100	Equipment									
7120	Equipment support	Q-5	20	.800	Ea.	70	46		116	146
7300	Pipe									
7310	Roller type									
7320	Up to 2-1/2" diam. pipe									

22 05 29.10 Hangers & Supp. for Plumb'g/HVAC Pipe/Equip.		Crew	Daily Output	Labor-Hours	Unit	2019 Bare Costs Material	Labor	Equipment	Total	Total Incl O&P
7324	3-1/2" off roof	Q-5	24	.667	Ea.	22	38.50		60.50	81.50
7326	Up to 10" off roof	"	20	.800	"	27	46		73	99
7340	2-1/2" to 3-1/2" diam. pipe									
7342	Up to 16" off roof	Q-5	18	.889	Ea.	48.50	51		99.50	131
7360	4" to 5" diam. pipe									
7362	Up to 12" off roof	Q-5	16	1	Ea.	69	57.50		126.50	163
7400	Strut/channel type									
7410	Up to 2-1/2" diam. pipe									
7424	3-1/2" off roof	Q-5	24	.667	Ea.	19.85	38.50		58.35	79.50
7426	Up to 10" off roof	"	20	.800	"	25	46		71	96.50
7440	Strut and roller type									
7452	2-1/2" to 3-1/2" diam. pipe									
7454	Up to 16" off roof	Q-5	18	.889	Ea.	70	51		121	154
7460	Strut and hanger type									
7470	Up to 3" diam. pipe									
7474	Up to 8" off roof	Q-5	19	.842	Ea.	44.50	48.50		93	121
8000	Pipe clamp, plastic, 1/2" CTS	1 Plum	80	.100		.23	6.30		6.53	9.75
8010	3/4" CTS		73	.110		.24	6.90		7.14	10.65
8020	1" CTS		68	.118		.53	7.45		7.98	11.75
8080	Economy clamp, 1/4" CTS		175	.046		.05	2.89		2.94	4.39
8090	3/8" CTS		168	.048		.05	3.01		3.06	4.57
8100	1/2" CTS		160	.050		.05	3.16		3.21	4.80
8110	3/4" CTS		145	.055		.05	3.48		3.53	5.30
8200	Half clamp, 1/2" CTS		80	.100		.07	6.30		6.37	9.60
8210	3/4" CTS		73	.110		.10	6.90		7	10.50
8300	Suspension clamp, 1/2" CTS		80	.100		.24	6.30		6.54	9.75
8310	3/4" CTS		73	.110		.25	6.90		7.15	10.70
8320	1" CTS		68	.118		.53	7.45		7.98	11.75
8400	Insulator, 1/2" CTS		80	.100		.35	6.30		6.65	9.90
8410	3/4" CTS		73	.110		.35	6.90		7.25	10.80
8420	1" CTS		68	.118		.38	7.45		7.83	11.55
8500	J hook clamp with nail, 1/2" CTS		240	.033		.10	2.10		2.20	3.27
8501	3/4" CTS		240	.033		.10	2.10		2.20	3.27
8800	Wire cable support system									
8810	Cable with hook terminal and locking device									
8830	2 mm (.079") diam. cable (100 lb. cap.)									
8840	1 m (3.3') length, with hook	1 Shee	96	.083	Ea.	4.05	5.10		9.15	12.15
8850	2 m (6.6') length, with hook		84	.095		4.59	5.80		10.39	13.85
8860	3 m (9.9') length, with hook		72	.111		3.84	6.75		10.59	14.50
8870	5 m (16.4') length, with hook	Q-9	60	.267		5.05	14.65		19.70	27.50
8880	10 m (32.8') length, with hook	"	30	.533		7.15	29.50		36.65	52.50
8900	3 mm (.118") diam. cable (200 lb. cap.)									
8910	1 m (3.3') length, with hook	1 Shee	96	.083	Ea.	8.20	5.10		13.30	16.75
8920	2 m (6.6') length, with hook		84	.095		9.10	5.80		14.90	18.85
8930	3 m (9.9') length, with hook		72	.111		9.90	6.75		16.65	21
8940	5 m (16.4') length, with hook	Q-9	60	.267		12.35	14.65		27	35.50
8950	10 m (32.8') length, with hook	"	30	.533		16.35	29.50		45.85	62.50
9000	Cable system accessories									
9010	Anchor bolt, 3/8", with nut	1 Shee	140	.057	Ea.	1.10	3.48		4.58	6.50
9020	Air duct corner protector		160	.050		1.26	3.05		4.31	6
9030	Air duct support attachment		140	.057		2.03	3.48		5.51	7.55
9040	Flange clip, hammer-on style									
9044	For flange thickness 3/32"-9/64", 160 lb. cap.	1 Shee	180	.044	Ea.	.38	2.71		3.09	4.53

22 05 29 – Hangers and Supports for Plumbing Piping and Equipment

22 05 29.10 Hangers & Supp. for Plumb'g/HVAC Pipe/Equip.		Crew	Daily Output	Labor-Hours	Unit	2019 Bare Costs Material	Labor	Equipment	Total	Total Incl O&P
9048	For flange thickness 1/8"-1/4", 200 lb. cap.	1 Shee	160	.050	Ea.	.31	3.05		3.36	4.97
9052	For flange thickness 5/16"-1/2", 200 lb. cap.		150	.053		.64	3.25		3.89	5.65
9056	For flange thickness 9/16"-3/4", 200 lb. cap.		140	.057		.86	3.48		4.34	6.25
9060	Wire insulation protection tube		180	.044	L.F.	.46	2.71		3.17	4.62
9070	Wire cutter				Ea.	43			43	47

22 05 33 – Heat Tracing for Plumbing Piping

22 05 33.20 Temperature Maintenance Cable

		Crew	Daily Output	Labor-Hours	Unit	Material	Labor	Equipment	Total	Total Incl O&P
0010	TEMPERATURE MAINTENANCE CABLE									
0040	Components									
0080	Heating cable									
0100	208 V									
0150	140°F	Q-1	1060.80	.015	L.F.	9.90	.86		10.76	12.20
0200	120 V									
0220	125°F	Q-1	1060.80	.015	L.F.	8.85	.86		9.71	11.05
0300	Power kit w/1 end seal	1 Elec	48.80	.164	Ea.	115	9.85		124.85	142
0310	Splice kit		35.50	.225		136	13.55		149.55	170
0320	End seal		160	.050		12.50	3		15.50	18.25
0330	Tee kit w/1 end seal		26.80	.299		140	17.95		157.95	181
0340	Powered splice w/2 end seals		20	.400		110	24		134	157
0350	Powered tee kit w/3 end seals		18	.444		142	26.50		168.50	196
0360	Cross kit w/2 end seals		18.60	.430		146	26		172	200
0500	Recommended thickness of fiberglass insulation									
0510	Pipe size									
0520	1/2" to 1" use 1" insulation									
0530	1-1/4" to 2" use 1-1/2" insulation									
0540	2-1/2" to 6" use 2" insulation									
0560	NOTE: For pipe sizes 1-1/4" and smaller use 1/4" larger diameter									
0570	insulation to allow room for installation over cable.									

22 05 48 – Vibration and Seismic Controls for Plumbing Piping and Equipment

22 05 48.40 Vibration Absorbers

		Crew	Daily Output	Labor-Hours	Unit	Material	Labor	Equipment	Total	Total Incl O&P
0010	VIBRATION ABSORBERS									
0100	Hangers, neoprene flex									
0200	10-120 lb. capacity				Ea.	25.50			25.50	28.50
0220	75-550 lb. capacity					41			41	45.50
0240	250-1,100 lb. capacity					82.50			82.50	90.50
0260	1,000-4,000 lb. capacity					152			152	167
0500	Spring flex, 60 lb. capacity					37			37	41
0520	450 lb. capacity					58			58	64
0540	900 lb. capacity					82.50			82.50	91
0560	1,100-1,300 lb. capacity					82.50			82.50	90.50
0600	Rubber in shear									
0610	45-340 lb., up to 1/2" rod size	1 Stpi	22	.364	Ea.	23	23.50		46.50	60.50
0620	130-700 lb., up to 3/4" rod size		20	.400		50	25.50		75.50	93.50
0630	50-1,000 lb., up to 3/4" rod size		18	.444		66	28.50		94.50	116
1000	Mounts, neoprene, 45-380 lb. capacity					18.55			18.55	20.50
1020	250-1,100 lb. capacity					61.50			61.50	68
1040	1,000-4,000 lb. capacity					132			132	146
1100	Spring flex, 60 lb. capacity					85			85	93.50
1120	165 lb. capacity					87.50			87.50	96
1140	260 lb. capacity					79.50			79.50	87.50
1160	450 lb. capacity					105			105	116
1180	600 lb. capacity					109			109	120

22 05 48 – Vibration and Seismic Controls for Plumbing Piping and Equipment

22 05 48.40 Vibration Absorbers		Crew	Daily Output	Labor-Hours	Unit	2019 Bare Costs Material	Labor	Equipment	Total	Total Incl O&P
1200	750 lb. capacity				Ea.	145			145	159
1220	900 lb. capacity					152			152	167
1240	1,100 lb. capacity					177			177	194
1260	1,300 lb. capacity					177			177	194
1280	1,500 lb. capacity					186			186	204
1300	1,800 lb. capacity					248			248	272
1320	2,200 lb. capacity					310			310	340
1340	2,600 lb. capacity					360			360	395
1399	Spring type									
1400	2 piece									
1410	50-1,000 lb.	1 Stpi	12	.667	Ea.	209	42.50		251.50	294
1420	1,100-1,600 lb.	"	12	.667	"	224	42.50		266.50	310
1500	Double spring open									
1510	150-450 lb.	1 Stpi	24	.333	Ea.	143	21.50		164.50	190
1520	500-1,000 lb.		24	.333		143	21.50		164.50	189
1530	1,100-1,600 lb.		24	.333		143	21.50		164.50	189
1540	1,700-2,400 lb.		24	.333		156	21.50		177.50	204
1550	2,500-3,400 lb.		24	.333		266	21.50		287.50	325
2000	Pads, cork rib, 18" x 18" x 1", 10-50 psi					193			193	213
2020	18" x 36" x 1", 10-50 psi					385			385	425
2100	Shear flexible pads, 18" x 18" x 3/8", 20-70 psi					84			84	92.50
2120	18" x 36" x 3/8", 20-70 psi					193			193	212
2150	Laminated neoprene and cork									
2160	1" thick	1 Stpi	16	.500	S.F.	88	32		120	145
2200	Neoprene elastomer isolation bearing pad									
2230	464 psi, 5/8" x 19-11/16" x 39-3/8"	1 Stpi	16	.500	Ea.	340	32		372	420
3000	Note overlap in capacities due to deflections									

22 05 53 – Identification for Plumbing Piping and Equipment

22 05 53.10 Piping System Identification Labels		Crew	Daily Output	Labor-Hours	Unit	Material	Labor	Equipment	Total	Total Incl O&P
0010	**PIPING SYSTEM IDENTIFICATION LABELS**									
0100	Indicate contents and flow direction									
0106	Pipe markers									
0110	Plastic snap around									
0114	1/2" pipe	1 Plum	80	.100	Ea.	6.80	6.30		13.10	16.95
0116	3/4" pipe		80	.100		7.10	6.30		13.40	17.30
0118	1" pipe		80	.100		7.10	6.30		13.40	17.30
0120	2" pipe		75	.107		8.40	6.75		15.15	19.35
0122	3" pipe		70	.114		13	7.20		20.20	25
0124	4" pipe		60	.133		12.95	8.40		21.35	27
0126	6" pipe		60	.133		13.80	8.40		22.20	28
0128	8" pipe		56	.143		18.40	9		27.40	34
0130	10" pipe		56	.143		18.40	9		27.40	34
0200	Over 10" pipe size		50	.160		22.50	10.10		32.60	39.50
1110	Self adhesive									
1114	1" pipe	1 Plum	80	.100	Ea.	3.58	6.30		9.88	13.45
1116	2" pipe		75	.107		3.49	6.75		10.24	13.95
1118	3" pipe		70	.114		5.20	7.20		12.40	16.60
1120	4" pipe		60	.133		5.20	8.40		13.60	18.40
1122	6" pipe		60	.133		5.20	8.40		13.60	18.40
1124	8" pipe		56	.143		7.50	9		16.50	22
1126	10" pipe		56	.143		7.45	9		16.45	22
1200	Over 10" pipe size		50	.160		10.50	10.10		20.60	26.50

22 05 Common Work Results for Plumbing

22 05 53 – Identification for Plumbing Piping and Equipment

22 05 53.10 Piping System Identification Labels		Crew	Daily Output	Labor-Hours	Unit	Material	2019 Bare Costs Labor	Equipment	Total	Total Incl O&P
2000	Valve tags									
2010	Numbered plus identifying legend									
2100	Brass, 2" diameter	1 Plum	40	.200	Ea.	3.14	12.65		15.79	22.50
2200	Plastic, 1-1/2" diameter	"	40	.200	"	3.08	12.65		15.73	22.50

22 07 Plumbing Insulation

22 07 16 – Plumbing Equipment Insulation

22 07 16.10 Insulation for Plumbing Equipment

		Crew	Daily Output	Labor-Hours	Unit	Material	Labor	Equipment	Total	Total Incl O&P
0010	**INSULATION FOR PLUMBING EQUIPMENT**									
2900	Domestic water heater wrap kit									
2920	1-1/2" with vinyl jacket, 20 to 60 gal. G	1 Plum	8	1	Ea.	17.05	63		80.05	114

22 07 19 – Plumbing Piping Insulation

22 07 19.10 Piping Insulation

		Crew	Daily Output	Labor-Hours	Unit	Material	Labor	Equipment	Total	Total Incl O&P
0010	**PIPING INSULATION**									
0110	Insulation req'd. is based on the surface size/area to be covered									
0230	Insulated protectors (ADA)									
0235	For exposed piping under sinks or lavatories									
0240	Vinyl coated foam, velcro tabs									
0245	P Trap, 1-1/4" or 1-1/2"	1 Plum	32	.250	Ea.	16.70	15.80		32.50	42
0260	Valve and supply cover									
0265	1/2", 3/8", and 7/16" pipe size	1 Plum	32	.250	Ea.	16.85	15.80		32.65	42
0280	Tailpiece offset (wheelchair)									
0285	1-1/4" pipe size	1 Plum	32	.250	Ea.	13.20	15.80		29	38
0600	Pipe covering (price copper tube one size less than IPS)									
1000	Mineral wool									
1010	Preformed, 1200°F, plain									
1014	1" wall									
1016	1/2" iron pipe size G	Q-14	230	.070	L.F.	1.88	3.59		5.47	7.60
1018	3/4" iron pipe size G		220	.073		1.91	3.76		5.67	7.90
1022	1" iron pipe size G		210	.076		2.01	3.93		5.94	8.25
1024	1-1/4" iron pipe size G		205	.078		2.07	4.03		6.10	8.50
1026	1-1/2" iron pipe size G		205	.078		1.95	4.03		5.98	8.35
1028	2" iron pipe size G		200	.080		2.64	4.13		6.77	9.30
1030	2-1/2" iron pipe size G		190	.084		2.71	4.35		7.06	9.70
1032	3" iron pipe size G		180	.089		2.90	4.59		7.49	10.30
1034	4" iron pipe size G		150	.107		3.57	5.50		9.07	12.45
1036	5" iron pipe size G		140	.114		4.04	5.90		9.94	13.55
1038	6" iron pipe size G		120	.133		4.31	6.90		11.21	15.40
1040	7" iron pipe size G		110	.145		5.05	7.50		12.55	17.15
1042	8" iron pipe size G		100	.160		6.95	8.25		15.20	20.50
1044	9" iron pipe size G		90	.178		7.30	9.20		16.50	22
1046	10" iron pipe size G		90	.178		8.25	9.20		17.45	23.50
1050	1-1/2" wall									
1052	1/2" iron pipe size G	Q-14	225	.071	L.F.	3.21	3.67		6.88	9.20
1054	3/4" iron pipe size G		215	.074		3.29	3.84		7.13	9.55
1056	1" iron pipe size G		205	.078		3.33	4.03		7.36	9.85
1058	1-1/4" iron pipe size G		200	.080		3.40	4.13		7.53	10.15
1060	1-1/2" iron pipe size G		200	.080		3.57	4.13		7.70	10.35
1062	2" iron pipe size G		190	.084		3.78	4.35		8.13	10.85
1064	2-1/2" iron pipe size G		180	.089		4.12	4.59		8.71	11.65

22 07 19 – Plumbing Piping Insulation

22 07 19.10 Piping Insulation		Crew	Daily Output	Labor-Hours	Unit	Material	Labor	Equipment	Total	Total Incl O&P
						2019 Bare Costs				
1066	3″ iron pipe size	G	Q-14	165	.097	L.F.	4.29	5		9.29	12.45
1068	4″ iron pipe size	G		140	.114		5.15	5.90		11.05	14.75
1070	5″ iron pipe size	G		130	.123		5.50	6.35		11.85	15.85
1072	6″ iron pipe size	G		110	.145		5.70	7.50		13.20	17.85
1074	7″ iron pipe size	G		100	.160		6.70	8.25		14.95	20
1076	8″ iron pipe size	G		90	.178		7.45	9.20		16.65	22.50
1078	9″ iron pipe size	G		85	.188		8.60	9.70		18.30	24.50
1080	10″ iron pipe size	G		80	.200		9.40	10.35		19.75	26.50
1082	12″ iron pipe size	G		75	.213		10.75	11		21.75	29
1084	14″ iron pipe size	G		70	.229		12.30	11.80		24.10	32
1086	16″ iron pipe size	G		65	.246		13.55	12.70		26.25	34.50
1088	18″ iron pipe size	G		60	.267		15.75	13.75		29.50	39
1090	20″ iron pipe size	G		55	.291		16.80	15		31.80	41.50
1092	22″ iron pipe size	G		50	.320		19.20	16.50		35.70	46.50
1094	24″ iron pipe size	G	↓	45	.356	↓	20.50	18.35		38.85	51
1100	2″ wall										
1102	1/2″ iron pipe size	G	Q-14	220	.073	L.F.	4.15	3.76		7.91	10.35
1104	3/4″ iron pipe size	G		210	.076		4.30	3.93		8.23	10.80
1106	1″ iron pipe size	G		200	.080		4.56	4.13		8.69	11.40
1108	1-1/4″ iron pipe size	G		190	.084		4.89	4.35		9.24	12.10
1110	1-1/2″ iron pipe size	G		190	.084		5.10	4.35		9.45	12.30
1112	2″ iron pipe size	G		180	.089		5.35	4.59		9.94	13
1114	2-1/2″ iron pipe size	G		170	.094		6.05	4.86		10.91	14.15
1116	3″ iron pipe size	G		160	.100		6.45	5.15		11.60	15.05
1118	4″ iron pipe size	G		130	.123		7.55	6.35		13.90	18.10
1120	5″ iron pipe size	G		120	.133		8.35	6.90		15.25	19.85
1122	6″ iron pipe size	G		100	.160		8.70	8.25		16.95	22.50
1124	8″ iron pipe size	G		80	.200		10.55	10.35		20.90	27.50
1126	10″ iron pipe size	G		70	.229		12.90	11.80		24.70	32.50
1128	12″ iron pipe size	G		65	.246		14.30	12.70		27	35.50
1130	14″ iron pipe size	G		60	.267		15.95	13.75		29.70	39
1132	16″ iron pipe size	G		55	.291		18.10	15		33.10	43
1134	18″ iron pipe size	G		50	.320		19.55	16.50		36.05	47
1136	20″ iron pipe size	G		45	.356		22.50	18.35		40.85	53.50
1138	22″ iron pipe size	G		45	.356		24.50	18.35		42.85	55.50
1140	24″ iron pipe size	G	↓	40	.400	↓	26	20.50		46.50	60.50
1150	4″ wall										
1152	1-1/4″ iron pipe size	G	Q-14	170	.094	L.F.	13.20	4.86		18.06	22
1154	1-1/2″ iron pipe size	G		165	.097		13.70	5		18.70	23
1156	2″ iron pipe size	G		155	.103		13.95	5.35		19.30	23.50
1158	4″ iron pipe size	G		105	.152		17.65	7.85		25.50	31.50
1160	6″ iron pipe size	G		75	.213		21	11		32	40
1162	8″ iron pipe size	G		60	.267		23.50	13.75		37.25	47.50
1164	10″ iron pipe size	G		50	.320		28.50	16.50		45	57
1166	12″ iron pipe size	G		45	.356		31.50	18.35		49.85	63
1168	14″ iron pipe size	G		40	.400		33.50	20.50		54	69
1170	16″ iron pipe size	G		35	.457		37.50	23.50		61	78
1172	18″ iron pipe size	G		32	.500		40.50	26		66.50	85
1174	20″ iron pipe size	G		30	.533		44	27.50		71.50	90.50
1176	22″ iron pipe size	G		28	.571		49	29.50		78.50	99.50
1178	24″ iron pipe size	G	↓	26	.615	↓	52	32		84	106
4280	Cellular glass, closed cell foam, all service jacket, sealant,										
4281	working temp. (-450°F to +900°F), 0 water vapor transmission										

22 07 19 – Plumbing Piping Insulation

22 07 19.10 Piping Insulation		Crew	Daily Output	Labor-Hours	Unit	Material	Labor	Equipment	Total	Total Incl O&P
						2019 Bare Costs				
4284	1" wall									
4286	1/2" iron pipe size G	Q-14	120	.133	L.F.	9.15	6.90		16.05	20.50
4300	1-1/2" wall									
4301	1" iron pipe size G	Q-14	105	.152	L.F.	10.45	7.85		18.30	23.50
4304	2-1/2" iron pipe size G		90	.178		13.65	9.20		22.85	29
4306	3" iron pipe size G		85	.188		17.50	9.70		27.20	34.50
4308	4" iron pipe size G		70	.229		21.50	11.80		33.30	41.50
4310	5" iron pipe size G	↓	65	.246	↓	22	12.70		34.70	43.50
4320	2" wall									
4322	1" iron pipe size G	Q-14	100	.160	L.F.	13.90	8.25		22.15	28
4324	2-1/2" iron pipe size G		85	.188		21	9.70		30.70	38.50
4326	3" iron pipe size G		80	.200		21	10.35		31.35	39.50
4328	4" iron pipe size G		65	.246		26	12.70		38.70	48
4330	5" iron pipe size G		60	.267		27	13.75		40.75	51
4332	6" iron pipe size G		50	.320		31	16.50		47.50	59.50
4336	8" iron pipe size G		40	.400		36.50	20.50		57	72
4338	10" iron pipe size G	↓	35	.457	↓	45	23.50		68.50	86
4350	2-1/2" wall									
4360	12" iron pipe size G	Q-14	32	.500	L.F.	65	26		91	112
4362	14" iron pipe size G	"	28	.571	"	71.50	29.50		101	124
4370	3" wall									
4378	6" iron pipe size G	Q-14	48	.333	L.F.	46.50	17.20		63.70	78
4380	8" iron pipe size G		38	.421		58.50	21.50		80	98
4382	10" iron pipe size G		33	.485		59.50	25		84.50	104
4384	16" iron pipe size G		25	.640		84	33		117	144
4386	18" iron pipe size G		22	.727		107	37.50		144.50	176
4388	20" iron pipe size G	↓	20	.800	↓	116	41.50		157.50	192
4400	3-1/2" wall									
4412	12" iron pipe size G	Q-14	27	.593	L.F.	79.50	30.50		110	134
4414	14" iron pipe size G	"	25	.640	"	81.50	33		114.50	141
4430	4" wall									
4446	16" iron pipe size G	Q-14	22	.727	L.F.	92.50	37.50		130	160
4448	18" iron pipe size G		20	.800		118	41.50		159.50	194
4450	20" iron pipe size G	↓	18	.889	↓	135	46		181	219
4480	Fittings, average with fabric and mastic									
4484	1" wall									
4486	1/2" iron pipe size G	1 Asbe	40	.200	Ea.	6.60	11.45		18.05	25
4500	1-1/2" wall									
4502	1" iron pipe size G	1 Asbe	38	.211	Ea.	8	12.05		20.05	27.50
4504	2-1/2" iron pipe size G		32	.250		15.90	14.35		30.25	39.50
4506	3" iron pipe size G		30	.267		16.10	15.30		31.40	41.50
4508	4" iron pipe size G		28	.286		18.80	16.40		35.20	46
4510	5" iron pipe size G	↓	24	.333	↓	24	19.10		43.10	56
4520	2" wall									
4522	1" iron pipe size G	1 Asbe	36	.222	Ea.	12.50	12.75		25.25	33.50
4524	2-1/2" iron pipe size G		30	.267		20	15.30		35.30	45.50
4526	3" iron pipe size G		28	.286		19.75	16.40		36.15	47
4528	4" iron pipe size G		24	.333		24.50	19.10		43.60	56
4530	5" iron pipe size G		22	.364		36.50	21		57.50	72
4532	6" iron pipe size G		20	.400		51.50	23		74.50	92
4536	8" iron pipe size G		12	.667		88.50	38		126.50	157
4538	10" iron pipe size G	↓	8	1	↓	101	57.50		158.50	200
4550	2-1/2" wall									

22 07 19.10 Piping Insulation			Crew	Daily Output	Labor-Hours	Unit	2019 Bare Costs Material	Labor	Equipment	Total	Total Incl O&P
4560	12″ iron pipe size	G	1 Asbe	6	1.333	Ea.	189	76.50		265.50	325
4562	14″ iron pipe size	G	″	4	2	″	207	115		322	405
4570	3″ wall										
4578	6″ iron pipe size	G	1 Asbe	16	.500	Ea.	63.50	28.50		92	114
4580	8″ iron pipe size	G		10	.800		94.50	46		140.50	175
4582	10″ iron pipe size	G		6	1.333		128	76.50		204.50	259
4900	Calcium silicate, with 8 oz. canvas cover										
5100	1″ wall, 1/2″ iron pipe size	G	Q-14	170	.094	L.F.	4.21	4.86		9.07	12.15
5130	3/4″ iron pipe size	G		170	.094		4.24	4.86		9.10	12.15
5140	1″ iron pipe size	G		170	.094		4.15	4.86		9.01	12.05
5150	1-1/4″ iron pipe size	G		165	.097		4.21	5		9.21	12.40
5160	1-1/2″ iron pipe size	G		165	.097		4.26	5		9.26	12.45
5170	2″ iron pipe size	G		160	.100		4.93	5.15		10.08	13.35
5180	2-1/2″ iron pipe size	G		160	.100		5.20	5.15		10.35	13.70
5190	3″ iron pipe size	G		150	.107		5.65	5.50		11.15	14.75
5200	4″ iron pipe size	G		140	.114		7	5.90		12.90	16.80
5210	5″ iron pipe size	G		135	.119		7.40	6.10		13.50	17.60
5220	6″ iron pipe size	G		130	.123		7.85	6.35		14.20	18.40
5280	1-1/2″ wall, 1/2″ iron pipe size	G		150	.107		4.56	5.50		10.06	13.50
5310	3/4″ iron pipe size	G		150	.107		4.65	5.50		10.15	13.60
5320	1″ iron pipe size	G		150	.107		5.05	5.50		10.55	14.05
5330	1-1/4″ iron pipe size	G		145	.110		5.45	5.70		11.15	14.80
5340	1-1/2″ iron pipe size	G		145	.110		5.85	5.70		11.55	15.20
5350	2″ iron pipe size	G		140	.114		6.45	5.90		12.35	16.15
5360	2-1/2″ iron pipe size	G		140	.114		7	5.90		12.90	16.80
5370	3″ iron pipe size	G		135	.119		7.35	6.10		13.45	17.55
5380	4″ iron pipe size	G		125	.128		8.50	6.60		15.10	19.55
5390	5″ iron pipe size	G		120	.133		9.60	6.90		16.50	21
5400	6″ iron pipe size	G		110	.145		9.95	7.50		17.45	22.50
5402	8″ iron pipe size	G		95	.168		13	8.70		21.70	27.50
5404	10″ iron pipe size	G		85	.188		17	9.70		26.70	33.50
5406	12″ iron pipe size	G		80	.200		20.50	10.35		30.85	38.50
5408	14″ iron pipe size	G		75	.213		22.50	11		33.50	42
5410	16″ iron pipe size	G		70	.229		25.50	11.80		37.30	46
5412	18″ iron pipe size	G		65	.246		28	12.70		40.70	50.50
5460	2″ wall, 1/2″ iron pipe size	G		135	.119		7.05	6.10		13.15	17.20
5490	3/4″ iron pipe size	G		135	.119		7.40	6.10		13.50	17.60
5500	1″ iron pipe size	G		135	.119		7.80	6.10		13.90	18
5510	1-1/4″ iron pipe size	G		130	.123		8.35	6.35		14.70	19
5520	1-1/2″ iron pipe size	G		130	.123		8.70	6.35		15.05	19.40
5530	2″ iron pipe size	G		125	.128		9.25	6.60		15.85	20.50
5540	2-1/2″ iron pipe size	G		125	.128		10.90	6.60		17.50	22
5550	3″ iron pipe size	G		120	.133		11	6.90		17.90	23
5560	4″ iron pipe size	G		115	.139		12.75	7.20		19.95	25
5570	5″ iron pipe size	G		110	.145		14.50	7.50		22	27.50
5580	6″ iron pipe size	G		105	.152		15.85	7.85		23.70	29.50
5581	8″ iron pipe size	G		95	.168		18.90	8.70		27.60	34.50
5582	10″ iron pipe size	G		85	.188		23.50	9.70		33.20	41
5583	12″ iron pipe size	G		80	.200		26.50	10.35		36.85	45
5584	14″ iron pipe size	G		75	.213		29	11		40	48.50
5585	16″ iron pipe size	G		70	.229		32	11.80		43.80	53
5586	18″ iron pipe size	G		65	.246		34.50	12.70		47.20	57.50
5587	3″ wall, 1-1/4″ iron pipe size	G		100	.160		12.90	8.25		21.15	27

22 07 19.10 Piping Insulation		Crew	Daily Output	Labor-Hours	Unit	Material	Labor	Equipment	Total	Total Incl O&P
						2019 Bare Costs				
5588	1-1/2" iron pipe size	G	Q-14	100	.160	L.F.	13.05	8.25		21.30	27
5589	2" iron pipe size	G		95	.168		13.50	8.70		22.20	28.50
5590	2-1/2" iron pipe size	G		95	.168		15.70	8.70		24.40	30.50
5591	3" iron pipe size	G		90	.178		15.90	9.20		25.10	31.50
5592	4" iron pipe size	G		85	.188		20	9.70		29.70	37
5593	6" iron pipe size	G		80	.200		24.50	10.35		34.85	43
5594	8" iron pipe size	G		75	.213		29	11		40	49
5595	10" iron pipe size	G		65	.246		35	12.70		47.70	58
5596	12" iron pipe size	G		60	.267		39	13.75		52.75	64
5597	14" iron pipe size	G		55	.291		43.50	15		58.50	71
5598	16" iron pipe size	G		50	.320		48	16.50		64.50	78.50
5599	18" iron pipe size	G		45	.356		52.50	18.35		70.85	86.50
5600	Calcium silicate, no cover										
5720	1" wall, 1/2" iron pipe size	G	Q-14	180	.089	L.F.	3.84	4.59		8.43	11.30
5740	3/4" iron pipe size	G		180	.089		3.84	4.59		8.43	11.30
5750	1" iron pipe size	G		180	.089		3.67	4.59		8.26	11.15
5760	1-1/4" iron pipe size	G		175	.091		3.75	4.72		8.47	11.45
5770	1-1/2" iron pipe size	G		175	.091		3.78	4.72		8.50	11.45
5780	2" iron pipe size	G		170	.094		4.34	4.86		9.20	12.25
5790	2-1/2" iron pipe size	G		170	.094		4.60	4.86		9.46	12.55
5800	3" iron pipe size	G		160	.100		4.94	5.15		10.09	13.40
5810	4" iron pipe size	G		150	.107		6.20	5.50		11.70	15.30
5820	5" iron pipe size	G		145	.110		6.45	5.70		12.15	15.85
5830	6" iron pipe size	G		140	.114		6.75	5.90		12.65	16.55
5900	1-1/2" wall, 1/2" iron pipe size	G		160	.100		4.06	5.15		9.21	12.40
5920	3/4" iron pipe size	G		160	.100		4.12	5.15		9.27	12.50
5930	1" iron pipe size	G		160	.100		4.49	5.15		9.64	12.90
5940	1-1/4" iron pipe size	G		155	.103		4.85	5.35		10.20	13.60
5950	1-1/2" iron pipe size	G		155	.103		5.20	5.35		10.55	14
5960	2" iron pipe size	G		150	.107		5.75	5.50		11.25	14.85
5970	2-1/2" iron pipe size	G		150	.107		6.25	5.50		11.75	15.40
5980	3" iron pipe size	G		145	.110		6.55	5.70		12.25	16
5990	4" iron pipe size	G		135	.119		7.55	6.10		13.65	17.80
6000	5" iron pipe size	G		130	.123		8.55	6.35		14.90	19.20
6010	6" iron pipe size	G		120	.133		8.70	6.90		15.60	20.50
6020	7" iron pipe size	G		115	.139		10.25	7.20		17.45	22.50
6030	8" iron pipe size	G		105	.152		11.55	7.85		19.40	25
6040	9" iron pipe size	G		100	.160		13.85	8.25		22.10	28
6050	10" iron pipe size	G		95	.168		15.30	8.70		24	30
6060	12" iron pipe size	G		90	.178		18.15	9.20		27.35	34
6070	14" iron pipe size	G		85	.188		20.50	9.70		30.20	37.50
6080	16" iron pipe size	G		80	.200		23	10.35		33.35	41.50
6090	18" iron pipe size	G		75	.213		25.50	11		36.50	45
6120	2" wall, 1/2" iron pipe size	G		145	.110		6.40	5.70		12.10	15.85
6140	3/4" iron pipe size	G		145	.110		6.75	5.70		12.45	16.25
6150	1" iron pipe size	G		145	.110		7.15	5.70		12.85	16.65
6160	1-1/4" iron pipe size	G		140	.114		7.60	5.90		13.50	17.50
6170	1-1/2" iron pipe size	G		140	.114		7.95	5.90		13.85	17.85
6180	2" iron pipe size	G		135	.119		8.35	6.10		14.45	18.65
6190	2-1/2" iron pipe size	G		135	.119		10.05	6.10		16.15	20.50
6200	3" iron pipe size	G		130	.123		10.10	6.35		16.45	21
6210	4" iron pipe size	G		125	.128		11.65	6.60		18.25	23
6220	5" iron pipe size	G		120	.133		13.30	6.90		20.20	25.50

22 07 19.10 Piping Insulation			Crew	Daily Output	Labor-Hours	Unit	2019 Bare Costs Material	Labor	Equipment	Total	Total Incl O&P
6230	6″ iron pipe size	G	Q-14	115	.139	L.F.	14.55	7.20		21.75	27
6240	7″ iron pipe size	G		110	.145		15.75	7.50		23.25	29
6250	8″ iron pipe size	G		105	.152		17.75	7.85		25.60	31.50
6260	9″ iron pipe size	G		100	.160		19.60	8.25		27.85	34.50
6270	10″ iron pipe size	G		95	.168		22	8.70		30.70	37.50
6280	12″ iron pipe size	G		90	.178		24	9.20		33.20	40.50
6290	14″ iron pipe size	G		85	.188		26.50	9.70		36.20	44.50
6300	16″ iron pipe size	G		80	.200		29.50	10.35		39.85	48
6310	18″ iron pipe size	G		75	.213		32	11		43	52
6320	20″ iron pipe size	G		65	.246		39.50	12.70		52.20	63
6330	22″ iron pipe size	G		60	.267		43.50	13.75		57.25	69.50
6340	24″ iron pipe size	G		55	.291		44.50	15		59.50	72
6360	3″ wall, 1/2″ iron pipe size	G		115	.139		11.60	7.20		18.80	24
6380	3/4″ iron pipe size	G		115	.139		11.70	7.20		18.90	24
6390	1″ iron pipe size	G		115	.139		11.80	7.20		19	24
6400	1-1/4″ iron pipe size	G		110	.145		11.95	7.50		19.45	25
6410	1-1/2″ iron pipe size	G		110	.145		12	7.50		19.50	25
6420	2″ iron pipe size	G		105	.152		12.40	7.85		20.25	26
6430	2-1/2″ iron pipe size	G		105	.152		14.60	7.85		22.45	28
6440	3″ iron pipe size	G		100	.160		14.75	8.25		23	29
6450	4″ iron pipe size	G		95	.168		18.80	8.70		27.50	34
6460	5″ iron pipe size	G		90	.178		20.50	9.20		29.70	36.50
6470	6″ iron pipe size	G		90	.178		23	9.20		32.20	39.50
6480	7″ iron pipe size	G		85	.188		25.50	9.70		35.20	43
6490	8″ iron pipe size	G		85	.188		27.50	9.70		37.20	45
6500	9″ iron pipe size	G		80	.200		31	10.35		41.35	50
6510	10″ iron pipe size	G		75	.213		33	11		44	53.50
6520	12″ iron pipe size	G		70	.229		36.50	11.80		48.30	58
6530	14″ iron pipe size	G		65	.246		41	12.70		53.70	64.50
6540	16″ iron pipe size	G		60	.267		45.50	13.75		59.25	71.50
6550	18″ iron pipe size	G		55	.291		49.50	15		64.50	77.50
6560	20″ iron pipe size	G		50	.320		59	16.50		75.50	90.50
6570	22″ iron pipe size	G		45	.356		64	18.35		82.35	99
6580	24″ iron pipe size	G		40	.400		69	20.50		89.50	108
6600	Fiberglass, with all service jacket										
6640	1/2″ wall, 1/2″ iron pipe size	G	Q-14	250	.064	L.F.	.72	3.30		4.02	5.90
6660	3/4″ iron pipe size	G		240	.067		.82	3.44		4.26	6.20
6670	1″ iron pipe size	G		230	.070		.85	3.59		4.44	6.50
6680	1-1/4″ iron pipe size	G		220	.073		.91	3.76		4.67	6.80
6690	1-1/2″ iron pipe size	G		220	.073		1.04	3.76		4.80	6.95
6700	2″ iron pipe size	G		210	.076		1.11	3.93		5.04	7.25
6710	2-1/2″ iron pipe size	G		200	.080		1.16	4.13		5.29	7.70
6840	1″ wall, 1/2″ iron pipe size	G		240	.067		.88	3.44		4.32	6.25
6860	3/4″ iron pipe size	G		230	.070		.96	3.59		4.55	6.60
6870	1″ iron pipe size	G		220	.073		1.03	3.76		4.79	6.95
6880	1-1/4″ iron pipe size	G		210	.076		1.11	3.93		5.04	7.25
6890	1-1/2″ iron pipe size	G		210	.076		1.20	3.93		5.13	7.35
6900	2″ iron pipe size	G		200	.080		1.61	4.13		5.74	8.15
6910	2-1/2″ iron pipe size	G		190	.084		1.65	4.35		6	8.50
6920	3″ iron pipe size	G		180	.089		1.78	4.59		6.37	9.05
6930	3-1/2″ iron pipe size	G		170	.094		2.11	4.86		6.97	9.80
6940	4″ iron pipe size	G		150	.107		2.36	5.50		7.86	11.10
6950	5″ iron pipe size	G		140	.114		2.62	5.90		8.52	12

22 07 19 – Plumbing Piping Insulation

22 07 19.10 Piping Insulation		Crew	Daily Output	Labor-Hours	Unit	2019 Bare Costs Material	Labor	Equipment	Total	Total Incl O&P
6960	6" iron pipe size	G	Q-14	120	.133	L.F.	2.82	6.90		9.72	13.75
6970	7" iron pipe size	G		110	.145		3.35	7.50		10.85	15.30
6980	8" iron pipe size	G		100	.160		4.59	8.25		12.84	17.80
6990	9" iron pipe size	G		90	.178		4.84	9.20		14.04	19.45
7000	10" iron pipe size	G		90	.178		4.92	9.20		14.12	19.55
7010	12" iron pipe size	G		80	.200		5.35	10.35		15.70	22
7020	14" iron pipe size	G		80	.200		6.40	10.35		16.75	23
7030	16" iron pipe size	G		70	.229		8.15	11.80		19.95	27
7040	18" iron pipe size	G		70	.229		9.05	11.80		20.85	28
7050	20" iron pipe size	G		60	.267		10.10	13.75		23.85	32.50
7060	24" iron pipe size	G		60	.267		12.30	13.75		26.05	35
7080	1-1/2" wall, 1/2" iron pipe size	G		230	.070		1.89	3.59		5.48	7.65
7100	3/4" iron pipe size	G		220	.073		1.89	3.76		5.65	7.90
7110	1" iron pipe size	G		210	.076		2.02	3.93		5.95	8.25
7120	1-1/4" iron pipe size	G		200	.080		2.20	4.13		6.33	8.80
7130	1-1/2" iron pipe size	G		200	.080		2.55	4.13		6.68	9.20
7140	2" iron pipe size	G		190	.084		2.54	4.35		6.89	9.50
7150	2-1/2" iron pipe size	G		180	.089		2.73	4.59		7.32	10.10
7160	3" iron pipe size	G		170	.094		2.85	4.86		7.71	10.65
7170	3-1/2" iron pipe size	G		160	.100		3.13	5.15		8.28	11.40
7180	4" iron pipe size	G		140	.114		3.58	5.90		9.48	13.05
7190	5" iron pipe size	G		130	.123		3.63	6.35		9.98	13.80
7200	6" iron pipe size	G		110	.145		3.84	7.50		11.34	15.80
7210	7" iron pipe size	G		100	.160		4.26	8.25		12.51	17.45
7220	8" iron pipe size	G		90	.178		5.40	9.20		14.60	20
7230	9" iron pipe size	G		85	.188		5.60	9.70		15.30	21
7240	10" iron pipe size	G		80	.200		5.80	10.35		16.15	22.50
7250	12" iron pipe size	G		75	.213		6.60	11		17.60	24.50
7260	14" iron pipe size	G		70	.229		7.85	11.80		19.65	27
7270	16" iron pipe size	G		65	.246		10.20	12.70		22.90	31
7280	18" iron pipe size	G		60	.267		11.50	13.75		25.25	34
7290	20" iron pipe size	G		55	.291		11.70	15		26.70	36
7300	24" iron pipe size	G		50	.320		14.35	16.50		30.85	41.50
7320	2" wall, 1/2" iron pipe size	G		220	.073		2.93	3.76		6.69	9
7340	3/4" iron pipe size	G		210	.076		3.02	3.93		6.95	9.35
7350	1" iron pipe size	G		200	.080		3.21	4.13		7.34	9.95
7360	1-1/4" iron pipe size	G		190	.084		3.39	4.35		7.74	10.45
7370	1-1/2" iron pipe size	G		190	.084		4.19	4.35		8.54	11.30
7380	2" iron pipe size	G		180	.089		3.73	4.59		8.32	11.20
7390	2-1/2" iron pipe size	G		170	.094		4.01	4.86		8.87	11.90
7400	3" iron pipe size	G		160	.100		4.26	5.15		9.41	12.65
7410	3-1/2" iron pipe size	G		150	.107		4.62	5.50		10.12	13.60
7420	4" iron pipe size	G		130	.123		5.85	6.35		12.20	16.25
7430	5" iron pipe size	G		120	.133		5.70	6.90		12.60	16.90
7440	6" iron pipe size	G		100	.160		6.70	8.25		14.95	20
7450	7" iron pipe size	G		90	.178		6.65	9.20		15.85	21.50
7460	8" iron pipe size	G		80	.200		7.15	10.35		17.50	24
7470	9" iron pipe size	G		75	.213		7.85	11		18.85	25.50
7480	10" iron pipe size	G		70	.229		8.55	11.80		20.35	27.50
7490	12" iron pipe size	G		65	.246		9.55	12.70		22.25	30
7500	14" iron pipe size	G		60	.267		12.30	13.75		26.05	35
7510	16" iron pipe size	G		55	.291		13.50	15		28.50	38
7520	18" iron pipe size	G		50	.320		15.05	16.50		31.55	42

22 07 19.10 Piping Insulation			Crew	Daily Output	Labor-Hours	Unit	2019 Bare Costs Material	Labor	Equipment	Total	Total Incl O&P
7530	20″ iron pipe size	G	Q-14	45	.356	L.F.	17.10	18.35		35.45	47.50
7540	24″ iron pipe size	G		40	.400		18.35	20.50		38.85	52
7560	2-1/2″ wall, 1/2″ iron pipe size	G		210	.076		3.46	3.93		7.39	9.85
7562	3/4″ iron pipe size	G		200	.080		3.62	4.13		7.75	10.40
7564	1″ iron pipe size	G		190	.084		3.77	4.35		8.12	10.85
7566	1-1/4″ iron pipe size	G		185	.086		3.91	4.47		8.38	11.20
7568	1-1/2″ iron pipe size	G		180	.089		4.11	4.59		8.70	11.60
7570	2″ iron pipe size	G		170	.094		4.32	4.86		9.18	12.25
7572	2-1/2″ iron pipe size	G		160	.100		4.97	5.15		10.12	13.40
7574	3″ iron pipe size	G		150	.107		5.25	5.50		10.75	14.25
7576	3-1/2″ iron pipe size	G		140	.114		5.70	5.90		11.60	15.35
7578	4″ iron pipe size	G		120	.133		6	6.90		12.90	17.25
7580	5″ iron pipe size	G		110	.145		5.60	7.50		13.10	17.75
7582	6″ iron pipe size	G		90	.178		8.60	9.20		17.80	23.50
7584	7″ iron pipe size	G		80	.200		8.65	10.35		19	25.50
7586	8″ iron pipe size	G		70	.229		9.20	11.80		21	28.50
7588	9″ iron pipe size	G		65	.246		10.05	12.70		22.75	30.50
7590	10″ iron pipe size	G		60	.267		10.95	13.75		24.70	33.50
7592	12″ iron pipe size	G		55	.291		13.30	15		28.30	37.50
7594	14″ iron pipe size	G		50	.320		15.50	16.50		32	42.50
7596	16″ iron pipe size	G		45	.356		17.80	18.35		36.15	48
7598	18″ iron pipe size	G		40	.400		19.25	20.50		39.75	53
7602	24″ iron pipe size	G		30	.533		25.50	27.50		53	70.50
7620	3″ wall, 1/2″ iron pipe size	G		200	.080		4.43	4.13		8.56	11.25
7622	3/4″ iron pipe size	G		190	.084		4.69	4.35		9.04	11.85
7624	1″ iron pipe size	G		180	.089		4.97	4.59		9.56	12.55
7626	1-1/4″ iron pipe size	G		175	.091		5.05	4.72		9.77	12.85
7628	1-1/2″ iron pipe size	G		170	.094		5.30	4.86		10.16	13.30
7630	2″ iron pipe size	G		160	.100		5.70	5.15		10.85	14.25
7632	2-1/2″ iron pipe size	G		150	.107		5.95	5.50		11.45	15.05
7634	3″ iron pipe size	G		140	.114		6.35	5.90		12.25	16.05
7636	3-1/2″ iron pipe size	G		130	.123		7	6.35		13.35	17.50
7638	4″ iron pipe size	G		110	.145		7.50	7.50		15	19.85
7640	5″ iron pipe size	G		100	.160		8.50	8.25		16.75	22
7642	6″ iron pipe size	G		80	.200		9.10	10.35		19.45	26
7644	7″ iron pipe size	G		70	.229		10.40	11.80		22.20	29.50
7646	8″ iron pipe size	G		60	.267		11.30	13.75		25.05	34
7648	9″ iron pipe size	G		55	.291		12.20	15		27.20	36.50
7650	10″ iron pipe size	G		50	.320		13.10	16.50		29.60	40
7652	12″ iron pipe size	G		45	.356		16.30	18.35		34.65	46.50
7654	14″ iron pipe size	G		40	.400		18.75	20.50		39.25	52.50
7656	16″ iron pipe size	G		35	.457		21.50	23.50		45	60
7658	18″ iron pipe size	G		32	.500		23	26		49	65
7660	20″ iron pipe size	G		30	.533		25	27.50		52.50	70
7662	24″ iron pipe size	G		28	.571		31.50	29.50		61	80
7664	26″ iron pipe size	G		26	.615		34.50	32		66.50	87
7666	30″ iron pipe size	G	▼	24	.667	▼	45.50	34.50		80	103
7800	For fiberglass with standard canvas jacket, deduct						5%				
7802	For fittings, add 3 L.F. for each fitting										
7804	plus 4 L.F. for each flange of the fitting										
7820	Polyethylene tubing flexible closed cell foam, UV resistant										
7828	Standard temperature (-90°F to +212°F)										
7830	3/8″ wall, 1/8″ iron pipe size	G	1 Asbe	130	.062	L.F.	.28	3.53		3.81	5.75

22 07 19.10 Piping Insulation			Crew	Daily Output	Labor-Hours	Unit	2019 Bare Costs Material	Labor	Equipment	Total	Total Incl O&P
7831	1/4" iron pipe size	G	1 Asbe	130	.062	L.F.	.29	3.53		3.82	5.75
7832	3/8" iron pipe size	G		130	.062		.33	3.53		3.86	5.80
7833	1/2" iron pipe size	G		126	.063		.37	3.64		4.01	6
7834	3/4" iron pipe size	G		122	.066		.44	3.76		4.20	6.30
7835	1" iron pipe size	G		120	.067		.48	3.82		4.30	6.45
7836	1-1/4" iron pipe size	G		118	.068		.61	3.89		4.50	6.65
7837	1-1/2" iron pipe size	G		118	.068		.72	3.89		4.61	6.80
7838	2" iron pipe size	G		116	.069		1.01	3.96		4.97	7.20
7839	2-1/2" iron pipe size	G		114	.070		1.27	4.02		5.29	7.60
7840	3" iron pipe size	G		112	.071		1.97	4.10		6.07	8.50
7842	1/2" wall, 1/8" iron pipe size	G		120	.067		.41	3.82		4.23	6.35
7843	1/4" iron pipe size	G		120	.067		.45	3.82		4.27	6.40
7844	3/8" iron pipe size	G		120	.067		.48	3.82		4.30	6.45
7845	1/2" iron pipe size	G		118	.068		.55	3.89		4.44	6.60
7846	3/4" iron pipe size	G		116	.069		.63	3.96		4.59	6.80
7847	1" iron pipe size	G		114	.070		.70	4.02		4.72	6.95
7848	1-1/4" iron pipe size	G		112	.071		.81	4.10		4.91	7.25
7849	1-1/2" iron pipe size	G		110	.073		.96	4.17		5.13	7.50
7850	2" iron pipe size	G		108	.074		1.40	4.25		5.65	8.10
7851	2-1/2" iron pipe size	G		106	.075		1.73	4.33		6.06	8.60
7852	3" iron pipe size	G		104	.077		2.35	4.41		6.76	9.40
7853	3-1/2" iron pipe size	G		102	.078		2.66	4.50		7.16	9.90
7854	4" iron pipe size	G		100	.080		3.02	4.59		7.61	10.40
7855	3/4" wall, 1/8" iron pipe size	G		110	.073		.62	4.17		4.79	7.15
7856	1/4" iron pipe size	G		110	.073		.67	4.17		4.84	7.20
7857	3/8" iron pipe size	G		108	.074		.78	4.25		5.03	7.40
7858	1/2" iron pipe size	G		106	.075		.93	4.33		5.26	7.70
7859	3/4" iron pipe size	G		104	.077		1.11	4.41		5.52	8
7860	1" iron pipe size	G		102	.078		1.29	4.50		5.79	8.35
7861	1-1/4" iron pipe size	G		100	.080		1.74	4.59		6.33	9
7862	1-1/2" iron pipe size	G		100	.080		1.91	4.59		6.50	9.20
7863	2" iron pipe size	G		98	.082		2.50	4.68		7.18	10
7864	2-1/2" iron pipe size	G		96	.083		3.03	4.78		7.81	10.75
7865	3" iron pipe size	G		94	.085		3.78	4.88		8.66	11.70
7866	3-1/2" iron pipe size	G		92	.087		4.34	4.99		9.33	12.45
7867	4" iron pipe size	G		90	.089		4.85	5.10		9.95	13.20
7868	1" wall, 1/4" iron pipe size	G		100	.080		1.35	4.59		5.94	8.60
7869	3/8" iron pipe size	G		98	.082		1.36	4.68		6.04	8.75
7870	1/2" iron pipe size	G		96	.083		1.71	4.78		6.49	9.30
7871	3/4" iron pipe size	G		94	.085		1.98	4.88		6.86	9.75
7872	1" iron pipe size	G		92	.087		2.41	4.99		7.40	10.35
7873	1-1/4" iron pipe size	G		90	.089		2.74	5.10		7.84	10.85
7874	1-1/2" iron pipe size	G		90	.089		3.12	5.10		8.22	11.30
7875	2" iron pipe size	G		88	.091		4.07	5.20		9.27	12.55
7876	2-1/2" iron pipe size	G		86	.093		5.35	5.35		10.70	14.15
7877	3" iron pipe size	G		84	.095		6	5.45		11.45	15.05
7878	Contact cement, quart can	G				Ea.	13.05			13.05	14.40
7879	Rubber tubing, flexible closed cell foam										
7880	3/8" wall, 1/4" iron pipe size	G	1 Asbe	120	.067	L.F.	.32	3.82		4.14	6.25
7900	3/8" iron pipe size	G		120	.067		.47	3.82		4.29	6.40
7910	1/2" iron pipe size	G		115	.070		.50	3.99		4.49	6.70
7920	3/4" iron pipe size	G		115	.070		.58	3.99		4.57	6.80
7930	1" iron pipe size	G		110	.073		.59	4.17		4.76	7.10

22 07 19.10 Piping Insulation			Crew	Daily Output	Labor-Hours	Unit	Material	2019 Bare Costs Labor	Equipment	Total	Total Incl O&P
7940	1-1/4" iron pipe size	G	1 Asbe	110	.073	L.F.	.67	4.17		4.84	7.20
7950	1-1/2" iron pipe size	G		110	.073		.77	4.17		4.94	7.30
8100	1/2" wall, 1/4" iron pipe size	G		90	.089		.83	5.10		5.93	8.75
8120	3/8" iron pipe size	G		90	.089		.87	5.10		5.97	8.80
8130	1/2" iron pipe size	G		89	.090		.91	5.15		6.06	8.95
8140	3/4" iron pipe size	G		89	.090		1.04	5.15		6.19	9.10
8150	1" iron pipe size	G		88	.091		.84	5.20		6.04	8.95
8160	1-1/4" iron pipe size	G		87	.092		1.19	5.25		6.44	9.45
8170	1-1/2" iron pipe size	G		87	.092		1.70	5.25		6.95	10
8180	2" iron pipe size	G		86	.093		2.06	5.35		7.41	10.50
8190	2-1/2" iron pipe size	G		86	.093		2.25	5.35		7.60	10.75
8200	3" iron pipe size	G		85	.094		2.37	5.40		7.77	10.95
8210	3-1/2" iron pipe size	G		85	.094		3.35	5.40		8.75	12.05
8220	4" iron pipe size	G		80	.100		3.61	5.75		9.36	12.80
8230	5" iron pipe size	G		80	.100		4.92	5.75		10.67	14.25
8240	6" iron pipe size	G		75	.107		4.92	6.10		11.02	14.85
8300	3/4" wall, 1/4" iron pipe size	G		90	.089		.92	5.10		6.02	8.85
8320	3/8" iron pipe size	G		90	.089		1.09	5.10		6.19	9.05
8330	1/2" iron pipe size	G		89	.090		1.11	5.15		6.26	9.15
8340	3/4" iron pipe size	G		89	.090		1.81	5.15		6.96	9.95
8350	1" iron pipe size	G		88	.091		2.08	5.20		7.28	10.35
8360	1-1/4" iron pipe size	G		87	.092		2.42	5.25		7.67	10.80
8370	1-1/2" iron pipe size	G		87	.092		3.04	5.25		8.29	11.50
8380	2" iron pipe size	G		86	.093		3.57	5.35		8.92	12.20
8390	2-1/2" iron pipe size	G		86	.093		4.27	5.35		9.62	12.95
8400	3" iron pipe size	G		85	.094		4.82	5.40		10.22	13.65
8410	3-1/2" iron pipe size	G		85	.094		4.93	5.40		10.33	13.75
8420	4" iron pipe size	G		80	.100		6	5.75		11.75	15.45
8430	5" iron pipe size	G		80	.100		7.15	5.75		12.90	16.75
8440	6" iron pipe size	G		80	.100		8.70	5.75		14.45	18.45
8444	1" wall, 1/2" iron pipe size	G		86	.093		2.71	5.35		8.06	11.25
8445	3/4" iron pipe size	G		84	.095		3.29	5.45		8.74	12.05
8446	1" iron pipe size	G		84	.095		3.43	5.45		8.88	12.20
8447	1-1/4" iron pipe size	G		82	.098		3.76	5.60		9.36	12.80
8448	1-1/2" iron pipe size	G		82	.098		5.25	5.60		10.85	14.45
8449	2" iron pipe size	G		80	.100		6.50	5.75		12.25	16
8450	2-1/2" iron pipe size	G	↓	80	.100	↓	7.50	5.75		13.25	17.10
8456	Rubber insulation tape, 1/8" x 2" x 30'	G				Ea.	21.50			21.50	23.50
8460	Polyolefin tubing, flexible closed cell foam, UV stabilized, work										
8462	temp. -165°F to +210°F, 0 water vapor transmission										
8464	3/8" wall, 1/8" iron pipe size	G	1 Asbe	140	.057	L.F.	.50	3.28		3.78	5.60
8466	1/4" iron pipe size	G		140	.057		.52	3.28		3.80	5.60
8468	3/8" iron pipe size	G		140	.057		.57	3.28		3.85	5.70
8470	1/2" iron pipe size	G		136	.059		.64	3.37		4.01	5.90
8472	3/4" iron pipe size	G		132	.061		.64	3.48		4.12	6.05
8474	1" iron pipe size	G		130	.062		.85	3.53		4.38	6.40
8476	1-1/4" iron pipe size	G		128	.063		1.03	3.58		4.61	6.70
8478	1-1/2" iron pipe size	G		128	.063		1.26	3.58		4.84	6.95
8480	2" iron pipe size	G		126	.063		1.53	3.64		5.17	7.30
8482	2-1/2" iron pipe size	G		123	.065		2.22	3.73		5.95	8.20
8484	3" iron pipe size	G		121	.066		2.49	3.79		6.28	8.60
8486	4" iron pipe size	G		118	.068		4.68	3.89		8.57	11.15
8500	1/2" wall, 1/8" iron pipe size	G	↓	130	.062	↓	.73	3.53		4.26	6.25

22 07 19 – Plumbing Piping Insulation

22 07 19.10 Piping Insulation			Crew	Daily Output	Labor-Hours	Unit	2019 Bare Costs Material	Labor	Equipment	Total	Total Incl O&P
8502	1/4" iron pipe size	G	1 Asbe	130	.062	L.F.	.78	3.53		4.31	6.30
8504	3/8" iron pipe size	G		130	.062		.84	3.53		4.37	6.35
8506	1/2" iron pipe size	G		128	.063		.90	3.58		4.48	6.55
8508	3/4" iron pipe size	G		126	.063		1.05	3.64		4.69	6.75
8510	1" iron pipe size	G		123	.065		1.18	3.73		4.91	7.05
8512	1-1/4" iron pipe size	G		121	.066		1.40	3.79		5.19	7.40
8514	1-1/2" iron pipe size	G		119	.067		1.71	3.86		5.57	7.85
8516	2" iron pipe size	G		117	.068		2.15	3.92		6.07	8.40
8518	2-1/2" iron pipe size	G		114	.070		2.90	4.02		6.92	9.40
8520	3" iron pipe size	G		112	.071		3.73	4.10		7.83	10.45
8522	4" iron pipe size	G		110	.073		5.25	4.17		9.42	12.20
8534	3/4" wall, 1/8" iron pipe size	G		120	.067		1.11	3.82		4.93	7.10
8536	1/4" iron pipe size	G		120	.067		1.17	3.82		4.99	7.20
8538	3/8" iron pipe size	G		117	.068		1.35	3.92		5.27	7.55
8540	1/2" iron pipe size	G		114	.070		1.49	4.02		5.51	7.85
8542	3/4" iron pipe size	G		112	.071		1.58	4.10		5.68	8.10
8544	1" iron pipe size	G		110	.073		2.26	4.17		6.43	8.95
8546	1-1/4" iron pipe size	G		108	.074		2.88	4.25		7.13	9.70
8548	1-1/2" iron pipe size	G		108	.074		3.54	4.25		7.79	10.45
8550	2" iron pipe size	G		106	.075		4.43	4.33		8.76	11.55
8552	2-1/2" iron pipe size	G		104	.077		5.65	4.41		10.06	13
8554	3" iron pipe size	G		102	.078		7.65	4.50		12.15	15.40
8556	4" iron pipe size	G		100	.080		9.20	4.59		13.79	17.20
8570	1" wall, 1/8" iron pipe size	G		110	.073		2	4.17		6.17	8.65
8572	1/4" iron pipe size	G		108	.074		2.25	4.25		6.50	9.05
8574	3/8" iron pipe size	G		106	.075		2.11	4.33		6.44	9
8576	1/2" iron pipe size	G		104	.077		1.97	4.41		6.38	8.95
8578	3/4" iron pipe size	G		102	.078		2.98	4.50		7.48	10.25
8580	1" iron pipe size	G		100	.080		2.92	4.59		7.51	10.30
8582	1-1/4" iron pipe size	G		97	.082		3.89	4.73		8.62	11.60
8584	1-1/2" iron pipe size	G		97	.082		4.38	4.73		9.11	12.10
8586	2" iron pipe size	G		95	.084		5.10	4.83		9.93	13.10
8588	2-1/2" iron pipe size	G		93	.086		8	4.93		12.93	16.40
8590	3" iron pipe size	G	↓	91	.088	↓	9.75	5.05		14.80	18.55
8606	Contact adhesive (R-320)	G				Qt.	19.25			19.25	21
8608	Contact adhesive (R-320)	G				Gal.	76.50			76.50	84
8610	NOTE: Preslit/preglued vs. unslit, same price										

22 07 19.30 Piping Insulation Protective Jacketing, PVC		Crew	Daily Output	Labor-Hours	Unit	Material	Labor	Equipment	Total	Total Incl O&P
0010	**PIPING INSULATION PROTECTIVE JACKETING, PVC**									
0100	PVC, white, 48" lengths cut from roll goods									
0120	20 mil thick									
0140	Size based on OD of insulation									
0150	1-1/2" ID	Q-14	270	.059	L.F.	.25	3.06		3.31	5
0152	2" ID		260	.062		.33	3.18		3.51	5.25
0154	2-1/2" ID		250	.064		.39	3.30		3.69	5.55
0156	3" ID		240	.067		.47	3.44		3.91	5.80
0158	3-1/2" ID		230	.070		.53	3.59		4.12	6.15
0160	4" ID		220	.073		.60	3.76		4.36	6.45
0162	4-1/2" ID		210	.076		.68	3.93		4.61	6.80
0164	5" ID		200	.080		.74	4.13		4.87	7.20
0166	5-1/2" ID		190	.084		.81	4.35		5.16	7.60
0168	6" ID	↓	180	.089	↓	.88	4.59		5.47	8.05

22 07 19 – Plumbing Piping Insulation

22 07 19.30 Piping Insulation Protective Jacketing, PVC		Crew	Daily Output	Labor-Hours	Unit	Material	2019 Bare Costs Labor	Equipment	Total	Total Incl O&P
0170	6-1/2" ID	Q-14	175	.091	L.F.	.95	4.72		5.67	8.35
0172	7" ID		170	.094		1.01	4.86		5.87	8.60
0174	7-1/2" ID		164	.098		1.10	5.05		6.15	9
0176	8" ID		161	.099		1.16	5.15		6.31	9.20
0178	8-1/2" ID		158	.101		1.23	5.25		6.48	9.40
0180	9" ID		155	.103		1.30	5.35		6.65	9.70
0182	9-1/2" ID		152	.105		1.37	5.45		6.82	9.90
0184	10" ID		149	.107		1.43	5.55		6.98	10.10
0186	10-1/2" ID		146	.110		1.51	5.65		7.16	10.40
0188	11" ID		143	.112		1.58	5.80		7.38	10.65
0190	11-1/2" ID		140	.114		1.65	5.90		7.55	10.90
0192	12" ID		137	.117		1.71	6.05		7.76	11.20
0194	12-1/2" ID		134	.119		1.79	6.15		7.94	11.45
0195	13" ID		132	.121		1.85	6.25		8.10	11.70
0196	13-1/2" ID		132	.121		1.93	6.25		8.18	11.75
0198	14" ID		130	.123		2	6.35		8.35	12
0200	15" ID		128	.125		2.14	6.45		8.59	12.30
0202	16" ID		126	.127		2.27	6.55		8.82	12.60
0204	17" ID		124	.129		2.42	6.65		9.07	12.95
0206	18" ID		122	.131		2.56	6.75		9.31	13.25
0208	19" ID		120	.133		2.69	6.90		9.59	13.60
0210	20" ID		118	.136		2.84	7		9.84	13.90
0212	21" ID		116	.138		2.97	7.10		10.07	14.25
0214	22" ID		114	.140		3.11	7.25		10.36	14.60
0216	23" ID		112	.143		3.25	7.40		10.65	15
0218	24" ID		110	.145		3.39	7.50		10.89	15.35
0220	25" ID		108	.148		3.53	7.65		11.18	15.70
0222	26" ID		106	.151		3.67	7.80		11.47	16.10
0224	27" ID		104	.154		3.81	7.95		11.76	16.45
0226	28" ID		102	.157		3.94	8.10		12.04	16.85
0228	29" ID		100	.160		4.09	8.25		12.34	17.25
0230	30" ID		98	.163		4.23	8.45		12.68	17.65
0300	For colors, add				Ea.	10%				
1000	30 mil thick									
1010	Size based on OD of insulation									
1020	2" ID	Q-14	260	.062	L.F.	.48	3.18		3.66	5.45
1022	2-1/2" ID		250	.064		.58	3.30		3.88	5.75
1024	3" ID		240	.067		.70	3.44		4.14	6.05
1026	3-1/2" ID		230	.070		.79	3.59		4.38	6.40
1028	4" ID		220	.073		.89	3.76		4.65	6.80
1030	4-1/2" ID		210	.076		.99	3.93		4.92	7.15
1032	5" ID		200	.080		1.10	4.13		5.23	7.60
1034	5-1/2" ID		190	.084		1.20	4.35		5.55	8
1036	6" ID		180	.089		1.30	4.59		5.89	8.55
1038	6-1/2" ID		175	.091		1.40	4.72		6.12	8.85
1040	7" ID		170	.094		1.51	4.86		6.37	9.15
1042	7-1/2" ID		164	.098		1.62	5.05		6.67	9.60
1044	8" ID		161	.099		1.72	5.15		6.87	9.80
1046	8-1/2" ID		158	.101		1.82	5.25		7.07	10.05
1048	9" ID		155	.103		1.93	5.35		7.28	10.35
1050	9-1/2" ID		152	.105		2.03	5.45		7.48	10.65
1052	10" ID		149	.107		2.13	5.55		7.68	10.90
1054	10-1/2" ID		146	.110		2.23	5.65		7.88	11.20

22 07 19.30 Piping Insulation Protective Jacketing, PVC		Crew	Daily Output	Labor-Hours	Unit	2019 Bare Costs Material	Labor	Equipment	Total	Total Incl O&P
1056	11" ID	Q-14	143	.112	L.F.	2.34	5.80		8.14	11.45
1058	11-1/2" ID		140	.114		2.44	5.90		8.34	11.80
1060	12" ID		137	.117		2.54	6.05		8.59	12.10
1062	12-1/2" ID		134	.119		2.65	6.15		8.80	12.40
1063	13" ID		132	.121		2.76	6.25		9.01	12.70
1064	13-1/2" ID		132	.121		2.86	6.25		9.11	12.80
1066	14" ID		130	.123		2.96	6.35		9.31	13.05
1068	15" ID		128	.125		3.17	6.45		9.62	13.45
1070	16" ID		126	.127		3.37	6.55		9.92	13.80
1072	17" ID		124	.129		3.58	6.65		10.23	14.25
1074	18" ID		122	.131		3.79	6.75		10.54	14.60
1076	19" ID		120	.133		4	6.90		10.90	15.05
1078	20" ID		118	.136		4.20	7		11.20	15.40
1080	21" ID		116	.138		4.41	7.10		11.51	15.85
1082	22" ID		114	.140		4.61	7.25		11.86	16.25
1084	23" ID		112	.143		4.82	7.40		12.22	16.70
1086	24" ID		110	.145		5	7.50		12.50	17.10
1088	25" ID		108	.148		5.25	7.65		12.90	17.55
1090	26" ID		106	.151		5.45	7.80		13.25	18.05
1092	27" ID		104	.154		5.65	7.95		13.60	18.45
1094	28" ID		102	.157		5.85	8.10		13.95	18.95
1096	29" ID		100	.160		6.05	8.25		14.30	19.40
1098	30" ID		98	.163		6.25	8.45		14.70	19.90
1300	For colors, add				Ea.	10%				
2000	PVC, white, fitting covers									
2020	Fiberglass insulation inserts included with sizes 1-3/4" thru 9-3/4"									
2030	Size is based on OD of insulation									
2040	90° elbow fitting									
2060	1-3/4"	Q-14	135	.119	Ea.	.54	6.10		6.64	10.05
2062	2"		130	.123		.67	6.35		7.02	10.55
2064	2-1/4"		128	.125		.78	6.45		7.23	10.80
2068	2-1/2"		126	.127		.83	6.55		7.38	11
2070	2-3/4"		123	.130		.98	6.70		7.68	11.45
2072	3"		120	.133		.95	6.90		7.85	11.70
2074	3-3/8"		116	.138		1.12	7.10		8.22	12.25
2076	3-3/4"		113	.142		1.20	7.30		8.50	12.60
2078	4-1/8"		110	.145		1.58	7.50		9.08	13.35
2080	4-3/4"		105	.152		1.91	7.85		9.76	14.25
2082	5-1/4"		100	.160		2.22	8.25		10.47	15.20
2084	5-3/4"		95	.168		2.81	8.70		11.51	16.50
2086	6-1/4"		90	.178		4.69	9.20		13.89	19.30
2088	6-3/4"		87	.184		4.96	9.50		14.46	20
2090	7-1/4"		85	.188		6.20	9.70		15.90	22
2092	7-3/4"		83	.193		6.55	9.95		16.50	22.50
2094	8-3/4"		80	.200		8.40	10.35		18.75	25
2096	9-3/4"		77	.208		11.20	10.75		21.95	29
2098	10-7/8"		74	.216		12.50	11.15		23.65	31
2100	11-7/8"		71	.225		14.35	11.65		26	33.50
2102	12-7/8"		68	.235		19.75	12.15		31.90	40.50
2104	14-1/8"		66	.242		20.50	12.50		33	42
2106	15-1/8"		64	.250		22	12.90		34.90	44.50
2108	16-1/8"		63	.254		24.50	13.10		37.60	46.50
2110	17-1/8"		62	.258		27	13.30		40.30	50

22 07 19.30 Piping Insulation Protective Jacketing, PVC		Crew	Daily Output	Labor-Hours	Unit	2019 Bare Costs Material	Labor	Equipment	Total	Total Incl O&P
2112	18-1/8"	Q-14	61	.262	Ea.	36	13.55		49.55	60.50
2114	19-1/8"		60	.267		47	13.75		60.75	73.50
2116	20-1/8"		59	.271		60.50	14		74.50	88
2200	45° elbow fitting									
2220	1-3/4" thru 9-3/4" same price as 90° elbow fitting									
2320	10-7/8"	Q-14	74	.216	Ea.	12.50	11.15		23.65	31
2322	11-7/8"		71	.225		13.75	11.65		25.40	33
2324	12-7/8"		68	.235		15.35	12.15		27.50	35.50
2326	14-1/8"		66	.242		17.50	12.50		30	38.50
2328	15-1/8"		64	.250		18.85	12.90		31.75	40.50
2330	16-1/8"		63	.254		21.50	13.10		34.60	43.50
2332	17-1/8"		62	.258		24	13.30		37.30	47
2334	18-1/8"		61	.262		29.50	13.55		43.05	53.50
2336	19-1/8"		60	.267		40.50	13.75		54.25	66
2338	20-1/8"		59	.271		46	14		60	72
2400	Tee fitting									
2410	1-3/4"	Q-14	96	.167	Ea.	1.03	8.60		9.63	14.45
2412	2"		94	.170		1.17	8.80		9.97	14.85
2414	2-1/4"		91	.176		1.26	9.10		10.36	15.40
2416	2-1/2"		88	.182		1.38	9.40		10.78	16
2418	2-3/4"		85	.188		1.52	9.70		11.22	16.65
2420	3"		82	.195		1.65	10.05		11.70	17.35
2422	3-3/8"		79	.203		1.90	10.45		12.35	18.25
2424	3-3/4"		76	.211		2.35	10.85		13.20	19.40
2426	4-1/8"		73	.219		2.55	11.30		13.85	20.50
2428	4-3/4"		70	.229		3.17	11.80		14.97	21.50
2430	5-1/4"		67	.239		3.81	12.35		16.16	23
2432	5-3/4"		63	.254		5.05	13.10		18.15	25.50
2434	6-1/4"		60	.267		6.65	13.75		20.40	29
2436	6-3/4"		59	.271		8.25	14		22.25	30.50
2438	7-1/4"		57	.281		13.30	14.50		27.80	37
2440	7-3/4"		54	.296		14.60	15.30		29.90	39.50
2442	8-3/4"		52	.308		17.75	15.90		33.65	44
2444	9-3/4"		50	.320		21	16.50		37.50	48.50
2446	10-7/8"		48	.333		21	17.20		38.20	50
2448	11-7/8"		47	.340		23.50	17.60		41.10	53
2450	12-7/8"		46	.348		26	17.95		43.95	56
2452	14-1/8"		45	.356		28.50	18.35		46.85	59.50
2454	15-1/8"		44	.364		30.50	18.75		49.25	62.50
2456	16-1/8"		43	.372		33	19.20		52.20	66
2458	17-1/8"		42	.381		35.50	19.65		55.15	69.50
2460	18-1/8"		41	.390		39	20		59	73.50
2462	19-1/8"		40	.400		42.50	20.50		63	78.50
2464	20-1/8"		39	.410		47	21		68	84
4000	Mechanical grooved fitting cover, including insert									
4020	90° elbow fitting									
4030	3/4" & 1"	Q-14	140	.114	Ea.	4.30	5.90		10.20	13.85
4040	1-1/4" & 1-1/2"		135	.119		5.45	6.10		11.55	15.45
4042	2"		130	.123		8.55	6.35		14.90	19.20
4044	2-1/2"		125	.128		9.55	6.60		16.15	20.50
4046	3"		120	.133		10.65	6.90		17.55	22.50
4048	3-1/2"		115	.139		12.35	7.20		19.55	24.50
4050	4"		110	.145		13.75	7.50		21.25	27

22 07 19.30 Piping Insulation Protective Jacketing, PVC		Crew	Daily Output	Labor-Hours	Unit	Material	2019 Bare Costs Labor	Equipment	Total	Total Incl O&P
4052	5"	Q-14	100	.160	Ea.	17.10	8.25		25.35	31.50
4054	6"		90	.178		25.50	9.20		34.70	42
4056	8"		80	.200		27.50	10.35		37.85	46.50
4058	10"		75	.213		35	11		46	55.50
4060	12"		68	.235		50.50	12.15		62.65	74.50
4062	14"		65	.246		52.50	12.70		65.20	77.50
4064	16"		63	.254		71.50	13.10		84.60	99
4066	18"		61	.262		98.50	13.55		112.05	129
4100	45° elbow fitting									
4120	3/4" & 1"	Q-14	140	.114	Ea.	3.82	5.90		9.72	13.30
4130	1-1/4" & 1-1/2"		135	.119		4.86	6.10		10.96	14.80
4140	2"		130	.123		8.35	6.35		14.70	18.95
4142	2-1/2"		125	.128		9.25	6.60		15.85	20.50
4144	3"		120	.133		9.55	6.90		16.45	21
4146	3-1/2"		115	.139		10.45	7.20		17.65	22.50
4148	4"		110	.145		14.40	7.50		21.90	27.50
4150	5"		100	.160		16.60	8.25		24.85	31
4152	6"		90	.178		23.50	9.20		32.70	39.50
4154	8"		80	.200		24.50	10.35		34.85	43
4156	10"		75	.213		31.50	11		42.50	51.50
4158	12"		68	.235		47	12.15		59.15	70.50
4160	14"		65	.246		61	12.70		73.70	86.50
4162	16"		63	.254		77.50	13.10		90.60	106
4164	18"		61	.262		87	13.55		100.55	117
4200	Tee fitting									
4220	3/4" & 1"	Q-14	93	.172	Ea.	5.60	8.90		14.50	19.85
4230	1-1/4" & 1-1/2"		90	.178		7.05	9.20		16.25	22
4240	2"		87	.184		11.55	9.50		21.05	27.50
4242	2-1/2"		84	.190		12.85	9.85		22.70	29.50
4244	3"		80	.200		14.25	10.35		24.60	31.50
4246	3-1/2"		77	.208		15.25	10.75		26	33.50
4248	4"		73	.219		18.65	11.30		29.95	38
4250	5"		67	.239		20.50	12.35		32.85	41.50
4252	6"		60	.267		34.50	13.75		48.25	59.50
4254	8"		54	.296		33	15.30		48.30	59.50
4256	10"		50	.320		46	16.50		62.50	76
4258	12"		46	.348		66.50	17.95		84.45	101
4260	14"		43	.372		94	19.20		113.20	133
4262	16"		42	.381		104	19.65		123.65	146
4264	18"		41	.390		109	20		129	151

22 07 19.40 Pipe Insulation Protective Jacketing, Aluminum		Crew	Daily Output	Labor-Hours	Unit	Material	Labor	Equipment	Total	Total Incl O&P
0010	**PIPE INSULATION PROTECTIVE JACKETING, ALUMINUM**									
0100	Metal roll jacketing									
0120	Aluminum with polykraft moisture barrier									
0140	Smooth, based on OD of insulation, .016" thick									
0180	1/2" ID	Q-14	220	.073	L.F.	.29	3.76		4.05	6.10
0190	3/4" ID		215	.074		.37	3.84		4.21	6.35
0200	1" ID		210	.076		.45	3.93		4.38	6.55
0210	1-1/4" ID		205	.078		.54	4.03		4.57	6.80
0220	1-1/2" ID		202	.079		.62	4.09		4.71	7
0230	1-3/4" ID		199	.080		.78	4.15		4.93	7.25
0240	2" ID		195	.082		.87	4.24		5.11	7.50

22 07 Plumbing Insulation

22 07 19 – Plumbing Piping Insulation

22 07 19.40 Pipe Insulation Protective Jacketing, Aluminum		Crew	Daily Output	Labor-Hours	Unit	2019 Bare Costs Material	Labor	Equipment	Total	Total Incl O&P
0250	2-1/4" ID	Q-14	191	.084	L.F.	.97	4.33		5.30	7.75
0260	2-1/2" ID		187	.086		.97	4.42		5.39	7.85
0270	2-3/4" ID		184	.087		1.06	4.49		5.55	8.10
0280	3" ID		180	.089		1.26	4.59		5.85	8.50
0290	3-1/4" ID		176	.091		1.36	4.69		6.05	8.75
0300	3-1/2" ID		172	.093		1.32	4.80		6.12	8.85
0310	3-3/4" ID		169	.095		1.40	4.89		6.29	9.10
0320	4" ID		165	.097		1.49	5		6.49	9.40
0330	4-1/4" ID		161	.099		1.73	5.15		6.88	9.80
0340	4-1/2" ID		157	.102		1.67	5.25		6.92	9.95
0350	4-3/4" ID		154	.104		1.92	5.35		7.27	10.40
0360	5" ID		150	.107		2.02	5.50		7.52	10.70
0370	5-1/4" ID		146	.110		2.11	5.65		7.76	11.05
0380	5-1/2" ID		143	.112		2.01	5.80		7.81	11.10
0390	5-3/4" ID		139	.115		2.32	5.95		8.27	11.70
0400	6" ID		135	.119		2.41	6.10		8.51	12.10
0410	6-1/4" ID		133	.120		2.50	6.20		8.70	12.35
0420	6-1/2" ID		131	.122		2.36	6.30		8.66	12.35
0430	7" ID		128	.125		2.51	6.45		8.96	12.70
0440	7-1/4" ID		125	.128		2.60	6.60		9.20	13.05
0450	7-1/2" ID		123	.130		2.68	6.70		9.38	13.30
0460	8" ID		121	.132		2.85	6.85		9.70	13.70
0470	8-1/2" ID		119	.134		3.02	6.95		9.97	14
0480	9" ID		116	.138		3.19	7.10		10.29	14.50
0490	9-1/2" ID		114	.140		3.37	7.25		10.62	14.90
0500	10" ID		112	.143		3.54	7.40		10.94	15.30
0510	10-1/2" ID		110	.145		4.07	7.50		11.57	16.10
0520	11" ID		107	.150		3.87	7.70		11.57	16.15
0530	11-1/2" ID		105	.152		4.05	7.85		11.90	16.60
0540	12" ID		103	.155		4.22	8		12.22	17.05
0550	12-1/2" ID		100	.160		4.39	8.25		12.64	17.60
0560	13" ID		99	.162		4.56	8.35		12.91	17.90
0570	14" ID		98	.163		4.91	8.45		13.36	18.40
0580	15" ID		96	.167		5.25	8.60		13.85	19.10
0590	16" ID		95	.168		5.60	8.70		14.30	19.55
0600	17" ID		93	.172		5.95	8.90		14.85	20
0610	18" ID		92	.174		6.15	9		15.15	20.50
0620	19" ID		90	.178		6.60	9.20		15.80	21.50
0630	20" ID		89	.180		6.95	9.30		16.25	22
0640	21" ID		87	.184		7.30	9.50		16.80	22.50
0650	22" ID		86	.186		7.65	9.60		17.25	23.50
0660	23" ID		84	.190		8	9.85		17.85	24
0670	24" ID	↓	83	.193		8.35	9.95		18.30	24.50
0710	For smooth .020" thick, add					27%	10%			
0720	For smooth .024" thick, add					52%	20%			
0730	For smooth .032" thick, add					104%	33%			
0800	For stucco embossed, add					1%				
0820	For corrugated, add				↓	2.50%				
0900	White aluminum with polysurlyn moisture barrier									
0910	Smooth, % is an add to polykraft lines of same thickness									
0940	For smooth .016" thick, add				L.F.	35%				
0960	For smooth .024" thick, add				"	22%				
1000	Aluminum fitting covers									

22 07 19.40 Pipe Insulation Protective Jacketing, Aluminum		Crew	Daily Output	Labor-Hours	Unit	2019 Bare Costs Material	Labor	Equipment	Total	Total Incl O&P
1010	Size is based on OD of insulation									
1020	90° LR elbow, 2 piece									
1100	1-1/2"	Q-14	140	.114	Ea.	4.16	5.90		10.06	13.70
1110	1-3/4"		135	.119		4.16	6.10		10.26	14.05
1120	2"		130	.123		5.25	6.35		11.60	15.55
1130	2-1/4"		128	.125		5.25	6.45		11.70	15.70
1140	2-1/2"		126	.127		5.25	6.55		11.80	15.85
1150	2-3/4"		123	.130		5.25	6.70		11.95	16.10
1160	3"		120	.133		5.70	6.90		12.60	16.90
1170	3-1/4"		117	.137		5.70	7.05		12.75	17.15
1180	3-1/2"		115	.139		6.70	7.20		13.90	18.50
1190	3-3/4"		113	.142		6.90	7.30		14.20	18.85
1200	4"		110	.145		7.20	7.50		14.70	19.50
1210	4-1/4"		108	.148		7.40	7.65		15.05	19.95
1220	4-1/2"		106	.151		7.40	7.80		15.20	20
1230	4-3/4"		104	.154		7.40	7.95		15.35	20.50
1240	5"		102	.157		8.35	8.10		16.45	21.50
1250	5-1/4"		100	.160		9.65	8.25		17.90	23.50
1260	5-1/2"		97	.165		12	8.50		20.50	26.50
1270	5-3/4"		95	.168		12	8.70		20.70	26.50
1280	6"		92	.174		10.25	9		19.25	25
1290	6-1/4"		90	.178		10.25	9.20		19.45	25.50
1300	6-1/2"		87	.184		12.30	9.50		21.80	28
1310	7"		85	.188		14.95	9.70		24.65	31.50
1320	7-1/4"		84	.190		14.95	9.85		24.80	31.50
1330	7-1/2"		83	.193		22	9.95		31.95	39.50
1340	8"		82	.195		16.40	10.05		26.45	33.50
1350	8-1/2"		80	.200		30.50	10.35		40.85	49.50
1360	9"		78	.205		30.50	10.60		41.10	50
1370	9-1/2"		77	.208		22.50	10.75		33.25	41
1380	10"		76	.211		22.50	10.85		33.35	41.50
1390	10-1/2"		75	.213		23	11		34	42
1400	11"		74	.216		23	11.15		34.15	42.50
1410	11-1/2"		72	.222		26	11.45		37.45	46
1420	12"		71	.225		26	11.65		37.65	46.50
1430	12-1/2"		69	.232		46	11.95		57.95	69
1440	13"		68	.235		46	12.15		58.15	69.50
1450	14"		66	.242		62	12.50		74.50	87.50
1460	15"		64	.250		64.50	12.90		77.40	91
1470	16"		63	.254		70.50	13.10		83.60	97.50
2000	45° elbow, 2 piece									
2010	2-1/2"	Q-14	126	.127	Ea.	4.32	6.55		10.87	14.85
2020	2-3/4"		123	.130		4.32	6.70		11.02	15.10
2030	3"		120	.133		4.89	6.90		11.79	16.05
2040	3-1/4"		117	.137		4.89	7.05		11.94	16.30
2050	3-1/2"		115	.139		5.55	7.20		12.75	17.25
2060	3-3/4"		113	.142		5.55	7.30		12.85	17.45
2070	4"		110	.145		5.65	7.50		13.15	17.85
2080	4-1/4"		108	.148		6.45	7.65		14.10	18.90
2090	4-1/2"		106	.151		6.45	7.80		14.25	19.15
2100	4-3/4"		104	.154		6.45	7.95		14.40	19.35
2110	5"		102	.157		7.60	8.10		15.70	21
2120	5-1/4"		100	.160		7.60	8.25		15.85	21

22 07 19.40 Pipe Insulation Protective Jacketing, Aluminum		Crew	Daily Output	Labor-Hours	Unit	Material	Labor	Equipment	Total	Total Incl O&P
						2019 Bare Costs				
2130	5-1/2"	Q-14	97	.165	Ea.	7.95	8.50		16.45	22
2140	6"		92	.174		7.95	9		16.95	22.50
2150	6-1/2"		87	.184		11	9.50		20.50	27
2160	7"		85	.188		11	9.70		20.70	27
2170	7-1/2"		83	.193		11.10	9.95		21.05	27.50
2180	8"		82	.195		11.10	10.05		21.15	28
2190	8-1/2"		80	.200		13.80	10.35		24.15	31
2200	9"		78	.205		13.80	10.60		24.40	31.50
2210	9-1/2"		77	.208		19.15	10.75		29.90	37.50
2220	10"		76	.211		19.15	10.85		30	38
2230	10-1/2"		75	.213		17.90	11		28.90	36.50
2240	11"		74	.216		17.90	11.15		29.05	37
2250	11-1/2"		72	.222		21.50	11.45		32.95	41
2260	12"		71	.225		21.50	11.65		33.15	41.50
2270	13"		68	.235		25.50	12.15		37.65	47
2280	14"		66	.242		31	12.50		43.50	53.50
2290	15"		64	.250		48	12.90		60.90	72.50
2300	16"		63	.254		52	13.10		65.10	77
2310	17"		62	.258		50.50	13.30		63.80	76
2320	18"		61	.262		57.50	13.55		71.05	84.50
2330	19"		60	.267		70.50	13.75		84.25	99
2340	20"		59	.271		68	14		82	96.50
2350	21"		58	.276		73.50	14.25		87.75	103
3000	Tee, 4 piece									
3010	2-1/2"	Q-14	88	.182	Ea.	29.50	9.40		38.90	47
3020	2-3/4"		86	.186		29.50	9.60		39.10	47.50
3030	3"		84	.190		33.50	9.85		43.35	52
3040	3-1/4"		82	.195		33.50	10.05		43.55	52.50
3050	3-1/2"		80	.200		35	10.35		45.35	54.50
3060	4"		78	.205		35.50	10.60		46.10	55.50
3070	4-1/4"		76	.211		37.50	10.85		48.35	58
3080	4-1/2"		74	.216		37.50	11.15		48.65	58.50
3090	4-3/4"		72	.222		37.50	11.45		48.95	58.50
3100	5"		70	.229		38.50	11.80		50.30	60.50
3110	5-1/4"		68	.235		38.50	12.15		50.65	61.50
3120	5-1/2"		66	.242		40.50	12.50		53	64
3130	6"		64	.250		40.50	12.90		53.40	64.50
3140	6-1/2"		60	.267		44.50	13.75		58.25	70.50
3150	7"		58	.276		44.50	14.25		58.75	71
3160	7-1/2"		56	.286		49.50	14.75		64.25	77.50
3170	8"		54	.296		49.50	15.30		64.80	78
3180	8-1/2"		52	.308		50.50	15.90		66.40	80
3190	9"		50	.320		50.50	16.50		67	81
3200	9-1/2"		49	.327		37.50	16.85		54.35	67.50
3210	10"		48	.333		37.50	17.20		54.70	68
3220	10-1/2"		47	.340		40	17.60		57.60	71
3230	11"		46	.348		40	17.95		57.95	71.50
3240	11-1/2"		45	.356		42	18.35		60.35	75
3250	12"		44	.364		42	18.75		60.75	75.50
3260	13"		43	.372		44.50	19.20		63.70	78
3270	14"		42	.381		46.50	19.65		66.15	81.50
3280	15"		41	.390		50.50	20		70.50	86.50
3290	16"		40	.400		51.50	20.50		72	88.50

22 07 19 – Plumbing Piping Insulation

22 07 19.40 Pipe Insulation Protective Jacketing, Aluminum		Crew	Daily Output	Labor-Hours	Unit	Material	2019 Bare Costs Labor	Equipment	Total	Total Incl O&P
3300	17"	Q-14	39	.410	Ea.	59	21		80	97.50
3310	18"		38	.421		61.50	21.50		83	101
3320	19"		37	.432		68	22.50		90.50	110
3330	20"		36	.444		69.50	23		92.50	112
3340	22"		35	.457		88.50	23.50		112	134
3350	23"		34	.471		91	24.50		115.50	138
3360	24"		31	.516		93.50	26.50		120	144

22 07 19.50 Pipe Insulation Protective Jacketing, St. Stl.		Crew	Daily Output	Labor-Hours	Unit	Material	Labor	Equipment	Total	Total Incl O&P
0010	**PIPE INSULATION PROTECTIVE JACKETING, STAINLESS STEEL**									
0100	Metal roll jacketing									
0120	Type 304 with moisture barrier									
0140	Smooth, based on OD of insulation, .010" thick									
0260	2-1/2" ID	Q-14	250	.064	L.F.	3.10	3.30		6.40	8.50
0270	2-3/4" ID		245	.065		3.37	3.37		6.74	8.90
0280	3" ID		240	.067		3.65	3.44		7.09	9.30
0290	3-1/4" ID		235	.068		3.93	3.52		7.45	9.75
0300	3-1/2" ID		230	.070		4.20	3.59		7.79	10.15
0310	3-3/4" ID		225	.071		4.48	3.67		8.15	10.60
0320	4" ID		220	.073		4.75	3.76		8.51	11.05
0330	4-1/4" ID		215	.074		5	3.84		8.84	11.45
0340	4-1/2" ID		210	.076		5.30	3.93		9.23	11.90
0350	5" ID		200	.080		5.85	4.13		9.98	12.85
0360	5-1/2" ID		190	.084		6.40	4.35		10.75	13.75
0370	6" ID		180	.089		6.95	4.59		11.54	14.75
0380	6-1/2" ID		175	.091		7.50	4.72		12.22	15.55
0390	7" ID		170	.094		8.05	4.86		12.91	16.35
0400	7-1/2" ID		164	.098		8.60	5.05		13.65	17.25
0410	8" ID		161	.099		9.15	5.15		14.30	17.95
0420	8-1/2" ID		158	.101		9.70	5.25		14.95	18.70
0430	9" ID		155	.103		10.25	5.35		15.60	19.55
0440	9-1/2" ID		152	.105		10.80	5.45		16.25	20.50
0450	10" ID		149	.107		11.35	5.55		16.90	21
0460	10-1/2" ID		146	.110		11.90	5.65		17.55	22
0470	11" ID		143	.112		12.45	5.80		18.25	22.50
0480	12" ID		137	.117		13.55	6.05		19.60	24
0490	13" ID		132	.121		14.65	6.25		20.90	26
0500	14" ID		130	.123		15.75	6.35		22.10	27
0700	For smooth .016" thick, add					45%	33%			
1000	Stainless steel, Type 316, fitting covers									
1010	Size is based on OD of insulation									
1020	90° LR elbow, 2 piece									
1100	1-1/2"	Q-14	126	.127	Ea.	12.80	6.55		19.35	24
1110	2-3/4"		123	.130		12.80	6.70		19.50	24.50
1120	3"		120	.133		13.40	6.90		20.30	25.50
1130	3-1/4"		117	.137		13.40	7.05		20.45	25.50
1140	3-1/2"		115	.139		14.05	7.20		21.25	26.50
1150	3-3/4"		113	.142		14.85	7.30		22.15	27.50
1160	4"		110	.145		16.35	7.50		23.85	29.50
1170	4-1/4"		108	.148		21.50	7.65		29.15	36
1180	4-1/2"		106	.151		21.50	7.80		29.30	36
1190	5"		102	.157		22	8.10		30.10	37
1200	5-1/2"		97	.165		33	8.50		41.50	49.50

22 07 19.50 Pipe Insulation Protective Jacketing, St. Stl.		Crew	Daily Output	Labor-Hours	Unit	2019 Bare Costs Material	Labor	Equipment	Total	Total Incl O&P
1210	6″	Q-14	92	.174	Ea.	36.50	9		45.50	54
1220	6-1/2″		87	.184		50	9.50		59.50	69.50
1230	7″		85	.188		50	9.70		59.70	70
1240	7-1/2″		83	.193		58.50	9.95		68.45	80
1250	8″		80	.200		58.50	10.35		68.85	80.50
1260	8-1/2″		80	.200		60.50	10.35		70.85	82.50
1270	9″		78	.205		92	10.60		102.60	117
1280	9-1/2″		77	.208		90.50	10.75		101.25	116
1290	10″		76	.211		90.50	10.85		101.35	116
1300	10-1/2″		75	.213		108	11		119	136
1310	11″		74	.216		103	11.15		114.15	130
1320	12″		71	.225		117	11.65		128.65	147
1330	13″		68	.235		162	12.15		174.15	197
1340	14″		66	.242		163	12.50		175.50	199
2000	45° elbow, 2 piece									
2010	2-1/2″	Q-14	126	.127	Ea.	10.70	6.55		17.25	22
2020	2-3/4″		123	.130		10.70	6.70		17.40	22
2030	3″		120	.133		11.50	6.90		18.40	23.50
2040	3-1/4″		117	.137		11.50	7.05		18.55	23.50
2050	3-1/2″		115	.139		11.65	7.20		18.85	24
2060	3-3/4″		113	.142		11.65	7.30		18.95	24
2070	4″		110	.145		14.95	7.50		22.45	28
2080	4-1/4″		108	.148		21	7.65		28.65	35
2090	4-1/2″		106	.151		21	7.80		28.80	35
2100	4-3/4″		104	.154		21	7.95		28.95	35.50
2110	5″		102	.157		21.50	8.10		29.60	36
2120	5-1/2″		97	.165		21.50	8.50		30	36.50
2130	6″		92	.174		25.50	9		34.50	42
2140	6-1/2″		87	.184		25.50	9.50		35	42.50
2150	7″		85	.188		43	9.70		52.70	62
2160	7-1/2″		83	.193		43.50	9.95		53.45	63.50
2170	8″		82	.195		43.50	10.05		53.55	63.50
2180	8-1/2″		80	.200		51	10.35		61.35	72.50
2190	9″		78	.205		51	10.60		61.60	73
2200	9-1/2″		77	.208		60	10.75		70.75	82.50
2210	10″		76	.211		60	10.85		70.85	83
2220	10-1/2″		75	.213		72	11		83	96.50
2230	11″		74	.216		72	11.15		83.15	97
2240	12″		71	.225		78.50	11.65		90.15	104
2250	13″		68	.235		92	12.15		104.15	120

22 11 13.14 Pipe, Brass

22 11 13.14 Pipe, Brass		Crew	Daily Output	Labor-Hours	Unit	2019 Bare Costs Material	Labor	Equipment	Total	Total Incl O&P
0010	**PIPE, BRASS**, Plain end									
0900	Field threaded, coupling & clevis hanger assembly 10′ OC									
0920	Regular weight									
1120	1/2″ diameter	1 Plum	48	.167	L.F.	7.80	10.55		18.35	24.50
1140	3/4″ diameter		46	.174		10.25	11		21.25	28
1160	1″ diameter		43	.186		7.20	11.75		18.95	25.50
1180	1-1/4″ diameter	Q-1	72	.222		23.50	12.65		36.15	45
1200	1-1/2″ diameter		65	.246		28	14		42	51.50
1220	2″ diameter		53	.302		19.25	17.15		36.40	47
1240	2-1/2″ diameter		41	.390		31	22		53	67.50
1260	3″ diameter		31	.516		42	29.50		71.50	90.50
1300	4″ diameter	Q-2	37	.649		149	38		187	222
1930	To delete coupling & hanger, subtract									
1940	1/2″ diam.					40%	46%			
1950	3/4″ diam. to 1-1/2″ diam.					39%	39%			
1960	2″ diam. to 4″ diam.					48%	35%			

22 11 13.16 Pipe Fittings, Brass

22 11 13.16 Pipe Fittings, Brass		Crew	Daily Output	Labor-Hours	Unit	Material	Labor	Equipment	Total	Total Incl O&P
0010	**PIPE FITTINGS, BRASS**, Rough bronze, threaded, lead free.									
1000	Standard wt., 90° elbow									
1040	1/8″	1 Plum	13	.615	Ea.	23	39		62	84
1060	1/4″		13	.615		23	39		62	84
1080	3/8″		13	.615		23	39		62	84
1100	1/2″		12	.667		23	42		65	88.50
1120	3/4″		11	.727		31	46		77	103
1140	1″		10	.800		50	50.50		100.50	131
1160	1-1/4″	Q-1	17	.941		81	53.50		134.50	170
1180	1-1/2″		16	1		100	57		157	196
1200	2″		14	1.143		162	65		227	276
1220	2-1/2″		11	1.455		390	82.50		472.50	555
1240	3″		8	2		600	114		714	830
1260	4″	Q-2	11	2.182		1,200	129		1,329	1,525
1280	5″		8	3		2,700	177		2,877	3,225
1300	6″		7	3.429		4,950	202		5,152	5,725
1500	45° elbow, 1/8″	1 Plum	13	.615		28.50	39		67.50	89.50
1540	1/4″		13	.615		28.50	39		67.50	89.50
1560	3/8″		13	.615		28.50	39		67.50	89.50
1580	1/2″		12	.667		28.50	42		70.50	94
1600	3/4″		11	.727		40	46		86	113
1620	1″		10	.800		68.50	50.50		119	152
1640	1-1/4″	Q-1	17	.941		109	53.50		162.50	201
1660	1-1/2″		16	1		137	57		194	237
1680	2″		14	1.143		222	65		287	340
1700	2-1/2″		11	1.455		425	82.50		507.50	590
1720	3″		8	2		525	114		639	750
1740	4″	Q-2	11	2.182		1,175	129		1,304	1,475
1760	5″		8	3		2,200	177		2,377	2,675
1780	6″		7	3.429		3,025	202		3,227	3,625
2000	Tee, 1/8″	1 Plum	9	.889		27	56		83	114
2040	1/4″		9	.889		27	56		83	114
2060	3/8″		9	.889		27	56		83	114
2080	1/2″		8	1		27	63		90	125
2100	3/4″		7	1.143		38.50	72		110.50	151

22 11 13.16 Pipe Fittings, Brass

22 11 13.16 Pipe Fittings, Brass		Crew	Daily Output	Labor-Hours	Unit	Material	Labor	Equipment	Total	Total Incl O&P
						2019 Bare Costs				
2120	1"	1 Plum	6	1.333	Ea.	69.50	84		153.50	203
2140	1-1/4"	Q-1	10	1.600		120	91		211	267
2160	1-1/2"		9	1.778		135	101		236	300
2180	2"		8	2		224	114		338	415
2200	2-1/2"		7	2.286		535	130		665	780
2220	3"		5	3.200		815	182		997	1,175
2240	4"	Q-2	7	3.429		2,025	202		2,227	2,525
2260	5"		5	4.800		3,225	283		3,508	3,975
2280	6"		4	6		5,300	355		5,655	6,350
2500	Coupling, 1/8"	1 Plum	26	.308		19.30	19.45		38.75	50
2540	1/4"		22	.364		19.30	23		42.30	55.50
2560	3/8"		18	.444		19.30	28		47.30	63
2580	1/2"		15	.533		20	33.50		53.50	72.50
2600	3/4"		14	.571		27	36		63	84
2620	1"		13	.615		46	39		85	110
2640	1-1/4"	Q-1	22	.727		77	41.50		118.50	147
2660	1-1/2"		20	.800		100	45.50		145.50	178
2680	2"		18	.889		166	50.50		216.50	258
2700	2-1/2"		14	1.143		286	65		351	415
2720	3"		10	1.600		395	91		486	570
2740	4"	Q-2	12	2		790	118		908	1,050
2760	5"		10	2.400		1,450	141		1,591	1,800
2780	6"		9	2.667		2,075	157		2,232	2,500
3000	Union, 125 lb.									
3020	1/8"	1 Plum	12	.667	Ea.	42	42		84	109
3040	1/4"		12	.667		42	42		84	109
3060	3/8"		12	.667		42	42		84	109
3080	1/2"		11	.727		42	46		88	115
3100	3/4"		10	.800		57.50	50.50		108	140
3120	1"		9	.889		87	56		143	180
3140	1-1/4"	Q-1	16	1		126	57		183	224
3160	1-1/2"		15	1.067		150	60.50		210.50	256
3180	2"		13	1.231		202	70		272	325
3200	2-1/2"		10	1.600		700	91		791	905
3220	3"		7	2.286		945	130		1,075	1,225
3240	4"	Q-2	10	2.400		2,525	141		2,666	2,975
3320	For 250 lb. (navy pattern), add					100%				

22 11 13.23 Pipe/Tube, Copper

22 11 13.23 Pipe/Tube, Copper		Crew	Daily Output	Labor-Hours	Unit	Material	Labor	Equipment	Total	Total Incl O&P
0010	**PIPE/TUBE, COPPER**, Solder joints R221113-50									
0100	Solder									
0120	Solder, lead free, roll				Lb.	13.65			13.65	15
1000	Type K tubing, couplings & clevis hanger assemblies 10' OC									
1100	1/4" diameter	1 Plum	84	.095	L.F.	4.12	6		10.12	13.55
1120	3/8" diameter		82	.098		4.69	6.15		10.84	14.40
1140	1/2" diameter		78	.103		5.25	6.50		11.75	15.45
1160	5/8" diameter		77	.104		6.65	6.55		13.20	17.20
1180	3/4" diameter		74	.108		8.55	6.85		15.40	19.65
1200	1" diameter		66	.121		12.70	7.65		20.35	25.50
1220	1-1/4" diameter		56	.143		15.20	9		24.20	30.50
1240	1-1/2" diameter		50	.160		18.15	10.10		28.25	35
1260	2" diameter		40	.200		27	12.65		39.65	48.50
1280	2-1/2" diameter	Q-1	60	.267		41	15.15		56.15	67.50

22 11 13.23 Pipe/Tube, Copper		Crew	Daily Output	Labor-Hours	Unit	2019 Bare Costs Material	Labor	Equipment	Total	Total Incl O&P
1300	3″ diameter	Q-1	54	.296	L.F.	55	16.85		71.85	86
1320	3-1/2″ diameter		42	.381		75	21.50		96.50	115
1330	4″ diameter		38	.421		96	24		120	142
1340	5″ diameter		32	.500		139	28.50		167.50	196
1360	6″ diameter	Q-2	38	.632		200	37		237	276
1380	8″ diameter	"	34	.706		350	41.50		391.50	450
1390	For other than full hard temper, add					13%				
1440	For silver solder, add						15%			
1800	For medical clean (oxygen class), add					12%				
1950	To delete cplgs. & hngrs., 1/4″-1″ pipe, subtract					27%	60%			
1960	1-1/4″-3″ pipe, subtract					14%	52%			
1970	3-1/2″-5″ pipe, subtract					10%	60%			
1980	6″-8″ pipe, subtract					19%	53%			
2000	Type L tubing, couplings & clevis hanger assemblies 10′ OC									
2100	1/4″ diameter	1 Plum	88	.091	L.F.	2.53	5.75		8.28	11.40
2120	3/8″ diameter		84	.095		3.16	6		9.16	12.45
2140	1/2″ diameter		81	.099		3.38	6.25		9.63	13.05
2160	5/8″ diameter		79	.101		5.55	6.40		11.95	15.75
2180	3/4″ diameter		76	.105		4.39	6.65		11.04	14.75
2200	1″ diameter		68	.118		7	7.45		14.45	18.90
2220	1-1/4″ diameter		58	.138		11.65	8.70		20.35	26
2240	1-1/2″ diameter		52	.154		10.45	9.70		20.15	26
2260	2″ diameter		42	.190		18.05	12.05		30.10	38
2280	2-1/2″ diameter	Q-1	62	.258		25.50	14.65		40.15	50
2300	3″ diameter		56	.286		43.50	16.25		59.75	72
2320	3-1/2″ diameter		43	.372		54	21		75	90.50
2340	4″ diameter		39	.410		62	23.50		85.50	103
2360	5″ diameter		34	.471		151	26.50		177.50	206
2380	6″ diameter	Q-2	40	.600		145	35.50		180.50	213
2400	8″ diameter	"	36	.667		242	39.50		281.50	325
2410	For other than full hard temper, add					21%				
2590	For silver solder, add						15%			
2900	For medical clean (oxygen class), add					12%				
2940	To delete cplgs. & hngrs., 1/4″-1″ pipe, subtract					37%	63%			
2960	1-1/4″-3″ pipe, subtract					12%	53%			
2970	3-1/2″-5″ pipe, subtract					12%	63%			
2980	6″-8″ pipe, subtract					24%	55%			
3000	Type M tubing, couplings & clevis hanger assemblies 10′ OC									
3100	1/4″ diameter	1 Plum	90	.089	L.F.	3.46	5.60		9.06	12.20
3120	3/8″ diameter		87	.092		3.79	5.80		9.59	12.85
3140	1/2″ diameter		84	.095		3.64	6		9.64	13
3160	5/8″ diameter		81	.099		4.81	6.25		11.06	14.65
3180	3/4″ diameter		78	.103		5.20	6.50		11.70	15.40
3200	1″ diameter		70	.114		8.40	7.20		15.60	20
3220	1-1/4″ diameter		60	.133		11.10	8.40		19.50	25
3240	1-1/2″ diameter		54	.148		14.20	9.35		23.55	29.50
3260	2″ diameter		44	.182		21	11.50		32.50	41
3280	2-1/2″ diameter	Q-1	64	.250		31.50	14.20		45.70	56
3300	3″ diameter		58	.276		39	15.70		54.70	66.50
3320	3-1/2″ diameter		45	.356		56	20		76	92
3340	4″ diameter		40	.400		74	22.50		96.50	115
3360	5″ diameter		36	.444		141	25.50		166.50	193
3370	6″ diameter	Q-2	42	.571		195	33.50		228.50	265

22 11 13 – Facility Water Distribution Piping

22 11 13.23 Pipe/Tube, Copper		Crew	Daily Output	Labor-Hours	Unit	2019 Bare Costs Material	Labor	Equipment	Total	Total Incl O&P
3380	8″ diameter	Q-2	38	.632	L.F.	325	37		362	410
3440	For silver solder, add						15%			
3960	To delete cplgs. & hngrs., 1/4″-1″ pipe, subtract					35%	65%			
3970	1-1/4″-3″ pipe, subtract					19%	56%			
3980	3-1/2″-5″ pipe, subtract					13%	65%			
3990	6″-8″ pipe, subtract					28%	58%			
4000	Type DWV tubing, couplings & clevis hanger assemblies 10′ OC									
4100	1-1/4″ diameter	1 Plum	60	.133	L.F.	12.45	8.40		20.85	26.50
4120	1-1/2″ diameter		54	.148		12	9.35		21.35	27.50
4140	2″ diameter		44	.182		18.55	11.50		30.05	38
4160	3″ diameter	Q-1	58	.276		24.50	15.70		40.20	50
4180	4″ diameter		40	.400		55.50	22.50		78	95
4200	5″ diameter		36	.444		103	25.50		128.50	151
4220	6″ diameter	Q-2	42	.571		148	33.50		181.50	214
4240	8″ diameter	″	38	.632		460	37		497	565
4730	To delete cplgs. & hngrs., 1-1/4″-2″ pipe, subtract					16%	53%			
4740	3″-4″ pipe, subtract					13%	60%			
4750	5″-8″ pipe, subtract					23%	58%			
5200	ACR tubing, type L, hard temper, cleaned and									
5220	capped, no couplings or hangers									
5240	3/8″ OD				L.F.	1.69			1.69	1.86
5250	1/2″ OD					2.53			2.53	2.78
5260	5/8″ OD					3.17			3.17	3.49
5270	3/4″ OD					4.32			4.32	4.75
5280	7/8″ OD					4.97			4.97	5.45
5290	1-1/8″ OD					7			7	7.70
5300	1-3/8″ OD					9.35			9.35	10.30
5310	1-5/8″ OD					12.40			12.40	13.65
5320	2-1/8″ OD					19.60			19.60	21.50
5330	2-5/8″ OD					27.50			27.50	30
5340	3-1/8″ OD					33			33	36.50
5350	3-5/8″ OD					57			57	63
5360	4-1/8″ OD					65			65	71.50
5380	ACR tubing, type L, hard, cleaned and capped									
5381	No couplings or hangers									
5384	3/8″	1 Stpi	160	.050	L.F.	1.69	3.20		4.89	6.65
5385	1/2″		160	.050		2.53	3.20		5.73	7.60
5386	5/8″		160	.050		3.17	3.20		6.37	8.30
5387	3/4″		130	.062		4.32	3.94		8.26	10.65
5388	7/8″		130	.062		4.97	3.94		8.91	11.35
5389	1-1/8″		115	.070		7	4.45		11.45	14.40
5390	1-3/8″		100	.080		9.35	5.10		14.45	18
5391	1-5/8″		90	.089		12.40	5.70		18.10	22
5392	2-1/8″		80	.100		19.60	6.40		26	31
5393	2-5/8″	Q-5	125	.128		27.50	7.35		34.85	41
5394	3-1/8″		105	.152		33	8.75		41.75	49.50
5395	4-1/8″		95	.168		65	9.70		74.70	86
5800	Refrigeration tubing, dryseal, 50′ coils									
5840	1/8″ OD				Coil	33.50			33.50	37
5850	3/16″ OD					41.50			41.50	46
5860	1/4″ OD					45			45	49.50
5870	5/16″ OD					64.50			64.50	71
5880	3/8″ OD					64			64	70.50

22 11 13.23 Pipe/Tube, Copper		Crew	Daily Output	Labor-Hours	Unit	Material	2019 Bare Costs Labor	Equipment	Total	Total Incl O&P
5890	1/2″ OD				Coil	91			91	100
5900	5/8″ OD					122			122	134
5910	3/4″ OD					142			142	156
5920	7/8″ OD					142			142	156
5930	1-1/8″ OD					315			315	345
5940	1-3/8″ OD					545			545	600
5950	1-5/8″ OD					690			690	760
9400	Sub assemblies used in assembly systems									
9410	Chilled water unit, coil connections per unit under 10 ton	Q-5	.80	20	System	1,425	1,150		2,575	3,275
9420	Chilled water unit, coil connections per unit 10 ton and up		1	16		2,200	920		3,120	3,800
9430	Chilled water dist. piping per ton, less than 61 ton systems		26	.615		22.50	35.50		58	78
9440	Chilled water dist. piping per ton, 61 through 120 ton systems	Q-6	31	.774		49.50	46		95.50	124
9450	Chilled water dist. piping/ton, 135 ton systems and up	Q-8	25.40	1.260		65.50	76.50	2.20	144.20	189
9510	Refrigerant piping/ton of cooling for remote condensers	Q-5	2	8		360	460		820	1,075
9520	Refrigerant piping per ton up to 10 ton w/remote condensing unit		2.40	6.667		191	385		576	785
9530	Refrigerant piping per ton, 20 ton w/remote condensing unit		2	8		222	460		682	935
9540	Refrigerant piping per ton, 40 ton w/remote condensing unit		1.90	8.421		315	485		800	1,075
9550	Refrigerant piping per ton, 75-80 ton w/remote condensing unit	Q-6	2.40	10		455	595		1,050	1,400
9560	Refrigerant piping per ton, 100 ton w/remote condensing unit	″	2.20	10.909		660	650		1,310	1,700

22 11 13.25 Pipe/Tube Fittings, Copper		Crew	Daily Output	Labor-Hours	Unit	Material	Labor	Equipment	Total	Total Incl O&P
0010	**PIPE/TUBE FITTINGS, COPPER**, Wrought unless otherwise noted									
0020	For silver solder, add						15%			
0040	Solder joints, copper x copper									
0070	90° elbow, 1/4″	1 Plum	22	.364	Ea.	3.68	23		26.68	38.50
0090	3/8″		22	.364		3.93	23		26.93	39
0100	1/2″		20	.400		1.20	25.50		26.70	39.50
0110	5/8″		19	.421		2.53	26.50		29.03	43
0120	3/4″		19	.421		2.40	26.50		28.90	42.50
0130	1″		16	.500		6.55	31.50		38.05	54.50
0140	1-1/4″		15	.533		11.05	33.50		44.55	62.50
0150	1-1/2″		13	.615		15.35	39		54.35	75.50
0160	2″		11	.727		28	46		74	99.50
0170	2-1/2″	Q-1	13	1.231		61.50	70		131.50	173
0180	3″		11	1.455		82.50	82.50		165	215
0190	3-1/2″		10	1.600		315	91		406	485
0200	4″		9	1.778		215	101		316	390
0210	5″		6	2.667		845	152		997	1,150
0220	6″	Q-2	9	2.667		1,125	157		1,282	1,475
0230	8″	″	8	3		4,175	177		4,352	4,850
0250	45° elbow, 1/4″	1 Plum	22	.364		8.35	23		31.35	43.50
0270	3/8″		22	.364		8	23		31	43.50
0280	1/2″		20	.400		3.02	25.50		28.52	41.50
0290	5/8″		19	.421		11.45	26.50		37.95	52.50
0300	3/4″		19	.421		4.15	26.50		30.65	44.50
0310	1″		16	.500		10.40	31.50		41.90	59
0320	1-1/4″		15	.533		18.60	33.50		52.10	71
0330	1-1/2″		13	.615		16	39		55	76
0340	2″		11	.727		27	46		73	98.50
0350	2-1/2″	Q-1	13	1.231		56.50	70		126.50	168
0360	3″		13	1.231		99.50	70		169.50	214
0370	3-1/2″		10	1.600		135	91		226	285
0380	4″		9	1.778		166	101		267	335

22 11 13.25 Pipe/Tube Fittings, Copper		Crew	Daily Output	Labor-Hours	Unit	Material	Labor	Equipment	Total	Total Incl O&P
						2019 Bare Costs				
0390	5"	Q-1	6	2.667	Ea.	625	152		777	915
0400	6"	Q-2	9	2.667		955	157		1,112	1,275
0410	8"	"	8	3		4,025	177		4,202	4,700
0450	Tee, 1/4"	1 Plum	14	.571		8	36		44	63
0470	3/8"		14	.571		7.10	36		43.10	62
0480	1/2"		13	.615		2.27	39		41.27	61
0490	5/8"		12	.667		18.90	42		60.90	84
0500	3/4"		12	.667		6.05	42		48.05	69.50
0510	1"		10	.800		16.20	50.50		66.70	94
0520	1-1/4"		9	.889		23.50	56		79.50	110
0530	1-1/2"		8	1		31.50	63		94.50	130
0540	2"		7	1.143		51	72		123	164
0550	2-1/2"	Q-1	8	2		115	114		229	297
0560	3"		7	2.286		160	130		290	370
0570	3-1/2"		6	2.667		535	152		687	815
0580	4"		5	3.200		350	182		532	660
0590	5"		4	4		900	227		1,127	1,325
0600	6"	Q-2	6	4		1,250	236		1,486	1,725
0610	8"	"	5	4.800		5,450	283		5,733	6,425
0612	Tee, reducing on the outlet, 1/4"	1 Plum	15	.533		15.60	33.50		49.10	67.50
0613	3/8"		15	.533		14.90	33.50		48.40	67
0614	1/2"		14	.571		13.95	36		49.95	69.50
0615	5/8"		13	.615		28	39		67	89.50
0616	3/4"		12	.667		9	42		51	73
0617	1"		11	.727		25	46		71	96.50
0618	1-1/4"		10	.800		30	50.50		80.50	109
0619	1-1/2"		9	.889		30.50	56		86.50	118
0620	2"		8	1		53	63		116	154
0621	2-1/2"	Q-1	9	1.778		129	101		230	294
0622	3"		8	2		149	114		263	335
0623	4"		6	2.667		287	152		439	540
0624	5"		5	3.200		1,775	182		1,957	2,225
0625	6"	Q-2	7	3.429		2,075	202		2,277	2,600
0626	8"	"	6	4		9,325	236		9,561	10,700
0630	Tee, reducing on the run, 1/4"	1 Plum	15	.533		22	33.50		55.50	75
0631	3/8"		15	.533		31	33.50		64.50	84.50
0632	1/2"		14	.571		19.20	36		55.20	75
0633	5/8"		13	.615		28	39		67	89.50
0634	3/4"		12	.667		21	42		63	86
0635	1"		11	.727		24.50	46		70.50	96
0636	1-1/4"		10	.800		38.50	50.50		89	119
0637	1-1/2"		9	.889		68	56		124	159
0638	2"		8	1		78.50	63		141.50	182
0639	2-1/2"	Q-1	9	1.778		180	101		281	350
0640	3"		8	2		262	114		376	460
0641	4"		6	2.667		555	152		707	835
0642	5"		5	3.200		1,675	182		1,857	2,125
0643	6"	Q-2	7	3.429		2,550	202		2,752	3,100
0644	8"	"	6	4		9,575	236		9,811	10,900
0650	Coupling, 1/4"	1 Plum	24	.333		1.01	21		22.01	32.50
0670	3/8"		24	.333		1.45	21		22.45	33
0680	1/2"		22	.364		.92	23		23.92	35.50
0690	5/8"		21	.381		3.94	24		27.94	40.50

22 11 13.25	Pipe/Tube Fittings, Copper	Crew	Daily Output	Labor-Hours	Unit	2019 Bare Costs Material	Labor	Equipment	Total	Total Incl O&P
0700	3/4″	1 Plum	21	.381	Ea.	2.49	24		26.49	38.50
0710	1″		18	.444		4.94	28		32.94	47.50
0715	1-1/4″		17	.471		7	29.50		36.50	52
0716	1-1/2″		15	.533		8.70	33.50		42.20	60
0718	2″		13	.615		14.50	39		53.50	74.50
0721	2-1/2″	Q-1	15	1.067		38.50	60.50		99	133
0722	3″		13	1.231		45	70		115	155
0724	3-1/2″		8	2		104	114		218	285
0726	4″		7	2.286		89.50	130		219.50	294
0728	5″		6	2.667		215	152		367	465
0731	6″	Q-2	8	3		335	177		512	635
0732	8″	″	7	3.429		1,200	202		1,402	1,600
0741	Coupling, reducing, concentric									
0743	1/2″	1 Plum	23	.348	Ea.	2.71	22		24.71	36
0745	3/4″		21.50	.372		5.40	23.50		28.90	41.50
0747	1″		19.50	.410		8	26		34	48
0748	1-1/4″		18	.444		11.65	28		39.65	55
0749	1-1/2″		16	.500		15.70	31.50		47.20	65
0751	2″		14	.571		28	36		64	85
0752	2-1/2″		13	.615		53	39		92	117
0753	3″	Q-1	14	1.143		50.50	65		115.50	153
0755	4″	″	8	2		134	114		248	320
0757	5″	Q-2	7.50	3.200		725	189		914	1,075
0759	6″		7	3.429		1,150	202		1,352	1,550
0761	8″		6.50	3.692		3,200	218		3,418	3,850
0771	Cap, sweat									
0773	1/2″	1 Plum	40	.200	Ea.	1.20	12.65		13.85	20.50
0775	3/4″		38	.211		2.23	13.30		15.53	22.50
0777	1″		32	.250		5.20	15.80		21	29.50
0778	1-1/4″		29	.276		6.65	17.40		24.05	33.50
0779	1-1/2″		26	.308		11.25	19.45		30.70	41.50
0781	2″		22	.364		18.60	23		41.60	55
0791	Flange, sweat									
0793	3″	Q-1	22	.727	Ea.	252	41.50		293.50	340
0795	4″		18	.889		350	50.50		400.50	460
0797	5″		12	1.333		660	76		736	840
0799	6″	Q-2	18	1.333		690	78.50		768.50	880
0801	8″	″	16	1.500		1,200	88.50		1,288.50	1,425
0850	Unions, 1/4″	1 Plum	21	.381		38	24		62	78
0870	3/8″		21	.381		38.50	24		62.50	78
0880	1/2″		19	.421		20.50	26.50		47	62.50
0890	5/8″		18	.444		78	28		106	128
0900	3/4″		18	.444		25.50	28		53.50	70
0910	1″		15	.533		44	33.50		77.50	99
0920	1-1/4″		14	.571		72	36		108	134
0930	1-1/2″		12	.667		99.50	42		141.50	172
0940	2″		10	.800		162	50.50		212.50	254
0950	2-1/2″	Q-1	12	1.333		355	76		431	505
0960	3″	″	10	1.600		920	91		1,011	1,125
0980	Adapter, copper x male IPS, 1/4″	1 Plum	20	.400		12.40	25.50		37.90	51.50
0990	3/8″		20	.400		8	25.50		33.50	47
1000	1/2″		18	.444		3.44	28		31.44	46
1010	3/4″		17	.471		5.75	29.50		35.25	51

22 11 13.25 Pipe/Tube Fittings, Copper		Crew	Daily Output	Labor-Hours	Unit	Material	Labor	Equipment	Total	Total Incl O&P
						2019 Bare Costs				
1020	1"	1 Plum	15	.533	Ea.	14.70	33.50		48.20	66.50
1030	1-1/4"		13	.615		23	39		62	83.50
1040	1-1/2"		12	.667		25	42		67	90.50
1050	2"		11	.727		42	46		88	116
1060	2-1/2"	Q-1	10.50	1.524		152	86.50		238.50	297
1070	3"		10	1.600		189	91		280	345
1080	3-1/2"		9	1.778		199	101		300	370
1090	4"		8	2		230	114		344	425
1200	5", cast		6	2.667		1,975	152		2,127	2,400
1210	6", cast	Q-2	8.50	2.824		2,225	166		2,391	2,675
1214	Adapter, copper x female IPS									
1216	1/2"	1 Plum	18	.444	Ea.	5.45	28		33.45	48
1218	3/4"		17	.471		7.45	29.50		36.95	52.50
1220	1"		15	.533		17.15	33.50		50.65	69.50
1221	1-1/4"		13	.615		25	39		64	86
1222	1-1/2"		12	.667		39	42		81	106
1224	2"		11	.727		53	46		99	127
1250	Cross, 1/2"		10	.800		26	50.50		76.50	105
1260	3/4"		9.50	.842		51	53		104	136
1270	1"		8	1		86.50	63		149.50	190
1280	1-1/4"		7.50	1.067		124	67.50		191.50	237
1290	1-1/2"		6.50	1.231		177	77.50		254.50	310
1300	2"		5.50	1.455		335	92		427	505
1310	2-1/2"	Q-1	6.50	2.462		770	140		910	1,050
1320	3"	"	5.50	2.909		740	165		905	1,075
1500	Tee fitting, mechanically formed (Type 1, 'branch sizes up to 2 in.')									
1520	1/2" run size, 3/8" to 1/2" branch size	1 Plum	80	.100	Ea.		6.30		6.30	9.50
1530	3/4" run size, 3/8" to 3/4" branch size		60	.133			8.40		8.40	12.65
1540	1" run size, 3/8" to 1" branch size		54	.148			9.35		9.35	14.05
1550	1-1/4" run size, 3/8" to 1-1/4" branch size		48	.167			10.55		10.55	15.80
1560	1-1/2" run size, 3/8" to 1-1/2" branch size		40	.200			12.65		12.65	18.95
1570	2" run size, 3/8" to 2" branch size		35	.229			14.45		14.45	21.50
1580	2-1/2" run size, 1/2" to 2" branch size		32	.250			15.80		15.80	23.50
1590	3" run size, 1" to 2" branch size		26	.308			19.45		19.45	29
1600	4" run size, 1" to 2" branch size		24	.333			21		21	31.50
1640	Tee fitting, mechanically formed (Type 2, branches 2-1/2" thru 4")									
1650	2-1/2" run size, 2-1/2" branch size	1 Plum	12.50	.640	Ea.		40.50		40.50	60.50
1660	3" run size, 2-1/2" to 3" branch size		12	.667			42		42	63
1670	3-1/2" run size, 2-1/2" to 3-1/2" branch size		11	.727			46		46	69
1680	4" run size, 2-1/2" to 4" branch size		10.50	.762			48		48	72
1698	5" run size, 2" to 4" branch size		9.50	.842			53		53	80
1700	6" run size, 2" to 4" branch size		8.50	.941			59.50		59.50	89
1710	8" run size, 2" to 4" branch size		7	1.143			72		72	108
1800	ACR fittings, OD size									
1802	Tee, straight									
1808	5/8"	1 Stpi	12	.667	Ea.	2.80	42.50		45.30	67
1810	3/4"		12	.667		18.90	42.50		61.40	85
1812	7/8"		10	.800		6.75	51		57.75	84.50
1813	1"		10	.800		52	51		103	134
1814	1-1/8"		10	.800		22	51		73	101
1816	1-3/8"		9	.889		28	57		85	117
1818	1-5/8"		8	1		45.50	64		109.50	146
1820	2-1/8"		7	1.143		67	73		140	184

22 11 13.25 Pipe/Tube Fittings, Copper		Crew	Daily Output	Labor-Hours	Unit	Material	2019 Bare Costs Labor	Equipment	Total	Total Incl O&P
1822	2-5/8"	Q-5	8	2	Ea.	128	115		243	315
1824	3-1/8"		7	2.286		184	132		316	400
1826	4-1/8"		5	3.200		395	184		579	710
1830	90° elbow									
1836	5/8"	1 Stpi	19	.421	Ea.	5.75	27		32.75	47
1838	3/4"		19	.421		10.45	27		37.45	52
1840	7/8"		16	.500		10.40	32		42.40	59.50
1842	1-1/8"		16	.500		13.90	32		45.90	63.50
1844	1-3/8"		15	.533		13.45	34		47.45	66
1846	1-5/8"		13	.615		21	39.50		60.50	82
1848	2-1/8"		11	.727		38	46.50		84.50	112
1850	2-5/8"	Q-5	13	1.231		72	71		143	185
1852	3-1/8"		11	1.455		98.50	83.50		182	234
1854	4-1/8"		9	1.778		215	102		317	390
1860	Coupling									
1866	5/8"	1 Stpi	21	.381	Ea.	1.22	24.50		25.72	38
1868	3/4"		21	.381		3.19	24.50		27.69	40
1870	7/8"		18	.444		2.36	28.50		30.86	45
1871	1"		18	.444		8.15	28.50		36.65	51.50
1872	1-1/8"		18	.444		4.97	28.50		33.47	48
1874	1-3/8"		17	.471		8.70	30		38.70	54.50
1876	1-5/8"		15	.533		11.50	34		45.50	63.50
1878	2-1/8"		13	.615		19.15	39.50		58.65	80
1880	2-5/8"	Q-5	15	1.067		37.50	61.50		99	134
1882	3-1/8"		13	1.231		51	71		122	162
1884	4-1/8"		7	2.286		113	132		245	320
2000	DWV, solder joints, copper x copper									
2030	90° elbow, 1-1/4"	1 Plum	13	.615	Ea.	17.50	39		56.50	78
2050	1-1/2"		12	.667		23	42		65	88.50
2070	2"		10	.800		38	50.50		88.50	118
2090	3"	Q-1	10	1.600		85.50	91		176.50	230
2100	4"	"	9	1.778		460	101		561	660
2150	45° elbow, 1-1/4"	1 Plum	13	.615		14.30	39		53.30	74
2170	1-1/2"		12	.667		13.30	42		55.30	77.50
2180	2"		10	.800		27.50	50.50		78	106
2190	3"	Q-1	10	1.600		61	91		152	203
2200	4"	"	9	1.778		97.50	101		198.50	259
2250	Tee, sanitary, 1-1/4"	1 Plum	9	.889		27	56		83	114
2270	1-1/2"		8	1		33.50	63		96.50	132
2290	2"		7	1.143		52.50	72		124.50	166
2310	3"	Q-1	7	2.286		200	130		330	415
2330	4"	"	6	2.667		485	152		637	760
2400	Coupling, 1-1/4"	1 Plum	14	.571		7.30	36		43.30	62
2420	1-1/2"		13	.615		9.05	39		48.05	68.50
2440	2"		11	.727		12.60	46		58.60	83
2460	3"	Q-1	11	1.455		29	82.50		111.50	156
2480	4"	"	10	1.600		64	91		155	207
2602	Traps, see Section 22 13 16.60									
3500	Compression joint fittings									
3510	As used for plumbing and oil burner work									
3520	Fitting price includes nuts and sleeves									
3540	Sleeve, 1/8"				Ea.	.15			.15	.17
3550	3/16"					.15			.15	.17

22 11 13.25	Pipe/Tube Fittings, Copper	Crew	Daily Output	Labor-Hours	Unit	Material	Labor (2019 Bare Costs)	Equipment	Total	Total Incl O&P
3560	1/4″				Ea.	.05			.05	.06
3570	5/16″					.15			.15	.17
3580	3/8″					.30			.30	.33
3600	1/2″					.40			.40	.44
3620	Nut, 1/8″					.20			.20	.22
3630	3/16″					.23			.23	.25
3640	1/4″					.21			.21	.23
3650	5/16″					.31			.31	.34
3660	3/8″					.36			.36	.40
3670	1/2″					.63			.63	.69
3710	Union, 1/8″	1 Plum	26	.308		1.28	19.45		20.73	30.50
3720	3/16″		24	.333		1.22	21		22.22	33
3730	1/4″		24	.333		1.73	21		22.73	33.50
3740	5/16″		23	.348		2.22	22		24.22	35.50
3750	3/8″		22	.364		1.98	23		24.98	36.50
3760	1/2″		22	.364		3.13	23		26.13	38
3780	5/8″		21	.381		2.92	24		26.92	39
3820	Union tee, 1/8″		17	.471		4.87	29.50		34.37	50
3830	3/16″		16	.500		2.81	31.50		34.31	50.50
3840	1/4″		15	.533		3.43	33.50		36.93	54.50
3850	5/16″		15	.533		3.58	33.50		37.08	54.50
3860	3/8″		15	.533		3.58	33.50		37.08	54.50
3870	1/2″		15	.533		7.65	33.50		41.15	59
3910	Union elbow, 1/4″		24	.333		1.89	21		22.89	33.50
3920	5/16″		23	.348		2.77	22		24.77	36
3930	3/8″		22	.364		3.59	23		26.59	38.50
3940	1/2″		22	.364		5.40	23		28.40	40.50
3980	Female connector, 1/8″		26	.308		.98	19.45		20.43	30
4000	3/16″ x 1/8″		24	.333		1.26	21		22.26	33
4010	1/4″ x 1/8″		24	.333		1.50	21		22.50	33
4020	1/4″		24	.333		2.02	21		23.02	33.50
4030	3/8″ x 1/4″		22	.364		2.02	23		25.02	36.50
4040	1/2″ x 3/8″		22	.364		3.21	23		26.21	38
4050	5/8″ x 1/2″		21	.381		3.96	24		27.96	40.50
4090	Male connector, 1/8″		26	.308		1.22	19.45		20.67	30.50
4100	3/16″ x 1/8″		24	.333		.83	21		21.83	32.50
4110	1/4″ x 1/8″		24	.333		1.15	21		22.15	33
4120	1/4″		24	.333		1.15	21		22.15	33
4130	5/16″ x 1/8″		23	.348		1.36	22		23.36	34.50
4140	5/16″ x 1/4″		23	.348		1.14	22		23.14	34.50
4150	3/8″ x 1/8″		22	.364		1.61	23		24.61	36.50
4160	3/8″ x 1/4″		22	.364		1.85	23		24.85	36.50
4170	3/8″		22	.364		1.85	23		24.85	36.50
4180	3/8″ x 1/2″		22	.364		2.55	23		25.55	37.50
4190	1/2″ x 3/8″		22	.364		2.46	23		25.46	37
4200	1/2″		22	.364		2.75	23		25.75	37.50
4210	5/8″ x 1/2″		21	.381		3.18	24		27.18	39.50
4240	Male elbow, 1/8″		26	.308		2.22	19.45		21.67	31.50
4250	3/16″ x 1/8″		24	.333		1.81	21		22.81	33.50
4260	1/4″ x 1/8″		24	.333		1.85	21		22.85	33.50
4270	1/4″		24	.333		2.10	21		23.10	34
4280	3/8″ x 1/4″		22	.364		2.66	23		25.66	37.50
4290	3/8″		22	.364		4.57	23		27.57	39.50

22 11 13.25 Pipe/Tube Fittings, Copper		Crew	Daily Output	Labor-Hours	Unit	Material	Labor	Equipment	Total	Total Incl O&P
						2019 Bare Costs				
4300	1/2" x 1/4"	1 Plum	22	.364	Ea.	4.42	23		27.42	39.50
4310	1/2" x 3/8"		22	.364		6.95	23		29.95	42
4340	Female elbow, 1/8"		26	.308		3.54	19.45		22.99	33
4350	1/4" x 1/8"		24	.333		2.71	21		23.71	34.50
4360	1/4"		24	.333		3.55	21		24.55	35.50
4370	3/8" x 1/4"		22	.364		3.49	23		26.49	38.50
4380	1/2" x 3/8"		22	.364		4.44	23		27.44	39.50
4390	1/2"		22	.364		6	23		29	41
4420	Male run tee, 1/4" x 1/8"		15	.533		3.31	33.50		36.81	54
4430	5/16" x 1/8"		15	.533		4.73	33.50		38.23	55.50
4440	3/8" x 1/4"		15	.533		5.10	33.50		38.60	56
4480	Male branch tee, 1/4" x 1/8"		15	.533		3.84	33.50		37.34	54.50
4490	1/4"		15	.533		4.13	33.50		37.63	55
4500	3/8" x 1/4"		15	.533		3.58	33.50		37.08	54.50
4510	1/2" x 3/8"		15	.533		5.55	33.50		39.05	56.50
4520	1/2"	↓	15	.533	↓	8.90	33.50		42.40	60.50
4800	Flare joint fittings									
4810	Refrigeration fittings									
4820	Flare joint nuts and labor not incl. in price. Add 1 nut per jnt.									
4830	90° elbow, 1/4"				Ea.	1.50			1.50	1.65
4840	3/8"					2.01			2.01	2.21
4850	1/2"					2.58			2.58	2.84
4860	5/8"					3.55			3.55	3.91
4870	3/4"					10.75			10.75	11.85
5030	Tee, 1/4"					2.21			2.21	2.43
5040	5/16"					2.64			2.64	2.90
5050	3/8"					2.36			2.36	2.60
5060	1/2"					3.56			3.56	3.92
5070	5/8"					5.10			5.10	5.65
5080	3/4"					5.75			5.75	6.35
5140	Union, 3/16"					.90			.90	.99
5150	1/4"					.67			.67	.74
5160	5/16"					1.25			1.25	1.38
5170	3/8"					.96			.96	1.06
5180	1/2"					1.43			1.43	1.57
5190	5/8"					2.29			2.29	2.52
5200	3/4"					7.75			7.75	8.50
5260	Long flare nut, 3/16"	1 Stpi	42	.190		1.15	12.20		13.35	19.55
5270	1/4"		41	.195		.70	12.50		13.20	19.45
5280	5/16"		40	.200		1.27	12.80		14.07	20.50
5290	3/8"		39	.205		1.44	13.10		14.54	21.50
5300	1/2"		38	.211		2.10	13.45		15.55	22.50
5310	5/8"		37	.216		3.76	13.85		17.61	25
5320	3/4"		34	.235		14.95	15.05		30	39
5380	Short flare nut, 3/16"		42	.190		4.05	12.20		16.25	23
5390	1/4"		41	.195		.49	12.50		12.99	19.25
5400	5/16"		40	.200		.59	12.80		13.39	19.85
5410	3/8"		39	.205		.67	13.10		13.77	20.50
5420	1/2"		38	.211		1.15	13.45		14.60	21.50
5430	5/8"		36	.222		1.61	14.20		15.81	23.50
5440	3/4"	↓	34	.235		5.50	15.05		20.55	28.50
5500	90° elbow flare by MIPS, 1/4"					2.01			2.01	2.21
5510	3/8"				↓	1.97			1.97	2.17

22 11 13.25 Pipe/Tube Fittings, Copper		Crew	Daily Output	Labor-Hours	Unit	2019 Bare Costs Material	Labor	Equipment	Total	Total Incl O&P
5520	1/2"				Ea.	3.64			3.64	4
5530	5/8"					7.45			7.45	8.20
5540	3/4"					12.45			12.45	13.70
5600	Flare by FIPS, 1/4"					3.42			3.42	3.76
5610	3/8"					3.33			3.33	3.66
5620	1/2"					4.87			4.87	5.35
5670	Flare by sweat, 1/4"					2.20			2.20	2.42
5680	3/8"					5.70			5.70	6.25
5690	1/2"					5.50			5.50	6.05
5700	5/8"					16.70			16.70	18.35
5760	Tee flare by IPS, 1/4"					3.27			3.27	3.60
5770	3/8"					4.75			4.75	5.25
5780	1/2"					8			8	8.80
5790	5/8"					5.45			5.45	6
5850	Connector, 1/4"					.76			.76	.84
5860	3/8"					1.12			1.12	1.23
5870	1/2"					1.82			1.82	2
5880	5/8"					3.41			3.41	3.75
5890	3/4"					5.50			5.50	6.05
5950	Seal cap, 1/4"					.47			.47	.52
5960	3/8"					.78			.78	.86
5970	1/2"					.98			.98	1.08
5980	5/8"					2.04			2.04	2.24
5990	3/4"					5.25			5.25	5.80
6000	Water service fittings									
6010	Flare joints nut and labor are included in the fitting price.									
6020	90° elbow, C x C, 3/8"	1 Plum	19	.421	Ea.	59.50	26.50		86	106
6030	1/2"		18	.444		64.50	28		92.50	113
6040	3/4"		16	.500		82.50	31.50		114	138
6050	1"		15	.533		156	33.50		189.50	223
6080	2"		10	.800		570	50.50		620.50	700
6090	90° elbow, C x MPT, 3/8"		19	.421		45	26.50		71.50	90
6100	1/2"		18	.444		49	28		77	96
6110	3/4"		16	.500		57	31.50		88.50	110
6120	1"		15	.533		155	33.50		188.50	222
6130	1-1/4"		13	.615		325	39		364	415
6140	1-1/2"		12	.667		325	42		367	420
6150	2"		10	.800		320	50.50		370.50	425
6160	90° elbow, C x FPT, 3/8"		19	.421		48.50	26.50		75	93.50
6170	1/2"		18	.444		52.50	28		80.50	100
6180	3/4"		16	.500		65.50	31.50		97	120
6190	1"		15	.533		151	33.50		184.50	217
6200	1-1/4"		13	.615		129	39		168	201
6210	1-1/2"		12	.667		495	42		537	610
6220	2"		10	.800		715	50.50		765.50	860
6230	Tee, C x C x C, 3/8"		13	.615		94.50	39		133.50	163
6240	1/2"		12	.667		94.50	42		136.50	167
6250	3/4"		11	.727		111	46		157	191
6260	1"		10	.800		196	50.50		246.50	292
6330	Tube nut, C x nut seat, 3/8"		40	.200		18.40	12.65		31.05	39.50
6340	1/2"		38	.211		18.40	13.30		31.70	40.50
6350	3/4"		34	.235		17.90	14.85		32.75	42
6360	1"		32	.250		35	15.80		50.80	62

22 11 13.25 Pipe/Tube Fittings, Copper

	22 11 13.25 Pipe/Tube Fittings, Copper	Crew	Daily Output	Labor-Hours	Unit	Material	Labor	Equipment	Total	Total Incl O&P
						2019 Bare Costs				
6380	Coupling, C x C, 3/8"	1 Plum	19	.421	Ea.	58	26.50		84.50	104
6390	1/2"		18	.444		58	28		86	106
6400	3/4"		16	.500		73.50	31.50		105	129
6410	1"		15	.533		136	33.50		169.50	200
6420	1-1/4"		12	.667		220	42		262	305
6430	1-1/2"		12	.667		335	42		377	430
6440	2"		10	.800		500	50.50		550.50	625
6450	Adapter, C x FPT, 3/8"		19	.421		44	26.50		70.50	88
6460	1/2"		18	.444		44	28		72	90
6470	3/4"		16	.500		53	31.50		84.50	106
6480	1"		15	.533		113	33.50		146.50	175
6490	1-1/4"		13	.615		100	39		139	169
6500	1-1/2"		12	.667		240	42		282	325
6510	2"		10	.800		310	50.50		360.50	420
6520	Adapter, C x MPT, 3/8"		19	.421		36	26.50		62.50	79.50
6530	1/2"		18	.444		39	28		67	85
6540	3/4"		16	.500		54	31.50		85.50	107
6550	1"		15	.533		95.50	33.50		129	156
6560	1-1/4"		13	.615		238	39		277	320
6570	1-1/2"		12	.667		239	42		281	325
6580	2"	↓	10	.800	↓	289	50.50		339.50	395
6992	Tube connector fittings, See Section 22 11 13.76 for plastic ftng.									
7000	Insert type brass/copper, 100 psi @ 180°F, CTS									
7010	Adapter MPT 3/8" x 1/2" CTS	1 Plum	29	.276	Ea.	2.70	17.40		20.10	29
7020	1/2" x 1/2"		26	.308		3.13	19.45		22.58	32.50
7030	3/4" x 1/2"		26	.308		4.21	19.45		23.66	33.50
7040	3/4" x 3/4"		25	.320		4.61	20		24.61	35.50
7050	Adapter CTS 1/2" x 1/2" sweat		24	.333		3.99	21		24.99	36
7060	3/4" x 3/4" sweat		22	.364		1.47	23		24.47	36
7070	Coupler center set 3/8" CTS		25	.320		1.44	20		21.44	32
7080	1/2" CTS		23	.348		4.04	22		26.04	37.50
7090	3/4" CTS		22	.364		1.63	23		24.63	36.50
7100	Elbow 90°, copper 3/8"		25	.320		3.26	20		23.26	34
7110	1/2" CTS		23	.348		2.23	22		24.23	35.50
7120	3/4" CTS		22	.364		3.03	23		26.03	38
7130	Tee copper 3/8" CTS		17	.471		4.13	29.50		33.63	49
7140	1/2" CTS		15	.533		3.02	33.50		36.52	54
7150	3/4" CTS		14	.571		4.62	36		40.62	59
7160	3/8" x 3/8" x 1/2"		16	.500		4.18	31.50		35.68	52
7170	1/2" x 3/8" x 1/2"		15	.533		2.55	33.50		36.05	53.50
7180	3/4" x 1/2" x 3/4"	↓	14	.571	↓	4.57	36		40.57	59

22 11 13.27 Pipe/Tube, Grooved Joint for Copper

	22 11 13.27 Pipe/Tube, Grooved Joint for Copper	Crew	Daily Output	Labor-Hours	Unit	Material	Labor	Equipment	Total	Total Incl O&P
0010	**PIPE/TUBE, GROOVED JOINT FOR COPPER**									
4000	Fittings: coupling material required at joints not incl. in fitting price.									
4001	Add 1 selected coupling, material only, per joint for installed price.									
4010	Coupling, rigid style									
4018	2" diameter	1 Plum	50	.160	Ea.	27.50	10.10		37.60	45.50
4020	2-1/2" diameter	Q-1	80	.200		31	11.35		42.35	51
4022	3" diameter		67	.239		34.50	13.55		48.05	58.50
4024	4" diameter		50	.320		52	18.20		70.20	85
4026	5" diameter	↓	40	.400		87.50	22.50		110	130
4028	6" diameter	Q-2	50	.480	↓	115	28.50		143.50	170

22 11 13.27 Pipe/Tube, Grooved Joint for Copper

	22 11 13.27 Pipe/Tube, Grooved Joint for Copper	Crew	Daily Output	Labor-Hours	Unit	2019 Bare Costs Material	Labor	Equipment	Total	Total Incl O&P
4100	Elbow, 90° or 45°									
4108	2" diameter	1 Plum	25	.320	Ea.	50.50	20		70.50	86
4110	2-1/2" diameter	Q-1	40	.400		54.50	22.50		77	94
4112	3" diameter		33	.485		76.50	27.50		104	126
4114	4" diameter		25	.640		175	36.50		211.50	247
4116	5" diameter		20	.800		495	45.50		540.50	615
4118	6" diameter	Q-2	25	.960		790	56.50		846.50	955
4200	Tee									
4208	2" diameter	1 Plum	17	.471	Ea.	82.50	29.50		112	135
4210	2-1/2" diameter	Q-1	27	.593		87.50	33.50		121	147
4212	3" diameter		22	.727		130	41.50		171.50	205
4214	4" diameter		17	.941		285	53.50		338.50	395
4216	5" diameter		13	1.231		800	70		870	980
4218	6" diameter	Q-2	17	1.412		985	83		1,068	1,200
4300	Reducer, concentric									
4310	3" x 2-1/2" diameter	Q-1	35	.457	Ea.	72.50	26		98.50	119
4312	4" x 2-1/2" diameter		32	.500		148	28.50		176.50	205
4314	4" x 3" diameter		29	.552		148	31.50		179.50	209
4316	5" x 3" diameter		25	.640		415	36.50		451.50	510
4318	5" x 4" diameter		22	.727		415	41.50		456.50	515
4320	6" x 3" diameter	Q-2	28	.857		450	50.50		500.50	570
4322	6" x 4" diameter		26	.923		450	54.50		504.50	575
4324	6" x 5" diameter		24	1		450	59		509	585
4350	Flange, w/groove gasket									
4351	ANSI class 125 and 150									
4355	2" diameter	1 Plum	23	.348	Ea.	208	22		230	262
4356	2-1/2" diameter	Q-1	37	.432		217	24.50		241.50	276
4358	3" diameter		31	.516		227	29.50		256.50	294
4360	4" diameter		23	.696		249	39.50		288.50	335
4362	5" diameter		19	.842		335	48		383	440
4364	6" diameter	Q-2	23	1.043		365	61.50		426.50	495

22 11 13.29 Pipe, Fittings and Valves, Copper, Pressed-Joint

	22 11 13.29 Pipe, Fittings and Valves, Copper, Pressed-Joint	Crew	Daily Output	Labor-Hours	Unit	Material	Labor	Equipment	Total	Total Incl O&P
0010	**PIPE, FITTINGS AND VALVES, COPPER, PRESSED-JOINT**									
0040	Pipe/tube includes coupling & clevis type hanger assy's, 10' OC									
0120	Type K									
0130	1/2" diameter	1 Plum	78	.103	L.F.	5.45	6.50		11.95	15.70
0134	3/4" diameter		74	.108		8.75	6.85		15.60	19.90
0138	1" diameter		66	.121		13.15	7.65		20.80	26
0142	1-1/4" diameter		56	.143		15.65	9		24.65	31
0146	1-1/2" diameter		50	.160		19.45	10.10		29.55	36.50
0150	2" diameter		40	.200		28	12.65		40.65	50
0154	2-1/2" diameter	Q-1	60	.267		46	15.15		61.15	73
0158	3" diameter		54	.296		61.50	16.85		78.35	93
0162	4" diameter		38	.421		103	24		127	149
0180	To delete cplgs. & hngrs., 1/2" pipe, subtract					19%	48%			
0184	3/4"-2" pipe, subtract					14%	46%			
0186	2-1/2"-4" pipe, subtract					24%	34%			
0220	Type L									
0230	1/2" diameter	1 Plum	81	.099	L.F.	3.59	6.25		9.84	13.30
0234	3/4" diameter		76	.105		4.62	6.65		11.27	15.05
0238	1" diameter		68	.118		7.50	7.45		14.95	19.40
0242	1-1/4" diameter		58	.138		12.10	8.70		20.80	26.50

22 11 13.29 Pipe, Fittings and Valves, Copper, Pressed-Joint		Crew	Daily Output	Labor-Hours	Unit	2019 Bare Costs Material	Labor	Equipment	Total	Total Incl O&P
0246	1-1/2" diameter	1 Plum	52	.154	L.F.	11.70	9.70		21.40	27.50
0250	2" diameter	↓	42	.190		19.35	12.05		31.40	39.50
0254	2-1/2" diameter	Q-1	62	.258		30.50	14.65		45.15	55.50
0258	3" diameter		56	.286		50	16.25		66.25	79.50
0262	4" diameter	↓	39	.410	↓	68.50	23.50		92	111
0280	To delete cplgs. & hngrs., 1/2" pipe, subtract					21%	52%			
0284	3/4"-2" pipe, subtract					17%	46%			
0286	2-1/2"-4" pipe, subtract					23%	35%			
0320	Type M									
0330	1/2" diameter	1 Plum	84	.095	L.F.	3.85	6		9.85	13.25
0334	3/4" diameter		78	.103		5.40	6.50		11.90	15.65
0338	1" diameter		70	.114		8.90	7.20		16.10	20.50
0342	1-1/4" diameter		60	.133		11.55	8.40		19.95	25.50
0346	1-1/2" diameter		54	.148		15.50	9.35		24.85	31
0350	2" diameter	↓	44	.182		22.50	11.50		34	42
0354	2-1/2" diameter	Q-1	64	.250		36	14.20		50.20	61.50
0358	3" diameter		58	.276		45.50	15.70		61.20	74
0362	4" diameter	↓	40	.400	↓	80.50	22.50		103	123
0380	To delete cplgs. & hngrs., 1/2" pipe, subtract					32%	49%			
0384	3/4"-2" pipe, subtract					21%	46%			
0386	2-1/2"-4" pipe, subtract					25%	36%			
1600	Fittings									
1610	Press joints, copper x copper									
1620	Note: Reducing fittings show most expensive size combination.									
1800	90° elbow, 1/2"	1 Plum	36.60	.219	Ea.	3.55	13.80		17.35	24.50
1810	3/4"		27.50	.291		5.75	18.35		24.10	34
1820	1"		25.90	.309		11.50	19.50		31	42
1830	1-1/4"		20.90	.383		22	24		46	60.50
1840	1-1/2"		18.30	.437		44	27.50		71.50	90
1850	2"	↓	15.70	.510		61.50	32		93.50	117
1860	2-1/2"	Q-1	25.90	.618		174	35		209	245
1870	3"		22	.727		219	41.50		260.50	305
1880	4"	↓	16.30	.982		271	56		327	380
2000	45° elbow, 1/2"	1 Plum	36.60	.219		3.94	13.80		17.74	25
2010	3/4"		27.50	.291		4.64	18.35		22.99	32.50
2020	1"		25.90	.309		14.85	19.50		34.35	46
2030	1-1/4"		20.90	.383		21.50	24		45.50	60
2040	1-1/2"		18.30	.437		34.50	27.50		62	79.50
2050	2"	↓	15.70	.510		48.50	32		80.50	102
2060	2-1/2"	Q-1	25.90	.618		116	35		151	181
2070	3"		22	.727		165	41.50		206.50	243
2080	4"	↓	16.30	.982		231	56		287	340
2200	Tee, 1/2"	1 Plum	27.50	.291		5.45	18.35		23.80	33.50
2210	3/4"		20.70	.386		8.90	24.50		33.40	46.50
2220	1"		19.40	.412		16.20	26		42.20	57
2230	1-1/4"		15.70	.510		27.50	32		59.50	79
2240	1-1/2"		13.80	.580		53	36.50		89.50	114
2250	2"	↓	11.80	.678		65.50	43		108.50	137
2260	2-1/2"	Q-1	19.40	.825		217	47		264	310
2270	3"		16.50	.970		269	55		324	380
2280	4"	↓	12.20	1.311	↓	380	74.50		454.50	530
2400	Tee, reducing on the outlet									
2410	3/4"	1 Plum	20.70	.386	Ea.	7.65	24.50		32.15	45

22 11 13.29 Pipe, Fittings and Valves, Copper, Pressed-Joint		Crew	Daily Output	Labor-Hours	Unit	Material	2019 Bare Costs Labor	Equipment	Total	Total Incl O&P
2420	1"	1 Plum	19.40	.412	Ea.	18.70	26		44.70	59.50
2430	1-1/4"		15.70	.510		27	32		59	78
2440	1-1/2"		13.80	.580		57	36.50		93.50	118
2450	2"		11.80	.678		91	43		134	164
2460	2-1/2"	Q-1	19.40	.825		310	47		357	410
2470	3"		16.50	.970		340	55		395	460
2480	4"		12.20	1.311		470	74.50		544.50	625
2600	Tee, reducing on the run									
2610	3/4"	1 Plum	20.70	.386	Ea.	14.85	24.50		39.35	53
2620	1"		19.40	.412		27	26		53	69
2630	1-1/4"		15.70	.510		56	32		88	110
2640	1-1/2"		13.80	.580		86	36.50		122.50	150
2650	2"		11.80	.678		94.50	43		137.50	168
2660	2-1/2"	Q-1	19.40	.825		320	47		367	420
2670	3"		16.50	.970		385	55		440	505
2680	4"		12.20	1.311		545	74.50		619.50	710
2800	Coupling, 1/2"	1 Plum	36.60	.219		3	13.80		16.80	24
2810	3/4"		27.50	.291		4.85	18.35		23.20	33
2820	1"		25.90	.309		9.75	19.50		29.25	40.50
2830	1-1/4"		20.90	.383		11.55	24		35.55	49
2840	1-1/2"		18.30	.437		21.50	27.50		49	65
2850	2"		15.70	.510		27	32		59	78.50
2860	2-1/2"	Q-1	25.90	.618		87	35		122	148
2870	3"		22	.727		111	41.50		152.50	184
2880	4"		16.30	.982		155	56		211	255
3000	Union, 1/2"	1 Plum	36.60	.219		26	13.80		39.80	49
3010	3/4"		27.50	.291		33	18.35		51.35	63.50
3020	1"		25.90	.309		53	19.50		72.50	88
3030	1-1/4"		20.90	.383		77.50	24		101.50	122
3040	1-1/2"		18.30	.437		103	27.50		130.50	155
3050	2"		15.70	.510		166	32		198	232
3200	Adapter, tube to MPT									
3210	1/2"	1 Plum	15.10	.530	Ea.	3.86	33.50		37.36	54.50
3220	3/4"		13.60	.588		7.55	37		44.55	64
3230	1"		11.60	.690		12.75	43.50		56.25	79.50
3240	1-1/4"		9.90	.808		28	51		79	107
3250	1-1/2"		9	.889		39	56		95	127
3260	2"		7.90	1.013		75.50	64		139.50	179
3270	2-1/2"	Q-1	13.40	1.194		176	68		244	296
3280	3"		10.60	1.509		221	86		307	375
3290	4"		7.70	2.078		264	118		382	465
3400	Adapter, tube to FPT									
3410	1/2"	1 Plum	15.10	.530	Ea.	4.65	33.50		38.15	55
3420	3/4"		13.60	.588		7.55	37		44.55	64
3430	1"		11.60	.690		13.90	43.50		57.40	81
3440	1-1/4"		9.90	.808		31.50	51		82.50	111
3450	1-1/2"		9	.889		45.50	56		101.50	134
3460	2"		7.90	1.013		78	64		142	182
3470	2-1/2"	Q-1	13.40	1.194		194	68		262	315
3480	3"		10.60	1.509		315	86		401	480
3490	4"		7.70	2.078		410	118		528	625
3600	Flange									
3620	1"	1 Plum	36.20	.221	Ea.	130	13.95		143.95	164

22 11 13.29 Pipe, Fittings and Valves, Copper, Pressed-Joint		Crew	Daily Output	Labor-Hours	Unit	2019 Bare Costs Material	Labor	Equipment	Total	Total Incl O&P
3630	1-1/4"	1 Plum	29.30	.273	Ea.	188	17.25		205.25	233
3640	1-1/2"		25.60	.313		208	19.75		227.75	259
3650	2"		22	.364		234	23		257	292
3660	2-1/2"	Q-1	36.20	.442		267	25		292	330
3670	3"		30.80	.519		325	29.50		354.50	400
3680	4"		22.80	.702		360	40		400	460
3800	Cap, 1/2"	1 Plum	53.10	.151		6.80	9.50		16.30	22
3810	3/4"		39.80	.201		10.85	12.70		23.55	31
3820	1"		37.50	.213		16.75	13.45		30.20	38.50
3830	1-1/4"		30.40	.263		19.50	16.60		36.10	46.50
3840	1-1/2"		26.60	.301		30	19		49	62
3850	2"		22.80	.351		36.50	22		58.50	74
3860	2-1/2"	Q-1	37.50	.427		118	24.50		142.50	167
3870	3"		31.90	.502		150	28.50		178.50	208
3880	4"		23.60	.678		199	38.50		237.50	277
4000	Reducer									
4010	3/4"	1 Plum	27.50	.291	Ea.	13.65	18.35		32	42.50
4020	1"		25.90	.309		24.50	19.50		44	56.50
4030	1-1/4"		20.90	.383		36.50	24		60.50	76.50
4040	1-1/2"		18.30	.437		43.50	27.50		71	89
4050	2"		15.70	.510		59.50	32		91.50	114
4060	2-1/2"	Q-1	25.90	.618		174	35		209	244
4070	3"		22	.727		231	41.50		272.50	315
4080	4"		16.30	.982		298	56		354	410
4100	Stub out, 1/2"	1 Plum	50	.160		8.80	10.10		18.90	25
4110	3/4"		35	.229		14.45	14.45		28.90	37.50
4120	1"		30	.267		19.80	16.85		36.65	47.50
6000	Valves									
6200	Ball valve									
6210	1/2"	1 Plum	25.60	.313	Ea.	41.50	19.75		61.25	75
6220	3/4"		19.20	.417		55	26.50		81.50	100
6230	1"		18.10	.442		61	28		89	110
6240	1-1/4"		14.70	.544		99.50	34.50		134	161
6250	1-1/2"		12.80	.625		244	39.50		283.50	330
6260	2"		11	.727		355	46		401	460
6400	Check valve									
6410	1/2"	1 Plum	30	.267	Ea.	36	16.85		52.85	65.50
6420	3/4"		22.50	.356		46.50	22.50		69	84.50
6430	1"		21.20	.377		53.50	24		77.50	94.50
6440	1-1/4"		17.20	.465		77.50	29.50		107	130
6450	1-1/2"		15	.533		110	33.50		143.50	172
6460	2"		12.90	.620		199	39		238	278
6600	Butterfly valve, lug type									
6660	2-1/2"	Q-1	9	1.778	Ea.	170	101		271	340
6670	3"		8	2		209	114		323	400
6680	4"		5	3.200		261	182		443	560

22 11 13.44 Pipe, Steel

		Crew	Daily Output	Labor-Hours	Unit	Material	Labor	Equipment	Total	Total Incl O&P
0010	**PIPE, STEEL** R221113-50									
0020	All pipe sizes are to Spec. A-53 unless noted otherwise R221113-70									
0032	Schedule 10, see Line 22 11 13.48 0500									
0050	Schedule 40, threaded, with couplings, and clevis hanger									
0060	assemblies sized for covering, 10' OC									

22 11 13.44 Pipe, Steel		Crew	Daily Output	Labor-Hours	Unit	Material	2019 Bare Costs Labor	Equipment	Total	Total Incl O&P
0540	Black, 1/4" diameter	1 Plum	66	.121	L.F.	5.80	7.65		13.45	17.85
0550	3/8" diameter		65	.123		6.45	7.75		14.20	18.75
0560	1/2" diameter		63	.127		3.91	8		11.91	16.35
0570	3/4" diameter		61	.131		4.25	8.30		12.55	17.15
0580	1" diameter		53	.151		4.42	9.55		13.97	19.15
0590	1-1/4" diameter	Q-1	89	.180		5.20	10.20		15.40	21
0600	1-1/2" diameter		80	.200		5.70	11.35		17.05	23.50
0610	2" diameter		64	.250		11.55	14.20		25.75	34
0620	2-1/2" diameter		50	.320		15.65	18.20		33.85	45
0630	3" diameter		43	.372		18.45	21		39.45	52
0640	3-1/2" diameter		40	.400		25	22.50		47.50	61.50
0650	4" diameter		36	.444		21.50	25.50		47	62
0809	A-106, gr. A/B, seamless w/cplgs. & clevis hanger assemblies									
0811	1/4" diameter	1 Plum	66	.121	L.F.	6.60	7.65		14.25	18.75
0812	3/8" diameter		65	.123		6.85	7.75		14.60	19.20
0813	1/2" diameter		63	.127		7.05	8		15.05	19.85
0814	3/4" diameter		61	.131		8.55	8.30		16.85	22
0815	1" diameter		53	.151		9.05	9.55		18.60	24.50
0816	1-1/4" diameter	Q-1	89	.180		10.95	10.20		21.15	27.50
0817	1-1/2" diameter		80	.200		15.40	11.35		26.75	34
0819	2" diameter		64	.250		13.70	14.20		27.90	36.50
0821	2-1/2" diameter		50	.320		19.25	18.20		37.45	48.50
0822	3" diameter		43	.372		26	21		47	60
0823	4" diameter		36	.444		42	25.50		67.50	84
1220	To delete coupling & hanger, subtract									
1230	1/4" diam. to 3/4" diam.					31%	56%			
1240	1" diam. to 1-1/2" diam.					23%	51%			
1250	2" diam. to 4" diam.					23%	41%			
1280	All pipe sizes are to Spec. A-53 unless noted otherwise									
1281	Schedule 40, threaded, with couplings and clevis hanger									
1282	assemblies sized for covering, 10' OC									
1290	Galvanized, 1/4" diameter	1 Plum	66	.121	L.F.	7.65	7.65		15.30	19.90
1300	3/8" diameter		65	.123		7.85	7.75		15.60	20.50
1310	1/2" diameter		63	.127		3.99	8		11.99	16.45
1320	3/4" diameter		61	.131		4.51	8.30		12.81	17.40
1330	1" diameter		53	.151		4.91	9.55		14.46	19.70
1340	1-1/4" diameter	Q-1	89	.180		5.85	10.20		16.05	22
1350	1-1/2" diameter		80	.200		6.45	11.35		17.80	24
1360	2" diameter		64	.250		12.55	14.20		26.75	35.50
1370	2-1/2" diameter		50	.320		17.90	18.20		36.10	47
1380	3" diameter		43	.372		20.50	21		41.50	54.50
1390	3-1/2" diameter		40	.400		26.50	22.50		49	63
1400	4" diameter		36	.444		24	25.50		49.50	64.50
1750	To delete coupling & hanger, subtract									
1760	1/4" diam. to 3/4" diam.					31%	56%			
1770	1" diam. to 1-1/2" diam.					23%	51%			
1780	2" diam. to 4" diam.					23%	41%			
2000	Welded, sch. 40, on yoke & roll hanger assy's, sized for covering, 10' OC									
2040	Black, 1" diameter	Q-15	93	.172	L.F.	4.23	9.80	.60	14.63	19.95
2050	1-1/4" diameter		84	.190		5.30	10.85	.67	16.82	23
2060	1-1/2" diameter		76	.211		5.70	11.95	.74	18.39	25
2070	2" diameter		61	.262		11.25	14.90	.92	27.07	36
2080	2-1/2" diameter		47	.340		15	19.35	1.19	35.54	47

22 11 13.44 Pipe, Steel		Crew	Daily Output	Labor-Hours	Unit	Material	Labor	Equipment	Total	Total Incl O&P
						2019 Bare Costs				
2090	3″ diameter	Q-15	43	.372	L.F.	16.55	21	1.31	38.86	51
2100	3-1/2″ diameter		39	.410		19.50	23.50	1.44	44.44	58
2110	4″ diameter		37	.432		17	24.50	1.52	43.02	57.50
2120	5″ diameter		32	.500		34.50	28.50	1.76	64.76	82.50
2130	6″ diameter	Q-16	36	.667		42.50	39.50	1.56	83.56	108
2140	8″ diameter		29	.828		69	49	1.94	119.94	151
2150	10″ diameter		24	1		88	59	2.34	149.34	188
2160	12″ diameter		19	1.263		105	74.50	2.96	182.46	230
2170	14″ diameter (two rod roll type hanger for 14″ diam. and up)		15	1.600		99.50	94.50	3.74	197.74	254
2180	16″ diameter (two rod roll type hanger)		13	1.846		155	109	4.32	268.32	340
2190	18″ diameter (two rod roll type hanger)		11	2.182		143	129	5.10	277.10	355
2200	20″ diameter (two rod roll type hanger)		9	2.667		147	157	6.25	310.25	405
2220	24″ diameter (two rod roll type hanger)		8	3		194	177	7	378	485
2345	Sch. 40, A-53, gr. A/B, ERW, welded w/hngrs.									
2346	2″ diameter	Q-15	61	.262	L.F.	7.10	14.90	.92	22.92	31.50
2347	2-1/2″ diameter		47	.340		9.10	19.35	1.19	29.64	40.50
2348	3″ diameter		43	.372		11.15	21	1.31	33.46	45
2349	4″ diameter		38	.421		15.50	24	1.48	40.98	54.50
2350	6″ diameter	Q-16	37	.649		26.50	38	1.52	66.02	88
2351	8″ diameter		29	.828		67.50	49	1.94	118.44	149
2352	10″ diameter		24	1		88.50	59	2.34	149.84	189
2560	To delete hanger, subtract									
2570	1″ diam. to 1-1/2″ diam.					15%	34%			
2580	2″ diam. to 3-1/2″ diam.					9%	21%			
2590	4″ diam. to 12″ diam.					5%	12%			
2596	14″ diam. to 24″ diam.					3%	10%			
2640	Galvanized, 1″ diameter	Q-15	93	.172	L.F.	4.99	9.80	.60	15.39	21
2650	1-1/4″ diameter		84	.190		6.20	10.85	.67	17.72	24
2660	1-1/2″ diameter		76	.211		6.70	11.95	.74	19.39	26
2670	2″ diameter		61	.262		12.65	14.90	.92	28.47	37.50
2680	2-1/2″ diameter		47	.340		15.55	19.35	1.19	36.09	47.50
2690	3″ diameter		43	.372		18.75	21	1.31	41.06	53.50
2700	3-1/2″ diameter		39	.410		21	23.50	1.44	45.94	60
2710	4″ diameter		37	.432		20	24.50	1.52	46.02	60.50
2720	5″ diameter		32	.500		23	28.50	1.76	53.26	69.50
2730	6″ diameter	Q-16	36	.667		28	39.50	1.56	69.06	91
2740	8″ diameter		29	.828		39	49	1.94	89.94	118
2750	10″ diameter		24	1		86	59	2.34	147.34	186
2760	12″ diameter		19	1.263		113	74.50	2.96	190.46	240
3160	To delete hanger, subtract									
3170	1″ diam. to 1-1/2″ diam.					30%	34%			
3180	2″ diam. to 3-1/2″ diam.					19%	21%			
3190	4″ diam. to 12″ diam.					10%	12%			
3250	Flanged, 150 lb. weld neck, on yoke & roll hangers									
3260	sized for covering, 10′ OC									
3290	Black, 1″ diameter	Q-15	70	.229	L.F.	10.35	13	.80	24.15	32
3300	1-1/4″ diameter		64	.250		11.40	14.20	.88	26.48	35
3310	1-1/2″ diameter		58	.276		11.85	15.70	.97	28.52	37.50
3320	2″ diameter		45	.356		14.85	20	1.25	36.10	48
3330	2-1/2″ diameter		36	.444		18.50	25.50	1.56	45.56	60
3340	3″ diameter		32	.500		21	28.50	1.76	51.26	67.50
3350	3-1/2″ diameter		29	.552		25	31.50	1.94	58.44	76.50
3360	4″ diameter		26	.615		30.50	35	2.16	67.66	88.50

	22 11 13.44 Pipe, Steel	Crew	Daily Output	Labor-Hours	Unit	Material	2019 Bare Costs Labor	Equipment	Total	Total Incl O&P
3370	5" diameter	Q-15	21	.762	L.F.	54	43.50	2.67	100.17	127
3380	6" diameter	Q-16	25	.960		60	56.50	2.25	118.75	153
3390	8" diameter		19	1.263		98.50	74.50	2.96	175.96	223
3400	10" diameter		16	1.500		139	88.50	3.51	231.01	290
3410	12" diameter		14	1.714		172	101	4.01	277.01	345
3470	For 300 lb. flanges, add					63%				
3480	For 600 lb. flanges, add					310%				
3960	To delete flanges & hanger, subtract									
3970	1" diam. to 2" diam.					76%	65%			
3980	2-1/2" diam. to 4" diam.					62%	59%			
3990	5" diam. to 12" diam.					60%	46%			
4040	Galvanized, 1" diameter	Q-15	70	.229	L.F.	11.10	13	.80	24.90	32.50
4050	1-1/4" diameter		64	.250		12.30	14.20	.88	27.38	36
4060	1-1/2" diameter		58	.276		12.80	15.70	.97	29.47	38.50
4070	2" diameter		45	.356		14.55	20	1.25	35.80	48
4080	2-1/2" diameter		36	.444		19.30	25.50	1.56	46.36	60.50
4090	3" diameter		32	.500		22.50	28.50	1.76	52.76	69
4100	3-1/2" diameter		29	.552		26.50	31.50	1.94	59.94	78
4110	4" diameter		26	.615		34	35	2.16	71.16	92
4120	5" diameter		21	.762		42	43.50	2.67	88.17	114
4130	6" diameter	Q-16	25	.960		45	56.50	2.25	103.75	137
4140	8" diameter		19	1.263		59	74.50	2.96	136.46	180
4150	10" diameter		16	1.500		123	88.50	3.51	215.01	272
4160	12" diameter		14	1.714		146	101	4.01	251.01	315
4220	For 300 lb. flanges, add					65%				
4240	For 600 lb. flanges, add					350%				
4660	To delete flanges & hanger, subtract									
4670	1" diam. to 2" diam.					73%	65%			
4680	2-1/2" diam. to 4" diam.					58%	59%			
4690	5" diam. to 12" diam.					43%	46%			
4750	Schedule 80, threaded, with couplings, and clevis hanger assemblies									
4760	sized for covering, 10' OC									
4790	Black, 1/4" diameter	1 Plum	54	.148	L.F.	11.30	9.35		20.65	26.50
4800	3/8" diameter		53	.151		15.70	9.55		25.25	31.50
4810	1/2" diameter		52	.154		4.70	9.70		14.40	19.75
4820	3/4" diameter		50	.160		5.80	10.10		15.90	21.50
4830	1" diameter		45	.178		6.90	11.25		18.15	24.50
4840	1-1/4" diameter	Q-1	75	.213		8.85	12.10		20.95	28
4850	1-1/2" diameter		69	.232		9.75	13.20		22.95	30.50
4860	2" diameter		56	.286		12.10	16.25		28.35	38
4870	2-1/2" diameter		44	.364		16.30	20.50		36.80	49
4880	3" diameter		38	.421		21.50	24		45.50	60
4890	3-1/2" diameter		35	.457		26.50	26		52.50	68
4900	4" diameter		32	.500		34.50	28.50		63	80.50
5061	A-106, gr. A/B seamless with cplgs. & clevis hanger assemblies, 1/4" diam.	1 Plum	63	.127		7.50	8		15.50	20.50
5062	3/8" diameter		62	.129		8.65	8.15		16.80	22
5063	1/2" diameter		61	.131		7.25	8.30		15.55	20.50
5064	3/4" diameter		57	.140		8.45	8.85		17.30	22.50
5065	1" diameter		51	.157		9.80	9.90		19.70	25.50
5066	1-1/4" diameter	Q-1	85	.188		12.85	10.70		23.55	30
5067	1-1/2" diameter		77	.208		16.35	11.80		28.15	35.50
5071	2" diameter		61	.262		16.75	14.90		31.65	41
5072	2-1/2" diameter		48	.333		19.15	18.95		38.10	49.50

22 11 13 – Facility Water Distribution Piping

22 11 13.44 Pipe, Steel		Crew	Daily Output	Labor-Hours	Unit	Material	Labor	Equipment	Total	Total Incl O&P
						2019 Bare Costs				
5073	3″ diameter	Q-1	41	.390	L.F.	26.50	22		48.50	62.50
5074	4″ diameter		35	.457		38.50	26		64.50	81
5430	To delete coupling & hanger, subtract									
5440	1/4″ diam. to 1/2″ diam.					31%	54%			
5450	3/4″ diam. to 1-1/2″ diam.					28%	49%			
5460	2″ diam. to 4″ diam.					21%	40%			
5510	Galvanized, 1/4″ diameter	1 Plum	54	.148	L.F.	8.95	9.35		18.30	24
5520	3/8″ diameter		53	.151		9.75	9.55		19.30	25
5530	1/2″ diameter		52	.154		5.45	9.70		15.15	20.50
5540	3/4″ diameter		50	.160		6.50	10.10		16.60	22.50
5550	1″ diameter		45	.178		7.95	11.25		19.20	25.50
5560	1-1/4″ diameter	Q-1	75	.213		10.45	12.10		22.55	29.50
5570	1-1/2″ diameter		69	.232		11.55	13.20		24.75	32.50
5580	2″ diameter		56	.286		15	16.25		31.25	41
5590	2-1/2″ diameter		44	.364		20.50	20.50		41	54
5600	3″ diameter		38	.421		26.50	24		50.50	65
5610	3-1/2″ diameter		35	.457		38	26		64	81
5620	4″ diameter		32	.500		38.50	28.50		67	85
5930	To delete coupling & hanger, subtract									
5940	1/4″ diam. to 1/2″ diam.					31%	54%			
5950	3/4″ diam. to 1-1/2″ diam.					28%	49%			
5960	2″ diam. to 4″ diam.					21%	40%			
6000	Welded, on yoke & roller hangers									
6010	sized for covering, 10′ OC									
6040	Black, 1″ diameter	Q-15	85	.188	L.F.	6.85	10.70	.66	18.21	24.50
6050	1-1/4″ diameter		79	.203		8.65	11.50	.71	20.86	27.50
6060	1-1/2″ diameter		72	.222		9.55	12.65	.78	22.98	30.50
6070	2″ diameter		57	.281		11.20	15.95	.99	28.14	37.50
6080	2-1/2″ diameter		44	.364		14.85	20.50	1.28	36.63	49
6090	3″ diameter		40	.400		19.50	22.50	1.40	43.40	57
6100	3-1/2″ diameter		34	.471		24	26.50	1.65	52.15	68
6110	4″ diameter		33	.485		31	27.50	1.70	60.20	77.50
6120	5″ diameter, A-106B		26	.615		68.50	35	2.16	105.66	130
6130	6″ diameter, A-106B	Q-16	30	.800		64.50	47	1.87	113.37	144
6140	8″ diameter, A-106B		25	.960		96	56.50	2.25	154.75	192
6150	10″ diameter, A-106B		20	1.200		149	70.50	2.81	222.31	273
6160	12″ diameter, A-106B		15	1.600		320	94.50	3.74	418.24	500
6540	To delete hanger, subtract									
6550	1″ diam. to 1-1/2″ diam.					30%	14%			
6560	2″ diam. to 3″ diam.					23%	9%			
6570	3-1/2″ diam. to 5″ diam.					12%	6%			
6580	6″ diam. to 12″ diam.					10%	4%			
6610	Galvanized, 1″ diameter	Q-15	85	.188	L.F.	7.90	10.70	.66	19.26	25.50
6650	1-1/4″ diameter		79	.203		10.25	11.50	.71	22.46	29.50
6660	1-1/2″ diameter		72	.222		11.35	12.65	.78	24.78	32.50
6670	2″ diameter		57	.281		14.05	15.95	.99	30.99	40.50
6680	2-1/2″ diameter		44	.364		19.30	20.50	1.28	41.08	53.50
6690	3″ diameter		40	.400		24	22.50	1.40	47.90	62
6700	3-1/2″ diameter		34	.471		35	26.50	1.65	63.15	81
6710	4″ diameter		33	.485		35	27.50	1.70	64.20	82
6720	5″ diameter, A-106B		26	.615		49.50	35	2.16	86.66	109
6730	6″ diameter, A-106B	Q-16	30	.800		60.50	47	1.87	109.37	140
6740	8″ diameter, A-106B		25	.960		115	56.50	2.25	173.75	213

22 11 Facility Water Distribution

22 11 13 – Facility Water Distribution Piping

22 11 13.44 Pipe, Steel		Crew	Daily Output	Labor-Hours	Unit	Material	2019 Bare Costs Labor	Equipment	Total	Total Incl O&P
6750	10" diameter, A-106B	Q-16	20	1.200	L.F.	260	70.50	2.81	333.31	395
6760	12" diameter, A-106B		15	1.600		495	94.50	3.74	593.24	690
7150	To delete hanger, subtract									
7160	1" diam. to 1-1/2" diam.					26%	14%			
7170	2" diam. to 3" diam.					16%	9%			
7180	3-1/2" diam. to 5" diam.					10%	6%			
7190	6" diam. to 12" diam.					7%	4%			
7250	Flanged, 300 lb. weld neck, on yoke & roll hangers									
7260	sized for covering, 10' OC									
7290	Black, 1" diameter	Q-15	66	.242	L.F.	14.05	13.80	.85	28.70	37
7300	1-1/4" diameter		61	.262		15.80	14.90	.92	31.62	41
7310	1-1/2" diameter		54	.296		16.75	16.85	1.04	34.64	45
7320	2" diameter		42	.381		21	21.50	1.34	43.84	57
7330	2-1/2" diameter		33	.485		26	27.50	1.70	55.20	72
7340	3" diameter		29	.552		29	31.50	1.94	62.44	81
7350	3-1/2" diameter		24	.667		40.50	38	2.34	80.84	104
7360	4" diameter		23	.696		47.50	39.50	2.44	89.44	115
7370	5" diameter		19	.842		92.50	48	2.96	143.46	177
7380	6" diameter	Q-16	23	1.043		89.50	61.50	2.44	153.44	194
7390	8" diameter		17	1.412		141	83	3.30	227.30	284
7400	10" diameter		14	1.714		236	101	4.01	341.01	415
7410	12" diameter		12	2		415	118	4.68	537.68	640
7470	For 600 lb. flanges, add					100%				
7940	To delete flanges & hanger, subtract									
7950	1" diam. to 1-1/2" diam.					75%	66%			
7960	2" diam. to 3" diam.					62%	60%			
7970	3-1/2" diam. to 5" diam.					54%	66%			
7980	6" diam. to 12" diam.					55%	62%			
8040	Galvanized, 1" diameter	Q-15	66	.242	L.F.	15.10	13.80	.85	29.75	38
8050	1-1/4" diameter		61	.262		17.45	14.90	.92	33.27	42.50
8060	1-1/2" diameter		54	.296		18.55	16.85	1.04	36.44	47
8070	2" diameter		42	.381		24	21.50	1.34	46.84	60.50
8080	2-1/2" diameter		33	.485		30.50	27.50	1.70	59.70	77
8090	3" diameter		29	.552		34	31.50	1.94	67.44	86
8100	3-1/2" diameter		24	.667		52	38	2.34	92.34	117
8110	4" diameter		23	.696		51.50	39.50	2.44	93.44	119
8120	5" diameter, A-106B		19	.842		74	48	2.96	124.96	157
8130	6" diameter, A-106B	Q-16	23	1.043		85.50	61.50	2.44	149.44	189
8140	8" diameter, A-106B		17	1.412		160	83	3.30	246.30	305
8150	10" diameter, A-106B		14	1.714		345	101	4.01	450.01	535
8160	12" diameter, A-106B		12	2		590	118	4.68	712.68	830
8240	For 600 lb. flanges, add					100%				
8300	To delete flanges & hangers, subtract									
8310	1" diam. to 1-1/2" diam.					72%	66%			
8320	2" diam. to 3" diam.					59%	60%			
8330	3-1/2" diam. to 5" diam.					51%	66%			
8340	6" diam. to 12" diam.					49%	62%			
9000	Threading pipe labor, one end, all schedules through 80									
9010	1/4" through 3/4" pipe size	1 Plum	80	.100	Ea.		6.30		6.30	9.50
9020	1" through 2" pipe size		73	.110			6.90		6.90	10.40
9030	2-1/2" pipe size		53	.151			9.55		9.55	14.30
9040	3" pipe size		50	.160			10.10		10.10	15.15
9050	3-1/2" pipe size	Q-1	89	.180			10.20		10.20	15.35

22 11 13.44 Pipe, Steel		Crew	Daily Output	Labor-Hours	Unit	Material	2019 Bare Costs Labor	Equipment	Total	Total Incl O&P
9060	4" pipe size	Q-1	73	.219	Ea.		12.45		12.45	18.70
9070	5" pipe size		53	.302			17.15		17.15	26
9080	6" pipe size		46	.348			19.75		19.75	29.50
9090	8" pipe size		29	.552			31.50		31.50	47
9100	10" pipe size		21	.762			43.50		43.50	65
9110	12" pipe size		13	1.231			70		70	105
9120	Cutting pipe labor, one cut									
9124	Shop fabrication, machine cut									
9126	Schedule 40, straight pipe									
9128	2" pipe size or less	1 Stpi	62	.129	Ea.		8.25		8.25	12.40
9130	2-1/2" pipe size		56	.143			9.15		9.15	13.70
9132	3" pipe size		42	.190			12.20		12.20	18.30
9134	4" pipe size		31	.258			16.50		16.50	25
9136	5" pipe size		26	.308			19.70		19.70	29.50
9138	6" pipe size		19	.421			27		27	40.50
9140	8" pipe size		14	.571			36.50		36.50	55
9142	10" pipe size		10	.800			51		51	77
9144	12" pipe size		7	1.143			73		73	110
9146	14" pipe size	Q-5	10.50	1.524			87.50		87.50	132
9148	16" pipe size		8.60	1.860			107		107	161
9150	18" pipe size		7	2.286			132		132	197
9152	20" pipe size		5.80	2.759			159		159	238
9154	24" pipe size		4	4			230		230	345
9160	Schedule 80, straight pipe									
9164	2" pipe size or less	1 Stpi	42	.190	Ea.		12.20		12.20	18.30
9166	2-1/2" pipe size		37	.216			13.85		13.85	21
9168	3" pipe size		31	.258			16.50		16.50	25
9170	4" pipe size		23	.348			22		22	33.50
9172	5" pipe size		18	.444			28.50		28.50	42.50
9174	6" pipe size		14.60	.548			35		35	52.50
9176	8" pipe size		10	.800			51		51	77
9178	10" pipe size	Q-5	14	1.143			66		66	98.50
9180	12" pipe size	"	10	1.600			92		92	138
9200	Welding labor per joint									
9210	Schedule 40									
9230	1/2" pipe size	Q-15	32	.500	Ea.		28.50	1.76	30.26	44.50
9240	3/4" pipe size		27	.593			33.50	2.08	35.58	53
9250	1" pipe size		23	.696			39.50	2.44	41.94	62
9260	1-1/4" pipe size		20	.800			45.50	2.81	48.31	71
9270	1-1/2" pipe size		19	.842			48	2.96	50.96	75.50
9280	2" pipe size		16	1			57	3.51	60.51	89.50
9290	2-1/2" pipe size		13	1.231			70	4.32	74.32	110
9300	3" pipe size		12	1.333			76	4.68	80.68	119
9310	4" pipe size		10	1.600			91	5.60	96.60	142
9320	5" pipe size		9	1.778			101	6.25	107.25	159
9330	6" pipe size		8	2			114	7	121	179
9340	8" pipe size		5	3.200			182	11.25	193.25	285
9350	10" pipe size		4	4			227	14.05	241.05	355
9360	12" pipe size		3	5.333			305	18.70	323.70	475
9370	14" pipe size		2.60	6.154			350	21.50	371.50	550
9380	16" pipe size		2.20	7.273			415	25.50	440.50	650
9390	18" pipe size		2	8			455	28	483	710
9400	20" pipe size		1.80	8.889			505	31	536	795

22 11 13.44 Pipe, Steel		Crew	Daily Output	Labor-Hours	Unit	Material	2019 Bare Costs Labor	Equipment	Total	Total Incl O&P
9410	22" pipe size	Q-15	1.70	9.412	Ea.		535	33	568	840
9420	24" pipe size		1.50	10.667			605	37.50	642.50	950
9450	Schedule 80									
9460	1/2" pipe size	Q-15	27	.593	Ea.		33.50	2.08	35.58	53
9470	3/4" pipe size		23	.696			39.50	2.44	41.94	62
9480	1" pipe size		20	.800			45.50	2.81	48.31	71
9490	1-1/4" pipe size		19	.842			48	2.96	50.96	75.50
9500	1-1/2" pipe size		18	.889			50.50	3.12	53.62	79.50
9510	2" pipe size		15	1.067			60.50	3.74	64.24	95
9520	2-1/2" pipe size		12	1.333			76	4.68	80.68	119
9530	3" pipe size		11	1.455			82.50	5.10	87.60	130
9540	4" pipe size		8	2			114	7	121	179
9550	5" pipe size		6	2.667			152	9.35	161.35	237
9560	6" pipe size		5	3.200			182	11.25	193.25	285
9570	8" pipe size		4	4			227	14.05	241.05	355
9580	10" pipe size		3	5.333			305	18.70	323.70	475
9590	12" pipe size		2	8			455	28	483	710
9600	14" pipe size	Q-16	2.60	9.231			545	21.50	566.50	840
9610	16" pipe size		2.30	10.435			615	24.50	639.50	950
9620	18" pipe size		2	12			705	28	733	1,075
9630	20" pipe size		1.80	13.333			785	31	816	1,200
9640	22" pipe size		1.60	15			885	35	920	1,375
9650	24" pipe size		1.50	16			945	37.50	982.50	1,475

22 11 13.45 Pipe Fittings, Steel, Threaded		Crew	Daily Output	Labor-Hours	Unit	Material	Labor	Equipment	Total	Total Incl O&P
0010	**PIPE FITTINGS, STEEL, THREADED**									
0020	Cast iron									
0040	Standard weight, black									
0060	90° elbow, straight									
0070	1/4"	1 Plum	16	.500	Ea.	11.15	31.50		42.65	60
0080	3/8"		16	.500		16.20	31.50		47.70	65.50
0090	1/2"		15	.533		7.10	33.50		40.60	58.50
0100	3/4"		14	.571		7.40	36		43.40	62
0110	1"		13	.615		8.75	39		47.75	68
0120	1-1/4"	Q-1	22	.727		12.40	41.50		53.90	75.50
0130	1-1/2"		20	.800		17.15	45.50		62.65	87
0140	2"		18	.889		27	50.50		77.50	106
0150	2-1/2"		14	1.143		64.50	65		129.50	168
0160	3"		10	1.600		105	91		196	252
0170	3-1/2"		8	2		286	114		400	485
0180	4"		6	2.667		195	152		347	440
0250	45° elbow, straight									
0260	1/4"	1 Plum	16	.500	Ea.	13.80	31.50		45.30	62.50
0270	3/8"		16	.500		14.85	31.50		46.35	64
0280	1/2"		15	.533		10.85	33.50		44.35	62.50
0300	3/4"		14	.571		10.90	36		46.90	66
0320	1"		13	.615		12.80	39		51.80	72.50
0330	1-1/4"	Q-1	22	.727		17.20	41.50		58.70	81
0340	1-1/2"		20	.800		28.50	45.50		74	99.50
0350	2"		18	.889		33	50.50		83.50	112
0360	2-1/2"		14	1.143		85.50	65		150.50	192
0370	3"		10	1.600		135	91		226	285
0380	3-1/2"		8	2		315	114		429	520

22 11 13.45 Pipe Fittings, Steel, Threaded		Crew	Daily Output	Labor-Hours	Unit	2019 Bare Costs Material	Labor	Equipment	Total	Total Incl O&P
0400	4"	Q-1	6	2.667	Ea.	282	152		434	535
0500	Tee, straight									
0510	1/4"	1 Plum	10	.800	Ea.	17.50	50.50		68	95.50
0520	3/8"		10	.800		17	50.50		67.50	94.50
0530	1/2"		9	.889		11.05	56		67.05	96
0540	3/4"		9	.889		12.85	56		68.85	98
0550	1"		8	1		11.45	63		74.45	108
0560	1-1/4"	Q-1	14	1.143		21	65		86	121
0570	1-1/2"		13	1.231		27	70		97	135
0580	2"		11	1.455		38	82.50		120.50	166
0590	2-1/2"		9	1.778		98.50	101		199.50	260
0600	3"		6	2.667		151	152		303	395
0610	3-1/2"		5	3.200		305	182		487	610
0620	4"		4	4		291	227		518	660
0660	Tee, reducing, run or outlet									
0661	1/2"	1 Plum	9	.889	Ea.	27	56		83	114
0662	3/4"		9	.889		24.50	56		80.50	111
0663	1"		8	1		27	63		90	125
0664	1-1/4"	Q-1	14	1.143		38	65		103	140
0665	1-1/2"		13	1.231		38	70		108	147
0666	2"		11	1.455		54.50	82.50		137	184
0667	2-1/2"		9	1.778		111	101		212	274
0668	3"		6	2.667		202	152		354	450
0669	3-1/2"		5	3.200		480	182		662	800
0670	4"		4	4		520	227		747	915
0674	Reducer, concentric									
0675	3/4"	1 Plum	18	.444	Ea.	19.80	28		47.80	64
0676	1"	"	15	.533		13.40	33.50		46.90	65.50
0677	1-1/4"	Q-1	26	.615		34.50	35		69.50	90.50
0678	1-1/2"		24	.667		63	38		101	127
0679	2"		21	.762		71.50	43.50		115	144
0680	2-1/2"		18	.889		102	50.50		152.50	188
0681	3"		14	1.143		164	65		229	278
0682	3-1/2"		12	1.333		296	76		372	440
0683	4"		10	1.600		298	91		389	465
0687	Reducer, eccentric									
0688	3/4"	1 Plum	16	.500	Ea.	41.50	31.50		73	93.50
0689	1"	"	14	.571		45	36		81	104
0690	1-1/4"	Q-1	25	.640		72.50	36.50		109	135
0691	1-1/2"		22	.727		90.50	41.50		132	162
0692	2"		20	.800		137	45.50		182.50	219
0693	2-1/2"		16	1		189	57		246	294
0694	3"		12	1.333		298	76		374	445
0695	3-1/2"		10	1.600		425	91		516	600
0696	4"		9	1.778		540	101		641	740
0700	Standard weight, galvanized cast iron									
0720	90° elbow, straight									
0730	1/4"	1 Plum	16	.500	Ea.	18.40	31.50		49.90	68
0740	3/8"		16	.500		18.40	31.50		49.90	68
0750	1/2"		15	.533		21	33.50		54.50	73.50
0760	3/4"		14	.571		21	36		57	77
0770	1"		13	.615		24	39		63	84.50
0780	1-1/4"	Q-1	22	.727		37	41.50		78.50	103

22 11 13.45 Pipe Fittings, Steel, Threaded		Crew	Daily Output	Labor-Hours	Unit	2019 Bare Costs Material	Labor	Equipment	Total	Total Incl O&P
0790	1-1/2"	Q-1	20	.800	Ea.	51	45.50		96.50	124
0800	2"		18	.889		75	50.50		125.50	159
0810	2-1/2"		14	1.143		154	65		219	268
0820	3"		10	1.600		234	91		325	395
0830	3-1/2"		8	2		375	114		489	580
0840	4"		6	2.667		430	152		582	700
0900	45° elbow, straight									
0910	1/4"	1 Plum	16	.500	Ea.	21.50	31.50		53	71
0920	3/8"		16	.500		21	31.50		52.50	71
0930	1/2"		15	.533		21	33.50		54.50	73.50
0940	3/4"		14	.571		24	36		60	80.50
0950	1"		13	.615		31	39		70	92.50
0960	1-1/4"	Q-1	22	.727		46	41.50		87.50	113
0970	1-1/2"		20	.800		65.50	45.50		111	141
0980	2"		18	.889		92	50.50		142.50	177
0990	2-1/2"		14	1.143		145	65		210	257
1000	3"		10	1.600		290	91		381	455
1010	3-1/2"		8	2		530	114		644	750
1020	4"		6	2.667		475	152		627	750
1100	Tee, straight									
1110	1/4"	1 Plum	10	.800	Ea.	20	50.50		70.50	98
1120	3/8"		10	.800		22.50	50.50		73	101
1130	1/2"		9	.889		22.50	56		78.50	109
1140	3/4"		9	.889		29.50	56		85.50	117
1150	1"		8	1		32	63		95	131
1160	1-1/4"	Q-1	14	1.143		56.50	65		121.50	160
1170	1-1/2"		13	1.231		74.50	70		144.50	187
1180	2"		11	1.455		93	82.50		175.50	226
1190	2-1/2"		9	1.778		189	101		290	360
1200	3"		6	2.667		425	152		577	695
1210	3-1/2"		5	3.200		530	182		712	860
1220	4"		4	4		585	227		812	985
1300	Extra heavy weight, black									
1310	Couplings, steel straight									
1320	1/4"	1 Plum	19	.421	Ea.	4.96	26.50		31.46	45.50
1330	3/8"		19	.421		5.45	26.50		31.95	46
1340	1/2"		19	.421		7.35	26.50		33.85	48
1350	3/4"		18	.444		7.85	28		35.85	50.50
1360	1"		15	.533		10	33.50		43.50	61.50
1370	1-1/4"	Q-1	26	.615		16.05	35		51.05	70
1380	1-1/2"		24	.667		16.05	38		54.05	74.50
1390	2"		21	.762		24.50	43.50		68	92
1400	2-1/2"		18	.889		36	50.50		86.50	116
1410	3"		14	1.143		43	65		108	145
1420	3-1/2"		12	1.333		58	76		134	178
1430	4"		10	1.600		67.50	91		158.50	211
1510	90° elbow, straight									
1520	1/2"	1 Plum	15	.533	Ea.	36.50	33.50		70	90.50
1530	3/4"		14	.571		37	36		73	95
1540	1"		13	.615		45	39		84	108
1550	1-1/4"	Q-1	22	.727		67	41.50		108.50	136
1560	1-1/2"		20	.800		83	45.50		128.50	159
1580	2"		18	.889		102	50.50		152.50	189

22 11 13.45 Pipe Fittings, Steel, Threaded		Crew	Daily Output	Labor-Hours	Unit	2019 Bare Costs Material	Labor	Equipment	Total	Total Incl O&P
1590	2-1/2"	Q-1	14	1.143	Ea.	249	65		314	370
1600	3"		10	1.600		335	91		426	500
1610	4"		6	2.667		725	152		877	1,025
1650	45° elbow, straight									
1660	1/2"	1 Plum	15	.533	Ea.	52	33.50		85.50	108
1670	3/4"		14	.571		50.50	36		86.50	110
1680	1"		13	.615		60.50	39		99.50	125
1690	1-1/4"	Q-1	22	.727		99.50	41.50		141	171
1700	1-1/2"		20	.800		109	45.50		154.50	188
1710	2"		18	.889		156	50.50		206.50	247
1720	2-1/2"		14	1.143		269	65		334	395
1800	Tee, straight									
1810	1/2"	1 Plum	9	.889	Ea.	57	56		113	147
1820	3/4"		9	.889		57	56		113	147
1830	1"		8	1		69	63		132	171
1840	1-1/4"	Q-1	14	1.143		103	65		168	211
1850	1-1/2"		13	1.231		132	70		202	250
1860	2"		11	1.455		164	82.50		246.50	305
1870	2-1/2"		9	1.778		345	101		446	530
1880	3"		6	2.667		490	152		642	760
1890	4"		4	4		950	227		1,177	1,400
4000	Standard weight, black									
4010	Couplings, steel straight, merchants									
4030	1/4"	1 Plum	19	.421	Ea.	1.31	26.50		27.81	41.50
4040	3/8"		19	.421		1.60	26.50		28.10	42
4050	1/2"		19	.421		1.69	26.50		28.19	42
4060	3/4"		18	.444		2.12	28		30.12	44.50
4070	1"		15	.533		2.97	33.50		36.47	54
4080	1-1/4"	Q-1	26	.615		3.79	35		38.79	56.50
4090	1-1/2"		24	.667		4.80	38		42.80	62.50
4100	2"		21	.762		6.85	43.50		50.35	72.50
4110	2-1/2"		18	.889		21.50	50.50		72	99.50
4120	3"		14	1.143		30	65		95	131
4130	3-1/2"		12	1.333		53.50	76		129.50	173
4140	4"		10	1.600		53.50	91		144.50	195
4166	Plug, 1/4"	1 Plum	38	.211		3.22	13.30		16.52	23.50
4167	3/8"		38	.211		3.05	13.30		16.35	23.50
4168	1/2"		38	.211		3.15	13.30		16.45	23.50
4169	3/4"		32	.250		9.50	15.80		25.30	34
4170	1"		30	.267		10.10	16.85		26.95	36.50
4171	1-1/4"	Q-1	52	.308		11.65	17.50		29.15	39
4172	1-1/2"		48	.333		16.50	18.95		35.45	46.50
4173	2"		42	.381		21.50	21.50		43	56
4176	2-1/2"		36	.444		31.50	25.50		57	72.50
4180	4"		20	.800		58	45.50		103.50	132
4200	Standard weight, galvanized									
4210	Couplings, steel straight, merchants									
4230	1/4"	1 Plum	19	.421	Ea.	1.53	26.50		28.03	41.50
4240	3/8"		19	.421		1.95	26.50		28.45	42
4250	1/2"		19	.421		2.06	26.50		28.56	42.50
4260	3/4"		18	.444		2.55	28		30.55	45
4270	1"		15	.533		3.57	33.50		37.07	54.50
4280	1-1/4"	Q-1	26	.615		4.57	35		39.57	57.50

22 11 13.45 Pipe Fittings, Steel, Threaded		Crew	Daily Output	Labor-Hours	Unit	Material	Labor	Equipment	Total	Total Incl O&P
						2019 Bare Costs				
4290	1-1/2"	Q-1	24	.667	Ea.	5.65	38		43.65	63.50
4300	2"		21	.762		8.45	43.50		51.95	74.50
4310	2-1/2"		18	.889		27	50.50		77.50	106
4320	3"		14	1.143		35.50	65		100.50	137
4330	3-1/2"		12	1.333		61.50	76		137.50	182
4340	4"		10	1.600		61.50	91		152.50	204
4370	Plug, galvanized, square head									
4374	1/2"	1 Plum	38	.211	Ea.	8.10	13.30		21.40	29
4375	3/4"		32	.250		7.60	15.80		23.40	32
4376	1"		30	.267		7.60	16.85		24.45	34
4377	1-1/4"	Q-1	52	.308		12.70	17.50		30.20	40
4378	1-1/2"		48	.333		17.15	18.95		36.10	47.50
4379	2"		42	.381		21.50	21.50		43	56
4380	2-1/2"		36	.444		45	25.50		70.50	87.50
4381	3"		28	.571		57.50	32.50		90	112
4382	4"		20	.800		158	45.50		203.50	242
4700	Nipple, black									
4710	1/2" x 4" long	1 Plum	19	.421	Ea.	3.07	26.50		29.57	43.50
4712	3/4" x 4" long		18	.444		3.70	28		31.70	46
4714	1" x 4" long		15	.533		5.15	33.50		38.65	56
4716	1-1/4" x 4" long	Q-1	26	.615		6.45	35		41.45	59.50
4718	1-1/2" x 4" long		24	.667		7.60	38		45.60	65.50
4720	2" x 4" long		21	.762		10.60	43.50		54.10	76.50
4722	2-1/2" x 4" long		18	.889		29	50.50		79.50	108
4724	3" x 4" long		14	1.143		36.50	65		101.50	138
4726	4" x 4" long		10	1.600		49.50	91		140.50	190
4800	Nipple, galvanized									
4810	1/2" x 4" long	1 Plum	19	.421	Ea.	3.76	26.50		30.26	44
4812	3/4" x 4" long		18	.444		4.63	28		32.63	47
4814	1" x 4" long		15	.533		6.25	33.50		39.75	57.50
4816	1-1/4" x 4" long	Q-1	26	.615		7.65	35		42.65	61
4818	1-1/2" x 4" long		24	.667		9.70	38		47.70	67.50
4820	2" x 4" long		21	.762		12.35	43.50		55.85	78.50
4822	2-1/2" x 4" long		18	.889		33	50.50		83.50	113
4824	3" x 4" long		14	1.143		43	65		108	145
4826	4" x 4" long		10	1.600		58	91		149	200
5000	Malleable iron, 150 lb.									
5020	Black									
5040	90° elbow, straight									
5060	1/4"	1 Plum	16	.500	Ea.	5.50	31.50		37	53.50
5070	3/8"		16	.500		5.50	31.50		37	53.50
5080	1/2"		15	.533		3.69	33.50		37.19	54.50
5090	3/4"		14	.571		4.46	36		40.46	59
5100	1"		13	.615		7.80	39		46.80	67
5110	1-1/4"	Q-1	22	.727		12.90	41.50		54.40	76
5120	1-1/2"		20	.800		17	45.50		62.50	86.50
5130	2"		18	.889		29.50	50.50		80	109
5140	2-1/2"		14	1.143		66.50	65		131.50	171
5150	3"		10	1.600		97.50	91		188.50	243
5160	3-1/2"		8	2		269	114		383	465
5170	4"		6	2.667		210	152		362	455
5250	45° elbow, straight									
5270	1/4"	1 Plum	16	.500	Ea.	8.30	31.50		39.80	56.50

22 11 13.45 Pipe Fittings, Steel, Threaded		Crew	Daily Output	Labor-Hours	Unit	Material	2019 Bare Costs Labor	Equipment	Total	Total Incl O&P
5280	3/8"	1 Plum	16	.500	Ea.	8.30	31.50		39.80	56.50
5290	1/2"		15	.533		6.30	33.50		39.80	57.50
5300	3/4"		14	.571		7.80	36		43.80	62.50
5310	1"		13	.615		9.80	39		48.80	69.50
5320	1-1/4"	Q-1	22	.727		17.35	41.50		58.85	81
5330	1-1/2"		20	.800		21.50	45.50		67	91.50
5340	2"		18	.889		32.50	50.50		83	112
5350	2-1/2"		14	1.143		94	65		159	201
5360	3"		10	1.600		124	91		215	273
5370	3-1/2"		8	2		254	114		368	450
5380	4"		6	2.667		239	152		391	490
5450	Tee, straight									
5470	1/4"	1 Plum	10	.800	Ea.	8	50.50		58.50	85
5480	3/8"		10	.800		8	50.50		58.50	85
5490	1/2"		9	.889		5.10	56		61.10	89.50
5500	3/4"		9	.889		7.35	56		63.35	92
5510	1"		8	1		12.55	63		75.55	109
5520	1-1/4"	Q-1	14	1.143		20.50	65		85.50	120
5530	1-1/2"		13	1.231		25.50	70		95.50	133
5540	2"		11	1.455		43	82.50		125.50	172
5550	2-1/2"		9	1.778		93	101		194	254
5560	3"		6	2.667		137	152		289	375
5570	3-1/2"		5	3.200		320	182		502	625
5580	4"		4	4		330	227		557	705
5601	Tee, reducing, on outlet									
5602	1/2"	1 Plum	9	.889	Ea.	13.30	56		69.30	98.50
5603	3/4"		9	.889		14.25	56		70.25	99.50
5604	1"		8	1		20	63		83	117
5605	1-1/4"	Q-1	14	1.143		35.50	65		100.50	137
5606	1-1/2"		13	1.231		35.50	70		105.50	144
5607	2"		11	1.455		48.50	82.50		131	177
5608	2-1/2"		9	1.778		141	101		242	305
5609	3"		6	2.667		192	152		344	440
5610	3-1/2"		5	3.200		475	182		657	800
5611	4"		4	4		385	227		612	765
5650	Coupling									
5670	1/4"	1 Plum	19	.421	Ea.	6.85	26.50		33.35	47.50
5680	3/8"		19	.421		6.85	26.50		33.35	47.50
5690	1/2"		19	.421		5.25	26.50		31.75	46
5700	3/4"		18	.444		6.20	28		34.20	49
5710	1"		15	.533		9.25	33.50		42.75	60.50
5720	1-1/4"	Q-1	26	.615		12.25	35		47.25	66
5730	1-1/2"		24	.667		16.20	38		54.20	75
5740	2"		21	.762		24	43.50		67.50	91.50
5750	2-1/2"		18	.889		66	50.50		116.50	149
5760	3"		14	1.143		89.50	65		154.50	196
5770	3-1/2"		12	1.333		153	76		229	282
5780	4"		10	1.600		180	91		271	335
5840	Reducer, concentric, 1/4"	1 Plum	19	.421		7.20	26.50		33.70	48
5850	3/8"		19	.421		9.25	26.50		35.75	50
5860	1/2"		19	.421		7.30	26.50		33.80	48
5870	3/4"		16	.500		8.65	31.50		40.15	57
5880	1"		15	.533		14.50	33.50		48	66.50

22 11 13.45 Pipe Fittings, Steel, Threaded		Crew	Daily Output	Labor-Hours	Unit	Material	2019 Bare Costs Labor	Equipment	Total	Total Incl O&P
5890	1-1/4"	Q-1	26	.615	Ea.	16.30	35		51.30	70.50
5900	1-1/2"		24	.667		23.50	38		61.50	82.50
5910	2"		21	.762		33.50	43.50		77	102
5911	2-1/2"		18	.889		75	50.50		125.50	159
5912	3"		14	1.143		91.50	65		156.50	199
5913	3-1/2"		12	1.333		289	76		365	435
5914	4"		10	1.600		220	91		311	380
5981	Bushing, 1/4"	1 Plum	19	.421		1.07	26.50		27.57	41
5982	3/8"		19	.421		5.65	26.50		32.15	46
5983	1/2"		19	.421		6.05	26.50		32.55	46.50
5984	3/4"		16	.500		7.15	31.50		38.65	55.50
5985	1"		15	.533		10.15	33.50		43.65	61.50
5986	1-1/4"	Q-1	26	.615		12.60	35		47.60	66.50
5987	1-1/2"		24	.667		10.65	38		48.65	68.50
5988	2"		21	.762		13.35	43.50		56.85	79.50
5989	Cap, 1/4"	1 Plum	38	.211		7.15	13.30		20.45	28
5991	3/8"		38	.211		6	13.30		19.30	26.50
5992	1/2"		38	.211		3.89	13.30		17.19	24
5993	3/4"		32	.250		5.20	15.80		21	29.50
5994	1"		30	.267		6.30	16.85		23.15	32.50
5995	1-1/4"	Q-1	52	.308		8.35	17.50		25.85	35
5996	1-1/2"		48	.333		11.45	18.95		30.40	41
5997	2"		42	.381		16.70	21.50		38.20	51
6000	For galvanized elbows, tees, and couplings, add					20%				
6058	For galvanized reducers, caps and bushings, add					20%				
6100	90° elbow, galvanized, 150 lb., reducing									
6110	3/4" x 1/2"	1 Plum	15.40	.519	Ea.	10.95	33		43.95	61
6112	1" x 3/4"		14	.571		14.05	36		50.05	69.50
6114	1" x 1/2"		14.50	.552		14.95	35		49.95	69
6116	1-1/4" x 1"	Q-1	24.20	.661		23.50	37.50		61	82.50
6118	1-1/4" x 3/4"		25.40	.630		28.50	36		64.50	84.50
6120	1-1/4" x 1/2"		26.20	.611		30	34.50		64.50	85
6122	1-1/2" x 1-1/4"		21.60	.741		37.50	42		79.50	105
6124	1-1/2" x 1"		23.50	.681		37.50	38.50		76	99.50
6126	1-1/2" x 3/4"		24.60	.650		37.50	37		74.50	97
6128	2" x 1-1/2"		20.50	.780		44	44.50		88.50	115
6130	2" x 1-1/4"		21	.762		50.50	43.50		94	121
6132	2" x 1"		22.80	.702		52.50	40		92.50	118
6134	2" x 3/4"		23.90	.669		53.50	38		91.50	116
6136	2-1/2" x 2"		12.30	1.301		149	74		223	275
6138	2-1/2" x 1-1/2"		12.50	1.280		166	72.50		238.50	292
6140	3" x 2-1/2"		8.60	1.860		272	106		378	460
6142	3" x 2"		11.80	1.356		236	77		313	375
6144	4" x 3"		8.20	1.951		615	111		726	840
6160	90° elbow, black, 150 lb., reducing									
6170	1" x 3/4"	1 Plum	14	.571	Ea.	9.60	36		45.60	64.50
6174	1-1/2" x 1"	Q-1	23.50	.681		23	38.50		61.50	83
6178	1-1/2" x 3/4"		24.60	.650		26	37		63	84
6182	2" x 1-1/2"		20.50	.780		33	44.50		77.50	103
6186	2" x 1"		22.80	.702		37.50	40		77.50	102
6190	2" x 3/4"		23.90	.669		40	38		78	101
6194	2-1/2" x 2"		12.30	1.301		90.50	74		164.50	211
7000	Union, with brass seat									

22 11 13.45 Pipe Fittings, Steel, Threaded		Crew	Daily Output	Labor-Hours	Unit	2019 Bare Costs Material	Labor	Equipment	Total	Total Incl O&P
7010	1/4"	1 Plum	15	.533	Ea.	28.50	33.50		62	82
7020	3/8"		15	.533		19.90	33.50		53.40	72.50
7030	1/2"		14	.571		18.05	36		54.05	74
7040	3/4"		13	.615		20.50	39		59.50	81.50
7050	1"		12	.667		27	42		69	92.50
7060	1-1/4"	Q-1	21	.762		39	43.50		82.50	108
7070	1-1/2"		19	.842		48.50	48		96.50	125
7080	2"		17	.941		56	53.50		109.50	143
7090	2-1/2"		13	1.231		165	70		235	287
7100	3"		9	1.778		202	101		303	375
7120	Union, galvanized									
7124	1/2"	1 Plum	14	.571	Ea.	24	36		60	80.50
7125	3/4"		13	.615		27.50	39		66.50	89
7126	1"		12	.667		36	42		78	103
7127	1-1/4"	Q-1	21	.762		52	43.50		95.50	123
7128	1-1/2"		19	.842		63	48		111	142
7129	2"		17	.941		72.50	53.50		126	161
7130	2-1/2"		13	1.231		252	70		322	385
7131	3"		9	1.778		350	101		451	535
7500	Malleable iron, 300 lb.									
7520	Black									
7540	90° elbow, straight, 1/4"	1 Plum	16	.500	Ea.	19.35	31.50		50.85	69
7560	3/8"		16	.500		17.15	31.50		48.65	66.50
7570	1/2"		15	.533		22.50	33.50		56	75
7580	3/4"		14	.571		25	36		61	81.50
7590	1"		13	.615		32	39		71	94
7600	1-1/4"	Q-1	22	.727		47	41.50		88.50	114
7610	1-1/2"		20	.800		55.50	45.50		101	129
7620	2"		18	.889		79.50	50.50		130	164
7630	2-1/2"		14	1.143		206	65		271	325
7640	3"		10	1.600		237	91		328	395
7650	4"		6	2.667		560	152		712	840
7700	45° elbow, straight, 1/4"	1 Plum	16	.500		29	31.50		60.50	79
7720	3/8"		16	.500		28.50	31.50		60	79
7730	1/2"		15	.533		32	33.50		65.50	85.50
7740	3/4"		14	.571		35.50	36		71.50	93
7750	1"		13	.615		39.50	39		78.50	102
7760	1-1/4"	Q-1	22	.727		63	41.50		104.50	132
7770	1-1/2"		20	.800		82.50	45.50		128	159
7780	2"		18	.889		124	50.50		174.50	213
7790	2-1/2"		14	1.143		278	65		343	405
7800	3"		10	1.600		365	91		456	540
7810	4"		6	2.667		835	152		987	1,150
7850	Tee, straight, 1/4"	1 Plum	10	.800		24.50	50.50		75	103
7870	3/8"		10	.800		26	50.50		76.50	105
7880	1/2"		9	.889		33	56		89	120
7890	3/4"		9	.889		35.50	56		91.50	123
7900	1"		8	1		42.50	63		105.50	142
7910	1-1/4"	Q-1	14	1.143		61	65		126	165
7920	1-1/2"		13	1.231		71	70		141	184
7930	2"		11	1.455		105	82.50		187.50	240
7940	2-1/2"		9	1.778		256	101		357	435
7950	3"		6	2.667		365	152		517	625

22 11 13.45 Pipe Fittings, Steel, Threaded		Crew	Daily Output	Labor-Hours	Unit	Material	2019 Bare Costs Labor	Equipment	Total	Total Incl O&P
7960	4"	Q-1	4	4	Ea.	1,050	227		1,277	1,500
8050	Couplings, straight, 1/4"	1 Plum	19	.421		17.45	26.50		43.95	59
8070	3/8"		19	.421		17.45	26.50		43.95	59
8080	1/2"		19	.421		19.45	26.50		45.95	61.50
8090	3/4"		18	.444		22.50	28		50.50	66.50
8100	1"		15	.533		25.50	33.50		59	78.50
8110	1-1/4"	Q-1	26	.615		30.50	35		65.50	86
8120	1-1/2"		24	.667		45.50	38		83.50	107
8130	2"		21	.762		65.50	43.50		109	137
8140	2-1/2"		18	.889		121	50.50		171.50	209
8150	3"		14	1.143		175	65		240	290
8160	4"		10	1.600		340	91		431	505
8162	6"		10	1.600		340	91		431	505
8200	Galvanized									
8220	90° elbow, straight, 1/4"	1 Plum	16	.500	Ea.	34.50	31.50		66	85.50
8222	3/8"		16	.500		36	31.50		67.50	87
8224	1/2"		15	.533		42.50	33.50		76	97.50
8226	3/4"		14	.571		48	36		84	107
8228	1"		13	.615		61.50	39		100.50	126
8230	1-1/4"	Q-1	22	.727		99	41.50		140.50	171
8232	1-1/2"		20	.800		104	45.50		149.50	183
8234	2"		18	.889		178	50.50		228.50	272
8236	2-1/2"		14	1.143		385	65		450	525
8238	3"		10	1.600		455	91		546	635
8240	4"		6	2.667		1,350	152		1,502	1,725
8280	45° elbow, straight									
8282	1/2"	1 Plum	15	.533	Ea.	62.50	33.50		96	119
8284	3/4"		14	.571		72	36		108	134
8286	1"		13	.615		79	39		118	146
8288	1-1/4"	Q-1	22	.727		122	41.50		163.50	197
8290	1-1/2"		20	.800		158	45.50		203.50	242
8292	2"		18	.889		219	50.50		269.50	315
8310	Tee, straight, 1/4"	1 Plum	10	.800		49.50	50.50		100	131
8312	3/8"		10	.800		50	50.50		100.50	131
8314	1/2"		9	.889		63	56		119	153
8316	3/4"		9	.889		69	56		125	160
8318	1"		8	1		85.50	63		148.50	189
8320	1-1/4"	Q-1	14	1.143		122	65		187	232
8322	1-1/2"		13	1.231		127	70		197	244
8324	2"		11	1.455		202	82.50		284.50	345
8326	2-1/2"		9	1.778		640	101		741	855
8328	3"		6	2.667		725	152		877	1,025
8330	4"		4	4		1,875	227		2,102	2,400
8380	Couplings, straight, 1/4"	1 Plum	19	.421		22.50	26.50		49	65
8382	3/8"		19	.421		32.50	26.50		59	76
8384	1/2"		19	.421		33.50	26.50		60	77
8386	3/4"		18	.444		38	28		66	83.50
8388	1"		15	.533		50	33.50		83.50	106
8390	1-1/4"	Q-1	26	.615		65.50	35		100.50	125
8392	1-1/2"		24	.667		92	38		130	158
8394	2"		21	.762		112	43.50		155.50	188
8396	2-1/2"		18	.889		294	50.50		344.50	400
8398	3"		14	1.143		350	65		415	485

22 11 13.45 Pipe Fittings, Steel, Threaded		Crew	Daily Output	Labor-Hours	Unit	Material	Labor	Equipment	Total	Total Incl O&P
						2019 Bare Costs				
8399	4"	Q-1	10	1.600	Ea.	540	91		631	730
8529	Black									
8530	Reducer, concentric, 1/4"	1 Plum	19	.421	Ea.	23	26.50		49.50	65.50
8531	3/8"		19	.421		23	26.50		49.50	65.50
8532	1/2"		17	.471		28	29.50		57.50	75
8533	3/4"		16	.500		39.50	31.50		71	91
8534	1"		15	.533		51.50	33.50		85	108
8535	1-1/4"	Q-1	26	.615		71	35		106	131
8536	1-1/2"		24	.667		74	38		112	138
8537	2"		21	.762		107	43.50		150.50	182
8550	Cap, 1/4"	1 Plum	38	.211		17.90	13.30		31.20	39.50
8551	3/8"		38	.211		17.90	13.30		31.20	39.50
8552	1/2"		34	.235		17.75	14.85		32.60	42
8553	3/4"		32	.250		22.50	15.80		38.30	48
8554	1"		30	.267		29.50	16.85		46.35	58
8555	1-1/4"	Q-1	52	.308		32.50	17.50		50	62
8556	1-1/2"		48	.333		47.50	18.95		66.45	81
8557	2"		42	.381		66	21.50		87.50	106
8570	Plug, 1/4"	1 Plum	38	.211		3.48	13.30		16.78	24
8571	3/8"		38	.211		3.62	13.30		16.92	24
8572	1/2"		34	.235		3.67	14.85		18.52	26.50
8573	3/4"		32	.250		4.58	15.80		20.38	28.50
8574	1"		30	.267		7	16.85		23.85	33
8575	1-1/4"	Q-1	52	.308		14.45	17.50		31.95	42
8576	1-1/2"		48	.333		16.50	18.95		35.45	46.50
8577	2"		42	.381		26	21.50		47.50	61
9500	Union with brass seat, 1/4"	1 Plum	15	.533		42	33.50		75.50	97
9530	3/8"		15	.533		42.50	33.50		76	97.50
9540	1/2"		14	.571		34.50	36		70.50	92
9550	3/4"		13	.615		38.50	39		77.50	101
9560	1"		12	.667		50	42		92	118
9570	1-1/4"	Q-1	21	.762		81.50	43.50		125	155
9580	1-1/2"		19	.842		85.50	48		133.50	166
9590	2"		17	.941		106	53.50		159.50	197
9600	2-1/2"		13	1.231		340	70		410	480
9610	3"		9	1.778		440	101		541	635
9620	4"		5	3.200		1,400	182		1,582	1,825
9630	Union, all iron, 1/4"	1 Plum	15	.533		65.50	33.50		99	123
9650	3/8"		15	.533		65.50	33.50		99	123
9660	1/2"		14	.571		68.50	36		104.50	130
9670	3/4"		13	.615		74	39		113	140
9680	1"		12	.667		91.50	42		133.50	164
9690	1-1/4"	Q-1	21	.762		139	43.50		182.50	218
9700	1-1/2"		19	.842		170	48		218	259
9710	2"		17	.941		215	53.50		268.50	315
9720	2-1/2"		13	1.231		380	70		450	525
9730	3"		9	1.778		725	101		826	945
9750	For galvanized unions, add					15%				
9757	Forged steel, 3000 lb.									
9758	Black									
9760	90° elbow, 1/4"	1 Plum	16	.500	Ea.	19.75	31.50		51.25	69
9761	3/8"		16	.500		19.75	31.50		51.25	69
9762	1/2"		15	.533		15.20	33.50		48.70	67

22 11 13.45 Pipe Fittings, Steel, Threaded		Crew	Daily Output	Labor-Hours	Unit	2019 Bare Costs Material	Labor	Equipment	Total	Total Incl O&P
9763	3/4"	1 Plum	14	.571	Ea.	18.95	36		54.95	75
9764	1"		13	.615		28	39		67	89.50
9765	1-1/4"	Q-1	22	.727		53.50	41.50		95	121
9766	1-1/2"		20	.800		69	45.50		114.50	144
9767	2"		18	.889		84	50.50		134.50	169
9780	45° elbow, 1/4"	1 Plum	16	.500		25	31.50		56.50	75
9781	3/8"		16	.500		25	31.50		56.50	75
9782	1/2"		15	.533		24.50	33.50		58	77.50
9783	3/4"		14	.571		28.50	36		64.50	85.50
9784	1"		13	.615		39	39		78	102
9785	1-1/4"	Q-1	22	.727		53.50	41.50		95	121
9786	1-1/2"		20	.800		77	45.50		122.50	153
9787	2"		18	.889		106	50.50		156.50	193
9800	Tee, 1/4"	1 Plum	10	.800		24	50.50		74.50	103
9801	3/8"		10	.800		24	50.50		74.50	103
9802	1/2"		9	.889		21.50	56		77.50	108
9803	3/4"		9	.889		29	56		85	116
9804	1"		8	1		38.50	63		101.50	137
9805	1-1/4"	Q-1	14	1.143		74.50	65		139.50	180
9806	1-1/2"		13	1.231		87	70		157	201
9807	2"		11	1.455		106	82.50		188.50	240
9820	Reducer, concentric, 1/4"	1 Plum	19	.421		12.25	26.50		38.75	53.50
9821	3/8"		19	.421		12.75	26.50		39.25	54
9822	1/2"		17	.471		12.75	29.50		42.25	58.50
9823	3/4"		16	.500		15.15	31.50		46.65	64
9824	1"		15	.533		19.70	33.50		53.20	72
9825	1-1/4"	Q-1	26	.615		34	35		69	90
9826	1-1/2"		24	.667		37	38		75	98
9827	2"		21	.762		54	43.50		97.50	124
9840	Cap, 1/4"	1 Plum	38	.211		7.95	13.30		21.25	28.50
9841	3/8"		38	.211		7.40	13.30		20.70	28
9842	1/2"		34	.235		7.20	14.85		22.05	30.50
9843	3/4"		32	.250		10.05	15.80		25.85	34.50
9844	1"		30	.267		15.45	16.85		32.30	42.50
9845	1-1/4"	Q-1	52	.308		25	17.50		42.50	53.50
9846	1-1/2"		48	.333		29.50	18.95		48.45	61
9847	2"		42	.381		42.50	21.50		64	79.50
9860	Plug, 1/4"	1 Plum	38	.211		3.49	13.30		16.79	24
9861	3/8"		38	.211		3.57	13.30		16.87	24
9862	1/2"		34	.235		3.66	14.85		18.51	26.50
9863	3/4"		32	.250		4.75	15.80		20.55	29
9864	1"		30	.267		7.30	16.85		24.15	33.50
9865	1-1/4"	Q-1	52	.308		15.60	17.50		33.10	43
9866	1-1/2"		48	.333		17.85	18.95		36.80	48
9867	2"		42	.381		28	21.50		49.50	63.50
9880	Union, bronze seat, 1/4"	1 Plum	15	.533		56	33.50		89.50	112
9881	3/8"		15	.533		56	33.50		89.50	112
9882	1/2"		14	.571		53.50	36		89.50	113
9883	3/4"		13	.615		72.50	39		111.50	139
9884	1"		12	.667		81	42		123	153
9885	1-1/4"	Q-1	21	.762		153	43.50		196.50	233
9886	1-1/2"		19	.842		162	48		210	250
9887	2"		17	.941		192	53.50		245.50	293

22 11 13.45 Pipe Fittings, Steel, Threaded

	22 11 13.45 Pipe Fittings, Steel, Threaded	Crew	Daily Output	Labor-Hours	Unit	Material	2019 Bare Costs Labor	Equipment	Total	Total Incl O&P
9900	Coupling, 1/4"	1 Plum	19	.421	Ea.	7.70	26.50		34.20	48.50
9901	3/8"		19	.421		7.70	26.50		34.20	48.50
9902	1/2"		17	.471		6.30	29.50		35.80	51.50
9903	3/4"		16	.500		8.20	31.50		39.70	56.50
9904	1"		15	.533		14.30	33.50		47.80	66.50
9905	1-1/4"	Q-1	26	.615		24	35		59	78.50
9906	1-1/2"		24	.667		31	38		69	91
9907	2"		21	.762		38.50	43.50		82	107

22 11 13.47 Pipe Fittings, Steel

	22 11 13.47 Pipe Fittings, Steel	Crew	Daily Output	Labor-Hours	Unit	Material	Labor	Equipment	Total	Total Incl O&P
0010	**PIPE FITTINGS, STEEL**, flanged, welded & special									
0020	Flanged joints, CI, standard weight, black. One gasket & bolt									
0040	set, mat'l only, required at each joint, not included (see line 0620)									
0060	90° elbow, straight, 1-1/2" pipe size	Q-1	14	1.143	Ea.	645	65		710	810
0080	2" pipe size		13	1.231		375	70		445	515
0090	2-1/2" pipe size		12	1.333		405	76		481	560
0100	3" pipe size		11	1.455		335	82.50		417.50	495
0110	4" pipe size		8	2		415	114		529	630
0120	5" pipe size		7	2.286		985	130		1,115	1,275
0130	6" pipe size	Q-2	9	2.667		650	157		807	950
0140	8" pipe size		8	3		1,125	177		1,302	1,500
0150	10" pipe size		7	3.429		2,475	202		2,677	3,000
0160	12" pipe size		6	4		4,950	236		5,186	5,800
0171	90° elbow, reducing									
0172	2-1/2" by 2" pipe size	Q-1	12	1.333	Ea.	1,250	76		1,326	1,500
0173	3" by 2-1/2" pipe size		11	1.455		1,250	82.50		1,332.50	1,500
0174	4" by 3" pipe size		8	2		925	114		1,039	1,200
0175	5" by 3" pipe size		7	2.286		2,000	130		2,130	2,425
0176	6" by 4" pipe size	Q-2	9	2.667		1,150	157		1,307	1,475
0177	8" by 6" pipe size		8	3		1,675	177		1,852	2,100
0178	10" by 8" pipe size		7	3.429		3,175	202		3,377	3,800
0179	12" by 10" pipe size		6	4		6,100	236		6,336	7,050
0200	45° elbow, straight, 1-1/2" pipe size	Q-1	14	1.143		785	65		850	960
0220	2" pipe size		13	1.231		540	70		610	700
0230	2-1/2" pipe size		12	1.333		575	76		651	750
0240	3" pipe size		11	1.455		560	82.50		642.50	740
0250	4" pipe size		8	2		630	114		744	865
0260	5" pipe size		7	2.286		1,500	130		1,630	1,850
0270	6" pipe size	Q-2	9	2.667		1,025	157		1,182	1,350
0280	8" pipe size		8	3		1,500	177		1,677	1,925
0290	10" pipe size		7	3.429		3,175	202		3,377	3,800
0300	12" pipe size		6	4		4,875	236		5,111	5,700
0310	Cross, straight									
0311	2-1/2" pipe size	Q-1	6	2.667	Ea.	1,175	152		1,327	1,525
0312	3" pipe size		5	3.200		1,250	182		1,432	1,650
0313	4" pipe size		4	4		1,625	227		1,852	2,125
0314	5" pipe size		3	5.333		3,650	305		3,955	4,475
0315	6" pipe size	Q-2	5	4.800		3,400	283		3,683	4,175
0316	8" pipe size		4	6		5,475	355		5,830	6,550
0317	10" pipe size		3	8		6,500	470		6,970	7,850
0318	12" pipe size		2	12		10,600	705		11,305	12,800
0350	Tee, straight, 1-1/2" pipe size	Q-1	10	1.600		745	91		836	955
0370	2" pipe size		9	1.778		410	101		511	600

22 11 13.47 Pipe Fittings, Steel		Crew	Daily Output	Labor-Hours	Unit	Material (2019 Bare Costs)	Labor (2019 Bare Costs)	Equipment (2019 Bare Costs)	Total (2019 Bare Costs)	Total Incl O&P
0380	2-1/2" pipe size	Q-1	8	2	Ea.	595	114		709	825
0390	3" pipe size		7	2.286		415	130		545	655
0400	4" pipe size		5	3.200		635	182		817	970
0410	5" pipe size	↓	4	4		1,700	227		1,927	2,225
0420	6" pipe size	Q-2	6	4		920	236		1,156	1,350
0430	8" pipe size		5	4.800		1,575	283		1,858	2,150
0440	10" pipe size		4	6		4,225	355		4,580	5,175
0450	12" pipe size	↓	3	8	↓	6,650	470		7,120	8,000
0459	Tee, reducing on outlet									
0460	2-1/2" by 2" pipe size	Q-1	8	2	Ea.	1,275	114		1,389	1,575
0461	3" by 2-1/2" pipe size		7	2.286		1,225	130		1,355	1,550
0462	4" by 3" pipe size		5	3.200		1,300	182		1,482	1,700
0463	5" by 4" pipe size	↓	4	4		2,900	227		3,127	3,525
0464	6" by 4" pipe size	Q-2	6	4		1,250	236		1,486	1,725
0465	8" by 6" pipe size		5	4.800		1,225	283		1,508	1,775
0466	10" by 8" pipe size		4	6		4,475	355		4,830	5,425
0467	12" by 10" pipe size	↓	3	8	↓	7,750	470		8,220	9,225
0476	Reducer, concentric									
0477	3" by 2-1/2"	Q-1	12	1.333	Ea.	745	76		821	935
0478	4" by 3"		9	1.778		840	101		941	1,075
0479	5" by 4"	↓	8	2		1,300	114		1,414	1,600
0480	6" by 4"	Q-2	10	2.400		1,125	141		1,266	1,425
0481	8" by 6"		9	2.667		1,425	157		1,582	1,775
0482	10" by 8"		8	3		2,950	177		3,127	3,525
0483	12" by 10"	↓	7	3.429	↓	5,025	202		5,227	5,825
0492	Reducer, eccentric									
0493	4" by 3"	Q-1	8	2	Ea.	1,350	114		1,464	1,675
0494	5" by 4"	"	7	2.286		2,125	130		2,255	2,525
0495	6" by 4"	Q-2	9	2.667		1,275	157		1,432	1,650
0496	8" by 6"		8	3		1,625	177		1,802	2,050
0497	10" by 8"		7	3.429		4,150	202		4,352	4,875
0498	12" by 10"	↓	6	4		5,650	236		5,886	6,550
0500	For galvanized elbows and tees, add					100%				
0520	For extra heavy weight elbows and tees, add					140%				
0620	Gasket and bolt set, 150 lb., 1/2" pipe size	1 Plum	20	.400		3.81	25.50		29.31	42
0622	3/4" pipe size		19	.421		4.06	26.50		30.56	44.50
0624	1" pipe size		18	.444		4.13	28		32.13	46.50
0626	1-1/4" pipe size		17	.471		4.32	29.50		33.82	49.50
0628	1-1/2" pipe size		15	.533		4.38	33.50		37.88	55.50
0630	2" pipe size		13	.615		8.50	39		47.50	68
0640	2-1/2" pipe size		12	.667		8.60	42		50.60	72.50
0650	3" pipe size		11	.727		8.60	46		54.60	78.50
0660	3-1/2" pipe size		9	.889		18.20	56		74.20	104
0670	4" pipe size		8	1		16.30	63		79.30	113
0680	5" pipe size		7	1.143		25.50	72		97.50	136
0690	6" pipe size		6	1.333		27	84		111	156
0700	8" pipe size		5	1.600		30.50	101		131.50	186
0710	10" pipe size		4.50	1.778		52	112		164	226
0720	12" pipe size		4.20	1.905		56	120		176	242
0730	14" pipe size		4	2		55.50	126		181.50	251
0740	16" pipe size		3	2.667		62	168		230	320
0750	18" pipe size		2.70	2.963		120	187		307	415
0760	20" pipe size	↓	2.30	3.478	↓	195	220		415	545

22 11 13.47 Pipe Fittings, Steel		Crew	Daily Output	Labor-Hours	Unit	2019 Bare Costs Material	Labor	Equipment	Total	Total Incl O&P
0780	24″ pipe size	1 Plum	1.90	4.211	Ea.	244	266		510	670
0790	26″ pipe size		1.60	5		330	315		645	840
0810	30″ pipe size		1.40	5.714		645	360		1,005	1,250
0830	36″ pipe size		1.10	7.273		1,200	460		1,660	2,025
0850	For 300 lb. gasket set, add					40%				
2000	Flanged unions, 125 lb., black, 1/2″ pipe size	1 Plum	17	.471	Ea.	97.50	29.50		127	152
2040	3/4″ pipe size		17	.471		135	29.50		164.50	194
2050	1″ pipe size		16	.500		131	31.50		162.50	192
2060	1-1/4″ pipe size	Q-1	28	.571		155	32.50		187.50	219
2070	1-1/2″ pipe size		27	.593		143	33.50		176.50	208
2080	2″ pipe size		26	.615		168	35		203	238
2090	2-1/2″ pipe size		24	.667		228	38		266	310
2100	3″ pipe size		22	.727		258	41.50		299.50	345
2110	3-1/2″ pipe size		18	.889		435	50.50		485.50	555
2120	4″ pipe size		16	1		350	57		407	470
2130	5″ pipe size		14	1.143		815	65		880	1,000
2140	6″ pipe size	Q-2	19	1.263		765	74.50		839.50	955
2150	8″ pipe size	″	16	1.500		1,775	88.50		1,863.50	2,075
2200	For galvanized unions, add					150%				
2290	Threaded flange									
2300	Cast iron									
2310	Black, 125 lb., per flange									
2320	1″ pipe size	1 Plum	27	.296	Ea.	54.50	18.70		73.20	88
2330	1-1/4″ pipe size	Q-1	44	.364		65.50	20.50		86	103
2340	1-1/2″ pipe size		40	.400		60.50	22.50		83	101
2350	2″ pipe size		36	.444		60.50	25.50		86	105
2360	2-1/2″ pipe size		28	.571		70.50	32.50		103	126
2370	3″ pipe size		20	.800		91	45.50		136.50	168
2380	3-1/2″ pipe size		16	1		129	57		186	228
2390	4″ pipe size		12	1.333		122	76		198	249
2400	5″ pipe size		10	1.600		173	91		264	325
2410	6″ pipe size	Q-2	14	1.714		196	101		297	365
2420	8″ pipe size		12	2		310	118		428	515
2430	10″ pipe size		10	2.400		550	141		691	810
2440	12″ pipe size		8	3		1,225	177		1,402	1,625
2460	For galvanized flanges, add					95%				
2490	Blind flange									
2492	Cast iron									
2494	Black, 125 lb., per flange									
2496	1″ pipe size	1 Plum	27	.296	Ea.	87	18.70		105.70	124
2500	1-1/2″ pipe size	Q-1	40	.400		89	22.50		111.50	132
2502	2″ pipe size		36	.444		110	25.50		135.50	159
2504	2-1/2″ pipe size		28	.571		111	32.50		143.50	171
2506	3″ pipe size		20	.800		145	45.50		190.50	228
2508	4″ pipe size		12	1.333		186	76		262	320
2510	5″ pipe size		10	1.600		300	91		391	465
2512	6″ pipe size	Q-2	14	1.714		330	101		431	510
2514	8″ pipe size		12	2		515	118		633	740
2516	10″ pipe size		10	2.400		765	141		906	1,050
2518	12″ pipe size		8	3		1,475	177		1,652	1,875
2520	For galvanized flanges, add					80%				
2570	Threaded flange									
2580	Forged steel									

22 11 13.47 Pipe Fittings, Steel		Crew	Daily Output	Labor-Hours	Unit	Material	Labor	Equipment	Total	Total Incl O&P
						2019 Bare Costs				
2590	Black, 150 lb., per flange									
2600	1/2" pipe size	1 Plum	30	.267	Ea.	27	16.85		43.85	55.50
2610	3/4" pipe size		28	.286		27	18.05		45.05	57
2620	1" pipe size		27	.296		27	18.70		45.70	58
2630	1-1/4" pipe size	Q-1	44	.364		27	20.50		47.50	61
2640	1-1/2" pipe size		40	.400		27	22.50		49.50	64
2650	2" pipe size		36	.444		30.50	25.50		56	71.50
2660	2-1/2" pipe size		28	.571		42	32.50		74.50	94.50
2670	3" pipe size		20	.800		42.50	45.50		88	115
2690	4" pipe size		12	1.333		47.50	76		123.50	167
2700	5" pipe size		10	1.600		78	91		169	222
2710	6" pipe size	Q-2	14	1.714		84	101		185	245
2720	8" pipe size		12	2		147	118		265	340
2730	10" pipe size		10	2.400		263	141		404	500
2860	Black, 300 lb., per flange									
2870	1/2" pipe size	1 Plum	30	.267	Ea.	30	16.85		46.85	58.50
2880	3/4" pipe size		28	.286		30	18.05		48.05	60
2890	1" pipe size		27	.296		30	18.70		48.70	61
2900	1-1/4" pipe size	Q-1	44	.364		30	20.50		50.50	64
2910	1-1/2" pipe size		40	.400		30	22.50		52.50	67
2920	2" pipe size		36	.444		39	25.50		64.50	81
2930	2-1/2" pipe size		28	.571		49	32.50		81.50	103
2940	3" pipe size		20	.800		49.50	45.50		95	123
2960	4" pipe size		12	1.333		71.50	76		147.50	193
2970	6" pipe size	Q-2	14	1.714		155	101		256	320
3000	Weld joint, butt, carbon steel, standard weight									
3040	90° elbow, long radius									
3050	1/2" pipe size	Q-15	16	1	Ea.	43.50	57	3.51	104.01	137
3060	3/4" pipe size		16	1		43.50	57	3.51	104.01	137
3070	1" pipe size		16	1		26	57	3.51	86.51	118
3080	1-1/4" pipe size		14	1.143		26	65	4.01	95.01	130
3090	1-1/2" pipe size		13	1.231		26	70	4.32	100.32	138
3100	2" pipe size		10	1.600		22	91	5.60	118.60	166
3110	2-1/2" pipe size		8	2		34.50	114	7	155.50	217
3120	3" pipe size		7	2.286		30	130	8	168	237
3130	4" pipe size		5	3.200		55.50	182	11.25	248.75	345
3136	5" pipe size		4	4		107	227	14.05	348.05	470
3140	6" pipe size	Q-16	5	4.800		124	283	11.25	418.25	575
3150	8" pipe size		3.75	6.400		234	375	15	624	840
3160	10" pipe size		3	8		460	470	18.70	948.70	1,225
3170	12" pipe size		2.50	9.600		605	565	22.50	1,192.50	1,550
3180	14" pipe size		2	12		1,000	705	28	1,733	2,175
3190	16" pipe size		1.50	16		1,375	945	37.50	2,357.50	3,000
3191	18" pipe size		1.25	19.200		1,625	1,125	45	2,795	3,550
3192	20" pipe size		1.15	20.870		2,375	1,225	49	3,649	4,500
3194	24" pipe size		1.02	23.529		3,575	1,375	55	5,005	6,050
3200	45° elbow, long									
3210	1/2" pipe size	Q-15	16	1	Ea.	64.50	57	3.51	125.01	160
3220	3/4" pipe size		16	1		64.50	57	3.51	125.01	160
3230	1" pipe size		16	1		23	57	3.51	83.51	114
3240	1-1/4" pipe size		14	1.143		23	65	4.01	92.01	127
3250	1-1/2" pipe size		13	1.231		23	70	4.32	97.32	135
3260	2" pipe size		10	1.600		23	91	5.60	119.60	167

22 11 13.47 Pipe Fittings, Steel		Crew	Daily Output	Labor-Hours	Unit	Material	Labor	Equipment	Total	Total Incl O&P
						2019 Bare Costs				
3270	2-1/2" pipe size	Q-15	8	2	Ea.	27.50	114	7	148.50	209
3280	3" pipe size		7	2.286		28.50	130	8	166.50	235
3290	4" pipe size		5	3.200		51	182	11.25	244.25	340
3296	5" pipe size		4	4		77.50	227	14.05	318.55	440
3300	6" pipe size	Q-16	5	4.800		101	283	11.25	395.25	550
3310	8" pipe size		3.75	6.400		168	375	15	558	765
3320	10" pipe size		3	8		335	470	18.70	823.70	1,100
3330	12" pipe size		2.50	9.600		475	565	22.50	1,062.50	1,400
3340	14" pipe size		2	12		635	705	28	1,368	1,775
3341	16" pipe size		1.50	16		1,150	945	37.50	2,132.50	2,725
3342	18" pipe size		1.25	19.200		1,600	1,125	45	2,770	3,525
3343	20" pipe size		1.15	20.870		1,675	1,225	49	2,949	3,725
3345	24" pipe size		1.05	22.857		2,525	1,350	53.50	3,928.50	4,850
3346	26" pipe size		.85	28.235		2,775	1,675	66	4,516	5,625
3347	30" pipe size		.45	53.333		3,050	3,150	125	6,325	8,225
3349	36" pipe size		.38	63.158		3,375	3,725	148	7,248	9,425
3350	Tee, straight									
3352	For reducing tees and concentrics see starting line 4600									
3360	1/2" pipe size	Q-15	10	1.600	Ea.	114	91	5.60	210.60	267
3370	3/4" pipe size		10	1.600		114	91	5.60	210.60	267
3380	1" pipe size		10	1.600		56.50	91	5.60	153.10	204
3390	1-1/4" pipe size		9	1.778		70.50	101	6.25	177.75	236
3400	1-1/2" pipe size		8	2		70.50	114	7	191.50	256
3410	2" pipe size		6	2.667		56.50	152	9.35	217.85	299
3420	2-1/2" pipe size		5	3.200		78	182	11.25	271.25	370
3430	3" pipe size		4	4		86.50	227	14.05	327.55	450
3440	4" pipe size		3	5.333		122	305	18.70	445.70	610
3446	5" pipe size		2.50	6.400		201	365	22.50	588.50	790
3450	6" pipe size	Q-16	3	8		209	470	18.70	697.70	955
3460	8" pipe size		2.50	9.600		365	565	22.50	952.50	1,275
3470	10" pipe size		2	12		715	705	28	1,448	1,875
3480	12" pipe size		1.60	15		1,000	885	35	1,920	2,475
3481	14" pipe size		1.30	18.462		1,750	1,100	43	2,893	3,600
3482	16" pipe size		1	24		1,975	1,425	56	3,456	4,350
3483	18" pipe size		.80	30		3,100	1,775	70	4,945	6,150
3484	20" pipe size		.75	32		4,900	1,875	75	6,850	8,300
3486	24" pipe size		.70	34.286		6,325	2,025	80	8,430	10,100
3487	26" pipe size		.55	43.636		6,950	2,575	102	9,627	11,600
3488	30" pipe size		.30	80		7,600	4,725	187	12,512	15,700
3490	36" pipe size		.25	96		8,400	5,650	225	14,275	18,000
3491	Eccentric reducer, 1-1/2" pipe size	Q-15	14	1.143		44.50	65	4.01	113.51	151
3492	2" pipe size		11	1.455		50.50	82.50	5.10	138.10	185
3493	2-1/2" pipe size		9	1.778		55.50	101	6.25	162.75	220
3494	3" pipe size		8	2		72.50	114	7	193.50	259
3495	4" pipe size		6	2.667		147	152	9.35	308.35	400
3496	6" pipe size	Q-16	5	4.800		189	283	11.25	483.25	645
3497	8" pipe size		4	6		247	355	14.05	616.05	815
3498	10" pipe size		3	8		435	470	18.70	923.70	1,200
3499	12" pipe size		2.50	9.600		625	565	22.50	1,212.50	1,550
3501	Cap, 1-1/2" pipe size	Q-15	28	.571		20	32.50	2.01	54.51	72.50
3502	2" pipe size		22	.727		22.50	41.50	2.55	66.55	89.50
3503	2-1/2" pipe size		18	.889		23.50	50.50	3.12	77.12	105
3504	3" pipe size		16	1		23.50	57	3.51	84.01	115

22 11 13.47 Pipe Fittings, Steel		Crew	Daily Output	Labor-Hours	Unit	Material	2019 Bare Costs Labor	Equipment	Total	Total Incl O&P
3505	4" pipe size	Q-15	12	1.333	Ea.	34.50	76	4.68	115.18	157
3506	6" pipe size	Q-16	10	2.400		58.50	141	5.60	205.10	282
3507	8" pipe size		8	3		88.50	177	7	272.50	370
3508	10" pipe size		6	4		161	236	9.35	406.35	540
3509	12" pipe size		5	4.800		242	283	11.25	536.25	705
3511	14" pipe size		4	6		420	355	14.05	789.05	1,000
3512	16" pipe size		4	6		470	355	14.05	839.05	1,050
3513	18" pipe size		3	8		515	470	18.70	1,003.70	1,300
3517	Weld joint, butt, carbon steel, extra strong									
3519	90° elbow, long									
3520	1/2" pipe size	Q-15	13	1.231	Ea.	57.50	70	4.32	131.82	173
3530	3/4" pipe size		12	1.333		57.50	76	4.68	138.18	182
3540	1" pipe size		11	1.455		28	82.50	5.10	115.60	161
3550	1-1/4" pipe size		10	1.600		28	91	5.60	124.60	173
3560	1-1/2" pipe size		9	1.778		28	101	6.25	135.25	190
3570	2" pipe size		8	2		28.50	114	7	149.50	210
3580	2-1/2" pipe size		7	2.286		40	130	8	178	248
3590	3" pipe size		6	2.667		51	152	9.35	212.35	293
3600	4" pipe size		4	4		84	227	14.05	325.05	450
3606	5" pipe size		3.50	4.571		201	260	16.05	477.05	630
3610	6" pipe size	Q-16	4.50	5.333		213	315	12.50	540.50	720
3620	8" pipe size		3.50	6.857		405	405	16.05	826.05	1,075
3630	10" pipe size		2.50	9.600		855	565	22.50	1,442.50	1,825
3640	12" pipe size		2.25	10.667		1,050	630	25	1,705	2,125
3650	45° elbow, long									
3660	1/2" pipe size	Q-15	13	1.231	Ea.	63.50	70	4.32	137.82	180
3670	3/4" pipe size		12	1.333		63.50	76	4.68	144.18	189
3680	1" pipe size		11	1.455		29.50	82.50	5.10	117.10	162
3690	1-1/4" pipe size		10	1.600		29.50	91	5.60	126.10	175
3700	1-1/2" pipe size		9	1.778		29.50	101	6.25	136.75	191
3710	2" pipe size		8	2		29.50	114	7	150.50	211
3720	2-1/2" pipe size		7	2.286		66	130	8	204	276
3730	3" pipe size		6	2.667		38.50	152	9.35	199.85	279
3740	4" pipe size		4	4		60.50	227	14.05	301.55	420
3746	5" pipe size		3.50	4.571		143	260	16.05	419.05	565
3750	6" pipe size	Q-16	4.50	5.333		165	315	12.50	492.50	665
3760	8" pipe size		3.50	6.857		286	405	16.05	707.05	940
3770	10" pipe size		2.50	9.600		550	565	22.50	1,137.50	1,475
3780	12" pipe size		2.25	10.667		815	630	25	1,470	1,875
3800	Tee, straight									
3810	1/2" pipe size	Q-15	9	1.778	Ea.	140	101	6.25	247.25	315
3820	3/4" pipe size		8.50	1.882		135	107	6.60	248.60	315
3830	1" pipe size		8	2		54	114	7	175	238
3840	1-1/4" pipe size		7	2.286		54	130	8	192	263
3850	1-1/2" pipe size		6	2.667		54	152	9.35	215.35	296
3860	2" pipe size		5	3.200		61.50	182	11.25	254.75	355
3870	2-1/2" pipe size		4	4		102	227	14.05	343.05	465
3880	3" pipe size		3.50	4.571		128	260	16.05	404.05	550
3890	4" pipe size		2.50	6.400		154	365	22.50	541.50	740
3896	5" pipe size		2.25	7.111		385	405	25	815	1,050
3900	6" pipe size	Q-16	2.25	10.667		295	630	25	950	1,300
3910	8" pipe size		2	12		565	705	28	1,298	1,700
3920	10" pipe size		1.75	13.714		870	810	32	1,712	2,225

22 11 13 – Facility Water Distribution Piping

22 11 13.47 Pipe Fittings, Steel		Crew	Daily Output	Labor-Hours	Unit	2019 Bare Costs Material	Labor	Equipment	Total	Total Incl O&P
3930	12″ pipe size	Q-16	1.50	16	Ea.	1,275	945	37.50	2,257.50	2,875
4000	Eccentric reducer, 1-1/2″ pipe size	Q-15	10	1.600		17.65	91	5.60	114.25	162
4010	2″ pipe size		9	1.778		48.50	101	6.25	155.75	212
4020	2-1/2″ pipe size		8	2		73.50	114	7	194.50	260
4030	3″ pipe size		7	2.286		59.50	130	8	197.50	269
4040	4″ pipe size		5	3.200		97	182	11.25	290.25	390
4046	5″ pipe size		4.70	3.404		265	193	11.95	469.95	595
4050	6″ pipe size	Q-16	4.50	5.333		267	315	12.50	594.50	775
4060	8″ pipe size		3.50	6.857		400	405	16.05	821.05	1,075
4070	10″ pipe size		2.50	9.600		685	565	22.50	1,272.50	1,625
4080	12″ pipe size		2.25	10.667		870	630	25	1,525	1,925
4090	14″ pipe size		2.10	11.429		1,625	675	26.50	2,326.50	2,800
4100	16″ pipe size		1.90	12.632		2,075	745	29.50	2,849.50	3,425
4151	Cap, 1-1/2″ pipe size	Q-15	24	.667		25	38	2.34	65.34	87
4152	2″ pipe size		18	.889		22.50	50.50	3.12	76.12	104
4153	2-1/2″ pipe size		16	1		30.50	57	3.51	91.01	123
4154	3″ pipe size		14	1.143		34.50	65	4.01	103.51	139
4155	4″ pipe size		10	1.600		45	91	5.60	141.60	192
4156	6″ pipe size	Q-16	9	2.667		94.50	157	6.25	257.75	345
4157	8″ pipe size		7	3.429		142	202	8	352	470
4158	10″ pipe size		5	4.800		218	283	11.25	512.25	675
4159	12″ pipe size		4	6		291	355	14.05	660.05	865
4190	Weld fittings, reducing, standard weight									
4200	Welding ring w/spacer pins, 2″ pipe size				Ea.	2.02			2.02	2.22
4210	2-1/2″ pipe size					2.23			2.23	2.45
4220	3″ pipe size					2.32			2.32	2.55
4230	4″ pipe size					2.56			2.56	2.82
4236	5″ pipe size					3.20			3.20	3.52
4240	6″ pipe size					3.37			3.37	3.71
4250	8″ pipe size					3.75			3.75	4.13
4260	10″ pipe size					4.34			4.34	4.77
4270	12″ pipe size					5.05			5.05	5.60
4280	14″ pipe size					5.85			5.85	6.40
4290	16″ pipe size					6.80			6.80	7.50
4300	18″ pipe size					7.50			7.50	8.25
4310	20″ pipe size					8.55			8.55	9.45
4330	24″ pipe size					11.75			11.75	12.95
4340	26″ pipe size					14.50			14.50	15.95
4350	30″ pipe size					17.15			17.15	18.85
4370	36″ pipe size					21.50			21.50	24
4600	Tee, reducing on outlet									
4601	2-1/2″ x 2″ pipe size	Q-15	5	3.200	Ea.	104	182	11.25	297.25	400
4602	3″ x 2-1/2″ pipe size		4	4		148	227	14.05	389.05	520
4604	4″ x 3″ pipe size		3	5.333		120	305	18.70	443.70	610
4605	5″ x 4″ pipe size		2.50	6.400		335	365	22.50	722.50	940
4606	6″ x 5″ pipe size	Q-16	3	8		465	470	18.70	953.70	1,225
4607	8″ x 6″ pipe size		2.50	9.600		575	565	22.50	1,162.50	1,500
4608	10″ x 8″ pipe size		2	12		820	705	28	1,553	1,975
4609	12″ x 10″ pipe size		1.60	15		1,300	885	35	2,220	2,800
4610	16″ x 12″ pipe size		1.50	16		2,425	945	37.50	3,407.50	4,150
4611	14″ x 12″ pipe size		1.52	15.789		2,350	930	37	3,317	4,025
4618	Reducer, concentric									
4619	2-1/2″ by 2″ pipe size	Q-15	10	1.600	Ea.	41	91	5.60	137.60	188

22 11 13.47 Pipe Fittings, Steel		Crew	Daily Output	Labor-Hours	Unit	Material	Labor	Equipment	Total	Total Incl O&P
						2019 Bare Costs				
4620	3" by 2-1/2" pipe size	Q-15	9	1.778	Ea.	34	101	6.25	141.25	196
4621	3-1/2" by 3" pipe size		8	2		110	114	7	231	300
4622	4" by 2-1/2" pipe size		7	2.286		56.50	130	8	194.50	266
4623	5" by 3" pipe size	↓	7	2.286		142	130	8	280	360
4624	6" by 4" pipe size	Q-16	6	4		117	236	9.35	362.35	495
4625	8" by 6" pipe size		5	4.800		120	283	11.25	414.25	570
4626	10" by 8" pipe size		4	6		247	355	14.05	616.05	815
4627	12" by 10" pipe size	↓	3	8	↓	287	470	18.70	775.70	1,050
4660	Reducer, eccentric									
4662	3" x 2" pipe size	Q-15	8	2	Ea.	73.50	114	7	194.50	259
4664	4" x 3" pipe size		6	2.667		89.50	152	9.35	250.85	335
4666	4" x 2" pipe size	↓	6	2.667		109	152	9.35	270.35	355
4670	6" x 4" pipe size	Q-16	5	4.800		184	283	11.25	478.25	640
4672	6" x 3" pipe size		5	4.800		289	283	11.25	583.25	755
4676	8" x 6" pipe size		4	6		235	355	14.05	604.05	805
4678	8" x 4" pipe size		4	6		430	355	14.05	799.05	1,025
4682	10" x 8" pipe size		3	8		310	470	18.70	798.70	1,075
4684	10" x 6" pipe size		3	8		420	470	18.70	908.70	1,175
4688	12" x 10" pipe size		2.50	9.600		480	565	22.50	1,067.50	1,400
4690	12" x 8" pipe size		2.50	9.600		755	565	22.50	1,342.50	1,700
4691	14" x 12" pipe size		2.20	10.909		840	645	25.50	1,510.50	1,925
4693	16" x 14" pipe size		1.80	13.333		1,000	785	31	1,816	2,300
4694	16" x 12" pipe size		2	12		1,700	705	28	2,433	2,950
4696	18" x 16" pipe size	↓	1.60	15	↓	1,075	885	35	1,995	2,550
5000	Weld joint, socket, forged steel, 3000 lb., schedule 40 pipe									
5010	90° elbow, straight									
5020	1/4" pipe size	Q-15	22	.727	Ea.	25.50	41.50	2.55	69.55	93
5030	3/8" pipe size		22	.727		25.50	41.50	2.55	69.55	93
5040	1/2" pipe size		20	.800		14.20	45.50	2.81	62.51	86.50
5050	3/4" pipe size		20	.800		13.15	45.50	2.81	61.46	85.50
5060	1" pipe size		20	.800		19	45.50	2.81	67.31	92
5070	1-1/4" pipe size		18	.889		36.50	50.50	3.12	90.12	119
5080	1-1/2" pipe size		16	1		42	57	3.51	102.51	135
5090	2" pipe size		12	1.333		62	76	4.68	142.68	188
5100	2-1/2" pipe size		10	1.600		177	91	5.60	273.60	335
5110	3" pipe size		8	2		305	114	7	426	520
5120	4" pipe size	↓	6	2.667	↓	805	152	9.35	966.35	1,125
5130	45° elbow, straight									
5134	1/4" pipe size	Q-15	22	.727	Ea.	25.50	41.50	2.55	69.55	93
5135	3/8" pipe size		22	.727		25.50	41.50	2.55	69.55	93
5136	1/2" pipe size		20	.800		18.90	45.50	2.81	67.21	92
5137	3/4" pipe size		20	.800		22	45.50	2.81	70.31	95
5140	1" pipe size		20	.800		28.50	45.50	2.81	76.81	103
5150	1-1/4" pipe size		18	.889		39.50	50.50	3.12	93.12	123
5160	1-1/2" pipe size		16	1		48.50	57	3.51	109.01	142
5170	2" pipe size		12	1.333		77.50	76	4.68	158.18	205
5180	2-1/2" pipe size		10	1.600		204	91	5.60	300.60	365
5190	3" pipe size		8	2		340	114	7	461	555
5200	4" pipe size	↓	6	2.667	↓	665	152	9.35	826.35	970
5250	Tee, straight									
5254	1/4" pipe size	Q-15	15	1.067	Ea.	28	60.50	3.74	92.24	126
5255	3/8" pipe size		15	1.067		28	60.50	3.74	92.24	126
5256	1/2" pipe size	↓	13	1.231	↓	17.55	70	4.32	91.87	129

22 11 13.47 Pipe Fittings, Steel		Crew	Daily Output	Labor-Hours	Unit	2019 Bare Costs Material	Labor	Equipment	Total	Total Incl O&P
5257	3/4" pipe size	Q-15	13	1.231	Ea.	21.50	70	4.32	95.82	133
5260	1" pipe size		13	1.231		29	70	4.32	103.32	142
5270	1-1/4" pipe size		12	1.333		45	76	4.68	125.68	169
5280	1-1/2" pipe size		11	1.455		59	82.50	5.10	146.60	195
5290	2" pipe size		8	2		86	114	7	207	273
5300	2-1/2" pipe size		6	2.667		246	152	9.35	407.35	510
5310	3" pipe size		5	3.200		590	182	11.25	783.25	935
5320	4" pipe size		4	4		950	227	14.05	1,191.05	1,400
5350	For reducing sizes, add					60%				
5450	Couplings									
5451	1/4" pipe size	Q-15	23	.696	Ea.	16.35	39.50	2.44	58.29	80
5452	3/8" pipe size		23	.696		16.35	39.50	2.44	58.29	80
5453	1/2" pipe size		21	.762		7.45	43.50	2.67	53.62	76
5454	3/4" pipe size		21	.762		9.60	43.50	2.67	55.77	78.50
5460	1" pipe size		20	.800		10.60	45.50	2.81	58.91	83
5470	1-1/4" pipe size		20	.800		18.85	45.50	2.81	67.16	91.50
5480	1-1/2" pipe size		18	.889		21	50.50	3.12	74.62	102
5490	2" pipe size		14	1.143		33.50	65	4.01	102.51	138
5500	2-1/2" pipe size		12	1.333		74.50	76	4.68	155.18	201
5510	3" pipe size		9	1.778		167	101	6.25	274.25	345
5520	4" pipe size		7	2.286		251	130	8	389	480
5570	Union, 1/4" pipe size		21	.762		34.50	43.50	2.67	80.67	106
5571	3/8" pipe size		21	.762		34.50	43.50	2.67	80.67	106
5572	1/2" pipe size		19	.842		29.50	48	2.96	80.46	108
5573	3/4" pipe size		19	.842		34	48	2.96	84.96	112
5574	1" pipe size		19	.842		43.50	48	2.96	94.46	123
5575	1-1/4" pipe size		17	.941		70.50	53.50	3.30	127.30	162
5576	1-1/2" pipe size		15	1.067		76	60.50	3.74	140.24	179
5577	2" pipe size		11	1.455		109	82.50	5.10	196.60	250
5600	Reducer, 1/4" pipe size		23	.696		41	39.50	2.44	82.94	107
5601	3/8" pipe size		23	.696		43	39.50	2.44	84.94	110
5602	1/2" pipe size		21	.762		26.50	43.50	2.67	72.67	97
5603	3/4" pipe size		21	.762		26.50	43.50	2.67	72.67	97
5604	1" pipe size		21	.762		36.50	43.50	2.67	82.67	108
5605	1-1/4" pipe size		19	.842		47.50	48	2.96	98.46	128
5607	1-1/2" pipe size		17	.941		49	53.50	3.30	105.80	138
5608	2" pipe size		13	1.231		54	70	4.32	128.32	169
5612	Cap, 1/4" pipe size		46	.348		15.45	19.75	1.22	36.42	48
5613	3/8" pipe size		46	.348		15.45	19.75	1.22	36.42	48
5614	1/2" pipe size		42	.381		9.40	21.50	1.34	32.24	44.50
5615	3/4" pipe size		42	.381		10.95	21.50	1.34	33.79	46
5616	1" pipe size		42	.381		16.65	21.50	1.34	39.49	52.50
5617	1-1/4" pipe size		38	.421		19.55	24	1.48	45.03	59
5618	1-1/2" pipe size		34	.471		27	26.50	1.65	55.15	72
5619	2" pipe size		26	.615		42	35	2.16	79.16	101
5630	T-O-L, 1/4" pipe size, nozzle		23	.696		7.80	39.50	2.44	49.74	71
5631	3/8" pipe size, nozzle		23	.696		7.90	39.50	2.44	49.84	71
5632	1/2" pipe size, nozzle		22	.727		7.80	41.50	2.55	51.85	73.50
5633	3/4" pipe size, nozzle		21	.762		9	43.50	2.67	55.17	78
5634	1" pipe size, nozzle		20	.800		10.50	45.50	2.81	58.81	82.50
5635	1-1/4" pipe size, nozzle		18	.889		16.55	50.50	3.12	70.17	97.50
5636	1-1/2" pipe size, nozzle		16	1		16.60	57	3.51	77.11	108
5637	2" pipe size, nozzle		12	1.333		18.40	76	4.68	99.08	139

22 11 13.47 Pipe Fittings, Steel		Crew	Daily Output	Labor-Hours	Unit	2019 Bare Costs Material	Labor	Equipment	Total	Total Incl O&P
5638	2-1/2" pipe size, nozzle	Q-15	10	1.600	Ea.	63.50	91	5.60	160.10	212
5639	4" pipe size, nozzle		6	2.667		143	152	9.35	304.35	395
5640	W-O-L, 1/4" pipe size, nozzle		23	.696		18.45	39.50	2.44	60.39	82.50
5641	3/8" pipe size, nozzle		23	.696		17.75	39.50	2.44	59.69	81.50
5642	1/2" pipe size, nozzle		22	.727		16.15	41.50	2.55	60.20	82.50
5643	3/4" pipe size, nozzle		21	.762		17.05	43.50	2.67	63.22	86.50
5644	1" pipe size, nozzle		20	.800		17.80	45.50	2.81	66.11	90.50
5645	1-1/4" pipe size, nozzle		18	.889		20.50	50.50	3.12	74.12	102
5646	1-1/2" pipe size, nozzle		16	1		21	57	3.51	81.51	113
5647	2" pipe size, nozzle		12	1.333		21.50	76	4.68	102.18	143
5648	2-1/2" pipe size, nozzle		10	1.600		49	91	5.60	145.60	196
5649	3" pipe size, nozzle		8	2		53.50	114	7	174.50	238
5650	4" pipe size, nozzle		6	2.667		67.50	152	9.35	228.85	310
5651	5" pipe size, nozzle		5	3.200		167	182	11.25	360.25	470
5652	6" pipe size, nozzle		4	4		187	227	14.05	428.05	560
5653	8" pipe size, nozzle		3	5.333		360	305	18.70	683.70	870
5654	10" pipe size, nozzle		2.60	6.154		520	350	21.50	891.50	1,125
5655	12" pipe size, nozzle		2.20	7.273		985	415	25.50	1,425.50	1,725
5674	S-O-L, 1/4" pipe size, outlet		23	.696		10.40	39.50	2.44	52.34	73.50
5675	3/8" pipe size, outlet		23	.696		10.40	39.50	2.44	52.34	73.50
5676	1/2" pipe size, outlet		22	.727		9.70	41.50	2.55	53.75	75.50
5677	3/4" pipe size, outlet		21	.762		9.80	43.50	2.67	55.97	78.50
5678	1" pipe size, outlet		20	.800		10.90	45.50	2.81	59.21	83
5679	1-1/4" pipe size, outlet		18	.889		18.20	50.50	3.12	71.82	99.50
5680	1-1/2" pipe size, outlet		16	1		18.20	57	3.51	78.71	109
5681	2" pipe size, outlet		12	1.333		20.50	76	4.68	101.18	142
6000	Weld-on flange, forged steel									
6020	Slip-on, 150 lb. flange (welded front and back)									
6050	1/2" pipe size	Q-15	18	.889	Ea.	17.55	50.50	3.12	71.17	98.50
6060	3/4" pipe size		18	.889		17.55	50.50	3.12	71.17	98.50
6070	1" pipe size		17	.941		17.55	53.50	3.30	74.35	103
6080	1-1/4" pipe size		16	1		17.55	57	3.51	78.06	109
6090	1-1/2" pipe size		15	1.067		17.55	60.50	3.74	81.79	114
6100	2" pipe size		12	1.333		23	76	4.68	103.68	144
6110	2-1/2" pipe size		10	1.600		34.50	91	5.60	131.10	180
6120	3" pipe size		9	1.778		31	101	6.25	138.25	193
6130	3-1/2" pipe size		7	2.286		34	130	8	172	241
6140	4" pipe size		6	2.667		32.50	152	9.35	193.85	273
6150	5" pipe size		5	3.200		66.50	182	11.25	259.75	360
6160	6" pipe size	Q-16	6	4		53.50	236	9.35	298.85	425
6170	8" pipe size		5	4.800		93	283	11.25	387.25	540
6180	10" pipe size		4	6		164	355	14.05	533.05	725
6190	12" pipe size		3	8		245	470	18.70	733.70	995
6191	14" pipe size		2.50	9.600		276	565	22.50	863.50	1,175
6192	16" pipe size		1.80	13.333		485	785	31	1,301	1,750
6200	300 lb. flange									
6210	1/2" pipe size	Q-15	17	.941	Ea.	24	53.50	3.30	80.80	111
6220	3/4" pipe size		17	.941		24	53.50	3.30	80.80	111
6230	1" pipe size		16	1		24	57	3.51	84.51	116
6240	1-1/4" pipe size		13	1.231		24	70	4.32	98.32	136
6250	1-1/2" pipe size		12	1.333		24	76	4.68	104.68	146
6260	2" pipe size		11	1.455		36	82.50	5.10	123.60	169
6270	2-1/2" pipe size		9	1.778		36	101	6.25	143.25	198

22 11 13.47 Pipe Fittings, Steel		Crew	Daily Output	Labor-Hours	Unit	2019 Bare Costs Material	Labor	Equipment	Total	Total Incl O&P
6280	3" pipe size	Q-15	7	2.286	Ea.	37.50	130	8	175.50	245
6290	4" pipe size		6	2.667		56.50	152	9.35	217.85	299
6300	5" pipe size		4	4		109	227	14.05	350.05	475
6310	6" pipe size	Q-16	5	4.800		106	283	11.25	400.25	555
6320	8" pipe size		4	6		182	355	14.05	551.05	745
6330	10" pipe size		3.40	7.059		320	415	16.50	751.50	995
6340	12" pipe size		2.80	8.571		335	505	20	860	1,150
6400	Welding neck, 150 lb. flange									
6410	1/2" pipe size	Q-15	40	.400	Ea.	25	22.50	1.40	48.90	63
6420	3/4" pipe size		36	.444		25	25.50	1.56	52.06	67
6430	1" pipe size		32	.500		25	28.50	1.76	55.26	72
6440	1-1/4" pipe size		29	.552		25	31.50	1.94	58.44	76.50
6450	1-1/2" pipe size		26	.615		25	35	2.16	62.16	82.50
6460	2" pipe size		20	.800		33.50	45.50	2.81	81.81	108
6470	2-1/2" pipe size		16	1		36.50	57	3.51	97.01	130
6480	3" pipe size		14	1.143		40.50	65	4.01	109.51	146
6500	4" pipe size		10	1.600		41.50	91	5.60	138.10	188
6510	5" pipe size		8	2		65.50	114	7	186.50	251
6520	6" pipe size	Q-16	10	2.400		68	141	5.60	214.60	293
6530	8" pipe size		7	3.429		116	202	8	326	440
6540	10" pipe size		6	4		186	236	9.35	431.35	570
6550	12" pipe size		5	4.800		292	283	11.25	586.25	755
6551	14" pipe size		4.50	5.333		435	315	12.50	762.50	960
6552	16" pipe size		3	8		585	470	18.70	1,073.70	1,375
6553	18" pipe size		2.50	9.600		785	565	22.50	1,372.50	1,750
6554	20" pipe size		2.30	10.435		955	615	24.50	1,594.50	2,000
6556	24" pipe size		2	12		1,300	705	28	2,033	2,500
6557	26" pipe size		1.70	14.118		1,400	830	33	2,263	2,825
6558	30" pipe size		.90	26.667		1,625	1,575	62.50	3,262.50	4,200
6559	36" pipe size		.75	32		1,850	1,875	75	3,800	4,950
6560	300 lb. flange									
6570	1/2" pipe size	Q-15	36	.444	Ea.	34	25.50	1.56	61.06	77
6580	3/4" pipe size		34	.471		34	26.50	1.65	62.15	79.50
6590	1" pipe size		30	.533		34	30.50	1.87	66.37	85
6600	1-1/4" pipe size		28	.571		34	32.50	2.01	68.51	88
6610	1-1/2" pipe size		24	.667		34	38	2.34	74.34	97
6620	2" pipe size		18	.889		45	50.50	3.12	98.62	129
6630	2-1/2" pipe size		14	1.143		51	65	4.01	120.01	158
6640	3" pipe size		12	1.333		44	76	4.68	124.68	167
6650	4" pipe size		8	2		75	114	7	196	261
6660	5" pipe size		7	2.286		109	130	8	247	325
6670	6" pipe size	Q-16	9	2.667		112	157	6.25	275.25	365
6680	8" pipe size		6	4		213	236	9.35	458.35	600
6690	10" pipe size		5	4.800		405	283	11.25	699.25	880
6700	12" pipe size		4	6		445	355	14.05	814.05	1,025
6710	14" pipe size		3.50	6.857		960	405	16.05	1,381.05	1,675
6720	16" pipe size		2	12		1,125	705	28	1,858	2,325
7740	Plain ends for plain end pipe, mechanically coupled									
7750	Cplg. & labor required at joints not included, add 1 per									
7760	joint for installed price, see line 9180									
7770	Malleable iron, painted, unless noted otherwise									
7800	90° elbow, 1"				Ea.	143			143	157
7810	1-1/2"					169			169	186

22 11 13.47 Pipe Fittings, Steel		Crew	Daily Output	Labor-Hours	Unit	2019 Bare Costs Material	Labor	Equipment	Total	Total Incl O&P
7820	2"				Ea.	252			252	277
7830	2-1/2"					295			295	325
7840	3"					305			305	335
7860	4"					345			345	380
7870	5" welded steel					400			400	440
7880	6"					505			505	555
7890	8" welded steel					945			945	1,050
7900	10" welded steel					1,200			1,200	1,325
7910	12" welded steel					1,350			1,350	1,475
7970	45° elbow, 1"					85.50			85.50	94
7980	1-1/2"					127			127	140
7990	2"					267			267	294
8000	2-1/2"					267			267	294
8010	3"					305			305	335
8030	4"					320			320	350
8040	5" welded steel					400			400	440
8050	6"					455			455	500
8060	8"					535			535	590
8070	10" welded steel					555			555	610
8080	12" welded steel					1,000			1,000	1,100
8140	Tee, straight 1"					160			160	176
8150	1-1/2"					206			206	226
8160	2"					206			206	226
8170	2-1/2"					268			268	295
8180	3"					410			410	455
8200	4"					590			590	650
8210	5" welded steel					805			805	885
8220	6"					700			700	770
8230	8" welded steel					1,000			1,000	1,100
8240	10" welded steel					1,600			1,600	1,750
8250	12" welded steel					1,850			1,850	2,050
8340	Segmentally welded steel, painted									
8390	Wye 2"				Ea.	230			230	253
8400	2-1/2"					230			230	253
8410	3"					280			280	310
8430	4"					355			355	390
8440	5"					530			530	580
8450	6"					635			635	700
8460	8"					940			940	1,025
8470	10"					1,575			1,575	1,725
8480	12"					1,650			1,650	1,825
8540	Wye, lateral 2"					278			278	305
8550	2-1/2"					320			320	350
8560	3"					380			380	415
8580	4"					520			520	575
8590	5"					885			885	975
8600	6"					915			915	1,000
8610	8"					1,525			1,525	1,675
8620	10"					2,275			2,275	2,500
8630	12"					2,850			2,850	3,125
8690	Cross, 2"					188			188	207
8700	2-1/2"					188			188	207
8710	3"					224			224	246

22 11 13.47 Pipe Fittings, Steel		Crew	Daily Output	Labor-Hours	Unit	Material	2019 Bare Costs Labor	Equipment	Total	Total Incl O&P
8730	4"				Ea.	450			450	495
8740	5"					520			520	570
8750	6"					680			680	750
8760	8"					1,025			1,025	1,125
8770	10"					1,325			1,325	1,450
8780	12"					1,775			1,775	1,950
8800	Tees, reducing 2" x 1"					221			221	243
8810	2" x 1-1/2"					221			221	243
8820	3" x 1"					204			204	225
8830	3" x 1-1/2"					297			297	325
8840	3" x 2"					202			202	222
8850	4" x 1"					296			296	325
8860	4" x 1-1/2"					296			296	325
8870	4" x 2"					440			440	485
8880	4" x 2-1/2"					450			450	495
8890	4" x 3"					450			450	495
8900	6" x 2"					370			370	405
8910	6" x 3"					298			298	330
8920	6" x 4"					370			370	405
8930	8" x 2"					550			550	605
8940	8" x 3"					590			590	645
8950	8" x 4"					625			625	685
8960	8" x 5"					450			450	495
8970	8" x 6"					595			595	655
8980	10" x 4"					550			550	605
8990	10" x 6"					570			570	625
9000	10" x 8"					870			870	960
9010	12" x 6"					1,225			1,225	1,350
9020	12" x 8"					1,000			1,000	1,100
9030	12" x 10"				↓	995			995	1,100
9080	Adapter nipples 3" long									
9090	1"				Ea.	25			25	27.50
9100	1-1/2"					25			25	27.50
9110	2"					25			25	27.50
9120	2-1/2"					29			29	32
9130	3"					35			35	38.50
9140	4"					57			57	63
9150	6"				↓	143			143	157
9180	Coupling, mechanical, plain end pipe to plain end pipe or fitting									
9190	1"	Q-1	29	.552	Ea.	86.50	31.50		118	142
9200	1-1/2"		28	.571		86.50	32.50		119	144
9210	2"		27	.593		86.50	33.50		120	146
9220	2-1/2"		26	.615		86.50	35		121.50	148
9230	3"		25	.640		126	36.50		162.50	194
9240	3-1/2"		24	.667		146	38		184	217
9250	4"	↓	22	.727		146	41.50		187.50	222
9260	5"	Q-2	28	.857		207	50.50		257.50	305
9270	6"		24	1		253	59		312	370
9280	8"		19	1.263		445	74.50		519.50	600
9290	10"		16	1.500		580	88.50		668.50	770
9300	12"	↓	12	2	↓	725	118		843	975
9310	Outlets for precut holes through pipe wall									
9331	Strapless type, with gasket									

22 11 13.47 Pipe Fittings, Steel		Crew	Daily Output	Labor-Hours	Unit	2019 Bare Costs Material	Labor	Equipment	Total	Total Incl O&P
9332	4" to 8" pipe x 1/2"	1 Plum	13	.615	Ea.	94.50	39		133.50	163
9333	4" to 8" pipe x 3/4"		13	.615		101	39		140	170
9334	10" pipe and larger x 1/2"		11	.727		94.50	46		140.50	173
9335	10" pipe and larger x 3/4"	↓	11	.727	↓	101	46		147	180
9341	Thermometer wells with gasket									
9342	4" to 8" pipe, 6" stem	1 Plum	14	.571	Ea.	142	36		178	210
9343	8" pipe and larger, 6" stem	"	13	.615	"	142	39		181	215
9800	Mech. tee, cast iron, grooved or threaded, w/nuts and bolts									
9940	For galvanized fittings for plain end pipe, add					20%				

22 11 13.48 Pipe, Fittings and Valves, Steel, Grooved-Joint		Crew	Daily Output	Labor-Hours	Unit	Material	Labor	Equipment	Total	Total Incl O&P
0010	**PIPE, FITTINGS AND VALVES, STEEL, GROOVED-JOINT**									
0012	Fittings are ductile iron. Steel fittings noted.									
0020	Pipe includes coupling & clevis type hanger assemblies, 10' OC									
0500	Schedule 10, black									
0550	2" diameter	1 Plum	43	.186	L.F.	6.10	11.75		17.85	24.50
0560	2-1/2" diameter	Q-1	61	.262		9.95	14.90		24.85	33.50
0570	3" diameter		55	.291		6.55	16.55		23.10	32
0580	3-1/2" diameter		53	.302		14.50	17.15		31.65	42
0590	4" diameter		49	.327		13.15	18.55		31.70	42.50
0600	5" diameter	↓	40	.400		16.50	22.50		39	52
0610	6" diameter	Q-2	46	.522		14.75	31		45.75	62.50
0620	8" diameter	"	41	.585	↓	33	34.50		67.50	88
0700	To delete couplings & hangers, subtract									
0710	2" diam. to 5" diam.					25%	20%			
0720	6" diam. to 8" diam.					27%	15%			
1000	Schedule 40, black									
1040	3/4" diameter	1 Plum	71	.113	L.F.	6.25	7.10		13.35	17.60
1050	1" diameter		63	.127		6.05	8		14.05	18.70
1060	1-1/4" diameter		58	.138		7.30	8.70		16	21
1070	1-1/2" diameter		51	.157		7.95	9.90		17.85	23.50
1080	2" diameter	↓	40	.200		9.20	12.65		21.85	29
1090	2-1/2" diameter	Q-1	57	.281		12.10	15.95		28.05	37.50
1100	3" diameter		50	.320		14.20	18.20		32.40	43
1110	4" diameter		45	.356		24	20		44	57
1120	5" diameter	↓	37	.432		43.50	24.50		68	85
1130	6" diameter	Q-2	42	.571		49.50	33.50		83	105
1140	8" diameter		37	.649		82	38		120	148
1150	10" diameter		31	.774		110	45.50		155.50	190
1160	12" diameter		27	.889		123	52.50		175.50	214
1170	14" diameter		20	1.200		131	70.50		201.50	250
1180	16" diameter		17	1.412		196	83		279	340
1190	18" diameter		14	1.714		199	101		300	370
1200	20" diameter		12	2		236	118		354	435
1210	24" diameter	↓	10	2.400	↓	269	141		410	505
1740	To delete coupling & hanger, subtract									
1750	3/4" diam. to 2" diam.					65%	27%			
1760	2-1/2" diam. to 5" diam.					41%	18%			
1770	6" diam. to 12" diam.					31%	13%			
1780	14" diam. to 24" diam.					35%	10%			
1800	Galvanized									
1840	3/4" diameter	1 Plum	71	.113	L.F.	6.50	7.10		13.60	17.85
1850	1" diameter	↓	63	.127	↓	6.85	8		14.85	19.55

22 11 13.48 Pipe, Fittings and Valves, Steel, Grooved-Joint		Crew	Daily Output	Labor-Hours	Unit	Material	2019 Bare Costs Labor	Equipment	Total	Total Incl O&P
1860	1-1/4" diameter	1 Plum	58	.138	L.F.	8.25	8.70		16.95	22
1870	1-1/2" diameter		51	.157		8.95	9.90		18.85	24.50
1880	2" diameter	↓	40	.200		10.60	12.65		23.25	30.50
1890	2-1/2" diameter	Q-1	57	.281		13.45	15.95		29.40	39
1900	3" diameter		50	.320		16.40	18.20		34.60	45.50
1910	4" diameter		45	.356		27	20		47	60.50
1920	5" diameter	↓	37	.432		31.50	24.50		56	72
1930	6" diameter	Q-2	42	.571		35	33.50		68.50	89
1940	8" diameter		37	.649		52	38		90	115
1950	10" diameter		31	.774		109	45.50		154.50	188
1960	12" diameter	↓	27	.889	↓	132	52.50		184.50	224
2540	To delete coupling & hanger, subtract									
2550	3/4" diam. to 2" diam.					36%	27%			
2560	2-1/2" diam. to 5" diam.					19%	18%			
2570	6" diam. to 12" diam.					14%	13%			
2600	Schedule 80, black									
2610	3/4" diameter	1 Plum	65	.123	L.F.	7.25	7.75		15	19.65
2650	1" diameter		61	.131		8	8.30		16.30	21.50
2660	1-1/4" diameter		55	.145		10	9.20		19.20	25
2670	1-1/2" diameter		49	.163		11.10	10.30		21.40	27.50
2680	2" diameter	↓	38	.211		13.10	13.30		26.40	34.50
2690	2-1/2" diameter	Q-1	54	.296		16.65	16.85		33.50	44
2700	3" diameter		48	.333		22	18.95		40.95	52.50
2710	4" diameter		44	.364		34	20.50		54.50	68.50
2720	5" diameter	↓	35	.457		73.50	26		99.50	120
2730	6" diameter	Q-2	40	.600		70.50	35.50		106	131
2740	8" diameter		35	.686		106	40.50		146.50	177
2750	10" diameter		29	.828		163	49		212	253
2760	12" diameter	↓	24	1	↓	335	59		394	460
3240	To delete coupling & hanger, subtract									
3250	3/4" diam. to 2" diam.					30%	25%			
3260	2-1/2" diam. to 5" diam.					14%	17%			
3270	6" diam. to 12" diam.					12%	12%			
3300	Galvanized									
3310	3/4" diameter	1 Plum	65	.123	L.F.	7.95	7.75		15.70	20.50
3350	1" diameter		61	.131		9.05	8.30		17.35	22.50
3360	1-1/4" diameter		55	.145		11.60	9.20		20.80	26.50
3370	1-1/2" diameter		46	.174		12.90	11		23.90	30.50
3380	2" diameter	↓	38	.211		16.50	13.30		29.80	38
3390	2-1/2" diameter	Q-1	54	.296		21.50	16.85		38.35	49.50
3400	3" diameter		48	.333		27.50	18.95		46.45	58.50
3410	4" diameter		44	.364		39	20.50		59.50	74
3420	5" diameter	↓	35	.457		54.50	26		80.50	99
3430	6" diameter	Q-2	40	.600		67.50	35.50		103	127
3440	8" diameter		35	.686		125	40.50		165.50	199
3450	10" diameter		29	.828		281	49		330	385
3460	12" diameter	↓	24	1	↓	520	59		579	660
3920	To delete coupling & hanger, subtract									
3930	3/4" diam. to 2" diam.					30%	25%			
3940	2-1/2" diam. to 5" diam.					15%	17%			
3950	6" diam. to 12" diam.					11%	12%			
3990	Fittings: coupling material required at joints not incl. in fitting price.									
3994	Add 1 selected coupling, material only, per joint for installed price.									

22 11 13.48 Pipe, Fittings and Valves, Steel, Grooved-Joint		Crew	Daily Output	Labor-Hours	Unit	2019 Bare Costs Material	Labor	Equipment	Total	Total Incl O&P
4000	Elbow, 90° or 45°, painted									
4030	3/4" diameter	1 Plum	50	.160	Ea.	77	10.10		87.10	99.50
4040	1" diameter		50	.160		41	10.10		51.10	60
4050	1-1/4" diameter		40	.200		41	12.65		53.65	64
4060	1-1/2" diameter		33	.242		41	15.30		56.30	68
4070	2" diameter		25	.320		41	20		61	75.50
4080	2-1/2" diameter	Q-1	40	.400		41	22.50		63.50	79
4090	3" diameter		33	.485		72	27.50		99.50	121
4100	4" diameter		25	.640		78.50	36.50		115	141
4110	5" diameter		20	.800		186	45.50		231.50	273
4120	6" diameter	Q-2	25	.960		219	56.50		275.50	325
4130	8" diameter		21	1.143		455	67.50		522.50	600
4140	10" diameter		18	1.333		825	78.50		903.50	1,025
4150	12" diameter		15	1.600		895	94.50		989.50	1,125
4170	14" diameter		12	2		940	118		1,058	1,200
4180	16" diameter		11	2.182		1,225	129		1,354	1,550
4190	18" diameter	Q-3	14	2.286		1,550	137		1,687	1,900
4200	20" diameter		12	2.667		2,050	160		2,210	2,500
4210	24" diameter		10	3.200		2,950	192		3,142	3,550
4250	For galvanized elbows, add					26%				
4690	Tee, painted									
4700	3/4" diameter	1 Plum	38	.211	Ea.	82	13.30		95.30	110
4740	1" diameter		33	.242		63.50	15.30		78.80	93
4750	1-1/4" diameter		27	.296		63.50	18.70		82.20	98
4760	1-1/2" diameter		22	.364		63.50	23		86.50	105
4770	2" diameter		17	.471		63.50	29.50		93	115
4780	2-1/2" diameter	Q-1	27	.593		63.50	33.50		97	121
4790	3" diameter		22	.727		86.50	41.50		128	158
4800	4" diameter		17	.941		132	53.50		185.50	226
4810	5" diameter		13	1.231		305	70		375	440
4820	6" diameter	Q-2	17	1.412		355	83		438	515
4830	8" diameter		14	1.714		775	101		876	1,000
4840	10" diameter		12	2		975	118		1,093	1,250
4850	12" diameter		10	2.400		1,300	141		1,441	1,625
4851	14" diameter		9	2.667		1,275	157		1,432	1,625
4852	16" diameter		8	3		1,425	177		1,602	1,825
4853	18" diameter	Q-3	10	3.200		1,775	192		1,967	2,250
4854	20" diameter		9	3.556		2,550	214		2,764	3,125
4855	24" diameter		8	4		3,875	240		4,115	4,625
4900	For galvanized tees, add					24%				
4906	Couplings, rigid style, painted									
4908	1" diameter	1 Plum	100	.080	Ea.	31.50	5.05		36.55	42.50
4909	1-1/4" diameter		100	.080		29	5.05		34.05	39.50
4910	1-1/2" diameter		67	.119		31.50	7.55		39.05	46.50
4912	2" diameter		50	.160		40	10.10		50.10	59
4914	2-1/2" diameter	Q-1	80	.200		45.50	11.35		56.85	67
4916	3" diameter		67	.239		53	13.55		66.55	79
4918	4" diameter		50	.320		73.50	18.20		91.70	109
4920	5" diameter		40	.400		95	22.50		117.50	138
4922	6" diameter	Q-2	50	.480		125	28.50		153.50	181
4924	8" diameter		42	.571		197	33.50		230.50	268
4926	10" diameter		35	.686		380	40.50		420.50	475
4928	12" diameter		32	.750		425	44		469	535

22 11 13.48 Pipe, Fittings and Valves, Steel, Grooved-Joint		Crew	Daily Output	Labor-Hours	Unit	2019 Bare Costs Material	Labor	Equipment	Total	Total Incl O&P
4930	14" diameter	Q-2	24	1	Ea.	555	59		614	700
4931	16" diameter		20	1.200		405	70.50		475.50	550
4932	18" diameter		18	1.333		505	78.50		583.50	675
4933	20" diameter		16	1.500		685	88.50		773.50	890
4934	24" diameter		13	1.846		885	109		994	1,125
4940	Flexible, standard, painted									
4950	3/4" diameter	1 Plum	100	.080	Ea.	23	5.05		28.05	33
4960	1" diameter		100	.080		23	5.05		28.05	33
4970	1-1/4" diameter		80	.100		30	6.30		36.30	42.50
4980	1-1/2" diameter		67	.119		32.50	7.55		40.05	47.50
4990	2" diameter		50	.160		35	10.10		45.10	53.50
5000	2-1/2" diameter	Q-1	80	.200		40.50	11.35		51.85	61.50
5010	3" diameter		67	.239		45	13.55		58.55	70
5020	3-1/2" diameter		57	.281		64.50	15.95		80.45	95
5030	4" diameter		50	.320		65	18.20		83.20	99
5040	5" diameter		40	.400		97.50	22.50		120	141
5050	6" diameter	Q-2	50	.480		115	28.50		143.50	170
5070	8" diameter		42	.571		187	33.50		220.50	257
5090	10" diameter		35	.686		305	40.50		345.50	395
5110	12" diameter		32	.750		345	44		389	445
5120	14" diameter		24	1		490	59		549	630
5130	16" diameter		20	1.200		640	70.50		710.50	810
5140	18" diameter		18	1.333		750	78.50		828.50	945
5150	20" diameter		16	1.500		1,175	88.50		1,263.50	1,425
5160	24" diameter		13	1.846		1,300	109		1,409	1,600
5176	Lightweight style, painted									
5178	1-1/2" diameter	1 Plum	67	.119	Ea.	29	7.55		36.55	43
5180	2" diameter	"	50	.160		34.50	10.10		44.60	53
5182	2-1/2" diameter	Q-1	80	.200		40	11.35		51.35	61
5184	3" diameter		67	.239		44.50	13.55		58.05	69.50
5186	3-1/2" diameter		57	.281		55	15.95		70.95	84.50
5188	4" diameter		50	.320		64	18.20		82.20	98
5190	5" diameter		40	.400		95.50	22.50		118	139
5192	6" diameter	Q-2	50	.480		113	28.50		141.50	167
5194	8" diameter		42	.571		183	33.50		216.50	252
5196	10" diameter		35	.686		380	40.50		420.50	475
5198	12" diameter		32	.750		425	44		469	530
5200	For galvanized couplings, add					33%				
5220	Tee, reducing, painted									
5225	2" x 1-1/2" diameter	Q-1	38	.421	Ea.	133	24		157	183
5226	2-1/2" x 2" diameter		28	.571		133	32.50		165.50	196
5227	3" x 2-1/2" diameter		23	.696		118	39.50		157.50	190
5228	4" x 3" diameter		18	.889		158	50.50		208.50	250
5229	5" x 4" diameter		15	1.067		340	60.50		400.50	460
5230	6" x 4" diameter	Q-2	18	1.333		370	78.50		448.50	530
5231	8" x 6" diameter		15	1.600		775	94.50		869.50	990
5232	10" x 8" diameter		13	1.846		855	109		964	1,100
5233	12" x 10" diameter		11	2.182		1,050	129		1,179	1,350
5234	14" x 12" diameter		10	2.400		1,375	141		1,516	1,700
5235	16" x 12" diameter		9	2.667		1,125	157		1,282	1,450
5236	18" x 12" diameter	Q-3	12	2.667		1,350	160		1,510	1,750
5237	18" x 16" diameter		11	2.909		1,725	175		1,900	2,150
5238	20" x 16" diameter		10	3.200		2,250	192		2,442	2,775

22 11 13.48 Pipe, Fittings and Valves, Steel, Grooved-Joint		Crew	Daily Output	Labor-Hours	Unit	2019 Bare Costs Material	Labor	Equipment	Total	Total Incl O&P
5239	24" x 20" diameter	Q-3	9	3.556	Ea.	3,550	214		3,764	4,225
5240	Reducer, concentric, painted									
5241	2-1/2" x 2" diameter	Q-1	43	.372	Ea.	48	21		69	84.50
5242	3" x 2-1/2" diameter		35	.457		57.50	26		83.50	103
5243	4" x 3" diameter		29	.552		69.50	31.50		101	124
5244	5" x 4" diameter		22	.727		95.50	41.50		137	167
5245	6" x 4" diameter	Q-2	26	.923		111	54.50		165.50	204
5246	8" x 6" diameter		23	1.043		286	61.50		347.50	410
5247	10" x 8" diameter		20	1.200		565	70.50		635.50	725
5248	12" x 10" diameter		16	1.500		1,000	88.50		1,088.50	1,225
5255	Eccentric, painted									
5256	2-1/2" x 2" diameter	Q-1	42	.381	Ea.	99.50	21.50		121	142
5257	3" x 2-1/2" diameter		34	.471		114	26.50		140.50	165
5258	4" x 3" diameter		28	.571		138	32.50		170.50	201
5259	5" x 4" diameter		21	.762		188	43.50		231.50	271
5260	6" x 4" diameter	Q-2	25	.960		218	56.50		274.50	325
5261	8" x 6" diameter		22	1.091		440	64.50		504.50	580
5262	10" x 8" diameter		19	1.263		1,175	74.50		1,249.50	1,375
5263	12" x 10" diameter		15	1.600		1,600	94.50		1,694.50	1,925
5270	Coupling, reducing, painted									
5272	2" x 1-1/2" diameter	1 Plum	52	.154	Ea.	46	9.70		55.70	65
5274	2-1/2" x 2" diameter	Q-1	82	.195		59.50	11.10		70.60	82
5276	3" x 2" diameter		69	.232		67.50	13.20		80.70	94.50
5278	4" x 2" diameter		52	.308		107	17.50		124.50	143
5280	5" x 4" diameter		42	.381		120	21.50		141.50	165
5282	6" x 4" diameter	Q-2	52	.462		181	27		208	240
5284	8" x 6" diameter	"	44	.545		270	32		302	345
5290	Outlet coupling, painted									
5294	1-1/2" x 1" pipe size	1 Plum	65	.123	Ea.	61	7.75		68.75	78.50
5296	2" x 1" pipe size	"	48	.167		62	10.55		72.55	84.50
5298	2-1/2" x 1" pipe size	Q-1	78	.205		95.50	11.65		107.15	123
5300	2-1/2" x 1-1/4" pipe size		70	.229		107	13		120	138
5302	3" x 1" pipe size		65	.246		122	14		136	155
5304	4" x 3/4" pipe size		48	.333		135	18.95		153.95	177
5306	4" x 1-1/2" pipe size		46	.348		191	19.75		210.75	240
5308	6" x 1-1/2" pipe size	Q-2	44	.545		269	32		301	345
5320	Outlet, strap-on T, painted									
5324	Threaded female branch									
5330	2" pipe x 1-1/2" branch size	1 Plum	50	.160	Ea.	43.50	10.10		53.60	62.50
5334	2-1/2" pipe x 1-1/2" branch size	Q-1	80	.200		53	11.35		64.35	75
5338	3" pipe x 1-1/2" branch size		67	.239		54.50	13.55		68.05	80.50
5342	3" pipe x 2" branch size		59	.271		63	15.40		78.40	92
5346	4" pipe x 2" branch size		50	.320		66	18.20		84.20	100
5350	4" pipe x 2-1/2" branch size		47	.340		69	19.35		88.35	105
5354	4" pipe x 3" branch size		43	.372		73.50	21		94.50	113
5358	5" pipe x 3" branch size		40	.400		90.50	22.50		113	134
5362	6" pipe x 2" branch size	Q-2	50	.480		83	28.50		111.50	134
5366	6" pipe x 3" branch size		47	.511		94.50	30		124.50	149
5370	6" pipe x 4" branch size		44	.545		105	32		137	163
5374	8" pipe x 2-1/2" branch size		42	.571		144	33.50		177.50	210
5378	8" pipe x 4" branch size		38	.632		157	37		194	228
5390	Grooved branch									
5394	2" pipe x 1-1/4" branch size	1 Plum	78	.103	Ea.	43.50	6.50		50	57

22 11 13.48 Pipe, Fittings and Valves, Steel, Grooved-Joint		Crew	Daily Output	Labor-Hours	Unit	2019 Bare Costs Material	Labor	Equipment	Total	Total Incl O&P
5398	2" pipe x 1-1/2" branch size	1 Plum	50	.160	Ea.	43.50	10.10		53.60	62.50
5402	2-1/2" pipe x 1-1/2" branch size	Q-1	80	.200		53	11.35		64.35	75
5406	3" pipe x 1-1/2" branch size		67	.239		54.50	13.55		68.05	80.50
5410	3" pipe x 2" branch size		59	.271		63	15.40		78.40	92
5414	4" pipe x 2" branch size		50	.320		66	18.20		84.20	100
5418	4" pipe x 2-1/2" branch size		47	.340		69	19.35		88.35	105
5422	4" pipe x 3" branch size		43	.372		73.50	21		94.50	113
5426	5" pipe x 3" branch size		40	.400		90.50	22.50		113	134
5430	6" pipe x 2" branch size	Q-2	50	.480		83	28.50		111.50	134
5434	6" pipe x 3" branch size		47	.511		94.50	30		124.50	149
5438	6" pipe x 4" branch size		44	.545		105	32		137	163
5442	8" pipe x 2-1/2" branch size		42	.571		144	33.50		177.50	210
5446	8" pipe x 4" branch size		38	.632		157	37		194	228
5750	Flange, w/groove gasket, black steel									
5754	See Line 22 11 13.47 0620 for gasket & bolt set									
5760	ANSI class 125 and 150, painted									
5780	2" pipe size	1 Plum	23	.348	Ea.	135	22		157	181
5790	2-1/2" pipe size	Q-1	37	.432		168	24.50		192.50	221
5800	3" pipe size		31	.516		181	29.50		210.50	243
5820	4" pipe size		23	.696		241	39.50		280.50	325
5830	5" pipe size		19	.842		280	48		328	380
5840	6" pipe size	Q-2	23	1.043		305	61.50		366.50	430
5850	8" pipe size		17	1.412		345	83		428	505
5860	10" pipe size		14	1.714		545	101		646	750
5870	12" pipe size		12	2		710	118		828	955
5880	14" pipe size		10	2.400		1,350	141		1,491	1,675
5890	16" pipe size		9	2.667		1,550	157		1,707	1,925
5900	18" pipe size		6	4		1,925	236		2,161	2,450
5910	20" pipe size		5	4.800		2,300	283		2,583	2,950
5920	24" pipe size		4.50	5.333		2,950	315		3,265	3,725
5940	ANSI class 350, painted									
5946	2" pipe size	1 Plum	23	.348	Ea.	169	22		191	219
5948	2-1/2" pipe size	Q-1	37	.432		195	24.50		219.50	251
5950	3" pipe size		31	.516		266	29.50		295.50	335
5952	4" pipe size		23	.696		355	39.50		394.50	450
5954	5" pipe size		19	.842		405	48		453	515
5956	6" pipe size	Q-2	23	1.043		470	61.50		531.50	610
5958	8" pipe size		17	1.412		535	83		618	715
5960	10" pipe size		14	1.714		855	101		956	1,100
5962	12" pipe size		12	2		910	118		1,028	1,175
6100	Cross, painted									
6110	2" diameter	1 Plum	12.50	.640	Ea.	114	40.50		154.50	186
6112	2-1/2" diameter	Q-1	20	.800		114	45.50		159.50	193
6114	3" diameter		15.50	1.032		203	58.50		261.50	310
6116	4" diameter		12.50	1.280		335	72.50		407.50	480
6118	6" diameter	Q-2	12.50	1.920		880	113		993	1,150
6120	8" diameter		10.50	2.286		1,150	135		1,285	1,475
6122	10" diameter		9	2.667		1,900	157		2,057	2,325
6124	12" diameter		7.50	3.200		2,750	189		2,939	3,300
7400	Suction diffuser									
7402	Grooved end inlet x flanged outlet									
7410	3" x 3"	Q-1	27	.593	Ea.	1,100	33.50		1,133.50	1,250
7412	4" x 4"		19	.842		1,475	48		1,523	1,700

22 11 13 – Facility Water Distribution Piping

22 11 13.48	Pipe, Fittings and Valves, Steel, Grooved-Joint	Crew	Daily Output	Labor-Hours	Unit	Material	2019 Bare Costs Labor	Equipment	Total	Total Incl O&P
7414	5″ x 5″	Q-1	14	1.143	Ea.	1,725	65		1,790	2,000
7416	6″ x 6″	Q-2	20	1.200		2,175	70.50		2,245.50	2,500
7418	8″ x 8″		15	1.600		4,050	94.50		4,144.50	4,600
7420	10″ x 10″		12	2		5,500	118		5,618	6,225
7422	12″ x 12″		9	2.667		9,050	157		9,207	10,200
7424	14″ x 14″		7	3.429		8,675	202		8,877	9,825
7426	16″ x 14″		6	4		9,375	236		9,611	10,700
7500	Strainer, tee type, painted									
7506	2″ pipe size	1 Plum	21	.381	Ea.	730	24		754	835
7508	2-1/2″ pipe size	Q-1	30	.533		765	30.50		795.50	885
7510	3″ pipe size		28	.571		860	32.50		892.50	995
7512	4″ pipe size		20	.800		970	45.50		1,015.50	1,150
7514	5″ pipe size		15	1.067		1,400	60.50		1,460.50	1,650
7516	6″ pipe size	Q-2	23	1.043		1,525	61.50		1,586.50	1,775
7518	8″ pipe size		16	1.500		2,350	88.50		2,438.50	2,700
7520	10″ pipe size		13	1.846		3,425	109		3,534	3,950
7522	12″ pipe size		10	2.400		4,425	141		4,566	5,050
7524	14″ pipe size		8	3		11,900	177		12,077	13,400
7526	16″ pipe size		7	3.429		15,200	202		15,402	17,000
7570	Expansion joint, max. 3″ travel									
7572	2″ diameter	1 Plum	24	.333	Ea.	920	21		941	1,025
7574	3″ diameter	Q-1	31	.516		1,050	29.50		1,079.50	1,200
7576	4″ diameter	″	23	.696		1,400	39.50		1,439.50	1,575
7578	6″ diameter	Q-2	38	.632		2,150	37		2,187	2,425
7790	Valves: coupling material required at joints not incl. in valve price.									
7794	Add 1 selected coupling, material only, per joint for installed price.									
7800	Ball valve w/handle, carbon steel trim									
7810	1-1/2″ pipe size	1 Plum	31	.258	Ea.	165	16.30		181.30	206
7812	2″ pipe size	″	24	.333		183	21		204	233
7814	2-1/2″ pipe size	Q-1	38	.421		400	24		424	475
7816	3″ pipe size		31	.516		645	29.50		674.50	750
7818	4″ pipe size		23	.696		995	39.50		1,034.50	1,150
7820	6″ pipe size	Q-2	24	1		3,050	59		3,109	3,450
7830	With gear operator									
7834	2-1/2″ pipe size	Q-1	38	.421	Ea.	805	24		829	920
7836	3″ pipe size		31	.516		1,150	29.50		1,179.50	1,300
7838	4″ pipe size		23	.696		1,450	39.50		1,489.50	1,650
7840	6″ pipe size	Q-2	24	1		3,425	59		3,484	3,875
7870	Check valve									
7874	2-1/2″ pipe size	Q-1	38	.421	Ea.	350	24		374	420
7876	3″ pipe size		31	.516		410	29.50		439.50	495
7878	4″ pipe size		23	.696		435	39.50		474.50	535
7880	5″ pipe size		18	.889		720	50.50		770.50	870
7882	6″ pipe size	Q-2	24	1		855	59		914	1,025
7884	8″ pipe size		19	1.263		1,175	74.50		1,249.50	1,375
7886	10″ pipe size		16	1.500		3,225	88.50		3,313.50	3,675
7888	12″ pipe size		13	1.846		3,825	109		3,934	4,375
7900	Plug valve, balancing, w/lever operator									
7906	3″ pipe size	Q-1	31	.516	Ea.	855	29.50		884.50	985
7908	4″ pipe size	″	23	.696		1,050	39.50		1,089.50	1,200
7909	6″ pipe size	Q-2	24	1		1,625	59		1,684	1,875
7916	With gear operator									
7920	3″ pipe size	Q-1	31	.516	Ea.	1,625	29.50		1,654.50	1,8[illegible]

22 11 13.48 Pipe, Fittings and Valves, Steel, Grooved-Joint		Crew	Daily Output	Labor-Hours	Unit	Material	2019 Bare Costs Labor	Equipment	Total	Total Incl O&P
7922	4″ pipe size	Q-1	23	.696	Ea.	1,675	39.50		1,714.50	1,900
7924	6″ pipe size	Q-2	24	1		2,275	59		2,334	2,600
7926	8″ pipe size		19	1.263		3,500	74.50		3,574.50	3,950
7928	10″ pipe size		16	1.500		4,475	88.50		4,563.50	5,050
7930	12″ pipe size		13	1.846		6,825	109		6,934	7,700
8000	Butterfly valve, 2 position handle, with standard trim									
8010	1-1/2″ pipe size	1 Plum	30	.267	Ea.	325	16.85		341.85	380
8020	2″ pipe size	″	23	.348		325	22		347	390
8030	3″ pipe size	Q-1	30	.533		460	30.50		490.50	550
8050	4″ pipe size	″	22	.727		515	41.50		556.50	625
8070	6″ pipe size	Q-2	23	1.043		1,025	61.50		1,086.50	1,225
8080	8″ pipe size		18	1.333		1,275	78.50		1,353.50	1,525
8090	10″ pipe size		15	1.600		1,825	94.50		1,919.50	2,150
8200	With stainless steel trim									
8240	1-1/2″ pipe size	1 Plum	30	.267	Ea.	410	16.85		426.85	480
8250	2″ pipe size	″	23	.348		410	22		432	490
8270	3″ pipe size	Q-1	30	.533		555	30.50		585.50	655
8280	4″ pipe size	″	22	.727		600	41.50		641.50	720
8300	6″ pipe size	Q-2	23	1.043		1,125	61.50		1,186.50	1,325
8310	8″ pipe size		18	1.333		2,800	78.50		2,878.50	3,200
8320	10″ pipe size		15	1.600		4,075	94.50		4,169.50	4,650
8322	12″ pipe size		12	2		5,475	118		5,593	6,200
8324	14″ pipe size		10	2.400		6,850	141		6,991	7,725
8326	16″ pipe size		9	2.667		11,800	157		11,957	13,200
8328	18″ pipe size	Q-3	12	2.667		12,200	160		12,360	13,600
8330	20″ pipe size		10	3.200		15,000	192		15,192	16,800
8332	24″ pipe size		9	3.556		20,100	214		20,314	22,400
8336	Note: sizes 8″ up w/manual gear operator									
9000	Cut one groove, labor									
9010	3/4″ pipe size	Q-1	152	.105	Ea.		6		6	9
9020	1″ pipe size		140	.114			6.50		6.50	9.75
9030	1-1/4″ pipe size		124	.129			7.35		7.35	11
9040	1-1/2″ pipe size		114	.140			8		8	11.95
9050	2″ pipe size		104	.154			8.75		8.75	13.10
9060	2-1/2″ pipe size		96	.167			9.45		9.45	14.20
9070	3″ pipe size		88	.182			10.35		10.35	15.50
9080	3-1/2″ pipe size		83	.193			10.95		10.95	16.45
9090	4″ pipe size		78	.205			11.65		11.65	17.50
9100	5″ pipe size		72	.222			12.65		12.65	18.95
9110	6″ pipe size		70	.229			13		13	19.50
9120	8″ pipe size		54	.296			16.85		16.85	25.50
9130	10″ pipe size		38	.421			24		24	36
9140	12″ pipe size		30	.533			30.50		30.50	45.50
9150	14″ pipe size		20	.800			45.50		45.50	68
9160	16″ pipe size		19	.842			48		48	72
9170	18″ pipe size		18	.889			50.50		50.50	76
9180	20″ pipe size		17	.941			53.50		53.50	80.50
9190	24″ pipe size		15	1.067			60.50		60.50	91
9210	Roll one groove									
9220	3/4″ pipe size	Q-1	266	.060	Ea.		3.42		3.42	5.15
9230	1″ pipe size		228	.070			3.99		3.99	6
9240	1-1/4″ pipe size		200	.080			4.55		4.55	6.80
9250	1-1/2″ pipe size		178	.090			5.10		5.10	7.65

22 11 13.48 Pipe, Fittings and Valves, Steel, Grooved-Joint		Crew	Daily Output	Labor-Hours	Unit	Material	2019 Bare Costs Labor	Equipment	Total	Total Incl O&P
9260	2" pipe size	Q-1	116	.138	Ea.		7.85		7.85	11.75
9270	2-1/2" pipe size		110	.145			8.25		8.25	12.40
9280	3" pipe size		100	.160			9.10		9.10	13.65
9290	3-1/2" pipe size		94	.170			9.65		9.65	14.50
9300	4" pipe size		86	.186			10.55		10.55	15.85
9310	5" pipe size		84	.190			10.85		10.85	16.25
9320	6" pipe size		80	.200			11.35		11.35	17.05
9330	8" pipe size		66	.242			13.80		13.80	20.50
9340	10" pipe size		58	.276			15.70		15.70	23.50
9350	12" pipe size		46	.348			19.75		19.75	29.50
9360	14" pipe size		30	.533			30.50		30.50	45.50
9370	16" pipe size		28	.571			32.50		32.50	48.50
9380	18" pipe size		27	.593			33.50		33.50	50.50
9390	20" pipe size		25	.640			36.50		36.50	54.50
9400	24" pipe size		23	.696			39.50		39.50	59.50

22 11 13.60 Tubing, Stainless Steel		Crew	Daily Output	Labor-Hours	Unit	Material	Labor	Equipment	Total	Total Incl O&P
0010	**TUBING, STAINLESS STEEL**									
5000	Tubing									
5010	Type 304, no joints, no hangers									
5020	.035 wall									
5021	1/4"	1 Plum	160	.050	L.F.	5.95	3.16		9.11	11.25
5022	3/8"		160	.050		5.50	3.16		8.66	10.80
5023	1/2"		160	.050		9.35	3.16		12.51	15.05
5024	5/8"		160	.050		8.95	3.16		12.11	14.60
5025	3/4"		133	.060		9.35	3.80		13.15	16
5026	7/8"		133	.060		11.05	3.80		14.85	17.85
5027	1"		114	.070		12.70	4.43		17.13	20.50
5040	.049 wall									
5041	1/4"	1 Plum	160	.050	L.F.	6.25	3.16		9.41	11.60
5042	3/8"		160	.050		9.35	3.16		12.51	15.05
5043	1/2"		160	.050		9.15	3.16		12.31	14.80
5044	5/8"		160	.050		10.20	3.16		13.36	15.95
5045	3/4"		133	.060		10.85	3.80		14.65	17.60
5046	7/8"		133	.060		12.35	3.80		16.15	19.30
5047	1"		114	.070		12.65	4.43		17.08	20.50
5060	.065 wall									
5061	1/4"	1 Plum	160	.050	L.F.	6.70	3.16		9.86	12.10
5062	3/8"		160	.050		9.35	3.16		12.51	15
5063	1/2"		160	.050		8.85	3.16		12.01	14.45
5064	5/8"		160	.050		11.30	3.16		14.46	17.20
5065	3/4"		133	.060		12.75	3.80		16.55	19.70
5066	7/8"		133	.060		13.75	3.80		17.55	21
5067	1"		114	.070		13.20	4.43		17.63	21
5210	Type 316									
5220	.035 wall									
5221	1/4"	1 Plum	160	.050	L.F.	5.90	3.16		9.06	11.20
5222	3/8"		160	.050		6	3.16		9.16	11.35
5223	1/2"		160	.050		9.55	3.16		12.71	15.30
5224	5/8"		160	.050		9.70	3.16		12.86	15.40
5225	3/4"		133	.060		12.15	3.80		15.95	19.10
5226	7/8"		133	.060		18.60	3.80		22.40	26
5227	1"		114	.070		16.65	4.43		21.08	25

22 11 13.60 Tubing, Stainless Steel		Crew	Daily Output	Labor-Hours	Unit	Material	Labor	Equipment	Total	Total Incl O&P
						2019 Bare Costs				
5240	.049 wall									
5241	1/4"	1 Plum	160	.050	L.F.	5.20	3.16		8.36	10.45
5242	3/8"		160	.050		9.80	3.16		12.96	15.55
5243	1/2"		160	.050		9.30	3.16		12.46	15
5244	5/8"		160	.050		10.15	3.16		13.31	15.95
5245	3/4"		133	.060		11.30	3.80		15.10	18.15
5246	7/8"		133	.060		18.05	3.80		21.85	25.50
5247	1"		114	.070		13.70	4.43		18.13	22
5260	.065 wall									
5261	1/4"	1 Plum	160	.050	L.F.	6.50	3.16		9.66	11.90
5262	3/8"		160	.050		10.40	3.16		13.56	16.20
5263	1/2"		160	.050		9.55	3.16		12.71	15.25
5264	5/8"		160	.050		12.05	3.16		15.21	18
5265	3/4"		133	.060		13.25	3.80		17.05	20.50
5266	7/8"		133	.060		18.50	3.80		22.30	26
5267	1"		114	.070		18.65	4.43		23.08	27

22 11 13.61 Tubing Fittings, Stainless Steel

		Crew	Daily Output	Labor-Hours	Unit	Material	Labor	Equipment	Total	Total Incl O&P
0010	**TUBING FITTINGS, STAINLESS STEEL**									
8200	Tube fittings, compression type									
8202	Type 316									
8204	90° elbow									
8206	1/4"	1 Plum	24	.333	Ea.	16.30	21		37.30	49.50
8207	3/8"		22	.364		20	23		43	56.50
8208	1/2"		22	.364		33	23		56	71
8209	5/8"		21	.381		37	24		61	77
8210	3/4"		21	.381		58.50	24		82.50	100
8211	7/8"		20	.400		90	25.50		115.50	137
8212	1"		20	.400		112	25.50		137.50	161
8220	Union tee									
8222	1/4"	1 Plum	15	.533	Ea.	23	33.50		56.50	76
8224	3/8"		15	.533		30	33.50		63.50	83.50
8225	1/2"		15	.533		46	33.50		79.50	101
8226	5/8"		14	.571		51.50	36		87.50	111
8227	3/4"		14	.571		69	36		105	130
8228	7/8"		13	.615		137	39		176	210
8229	1"		13	.615		147	39		186	221
8234	Union									
8236	1/4"	1 Plum	24	.333	Ea.	11.30	21		32.30	44
8237	3/8"		22	.364		16.10	23		39.10	52.50
8238	1/2"		22	.364		24	23		47	61
8239	5/8"		21	.381		31.50	24		55.50	70.50
8240	3/4"		21	.381		39	24		63	79
8241	7/8"		20	.400		64	25.50		89.50	108
8242	1"		20	.400		67.50	25.50		93	112
8250	Male connector									
8252	1/4" x 1/4"	1 Plum	24	.333	Ea.	7.25	21		28.25	39.50
8253	3/8" x 3/8"		22	.364		11.30	23		34.30	47
8254	1/2" x 1/2"		22	.364		16.75	23		39.75	53
8256	3/4" x 3/4"		21	.381		25.50	24		49.50	64.50
8258	1" x 1"		20	.400		45	25.50		70.50	87.50

22 11 13 – Facility Water Distribution Piping

22 11 13.64 Pipe, Stainless Steel		Crew	Daily Output	Labor-Hours	Unit	Material	Labor	Equipment	Total	Total Incl O&P
						2019 Bare Costs				
0010	**PIPE, STAINLESS STEEL**									
0020	Welded, with clevis type hanger assemblies, 10′ OC									
0500	Schedule 5, type 304									
0540	1/2″ diameter	Q-15	128	.125	L.F.	12.10	7.10	.44	19.64	24.50
0550	3/4″ diameter		116	.138		16.65	7.85	.48	24.98	30.50
0560	1″ diameter		103	.155		17.35	8.85	.55	26.75	33
0570	1-1/4″ diameter		93	.172		21	9.80	.60	31.40	38.50
0580	1-1/2″ diameter		85	.188		23.50	10.70	.66	34.86	43
0590	2″ diameter		69	.232		34	13.20	.81	48.01	58
0600	2-1/2″ diameter		53	.302		75	17.15	1.06	93.21	110
0610	3″ diameter		48	.333		91	18.95	1.17	111.12	130
0620	4″ diameter		44	.364		127	20.50	1.28	148.78	172
0630	5″ diameter		36	.444		146	25.50	1.56	173.06	201
0640	6″ diameter	Q-16	42	.571		310	33.50	1.34	344.84	395
0650	8″ diameter		34	.706		480	41.50	1.65	523.15	595
0660	10″ diameter		26	.923		277	54.50	2.16	333.66	390
0670	12″ diameter		21	1.143		365	67.50	2.67	435.17	505
0700	To delete hangers, subtract									
0710	1/2″ diam. to 1-1/2″ diam.					8%	19%			
0720	2″ diam. to 5″ diam.					4%	9%			
0730	6″ diam. to 12″ diam.					3%	4%			
0750	For small quantities, add				L.F.	10%				
1250	Schedule 5, type 316									
1290	1/2″ diameter	Q-15	128	.125	L.F.	12.45	7.10	.44	19.99	25
1300	3/4″ diameter		116	.138		20.50	7.85	.48	28.83	35
1310	1″ diameter		103	.155		22	8.85	.55	31.40	38
1320	1-1/4″ diameter		93	.172		25.50	9.80	.60	35.90	43.50
1330	1-1/2″ diameter		85	.188		39.50	10.70	.66	50.86	60.50
1340	2″ diameter		69	.232		56	13.20	.81	70.01	82
1350	2-1/2″ diameter		53	.302		82.50	17.15	1.06	100.71	118
1360	3″ diameter		48	.333		104	18.95	1.17	124.12	145
1370	4″ diameter		44	.364		134	20.50	1.28	155.78	179
1380	5″ diameter		36	.444		152	25.50	1.56	179.06	207
1390	6″ diameter	Q-16	42	.571		173	33.50	1.34	207.84	243
1400	8″ diameter		34	.706		261	41.50	1.65	304.15	350
1410	10″ diameter		26	.923		310	54.50	2.16	366.66	425
1420	12″ diameter		21	1.143		395	67.50	2.67	465.17	540
1490	For small quantities, add					10%				
1940	To delete hanger, subtract									
1950	1/2″ diam. to 1-1/2″ diam.					5%	19%			
1960	2″ diam. to 5″ diam.					3%	9%			
1970	6″ diam. to 12″ diam.					2%	4%			
2000	Schedule 10, type 304									
2040	1/4″ diameter	Q-15	131	.122	L.F.	7.95	6.95	.43	15.33	19.60
2050	3/8″ diameter		128	.125		8.15	7.10	.44	15.69	20
2060	1/2″ diameter		125	.128		13.80	7.25	.45	21.50	26.50
2070	3/4″ diameter		113	.142		16	8.05	.50	24.55	30.50
2080	1″ diameter		100	.160		14.75	9.10	.56	24.41	30.50
2090	1-1/4″ diameter		91	.176		23	10	.62	33.62	40.50
2100	1-1/2″ diameter		83	.193		21	10.95	.68	32.63	40
2110	2″ diameter		67	.239		22.50	13.55	.84	36.89	46.50
2120	2-1/2″ diameter		51	.314		34.50	17.85	1.10	53.45	66

22 11 13 – Facility Water Distribution Piping

22 11 13.64 Pipe, Stainless Steel		Crew	Daily Output	Labor-Hours	Unit	Material	2019 Bare Costs Labor	Equipment	Total	Total Incl O&P
2130	3" diameter	Q-15	46	.348	L.F.	42	19.75	1.22	62.97	77
2140	4" diameter		42	.381		48	21.50	1.34	70.84	87
2150	5" diameter		35	.457		61	26	1.60	88.60	108
2160	6" diameter	Q-16	40	.600		57	35.50	1.40	93.90	117
2170	8" diameter		33	.727		97	43	1.70	141.70	173
2180	10" diameter		25	.960		133	56.50	2.25	191.75	233
2190	12" diameter		21	1.143		165	67.50	2.67	235.17	285
2250	For small quantities, add					10%				
2650	To delete hanger, subtract									
2660	1/4" diam. to 3/4" diam.					9%	22%			
2670	1" diam. to 2" diam.					4%	15%			
2680	2-1/2" diam. to 5" diam.					3%	8%			
2690	6" diam. to 12" diam.					3%	4%			
2750	Schedule 10, type 316									
2790	1/4" diameter	Q-15	131	.122	L.F.	6.95	6.95	.43	14.33	18.50
2800	3/8" diameter		128	.125		8.60	7.10	.44	16.14	20.50
2810	1/2" diameter		125	.128		10.90	7.25	.45	18.60	23.50
2820	3/4" diameter		113	.142		19.30	8.05	.50	27.85	33.50
2830	1" diameter		100	.160		23	9.10	.56	32.66	40
2840	1-1/4" diameter		91	.176		22	10	.62	32.62	40
2850	1-1/2" diameter		83	.193		30	10.95	.68	41.63	50
2860	2" diameter		67	.239		36.50	13.55	.84	50.89	61.50
2870	2-1/2" diameter		51	.314		35.50	17.85	1.10	54.45	67
2880	3" diameter		46	.348		54.50	19.75	1.22	75.47	91
2890	4" diameter		42	.381		62	21.50	1.34	84.84	102
2900	5" diameter		35	.457		58.50	26	1.60	86.10	105
2910	6" diameter	Q-16	40	.600		85.50	35.50	1.40	122.40	149
2920	8" diameter		33	.727		126	43	1.70	170.70	205
2930	10" diameter		25	.960		166	56.50	2.25	224.75	270
2940	12" diameter		21	1.143		208	67.50	2.67	278.17	330
2990	For small quantities, add					10%				
3430	To delete hanger, subtract									
3440	1/4" diam. to 3/4" diam.					6%	22%			
3450	1" diam. to 2" diam.					3%	15%			
3460	2-1/2" diam. to 5" diam.					2%	8%			
3470	6" diam. to 12" diam.					2%	4%			
3500	Threaded, couplings and clevis hanger assemblies, 10' OC									
3520	Schedule 40, type 304									
3540	1/4" diameter	1 Plum	54	.148	L.F.	9.70	9.35		19.05	25
3550	3/8" diameter		53	.151		11.05	9.55		20.60	26.50
3560	1/2" diameter		52	.154		12	9.70		21.70	28
3570	3/4" diameter		51	.157		18.05	9.90		27.95	34.50
3580	1" diameter		45	.178		23.50	11.25		34.75	42.50
3590	1-1/4" diameter	Q-1	76	.211		39.50	11.95		51.45	61.50
3600	1-1/2" diameter		69	.232		38	13.20		51.20	62
3610	2" diameter		57	.281		48	15.95		63.95	77
3620	2-1/2" diameter		44	.364		87	20.50		107.50	127
3630	3" diameter		38	.421		112	24		136	159
3640	4" diameter	Q-2	51	.471		129	27.50		156.50	184
3740	For small quantities, add					10%				
4200	To delete couplings & hangers, subtract									
4210	1/4" diam. to 3/4" diam.					15%	56%			
4220	1" diam. to 2" diam.					18%	49%			

22 11 13.64 Pipe, Stainless Steel		Crew	Daily Output	Labor-Hours	Unit	2019 Bare Costs Material	Labor	Equipment	Total	Total Incl O&P
4230	2-1/2" diam. to 4" diam.					34%	40%			
4250	Schedule 40, type 316									
4290	1/4" diameter	1 Plum	54	.148	L.F.	22	9.35		31.35	38.50
4300	3/8" diameter		53	.151		24.50	9.55		34.05	41
4310	1/2" diameter		52	.154		29.50	9.70		39.20	47
4320	3/4" diameter		51	.157		34	9.90		43.90	52.50
4330	1" diameter		45	.178		47.50	11.25		58.75	69
4340	1-1/4" diameter	Q-1	76	.211		69.50	11.95		81.45	94.50
4350	1-1/2" diameter		69	.232		71	13.20		84.20	98
4360	2" diameter		57	.281		95.50	15.95		111.45	129
4370	2-1/2" diameter		44	.364		135	20.50		155.50	180
4380	3" diameter		38	.421		149	24		173	200
4390	4" diameter	Q-2	51	.471		220	27.50		247.50	284
4490	For small quantities, add					10%				
4900	To delete couplings & hangers, subtract									
4910	1/4" diam. to 3/4" diam.					12%	56%			
4920	1" diam. to 2" diam.					14%	49%			
4930	2-1/2" diam. to 4" diam.					27%	40%			
5000	Schedule 80, type 304									
5040	1/4" diameter	1 Plum	53	.151	L.F.	20.50	9.55		30.05	37
5050	3/8" diameter		52	.154		36.50	9.70		46.20	55
5060	1/2" diameter		51	.157		30.50	9.90		40.40	49
5070	3/4" diameter		48	.167		35	10.55		45.55	54.50
5080	1" diameter		43	.186		51.50	11.75		63.25	74.50
5090	1-1/4" diameter	Q-1	73	.219		55	12.45		67.45	79
5100	1-1/2" diameter		67	.239		66	13.55		79.55	93
5110	2" diameter		54	.296		104	16.85		120.85	140
5190	For small quantities, add					10%				
5700	To delete couplings & hangers, subtract									
5710	1/4" diam. to 3/4" diam.					10%	53%			
5720	1" diam. to 2" diam.					14%	47%			
5750	Schedule 80, type 316									
5790	1/4" diameter	1 Plum	53	.151	L.F.	42	9.55		51.55	60.50
5800	3/8" diameter		52	.154		43	9.70		52.70	62
5810	1/2" diameter		51	.157		49	9.90		58.90	68.50
5820	3/4" diameter		48	.167		52.50	10.55		63.05	74
5830	1" diameter		43	.186		50.50	11.75		62.25	73
5840	1-1/4" diameter	Q-1	73	.219		75.50	12.45		87.95	102
5850	1-1/2" diameter		67	.239		103	13.55		116.55	134
5860	2" diameter		54	.296		112	16.85		128.85	149
5950	For small quantities, add					10%				
7000	To delete couplings & hangers, subtract									
7010	1/4" diam. to 3/4" diam.					9%	53%			
7020	1" diam. to 2" diam.					14%	47%			
8000	Weld joints with clevis type hanger assemblies, 10' OC									
8010	Schedule 40, type 304									
8050	1/8" pipe size	Q-15	126	.127	L.F.	7.30	7.20	.45	14.95	19.40
8060	1/4" pipe size		125	.128		8.45	7.25	.45	16.15	20.50
8070	3/8" pipe size		122	.131		9.55	7.45	.46	17.46	22
8080	1/2" pipe size		118	.136		10	7.70	.48	18.18	23
8090	3/4" pipe size		109	.147		15.35	8.35	.52	24.22	30
8100	1" pipe size		95	.168		19.05	9.55	.59	29.19	36
8110	1-1/4" pipe size		86	.186		29	10.55	.65	40.20	48.50

22 11 13.64 Pipe, Stainless Steel		Crew	Daily Output	Labor-Hours	Unit	2019 Bare Costs Material	Labor	Equipment	Total	Total Incl O&P
8120	1-1/2" pipe size	Q-15	78	.205	L.F.	30.50	11.65	.72	42.87	52
8130	2" pipe size		62	.258		35	14.65	.91	50.56	61.50
8140	2-1/2" pipe size		49	.327		56.50	18.55	1.15	76.20	91.50
8150	3" pipe size		44	.364		70.50	20.50	1.28	92.28	110
8160	3-1/2" pipe size		44	.364		93.50	20.50	1.28	115.28	135
8170	4" pipe size		39	.410		71	23.50	1.44	95.94	115
8180	5" pipe size		32	.500		123	28.50	1.76	153.26	179
8190	6" pipe size	Q-16	37	.649		141	38	1.52	180.52	214
8200	8" pipe size		29	.828		157	49	1.94	207.94	247
8210	10" pipe size		24	1		261	59	2.34	322.34	380
8220	12" pipe size		20	1.200		505	70.50	2.81	578.31	665
8300	Schedule 40, type 316									
8310	1/8" pipe size	Q-15	126	.127	L.F.	18.95	7.20	.45	26.60	32.50
8320	1/4" pipe size		125	.128		20.50	7.25	.45	28.20	34
8330	3/8" pipe size		122	.131		22.50	7.45	.46	30.41	36
8340	1/2" pipe size		118	.136		27	7.70	.48	35.18	42
8350	3/4" pipe size		109	.147		31	8.35	.52	39.87	47
8360	1" pipe size		95	.168		42	9.55	.59	52.14	61
8370	1-1/4" pipe size		86	.186		57	10.55	.65	68.20	79
8380	1-1/2" pipe size		78	.205		61.50	11.65	.72	73.87	86.50
8390	2" pipe size		62	.258		80	14.65	.91	95.56	111
8400	2-1/2" pipe size		49	.327		99	18.55	1.15	118.70	138
8410	3" pipe size		44	.364		99	20.50	1.28	120.78	141
8420	3-1/2" pipe size		44	.364		108	20.50	1.28	129.78	151
8430	4" pipe size		39	.410		151	23.50	1.44	175.94	203
8440	5" pipe size		32	.500		162	28.50	1.76	192.26	222
8450	6" pipe size	Q-16	37	.649		244	38	1.52	283.52	330
8460	8" pipe size		29	.828		266	49	1.94	316.94	370
8470	10" pipe size		24	1		320	59	2.34	381.34	445
8480	12" pipe size		20	1.200		435	70.50	2.81	508.31	590
8500	Schedule 80, type 304									
8510	1/4" pipe size	Q-15	110	.145	L.F.	19.05	8.25	.51	27.81	34
8520	3/8" pipe size		109	.147		35	8.35	.52	43.87	51.50
8530	1/2" pipe size		106	.151		29	8.60	.53	38.13	45
8540	3/4" pipe size		96	.167		32.50	9.45	.59	42.54	50.50
8550	1" pipe size		87	.184		47.50	10.45	.65	58.60	68.50
8560	1-1/4" pipe size		81	.198		44	11.25	.69	55.94	66
8570	1-1/2" pipe size		74	.216		53.50	12.30	.76	66.56	78.50
8580	2" pipe size		58	.276		87.50	15.70	.97	104.17	121
8590	2-1/2" pipe size		46	.348		95.50	19.75	1.22	116.47	136
8600	3" pipe size		41	.390		113	22	1.37	136.37	159
8610	4" pipe size		33	.485		144	27.50	1.70	173.20	201
8630	6" pipe size	Q-16	30	.800		249	47	1.87	297.87	345
8640	Schedule 80, type 316									
8650	1/4" pipe size	Q-15	110	.145	L.F.	40.50	8.25	.51	49.26	57.50
8660	3/8" pipe size		109	.147		41.50	8.35	.52	50.37	58.50
8670	1/2" pipe size		106	.151		46.50	8.60	.53	55.63	64.50
8680	3/4" pipe size		96	.167		49	9.45	.59	59.04	69
8690	1" pipe size		87	.184		44.50	10.45	.65	55.60	65.50
8700	1-1/4" pipe size		81	.198		62	11.25	.69	73.94	85.50
8710	1-1/2" pipe size		74	.216		87	12.30	.76	100.06	115
8720	2" pipe size		58	.276		90	15.70	.97	106.67	124
8730	2-1/2" pipe size		46	.348		83.50	19.75	1.22	104.47	123

22 11 13.64 Pipe, Stainless Steel		Crew	Daily Output	Labor-Hours	Unit	Material	2019 Bare Costs Labor	Equipment	Total	Total Incl O&P
8740	3" pipe size	Q-15	41	.390	L.F.	135	22	1.37	158.37	183
8760	4" pipe size		33	.485		183	27.50	1.70	212.20	244
8770	6" pipe size	Q-16	30	.800		315	47	1.87	363.87	420
9100	Threading pipe labor, sst, one end, schedules 40 & 80									
9110	1/4" through 3/4" pipe size	1 Plum	61.50	.130	Ea.		8.20		8.20	12.35
9120	1" through 2" pipe size		55.90	.143			9.05		9.05	13.55
9130	2-1/2" pipe size		41.50	.193			12.15		12.15	18.25
9140	3" pipe size		38.50	.208			13.10		13.10	19.70
9150	3-1/2" pipe size	Q-1	68.40	.234			13.30		13.30	19.95
9160	4" pipe size		73	.219			12.45		12.45	18.70
9170	5" pipe size		40.70	.393			22.50		22.50	33.50
9180	6" pipe size		35.40	.452			25.50		25.50	38.50
9190	8" pipe size		22.30	.717			41		41	61
9200	10" pipe size		16.10	.994			56.50		56.50	85
9210	12" pipe size		12.30	1.301			74		74	111
9250	Welding labor per joint for stainless steel									
9260	Schedule 5 and 10									
9270	1/4" pipe size	Q-15	36	.444	Ea.		25.50	1.56	27.06	39.50
9280	3/8" pipe size		35	.457			26	1.60	27.60	41
9290	1/2" pipe size		35	.457			26	1.60	27.60	41
9300	3/4" pipe size		28	.571			32.50	2.01	34.51	50.50
9310	1" pipe size		25	.640			36.50	2.25	38.75	57
9320	1-1/4" pipe size		22	.727			41.50	2.55	44.05	65
9330	1-1/2" pipe size		21	.762			43.50	2.67	46.17	68
9340	2" pipe size		18	.889			50.50	3.12	53.62	79.50
9350	2-1/2" pipe size		12	1.333			76	4.68	80.68	119
9360	3" pipe size		9.73	1.644			93.50	5.75	99.25	146
9370	4" pipe size		7.37	2.171			123	7.60	130.60	193
9380	5" pipe size		6.15	2.602			148	9.15	157.15	232
9390	6" pipe size		5.71	2.802			159	9.85	168.85	250
9400	8" pipe size		3.69	4.336			246	15.20	261.20	385
9410	10" pipe size		2.91	5.498			310	19.30	329.30	490
9420	12" pipe size		2.31	6.926			395	24.50	419.50	615
9500	Schedule 40									
9510	1/4" pipe size	Q-15	28	.571	Ea.		32.50	2.01	34.51	50.50
9520	3/8" pipe size		27	.593			33.50	2.08	35.58	53
9530	1/2" pipe size		25.40	.630			36	2.21	38.21	56
9540	3/4" pipe size		22.22	.720			41	2.53	43.53	64.50
9550	1" pipe size		20.25	.790			45	2.77	47.77	70.50
9560	1-1/4" pipe size		18.82	.850			48.50	2.98	51.48	76
9570	1-1/2" pipe size		17.78	.900			51	3.16	54.16	80
9580	2" pipe size		15.09	1.060			60.50	3.72	64.22	94.50
9590	2-1/2" pipe size		7.96	2.010			114	7.05	121.05	179
9600	3" pipe size		6.43	2.488			141	8.75	149.75	222
9610	4" pipe size		4.88	3.279			186	11.50	197.50	293
9620	5" pipe size		4.26	3.756			213	13.20	226.20	335
9630	6" pipe size		3.77	4.244			241	14.90	255.90	375
9640	8" pipe size		2.44	6.557			375	23	398	585
9650	10" pipe size		1.92	8.333			475	29.50	504.50	740
9660	12" pipe size		1.52	10.526			600	37	637	940
9750	Schedule 80									
9760	1/4" pipe size	Q-15	21.55	.742	Ea.		42	2.61	44.61	66.50
9770	3/8" pipe size		20.75	.771			44	2.71	46.71	69

22 11 13.64 Pipe, Stainless Steel		Crew	Daily Output	Labor-Hours	Unit	Material	2019 Bare Costs Labor	Equipment	Total	Total Incl O&P
9780	1/2" pipe size	Q-15	19.54	.819	Ea.		46.50	2.87	49.37	73
9790	3/4" pipe size		17.09	.936			53	3.29	56.29	83.50
9800	1" pipe size		15.58	1.027			58.50	3.60	62.10	91.50
9810	1-1/4" pipe size		14.48	1.105			63	3.88	66.88	98.50
9820	1-1/2" pipe size		13.68	1.170			66.50	4.11	70.61	104
9830	2" pipe size		11.61	1.378			78.50	4.84	83.34	123
9840	2-1/2" pipe size		6.12	2.614			149	9.20	158.20	233
9850	3" pipe size		4.94	3.239			184	11.35	195.35	289
9860	4" pipe size		3.75	4.267			242	15	257	380
9870	5" pipe size		3.27	4.893			278	17.15	295.15	435
9880	6" pipe size		2.90	5.517			315	19.35	334.35	490
9890	8" pipe size		1.87	8.556			485	30	515	765
9900	10" pipe size		1.48	10.811			615	38	653	960
9910	12" pipe size		1.17	13.675			775	48	823	1,225
9920	Schedule 160, 1/2" pipe size		17	.941			53.50	3.30	56.80	84
9930	3/4" pipe size		14.81	1.080			61.50	3.79	65.29	96
9940	1" pipe size		13.50	1.185			67.50	4.16	71.66	106
9950	1-1/4" pipe size		12.55	1.275			72.50	4.47	76.97	114
9960	1-1/2" pipe size		11.85	1.350			76.50	4.74	81.24	120
9970	2" pipe size		10	1.600			91	5.60	96.60	142
9980	3" pipe size		4.28	3.738			212	13.10	225.10	335
9990	4" pipe size	▼	3.25	4.923	▼		280	17.30	297.30	440

22 11 13.66 Pipe Fittings, Stainless Steel		Crew	Daily Output	Labor-Hours	Unit	Material	Labor	Equipment	Total	Total Incl O&P
0010	**PIPE FITTINGS, STAINLESS STEEL**									
0100	Butt weld joint, schedule 5, type 304									
0120	90° elbow, long									
0140	1/2"	Q-15	17.50	.914	Ea.	18.75	52	3.21	73.96	102
0150	3/4"		14	1.143		18.75	65	4.01	87.76	122
0160	1"		12.50	1.280		19.75	72.50	4.49	96.74	136
0170	1-1/4"		11	1.455		27	82.50	5.10	114.60	160
0180	1-1/2"		10.50	1.524		23	86.50	5.35	114.85	161
0190	2"		9	1.778		27	101	6.25	134.25	189
0200	2-1/2"		6	2.667		62.50	152	9.35	223.85	305
0210	3"		4.86	3.292		58.50	187	11.55	257.05	360
0220	3-1/2"		4.27	3.747		172	213	13.15	398.15	525
0230	4"		3.69	4.336		98	246	15.20	359.20	495
0240	5"	▼	3.08	5.195		365	295	18.25	678.25	865
0250	6"	Q-16	4.29	5.594		281	330	13.10	624.10	820
0260	8"		2.76	8.696		595	510	20.50	1,125.50	1,450
0270	10"		2.18	11.009		915	650	26	1,591	2,000
0280	12"	▼	1.73	13.873		1,300	820	32.50	2,152.50	2,675
0320	For schedule 5, type 316, add				▼	30%				
0600	45° elbow, long									
0620	1/2"	Q-15	17.50	.914	Ea.	18.75	52	3.21	73.96	102
0630	3/4"		14	1.143		18.75	65	4.01	87.76	122
0640	1"		12.50	1.280		19.75	72.50	4.49	96.74	136
0650	1-1/4"		11	1.455		27	82.50	5.10	114.60	160
0660	1-1/2"		10.50	1.524		23	86.50	5.35	114.85	161
0670	2"		9	1.778		27	101	6.25	134.25	189
0680	2-1/2"		6	2.667		62.50	152	9.35	223.85	305
0690	3"		4.86	3.292		47	187	11.55	245.55	345
0700	3-1/2"	▼	4.27	3.747	▼	172	213	13.15	398.15	525

22 11 13.66 Pipe Fittings, Stainless Steel		Crew	Daily Output	Labor-Hours	Unit	Material	2019 Bare Costs Labor	Equipment	Total	Total Incl O&P
0710	4"	Q-15	3.69	4.336	Ea.	79	246	15.20	340.20	475
0720	5"		3.08	5.195		291	295	18.25	604.25	785
0730	6"	Q-16	4.29	5.594		198	330	13.10	541.10	725
0740	8"		2.76	8.696		415	510	20.50	945.50	1,250
0750	10"		2.18	11.009		730	650	26	1,406	1,800
0760	12"		1.73	13.873		910	820	32.50	1,762.50	2,250
0800	For schedule 5, type 316, add					25%				
1100	Tee, straight									
1130	1/2"	Q-15	11.66	1.372	Ea.	56	78	4.82	138.82	184
1140	3/4"		9.33	1.715		56	97.50	6	159.50	215
1150	1"		8.33	1.921		59.50	109	6.75	175.25	236
1160	1-1/4"		7.33	2.183		48	124	7.65	179.65	247
1170	1-1/2"		7	2.286		47	130	8	185	255
1180	2"		6	2.667		49	152	9.35	210.35	291
1190	2-1/2"		4	4		118	227	14.05	359.05	485
1200	3"		3.24	4.938		83	281	17.35	381.35	530
1210	3-1/2"		2.85	5.614		246	320	19.70	585.70	775
1220	4"		2.46	6.504		133	370	23	526	725
1230	5"		2	8		420	455	28	903	1,175
1240	6"	Q-16	2.85	8.421		335	495	19.70	849.70	1,125
1250	8"		1.84	13.043		705	770	30.50	1,505.50	1,975
1260	10"		1.45	16.552		1,150	975	38.50	2,163.50	2,775
1270	12"		1.15	20.870		1,600	1,225	49	2,874	3,650
1320	For schedule 5, type 316, add					25%				
2000	Butt weld joint, schedule 10, type 304									
2020	90° elbow, long									
2040	1/2"	Q-15	17	.941	Ea.	13.90	53.50	3.30	70.70	99.50
2050	3/4"		14	1.143		13.20	65	4.01	82.21	116
2060	1"		12.50	1.280		14.25	72.50	4.49	91.24	130
2070	1-1/4"		11	1.455		20	82.50	5.10	107.60	152
2080	1-1/2"		10.50	1.524		16.50	86.50	5.35	108.35	154
2090	2"		9	1.778		19.50	101	6.25	126.75	180
2100	2-1/2"		6	2.667		46.50	152	9.35	207.85	288
2110	3"		4.86	3.292		43.50	187	11.55	242.05	340
2120	3-1/2"		4.27	3.747		128	213	13.15	354.15	475
2130	4"		3.69	4.336		70.50	246	15.20	331.70	465
2140	5"		3.08	5.195		270	295	18.25	583.25	765
2150	6"	Q-16	4.29	5.594		203	330	13.10	546.10	730
2160	8"		2.76	8.696		430	510	20.50	960.50	1,275
2170	10"		2.18	11.009		660	650	26	1,336	1,725
2180	12"		1.73	13.873		940	820	32.50	1,792.50	2,275
2500	45° elbow, long									
2520	1/2"	Q-15	17.50	.914	Ea.	13.90	52	3.21	69.11	97
2530	3/4"		14	1.143		13.90	65	4.01	82.91	117
2540	1"		12.50	1.280		14.25	72.50	4.49	91.24	130
2550	1-1/4"		11	1.455		20	82.50	5.10	107.60	152
2560	1-1/2"		10.50	1.524		16.50	86.50	5.35	108.35	154
2570	2"		9	1.778		19.50	101	6.25	126.75	180
2580	2-1/2"		6	2.667		46.50	152	9.35	207.85	288
2590	3"		4.86	3.292		35	187	11.55	233.55	330
2600	3-1/2"		4.27	3.747		128	213	13.15	354.15	475
2610	4"		3.69	4.336		57	246	15.20	318.20	450
2620	5"		3.08	5.195		216	295	18.25	529.25	705

22 11 13.66 Pipe Fittings, Stainless Steel		Crew	Daily Output	Labor-Hours	Unit	Material	Labor	Equipment	Total	Total Incl O&P
						2019 Bare Costs				
2630	6"	Q-16	4.29	5.594	Ea.	143	330	13.10	486.10	665
2640	8"		2.76	8.696		300	510	20.50	830.50	1,125
2650	10"		2.18	11.009		525	650	26	1,201	1,575
2660	12"		1.73	13.873		655	820	32.50	1,507.50	1,975
2670	Reducer, concentric									
2674	1" x 3/4"	Q-15	13.25	1.208	Ea.	29.50	68.50	4.24	102.24	140
2676	2" x 1-1/2"	"	9.75	1.641		24.50	93.50	5.75	123.75	173
2678	6" x 4"	Q-16	4.91	4.888		73	288	11.45	372.45	525
2680	Caps									
2682	1"	Q-15	25	.640	Ea.	30.50	36.50	2.25	69.25	90.50
2684	1-1/2"		21	.762		36.50	43.50	2.67	82.67	108
2685	2"		18	.889		34.50	50.50	3.12	88.12	117
2686	4"		7.38	2.168		55	123	7.60	185.60	254
2687	6"	Q-16	8.58	2.797		87.50	165	6.55	259.05	350
3000	Tee, straight									
3030	1/2"	Q-15	11.66	1.372	Ea.	41.50	78	4.82	124.32	168
3040	3/4"		9.33	1.715		41.50	97.50	6	145	199
3050	1"		8.33	1.921		43	109	6.75	158.75	218
3060	1-1/4"		7.33	2.183		47.50	124	7.65	179.15	246
3070	1-1/2"		7	2.286		45	130	8	183	253
3080	2"		6	2.667		47	152	9.35	208.35	289
3090	2-1/2"		4	4		58	227	14.05	299.05	420
3100	3"		3.24	4.938		69	281	17.35	367.35	515
3110	3-1/2"		2.85	5.614		101	320	19.70	440.70	615
3120	4"		2.46	6.504		96	370	23	489	685
3130	5"		2	8		315	455	28	798	1,050
3140	6"	Q-16	2.85	8.421		240	495	19.70	754.70	1,025
3150	8"		1.84	13.043		510	770	30.50	1,310.50	1,750
3151	10"		1.45	16.552		825	975	38.50	1,838.50	2,425
3152	12"		1.15	20.870		1,150	1,225	49	2,424	3,150
3154	For schedule 10, type 316, add					25%				
3281	Butt weld joint, schedule 40, type 304									
3284	90° elbow, long, 1/2"	Q-15	12.70	1.260	Ea.	16.25	71.50	4.42	92.17	130
3288	3/4"		11.10	1.441		16.25	82	5.05	103.30	146
3289	1"		10.13	1.579		17	90	5.55	112.55	160
3290	1-1/4"		9.40	1.702		22.50	96.50	5.95	124.95	176
3300	1-1/2"		8.89	1.800		17.75	102	6.30	126.05	180
3310	2"		7.55	2.119		25.50	120	7.45	152.95	217
3320	2-1/2"		3.98	4.020		48.50	228	14.10	290.60	415
3330	3"		3.21	4.984		62.50	283	17.50	363	515
3340	3-1/2"		2.83	5.654		232	320	19.85	571.85	755
3350	4"		2.44	6.557		108	375	23	506	705
3360	5"		2.13	7.512		365	425	26.50	816.50	1,075
3370	6"	Q-16	2.83	8.481		315	500	19.85	834.85	1,125
3380	8"		1.83	13.115		615	775	30.50	1,420.50	1,850
3390	10"		1.44	16.667		1,300	980	39	2,319	2,950
3400	12"		1.14	21.053		1,650	1,250	49.50	2,949.50	3,725
3410	For schedule 40, type 316, add					25%				
3460	45° elbow, long, 1/2"	Q-15	12.70	1.260	Ea.	16.25	71.50	4.42	92.17	130
3470	3/4"		11.10	1.441		16.25	82	5.05	103.30	146
3480	1"		10.13	1.579		17	90	5.55	112.55	160
3490	1-1/4"		9.40	1.702		22.50	96.50	5.95	124.95	176
3500	1-1/2"		8.89	1.800		17.75	102	6.30	126.05	180

22 11 13 – Facility Water Distribution Piping

22 11 13.66 Pipe Fittings, Stainless Steel		Crew	Daily Output	Labor-Hours	Unit	Material	Labor	Equipment	Total	Total Incl O&P
						2019 Bare Costs				
3510	2"	Q-15	7.55	2.119	Ea.	25.50	120	7.45	152.95	217
3520	2-1/2"		3.98	4.020		48.50	228	14.10	290.60	415
3530	3"		3.21	4.984		49.50	283	17.50	350	500
3540	3-1/2"		2.83	5.654		232	320	19.85	571.85	755
3550	4"		2.44	6.557		77.50	375	23	475.50	670
3560	5"		2.13	7.512		255	425	26.50	706.50	950
3570	6"	Q-16	2.83	8.481		224	500	19.85	743.85	1,025
3580	8"		1.83	13.115		430	775	30.50	1,235.50	1,650
3590	10"		1.44	16.667		910	980	39	1,929	2,525
3600	12"		1.14	21.053		1,150	1,250	49.50	2,449.50	3,175
3610	For schedule 40, type 316, add					25%				
3660	Tee, straight 1/2"	Q-15	8.46	1.891	Ea.	41.50	107	6.65	155.15	214
3670	3/4"		7.40	2.162		41.50	123	7.60	172.10	238
3680	1"		6.74	2.374		44	135	8.35	187.35	260
3690	1-1/4"		6.27	2.552		103	145	8.95	256.95	340
3700	1-1/2"		5.92	2.703		47	154	9.50	210.50	292
3710	2"		5.03	3.181		70.50	181	11.15	262.65	360
3720	2-1/2"		2.65	6.038		89	345	21	455	635
3730	3"		2.14	7.477		89.50	425	26	540.50	770
3740	3-1/2"		1.88	8.511		169	485	30	684	945
3750	4"		1.62	9.877		169	560	34.50	763.50	1,075
3760	5"		1.42	11.268		390	640	39.50	1,069.50	1,425
3770	6"	Q-16	1.88	12.766		375	750	30	1,155	1,575
3780	8"		1.22	19.672		735	1,150	46	1,931	2,600
3790	10"		.96	25		1,425	1,475	58.50	2,958.50	3,850
3800	12"		.76	31.579		1,900	1,850	74	3,824	4,975
3810	For schedule 40, type 316, add					25%				
3820	Tee, reducing on outlet, 3/4" x 1/2"	Q-15	7.73	2.070	Ea.	55.50	118	7.25	180.75	246
3822	1" x 1/2"		7.24	2.210		60.50	126	7.75	194.25	263
3824	1" x 3/4"		6.96	2.299		55.50	131	8.05	194.55	266
3826	1-1/4" x 1"		6.43	2.488		136	141	8.75	285.75	370
3828	1-1/2" x 1/2"		6.58	2.432		82	138	8.55	228.55	305
3830	1-1/2" x 3/4"		6.35	2.520		76.50	143	8.85	228.35	310
3832	1-1/2" x 1"		6.18	2.589		58.50	147	9.10	214.60	296
3834	2" x 1"		5.50	2.909		105	165	10.20	280.20	375
3836	2" x 1-1/2"		5.30	3.019		88	172	10.60	270.60	365
3838	2-1/2" x 2"		3.15	5.079		133	289	17.85	439.85	600
3840	3" x 1-1/2"		2.72	5.882		134	335	20.50	489.50	670
3842	3" x 2"		2.65	6.038		112	345	21	478	660
3844	4" x 2"		2.10	7.619		280	435	26.50	741.50	990
3846	4" x 3"		1.77	9.040		203	515	31.50	749.50	1,025
3848	5" x 4"		1.48	10.811		470	615	38	1,123	1,475
3850	6" x 3"	Q-16	2.19	10.959		520	645	25.50	1,190.50	1,575
3852	6" x 4"		2.04	11.765		450	695	27.50	1,172.50	1,575
3854	8" x 4"		1.46	16.438		1,050	970	38.50	2,058.50	2,650
3856	10" x 8"		.69	34.783		1,775	2,050	81.50	3,906.50	5,125
3858	12" x 10"		.55	43.636		2,350	2,575	102	5,027	6,550
3950	Reducer, concentric, 3/4" x 1/2"	Q-15	11.85	1.350		31	76.50	4.74	112.24	154
3952	1" x 3/4"		10.60	1.509		32.50	86	5.30	123.80	170
3954	1-1/4" x 3/4"		10.19	1.570		79.50	89	5.50	174	228
3956	1-1/4" x 1"		9.76	1.639		40	93	5.75	138.75	190
3958	1-1/2" x 3/4"		9.88	1.619		67	92	5.70	164.70	218
3960	1-1/2" x 1"		9.47	1.690		54	96	5.95	155.95	210

22 11 13.66 Pipe Fittings, Stainless Steel		Crew	Daily Output	Labor-Hours	Unit	2019 Bare Costs Material	Labor	Equipment	Total	Total Incl O&P
3962	2" x 1"	Q-15	8.65	1.850	Ea.	31	105	6.50	142.50	199
3964	2" x 1-1/2"		8.16	1.961		27	111	6.90	144.90	205
3966	2-1/2" x 1"		5.71	2.802		117	159	9.85	285.85	380
3968	2-1/2" x 2"		5.21	3.071		59.50	175	10.80	245.30	340
3970	3" x 1"		4.88	3.279		80.50	186	11.50	278	380
3972	3" x 1-1/2"		4.72	3.390		41	193	11.90	245.90	345
3974	3" x 2"		4.51	3.548		38.50	202	12.45	252.95	360
3976	4" x 2"		3.69	4.336		55	246	15.20	316.20	445
3978	4" x 3"		2.77	5.776		41	330	20.50	391.50	565
3980	5" x 3"		2.56	6.250		244	355	22	621	830
3982	5" x 4"		2.27	7.048		197	400	24.50	621.50	845
3984	6" x 3"	Q-16	3.57	6.723		126	395	15.75	536.75	750
3986	6" x 4"		3.19	7.524		104	445	17.60	566.60	800
3988	8" x 4"		2.44	9.836		315	580	23	918	1,250
3990	8" x 6"		2.22	10.811		236	635	25.50	896.50	1,250
3992	10" x 6"		1.91	12.565		440	740	29.50	1,209.50	1,625
3994	10" x 8"		1.61	14.907		365	880	35	1,280	1,775
3995	12" x 6"		1.63	14.724		695	870	34.50	1,599.50	2,100
3996	12" x 8"		1.41	17.021		620	1,000	40	1,660	2,225
3997	12" x 10"		1.27	18.898		435	1,125	44	1,604	2,200
4000	Socket weld joint, 3,000 lb., type 304									
4100	90° elbow									
4140	1/4"	Q-15	13.47	1.188	Ea.	44	67.50	4.17	115.67	154
4150	3/8"		12.97	1.234		57.50	70	4.33	131.83	173
4160	1/2"		12.21	1.310		65	74.50	4.60	144.10	189
4170	3/4"		10.68	1.498		72	85	5.25	162.25	213
4180	1"		9.74	1.643		109	93.50	5.75	208.25	266
4190	1-1/4"		9.05	1.768		190	100	6.20	296.20	365
4200	1-1/2"		8.55	1.871		230	106	6.55	342.55	420
4210	2"		7.26	2.204		370	125	7.75	502.75	605
4300	45° elbow									
4340	1/4"	Q-15	13.47	1.188	Ea.	83	67.50	4.17	154.67	197
4350	3/8"		12.97	1.234		83	70	4.33	157.33	201
4360	1/2"		12.21	1.310		83	74.50	4.60	162.10	209
4370	3/4"		10.68	1.498		94.50	85	5.25	184.75	238
4380	1"		9.74	1.643		137	93.50	5.75	236.25	296
4390	1-1/4"		9.05	1.768		225	100	6.20	331.20	405
4400	1-1/2"		8.55	1.871		226	106	6.55	338.55	415
4410	2"		7.26	2.204		410	125	7.75	542.75	645
4500	Tee									
4540	1/4"	Q-15	8.97	1.784	Ea.	59	101	6.25	166.25	223
4550	3/8"		8.64	1.852		71.50	105	6.50	183	244
4560	1/2"		8.13	1.968		87.50	112	6.90	206.40	272
4570	3/4"		7.12	2.247		101	128	7.90	236.90	310
4580	1"		6.48	2.469		136	140	8.65	284.65	370
4590	1-1/4"		6.03	2.653		241	151	9.30	401.30	500
4600	1-1/2"		5.69	2.812		345	160	9.85	514.85	630
4610	2"		4.83	3.313		525	188	11.65	724.65	870
5000	Socket weld joint, 3,000 lb., type 316									
5100	90° elbow									
5140	1/4"	Q-15	13.47	1.188	Ea.	54	67.50	4.17	125.67	165
5150	3/8"		12.97	1.234		63	70	4.33	137.33	179
5160	1/2"		12.21	1.310		76	74.50	4.60	155.10	201

22 11 13.66 Pipe Fittings, Stainless Steel		Crew	Daily Output	Labor-Hours	Unit	2019 Bare Costs Material	Labor	Equipment	Total	Total Incl O&P
5170	3/4"	Q-15	10.68	1.498	Ea.	100	85	5.25	190.25	244
5180	1"		9.74	1.643		142	93.50	5.75	241.25	305
5190	1-1/4"		9.05	1.768		254	100	6.20	360.20	435
5200	1-1/2"		8.55	1.871		287	106	6.55	399.55	480
5210	2"		7.26	2.204		490	125	7.75	622.75	730
5300	45° elbow									
5340	1/4"	Q-15	13.47	1.188	Ea.	106	67.50	4.17	177.67	222
5350	3/8"		12.97	1.234		106	70	4.33	180.33	226
5360	1/2"		12.21	1.310		111	74.50	4.60	190.10	239
5370	3/4"		10.68	1.498		121	85	5.25	211.25	267
5380	1"		9.74	1.643		183	93.50	5.75	282.25	345
5390	1-1/4"		9.05	1.768		267	100	6.20	373.20	450
5400	1-1/2"		8.55	1.871		293	106	6.55	405.55	485
5410	2"		7.26	2.204		435	125	7.75	567.75	675
5500	Tee									
5540	1/4"	Q-15	8.97	1.784	Ea.	73.50	101	6.25	180.75	240
5550	3/8"		8.64	1.852		89.50	105	6.50	201	264
5560	1/2"		8.13	1.968		100	112	6.90	218.90	286
5570	3/4"		7.12	2.247		123	128	7.90	258.90	335
5580	1"		6.48	2.469		189	140	8.65	337.65	430
5590	1-1/4"		6.03	2.653		300	151	9.30	460.30	565
5600	1-1/2"		5.69	2.812		425	160	9.85	594.85	715
5610	2"		4.83	3.313		675	188	11.65	874.65	1,050
5700	For socket weld joint, 6,000 lb., type 304 and 316, add					100%				
6000	Threaded companion flange									
6010	Stainless steel, 150 lb., type 304									
6020	1/2" diam.	1 Plum	30	.267	Ea.	46.50	16.85		63.35	76.50
6030	3/4" diam.		28	.286		51.50	18.05		69.55	83.50
6040	1" diam.		27	.296		56.50	18.70		75.20	90.50
6050	1-1/4" diam.	Q-1	44	.364		73	20.50		93.50	112
6060	1-1/2" diam.		40	.400		73	22.50		95.50	115
6070	2" diam.		36	.444		96	25.50		121.50	143
6080	2-1/2" diam.		28	.571		134	32.50		166.50	196
6090	3" diam.		20	.800		140	45.50		185.50	222
6110	4" diam.		12	1.333		192	76		268	325
6130	6" diam.	Q-2	14	1.714		345	101		446	530
6140	8" diam.	"	12	2		665	118		783	905
6150	For type 316, add					40%				
6260	Weld flanges, stainless steel, type 304									
6270	Slip on, 150 lb. (welded, front and back)									
6280	1/2" diam.	Q-15	12.70	1.260	Ea.	41	71.50	4.42	116.92	157
6290	3/4" diam.		11.11	1.440		42.50	82	5.05	129.55	175
6300	1" diam.		10.13	1.579		46.50	90	5.55	142.05	192
6310	1-1/4" diam.		9.41	1.700		63	96.50	5.95	165.45	221
6320	1-1/2" diam.		8.89	1.800		63	102	6.30	171.30	229
6330	2" diam.		7.55	2.119		81.50	120	7.45	208.95	279
6340	2-1/2" diam.		3.98	4.020		114	228	14.10	356.10	485
6350	3" diam.		3.21	4.984		123	283	17.50	423.50	580
6370	4" diam.		2.44	6.557		167	375	23	565	770
6390	6" diam.	Q-16	1.89	12.698		255	750	29.50	1,034.50	1,450
6400	8" diam.	"	1.22	19.672		480	1,150	46	1,676	2,325
6410	For type 316, add					40%				
6530	Weld neck 150 lb.									

22 11 13.66 Pipe Fittings, Stainless Steel		Crew	Daily Output	Labor-Hours	Unit	Material	2019 Bare Costs Labor	Equipment	Total	Total Incl O&P
6540	1/2" diam.	Q-15	25.40	.630	Ea.	21	36	2.21	59.21	79
6550	3/4" diam.		22.22	.720		25	41	2.53	68.53	92
6560	1" diam.		20.25	.790		28.50	45	2.77	76.27	102
6570	1-1/4" diam.		18.82	.850		43.50	48.50	2.98	94.98	124
6580	1-1/2" diam.		17.78	.900		40.50	51	3.16	94.66	124
6590	2" diam.		15.09	1.060		42	60.50	3.72	106.22	141
6600	2-1/2" diam.		7.96	2.010		67.50	114	7.05	188.55	253
6610	3" diam.		6.43	2.488		69	141	8.75	218.75	298
6630	4" diam.		4.88	3.279		108	186	11.50	305.50	410
6640	5" diam.		4.26	3.756		134	213	13.20	360.20	485
6650	6" diam.	Q-16	5.66	4.240		155	250	9.90	414.90	555
6652	8" diam.		3.65	6.575		277	385	15.40	677.40	900
6654	10" diam.		2.88	8.333		390	490	19.50	899.50	1,175
6656	12" diam.		2.28	10.526		720	620	24.50	1,364.50	1,750
6670	For type 316, add					23%				
7000	Threaded joint, 150 lb., type 304									
7030	90° elbow									
7040	1/8"	1 Plum	13	.615	Ea.	28	39		67	89.50
7050	1/4"		13	.615		28	39		67	89.50
7070	3/8"		13	.615		33.50	39		72.50	95
7080	1/2"		12	.667		28.50	42		70.50	94.50
7090	3/4"		11	.727		34	46		80	107
7100	1"		10	.800		46.50	50.50		97	128
7110	1-1/4"	Q-1	17	.941		75.50	53.50		129	164
7120	1-1/2"		16	1		83.50	57		140.50	177
7130	2"		14	1.143		120	65		185	230
7140	2-1/2"		11	1.455		305	82.50		387.50	460
7150	3"		8	2		430	114		544	645
7160	4"	Q-2	11	2.182		725	129		854	990
7180	45° elbow									
7190	1/8"	1 Plum	13	.615	Ea.	41	39		80	104
7200	1/4"		13	.615		41	39		80	104
7210	3/8"		13	.615		42	39		81	105
7220	1/2"		12	.667		42.50	42		84.50	110
7230	3/4"		11	.727		45.50	46		91.50	119
7240	1"		10	.800		52	50.50		102.50	133
7250	1-1/4"	Q-1	17	.941		75	53.50		128.50	163
7260	1-1/2"		16	1		97.50	57		154.50	194
7270	2"		14	1.143		139	65		204	251
7280	2-1/2"		11	1.455		430	82.50		512.50	600
7290	3"		8	2		630	114		744	865
7300	4"	Q-2	11	2.182		1,150	129		1,279	1,450
7320	Tee, straight									
7330	1/8"	1 Plum	9	.889	Ea.	43.50	56		99.50	132
7340	1/4"		9	.889		43.50	56		99.50	132
7350	3/8"		9	.889		47	56		103	136
7360	1/2"		8	1		44.50	63		107.50	144
7370	3/4"		7	1.143		47.50	72		119.50	161
7380	1"		6.50	1.231		62	77.50		139.50	185
7390	1-1/4"	Q-1	11	1.455		103	82.50		185.50	237
7400	1-1/2"		10	1.600		133	91		224	283
7410	2"		9	1.778		166	101		267	335
7420	2-1/2"		7	2.286		440	130		570	675

22 11 13.66 Pipe Fittings, Stainless Steel		Crew	Daily Output	Labor-Hours	Unit	Material	2019 Bare Costs Labor	Equipment	Total	Total Incl O&P
7430	3"	Q-1	5	3.200	Ea.	665	182		847	1,000
7440	4"	Q-2	7	3.429		1,650	202		1,852	2,125
7460	Coupling, straight									
7470	1/8"	1 Plum	19	.421	Ea.	10.75	26.50		37.25	52
7480	1/4"		19	.421		12.75	26.50		39.25	54
7490	3/8"		19	.421		15.25	26.50		41.75	57
7500	1/2"		19	.421		20	26.50		46.50	62.50
7510	3/4"		18	.444		27	28		55	71.50
7520	1"		15	.533		43.50	33.50		77	98.50
7530	1-1/4"	Q-1	26	.615		105	35		140	168
7540	1-1/2"		24	.667		78	38		116	143
7550	2"		21	.762		131	43.50		174.50	209
7560	2-1/2"		18	.889		305	50.50		355.50	410
7570	3"		14	1.143		415	65		480	555
7580	4"	Q-2	16	1.500		575	88.50		663.50	770
7600	Reducer, concentric, 1/2"	1 Plum	12	.667		21.50	42		63.50	86.50
7610	3/4"		11	.727		27.50	46		73.50	99.50
7612	1"		10	.800		45.50	50.50		96	127
7614	1-1/4"	Q-1	17	.941		99	53.50		152.50	190
7616	1-1/2"		16	1		108	57		165	205
7618	2"		14	1.143		165	65		230	280
7620	2-1/2"		11	1.455		450	82.50		532.50	620
7622	3"		8	2		500	114		614	720
7624	4"	Q-2	11	2.182		825	129		954	1,100
7710	Union									
7720	1/8"	1 Plum	12	.667	Ea.	51.50	42		93.50	120
7730	1/4"		12	.667		51.50	42		93.50	120
7740	3/8"		12	.667		60	42		102	129
7750	1/2"		11	.727		73	46		119	149
7760	3/4"		10	.800		100	50.50		150.50	186
7770	1"		9	.889		145	56		201	243
7780	1-1/4"	Q-1	16	1		370	57		427	495
7790	1-1/2"		15	1.067		390	60.50		450.50	520
7800	2"		13	1.231		490	70		560	645
7810	2-1/2"		10	1.600		960	91		1,051	1,175
7820	3"		7	2.286		1,275	130		1,405	1,600
7830	4"	Q-2	10	2.400		1,775	141		1,916	2,150
7838	Caps									
7840	1/2"	1 Plum	24	.333	Ea.	13.30	21		34.30	46
7841	3/4"		22	.364		19.35	23		42.35	56
7842	1"		20	.400		31	25.50		56.50	72
7843	1-1/2"	Q-1	32	.500		78	28.50		106.50	129
7844	2"	"	28	.571		99	32.50		131.50	158
7845	4"	Q-2	22	1.091		390	64.50		454.50	525
7850	For 150 lb., type 316, add					25%				
8750	Threaded joint, 2,000 lb., type 304									
8770	90° elbow									
8780	1/8"	1 Plum	13	.615	Ea.	32.50	39		71.50	94.50
8790	1/4"		13	.615		32.50	39		71.50	94.50
8800	3/8"		13	.615		40.50	39		79.50	103
8810	1/2"		12	.667		54.50	42		96.50	123
8820	3/4"		11	.727		63.50	46		109.50	139
8830	1"		10	.800		84	50.50		134.50	168

22 11 13.66 Pipe Fittings, Stainless Steel		Crew	Daily Output	Labor-Hours	Unit	2019 Bare Costs Material	Labor	Equipment	Total	Total Incl O&P
8840	1-1/4"	Q-1	17	.941	Ea.	137	53.50		190.50	231
8850	1-1/2"		16	1		219	57		276	325
8860	2"		14	1.143		310	65		375	440
8880	45° elbow									
8890	1/8"	1 Plum	13	.615	Ea.	68	39		107	133
8900	1/4"		13	.615		68	39		107	133
8910	3/8"		13	.615		85	39		124	152
8920	1/2"		12	.667		85	42		127	157
8930	3/4"		11	.727		93.50	46		139.50	172
8940	1"		10	.800		111	50.50		161.50	199
8950	1-1/4"	Q-1	17	.941		195	53.50		248.50	295
8960	1-1/2"		16	1		261	57		318	375
8970	2"		14	1.143		300	65		365	430
8990	Tee, straight									
9000	1/8"	1 Plum	9	.889	Ea.	42	56		98	130
9010	1/4"		9	.889		42	56		98	130
9020	3/8"		9	.889		54	56		110	144
9030	1/2"		8	1		69.50	63		132.50	172
9040	3/4"		7	1.143		78.50	72		150.50	195
9050	1"		6.50	1.231		109	77.50		186.50	236
9060	1-1/4"	Q-1	11	1.455		183	82.50		265.50	325
9070	1-1/2"		10	1.600		305	91		396	470
9080	2"		9	1.778		425	101		526	615
9100	For couplings and unions use 3,000 lb., type 304									
9120	2,000 lb., type 316									
9130	90° elbow									
9140	1/8"	1 Plum	13	.615	Ea.	40.50	39		79.50	103
9150	1/4"		13	.615		40.50	39		79.50	103
9160	3/8"		13	.615		47.50	39		86.50	111
9170	1/2"		12	.667		61.50	42		103.50	131
9180	3/4"		11	.727		75	46		121	152
9190	1"		10	.800		115	50.50		165.50	202
9200	1-1/4"	Q-1	17	.941		205	53.50		258.50	305
9210	1-1/2"		16	1		234	57		291	345
9220	2"		14	1.143		395	65		460	535
9240	45° elbow									
9250	1/8"	1 Plum	13	.615	Ea.	86	39		125	153
9260	1/4"		13	.615		86	39		125	153
9270	3/8"		13	.615		86	39		125	153
9280	1/2"		12	.667		86	42		128	158
9300	3/4"		11	.727		97	46		143	176
9310	1"		10	.800		146	50.50		196.50	237
9320	1-1/4"	Q-1	17	.941		209	53.50		262.50	310
9330	1-1/2"		16	1		286	57		343	400
9340	2"		14	1.143		350	65		415	485
9360	Tee, straight									
9370	1/8"	1 Plum	9	.889	Ea.	51.50	56		107.50	141
9380	1/4"		9	.889		51.50	56		107.50	141
9390	3/8"		9	.889		63.50	56		119.50	154
9400	1/2"		8	1		78.50	63		141.50	182
9410	3/4"		7	1.143		97	72		169	215
9420	1"		6.50	1.231		152	77.50		229.50	284
9430	1-1/4"	Q-1	11	1.455		240	82.50		322.50	390

22 11 13 – Facility Water Distribution Piping

22 11 13.66 Pipe Fittings, Stainless Steel		Crew	Daily Output	Labor-Hours	Unit	2019 Bare Costs Material	Labor	Equipment	Total	Total Incl O&P
9440	1-1/2"	Q-1	10	1.600	Ea.	345	91		436	515
9450	2"	↓	9	1.778	↓	550	101		651	755
9470	For couplings and unions use 3,000 lb., type 316									
9490	3,000 lb., type 304									
9510	Coupling									
9520	1/8"	1 Plum	19	.421	Ea.	12.85	26.50		39.35	54
9530	1/4"		19	.421		13.95	26.50		40.45	55.50
9540	3/8"		19	.421		16.25	26.50		42.75	58
9550	1/2"		19	.421		19.60	26.50		46.10	61.50
9560	3/4"		18	.444		26	28		54	71
9570	1"	↓	15	.533		44.50	33.50		78	99
9580	1-1/4"	Q-1	26	.615		107	35		142	171
9590	1-1/2"		24	.667		126	38		164	195
9600	2"	↓	21	.762	↓	165	43.50		208.50	247
9620	Union									
9630	1/8"	1 Plum	12	.667	Ea.	98.50	42		140.50	171
9640	1/4"		12	.667		98.50	42		140.50	171
9650	3/8"		12	.667		105	42		147	179
9660	1/2"		11	.727		105	46		151	185
9670	3/4"		10	.800		129	50.50		179.50	218
9680	1"	↓	9	.889		198	56		254	300
9690	1-1/4"	Q-1	16	1		370	57		427	490
9700	1-1/2"		15	1.067		415	60.50		475.50	545
9710	2"	↓	13	1.231	↓	555	70		625	715
9730	3,000 lb., type 316									
9750	Coupling									
9770	1/8"	1 Plum	19	.421	Ea.	14.50	26.50		41	56
9780	1/4"		19	.421		16.30	26.50		42.80	58
9790	3/8"		19	.421		17.45	26.50		43.95	59
9800	1/2"		19	.421		25	26.50		51.50	67.50
9810	3/4"		18	.444		36.50	28		64.50	82
9820	1"	↓	15	.533		61	33.50		94.50	118
9830	1-1/4"	Q-1	26	.615		137	35		172	203
9840	1-1/2"		24	.667		159	38		197	232
9850	2"	↓	21	.762	↓	220	43.50		263.50	305
9870	Union									
9880	1/8"	1 Plum	12	.667	Ea.	108	42		150	182
9890	1/4"		12	.667		108	42		150	182
9900	3/8"		12	.667		125	42		167	201
9910	1/2"		11	.727		125	46		171	207
9920	3/4"		10	.800		163	50.50		213.50	256
9930	1"	↓	9	.889		258	56		314	365
9940	1-1/4"	Q-1	16	1		460	57		517	595
9950	1-1/2"		15	1.067		555	60.50		615.50	700
9960	2"	↓	13	1.231	↓	705	70		775	885

22 11 13.74 Pipe, Plastic		Crew	Daily Output	Labor-Hours	Unit	Material	Labor	Equipment	Total	Total Incl O&P
0010	**PIPE, PLASTIC**									
0020	Fiberglass reinforced, couplings 10' OC, clevis hanger assy's, 3 per 10'									
0080	General service									
0120	2" diameter	Q-1	59	.271	L.F.	16.90	15.40		32.30	41.50
0140	3" diameter		52	.308		18.15	17.50		35.65	46
0150	4" diameter	↓	48	.333	↓	36.50	18.95		55.45	68.50

22 11 13.74 Pipe, Plastic		Crew	Daily Output	Labor-Hours	Unit	Material	Labor	Equipment	Total	Total Incl O&P
						2019 Bare Costs				
0160	6″ diameter	Q-1	39	.410	L.F.	40	23.50		63.50	79
0170	8″ diameter	Q-2	49	.490		63	29		92	113
0180	10″ diameter		41	.585		90.50	34.50		125	152
0190	12″ diameter		36	.667		116	39.50		155.50	187
0600	PVC, high impact/pressure, cplgs. 10′ OC, clevis hanger assy's, 3 per 10′									
1020	Schedule 80									
1070	1/2″ diameter	1 Plum	50	.160	L.F.	6.40	10.10		16.50	22
1080	3/4″ diameter		47	.170		7.55	10.75		18.30	24.50
1090	1″ diameter		43	.186		12.30	11.75		24.05	31
1100	1-1/4″ diameter		39	.205		14.30	12.95		27.25	35
1110	1-1/2″ diameter		34	.235		15.50	14.85		30.35	39.50
1120	2″ diameter	Q-1	55	.291		18.55	16.55		35.10	45.50
1140	3″ diameter		50	.320		29.50	18.20		47.70	59.50
1150	4″ diameter		46	.348		53	19.75		72.75	88
1170	6″ diameter		38	.421		62	24		86	104
1730	To delete coupling & hangers, subtract									
1740	1/2″ diam.					62%	80%			
1750	3/4″ diam. to 1-1/4″ diam.					58%	73%			
1760	1-1/2″ diam. to 6″ diam.					40%	57%			
1800	PVC, couplings 10′ OC, clevis hanger assemblies, 3 per 10′									
1820	Schedule 40									
1860	1/2″ diameter	1 Plum	54	.148	L.F.	4.76	9.35		14.11	19.30
1870	3/4″ diameter		51	.157		5.30	9.90		15.20	20.50
1880	1″ diameter		46	.174		9	11		20	26.50
1890	1-1/4″ diameter		42	.190		9.80	12.05		21.85	29
1900	1-1/2″ diameter		36	.222		9.85	14.05		23.90	32
1910	2″ diameter	Q-1	59	.271		11.35	15.40		26.75	35.50
1920	2-1/2″ diameter		56	.286		10.90	16.25		27.15	36.50
1930	3″ diameter		53	.302		13.10	17.15		30.25	40.50
1940	4″ diameter		48	.333		29.50	18.95		48.45	60.50
1950	5″ diameter		43	.372		38	21		59	73
1960	6″ diameter		39	.410		27.50	23.50		51	65.50
1970	8″ diameter	Q-2	48	.500		35.50	29.50		65	83
1980	10″ diameter		43	.558		76	33		109	133
1990	12″ diameter		42	.571		91.50	33.50		125	151
2000	14″ diameter		31	.774		150	45.50		195.50	234
2010	16″ diameter		23	1.043		214	61.50		275.50	330
2340	To delete coupling & hangers, subtract									
2360	1/2″ diam. to 1-1/4″ diam.					65%	74%			
2370	1-1/2″ diam. to 6″ diam.					44%	57%			
2380	8″ diam. to 12″ diam.					41%	53%			
2390	14″ diam. to 16″ diam.					48%	45%			
2420	Schedule 80									
2440	1/4″ diameter	1 Plum	58	.138	L.F.	4.29	8.70		12.99	17.75
2450	3/8″ diameter		55	.145		4.29	9.20		13.49	18.50
2460	1/2″ diameter		50	.160		4.86	10.10		14.96	20.50
2470	3/4″ diameter		47	.170		5.45	10.75		16.20	22
2480	1″ diameter		43	.186		9.25	11.75		21	28
2490	1-1/4″ diameter		39	.205		10.05	12.95		23	30.50
2500	1-1/2″ diameter		34	.235		10.20	14.85		25.05	34
2510	2″ diameter	Q-1	55	.291		11.25	16.55		27.80	37.50
2520	2-1/2″ diameter		52	.308		10.30	17.50		27.80	37.50
2530	3″ diameter		50	.320		14.60	18.20		32.80	43.50

22 11 13.74 Pipe, Plastic		Crew	Daily Output	Labor-Hours	Unit	Material	2019 Bare Costs Labor	Equipment	Total	Total Incl O&P
2540	4″ diameter	Q-1	46	.348	L.F.	32.50	19.75		52.25	65
2550	5″ diameter		42	.381		41	21.50		62.50	77.50
2560	6″ diameter	↓	38	.421		35	24		59	74.50
2570	8″ diameter	Q-2	47	.511		45	30		75	94.50
2580	10″ diameter		42	.571		101	33.50		134.50	162
2590	12″ diameter	↓	38	.632	↓	122	37		159	190
2830	To delete coupling & hangers, subtract									
2840	1/4″ diam. to 1/2″ diam.					66%	80%			
2850	3/4″ diam. to 1-1/4″ diam.					61%	73%			
2860	1-1/2″ diam. to 6″ diam.					41%	57%			
2870	8″ diam. to 12″ diam.					31%	50%			
2900	Schedule 120									
2910	1/2″ diameter	1 Plum	50	.160	L.F.	5.25	10.10		15.35	21
2950	3/4″ diameter		47	.170		5.95	10.75		16.70	22.50
2960	1″ diameter		43	.186		9.95	11.75		21.70	28.50
2970	1-1/4″ diameter		39	.205		11.15	12.95		24.10	32
2980	1-1/2″ diameter	↓	33	.242		11.65	15.30		26.95	36
2990	2″ diameter	Q-1	54	.296		13.20	16.85		30.05	40
3000	2-1/2″ diameter		52	.308		15.45	17.50		32.95	43
3010	3″ diameter		49	.327		25	18.55		43.55	55.50
3020	4″ diameter		45	.356		38	20		58	72.50
3030	6″ diameter	↓	37	.432	↓	62	24.50		86.50	105
3240	To delete coupling & hangers, subtract									
3250	1/2″ diam. to 1-1/4″ diam.					52%	74%			
3260	1-1/2″ diam. to 4″ diam.					30%	57%			
3270	6″ diam.					17%	50%			
3300	PVC, pressure, couplings 10′ OC, clevis hanger assy's, 3 per 10′									
3310	SDR 26, 160 psi									
3350	1-1/4″ diameter	1 Plum	42	.190	L.F.	9.40	12.05		21.45	28.50
3360	1-1/2″ diameter	″	36	.222		10.15	14.05		24.20	32
3370	2″ diameter	Q-1	59	.271		10.45	15.40		25.85	34.50
3380	2-1/2″ diameter		56	.286		10.15	16.25		26.40	35.50
3390	3″ diameter		53	.302		12.75	17.15		29.90	40
3400	4″ diameter		48	.333		29.50	18.95		48.45	61
3420	6″ diameter	↓	39	.410		30	23.50		53.50	68.50
3430	8″ diameter	Q-2	48	.500	↓	44.50	29.50		74	93
3660	To delete coupling & clevis hanger assy's, subtract									
3670	1-1/4″ diam.					63%	68%			
3680	1-1/2″ diam. to 4″ diam.					48%	57%			
3690	6″ diam. to 8″ diam.					60%	54%			
3720	SDR 21, 200 psi, 1/2″ diameter	1 Plum	54	.148	L.F.	4.64	9.35		13.99	19.15
3740	3/4″ diameter		51	.157		5.05	9.90		14.95	20.50
3750	1″ diameter		46	.174		8.65	11		19.65	26
3760	1-1/4″ diameter		42	.190		9.55	12.05		21.60	28.50
3770	1-1/2″ diameter	↓	36	.222		10.10	14.05		24.15	32
3780	2″ diameter	Q-1	59	.271		10.35	15.40		25.75	34.50
3790	2-1/2″ diameter		56	.286		11.45	16.25		27.70	37
3800	3″ diameter		53	.302		12.40	17.15		29.55	39.50
3810	4″ diameter		48	.333		31	18.95		49.95	62.50
3830	6″ diameter	↓	39	.410		30	23.50		53.50	68
3840	8″ diameter	Q-2	48	.500	↓	46.50	29.50		76	95
4000	To delete coupling & hangers, subtract									
4010	1/2″ diam. to 3/4″ diam.					71%	77%			

22 11 13.74 Pipe, Plastic		Crew	Daily Output	Labor-Hours	Unit	Material	2019 Bare Costs Labor	Equipment	Total	Total Incl O&P
4020	1″ diam. to 1-1/4″ diam.					63%	70%			
4030	1-1/2″ diam. to 6″ diam.					44%	57%			
4040	8″ diam.					46%	54%			
4100	DWV type, schedule 40, couplings 10′ OC, clevis hanger assy's, 3 per 10′									
4210	ABS, schedule 40, foam core type									
4212	Plain end black									
4214	1-1/2″ diameter	1 Plum	39	.205	L.F.	8.25	12.95		21.20	28.50
4216	2″ diameter	Q-1	62	.258		8.80	14.65		23.45	31.50
4218	3″ diameter		56	.286		8.55	16.25		24.80	34
4220	4″ diameter		51	.314		23.50	17.85		41.35	53
4222	6″ diameter		42	.381		19.80	21.50		41.30	54.50
4240	To delete coupling & hangers, subtract									
4244	1-1/2″ diam. to 6″ diam.					43%	48%			
4400	PVC									
4410	1-1/4″ diameter	1 Plum	42	.190	L.F.	8.80	12.05		20.85	27.50
4420	1-1/2″ diameter	″	36	.222		8	14.05		22.05	30
4460	2″ diameter	Q-1	59	.271		8.85	15.40		24.25	33
4470	3″ diameter		53	.302		8.65	17.15		25.80	35.50
4480	4″ diameter		48	.333		10.45	18.95		29.40	40
4490	6″ diameter		39	.410		17.95	23.50		41.45	54.50
4500	8″ diameter	Q-2	48	.500		24.50	29.50		54	71
4510	To delete coupling & hangers, subtract									
4520	1-1/4″ diam. to 1-1/2″ diam.					48%	60%			
4530	2″ diam. to 8″ diam.					42%	54%			
4532	to delete hangers, 2″ diam. to 8″ diam.	Q-1	50	.320	L.F.	19.80	18.20		38	49.50
4550	PVC, schedule 40, foam core type									
4552	Plain end, white									
4554	1-1/2″ diameter	1 Plum	39	.205	L.F.	8.05	12.95		21	28.50
4556	2″ diameter	Q-1	62	.258		8.55	14.65		23.20	31.50
4558	3″ diameter		56	.286		8.05	16.25		24.30	33.50
4560	4″ diameter		51	.314		9.55	17.85		27.40	37.50
4562	6″ diameter		42	.381		16.10	21.50		37.60	50.50
4564	8″ diameter	Q-2	51	.471		22	27.50		49.50	65.50
4568	10″ diameter		48	.500		25.50	29.50		55	72
4570	12″ diameter		46	.522		29	31		60	78
4580	To delete coupling & hangers, subtract									
4582	1-1/2″ diam. to 2″ diam.					58%	54%			
4584	3″ diam. to 12″ diam.					46%	42%			
4800	PVC, clear pipe, cplgs. 10′ OC, clevis hanger assy's 3 per 10′, Sched. 40									
4840	1/4″ diameter	1 Plum	59	.136	L.F.	4.67	8.55		13.22	18
4850	3/8″ diameter		56	.143		4.96	9		13.96	19
4860	1/2″ diameter		54	.148		5.50	9.35		14.85	20
4870	3/4″ diameter		51	.157		6.15	9.90		16.05	21.50
4880	1″ diameter		46	.174		10.85	11		21.85	28.50
4890	1-1/4″ diameter		42	.190		12.15	12.05		24.20	31.50
4900	1-1/2″ diameter		36	.222		12.95	14.05		27	35.50
4910	2″ diameter	Q-1	59	.271		15.15	15.40		30.55	39.50
4920	2-1/2″ diameter		56	.286		17.85	16.25		34.10	44
4930	3″ diameter		53	.302		21.50	17.15		38.65	50
4940	3-1/2″ diameter		50	.320		30	18.20		48.20	60.50
4950	4″ diameter		48	.333		40.50	18.95		59.45	73
5250	To delete coupling & hangers, subtract									
5260	1/4″ diam. to 3/8″ diam.					60%	81%			

22 11 13 – Facility Water Distribution Piping

22 11 13.74 Pipe, Plastic		Crew	Daily Output	Labor-Hours	Unit	2019 Bare Costs Material	Labor	Equipment	Total	Total Incl O&P
5270	1/2" diam. to 3/4" diam.					41%	77%			
5280	1" diam. to 1-1/2" diam.					26%	67%			
5290	2" diam. to 4" diam.					16%	58%			
5300	CPVC, socket joint, couplings 10' OC, clevis hanger assemblies, 3 per 10'									
5302	Schedule 40									
5304	1/2" diameter	1 Plum	54	.148	L.F.	5.60	9.35		14.95	20
5305	3/4" diameter		51	.157		6.60	9.90		16.50	22
5306	1" diameter		46	.174		10.85	11		21.85	28.50
5307	1-1/4" diameter		42	.190		12.25	12.05		24.30	31.50
5308	1-1/2" diameter		36	.222		12.15	14.05		26.20	34.50
5309	2" diameter	Q-1	59	.271		15.40	15.40		30.80	40
5310	2-1/2" diameter		56	.286		19.40	16.25		35.65	46
5311	3" diameter		53	.302		22	17.15		39.15	50.50
5312	4" diameter		48	.333		45	18.95		63.95	78
5314	6" diameter		43	.372		54.50	21		75.50	91.50
5318	To delete coupling & hangers, subtract									
5319	1/2" diam. to 3/4" diam.					37%	77%			
5320	1" diam. to 1-1/4" diam.					27%	70%			
5321	1-1/2" diam. to 3" diam.					21%	57%			
5322	4" diam. to 6" diam.					16%	57%			
5324	Schedule 80									
5325	1/2" diameter	1 Plum	50	.160	L.F.	6.25	10.10		16.35	22
5326	3/4" diameter		47	.170		6.65	10.75		17.40	23.50
5327	1" diameter		43	.186		11.50	11.75		23.25	30.50
5328	1-1/4" diameter		39	.205		13.25	12.95		26.20	34
5329	1-1/2" diameter		34	.235		14.40	14.85		29.25	38.50
5330	2" diameter	Q-1	55	.291		17	16.55		33.55	43.50
5331	2-1/2" diameter		52	.308		21.50	17.50		39	49.50
5332	3" diameter		50	.320		23.50	18.20		41.70	53.50
5333	4" diameter		46	.348		46	19.75		65.75	80
5334	6" diameter		38	.421		60.50	24		84.50	103
5335	8" diameter	Q-2	47	.511		162	30		192	223
5339	To delete couplings & hangers, subtract									
5340	1/2" diam. to 3/4" diam.					44%	77%			
5341	1" diam. to 1-1/4" diam.					32%	71%			
5342	1-1/2" diam. to 4" diam.					25%	58%			
5343	6" diam. to 8" diam.					20%	53%			
5360	CPVC, threaded, couplings 10' OC, clevis hanger assemblies, 3 per 10'									
5380	Schedule 40									
5460	1/2" diameter	1 Plum	54	.148	L.F.	6.45	9.35		15.80	21
5470	3/4" diameter		51	.157		8.15	9.90		18.05	24
5480	1" diameter		46	.174		12.40	11		23.40	30
5490	1-1/4" diameter		42	.190		13.45	12.05		25.50	33
5500	1-1/2" diameter		36	.222		13.20	14.05		27.25	35.50
5510	2" diameter	Q-1	59	.271		16.60	15.40		32	41.50
5520	2-1/2" diameter		56	.286		20.50	16.25		36.75	47.50
5530	3" diameter		53	.302		24	17.15		41.15	52.50
5540	4" diameter		48	.333		52.50	18.95		71.45	86.50
5550	6" diameter		43	.372		58	21		79	95.50
5730	To delete coupling & hangers, subtract									
5740	1/2" diam. to 3/4" diam.					37%	77%			
5750	1" diam. to 1-1/4" diam.					27%	70%			
5760	1-1/2" diam. to 3" diam.					21%	57%			

22 11 13.74 Pipe, Plastic		Crew	Daily Output	Labor-Hours	Unit	Material	2019 Bare Costs Labor	Equipment	Total	Total Incl O&P
5770	4" diam. to 6" diam.					16%	57%			
5800	Schedule 80									
5860	1/2" diameter	1 Plum	50	.160	L.F.	7.10	10.10		17.20	23
5870	3/4" diameter		47	.170		8.15	10.75		18.90	25
5880	1" diameter		43	.186		13.05	11.75		24.80	32
5890	1-1/4" diameter		39	.205		14.45	12.95		27.40	35.50
5900	1-1/2" diameter		34	.235		15.45	14.85		30.30	39.50
5910	2" diameter	Q-1	55	.291		18.20	16.55		34.75	45
5920	2-1/2" diameter		52	.308		23	17.50		40.50	51
5930	3" diameter		50	.320		25.50	18.20		43.70	55.50
5940	4" diameter		46	.348		53.50	19.75		73.25	88.50
5950	6" diameter		38	.421		64	24		88	107
5960	8" diameter	Q-2	47	.511		160	30		190	221
6060	To delete couplings & hangers, subtract									
6070	1/2" diam. to 3/4" diam.					44%	77%			
6080	1" diam. to 1-1/4" diam.					32%	71%			
6090	1-1/2" diam. to 4" diam.					25%	58%			
6100	6" diam. to 8" diam.					20%	53%			
6240	CTS, 1/2" diameter	1 Plum	54	.148	L.F.	4.59	9.35		13.94	19.10
6250	3/4" diameter		51	.157		9.45	9.90		19.35	25.50
6260	1" diameter		46	.174		13.40	11		24.40	31.50
6270	1-1/4" diameter		42	.190		18.10	12.05		30.15	38
6280	1-1/2" diameter		36	.222		21.50	14.05		35.55	45
6290	2" diameter	Q-1	59	.271		32	15.40		47.40	58
6370	To delete coupling & hangers, subtract									
6380	1/2" diam.					51%	79%			
6390	3/4" diam.					40%	76%			
6392	1" thru 2" diam.					72%	68%			
7280	PEX, flexible, no couplings or hangers									
7282	Note: For labor costs add 25% to the couplings and fittings labor total.									
7285	For fittings see section 23 83 16.10 7000									
7300	Non-barrier type, hot/cold tubing rolls									
7310	1/4" diameter x 100'				L.F.	.52			.52	.57
7350	3/8" diameter x 100'					.55			.55	.61
7360	1/2" diameter x 100'					.64			.64	.70
7370	1/2" diameter x 500'					.64			.64	.70
7380	1/2" diameter x 1000'					.63			.63	.69
7400	3/4" diameter x 100'					1.01			1.01	1.11
7410	3/4" diameter x 500'					1.14			1.14	1.25
7420	3/4" diameter x 1000'					1.14			1.14	1.25
7460	1" diameter x 100'					1.96			1.96	2.16
7470	1" diameter x 300'					1.96			1.96	2.16
7480	1" diameter x 500'					1.96			1.96	2.16
7500	1-1/4" diameter x 100'					3.33			3.33	3.66
7510	1-1/4" diameter x 300'					3.33			3.33	3.66
7540	1-1/2" diameter x 100'					4.52			4.52	4.97
7550	1-1/2" diameter x 300'					4.50			4.50	4.95
7596	Most sizes available in red or blue									
7700	Non-barrier type, hot/cold tubing straight lengths									
7710	1/2" diameter x 20'				L.F.	.64			.64	.70
7750	3/4" diameter x 20'					1.14			1.14	1.25
7760	1" diameter x 20'					1.96			1.96	2.16
7770	1-1/4" diameter x 20'					3.44			3.44	3.78

22 11 13.74 Pipe, Plastic		Crew	Daily Output	Labor-Hours	Unit	2019 Bare Costs Material	Labor	Equipment	Total	Total Incl O&P
7780	1-1/2" diameter x 20'				L.F.	4.54			4.54	4.99
7790	2" diameter					8.85			8.85	9.75
7796	Most sizes available in red or blue									
9000	Polypropylene pipe									
9002	For fusion weld fittings and accessories see line 22 11 13.76 9400									
9004	Note: sizes 1/2" thru 4" use socket fusion									
9005	Sizes 6" thru 10" use butt fusion									
9010	SDR 7.4 (domestic hot water piping)									
9011	Enhanced to minimize thermal expansion and high temperature life									
9016	13' lengths, size is ID, includes joints 13' OC and hangers 3 per 10'									
9020	3/8" diameter	1 Plum	53	.151	L.F.	3.31	9.55		12.86	17.95
9022	1/2" diameter		52	.154		3.77	9.70		13.47	18.75
9024	3/4" diameter		50	.160		4.32	10.10		14.42	19.90
9026	1" diameter		45	.178		5.40	11.25		16.65	23
9028	1-1/4" diameter		40	.200		7.65	12.65		20.30	27.50
9030	1-1/2" diameter		35	.229		10.15	14.45		24.60	32.50
9032	2" diameter	Q-1	58	.276		14.10	15.70		29.80	39
9034	2-1/2" diameter		55	.291		18.90	16.55		35.45	46
9036	3" diameter		52	.308		24.50	17.50		42	53
9038	3-1/2" diameter		49	.327		33.50	18.55		52.05	65
9040	4" diameter		46	.348		36	19.75		55.75	69
9042	6" diameter		39	.410		48.50	23.50		72	88
9044	8" diameter	Q-2	48	.500		73	29.50		102.50	124
9046	10" diameter	"	43	.558		115	33		148	176
9050	To delete joint & hangers, subtract									
9052	3/8" diam. to 1" diam.					45%	65%			
9054	1-1/4" diam. to 4" diam.					15%	45%			
9056	6" diam. to 10" diam.					5%	24%			
9060	SDR 11 (domestic cold water piping)									
9062	13' lengths, size is ID, includes joints 13' OC and hangers 3 per 10'									
9064	1/2" diameter	1 Plum	57	.140	L.F.	3.44	8.85		12.29	17.10
9066	3/4" diameter		54	.148		3.79	9.35		13.14	18.20
9068	1" diameter		49	.163		4.42	10.30		14.72	20.50
9070	1-1/4" diameter		45	.178		6.25	11.25		17.50	23.50
9072	1-1/2" diameter		40	.200		7.70	12.65		20.35	27.50
9074	2" diameter	Q-1	62	.258		10.60	14.65		25.25	33.50
9076	2-1/2" diameter		59	.271		13.55	15.40		28.95	38
9078	3" diameter		56	.286		17.90	16.25		34.15	44
9080	3-1/2" diameter		53	.302		25.50	17.15		42.65	54
9082	4" diameter		50	.320		28.50	18.20		46.70	58.50
9084	6" diameter		47	.340		33	19.35		52.35	65.50
9086	8" diameter	Q-2	51	.471		51	27.50		78.50	97.50
9088	10" diameter	"	46	.522		76	31		107	130
9090	To delete joint & hangers, subtract									
9092	1/2" diam. to 1" diam.					45%	65%			
9094	1-1/4" diam. to 4" diam.					15%	45%			
9096	6" diam. to 10" diam.					5%	24%			

22 11 13.76 Pipe Fittings, Plastic		Crew	Daily Output	Labor-Hours	Unit	Material	Labor	Equipment	Total	Total Incl O&P
0010	PIPE FITTINGS, PLASTIC									
0030	Epoxy resin, fiberglass reinforced, general service									
0100	3"	Q-1	20.80	.769	Ea.	117	43.50		160.50	195
0110	4"		16.50	.970		120	55		175	215

22 11 13 – Facility Water Distribution Piping

22 11 13.76 Pipe Fittings, Plastic		Crew	Daily Output	Labor-Hours	Unit	Material	Labor	Equipment	Total	Total Incl O&P
						2019 Bare Costs				
0120	6″	Q-1	10.10	1.584	Ea.	234	90		324	390
0130	8″	Q-2	9.30	2.581		430	152		582	705
0140	10″		8.50	2.824		540	166		706	845
0150	12″		7.60	3.158		775	186		961	1,125
0170	Elbow, 90°, flanged									
0172	2″	Q-1	23	.696	Ea.	226	39.50		265.50	310
0173	3″		16	1		259	57		316	370
0174	4″		13	1.231		340	70		410	475
0176	6″		8	2		615	114		729	845
0177	8″	Q-2	9	2.667		1,100	157		1,257	1,450
0178	10″		7	3.429		1,525	202		1,727	1,975
0179	12″		5	4.800		2,050	283		2,333	2,675
0186	Elbow, 45°, flanged									
0188	2″	Q-1	23	.696	Ea.	226	39.50		265.50	310
0189	3″		16	1		259	57		316	370
0190	4″		13	1.231		340	70		410	475
0192	6″		8	2		615	114		729	845
0193	8″	Q-2	9	2.667		1,125	157		1,282	1,450
0194	10″		7	3.429		1,550	202		1,752	2,025
0195	12″		5	4.800		2,125	283		2,408	2,775
0352	Tee, flanged									
0354	2″	Q-1	17	.941	Ea.	305	53.50		358.50	415
0355	3″		10	1.600		405	91		496	580
0356	4″		8	2		455	114		569	670
0358	6″		5	3.200		780	182		962	1,125
0359	8″	Q-2	6	4		1,475	236		1,711	1,975
0360	10″		5	4.800		2,150	283		2,433	2,800
0361	12″		4	6		2,975	355		3,330	3,775
0365	Wye, flanged									
0367	2″	Q-1	17	.941	Ea.	605	53.50		658.50	745
0368	3″		10	1.600		835	91		926	1,050
0369	4″		8	2		1,125	114		1,239	1,400
0371	6″		5	3.200		1,575	182		1,757	2,000
0372	8″	Q-2	6	4		2,575	236		2,811	3,175
0373	10″		5	4.800		4,000	283		4,283	4,825
0374	12″		4	6		5,825	355		6,180	6,925
0380	Couplings									
0410	2″	Q-1	33.10	.483	Ea.	26	27.50		53.50	69.50
0420	3″		20.80	.769		28.50	43.50		72	96.50
0430	4″		16.50	.970		38.50	55		93.50	125
0440	6″		10.10	1.584		90	90		180	234
0450	8″	Q-2	9.30	2.581		153	152		305	395
0460	10″		8.50	2.824		224	166		390	495
0470	12″		7.60	3.158		300	186		486	610
0473	High corrosion resistant couplings, add					30%				
0474	Reducer, concentric, flanged									
0475	2″ x 1-1/2″	Q-1	30	.533	Ea.	590	30.50		620.50	690
0476	3″ x 2″		24	.667		640	38		678	760
0477	4″ x 3″		19	.842		720	48		768	860
0479	6″ x 4″		15	1.067		720	60.50		780.50	880
0480	8″ x 6″	Q-2	16	1.500		1,175	88.50		1,263.50	1,425
0481	10″ x 8″		13	1.846		1,100	109		1,209	1,375
0482	12″ x 10″		11	2.182		1,700	129		1,829	2,075

22 11 13.76 Pipe Fittings, Plastic		Crew	Daily Output	Labor-Hours	Unit	2019 Bare Costs Material	Labor	Equipment	Total	Total Incl O&P
0486	Adapter, bell x male or female									
0488	2"	Q-1	28	.571	Ea.	29.50	32.50		62	81
0489	3"		20	.800		44.50	45.50		90	117
0491	4"		17	.941		57	53.50		110.50	143
0492	6"		12	1.333		123	76		199	249
0493	8"	Q-2	15	1.600		177	94.50		271.50	335
0494	10"	"	11	2.182		254	129		383	475
0528	Flange									
0532	2"	Q-1	46	.348	Ea.	42	19.75		61.75	75.50
0533	3"		32	.500		54	28.50		82.50	102
0534	4"		26	.615		69	35		104	129
0536	6"		16	1		131	57		188	231
0537	8"	Q-2	18	1.333		201	78.50		279.50	340
0538	10"		14	1.714		285	101		386	465
0539	12"		10	2.400		355	141		496	605
2100	PVC schedule 80, socket joint									
2110	90° elbow, 1/2"	1 Plum	30.30	.264	Ea.	2.64	16.65		19.29	28
2130	3/4"		26	.308		3.38	19.45		22.83	32.50
2140	1"		22.70	.352		5.60	22.50		28.10	39.50
2150	1-1/4"		20.20	.396		7.30	25		32.30	45.50
2160	1-1/2"		18.20	.440		7.80	28		35.80	50
2170	2"	Q-1	33.10	.483		9.40	27.50		36.90	51.50
2180	3"		20.80	.769		25	43.50		68.50	93
2190	4"		16.50	.970		37.50	55		92.50	124
2200	6"		10.10	1.584		107	90		197	253
2210	8"	Q-2	9.30	2.581		296	152		448	555
2250	45° elbow, 1/2"	1 Plum	30.30	.264		4.97	16.65		21.62	30.50
2270	3/4"		26	.308		7.60	19.45		27.05	37.50
2280	1"		22.70	.352		11.40	22.50		33.90	46
2290	1-1/4"		20.20	.396		14.50	25		39.50	53.50
2300	1-1/2"		18.20	.440		17.20	28		45.20	60.50
2310	2"	Q-1	33.10	.483		22	27.50		49.50	65.50
2320	3"		20.80	.769		56.50	43.50		100	128
2330	4"		16.50	.970		102	55		157	196
2340	6"		10.10	1.584		129	90		219	277
2350	8"	Q-2	9.30	2.581		279	152		431	535
2400	Tee, 1/2"	1 Plum	20.20	.396		7.45	25		32.45	45.50
2420	3/4"		17.30	.462		7.80	29		36.80	52.50
2430	1"		15.20	.526		9.75	33		42.75	61
2440	1-1/4"		13.50	.593		27	37.50		64.50	85.50
2450	1-1/2"		12.10	.661		27	42		69	92
2460	2"	Q-1	20	.800		33.50	45.50		79	105
2470	3"		13.90	1.151		45.50	65.50		111	148
2480	4"		11	1.455		53	82.50		135.50	182
2490	6"		6.70	2.388		181	136		317	405
2500	8"	Q-2	6.20	3.871		400	228		628	780
2510	Flange, socket, 150 lb., 1/2"	1 Plum	55.60	.144		14.50	9.10		23.60	29.50
2514	3/4"		47.60	.168		15.45	10.60		26.05	33
2518	1"		41.70	.192		17.20	12.10		29.30	37
2522	1-1/2"		33.30	.240		18.35	15.15		33.50	43
2526	2"	Q-1	60.60	.264		24	15		39	49
2530	4"		30.30	.528		52	30		82	103
2534	6"		18.50	.865		82	49		131	164

22 11 13.76 Pipe Fittings, Plastic		Crew	Daily Output	Labor-Hours	Unit	2019 Bare Costs Material	Labor	Equipment	Total	Total Incl O&P
2538	8"	Q-2	17.10	1.404	Ea.	147	82.50		229.50	285
2550	Coupling, 1/2"	1 Plum	30.30	.264		4.83	16.65		21.48	30.50
2570	3/4"		26	.308		8.50	19.45		27.95	38.50
2580	1"		22.70	.352		6.95	22.50		29.45	41
2590	1-1/4"		20.20	.396		10.10	25		35.10	48.50
2600	1-1/2"		18.20	.440		10.90	28		38.90	53.50
2610	2"	Q-1	33.10	.483		11.75	27.50		39.25	54
2620	3"		20.80	.769		33.50	43.50		77	103
2630	4"		16.50	.970		41.50	55		96.50	128
2640	6"		10.10	1.584		89.50	90		179.50	234
2650	8"	Q-2	9.30	2.581		122	152		274	360
2660	10"		8.50	2.824		415	166		581	705
2670	12"		7.60	3.158		480	186		666	810
2700	PVC (white), schedule 40, socket joints									
2760	90° elbow, 1/2"	1 Plum	33.30	.240	Ea.	.51	15.15		15.66	23.50
2770	3/4"		28.60	.280		.59	17.65		18.24	27
2780	1"		25	.320		1.04	20		21.04	31.50
2790	1-1/4"		22.20	.360		1.82	23		24.82	36
2800	1-1/2"		20	.400		1.97	25.50		27.47	40
2810	2"	Q-1	36.40	.440		3.08	25		28.08	41
2820	2-1/2"		26.70	.599		9.90	34		43.90	62
2830	3"		22.90	.699		11.20	39.50		50.70	72
2840	4"		18.20	.879		20	50		70	97
2850	5"		12.10	1.322		52	75		127	170
2860	6"		11.10	1.441		64	82		146	193
2870	8"	Q-2	10.30	2.330		164	137		301	385
2980	45° elbow, 1/2"	1 Plum	33.30	.240		.86	15.15		16.01	24
2990	3/4"		28.60	.280		1.32	17.65		18.97	28
3000	1"		25	.320		1.59	20		21.59	32.50
3010	1-1/4"		22.20	.360		2.23	23		25.23	36.50
3020	1-1/2"		20	.400		2.77	25.50		28.27	41
3030	2"	Q-1	36.40	.440		3.60	25		28.60	41.50
3040	2-1/2"		26.70	.599		9.40	34		43.40	61.50
3050	3"		22.90	.699		14.55	39.50		54.05	75.50
3060	4"		18.20	.879		26	50		76	104
3070	5"		12.10	1.322		52	75		127	170
3080	6"		11.10	1.441		64.50	82		146.50	194
3090	8"	Q-2	10.30	2.330		155	137		292	375
3180	Tee, 1/2"	1 Plum	22.20	.360		.64	23		23.64	34.50
3190	3/4"		19	.421		.76	26.50		27.26	41
3200	1"		16.70	.479		1.38	30.50		31.88	47
3210	1-1/4"		14.80	.541		2.15	34		36.15	53.50
3220	1-1/2"		13.30	.602		2.61	38		40.61	60
3230	2"	Q-1	24.20	.661		3.80	37.50		41.30	60.50
3240	2-1/2"		17.80	.899		12.50	51		63.50	90.50
3250	3"		15.20	1.053		16.45	60		76.45	108
3260	4"		12.10	1.322		30	75		105	146
3270	5"		8.10	1.975		72	112		184	247
3280	6"		7.40	2.162		100	123		223	294
3290	8"	Q-2	6.80	3.529		238	208		446	570
3380	Coupling, 1/2"	1 Plum	33.30	.240		.34	15.15		15.49	23.50
3390	3/4"		28.60	.280		.47	17.65		18.12	27
3400	1"		25	.320		.82	20		20.82	31.50

22 11 13.76 Pipe Fittings, Plastic		Crew	Daily Output	Labor-Hours	Unit	Material	Labor	Equipment	Total	Total Incl O&P
3410	1-1/4"	1 Plum	22.20	.360	Ea.	1.13	23		24.13	35
3420	1-1/2"		20	.400		1.20	25.50		26.70	39.50
3430	2"	Q-1	36.40	.440		1.83	25		26.83	39.50
3440	2-1/2"		26.70	.599		4.06	34		38.06	55.50
3450	3"		22.90	.699		6.35	39.50		45.85	66.50
3460	4"		18.20	.879		9.20	50		59.20	85
3470	5"		12.10	1.322		16.85	75		91.85	132
3480	6"		11.10	1.441		29	82		111	155
3490	8"	Q-2	10.30	2.330		54.50	137		191.50	266
3600	Cap, schedule 40, PVC socket 1/2"	1 Plum	60.60	.132		.47	8.35		8.82	13
3610	3/4"		51.90	.154		.54	9.75		10.29	15.20
3620	1"		45.50	.176		.86	11.10		11.96	17.60
3630	1-1/4"		40.40	.198		1.20	12.50		13.70	20
3640	1-1/2"		36.40	.220		1.32	13.90		15.22	22.50
3650	2"	Q-1	66.10	.242		1.58	13.75		15.33	22
3660	2-1/2"		48.50	.330		5.05	18.75		23.80	33.50
3670	3"		41.60	.385		5.55	22		27.55	39
3680	4"		33.10	.483		12.50	27.50		40	55
3690	6"		20.20	.792		30	45		75	101
3700	8"	Q-2	18.60	1.290		75.50	76		151.50	197
3710	Reducing insert, schedule 40, socket weld									
3712	3/4"	1 Plum	31.50	.254	Ea.	.54	16.05		16.59	24.50
3713	1"		27.50	.291		.98	18.35		19.33	28.50
3715	1-1/2"		22	.364		1.39	23		24.39	36
3716	2"	Q-1	40	.400		2.29	22.50		24.79	36.50
3717	4"		20	.800		12.15	45.50		57.65	81.50
3718	6"		12.20	1.311		30	74.50		104.50	145
3719	8"	Q-2	11.30	2.124		119	125		244	320
3730	Reducing insert, socket weld x female/male thread									
3732	1/2"	1 Plum	38.30	.209	Ea.	2.43	13.20		15.63	22.50
3733	3/4"		32.90	.243		1.51	15.35		16.86	24.50
3734	1"		28.80	.278		2.12	17.55		19.67	29
3736	1-1/2"		23	.348		3.80	22		25.80	37
3737	2"	Q-1	41.90	.382		4.06	21.50		25.56	37
3738	4"	"	20.90	.766		39.50	43.50		83	109
3742	Male adapter, socket weld x male thread									
3744	1/2"	1 Plum	38.30	.209	Ea.	.39	13.20		13.59	20
3745	3/4"		32.90	.243		.52	15.35		15.87	23.50
3746	1"		28.80	.278		.95	17.55		18.50	27.50
3748	1-1/2"		23	.348		1.54	22		23.54	34.50
3749	2"	Q-1	41.90	.382		2	21.50		23.50	34.50
3750	4"	"	20.90	.766		10.90	43.50		54.40	77.50
3754	Female adapter, socket weld x female thread									
3756	1/2"	1 Plum	38.30	.209	Ea.	.59	13.20		13.79	20.50
3757	3/4"		32.90	.243		.76	15.35		16.11	24
3758	1"		28.80	.278		.86	17.55		18.41	27.50
3760	1-1/2"		23	.348		1.51	22		23.51	34.50
3761	2"	Q-1	41.90	.382		2.02	21.50		23.52	34.50
3762	4"	"	20.90	.766		11.40	43.50		54.90	78
3800	PVC, schedule 80, socket joints									
3810	Reducing insert									
3812	3/4"	1 Plum	28.60	.280	Ea.	1.54	17.65		19.19	28
3813	1"		25	.320		4.41	20		24.41	35.50

22 11 13.76 Pipe Fittings, Plastic		Crew	Daily Output	Labor-Hours	Unit	Material	Labor	Equipment	Total	Total Incl O&P
						2019 Bare Costs				
3815	1-1/2"	1 Plum	20	.400	Ea.	9.65	25.50		35.15	48.50
3816	2"	Q-1	36.40	.440		13.45	25		38.45	52.50
3817	4"		18.20	.879		51.50	50		101.50	132
3818	6"		11.10	1.441		72	82		154	203
3819	8"	Q-2	10.20	2.353		410	139		549	660
3830	Reducing insert, socket weld x female/male thread									
3832	1/2"	1 Plum	34.80	.230	Ea.	9.65	14.50		24.15	32.50
3833	3/4"		29.90	.268		5.90	16.90		22.80	32
3834	1"		26.10	.307		9.25	19.35		28.60	39
3836	1-1/2"		20.90	.383		11.65	24		35.65	49.50
3837	2"	Q-1	38	.421		17.05	24		41.05	55
3838	4"	"	19	.842		82.50	48		130.50	163
3844	Adapter, male socket x male thread									
3846	1/2"	1 Plum	34.80	.230	Ea.	3.73	14.50		18.23	26
3847	3/4"		29.90	.268		4.11	16.90		21.01	30
3848	1"		26.10	.307		7.10	19.35		26.45	37
3850	1-1/2"		20.90	.383		11.95	24		35.95	49.50
3851	2"	Q-1	38	.421		17.30	24		41.30	55
3852	4"	"	19	.842		39	48		87	115
3860	Adapter, female socket x female thread									
3862	1/2"	1 Plum	34.80	.230	Ea.	4.48	14.50		18.98	27
3863	3/4"		29.90	.268		6.65	16.90		23.55	33
3864	1"		26.10	.307		9.80	19.35		29.15	40
3866	1-1/2"		20.90	.383		19.45	24		43.45	58
3867	2"	Q-1	38	.421		34	24		58	73.50
3868	4"	"	19	.842		104	48		152	186
3872	Union, socket joints									
3874	1/2"	1 Plum	25.80	.310	Ea.	9.80	19.60		29.40	40.50
3875	3/4"		22.10	.362		12.45	23		35.45	48
3876	1"		19.30	.415		14.20	26		40.20	55
3878	1-1/2"		15.50	.516		32	32.50		64.50	84
3879	2"	Q-1	28.10	.569		43.50	32.50		76	96
3888	Cap									
3890	1/2"	1 Plum	54.50	.147	Ea.	4.68	9.25		13.93	19.05
3891	3/4"		46.70	.171		4.93	10.80		15.73	21.50
3892	1"		41	.195		8.75	12.30		21.05	28
3894	1-1/2"		32.80	.244		10.60	15.40		26	34.50
3895	2"	Q-1	59.50	.269		28	15.30		43.30	53.50
3896	4"		30	.533		84	30.50		114.50	138
3897	6"		18.20	.879		209	50		259	305
3898	8"	Q-2	16.70	1.437		269	84.50		353.50	420
4500	DWV, ABS, non pressure, socket joints									
4540	1/4 bend, 1-1/4"	1 Plum	20.20	.396	Ea.	4.73	25		29.73	42.50
4560	1-1/2"	"	18.20	.440		3.65	28		31.65	45.50
4570	2"	Q-1	33.10	.483		5.60	27.50		33.10	47
4580	3"		20.80	.769		14.15	43.50		57.65	81
4590	4"		16.50	.970		29	55		84	115
4600	6"		10.10	1.584		124	90		214	272
4650	1/8 bend, same as 1/4 bend									
4800	Tee, sanitary									
4820	1-1/4"	1 Plum	13.50	.593	Ea.	6.30	37.50		43.80	63
4830	1-1/2"	"	12.10	.661		5.40	42		47.40	68.50
4840	2"	Q-1	20	.800		8.30	45.50		53.80	77

22 11 13.76 Pipe Fittings, Plastic		Crew	Daily Output	Labor-Hours	Unit	Material	Labor	Equipment	Total	Total Incl O&P
						2019 Bare Costs				
4850	3"	Q-1	13.90	1.151	Ea.	22.50	65.50		88	123
4860	4"		11	1.455		40.50	82.50		123	169
4862	Tee, sanitary, reducing, 2" x 1-1/2"		22	.727		7.20	41.50		48.70	70
4864	3" x 2"		15.30	1.046		13.60	59.50		73.10	104
4868	4" x 3"		12.10	1.322		39.50	75		114.50	157
4870	Combination Y and 1/8 bend									
4872	1-1/2"	1 Plum	12.10	.661	Ea.	12.90	42		54.90	76.50
4874	2"	Q-1	20	.800		14.35	45.50		59.85	84
4876	3"		13.90	1.151		34	65.50		99.50	136
4878	4"		11	1.455		65	82.50		147.50	196
4880	3" x 1-1/2"		15.50	1.032		34.50	58.50		93	126
4882	4" x 3"		12.10	1.322		51	75		126	170
4900	Wye, 1-1/4"	1 Plum	13.50	.593		7.25	37.50		44.75	64
4902	1-1/2"	"	12.10	.661		8.20	42		50.20	71.50
4904	2"	Q-1	20	.800		10.70	45.50		56.20	80
4906	3"		13.90	1.151		25.50	65.50		91	126
4908	4"		11	1.455		51.50	82.50		134	181
4910	6"		6.70	2.388		155	136		291	375
4918	3" x 1-1/2"		15.50	1.032		20.50	58.50		79	111
4920	4" x 3"		12.10	1.322		40.50	75		115.50	158
4922	6" x 4"		6.90	2.319		123	132		255	335
4930	Double wye, 1-1/2"	1 Plum	9.10	.879		26.50	55.50		82	113
4932	2"	Q-1	16.60	.964		31.50	55		86.50	117
4934	3"		10.40	1.538		74	87.50		161.50	213
4936	4"		8.25	1.939		145	110		255	325
4940	2" x 1-1/2"		16.80	.952		29	54		83	113
4942	3" x 2"		10.60	1.509		51.50	86		137.50	186
4944	4" x 3"		8.45	1.893		115	108		223	287
4946	6" x 4"		7.25	2.207		165	125		290	370
4950	Reducer bushing, 2" x 1-1/2"		36.40	.440		2.86	25		27.86	40.50
4952	3" x 1-1/2"		27.30	.586		12.60	33.50		46.10	64
4954	4" x 2"		18.20	.879		23.50	50		73.50	101
4956	6" x 4"		11.10	1.441		66	82		148	196
4960	Couplings, 1-1/2"	1 Plum	18.20	.440		1.75	28		29.75	43.50
4962	2"	Q-1	33.10	.483		2.34	27.50		29.84	43.50
4963	3"		20.80	.769		6.60	43.50		50.10	73
4964	4"		16.50	.970		11.85	55		66.85	95.50
4966	6"		10.10	1.584		49.50	90		139.50	190
4970	2" x 1-1/2"		33.30	.480		5.05	27.50		32.55	46.50
4972	3" x 1-1/2"		21	.762		14.95	43.50		58.45	81.50
4974	4" x 3"		16.70	.958		22.50	54.50		77	106
4978	Closet flange, 4"	1 Plum	32	.250		12.30	15.80		28.10	37
4980	4" x 3"	"	34	.235		14.70	14.85		29.55	38.50
5000	DWV, PVC, schedule 40, socket joints									
5040	1/4 bend, 1-1/4"	1 Plum	20.20	.396	Ea.	9.30	25		34.30	48
5060	1-1/2"	"	18.20	.440		2.66	28		30.66	44.50
5070	2"	Q-1	33.10	.483		4.19	27.50		31.69	45.50
5080	3"		20.80	.769		12.30	43.50		55.80	79
5090	4"		16.50	.970		24.50	55		79.50	110
5100	6"		10.10	1.584		86	90		176	230
5105	8"	Q-2	9.30	2.581		112	152		264	350
5106	10"	"	8.50	2.824		360	166		526	645
5110	1/4 bend, long sweep, 1-1/2"	1 Plum	18.20	.440		6.15	28		34.15	48.50

22 11 13.76 Pipe Fittings, Plastic		Crew	Daily Output	Labor-Hours	Unit	Material	2019 Bare Costs Labor	Equipment	Total	Total Incl O&P
5112	2"	Q-1	33.10	.483	Ea.	6.85	27.50		34.35	48.50
5114	3"		20.80	.769		15.85	43.50		59.35	83
5116	4"		16.50	.970		30	55		85	116
5150	1/8 bend, 1-1/4"	1 Plum	20.20	.396		6.40	25		31.40	44.50
5170	1-1/2"	"	18.20	.440		2.61	28		30.61	44.50
5180	2"	Q-1	33.10	.483		3.88	27.50		31.38	45.50
5190	3"		20.80	.769		11.05	43.50		54.55	77.50
5200	4"		16.50	.970		20	55		75	105
5210	6"		10.10	1.584		75	90		165	218
5215	8"	Q-2	9.30	2.581		121	152		273	360
5216	10"		8.50	2.824		240	166		406	515
5217	12"		7.60	3.158		315	186		501	630
5250	Tee, sanitary 1-1/4"	1 Plum	13.50	.593		10	37.50		47.50	67
5254	1-1/2"	"	12.10	.661		4.64	42		46.64	67.50
5255	2"	Q-1	20	.800		6.85	45.50		52.35	75.50
5256	3"		13.90	1.151		18	65.50		83.50	118
5257	4"		11	1.455		33	82.50		115.50	161
5259	6"		6.70	2.388		133	136		269	350
5261	8"	Q-2	6.20	3.871		289	228		517	660
5276	Tee, sanitary, reducing									
5281	2" x 1-1/2" x 1-1/2"	Q-1	23	.696	Ea.	6	39.50		45.50	66
5282	2" x 1-1/2" x 2"		22	.727		7.30	41.50		48.80	70
5283	2" x 2" x 1-1/2"		22	.727		6.05	41.50		47.55	68.50
5284	3" x 3" x 1-1/2"		15.50	1.032		13.15	58.50		71.65	102
5285	3" x 3" x 2"		15.30	1.046		13.55	59.50		73.05	104
5286	4" x 4" x 1-1/2"		12.30	1.301		34.50	74		108.50	149
5287	4" x 4" x 2"		12.20	1.311		28.50	74.50		103	144
5288	4" x 4" x 3"		12.10	1.322		38.50	75		113.50	156
5291	6" x 6" x 4"		6.90	2.319		128	132		260	340
5294	Tee, double sanitary									
5295	1-1/2"	1 Plum	9.10	.879	Ea.	10.30	55.50		65.80	95
5296	2"	Q-1	16.60	.964		13.90	55		68.90	97.50
5297	3"		10.40	1.538		39	87.50		126.50	174
5298	4"		8.25	1.939		62.50	110		172.50	234
5303	Wye, reducing									
5304	2" x 1-1/2" x 1-1/2"	Q-1	23	.696	Ea.	11.70	39.50		51.20	72.50
5305	2" x 2" x 1-1/2"		22	.727		10.25	41.50		51.75	73.50
5306	3" x 3" x 2"		15.30	1.046		33	59.50		92.50	126
5307	4" x 4" x 2"		12.20	1.311		24.50	74.50		99	139
5309	4" x 4" x 3"		12.10	1.322		33	75		108	150
5314	Combination Y & 1/8 bend, 1-1/2"	1 Plum	12.10	.661		11.30	42		53.30	75
5315	2"	Q-1	20	.800		14.15	45.50		59.65	83.50
5317	3"		13.90	1.151		31	65.50		96.50	132
5318	4"		11	1.455		61.50	82.50		144	192
5319	6"		6.70	2.388		222	136		358	450
5320	8"	Q-2	6.20	3.871		320	228		548	695
5321	10"		5.70	4.211		890	248		1,138	1,350
5322	12"		5.10	4.706		1,200	277		1,477	1,750
5324	Combination Y & 1/8 bend, reducing									
5325	2" x 2" x 1-1/2"	Q-1	22	.727	Ea.	15.90	41.50		57.40	79.50
5327	3" x 3" x 1-1/2"		15.50	1.032		28.50	58.50		87	119
5328	3" x 3" x 2"		15.30	1.046		21	59.50		80.50	113
5329	4" x 4" x 2"		12.20	1.311		32	74.50		106.50	147

22 11 13.76 Pipe Fittings, Plastic		Crew	Daily Output	Labor-Hours	Unit	2019 Bare Costs Material	Labor	Equipment	Total	Total Incl O&P
5331	Wye, 1-1/4"	1 Plum	13.50	.593	Ea.	12.80	37.50		50.30	70
5332	1-1/2"	"	12.10	.661		8.50	42		50.50	72
5333	2"	Q-1	20	.800		8.35	45.50		53.85	77
5334	3"		13.90	1.151		22.50	65.50		88	123
5335	4"		11	1.455		41	82.50		123.50	169
5336	6"		6.70	2.388		119	136		255	335
5337	8"	Q-2	6.20	3.871		211	228		439	570
5338	10"		5.70	4.211		185	248		433	575
5339	12"		5.10	4.706		300	277		577	745
5341	2" x 1-1/2"	Q-1	22	.727		10.25	41.50		51.75	73.50
5342	3" x 1-1/2"		15.50	1.032		15.15	58.50		73.65	105
5343	4" x 3"		12.10	1.322		33	75		108	150
5344	6" x 4"		6.90	2.319		90	132		222	297
5345	8" x 6"	Q-2	6.40	3.750		197	221		418	545
5347	Double wye, 1-1/2"	1 Plum	9.10	.879		19.15	55.50		74.65	105
5348	2"	Q-1	16.60	.964		21.50	55		76.50	106
5349	3"		10.40	1.538		44.50	87.50		132	180
5350	4"		8.25	1.939		90	110		200	264
5353	Double wye, reducing									
5354	2" x 2" x 1-1/2" x 1-1/2"	Q-1	16.80	.952	Ea.	19.50	54		73.50	103
5355	3" x 3" x 2" x 2"		10.60	1.509		33	86		119	166
5356	4" x 4" x 3" x 3"		8.45	1.893		71.50	108		179.50	240
5357	6" x 6" x 4" x 4"		7.25	2.207		250	125		375	465
5374	Coupling, 1-1/4"	1 Plum	20.20	.396		6.05	25		31.05	44
5376	1-1/2"	"	18.20	.440		1.24	28		29.24	43
5378	2"	Q-1	33.10	.483		1.71	27.50		29.21	43
5380	3"		20.80	.769		5.95	43.50		49.45	72
5390	4"		16.50	.970		10.15	55		65.15	93.50
5400	6"		10.10	1.584		33.50	90		123.50	172
5402	8"	Q-2	9.30	2.581		56	152		208	290
5404	2" x 1-1/2"	Q-1	33.30	.480		3.82	27.50		31.32	45
5406	3" x 1-1/2"		21	.762		11.20	43.50		54.70	77.50
5408	4" x 3"		16.70	.958		18.30	54.50		72.80	102
5410	Reducer bushing, 2" x 1-1/4"		36.50	.438		3.50	25		28.50	41.50
5411	2" x 1-1/2"		36.40	.440		2.17	25		27.17	40
5412	3" x 1-1/2"		27.30	.586		10.50	33.50		44	61.50
5413	3" x 2"		27.10	.590		5.50	33.50		39	56.50
5414	4" x 2"		18.20	.879		18.25	50		68.25	95
5415	4" x 3"		16.70	.958		9.40	54.50		63.90	92
5416	6" x 4"		11.10	1.441		47	82		129	175
5418	8" x 6"	Q-2	10.20	2.353		92	139		231	310
5425	Closet flange 4"	Q-1	32	.500		14.80	28.50		43.30	59
5426	4" x 3"	"	34	.471		14.65	26.50		41.15	56
5450	Solvent cement for PVC, industrial grade, per quart				Qt.	30.50			30.50	33.50
5500	CPVC, Schedule 80, threaded joints									
5540	90° elbow, 1/4"	1 Plum	32	.250	Ea.	12.95	15.80		28.75	38
5560	1/2"		30.30	.264		7.75	16.65		24.40	33.50
5570	3/4"		26	.308		11.25	19.45		30.70	41.50
5580	1"		22.70	.352		15.80	22.50		38.30	51
5590	1-1/4"		20.20	.396		30.50	25		55.50	71
5600	1-1/2"		18.20	.440		32.50	28		60.50	77.50
5610	2"	Q-1	33.10	.483		44	27.50		71.50	89
5620	2-1/2"		24.20	.661		136	37.50		173.50	207

22 11 13.76	Pipe Fittings, Plastic	Crew	Daily Output	Labor-Hours	Unit	2019 Bare Costs Material	Labor	Equipment	Total	Total Incl O&P
5630	3"	Q-1	20.80	.769	Ea.	146	43.50		189.50	226
5640	4"		16.50	.970		228	55		283	335
5650	6"		10.10	1.584		264	90		354	425
5660	45° elbow same as 90° elbow									
5700	Tee, 1/4"	1 Plum	22	.364	Ea.	25	23		48	62
5702	1/2"		20.20	.396		25	25		50	65
5704	3/4"		17.30	.462		36	29		65	84
5706	1"		15.20	.526		39	33		72	93
5708	1-1/4"		13.50	.593		39	37.50		76.50	99
5710	1-1/2"		12.10	.661		41	42		83	108
5712	2"	Q-1	20	.800		46	45.50		91.50	119
5714	2-1/2"		16.20	.988		228	56		284	335
5716	3"		13.90	1.151		261	65.50		326.50	385
5718	4"		11	1.455		620	82.50		702.50	805
5720	6"		6.70	2.388		700	136		836	975
5730	Coupling, 1/4"	1 Plum	32	.250		16.50	15.80		32.30	41.50
5732	1/2"		30.30	.264		13.95	16.65		30.60	40.50
5734	3/4"		26	.308		22.50	19.45		41.95	53.50
5736	1"		22.70	.352		25	22.50		47.50	61
5738	1-1/4"		20.20	.396		26.50	25		51.50	66.50
5740	1-1/2"		18.20	.440		28.50	28		56.50	72.50
5742	2"	Q-1	33.10	.483		33.50	27.50		61	78
5744	2-1/2"		24.20	.661		60	37.50		97.50	123
5746	3"		20.80	.769		70	43.50		113.50	143
5748	4"		16.50	.970		142	55		197	240
5750	6"		10.10	1.584		194	90		284	350
5752	8"	Q-2	9.30	2.581		405	152		557	675
5900	CPVC, Schedule 80, socket joints									
5904	90° elbow, 1/4"	1 Plum	32	.250	Ea.	12.35	15.80		28.15	37
5906	1/2"		30.30	.264		4.83	16.65		21.48	30.50
5908	3/4"		26	.308		6.15	19.45		25.60	36
5910	1"		22.70	.352		9.80	22.50		32.30	44.50
5912	1-1/4"		20.20	.396		21	25		46	61
5914	1-1/2"		18.20	.440		23.50	28		51.50	67.50
5916	2"	Q-1	33.10	.483		28.50	27.50		56	72.50
5918	2-1/2"		24.20	.661		65.50	37.50		103	129
5920	3"		20.80	.769		74.50	43.50		118	148
5922	4"		16.50	.970		134	55		189	230
5924	6"		10.10	1.584		269	90		359	430
5926	8"		9.30	1.720		660	98		758	870
5930	45° elbow, 1/4"	1 Plum	32	.250		18.35	15.80		34.15	43.50
5932	1/2"		30.30	.264		5.90	16.65		22.55	31.50
5934	3/4"		26	.308		8.55	19.45		28	38.50
5936	1"		22.70	.352		13.60	22.50		36.10	48.50
5938	1-1/4"		20.20	.396		26.50	25		51.50	67
5940	1-1/2"		18.20	.440		27.50	28		55.50	71.50
5942	2"	Q-1	33.10	.483		30.50	27.50		58	74.50
5944	2-1/2"		24.20	.661		63	37.50		100.50	126
5946	3"		20.80	.769		80.50	43.50		124	154
5948	4"		16.50	.970		111	55		166	205
5950	6"		10.10	1.584		350	90		440	520
5952	8"		9.30	1.720		710	98		808	925
5960	Tee, 1/4"	1 Plum	22	.364		11.35	23		34.35	47

22 11 13.76 Pipe Fittings, Plastic		Crew	Daily Output	Labor-Hours	Unit	Material	Labor	Equipment	Total	Total Incl O&P
						2019 Bare Costs				
5962	1/2"	1 Plum	20.20	.396	Ea.	11.35	25		36.35	50
5964	3/4"		17.30	.462		11.55	29		40.55	56.50
5966	1"		15.20	.526		14.15	33		47.15	65.50
5968	1-1/4"		13.50	.593		30	37.50		67.50	89
5970	1-1/2"		12.10	.661		34	42		76	100
5972	2"	Q-1	20	.800		38	45.50		83.50	110
5974	2-1/2"		16.20	.988		96.50	56		152.50	190
5976	3"		13.90	1.151		96.50	65.50		162	204
5978	4"		11	1.455		129	82.50		211.50	266
5980	6"		6.70	2.388		335	136		471	575
5982	8"	Q-2	6.20	3.871		960	228		1,188	1,400
5990	Coupling, 1/4"	1 Plum	32	.250		13.15	15.80		28.95	38
5992	1/2"		30.30	.264		5.10	16.65		21.75	30.50
5994	3/4"		26	.308		7.15	19.45		26.60	37
5996	1"		22.70	.352		9.60	22.50		32.10	44
5998	1-1/4"		20.20	.396		14.40	25		39.40	53.50
6000	1-1/2"		18.20	.440		18.10	28		46.10	61.50
6002	2"	Q-1	33.10	.483		21	27.50		48.50	64.50
6004	2-1/2"		24.20	.661		47	37.50		84.50	108
6006	3"		20.80	.769		51	43.50		94.50	122
6008	4"		16.50	.970		66.50	55		121.50	156
6010	6"		10.10	1.584		157	90		247	310
6012	8"	Q-2	9.30	2.581		425	152		577	695
6200	CTS, 100 psi at 180°F, hot and cold water									
6230	90° elbow, 1/2"	1 Plum	20	.400	Ea.	.34	25.50		25.84	38.50
6250	3/4"		19	.421		.55	26.50		27.05	40.50
6251	1"		16	.500		2	31.50		33.50	49.50
6252	1-1/4"		15	.533		3.10	33.50		36.60	54
6253	1-1/2"		14	.571		5.60	36		41.60	60
6254	2"	Q-1	23	.696		10.80	39.50		50.30	71.50
6260	45° elbow, 1/2"	1 Plum	20	.400		.42	25.50		25.92	38.50
6280	3/4"		19	.421		.74	26.50		27.24	41
6281	1"		16	.500		1.78	31.50		33.28	49.50
6282	1-1/4"		15	.533		3.83	33.50		37.33	54.50
6283	1-1/2"		14	.571		5.55	36		41.55	60
6284	2"	Q-1	23	.696		12.05	39.50		51.55	73
6290	Tee, 1/2"	1 Plum	13	.615		.43	39		39.43	59
6310	3/4"		12	.667		.80	42		42.80	64
6311	1"		11	.727		3.90	46		49.90	73.50
6312	1-1/4"		10	.800		6.05	50.50		56.55	82.50
6313	1-1/2"		10	.800		7.80	50.50		58.30	84.50
6314	2"	Q-1	17	.941		12.60	53.50		66.10	94.50
6320	Coupling, 1/2"	1 Plum	22	.364		.27	23		23.27	35
6340	3/4"		21	.381		.36	24		24.36	36.50
6341	1"		18	.444		1.58	28		29.58	43.50
6342	1-1/4"		17	.471		2.04	29.50		31.54	46.50
6343	1-1/2"		16	.500		3	31.50		34.50	51
6344	2"	Q-1	28	.571		5.50	32.50		38	54.50
6360	Solvent cement for CPVC, commercial grade, per quart				Qt.	42.50			42.50	47
7340	PVC flange, slip-on, Sch 80 std., 1/2"	1 Plum	22	.364	Ea.	14.50	23		37.50	50.50
7350	3/4"		21	.381		15.45	24		39.45	53
7360	1"		18	.444		17.20	28		45.20	61
7370	1-1/4"		17	.471		17.75	29.50		47.25	64

22 11 13.76 Pipe Fittings, Plastic		Crew	Daily Output	Labor-Hours	Unit	Material	Labor	Equipment	Total	Total Incl O&P
						2019 Bare Costs				
7380	1-1/2"	1 Plum	16	.500	Ea.	18.10	31.50		49.60	67.50
7390	2"	Q-1	26	.615		24	35		59	79
7400	2-1/2"		24	.667		37.50	38		75.50	98
7410	3"		18	.889		41	50.50		91.50	122
7420	4"		15	1.067		52	60.50		112.50	149
7430	6"		10	1.600		82	91		173	226
7440	8"	Q-2	11	2.182		147	129		276	355
7550	Union, schedule 40, socket joints, 1/2"	1 Plum	19	.421		5.60	26.50		32.10	46
7560	3/4"		18	.444		5.75	28		33.75	48.50
7570	1"		15	.533		5.90	33.50		39.40	57
7580	1-1/4"		14	.571		17.75	36		53.75	73.50
7590	1-1/2"		13	.615		19.95	39		58.95	80.50
7600	2"	Q-1	20	.800		26.50	45.50		72	97.50
7992	Polybutyl/polyethyl pipe, for copper fittings see Line 22 11 13.25 7000									
8000	Compression type, PVC, 160 psi cold water									
8010	Coupling, 3/4" CTS	1 Plum	21	.381	Ea.	4.45	24		28.45	41
8020	1" CTS		18	.444		5.70	28		33.70	48.50
8030	1-1/4" CTS		17	.471		7.95	29.50		37.45	53.50
8040	1-1/2" CTS		16	.500		10.35	31.50		41.85	59
8050	2" CTS		15	.533		15.05	33.50		48.55	67
8060	Female adapter, 3/4" FPT x 3/4" CTS		23	.348		5.95	22		27.95	39.50
8070	3/4" FPT x 1" CTS		21	.381		6.90	24		30.90	43.50
8080	1" FPT x 1" CTS		20	.400		6.75	25.50		32.25	45.50
8090	1-1/4" FPT x 1-1/4" CTS		18	.444		8.85	28		36.85	51.50
8100	1-1/2" FPT x 1-1/2" CTS		16	.500		10.10	31.50		41.60	58.50
8110	2" FPT x 2" CTS		13	.615		15.05	39		54.05	75
8130	Male adapter, 3/4" MPT x 3/4" CTS		23	.348		4.96	22		26.96	38.50
8140	3/4" MPT x 1" CTS		21	.381		5.90	24		29.90	42.50
8150	1" MPT x 1" CTS		20	.400		5.75	25.50		31.25	44.50
8160	1-1/4" MPT x 1-1/4" CTS		18	.444		7.80	28		35.80	50.50
8170	1-1/2" MPT x 1-1/2" CTS		16	.500		9.35	31.50		40.85	58
8180	2" MPT x 2" CTS		13	.615		12.10	39		51.10	72
8200	Spigot adapter, 3/4" IPS x 3/4" CTS		23	.348		2.24	22		24.24	35.50
8210	3/4" IPS x 1" CTS		21	.381		2.72	24		26.72	39
8220	1" IPS x 1" CTS		20	.400		2.70	25.50		28.20	41
8230	1-1/4" IPS x 1-1/4" CTS		18	.444		4.07	28		32.07	46.50
8240	1-1/2" IPS x 1-1/2" CTS		16	.500		4.27	31.50		35.77	52
8250	2" IPS x 2" CTS		13	.615		5.30	39		44.30	64.50
8270	Price includes insert stiffeners									
8280	250 psi is same price as 160 psi									
8300	Insert type, nylon, 160 & 250 psi, cold water									
8310	Clamp ring stainless steel, 3/4" IPS	1 Plum	115	.070	Ea.	2.87	4.39		7.26	9.75
8320	1" IPS		107	.075		2.92	4.72		7.64	10.30
8330	1-1/4" IPS		101	.079		2.95	5		7.95	10.75
8340	1-1/2" IPS		95	.084		3.95	5.30		9.25	12.35
8350	2" IPS		85	.094		4.51	5.95		10.46	13.85
8370	Coupling, 3/4" IPS		22	.364		.94	23		23.94	35.50
8380	1" IPS		19	.421		.98	26.50		27.48	41
8390	1-1/4" IPS		18	.444		1.46	28		29.46	43.50
8400	1-1/2" IPS		17	.471		1.72	29.50		31.22	46.50
8410	2" IPS		16	.500		3.42	31.50		34.92	51.50
8430	Elbow, 90°, 3/4" IPS		22	.364		1.88	23		24.88	36.50
8440	1" IPS		19	.421		2.09	26.50		28.59	42.50

22 11 13.76 Pipe Fittings, Plastic		Crew	Daily Output	Labor-Hours	Unit	2019 Bare Costs Material	Labor	Equipment	Total	Total Incl O&P
8450	1-1/4" IPS	1 Plum	18	.444	Ea.	2.33	28		30.33	44.50
8460	1-1/2" IPS		17	.471		2.75	29.50		32.25	47.50
8470	2" IPS		16	.500		3.84	31.50		35.34	51.50
8490	Male adapter, 3/4" IPS x 3/4" MPT		25	.320		.94	20		20.94	31.50
8500	1" IPS x 1" MPT		21	.381		.98	24		24.98	37
8510	1-1/4" IPS x 1-1/4" MPT		20	.400		1.54	25.50		27.04	39.50
8520	1-1/2" IPS x 1-1/2" MPT		18	.444		1.72	28		29.72	44
8530	2" IPS x 2" MPT		15	.533		3.31	33.50		36.81	54
8550	Tee, 3/4" IPS		14	.571		1.82	36		37.82	56
8560	1" IPS		13	.615		2.38	39		41.38	61
8570	1-1/4" IPS		12	.667		3.71	42		45.71	67
8580	1-1/2" IPS		11	.727		4.22	46		50.22	73.50
8590	2" IPS	↓	10	.800	↓	8.30	50.50		58.80	85
8610	Insert type, PVC, 100 psi @ 180°F, hot & cold water									
8620	Coupler, male, 3/8" CTS x 3/8" MPT	1 Plum	29	.276	Ea.	.77	17.40		18.17	27
8630	3/8" CTS x 1/2" MPT		28	.286		.77	18.05		18.82	28
8640	1/2" CTS x 1/2" MPT		27	.296		.79	18.70		19.49	29
8650	1/2" CTS x 3/4" MPT		26	.308		2.44	19.45		21.89	31.50
8660	3/4" CTS x 1/2" MPT		25	.320		2.21	20		22.21	33
8670	3/4" CTS x 3/4" MPT		25	.320		.94	20		20.94	31.50
8700	Coupling, 3/8" CTS x 1/2" CTS		25	.320		4.75	20		24.75	36
8710	1/2" CTS		23	.348		5.70	22		27.70	39.50
8730	3/4" CTS		22	.364		12.05	23		35.05	48
8750	Elbow 90°, 3/8" CTS		25	.320		4.12	20		24.12	35
8760	1/2" CTS		23	.348		4.95	22		26.95	38.50
8770	3/4" CTS		22	.364		7.45	23		30.45	42.50
8800	Rings, crimp, copper, 3/8" CTS		120	.067		.18	4.21		4.39	6.50
8810	1/2" CTS		117	.068		.19	4.32		4.51	6.70
8820	3/4" CTS		115	.070		.24	4.39		4.63	6.85
8850	Reducer tee, bronze, 3/8" x 3/8" x 1/2" CTS		17	.471		5.55	29.50		35.05	50.50
8860	1/2" x 1/2" x 3/4" CTS		15	.533		7.10	33.50		40.60	58.50
8870	3/4" x 1/2" x 1/2" CTS		14	.571		7.10	36		43.10	62
8890	3/4" x 3/4" x 1/2" CTS		14	.571		7.20	36		43.20	62
8900	1" x 1/2" x 1/2" CTS		14	.571		11.85	36		47.85	67
8930	Tee, 3/8" CTS		17	.471		3.41	29.50		32.91	48.50
8940	1/2" CTS		15	.533		2.44	33.50		35.94	53
8950	3/4" CTS	↓	14	.571	↓	3.61	36		39.61	58
8960	Copper rings included in fitting price									
9000	Flare type, assembled, acetal, hot & cold water									
9010	Coupling, 1/4" & 3/8" CTS	1 Plum	24	.333	Ea.	3.46	21		24.46	35.50
9020	1/2" CTS		22	.364		3.96	23		26.96	39
9030	3/4" CTS		21	.381		5.85	24		29.85	42.50
9040	1" CTS		18	.444		7.45	28		35.45	50
9050	Elbow 90°, 1/4" CTS		26	.308		3.84	19.45		23.29	33
9060	3/8" CTS		24	.333		4.12	21		25.12	36
9070	1/2" CTS		22	.364		4.87	23		27.87	40
9080	3/4" CTS		21	.381		7.45	24		31.45	44
9090	1" CTS		18	.444		9.80	28		37.80	53
9110	Tee, 1/4" CTS		16	.500		4.21	31.50		35.71	52
9114	3/8" CTS		15	.533		4.26	33.50		37.76	55
9120	1/2" CTS		14	.571		5.40	36		41.40	60
9130	3/4" CTS		13	.615		8.50	39		47.50	68
9140	1" CTS	↓	12	.667	↓	11.40	42		53.40	75.50

	22 11 13.76 Pipe Fittings, Plastic	Crew	Daily Output	Labor-Hours	Unit	Material	2019 Bare Costs Labor	Equipment	Total	Total Incl O&P
9400	Polypropylene, fittings and accessories									
9404	Fittings fusion welded, sizes are ID									
9408	Note: sizes 1/2" thru 4" use socket fusion									
9410	Sizes 6" thru 10" use butt fusion									
9416	Coupling									
9420	3/8"	1 Plum	39	.205	Ea.	.89	12.95		13.84	20.50
9422	1/2"		37.40	.214		1.19	13.50		14.69	22
9424	3/4"		35.40	.226		1.32	14.25		15.57	23
9426	1"		29.70	.269		1.74	17		18.74	27.50
9428	1-1/4"		27.60	.290		2.08	18.30		20.38	30
9430	1-1/2"		24.80	.323		4.38	20.50		24.88	35.50
9432	2"	Q-1	43	.372		8.75	21		29.75	41
9434	2-1/2"		35.60	.449		9.80	25.50		35.30	49.50
9436	3"		30.90	.518		21.50	29.50		51	67.50
9438	3-1/2"		27.80	.576		35	32.50		67.50	87.50
9440	4"		25	.640		46	36.50		82.50	105
9442	Reducing coupling, female to female									
9446	2" to 1-1/2"	Q-1	49	.327	Ea.	12.60	18.55		31.15	42
9448	2-1/2" to 2"		41.20	.388		13.85	22		35.85	48.50
9450	3" to 2-1/2"		33.10	.483		18.70	27.50		46.20	61.50
9470	Reducing bushing, female to female									
9472	1/2" to 3/8"	1 Plum	38.20	.209	Ea.	1.19	13.20		14.39	21
9474	3/4" to 3/8" or 1/2"		36.50	.219		1.32	13.85		15.17	22.50
9476	1" to 3/4" or 1/2"		33.10	.242		1.76	15.25		17.01	25
9478	1-1/4" to 3/4" or 1"		28.70	.279		2.71	17.60		20.31	29.50
9480	1-1/2" to 1/2" thru 1-1/4"		26.20	.305		4.49	19.30		23.79	34
9482	2" to 1/2" thru 1-1/2"	Q-1	43	.372		8.95	21		29.95	41.50
9484	2-1/2" to 1/2" thru 2"		41.20	.388		10.05	22		32.05	44
9486	3" to 1-1/2" thru 2-1/2"		33.10	.483		22.50	27.50		50	65.50
9488	3-1/2" to 2" thru 3"		29.20	.548		35.50	31		66.50	86
9490	4" to 2-1/2" thru 3-1/2"		26.20	.611		56	34.50		90.50	114
9491	6" to 4" SDR 7.4		16.50	.970		67	55		122	156
9492	6" to 4" SDR 11		16.50	.970		67	55		122	156
9493	8" to 6" SDR 7.4		10.10	1.584		98.50	90		188.50	244
9494	8" to 6" SDR 11		10.10	1.584		63	90		153	204
9495	10" to 8" SDR 7.4		7.80	2.051		135	117		252	325
9496	10" to 8" SDR 11		7.80	2.051		93	117		210	277
9500	90° elbow									
9504	3/8"	1 Plum	39	.205	Ea.	1.27	12.95		14.22	21
9506	1/2"		37.40	.214		1.27	13.50		14.77	22
9508	3/4"		35.40	.226		1.63	14.25		15.88	23.50
9510	1"		29.70	.269		2.35	17		19.35	28
9512	1-1/4"		27.60	.290		3.62	18.30		21.92	31.50
9514	1-1/2"		24.80	.323		7.80	20.50		28.30	39
9516	2"	Q-1	43	.372		11.95	21		32.95	44.50
9518	2-1/2"		35.60	.449		26.50	25.50		52	67.50
9520	3"		30.90	.518		44	29.50		73.50	92.50
9522	3-1/2"		27.80	.576		62.50	32.50		95	118
9524	4"		25	.640		96.50	36.50		133	161
9526	6" SDR 7.4		5.55	2.883		110	164		274	365
9528	6" SDR 11		5.55	2.883		86	164		250	340
9530	8" SDR 7.4	Q-2	8.10	2.963		258	175		433	545
9532	8" SDR 11		8.10	2.963		231	175		406	515

22 11 13.76 Pipe Fittings, Plastic		Crew	Daily Output	Labor-Hours	Unit	2019 Bare Costs Material	Labor	Equipment	Total	Total Incl O&P
9534	10″ SDR 7.4	Q-2	7.50	3.200	Ea.	395	189		584	715
9536	10″ SDR 11		7.50	3.200		355	189		544	675
9551	45° elbow									
9554	3/8″	1 Plum	39	.205	Ea.	1.27	12.95		14.22	21
9556	1/2″		37.40	.214		1.27	13.50		14.77	22
9558	3/4″		35.40	.226		1.63	14.25		15.88	23.50
9564	1″		29.70	.269		2.35	17		19.35	28
9566	1-1/4″		27.60	.290		3.62	18.30		21.92	31.50
9568	1-1/2″		24.80	.323		7.80	20.50		28.30	39
9570	2″	Q-1	43	.372		11.80	21		32.80	44.50
9572	2-1/2″		35.60	.449		26	25.50		51.50	67.50
9574	3″		30.90	.518		48.50	29.50		78	97
9576	3-1/2″		27.80	.576		69	32.50		101.50	125
9578	4″		25	.640		106	36.50		142.50	172
9580	6″ SDR 7.4		5.55	2.883		122	164		286	380
9582	6″ SDR 11		5.55	2.883		101	164		265	355
9584	8″ SDR 7.4	Q-2	8.10	2.963		252	175		427	540
9586	8″ SDR 11		8.10	2.963		220	175		395	505
9588	10″ SDR 7.4		7.50	3.200		405	189		594	730
9590	10″ SDR 11		7.50	3.200		335	189		524	655
9600	Tee									
9604	3/8″	1 Plum	26	.308	Ea.	1.71	19.45		21.16	31
9606	1/2″		24.90	.321		2.35	20.50		22.85	33
9608	3/4″		23.70	.338		2.35	21.50		23.85	34.50
9610	1″		19.90	.402		2.96	25.50		28.46	41.50
9612	1-1/4″		18.50	.432		4.56	27.50		32.06	46
9614	1-1/2″		16.60	.482		13	30.50		43.50	60
9616	2″	Q-1	26.80	.597		17.50	34		51.50	70.50
9618	2-1/2″		23.80	.672		28.50	38		66.50	89
9620	3″		20.60	.777		57.50	44		101.50	129
9622	3-1/2″		18.40	.870		89.50	49.50		139	173
9624	4″		16.70	.958		107	54.50		161.50	200
9626	6″ SDR 7.4		3.70	4.324		133	246		379	515
9628	6″ SDR 11		3.70	4.324		174	246		420	560
9630	8″ SDR 7.4	Q-2	5.40	4.444		365	262		627	795
9632	8″ SDR 11		5.40	4.444		320	262		582	750
9634	10″ SDR 7.4		5	4.800		630	283		913	1,125
9636	10″ SDR 11		5	4.800		545	283		828	1,025
9638	For reducing tee use same tee price									
9660	End cap									
9662	3/8″	1 Plum	78	.103	Ea.	1.85	6.50		8.35	11.75
9664	1/2″		74.60	.107		1.85	6.75		8.60	12.20
9666	3/4″		71.40	.112		2.35	7.10		9.45	13.20
9668	1″		59.50	.134		2.84	8.50		11.34	15.85
9670	1-1/4″		55.60	.144		4.49	9.10		13.59	18.60
9672	1-1/2″		49.50	.162		6.20	10.20		16.40	22
9674	2″	Q-1	80	.200		10.35	11.35		21.70	28.50
9676	2-1/2″		71.40	.224		15	12.75		27.75	35.50
9678	3″		61.70	.259		34	14.75		48.75	59.50
9680	3-1/2″		55.20	.290		41	16.45		57.45	69.50
9682	4″		50	.320		62	18.20		80.20	96
9684	6″ SDR 7.4		26.50	.604		85.50	34.50		120	146
9686	6″ SDR 11		26.50	.604		66	34.50		100.50	124

22 11 13.76 Pipe Fittings, Plastic		Crew	Daily Output	Labor-Hours	Unit	Material	Labor	Equipment	Total	Total Incl O&P
						2019 Bare Costs				
9688	8″ SDR 7.4	Q-2	16.30	1.472	Ea.	85.50	87		172.50	224
9690	8″ SDR 11		16.30	1.472		75.50	87		162.50	213
9692	10″ SDR 7.4		14.90	1.611		128	95		223	283
9694	10″ SDR 11	↓	14.90	1.611	↓	87.50	95		182.50	239
9800	Accessories and tools									
9802	Pipe clamps for suspension, not including rod or beam clamp									
9804	3/8″	1 Plum	74	.108	Ea.	2.19	6.85		9.04	12.65
9805	1/2″		70	.114		2.75	7.20		9.95	13.90
9806	3/4″		68	.118		3.16	7.45		10.61	14.65
9807	1″		66	.121		3.45	7.65		11.10	15.30
9808	1-1/4″		64	.125		3.54	7.90		11.44	15.75
9809	1-1/2″	↓	62	.129		3.87	8.15		12.02	16.50
9810	2″	Q-1	110	.145		4.80	8.25		13.05	17.70
9811	2-1/2″		104	.154		6.15	8.75		14.90	19.85
9812	3″		98	.163		6.65	9.30		15.95	21
9813	3-1/2″		92	.174		7.25	9.90		17.15	23
9814	4″		86	.186		8.15	10.55		18.70	25
9815	6″	↓	70	.229		9.95	13		22.95	30.50
9816	8″	Q-2	100	.240		35.50	14.15		49.65	60
9817	10″	″	94	.255	↓	40.50	15.05		55.55	67
9820	Pipe cutter									
9822	For 3/8″ thru 1-1/4″				Ea.	110			110	121
9824	For 1-1/2″ thru 4″				″	288			288	315
9826	Note: Pipes may be cut with standard									
9827	iron saw with blades for plastic.									
9982	For plastic hangers see Line 22 05 29.10 8000									
9986	For copper/brass fittings see Line 22 11 13.25 7000									

22 11 13.78 Pipe, High Density Polyethylene Plastic (HDPE)		Crew	Daily Output	Labor-Hours	Unit	Material	Labor	Equipment	Total	Total Incl O&P
0010	**PIPE, HIGH DENSITY POLYETHYLENE PLASTIC (HDPE)**									
0020	Not incl. hangers, trenching, backfill, hoisting or digging equipment.									
0030	Standard length is 40′, add a weld for each joint									
0040	Single wall									
0050	Straight									
0054	1″ diameter DR 11				L.F.	.80			.80	.88
0058	1-1/2″ diameter DR 11					1.02			1.02	1.12
0062	2″ diameter DR 11					1.70			1.70	1.87
0066	3″ diameter DR 11					2.05			2.05	2.26
0070	3″ diameter DR 17					1.64			1.64	1.80
0074	4″ diameter DR 11					3.44			3.44	3.78
0078	4″ diameter DR 17					3.44			3.44	3.78
0082	6″ diameter DR 11					8.55			8.55	9.40
0086	6″ diameter DR 17					5.65			5.65	6.20
0090	8″ diameter DR 11					14.20			14.20	15.60
0094	8″ diameter DR 26					6.65			6.65	7.30
0098	10″ diameter DR 11					22			22	24.50
0102	10″ diameter DR 26					10.25			10.25	11.30
0106	12″ diameter DR 11					32.50			32.50	35.50
0110	12″ diameter DR 26					15.40			15.40	16.95
0114	16″ diameter DR 11					49.50			49.50	54.50
0118	16″ diameter DR 26					22			22	24.50
0122	18″ diameter DR 11					63.50			63.50	69.50
0126	18″ diameter DR 26				↓	29			29	32

22 11 13 – Facility Water Distribution Piping

22 11 13.78 Pipe, High Density Polyethylene Plastic (HDPE)		Crew	Daily Output	Labor-Hours	Unit	2019 Bare Costs Material	Labor	Equipment	Total	Total Incl O&P
0130	20" diameter DR 11				L.F.	77			77	84.50
0134	20" diameter DR 26					34			34	37.50
0138	22" diameter DR 11					94			94	104
0142	22" diameter DR 26					43			43	47
0146	24" diameter DR 11					111			111	122
0150	24" diameter DR 26					49.50			49.50	54.50
0154	28" diameter DR 17					103			103	113
0158	28" diameter DR 26					68.50			68.50	75.50
0162	30" diameter DR 21					96			96	105
0166	30" diameter DR 26					78.50			78.50	86.50
0170	36" diameter DR 26					78.50			78.50	86.50
0174	42" diameter DR 26					152			152	167
0178	48" diameter DR 26					200			200	220
0182	54" diameter DR 26				↓	251			251	277
0300	90° elbow									
0304	1" diameter DR 11				Ea.	5.10			5.10	5.60
0308	1-1/2" diameter DR 11					7.20			7.20	7.90
0312	2" diameter DR 11					7.20			7.20	7.90
0316	3" diameter DR 11					14.35			14.35	15.80
0320	3" diameter DR 17					14.35			14.35	15.80
0324	4" diameter DR 11					20			20	22
0328	4" diameter DR 17					20			20	22
0332	6" diameter DR 11					46			46	50.50
0336	6" diameter DR 17					46			46	50.50
0340	8" diameter DR 11					114			114	125
0344	8" diameter DR 26					101			101	111
0348	10" diameter DR 11					425			425	465
0352	10" diameter DR 26					390			390	430
0356	12" diameter DR 11					450			450	495
0360	12" diameter DR 26					405			405	445
0364	16" diameter DR 11					530			530	585
0368	16" diameter DR 26					525			525	575
0372	18" diameter DR 11					665			665	730
0376	18" diameter DR 26					625			625	690
0380	20" diameter DR 11					785			785	860
0384	20" diameter DR 26					760			760	835
0388	22" diameter DR 11					810			810	890
0392	22" diameter DR 26					785			785	860
0396	24" diameter DR 11					915			915	1,000
0400	24" diameter DR 26					890			890	975
0404	28" diameter DR 17					1,100			1,100	1,225
0408	28" diameter DR 26					1,050			1,050	1,150
0412	30" diameter DR 17					1,575			1,575	1,725
0416	30" diameter DR 26					1,425			1,425	1,575
0420	36" diameter DR 26					1,825			1,825	2,000
0424	42" diameter DR 26					2,350			2,350	2,575
0428	48" diameter DR 26					2,750			2,750	3,025
0432	54" diameter DR 26				↓	6,525			6,525	7,175
0500	45° elbow									
0512	2" diameter DR 11				Ea.	5.75			5.75	6.35
0516	3" diameter DR 11					14.35			14.35	15.80
0520	3" diameter DR 17					14.35			14.35	15.80
0524	4" diameter DR 11				↓	20			20	22

22 11 13 – Facility Water Distribution Piping

22 11 13.78 Pipe, High Density Polyethylene Plastic (HDPE)		Crew	Daily Output	Labor-Hours	Unit	Material	Labor	Equipment	Total	Total Incl O&P
0528	4" diameter DR 17				Ea.	20			20	22
0532	6" diameter DR 11					46			46	50.50
0536	6" diameter DR 17					46			46	50.50
0540	8" diameter DR 11					114			114	125
0544	8" diameter DR 26					66			66	72.50
0548	10" diameter DR 11					425			425	465
0552	10" diameter DR 26					390			390	430
0556	12" diameter DR 11					450			450	495
0560	12" diameter DR 26					405			405	445
0564	16" diameter DR 11					235			235	259
0568	16" diameter DR 26					222			222	244
0572	18" diameter DR 11					250			250	275
0576	18" diameter DR 26					229			229	252
0580	20" diameter DR 11					380			380	415
0584	20" diameter DR 26					360			360	395
0588	22" diameter DR 11					525			525	575
0592	22" diameter DR 26					495			495	545
0596	24" diameter DR 11					640			640	705
0600	24" diameter DR 26					615			615	675
0604	28" diameter DR 17					725			725	800
0608	28" diameter DR 26					705			705	775
0612	30" diameter DR 17					900			900	990
0616	30" diameter DR 26					875			875	965
0620	36" diameter DR 26					1,125			1,125	1,225
0624	42" diameter DR 26					1,425			1,425	1,575
0628	48" diameter DR 26					1,550			1,550	1,700
0632	54" diameter DR 26				▼	2,075			2,075	2,275
0700	Tee									
0704	1" diameter DR 11				Ea.	7.50			7.50	8.25
0708	1-1/2" diameter DR 11					10.50			10.50	11.55
0712	2" diameter DR 11					9			9	9.90
0716	3" diameter DR 11					16.50			16.50	18.15
0720	3" diameter DR 17					16.50			16.50	18.15
0724	4" diameter DR 11					24			24	26.50
0728	4" diameter DR 17					24			24	26.50
0732	6" diameter DR 11					60			60	66
0736	6" diameter DR 17					60			60	66
0740	8" diameter DR 11					149			149	164
0744	8" diameter DR 17					149			149	164
0748	10" diameter DR 11					445			445	485
0752	10" diameter DR 17					445			445	485
0756	12" diameter DR 11					590			590	650
0760	12" diameter DR 17					590			590	650
0764	16" diameter DR 11					310			310	345
0768	16" diameter DR 17					257			257	283
0772	18" diameter DR 11					440			440	485
0776	18" diameter DR 17					360			360	400
0780	20" diameter DR 11					540			540	590
0784	20" diameter DR 17					440			440	485
0788	22" diameter DR 11					690			690	760
0792	22" diameter DR 17					545			545	600
0796	24" diameter DR 11					855			855	940
0800	24" diameter DR 17				▼	720			720	795

Labor, Equipment and Total columns are under "2019 Bare Costs".

	22 11 13.78 Pipe, High Density Polyethylene Plastic (HDPE)	Crew	Daily Output	Labor-Hours	Unit	2019 Bare Costs Material	Labor	Equipment	Total	Total Incl O&P
0804	28" diameter DR 17				Ea.	1,400			1,400	1,525
0812	30" diameter DR 17					1,600			1,600	1,750
0820	36" diameter DR 17					2,650			2,650	2,900
0824	42" diameter DR 26					2,950			2,950	3,225
0828	48" diameter DR 26				↓	3,150			3,150	3,475
1000	Flange adptr, w/back-up ring and 1/2 cost of plated bolt set									
1004	1" diameter DR 11				Ea.	28.50			28.50	31.50
1008	1-1/2" diameter DR 11					28.50			28.50	31.50
1012	2" diameter DR 11					18.05			18.05	19.85
1016	3" diameter DR 11					21			21	23
1020	3" diameter DR 17					21			21	23
1024	4" diameter DR 11					28.50			28.50	31.50
1028	4" diameter DR 17					28.50			28.50	31.50
1032	6" diameter DR 11					40.50			40.50	44.50
1036	6" diameter DR 17					40.50			40.50	44.50
1040	8" diameter DR 11					58.50			58.50	64.50
1044	8" diameter DR 26					58.50			58.50	64.50
1048	10" diameter DR 11					93			93	102
1052	10" diameter DR 26					93			93	102
1056	12" diameter DR 11					137			137	150
1060	12" diameter DR 26					137			137	150
1064	16" diameter DR 11					293			293	320
1068	16" diameter DR 26					293			293	320
1072	18" diameter DR 11					380			380	420
1076	18" diameter DR 26					380			380	420
1080	20" diameter DR 11					525			525	580
1084	20" diameter DR 26					525			525	580
1088	22" diameter DR 11					570			570	630
1092	22" diameter DR 26					570			570	630
1096	24" diameter DR 17					620			620	685
1100	24" diameter DR 32.5					620			620	685
1104	28" diameter DR 15.5					840			840	925
1108	28" diameter DR 32.5					840			840	925
1112	30" diameter DR 11					975			975	1,075
1116	30" diameter DR 21					975			975	1,075
1120	36" diameter DR 26					1,075			1,075	1,175
1124	42" diameter DR 26					1,225			1,225	1,350
1128	48" diameter DR 26					1,500			1,500	1,625
1132	54" diameter DR 26				↓	1,800			1,800	1,975
1200	Reducer									
1208	2" x 1-1/2" diameter DR 11				Ea.	8.65			8.65	9.50
1212	3" x 2" diameter DR 11					8.65			8.65	9.50
1216	4" x 2" diameter DR 11					10.05			10.05	11.05
1220	4" x 3" diameter DR 11					12.95			12.95	14.25
1224	6" x 4" diameter DR 11					30			30	33
1228	8" x 6" diameter DR 11					46			46	50.50
1232	10" x 8" diameter DR 11					79			79	87
1236	12" x 8" diameter DR 11					129			129	142
1240	12" x 10" diameter DR 11					103			103	114
1244	14" x 12" diameter DR 11					115			115	126
1248	16" x 14" diameter DR 11					147			147	161
1252	18" x 16" diameter DR 11					178			178	196
1256	20" x 18" diameter DR 11				↓	355			355	390

22 11 13.78 Pipe, High Density Polyethylene Plastic (HDPE)		Crew	Daily Output	Labor-Hours	Unit	Material	2019 Bare Costs Labor	Equipment	Total	Total Incl O&P
1260	22" x 20" diameter DR 11				Ea.	435			435	475
1264	24" x 22" diameter DR 11					490			490	540
1268	26" x 24" diameter DR 11					575			575	630
1272	28" x 24" diameter DR 11					735			735	805
1276	32" x 28" diameter DR 17					950			950	1,050
1280	36" x 32" diameter DR 17					1,300			1,300	1,425
4000	Welding labor per joint, not including welding machine									
4010	Pipe joint size (cost based on thickest wall for each diam.)									
4030	1" pipe size	4 Skwk	273	.117	Ea.		6.25		6.25	9.50
4040	1-1/2" pipe size		175	.183			9.75		9.75	14.85
4050	2" pipe size		128	.250			13.35		13.35	20.50
4060	3" pipe size		100	.320			17.10		17.10	26
4070	4" pipe size		77	.416			22		22	34
4080	6" pipe size	5 Skwk	63	.635			34		34	51.50
4090	8" pipe size		48	.833			44.50		44.50	67.50
4100	10" pipe size		40	1			53.50		53.50	81.50
4110	12" pipe size	6 Skwk	41	1.171			62.50		62.50	95
4120	16" pipe size		34	1.412			75.50		75.50	115
4130	18" pipe size		32	1.500			80		80	122
4140	20" pipe size	8 Skwk	37	1.730			92.50		92.50	141
4150	22" pipe size		35	1.829			97.50		97.50	149
4160	24" pipe size		34	1.882			101		101	153
4170	28" pipe size		33	1.939			104		104	158
4180	30" pipe size		32	2			107		107	163
4190	36" pipe size		31	2.065			110		110	168
4200	42" pipe size		30	2.133			114		114	173
4210	48" pipe size	9 Skwk	33	2.182			117		117	177
4220	54" pipe size	"	31	2.323			124		124	189
4300	Note: Cost for set up each time welder is moved.									
4301	Add 50% of a weld cost									
4310	Welder usually remains stationary with pipe moved through it.									
4340	Weld machine, rental per day based on diam. capacity									
4350	1" thru 2" diameter				Ea.			40.50	40.50	44.50
4360	3" thru 4" diameter							46	46	50.50
4370	6" thru 8" diameter							103	103	113
4380	10" thru 12" diameter							178	178	196
4390	16" thru 18" diameter							259	259	285
4400	20" thru 24" diameter							500	500	550
4410	28" thru 32" diameter							545	545	600
4420	36" diameter							570	570	625
4430	42" thru 54" diameter							890	890	980
5000	Dual wall contained pipe									
5040	Straight									
5054	1" DR 11 x 3" DR 11				L.F.	5.35			5.35	5.85
5058	1" DR 11 x 4" DR 11					5.60			5.60	6.20
5062	1-1/2" DR 11 x 4" DR 17					6.25			6.25	6.90
5066	2" DR 11 x 4" DR 17					6.60			6.60	7.25
5070	2" DR 11 x 6" DR 17					9.95			9.95	10.95
5074	3" DR 11 x 6" DR 17					11.20			11.20	12.30
5078	3" DR 11 x 6" DR 26					9.55			9.55	10.50
5086	3" DR 17 x 8" DR 17					11.95			11.95	13.15
5090	4" DR 11 x 8" DR 17					16.90			16.90	18.60
5094	4" DR 17 x 8" DR 26					13.20			13.20	14.50

22 11 13.78 Pipe, High Density Polyethylene Plastic (HDPE)		Crew	Daily Output	Labor-Hours	Unit	2019 Bare Costs Material	Labor	Equipment	Total	Total Incl O&P
5098	6" DR 11 x 10" DR 17				L.F.	26			26	29
5102	6" DR 17 x 10" DR 26					17.65			17.65	19.40
5106	6" DR 26 x 10" DR 26					16.50			16.50	18.15
5110	8" DR 17 x 12" DR 26					25.50			25.50	28
5114	8" DR 26 x 12" DR 32.5					23.50			23.50	26
5118	10" DR 17 x 14" DR 26					38.50			38.50	42
5122	10" DR 17 x 16" DR 26					28.50			28.50	31.50
5126	10" DR 26 x 16" DR 26					36.50			36.50	40
5130	12" DR 26 x 16" DR 26					38			38	42
5134	12" DR 17 x 18" DR 26					53			53	58.50
5138	12" DR 26 x 18" DR 26					51			51	56
5142	14" DR 26 x 20" DR 32.5					51.50			51.50	56.50
5146	16" DR 26 x 22" DR 32.5					59.50			59.50	65.50
5150	18" DR 26 x 24" DR 32.5					70.50			70.50	78
5154	20" DR 32.5 x 28" DR 32.5					78.50			78.50	86.50
5158	22" DR 32.5 x 30" DR 32.5					78.50			78.50	86.50
5162	24" DR 32.5 x 32" DR 32.5					87			87	96
5166	36" DR 32.5 x 42" DR 32.5					102			102	112
5300	Force transfer coupling									
5354	1" DR 11 x 3" DR 11				Ea.	183			183	201
5358	1" DR 11 x 4" DR 17					206			206	227
5362	1-1/2" DR 11 x 4" DR 17					206			206	227
5366	2" DR 11 x 4" DR 17					206			206	227
5370	2" DR 11 x 6" DR 17					340			340	375
5374	3" DR 11 x 6" DR 17					365			365	405
5378	3" DR 11 x 6" DR 26					365			365	405
5382	3" DR 11 x 8" DR 11					365			365	400
5386	3" DR 11 x 8" DR 17					405			405	445
5390	4" DR 11 x 8" DR 17					415			415	460
5394	4" DR 17 x 8" DR 26					425			425	470
5398	6" DR 11 x 10" DR 17					490			490	540
5402	6" DR 17 x 10" DR 26					480			480	525
5406	6" DR 26 x 10" DR 26					670			670	740
5410	8" DR 17 x 12" DR 26					700			700	770
5414	8" DR 26 x 12" DR 32.5					830			830	910
5418	10" DR 17 x 14" DR 26					960			960	1,050
5422	10" DR 17 x 16" DR 26					970			970	1,075
5426	10" DR 26 x 16" DR 26					1,100			1,100	1,200
5430	12" DR 26 x 16" DR 26					1,325			1,325	1,450
5434	12" DR 17 x 18" DR 26					1,325			1,325	1,450
5438	12" DR 26 x 18" DR 26					1,475			1,475	1,625
5442	14" DR 26 x 20" DR 32.5					1,625			1,625	1,800
5446	16" DR 26 x 22" DR 32.5					1,800			1,800	2,000
5450	18" DR 26 x 24" DR 32.5					2,000			2,000	2,200
5454	20" DR 32.5 x 28" DR 32.5					2,175			2,175	2,400
5458	22" DR 32.5 x 30" DR 32.5					2,400			2,400	2,650
5462	24" DR 32.5 x 32" DR 32.5					2,675			2,675	2,925
5466	36" DR 32.5 x 42" DR 32.5					2,750			2,750	3,025
5600	90° elbow									
5654	1" DR 11 x 3" DR 11				Ea.	196			196	215
5658	1" DR 11 x 4" DR 17					238			238	262
5662	1-1/2" DR 11 x 4" DR 17					243			243	268
5666	2" DR 11 x 4" DR 17					236			236	260

22 11 13.78 Pipe, High Density Polyethylene Plastic (HDPE)		Crew	Daily Output	Labor-Hours	Unit	2019 Bare Costs Material	Labor	Equipment	Total	Total Incl O&P
5670	2" DR 11 x 6" DR 17				Ea.	345			345	380
5674	3" DR 11 x 6" DR 17					370			370	410
5678	3" DR 17 x 6" DR 26					385			385	425
5682	3" DR 17 x 8" DR 11					495			495	545
5686	3" DR 17 x 8" DR 17					545			545	600
5690	4" DR 11 x 8" DR 17					445			445	490
5694	4" DR 17 x 8" DR 26					440			440	485
5698	6" DR 11 x 10" DR 17					805			805	890
5702	6" DR 17 x 10" DR 26					835			835	920
5706	6" DR 26 x 10" DR 26					820			820	900
5710	8" DR 17 x 12" DR 26					1,075			1,075	1,175
5714	8" DR 26 x 12" DR 32.5					890			890	980
5718	10" DR 17 x 14" DR 26					1,000			1,000	1,125
5722	10" DR 17 x 16" DR 26					985			985	1,075
5726	10" DR 26 x 16" DR 26					995			995	1,100
5730	12" DR 26 x 16" DR 26					1,000			1,000	1,100
5734	12" DR 17 x 18" DR 26					1,075			1,075	1,200
5738	12" DR 26 x 18" DR 26					1,125			1,125	1,225
5742	14" DR 26 x 20" DR 32.5					1,225			1,225	1,350
5746	16" DR 26 x 22" DR 32.5					1,275			1,275	1,400
5750	18" DR 26 x 24" DR 32.5					1,325			1,325	1,450
5754	20" DR 32.5 x 28" DR 32.5					1,375			1,375	1,525
5758	22" DR 32.5 x 30" DR 32.5					1,425			1,425	1,575
5762	24" DR 32.5 x 32" DR 32.5					1,475			1,475	1,625
5766	36" DR 32.5 x 42" DR 32.5				▼	1,525			1,525	1,675
5800	45° elbow									
5804	1" DR 11 x 3" DR 11				Ea.	189			189	207
5808	1" DR 11 x 4" DR 17					195			195	214
5812	1-1/2" DR 11 x 4" DR 17					198			198	217
5816	2" DR 11 x 4" DR 17					187			187	206
5820	2" DR 11 x 6" DR 17					271			271	298
5824	3" DR 11 x 6" DR 17					289			289	320
5828	3" DR 17 x 6" DR 26					296			296	325
5832	3" DR 17 x 8" DR 11					365			365	405
5836	3" DR 17 x 8" DR 17					420			420	465
5840	4" DR 11 x 8" DR 17					340			340	375
5844	4" DR 17 x 8" DR 26					350			350	385
5848	6" DR 11 x 10" DR 17					540			540	595
5852	6" DR 17 x 10" DR 26					555			555	615
5856	6" DR 26 x 10" DR 26					635			635	695
5860	8" DR 17 x 12" DR 26					740			740	815
5864	8" DR 26 x 12" DR 32.5					695			695	765
5868	10" DR 17 x 14" DR 26					760			760	835
5872	10" DR 17 x 16" DR 26					730			730	805
5876	10" DR 26 x 16" DR 26					785			785	860
5880	12" DR 26 x 16" DR 26					690			690	760
5884	12" DR 17 x 18" DR 26					1,000			1,000	1,100
5888	12" DR 26 x 18" DR 26					885			885	975
5892	14" DR 26 x 20" DR 32.5					940			940	1,025
5896	16" DR 26 x 22" DR 32.5					975			975	1,075
5900	18" DR 26 x 24" DR 32.5					1,025			1,025	1,125
5904	20" DR 32.5 x 28" DR 32.5					1,050			1,050	1,150
5908	22" DR 32.5 x 30" DR 32.5				▼	1,100			1,100	1,200

22 11 13.78 Pipe, High Density Polyethylene Plastic (HDPE)		Crew	Daily Output	Labor-Hours	Unit	Material	2019 Bare Costs Labor	Equipment	Total	Total Incl O&P
5912	24" DR 32.5 x 32" DR 32.5				Ea.	1,125			1,125	1,250
5916	36" DR 32.5 x 42" DR 32.5				↓	1,175			1,175	1,275
6000	Access port with 4" riser									
6050	1" DR 11 x 4" DR 17				Ea.	274			274	300
6054	1-1/2" DR 11 x 4" DR 17					276			276	305
6058	2" DR 11 x 6" DR 17					335			335	370
6062	3" DR 11 x 6" DR 17					340			340	375
6066	3" DR 17 x 6" DR 26					330			330	365
6070	3" DR 17 x 8" DR 11					340			340	370
6074	3" DR 17 x 8" DR 17					355			355	390
6078	4" DR 11 x 8" DR 17					365			365	400
6082	4" DR 17 x 8" DR 26					350			350	385
6086	6" DR 11 x 10" DR 17					420			420	460
6090	6" DR 17 x 10" DR 26					380			380	420
6094	6" DR 26 x 10" DR 26					405			405	445
6098	8" DR 17 x 12" DR 26					435			435	480
6102	8" DR 26 x 12" DR 32.5				↓	430			430	475
6200	End termination with vent plug									
6204	1" DR 11 x 3" DR 11				Ea.	105			105	115
6208	1" DR 11 x 4" DR 17					137			137	151
6212	1-1/2" DR 11 x 4" DR 17					137			137	151
6216	2" DR 11 x 4" DR 17					137			137	151
6220	2" DR 11 x 6" DR 17					272			272	299
6224	3" DR 11 x 6" DR 17					272			272	299
6228	3" DR 17 x 6" DR 26					272			272	299
6232	3" DR 17 x 8" DR 11					269			269	296
6236	3" DR 17 x 8" DR 17					300			300	330
6240	4" DR 11 x 8" DR 17					345			345	380
6244	4" DR 17 x 8" DR 26					330			330	365
6248	6" DR 11 x 10" DR 17					375			375	410
6252	6" DR 17 x 10" DR 26					355			355	390
6256	6" DR 26 x 10" DR 26					535			535	585
6260	8" DR 17 x 12" DR 26					545			545	595
6264	8" DR 26 x 12" DR 32.5					675			675	740
6268	10" DR 17 x 14" DR 26					895			895	980
6272	10" DR 17 x 16" DR 26					840			840	925
6276	10" DR 26 x 16" DR 26					955			955	1,050
6280	12" DR 26 x 16" DR 26					1,050			1,050	1,175
6284	12" DR 17 x 18" DR 26					1,225			1,225	1,350
6288	12" DR 26 x 18" DR 26					1,350			1,350	1,475
6292	14" DR 26 x 20" DR 32.5					1,525			1,525	1,675
6296	16" DR 26 x 22" DR 32.5					1,700			1,700	1,875
6300	18" DR 26 x 24" DR 32.5					1,925			1,925	2,100
6304	20" DR 32.5 x 28" DR 32.5					2,125			2,125	2,325
6308	22" DR 32.5 x 30" DR 32.5					2,375			2,375	2,600
6312	24" DR 32.5 x 32" DR 32.5					2,650			2,650	2,925
6316	36" DR 32.5 x 42" DR 32.5				↓	2,875			2,875	3,175
6600	Tee									
6604	1" DR 11 x 3" DR 11				Ea.	228			228	251
6608	1" DR 11 x 4" DR 17					246			246	270
6612	1-1/2" DR 11 x 4" DR 17					251			251	276
6616	2" DR 11 x 4" DR 17					238			238	261
6620	2" DR 11 x 6" DR 17				↓	375			375	410

22 11 13.78 Pipe, High Density Polyethylene Plastic (HDPE)		Crew	Daily Output	Labor-Hours	Unit	2019 Bare Costs Material	Labor	Equipment	Total	Total Incl O&P
6624	3" DR 11 x 6" DR 17				Ea.	440			440	480
6628	3" DR 17 x 6" DR 26					415			415	460
6632	3" DR 17 x 8" DR 11					510			510	560
6636	3" DR 17 x 8" DR 17					565			565	620
6640	4" DR 11 x 8" DR 17					685			685	750
6644	4" DR 17 x 8" DR 26					690			690	755
6648	6" DR 11 x 10" DR 17					890			890	980
6652	6" DR 17 x 10" DR 26					900			900	990
6656	6" DR 26 x 10" DR 26					1,125			1,125	1,225
6660	8" DR 17 x 12" DR 26					1,125			1,125	1,225
6664	8" DR 26 x 12" DR 32.5					1,450			1,450	1,600
6668	10" DR 17 x 14" DR 26					1,675			1,675	1,850
6672	10" DR 17 x 16" DR 26					1,875			1,875	2,050
6676	10" DR 26 x 16" DR 26					1,875			1,875	2,050
6680	12" DR 26 x 16" DR 26					2,150			2,150	2,350
6684	12" DR 17 x 18" DR 26					2,150			2,150	2,350
6688	12" DR 26 x 18" DR 26					2,425			2,425	2,675
6692	14" DR 26 x 20" DR 32.5					2,725			2,725	3,000
6696	16" DR 26 x 22" DR 32.5					3,050			3,050	3,375
6700	18" DR 26 x 24" DR 32.5					3,500			3,500	3,850
6704	20" DR 32.5 x 28" DR 32.5					3,900			3,900	4,300
6708	22" DR 32.5 x 30" DR 32.5					4,400			4,400	4,850
6712	24" DR 32.5 x 32" DR 32.5					4,975			4,975	5,475
6716	36" DR 32.5 x 42" DR 32.5				↓	5,675			5,675	6,225
6800	Wye									
6816	2" DR 11 x 4" DR 17				Ea.	585			585	645
6820	2" DR 11 x 6" DR 17					730			730	805
6824	3" DR 11 x 6" DR 17					775			775	855
6828	3" DR 17 x 6" DR 26					780			780	855
6832	3" DR 17 x 8" DR 11					820			820	900
6836	3" DR 17 x 8" DR 17					900			900	985
6840	4" DR 11 x 8" DR 17					990			990	1,075
6844	4" DR 17 x 8" DR 26					605			605	665
6848	6" DR 11 x 10" DR 17					1,550			1,550	1,700
6852	6" DR 17 x 10" DR 26					940			940	1,025
6856	6" DR 26 x 10" DR 26					1,325			1,325	1,475
6860	8" DR 17 x 12" DR 26					1,850			1,850	2,050
6864	8" DR 26 x 12" DR 32.5					1,525			1,525	1,675
6868	10" DR 17 x 14" DR 26					1,600			1,600	1,775
6872	10" DR 17 x 16" DR 26					1,750			1,750	1,900
6876	10" DR 26 x 16" DR 26					1,900			1,900	2,075
6880	12" DR 26 x 16" DR 26					2,025			2,025	2,250
6884	12" DR 17 x 18" DR 26					2,175			2,175	2,400
6888	12" DR 26 x 18" DR 26					2,375			2,375	2,600
6892	14" DR 26 x 20" DR 32.5					2,525			2,525	2,775
6896	16" DR 26 x 22" DR 32.5					2,575			2,575	2,850
6900	18" DR 26 x 24" DR 32.5					2,925			2,925	3,225
6904	20" DR 32.5 x 28" DR 32.5					3,150			3,150	3,475
6908	22" DR 32.5 x 30" DR 32.5					3,400			3,400	3,750
6912	24" DR 32.5 x 32" DR 32.5				↓	3,675			3,675	4,025
9000	Welding labor per joint, not including welding machine									
9010	Pipe joint size, outer pipe (cost based on the thickest walls)									
9020	Straight pipe									

22 11 13 – Facility Water Distribution Piping

22 11 13.78 Pipe, High Density Polyethylene Plastic (HDPE)		Crew	Daily Output	Labor-Hours	Unit	Material	2019 Bare Costs Labor	Equipment	Total	Total Incl O&P
9050	3″ pipe size	4 Skwk	96	.333	Ea.		17.80		17.80	27
9060	4″ pipe size	″	77	.416			22		22	34
9070	6″ pipe size	5 Skwk	60	.667			35.50		35.50	54
9080	8″ pipe size	″	40	1			53.50		53.50	81.50
9090	10″ pipe size	6 Skwk	41	1.171			62.50		62.50	95
9100	12″ pipe size		39	1.231			65.50		65.50	100
9110	14″ pipe size		38	1.263			67.50		67.50	103
9120	16″ pipe size		35	1.371			73		73	111
9130	18″ pipe size	8 Skwk	45	1.422			76		76	116
9140	20″ pipe size		42	1.524			81.50		81.50	124
9150	22″ pipe size		40	1.600			85.50		85.50	130
9160	24″ pipe size		38	1.684			90		90	137
9170	28″ pipe size		37	1.730			92.50		92.50	141
9180	30″ pipe size		36	1.778			95		95	144
9190	32″ pipe size		35	1.829			97.50		97.50	149
9200	42″ pipe size		32	2			107		107	163
9300	Note: Cost for set up each time welder is moved.									
9301	Add 100% of weld labor cost									
9310	For handling between fitting welds add 50% of weld labor cost									
9320	Welder usually remains stationary with pipe moved through it									
9360	Weld machine, rental per day based on diam. capacity									
9380	3″ thru 4″ diameter				Ea.			63.50	63.50	70
9390	6″ thru 8″ diameter							207	207	228
9400	10″ thru 12″ diameter							310	310	340
9410	14″ thru 18″ diameter							420	420	460
9420	20″ thru 24″ diameter							750	750	825
9430	28″ thru 32″ diameter							880	880	970
9440	42″ thru 58″ diameter							910	910	1,000

22 11 19 – Domestic Water Piping Specialties

22 11 19.10 Flexible Connectors

Line	Description	Crew	Daily Output	Labor-Hours	Unit	Material	Labor	Equipment	Total	Total Incl O&P
0010	**FLEXIBLE CONNECTORS**, Corrugated, 7/8″ OD, 1/2″ ID									
0050	Gas, seamless brass, steel fittings									
0200	12″ long	1 Plum	36	.222	Ea.	6.60	14.05		20.65	28.50
0220	18″ long		36	.222		8.35	14.05		22.40	30
0240	24″ long		34	.235		9.70	14.85		24.55	33
0260	30″ long		34	.235		10.75	14.85		25.60	34.50
0280	36″ long		32	.250		11.55	15.80		27.35	36.50
0320	48″ long		30	.267		14.70	16.85		31.55	41.50
0340	60″ long		30	.267		17.45	16.85		34.30	44.50
0360	72″ long		30	.267		20	16.85		36.85	47.50
2000	Water, copper tubing, dielectric separators									
2100	12″ long	1 Plum	36	.222	Ea.	6.80	14.05		20.85	28.50
2220	15″ long		36	.222		7.55	14.05		21.60	29.50
2240	18″ long		36	.222		8.20	14.05		22.25	30
2260	24″ long		34	.235		10.05	14.85		24.90	33.50

22 11 19.14 Flexible Metal Hose

Line	Description	Crew	Daily Output	Labor-Hours	Unit	Material	Labor	Equipment	Total	Total Incl O&P
0010	**FLEXIBLE METAL HOSE**, Connectors, standard lengths									
0100	Bronze braided, bronze ends									
0120	3/8″ diameter x 12″	1 Stpi	26	.308	Ea.	28	19.70		47.70	60.50
0140	1/2″ diameter x 12″		24	.333		28.50	21.50		50	63.50
0160	3/4″ diameter x 12″		20	.400		39	25.50		64.50	81.50
0180	1″ diameter x 18″		19	.421		50.50	27		77.50	96

22 11 19 – Domestic Water Piping Specialties

22 11 19.14 Flexible Metal Hose		Crew	Daily Output	Labor-Hours	Unit	Material	Labor	Equipment	Total	Total Incl O&P
						2019 Bare Costs				
0200	1-1/2" diameter x 18"	1 Stpi	13	.615	Ea.	57	39.50		96.50	122
0220	2" diameter x 18"		11	.727		86	46.50		132.50	165
1000	Carbon steel ends									
1020	1/4" diameter x 12"	1 Stpi	28	.286	Ea.	31.50	18.25		49.75	62.50
1040	3/8" diameter x 12"		26	.308		33	19.70		52.70	66
1060	1/2" diameter x 12"		24	.333		28.50	21.50		50	63
1080	1/2" diameter x 24"		24	.333		58.50	21.50		80	96.50
1120	3/4" diameter x 12"		20	.400		45.50	25.50		71	88.50
1140	3/4" diameter x 24"		20	.400		70	25.50		95.50	116
1160	3/4" diameter x 36"		20	.400		78.50	25.50		104	125
1180	1" diameter x 18"		19	.421		62.50	27		89.50	110
1200	1" diameter x 30"		19	.421		74.50	27		101.50	123
1220	1" diameter x 36"		19	.421		98.50	27		125.50	149
1240	1-1/4" diameter x 18"		15	.533		85	34		119	145
1260	1-1/4" diameter x 36"		15	.533		134	34		168	198
1280	1-1/2" diameter x 18"		13	.615		106	39.50		145.50	176
1300	1-1/2" diameter x 36"		13	.615		138	39.50		177.50	211
1320	2" diameter x 24"		11	.727		189	46.50		235.50	278
1340	2" diameter x 36"		11	.727		176	46.50		222.50	263
1360	2-1/2" diameter x 24"		9	.889		360	57		417	480
1380	2-1/2" diameter x 36"		9	.889		124	57		181	222
1400	3" diameter x 24"		7	1.143		480	73		553	640
1420	3" diameter x 36"		7	1.143		930	73		1,003	1,125
2000	Carbon steel braid, carbon steel solid ends									
2100	1/2" diameter x 12"	1 Stpi	24	.333	Ea.	49	21.50		70.50	85.50
2120	3/4" diameter x 12"		20	.400		69.50	25.50		95	115
2140	1" diameter x 12"		19	.421		109	27		136	161
2160	1-1/4" diameter x 12"		15	.533		50	34		84	106
2180	1-1/2" diameter x 12"		13	.615		55.50	39.50		95	120
3000	Stainless steel braid, welded on carbon steel ends									
3100	1/2" diameter x 12"	1 Stpi	24	.333	Ea.	71	21.50		92.50	110
3120	3/4" diameter x 12"		20	.400		86.50	25.50		112	134
3140	3/4" diameter x 24"		20	.400		99.50	25.50		125	148
3160	3/4" diameter x 36"		20	.400		114	25.50		139.50	164
3180	1" diameter x 12"		19	.421		100	27		127	151
3200	1" diameter x 24"		19	.421		112	27		139	164
3220	1" diameter x 36"		19	.421		141	27		168	196
3240	1-1/4" diameter x 12"		15	.533		142	34		176	208
3260	1-1/4" diameter x 24"		15	.533		169	34		203	237
3280	1-1/4" diameter x 36"		15	.533		191	34		225	261
3300	1-1/2" diameter x 12"		13	.615		56	39.50		95.50	121
3320	1-1/2" diameter x 24"		13	.615		177	39.50		216.50	254
3340	1-1/2" diameter x 36"		13	.615		216	39.50		255.50	297
3400	Metal stainless steel braid, over corrugated stainless steel, flanged ends									
3410	150 psi									
3420	1/2" diameter x 12"	1 Stpi	24	.333	Ea.	79.50	21.50		101	120
3430	1" diameter x 12"		20	.400		130	25.50		155.50	182
3440	1-1/2" diameter x 12"		15	.533		140	34		174	206
3450	2-1/2" diameter x 9"		12	.667		61	42.50		103.50	132
3460	3" diameter x 9"		9	.889		31.50	57		88.50	121
3470	4" diameter x 9"		7	1.143		40.50	73		113.50	155
3480	4" diameter x 30"		5	1.600		400	102		502	595
3490	4" diameter x 36"		4.80	1.667		420	107		527	620

22 11 19 – Domestic Water Piping Specialties

22 11 19.14 Flexible Metal Hose		Crew	Daily Output	Labor-Hours	Unit	Material	Labor	Equipment	Total	Total Incl O&P
						2019 Bare Costs				
3500	6″ diameter x 11″	1 Stpi	5	1.600	Ea.	73	102		175	234
3510	6″ diameter x 36″		3.80	2.105		500	135		635	750
3520	8″ diameter x 12″		4	2		292	128		420	510
3530	10″ diameter x 13″		3	2.667		208	171		379	485
3540	12″ diameter x 14″	Q-5	4	4		315	230		545	695
6000	Molded rubber with helical wire reinforcement									
6010	150 psi									
6020	1-1/2″ diameter x 12″	1 Stpi	15	.533	Ea.	31	34		65	85
6030	2″ diameter x 12″		12	.667		77.50	42.50		120	149
6040	3″ diameter x 12″		8	1		81	64		145	185
6050	4″ diameter x 12″		6	1.333		103	85.50		188.50	241
6060	6″ diameter x 18″		4	2		154	128		282	360
6070	8″ diameter x 24″		3	2.667		215	171		386	495
6080	10″ diameter x 24″		2	4		261	256		517	675
6090	12″ diameter x 24″	Q-5	3	5.333		298	305		603	790
7000	Molded teflon with stainless steel flanges									
7010	150 psi									
7020	2-1/2″ diameter x 3-3/16″	Q-1	7.80	2.051	Ea.	755	117		872	1,000
7030	3″ diameter x 3-5/8″		6.50	2.462		535	140		675	800
7040	4″ diameter x 3-5/8″		5	3.200		690	182		872	1,025
7050	6″ diameter x 4″		4.30	3.721		960	211		1,171	1,375
7060	8″ diameter x 6″		3.80	4.211		1,525	239		1,764	2,025

22 11 19.18 Mixing Valve		Crew	Daily Output	Labor-Hours	Unit	Material	Labor	Equipment	Total	Total Incl O&P
0010	**MIXING VALVE**, Automatic, water tempering.									
0040	1/2″ size	1 Stpi	19	.421	Ea.	600	27		627	705
0050	3/4″ size		18	.444		600	28.50		628.50	705
0100	1″ size		16	.500		935	32		967	1,075
0120	1-1/4″ size		13	.615		1,250	39.50		1,289.50	1,450
0140	1-1/2″ size		10	.800		1,525	51		1,576	1,750
0160	2″ size		8	1		1,925	64		1,989	2,200
0170	2-1/2″ size		6	1.333		1,925	85.50		2,010.50	2,225
0180	3″ size		4	2		4,550	128		4,678	5,200
0190	4″ size		3	2.667		4,550	171		4,721	5,250

22 11 19.22 Pressure Reducing Valve		Crew	Daily Output	Labor-Hours	Unit	Material	Labor	Equipment	Total	Total Incl O&P
0010	**PRESSURE REDUCING VALVE**, Steam, pilot operated.									
0100	Threaded, iron body									
0200	1-1/2″ size	1 Stpi	8	1	Ea.	2,200	64		2,264	2,525
0220	2″ size	″	5	1.600	″	2,500	102		2,602	2,900
1000	Flanged, iron body, 125 lb. flanges									
1020	2″ size	1 Stpi	8	1	Ea.	3,425	64		3,489	3,850
1040	2-1/2″ size	″	4	2		2,475	128		2,603	2,925
1060	3″ size	Q-5	4.50	3.556		2,725	205		2,930	3,300
1080	4″ size	″	3	5.333		3,950	305		4,255	4,800
1500	For 250 lb. flanges, add					5%				

22 11 19.26 Pressure Regulators		Crew	Daily Output	Labor-Hours	Unit	Material	Labor	Equipment	Total	Total Incl O&P
0010	**PRESSURE REGULATORS**									
0100	Gas appliance regulators									
0106	Main burner and pilot applications									
0108	Rubber seat poppet type									
0109	1/8″ pipe size	1 Stpi	24	.333	Ea.	19.70	21.50		41.20	53.50
0110	1/4″ pipe size		24	.333		23.50	21.50		45	57.50
0112	3/8″ pipe size		24	.333		24	21.50		45.50	58

22 11 19.26 Pressure Regulators		Crew	Daily Output	Labor-Hours	Unit	Material	2019 Bare Costs Labor	Equipment	Total	Total Incl O&P
0113	1/2" pipe size	1 Stpi	24	.333	Ea.	26.50	21.50		48	61
0114	3/4" pipe size		20	.400		31	25.50		56.50	72.50
0122	Lever action type									
0123	3/8" pipe size	1 Stpi	24	.333	Ea.	29	21.50		50.50	63.50
0124	1/2" pipe size		24	.333		29	21.50		50.50	63.50
0125	3/4" pipe size		20	.400		56.50	25.50		82	101
0126	1" pipe size		19	.421		56.50	27		83.50	103
0132	Double diaphragm type									
0133	3/8" pipe size	1 Stpi	24	.333	Ea.	45	21.50		66.50	81.50
0134	1/2" pipe size		24	.333		60	21.50		81.50	97.50
0135	3/4" pipe size		20	.400		93	25.50		118.50	141
0136	1" pipe size		19	.421		115	27		142	167
0137	1-1/4" pipe size		15	.533		395	34		429	485
0138	1-1/2" pipe size		13	.615		730	39.50		769.50	860
0139	2" pipe size		11	.727		730	46.50		776.50	870
0140	2-1/2" pipe size	Q-5	15	1.067		1,425	61.50		1,486.50	1,650
0141	3" pipe size		13	1.231		1,425	71		1,496	1,650
0142	4" pipe size (flanged)		8	2		2,525	115		2,640	2,950
0160	Main burner only									
0162	Straight-thru-flow design									
0163	1/2" pipe size	1 Stpi	24	.333	Ea.	48.50	21.50		70	85.50
0164	3/4" pipe size		20	.400		66.50	25.50		92	112
0165	1" pipe size		19	.421		88	27		115	138
0166	1-1/4" pipe size		15	.533		88	34		122	148
0200	Oil, light, hot water, ordinary steam, threaded									
0220	Bronze body, 1/4" size	1 Stpi	24	.333	Ea.	168	21.50		189.50	217
0230	3/8" size		24	.333		180	21.50		201.50	230
0240	1/2" size		24	.333		227	21.50		248.50	281
0250	3/4" size		20	.400		270	25.50		295.50	335
0260	1" size		19	.421		410	27		437	495
0270	1-1/4" size		15	.533		555	34		589	660
0280	1-1/2" size		13	.615		575	39.50		614.50	695
0290	2" size		11	.727		1,075	46.50		1,121.50	1,275
0320	Iron body, 1/4" size		24	.333		126	21.50		147.50	170
0330	3/8" size		24	.333		148	21.50		169.50	195
0340	1/2" size		24	.333		156	21.50		177.50	204
0350	3/4" size		20	.400		190	25.50		215.50	248
0360	1" size		19	.421		246	27		273	310
0370	1-1/4" size		15	.533		340	34		374	425
0380	1-1/2" size		13	.615		380	39.50		419.50	480
0390	2" size		11	.727		550	46.50		596.50	675
0500	Oil, heavy, viscous fluids, threaded									
0520	Bronze body, 3/8" size	1 Stpi	24	.333	Ea.	278	21.50		299.50	335
0530	1/2" size		24	.333		355	21.50		376.50	425
0540	3/4" size		20	.400		400	25.50		425.50	475
0550	1" size		19	.421		495	27		522	585
0560	1-1/4" size		15	.533		700	34		734	820
0570	1-1/2" size		13	.615		800	39.50		839.50	945
0600	Iron body, 3/8" size		24	.333		221	21.50		242.50	275
0620	1/2" size		24	.333		268	21.50		289.50	325
0630	3/4" size		20	.400		305	25.50		330.50	375
0640	1" size		19	.421		360	27		387	435
0650	1-1/4" size		15	.533		520	34		554	620

22 11 19.26 Pressure Regulators		Crew	Daily Output	Labor-Hours	Unit	Material	2019 Bare Costs Labor	Equipment	Total	Total Incl O&P
0660	1-1/2" size	1 Stpi	13	.615	Ea.	555	39.50		594.50	670
0800	Process steam, wet or super heated, monel trim, threaded									
0820	Bronze body, 1/4" size	1 Stpi	24	.333	Ea.	635	21.50		656.50	725
0830	3/8" size		24	.333		635	21.50		656.50	725
0840	1/2" size		24	.333		725	21.50		746.50	830
0850	3/4" size		20	.400		855	25.50		880.50	980
0860	1" size		19	.421		1,075	27		1,102	1,225
0870	1-1/4" size		15	.533		1,325	34		1,359	1,525
0880	1-1/2" size		13	.615		1,600	39.50		1,639.50	1,825
0920	Iron body, max 125 PSIG press out, 1/4" size		24	.333		830	21.50		851.50	940
0930	3/8" size		24	.333		850	21.50		871.50	965
0940	1/2" size		24	.333		1,050	21.50		1,071.50	1,175
0950	3/4" size		20	.400		1,275	25.50		1,300.50	1,450
0960	1" size		19	.421		1,625	27		1,652	1,850
0970	1-1/4" size		15	.533		1,975	34		2,009	2,200
0980	1-1/2" size		13	.615		2,350	39.50		2,389.50	2,650
3000	Steam, high capacity, bronze body, stainless steel trim									
3020	Threaded, 1/2" diameter	1 Stpi	24	.333	Ea.	2,675	21.50		2,696.50	2,975
3030	3/4" diameter		24	.333		2,675	21.50		2,696.50	2,975
3040	1" diameter		19	.421		3,000	27		3,027	3,350
3060	1-1/4" diameter		15	.533		3,325	34		3,359	3,700
3080	1-1/2" diameter		13	.615		3,800	39.50		3,839.50	4,225
3100	2" diameter		11	.727		4,650	46.50		4,696.50	5,175
3120	2-1/2" diameter	Q-5	12	1.333		5,775	76.50		5,851.50	6,475
3140	3" diameter	"	11	1.455		6,625	83.50		6,708.50	7,400
3500	Flanged connection, iron body, 125 lb. W.S.P.									
3520	3" diameter	Q-5	11	1.455	Ea.	7,250	83.50		7,333.50	8,100
3540	4" diameter	"	5	3.200	"	9,150	184		9,334	10,400
9002	For water pressure regulators, see Section 22 05 23.20									

22 11 19.30 Pressure and Temperature Safety Plug		Crew	Daily Output	Labor-Hours	Unit	Material	Labor	Equipment	Total	Total Incl O&P
0010	**PRESSURE & TEMPERATURE SAFETY PLUG**									
1000	3/4" external thread, 3/8" diam. element									
1020	Carbon steel									
1050	7-1/2" insertion	1 Stpi	32	.250	Ea.	48.50	16		64.50	77.50
1120	304 stainless steel									
1150	7-1/2" insertion	1 Stpi	32	.250	Ea.	47.50	16		63.50	76
1220	316 stainless steel									
1250	7-1/2" insertion	1 Stpi	32	.250	Ea.	51	16		67	80

22 11 19.32 Pressure and Temperature Measurement Plug		Crew	Daily Output	Labor-Hours	Unit	Material	Labor	Equipment	Total	Total Incl O&P
0010	**PRESSURE & TEMPERATURE MEASUREMENT PLUG**									
0020	A permanent access port for insertion of									
0030	a pressure or temperature measuring probe									
0100	Plug, brass									
0110	1/4" MNPT, 1-1/2" long	1 Stpi	32	.250	Ea.	6.35	16		22.35	31
0120	3" long		31	.258		11.85	16.50		28.35	38
0140	1/2" MNPT, 1-1/2" long		30	.267		8.70	17.05		25.75	35
0150	3" long		29	.276		14.40	17.65		32.05	42.50
0200	Pressure gauge probe adapter									
0210	1/8" diameter, 1-1/2" probe				Ea.	20.50			20.50	22.50
0220	3" probe				"	33.50			33.50	37
0300	Temperature gauge, 5" stem									
0310	Analog				Ea.	10.60			10.60	11.65

22 11 19.32 Pressure and Temperature Measurement Plug		Crew	Daily Output	Labor-Hours	Unit	2019 Bare Costs Material	Labor	Equipment	Total	Total Incl O&P
0330	Digital				Ea.	29			29	31.50
0400	Pressure gauge, compound									
0410	1/4" MNPT				Ea.	32.50			32.50	36
0500	Pressure and temperature test kit									
0510	Contains 2 thermometers and 2 pressure gauges									
0520	Kit				Ea.	345			345	380

22 11 19.34 Sleeves and Escutcheons

Line	Description	Crew	Daily Output	Labor-Hours	Unit	Material	Labor	Equipment	Total	Total Incl O&P
0010	**SLEEVES & ESCUTCHEONS**									
0100	Pipe sleeve									
0110	Steel, w/water stop, 12" long, with link seal									
0120	2" diam. for 1/2" carrier pipe	1 Plum	8.40	.952	Ea.	62.50	60		122.50	159
0130	2-1/2" diam. for 3/4" carrier pipe		8	1		71	63		134	173
0140	2-1/2" diam. for 1" carrier pipe		8	1		67	63		130	169
0150	3" diam. for 1-1/4" carrier pipe		7.20	1.111		86.50	70		156.50	200
0160	3-1/2" diam. for 1-1/2" carrier pipe		6.80	1.176		87	74.50		161.50	207
0170	4" diam. for 2" carrier pipe		6	1.333		93.50	84		177.50	229
0180	4" diam. for 2-1/2" carrier pipe		6	1.333		95	84		179	230
0190	5" diam. for 3" carrier pipe		5.40	1.481		108	93.50		201.50	258
0200	6" diam. for 4" carrier pipe		4.80	1.667		130	105		235	300
0210	10" diam. for 6" carrier pipe	Q-1	8	2		254	114		368	450
0220	12" diam. for 8" carrier pipe		7.20	2.222		345	126		471	570
0230	14" diam. for 10" carrier pipe		6.40	2.500		360	142		502	610
0240	16" diam. for 12" carrier pipe		5.80	2.759		400	157		557	675
0250	18" diam. for 14" carrier pipe		5.20	3.077		745	175		920	1,075
0260	24" diam. for 18" carrier pipe		4	4		1,125	227		1,352	1,575
0270	24" diam. for 20" carrier pipe		4	4		935	227		1,162	1,375
0280	30" diam. for 24" carrier pipe		3.20	5		1,375	284		1,659	1,925
0500	Wall sleeve									
0510	Ductile iron with rubber gasket seal									
0520	3"	1 Plum	8.40	.952	Ea.	211	60		271	320
0530	4"		7.20	1.111		225	70		295	355
0540	6"		6	1.333		274	84		358	425
0550	8"		4	2		1,200	126		1,326	1,525
0560	10"		3	2.667		1,475	168		1,643	1,875
0570	12"		2.40	3.333		1,750	211		1,961	2,225
5000	Escutcheon									
5100	Split ring, pipe									
5110	Chrome plated									
5120	1/2"	1 Plum	160	.050	Ea.	1.15	3.16		4.31	6
5130	3/4"		160	.050		1.29	3.16		4.45	6.15
5140	1"		135	.059		1.38	3.74		5.12	7.10
5150	1-1/2"		115	.070		1.94	4.39		6.33	8.75
5160	2"		100	.080		2.15	5.05		7.20	9.95
5170	4"		80	.100		4.95	6.30		11.25	14.95
5180	6"		68	.118		7.80	7.45		15.25	19.70
5400	Shallow flange type									
5410	Chrome plated steel									
5420	1/2" CTS	1 Plum	180	.044	Ea.	.33	2.81		3.14	4.57
5430	3/4" CTS		180	.044		.36	2.81		3.17	4.61
5440	1/2" IPS		180	.044		.38	2.81		3.19	4.63
5450	3/4" IPS		180	.044		.42	2.81		3.23	4.67
5460	1" IPS		175	.046		.47	2.89		3.36	4.85

22 11 19.34 Sleeves and Escutcheons		Crew	Daily Output	Labor-Hours	Unit	Material	Labor (2019 Bare Costs)	Equipment	Total	Total Incl O&P
5470	1-1/2″ IPS	1 Plum	170	.047	Ea.	.80	2.97		3.77	5.35
5480	2″ IPS	↓	160	.050	↓	.97	3.16		4.13	5.80

22 11 19.38 Water Supply Meters

		Crew	Daily Output	Labor-Hours	Unit	Material	Labor	Equipment	Total	Total Incl O&P
0010	**WATER SUPPLY METERS**									
1000	Detector, serves dual systems such as fire and domestic or									
1020	process water, wide range cap., UL and FM approved									
1100	3″ mainline x 2″ by-pass, 400 GPM	Q-1	3.60	4.444	Ea.	7,675	253		7,928	8,825
1140	4″ mainline x 2″ by-pass, 700 GPM	″	2.50	6.400		7,675	365		8,040	9,000
1180	6″ mainline x 3″ by-pass, 1,600 GPM	Q-2	2.60	9.231		11,700	545		12,245	13,700
1220	8″ mainline x 4″ by-pass, 2,800 GPM		2.10	11.429		17,400	675		18,075	20,100
1260	10″ mainline x 6″ by-pass, 4,400 GPM		2	12		24,900	705		25,605	28,500
1300	10″ x 12″ mainlines x 6″ by-pass, 5,400 GPM	↓	1.70	14.118	↓	33,700	830		34,530	38,400
2000	Domestic/commercial, bronze									
2020	Threaded									
2060	5/8″ diameter, to 20 GPM	1 Plum	16	.500	Ea.	51.50	31.50		83	105
2080	3/4″ diameter, to 30 GPM		14	.571		94	36		130	157
2100	1″ diameter, to 50 GPM	↓	12	.667	↓	143	42		185	220
2300	Threaded/flanged									
2340	1-1/2″ diameter, to 100 GPM	1 Plum	8	1	Ea.	350	63		413	480
2360	2″ diameter, to 160 GPM	″	6	1.333	″	475	84		559	645
2600	Flanged, compound									
2640	3″ diameter, 320 GPM	Q-1	3	5.333	Ea.	2,425	305		2,730	3,100
2660	4″ diameter, to 500 GPM		1.50	10.667		3,875	605		4,480	5,150
2680	6″ diameter, to 1,000 GPM		1	16		6,275	910		7,185	8,275
2700	8″ diameter, to 1,800 GPM	↓	.80	20	↓	9,800	1,125		10,925	12,500
7000	Turbine									
7260	Flanged									
7300	2″ diameter, to 160 GPM	1 Plum	7	1.143	Ea.	605	72		677	775
7320	3″ diameter, to 450 GPM	Q-1	3.60	4.444		1,000	253		1,253	1,475
7340	4″ diameter, to 650 GPM	″	2.50	6.400		1,675	365		2,040	2,400
7360	6″ diameter, to 1,800 GPM	Q-2	2.60	9.231		3,050	545		3,595	4,175
7380	8″ diameter, to 2,500 GPM		2.10	11.429		4,725	675		5,400	6,200
7400	10″ diameter, to 5,500 GPM	↓	1.70	14.118	↓	6,350	830		7,180	8,250

22 11 19.42 Backflow Preventers

		Crew	Daily Output	Labor-Hours	Unit	Material	Labor	Equipment	Total	Total Incl O&P
0010	**BACKFLOW PREVENTERS**, Includes valves									
0020	and four test cocks, corrosion resistant, automatic operation									
1000	Double check principle									
1010	Threaded, with ball valves									
1020	3/4″ pipe size	1 Plum	16	.500	Ea.	224	31.50		255.50	294
1030	1″ pipe size		14	.571		250	36		286	330
1040	1-1/2″ pipe size		10	.800		580	50.50		630.50	715
1050	2″ pipe size	↓	7	1.143	↓	675	72		747	855
1080	Threaded, with gate valves									
1100	3/4″ pipe size	1 Plum	16	.500	Ea.	1,050	31.50		1,081.50	1,200
1120	1″ pipe size		14	.571		1,050	36		1,086	1,225
1140	1-1/2″ pipe size		10	.800		1,200	50.50		1,250.50	1,400
1160	2″ pipe size	↓	7	1.143	↓	1,675	72		1,747	1,925
1200	Flanged, valves are gate									
1210	3″ pipe size	Q-1	4.50	3.556	Ea.	2,950	202		3,152	3,550
1220	4″ pipe size	″	3	5.333		2,975	305		3,280	3,725
1230	6″ pipe size	Q-2	3	8		4,525	470		4,995	5,675
1240	8″ pipe size	↓	2	12	↓	8,050	705		8,755	9,900

22 11 19 – Domestic Water Piping Specialties

22 11 19.42 Backflow Preventers		Crew	Daily Output	Labor-Hours	Unit	Material	2019 Bare Costs Labor	Equipment	Total	Total Incl O&P
1250	10" pipe size	Q-2	1	24	Ea.	11,900	1,425		13,325	15,200
1300	Flanged, valves are OS&Y									
1370	1" pipe size	1 Plum	5	1.600	Ea.	1,175	101		1,276	1,425
1374	1-1/2" pipe size		5	1.600		1,325	101		1,426	1,625
1378	2" pipe size		4.80	1.667		1,800	105		1,905	2,125
1380	3" pipe size	Q-1	4.50	3.556		2,775	202		2,977	3,350
1400	4" pipe size	"	3	5.333		4,625	305		4,930	5,525
1420	6" pipe size	Q-2	3	8		6,650	470		7,120	8,025
1430	8" pipe size	"	2	12		14,000	705		14,705	16,500
4000	Reduced pressure principle									
4100	Threaded, bronze, valves are ball									
4120	3/4" pipe size	1 Plum	16	.500	Ea.	465	31.50		496.50	565
4140	1" pipe size		14	.571		495	36		531	600
4150	1-1/4" pipe size		12	.667		950	42		992	1,125
4160	1-1/2" pipe size		10	.800		1,000	50.50		1,050.50	1,175
4180	2" pipe size		7	1.143		1,175	72		1,247	1,400
5000	Flanged, bronze, valves are OS&Y									
5060	2-1/2" pipe size	Q-1	5	3.200	Ea.	5,650	182		5,832	6,475
5080	3" pipe size		4.50	3.556		6,425	202		6,627	7,350
5100	4" pipe size		3	5.333		8,025	305		8,330	9,275
5120	6" pipe size	Q-2	3	8		11,700	470		12,170	13,500
5200	Flanged, iron, valves are gate									
5210	2-1/2" pipe size	Q-1	5	3.200	Ea.	2,825	182		3,007	3,375
5220	3" pipe size		4.50	3.556		2,950	202		3,152	3,550
5230	4" pipe size		3	5.333		3,825	305		4,130	4,650
5240	6" pipe size	Q-2	3	8		5,625	470		6,095	6,875
5250	8" pipe size		2	12		9,650	705		10,355	11,700
5260	10" pipe size		1	24		15,600	1,425		17,025	19,300
5600	Flanged, iron, valves are OS&Y									
5660	2-1/2" pipe size	Q-1	5	3.200	Ea.	3,225	182		3,407	3,800
5680	3" pipe size		4.50	3.556		3,250	202		3,452	3,875
5700	4" pipe size		3	5.333		4,250	305		4,555	5,125
5720	6" pipe size	Q-2	3	8		5,850	470		6,320	7,125
5740	8" pipe size		2	12		10,700	705		11,405	12,800
5760	10" pipe size		1	24		19,700	1,425		21,125	23,800

22 11 19.50 Vacuum Breakers

		Crew	Daily Output	Labor-Hours	Unit	Material	Labor	Equipment	Total	Total Incl O&P
0010	**VACUUM BREAKERS**									
0013	See also backflow preventers Section 22 11 19.42									
1000	Anti-siphon continuous pressure type									
1010	Max. 150 psi - 210°F									
1020	Bronze body									
1030	1/2" size	1 Stpi	24	.333	Ea.	183	21.50		204.50	233
1040	3/4" size		20	.400		183	25.50		208.50	240
1050	1" size		19	.421		187	27		214	247
1060	1-1/4" size		15	.533		370	34		404	455
1070	1-1/2" size		13	.615		455	39.50		494.50	560
1080	2" size		11	.727		465	46.50		511.50	585
1200	Max. 125 psi with atmospheric vent									
1210	Brass, in-line construction									
1220	1/4" size	1 Stpi	24	.333	Ea.	138	21.50		159.50	184
1230	3/8" size	"	24	.333		138	21.50		159.50	184
1260	For polished chrome finish, add					13%				

22 11 19 – Domestic Water Piping Specialties

22 11 19.50 Vacuum Breakers		Crew	Daily Output	Labor-Hours	Unit	Material	2019 Bare Costs Labor	Equipment	Total	Total Incl O&P
2000	Anti-siphon, non-continuous pressure type									
2010	Hot or cold water 125 psi - 210°F									
2020	Bronze body									
2030	1/4" size	1 Stpi	24	.333	Ea.	77.50	21.50		99	117
2040	3/8" size		24	.333		77.50	21.50		99	117
2050	1/2" size		24	.333		87	21.50		108.50	128
2060	3/4" size		20	.400		104	25.50		129.50	153
2070	1" size		19	.421		161	27		188	218
2080	1-1/4" size		15	.533		282	34		316	360
2090	1-1/2" size		13	.615		330	39.50		369.50	425
2100	2" size		11	.727		515	46.50		561.50	635
2110	2-1/2" size		8	1		1,475	64		1,539	1,725
2120	3" size		6	1.333		1,975	85.50		2,060.50	2,275
2150	For polished chrome finish, add					50%				
3000	Air gap fitting									
3020	1/2" NPT size	1 Plum	19	.421	Ea.	60	26.50		86.50	106
3030	1" NPT size	"	15	.533		69	33.50		102.50	126
3040	2" NPT size	Q-1	21	.762		134	43.50		177.50	213
3050	3" NPT size		14	1.143		266	65		331	390
3060	4" NPT size		10	1.600		266	91		357	430

22 11 19.54 Water Hammer Arresters/Shock Absorbers		Crew	Daily Output	Labor-Hours	Unit	Material	Labor	Equipment	Total	Total Incl O&P
0010	**WATER HAMMER ARRESTERS/SHOCK ABSORBERS**									
0490	Copper									
0500	3/4" male IPS for 1 to 11 fixtures	1 Plum	12	.667	Ea.	29	42		71	95
0600	1" male IPS for 12 to 32 fixtures		8	1		49	63		112	149
0700	1-1/4" male IPS for 33 to 60 fixtures		8	1		49.50	63		112.50	150
0800	1-1/2" male IPS for 61 to 113 fixtures		8	1		71	63		134	173
0900	2" male IPS for 114 to 154 fixtures		8	1		104	63		167	209
1000	2-1/2" male IPS for 155 to 330 fixtures		4	2		315	126		441	535
4000	Bellows type									
4010	3/4" FNPT, to 11 fixture units	1 Plum	10	.800	Ea.	305	50.50		355.50	415
4020	1" FNPT, to 32 fixture units		8.80	.909		620	57.50		677.50	770
4030	1" FNPT, to 60 fixture units		8.80	.909		935	57.50		992.50	1,100
4040	1" FNPT, to 113 fixture units		8.80	.909		2,350	57.50		2,407.50	2,650
4050	1" FNPT, to 154 fixture units		6.60	1.212		2,650	76.50		2,726.50	3,025
4060	1-1/2" FNPT, to 300 fixture units		5	1.600		3,250	101		3,351	3,725

22 11 19.64 Hydrants		Crew	Daily Output	Labor-Hours	Unit	Material	Labor	Equipment	Total	Total Incl O&P
0010	**HYDRANTS**									
0050	Wall type, moderate climate, bronze, encased									
0200	3/4" IPS connection	1 Plum	16	.500	Ea.	900	31.50		931.50	1,050
0300	1" IPS connection		14	.571		975	36		1,011	1,125
0500	Anti-siphon type, 3/4" connection		16	.500		775	31.50		806.50	900
1000	Non-freeze, bronze, exposed									
1100	3/4" IPS connection, 4" to 9" thick wall	1 Plum	14	.571	Ea.	420	36		456	515
1120	10" to 14" thick wall		12	.667		435	42		477	540
1140	15" to 19" thick wall		12	.667		490	42		532	605
1160	20" to 24" thick wall		10	.800		760	50.50		810.50	915
1200	For 1" IPS connection, add					15%	10%			
1240	For 3/4" adapter type vacuum breaker, add				Ea.	73.50			73.50	81
1280	For anti-siphon type, add				"	156			156	172
2000	Non-freeze bronze, encased, anti-siphon type									
2100	3/4" IPS connection, 5" to 9" thick wall	1 Plum	14	.571	Ea.	1,225	36		1,261	1,400

22 11 19.64 Hydrants		Crew	Daily Output	Labor-Hours	Unit	Material	2019 Bare Costs Labor	Equipment	Total	Total Incl O&P
2120	10" to 14" thick wall	1 Plum	12	.667	Ea.	1,550	42		1,592	1,775
2140	15" to 19" thick wall		12	.667		1,650	42		1,692	1,900
2160	20" to 24" thick wall		10	.800		1,700	50.50		1,750.50	1,950
2200	For 1" IPS connection, add					10%	10%			
3000	Ground box type, bronze frame, 3/4" IPS connection									
3080	Non-freeze, all bronze, polished face, set flush									
3100	2' depth of bury	1 Plum	8	1	Ea.	1,175	63		1,238	1,400
3120	3' depth of bury		8	1		1,250	63		1,313	1,475
3140	4' depth of bury		8	1		1,350	63		1,413	1,575
3160	5' depth of bury		7	1.143		1,450	72		1,522	1,700
3180	6' depth of bury		7	1.143		1,525	72		1,597	1,800
3200	7' depth of bury		6	1.333		1,600	84		1,684	1,900
3220	8' depth of bury		5	1.600		1,700	101		1,801	2,025
3240	9' depth of bury		4	2		1,775	126		1,901	2,175
3260	10' depth of bury		4	2		1,875	126		2,001	2,250
3400	For 1" IPS connection, add					15%	10%			
3450	For 1-1/4" IPS connection, add					325%	14%			
3500	For 1-1/2" IPS connection, add					370%	18%			
3550	For 2" IPS connection, add					445%	24%			
3600	For tapped drain port in box, add					99			99	109
4000	Non-freeze, CI body, bronze frame & scoriated cover									
4010	with hose storage									
4100	2' depth of bury	1 Plum	7	1.143	Ea.	2,100	72		2,172	2,400
4120	3' depth of bury		7	1.143		2,200	72		2,272	2,525
4140	4' depth of bury		7	1.143		2,275	72		2,347	2,600
4160	5' depth of bury		6.50	1.231		2,300	77.50		2,377.50	2,675
4180	6' depth of bury		6	1.333		2,350	84		2,434	2,700
4200	7' depth of bury		5.50	1.455		2,425	92		2,517	2,825
4220	8' depth of bury		5	1.600		2,500	101		2,601	2,900
4240	9' depth of bury		4.50	1.778		2,500	112		2,612	2,925
4260	10' depth of bury		4	2		2,675	126		2,801	3,150
4280	For 1" IPS connection, add					465			465	515
4300	For tapped drain port in box, add					105			105	115
5000	Moderate climate, all bronze, polished face									
5020	and scoriated cover, set flush									
5100	3/4" IPS connection	1 Plum	16	.500	Ea.	790	31.50		821.50	920
5120	1" IPS connection	"	14	.571		1,950	36		1,986	2,200
5200	For tapped drain port in box, add					105			105	115
6000	Ground post type, all non-freeze, all bronze, aluminum casing									
6010	guard, exposed head, 3/4" IPS connection									
6100	2' depth of bury	1 Plum	8	1	Ea.	1,125	63		1,188	1,350
6120	3' depth of bury		8	1		1,225	63		1,288	1,450
6140	4' depth of bury		8	1		1,300	63		1,363	1,550
6160	5' depth of bury		7	1.143		1,400	72		1,472	1,625
6180	6' depth of bury		7	1.143		1,475	72		1,547	1,725
6200	7' depth of bury		6	1.333		1,600	84		1,684	1,875
6220	8' depth of bury		5	1.600		1,700	101		1,801	2,025
6240	9' depth of bury		4	2		1,800	126		1,926	2,175
6260	10' depth of bury		4	2		1,900	126		2,026	2,275
6300	For 1" IPS connection, add					40%	10%			
6350	For 1-1/4" IPS connection, add					140%	14%			
6400	For 1-1/2" IPS connection, add					225%	18%			
6450	For 2" IPS connection, add					315%	24%			

22 11 23 – Domestic Water Pumps

22 11 23.10 General Utility Pumps

	22 11 23.10 General Utility Pumps	Crew	Daily Output	Labor-Hours	Unit	2019 Bare Costs Material	Labor	Equipment	Total	Total Incl O&P
0010	**GENERAL UTILITY PUMPS**									
2000	Single stage									
3000	Double suction,									
3140	50 HP, 5″ D x 6″ S	Q-2	.33	72.727	Ea.	10,200	4,275		14,475	17,600
3180	60 HP, 6″ D x 8″ S	Q-3	.30	107		16,700	6,400		23,100	28,000
3190	75 HP, to 2,500 GPM		.28	114		15,600	6,875		22,475	27,500
3220	100 HP, to 3,000 GPM		.26	123		19,100	7,400		26,500	32,100
3240	150 HP, to 4,000 GPM		.24	133		22,000	8,025		30,025	36,200
4000	Centrifugal, end suction, mounted on base									
4010	Horizontal mounted, with drip proof motor, rated @ 100′ head									
4020	Vertical split case, single stage									
4040	100 GPM, 5 HP, 1-1/2″ discharge	Q-1	1.70	9.412	Ea.	4,425	535		4,960	5,675
4050	200 GPM, 10 HP, 2″ discharge		1.30	12.308		4,500	700		5,200	6,000
4060	250 GPM, 10 HP, 3″ discharge		1.28	12.500		5,025	710		5,735	6,600
4070	300 GPM, 15 HP, 2″ discharge	Q-2	1.56	15.385		6,050	905		6,955	8,000
4080	500 GPM, 20 HP, 4″ discharge		1.44	16.667		6,925	980		7,905	9,100
4090	750 GPM, 30 HP, 4″ discharge		1.20	20		7,950	1,175		9,125	10,500
4100	1,050 GPM, 40 HP, 5″ discharge		1	24		10,200	1,425		11,625	13,300
4110	1,500 GPM, 60 HP, 6″ discharge		.60	40		16,700	2,350		19,050	21,900
4120	2,000 GPM, 75 HP, 6″ discharge		.50	48		17,800	2,825		20,625	23,900
4130	3,000 GPM, 100 HP, 8″ discharge		.40	60		23,500	3,525		27,025	31,100
4200	Horizontal split case, single stage									
4210	100 GPM, 7.5 HP, 1-1/2″ discharge	Q-1	1.70	9.412	Ea.	4,450	535		4,985	5,700
4220	250 GPM, 15 HP, 2-1/2″ discharge	″	1.30	12.308		7,325	700		8,025	9,100
4230	500 GPM, 20 HP, 4″ discharge	Q-2	1.60	15		12,200	885		13,085	14,800
4240	750 GPM, 25 HP, 5″ discharge		1.54	15.584		12,200	920		13,120	14,800
4250	1,000 GPM, 40 HP, 5″ discharge		1.20	20		15,500	1,175		16,675	18,900
4260	1,500 GPM, 50 HP, 6″ discharge	Q-3	1.42	22.535		20,000	1,350		21,350	24,000
4270	2,000 GPM, 75 HP, 8″ discharge		1.14	28.070		25,300	1,675		26,975	30,300
4280	3,000 GPM, 100 HP, 10″ discharge		.96	33.333		26,800	2,000		28,800	32,500
4290	3,500 GPM, 150 HP, 10″ discharge		.86	37.209		36,400	2,225		38,625	43,500
4300	4,000 GPM, 200 HP, 10″ discharge		.66	48.485		32,800	2,925		35,725	40,400
4330	Horizontal split case, two stage, 500′ head									
4340	100 GPM, 40 HP, 1-1/2″ discharge	Q-2	1.70	14.118	Ea.	17,900	830		18,730	21,000
4350	200 GPM, 50 HP, 1-1/2″ discharge	″	1.44	16.667		18,800	980		19,780	22,100
4360	300 GPM, 75 HP, 2″ discharge	Q-3	1.57	20.382		25,800	1,225		27,025	30,200
4370	400 GPM, 100 HP, 3″ discharge		1.14	28.070		30,600	1,675		32,275	36,200
4380	800 GPM, 200 HP, 4″ discharge		.86	37.209		40,500	2,225		42,725	48,000
5000	Centrifugal, in-line									
5006	Vertical mount, iron body, 125 lb. flgd, 1,800 RPM TEFC mtr									
5010	Single stage									
5012	.5 HP, 1-1/2″ suction & discharge	Q-1	3.20	5	Ea.	1,475	284		1,759	2,050
5014	.75 HP, 2″ suction & discharge		2.80	5.714		2,850	325		3,175	3,600
5015	1 HP, 3″ suction & discharge		2.50	6.400		3,200	365		3,565	4,075
5016	1.5 HP, 3″ suction & discharge		2.30	6.957		3,325	395		3,720	4,250
5018	2 HP, 4″ suction & discharge		2.10	7.619		4,250	435		4,685	5,325
5020	3 HP, 5″ suction & discharge		1.90	8.421		4,900	480		5,380	6,125
5030	5 HP, 6″ suction & discharge		1.60	10		5,575	570		6,145	6,975
5040	7.5 HP, 6″ suction & discharge		1.30	12.308		6,400	700		7,100	8,100
5050	10 HP, 8″ suction & discharge	Q-2	1.80	13.333		8,550	785		9,335	10,600
5060	15 HP, 8″ suction & discharge		1.70	14.118		10,000	830		10,830	12,300
5064	20 HP, 8″ suction & discharge		1.60	15		11,800	885		12,685	14,200

22 11 23 – Domestic Water Pumps

22 11 23.10 General Utility Pumps		Crew	Daily Output	Labor-Hours	Unit	2019 Bare Costs Material	Labor	Equipment	Total	Total Incl O&P
5070	30 HP, 8" suction & discharge	Q-2	1.50	16	Ea.	14,700	945		15,645	17,600
5080	40 HP, 8" suction & discharge		1.40	17.143		17,700	1,000		18,700	21,000
5090	50 HP, 8" suction & discharge		1.30	18.462		18,000	1,100		19,100	21,400
5094	60 HP, 8" suction & discharge		.90	26.667		19,200	1,575		20,775	23,500
5100	75 HP, 8" suction & discharge	Q-3	.60	53.333		21,600	3,200		24,800	28,600
5110	100 HP, 8" suction & discharge	"	.50	64		24,600	3,850		28,450	32,900

22 11 23.11 Miscellaneous Pumps		Crew	Daily Output	Labor-Hours	Unit	Material	Labor	Equipment	Total	Total Incl O&P
0010	**MISCELLANEOUS PUMPS**									
0020	Water pump, portable, gasoline powered									
0100	170 GPH, 2" discharge	Q-1	11	1.455	Ea.	715	82.50		797.50	910
0110	343 GPH, 3" discharge		10.50	1.524		905	86.50		991.50	1,125
0120	608 GPH, 4" discharge		10	1.600		1,600	91		1,691	1,900
0500	Pump, propylene body, housing and impeller									
0510	22 GPM, 1/3 HP, 40' HD	1 Plum	5	1.600	Ea.	930	101		1,031	1,175
0520	33 GPM, 1/2 HP, 40' HD	Q-1	5	3.200		855	182		1,037	1,225
0530	53 GPM, 3/4 HP, 40' HD	"	4	4		1,025	227		1,252	1,475
0600	Rotary pump, CI									
0614	282 GPH, 3/4 HP, 1" discharge	Q-1	4.50	3.556	Ea.	1,200	202		1,402	1,625
0618	277 GPH, 1 HP, 1" discharge		4	4		1,200	227		1,427	1,675
0624	1,100 GPH, 1.5 HP, 1-1/4" discharge		3.60	4.444		2,950	253		3,203	3,625
0628	1,900 GPH, 2 HP, 1-1/4" discharge		3.20	5		4,175	284		4,459	5,000
1000	Turbine pump, CI									
1010	50 GPM, 2 HP, 3" discharge	Q-1	.80	20	Ea.	3,925	1,125		5,050	6,025
1020	100 GPM, 3 HP, 4" discharge	Q-2	.96	25		6,625	1,475		8,100	9,475
1030	250 GPM, 15 HP, 6" discharge	"	.94	25.532		8,200	1,500		9,700	11,300
1040	500 GPM, 25 HP, 6" discharge	Q-3	1.22	26.230		8,275	1,575		9,850	11,500
1050	1,000 GPM, 50 HP, 8" discharge		1.14	28.070		11,400	1,675		13,075	15,100
1060	2,000 GPM, 100 HP, 10" discharge		1	32		13,500	1,925		15,425	17,800
1070	3,000 GPM, 150 HP, 10" discharge		.80	40		21,800	2,400		24,200	27,600
1080	4,000 GPM, 200 HP, 12" discharge		.70	45.714		23,800	2,750		26,550	30,300
1090	6,000 GPM, 300 HP, 14" discharge		.60	53.333		36,300	3,200		39,500	44,800
1100	10,000 GPM, 300 HP, 18" discharge		.58	55.172		42,000	3,325		45,325	51,000
2000	Centrifugal stainless steel pumps									
2100	100 GPM, 100' TDH, 3 HP	Q-1	1.80	8.889	Ea.	11,900	505		12,405	13,900
2130	250 GPM, 100' TDH, 10 HP	"	1.28	12.500		16,600	710		17,310	19,400
2160	500 GPM, 100' TDH, 20 HP	Q-2	1.44	16.667		19,000	980		19,980	22,400
2200	Vertical turbine stainless steel pumps									
2220	100 GPM, 100' TDH, 7.5 HP	Q-2	.95	25.263	Ea.	66,500	1,500		68,000	75,000
2240	250 GPM, 100' TDH, 15 HP		.94	25.532		73,500	1,500		75,000	83,500
2260	500 GPM, 100' TDH, 20 HP		.93	25.806		81,000	1,525		82,525	91,500
2280	750 GPM, 100' TDH, 30 HP	Q-3	1.19	26.891		88,000	1,625		89,625	99,000
2300	1,000 GPM, 100' TDH, 40 HP	"	1.16	27.586		95,000	1,650		96,650	107,000
2320	100 GPM, 200' TDH, 15 HP	Q-2	.94	25.532		102,000	1,500		103,500	115,000
2340	250 GPM, 200' TDH, 25 HP	Q-3	1.22	26.230		109,500	1,575		111,075	122,500
2360	500 GPM, 200' TDH, 40 HP		1.16	27.586		116,500	1,650		118,150	130,500
2380	750 GPM, 200' TDH, 60 HP		1.13	28.319		117,500	1,700		119,200	132,000
2400	1,000 GPM, 200' TDH, 75 HP		1.12	28.571		121,000	1,725		122,725	135,500

22 11 23.13 Domestic-Water Packaged Booster Pumps		Crew	Daily Output	Labor-Hours	Unit	Material	Labor	Equipment	Total	Total Incl O&P
0010	**DOMESTIC-WATER PACKAGED BOOSTER PUMPS**									
0200	Pump system, with diaphragm tank, control, press. switch									
0300	1 HP pump	Q-1	1.30	12.308	Ea.	7,075	700		7,775	8,825
0400	1-1/2 HP pump		1.25	12.800		7,150	725		7,875	8,950

22 11 Facility Water Distribution

22 11 23 – Domestic Water Pumps

22 11 23.13 Domestic-Water Packaged Booster Pumps		Crew	Daily Output	Labor-Hours	Unit	2019 Bare Costs Material	Labor	Equipment	Total	Total Incl O&P
0420	2 HP pump	Q-1	1.20	13.333	Ea.	7,325	760		8,085	9,175
0440	3 HP pump		1.10	14.545		7,425	825		8,250	9,425
0460	5 HP pump	Q-2	1.50	16		8,225	945		9,170	10,500
0480	7-1/2 HP pump		1.42	16.901		9,150	995		10,145	11,600
0500	10 HP pump		1.34	17.910		9,575	1,050		10,625	12,100
2000	Pump system, variable speed, base, controls, starter									
2010	Duplex, 100' head									
2020	400 GPM, 7-1/2 HP, 4" discharge	Q-2	.70	34.286	Ea.	40,900	2,025		42,925	47,900
2025	Triplex, 100' head									
2030	1,000 GPM, 15 HP, 6" discharge	Q-2	.50	48	Ea.	48,000	2,825		50,825	57,500
2040	1,700 GPM, 30 HP, 6" discharge	"	.30	80	"	71,000	4,725		75,725	85,000

22 12 Facility Potable-Water Storage Tanks

22 12 21 – Facility Underground Potable-Water Storage Tanks

22 12 21.13 Fiberglass, Undrgrnd Pot.-Water Storage Tanks		Crew	Daily Output	Labor-Hours	Unit	Material	Labor	Equipment	Total	Total Incl O&P
0010	FIBERGLASS, UNDERGROUND POTABLE-WATER STORAGE TANKS									
0020	Excludes excavation, backfill & piping									
0030	Single wall									
2000	600 gallon capacity	B-21B	3.75	10.667	Ea.	3,975	475	126	4,576	5,225
2010	1,000 gallon capacity		3.50	11.429		5,100	510	135	5,745	6,550
2020	2,000 gallon capacity		3.25	12.308		7,375	550	145	8,070	9,125
2030	4,000 gallon capacity		3	13.333		10,100	595	157	10,852	12,200
2040	6,000 gallon capacity		2.65	15.094		11,400	675	178	12,253	13,700
2050	8,000 gallon capacity		2.30	17.391		13,700	780	205	14,685	16,500
2060	10,000 gallon capacity		2	20		15,600	895	236	16,731	18,700
2070	12,000 gallon capacity		1.50	26.667		21,600	1,200	315	23,115	25,800
2080	15,000 gallon capacity		1	40		25,700	1,800	470	27,970	31,400
2090	20,000 gallon capacity		.75	53.333		33,000	2,375	630	36,005	40,600
2100	25,000 gallon capacity		.50	80		47,900	3,575	945	52,420	59,000
2110	30,000 gallon capacity		.35	114		56,500	5,100	1,350	62,950	71,000
2120	40,000 gallon capacity		.30	133		86,500	5,975	1,575	94,050	106,000

22 12 23 – Facility Indoor Potable-Water Storage Tanks

22 12 23.13 Facility Steel, Indoor Pot.-Water Storage Tanks		Crew	Daily Output	Labor-Hours	Unit	Material	Labor	Equipment	Total	Total Incl O&P
0010	FACILITY STEEL, INDOOR POT.-WATER STORAGE TANKS									
2000	Galvanized steel, 15 gal., 14" diam. x 26" LOA	1 Plum	12	.667	Ea.	1,450	42		1,492	1,675
2060	30 gal., 14" diam. x 49" LOA		11	.727		1,725	46		1,771	1,975
2080	80 gal., 20" diam. x 64" LOA		9	.889		2,675	56		2,731	3,025
2100	135 gal., 24" diam. x 75" LOA		6	1.333		3,925	84		4,009	4,450
2120	240 gal., 30" diam. x 86" LOA		4	2		7,675	126		7,801	8,650
2140	300 gal., 36" diam. x 76" LOA		3	2.667		10,900	168		11,068	12,300
2160	400 gal., 36" diam. x 100" LOA	Q-1	4	4		13,400	227		13,627	15,100
2180	500 gal., 36" diam. x 126" LOA	"	3	5.333		16,600	305		16,905	18,800
3000	Glass lined, P.E., 80 gal., 20" diam. x 60" LOA	1 Plum	9	.889		4,225	56		4,281	4,725
3060	140 gal., 24" diam. x 75" LOA		6	1.333		5,800	84		5,884	6,500
3080	200 gal., 30" diam. x 71" LOA		4	2		8,050	126		8,176	9,050
3100	350 gal., 36" diam. x 86" LOA		3	2.667		9,700	168		9,868	11,000
3120	450 gal., 42" diam. x 79" LOA	Q-1	4	4		13,600	227		13,827	15,300
3140	600 gal., 48" diam. x 81" LOA		3	5.333		16,000	305		16,305	18,100
3160	750 gal., 48" diam. x 105" LOA		3	5.333		17,900	305		18,205	20,200
3180	900 gal., 54" diam. x 95" LOA		2.50	6.400		26,100	365		26,465	29,200

22 12 Facility Potable-Water Storage Tanks

22 12 23 – Facility Indoor Potable-Water Storage Tanks

22 12 23.13 Facility Steel, Indoor Pot.-Water Storage Tanks		Crew	Daily Output	Labor-Hours	Unit	2019 Bare Costs Material	Labor	Equipment	Total	Total Incl O&P
3200	1,500 gal., 54″ diam. x 153″ LOA	Q-1	2	8	Ea.	32,000	455		32,455	35,900
3220	1,800 gal., 54″ diam. x 181″ LOA		1.50	10.667		35,700	605		36,305	40,200
3240	2,000 gal., 60″ diam. x 165″ LOA		1	16		41,000	910		41,910	46,500
3260	3,500 gal., 72″ diam. x 201″ LOA	Q-2	1.50	16		52,000	945		52,945	59,000

22 13 Facility Sanitary Sewerage

22 13 16 – Sanitary Waste and Vent Piping

22 13 16.40 Pipe Fittings, Cast Iron for Drainage		Crew	Daily Output	Labor-Hours	Unit	Material	Labor	Equipment	Total	Total Incl O&P
0010	**PIPE FITTINGS, CAST IRON FOR DRAINAGE**, Special									
1000	Drip pan elbow (safety valve discharge elbow)									
1010	Cast iron, threaded inlet									
1014	2-1/2″	Q-1	8	2	Ea.	1,475	114		1,589	1,800
1015	3″		6.40	2.500		1,600	142		1,742	1,975
1017	4″		4.80	3.333		2,150	189		2,339	2,625
1018	Cast iron, flanged inlet									
1019	6″	Q-2	3.60	6.667	Ea.	2,425	395		2,820	3,275
1020	8″	″	2.60	9.231	″	2,450	545		2,995	3,525

22 13 16.60 Traps		Crew	Daily Output	Labor-Hours	Unit	Material	Labor	Equipment	Total	Total Incl O&P
0010	**TRAPS**									
4700	Copper, drainage, drum trap									
4800	3″ x 5″ solid, 1-1/2″ pipe size	1 Plum	16	.500	Ea.	142	31.50		173.50	205
4840	3″ x 6″ swivel, 1-1/2″ pipe size	″	16	.500	″	287	31.50		318.50	365
5100	P trap, standard pattern									
5200	1-1/4″ pipe size	1 Plum	18	.444	Ea.	92.50	28		120.50	144
5240	1-1/2″ pipe size		17	.471		103	29.50		132.50	158
5260	2″ pipe size		15	.533		158	33.50		191.50	225
5280	3″ pipe size		11	.727		495	46		541	615
5340	With cleanout, swivel joint and slip joint									
5360	1-1/4″ pipe size	1 Plum	18	.444	Ea.	119	28		147	173
5400	1-1/2″ pipe size		17	.471		186	29.50		215.50	250
5420	2″ pipe size		15	.533		160	33.50		193.50	226
6710	ABS DWV P trap, solvent weld joint									
6720	1-1/2″ pipe size	1 Plum	18	.444	Ea.	11.40	28		39.40	54.50
6722	2″ pipe size		17	.471		15	29.50		44.50	61
6724	3″ pipe size		15	.533		59.50	33.50		93	116
6726	4″ pipe size		14	.571		119	36		155	185
6732	PVC DWV P trap, solvent weld joint									
6733	1-1/2″ pipe size	1 Plum	18	.444	Ea.	9.70	28		37.70	52.50
6734	2″ pipe size		17	.471		11.85	29.50		41.35	57.50
6735	3″ pipe size		15	.533		40	33.50		73.50	94.50
6736	4″ pipe size		14	.571		91	36		127	154
6760	PP DWV, dilution trap, 1-1/2″ pipe size		16	.500		265	31.50		296.50	340
6770	P trap, 1-1/2″ pipe size		17	.471		66	29.50		95.50	117
6780	2″ pipe size		16	.500		97.50	31.50		129	155
6790	3″ pipe size		14	.571		179	36		215	251
6800	4″ pipe size		13	.615		300	39		339	390
6830	S trap, 1-1/2″ pipe size		16	.500		50	31.50		81.50	103
6840	2″ pipe size		15	.533		75	33.50		108.50	133
6850	Universal trap, 1-1/2″ pipe size		14	.571		107	36		143	172
6860	PVC DWV hub x hub, basin trap, 1-1/4″ pipe size		18	.444		52	28		80	99
6870	Sink P trap, 1-1/2″ pipe size		18	.444		14.90	28		42.90	58.50

22 13 16 – Sanitary Waste and Vent Piping

22 13 16.60 Traps		Crew	Daily Output	Labor-Hours	Unit	2019 Bare Costs Material	Labor	Equipment	Total	Total Incl O&P
6880	Tubular S trap, 1-1/2" pipe size	1 Plum	17	.471	Ea.	25.50	29.50		55	73
6890	PVC sch. 40 DWV, drum trap									
6900	1-1/2" pipe size	1 Plum	16	.500	Ea.	37	31.50		68.50	88
6910	P trap, 1-1/2" pipe size		18	.444		9.20	28		37.20	52
6920	2" pipe size		17	.471		12.35	29.50		41.85	58
6930	3" pipe size		15	.533		42	33.50		75.50	96.50
6940	4" pipe size		14	.571		95	36		131	159
6950	P trap w/clean out, 1-1/2" pipe size		18	.444		15.60	28		43.60	59
6960	2" pipe size		17	.471		25.50	29.50		55	72.50
6970	P trap adjustable, 1-1/2" pipe size		17	.471		11.75	29.50		41.25	57.50
6980	P trap adj. w/union & cleanout, 1-1/2" pipe size		16	.500		47.50	31.50		79	100

22 13 16.80 Vent Flashing and Caps

Line	Description	Crew	Daily Output	Labor-Hours	Unit	Material	Labor	Equipment	Total	Total Incl O&P
0010	**VENT FLASHING AND CAPS**									
0120	Vent caps									
0140	Cast iron									
0180	2-1/2" to 3-5/8" pipe	1 Plum	21	.381	Ea.	51.50	24		75.50	92.50
0190	4" to 4-1/8" pipe	"	19	.421	"	75	26.50		101.50	123
0900	Vent flashing									
1000	Aluminum with lead ring									
1020	1-1/4" pipe	1 Plum	20	.400	Ea.	5.50	25.50		31	44
1030	1-1/2" pipe		20	.400		5.15	25.50		30.65	43.50
1040	2" pipe		18	.444		5.20	28		33.20	47.50
1050	3" pipe		17	.471		5.75	29.50		35.25	51
1060	4" pipe		16	.500		6.95	31.50		38.45	55
1350	Copper with neoprene ring									
1400	1-1/4" pipe	1 Plum	20	.400	Ea.	72.50	25.50		98	118
1430	1-1/2" pipe		20	.400		72.50	25.50		98	118
1440	2" pipe		18	.444		72.50	28		100.50	122
1450	3" pipe		17	.471		87.50	29.50		117	141
1460	4" pipe		16	.500		87.50	31.50		119	144
2000	Galvanized with neoprene ring									
2020	1-1/4" pipe	1 Plum	20	.400	Ea.	14.75	25.50		40.25	54
2030	1-1/2" pipe		20	.400		15.85	25.50		41.35	55.50
2040	2" pipe		18	.444		15.70	28		43.70	59.50
2050	3" pipe		17	.471		19	29.50		48.50	65.50
2060	4" pipe		16	.500		16.95	31.50		48.45	66
2980	Neoprene, one piece									
3000	1-1/4" pipe	1 Plum	24	.333	Ea.	3.06	21		24.06	35
3030	1-1/2" pipe		24	.333		3.06	21		24.06	35
3040	2" pipe		23	.348		5.70	22		27.70	39.50
3050	3" pipe		21	.381		6.75	24		30.75	43.50
3060	4" pipe		20	.400		9.95	25.50		35.45	49
4000	Lead, 4#, 8" skirt, vent through roof									
4100	2" pipe	1 Plum	18	.444	Ea.	43	28		71	89
4110	3" pipe		17	.471		46.50	29.50		76	96
4120	4" pipe		16	.500		55.50	31.50		87	109
4130	6" pipe		14	.571		83	36		119	146

22 13 26 – Sanitary Waste Separators

22 13 26.10 Separators

Line	Description	Crew	Daily Output	Labor-Hours	Unit	Material	Labor	Equipment	Total	Total Incl O&P
0010	**SEPARATORS**, Entrainment eliminator, steel body, 150 PSIG									
0100	1/4" size	1 Stpi	24	.333	Ea.	271	21.50		292.50	330
0120	1/2" size		24	.333		279	21.50		300.50	335

22 13 26 – Sanitary Waste Separators

22 13 26.10 Separators		Crew	Daily Output	Labor-Hours	Unit	Material	2019 Bare Costs Labor	Equipment	Total	Total Incl O&P
0140	3/4" size	1 Stpi	20	.400	Ea.	280	25.50		305.50	345
0160	1" size		19	.421		299	27		326	370
0180	1-1/4" size		15	.533		315	34		349	400
0200	1-1/2" size		13	.615		350	39.50		389.50	445
0220	2" size		11	.727		390	46.50		436.50	495
0240	2-1/2" size	Q-5	15	1.067		1,700	61.50		1,761.50	1,950
0260	3" size		13	1.231		1,925	71		1,996	2,200
0280	4" size		10	1.600		2,175	92		2,267	2,525
0300	5" size		6	2.667		2,575	153		2,728	3,050
0320	6" size		3	5.333		2,800	305		3,105	3,525
0340	8" size	Q-6	4.40	5.455		3,550	325		3,875	4,400
0360	10" size	"	4	6		5,075	360		5,435	6,100
1000	For 300 PSIG, add					15%				

22 13 29 – Sanitary Sewerage Pumps

22 13 29.13 Wet-Pit-Mounted, Vertical Sewerage Pumps

Line	Description	Crew	Daily Output	Labor-Hours	Unit	Material	Labor	Equipment	Total	Total Incl O&P
0010	**WET-PIT-MOUNTED, VERTICAL SEWERAGE PUMPS**									
0020	Controls incl. alarm/disconnect panel w/wire. Excavation not included									
0260	Simplex, 9 GPM at 60 PSIG, 91 gal. tank				Ea.	3,450			3,450	3,800
0300	Unit with manway, 26" ID, 18" high					4,100			4,100	4,500
0340	26" ID, 36" high					4,075			4,075	4,475
0380	43" ID, 4' high					4,150			4,150	4,575
0600	Simplex, 9 GPM at 60 PSIG, 150 gal. tank, indoor					3,975			3,975	4,375
0700	Unit with manway, 26" ID, 36" high					4,675			4,675	5,125
0740	26" ID, 4' high					4,900			4,900	5,375
2000	Duplex, 18 GPM at 60 PSIG, 150 gal. tank, indoor					7,725			7,725	8,500
2060	Unit with manway, 43" ID, 4' high					8,525			8,525	9,375
2400	For core only					2,025			2,025	2,225
3000	Indoor residential type installation									
3020	Simplex, 9 GPM at 60 PSIG, 91 gal. HDPE tank				Ea.	3,500			3,500	3,850

22 13 29.14 Sewage Ejector Pumps

Line	Description	Crew	Daily Output	Labor-Hours	Unit	Material	Labor	Equipment	Total	Total Incl O&P
0010	**SEWAGE EJECTOR PUMPS**, With operating and level controls									
0100	Simplex system incl. tank, cover, pump 15' head									
0500	37 gal. PE tank, 12 GPM, 1/2 HP, 2" discharge	Q-1	3.20	5	Ea.	500	284		784	975
0510	3" discharge		3.10	5.161		545	293		838	1,050
0530	87 GPM, .7 HP, 2" discharge		3.20	5		770	284		1,054	1,275
0540	3" discharge		3.10	5.161		835	293		1,128	1,350
0600	45 gal. coated stl. tank, 12 GPM, 1/2 HP, 2" discharge		3	5.333		895	305		1,200	1,450
0610	3" discharge		2.90	5.517		930	315		1,245	1,500
0630	87 GPM, .7 HP, 2" discharge		3	5.333		1,150	305		1,455	1,725
0640	3" discharge		2.90	5.517		1,225	315		1,540	1,800
0660	134 GPM, 1 HP, 2" discharge		2.80	5.714		1,250	325		1,575	1,850
0680	3" discharge		2.70	5.926		1,300	335		1,635	1,950
0700	70 gal. PE tank, 12 GPM, 1/2 HP, 2" discharge		2.60	6.154		965	350		1,315	1,600
0710	3" discharge		2.40	6.667		1,025	380		1,405	1,700
0730	87 GPM, .7 HP, 2" discharge		2.50	6.400		1,250	365		1,615	1,925
0740	3" discharge		2.30	6.957		1,325	395		1,720	2,050
0760	134 GPM, 1 HP, 2" discharge		2.20	7.273		1,350	415		1,765	2,125
0770	3" discharge		2	8		1,450	455		1,905	2,275
0800	75 gal. coated stl. tank, 12 GPM, 1/2 HP, 2" discharge		2.40	6.667		1,075	380		1,455	1,750
0810	3" discharge		2.20	7.273		1,125	415		1,540	1,850
0830	87 GPM, .7 HP, 2" discharge		2.30	6.957		1,350	395		1,745	2,075
0840	3" discharge		2.10	7.619		1,425	435		1,860	2,200

22 13 Facility Sanitary Sewerage

22 13 29 – Sanitary Sewerage Pumps

22 13 29.14 Sewage Ejector Pumps		Crew	Daily Output	Labor-Hours	Unit	Material	2019 Bare Costs Labor	Equipment	Total	Total Incl O&P
0860	134 GPM, 1 HP, 2" discharge	Q-1	2	8	Ea.	1,450	455		1,905	2,275
0880	3" discharge	↓	1.80	8.889	↓	1,525	505		2,030	2,425
1040	Duplex system incl. tank, covers, pumps									
1060	110 gal. fiberglass tank, 24 GPM, 1/2 HP, 2" discharge	Q-1	1.60	10	Ea.	1,950	570		2,520	2,975
1080	3" discharge		1.40	11.429		2,025	650		2,675	3,200
1100	174 GPM, .7 HP, 2" discharge		1.50	10.667		2,500	605		3,105	3,650
1120	3" discharge		1.30	12.308		2,600	700		3,300	3,925
1140	268 GPM, 1 HP, 2" discharge		1.20	13.333		2,725	760		3,485	4,125
1160	3" discharge	↓	1	16		2,825	910		3,735	4,475
1260	135 gal. coated stl. tank, 24 GPM, 1/2 HP, 2" discharge	Q-2	1.70	14.118		2,000	830		2,830	3,450
2000	3" discharge		1.60	15		2,125	885		3,010	3,650
2640	174 GPM, .7 HP, 2" discharge		1.60	15		2,625	885		3,510	4,200
2660	3" discharge		1.50	16		2,750	945		3,695	4,450
2700	268 GPM, 1 HP, 2" discharge		1.30	18.462		2,850	1,100		3,950	4,750
3040	3" discharge		1.10	21.818		3,025	1,275		4,300	5,250
3060	275 gal. coated stl. tank, 24 GPM, 1/2 HP, 2" discharge		1.50	16		2,500	945		3,445	4,175
3080	3" discharge		1.40	17.143		2,525	1,000		3,525	4,300
3100	174 GPM, .7 HP, 2" discharge		1.40	17.143		3,225	1,000		4,225	5,075
3120	3" discharge		1.30	18.462		3,500	1,100		4,600	5,475
3140	268 GPM, 1 HP, 2" discharge		1.10	21.818		3,525	1,275		4,800	5,800
3160	3" discharge	↓	.90	26.667	↓	3,725	1,575		5,300	6,450
3260	Pump system accessories, add									
3300	Alarm horn and lights, 115 V mercury switch	Q-1	8	2	Ea.	102	114		216	283
3340	Switch, mag. contactor, alarm bell, light, 3 level control		5	3.200		495	182		677	820
3380	Alternator, mercury switch activated	↓	4	4	↓	890	227		1,117	1,325

22 14 Facility Storm Drainage

22 14 29 – Sump Pumps

22 14 29.13 Wet-Pit-Mounted, Vertical Sump Pumps

		Crew	Daily Output	Labor-Hours	Unit	Material	Labor	Equipment	Total	Total Incl O&P
0010	**WET-PIT-MOUNTED, VERTICAL SUMP PUMPS**									
0400	Molded PVC base, 21 GPM at 15' head, 1/3 HP	1 Plum	5	1.600	Ea.	141	101		242	305
0800	Iron base, 21 GPM at 15' head, 1/3 HP		5	1.600		154	101		255	320
1200	Solid brass, 21 GPM at 15' head, 1/3 HP	↓	5	1.600	↓	239	101		340	415
2000	Sump pump, single stage									
2010	25 GPM, 1 HP, 1-1/2" discharge	Q-1	1.80	8.889	Ea.	3,800	505		4,305	4,950
2020	75 GPM, 1-1/2 HP, 2" discharge		1.50	10.667		4,025	605		4,630	5,325
2030	100 GPM, 2 HP, 2-1/2" discharge		1.30	12.308		4,100	700		4,800	5,550
2040	150 GPM, 3 HP, 3" discharge		1.10	14.545		4,100	825		4,925	5,750
2050	200 GPM, 3 HP, 3" discharge	↓	1	16		4,350	910		5,260	6,150
2060	300 GPM, 10 HP, 4" discharge	Q-2	1.20	20		4,700	1,175		5,875	6,925
2070	500 GPM, 15 HP, 5" discharge		1.10	21.818		5,350	1,275		6,625	7,800
2080	800 GPM, 20 HP, 6" discharge		1	24		6,300	1,425		7,725	9,050
2090	1,000 GPM, 30 HP, 6" discharge		.85	28.235		6,925	1,675		8,600	10,100
2100	1,600 GPM, 50 HP, 8" discharge	↓	.72	33.333		10,800	1,975		12,775	14,900
2110	2,000 GPM, 60 HP, 8" discharge	Q-3	.85	37.647		11,100	2,275		13,375	15,600
2202	For general purpose float switch, copper coated float, add	Q-1	5	3.200	↓	103	182		285	385

22 14 29 – Sump Pumps

22 14 29.16 Submersible Sump Pumps		Crew	Daily Output	Labor-Hours	Unit	Material	Labor	Equipment	Total	Total Incl O&P
						2019 Bare Costs				
0010	**SUBMERSIBLE SUMP PUMPS**									
1000	Elevator sump pumps, automatic									
1010	Complete systems, pump, oil detector, controls and alarm									
1020	1-1/2" discharge, does not include the sump pit/tank									
1040	1/3 HP, 115 V	1 Plum	4.40	1.818	Ea.	1,850	115		1,965	2,200
1050	1/2 HP, 115 V		4	2		1,900	126		2,026	2,300
1060	1/2 HP, 230 V		4	2		1,950	126		2,076	2,350
1070	3/4 HP, 115 V		3.60	2.222		2,050	140		2,190	2,450
1080	3/4 HP, 230 V		3.60	2.222		2,025	140		2,165	2,450
1100	Sump pump only									
1110	1/3 HP, 115 V	1 Plum	6.40	1.250	Ea.	170	79		249	305
1120	1/2 HP, 115 V		5.80	1.379		242	87		329	395
1130	1/2 HP, 230 V		5.80	1.379		286	87		373	445
1140	3/4 HP, 115 V		5.40	1.481		325	93.50		418.50	495
1150	3/4 HP, 230 V		5.40	1.481		370	93.50		463.50	545
1200	Oil detector, control and alarm only									
1210	115 V	1 Plum	8	1	Ea.	1,675	63		1,738	1,950
1220	230 V	"	8	1	"	1,675	63		1,738	1,950
7000	Sump pump, automatic									
7100	Plastic, 1-1/4" discharge, 1/4 HP	1 Plum	6.40	1.250	Ea.	163	79		242	297
7140	1/3 HP		6	1.333		221	84		305	370
7160	1/2 HP		5.40	1.481		255	93.50		348.50	420
7180	1-1/2" discharge, 1/2 HP		5.20	1.538		310	97		407	485
7500	Cast iron, 1-1/4" discharge, 1/4 HP		6	1.333		215	84		299	365
7540	1/3 HP		6	1.333		253	84		337	405
7560	1/2 HP		5	1.600		305	101		406	485

22 14 53 – Rainwater Storage Tanks

22 14 53.13 Fiberglass, Rainwater Storage Tank		Crew	Daily Output	Labor-Hours	Unit	Material	Labor	Equipment	Total	Total Incl O&P
0010	**FIBERGLASS, RAINWATER STORAGE TANK**									
2000	600 gallon	B-21B	3.75	10.667	Ea.	3,975	475	126	4,576	5,225
2010	1,000 gallon		3.50	11.429		5,100	510	135	5,745	6,550
2020	2,000 gallon		3.25	12.308		7,375	550	145	8,070	9,125
2030	4,000 gallon		3	13.333		10,100	595	157	10,852	12,200
2040	6,000 gallon		2.65	15.094		11,400	675	178	12,253	13,700
2050	8,000 gallon		2.30	17.391		13,700	780	205	14,685	16,500
2060	10,000 gallon		2	20		15,600	895	236	16,731	18,700
2070	12,000 gallon		1.50	26.667		21,600	1,200	315	23,115	25,800
2080	15,000 gallon		1	40		25,700	1,800	470	27,970	31,400
2090	20,000 gallon		.75	53.333		33,000	2,375	630	36,005	40,600
2100	25,000 gallon		.50	80		47,900	3,575	945	52,420	59,000
2110	30,000 gallon		.35	114		56,500	5,100	1,350	62,950	71,000
2120	40,000 gallon		.30	133		86,500	5,975	1,575	94,050	106,000

22 15 General Service Compressed-Air Systems

22 15 13 – General Service Compressed-Air Piping

22 15 13.10 Compressor Accessories		Crew	Daily Output	Labor-Hours	Unit	2019 Bare Costs Material	Labor	Equipment	Total	Total Incl O&P
0010	COMPRESSOR ACCESSORIES									
1700	Refrigerated air dryers with ambient air filters									
1710	10 CFM	Q-5	8	2	Ea.	745	115		860	995
1715	15 CFM		7.30	2.192		1,125	126		1,251	1,425
1716	20 CFM		6.90	2.319		2,250	133		2,383	2,700
1720	25 CFM		6.60	2.424		1,400	140		1,540	1,725
1730	50 CFM		6.20	2.581		1,875	149		2,024	2,300
1740	75 CFM		5.80	2.759		1,775	159		1,934	2,200
1750	100 CFM		5.60	2.857		3,000	164		3,164	3,550
1751	125 CFM		5.40	2.963		3,700	171		3,871	4,300
1752	175 CFM		5	3.200		3,700	184		3,884	4,325
1753	200 CFM		4.80	3.333		5,350	192		5,542	6,175
1754	300 CFM		4.40	3.636		6,000	209		6,209	6,925
1755	400 CFM		4	4		7,950	230		8,180	9,100
1756	500 CFM		3.80	4.211		8,975	242		9,217	10,200
1757	600 CFM		3.40	4.706		11,500	271		11,771	13,100
1758	800 CFM		3	5.333		19,000	305		19,305	21,400
1759	1,000 CFM		2.60	6.154		22,700	355		23,055	25,500
3460	Air filter, regulator, lubricator combination									
3470	Flush mount									
3480	Adjustable range 0-140 psi									
3500	1/8″ NPT, 34 SCFM	1 Stpi	17	.471	Ea.	105	30		135	160
3510	1/4″ NPT, 61 SCFM		17	.471		188	30		218	252
3520	3/8″ NPT, 85 SCFM		16	.500		191	32		223	259
3530	1/2″ NPT, 150 SCFM		15	.533		205	34		239	277
3540	3/4″ NPT, 171 SCFM		14	.571		237	36.50		273.50	315
3550	1″ NPT, 150 SCFM		13	.615		405	39.50		444.50	505
4000	Couplers, air line, sleeve type									
4010	Female, connection size NPT									
4020	1/4″	1 Stpi	38	.211	Ea.	7.50	13.45		20.95	28.50
4030	3/8″		36	.222		10.85	14.20		25.05	33.50
4040	1/2″		35	.229		23	14.60		37.60	47.50
4050	3/4″		34	.235		22.50	15.05		37.55	47
4100	Male									
4110	1/4″	1 Stpi	38	.211	Ea.	9.10	13.45		22.55	30
4120	3/8″		36	.222		10.80	14.20		25	33.50
4130	1/2″		35	.229		19.65	14.60		34.25	43.50
4140	3/4″		34	.235		20.50	15.05		35.55	45.50
4150	Coupler, combined male and female halves									
4160	1/2″	1 Stpi	17	.471	Ea.	42.50	30		72.50	92
4170	3/4″	″	15	.533	″	43	34		77	98.50

22 15 19 – General Service Packaged Air Compressors and Receivers

22 15 19.10 Air Compressors		Crew	Daily Output	Labor-Hours	Unit	Material	Labor	Equipment	Total	Total Incl O&P
0010	AIR COMPRESSORS									
5250	Air, reciprocating air cooled, splash lubricated, tank mounted									
5300	Single stage, 1 phase, 140 psi									
5303	1/2 HP, 17 gal. tank	1 Stpi	3	2.667	Ea.	1,975	171		2,146	2,425
5305	3/4 HP, 30 gal. tank		2.60	3.077		1,800	197		1,997	2,275
5307	1 HP, 30 gal. tank		2.20	3.636		2,050	233		2,283	2,600
5309	2 HP, 30 gal. tank	Q-5	4	4		2,475	230		2,705	3,075
5310	3 HP, 30 gal. tank		3.60	4.444		2,425	256		2,681	3,050
5314	3 HP, 60 gal. tank		3.50	4.571		3,075	263		3,338	3,775

22 15 General Service Compressed-Air Systems

22 15 19 – General Service Packaged Air Compressors and Receivers

22 15 19.10 Air Compressors		Crew	Daily Output	Labor-Hours	Unit	Material	2019 Bare Costs Labor	Equipment	Total	Total Incl O&P
5320	5 HP, 60 gal. tank	Q-5	3.20	5	Ea.	3,300	288		3,588	4,050
5330	5 HP, 80 gal. tank		3	5.333		3,300	305		3,605	4,075
5340	7.5 HP, 80 gal. tank		2.60	6.154		4,200	355		4,555	5,150
5600	2 stage pkg., 3 phase									
5650	6 CFM at 125 psi, 1-1/2 HP, 60 gal. tank	Q-5	3	5.333	Ea.	2,575	305		2,880	3,275
5670	10.9 CFM at 125 psi, 3 HP, 80 gal. tank		1.50	10.667		3,275	615		3,890	4,525
5680	38.7 CFM at 125 psi, 10 HP, 120 gal. tank		.60	26.667		6,225	1,525		7,750	9,150
5690	105 CFM at 125 psi, 25 HP, 250 gal. tank	Q-6	.60	40		14,000	2,375		16,375	19,000
5800	With single stage pump									
5850	8.3 CFM at 125 psi, 2 HP, 80 gal. tank	Q-6	3.50	6.857	Ea.	3,150	410		3,560	4,075
5860	38.7 CFM at 125 psi, 10 HP, 120 gal. tank	"	.90	26.667	"	5,525	1,600		7,125	8,475
6000	Reciprocating, 2 stage, tank mtd, 3 ph, cap rated @ 175 PSIG									
6050	Pressure lubricated, hvy. duty, 9.7 CFM, 3 HP, 120 gal. tank	Q-5	1.30	12.308	Ea.	4,900	710		5,610	6,450
6054	5 CFM, 1-1/2 HP, 80 gal. tank		2.80	5.714		3,050	330		3,380	3,850
6056	6.4 CFM, 2 HP, 80 gal. tank		2	8		3,150	460		3,610	4,150
6058	8.1 CFM, 3 HP, 80 gal. tank		1.70	9.412		3,325	540		3,865	4,500
6059	14.8 CFM, 5 HP, 80 gal. tank		1	16		3,375	920		4,295	5,075
6060	16.5 CFM, 5 HP, 120 gal. tank		1	16		5,175	920		6,095	7,050
6063	13 CFM, 6 HP, 80 gal. tank		.90	17.778		4,750	1,025		5,775	6,750
6066	19.8 CFM, 7.5 HP, 80 gal. tank		.80	20		4,750	1,150		5,900	6,950
6070	25.8 CFM, 7-1/2 HP, 120 gal. tank		.80	20		5,100	1,150		6,250	7,325
6078	34.8 CFM, 10 HP, 80 gal. tank		.70	22.857		7,325	1,325		8,650	10,000
6080	34.8 CFM, 10 HP, 120 gal. tank		.60	26.667		7,650	1,525		9,175	10,700
6090	53.7 CFM, 15 HP, 120 gal. tank	Q-6	.80	30		8,975	1,800		10,775	12,600
6100	76.7 CFM, 20 HP, 120 gal. tank		.70	34.286		11,300	2,050		13,350	15,500
6104	76.7 CFM, 20 HP, 240 gal. tank		.68	35.294		11,900	2,100		14,000	16,300
6110	90.1 CFM, 25 HP, 120 gal. tank		.63	38.095		12,800	2,275		15,075	17,500
6120	101 CFM, 30 HP, 120 gal. tank		.57	42.105		11,800	2,525		14,325	16,800
6130	101 CFM, 30 HP, 250 gal. tank		.52	46.154		12,500	2,750		15,250	17,800
6200	Oil-less, 13.6 CFM, 5 HP, 120 gal. tank	Q-5	.88	18.182		13,400	1,050		14,450	16,400
6210	13.6 CFM, 5 HP, 250 gal. tank		.80	20		14,400	1,150		15,550	17,600
6220	18.2 CFM, 7.5 HP, 120 gal. tank		.73	21.918		13,400	1,250		14,650	16,700
6230	18.2 CFM, 7.5 HP, 250 gal. tank		.67	23.881		14,400	1,375		15,775	18,000
6250	30.5 CFM, 10 HP, 120 gal. tank		.57	28.070		15,900	1,625		17,525	19,800
6260	30.5 CFM, 10 HP, 250 gal. tank		.53	30.189		16,800	1,725		18,525	21,100
6270	41.3 CFM, 15 HP, 120 gal. tank	Q-6	.70	34.286		17,200	2,050		19,250	22,100
6280	41.3 CFM, 15 HP, 250 gal. tank	"	.67	35.821		18,200	2,150		20,350	23,300

22 31 Domestic Water Softeners

22 31 13 – Residential Domestic Water Softeners

22 31 13.10 Residential Water Softeners		Crew	Daily Output	Labor-Hours	Unit	Material	Labor	Equipment	Total	Total Incl O&P
0010	**RESIDENTIAL WATER SOFTENERS**									
7350	Water softener, automatic, to 30 grains per gallon	2 Plum	5	3.200	Ea.	375	202		577	715
7400	To 100 grains per gallon	"	4	4	"	855	253		1,108	1,325

22 35 Domestic Water Heat Exchangers

22 35 30 – Water Heating by Steam

22 35 30.10 Water Heating Transfer Package		Crew	Daily Output	Labor-Hours	Unit	Material	2019 Bare Costs Labor	Equipment	Total	Total Incl O&P
0010	**WATER HEATING TRANSFER PACKAGE**, Complete controls,									
0020	expansion tank, converter, air separator									
1000	Hot water, 180°F enter, 200°F leaving, 15# steam									
1010	One pump system, 28 GPM	Q-6	.75	32	Ea.	21,500	1,900		23,400	26,600
1020	35 GPM		.70	34.286		23,400	2,050		25,450	28,900
1040	55 GPM		.65	36.923		27,600	2,200		29,800	33,700
1060	130 GPM		.55	43.636		34,900	2,600		37,500	42,300
1080	255 GPM		.40	60		45,900	3,575		49,475	56,000
1100	550 GPM		.30	80		62,000	4,775		66,775	75,000
1120	800 GPM		.25	96		74,500	5,725		80,225	90,500
1220	Two pump system, 28 GPM		.70	34.286		29,200	2,050		31,250	35,300
1240	35 GPM		.65	36.923		34,200	2,200		36,400	41,000
1260	55 GPM		.60	40		37,200	2,375		39,575	44,500
1280	130 GPM		.50	48		48,500	2,875		51,375	58,000
1300	255 GPM		.35	68.571		62,000	4,100		66,100	74,500
1320	550 GPM		.25	96		78,000	5,725		83,725	94,500
1340	800 GPM	▼	.20	120	▼	101,000	7,150		108,150	121,500

22 41 Residential Plumbing Fixtures

22 41 39 – Residential Faucets, Supplies and Trim

22 41 39.10 Faucets and Fittings		Crew	Daily Output	Labor-Hours	Unit	Material	Labor	Equipment	Total	Total Incl O&P
0010	**FAUCETS AND FITTINGS**									
5000	Sillcock, compact, brass, IPS or copper to hose	1 Plum	24	.333	Ea.	10.60	21		31.60	43
6000	Stop and waste valves, bronze									
6100	Angle, solder end 1/2"	1 Plum	24	.333	Ea.	24.50	21		45.50	58.50
6110	3/4"		20	.400		31.50	25.50		57	73
6300	Straightway, solder end 3/8"		24	.333		17.50	21		38.50	51
6310	1/2"		24	.333		17.85	21		38.85	51
6320	3/4"		20	.400		20.50	25.50		46	60.50
6410	Straightway, threaded 1/2"		24	.333		21.50	21		42.50	55
6420	3/4"		20	.400		24	25.50		49.50	64.50
6430	1"	▼	19	.421	▼	27	26.50		53.50	70

22 51 Swimming Pool Plumbing Systems

22 51 19 – Swimming Pool Water Treatment Equipment

22 51 19.50 Swimming Pool Filtration Equipment		Crew	Daily Output	Labor-Hours	Unit	Material	Labor	Equipment	Total	Total Incl O&P
0010	**SWIMMING POOL FILTRATION EQUIPMENT**									
0900	Filter system, sand or diatomite type, incl. pump, 6,000 gal./hr.	2 Plum	1.80	8.889	Total	2,200	560		2,760	3,275
1020	Add for chlorination system, 800 S.F. pool		3	5.333	Ea.	220	335		555	745
1040	5,000 S.F. pool	▼	3	5.333	"	1,875	335		2,210	2,550

22 52 Fountain Plumbing Systems

22 52 16 – Fountain Pumps

22 52 16.10 Fountain Water Pumps		Crew	Daily Output	Labor-Hours	Unit	Material	2019 Bare Costs Labor	Equipment	Total	Total Incl O&P
0010	**FOUNTAIN WATER PUMPS**									
0100	Pump w/controls									
0200	Single phase, 100′ cord, 1/2 HP pump	2 Skwk	4.40	3.636	Ea.	1,275	194		1,469	1,700
0300	3/4 HP pump		4.30	3.721		1,275	199		1,474	1,725
0400	1 HP pump		4.20	3.810		1,900	203		2,103	2,400
0500	1-1/2 HP pump		4.10	3.902		2,575	208		2,783	3,150
0600	2 HP pump		4	4		3,975	214		4,189	4,675
0700	Three phase, 200′ cord, 5 HP pump		3.90	4.103		5,575	219		5,794	6,450
0800	7-1/2 HP pump		3.80	4.211		11,300	225		11,525	12,800
0900	10 HP pump		3.70	4.324		14,100	231		14,331	15,900
1000	15 HP pump		3.60	4.444		21,200	237		21,437	23,700
2000	DESIGN NOTE: Use two horsepower per surface acre.									

22 52 33 – Fountain Ancillary

22 52 33.10 Fountain Miscellaneous		Crew	Daily Output	Labor-Hours	Unit	Material	Labor	Equipment	Total	Total Incl O&P
0010	**FOUNTAIN MISCELLANEOUS**									
1300	Lights w/mounting kits, 200 watt	2 Skwk	18	.889	Ea.	1,200	47.50		1,247.50	1,375
1400	300 watt		18	.889		1,350	47.50		1,397.50	1,575
1500	500 watt		18	.889		1,500	47.50		1,547.50	1,750
1600	Color blender		12	1.333		585	71		656	755

22 62 Vacuum Systems for Laboratory and Healthcare Facilities

22 62 19 – Vacuum Equipment for Laboratory and Healthcare Facilities

22 62 19.70 Healthcare Vacuum Equipment		Crew	Daily Output	Labor-Hours	Unit	Material	Labor	Equipment	Total	Total Incl O&P
0010	**HEALTHCARE VACUUM EQUIPMENT**									
0300	Dental oral									
0310	Duplex									
0330	165 SCFM with 77 gal. separator	Q-2	1.30	18.462	Ea.	53,000	1,100		54,100	59,500
1100	Vacuum system									
1110	Vacuum outlet alarm panel	1 Plum	3.20	2.500	Ea.	1,125	158		1,283	1,450
2000	Medical, with receiver									
2100	Rotary vane type, lubricated, with controls									
2110	Simplex									
2120	1.5 HP, 80 gal. tank	Q-1	8	2	Ea.	6,625	114		6,739	7,475
2130	2 HP, 80 gal. tank		7	2.286		6,875	130		7,005	7,750
2140	3 HP, 80 gal. tank		6.60	2.424		7,225	138		7,363	8,150
2150	5 HP, 80 gal. tank		6	2.667		7,900	152		8,052	8,900
2200	Duplex									
2210	1 HP, 80 gal. tank	Q-1	8.40	1.905	Ea.	10,600	108		10,708	11,900
2220	1.5 HP, 80 gal. tank		7.80	2.051		10,900	117		11,017	12,200
2230	2 HP, 80 gal. tank		6.80	2.353		11,700	134		11,834	13,100
2240	3 HP, 120 gal. tank		6	2.667		12,800	152		12,952	14,300
2250	5 HP, 120 gal. tank		5.40	2.963		14,500	168		14,668	16,300
2260	7.5 HP, 200 gal. tank	Q-2	7	3.429		20,800	202		21,002	23,100
2270	10 HP, 200 gal. tank		6.40	3.750		24,000	221		24,221	26,700
2280	15 HP, 200 gal. tank		5.80	4.138		42,900	244		43,144	47,600
2290	20 HP, 200 gal. tank		5	4.800		48,900	283		49,183	54,500
2300	25 HP, 200 gal. tank		4	6		57,000	355		57,355	63,000
2400	Triplex									
2410	7.5 HP, 200 gal. tank	Q-2	6.40	3.750	Ea.	32,900	221		33,121	36,500
2420	10 HP, 200 gal. tank		5.70	4.211		38,400	248		38,648	42,600

22 62 Vacuum Systems for Laboratory and Healthcare Facilities

22 62 19 – Vacuum Equipment for Laboratory and Healthcare Facilities

22 62 19.70 Healthcare Vacuum Equipment		Crew	Daily Output	Labor-Hours	Unit	2019 Bare Costs Material	Labor	Equipment	Total	Total Incl O&P
2430	15 HP, 200 gal. tank	Q-2	4.90	4.898	Ea.	65,000	289		65,289	72,000
2440	20 HP, 200 gal. tank		4	6		74,000	355		74,355	82,000
2450	25 HP, 200 gal. tank		3.70	6.486		85,500	380		85,880	94,500
2500	Quadruplex									
2510	7.5 HP, 200 gal. tank	Q-2	5.30	4.528	Ea.	41,000	267		41,267	45,500
2520	10 HP, 200 gal. tank		4.70	5.106		47,300	300		47,600	52,500
2530	15 HP, 200 gal. tank		4	6		85,500	355		85,855	94,500
2540	20 HP, 200 gal. tank		3.60	6.667		97,500	395		97,895	107,500
2550	25 HP, 200 gal. tank		3.20	7.500		113,000	440		113,440	125,000
4000	Liquid ring type, water sealed, with controls									
4200	Duplex									
4210	1.5 HP, 120 gal. tank	Q-1	6	2.667	Ea.	21,000	152		21,152	23,200
4220	3 HP, 120 gal. tank	"	5.50	2.909		21,800	165		21,965	24,200
4230	4 HP, 120 gal. tank	Q-2	6.80	3.529		23,300	208		23,508	25,900
4240	5 HP, 120 gal. tank		6.50	3.692		24,600	218		24,818	27,300
4250	7.5 HP, 200 gal. tank		6.20	3.871		29,900	228		30,128	33,200
4260	10 HP, 200 gal. tank		5.80	4.138		37,300	244		37,544	41,400
4270	15 HP, 200 gal. tank		5.10	4.706		45,200	277		45,477	50,000
4280	20 HP, 200 gal. tank		4.60	5.217		58,500	305		58,805	65,000
4290	30 HP, 200 gal. tank		4	6		70,000	355		70,355	77,500

22 63 Gas Systems for Laboratory and Healthcare Facilities

22 63 13 – Gas Piping for Laboratory and Healthcare Facilities

22 63 13.70 Healthcare Gas Piping		Crew	Daily Output	Labor-Hours	Unit	Material	Labor	Equipment	Total	Total Incl O&P
0010	**HEALTHCARE GAS PIPING**									
1000	Nitrogen or oxygen system									
1010	Cylinder manifold									
1020	5 cylinder	1 Plum	.80	10	Ea.	6,050	630		6,680	7,600
1026	10 cylinder	"	.40	20		7,275	1,275		8,550	9,900
1050	Nitrogen generator, 30 LPM	Q-5	.40	40		28,800	2,300		31,100	35,200
3000	Outlets and valves									
3010	Recessed, wall mounted									
3012	Single outlet	1 Plum	3.20	2.500	Ea.	73	158		231	315
3100	Ceiling outlet									
3190	Zone valve with box									
3210	2" valve size	1 Plum	3.20	2.500	Ea.	425	158		583	700
4000	Alarm panel, medical gases and vacuum									
4010	Alarm panel	1 Plum	3.20	2.500	Ea.	1,125	158		1,283	1,450

Estimating Tips

The labor adjustment factors listed in Subdivision 22 01 02.20 also apply to Division 23.

23 10 00 Facility Fuel Systems

- The prices in this subdivision for above- and below-ground storage tanks do not include foundations or hold-down slabs, unless noted. The estimator should refer to Divisions 3 and 31 for foundation system pricing. In addition to the foundations, required tank accessories, such as tank gauges, leak detection devices, and additional manholes and piping, must be added to the tank prices.

23 50 00 Central Heating Equipment

- When estimating the cost of an HVAC system, check to see who is responsible for providing and installing the temperature control system. It is possible to overlook controls, assuming that they would be included in the electrical estimate.
- When looking up a boiler, be careful on specified capacity. Some manufacturers rate their products on output while others use input.
- Include HVAC insulation for pipe, boiler, and duct (wrap and liner).
- Be careful when looking up mechanical items to get the correct pressure rating and connection type (thread, weld, flange).

23 70 00 Central HVAC Equipment

- Combination heating and cooling units are sized by the air conditioning requirements. (See Reference No. R236000-20 for the preliminary sizing guide.)
- A ton of air conditioning is nominally 400 CFM.
- Rectangular duct is taken off by the linear foot for each size, but its cost is usually estimated by the pound. Remember that SMACNA standards now base duct on internal pressure.
- Prefabricated duct is estimated and purchased like pipe: straight sections and fittings.
- Note that cranes or other lifting equipment are not included on any lines in Division 23. For example, if a crane is required to lift a heavy piece of pipe into place high above a gym floor, or to put a rooftop unit on the roof of a four-story building, etc., it must be added. Due to the potential for extreme variation—from nothing additional required to a major crane or helicopter—we feel that including a nominal amount for "lifting contingency" would be useless and detract from the accuracy of the estimate. When using equipment rental cost data from RSMeans, do not forget to include the cost of the operator(s).

Reference Numbers

Reference numbers are shown at the beginning of some major classifications. These numbers refer to related items in the Reference Section. The reference information may be an estimating procedure, an alternate pricing method, or technical information.

Note: Not all subdivisions listed here necessarily appear. ■

Note: "Powered in part by CINX™, based on liscensed proprietary information of Harrison Publishing House, Inc."

Note: Trade Service, in part, has been used as a reference source for some of the material prices used in Division 23.

23 05 02 – HVAC General

23 05 02.10 Air Conditioning, General	Crew	Daily Output	Labor-Hours	Unit	Material	2019 Bare Costs Labor	Equipment	Total	Total Incl O&P
0010 **AIR CONDITIONING, GENERAL** Prices are for standard efficiencies (SEER 13)									
0020 for upgrade to SEER 14 add					10%				

23 05 05 – Selective Demolition for HVAC

23 05 05.10 HVAC Demolition	Crew	Daily Output	Labor-Hours	Unit	Material	Labor	Equipment	Total	Total Incl O&P
0010 **HVAC DEMOLITION** R220105-10									
0100 Air conditioner, split unit, 3 ton	Q-5	2	8	Ea.		460		460	690
0150 Package unit, 3 ton R024119-10	Q-6	3	8			475		475	715
0190 Rooftop, self contained, up to 5 ton	1 Plum	1.20	6.667			420		420	630
0250 Air curtain	Q-9	20	.800	L.F.		44		44	66.50
0254 Air filters, up thru 16,000 CFM		20	.800	Ea.		44		44	66.50
0256 20,000 thru 60,000 CFM		16	1			55		55	83.50
0297 Boiler blowdown	Q-5	8	2			115		115	173
0298 Boilers									
0300 Electric, up thru 148 kW	Q-19	2	12	Ea.		700		700	1,050
0310 150 thru 518 kW	"	1	24			1,400		1,400	2,100
0320 550 thru 2,000 kW	Q-21	.40	80			4,775		4,775	7,175
0330 2,070 kW and up	"	.30	107			6,375		6,375	9,550
0340 Gas and/or oil, up thru 150 MBH	Q-7	2.20	14.545			885		885	1,325
0350 160 thru 2,000 MBH		.80	40			2,425		2,425	3,650
0360 2,100 thru 4,500 MBH		.50	64			3,900		3,900	5,850
0370 4,600 thru 7,000 MBH		.30	107			6,500		6,500	9,750
0380 7,100 thru 12,000 MBH		.16	200			12,200		12,200	18,300
0390 12,200 thru 25,000 MBH		.12	267			16,200		16,200	24,400
0400 Central station air handler unit, up thru 15 ton	Q-5	1.60	10			575		575	865
0410 17.5 thru 30 ton	"	.80	20			1,150		1,150	1,725
0430 Computer room unit									
0434 Air cooled split, up thru 10 ton	Q-5	.67	23.881	Ea.		1,375		1,375	2,050
0436 12 thru 23 ton		.53	30.189			1,725		1,725	2,600
0440 Chilled water, up thru 10 ton		1.30	12.308			710		710	1,075
0444 12 thru 23 ton		1	16			920		920	1,375
0450 Glycol system, up thru 10 ton		.53	30.189			1,725		1,725	2,600
0454 12 thru 23 ton		.40	40			2,300		2,300	3,450
0460 Water cooled, not including condenser, up thru 10 ton		.80	20			1,150		1,150	1,725
0464 12 thru 23 ton		.60	26.667			1,525		1,525	2,300
0600 Condenser, up thru 50 ton		1	16			920		920	1,375
0610 51 thru 100 ton	Q-6	.90	26.667			1,600		1,600	2,400
0620 101 thru 1,000 ton	"	.70	34.286			2,050		2,050	3,075
0660 Condensing unit, up thru 10 ton	Q-5	1.25	12.800			735		735	1,100
0670 11 thru 50 ton	"	.40	40			2,300		2,300	3,450
0680 60 thru 100 ton	Q-6	.30	80			4,775		4,775	7,175
0700 Cooling tower, up thru 400 ton		.80	30			1,800		1,800	2,675
0710 450 thru 600 ton		.53	45.283			2,700		2,700	4,050
0720 700 thru 1,300 ton		.40	60			3,575		3,575	5,375
0780 Dehumidifier, up thru 155 lb./hr.	Q-1	8	2			114		114	171
0790 240 lb./hr. and up	"	2	8			455		455	680
1560 Ductwork									
1570 Metal, steel, sst, fabricated	Q-9	1000	.016	Lb.		.88		.88	1.33
1580 Aluminum, fabricated		485	.033	"		1.81		1.81	2.75
1590 Spiral, prefabricated		400	.040	L.F.		2.19		2.19	3.33
1600 Fiberglass, prefabricated		400	.040			2.19		2.19	3.33
1610 Flex, prefabricated		500	.032			1.76		1.76	2.67
1620 Glass fiber reinforced plastic, prefabricated		280	.057			3.13		3.13	4.76

	23 05 05.10 HVAC Demolition	Crew	Daily Output	Labor-Hours	Unit	Material	2019 Bare Costs Labor	Equipment	Total	Total Incl O&P
1630	Diffusers, registers or grills, up thru 20" max dimension	1 Shee	50	.160	Ea.		9.75		9.75	14.80
1640	21 thru 36" max dimension		36	.222			13.55		13.55	20.50
1650	Above 36" max dimension		30	.267			16.25		16.25	24.50
1700	Evaporator, up thru 12,000 BTUH	Q-5	5.30	3.019			174		174	261
1710	12,500 thru 30,000 BTUH	"	2.70	5.926			340		340	510
1720	31,000 BTUH and up	Q-6	1.50	16			955		955	1,425
1730	Evaporative cooler, up thru 5 HP	Q-9	2.70	5.926			325		325	495
1740	10 thru 30 HP	"	.67	23.881			1,300		1,300	2,000
1750	Exhaust systems									
1760	Exhaust components	1 Shee	8	1	System		61		61	92.50
1770	Weld fume hoods	"	20	.400	Ea.		24.50		24.50	37
2120	Fans, up thru 1 HP or 2,000 CFM	Q-9	8	2			110		110	167
2124	1-1/2 thru 10 HP or 20,000 CFM		5.30	3.019			166		166	251
2128	15 thru 30 HP or above 20,000 CFM		4	4			219		219	335
2150	Fan coil air conditioner, chilled water, up thru 7.5 ton	Q-5	14	1.143			66		66	98.50
2154	Direct expansion, up thru 10 ton		8	2			115		115	173
2158	11 thru 30 ton		2	8			460		460	690
2170	Flue shutter damper	Q-9	8	2			110		110	167
2200	Furnace, electric	Q-20	2	10			560		560	845
2300	Gas or oil, under 120 MBH	Q-9	4	4			219		219	335
2340	Over 120 MBH	"	3	5.333			293		293	445
2730	Heating and ventilating unit	Q-5	2.70	5.926			340		340	510
2740	Heater, electric, wall, baseboard and quartz	1 Elec	10	.800			48		48	71.50
2750	Heater, electric, unit, cabinet, fan and convector	"	8	1			60		60	89.50
2760	Heat exchanger, shell and tube type	Q-5	1.60	10			575		575	865
2770	Plate type	Q-6	.60	40			2,375		2,375	3,575
2810	Heat pump									
2820	Air source, split, 4 thru 10 ton	Q-5	.90	17.778	Ea.		1,025		1,025	1,525
2830	15 thru 25 ton	Q-6	.80	30			1,800		1,800	2,675
2850	Single package, up thru 12 ton	Q-5	1	16			920		920	1,375
2860	Water source, up thru 15 ton	"	.90	17.778			1,025		1,025	1,525
2870	20 thru 50 ton	Q-6	.80	30			1,800		1,800	2,675
2910	Heat recovery package, up thru 20,000 CFM	Q-5	2	8			460		460	690
2920	25,000 CFM and up		1.20	13.333			765		765	1,150
2930	Heat transfer package, up thru 130 GPM		.80	20			1,150		1,150	1,725
2934	255 thru 800 GPM		.42	38.095			2,200		2,200	3,300
2940	Humidifier		10.60	1.509			87		87	130
2961	Hydronic unit heaters, up thru 200 MBH		14	1.143			66		66	98.50
2962	Above 200 MBH		8	2			115		115	173
2964	Valance units		32	.500			29		29	43
2966	Radiant floor heating									
2967	System valves, controls, manifolds	Q-5	16	1	Ea.		57.50		57.50	86.50
2968	Per room distribution		8	2			115		115	173
2970	Hydronic heating, baseboard radiation		16	1			57.50		57.50	86.50
2976	Convectors and free standing radiators		18	.889			51		51	77
2980	Induced draft fan, up thru 1 HP	Q-9	4.60	3.478			191		191	290
2984	1-1/2 HP thru 7-1/2 HP	"	2.20	7.273			400		400	605
2988	Infrared unit	Q-5	16	1			57.50		57.50	86.50
2992	Louvers	1 Shee	46	.174	S.F.		10.60		10.60	16.10
3000	Mechanical equipment, light items. Unit is weight, not cooling.	Q-5	.90	17.778	Ton		1,025		1,025	1,525
3600	Heavy items		1.10	14.545	"		835		835	1,250
3720	Make-up air unit, up thru 6,000 CFM		3	5.333	Ea.		305		305	460
3730	6,500 thru 30,000 CFM		1.60	10			575		575	865

23 05 05 – Selective Demolition for HVAC

23 05 05.10 HVAC Demolition		Crew	Daily Output	Labor-Hours	Unit	Material	2019 Bare Costs Labor	Equipment	Total	Total Incl O&P
3740	35,000 thru 75,000 CFM	Q-6	1	24	Ea.		1,425		1,425	2,150
3800	Mixing boxes, constant and VAV	Q-9	18	.889			49		49	74
4000	Packaged terminal air conditioner, up thru 18,000 BTUH	Q-5	8	2			115		115	173
4010	24,000 thru 48,000 BTUH	″	2.80	5.714			330		330	495
5000	Refrigerant compressor, reciprocating or scroll									
5010	Up thru 5 ton	1 Stpi	6	1.333	Ea.		85.50		85.50	128
5020	5.08 thru 10 ton	Q-5	6	2.667			153		153	230
5030	15 thru 50 ton	″	3	5.333			305		305	460
5040	60 thru 130 ton	Q-6	2.80	8.571			510		510	770
5090	Remove refrigerant from system	1 Stpi	40	.200	Lb.		12.80		12.80	19.20
5100	Rooftop air conditioner, up thru 10 ton	Q-5	1.40	11.429	Ea.		660		660	985
5110	12 thru 40 ton	Q-6	1	24			1,425		1,425	2,150
5120	50 thru 140 ton		.50	48			2,875		2,875	4,300
5130	150 thru 300 ton		.30	80			4,775		4,775	7,175
6000	Self contained single package air conditioner, up thru 10 ton	Q-5	1.60	10			575		575	865
6010	15 thru 60 ton	Q-6	1.20	20			1,200		1,200	1,800
6100	Space heaters, up thru 200 MBH	Q-5	10	1.600			92		92	138
6110	Over 200 MBH		5	3.200			184		184	276
6200	Split ductless, both sections		8	2			115		115	173
6300	Steam condensate meter	1 Stpi	11	.727			46.50		46.50	70
6600	Thru-the-wall air conditioner	L-2	8	2			90.50		90.50	138
7000	Vent chimney, prefabricated, up thru 12″ diameter	Q-9	94	.170	V.L.F.		9.35		9.35	14.20
7010	14″ thru 36″ diameter		40	.400			22		22	33.50
7020	38″ thru 48″ diameter		32	.500			27.50		27.50	41.50
7030	54″ thru 60″ diameter	Q-10	14	1.714			97.50		97.50	148
7400	Ventilators, up thru 14″ neck diameter	Q-9	58	.276	Ea.		15.15		15.15	23
7410	16″ thru 50″ neck diameter		40	.400			22		22	33.50
7450	Relief vent, up thru 24″ x 96″		22	.727			40		40	60.50
7460	48″ x 60″ thru 96″ x 144″		10	1.600			88		88	133
8000	Water chiller up thru 10 ton	Q-5	2.50	6.400			370		370	555
8010	15 thru 100 ton	Q-6	.48	50			2,975		2,975	4,475
8020	110 thru 500 ton	Q-7	.29	110			6,725		6,725	10,100
8030	600 thru 1000 ton		.23	139			8,475		8,475	12,700
8040	1100 ton and up		.20	160			9,750		9,750	14,600
8400	Window air conditioner	1 Carp	16	.500			26		26	39

23 05 23 – General-Duty Valves for HVAC Piping

23 05 23.20 Valves, Bronze/Brass		Crew	Daily Output	Labor-Hours	Unit	Material	Labor	Equipment	Total	Total Incl O&P
0010	**VALVES, BRONZE/BRASS**									
0020	Brass									
1300	Ball combination valves, shut-off and union									
1310	Solder, with strainer, drain and PT ports									
1320	1/2″	1 Stpi	17	.471	Ea.	57	30		87	108
1330	3/4″		16	.500		62	32		94	117
1340	1″		14	.571		91	36.50		127.50	155
1350	1-1/4″		12	.667		108	42.50		150.50	183
1360	1-1/2″		10	.800		159	51		210	252
1370	2″		8	1		199	64		263	315
1410	Threaded, with strainer, drain and PT ports									
1420	1/2″	1 Stpi	20	.400	Ea.	57	25.50		82.50	101
1430	3/4″		18	.444		62	28.50		90.50	111
1440	1″		17	.471		91	30		121	145
1450	1-1/4″		12	.667		108	42.50		150.50	183

23 05 23 – General-Duty Valves for HVAC Piping

23 05 23.20 Valves, Bronze/Brass		Crew	Daily Output	Labor-Hours	Unit	Material	2019 Bare Costs Labor	Equipment	Total	Total Incl O&P
1460	1-1/2"	1 Stpi	11	.727	Ea.	159	46.50		205.50	245
1470	2"	↓	9	.889	↓	199	57		256	305

23 05 23.30 Valves, Iron Body		Crew	Daily Output	Labor-Hours	Unit	Material	Labor	Equipment	Total	Total Incl O&P
0010	**VALVES, IRON BODY** R220523-90									
0022	For grooved joint, see Section 22 11 13.48									
0100	Angle, 125 lb.									
0110	Flanged									
0116	2"	1 Plum	5	1.600	Ea.	1,300	101		1,401	1,575
0118	4"	Q-1	3	5.333		2,475	305		2,780	3,175
0120	6"	Q-2	3	8		4,825	470		5,295	6,000
0122	8"	"	2.50	9.600		6,475	565		7,040	7,975
0560	Butterfly, lug type, pneumatic operator, 2" size	1 Stpi	14	.571		605	36.50		641.50	720
0570	3"	Q-1	8	2		655	114		769	890
0580	4"	"	5	3.200		775	182		957	1,125
0590	6"	Q-2	5	4.800		995	283		1,278	1,525
0600	8"		4.50	5.333		1,225	315		1,540	1,825
0610	10"		4	6		1,575	355		1,930	2,250
0620	12"		3	8		1,900	470		2,370	2,800
0630	14"		2.30	10.435		2,525	615		3,140	3,700
0640	18"		1.50	16		4,325	945		5,270	6,175
0650	20"	↓	1	24		5,525	1,425		6,950	8,200
0790	Butterfly, lug type, gear operated, 2" size, 200 lb. except noted	1 Plum	14	.571		355	36		391	445
0800	2-1/2"	Q-1	9	1.778		355	101		456	545
0810	3"		8	2		375	114		489	585
0820	4"	↓	5	3.200		440	182		622	760
0830	5"	Q-2	5	4.800		585	283		868	1,075
0840	6"		5	4.800		605	283		888	1,100
0850	8"		4.50	5.333		785	315		1,100	1,325
0860	10"		4	6		1,100	355		1,455	1,750
0870	12"		3	8		1,500	470		1,970	2,350
0880	14", 150 lb.		2.30	10.435		2,575	615		3,190	3,750
0890	16", 150 lb.		1.75	13.714		3,825	810		4,635	5,425
0900	18", 150 lb.		1.50	16		4,700	945		5,645	6,600
0910	20", 150 lb.		1	24		6,275	1,425		7,700	9,025
0930	24", 150 lb.	↓	.75	32	↓	10,700	1,875		12,575	14,500
1020	Butterfly, wafer type, gear actuator, 200 lb.									
1030	2"	1 Plum	14	.571	Ea.	114	36		150	180
1040	2-1/2"	Q-1	9	1.778		131	101		232	296
1050	3"		8	2		121	114		235	305
1060	4"	↓	5	3.200		135	182		317	420
1070	5"	Q-2	5	4.800		153	283		436	595
1080	6"		5	4.800		174	283		457	615
1090	8"		4.50	5.333		219	315		534	710
1100	10"		4	6		289	355		644	845
1110	12"	↓	3	8	↓	585	470		1,055	1,350
1200	Wafer type, lever actuator, 200 lb.									
1220	2"	1 Plum	14	.571	Ea.	189	36		225	262
1230	2-1/2"	Q-1	9	1.778		180	101		281	350
1240	3"		8	2		212	114		326	405
1250	4"	↓	5	3.200		237	182		419	535
1260	5"	Q-2	5	4.800		340	283		623	800
1270	6"	↓	5	4.800	↓	390	283		673	855

23 05 23.30 Valves, Iron Body		Crew	Daily Output	Labor-Hours	Unit	Material	Labor	Equipment	Total	Total Incl O&P
						2019 Bare Costs				
1280	8"	Q-2	4.50	5.333	Ea.	665	315		980	1,200
1290	10"		4	6		775	355		1,130	1,375
1300	12"		3	8		1,300	470		1,770	2,125
1650	Gate, 125 lb., N.R.S.									
2150	Flanged									
2200	2"	1 Plum	5	1.600	Ea.	600	101		701	810
2240	2-1/2"	Q-1	5	3.200		615	182		797	955
2260	3"		4.50	3.556		690	202		892	1,075
2280	4"		3	5.333		990	305		1,295	1,550
2290	5"	Q-2	3.40	7.059		1,675	415		2,090	2,475
2300	6"		3	8		1,675	470		2,145	2,550
2320	8"		2.50	9.600		2,900	565		3,465	4,025
2340	10"		2.20	10.909		5,100	645		5,745	6,575
2360	12"		1.70	14.118		7,000	830		7,830	8,950
2370	14"		1.30	18.462		10,500	1,100		11,600	13,100
2380	16"		1	24		14,200	1,425		15,625	17,700
2420	For 250 lb. flanged, add					200%	10%			
3550	OS&Y, 125 lb., flanged									
3600	2"	1 Plum	5	1.600	Ea.	400	101		501	590
3640	2-1/2"	Q-1	5	3.200		420	182		602	740
3660	3"		4.50	3.556		450	202		652	800
3670	3-1/2"		3	5.333		745	305		1,050	1,275
3680	4"		3	5.333		675	305		980	1,200
3690	5"	Q-2	3.40	7.059		1,075	415		1,490	1,800
3700	6"		3	8		1,075	470		1,545	1,875
3720	8"		2.50	9.600		1,900	565		2,465	2,925
3740	10"		2.20	10.909		3,450	645		4,095	4,775
3760	12"		1.70	14.118		4,725	830		5,555	6,425
3770	14"		1.30	18.462		10,300	1,100		11,400	12,900
3780	16"		1	24		14,500	1,425		15,925	18,100
3790	18"		.80	30		20,600	1,775		22,375	25,400
3800	20"		.60	40		24,500	2,350		26,850	30,500
3830	24"		.50	48		36,200	2,825		39,025	44,200
3900	For 175 lb., flanged, add					200%	10%			
4350	Globe, OS&Y									
4540	Class 125, flanged									
4550	2"	1 Plum	5	1.600	Ea.	910	101		1,011	1,150
4560	2-1/2"	Q-1	5	3.200		910	182		1,092	1,275
4570	3"		4.50	3.556		1,100	202		1,302	1,525
4580	4"		3	5.333		1,600	305		1,905	2,200
4590	5"	Q-2	3.40	7.059		2,900	415		3,315	3,800
4600	6"		3	8		2,900	470		3,370	3,875
4610	8"		2.50	9.600		5,700	565		6,265	7,125
4612	10"		2.20	10.909		10,300	645		10,945	12,300
4614	12"		1.70	14.118		10,800	830		11,630	13,200
5040	Class 250, flanged									
5050	2"	1 Plum	4.50	1.778	Ea.	1,475	112		1,587	1,775
5060	2-1/2"	Q-1	4.50	3.556		1,900	202		2,102	2,400
5070	3"		4	4		1,975	227		2,202	2,525
5080	4"		2.70	5.926		2,875	335		3,210	3,675
5090	5"	Q-2	3	8		6,525	470		6,995	7,875
5100	6"		2.70	8.889		5,275	525		5,800	6,600
5110	8"		2.20	10.909		8,775	645		9,420	10,600

23 05 23 – General-Duty Valves for HVAC Piping

23 05 23.30 Valves, Iron Body		Crew	Daily Output	Labor-Hours	Unit	Material	2019 Bare Costs Labor	Equipment	Total	Total Incl O&P
5120	10"	Q-2	2	12	Ea.	15,300	705		16,005	17,900
5130	12"		1.60	15		23,500	885		24,385	27,200
5240	Valve sprocket rim w/chain, for 2" valve	1 Stpi	30	.267		179	17.05		196.05	223
5250	2-1/2" valve		27	.296		179	18.95		197.95	226
5260	3-1/2" valve		25	.320		188	20.50		208.50	238
5270	6" valve		20	.400		188	25.50		213.50	246
5280	8" valve		18	.444		223	28.50		251.50	288
5290	12" valve		16	.500		235	32		267	305
5300	16" valve		12	.667		252	42.50		294.50	340
5310	20" valve		10	.800		252	51		303	355
5320	36" valve		8	1		390	64		454	520
5450	Swing check, 125 lb., threaded									
5500	2"	1 Plum	11	.727	Ea.	540	46		586	665
5540	2-1/2"	Q-1	15	1.067		615	60.50		675.50	770
5550	3"		13	1.231		670	70		740	840
5560	4"		10	1.600		1,050	91		1,141	1,300
5950	Flanged									
5994	1"	1 Plum	7	1.143	Ea.	279	72		351	415
5998	1-1/2"		6	1.333		440	84		524	610
6000	2"		5	1.600		475	101		576	670
6040	2-1/2"	Q-1	5	3.200		360	182		542	670
6050	3"		4.50	3.556		415	202		617	765
6060	4"		3	5.333		730	305		1,035	1,250
6070	6"	Q-2	3	8		1,200	470		1,670	2,025
6080	8"		2.50	9.600		2,350	565		2,915	3,425
6090	10"		2.20	10.909		3,975	645		4,620	5,350
6100	12"		1.70	14.118		5,950	830		6,780	7,800
6102	14"		1.55	15.484		13,200	910		14,110	15,900
6104	16"		1.40	17.143		16,700	1,000		17,700	19,900
6110	18"		1.30	18.462		25,400	1,100		26,500	29,600
6112	20"		1	24		40,400	1,425		41,825	46,600
6114	24"		.75	32		58,500	1,875		60,375	67,500
6160	For 250 lb. flanged, add					200%	20%			
6600	Silent check, bronze trim									
6610	Compact wafer type, for 125 or 150 lb. flanges									
6630	1-1/2"	1 Plum	11	.727	Ea.	155	46		201	240
6640	2"	"	9	.889		190	56		246	293
6650	2-1/2"	Q-1	9	1.778		207	101		308	380
6660	3"		8	2		240	114		354	435
6670	4"		5	3.200		271	182		453	570
6680	5"	Q-2	6	4		375	236		611	765
6690	6"		6	4		525	236		761	935
6700	8"		4.50	5.333		890	315		1,205	1,450
6710	10"		4	6		1,525	355		1,880	2,200
6720	12"		3	8		2,750	470		3,220	3,725
6740	For 250 or 300 lb. flanges, thru 6" no change									
6741	For 8" and 10", add				Ea.	11%	10%			
6750	Twin disc									
6752	2"	1 Plum	9	.889	Ea.	355	56		411	475
6754	4"	Q-1	5	3.200		640	182		822	980
6756	6"	Q-2	5	4.800		955	283		1,238	1,475
6758	8"		4.50	5.333		1,450	315		1,765	2,075
6760	10"		4	6		2,325	355		2,680	3,075

23 05 23.30 Valves, Iron Body		Crew	Daily Output	Labor-Hours	Unit	2019 Bare Costs Material	Labor	Equipment	Total	Total Incl O&P
6762	12"	Q-2	3	8	Ea.	3,025	470		3,495	4,025
6764	18"		1.50	16		13,000	945		13,945	15,700
6766	24"		.75	32		18,400	1,875		20,275	23,000
6800	Full flange type, 150 lb.									
6810	1"	1 Plum	14	.571	Ea.	181	36		217	253
6811	1-1/4"		12	.667		190	42		232	273
6812	1-1/2"		11	.727		221	46		267	310
6813	2"		9	.889		253	56		309	365
6814	2-1/2"	Q-1	9	1.778		286	101		387	465
6815	3"		8	2		350	114		464	550
6816	4"		5	3.200		395	182		577	710
6817	5"	Q-2	6	4		620	236		856	1,025
6818	6"		5	4.800		815	283		1,098	1,325
6819	8"		4.50	5.333		1,400	315		1,715	2,025
6820	10"		4	6		1,725	355		2,080	2,425
6840	For 250 lb., add					30%	10%			
6900	Globe type, 125 lb.									
6910	2"	1 Plum	9	.889	Ea.	370	56		426	490
6911	2-1/2"	Q-1	9	1.778		490	101		591	690
6912	3"		8	2		435	114		549	645
6913	4"		5	3.200		740	182		922	1,075
6914	5"	Q-2	6	4		1,100	236		1,336	1,550
6915	6"		5	4.800		1,275	283		1,558	1,850
6916	8"		4.50	5.333		2,350	315		2,665	3,050
6917	10"		4	6		2,950	355		3,305	3,775
6918	12"		3	8		5,125	470		5,595	6,325
6919	14"		2.30	10.435		6,700	615		7,315	8,300
6920	16"		1.75	13.714		7,275	810		8,085	9,225
6921	18"		1.50	16		11,800	945		12,745	14,400
6922	20"		1	24		13,700	1,425		15,125	17,200
6923	24"		.75	32		15,000	1,875		16,875	19,300
6940	For 250 lb., add					40%	10%			
6980	Screwed end type, 125 lb.									
6981	1"	1 Plum	19	.421	Ea.	93	26.50		119.50	142
6982	1-1/4"		15	.533		112	33.50		145.50	174
6983	1-1/2"		13	.615		136	39		175	209
6984	2"		11	.727		186	46		232	274

23 05 23.50 Valves, Multipurpose		Crew	Daily Output	Labor-Hours	Unit	Material	Labor	Equipment	Total	Total Incl O&P
0010	**VALVES, MULTIPURPOSE**									
0100	Functions as a shut off, balancing, check & metering valve									
1000	Cast iron body									
1010	Threaded									
1020	1-1/2" size	1 Stpi	11	.727	Ea.	425	46.50		471.50	540
1030	2" size	"	8	1		520	64		584	665
1040	2-1/2" size	Q-5	5	3.200		560	184		744	890
1200	Flanged									
1210	3" size	Q-5	4.20	3.810	Ea.	645	219		864	1,050
1220	4" size	"	3	5.333		1,300	305		1,605	1,875
1230	5" size	Q-6	3.80	6.316		1,550	375		1,925	2,275
1240	6" size		3	8		1,925	475		2,400	2,850
1250	8" size		2.50	9.600		3,375	575		3,950	4,550
1260	10" size		2.20	10.909		4,825	650		5,475	6,275

23 05 23 – General-Duty Valves for HVAC Piping

23 05 23.50 Valves, Multipurpose		Crew	Daily Output	Labor-Hours	Unit	2019 Bare Costs Material	Labor	Equipment	Total	Total Incl O&P
1270	12″ size	Q-6	2.10	11.429	Ea.	8,350	680		9,030	10,200
1280	14″ size		2	12		12,400	715		13,115	14,800

23 05 23.70 Valves, Semi-Steel		Crew	Daily Output	Labor-Hours	Unit	Material	Labor	Equipment	Total	Total Incl O&P
0010	**VALVES, SEMI-STEEL** R220523-90									
1020	Lubricated plug valve, threaded, 200 psi									
1030	1/2″	1 Plum	18	.444	Ea.	106	28		134	159
1040	3/4″		16	.500		106	31.50		137.50	165
1050	1″		14	.571		112	36		148	177
1060	1-1/4″		12	.667		132	42		174	208
1070	1-1/2″		11	.727		145	46		191	228
1080	2″		8	1		233	63		296	350
1090	2-1/2″	Q-1	5	3.200		350	182		532	660
1100	3″	″	4.50	3.556		425	202		627	770
6990	Flanged, 200 psi									
7000	2″	1 Plum	8	1	Ea.	270	63		333	390
7010	2-1/2″	Q-1	5	3.200		380	182		562	690
7020	3″		4.50	3.556		455	202		657	810
7030	4″		3	5.333		570	305		875	1,075
7036	5″		2.50	6.400		1,600	365		1,965	2,325
7040	6″	Q-2	3	8		1,925	470		2,395	2,825
7050	8″		2.50	9.600		2,050	565		2,615	3,100
7060	10″		2.20	10.909		3,275	645		3,920	4,600
7070	12″		1.70	14.118		5,075	830		5,905	6,825

23 05 23.80 Valves, Steel		Crew	Daily Output	Labor-Hours	Unit	Material	Labor	Equipment	Total	Total Incl O&P
0010	**VALVES, STEEL** R220523-90									
0800	Cast									
1350	Check valve, swing type, 150 lb., flanged									
1370	1″	1 Plum	10	.800	Ea.	360	50.50		410.50	475
1400	2″	″	8	1		635	63		698	795
1440	2-1/2″	Q-1	5	3.200		740	182		922	1,075
1450	3″		4.50	3.556		750	202		952	1,125
1460	4″		3	5.333		1,050	305		1,355	1,600
1470	6″	Q-2	3	8		1,625	470		2,095	2,500
1480	8″		2.50	9.600		2,875	565		3,440	4,025
1490	10″		2.20	10.909		4,275	645		4,920	5,675
1500	12″		1.70	14.118		6,200	830		7,030	8,075
1510	14″		1.30	18.462		9,300	1,100		10,400	11,800
1520	16″		1	24		10,300	1,425		11,725	13,400
1540	For 300 lb., flanged, add					50%	15%			
1548	For 600 lb., flanged, add					110%	20%			
1571	300 lb., 2″	1 Plum	7.40	1.081		660	68.50		728.50	825
1572	2-1/2″	Q-1	4.20	3.810		990	217		1,207	1,400
1573	3″		4	4		990	227		1,217	1,425
1574	4″		2.80	5.714		1,375	325		1,700	2,000
1575	6″	Q-2	2.90	8.276		2,675	490		3,165	3,675
1576	8″		2.40	10		3,950	590		4,540	5,225
1577	10″		2.10	11.429		5,950	675		6,625	7,550
1578	12″		1.60	15		8,650	885		9,535	10,800
1579	14″		1.20	20		12,800	1,175		13,975	15,900
1581	16″		.90	26.667		16,400	1,575		17,975	20,400
1950	Gate valve, 150 lb., flanged									
2000	2″	1 Plum	8	1	Ea.	635	63		698	795

23 05 23.80 Valves, Steel		Crew	Daily Output	Labor-Hours	Unit	2019 Bare Costs Material	Labor	Equipment	Total	Total Incl O&P
2040	2-1/2"	Q-1	5	3.200	Ea.	900	182		1,082	1,275
2050	3"		4.50	3.556		900	202		1,102	1,300
2060	4"		3	5.333		1,150	305		1,455	1,700
2070	6"	Q-2	3	8		1,825	470		2,295	2,725
2080	8"		2.50	9.600		2,850	565		3,415	4,000
2090	10"		2.20	10.909		4,125	645		4,770	5,525
2100	12"		1.70	14.118		5,650	830		6,480	7,450
2110	14"		1.30	18.462		9,225	1,100		10,325	11,700
2120	16"		1	24		11,900	1,425		13,325	15,200
2130	18"		.80	30		15,600	1,775		17,375	19,900
2140	20"		.60	40		18,000	2,350		20,350	23,300
2650	300 lb., flanged									
2700	2"	1 Plum	7.40	1.081	Ea.	870	68.50		938.50	1,050
2740	2-1/2"	Q-1	4.20	3.810		1,200	217		1,417	1,650
2750	3"		4	4		1,200	227		1,427	1,675
2760	4"		2.80	5.714		1,675	325		2,000	2,300
2770	6"	Q-2	2.90	8.276		2,825	490		3,315	3,825
2780	8"		2.40	10		4,400	590		4,990	5,725
2790	10"		2.10	11.429		5,925	675		6,600	7,500
2800	12"		1.60	15		8,550	885		9,435	10,700
2810	14"		1.20	20		17,100	1,175		18,275	20,600
2820	16"		.90	26.667		21,500	1,575		23,075	26,000
2830	18"		.70	34.286		30,600	2,025		32,625	36,700
2840	20"		.50	48		34,100	2,825		36,925	41,800
3650	Globe valve, 150 lb., flanged									
3700	2"	1 Plum	8	1	Ea.	800	63		863	975
3740	2-1/2"	Q-1	5	3.200		1,025	182		1,207	1,400
3750	3"		4.50	3.556		1,025	202		1,227	1,425
3760	4"		3	5.333		1,475	305		1,780	2,075
3770	6"	Q-2	3	8		2,325	470		2,795	3,275
3780	8"		2.50	9.600		4,375	565		4,940	5,650
3790	10"		2.20	10.909		8,050	645		8,695	9,825
3800	12"		1.70	14.118		10,700	830		11,530	13,100
4080	300 lb., flanged									
4100	2"	1 Plum	7.40	1.081	Ea.	1,075	68.50		1,143.50	1,275
4140	2-1/2"	Q-1	4.20	3.810		1,450	217		1,667	1,925
4150	3"		4	4		1,450	227		1,677	1,950
4160	4"		2.80	5.714		2,025	325		2,350	2,700
4170	6"	Q-2	2.90	8.276		3,650	490		4,140	4,725
4180	8"		2.40	10		6,200	590		6,790	7,700
4190	10"		2.10	11.429		11,900	675		12,575	14,100
4200	12"		1.60	15		14,300	885		15,185	17,000
4680	600 lb., flanged									
4700	2"	1 Plum	7	1.143	Ea.	1,500	72		1,572	1,750
4740	2-1/2"	Q-1	4	4		2,400	227		2,627	2,975
4750	3"		3.60	4.444		2,400	253		2,653	3,000
4760	4"		2.50	6.400		3,675	365		4,040	4,600
4770	6"	Q-2	2.60	9.231		7,775	545		8,320	9,375
4780	8"	"	2.10	11.429		12,900	675		13,575	15,200
4800	Silent check, 316 S.S. trim									
4810	Full flange type, 150 lb.									
4811	1"	1 Plum	14	.571	Ea.	585	36		621	700
4812	1-1/4"		12	.667		585	42		627	705

23 05 23.80 Valves, Steel		Crew	Daily Output	Labor-Hours	Unit	2019 Bare Costs Material	Labor	Equipment	Total	Total Incl O&P
4813	1-1/2"	1 Plum	11	.727	Ea.	640	46		686	775
4814	2"		9	.889		740	56		796	900
4815	2-1/2"	Q-1	9	1.778		930	101		1,031	1,175
4816	3"		8	2		1,050	114		1,164	1,325
4817	4"		5	3.200		1,450	182		1,632	1,850
4818	5"	Q-2	6	4		1,850	236		2,086	2,375
4819	6"		5	4.800		2,300	283		2,583	2,950
4820	8"		4.50	5.333		3,450	315		3,765	4,275
4821	10"		4	6		4,875	355		5,230	5,900
4840	For 300 lb., add					20%	10%			
4860	For 600 lb., add					50%	15%			
4900	Globe type, 150 lb.									
4910	2"	1 Plum	9	.889	Ea.	880	56		936	1,050
4911	2-1/2"	Q-1	9	1.778		1,100	101		1,201	1,350
4912	3"		8	2		1,175	114		1,289	1,475
4913	4"		5	3.200		1,575	182		1,757	2,000
4914	5"	Q-2	6	4		2,050	236		2,286	2,600
4915	6"		5	4.800		2,425	283		2,708	3,100
4916	8"		4.50	5.333		3,525	315		3,840	4,375
4917	10"		4	6		6,100	355		6,455	7,250
4918	12"		3	8		7,525	470		7,995	8,975
4919	14"		2.30	10.435		10,600	615		11,215	12,600
4920	16"		1.75	13.714		14,800	810		15,610	17,500
4921	18"		1.50	16		15,600	945		16,545	18,600
4940	For 300 lb., add					30%	10%			
4960	For 600 lb., add					60%	15%			
5150	Forged									
5340	Ball valve, 1,500 psi, threaded, 1/4" size	1 Plum	24	.333	Ea.	67.50	21		88.50	106
5350	3/8"		24	.333		65	21		86	103
5360	1/2"		24	.333		86	21		107	126
5370	3/4"		20	.400		102	25.50		127.50	150
5380	1"		19	.421		138	26.50		164.50	192
5390	1-1/4"		15	.533		178	33.50		211.50	247
5400	1-1/2"		13	.615		241	39		280	325
5410	2"		11	.727		315	46		361	420
5460	Ball valve, 800 lb., socket weld, 1/4" size	Q-15	19	.842		69	48	2.96	119.96	151
5470	3/8"		19	.842		69	48	2.96	119.96	151
5480	1/2"		19	.842		90	48	2.96	140.96	174
5490	3/4"		19	.842		119	48	2.96	169.96	206
5500	1"		15	1.067		149	60.50	3.74	213.24	258
5510	1-1/4"		13	1.231		198	70	4.32	272.32	330
5520	1-1/2"		11	1.455		248	82.50	5.10	335.60	400
5530	2"		8.50	1.882		320	107	6.60	433.60	520
5550	Ball valve, 150 lb., flanged									
5560	4"	Q-1	3	5.333	Ea.	1,575	305		1,880	2,175
5570	6"	Q-2	3	8		2,400	470		2,870	3,325
5580	8"	"	2.50	9.600		5,725	565		6,290	7,150
5650	Check valve, class 800, horizontal									
5651	Socket									
5652	1/4"	Q-15	19	.842	Ea.	83	48	2.96	133.96	167
5654	3/8"		19	.842		83	48	2.96	133.96	167
5656	1/2"		19	.842		83	48	2.96	133.96	167
5658	3/4"		19	.842		89	48	2.96	139.96	173

23 05 23.80 Valves, Steel		Crew	Daily Output	Labor-Hours	Unit	Material	Labor	Equipment	Total	Total Incl O&P
						2019 Bare Costs				
5660	1"	Q-15	15	1.067	Ea.	105	60.50	3.74	169.24	210
5662	1-1/4"		13	1.231		205	70	4.32	279.32	335
5664	1-1/2"		11	1.455		205	82.50	5.10	292.60	355
5666	2"		8.50	1.882		288	107	6.60	401.60	485
5698	Threaded									
5700	1/4"	1 Plum	24	.333	Ea.	83	21		104	123
5720	3/8"		24	.333		83	21		104	123
5730	1/2"		24	.333		83	21		104	123
5740	3/4"		20	.400		89	25.50		114.50	136
5750	1"		19	.421		105	26.50		131.50	155
5760	1-1/4"		15	.533		205	33.50		238.50	277
5770	1-1/2"		13	.615		205	39		244	285
5780	2"		11	.727		288	46		334	385
5840	For class 150, flanged, add					100%	15%			
5860	For class 300, flanged, add					120%	20%			
6100	Gate, class 800, OS&Y, socket									
6102	3/8"	Q-15	19	.842	Ea.	57.50	48	2.96	108.46	139
6103	1/2"		19	.842		57.50	48	2.96	108.46	139
6104	3/4"		19	.842		63	48	2.96	113.96	145
6105	1"		15	1.067		76.50	60.50	3.74	140.74	180
6106	1-1/4"		13	1.231		145	70	4.32	219.32	269
6107	1-1/2"		11	1.455		145	82.50	5.10	232.60	289
6108	2"		8.50	1.882		199	107	6.60	312.60	385
6118	Threaded									
6120	3/8"	1 Plum	24	.333	Ea.	57.50	21		78.50	95
6130	1/2"		24	.333		57.50	21		78.50	95
6140	3/4"		20	.400		63	25.50		88.50	108
6150	1"		19	.421		76.50	26.50		103	125
6160	1-1/4"		15	.533		145	33.50		178.50	210
6170	1-1/2"		13	.615		145	39		184	218
6180	2"		11	.727		199	46		245	288
6260	For OS&Y, flanged, add					100%	20%			
6700	Globe, OS&Y, class 800, socket									
6710	1/4"	Q-15	19	.842	Ea.	88.50	48	2.96	139.46	173
6720	3/8"		19	.842		88.50	48	2.96	139.46	173
6730	1/2"		19	.842		88.50	48	2.96	139.46	173
6740	3/4"		19	.842		102	48	2.96	152.96	187
6750	1"		15	1.067		133	60.50	3.74	197.24	241
6760	1-1/4"		13	1.231		260	70	4.32	334.32	395
6770	1-1/2"		11	1.455		260	82.50	5.10	347.60	415
6780	2"		8.50	1.882		330	107	6.60	443.60	535
6860	For OS&Y, flanged, add					300%	20%			
6880	Threaded									
6882	1/4"	1 Plum	24	.333	Ea.	88.50	21		109.50	129
6884	3/8"		24	.333		88.50	21		109.50	129
6886	1/2"		24	.333		83	21		104	123
6888	3/4"		20	.400		102	25.50		127.50	150
6890	1"		19	.421		133	26.50		159.50	186
6892	1-1/4"		15	.533		260	33.50		293.50	335
6894	1-1/2"		13	.615		260	39		299	345
6896	2"		11	.727		330	46		376	435

23 05 23.90 Valves, Stainless Steel		Crew	Daily Output	Labor-Hours	Unit	Material	Labor	Equipment	Total	Total Incl O&P
						2019 Bare Costs				
0010	**VALVES, STAINLESS STEEL** R220523-90									
1610	Ball, threaded 1/4"	1 Stpi	24	.333	Ea.	43	21.50		64.50	79
1620	3/8"		24	.333		43	21.50		64.50	79
1630	1/2"		22	.364		43	23.50		66.50	82
1640	3/4"		20	.400		71	25.50		96.50	117
1650	1"		19	.421		86.50	27		113.50	136
1660	1-1/4"		15	.533		211	34		245	283
1670	1-1/2"		13	.615		217	39.50		256.50	298
1680	2"	↓	11	.727	↓	278	46.50		324.50	375
1700	Check, 200 lb., threaded									
1710	1/4"	1 Plum	24	.333	Ea.	143	21		164	189
1720	1/2"		22	.364		143	23		166	192
1730	3/4"		20	.400		148	25.50		173.50	201
1750	1"		19	.421		193	26.50		219.50	252
1760	1-1/2"		13	.615		360	39		399	460
1770	2"	↓	11	.727	↓	615	46		661	745
1800	150 lb., flanged									
1810	2-1/2"	Q-1	5	3.200	Ea.	1,900	182		2,082	2,350
1820	3"		4.50	3.556		1,875	202		2,077	2,375
1830	4"	↓	3	5.333		2,800	305		3,105	3,525
1840	6"	Q-2	3	8		4,950	470		5,420	6,125
1850	8"	"	2.50	9.600	↓	10,100	565		10,665	12,000
2100	Gate, OS&Y, 150 lb., flanged									
2120	1/2"	1 Plum	18	.444	Ea.	520	28		548	610
2140	3/4"		16	.500		505	31.50		536.50	605
2150	1"		14	.571		635	36		671	755
2160	1-1/2"		11	.727		1,225	46		1,271	1,425
2170	2"	↓	8	1		1,450	63		1,513	1,700
2180	2-1/2"	Q-1	5	3.200		1,750	182		1,932	2,200
2190	3"		4.50	3.556		2,275	202		2,477	2,800
2200	4"		3	5.333		2,575	305		2,880	3,275
2205	5"	↓	2.80	5.714		4,850	325		5,175	5,800
2210	6"	Q-2	3	8		4,800	470		5,270	5,975
2220	8"		2.50	9.600		8,350	565		8,915	10,100
2230	10"		2.30	10.435		14,900	615		15,515	17,200
2240	12"	↓	1.90	12.632		19,700	745		20,445	22,800
2260	For 300 lb., flanged, add				↓	120%	15%			
2600	600 lb., flanged									
2620	1/2"	1 Plum	16	.500	Ea.	187	31.50		218.50	254
2640	3/4"		14	.571		203	36		239	277
2650	1"		12	.667		244	42		286	330
2660	1-1/2"		10	.800		390	50.50		440.50	505
2670	2"	↓	7	1.143		535	72		607	700
2680	2-1/2"	Q-1	4	4		6,475	227		6,702	7,475
2690	3"	"	3.60	4.444	↓	6,475	253		6,728	7,500
3100	Globe, OS&Y, 150 lb., flanged									
3120	1/2"	1 Plum	18	.444	Ea.	500	28		528	590
3140	3/4"		16	.500		540	31.50		571.50	645
3150	1"		14	.571		705	36		741	835
3160	1-1/2"		11	.727		1,125	46		1,171	1,300
3170	2"	↓	8	1		1,450	63		1,513	1,675
3180	2-1/2"	Q-1	5	3.200	↓	2,975	182		3,157	3,550

23 05 23 – General-Duty Valves for HVAC Piping

23 05 23.90 Valves, Stainless Steel		Crew	Daily Output	Labor-Hours	Unit	2019 Bare Costs Material	Labor	Equipment	Total	Total Incl O&P
3190	3"	Q-1	4.50	3.556	Ea.	2,975	202		3,177	3,575
3200	4"		3	5.333		4,775	305		5,080	5,700
3210	6"	Q-2	3	8		8,050	470		8,520	9,550
5000	Silent check, 316 S.S. body and trim									
5010	Compact wafer type, 300 lb.									
5020	1"	1 Plum	14	.571	Ea.	1,075	36		1,111	1,250
5021	1-1/4"		12	.667		1,200	42		1,242	1,375
5022	1-1/2"		11	.727		1,225	46		1,271	1,400
5023	2"		9	.889		1,375	56		1,431	1,600
5024	2-1/2"	Q-1	9	1.778		1,900	101		2,001	2,250
5025	3"	"	8	2		2,200	114		2,314	2,600
5100	Full flange wafer type, 150 lb.									
5110	1"	1 Plum	14	.571	Ea.	765	36		801	895
5111	1-1/4"		12	.667		820	42		862	965
5112	1-1/2"		11	.727		1,025	46		1,071	1,200
5113	2"		9	.889		1,225	56		1,281	1,425
5114	2-1/2"	Q-1	9	1.778		1,575	101		1,676	1,875
5115	3"		8	2		1,650	114		1,764	1,975
5116	4"		5	3.200		2,450	182		2,632	2,975
5117	5"	Q-2	6	4		2,850	236		3,086	3,475
5118	6"		5	4.800		3,575	283		3,858	4,350
5119	8"		4.50	5.333		4,950	315		5,265	5,925
5120	10"		4	6		9,300	355		9,655	10,700
5200	Globe type, 300 lb.									
5210	2"	1 Plum	9	.889	Ea.	1,275	56		1,331	1,475
5211	2-1/2"	Q-1	9	1.778		1,575	101		1,676	1,875
5212	3"		8	2		1,950	114		2,064	2,300
5213	4"		5	3.200		2,700	182		2,882	3,250
5214	5"	Q-2	6	4		4,175	236		4,411	4,925
5215	6"		5	4.800		4,275	283		4,558	5,125
5216	8"		4.50	5.333		6,025	315		6,340	7,100
5217	10"		4	6		8,550	355		8,905	9,925
5300	Screwed end type, 300 lb.									
5310	1/2"	1 Plum	24	.333	Ea.	82	21		103	122
5311	3/4"		20	.400		112	25.50		137.50	161
5312	1"		19	.421		129	26.50		155.50	182
5313	1-1/4"		15	.533		151	33.50		184.50	217
5314	1-1/2"		13	.615		162	39		201	238
5315	2"		11	.727		200	46		246	289

23 05 93 – Testing, Adjusting, and Balancing for HVAC

23 05 93.10 Balancing, Air		Crew	Daily Output	Labor-Hours	Unit	Material	Labor	Equipment	Total	Total Incl O&P
0010	**BALANCING, AIR** (Subcontractor's quote incl. material and labor) R230500-10									
0900	Heating and ventilating equipment									
1000	Centrifugal fans, utility sets				Ea.				420	420
1100	Heating and ventilating unit								630	630
1200	In-line fan								630	630
1300	Propeller and wall fan								119	119
1400	Roof exhaust fan								280	280
2000	Air conditioning equipment, central station								910	910
2100	Built-up low pressure unit								840	840
2200	Built-up high pressure unit								980	980
2300	Built-up high pressure dual duct								1,550	1,550

23 05 93 – Testing, Adjusting, and Balancing for HVAC

23 05 93.10 Balancing, Air		Crew	Daily Output	Labor-Hours	Unit	2019 Bare Costs Material	Labor	Equipment	Total	Total Incl O&P
2400	Built-up variable volume				Ea.				1,825	1,825
2500	Multi-zone A.C. and heating unit								630	630
2600	For each zone over one, add								140	140
2700	Package A.C. unit								350	350
2800	Rooftop heating and cooling unit								490	490
3000	Supply, return, exhaust, registers & diffusers, avg. height ceiling								84	84
3100	High ceiling								126	126
3200	Floor height								70	70
3300	Off mixing box								56	56
3500	Induction unit								91	91
3600	Lab fume hood								420	420
3700	Linear supply								210	210
3800	Linear supply high								245	245
4000	Linear return								70	70
4100	Light troffers								84	84
4200	Moduline - master								84	84
4300	Moduline - slaves								42	42
4400	Regenerators								560	560
4500	Taps into ceiling plenums								105	105
4600	Variable volume boxes								84	84

23 05 93.20 Balancing, Water		Crew	Daily Output	Labor-Hours	Unit	Material	Labor	Equipment	Total	Total Incl O&P
0010	**BALANCING, WATER** (Subcontractor's quote incl. material and labor) R230500-10									
0050	Air cooled condenser				Ea.				256	256
0080	Boiler								515	515
0100	Cabinet unit heater								88	88
0200	Chiller								620	620
0300	Convector								73	73
0400	Converter								365	365
0500	Cooling tower								475	475
0600	Fan coil unit, unit ventilator								132	132
0700	Fin tube and radiant panels								146	146
0800	Main and duct re-heat coils								135	135
0810	Heat exchanger								135	135
0900	Main balancing cocks								110	110
1000	Pumps								320	320
1100	Unit heater								102	102

23 05 93.50 Piping, Testing		Crew	Daily Output	Labor-Hours	Unit	Material	Labor	Equipment	Total	Total Incl O&P
0010	**PIPING, TESTING**									
0100	Nondestructive testing									
0110	Nondestructive hydraulic pressure test, isolate & 1 hr. hold									
0120	1" - 4" pipe									
0140	0-250 L.F.	1 Stpi	1.33	6.015	Ea.		385		385	575
0160	250-500 L.F.	"	.80	10			640		640	960
0180	500-1000 L.F.	Q-5	1.14	14.035			810		810	1,200
0200	1000-2000 L.F.	"	.80	20			1,150		1,150	1,725
0300	6" - 10" pipe									
0320	0-250 L.F.	Q-5	1	16	Ea.		920		920	1,375
0340	250-500 L.F.		.73	21.918			1,250		1,250	1,900
0360	500-1000 L.F.		.53	30.189			1,725		1,725	2,600
0380	1000-2000 L.F.		.38	42.105			2,425		2,425	3,625
1000	Pneumatic pressure test, includes soaping joints									
1120	1" - 4" pipe									

23 05 Common Work Results for HVAC

23 05 93 – Testing, Adjusting, and Balancing for HVAC

23 05 93.50 Piping, Testing		Crew	Daily Output	Labor-Hours	Unit	Material	2019 Bare Costs Labor	Equipment	Total	Total Incl O&P
1140	0-250 L.F.	Q-5	2.67	5.993	Ea.	12	345		357	530
1160	250-500 L.F.		1.33	12.030		24	690		714	1,075
1180	500-1000 L.F.		.80	20		36	1,150		1,186	1,775
1200	1000-2000 L.F.		.50	32		48	1,850		1,898	2,825
1300	6" - 10" pipe									
1320	0-250 L.F.	Q-5	1.33	12.030	Ea.	12	690		702	1,075
1340	250-500 L.F.		.67	23.881		24	1,375		1,399	2,075
1360	500-1000 L.F.		.40	40		48	2,300		2,348	3,500
1380	1000-2000 L.F.		.25	64		60	3,675		3,735	5,600
2000	X-Ray of welds									
2110	2" diam.	1 Stpi	8	1	Ea.	16.05	64		80.05	114
2120	3" diam.		8	1		16.05	64		80.05	114
2130	4" diam.		8	1		24	64		88	123
2140	6" diam.		8	1		24	64		88	123
2150	8" diam.		6.60	1.212		24	77.50		101.50	143
2160	10" diam.		6	1.333		32	85.50		117.50	164
3000	Liquid penetration of welds									
3110	2" diam.	1 Stpi	14	.571	Ea.	12.40	36.50		48.90	68.50
3120	3" diam.		13.60	.588		12.40	37.50		49.90	70
3130	4" diam.		13.40	.597		12.40	38		50.40	71
3140	6" diam.		13.20	.606		12.40	39		51.40	71.50
3150	8" diam.		13	.615		18.55	39.50		58.05	79.50
3160	10" diam.		12.80	.625		18.55	40		58.55	80.50

23 07 HVAC Insulation

23 07 13 – Duct Insulation

23 07 13.10 Duct Thermal Insulation		Crew	Daily Output	Labor-Hours	Unit	Material	Labor	Equipment	Total	Total Incl O&P
0010	**DUCT THERMAL INSULATION**									
0110	Insulation req'd. is based on the surface size/area to be covered									
3000	Ductwork									
3020	Blanket type, fiberglass, flexible									
3030	Fire rated for grease and hazardous exhaust ducts									
3060	1-1/2" thick	Q-14	84	.190	S.F.	4.62	9.85		14.47	20.50
3090	Fire rated for plenums									
3100	1/2" x 24" x 25'	Q-14	1.94	8.247	Roll	172	425		597	845
3110	1/2" x 24" x 25'		98	.163	S.F.	3.43	8.45		11.88	16.75
3120	1/2" x 48" x 25'		1.04	15.385	Roll	340	795		1,135	1,600
3126	1/2" x 48" x 25'		104	.154	S.F.	3.39	7.95		11.34	16
3140	FSK vapor barrier wrap, .75 lb. density									
3160	1" thick G	Q-14	350	.046	S.F.	.23	2.36		2.59	3.89
3170	1-1/2" thick G		320	.050		.27	2.58		2.85	4.29
3180	2" thick G		300	.053		.32	2.75		3.07	4.60
3190	3" thick G		260	.062		.47	3.18		3.65	5.40
3200	4" thick G		242	.066		.67	3.41		4.08	6
3210	Vinyl jacket, same as FSK									
3280	Unfaced, 1 lb. density									
3310	1" thick G	Q-14	360	.044	S.F.	.25	2.29		2.54	3.82
3320	1-1/2" thick G		330	.048		.40	2.50		2.90	4.30
3330	2" thick G		310	.052		.45	2.66		3.11	4.61
3400	FSK facing, 1 lb. density									
3420	1-1/2" thick G	Q-14	310	.052	S.F.	.38	2.66		3.04	4.53

23 07 13 – Duct Insulation

23 07 13.10 Duct Thermal Insulation			Crew	Daily Output	Labor-Hours	Unit	2019 Bare Costs Material	Labor	Equipment	Total	Total Incl O&P
3430	2" thick	G	Q-14	300	.053	S.F.	.44	2.75		3.19	4.73
3450	FSK facing, 1.5 lb. density										
3470	1-1/2" thick	G	Q-14	300	.053	S.F.	.46	2.75		3.21	4.76
3480	2" thick	G	"	290	.055	"	.57	2.85		3.42	5.05
3730	Sheet insulation										
3760	Polyethylene foam, closed cell, UV resistant										
3770	Standard temperature (-90°F to +212°F)										
3771	1/4" thick	G	Q-14	450	.036	S.F.	1.78	1.84		3.62	4.79
3772	3/8" thick	G		440	.036		2.55	1.88		4.43	5.70
3773	1/2" thick	G		420	.038		3.13	1.97		5.10	6.50
3774	3/4" thick	G		400	.040		4.47	2.07		6.54	8.10
3775	1" thick	G		380	.042		6.05	2.17		8.22	10
3776	1-1/2" thick	G		360	.044		9.55	2.29		11.84	14.05
3777	2" thick	G		340	.047		12.65	2.43		15.08	17.65
3778	2-1/2" thick	G	↓	320	.050	↓	16.20	2.58		18.78	22
3779	Adhesive (see line 7878)										
3780	Foam, rubber										
3782	1" thick	G	1 Stpi	50	.160	S.F.	2.98	10.25		13.23	18.65
3795	Finishes										
3800	Stainless steel woven mesh		Q-14	100	.160	S.F.	.95	8.25		9.20	13.80
3810	For .010" stainless steel, add			160	.100		3.32	5.15		8.47	11.60
3820	18 oz. fiberglass cloth, pasted on			170	.094		.86	4.86		5.72	8.45
3900	8 oz. canvas, pasted on			180	.089		.27	4.59		4.86	7.40
3940	For .016" aluminum jacket, add		↓	200	.080	↓	1.06	4.13		5.19	7.55
7000	Board insulation										
7020	Mineral wool, 1200° F										
7022	6 lb. density, plain										
7024	1" thick	G	Q-14	370	.043	S.F.	.34	2.23		2.57	3.82
7026	1-1/2" thick	G		350	.046		.51	2.36		2.87	4.20
7028	2" thick	G		330	.048		.68	2.50		3.18	4.61
7030	3" thick	G		300	.053		.84	2.75		3.59	5.15
7032	4" thick	G	↓	280	.057	↓	1.37	2.95		4.32	6.05
7038	8 lb. density, plain										
7040	1" thick	G	Q-14	360	.044	S.F.	.42	2.29		2.71	4
7042	1-1/2" thick	G		340	.047		.61	2.43		3.04	4.42
7044	2" thick	G		320	.050		.83	2.58		3.41	4.90
7046	3" thick	G		290	.055		1.25	2.85		4.10	5.80
7048	4" thick	G	↓	270	.059	↓	1.62	3.06		4.68	6.50
7060	10 lb. density, plain										
7062	1" thick	G	Q-14	350	.046	S.F.	.58	2.36		2.94	4.28
7064	1-1/2" thick	G		330	.048		.87	2.50		3.37	4.82
7066	2" thick	G		310	.052		1.16	2.66		3.82	5.40
7068	3" thick	G		280	.057		1.75	2.95		4.70	6.50
7070	4" thick	G	↓	260	.062	↓	2.22	3.18		5.40	7.35
7878	Contact cement, quart can					Ea.	13.05			13.05	14.40

23 07 16 – HVAC Equipment Insulation

23 07 16.10 HVAC Equipment Thermal Insulation			Crew	Daily Output	Labor-Hours	Unit	Material	Labor	Equipment	Total	Total Incl O&P
0010	**HVAC EQUIPMENT THERMAL INSULATION**										
0110	Insulation req'd. is based on the surface size/area to be covered										
1000	Boiler, 1-1/2" calcium silicate only	G	Q-14	110	.145	S.F.	4.02	7.50		11.52	16
1020	Plus 2" fiberglass	G	"	80	.200	"	5.55	10.35		15.90	22
2000	Breeching, 2" calcium silicate										

23 07 HVAC Insulation

23 07 16 – HVAC Equipment Insulation

23 07 16.10 HVAC Equipment Thermal Insulation			Crew	Daily Output	Labor-Hours	Unit	2019 Bare Costs Material	Labor	Equipment	Total	Total Incl O&P
2020	Rectangular	G	Q-14	42	.381	S.F.	7.75	19.65		27.40	39
2040	Round	G	"	38.70	.413	"	8.05	21.50		29.55	42
2300	Calcium silicate block, +200°F to +1,200°F										
2310	On irregular surfaces, valves and fittings										
2340	1" thick	G	Q-14	30	.533	S.F.	3.50	27.50		31	46.50
2360	1-1/2" thick	G		25	.640		3.95	33		36.95	55.50
2380	2" thick	G		22	.727		4.02	37.50		41.52	62.50
2400	3" thick	G		18	.889		6.20	46		52.20	78
2410	On plane surfaces										
2420	1" thick	G	Q-14	126	.127	S.F.	3.50	6.55		10.05	13.95
2430	1-1/2" thick	G		120	.133		3.95	6.90		10.85	15
2440	2" thick	G		100	.160		4.02	8.25		12.27	17.15
2450	3" thick	G		70	.229		6.20	11.80		18	25

23 09 Instrumentation and Control for HVAC

23 09 13 – Instrumentation and Control Devices for HVAC

23 09 13.60 Water Level Controls

		Crew	Daily Output	Labor-Hours	Unit	Material	Labor	Equipment	Total	Total Incl O&P
0010	**WATER LEVEL CONTROLS**									
1000	Electric water feeder	1 Stpi	12	.667	Ea.	340	42.50		382.50	440
2000	Feeder cut-off combination									
2100	Steam system up to 5,000 sq. ft.	1 Stpi	12	.667	Ea.	680	42.50		722.50	815
2200	Steam system above 5,000 sq. ft.		12	.667		925	42.50		967.50	1,100
2300	Steam and hot water, high pressure		10	.800		1,100	51		1,151	1,275
3000	Low water cut-off for hot water boiler, 50 psi maximum									
3100	1" top & bottom equalizing pipes, manual reset	1 Stpi	14	.571	Ea.	415	36.50		451.50	510
3200	1" top & bottom equalizing pipes		14	.571		380	36.50		416.50	470
3300	2-1/2" side connection for nipple-to-boiler		14	.571		380	36.50		416.50	470
4000	Low water cut-off for low pressure steam with quick hook-up ftgs.									
4100	For installation in gauge glass tappings	1 Stpi	16	.500	Ea.	310	32		342	390
4200	Built-in type, 2-1/2" tap, 3-1/8" insertion		16	.500		248	32		280	320
4300	Built-in type, 2-1/2" tap, 1-3/4" insertion		16	.500		273	32		305	350
4400	Side connection to 2-1/2" tapping		16	.500		290	32		322	370
5000	Pump control, low water cut-off and alarm switch		14	.571		750	36.50		786.50	880
9000	Water gauges, complete									
9010	Rough brass, wheel type									
9020	125 psi at 350°F									
9030	3/8" pipe size	1 Stpi	11	.727	Ea.	67.50	46.50		114	145
9040	1/2" pipe size	"	10	.800	"	69.50	51		120.50	154
9060	200 psi at 400°F									
9070	3/8" pipe size	1 Stpi	11	.727	Ea.	84	46.50		130.50	163
9080	1/2" pipe size		10	.800		84	51		135	170
9090	3/4" pipe size		9	.889		107	57		164	204
9130	Rough brass, chain lever type									
9140	250 psi at 400°F									
9200	Polished brass, wheel type									
9210	200 psi at 400°F									
9220	3/8" pipe size	1 Stpi	11	.727	Ea.	159	46.50		205.50	245
9230	1/2" pipe size		10	.800		162	51		213	255
9240	3/4" pipe size		9	.889		171	57		228	274
9260	Polished brass, chain lever type									
9270	250 psi at 400°F									

23 09 13 – Instrumentation and Control Devices for HVAC

23 09 13.60 Water Level Controls		Crew	Daily Output	Labor-Hours	Unit	Material	Labor	Equipment	Total	Total Incl O&P
						2019 Bare Costs				
9280	1/2" pipe size	1 Stpi	10	.800	Ea.	234	51		285	335
9290	3/4" pipe size	"	9	.889	"	254	57		311	365
9400	Bronze, high pressure, ASME									
9410	1/2" pipe size	1 Stpi	10	.800	Ea.	274	51		325	375
9420	3/4" pipe size	"	9	.889	"	305	57		362	420
9460	Chain lever type									
9470	1/2" pipe size	1 Stpi	10	.800	Ea.	375	51		426	485
9480	3/4" pipe size	"	9	.889	"	445	57		502	575
9500	316 stainless steel, high pressure, ASME									
9510	500 psi at 450°F									
9520	1/2" pipe size	1 Stpi	10	.800	Ea.	830	51		881	985
9530	3/4" pipe size	"	9	.889	"	805	57		862	970

23 09 23 – Direct-Digital Control System for HVAC

23 09 23.10 Control Components/DDC Systems		Crew	Daily Output	Labor-Hours	Unit	Material	Labor	Equipment	Total	Total Incl O&P
0010	**CONTROL COMPONENTS/DDC SYSTEMS** (Sub's quote incl. M & L)									
0100	Analog inputs									
0110	Sensors (avg. 50' run in 1/2" EMT)									
0120	Duct temperature				Ea.				415	415
0130	Space temperature								665	665
0140	Duct humidity, +/- 3%								695	695
0150	Space humidity, +/- 2%								1,075	1,075
0160	Duct static pressure								565	565
0170	CFM/transducer								765	765
0172	Water temperature								655	655
0174	Water flow								2,400	2,400
0176	Water pressure differential								975	975
0177	Steam flow								2,400	2,400
0178	Steam pressure								1,025	1,025
0180	KW/transducer								1,350	1,350
0182	KWH totalization (not incl. elec. meter pulse xmtr.)								625	625
0190	Space static pressure				↓				1,075	1,075
1000	Analog outputs (avg. 50' run in 1/2" EMT)									
1010	P/I transducer				Ea.				635	635
1020	Analog output, matl. in MUX								305	305
1030	Pneumatic (not incl. control device)								645	645
1040	Electric (not incl. control device)				↓				380	380
2000	Status (alarms)									
2100	Digital inputs (avg. 50' run in 1/2" EMT)									
2110	Freeze				Ea.				435	435
2120	Fire								395	395
2130	Differential pressure (air)								595	595
2140	Differential pressure (water)								975	975
2150	Current sensor								435	435
2160	Duct high temperature thermostat								570	570
2170	Duct smoke detector				↓				705	705
2200	Digital output (avg. 50' run in 1/2" EMT)									
2210	Start/stop				Ea.				340	340
2220	On/off (maintained contact)				"				585	585
3000	Controller MUX panel, incl. function boards									
3100	48 point				Ea.				5,275	5,275
3110	128 point				"				7,225	7,225
3200	DDC controller (avg. 50' run in conduit)									

23 09 Instrumentation and Control for HVAC

23 09 23 – Direct-Digital Control System for HVAC

23 09 23.10 Control Components/DDC Systems		Crew	Daily Output	Labor-Hours	Unit	Material	2019 Bare Costs Labor	Equipment	Total	Total Incl O&P
3210	Mechanical room									
3214	16 point controller (incl. 120 volt/1 phase power supply)				Ea.				3,275	3,275
3229	32 point controller (incl. 120 volt/1 phase power supply)				"				5,425	5,425
3230	Includes software programming and checkout									
3260	Space									
3266	VAV terminal box (incl. space temp. sensor)				Ea.				840	840
3280	Host computer (avg. 50′ run in conduit)									
3281	Package complete with PC, keyboard,									
3282	printer, monitor, basic software				Ea.				3,150	3,150
4000	Front end costs									
4100	Computer (P.C.) with software program				Ea.				6,350	6,350
4200	Color graphics software								3,925	3,925
4300	Color graphics slides								490	490
4350	Additional printer				↓				980	980
4400	Communications trunk cable				L.F.				3.80	3.80
4500	Engineering labor (not incl. dftg.)				Point				94	94
4600	Calibration labor								120	120
4700	Start-up, checkout labor				↓				120	120
4800	Programming labor, as req'd									
5000	Communications bus (data transmission cable)									
5010	#18 twisted shielded pair in 1/2″ EMT conduit				C.L.F.				380	380
8000	Applications software									
8050	Basic maintenance manager software (not incl. data base entry)				Ea.				1,950	1,950
8100	Time program				Point				6.85	6.85
8120	Duty cycle								13.65	13.65
8140	Optimum start/stop								41.50	41.50
8160	Demand limiting								20.50	20.50
8180	Enthalpy program				↓				41.50	41.50
8200	Boiler optimization				Ea.				1,225	1,225
8220	Chiller optimization				"				1,625	1,625
8240	Custom applications									
8260	Cost varies with complexity									

23 09 33 – Electric and Electronic Control System for HVAC

23 09 33.10 Electronic Control Systems

		Crew	Daily Output	Labor-Hours	Unit	Material	Labor	Equipment	Total	Total Incl O&P
0010	**ELECTRONIC CONTROL SYSTEMS** R230500-10									
0020	For electronic costs, add to Section 23 09 43.10				Ea.				15%	15%

23 09 43 – Pneumatic Control System for HVAC

23 09 43.10 Pneumatic Control Systems

		Crew	Daily Output	Labor-Hours	Unit	Material	Labor	Equipment	Total	Total Incl O&P
0010	**PNEUMATIC CONTROL SYSTEMS** R230500-10									
0011	Including a nominal 50′ of tubing. Add control panelboard if req'd.									
0100	Heating and ventilating, split system									
0200	Mixed air control, economizer cycle, panel readout, tubing									
0220	Up to 10 tons G	Q-19	.68	35.294	Ea.	4,500	2,050		6,550	8,050
0240	For 10 to 20 tons G		.63	37.915		4,775	2,225		7,000	8,575
0260	For over 20 tons G		.58	41.096		5,175	2,400		7,575	9,275
0270	Enthalpy cycle, up to 10 tons		.50	48.387		4,925	2,825		7,750	9,650
0280	For 10 to 20 tons		.46	52.174		5,325	3,050		8,375	10,400
0290	For over 20 tons	↓	.42	56.604	↓	5,825	3,300		9,125	11,400
0300	Heating coil, hot water, 3 way valve,									
0320	Freezestat, limit control on discharge, readout	Q-5	.69	23.088	Ea.	3,350	1,325		4,675	5,675
0500	Cooling coil, chilled water, room									
0520	Thermostat, 3 way valve	Q-5	2	8	Ea.	1,500	460		1,960	2,350

23 09 43 – Pneumatic Control System for HVAC

23 09 43.10 Pneumatic Control Systems		Crew	Daily Output	Labor-Hours	Unit	2019 Bare Costs Material	Labor	Equipment	Total	Total Incl O&P
0600	Cooling tower, fan cycle, damper control,									
0620	Control system including water readout in/out at panel	Q-19	.67	35.821	Ea.	5,925	2,100		8,025	9,650
1000	Unit ventilator, day/night operation,									
1100	freezestat, ASHRAE, cycle 2	Q-19	.91	26.374	Ea.	3,275	1,550		4,825	5,925
2000	Compensated hot water from boiler, valve control,									
2100	readout and reset at panel, up to 60 GPM	Q-19	.55	43.956	Ea.	6,075	2,575		8,650	10,500
2120	For 120 GPM		.51	47.059		6,575	2,750		9,325	11,400
2140	For 240 GPM		.49	49.180		6,800	2,875		9,675	11,800
3000	Boiler room combustion air, damper to 5 S.F., controls		1.37	17.582		2,975	1,025		4,000	4,775
3500	Fan coil, heating and cooling valves, 4 pipe control system		3	8		1,325	465		1,790	2,150
3600	Heat exchanger system controls		.86	27.907		2,875	1,625		4,500	5,625
3900	Multizone control (one per zone), includes thermostat, damper									
3910	motor and reset of discharge temperature	Q-5	.51	31.373	Ea.	2,975	1,800		4,775	5,950
4000	Pneumatic thermostat, including controlling room radiator valve	"	2.43	6.593		890	380		1,270	1,550
4040	Program energy saving optimizer G	Q-19	1.21	19.786		7,400	1,150		8,550	9,875
4060	Pump control system	"	3	8		1,375	465		1,840	2,200
4080	Reheat coil control system, not incl. coil	Q-5	2.43	6.593		1,150	380		1,530	1,850
4500	Air supply for pneumatic control system									
4600	Tank mounted duplex compressor, starter, alternator,									
4620	piping, dryer, PRV station and filter									
4630	1/2 HP	Q-19	.68	35.139	Ea.	10,900	2,050		12,950	15,100
4640	3/4 HP		.64	37.383		11,500	2,175		13,675	16,000
4650	1 HP		.61	39.539		12,600	2,300		14,900	17,400
4660	1-1/2 HP		.58	41.739		13,500	2,425		15,925	18,500
4680	3 HP		.55	43.956		18,300	2,575		20,875	24,100
4690	5 HP		.42	57.143		31,600	3,325		34,925	39,800
4800	Main air supply, includes 3/8" copper main and labor	Q-5	1.82	8.791	C.L.F.	365	505		870	1,175
4810	If poly tubing used, deduct								30%	30%
7000	Static pressure control for air handling unit, includes pressure									
7010	sensor, receiver controller, readout and damper motors	Q-19	.64	37.383	Ea.	8,675	2,175		10,850	12,800
7020	If return air fan requires control, add								70%	70%
8600	VAV boxes, incl. thermostat, damper motor, reheat coil & tubing	Q-5	1.46	10.989		1,375	630		2,005	2,475
8610	If no reheat coil, deduct								204	204

23 09 53 – Pneumatic and Electric Control System for HVAC

23 09 53.10 Control Components		Crew	Daily Output	Labor-Hours	Unit	Material	Labor	Equipment	Total	Total Incl O&P
0010	**CONTROL COMPONENTS**									
0500	Aquastats									
0508	Immersion type									
0510	High/low limit, breaks contact w/temp rise	1 Stpi	4	2	Ea.	365	128		493	590
0514	Sequencing, break 2 switches w/temp rise		4	2		335	128		463	555
0518	Circulating, makes contact w/temp rise		4	2		238	128		366	455
0600	Carbon monoxide detector system									
0606	Panel	1 Stpi	4	2	Ea.	980	128		1,108	1,275
0610	Sensor		7.30	1.096		655	70		725	825
0680	Controller for VAV box, includes actuator		7.30	1.096		279	70		349	410
0700	Controller, receiver									
0730	Pneumatic, panel mount, single input	1 Plum	8	1	Ea.	460	63		523	605
0740	With conversion mounting bracket		8	1		460	63		523	605
0750	Dual input, with control point adjustment		7	1.143		640	72		712	810
0850	Electric, single snap switch	1 Elec	4	2		490	120		610	715
0860	Dual snap switches		3	2.667		655	160		815	960
0870	Humidity controller		8	1		266	60		326	385

23 09 53.10 Control Components		Crew	Daily Output	Labor-Hours	Unit	2019 Bare Costs Material	Labor	Equipment	Total	Total Incl O&P
0880	Load limiting controller	1 Elec	8	1	Ea.	875	60		935	1,050
0890	Temperature controller		8	1		490	60		550	630
0900	Control panel readout									
0910	Panel up to 12 indicators	1 Stpi	1.20	6.667	Ea.	315	425		740	985
0914	Panel up to 24 indicators		.86	9.302		330	595		925	1,250
0918	Panel up to 48 indicators		.50	16		800	1,025		1,825	2,400
1000	Enthalpy control, boiler water temperature control									
1010	governed by outdoor temperature, with timer	1 Elec	3	2.667	Ea.	360	160		520	635
1300	Energy control/monitor									
1320	BTU computer meter and controller	1 Stpi	1	8	Ea.	1,075	510		1,585	1,950
1600	Flow meters									
1610	Gas	1 Stpi	4	2	Ea.	98.50	128		226.50	300
1620	Liquid	"	4	2	"	102	128		230	305
1640	Freezestat									
1644	20′ sensing element, adjustable	1 Stpi	3.50	2.286	Ea.	106	146		252	335
2000	Gauges, pressure or vacuum									
2100	2″ diameter dial	1 Stpi	32	.250	Ea.	9.80	16		25.80	35
2200	2-1/2″ diameter dial		32	.250		11.90	16		27.90	37
2300	3-1/2″ diameter dial		32	.250		18.50	16		34.50	44.50
2400	4-1/2″ diameter dial		32	.250		21	16		37	47
2700	Flanged iron case, black ring									
2800	3-1/2″ diameter dial	1 Stpi	32	.250	Ea.	101	16		117	135
2900	4-1/2″ diameter dial		32	.250		105	16		121	139
3000	6″ diameter dial		32	.250		167	16		183	208
3010	Steel case, 0-300 psi									
3012	2″ diameter dial	1 Stpi	16	.500	Ea.	8.05	32		40.05	57
3014	4″ diameter dial	"	16	.500	"	20.50	32		52.50	70.50
3020	Aluminum case, 0-300 psi									
3022	3-1/2″ diameter dial	1 Stpi	16	.500	Ea.	12.80	32		44.80	62
3024	4-1/2″ diameter dial		16	.500		122	32		154	182
3026	6″ diameter dial		16	.500		191	32		223	258
3028	8-1/2″ diameter dial		16	.500		330	32		362	410
3030	Brass case, 0-300 psi									
3032	2″ diameter dial	1 Stpi	16	.500	Ea.	50	32		82	103
3034	4-1/2″ diameter dial	"	16	.500	"	100	32		132	158
3040	Steel case, high pressure, 0-10,000 psi									
3042	4-1/2″ diameter dial	1 Stpi	16	.500	Ea.	269	32		301	345
3044	6-1/2″ diameter dial		16	.500		325	32		357	405
3046	8-1/2″ diameter dial		16	.500		420	32		452	510
3080	Pressure gauge, differential, magnehelic									
3084	0-2″ W.C., with air filter kit	1 Stpi	6	1.333	Ea.	73	85.50		158.50	209
3300	For compound pressure-vacuum, add					18%				
3350	Humidistat									
3360	Pneumatic operation									
3361	Room humidistat, direct acting	1 Stpi	12	.667	Ea.	325	42.50		367.50	425
3362	Room humidistat, reverse acting		12	.667		325	42.50		367.50	425
3363	Room humidity transmitter		17	.471		350	30		380	425
3364	Duct mounted controller		12	.667		380	42.50		422.50	485
3365	Duct mounted transmitter		12	.667		345	42.50		387.50	445
3366	Humidity indicator, 3-1/2″		28	.286		134	18.25		152.25	175
3390	Electric operated	1 Shee	8	1		46	61		107	144
3400	Relays									
3430	Pneumatic/electric	1 Plum	16	.500	Ea.	310	31.50		341.50	390

23 09 53 – Pneumatic and Electric Control System for HVAC

23 09 53.10 Control Components		Crew	Daily Output	Labor-Hours	Unit	2019 Bare Costs Material	Labor	Equipment	Total	Total Incl O&P
3440	Pneumatic proportioning	1 Plum	8	1	Ea.	242	63		305	360
3450	Pneumatic switching		12	.667		154	42		196	232
3460	Selector, 3 point		6	1.333		109	84		193	246
3470	Pneumatic time delay		8	1		297	63		360	420
3500	Sensor, air operated									
3520	Humidity	1 Plum	16	.500	Ea.	380	31.50		411.50	470
3540	Pressure		16	.500		68.50	31.50		100	123
3560	Temperature		12	.667		165	42		207	244
3600	Electric operated									
3620	Humidity	1 Elec	8	1	Ea.	269	60		329	385
3650	Pressure		8	1		305	60		365	425
3680	Temperature		10	.800		118	48		166	202
3700	Switches									
3710	Minimum position									
3720	Electrical	1 Stpi	4	2	Ea.	21	128		149	216
3730	Pneumatic	"	4	2	"	162	128		290	370
4000	Thermometers									
4100	Dial type, 3-1/2" diameter, vapor type, union connection	1 Stpi	32	.250	Ea.	212	16		228	257
4120	Liquid type, union connection		32	.250		410	16		426	475
4130	Remote reading, 15′ capillary		32	.250		228	16		244	275
4500	Stem type, 6-1/2" case, 2" stem, 1/2" NPT		32	.250		62	16		78	92
4520	4" stem, 1/2" NPT		32	.250		76	16		92	108
4600	9" case, 3-1/2" stem, 3/4" NPT		28	.286		92.50	18.25		110.75	130
4620	6" stem, 3/4" NPT		28	.286		113	18.25		131.25	152
4640	8" stem, 3/4" NPT		28	.286		178	18.25		196.25	224
4660	12" stem, 1" NPT		26	.308		153	19.70		172.70	199
4670	Bi-metal, dial type, steel case brass stem									
4672	2" dial, 4" - 9" stem	1 Stpi	16	.500	Ea.	39.50	32		71.50	91.50
4673	2-1/2" dial, 4" - 9" stem		16	.500		39.50	32		71.50	91.50
4674	3-1/2" dial, 4" - 9" stem		16	.500		48	32		80	101
4680	Mercury filled, industrial, union connection type									
4682	Angle stem, 7" scale	1 Stpi	16	.500	Ea.	59	32		91	113
4683	9" scale		16	.500		160	32		192	224
4684	12" scale		16	.500		206	32		238	275
4686	Straight stem, 7" scale		16	.500		158	32		190	221
4687	9" scale		16	.500		147	32		179	209
4688	12" scale		16	.500		181	32		213	247
4690	Mercury filled, industrial, separable socket type, with well									
4692	Angle stem, with socket, 7" scale	1 Stpi	16	.500	Ea.	190	32		222	257
4693	9" scale		16	.500		190	32		222	257
4694	12" scale		16	.500		225	32		257	296
4696	Straight stem, with socket, 7" scale		16	.500		148	32		180	210
4697	9" scale		16	.500		154	32		186	217
4698	12" scale		16	.500		182	32		214	248
5000	Thermostats									
5030	Manual	1 Stpi	8	1	Ea.	42	64		106	143
5040	1 set back, electric, timed G		8	1		75.50	64		139.50	179
5050	2 set back, electric, timed G		8	1		224	64		288	345
5100	Locking cover		20	.400		20.50	25.50		46	61
5200	24 hour, automatic, clock G	1 Shee	8	1		236	61		297	350
5220	Electric, low voltage, 2 wire	1 Elec	13	.615		62	37		99	123
5230	3 wire		10	.800		53	48		101	130
5236	Heating/cooling, low voltage, with clock		8	1		226	60		286	340

23 09 53 – Pneumatic and Electric Control System for HVAC

23 09 53.10 Control Components		Crew	Daily Output	Labor-Hours	Unit	2019 Bare Costs Material	Labor	Equipment	Total	Total Incl O&P
5240	Pneumatic									
5250	Single temp., single pressure	1 Stpi	8	1	Ea.	216	64		280	335
5251	Dual pressure		8	1		325	64		389	450
5252	Dual temp., dual pressure		8	1		287	64		351	410
5253	Reverse acting w/averaging element		8	1		224	64		288	345
5254	Heating-cooling w/deadband		8	1		485	64		549	625
5255	Integral w/piston top valve actuator		8	1		199	64		263	315
5256	Dual temp., dual pressure		8	1		204	64		268	320
5257	Low limit, 8′ averaging element		8	1		185	64		249	300
5258	Room single temp. proportional		8	1		76	64		140	180
5260	Dual temp, direct acting for VAV fan		8	1		252	64		316	375
5262	Capillary tube type, 20′, 30°F to 100°F		8	1		345	64		409	475
5300	Transmitter, pneumatic									
5320	Temperature averaging element	Q-1	8	2	Ea.	130	114		244	315
5350	Pressure differential	1 Plum	7	1.143		1,100	72		1,172	1,300
5370	Humidity, duct		8	1		345	63		408	475
5380	Room		12	.667		350	42		392	445
5390	Temperature, with averaging element		6	1.333		178	84		262	320
5500	Timer/time clocks									
5510	7 day, 12 hr. battery	1 Stpi	2.60	3.077	Ea.	213	197		410	530
5600	Recorders									
5610	Hydrograph humidity, wall mtd., aluminum case, grad. chart									
5612	8″	1 Stpi	1.60	5	Ea.	955	320		1,275	1,525
5614	10″	″	1.60	5	″	1,000	320		1,320	1,575
5620	Thermo-hydrograph, wall mtd., grad. chart, 5′ SS probe									
5622	8″	1 Stpi	1.60	5	Ea.	1,050	320		1,370	1,625
5624	10″	″	1.60	5	″	1,625	320		1,945	2,275
5630	Time of operation, wall mtd., 1 pen cap, 24 hr. clock									
5632	6″	1 Stpi	1.60	5	Ea.	435	320		755	960
5634	8″	″	1.60	5	″	655	320		975	1,200
5640	Pressure & vacuum, bourdon tube type									
5642	Flush mount, 1 pen									
5644	4″	1 Stpi	1.60	5	Ea.	465	320		785	995
5645	6″		1.60	5		405	320		725	925
5646	8″		1.60	5		675	320		995	1,225
5648	10″		1.60	5		1,450	320		1,770	2,075
6000	Valves, motorized zone									
6100	Sweat connections, 1/2″ C x C	1 Stpi	20	.400	Ea.	144	25.50		169.50	197
6110	3/4″ C x C		20	.400		139	25.50		164.50	192
6120	1″ C x C		19	.421		188	27		215	248
6140	1/2″ C x C, with end switch, 2 wire		20	.400		137	25.50		162.50	189
6150	3/4″ C x C, with end switch, 2 wire		20	.400		120	25.50		145.50	171
6160	1″ C x C, with end switch, 2 wire		19	.421		176	27		203	234
7090	Valves, motor controlled, including actuator									
7100	Electric motor actuated									
7200	Brass, two way, screwed									
7210	1/2″ pipe size	L-6	36	.333	Ea.	365	20.50		385.50	430
7220	3/4″ pipe size		30	.400		615	25		640	710
7230	1″ pipe size		28	.429		730	26.50		756.50	845
7240	1-1/2″ pipe size		19	.632		825	39		864	965
7250	2″ pipe size		16	.750		1,175	46.50		1,221.50	1,375
7350	Brass, three way, screwed									
7360	1/2″ pipe size	L-6	33	.364	Ea.	320	22.50		342.50	385

23 09 53 – Pneumatic and Electric Control System for HVAC

23 09 53.10 Control Components		Crew	Daily Output	Labor-Hours	Unit	2019 Bare Costs Material	Labor	Equipment	Total	Total Incl O&P
7370	3/4" pipe size	L-6	27	.444	Ea.	350	27.50		377.50	425
7380	1" pipe size		25.50	.471		405	29		434	490
7384	1-1/4" pipe size		21	.571		475	35.50		510.50	575
7390	1-1/2" pipe size		17	.706		525	44		569	640
7400	2" pipe size		14	.857		765	53.50		818.50	925
7550	Iron body, two way, flanged									
7560	2-1/2" pipe size	L-6	4	3	Ea.	1,150	186		1,336	1,550
7570	3" pipe size		3	4		1,175	248		1,423	1,675
7580	4" pipe size		2	6		2,200	375		2,575	2,975
7850	Iron body, three way, flanged									
7860	2-1/2" pipe size	L-6	3	4	Ea.	1,150	248		1,398	1,650
7870	3" pipe size		2.50	4.800		1,325	298		1,623	1,900
7880	4" pipe size		2	6		1,675	375		2,050	2,400
8000	Pneumatic, air operated									
8050	Brass, two way, screwed									
8060	1/2" pipe size, class 250	1 Plum	24	.333	Ea.	176	21		197	226
8070	3/4" pipe size, class 250		20	.400		211	25.50		236.50	270
8080	1" pipe size, class 250		19	.421		246	26.50		272.50	310
8090	1-1/4" pipe size, class 125		15	.533		305	33.50		338.50	385
8100	1-1/2" pipe size, class 125		13	.615		390	39		429	490
8110	2" pipe size, class 125		11	.727		455	46		501	570
8180	Brass, three way, screwed									
8190	1/2" pipe size, class 250	1 Plum	22	.364	Ea.	181	23		204	234
8200	3/4" pipe size, class 250		18	.444		218	28		246	282
8210	1" pipe size, class 250		17	.471		255	29.50		284.50	325
8214	1-1/4" pipe size, class 250		14	.571		325	36		361	415
8220	1-1/2" pipe size, class 125		11	.727		410	46		456	520
8230	2" pipe size, class 125		9	.889		475	56		531	610
8450	Iron body, two way, flanged									
8510	2-1/2" pipe size, class 125	Q-1	5	3.200	Ea.	920	182		1,102	1,275
8520	3" pipe size, class 125		4.50	3.556		1,000	202		1,202	1,400
8530	4" pipe size, class 125		3	5.333		1,700	305		2,005	2,325
8540	5" pipe size, class 125	Q-2	3.40	7.059		3,725	415		4,140	4,725
8550	6" pipe size, class 125	"	3	8		3,375	470		3,845	4,425
8560	Iron body, three way, flanged									
8570	2-1/2" pipe size, class 125	Q-1	4.50	3.556	Ea.	990	202		1,192	1,400
8580	3" pipe size, class 125		4	4		1,200	227		1,427	1,675
8590	4" pipe size, class 125		2.50	6.400		2,300	365		2,665	3,100
8600	6" pipe size, class 125	Q-2	3	8		3,500	470		3,970	4,550
9005	Pneumatic system misc. components									
9010	Adding/subtracting repeater	1 Stpi	20	.400	Ea.	169	25.50		194.50	225
9020	Adjustable ratio network		16	.500		160	32		192	224
9030	Comparator		20	.400		350	25.50		375.50	420
9040	Cumulator									
9041	Air switching	1 Stpi	20	.400	Ea.	154	25.50		179.50	208
9042	Averaging		20	.400		195	25.50		220.50	254
9043	2:1 ratio		20	.400		325	25.50		350.50	395
9044	Sequencing		20	.400		98	25.50		123.50	147
9045	Two-position		16	.500		192	32		224	259
9046	Two-position pilot		16	.500		365	32		397	455
9050	Damper actuator									
9051	For smoke control	1 Stpi	8	1	Ea.	272	64		336	395
9052	For ventilation	"	8	1	"	240	64		304	360

23 09 53.10 Control Components		Crew	Daily Output	Labor-Hours	Unit	Material	2019 Bare Costs Labor	Equipment	Total	Total Incl O&P
9053	Series duplex pedestal mounted									
9054	For inlet vanes on fans, compressors	1 Stpi	6	1.333	Ea.	1,950	85.50		2,035.50	2,250
9060	Enthalpy logic center		16	.500		325	32		357	405
9070	High/low pressure selector		22	.364		62	23.50		85.50	103
9080	Signal limiter		20	.400		98	25.50		123.50	147
9081	Transmitter		28	.286		37.50	18.25		55.75	68.50
9090	Optimal start									
9092	Mass temperature transmitter	1 Stpi	18	.444	Ea.	147	28.50		175.50	204
9100	Step controller with positioner, time delay restrictor									
9110	recycler air valve, switches and cam settings									
9112	With cabinet									
9113	6 points	1 Stpi	4	2	Ea.	990	128		1,118	1,300
9114	6 points, 6 manual sequences		2	4		870	256		1,126	1,350
9115	8 points		3	2.667		1,100	171		1,271	1,450
9200	Pressure controller and switches									
9210	High static pressure limit	1 Stpi	8	1	Ea.	281	64		345	405
9220	Pressure transmitter		8	1		232	64		296	350
9230	Differential pressure transmitter		8	1		910	64		974	1,100
9240	Static pressure transmitter		8	1		940	64		1,004	1,125
9250	Proportional-only control, single input		6	1.333		460	85.50		545.50	640
9260	Proportional plus integral, single input		6	1.333		660	85.50		745.50	855
9270	Proportional-only, dual input		6	1.333		640	85.50		725.50	830
9280	Proportional plus integral, dual input		6	1.333		805	85.50		890.50	1,025
9281	Time delay for above proportional units		16	.500		162	32		194	226
9290	Proportional/2 position controller units		6	1.333		455	85.50		540.50	630
9300	Differential pressure control direct/reverse		6	1.333		410	85.50		495.50	585
9310	Air pressure reducing valve									
9311	1/8" size	1 Stpi	18	.444	Ea.	38	28.50		66.50	84.50
9315	Precision valve, 1/4" size		17	.471		214	30		244	280
9320	Air flow controller		12	.667		181	42.50		223.50	263
9330	Booster relay volume amplifier		18	.444		105	28.50		133.50	159
9340	Series restrictor, straight		60	.133		5.35	8.55		13.90	18.70
9341	T-fitting		40	.200		7.85	12.80		20.65	28
9342	In-line adjustable		48	.167		35	10.65		45.65	54.50
9350	Diode tee		40	.200		18.10	12.80		30.90	39
9351	Restrictor tee		40	.200		18.10	12.80		30.90	39
9360	Pneumatic gradual switch		8	1		305	64		369	430
9361	Selector switch		8	1		109	64		173	216
9370	Electro-pneumatic motor driven servo		10	.800		990	51		1,041	1,175
9390	Fan control switch and mounting base		16	.500		110	32		142	169
9400	Circulating pump sequencer		14	.571		1,275	36.50		1,311.50	1,450
9410	Pneumatic tubing, fittings and accessories									
9414	Tubing, urethane									
9415	1/8" OD x 1/16" ID	1 Stpi	120	.067	L.F.	.71	4.26		4.97	7.20
9416	1/4" OD x 1/8" ID		115	.070		.73	4.45		5.18	7.50
9417	5/32" OD x 3/32" ID		110	.073		.41	4.65		5.06	7.45
9420	Coupling, straight									
9422	Barb x barb									
9423	1/4" x 5/32"	1 Stpi	160	.050	Ea.	.81	3.20		4.01	5.70
9424	1/4" x 1/4"		158	.051		.61	3.24		3.85	5.55
9425	3/8" x 3/8"		154	.052		.60	3.32		3.92	5.65
9426	3/8" x 1/4"		150	.053		.55	3.41		3.96	5.70
9427	1/2" x 1/4"		148	.054		5.25	3.46		8.71	11

23 09 53.10 Control Components		Crew	Daily Output	Labor-Hours	Unit	2019 Bare Costs Material	Labor	Equipment	Total	Total Incl O&P
9428	1/2″ x 3/8″	1 Stpi	144	.056	Ea.	1.06	3.55		4.61	6.50
9429	1/2″ x 1/2″		140	.057		.71	3.65		4.36	6.30
9440	Tube x tube									
9441	1/4″ x 1/4″	1 Stpi	100	.080	Ea.	2.20	5.10		7.30	10.10
9442	3/8″ x 1/4″		96	.083		4.36	5.35		9.71	12.80
9443	3/8″ x 3/8″		92	.087		2.73	5.55		8.28	11.35
9444	1/2″ x 3/8″		88	.091		4.53	5.80		10.33	13.70
9445	1/2″ x 1/2″		84	.095		5	6.10		11.10	14.65
9450	Elbow coupling									
9452	Barb x barb									
9454	1/4″ x 1/4″	1 Stpi	158	.051	Ea.	1.29	3.24		4.53	6.30
9455	3/8″ x 3/8″		154	.052		3.50	3.32		6.82	8.85
9456	1/2″ x 1/2″		140	.057		5.30	3.65		8.95	11.30
9460	Tube x tube									
9462	1/2″ x 1/2″	1 Stpi	84	.095	Ea.	7.25	6.10		13.35	17.10
9470	Tee coupling									
9472	Barb x barb x barb									
9474	1/4″ x 1/4″ x 5/32″	1 Stpi	108	.074	Ea.	1.51	4.74		6.25	8.75
9475	1/4″ x 1/4″ x 1/4″		104	.077		.99	4.92		5.91	8.50
9476	3/8″ x 3/8″ x 5/32″		102	.078		3.08	5		8.08	10.95
9477	3/8″ x 3/8″ x 1/4″		99	.081		1.51	5.15		6.66	9.40
9478	3/8″ x 3/8″ x 3/8″		98	.082		4.56	5.20		9.76	12.85
9479	1/2″ x 1/2″ x 1/4″		96	.083		3.98	5.35		9.33	12.40
9480	1/2″ x 1/2″ x 3/8″		95	.084		2.76	5.40		8.16	11.15
9481	1/2″ x 1/2″ x 1/2″		92	.087		4.46	5.55		10.01	13.25
9484	Tube x tube x tube									
9485	1/4″ x 1/4″ x 1/4″	1 Stpi	66	.121	Ea.	4.64	7.75		12.39	16.75
9486	3/8″ x 1/4″ x 1/4″		64	.125		9.60	8		17.60	22.50
9487	3/8″ x 3/8″ x 1/4″		61	.131		6.35	8.40		14.75	19.55
9488	3/8″ x 3/8″ x 3/8″		60	.133		6.65	8.55		15.20	20
9489	1/2″ x 1/2″ x 3/8″		58	.138		15.05	8.80		23.85	30
9490	1/2″ x 1/2″ x 1/2″		56	.143		10.65	9.15		19.80	25.50
9492	Needle valve									
9494	Tube x tube									
9495	1/4″ x 1/4″	1 Stpi	86	.093	Ea.	3.99	5.95		9.94	13.35
9496	3/8″ x 3/8″	″	82	.098	″	9.35	6.25		15.60	19.60
9600	Electronic system misc. components									
9610	Electric motor damper actuator	1 Elec	8	1	Ea.	525	60		585	665
9620	Damper position indicator		10	.800		121	48		169	206
9630	Pneumatic-electronic transducer		12	.667		60	40		100	126
9700	Modulating step controller									
9701	For 2-5 steps	1 Elec	5.30	1.509	Ea.	805	90.50		895.50	1,025
9702	6-10 steps		3.20	2.500		238	150		388	485
9703	11-15 steps		2.70	2.963		465	178		643	775
9704	16-20 steps		1.60	5		690	300		990	1,200
9710	Staging network									
9740	Accessory unit regulator									
9750	Accessory power supply	1 Elec	18	.444	Ea.	197	26.50		223.50	257
9760	Step down transformer	″	16	.500	″	184	30		214	248

23 11 13 – Facility Fuel-Oil Piping

23 11 13.10 Fuel Oil Specialties		Crew	Daily Output	Labor-Hours	Unit	2019 Bare Costs Material	Labor	Equipment	Total	Total Incl O&P
0010	**FUEL OIL SPECIALTIES**									
0020	Foot valve, single poppet, metal to metal construction									
0040	Bevel seat, 1/2" diameter	1 Stpi	20	.400	Ea.	80	25.50		105.50	127
0060	3/4" diameter		18	.444		95.50	28.50		124	148
0080	1" diameter		16	.500		105	32		137	164
0100	1-1/4" diameter		15	.533		167	34		201	235
0120	1-1/2" diameter		13	.615		192	39.50		231.50	271
0140	2" diameter		11	.727		242	46.50		288.50	335
0160	Foot valve, double poppet, metal to metal construction									
0164	1" diameter	1 Stpi	15	.533	Ea.	167	34		201	235
0166	1-1/2" diameter	"	12	.667	"	268	42.50		310.50	360
0400	Fuel fill box, flush type									
0408	Nonlocking, watertight									
0410	1-1/2" diameter	1 Stpi	12	.667	Ea.	22.50	42.50		65	89
0440	2" diameter		10	.800		18.95	51		69.95	98
0450	2-1/2" diameter		9	.889		65.50	57		122.50	158
0460	3" diameter		7	1.143		53	73		126	169
0470	4" diameter		5	1.600		77	102		179	239
0500	Locking inner cover									
0510	2" diameter	1 Stpi	8	1	Ea.	82.50	64		146.50	187
0520	2-1/2" diameter		7	1.143		110	73		183	231
0530	3" diameter		5	1.600		118	102		220	284
0540	4" diameter		4	2		152	128		280	360
0600	Fuel system components									
0620	Spill container	1 Stpi	4	2	Ea.	950	128		1,078	1,250
0640	Fill adapter, 4", straight drop		8	1		60	64		124	162
0680	Fill cap, 4"		30	.267		36.50	17.05		53.55	65.50
0700	Extractor fitting, 4" x 1-1/2"		8	1		510	64		574	660
0740	Vapor hose adapter, 4"		8	1		107	64		171	214
0760	Wood gage stick, 10'					12.05			12.05	13.25
1000	Oil filters, 3/8" IPT, 20 gal. per hour	1 Stpi	20	.400		40.50	25.50		66	83.50
1020	32 gal. per hour		18	.444		55.50	28.50		84	104
1040	40 gal. per hour		16	.500		54.50	32		86.50	108
1060	50 gal. per hour		14	.571		61.50	36.50		98	123
2000	Remote tank gauging system, self contained									
2100	Single tank kit/8 sensors inputs w/printer	1 Stpi	2.50	3.200	Ea.	4,100	205		4,305	4,800
2120	Two tank kit/8 sensors inputs w/printer		2	4		5,750	256		6,006	6,700
3000	Valve, ball check, globe type, 3/8" diameter		24	.333		12.50	21.50		34	46
3500	Fusible, 3/8" diameter		24	.333		15.45	21.50		36.95	49
3600	1/2" diameter		24	.333		38	21.50		59.50	74
3610	3/4" diameter		20	.400		84	25.50		109.50	131
3620	1" diameter		19	.421		243	27		270	310
4000	Nonfusible, 3/8" diameter		24	.333		26	21.50		47.50	61
4500	Shutoff, gate type, lever handle, spring-fusible kit									
4520	1/4" diameter	1 Stpi	14	.571	Ea.	42	36.50		78.50	101
4540	3/8" diameter		12	.667		40.50	42.50		83	109
4560	1/2" diameter		10	.800		57.50	51		108.50	141
4570	3/4" diameter		8	1		105	64		169	211
4580	Lever handle, requires weight and fusible kit									
4600	1" diameter	1 Stpi	9	.889	Ea.	390	57		447	515
4620	1-1/4" diameter		8	1		460	64		524	600
4640	1-1/2" diameter		7	1.143		420	73		493	575
4660	2" diameter		6	1.333		780	85.50		865.50	985

23 11 Facility Fuel Piping

23 11 13 – Facility Fuel-Oil Piping

23 11 13.10 Fuel Oil Specialties		Crew	Daily Output	Labor-Hours	Unit	2019 Bare Costs Material	Labor	Equipment	Total	Total Incl O&P
4680	For fusible link, weight and braided wire, add				Ea.	5%				
5000	Vent alarm, whistling signal					37.50			37.50	41.50
5500	Vent protector/breather, 1-1/4" diameter	1 Stpi	32	.250		20.50	16		36.50	46.50
5520	1-1/2" diameter		32	.250		23.50	16		39.50	50
5540	2" diameter		32	.250		48.50	16		64.50	77
5560	3" diameter		28	.286		68.50	18.25		86.75	103
5580	4" diameter		24	.333		79	21.50		100.50	119
5600	Dust cap, breather, 2"		40	.200		12.35	12.80		25.15	33
8000	Fuel oil and tank heaters									
8020	Electric, capacity rated at 230 volts									
8040	Immersion element in steel manifold									
8060	96 GPH at 50°F rise	Q-5	6.40	2.500	Ea.	1,350	144		1,494	1,700
8070	128 GPH at 50°F rise		6.20	2.581		1,350	149		1,499	1,700
8080	160 GPH at 50°F rise		5.90	2.712		1,550	156		1,706	1,925
8090	192 GPH at 50°F rise		5.50	2.909		1,675	167		1,842	2,100
8100	240 GPH at 50°F rise		5.10	3.137		1,950	181		2,131	2,425
8110	288 GPH at 50°F rise		4.60	3.478		2,125	200		2,325	2,625
8120	384 GPH at 50°F rise		3.10	5.161		2,475	297		2,772	3,175
8130	480 GPH at 50°F rise		2.30	6.957		3,100	400		3,500	4,025
8140	576 GPH at 50°F rise		2.10	7.619		3,100	440		3,540	4,075
8300	Suction stub, immersion type									
8320	75" long, 750 watts	1 Stpi	14	.571	Ea.	885	36.50		921.50	1,025
8330	99" long, 2000 watts		12	.667		1,275	42.50		1,317.50	1,475
8340	123" long, 3000 watts		10	.800		1,375	51		1,426	1,600
8660	Steam, cross flow, rated at 5 PSIG									
8680	42 GPH	Q-5	7	2.286	Ea.	2,050	132		2,182	2,450
8690	73 GPH		6.70	2.388		2,450	137		2,587	2,900
8700	112 GPH		6.20	2.581		2,475	149		2,624	2,950
8710	158 GPH		5.80	2.759		2,700	159		2,859	3,225
8720	187 GPH		4	4		3,825	230		4,055	4,550
8730	270 GPH		3.60	4.444		3,975	256		4,231	4,750
8740	365 GPH	Q-6	4.90	4.898		4,825	292		5,117	5,750
8750	635 GPH		3.70	6.486		5,900	385		6,285	7,075
8760	845 GPH		2.50	9.600		8,950	575		9,525	10,700
8770	1420 GPH		1.60	15		14,600	895		15,495	17,400
8780	2100 GPH		1.10	21.818		20,000	1,300		21,300	24,000

23 11 23 – Facility Natural-Gas Piping

23 11 23.10 Gas Meters		Crew	Daily Output	Labor-Hours	Unit	Material	Labor	Equipment	Total	Total Incl O&P
0010	**GAS METERS**									
4000	Residential									
4010	Gas meter, residential, 3/4" pipe size	1 Plum	14	.571	Ea.	320	36		356	410
4020	Gas meter, residential, 1" pipe size		12	.667		258	42		300	345
4030	Gas meter, residential, 1-1/4" pipe size		10	.800		261	50.50		311.50	365
4032	Gas meter, 2"		8	1		1,675	63		1,738	1,950

23 11 23.20 Gas Piping, Flexible (Csst)		Crew	Daily Output	Labor-Hours	Unit	Material	Labor	Equipment	Total	Total Incl O&P
0010	**GAS PIPING, FLEXIBLE (CSST)**									
0100	Tubing with lightning protection									
0110	3/8"	1 Stpi	65	.123	L.F.	3.06	7.85		10.91	15.15
0120	1/2"		62	.129		3.44	8.25		11.69	16.20
0130	3/4"		60	.133		4.48	8.55		13.03	17.75
0140	1"		55	.145		6.90	9.30		16.20	21.50
0150	1-1/4"		50	.160		8	10.25		18.25	24

23 11 Facility Fuel Piping

23 11 23 – Facility Natural-Gas Piping

23 11 23.20 Gas Piping, Flexible (Csst)		Crew	Daily Output	Labor-Hours	Unit	Material	2019 Bare Costs Labor	Equipment	Total	Total Incl O&P
0160	1-1/2"	1 Stpi	45	.178	L.F.	14	11.35		25.35	32.50
0170	2"		40	.200		20	12.80		32.80	41
0200	Tubing for underground/underslab burial									
0210	3/8"	1 Stpi	65	.123	L.F.	4.26	7.85		12.11	16.50
0220	1/2"		62	.129		4.86	8.25		13.11	17.75
0230	3/4"		60	.133		6.15	8.55		14.70	19.60
0240	1"		55	.145		8.30	9.30		17.60	23
0250	1-1/4"		50	.160		11.15	10.25		21.40	27.50
0260	1-1/2"		45	.178		21	11.35		32.35	40
0270	2"		40	.200		25	12.80		37.80	46.50
3000	Fittings									
3010	Straight									
3100	Tube to NPT									
3110	3/8"	1 Stpi	29	.276	Ea.	12.80	17.65		30.45	40.50
3120	1/2"		27	.296		13.80	18.95		32.75	43.50
3130	3/4"		25	.320		18.80	20.50		39.30	51
3140	1"		23	.348		29	22		51	65.50
3150	1-1/4"		20	.400		66	25.50		91.50	111
3160	1-1/2"		17	.471		133	30		163	191
3170	2"		15	.533		226	34		260	300
3200	Coupling									
3210	3/8"	1 Stpi	29	.276	Ea.	22.50	17.65		40.15	51
3220	1/2"		27	.296		25.50	18.95		44.45	56.50
3230	3/4"		25	.320		35	20.50		55.50	69
3240	1"		23	.348		54	22		76	93
3250	1-1/4"		20	.400		120	25.50		145.50	171
3260	1-1/2"		17	.471		252	30		282	320
3270	2"		15	.533		430	34		464	525
3300	Flange fitting									
3310	3/8"	1 Stpi	25	.320	Ea.	19.60	20.50		40.10	52
3320	1/2"		22	.364		18.75	23.50		42.25	55.50
3330	3/4"		19	.421		24.50	27		51.50	67
3340	1"		16	.500		33.50	32		65.50	85
3350	1-1/4"		12	.667		77.50	42.50		120	150
3400	90° flange valve									
3410	3/8"	1 Stpi	25	.320	Ea.	37.50	20.50		58	71.50
3420	1/2"		22	.364		38	23.50		61.50	77
3430	3/4"		19	.421		47	27		74	92
4000	Tee									
4120	1/2"	1 Stpi	20.50	.390	Ea.	40	25		65	81.50
4130	3/4"		19	.421		53	27		80	99
4140	1"		17.50	.457		96	29		125	149
5000	Reducing									
5110	Tube to NPT									
5120	3/4" to 1/2" NPT	1 Stpi	26	.308	Ea.	20	19.70		39.70	51.50
5130	1" to 3/4" NPT	"	24	.333	"	31.50	21.50		53	66.50
5200	Reducing tee									
5210	1/2" x 3/8" x 3/8"	1 Stpi	21	.381	Ea.	53	24.50		77.50	94.50
5220	3/4" x 1/2" x 1/2"		20	.400		49.50	25.50		75	93
5230	1" x 3/4" x 1/2"		18	.444		82	28.50		110.50	133
5240	1-1/4" x 1-1/4" x 1"		15.60	.513		219	33		252	290
5250	1-1/2" x 1-1/2" x 1-1/4"		13.30	.602		335	38.50		373.50	430
5260	2" x 2" x 1-1/2"		11.80	.678		505	43.50		548.50	620

23 11 Facility Fuel Piping

23 11 23 – Facility Natural-Gas Piping

23 11 23.20 Gas Piping, Flexible (Csst)		Crew	Daily Output	Labor-Hours	Unit	2019 Bare Costs Material	Labor	Equipment	Total	Total Incl O&P
5300	Manifold with four ports and mounting bracket									
5302	Labor to mount manifold does not include making pipe									
5304	connections which are included in fitting labor.									
5310	3/4" x 1/2" x 1/2"(4)	1 Stpi	76	.105	Ea.	40.50	6.75		47.25	54.50
5330	1-1/4" x 1" x 3/4"(4)	↓	72	.111	↓	58	7.10		65.10	74.50
5350	2" x 1-1/2" x 1"(4)	↓	68	.118	↓	73	7.50		80.50	91.50
5600	Protective striker plate									
5610	Quarter plate, 3" x 2"	1 Stpi	88	.091	Ea.	.92	5.80		6.72	9.70
5620	Half plate, 3" x 7"	↓	82	.098	↓	2.10	6.25		8.35	11.65
5630	Full plate, 3" x 12"	↓	78	.103	↓	3.76	6.55		10.31	14

23 12 Facility Fuel Pumps

23 12 13 – Facility Fuel-Oil Pumps

23 12 13.10 Pump and Motor Sets

Line	Description	Crew	Daily Output	Labor-Hours	Unit	Material	Labor	Equipment	Total	Total Incl O&P
0010	**PUMP AND MOTOR SETS**									
1810	Light fuel and diesel oils									
1820	20 GPH, 1/3 HP	Q-5	6	2.667	Ea.	1,475	153		1,628	1,850
1830	27 GPH, 1/3 HP	↓	6	2.667	↓	1,475	153		1,628	1,850
1840	80 GPH, 1/3 HP	↓	5	3.200	↓	1,475	184		1,659	1,900
1850	145 GPH, 1/2 HP	↓	4	4	↓	1,550	230		1,780	2,075
1860	277 GPH, 1 HP	↓	4	4	↓	1,750	230		1,980	2,275
1870	700 GPH, 1-1/2 HP	↓	3	5.333	↓	3,900	305		4,205	4,750
1880	1000 GPH, 2 HP	↓	3	5.333	↓	4,600	305		4,905	5,500
1890	1800 GPH, 5 HP	↓	1.80	8.889	↓	7,300	510		7,810	8,800

23 13 Facility Fuel-Storage Tanks

23 13 13 – Facility Underground Fuel-Oil, Storage Tanks

23 13 13.09 Single-Wall Steel Fuel-Oil Tanks

Line	Description	Crew	Daily Output	Labor-Hours	Unit	Material	Labor	Equipment	Total	Total Incl O&P
0010	**SINGLE-WALL STEEL FUEL-OIL TANKS**									
5000	Tanks, steel ugnd., sti-p3, not incl. hold-down bars									
5500	Excavation, pad, pumps and piping not included									
5510	Single wall, 500 gallon capacity, 7 ga. shell	Q-5	2.70	5.926	Ea.	1,850	340		2,190	2,525
5520	1,000 gallon capacity, 7 ga. shell	"	2.50	6.400	↓	2,725	370		3,095	3,550
5530	2,000 gallon capacity, 1/4" thick shell	Q-7	4.60	6.957	↓	3,500	425		3,925	4,475
5535	2,500 gallon capacity, 7 ga. shell	Q-5	3	5.333	↓	5,325	305		5,630	6,325
5540	5,000 gallon capacity, 1/4" thick shell	Q-7	3.20	10	↓	11,900	610		12,510	14,000
5560	10,000 gallon capacity, 1/4" thick shell	↓	2	16	↓	10,300	975		11,275	12,800
5580	15,000 gallon capacity, 5/16" thick shell	↓	1.70	18.824	↓	10,200	1,150		11,350	12,900
5600	20,000 gallon capacity, 5/16" thick shell	↓	1.50	21.333	↓	21,000	1,300		22,300	25,100
5610	25,000 gallon capacity, 3/8" thick shell	↓	1.30	24.615	↓	27,500	1,500		29,000	32,600
5620	30,000 gallon capacity, 3/8" thick shell	↓	1.10	29.091	↓	33,500	1,775		35,275	39,600
5630	40,000 gallon capacity, 3/8" thick shell	↓	.90	35.556	↓	43,200	2,175		45,375	51,000
5640	50,000 gallon capacity, 3/8" thick shell	↓	.80	40	↓	48,000	2,425		50,425	56,500

23 13 13.13 Dbl-Wall Steel, Undrgrnd Fuel-Oil, Stor. Tanks		Crew	Daily Output	Labor-Hours	Unit	2019 Bare Costs Material	Labor	Equipment	Total	Total Incl O&P
0010	**DOUBLE-WALL STEEL, UNDERGROUND FUEL-OIL, STORAGE TANKS**									
6200	Steel, underground, 360°, double wall, UL listed,									
6210	with sti-P3 corrosion protection,									
6220	(dielectric coating, cathodic protection, electrical									
6230	isolation) 30 year warranty,									
6240	not incl. manholes or hold-downs.									
6250	500 gallon capacity	Q-5	2.40	6.667	Ea.	3,000	385		3,385	3,875
6260	1,000 gallon capacity	″	2.25	7.111		3,650	410		4,060	4,650
6270	2,000 gallon capacity	Q-7	4.16	7.692		4,825	470		5,295	6,025
6280	3,000 gallon capacity		3.90	8.205		6,725	500		7,225	8,150
6290	4,000 gallon capacity		3.64	8.791		7,925	535		8,460	9,525
6300	5,000 gallon capacity		2.91	10.997		8,575	670		9,245	10,400
6310	6,000 gallon capacity		2.42	13.223		12,100	805		12,905	14,500
6320	8,000 gallon capacity		2.08	15.385		10,900	935		11,835	13,400
6330	10,000 gallon capacity		1.82	17.582		13,400	1,075		14,475	16,300
6340	12,000 gallon capacity		1.70	18.824		15,000	1,150		16,150	18,200
6350	15,000 gallon capacity		1.33	24.060		20,000	1,475		21,475	24,200
6360	20,000 gallon capacity		1.33	24.060		33,100	1,475		34,575	38,600
6370	25,000 gallon capacity		1.16	27.586		35,600	1,675		37,275	41,700
6380	30,000 gallon capacity		1.03	31.068		36,600	1,900		38,500	43,200
6390	40,000 gallon capacity		.80	40		114,000	2,425		116,425	129,000
6395	50,000 gallon capacity		.73	43.836		138,500	2,675		141,175	156,000
6400	For hold-downs 500-2,000 gal., add		16	2	Set	219	122		341	425
6410	For hold-downs 3,000-6,000 gal., add		12	2.667		380	162		542	665
6420	For hold-downs 8,000-12,000 gal., add		11	2.909		450	177		627	760
6430	For hold-downs 15,000 gal., add		9	3.556		645	216		861	1,025
6440	For hold-downs 20,000 gal., add		8	4		745	244		989	1,175
6450	For hold-downs 20,000 gal. plus, add		6	5.333		985	325		1,310	1,550
6500	For manways, add				Ea.	995			995	1,100
6600	In place with hold-downs									
6652	550 gallon capacity	Q-5	1.84	8.696	Ea.	3,225	500		3,725	4,300

23 13 13.23 Glass-Fiber-Reinfcd-Plastic, Fuel-Oil, Storage		Crew	Daily Output	Labor-Hours	Unit	Material	Labor	Equipment	Total	Total Incl O&P
0010	**GLASS-FIBER-REINFCD-PLASTIC, UNDERGRND. FUEL-OIL, STORAGE**									
0210	Fiberglass, underground, single wall, UL listed, not including									
0220	manway or hold-down strap									
0225	550 gallon capacity	Q-5	2.67	5.993	Ea.	3,950	345		4,295	4,875
0230	1,000 gallon capacity	″	2.46	6.504		5,075	375		5,450	6,125
0240	2,000 gallon capacity	Q-7	4.57	7.002		7,475	425		7,900	8,875
0245	3,000 gallon capacity		3.90	8.205		8,250	500		8,750	9,825
0250	4,000 gallon capacity		3.55	9.014		9,950	550		10,500	11,700
0255	5,000 gallon capacity		3.20	10		10,600	610		11,210	12,600
0260	6,000 gallon capacity		2.67	11.985		10,900	730		11,630	13,100
0270	8,000 gallon capacity		2.29	13.974		13,500	850		14,350	16,100
0280	10,000 gallon capacity		2	16		15,400	975		16,375	18,400
0282	12,000 gallon capacity		1.88	17.021		17,500	1,025		18,525	20,800
0284	15,000 gallon capacity		1.68	19.048		23,800	1,150		24,950	27,900
0290	20,000 gallon capacity		1.45	22.069		32,400	1,350		33,750	37,700
0300	25,000 gallon capacity		1.28	25		65,000	1,525		66,525	74,000
0320	30,000 gallon capacity		1.14	28.070		90,500	1,700		92,200	102,500
0340	40,000 gallon capacity		.89	35.955		108,000	2,200		110,200	122,500
0360	48,000 gallon capacity		.81	39.506		153,000	2,400		155,400	172,000
0500	For manway, fittings and hold-downs, add					20%	15%			

23 13 13 – Facility Underground Fuel-Oil, Storage Tanks

23 13 13.23 Glass-Fiber-Reinfcd-Plastic, Fuel-Oil, Storage		Crew	Daily Output	Labor-Hours	Unit	2019 Bare Costs Material	Labor	Equipment	Total	Total Incl O&P
0600	For manways, add				Ea.	2,375			2,375	2,600
1000	For helical heating coil, add	Q-5	2.50	6.400		5,500	370		5,870	6,600
1020	Fiberglass, underground, double wall, UL listed									
1030	includes manways, not incl. hold-down straps									
1040	600 gallon capacity	Q-5	2.42	6.612	Ea.	8,875	380		9,255	10,300
1050	1,000 gallon capacity	"	2.25	7.111		11,900	410		12,310	13,700
1060	2,500 gallon capacity	Q-7	4.16	7.692		16,900	470		17,370	19,200
1070	3,000 gallon capacity		3.90	8.205		18,800	500		19,300	21,500
1080	4,000 gallon capacity		3.64	8.791		19,100	535		19,635	21,800
1090	6,000 gallon capacity		2.42	13.223		25,200	805		26,005	28,900
1100	8,000 gallon capacity		2.08	15.385		32,000	935		32,935	36,600
1110	10,000 gallon capacity		1.82	17.582		36,000	1,075		37,075	41,200
1120	12,000 gallon capacity		1.70	18.824		44,900	1,150		46,050	51,000
1122	15,000 gallon capacity		1.52	21.053		63,500	1,275		64,775	72,000
1124	20,000 gallon capacity		1.33	24.060		75,000	1,475		76,475	84,500
1126	25,000 gallon capacity		1.16	27.586		92,000	1,675		93,675	104,000
1128	30,000 gallon capacity		1.03	31.068		110,500	1,900		112,400	124,500
1140	For hold-down straps, add					2%	10%			
1150	For hold-downs 500-4,000 gal., add	Q-7	16	2	Set	505	122		627	740
1160	For hold-downs 5,000-15,000 gal., add		8	4		1,025	244		1,269	1,500
1170	For hold-downs 20,000 gal., add		5.33	6.004		1,525	365		1,890	2,225
1180	For hold-downs 25,000 gal., add		4	8		2,025	485		2,510	2,950
1190	For hold-downs 30,000 gal., add		2.60	12.308		3,050	750		3,800	4,475
2210	Fiberglass, underground, single wall, UL listed, including									
2220	hold-down straps, no manways									
2225	550 gallon capacity	Q-5	2	8	Ea.	4,450	460		4,910	5,600
2230	1,000 gallon capacity	"	1.88	8.511		5,575	490		6,065	6,850
2240	2,000 gallon capacity	Q-7	3.55	9.014		7,975	550		8,525	9,600
2250	4,000 gallon capacity		2.90	11.034		10,400	670		11,070	12,500
2260	6,000 gallon capacity		2	16		11,900	975		12,875	14,600
2270	8,000 gallon capacity		1.78	17.978		14,500	1,100		15,600	17,700
2280	10,000 gallon capacity		1.60	20		16,400	1,225		17,625	19,800
2282	12,000 gallon capacity		1.52	21.053		18,500	1,275		19,775	22,200
2284	15,000 gallon capacity		1.39	23.022		24,800	1,400		26,200	29,300
2290	20,000 gallon capacity		1.14	28.070		34,000	1,700		35,700	39,900
2300	25,000 gallon capacity		.96	33.333		67,000	2,025		69,025	77,000
2320	30,000 gallon capacity		.80	40		94,000	2,425		96,425	106,500
3020	Fiberglass, underground, double wall, UL listed									
3030	includes manways and hold-down straps									
3040	600 gallon capacity	Q-5	1.86	8.602	Ea.	9,375	495		9,870	11,000
3050	1,000 gallon capacity	"	1.70	9.412		12,400	540		12,940	14,500
3060	2,500 gallon capacity	Q-7	3.29	9.726		17,400	590		17,990	20,000
3070	3,000 gallon capacity		3.13	10.224		19,300	620		19,920	22,100
3080	4,000 gallon capacity		2.93	10.922		19,600	665		20,265	22,500
3090	6,000 gallon capacity		1.86	17.204		26,200	1,050		27,250	30,400
3100	8,000 gallon capacity		1.65	19.394		33,100	1,175		34,275	38,200
3110	10,000 gallon capacity		1.48	21.622		37,000	1,325		38,325	42,700
3120	12,000 gallon capacity		1.40	22.857		46,000	1,400		47,400	52,500
3122	15,000 gallon capacity		1.28	25		64,500	1,525		66,025	73,500
3124	20,000 gallon capacity		1.06	30.189		76,500	1,850		78,350	87,000
3126	25,000 gallon capacity		.90	35.556		94,000	2,175		96,175	107,000
3128	30,000 gallon capacity		.74	43.243		113,500	2,625		116,125	129,000

23 13 23.13 Vertical, Steel, Abvground Fuel-Oil, Stor. Tanks		Crew	Daily Output	Labor-Hours	Unit	2019 Bare Costs Material	Labor	Equipment	Total	Total Incl O&P
0010	**VERTICAL, STEEL, ABOVEGROUND FUEL-OIL, STORAGE TANKS**									
4000	Fixed roof oil storage tanks, steel (1 BBL=42 gal. w/foundation 3'D x 1'W)									
4200	5,000 barrels				Ea.				194,000	213,500
4300	24,000 barrels								333,500	367,000
4500	56,000 barrels								729,000	802,000
4600	110,000 barrels								1,060,000	1,166,000
4800	143,000 barrels								1,250,000	1,375,000
4900	225,000 barrels								1,360,000	1,496,000
5100	Floating roof gasoline tanks, steel, 5,000 barrels (w/foundation 3'D x 1'W)								204,000	225,000
5200	25,000 barrels								381,000	419,000
5400	55,000 barrels								839,000	923,000
5500	100,000 barrels								1,253,000	1,379,000
5700	150,000 barrels								1,532,000	1,685,000
5800	225,000 barrels								2,300,000	2,783,000

23 13 23.16 Horizontal, Stl, Abvgrd Fuel-Oil, Storage Tanks		Crew	Daily Output	Labor-Hours	Unit	Material	Labor	Equipment	Total	Total Incl O&P
0010	**HORIZONTAL, STEEL, ABOVEGROUND FUEL-OIL, STORAGE TANKS**									
3000	Steel, storage, aboveground, including cradles, coating,									
3020	fittings, not including foundation, pumps or piping									
3040	Single wall, 275 gallon	Q-5	5	3.200	Ea.	520	184		704	845
3060	550 gallon	″	2.70	5.926		4,425	340		4,765	5,375
3080	1,000 gallon	Q-7	5	6.400		7,400	390		7,790	8,700
3100	1,500 gallon		4.75	6.737		10,700	410		11,110	12,300
3120	2,000 gallon		4.60	6.957		12,900	425		13,325	14,800
3140	5,000 gallon		3.20	10		21,400	610		22,010	24,500
3150	10,000 gallon		2	16		41,400	975		42,375	47,000
3160	15,000 gallon		1.70	18.824		54,500	1,150		55,650	61,500
3170	20,000 gallon		1.45	22.069		71,000	1,350		72,350	80,000
3180	25,000 gallon		1.30	24.615		82,500	1,500		84,000	93,500
3190	30,000 gallon		1.10	29.091		99,000	1,775		100,775	111,500
3320	Double wall, 500 gallon capacity	Q-5	2.40	6.667		1,850	385		2,235	2,625
3330	2,000 gallon capacity	Q-7	4.15	7.711		6,125	470		6,595	7,425
3340	4,000 gallon capacity		3.60	8.889		13,600	540		14,140	15,700
3350	6,000 gallon capacity		2.40	13.333		15,400	810		16,210	18,100
3360	8,000 gallon capacity		2	16		18,100	975		19,075	21,400
3370	10,000 gallon capacity		1.80	17.778		30,300	1,075		31,375	34,900
3380	15,000 gallon capacity		1.50	21.333		40,000	1,300		41,300	46,000
3390	20,000 gallon capacity		1.30	24.615		46,500	1,500		48,000	53,500
3400	25,000 gallon capacity		1.15	27.826		57,500	1,700		59,200	66,000
3410	30,000 gallon capacity		1	32		66,500	1,950		68,450	76,500

23 13 23.26 Horizontal, Conc., Abvgrd Fuel-Oil, Stor. Tanks		Crew	Daily Output	Labor-Hours	Unit	Material	Labor	Equipment	Total	Total Incl O&P
0010	**HORIZONTAL, CONCRETE, ABOVEGROUND FUEL-OIL, STORAGE TANKS**									
0050	Concrete, storage, aboveground, including pad & pump									
0100	500 gallon	F-3	2	20	Ea.	10,100	1,050	236	11,386	13,000
0200	1,000 gallon	″	2	20		14,200	1,050	236	15,486	17,500
0300	2,000 gallon	F-4	2	24		18,200	1,250	490	19,940	22,500
0400	4,000 gallon		2	24		23,300	1,250	490	25,040	28,000
0500	8,000 gallon		2	24		36,500	1,250	490	38,240	42,500
0600	12,000 gallon		2	24		48,600	1,250	490	50,340	56,000

23 21 20 – Hydronic HVAC Piping Specialties

23 21 20.10 Air Control		Crew	Daily Output	Labor-Hours	Unit	Material	2019 Bare Costs Labor	Equipment	Total	Total Incl O&P
0010	**AIR CONTROL**									
0030	Air separator, with strainer									
0040	2″ diameter	Q-5	6	2.667	Ea.	1,475	153		1,628	1,850
0080	2-1/2″ diameter		5	3.200		1,650	184		1,834	2,075
0100	3″ diameter		4	4		2,550	230		2,780	3,150
0120	4″ diameter		3	5.333		3,650	305		3,955	4,475
0130	5″ diameter	Q-6	3.60	6.667		4,650	400		5,050	5,725
0140	6″ diameter		3.40	7.059		5,575	420		5,995	6,775
0160	8″ diameter		3	8		8,325	475		8,800	9,875
0180	10″ diameter		2.20	10.909		12,900	650		13,550	15,200
0200	12″ diameter		1.70	14.118		21,400	845		22,245	24,900
0210	14″ diameter		1.30	18.462		25,500	1,100		26,600	29,800
0220	16″ diameter		1	24		38,000	1,425		39,425	44,000
0230	18″ diameter		.80	30		49,900	1,800		51,700	57,500
0240	20″ diameter		.60	40		56,000	2,375		58,375	65,000
0300	Without strainer									
0310	2″ diameter	Q-5	6	2.667	Ea.	910	153		1,063	1,225
0320	2-1/2″ diameter		5	3.200		1,100	184		1,284	1,475
0330	3″ diameter		4	4		1,525	230		1,755	2,025
0340	4″ diameter		3	5.333		2,425	305		2,730	3,125
0350	5″ diameter		2.40	6.667		3,275	385		3,660	4,175
0360	6″ diameter	Q-6	3.40	7.059		3,725	420		4,145	4,725
0370	8″ diameter		3	8		5,125	475		5,600	6,375
0380	10″ diameter		2.20	10.909		7,575	650		8,225	9,300
0390	12″ diameter		1.70	14.118		11,600	845		12,445	14,100
0400	14″ diameter		1.30	18.462		17,400	1,100		18,500	20,900
0410	16″ diameter		1	24		24,400	1,425		25,825	29,000
0420	18″ diameter		.80	30		31,800	1,800		33,600	37,600
0430	20″ diameter		.60	40		38,100	2,375		40,475	45,500
1000	Micro-bubble separator for total air removal, closed loop system									
1010	Requires bladder type tank in system.									
1020	Water (hot or chilled) or glycol system									
1030	Threaded									
1040	3/4″ diameter	1 Stpi	20	.400	Ea.	89.50	25.50		115	137
1050	1″ diameter		19	.421		96	27		123	146
1060	1-1/4″ diameter		16	.500		139	32		171	201
1070	1-1/2″ diameter		13	.615		181	39.50		220.50	259
1080	2″ diameter		11	.727		1,125	46.50		1,171.50	1,300
1090	2-1/2″ diameter	Q-5	15	1.067		1,325	61.50		1,386.50	1,550
1100	3″ diameter		13	1.231		1,825	71		1,896	2,100
1110	4″ diameter		10	1.600		1,925	92		2,017	2,275
1230	Flanged									
1250	2″ diameter	1 Stpi	8	1	Ea.	1,650	64		1,714	1,900
1260	2-1/2″ diameter	Q-5	5	3.200		1,725	184		1,909	2,175
1270	3″ diameter		4.50	3.556		2,350	205		2,555	2,875
1280	4″ diameter		3	5.333		2,550	305		2,855	3,250
1290	5″ diameter	Q-6	3.40	7.059		4,050	420		4,470	5,100
1300	6″ diameter	″	3	8		5,275	475		5,750	6,525
1400	Larger sizes available									
1590	With extended tank dirt catcher									
1600	Flanged									
1620	2″ diameter	1 Stpi	7.40	1.081	Ea.	3,500	69		3,569	3,950
1630	2-1/2″ diameter	Q-5	4.20	3.810		3,800	219		4,019	4,500

23 21 20 – Hydronic HVAC Piping Specialties

23 21 20.10 Air Control		Crew	Daily Output	Labor-Hours	Unit	Material	Labor	Equipment	Total	Total Incl O&P
						2019 Bare Costs				
1640	3″ diameter	Q-5	4	4	Ea.	5,025	230		5,255	5,875
1650	4″ diameter		2.80	5.714		5,550	330		5,880	6,600
1660	5″ diameter	Q-6	3.10	7.742		9,200	460		9,660	10,800
1670	6″ diameter	″	2.90	8.276		11,300	495		11,795	13,100
1800	With drain/dismantle/cleaning access flange									
1810	Flanged									
1820	2″ diameter	Q-5	4.60	3.478	Ea.	2,950	200		3,150	3,550
1830	2-1/2″ diameter		4	4		3,150	230		3,380	3,825
1840	3″ diameter		3.60	4.444		4,250	256		4,506	5,050
1850	4″ diameter		2.50	6.400		4,750	370		5,120	5,775
1860	5″ diameter	Q-6	2.80	8.571		7,625	510		8,135	9,150
1870	6″ diameter	″	2.60	9.231		9,850	550		10,400	11,600
2000	Boiler air fitting, separator									
2010	1-1/4″	Q-5	22	.727	Ea.	256	42		298	345
2020	1-1/2″		20	.800		485	46		531	605
2021	2″		18	.889		855	51		906	1,025
2400	Compression tank air fitting									
2410	For tanks 9″ to 24″ diameter	1 Stpi	15	.533	Ea.	93.50	34		127.50	154
2420	For tanks 100 gallon or larger	″	12	.667	″	370	42.50		412.50	470

23 21 20.14 Air Purging Scoop		Crew	Daily Output	Labor-Hours	Unit	Material	Labor	Equipment	Total	Total Incl O&P
0010	**AIR PURGING SCOOP**, With tappings.									
0020	For air vent and expansion tank connection									
0100	1″ pipe size, threaded	1 Stpi	19	.421	Ea.	29	27		56	72.50
0110	1-1/4″ pipe size, threaded		15	.533		29	34		63	83
0120	1-1/2″ pipe size, threaded		13	.615		65	39.50		104.50	131
0130	2″ pipe size, threaded		11	.727		74.50	46.50		121	152
0140	2-1/2″ pipe size, threaded	Q-5	15	1.067		166	61.50		227.50	274
0150	3″ pipe size, threaded		13	1.231		219	71		290	345
0160	4″ 150 lb. flanges		3	5.333		485	305		790	995

23 21 20.18 Automatic Air Vent		Crew	Daily Output	Labor-Hours	Unit	Material	Labor	Equipment	Total	Total Incl O&P
0010	**AUTOMATIC AIR VENT**									
0020	Cast iron body, stainless steel internals, float type									
0060	1/2″ NPT inlet, 300 psi	1 Stpi	12	.667	Ea.	171	42.50		213.50	252
0140	3/4″ NPT inlet, 300 psi		12	.667		171	42.50		213.50	252
0180	1/2″ NPT inlet, 250 psi		10	.800		380	51		431	490
0220	3/4″ NPT inlet, 250 psi		10	.800		380	51		431	490
0260	1″ NPT inlet, 250 psi		10	.800		555	51		606	685
0340	1-1/2″ NPT inlet, 250 psi	Q-5	12	1.333		1,175	76.50		1,251.50	1,425
0380	2″ NPT inlet, 250 psi	″	12	1.333		1,175	76.50		1,251.50	1,425
0600	Forged steel body, stainless steel internals, float type									
0640	1/2″ NPT inlet, 750 psi	1 Stpi	12	.667	Ea.	1,250	42.50		1,292.50	1,450
0680	3/4″ NPT inlet, 750 psi		12	.667		1,250	42.50		1,292.50	1,450
0760	3/4″ NPT inlet, 1,000 psi		10	.800		1,850	51		1,901	2,125
0800	1″ NPT inlet, 1,000 psi	Q-5	12	1.333		1,850	76.50		1,926.50	2,175
0880	1-1/2″ NPT inlet, 1,000 psi		10	1.600		5,275	92		5,367	5,950
0920	2″ NPT inlet, 1,000 psi		10	1.600		5,275	92		5,367	5,950
1100	Formed steel body, noncorrosive									
1110	1/8″ NPT inlet, 150 psi	1 Stpi	32	.250	Ea.	14.65	16		30.65	40
1120	1/4″ NPT inlet, 150 psi		32	.250		52	16		68	81.50
1130	3/4″ NPT inlet, 150 psi		32	.250		52	16		68	81.50
1300	Chrome plated brass, automatic/manual, for radiators									
1310	1/8″ NPT inlet, nickel plated brass	1 Stpi	32	.250	Ea.	9	16		25	34

23 21 20.22 Circuit Sensor

	23 21 20.22 Circuit Sensor	Crew	Daily Output	Labor-Hours	Unit	Material	2019 Bare Costs Labor	Equipment	Total	Total Incl O&P
0010	**CIRCUIT SENSOR**, Flow meter									
0020	Metering stations									
0040	Wafer orifice insert type									
0060	2-1/2" pipe size	Q-5	12	1.333	Ea.	270	76.50		346.50	410
0100	3" pipe size		11	1.455		305	83.50		388.50	460
0140	4" pipe size		8	2		350	115		465	560
0180	5" pipe size		7.30	2.192		445	126		571	680
0220	6" pipe size		6.40	2.500		535	144		679	800
0260	8" pipe size	Q-6	5.30	4.528		740	270		1,010	1,225
0280	10" pipe size		4.60	5.217		840	310		1,150	1,400
0360	12" pipe size		4.20	5.714		1,400	340		1,740	2,050

23 21 20.26 Circuit Setter

	23 21 20.26 Circuit Setter	Crew	Daily Output	Labor-Hours	Unit	Material	Labor	Equipment	Total	Total Incl O&P
0010	**CIRCUIT SETTER**, Balance valve									
0012	Bronze body, soldered									
0013	1/2" size	1 Stpi	24	.333	Ea.	80.50	21.50		102	121
0014	3/4" size	"	20	.400	"	85.50	25.50		111	133
0018	Threaded									
0019	1/2" pipe size	1 Stpi	22	.364	Ea.	83.50	23.50		107	127
0020	3/4" pipe size		20	.400		89.50	25.50		115	137
0040	1" pipe size		18	.444		116	28.50		144.50	170
0050	1-1/4" pipe size		15	.533		170	34		204	238
0060	1-1/2" pipe size		12	.667		200	42.50		242.50	284
0080	2" pipe size		10	.800		287	51		338	390
0100	2-1/2" pipe size	Q-5	15	1.067		650	61.50		711.50	805
0120	3" pipe size	"	10	1.600		915	92		1,007	1,150
0130	Cast iron body, flanged									
0136	3" pipe size, flanged	Q-5	4	4	Ea.	915	230		1,145	1,350
0140	4" pipe size	"	3	5.333		1,375	305		1,680	1,950
0200	For differential meter, accurate to 1%, add					960			960	1,050
0300	For differential meter 400 psi/200°F continuous, add					2,400			2,400	2,650

23 21 20.30 Cocks, Drains and Specialties

	23 21 20.30 Cocks, Drains and Specialties	Crew	Daily Output	Labor-Hours	Unit	Material	Labor	Equipment	Total	Total Incl O&P
0010	**COCKS, DRAINS AND SPECIALTIES**									
1000	Boiler drain									
1010	Pipe thread to hose									
1020	Bronze									
1030	1/2" size	1 Stpi	36	.222	Ea.	12.40	14.20		26.60	35
1040	3/4" size	"	34	.235	"	13.40	15.05		28.45	37
1100	Solder to hose									
1110	Bronze									
1120	1/2" size	1 Stpi	46	.174	Ea.	12.30	11.10		23.40	30.50
1130	3/4" size	"	44	.182	"	13.40	11.65		25.05	32
1600	With built-in vacuum breaker									
1610	1/2" I.P. or solder	1 Stpi	36	.222	Ea.	42.50	14.20		56.70	68
1630	With tamper proof vacuum breaker									
1640	1/2" I.P. or solder	1 Stpi	36	.222	Ea.	46.50	14.20		60.70	72.50
1650	3/4" I.P. or solder	"	34	.235	"	53.50	15.05		68.55	81
3000	Cocks									
3010	Air, lever or tee handle									
3020	Bronze, single thread									
3030	1/8" size	1 Stpi	52	.154	Ea.	11.85	9.85		21.70	28
3040	1/4" size		46	.174		12.50	11.10		23.60	30.50
3050	3/8" size		40	.200		12.55	12.80		25.35	33

23 21 20.30 Cocks, Drains and Specialties		Crew	Daily Output	Labor-Hours	Unit	Material	2019 Bare Costs Labor	Equipment	Total	Total Incl O&P
3060	1/2" size	1 Stpi	36	.222	Ea.	14.95	14.20		29.15	38
3100	Bronze, double thread									
3110	1/8" size	1 Stpi	26	.308	Ea.	14	19.70		33.70	45
3120	1/4" size		22	.364		14.35	23.50		37.85	51
3130	3/8" size		18	.444		16.15	28.50		44.65	60.50
3140	1/2" size	↓	15	.533	↓	19.80	34		53.80	73
4000	Steam									
4010	Bronze									
4020	1/8" size	1 Stpi	26	.308	Ea.	24	19.70		43.70	56
4030	1/4" size		22	.364		23	23.50		46.50	60.50
4040	3/8" size		18	.444		26.50	28.50		55	72
4050	1/2" size		15	.533		27	34		61	80.50
4060	3/4" size		14	.571		33	36.50		69.50	91.50
4070	1" size		13	.615		48	39.50		87.50	112
4080	1-1/4" size		11	.727		66	46.50		112.50	143
4090	1-1/2" size		10	.800		99.50	51		150.50	187
4100	2" size	↓	9	.889	↓	144	57		201	244
4300	Bronze, 3 way									
4320	1/4" size	1 Stpi	15	.533	Ea.	50	34		84	106
4330	3/8" size		12	.667		50	42.50		92.50	119
4340	1/2" size		10	.800		44.50	51		95.50	126
4350	3/4" size		9.50	.842		76	54		130	165
4360	1" size		8.50	.941		92	60		152	192
4370	1-1/4" size		7.50	1.067		138	68		206	254
4380	1-1/2" size		6.50	1.231		160	78.50		238.50	294
4390	2" size	↓	6	1.333	↓	229	85.50		314.50	380
4500	Gauge cock, brass									
4510	1/4" FPT	1 Stpi	24	.333	Ea.	12.45	21.50		33.95	45.50
4512	1/4" MPT	"	24	.333	"	14.55	21.50		36.05	48
4600	Pigtail, steam syphon									
4604	1/4"	1 Stpi	24	.333	Ea.	20.50	21.50		42	54.50
4650	Snubber valve									
4654	1/4"	1 Stpi	22	.364	Ea.	9.45	23.50		32.95	45.50
4660	Nipple, black steel									
4664	1/4" x 3"	1 Stpi	37	.216	Ea.	3.05	13.85		16.90	24.50

23 21 20.34 Dielectric Unions		Crew	Daily Output	Labor-Hours	Unit	Material	Labor	Equipment	Total	Total Incl O&P
0010	**DIELECTRIC UNIONS**, Standard gaskets for water and air									
0020	250 psi maximum pressure									
0280	Female IPT to sweat, straight									
0300	1/2" pipe size	1 Plum	24	.333	Ea.	7.30	21		28.30	39.50
0340	3/4" pipe size		20	.400		8.15	25.50		33.65	47
0360	1" pipe size		19	.421		11.10	26.50		37.60	52
0380	1-1/4" pipe size		15	.533		15.35	33.50		48.85	67.50
0400	1-1/2" pipe size		13	.615		26	39		65	87
0420	2" pipe size	↓	11	.727	↓	35	46		81	108
0580	Female IPT to brass pipe thread, straight									
0600	1/2" pipe size	1 Plum	24	.333	Ea.	15.85	21		36.85	49
0640	3/4" pipe size		20	.400		17.45	25.50		42.95	57
0660	1" pipe size		19	.421		32.50	26.50		59	75.50
0680	1-1/4" pipe size		15	.533		34	33.50		67.50	88
0700	1-1/2" pipe size		13	.615		49.50	39		88.50	113
0720	2" pipe size	↓	11	.727	↓	99	46		145	178

23 21 20 – Hydronic HVAC Piping Specialties

23 21 20.34 Dielectric Unions		Crew	Daily Output	Labor-Hours	Unit	Material	2019 Bare Costs Labor	Equipment	Total	Total Incl O&P
0780	Female IPT to female IPT, straight									
0800	1/2" pipe size	1 Plum	24	.333	Ea.	14.60	21		35.60	47.50
0840	3/4" pipe size		20	.400		16.90	25.50		42.40	56.50
0860	1" pipe size		19	.421		22	26.50		48.50	64
0880	1-1/4" pipe size		15	.533		29	33.50		62.50	82
0900	1-1/2" pipe size		13	.615		56.50	39		95.50	121
0920	2" pipe size		11	.727		83.50	46		129.50	161
2000	175 psi maximum pressure									
2180	Female IPT to sweat									
2240	2" pipe size	1 Plum	9	.889	Ea.	173	56		229	275
2260	2-1/2" pipe size	Q-1	15	1.067		185	60.50		245.50	295
2280	3" pipe size		14	1.143		254	65		319	380
2300	4" pipe size		11	1.455		675	82.50		757.50	865
2480	Female IPT to brass pipe									
2500	1-1/2" pipe size	1 Plum	11	.727	Ea.	211	46		257	300
2540	2" pipe size	"	9	.889		223	56		279	330
2560	2-1/2" pipe size	Q-1	15	1.067		315	60.50		375.50	435
2580	3" pipe size		14	1.143		350	65		415	485
2600	4" pipe size		11	1.455		495	82.50		577.50	670

23 21 20.38 Expansion Couplings		Crew	Daily Output	Labor-Hours	Unit	Material	Labor	Equipment	Total	Total Incl O&P
0010	**EXPANSION COUPLINGS**, Hydronic									
0100	Copper to copper, sweat									
1000	Baseboard riser fitting, 5" stub by coupling 12" long									
1020	1/2" diameter	1 Stpi	24	.333	Ea.	24.50	21.50		46	59
1040	3/4" diameter		20	.400		35	25.50		60.50	77
1060	1" diameter		19	.421		48.50	27		75.50	93.50
1080	1-1/4" diameter		15	.533		63.50	34		97.50	121
1180	9" stub by tubing 8" long									
1200	1/2" diameter	1 Stpi	24	.333	Ea.	22	21.50		43.50	56
1220	3/4" diameter		20	.400		29	25.50		54.50	70.50
1240	1" diameter		19	.421		40.50	27		67.50	85
1260	1-1/4" diameter		15	.533		56.50	34		90.50	114

23 21 20.42 Expansion Joints		Crew	Daily Output	Labor-Hours	Unit	Material	Labor	Equipment	Total	Total Incl O&P
0010	**EXPANSION JOINTS**									
0100	Bellows type, neoprene cover, flanged spool									
0140	6" face to face, 1-1/4" diameter	1 Stpi	11	.727	Ea.	262	46.50		308.50	360
0160	1-1/2" diameter	"	10.60	.755		262	48.50		310.50	360
0180	2" diameter	Q-5	13.30	1.203		265	69		334	395
0190	2-1/2" diameter		12.40	1.290		275	74.50		349.50	410
0200	3" diameter		11.40	1.404		305	81		386	460
0210	4" diameter		8.40	1.905		330	110		440	530
0220	5" diameter		7.60	2.105		405	121		526	625
0230	6" diameter		6.80	2.353		420	135		555	665
0240	8" diameter		5.40	2.963		485	171		656	790
0250	10" diameter		5	3.200		670	184		854	1,000
0260	12" diameter		4.60	3.478		770	200		970	1,150
0480	10" face to face, 2" diameter		13	1.231		380	71		451	525
0500	2-1/2" diameter		12	1.333		400	76.50		476.50	555
0520	3" diameter		11	1.455		410	83.50		493.50	575
0540	4" diameter		8	2		465	115		580	685
0560	5" diameter		7	2.286		555	132		687	805
0580	6" diameter		6	2.667		575	153		728	860

23 21 20.42 Expansion Joints

	23 21 20.42 Expansion Joints	Crew	Daily Output	Labor-Hours	Unit	Material	Labor	Equipment	Total	Total Incl O&P
0600	8" diameter	Q-5	5	3.200	Ea.	685	184		869	1,025
0620	10" diameter		4.60	3.478		755	200		955	1,125
0640	12" diameter		4	4		940	230		1,170	1,375
0660	14" diameter		3.80	4.211		1,150	242		1,392	1,650
0680	16" diameter		2.90	5.517		1,350	320		1,670	1,950
0700	18" diameter		2.50	6.400		1,500	370		1,870	2,200
0720	20" diameter		2.10	7.619		1,575	440		2,015	2,400
0740	24" diameter		1.80	8.889		1,850	510		2,360	2,825
0760	26" diameter		1.40	11.429		2,075	660		2,735	3,250
0780	30" diameter		1.20	13.333		2,325	765		3,090	3,700
0800	36" diameter	↓	1	16	↓	2,875	920		3,795	4,525
1000	Bellows with internal sleeves and external covers									
1010	Stainless steel, 150 lb.									
1020	With male threads									
1040	3/4" diameter	1 Stpi	20	.400	Ea.	111	25.50		136.50	161
1050	1" diameter		19	.421		120	27		147	173
1060	1-1/4" diameter		16	.500		130	32		162	191
1070	1-1/2" diameter		13	.615		147	39.50		186.50	221
1080	2" diameter	↓	11	.727	↓	157	46.50		203.50	243
1110	With flanged ends									
1120	1-1/4" diameter	1 Stpi	12	.667	Ea.	138	42.50		180.50	216
1130	1-1/2" diameter		11	.727		156	46.50		202.50	242
1140	2" diameter	↓	9	.889		214	57		271	320
1150	3" diameter	Q-5	8	2		375	115		490	585
1160	4" diameter	"	5	3.200		1,025	184		1,209	1,400
1170	5" diameter	Q-6	6	4		1,225	239		1,464	1,700
1180	6" diameter		5	4.800		1,450	286		1,736	2,025
1190	8" diameter		4.50	5.333		1,975	320		2,295	2,625
1200	10" diameter		4	6		2,475	360		2,835	3,250
1210	12" diameter	↓	3.60	6.667	↓	3,625	400		4,025	4,600

2019 Bare Costs: Material, Labor, Equipment, Total

23 21 20.46 Expansion Tanks

	23 21 20.46 Expansion Tanks	Crew	Daily Output	Labor-Hours	Unit	Material	Labor	Equipment	Total	Total Incl O&P
0010	**EXPANSION TANKS**									
1400	Plastic, corrosion resistant, see Plumbing Costs									
1507	Underground fuel-oil storage tanks, see Section 23 13 13									
1512	Tank leak detection systems, see Section 28 33 33.50									
2000	Steel, liquid expansion, ASME, painted, 15 gallon capacity	Q-5	17	.941	Ea.	790	54		844	950
2020	24 gallon capacity		14	1.143		790	66		856	970
2040	30 gallon capacity		12	1.333		900	76.50		976.50	1,100
2060	40 gallon capacity		10	1.600		1,050	92		1,142	1,300
2080	60 gallon capacity		8	2		1,275	115		1,390	1,575
2100	80 gallon capacity		7	2.286		1,350	132		1,482	1,700
2120	100 gallon capacity		6	2.667		1,825	153		1,978	2,250
2130	120 gallon capacity		5	3.200		1,950	184		2,134	2,425
2140	135 gallon capacity		4.50	3.556		2,050	205		2,255	2,550
2150	175 gallon capacity		4	4		3,200	230		3,430	3,875
2160	220 gallon capacity		3.60	4.444		3,650	256		3,906	4,375
2170	240 gallon capacity		3.30	4.848		3,800	279		4,079	4,600
2180	305 gallon capacity		3	5.333		5,350	305		5,655	6,325
2190	400 gallon capacity	↓	2.80	5.714	↓	6,550	330		6,880	7,725
2360	Galvanized									
2370	15 gallon capacity	Q-5	17	.941	Ea.	1,250	54		1,304	1,450
2380	24 gallon capacity	↓	14	1.143	↓	1,475	66		1,541	1,725

23 21 20.46 Expansion Tanks		Crew	Daily Output	Labor-Hours	Unit	2019 Bare Costs Material	Labor	Equipment	Total	Total Incl O&P
2390	30 gallon capacity	Q-5	12	1.333	Ea.	1,525	76.50		1,601.50	1,825
2400	40 gallon capacity		10	1.600		1,800	92		1,892	2,150
2410	60 gallon capacity		8	2		2,075	115		2,190	2,450
2420	80 gallon capacity		7	2.286		2,525	132		2,657	2,975
2430	100 gallon capacity		6	2.667		3,375	153		3,528	3,925
2440	120 gallon capacity		5	3.200		3,475	184		3,659	4,100
2450	135 gallon capacity		4.50	3.556		3,800	205		4,005	4,475
2460	175 gallon capacity		4	4		5,150	230		5,380	6,000
2470	220 gallon capacity		3.60	4.444		6,900	256		7,156	7,975
2480	240 gallon capacity		3.30	4.848		7,225	279		7,504	8,350
2490	305 gallon capacity		3	5.333		8,825	305		9,130	10,200
2500	400 gallon capacity		2.80	5.714		12,900	330		13,230	14,700
3000	Steel ASME expansion, rubber diaphragm, 19 gal. cap. accept.		12	1.333		2,775	76.50		2,851.50	3,200
3020	31 gallon capacity		8	2		3,275	115		3,390	3,775
3040	61 gallon capacity		6	2.667		4,750	153		4,903	5,450
3060	79 gallon capacity		5	3.200		4,675	184		4,859	5,425
3080	119 gallon capacity		4	4		4,850	230		5,080	5,675
3100	158 gallon capacity		3.80	4.211		6,725	242		6,967	7,775
3120	211 gallon capacity		3.30	4.848		7,775	279		8,054	9,000
3140	317 gallon capacity		2.80	5.714		10,200	330		10,530	11,700
3160	422 gallon capacity		2.60	6.154		15,100	355		15,455	17,100
3180	528 gallon capacity		2.40	6.667		16,500	385		16,885	18,800

23 21 20.50 Float Valves		Crew	Daily Output	Labor-Hours	Unit	Material	Labor	Equipment	Total	Total Incl O&P
0010	**FLOAT VALVES**									
0020	With ball and bracket									
0030	Single seat, threaded									
0040	Brass body									
0050	1/2"	1 Stpi	11	.727	Ea.	99.50	46.50		146	180
0060	3/4"		9	.889		114	57		171	211
0070	1"		7	1.143		150	73		223	275
0080	1-1/2"		4.50	1.778		228	114		342	420
0090	2"		3.60	2.222		224	142		366	460
0300	For condensate receivers, CI, in-line mount									
0320	1" inlet	1 Stpi	7	1.143	Ea.	150	73		223	275
0360	For condensate receiver, CI, external float, flanged tank mount									
0370	3/4" inlet	1 Stpi	5	1.600	Ea.	112	102		214	277

23 21 20.54 Flow Check Control		Crew	Daily Output	Labor-Hours	Unit	Material	Labor	Equipment	Total	Total Incl O&P
0010	**FLOW CHECK CONTROL**									
0100	Bronze body, soldered									
0110	3/4" size	1 Stpi	20	.400	Ea.	80	25.50		105.50	127
0120	1" size	"	19	.421	"	96.50	27		123.50	147
0200	Cast iron body, threaded									
0210	3/4" size	1 Stpi	20	.400	Ea.	62	25.50		87.50	107
0220	1" size		19	.421		70	27		97	118
0230	1-1/4" size		15	.533		85	34		119	145
0240	1-1/2" size		13	.615		130	39.50		169.50	202
0250	2" size		11	.727		186	46.50		232.50	274
0300	Flanged inlet, threaded outlet									
0310	2-1/2" size	1 Stpi	8	1	Ea.	350	64		414	480
0320	3" size	Q-5	9	1.778		420	102		522	615

23 21 Hydronic Piping and Pumps

23 21 20 – Hydronic HVAC Piping Specialties

23 21 20.54 Flow Check Control		Crew	Daily Output	Labor-Hours	Unit	Material	2019 Bare Costs Labor	Equipment	Total	Total Incl O&P
0330	4″ size	Q-5	7	2.286	Ea.	815	132		947	1,100

23 21 20.58 Hydronic Heating Control Valves

		Crew	Daily Output	Labor-Hours	Unit	Material	Labor	Equipment	Total	Total Incl O&P
0010	**HYDRONIC HEATING CONTROL VALVES**									
0050	Hot water, nonelectric, thermostatic									
0100	Radiator supply, 1/2″ diameter	1 Stpi	24	.333	Ea.	70	21.50		91.50	109
0120	3/4″ diameter		20	.400		72.50	25.50		98	118
0140	1″ diameter		19	.421		89.50	27		116.50	139
0160	1-1/4″ diameter		15	.533		126	34		160	189
0500	For low pressure steam, add					25%				
1000	Manual, radiator supply									
1010	1/2″ pipe size, angle union	1 Stpi	24	.333	Ea.	65	21.50		86.50	104
1020	3/4″ pipe size, angle union		20	.400		81.50	25.50		107	128
1030	1″ pipe size, angle union		19	.421		97	27		124	148
1100	Radiator, balancing, straight, sweat connections									
1110	1/2″ pipe size	1 Stpi	24	.333	Ea.	21	21.50		42.50	55.50
1120	3/4″ pipe size		20	.400		29.50	25.50		55	71
1130	1″ pipe size		19	.421		48.50	27		75.50	93.50
1200	Steam, radiator, supply									
1210	1/2″ pipe size, angle union	1 Stpi	24	.333	Ea.	61	21.50		82.50	99.50
1220	3/4″ pipe size, angle union		20	.400		68.50	25.50		94	114
1230	1″ pipe size, angle union		19	.421		77	27		104	125
1240	1-1/4″ pipe size, angle union		15	.533		101	34		135	163
8000	System balancing and shut-off									
8020	Butterfly, quarter turn, calibrated, threaded or solder									
8040	Bronze, -30°F to +350°F, pressure to 175 psi									
8060	1/2″ size	1 Stpi	22	.364	Ea.	17.30	23.50		40.80	54
8070	3/4″ size		20	.400		28	25.50		53.50	69
8080	1″ size		19	.421		34	27		61	78
8090	1-1/4″ size		15	.533		55	34		89	112
8100	1-1/2″ size		13	.615		71	39.50		110.50	137
8110	2″ size		11	.727		90	46.50		136.50	169

23 21 20.62 Liquid Drainers

		Crew	Daily Output	Labor-Hours	Unit	Material	Labor	Equipment	Total	Total Incl O&P
0010	**LIQUID DRAINERS**									
0100	Guided lever type									
0110	Cast iron body, threaded									
0120	1/2″ pipe size	1 Stpi	22	.364	Ea.	135	23.50		158.50	184
0130	3/4″ pipe size		20	.400		355	25.50		380.50	430
0140	1″ pipe size		19	.421		565	27		592	665
0150	1-1/2″ pipe size		11	.727		1,075	46.50		1,121.50	1,250
0160	2″ pipe size		8	1		1,075	64		1,139	1,275
0200	Forged steel body, threaded									
0210	1/2″ pipe size	1 Stpi	18	.444	Ea.	1,175	28.50		1,203.50	1,325
0220	3/4″ pipe size		16	.500		1,175	32		1,207	1,325
0230	1″ pipe size		14	.571		1,750	36.50		1,786.50	1,975
0240	1-1/2″ pipe size	Q-5	10	1.600		5,875	92		5,967	6,600
0250	2″ pipe size	″	8	2		5,875	115		5,990	6,625
0300	Socket weld									
0310	1/2″ pipe size	Q-17	16	1	Ea.	1,475	57.50	3.51	1,536.01	1,725
0320	3/4″ pipe size		14	1.143		1,475	66	4.01	1,545.01	1,725
0330	1″ pipe size		12	1.333		2,225	76.50	4.68	2,306.18	2,575
0340	1-1/2″ pipe size		9	1.778		6,075	102	6.25	6,183.25	6,825
0350	2″ pipe size		7	2.286		5,850	132	8	5,990	6,650

23 21 20 – Hydronic HVAC Piping Specialties

23 21 20.62 Liquid Drainers		Crew	Daily Output	Labor-Hours	Unit	Material	2019 Bare Costs Labor	Equipment	Total	Total Incl O&P
0400	Flanged									
0410	1/2" pipe size	1 Stpi	12	.667	Ea.	3,050	42.50		3,092.50	3,425
0420	3/4" pipe size		10	.800		3,050	51		3,101	3,425
0430	1" pipe size		7	1.143		3,825	73		3,898	4,325
0440	1-1/2" pipe size	Q-5	9	1.778		8,075	102		8,177	9,025
0450	2" pipe size	"	7	2.286		8,075	132		8,207	9,075
1000	Fixed lever and snap action type									
1100	Cast iron body, threaded									
1110	1/2" pipe size	1 Stpi	22	.364	Ea.	209	23.50		232.50	264
1120	3/4" pipe size		20	.400		550	25.50		575.50	645
1130	1" pipe size		19	.421		550	27		577	645
1200	Forged steel body, threaded									
1210	1/2" pipe size	1 Stpi	18	.444	Ea.	1,450	28.50		1,478.50	1,650
1220	3/4" pipe size		16	.500		1,450	32		1,482	1,650
1230	1" pipe size		14	.571		3,450	36.50		3,486.50	3,850
1240	1-1/4" pipe size		12	.667		2,950	42.50		2,992.50	3,300
2000	High pressure high leverage type									
2100	Forged steel body, threaded									
2110	1/2" pipe size	1 Stpi	18	.444	Ea.	2,350	28.50		2,378.50	2,625
2120	3/4" pipe size		16	.500		2,750	32		2,782	3,075
2130	1" pipe size		14	.571		3,925	36.50		3,961.50	4,350
2140	1-1/4" pipe size		12	.667		4,650	42.50		4,692.50	5,175
2150	1-1/2" pipe size	Q-5	10	1.600		5,900	92		5,992	6,650
2160	2" pipe size	"	8	2		5,900	115		6,015	6,675
2200	Socket weld									
2210	1/2" pipe size	Q-17	16	1	Ea.	2,975	57.50	3.51	3,036.01	3,350
2220	3/4" pipe size		14	1.143		2,525	66	4.01	2,595.01	2,875
2230	1" pipe size		12	1.333		4,125	76.50	4.68	4,206.18	4,650
2240	1-1/4" pipe size		11	1.455		4,825	83.50	5.10	4,913.60	5,425
2250	1-1/2" pipe size		9	1.778		6,400	102	6.25	6,508.25	7,175
2260	2" pipe size		7	2.286		6,400	132	8	6,540	7,225
2300	Flanged									
2310	1/2" pipe size	1 Stpi	12	.667	Ea.	3,400	42.50		3,442.50	3,800
2320	3/4" pipe size		10	.800		3,975	51		4,026	4,450
2330	1" pipe size		7	1.143		5,350	73		5,423	6,000
2340	1-1/4" pipe size		6	1.333		5,350	85.50		5,435.50	6,025
2350	1-1/2" pipe size	Q-5	9	1.778		7,325	102		7,427	8,200
2360	2" pipe size	"	7	2.286		7,325	132		7,457	8,250
3000	Guided lever dual gravity type									
3100	Cast iron body, threaded									
3110	1/2" pipe size	1 Stpi	22	.364	Ea.	500	23.50		523.50	585
3120	3/4" pipe size		20	.400		425	25.50		450.50	510
3130	1" pipe size		19	.421		620	27		647	725
3140	1-1/2" pipe size		11	.727		1,200	46.50		1,246.50	1,400
3150	2" pipe size		8	1		1,450	64		1,514	1,700
3200	Forged steel body, threaded									
3210	1/2" pipe size	1 Stpi	18	.444	Ea.	1,275	28.50		1,303.50	1,450
3220	3/4" pipe size		16	.500		1,275	32		1,307	1,450
3230	1" pipe size		14	.571		1,900	36.50		1,936.50	2,125
3240	1-1/2" pipe size	Q-5	10	1.600		5,050	92		5,142	5,700
3250	2" pipe size	"	8	2		5,050	115		5,165	5,725
3300	Socket weld									
3310	1/2" pipe size	Q-17	16	1	Ea.	1,450	57.50	3.51	1,511.01	1,700

23 21 20.62 Liquid Drainers		Crew	Daily Output	Labor-Hours	Unit	Material	2019 Bare Costs Labor	Equipment	Total	Total Incl O&P
3320	3/4" pipe size	Q-17	14	1.143	Ea.	1,450	66	4.01	1,520.01	1,700
3330	1" pipe size		12	1.333		2,075	76.50	4.68	2,156.18	2,425
3340	1-1/2" pipe size		9	1.778		6,150	102	6.25	6,258.25	6,900
3350	2" pipe size		7	2.286		5,225	132	8	5,365	5,950
3400	Flanged									
3410	1/2" pipe size	1 Stpi	12	.667	Ea.	2,325	42.50		2,367.50	2,625
3420	3/4" pipe size		10	.800		2,325	51		2,376	2,625
3430	1" pipe size		7	1.143		3,575	73		3,648	4,025
3440	1-1/2" pipe size	Q-5	9	1.778		6,450	102		6,552	7,250
3450	2" pipe size	"	7	2.286		6,450	132		6,582	7,300

23 21 20.66 Monoflow Tee Fitting		Crew	Daily Output	Labor-Hours	Unit	Material	Labor	Equipment	Total	Total Incl O&P
0010	**MONOFLOW TEE FITTING**									
1100	For one pipe hydronic, supply and return									
1110	Copper, soldered									
1120	3/4" x 1/2" size	1 Stpi	13	.615	Ea.	22	39.50		61.50	83.50
1130	1" x 1/2" size		12	.667		22	42.50		64.50	88.50
1140	1" x 3/4" size		11	.727		22	46.50		68.50	94.50
1150	1-1/4" x 1/2" size		11	.727		23	46.50		69.50	95.50
1160	1-1/4" x 3/4" size		10	.800		23	51		74	103
1170	1-1/2" x 3/4" size		10	.800		33	51		84	113
1180	1-1/2" x 1" size		9	.889		33	57		90	122
1190	2" x 3/4" size		9	.889		43.50	57		100.50	133
1200	2" x 1" size		8	1		43.50	64		107.50	144

23 21 20.70 Steam Traps		Crew	Daily Output	Labor-Hours	Unit	Material	Labor	Equipment	Total	Total Incl O&P
0010	**STEAM TRAPS**									
0030	Cast iron body, threaded									
0040	Inverted bucket									
0050	1/2" pipe size	1 Stpi	12	.667	Ea.	167	42.50		209.50	248
0070	3/4" pipe size		10	.800		243	51		294	345
0100	1" pipe size		9	.889		370	57		427	495
0120	1-1/4" pipe size		8	1		560	64		624	715
0130	1-1/2" pipe size		7	1.143		705	73		778	890
0140	2" pipe size		6	1.333		1,350	85.50		1,435.50	1,600
0200	With thermic vent & check valve									
0202	1/2" pipe size	1 Stpi	12	.667	Ea.	215	42.50		257.50	300
0204	3/4" pipe size		10	.800		279	51		330	380
0206	1" pipe size		9	.889		490	57		547	620
0208	1-1/4" pipe size		8	1		740	64		804	910
0210	1-1/2" pipe size		7	1.143		945	73		1,018	1,150
0212	2" pipe size		6	1.333		1,450	85.50		1,535.50	1,700
1000	Float & thermostatic, 15 psi									
1010	3/4" pipe size	1 Stpi	16	.500	Ea.	175	32		207	240
1020	1" pipe size		15	.533		183	34		217	253
1030	1-1/4" pipe size		13	.615		231	39.50		270.50	315
1040	1-1/2" pipe size		9	.889		310	57		367	425
1060	2" pipe size		6	1.333		965	85.50		1,050.50	1,175
1100	For vacuum breaker, add					74			74	81.50
1110	Differential condensate controller									
1120	With union and manual metering valve									
1130	1/2" pipe size	1 Stpi	24	.333	Ea.	695	21.50		716.50	795
1140	3/4" pipe size		20	.400		850	25.50		875.50	970
1150	1" pipe size		19	.421		1,150	27		1,177	1,300

23 21 20.70 Steam Traps		Crew	Daily Output	Labor-Hours	Unit	Material	2019 Bare Costs Labor	Equipment	Total	Total Incl O&P
1160	1-1/4" pipe size	1 Stpi	15	.533	Ea.	1,550	34		1,584	1,750
1170	1-1/2" pipe size		13	.615		1,875	39.50		1,914.50	2,100
1180	2" pipe size		11	.727		2,425	46.50		2,471.50	2,750
1290	Brass body, threaded									
1300	Thermostatic, angle union, 25 psi									
1310	1/2" pipe size	1 Stpi	24	.333	Ea.	71.50	21.50		93	111
1320	3/4" pipe size		20	.400		103	25.50		128.50	152
1330	1" pipe size		19	.421		165	27		192	223
1990	Carbon steel body, threaded									
2000	Controlled disc, integral strainer & blow down									
2010	Threaded									
2020	3/8" pipe size	1 Stpi	24	.333	Ea.	325	21.50		346.50	385
2030	1/2" pipe size		24	.333		325	21.50		346.50	385
2040	3/4" pipe size		20	.400		435	25.50		460.50	515
2050	1" pipe size		19	.421		560	27		587	655
2100	Socket weld									
2110	3/8" pipe size	Q-17	17	.941	Ea.	390	54	3.30	447.30	510
2120	1/2" pipe size		16	1		390	57.50	3.51	451.01	515
2130	3/4" pipe size		14	1.143		505	66	4.01	575.01	660
2140	1" pipe size		12	1.333		635	76.50	4.68	716.18	815
2170	Flanged, 600 lb. ASA									
2180	3/8" pipe size	1 Stpi	12	.667	Ea.	875	42.50		917.50	1,025
2190	1/2" pipe size		12	.667		875	42.50		917.50	1,025
2200	3/4" pipe size		10	.800		1,050	51		1,101	1,225
2210	1" pipe size		7	1.143		1,200	73		1,273	1,425
5000	Forged steel body									
5010	Inverted bucket									
5020	Threaded									
5030	1/2" pipe size	1 Stpi	12	.667	Ea.	595	42.50		637.50	715
5040	3/4" pipe size		10	.800		1,050	51		1,101	1,225
5050	1" pipe size		9	.889		1,550	57		1,607	1,775
5060	1-1/4" pipe size		8	1		2,250	64		2,314	2,575
5070	1-1/2" pipe size		7	1.143		3,175	73		3,248	3,600
5080	2" pipe size		6	1.333		5,575	85.50		5,660.50	6,250
5100	Socket weld									
5110	1/2" pipe size	Q-17	16	1	Ea.	685	57.50	3.51	746.01	845
5120	3/4" pipe size		14	1.143		1,100	66	4.01	1,170.01	1,325
5130	1" pipe size		12	1.333		1,600	76.50	4.68	1,681.18	1,875
5140	1-1/4" pipe size		10	1.600		2,300	92	5.60	2,397.60	2,675
5150	1-1/2" pipe size		9	1.778		3,175	102	6.25	3,283.25	3,625
5160	2" pipe size		7	2.286		5,500	132	8	5,640	6,250
5200	Flanged									
5210	1/2" pipe size	1 Stpi	12	.667	Ea.	1,050	42.50		1,092.50	1,250
5220	3/4" pipe size		10	.800		1,725	51		1,776	1,975
5230	1" pipe size		7	1.143		2,350	73		2,423	2,700
5240	1-1/4" pipe size		6	1.333		3,250	85.50		3,335.50	3,700
5250	1-1/2" pipe size	Q-5	9	1.778		4,075	102		4,177	4,650
5260	2" pipe size	"	7	2.286		6,425	132		6,557	7,275
6000	Stainless steel body									
6010	Inverted bucket									
6020	Threaded									
6030	1/2" pipe size	1 Stpi	12	.667	Ea.	196	42.50		238.50	280
6040	3/4" pipe size		10	.800		256	51		307	360

23 21 Hydronic Piping and Pumps

23 21 20 – Hydronic HVAC Piping Specialties

23 21 20.70 Steam Traps		Crew	Daily Output	Labor-Hours	Unit	2019 Bare Costs Material	Labor	Equipment	Total	Total Incl O&P
6050	1″ pipe size	1 Stpi	9	.889	Ea.	910	57		967	1,075
6100	Threaded with thermic vent									
6110	1/2″ pipe size	1 Stpi	12	.667	Ea.	192	42.50		234.50	275
6120	3/4″ pipe size		10	.800		253	51		304	355
6130	1″ pipe size		9	.889		900	57		957	1,075
6200	Threaded with check valve									
6210	1/2″ pipe size	1 Stpi	12	.667	Ea.	280	42.50		322.50	375
6220	3/4″ pipe size	″	10	.800	″	385	51		436	500
6300	Socket weld									
6310	1/2″ pipe size	Q-17	16	1	Ea.	226	57.50	3.51	287.01	340
6320	3/4″ pipe size	″	14	1.143	″	242	66	4.01	312.01	370

23 21 20.74 Strainers, Basket Type

Line	Description	Crew	Daily Output	Labor-Hours	Unit	Material	Labor	Equipment	Total	Total Incl O&P
0010	**STRAINERS, BASKET TYPE**, Perforated stainless steel basket									
0100	Brass or monel available									
2000	Simplex style									
2300	Bronze body									
2320	Screwed, 3/8″ pipe size	1 Stpi	22	.364	Ea.	179	23.50		202.50	232
2340	1/2″ pipe size		20	.400		214	25.50		239.50	274
2360	3/4″ pipe size		17	.471		291	30		321	365
2380	1″ pipe size		15	.533		375	34		409	465
2400	1-1/4″ pipe size		13	.615		450	39.50		489.50	555
2420	1-1/2″ pipe size		12	.667		465	42.50		507.50	575
2440	2″ pipe size		10	.800		690	51		741	835
2460	2-1/2″ pipe size	Q-5	15	1.067		1,000	61.50		1,061.50	1,200
2480	3″ pipe size	″	14	1.143		1,400	66		1,466	1,650
2600	Flanged, 2″ pipe size	1 Stpi	6	1.333		760	85.50		845.50	965
2620	2-1/2″ pipe size	Q-5	4.50	3.556		1,425	205		1,630	1,850
2640	3″ pipe size		3.50	4.571		1,600	263		1,863	2,150
2660	4″ pipe size		3	5.333		2,275	305		2,580	2,950
2680	5″ pipe size	Q-6	3.40	7.059		3,925	420		4,345	4,950
2700	6″ pipe size		3	8		4,400	475		4,875	5,575
2710	8″ pipe size		2.50	9.600		7,175	575		7,750	8,750
3600	Iron body									
3700	Screwed, 3/8″ pipe size	1 Stpi	22	.364	Ea.	155	23.50		178.50	206
3720	1/2″ pipe size		20	.400		161	25.50		186.50	216
3740	3/4″ pipe size		17	.471		204	30		234	270
3760	1″ pipe size		15	.533		209	34		243	280
3780	1-1/4″ pipe size		13	.615		272	39.50		311.50	360
3800	1-1/2″ pipe size		12	.667		300	42.50		342.50	395
3820	2″ pipe size		10	.800		360	51		411	470
3840	2-1/2″ pipe size	Q-5	15	1.067		480	61.50		541.50	615
3860	3″ pipe size	″	14	1.143		580	66		646	740
4000	Flanged, 2″ pipe size	1 Stpi	6	1.333		470	85.50		555.50	645
4020	2-1/2″ pipe size	Q-5	4.50	3.556		635	205		840	1,000
4040	3″ pipe size		3.50	4.571		665	263		928	1,125
4060	4″ pipe size		3	5.333		1,050	305		1,355	1,600
4080	5″ pipe size	Q-6	3.40	7.059		1,550	420		1,970	2,325
4100	6″ pipe size		3	8		1,975	475		2,450	2,900
4120	8″ pipe size		2.50	9.600		3,550	575		4,125	4,775
4140	10″ pipe size		2.20	10.909		7,825	650		8,475	9,575
6000	Cast steel body									
6400	Screwed, 1″ pipe size	1 Stpi	15	.533	Ea.	295	34		329	375

23 21 20 – Hydronic HVAC Piping Specialties

23 21 20.74 Strainers, Basket Type		Crew	Daily Output	Labor-Hours	Unit	2019 Bare Costs Material	Labor	Equipment	Total	Total Incl O&P
6410	1-1/4" pipe size	1 Stpi	13	.615	Ea.	455	39.50		494.50	560
6420	1-1/2" pipe size		12	.667		520	42.50		562.50	635
6440	2" pipe size		10	.800		610	51		661	745
6460	2-1/2" pipe size	Q-5	15	1.067		880	61.50		941.50	1,050
6480	3" pipe size	"	14	1.143		1,125	66		1,191	1,350
6560	Flanged, 2" pipe size	1 Stpi	6	1.333		1,300	85.50		1,385.50	1,575
6580	2-1/2" pipe size	Q-5	4.50	3.556		1,950	205		2,155	2,450
6600	3" pipe size		3.50	4.571		2,350	263		2,613	2,975
6620	4" pipe size		3	5.333		2,475	305		2,780	3,175
6640	6" pipe size	Q-6	3	8		4,550	475		5,025	5,725
6660	8" pipe size	"	2.50	9.600		8,650	575		9,225	10,400
7000	Stainless steel body									
7200	Screwed, 1" pipe size	1 Stpi	15	.533	Ea.	430	34		464	520
7210	1-1/4" pipe size		13	.615		670	39.50		709.50	795
7220	1-1/2" pipe size		12	.667		670	42.50		712.50	800
7240	2" pipe size		10	.800		995	51		1,046	1,175
7260	2-1/2" pipe size	Q-5	15	1.067		1,400	61.50		1,461.50	1,650
7280	3" pipe size	"	14	1.143		1,950	66		2,016	2,225
7400	Flanged, 2" pipe size	1 Stpi	6	1.333		1,475	85.50		1,560.50	1,725
7420	2-1/2" pipe size	Q-5	4.50	3.556		2,650	205		2,855	3,225
7440	3" pipe size		3.50	4.571		2,725	263		2,988	3,400
7460	4" pipe size		3	5.333		4,275	305		4,580	5,150
7480	6" pipe size	Q-6	3	8		9,700	475		10,175	11,400
7500	8" pipe size	"	2.50	9.600		14,900	575		15,475	17,300
8100	Duplex style									
8200	Bronze body									
8240	Screwed, 3/4" pipe size	1 Stpi	16	.500	Ea.	1,250	32		1,282	1,450
8260	1" pipe size		14	.571		1,250	36.50		1,286.50	1,450
8280	1-1/4" pipe size		12	.667		2,750	42.50		2,792.50	3,100
8300	1-1/2" pipe size		11	.727		2,750	46.50		2,796.50	3,100
8320	2" pipe size		9	.889		4,000	57		4,057	4,475
8340	2-1/2" pipe size	Q-5	14	1.143		5,175	66		5,241	5,775
8420	Flanged, 2" pipe size	1 Stpi	6	1.333		4,775	85.50		4,860.50	5,375
8440	2-1/2" pipe size	Q-5	4.50	3.556		6,600	205		6,805	7,550
8460	3" pipe size		3.50	4.571		7,200	263		7,463	8,300
8480	4" pipe size		3	5.333		10,500	305		10,805	12,000
8500	5" pipe size	Q-6	3.40	7.059		22,900	420		23,320	25,700
8520	6" pipe size	"	3	8		22,900	475		23,375	25,900
8700	Iron body									
8740	Screwed, 3/4" pipe size	1 Stpi	16	.500	Ea.	1,425	32		1,457	1,600
8760	1" pipe size		14	.571		1,425	36.50		1,461.50	1,600
8780	1-1/4" pipe size		12	.667		1,525	42.50		1,567.50	1,750
8800	1-1/2" pipe size		11	.727		1,525	46.50		1,571.50	1,750
8820	2" pipe size		9	.889		2,600	57		2,657	2,925
8840	2-1/2" pipe size	Q-5	14	1.143		2,850	66		2,916	3,250
9000	Flanged, 2" pipe size	1 Stpi	6	1.333		2,750	85.50		2,835.50	3,125
9020	2-1/2" pipe size	Q-5	4.50	3.556		2,950	205		3,155	3,550
9040	3" pipe size		3.50	4.571		3,225	263		3,488	3,950
9060	4" pipe size		3	5.333		5,425	305		5,730	6,425
9080	5" pipe size	Q-6	3.40	7.059		11,900	420		12,320	13,700
9100	6" pipe size		3	8		11,900	475		12,375	13,800
9120	8" pipe size		2.50	9.600		23,700	575		24,275	27,000
9140	10" pipe size		2.20	10.909		34,000	650		34,650	38,400

23 21 20.74 Strainers, Basket Type		Crew	Daily Output	Labor-Hours	Unit	Material	2019 Bare Costs Labor	Equipment	Total	Total Incl O&P
9160	12″ pipe size	Q-6	1.70	14.118	Ea.	37,500	845		38,345	42,500
9170	14″ pipe size		1.40	17.143		43,800	1,025		44,825	49,700
9180	16″ pipe size		1	24		53,500	1,425		54,925	61,000
9300	Cast steel body									
9340	Screwed, 1″ pipe size	1 Stpi	14	.571	Ea.	1,775	36.50		1,811.50	2,000
9360	1-1/2″ pipe size		11	.727		2,850	46.50		2,896.50	3,225
9380	2″ pipe size		9	.889		3,775	57		3,832	4,250
9460	Flanged, 2″ pipe size		6	1.333		3,900	85.50		3,985.50	4,400
9480	2-1/2″ pipe size	Q-5	4.50	3.556		5,325	205		5,530	6,175
9500	3″ pipe size		3.50	4.571		5,825	263		6,088	6,800
9520	4″ pipe size		3	5.333		7,175	305		7,480	8,350
9540	6″ pipe size	Q-6	3	8		13,400	475		13,875	15,500
9560	8″ pipe size	″	2.50	9.600		32,800	575		33,375	37,000
9700	Stainless steel body									
9740	Screwed, 1″ pipe size	1 Stpi	14	.571	Ea.	3,200	36.50		3,236.50	3,550
9760	1-1/2″ pipe size		11	.727		4,775	46.50		4,821.50	5,325
9780	2″ pipe size		9	.889		6,650	57		6,707	7,375
9860	Flanged, 2″ pipe size		6	1.333		7,275	85.50		7,360.50	8,125
9880	2-1/2″ pipe size	Q-5	4.50	3.556		12,200	205		12,405	13,700
9900	3″ pipe size		3.50	4.571		13,300	263		13,563	15,000
9920	4″ pipe size		3	5.333		17,300	305		17,605	19,500
9940	6″ pipe size	Q-6	3	8		26,400	475		26,875	29,800
9960	8″ pipe size	″	2.50	9.600		74,000	575		74,575	82,000

23 21 20.76 Strainers, Y Type, Bronze Body		Crew	Daily Output	Labor-Hours	Unit	Material	Labor	Equipment	Total	Total Incl O&P
0010	**STRAINERS, Y TYPE, BRONZE BODY**									
0050	Screwed, 125 lb., 1/4″ pipe size	1 Stpi	24	.333	Ea.	39.50	21.50		61	75.50
0070	3/8″ pipe size		24	.333		43	21.50		64.50	79.50
0100	1/2″ pipe size		20	.400		43	25.50		68.50	86
0120	3/4″ pipe size		19	.421		53	27		80	99
0140	1″ pipe size		17	.471		62	30		92	113
0150	1-1/4″ pipe size		15	.533		126	34		160	189
0160	1-1/2″ pipe size		14	.571		134	36.50		170.50	202
0180	2″ pipe size		13	.615		176	39.50		215.50	253
0182	3″ pipe size		12	.667		1,150	42.50		1,192.50	1,350
0200	300 lb., 2-1/2″ pipe size	Q-5	17	.941		760	54		814	915
0220	3″ pipe size		16	1		1,275	57.50		1,332.50	1,500
0240	4″ pipe size		15	1.067		2,400	61.50		2,461.50	2,750
0500	For 300 lb. rating 1/4″ thru 2″, add					15%				
1000	Flanged, 150 lb., 1-1/2″ pipe size	1 Stpi	11	.727	Ea.	490	46.50		536.50	610
1020	2″ pipe size	″	8	1		665	64		729	825
1030	2-1/2″ pipe size	Q-5	5	3.200		1,075	184		1,259	1,450
1040	3″ pipe size		4.50	3.556		1,325	205		1,530	1,775
1060	4″ pipe size		3	5.333		2,050	305		2,355	2,700
1080	5″ pipe size	Q-6	3.40	7.059		2,500	420		2,920	3,375
1100	6″ pipe size		3	8		3,950	475		4,425	5,075
1106	8″ pipe size		2.60	9.231		4,300	550		4,850	5,550
1500	For 300 lb. rating, add					40%				

23 21 20.78 Strainers, Y Type, Iron Body		Crew	Daily Output	Labor-Hours	Unit	Material	Labor	Equipment	Total	Total Incl O&P
0010	**STRAINERS, Y TYPE, IRON BODY**									
0050	Screwed, 250 lb., 1/4″ pipe size	1 Stpi	20	.400	Ea.	17.40	25.50		42.90	57.50
0070	3/8″ pipe size		20	.400		17.40	25.50		42.90	57.50
0100	1/2″ pipe size		20	.400		17.40	25.50		42.90	57.50

23 21 20 – Hydronic HVAC Piping Specialties

23 21 20.78 Strainers, Y Type, Iron Body		Crew	Daily Output	Labor-Hours	Unit	Material	2019 Bare Costs Labor	Equipment	Total	Total Incl O&P
0120	3/4" pipe size	1 Stpi	18	.444	Ea.	20	28.50		48.50	64.50
0140	1" pipe size		16	.500		24.50	32		56.50	75
0150	1-1/4" pipe size		15	.533		37.50	34		71.50	92.50
0160	1-1/2" pipe size		12	.667		51.50	42.50		94	121
0180	2" pipe size		8	1		76.50	64		140.50	181
0200	2-1/2" pipe size	Q-5	12	1.333		278	76.50		354.50	420
0220	3" pipe size		11	1.455		300	83.50		383.50	455
0240	4" pipe size		5	3.200		510	184		694	835
0500	For galvanized body, add					50%				
1000	Flanged, 125 lb., 1-1/2" pipe size	1 Stpi	11	.727	Ea.	173	46.50		219.50	260
1020	2" pipe size	"	8	1		156	64		220	268
1030	2-1/2" pipe size	Q-5	5	3.200		167	184		351	460
1040	3" pipe size		4.50	3.556		182	205		387	505
1060	4" pipe size		3	5.333		310	305		615	805
1080	5" pipe size	Q-6	3.40	7.059		400	420		820	1,075
1100	6" pipe size		3	8		630	475		1,105	1,400
1120	8" pipe size		2.50	9.600		1,150	575		1,725	2,100
1140	10" pipe size		2	12		1,900	715		2,615	3,175
1160	12" pipe size		1.70	14.118		3,125	845		3,970	4,725
1170	14" pipe size		1.30	18.462		5,800	1,100		6,900	8,025
1180	16" pipe size		1	24		8,225	1,425		9,650	11,200
1500	For 250 lb. rating, add					20%				
2000	For galvanized body, add					50%				
2500	For steel body, add					40%				

23 21 20.80 Suction Diffusers		Crew	Daily Output	Labor-Hours	Unit	Material	Labor	Equipment	Total	Total Incl O&P
0010	**SUCTION DIFFUSERS**									
0100	Cast iron body with integral straightening vanes, strainer									
1000	Flanged									
1010	2" inlet, 1-1/2" pump side	1 Stpi	6	1.333	Ea.	425	85.50		510.50	600
1020	2" pump side	"	5	1.600		510	102		612	715
1030	3" inlet, 2" pump side	Q-5	6.50	2.462		530	142		672	800
1040	2-1/2" pump side		5.30	3.019		785	174		959	1,125
1050	3" pump side		4.50	3.556		790	205		995	1,175
1060	4" inlet, 3" pump side		3.50	4.571		910	263		1,173	1,400
1070	4" pump side		3	5.333		1,050	305		1,355	1,600
1080	5" inlet, 4" pump side	Q-6	3.80	6.316		1,200	375		1,575	1,900
1090	5" pump side		3.40	7.059		1,425	420		1,845	2,200
1100	6" inlet, 4" pump side		3.30	7.273		1,250	435		1,685	2,025
1110	5" pump side		3.10	7.742		1,500	460		1,960	2,350
1120	6" pump side		3	8		1,575	475		2,050	2,475
1130	8" inlet, 6" pump side		2.70	8.889		1,750	530		2,280	2,725
1140	8" pump side		2.50	9.600		3,050	575		3,625	4,200
1150	10" inlet, 6" pump side		2.40	10		4,275	595		4,870	5,600
1160	8" pump side		2.30	10.435		3,275	625		3,900	4,550
1170	10" pump side		2.20	10.909		5,100	650		5,750	6,600
1180	12" inlet, 8" pump side		2	12		5,400	715		6,115	7,000
1190	10" pump side		1.80	13.333		6,100	795		6,895	7,900
1200	12" pump side		1.70	14.118		6,725	845		7,570	8,675
1210	14" inlet, 12" pump side		1.50	16		7,575	955		8,530	9,750
1220	14" pump side		1.30	18.462		9,750	1,100		10,850	12,400

23 21 20.84 Thermoflo Indicator

	23 21 20.84 Thermoflo Indicator	Crew	Daily Output	Labor-Hours	Unit	Material	Labor (2019 Bare Costs)	Equipment	Total	Total Incl O&P
0010	**THERMOFLO INDICATOR**, For balancing									
1000	Sweat connections, 1-1/4" pipe size	1 Stpi	12	.667	Ea.	755	42.50		797.50	895
1020	1-1/2" pipe size		10	.800		765	51		816	920
1040	2" pipe size		8	1		805	64		869	980
1060	2-1/2" pipe size		7	1.143		1,225	73		1,298	1,450
2000	Flange connections, 3" pipe size	Q-5	5	3.200		1,475	184		1,659	1,900
2020	4" pipe size		4	4		1,775	230		2,005	2,300
2030	5" pipe size		3.50	4.571		2,200	263		2,463	2,825
2040	6" pipe size		3	5.333		2,350	305		2,655	3,050
2060	8" pipe size		2	8		2,875	460		3,335	3,850

23 21 20.88 Venturi Flow

	23 21 20.88 Venturi Flow	Crew	Daily Output	Labor-Hours	Unit	Material	Labor	Equipment	Total	Total Incl O&P
0010	**VENTURI FLOW**, Measuring device									
0050	1/2" diameter	1 Stpi	24	.333	Ea.	300	21.50		321.50	360
0100	3/4" diameter		20	.400		275	25.50		300.50	340
0120	1" diameter		19	.421		294	27		321	365
0140	1-1/4" diameter		15	.533		365	34		399	450
0160	1-1/2" diameter		13	.615		380	39.50		419.50	480
0180	2" diameter		11	.727		390	46.50		436.50	500
0200	2-1/2" diameter	Q-5	16	1		535	57.50		592.50	675
0220	3" diameter		14	1.143		550	66		616	705
0240	4" diameter		11	1.455		825	83.50		908.50	1,025
0260	5" diameter	Q-6	4	6		1,075	360		1,435	1,725
0280	6" diameter		3.50	6.857		1,200	410		1,610	1,950
0300	8" diameter		3	8		1,550	475		2,025	2,425
0320	10" diameter		2	12		3,650	715		4,365	5,100
0330	12" diameter		1.80	13.333		5,150	795		5,945	6,875
0340	14" diameter		1.60	15		5,850	895		6,745	7,800
0350	16" diameter		1.40	17.143		6,675	1,025		7,700	8,875
0500	For meter, add					2,325			2,325	2,550

23 21 20.94 Weld End Ball Joints

	23 21 20.94 Weld End Ball Joints	Crew	Daily Output	Labor-Hours	Unit	Material	Labor	Equipment	Total	Total Incl O&P
0010	**WELD END BALL JOINTS**, Steel									
0050	2-1/2" diameter	Q-17	13	1.231	Ea.	835	71	4.32	910.32	1,025
0100	3" diameter		12	1.333		970	76.50	4.68	1,051.18	1,200
0120	4" diameter		11	1.455		1,425	83.50	5.10	1,513.60	1,700
0140	5" diameter	Q-18	14	1.714		1,925	102	4.01	2,031.01	2,275
0160	6" diameter		12	2		2,275	119	4.68	2,398.68	2,675
0180	8" diameter		9	2.667		3,350	159	6.25	3,515.25	3,950
0200	10" diameter		8	3		4,600	179	7	4,786	5,350
0220	12" diameter		6	4		6,900	239	9.35	7,148.35	7,950

23 21 20.98 Zone Valves

	23 21 20.98 Zone Valves	Crew	Daily Output	Labor-Hours	Unit	Material	Labor	Equipment	Total	Total Incl O&P
0010	**ZONE VALVES**									
1000	Bronze body									
1010	2 way, 125 psi									
1020	Standard, 65' pump head									
1030	1/2" soldered	1 Stpi	24	.333	Ea.	149	21.50		170.50	196
1040	3/4" soldered		20	.400		149	25.50		174.50	203
1050	1" soldered		19	.421		162	27		189	219
1060	1-1/4" soldered		15	.533		191	34		225	261
1200	High head, 125' pump head									
1210	1/2" soldered	1 Stpi	24	.333	Ea.	220	21.50		241.50	274
1220	3/4" soldered		20	.400		228	25.50		253.50	289

23 21 20 – Hydronic HVAC Piping Specialties

23 21 20.98 Zone Valves		Crew	Daily Output	Labor-Hours	Unit	Material	2019 Bare Costs Labor	Equipment	Total	Total Incl O&P
1230	1" soldered	1 Stpi	19	.421	Ea.	247	27		274	310
1300	Geothermal, 150' pump head									
1310	3/4" soldered	1 Stpi	20	.400	Ea.	225	25.50		250.50	286
1320	1" soldered		19	.421		275	27		302	345
1330	3/4" threaded	↓	20	.400	↓	223	25.50		248.50	285
3000	3 way, 125 psi									
3200	Bypass, 65' pump head									
3210	1/2" soldered	1 Stpi	24	.333	Ea.	201	21.50		222.50	254
3220	3/4" soldered		20	.400		237	25.50		262.50	300
3230	1" soldered		19	.421		268	27		295	335
9000	Transformer, for up to 3 zone valves	↓	20	.400	↓	32.50	25.50		58	74.50

23 21 23 – Hydronic Pumps

23 21 23.13 In-Line Centrifugal Hydronic Pumps

23 21 23.13	In-Line Centrifugal Hydronic Pumps	Crew	Daily Output	Labor-Hours	Unit	Material	Labor	Equipment	Total	Total Incl O&P
0010	**IN-LINE CENTRIFUGAL HYDRONIC PUMPS**									
0600	Bronze, sweat connections, 1/40 HP, in line									
0640	3/4" size	Q-1	16	1	Ea.	288	57		345	400
1000	Flange connection, 3/4" to 1-1/2" size									
1040	1/12 HP	Q-1	6	2.667	Ea.	755	152		907	1,050
1060	1/8 HP		6	2.667		1,325	152		1,477	1,675
1100	1/3 HP		6	2.667		1,450	152		1,602	1,825
1140	2" size, 1/6 HP		5	3.200		1,650	182		1,832	2,075
1180	2-1/2" size, 1/4 HP		5	3.200		2,125	182		2,307	2,625
1220	3" size, 1/4 HP		4	4		2,300	227		2,527	2,875
1260	1/3 HP		4	4		2,725	227		2,952	3,350
1300	1/2 HP		4	4		3,025	227		3,252	3,675
1340	3/4 HP		4	4		2,975	227		3,202	3,625
1380	1 HP	↓	4	4	↓	4,800	227		5,027	5,625
2000	Cast iron, flange connection									
2040	3/4" to 1-1/2" size, in line, 1/12 HP	Q-1	6	2.667	Ea.	460	152		612	730
2060	1/8 HP		6	2.667		815	152		967	1,125
2100	1/3 HP		6	2.667		910	152		1,062	1,225
2140	2" size, 1/6 HP		5	3.200		1,000	182		1,182	1,375
2180	2-1/2" size, 1/4 HP		5	3.200		1,125	182		1,307	1,500
2220	3" size, 1/4 HP		4	4		1,150	227		1,377	1,625
2260	1/3 HP		4	4		1,525	227		1,752	2,025
2300	1/2 HP		4	4		1,575	227		1,802	2,075
2340	3/4 HP		4	4		1,825	227		2,052	2,350
2380	1 HP	↓	4	4	↓	2,625	227		2,852	3,250
2600	For nonferrous impeller, add					3%				
3000	High head, bronze impeller									
3030	1-1/2" size, 1/2 HP	Q-1	5	3.200	Ea.	1,350	182		1,532	1,750
3040	1-1/2" size, 3/4 HP		5	3.200		1,475	182		1,657	1,900
3050	2" size, 1 HP		4	4		1,800	227		2,027	2,350
3090	2" size, 1-1/2 HP	↓	4	4	↓	2,250	227		2,477	2,825
4000	Close coupled, end suction, bronze impeller									
4040	1-1/2" size, 1-1/2 HP, to 40 GPM	Q-1	3	5.333	Ea.	2,600	305		2,905	3,325
4090	2" size, 2 HP, to 50 GPM		3	5.333		3,025	305		3,330	3,775
4100	2" size, 3 HP, to 90 GPM		2.30	6.957		3,050	395		3,445	3,950
4190	2-1/2" size, 3 HP, to 150 GPM		2	8		3,500	455		3,955	4,525
4300	3" size, 5 HP, to 225 GPM		1.80	8.889		4,000	505		4,505	5,175
4410	3" size, 10 HP, to 350 GPM		1.60	10		5,825	570		6,395	7,250
4420	4" size, 7-1/2 HP, to 350 GPM	↓	1.60	10	↓	5,975	570		6,545	7,400

23 21 Hydronic Piping and Pumps

23 21 23 – Hydronic Pumps

23 21 23.13 In-Line Centrifugal Hydronic Pumps		Crew	Daily Output	Labor-Hours	Unit	Material	2019 Bare Costs Labor	Equipment	Total	Total Incl O&P
4520	4" size, 10 HP, to 600 GPM	Q-2	1.70	14.118	Ea.	5,950	830		6,780	7,800
4530	5" size, 15 HP, to 1,000 GPM		1.70	14.118		5,975	830		6,805	7,825
4610	5" size, 20 HP, to 1,350 GPM		1.50	16		6,350	945		7,295	8,400
4620	5" size, 25 HP, to 1,550 GPM		1.50	16		8,675	945		9,620	11,000
5000	Base mounted, bronze impeller, coupling guard									
5040	1-1/2" size, 1-1/2 HP, to 40 GPM	Q-1	2.30	6.957	Ea.	6,925	395		7,320	8,200
5090	2" size, 2 HP, to 50 GPM		2.30	6.957		7,750	395		8,145	9,125
5100	2" size, 3 HP, to 90 GPM		2	8		8,325	455		8,780	9,850
5190	2-1/2" size, 3 HP, to 150 GPM		1.80	8.889		9,125	505		9,630	10,800
5300	3" size, 5 HP, to 225 GPM		1.60	10		10,500	570		11,070	12,500
5410	4" size, 5 HP, to 350 GPM		1.50	10.667		11,300	605		11,905	13,400
5420	4" size, 7-1/2 HP, to 350 GPM		1.50	10.667		13,100	605		13,705	15,300
5520	5" size, 10 HP, to 600 GPM	Q-2	1.60	15		17,100	885		17,985	20,100
5530	5" size, 15 HP, to 1,000 GPM		1.60	15		18,300	885		19,185	21,400
5610	6" size, 20 HP, to 1,350 GPM		1.40	17.143		20,700	1,000		21,700	24,300
5620	6" size, 25 HP, to 1,550 GPM		1.40	17.143		23,400	1,000		24,400	27,200
5800	The above pump capacities are based on 1,800 RPM,									
5810	at a 60' head. Increasing the RPM									
5820	or decreasing the head will increase the GPM.									

23 21 29 – Automatic Condensate Pump Units

23 21 29.10 Condensate Removal Pump System		Crew	Daily Output	Labor-Hours	Unit	Material	Labor	Equipment	Total	Total Incl O&P
0010	**CONDENSATE REMOVAL PUMP SYSTEM**									
0020	Pump with 1 gal. ABS tank									
0100	115 V									
0120	1/50 HP, 200 GPH G	1 Stpi	12	.667	Ea.	183	42.50		225.50	266
0140	1/18 HP, 270 GPH G		10	.800		187	51		238	282
0160	1/5 HP, 450 GPH G		8	1		420	64		484	560
0200	230 V									
0240	1/18 HP, 270 GPH	1 Stpi	10	.800	Ea.	202	51		253	299
0260	1/5 HP, 450 GPH G	"	8	1	"	485	64		549	625

23 22 Steam and Condensate Piping and Pumps

23 22 13 – Steam and Condensate Heating Piping

23 22 13.23 Aboveground Steam and Condensate Piping		Crew	Daily Output	Labor-Hours	Unit	Material	Labor	Equipment	Total	Total Incl O&P
0010	**ABOVEGROUND STEAM AND CONDENSATE HEATING PIPING**									
0020	Condensate meter									
0100	500 lb. per hour	1 Stpi	14	.571	Ea.	3,725	36.50		3,761.50	4,150
0140	1500 lb. per hour		7	1.143		4,350	73		4,423	4,900
0160	3000 lb. per hour		5	1.600		5,025	102		5,127	5,700
0200	12,000 lb. per hour	Q-5	3.50	4.571		6,000	263		6,263	7,000

23 22 23 – Steam Condensate Pumps

23 22 23.10 Condensate Return System		Crew	Daily Output	Labor-Hours	Unit	Material	Labor	Equipment	Total	Total Incl O&P
0010	**CONDENSATE RETURN SYSTEM**									
2000	Simplex									
2010	With pump, motor, CI receiver, float switch									
2020	3/4 HP, 15 GPM	Q-1	1.80	8.889	Ea.	7,025	505		7,530	8,500
2100	Duplex									
2110	With 2 pumps and motors, CI receiver, float switch, alternator									
2120	3/4 HP, 15 GPM, 15 gal. CI receiver	Q-1	1.40	11.429	Ea.	7,675	650		8,325	9,425
2130	1 HP, 25 GPM		1.20	13.333		9,225	760		9,985	11,200

23 22 Steam and Condensate Piping and Pumps

23 22 23 – Steam Condensate Pumps

23 22 23.10 Condensate Return System		Crew	Daily Output	Labor-Hours	Unit	Material	2019 Bare Costs Labor	Equipment	Total	Total Incl O&P
2140	1-1/2 HP, 45 GPM	Q-1	1	16	Ea.	10,700	910		11,610	13,100
2150	1-1/2 HP, 60 GPM	↓	1	16	↓	12,000	910		12,910	14,600

23 23 Refrigerant Piping

23 23 13 – Refrigerant Piping Valves

23 23 13.10 Valves		Crew	Daily Output	Labor-Hours	Unit	Material	Labor	Equipment	Total	Total Incl O&P
0010	**VALVES**									
8100	Check valve, soldered									
8110	5/8"	1 Stpi	36	.222	Ea.	93	14.20		107.20	124
8114	7/8"		26	.308		115	19.70		134.70	156
8118	1-1/8"		18	.444		124	28.50		152.50	180
8122	1-3/8"		14	.571		201	36.50		237.50	276
8126	1-5/8"		13	.615		237	39.50		276.50	320
8130	2-1/8"	↓	12	.667		287	42.50		329.50	380
8134	2-5/8"	Q-5	22	.727		405	42		447	510
8138	3-1/8"	"	20	.800	↓	540	46		586	665
8500	Refrigeration valve, packless, soldered									
8510	1/2"	1 Stpi	38	.211	Ea.	46	13.45		59.45	70.50
8514	5/8"		36	.222		48	14.20		62.20	74
8518	7/8"	↓	26	.308	↓	180	19.70		199.70	228
8520	Packed, soldered									
8522	1-1/8"	1 Stpi	18	.444	Ea.	215	28.50		243.50	280
8526	1-3/8"		14	.571		360	36.50		396.50	450
8530	1-5/8"		13	.615		350	39.50		389.50	445
8534	2-1/8"	↓	12	.667		720	42.50		762.50	860
8538	2-5/8"	Q-5	22	.727		1,050	42		1,092	1,250
8542	3-1/8"		20	.800		1,250	46		1,296	1,450
8546	4-1/8"	↓	18	.889	↓	1,825	51		1,876	2,075
8600	Solenoid valve, flange/solder									
8610	1/2"	1 Stpi	38	.211	Ea.	254	13.45		267.45	300
8614	5/8"		36	.222		325	14.20		339.20	380
8618	3/4"		30	.267		400	17.05		417.05	465
8622	7/8"		26	.308		485	19.70		504.70	565
8626	1-1/8"		18	.444		595	28.50		623.50	700
8630	1-3/8"		14	.571		730	36.50		766.50	860
8634	1-5/8"		13	.615		1,025	39.50		1,064.50	1,175
8638	2-1/8"	↓	12	.667	↓	1,125	42.50		1,167.50	1,300
8800	Thermostatic valve, flange/solder									
8810	1/2 – 3 ton, 3/8" x 5/8"	1 Stpi	9	.889	Ea.	126	57		183	225
8814	4 – 5 ton, 1/2" x 7/8"		7	1.143		150	73		223	275
8818	6 – 8 ton, 5/8" x 7/8"		5	1.600		256	102		358	435
8822	7 – 12 ton, 7/8" x 1-1/8"		4	2		300	128		428	520
8826	15 – 20 ton, 7/8" x 1-3/8"	↓	3.20	2.500	↓	300	160		460	570

23 23 16 – Refrigerant Piping Specialties

23 23 16.10 Refrigerant Piping Component Specialties		Crew	Daily Output	Labor-Hours	Unit	Material	Labor	Equipment	Total	Total Incl O&P
0010	**REFRIGERANT PIPING COMPONENT SPECIALTIES**									
0600	Accumulator									
0610	3/4"	1 Stpi	8.80	.909	Ea.	52.50	58		110.50	145
0614	7/8"		6.40	1.250		67.50	80		147.50	194
0618	1-1/8"	↓	4.80	1.667	↓	140	107		247	315

23 23 16.10 Refrigerant Piping Component Specialties		Crew	Daily Output	Labor-Hours	Unit	Material	Labor	Equipment	Total	Total Incl O&P
						2019 Bare Costs				
0622	1-3/8"	1 Stpi	4	2	Ea.	122	128		250	325
0626	1-5/8"		3.20	2.500		129	160		289	380
0630	2-1/8"		2.40	3.333		330	213		543	680
0700	Condensate drip/drain pan, electric, 10" x 9" x 2-3/4"		24	.333		105	21.50		126.50	147
1000	Filter dryer									
1010	Replaceable core type, solder									
1020	1/2"	1 Stpi	20	.400	Ea.	66	25.50		91.50	111
1030	5/8"		19	.421		66.50	27		93.50	114
1040	7/8"		18	.444		68	28.50		96.50	117
1050	1-1/8"		15	.533		98.50	34		132.50	159
1060	1-3/8"		14	.571		152	36.50		188.50	222
1070	1-5/8"		12	.667		152	42.50		194.50	231
1080	2-1/8"		10	.800		405	51		456	520
1090	2-5/8"		9	.889		745	57		802	905
1100	3-1/8"		8	1		800	64		864	975
1200	Sealed in-line, solder									
1210	1/4"	1 Stpi	22	.364	Ea.	30.50	23.50		54	68.50
1220	3/8"		21	.381		25.50	24.50		50	64.50
1230	1/2"		20	.400		30.50	25.50		56	72
1260	5/8"		19	.421		41.50	27		68.50	86
1270	7/8"		18	.444		45.50	28.50		74	92.50
1290	1-1/8"		15	.533		68.50	34		102.50	127
4000	P-trap, suction line, solder									
4010	5/8"	1 Stpi	19	.421	Ea.	40	27		67	84.50
4020	3/4"		19	.421		66	27		93	113
4030	7/8"		18	.444		51	28.50		79.50	98.50
4040	1-1/8"		15	.533		77	34		111	136
4050	1-3/8"		14	.571		139	36.50		175.50	208
4060	1-5/8"		12	.667		212	42.50		254.50	297
4070	2-1/8"		10	.800		272	51		323	375
5000	Sightglass									
5010	Moisture and liquid indicator, solder									
5020	1/4"	1 Stpi	22	.364	Ea.	18.25	23.50		41.75	55
5030	3/8"		21	.381		18.55	24.50		43.05	57
5040	1/2"		20	.400		23	25.50		48.50	64
5050	5/8"		19	.421		24	27		51	66.50
5060	7/8"		18	.444		35.50	28.50		64	81.50
5070	1-1/8"		15	.533		38.50	34		72.50	93
5080	1-3/8"		14	.571		75.50	36.50		112	138
5090	1-5/8"		12	.667		85	42.50		127.50	158
5100	2-1/8"		10	.800		90	51		141	176
7400	Vacuum pump set									
7410	Two stage, high vacuum continuous duty	1 Stpi	8	1	Ea.	3,975	64		4,039	4,475

23 23 16.16 Refrigerant Line Sets

		Crew	Daily Output	Labor-Hours	Unit	Material	Labor	Equipment	Total	Total Incl O&P
0010	**REFRIGERANT LINE SETS**, Standard									
0100	Copper tube									
0110	1/2" insulation, both tubes									
0120	Combination 1/4" and 1/2" tubes									
0130	10' set	Q-5	42	.381	Ea.	48	22		70	85.50
0135	15' set		42	.381		69	22		91	109
0140	20' set		40	.400		74.50	23		97.50	117
0150	30' set		37	.432		105	25		130	154

23 23 16.16 Refrigerant Line Sets		Crew	Daily Output	Labor-Hours	Unit	2019 Bare Costs Material	Labor	Equipment	Total	Total Incl O&P
0160	40′ set	Q-5	35	.457	Ea.	132	26.50		158.50	185
0170	50′ set		32	.500		163	29		192	223
0180	100′ set		22	.727		355	42		397	455
0300	Combination 1/4″ and 3/4″ tubes									
0310	10′ set	Q-5	40	.400	Ea.	53.50	23		76.50	93.50
0320	20′ set		38	.421		94.50	24		118.50	141
0330	30′ set		35	.457		142	26.50		168.50	196
0340	40′ set		33	.485		185	28		213	246
0350	50′ set		30	.533		233	30.50		263.50	300
0380	100′ set		20	.800		530	46		576	655
0500	Combination 3/8″ & 3/4″ tubes									
0510	10′ set	Q-5	28	.571	Ea.	61.50	33		94.50	117
0520	20′ set		36	.444		95.50	25.50		121	144
0530	30′ set		34	.471		130	27		157	184
0540	40′ set		31	.516		171	29.50		200.50	233
0550	50′ set		28	.571		201	33		234	271
0580	100′ set		18	.889		595	51		646	730
0700	Combination 3/8″ & 1-1/8″ tubes									
0710	10′ set	Q-5	36	.444	Ea.	108	25.50		133.50	157
0720	20′ set		33	.485		180	28		208	240
0730	30′ set		31	.516		242	29.50		271.50	310
0740	40′ set		28	.571		365	33		398	450
0750	50′ set		26	.615		350	35.50		385.50	440
0900	Combination 1/2″ & 3/4″ tubes									
0910	10′ set	Q-5	37	.432	Ea.	65	25		90	109
0920	20′ set		35	.457		115	26.50		141.50	167
0930	30′ set		33	.485		173	28		201	233
0940	40′ set		30	.533		228	30.50		258.50	297
0950	50′ set		27	.593		287	34		321	365
0980	100′ set		17	.941		645	54		699	790
2100	Combination 1/2″ & 1-1/8″ tubes									
2110	10′ set	Q-5	35	.457	Ea.	111	26.50		137.50	162
2120	20′ set		31	.516		190	29.50		219.50	254
2130	30′ set		29	.552		285	32		317	365
2140	40′ set		25	.640		385	37		422	475
2150	50′ set		14	1.143		465	66		531	615
2300	For 1″ thick insulation add					30%	15%			
3000	Refrigerant line sets, min-split, flared									
3100	Combination 1/4″ & 3/8″ tubes									
3120	15′ set	Q-5	41	.390	Ea.	68.50	22.50		91	109
3140	25′ set		38.50	.416		97	24		121	143
3160	35′ set		37	.432		122	25		147	172
3180	50′ set		30	.533		166	30.50		196.50	229
3200	Combination 1/4″ & 1/2″ tubes									
3220	15′ set	Q-5	41	.390	Ea.	73	22.50		95.50	114
3240	25′ set		38.50	.416		99.50	24		123.50	145
3260	35′ set		37	.432		128	25		153	179
3280	50′ set		30	.533		173	30.50		203.50	236

23 23 Refrigerant Piping

23 23 23 – Refrigerants

23 23 23.10 Anti-Freeze		Crew	Daily Output	Labor-Hours	Unit	Material	Labor	Equipment	Total	Total Incl O&P
						2019 Bare Costs				
0010	**ANTI-FREEZE**, Inhibited									
0900	Ethylene glycol concentrated									
1000	55 gallon drums, small quantities				Gal.	12.70			12.70	13.95
1200	Large quantities					10.80			10.80	11.85
2000	Propylene glycol, for solar heat, small quantities					23.50			23.50	26
2100	Large quantities				↓	10.95			10.95	12.05

23 23 23.20 Refrigerant		Crew	Daily Output	Labor-Hours	Unit	Material	Labor	Equipment	Total	Total Incl O&P
0010	**REFRIGERANT**									
4420	R-22, 30 lb. disposable cylinder				Lb.	21			21	23
4428	R-134A, 30 lb. disposable cylinder					15.50			15.50	17.05
4434	R-407C, 30 lb. disposable cylinder					11.85			11.85	13.05
4440	R-408A, 25 lb. disposable cylinder					27.50			27.50	30
4450	R-410A, 25 lb. disposable cylinder					11.65			11.65	12.80
4470	R-507, 25 lb. disposable cylinder				↓	7.40			7.40	8.15

23 31 HVAC Ducts and Casings

23 31 13 – Metal Ducts

23 31 13.13 Rectangular Metal Ducts		Crew	Daily Output	Labor-Hours	Unit	Material	Labor	Equipment	Total	Total Incl O&P
0010	**RECTANGULAR METAL DUCTS**									
0020	Fabricated rectangular, includes fittings, joints, supports,									
0021	allowance for flexible connections and field sketches.									
0030	Does not include "as-built dwgs." or insulation. R233100-30									
0031	NOTE: Fabrication and installation are combined									
0040	as LABOR cost. Approx. 25% fittings assumed. R233100-40									
0042	Fabrication/Inst. is to commercial quality standards									
0043	(SMACNA or equiv.) for structure, sealing, leak testing, etc. R233100-50									
0050	Add to labor for elevated installation									
0051	of fabricated ductwork									
0052	10' to 15' high						6%			
0053	15' to 20' high						12%			
0054	20' to 25' high						15%			
0055	25' to 30' high						21%			
0056	30' to 35' high						24%			
0057	35' to 40' high						30%			
0058	Over 40' high						33%			
0072	For duct insulation see Line 23 07 13.10 3000									
0100	Aluminum, alloy 3003-H14, under 100 lb.	Q-10	75	.320	Lb.	3.20	18.20		21.40	31
0110	100 to 500 lb.		80	.300		1.89	17.05		18.94	28
0120	500 to 1,000 lb.		95	.253		1.82	14.35		16.17	24
0140	1,000 to 2,000 lb.		120	.200		1.78	11.40		13.18	19.25
0150	2,000 to 5,000 lb.		130	.185		1.78	10.50		12.28	17.90
0160	Over 5,000 lb.		145	.166		1.78	9.40		11.18	16.25
0500	Galvanized steel, under 200 lb.		235	.102		.57	5.80		6.37	9.45
0520	200 to 500 lb.		245	.098		.58	5.55		6.13	9.10
0540	500 to 1,000 lb.		255	.094		.56	5.35		5.91	8.75
0560	1,000 to 2,000 lb.		265	.091		.57	5.15		5.72	8.45
0570	2,000 to 5,000 lb.		275	.087		.56	4.96		5.52	8.15
0580	Over 5,000 lb.	↓	285	.084	↓	.57	4.79		5.36	7.90
0600	For large quantities special prices available from supplier									
1000	Stainless steel, type 304, under 100 lb.	Q-10	165	.145	Lb.	4.71	8.25		12.96	17.75

23 31 13.13 Rectangular Metal Ducts		Crew	Daily Output	Labor-Hours	Unit	2019 Bare Costs Material	Labor	Equipment	Total	Total Incl O&P
1020	100 to 500 lb.	Q-10	175	.137	Lb.	3.83	7.80		11.63	16.05
1030	500 to 1,000 lb.		190	.126		2.62	7.20		9.82	13.80
1040	1,000 to 2,000 lb.		200	.120		2.30	6.85		9.15	12.90
1050	2,000 to 5,000 lb.		225	.107		2.16	6.05		8.21	11.60
1060	Over 5,000 lb.		235	.102		2.51	5.80		8.31	11.55
1080	Note: Minimum order cost exceeds per lb. cost for min. wt.									
1100	For medium pressure ductwork, add				Lb.		15%			
1200	For high pressure ductwork, add						40%			
1210	For welded ductwork, add						85%			
1220	For 30% fittings, add						11%			
1224	For 40% fittings, add						34%			
1228	For 50% fittings, add						56%			
1232	For 60% fittings, add						79%			
1236	For 70% fittings, add						101%			
1240	For 80% fittings, add						124%			
1244	For 90% fittings, add						147%			
1248	For 100% fittings, add						169%			
1252	Note: Fittings add includes time for detailing and installation.									
1276	Duct tape, 2" x 60 yd roll				Ea.	6.90			6.90	7.55

23 31 13.16 Round and Flat-Oval Spiral Ducts		Crew	Daily Output	Labor-Hours	Unit	Material	Labor	Equipment	Total	Total Incl O&P
0010	**ROUND AND FLAT-OVAL SPIRAL DUCTS**									
0020	Fabricated round and flat oval spiral, includes hangers,									
0021	supports and field sketches.									
1280	Add to labor for elevated installation									
1282	of prefabricated (purchased) ductwork									
1283	10′ to 15′ high						10%			
1284	15′ to 20′ high						20%			
1285	20′ to 25′ high						25%			
1286	25′ to 30′ high						35%			
1287	30′ to 35′ high						40%			
1288	35′ to 40′ high						50%			
1289	Over 40′ high						55%			
5400	Spiral preformed, steel, galv., straight lengths, max 10" spwg.									
5410	4" diameter, 26 ga.	Q-9	360	.044	L.F.	1.89	2.44		4.33	5.80
5416	5" diameter, 26 ga.		320	.050		1.98	2.74		4.72	6.35
5420	6" diameter, 26 ga.		280	.057		2.03	3.13		5.16	7
5425	7" diameter, 26 ga.		240	.067		2.18	3.66		5.84	7.95
5430	8" diameter, 26 ga.		200	.080		2.72	4.39		7.11	9.65
5440	10" diameter, 26 ga.		160	.100		3.38	5.50		8.88	12.05
5450	12" diameter, 26 ga.		120	.133		3.97	7.30		11.27	15.45
5460	14" diameter, 26 ga.		80	.200		4.35	10.95		15.30	21.50
5480	16" diameter, 24 ga.		60	.267		6.40	14.65		21.05	29
5490	18" diameter, 24 ga.		50	.320		7.30	17.55		24.85	34.50
5500	20" diameter, 24 ga.	Q-10	65	.369		8.10	21		29.10	41
5510	22" diameter, 24 ga.		60	.400		8.90	23		31.90	44.50
5520	24" diameter, 24 ga.		55	.436		9.70	25		34.70	48
5540	30" diameter, 22 ga.		45	.533		14.40	30.50		44.90	62
5600	36" diameter, 22 ga.		40	.600		17.55	34		51.55	71.50
5800	Connector, 4" diameter	Q-9	100	.160	Ea.	3.24	8.80		12.04	16.90
5810	5" diameter		94	.170		3.25	9.35		12.60	17.80
5820	6" diameter		88	.182		3.37	9.95		13.32	18.85
5840	8" diameter		78	.205		3.82	11.25		15.07	21.50

23 31 13.16	Round and Flat-Oval Spiral Ducts	Crew	Daily Output	Labor-Hours	Unit	Material	Labor	Equipment	Total	Total Incl O&P
						2019 Bare Costs				
5860	10″ diameter	Q-9	70	.229	Ea.	4.39	12.55		16.94	24
5880	12″ diameter		50	.320		4.95	17.55		22.50	32
5900	14″ diameter		44	.364		5.35	19.95		25.30	36.50
5920	16″ diameter		40	.400		5.85	22		27.85	40
5930	18″ diameter		37	.432		6.25	23.50		29.75	43
5940	20″ diameter		34	.471		7.10	26		33.10	47
5950	22″ diameter		31	.516		8.30	28.50		36.80	52
5960	24″ diameter		28	.571		8.80	31.50		40.30	57
5980	30″ diameter		22	.727		14.80	40		54.80	77
6000	36″ diameter		18	.889		18	49		67	94
6300	Elbow, 45°, 4″ diameter		60	.267		6.15	14.65		20.80	29
6310	5″ diameter		52	.308		6.70	16.90		23.60	33
6320	6″ diameter		44	.364		6.70	19.95		26.65	38
6340	8″ diameter		28	.571		8.05	31.50		39.55	56.50
6360	10″ diameter		18	.889		8.50	49		57.50	83.50
6380	12″ diameter		13	1.231		9.25	67.50		76.75	113
6400	14″ diameter		11	1.455		12.80	80		92.80	135
6420	16″ diameter		10	1.600		20	88		108	155
6430	18″ diameter		9.60	1.667		52.50	91.50		144	197
6440	20″ diameter	Q-10	14	1.714		55	97.50		152.50	209
6450	22″ diameter		13	1.846		72.50	105		177.50	239
6460	24″ diameter		12	2		73	114		187	254
6480	30″ diameter		9	2.667		119	152		271	360
6500	36″ diameter		7	3.429		147	195		342	455
6600	Elbow, 90°, 4″ diameter	Q-9	60	.267		4.62	14.65		19.27	27
6610	5″ diameter		52	.308		4.60	16.90		21.50	30.50
6620	6″ diameter		44	.364		4.61	19.95		24.56	35.50
6625	7″ diameter		36	.444		4.98	24.50		29.48	42.50
6630	8″ diameter		28	.571		5.65	31.50		37.15	54
6640	10″ diameter		18	.889		7.40	49		56.40	82
6650	12″ diameter		13	1.231		9.25	67.50		76.75	113
6660	14″ diameter		11	1.455		15.90	80		95.90	139
6670	16″ diameter		10	1.600		18.10	88		106.10	153
6676	18″ diameter		9.60	1.667		68.50	91.50		160	214
6680	20″ diameter	Q-10	14	1.714		73	97.50		170.50	229
6684	22″ diameter		13	1.846		75	105		180	242
6690	24″ diameter		12	2		80.50	114		194.50	262
6700	30″ diameter		9	2.667		152	152		304	400
6710	36″ diameter		7	3.429		223	195		418	540
6720	End cap									
6722	4″	Q-9	110	.145	Ea.	4.82	8		12.82	17.40
6724	6″		94	.170		4.82	9.35		14.17	19.50
6725	8″		88	.182		6.05	9.95		16	22
6726	10″		78	.205		7.55	11.25		18.80	25.50
6727	12″		70	.229		8.10	12.55		20.65	28
6728	14″		50	.320		11.30	17.55		28.85	39
6729	16″		44	.364		12.75	19.95		32.70	44.50
6730	18″		40	.400		17.20	22		39.20	52.50
6731	20″		37	.432		36	23.50		59.50	75.50
6732	22″		34	.471		40.50	26		66.50	83.50
6733	24″		30	.533		37	29.50		66.50	85
6800	Reducing coupling, 6″ x 4″		46	.348		15.15	19.10		34.25	45.50
6820	8″ x 6″		40	.400		19.30	22		41.30	54.50

23 31 13.16 Round and Flat-Oval Spiral Ducts		Crew	Daily Output	Labor-Hours	Unit	Material	Labor	Equipment	Total	Total Incl O&P
						2019 Bare Costs				
6840	10″ x 8″	Q-9	32	.500	Ea.	21	27.50		48.50	65
6860	12″ x 10″		24	.667		21	36.50		57.50	78.50
6880	14″ x 12″		20	.800		22.50	44		66.50	91
6900	16″ x 14″		18	.889		28	49		77	105
6920	18″ x 16″		16	1		39	55		94	127
6940	20″ x 18″	Q-10	24	1		42	57		99	133
6950	22″ x 20″		23	1.043		49.50	59.50		109	145
6960	24″ x 22″		22	1.091		54	62		116	154
6980	30″ x 28″		18	1.333		105	76		181	230
7000	36″ x 34″		16	1.500		144	85.50		229.50	289
7100	Tee, 90°, 4″ diameter	Q-9	40	.400		13.65	22		35.65	48.50
7110	5″ diameter		34.60	.462		14.25	25.50		39.75	54
7120	6″ diameter		30	.533		14.85	29.50		44.35	61
7130	8″ diameter		19	.842		18.70	46		64.70	90.50
7140	10″ diameter		12	1.333		28	73		101	142
7150	12″ diameter		9	1.778		33.50	97.50		131	185
7160	14″ diameter		7	2.286		45	125		170	240
7170	16″ diameter		6.70	2.388		72	131		203	278
7176	18″ diameter		6	2.667		105	146		251	335
7180	20″ diameter	Q-10	8.70	2.759		125	157		282	375
7190	24″ diameter		7	3.429		195	195		390	510
7200	30″ diameter		6	4		251	228		479	620
7210	36″ diameter		5	4.800		320	273		593	765
7250	Saddle tap									
7254	4″ diameter on 6″ diameter	Q-9	36	.444	Ea.	23	24.50		47.50	62.50
7256	5″ diameter on 7″ diameter		35.80	.447		23	24.50		47.50	62.50
7258	6″ diameter on 8″ diameter		35.40	.452		23	25		48	63
7260	6″ diameter on 10″ diameter		35.20	.455		23	25		48	63.50
7262	6″ diameter on 12″ diameter		35	.457		23	25		48	63.50
7264	7″ diameter on 9″ diameter		35.20	.455		25.50	25		50.50	66
7266	8″ diameter on 10″ diameter		35	.457		25.50	25		50.50	66
7268	8″ diameter on 12″ diameter		34.80	.460		25.50	25		50.50	66.50
7270	8″ diameter on 14″ diameter		34.50	.464		26	25.50		51.50	67
7272	10″ diameter on 12″ diameter		34.20	.468		31	25.50		56.50	73
7274	10″ diameter on 14″ diameter		33.80	.473		31	26		57	73.50
7276	10″ diameter on 16″ diameter		33.40	.479		31	26.50		57.50	74
7278	10″ diameter on 18″ diameter		33	.485		31	26.50		57.50	74.50
7280	12″ diameter on 14″ diameter		32.60	.491		36	27		63	80.50
7282	12″ diameter on 18″ diameter		32.20	.497		36	27.50		63.50	81
7284	14″ diameter on 16″ diameter		31.60	.506		50.50	28		78.50	98
7286	14″ diameter on 20″ diameter		31	.516		50.50	28.50		79	99
7288	16″ diameter on 20″ diameter		30	.533		68.50	29.50		98	120
7290	18″ diameter on 20″ diameter		28	.571		84.50	31.50		116	140
7292	20″ diameter on 22″ diameter		26	.615		109	34		143	171
7294	22″ diameter on 24″ diameter		24	.667		121	36.50		157.50	189
7400	Steel, PVC coated both sides, straight lengths									
7410	4″ diameter, 26 ga.	Q-9	360	.044	L.F.	4.49	2.44		6.93	8.65
7416	5″ diameter, 26 ga.		240	.067		4.38	3.66		8.04	10.35
7420	6″ diameter, 26 ga.		280	.057		4.38	3.13		7.51	9.60
7440	8″ diameter, 26 ga.		200	.080		6.15	4.39		10.54	13.45
7460	10″ diameter, 26 ga.		160	.100		7.70	5.50		13.20	16.80
7480	12″ diameter, 26 ga.		120	.133		8.85	7.30		16.15	21
7500	14″ diameter, 26 ga.		80	.200		10.35	10.95		21.30	28

23 31 13.16 Round and Flat-Oval Spiral Ducts		Crew	Daily Output	Labor-Hours	Unit	2019 Bare Costs Material	Labor	Equipment	Total	Total Incl O&P
7520	16" diameter, 24 ga.	Q-9	60	.267	L.F.	11.90	14.65		26.55	35
7540	18" diameter, 24 ga.	↓	45	.356		13.40	19.50		32.90	44.50
7560	20" diameter, 24 ga.	Q-10	65	.369		14.85	21		35.85	48.50
7580	24" diameter, 24 ga.		55	.436		17.85	25		42.85	57
7600	30" diameter, 22 ga.		45	.533		27	30.50		57.50	75.50
7620	36" diameter, 22 ga.	↓	40	.600	↓	32	34		66	87.50
7890	Connector, 4" diameter	Q-9	100	.160	Ea.	2.61	8.80		11.41	16.20
7896	5" diameter		83	.193		2.91	10.55		13.46	19.25
7900	6" diameter		88	.182		3.31	9.95		13.26	18.80
7920	8" diameter		78	.205		3.91	11.25		15.16	21.50
7940	10" diameter		70	.229		4.51	12.55		17.06	24
7960	12" diameter		50	.320		5.30	17.55		22.85	32.50
7980	14" diameter		44	.364		5.60	19.95		25.55	36.50
8000	16" diameter		40	.400		6.10	22		28.10	40.50
8020	18" diameter		37	.432		6.80	23.50		30.30	43.50
8040	20" diameter		34	.471		8.90	26		34.90	49
8060	24" diameter		28	.571		11.05	31.50		42.55	59.50
8080	30" diameter		22	.727		14.75	40		54.75	76.50
8100	36" diameter		18	.889		17.95	49		66.95	94
8390	Elbow, 45°, 4" diameter		60	.267		23	14.65		37.65	47.50
8396	5" diameter		36	.444		23	24.50		47.50	62.50
8400	6" diameter		44	.364		23	19.95		42.95	56
8420	8" diameter		28	.571		23	31.50		54.50	73
8440	10" diameter		18	.889		24	49		73	101
8460	12" diameter		13	1.231		24.50	67.50		92	130
8480	14" diameter		11	1.455		24	80		104	148
8500	16" diameter		10	1.600		23	88		111	159
8520	18" diameter	↓	9	1.778		25	97.50		122.50	176
8540	20" diameter	Q-10	13	1.846		77	105		182	244
8560	24" diameter		11	2.182		98	124		222	296
8580	30" diameter		9	2.667		180	152		332	430
8600	36" diameter	↓	7	3.429		211	195		406	530
8800	Elbow, 90°, 4" diameter	Q-9	60	.267		25	14.65		39.65	49.50
8804	5" diameter		52	.308		25	16.90		41.90	53
8806	6" diameter		44	.364		25	19.95		44.95	58
8808	8" diameter		28	.571		26	31.50		57.50	76
8810	10" diameter		18	.889		26	49		75	103
8812	12" diameter		13	1.231		26.50	67.50		94	132
8814	14" diameter		11	1.455		25	80		105	149
8816	16" diameter		10	1.600		29	88		117	165
8818	18" diameter	↓	9.60	1.667		35.50	91.50		127	178
8820	20" diameter	Q-10	14	1.714		45	97.50		142.50	198
8822	24" diameter		12	2		104	114		218	288
8824	30" diameter		9	2.667		228	152		380	480
8826	36" diameter	↓	7	3.429	↓	350	195		545	680
9000	Reducing coupling									
9040	6" x 4"	Q-9	46	.348	Ea.	19.55	19.10		38.65	50.50
9060	8" x 6"		40	.400		19.55	22		41.55	55
9080	10" x 8"		32	.500		20.50	27.50		48	64
9100	12" x 10"		24	.667		23	36.50		59.50	81
9120	14" x 12"		20	.800		26.50	44		70.50	95.50
9140	16" x 14"		18	.889		36.50	49		85.50	115
9160	18" x 16"	↓	16	1	↓	46	55		101	135

23 31 13.16 Round and Flat-Oval Spiral Ducts		Crew	Daily Output	Labor-Hours	Unit	2019 Bare Costs Material	Labor	Equipment	Total	Total Incl O&P
9180	20" x 18"	Q-10	24	1	Ea.	52	57		109	144
9200	24" x 22"		22	1.091		70	62		132	171
9220	30" x 28"		18	1.333		104	76		180	230
9240	36" x 34"		16	1.500		160	85.50		245.50	305
9800	Steel, stainless, straight lengths									
9810	3" diameter, 26 ga.	Q-9	400	.040	L.F.	15.65	2.19		17.84	20.50
9820	4" diameter, 26 ga.		360	.044		18.15	2.44		20.59	23.50
9830	5" diameter, 26 ga.		320	.050		19.60	2.74		22.34	25.50
9840	6" diameter, 26 ga.		280	.057		19.60	3.13		22.73	26.50
9850	7" diameter, 26 ga.		240	.067		36	3.66		39.66	45.50
9860	8" diameter, 26 ga.		200	.080		25.50	4.39		29.89	34.50
9870	9" diameter, 26 ga.		180	.089		37	4.88		41.88	48.50
9880	10" diameter, 26 ga.		160	.100		48	5.50		53.50	61.50
9890	12" diameter, 26 ga.		120	.133		68	7.30		75.30	86
9900	14" diameter, 26 ga.		80	.200		71.50	10.95		82.45	95

23 31 13.17 Round Grease Duct		Crew	Daily Output	Labor-Hours	Unit	Material	Labor	Equipment	Total	Total Incl O&P
0010	**ROUND GREASE DUCT**									
1280	Add to labor for elevated installation									
1282	of prefabricated (purchased) ductwork									
1283	10' to 15' high						10%			
1284	15' to 20' high						20%			
1285	20' to 25' high						25%			
1286	25' to 30' high						35%			
1287	30' to 35' high						40%			
1288	35' to 40' high						50%			
1289	Over 40' high						55%			
4020	Zero clearance, 2 hr. fire rated, UL listed, 3" double wall									
4030	304 interior, aluminized exterior, all sizes are ID									
4032	For 316 interior and aluminized exterior, add					6%				
4034	For 304 interior and 304 exterior, add					21%				
4036	For 316 interior and 316 exterior, add					40%				
4040	Straight									
4052	6"	Q-9	34.20	.468	L.F.	75	25.50		100.50	122
4054	8"		29.64	.540		86	29.50		115.50	140
4056	10"		27.36	.585		96	32		128	154
4058	12"		25.10	.637		110	35		145	174
4060	14"		23.94	.668		123	36.50		159.50	192
4062	16"		22.80	.702		139	38.50		177.50	212
4064	18"		21.66	.739		157	40.50		197.50	235
4066	20"	Q-10	20.52	1.170		179	66.50		245.50	298
4068	22"		19.38	1.238		201	70.50		271.50	330
4070	24"		18.24	1.316		229	75		304	365
4072	26"		17.67	1.358		246	77.50		323.50	385
4074	28"		17.10	1.404		262	80		342	410
4076	30"		15.96	1.504		275	85.50		360.50	435
4078	32"		15.39	1.559		297	88.50		385.50	460
4080	36"		14.25	1.684		330	96		426	510
4100	Adjustable section, 30" long									
4104	6"	Q-9	17.10	.936	Ea.	226	51.50		277.50	325
4106	8"		14.82	1.080		283	59		342	400
4108	10"		13.68	1.170		320	64		384	455
4110	12"		12.54	1.276		360	70		430	500

23 31 13.17 Round Grease Duct		Crew	Daily Output	Labor-Hours	Unit	Material	2019 Bare Costs Labor	Equipment	Total	Total Incl O&P
4112	14"	Q-9	11.97	1.337	Ea.	405	73.50		478.50	555
4114	16"		11.40	1.404		455	77		532	615
4116	18"		10.83	1.477		515	81		596	695
4118	20"	Q-10	10.26	2.339		585	133		718	845
4120	22"		9.69	2.477		660	141		801	945
4122	24"		9.12	2.632		750	150		900	1,050
4124	26"		8.83	2.718		805	155		960	1,125
4126	28"		8.55	2.807		865	160		1,025	1,200
4128	30"		8.26	2.906		920	165		1,085	1,250
4130	32"		7.98	3.008		975	171		1,146	1,325
4132	36"		6.84	3.509		1,075	200		1,275	1,500
4140	Tee, 90°, grease									
4144	6"	Q-9	13.68	1.170	Ea.	400	64		464	545
4146	8"		12.54	1.276		440	70		510	590
4148	10"		11.97	1.337		490	73.50		563.50	650
4150	12"		11.40	1.404		570	77		647	740
4152	14"		10.26	1.559		650	85.50		735.50	845
4154	16"		9.12	1.754		715	96		811	930
4156	18"		7.98	2.005		840	110		950	1,100
4158	20"		9.69	1.651		970	90.50		1,060.50	1,225
4160	22"		8.26	1.937		1,075	106		1,181	1,350
4162	24"		6.84	2.339		1,200	128		1,328	1,525
4164	26"		6.55	2.443		1,325	134		1,459	1,650
4166	28"	Q-10	6.27	3.828		1,450	218		1,668	1,925
4168	30"		5.98	4.013		1,725	228		1,953	2,250
4170	32"		5.70	4.211		2,025	239		2,264	2,600
4172	36"		5.13	4.678		2,250	266		2,516	2,875
4180	Cleanout tee cap									
4184	6"	Q-9	21.09	.759	Ea.	81.50	41.50		123	153
4186	8"		19.38	.826		86	45.50		131.50	164
4188	10"		18.24	.877		92.50	48		140.50	175
4190	12"		17.10	.936		99	51.50		150.50	187
4192	14"		15.96	1.003		105	55		160	199
4194	16"		14.25	1.123		116	61.50		177.50	221
4196	18"		13.68	1.170		122	64		186	233
4198	20"	Q-10	15.39	1.559		136	88.50		224.50	284
4200	22"		12.54	1.914		147	109		256	325
4202	24"		11.97	2.005		160	114		274	350
4204	26"		11.40	2.105		174	120		294	375
4206	28"		10.83	2.216		189	126		315	400
4208	30"		10.26	2.339		206	133		339	430
4210	32"		9.69	2.477		225	141		366	460
4212	36"		8.55	2.807		263	160		423	530
4220	Elbow, 45°									
4224	6"	Q-9	17.10	.936	Ea.	283	51.50		334.50	390
4226	8"		14.82	1.080		320	59		379	440
4228	10"		13.68	1.170		365	64		429	500
4230	12"		12.54	1.276		415	70		485	560
4232	14"		11.97	1.337		465	73.50		538.50	620
4234	16"		11.40	1.404		525	77		602	690
4236	18"		10.83	1.477		595	81		676	775
4238	20"	Q-10	10.26	2.339		670	133		803	940
4240	22"		9.69	2.477		750	141		891	1,050

23 31 13.17 Round Grease Duct		Crew	Daily Output	Labor-Hours	Unit	Material	Labor	Equipment	Total	Total Incl O&P
						2019 Bare Costs				
4242	24"	Q-10	9.12	2.632	Ea.	860	150		1,010	1,175
4244	26"		8.83	2.718		910	155		1,065	1,225
4246	28"		8.55	2.807		1,075	160		1,235	1,425
4248	30"		8.26	2.906		1,075	165		1,240	1,450
4250	32"		7.98	3.008		1,300	171		1,471	1,675
4252	36"		6.84	3.509		1,750	200		1,950	2,225
4260	Elbow, 90°									
4264	6"	Q-9	17.10	.936	Ea.	565	51.50		616.50	700
4266	8"		14.82	1.080		635	59		694	790
4268	10"		13.68	1.170		775	64		839	955
4270	12"		12.54	1.276		880	70		950	1,075
4272	14"		11.97	1.337		925	73.50		998.50	1,125
4274	16"		11.40	1.404		1,050	77		1,127	1,275
4276	18"		10.83	1.477		1,275	81		1,356	1,525
4278	20"	Q-10	10.26	2.339		1,325	133		1,458	1,675
4280	22"		9.69	2.477		1,500	141		1,641	1,875
4282	24"		9.12	2.632		1,700	150		1,850	2,100
4284	26"		8.83	2.718		1,825	155		1,980	2,225
4286	28"		8.55	2.807		1,925	160		2,085	2,375
4288	30"		8.26	2.906		2,150	165		2,315	2,625
4290	32"		7.98	3.008		2,600	171		2,771	3,100
4292	36"		6.84	3.509		3,475	200		3,675	4,125
4300	Support strap									
4304	6"	Q-9	25.50	.627	Ea.	99	34.50		133.50	162
4306	8"		22	.727		116	40		156	188
4308	10"		20.50	.780		119	43		162	196
4310	12"		18.80	.851		128	46.50		174.50	212
4312	14"		17.90	.894		138	49		187	226
4314	16"		17	.941		139	51.50		190.50	232
4316	18"		16.20	.988		141	54		195	238
4318	20"	Q-10	15.40	1.558		156	88.50		244.50	305
4320	22"	"	14.50	1.655		173	94		267	335
4340	Plate support assembly									
4344	6"	Q-9	14.82	1.080	Ea.	139	59		198	243
4346	8"		12.54	1.276		165	70		235	287
4348	10"		11.40	1.404		177	77		254	310
4350	12"		10.26	1.559		188	85.50		273.50	335
4352	14"		9.69	1.651		220	90.50		310.50	380
4354	16"		9.12	1.754		232	96		328	400
4356	18"		8.55	1.871		244	103		347	425
4358	20"	Q-10	9.12	2.632		256	150		406	510
4360	22"		8.55	2.807		262	160		422	530
4362	24"		7.98	3.008		266	171		437	550
4364	26"		7.69	3.121		299	178		477	600
4366	28"		7.41	3.239		325	184		509	640
4368	30"		7.12	3.371		365	192		557	690
4370	32"		6.84	3.509		400	200		600	745
4372	36"		5.70	4.211		480	239		719	895
4380	Wall guide assembly									
4384	6"	Q-9	19.10	.838	Ea.	184	46		230	273
4386	8"		16.50	.970		209	53		262	310
4388	10"		15.30	1.046		221	57.50		278.50	330
4390	12"		14.10	1.135		231	62		293	350

23 31 13.17 Round Grease Duct		Crew	Daily Output	Labor-Hours	Unit	Material	2019 Bare Costs Labor	Equipment	Total	Total Incl O&P
4392	14″	Q-9	13.40	1.194	Ea.	250	65.50		315.50	375
4394	16″		12.75	1.255		253	69		322	385
4396	18″		12.10	1.322		256	72.50		328.50	390
4398	20″	Q-10	11.50	2.087		286	119		405	495
4400	22″		10.90	2.202		310	125		435	530
4402	24″		10.20	2.353		345	134		479	585
4404	26″		9.90	2.424		380	138		518	630
4406	28″		9.60	2.500		395	142		537	650
4408	30″		9.30	2.581		450	147		597	720
4410	32″		9	2.667		480	152		632	755
4412	36″		7.60	3.158		525	180		705	850
4420	Hood transition, flanged or unflanged									
4424	6″	Q-9	15.60	1.026	Ea.	78.50	56.50		135	172
4426	8″		13.50	1.185		89.50	65		154.50	197
4428	10″		12.50	1.280		99.50	70		169.50	216
4430	12″		11.40	1.404		119	77		196	248
4432	14″		10.90	1.468		128	80.50		208.50	263
4434	16″		10.40	1.538		138	84.50		222.50	280
4436	18″		9.80	1.633		158	89.50		247.50	310
4438	20″	Q-10	9.40	2.553		168	145		313	405
4440	22″		8.80	2.727		179	155		334	435
4442	24″		8.30	2.892		201	164		365	470
4444	26″		8.10	2.963		206	169		375	485
4446	28″		7.80	3.077		218	175		393	505
4448	30″		7.50	3.200		237	182		419	535
4450	32″		7.30	3.288		248	187		435	555
4452	36″		6.20	3.871		274	220		494	635
4460	Fan adapter									
4464	6″	Q-9	15	1.067	Ea.	222	58.50		280.50	335
4466	8″		13	1.231		252	67.50		319.50	380
4468	10″		12	1.333		275	73		348	415
4470	12″		11	1.455		310	80		390	460
4472	14″		10.50	1.524		330	83.50		413.50	490
4474	16″		10	1.600		360	88		448	535
4476	18″		9.50	1.684		390	92.50		482.50	570
4478	20″	Q-10	9	2.667		415	152		567	690
4480	22″		8.50	2.824		445	161		606	735
4482	24″		8	3		470	171		641	775
4484	26″		7.70	3.117		500	177		677	820
4486	28″		7.50	3.200		525	182		707	855
4488	30″		7.20	3.333		555	190		745	900
4490	32″		7	3.429		585	195		780	935
4492	36″		6	4		650	228		878	1,050
4500	Tapered increaser/reducer									
4530	6″ x 10″ diam.	Q-9	13.40	1.194	Ea.	282	65.50		347.50	410
4540	6″ x 20″ diam.		10.10	1.584		282	87		369	440
4550	8″ x 10″ diam.		13.20	1.212		345	66.50		411.50	475
4560	8″ x 20″ diam.		9.90	1.616		340	88.50		428.50	510
4570	10″ x 20″ diam.		10.60	1.509		395	83		478	560
4580	10″ x 28″ diam.	Q-10	8.60	2.791		395	159		554	675
4590	12″ x 20″ diam.		8.80	2.727		455	155		610	735
4600	14″ x 28″ diam.		8.40	2.857		515	163		678	810
4610	16″ x 20″ diam.		8.60	2.791		575	159		734	875

23 31 13 – Metal Ducts

23 31 13.17 Round Grease Duct		Crew	Daily Output	Labor-Hours	Unit	Material	2019 Bare Costs Labor	Equipment	Total	Total Incl O&P
4620	16" x 30" diam.	Q-10	8.30	2.892	Ea.	570	164		734	875
4630	18" x 24" diam.		8.80	2.727		630	155		785	925
4640	18" x 30" diam.		8.10	2.963		630	169		799	945
4650	20" x 24" diam.		8.30	2.892		685	164		849	1,000
4660	20" x 32" diam.		8	3		685	171		856	1,000
4670	24" x 26" diam.		8	3		805	171		976	1,150
4680	24" x 36" diam.		7.30	3.288		800	187		987	1,175
4690	28" x 30" diam.		7.90	3.038		925	173		1,098	1,275
4700	28" x 36" diam.		7.10	3.380		915	192		1,107	1,300
4710	30" x 36" diam.	↓	6.90	3.478	↓	970	198		1,168	1,375
4730	Many intermediate standard sizes of tapered fittings are available									

23 31 13.19 Metal Duct Fittings		Crew	Daily Output	Labor-Hours	Unit	Material	Labor	Equipment	Total	Total Incl O&P
0010	**METAL DUCT FITTINGS**									
0350	Duct collar, spin-in type									
0352	Without damper									
0354	Round									
0356	4"	1 Shee	32	.250	Ea.	3.66	15.25		18.91	27
0357	5"		30	.267		4.44	16.25		20.69	29.50
0358	6"		28	.286		4.90	17.40		22.30	32
0359	7"		26	.308		5.35	18.75		24.10	34.50
0360	8"		24	.333		5.35	20.50		25.85	37
0361	9"		22	.364		5.60	22		27.60	39.50
0362	10"		20	.400		5.85	24.50		30.35	43.50
0363	12"		17	.471	↓	7.40	28.50		35.90	51.50
2000	Fabrics for flexible connections, with metal edge		100	.080	L.F.	3.92	4.88		8.80	11.70
2100	Without metal edge	↓	160	.050	"	3.51	3.05		6.56	8.50

23 31 16 – Nonmetal Ducts

23 31 16.13 Fibrous-Glass Ducts		Crew	Daily Output	Labor-Hours	Unit	Material	Labor	Equipment	Total	Total Incl O&P
0010	**FIBROUS-GLASS DUCTS**									
1280	Add to labor for elevated installation									
1282	of prefabricated (purchased) ductwork									
1283	10′ to 15′ high						10%			
1284	15′ to 20′ high						20%			
1285	20′ to 25′ high						25%			
1286	25′ to 30′ high						35%			
1287	30′ to 35′ high						40%			
1288	35′ to 40′ high						50%			
1289	Over 40′ high						55%			
3490	Rigid fiberglass duct board, foil reinf. kraft facing									
3500	Rectangular, 1" thick, alum. faced, (FRK), std. weight	Q-10	350	.069	SF Surf	.90	3.90		4.80	6.90

23 31 16.16 Thermoset Fiberglass-Reinforced Plastic Ducts		Crew	Daily Output	Labor-Hours	Unit	Material	Labor	Equipment	Total	Total Incl O&P
0010	**THERMOSET FIBERGLASS-REINFORCED PLASTIC DUCTS**									
1280	Add to labor for elevated installation									
1282	of prefabricated (purchased) ductwork									
1283	10′ to 15′ high						10%			
1284	15′ to 20′ high						20%			
1285	20′ to 25′ high						25%			
1286	25′ to 30′ high						35%			
1287	30′ to 35′ high						40%			
1288	35′ to 40′ high						50%			
1289	Over 40′ high						55%			
3550	Rigid fiberglass reinforced plastic, FM approved									

23 31 16 – Nonmetal Ducts

23 31 16.16 Thermoset Fiberglass-Reinforced Plastic Ducts		Crew	Daily Output	Labor-Hours	Unit	2019 Bare Costs Material	Labor	Equipment	Total	Total Incl O&P
3552	for acid fume and smoke exhaust system, nonflammable									
3554	Straight, 4" diameter	Q-9	200	.080	L.F.	10.50	4.39		14.89	18.20
3555	6" diameter		146	.110		13.80	6		19.80	24.50
3556	8" diameter		106	.151		16.70	8.30		25	31
3557	10" diameter		85	.188		20	10.35		30.35	37.50
3558	12" diameter		68	.235		23.50	12.90		36.40	45.50
3559	14" diameter		50	.320		34.50	17.55		52.05	64.50
3560	16" diameter		40	.400		38.50	22		60.50	76
3561	18" diameter		31	.516		43	28.50		71.50	90.50
3562	20" diameter	Q-10	44	.545		47.50	31		78.50	99
3563	22" diameter		40	.600		51.50	34		85.50	109
3564	24" diameter		37	.649		55	37		92	117
3565	26" diameter		34	.706		59.50	40		99.50	127
3566	28" diameter		28	.857		64	49		113	145
3567	30" diameter		24	1		68	57		125	161
3568	32" diameter		22.60	1.062		72.50	60.50		133	171
3569	34" diameter		21.60	1.111		76	63		139	180
3570	36" diameter		20.60	1.165		80.50	66.50		147	190
3571	38" diameter		19.80	1.212		124	69		193	242
3572	42" diameter		18.30	1.311		139	74.50		213.50	266
3573	46" diameter		17	1.412		153	80.50		233.50	291
3574	50" diameter		15.90	1.509		167	86		253	315
3575	54" diameter		14.90	1.611		184	91.50		275.50	340
3576	60" diameter		13.60	1.765		210	100		310	385
3577	64" diameter		12.80	1.875		310	107		417	500
3578	68" diameter		12.20	1.967		325	112		437	530
3579	72" diameter		11.90	2.017		350	115		465	560
3584	Note: joints are cemented with									
3586	fiberglass resin, included in material cost.									
3590	Elbow, 90°, 4" diameter	Q-9	20.30	.788	Ea.	51	43		94	122
3591	6" diameter		13.80	1.159		71	63.50		134.50	175
3592	8" diameter		10	1.600		78	88		166	219
3593	10" diameter		7.80	2.051		103	113		216	284
3594	12" diameter		6.50	2.462		130	135		265	350
3595	14" diameter		5.50	2.909		159	160		319	415
3596	16" diameter		4.80	3.333		190	183		373	485
3597	18" diameter		4.40	3.636		272	199		471	605
3598	20" diameter	Q-10	5.90	4.068		286	231		517	665
3599	22" diameter		5.40	4.444		350	253		603	770
3600	24" diameter		4.90	4.898		405	279		684	870
3601	26" diameter		4.60	5.217		410	297		707	900
3602	28" diameter		4.20	5.714		515	325		840	1,050
3603	30" diameter		3.90	6.154		570	350		920	1,150
3604	32" diameter		3.70	6.486		635	370		1,005	1,250
3605	34" diameter		3.45	6.957		750	395		1,145	1,425
3606	36" diameter		3.20	7.500		860	425		1,285	1,600
3607	38" diameter		3.07	7.818		1,225	445		1,670	2,025
3608	42" diameter		2.70	8.889		1,475	505		1,980	2,400
3609	46" diameter		2.54	9.449		1,700	535		2,235	2,700
3610	50" diameter		2.33	10.300		1,975	585		2,560	3,075
3611	54" diameter		2.14	11.215		2,275	640		2,915	3,475
3612	60" diameter		1.94	12.371		3,025	705		3,730	4,400
3613	64" diameter		1.82	13.187		4,400	750		5,150	6,000

23 31 16 – Nonmetal Ducts

23 31 16.16 Thermoset Fiberglass-Reinforced Plastic Ducts		Crew	Daily Output	Labor-Hours	Unit	Material	2019 Bare Costs Labor	Equipment	Total	Total Incl O&P
3614	68" diameter	Q-10	1.71	14.035	Ea.	4,875	800		5,675	6,575
3615	72" diameter		1.62	14.815		5,400	845		6,245	7,225
3626	Elbow, 45°, 4" diameter	Q-9	22.30	.717		49	39.50		88.50	114
3627	6" diameter		15.20	1.053		67.50	57.50		125	162
3628	8" diameter		11	1.455		72	80		152	200
3629	10" diameter		8.60	1.860		75.50	102		177.50	239
3630	12" diameter		7.20	2.222		83	122		205	277
3631	14" diameter		6.10	2.623		102	144		246	330
3632	16" diameter		5.28	3.030		121	166		287	385
3633	18" diameter		4.84	3.306		142	181		323	430
3634	20" diameter	Q-10	6.50	3.692		164	210		374	500
3635	22" diameter		5.94	4.040		224	230		454	595
3636	24" diameter		5.40	4.444		255	253		508	665
3637	26" diameter		5.06	4.743		289	270		559	730
3638	28" diameter		4.62	5.195		315	295		610	795
3639	30" diameter		4.30	5.581		360	315		675	880
3640	32" diameter		4.07	5.897		400	335		735	950
3641	34" diameter		3.80	6.316		475	360		835	1,075
3642	36" diameter		3.52	6.818		545	390		935	1,200
3643	38" diameter		3.38	7.101		765	405		1,170	1,450
3644	42" diameter		2.97	8.081		920	460		1,380	1,700
3645	46" diameter		2.80	8.571		1,075	490		1,565	1,925
3646	50" diameter		2.56	9.375		1,250	535		1,785	2,175
3647	54" diameter		2.35	10.213		1,425	580		2,005	2,425
3648	60" diameter		2.13	11.268		1,900	640		2,540	3,075
3649	64" diameter		2	12		2,750	685		3,435	4,050
3650	68" diameter		1.88	12.766		3,050	725		3,775	4,450
3651	72" diameter		1.78	13.483		3,375	765		4,140	4,875
3660	Tee, 90°, 4" diameter	Q-9	14.06	1.138		28	62.50		90.50	126
3661	6" diameter		9.52	1.681		42.50	92		134.50	187
3662	8" diameter		6.98	2.292		58	126		184	255
3663	10" diameter		5.45	2.936		75.50	161		236.50	330
3664	12" diameter		4.60	3.478		95	191		286	395
3665	14" diameter		3.86	4.145		138	227		365	495
3666	16" diameter		3.40	4.706		164	258		422	570
3667	18" diameter		3.05	5.246		194	288		482	650
3668	20" diameter	Q-10	4.12	5.825		228	330		558	755
3669	22" diameter		3.78	6.349		260	360		620	835
3670	24" diameter		3.46	6.936		293	395		688	925
3671	26" diameter		3.20	7.500		330	425		755	1,025
3672	28" diameter		2.96	8.108		370	460		830	1,100
3673	30" diameter		2.75	8.727		415	495		910	1,200
3674	32" diameter		2.54	9.449		460	535		995	1,325
3675	34" diameter		2.40	10		505	570		1,075	1,425
3676	36" diameter		2.27	10.573		550	600		1,150	1,525
3677	38" diameter		2.15	11.163		830	635		1,465	1,875
3678	42" diameter		1.91	12.565		1,000	715		1,715	2,175
3679	46" diameter		1.78	13.483		1,175	765		1,940	2,475
3680	50" diameter		1.63	14.724		1,375	840		2,215	2,775
3681	54" diameter		1.50	16		1,600	910		2,510	3,125
3682	60" diameter		1.35	17.778		1,975	1,000		2,975	3,700
3683	64" diameter		1.27	18.898		2,975	1,075		4,050	4,900
3684	68" diameter		1.20	20		3,325	1,150		4,475	5,375

23 31 16.16 Thermoset Fiberglass-Reinforced Plastic Ducts		Crew	Daily Output	Labor-Hours	Unit	2019 Bare Costs Material	Labor	Equipment	Total	Total Incl O&P
3685	72" diameter	Q-10	1.13	21.239	Ea.	3,725	1,200		4,925	5,900
3690	For Y @ 45°, add					22%				
3700	Blast gate, 4" diameter	Q-9	22.30	.717		135	39.50		174.50	208
3701	6" diameter		15.20	1.053		152	57.50		209.50	256
3702	8" diameter		11	1.455		205	80		285	345
3703	10" diameter		8.60	1.860		235	102		337	415
3704	12" diameter		7.20	2.222		286	122		408	500
3705	14" diameter		6.10	2.623		320	144		464	570
3706	16" diameter		5.28	3.030		350	166		516	635
3707	18" diameter		4.84	3.306		350	181		531	660
3708	20" diameter	Q-10	6.50	3.692		415	210		625	775
3709	22" diameter		5.94	4.040		450	230		680	845
3710	24" diameter		5.40	4.444		485	253		738	915
3711	26" diameter		5.06	4.743		520	270		790	980
3712	28" diameter		4.62	5.195		555	295		850	1,075
3713	30" diameter		4.30	5.581		590	315		905	1,125
3714	32" diameter		4.07	5.897		645	335		980	1,225
3715	34" diameter		3.80	6.316		705	360		1,065	1,325
3716	36" diameter		3.52	6.818		755	390		1,145	1,425
3717	38" diameter		3.38	7.101		1,025	405		1,430	1,750
3718	42" diameter		2.97	8.081		1,175	460		1,635	2,000
3719	46" diameter		2.80	8.571		1,350	490		1,840	2,225
3720	50" diameter		2.56	9.375		1,550	535		2,085	2,500
3721	54" diameter		2.35	10.213		1,725	580		2,305	2,775
3722	60" diameter		2.13	11.268		2,025	640		2,665	3,200
3723	64" diameter		2	12		2,250	685		2,935	3,500
3724	68" diameter		1.88	12.766		2,500	725		3,225	3,850
3725	72" diameter		1.78	13.483		2,725	765		3,490	4,175
3740	Reducers, 4" diameter	Q-9	24.50	.653		33.50	36		69.50	91.50
3741	6" diameter		16.70	.958		50.50	52.50		103	136
3742	8" diameter		12	1.333		68	73		141	186
3743	10" diameter		9.50	1.684		84	92.50		176.50	233
3744	12" diameter		7.90	2.025		101	111		212	280
3745	14" diameter		6.70	2.388		117	131		248	330
3746	16" diameter		5.80	2.759		134	151		285	375
3747	18" diameter		5.30	3.019		152	166		318	420
3748	20" diameter	Q-10	7.20	3.333		168	190		358	475
3749	22" diameter		6.50	3.692		271	210		481	620
3750	24" diameter		5.90	4.068		293	231		524	675
3751	26" diameter		5.60	4.286		320	244		564	720
3752	28" diameter		5.10	4.706		345	268		613	785
3753	30" diameter		4.70	5.106		370	290		660	850
3754	32" diameter		4.50	5.333		395	305		700	895
3755	34" diameter		4.20	5.714		420	325		745	955
3756	36" diameter		3.90	6.154		445	350		795	1,025
3757	38" diameter		3.70	6.486		675	370		1,045	1,300
3758	42" diameter		3.30	7.273		745	415		1,160	1,450
3759	46" diameter		3.10	7.742		815	440		1,255	1,575
3760	50" diameter		2.80	8.571		885	490		1,375	1,725
3761	54" diameter		2.60	9.231		960	525		1,485	1,850
3762	60" diameter		2.34	10.256		1,075	585		1,660	2,050
3763	64" diameter		2.20	10.909		1,650	620		2,270	2,775
3764	68" diameter		2.06	11.650		1,750	665		2,415	2,925

23 31 16.16 Thermoset Fiberglass-Reinforced Plastic Ducts		Crew	Daily Output	Labor-Hours	Unit	2019 Bare Costs Material	Labor	Equipment	Total	Total Incl O&P
3765	72" diameter	Q-10	1.96	12.245	Ea.	1,875	695		2,570	3,125
3780	Flange, per each, 4" diameter	Q-9	40.60	.394		81.50	21.50		103	123
3781	6" diameter		27.70	.578		95.50	31.50		127	153
3782	8" diameter		20.50	.780		119	43		162	196
3783	10" diameter		16.30	.982		143	54		197	239
3784	12" diameter		13.90	1.151		164	63		227	277
3785	14" diameter		11.30	1.416		190	77.50		267.50	325
3786	16" diameter		9.70	1.649		213	90.50		303.50	370
3787	18" diameter		8.30	1.928		237	106		343	420
3788	20" diameter	Q-10	11.30	2.124		260	121		381	470
3789	22" diameter		10.40	2.308		283	131		414	510
3790	24" diameter		9.50	2.526		310	144		454	560
3791	26" diameter		8.80	2.727		330	155		485	600
3792	28" diameter		8.20	2.927		355	166		521	645
3793	30" diameter		7.60	3.158		375	180		555	690
3794	32" diameter		7.20	3.333		400	190		590	730
3795	34" diameter		6.80	3.529		425	201		626	775
3796	36" diameter		6.40	3.750		450	213		663	820
3797	38" diameter		6.10	3.934		475	224		699	860
3798	42" diameter		5.50	4.364		520	248		768	945
3799	46" diameter		5	4.800		565	273		838	1,050
3800	50" diameter		4.60	5.217		615	297		912	1,125
3801	54" diameter		4.20	5.714		660	325		985	1,225
3802	60" diameter		3.80	6.316		735	360		1,095	1,350
3803	64" diameter		3.60	6.667		775	380		1,155	1,425
3804	68" diameter		3.40	7.059		825	400		1,225	1,525
3805	72" diameter		3.20	7.500		875	425		1,300	1,600

23 31 16.19 PVC Ducts		Crew	Daily Output	Labor-Hours	Unit	Material	Labor	Equipment	Total	Total Incl O&P
0010	**PVC DUCTS**									
4000	Rigid plastic, corrosive fume resistant PVC									
4020	Straight, 6" diameter	Q-9	220	.073	L.F.	13.65	3.99		17.64	21
4040	8" diameter		160	.100		20	5.50		25.50	30.50
4060	10" diameter		120	.133		25.50	7.30		32.80	39
4070	12" diameter		100	.160		29.50	8.80		38.30	46
4080	14" diameter		70	.229		37	12.55		49.55	59.50
4090	16" diameter		62	.258		43	14.15		57.15	68.50
4100	18" diameter		58	.276		59.50	15.15		74.65	88.50
4110	20" diameter	Q-10	75	.320		51.50	18.20		69.70	84
4130	24" diameter	"	55	.436		72.50	25		97.50	118
4250	Coupling, 6" diameter	Q-9	88	.182	Ea.	28	9.95		37.95	46
4270	8" diameter		78	.205		36.50	11.25		47.75	57
4290	10" diameter		70	.229		42	12.55		54.55	65
4300	12" diameter		55	.291		45.50	15.95		61.45	74
4310	14" diameter		44	.364		58	19.95		77.95	94.50
4320	16" diameter		40	.400		61.50	22		83.50	101
4330	18" diameter		38	.421		80	23		103	123
4340	20" diameter	Q-10	55	.436		96	25		121	144
4360	24" diameter	"	45	.533		122	30.50		152.50	180
4470	Elbow, 90°, 6" diameter	Q-9	44	.364		128	19.95		147.95	171
4490	8" diameter		28	.571		148	31.50		179.50	210
4510	10" diameter		18	.889		195	49		244	288
4520	12" diameter		15	1.067		195	58.50		253.50	305

23 31 16.19 PVC Ducts		Crew	Daily Output	Labor-Hours	Unit	2019 Bare Costs Material	Labor	Equipment	Total	Total Incl O&P
4530	14" diameter	Q-9	11	1.455	Ea.	220	80		300	365
4540	16" diameter		10	1.600		293	88		381	460
4550	18" diameter	Q-10	15	1.600		400	91		491	580
4560	20" diameter		14	1.714		640	97.50		737.50	855
4580	24" diameter		12	2		745	114		859	995
4750	Elbow 45°, use 90° and deduct					25%				

23 31 16.21 Polypropylene Ducts		Crew	Daily Output	Labor-Hours	Unit	Material	Labor	Equipment	Total	Total Incl O&P
0010	**POLYPROPYLENE DUCTS**									
0020	Rigid plastic, corrosive fume resistant, flame retardant, PP									
0110	Straight, 4" diameter	Q-9	200	.080	L.F.	7.60	4.39		11.99	15
0120	6" diameter		146	.110		10.75	6		16.75	21
0140	8" diameter		106	.151		13.80	8.30		22.10	28
0160	10" diameter		85	.188		20.50	10.35		30.85	38
0170	12" diameter		68	.235		39	12.90		51.90	62.50
0180	14" diameter		50	.320		44.50	17.55		62.05	75
0190	16" diameter		40	.400		59.50	22		81.50	99
0200	18" diameter		31	.516		97	28.50		125.50	150
0210	20" diameter	Q-10	44	.545		123	31		154	182
0220	22" diameter		40	.600		138	34		172	203
0230	24" diameter		37	.649		187	37		224	261
0250	28" diameter		28	.857		251	49		300	350
0270	32" diameter		24	1		283	57		340	395
0400	Coupling									
0410	4" diameter	Q-9	23	.696	Ea.	24	38		62	84.50
0420	6" diameter		16	1		25	55		80	111
0430	8" diameter		11.40	1.404		31.50	77		108.50	152
0440	10" diameter		9	1.778		44	97.50		141.50	197
0450	12" diameter		8	2		52	110		162	225
0460	14" diameter		6.60	2.424		94.50	133		227.50	305
0470	16" diameter		5.70	2.807		102	154		256	345
0480	18" diameter		5.20	3.077		125	169		294	395
0490	20" diameter	Q-10	6.80	3.529		136	201		337	455
0500	22" diameter		6.20	3.871		345	220		565	715
0510	24" diameter		5.80	4.138		430	235		665	825
0520	28" diameter		5.50	4.364		440	248		688	860
0530	32" diameter		5.20	4.615		490	263		753	940
0800	Elbow, 90°									
0810	4" diameter	Q-9	22	.727	Ea.	35	40		75	99
0820	6" diameter		15.40	1.039		49	57		106	141
0830	8" diameter		11	1.455		65	80		145	193
0840	10" diameter		8.60	1.860		94	102		196	258
0850	12" diameter		7.70	2.078		242	114		356	440
0860	14" diameter		6.40	2.500		280	137		417	520
0870	16" diameter		5.40	2.963		297	163		460	570
0880	18" diameter	Q-10	7.50	3.200		815	182		997	1,175
0890	20" diameter		6.60	3.636		700	207		907	1,075
0896	22" diameter		6.10	3.934		1,250	224		1,474	1,725
0900	24" diameter		5.60	4.286		1,200	244		1,444	1,700
0910	28" diameter		5.40	4.444		2,300	253		2,553	2,925
0920	32" diameter		5.20	4.615		3,025	263		3,288	3,725
1000	Elbow, 45°									
1020	4" diameter	Q-9	22	.727	Ea.	25	40		65	88

23 31 HVAC Ducts and Casings

23 31 16 – Nonmetal Ducts

23 31 16.21 Polypropylene Ducts		Crew	Daily Output	Labor-Hours	Unit	Material	Labor (2019 Bare Costs)	Equipment	Total	Total Incl O&P
1030	6″ diameter	Q-9	15.40	1.039	Ea.	45.50	57		102.50	137
1040	8″ diameter		11	1.455		48	80		128	174
1050	10″ diameter		8.60	1.860		63.50	102		165.50	225
1060	12″ diameter		7.70	2.078		187	114		301	380
1070	14″ diameter		6.40	2.500		211	137		348	440
1080	16″ diameter		5.40	2.963		227	163		390	495
1090	18″ diameter	Q-10	7.50	3.200		510	182		692	835
1100	20″ diameter		6.60	3.636		575	207		782	950
1106	22″ diameter		6.10	3.934		800	224		1,024	1,225
1110	24″ diameter		5.60	4.286		855	244		1,099	1,300
1120	28″ diameter		5.40	4.444		1,550	253		1,803	2,075
1130	32″ diameter		5.20	4.615		1,950	263		2,213	2,550
1200	Tee									
1210	4″ diameter	Q-9	16.50	.970	Ea.	150	53		203	246
1220	6″ diameter		11.60	1.379		179	75.50		254.50	310
1230	8″ diameter		8.30	1.928		223	106		329	405
1240	10″ diameter		6.50	2.462		310	135		445	550
1250	12″ diameter		5.80	2.759		425	151		576	700
1260	14″ diameter		4.80	3.333		545	183		728	880
1270	16″ diameter		4.05	3.951		615	217		832	1,000
1280	18″ diameter	Q-10	5.60	4.286		790	244		1,034	1,250
1290	20″ diameter		4.90	4.898		860	279		1,139	1,375
1300	22″ diameter		4.40	5.455		1,050	310		1,360	1,625
1310	24″ diameter		4.20	5.714		1,525	325		1,850	2,175
1320	28″ diameter		4.10	5.854		2,400	335		2,735	3,125
1330	32″ diameter		4	6		2,900	340		3,240	3,725

23 31 16.22 Duct Board		Crew	Daily Output	Labor-Hours	Unit	Material	Labor	Equipment	Total	Total Incl O&P
0010	**DUCT BOARD**									
3800	Rigid, resin bonded fibrous glass, FSK									
3810	Temperature, bacteria and fungi resistant									
3820	1″ thick	Q-14	150	.107	S.F.	1.63	5.50		7.13	10.30
3830	1-1/2″ thick		130	.123		2.15	6.35		8.50	12.15
3840	2″ thick		120	.133		3.71	6.90		10.61	14.75

23 33 Air Duct Accessories

23 33 13 – Dampers

23 33 13.13 Volume-Control Dampers		Crew	Daily Output	Labor-Hours	Unit	Material	Labor	Equipment	Total	Total Incl O&P
0010	**VOLUME-CONTROL DAMPERS**									
5990	Multi-blade dampers, opposed blade, 8″ x 6″	1 Shee	24	.333	Ea.	32.50	20.50		53	66.50
5994	8″ x 8″		22	.364		33	22		55	70
5996	10″ x 10″		21	.381		36	23		59	75
6000	12″ x 12″		21	.381		45	23		68	85
6020	12″ x 18″		18	.444		60.50	27		87.50	108
6030	14″ x 10″		20	.400		44	24.50		68.50	85.50
6031	14″ x 14″		17	.471		47.50	28.50		76	95.50
6032	16″ x 10″		18	.444		46.50	27		73.50	92.50
6033	16″ x 12″		17	.471		47.50	28.50		76	95.50
6035	16″ x 16″		16	.500		59	30.50		89.50	112
6036	18″ x 14″		16	.500		59	30.50		89.50	112
6037	18″ x 16″		15	.533		74	32.50		106.50	131
6038	18″ x 18″		15	.533		78	32.50		110.50	136

23 33 13.13 Volume-Control Dampers		Crew	Daily Output	Labor-Hours	Unit	Material	Labor	Equipment	Total	Total Incl O&P
						2019 Bare Costs				
6040	18" x 24"	1 Shee	12	.667	Ea.	103	40.50		143.50	175
6060	18" x 28"		10	.800		119	49		168	205
6068	20" x 6"		18	.444		43.50	27		70.50	88.50
6069	20" x 8"		16	.500		48.50	30.50		79	99.50
6070	20" x 16"		14	.571		78	35		113	139
6071	20" x 18"		13	.615		86.50	37.50		124	153
6072	20" x 20"		13	.615		84.50	37.50		122	150
6073	22" x 6"		20	.400		44	24.50		68.50	85.50
6074	22" x 18"		14	.571		84.50	35		119.50	146
6075	22" x 20"		12	.667		104	40.50		144.50	176
6076	24" x 16"		11	.727		92.50	44.50		137	170
6077	22" x 22"		10	.800		113	49		162	198
6078	24" x 20"		8	1		113	61		174	217
6080	24" x 24"		8	1		122	61		183	228
6100	24" x 28"		6	1.333		143	81.50		224.50	280
6110	26" x 26"	▼	6	1.333		143	81.50		224.50	280
6120	28" x 28"	Q-9	11	1.455		143	80		223	279
6130	30" x 18"		10	1.600		126	88		214	272
6132	30" x 24"		7	2.286		160	125		285	365
6133	30" x 30"		6.60	2.424		212	133		345	435
6135	32" x 32"		6.40	2.500		229	137		366	460
6150	34" x 34"		6	2.667		257	146		403	505
6151	36" x 12"		10	1.600		103	88		191	246
6152	36" x 16"		8	2		143	110		253	325
6156	36" x 32"		6.30	2.540		257	139		396	495
6157	36" x 34"		6.20	2.581		271	142		413	515
6158	36" x 36"		6	2.667		280	146		426	530
6160	44" x 28"		5.80	2.759		290	151		441	550
6180	48" x 36"		5.60	2.857		375	157		532	655
6200	56" x 36"		5.40	2.963		480	163		643	770
6220	60" x 36"		5.20	3.077		515	169		684	820
6240	60" x 44"	▼	5	3.200	▼	595	176		771	920
7500	Variable volume modulating motorized damper, incl. elect. mtr.									
7504	8" x 6"	1 Shee	15	.533	Ea.	180	32.50		212.50	248
7506	10" x 6"		14	.571		180	35		215	251
7510	10" x 10"		13	.615		188	37.50		225.50	264
7520	12" x 12"		12	.667		194	40.50		234.50	275
7522	12" x 16"		11	.727		195	44.50		239.50	282
7524	16" x 10"		12	.667		191	40.50		231.50	272
7526	16" x 14"		10	.800		200	49		249	295
7528	16" x 18"		9	.889		206	54		260	310
7540	18" x 12"		10	.800		201	49		250	296
7542	18" x 18"		8	1		210	61		271	325
7544	20" x 14"		8	1		218	61		279	330
7546	20" x 18"		7	1.143		224	69.50		293.50	355
7560	24" x 12"		8	1		219	61		280	335
7562	24" x 18"		7	1.143		236	69.50		305.50	365
7568	28" x 10"		7	1.143		219	69.50		288.50	345
7580	28" x 16"		6	1.333		219	81.50		300.50	365
7590	30" x 14"		5	1.600		227	97.50		324.50	400
7600	30" x 18"		4	2		297	122		419	510
7610	30" x 24"	▼	3.80	2.105		395	128		523	630
7690	48" x 48"	Q-9	6	2.667	▼	1,275	146		1,421	1,625

23 33 Air Duct Accessories

23 33 13 – Dampers

23 33 13.13 Volume-Control Dampers

	23 33 13.13 Volume-Control Dampers	Crew	Daily Output	Labor-Hours	Unit	2019 Bare Costs Material	Labor	Equipment	Total	Total Incl O&P
7694	6′ x 14′ w/2 motors	Q-9	3	5.333	Ea.	4,550	293		4,843	5,450
7700	For thermostat, add	1 Shee	8	1		43	61		104	140
7800	For transformer 40 VA capacity, add	″	16	.500		30	30.50		60.50	79.50
8000	Multi-blade dampers, parallel blade									
8100	8″ x 8″	1 Shee	24	.333	Ea.	116	20.50		136.50	159
8120	12″ x 8″		22	.364		116	22		138	162
8140	16″ x 10″		20	.400		147	24.50		171.50	199
8160	18″ x 12″		18	.444		160	27		187	217
8180	22″ x 12″		15	.533		173	32.50		205.50	240
8200	24″ x 16″		11	.727		191	44.50		235.50	278
8220	28″ x 16″		10	.800		213	49		262	310
8240	30″ x 16″		8	1		216	61		277	330
8260	30″ x 18″		7	1.143		264	69.50		333.50	395
8400	Round damper, butterfly, vol. control w/lever lock reg.									
8410	6″ diam.	1 Shee	22	.364	Ea.	34	22		56	71
8412	7″ diam.		21	.381		38.50	23		61.50	78
8414	8″ diam.		20	.400		41.50	24.50		66	82.50
8416	9″ diam.		19	.421		42	25.50		67.50	85.50
8418	10″ diam.		18	.444		46	27		73	91.50
8420	12″ diam.		16	.500		46.50	30.50		77	98
8422	14″ diam.		14	.571		55	35		90	114
8424	16″ diam.		13	.615		59	37.50		96.50	122
8426	18″ diam.		12	.667		72.50	40.50		113	142
8428	20″ diam.		11	.727		101	44.50		145.50	179
8430	24″ diam.		10	.800		125	49		174	212
8432	30″ diam.		9	.889		175	54		229	276
8434	36″ diam.		8	1		234	61		295	350
8500	Round motor operated damper									
8510	6″ diam.	1 Shee	22	.364	Ea.	92.50	22		114.50	136
8512	7″ diam.		21	.381		100	23		123	146
8514	8″ diam.		20	.400		103	24.50		127.50	150
8516	10″ diam.		18	.444		115	27		142	167
8518	12″ diam.		16	.500		121	30.50		151.50	180
8520	14″ diam.		14	.571		137	35		172	203

23 33 13.16 Fire Dampers

	23 33 13.16 Fire Dampers	Crew	Daily Output	Labor-Hours	Unit	Material	Labor	Equipment	Total	Total Incl O&P
0010	**FIRE DAMPERS**									
3000	Fire damper, curtain type, 1-1/2 hr. rated, vertical, 6″ x 6″	1 Shee	24	.333	Ea.	34.50	20.50		55	69
3020	8″ x 6″		22	.364		34.50	22		56.50	71.50
3021	8″ x 8″		22	.364		29.50	22		51.50	66
3030	10″ x 10″		22	.364		34.50	22		56.50	71.50
3040	12″ x 6″		22	.364		34.50	22		56.50	71.50
3044	12″ x 12″		20	.400		32	24.50		56.50	72
3060	20″ x 6″		18	.444		36	27		63	80.50
3080	12″ x 8″		22	.364		27.50	22		49.50	64
3100	24″ x 8″		16	.500		41	30.50		71.50	91.50
3120	12″ x 10″		21	.381		34.50	23		57.50	73.50
3140	24″ x 10″		15	.533		55.50	32.50		88	111
3160	36″ x 10″		12	.667		72	40.50		112.50	141
3180	16″ x 12″		20	.400		47	24.50		71.50	88.50
3200	24″ x 12″		13	.615		46	37.50		83.50	108
3220	48″ x 12″		10	.800		71	49		120	152
3238	14″ x 14″		19	.421		51	25.50		76.50	95

23 33 13.16 Fire Dampers		Crew	Daily Output	Labor-Hours	Unit	Material	2019 Bare Costs Labor	Equipment	Total	Total Incl O&P
3240	16" x 14"	1 Shee	18	.444	Ea.	62.50	27		89.50	110
3260	24" x 14"		12	.667		75	40.50		115.50	144
3280	30" x 14"		11	.727		72	44.50		116.50	147
3298	16" x 16"		18	.444		45	27		72	90.50
3300	18" x 16"		17	.471		61	28.50		89.50	111
3320	24" x 16"		11	.727		51.50	44.50		96	125
3340	36" x 16"		10	.800		63.50	49		112.50	144
3356	18" x 18"		16	.500		63	30.50		93.50	116
3360	24" x 18"		10	.800		76.50	49		125.50	158
3380	48" x 18"		8	1		110	61		171	214
3398	20" x 20"		10	.800		53	49		102	132
3400	24" x 20"		8	1		60	61		121	159
3420	36" x 20"		7	1.143		72	69.50		141.50	186
3440	24" x 22"		7	1.143		79	69.50		148.50	193
3460	30" x 22"		6	1.333		86	81.50		167.50	218
3478	24" x 24"		8	1		58.50	61		119.50	157
3480	26" x 24"		7	1.143		80	69.50		149.50	194
3484	32" x 24"	Q-9	13	1.231		75	67.50		142.50	186
3500	48" x 24"		12	1.333		97	73		170	218
3520	28" x 26"		13	1.231		90	67.50		157.50	202
3540	30" x 28"		12	1.333		99.50	73		172.50	220
3560	48" x 30"		11	1.455		145	80		225	280
3580	48" x 36"		10	1.600		151	88		239	300
3590	32" x 40"		10	1.600		117	88		205	262
3600	44" x 44"		10	1.600		161	88		249	310
3620	48" x 48"		8	2		179	110		289	365
3700	UL label included in above									
3800	For horizontal operation, add				Ea.	20%				
3900	For cap for blades out of air stream, add					20%				
4000	For 10" 22 ga., UL approved sleeve, add					35%				
4100	For 10" 22 ga., UL sleeve, 100% free area, add					65%				
4200	For oversize openings group dampers									
4210	Fire damper, round, Type A, 1-1/2 hour									
4214	4" diam.	1 Shee	20	.400	Ea.	55	24.50		79.50	97.50
4216	5" diam.		16	.500		48	30.50		78.50	99
4218	6" diam.		14.30	.559		49	34		83	106
4220	7" diam.		13.30	.602		69	36.50		105.50	132
4222	8" diam.		11.40	.702		70.50	43		113.50	143
4224	9" diam.		10.80	.741		55	45		100	129
4226	10" diam.		10	.800		55	49		104	135
4228	12" diam.		9.50	.842		59.50	51.50		111	143
4230	14" diam.		8.20	.976		80.50	59.50		140	179
4232	16" diam.		7.60	1.053		90	64		154	197
4234	18" diam.		7.20	1.111		108	67.50		175.50	222
4236	20" diam.		6.40	1.250		137	76		213	267
4238	22" diam.		6.10	1.311		170	80		250	310
4240	24" diam.		5.70	1.404		178	85.50		263.50	325
4242	26" diam.		5.50	1.455		201	88.50		289.50	355
4244	28" diam.		5.40	1.481		185	90.50		275.50	340
4246	30" diam.		5.30	1.509		220	92		312	380
4502	Power open-spring close, sleeve and actuator motor mounted, 1-1/2 hour									
4506	6" x 6"	1 Shee	24	.333	Ea.	82.50	20.50		103	122
4510	8" x 8"		22	.364		290	22		312	355

23 33 13.16 Fire Dampers

23 33 13.16 Fire Dampers		Crew	Daily Output	Labor-Hours	Unit	Material	Labor	Equipment	Total	Total Incl O&P
4520	16" x 8"	1 Shee	20	.400	Ea.	350	24.50		374.50	420
4540	18" x 8"		18	.444		355	27		382	430
4560	20" x 8"		16	.500		355	30.50		385.50	435
4580	10" x 10"		21	.381		335	23		358	400
4600	24" x 10"		15	.533		390	32.50		422.50	480
4620	30" x 10"		12	.667		440	40.50		480.50	545
4640	12" x 12"		20	.400		375	24.50		399.50	445
4660	18" x 12"		18	.444		365	27		392	440
4680	24" x 12"		13	.615		400	37.50		437.50	495
4700	30" x 12"		11	.727		405	44.50		449.50	515
4720	14" x 14"		17	.471		400	28.50		428.50	485
4740	16" x 14"		18	.444		400	27		427	480
4760	20" x 14"		14	.571		390	35		425	480
4780	24" x 14"		12	.667		410	40.50		450.50	510
4800	30" x 14"		10	.800		415	49		464	530
4820	16" x 16"		16	.500		395	30.50		425.50	480
4840	20" x 16"		14	.571		405	35		440	500
4860	24" x 16"		11	.727		415	44.50		459.50	525
4880	30" x 16"		8	1		445	61		506	585
4900	18" x 18"		15	.533		435	32.50		467.50	530
5000	24" x 18"		10	.800		435	49		484	555
5020	36" x 18"		7	1.143		450	69.50		519.50	600
5040	20" x 20"		13	.615		435	37.50		472.50	535
5060	24" x 20"		8	1		440	61		501	575
5080	30" x 20"		8	1		455	61		516	595
5100	36" x 20"		7	1.143		495	69.50		564.50	645
5120	24" x 24"		8	1		450	61		511	590
5130	30" x 24"		7.60	1.053		470	64		534	620
5140	36" x 24"	Q-9	12	1.333		505	73		578	665
5141	36" x 30"		10	1.600		585	88		673	780
5142	40" x 36"		8	2		985	110		1,095	1,250
5143	48" x 48"		4	4		1,800	219		2,019	2,300
5150	Damper operator motor, 24 or 120 volt	1 Shee	16	.500		505	30.50		535.50	600

23 33 13.25 Exhaust Vent Damper

23 33 13.25 Exhaust Vent Damper		Crew	Daily Output	Labor-Hours	Unit	Material	Labor	Equipment	Total	Total Incl O&P
0010	**EXHAUST VENT DAMPER**, Auto, OB with elect. actuator.									
1110	6" x 6"	1 Shee	24	.333	Ea.	99.50	20.50		120	140
1114	8" x 6"		23	.348		101	21		122	143
1116	8" x 8"		22	.364		104	22		126	148
1118	10" x 6"		23	.348		104	21		125	146
1120	10" x 10"		22	.364		113	22		135	158
1122	12" x 6"		22	.364		104	22		126	149
1124	12" x 10"		21	.381		117	23		140	165
1126	12" x 12"		20	.400		113	24.50		137.50	162
1128	14" x 6"		21	.381		107	23		130	154
1130	14" x 10"		19	.421		121	25.50		146.50	173
1132	16" x 6"		21	.381		108	23		131	155
1134	16" x 8"		20	.400		121	24.50		145.50	170
1136	18" x 6"		20	.400		116	24.50		140.50	165
1138	18" x 8"		19	.421		121	25.50		146.50	173

23 33 13 – Dampers

23 33 13.28 Splitter Damper Assembly

		Crew	Daily Output	Labor-Hours	Unit	Material	2019 Bare Costs Labor	Equipment	Total	Total Incl O&P
0010	**SPLITTER DAMPER ASSEMBLY**									
7000	Self locking, 1′ rod	1 Shee	24	.333	Ea.	29.50	20.50		50	63.50
7020	3′ rod		22	.364		40	22		62	77.50
7040	4′ rod		20	.400		44.50	24.50		69	86
7060	6′ rod	↓	18	.444	↓	54.50	27		81.50	101

23 33 13.32 Relief Damper

		Crew	Daily Output	Labor-Hours	Unit	Material	Labor	Equipment	Total	Total Incl O&P
0010	**RELIEF DAMPER**, Electronic bypass with tight seal									
8310	8″ x 6″	1 Shee	22	.364	Ea.	270	22		292	330
8314	10″ x 6″		22	.364		270	22		292	330
8318	10″ x 10″		21	.381		277	23		300	340
8322	12″ x 12″		20	.400		289	24.50		313.50	355
8326	12″ x 16″		19	.421		297	25.50		322.50	365
8330	16″ x 10″		20	.400		315	24.50		339.50	380
8334	16″ x 14″		18	.444		315	27		342	390
8338	16″ x 18″		17	.471		350	28.50		378.50	430
8342	18″ x 12″		18	.444		350	27		377	425
8346	18″ x 18″		15	.533		310	32.50		342.50	390
8350	20″ x 14″		14	.571		325	35		360	415
8354	20″ x 18″		13	.615		330	37.50		367.50	420
8358	24″ x 12″		13	.615		335	37.50		372.50	420
8362	24″ x 18″		10	.800		325	49		374	430
8363	24″ x 24″		10	.800		360	49		409	475
8364	24″ x 36″		9	.889		430	54		484	555
8365	24″ x 48″		6	1.333		480	81.50		561.50	655
8366	28″ x 10″		13	.615		235	37.50		272.50	315
8370	28″ x 16″		10	.800		370	49		419	480
8374	30″ x 14″		11	.727		246	44.50		290.50	340
8378	30″ x 18″		8	1		375	61		436	510
8382	30″ x 24″		6	1.333		395	81.50		476.50	560
8390	46″ x 36″		4	2		530	122		652	770
8394	48″ x 48″		3	2.667		580	163		743	885
8396	54″ x 36″	↓	2	4	↓	615	244		859	1,050

23 33 19 – Duct Silencers

23 33 19.10 Duct Silencers

		Crew	Daily Output	Labor-Hours	Unit	Material	Labor	Equipment	Total	Total Incl O&P
0010	**DUCT SILENCERS**									
9000	Silencers, noise control for air flow, duct				MCFM	81.50			81.50	89.50
9004	Duct sound trap, packaged									
9010	12″ x 18″ x 36″, 2,000 CFM	1 Shee	12	.667	Ea.	655	40.50		695.50	780
9011	24″ x 18″ x 36″, 9,000 CFM	Q-9	80	.200		750	10.95		760.95	840
9012	24″ x 24″ x 36″, 9,000 CFM		60	.267		1,125	14.65		1,139.65	1,275
9013	24″ x 30″ x 36″, 9,000 CFM		50	.320		1,200	17.55		1,217.55	1,350
9014	24″ x 36″ x 36″, 9,000 CFM		40	.400		1,500	22		1,522	1,675
9015	24″ x 48″ x 36″, 9,000 CFM		28	.571		1,950	31.50		1,981.50	2,200
9018	36″ x 18″ x 36″, 6,300 CFM		42	.381		1,500	21		1,521	1,675
9019	36″ x 36″ x 36″, 6,300 CFM		21	.762		3,000	42		3,042	3,375
9020	36″ x 48″ x 36″, 6,300 CFM		14	1.143		3,900	62.50		3,962.50	4,400
9021	36″ x 60″ x 36″, 6,300 CFM		13	1.231		4,950	67.50		5,017.50	5,550
9022	48″ x 48″ x 36″, 6,300 CFM		12	1.333		5,275	73		5,348	5,900
9023	24″ x 24″ x 60″, 6,000 CFM		3.60	4.444		1,725	244		1,969	2,275
9026	24″ x 30″ x 60″, 7,000 CFM		3.50	4.571		1,950	251		2,201	2,500
9030	24″ x 36″ x 60″, 8,000 CFM	↓	3.40	4.706	↓	2,150	258		2,408	2,750

23 33 19 – Duct Silencers

23 33 19.10 Duct Silencers		Crew	Daily Output	Labor-Hours	Unit	Material	Labor	Equipment	Total	Total Incl O&P
						2019 Bare Costs				
9034	24" x 48" x 60", 11,000 CFM	Q-9	3.30	4.848	Ea.	1,825	266		2,091	2,425
9038	36" x 18" x 60", 6,300 CFM		3.50	4.571		1,450	251		1,701	1,950
9042	36" x 36" x 60", 12,000 CFM		3.30	4.848		2,325	266		2,591	2,950
9046	36" x 48" x 60", 17,000 CFM		3.10	5.161		3,025	283		3,308	3,775
9050	36" x 60" x 60", 20,000 CFM		2.80	5.714		3,700	315		4,015	4,550
9054	48" x 48" x 60", 23,000 CFM		2.60	6.154		4,275	340		4,615	5,225
9056	48" x 60" x 60", 23,000 CFM		2.10	7.619		5,350	420		5,770	6,500

23 33 23 – Turning Vanes

23 33 23.13 Air Turning Vanes

		Crew	Daily Output	Labor-Hours	Unit	Material	Labor	Equipment	Total	Total Incl O&P
0010	**AIR TURNING VANES**									
9400	Turning vane components									
9410	Turning vane rail	1 Shee	160	.050	L.F.	.84	3.05		3.89	5.55
9420	Double thick, factory fab. vane		300	.027		1.25	1.63		2.88	3.85
9428	12" high set		170	.047		1.80	2.87		4.67	6.35
9432	14" high set		160	.050		2.10	3.05		5.15	6.95
9434	16" high set		150	.053		2.40	3.25		5.65	7.60
9436	18" high set		144	.056		2.70	3.39		6.09	8.10
9438	20" high set		138	.058		3	3.53		6.53	8.65
9440	22" high set		130	.062		3.30	3.75		7.05	9.35
9442	24" high set		124	.065		3.60	3.93		7.53	9.90
9444	26" high set		116	.069		3.90	4.20		8.10	10.70
9446	30" high set		112	.071		4.50	4.35		8.85	11.55

23 33 33 – Duct-Mounting Access Doors

23 33 33.13 Duct Access Doors

		Crew	Daily Output	Labor-Hours	Unit	Material	Labor	Equipment	Total	Total Incl O&P
0010	**DUCT ACCESS DOORS**									
1000	Duct access door, insulated, 6" x 6"	1 Shee	14	.571	Ea.	25	35		60	80.50
1020	10" x 10"		11	.727		22	44.50		66.50	91.50
1040	12" x 12"		10	.800		27.50	49		76.50	105
1050	12" x 18"		9	.889		52	54		106	140
1060	16" x 12"		9	.889		42	54		96	129
1070	18" x 18"		8	1		37.50	61		98.50	134
1074	24" x 18"		8	1		62	61		123	161
1080	24" x 24"		8	1		67.50	61		128.50	167

23 33 46 – Flexible Ducts

23 33 46.10 Flexible Air Ducts

		Crew	Daily Output	Labor-Hours	Unit	Material	Labor	Equipment	Total	Total Incl O&P
0010	**FLEXIBLE AIR DUCTS**									
1280	Add to labor for elevated installation									
1282	of prefabricated (purchased) ductwork									
1283	10' to 15' high						10%			
1284	15' to 20' high						20%			
1285	20' to 25' high						25%			
1286	25' to 30' high						35%			
1287	30' to 35' high						40%			
1288	35' to 40' high						50%			
1289	Over 40' high						55%			
1300	Flexible, coated fiberglass fabric on corr. resist. metal helix									
1400	pressure to 12" (WG) UL-181									
1500	Noninsulated, 3" diameter	Q-9	400	.040	L.F.	1.21	2.19		3.40	4.66
1520	4" diameter		360	.044		1.24	2.44		3.68	5.05
1540	5" diameter		320	.050		1.38	2.74		4.12	5.70
1560	6" diameter		280	.057		1.61	3.13		4.74	6.55

23 33 46.10 Flexible Air Ducts			Crew	Daily Output	Labor-Hours	Unit	2019 Bare Costs Material	Labor	Equipment	Total	Total Incl O&P
1580	7″ diameter		Q-9	240	.067	L.F.	1.75	3.66		5.41	7.50
1600	8″ diameter			200	.080		2.01	4.39		6.40	8.85
1620	9″ diameter			180	.089		2.22	4.88		7.10	9.85
1640	10″ diameter			160	.100		2.60	5.50		8.10	11.20
1660	12″ diameter			120	.133		3.09	7.30		10.39	14.50
1680	14″ diameter			80	.200		3.76	10.95		14.71	21
1700	16″ diameter			60	.267		5.40	14.65		20.05	28
1800	For plastic cable tie, add					Ea.	.60			.60	.66
1900	Insulated, 1″ thick, PE jacket, 3″ diameter	G	Q-9	380	.042	L.F.	2.66	2.31		4.97	6.45
1910	4″ diameter	G		340	.047		2.83	2.58		5.41	7.05
1920	5″ diameter	G		300	.053		3.01	2.93		5.94	7.75
1940	6″ diameter	G		260	.062		3.14	3.38		6.52	8.60
1960	7″ diameter	G		220	.073		3.59	3.99		7.58	10
1980	8″ diameter	G		180	.089		3.96	4.88		8.84	11.75
2000	9″ diameter	G		160	.100		4.18	5.50		9.68	12.95
2020	10″ diameter	G		140	.114		4.59	6.25		10.84	14.55
2040	12″ diameter	G		100	.160		5.55	8.80		14.35	19.45
2060	14″ diameter	G		80	.200		6.90	10.95		17.85	24
2080	16″ diameter	G		60	.267		9.25	14.65		23.90	32
2100	18″ diameter	G		45	.356		10.95	19.50		30.45	41.50
2120	20″ diameter	G	Q-10	65	.369		10.80	21		31.80	44
2500	Insulated, heavy duty, coated fiberglass fabric										
2520	4″ diameter	G	Q-9	340	.047	L.F.	3.80	2.58		6.38	8.10
2540	5″ diameter	G		300	.053		4.53	2.93		7.46	9.40
2560	6″ diameter	G		260	.062		4.48	3.38		7.86	10.10
2580	7″ diameter	G		220	.073		5.70	3.99		9.69	12.35
2600	8″ diameter	G		180	.089		5.70	4.88		10.58	13.70
2620	9″ diameter	G		160	.100		7.10	5.50		12.60	16.15
2640	10″ diameter	G		140	.114		7.05	6.25		13.30	17.30
2660	12″ diameter	G		100	.160		8.20	8.80		17	22.50
2680	14″ diameter	G		80	.200		10.05	10.95		21	27.50
2700	16″ diameter	G		60	.267		13.15	14.65		27.80	36.50
2720	18″ diameter	G		45	.356		15.75	19.50		35.25	47
2800	Flexible, aluminum, pressure to 12″ (WG) UL-181										
2820	Non-insulated										
2830	3″ diameter		Q-9	400	.040	L.F.	1.27	2.19		3.46	4.73
2831	4″ diameter			360	.044		1.40	2.44		3.84	5.25
2832	5″ diameter			320	.050		1.80	2.74		4.54	6.15
2833	6″ diameter			280	.057		2.28	3.13		5.41	7.25
2834	7″ diameter			240	.067		2.83	3.66		6.49	8.65
2835	8″ diameter			200	.080		3.17	4.39		7.56	10.15
2836	9″ diameter			180	.089		3.71	4.88		8.59	11.50
2837	10″ diameter			160	.100		4.49	5.50		9.99	13.30
2838	12″ diameter			120	.133		5.80	7.30		13.10	17.45
2839	14″ diameter			80	.200		6.60	10.95		17.55	24
2840	15″ diameter			70	.229		7.80	12.55		20.35	27.50
2841	16″ diameter			60	.267		10.35	14.65		25	33.50
2842	18″ diameter			45	.356		11.75	19.50		31.25	42.50
2843	20″ diameter		Q-10	65	.369		13.05	21		34.05	46.50
2880	Insulated, 1″ thick with 3/4 lb., PE jacket										
2890	3″ diameter	G	Q-9	380	.042	L.F.	3.18	2.31		5.49	7
2891	4″ diameter	G		340	.047		3.58	2.58		6.16	7.85
2892	5″ diameter	G		300	.053		4.06	2.93		6.99	8.90

23 33 46 – Flexible Ducts

23 33 46.10 Flexible Air Ducts			Crew	Daily Output	Labor-Hours	Unit	2019 Bare Costs Material	Labor	Equipment	Total	Total Incl O&P
2893	6″ diameter	G	Q-9	260	.062	L.F.	4.75	3.38		8.13	10.40
2894	7″ diameter	G		220	.073		5.85	3.99		9.84	12.45
2895	8″ diameter	G		180	.089		5.95	4.88		10.83	13.90
2896	9″ diameter	G		160	.100		6.90	5.50		12.40	15.95
2897	10″ diameter	G		140	.114		8.15	6.25		14.40	18.45
2898	12″ diameter	G		100	.160		9.40	8.80		18.20	23.50
2899	14″ diameter	G		80	.200		10.70	10.95		21.65	28.50
2900	15″ diameter	G		70	.229		11.75	12.55		24.30	32
2901	16″ diameter	G		60	.267		14.90	14.65		29.55	38.50
2902	18″ diameter	G		45	.356		17.45	19.50		36.95	48.50
3000	Transitions, 3″ deep, square to round										
3010	6″ x 6″ to round		Q-9	44	.364	Ea.	11.05	19.95		31	42.50
3014	6″ x 12″ to round			42	.381		20	21		41	53.50
3018	8″ x 8″ to round			40	.400		14.25	22		36.25	49
3022	9″ x 9″ to round			38	.421		14.25	23		37.25	50.50
3026	9″ x 12″ to round			37	.432		25.50	23.50		49	64
3030	9″ x 15″ to round			36	.444		32.50	24.50		57	73
3034	10″ x 10″ to round			34	.471		15.60	26		41.60	56
3038	10″ x 22″ to round			32	.500		36.50	27.50		64	82
3042	12″ x 12″ to round			30	.533		15.60	29.50		45.10	61.50
3046	12″ x 18″ to round			28	.571		36.50	31.50		68	88
3050	12″ x 24″ to round			26	.615		36.50	34		70.50	92
3054	15″ x 15″ to round			25	.640		21	35		56	76.50
3058	15″ x 18″ to round			24	.667		39.50	36.50		76	99
3062	18″ x 18″ to round			22	.727		30	40		70	93.50
3066	21″ x 21″ to round			21	.762		32.50	42		74.50	99
3070	22″ x 22″ to round			20	.800		32	44		76	102
3074	24″ x 24″ to round			18	.889		34	49		83	112
5000	Flexible, aluminum, acoustical, pressure to 2″ (WG), NFPA-90A										
5010	Fiberglass insulation 1-1/2″ thick, 1/2 lb. density										
5020	Polyethylene jacket, UL approved										
5026	5″ diameter	G	Q-9	300	.053	L.F.	4.79	2.93		7.72	9.70
5030	6″ diameter	G		260	.062		5.80	3.38		9.18	11.50
5034	7″ diameter	G		220	.073		7.05	3.99		11.04	13.80
5038	8″ diameter	G		180	.089		7.10	4.88		11.98	15.25
5042	9″ diameter	G		160	.100		8.45	5.50		13.95	17.65
5046	10″ diameter	G		140	.114		9.65	6.25		15.90	20
5050	12″ diameter	G		100	.160		11.35	8.80		20.15	26
5054	14″ diameter	G		80	.200		13.65	10.95		24.60	31.50
5058	16″ diameter	G		60	.267		17.80	14.65		32.45	41.50
5072	Hospital grade, PE jacket, UL approved										
5076	5″ diameter	G	Q-9	300	.053	L.F.	5.20	2.93		8.13	10.15
5080	6″ diameter	G		260	.062		6.20	3.38		9.58	11.95
5084	7″ diameter	G		220	.073		7.60	3.99		11.59	14.40
5088	8″ diameter	G		180	.089		7.70	4.88		12.58	15.85
5092	9″ diameter	G		160	.100		8.80	5.50		14.30	18.05
5096	10″ diameter	G		140	.114		10.35	6.25		16.60	21
5100	12″ diameter	G		100	.160		12.20	8.80		21	27
5104	14″ diameter	G		80	.200		14.65	10.95		25.60	33
5108	16″ diameter	G		60	.267		19.25	14.65		33.90	43
5140	Flexible, aluminum, silencer, pressure to 12″ (WG), NFPA-90A										
5144	Fiberglass insulation 1-1/2″ thick, 1/2 lb. density										
5148	Aluminum outer shell, UL approved										

23 33 46 – Flexible Ducts

23 33 46.10 Flexible Air Ducts		Crew	Daily Output	Labor-Hours	Unit	Material	2019 Bare Costs Labor	Equipment	Total	Total Incl O&P
5152	5" diameter	G	Q-9	290	.055	L.F.	14.10	3.03		17.13	20
5156	6" diameter	G		250	.064		13.70	3.51		17.21	20.50
5160	7" diameter	G		210	.076		18	4.18		22.18	26
5164	8" diameter	G		170	.094		17.70	5.15		22.85	27.50
5168	9" diameter	G		150	.107		21.50	5.85		27.35	33
5172	10" diameter	G		130	.123		21	6.75		27.75	33.50
5176	12" diameter	G		90	.178		25	9.75		34.75	42.50
5180	14" diameter	G		70	.229		30.50	12.55		43.05	52.50
5184	16" diameter	G	↓	50	.320	↓	39.50	17.55		57.05	70
5200	Hospital grade, aluminum shell, UL approved										
5204	5" diameter	G	Q-9	280	.057	L.F.	14.70	3.13		17.83	21
5208	6" diameter	G		240	.067		14.30	3.66		17.96	21.50
5212	7" diameter	G		200	.080		18.75	4.39		23.14	27
5216	8" diameter	G		160	.100		18.40	5.50		23.90	28.50
5220	9" diameter	G		140	.114		22.50	6.25		28.75	34.50
5224	10" diameter	G		130	.123		21.50	6.75		28.25	34
5228	12" diameter	G		80	.200		25.50	10.95		36.45	44.50
5232	14" diameter	G		60	.267		32	14.65		46.65	57
5236	16" diameter	G	↓	40	.400	↓	41	22		63	78.50

23 33 53 – Duct Liners

23 33 53.10 Duct Liner Board			Crew	Daily Output	Labor-Hours	Unit	Material	Labor	Equipment	Total	Total Incl O&P
0010	**DUCT LINER BOARD**										
3340	Board type fiberglass liner, FSK, 1-1/2 lb. density										
3344	1" thick	G	Q-14	150	.107	S.F.	1.04	5.50		6.54	9.65
3345	1-1/2" thick	G		130	.123		1.13	6.35		7.48	11.05
3346	2" thick	G		120	.133		1.31	6.90		8.21	12.10
3348	3" thick	G		110	.145		1.71	7.50		9.21	13.50
3350	4" thick	G		100	.160		2.08	8.25		10.33	15.05
3356	3 lb. density, 1" thick	G		150	.107		1.31	5.50		6.81	9.95
3358	1-1/2" thick	G		130	.123		1.66	6.35		8.01	11.65
3360	2" thick	G		120	.133		2.03	6.90		8.93	12.90
3362	2-1/2" thick	G		110	.145		2.39	7.50		9.89	14.25
3364	3" thick	G		100	.160		2.74	8.25		10.99	15.75
3366	4" thick	G		90	.178		3.43	9.20		12.63	17.90
3370	6 lb. density, 1" thick	G		140	.114		1.86	5.90		7.76	11.15
3374	1-1/2" thick	G		120	.133		2.49	6.90		9.39	13.40
3378	2" thick	G	↓	100	.160	↓	3.14	8.25		11.39	16.20
3490	Board type, fiberglass liner, 3 lb. density										
3680	No finish										
3700	1" thick	G	Q-14	170	.094	S.F.	.73	4.86		5.59	8.30
3710	1-1/2" thick	G		140	.114		1.10	5.90		7	10.30
3720	2" thick	G	↓	130	.123	↓	1.46	6.35		7.81	11.40
3940	Board type, non-fibrous foam										
3950	Temperature, bacteria and fungi resistant										
3960	1" thick	G	Q-14	150	.107	S.F.	2.47	5.50		7.97	11.20
3970	1-1/2" thick	G		130	.123		4.23	6.35		10.58	14.45
3980	2" thick	G	↓	120	.133	↓	4.12	6.90		11.02	15.20

23 34 HVAC Fans

23 34 13 – Axial HVAC Fans

23 34 13.10 Axial Flow HVAC Fans		Crew	Daily Output	Labor-Hours	Unit	Material	2019 Bare Costs Labor	Equipment	Total	Total Incl O&P
0010	**AXIAL FLOW HVAC FANS**									
0020	Air conditioning and process air handling									
0500	Axial flow, constant speed									
0505	Direct drive, 1/8" S.P.									
0510	12", 1,060 CFM, 1/6 HP	Q-20	3	6.667	Ea.	755	375		1,130	1,400
0514	12", 2,095 CFM, 1/2 HP		3	6.667		725	375		1,100	1,375
0518	16", 2,490 CFM, 1/3 HP		2.80	7.143		730	400		1,130	1,400
0522	20", 4,130 CFM, 3/4 HP		2.60	7.692		1,050	430		1,480	1,800
0526	22", 4,700 CFM, 3/4 HP		2.60	7.692		1,550	430		1,980	2,350
0530	24", 5,850 CFM, 1 HP		2.50	8		1,750	445		2,195	2,600
0534	24", 7,925 CFM, 1-1/2 HP		2.40	8.333		1,750	465		2,215	2,625
0538	30", 10,640 CFM, 2 HP		2.20	9.091		2,575	510		3,085	3,600
0542	30", 14,765 CFM, 2-1/2 HP		2.10	9.524		2,975	530		3,505	4,075
0546	36", 16,780 CFM, 2 HP		2.10	9.524		3,100	530		3,630	4,200
0550	36", 22,920 CFM, 5 HP	↓	1.80	11.111	↓	3,375	620		3,995	4,675
0560	Belt drive, 1/8" S.P.									
0562	15", 2,800 CFM, 1/3 HP	Q-20	3.20	6.250	Ea.	1,300	350		1,650	1,950
0564	15", 3,400 CFM, 1/2 HP		3	6.667		1,325	375		1,700	2,025
0568	18", 3,280 CFM, 1/3 HP		2.80	7.143		1,400	400		1,800	2,125
0572	18", 3,900 CFM, 1/2 HP		2.80	7.143		1,425	400		1,825	2,150
0576	18", 5,250 CFM, 1 HP		2.70	7.407		1,525	415		1,940	2,300
0584	24", 6,430 CFM, 1 HP		2.50	8		2,000	445		2,445	2,900
0588	24", 8,860 CFM, 2 HP		2.40	8.333		2,100	465		2,565	3,025
0592	30", 9,250 CFM, 1 HP		2.20	9.091		2,600	510		3,110	3,625
0596	30", 16,900 CFM, 5 HP		2	10		2,775	560		3,335	3,900
0604	36", 14,475 CFM, 2 HP		2.30	8.696		3,075	485		3,560	4,125
0608	36", 20,080 CFM, 5 HP		1.80	11.111		3,300	620		3,920	4,575
0616	42", 29,000 CFM, 7-1/2 HP		1.40	14.286		4,225	800		5,025	5,850
1500	Vaneaxial, low pressure, 2,000 CFM, 1/2 HP		3.60	5.556		2,450	310		2,760	3,175
1520	4,000 CFM, 1 HP		3.20	6.250		2,575	350		2,925	3,350
1540	8,000 CFM, 2 HP		2.80	7.143		3,650	400		4,050	4,625
1560	16,000 CFM, 5 HP	↓	2.40	8.333	↓	5,100	465		5,565	6,300

23 34 14 – Blower HVAC Fans

23 34 14.10 Blower Type HVAC Fans		Crew	Daily Output	Labor-Hours	Unit	Material	Labor	Equipment	Total	Total Incl O&P
0010	**BLOWER TYPE HVAC FANS**									
2000	Blowers, direct drive with motor, complete									
2020	1,045 CFM @ 0.5" S.P., 1/5 HP	Q-20	18	1.111	Ea.	345	62		407	475
2040	1,385 CFM @ 0.5" S.P., 1/4 HP		18	1.111		360	62		422	490
2060	1,640 CFM @ 0.5" S.P., 1/3 HP		18	1.111		330	62		392	455
2080	1,760 CFM @ 0.5" S.P., 1/2 HP	↓	18	1.111	↓	340	62		402	470
2090	4 speed									
2100	1,164 to 1,739 CFM @ 0.5" S.P., 1/3 HP	Q-20	16	1.250	Ea.	345	70		415	485
2120	1,467 to 2,218 CFM @ 1.0" S.P., 3/4 HP	"	14	1.429	"	395	80		475	555
2500	Ceiling fan, right angle, extra quiet, 0.10" S.P.									
2520	95 CFM	Q-20	20	1	Ea.	297	56		353	410
2540	210 CFM		19	1.053		350	59		409	475
2560	385 CFM		18	1.111		445	62		507	585
2580	885 CFM		16	1.250		880	70		950	1,075
2600	1,650 CFM		13	1.538		1,225	86		1,311	1,475
2620	2,960 CFM	↓	11	1.818		1,625	102		1,727	1,925
2640	For wall or roof cap, add	1 Shee	16	.500		297	30.50		327.50	370
2660	For straight thru fan, add				↓	10%				

23 34 14 – Blower HVAC Fans

23 34 14.10 Blower Type HVAC Fans		Crew	Daily Output	Labor-Hours	Unit	2019 Bare Costs Material	Labor	Equipment	Total	Total Incl O&P
2680	For speed control switch, add	1 Elec	16	.500	Ea.	162	30		192	223
7500	Utility set, steel construction, pedestal, 1/4" S.P.									
7520	Direct drive, 150 CFM, 1/8 HP	Q-20	6.40	3.125	Ea.	940	175		1,115	1,300
7540	485 CFM, 1/6 HP		5.80	3.448		1,075	193		1,268	1,475
7560	1,950 CFM, 1/2 HP		4.80	4.167		1,400	233		1,633	1,875
7580	2,410 CFM, 3/4 HP		4.40	4.545		2,575	254		2,829	3,200
7600	3,328 CFM, 1-1/2 HP		3	6.667		2,575	375		2,950	3,400
7680	V-belt drive, drive cover, 3 phase									
7700	800 CFM, 1/4 HP	Q-20	6	3.333	Ea.	1,025	186		1,211	1,400
7720	1,300 CFM, 1/3 HP		5	4		1,075	224		1,299	1,525
7740	2,000 CFM, 1 HP		4.60	4.348		1,275	243		1,518	1,775
7760	2,900 CFM, 3/4 HP		4.20	4.762		1,700	266		1,966	2,275
7780	3,600 CFM, 3/4 HP		4	5		2,100	279		2,379	2,725
7800	4,800 CFM, 1 HP		3.50	5.714		2,475	320		2,795	3,200
7820	6,700 CFM, 1-1/2 HP		3	6.667		3,050	375		3,425	3,950
7830	7,500 CFM, 2 HP		2.50	8		4,150	445		4,595	5,250
7840	11,000 CFM, 3 HP		2	10		5,550	560		6,110	6,950
7860	13,000 CFM, 3 HP		1.60	12.500		5,625	700		6,325	7,250
7880	15,000 CFM, 5 HP		1	20		5,825	1,125		6,950	8,125
7900	17,000 CFM, 7-1/2 HP		.80	25		6,250	1,400		7,650	9,000
7920	20,000 CFM, 7-1/2 HP		.80	25		7,450	1,400		8,850	10,300

23 34 16 – Centrifugal HVAC Fans

23 34 16.10 Centrifugal Type HVAC Fans		Crew	Daily Output	Labor-Hours	Unit	Material	Labor	Equipment	Total	Total Incl O&P
0010	**CENTRIFUGAL TYPE HVAC FANS**									
0200	In-line centrifugal, supply/exhaust booster									
0220	aluminum wheel/hub, disconnect switch, 1/4" S.P.									
0240	500 CFM, 10" diameter connection	Q-20	3	6.667	Ea.	1,625	375		2,000	2,350
0260	1,380 CFM, 12" diameter connection		2	10		1,675	560		2,235	2,700
0280	1,520 CFM, 16" diameter connection		2	10		1,700	560		2,260	2,700
0300	2,560 CFM, 18" diameter connection		1	20		1,850	1,125		2,975	3,725
0320	3,480 CFM, 20" diameter connection		.80	25		2,150	1,400		3,550	4,500
0326	5,080 CFM, 20" diameter connection		.75	26.667		2,200	1,500		3,700	4,675
0340	7,500 CFM, 22" diameter connection		.70	28.571		2,600	1,600		4,200	5,300
0350	10,000 CFM, 27" diameter connection		.65	30.769		3,100	1,725		4,825	6,000
3500	Centrifugal, airfoil, motor and drive, complete									
3520	1,000 CFM, 1/2 HP	Q-20	2.50	8	Ea.	2,150	445		2,595	3,025
3540	2,000 CFM, 1 HP		2	10		2,325	560		2,885	3,400
3560	4,000 CFM, 3 HP		1.80	11.111		2,975	620		3,595	4,225
3580	8,000 CFM, 7-1/2 HP		1.40	14.286		4,300	800		5,100	5,925
3600	12,000 CFM, 10 HP		1	20		5,925	1,125		7,050	8,225
4000	Single width, belt drive, not incl. motor, capacities									
4020	at 2,000 fpm, 2.5" S.P. for indicated motor									
4040	6,900 CFM, 5 HP	Q-9	2.40	6.667	Ea.	4,100	365		4,465	5,075
4060	10,340 CFM, 7-1/2 HP		2.20	7.273		5,700	400		6,100	6,875
4080	15,320 CFM, 10 HP		2	8		6,350	440		6,790	7,675
4100	22,780 CFM, 15 HP		1.80	8.889		10,000	490		10,490	11,700
4120	33,840 CFM, 20 HP		1.60	10		13,300	550		13,850	15,400
4140	41,400 CFM, 25 HP		1.40	11.429		15,800	625		16,425	18,400
4160	50,100 CFM, 30 HP		.80	20		21,700	1,100		22,800	25,600
4200	Double width wheel, 12,420 CFM, 7-1/2 HP		2.20	7.273		5,550	400		5,950	6,700
4220	18,620 CFM, 15 HP		2	8		8,400	440		8,840	9,900
4240	27,580 CFM, 20 HP		1.80	8.889		9,700	490		10,190	11,400

23 34 16 – Centrifugal HVAC Fans

23 34 16.10 Centrifugal Type HVAC Fans		Crew	Daily Output	Labor-Hours	Unit	2019 Bare Costs Material	Labor	Equipment	Total	Total Incl O&P
4260	40,980 CFM, 25 HP	Q-9	1.50	10.667	Ea.	15,200	585		15,785	17,600
4280	60,920 CFM, 40 HP		1	16		21,600	880		22,480	25,100
4300	74,520 CFM, 50 HP		.80	20		24,300	1,100		25,400	28,400
4320	90,160 CFM, 50 HP		.70	22.857		35,100	1,250		36,350	40,500
4340	110,300 CFM, 60 HP		.50	32		47,500	1,750		49,250	55,000
4360	134,960 CFM, 75 HP		.40	40		63,000	2,200		65,200	73,000
5000	Utility set, centrifugal, V belt drive, motor									
5020	1/4" S.P., 1,200 CFM, 1/4 HP	Q-20	6	3.333	Ea.	1,900	186		2,086	2,375
5040	1,520 CFM, 1/3 HP		5	4		2,475	224		2,699	3,050
5060	1,850 CFM, 1/2 HP		4	5		2,450	279		2,729	3,125
5080	2,180 CFM, 3/4 HP		3	6.667		2,850	375		3,225	3,700
5100	1/2" S.P., 3,600 CFM, 1 HP		2	10		3,050	560		3,610	4,225
5120	4,250 CFM, 1-1/2 HP		1.60	12.500		3,650	700		4,350	5,050
5140	4,800 CFM, 2 HP		1.40	14.286		4,475	800		5,275	6,125
5160	6,920 CFM, 5 HP		1.30	15.385		5,400	860		6,260	7,225
5180	7,700 CFM, 7-1/2 HP		1.20	16.667		6,550	930		7,480	8,600
5200	For explosion proof motor, add					15%				
5300	Fume exhauster without hose and intake nozzle									
5310	630 CFM, 1-1/2 HP	Q-20	2	10	Ea.	905	560		1,465	1,850
5320	Hose extension kit 5'	"	20	1	"	36.50	56		92.50	125
5500	Fans, industrial exhauster, for air which may contain granular matl.									
5520	1,000 CFM, 1-1/2 HP	Q-20	2.50	8	Ea.	3,150	445		3,595	4,150
5540	2,000 CFM, 3 HP		2	10		3,850	560		4,410	5,075
5560	4,000 CFM, 7-1/2 HP		1.80	11.111		5,400	620		6,020	6,875
5580	8,000 CFM, 15 HP		1.40	14.286		6,950	800		7,750	8,850
5600	12,000 CFM, 30 HP		1	20		10,700	1,125		11,825	13,500
7000	Roof exhauster, centrifugal, aluminum housing, 12" galvanized									
7020	curb, bird screen, back draft damper, 1/4" S.P.									
7100	Direct drive, 320 CFM, 11" sq. damper	Q-20	7	2.857	Ea.	805	160		965	1,125
7120	600 CFM, 11" sq. damper		6	3.333		950	186		1,136	1,325
7140	815 CFM, 13" sq. damper		5	4		1,025	224		1,249	1,475
7160	1,450 CFM, 13" sq. damper		4.20	4.762		1,600	266		1,866	2,150
7180	2,050 CFM, 16" sq. damper		4	5		1,825	279		2,104	2,425
7200	V-belt drive, 1,650 CFM, 12" sq. damper		6	3.333		1,475	186		1,661	1,900
7220	2,750 CFM, 21" sq. damper		5	4		1,750	224		1,974	2,275
7230	3,500 CFM, 21" sq. damper		4.50	4.444		1,975	248		2,223	2,550
7240	4,910 CFM, 23" sq. damper		4	5		2,400	279		2,679	3,075
7260	8,525 CFM, 28" sq. damper		3	6.667		3,200	375		3,575	4,100
7280	13,760 CFM, 35" sq. damper		2	10		4,450	560		5,010	5,750
7300	20,558 CFM, 43" sq. damper		1	20		8,825	1,125		9,950	11,400
7320	For 2 speed winding, add					15%				
7340	For explosion proof motor, add					765			765	840
7360	For belt driven, top discharge, add					15%				
7400	Roof mounted kitchen exhaust, aluminum, centrifugal									
7410	Direct drive, 1/4" S.P., 2 speed, temp to 200°F									
7414	1/3 HP, 2,038 CFM	Q-20	7	2.857	Ea.	610	160		770	915
7416	1/2 HP, 2,393 CFM		6	3.333		595	186		781	935
7418	3/4 HP, 4,329 CFM		5	4		920	224		1,144	1,350
7424	Belt drive, 1/4" S.P., temp to 250°F									
7426	3/4 HP, 2,757 CFM	Q-20	5	4	Ea.	940	224		1,164	1,375
7428	1-1/2 HP, 4,913 CFM		4	5		1,325	279		1,604	1,875
7430	1-1/2 HP, 5,798 CFM		4	5		1,375	279		1,654	1,950
7450	Upblast, propeller, w/backdraft damper, bird screen & curb									

23 34 16 – Centrifugal HVAC Fans

23 34 16.10 Centrifugal Type HVAC Fans		Crew	Daily Output	Labor-Hours	Unit	Material	2019 Bare Costs Labor	Equipment	Total	Total Incl O&P
7454	30,300 CFM @ 3/8" S.P., 5 HP	Q-20	1.60	12.500	Ea.	4,800	700		5,500	6,350
7458	36,000 CFM @ 1/2" S.P., 15 HP	"	1.60	12.500	"	6,825	700		7,525	8,575
8500	Wall exhausters, centrifugal, auto damper, 1/8" S.P.									
8520	Direct drive, 610 CFM, 1/20 HP	Q-20	14	1.429	Ea.	420	80		500	580
8540	796 CFM, 1/12 HP		13	1.538		1,025	86		1,111	1,250
8560	822 CFM, 1/6 HP		12	1.667		1,050	93		1,143	1,325
8580	1,320 CFM, 1/4 HP		12	1.667		1,400	93		1,493	1,675
8600	1,756 CFM, 1/4 HP		11	1.818		1,250	102		1,352	1,525
8620	1,983 CFM, 1/4 HP		10	2		1,275	112		1,387	1,575
8640	2,900 CFM, 1/2 HP		9	2.222		1,350	124		1,474	1,700
8660	3,307 CFM, 3/4 HP		8	2.500		1,475	140		1,615	1,825
8670	5,940 CFM, 1 HP	↓	7	2.857	↓	1,600	160		1,760	2,000
9500	V-belt drive, 3 phase									
9520	2,800 CFM, 1/4 HP	Q-20	9	2.222	Ea.	1,900	124		2,024	2,275
9540	3,740 CFM, 1/2 HP		8	2.500		1,975	140		2,115	2,375
9560	4,400 CFM, 3/4 HP		7	2.857		2,000	160		2,160	2,450
9580	5,700 CFM, 1-1/2 HP	↓	6	3.333	↓	2,075	186		2,261	2,550

23 34 23 – HVAC Power Ventilators

23 34 23.10 HVAC Power Circulators and Ventilators		Crew	Daily Output	Labor-Hours	Unit	Material	Labor	Equipment	Total	Total Incl O&P
0010	**HVAC POWER CIRCULATORS AND VENTILATORS**									
3000	Paddle blade air circulator, 3 speed switch									
3020	42", 5,000 CFM high, 3,000 CFM low G	1 Elec	2.40	3.333	Ea.	116	200		316	425
3040	52", 6,500 CFM high, 4,000 CFM low G	"	2.20	3.636	"	151	218		369	490
3100	For antique white motor, same cost									
3200	For brass plated motor, same cost									
3300	For light adaptor kit, add G				Ea.	40			40	44
3310	Industrial grade, reversible, 4 blade									
3312	5,500 CFM	Q-20	5	4	Ea.	137	224		361	490
3314	7,600 CFM		5	4		154	224		378	510
3316	21,015 CFM	↓	4	5	↓	172	279		451	615
4000	High volume, low speed (HVLS), paddle blade air circulator									
4010	Variable speed, reversible, 1 HP motor, motor control panel,									
4020	motor drive cable, control cable with remote, and safety cable.									
4140	8' diameter	Q-5	1.50	10.667	Ea.	4,825	615		5,440	6,225
4150	10' diameter		1.45	11.034		4,875	635		5,510	6,300
4160	12' diameter		1.40	11.429		5,075	660		5,735	6,575
4170	14' diameter		1.35	11.852		5,175	680		5,855	6,700
4180	16' diameter		1.30	12.308		5,200	710		5,910	6,800
4190	18' diameter		1.25	12.800		5,250	735		5,985	6,875
4200	20' diameter		1.20	13.333		5,600	765		6,365	7,300
4210	24' diameter	↓	1	16	↓	5,725	920		6,645	7,675
6000	Propeller exhaust, wall shutter									
6020	Direct drive, one speed, 0.075" S.P.									
6100	653 CFM, 1/30 HP	Q-20	10	2	Ea.	209	112		321	400
6120	1,033 CFM, 1/20 HP		9	2.222		310	124		434	530
6140	1,323 CFM, 1/15 HP		8	2.500		350	140		490	590
6160	2,444 CFM, 1/4 HP	↓	7	2.857	↓	435	160		595	715
6300	V-belt drive, 3 phase									
6320	6,175 CFM, 3/4 HP	Q-20	5	4	Ea.	1,075	224		1,299	1,550
6340	7,500 CFM, 3/4 HP		5	4		1,150	224		1,374	1,625
6360	10,100 CFM, 1 HP		4.50	4.444		1,375	248		1,623	1,875
6380	14,300 CFM, 1-1/2 HP	↓	4	5	↓	1,625	279		1,904	2,225

23 34 23.10 HVAC Power Circulators and Ventilators		Crew	Daily Output	Labor-Hours	Unit	Material	2019 Bare Costs Labor	Equipment	Total	Total Incl O&P
6400	19,800 CFM, 2 HP	Q-20	3	6.667	Ea.	2,175	375		2,550	2,950
6420	26,250 CFM, 3 HP		2.60	7.692		2,300	430		2,730	3,175
6440	38,500 CFM, 5 HP		2.20	9.091		2,825	510		3,335	3,900
6460	46,000 CFM, 7-1/2 HP		2	10		2,900	560		3,460	4,050
6480	51,500 CFM, 10 HP		1.80	11.111		3,000	620		3,620	4,250
6490	V-belt drive, 115 V, residential, whole house									
6500	Ceiling-wall, 5,200 CFM, 1/4 HP, 30" x 30"	1 Shee	6	1.333	Ea.	575	81.50		656.50	760
6510	7,500 CFM, 1/3 HP, 36" x 36"		5	1.600		605	97.50		702.50	815
6520	10,500 CFM, 1/3 HP, 42" x 42"		5	1.600		630	97.50		727.50	845
6530	13,200 CFM, 1/3 HP, 48" x 48"		4	2		695	122		817	950
6540	15,445 CFM, 1/2 HP, 48" x 48"		4	2		725	122		847	980
6550	17,025 CFM, 1/2 HP, 54" x 54"		4	2		1,400	122		1,522	1,700
6560	For two speed motor, add					20%				
6570	Shutter, automatic, ceiling/wall									
6580	30" x 30"	1 Shee	8	1	Ea.	188	61		249	299
6590	36" x 36"		8	1		214	61		275	330
6600	42" x 42"		8	1		268	61		329	390
6610	48" x 48"		7	1.143		291	69.50		360.50	425
6620	54" x 54"		6	1.333		385	81.50		466.50	545
6630	Timer, shut off, to 12 hour		20	.400		61.50	24.50		86	105
6650	Residential, bath exhaust, grille, back draft damper									
6660	50 CFM	Q-20	24	.833	Ea.	53	46.50		99.50	129
6670	110 CFM		22	.909		106	51		157	193
6672	180 CFM		22	.909		141	51		192	232
6673	210 CFM		22	.909		147	51		198	238
6674	260 CFM		22	.909		229	51		280	330
6675	300 CFM		22	.909		153	51		204	246
6680	Light combination, squirrel cage, 100 watt, 70 CFM		24	.833		117	46.50		163.50	199
6700	Light/heater combination, ceiling mounted									
6710	70 CFM, 1,450 watt	Q-20	24	.833	Ea.	144	46.50		190.50	229
6800	Heater combination, recessed, 70 CFM		24	.833		75.50	46.50		122	154
6820	With 2 infrared bulbs		23	.870		96	48.50		144.50	180
6840	Wall mount, 170 CFM		22	.909		273	51		324	375
6846	Ceiling mount, 180 CFM		22	.909		275	51		326	380
6900	Kitchen exhaust, grille, complete, 160 CFM		22	.909		107	51		158	195
6910	180 CFM		20	1		97	56		153	192
6920	270 CFM		18	1.111		235	62		297	350
6930	350 CFM		16	1.250		150	70		220	271
6940	Residential roof jacks and wall caps									
6944	Wall cap with back draft damper									
6946	3" & 4" diam. round duct	1 Shee	11	.727	Ea.	25	44.50		69.50	95
6948	6" diam. round duct	"	11	.727	"	77	44.50		121.50	153
6958	Roof jack with bird screen and back draft damper									
6960	3" & 4" diam. round duct	1 Shee	11	.727	Ea.	20.50	44.50		65	90
6962	3-1/4" x 10" rectangular duct	"	10	.800	"	37	49		86	115
6980	Transition									
6982	3-1/4" x 10" to 6" diam. round	1 Shee	20	.400	Ea.	35	24.50		59.50	75.50
8020	Attic, roof type									
8030	Aluminum dome, damper & curb									
8080	12" diameter, 1,000 CFM (gravity)	1 Elec	10	.800	Ea.	610	48		658	740
8090	16" diameter, 1,500 CFM (gravity)		9	.889		735	53.50		788.50	890
8100	20" diameter, 2,500 CFM (gravity)		8	1		900	60		960	1,075
8110	26" diameter, 4,000 CFM (gravity)		7	1.143		1,100	68.50		1,168.50	1,300

23 34 23 – HVAC Power Ventilators

23 34 23.10 HVAC Power Circulators and Ventilators		Crew	Daily Output	Labor-Hours	Unit	Material	Labor	Equipment	Total	Total Incl O&P
						2019 Bare Costs				
8120	32" diameter, 6,500 CFM (gravity)	1 Elec	6	1.333	Ea.	1,500	80		1,580	1,775
8130	38" diameter, 8,000 CFM (gravity)		5	1.600		2,225	96		2,321	2,600
8140	50" diameter, 13,000 CFM (gravity)		4	2		3,225	120		3,345	3,725
8160	Plastic, ABS dome									
8180	1,050 CFM	1 Elec	14	.571	Ea.	179	34.50		213.50	248
8200	1,600 CFM	"	12	.667	"	268	40		308	355
8240	Attic, wall type, with shutter, one speed									
8250	12" diameter, 1,000 CFM	1 Elec	14	.571	Ea.	415	34.50		449.50	505
8260	14" diameter, 1,500 CFM		12	.667		450	40		490	550
8270	16" diameter, 2,000 CFM		9	.889		505	53.50		558.50	635
8290	Whole house, wall type, with shutter, one speed									
8300	30" diameter, 4,800 CFM	1 Elec	7	1.143	Ea.	1,075	68.50		1,143.50	1,300
8310	36" diameter, 7,000 CFM		6	1.333		1,175	80		1,255	1,425
8320	42" diameter, 10,000 CFM		5	1.600		1,325	96		1,421	1,600
8330	48" diameter, 16,000 CFM		4	2		1,650	120		1,770	1,975
8340	For two speed, add					99			99	109
8350	Whole house, lay-down type, with shutter, one speed									
8360	30" diameter, 4,500 CFM	1 Elec	8	1	Ea.	1,150	60		1,210	1,375
8370	36" diameter, 6,500 CFM		7	1.143		1,250	68.50		1,318.50	1,475
8380	42" diameter, 9,000 CFM		6	1.333		1,375	80		1,455	1,625
8390	48" diameter, 12,000 CFM		5	1.600		1,550	96		1,646	1,850
8440	For two speed, add					74.50			74.50	82
8450	For 12 hour timer switch, add	1 Elec	32	.250		74.50	15		89.50	105

23 34 33 – Air Curtains

23 34 33.10 Air Barrier Curtains		Crew	Daily Output	Labor-Hours	Unit	Material	Labor	Equipment	Total	Total Incl O&P
0010	**AIR BARRIER CURTAINS**, Incl. motor starters, transformers,									
0050	and door switches									
2450	Conveyor openings or service windows									
3000	Service window, 5' high x 25" wide	2 Shee	5	3.200	Ea.	355	195		550	690
3100	Environmental separation									
3110	Door heights up to 8', low profile, super quiet									
3120	Unheated, variable speed									
3130	36" wide	2 Shee	4	4	Ea.	735	244		979	1,175
3134	42" wide		3.80	4.211		755	257		1,012	1,225
3138	48" wide		3.60	4.444		780	271		1,051	1,275
3142	60" wide		3.40	4.706		830	287		1,117	1,350
3146	72" wide	Q-3	4.60	6.957		1,000	420		1,420	1,725
3150	96" wide		4.40	7.273		1,700	435		2,135	2,525
3154	120" wide		4.20	7.619		1,475	460		1,935	2,300
3158	144" wide		4	8		1,925	480		2,405	2,850
3200	Door heights up to 10'									
3210	Unheated									
3230	36" wide	2 Shee	3.80	4.211	Ea.	715	257		972	1,175
3234	42" wide		3.60	4.444		710	271		981	1,200
3238	48" wide		3.40	4.706		725	287		1,012	1,225
3242	60" wide		3.20	5		1,100	305		1,405	1,675
3246	72" wide	Q-3	4.40	7.273		1,250	435		1,685	2,050
3250	96" wide		4.20	7.619		1,475	460		1,935	2,300
3254	120" wide		4	8		2,000	480		2,480	2,925
3258	144" wide		3.80	8.421		2,175	505		2,680	3,125
3300	Door heights up to 12'									
3310	Unheated									

23 34 HVAC Fans

23 34 33 – Air Curtains

23 34 33.10 Air Barrier Curtains		Crew	Daily Output	Labor-Hours	Unit	Material	Labor	Equipment	Total	Total Incl O&P
						2019 Bare Costs				
3334	42″ wide	2 Shee	3.40	4.706	Ea.	1,100	287		1,387	1,625
3338	48″ wide		3.20	5		1,100	305		1,405	1,675
3342	60″ wide		3	5.333		1,125	325		1,450	1,725
3346	72″ wide	Q-3	4.20	7.619		1,875	460		2,335	2,725
3350	96″ wide		4	8		2,175	480		2,655	3,125
3354	120″ wide		3.80	8.421		2,350	505		2,855	3,350
3358	144″ wide		3.60	8.889		3,125	535		3,660	4,250
3400	Door heights up to 16′									
3410	Unheated									
3438	48″ wide	2 Shee	3	5.333	Ea.	1,475	325		1,800	2,125
3442	60″ wide	″	2.80	5.714		1,575	350		1,925	2,250
3446	72″ wide	Q-3	3.80	8.421		2,525	505		3,030	3,525
3450	96″ wide		3.60	8.889		2,800	535		3,335	3,875
3454	120″ wide		3.40	9.412		3,400	565		3,965	4,575
3458	144″ wide		3.20	10		3,500	600		4,100	4,775
3470	Heated, electric									
3474	48″ wide	2 Shee	2.90	5.517	Ea.	2,375	335		2,710	3,100
3478	60″ wide	″	2.70	5.926		2,425	360		2,785	3,225
3482	72″ wide	Q-3	3.70	8.649		4,125	520		4,645	5,325
3486	96″ wide		3.50	9.143		4,350	550		4,900	5,625
3490	120″ wide		3.30	9.697		5,825	585		6,410	7,300
3494	144″ wide		3.10	10.323		5,825	620		6,445	7,325

23 35 Special Exhaust Systems

23 35 16 – Engine Exhaust Systems

23 35 16.10 Engine Exhaust Removal Systems

		Crew	Daily Output	Labor-Hours	Unit	Material	Labor	Equipment	Total	Total Incl O&P
0010	**ENGINE EXHAUST REMOVAL SYSTEMS** D3090–320									
0500	Engine exhaust, garage, in-floor system									
0510	Single tube outlet assemblies									
0520	For transite pipe ducting, self-storing tube									
0530	3″ tubing adapter plate	1 Shee	16	.500	Ea.	221	30.50		251.50	290
0540	4″ tubing adapter plate		16	.500		224	30.50		254.50	294
0550	5″ tubing adapter plate		16	.500		225	30.50		255.50	294
0600	For vitrified tile ducting									
0610	3″ tubing adapter plate, self-storing tube	1 Shee	16	.500	Ea.	221	30.50		251.50	290
0620	4″ tubing adapter plate, self-storing tube		16	.500		224	30.50		254.50	294
0660	5″ tubing adapter plate, self-storing tube		16	.500		224	30.50		254.50	294
0800	Two tube outlet assemblies									
0810	For transite pipe ducting, self-storing tube									
0820	3″ tubing, dual exhaust adapter plate	1 Shee	16	.500	Ea.	221	30.50		251.50	290
0850	For vitrified tile ducting									
0860	3″ tubing, dual exhaust, self-storing tube	1 Shee	16	.500	Ea.	281	30.50		311.50	355
0870	3″ tubing, double outlet, non-storing tubes	″	16	.500	″	281	30.50		311.50	355
0900	Accessories for metal tubing (overhead systems also)									
0910	Adapters, for metal tubing end									
0920	3″ tail pipe type				Ea.	44.50			44.50	49
0930	4″ tail pipe type					48.50			48.50	53.50
0940	5″ tail pipe type					49.50			49.50	54.50
0990	5″ diesel stack type					284			284	310
1000	6″ diesel stack type					293			293	325
1100	Bullnose (guide) required for in-floor assemblies									

23 35 16.10 Engine Exhaust Removal Systems		Crew	Daily Output	Labor-Hours	Unit	Material	2019 Bare Costs Labor	Equipment	Total	Total Incl O&P
1110	3″ tubing size				Ea.	25.50			25.50	28
1120	4″ tubing size					27			27	30
1130	5″ tubing size					29			29	32
1150	Plain rings, for tubing end									
1160	3″ tubing size				Ea.	22.50			22.50	24.50
1170	4″ tubing size				"	37			37	40.50
1200	Tubing, galvanized, flexible (for overhead systems also)									
1210	3″ ID				L.F.	10.55			10.55	11.60
1220	4″ ID					12.95			12.95	14.25
1230	5″ ID					15.25			15.25	16.80
1240	6″ ID					17.65			17.65	19.40
1250	Stainless steel, flexible (for overhead system also)									
1260	3″ ID				L.F.	24			24	26.50
1270	4″ ID					32.50			32.50	36
1280	5″ ID					37			37	40.50
1290	6″ ID					43			43	47
1500	Engine exhaust, garage, overhead components, for neoprene tubing									
1510	Alternate metal tubing & accessories see above									
1550	Adapters, for neoprene tubing end									
1560	3″ tail pipe, adjustable, neoprene				Ea.	51			51	56
1570	3″ tail pipe, heavy wall neoprene					56.50			56.50	62
1580	4″ tail pipe, heavy wall neoprene					94			94	103
1590	5″ tail pipe, heavy wall neoprene					97			97	107
1650	Connectors, tubing									
1660	3″ interior, aluminum				Ea.	22			22	24.50
1670	4″ interior, aluminum					37			37	40.50
1710	5″ interior, neoprene					55.50			55.50	61
1750	3″ spiralock, neoprene					22.50			22.50	24.50
1760	4″ spiralock, neoprene					37			37	40.50
1780	Y for 3″ ID tubing, neoprene, dual exhaust					180			180	198
1790	Y for 4″ ID tubing, aluminum, dual exhaust					172			172	189
1850	Elbows, aluminum, splice into tubing for strap									
1860	3″ neoprene tubing size				Ea.	45			45	49.50
1870	4″ neoprene tubing size					36.50			36.50	40
1900	Flange assemblies, connect tubing to overhead duct					66.50			66.50	73
2000	Hardware and accessories									
2020	Cable, galvanized, 1/8″ diameter				L.F.	.45			.45	.50
2040	Cleat, tie down cable or rope				Ea.	5.05			5.05	5.55
2060	Pulley					6.95			6.95	7.65
2080	Pulley hook, universal					5.05			5.05	5.55
2100	Rope, nylon, 1/4″ diameter				L.F.	.38			.38	.42
2120	Winch, 1″ diameter				Ea.	102			102	112
2150	Lifting strap, mounts on neoprene									
2160	3″ tubing size				Ea.	25			25	27.50
2170	4″ tubing size					25			25	27.50
2180	5″ tubing size					25			25	27.50
2190	6″ tubing size					25			25	27.50
2200	Tubing, neoprene, 11′ lengths									
2210	3″ ID				L.F.	10.30			10.30	11.35
2220	4″ ID					16.50			16.50	18.15
2230	5″ ID					24.50			24.50	27
2500	Engine exhaust, thru-door outlet									
2510	3″ tube size	1 Carp	16	.500	Ea.	54.50	26		80.50	99

23 35 Special Exhaust Systems

23 35 16 – Engine Exhaust Systems

23 35 16.10 Engine Exhaust Removal Systems		Crew	Daily Output	Labor-Hours	Unit	Material	2019 Bare Costs Labor	Equipment	Total	Total Incl O&P
2530	4" tube size	1 Carp	16	.500	Ea.	65.50	26		91.50	112
3000	Tubing, exhaust, flex hose, with									
3010	coupler, damper and tail pipe adapter									
3020	Neoprene									
3040	3" x 20'	1 Shee	6	1.333	Ea.	310	81.50		391.50	465
3050	4" x 15'		5.40	1.481		400	90.50		490.50	580
3060	4" x 20'		5	1.600		485	97.50		582.50	685
3070	5" x 15'		4.40	1.818		545	111		656	770
3100	Galvanized									
3110	3" x 20'	1 Shee	6	1.333	Ea.	300	81.50		381.50	455
3120	4" x 17'		5.60	1.429		330	87		417	495
3130	4" x 20'		5	1.600		370	97.50		467.50	555
3140	5" x 17'		4.60	1.739		395	106		501	590
8000	Blower, for tailpipe exhaust system									
8010	Direct drive									
8012	495 CFM, 1/3 HP	Q-9	5	3.200	Ea.	2,400	176		2,576	2,900
8014	1,445 CFM, 3/4 HP		4	4		2,975	219		3,194	3,600
8016	1,840 CFM, 1-1/2 HP		3	5.333		3,575	293		3,868	4,375
8030	Beltdrive									
8032	1,400 CFM, 1 HP	Q-9	4	4	Ea.	2,975	219		3,194	3,600
8034	2,023 CFM, 1-1/2 HP		3	5.333		4,125	293		4,418	5,000
8036	2,750 CFM, 2 HP		2	8		4,875	440		5,315	6,050
8038	4,400 CFM, 3 HP		1.80	8.889		7,200	490		7,690	8,675
8040	7,060 CFM, 5 HP		1.40	11.429		10,200	625		10,825	12,200

23 35 43 – Welding Fume Elimination Systems

23 35 43.10 Welding Fume Elimination System Components

Line	Description	Crew	Daily Output	Labor-Hours	Unit	Material	Labor	Equipment	Total	Total Incl O&P
0010	**WELDING FUME ELIMINATION SYSTEM COMPONENTS**									
7500	Welding fume elimination accessories for garage exhaust systems									
7600	Cut off (blast gate)									
7610	3" tubing size, 3" x 6" opening	1 Shee	24	.333	Ea.	24	20.50		44.50	57
7620	4" tubing size, 4" x 8" opening		24	.333		24	20.50		44.50	57.50
7630	5" tubing size, 5" x 10" opening		24	.333		28	20.50		48.50	62
7640	6" tubing size		24	.333		29	20.50		49.50	63
7650	8" tubing size		24	.333		40	20.50		60.50	75
7700	Hoods, magnetic, with handle & screen									
7710	3" tubing size, 3" x 6" opening	1 Shee	24	.333	Ea.	92.50	20.50		113	133
7720	4" tubing size, 4" x 8" opening		24	.333		86.50	20.50		107	127
7730	5" tubing size, 5" x 10" opening		24	.333		86.50	20.50		107	127

23 36 Air Terminal Units

23 36 13 – Constant-Air-Volume Units

23 36 13.10 Constant Volume Mixing Boxes

Line	Description	Crew	Daily Output	Labor-Hours	Unit	Material	Labor	Equipment	Total	Total Incl O&P
0010	**CONSTANT VOLUME MIXING BOXES**									
5180	Mixing box, includes electric or pneumatic motor									
5192	Recommend use with silencer, see Line 23 33 19.10 0010									
5200	Constant volume, 150 to 270 CFM	Q-9	12	1.333	Ea.	1,050	73		1,123	1,275
5210	270 to 600 CFM		11	1.455		1,100	80		1,180	1,325
5230	550 to 1,000 CFM		9	1.778		1,100	97.50		1,197.50	1,350
5240	1,000 to 1,600 CFM		8	2		1,125	110		1,235	1,400
5250	1,300 to 1,900 CFM		6	2.667		1,150	146		1,296	1,475

23 36 Air Terminal Units

23 36 13 – Constant-Air-Volume Units

23 36 13.10	Constant Volume Mixing Boxes	Crew	Daily Output	Labor-Hours	Unit	Material	2019 Bare Costs Labor	Equipment	Total	Total Incl O&P
5260	550 to 2,640 CFM	Q-9	5.60	2.857	Ea.	1,250	157		1,407	1,625
5270	650 to 3,120 CFM		5.20	3.077		1,350	169		1,519	1,725

23 36 16 – Variable-Air-Volume Units

23 36 16.10	Variable Volume Mixing Boxes	Crew	Daily Output	Labor-Hours	Unit	Material	Labor	Equipment	Total	Total Incl O&P
0010	**VARIABLE VOLUME MIXING BOXES**									
5180	Mixing box, includes electric or pneumatic motor									
5192	Recommend use with attenuator, see Line 23 33 19.10 0010									
5500	VAV cool only, pneumatic, pressure independent 300 to 600 CFM	Q-9	11	1.455	Ea.	810	80		890	1,025
5510	500 to 1,000 CFM		9	1.778		835	97.50		932.50	1,075
5520	800 to 1,600 CFM		9	1.778		860	97.50		957.50	1,100
5530	1,100 to 2,000 CFM		8	2		880	110		990	1,125
5540	1,500 to 3,000 CFM		7	2.286		940	125		1,065	1,225
5550	2,000 to 4,000 CFM		6	2.667		950	146		1,096	1,275
5560	For electric, w/thermostat, pressure dependent, add					23			23	25.50
5600	VAV, HW coils, damper, actuator and thermostat									
5610	200 CFM	Q-9	11	1.455	Ea.	1,200	80		1,280	1,450
5620	400 CFM		10	1.600		1,200	88		1,288	1,450
5630	600 CFM		10	1.600		1,200	88		1,288	1,450
5640	800 CFM		8	2		1,250	110		1,360	1,550
5650	1,000 CFM		8	2		1,250	110		1,360	1,550
5660	1,250 CFM		6	2.667		1,350	146		1,496	1,725
5670	1,500 CFM		6	2.667		1,350	146		1,496	1,725
5680	2,000 CFM		4	4		1,475	219		1,694	1,950
5684	3,000 CFM		3.80	4.211		1,600	231		1,831	2,125
5700	VAV cool only, fan powered, damper, actuator, thermostat									
5710	200 CFM	Q-9	10	1.600	Ea.	1,650	88		1,738	1,925
5720	400 CFM		9	1.778		1,800	97.50		1,897.50	2,125
5730	600 CFM		9	1.778		1,800	97.50		1,897.50	2,125
5740	800 CFM		7	2.286		1,925	125		2,050	2,300
5750	1,000 CFM		7	2.286		1,925	125		2,050	2,300
5760	1,250 CFM		5	3.200		2,075	176		2,251	2,550
5770	1,500 CFM		5	3.200		2,075	176		2,251	2,550
5780	2,000 CFM		4	4		2,275	219		2,494	2,825
5800	VAV fan powr'd, with HW coils, dampers, actuators, thermostat									
5810	200 CFM	Q-9	10	1.600	Ea.	2,075	88		2,163	2,400
5820	400 CFM		9	1.778		2,150	97.50		2,247.50	2,500
5830	600 CFM		9	1.778		2,150	97.50		2,247.50	2,500
5840	800 CFM		7	2.286		2,275	125		2,400	2,700
5850	1,000 CFM		7	2.286		2,275	125		2,400	2,700
5860	1,250 CFM		5	3.200		2,575	176		2,751	3,125
5870	1,500 CFM		5	3.200		2,575	176		2,751	3,125
5880	2,000 CFM		3.50	4.571		2,775	251		3,026	3,425

23 37 Air Outlets and Inlets

23 37 13 – Diffusers, Registers, and Grilles

23 37 13.10 Diffusers		Crew	Daily Output	Labor-Hours	Unit	Material	2019 Bare Costs Labor	Equipment	Total	Total Incl O&P
0010	**DIFFUSERS**, Aluminum, opposed blade damper unless noted									
0100	Ceiling, linear, also for sidewall									
0120	2″ wide R233100-30	1 Shee	32	.250	L.F.	18.85	15.25		34.10	43.50
0140	3″ wide		30	.267		21.50	16.25		37.75	48
0160	4″ wide R233700-60		26	.308		25.50	18.75		44.25	56.50
0180	6″ wide		24	.333		31	20.50		51.50	65.50
0200	8″ wide		22	.364		36.50	22		58.50	73.50
0220	10″ wide		20	.400		41.50	24.50		66	82.50
0240	12″ wide		18	.444		47.50	27		74.50	93.50
0260	For floor or sill application, add					15%				
0500	Perforated, 24″ x 24″ lay-in panel size, 6″ x 6″	1 Shee	16	.500	Ea.	165	30.50		195.50	228
0520	8″ x 8″		15	.533		173	32.50		205.50	240
0530	9″ x 9″		14	.571		176	35		211	246
0540	10″ x 10″		14	.571		176	35		211	247
0560	12″ x 12″		12	.667		183	40.50		223.50	264
0580	15″ x 15″		11	.727		185	44.50		229.50	272
0590	16″ x 16″		11	.727		205	44.50		249.50	294
0600	18″ x 18″		10	.800		220	49		269	315
0610	20″ x 20″		10	.800		237	49		286	335
0620	24″ x 24″		9	.889		260	54		314	370
1000	Rectangular, 1 to 4 way blow, 6″ x 6″		16	.500		43	30.50		73.50	93.50
1010	8″ x 8″		15	.533		61	32.50		93.50	117
1014	9″ x 9″		15	.533		53.50	32.50		86	108
1016	10″ x 10″		15	.533		83	32.50		115.50	141
1020	12″ x 6″		15	.533		75	32.50		107.50	132
1040	12″ x 9″		14	.571		79.50	35		114.50	141
1060	12″ x 12″		12	.667		74.50	40.50		115	143
1070	14″ x 6″		13	.615		81.50	37.50		119	147
1074	14″ x 14″		12	.667		133	40.50		173.50	209
1080	18″ x 12″		11	.727		144	44.50		188.50	226
1120	15″ x 15″		10	.800		93.50	49		142.50	177
1140	18″ x 15″		9	.889		163	54		217	263
1150	18″ x 18″		9	.889		117	54		171	212
1160	21″ x 21″		8	1		225	61		286	340
1170	24″ x 12″		10	.800		171	49		220	263
1180	24″ x 24″		7	1.143		284	69.50		353.50	415
1500	Round, butterfly damper, steel, diffuser size, 6″ diameter		18	.444		12.55	27		39.55	55
1520	8″ diameter		16	.500		13.25	30.50		43.75	61
1540	10″ diameter		14	.571		17.20	35		52.20	72
1560	12″ diameter		12	.667		22.50	40.50		63	86.50
1580	14″ diameter		10	.800		28.50	49		77.50	105
1600	18″ diameter		9	.889		66.50	54		120.50	156
1700	Round, steel, adjustable core, annular segmented damper, diffuser size									
1704	6″ diameter	1 Shee	18	.444	Ea.	82.50	27		109.50	132
1708	8″ diameter		16	.500		97	30.50		127.50	154
1712	10″ diameter		14	.571		115	35		150	179
1716	12″ diameter		12	.667		140	40.50		180.50	216
1720	14″ diameter		10	.800		172	49		221	263
1724	18″ diameter		9	.889		300	54		354	415
1728	20″ diameter		9	.889		345	54		399	465
1732	24″ diameter		8	1		355	61		416	490
1736	30″ diameter		7	1.143		895	69.50		964.50	1,100
1740	36″ diameter		6	1.333		1,100	81.50		1,181.50	1,325

23 37 13.10 Diffusers		Crew	Daily Output	Labor-Hours	Unit	Material	2019 Bare Costs Labor	Equipment	Total	Total Incl O&P
2000	T-bar mounting, 24" x 24" lay-in frame, 6" x 6"	1 Shee	16	.500	Ea.	72.50	30.50		103	126
2020	8" x 8"		14	.571		72	35		107	133
2040	12" x 12"		12	.667		88.50	40.50		129	159
2060	16" x 16"		11	.727		115	44.50		159.50	194
2080	18" x 18"		10	.800		121	49		170	207
2500	Combination supply and return									
2520	21" x 21" supply, 15" x 15" return	Q-9	10	1.600	Ea.	164	88		252	315
2540	24" x 24" supply, 18" x 18" return		9.50	1.684		219	92.50		311.50	380
2560	27" x 27" supply, 18" x 18" return		9	1.778		243	97.50		340.50	415
2580	30" x 30" supply, 21" x 21" return		8.50	1.882		325	103		428	515
2600	33" x 33" supply, 21" x 21" return		8	2		365	110		475	570
2620	36" x 36" supply, 24" x 24" return		7.50	2.133		415	117		532	640
3000	Baseboard, white enameled steel									
3100	18" long	1 Shee	20	.400	Ea.	7	24.50		31.50	44.50
3120	24" long		18	.444		13.70	27		40.70	56
3140	48" long		16	.500		25	30.50		55.50	74
3400	For matching return, deduct					10%				
4000	Floor, steel, adjustable pattern									
4100	2" x 10"	1 Shee	34	.235	Ea.	9.45	14.35		23.80	32.50
4120	2" x 12"		32	.250		10.35	15.25		25.60	34.50
4140	2" x 14"		30	.267		11.75	16.25		28	37.50
4200	4" x 10"		28	.286		10.95	17.40		28.35	38.50
4220	4" x 12"		26	.308		12.10	18.75		30.85	42
4240	4" x 14"		25	.320		13.65	19.50		33.15	44.50
4260	6" x 10"		26	.308		16.50	18.75		35.25	46.50
4280	6" x 12"		24	.333		17.20	20.50		37.70	50
4300	6" x 14"		22	.364		17.95	22		39.95	53.50
4500	Linear bar diffuser in wall, sill or ceiling									
4510	2" wide	1 Shee	32	.250	L.F.	64	15.25		79.25	93.50
4512	3" wide		30	.267		69	16.25		85.25	101
4514	4" wide		26	.308		85.50	18.75		104.25	123
4516	6" wide		24	.333		106	20.50		126.50	147
4520	10" wide		20	.400		141	24.50		165.50	192
4522	12" wide		18	.444		158	27		185	215
5000	Sidewall, aluminum, 3 way dispersion									
5100	8" x 4"	1 Shee	24	.333	Ea.	5.95	20.50		26.45	37.50
5110	8" x 6"		24	.333		6.35	20.50		26.85	38
5120	10" x 4"		22	.364		6.25	22		28.25	40.50
5130	10" x 6"		22	.364		6.70	22		28.70	41
5140	10" x 8"		20	.400		9.15	24.50		33.65	47
5160	12" x 4"		18	.444		6.75	27		33.75	48.50
5170	12" x 6"		17	.471		7.45	28.50		35.95	51.50
5180	12" x 8"		16	.500		9.65	30.50		40.15	57
5200	14" x 4"		15	.533		7.50	32.50		40	58
5220	14" x 6"		14	.571		8.30	35		43.30	62
5240	14" x 8"		13	.615		10.75	37.50		48.25	69
5260	16" x 6"		12.50	.640		11.80	39		50.80	72
6000	For steel diffusers instead of aluminum, deduct					10%				

23 37 13 – Diffusers, Registers, and Grilles

23 37 13.30 Grilles		Crew	Daily Output	Labor-Hours	Unit	Material	Labor	Equipment	Total	Total Incl O&P
						2019 Bare Costs				
0010	**GRILLES**									
0020	Aluminum, unless noted otherwise									
0100	Air supply, single deflection, adjustable									
0120	8" x 4"	1 Shee	30	.267	Ea.	13.70	16.25		29.95	39.50
0140	8" x 8"		28	.286		16.65	17.40		34.05	45
0160	10" x 4"		24	.333		15.10	20.50		35.60	47.50
0180	10" x 10"		23	.348		18.90	21		39.90	53
0200	12" x 6"		23	.348		10.85	21		31.85	44
0220	12" x 12"		22	.364		20.50	22		42.50	56
0230	14" x 6"		23	.348		13.50	21		34.50	47
0240	14" x 8"		23	.348		14.65	21		35.65	48
0260	14" x 14"		22	.364		29	22		51	65
0270	18" x 8"		23	.348		24.50	21		45.50	59
0280	18" x 10"		23	.348		24.50	21		45.50	59
0300	18" x 18"		21	.381		30	23		53	68.50
0320	20" x 12"		22	.364		19.45	22		41.45	55
0340	20" x 20"		21	.381		36.50	23		59.50	75.50
0350	24" x 8"		20	.400		18.80	24.50		43.30	57.50
0360	24" x 14"		18	.444		29.50	27		56.50	73.50
0380	24" x 24"		15	.533		45.50	32.50		78	100
0400	30" x 8"		20	.400		39	24.50		63.50	80
0420	30" x 10"		19	.421		26.50	25.50		52	68
0440	30" x 12"		18	.444		29	27		56	73
0460	30" x 16"		17	.471		39.50	28.50		68	87
0480	30" x 18"		17	.471		48	28.50		76.50	96
0500	30" x 30"		14	.571		84	35		119	145
0520	36" x 12"		17	.471		41	28.50		69.50	88.50
0540	36" x 16"		16	.500		50	30.50		80.50	102
0560	36" x 18"		15	.533		57.50	32.50		90	113
0570	36" x 20"		15	.533		62	32.50		94.50	118
0580	36" x 24"		14	.571		82	35		117	143
0600	36" x 28"		13	.615		88	37.50		125.50	154
0620	36" x 30"		12	.667		110	40.50		150.50	184
0640	36" x 32"		12	.667		107	40.50		147.50	179
0660	36" x 34"		11	.727		130	44.50		174.50	211
0680	36" x 36"		11	.727		143	44.50		187.50	226
0700	For double deflecting, add					70%				
1000	Air return, steel, 6" x 6"	1 Shee	26	.308		21	18.75		39.75	51.50
1020	10" x 6"		24	.333		21	20.50		41.50	54
1040	14" x 6"		23	.348		24.50	21		45.50	59
1060	10" x 8"		23	.348		24.50	21		45.50	59
1080	16" x 8"		22	.364		28.50	22		50.50	65
1100	12" x 12"		22	.364		28	22		50	64.50
1120	24" x 12"		18	.444		37.50	27		64.50	82.50
1140	30" x 12"		16	.500		49	30.50		79.50	101
1160	14" x 14"		22	.364		33	22		55	70
1180	16" x 16"		22	.364		36.50	22		58.50	73.50
1200	18" x 18"		21	.381		42.50	23		65.50	82.50
1220	24" x 18"		16	.500		48	30.50		78.50	99
1240	36" x 18"		15	.533		72.50	32.50		105	129
1260	24" x 24"		15	.533		56	32.50		88.50	112
1280	36" x 24"		14	.571		82	35		117	143

23 37 13.30 Grilles		Crew	Daily Output	Labor-Hours	Unit	Material	2019 Bare Costs Labor	Equipment	Total	Total Incl O&P
1300	48" x 24"	1 Shee	12	.667	Ea.	134	40.50		174.50	209
1320	48" x 30"		11	.727		173	44.50		217.50	258
1340	36" x 36"		13	.615		106	37.50		143.50	173
1360	48" x 36"		11	.727		215	44.50		259.50	305
1380	48" x 48"		8	1		264	61		325	385
2000	Door grilles, 12" x 12"		22	.364		71.50	22		93.50	113
2020	18" x 12"		22	.364		82.50	22		104.50	125
2040	24" x 12"		18	.444		93	27		120	143
2060	18" x 18"		18	.444		115	27		142	168
2080	24" x 18"		16	.500		130	30.50		160.50	190
2100	24" x 24"		15	.533		184	32.50		216.50	253
3000	Filter grille with filter, 12" x 12"		24	.333		55.50	20.50		76	92
3020	18" x 12"		20	.400		74.50	24.50		99	119
3040	24" x 18"		18	.444		89	27		116	139
3060	24" x 24"		16	.500		105	30.50		135.50	163
3080	30" x 24"		14	.571		126	35		161	191
3100	30" x 30"		13	.615		149	37.50		186.50	221
3950	Eggcrate, framed, 6" x 6" opening		26	.308		16	18.75		34.75	46
3954	8" x 8" opening		24	.333		19.90	20.50		40.40	53
3960	10" x 10" opening		23	.348		24.50	21		45.50	59
3970	12" x 12" opening		22	.364		28	22		50	64
3980	14" x 14" opening		22	.364		35.50	22		57.50	72.50
3984	16" x 16" opening		21	.381		39.50	23		62.50	79
3990	18" x 18" opening		21	.381		46.50	23		69.50	86.50
4020	22" x 22" opening		17	.471		55.50	28.50		84	105
4040	24" x 24" opening		15	.533		71	32.50		103.50	128
4044	28" x 28" opening		14	.571		109	35		144	172
4048	36" x 36" opening		12	.667		124	40.50		164.50	199
4050	48" x 24" opening		12	.667		158	40.50		198.50	235
4060	Eggcrate, lay-in, T-bar system									
4070	48" x 24" sheet	1 Shee	40	.200	Ea.	68.50	12.20		80.70	93.50
5000	Transfer grille, vision proof, 8" x 4"		30	.267		13.95	16.25		30.20	40
5020	8" x 6"		28	.286		22	17.40		39.40	50.50
5040	8" x 8"		26	.308		24	18.75		42.75	55
5060	10" x 6"		24	.333		23	20.50		43.50	56
5080	10" x 10"		23	.348		22	21		43	56
5090	12" x 6"		23	.348		16.80	21		37.80	50.50
5100	12" x 10"		22	.364		24	22		46	60
5110	12" x 12"		22	.364		26.50	22		48.50	62.50
5120	14" x 10"		22	.364		34.50	22		56.50	71
5140	16" x 8"		21	.381		34.50	23		57.50	73
5160	18" x 12"		20	.400		34	24.50		58.50	74.50
5170	18" x 18"		21	.381		48	23		71	88.50
5180	20" x 12"		20	.400		37.50	24.50		62	78
5200	20" x 20"		19	.421		58	25.50		83.50	103
5220	24" x 12"		17	.471		36.50	28.50		65	83.50
5240	24" x 24"		16	.500		81	30.50		111.50	136
5260	30" x 6"		15	.533		37.50	32.50		70	90.50
5280	30" x 8"		15	.533		38.50	32.50		71	92
5300	30" x 12"		14	.571		54.50	35		89.50	113
5320	30" x 16"		13	.615		63.50	37.50		101	127
5340	30" x 20"		12	.667		80	40.50		120.50	150
5360	30" x 24"		11	.727		101	44.50		145.50	179

23 37 Air Outlets and Inlets

23 37 13 – Diffusers, Registers, and Grilles

23 37 13.30 Grilles		Crew	Daily Output	Labor-Hours	Unit	2019 Bare Costs Material	Labor	Equipment	Total	Total Incl O&P
5380	30″ x 30″	1 Shee	10	.800	Ea.	117	49		166	203
6000	For steel grilles instead of aluminum in above, deduct					10%				
6200	Plastic, eggcrate, lay-in, T-bar system									
6210	48″ x 24″ sheet	1 Shee	50	.160	Ea.	20	9.75		29.75	37
6250	Steel door louver									
6270	With fire link, steel only									
6350	12″ x 18″	1 Shee	22	.364	Ea.	222	22		244	279
6410	12″ x 24″		21	.381		242	23		265	300
6470	18″ x 18″		20	.400		247	24.50		271.50	310
6530	18″ x 24″		19	.421		284	25.50		309.50	355
6660	24″ x 24″		15	.533		288	32.50		320.50	365

23 37 13.60 Registers		Crew	Daily Output	Labor-Hours	Unit	Material	Labor	Equipment	Total	Total Incl O&P
0010	**REGISTERS**									
0980	Air supply									
1000	Ceiling/wall, O.B. damper, anodized aluminum									
1010	One or two way deflection, adj. curved face bars									
1014	6″ x 6″	1 Shee	24	.333	Ea.	25.50	20.50		46	59
1020	8″ x 4″		26	.308		28	18.75		46.75	59
1040	8″ x 8″		24	.333		32	20.50		52.50	66.50
1060	10″ x 6″		20	.400		19.90	24.50		44.40	59
1080	10″ x 10″		19	.421		27	25.50		52.50	68.50
1100	12″ x 6″		19	.421		30	25.50		55.50	72
1120	12″ x 12″		18	.444		34.50	27		61.50	79
1140	14″ x 8″		17	.471		20.50	28.50		49	66
1160	14″ x 14″		18	.444		37	27		64	81.50
1170	16″ x 16″		17	.471		41	28.50		69.50	88.50
1180	18″ x 8″		18	.444		45	27		72	90.50
1200	18″ x 18″		17	.471		82	28.50		110.50	134
1220	20″ x 4″		19	.421		17.95	25.50		43.45	59
1240	20″ x 6″		18	.444		28.50	27		55.50	72
1260	20″ x 8″		18	.444		45.50	27		72.50	91
1280	20″ x 20″		17	.471		100	28.50		128.50	154
1290	22″ x 22″		15	.533		115	32.50		147.50	176
1300	24″ x 4″		17	.471		22	28.50		50.50	68
1320	24″ x 6″		16	.500		51	30.50		81.50	103
1340	24″ x 8″		13	.615		56.50	37.50		94	119
1350	24″ x 18″		12	.667		101	40.50		141.50	173
1360	24″ x 24″		11	.727		96.50	44.50		141	174
1380	30″ x 4″		16	.500		28	30.50		58.50	77
1400	30″ x 6″		15	.533		65	32.50		97.50	121
1420	30″ x 8″		14	.571		73	35		108	134
1440	30″ x 24″		12	.667		182	40.50		222.50	262
1460	30″ x 30″		10	.800		233	49		282	330
1504	4 way deflection, adjustable curved face bars									
1510	6″ x 6″	1 Shee	26	.308	Ea.	30.50	18.75		49.25	62.50
1514	8″ x 8″		24	.333		38.50	20.50		59	73.50
1518	10″ x 10″		19	.421		32	25.50		57.50	74.50
1522	12″ x 6″		19	.421		36	25.50		61.50	78.50
1526	12″ x 12″		18	.444		41.50	27		68.50	86.50
1530	14″ x 14″		18	.444		44	27		71	89.50
1534	16″ x 16″		17	.471		49	28.50		77.50	97.50
1538	18″ x 18″		17	.471		98.50	28.50		127	152

23 37 13.60 Registers		Crew	Daily Output	Labor-Hours	Unit	Material	2019 Bare Costs Labor	Equipment	Total	Total Incl O&P
1542	22" x 22"	1 Shee	16	.500	Ea.	138	30.50		168.50	199
1980	One way deflection, adj. vert. or horiz. face bars									
1990	6" x 6"	1 Shee	26	.308	Ea.	23	18.75		41.75	54
2000	8" x 4"		26	.308		21.50	18.75		40.25	52.50
2020	8" x 8"		24	.333		25.50	20.50		46	59.50
2040	10" x 6"		20	.400		21.50	24.50		46	60.50
2060	10" x 10"		19	.421		25	25.50		50.50	66.50
2080	12" x 6"		19	.421		23.50	25.50		49	64.50
2100	12" x 12"		18	.444		30.50	27		57.50	74.50
2120	14" x 8"		17	.471		28	28.50		56.50	74
2140	14" x 12"		18	.444		48	27		75	94
2160	14" x 14"		18	.444		55	27		82	102
2180	16" x 6"		18	.444		27.50	27		54.50	71
2200	16" x 12"		18	.444		53	27		80	99.50
2220	16" x 16"		17	.471		58	28.50		86.50	108
2240	18" x 8"		18	.444		46.50	27		73.50	92.50
2260	18" x 12"		17	.471		39.50	28.50		68	87
2280	18" x 18"		16	.500		82.50	30.50		113	137
2300	20" x 10"		18	.444		55.50	27		82.50	102
2320	20" x 16"		16	.500		77	30.50		107.50	132
2340	20" x 20"		16	.500		101	30.50		131.50	158
2360	24" x 12"		15	.533		70	32.50		102.50	127
2380	24" x 16"		14	.571		93	35		128	155
2400	24" x 20"		12	.667		116	40.50		156.50	190
2420	24" x 24"		10	.800		98.50	49		147.50	182
2440	30" x 12"		13	.615		90.50	37.50		128	157
2460	30" x 16"		13	.615		86	37.50		123.50	152
2480	30" x 24"		11	.727		142	44.50		186.50	224
2500	30" x 30"		9	.889		235	54		289	340
2520	36" x 12"		11	.727		112	44.50		156.50	191
2540	36" x 24"		10	.800		229	49		278	325
2560	36" x 36"		8	1		385	61		446	515
2600	For 2 way deflect., adj. vert. or horiz. face bars, add					40%				
2700	Above registers in steel instead of aluminum, deduct					10%				
3000	Baseboard, hand adj. damper, enameled steel									
3012	8" x 6"	1 Shee	26	.308	Ea.	7.45	18.75		26.20	36.50
3020	10" x 6"		24	.333		8.10	20.50		28.60	40
3040	12" x 5"		23	.348		6.85	21		27.85	39.50
3060	12" x 6"		23	.348		9.75	21		30.75	43
3080	12" x 8"		22	.364		12.75	22		34.75	47.50
3100	14" x 6"		20	.400		9.55	24.50		34.05	47.50
4000	Floor, toe operated damper, enameled steel									
4020	4" x 8"	1 Shee	32	.250	Ea.	13.60	15.25		28.85	38
4040	4" x 12"		26	.308		16	18.75		34.75	46
4060	6" x 8"		28	.286		14.65	17.40		32.05	42.50
4080	6" x 14"		22	.364		18.50	22		40.50	54
4100	8" x 10"		22	.364		16.65	22		38.65	52
4120	8" x 16"		20	.400		13.95	24.50		38.45	52.50
4140	10" x 10"		20	.400		19.95	24.50		44.45	59
4160	10" x 16"		18	.444		23.50	27		50.50	67
4180	12" x 12"		18	.444		24.50	27		51.50	68
4200	12" x 24"		16	.500		25	30.50		55.50	74
4220	14" x 14"		16	.500		50.50	30.50		81	102

23 37 13.60 Registers		Crew	Daily Output	Labor-Hours	Unit	Material	2019 Bare Costs Labor	Equipment	Total	Total Incl O&P
4240	14" x 20"	1 Shee	15	.533	Ea.	46.50	32.50		79	101
4300	Spiral pipe supply register									
4310	Steel, with air scoop									
4320	4" x 12", for 8" thru 13" diameter duct	1 Shee	25	.320	Ea.	81	19.50		100.50	119
4330	4" x 18", for 8" thru 13" diameter duct		18	.444		93.50	27		120.50	144
4340	6" x 12", for 14" thru 21" diameter duct		19	.421		85.50	25.50		111	133
4350	6" x 16", for 14" thru 21" diameter duct		18	.444		97	27		124	148
4360	6" x 20", for 14" thru 21" diameter duct		17	.471		107	28.50		135.50	162
4370	6" x 24", for 14" thru 21" diameter duct		16	.500		129	30.50		159.50	189
4380	8" x 16", for 22" thru 31" diameter duct		19	.421		101	25.50		126.50	151
4390	8" x 18", for 22" thru 31" diameter duct		18	.444		109	27		136	161
4400	8" x 24", for 22" thru 31" diameter duct		15	.533		136	32.50		168.50	200
4980	Air return									
5000	Ceiling or wall, fixed 45° face blades									
5010	Adjustable O.B. damper, anodized aluminum									
5020	4" x 8"	1 Shee	26	.308	Ea.	19	18.75		37.75	49.50
5040	6" x 8"		24	.333		24	20.50		44.50	57.50
5060	6" x 10"		19	.421		24.50	25.50		50	66
5070	6" x 12"		19	.421		28	25.50		53.50	70
5080	6" x 16"		18	.444		29	27		56	73
5100	8" x 10"		19	.421		28.50	25.50		54	70
5120	8" x 12"		16	.500		31.50	30.50		62	81
5140	10" x 10"		18	.444		26.50	27		53.50	70
5160	10" x 16"		17	.471		45.50	28.50		74	93.50
5170	12" x 12"		18	.444		31.50	27		58.50	75.50
5180	12" x 18"		18	.444		41	27		68	86
5184	12" x 24"		14	.571		66.50	35		101.50	126
5200	12" x 30"		12	.667		65	40.50		105.50	133
5210	12" x 36"		10	.800		89	49		138	172
5220	16" x 16"		17	.471		65.50	28.50		94	116
5240	18" x 18"		16	.500		59	30.50		89.50	111
5250	18" x 24"		13	.615		85.50	37.50		123	151
5254	18" x 30"		11	.727		95	44.50		139.50	172
5260	18" x 36"		10	.800		149	49		198	238
5270	20" x 20"		10	.800		82.50	49		131.50	165
5274	20" x 24"		12	.667		84	40.50		124.50	154
5280	24" x 24"		11	.727		117	44.50		161.50	196
5290	24" x 30"		10	.800		146	49		195	235
5300	24" x 36"		8	1		186	61		247	297
5320	24" x 48"		6	1.333		221	81.50		302.50	365
5340	30" x 30"		6.50	1.231		170	75		245	300
5344	30" x 120"	Q-9	6	2.667		1,050	146		1,196	1,375
5360	36" x 36"	1 Shee	6	1.333		275	81.50		356.50	425
5370	40" x 32"	"	6	1.333		270	81.50		351.50	420
5400	Ceiling or wall, removable-reversible core, single deflection									
5402	Adjustable O.B. damper, radiused frame with border, aluminum									
5410	8" x 4"	1 Shee	26	.308	Ea.	20	18.75		38.75	50.50
5414	8" x 6"		24	.333		19.95	20.50		40.45	53
5418	10" x 6"		19	.421		25.50	25.50		51	67
5422	10" x 10"		18	.444		29	27		56	73
5426	12" x 6"		19	.421		30.50	25.50		56	72.50
5430	12" x 12"		18	.444		41	27		68	86.50
5434	16" x 16"		16	.500		48.50	30.50		79	100

23 37 13 – Diffusers, Registers, and Grilles

23 37 13.60 Registers		Crew	Daily Output	Labor-Hours	Unit	Material	2019 Bare Costs Labor	Equipment	Total	Total Incl O&P
5438	18" x 18"	1 Shee	17	.471	Ea.	65.50	28.50		94	116
5442	20" x 20"		16	.500		74.50	30.50		105	129
5446	24" x 12"		15	.533		49	32.50		81.50	104
5450	24" x 18"		13	.615		74.50	37.50		112	139
5454	24" x 24"		11	.727		104	44.50		148.50	182
5458	30" x 12"		13	.615		65.50	37.50		103	129
5462	30" x 18"		12.50	.640		94.50	39		133.50	163
5466	30" x 24"		11	.727		214	44.50		258.50	305
5470	30" x 30"		9	.889		234	54		288	340
5474	36" x 12"		11	.727		81	44.50		125.50	157
5478	36" x 18"		10.50	.762		113	46.50		159.50	196
5482	36" x 24"		10	.800		227	49		276	325
5486	36" x 30"		9	.889		245	54		299	350
5490	36" x 36"	↓	8	1	↓	276	61		337	400
6000	For steel construction instead of aluminum, deduct					10%				

23 37 15 – Louvers

23 37 15.40 HVAC Louvers		Crew	Daily Output	Labor-Hours	Unit	Material	Labor	Equipment	Total	Total Incl O&P
0010	**HVAC LOUVERS**									
0100	Aluminum, extruded, with screen, mill finish									
1002	Brick vent, see also Section 04 05 23.19									
1100	Standard, 4" deep, 8" wide, 5" high	1 Shee	24	.333	Ea.	37	20.50		57.50	71.50
1200	Modular, 4" deep, 7-3/4" wide, 5" high		24	.333		38	20.50		58.50	72.50
1300	Speed brick, 4" deep, 11-5/8" wide, 3-7/8" high		24	.333		38	20.50		58.50	72.50
1400	Fuel oil brick, 4" deep, 8" wide, 5" high		24	.333	↓	65.50	20.50		86	103
2000	Cooling tower and mechanical equip., screens, light weight		40	.200	S.F.	16.30	12.20		28.50	36.50
2020	Standard weight		35	.229		43.50	13.95		57.45	68.50
2500	Dual combination, automatic, intake or exhaust		20	.400		59.50	24.50		84	102
2520	Manual operation		20	.400		44	24.50		68.50	85.50
2540	Electric or pneumatic operation		20	.400	↓	44	24.50		68.50	85.50
2560	Motor, for electric or pneumatic	↓	14	.571	Ea.	510	35		545	615
3000	Fixed blade, continuous line									
3100	Mullion type, stormproof	1 Shee	28	.286	S.F.	44	17.40		61.40	75
3200	Stormproof		28	.286		44	17.40		61.40	75
3300	Vertical line	↓	28	.286	↓	52	17.40		69.40	84
3500	For damper to use with above, add					50%	30%			
3520	Motor, for damper, electric or pneumatic	1 Shee	14	.571	Ea.	510	35		545	615
4000	Operating, 45°, manual, electric or pneumatic		24	.333	S.F.	55.50	20.50		76	92
4100	Motor, for electric or pneumatic		14	.571	Ea.	510	35		545	615
4200	Penthouse, roof		56	.143	S.F.	26	8.70		34.70	41.50
4300	Walls		40	.200		61.50	12.20		73.70	86
5000	Thinline, under 4" thick, fixed blade	↓	40	.200	↓	25.50	12.20		37.70	46.50
5010	Finishes, applied by mfr. at additional cost, available in colors									
5020	Prime coat only, add				S.F.	3.50			3.50	3.85
5040	Baked enamel finish coating, add					6.45			6.45	7.10
5060	Anodized finish, add					7			7	7.70
5080	Duranodic finish, add					12.70			12.70	14
5100	Fluoropolymer finish coating, add				↓	20			20	22
9000	Stainless steel, fixed blade, continuous line									
9010	Hospital grade									
9110	20 ga.	1 Shee	22	.364	S.F.	41	22		63	78.50
9980	For small orders (under 10 pieces), add				"	25%				

23 37 23 – HVAC Gravity Ventilators

23 37 23.10 HVAC Gravity Air Ventilators		Crew	Daily Output	Labor-Hours	Unit	Material	2019 Bare Costs Labor	Equipment	Total	Total Incl O&P
0010	**HVAC GRAVITY AIR VENTILATORS**, Includes base									
1280	Rotary ventilators, wind driven, galvanized									
1300	4" neck diameter	Q-9	20	.800	Ea.	49.50	44		93.50	121
1320	5" neck diameter		18	.889		50	49		99	129
1340	6" neck diameter		16	1		50	55		105	139
1360	8" neck diameter		14	1.143		58	62.50		120.50	159
1380	10" neck diameter		12	1.333		64	73		137	181
1400	12" neck diameter		10	1.600		69.50	88		157.50	210
1420	14" neck diameter		10	1.600		118	88		206	263
1440	16" neck diameter		9	1.778		138	97.50		235.50	300
1460	18" neck diameter		9	1.778		166	97.50		263.50	330
1480	20" neck diameter		8	2		199	110		309	385
1500	24" neck diameter		8	2		310	110		420	505
1520	30" neck diameter		7	2.286		375	125		500	600
1540	36" neck diameter		6	2.667		595	146		741	875
1600	For aluminum, add					30%				
2000	Stationary, gravity, syphon, galvanized									
2100	3" neck diameter, 40 CFM	Q-9	24	.667	Ea.	29.50	36.50		66	87.50
2120	4" neck diameter, 50 CFM		20	.800		31	44		75	101
2140	5" neck diameter, 58 CFM		18	.889		34	49		83	112
2160	6" neck diameter, 66 CFM		16	1		35	55		90	122
2180	7" neck diameter, 86 CFM		15	1.067		46.50	58.50		105	141
2200	8" neck diameter, 110 CFM		14	1.143		48	62.50		110.50	148
2220	10" neck diameter, 140 CFM		12	1.333		65.50	73		138.50	183
2240	12" neck diameter, 160 CFM		10	1.600		81.50	88		169.50	223
2260	14" neck diameter, 250 CFM		10	1.600		130	88		218	276
2280	16" neck diameter, 380 CFM		9	1.778		167	97.50		264.50	330
2300	18" neck diameter, 500 CFM		9	1.778		188	97.50		285.50	355
2320	20" neck diameter, 625 CFM		8	2		224	110		334	415
2340	24" neck diameter, 900 CFM		8	2		277	110		387	470
2360	30" neck diameter, 1,375 CFM		7	2.286		345	125		470	570
2380	36" neck diameter, 2,000 CFM		6	2.667		430	146		576	695
2400	42" neck diameter, 3,000 CFM		4	4		640	219		859	1,025
2410	48" neck diameter, 4,000 CFM		3	5.333		930	293		1,223	1,475
2500	For aluminum, add					30%				
2520	For stainless steel, add					60%				
2530	For copper, add					100%				
3000	Rotating chimney cap, galvanized, 4" neck diameter	Q-9	20	.800		20	44		64	89
3020	5" neck diameter		18	.889		20.50	49		69.50	96.50
3040	6" neck diameter		16	1		20.50	55		75.50	106
3060	7" neck diameter		15	1.067		26	58.50		84.50	118
3080	8" neck diameter		14	1.143		26	62.50		88.50	124
3100	10" neck diameter		12	1.333		34.50	73		107.50	149
3600	Stationary chimney rain cap, galvanized, 3" neck diameter		24	.667		3.34	36.50		39.84	59
3620	4" neck diameter		20	.800		3.65	44		47.65	70.50
3640	6" neck diameter		16	1		4.34	55		59.34	88.50
3680	8" neck diameter		14	1.143		5.45	62.50		67.95	101
3700	10" neck diameter		12	1.333		6.55	73		79.55	118
3720	12" neck diameter		10	1.600		9.65	88		97.65	144
3740	14" neck diameter		10	1.600		15.35	88		103.35	150
3760	16" neck diameter		9	1.778		16.15	97.50		113.65	166
3770	18" neck diameter		9	1.778		25.50	97.50		123	176

23 37 Air Outlets and Inlets

23 37 23 – HVAC Gravity Ventilators

23 37 23.10 HVAC Gravity Air Ventilators		Crew	Daily Output	Labor-Hours	Unit	2019 Bare Costs Material	Labor	Equipment	Total	Total Incl O&P
3780	20" neck diameter	Q-9	8	2	Ea.	34.50	110		144.50	205
3790	24" neck diameter		8	2		43.50	110		153.50	215
4200	Stationary mushroom, aluminum, 16" orifice diameter		10	1.600		675	88		763	880
4220	26" orifice diameter		6.15	2.602		1,000	143		1,143	1,325
4230	30" orifice diameter		5.71	2.802		1,475	154		1,629	1,825
4240	38" orifice diameter		5	3.200		2,100	176		2,276	2,575
4250	42" orifice diameter		4.70	3.404		2,775	187		2,962	3,325
4260	50" orifice diameter		4.44	3.604		3,300	198		3,498	3,925
5000	Relief vent									
5500	Rectangular, aluminum, galvanized curb									
5510	intake/exhaust, 0.033" SP									
5580	500 CFM, 12" x 12"	Q-9	8.60	1.860	Ea.	745	102		847	975
5600	600 CFM, 12" x 16"		8	2		835	110		945	1,075
5620	750 CFM, 12" x 20"		7.20	2.222		895	122		1,017	1,175
5640	1,000 CFM, 12" x 24"		6.60	2.424		935	133		1,068	1,225
5660	1,500 CFM, 12" x 36"		5.80	2.759		1,250	151		1,401	1,600
5680	3,000 CFM, 20" x 42"		4	4		1,650	219		1,869	2,125
5700	5,000 CFM, 20" x 72"		3	5.333		2,375	293		2,668	3,050
5720	6,000 CFM, 24" x 72"		2.30	6.957		2,550	380		2,930	3,400
5740	10,000 CFM, 48" x 60"		1.80	8.889		3,225	490		3,715	4,300
5760	12,000 CFM, 48" x 72"		1.60	10		3,825	550		4,375	5,025
5780	13,750 CFM, 60" x 66"		1.40	11.429		4,225	625		4,850	5,600
5800	15,000 CFM, 60" x 72"		1.30	12.308		4,525	675		5,200	6,000
5820	18,000 CFM, 72" x 72"		1.20	13.333		5,125	730		5,855	6,750
5880	Size is throat area, volume is at 500 fpm									
7000	Note: sizes based on exhaust. Intake, with 0.125" SP									
7100	loss, approximately twice listed capacity.									

23 38 Ventilation Hoods

23 38 13 – Commercial-Kitchen Hoods

23 38 13.10 Hood and Ventilation Equipment		Crew	Daily Output	Labor-Hours	Unit	Material	Labor	Equipment	Total	Total Incl O&P
0010	**HOOD AND VENTILATION EQUIPMENT**									
2970	Exhaust hood, sst, gutter on all sides, 4' x 4' x 2'	1 Carp	1.80	4.444	Ea.	4,225	230		4,455	5,000
2980	4' x 4' x 7'	"	1.60	5		6,725	258		6,983	7,800
7800	Vent hood, wall canopy with fire protection, 30"	L-3A	9	1.333		395	75		470	550
7810	Without fire protection, 36"		10	1.200		420	67.50		487.50	560
7820	Island canopy with fire protection, 30"		7	1.714		775	96		871	995
7830	Without fire protection, 36"		8	1.500		805	84		889	1,000
7840	Back shelf with fire protection, 30"		11	1.091		400	61		461	535
7850	Without fire protection, black, 36"		12	1		450	56		506	580
7852	Without fire protection, stainless steel, 36"		12	1		470	56		526	605
7860	Range hood & CO_2 system, 30"	1 Carp	2.50	3.200		4,350	165		4,515	5,025
7950	Hood fire protection system, electric stove	Q-1	3	5.333		2,075	305		2,380	2,725
7952	Hood fire protection system, gas stove	"	3	5.333		2,375	305		2,680	3,050

23 38 13.20 Kitchen Ventilation		Crew	Daily Output	Labor-Hours	Unit	Material	Labor	Equipment	Total	Total Incl O&P
0010	**KITCHEN VENTILATION**, Commercial									
6000	Exhaust hoods and fans									
6100	Centrifugal grease extraction									
6110	Water wash type									
6120	Per foot of length	Q-10	7.10	3.380	L.F.	1,900	192		2,092	2,375
6200	Non water wash type									

23 38 Ventilation Hoods

23 38 13 – Commercial-Kitchen Hoods

23 38 13.20 Kitchen Ventilation		Crew	Daily Output	Labor-Hours	Unit	Material	2019 Bare Costs Labor	Equipment	Total	Total Incl O&P
6210	Per foot of length	Q-10	13.40	1.791	L.F.	1,375	102		1,477	1,675
6400	Island style ventilator									
6410	Water wash type									
6420	Per foot of length	Q-10	7.10	3.380	L.F.	2,525	192		2,717	3,075
6460	Non water wash type									
6470	Per foot of length	Q-10	13.40	1.791	L.F.	885	102		987	1,125

23 41 Particulate Air Filtration

23 41 13 – Panel Air Filters

23 41 13.10 Panel Type Air Filters

		Crew	Daily Output	Labor-Hours	Unit	Material	Labor	Equipment	Total	Total Incl O&P
0010	**PANEL TYPE AIR FILTERS**									
2950	Mechanical media filtration units									
3000	High efficiency type, with frame, non-supported G				MCFM	35			35	38.50
3100	Supported type G				"	54			54	59.50
5500	Throwaway glass or paper media type, 12" x 36" x 1"				Ea.	2.27			2.27	2.50

23 41 16 – Renewable-Media Air Filters

23 41 16.10 Disposable Media Air Filters

		Crew	Daily Output	Labor-Hours	Unit	Material	Labor	Equipment	Total	Total Incl O&P
0010	**DISPOSABLE MEDIA AIR FILTERS**									
5000	Renewable disposable roll				C.S.F.	5.25			5.25	5.75
5800	Filter, bag type									
5810	90-95% efficiency									
5820	24" x 12" x 29", 0.75-1.25 MCFM	1 Shee	1.90	4.211	Ea.	61	257		318	455
5830	24" x 24" x 29", 1.5-2.5 MCFM	"	1.90	4.211	"	100	257		357	500
5850	80-85% efficiency									
5860	24" x 12" x 29", 0.75-1.25 MCFM	1 Shee	1.90	4.211	Ea.	50.50	257		307.50	445
5870	24" x 24" x 29", 1.5-2.5 MCFM	"	1.90	4.211	"	93	257		350	495

23 41 19 – Washable Air Filters

23 41 19.10 Permanent Air Filters

		Crew	Daily Output	Labor-Hours	Unit	Material	Labor	Equipment	Total	Total Incl O&P
0010	**PERMANENT AIR FILTERS**									
4500	Permanent washable G				MCFM	25			25	27.50

23 41 23 – Extended Surface Filters

23 41 23.10 Expanded Surface Filters

		Crew	Daily Output	Labor-Hours	Unit	Material	Labor	Equipment	Total	Total Incl O&P
0010	**EXPANDED SURFACE FILTERS**									
4000	Medium efficiency, extended surface G				MCFM	6.15			6.15	6.75

23 41 33 – High-Efficiency Particulate Filtration

23 41 33.10 HEPA Filters

		Crew	Daily Output	Labor-Hours	Unit	Material	Labor	Equipment	Total	Total Incl O&P
0010	**HEPA FILTERS**									
6000	HEPA filter complete w/particle board,									
6010	kraft paper frame, separator material									
6020	95% DOP efficiency									
6030	12" x 12" x 6", 150 CFM	1 Shee	3.70	2.162	Ea.	44.50	132		176.50	249
6034	24" x 12" x 6", 375 CFM		1.90	4.211		74	257		331	470
6038	24" x 18" x 6", 450 CFM		1.90	4.211		91	257		348	490
6042	24" x 24" x 6", 700 CFM		1.90	4.211		102	257		359	505
6046	12" x 12" x 12", 250 CFM		3.70	2.162		75	132		207	283
6050	24" x 12" x 12", 500 CFM		1.90	4.211		106	257		363	505
6054	24" x 18" x 12", 875 CFM		1.90	4.211		148	257		405	555
6058	24" x 24" x 12", 1,000 CFM	↓	1.90	4.211	↓	160	257		417	565
6100	99% DOP efficiency									

23 41 Particulate Air Filtration

23 41 33 – High-Efficiency Particulate Filtration

23 41 33.10 HEPA Filters		Crew	Daily Output	Labor-Hours	Unit	Material	2019 Bare Costs Labor	Equipment	Total	Total Incl O&P
6110	12" x 12" x 6", 150 CFM	1 Shee	3.70	2.162	Ea.	46.50	132		178.50	251
6114	24" x 12" x 6", 325 CFM		1.90	4.211		77	257		334	475
6118	24" x 18" x 6", 550 CFM		1.90	4.211		94.50	257		351.50	495
6122	24" x 24" x 6", 775 CFM		1.90	4.211		106	257		363	505
6124	24" x 48" x 6", 775 CFM		1.30	6.154		206	375		581	795
6126	12" x 12" x 12", 250 CFM		3.60	2.222		78	135		213	292
6130	24" x 12" x 12", 500 CFM		1.90	4.211		110	257		367	510
6134	24" x 18" x 12", 775 CFM		1.90	4.211		154	257		411	560
6138	24" x 24" x 12", 1,100 CFM		1.90	4.211		166	257		423	570
6500	HEPA filter housing, 14 ga. galv. sheet metal									
6510	12" x 12" x 6"	1 Shee	2.50	3.200	Ea.	670	195		865	1,025
6514	24" x 12" x 6"		2	4		745	244		989	1,200
6518	12" x 12" x 12"		2.40	3.333		670	203		873	1,050
6522	14" x 12" x 12"		2.30	3.478		670	212		882	1,050
6526	24" x 18" x 6"		1.90	4.211		950	257		1,207	1,450
6530	24" x 24" x 6"		1.80	4.444		865	271		1,136	1,375
6534	24" x 48" x 6"		1.70	4.706		1,150	287		1,437	1,700
6538	24" x 72" x 6"		1.60	5		1,475	305		1,780	2,100
6542	24" x 18" x 12"		1.80	4.444		950	271		1,221	1,450
6546	24" x 24" x 12"		1.70	4.706		1,075	287		1,362	1,600
6550	24" x 48" x 12"		1.60	5		1,250	305		1,555	1,850
6554	24" x 72" x 12"		1.50	5.333		1,550	325		1,875	2,200
6558	48" x 48" x 6"	Q-9	2.80	5.714		1,700	315		2,015	2,350
6562	48" x 72" x 6"		2.60	6.154		2,050	340		2,390	2,775
6566	48" x 96" x 6"		2.40	6.667		2,600	365		2,965	3,400
6570	48" x 48" x 12"		2.70	5.926		1,700	325		2,025	2,375
6574	48" x 72" x 12"		2.50	6.400		2,050	350		2,400	2,775
6578	48" x 96" x 12"		2.30	6.957		2,600	380		2,980	3,425
6582	114" x 72" x 12"		2	8		3,600	440		4,040	4,625

23 42 Gas-Phase Air Filtration

23 42 13 – Activated-Carbon Air Filtration

23 42 13.10 Charcoal Type Air Filtration

		Crew	Daily Output	Labor-Hours	Unit	Material	Labor	Equipment	Total	Total Incl O&P
0010	**CHARCOAL TYPE AIR FILTRATION**									
0050	Activated charcoal type, full flow				MCFM	650			650	715
0060	Full flow, impregnated media 12" deep					225			225	248
0070	HEPA filter & frame for field erection					440			440	485
0080	HEPA filter-diffuser, ceiling install.					350			350	385

23 43 Electronic Air Cleaners

23 43 13 – Washable Electronic Air Cleaners

23 43 13.10 Electronic Air Cleaners

		Crew	Daily Output	Labor-Hours	Unit	Material	Labor	Equipment	Total	Total Incl O&P
0010	**ELECTRONIC AIR CLEANERS**									
2000	Electronic air cleaner, duct mounted									
2150	1,000 CFM	1 Shee	4	2	Ea.	430	122		552	660
2200	1,200 CFM		3.80	2.105		505	128		633	750
2250	1,400 CFM		3.60	2.222		530	135		665	785
2260	2,000 CFM		3.20	2.500		585	152		737	875

23 51 13 – Draft Control Devices

23 51 13.13 Draft-Induction Fans

	23 51 13.13 Draft-Induction Fans	Crew	Daily Output	Labor-Hours	Unit	Material	2019 Bare Costs Labor	Equipment	Total	Total Incl O&P
0010	**DRAFT-INDUCTION FANS**									
1000	Breeching installation									
1800	Hot gas, 600°F, variable pitch pulley and motor									
1840	6″ diam. inlet, 1/4 HP, 1 phase, 400 CFM	Q-9	6	2.667	Ea.	1,500	146		1,646	1,875
1860	8″ diam. inlet, 1/4 HP, 1 phase, 1,120 CFM		4	4		2,150	219		2,369	2,675
1870	9″ diam. inlet, 3/4 HP, 1 phase, 1,440 CFM		3.60	4.444		2,650	244		2,894	3,300
1880	10″ diam. inlet, 3/4 HP, 1 phase, 2,000 CFM		3.30	4.848		2,625	266		2,891	3,300
1900	12″ diam. inlet, 3/4 HP, 3 phase, 2,960 CFM		3	5.333		2,925	293		3,218	3,675
1910	14″ diam. inlet, 1 HP, 3 phase, 4,160 CFM		2.60	6.154		2,900	340		3,240	3,725
1920	16″ diam. inlet, 2 HP, 3 phase, 5,500 CFM		2.30	6.957		3,225	380		3,605	4,125
1950	20″ diam. inlet, 3 HP, 3 phase, 9,760 CFM		1.50	10.667		4,675	585		5,260	6,050
1960	22″ diam. inlet, 5 HP, 3 phase, 13,360 CFM		1	16		7,650	880		8,530	9,725
1980	24″ diam. inlet, 7-1/2 HP, 3 phase, 17,760 CFM		.80	20		8,400	1,100		9,500	10,900
2300	For multi-blade damper at fan inlet, add					20%				
3600	Chimney-top installation									
3700	6″ size	1 Shee	8	1	Ea.	1,450	61		1,511	1,700
3740	8″ size		7	1.143		1,450	69.50		1,519.50	1,700
3750	10″ size		6.50	1.231		2,125	75		2,200	2,450
3780	13″ size		6	1.333		1,950	81.50		2,031.50	2,275
3880	For speed control switch, add					136			136	150
3920	For thermal fan control, add					136			136	150
5500	Flue blade style damper for draft control,									
5510	locking quadrant blade									
5550	8″ size	Q-9	8	2	Ea.	405	110		515	610
5560	9″ size		7.50	2.133		405	117		522	625
5570	10″ size		7	2.286		415	125		540	645
5580	12″ size		6.50	2.462		425	135		560	670
5590	14″ size		6	2.667		435	146		581	695
5600	16″ size		5.50	2.909		445	160		605	730
5610	18″ size		5	3.200		465	176		641	775
5620	20″ size		4.50	3.556		490	195		685	835
5630	22″ size		4	4		515	219		734	900
5640	24″ size		3.50	4.571		545	251		796	980
5650	27″ size		3	5.333		575	293		868	1,075
5660	30″ size		2.50	6.400		625	350		975	1,225
5670	32″ size		2	8		655	440		1,095	1,375
5680	36″ size		1.50	10.667		725	585		1,310	1,675

23 51 13.16 Vent Dampers

	23 51 13.16 Vent Dampers	Crew	Daily Output	Labor-Hours	Unit	Material	Labor	Equipment	Total	Total Incl O&P
0010	**VENT DAMPERS**									
5000	Vent damper, bi-metal, gas, 3″ diameter	Q-9	24	.667	Ea.	62	36.50		98.50	124
5010	4″ diameter		24	.667		62	36.50		98.50	124
5020	5″ diameter		23	.696		62	38		100	127
5030	6″ diameter		22	.727		62	40		102	129
5040	7″ diameter		21	.762		66	42		108	136
5050	8″ diameter		20	.800		68	44		112	142

23 51 13.19 Barometric Dampers

	23 51 13.19 Barometric Dampers	Crew	Daily Output	Labor-Hours	Unit	Material	Labor	Equipment	Total	Total Incl O&P
0010	**BAROMETRIC DAMPERS**									
1000	Barometric, gas fired system only, 6″ size for 5″ and 6″ pipes	1 Shee	20	.400	Ea.	105	24.50		129.50	152
1020	7″ size, for 6″ and 7″ pipes		19	.421		114	25.50		139.50	165
1040	8″ size, for 7″ and 8″ pipes		18	.444		147	27		174	203
1060	9″ size, for 8″ and 9″ pipes		16	.500		163	30.50		193.50	226
2000	All fuel, oil, oil/gas, coal									

23 51 Breechings, Chimneys, and Stacks

23 51 13 – Draft Control Devices

23 51 13.19 Barometric Dampers		Crew	Daily Output	Labor-Hours	Unit	Material	2019 Bare Costs Labor	Equipment	Total	Total Incl O&P
2020	10″ for 9″ and 10″ pipes	1 Shee	15	.533	Ea.	254	32.50		286.50	330
2040	12″ for 11″ and 12″ pipes		15	.533		330	32.50		362.50	415
2060	14″ for 13″ and 14″ pipes		14	.571		430	35		465	530
2080	16″ for 15″ and 16″ pipes		13	.615		605	37.50		642.50	720
2100	18″ for 17″ and 18″ pipes		12	.667		800	40.50		840.50	940
2120	20″ for 19″ and 21″ pipes		10	.800		960	49		1,009	1,125
2140	24″ for 22″ and 25″ pipes	Q-9	12	1.333		1,175	73		1,248	1,400
2160	28″ for 26″ and 30″ pipes		10	1.600		1,450	88		1,538	1,725
2180	32″ for 31″ and 34″ pipes		8	2		1,875	110		1,985	2,225
3260	For thermal switch for above, add	1 Shee	24	.333		105	20.50		125.50	146

23 51 23 – Gas Vents

23 51 23.10 Gas Chimney Vents		Crew	Daily Output	Labor-Hours	Unit	Material	Labor	Equipment	Total	Total Incl O&P
0010	**GAS CHIMNEY VENTS**, Prefab metal, UL listed									
0020	Gas, double wall, galvanized steel									
0080	3″ diameter	Q-9	72	.222	V.L.F.	6.80	12.20		19	26
0100	4″ diameter		68	.235		8.35	12.90		21.25	29
0120	5″ diameter		64	.250		9.30	13.70		23	31.50
0140	6″ diameter		60	.267		11.20	14.65		25.85	34.50
0160	7″ diameter		56	.286		22.50	15.65		38.15	49
0180	8″ diameter		52	.308		23	16.90		39.90	51
0200	10″ diameter		48	.333		44.50	18.30		62.80	77
0220	12″ diameter		44	.364		52	19.95		71.95	88
0240	14″ diameter		42	.381		87	21		108	128
0260	16″ diameter		40	.400		125	22		147	172
0280	18″ diameter		38	.421		154	23		177	205
0300	20″ diameter	Q-10	36	.667		184	38		222	261
0320	22″ diameter		34	.706		236	40		276	320
0340	24″ diameter		32	.750		285	42.50		327.50	380
0600	For 4″, 5″ and 6″ oval, add					50%				
0650	Gas, double wall, galvanized steel, fittings									
0660	Elbow 45°, 3″ diameter	Q-9	36	.444	Ea.	12.10	24.50		36.60	50.50
0670	4″ diameter		34	.471		14.85	26		40.85	55.50
0680	5″ diameter		32	.500		17.10	27.50		44.60	60.50
0690	6″ diameter		30	.533		21.50	29.50		51	68
0700	7″ diameter		28	.571		34	31.50		65.50	84.50
0710	8″ diameter		26	.615		44.50	34		78.50	101
0720	10″ diameter		24	.667		92	36.50		128.50	157
0730	12″ diameter		22	.727		101	40		141	172
0740	14″ diameter		21	.762		154	42		196	233
0750	16″ diameter		20	.800		200	44		244	287
0760	18″ diameter		19	.842		262	46		308	360
0770	20″ diameter	Q-10	18	1.333		293	76		369	435
0780	22″ diameter		17	1.412		490	80.50		570.50	660
0790	24″ diameter		16	1.500		625	85.50		710.50	820
0916	Adjustable length									
0918	3″ diameter, to 12″	Q-9	36	.444	Ea.	15	24.50		39.50	53.50
0920	4″ diameter, to 12″		34	.471		17.45	26		43.45	58
0924	6″ diameter, to 12″		30	.533		22.50	29.50		52	69.50
0928	8″ diameter, to 12″		26	.615		42	34		76	98
0930	10″ diameter, to 18″		24	.667		132	36.50		168.50	201
0932	12″ diameter, to 18″		22	.727		135	40		175	209
0936	16″ diameter, to 18″		20	.800		272	44		316	365

23 51 23.10 Gas Chimney Vents		Crew	Daily Output	Labor-Hours	Unit	2019 Bare Costs Material	Labor	Equipment	Total	Total Incl O&P
0938	18″ diameter, to 18″	Q-9	19	.842	Ea.	315	46		361	415
0944	24″ diameter, to 18″	Q-10	16	1.500		585	85.50		670.50	775
0950	Elbow 90°, adjustable, 3″ diameter	Q-9	36	.444		20.50	24.50		45	59.50
0960	4″ diameter		34	.471		24	26		50	65.50
0970	5″ diameter		32	.500		29.50	27.50		57	74
0980	6″ diameter		30	.533		35	29.50		64.50	83
0990	7″ diameter		28	.571		62	31.50		93.50	116
1010	8″ diameter		26	.615		61.50	34		95.50	120
1020	Wall thimble, 4 to 7″ adjustable, 3″ diameter		36	.444		12.90	24.50		37.40	51
1022	4″ diameter		34	.471		14.50	26		40.50	55
1024	5″ diameter		32	.500		17.10	27.50		44.60	60.50
1026	6″ diameter		30	.533		17.85	29.50		47.35	64
1028	7″ diameter		28	.571		37.50	31.50		69	88.50
1030	8″ diameter		26	.615		44.50	34		78.50	101
1040	Roof flashing, 3″ diameter		36	.444		7.05	24.50		31.55	45
1050	4″ diameter		34	.471		8.20	26		34.20	48
1060	5″ diameter		32	.500		24.50	27.50		52	68
1070	6″ diameter		30	.533		19.55	29.50		49.05	66
1080	7″ diameter		28	.571		26	31.50		57.50	76
1090	8″ diameter		26	.615		27.50	34		61.50	82
1100	10″ diameter		24	.667		37.50	36.50		74	96.50
1110	12″ diameter		22	.727		52	40		92	118
1120	14″ diameter		20	.800		118	44		162	197
1130	16″ diameter		18	.889		144	49		193	232
1140	18″ diameter		16	1		201	55		256	305
1150	20″ diameter	Q-10	18	1.333		259	76		335	400
1160	22″ diameter		14	1.714		320	97.50		417.50	505
1170	24″ diameter		12	2		370	114		484	580
1200	Tee, 3″ diameter	Q-9	27	.593		32	32.50		64.50	84.50
1210	4″ diameter		26	.615		34	34		68	89
1220	5″ diameter		25	.640		36	35		71	93
1230	6″ diameter		24	.667		40.50	36.50		77	101
1240	7″ diameter		23	.696		58	38		96	122
1250	8″ diameter		22	.727		64	40		104	131
1260	10″ diameter		21	.762		172	42		214	253
1270	12″ diameter		20	.800		179	44		223	264
1280	14″ diameter		18	.889		310	49		359	420
1290	16″ diameter		16	1		470	55		525	600
1300	18″ diameter		14	1.143		565	62.50		627.50	715
1310	20″ diameter	Q-10	17	1.412		765	80.50		845.50	960
1320	22″ diameter		13	1.846		990	105		1,095	1,250
1330	24″ diameter		12	2		1,150	114		1,264	1,425
1460	Tee cap, 3″ diameter	Q-9	45	.356		2.15	19.50		21.65	32
1470	4″ diameter		42	.381		2.33	21		23.33	34
1480	5″ diameter		40	.400		3	22		25	37
1490	6″ diameter		37	.432		4.18	23.50		27.68	40.50
1500	7″ diameter		35	.457		6.95	25		31.95	45.50
1510	8″ diameter		34	.471		7.70	26		33.70	47.50
1520	10″ diameter		32	.500		62	27.50		89.50	110
1530	12″ diameter		30	.533		61.50	29.50		91	112
1540	14″ diameter		28	.571		65	31.50		96.50	119
1550	16″ diameter		25	.640		66.50	35		101.50	127
1560	18″ diameter		24	.667		78.50	36.50		115	142

23 51 23.10 Gas Chimney Vents		Crew	Daily Output	Labor-Hours	Unit	Material	2019 Bare Costs Labor	Equipment	Total	Total Incl O&P
1570	20″ diameter	Q-10	27	.889	Ea.	85.50	50.50		136	171
1580	22″ diameter		22	1.091		149	62		211	258
1590	24″ diameter		21	1.143		176	65		241	293
1750	Top, 3″ diameter	Q-9	46	.348		15.80	19.10		34.90	46.50
1760	4″ diameter		44	.364		16.40	19.95		36.35	48.50
1770	5″ diameter		42	.381		16.70	21		37.70	50
1780	6″ diameter		40	.400		21.50	22		43.50	57
1790	7″ diameter		38	.421		47	23		70	86.50
1800	8″ diameter		36	.444		56.50	24.50		81	99
1810	10″ diameter		34	.471		94	26		120	142
1820	12″ diameter		32	.500		122	27.50		149.50	176
1830	14″ diameter		30	.533		190	29.50		219.50	254
1840	16″ diameter		28	.571		233	31.50		264.50	305
1850	18″ diameter		26	.615		325	34		359	405
1860	20″ diameter	Q-10	28	.857		500	49		549	625
1870	22″ diameter		22	1.091		760	62		822	930
1880	24″ diameter		20	1.200		960	68.50		1,028.50	1,150
1900	Gas, double wall, galvanized steel, oval									
1904	4″ x 1′	Q-9	68	.235	V.L.F.	16.40	12.90		29.30	37.50
1906	5″ x 1′		64	.250		84	13.70		97.70	114
1908	5″/6″ x 1′		60	.267		42	14.65		56.65	68
1910	Oval fittings									
1912	Adjustable length									
1914	4″ diameter to 12″ long	Q-9	34	.471	Ea.	17.80	26		43.80	58.50
1916	5″ diameter to 12″ long		32	.500		18	27.50		45.50	61.50
1918	5″/6″ diameter to 12″ long		30	.533		22.50	29.50		52	69.50
1920	Elbow 45°									
1922	4″	Q-9	34	.471	Ea.	28.50	26		54.50	70.50
1924	5″		32	.500		55	27.50		82.50	102
1926	5″/6″		30	.533		57.50	29.50		87	108
1930	Elbow 45°, flat									
1932	4″	Q-9	34	.471	Ea.	27.50	26		53.50	69
1934	5″		32	.500		55	27.50		82.50	102
1936	5″/6″		30	.533		60	29.50		89.50	111
1940	Top									
1942	4″	Q-9	44	.364	Ea.	35.50	19.95		55.45	69.50
1944	5″		42	.381		35.50	21		56.50	71
1946	5″/6″		40	.400		46.50	22		68.50	84.50
1950	Adjustable flashing									
1952	4″	Q-9	34	.471	Ea.	12	26		38	52
1954	5″		32	.500		35	27.50		62.50	80
1956	5″/6″		30	.533		39	29.50		68.50	87.50
1960	Tee									
1962	4″	Q-9	26	.615	Ea.	40.50	34		74.50	96
1964	5″		25	.640		88	35		123	151
1966	5″/6″		24	.667		87.50	36.50		124	152
1970	Tee with short snout									
1972	4″	Q-9	26	.615	Ea.	43	34		77	98.50

23 51 26.10 All-Fuel Vent Chimneys, Press. Tight, Dbl. Wall		Crew	Daily Output	Labor-Hours	Unit	Material	Labor	Equipment	Total	Total Incl O&P
						2019 Bare Costs				
0010	ALL-FUEL VENT CHIMNEYS, PRESSURE TIGHT, DOUBLE WALL									
3200	All fuel, pressure tight, double wall, 1" insulation, UL listed, 1,400°F.									
3210	304 stainless steel liner, aluminized steel outer jacket									
3220	6" diameter	Q-9	60	.267	L.F.	55.50	14.65		70.15	83
3221	8" diameter		52	.308		63.50	16.90		80.40	95.50
3222	10" diameter		48	.333		71	18.30		89.30	106
3223	12" diameter		44	.364		80	19.95		99.95	119
3224	14" diameter		42	.381		91.50	21		112.50	133
3225	16" diameter		40	.400		103	22		125	148
3226	18" diameter		38	.421		118	23		141	164
3227	20" diameter	Q-10	36	.667		133	38		171	204
3228	24" diameter		32	.750		170	42.50		212.50	252
3229	28" diameter		30	.800		195	45.50		240.50	284
3230	32" diameter		27	.889		221	50.50		271.50	320
3231	36" diameter		25	.960		245	54.50		299.50	355
3232	42" diameter		22	1.091		288	62		350	410
3233	48" diameter		19	1.263		325	72		397	470
3260	For 316 stainless steel liner, add					30%				
3280	All fuel, pressure tight, double wall fittings									
3284	304 stainless steel inner, aluminized steel jacket									
3288	Adjustable 20"/29" section									
3292	6" diameter	Q-9	30	.533	Ea.	201	29.50		230.50	266
3293	8" diameter		26	.615		210	34		244	283
3294	10" diameter		24	.667		238	36.50		274.50	315
3295	12" diameter		22	.727		268	40		308	355
3296	14" diameter		21	.762		305	42		347	400
3297	16" diameter		20	.800		340	44		384	440
3298	18" diameter		19	.842		385	46		431	495
3299	20" diameter	Q-10	18	1.333		435	76		511	595
3300	24" diameter		16	1.500		560	85.50		645.50	745
3301	28" diameter		15	1.600		600	91		691	800
3302	32" diameter		14	1.714		725	97.50		822.50	950
3303	36" diameter		12	2		810	114		924	1,075
3304	42" diameter		11	2.182		950	124		1,074	1,250
3305	48" diameter		10	2.400		1,075	137		1,212	1,400
3350	Elbow 90° fixed									
3354	6" diameter	Q-9	30	.533	Ea.	420	29.50		449.50	510
3355	8" diameter		26	.615		475	34		509	570
3356	10" diameter		24	.667		535	36.50		571.50	645
3357	12" diameter		22	.727		610	40		650	730
3358	14" diameter		21	.762		690	42		732	825
3359	16" diameter		20	.800		780	44		824	920
3360	18" diameter		19	.842		880	46		926	1,025
3361	20" diameter	Q-10	18	1.333		995	76		1,071	1,225
3362	24" diameter		16	1.500		1,275	85.50		1,360.50	1,525
3363	28" diameter		15	1.600		1,425	91		1,516	1,725
3364	32" diameter		14	1.714		1,925	97.50		2,022.50	2,275
3365	36" diameter		12	2		2,600	114		2,714	3,025
3366	42" diameter		11	2.182		3,100	124		3,224	3,600
3367	48" diameter		10	2.400		3,450	137		3,587	3,975
3380	For 316 stainless steel liner, add					30%				
3400	Elbow 45°									

23 51 26 – All-Fuel Vent Chimneys

	23 51 26.10 All-Fuel Vent Chimneys, Press. Tight, Dbl. Wall	Crew	Daily Output	Labor-Hours	Unit	Material	Labor	Equipment	Total	Total Incl O&P
						2019 Bare Costs				
3404	6″ diameter	Q-9	30	.533	Ea.	210	29.50		239.50	276
3405	8″ diameter		26	.615		236	34		270	310
3406	10″ diameter		24	.667		269	36.50		305.50	350
3407	12″ diameter		22	.727		305	40		345	400
3408	14″ diameter		21	.762		345	42		387	445
3409	16″ diameter		20	.800		390	44		434	490
3410	18″ diameter		19	.842		440	46		486	555
3411	20″ diameter	Q-10	18	1.333		500	76		576	665
3412	24″ diameter		16	1.500		640	85.50		725.50	835
3413	28″ diameter		15	1.600		715	91		806	925
3414	32″ diameter		14	1.714		965	97.50		1,062.50	1,200
3415	36″ diameter		12	2		1,300	114		1,414	1,600
3416	42″ diameter		11	2.182		1,500	124		1,624	1,850
3417	48″ diameter		10	2.400		1,725	137		1,862	2,100
3430	For 316 stainless steel liner, add					30%				
3450	Tee 90°									
3454	6″ diameter	Q-9	24	.667	Ea.	248	36.50		284.50	330
3455	8″ diameter		22	.727		271	40		311	360
3456	10″ diameter		21	.762		305	42		347	400
3457	12″ diameter		20	.800		355	44		399	455
3458	14″ diameter		18	.889		405	49		454	520
3459	16″ diameter		16	1		445	55		500	570
3460	18″ diameter		14	1.143		520	62.50		582.50	670
3461	20″ diameter	Q-10	17	1.412		600	80.50		680.50	780
3462	24″ diameter		12	2		755	114		869	1,000
3463	28″ diameter		11	2.182		895	124		1,019	1,175
3464	32″ diameter		10	2.400		1,250	137		1,387	1,575
3465	36″ diameter		9	2.667		1,400	152		1,552	1,775
3466	42″ diameter		6	4		1,650	228		1,878	2,150
3467	48″ diameter		5	4.800		1,875	273		2,148	2,475
3480	For tee cap, add					35%	20%			
3500	For 316 stainless steel liner, add					30%				
3520	Plate support, galvanized									
3524	6″ diameter	Q-9	26	.615	Ea.	124	34		158	188
3525	8″ diameter		22	.727		144	40		184	220
3526	10″ diameter		20	.800		157	44		201	240
3527	12″ diameter		18	.889		165	49		214	256
3528	14″ diameter		17	.941		195	51.50		246.50	294
3529	16″ diameter		16	1		206	55		261	310
3530	18″ diameter		15	1.067		217	58.50		275.50	330
3531	20″ diameter	Q-10	16	1.500		228	85.50		313.50	380
3532	24″ diameter		14	1.714		237	97.50		334.50	410
3533	28″ diameter		13	1.846		291	105		396	480
3534	32″ diameter		12	2		355	114		469	565
3535	36″ diameter		10	2.400		430	137		567	675
3536	42″ diameter		9	2.667		500	152		652	780
3537	48″ diameter		8	3		570	171		741	890
3570	Bellows, lined									
3574	6″ diameter	Q-9	30	.533	Ea.	1,350	29.50		1,379.50	1,525
3575	8″ diameter		26	.615		1,425	34		1,459	1,600
3576	10″ diameter		24	.667		1,425	36.50		1,461.50	1,625
3577	12″ diameter		22	.727		1,475	40		1,515	1,675
3578	14″ diameter		21	.762		1,525	42		1,567	1,750

23 51 26 – All-Fuel Vent Chimneys

23 51 26.10 All-Fuel Vent Chimneys, Press. Tight, Dbl. Wall		Crew	Daily Output	Labor-Hours	Unit	2019 Bare Costs Material	Labor	Equipment	Total	Total Incl O&P
3579	16" diameter	Q-9	20	.800	Ea.	1,575	44		1,619	1,825
3580	18" diameter		19	.842		1,650	46		1,696	1,875
3581	20" diameter	Q-10	18	1.333		1,675	76		1,751	1,975
3582	24" diameter		16	1.500		1,975	85.50		2,060.50	2,300
3583	28" diameter		15	1.600		2,100	91		2,191	2,450
3584	32" diameter		14	1.714		2,325	97.50		2,422.50	2,725
3585	36" diameter		12	2		2,575	114		2,689	3,000
3586	42" diameter		11	2.182		2,875	124		2,999	3,375
3587	48" diameter		10	2.400		3,300	137		3,437	3,850
3590	For all 316 stainless steel construction, add					55%				
3600	Ventilated roof thimble, 304 stainless steel									
3620	6" diameter	Q-9	26	.615	Ea.	258	34		292	335
3624	8" diameter		22	.727		263	40		303	350
3625	10" diameter		20	.800		274	44		318	365
3626	12" diameter		18	.889		283	49		332	385
3627	14" diameter		17	.941		294	51.50		345.50	405
3628	16" diameter		16	1		315	55		370	435
3629	18" diameter		15	1.067		340	58.50		398.50	465
3630	20" diameter	Q-10	16	1.500		360	85.50		445.50	525
3631	24" diameter		14	1.714		395	97.50		492.50	585
3632	28" diameter		13	1.846		425	105		530	630
3633	32" diameter		12	2		460	114		574	685
3634	36" diameter		10	2.400		495	137		632	750
3635	42" diameter		9	2.667		550	152		702	835
3636	48" diameter		8	3		605	171		776	925
3650	For 316 stainless steel, add					30%				
3670	Exit cone, 316 stainless steel only									
3674	6" diameter	Q-9	46	.348	Ea.	195	19.10		214.10	243
3675	8" diameter		42	.381		200	21		221	252
3676	10" diameter		40	.400		212	22		234	267
3677	12" diameter		38	.421		225	23		248	283
3678	14" diameter		37	.432		241	23.50		264.50	300
3679	16" diameter		36	.444		305	24.50		329.50	370
3680	18" diameter		35	.457		325	25		350	400
3681	20" diameter	Q-10	28	.857		390	49		439	505
3682	24" diameter		26	.923		515	52.50		567.50	645
3683	28" diameter		25	.960		585	54.50		639.50	730
3684	32" diameter		24	1		720	57		777	880
3685	36" diameter		22	1.091		880	62		942	1,050
3686	42" diameter		21	1.143		1,025	65		1,090	1,225
3687	48" diameter		20	1.200		1,175	68.50		1,243.50	1,375
3720	Roof guide, 304 stainless steel									
3724	6" diameter	Q-9	25	.640	Ea.	89.50	35		124.50	152
3725	8" diameter		21	.762		104	42		146	178
3726	10" diameter		19	.842		114	46		160	195
3727	12" diameter		17	.941		118	51.50		169.50	209
3728	14" diameter		16	1		137	55		192	235
3729	16" diameter		15	1.067		145	58.50		203.50	249
3730	18" diameter		14	1.143		154	62.50		216.50	264
3731	20" diameter	Q-10	15	1.600		162	91		253	315
3732	24" diameter		13	1.846		169	105		274	345
3733	28" diameter		12	2		206	114		320	400
3734	32" diameter		11	2.182		253	124		377	465

23 51 26.10 All-Fuel Vent Chimneys, Press. Tight, Dbl. Wall		Crew	Daily Output	Labor-Hours	Unit	2019 Bare Costs Material	Labor	Equipment	Total	Total Incl O&P
3735	36″ diameter	Q-10	9	2.667	Ea.	300	152		452	565
3736	42″ diameter		8	3		350	171		521	645
3737	48″ diameter		7	3.429		400	195		595	740
3750	For 316 stainless steel, add					30%				
3770	Rain cap with bird screen									
3774	6″ diameter	Q-9	46	.348	Ea.	300	19.10		319.10	365
3775	8″ diameter		42	.381		350	21		371	415
3776	10″ diameter		40	.400		410	22		432	485
3777	12″ diameter		38	.421		470	23		493	555
3778	14″ diameter		37	.432		535	23.50		558.50	625
3779	16″ diameter		36	.444		610	24.50		634.50	705
3780	18″ diameter		35	.457		705	25		730	815
3781	20″ diameter	Q-10	28	.857		800	49		849	955
3782	24″ diameter		26	.923		960	52.50		1,012.50	1,125
3783	28″ diameter		25	.960		1,150	54.50		1,204.50	1,325
3784	32″ diameter		24	1		1,300	57		1,357	1,500
3785	36″ diameter		22	1.091		1,525	62		1,587	1,775
3786	42″ diameter		21	1.143		1,800	65		1,865	2,075
3787	48″ diameter		20	1.200		2,050	68.50		2,118.50	2,350

23 51 26.30 All-Fuel Vent Chimneys, Double Wall, St. Stl.

		Crew	Daily Output	Labor-Hours	Unit	Material	Labor	Equipment	Total	Total Incl O&P
0010	**ALL-FUEL VENT CHIMNEYS, DOUBLE WALL, STAINLESS STEEL**									
7780	All fuel, pressure tight, double wall, 4″ insulation, UL listed, 1,400°F.									
7790	304 stainless steel liner, aluminized steel outer jacket									
7800	6″ diameter	Q-9	60	.267	V.L.F.	66.50	14.65		81.15	95
7804	8″ diameter		52	.308		76	16.90		92.90	110
7806	10″ diameter		48	.333		85	18.30		103.30	122
7808	12″ diameter		44	.364		98	19.95		117.95	139
7810	14″ diameter		42	.381		109	21		130	152
7880	For 316 stainless steel liner, add				L.F.	30%				
8000	All fuel, double wall, stainless steel fittings									
8010	Roof support, 6″ diameter	Q-9	30	.533	Ea.	118	29.50		147.50	174
8030	8″ diameter		26	.615		137	34		171	202
8040	10″ diameter		24	.667		145	36.50		181.50	216
8050	12″ diameter		22	.727		154	40		194	230
8060	14″ diameter		21	.762		163	42		205	243
8100	Elbow 45°, 6″ diameter		30	.533		251	29.50		280.50	320
8140	8″ diameter		26	.615		282	34		316	360
8160	10″ diameter		24	.667		320	36.50		356.50	410
8180	12″ diameter		22	.727		365	40		405	460
8200	14″ diameter		21	.762		410	42		452	515
8300	Insulated tee, 6″ diameter		30	.533		295	29.50		324.50	370
8360	8″ diameter		26	.615		325	34		359	405
8380	10″ diameter		24	.667		355	36.50		391.50	445
8400	12″ diameter		22	.727		415	40		455	515
8420	14″ diameter		20	.800		480	44		524	595
8500	Boot tee, 6″ diameter		28	.571		555	31.50		586.50	660
8520	8″ diameter		24	.667		620	36.50		656.50	735
8530	10″ diameter		22	.727		720	40		760	850
8540	12″ diameter		20	.800		850	44		894	1,000
8550	14″ diameter		18	.889		965	49		1,014	1,125
8600	Rain cap with bird screen, 6″ diameter		30	.533		300	29.50		329.50	375
8640	8″ diameter		26	.615		350	34		384	440

23 51 Breechings, Chimneys, and Stacks

23 51 26 – All-Fuel Vent Chimneys

23 51 26.30 All-Fuel Vent Chimneys, Double Wall, St. Stl.		Crew	Daily Output	Labor-Hours	Unit	Material	2019 Bare Costs Labor	Equipment	Total	Total Incl O&P
8660	10" diameter	Q-9	24	.667	Ea.	410	36.50		446.50	505
8680	12" diameter		22	.727		470	40		510	580
8700	14" diameter		21	.762		535	42		577	655
8800	Flat roof flashing, 6" diameter		30	.533		111	29.50		140.50	167
8840	8" diameter		26	.615		121	34		155	185
8860	10" diameter		24	.667		130	36.50		166.50	199
8880	12" diameter		22	.727		142	40		182	217
8900	14" diameter		21	.762		145	42		187	223

23 52 Heating Boilers

23 52 13 – Electric Boilers

23 52 13.10 Electric Boilers, ASME		Crew	Daily Output	Labor-Hours	Unit	Material	Labor	Equipment	Total	Total Incl O&P
0010	**ELECTRIC BOILERS, ASME**, Standard controls and trim R235000-30									
1000	Steam, 6 KW, 20.5 MBH	Q-19	1.20	20	Ea.	4,300	1,175		5,475	6,475
1040	9 KW, 30.7 MBH D3020-102		1.20	20		4,275	1,175		5,450	6,450
1060	18 KW, 61.4 MBH		1.20	20		4,400	1,175		5,575	6,600
1080	24 KW, 81.8 MBH D3020-104		1.10	21.818		5,100	1,275		6,375	7,525
1120	36 KW, 123 MBH		1.10	21.818		5,650	1,275		6,925	8,125
1160	60 KW, 205 MBH		1	24		7,075	1,400		8,475	9,875
1220	112 KW, 382 MBH		.75	32		9,600	1,875		11,475	13,300
1240	148 KW, 505 MBH		.65	36.923		10,200	2,150		12,350	14,400
1260	168 KW, 573 MBH		.60	40		20,500	2,325		22,825	26,000
1280	222 KW, 758 MBH		.55	43.636		23,900	2,550		26,450	30,100
1300	296 KW, 1,010 MBH		.45	53.333		25,100	3,125		28,225	32,300
1320	300 KW, 1,023 MBH		.40	60		26,400	3,500		29,900	34,400
1340	370 KW, 1,263 MBH		.35	68.571		29,300	4,000		33,300	38,200
1360	444 KW, 1,515 MBH		.30	80		31,400	4,675		36,075	41,600
1380	518 KW, 1,768 MBH	Q-21	.36	88.889		32,700	5,325		38,025	44,000
1400	592 KW, 2,020 MBH		.34	94.118		35,500	5,625		41,125	47,400
1420	666 KW, 2,273 MBH		.32	100		36,900	5,975		42,875	49,600
1460	740 KW, 2,526 MBH		.28	114		38,400	6,825		45,225	52,500
1480	814 KW, 2,778 MBH		.25	128		40,800	7,650		48,450	56,500
1500	962 KW, 3,283 MBH		.22	145		45,000	8,700		53,700	62,500
1520	1,036 KW, 3,536 MBH		.20	160		47,800	9,575		57,375	67,000
1540	1,110 KW, 3,788 MBH		.19	168		49,700	10,100		59,800	69,500
1560	2,070 KW, 7,063 MBH		.18	178		68,500	10,600		79,100	91,000
1580	2,250 KW, 7,677 MBH		.17	188		80,000	11,300		91,300	105,000
1600	2,340 KW, 7,984 MBH		.16	200		86,500	12,000		98,500	113,500
2000	Hot water, 7.5 KW, 25.6 MBH	Q-19	1.30	18.462		5,200	1,075		6,275	7,350
2020	15 KW, 51.2 MBH		1.30	18.462		5,250	1,075		6,325	7,400
2040	30 KW, 102 MBH		1.20	20		5,600	1,175		6,775	7,900
2060	45 KW, 164 MBH		1.20	20		5,700	1,175		6,875	8,025
2070	60 KW, 205 MBH		1.20	20		5,800	1,175		6,975	8,150
2080	75 KW, 256 MBH		1.10	21.818		6,150	1,275		7,425	8,675
2100	90 KW, 307 MBH		1.10	21.818		6,150	1,275		7,425	8,675
2120	105 KW, 358 MBH		1	24		6,650	1,400		8,050	9,400
2140	120 KW, 410 MBH		.90	26.667		6,675	1,550		8,225	9,650
2160	135 KW, 461 MBH		.75	32		7,550	1,875		9,425	11,100
2180	150 KW, 512 MBH		.65	36.923		7,950	2,150		10,100	12,000
2200	165 KW, 563 MBH		.60	40		8,025	2,325		10,350	12,300
2220	296 KW, 1,010 MBH		.55	43.636		15,600	2,550		18,150	21,000

23 52 Heating Boilers

23 52 13 – Electric Boilers

23 52 13.10 Electric Boilers, ASME		Crew	Daily Output	Labor-Hours	Unit	Material	2019 Bare Costs Labor	Equipment	Total	Total Incl O&P
2280	370 KW, 1,263 MBH	Q-19	.40	60	Ea.	17,300	3,500		20,800	24,300
2300	444 KW, 1,515 MBH	↓	.35	68.571		20,100	4,000		24,100	28,100
2340	518 KW, 1,768 MBH	Q-21	.44	72.727		22,300	4,350		26,650	31,000
2360	592 KW, 2,020 MBH		.43	74.419		25,300	4,450		29,750	34,500
2400	666 KW, 2,273 MBH		.40	80		26,400	4,775		31,175	36,200
2420	740 KW, 2,526 MBH		.39	82.051		27,500	4,900		32,400	37,700
2440	814 KW, 2,778 MBH		.38	84.211		29,500	5,025		34,525	40,100
2460	888 KW, 3,031 MBH		.37	86.486		32,000	5,175		37,175	43,000
2480	962 KW, 3,283 MBH		.36	88.889		32,800	5,325		38,125	44,000
2500	1,036 KW, 3,536 MBH		.34	94.118		35,200	5,625		40,825	47,200
2520	1,110 KW, 3,788 MBH		.33	96.970		38,500	5,800		44,300	51,000
2540	1,440 KW, 4,915 MBH		.32	100		42,900	5,975		48,875	56,000
2560	1,560 KW, 5,323 MBH		.31	103		46,900	6,175		53,075	61,000
2580	1,680 KW, 5,733 MBH		.30	107		50,000	6,375		56,375	64,500
2600	1,800 KW, 6,143 MBH		.29	110		53,500	6,600		60,100	68,500
2620	1,980 KW, 6,757 MBH		.28	114		58,500	6,825		65,325	74,000
2640	2,100 KW, 7,167 MBH		.27	119		64,500	7,075		71,575	81,500
2660	2,220 KW, 7,576 MBH		.26	123		65,000	7,350		72,350	82,500
2680	2,400 KW, 8,191 MBH		.25	128		68,000	7,650		75,650	86,500
2700	2,610 KW, 8,905 MBH		.24	133		72,500	7,975		80,475	92,000
2720	2,790 KW, 9,519 MBH		.23	139		76,000	8,325		84,325	96,000
2740	2,970 KW, 10,133 MBH		.21	152		77,500	9,100		86,600	99,000
2760	3,150 KW, 10,748 MBH		.19	168		83,000	10,100		93,100	106,000
2780	3,240 KW, 11,055 MBH		.18	178		84,500	10,600		95,100	109,000
2800	3,420 KW, 11,669 MBH		.17	188		90,500	11,300		101,800	116,500
2820	3,600 KW, 12,283 MBH	↓	.16	200	↓	94,500	12,000		106,500	122,000

23 52 16 – Condensing Boilers

23 52 16.24 Condensing Boilers		Crew	Daily Output	Labor-Hours	Unit	Material	Labor	Equipment	Total	Total Incl O&P
0010	**CONDENSING BOILERS**, Cast iron, high efficiency									
0020	Packaged with standard controls, circulator and trim									
0030	Intermittent (spark) pilot, natural or LP gas									
0040	Hot water, DOE MBH output (AFUE %)									
0100	42 MBH (84.0%) G	Q-5	1.80	8.889	Ea.	1,525	510		2,035	2,475
0120	57 MBH (84.3%) G		1.60	10		1,675	575		2,250	2,725
0140	85 MBH (84.0%) G		1.40	11.429		1,875	660		2,535	3,025
0160	112 MBH (83.7%) G	↓	1.20	13.333		2,125	765		2,890	3,475
0180	140 MBH (83.3%) G	Q-6	1.60	15		2,350	895		3,245	3,950
0200	167 MBH (83.0%) G		1.40	17.143		2,650	1,025		3,675	4,450
0220	194 MBH (82.7%) G	↓	1.20	20	↓	2,950	1,200		4,150	5,025

23 52 19 – Pulse Combustion Boilers

23 52 19.20 Pulse Type Combustion Boilers		Crew	Daily Output	Labor-Hours	Unit	Material	Labor	Equipment	Total	Total Incl O&P
0010	**PULSE TYPE COMBUSTION BOILERS**, High efficiency									
7990	Special feature gas fired boilers									
8000	Pulse combustion, standard controls/trim									
8010	Hot water, DOE MBH output (AFUE %)									
8030	71 MBH (95.2%)	Q-5	1.60	10	Ea.	3,825	575		4,400	5,075
8050	94 MBH (95.3%) G		1.40	11.429		3,825	660		4,485	5,200
8080	139 MBH (95.6%) G		1.20	13.333		4,725	765		5,490	6,350
8090	207 MBH (95.4%)		1.16	13.793		5,425	795		6,220	7,150
8120	270 MBH (96.4%)		1.12	14.286		7,425	820		8,245	9,400
8130	365 MBH (91.7%)	↓	1.07	14.953	↓	8,375	860		9,235	10,500

23 52 23 – Cast-Iron Boilers

23 52 23.20 Gas-Fired Boilers		Crew	Daily Output	Labor-Hours	Unit	Material	Labor	Equipment	Total	Total Incl O&P
						2019 Bare Costs				
0010	**GAS-FIRED BOILERS**, Natural or propane, standard controls, packaged R235000-10									
1000	Cast iron, with insulated jacket									
2000	Steam, gross output, 81 MBH R235000-20	Q-7	1.40	22.857	Ea.	2,525	1,400		3,925	4,875
2020	102 MBH		1.30	24.615		2,625	1,500		4,125	5,125
2040	122 MBH R235000-30		1	32		3,050	1,950		5,000	6,275
2060	163 MBH		.90	35.556		3,200	2,175		5,375	6,775
2080	203 MBH		.90	35.556		3,725	2,175		5,900	7,350
2100	240 MBH		.85	37.647		3,775	2,300		6,075	7,625
2120	280 MBH		.80	40		4,600	2,425		7,025	8,700
2140	320 MBH		.70	45.714		4,850	2,775		7,625	9,525
2160	360 MBH		.63	51.200		5,475	3,125		8,600	10,700
2180	400 MBH		.56	56.838		5,800	3,450		9,250	11,600
2200	440 MBH		.51	62.500		6,250	3,800		10,050	12,600
2220	544 MBH		.45	71.588		8,925	4,350		13,275	16,400
2240	765 MBH		.43	74.419		10,800	4,525		15,325	18,600
2260	892 MBH		.38	84.211		12,400	5,125		17,525	21,400
2280	1,275 MBH		.34	94.118		15,200	5,725		20,925	25,300
2300	1,530 MBH		.32	100		16,300	6,100		22,400	27,100
2320	1,875 MBH		.30	107		24,400	6,500		30,900	36,600
2340	2,170 MBH		.26	122		26,200	7,425		33,625	40,000
2360	2,675 MBH		.20	163		27,900	9,950		37,850	45,600
2380	3,060 MBH		.19	172		28,600	10,500		39,100	47,100
2400	3,570 MBH		.18	182		34,800	11,100		45,900	55,000
2420	4,207 MBH		.16	205		38,000	12,500		50,500	60,500
2440	4,720 MBH		.15	208		65,500	12,700		78,200	91,000
2460	5,660 MBH		.15	221		74,500	13,400		87,900	101,500
2480	6,100 MBH		.13	246		84,500	15,000		99,500	115,000
2500	6,390 MBH		.12	267		88,000	16,200		104,200	121,500
2520	6,680 MBH		.11	291		90,000	17,700		107,700	125,500
2540	6,970 MBH		.10	320		94,500	19,500		114,000	132,500
3000	Hot water, gross output, 80 MBH		1.46	21.918		1,950	1,325		3,275	4,125
3020	100 MBH		1.35	23.704		2,275	1,450		3,725	4,675
3040	122 MBH		1.10	29.091		2,575	1,775		4,350	5,475
3060	163 MBH		1	32		2,900	1,950		4,850	6,100
3080	203 MBH		1	32		3,725	1,950		5,675	7,025
3100	240 MBH		.95	33.684		3,825	2,050		5,875	7,275
3120	280 MBH		.90	35.556		4,000	2,175		6,175	7,625
3140	320 MBH		.80	40		4,350	2,425		6,775	8,425
3160	360 MBH		.71	45.070		4,975	2,750		7,725	9,575
3180	400 MBH		.64	50		5,150	3,050		8,200	10,300
3200	440 MBH		.58	54.983		5,525	3,350		8,875	11,100
3220	544 MBH		.51	62.992		8,600	3,825		12,425	15,200
3240	765 MBH		.46	70.022		10,400	4,275		14,675	17,800
3260	1,088 MBH		.40	80		13,000	4,875		17,875	21,600
3280	1,275 MBH		.36	89.888		15,000	5,475		20,475	24,700
3300	1,530 MBH		.31	105		16,100	6,375		22,475	27,300
3320	2,000 MBH		.26	125		20,600	7,600		28,200	34,100
3340	2,312 MBH		.22	148		25,800	9,025		34,825	41,900
3360	2,856 MBH		.20	160		29,600	9,750		39,350	47,200
3380	3,264 MBH		.18	180		31,400	10,900		42,300	51,000
3400	3,996 MBH		.16	195		34,400	11,900		46,300	55,500
3420	4,488 MBH		.15	211		37,800	12,800		50,600	61,000

23 52 23.20 Gas-Fired Boilers		Crew	Daily Output	Labor-Hours	Unit	Material	2019 Bare Costs Labor	Equipment	Total	Total Incl O&P
3440	4,720 MBH	Q-7	.15	221	Ea.	71,500	13,400		84,900	98,500
3460	5,520 MBH		.14	229		90,500	13,900		104,400	120,500
3480	6,100 MBH		.13	250		112,500	15,200		127,700	147,000
3500	6,390 MBH		.11	286		115,500	17,400		132,900	153,500
3520	6,680 MBH		.10	311		120,500	18,900		139,400	161,000
3540	6,970 MBH		.09	360		124,500	21,900		146,400	170,000
7000	For tankless water heater, add					10%				
7050	For additional zone valves up to 312 MBH, add					197			197	217

23 52 23.30 Gas/Oil Fired Boilers		Crew	Daily Output	Labor-Hours	Unit	Material	Labor	Equipment	Total	Total Incl O&P
0010	**GAS/OIL FIRED BOILERS**, Combination with burners and controls, packaged									
1000	Cast iron with insulated jacket									
2000	Steam, gross output, 720 MBH	Q-7	.43	74.074	Ea.	14,000	4,500		18,500	22,200
2020	810 MBH		.38	83.990		14,000	5,125		19,125	23,100
2040	1,084 MBH		.34	93.023		16,400	5,675		22,075	26,500
2060	1,360 MBH		.33	98.160		18,700	5,975		24,675	29,600
2080	1,600 MBH		.30	107		20,000	6,525		26,525	31,800
2100	2,040 MBH		.25	130		24,300	7,925		32,225	38,700
2120	2,450 MBH		.21	156		25,700	9,500		35,200	42,600
2140	2,700 MBH		.19	166		27,100	10,100		37,200	45,000
2160	3,000 MBH		.18	176		29,400	10,700		40,100	48,400
2180	3,270 MBH		.17	184		31,500	11,200		42,700	51,500
2200	3,770 MBH		.17	192		64,500	11,700		76,200	88,500
2220	4,070 MBH		.16	200		68,000	12,200		80,200	93,500
2240	4,650 MBH		.15	211		71,500	12,800		84,300	98,000
2260	5,230 MBH		.14	224		79,500	13,600		93,100	108,000
2280	5,520 MBH		.14	235		86,500	14,300		100,800	116,500
2300	5,810 MBH		.13	248		88,000	15,100		103,100	119,500
2320	6,100 MBH		.12	260		91,500	15,800		107,300	124,500
2340	6,390 MBH		.11	296		92,500	18,000		110,500	129,000
2360	6,680 MBH		.10	320		96,000	19,500		115,500	134,500
2380	6,970 MBH		.09	372		98,000	22,700		120,700	142,000
2900	Hot water, gross output									
2910	200 MBH	Q-6	.62	39.024	Ea.	10,400	2,325		12,725	14,900
2920	300 MBH		.49	49.080		10,400	2,925		13,325	15,800
2930	400 MBH		.41	57.971		12,200	3,450		15,650	18,600
2940	500 MBH		.36	67.039		13,100	4,000		17,100	20,400
3000	584 MBH	Q-7	.44	72.072		14,300	4,400		18,700	22,300
3020	876 MBH		.41	79.012		19,100	4,800		23,900	28,300
3040	1,168 MBH		.31	104		27,900	6,325		34,225	40,200
3060	1,460 MBH		.28	113		35,500	6,875		42,375	49,300
3080	2,044 MBH		.26	122		40,800	7,425		48,225	56,000
3100	2,628 MBH		.21	150		44,000	9,150		53,150	62,000
3120	3,210 MBH		.18	175		47,800	10,600		58,400	68,500
3140	3,796 MBH		.17	186		53,500	11,300		64,800	76,000
3160	4,088 MBH		.16	195		61,000	11,900		72,900	85,000
3180	4,672 MBH		.16	204		66,500	12,400		78,900	91,500
3200	5,256 MBH		.15	218		75,000	13,300		88,300	102,500
3220	6,000 MBH, 179 BHP		.13	256		106,500	15,600		122,100	140,500
3240	7,130 MBH, 213 BHP		.08	386		111,500	23,500		135,000	157,500
3260	9,800 MBH, 286 BHP		.06	533		128,500	32,500		161,000	189,500
3280	10,900 MBH, 325.6 BHP		.05	593		155,500	36,100		191,600	225,000
3290	12,200 MBH, 364.5 BHP		.05	667		170,000	40,600		210,600	248,000

23 52 Heating Boilers

23 52 23 – Cast-Iron Boilers

23 52 23.30 Gas/Oil Fired Boilers		Crew	Daily Output	Labor-Hours	Unit	Material	2019 Bare Costs Labor	Equipment	Total	Total Incl O&P
3300	13,500 MBH, 403.3 BHP	Q-7	.04	727	Ea.	188,000	44,300		232,300	273,500

23 52 23.40 Oil-Fired Boilers

		Crew	Daily Output	Labor-Hours	Unit	Material	Labor	Equipment	Total	Total Incl O&P
0010	**OIL-FIRED BOILERS**, Standard controls, flame retention burner, packaged									
1000	Cast iron, with insulated flush jacket									
2000	Steam, gross output, 109 MBH	Q-7	1.20	26.667	Ea.	2,225	1,625		3,850	4,875
2020	144 MBH		1.10	29.091		2,500	1,775		4,275	5,400
2040	173 MBH		1	32		2,825	1,950		4,775	6,050
2060	207 MBH		.90	35.556		3,025	2,175		5,200	6,575
2080	236 MBH		.85	37.647		3,625	2,300		5,925	7,425
2100	300 MBH		.70	45.714		4,525	2,775		7,300	9,150
2120	480 MBH		.50	64		5,875	3,900		9,775	12,300
2140	665 MBH		.45	71.111		8,350	4,325		12,675	15,700
2160	794 MBH		.41	78.049		9,300	4,750		14,050	17,300
2180	1,084 MBH		.38	85.106		10,700	5,175		15,875	19,500
2200	1,360 MBH		.33	98.160		12,200	5,975		18,175	22,400
2220	1,600 MBH		.26	122		16,800	7,425		24,225	29,700
2240	2,175 MBH		.24	134		17,500	8,150		25,650	31,400
2260	2,480 MBH		.21	156		20,000	9,500		29,500	36,300
2280	3,000 MBH		.19	170		23,000	10,400		33,400	40,800
2300	3,550 MBH		.17	187		26,800	11,400		38,200	46,600
2320	3,820 MBH		.16	200		37,700	12,200		49,900	60,000
2340	4,360 MBH		.15	215		41,400	13,100		54,500	65,000
2360	4,940 MBH		.14	225		65,500	13,700		79,200	92,500
2380	5,520 MBH		.14	235		77,000	14,300		91,300	106,500
2400	6,100 MBH		.13	256		88,500	15,600		104,100	120,500
2420	6,390 MBH		.11	291		93,000	17,700		110,700	129,000
2440	6,680 MBH		.10	314		95,500	19,100		114,600	133,500
2460	6,970 MBH		.09	364		99,000	22,100		121,100	142,000
3000	Hot water, same price as steam									
4000	For tankless coil in smaller sizes, add				Ea.	15%				

23 52 23.60 Solid-Fuel Boilers

		Crew	Daily Output	Labor-Hours	Unit	Material	Labor	Equipment	Total	Total Incl O&P
0010	**SOLID-FUEL BOILERS**									
3000	Stoker fired (coal/wood/biomass) cast iron with flush jacket and									
3400	insulation, steam or water, gross output, 1,280 MBH	Q-6	.36	66.667	Ea.	175,000	3,975		178,975	198,500
3420	1,460 MBH		.30	80		180,500	4,775		185,275	205,500
3440	1,640 MBH		.28	85.714		193,500	5,125		198,625	220,500
3460	1,820 MBH		.26	92.308		209,000	5,500		214,500	238,500
3480	2,000 MBH		.25	96		226,000	5,725		231,725	257,000
3500	2,360 MBH		.23	104		233,000	6,225		239,225	266,000
3540	2,725 MBH		.20	120		240,000	7,150		247,150	275,000
3800	2,950 MBH	Q-7	.16	200		249,000	12,200		261,200	292,500
3820	3,210 MBH		.15	213		263,000	13,000		276,000	308,500
3840	3,480 MBH		.14	229		264,500	13,900		278,400	312,000
3860	3,745 MBH		.14	229		266,500	13,900		280,400	314,000
3880	4,000 MBH		.13	246		268,000	15,000		283,000	317,500
3900	4,200 MBH		.13	246		270,000	15,000		285,000	319,000
3920	4,400 MBH		.12	267		271,500	16,200		287,700	323,000
3940	4,600 MBH		.12	267		274,500	16,200		290,700	326,500

23 52 26 – Steel Boilers

23 52 26.40 Oil-Fired Boilers		Crew	Daily Output	Labor-Hours	Unit	Material	2019 Bare Costs Labor	Equipment	Total	Total Incl O&P
0010	**OIL-FIRED BOILERS**, Standard controls, flame retention burner									
5000	Steel, with insulated flush jacket									
7000	Hot water, gross output, 103 MBH	Q-6	1.60	15	Ea.	1,875	895		2,770	3,400
7020	122 MBH		1.45	16.506		2,025	985		3,010	3,700
7040	137 MBH		1.36	17.595		2,300	1,050		3,350	4,125
7060	168 MBH		1.30	18.405		2,275	1,100		3,375	4,150
7080	225 MBH		1.22	19.704		3,525	1,175		4,700	5,650
7100	315 MBH		.96	25.105		4,475	1,500		5,975	7,175
7120	420 MBH		.70	34.483		5,825	2,050		7,875	9,500
7140	525 MBH		.57	42.403		7,475	2,525		10,000	12,000
7180	735 MBH		.48	50.104		7,175	3,000		10,175	12,400
7220	1,050 MBH		.37	65.753		20,400	3,925		24,325	28,400
7280	2,310 MBH		.21	115		26,400	6,850		33,250	39,300
7320	3,150 MBH	↓	.13	185		43,100	11,000		54,100	64,000
7340	For tankless coil in steam or hot water, add				↓	7%				

23 52 26.70 Packaged Water Tube Boilers		Crew	Daily Output	Labor-Hours	Unit	Material	Labor	Equipment	Total	Total Incl O&P
0010	**PACKAGED WATER TUBE BOILERS**									
2000	Packaged water tube, #2 oil, steam or hot water, gross output									
2010	200 MBH	Q-6	1.50	16	Ea.	4,675	955		5,630	6,575
2014	275 MBH		1.40	17.143		5,250	1,025		6,275	7,300
2018	360 MBH		1.10	21.818		6,925	1,300		8,225	9,575
2022	520 MBH		.65	36.923		8,200	2,200		10,400	12,300
2026	600 MBH		.60	40		11,100	2,375		13,475	15,800
2030	720 MBH		.55	43.636		12,200	2,600		14,800	17,300
2034	960 MBH	↓	.48	50		14,200	2,975		17,175	20,100
2040	1,200 MBH	Q-7	.50	64		25,000	3,900		28,900	33,400
2044	1,440 MBH		.45	71.111		34,000	4,325		38,325	43,900
2060	1,600 MBH		.40	80		40,000	4,875		44,875	51,500
2068	1,920 MBH		.35	91.429		43,300	5,575		48,875	56,000
2072	2,160 MBH		.33	96.970		48,100	5,900		54,000	62,000
2080	2,400 MBH		.30	107		54,000	6,500		60,500	69,000
2100	3,200 MBH		.25	128		61,500	7,800		69,300	79,000
2120	4,800 MBH	↓	.20	160	↓	80,500	9,750		90,250	103,000
2200	Gas fired									
2204	200 MBH	Q-6	1.50	16	Ea.	8,600	955		9,555	10,900
2208	275 MBH		1.40	17.143		9,000	1,025		10,025	11,400
2212	360 MBH		1.10	21.818		9,225	1,300		10,525	12,200
2216	520 MBH		.65	36.923		8,475	2,200		10,675	12,600
2220	600 MBH		.60	40		11,000	2,375		13,375	15,700
2224	720 MBH		.55	43.636		10,000	2,600		12,600	14,900
2228	960 MBH	↓	.48	50		18,400	2,975		21,375	24,800
2232	1,220 MBH	Q-7	.50	64		18,900	3,900		22,800	26,700
2236	1,440 MBH		.45	71.111		20,100	4,325		24,425	28,600
2240	1,680 MBH		.40	80		29,700	4,875		34,575	40,000
2244	1,920 MBH		.35	91.429		28,700	5,575		34,275	39,900
2248	2,160 MBH		.33	96.970		30,400	5,900		36,300	42,400
2252	2,400 MBH	↓	.30	107	↓	34,200	6,500		40,700	47,400

23 52 Heating Boilers

23 52 28 – Swimming Pool Boilers

23 52 28.10 Swimming Pool Heaters		Crew	Daily Output	Labor-Hours	Unit	2019 Bare Costs Material	Labor	Equipment	Total	Total Incl O&P
0010	**SWIMMING POOL HEATERS**, Not including wiring, external									
0020	piping, base or pad									
0160	Gas fired, input, 155 MBH	Q-6	1.50	16	Ea.	1,975	955		2,930	3,600
0200	199 MBH		1	24		2,125	1,425		3,550	4,500
0220	250 MBH		.70	34.286		2,325	2,050		4,375	5,625
0240	300 MBH		.60	40		2,425	2,375		4,800	6,250
0260	399 MBH		.50	48		2,725	2,875		5,600	7,300
0280	500 MBH		.40	60		8,625	3,575		12,200	14,900
0300	650 MBH		.35	68.571		9,200	4,100		13,300	16,300
0320	750 MBH		.33	72.727		10,000	4,350		14,350	17,500
0360	990 MBH		.22	109		13,500	6,500		20,000	24,600
0370	1,260 MBH		.21	114		17,500	6,825		24,325	29,500
0380	1,440 MBH		.19	126		19,200	7,550		26,750	32,400
0400	1,800 MBH		.14	171		20,800	10,200		31,000	38,300
0410	2,070 MBH		.13	185		24,600	11,000		35,600	43,500
2000	Electric, 12 KW, 4,800 gallon pool	Q-19	3	8		2,125	465		2,590	3,050
2020	15 KW, 7,200 gallon pool		2.80	8.571		2,150	500		2,650	3,100
2040	24 KW, 9,600 gallon pool		2.40	10		2,500	585		3,085	3,625
2060	30 KW, 12,000 gallon pool		2	12		2,550	700		3,250	3,875
2080	36 KW, 14,400 gallon pool		1.60	15		2,950	875		3,825	4,550
2100	57 KW, 24,000 gallon pool		1.20	20		3,700	1,175		4,875	5,800
9000	To select pool heater: 12 BTUH x S.F. pool area									
9010	X temperature differential = required output									
9050	For electric, KW = gallons x 2.5 divided by 1,000									
9100	For family home type pool, double the									
9110	Rated gallon capacity = 1/2°F rise per hour									

23 52 39 – Fire-Tube Boilers

23 52 39.13 Scotch Marine Boilers		Crew	Daily Output	Labor-Hours	Unit	Material	Labor	Equipment	Total	Total Incl O&P
0010	**SCOTCH MARINE BOILERS**									
1000	Packaged fire tube, #2 oil, gross output									
1006	15 psi steam									
1010	1,005 MBH, 30 HP	Q-6	.33	72.072	Ea.	56,000	4,300		60,300	68,500
1014	1,675 MBH, 50 HP	"	.26	93.023		70,500	5,550		76,050	86,000
1020	3,348 MBH, 100 HP	Q-7	.21	152		94,500	9,275		103,775	118,000
1030	5,025 MBH, 150 HP		.17	193		119,500	11,700		131,200	148,500
1040	6,696 MBH, 200 HP		.14	224		138,500	13,600		152,100	173,000
1050	8,375 MBH, 250 HP		.14	232		164,000	14,100		178,100	201,000
1060	10,044 MBH, 300 HP		.13	252		181,000	15,300		196,300	222,000
1080	16,740 MBH, 500 HP		.08	381		246,000	23,200		269,200	305,500
1100	23,435 MBH, 700 HP		.07	485		301,500	29,500		331,000	376,000
1102	26,780 MBH, 800 HP		.07	485		324,000	29,500		353,500	401,000
1104	30,125 MBH, 900 HP		.06	500		591,000	30,400		621,400	695,500
1106	33,475 MBH, 1,000 HP		.06	525		625,500	31,900		657,400	736,000
1107	36,825 MBH, 1,100 HP		.06	552		671,500	33,600		705,100	789,000
1108	40,170 MBH, 1,200 HP		.06	582		704,000	35,400		739,400	827,500
1110	46,865 MBH, 1,400 HP		.05	604		741,500	36,800		778,300	871,000
1112	53,560 MBH, 1,600 HP		.05	627		806,000	38,200		844,200	944,000
1114	60,250 MBH, 1,800 HP		.05	667		832,500	40,600		873,100	976,500
1116	66,950 MBH, 2,000 HP		.05	696		881,500	42,400		923,900	1,033,000
1118	73,650 MBH, 2,200 HP		.04	744		952,500	45,300		997,800	1,115,500
1120	To fire #6, add		.83	38.554		18,200	2,350		20,550	23,500
1140	To fire #6, and gas, add		.42	76.190		26,800	4,650		31,450	36,500

23 52 Heating Boilers

23 52 39 – Fire-Tube Boilers

23 52 39.13 Scotch Marine Boilers		Crew	Daily Output	Labor-Hours	Unit	Material	2019 Bare Costs Labor	Equipment	Total	Total Incl O&P
1141	1,675 MBH, 50 HP	Q-6	.26	93.023	Ea.	91,500	5,550		97,050	109,000
1142	6,696 MBH, 200 HP	Q-7	.14	224		159,000	13,600		172,600	195,000
1143	10,044 MBH, 300 HP		.13	252		195,000	15,300		210,300	237,500
1144	16,740 MBH, 500 HP		.08	381		256,500	23,200		279,700	317,000
1145	23,435 MBH, 700 HP		.07	485		308,500	29,500		338,000	384,000
1146	30,125 MBH, 900 HP		.06	500		542,000	30,400		572,400	641,500
1160	For high pressure, add					29,800			29,800	32,700
1180	For duplex package feed system									
1200	To 3,348 MBH boiler, add	Q-7	.54	59.259	Ea.	14,400	3,600		18,000	21,300
1220	To 6,696 MBH boiler, add		.41	78.049		17,100	4,750		21,850	25,900
1240	To 10,044 MBH boiler, add		.38	84.211		19,600	5,125		24,725	29,300
1260	To 16,740 MBH boiler, add		.28	114		22,300	6,950		29,250	35,000
1280	To 23,435 MBH boiler, add		.25	128		26,400	7,800		34,200	40,700
1300	150 psi steam									
1310	1,005 MBH, 30 HP	Q-6	.33	72.072	Ea.	64,000	4,300		68,300	77,000
1320	1,675 MBH, 50 HP	"	.26	93.023		79,500	5,550		85,050	96,000
1330	3,350 MBH, 100 HP	Q-7	.21	152		113,500	9,275		122,775	139,000
1340	5,025 MBH, 150 HP		.17	193		145,000	11,700		156,700	177,000
1350	6,700 MBH, 200 HP		.14	224		160,500	13,600		174,100	197,000
1360	10,050 MBH, 300 HP		.13	252		226,500	15,300		241,800	272,000
1400	Packaged Scotch Marine, #6 oil									
1410	15 psi steam									
1420	1,675 MBH, 50 HP	Q-6	.25	97.166	Ea.	86,500	5,800		92,300	103,500
1430	3,350 MBH, 100 HP	Q-7	.20	159		113,000	9,700		122,700	138,500
1440	5,025 MBH, 150 HP		.16	203		138,000	12,300		150,300	170,000
1450	6,700 MBH, 200 HP		.14	234		157,500	14,200		171,700	195,000
1460	8,375 MBH, 250 HP		.13	244		180,000	14,900		194,900	220,500
1470	10,050 MBH, 300 HP		.12	262		202,000	16,000		218,000	246,500
1480	13,400 MBH, 400 HP		.09	352		249,500	21,400		270,900	306,000
1500	150 psi steam									
1510	1,675 MBH, 50 HP	Q-6	.25	97.166	Ea.	102,500	5,800		108,300	121,000
1520	3,350 MBH, 100 HP	Q-7	.20	159		130,500	9,700		140,200	158,000
1530	5,025 MBH, 150 HP		.16	203		160,500	12,300		172,800	195,000
1540	6,700 MBH, 200 HP		.14	234		176,000	14,200		190,200	215,000
1550	8,375 MBH, 250 HP		.13	244		214,000	14,900		228,900	258,000
1560	10,050 MBH, 300 HP		.12	262		238,000	16,000		254,000	286,000
1570	13,400 MBH, 400 HP		.09	352		276,000	21,400		297,400	336,000

23 52 84 – Boiler Blowdown

23 52 84.10 Boiler Blowdown Systems		Crew	Daily Output	Labor-Hours	Unit	Material	Labor	Equipment	Total	Total Incl O&P
0010	**BOILER BLOWDOWN SYSTEMS**									
1010	Boiler blowdown, auto/manual to 2,000 MBH	Q-5	3.75	4.267	Ea.	6,550	246		6,796	7,575
1020	7,300 MBH	"	3	5.333	"	7,525	305		7,830	8,725

23 52 88 – Burners

23 52 88.10 Replacement Type Burners		Crew	Daily Output	Labor-Hours	Unit	Material	Labor	Equipment	Total	Total Incl O&P
0010	**REPLACEMENT TYPE BURNERS**									
0990	Residential, conversion, gas fired, LP or natural									
1000	Gun type, atmospheric input 50 to 225 MBH	Q-1	2.50	6.400	Ea.	1,000	365		1,365	1,650
1020	100 to 400 MBH		2	8		1,675	455		2,130	2,525
1040	300 to 1,000 MBH		1.70	9.412		5,050	535		5,585	6,350
2000	Commercial and industrial, gas/oil, input									
2050	400 MBH	Q-1	1.50	10.667	Ea.	4,450	605		5,055	5,800
2090	670 MBH		1.40	11.429		4,450	650		5,100	5,850

23 52 Heating Boilers

23 52 88 – Burners

23 52 88.10 Replacement Type Burners		Crew	Daily Output	Labor-Hours	Unit	2019 Bare Costs Material	Labor	Equipment	Total	Total Incl O&P
2140	1,155 MBH	Q-1	1.30	12.308	Ea.	4,950	700		5,650	6,500
2200	1,800 MBH		1.20	13.333		4,975	760		5,735	6,575
2260	3,000 MBH		1.10	14.545		5,050	825		5,875	6,825
2320	4,100 MBH		1	16		6,950	910		7,860	9,025
3000	Flame retention oil fired assembly, input									
3020	.50 to 2.25 GPH	Q-1	2.40	6.667	Ea.	360	380		740	965
3040	2.0 to 5.0 GPH		2	8		340	455		795	1,050
3060	3.0 to 7.0 GPH		1.80	8.889		580	505		1,085	1,400
3080	6.0 to 12.0 GPH		1.60	10		930	570		1,500	1,875
4600	Gas safety, shut off valve, 3/4" threaded	1 Stpi	20	.400		187	25.50		212.50	245
4610	1" threaded		19	.421		181	27		208	240
4620	1-1/4" threaded		15	.533		204	34		238	275
4630	1-1/2" threaded		13	.615		220	39.50		259.50	300
4640	2" threaded		11	.727		247	46.50		293.50	340
4650	2-1/2" threaded	Q-1	15	1.067		282	60.50		342.50	400
4660	3" threaded		13	1.231		385	70		455	530
4670	4" flanged		3	5.333		2,775	305		3,080	3,500
4680	6" flanged	Q-2	3	8		6,175	470		6,645	7,500

23 53 Heating Boiler Feedwater Equipment

23 53 24 – Shot Chemical Feeder

23 53 24.10 Chemical Feeder		Crew	Daily Output	Labor-Hours	Unit	Material	Labor	Equipment	Total	Total Incl O&P
0010	**CHEMICAL FEEDER**									
0200	Shot chem feeder, by pass, in line mount, 300 PSIG, 1.7 gallon	2 Stpi	1.80	8.889	Ea.	284	570		854	1,175
0230	12 gallon		.93	17.204		500	1,100		1,600	2,200
0400	300 lb., ASME, 5 gallon		1.24	12.903		310	825		1,135	1,600
0420	10 gallon		.93	17.204		460	1,100		1,560	2,150

23 54 Furnaces

23 54 13 – Electric-Resistance Furnaces

23 54 13.10 Electric Furnaces		Crew	Daily Output	Labor-Hours	Unit	Material	Labor	Equipment	Total	Total Incl O&P
0010	**ELECTRIC FURNACES**, Hot air, blowers, std. controls									
0011	not including gas, oil or flue piping									
1000	Electric, UL listed									
1070	10.2 MBH	Q-20	4.40	4.545	Ea.	350	254		604	770
1080	17.1 MBH		4.60	4.348		430	243		673	845
1100	34.1 MBH		4.40	4.545		540	254		794	975
1120	51.6 MBH		4.20	4.762		590	266		856	1,050
1140	68.3 MBH		4	5		610	279		889	1,100
1160	85.3 MBH		3.80	5.263		635	294		929	1,150

23 54 16 – Fuel-Fired Furnaces

23 54 16.13 Gas-Fired Furnaces		Crew	Daily Output	Labor-Hours	Unit	Material	Labor	Equipment	Total	Total Incl O&P
0010	**GAS-FIRED FURNACES**									
3000	Gas, AGA certified, upflow, direct drive models									
3020	45 MBH input	Q-9	4	4	Ea.	670	219		889	1,075
3040	60 MBH input		3.80	4.211		695	231		926	1,100
3060	75 MBH input		3.60	4.444		755	244		999	1,200
3100	100 MBH input		3.20	5		785	274		1,059	1,275
3120	125 MBH input		3	5.333		815	293		1,108	1,350

23 54 16 – Fuel-Fired Furnaces

23 54 16.13 Gas-Fired Furnaces		Crew	Daily Output	Labor-Hours	Unit	2019 Bare Costs Material	Labor	Equipment	Total	Total Incl O&P
3130	150 MBH input	Q-9	2.80	5.714	Ea.	1,825	315		2,140	2,475
3140	200 MBH input		2.60	6.154		3,400	340		3,740	4,275
3160	300 MBH input		2.30	6.957		3,925	380		4,305	4,900
3180	400 MBH input		2	8		5,075	440		5,515	6,275
3200	Gas fired wall furnace									
3204	Horizontal flow									
3208	7.7 MBH	Q-9	7	2.286	Ea.	755	125		880	1,025
3212	14 MBH		6.50	2.462		755	135		890	1,025
3216	24 MBH		5	3.200		785	176		961	1,125
3220	49 MBH		4	4		995	219		1,214	1,425
3224	65 MBH		3.60	4.444		1,175	244		1,419	1,675
3240	Up-flow									
3244	7.7 MBH	Q-9	7	2.286	Ea.	720	125		845	980
3248	14 MBH		6.50	2.462		720	135		855	995
3252	24 MBH		5	3.200		840	176		1,016	1,200
3254	35 MBH		4.50	3.556		735	195		930	1,100
3256	49 MBH		4	4		1,500	219		1,719	1,975
3260	65 MBH		3.60	4.444		1,350	244		1,594	1,875
3500	Gas furnace									
3510	Up-flow									
3514	1,200 CFM, 45 MBH	Q-9	4	4	Ea.	1,150	219		1,369	1,600
3518	75 MBH		3.60	4.444		1,325	244		1,569	1,825
3522	100 MBH		3.20	5		1,475	274		1,749	2,025
3526	125 MBH		3	5.333		1,425	293		1,718	2,025
3530	150 MBH		2.80	5.714		1,625	315		1,940	2,250
3540	2,000 CFM, 75 MBH		3.60	4.444		1,375	244		1,619	1,900
3544	100 MBH		3.20	5		1,550	274		1,824	2,150
3548	120 MBH		3	5.333		1,650	293		1,943	2,275
3560	Horizontal flow									
3564	1,200 CFM, 45 MBH	Q-9	4	4	Ea.	1,325	219		1,544	1,775
3568	60 MBH		3.80	4.211		1,400	231		1,631	1,875
3572	70 MBH		3.60	4.444		1,425	244		1,669	1,950
3576	2,000 CFM, 90 MBH		3.30	4.848		1,575	266		1,841	2,125
3580	110 MBH		3.10	5.161		1,850	283		2,133	2,475

23 54 16.14 Condensing Furnaces		Crew	Daily Output	Labor-Hours	Unit	Material	Labor	Equipment	Total	Total Incl O&P
0010	**CONDENSING FURNACES**, High efficiency									
0020	Oil fired, packaged, complete									
0030	Upflow									
0040	Output @ 95% A.F.U.E.									
0100	49 MBH @ 1,000 CFM	Q-9	3.70	4.324	Ea.	6,750	237		6,987	7,775
0110	73.5 MBH @ 2,000 CFM		3.60	4.444		7,025	244		7,269	8,100
0120	96 MBH @ 2,000 CFM		3.40	4.706		7,025	258		7,283	8,125
0130	115.6 MBH @ 2,000 CFM		3.40	4.706		7,025	258		7,283	8,125
0140	147 MBH @ 2,000 CFM		3.30	4.848		18,400	266		18,666	20,600
0150	192 MBH @ 4,000 CFM		2.60	6.154		19,400	340		19,740	21,800
0170	231.5 MBH @ 4,000 CFM		2.30	6.957		19,400	380		19,780	21,900
0260	For variable speed motor, add					785			785	865
0270	Note: Also available in horizontal, counterflow and lowboy configurations.									

23 54 16 – Fuel-Fired Furnaces

23 54 16.16 Oil-Fired Furnaces

	23 54 16.16 Oil-Fired Furnaces	Crew	Daily Output	Labor-Hours	Unit	Material	2019 Bare Costs Labor	Equipment	Total	Total Incl O&P
0010	**OIL-FIRED FURNACES**									
6000	Oil, UL listed, atomizing gun type burner									
6020	56 MBH output	Q-9	3.60	4.444	Ea.	3,125	244		3,369	3,800
6030	84 MBH output		3.50	4.571		2,975	251		3,226	3,625
6040	95 MBH output		3.40	4.706		2,900	258		3,158	3,575
6060	134 MBH output		3.20	5		2,900	274		3,174	3,625
6080	151 MBH output		3	5.333		3,225	293		3,518	4,000
6100	200 MBH input		2.60	6.154		3,775	340		4,115	4,675
6120	300 MBH input		2.30	6.957		4,325	380		4,705	5,325
6140	400 MBH input		2	8		4,950	440		5,390	6,125
6200	Up flow									
6210	105 MBH	Q-9	3.40	4.706	Ea.	2,050	258		2,308	2,675
6220	140 MBH		3.30	4.848		2,775	266		3,041	3,450
6230	168 MBH		2.90	5.517		4,250	305		4,555	5,125
6260	Lo-boy style									
6270	105 MBH	Q-9	3.40	4.706	Ea.	2,800	258		3,058	3,475
6280	140 MBH		3.30	4.848		3,025	266		3,291	3,725
6290	189 MBH		2.80	5.714		4,825	315		5,140	5,775
6320	Down flow									
6330	105 MBH	Q-9	3.40	4.706	Ea.	2,800	258		3,058	3,475
6340	140 MBH		3.30	4.848		3,025	266		3,291	3,725
6350	168 MBH		2.90	5.517		5,200	305		5,505	6,150
6380	Horizontal									
6390	105 MBH	Q-9	3.40	4.706	Ea.	2,150	258		2,408	2,750
6400	140 MBH		3.30	4.848		3,025	266		3,291	3,725
6410	189 MBH		2.80	5.714		4,600	315		4,915	5,525

23 54 16.21 Solid Fuel-Fired Furnaces

	23 54 16.21 Solid Fuel-Fired Furnaces		Crew	Daily Output	Labor-Hours	Unit	Material	Labor	Equipment	Total	Total Incl O&P
0010	**SOLID FUEL-FIRED FURNACES**										
6020	Wood fired furnaces										
6030	Includes hot water coil, thermostat, and auto draft control										
6040	24" long firebox	G	Q-9	4	4	Ea.	4,550	219		4,769	5,350
6050	30" long firebox	G		3.60	4.444		5,300	244		5,544	6,200
6060	With fireplace glass doors	G		3.20	5		6,675	274		6,949	7,775
6200	Wood/oil fired furnaces, includes two thermostats										
6210	Includes hot water coil and auto draft control										
6240	24" long firebox	G	Q-9	3.40	4.706	Ea.	5,600	258		5,858	6,550
6250	30" long firebox	G		3	5.333		6,225	293		6,518	7,300
6260	With fireplace glass doors	G		2.80	5.714		7,625	315		7,940	8,850
6400	Wood/gas fired furnaces, includes two thermostats										
6410	Includes hot water coil and auto draft control										
6440	24" long firebox	G	Q-9	2.80	5.714	Ea.	6,200	315		6,515	7,300
6450	30" long firebox	G		2.40	6.667		6,950	365		7,315	8,200
6460	With fireplace glass doors	G		2	8		8,425	440		8,865	9,950
6600	Wood/oil/gas fired furnaces, optional accessories										
6610	Hot air plenum		Q-9	16	1	Ea.	112	55		167	207
6620	Safety heat dump			24	.667		97.50	36.50		134	163
6630	Auto air intake			18	.889		153	49		202	243
6640	Cold air return package			14	1.143		165	62.50		227.50	276
6650	Wood fork						53			53	58
6700	Wood fired outdoor furnace										
6740	24" long firebox	G	Q-9	3.80	4.211	Ea.	4,100	231		4,331	4,875
6760	Wood fired outdoor furnace, optional accessories										

23 54 Furnaces

23 54 16 – Fuel-Fired Furnaces

23 54 16.21 Solid Fuel-Fired Furnaces			Crew	Daily Output	Labor-Hours	Unit	2019 Bare Costs Material	Labor	Equipment	Total	Total Incl O&P
6770	Chimney section, stainless steel, 6" ID x 3' long	G	Q-9	36	.444	Ea.	194	24.50		218.50	250
6780	Chimney cap, stainless steel	G	"	40	.400	"	116	22		138	162
6800	Wood fired hot water furnace										
6820	Includes 200 gal. hot water storage, thermostat, and auto draft control										
6840	30" long firebox	G	Q-9	2.10	7.619	Ea.	8,675	420		9,095	10,200
6850	Water to air heat exchanger										
6870	Includes mounting kit and blower relay										
6880	140 MBH, 18.75" W x 18.75" L		Q-9	7.50	2.133	Ea.	385	117		502	605
6890	200 MBH, 24" W x 24" L		"	7	2.286	"	550	125		675	795
6900	Water to water heat exchanger										
6940	100 MBH		Q-9	6.50	2.462	Ea.	390	135		525	630
6960	290 MBH		"	6	2.667	"	560	146		706	835
7000	Optional accessories										
7010	Large volume circulation pump (2 included)		Q-9	14	1.143	Ea.	253	62.50		315.50	375
7020	Air bleed fittings (package)			24	.667		53	36.50		89.50	114
7030	Domestic water preheater			6	2.667		229	146		375	475
7040	Smoke pipe kit			4	4		96	219		315	440

23 54 24 – Furnace Components for Cooling

23 54 24.10 Furnace Components and Combinations		Crew	Daily Output	Labor-Hours	Unit	Material	Labor	Equipment	Total	Total Incl O&P
0010	**FURNACE COMPONENTS AND COMBINATIONS**									
0080	Coils, A.C. evaporator, for gas or oil furnaces									
0090	Add-on, with holding charge									
0100	Upflow									
0120	1-1/2 ton cooling	Q-5	4	4	Ea.	217	230		447	585
0130	2 ton cooling		3.70	4.324		281	249		530	685
0140	3 ton cooling		3.30	4.848		430	279		709	890
0150	4 ton cooling		3	5.333		585	305		890	1,100
0160	5 ton cooling		2.70	5.926		630	340		970	1,200
0300	Downflow									
0330	2-1/2 ton cooling	Q-5	3	5.333	Ea.	335	305		640	830
0340	3-1/2 ton cooling		2.60	6.154		495	355		850	1,075
0350	5 ton cooling		2.20	7.273		520	420		940	1,200
0600	Horizontal									
0630	2 ton cooling	Q-5	3.90	4.103	Ea.	410	236		646	810
0640	3 ton cooling		3.50	4.571		445	263		708	885
0650	4 ton cooling		3.20	5		490	288		778	970
0660	5 ton cooling		2.90	5.517		490	320		810	1,025
2000	Cased evaporator coils for air handlers									
2100	1-1/2 ton cooling	Q-5	4.40	3.636	Ea.	315	209		524	665
2110	2 ton cooling		4.10	3.902		365	225		590	735
2120	2-1/2 ton cooling		3.90	4.103		385	236		621	780
2130	3 ton cooling		3.70	4.324		430	249		679	850
2140	3-1/2 ton cooling		3.50	4.571		540	263		803	985
2150	4 ton cooling		3.20	5		650	288		938	1,150
2160	5 ton cooling		2.90	5.517		640	320		960	1,175
3010	Air handler, modular									
3100	With cased evaporator cooling coil									
3120	1-1/2 ton cooling	Q-5	3.80	4.211	Ea.	985	242		1,227	1,450
3130	2 ton cooling		3.50	4.571		1,025	263		1,288	1,525
3140	2-1/2 ton cooling		3.30	4.848		1,100	279		1,379	1,625
3150	3 ton cooling		3.10	5.161		1,250	297		1,547	1,825
3160	3-1/2 ton cooling		2.90	5.517		1,500	320		1,820	2,125

23 54 Furnaces

23 54 24 – Furnace Components for Cooling

23 54 24.10 Furnace Components and Combinations		Crew	Daily Output	Labor-Hours	Unit	Material	2019 Bare Costs Labor	Equipment	Total	Total Incl O&P
3170	4 ton cooling	Q-5	2.50	6.400	Ea.	1,550	370		1,920	2,275
3180	5 ton cooling		2.10	7.619		1,850	440		2,290	2,700
3500	With no cooling coil									
3520	1-1/2 ton coil size	Q-5	12	1.333	Ea.	1,175	76.50		1,251.50	1,425
3530	2 ton coil size		10	1.600		1,275	92		1,367	1,550
3540	2-1/2 ton coil size		10	1.600		1,300	92		1,392	1,575
3554	3 ton coil size		9	1.778		1,275	102		1,377	1,575
3560	3-1/2 ton coil size		9	1.778		1,325	102		1,427	1,625
3570	4 ton coil size		8.50	1.882		1,400	108		1,508	1,725
3580	5 ton coil size		8	2		1,525	115		1,640	1,850
4000	With heater									
4120	5 kW, 17.1 MBH	Q-5	16	1	Ea.	470	57.50		527.50	605
4130	7.5 kW, 25.6 MBH		15.60	1.026		775	59		834	940
4140	10 kW, 34.2 MBH		15.20	1.053		1,225	60.50		1,285.50	1,450
4150	12.5 kW, 42.7 MBH		14.80	1.081		1,625	62		1,687	1,875
4160	15 kW, 51.2 MBH		14.40	1.111		1,775	64		1,839	2,050
4170	25 kW, 85.4 MBH		14	1.143		2,400	66		2,466	2,725
4180	30 kW, 102 MBH		13	1.231		2,775	71		2,846	3,150

23 55 Fuel-Fired Heaters

23 55 13 – Fuel-Fired Duct Heaters

23 55 13.16 Gas-Fired Duct Heaters		Crew	Daily Output	Labor-Hours	Unit	Material	2019 Bare Costs Labor	Equipment	Total	Total Incl O&P
0010	**GAS-FIRED DUCT HEATERS**, Includes burner, controls, stainless steel									
0020	heat exchanger. Gas fired, electric ignition									
0030	Indoor installation									
0080	100 MBH output	Q-5	5	3.200	Ea.	3,050	184		3,234	3,625
0100	120 MBH output		4	4		3,350	230		3,580	4,025
0130	200 MBH output		2.70	5.926		4,250	340		4,590	5,175
0140	240 MBH output		2.30	6.957		4,425	400		4,825	5,450
0160	280 MBH output		2	8		4,850	460		5,310	6,050
0180	320 MBH output		1.60	10		5,200	575		5,775	6,600
0300	For powered venter and adapter, add					535			535	585
0502	For required flue pipe, see Section 23 51 23.10									
1000	Outdoor installation, with power venter									
1020	75 MBH output	Q-5	4	4	Ea.	3,700	230		3,930	4,425
1040	94 MBH output		4	4		3,750	230		3,980	4,475
1060	120 MBH output		4	4		4,075	230		4,305	4,825
1080	157 MBH output		3.50	4.571		4,375	263		4,638	5,200
1100	187 MBH output		3	5.333		4,925	305		5,230	5,850
1120	225 MBH output		2.50	6.400		5,125	370		5,495	6,200
1140	300 MBH output		1.80	8.889		6,400	510		6,910	7,800
1160	375 MBH output		1.60	10		7,200	575		7,775	8,800
1180	450 MBH output		1.40	11.429		9,575	660		10,235	11,500
1200	600 MBH output		1	16		10,900	920		11,820	13,300
1300	Aluminized exchanger, subtract					15%				
1500	For two stage gas valve, add					800			800	880

23 55 Fuel-Fired Heaters

23 55 23 – Gas-Fired Radiant Heaters

23 55 23.10 Infrared Type Heating Units		Crew	Daily Output	Labor-Hours	Unit	Material	Labor (2019 Bare Costs)	Equipment	Total	Total Incl O&P
0010	**INFRARED TYPE HEATING UNITS**									
0020	Gas fired, unvented, electric ignition, 100% shutoff.									
0030	Piping and wiring not included									
0100	Input, 30 MBH	Q-5	6	2.667	Ea.	1,050	153		1,203	1,400
0120	45 MBH		5	3.200		995	184		1,179	1,375
0140	50 MBH		4.50	3.556		1,000	205		1,205	1,400
0160	60 MBH		4	4		1,025	230		1,255	1,475
0180	75 MBH		3	5.333		1,025	305		1,330	1,575
0200	90 MBH		2.50	6.400		1,050	370		1,420	1,700
0220	105 MBH		2	8		1,050	460		1,510	1,850
0240	120 MBH		2	8		1,625	460		2,085	2,475
1000	Gas fired, vented, electric ignition, tubular									
1020	Piping and wiring not included, 20′ to 80′ lengths									
1030	Single stage, input, 60 MBH	Q-6	4.50	5.333	Ea.	1,500	320		1,820	2,125
1040	80 MBH		3.90	6.154		1,500	365		1,865	2,200
1050	100 MBH		3.40	7.059		1,500	420		1,920	2,275
1060	125 MBH		2.90	8.276		1,500	495		1,995	2,400
1070	150 MBH		2.70	8.889		1,500	530		2,030	2,450
1080	170 MBH		2.50	9.600		1,500	575		2,075	2,500
1090	200 MBH		2.20	10.909		1,700	650		2,350	2,850
1100	Note: Final pricing may vary due to									
1110	tube length and configuration package selected									
1130	Two stage, input, 60 MBH high, 45 MBH low	Q-6	4.50	5.333	Ea.	1,825	320		2,145	2,500
1140	80 MBH high, 60 MBH low		3.90	6.154		1,825	365		2,190	2,575
1150	100 MBH high, 65 MBH low		3.40	7.059		1,825	420		2,245	2,650
1160	125 MBH high, 95 MBH low		2.90	8.276		1,850	495		2,345	2,775
1170	150 MBH high, 100 MBH low		2.70	8.889		1,850	530		2,380	2,825
1180	170 MBH high, 125 MBH low		2.50	9.600		2,050	575		2,625	3,100
1190	200 MBH high, 150 MBH low		2.20	10.909		2,050	650		2,700	3,225
1220	Note: Final pricing may vary due to									
1230	tube length and configuration package selected									

23 55 33 – Fuel-Fired Unit Heaters

23 55 33.13 Oil-Fired Unit Heaters

Line	Description	Crew	Daily Output	Labor-Hours	Unit	Material	Labor	Equipment	Total	Total Incl O&P
0010	**OIL-FIRED UNIT HEATERS**, Cabinet, grilles, fan, ctrl., burner, no piping									
6000	Oil fired, suspension mounted, 94 MBH output	Q-5	4	4	Ea.	5,125	230		5,355	6,000
6040	140 MBH output		3	5.333		5,450	305		5,755	6,450
6060	184 MBH output		3	5.333		5,800	305		6,105	6,825

23 55 33.16 Gas-Fired Unit Heaters

Line	Description	Crew	Daily Output	Labor-Hours	Unit	Material	Labor	Equipment	Total	Total Incl O&P
0010	**GAS-FIRED UNIT HEATERS**, Cabinet, grilles, fan, ctrls., burner, no piping									
0022	thermostat, no piping. For flue see Section 23 51 23.10									
1000	Gas fired, floor mounted									
1100	60 MBH output	Q-5	10	1.600	Ea.	720	92		812	935
1120	80 MBH output		9	1.778		730	102		832	960
1140	100 MBH output		8	2		795	115		910	1,050
1160	120 MBH output		7	2.286		950	132		1,082	1,250
1180	180 MBH output		6	2.667		1,150	153		1,303	1,475
1500	Rooftop mounted, power vent, stainless steel exchanger									
1520	75 MBH input	Q-6	4	6	Ea.	5,900	360		6,260	7,025
1540	100 MBH input		3.60	6.667		6,175	400		6,575	7,375
1560	125 MBH input		3.30	7.273		6,425	435		6,860	7,725
1580	150 MBH input		3	8		6,925	475		7,400	8,350

23 55 33 – Fuel-Fired Unit Heaters

23 55 33.16 Gas-Fired Unit Heaters		Crew	Daily Output	Labor-Hours	Unit	Material	2019 Bare Costs Labor	Equipment	Total	Total Incl O&P
1600	175 MBH input	Q-6	2.60	9.231	Ea.	7,100	550		7,650	8,625
1620	225 MBH input		2.30	10.435		8,100	625		8,725	9,825
1640	300 MBH input		1.90	12.632		9,350	755		10,105	11,400
1660	350 MBH input		1.40	17.143		10,000	1,025		11,025	12,500
1680	450 MBH input		1.20	20		10,700	1,200		11,900	13,600
1720	700 MBH input		.80	30		13,000	1,800		14,800	17,000
1760	1,200 MBH input		.30	80		13,800	4,775		18,575	22,300
1900	For aluminized steel exchanger, subtract					10%				
2000	Suspension mounted, propeller fan, 20 MBH output	Q-5	8.50	1.882		1,725	108		1,833	2,050
2020	40 MBH output		7.50	2.133		1,650	123		1,773	1,975
2040	60 MBH output		7	2.286		1,800	132		1,932	2,200
2060	80 MBH output		6	2.667		2,000	153		2,153	2,425
2080	100 MBH output		5.50	2.909		2,125	167		2,292	2,600
2100	130 MBH output		5	3.200		2,350	184		2,534	2,875
2120	140 MBH output		4.50	3.556		2,600	205		2,805	3,150
2140	160 MBH output		4	4		2,575	230		2,805	3,175
2160	180 MBH output		3.50	4.571		3,175	263		3,438	3,900
2180	200 MBH output		3	5.333		3,075	305		3,380	3,850
2200	240 MBH output		2.70	5.926		3,400	340		3,740	4,225
2220	280 MBH output		2.30	6.957		3,825	400		4,225	4,825
2240	320 MBH output		2	8		4,375	460		4,835	5,525
2500	For powered venter and adapter, add					400			400	440
3000	Suspension mounted, blower type, 40 MBH output	Q-5	6.80	2.353		765	135		900	1,050
3020	60 MBH output		6.60	2.424		910	140		1,050	1,200
3040	84 MBH output		5.80	2.759		860	159		1,019	1,175
3060	104 MBH output		5.20	3.077		850	177		1,027	1,200
3080	140 MBH output		4.30	3.721		1,100	214		1,314	1,550
3100	180 MBH output		3.30	4.848		1,450	279		1,729	2,000
3120	240 MBH output		2.50	6.400		1,600	370		1,970	2,300
3140	280 MBH output		2	8		1,400	460		1,860	2,250
4000	Suspension mounted, sealed combustion system,									
4020	Aluminized steel exchanger, powered vent									
4040	100 MBH output	Q-5	5	3.200	Ea.	3,100	184		3,284	3,675
4060	120 MBH output		4.70	3.404		3,700	196		3,896	4,350
4080	160 MBH output		3.70	4.324		4,300	249		4,549	5,100
4100	200 MBH output		2.90	5.517		5,200	320		5,520	6,200
4120	240 MBH output		2.50	6.400		5,625	370		5,995	6,750
4140	320 MBH output		1.70	9.412		6,750	540		7,290	8,250
5000	Wall furnace, 17.5 MBH output		6	2.667		630	153		783	925
5020	24 MBH output		5	3.200		690	184		874	1,025
5040	35 MBH output		4	4		990	230		1,220	1,425

23 56 Solar Energy Heating Equipment

23 56 16 – Packaged Solar Heating Equipment

23 56 16.40 Solar Heating Systems			Crew	Daily Output	Labor-Hours	Unit	2019 Bare Costs Material	Labor	Equipment	Total	Total Incl O&P
0010	SOLAR HEATING SYSTEMS R235616-60										
0020	System/package prices, not including connecting										
0030	pipe, insulation, or special heating/plumbing fixtures										
0152	For solar ultraviolet pipe insulation see Section 22 07 19.10										
0500	Hot water, standard package, low temperature										
0540	1 collector, circulator, fittings, 65 gal. tank	G	Q-1	.50	32	Ea.	4,050	1,825		5,875	7,175
0580	2 collectors, circulator, fittings, 120 gal. tank	G		.40	40		5,350	2,275		7,625	9,275
0620	3 collectors, circulator, fittings, 120 gal. tank	G		.34	47.059		7,300	2,675		9,975	12,100
0700	Medium temperature package										
0720	1 collector, circulator, fittings, 80 gal. tank	G	Q-1	.50	32	Ea.	5,300	1,825		7,125	8,550
0740	2 collectors, circulator, fittings, 120 gal. tank	G		.40	40		6,975	2,275		9,250	11,100
0780	3 collectors, circulator, fittings, 120 gal. tank	G		.30	53.333		7,900	3,025		10,925	13,200
0980	For each additional 120 gal. tank, add	G					1,875			1,875	2,075

23 56 19 – Solar Heating Components

23 56 19.50 Solar Heating Ancillary			Crew	Daily Output	Labor-Hours	Unit	Material	Labor	Equipment	Total	Total Incl O&P
0010	SOLAR HEATING ANCILLARY										
2300	Circulators, air										
2310	Blowers										
2330	100-300 S.F. system, 1/10 HP	G	Q-9	16	1	Ea.	251	55		306	360
2340	300-500 S.F. system, 1/5 HP	G		15	1.067		330	58.50		388.50	455
2350	Two speed, 100-300 S.F., 1/10 HP	G		14	1.143		145	62.50		207.50	255
2400	Reversible fan, 20" diameter, 2 speed	G		18	.889		115	49		164	200
2550	Booster fan 6" diameter, 120 CFM	G		16	1		37.50	55		92.50	125
2570	6" diameter, 225 CFM	G		16	1		46	55		101	134
2580	8" diameter, 150 CFM	G		16	1		42	55		97	130
2590	8" diameter, 310 CFM	G		14	1.143		64.50	62.50		127	166
2600	8" diameter, 425 CFM	G		14	1.143		72.50	62.50		135	175
2650	Rheostat	G		32	.500		15.80	27.50		43.30	59
2660	Shutter/damper	G		12	1.333		57	73		130	174
2670	Shutter motor	G		16	1		142	55		197	240
2800	Circulators, liquid, 1/25 HP, 5.3 GPM	G	Q-1	14	1.143		200	65		265	315
2820	1/20 HP, 17 GPM	G		12	1.333		268	76		344	410
2850	1/20 HP, 17 GPM, stainless steel	G		12	1.333		259	76		335	400
2870	1/12 HP, 30 GPM	G		10	1.600		355	91		446	525
3000	Collector panels, air with aluminum absorber plate										
3010	Wall or roof mount										
3040	Flat black, plastic glazing										
3080	4' x 8'	G	Q-9	6	2.667	Ea.	675	146		821	965
3100	4' x 10'	G		5	3.200	"	830	176		1,006	1,175
3200	Flush roof mount, 10' to 16' x 22" wide	G		96	.167	L.F.	143	9.15		152.15	171
3210	Manifold, by L.F. width of collectors	G		160	.100	"	155	5.50		160.50	179
3300	Collector panels, liquid with copper absorber plate										
3320	Black chrome, tempered glass glazing										
3330	Alum. frame, 4' x 8', 5/32" single glazing	G	Q-1	9.50	1.684	Ea.	1,025	95.50		1,120.50	1,275
3390	Alum. frame, 4' x 10', 5/32" single glazing	G		6	2.667		1,175	152		1,327	1,525
3450	Flat black, alum. frame, 3.5' x 7.5'	G		9	1.778		835	101		936	1,075
3500	4' x 8'	G		5.50	2.909		1,075	165		1,240	1,425
3520	4' x 10'	G		10	1.600		1,275	91		1,366	1,525
3540	4' x 12.5'	G		5	3.200		1,300	182		1,482	1,700
3550	Liquid with fin tube absorber plate										
3560	Alum. frame 4' x 8' tempered glass	G	Q-1	10	1.600	Ea.	600	91		691	795
3580	Liquid with vacuum tubes, 4' x 6'-10"	G		9	1.778		925	101		1,026	1,175

23 56 Solar Energy Heating Equipment

23 56 19 – Solar Heating Components

23 56 19.50 Solar Heating Ancillary			Crew	Daily Output	Labor-Hours	Unit	Material	Labor (2019 Bare Costs)	Equipment	Total	Total Incl O&P
3600	Liquid, full wetted, plastic, alum. frame, 4′ x 10′	G	Q-1	5	3.200	Ea.	330	182		512	640
3650	Collector panel mounting, flat roof or ground rack	G		7	2.286		250	130		380	470
3670	Roof clamps	G		70	.229	Set	2.89	13		15.89	22.50
3700	Roof strap, teflon	G	1 Plum	205	.039	L.F.	25	2.46		27.46	31
3900	Differential controller with two sensors										
3930	Thermostat, hard wired	G	1 Plum	8	1	Ea.	103	63		166	209
3950	Line cord and receptacle	G		12	.667		131	42		173	208
4050	Pool valve system	G		2.50	3.200		168	202		370	490
4070	With 12 VAC actuator	G		2	4		335	253		588	745
4080	Pool pump system, 2″ pipe size	G		6	1.333		200	84		284	345
4100	Five station with digital read-out	G		3	2.667		271	168		439	550
4150	Sensors										
4200	Brass plug, 1/2″ MPT	G	1 Plum	32	.250	Ea.	16.65	15.80		32.45	42
4210	Brass plug, reversed	G		32	.250		27	15.80		42.80	53
4220	Freeze prevention	G		32	.250		23	15.80		38.80	49
4240	Screw attached	G		32	.250		10	15.80		25.80	34.50
4250	Brass, immersion	G		32	.250		14.10	15.80		29.90	39
4300	Heat exchanger										
4315	includes coil, blower, circulator										
4316	and controller for DHW and space hot air										
4330	Fluid to air coil, up flow, 45 MBH	G	Q-1	4	4	Ea.	325	227		552	695
4380	70 MBH	G		3.50	4.571		365	260		625	790
4400	80 MBH	G		3	5.333		485	305		790	990
4580	Fluid to fluid package includes two circulating pumps										
4590	expansion tank, check valve, relief valve										
4600	controller, high temperature cutoff and sensors	G	Q-1	2.50	6.400	Ea.	820	365		1,185	1,450
4650	Heat transfer fluid										
4700	Propylene glycol, inhibited anti-freeze	G	1 Plum	28	.286	Gal.	23.50	18.05		41.55	53
4800	Solar storage tanks, knocked down										
4810	Air, galvanized steel clad, double wall, 4″ fiberglass insulation										
5120	45 mil reinforced polypropylene lining,										
5140	4′ high, 4′ x 4′ = 64 C.F./450 gallons	G	Q-9	2	8	Ea.	3,725	440		4,165	4,750
5150	4′ x 8′ = 128 C.F./900 gallons	G		1.50	10.667		5,575	585		6,160	7,025
5160	4′ x 12′ = 190 C.F./1,300 gallons	G		1.30	12.308		7,425	675		8,100	9,200
5170	8′ x 8′ = 250 C.F./1,700 gallons	G		1	16		7,425	880		8,305	9,500
5190	6′-3″ high, 7′ x 7′ = 306 C.F./2,000 gallons	G	Q-10	1.20	20		14,500	1,150		15,650	17,700
5200	7′ x 10′-6″ = 459 C.F./3,000 gallons	G		.80	30		18,200	1,700		19,900	22,600
5210	7′ x 14′ = 613 C.F./4,000 gallons	G		.60	40		21,800	2,275		24,075	27,500
5220	10′-6″ x 10′-6″ = 689 C.F./4,500 gallons	G		.50	48		21,800	2,725		24,525	28,200
5230	10′-6″ x 14′ = 919 C.F./6,000 gallons	G		.40	60		25,400	3,425		28,825	33,200
5240	14′ x 14′ = 1,225 C.F./8,000 gallons	G	Q-11	.40	80		29,000	4,650		33,650	39,000
5250	14′ x 17′-6″ = 1,531 C.F./10,000 gallons	G		.30	107		32,700	6,200		38,900	45,400
5260	17′-6″ x 17′-6″ = 1,914 C.F./12,500 gallons	G		.25	128		36,200	7,425		43,625	51,000
5270	17′-6″ x 21′ = 2,297 C.F./15,000 gallons	G		.20	160		39,900	9,275		49,175	58,000
5280	21′ x 21′ = 2,756 C.F./18,000 gallons	G		.18	178		43,200	10,300		53,500	63,000
5290	30 mil reinforced Hypalon lining, add						.02%				
7000	Solar control valves and vents										
7050	Air purger, 1″ pipe size	G	1 Plum	12	.667	Ea.	47.50	42		89.50	116
7070	Air eliminator, automatic 3/4″ size	G		32	.250		30.50	15.80		46.30	57
7090	Air vent, automatic, 1/8″ fitting	G		32	.250		18.45	15.80		34.25	44
7100	Manual, 1/8″ NPT	G		32	.250		2.66	15.80		18.46	26.50
7120	Backflow preventer, 1/2″ pipe size	G		16	.500		84.50	31.50		116	140
7130	3/4″ pipe size	G		16	.500		88	31.50		119.50	145

23 56 Solar Energy Heating Equipment

23 56 19 – Solar Heating Components

23 56 19.50 Solar Heating Ancillary			Crew	Daily Output	Labor-Hours	Unit	Material	Labor	Equipment	Total	Total Incl O&P
							2019 Bare Costs				
7150	Balancing valve, 3/4" pipe size	G	1 Plum	20	.400	Ea.	64	25.50		89.50	109
7180	Draindown valve, 1/2" copper tube	G		9	.889		216	56		272	320
7200	Flow control valve, 1/2" pipe size	G		22	.364		134	23		157	183
7220	Expansion tank, up to 5 gal.	G		32	.250		69.50	15.80		85.30	100
7250	Hydronic controller (aquastat)	G		8	1		220	63		283	335
7400	Pressure gauge, 2" dial	G		32	.250		24.50	15.80		40.30	50.50
7450	Relief valve, temp. and pressure 3/4" pipe size	G		30	.267		24	16.85		40.85	52
7500	Solenoid valve, normally closed										
7520	Brass, 3/4" NPT, 24V	G	1 Plum	9	.889	Ea.	139	56		195	237
7530	1" NPT, 24V	G		9	.889		222	56		278	330
7750	Vacuum relief valve, 3/4" pipe size	G		32	.250		30	15.80		45.80	56.50
7800	Thermometers										
7820	Digital temperature monitoring, 4 locations	G	1 Plum	2.50	3.200	Ea.	140	202		342	460
7900	Upright, 1/2" NPT	G		8	1		39.50	63		102.50	139
7970	Remote probe, 2" dial	G		8	1		34	63		97	132
7990	Stem, 2" dial, 9" stem	G		16	.500		22	31.50		53.50	72
8250	Water storage tank with heat exchanger and electric element										
8270	66 gal. with 2" x 2 lb. density insulation	G	1 Plum	1.60	5	Ea.	2,000	315		2,315	2,675
8300	80 gal. with 2" x 2 lb. density insulation	G		1.60	5		2,000	315		2,315	2,675
8380	120 gal. with 2" x 2 lb. density insulation	G		1.40	5.714		1,950	360		2,310	2,675
8400	120 gal. with 2" x 2 lb. density insul., 40 S.F. heat coil	G		1.40	5.714		2,875	360		3,235	3,725
8500	Water storage module, plastic										
8600	Tubular, 12" diameter, 4' high	G	1 Carp	48	.167	Ea.	89.50	8.60		98.10	112
8610	12" diameter, 8' high	G		40	.200		144	10.35		154.35	174
8620	18" diameter, 5' high	G		38	.211		162	10.85		172.85	194
8630	18" diameter, 10' high	G		32	.250		264	12.90		276.90	310
8640	58" diameter, 5' high	G	2 Carp	32	.500		700	26		726	810
8650	Cap, 12" diameter	G					19.95			19.95	22
8660	18" diameter	G					25.50			25.50	28

23 57 Heat Exchangers for HVAC

23 57 16 – Steam-to-Water Heat Exchangers

23 57 16.10 Shell/Tube Type Steam-to-Water Heat Exch.		Crew	Daily Output	Labor-Hours	Unit	Material	Labor	Equipment	Total	Total Incl O&P
0010	**SHELL AND TUBE TYPE STEAM-TO-WATER HEAT EXCHANGERS**									
0016	Shell & tube type, 2 or 4 pass, 3/4" OD copper tubes,									
0020	C.I. heads, C.I. tube sheet, steel shell									
0100	Hot water 40°F to 180°F, by steam at 10 psi									
0120	8 GPM	Q-5	6	2.667	Ea.	2,525	153		2,678	3,000
0140	10 GPM		5	3.200		3,800	184		3,984	4,450
0160	40 GPM		4	4		5,850	230		6,080	6,800
0180	64 GPM		2	8		8,975	460		9,435	10,600
0200	96 GPM		1	16		12,000	920		12,920	14,600
0220	120 GPM	Q-6	1.50	16		15,400	955		16,355	18,400
0240	168 GPM		1	24		19,300	1,425		20,725	23,500
0260	240 GPM		.80	30		30,200	1,800		32,000	35,900
0300	600 GPM		.70	34.286		62,500	2,050		64,550	71,500
0500	For bronze head and tube sheet, add					50%				

23 57 Heat Exchangers for HVAC

23 57 19 – Liquid-to-Liquid Heat Exchangers

23 57 19.13 Plate-Type, Liquid-to-Liquid Heat Exchangers		Crew	Daily Output	Labor-Hours	Unit	2019 Bare Costs Material	Labor	Equipment	Total	Total Incl O&P
0010	PLATE-TYPE, LIQUID-TO-LIQUID HEAT EXCHANGERS									
3000	Plate type,									
3100	400 GPM	Q-6	.80	30	Ea.	44,900	1,800		46,700	52,000
3120	800 GPM	"	.50	48		77,500	2,875		80,375	89,500
3140	1,200 GPM	Q-7	.34	94.118		115,000	5,725		120,725	135,000
3160	1,800 GPM	"	.24	133		152,500	8,125		160,625	180,000

23 57 19.16 Shell-Type, Liquid-to-Liquid Heat Exchangers		Crew	Daily Output	Labor-Hours	Unit	Material	Labor	Equipment	Total	Total Incl O&P
0010	SHELL-TYPE, LIQUID-TO-LIQUID HEAT EXCHANGERS									
1000	Hot water 40°F to 140°F, by water at 200°F									
1020	7 GPM	Q-5	6	2.667	Ea.	3,075	153		3,228	3,625
1040	16 GPM		5	3.200		4,375	184		4,559	5,075
1060	34 GPM		4	4		6,600	230		6,830	7,625
1080	55 GPM		3	5.333		9,400	305		9,705	10,800
1100	74 GPM		1.50	10.667		12,000	615		12,615	14,100
1120	86 GPM		1.40	11.429		16,000	660		16,660	18,600
1140	112 GPM	Q-6	2	12		20,000	715		20,715	23,100
1160	126 GPM		1.80	13.333		24,900	795		25,695	28,600
1180	152 GPM		1	24		31,200	1,425		32,625	36,500

23 61 Refrigerant Compressors

23 61 13 – Centrifugal Refrigerant Compressors

23 61 13.10 Centrifugal Compressors		Crew	Daily Output	Labor-Hours	Unit	Material	Labor	Equipment	Total	Total Incl O&P
0010	CENTRIFUGAL COMPRESSORS									
0100	Refrigeration, centrifugal including switches & protective devices									
0210	200 ton	Q-7	.23	137	Ea.	35,300	8,350		43,650	51,500
0220	300 ton		.17	187		38,900	11,400		50,300	60,000
0230	400 ton		.14	227		39,800	13,800		53,600	64,500
0240	500 ton		.14	237		40,200	14,400		54,600	66,000
0250	600 ton		.13	248		48,300	15,100		63,400	75,500
0260	700 ton		.12	260		48,600	15,800		64,400	77,500
0270	800 ton		.12	271		61,000	16,500		77,500	92,000
0280	900 ton		.11	288		68,000	17,600		85,600	101,000
0290	1,000 ton		.11	296		77,500	18,000		95,500	112,500
0300	1,100 ton		.10	308		81,000	18,700		99,700	117,000
0310	1,200 ton		.10	317		82,000	19,300		101,300	119,500
0320	1,300 ton		.10	327		82,500	19,900		102,400	120,500
0330	1,400 ton		.09	340		85,000	20,700		105,700	124,500
0340	1,500 ton		.09	352		88,500	21,400		109,900	129,500

23 61 15 – Rotary Refrigerant Compressors

23 61 15.10 Rotary Compressors		Crew	Daily Output	Labor-Hours	Unit	Material	Labor	Equipment	Total	Total Incl O&P
0010	ROTARY COMPRESSORS									
0100	Refrigeration, hermetic, switches and protective devices									
0210	1.25 ton	1 Stpi	3	2.667	Ea.	251	171		422	535
0220	1.42 ton		3	2.667		254	171		425	535
0230	1.68 ton		2.80	2.857		276	183		459	580
0240	2.00 ton		2.60	3.077		285	197		482	610
0250	2.37 ton		2.50	3.200		320	205		525	660
0260	2.67 ton		2.40	3.333		335	213		548	690
0270	3.53 ton		2.30	3.478		370	222		592	740
0280	4.43 ton		2.20	3.636		470	233		703	865

23 61 Refrigerant Compressors

23 61 15 – Rotary Refrigerant Compressors

23 61 15.10 Rotary Compressors		Crew	Daily Output	Labor-Hours	Unit	Material	2019 Bare Costs Labor	Equipment	Total	Total Incl O&P
0290	5.08 ton	1 Stpi	2.10	3.810	Ea.	520	244		764	935
0300	5.22 ton		2	4		730	256		986	1,200
0310	7.31 ton	Q-5	3	5.333		1,175	305		1,480	1,725
0320	9.95 ton		2.60	6.154		1,375	355		1,730	2,025
0330	11.6 ton		2.50	6.400		1,750	370		2,120	2,475
0340	15.25 ton		2.30	6.957		2,425	400		2,825	3,250
0350	17.7 ton		2.10	7.619		2,575	440		3,015	3,500

23 61 16 – Reciprocating Refrigerant Compressors

23 61 16.10 Reciprocating Compressors		Crew	Daily Output	Labor-Hours	Unit	Material	Labor	Equipment	Total	Total Incl O&P
0010	RECIPROCATING COMPRESSORS									
0990	Refrigeration, recip. hermetic, switches & protective devices									
1000	10 ton	Q-5	1	16	Ea.	16,200	920		17,120	19,200
1100	20 ton	Q-6	.72	33.333		23,500	2,000		25,500	28,800
1200	30 ton		.64	37.500		28,400	2,250		30,650	34,700
1300	40 ton		.44	54.545		32,700	3,250		35,950	40,800
1400	50 ton		.20	120		34,000	7,150		41,150	48,100
1500	75 ton	Q-7	.27	119		37,600	7,225		44,825	52,000
1600	130 ton	"	.21	152		42,900	9,275		52,175	61,000

23 61 19 – Scroll Refrigerant Compressors

23 61 19.10 Scroll Compressors		Crew	Daily Output	Labor-Hours	Unit	Material	Labor	Equipment	Total	Total Incl O&P
0010	SCROLL COMPRESSORS									
1800	Refrigeration, scroll type									
1810	1.5 ton	1 Stpi	2.70	2.963	Ea.	885	189		1,074	1,250
1820	2.5 ton		2.50	3.200		885	205		1,090	1,275
1830	2.75 ton		2.35	3.404		730	218		948	1,125
1840	3 ton		2.30	3.478		940	222		1,162	1,350
1850	3.5 ton		2.25	3.556		970	227		1,197	1,425
1860	4 ton		2.20	3.636		1,200	233		1,433	1,675
1870	5 ton		2.10	3.810		1,275	244		1,519	1,775
1880	6 ton	Q-5	3.70	4.324		1,450	249		1,699	1,975
1900	7.5 ton		3.20	5		2,575	288		2,863	3,275
1910	10 ton		3.10	5.161		2,750	297		3,047	3,475
1920	13 ton		2.90	5.517		2,625	320		2,945	3,375
1960	25 ton		2.20	7.273		5,075	420		5,495	6,225

23 62 Packaged Compressor and Condenser Units

23 62 13 – Packaged Air-Cooled Refrigerant Compressor and Condenser Units

23 62 13.10 Packaged Air-Cooled Refrig. Condensing Units		Crew	Daily Output	Labor-Hours	Unit	Material	Labor	Equipment	Total	Total Incl O&P
0010	PACKAGED AIR-COOLED REFRIGERANT CONDENSING UNITS									
0020	Condensing unit									
0030	Air cooled, compressor, standard controls									
0050	1.5 ton	Q-5	2.50	6.400	Ea.	1,075	370		1,445	1,725
0100	2 ton		2.10	7.619		1,050	440		1,490	1,800
0200	2.5 ton		1.70	9.412		1,250	540		1,790	2,200
0300	3 ton		1.30	12.308		1,325	710		2,035	2,525
0350	3.5 ton		1.10	14.545		1,550	835		2,385	2,950
0400	4 ton		.90	17.778		1,775	1,025		2,800	3,475
0500	5 ton		.60	26.667		2,125	1,525		3,650	4,650
0550	7.5 ton		.55	29.091		3,325	1,675		5,000	6,175
0560	8.5 ton		.53	30.189		3,500	1,725		5,225	6,450

23 62 Packaged Compressor and Condenser Units

23 62 13 – Packaged Air-Cooled Refrigerant Compressor and Condenser Units

23 62 13.10 Packaged Air-Cooled Refrig. Condensing Units		Crew	Daily Output	Labor-Hours	Unit	2019 Bare Costs Material	Labor	Equipment	Total	Total Incl O&P
0600	10 ton	Q-5	.50	32	Ea.	3,650	1,850		5,500	6,775
0620	12.5 ton		.48	33.333		4,275	1,925		6,200	7,575
0650	15 ton		.40	40		7,175	2,300		9,475	11,400
0700	20 ton	Q-6	.40	60		8,350	3,575		11,925	14,600
0720	25 ton		.35	68.571		14,200	4,100		18,300	21,800
0750	30 ton		.30	80		15,600	4,775		20,375	24,300
0800	40 ton		.20	120		20,000	7,150		27,150	32,700
0840	50 ton		.18	133		22,000	7,950		29,950	36,100
0860	60 ton		.16	150		23,800	8,950		32,750	39,500
0900	70 ton		.14	171		29,100	10,200		39,300	47,500
1000	80 ton		.12	200		32,400	11,900		44,300	53,500
1010	90 ton		.09	267		34,600	15,900		50,500	62,000
1100	100 ton		.09	267		40,600	15,900		56,500	68,500

23 62 23 – Packaged Water-Cooled Refrigerant Compressor and Condenser Units

23 62 23.10 Pckgd. Water-Cooled Refrigerant Condns. Units		Crew	Daily Output	Labor-Hours	Unit	Material	Labor	Equipment	Total	Total Incl O&P
0010	**PACKAGED WATER-COOLED REFRIGERANT CONDENSING UNITS**									
2000	Water cooled, compressor, heat exchanger, controls									
2100	5 ton	Q-5	.70	22.857	Ea.	16,800	1,325		18,125	20,500
2110	10 ton		.60	26.667		20,700	1,525		22,225	25,100
2200	15 ton		.50	32		27,700	1,850		29,550	33,300
2300	20 ton	Q-6	.40	60		31,200	3,575		34,775	39,700
2310	30 ton		.30	80		32,500	4,775		37,275	43,000
2400	40 ton		.20	120		35,100	7,150		42,250	49,300

23 63 Refrigerant Condensers

23 63 13 – Air-Cooled Refrigerant Condensers

23 63 13.10 Air-Cooled Refrig. Condensers		Crew	Daily Output	Labor-Hours	Unit	Material	Labor	Equipment	Total	Total Incl O&P
0010	**AIR-COOLED REFRIG. CONDENSERS**									
0080	Air cooled, belt drive, propeller fan									
0220	45 ton	Q-6	.70	34.286	Ea.	10,200	2,050		12,250	14,300
0240	50 ton		.69	34.985		11,000	2,100		13,100	15,100
0260	54 ton		.64	37.795		11,900	2,250		14,150	16,500
0280	59 ton		.58	41.308		13,000	2,475		15,475	18,100
0300	65 ton		.53	45.541		15,500	2,725		18,225	21,100
0320	73 ton		.47	51.173		17,000	3,050		20,050	23,300
0340	81 ton		.42	56.738		18,600	3,375		21,975	25,500
0360	86 ton		.40	60.302		19,500	3,600		23,100	26,900
0380	88 ton		.39	61.697		21,000	3,675		24,675	28,600
0400	101 ton	Q-7	.45	70.640		23,900	4,300		28,200	32,700
0500	159 ton		.31	103		24,900	6,275		31,175	36,800
0600	228 ton		.22	148		52,000	9,025		61,025	70,500
0650	320 ton		.16	206		49,900	12,600		62,500	74,000
0660	475 ton		.11	296		108,000	18,000		126,000	145,500
1500	May be specified single or multi-circuit									
1550	Air cooled, direct drive, propeller fan									
1590	1 ton	Q-5	3.80	4.211	Ea.	1,700	242		1,942	2,250
1600	1-1/2 ton		3.60	4.444		2,075	256		2,331	2,650
1620	2 ton		3.20	5		2,250	288		2,538	2,900
1630	3 ton		2.40	6.667		2,500	385		2,885	3,325
1640	5 ton		2	8		5,225	460		5,685	6,450

23 63 Refrigerant Condensers

23 63 13 – Air-Cooled Refrigerant Condensers

23 63 13.10 Air-Cooled Refrig. Condensers		Crew	Daily Output	Labor-Hours	Unit	Material	Labor	Equipment	Total	Total Incl O&P
						2019 Bare Costs				
1650	8 ton	Q-5	1.80	8.889	Ea.	5,775	510		6,285	7,150
1660	10 ton		1.40	11.429		6,950	660		7,610	8,600
1670	12 ton		1.30	12.308		7,950	710		8,660	9,800
1680	14 ton		1.20	13.333		9,075	765		9,840	11,200
1690	16 ton		1.10	14.545		10,500	835		11,335	12,900
1700	21 ton		1	16		11,500	920		12,420	14,000
1720	26 ton		.84	19.002		12,900	1,100		14,000	15,900
1740	30 ton		.70	22.792		16,300	1,300		17,600	19,900
1760	41 ton	Q-6	.77	31.008		18,000	1,850		19,850	22,600
1780	52 ton		.66	36.419		25,600	2,175		27,775	31,500
1800	63 ton		.55	44.037		28,300	2,625		30,925	35,100
1820	76 ton		.45	52.980		32,100	3,150		35,250	40,100
1840	86 ton		.40	60		39,600	3,575		43,175	49,000
1860	97 ton		.35	67.989		45,100	4,050		49,150	55,500
1880	105 ton	Q-7	.44	73.563		52,000	4,475		56,475	63,500
1890	118 ton		.39	82.687		56,000	5,025		61,025	69,000
1900	126 ton		.36	88.154		65,000	5,375		70,375	79,500
1910	136 ton		.34	95.238		73,000	5,800		78,800	88,500
1920	142 ton		.32	99.379		74,500	6,050		80,550	91,000

23 63 30 – Lubricants

23 63 30.10 Lubricant Oils

		Crew	Daily Output	Labor-Hours	Unit	Material	Labor	Equipment	Total	Total Incl O&P
0010	**LUBRICANT OILS**									
8000	Oils									
8100	Lubricating									
8120	Oil, lubricating				Oz.	.30			.30	.33
8500	Refrigeration									
8520	Oil, refrigeration				Gal.	48.50			48.50	53.50
8525	Oil, refrigeration				Qt.	12.15			12.15	13.35

23 63 33 – Evaporative Refrigerant Condensers

23 63 33.10 Evaporative Condensers

		Crew	Daily Output	Labor-Hours	Unit	Material	Labor	Equipment	Total	Total Incl O&P
0010	**EVAPORATIVE CONDENSERS**									
3400	Evaporative, copper coil, pump, fan motor									
3440	10 ton	Q-5	.54	29.630	Ea.	12,400	1,700		14,100	16,200
3460	15 ton		.50	32		12,500	1,850		14,350	16,600
3480	20 ton		.47	34.043		12,800	1,950		14,750	17,100
3500	25 ton		.45	35.556		10,100	2,050		12,150	14,200
3520	30 ton		.42	38.095		10,600	2,200		12,800	14,900
3540	40 ton	Q-6	.49	48.980		11,700	2,925		14,625	17,200
3560	50 ton		.39	61.538		13,000	3,675		16,675	19,800
3580	65 ton		.35	68.571		13,700	4,100		17,800	21,300
3600	80 ton		.33	72.727		15,200	4,350		19,550	23,300
3620	90 ton		.29	82.759		16,700	4,950		21,650	25,700
3640	100 ton	Q-7	.36	88.889		18,400	5,400		23,800	28,400
3660	110 ton		.33	96.970		25,200	5,900		31,100	36,600
3680	125 ton		.30	107		27,300	6,500		33,800	39,800
3700	135 ton		.28	114		29,100	6,950		36,050	42,400
3720	150 ton		.25	128		31,700	7,800		39,500	46,600
3740	165 ton		.23	139		34,200	8,475		42,675	50,500
3760	185 ton		.22	145		37,400	8,850		46,250	54,500
3860	For fan damper control, add	Q-5	2	8		880	460		1,340	1,650

23 64 Packaged Water Chillers

23 64 13 – Absorption Water Chillers

23 64 13.13 Direct-Fired Absorption Water Chillers

	23 64 13.13 Direct-Fired Absorption Water Chillers	Crew	Daily Output	Labor-Hours	Unit	Material	Labor	Equipment	Total	Total Incl O&P
						2019 Bare Costs				
0010	**DIRECT-FIRED ABSORPTION WATER CHILLERS**									
3000	Gas fired, air cooled									
3220	5 ton	Q-5	.60	26.667	Ea.	8,900	1,525		10,425	12,100
3270	10 ton	″	.40	40	″	22,700	2,300		25,000	28,500
4000	Water cooled, duplex									
4130	100 ton	Q-7	.13	246	Ea.	143,500	15,000		158,500	180,500
4140	200 ton		.11	283		186,500	17,200		203,700	231,000
4150	300 ton		.11	299		232,500	18,200		250,700	283,000
4160	400 ton		.10	317		302,000	19,300		321,300	361,000
4170	500 ton		.10	330		380,500	20,100		400,600	448,500
4180	600 ton		.09	340		449,000	20,700		469,700	525,000
4190	700 ton		.09	360		498,000	21,900		519,900	581,000
4200	800 ton		.08	381		601,500	23,200		624,700	696,500
4210	900 ton		.08	400		697,000	24,400		721,400	803,000
4220	1,000 ton		.08	421		774,000	25,600		799,600	890,000

23 64 13.16 Indirect-Fired Absorption Water Chillers

	23 64 13.16 Indirect-Fired Absorption Water Chillers	Crew	Daily Output	Labor-Hours	Unit	Material	Labor	Equipment	Total	Total Incl O&P
0010	**INDIRECT-FIRED ABSORPTION WATER CHILLERS**									
0020	Steam or hot water, water cooled									
0050	100 ton	Q-7	.13	240	Ea.	132,500	14,600		147,100	168,000
0100	148 ton		.12	258		240,500	15,700		256,200	288,000
0200	200 ton		.12	276		267,500	16,800		284,300	319,500
0240	250 ton		.11	291		293,500	17,700		311,200	349,000
0300	354 ton		.11	305		368,000	18,600		386,600	432,500
0400	420 ton		.10	323		400,500	19,700		420,200	470,000
0500	665 ton		.09	340		564,500	20,700		585,200	652,000
0600	750 ton		.09	364		657,000	22,100		679,100	756,000
0700	850 ton		.08	386		704,000	23,500		727,500	809,500
0800	955 ton		.08	410		752,500	25,000		777,500	865,500
0900	1,125 ton		.08	421		869,000	25,600		894,600	994,500
1000	1,250 ton		.07	444		931,000	27,100		958,100	1,064,500
1100	1,465 ton		.07	464		1,108,000	28,200		1,136,200	1,261,000
1200	1,660 ton		.07	478		1,303,000	29,100		1,332,100	1,477,000
2000	For two stage unit, add					80%	25%			

23 64 16 – Centrifugal Water Chillers

23 64 16.10 Centrifugal Type Water Chillers

	23 64 16.10 Centrifugal Type Water Chillers	Crew	Daily Output	Labor-Hours	Unit	Material	Labor	Equipment	Total	Total Incl O&P
0010	**CENTRIFUGAL TYPE WATER CHILLERS**, With standard controls									
0020	Centrifugal liquid chiller, water cooled									
0030	not including water tower									
0100	2,000 ton (twin 1,000 ton units)	Q-7	.07	478	Ea.	750,000	29,100		779,100	868,500
0274	Centrifugal, packaged unit, water cooled, not incl. tower									
0278	200 ton	Q-7	.19	172	Ea.	132,500	10,500		143,000	161,000
0279	300 ton		.14	234		151,500	14,200		165,700	188,500
0280	400 ton		.11	283		154,500	17,200		171,700	196,000
0282	450 ton		.11	291		168,500	17,700		186,200	212,000
0290	500 ton		.11	296		189,500	18,000		207,500	235,000
0292	550 ton		.11	305		217,500	18,600		236,100	267,500
0300	600 ton		.10	311		232,500	18,900		251,400	284,500
0302	650 ton		.10	320		245,000	19,500		264,500	298,500
0304	700 ton		.10	327		262,500	19,900		282,400	318,500
0306	750 ton		.10	333		281,000	20,300		301,300	339,500
0310	800 ton		.09	340		299,500	20,700		320,200	360,500
0312	850 ton		.09	352		318,500	21,400		339,900	382,500

23 64 Packaged Water Chillers

23 64 16 – Centrifugal Water Chillers

23 64 16.10 Centrifugal Type Water Chillers		Crew	Daily Output	Labor-Hours	Unit	Material	Labor	Equipment	Total	Total Incl O&P
						2019 Bare Costs				
0316	900 ton	Q-7	.09	360	Ea.	337,500	21,900		359,400	404,500
0318	950 ton		.09	364		356,500	22,100		378,600	425,000
0320	1,000 ton		.09	372		375,000	22,700		397,700	446,500
0324	1,100 ton		.08	386		340,000	23,500		363,500	408,500
0330	1,200 ton		.08	395		406,500	24,100		430,600	483,000
0340	1,300 ton		.08	410		430,000	25,000		455,000	510,500
0350	1,400 ton		.08	421		474,000	25,600		499,600	560,000
0360	1,500 ton		.08	427		499,500	26,000		525,500	588,500
0370	1,600 ton		.07	438		532,000	26,700		558,700	625,500
0380	1,700 ton		.07	451		566,000	27,400		593,400	663,500
0390	1,800 ton		.07	457		599,500	27,800		627,300	701,500
0400	1,900 ton		.07	471		632,000	28,600		660,600	738,000
0430	2,000 ton		.07	492		667,000	30,000		697,000	778,500
0440	2,500 ton	↓	.06	525	↓	835,000	31,900		866,900	966,500

23 64 19 – Reciprocating Water Chillers

23 64 19.10 Reciprocating Type Water Chillers		Crew	Daily Output	Labor-Hours	Unit	Material	Labor	Equipment	Total	Total Incl O&P
0010	**RECIPROCATING TYPE WATER CHILLERS**, With standard controls									
0494	Water chillers, integral air cooled condenser									
0538	60 ton cooling	Q-7	.28	115	Ea.	45,300	7,000		52,300	60,500
0552	80 ton cooling		.26	123		54,500	7,500		62,000	71,000
0554	90 ton cooling		.25	126		61,500	7,675		69,175	79,000
0640	150 ton cooling		.23	138		93,500	8,400		101,900	115,500
0654	190 ton cooling		.22	144		116,000	8,775		124,775	140,500
0660	210 ton cooling		.22	148		123,000	9,025		132,025	149,000
0664	275 ton cooling		.21	156		154,500	9,500		164,000	184,500
0666	300 ton cooling		.20	160		167,000	9,750		176,750	198,500
0668	330 ton cooling		.20	164		187,000	10,000		197,000	221,000
0670	360 ton cooling		.19	168		207,000	10,300		217,300	243,500
0672	390 ton cooling	↓	.19	173	↓	229,500	10,500		240,000	268,000
0980	Water cooled, multiple compressor, semi-hermetic, tower not incl.									
1090	45 ton cooling	Q-7	.29	111	Ea.	25,100	6,775		31,875	37,700
1130	75 ton cooling		.23	139		46,600	8,475		55,075	64,000
1140	85 ton cooling		.21	152		49,400	9,225		58,625	68,500
1150	95 ton cooling		.19	165		52,500	10,000		62,500	73,000
1160	100 ton cooling		.18	180		50,500	10,900		61,400	72,000
1170	115 ton cooling		.17	190		51,500	11,600		63,100	74,500
1180	125 ton cooling		.16	196		52,500	12,000		64,500	76,000
1200	145 ton cooling		.16	203		59,500	12,300		71,800	84,000
1210	155 ton cooling	↓	.15	211	↓	63,000	12,800		75,800	88,000
1300	Water cooled, single compressor, semi-hermetic, tower not incl.									
1320	40 ton cooling	Q-7	.30	108	Ea.	17,600	6,575		24,175	29,200
1340	50 ton cooling		.28	114		21,800	6,925		28,725	34,400
1360	60 ton cooling	↓	.25	126	↓	24,200	7,675		31,875	38,200
4000	Packaged chiller, remote air cooled condensers not incl.									
4020	15 ton cooling	Q-7	.30	108	Ea.	20,500	6,575		27,075	32,400
4030	20 ton cooling		.28	116		21,200	7,050		28,250	33,900
4040	25 ton cooling		.25	126		23,200	7,675		30,875	37,000
4050	35 ton cooling		.24	134		28,700	8,150		36,850	43,800
4060	40 ton cooling		.22	144		27,500	8,775		36,275	43,500
4066	45 ton cooling		.21	150		30,600	9,100		39,700	47,300
4070	50 ton cooling	↓	.21	154	↓	30,900	9,375		40,275	48,100

23 64 Packaged Water Chillers

23 64 19 – Reciprocating Water Chillers

23 64 19.10 Reciprocating Type Water Chillers		Crew	Daily Output	Labor-Hours	Unit	Material	Labor (2019 Bare Costs)	Equipment	Total	Total Incl O&P
4080	60 ton cooling	Q-7	.20	164	Ea.	33,200	10,000		43,200	51,500
4090	75 ton cooling		.18	174		54,000	10,600		64,600	75,500
4100	85 ton cooling		.17	184		56,000	11,200		67,200	78,500
4110	95 ton cooling		.17	194		56,500	11,800		68,300	79,500
4120	105 ton cooling		.16	204		61,500	12,400		73,900	86,500
4130	115 ton cooling		.15	213		63,500	13,000		76,500	89,500
4140	125 ton cooling		.14	224		64,000	13,600		77,600	91,000
4150	145 ton cooling		.14	234		71,500	14,200		85,700	100,000

23 64 23 – Scroll Water Chillers

23 64 23.10 Scroll Water Chillers

Line	Description	Crew	Daily Output	Labor-Hours	Unit	Material	Labor	Equipment	Total	Total Incl O&P
0010	**SCROLL WATER CHILLERS**, With standard controls									
0480	Packaged w/integral air cooled condenser									
0482	10 ton cooling	Q-7	.34	94.118	Ea.	15,100	5,725		20,825	25,200
0490	15 ton cooling		.37	86.486		18,000	5,275		23,275	27,700
0500	20 ton cooling		.34	94.118		22,200	5,725		27,925	33,100
0510	25 ton cooling		.34	94.118		25,900	5,725		31,625	37,100
0515	30 ton cooling		.31	102		28,000	6,200		34,200	40,200
0517	35 ton cooling		.31	105		31,000	6,375		37,375	43,700
0520	40 ton cooling		.30	108		33,100	6,575		39,675	46,300
0528	45 ton cooling		.29	110		36,300	6,700		43,000	49,900
0536	50 ton cooling		.28	113		39,000	6,875		45,875	53,500
0540	70 ton cooling		.27	119		51,000	7,225		58,225	67,000
0590	100 ton cooling		.25	128		69,000	7,800		76,800	87,500
0600	110 ton cooling		.24	133		74,500	8,125		82,625	94,000
0630	130 ton cooling		.24	133		83,500	8,125		91,625	104,000
0670	170 ton cooling		.23	139		104,500	8,475		112,975	127,500
0671	210 ton cooling		.22	145		123,000	8,850		131,850	149,000
0675	250 ton cooling		.21	152		143,000	9,275		152,275	171,500
0676	275 ton cooling		.21	152		154,500	9,275		163,775	184,000
0677	300 ton cooling		.20	160		167,000	9,750		176,750	198,500
0678	330 ton cooling		.20	160		187,000	9,750		196,750	220,500
0679	390 ton cooling		.19	168		229,500	10,300		239,800	267,500
0680	Scroll water cooled, single compressor, hermetic, tower not incl.									
0700	2 ton cooling	Q-5	.57	28.070	Ea.	3,550	1,625		5,175	6,325
0710	5 ton cooling		.57	28.070		4,250	1,625		5,875	7,100
0720	6 ton cooling		.42	38.005		5,350	2,175		7,525	9,150
0740	8 ton cooling		.31	52.117		5,850	3,000		8,850	11,000
0760	10 ton cooling	Q-6	.36	67.039		6,725	4,000		10,725	13,400
0780	15 ton cooling	"	.36	66.667		10,500	3,975		14,475	17,500
0800	20 ton cooling	Q-7	.38	83.990		12,000	5,125		17,125	20,900
0820	30 ton cooling	"	.33	96.096		13,400	5,850		19,250	23,600
0900	Scroll water cooled, multiple compressor, hermetic, tower not incl.									
0910	15 ton cooling	Q-6	.36	66.667	Ea.	20,500	3,975		24,475	28,500
0930	30 ton cooling	Q-7	.36	88.889		23,200	5,400		28,600	33,600
0935	35 ton cooling		.31	103		28,700	6,275		34,975	41,000
0940	40 ton cooling		.30	107		27,500	6,500		34,000	40,100
0950	50 ton cooling		.28	114		30,900	6,950		37,850	44,400
0960	60 ton cooling		.25	128		33,200	7,800		41,000	48,200
0962	75 ton cooling		.23	139		54,000	8,475		62,475	72,000
0964	85 ton cooling		.23	139		56,000	8,475		64,475	74,000
0970	100 ton cooling		.18	178		61,500	10,800		72,300	84,000
0980	115 ton cooling		.17	188		63,500	11,500		75,000	87,000

23 64 Packaged Water Chillers

23 64 23 – Scroll Water Chillers

23 64 23.10 Scroll Water Chillers		Crew	Daily Output	Labor-Hours	Unit	Material	Labor	Equipment	Total	Total Incl O&P
						2019 Bare Costs				
0990	125 ton cooling	Q-7	.16	200	Ea.	64,000	12,200		76,200	89,000
0995	145 ton cooling	"	.16	200	"	71,500	12,200		83,700	97,000

23 64 26 – Rotary-Screw Water Chillers

23 64 26.10 Rotary-Screw Type Water Chillers		Crew	Daily Output	Labor-Hours	Unit	Material	Labor	Equipment	Total	Total Incl O&P
0010	**ROTARY-SCREW TYPE WATER CHILLERS**, With standard controls									
0110	Screw, liquid chiller, air cooled, insulated evaporator									
0120	130 ton	Q-7	.14	229	Ea.	95,500	13,900		109,400	126,000
0124	160 ton		.13	246		122,500	15,000		137,500	157,500
0128	180 ton		.13	250		138,000	15,200		153,200	174,500
0132	210 ton		.12	258		146,000	15,700		161,700	184,000
0136	270 ton		.12	267		173,500	16,200		189,700	215,500
0140	320 ton	↓	.12	276	↓	217,500	16,800		234,300	264,000
0200	Packaged unit, water cooled, not incl. tower									
0210	80 ton	Q-7	.14	224	Ea.	47,800	13,600		61,400	73,000
0220	100 ton		.14	230		58,500	14,000		72,500	85,500
0230	150 ton		.13	241		77,000	14,600		91,600	106,500
0240	200 ton		.13	252		83,500	15,300		98,800	115,000
0250	250 ton		.12	260		107,000	15,800		122,800	141,500
0260	300 ton		.12	267		130,000	16,200		146,200	168,000
0270	350 ton	↓	.12	276	↓	143,000	16,800		159,800	182,000
1450	Water cooled, tower not included									
1560	135 ton cooling, screw compressors	Q-7	.14	235	Ea.	60,500	14,300		74,800	88,000
1580	150 ton cooling, screw compressors		.13	241		70,000	14,600		84,600	99,000
1620	200 ton cooling, screw compressors		.13	250		96,500	15,200		111,700	129,000
1660	291 ton cooling, screw compressors	↓	.12	260	↓	100,500	15,800		116,300	135,000

23 64 33 – Modular Water Chillers

23 64 33.10 Direct Expansion Type Water Chillers		Crew	Daily Output	Labor-Hours	Unit	Material	Labor	Equipment	Total	Total Incl O&P
0010	**DIRECT EXPANSION TYPE WATER CHILLERS**, With standard controls									
8000	Direct expansion, shell and tube type, for built up systems									
8020	1 ton	Q-5	2	8	Ea.	7,975	460		8,435	9,450
8030	5 ton		1.90	8.421		13,200	485		13,685	15,200
8040	10 ton		1.70	9.412		16,400	540		16,940	18,800
8050	20 ton		1.50	10.667		18,700	615		19,315	21,500
8060	30 ton		1	16		23,800	920		24,720	27,600
8070	50 ton	↓	.90	17.778		38,300	1,025		39,325	43,600
8080	100 ton	Q-6	.90	26.667	↓	68,000	1,600		69,600	77,000

23 65 Cooling Towers

23 65 13 – Forced-Draft Cooling Towers

23 65 13.10 Forced-Draft Type Cooling Towers		Crew	Daily Output	Labor-Hours	Unit	Material	Labor	Equipment	Total	Total Incl O&P
0010	**FORCED-DRAFT TYPE COOLING TOWERS**, Packaged units D3030–310									
0070	Galvanized steel									
0080	Induced draft, crossflow									
0100	Vertical, belt drive, 61 tons	Q-6	90	.267	TonAC	198	15.90		213.90	242
0150	100 ton		100	.240		191	14.30		205.30	232
0200	115 ton		109	.220		166	13.15		179.15	203
0250	131 ton		120	.200		199	11.95		210.95	237
0260	162 ton	↓	132	.182	↓	161	10.85		171.85	193
1000	For higher capacities, use multiples									
1500	Induced air, double flow									

23 65 13 – Forced-Draft Cooling Towers

	23 65 13.10 Forced-Draft Type Cooling Towers	Crew	Daily Output	Labor-Hours	Unit	2019 Bare Costs Material	Labor	Equipment	Total	Total Incl O&P
1900	Vertical, gear drive, 167 ton	Q-6	126	.190	TonAC	156	11.35		167.35	189
2000	297 ton		129	.186		96.50	11.10		107.60	123
2100	582 ton		132	.182		53.50	10.85		64.35	75.50
2150	849 ton		142	.169		72	10.10		82.10	94.50
2200	1,016 ton		150	.160		73	9.55		82.55	94.50
2500	Blow through, centrifugal type									
2510	50 ton	Q-6	2.28	10.526	Ea.	13,300	630		13,930	15,500
2520	75 ton		1.52	15.789		17,100	940		18,040	20,300
2524	100 ton		1.22	19.672		20,500	1,175		21,675	24,400
2528	125 ton	Q-7	1.31	24.427		21,700	1,475		23,175	26,100
2532	200 ton		.82	39.216		32,800	2,375		35,175	39,700
2536	250 ton		.65	49.231		37,400	3,000		40,400	45,700
2540	300 ton		.54	59.259		43,100	3,600		46,700	53,000
2544	350 ton		.47	68.085		48,700	4,150		52,850	59,500
2548	400 ton		.41	78.049		50,500	4,750		55,250	62,500
2552	450 ton		.36	88.154		61,000	5,375		66,375	75,000
2556	500 ton		.32	100		65,000	6,100		71,100	80,500
2560	550 ton		.30	108		70,000	6,550		76,550	87,000
2564	600 ton		.27	118		73,500	7,150		80,650	91,000
2568	650 ton		.25	127		81,500	7,750		89,250	101,500
2572	700 ton		.23	137		95,500	8,350		103,850	118,000
2576	750 ton		.22	147		102,500	8,925		111,425	126,500
2580	800 ton		.21	152		111,000	9,275		120,275	136,500
2584	850 ton		.20	157		118,000	9,550		127,550	144,500
2588	900 ton		.20	162		125,000	9,850		134,850	152,500
2592	950 ton		.19	170		132,000	10,400		142,400	161,000
2596	1,000 ton		.18	181		136,500	11,000		147,500	166,500
2700	Axial fan, induced draft									
2710	50 ton	Q-6	2.28	10.526	Ea.	11,700	630		12,330	13,700
2720	75 ton		1.52	15.789		13,700	940		14,640	16,500
2724	100 ton		1.22	19.672		14,500	1,175		15,675	17,800
2728	125 ton	Q-7	1.31	24.427		17,900	1,475		19,375	21,900
2732	200 ton		.82	39.216		30,400	2,375		32,775	37,000
2736	250 ton		.65	49.231		34,200	3,000		37,200	42,200
2740	300 ton		.54	59.259		40,000	3,600		43,600	49,400
2744	350 ton		.47	68.085		42,700	4,150		46,850	53,000
2748	400 ton		.41	78.049		49,100	4,750		53,850	61,000
2752	450 ton		.36	88.154		54,500	5,375		59,875	68,000
2756	500 ton		.32	100		58,000	6,100		64,100	72,500
2760	550 ton		.30	108		64,500	6,550		71,050	81,000
2764	600 ton		.27	118		68,000	7,150		75,150	85,500
2768	650 ton		.25	127		74,500	7,750		82,250	93,500
2772	700 ton		.23	137		84,500	8,350		92,850	105,500
2776	750 ton		.22	147		88,000	8,925		96,925	110,500
2780	800 ton		.21	152		92,000	9,275		101,275	115,000
2784	850 ton		.20	157		92,500	9,550		102,050	116,000
2788	900 ton		.20	162		95,500	9,850		105,350	120,500
2792	950 ton		.19	170		100,500	10,400		110,900	126,000
2796	1,000 ton		.18	181		107,000	11,000		118,000	134,000
3000	For higher capacities, use multiples									
3500	For pumps and piping, add	Q-6	38	.632	TonAC	99	37.50		136.50	166
4000	For absorption systems, add				"	75%	75%			
4100	Cooling water chemical feeder	Q-5	3	5.333	Ea.	415	305		720	920

23 65 Cooling Towers

23 65 13 – Forced-Draft Cooling Towers

23 65 13.10 Forced-Draft Type Cooling Towers		Crew	Daily Output	Labor-Hours	Unit	2019 Bare Costs Material	Labor	Equipment	Total	Total Incl O&P
5000	Fiberglass tower on galvanized steel support structure									
5010	Draw thru									
5100	100 ton	Q-6	1.40	17.143	Ea.	14,400	1,025		15,425	17,300
5120	120 ton		1.20	20		16,700	1,200		17,900	20,200
5140	140 ton		1	24		18,100	1,425		19,525	22,100
5160	160 ton		.80	30		20,000	1,800		21,800	24,700
5180	180 ton		.65	36.923		22,700	2,200		24,900	28,300
5251	600 ton		.16	150		76,500	8,950		85,450	97,500
5252	1,000 ton		.10	250		127,000	14,900		141,900	162,500
5300	For stainless steel support structure, add					30%				
5360	For higher capacities, use multiples of each size									
6000	Stainless steel									
6010	Induced draft, crossflow, horizontal, belt drive									
6100	57 ton	Q-6	1.50	16	Ea.	30,600	955		31,555	35,000
6120	91 ton		.99	24.242		37,400	1,450		38,850	43,400
6140	111 ton		.43	55.814		48,300	3,325		51,625	58,000
6160	126 ton		.22	109		50,500	6,500		57,000	65,500
6170	Induced draft, crossflow, vertical, gear drive									
6172	167 ton	Q-6	.75	32	Ea.	58,500	1,900		60,400	67,500
6174	297 ton		.43	55.814		65,000	3,325		68,325	76,500
6176	582 ton		.23	104		104,500	6,225		110,725	124,500
6178	849 ton		.17	141		144,500	8,425		152,925	171,500
6180	1,016 ton		.15	160		180,000	9,550		189,550	212,500

23 65 33 – Liquid Coolers

23 65 33.10 Liquid Cooler, Closed Circuit Type Cooling Towers

Line	Description	Crew	Daily Output	Labor-Hours	Unit	Material	Labor	Equipment	Total	Total Incl O&P
0010	**LIQUID COOLER, CLOSED CIRCUIT TYPE COOLING TOWERS**									
0070	Packaged options include: vibration switch, sump sweeper piping,									
0076	grooved coil connections, fan motor inverter capable,									
0080	304 st. stl. cold water basin, external platform w/ladder.									
0100	50 ton	Q-6	2	12	Ea.	35,800	715		36,515	40,500
0150	100 ton	"	1.13	21.239		71,500	1,275		72,775	81,000
0200	150 ton	Q-7	.98	32.653		107,500	2,000		109,500	121,500
0250	200 ton		.78	41.026		143,500	2,500		146,000	161,500
0300	250 ton		.63	50.794		179,000	3,100		182,100	201,500

23 72 Air-to-Air Energy Recovery Equipment

23 72 13 – Heat-Wheel Air-to-Air Energy-Recovery Equipment

23 72 13.10 Heat-Wheel Air-to-Air Energy Recovery Equip.

Line	Description		Crew	Daily Output	Labor-Hours	Unit	Material	Labor	Equipment	Total	Total Incl O&P
0010	**HEAT-WHEEL AIR-TO-AIR ENERGY RECOVERY EQUIPMENT**										
0100	Air to air										
4000	Enthalpy recovery wheel										
4010	1,000 max CFM	G	Q-9	1.20	13.333	Ea.	7,225	730		7,955	9,050
4020	2,000 max CFM	G		1	16		8,450	880		9,330	10,600
4030	4,000 max CFM	G		.80	20		9,775	1,100		10,875	12,500
4040	6,000 max CFM	G		.70	22.857		11,400	1,250		12,650	14,500
4050	8,000 max CFM	G	Q-10	1	24		12,600	1,375		13,975	16,000
4060	10,000 max CFM	G		.90	26.667		15,100	1,525		16,625	18,900
4070	20,000 max CFM	G		.80	30		27,400	1,700		29,100	32,700
4080	25,000 max CFM	G		.70	34.286		33,400	1,950		35,350	39,800
4090	30,000 max CFM	G		.50	48		37,100	2,725		39,825	45,000

23 72 Air-to-Air Energy Recovery Equipment

23 72 13 – Heat-Wheel Air-to-Air Energy-Recovery Equipment

23 72 13.10 Heat-Wheel Air-to-Air Energy Recovery Equip.		Crew	Daily Output	Labor-Hours	Unit	2019 Bare Costs Material	Labor	Equipment	Total	Total Incl O&P
4100	40,000 max CFM G	Q-10	.45	53.333	Ea.	51,000	3,025		54,025	61,000
4110	50,000 max CFM G	↓	.40	60	↓	59,500	3,425		62,925	70,500

23 72 16 – Heat-Pipe Air-To-Air Energy-Recovery Equipment

23 72 16.10 Heat Pipes		Crew	Daily Output	Labor-Hours	Unit	Material	Labor	Equipment	Total	Total Incl O&P
0010	**HEAT PIPES**									
8000	Heat pipe type, glycol, 50% efficient									
8010	100 MBH, 1,700 CFM	1 Stpi	.80	10	Ea.	4,675	640		5,315	6,100
8020	160 MBH, 2,700 CFM		.60	13.333		6,300	855		7,155	8,200
8030	620 MBH, 4,000 CFM	↓	.40	20	↓	9,500	1,275		10,775	12,300

23 73 Indoor Central-Station Air-Handling Units

23 73 13 – Modular Indoor Central-Station Air-Handling Units

23 73 13.10 Air-Handling Units		Crew	Daily Output	Labor-Hours	Unit	Material	Labor	Equipment	Total	Total Incl O&P
0010	**AIR-HANDLING UNITS**, Built-Up									
0100	With cooling/heating coil section, filters, mixing box									
0880	Single zone, horizontal/vertical									
0890	Constant volume									
0900	1,600 CFM	Q-5	1.20	13.333	Ea.	5,550	765		6,315	7,275
0906	2,000 CFM		1.10	14.545		6,675	835		7,510	8,600
0910	3,000 CFM		1	16		9,775	920		10,695	12,200
0916	4,000 CFM	↓	.95	16.842		12,800	970		13,770	15,600
0920	5,000 CFM	Q-6	1.40	17.143		15,600	1,025		16,625	18,600
0926	6,500 CFM		1.30	18.462		19,500	1,100		20,600	23,100
0930	7,500 CFM		1.20	20		21,100	1,200		22,300	25,000
0936	9,200 CFM		1.10	21.818		24,500	1,300		25,800	28,900
0940	11,500 CFM		1	24		27,800	1,425		29,225	32,800
0946	13,200 CFM		.90	26.667		33,400	1,600		35,000	39,100
0950	16,500 CFM		.80	30		41,200	1,800		43,000	48,000
0960	19,500 CFM		.70	34.286		48,400	2,050		50,450	56,000
0970	22,000 CFM		.60	40		54,500	2,375		56,875	63,500
0980	27,000 CFM		.50	48		65,500	2,875		68,375	76,500
0990	34,000 CFM		.40	60		83,500	3,575		87,075	97,000
1000	40,000 CFM		.30	80		100,000	4,775		104,775	117,000
1010	47,000 CFM	↓	.20	120	↓	109,500	7,150		116,650	131,000
2300	Multi-zone, horizontal or vertical, blow-thru fan									
2310	3,000 CFM	Q-5	1	16	Ea.	7,475	920		8,395	9,600
2314	4,000 CFM	"	.95	16.842		10,400	970		11,370	13,000
2318	5,000 CFM	Q-6	1.40	17.143		13,000	1,025		14,025	15,800
2322	6,500 CFM		1.30	18.462		16,900	1,100		18,000	20,200
2326	7,500 CFM		1.20	20		18,400	1,200		19,600	22,100
2330	9,200 CFM		1.10	21.818		24,000	1,300		25,300	28,400
2334	11,500 CFM		1	24		26,400	1,425		27,825	31,300
2336	13,200 CFM		.90	26.667		30,200	1,600		31,800	35,700
2340	16,500 CFM		.80	30		37,700	1,800		39,500	44,200
2344	19,500 CFM		.70	34.286		44,500	2,050		46,550	52,000
2348	22,000 CFM		.60	40		50,000	2,375		52,375	58,500
2350	22,500 CFM		.55	43.636		50,500	2,600		53,100	59,500
2352	27,000 CFM		.50	48		53,000	2,875		55,875	63,000
2356	34,000 CFM		.40	60		67,000	3,575		70,575	79,000
2360	40,000 CFM	↓	.30	80	↓	72,500	4,775		77,275	87,000

23 73 13 – Modular Indoor Central-Station Air-Handling Units

23 73 13.10 Air-Handling Units		Crew	Daily Output	Labor-Hours	Unit	Material	2019 Bare Costs Labor	Equipment	Total	Total Incl O&P
2364	47,000 CFM	Q-6	.20	120	Ea.	92,500	7,150		99,650	112,000

23 73 13.20 Air-Handling Units, Packaged Indoor Type

		Crew	Daily Output	Labor-Hours	Unit	Material	Labor	Equipment	Total	Total Incl O&P
0010	**AIR-HANDLING UNITS, PACKAGED INDOOR TYPE**									
0040	Cooling coils may be chilled water or DX									
0050	Heating coils may be hot water, steam or electric									
1090	Constant volume									
1200	2,000 CFM	Q-5	1.10	14.545	Ea.	10,900	835		11,735	13,300
1400	5,000 CFM	Q-6	.80	30		17,500	1,800		19,300	22,000
1550	10,000 CFM		.54	44.444		25,800	2,650		28,450	32,400
1670	15,000 CFM		.38	63.158		52,500	3,775		56,275	63,500
1700	20,000 CFM		.31	77.419		67,000	4,625		71,625	80,500
1710	30,000 CFM		.27	88.889		91,000	5,300		96,300	108,000
2300	Variable air volume									
2330	2,000 CFM	Q-5	1	16	Ea.	15,500	920		16,420	18,400
2340	5,000 CFM	Q-6	.70	34.286		21,400	2,050		23,450	26,600
2350	10,000 CFM		.50	48		41,500	2,875		44,375	50,000
2360	15,000 CFM		.40	60		60,500	3,575		64,075	72,000
2370	20,000 CFM		.29	82.759		77,500	4,950		82,450	93,000
2380	30,000 CFM		.20	120		93,500	7,150		100,650	113,500

23 73 39 – Indoor, Direct Gas-Fired Heating and Ventilating Units

23 73 39.10 Make-Up Air Unit

		Crew	Daily Output	Labor-Hours	Unit	Material	Labor	Equipment	Total	Total Incl O&P
0010	**MAKE-UP AIR UNIT**									
0020	Indoor suspension, natural/LP gas, direct fired,									
0032	standard control. For flue see Section 23 51 23.10									
0040	70°F temperature rise, MBH is input									
0100	75 MBH input	Q-6	3.60	6.667	Ea.	5,250	400		5,650	6,375
0120	100 MBH input		3.40	7.059		5,450	420		5,870	6,625
0140	125 MBH input		3.20	7.500		5,800	450		6,250	7,050
0160	150 MBH input		3	8		6,375	475		6,850	7,725
0180	175 MBH input		2.80	8.571		6,525	510		7,035	7,950
0200	200 MBH input		2.60	9.231		6,875	550		7,425	8,400
0220	225 MBH input		2.40	10		7,125	595		7,720	8,750
0240	250 MBH input		2	12		8,225	715		8,940	10,100
0260	300 MBH input		1.90	12.632		8,425	755		9,180	10,400
0280	350 MBH input		1.75	13.714		9,325	820		10,145	11,500
0300	400 MBH input		1.60	15		16,700	895		17,595	19,800
0320	840 MBH input		1.20	20		22,900	1,200		24,100	27,000
0340	1,200 MBH input		.88	27.273		33,800	1,625		35,425	39,600
0600	For discharge louver assembly, add					5%				
0700	For filters, add					10%				
0800	For air shut-off damper section, add					30%				

23 74 Packaged Outdoor HVAC Equipment

23 74 13 – Packaged, Outdoor, Central-Station Air-Handling Units

23 74 13.10 Packaged, Outdoor Type, Central-Station AHU		Crew	Daily Output	Labor-Hours	Unit	Material	Labor	Equipment	Total	Total Incl O&P
						2019 Bare Costs				
0010	**PACKAGED, OUTDOOR TYPE, CENTRAL-STATION AIR-HANDLING UNITS**									
2900	Cooling coils may be chilled water or DX									
2910	Heating coils may be hot water, steam or electric									
3000	Weathertight, ground or rooftop									
3010	Constant volume									
3100	2,000 CFM	Q-6	1	24	Ea.	23,900	1,425		25,325	28,400
3140	5,000 CFM		.79	30.380		27,700	1,825		29,525	33,200
3150	10,000 CFM		.52	46.154		58,000	2,750		60,750	67,500
3160	15,000 CFM		.41	58.537		84,000	3,500		87,500	98,000
3170	20,000 CFM		.30	80		115,000	4,775		119,775	133,500
3180	30,000 CFM		.24	100		144,000	5,975		149,975	167,500
3200	Variable air volume									
3210	2,000 CFM	Q-6	.96	25	Ea.	15,100	1,500		16,600	18,900
3240	5,000 CFM		.73	32.877		30,500	1,950		32,450	36,500
3250	10,000 CFM		.50	48		65,000	2,875		67,875	75,500
3260	15,000 CFM		.40	60		99,000	3,575		102,575	114,500
3270	20,000 CFM		.29	82.759		128,500	4,950		133,450	149,000
3280	30,000 CFM		.20	120		151,500	7,150		158,650	177,000

23 74 23 – Packaged, Outdoor, Heating-Only Makeup-Air Units

23 74 23.16 Packaged, Ind.-Fired, In/Outdoor		Crew	Daily Output	Labor-Hours	Unit	Material	Labor	Equipment	Total	Total Incl O&P
0010	**PACKAGED, IND.-FIRED, IN/OUTDOOR**, Heat-Only Makeup-Air Units									
1000	Rooftop unit, natural gas, gravity vent, S.S. exchanger									
1010	70°F temperature rise, MBH is input									
1020	250 MBH	Q-6	4	6	Ea.	16,100	360		16,460	18,300
1040	400 MBH		3.60	6.667		18,600	400		19,000	21,100
1060	550 MBH		3.30	7.273		22,100	435		22,535	25,000
1080	750 MBH		3	8		23,500	475		23,975	26,500
1100	1,000 MBH		2.60	9.231		26,500	550		27,050	29,900
1120	1,750 MBH		2.30	10.435		31,700	625		32,325	35,800
1140	2,500 MBH		1.90	12.632		46,800	755		47,555	52,500
1160	3,250 MBH		1.40	17.143		52,000	1,025		53,025	59,000
1180	4,000 MBH		1.20	20		70,000	1,200		71,200	79,000
1200	6,000 MBH		1	24		89,500	1,425		90,925	100,000
1600	For cleanable filters, add					5%				
1700	For electric modulating gas control, add					10%				

23 74 33 – Dedicated Outdoor-Air Units

23 74 33.10 Rooftop Air Conditioners		Crew	Daily Output	Labor-Hours	Unit	Material	Labor	Equipment	Total	Total Incl O&P
0010	**ROOFTOP AIR CONDITIONERS**, Standard controls, curb, economizer									
1000	Single zone, electric cool, gas heat									
1100	3 ton cooling, 60 MBH heating D3050–150	Q-5	.70	22.857	Ea.	3,025	1,325		4,350	5,300
1120	4 ton cooling, 95 MBH heating D3050–155		.61	26.403		3,575	1,525		5,100	6,200
1140	5 ton cooling, 112 MBH heating R236000-10		.56	28.521		4,425	1,650		6,075	7,325
1145	6 ton cooling, 140 MBH heating R236000-20		.52	30.769		5,050	1,775		6,825	8,200
1150	7.5 ton cooling, 170 MBH heating		.50	32.258		5,825	1,850		7,675	9,200
1156	8.5 ton cooling, 170 MBH heating		.46	34.783		7,100	2,000		9,100	10,800
1160	10 ton cooling, 200 MBH heating	Q-6	.67	35.982		9,200	2,150		11,350	13,300
1170	12.5 ton cooling, 230 MBH heating		.63	37.975		10,500	2,275		12,775	15,000
1180	15 ton cooling, 270 MBH heating		.57	42.032		13,100	2,500		15,600	18,200
1190	17.5 ton cooling, 330 MBH heating		.52	45.889		14,000	2,750		16,750	19,500
1200	20 ton cooling, 360 MBH heating	Q-7	.67	47.976		29,500	2,925		32,425	36,900
1210	25 ton cooling, 450 MBH heating		.56	57.554		32,800	3,500		36,300	41,300
1220	30 ton cooling, 540 MBH heating		.47	68.376		35,700	4,175		39,875	45,600

23 74 33.10 Rooftop Air Conditioners		Crew	Daily Output	Labor-Hours	Unit	2019 Bare Costs Material	Labor	Equipment	Total	Total Incl O&P
1240	40 ton cooling, 675 MBH heating	Q-7	.35	91.168	Ea.	43,900	5,550		49,450	56,500
1260	50 ton cooling, 810 MBH heating		.28	114		49,900	6,925		56,825	65,500
1265	60 ton cooling, 900 MBH heating		.23	137		59,500	8,325		67,825	78,000
1270	80 ton cooling, 1,000 MBH heating		.20	160		77,500	9,750		87,250	99,500
1275	90 ton cooling, 1,200 MBH heating		.17	188		107,000	11,500		118,500	135,000
1280	100 ton cooling, 1,350 MBH heating		.14	229		125,500	13,900		139,400	159,000
1300	Electric cool, electric heat									
1310	5 ton	Q-5	.63	25.600	Ea.	5,275	1,475		6,750	8,000
1314	10 ton	Q-6	.74	32.432		8,275	1,925		10,200	12,000
1318	15 ton	"	.64	37.795		12,500	2,250		14,750	17,100
1322	20 ton	Q-7	.74	43.185		16,600	2,625		19,225	22,200
1326	25 ton		.62	51.696		20,700	3,150		23,850	27,500
1330	30 ton		.52	61.657		24,900	3,750		28,650	33,000
1334	40 ton		.39	82.051		33,200	5,000		38,200	44,000
1338	50 ton		.31	103		39,300	6,250		45,550	52,500
1342	60 ton		.26	123		47,300	7,500		54,800	63,000
1346	75 ton		.21	154		59,000	9,375		68,375	79,000
2000	Multizone, electric cool, gas heat, economizer									
2100	15 ton cooling, 360 MBH heating	Q-7	.61	52.545	Ea.	65,000	3,200		68,200	76,500
2120	20 ton cooling, 360 MBH heating		.53	60.038		69,500	3,650		73,150	82,000
2140	25 ton cooling, 450 MBH heating		.45	71.910		85,500	4,375		89,875	100,500
2160	28 ton cooling, 450 MBH heating		.41	79.012		103,500	4,800		108,300	121,000
2180	30 ton cooling, 540 MBH heating		.37	85.562		108,000	5,200		113,200	127,000
2200	40 ton cooling, 540 MBH heating		.28	114		123,500	6,925		130,425	146,500
2210	50 ton cooling, 540 MBH heating		.23	142		155,500	8,650		164,150	184,000
2220	70 ton cooling, 1,500 MBH heating		.16	199		165,000	12,100		177,100	199,500
2240	80 ton cooling, 1,500 MBH heating		.14	229		188,500	13,900		202,400	228,500
2260	90 ton cooling, 1,500 MBH heating		.13	256		193,000	15,600		208,600	236,000
2280	105 ton cooling, 1,500 MBH heating		.11	291		215,000	17,700		232,700	262,500
2400	For hot water heat coil, deduct					5%				
2500	For steam heat coil, deduct					2%				
2600	For electric heat, deduct					3%	5%			
4000	Single zone, electric cool, gas heat, variable air volume									
4100	12.5 ton cooling, 230 MBH heating	Q-6	.55	43.716	Ea.	13,300	2,600		15,900	18,500
4120	18 ton cooling, 330 MBH heating	"	.46	52.747		17,600	3,150		20,750	24,100
4140	25 ton cooling, 450 MBH heating	Q-7	.48	66.116		37,200	4,025		41,225	47,000
4160	40 ton cooling, 675 MBH heating		.31	105		51,500	6,375		57,875	66,000
4180	60 ton cooling, 900 MBH heating		.20	158		73,500	9,600		83,100	95,000
4200	80 ton cooling, 1,000 MBH heating		.15	209		94,500	12,700		107,200	123,000
5000	Single zone, electric cool only									
5050	3 ton cooling	Q-5	.88	18.203	Ea.	2,700	1,050		3,750	4,550
5060	4 ton cooling		.76	21.108		3,350	1,225		4,575	5,500
5070	5 ton cooling		.70	22.792		3,600	1,300		4,900	5,925
5080	6 ton cooling		.65	24.502		4,375	1,400		5,775	6,950
5090	7.5 ton cooling		.62	25.806		6,500	1,475		7,975	9,400
5100	8.5 ton cooling		.59	26.981		7,150	1,550		8,700	10,200
5110	10 ton cooling	Q-6	.83	28.812		8,475	1,725		10,200	11,900
5120	12.5 ton cooling		.80	30		11,500	1,800		13,300	15,400
5130	15 ton cooling		.71	33.613		15,500	2,000		17,500	20,100
5140	17.5 ton cooling		.65	36.697		16,600	2,200		18,800	21,600
5150	20 ton cooling	Q-7	.83	38.415		21,700	2,350		24,050	27,300
5160	25 ton cooling		.70	45.977		22,600	2,800		25,400	29,100
5170	30 ton cooling		.59	54.701		24,600	3,325		27,925	32,100

23 74 33 – Dedicated Outdoor-Air Units

23 74 33.10 Rooftop Air Conditioners		Crew	Daily Output	Labor-Hours	Unit	Material	2019 Bare Costs Labor	Equipment	Total	Total Incl O&P
5180	40 ton cooling	Q-7	.44	73.059	Ea.	33,100	4,450		37,550	43,100
5190	50 ton cooling		.38	84.211		38,400	5,125		43,525	49,900
5200	60 ton cooling		.28	114		47,100	6,950		54,050	62,500
5400	For low heat, add					7%				
5410	For high heat, add					10%				
6000	Single zone electric cooling with variable volume distribution									
6020	20 ton cooling	Q-7	.72	44.199	Ea.	23,500	2,700		26,200	29,800
6030	25 ton cooling		.61	52.893		25,400	3,225		28,625	32,700
6040	30 ton cooling		.51	62.868		29,700	3,825		33,525	38,400
6050	40 ton cooling		.38	83.990		34,100	5,125		39,225	45,200
6060	50 ton cooling		.31	105		35,300	6,375		41,675	48,400
6070	60 ton cooling		.25	126		48,700	7,675		56,375	65,000
6200	For low heat, add					7%				
6210	For high heat, add					10%				
7000	Multizone, cool/heat, variable volume distribution									
7100	50 ton cooling	Q-7	.20	164	Ea.	115,000	10,000		125,000	141,500
7110	70 ton cooling		.14	229		161,000	13,900		174,900	198,000
7120	90 ton cooling		.11	296		187,500	18,000		205,500	233,000
7130	105 ton cooling		.10	333		206,500	20,300		226,800	257,500
7140	120 ton cooling		.08	381		236,000	23,200		259,200	294,500
7150	140 ton cooling		.07	444		272,500	27,100		299,600	340,000
7300	Penthouse unit, cool/heat, variable volume distribution									
7400	150 ton cooling	Q-7	.10	311	Ea.	426,500	18,900		445,400	498,000
7410	170 ton cooling		.10	320		440,000	19,500		459,500	513,000
7420	200 ton cooling		.10	327		518,000	19,900		537,900	599,500
7430	225 ton cooling		.09	340		558,500	20,700		579,200	645,500
7440	250 ton cooling		.09	360		621,000	21,900		642,900	716,000
7450	270 ton cooling		.09	364		670,500	22,100		692,600	770,500
7460	300 ton cooling		.09	372		745,000	22,700		767,700	853,500
9400	Sub assemblies for assembly systems									
9410	Ductwork, per ton, rooftop 1 zone units	Q-9	.96	16.667	Ton	266	915		1,181	1,700
9420	Ductwork, per ton, rooftop multizone units		.50	32		350	1,750		2,100	3,050
9430	Ductwork, VAV cooling/ton, rooftop multizone, 1 zone/ton		.32	50		460	2,750		3,210	4,675
9440	Ductwork per ton, packaged water & air cooled units		1.26	12.698		62.50	695		757.50	1,125
9450	Ductwork per ton, split system remote condensing units		1.32	12.121		59	665		724	1,075
9500	Ductwork package for residential units, 1 ton		.80	20	Ea.	370	1,100		1,470	2,075
9520	2 ton		.40	40		745	2,200		2,945	4,150
9530	3 ton		.27	59.259		1,125	3,250		4,375	6,150
9540	4 ton	Q-10	.30	80		1,475	4,550		6,025	8,525
9550	5 ton		.23	104		1,850	5,925		7,775	11,100
9560	6 ton		.20	120		2,225	6,825		9,050	12,900
9570	7 ton		.17	141		2,600	8,025		10,625	15,100
9580	8 ton		.15	160		2,975	9,100		12,075	17,100

23 76 13 – Direct Evaporative Air Coolers

23 76 13.10 Evaporative Coolers		Crew	Daily Output	Labor-Hours	Unit	Material	2019 Bare Costs Labor	Equipment	Total	Total Incl O&P
0010	EVAPORATIVE COOLERS (Swamp Coolers) ducted, not incl. duct.									
0100	Side discharge style, capacities at 0.25" S.P.									
0120	1,785 CFM, 1/3 HP, 115 V G	Q-9	5	3.200	Ea.	470	176		646	780
0140	2,740 CFM, 1/3 HP, 115 V G		4.50	3.556		560	195		755	915
0160	3,235 CFM, 1/2 HP, 115 V G		4	4		565	219		784	960
0180	3,615 CFM, 1/2 HP, 230 V G		3.60	4.444		695	244		939	1,125
0200	4,215 CFM, 3/4 HP, 230 V G		3.20	5		740	274		1,014	1,225
0220	5,255 CFM, 1 HP, 115/230 V G		3	5.333		1,150	293		1,443	1,700
0240	6,090 CFM, 1 HP, 230/460 V G		2.80	5.714		1,950	315		2,265	2,600
0260	8,300 CFM, 1-1/2 HP, 230/460 V G		2.60	6.154		1,950	340		2,290	2,650
0280	8,360 CFM, 1-1/2 HP, 230/460 V G		2.20	7.273		1,950	400		2,350	2,750
0300	9,725 CFM, 2 HP, 230/460 V G		1.80	8.889		2,200	490		2,690	3,175
0320	11,715 CFM, 3 HP, 230/460 V G		1.40	11.429		2,425	625		3,050	3,625
0340	14,410 CFM, 5 HP, 230/460 V G	↓	1	16	↓	2,650	880		3,530	4,225
0400	For two-speed motor, add					5%				
0500	For down discharge style, add					10%				

23 76 26 – Fan Coil Evaporators for Coolers/Freezers

23 76 26.10 Evaporators		Crew	Daily Output	Labor-Hours	Unit	Material	Labor	Equipment	Total	Total Incl O&P
0010	EVAPORATORS, DX coils, remote compressors not included.									
1000	Coolers, reach-in type, above freezing temperatures									
1300	Shallow depth, wall mount, 7 fins per inch, air defrost									
1310	600 BTUH, 8" fan, rust-proof core	Q-5	3.80	4.211	Ea.	700	242		942	1,125
1320	900 BTUH, 8" fan, rust-proof core		3.30	4.848		815	279		1,094	1,325
1330	1,200 BTUH, 8" fan, rust-proof core		3	5.333		1,025	305		1,330	1,575
1340	1,800 BTUH, 8" fan, rust-proof core		2.40	6.667		1,025	385		1,410	1,725
1350	2,500 BTUH, 10" fan, rust-proof core		2.20	7.273		1,100	420		1,520	1,825
1360	3,500 BTUH, 10" fan		2	8		1,150	460		1,610	1,975
1370	4,500 BTUH, 10" fan	↓	1.90	8.421		1,225	485		1,710	2,075
1390	For pan drain, add				↓	62.50			62.50	69
1600	Undercounter refrigerators, ceiling or wall mount,									
1610	8 fins per inch, air defrost, rust-proof core									
1630	800 BTUH, one 6" fan	Q-5	5	3.200	Ea.	286	184		470	590
1640	1,300 BTUH, two 6" fans		4	4		365	230		595	750
1650	1,700 BTUH, two 6" fans	↓	3.60	4.444	↓	410	256		666	835
2000	Coolers, reach-in and walk-in types, above freezing temperatures									
2600	Two-way discharge, ceiling mount, 150 to 4,100 CFM,									
2610	above 34°F applications, air defrost									
2630	900 BTUH, 7 fins per inch, 8" fan	Q-5	4	4	Ea.	560	230		790	960
2660	2,500 BTUH, 7 fins per inch, 10" fan		2.60	6.154		855	355		1,210	1,475
2690	5,500 BTUH, 7 fins per inch, 12" fan		1.70	9.412		1,525	540		2,065	2,500
2720	8,500 BTUH, 8 fins per inch, 16" fan		1.20	13.333		1,600	765		2,365	2,925
2750	15,000 BTUH, 7 fins per inch, 18" fan		1.10	14.545		2,575	835		3,410	4,100
2770	24,000 BTUH, 7 fins per inch, two 16" fans		1	16		3,850	920		4,770	5,625
2790	30,000 BTUH, 7 fins per inch, two 18" fans	↓	.90	17.778	↓	4,800	1,025		5,825	6,800
2850	Two-way discharge, low profile, ceiling mount,									
2860	8 fins per inch, 200 to 570 CFM, air defrost									
2880	800 BTUH, one 6" fan	Q-5	4	4	Ea.	335	230		565	715
2890	1,300 BTUH, two 6" fans		3.50	4.571		405	263		668	840
2900	1,800 BTUH, two 6" fans		3.30	4.848		485	279		764	955
2910	2,700 BTUH, three 6" fans	↓	2.70	5.926	↓	585	340		925	1,150
3000	Coolers, walk-in type, above freezing temperatures									
3300	General use, ceiling mount, 108 to 2,080 CFM									

23 76 26 – Fan Coil Evaporators for Coolers/Freezers

	23 76 26.10 Evaporators	Crew	Daily Output	Labor-Hours	Unit	Material	2019 Bare Costs Labor	Equipment	Total	Total Incl O&P
3320	600 BTUH, 7 fins per inch, 6" fan	Q-5	5.20	3.077	Ea.	335	177		512	635
3340	1,200 BTUH, 7 fins per inch, 8" fan		3.70	4.324		465	249		714	885
3360	1,800 BTUH, 7 fins per inch, 10" fan		2.70	5.926		655	340		995	1,225
3380	3,500 BTUH, 7 fins per inch, 12" fan		2.40	6.667		960	385		1,345	1,625
3400	5,500 BTUH, 8 fins per inch, 12" fan		2	8		1,075	460		1,535	1,875
3430	8,500 BTUH, 8 fins per inch, 16" fan		1.40	11.429		1,600	660		2,260	2,725
3460	15,000 BTUH, 7 fins per inch, two 16" fans		1.10	14.545		2,550	835		3,385	4,050
3640	Low velocity, high latent load, ceiling mount,									
3650	1,050 to 2,420 CFM, air defrost, 6 fins per inch									
3670	6,700 BTUH, two 10" fans	Q-5	1.30	12.308	Ea.	1,775	710		2,485	3,025
3680	10,000 BTUH, three 10" fans		1	16		2,250	920		3,170	3,850
3690	13,500 BTUH, three 10" fans		1	16		2,825	920		3,745	4,475
3700	18,000 BTUH, four 10" fans		1	16		3,550	920		4,470	5,275
3710	26,500 BTUH, four 10" fans		.80	20		4,450	1,150		5,600	6,625
3730	For electric defrost, add					17%				
5000	Freezers and coolers, reach-in type, above 34°F									
5030	to sub-freezing temperature range, low latent load,									
5050	air defrost, 7 fins per inch									
5070	1,200 BTUH, 8" fan, rustproof core	Q-5	4.40	3.636	Ea.	690	209		899	1,075
5080	1,500 BTUH, 8" fan, rustproof core		4.20	3.810		705	219		924	1,100
5090	1,800 BTUH, 8" fan, rustproof core		3.70	4.324		720	249		969	1,175
5100	2,500 BTUH, 10" fan, rustproof core		2.90	5.517		690	320		1,010	1,225
5110	3,500 BTUH, 10" fan, rustproof core		2.50	6.400		860	370		1,230	1,500
5120	4,500 BTUH, 12" fan, rustproof core		2.10	7.619		1,075	440		1,515	1,825
6000	Freezers and coolers, walk-in type									
6050	1,960 to 18,000 CFM, medium profile									
6060	Standard motor, 6 fins per inch, to -30°F, all aluminum									
6080	10,500 BTUH, air defrost, one 18" fan	Q-5	1.40	11.429	Ea.	1,700	660		2,360	2,850
6090	12,500 BTUH, air defrost, one 18" fan		1.40	11.429		1,825	660		2,485	3,000
6100	16,400 BTUH, air defrost, one 18" fan		1.30	12.308		1,850	710		2,560	3,125
6110	20,900 BTUH, air defrost, one 18" fan		1	16		2,175	920		3,095	3,750
6120	27,000 BTUH, air defrost, two 18" fans		1	16		3,075	920		3,995	4,750
6130	32,900 BTUH, air defrost, two 20" fans	Q-6	1.30	18.462		3,400	1,100		4,500	5,400
6140	39,000 BTUH, air defrost, two 20" fans		1.25	19.200		3,950	1,150		5,100	6,050
6150	44,100 BTUH, air defrost, three 20" fans		1	24		4,550	1,425		5,975	7,150
6155	53,000 BTUH, air defrost, three 20" fans		1	24		4,375	1,425		5,800	6,975
6160	66,200 BTUH, air defrost, four 20" fans		1	24		5,025	1,425		6,450	7,700
6170	78,000 BTUH, air defrost, four 20" fans		.80	30		5,800	1,800		7,600	9,050
6180	88,200 BTUH, air defrost, four 24" fans		.60	40		6,250	2,375		8,625	10,500
6190	110,000 BTUH, air defrost, four 24" fans		.50	48		7,550	2,875		10,425	12,600
6330	Hot gas defrost, standard motor units, add					35%				
6370	Electric defrost, 230V, standard motor, add					16%				
6410	460V, standard or high capacity motor, add					25%				
6800	Eight fins per inch, increases BTUH 75%									
6810	Standard capacity, add				Ea.	5%				
6830	Hot gas & electric defrost not recommended									
7000	800 to 4,150 CFM, 12" fans									
7010	Four fins per inch									
7030	3,400 BTUH, 1 fan	Q-5	2.80	5.714	Ea.	570	330		900	1,125
7040	4,200 BTUH, 1 fan		2.40	6.667		665	385		1,050	1,300
7050	5,300 BTUH, 1 fan		1.90	8.421		740	485		1,225	1,550
7060	6,800 BTUH, 1 fan		1.70	9.412		840	540		1,380	1,725
7070	8,400 BTUH, 2 fans		1.60	10		845	575		1,420	1,800

23 76 Evaporative Air-Cooling Equipment

23 76 26 – Fan Coil Evaporators for Coolers/Freezers

23 76 26.10 Evaporators		Crew	Daily Output	Labor-Hours	Unit	Material	2019 Bare Costs Labor	Equipment	Total	Total Incl O&P
7080	10,500 BTUH, 2 fans	Q-5	1.40	11.429	Ea.	1,050	660		1,710	2,125
7090	13,000 BTUH, 3 fans		1.30	12.308		1,650	710		2,360	2,900
7100	17,000 BTUH, 4 fans		1.20	13.333		2,050	765		2,815	3,400
7110	21,500 BTUH, 5 fans		1	16		2,350	920		3,270	3,975
7160	Six fins per inch increases BTUH 35%, add					5%				
7180	Eight fins per inch increases BTUH 53%,									
7190	not recommended for below 35°F, add				Ea.	8%				
7210	For 12°F temperature differential, add					20%				
7230	For adjustable thermostat control, add					126			126	139
7240	For hot gas defrost, except 8 fin models									
7250	on applications below 35°F, add				Ea.	58%				
7260	For electric defrost, except 8 fin models									
7270	on applications below 35°F, add				Ea.	36%				
8000	Freezers, pass-thru door uprights, walk-in storage									
8300	Low temperature, thin profile, electric defrost									
8310	138 to 1,800 CFM, 5 fins per inch									
8320	900 BTUH, one 6" fan	Q-5	3.30	4.848	Ea.	435	279		714	900
8340	1,500 BTUH, two 6" fans		2.20	7.273		660	420		1,080	1,350
8360	2,600 BTUH, three 6" fans		1.90	8.421		920	485		1,405	1,725
8370	3,300 BTUH, four 6" fans		1.70	9.412		1,025	540		1,565	1,975
8380	4,400 BTUH, five 6" fans		1.40	11.429		1,450	660		2,110	2,550
8400	7,000 BTUH, three 10" fans		1.30	12.308		1,775	710		2,485	3,025
8410	8,700 BTUH, four 10" fans		1.10	14.545		2,100	835		2,935	3,575
8450	For air defrost, deduct					263			263	290

23 81 Decentralized Unitary HVAC Equipment

23 81 13 – Packaged Terminal Air-Conditioners

23 81 13.10 Packaged Cabinet Type Air-Conditioners		Crew	Daily Output	Labor-Hours	Unit	Material	Labor	Equipment	Total	Total Incl O&P
0010	**PACKAGED CABINET TYPE AIR-CONDITIONERS**, Cabinet, wall sleeve,									
0100	louver, electric heat, thermostat, manual changeover, 208 V									
0200	6,000 BTUH cooling, 8,800 BTU heat	Q-5	6	2.667	Ea.	705	153		858	1,000
0220	9,000 BTUH cooling, 13,900 BTU heat		5	3.200		970	184		1,154	1,350
0240	12,000 BTUH cooling, 13,900 BTU heat		4	4		1,475	230		1,705	1,975
0260	15,000 BTUH cooling, 13,900 BTU heat		3	5.333		1,450	305		1,755	2,050
0280	18,000 BTUH cooling, 10 KW heat		2.40	6.667		2,450	385		2,835	3,250
0300	24,000 BTUH cooling, 10 KW heat		1.90	8.421		2,475	485		2,960	3,450
0320	30,000 BTUH cooling, 10 KW heat		1.40	11.429		2,875	660		3,535	4,125
0340	36,000 BTUH cooling, 10 KW heat		1.25	12.800		2,875	735		3,610	4,250
0360	42,000 BTUH cooling, 10 KW heat		1	16		3,125	920		4,045	4,825
0380	48,000 BTUH cooling, 10 KW heat		.90	17.778		3,325	1,025		4,350	5,175
0500	For hot water coil, increase heat by 10%, add					5%	10%			
1000	For steam, increase heat output by 30%, add					8%	10%			

23 81 19 – Self-Contained Air-Conditioners

23 81 19.10 Window Unit Air Conditioners		Crew	Daily Output	Labor-Hours	Unit	Material	Labor	Equipment	Total	Total Incl O&P
0010	**WINDOW UNIT AIR CONDITIONERS**									
4000	Portable/window, 15 amp, 125 V grounded receptacle required									
4060	5,000 BTUH	1 Carp	8	1	Ea.	305	51.50		356.50	415
4340	6,000 BTUH		8	1		252	51.50		303.50	355
4480	8,000 BTUH		6	1.333		355	69		424	495
4500	10,000 BTUH		6	1.333		600	69		669	765

23 81 19 – Self-Contained Air-Conditioners

23 81 19.10 Window Unit Air Conditioners		Crew	Daily Output	Labor-Hours	Unit	2019 Bare Costs Material	Labor	Equipment	Total	Total Incl O&P
4520	12,000 BTUH	L-2	8	2	Ea.	1,950	90.50		2,040.50	2,300
4600	Window/thru-the-wall, 15 amp, 230 V grounded receptacle required									
4780	18,000 BTUH	L-2	6	2.667	Ea.	680	121		801	935
4940	25,000 BTUH		4	4		930	181		1,111	1,300
4960	29,000 BTUH	↓	4	4	↓	955	181		1,136	1,325

23 81 19.20 Self-Contained Single Package		Crew	Daily Output	Labor-Hours	Unit	Material	Labor	Equipment	Total	Total Incl O&P
0010	**SELF-CONTAINED SINGLE PACKAGE**									
0100	Air cooled, for free blow or duct, not incl. remote condenser									
0110	Constant volume									
0200	3 ton cooling	Q-5	1	16	Ea.	3,825	920		4,745	5,575
0210	4 ton cooling	"	.80	20		4,175	1,150		5,325	6,300
0220	5 ton cooling	Q-6	1.20	20		4,400	1,200		5,600	6,650
0230	7.5 ton cooling	"	.90	26.667		5,650	1,600		7,250	8,600
0240	10 ton cooling	Q-7	1	32		7,925	1,950		9,875	11,600
0250	15 ton cooling		.95	33.684		10,300	2,050		12,350	14,400
0260	20 ton cooling		.90	35.556		13,100	2,175		15,275	17,700
0270	25 ton cooling		.85	37.647		19,900	2,300		22,200	25,400
0280	30 ton cooling		.80	40		25,800	2,425		28,225	32,100
0300	40 ton cooling		.60	53.333		37,200	3,250		40,450	45,900
0320	50 ton cooling	↓	.50	64		50,500	3,900		54,400	61,500
0340	60 ton cooling	Q-8	.40	80	↓	58,000	4,875	140	63,015	71,000
0380	With hot water heat, variable air volume									
0390	10 ton cooling	Q-7	.96	33.333	Ea.	11,300	2,025		13,325	15,500
0400	20 ton cooling		.86	37.209		18,800	2,275		21,075	24,100
0410	30 ton cooling		.76	42.105		35,000	2,575		37,575	42,400
0420	40 ton cooling		.58	55.172		48,500	3,350		51,850	58,500
0430	50 ton cooling	↓	.48	66.667		65,500	4,050		69,550	78,000
0440	60 ton cooling	Q-8	.38	84.211	↓	77,500	5,125	147	82,772	93,000
0490	For duct mounting, no price change									
0500	For steam heating coils, add				Ea.	10%	10%			
0550	Packaged, with electric heat									
0552	Constant volume									
0560	5 ton cooling	Q-6	1.20	20	Ea.	8,150	1,200		9,350	10,800
0570	10 ton cooling	Q-7	1	32		13,700	1,950		15,650	18,000
0580	20 ton cooling		.90	35.556		21,300	2,175		23,475	26,700
0590	30 ton cooling		.80	40		39,600	2,425		42,025	47,300
0600	40 ton cooling		.60	53.333		52,000	3,250		55,250	62,500
0610	50 ton cooling	↓	.50	64	↓	71,000	3,900		74,900	84,000
0700	Variable air volume									
0710	10 ton cooling	Q-7	1	32	Ea.	16,200	1,950		18,150	20,800
0720	20 ton cooling		.90	35.556		25,800	2,175		27,975	31,600
0730	30 ton cooling		.80	40		46,200	2,425		48,625	54,500
0740	40 ton cooling		.60	53.333		59,500	3,250		62,750	70,500
0750	50 ton cooling	↓	.50	64		81,000	3,900		84,900	95,000
0760	60 ton cooling	Q-8	.40	80	↓	90,500	4,875	140	95,515	107,000
1000	Water cooled for free blow or duct, not including tower									
1010	Constant volume									
1100	3 ton cooling	Q-6	1	24	Ea.	3,775	1,425		5,200	6,300
1120	5 ton cooling		1	24		4,600	1,425		6,025	7,200
1130	7.5 ton cooling	↓	.80	30		6,225	1,800		8,025	9,525
1140	10 ton cooling	Q-7	.90	35.556		8,725	2,175		10,900	12,900
1150	15 ton cooling	↓	.85	37.647	↓	12,600	2,300		14,900	17,400

23 81 19 – Self-Contained Air-Conditioners

23 81 19.20 Self-Contained Single Package		Crew	Daily Output	Labor-Hours	Unit	2019 Bare Costs Material	Labor	Equipment	Total	Total Incl O&P
1160	20 ton cooling	Q-7	.80	40	Ea.	28,100	2,425		30,525	34,600
1170	25 ton cooling		.75	42.667		33,000	2,600		35,600	40,200
1180	30 ton cooling		.70	45.714		37,200	2,775		39,975	45,100
1200	40 ton cooling		.40	80		42,700	4,875		47,575	54,500
1220	50 ton cooling		.30	107		50,500	6,500		57,000	66,000
1240	60 ton cooling	Q-8	.30	107		59,000	6,500	187	65,687	75,000
1300	For hot water or steam heat coils, add					12%	10%			
2000	Water cooled with electric heat, not including tower									
2010	Constant volume									
2100	5 ton cooling	Q-6	1.20	20	Ea.	8,350	1,200		9,550	11,000
2120	10 ton cooling	Q-7	1	32		14,500	1,950		16,450	18,800
2130	15 ton cooling	"	.90	35.556		20,800	2,175		22,975	26,200
2400	Variable air volume									
2440	10 ton cooling	Q-7	1	32	Ea.	17,000	1,950		18,950	21,700
2450	15 ton cooling	"	.90	35.556	"	25,300	2,175		27,475	31,100
2600	Water cooled, hot water coils, not including tower									
2610	Variable air volume									
2640	10 ton cooling	Q-7	1	32	Ea.	12,200	1,950		14,150	16,300
2650	15 ton cooling	"	.80	40	"	18,300	2,425		20,725	23,900

23 81 23 – Computer-Room Air-Conditioners

23 81 23.10 Computer Room Units		Crew	Daily Output	Labor-Hours	Unit	Material	Labor	Equipment	Total	Total Incl O&P
0010	**COMPUTER ROOM UNITS** D3050–185									
1000	Air cooled, includes remote condenser but not									
1020	interconnecting tubing or refrigerant									
1080	3 ton	Q-5	.50	32	Ea.	26,600	1,850		28,450	32,100
1120	5 ton		.45	35.556		28,400	2,050		30,450	34,400
1160	6 ton		.30	53.333		54,500	3,075		57,575	64,000
1200	8 ton		.27	59.259		53,000	3,400		56,400	63,500
1240	10 ton		.25	64		55,500	3,675		59,175	66,500
1280	15 ton		.22	72.727		61,000	4,175		65,175	73,500
1290	18 ton		.20	80		70,500	4,600		75,100	84,500
1300	20 ton	Q-6	.26	92.308		73,000	5,500		78,500	89,000
1320	22 ton		.24	100		74,000	5,975		79,975	90,500
1360	30 ton		.21	114		91,000	6,825		97,825	110,500
2200	Chilled water, for connection to									
2220	existing chiller system of adequate capacity									
2260	5 ton	Q-5	.74	21.622	Ea.	19,200	1,250		20,450	23,100
2280	6 ton		.52	30.769		19,500	1,775		21,275	24,100
2300	8 ton		.50	32		19,800	1,850		21,650	24,600
2320	10 ton		.49	32.653		19,900	1,875		21,775	24,700
2330	12 ton		.49	32.990		21,400	1,900		23,300	26,500
2360	15 ton		.48	33.333		21,800	1,925		23,725	26,900
2400	20 ton		.46	34.783		23,200	2,000		25,200	28,500
2440	25 ton	Q-6	.63	38.095		24,700	2,275		26,975	30,500
2460	30 ton		.57	42.105		26,800	2,525		29,325	33,300
2480	35 ton		.52	46.154		27,100	2,750		29,850	33,900
2500	45 ton		.46	52.174		28,600	3,125		31,725	36,100
2520	50 ton		.42	57.143		32,800	3,400		36,200	41,200
2540	60 ton		.38	63.158		34,900	3,775		38,675	44,100
2560	77 ton		.35	68.571		54,500	4,100		58,600	66,000
2580	95 ton		.32	75		63,500	4,475		67,975	76,500
4000	Glycol system, complete except for interconnecting tubing									

23 81 Decentralized Unitary HVAC Equipment

23 81 23 – Computer-Room Air-Conditioners

23 81 23.10 Computer Room Units		Crew	Daily Output	Labor-Hours	Unit	Material	Labor	Equipment	Total	Total Incl O&P
						2019 Bare Costs				
4060	3 ton	Q-5	.40	40	Ea.	33,400	2,300		35,700	40,200
4100	5 ton		.38	42.105		36,600	2,425		39,025	43,800
4120	6 ton		.25	64		51,500	3,675		55,175	62,000
4140	8 ton		.23	69.565		58,500	4,000		62,500	70,500
4160	10 ton		.21	76.190		62,500	4,375		66,875	75,000
4180	12 ton		.19	84.211		62,500	4,850		67,350	76,000
4200	15 ton	Q-6	.26	92.308		77,500	5,500		83,000	93,500
4240	20 ton		.24	100		85,000	5,975		90,975	102,500
4280	22 ton		.23	104		89,500	6,225		95,725	108,000
4430	30 ton		.22	109		110,500	6,500		117,000	132,000
8000	Water cooled system, not including condenser,									
8020	water supply or cooling tower									
8060	3 ton	Q-5	.62	25.806	Ea.	27,100	1,475		28,575	32,000
8100	5 ton		.54	29.630		29,500	1,700		31,200	35,000
8120	6 ton		.35	45.714		41,800	2,625		44,425	50,000
8140	8 ton		.33	48.485		48,300	2,800		51,100	57,000
8160	10 ton		.31	51.613		50,000	2,975		52,975	59,500
8180	12 ton		.29	55.172		50,000	3,175		53,175	60,000
8200	15 ton		.27	59.259		58,500	3,400		61,900	69,500
8240	20 ton	Q-6	.38	63.158		63,000	3,775		66,775	74,500
8280	22 ton		.35	68.571		66,000	4,100		70,100	78,500
8300	30 ton		.30	80		82,000	4,775		86,775	97,500

23 81 26 – Split-System Air-Conditioners

23 81 26.10 Split Ductless Systems		Crew	Daily Output	Labor-Hours	Unit	Material	Labor	Equipment	Total	Total Incl O&P
0010	**SPLIT DUCTLESS SYSTEMS**									
0100	Cooling only, single zone									
0110	Wall mount									
0120	3/4 ton cooling	Q-5	2	8	Ea.	1,100	460		1,560	1,925
0130	1 ton cooling		1.80	8.889		1,225	510		1,735	2,125
0140	1-1/2 ton cooling		1.60	10		1,975	575		2,550	3,050
0150	2 ton cooling		1.40	11.429		2,250	660		2,910	3,450
1000	Ceiling mount									
1020	2 ton cooling	Q-5	1.40	11.429	Ea.	2,000	660		2,660	3,175
1030	3 ton cooling	"	1.20	13.333	"	2,600	765		3,365	4,000
3000	Multizone									
3010	Wall mount									
3020	2 @ 3/4 ton cooling	Q-5	1.80	8.889	Ea.	3,475	510		3,985	4,600
5000	Cooling/Heating									
5010	Wall mount									
5110	1 ton cooling	Q-5	1.70	9.412	Ea.	1,300	540		1,840	2,275
5120	1-1/2 ton cooling	"	1.50	10.667	"	2,000	615		2,615	3,125
7000	Accessories for all split ductless systems									
7010	Add for ambient frost control	Q-5	8	2	Ea.	127	115		242	315
7020	Add for tube/wiring kit (line sets)									
7030	15′ kit	Q-5	32	.500	Ea.	89	29		118	141
7036	25′ kit		28	.571		138	33		171	201
7040	35′ kit		24	.667		200	38.50		238.50	278
7050	50′ kit		20	.800		203	46		249	292

23 81 29 – Variable Refrigerant Flow HVAC Systems

23 81 29.10 Heat Pump, Gas Driven		Crew	Daily Output	Labor-Hours	Unit	Material	2019 Bare Costs Labor	Equipment	Total	Total Incl O&P
0010	**HEAT PUMP, GAS DRIVEN,** Variable refrigerant volume (VRV) type									
0020	Not including interconnecting tubing or multi-zone controls									
1000	For indoor fan VRV type AHU see 23 82 19.40									
1010	Multi-zone split									
1020	Outdoor unit									
1100	8 tons cooling, for up to 17 zones	Q-5	1.30	12.308	Ea.	29,700	710		30,410	33,700
1110	Isolation rails		2.60	6.154	Pair	1,050	355		1,405	1,675
1160	15 tons cooling, for up to 33 zones		1	16	Ea.	35,900	920		36,820	40,900
1170	Isolation rails		2.60	6.154	Pair	1,225	355		1,580	1,875
2000	Packaged unit									
2020	Outdoor unit									
2200	11 tons cooling	Q-5	1.30	12.308	Ea.	38,200	710		38,910	43,100
2210	Roof curb adapter	"	2.60	6.154		660	355		1,015	1,250
2220	Thermostat	1 Stpi	1.30	6.154		395	395		790	1,025

23 81 43 – Air-Source Unitary Heat Pumps

23 81 43.10 Air-Source Heat Pumps		Crew	Daily Output	Labor-Hours	Unit	Material	2019 Bare Costs Labor	Equipment	Total	Total Incl O&P
0010	**AIR-SOURCE HEAT PUMPS,** Not including interconnecting tubing									
1000	Air to air, split system, not including curbs, pads, fan coil and ductwork									
1010	For curbs/pads see Section 23 91 10									
1012	Outside condensing unit only, for fan coil see Section 23 82 19.10									
1015	1.5 ton cooling, 7 MBH heat @ 0°F	Q-5	2.40	6.667	Ea.	1,625	385		2,010	2,375
1020	2 ton cooling, 8.5 MBH heat @ 0°F		2	8		1,775	460		2,235	2,650
1030	2.5 ton cooling, 10 MBH heat @ 0°F		1.60	10		1,925	575		2,500	3,000
1040	3 ton cooling, 13 MBH heat @ 0°F		1.20	13.333		2,150	765		2,915	3,500
1050	3.5 ton cooling, 18 MBH heat @ 0°F		1	16		2,275	920		3,195	3,875
1054	4 ton cooling, 24 MBH heat @ 0°F		.80	20		2,475	1,150		3,625	4,450
1060	5 ton cooling, 27 MBH heat @ 0°F		.50	32		2,700	1,850		4,550	5,750
1080	7.5 ton cooling, 33 MBH heat @ 0°F		.45	35.556		4,000	2,050		6,050	7,475
1100	10 ton cooling, 50 MBH heat @ 0°F	Q-6	.64	37.500		6,350	2,250		8,600	10,300
1120	15 ton cooling, 64 MBH heat @ 0°F		.50	48		9,650	2,875		12,525	14,900
1130	20 ton cooling, 85 MBH heat @ 0°F		.35	68.571		18,100	4,100		22,200	26,100
1140	25 ton cooling, 119 MBH heat @ 0°F		.25	96		21,500	5,725		27,225	32,300
1500	Single package, not including curbs, pads, or plenums									
1502	0.5 ton cooling, supplementary heat included	Q-5	8	2	Ea.	3,375	115		3,490	3,900
1504	0.75 ton cooling, supplementary heat included		6	2.667		3,825	153		3,978	4,425
1506	1 ton cooling, supplementary heat included		4	4		3,275	230		3,505	3,950
1510	1.5 ton cooling, 5 MBH heat @ 0°F		1.55	10.323		3,325	595		3,920	4,550
1520	2 ton cooling, 6.5 MBH heat @ 0°F		1.50	10.667		3,300	615		3,915	4,550
1540	2.5 ton cooling, 8 MBH heat @ 0°F		1.40	11.429		3,325	660		3,985	4,625
1560	3 ton cooling, 10 MBH heat @ 0°F		1.20	13.333		3,750	765		4,515	5,275
1570	3.5 ton cooling, 11 MBH heat @ 0°F		1	16		4,075	920		4,995	5,875
1580	4 ton cooling, 13 MBH heat @ 0°F		.96	16.667		4,375	960		5,335	6,275
1620	5 ton cooling, 27 MBH heat @ 0°F		.65	24.615		5,050	1,425		6,475	7,675
1640	7.5 ton cooling, 35 MBH heat @ 0°F		.40	40		7,350	2,300		9,650	11,500
1648	10 ton cooling, 45 MBH heat @ 0°F	Q-6	.40	60		10,400	3,575		13,975	16,800
1652	12 ton cooling, 50 MBH heat @ 0°F	"	.36	66.667		10,800	3,975		14,775	17,900
1696	Supplementary electric heat coil incl., unless noted otherwise									
6000	Air to water, single package, excluding storage tank and ductwork									
6010	Includes circulating water pump, air duct connections, digital temperature									
6020	controller with remote tank temp. probe and sensor for storage tank.									
6040	Water heating - air cooling capacity									
6110	35.5 MBH heat water, 2.3 ton cool air	Q-5	1.60	10	Ea.	17,400	575		17,975	20,100

23 81 Decentralized Unitary HVAC Equipment

23 81 43 – Air-Source Unitary Heat Pumps

23 81 43.10 Air-Source Heat Pumps		Crew	Daily Output	Labor-Hours	Unit	Material	2019 Bare Costs Labor	Equipment	Total	Total Incl O&P
6120	58 MBH heat water, 3.8 ton cool air	Q-5	1.10	14.545	Ea.	19,900	835		20,735	23,100
6130	76 MBH heat water, 4.9 ton cool air		.87	18.391		23,700	1,050		24,750	27,700
6140	98 MBH heat water, 6.5 ton cool air		.62	25.806		30,600	1,475		32,075	35,900
6150	113 MBH heat water, 7.4 ton cool air		.59	27.119		33,200	1,550		34,750	38,900
6160	142 MBH heat water, 9.2 ton cool air		.52	30.769		40,700	1,775		42,475	47,500
6170	171 MBH heat water, 11.1 ton cool air		.49	32.653		47,300	1,875		49,175	55,000

23 81 46 – Water-Source Unitary Heat Pumps

23 81 46.10 Water Source Heat Pumps		Crew	Daily Output	Labor-Hours	Unit	Material	Labor	Equipment	Total	Total Incl O&P
0010	**WATER SOURCE HEAT PUMPS**, Not incl. connecting tubing or water source									
2000	Water source to air, single package									
2100	1 ton cooling, 13 MBH heat @ 75°F	Q-5	2	8	Ea.	1,925	460		2,385	2,800
2120	1.5 ton cooling, 17 MBH heat @ 75°F		1.80	8.889		1,900	510		2,410	2,875
2140	2 ton cooling, 19 MBH heat @ 75°F		1.70	9.412		2,275	540		2,815	3,325
2160	2.5 ton cooling, 25 MBH heat @ 75°F		1.60	10		2,475	575		3,050	3,600
2180	3 ton cooling, 27 MBH heat @ 75°F		1.40	11.429		2,525	660		3,185	3,750
2190	3.5 ton cooling, 29 MBH heat @ 75°F		1.30	12.308		2,750	710		3,460	4,100
2200	4 ton cooling, 31 MBH heat @ 75°F		1.20	13.333		3,075	765		3,840	4,550
2220	5 ton cooling, 29 MBH heat @ 75°F		.90	17.778		3,375	1,025		4,400	5,225
2240	7.5 ton cooling, 35 MBH heat @ 75°F		.60	26.667		6,850	1,525		8,375	9,850
2250	8 ton cooling, 40 MBH heat @ 75°F		.58	27.586		7,000	1,600		8,600	10,100
2260	10 ton cooling, 50 MBH heat @ 75°F		.53	30.189		7,675	1,725		9,400	11,000
2280	15 ton cooling, 64 MBH heat @ 75°F	Q-6	.47	51.064		14,800	3,050		17,850	20,900
2300	20 ton cooling, 100 MBH heat @ 75°F		.41	58.537		16,100	3,500		19,600	23,000
2310	25 ton cooling, 100 MBH heat @ 75°F		.32	75		22,000	4,475		26,475	30,900
2320	30 ton cooling, 128 MBH heat @ 75°F		.24	102		24,100	6,100		30,200	35,700
2340	40 ton cooling, 200 MBH heat @ 75°F		.21	117		34,000	6,975		40,975	47,900
2360	50 ton cooling, 200 MBH heat @ 75°F		.15	160		38,400	9,550		47,950	56,500
3960	For supplementary heat coil, add					10%				
4000	For increase in capacity thru use									
4020	of solar collector, size boiler at 60%									

23 82 Convection Heating and Cooling Units

23 82 13 – Valance Heating and Cooling Units

23 82 13.16 Valance Units		Crew	Daily Output	Labor-Hours	Unit	Material	Labor	Equipment	Total	Total Incl O&P
0010	**VALANCE UNITS**									
6000	Valance units, complete with 1/2" cooling coil, enclosure									
6020	2 tube	Q-5	18	.889	L.F.	35	51		86	116
6040	3 tube		16	1		42	57.50		99.50	133
6060	4 tube		16	1		52	57.50		109.50	144
6080	5 tube		15	1.067		54	61.50		115.50	152
6100	6 tube		15	1.067		59.50	61.50		121	158
6120	8 tube		14	1.143		87	66		153	194
6200	For 3/4" cooling coil, add					10%				

23 82 16 – Air Coils

23 82 16.10 Flanged Coils		Crew	Daily Output	Labor-Hours	Unit	Material	Labor	Equipment	Total	Total Incl O&P
0010	**FLANGED COILS**									
0100	Basic water, DX, or condenser coils									
0110	Copper tubes, alum. fins, galv. end sheets									
0112	H is finned height, L is finned length									
0120	3/8" x 0.016" tubing, 0.0065 aluminum fins									

23 82 16 – Air Coils

23 82 16.10 Flanged Coils		Crew	Daily Output	Labor-Hours	Unit	Material	2019 Bare Costs Labor	Equipment	Total	Total Incl O&P
0130	2 row, 8 fins per inch									
0140	4″ H x 12″ L	Q-5	48	.333	Ea.	1,225	19.20		1,244.20	1,375
0150	4″ H x 24″ L		38.38	.417		1,275	24		1,299	1,425
0160	4″ H x 48″ L		19.25	.831		1,325	48		1,373	1,525
0170	4″ H x 72″ L		12.80	1.250		1,475	72		1,547	1,725
0180	6″ H x 12″ L		48	.333		1,200	19.20		1,219.20	1,350
0190	6″ H x 24″ L		25.60	.625		1,250	36		1,286	1,425
0200	6″ H x 48″ L		12.80	1.250		1,375	72		1,447	1,625
0210	6″ H x 72″ L		8.53	1.876		1,575	108		1,683	1,875
0220	10″ H x 12″ L		30.73	.521		1,300	30		1,330	1,500
0230	10″ H x 24″ L		15.33	1.044		1,400	60		1,460	1,650
0240	10″ H x 48″ L		7.69	2.081		1,500	120		1,620	1,825
0250	10″ H x 72″ L		5.12	3.125		1,650	180		1,830	2,100
0260	12″ H x 12″ L		25.60	.625		1,400	36		1,436	1,575
0270	12″ H x 24″ L		12.80	1.250		1,450	72		1,522	1,675
0280	12″ H x 48″ L		6.40	2.500		1,550	144		1,694	1,950
0290	12″ H x 72″ L		4.27	3.747		1,775	216		1,991	2,275
0300	16″ H x 16″ L		14.38	1.113		1,425	64		1,489	1,675
0310	16″ H x 24″ L		9.59	1.668		1,525	96		1,621	1,825
0320	16″ H x 48″ L		4.80	3.333		1,675	192		1,867	2,150
0330	16″ H x 72″ L		3.20	5		1,900	288		2,188	2,500
0340	20″ H x 24″ L		7.69	2.081		1,625	120		1,745	1,950
0350	20″ H x 48″ L		3.84	4.167		1,775	240		2,015	2,300
0360	20″ H x 72″ L		2.56	6.250		2,000	360		2,360	2,750
0370	24″ H x 24″ L		6.40	2.500		1,700	144		1,844	2,100
0380	24″ H x 48″ L		3.20	5		1,900	288		2,188	2,500
0390	24″ H x 72″ L		2.13	7.512		2,150	430		2,580	3,000
0400	30″ H x 28″ L		4.39	3.645		1,850	210		2,060	2,375
0410	30″ H x 48″ L		2.56	6.250		2,050	360		2,410	2,825
0420	30″ H x 72″ L		1.71	9.357		2,375	540		2,915	3,400
0430	30″ H x 84″ L		1.46	10.959		2,575	630		3,205	3,775
0440	34″ H x 32″ L		3.39	4.720		1,975	272		2,247	2,575
0450	34″ H x 48″ L		2.26	7.080		2,175	405		2,580	3,000
0460	34″ H x 72″ L		1.51	10.596		2,500	610		3,110	3,675
0470	34″ H x 84″ L	↓	1.29	12.403	↓	2,725	715		3,440	4,075
0600	For 10 fins per inch, add					1%				
0610	For 12 fins per inch, add					3%				
0620	For 4 row, add				Ea.	30%	50%			
0630	For 6 row, add					60%	100%			
0640	For 8 row, add				↓	80%	150%			
1100	1/2″ x 0.017″ tubing, 0.0065 aluminum fins									
1110	2 row, 8 fins per inch									
1120	5″ H x 6″ L	Q-5	40	.400	Ea.	1,225	23		1,248	1,375
1130	5″ H x 20″ L		37.10	.431		1,300	25		1,325	1,450
1140	5″ H x 45″ L		16.41	.975		1,325	56		1,381	1,525
1150	5″ H x 90″ L		8.21	1.949		1,550	112		1,662	1,875
1160	7.5″ H x 5″ L		38.50	.416		1,175	24		1,199	1,325
1170	7.5″ H x 20″ L		24.62	.650		1,275	37.50		1,312.50	1,450
1180	7.5″ H x 45″ L		10.94	1.463		1,400	84		1,484	1,675
1190	7.5″ H x 90″ L		5.46	2.930		1,750	169		1,919	2,175
1200	12.5″ H x 10″ L		29.49	.543		1,300	31		1,331	1,475
1210	12.5″ H x 20″ L		14.75	1.085		1,400	62.50		1,462.50	1,650
1220	12.5″ H x 45″ L	↓	6.55	2.443	↓	1,575	141		1,716	1,925

23 82 16.10 Flanged Coils		Crew	Daily Output	Labor-Hours	Unit	2019 Bare Costs Material	Labor	Equipment	Total	Total Incl O&P
1230	12.5" H x 90" L	Q-5	3.28	4.878	Ea.	1,875	281		2,156	2,500
1240	15" H x 15" L		16.41	.975		1,350	56		1,406	1,575
1250	15" H x 30" L		8.21	1.949		1,550	112		1,662	1,900
1260	15" H x 45" L		5.46	2.930		1,650	169		1,819	2,075
1270	15" H x 90" L		2.73	5.861		1,925	335		2,260	2,625
1280	20" H x 20" L		9.22	1.735		1,600	100		1,700	1,900
1290	20" H x 45" L		4.10	3.902		1,725	225		1,950	2,225
1300	20" H x 90" L		2.05	7.805		2,250	450		2,700	3,150
1310	25" H x 25" L		5.90	2.712		1,700	156		1,856	2,100
1320	25" H x 45" L		3.28	4.878		1,775	281		2,056	2,400
1330	25" H x 90" L		1.64	9.756		2,375	560		2,935	3,475
1340	30" H x 30" L		4.10	3.902		1,850	225		2,075	2,350
1350	30" H x 45" L		2.73	5.861		1,925	335		2,260	2,600
1360	30" H x 90" L		1.37	11.679		2,600	670		3,270	3,850
1370	35" H x 35" L		3.01	5.316		2,525	305		2,830	3,225
1380	35" H x 45" L		2.34	6.838		2,050	395		2,445	2,850
1390	35" H x 90" L		1.17	13.675		2,750	785		3,535	4,200
1400	42.5" H x 45" L		1.93	8.290		2,300	475		2,775	3,250
1410	42.5" H x 75" L		1.16	13.793		2,750	795		3,545	4,225
1420	42.5" H x 105" L		.83	19.277		3,250	1,100		4,350	5,250
1600	For 10 fins per inch, add					2%				
1610	For 12 fins per inch, add					5%				
1620	For 4 row, add					35%	50%			
1630	For 6 row, add					65%	100%			
1640	For 8 row, add					90%	150%			
2010	5/8" x 0.020" tubing, 0.0065 aluminum fins									
2020	2 row, 8 fins per inch									
2030	6" H x 6" L	Q-5	38	.421	Ea.	1,075	24		1,099	1,200
2040	6" H x 24" L		24.96	.641		1,150	37		1,187	1,300
2050	6" H x 48" L		12.48	1.282		1,200	74		1,274	1,425
2060	6" H x 96" L		6.24	2.564		1,475	148		1,623	1,850
2070	12" H x 12" L		24.96	.641		1,150	37		1,187	1,325
2080	12" H x 24" L		12.48	1.282		1,350	74		1,424	1,575
2090	12" H x 48" L		6.24	2.564		1,325	148		1,473	1,700
2100	12" H x 96" L		3.12	5.128		2,675	295		2,970	3,400
2110	18" H x 24" L		8.32	1.923		1,325	111		1,436	1,650
2120	18" H x 48" L		4.16	3.846		1,475	221		1,696	1,950
2130	18" H x 96" L		2.08	7.692		1,925	445		2,370	2,800
2140	24" H x 24" L		6.24	2.564		1,425	148		1,573	1,800
2150	24" H x 48" L		3.12	5.128		1,675	295		1,970	2,300
2160	24" H x 96" L		1.56	10.256		2,150	590		2,740	3,250
2170	30" H x 30" L		3.99	4.010		1,750	231		1,981	2,250
2180	30" H x 48" L		2.50	6.400		1,800	370		2,170	2,525
2190	30" H x 96" L		1.25	12.800		2,425	735		3,160	3,775
2200	30" H x 120" L		1	16		2,725	920		3,645	4,375
2210	36" H x 36" L		2.77	5.776		1,800	330		2,130	2,475
2220	36" H x 48" L		2.08	7.692		1,925	445		2,370	2,800
2230	36" H x 90" L		1.11	14.414		2,600	830		3,430	4,100
2240	36" H x 120" L		.83	19.277		2,975	1,100		4,075	4,950
2250	42" H x 48" L		1.78	8.989		2,075	515		2,590	3,050
2260	42" H x 90" L		.95	16.842		2,875	970		3,845	4,625
2270	42" H x 120" L		.71	22.535		3,550	1,300		4,850	5,850
2280	51" H x 48" L		1.47	10.884		2,325	625		2,950	3,500

23 82 16.10 Flanged Coils		Crew	Daily Output	Labor-Hours	Unit	Material	Labor	Equipment	Total	Total Incl O&P
						2019 Bare Costs				
2290	51" H x 96" L	Q-5	.73	21.918	Ea.	3,550	1,250		4,800	5,800
2300	51" H x 120" L		.59	27.119		3,550	1,550		5,100	6,275
2500	For 10 fins per inch, add					3%				
2510	For 12 fins per inch, add					6%				
2520	For 4 row, add				Ea.	40%	50%			
2530	For 6 row, add					80%	100%			
2540	For 8 row, add					110%	150%			
3000	Hot water booster coils									
3010	Copper tubes, alum. fins, galv. end sheets									
3012	H is finned height, L is finned length									
3020	1/2" x 0.017" tubing, 0.0065 aluminum fins									
3030	1 row, 10 fins per inch									
3040	5" H x 10" L	Q-5	40	.400	Ea.	885	23		908	1,000
3050	5" H x 25" L		33.18	.482		955	28		983	1,100
3060	10" H x 10" L		39	.410		920	23.50		943.50	1,050
3070	10" H x 20" L		20.73	.772		985	44.50		1,029.50	1,150
3080	10" H x 30" L		13.83	1.157		1,025	66.50		1,091.50	1,225
3090	15" H x 15" L		18.44	.868		935	50		985	1,100
3110	15" H x 20" L		13.83	1.157		1,025	66.50		1,091.50	1,225
3120	15" H x 30" L		9.22	1.735		1,075	100		1,175	1,325
3130	20" H x 20" L		10.37	1.543		1,075	89		1,164	1,300
3140	20" H x 30" L		6.91	2.315		1,125	133		1,258	1,450
3150	20" H x 40" L		5.18	3.089		1,175	178		1,353	1,575
3160	25" H x 25" L		6.64	2.410		1,150	139		1,289	1,475
3170	25" H x 35" L		4.74	3.376		1,200	194		1,394	1,625
3180	25" H x 45" L		3.69	4.336		1,250	250		1,500	1,750
3190	30" H x 30" L		4.61	3.471		1,225	200		1,425	1,650
3200	30" H x 40" L		3.46	4.624		1,300	266		1,566	1,825
3210	30" H x 50" L		2.76	5.797		1,350	335		1,685	2,000
3300	For 2 row, add					20%	50%			
3400	1/2" x 0.020" tubing, 0.008 aluminum fins									
3410	2 row, 10 fins per inch									
3420	20" H x 27" L	Q-5	7.27	2.201	Ea.	1,275	127		1,402	1,600
3430	20" H x 36" L		5.44	2.941		1,350	169		1,519	1,725
3440	20" H x 45" L		4.35	3.678		1,350	212		1,562	1,800
3450	25" H x 45" L		3.49	4.585		1,450	264		1,714	2,000
3460	25" H x 54" L		2.90	5.517		1,650	320		1,970	2,300
3470	30" H x 54" L		2.42	6.612		1,775	380		2,155	2,525
3480	35" H x 54" L		2.07	7.729		1,925	445		2,370	2,775
3490	35" H x 63" L		1.78	8.989		1,975	515		2,490	2,950
3500	40" H x 72" L		1.36	11.765		2,350	675		3,025	3,600
3510	55" H x 63" L		1.13	14.159		2,625	815		3,440	4,125
3520	55" H x 72" L		.99	16.162		2,925	930		3,855	4,625
3530	55" H x 81" L		.88	18.182		3,175	1,050		4,225	5,075
3540	75" H x 81" L		.64	25		3,925	1,450		5,375	6,475
3550	75" H x 90" L		.58	27.586		3,975	1,600		5,575	6,750
3560	75" H x 108" L		.48	33.333		4,425	1,925		6,350	7,750
3800	5/8" x 0.020" tubing, 0.0065 aluminum fins									
3810	1 row, 10 fins per inch									
3820	6" H x 12" L	Q-5	38	.421	Ea.	715	24		739	820
3830	6" H x 24" L		28.80	.556		780	32		812	910
3840	6" H x 30" L		23.04	.694		805	40		845	945
3850	12" H x 12" L		28.80	.556		810	32		842	940

23 82 16 – Air Coils

23 82 16.10 Flanged Coils		Crew	Daily Output	Labor-Hours	Unit	2019 Bare Costs Material	Labor	Equipment	Total	Total Incl O&P
3860	12″ H x 24″ L	Q-5	14.40	1.111	Ea.	810	64		874	985
3870	12″ H x 30″ L		11.52	1.389		870	80		950	1,075
3880	18″ H x 18″ L		12.80	1.250		1,000	72		1,072	1,200
3890	18″ H x 24″ L		9.60	1.667		1,050	96		1,146	1,300
3900	18″ H x 36″ L		6.40	2.500		1,100	144		1,244	1,425
3910	24″ H x 24″ L		7.20	2.222		1,100	128		1,228	1,400
3920	24″ H x 30″ L		4.80	3.333		1,150	192		1,342	1,575
3930	24″ H x 36″ L		5.05	3.168		1,175	182		1,357	1,575
3940	30″ H x 30″ L		4.61	3.471		1,225	200		1,425	1,650
3950	30″ H x 36″ L		3.85	4.159		1,250	239		1,489	1,725
3960	30″ H x 42″ L		3.29	4.863		1,275	280		1,555	1,850
3970	36″ H x 36″ L		3.20	5		1,325	288		1,613	1,875
3980	36″ H x 42″ L		2.74	5.839		1,350	335		1,685	2,000
4100	For 2 row, add					30%	50%			
4400	Steam and hot water heating coils									
4410	Copper tubes, alum. fins. galv. end sheets									
4412	H is finned height, L is finned length									
4420	5/8″ tubing, to 175 PSIG working pressure									
4430	1 row, 8 fins per inch									
4440	6″ H x 6″ L	Q-5	38	.421	Ea.	1,200	24		1,224	1,325
4450	6″ H x 12″ L		37	.432		1,200	25		1,225	1,375
4460	6″ H x 24″ L		24.48	.654		1,300	37.50		1,337.50	1,475
4470	6″ H x 36″ L		16.32	.980		1,350	56.50		1,406.50	1,550
4480	6″ H x 48″ L		12.24	1.307		1,375	75		1,450	1,650
4490	6″ H x 84″ L		6.99	2.289		2,050	132		2,182	2,450
4500	6″ H x 108″ L		5.44	2.941		2,250	169		2,419	2,725
4510	12″ H x 12″ L		24.48	.654		1,350	37.50		1,387.50	1,525
4520	12″ H x 24″ L		12.24	1.307		1,450	75		1,525	1,725
4530	12″ H x 36″ L		8.16	1.961		1,525	113		1,638	1,875
4540	12″ H x 48″ L		6.12	2.614		1,600	150		1,750	2,000
4550	12″ H x 84″ L		3.50	4.571		2,375	263		2,638	3,025
4560	12″ H x 108″ L		2.72	5.882		2,700	340		3,040	3,475
4570	18″ H x 18″ L		10.88	1.471		1,550	84.50		1,634.50	1,850
4580	18″ H x 24″ L		8.16	1.961		1,600	113		1,713	1,950
4590	18″ H x 48″ L		4.08	3.922		1,900	226		2,126	2,425
4600	18″ H x 84″ L		2.33	6.867		2,775	395		3,170	3,650
4610	18″ H x 108″ L		1.81	8.840		3,100	510		3,610	4,175
4620	24″ H x 24″ L		6.12	2.614		1,875	150		2,025	2,300
4630	24″ H x 48″ L		3.06	5.229		2,200	300		2,500	2,875
4640	24″ H x 84″ L		1.75	9.143		3,175	525		3,700	4,275
4650	24″ H x 108″ L		1.36	11.765		3,800	675		4,475	5,200
4660	36″ H x 36″ L		2.72	5.882		2,450	340		2,790	3,200
4670	36″ H x 48″ L		2.04	7.843		2,650	450		3,100	3,600
4680	36″ H x 84″ L		1.17	13.675		4,250	785		5,035	5,850
4690	36″ H x 108″ L		.91	17.582		5,025	1,000		6,025	7,050
4700	48″ H x 48″ L		1.53	10.458		3,175	600		3,775	4,400
4710	48″ H x 84″ L		.87	18.391		5,075	1,050		6,125	7,175
4720	48″ H x 108″ L		.68	23.529		6,075	1,350		7,425	8,700
4730	48″ H x 126″ L		.58	27.586		6,425	1,600		8,025	9,450
4910	For 10 fins per inch, add					2%				
4920	For 12 fins per inch, add					5%				
4930	For 2 row, add					30%	50%			
4940	For 3 row, add					50%	100%			

23 82 16 – Air Coils

23 82 16.10 Flanged Coils		Crew	Daily Output	Labor-Hours	Unit	Material	Labor	Equipment	Total	Total Incl O&P
						2019 Bare Costs				
4950	For 4 row, add					80%	123%			
4960	For 230 PSIG heavy duty, add					20%	30%			
6000	Steam, heavy duty, 200 PSIG									
6010	Copper tubes, alum. fins, galv. end sheets									
6012	H is finned height, L is finned length									
6020	1" x 0.035" tubing, 0.010 aluminum fins									
6030	1 row, 8 fins per inch									
6040	12" H x 12" L	Q-5	24.48	.654	Ea.	1,400	37.50		1,437.50	1,575
6050	12" H x 24" L		12.24	1.307		1,675	75		1,750	1,975
6060	12" H x 36" L		8.16	1.961		1,800	113		1,913	2,150
6070	12" H x 48" L		6.12	2.614		1,900	150		2,050	2,300
6080	12" H x 84" L		3.50	4.571		2,425	263		2,688	3,075
6090	12" H x 108" L		2.72	5.882		2,975	340		3,315	3,775
6100	18" H x 24" L		8.16	1.961		1,925	113		2,038	2,300
6110	18" H x 36" L		5.44	2.941		2,125	169		2,294	2,575
6120	18" H x 48" L		4.08	3.922		2,100	226		2,326	2,675
6130	18" H x 84" L		2.33	6.867		3,400	395		3,795	4,350
6140	18" H x 108" L		1.81	8.840		3,975	510		4,485	5,150
6150	24" H x 24" L		6.12	2.614		2,225	150		2,375	2,675
6160	24" H x 36" L		4.08	3.922		2,400	226		2,626	2,975
6170	24" H x 48" L		3.06	5.229		2,700	300		3,000	3,400
6180	24" H x 84" L		1.75	9.143		3,825	525		4,350	5,000
6190	24" H x 108" L		1.36	11.765		4,650	675		5,325	6,150
6200	30" H x 36" L		3.26	4.908		3,025	282		3,307	3,750
6210	30" H x 48" L		2.45	6.531		3,350	375		3,725	4,275
6220	30" H x 84" L		1.40	11.429		4,650	660		5,310	6,075
6230	30" H x 108" L		1.09	14.679		5,550	845		6,395	7,375
6240	36" H x 36" L		2.72	5.882		3,400	340		3,740	4,225
6250	36" H x 48" L		2.04	7.843		3,600	450		4,050	4,625
6260	36" H x 84" L		1.17	13.675		5,275	785		6,060	6,975
6270	36" H x 108" L		.91	17.582		6,275	1,000		7,275	8,425
6280	36" H x 120" L		.82	19.512		6,700	1,125		7,825	9,050
6290	48" H x 48" L		1.53	10.458		4,650	600		5,250	6,025
6300	48" H x 84" L		.87	18.391		6,425	1,050		7,475	8,675
6310	48" H x 108" L		.68	23.529		7,850	1,350		9,200	10,700
6320	48" H x 120" L		.61	26.230		8,400	1,500		9,900	11,500
6400	For 10 fins per inch, add					4%				
6410	For 12 fins per inch, add					10%				

23 82 16.20 Duct Heaters		Crew	Daily Output	Labor-Hours	Unit	Material	Labor	Equipment	Total	Total Incl O&P
0010	**DUCT HEATERS**, Electric, 480 V, 3 Ph.									
0020	Finned tubular insert, 500°F									
0100	8" wide x 6" high, 4.0 kW	Q-20	16	1.250	Ea.	825	70		895	1,000
0120	12" high, 8.0 kW		15	1.333		1,375	74.50		1,449.50	1,625
0140	18" high, 12.0 kW		14	1.429		1,900	80		1,980	2,225
0160	24" high, 16.0 kW		13	1.538		2,450	86		2,536	2,825
0180	30" high, 20.0 kW		12	1.667		3,000	93		3,093	3,450
0300	12" wide x 6" high, 6.7 kW		15	1.333		875	74.50		949.50	1,075
0320	12" high, 13.3 kW		14	1.429		1,400	80		1,480	1,675
0340	18" high, 20.0 kW		13	1.538		1,975	86		2,061	2,300
0360	24" high, 26.7 kW		12	1.667		2,550	93		2,643	2,950
0380	30" high, 33.3 kW		11	1.818		3,100	102		3,202	3,550
0500	18" wide x 6" high, 13.3 kW		14	1.429		940	80		1,020	1,150

23 82 16.20 Duct Heaters		Crew	Daily Output	Labor-Hours	Unit	Material	2019 Bare Costs Labor	Equipment	Total	Total Incl O&P
0520	12" high, 26.7 kW	Q-20	13	1.538	Ea.	1,625	86		1,711	1,900
0540	18" high, 40.0 kW		12	1.667		2,150	93		2,243	2,500
0560	24" high, 53.3 kW		11	1.818		2,850	102		2,952	3,300
0580	30" high, 66.7 kW		10	2		3,575	112		3,687	4,100
0700	24" wide x 6" high, 17.8 kW		13	1.538		1,025	86		1,111	1,275
0720	12" high, 35.6 kW		12	1.667		1,750	93		1,843	2,075
0740	18" high, 53.3 kW		11	1.818		2,450	102		2,552	2,850
0760	24" high, 71.1 kW		10	2		3,175	112		3,287	3,675
0780	30" high, 88.9 kW		9	2.222		3,900	124		4,024	4,475
0900	30" wide x 6" high, 22.2 kW		12	1.667		1,100	93		1,193	1,350
0920	12" high, 44.4 kW		11	1.818		1,850	102		1,952	2,175
0940	18" high, 66.7 kW		10	2		2,600	112		2,712	3,050
0960	24" high, 88.9 kW		9	2.222		3,375	124		3,499	3,900
0980	30" high, 111.0 kW		8	2.500		4,125	140		4,265	4,725
1400	Note decreased kW available for									
1410	each duct size at same cost									
1420	See line 5000 for modifications and accessories									
2000	Finned tubular flange with insulated									
2020	terminal box, 500°F									
2100	12" wide x 36" high, 54 kW	Q-20	10	2	Ea.	4,050	112		4,162	4,625
2120	40" high, 60 kW		9	2.222		4,650	124		4,774	5,325
2200	24" wide x 36" high, 118.8 kW		9	2.222		5,125	124		5,249	5,825
2220	40" high, 132 kW		8	2.500		5,125	140		5,265	5,825
2400	36" wide x 8" high, 40 kW		11	1.818		2,050	102		2,152	2,400
2420	16" high, 80 kW		10	2		2,750	112		2,862	3,200
2440	24" high, 120 kW		9	2.222		3,600	124		3,724	4,150
2460	32" high, 160 kW		8	2.500		4,600	140		4,740	5,275
2480	36" high, 180 kW		7	2.857		5,675	160		5,835	6,500
2500	40" high, 200 kW		6	3.333		6,300	186		6,486	7,200
2600	40" wide x 8" high, 45 kW		11	1.818		2,150	102		2,252	2,525
2620	16" high, 90 kW		10	2		2,975	112		3,087	3,425
2640	24" high, 135 kW		9	2.222		3,750	124		3,874	4,325
2660	32" high, 180 kW		8	2.500		5,000	140		5,140	5,700
2680	36" high, 202.5 kW		7	2.857		5,775	160		5,935	6,600
2700	40" high, 225 kW		6	3.333		6,575	186		6,761	7,500
2800	48" wide x 8" high, 54.8 kW		10	2		2,250	112		2,362	2,650
2820	16" high, 109.8 kW		9	2.222		3,125	124		3,249	3,650
2840	24" high, 164.4 kW		8	2.500		3,975	140		4,115	4,575
2860	32" high, 219.2 kW		7	2.857		5,300	160		5,460	6,100
2880	36" high, 246.6 kW		6	3.333		6,100	186		6,286	7,000
2900	40" high, 274 kW		5	4		6,900	224		7,124	7,950
3000	56" wide x 8" high, 64 kW		9	2.222		2,575	124		2,699	3,025
3020	16" high, 128 kW		8	2.500		3,550	140		3,690	4,125
3040	24" high, 192 kW		7	2.857		4,325	160		4,485	5,000
3060	32" high, 256 kW		6	3.333		6,025	186		6,211	6,900
3080	36" high, 288 kW		5	4		6,925	224		7,149	7,950
3100	40" high, 320 kW		4	5		7,650	279		7,929	8,850
3200	64" wide x 8" high, 74 kW		8	2.500		2,650	140		2,790	3,100
3220	16" high, 148 kW		7	2.857		3,650	160		3,810	4,275
3240	24" high, 222 kW		6	3.333		4,650	186		4,836	5,400
3260	32" high, 296 kW		5	4		6,250	224		6,474	7,225
3280	36" high, 333 kW		4	5		7,450	279		7,729	8,625
3300	40" high, 370 kW		3	6.667		8,225	375		8,600	9,625

23 82 16 – Air Coils

23 82 16.20 Duct Heaters		Crew	Daily Output	Labor-Hours	Unit	Material	2019 Bare Costs Labor	Equipment	Total	Total Incl O&P
3800	Note decreased kW available for									
3820	each duct size at same cost									
5000	Duct heater modifications and accessories									
5120	T.C.O. limit auto or manual reset	Q-20	42	.476	Ea.	135	26.50		161.50	189
5140	Thermostat		28	.714		575	40		615	695
5160	Overheat thermocouple (removable)		7	2.857		815	160		975	1,125
5180	Fan interlock relay		18	1.111		197	62		259	310
5200	Air flow switch		20	1		170	56		226	272
5220	Split terminal box cover		100	.200		56	11.20		67.20	78.50
8000	To obtain BTU multiply kW by 3413									

23 82 19 – Fan Coil Units

23 82 19.10 Fan Coil Air Conditioning		Crew	Daily Output	Labor-Hours	Unit	Material	Labor	Equipment	Total	Total Incl O&P
0010	**FAN COIL AIR CONDITIONING**									
0030	Fan coil AC, cabinet mounted, filters and controls									
0100	Chilled water, 1/2 ton cooling	Q-5	8	2	Ea.	585	115		700	820
0110	3/4 ton cooling		7	2.286		730	132		862	995
0120	1 ton cooling		6	2.667		855	153		1,008	1,175
0140	1-1/2 ton cooling		5.50	2.909		900	167		1,067	1,250
0150	2 ton cooling		5.25	3.048		1,225	175		1,400	1,625
0160	2-1/2 ton cooling		5	3.200		1,775	184		1,959	2,250
0180	3 ton cooling		4	4		1,975	230		2,205	2,525
0262	For hot water coil, add					40%	10%			
0300	Console, 2 pipe with electric heat									
0310	1/2 ton cooling	Q-5	8	2	Ea.	1,425	115		1,540	1,750
0315	3/4 ton cooling		7	2.286		1,550	132		1,682	1,925
0320	1 ton cooling		6	2.667		1,675	153		1,828	2,075
0330	1-1/2 ton cooling		5.50	2.909		2,075	167		2,242	2,525
0340	2 ton cooling		4.70	3.404		2,750	196		2,946	3,325
0345	2-1/2 ton cooling		4.70	3.404		3,975	196		4,171	4,675
0350	3 ton cooling		4	4		5,350	230		5,580	6,225
0360	4 ton cooling		4	4		5,575	230		5,805	6,475
0940	Direct expansion, for use w/air cooled condensing unit, 1-1/2 ton cooling		5	3.200		650	184		834	985
0950	2 ton cooling		4.80	3.333		700	192		892	1,050
0960	2-1/2 ton cooling		4.40	3.636		735	209		944	1,125
0970	3 ton cooling		3.80	4.211		890	242		1,132	1,350
0980	3-1/2 ton cooling		3.60	4.444		930	256		1,186	1,400
0990	4 ton cooling		3.40	4.706		1,025	271		1,296	1,550
1000	5 ton cooling		3	5.333		1,125	305		1,430	1,700
1020	7-1/2 ton cooling		3	5.333		1,625	305		1,930	2,225
1040	10 ton cooling	Q-6	2.60	9.231		2,125	550		2,675	3,150
1042	12-1/2 ton cooling		2.30	10.435		2,675	625		3,300	3,850
1050	15 ton cooling		1.60	15		2,975	895		3,870	4,600
1060	20 ton cooling		.70	34.286		4,075	2,050		6,125	7,550
1070	25 ton cooling		.65	36.923		4,875	2,200		7,075	8,650
1080	30 ton cooling		.60	40		5,300	2,375		7,675	9,400
1500	For hot water coil, add					40%	10%			
1512	For condensing unit add see Section 23 62									
3000	Chilled water, horizontal unit, housing, 2 pipe, fan control, no valves									
3100	1/2 ton cooling	Q-5	8	2	Ea.	1,375	115		1,490	1,700
3110	1 ton cooling		6	2.667		1,600	153		1,753	2,000
3120	1-1/2 ton cooling		5.50	2.909		1,850	167		2,017	2,275
3130	2 ton cooling		5.25	3.048		2,200	175		2,375	2,700

23 82 19.10 Fan Coil Air Conditioning			Crew	Daily Output	Labor-Hours	Unit	Material	2019 Bare Costs Labor	Equipment	Total	Total Incl O&P
3140	3 ton cooling		Q-5	4	4	Ea.	2,675	230		2,905	3,275
3150	3-1/2 ton cooling			3.80	4.211		2,925	242		3,167	3,600
3160	4 ton cooling			3.80	4.211		2,925	242		3,167	3,600
3170	5 ton cooling			3.40	4.706		3,400	271		3,671	4,150
3180	6 ton cooling			3.40	4.706		3,400	271		3,671	4,150
3190	7 ton cooling			2.90	5.517		3,725	320		4,045	4,550
3200	8 ton cooling			2.90	5.517		3,725	320		4,045	4,550
3210	10 ton cooling		Q-6	2.80	8.571		3,900	510		4,410	5,075
3212	12-1/2 ton cooling			2.50	9.600		3,225	575		3,800	4,400
3214	15 ton cooling			2.30	10.435		3,450	625		4,075	4,725
3216	20 ton cooling			2.10	11.429		3,975	680		4,655	5,400
3218	25 ton cooling			1.80	13.333		6,500	795		7,295	8,350
3220	30 ton cooling			1.60	15		7,125	895		8,020	9,200
3240	For hot water coil, add						40%	10%			
4000	With electric heat, 2 pipe										
4100	1/2 ton cooling		Q-5	8	2	Ea.	1,625	115		1,740	1,950
4105	3/4 ton cooling			7	2.286		1,800	132		1,932	2,175
4110	1 ton cooling			6	2.667		1,975	153		2,128	2,400
4120	1-1/2 ton cooling			5.50	2.909		2,175	167		2,342	2,650
4130	2 ton cooling			5.25	3.048		2,750	175		2,925	3,300
4135	2-1/2 ton cooling			4.80	3.333		3,800	192		3,992	4,475
4140	3 ton cooling			4	4		5,125	230		5,355	5,975
4150	3-1/2 ton cooling			3.80	4.211		6,100	242		6,342	7,100
4160	4 ton cooling			3.60	4.444		7,050	256		7,306	8,125
4170	5 ton cooling			3.40	4.706		8,525	271		8,796	9,775
4180	6 ton cooling			3.20	5		8,525	288		8,813	9,800
4190	7 ton cooling			2.90	5.517		8,775	320		9,095	10,200
4200	8 ton cooling			2.20	7.273		9,400	420		9,820	10,900
4210	10 ton cooling		Q-6	2.80	8.571		10,200	510		10,710	12,000

23 82 19.20 Heating and Ventilating Units			Crew	Daily Output	Labor-Hours	Unit	Material	Labor	Equipment	Total	Total Incl O&P
0010	**HEATING AND VENTILATING UNITS**, Classroom units										
0020	Includes filter, heating/cooling coils, standard controls										
0080	750 CFM, 2 tons cooling		Q-6	2	12	Ea.	4,575	715		5,290	6,125
0100	1,000 CFM, 2-1/2 tons cooling			1.60	15		5,375	895		6,270	7,275
0120	1,250 CFM, 3 tons cooling			1.40	17.143		5,575	1,025		6,600	7,650
0140	1,500 CFM, 4 tons cooling			.80	30		5,950	1,800		7,750	9,225
0160	2,000 CFM, 5 tons cooling			.50	48		7,000	2,875		9,875	12,000
0500	For electric heat, add						35%				
1000	For no cooling, deduct						25%	10%			

23 82 19.40 Fan Coil Air Conditioning			Crew	Daily Output	Labor-Hours	Unit	Material	Labor	Equipment	Total	Total Incl O&P
0010	**FAN COIL AIR CONDITIONING**, Variable refrigerant volume (VRV) type										
0020	Not including interconnecting tubing or multi-zone controls										
0030	For VRV condensing unit see section 23 81 29.10										
0050	Indoor type, ducted										
0100	Vertical concealed										
0130	1 ton cooling	G	Q-5	2.97	5.387	Ea.	2,550	310		2,860	3,275
0140	1.5 ton cooling	G		2.77	5.776		2,600	330		2,930	3,350
0150	2 ton cooling	G		2.70	5.926		2,800	340		3,140	3,575
0160	2.5 ton cooling	G		2.60	6.154		2,950	355		3,305	3,775
0170	3 ton cooling	G		2.50	6.400		3,000	370		3,370	3,850
0180	3.5 ton cooling	G		2.30	6.957		3,150	400		3,550	4,075
0190	4 ton cooling	G		2.19	7.306		3,175	420		3,595	4,125

23 82 19.40 Fan Coil Air Conditioning			Crew	Daily Output	Labor-Hours	Unit	2019 Bare Costs Material	Labor	Equipment	Total	Total Incl O&P
0200	4.5 ton cooling	G	Q-5	2.08	7.692	Ea.	3,500	445		3,945	4,525
0230	Outside air connection possible										
1100	Ceiling concealed										
1130	0.6 ton cooling	G	Q-5	2.60	6.154	Ea.	1,675	355		2,030	2,350
1140	0.75 ton cooling	G		2.60	6.154		1,725	355		2,080	2,425
1150	1 ton cooling	G		2.60	6.154		1,825	355		2,180	2,525
1152	Screening door	G	1 Stpi	5.20	1.538		102	98.50		200.50	260
1160	1.5 ton cooling	G	Q-5	2.60	6.154		1,875	355		2,230	2,600
1162	Screening door	G	1 Stpi	5.20	1.538		118	98.50		216.50	278
1170	2 ton cooling	G	Q-5	2.60	6.154		2,225	355		2,580	2,975
1172	Screening door	G	1 Stpi	5.20	1.538		102	98.50		200.50	260
1180	2.5 ton cooling	G	Q-5	2.60	6.154		2,575	355		2,930	3,350
1182	Screening door	G	1 Stpi	5.20	1.538		177	98.50		275.50	340
1190	3 ton cooling	G	Q-5	2.60	6.154		2,725	355		3,080	3,525
1192	Screening door	G	1 Stpi	5.20	1.538		177	98.50		275.50	340
1200	4 ton cooling	G	Q-5	2.30	6.957		2,875	400		3,275	3,750
1202	Screening door	G	1 Stpi	5.20	1.538		177	98.50		275.50	340
1220	6 ton cooling	G	Q-5	2.08	7.692		4,750	445		5,195	5,900
1230	8 ton cooling	G	"	1.89	8.466		5,350	485		5,835	6,600
1260	Outside air connection possible										
4050	Indoor type, duct-free										
4100	Ceiling mounted cassette										
4130	0.75 ton cooling	G	Q-5	3.46	4.624	Ea.	1,975	266		2,241	2,575
4140	1 ton cooling	G		2.97	5.387		2,100	310		2,410	2,800
4150	1.5 ton cooling	G		2.77	5.776		2,150	330		2,480	2,875
4160	2 ton cooling	G		2.60	6.154		2,300	355		2,655	3,050
4170	2.5 ton cooling	G		2.50	6.400		2,350	370		2,720	3,150
4180	3 ton cooling	G		2.30	6.957		2,525	400		2,925	3,375
4190	4 ton cooling	G		2.08	7.692		2,825	445		3,270	3,800
4198	For ceiling cassette decoration panel, add	G	1 Stpi	5.20	1.538		305	98.50		403.50	485
4230	Outside air connection possible										
4500	Wall mounted										
4530	0.6 ton cooling	G	Q-5	5.20	3.077	Ea.	1,025	177		1,202	1,400
4540	0.75 ton cooling	G		5.20	3.077		1,075	177		1,252	1,450
4550	1 ton cooling	G		4.16	3.846		1,225	221		1,446	1,675
4560	1.5 ton cooling	G		3.77	4.244		1,350	244		1,594	1,850
4570	2 ton cooling	G		3.46	4.624		1,450	266		1,716	2,000
4590	Condensate pump for the above air handlers	G	1 Stpi	6.46	1.238		219	79		298	360
4700	Floor standing unit										
4730	1 ton cooling	G	Q-5	2.60	6.154	Ea.	1,725	355		2,080	2,425
4740	1.5 ton cooling	G		2.60	6.154		1,950	355		2,305	2,675
4750	2 ton cooling	G		2.60	6.154		2,100	355		2,455	2,825
4800	Floor standing unit, concealed										
4830	1 ton cooling	G	Q-5	2.60	6.154	Ea.	1,650	355		2,005	2,350
4840	1.5 ton cooling	G		2.60	6.154		1,875	355		2,230	2,575
4850	2 ton cooling	G		2.60	6.154		2,000	355		2,355	2,725
4880	Outside air connection possible										
8000	Accessories										
8100	Branch divergence pipe fitting										
8110	Capacity under 76 MBH		1 Stpi	2.50	3.200	Ea.	155	205		360	475
8120	76 MBH to 112 MBH			2.50	3.200		183	205		388	505
8130	112 MBH to 234 MBH			2.50	3.200		520	205		725	875
8200	Header pipe fitting										

23 82 Convection Heating and Cooling Units

23 82 19 – Fan Coil Units

23 82 19.40 Fan Coil Air Conditioning		Crew	Daily Output	Labor-Hours	Unit	2019 Bare Costs Material	Labor	Equipment	Total	Total Incl O&P
8210	Max 4 branches									
8220	Capacity under 76 MBH	1 Stpi	1.70	4.706	Ea.	228	300		528	700
8260	Max 8 branches									
8270	76 MBH to 112 MBH	1 Stpi	1.70	4.706	Ea.	465	300		765	960
8280	112 MBH to 234 MBH	"	1.30	6.154	"	740	395		1,135	1,400

23 82 29 – Radiators

23 82 29.10 Hydronic Heating

Line	Description	Crew	Daily Output	Labor-Hours	Unit	Material	Labor	Equipment	Total	Total Incl O&P
0010	**HYDRONIC HEATING**, Terminal units, not incl. main supply pipe									
1000	Radiation									
1100	Panel, baseboard, C.I., including supports, no covers	Q-5	46	.348	L.F.	46	20		66	80.50
3000	Radiators, cast iron									
3100	Free standing or wall hung, 6 tube, 25" high	Q-5	96	.167	Section	56.50	9.60		66.10	76.50
3150	4 tube, 25" high		96	.167		49.50	9.60		59.10	69
3200	4 tube, 19" high		96	.167		41.50	9.60		51.10	60
3250	Adj. brackets, 2 per wall radiator up to 30 sections	1 Stpi	32	.250	Ea.	65	16		81	95.50
3500	Recessed, 20" high x 5" deep, without grille	Q-5	60	.267	Section	51.50	15.35		66.85	79.50
3525	Free standing or wall hung, 30" high	"	60	.267		46	15.35		61.35	73.50
3600	For inlet grille, add					6.30			6.30	6.95
9500	To convert SFR to BTU rating: Hot water, 150 x SFR									
9510	Forced hot water, 180 x SFR; steam, 240 x SFR									

23 82 33 – Convectors

23 82 33.10 Convector Units

Line	Description	Crew	Daily Output	Labor-Hours	Unit	Material	Labor	Equipment	Total	Total Incl O&P
0010	**CONVECTOR UNITS**, Terminal units, not incl. main supply pipe									
2204	Convector, multifin, 2 pipe w/cabinet									
2210	17" H x 24" L	Q-5	10	1.600	Ea.	108	92		200	257
2214	17" H x 36" L		8.60	1.860		162	107		269	340
2218	17" H x 48" L		7.40	2.162		216	124		340	425
2222	21" H x 24" L		9	1.778		116	102		218	282
2226	21" H x 36" L		8.20	1.951		174	112		286	360
2228	21" H x 48" L		6.80	2.353		232	135		367	460
2240	For knob operated damper, add					140%				
2241	For metal trim strips, add	Q-5	64	.250	Ea.	13.95	14.40		28.35	37
2243	For snap-on inlet grille, add					10%	10%			
2245	For hinged access door, add	Q-5	64	.250	Ea.	38.50	14.40		52.90	64
2246	For air chamber, auto-venting, add	"	58	.276	"	8.75	15.90		24.65	33.50

23 82 36 – Finned-Tube Radiation Heaters

23 82 36.10 Finned Tube Radiation

Line	Description	Crew	Daily Output	Labor-Hours	Unit	Material	Labor	Equipment	Total	Total Incl O&P
0010	**FINNED TUBE RADIATION**, Terminal units, not incl. main supply pipe									
1150	Fin tube, wall hung, 14" slope top cover, with damper									
1200	1-1/4" copper tube, 4-1/4" alum. fin	Q-5	38	.421	L.F.	49.50	24		73.50	90.50
1250	1-1/4" steel tube, 4-1/4" steel fin		36	.444		42.50	25.50		68	85
1255	2" steel tube, 4-1/4" steel fin		32	.500		45.50	29		74.50	93
1260	21", two tier, slope top, w/damper									
1262	1-1/4" copper tube, 4-1/4" alum. fin	Q-5	30	.533	L.F.	58	30.50		88.50	110
1264	1-1/4" steel tube, 4-1/4" steel fin		28	.571		50.50	33		83.50	105
1266	2" steel tube, 4-1/4" steel fin		25	.640		53.50	37		90.50	115
1270	10", flat top, with damper									
1272	1-1/4" copper tube, 4-1/4" alum. fin	Q-5	38	.421	L.F.	22.50	24		46.50	61
1274	1-1/4" steel tube, 4-1/4" steel fin		36	.444		42	25.50		67.50	85
1276	2" steel tube, 4-1/4" steel fin		32	.500		45	29		74	92.50
1280	17", two tier, flat top, w/damper									

23 82 36 – Finned-Tube Radiation Heaters

23 82 36.10 Finned Tube Radiation		Crew	Daily Output	Labor-Hours	Unit	2019 Bare Costs Material	Labor	Equipment	Total	Total Incl O&P
1282	1-1/4" copper tube, 4-1/4" alum. fin	Q-5	30	.533	L.F.	54	30.50		84.50	106
1284	1-1/4" steel tube, 4-1/4" steel fin		28	.571		50.50	33		83.50	105
1286	2" steel tube, 4-1/4" steel fin		25	.640		53.50	37		90.50	115
1301	Rough in wall hung, steel fin w/supply & balance valves	1 Stpi	.91	8.791	Ea.	330	560		890	1,200
1310	Baseboard, pkgd, 1/2" copper tube, alum. fin, 7" high	Q-5	60	.267	L.F.	11.75	15.35		27.10	36
1320	3/4" copper tube, alum. fin, 7" high		58	.276		8.10	15.90		24	33
1340	1" copper tube, alum. fin, 8-7/8" high		56	.286		21.50	16.45		37.95	48
1360	1-1/4" copper tube, alum. fin, 8-7/8" high		54	.296		32	17.05		49.05	60.50
1380	1-1/4" IPS steel tube with steel fins		52	.308		32	17.70		49.70	61.50
1381	Rough in baseboard panel & fin tube, supply & balance valves	1 Stpi	1.06	7.547	Ea.	258	485		743	1,000
1500	Note: fin tube may also require corners, caps, etc.									

23 82 39 – Unit Heaters

23 82 39.13 Cabinet Unit Heaters		Crew	Daily Output	Labor-Hours	Unit	Material	Labor	Equipment	Total	Total Incl O&P
0010	**CABINET UNIT HEATERS**									
5400	Cabinet, horizontal, hot water, blower type									
5420	20 MBH	Q-5	10	1.600	Ea.	1,100	92		1,192	1,375
5430	60 MBH		7	2.286		1,925	132		2,057	2,325
5440	100 MBH		5	3.200		2,075	184		2,259	2,550
5450	120 MBH		4	4		2,075	230		2,305	2,625

23 82 39.16 Propeller Unit Heaters		Crew	Daily Output	Labor-Hours	Unit	Material	Labor	Equipment	Total	Total Incl O&P
0010	**PROPELLER UNIT HEATERS**									
3950	Unit heaters, propeller, 115 V 2 psi steam, 60°F entering air									
4000	Horizontal, 12 MBH	Q-5	12	1.333	Ea.	400	76.50		476.50	550
4020	28.2 MBH		10	1.600		515	92		607	705
4040	36.5 MBH		8	2		600	115		715	840
4060	43.9 MBH		8	2		600	115		715	840
4080	56.5 MBH		7.50	2.133		665	123		788	915
4100	65.6 MBH		7	2.286		675	132		807	940
4120	87.6 MBH		6.50	2.462		715	142		857	1,000
4140	96.8 MBH		6	2.667		880	153		1,033	1,200
4160	133.3 MBH		5	3.200		950	184		1,134	1,325
4180	157.6 MBH		4	4		1,175	230		1,405	1,625
4200	197.7 MBH		3	5.333		1,325	305		1,630	1,900
4220	257.2 MBH		2.50	6.400		1,600	370		1,970	2,300
4240	286.9 MBH		2	8		1,775	460		2,235	2,650
4260	364 MBH		1.80	8.889		2,250	510		2,760	3,225
4270	404 MBH		1.60	10		2,350	575		2,925	3,450
4300	Vertical diffuser same price									
4310	Vertical flow, 40 MBH	Q-5	11	1.455	Ea.	605	83.50		688.50	790
4314	58.5 MBH		8	2		595	115		710	830
4318	92 MBH		7	2.286		760	132		892	1,025
4322	109.7 MBH		6	2.667		895	153		1,048	1,225
4326	131 MBH		4	4		895	230		1,125	1,325
4330	160 MBH		3	5.333		950	305		1,255	1,500
4334	194 MBH		2.20	7.273		1,100	420		1,520	1,825
4338	212 MBH		2.10	7.619		1,350	440		1,790	2,150
4342	247 MBH		1.96	8.163		1,600	470		2,070	2,475
4346	297 MBH		1.80	8.889		1,800	510		2,310	2,750
4350	333 MBH	Q-6	1.90	12.632		1,875	755		2,630	3,175
4354	420 MBH (460 V)		1.80	13.333		2,300	795		3,095	3,725
4358	500 MBH (460 V)		1.71	14.035		3,050	840		3,890	4,625
4362	570 MBH (460 V)		1.40	17.143		4,175	1,025		5,200	6,125

23 82 Convection Heating and Cooling Units

23 82 39 – Unit Heaters

23 82 39.16 Propeller Unit Heaters		Crew	Daily Output	Labor-Hours	Unit	Material	Labor	Equipment	Total	Total Incl O&P
						2019 Bare Costs				
4366	620 MBH (460 V)	Q-6	1.30	18.462	Ea.	4,350	1,100		5,450	6,450
4370	960 MBH (460 V)		1.10	21.818		8,125	1,300		9,425	10,900
4400	Unit heaters, propeller, hot water, 115 V, 60 degree F entering air									
4404	Horizontal									
4408	6 MBH	Q-5	11	1.455	Ea.	400	83.50		483.50	560
4412	12 MBH		11	1.455		400	83.50		483.50	560
4416	16 MBH		10	1.600		450	92		542	635
4420	24 MBH		9	1.778		515	102		617	720
4424	29 MBH		8	2		605	115		720	840
4428	47 MBH		7	2.286		660	132		792	920
4432	63 MBH		6	2.667		715	153		868	1,025
4436	81 MBH		5.50	2.909		875	167		1,042	1,225
4440	90 MBH		5	3.200		950	184		1,134	1,325
4460	133 MBH		4	4		1,175	230		1,405	1,625
4464	139 MBH		3.50	4.571		1,325	263		1,588	1,850
4468	198 MBH		2.50	6.400		1,600	370		1,970	2,300
4472	224 MBH		2	8		1,775	460		2,235	2,650
4476	273 MBH		1.50	10.667		2,250	615		2,865	3,400
4500	Vertical									
4504	8 MBH	Q-5	13	1.231	Ea.	1,225	71		1,296	1,450
4508	12 MBH		13	1.231		1,225	71		1,296	1,450
4512	16 MBH		13	1.231		1,225	71		1,296	1,450
4516	30 MBH		12	1.333		1,225	76.50		1,301.50	1,475
4520	43 MBH		11	1.455		1,225	83.50		1,308.50	1,475
4524	57 MBH		8	2		1,350	115		1,465	1,650
4528	68 MBH		7	2.286		1,375	132		1,507	1,700
4540	105 MBH		6	2.667		1,450	153		1,603	1,800
4544	123 MBH		5	3.200		1,525	184		1,709	1,950
4550	140 MBH		5	3.200		1,700	184		1,884	2,125
4554	156 MBH		4	4		1,900	230		2,130	2,450
4560	210 MBH		4	4		2,300	230		2,530	2,875
4564	223 MBH		3	5.333		2,875	305		3,180	3,600
4568	257 MBH		2	8		2,875	460		3,335	3,850

23 83 Radiant Heating Units

23 83 16 – Radiant-Heating Hydronic Piping

23 83 16.10 Radiant Floor Heating		Crew	Daily Output	Labor-Hours	Unit	Material	Labor	Equipment	Total	Total Incl O&P
0010	**RADIANT FLOOR HEATING**									
0100	Tubing, PEX (cross-linked polyethylene)									
0110	Oxygen barrier type for systems with ferrous materials									
0120	1/2"	Q-5	800	.020	L.F.	1.06	1.15		2.21	2.90
0130	3/4"		535	.030		1.65	1.72		3.37	4.40
0140	1"		400	.040		2.32	2.30		4.62	6
0200	Non barrier type for ferrous free systems									
0210	1/2"	Q-5	800	.020	L.F.	.53	1.15		1.68	2.31
0220	3/4"		535	.030		.96	1.72		2.68	3.64
0230	1"		400	.040		1.72	2.30		4.02	5.35
1000	Manifolds									
1110	Brass									
1120	With supply and return valves, flow meter, thermometer,									
1122	auto air vent and drain/fill valve.									

23 83 16 – Radiant-Heating Hydronic Piping

23 83 16.10 Radiant Floor Heating		Crew	Daily Output	Labor-Hours	Unit	Material	2019 Bare Costs Labor	Equipment	Total	Total Incl O&P
1130	1", 2 circuit	Q-5	14	1.143	Ea.	350	66		416	485
1140	1", 3 circuit		13.50	1.185		385	68		453	525
1150	1", 4 circuit		13	1.231		405	71		476	550
1154	1", 5 circuit		12.50	1.280		485	73.50		558.50	645
1158	1", 6 circuit		12	1.333		590	76.50		666.50	760
1162	1", 7 circuit		11.50	1.391		630	80		710	815
1166	1", 8 circuit		11	1.455		640	83.50		723.50	830
1172	1", 9 circuit		10.50	1.524		750	87.50		837.50	955
1174	1", 10 circuit		10	1.600		765	92		857	980
1178	1", 11 circuit		9.50	1.684		775	97		872	1,000
1182	1", 12 circuit		9	1.778		905	102		1,007	1,150
1610	Copper manifold header (cut to size)									
1620	1" header, 12 circuit 1/2" sweat outlets	Q-5	3.33	4.805	Ea.	110	277		387	535
1630	1-1/4" header, 12 circuit 1/2" sweat outlets		3.20	5		127	288		415	570
1640	1-1/4" header, 12 circuit 3/4" sweat outlets		3	5.333		129	305		434	600
1650	1-1/2" header, 12 circuit 3/4" sweat outlets		3.10	5.161		157	297		454	620
1660	2" header, 12 circuit 3/4" sweat outlets		2.90	5.517		235	320		555	735
3000	Valves									
3110	Thermostatic zone valve actuator with end switch	Q-5	40	.400	Ea.	49	23		72	88
3114	Thermostatic zone valve actuator	"	36	.444	"	91	25.50		116.50	139
3120	Motorized straight zone valve with operator complete									
3130	3/4"	Q-5	35	.457	Ea.	150	26.50		176.50	206
3140	1"		32	.500		163	29		192	223
3150	1-1/4"		29.60	.541		205	31		236	272
3500	4 way mixing valve, manual, brass									
3530	1"	Q-5	13.30	1.203	Ea.	218	69		287	345
3540	1-1/4"		11.40	1.404		237	81		318	380
3550	1-1/2"		11	1.455		292	83.50		375.50	445
3560	2"		10.60	1.509		430	87		517	600
3800	Mixing valve motor, 4 way for valves, 1" and 1-1/4"		34	.471		385	27		412	465
3810	Mixing valve motor, 4 way for valves, 1-1/2" and 2"		30	.533		410	30.50		440.50	495
5000	Radiant floor heating, zone control panel									
5120	4 zone actuator valve control, expandable	Q-5	20	.800	Ea.	167	46		213	253
5130	6 zone actuator valve control, expandable		18	.889		252	51		303	355
6070	Thermal track, straight panel for long continuous runs, 5.333 S.F.		40	.400		34	23		57	72
6080	Thermal track, utility panel, for direction reverse at run end, 5.333 S.F.		40	.400		34	23		57	72
6090	Combination panel, for direction reverse plus straight run, 5.333 S.F.		40	.400		34	23		57	72
7000	PEX tubing fittings									
7100	Compression type									
7116	Coupling									
7120	1/2" x 1/2"	1 Stpi	27	.296	Ea.	6.70	18.95		25.65	36
7124	3/4" x 3/4"	"	23	.348	"	15.05	22		37.05	50
7130	Adapter									
7132	1/2" x female sweat 1/2"	1 Stpi	27	.296	Ea.	4.35	18.95		23.30	33.50
7134	1/2" x female sweat 3/4"		26	.308		4.86	19.70		24.56	35
7136	5/8" x female sweat 3/4"		24	.333		6.95	21.50		28.45	39.50
7140	Elbow									
7142	1/2" x female sweat 1/2"	1 Stpi	27	.296	Ea.	7.85	18.95		26.80	37
7144	1/2" x female sweat 3/4"		26	.308		9.25	19.70		28.95	39.50
7146	5/8" x female sweat 3/4"		24	.333		10.40	21.50		31.90	43.50
7200	Insert type									
7206	PEX x male NPT									
7210	1/2" x 1/2"	1 Stpi	29	.276	Ea.	2.80	17.65		20.45	29.50

23 83 Radiant Heating Units

23 83 16 – Radiant-Heating Hydronic Piping

23 83 16.10 Radiant Floor Heating		Crew	Daily Output	Labor-Hours	Unit	Material	Labor (2019 Bare Costs)	Equipment	Total	Total Incl O&P
7220	3/4" x 3/4"	1 Stpi	27	.296	Ea.	4.10	18.95		23.05	33
7230	1" x 1"		26	.308		6.85	19.70		26.55	37
7300	PEX coupling									
7310	1/2" x 1/2"	1 Stpi	30	.267	Ea.	.60	17.05		17.65	26
7320	3/4" x 3/4"		29	.276		.79	17.65		18.44	27.50
7330	1" x 1"		28	.286		1.37	18.25		19.62	29
7400	PEX stainless crimp ring									
7410	1/2" x 1/2"	1 Stpi	86	.093	Ea.	.52	5.95		6.47	9.50
7420	3/4" x 3/4"		84	.095		.71	6.10		6.81	9.95
7430	1" x 1"		82	.098		1.05	6.25		7.30	10.50

23 83 33 – Electric Radiant Heaters

23 83 33.10 Electric Heating		Crew	Daily Output	Labor-Hours	Unit	Material	Labor (2019 Bare Costs)	Equipment	Total	Total Incl O&P
0010	**ELECTRIC HEATING**, not incl. conduit or feed wiring									
1100	Rule of thumb: Baseboard units, including control	1 Elec	4.40	1.818	kW	112	109		221	286
1300	Baseboard heaters, 2' long, 350 watt		8	1	Ea.	27.50	60		87.50	120
1400	3' long, 750 watt		8	1		32	60		92	125
1600	4' long, 1,000 watt		6.70	1.194		39	71.50		110.50	150
1800	5' long, 935 watt		5.70	1.404		46.50	84.50		131	177
2000	6' long, 1,500 watt		5	1.600		51.50	96		147.50	200
2200	7' long, 1,310 watt		4.40	1.818		56	109		165	225
2400	8' long, 2,000 watt		4	2		65.50	120		185.50	251
2600	9' long, 1,680 watt		3.60	2.222		86	133		219	294
2800	10' long, 1,875 watt		3.30	2.424		154	146		300	385
2950	Wall heaters with fan, 120 to 277 volt									
3160	Recessed, residential, 750 watt	1 Elec	6	1.333	Ea.	94.50	80		174.50	223
3170	1,000 watt		6	1.333		94.50	80		174.50	223
3180	1,250 watt		5	1.600		102	96		198	255
3190	1,500 watt		4	2		102	120		222	291
3210	2,000 watt		4	2		94.50	120		214.50	283
3230	2,500 watt		3.50	2.286		237	137		374	465
3240	3,000 watt		3	2.667		365	160		525	645
3250	4,000 watt		2.70	2.963		370	178		548	670
3260	Commercial, 750 watt		6	1.333		202	80		282	340
3270	1,000 watt		6	1.333		202	80		282	340
3280	1,250 watt		5	1.600		202	96		298	365
3290	1,500 watt		4	2		202	120		322	400
3300	2,000 watt		4	2		202	120		322	400
3310	2,500 watt		3.50	2.286		202	137		339	430
3320	3,000 watt		3	2.667		425	160		585	710
3330	4,000 watt		2.70	2.963		495	178		673	810
3600	Thermostats, integral		16	.500		28	30		58	75.50
3800	Line voltage, 1 pole		8	1		17.15	60		77.15	108
3810	2 pole		8	1		22	60		82	114
4000	Heat trace system, 400 degree									
4020	115 V, 2.5 watts/L.F.	1 Elec	530	.015	L.F.	11.35	.91		12.26	13.80
4030	5 watts/L.F.		530	.015		11.35	.91		12.26	13.80
4050	10 watts/L.F.		530	.015		11.35	.91		12.26	13.80
4060	208 V, 5 watts/L.F.		530	.015		11.35	.91		12.26	13.80
4080	480 V, 8 watts/L.F.		530	.015		11.35	.91		12.26	13.80
4200	Heater raceway									
4260	Heat transfer cement									
4280	1 gallon				Ea.	63.50			63.50	70

23 83 33 – Electric Radiant Heaters

23 83 33.10 Electric Heating		Crew	Daily Output	Labor-Hours	Unit	2019 Bare Costs Material	Labor	Equipment	Total	Total Incl O&P
4300	5 gallon				Ea.	246			246	271
4320	Cable tie									
4340	3/4" pipe size	1 Elec	470	.017	Ea.	.02	1.02		1.04	1.55
4360	1" pipe size		444	.018		.02	1.08		1.10	1.63
4380	1-1/4" pipe size		400	.020		.02	1.20		1.22	1.81
4400	1-1/2" pipe size		355	.023		.02	1.35		1.37	2.04
4420	2" pipe size		320	.025		.05	1.50		1.55	2.30
4440	3" pipe size		160	.050		.05	3		3.05	4.54
4460	4" pipe size		100	.080		.06	4.80		4.86	7.20
4480	Thermostat NEMA 3R, 22 amp, 0-150 degree, 10' cap.		8	1		201	60		261	310
4500	Thermostat NEMA 4X, 25 amp, 40 degree, 5-1/2' cap.		7	1.143		254	68.50		322.50	380
4520	Thermostat NEMA 4X, 22 amp, 25-325 degree, 10' cap.		7	1.143		650	68.50		718.50	815
4540	Thermostat NEMA 4X, 22 amp, 15-140 degree		6	1.333		575	80		655	750
4580	Thermostat NEMA 4, 7, 9, 22 amp, 25-325 degree, 10' cap.		3.60	2.222		845	133		978	1,125
4600	Thermostat NEMA 4, 7, 9, 22 amp, 15-140 degree		3	2.667		835	160		995	1,150
4720	Fiberglass application tape, 36 yard roll		11	.727		79	43.50		122.50	152
5000	Radiant heating ceiling panels, 2' x 4', 500 watt		16	.500		395	30		425	480
5050	750 watt		16	.500		500	30		530	595
5200	For recessed plaster frame, add		32	.250		134	15		149	170
5300	Infrared quartz heaters, 120 volts, 1,000 watt		6.70	1.194		345	71.50		416.50	485
5350	1,500 watt		5	1.600		345	96		441	525
5400	240 volts, 1,500 watt		5	1.600		395	96		491	580
5450	2,000 watt		4	2		345	120		465	560
5500	3,000 watt		3	2.667		390	160		550	670
5550	4,000 watt		2.60	3.077		390	185		575	705
5570	Modulating control		.80	10		140	600		740	1,050
5600	Unit heaters, heavy duty, with fan & mounting bracket									
5650	Single phase, 208-240-277 volt, 3 kW	1 Elec	6	1.333	Ea.	695	80		775	885
5750	5 kW		5.50	1.455		785	87.50		872.50	995
5800	7 kW		5	1.600		1,300	96		1,396	1,575
5850	10 kW		4	2		1,025	120		1,145	1,300
5950	15 kW		3.80	2.105		1,650	126		1,776	2,000
6000	480 volt, 3 kW		6	1.333		535	80		615	710
6020	4 kW		5.80	1.379		745	83		828	945
6040	5 kW		5.50	1.455		740	87.50		827.50	945
6060	7 kW		5	1.600		1,125	96		1,221	1,375
6080	10 kW		4	2		1,175	120		1,295	1,475
6100	13 kW		3.80	2.105		1,750	126		1,876	2,125
6120	15 kW		3.70	2.162		2,525	130		2,655	2,975
6140	20 kW		3.50	2.286		2,975	137		3,112	3,450
6300	3 phase, 208-240 volt, 5 kW		5.50	1.455		575	87.50		662.50	760
6320	7 kW		5	1.600		935	96		1,031	1,175
6340	10 kW		4	2		875	120		995	1,150
6360	15 kW		3.70	2.162		1,625	130		1,755	1,975
6380	20 kW		3.50	2.286		2,825	137		2,962	3,300
6400	25 kW		3.30	2.424		3,550	146		3,696	4,125
6500	480 volt, 5 kW		5.50	1.455		730	87.50		817.50	930
6520	7 kW		5	1.600		1,100	96		1,196	1,375
6540	10 kW		4	2		975	120		1,095	1,250
6560	13 kW		3.80	2.105		1,700	126		1,826	2,050
6580	15 kW		3.70	2.162		1,425	130		1,555	1,750
6600	20 kW		3.50	2.286		2,425	137		2,562	2,875
6620	25 kW		3.30	2.424		3,325	146		3,471	3,900

23 83 33.10 Electric Heating		Crew	Daily Output	Labor-Hours	Unit	Material	Labor	Equipment	Total	Total Incl O&P
						2019 Bare Costs				
6630	30 kW	1 Elec	3	2.667	Ea.	3,375	160		3,535	3,950
6640	40 kW		2	4		4,075	240		4,315	4,825
6650	50 kW		1.60	5		6,375	300		6,675	7,475
6800	Vertical discharge heaters, with fan									
6820	Single phase, 208-240-277 volt, 10 kW	1 Elec	4	2	Ea.	985	120		1,105	1,250
6840	15 kW		3.70	2.162		1,700	130		1,830	2,075
6900	3 phase, 208-240 volt, 10 kW		4	2		930	120		1,050	1,200
6920	15 kW		3.70	2.162		1,700	130		1,830	2,050
6940	20 kW		3.50	2.286		3,025	137		3,162	3,525
6960	25 kW		3.30	2.424		3,275	146		3,421	3,825
6980	30 kW		3	2.667		4,150	160		4,310	4,800
7000	40 kW		2	4		4,650	240		4,890	5,475
7020	50 kW		1.60	5		6,625	300		6,925	7,725
7100	480 volt, 10 kW		4	2		1,075	120		1,195	1,350
7120	15 kW		3.70	2.162		1,750	130		1,880	2,125
7140	20 kW		3.50	2.286		2,825	137		2,962	3,300
7160	25 kW		3.30	2.424		3,550	146		3,696	4,150
7180	30 kW		3	2.667		3,925	160		4,085	4,575
7200	40 kW		2	4		4,075	240		4,315	4,825
7222	Electric heating, vertical disch heaters, 3 phase, 480 volt, 80 kW		1.60	5		6,375	300		6,675	7,475
7900	Cabinet convector heaters, 240 volt, three phase,									
7920	3′ long, 2,000 watt	1 Elec	5.30	1.509	Ea.	2,300	90.50		2,390.50	2,650
7940	3,000 watt		5.30	1.509		2,400	90.50		2,490.50	2,750
7960	4,000 watt		5.30	1.509		2,450	90.50		2,540.50	2,825
7980	6,000 watt		4.60	1.739		2,725	104		2,829	3,150
8000	8,000 watt		4.60	1.739		2,775	104		2,879	3,225
8020	4′ long, 4,000 watt		4.60	1.739		2,500	104		2,604	2,900
8040	6,000 watt		4	2		2,575	120		2,695	3,025
8060	8,000 watt		4	2		2,650	120		2,770	3,100
8080	10,000 watt		4	2		2,700	120		2,820	3,150
8100	Available also in 208 or 277 volt									
8200	Cabinet unit heaters, 120 to 277 volt, 1 pole,									
8220	wall mounted, 2 kW	1 Elec	4.60	1.739	Ea.	2,150	104		2,254	2,525
8230	3 kW		4.60	1.739		2,200	104		2,304	2,575
8240	4 kW		4.40	1.818		2,325	109		2,434	2,725
8250	5 kW		4.40	1.818		2,350	109		2,459	2,775
8260	6 kW		4.20	1.905		2,400	114		2,514	2,825
8270	8 kW		4	2		2,450	120		2,570	2,875
8280	10 kW		3.80	2.105		2,500	126		2,626	2,950
8290	12 kW		3.50	2.286		2,625	137		2,762	3,100
8300	13.5 kW		2.90	2.759		3,175	166		3,341	3,750
8310	16 kW		2.70	2.963		3,200	178		3,378	3,800
8320	20 kW		2.30	3.478		4,000	209		4,209	4,700
8330	24 kW		1.90	4.211		4,200	253		4,453	4,975
8350	Recessed, 2 kW		4.40	1.818		2,175	109		2,284	2,575
8370	3 kW		4.40	1.818		2,275	109		2,384	2,675
8380	4 kW		4.20	1.905		2,350	114		2,464	2,775
8390	5 kW		4.20	1.905		2,375	114		2,489	2,800
8400	6 kW		4	2		2,425	120		2,545	2,850
8410	8 kW		3.80	2.105		2,550	126		2,676	3,000
8420	10 kW		3.50	2.286		3,200	137		3,337	3,725
8430	12 kW		2.90	2.759		3,200	166		3,366	3,775
8440	13.5 kW		2.70	2.963		3,175	178		3,353	3,750

23 83 33 – Electric Radiant Heaters

23 83 33.10 Electric Heating		Crew	Daily Output	Labor-Hours	Unit	Material	2019 Bare Costs Labor	Equipment	Total	Total Incl O&P
8450	16 kW	1 Elec	2.30	3.478	Ea.	3,325	209		3,534	3,950
8460	20 kW		1.90	4.211		3,850	253		4,103	4,625
8470	24 kW		1.60	5		3,825	300		4,125	4,650
8490	Ceiling mounted, 2 kW		3.20	2.500		2,150	150		2,300	2,575
8510	3 kW		3.20	2.500		2,250	150		2,400	2,700
8520	4 kW		3	2.667		2,300	160		2,460	2,775
8530	5 kW		3	2.667		2,375	160		2,535	2,875
8540	6 kW		2.80	2.857		2,400	172		2,572	2,875
8550	8 kW		2.40	3.333		2,525	200		2,725	3,075
8560	10 kW		2.20	3.636		2,650	218		2,868	3,250
8570	12 kW		2	4		2,700	240		2,940	3,325
8580	13.5 kW		1.50	5.333		2,675	320		2,995	3,425
8590	16 kW		1.30	6.154		2,750	370		3,120	3,575
8600	20 kW		.90	8.889		3,800	535		4,335	5,000
8610	24 kW		.60	13.333		3,825	800		4,625	5,400
8630	208 to 480 V, 3 pole									
8650	Wall mounted, 2 kW	1 Elec	4.60	1.739	Ea.	2,275	104		2,379	2,650
8670	3 kW		4.60	1.739		2,375	104		2,479	2,750
8680	4 kW		4.40	1.818		2,425	109		2,534	2,850
8690	5 kW		4.40	1.818		2,500	109		2,609	2,925
8700	6 kW		4.20	1.905		2,525	114		2,639	2,950
8710	8 kW		4	2		2,600	120		2,720	3,050
8720	10 kW		3.80	2.105		2,675	126		2,801	3,125
8730	12 kW		3.50	2.286		2,725	137		2,862	3,200
8740	13.5 kW		2.90	2.759		2,800	166		2,966	3,350
8750	16 kW		2.70	2.963		2,825	178		3,003	3,375
8760	20 kW		2.30	3.478		3,975	209		4,184	4,675
8770	24 kW		1.90	4.211		4,200	253		4,453	5,000
8790	Recessed, 2 kW		4.40	1.818		2,275	109		2,384	2,675
8810	3 kW		4.40	1.818		2,300	109		2,409	2,700
8820	4 kW		4.20	1.905		2,425	114		2,539	2,850
8830	5 kW		4.20	1.905		2,425	114		2,539	2,825
8840	6 kW		4	2		2,450	120		2,570	2,850
8850	8 kW		3.80	2.105		2,525	126		2,651	2,975
8860	10 kW		3.50	2.286		2,550	137		2,687	3,025
8870	12 kW		2.90	2.759		2,850	166		3,016	3,400
8880	13.5 kW		2.70	2.963		2,425	178		2,603	2,925
8890	16 kW		2.30	3.478		4,725	209		4,934	5,500
8900	20 kW		1.90	4.211		4,275	253		4,528	5,075
8920	24 kW		1.60	5		4,350	300		4,650	5,225
8940	Ceiling mount, 2 kW		3.20	2.500		2,400	150		2,550	2,875
8950	3 kW		3.20	2.500		2,150	150		2,300	2,575
8960	4 kW		3	2.667		2,575	160		2,735	3,075
8970	5 kW		3	2.667		2,475	160		2,635	2,975
8980	6 kW		2.80	2.857		2,500	172		2,672	3,000
8990	8 kW		2.40	3.333		2,575	200		2,775	3,150
9000	10 kW		2.20	3.636		2,350	218		2,568	2,925
9020	13.5 kW		1.50	5.333		2,950	320		3,270	3,725
9030	16 kW		1.30	6.154		4,425	370		4,795	5,425
9040	20 kW		.90	8.889		4,275	535		4,810	5,500
9060	24 kW		.60	13.333		4,750	800		5,550	6,425
9230	13.5 kW, 40,956 BTU		2.20	3.636		1,725	218		1,943	2,225
9250	24 kW, 81,912 BTU		2	4		1,750	240		1,990	2,275

23 84 Humidity Control Equipment

23 84 13 – Humidifiers

23 84 13.10 Humidifier Units		Crew	Daily Output	Labor-Hours	Unit	2019 Bare Costs Material	Labor	Equipment	Total	Total Incl O&P
0010	**HUMIDIFIER UNITS**									
0520	Steam, room or duct, filter, regulators, auto. controls, 220 V									
0540	11 lb./hr.	Q-5	6	2.667	Ea.	2,750	153		2,903	3,250
0560	22 lb./hr.		5	3.200		3,025	184		3,209	3,600
0580	33 lb./hr.		4	4		3,100	230		3,330	3,775
0600	50 lb./hr.		4	4		3,700	230		3,930	4,425
0620	100 lb./hr.		3	5.333		5,100	305		5,405	6,075
0640	150 lb./hr.		2.50	6.400		6,500	370		6,870	7,700
0660	200 lb./hr.		2	8		8,150	460		8,610	9,675
0700	With blower									
0720	11 lb./hr.	Q-5	5.50	2.909	Ea.	4,100	167		4,267	4,750
0740	22 lb./hr.		4.75	3.368		4,325	194		4,519	5,050
0760	33 lb./hr.		3.75	4.267		4,425	246		4,671	5,250
0780	50 lb./hr.		3.50	4.571		5,425	263		5,688	6,350
0800	100 lb./hr.		2.75	5.818		6,050	335		6,385	7,150
0820	150 lb./hr.		2	8		8,825	460		9,285	10,400
0840	200 lb./hr.		1.50	10.667		11,100	615		11,715	13,100
1000	Steam for duct installation, manifold with pneum. controls									
1010	3 -191 lb./hr., 24" x 24"	Q-5	5	3.200	Ea.	2,175	184		2,359	2,675
1020	3-191 lb./hr., 48" x 48"		4.60	3.478		4,200	200		4,400	4,900
1030	65-334 lb./hr., 36" x 36"		4.80	3.333		4,100	192		4,292	4,800
1040	65-334 lb./hr., 72" x 72"		4	4		4,425	230		4,655	5,225
1050	100-690 lb./hr., 48" x 48"		4	4		5,750	230		5,980	6,675
1060	100-690 lb./hr., 84" x 84"		3.60	4.444		5,950	256		6,206	6,900
1070	220-2000 lb./hr., 72" x 72"		3.80	4.211		6,025	242		6,267	7,000
1080	220-2000 lb./hr., 144" x 144"		3.40	4.706		7,025	271		7,296	8,150
1090	For electrically controlled unit, add					805			805	885
1092	For additional manifold, add					25%	15%			
1200	Electric steam, self generating, with manifold									
1210	1-96 lb./hr., 12" x 72"	Q-5	3.60	4.444	Ea.	10,600	256		10,856	12,000
1220	20-168 lb./hr., 12" x 72"		3.60	4.444		16,200	256		16,456	18,200
1290	For room mounting, add		16	1		265	57.50		322.50	380
5000	Furnace type, wheel bypass									
5020	10 GPD	1 Stpi	4	2	Ea.	188	128		316	400
5040	14 GPD		3.80	2.105		206	135		341	430
5060	19 GPD		3.60	2.222		330	142		472	580

23 84 16 – Mechanical Dehumidification Units

23 84 16.10 Dehumidifier Units		Crew	Daily Output	Labor-Hours	Unit	Material	Labor	Equipment	Total	Total Incl O&P
0010	**DEHUMIDIFIER UNITS**									
6000	Self contained with filters and standard controls									
6040	1.5 lb./hr., 50 CFM	1 Plum	8	1	Ea.	8,325	63		8,388	9,275
6060	3 lb./hr., 150 CFM	Q-1	12	1.333		8,950	76		9,026	9,975
6065	6 lb./hr., 150 CFM		9	1.778		14,600	101		14,701	16,200
6070	16 to 20 lb./hr., 600 CFM		5	3.200		27,700	182		27,882	30,800
6080	30 to 40 lb./hr., 1,125 CFM		4	4		43,100	227		43,327	47,800
6090	60 to 75 lb./hr., 2,250 CFM		3	5.333		56,500	305		56,805	62,500
6100	120 to 155 lb./hr., 4,500 CFM		2	8		99,500	455		99,955	110,000
6110	240 to 310 lb./hr., 9,000 CFM		1.50	10.667		143,000	605		143,605	158,000
6120	400 to 515 lb./hr., 15,000 CFM	Q-2	1.60	15		180,000	885		180,885	199,500
6130	530 to 690 lb./hr., 20,000 CFM		1.40	17.143		195,000	1,000		196,000	216,000
6140	800 to 1030 lb./hr., 30,000 CFM		1.20	20		222,000	1,175		223,175	246,000
6150	1060 to 1375 lb./hr., 40,000 CFM		1	24		309,000	1,425		310,425	341,500

23 91 Prefabricated Equipment Supports

23 91 10 – Prefabricated Curbs, Pads and Stands

23 91 10.10 Prefabricated Pads and Stands		Crew	Daily Output	Labor-Hours	Unit	2019 Bare Costs Material	Labor	Equipment	Total	Total Incl O&P
0010	**PREFABRICATED PADS AND STANDS**									
6000	Pad, fiberglass reinforced concrete with polystyrene foam core									
6070	24" x 24"	1 Shee	16	.500	Ea.	25	30.50		55.50	74
6090	24" x 36"		12	.667		40	40.50		80.50	106
6110	24" x 42"		8	1		46	61		107	144
6150	26" x 36"	Q-9	8	2		41.50	110		151.50	213
6170	28" x 38"	1 Shee	8	1		46.50	61		107.50	144
6190	30" x 30"		12	.667		42	40.50		82.50	108
6220	30" x 36"		8	1		49.50	61		110.50	147
6240	30" x 40"	Q-9	8	2		54	110		164	227
6260	32" x 32"		7	2.286		46.50	125		171.50	242
6280	36" x 36"		8	2		57.50	110		167.50	230
6300	36" x 40"		7	2.286		63.50	125		188.50	260
6320	36" x 48"		7	2.286		76.50	125		201.50	274
6340	36" x 54"		6	2.667		92.50	146		238.50	325
6360	36" x 60" x 3"		7	2.286		107	125		232	305
6400	41" x 47" x 3"		6	2.667		99	146		245	330
6490	26" round		10	1.600		24.50	88		112.50	160
6550	30" round		9	1.778		31.50	97.50		129	183
6600	36" round		8	2		57	110		167	230
6700	For 3" thick pad, add					20%	5%			
8800	Stand, corrosion-resistant plastic									
8850	Heat pump, 6" high	1 Shee	32	.250	Ea.	5.05	15.25		20.30	28.50
8870	12" high	"	32	.250	"	14.30	15.25		29.55	38.50

Estimating Tips

26 05 00 Common Work Results for Electrical

- Conduit should be taken off in three main categories—power distribution, branch power, and branch lighting—so the estimator can concentrate on systems and components, therefore making it easier to ensure all items have been accounted for.
- For cost modifications for elevated conduit installation, add the percentages to labor according to the height of installation and only to the quantities exceeding the different height levels, not to the total conduit quantities. Refer to subdivision 26 01 02.20 for labor adjustment factors.
- Remember that aluminum wiring of equal ampacity is larger in diameter than copper and may require larger conduit.
- If more than three wires at a time are being pulled, deduct percentages from the labor hours of that grouping of wires.
- When taking off grounding systems, identify separately the type and size of wire, and list each unique type of ground connection.
- The estimator should take the weights of materials into consideration when completing a takeoff. Topics to consider include: How will the materials be supported? What methods of support are available? How high will the support structure have to reach? Will the final support structure be able to withstand the total burden? Is the support material included or separate from the fixture, equipment, and material specified?
- Do not overlook the costs for equipment used in the installation. If scaffolding or highlifts are available in the field, contractors may use them in lieu of the proposed ladders and rolling staging.

26 20 00 Low-Voltage Electrical Transmission

- Supports and concrete pads may be shown on drawings for the larger equipment, or the support system may be only a piece of plywood for the back of a panelboard. In either case, they must be included in the costs.

26 40 00 Electrical and Cathodic Protection

- When taking off cathodic protection systems, identify the type and size of cable, and list each unique type of anode connection.

26 50 00 Lighting

- Fixtures should be taken off room by room using the fixture schedule, specifications, and the ceiling plan. For large concentrations of lighting fixtures in the same area, deduct the percentages from labor hours.

Reference Numbers

Reference numbers are shown at the beginning of some major classifications. These numbers refer to related items in the Reference Section. The reference information may be an estimating procedure, an alternate pricing method, or technical information.

Note: Not all subdivisions listed here necessarily appear. ■

Note: "Powered in part by CINX™, based on licensed proprietary information of Harrison Publishing House, Inc."

Note: The following companies, in part, have been used as a reference source for some of the material prices used in Division 26:

Electriflex

Trade Service

26 01 02.20 Labor Adjustment Factors		Crew	Daily Output	Labor-Hours	Unit	Material	2019 Bare Costs Labor	Equipment	Total	Total Incl O&P
0010	**LABOR ADJUSTMENT FACTORS** (For Div. 26, 27 and 28) R260519-90									
0100	Subtract from labor for Economy of Scale for Wire									
0110	4-5 wires						25%			
0120	6-10 wires						30%			
0130	11-15 wires						35%			
0140	over 15 wires						40%			
0150	Labor adjustment factors (For Div. 26, 27, 28 and 48)									
0200	Labor factors: The below are reasonable suggestions, but									
0210	each project must be evaluated for its own peculiarities, and									
0220	the adjustments be increased or decreased depending on the									
0230	severity of the special conditions.									
1000	Add to labor for elevated installation (above floor level)									
1010	10′ to 14.5′ high						10%			
1020	15′ to 19.5′ high						20%			
1030	20′ to 24.5′ high						25%			
1040	25′to 29.5′ high						35%			
1050	30′ to 34.5′ high						40%			
1060	35′ to 39.5′ high						50%			
1070	40′ and higher						55%			
2000	Add to labor for crawl space									
2010	3′ high						40%			
2020	4′ high						30%			
3000	Add to labor for multi-story building									
3100	For new construction (No elevator available)									
3110	Add for floors 3 thru 10						5%			
3120	Add for floors 11 thru 15						10%			
3130	Add for floors 16 thru 20						15%			
3140	Add for floors 21 thru 30						20%			
3150	Add for floors 31 and up						30%			
3200	For existing structure (Elevator available)									
3210	Add for work on floor 3 and above						2%			
4000	Add to labor for working in existing occupied buildings									
4010	Hospital						35%			
4020	Office building						25%			
4030	School						20%			
4040	Factory or warehouse						15%			
4050	Multi-dwelling						15%			
5000	Add to labor, miscellaneous									
5010	Cramped shaft						35%			
5020	Congested area						15%			
5030	Excessive heat or cold						30%			
6000	Labor factors: the above are reasonable suggestions, but									
6100	each project should be evaluated for its own peculiarities									
6200	Other factors to be considered are:									
6210	Movement of material and equipment through finished areas						10%			
6220	Equipment room min security direct access w/authorization						15%			
6230	Attic space						25%			
6240	No service road						25%			
6250	Poor unloading/storage area, no hydraulic lifts or jacks						20%			
6260	Congested site area/heavy traffic						20%			
7000	Correctional facilities (no compounding division 1 adjustment factors)									
7010	Minimum security w/facilities escort						30%			
7020	Medium security w/facilities and correctional officer escort						40%			

26 01 Operation and Maintenance of Electrical Systems

26 01 02 – Labor Adjustment

	26 01 02.20 Labor Adjustment Factors	Crew	Daily Output	Labor-Hours	Unit	Material	2019 Bare Costs Labor	Equipment	Total	Total Incl O&P
7030	Max security w/facilities & correctional officer escort (no inmate contact)						50%			

26 05 Common Work Results for Electrical

26 05 33 – Raceway and Boxes for Electrical Systems

	26 05 33.95 Cutting and Drilling	Crew	Daily Output	Labor-Hours	Unit	Material	Labor	Equipment	Total	Total Incl O&P
0010	**CUTTING AND DRILLING**									
0100	Hole drilling to 10′ high, concrete wall									
0110	8″ thick, 1/2″ pipe size	R-31	12	.667	Ea.	.24	40	3.88	44.12	64
0120	3/4″ pipe size		12	.667		.24	40	3.88	44.12	64
0130	1″ pipe size		9.50	.842		.39	50.50	4.90	55.79	81.50
0140	1-1/4″ pipe size		9.50	.842		.39	50.50	4.90	55.79	81.50
0150	1-1/2″ pipe size		9.50	.842		.39	50.50	4.90	55.79	81.50
0160	2″ pipe size		4.40	1.818		.52	109	10.60	120.12	175
0170	2-1/2″ pipe size		4.40	1.818		.52	109	10.60	120.12	175
0180	3″ pipe size		4.40	1.818		.52	109	10.60	120.12	175
0190	3-1/2″ pipe size		3.30	2.424		.68	146	14.10	160.78	233
0200	4″ pipe size		3.30	2.424		.68	146	14.10	160.78	233
0500	12″ thick, 1/2″ pipe size		9.40	.851		.37	51	4.95	56.32	82.50
0520	3/4″ pipe size		9.40	.851		.37	51	4.95	56.32	82.50
0540	1″ pipe size		7.30	1.096		.59	66	6.40	72.99	106
0560	1-1/4″ pipe size		7.30	1.096		.59	66	6.40	72.99	106
0570	1-1/2″ pipe size		7.30	1.096		.59	66	6.40	72.99	106
0580	2″ pipe size		3.60	2.222		.79	133	12.95	146.74	214
0590	2-1/2″ pipe size		3.60	2.222		.79	133	12.95	146.74	214
0600	3″ pipe size		3.60	2.222		.79	133	12.95	146.74	214
0610	3-1/2″ pipe size		2.80	2.857		1.02	172	16.65	189.67	275
0630	4″ pipe size		2.50	3.200		1.02	192	18.60	211.62	310
0650	16″ thick, 1/2″ pipe size		7.60	1.053		.49	63	6.15	69.64	102
0670	3/4″ pipe size		7	1.143		.49	68.50	6.65	75.64	110
0690	1″ pipe size		6	1.333		.79	80	7.75	88.54	128
0710	1-1/4″ pipe size		5.50	1.455		.79	87.50	8.45	96.74	140
0730	1-1/2″ pipe size		5.50	1.455		.79	87.50	8.45	96.74	140
0750	2″ pipe size		3	2.667		1.05	160	15.50	176.55	257
0770	2-1/2″ pipe size		2.70	2.963		1.05	178	17.25	196.30	285
0790	3″ pipe size		2.50	3.200		1.05	192	18.60	211.65	310
0810	3-1/2″ pipe size		2.30	3.478		1.36	209	20	230.36	335
0830	4″ pipe size		2	4		1.36	240	23.50	264.86	385
0850	20″ thick, 1/2″ pipe size		6.40	1.250		.61	75	7.30	82.91	121
0870	3/4″ pipe size		6	1.333		.61	80	7.75	88.36	128
0890	1″ pipe size		5	1.600		.98	96	9.30	106.28	154
0910	1-1/4″ pipe size		4.80	1.667		.98	100	9.70	110.68	161
0930	1-1/2″ pipe size		4.60	1.739		.98	104	10.10	115.08	168
0950	2″ pipe size		2.70	2.963		1.31	178	17.25	196.56	285
0970	2-1/2″ pipe size		2.40	3.333		1.31	200	19.40	220.71	320
0990	3″ pipe size		2.20	3.636		1.31	218	21	240.31	350
1010	3-1/2″ pipe size		2	4		1.70	240	23.50	265.20	385
1030	4″ pipe size		1.70	4.706		1.70	283	27.50	312.20	450
1050	24″ thick, 1/2″ pipe size		5.50	1.455		.73	87.50	8.45	96.68	140
1070	3/4″ pipe size		5.10	1.569		.73	94	9.15	103.88	152
1090	1″ pipe size		4.30	1.860		1.18	112	10.85	124.03	180
1110	1-1/4″ pipe size		4	2		1.18	120	11.65	132.83	193

26 05 33.95 Cutting and Drilling		Crew	Daily Output	Labor-Hours	Unit	Material	2019 Bare Costs Labor	Equipment	Total	Total Incl O&P
1130	1-1/2" pipe size	R-31	4	2	Ea.	1.18	120	11.65	132.83	193
1150	2" pipe size		2.40	3.333		1.57	200	19.40	220.97	320
1170	2-1/2" pipe size		2.20	3.636		1.57	218	21	240.57	350
1190	3" pipe size		2	4		1.57	240	23.50	265.07	385
1210	3-1/2" pipe size		1.80	4.444		2.04	267	26	295.04	430
1230	4" pipe size		1.50	5.333		2.04	320	31	353.04	515
1500	Brick wall, 8" thick, 1/2" pipe size		18	.444		.24	26.50	2.59	29.33	43
1520	3/4" pipe size		18	.444		.24	26.50	2.59	29.33	43
1540	1" pipe size		13.30	.602		.39	36	3.50	39.89	58.50
1560	1-1/4" pipe size		13.30	.602		.39	36	3.50	39.89	58.50
1580	1-1/2" pipe size		13.30	.602		.39	36	3.50	39.89	58.50
1600	2" pipe size		5.70	1.404		.52	84.50	8.15	93.17	136
1620	2-1/2" pipe size		5.70	1.404		.52	84.50	8.15	93.17	136
1640	3" pipe size		5.70	1.404		.52	84.50	8.15	93.17	136
1660	3-1/2" pipe size		4.40	1.818		.68	109	10.60	120.28	175
1680	4" pipe size		4	2		.68	120	11.65	132.33	193
1700	12" thick, 1/2" pipe size		14.50	.552		.37	33	3.21	36.58	53.50
1720	3/4" pipe size		14.50	.552		.37	33	3.21	36.58	53.50
1740	1" pipe size		11	.727		.59	43.50	4.23	48.32	70.50
1760	1-1/4" pipe size		11	.727		.59	43.50	4.23	48.32	70.50
1780	1-1/2" pipe size		11	.727		.59	43.50	4.23	48.32	70.50
1800	2" pipe size		5	1.600		.79	96	9.30	106.09	154
1820	2-1/2" pipe size		5	1.600		.79	96	9.30	106.09	154
1840	3" pipe size		5	1.600		.79	96	9.30	106.09	154
1860	3-1/2" pipe size		3.80	2.105		1.02	126	12.25	139.27	204
1880	4" pipe size		3.30	2.424		1.02	146	14.10	161.12	234
1900	16" thick, 1/2" pipe size		12.30	.650		.49	39	3.79	43.28	63
1920	3/4" pipe size		12.30	.650		.49	39	3.79	43.28	63
1940	1" pipe size		9.30	.860		.79	51.50	5	57.29	83.50
1960	1-1/4" pipe size		9.30	.860		.79	51.50	5	57.29	83.50
1980	1-1/2" pipe size		9.30	.860		.79	51.50	5	57.29	83.50
2000	2" pipe size		4.40	1.818		1.05	109	10.60	120.65	176
2010	2-1/2" pipe size		4.40	1.818		1.05	109	10.60	120.65	176
2030	3" pipe size		4.40	1.818		1.05	109	10.60	120.65	176
2050	3-1/2" pipe size		3.30	2.424		1.36	146	14.10	161.46	234
2070	4" pipe size		3	2.667		1.36	160	15.50	176.86	258
2090	20" thick, 1/2" pipe size		10.70	.748		.61	45	4.35	49.96	72.50
2110	3/4" pipe size		10.70	.748		.61	45	4.35	49.96	72.50
2130	1" pipe size		8	1		.98	60	5.80	66.78	97
2150	1-1/4" pipe size		8	1		.98	60	5.80	66.78	97
2170	1-1/2" pipe size		8	1		.98	60	5.80	66.78	97
2190	2" pipe size		4	2		1.31	120	11.65	132.96	193
2210	2-1/2" pipe size		4	2		1.31	120	11.65	132.96	193
2230	3" pipe size		4	2		1.31	120	11.65	132.96	193
2250	3-1/2" pipe size		3	2.667		1.70	160	15.50	177.20	258
2270	4" pipe size		2.70	2.963		1.70	178	17.25	196.95	286
2290	24" thick, 1/2" pipe size		9.40	.851		.73	51	4.95	56.68	83
2310	3/4" pipe size		9.40	.851		.73	51	4.95	56.68	83
2330	1" pipe size		7.10	1.127		1.18	67.50	6.55	75.23	110
2350	1-1/4" pipe size		7.10	1.127		1.18	67.50	6.55	75.23	110
2370	1-1/2" pipe size		7.10	1.127		1.18	67.50	6.55	75.23	110
2390	2" pipe size		3.60	2.222		1.57	133	12.95	147.52	215
2410	2-1/2" pipe size		3.60	2.222		1.57	133	12.95	147.52	215

26 05 33 – Raceway and Boxes for Electrical Systems

26 05 33.95 Cutting and Drilling		Crew	Daily Output	Labor-Hours	Unit	2019 Bare Costs Material	Labor	Equipment	Total	Total Incl O&P
2430	3" pipe size	R-31	3.60	2.222	Ea.	1.57	133	12.95	147.52	215
2450	3-1/2" pipe size		2.80	2.857		2.04	172	16.65	190.69	277
2470	4" pipe size		2.50	3.200		2.04	192	18.60	212.64	310
3000	Knockouts to 8' high, metal boxes & enclosures									
3020	With hole saw, 1/2" pipe size	1 Elec	53	.151	Ea.		9.05		9.05	13.50
3040	3/4" pipe size		47	.170			10.20		10.20	15.25
3050	1" pipe size		40	.200			12		12	17.90
3060	1-1/4" pipe size		36	.222			13.35		13.35	19.90
3070	1-1/2" pipe size		32	.250			15		15	22.50
3080	2" pipe size		27	.296			17.80		17.80	26.50
3090	2-1/2" pipe size		20	.400			24		24	36
4010	3" pipe size		16	.500			30		30	45
4030	3-1/2" pipe size		13	.615			37		37	55
4050	4" pipe size		11	.727			43.50		43.50	65

26 05 80 – Wiring Connections

26 05 80.10 Motor Connections

Line	Description	Crew	Daily Output	Labor-Hours	Unit	Material	Labor	Equipment	Total	Total Incl O&P
0010	**MOTOR CONNECTIONS**									
0020	Flexible conduit and fittings, 115 volt, 1 phase, up to 1 HP motor	1 Elec	8	1	Ea.	5.70	60		65.70	96
0050	2 HP motor		6.50	1.231		10.15	74		84.15	121
0100	3 HP motor		5.50	1.455		9.05	87.50		96.55	140
0110	230 volt, 3 phase, 3 HP motor		6.78	1.180		6.80	71		77.80	114
0112	5 HP motor		5.47	1.463		5.85	88		93.85	137
0114	7-1/2 HP motor		4.61	1.735		9.35	104		113.35	165
0120	10 HP motor		4.20	1.905		18.25	114		132.25	191
0150	15 HP motor		3.30	2.424		18.25	146		164.25	237
0200	25 HP motor		2.70	2.963		28	178		206	296
0400	50 HP motor		2.20	3.636		55.50	218		273.50	385
0600	100 HP motor		1.50	5.333		118	320		438	610
1500	460 volt, 5 HP motor, 3 phase		8	1		6.15	60		66.15	96.50
1520	10 HP motor		8	1		6.15	60		66.15	96.50
1530	25 HP motor		6	1.333		10.95	80		90.95	131
1540	30 HP motor		6	1.333		10.95	80		90.95	131
1550	40 HP motor		5	1.600		17.65	96		113.65	162
1560	50 HP motor		5	1.600		22.50	96		118.50	168
1570	60 HP motor		3.80	2.105		24	126		150	216
1580	75 HP motor		3.50	2.286		31	137		168	240
1590	100 HP motor		2.50	3.200		48	192		240	340
1600	125 HP motor		2	4		56	240		296	420
1610	150 HP motor		1.80	4.444		59	267		326	465
1620	200 HP motor		1.50	5.333		96.50	320		416.50	585
2005	460 volt, 5 HP motor, 3 phase, w/sealtite		8	1		8.55	60		68.55	99
2010	10 HP motor		8	1		8.55	60		68.55	99
2015	25 HP motor		6	1.333		16.95	80		96.95	138
2020	30 HP motor		6	1.333		16.95	80		96.95	138
2025	40 HP motor		5	1.600		29	96		125	175
2030	50 HP motor		5	1.600		32.50	96		128.50	179
2035	60 HP motor		3.80	2.105		36	126		162	229
2040	75 HP motor		3.50	2.286		42	137		179	251
2045	100 HP motor		2.50	3.200		75	192		267	370
2055	150 HP motor		1.80	4.444		78	267		345	485
2060	200 HP motor		1.50	5.333		269	320		589	775

26 05 90.10 Residential Wiring		Crew	Daily Output	Labor-Hours	Unit	Material	Labor	Equipment	Total	Total Incl O&P
						2019 Bare Costs				
0010	**RESIDENTIAL WIRING**									
0020	20' avg. runs and #14/2 wiring incl. unless otherwise noted									
1000	Service & panel, includes 24' SE-AL cable, service eye, meter,									
1010	Socket, panel board, main bkr., ground rod, 15 or 20 amp									
1020	1-pole circuit breakers, and misc. hardware									
1100	100 amp, with 10 branch breakers	1 Elec	1.19	6.723	Ea.	325	405		730	960
1110	With PVC conduit and wire		.92	8.696		360	520		880	1,175
1120	With RGS conduit and wire		.73	10.959		575	660		1,235	1,600
1150	150 amp, with 14 branch breakers		1.03	7.767		845	465		1,310	1,625
1170	With PVC conduit and wire		.82	9.756		915	585		1,500	1,875
1180	With RGS conduit and wire		.67	11.940		1,225	715		1,940	2,425
1200	200 amp, with 18 branch breakers	2 Elec	1.80	8.889		1,000	535		1,535	1,900
1220	With PVC conduit and wire		1.46	10.959		1,075	660		1,735	2,150
1230	With RGS conduit and wire		1.24	12.903		1,450	775		2,225	2,750
1800	Lightning surge suppressor	1 Elec	32	.250		75	15		90	105
2000	Switch devices									
2100	Single pole, 15 amp, ivory, with a 1-gang box, cover plate,									
2110	Type NM (Romex) cable	1 Elec	17.10	.468	Ea.	15.50	28		43.50	59
2120	Type MC cable		14.30	.559		26	33.50		59.50	78.50
2130	EMT & wire		5.71	1.401		37.50	84		121.50	167
2150	3-way, #14/3, type NM cable		14.55	.550		9.80	33		42.80	60.50
2170	Type MC cable		12.31	.650		23.50	39		62.50	84
2180	EMT & wire		5	1.600		31.50	96		127.50	178
2200	4-way, #14/3, type NM cable		14.55	.550		18.50	33		51.50	70
2220	Type MC cable		12.31	.650		32.50	39		71.50	93.50
2230	EMT & wire		5	1.600		40.50	96		136.50	188
2250	S.P., 20 amp, #12/2, type NM cable		13.33	.600		11.30	36		47.30	66.50
2270	Type MC cable		11.43	.700		21	42		63	85.50
2280	EMT & wire		4.85	1.649		35.50	99		134.50	187
2290	S.P. rotary dimmer, 600 W, no wiring		17	.471		31.50	28.50		60	76.50
2300	S.P. rotary dimmer, 600 W, type NM cable		14.55	.550		35.50	33		68.50	88.50
2320	Type MC cable		12.31	.650		46	39		85	109
2330	EMT & wire		5	1.600		58.50	96		154.50	207
2350	3-way rotary dimmer, type NM cable		13.33	.600		23.50	36		59.50	79.50
2370	Type MC cable		11.43	.700		34	42		76	100
2380	EMT & wire		4.85	1.649		46.50	99		145.50	199
2400	Interval timer wall switch, 20 amp, 1-30 min., #12/2									
2410	Type NM cable	1 Elec	14.55	.550	Ea.	58.50	33		91.50	114
2420	Type MC cable		12.31	.650		64.50	39		103.50	129
2430	EMT & wire		5	1.600		82.50	96		178.50	234
2500	Decorator style									
2510	S.P., 15 amp, type NM cable	1 Elec	17.10	.468	Ea.	20.50	28		48.50	65
2520	Type MC cable		14.30	.559		31.50	33.50		65	84.50
2530	EMT & wire		5.71	1.401		42.50	84		126.50	173
2550	3-way, #14/3, type NM cable		14.55	.550		15	33		48	66
2570	Type MC cable		12.31	.650		29	39		68	90
2580	EMT & wire		5	1.600		37	96		133	184
2600	4-way, #14/3, type NM cable		14.55	.550		23.50	33		56.50	75.50
2620	Type MC cable		12.31	.650		37.50	39		76.50	99.50
2630	EMT & wire		5	1.600		45.50	96		141.50	193
2650	S.P., 20 amp, #12/2, type NM cable		13.33	.600		16.55	36		52.55	72
2670	Type MC cable		11.43	.700		26	42		68	91

26 05 90.10 Residential Wiring		Crew	Daily Output	Labor-Hours	Unit	Material	Labor	Equipment	Total	Total Incl O&P
						2019 Bare Costs				
2680	EMT & wire	1 Elec	4.85	1.649	Ea.	41	99		140	193
2700	S.P., slide dimmer, type NM cable		17.10	.468		37	28		65	83
2720	Type MC cable		14.30	.559		47.50	33.50		81	103
2730	EMT & wire		5.71	1.401		60	84		144	192
2750	S.P., touch dimmer, type NM cable		17.10	.468		48.50	28		76.50	95
2770	Type MC cable		14.30	.559		59	33.50		92.50	115
2780	EMT & wire		5.71	1.401		71.50	84		155.50	205
2800	3-way touch dimmer, type NM cable		13.33	.600		49.50	36		85.50	109
2820	Type MC cable		11.43	.700		60.50	42		102.50	129
2830	EMT & wire		4.85	1.649		73	99		172	228
3000	Combination devices									
3100	S.P. switch/15 amp recpt., ivory, 1-gang box, plate									
3110	Type NM cable	1 Elec	11.43	.700	Ea.	23	42		65	87.50
3120	Type MC cable		10	.800		33.50	48		81.50	109
3130	EMT & wire		4.40	1.818		46	109		155	214
3150	S.P. switch/pilot light, type NM cable		11.43	.700		24	42		66	89
3170	Type MC cable		10	.800		35	48		83	110
3180	EMT & wire		4.43	1.806		47.50	108		155.50	214
3190	2-S.P. switches, 2-#14/2, no wiring		14	.571		13.25	34.50		47.75	65.50
3200	2-S.P. switches, 2-#14/2, type NM cables		10	.800		25	48		73	99
3220	Type MC cable		8.89	.900		40	54		94	125
3230	EMT & wire		4.10	1.951		48	117		165	228
3250	3-way switch/15 amp recpt., #14/3, type NM cable		10	.800		30.50	48		78.50	105
3270	Type MC cable		8.89	.900		44.50	54		98.50	130
3280	EMT & wire		4.10	1.951		52.50	117		169.50	233
3300	2-3 way switches, 2-#14/3, type NM cables		8.89	.900		39	54		93	124
3320	Type MC cable		8	1		60.50	60		120.50	156
3330	EMT & wire		4	2		59.50	120		179.50	244
3350	S.P. switch/20 amp recpt., #12/2, type NM cable		10	.800		39	48		87	114
3370	Type MC cable		8.89	.900		44.50	54		98.50	130
3380	EMT & wire		4.10	1.951		63	117		180	245
3400	Decorator style									
3410	S.P. switch/15 amp recpt., type NM cable	1 Elec	11.43	.700	Ea.	28	42		70	93.50
3420	Type MC cable		10	.800		38.50	48		86.50	114
3430	EMT & wire		4.40	1.818		51	109		160	220
3450	S.P. switch/pilot light, type NM cable		11.43	.700		29.50	42		71.50	95
3470	Type MC cable		10	.800		40	48		88	116
3480	EMT & wire		4.40	1.818		52.50	109		161.50	221
3500	2-S.P. switches, 2-#14/2, type NM cables		10	.800		30	48		78	105
3520	Type MC cable		8.89	.900		45	54		99	130
3530	EMT & wire		4.10	1.951		53	117		170	234
3550	3-way/15 amp recpt., #14/3, type NM cable		10	.800		36	48		84	111
3570	Type MC cable		8.89	.900		49.50	54		103.50	135
3580	EMT & wire		4.10	1.951		57.50	117		174.50	239
3650	2-3 way switches, 2-#14/3, type NM cables		8.89	.900		44.50	54		98.50	129
3670	Type MC cable		8	1		66	60		126	162
3680	EMT & wire		4	2		64.50	120		184.50	250
3700	S.P. switch/20 amp recpt., #12/2, type NM cable		10	.800		44	48		92	120
3720	Type MC cable		8.89	.900		50	54		104	136
3730	EMT & wire		4.10	1.951		68.50	117		185.50	250
4000	Receptacle devices									
4010	Duplex outlet, 15 amp recpt., ivory, 1-gang box, plate									
4015	Type NM cable	1 Elec	14.55	.550	Ea.	8.70	33		41.70	59

26 05 90.10 Residential Wiring		Crew	Daily Output	Labor-Hours	Unit	Material	Labor	Equipment	Total	Total Incl O&P
						2019 Bare Costs				
4020	Type MC cable	1 Elec	12.31	.650	Ea.	19.30	39		58.30	79
4030	EMT & wire		5.33	1.501		30.50	90		120.50	168
4050	With #12/2, type NM cable		12.31	.650		9.85	39		48.85	69
4070	Type MC cable		10.67	.750		19.40	45		64.40	88.50
4080	EMT & wire		4.71	1.699		34	102		136	190
4100	20 amp recpt., #12/2, type NM cable		12.31	.650		17.85	39		56.85	77.50
4120	Type MC cable		10.67	.750		27.50	45		72.50	97
4130	EMT & wire		4.71	1.699		42	102		144	199
4140	For GFI see Section 26 05 90.10 line 4300 below									
4150	Decorator style, 15 amp recpt., type NM cable	1 Elec	14.55	.550	Ea.	13.95	33		46.95	65
4170	Type MC cable		12.31	.650		24.50	39		63.50	85
4180	EMT & wire		5.33	1.501		36	90		126	174
4200	With #12/2, type NM cable		12.31	.650		15.10	39		54.10	74.50
4220	Type MC cable		10.67	.750		24.50	45		69.50	94
4230	EMT & wire		4.71	1.699		39.50	102		141.50	196
4250	20 amp recpt., #12/2, type NM cable		12.31	.650		23	39		62	83.50
4270	Type MC cable		10.67	.750		32.50	45		77.50	103
4280	EMT & wire		4.71	1.699		47.50	102		149.50	204
4300	GFI, 15 amp recpt., type NM cable		12.31	.650		21	39		60	81.50
4320	Type MC cable		10.67	.750		32	45		77	102
4330	EMT & wire		4.71	1.699		43	102		145	200
4350	GFI with #12/2, type NM cable		10.67	.750		22.50	45		67.50	91.50
4370	Type MC cable		9.20	.870		32	52		84	113
4380	EMT & wire		4.21	1.900		46.50	114		160.50	222
4400	20 amp recpt., #12/2, type NM cable		10.67	.750		49	45		94	121
4420	Type MC cable		9.20	.870		58.50	52		110.50	142
4430	EMT & wire		4.21	1.900		73	114		187	251
4500	Weather-proof cover for above receptacles, add		32	.250		2.06	15		17.06	25
4550	Air conditioner outlet, 20 amp-240 volt recpt.									
4560	30′ of #12/2, 2 pole circuit breaker									
4570	Type NM cable	1 Elec	10	.800	Ea.	60	48		108	138
4580	Type MC cable		9	.889		71	53.50		124.50	158
4590	EMT & wire		4	2		85	120		205	272
4600	Decorator style, type NM cable		10	.800		65	48		113	143
4620	Type MC cable		9	.889		76	53.50		129.50	163
4630	EMT & wire		4	2		89.50	120		209.50	278
4650	Dryer outlet, 30 amp-240 volt recpt., 20′ of #10/3									
4660	2 pole circuit breaker									
4670	Type NM cable	1 Elec	6.41	1.248	Ea.	54	75		129	172
4680	Type MC cable		5.71	1.401		63.50	84		147.50	196
4690	EMT & wire		3.48	2.299		76	138		214	290
4700	Range outlet, 50 amp-240 volt recpt., 30′ of #8/3									
4710	Type NM cable	1 Elec	4.21	1.900	Ea.	82.50	114		196.50	261
4720	Type MC cable		4	2		127	120		247	320
4730	EMT & wire		2.96	2.703		107	162		269	360
4750	Central vacuum outlet, type NM cable		6.40	1.250		53.50	75		128.50	171
4770	Type MC cable		5.71	1.401		67.50	84		151.50	201
4780	EMT & wire		3.48	2.299		88	138		226	305
4800	30 amp-110 volt locking recpt., #10/2 circ. bkr.									
4810	Type NM cable	1 Elec	6.20	1.290	Ea.	63	77.50		140.50	185
4820	Type MC cable		5.40	1.481		80	89		169	221
4830	EMT & wire		3.20	2.500		99	150		249	335
4900	Low voltage outlets									

26 05 90 – Residential Applications

26 05 90.10 Residential Wiring		Crew	Daily Output	Labor-Hours	Unit	2019 Bare Costs Material	Labor	Equipment	Total	Total Incl O&P
4910	Telephone recpt., 20′ of 4/C phone wire	1 Elec	26	.308	Ea.	8.55	18.50		27.05	37
4920	TV recpt., 20′ of RG59U coax wire, F type connector	"	16	.500	"	17.10	30		47.10	64
4950	Door bell chime, transformer, 2 buttons, 60′ of bellwire									
4970	Economy model	1 Elec	11.50	.696	Ea.	55	42		97	123
4980	Custom model		11.50	.696		105	42		147	179
4990	Luxury model, 3 buttons		9.50	.842		188	50.50		238.50	283
6000	Lighting outlets									
6050	Wire only (for fixture), type NM cable	1 Elec	32	.250	Ea.	5.80	15		20.80	29
6070	Type MC cable		24	.333		11.65	20		31.65	43
6080	EMT & wire		10	.800		22	48		70	96
6100	Box (4″), and wire (for fixture), type NM cable		25	.320		14.85	19.20		34.05	45
6120	Type MC cable		20	.400		20.50	24		44.50	59
6130	EMT & wire		11	.727		31.50	43.50		75	99.50
6200	Fixtures (use with line 6050 or 6100 above)									
6210	Canopy style, economy grade	1 Elec	40	.200	Ea.	22.50	12		34.50	43
6220	Custom grade		40	.200		52.50	12		64.50	75.50
6250	Dining room chandelier, economy grade		19	.421		83	25.50		108.50	129
6260	Custom grade		19	.421		335	25.50		360.50	405
6270	Luxury grade		15	.533		1,250	32		1,282	1,425
6310	Kitchen fixture (fluorescent), economy grade		30	.267		71.50	16		87.50	103
6320	Custom grade		25	.320		149	19.20		168.20	193
6350	Outdoor, wall mounted, economy grade		30	.267		29.50	16		45.50	56.50
6360	Custom grade		30	.267		120	16		136	156
6370	Luxury grade		25	.320		243	19.20		262.20	296
6410	Outdoor PAR floodlights, 1 lamp, 150 watt		20	.400		26.50	24		50.50	65.50
6420	2 lamp, 150 watt each		20	.400		45	24		69	85.50
6430	For infrared security sensor, add		32	.250		102	15		117	135
6450	Outdoor, quartz-halogen, 300 watt flood		20	.400		38.50	24		62.50	78
6600	Recessed downlight, round, pre-wired, 50 or 75 watt trim		30	.267		66	16		82	96.50
6610	With shower light trim		30	.267		90	16		106	123
6620	With wall washer trim		28	.286		92	17.15		109.15	127
6630	With eye-ball trim		28	.286		77	17.15		94.15	111
6700	Porcelain lamp holder		40	.200		3	12		15	21
6710	With pull switch		40	.200		11.30	12		23.30	30.50
6750	Fluorescent strip, 2-20 watt tube, wrap around diffuser, 24″		24	.333		48	20		68	82.50
6770	2-34 watt tubes, 48″		20	.400		158	24		182	209
6800	Bathroom heat lamp, 1-250 watt		28	.286		31.50	17.15		48.65	60.50
6810	2-250 watt lamps		28	.286		76.50	17.15		93.65	110
6820	For timer switch, see Section 26 05 90.10 line 2400									
6900	Outdoor post lamp, incl. post, fixture, 35′ of #14/2									
6910	Type NM cable	1 Elec	3.50	2.286	Ea.	299	137		436	535
6920	Photo-eye, add		27	.296		28.50	17.80		46.30	57.50
6950	Clock dial time switch, 24 hr., w/enclosure, type NM cable		11.43	.700		78.50	42		120.50	149
6970	Type MC cable		11	.727		89	43.50		132.50	163
6980	EMT & wire		4.85	1.649		100	99		199	259
7000	Alarm systems									
7050	Smoke detectors, box, #14/3, type NM cable	1 Elec	14.55	.550	Ea.	34.50	33		67.50	87.50
7070	Type MC cable		12.31	.650		44.50	39		83.50	107
7080	EMT & wire		5	1.600		52.50	96		148.50	201
7090	For relay output to security system, add					11.15			11.15	12.25
8000	Residential equipment									
8050	Disposal hook-up, incl. switch, outlet box, 3′ of flex									
8060	20 amp-1 pole circ. bkr., and 25′ of #12/2									

26 05 90 – Residential Applications

26 05 90.10 Residential Wiring		Crew	Daily Output	Labor-Hours	Unit	Material	Labor	Equipment	Total	Total Incl O&P
						2019 Bare Costs				
8070	Type NM cable	1 Elec	10	.800	Ea.	28	48		76	103
8080	Type MC cable		8	1		38.50	60		98.50	132
8090	EMT & wire	↓	5	1.600	↓	56	96		152	205
8100	Trash compactor or dishwasher hook-up, incl. outlet box,									
8110	3′ of flex, 15 amp-1 pole circ. bkr., and 25′ of #14/2									
8120	Type NM cable	1 Elec	10	.800	Ea.	15.60	48		63.60	88.50
8130	Type MC cable		8	1		27.50	60		87.50	120
8140	EMT & wire	↓	5	1.600	↓	42.50	96		138.50	190
8150	Hot water sink dispenser hook-up, use line 8100									
8200	Vent/exhaust fan hook-up, type NM cable	1 Elec	32	.250	Ea.	5.80	15		20.80	29
8220	Type MC cable		24	.333		11.65	20		31.65	43
8230	EMT & wire	↓	10	.800	↓	22	48		70	96
8250	Bathroom vent fan, 50 CFM (use with above hook-up)									
8260	Economy model	1 Elec	15	.533	Ea.	19.45	32		51.45	69.50
8270	Low noise model		15	.533		45.50	32		77.50	98
8280	Custom model	↓	12	.667	↓	120	40		160	192
8300	Bathroom or kitchen vent fan, 110 CFM									
8310	Economy model	1 Elec	15	.533	Ea.	67	32		99	122
8320	Low noise model	"	15	.533	"	93.50	32		125.50	151
8350	Paddle fan, variable speed (w/o lights)									
8360	Economy model (AC motor)	1 Elec	10	.800	Ea.	131	48		179	216
8362	With light kit		10	.800		171	48		219	260
8370	Custom model (AC motor)		10	.800		340	48		388	445
8372	With light kit		10	.800		380	48		428	490
8380	Luxury model (DC motor)		8	1		330	60		390	450
8382	With light kit		8	1		370	60		430	495
8390	Remote speed switch for above, add	↓	12	.667	↓	34	40		74	96.50
8500	Whole house exhaust fan, ceiling mount, 36″, variable speed									
8510	Remote switch, incl. shutters, 20 amp-1 pole circ. bkr.									
8520	30′ of #12/2, type NM cable	1 Elec	4	2	Ea.	1,325	120		1,445	1,650
8530	Type MC cable		3.50	2.286		1,350	137		1,487	1,675
8540	EMT & wire	↓	3	2.667	↓	1,375	160		1,535	1,750
8600	Whirlpool tub hook-up, incl. timer switch, outlet box									
8610	3′ of flex, 20 amp-1 pole GFI circ. bkr.									
8620	30′ of #12/2, type NM cable	1 Elec	5	1.600	Ea.	128	96		224	283
8630	Type MC cable		4.20	1.905		135	114		249	320
8640	EMT & wire	↓	3.40	2.353	↓	150	141		291	375
8650	Hot water heater hook-up, incl. 1-2 pole circ. bkr., box;									
8660	3′ of flex, 20′ of #10/2, type NM cable	1 Elec	5	1.600	Ea.	28.50	96		124.50	174
8670	Type MC cable		4.20	1.905		41.50	114		155.50	217
8680	EMT & wire	↓	3.40	2.353	↓	49.50	141		190.50	266
9000	Heating/air conditioning									
9050	Furnace/boiler hook-up, incl. firestat, local on-off switch									
9060	Emergency switch, and 40′ of type NM cable	1 Elec	4	2	Ea.	53	120		173	237
9070	Type MC cable		3.50	2.286		68	137		205	280
9080	EMT & wire	↓	1.50	5.333	↓	92	320		412	580
9100	Air conditioner hook-up, incl. local 60 amp disc. switch									
9110	3′ sealtite, 40 amp, 2 pole circuit breaker									
9130	40′ of #8/2, type NM cable	1 Elec	3.50	2.286	Ea.	170	137		307	390
9140	Type MC cable		3	2.667		235	160		395	495
9150	EMT & wire	↓	1.30	6.154	↓	218	370		588	790
9200	Heat pump hook-up, 1-40 & 1-100 amp 2 pole circ. bkr.									
9210	Local disconnect switch, 3′ sealtite									

26 05 Common Work Results for Electrical

26 05 90 – Residential Applications

26 05 90.10 Residential Wiring		Crew	Daily Output	Labor-Hours	Unit	Material	2019 Bare Costs Labor	Equipment	Total	Total Incl O&P
9220	40′ of #8/2 & 30′ of #3/2									
9230	Type NM cable	1 Elec	1.30	6.154	Ea.	505	370		875	1,100
9240	Type MC cable		1.08	7.407		555	445		1,000	1,275
9250	EMT & wire		.94	8.511		520	510		1,030	1,350
9500	Thermostat hook-up, using low voltage wire									
9520	Heating only, 25′ of #18-3	1 Elec	24	.333	Ea.	6.20	20		26.20	37
9530	Heating/cooling, 25′ of #18-4	"	20	.400	"	8.15	24		32.15	45

26 24 Switchboards and Panelboards

26 24 19 – Motor-Control Centers

26 24 19.40 Motor Starters and Controls		Crew	Daily Output	Labor-Hours	Unit	Material	2019 Bare Costs Labor	Equipment	Total	Total Incl O&P
0010	**MOTOR STARTERS AND CONTROLS**									
0050	Magnetic, FVNR, with enclosure and heaters, 480 volt									
0080	2 HP, size 00	1 Elec	3.50	2.286	Ea.	199	137		336	425
0100	5 HP, size 0		2.30	3.478		290	209		499	630
0200	10 HP, size 1		1.60	5		266	300		566	740
0300	25 HP, size 2	2 Elec	2.20	7.273		500	435		935	1,200
0400	50 HP, size 3		1.80	8.889		815	535		1,350	1,700
0500	100 HP, size 4		1.20	13.333		1,800	800		2,600	3,175
0600	200 HP, size 5		.90	17.778		4,225	1,075		5,300	6,250
0610	400 HP, size 6		.80	20		18,100	1,200		19,300	21,700
0620	NEMA 7, 5 HP, size 0	1 Elec	1.60	5		1,350	300		1,650	1,925
0630	10 HP, size 1	"	1.10	7.273		1,400	435		1,835	2,200
0640	25 HP, size 2	2 Elec	1.80	8.889		2,275	535		2,810	3,300
0650	50 HP, size 3		1.20	13.333		3,425	800		4,225	4,950
0660	100 HP, size 4		.90	17.778		5,500	1,075		6,575	7,675
0670	200 HP, size 5		.50	32		13,200	1,925		15,125	17,400
0700	Combination, with motor circuit protectors, 5 HP, size 0	1 Elec	1.80	4.444		990	267		1,257	1,500
0800	10 HP, size 1	"	1.30	6.154		1,025	370		1,395	1,675
0900	25 HP, size 2	2 Elec	2	8		1,425	480		1,905	2,300
1000	50 HP, size 3		1.32	12.121		2,100	730		2,830	3,375
1200	100 HP, size 4		.80	20		4,500	1,200		5,700	6,750
1220	NEMA 7, 5 HP, size 0	1 Elec	1.30	6.154		3,050	370		3,420	3,925
1230	10 HP, size 1	"	1	8		3,125	480		3,605	4,175
1240	25 HP, size 2	2 Elec	1.32	12.121		4,175	730		4,905	5,675
1250	50 HP, size 3		.80	20		6,900	1,200		8,100	9,400
1260	100 HP, size 4		.60	26.667		10,800	1,600		12,400	14,200
1270	200 HP, size 5		.40	40		23,400	2,400		25,800	29,300
1400	Combination, with fused switch, 5 HP, size 0	1 Elec	1.80	4.444		595	267		862	1,050
1600	10 HP, size 1	"	1.30	6.154		640	370		1,010	1,250
1800	25 HP, size 2	2 Elec	2	8		1,025	480		1,505	1,875
2000	50 HP, size 3		1.32	12.121		1,750	730		2,480	3,000
2200	100 HP, size 4		.80	20		3,050	1,200		4,250	5,175
3500	Magnetic FVNR with NEMA 12, enclosure & heaters, 480 volt									
3600	5 HP, size 0	1 Elec	2.20	3.636	Ea.	240	218		458	590
3700	10 HP, size 1	"	1.50	5.333		360	320		680	880
3800	25 HP, size 2	2 Elec	2	8		675	480		1,155	1,450
3900	50 HP, size 3		1.60	10		1,050	600		1,650	2,050
4000	100 HP, size 4		1	16		2,475	960		3,435	4,150
4100	200 HP, size 5		.80	20		5,975	1,200		7,175	8,375
4200	Combination, with motor circuit protectors, 5 HP, size 0	1 Elec	1.70	4.706		795	283		1,078	1,300

26 24 Switchboards and Panelboards

26 24 19 – Motor-Control Centers

26 24 19.40 Motor Starters and Controls		Crew	Daily Output	Labor-Hours	Unit	2019 Bare Costs Material	Labor	Equipment	Total	Total Incl O&P
4300	10 HP, size 1	1 Elec	1.20	6.667	Ea.	825	400		1,225	1,500
4400	25 HP, size 2	2 Elec	1.80	8.889		1,225	535		1,760	2,150
4500	50 HP, size 3		1.20	13.333		2,025	800		2,825	3,425
4600	100 HP, size 4		.74	21.622		4,550	1,300		5,850	6,925
4700	Combination, with fused switch, 5 HP, size 0	1 Elec	1.70	4.706		715	283		998	1,200
4800	10 HP, size 1	"	1.20	6.667		745	400		1,145	1,425
4900	25 HP, size 2	2 Elec	1.80	8.889		1,125	535		1,660	2,050
5000	50 HP, size 3		1.20	13.333		1,825	800		2,625	3,200
5100	100 HP, size 4		.74	21.622		3,700	1,300		5,000	5,975
5200	Factory installed controls, adders to size 0 thru 5									
5300	Start-stop push button	1 Elec	32	.250	Ea.	50.50	15		65.50	78
5400	Hand-off-auto-selector switch		32	.250		50.50	15		65.50	78
5500	Pilot light		32	.250		94.50	15		109.50	127
5600	Start-stop-pilot		32	.250		145	15		160	182
5700	Auxiliary contact, NO or NC		32	.250		69	15		84	98.50
5800	NO-NC		32	.250		138	15		153	175

26 29 Low-Voltage Controllers

26 29 13 – Enclosed Controllers

26 29 13.20 Control Stations

Line	Description	Crew	Daily Output	Labor-Hours	Unit	Material	Labor	Equipment	Total	Total Incl O&P
0010	**CONTROL STATIONS**									
0050	NEMA 1, heavy duty, stop/start	1 Elec	8	1	Ea.	146	60		206	250
0100	Stop/start, pilot light		6.20	1.290		199	77.50		276.50	335
0200	Hand/off/automatic		6.20	1.290		108	77.50		185.50	235
0400	Stop/start/reverse		5.30	1.509		196	90.50		286.50	350
0500	NEMA 7, heavy duty, stop/start		6	1.333		540	80		620	715
0600	Stop/start, pilot light		4	2		655	120		775	905

26 29 23 – Variable-Frequency Motor Controllers

26 29 23.10 Variable Frequency Drives/Adj. Frequency Drives

Line	Description		Crew	Daily Output	Labor-Hours	Unit	Material	Labor	Equipment	Total	Total Incl O&P
0010	**VARIABLE FREQUENCY DRIVES/ADJ. FREQUENCY DRIVES**										
0100	Enclosed (NEMA 1), 460 volt, for 3 HP motor size	G	1 Elec	.80	10	Ea.	1,975	600		2,575	3,075
0110	5 HP motor size	G		.80	10		2,175	600		2,775	3,275
0120	7.5 HP motor size	G		.67	11.940		2,550	715		3,265	3,875
0130	10 HP motor size	G		.67	11.940		2,800	715		3,515	4,175
0140	15 HP motor size	G	2 Elec	.89	17.978		3,625	1,075		4,700	5,600
0150	20 HP motor size	G		.89	17.978		4,450	1,075		5,525	6,500
0160	25 HP motor size	G		.67	23.881		5,250	1,425		6,675	7,925
0170	30 HP motor size	G		.67	23.881		6,550	1,425		7,975	9,350
0180	40 HP motor size	G		.67	23.881		7,550	1,425		8,975	10,500
0190	50 HP motor size	G		.53	30.189		10,100	1,825		11,925	13,800
0200	60 HP motor size	G	R-3	.56	35.714		11,800	2,125	232	14,157	16,300
0210	75 HP motor size	G		.56	35.714		13,900	2,125	232	16,257	18,700
0220	100 HP motor size	G		.50	40		17,400	2,400	260	20,060	23,000
0230	125 HP motor size	G		.50	40		18,900	2,400	260	21,560	24,700
0240	150 HP motor size	G		.50	40		23,900	2,400	260	26,560	30,200
0250	200 HP motor size	G		.42	47.619		26,100	2,850	310	29,260	33,300
1100	Custom-engineered, 460 volt, for 3 HP motor size	G	1 Elec	.56	14.286		3,100	860		3,960	4,675
1110	5 HP motor size	G		.56	14.286		3,075	860		3,935	4,650
1120	7.5 HP motor size	G		.47	17.021		3,250	1,025		4,275	5,100
1130	10 HP motor size	G		.47	17.021		3,400	1,025		4,425	5,275

26 29 23 – Variable-Frequency Motor Controllers

26 29 23.10 Variable Frequency Drives/Adj. Frequency Drives			Crew	Daily Output	Labor-Hours	Unit	2019 Bare Costs Material	Labor	Equipment	Total	Total Incl O&P
1140	15 HP motor size	G	2 Elec	.62	25.806	Ea.	4,250	1,550		5,800	6,975
1150	20 HP motor size	G		.62	25.806		4,500	1,550		6,050	7,250
1160	25 HP motor size	G		.47	34.043		5,625	2,050		7,675	9,225
1170	30 HP motor size	G		.47	34.043		6,975	2,050		9,025	10,700
1180	40 HP motor size	G		.47	34.043		8,250	2,050		10,300	12,200
1190	50 HP motor size	G		.37	43.243		9,450	2,600		12,050	14,300
1200	60 HP motor size	G	R-3	.39	51.282		14,300	3,075	335	17,710	20,600
1210	75 HP motor size	G		.39	51.282		15,100	3,075	335	18,510	21,500
1220	100 HP motor size	G		.35	57.143		16,600	3,425	370	20,395	23,700
1230	125 HP motor size	G		.35	57.143		17,800	3,425	370	21,595	25,000
1240	150 HP motor size	G		.35	57.143		20,400	3,425	370	24,195	27,900
1250	200 HP motor size	G		.29	68.966		26,500	4,125	450	31,075	35,700
2000	For complex & special design systems to meet specific										
2010	requirements, obtain quote from vendor.										

Division Notes

	CREW	DAILY OUTPUT	LABOR-HOURS	UNIT	BARE COSTS MAT.	LABOR	EQUIP.	TOTAL	TOTAL INCL O&P

Division 28 Electronic Safety & Security

Estimating Tips

- When estimating material costs for electronic safety and security systems, it is always prudent to obtain manufacturers' quotations for equipment prices and special installation requirements that may affect the total cost.
- Fire alarm systems consist of control panels, annunciator panels, batteries with rack, charger, and fire alarm actuating and indicating devices. Some fire alarm systems include speakers, telephone lines, door closer controls, and other components. Be careful not to overlook the costs related to installation for these items. Also be aware of costs for integrated automation instrumentation and terminal devices, control equipment, control wiring, and programming. Insurance underwriters may have specific requirements for the type of materials to be installed or design requirements based on the hazard to be protected. Local jurisdictions may have requirements not covered by code. It is advisable to be aware of any special conditions.
- Security equipment includes items such as CCTV, access control, and other detection and identification systems to perform alert and alarm functions. Be sure to consider the costs related to installation for this security equipment, such as for integrated automation instrumentation and terminal devices, control equipment, control wiring, and programming.

Reference Numbers

Reference numbers are shown at the beginning of some major classifications. These numbers refer to related items in the Reference Section. The reference information may be an estimating procedure, an alternate pricing method, or technical information.

Note: Trade Service, in part, has been used as a reference source for some of the material prices used in Division 28.

28 18 Security Access Detection Equipment

28 18 11 – Security Access Metal Detectors

28 18 11.13 Security Access Metal Detectors		Crew	Daily Output	Labor-Hours	Unit	2019 Bare Costs Material	Labor	Equipment	Total	Total Incl O&P
0010	**SECURITY ACCESS METAL DETECTORS**									
0240	Metal detector, hand-held, wand type, unit only				Ea.	129			129	138
0250	Metal detector, walk through portal type, single zone	1 Elec	2	4		3,300	240		3,540	4,000
0260	Multi-zone	″	2	4	↓	4,500	240		4,740	5,300

28 18 13 – Security Access X-Ray Equipment

28 18 13.16 Security Access X-Ray Equipment		Crew	Daily Output	Labor-Hours	Unit	Material	Labor	Equipment	Total	Total Incl O&P
0010	**SECURITY ACCESS X-RAY EQUIPMENT**									
0290	X-ray machine, desk top, for mail/small packages/letters	1 Elec	4	2	Ea.	3,450	120		3,570	3,950
0300	Conveyor type, incl. monitor		2	4		15,800	240		16,040	17,800
0310	Includes additional features	↓	2	4		27,700	240		27,940	30,800
0320	X-ray machine, large unit, for airports, incl. monitor	2 Elec	1	16		39,600	960		40,560	44,900
0330	Full console	″	.50	32	↓	68,000	1,925		69,925	77,500

28 18 15 – Security Access Explosive Detection Equipment

28 18 15.23 Security Access Explosive Detection Equipment		Crew	Daily Output	Labor-Hours	Unit	Material	Labor	Equipment	Total	Total Incl O&P
0010	**SECURITY ACCESS EXPLOSIVE DETECTION EQUIPMENT**									
0270	Explosives detector, walk through portal type	1 Elec	2	4	Ea.	43,700	240		43,940	48,500
0280	Hand-held, battery operated				″				25,500	28,100

28 23 Video Management System

28 23 23 – Video Surveillance Systems Infrastructure

28 23 23.50 Video Surveillance Equipments		Crew	Daily Output	Labor-Hours	Unit	Material	Labor	Equipment	Total	Total Incl O&P
0010	**VIDEO SURVEILLANCE EQUIPMENTS**									
0200	Video cameras, wireless, hidden in exit signs, clocks, etc., incl. receiver	1 Elec	3	2.667	Ea.	171	160		331	425
0210	Accessories for video recorder, single camera		3	2.667		175	160		335	430
0220	For multiple cameras		3	2.667		1,675	160		1,835	2,100
0230	Video cameras, wireless, for under vehicle searching, complete		2	4		15,100	240		15,340	17,000
0234	Master monitor station, 3 doors x 5 color monitor with tilt feature	↓	2	4	↓	935	240		1,175	1,375

28 31 Intrusion Detection

28 31 16 – Intrusion Detection Systems Infrastructure

28 31 16.50 Intrusion Detection		Crew	Daily Output	Labor-Hours	Unit	Material	Labor	Equipment	Total	Total Incl O&P
0010	**INTRUSION DETECTION**, not including wires & conduits									
0100	Burglar alarm, battery operated, mechanical trigger	1 Elec	4	2	Ea.	281	120		401	490
0200	Electrical trigger		4	2		335	120		455	550
0400	For outside key control, add		8	1		87	60		147	185
0600	For remote signaling circuitry, add		8	1		144	60		204	249
0800	Card reader, flush type, standard		2.70	2.963		815	178		993	1,150
1000	Multi-code	↓	2.70	2.963	↓	1,225	178		1,403	1,600

28 42 Gas Detection and Alarm

28 42 15 – Gas Detection Sensors

28 42 15.50 Tank Leak Detection Systems		Crew	Daily Output	Labor-Hours	Unit	Material	2019 Bare Costs Labor	Equipment	Total	Total Incl O&P
0010	**TANK LEAK DETECTION SYSTEMS** Liquid and vapor									
0100	For hydrocarbons and hazardous liquids/vapors									
0120	Controller, data acquisition, incl. printer, modem, RS232 port									
0140	24 channel, for use with all probes				Ea.	2,775			2,775	3,050
0160	9 channel, for external monitoring				"	1,125			1,125	1,250
0200	Probes									
0210	Well monitoring									
0220	Liquid phase detection				Ea.	630			630	695
0230	Hydrocarbon vapor, fixed position					785			785	865
0240	Hydrocarbon vapor, float mounted					560			560	615
0250	Both liquid and vapor hydrocarbon					550			550	605
0300	Secondary containment, liquid phase									
0310	Pipe trench/manway sump				Ea.	725			725	800
0320	Double wall pipe and manual sump					610			610	670
0330	Double wall fiberglass annular space					420			420	460
0340	Double wall steel tank annular space					295			295	325
0500	Accessories									
0510	Modem, non-dedicated phone line				Ea.	269			269	296
0600	Monitoring, internal									
0610	Automatic tank gauge, incl. overfill				Ea.	1,200			1,200	1,325
0620	Product line				"	1,250			1,250	1,375
0700	Monitoring, special									
0710	Cathodic protection				Ea.	675			675	745
0720	Annular space chemical monitor				"	920			920	1,025

28 46 Fire Detection and Alarm

28 46 11 – Fire Sensors and Detectors

28 46 11.21 Carbon-Monoxide Detection Sensors

		Crew	Daily Output	Labor-Hours	Unit	Material	Labor	Equipment	Total	Total Incl O&P
0010	**CARBON-MONOXIDE DETECTION SENSORS**									
8400	Smoke and carbon monoxide alarm battery operated photoelectric low profile	1 Elec	24	.333	Ea.	59	20		79	95
8410	low profile photoelectric battery powered		24	.333		36.50	20		56.50	70
8420	photoelectric low profile sealed lithium		24	.333		66.50	20		86.50	104
8430	Photoelectric low profile sealed lithium smoke and CO with voice combo		24	.333		41.50	20		61.50	75.50
8500	Carbon monoxide sensor, wall mount 1Mod 1 relay output smoke & heat		24	.333		505	20		525	585
8700	Carbon monoxide detector, battery operated, wall mounted		16	.500		46	30		76	95.50
8710	Hardwired, wall and ceiling mounted		8	1		90	60		150	189
8720	Duct mounted		8	1		350	60		410	475

28 46 11.27 Other Sensors

		Crew	Daily Output	Labor-Hours	Unit	Material	Labor	Equipment	Total	Total Incl O&P
0010	**OTHER SENSORS**									
5200	Smoke detector, ceiling type	1 Elec	6.20	1.290	Ea.	123	77.50		200.50	251
5240	Smoke detector, addressable type		6	1.333		227	80		307	370
5400	Duct type		3.20	2.500		294	150		444	550
5420	Duct addressable type		3.20	2.500		525	150		675	800

28 46 11.50 Fire and Heat Detectors

		Crew	Daily Output	Labor-Hours	Unit	Material	Labor	Equipment	Total	Total Incl O&P
0010	**FIRE & HEAT DETECTORS**									
5000	Detector, rate of rise	1 Elec	8	1	Ea.	52	60		112	147
5100	Fixed temp fire alarm	"	7	1.143	"	45.50	68.50		114	152

28 46 21 – Fire Alarm

28 46 21.50 Alarm Panels and Devices		Crew	Daily Output	Labor-Hours	Unit	Material	Labor	Equipment	Total	Total Incl O&P
						2019 Bare Costs				
0010	**ALARM PANELS AND DEVICES**, not including wires & conduits									
3600	4 zone	2 Elec	2	8	Ea.	380	480		860	1,125
3800	8 zone		1	16		960	960		1,920	2,475
3810	5 zone		1.50	10.667		660	640		1,300	1,675
3900	10 zone		1.25	12.800		910	770		1,680	2,150
4000	12 zone		.67	23.988		2,175	1,450		3,625	4,550
4020	Alarm device, tamper, flow	1 Elec	8	1		227	60		287	340
4050	Actuating device		8	1		345	60		405	470
4200	Battery and rack		4	2		390	120		510	610
4400	Automatic charger		8	1		525	60		585	665
4600	Signal bell		8	1		135	60		195	239
4610	Fire alarm signal bell 10" red 20-24 V P		8	1		149	60		209	254
4800	Trouble buzzer or manual station		8	1		87.50	60		147.50	186
5425	Duct smoke and heat detector 2 wire		8	1		118	60		178	220
5430	Fire alarm duct detector controller		3	2.667		255	160		415	520
5435	Fire alarm duct detector sensor kit		8	1		71	60		131	168
5440	Remote test station for smoke detector duct type		5.30	1.509		60	90.50		150.50	202
5460	Remote fire alarm indicator light		5.30	1.509		22.50	90.50		113	160
5600	Strobe and horn		5.30	1.509		140	90.50		230.50	289
5610	Strobe and horn (ADA type)		5.30	1.509		168	90.50		258.50	320
5620	Visual alarm (ADA type)		6.70	1.194		117	71.50		188.50	236
5800	Fire alarm horn		6.70	1.194		55.50	71.50		127	168
6000	Door holder, electro-magnetic		4	2		91.50	120		211.50	280
6200	Combination holder and closer		3.20	2.500		126	150		276	365
6600	Drill switch		8	1		380	60		440	510
6800	Master box		2.70	2.963		6,575	178		6,753	7,500
7000	Break glass station		8	1		58	60		118	153
7800	Remote annunciator, 8 zone lamp		1.80	4.444		213	267		480	635
8000	12 zone lamp	2 Elec	2.60	6.154		415	370		785	1,000
8200	16 zone lamp	"	2.20	7.273		375	435		810	1,050

Estimating Tips

31 05 00 Common Work Results for Earthwork

- Estimating the actual cost of performing earthwork requires careful consideration of the variables involved. This includes items such as type of soil, whether water will be encountered, dewatering, whether banks need bracing, disposal of excavated earth, and length of haul to fill or spoil sites, etc. If the project has large quantities of cut or fill, consider raising or lowering the site to reduce costs, while paying close attention to the effect on site drainage and utilities.
- If the project has large quantities of fill, creating a borrow pit on the site can significantly lower the costs.
- It is very important to consider what time of year the project is scheduled for completion. Bad weather can create large cost overruns from dewatering, site repair, and lost productivity from cold weather.

Reference Numbers

Reference numbers are shown at the beginning of some major classifications. These numbers refer to related items in the Reference Section. The reference information may be an estimating procedure, an alternate pricing method, or technical information.

Note: Not all subdivisions listed here necessarily appear. ■

31 05 23 – Cement and Concrete for Earthwork

31 05 23.30 Plant Mixed Bituminous Concrete		Crew	Daily Output	Labor-Hours	Unit	Material	2019 Bare Costs Labor	Equipment	Total	Total Incl O&P
0010	**PLANT MIXED BITUMINOUS CONCRETE**									
0020	Asphaltic concrete plant mix (145 lb./C.F.)				Ton	64			64	70.50
0040	Asphaltic concrete less than 300 tons add trucking costs									
0050	See Section 31 23 23.20 for hauling costs									
0200	All weather patching mix, hot				Ton	65.50			65.50	72
0250	Cold patch					72.50			72.50	79.50
0300	Berm mix				↓	68.50			68.50	75.50

31 23 Excavation and Fill

31 23 16 – Excavation

31 23 16.13 Excavating, Trench		Crew	Daily Output	Labor-Hours	Unit	Material	2019 Bare Costs Labor	Equipment	Total	Total Incl O&P
0010	**EXCAVATING, TRENCH** G1030–805									
0011	Or continuous footing									
0020	Common earth with no sheeting or dewatering included									
0050	1' to 4' deep, 3/8 C.Y. excavator	B-11C	150	.107	B.C.Y.		5.15	2.13	7.28	10.05
0060	1/2 C.Y. excavator	B-11M	200	.080			3.85	1.94	5.79	7.95
0090	4' to 6' deep, 1/2 C.Y. excavator	"	200	.080			3.85	1.94	5.79	7.95
0100	5/8 C.Y. excavator	B-12Q	250	.064			3.15	2.35	5.50	7.35
0110	3/4 C.Y. excavator	B-12F	300	.053			2.63	2.27	4.90	6.45
0300	1/2 C.Y. excavator, truck mounted	B-12J	200	.080			3.94	4.31	8.25	10.70
0500	6' to 10' deep, 3/4 C.Y. excavator	B-12F	225	.071			3.50	3.03	6.53	8.60
0600	1 C.Y. excavator, truck mounted	B-12K	400	.040			1.97	2.96	4.93	6.25
0900	10' to 14' deep, 3/4 C.Y. excavator	B-12F	200	.080			3.94	3.41	7.35	9.70
1000	1-1/2 C.Y. excavator	B-12B	540	.030			1.46	1.68	3.14	4.05
1300	14' to 20' deep, 1 C.Y. excavator	B-12A	320	.050			2.46	2.34	4.80	6.30
1340	20' to 24' deep, 1 C.Y. excavator	"	288	.056			2.74	2.60	5.34	7
1352	4' to 6' deep, 1/2 C.Y. excavator w/trench box	B-13H	188	.085			4.19	5	9.19	11.80
1354	5/8 C.Y. excavator	"	235	.068			3.35	4	7.35	9.45
1356	3/4 C.Y. excavator	B-13G	282	.057			2.79	2.69	5.48	7.15
1362	6' to 10' deep, 3/4 C.Y. excavator w/trench box		212	.075			3.72	3.58	7.30	9.55
1374	10' to 14' deep, 3/4 C.Y. excavator w/trench box	↓	188	.085			4.19	4.04	8.23	10.75
1376	1-1/2 C.Y. excavator	B-13E	508	.032			1.55	1.94	3.49	4.48
1381	14' to 20' deep, 1 C.Y. excavator w/trench box	B-13D	301	.053			2.62	2.75	5.37	6.95
1386	20' to 24' deep, 1 C.Y. excavator w/trench box	"	271	.059			2.91	3.05	5.96	7.75
1400	By hand with pick and shovel 2' to 6' deep, light soil	1 Clab	8	1			41		41	62
1500	Heavy soil	"	4	2	↓		82		82	124
1700	For tamping backfilled trenches, air tamp, add	A-1G	100	.080	E.C.Y.		3.28	.53	3.81	5.55
1900	Vibrating plate, add	B-18	180	.133	"		5.55	.23	5.78	8.65
2100	Trim sides and bottom for concrete pours, common earth		1500	.016	S.F.		.67	.03	.70	1.04
2300	Hardpan	↓	600	.040	"		1.67	.07	1.74	2.59
5020	Loam & sandy clay with no sheeting or dewatering included									
5050	1' to 4' deep, 3/8 C.Y. tractor loader/backhoe	B-11C	162	.099	B.C.Y.		4.75	1.98	6.73	9.30
5060	1/2 C.Y. excavator	B-11M	216	.074			3.56	1.80	5.36	7.30
5080	4' to 6' deep, 1/2 C.Y. excavator	"	216	.074			3.56	1.80	5.36	7.30
5090	5/8 C.Y. excavator	B-12Q	276	.058			2.86	2.13	4.99	6.65
5100	3/4 C.Y. excavator	B-12F	324	.049			2.43	2.10	4.53	5.95
5130	1/2 C.Y. excavator, truck mounted	B-12J	216	.074			3.65	3.99	7.64	9.90
5140	6' to 10' deep, 3/4 C.Y. excavator	B-12F	243	.066			3.24	2.80	6.04	7.95
5160	1 C.Y. excavator, truck mounted	B-12K	432	.037			1.82	2.74	4.56	5.75
5190	10' to 14' deep, 3/4 C.Y. excavator	B-12F	216	.074			3.65	3.15	6.80	8.95
5210	1-1/2 C.Y. excavator	B-12B	583	.027	↓		1.35	1.56	2.91	3.75

31 23 16.13 Excavating, Trench

	31 23 16.13 Excavating, Trench	Crew	Daily Output	Labor-Hours	Unit	Material	2019 Bare Costs Labor	Equipment	Total	Total Incl O&P
5250	14′ to 20′ deep, 1 C.Y. excavator	B-12A	346	.046	B.C.Y.		2.28	2.17	4.45	5.80
5300	20′ to 24′ deep, 1 C.Y. excavator	″	311	.051			2.53	2.41	4.94	6.45
5352	4′ to 6′ deep, 1/2 C.Y. excavator w/trench box	B-13H	205	.078			3.84	4.59	8.43	10.85
5354	5/8 C.Y. excavator	″	257	.062			3.07	3.66	6.73	8.65
5356	3/4 C.Y. excavator	B-13G	308	.052			2.56	2.46	5.02	6.55
5362	6′ to 10′ deep, 3/4 C.Y. excavator w/trench box		231	.069			3.41	3.29	6.70	8.75
5370	10′ to 14′ deep, 3/4 C.Y. excavator w/trench box		205	.078			3.84	3.70	7.54	9.85
5374	1-1/2 C.Y. excavator	B-13E	554	.029			1.42	1.78	3.20	4.10
5382	14′ to 20′ deep, 1 C.Y. excavator w/trench box	B-13D	329	.049			2.40	2.52	4.92	6.40
5392	20′ to 24′ deep, 1 C.Y. excavator w/trench box	″	295	.054			2.67	2.81	5.48	7.10
6020	Sand & gravel with no sheeting or dewatering included									
6050	1′ to 4′ deep, 3/8 C.Y. excavator	B-11C	165	.097	B.C.Y.		4.66	1.94	6.60	9.15
6060	1/2 C.Y. excavator	B-11M	220	.073			3.50	1.76	5.26	7.20
6080	4′ to 6′ deep, 1/2 C.Y. excavator	″	220	.073			3.50	1.76	5.26	7.20
6090	5/8 C.Y. excavator	B-12Q	275	.058			2.87	2.14	5.01	6.65
6100	3/4 C.Y. excavator	B-12F	330	.048			2.39	2.06	4.45	5.85
6130	1/2 C.Y. excavator, truck mounted	B-12J	220	.073			3.58	3.92	7.50	9.70
6140	6′ to 10′ deep, 3/4 C.Y. excavator	B-12F	248	.065			3.18	2.75	5.93	7.80
6160	1 C.Y. excavator, truck mounted	B-12K	440	.036			1.79	2.69	4.48	5.65
6190	10′ to 14′ deep, 3/4 C.Y. excavator	B-12F	220	.073			3.58	3.10	6.68	8.80
6210	1-1/2 C.Y. excavator	B-12B	594	.027			1.33	1.53	2.86	3.68
6250	14′ to 20′ deep, 1 C.Y. excavator	B-12A	352	.045			2.24	2.13	4.37	5.70
6300	20′ to 24′ deep, 1 C.Y. excavator	″	317	.050			2.49	2.37	4.86	6.35
6352	4′ to 6′ deep, 1/2 C.Y. excavator w/trench box	B-13H	209	.077			3.77	4.50	8.27	10.65
6362	6′ to 10′ deep, 3/4 C.Y. excavator w/trench box	B-13G	236	.068			3.34	3.22	6.56	8.60
6370	10′ to 14′ deep, 3/4 C.Y. excavator w/trench box	″	209	.077			3.77	3.63	7.40	9.70
6374	1-1/2 C.Y. excavator	B-13E	564	.028			1.40	1.75	3.15	4.02
6382	14′ to 20′ deep, 1 C.Y. excavator w/trench box	B-13D	334	.048			2.36	2.48	4.84	6.30
6392	20′ to 24′ deep, 1 C.Y. excavator w/trench box	″	301	.053			2.62	2.75	5.37	6.95
7020	Dense hard clay with no sheeting or dewatering included									
7050	1′ to 4′ deep, 3/8 C.Y. excavator	B-11C	132	.121	B.C.Y.		5.85	2.43	8.28	11.45
7060	1/2 C.Y. excavator	B-11M	176	.091			4.37	2.20	6.57	9
7080	4′ to 6′ deep, 1/2 C.Y. excavator	″	176	.091			4.37	2.20	6.57	9
7090	5/8 C.Y. excavator	B-12Q	220	.073			3.58	2.67	6.25	8.35
7100	3/4 C.Y. excavator	B-12F	264	.061			2.99	2.58	5.57	7.35
7130	1/2 C.Y. excavator, truck mounted	B-12J	176	.091			4.48	4.90	9.38	12.15
7140	6′ to 10′ deep, 3/4 C.Y. excavator	B-12F	198	.081			3.98	3.44	7.42	9.80
7160	1 C.Y. excavator, truck mounted	B-12K	352	.045			2.24	3.37	5.61	7.05
7190	10′ to 14′ deep, 3/4 C.Y. excavator	B-12F	176	.091			4.48	3.87	8.35	11
7210	1-1/2 C.Y. excavator	B-12B	475	.034			1.66	1.91	3.57	4.60
7250	14′ to 20′ deep, 1 C.Y. excavator	B-12A	282	.057			2.79	2.66	5.45	7.15
7300	20′ to 24′ deep, 1 C.Y. excavator	″	254	.063			3.10	2.95	6.05	7.90

31 23 16.14 Excavating, Utility Trench

	31 23 16.14 Excavating, Utility Trench	Crew	Daily Output	Labor-Hours	Unit	Material	Labor	Equipment	Total	Total Incl O&P
0010	**EXCAVATING, UTILITY TRENCH**									
0011	Common earth									
0050	Trenching with chain trencher, 12 HP, operator walking									
0100	4″ wide trench, 12″ deep	B-53	800	.010	L.F.		.52	.08	.60	.87
0150	18″ deep		750	.011			.55	.08	.63	.92
0200	24″ deep		700	.011			.59	.09	.68	.99
0300	6″ wide trench, 12″ deep		650	.012			.64	.10	.74	1.06
0350	18″ deep		600	.013			.69	.11	.80	1.15
0400	24″ deep		550	.015			.75	.12	.87	1.26

31 23 16.14 Excavating, Utility Trench		Crew	Daily Output	Labor-Hours	Unit	Material	2019 Bare Costs Labor	Equipment	Total	Total Incl O&P
0450	36" deep	B-53	450	.018	L.F.		.92	.14	1.06	1.54
0600	8" wide trench, 12" deep		475	.017			.87	.13	1	1.46
0650	18" deep		400	.020			1.03	.16	1.19	1.72
0700	24" deep		350	.023			1.18	.18	1.36	1.97
0750	36" deep	↓	300	.027	↓		1.38	.21	1.59	2.30
1000	Backfill by hand including compaction, add									
1050	4" wide trench, 12" deep	A-1G	800	.010	L.F.		.41	.07	.48	.69
1100	18" deep		530	.015			.62	.10	.72	1.05
1150	24" deep		400	.020			.82	.13	.95	1.39
1300	6" wide trench, 12" deep		540	.015			.61	.10	.71	1.03
1350	18" deep		405	.020			.81	.13	.94	1.37
1400	24" deep		270	.030			1.22	.20	1.42	2.06
1450	36" deep		180	.044			1.82	.29	2.11	3.08
1600	8" wide trench, 12" deep		400	.020			.82	.13	.95	1.39
1650	18" deep		265	.030			1.24	.20	1.44	2.09
1700	24" deep		200	.040			1.64	.27	1.91	2.77
1750	36" deep	↓	135	.059	↓		2.43	.39	2.82	4.11
2000	Chain trencher, 40 HP operator riding									
2050	6" wide trench and backfill, 12" deep	B-54	1200	.007	L.F.		.34	.29	.63	.84
2100	18" deep		1000	.008			.41	.35	.76	1
2150	24" deep		975	.008			.42	.36	.78	1.03
2200	36" deep		900	.009			.46	.39	.85	1.12
2250	48" deep		750	.011			.55	.46	1.01	1.34
2300	60" deep		650	.012			.64	.54	1.18	1.54
2400	8" wide trench and backfill, 12" deep		1000	.008			.41	.35	.76	1
2450	18" deep		950	.008			.43	.37	.80	1.05
2500	24" deep		900	.009			.46	.39	.85	1.12
2550	36" deep		800	.010			.52	.44	.96	1.26
2600	48" deep		650	.012			.64	.54	1.18	1.54
2700	12" wide trench and backfill, 12" deep		975	.008			.42	.36	.78	1.03
2750	18" deep		860	.009			.48	.41	.89	1.17
2800	24" deep		800	.010			.52	.44	.96	1.26
2850	36" deep		725	.011			.57	.48	1.05	1.39
3000	16" wide trench and backfill, 12" deep		835	.010			.49	.42	.91	1.20
3050	18" deep		750	.011			.55	.46	1.01	1.34
3100	24" deep	↓	700	.011	↓		.59	.50	1.09	1.44
3200	Compaction with vibratory plate, add								35%	35%
5100	Hand excavate and trim for pipe bells after trench excavation									
5200	8" pipe	1 Clab	155	.052	L.F.		2.12		2.12	3.21
5300	18" pipe	"	130	.062	"		2.53		2.53	3.82

31 23 16.16 Structural Excavation for Minor Structures		Crew	Daily Output	Labor-Hours	Unit	Material	Labor	Equipment	Total	Total Incl O&P
0010	**STRUCTURAL EXCAVATION FOR MINOR STRUCTURES**									
0015	Hand, pits to 6' deep, sandy soil	1 Clab	8	1	B.C.Y.		41		41	62
0100	Heavy soil or clay		4	2			82		82	124
0300	Pits 6' to 12' deep, sandy soil		5	1.600			65.50		65.50	99.50
0500	Heavy soil or clay		3	2.667			109		109	166
0700	Pits 12' to 18' deep, sandy soil		4	2			82		82	124
0900	Heavy soil or clay	↓	2	4	↓		164		164	248
1500	For wet or muck hand excavation, add to above								50%	50%

31 23 Excavation and Fill

31 23 16 – Excavation

31 23 16.30 Drilling and Blasting Rock		Crew	Daily Output	Labor-Hours	Unit	Material	Labor	Equipment	Total	Total Incl O&P
						2019 Bare Costs				
0010	**DRILLING AND BLASTING ROCK**									
0020	Rock, open face, under 1,500 C.Y.	B-47	225	.107	B.C.Y.	4.54	4.83	6.55	15.92	19.50
0100	Over 1,500 C.Y.		300	.080		4.54	3.62	4.91	13.07	15.85
0200	Areas where blasting mats are required, under 1,500 C.Y.		175	.137		4.54	6.20	8.40	19.14	23.50
0250	Over 1,500 C.Y.		250	.096		4.54	4.34	5.90	14.78	18.05
2200	Trenches, up to 1,500 C.Y.		22	1.091		13.15	49.50	67	129.65	163
2300	Over 1,500 C.Y.		26	.923		13.15	42	56.50	111.65	140

31 23 19 – Dewatering

31 23 19.20 Dewatering Systems

Line	Description	Crew	Daily Output	Labor-Hours	Unit	Material	Labor	Equipment	Total	Total Incl O&P
0010	**DEWATERING SYSTEMS**									
0020	Excavate drainage trench, 2′ wide, 2′ deep	B-11C	90	.178	C.Y.		8.55	3.56	12.11	16.75
0100	2′ wide, 3′ deep, with backhoe loader	"	135	.119			5.70	2.37	8.07	11.20
0200	Excavate sump pits by hand, light soil	1 Clab	7.10	1.127			46.50		46.50	70
0300	Heavy soil	"	3.50	2.286			94		94	142
0500	Pumping 8 hrs., attended 2 hrs./day, incl. 20 L.F.									
0550	of suction hose & 100 L.F. discharge hose									
0600	2″ diaphragm pump used for 8 hrs.	B-10H	4	3	Day		151	19.55	170.55	249
0650	4″ diaphragm pump used for 8 hrs.	B-10I	4	3			151	37.50	188.50	268
0800	8 hrs. attended, 2″ diaphragm pump	B-10H	1	12			605	78	683	995
0900	3″ centrifugal pump	B-10J	1	12			605	84.50	689.50	1,000
1000	4″ diaphragm pump	B-10I	1	12			605	149	754	1,075
1100	6″ centrifugal pump	B-10K	1	12			605	320	925	1,250
1300	CMP, incl. excavation 3′ deep, 12″ diameter	B-6	115	.209	L.F.	10.70	9.30	2.78	22.78	29
1400	18″ diameter		100	.240	"	18.95	10.70	3.20	32.85	40.50
1600	Sump hole construction, incl. excavation and gravel, pit		1250	.019	C.F.	1.10	.86	.26	2.22	2.78
1700	With 12″ gravel collar, 12″ pipe, corrugated, 16 ga.		70	.343	L.F.	21.50	15.30	4.57	41.37	51.50
1800	15″ pipe, corrugated, 16 ga.		55	.436		28	19.45	5.80	53.25	67
1900	18″ pipe, corrugated, 16 ga.		50	.480		32.50	21.50	6.40	60.40	75.50
2000	24″ pipe, corrugated, 14 ga.		40	.600		39	27	8	74	92.50
2200	Wood lining, up to 4′ x 4′, add		300	.080	SFCA	16.10	3.57	1.07	20.74	24.50
9950	See Section 31 23 19.40 for wellpoints									
9960	See Section 31 23 19.30 for deep well systems									

31 23 19.30 Wells

Line	Description	Crew	Daily Output	Labor-Hours	Unit	Material	Labor	Equipment	Total	Total Incl O&P
0010	**WELLS**									
0011	For dewatering 10′ to 20′ deep, 2′ diameter									
0020	with steel casing, minimum	B-6	165	.145	V.L.F.	37.50	6.50	1.94	45.94	53.50
0050	Average		98	.245		45	10.90	3.27	59.17	69.50
0100	Maximum		49	.490		47.50	22	6.55	76.05	92.50
0300	For dewatering pumps see 01 54 33 in Reference Section									
0500	For domestic water wells, see Section 33 21 13.10									

31 23 19.40 Wellpoints

Line	Description	Crew	Daily Output	Labor-Hours	Unit	Material	Labor	Equipment	Total	Total Incl O&P
0010	**WELLPOINTS** R312319-90									
0011	For equipment rental, see 01 54 33 in Reference Section									
0100	Installation and removal of single stage system									
0110	Labor only, 0.75 labor-hours per L.F.	1 Clab	10.70	.748	LF Hdr		30.50		30.50	46.50
0200	2.0 labor-hours per L.F.	"	4	2	"		82		82	124
0400	Pump operation, 4 @ 6 hr. shifts									
0410	Per 24 hr. day	4 Eqlt	1.27	25.197	Day		1,300		1,300	1,950
0500	Per 168 hr. week, 160 hr. straight, 8 hr. double time		.18	178	Week		9,175		9,175	13,800
0550	Per 4.3 week month		.04	800	Month		41,300		41,300	62,000
0600	Complete installation, operation, equipment rental, fuel &									

31 23 19 – Dewatering

31 23 19.40 Wellpoints		Crew	Daily Output	Labor-Hours	Unit	2019 Bare Costs Material	Labor	Equipment	Total	Total Incl O&P
0610	removal of system with 2″ wellpoints 5′ OC									
0700	100′ long header, 6″ diameter, first month	4 Eqlt	3.23	9.907	LF Hdr	160	510		670	945
0800	Thereafter, per month		4.13	7.748		128	400		528	740
1000	200′ long header, 8″ diameter, first month		6	5.333		146	275		421	575
1100	Thereafter, per month		8.39	3.814		72	197		269	375
1300	500′ long header, 8″ diameter, first month		10.63	3.010		56	155		211	295
1400	Thereafter, per month		20.91	1.530		40	79		119	163
1600	1,000′ long header, 10″ diameter, first month		11.62	2.754		48	142		190	267
1700	Thereafter, per month		41.81	.765		24	39.50		63.50	86
1900	Note: above figures include pumping 168 hrs. per week,									
1910	the pump operator, and one stand-by pump.									

31 23 23 – Fill

31 23 23.13 Backfill

Code	Description	Crew	Daily Output	Labor-Hours	Unit	Material	Labor	Equipment	Total	Total Incl O&P
0010	**BACKFILL** G1030–805									
0015	By hand, no compaction, light soil	1 Clab	14	.571	L.C.Y.		23.50		23.50	35.50
0100	Heavy soil		11	.727	"		30		30	45
0300	Compaction in 6″ layers, hand tamp, add to above		20.60	.388	E.C.Y.		15.95		15.95	24
0400	Roller compaction operator walking, add	B-10A	100	.120			6.05	1.82	7.87	11.10
0500	Air tamp, add	B-9D	190	.211			8.75	1.42	10.17	14.75
0600	Vibrating plate, add	A-1D	60	.133			5.45	.52	5.97	8.85
0800	Compaction in 12″ layers, hand tamp, add to above	1 Clab	34	.235			9.65		9.65	14.60
0900	Roller compaction operator walking, add	B-10A	150	.080			4.03	1.21	5.24	7.40
1000	Air tamp, add	B-9	285	.140			5.80	.82	6.62	9.70
1100	Vibrating plate, add	A-1E	90	.089			3.65	.45	4.10	6
1300	Dozer backfilling, bulk, up to 300′ haul, no compaction	B-10B	1200	.010	L.C.Y.		.50	1.08	1.58	1.94
1400	Air tamped, add	B-11B	80	.200	E.C.Y.		9.25	3.59	12.84	17.90
1600	Compacting backfill, 6″ to 12″ lifts, vibrating roller	B-10C	800	.015			.76	2.14	2.90	3.49
1700	Sheepsfoot roller	B-10D	750	.016			.81	2.29	3.10	3.73
1900	Dozer backfilling, trench, up to 300′ haul, no compaction	B-10B	900	.013	L.C.Y.		.67	1.43	2.10	2.59
2000	Air tamped, add	B-11B	80	.200	E.C.Y.		9.25	3.59	12.84	17.90
2200	Compacting backfill, 6″ to 12″ lifts, vibrating roller	B-10C	700	.017			.86	2.45	3.31	3.99
2300	Sheepsfoot roller	B-10D	650	.018			.93	2.65	3.58	4.31

31 23 23.16 Fill By Borrow and Utility Bedding

Code	Description	Crew	Daily Output	Labor-Hours	Unit	Material	Labor	Equipment	Total	Total Incl O&P
0010	**FILL BY BORROW AND UTILITY BEDDING**									
0049	Utility bedding, for pipe & conduit, not incl. compaction									
0050	Crushed or screened bank run gravel	B-6	150	.160	L.C.Y.	21	7.15	2.13	30.28	36.50
0100	Crushed stone 3/4″ to 1/2″		150	.160		27.50	7.15	2.13	36.78	43
0200	Sand, dead or bank		150	.160		18.10	7.15	2.13	27.38	33
0500	Compacting bedding in trench	A-1D	90	.089	E.C.Y.		3.65	.35	4	5.90
0600	If material source exceeds 2 miles, add for extra mileage.									
0610	See Section 31 23 23.20 for hauling mileage add.									

31 23 23.17 General Fill

Code	Description	Crew	Daily Output	Labor-Hours	Unit	Material	Labor	Equipment	Total	Total Incl O&P
0010	**GENERAL FILL**									
0011	Spread dumped material, no compaction									
0020	By dozer	B-10B	1000	.012	L.C.Y.		.61	1.29	1.90	2.33
0100	By hand	1 Clab	12	.667	"		27.50		27.50	41.50
0500	Gravel fill, compacted, under floor slabs, 4″ deep	B-37	10000	.005	S.F.	.40	.21	.02	.63	.77
0600	6″ deep		8600	.006		.61	.24	.02	.87	1.05
0700	9″ deep		7200	.007		1.01	.29	.02	1.32	1.56
0800	12″ deep		6000	.008		1.41	.35	.03	1.79	2.11
1000	Alternate pricing method, 4″ deep		120	.400	E.C.Y.	30.50	17.25	1.27	49.02	61
1100	6″ deep		160	.300		30.50	12.95	.95	44.40	54

31 23 23.17 General Fill		Crew	Daily Output	Labor-Hours	Unit	Material	2019 Bare Costs Labor	Equipment	Total	Total Incl O&P
1200	9" deep	B-37	200	.240	E.C.Y.	30.50	10.35	.76	41.61	50
1300	12" deep	↓	220	.218	↓	30.50	9.40	.69	40.59	48.50
1400	Granular fill				L.C.Y.	20			20	22

31 23 23.20 Hauling		Crew	Daily Output	Labor-Hours	Unit	Material	Labor	Equipment	Total	Total Incl O&P
0010	**HAULING**									
0011	Excavated or borrow, loose cubic yards									
0012	no loading equipment, including hauling, waiting, loading/dumping									
0013	time per cycle (wait, load, travel, unload or dump & return)									
0014	8 C.Y. truck, 15 MPH avg., cycle 0.5 miles, 10 min. wait/ld./uld.	B-34A	320	.025	L.C.Y.		1.18	1.06	2.24	2.94
0016	cycle 1 mile		272	.029			1.38	1.25	2.63	3.45
0018	cycle 2 miles		208	.038			1.81	1.63	3.44	4.52
0020	cycle 4 miles		144	.056			2.61	2.36	4.97	6.50
0022	cycle 6 miles		112	.071			3.36	3.03	6.39	8.40
0024	cycle 8 miles		88	.091			4.27	3.86	8.13	10.70
0026	20 MPH avg., cycle 0.5 mile		336	.024			1.12	1.01	2.13	2.79
0028	cycle 1 mile		296	.027			1.27	1.15	2.42	3.17
0030	cycle 2 miles		240	.033			1.57	1.41	2.98	3.92
0032	cycle 4 miles		176	.045			2.14	1.93	4.07	5.35
0034	cycle 6 miles		136	.059			2.76	2.50	5.26	6.90
0036	cycle 8 miles		112	.071			3.36	3.03	6.39	8.40
0044	25 MPH avg., cycle 4 miles		192	.042			1.96	1.77	3.73	4.90
0046	cycle 6 miles		160	.050			2.35	2.12	4.47	5.90
0048	cycle 8 miles		128	.063			2.94	2.65	5.59	7.35
0050	30 MPH avg., cycle 4 miles		216	.037			1.74	1.57	3.31	4.35
0052	cycle 6 miles		176	.045			2.14	1.93	4.07	5.35
0054	cycle 8 miles		144	.056			2.61	2.36	4.97	6.50
0114	15 MPH avg., cycle 0.5 mile, 15 min. wait/ld./uld.		224	.036			1.68	1.52	3.20	4.20
0116	cycle 1 mile		200	.040			1.88	1.70	3.58	4.70
0118	cycle 2 miles		168	.048			2.24	2.02	4.26	5.60
0120	cycle 4 miles		120	.067			3.13	2.83	5.96	7.85
0122	cycle 6 miles		96	.083			3.92	3.54	7.46	9.80
0124	cycle 8 miles		80	.100			4.70	4.25	8.95	11.75
0126	20 MPH avg., cycle 0.5 mile		232	.034			1.62	1.46	3.08	4.05
0128	cycle 1 mile		208	.038			1.81	1.63	3.44	4.52
0130	cycle 2 miles		184	.043			2.04	1.85	3.89	5.10
0132	cycle 4 miles		144	.056			2.61	2.36	4.97	6.50
0134	cycle 6 miles		112	.071			3.36	3.03	6.39	8.40
0136	cycle 8 miles		96	.083			3.92	3.54	7.46	9.80
0144	25 MPH avg., cycle 4 miles		152	.053			2.47	2.23	4.70	6.20
0146	cycle 6 miles		128	.063			2.94	2.65	5.59	7.35
0148	cycle 8 miles		112	.071			3.36	3.03	6.39	8.40
0150	30 MPH avg., cycle 4 miles		168	.048			2.24	2.02	4.26	5.60
0152	cycle 6 miles		144	.056			2.61	2.36	4.97	6.50
0154	cycle 8 miles		120	.067			3.13	2.83	5.96	7.85
0214	15 MPH avg., cycle 0.5 mile, 20 min. wait/ld./uld.		176	.045			2.14	1.93	4.07	5.35
0216	cycle 1 mile		160	.050			2.35	2.12	4.47	5.90
0218	cycle 2 miles		136	.059			2.76	2.50	5.26	6.90
0220	cycle 4 miles		104	.077			3.62	3.27	6.89	9.05
0222	cycle 6 miles		88	.091			4.27	3.86	8.13	10.70
0224	cycle 8 miles		72	.111			5.20	4.72	9.92	13.05
0226	20 MPH avg., cycle 0.5 mile		176	.045			2.14	1.93	4.07	5.35
0228	cycle 1 mile	↓	168	.048	↓		2.24	2.02	4.26	5.60

31 23 23.20 Hauling		Crew	Daily Output	Labor-Hours	Unit	Material	2019 Bare Costs Labor	Equipment	Total	Total Incl O&P
0230	cycle 2 miles	B-34A	144	.056	L.C.Y.		2.61	2.36	4.97	6.50
0232	cycle 4 miles		120	.067			3.13	2.83	5.96	7.85
0234	cycle 6 miles		96	.083			3.92	3.54	7.46	9.80
0236	cycle 8 miles		88	.091			4.27	3.86	8.13	10.70
0244	25 MPH avg., cycle 4 miles		128	.063			2.94	2.65	5.59	7.35
0246	cycle 6 miles		112	.071			3.36	3.03	6.39	8.40
0248	cycle 8 miles		96	.083			3.92	3.54	7.46	9.80
0250	30 MPH avg., cycle 4 miles		136	.059			2.76	2.50	5.26	6.90
0252	cycle 6 miles		120	.067			3.13	2.83	5.96	7.85
0254	cycle 8 miles		104	.077			3.62	3.27	6.89	9.05
0314	15 MPH avg., cycle 0.5 mile, 25 min. wait/ld./uld.		144	.056			2.61	2.36	4.97	6.50
0316	cycle 1 mile		128	.063			2.94	2.65	5.59	7.35
0318	cycle 2 miles		112	.071			3.36	3.03	6.39	8.40
0320	cycle 4 miles		96	.083			3.92	3.54	7.46	9.80
0322	cycle 6 miles		80	.100			4.70	4.25	8.95	11.75
0324	cycle 8 miles		64	.125			5.90	5.30	11.20	14.70
0326	20 MPH avg., cycle 0.5 mile		144	.056			2.61	2.36	4.97	6.50
0328	cycle 1 mile		136	.059			2.76	2.50	5.26	6.90
0330	cycle 2 miles		120	.067			3.13	2.83	5.96	7.85
0332	cycle 4 miles		104	.077			3.62	3.27	6.89	9.05
0334	cycle 6 miles		88	.091			4.27	3.86	8.13	10.70
0336	cycle 8 miles		80	.100			4.70	4.25	8.95	11.75
0344	25 MPH avg., cycle 4 miles		112	.071			3.36	3.03	6.39	8.40
0346	cycle 6 miles		96	.083			3.92	3.54	7.46	9.80
0348	cycle 8 miles		88	.091			4.27	3.86	8.13	10.70
0350	30 MPH avg., cycle 4 miles		112	.071			3.36	3.03	6.39	8.40
0352	cycle 6 miles		104	.077			3.62	3.27	6.89	9.05
0354	cycle 8 miles		96	.083			3.92	3.54	7.46	9.80
0414	15 MPH avg., cycle 0.5 mile, 30 min. wait/ld./uld.		120	.067			3.13	2.83	5.96	7.85
0416	cycle 1 mile		112	.071			3.36	3.03	6.39	8.40
0418	cycle 2 miles		96	.083			3.92	3.54	7.46	9.80
0420	cycle 4 miles		80	.100			4.70	4.25	8.95	11.75
0422	cycle 6 miles		72	.111			5.20	4.72	9.92	13.05
0424	cycle 8 miles		64	.125			5.90	5.30	11.20	14.70
0426	20 MPH avg., cycle 0.5 mile		120	.067			3.13	2.83	5.96	7.85
0428	cycle 1 mile		112	.071			3.36	3.03	6.39	8.40
0430	cycle 2 miles		104	.077			3.62	3.27	6.89	9.05
0432	cycle 4 miles		88	.091			4.27	3.86	8.13	10.70
0434	cycle 6 miles		80	.100			4.70	4.25	8.95	11.75
0436	cycle 8 miles		72	.111			5.20	4.72	9.92	13.05
0444	25 MPH avg., cycle 4 miles		96	.083			3.92	3.54	7.46	9.80
0446	cycle 6 miles		88	.091			4.27	3.86	8.13	10.70
0448	cycle 8 miles		80	.100			4.70	4.25	8.95	11.75
0450	30 MPH avg., cycle 4 miles		96	.083			3.92	3.54	7.46	9.80
0452	cycle 6 miles		88	.091			4.27	3.86	8.13	10.70
0454	cycle 8 miles		80	.100			4.70	4.25	8.95	11.75
0514	15 MPH avg., cycle 0.5 mile, 35 min. wait/ld./uld.		104	.077			3.62	3.27	6.89	9.05
0516	cycle 1 mile		96	.083			3.92	3.54	7.46	9.80
0518	cycle 2 miles		88	.091			4.27	3.86	8.13	10.70
0520	cycle 4 miles		72	.111			5.20	4.72	9.92	13.05
0522	cycle 6 miles		64	.125			5.90	5.30	11.20	14.70
0524	cycle 8 miles		56	.143			6.70	6.05	12.75	16.75
0526	20 MPH avg., cycle 0.5 mile		104	.077			3.62	3.27	6.89	9.05

31 23 23.20 Hauling		Crew	Daily Output	Labor-Hours	Unit	Material	2019 Bare Costs Labor	Equipment	Total	Total Incl O&P
0528	cycle 1 mile	B-34A	96	.083	L.C.Y.		3.92	3.54	7.46	9.80
0530	cycle 2 miles		96	.083			3.92	3.54	7.46	9.80
0532	cycle 4 miles		80	.100			4.70	4.25	8.95	11.75
0534	cycle 6 miles		72	.111			5.20	4.72	9.92	13.05
0536	cycle 8 miles		64	.125			5.90	5.30	11.20	14.70
0544	25 MPH avg., cycle 4 miles		88	.091			4.27	3.86	8.13	10.70
0546	cycle 6 miles		80	.100			4.70	4.25	8.95	11.75
0548	cycle 8 miles		72	.111			5.20	4.72	9.92	13.05
0550	30 MPH avg., cycle 4 miles		88	.091			4.27	3.86	8.13	10.70
0552	cycle 6 miles		80	.100			4.70	4.25	8.95	11.75
0554	cycle 8 miles		72	.111			5.20	4.72	9.92	13.05
1014	12 C.Y. truck, cycle 0.5 mile, 15 MPH avg., 15 min. wait/ld./uld.	B-34B	336	.024			1.12	1.69	2.81	3.54
1016	cycle 1 mile		300	.027			1.25	1.89	3.14	3.97
1018	cycle 2 miles		252	.032			1.49	2.25	3.74	4.73
1020	cycle 4 miles		180	.044			2.09	3.15	5.24	6.60
1022	cycle 6 miles		144	.056			2.61	3.94	6.55	8.25
1024	cycle 8 miles		120	.067			3.13	4.73	7.86	9.90
1025	cycle 10 miles		96	.083			3.92	5.90	9.82	12.40
1026	20 MPH avg., cycle 0.5 mile		348	.023			1.08	1.63	2.71	3.42
1028	cycle 1 mile		312	.026			1.21	1.82	3.03	3.81
1030	cycle 2 miles		276	.029			1.36	2.05	3.41	4.31
1032	cycle 4 miles		216	.037			1.74	2.63	4.37	5.50
1034	cycle 6 miles		168	.048			2.24	3.38	5.62	7.10
1036	cycle 8 miles		144	.056			2.61	3.94	6.55	8.25
1038	cycle 10 miles		120	.067			3.13	4.73	7.86	9.90
1040	25 MPH avg., cycle 4 miles		228	.035			1.65	2.49	4.14	5.20
1042	cycle 6 miles		192	.042			1.96	2.95	4.91	6.20
1044	cycle 8 miles		168	.048			2.24	3.38	5.62	7.10
1046	cycle 10 miles		144	.056			2.61	3.94	6.55	8.25
1050	30 MPH avg., cycle 4 miles		252	.032			1.49	2.25	3.74	4.73
1052	cycle 6 miles		216	.037			1.74	2.63	4.37	5.50
1054	cycle 8 miles		180	.044			2.09	3.15	5.24	6.60
1056	cycle 10 miles		156	.051			2.41	3.63	6.04	7.65
1060	35 MPH avg., cycle 4 miles		264	.030			1.42	2.15	3.57	4.50
1062	cycle 6 miles		228	.035			1.65	2.49	4.14	5.20
1064	cycle 8 miles		204	.039			1.84	2.78	4.62	5.85
1066	cycle 10 miles		180	.044			2.09	3.15	5.24	6.60
1068	cycle 20 miles		120	.067			3.13	4.73	7.86	9.90
1069	cycle 30 miles		84	.095			4.48	6.75	11.23	14.20
1070	cycle 40 miles		72	.111			5.20	7.90	13.10	16.50
1072	40 MPH avg., cycle 6 miles		240	.033			1.57	2.36	3.93	4.96
1074	cycle 8 miles		216	.037			1.74	2.63	4.37	5.50
1076	cycle 10 miles		192	.042			1.96	2.95	4.91	6.20
1078	cycle 20 miles		120	.067			3.13	4.73	7.86	9.90
1080	cycle 30 miles		96	.083			3.92	5.90	9.82	12.40
1082	cycle 40 miles		72	.111			5.20	7.90	13.10	16.50
1084	cycle 50 miles		60	.133			6.25	9.45	15.70	19.85
1094	45 MPH avg., cycle 8 miles		216	.037			1.74	2.63	4.37	5.50
1096	cycle 10 miles		204	.039			1.84	2.78	4.62	5.85
1098	cycle 20 miles		132	.061			2.85	4.30	7.15	9
1100	cycle 30 miles		108	.074			3.48	5.25	8.73	11.05
1102	cycle 40 miles		84	.095			4.48	6.75	11.23	14.20
1104	cycle 50 miles		72	.111			5.20	7.90	13.10	16.50

31 23 23.20 Hauling		Crew	Daily Output	Labor-Hours	Unit	Material	2019 Bare Costs Labor	Equipment	Total	Total Incl O&P
1106	50 MPH avg., cycle 10 miles	B-34B	216	.037	L.C.Y.		1.74	2.63	4.37	5.50
1108	cycle 20 miles		144	.056			2.61	3.94	6.55	8.25
1110	cycle 30 miles		108	.074			3.48	5.25	8.73	11.05
1112	cycle 40 miles		84	.095			4.48	6.75	11.23	14.20
1114	cycle 50 miles		72	.111			5.20	7.90	13.10	16.50
1214	15 MPH avg., cycle 0.5 mile, 20 min. wait/ld./uld.		264	.030			1.42	2.15	3.57	4.50
1216	cycle 1 mile		240	.033			1.57	2.36	3.93	4.96
1218	cycle 2 miles		204	.039			1.84	2.78	4.62	5.85
1220	cycle 4 miles		156	.051			2.41	3.63	6.04	7.65
1222	cycle 6 miles		132	.061			2.85	4.30	7.15	9
1224	cycle 8 miles		108	.074			3.48	5.25	8.73	11.05
1225	cycle 10 miles		96	.083			3.92	5.90	9.82	12.40
1226	20 MPH avg., cycle 0.5 mile		264	.030			1.42	2.15	3.57	4.50
1228	cycle 1 mile		252	.032			1.49	2.25	3.74	4.73
1230	cycle 2 miles		216	.037			1.74	2.63	4.37	5.50
1232	cycle 4 miles		180	.044			2.09	3.15	5.24	6.60
1234	cycle 6 miles		144	.056			2.61	3.94	6.55	8.25
1236	cycle 8 miles		132	.061			2.85	4.30	7.15	9
1238	cycle 10 miles		108	.074			3.48	5.25	8.73	11.05
1240	25 MPH avg., cycle 4 miles		192	.042			1.96	2.95	4.91	6.20
1242	cycle 6 miles		168	.048			2.24	3.38	5.62	7.10
1244	cycle 8 miles		144	.056			2.61	3.94	6.55	8.25
1246	cycle 10 miles		132	.061			2.85	4.30	7.15	9
1250	30 MPH avg., cycle 4 miles		204	.039			1.84	2.78	4.62	5.85
1252	cycle 6 miles		180	.044			2.09	3.15	5.24	6.60
1254	cycle 8 miles		156	.051			2.41	3.63	6.04	7.65
1256	cycle 10 miles		144	.056			2.61	3.94	6.55	8.25
1260	35 MPH avg., cycle 4 miles		216	.037			1.74	2.63	4.37	5.50
1262	cycle 6 miles		192	.042			1.96	2.95	4.91	6.20
1264	cycle 8 miles		168	.048			2.24	3.38	5.62	7.10
1266	cycle 10 miles		156	.051			2.41	3.63	6.04	7.65
1268	cycle 20 miles		108	.074			3.48	5.25	8.73	11.05
1269	cycle 30 miles		72	.111			5.20	7.90	13.10	16.50
1270	cycle 40 miles		60	.133			6.25	9.45	15.70	19.85
1272	40 MPH avg., cycle 6 miles		192	.042			1.96	2.95	4.91	6.20
1274	cycle 8 miles		180	.044			2.09	3.15	5.24	6.60
1276	cycle 10 miles		156	.051			2.41	3.63	6.04	7.65
1278	cycle 20 miles		108	.074			3.48	5.25	8.73	11.05
1280	cycle 30 miles		84	.095			4.48	6.75	11.23	14.20
1282	cycle 40 miles		72	.111			5.20	7.90	13.10	16.50
1284	cycle 50 miles		60	.133			6.25	9.45	15.70	19.85
1294	45 MPH avg., cycle 8 miles		180	.044			2.09	3.15	5.24	6.60
1296	cycle 10 miles		168	.048			2.24	3.38	5.62	7.10
1298	cycle 20 miles		120	.067			3.13	4.73	7.86	9.90
1300	cycle 30 miles		96	.083			3.92	5.90	9.82	12.40
1302	cycle 40 miles		72	.111			5.20	7.90	13.10	16.50
1304	cycle 50 miles		60	.133			6.25	9.45	15.70	19.85
1306	50 MPH avg., cycle 10 miles		180	.044			2.09	3.15	5.24	6.60
1308	cycle 20 miles		132	.061			2.85	4.30	7.15	9
1310	cycle 30 miles		96	.083			3.92	5.90	9.82	12.40
1312	cycle 40 miles		84	.095			4.48	6.75	11.23	14.20
1314	cycle 50 miles		72	.111			5.20	7.90	13.10	16.50
1414	15 MPH avg., cycle 0.5 mile, 25 min. wait/ld./uld.		204	.039			1.84	2.78	4.62	5.85

31 23 23.20 Hauling		Crew	Daily Output	Labor-Hours	Unit	Material	2019 Bare Costs Labor	Equipment	Total	Total Incl O&P
1416	cycle 1 mile	B-34B	192	.042	L.C.Y.		1.96	2.95	4.91	6.20
1418	cycle 2 miles		168	.048			2.24	3.38	5.62	7.10
1420	cycle 4 miles		132	.061			2.85	4.30	7.15	9
1422	cycle 6 miles		120	.067			3.13	4.73	7.86	9.90
1424	cycle 8 miles		96	.083			3.92	5.90	9.82	12.40
1425	cycle 10 miles		84	.095			4.48	6.75	11.23	14.20
1426	20 MPH avg., cycle 0.5 mile		216	.037			1.74	2.63	4.37	5.50
1428	cycle 1 mile		204	.039			1.84	2.78	4.62	5.85
1430	cycle 2 miles		180	.044			2.09	3.15	5.24	6.60
1432	cycle 4 miles		156	.051			2.41	3.63	6.04	7.65
1434	cycle 6 miles		132	.061			2.85	4.30	7.15	9
1436	cycle 8 miles		120	.067			3.13	4.73	7.86	9.90
1438	cycle 10 miles		96	.083			3.92	5.90	9.82	12.40
1440	25 MPH avg., cycle 4 miles		168	.048			2.24	3.38	5.62	7.10
1442	cycle 6 miles		144	.056			2.61	3.94	6.55	8.25
1444	cycle 8 miles		132	.061			2.85	4.30	7.15	9
1446	cycle 10 miles		108	.074			3.48	5.25	8.73	11.05
1450	30 MPH avg., cycle 4 miles		168	.048			2.24	3.38	5.62	7.10
1452	cycle 6 miles		156	.051			2.41	3.63	6.04	7.65
1454	cycle 8 miles		132	.061			2.85	4.30	7.15	9
1456	cycle 10 miles		120	.067			3.13	4.73	7.86	9.90
1460	35 MPH avg., cycle 4 miles		180	.044			2.09	3.15	5.24	6.60
1462	cycle 6 miles		156	.051			2.41	3.63	6.04	7.65
1464	cycle 8 miles		144	.056			2.61	3.94	6.55	8.25
1466	cycle 10 miles		132	.061			2.85	4.30	7.15	9
1468	cycle 20 miles		96	.083			3.92	5.90	9.82	12.40
1469	cycle 30 miles		72	.111			5.20	7.90	13.10	16.50
1470	cycle 40 miles		60	.133			6.25	9.45	15.70	19.85
1472	40 MPH avg., cycle 6 miles		168	.048			2.24	3.38	5.62	7.10
1474	cycle 8 miles		156	.051			2.41	3.63	6.04	7.65
1476	cycle 10 miles		144	.056			2.61	3.94	6.55	8.25
1478	cycle 20 miles		96	.083			3.92	5.90	9.82	12.40
1480	cycle 30 miles		84	.095			4.48	6.75	11.23	14.20
1482	cycle 40 miles		60	.133			6.25	9.45	15.70	19.85
1484	cycle 50 miles		60	.133			6.25	9.45	15.70	19.85
1494	45 MPH avg., cycle 8 miles		156	.051			2.41	3.63	6.04	7.65
1496	cycle 10 miles		144	.056			2.61	3.94	6.55	8.25
1498	cycle 20 miles		108	.074			3.48	5.25	8.73	11.05
1500	cycle 30 miles		84	.095			4.48	6.75	11.23	14.20
1502	cycle 40 miles		72	.111			5.20	7.90	13.10	16.50
1504	cycle 50 miles		60	.133			6.25	9.45	15.70	19.85
1506	50 MPH avg., cycle 10 miles		156	.051			2.41	3.63	6.04	7.65
1508	cycle 20 miles		120	.067			3.13	4.73	7.86	9.90
1510	cycle 30 miles		96	.083			3.92	5.90	9.82	12.40
1512	cycle 40 miles		72	.111			5.20	7.90	13.10	16.50
1514	cycle 50 miles		60	.133			6.25	9.45	15.70	19.85
1614	15 MPH avg., cycle 0.5 mile, 30 min. wait/ld./uld.		180	.044			2.09	3.15	5.24	6.60
1616	cycle 1 mile		168	.048			2.24	3.38	5.62	7.10
1618	cycle 2 miles		144	.056			2.61	3.94	6.55	8.25
1620	cycle 4 miles		120	.067			3.13	4.73	7.86	9.90
1622	cycle 6 miles		108	.074			3.48	5.25	8.73	11.05
1624	cycle 8 miles		84	.095			4.48	6.75	11.23	14.20
1625	cycle 10 miles		84	.095			4.48	6.75	11.23	14.20

31 23 23.20 Hauling		Crew	Daily Output	Labor-Hours	Unit	Material	2019 Bare Costs Labor	Equipment	Total	Total Incl O&P
1626	20 MPH avg., cycle 0.5 mile	B-34B	180	.044	L.C.Y.		2.09	3.15	5.24	6.60
1628	cycle 1 mile		168	.048			2.24	3.38	5.62	7.10
1630	cycle 2 miles		156	.051			2.41	3.63	6.04	7.65
1632	cycle 4 miles		132	.061			2.85	4.30	7.15	9
1634	cycle 6 miles		120	.067			3.13	4.73	7.86	9.90
1636	cycle 8 miles		108	.074			3.48	5.25	8.73	11.05
1638	cycle 10 miles		96	.083			3.92	5.90	9.82	12.40
1640	25 MPH avg., cycle 4 miles		144	.056			2.61	3.94	6.55	8.25
1642	cycle 6 miles		132	.061			2.85	4.30	7.15	9
1644	cycle 8 miles		108	.074			3.48	5.25	8.73	11.05
1646	cycle 10 miles		108	.074			3.48	5.25	8.73	11.05
1650	30 MPH avg., cycle 4 miles		144	.056			2.61	3.94	6.55	8.25
1652	cycle 6 miles		132	.061			2.85	4.30	7.15	9
1654	cycle 8 miles		120	.067			3.13	4.73	7.86	9.90
1656	cycle 10 miles		108	.074			3.48	5.25	8.73	11.05
1660	35 MPH avg., cycle 4 miles		156	.051			2.41	3.63	6.04	7.65
1662	cycle 6 miles		144	.056			2.61	3.94	6.55	8.25
1664	cycle 8 miles		132	.061			2.85	4.30	7.15	9
1666	cycle 10 miles		120	.067			3.13	4.73	7.86	9.90
1668	cycle 20 miles		84	.095			4.48	6.75	11.23	14.20
1669	cycle 30 miles		72	.111			5.20	7.90	13.10	16.50
1670	cycle 40 miles		60	.133			6.25	9.45	15.70	19.85
1672	40 MPH avg., cycle 6 miles		144	.056			2.61	3.94	6.55	8.25
1674	cycle 8 miles		132	.061			2.85	4.30	7.15	9
1676	cycle 10 miles		120	.067			3.13	4.73	7.86	9.90
1678	cycle 20 miles		96	.083			3.92	5.90	9.82	12.40
1680	cycle 30 miles		72	.111			5.20	7.90	13.10	16.50
1682	cycle 40 miles		60	.133			6.25	9.45	15.70	19.85
1684	cycle 50 miles		48	.167			7.85	11.80	19.65	25
1694	45 MPH avg., cycle 8 miles		144	.056			2.61	3.94	6.55	8.25
1696	cycle 10 miles		132	.061			2.85	4.30	7.15	9
1698	cycle 20 miles		96	.083			3.92	5.90	9.82	12.40
1700	cycle 30 miles		84	.095			4.48	6.75	11.23	14.20
1702	cycle 40 miles		60	.133			6.25	9.45	15.70	19.85
1704	cycle 50 miles		60	.133			6.25	9.45	15.70	19.85
1706	50 MPH avg., cycle 10 miles		132	.061			2.85	4.30	7.15	9
1708	cycle 20 miles		108	.074			3.48	5.25	8.73	11.05
1710	cycle 30 miles		84	.095			4.48	6.75	11.23	14.20
1712	cycle 40 miles		72	.111			5.20	7.90	13.10	16.50
1714	cycle 50 miles		60	.133			6.25	9.45	15.70	19.85
2000	Hauling, 8 C.Y. truck, small project cost per hour	B-34A	8	1	Hr.		47	42.50	89.50	118
2100	12 C.Y. truck	B-34B	8	1			47	71	118	149
2150	16.5 C.Y. truck	B-34C	8	1			47	77.50	124.50	157
2175	18 C.Y. 8 wheel truck	B-34I	8	1			47	88	135	168
2200	20 C.Y. truck	B-34D	8	1			47	79.50	126.50	159
2300	Grading at dump, or embankment if required, by dozer	B-10B	1000	.012	L.C.Y.		.61	1.29	1.90	2.33
2310	Spotter at fill or cut, if required	1 Clab	8	1	Hr.		41		41	62
9014	18 C.Y. truck, 8 wheels,15 min. wait/ld./uld.,15 MPH, cycle 0.5 mi.	B-34I	504	.016	L.C.Y.		.75	1.40	2.15	2.66
9016	cycle 1 mile		450	.018			.84	1.57	2.41	2.98
9018	cycle 2 miles		378	.021			.99	1.86	2.85	3.55
9020	cycle 4 miles		270	.030			1.39	2.61	4	4.97
9022	cycle 6 miles		216	.037			1.74	3.26	5	6.20
9024	cycle 8 miles		180	.044			2.09	3.92	6.01	7.45

31 23 23.20 Hauling		Crew	Daily Output	Labor-Hours	Unit	Material	2019 Bare Costs Labor	Equipment	Total	Total Incl O&P
9025	cycle 10 miles	B-34I	144	.056	L.C.Y.		2.61	4.90	7.51	9.35
9026	20 MPH avg., cycle 0.5 mile		522	.015			.72	1.35	2.07	2.57
9028	cycle 1 mile		468	.017			.80	1.51	2.31	2.87
9030	cycle 2 miles		414	.019			.91	1.70	2.61	3.24
9032	cycle 4 miles		324	.025			1.16	2.18	3.34	4.14
9034	cycle 6 miles		252	.032			1.49	2.80	4.29	5.35
9036	cycle 8 miles		216	.037			1.74	3.26	5	6.20
9038	cycle 10 miles		180	.044			2.09	3.92	6.01	7.45
9040	25 MPH avg., cycle 4 miles		342	.023			1.10	2.06	3.16	3.92
9042	cycle 6 miles		288	.028			1.31	2.45	3.76	4.66
9044	cycle 8 miles		252	.032			1.49	2.80	4.29	5.35
9046	cycle 10 miles		216	.037			1.74	3.26	5	6.20
9050	30 MPH avg., cycle 4 miles		378	.021			.99	1.86	2.85	3.55
9052	cycle 6 miles		324	.025			1.16	2.18	3.34	4.14
9054	cycle 8 miles		270	.030			1.39	2.61	4	4.97
9056	cycle 10 miles		234	.034			1.61	3.01	4.62	5.75
9060	35 MPH avg., cycle 4 miles		396	.020			.95	1.78	2.73	3.39
9062	cycle 6 miles		342	.023			1.10	2.06	3.16	3.92
9064	cycle 8 miles		288	.028			1.31	2.45	3.76	4.66
9066	cycle 10 miles		270	.030			1.39	2.61	4	4.97
9068	cycle 20 miles		162	.049			2.32	4.35	6.67	8.30
9070	cycle 30 miles		126	.063			2.98	5.60	8.58	10.65
9072	cycle 40 miles		90	.089			4.18	7.85	12.03	14.90
9074	40 MPH avg., cycle 6 miles		360	.022			1.04	1.96	3	3.72
9076	cycle 8 miles		324	.025			1.16	2.18	3.34	4.14
9078	cycle 10 miles		288	.028			1.31	2.45	3.76	4.66
9080	cycle 20 miles		180	.044			2.09	3.92	6.01	7.45
9082	cycle 30 miles		144	.056			2.61	4.90	7.51	9.35
9084	cycle 40 miles		108	.074			3.48	6.55	10.03	12.45
9086	cycle 50 miles		90	.089			4.18	7.85	12.03	14.90
9094	45 MPH avg., cycle 8 miles		324	.025			1.16	2.18	3.34	4.14
9096	cycle 10 miles		306	.026			1.23	2.30	3.53	4.38
9098	cycle 20 miles		198	.040			1.90	3.56	5.46	6.80
9100	cycle 30 miles		144	.056			2.61	4.90	7.51	9.35
9102	cycle 40 miles		126	.063			2.98	5.60	8.58	10.65
9104	cycle 50 miles		108	.074			3.48	6.55	10.03	12.45
9106	50 MPH avg., cycle 10 miles		324	.025			1.16	2.18	3.34	4.14
9108	cycle 20 miles		216	.037			1.74	3.26	5	6.20
9110	cycle 30 miles		162	.049			2.32	4.35	6.67	8.30
9112	cycle 40 miles		126	.063			2.98	5.60	8.58	10.65
9114	cycle 50 miles		108	.074			3.48	6.55	10.03	12.45
9214	20 min. wait/ld./uld.,15 MPH, cycle 0.5 mi.		396	.020			.95	1.78	2.73	3.39
9216	cycle 1 mile		360	.022			1.04	1.96	3	3.72
9218	cycle 2 miles		306	.026			1.23	2.30	3.53	4.38
9220	cycle 4 miles		234	.034			1.61	3.01	4.62	5.75
9222	cycle 6 miles		198	.040			1.90	3.56	5.46	6.80
9224	cycle 8 miles		162	.049			2.32	4.35	6.67	8.30
9225	cycle 10 miles		144	.056			2.61	4.90	7.51	9.35
9226	20 MPH avg., cycle 0.5 mile		396	.020			.95	1.78	2.73	3.39
9228	cycle 1 mile		378	.021			.99	1.86	2.85	3.55
9230	cycle 2 miles		324	.025			1.16	2.18	3.34	4.14
9232	cycle 4 miles		270	.030			1.39	2.61	4	4.97
9234	cycle 6 miles		216	.037			1.74	3.26	5	6.20

31 23 23.20 Hauling		Crew	Daily Output	Labor-Hours	Unit	Material	2019 Bare Costs Labor	Equipment	Total	Total Incl O&P
9236	cycle 8 miles	B-34I	198	.040	L.C.Y.		1.90	3.56	5.46	6.80
9238	cycle 10 miles		162	.049			2.32	4.35	6.67	8.30
9240	25 MPH avg., cycle 4 miles		288	.028			1.31	2.45	3.76	4.66
9242	cycle 6 miles		252	.032			1.49	2.80	4.29	5.35
9244	cycle 8 miles		216	.037			1.74	3.26	5	6.20
9246	cycle 10 miles		198	.040			1.90	3.56	5.46	6.80
9250	30 MPH avg., cycle 4 miles		306	.026			1.23	2.30	3.53	4.38
9252	cycle 6 miles		270	.030			1.39	2.61	4	4.97
9254	cycle 8 miles		234	.034			1.61	3.01	4.62	5.75
9256	cycle 10 miles		216	.037			1.74	3.26	5	6.20
9260	35 MPH avg., cycle 4 miles		324	.025			1.16	2.18	3.34	4.14
9262	cycle 6 miles		288	.028			1.31	2.45	3.76	4.66
9264	cycle 8 miles		252	.032			1.49	2.80	4.29	5.35
9266	cycle 10 miles		234	.034			1.61	3.01	4.62	5.75
9268	cycle 20 miles		162	.049			2.32	4.35	6.67	8.30
9270	cycle 30 miles		108	.074			3.48	6.55	10.03	12.45
9272	cycle 40 miles		90	.089			4.18	7.85	12.03	14.90
9274	40 MPH avg., cycle 6 miles		288	.028			1.31	2.45	3.76	4.66
9276	cycle 8 miles		270	.030			1.39	2.61	4	4.97
9278	cycle 10 miles		234	.034			1.61	3.01	4.62	5.75
9280	cycle 20 miles		162	.049			2.32	4.35	6.67	8.30
9282	cycle 30 miles		126	.063			2.98	5.60	8.58	10.65
9284	cycle 40 miles		108	.074			3.48	6.55	10.03	12.45
9286	cycle 50 miles		90	.089			4.18	7.85	12.03	14.90
9294	45 MPH avg., cycle 8 miles		270	.030			1.39	2.61	4	4.97
9296	cycle 10 miles		252	.032			1.49	2.80	4.29	5.35
9298	cycle 20 miles		180	.044			2.09	3.92	6.01	7.45
9300	cycle 30 miles		144	.056			2.61	4.90	7.51	9.35
9302	cycle 40 miles		108	.074			3.48	6.55	10.03	12.45
9304	cycle 50 miles		90	.089			4.18	7.85	12.03	14.90
9306	50 MPH avg., cycle 10 miles		270	.030			1.39	2.61	4	4.97
9308	cycle 20 miles		198	.040			1.90	3.56	5.46	6.80
9310	cycle 30 miles		144	.056			2.61	4.90	7.51	9.35
9312	cycle 40 miles		126	.063			2.98	5.60	8.58	10.65
9314	cycle 50 miles		108	.074			3.48	6.55	10.03	12.45
9414	25 min. wait/ld./uld.,15 MPH, cycle 0.5 mi.		306	.026			1.23	2.30	3.53	4.38
9416	cycle 1 mile		288	.028			1.31	2.45	3.76	4.66
9418	cycle 2 miles		252	.032			1.49	2.80	4.29	5.35
9420	cycle 4 miles		198	.040			1.90	3.56	5.46	6.80
9422	cycle 6 miles		180	.044			2.09	3.92	6.01	7.45
9424	cycle 8 miles		144	.056			2.61	4.90	7.51	9.35
9425	cycle 10 miles		126	.063			2.98	5.60	8.58	10.65
9426	20 MPH avg., cycle 0.5 mile		324	.025			1.16	2.18	3.34	4.14
9428	cycle 1 mile		306	.026			1.23	2.30	3.53	4.38
9430	cycle 2 miles		270	.030			1.39	2.61	4	4.97
9432	cycle 4 miles		234	.034			1.61	3.01	4.62	5.75
9434	cycle 6 miles		198	.040			1.90	3.56	5.46	6.80
9436	cycle 8 miles		180	.044			2.09	3.92	6.01	7.45
9438	cycle 10 miles		144	.056			2.61	4.90	7.51	9.35
9440	25 MPH avg., cycle 4 miles		252	.032			1.49	2.80	4.29	5.35
9442	cycle 6 miles		216	.037			1.74	3.26	5	6.20
9444	cycle 8 miles		198	.040			1.90	3.56	5.46	6.80
9446	cycle 10 miles		180	.044			2.09	3.92	6.01	7.45

31 23 23.20 Hauling		Crew	Daily Output	Labor-Hours	Unit	Material	2019 Bare Costs Labor	Equipment	Total	Total Incl O&P
9450	30 MPH avg., cycle 4 miles	B-34I	252	.032	L.C.Y.		1.49	2.80	4.29	5.35
9452	cycle 6 miles		234	.034			1.61	3.01	4.62	5.75
9454	cycle 8 miles		198	.040			1.90	3.56	5.46	6.80
9456	cycle 10 miles		180	.044			2.09	3.92	6.01	7.45
9460	35 MPH avg., cycle 4 miles		270	.030			1.39	2.61	4	4.97
9462	cycle 6 miles		234	.034			1.61	3.01	4.62	5.75
9464	cycle 8 miles		216	.037			1.74	3.26	5	6.20
9466	cycle 10 miles		198	.040			1.90	3.56	5.46	6.80
9468	cycle 20 miles		144	.056			2.61	4.90	7.51	9.35
9470	cycle 30 miles		108	.074			3.48	6.55	10.03	12.45
9472	cycle 40 miles		90	.089			4.18	7.85	12.03	14.90
9474	40 MPH avg., cycle 6 miles		252	.032			1.49	2.80	4.29	5.35
9476	cycle 8 miles		234	.034			1.61	3.01	4.62	5.75
9478	cycle 10 miles		216	.037			1.74	3.26	5	6.20
9480	cycle 20 miles		144	.056			2.61	4.90	7.51	9.35
9482	cycle 30 miles		126	.063			2.98	5.60	8.58	10.65
9484	cycle 40 miles		90	.089			4.18	7.85	12.03	14.90
9486	cycle 50 miles		90	.089			4.18	7.85	12.03	14.90
9494	45 MPH avg., cycle 8 miles		234	.034			1.61	3.01	4.62	5.75
9496	cycle 10 miles		216	.037			1.74	3.26	5	6.20
9498	cycle 20 miles		162	.049			2.32	4.35	6.67	8.30
9500	cycle 30 miles		126	.063			2.98	5.60	8.58	10.65
9502	cycle 40 miles		108	.074			3.48	6.55	10.03	12.45
9504	cycle 50 miles		90	.089			4.18	7.85	12.03	14.90
9506	50 MPH avg., cycle 10 miles		234	.034			1.61	3.01	4.62	5.75
9508	cycle 20 miles		180	.044			2.09	3.92	6.01	7.45
9510	cycle 30 miles		144	.056			2.61	4.90	7.51	9.35
9512	cycle 40 miles		108	.074			3.48	6.55	10.03	12.45
9514	cycle 50 miles		90	.089			4.18	7.85	12.03	14.90
9614	30 min. wait/ld./uld.,15 MPH, cycle 0.5 mi.		270	.030			1.39	2.61	4	4.97
9616	cycle 1 mile		252	.032			1.49	2.80	4.29	5.35
9618	cycle 2 miles		216	.037			1.74	3.26	5	6.20
9620	cycle 4 miles		180	.044			2.09	3.92	6.01	7.45
9622	cycle 6 miles		162	.049			2.32	4.35	6.67	8.30
9624	cycle 8 miles		126	.063			2.98	5.60	8.58	10.65
9625	cycle 10 miles		126	.063			2.98	5.60	8.58	10.65
9626	20 MPH avg., cycle 0.5 mile		270	.030			1.39	2.61	4	4.97
9628	cycle 1 mile		252	.032			1.49	2.80	4.29	5.35
9630	cycle 2 miles		234	.034			1.61	3.01	4.62	5.75
9632	cycle 4 miles		198	.040			1.90	3.56	5.46	6.80
9634	cycle 6 miles		180	.044			2.09	3.92	6.01	7.45
9636	cycle 8 miles		162	.049			2.32	4.35	6.67	8.30
9638	cycle 10 miles		144	.056			2.61	4.90	7.51	9.35
9640	25 MPH avg., cycle 4 miles		216	.037			1.74	3.26	5	6.20
9642	cycle 6 miles		198	.040			1.90	3.56	5.46	6.80
9644	cycle 8 miles		180	.044			2.09	3.92	6.01	7.45
9646	cycle 10 miles		162	.049			2.32	4.35	6.67	8.30
9650	30 MPH avg., cycle 4 miles		216	.037			1.74	3.26	5	6.20
9652	cycle 6 miles		198	.040			1.90	3.56	5.46	6.80
9654	cycle 8 miles		180	.044			2.09	3.92	6.01	7.45
9656	cycle 10 miles		162	.049			2.32	4.35	6.67	8.30
9660	35 MPH avg., cycle 4 miles		234	.034			1.61	3.01	4.62	5.75
9662	cycle 6 miles		216	.037			1.74	3.26	5	6.20

31 23 Excavation and Fill

31 23 23 – Fill

31 23 23.20 Hauling		Crew	Daily Output	Labor-Hours	Unit	Material	2019 Bare Costs Labor	Equipment	Total	Total Incl O&P
9664	cycle 8 miles	B-34I	198	.040	L.C.Y.		1.90	3.56	5.46	6.80
9666	cycle 10 miles		180	.044			2.09	3.92	6.01	7.45
9668	cycle 20 miles		126	.063			2.98	5.60	8.58	10.65
9670	cycle 30 miles		108	.074			3.48	6.55	10.03	12.45
9672	cycle 40 miles		90	.089			4.18	7.85	12.03	14.90
9674	40 MPH avg., cycle 6 miles		216	.037			1.74	3.26	5	6.20
9676	cycle 8 miles		198	.040			1.90	3.56	5.46	6.80
9678	cycle 10 miles		180	.044			2.09	3.92	6.01	7.45
9680	cycle 20 miles		144	.056			2.61	4.90	7.51	9.35
9682	cycle 30 miles		108	.074			3.48	6.55	10.03	12.45
9684	cycle 40 miles		90	.089			4.18	7.85	12.03	14.90
9686	cycle 50 miles		72	.111			5.20	9.80	15	18.60
9694	45 MPH avg., cycle 8 miles		216	.037			1.74	3.26	5	6.20
9696	cycle 10 miles		198	.040			1.90	3.56	5.46	6.80
9698	cycle 20 miles		144	.056			2.61	4.90	7.51	9.35
9700	cycle 30 miles		126	.063			2.98	5.60	8.58	10.65
9702	cycle 40 miles		108	.074			3.48	6.55	10.03	12.45
9704	cycle 50 miles		90	.089			4.18	7.85	12.03	14.90
9706	50 MPH avg., cycle 10 miles		198	.040			1.90	3.56	5.46	6.80
9708	cycle 20 miles		162	.049			2.32	4.35	6.67	8.30
9710	cycle 30 miles		126	.063			2.98	5.60	8.58	10.65
9712	cycle 40 miles		108	.074			3.48	6.55	10.03	12.45
9714	cycle 50 miles	↓	90	.089	↓		4.18	7.85	12.03	14.90

31 41 Shoring

31 41 16 – Sheet Piling

31 41 16.10 Sheet Piling Systems		Crew	Daily Output	Labor-Hours	Unit	Material	Labor	Equipment	Total	Total Incl O&P
0010	**SHEET PILING SYSTEMS**									
0020	Sheet piling, 50,000 psi steel, not incl. wales, 22 psf, left in place	B-40	10.81	5.920	Ton	1,600	315	335	2,250	2,600
0100	Drive, extract & salvage R314116-45		6	10.667		520	570	605	1,695	2,100
0300	20′ deep excavation, 27 psf, left in place		12.95	4.942		1,600	264	280	2,144	2,475
0400	Drive, extract & salvage		6.55	9.771	↓	520	520	555	1,595	1,975
1200	15′ deep excavation, 22 psf, left in place		983	.065	S.F.	18.55	3.47	3.69	25.71	30
1300	Drive, extract & salvage		545	.117		5.85	6.25	6.65	18.75	23.50
1500	20′ deep excavation, 27 psf, left in place		960	.067		23.50	3.56	3.78	30.84	35
1600	Drive, extract & salvage	↓	485	.132	↓	7.55	7.05	7.50	22.10	27.50
2100	Rent steel sheet piling and wales, first month				Ton	310			310	345
2200	Per added month					31			31	34.50
2300	Rental piling left in place, add to rental					1,150			1,150	1,250
2500	Wales, connections & struts, 2/3 salvage				↓	480			480	530
3900	Wood, solid sheeting, incl. wales, braces and spacers, R314116-40									
3910	drive, extract & salvage, 8′ deep excavation	B-31	330	.121	S.F.	2	5.30	.67	7.97	10.95
4000	10′ deep, 50 S.F./hr. in & 150 S.F./hr. out		300	.133		2.06	5.80	.74	8.60	11.85
4100	12′ deep, 45 S.F./hr. in & 135 S.F./hr. out		270	.148		2.12	6.45	.82	9.39	13
4200	14′ deep, 42 S.F./hr. in & 126 S.F./hr. out		250	.160		2.18	6.95	.88	10.01	13.90
4300	16′ deep, 40 S.F./hr. in & 120 S.F./hr. out		240	.167		2.25	7.25	.92	10.42	14.50
4400	18′ deep, 38 S.F./hr. in & 114 S.F./hr. out		230	.174		2.32	7.60	.96	10.88	15.05
4520	Left in place, 8′ deep, 55 S.F./hr.		440	.091		3.60	3.96	.50	8.06	10.50
4540	10′ deep, 50 S.F./hr.		400	.100		3.79	4.36	.55	8.70	11.40
4560	12′ deep, 45 S.F./hr.		360	.111		4	4.84	.61	9.45	12.35
4565	14′ deep, 42 S.F./hr.	↓	335	.119	↓	4.24	5.20	.66	10.10	13.25

31 41 Shoring

31 41 16 – Sheet Piling

31 41 16.10 Sheet Piling Systems		Crew	Daily Output	Labor-Hours	Unit	2019 Bare Costs Material	Labor	Equipment	Total	Total Incl O&P
4570	16′ deep, 40 S.F./hr.	B-31	320	.125	S.F.	4.50	5.45	.69	10.64	13.95
4700	Alternate pricing, left in place, 8′ deep		1.76	22.727	M.B.F.	810	990	125	1,925	2,525
4800	Drive, extract and salvage, 8′ deep	↓	1.32	30.303	"	720	1,325	167	2,212	2,975

31 52 Cofferdams

31 52 16 – Timber Cofferdams

31 52 16.10 Cofferdams		Crew	Daily Output	Labor-Hours	Unit	Material	Labor	Equipment	Total	Total Incl O&P
0010	**COFFERDAMS**									
0011	Incl. mobilization and temporary sheeting									
0080	Soldier beams & lagging H-piles with 3″ wood sheeting									
0090	horizontal between piles, including removal of wales & braces									
0100	No hydrostatic head, 15′ deep, 1 line of braces, minimum	B-50	545	.206	S.F.	9.05	10.40	4.79	24.24	31
0200	Maximum		495	.226		10.05	11.45	5.25	26.75	34.50
1300	No hydrostatic head, left in place, 15′ deep, 1 line of braces, min.		635	.176		12.10	8.90	4.11	25.11	31.50
1400	Maximum	↓	575	.195		12.95	9.85	4.54	27.34	34.50
2350	Lagging only, 3″ thick wood between piles 8′ OC, minimum	B-46	400	.120		2.01	5.65	.12	7.78	11.05
2370	Maximum		250	.192		3.02	9	.19	12.21	17.45
2400	Open sheeting no bracing, for trenches to 10′ deep, min.		1736	.028		.91	1.30	.03	2.24	3.03
2450	Maximum	↓	1510	.032		1.01	1.49	.03	2.53	3.44
2500	Tie-back method, add to open sheeting, add, minimum								20%	20%
2550	Maximum				↓				60%	60%
2700	Tie-backs only, based on tie-backs total length, minimum	B-46	86.80	.553	L.F.	15.80	26	.54	42.34	58
2750	Maximum		38.50	1.247	"	28	58.50	1.21	87.71	122
3500	Tie-backs only, typical average, 25′ long		2	24	Ea.	695	1,125	23.50	1,843.50	2,525
3600	35′ long	↓	1.58	30.380	"	925	1,425	29.50	2,379.50	3,250
6000	See also Section 31 41 16.10									

Division Notes

I	CREW	DAILY OUTPUT	LABOR-HOURS	UNIT	BARE COSTS MAT.	LABOR	EQUIP.	TOTAL	TOTAL INCL O&P

Estimating Tips

32 01 00 Operations and Maintenance of Exterior Improvements

- Recycling of asphalt pavement is becoming very popular and is an alternative to removal and replacement. It can be a good value engineering proposal if removed pavement can be recycled, either at the project site or at another site that is reasonably close to the project site. Sections on repair of flexible and rigid pavement are included.

32 10 00 Bases, Ballasts, and Paving

- When estimating paving, keep in mind the project schedule. Also note that prices for asphalt and concrete are generally higher in the cold seasons. Lines for pavement markings, including tactile warning systems and fence lines, are included.

32 90 00 Planting

- The timing of planting and guarantee specifications often dictate the costs for establishing tree and shrub growth and a stand of grass or ground cover. Establish the work performance schedule to coincide with the local planting season. Maintenance and growth guarantees can add 20–100% to the total landscaping cost and can be contractually cumbersome. The cost to replace trees and shrubs can be as high as 5% of the total cost, depending on the planting zone, soil conditions, and time of year.

Reference Numbers

Reference numbers are shown at the beginning of some major classifications. These numbers refer to related items in the Reference Section. The reference information may be an estimating procedure, an alternate pricing method, or technical information.

Note: Not all subdivisions listed here necessarily appear. ■

32 12 16 – Asphalt Paving

32 12 16.13 Plant-Mix Asphalt Paving		Crew	Daily Output	Labor-Hours	Unit	2019 Bare Costs Material	Labor	Equipment	Total	Total Incl O&P
0010	**PLANT-MIX ASPHALT PAVING**									
0020	And large paved areas with no hauling included									
0025	See Section 31 23 23.20 for hauling costs									
0080	Binder course, 1-1/2" thick	B-25	7725	.011	S.Y.	5.25	.51	.34	6.10	6.90
0120	2" thick		6345	.014		6.95	.63	.42	8	9.05
0130	2-1/2" thick		5620	.016		8.70	.71	.47	9.88	11.20
0160	3" thick		4905	.018		10.45	.81	.54	11.80	13.30
0170	3-1/2" thick		4520	.019		12.20	.88	.59	13.67	15.35
0200	4" thick		4140	.021		13.95	.96	.64	15.55	17.50
0300	Wearing course, 1" thick	B-25B	10575	.009		3.46	.42	.27	4.15	4.73
0340	1-1/2" thick		7725	.012		5.80	.57	.37	6.74	7.65
0380	2" thick		6345	.015		7.80	.69	.45	8.94	10.15
0420	2-1/2" thick		5480	.018		9.65	.80	.53	10.98	12.40
0460	3" thick		4900	.020		11.50	.90	.59	12.99	14.60
0470	3-1/2" thick		4520	.021		13.45	.97	.64	15.06	16.95
0480	4" thick		4140	.023		15.40	1.06	.70	17.16	19.30
0500	Open graded friction course	B-25C	5000	.010		2.47	.44	.46	3.37	3.90
0800	Alternate method of figuring paving costs									
0810	Binder course, 1-1/2" thick	B-25	630	.140	Ton	64	6.30	4.20	74.50	84.50
0811	2" thick		690	.128		64	5.75	3.84	73.59	83.50
0812	3" thick		800	.110		64	4.96	3.31	72.27	81.50
0813	4" thick		900	.098		64	4.41	2.94	71.35	80.50
0850	Wearing course, 1" thick	B-25B	575	.167		75.50	7.65	5	88.15	100
0851	1-1/2" thick		630	.152		75.50	7	4.57	87.07	98.50
0852	2" thick		690	.139		75.50	6.40	4.17	86.07	97
0853	2-1/2" thick		765	.125		75.50	5.75	3.76	85.01	96
0854	3" thick		800	.120		75.50	5.50	3.60	84.60	95.50
1000	Pavement replacement over trench, 2" thick	B-37	90	.533	S.Y.	7.20	23	1.70	31.90	45
1050	4" thick		70	.686		14.25	29.50	2.18	45.93	62.50
1080	6" thick		55	.873		22.50	37.50	2.78	62.78	85

32 12 16.14 Asphaltic Concrete Paving		Crew	Daily Output	Labor-Hours	Unit	Material	Labor	Equipment	Total	Total Incl O&P
0011	**ASPHALTIC CONCRETE PAVING**, parking lots & driveways									
0015	No asphalt hauling included									
0018	Use 6.05 C.Y. per inch per M.S.F. for hauling									
0020	6" stone base, 2" binder course, 1" topping	B-25C	9000	.005	S.F.	2.20	.25	.26	2.71	3.07
0025	2" binder course, 2" topping		9000	.005		2.65	.25	.26	3.16	3.57
0030	3" binder course, 2" topping		9000	.005		3.05	.25	.26	3.56	4
0035	4" binder course, 2" topping		9000	.005		3.43	.25	.26	3.94	4.42
0040	1-1/2" binder course, 1" topping		9000	.005		2.01	.25	.26	2.52	2.86
0042	3" binder course, 1" topping		9000	.005		2.59	.25	.26	3.10	3.50
0045	3" binder course, 3" topping		9000	.005		3.50	.25	.26	4.01	4.50
0050	4" binder course, 3" topping		9000	.005		3.89	.25	.26	4.40	4.93
0055	4" binder course, 4" topping		9000	.005		4.34	.25	.26	4.85	5.40
0300	Binder course, 1-1/2" thick		35000	.001		.58	.06	.07	.71	.81
0400	2" thick		25000	.002		.75	.09	.09	.93	1.06
0500	3" thick		15000	.003		1.17	.15	.15	1.47	1.67
0600	4" thick		10800	.004		1.53	.20	.21	1.94	2.22
0800	Sand finish course, 3/4" thick		41000	.001		.31	.05	.06	.42	.48
0900	1" thick		34000	.001		.39	.07	.07	.53	.60
1000	Fill pot holes, hot mix, 2" thick	B-16	4200	.008		.81	.33	.14	1.28	1.54
1100	4" thick		3500	.009		1.19	.39	.16	1.74	2.07
1120	6" thick		3100	.010		1.59	.44	.18	2.21	2.62

32 12 Flexible Paving

32 12 16 – Asphalt Paving

32 12 16.14 Asphaltic Concrete Paving		Crew	Daily Output	Labor-Hours	Unit	Material	Labor	Equipment	Total	Total Incl O&P
						2019 Bare Costs				
1140	Cold patch, 2" thick	B-51	3000	.016	S.F.	.90	.67	.07	1.64	2.08
1160	4" thick		2700	.018		1.72	.75	.07	2.54	3.10
1180	6" thick	↓	1900	.025	↓	2.68	1.06	.10	3.84	4.67

32 84 Planting Irrigation

32 84 23 – Underground Sprinklers

32 84 23.10 Sprinkler Irrigation System		Crew	Daily Output	Labor-Hours	Unit	Material	Labor	Equipment	Total	Total Incl O&P
0010	**SPRINKLER IRRIGATION SYSTEM**									
0011	For lawns									
0800	Residential system, custom, 1" supply	B-20	2000	.012	S.F.	.26	.55		.81	1.12
0900	1-1/2" supply	"	1800	.013	"	.43	.61		1.04	1.40

Division Notes

	CREW	DAILY OUTPUT	LABOR-HOURS	UNIT	BARE COSTS MAT.	BARE COSTS LABOR	BARE COSTS EQUIP.	BARE COSTS TOTAL	TOTAL INCL O&P

Estimating Tips

33 10 00 Water Utilities
33 30 00 Sanitary Sewerage Utilities
33 40 00 Storm Drainage Utilities

- Never assume that the water, sewer, and drainage lines will go in at the early stages of the project. Consider the site access needs before dividing the site in half with open trenches, loose pipe, and machinery obstructions. Always inspect the site to establish that the site drawings are complete. Check off all existing utilities on your drawings as you locate them. Be especially careful with underground utilities because appurtenances are sometimes buried during regrading or repaving operations. If you find any discrepancies, mark up the site plan for further research. Differing site conditions can be very costly if discovered later in the project.
- See also Section 33 01 00 for restoration of pipe where removal/ replacement may be undesirable. Use of new types of piping materials can reduce the overall project cost. Owners/design engineers should consider the installing contractor as a valuable source of current information on utility products and local conditions that could lead to significant cost savings.

Reference Numbers

Reference numbers are shown at the beginning of some major classifications. These numbers refer to related items in the Reference Section. The reference information may be an estimating procedure, an alternate pricing method, or technical information.

Note: Not all subdivisions listed here necessarily appear. ■

Note: "Powered in part by CINX™, based on licensed proprietary information of Harrison Publishing House, Inc."

Note: Trade Service, in part, has been used as a reference source for some of the material prices used in Division 33

33 01 10.10 Corrosion Resistance		Crew	Daily Output	Labor-Hours	Unit	2019 Bare Costs Material	Labor	Equipment	Total	Total Incl O&P
0010	**CORROSION RESISTANCE**									
0012	Wrap & coat, add to pipe, 4" diameter				L.F.	2.45			2.45	2.70
0020	5" diameter					3.09			3.09	3.40
0040	6" diameter					3.72			3.72	4.09
0060	8" diameter					5			5	5.50
0080	10" diameter					6.20			6.20	6.80
0100	12" diameter					7.40			7.40	8.15
0120	14" diameter					8.65			8.65	9.55
0140	16" diameter					10.70			10.70	11.75
0160	18" diameter					11.15			11.15	12.25
0180	20" diameter					12.35			12.35	13.60
0200	24" diameter					14.85			14.85	16.35
0220	Small diameter pipe, 1" diameter, add					.93			.93	1.02
0240	2" diameter					1.28			1.28	1.41
0260	2-1/2" diameter					1.61			1.61	1.77
0280	3" diameter					1.86			1.86	2.05
0300	Fittings, field covered, add				S.F.	9.10			9.10	10
0500	Coating, bituminous, per diameter inch, 1 coat, add				L.F.	.19			.19	.21
0540	3 coat					.48			.48	.53
0560	Coal tar epoxy, per diameter inch, 1 coat, add					.20			.20	.22
0600	3 coat					.56			.56	.62
1000	Polyethylene H.D. extruded, 0.025" thk., 1/2" diameter, add					.09			.09	.10
1020	3/4" diameter					.15			.15	.17
1040	1" diameter					.20			.20	.22
1060	1-1/4" diameter					.26			.26	.29
1080	1-1/2" diameter					.29			.29	.32
1100	0.030" thk., 2" diameter					.38			.38	.42
1120	2-1/2" diameter					.47			.47	.52
1140	0.035" thk., 3" diameter					.57			.57	.63
1160	3-1/2" diameter					.66			.66	.73
1180	4" diameter					.76			.76	.84
1200	0.040" thk., 5" diameter					.94			.94	1.03
1220	6" diameter					1.13			1.13	1.24
1240	8" diameter					1.45			1.45	1.60
1260	10" diameter					1.83			1.83	2.01
1280	12" diameter					2.20			2.20	2.42
1300	0.060" thk., 14" diameter					2.59			2.59	2.85
1320	16" diameter					2.96			2.96	3.26
1340	18" diameter					3.34			3.34	3.67
1360	20" diameter					3.72			3.72	4.09
1380	Fittings, field wrapped, add				S.F.	4.39			4.39	4.83

33 01 10.20 Pipe Repair		Crew	Daily Output	Labor-Hours	Unit	Material	Labor	Equipment	Total	Total Incl O&P
0010	**PIPE REPAIR**									
0020	Not including excavation or backfill									
0100	Clamp, stainless steel, lightweight, for steel pipe									
0110	3" long, 1/2" diameter pipe	1 Plum	34	.235	Ea.	10.80	14.85		25.65	34.50
0120	3/4" diameter pipe		32	.250		13.05	15.80		28.85	38
0130	1" diameter pipe		30	.267		13.90	16.85		30.75	41
0140	1-1/4" diameter pipe		28	.286		14.55	18.05		32.60	43
0150	1-1/2" diameter pipe		26	.308		15.15	19.45		34.60	45.50
0160	2" diameter pipe		24	.333		16.55	21		37.55	50
0170	2-1/2" diameter pipe		23	.348		19.35	22		41.35	54.50

33 01 10.20 Pipe Repair		Crew	Daily Output	Labor-Hours	Unit	2019 Bare Costs Material	Labor	Equipment	Total	Total Incl O&P
0180	3″ diameter pipe	1 Plum	22	.364	Ea.	21	23		44	57.50
0190	3-1/2″ diameter pipe		21	.381		22	24		46	60.50
0200	4″ diameter pipe	B-20	44	.545		22.50	25		47.50	62.50
0210	5″ diameter pipe		42	.571		26.50	26		52.50	68.50
0220	6″ diameter pipe		38	.632		30	29		59	77
0230	8″ diameter pipe		30	.800		36.50	36.50		73	96
0240	10″ diameter pipe		28	.857		126	39.50		165.50	199
0250	12″ diameter pipe		24	1		136	46		182	220
0260	14″ diameter pipe		22	1.091		153	50		203	244
0270	16″ diameter pipe		20	1.200		165	55		220	265
0280	18″ diameter pipe		18	1.333		190	61		251	300
0290	20″ diameter pipe		16	1.500		199	69		268	325
0300	24″ diameter pipe		14	1.714		221	78.50		299.50	360
0360	For 6″ long, add					100%	40%			
0370	For 9″ long, add					200%	100%			
0380	For 12″ long, add					300%	150%			
0390	For 18″ long, add					500%	200%			
0400	Pipe freezing for live repairs of systems 3/8″ to 6″									
0410	Note: Pipe freezing can also be used to install a valve into a live system									
0420	Pipe freezing each side 3/8″	2 Skwk	8	2	Ea.	575	107		682	800
0425	Pipe freezing each side 3/8″, second location same kit		8	2		30.50	107		137.50	197
0430	Pipe freezing each side 3/4″		8	2		510	107		617	725
0435	Pipe freezing each side 3/4″, second location same kit		8	2		30.50	107		137.50	197
0440	Pipe freezing each side 1-1/2″		6	2.667		510	142		652	775
0445	Pipe freezing each side 1-1/2″ , second location same kit		6	2.667		30.50	142		172.50	251
0450	Pipe freezing each side 2″		6	2.667		960	142		1,102	1,275
0455	Pipe freezing each side 2″, second location same kit		6	2.667		30.50	142		172.50	251
0460	Pipe freezing each side 2-1/2″ to 3″		6	2.667		955	142		1,097	1,275
0465	Pipe freezing each side 2-1/2″ to 3″, second location same kit		6	2.667		61	142		203	284
0470	Pipe freezing each side 4″		4	4		1,600	214		1,814	2,075
0475	Pipe freezing each side 4″, second location same kit		4	4		73.50	214		287.50	405
0480	Pipe freezing each side 5″ to 6″		4	4		4,250	214		4,464	5,000
0485	Pipe freezing each side 5″ to 6″, second location same kit		4	4		220	214		434	565
0490	Pipe freezing extra 20 lb. CO_2 cylinders (3/8″ to 2″ - 1 ea, 3″ - 2 ea)					227			227	250
0500	Pipe freezing extra 50 lb. CO_2 cylinders (4″ - 2 ea, 5″-6″ - 6 ea)					515			515	565
1000	Clamp, stainless steel, with threaded service tap									
1040	Full seal for iron, steel, PVC pipe									
1100	6″ long, 2″ diameter pipe	1 Plum	17	.471	Ea.	108	29.50		137.50	164
1110	2-1/2″ diameter pipe		16	.500		110	31.50		141.50	169
1120	3″ diameter pipe		15.60	.513		125	32.50		157.50	186
1130	3-1/2″ diameter pipe		15	.533		129	33.50		162.50	193
1140	4″ diameter pipe	B-20	32	.750		135	34.50		169.50	200
1150	6″ diameter pipe		28	.857		178	39.50		217.50	255
1160	8″ diameter pipe		21	1.143		187	52.50		239.50	285
1170	10″ diameter pipe		20	1.200		232	55		287	340
1180	12″ diameter pipe		17	1.412		268	64.50		332.50	390
1200	8″ long, 2″ diameter pipe	1 Plum	11.72	.683		136	43		179	214
1210	2-1/2″ diameter pipe		11	.727		139	46		185	222
1220	3″ diameter pipe		10.75	.744		145	47		192	230
1230	3-1/2″ diameter pipe		10.34	.774		149	49		198	238
1240	4″ diameter pipe	B-20	22	1.091		160	50		210	252
1250	6″ diameter pipe		19.31	1.243		187	57		244	293
1260	8″ diameter pipe		14.48	1.657		214	76		290	350

33 01 Operation and Maintenance of Utilities

33 01 10 – Operation and Maintenance of Water Utilities

33 01 10.20 Pipe Repair		Crew	Daily Output	Labor-Hours	Unit	2019 Bare Costs Material	Labor	Equipment	Total	Total Incl O&P
1270	10″ diameter pipe	B-20	13.80	1.739	Ea.	279	79.50		358.50	425
1280	12″ diameter pipe		11.72	2.048		310	94		404	485
1300	12″ long, 2″ diameter pipe	1 Plum	9.44	.847		217	53.50		270.50	320
1310	2-1/2″ diameter pipe		8.89	.900		212	57		269	320
1320	3″ diameter pipe		8.67	.923		219	58.50		277.50	330
1330	3-1/2″ diameter pipe		8.33	.960		229	60.50		289.50	340
1340	4″ diameter pipe	B-20	17.78	1.350		240	62		302	360
1350	6″ diameter pipe		15.56	1.542		279	70.50		349.50	410
1360	8″ diameter pipe		11.67	2.057		330	94.50		424.50	510
1370	10″ diameter pipe		11.11	2.160		420	99		519	615
1380	12″ diameter pipe		9.44	2.542		485	117		602	710
1400	20″ long, 2″ diameter pipe	1 Plum	8.10	.988		247	62.50		309.50	365
1410	2-1/2″ diameter pipe		7.62	1.050		270	66.50		336.50	395
1420	3″ diameter pipe		7.43	1.077		315	68		383	445
1430	3-1/2″ diameter pipe		7.14	1.120		325	71		396	460
1440	4″ diameter pipe	B-20	15.24	1.575		390	72		462	540
1450	6″ diameter pipe		13.33	1.800		460	82.50		542.50	630
1460	8″ diameter pipe		10	2.400		520	110		630	735
1470	10″ diameter pipe		9.52	2.521		635	116		751	875
1480	12″ diameter pipe		8.10	2.963		750	136		886	1,025
1600	Clamp, stainless steel, single section									
1640	Full seal for iron, steel, PVC pipe									
1700	6″ long, 2″ diameter pipe	1 Plum	17	.471	Ea.	81	29.50		110.50	134
1710	2-1/2″ diameter pipe		16	.500		85	31.50		116.50	141
1720	3″ diameter pipe		15.60	.513		84	32.50		116.50	141
1730	3-1/2″ diameter pipe		15	.533		102	33.50		135.50	163
1740	4″ diameter pipe	B-20	32	.750		109	34.50		143.50	172
1750	6″ diameter pipe		27	.889		136	40.50		176.50	211
1760	8″ diameter pipe		21	1.143		139	52.50		191.50	233
1770	10″ diameter pipe		20	1.200		208	55		263	315
1780	12″ diameter pipe		17	1.412		210	64.50		274.50	330
1800	8″ long, 2″ diameter pipe	1 Plum	11.72	.683		110	43		153	186
1805	2-1/2″ diameter pipe		11.03	.725		116	46		162	197
1810	3″ diameter pipe		10.76	.743		121	47		168	204
1815	3-1/2″ diameter pipe		10.34	.774		128	49		177	215
1820	4″ diameter pipe	B-20	22.07	1.087		136	50		186	225
1825	6″ diameter pipe		19.31	1.243		163	57		220	266
1830	8″ diameter pipe		14.48	1.657		190	76		266	325
1835	10″ diameter pipe		13.79	1.740		254	80		334	400
1840	12″ diameter pipe		11.72	2.048		288	94		382	455
1850	12″ long, 2″ diameter pipe	1 Plum	9.44	.847		183	53.50		236.50	282
1855	2-1/2″ diameter pipe		8.89	.900		188	57		245	292
1860	3″ diameter pipe		8.67	.923		194	58.50		252.50	300
1865	3-1/2″ diameter pipe		8.33	.960		204	60.50		264.50	315
1870	4″ diameter pipe	B-20	17.78	1.350		219	62		281	335
1875	6″ diameter pipe		15.56	1.542		253	70.50		323.50	385
1880	8″ diameter pipe		11.67	2.057		305	94.50		399.50	480
1885	10″ diameter pipe		11.11	2.160		400	99		499	590
1890	12″ diameter pipe		9.44	2.542		460	117		577	680
1900	20″ long, 2″ diameter pipe	1 Plum	8.10	.988		223	62.50		285.50	340
1905	2-1/2″ diameter pipe		7.62	1.050		246	66.50		312.50	370
1910	3″ diameter pipe		7.43	1.077		282	68		350	410
1915	3-1/2″ diameter pipe		7.14	1.120		300	71		371	435

33 01 10 – Operation and Maintenance of Water Utilities

33 01 10.20 Pipe Repair		Crew	Daily Output	Labor-Hours	Unit	2019 Bare Costs Material	Labor	Equipment	Total	Total Incl O&P
1920	4″ diameter pipe	B-20	15.24	1.575	Ea.	375	72		447	520
1925	6″ diameter pipe		13.33	1.800		430	82.50		512.50	600
1930	8″ diameter pipe		10	2.400		500	110		610	715
1935	10″ diameter pipe		9.52	2.521		610	116		726	845
1940	12″ diameter pipe	↓	8.10	2.963	↓	725	136		861	1,000
2000	Clamp, stainless steel, two section									
2040	Full seal, for iron, steel, PVC pipe									
2100	8″ long, 4″ diameter pipe	B-20	24	1	Ea.	214	46		260	305
2110	6″ diameter pipe		20	1.200		246	55		301	355
2120	8″ diameter pipe		13	1.846		283	84.50		367.50	440
2130	10″ diameter pipe		12	2		243	91.50		334.50	405
2140	12″ diameter pipe		10	2.400		365	110		475	565
2200	10″ long, 4″ diameter pipe		16	1.500		278	69		347	410
2210	6″ diameter pipe		13	1.846		315	84.50		399.50	475
2220	8″ diameter pipe		9	2.667		350	122		472	570
2230	10″ diameter pipe		8	3		455	137		592	710
2240	12″ diameter pipe	↓	7	3.429	↓	525	157		682	820
2242	Clamp, stainless steel, three section									
2244	Full seal, for iron, steel, PVC pipe									
2250	10″ long, 14″ diameter pipe	B-20	6.40	3.750	Ea.	670	172		842	995
2260	16″ diameter pipe		6	4		665	183		848	1,000
2270	18″ diameter pipe		5	4.800		730	220		950	1,125
2280	20″ diameter pipe		4.60	5.217		945	239		1,184	1,425
2290	24″ diameter pipe	↓	4	6		1,325	275		1,600	1,900
2320	For 12″ long, add to 10″					15%	25%			
2330	For 20″ long, add to 10″				↓	70%	55%			
8000	For internal cleaning and inspection, see Section 33 01 30.11									
8100	For pipe testing, see Section 23 05 93.50									

33 01 30 – Operation and Maintenance of Sewer Utilities

33 01 30.11 Television Inspection of Sewers		Crew	Daily Output	Labor-Hours	Unit	Material	Labor	Equipment	Total	Total Incl O&P
0010	**TELEVISION INSPECTION OF SEWERS**									
0100	Pipe internal cleaning & inspection, cleaning, pressure pipe systems									
0120	Pig method, lengths 1000′ to 10,000′									
0140	4″ diameter thru 24″ diameter, minimum				L.F.				3.60	4.14
0160	Maximum				″				18	21
6000	Sewage/sanitary systems									
6100	Power rodder with header & cutters									
6110	Mobilization charge, minimum				Total				695	800
6120	Mobilization charge, maximum				″				9,125	10,600
6140	Cleaning 4″-12″ diameter				L.F.				6	6.60
6190	14″-24″ diameter								8	8.80
6240	30″ diameter								9.60	10.50
6250	36″ diameter								12	13.20
6260	48″ diameter								16	17.60
6270	60″ diameter								24	26.50
6280	72″ diameter				↓				48	53
9000	Inspection, television camera with video									
9060	up to 500 linear feet				Total				715	820

33 01 Operation and Maintenance of Utilities

33 01 30 – Operation and Maintenance of Sewer Utilities

33 01 30.23 Pipe Bursting		Crew	Daily Output	Labor-Hours	Unit	2019 Bare Costs Material	Labor	Equipment	Total	Total Incl O&P
0010	PIPE BURSTING									
0011	300′ runs, replace with HDPE pipe									
0020	Not including excavation, backfill, shoring, or dewatering									
0100	6″ to 15″ diameter, minimum				L.F.				200	220
0200	Maximum								550	605
0300	18″ to 36″ diameter, minimum								650	715
0400	Maximum				↓				1,050	1,150
0500	Mobilize and demobilize, minimum				Job				3,000	3,300
0600	Maximum				″				32,100	35,400

33 01 30.74 Sliplining, Excludes Cleaning

		Crew	Daily Output	Labor-Hours	Unit	Material	Labor	Equipment	Total	Total Incl O&P
0010	**SLIPLINING, excludes cleaning** and video inspection									
0020	Pipe relined with one pipe size smaller than original (4″ for 6″)									
0100	6″ diameter, original size	B-6B	600	.080	L.F.	5.75	3.34	1.43	10.52	13
0150	8″ diameter, original size		600	.080		9	3.34	1.43	13.77	16.55
0200	10″ diameter, original size		600	.080		11.65	3.34	1.43	16.42	19.50
0250	12″ diameter, original size		400	.120		13.40	5	2.15	20.55	24.50
0300	14″ diameter, original size	↓	400	.120		15.60	5	2.15	22.75	27
0350	16″ diameter, original size	B-6C	300	.160		19.45	6.70	8.50	34.65	41
0400	18″ diameter, original size	″	300	.160	↓	28.50	6.70	8.50	43.70	51
1000	Pipe HDPE lining, make service line taps	B-6	4	6	Ea.	89	267	80	436	590

33 05 Common Work Results for Utilities

33 05 07 – Trenchless Installation of Utility Piping

33 05 07.23 Utility Boring and Jacking

		Crew	Daily Output	Labor-Hours	Unit	Material	Labor	Equipment	Total	Total Incl O&P
0010	UTILITY BORING AND JACKING									
0011	Casing only, 100′ minimum,									
0020	not incl. jacking pits or dewatering									
0100	Roadwork, 1/2″ thick wall, 24″ diameter casing	B-42	20	3.200	L.F.	119	148	56	323	420
0200	36″ diameter		16	4		214	185	70.50	469.50	595
0300	48″ diameter		15	4.267		287	197	75	559	700
0500	Railroad work, 24″ diameter		15	4.267		119	197	75	391	515
0600	36″ diameter		14	4.571		214	211	80.50	505.50	645
0700	48″ diameter	↓	12	5.333		287	247	93.50	627.50	795
0900	For ledge, add								20%	20%
1000	Small diameter boring, 3″, sandy soil	B-82	900	.018		21.50	.82	.09	22.41	25
1040	Rocky soil	″	500	.032	↓	21.50	1.48	.16	23.14	26
1100	Prepare jacking pits, incl. mobilization & demobilization, minimum				Ea.				3,225	3,700
1101	Maximum				″				22,000	25,500

33 05 07.36 Microtunneling

		Crew	Daily Output	Labor-Hours	Unit	Material	Labor	Equipment	Total	Total Incl O&P
0010	MICROTUNNELING									
0011	Not including excavation, backfill, shoring,									
0020	or dewatering, average 50′/day, slurry method									
0100	24″ to 48″ outside diameter, minimum				L.F.				965	965
0110	Adverse conditions, add				″				500	500
1000	Rent microtunneling machine, average monthly lease				Month				97,500	107,000
1010	Operating technician				Day				630	705
1100	Mobilization and demobilization, minimum				Job				41,200	45,900
1110	Maximum				″				445,500	490,500

33 05 61 – Concrete Manholes

33 05 61.10 Storm Drainage Manholes, Frames and Covers		Crew	Daily Output	Labor-Hours	Unit	2019 Bare Costs Material	Labor	Equipment	Total	Total Incl O&P
0010	**STORM DRAINAGE MANHOLES, FRAMES & COVERS**									
0020	Excludes footing, excavation, backfill (See line items for frame & cover)									
0050	Brick, 4' inside diameter, 4' deep	D-1	1	16	Ea.	590	730		1,320	1,775
0100	6' deep		.70	22.857		835	1,050		1,885	2,525
0150	8' deep		.50	32		1,075	1,475		2,550	3,400
0200	For depths over 8', add		4	4	V.L.F.	87.50	183		270.50	375
0400	Concrete blocks (radial), 4' ID, 4' deep		1.50	10.667	Ea.	415	490		905	1,200
0500	6' deep		1	16		550	730		1,280	1,725
0600	8' deep		.70	22.857		685	1,050		1,735	2,350
0700	For depths over 8', add		5.50	2.909	V.L.F.	70	133		203	280
0800	Concrete, cast in place, 4' x 4', 8" thick, 4' deep	C-14H	2	24	Ea.	555	1,200	13.20	1,768.20	2,450
0900	6' deep		1.50	32		795	1,600	17.60	2,412.60	3,325
1000	8' deep		1	48		1,150	2,425	26.50	3,601.50	4,950
1100	For depths over 8', add		8	6	V.L.F.	127	300	3.30	430.30	600
1110	Precast, 4' ID, 4' deep	B-22	4.10	7.317	Ea.	865	350	47.50	1,262.50	1,525
1120	6' deep		3	10		1,075	480	65	1,620	1,975
1130	8' deep		2	15		1,200	720	97.50	2,017.50	2,525
1140	For depths over 8', add		16	1.875	V.L.F.	128	90.50	12.15	230.65	291
1150	5' ID, 4' deep	B-6	3	8	Ea.	1,750	355	107	2,212	2,575
1160	6' deep		2	12		2,075	535	160	2,770	3,250
1170	8' deep		1.50	16		2,575	715	213	3,503	4,150
1180	For depths over 8', add		12	2	V.L.F.	290	89	26.50	405.50	485
1190	6' ID, 4' deep		2	12	Ea.	2,475	535	160	3,170	3,700
1200	6' deep		1.50	16		3,025	715	213	3,953	4,625
1210	8' deep		1	24		3,725	1,075	320	5,120	6,075
1220	For depths over 8', add		8	3	V.L.F.	405	134	40	579	690
1250	Slab tops, precast, 8" thick									
1300	4' diameter manhole	B-6	8	3	Ea.	282	134	40	456	555
1400	5' diameter manhole		7.50	3.200		460	143	42.50	645.50	765
1500	6' diameter manhole		7	3.429		715	153	45.50	913.50	1,075
3800	Steps, heavyweight cast iron, 7" x 9"	1 Bric	40	.200		18.75	10.20		28.95	36
3900	8" x 9"		40	.200		22.50	10.20		32.70	40.50
3928	12" x 10-1/2"		40	.200		27.50	10.20		37.70	45.50
4000	Standard sizes, galvanized steel		40	.200		25	10.20		35.20	43
4100	Aluminum		40	.200		29	10.20		39.20	47.50
4150	Polyethylene		40	.200		31	10.20		41.20	49.50

33 05 97 – Identification and Signage for Utilities

33 05 97.05 Utility Connection

		Crew	Daily Output	Labor-Hours	Unit	Material	Labor	Equipment	Total	Total Incl O&P
0010	**UTILITY CONNECTION**									
0020	Water, sanitary, stormwater, gas, single connection	B-14	1	48	Ea.	3,050	2,075	320	5,445	6,825
0030	Telecommunication	"	3	16	"	435	690	107	1,232	1,650

33 05 97.10 Utility Accessories

		Crew	Daily Output	Labor-Hours	Unit	Material	Labor	Equipment	Total	Total Incl O&P
0010	**UTILITY ACCESSORIES** G1030–805									
0400	Underground tape, detectable, reinforced, alum. foil core, 2"	1 Clab	150	.053	C.L.F.	9	2.19		11.19	13.20
0500	6"	"	140	.057	"	36	2.35		38.35	43

33 11 13 – Potable Water Supply Wells

	33 11 13.10 Wells and Accessories	Crew	Daily Output	Labor-Hours	Unit	2019 Bare Costs Material	Labor	Equipment	Total	Total Incl O&P
0010	**WELLS & ACCESSORIES**									
0011	Domestic									
0100	Drilled, 4″ to 6″ diameter	B-23	120	.333	L.F.		13.80	22.50	36.30	45.50
0200	8″ diameter	″	95.20	.420	″		17.40	28	45.40	57.50
0400	Gravel pack well, 40′ deep, incl. gravel & casing, complete									
0500	24″ diameter casing x 18″ diameter screen	B-23	.13	308	Total	39,000	12,800	20,600	72,400	85,000
0600	36″ diameter casing x 18″ diameter screen		.12	333	″	39,900	13,800	22,300	76,000	89,500
0800	Observation wells, 1-1/4″ riser pipe		163	.245	V.L.F.	20.50	10.15	16.40	47.05	56
0900	For flush Buffalo roadway box, add	1 Skwk	16.60	.482	Ea.	51.50	25.50		77	96
1200	Test well, 2-1/2″ diameter, up to 50′ deep (15 to 50 GPM)	B-23	1.51	26.490	″	810	1,100	1,775	3,685	4,500
1300	Over 50′ deep, add	″	121.80	.328	L.F.	21.50	13.60	22	57.10	68
1500	Pumps, installed in wells to 100′ deep, 4″ submersible									
1510	1/2 HP	Q-1	3.22	4.969	Ea.	690	282		972	1,175
1520	3/4 HP		2.66	6.015		860	340		1,200	1,450
1600	1 HP		2.29	6.987		925	395		1,320	1,625
1700	1-1/2 HP	Q-22	1.60	10		1,850	570	295	2,715	3,225
1800	2 HP		1.33	12.030		1,825	685	355	2,865	3,450
1900	3 HP		1.14	14.035		2,250	800	415	3,465	4,125
2000	5 HP		1.14	14.035		3,100	800	415	4,315	5,050
2050	Remove and install motor only, 4 HP		1.14	14.035		1,300	800	415	2,515	3,100
3000	Pump, 6″ submersible, 25′ to 150′ deep, 25 HP, 249 to 297 GPM		.89	17.978		2,000	1,025	530	3,555	4,300
3100	25′ to 500′ deep, 30 HP, 100 to 300 GPM		.73	21.918		2,300	1,250	645	4,195	5,100
8110	Well screen assembly, stainless steel, 2″ diameter	B-23A	273	.088	L.F.	82	4.08	9.30	95.38	106
8120	3″ diameter		253	.095		138	4.40	10.05	152.45	170
8130	4″ diameter		200	.120		166	5.55	12.70	184.25	204
8140	5″ diameter		168	.143		181	6.65	15.10	202.75	226
8150	6″ diameter		126	.190		208	8.85	20	236.85	263
8160	8″ diameter		98.50	.244		280	11.30	26	317.30	355
8170	10″ diameter		73	.329		350	15.25	35	400.25	445
8180	12″ diameter		62.50	.384		400	17.80	40.50	458.30	510
8190	14″ diameter		54.30	.442		450	20.50	46.50	517	580
8200	16″ diameter		48.30	.497		500	23	52.50	575.50	645
8210	18″ diameter		39.20	.612		630	28.50	64.50	723	810
8220	20″ diameter		31.20	.769		705	35.50	81.50	822	925
8230	24″ diameter		23.80	1.008		835	47	107	989	1,100
8240	26″ diameter		21	1.143		940	53	121	1,114	1,250
8244	Well casing or drop pipe, PVC, 1/2″ diameter		550	.044		1.29	2.02	4.61	7.92	9.55
8245	3/4″ diameter		550	.044		1.30	2.02	4.61	7.93	9.60
8246	1″ diameter		550	.044		1.34	2.02	4.61	7.97	9.60
8247	1-1/4″ diameter		520	.046		1.69	2.14	4.88	8.71	10.45
8248	1-1/2″ diameter		490	.049		1.79	2.27	5.20	9.26	11.10
8249	1-3/4″ diameter		380	.063		1.83	2.93	6.70	11.46	13.80
8250	2″ diameter		280	.086		2.23	3.98	9.05	15.26	18.40
8252	3″ diameter		260	.092		4.11	4.28	9.75	18.14	21.50
8254	4″ diameter		205	.117		4.40	5.45	12.40	22.25	26.50
8255	5″ diameter		170	.141		4.42	6.55	14.95	25.92	31
8256	6″ diameter		130	.185		8.65	8.55	19.50	36.70	44
8258	8″ diameter		100	.240		13.10	11.15	25.50	49.75	59
8260	10″ diameter		73	.329		19.65	15.25	35	69.90	82.50
8262	12″ diameter		62	.387		23.50	17.95	41	82.45	98
8300	Slotted PVC, 1-1/4″ diameter		521	.046		2.75	2.14	4.87	9.76	11.60
8310	1-1/2″ diameter		488	.049		3.30	2.28	5.20	10.78	12.75
8320	2″ diameter		273	.088		4.18	4.08	9.30	17.56	21

33 11 Groundwater Sources

33 11 13 – Potable Water Supply Wells

	33 11 13.10 Wells and Accessories	Crew	Daily Output	Labor-Hours	Unit	2019 Bare Costs Material	Labor	Equipment	Total	Total Incl O&P
8330	3" diameter	B-23A	253	.095	L.F.	6.30	4.40	10.05	20.75	24.50
8340	4" diameter		200	.120		8.50	5.55	12.70	26.75	31.50
8350	5" diameter		168	.143		14.65	6.65	15.10	36.40	42.50
8360	6" diameter		126	.190		16.55	8.85	20	45.40	53.50
8370	8" diameter		98.50	.244		27	11.30	26	64.30	75
8400	Artificial gravel pack, 2" screen, 6" casing	B-23B	174	.138		4.80	6.40	16.30	27.50	33
8405	8" casing		111	.216		6.10	10.05	25.50	41.65	50
8410	10" casing		74.50	.322		8.90	14.95	38	61.85	74.50
8415	12" casing		60	.400		12.35	18.55	47	77.90	93.50
8420	14" casing		50.20	.478		14.25	22	56.50	92.75	111
8425	16" casing		40.70	.590		17.95	27.50	69.50	114.95	138
8430	18" casing		36	.667		21.50	31	78.50	131	157
8435	20" casing		29.50	.814		25.50	38	96	159.50	191
8440	24" casing		25.70	.934		29	43.50	110	182.50	219
8445	26" casing		24.60	.976		32.50	45.50	115	193	231
8450	30" casing		20	1.200		37	55.50	142	234.50	281
8455	36" casing		16.40	1.463		42	68	173	283	340
8500	Develop well		8	3	Hr.	290	139	355	784	920
8550	Pump test well		8	3		94.50	139	355	588.50	705
8560	Standby well	B-23A	8	3		103	139	315	557	675
8570	Standby, drill rig		8	3			139	315	454	560
8580	Surface seal well, concrete filled		1	24	Ea.	1,025	1,125	2,550	4,700	5,600
8590	Well test pump, install & remove	B-23	1	40			1,650	2,675	4,325	5,450
8600	Well sterilization, chlorine	2 Clab	1	16		123	655		778	1,125
8610	Well water pressure switch	1 Clab	12	.667		103	27.50		130.50	155
8630	Well water pressure switch with manual reset	"	12	.667		129	27.50		156.50	184
9950	See Section 31 23 19.40 for wellpoints									
9960	See Section 31 23 19.30 for drainage wells									

	33 11 13.20 Water Supply Wells, Pumps	Crew	Daily Output	Labor-Hours	Unit	Material	Labor	Equipment	Total	Total Incl O&P
0010	**WATER SUPPLY WELLS, PUMPS**									
0011	With pressure control									
1000	Deep well, jet, 42 gal. galvanized tank									
1040	3/4 HP	1 Plum	.80	10	Ea.	1,125	630		1,755	2,200
3000	Shallow well, jet, 30 gal. galvanized tank									
3040	1/2 HP	1 Plum	2	4	Ea.	910	253		1,163	1,375

33 14 Water Utility Transmission and Distribution

33 14 13 – Public Water Utility Distribution Piping

	33 14 13.15 Water Supply, Ductile Iron Pipe	Crew	Daily Output	Labor-Hours	Unit	Material	Labor	Equipment	Total	Total Incl O&P
0010	**WATER SUPPLY, DUCTILE IRON PIPE** R331113-80									
0020	Not including excavation or backfill									
2000	Pipe, class 50 water piping, 18' lengths									
2020	Mechanical joint, 4" diameter	B-21A	200	.200	L.F.	36	10.20	1.86	48.06	57
2040	6" diameter		160	.250		43	12.75	2.33	58.08	69.50
2060	8" diameter		133.33	.300		53.50	15.30	2.79	71.59	84.50
2080	10" diameter		114.29	.350		70.50	17.85	3.25	91.60	109
2100	12" diameter		105.26	.380		90	19.40	3.53	112.93	132
2120	14" diameter		100	.400		96.50	20.50	3.72	120.72	141
2140	16" diameter		72.73	.550		104	28	5.10	137.10	162
2160	18" diameter		68.97	.580		128	29.50	5.40	162.90	190
2170	20" diameter		57.14	.700		139	35.50	6.50	181	214

33 14 13 – Public Water Utility Distribution Piping

33 14 13.15 Water Supply, Ductile Iron Pipe		Crew	Daily Output	Labor-Hours	Unit	Material	2019 Bare Costs Labor	Equipment	Total	Total Incl O&P
2180	24" diameter	B-21A	47.06	.850	L.F.	165	43.50	7.90	216.40	256
3000	Push-on joint, 4" diameter		400	.100		20.50	5.10	.93	26.53	31
3020	6" diameter		333.33	.120		23	6.10	1.12	30.22	35.50
3040	8" diameter		200	.200		31	10.20	1.86	43.06	52
3060	10" diameter		181.82	.220		46.50	11.25	2.05	59.80	70.50
3080	12" diameter		160	.250		49	12.75	2.33	64.08	76
3100	14" diameter		133.33	.300		49	15.30	2.79	67.09	80
3120	16" diameter		114.29	.350		57.50	17.85	3.25	78.60	93.50
3140	18" diameter		100	.400		64	20.50	3.72	88.22	105
3160	20" diameter		88.89	.450		66.50	23	4.18	93.68	112
3180	24" diameter		76.92	.520		91	26.50	4.84	122.34	145
8000	Piping, fittings, mechanical joint, AWWA C110									
8006	90° bend, 4" diameter	B-20A	16	2	Ea.	175	99		274	340
8020	6" diameter		12.80	2.500		256	124		380	470
8040	8" diameter		10.67	2.999		475	148		623	745
8060	10" diameter	B-21A	11.43	3.500		695	179	32.50	906.50	1,075
8080	12" diameter		10.53	3.799		885	194	35.50	1,114.50	1,300
8100	14" diameter		10	4		1,375	204	37	1,616	1,850
8120	16" diameter		7.27	5.502		1,750	281	51	2,082	2,400
8140	18" diameter		6.90	5.797		2,400	296	54	2,750	3,150
8160	20" diameter		5.71	7.005		2,925	360	65	3,350	3,825
8180	24" diameter		4.70	8.511		4,675	435	79	5,189	5,900
8200	Wye or tee, 4" diameter	B-20A	10.67	2.999		375	148		523	640
8220	6" diameter		8.53	3.751		570	185		755	905
8240	8" diameter		7.11	4.501		895	223		1,118	1,325
8260	10" diameter	B-21A	7.62	5.249		1,150	268	49	1,467	1,725
8280	12" diameter		7.02	5.698		2,000	291	53	2,344	2,700
8300	14" diameter		6.67	5.997		2,700	305	56	3,061	3,500
8320	16" diameter		4.85	8.247		3,125	420	76.50	3,621.50	4,150
8340	18" diameter		4.60	8.696		4,300	445	81	4,826	5,475
8360	20" diameter		3.81	10.499		6,100	535	97.50	6,732.50	7,600
8380	24" diameter		3.14	12.739		8,550	650	118	9,318	10,500
8398	45° bend, 4" diameter	B-20A	16	2		214	99		313	385
8400	6" diameter		12.80	2.500		271	124		395	485
8405	8" diameter		10.67	2.999		390	148		538	655
8410	12" diameter	B-21A	10.53	3.799		830	194	35.50	1,059.50	1,250
8420	16" diameter		7.27	5.502		1,575	281	51	1,907	2,200
8430	20" diameter		5.71	7.005		2,475	360	65	2,900	3,300
8440	24" diameter		4.70	8.511		3,450	435	79	3,964	4,550
8450	Decreaser, 6" x 4" diameter	B-20A	14.22	2.250		234	111		345	425
8460	8" x 6" diameter	"	11.64	2.749		360	136		496	600
8470	10" x 6" diameter	B-21A	13.33	3.001		440	153	28	621	745
8480	12" x 6" diameter		12.70	3.150		590	161	29.50	780.50	920
8490	16" x 6" diameter		10	4		965	204	37	1,206	1,400
8500	20" x 6" diameter		8.42	4.751		1,625	242	44	1,911	2,225
8552	For water utility valves see Section 33 14 19									
8700	Joint restraint, ductile iron mechanical joints									
8710	4" diameter	B-20A	32	1	Ea.	32	49.50		81.50	110
8720	6" diameter		25.60	1.250		38.50	62		100.50	136
8730	8" diameter		21.33	1.500		57	74		131	175
8740	10" diameter		18.28	1.751		88.50	86.50		175	228
8750	12" diameter		16.84	1.900		117	94		211	270
8760	14" diameter		16	2		145	99		244	310

33 14 13.15 Water Supply, Ductile Iron Pipe		Crew	Daily Output	Labor-Hours	Unit	Material	2019 Bare Costs Labor	Equipment	Total	Total Incl O&P
8770	16″ diameter	B-20A	11.64	2.749	Ea.	180	136		316	405
8780	18″ diameter		11.03	2.901		254	143		397	495
8785	20″ diameter		9.14	3.501		315	173		488	605
8790	24″ diameter		7.53	4.250		430	210		640	785
9600	Steel sleeve with tap, 4″ diameter	B-20	3	8		425	365		790	1,025
9620	6″ diameter		2	12		495	550		1,045	1,375
9630	8″ diameter		2	12		670	550		1,220	1,575

33 14 13.25 Water Supply, Polyvinyl Chloride Pipe		Crew	Daily Output	Labor-Hours	Unit	Material	Labor	Equipment	Total	Total Incl O&P
0010	**WATER SUPPLY, POLYVINYL CHLORIDE PIPE** R331113-80									
0020	Not including excavation or backfill, unless specified									
2100	PVC pipe, Class 150, 1-1/2″ diameter	Q-1A	750	.013	L.F.	.52	.85		1.37	1.84
2120	2″ diameter		686	.015		.88	.93		1.81	2.36
2140	2-1/2″ diameter		500	.020		1.19	1.27		2.46	3.22
2160	3″ diameter	B-20	430	.056		1.62	2.56		4.18	5.65
3010	AWWA C905, PR 100, DR 25									
3030	14″ diameter	B-20A	213	.150	L.F.	13.85	7.45		21.30	26.50
3040	16″ diameter		200	.160		18.65	7.90		26.55	32.50
3050	18″ diameter		160	.200		24.50	9.90		34.40	41.50
3060	20″ diameter		133	.241		29.50	11.90		41.40	50
3070	24″ diameter		107	.299		44	14.80		58.80	71
3080	30″ diameter		80	.400		72	19.80		91.80	109
3090	36″ diameter		80	.400		111	19.80		130.80	152
3100	42″ diameter		60	.533		146	26.50		172.50	201
3200	48″ diameter		60	.533		191	26.50		217.50	250
4520	Pressure pipe Class 150, SDR 18, AWWA C900, 4″ diameter		380	.084		2.80	4.16		6.96	9.35
4530	6″ diameter		316	.101		5.05	5		10.05	13.10
4540	8″ diameter		264	.121		8.50	6		14.50	18.35
4550	10″ diameter		220	.145		12.70	7.20		19.90	25
4560	12″ diameter		186	.172		17.75	8.50		26.25	32.50
8000	Fittings with rubber gasket									
8003	Class 150, DR 18									
8006	90° bend , 4″ diameter	B-20	100	.240	Ea.	41	11		52	61.50
8020	6″ diameter		90	.267		76.50	12.20		88.70	103
8040	8″ diameter		80	.300		140	13.75		153.75	175
8060	10″ diameter		50	.480		330	22		352	400
8080	12″ diameter		30	.800		415	36.50		451.50	510
8100	Tee, 4″ diameter		90	.267		56.50	12.20		68.70	81
8120	6″ diameter		80	.300		137	13.75		150.75	172
8140	8″ diameter		70	.343		201	15.70		216.70	245
8160	10″ diameter		40	.600		665	27.50		692.50	770
8180	12″ diameter		20	1.200		905	55		960	1,075
8200	45° bend, 4″ diameter		100	.240		41	11		52	61.50
8220	6″ diameter		90	.267		75	12.20		87.20	101
8240	8″ diameter		50	.480		136	22		158	183
8260	10″ diameter		50	.480		291	22		313	355
8280	12″ diameter		30	.800		380	36.50		416.50	470
8300	Reducing tee 6″ x 4″		100	.240		126	11		137	155
8320	8″ x 6″		90	.267		202	12.20		214.20	241
8330	10″ x 6″		90	.267		340	12.20		352.20	395
8340	10″ x 8″		90	.267		355	12.20		367.20	410
8350	12″ x 6″		90	.267		405	12.20		417.20	465
8360	12″ x 8″		90	.267		425	12.20		437.20	485

33 14 13.25 Water Supply, Polyvinyl Chloride Pipe		Crew	Daily Output	Labor-Hours	Unit	2019 Bare Costs Material	Labor	Equipment	Total	Total Incl O&P
8400	Tapped service tee (threaded type) 6" x 6" x 3/4"	B-20	100	.240	Ea.	88	11		99	114
8430	6" x 6" x 1"		90	.267		88	12.20		100.20	116
8440	6" x 6" x 1-1/2"		90	.267		88	12.20		100.20	116
8450	6" x 6" x 2"		90	.267		88	12.20		100.20	116
8460	8" x 8" x 3/4"		90	.267		126	12.20		138.20	158
8470	8" x 8" x 1"		90	.267		126	12.20		138.20	158
8480	8" x 8" x 1-1/2"		90	.267		126	12.20		138.20	158
8490	8" x 8" x 2"		90	.267		126	12.20		138.20	158
8500	Repair coupling 4"		100	.240		23	11		34	42
8520	6" diameter		90	.267		36.50	12.20		48.70	58.50
8540	8" diameter		50	.480		85.50	22		107.50	128
8560	10" diameter		50	.480		184	22		206	236
8580	12" diameter		50	.480		260	22		282	320
8600	Plug end 4"		100	.240		21	11		32	39.50
8620	6" diameter		90	.267		36	12.20		48.20	58
8640	8" diameter		50	.480		60.50	22		82.50	100
8660	10" diameter		50	.480		98	22		120	142
8680	12" diameter		50	.480		121	22		143	167
8700	PVC pipe, joint restraint									
8710	4" diameter	B-20A	32	1	Ea.	46	49.50		95.50	125
8720	6" diameter		25.60	1.250		56.50	62		118.50	156
8730	8" diameter		21.33	1.500		83.50	74		157.50	204
8740	10" diameter		18.28	1.751		149	86.50		235.50	294
8750	12" diameter		16.84	1.900		158	94		252	315
8760	14" diameter		16	2		223	99		322	395
8770	16" diameter		11.64	2.749		300	136		436	540
8780	18" diameter		11.03	2.901		370	143		513	620
8785	20" diameter		9.14	3.501		465	173		638	770
8790	24" diameter		7.53	4.250		540	210		750	905

33 14 13.35 Water Supply, HDPE		Crew	Daily Output	Labor-Hours	Unit	Material	Labor	Equipment	Total	Total Incl O&P
0010	**WATER SUPPLY, HDPE**									
0011	Butt fusion joints, SDR 21 40' lengths not including excavation or backfill									
0100	4" diameter	B-22A	400	.100	L.F.	2.54	4.72	1.66	8.92	11.75
0200	6" diameter		380	.105		5.75	4.97	1.75	12.47	15.75
0300	8" diameter		320	.125		9	5.90	2.08	16.98	21
0400	10" diameter		300	.133		11.65	6.30	2.22	20.17	25
0500	12" diameter		260	.154		13.40	7.25	2.56	23.21	28.50
0600	14" diameter	B-22B	220	.182		15.60	8.60	9.95	34.15	41
0700	16" diameter		180	.222		19.45	10.50	12.15	42.10	50.50
0800	18" diameter		140	.286		28.50	13.50	15.60	57.60	69
0900	24" diameter		100	.400		52.50	18.90	22	93.40	111
1000	Fittings									
1100	Elbows, 90 degrees									
1200	4" diameter	B-22A	32	1.250	Ea.	17	59	21	97	131
1300	6" diameter		28	1.429		45	67.50	23.50	136	178
1400	8" diameter		24	1.667		117	78.50	27.50	223	279
1500	10" diameter		18	2.222		249	105	37	391	475
1600	12" diameter		12	3.333		296	157	55.50	508.50	625
1700	14" diameter	B-22B	9	4.444		565	210	243	1,018	1,200
1800	16" diameter		6	6.667		845	315	365	1,525	1,800
1900	18" diameter		4	10		970	470	545	1,985	2,400
2000	24" diameter		3	13.333		1,725	630	730	3,085	3,650

33 14 13 – Public Water Utility Distribution Piping

33 14 13.35 Water Supply, HDPE		Crew	Daily Output	Labor-Hours	Unit	2019 Bare Costs Material	Labor	Equipment	Total	Total Incl O&P
2100	Tees									
2200	4″ diameter	B-22A	30	1.333	Ea.	22	63	22	107	144
2300	6″ diameter		26	1.538		58.50	72.50	25.50	156.50	203
2400	8″ diameter		22	1.818		132	86	30	248	310
2500	10″ diameter		15	2.667		175	126	44.50	345.50	430
2600	12″ diameter		10	4		350	189	66.50	605.50	740
2700	14″ diameter	B-22B	8	5		410	236	273	919	1,100
2800	16″ diameter		6	6.667		485	315	365	1,165	1,400
2900	18″ diameter		4	10		570	470	545	1,585	1,950
3000	24″ diameter		2	20		985	945	1,100	3,030	3,700
4100	Caps									
4110	4″ diameter	B-22A	34	1.176	Ea.	15.85	55.50	19.55	90.90	123
4120	6″ diameter		30	1.333		31	63	22	116	154
4130	8″ diameter		26	1.538		51.50	72.50	25.50	149.50	195
4150	10″ diameter		20	2		164	94.50	33	291.50	360
4160	12″ diameter		14	2.857		204	135	47.50	386.50	480

33 14 17 – Site Water Utility Service Laterals

33 14 17.15 Tapping, Crosses and Sleeves		Crew	Daily Output	Labor-Hours	Unit	2019 Bare Costs Material	Labor	Equipment	Total	Total Incl O&P
0010	**TAPPING, CROSSES AND SLEEVES**									
4000	Drill and tap pressurized main (labor only)									
4100	6″ main, 1″ to 2″ service	Q-1	3	5.333	Ea.		305		305	455
4150	8″ main, 1″ to 2″ service	"	2.75	5.818	"		330		330	495
4500	Tap and insert gate valve									
4600	8″ main, 4″ branch	B-21	3.20	8.750	Ea.		415	40.50	455.50	675
4650	6″ branch		2.70	10.370			490	48	538	800
4700	10″ main, 4″ branch		2.70	10.370			490	48	538	800
4750	6″ branch		2.35	11.915			565	55	620	915
4800	12″ main, 6″ branch		2.35	11.915			565	55	620	915
4850	8″ branch		2.35	11.915			565	55	620	915
7020	Crosses, 4″ x 4″		37	.757		1,450	36	3.50	1,489.50	1,650
7030	6″ x 4″		25	1.120		1,625	53	5.20	1,683.20	1,875
7040	6″ x 6″		25	1.120		1,700	53	5.20	1,758.20	1,950
7060	8″ x 6″		21	1.333		2,000	63.50	6.15	2,069.65	2,325
7080	8″ x 8″		21	1.333		2,100	63.50	6.15	2,169.65	2,400
7100	10″ x 6″		21	1.333		4,100	63.50	6.15	4,169.65	4,600
7120	10″ x 10″		21	1.333		4,400	63.50	6.15	4,469.65	4,925
7140	12″ x 6″		18	1.556		4,075	74	7.20	4,156.20	4,600
7160	12″ x 12″		18	1.556		5,325	74	7.20	5,406.20	5,975
7180	14″ x 6″		16	1.750		10,000	83	8.10	10,091.10	11,100
7200	14″ x 14″		16	1.750		10,600	83	8.10	10,691.10	11,800
7220	16″ x 6″		14	2		10,800	95	9.25	10,904.25	12,000
7240	16″ x 10″		14	2		10,900	95	9.25	11,004.25	12,200
7260	16″ x 16″		14	2		11,500	95	9.25	11,604.25	12,800
7280	18″ x 6″		10	2.800		16,200	133	12.95	16,345.95	18,100
7300	18″ x 12″		10	2.800		16,300	133	12.95	16,445.95	18,200
7320	18″ x 18″		10	2.800		16,800	133	12.95	16,945.95	18,700
7340	20″ x 6″		8	3.500		13,000	166	16.20	13,182.20	14,600
7360	20″ x 12″		8	3.500		14,000	166	16.20	14,182.20	15,700
7380	20″ x 20″		8	3.500		20,600	166	16.20	20,782.20	23,000
7400	24″ x 6″		6	4.667		16,700	222	21.50	16,943.50	18,800
7420	24″ x 12″		6	4.667		17,100	222	21.50	17,343.50	19,200
7440	24″ x 18″		6	4.667		26,600	222	21.50	26,843.50	29,600

33 14 17 – Site Water Utility Service Laterals

33 14 17.15 Tapping, Crosses and Sleeves		Crew	Daily Output	Labor-Hours	Unit	2019 Bare Costs Material	Labor	Equipment	Total	Total Incl O&P
7460	24" x 24"	B-21	6	4.667	Ea.	26,800	222	21.50	27,043.50	29,900
7600	Cut-in sleeves with rubber gaskets, 4"		18	1.556		480	74	7.20	561.20	650
7620	6"		12	2.333		615	111	10.80	736.80	855
7640	8"		10	2.800		825	133	12.95	970.95	1,125
7660	10"		10	2.800		1,150	133	12.95	1,295.95	1,475
7680	12"		9	3.111		1,500	148	14.40	1,662.40	1,900
7800	Cut-in valves with rubber gaskets, 4"		18	1.556		475	74	7.20	556.20	645
7820	6"		12	2.333		605	111	10.80	726.80	845
7840	8"		10	2.800		1,175	133	12.95	1,320.95	1,525
7860	10"		10	2.800		1,675	133	12.95	1,820.95	2,050
7880	12"		9	3.111		2,225	148	14.40	2,387.40	2,700
7900	Tapping valve 4", MJ, ductile iron		18	1.556		455	74	7.20	536.20	620
7920	6", MJ, ductile iron		12	2.333		595	111	10.80	716.80	835
8000	Sleeves with rubber gaskets, 4" x 4"		37	.757		875	36	3.50	914.50	1,025
8010	6" x 4"		25	1.120		945	53	5.20	1,003.20	1,125
8020	6" x 6"		25	1.120		1,050	53	5.20	1,108.20	1,225
8030	8" x 4"		21	1.333		970	63.50	6.15	1,039.65	1,175
8040	8" x 6"		21	1.333		965	63.50	6.15	1,034.65	1,175
8060	8" x 8"		21	1.333		1,300	63.50	6.15	1,369.65	1,525
8070	10" x 4"		21	1.333		935	63.50	6.15	1,004.65	1,125
8080	10" x 6"		21	1.333		1,075	63.50	6.15	1,144.65	1,300
8090	10" x 8"		21	1.333		1,325	63.50	6.15	1,394.65	1,575
8100	10" x 10"		21	1.333		1,975	63.50	6.15	2,044.65	2,275
8110	12" x 4"		18	1.556		1,475	74	7.20	1,556.20	1,750
8120	12" x 6"		18	1.556		1,225	74	7.20	1,306.20	1,450
8130	12" x 8"		18	1.556		1,400	74	7.20	1,481.20	1,675
8135	12" x 10"		18	1.556		2,475	74	7.20	2,556.20	2,850
8140	12" x 12"		18	1.556		2,675	74	7.20	2,756.20	3,050
8160	14" x 6"		16	1.750		1,450	83	8.10	1,541.10	1,725
8180	14" x 14"		16	1.750		2,300	83	8.10	2,391.10	2,650
8200	16" x 6"		14	2		1,700	95	9.25	1,804.25	2,025
8220	16" x 10"		14	2		3,175	95	9.25	3,279.25	3,650
8240	16" x 16"		14	2		9,850	95	9.25	9,954.25	11,000
8260	18" x 6"		10	2.800		1,725	133	12.95	1,870.95	2,100
8280	18" x 12"		10	2.800		3,225	133	12.95	3,370.95	3,775
8300	18" x 18"		10	2.800		8,600	133	12.95	8,745.95	9,700
8320	20" x 6"		8	3.500		2,350	166	16.20	2,532.20	2,850
8340	20" x 12"		8	3.500		4,500	166	16.20	4,682.20	5,225
8360	20" x 20"		8	3.500		10,800	166	16.20	10,982.20	12,100
8380	24" x 6"		6	4.667		2,275	222	21.50	2,518.50	2,850
8400	24" x 12"		6	4.667		6,250	222	21.50	6,493.50	7,225
8420	24" x 18"		6	4.667		11,100	222	21.50	11,343.50	12,600
8440	24" x 24"		6	4.667		13,200	222	21.50	13,443.50	15,000
8800	Hydrant valve box, 6' long	B-20	20	1.200		211	55		266	315
8820	8' long		18	1.333		290	61		351	415
8830	Valve box w/lid 4' deep		14	1.714		91	78.50		169.50	219
8840	Valve box and large base w/lid		14	1.714		315	78.50		393.50	470

33 14 19 – Valves and Hydrants for Water Utility Service

33 14 19.10 Valves		Crew	Daily Output	Labor-Hours	Unit	Material	2019 Bare Costs Labor	Equipment	Total	Total Incl O&P
0010	**VALVES**, water distribution									
0011	See Sections 22 05 23.20 and 22 05 23.60									
3000	Butterfly valves with boxes, cast iron, mechanical joint									
3100	4" diameter	B-6	6	4	Ea.	530	178	53.50	761.50	915
3140	6" diameter		6	4		695	178	53.50	926.50	1,100
3180	8" diameter		6	4		875	178	53.50	1,106.50	1,300
3300	10" diameter		6	4		1,200	178	53.50	1,431.50	1,650
3340	12" diameter		6	4		1,600	178	53.50	1,831.50	2,075
3400	14" diameter		4	6		2,900	267	80	3,247	3,675
3440	16" diameter		4	6		4,150	267	80	4,497	5,050
3460	18" diameter		4	6		5,025	267	80	5,372	6,025
3480	20" diameter		4	6		6,600	267	80	6,947	7,750
3500	24" diameter		4	6		11,000	267	80	11,347	12,600
3510	30" diameter		4	6		11,400	267	80	11,747	13,000
3520	36" diameter	↓	4	6	↓	14,100	267	80	14,447	16,000
3600	With lever operator									
3610	4" diameter	B-6	6	4	Ea.	440	178	53.50	671.50	815
3614	6" diameter		6	4		605	178	53.50	836.50	995
3616	8" diameter		6	4		785	178	53.50	1,016.50	1,200
3618	10" diameter		6	4		1,100	178	53.50	1,331.50	1,550
3620	12" diameter		6	4		1,500	178	53.50	1,731.50	1,975
3622	14" diameter		4	6		2,575	267	80	2,922	3,325
3624	16" diameter		4	6		3,825	267	80	4,172	4,700
3626	18" diameter		4	6		4,700	267	80	5,047	5,675
3628	20" diameter		4	6		6,275	267	80	6,622	7,400
3630	24" diameter	↓	4	6	↓	10,700	267	80	11,047	12,200
3700	Check valves, flanged									
3710	4" diameter	B-6	6	4	Ea.	730	178	53.50	961.50	1,125
3714	6" diameter		6	4		1,200	178	53.50	1,431.50	1,650
3716	8" diameter		6	4		2,350	178	53.50	2,581.50	2,900
3718	10" diameter		6	4		3,975	178	53.50	4,206.50	4,700
3720	12" diameter		6	4		5,950	178	53.50	6,181.50	6,875
3726	18" diameter		4	6		25,400	267	80	25,747	28,500
3730	24" diameter	↓	4	6	↓	58,500	267	80	58,847	65,000
3800	Gate valves, C.I., 125 psi, mechanical joint, w/boxes									
3810	4" diameter	B-6	6	4	Ea.	765	178	53.50	996.50	1,175
3814	6" diameter		6	4		1,150	178	53.50	1,381.50	1,600
3816	8" diameter		6	4		1,975	178	53.50	2,206.50	2,500
3818	10" diameter		6	4		3,550	178	53.50	3,781.50	4,225
3820	12" diameter		6	4		4,800	178	53.50	5,031.50	5,600
3822	14" diameter		4	6		10,600	267	80	10,947	12,200
3824	16" diameter		4	6		14,900	267	80	15,247	16,800
3826	18" diameter		4	6		20,900	267	80	21,247	23,500
3828	20" diameter		4	6		24,900	267	80	25,247	27,800
3830	24" diameter		4	6		36,600	267	80	36,947	40,700
3831	30" diameter		4	6		42,200	267	80	42,547	47,000
3832	36" diameter	↓	4	6	↓	59,500	267	80	59,847	66,000

33 14 19.20 Valves		Crew	Daily Output	Labor-Hours	Unit	Material	Labor	Equipment	Total	Total Incl O&P
0010	**VALVES**									
0011	Special trim or use									
9000	Valves, gate valve, N.R.S. PIV with post, 4" diameter	B-6	6	4	Ea.	2,025	178	53.50	2,256.50	2,550
9020	6" diameter	↓	6	4	↓	2,425	178	53.50	2,656.50	2,975

33 14 Water Utility Transmission and Distribution

33 14 19 – Valves and Hydrants for Water Utility Service

33 14 19.20 Valves		Crew	Daily Output	Labor-Hours	Unit	Material	Labor	Equipment	Total	Total Incl O&P
						2019 Bare Costs				
9040	8″ diameter	B-6	6	4	Ea.	3,250	178	53.50	3,481.50	3,900
9060	10″ diameter		6	4		4,800	178	53.50	5,031.50	5,625
9080	12″ diameter		6	4		6,075	178	53.50	6,306.50	7,000
9100	14″ diameter		6	4		11,600	178	53.50	11,831.50	13,100
9120	OS&Y, 4″ diameter		6	4		805	178	53.50	1,036.50	1,225
9140	6″ diameter		6	4		1,125	178	53.50	1,356.50	1,575
9160	8″ diameter		6	4		1,775	178	53.50	2,006.50	2,275
9180	10″ diameter		6	4		2,750	178	53.50	2,981.50	3,375
9200	12″ diameter		6	4		4,275	178	53.50	4,506.50	5,025
9220	14″ diameter		4	6		5,750	267	80	6,097	6,825
9400	Check valves, rubber disc, 2-1/2″ diameter		6	4		550	178	53.50	781.50	935
9420	3″ diameter		6	4		585	178	53.50	816.50	970
9440	4″ diameter		6	4		730	178	53.50	961.50	1,125
9480	6″ diameter		6	4		1,200	178	53.50	1,431.50	1,650
9500	8″ diameter		6	4		2,350	178	53.50	2,581.50	2,900
9520	10″ diameter		6	4		3,975	178	53.50	4,206.50	4,700
9540	12″ diameter		6	4		5,950	178	53.50	6,181.50	6,875
9542	14″ diameter		4	6		6,500	267	80	6,847	7,650
9700	Detector check valves, reducing, 4″ diameter		6	4		1,175	178	53.50	1,406.50	1,600
9720	6″ diameter		6	4		1,950	178	53.50	2,181.50	2,475
9740	8″ diameter		6	4		2,900	178	53.50	3,131.50	3,525
9760	10″ diameter		6	4		5,100	178	53.50	5,331.50	5,950
9800	Galvanized, 4″ diameter		6	4		1,875	178	53.50	2,106.50	2,400
9820	6″ diameter		6	4		2,525	178	53.50	2,756.50	3,100
9840	8″ diameter		6	4		3,775	178	53.50	4,006.50	4,475
9860	10″ diameter	↓	6	4	↓	5,200	178	53.50	5,431.50	6,050

33 16 Water Utility Storage Tanks

33 16 11 – Elevated Composite Water Storage Tanks

33 16 11.50 Elevated Water Storage Tanks		Crew	Daily Output	Labor-Hours	Unit	Material	Labor	Equipment	Total	Total Incl O&P
0010	**ELEVATED WATER STORAGE TANKS**									
0011	Not incl. pipe, pumps or foundation									
3000	Elevated water tanks, 100′ to bottom capacity line, incl. painting									
3010	50,000 gallons				Ea.				185,000	204,000
3300	100,000 gallons								280,000	307,500
3400	250,000 gallons								751,500	826,500
3600	500,000 gallons								1,336,000	1,470,000
3700	750,000 gallons								1,622,000	1,783,500
3900	1,000,000 gallons				↓				2,322,000	2,556,000

33 16 23 – Ground-Level Steel Water Storage Tanks

33 16 23.13 Steel Water Storage Tanks		Crew	Daily Output	Labor-Hours	Unit	Material	Labor	Equipment	Total	Total Incl O&P
0010	**STEEL WATER STORAGE TANKS**									
0910	Steel, ground level, ht./diam. less than 1, not incl. fdn., 100,000 gallons				Ea.				202,000	244,500
1000	250,000 gallons								295,500	324,000
1200	500,000 gallons								417,000	458,500
1250	750,000 gallons								538,000	591,500
1300	1,000,000 gallons								558,000	725,500
1500	2,000,000 gallons								1,043,000	1,148,000
1600	4,000,000 gallons								2,121,000	2,333,000
1800	6,000,000 gallons				↓				3,095,000	3,405,000

33 16 Water Utility Storage Tanks

33 16 23 – Ground-Level Steel Water Storage Tanks

	33 16 23.13 Steel Water Storage Tanks	Crew	Daily Output	Labor-Hours	Unit	2019 Bare Costs Material	Labor	Equipment	Total	Total Incl O&P
1850	8,000,000 gallons				Ea.				4,068,000	4,475,000
1910	10,000,000 gallons								5,050,000	5,554,500
2100	Steel standpipes, ht./diam. more than 1,100′ to overflow, no fdn.									
2200	500,000 gallons				Ea.				546,500	600,500
2400	750,000 gallons								722,500	794,500
2500	1,000,000 gallons								1,060,500	1,167,000
2700	1,500,000 gallons								1,749,000	1,923,000
2800	2,000,000 gallons								2,327,000	2,559,000

33 16 36 – Ground-Level Reinforced Concrete Water Storage Tanks

	33 16 36.16 Prestressed Conc. Water Storage Tanks	Crew	Daily Output	Labor-Hours	Unit	Material	Labor	Equipment	Total	Total Incl O&P
0010	**PRESTRESSED CONC. WATER STORAGE TANKS**									
0020	Not including fdn., pipe or pumps, 250,000 gallons				Ea.				299,000	329,500
0100	500,000 gallons								487,000	536,000
0300	1,000,000 gallons								707,000	807,500
0400	2,000,000 gallons								1,072,000	1,179,000
0600	4,000,000 gallons								1,706,000	1,877,000
0700	6,000,000 gallons								2,266,000	2,493,000
0750	8,000,000 gallons								2,924,000	3,216,000
0800	10,000,000 gallons								3,533,000	3,886,000

33 16 56 – Ground-Level Plastic Water Storage Cisterns

	33 16 56.23 Plastic-Coated Fabric Pillow Water Tanks	Crew	Daily Output	Labor-Hours	Unit	Material	Labor	Equipment	Total	Total Incl O&P
0010	**PLASTIC-COATED FABRIC PILLOW WATER TANKS**									
7000	Water tanks, vinyl coated fabric pillow tanks, freestanding, 5,000 gallons	4 Clab	4	8	Ea.	3,625	330		3,955	4,500
7100	Supporting embankment not included, 25,000 gallons	6 Clab	2	24		14,000	985		14,985	16,900
7200	50,000 gallons	8 Clab	1.50	42.667		20,500	1,750		22,250	25,300
7300	100,000 gallons	9 Clab	.90	80		39,800	3,275		43,075	48,800
7400	150,000 gallons		.50	144		46,600	5,900		52,500	60,500
7500	200,000 gallons		.40	180		83,000	7,400		90,400	102,000
7600	250,000 gallons		.30	240		93,500	9,850		103,350	118,000

33 31 Sanitary Sewerage Piping

33 31 11 – Public Sanitary Sewerage Gravity Piping

	33 31 11.15 Sewage Collection, Concrete Pipe	Crew	Daily Output	Labor-Hours	Unit	Material	Labor	Equipment	Total	Total Incl O&P
0010	**SEWAGE COLLECTION, CONCRETE PIPE**									
0020	See Section 33 42 11.60 for sewage/drainage collection, concrete pipe									

	33 31 11.25 Sewage Collection, Polyvinyl Chloride Pipe	Crew	Daily Output	Labor-Hours	Unit	Material	Labor	Equipment	Total	Total Incl O&P
0010	**SEWAGE COLLECTION, POLYVINYL CHLORIDE PIPE**									
0020	Not including excavation or backfill									
2000	20′ lengths, SDR 35, B&S, 4″ diameter	B-20	375	.064	L.F.	1.67	2.93		4.60	6.30
2040	6″ diameter		350	.069		3.67	3.14		6.81	8.80
2080	13′ lengths, SDR 35, B&S, 8″ diameter		335	.072		6.60	3.28		9.88	12.25
2120	10″ diameter	B-21	330	.085		11.40	4.03	.39	15.82	19.10
2160	12″ diameter		320	.088		13	4.16	.41	17.57	21
2200	15″ diameter		240	.117		15.10	5.55	.54	21.19	25.50
4000	Piping, DWV PVC, no exc./bkfill., 10′ L, Sch 40, 4″ diameter	B-20	375	.064		3.80	2.93		6.73	8.65
4010	6″ diameter		350	.069		8.30	3.14		11.44	13.85
4020	8″ diameter		335	.072		13.05	3.28		16.33	19.40

33 41 Subdrainage

33 41 16 – Subdrainage Piping

33 41 16.25 Piping, Subdrainage, Corrugated Metal		Crew	Daily Output	Labor-Hours	Unit	2019 Bare Costs Material	Labor	Equipment	Total	Total Incl O&P
0010	**PIPING, SUBDRAINAGE, CORRUGATED METAL**									
0021	Not including excavation and backfill									
2010	Aluminum, perforated									
2020	6" diameter, 18 ga.	B-20	380	.063	L.F.	6.70	2.89		9.59	11.80
2200	8" diameter, 16 ga.	"	370	.065		8.45	2.97		11.42	13.75
2220	10" diameter, 16 ga.	B-21	360	.078		10.55	3.69	.36	14.60	17.60
2240	12" diameter, 16 ga.		285	.098		11.80	4.67	.45	16.92	20.50
2260	18" diameter, 16 ga.		205	.137		17.70	6.50	.63	24.83	30
3000	Uncoated galvanized, perforated									
3020	6" diameter, 18 ga.	B-20	380	.063	L.F.	6.45	2.89		9.34	11.45
3200	8" diameter, 16 ga.	"	370	.065		7.95	2.97		10.92	13.20
3220	10" diameter, 16 ga.	B-21	360	.078		8.40	3.69	.36	12.45	15.25
3240	12" diameter, 16 ga.		285	.098		9.35	4.67	.45	14.47	17.85
3260	18" diameter, 16 ga.		205	.137		14.35	6.50	.63	21.48	26.50
4000	Steel, perforated, asphalt coated									
4020	6" diameter, 18 ga.	B-20	380	.063	L.F.	6.60	2.89		9.49	11.70
4030	8" diameter, 18 ga.	"	370	.065		8.35	2.97		11.32	13.70
4040	10" diameter, 16 ga.	B-21	360	.078		10.50	3.69	.36	14.55	17.55
4050	12" diameter, 16 ga.		285	.098		11.80	4.67	.45	16.92	20.50
4060	18" diameter, 16 ga.		205	.137		18.60	6.50	.63	25.73	31

33 41 16.30 Piping, Subdrainage, Plastic		Crew	Daily Output	Labor-Hours	Unit	Material	Labor	Equipment	Total	Total Incl O&P
0010	**PIPING, SUBDRAINAGE, PLASTIC**									
0020	Not including excavation and backfill									
2100	Perforated PVC, 4" diameter	B-14	314	.153	L.F.	1.67	6.60	1.02	9.29	12.90
2110	6" diameter		300	.160		3.67	6.90	1.07	11.64	15.65
2120	8" diameter		290	.166		6	7.15	1.10	14.25	18.60
2130	10" diameter		280	.171		9.15	7.40	1.14	17.69	22.50
2140	12" diameter		270	.178		13	7.65	1.19	21.84	27

33 42 Stormwater Conveyance

33 42 11 – Stormwater Gravity Piping

33 42 11.40 Piping, Storm Drainage, Corrugated Metal		Crew	Daily Output	Labor-Hours	Unit	Material	Labor	Equipment	Total	Total Incl O&P
0010	**PIPING, STORM DRAINAGE, CORRUGATED METAL**									
0020	Not including excavation or backfill									
2000	Corrugated metal pipe, galvanized									
2020	Bituminous coated with paved invert, 20' lengths									
2040	8" diameter, 16 ga.	B-14	330	.145	L.F.	8.25	6.30	.97	15.52	19.65
2060	10" diameter, 16 ga.		260	.185		8.60	7.95	1.23	17.78	23
2080	12" diameter, 16 ga.		210	.229		10.70	9.85	1.52	22.07	28.50
2100	15" diameter, 16 ga.		200	.240		14.45	10.35	1.60	26.40	33.50
2120	18" diameter, 16 ga.		190	.253		18.95	10.90	1.69	31.54	39.50
2140	24" diameter, 14 ga.		160	.300		23	12.95	2	37.95	47
2160	30" diameter, 14 ga.	B-13	120	.467		27.50	21	4.85	53.35	67.50
2180	36" diameter, 12 ga.		120	.467		37	21	4.85	62.85	77.50
2200	48" diameter, 12 ga.		100	.560		48.50	25	5.80	79.30	98
2220	60" diameter, 10 ga.	B-13B	75	.747		74.50	33.50	13.10	121.10	147
2240	72" diameter, 8 ga.	"	45	1.244		88.50	56	22	166.50	206
2500	Galvanized, uncoated, 20' lengths									
2520	8" diameter, 16 ga.	B-14	355	.135	L.F.	7.55	5.85	.90	14.30	18.15
2540	10" diameter, 16 ga.		280	.171		8.55	7.40	1.14	17.09	22
2560	12" diameter, 16 ga.		220	.218		9.60	9.40	1.46	20.46	26.50

33 42 11.40 Piping, Storm Drainage, Corrugated Metal

	33 42 11.40 Piping, Storm Drainage, Corrugated Metal	Crew	Daily Output	Labor-Hours	Unit	2019 Bare Costs Material	Labor	Equipment	Total	Total Incl O&P
2580	15″ diameter, 16 ga.	B-14	220	.218	L.F.	11.10	9.40	1.46	21.96	28
2600	18″ diameter, 16 ga.		205	.234		13.85	10.10	1.56	25.51	32
2620	24″ diameter, 14 ga.		175	.274		21.50	11.85	1.83	35.18	43.50
2640	30″ diameter, 14 ga.	B-13	130	.431		27.50	19.30	4.48	51.28	64
2660	36″ diameter, 12 ga.		130	.431		38	19.30	4.48	61.78	76
2680	48″ diameter, 12 ga.		110	.509		56	23	5.30	84.30	102
2690	60″ diameter, 10 ga.	B-13B	78	.718		87	32	12.60	131.60	158
2780	End sections, 8″ diameter	B-14	35	1.371	Ea.	70	59	9.15	138.15	177
2785	10″ diameter		35	1.371		73.50	59	9.15	141.65	181
2790	12″ diameter		35	1.371		105	59	9.15	173.15	215
2800	18″ diameter		30	1.600		110	69	10.65	189.65	237
2810	24″ diameter	B-13	25	2.240		207	100	23.50	330.50	405
2820	30″ diameter		25	2.240		320	100	23.50	443.50	530
2825	36″ diameter		20	2.800		470	126	29	625	740
2830	48″ diameter		10	5.600		935	251	58	1,244	1,475
2835	60″ diameter	B-13B	5	11.200		1,600	500	196	2,296	2,725
2840	72″ diameter	″	4	14		2,000	630	245	2,875	3,425

33 42 11.60 Sewage/Drainage Collection, Concrete Pipe

	33 42 11.60 Sewage/Drainage Collection, Concrete Pipe	Crew	Daily Output	Labor-Hours	Unit	Material	Labor	Equipment	Total	Total Incl O&P
0010	**SEWAGE/DRAINAGE COLLECTION, CONCRETE PIPE**									
0020	Not including excavation or backfill									
1000	Non-reinforced pipe, extra strength, B&S or T&G joints									
1010	6″ diameter	B-14	265.04	.181	L.F.	7.65	7.80	1.21	16.66	21.50
1020	8″ diameter		224	.214		8.45	9.25	1.43	19.13	25
1030	10″ diameter		216	.222		9.35	9.60	1.48	20.43	26.50
1040	12″ diameter		200	.240		10.90	10.35	1.60	22.85	29.50
1050	15″ diameter		180	.267		15.35	11.50	1.78	28.63	36
1060	18″ diameter		144	.333		18.50	14.40	2.22	35.12	44.50
1070	21″ diameter		112	.429		19.75	18.50	2.86	41.11	52.50
1080	24″ diameter		100	.480		22	20.50	3.20	45.70	59.50
2000	Reinforced culvert, class 3, no gaskets									
2010	12″ diameter	B-14	150	.320	L.F.	14.15	13.80	2.13	30.08	39
2020	15″ diameter		150	.320		19.10	13.80	2.13	35.03	44.50
2030	18″ diameter		132	.364		23	15.70	2.43	41.13	51.50
2035	21″ diameter		120	.400		25.50	17.25	2.67	45.42	57
2040	24″ diameter		100	.480		30.50	20.50	3.20	54.20	68.50
2045	27″ diameter	B-13	92	.609		46	27.50	6.30	79.80	98.50
2050	30″ diameter		88	.636		52.50	28.50	6.60	87.60	108
2060	36″ diameter		72	.778		77	35	8.10	120.10	146
2070	42″ diameter	B-13B	72	.778		98	35	13.65	146.65	176
2080	48″ diameter		64	.875		93.50	39	15.35	147.85	179
2085	54″ diameter		56	1		146	45	17.55	208.55	247
2090	60″ diameter		48	1.167		174	52.50	20.50	247	294
2100	72″ diameter		40	1.400		283	63	24.50	370.50	430
2120	84″ diameter		32	1.750		355	78.50	30.50	464	540
2140	96″ diameter		24	2.333		425	105	41	571	675
2200	With gaskets, class 3, 12″ diameter	B-21	168	.167		15.60	7.90	.77	24.27	30
2220	15″ diameter		160	.175		21	8.30	.81	30.11	36.50
2230	18″ diameter		152	.184		25.50	8.75	.85	35.10	42
2235	21″ diameter		152	.184		28	8.75	.85	37.60	45
2240	24″ diameter		136	.206		37.50	9.80	.95	48.25	57
2260	30″ diameter	B-13	88	.636		60.50	28.50	6.60	95.60	117
2270	36″ diameter	″	72	.778		86.50	35	8.10	129.60	156

33 42 11 – Stormwater Gravity Piping

33 42 11.60 Sewage/Drainage Collection, Concrete Pipe		Crew	Daily Output	Labor-Hours	Unit	2019 Bare Costs Material	Labor	Equipment	Total	Total Incl O&P
2290	48″ diameter	B-13B	64	.875	L.F.	106	39	15.35	160.35	193
2310	72″ diameter	″	40	1.400		300	63	24.50	387.50	455
2330	Flared ends, 12″ diameter	B-21	31	.903	Ea.	278	43	4.18	325.18	375
2340	15″ diameter		25	1.120		315	53	5.20	373.20	435
2400	18″ diameter		20	1.400		365	66.50	6.50	438	515
2420	24″ diameter		14	2		425	95	9.25	529.25	625
2440	36″ diameter	B-13	10	5.600		895	251	58	1,204	1,425
3080	Radius pipe, add to pipe prices, 12″ to 60″ diameter				L.F.	50%				
3090	Over 60″ diameter, add				″	20%				
3500	Reinforced elliptical, 8′ lengths, C507 class 3									
3520	14″ x 23″ inside, round equivalent 18″ diameter	B-21	82	.341	L.F.	42.50	16.20	1.58	60.28	73
3530	24″ x 38″ inside, round equivalent 30″ diameter	B-13	58	.966		65.50	43.50	10.05	119.05	149
3540	29″ x 45″ inside, round equivalent 36″ diameter		52	1.077		91	48.50	11.20	150.70	185
3550	38″ x 60″ inside, round equivalent 48″ diameter		38	1.474		154	66	15.30	235.30	285
3560	48″ x 76″ inside, round equivalent 60″ diameter		26	2.154		214	96.50	22.50	333	405
3570	58″ x 91″ inside, round equivalent 72″ diameter		22	2.545		305	114	26.50	445.50	535
3780	Concrete slotted pipe, class 4 mortar joint									
3800	12″ diameter	B-21	168	.167	L.F.	31	7.90	.77	39.67	47.50
3840	18″ diameter	″	152	.184	″	35.50	8.75	.85	45.10	53.50
3900	Concrete slotted pipe, Class 4 O-ring joint									
3940	12″ diameter	B-21	168	.167	L.F.	28	7.90	.77	36.67	44
3960	18″ diameter	″	152	.184	″	32	8.75	.85	41.60	49

33 42 13 – Stormwater Culverts

33 42 13.15 Oval Arch Culverts

		Crew	Daily Output	Labor-Hours	Unit	Material	Labor	Equipment	Total	Total Incl O&P
0010	**OVAL ARCH CULVERTS**									
3000	Corrugated galvanized or aluminum, coated & paved									
3020	17″ x 13″, 16 ga., 15″ equivalent	B-14	200	.240	L.F.	13.35	10.35	1.60	25.30	32
3040	21″ x 15″, 16 ga., 18″ equivalent		150	.320		17.45	13.80	2.13	33.38	42.50
3060	28″ x 20″, 14 ga., 24″ equivalent		125	.384		22.50	16.55	2.56	41.61	53
3080	35″ x 24″, 14 ga., 30″ equivalent		100	.480		27.50	20.50	3.20	51.20	65.50
3100	42″ x 29″, 12 ga., 36″ equivalent	B-13	100	.560		33	25	5.80	63.80	80.50
3120	49″ x 33″, 12 ga., 42″ equivalent		90	.622		40	28	6.45	74.45	93
3140	57″ x 38″, 12 ga., 48″ equivalent		75	.747		51.50	33.50	7.75	92.75	116
3160	Steel, plain oval arch culverts, plain									
3180	17″ x 13″, 16 ga., 15″ equivalent	B-14	225	.213	L.F.	12.95	9.20	1.42	23.57	29.50
3200	21″ x 15″, 16 ga., 18″ equivalent		175	.274		15.65	11.85	1.83	29.33	37
3220	28″ x 20″, 14 ga., 24″ equivalent		150	.320		23.50	13.80	2.13	39.43	49
3240	35″ x 24″, 14 ga., 30″ equivalent	B-13	108	.519		28.50	23.50	5.40	57.40	72.50
3260	42″ x 29″, 12 ga., 36″ equivalent		108	.519		38.50	23.50	5.40	67.40	83.50
3280	49″ x 33″, 12 ga., 42″ equivalent		92	.609		50.50	27.50	6.30	84.30	103
3300	57″ x 38″, 12 ga., 48″ equivalent		75	.747		62	33.50	7.75	103.25	127
3320	End sections, 17″ x 13″		22	2.545	Ea.	141	114	26.50	281.50	355
3340	42″ x 29″		17	3.294	″	385	148	34	567	685
3360	Multi-plate arch, steel	B-20	1690	.014	Lb.	1.30	.65		1.95	2.42

33 42 33 – Stormwater Curbside Drains and Inlets

33 42 33.13 Catch Basins

		Crew	Daily Output	Labor-Hours	Unit	Material	Labor	Equipment	Total	Total Incl O&P
0010	**CATCH BASINS**									
0011	Not including footing & excavation									
1600	Frames & grates, C.I., 24″ square, 500 lb.	B-6	7.80	3.077	Ea.	355	137	41	533	640
1700	26″ D shape, 600 lb.		7	3.429		640	153	45.50	838.50	985
1800	Light traffic, 18″ diameter, 100 lb.		10	2.400		211	107	32	350	430
1900	24″ diameter, 300 lb.		8.70	2.759		202	123	37	362	450

33 42 Stormwater Conveyance

33 42 33 – Stormwater Curbside Drains and Inlets

33 42 33.13 Catch Basins		Crew	Daily Output	Labor-Hours	Unit	Material	2019 Bare Costs Labor	Equipment	Total	Total Incl O&P
2000	36″ diameter, 900 lb.	B-6	5.80	4.138	Ea.	660	184	55	899	1,075
2100	Heavy traffic, 24″ diameter, 400 lb.		7.80	3.077		273	137	41	451	550
2200	36″ diameter, 1,150 lb.		3	8		885	355	107	1,347	1,625
2300	Mass. State standard, 26″ diameter, 475 lb.		7	3.429		680	153	45.50	878.50	1,025
2400	30″ diameter, 620 lb.		7	3.429		345	153	45.50	543.50	660
2500	Watertight, 24″ diameter, 350 lb.		7.80	3.077		365	137	41	543	655
2600	26″ diameter, 500 lb.		7	3.429		430	153	45.50	628.50	755
2700	32″ diameter, 575 lb.		6	4		950	178	53.50	1,181.50	1,375
2800	3 piece cover & frame, 10″ deep,									
2900	1,200 lb., for heavy equipment	B-6	3	8	Ea.	1,075	355	107	1,537	1,825
3000	Raised for paving 1-1/4″ to 2″ high									
3100	4 piece expansion ring									
3200	20″ to 26″ diameter	1 Clab	3	2.667	Ea.	184	109		293	370
3300	30″ to 36″ diameter	″	3	2.667	″	255	109		364	445
3320	Frames and covers, existing, raised for paving, 2″, including									
3340	row of brick, concrete collar, up to 12″ wide frame	B-6	18	1.333	Ea.	52	59.50	17.80	129.30	166
3360	20″ to 26″ wide frame		11	2.182		78.50	97.50	29	205	266
3380	30″ to 36″ wide frame		9	2.667		97.50	119	35.50	252	325
3400	Inverts, single channel brick	D-1	3	5.333		106	244		350	485
3500	Concrete		5	3.200		127	146		273	365
3600	Triple channel, brick		2	8		170	365		535	745
3700	Concrete		3	5.333		156	244		400	540

33 52 Hydrocarbon Transmission and Distribution

33 52 13 – Liquid Hydrocarbon Piping

33 52 13.16 Gasoline Piping		Crew	Daily Output	Labor-Hours	Unit	Material	Labor	Equipment	Total	Total Incl O&P
0010	**GASOLINE PIPING**									
0020	Primary containment pipe, fiberglass-reinforced									
0030	Plastic pipe 15′ & 30′ lengths									
0040	2″ diameter	Q-6	425	.056	L.F.	6.55	3.37		9.92	12.25
0050	3″ diameter		400	.060		10.20	3.58		13.78	16.55
0060	4″ diameter		375	.064		13.20	3.82		17.02	20.50
0100	Fittings									
0110	Elbows, 90° & 45°, bell ends, 2″	Q-6	24	1	Ea.	41.50	59.50		101	136
0120	3″ diameter		22	1.091		52.50	65		117.50	155
0130	4″ diameter		20	1.200		67.50	71.50		139	182
0200	Tees, bell ends, 2″		21	1.143		57.50	68		125.50	165
0210	3″ diameter		18	1.333		61	79.50		140.50	186
0220	4″ diameter		15	1.600		82	95.50		177.50	233
0230	Flanges bell ends, 2″		24	1		31	59.50		90.50	124
0240	3″ diameter		22	1.091		37.50	65		102.50	139
0250	4″ diameter		20	1.200		43.50	71.50		115	155
0260	Sleeve couplings, 2″		21	1.143		12.10	68		80.10	115
0270	3″ diameter		18	1.333		17.20	79.50		96.70	138
0280	4″ diameter		15	1.600		22	95.50		117.50	167
0290	Threaded adapters, 2″		21	1.143		17.30	68		85.30	121
0300	3″ diameter		18	1.333		32	79.50		111.50	154
0310	4″ diameter		15	1.600		36	95.50		131.50	183
0320	Reducers, 2″		27	.889		27	53		80	109
0330	3″ diameter		22	1.091		27	65		92	127
0340	4″ diameter		20	1.200		36.50	71.50		108	147

33 52 13 – Liquid Hydrocarbon Piping

33 52 13.16 Gasoline Piping		Crew	Daily Output	Labor-Hours	Unit	Material	2019 Bare Costs Labor	Equipment	Total	Total Incl O&P
1010	Gas station product line for secondary containment (double wall)									
1100	Fiberglass reinforced plastic pipe 25′ lengths									
1120	Pipe, plain end, 3″ diameter	Q-6	375	.064	L.F.	25.50	3.82		29.32	34.50
1130	4″ diameter		350	.069		30.50	4.09		34.59	39.50
1140	5″ diameter		325	.074		33.50	4.41		37.91	43
1150	6″ diameter	↓	300	.080	↓	37	4.77		41.77	48
1200	Fittings									
1230	Elbows, 90° & 45°, 3″ diameter	Q-6	18	1.333	Ea.	136	79.50		215.50	268
1240	4″ diameter		16	1.500		158	89.50		247.50	305
1250	5″ diameter		14	1.714		175	102		277	345
1260	6″ diameter		12	2		195	119		314	395
1270	Tees, 3″ diameter		15	1.600		161	95.50		256.50	320
1280	4″ diameter		12	2		188	119		307	385
1290	5″ diameter		9	2.667		298	159		457	570
1300	6″ diameter		6	4		355	239		594	750
1310	Couplings, 3″ diameter		18	1.333		55	79.50		134.50	180
1320	4″ diameter		16	1.500		117	89.50		206.50	263
1330	5″ diameter		14	1.714		206	102		308	380
1340	6″ diameter		12	2		299	119		418	510
1350	Cross-over nipples, 3″ diameter		18	1.333		10.30	79.50		89.80	130
1360	4″ diameter		16	1.500		12.55	89.50		102.05	148
1370	5″ diameter		14	1.714		15.35	102		117.35	171
1380	6″ diameter		12	2		18.70	119		137.70	200
1400	Telescoping, reducers, concentric 4″ x 3″		18	1.333		44	79.50		123.50	167
1410	5″ x 4″		17	1.412		94.50	84.50		179	230
1420	6″ x 5″	↓	16	1.500	↓	220	89.50		309.50	375

33 52 16 – Gas Hydrocarbon Piping

33 52 16.20 Piping, Gas Service and Distribution, P.E.		Crew	Daily Output	Labor-Hours	Unit	Material	Labor	Equipment	Total	Total Incl O&P
0010	**PIPING, GAS SERVICE AND DISTRIBUTION, POLYETHYLENE**									
0020	Not including excavation or backfill									
1000	60 psi coils, compression coupling @ 100′, 1/2″ diameter, SDR 11	B-20A	608	.053	L.F.	.48	2.60		3.08	4.45
1010	1″ diameter, SDR 11		544	.059		1.13	2.91		4.04	5.60
1040	1-1/4″ diameter, SDR 11		544	.059		1.64	2.91		4.55	6.20
1100	2″ diameter, SDR 11		488	.066		2.54	3.24		5.78	7.65
1160	3″ diameter, SDR 11	↓	408	.078		5.60	3.88		9.48	12
1500	60 psi 40′ joints with coupling, 3″ diameter, SDR 11	B-21A	408	.098		6.90	5	.91	12.81	16.10
1540	4″ diameter, SDR 11		352	.114		11.70	5.80	1.06	18.56	23
1600	6″ diameter, SDR 11		328	.122		31.50	6.20	1.13	38.83	45
1640	8″ diameter, SDR 11	↓	272	.147	↓	47	7.50	1.37	55.87	64.50

33 61 Hydronic Energy Distribution

33 61 13 – Underground Hydronic Energy Distribution

33 61 13.20 Pipe Conduit, Prefabricated/Preinsulated		Crew	Daily Output	Labor-Hours	Unit	Material	Labor	Equipment	Total	Total Incl O&P
0010	**PIPE CONDUIT, PREFABRICATED/PREINSULATED**									
0020	Does not include trenching, fittings or crane.									
0300	For cathodic protection, add 12 to 14%									
0310	of total built-up price (casing plus service pipe)									
0580	Polyurethane insulated system, 250°F max. temp.									
0620	Black steel service pipe, standard wt., 1/2″ insulation									
0660	3/4″ diam. pipe size	Q-17	54	.296	L.F.	93	17.05	1.04	111.09	129

33 61 13.20 Pipe Conduit, Prefabricated/Preinsulated		Crew	Daily Output	Labor-Hours	Unit	2019 Bare Costs Material	Labor	Equipment	Total	Total Incl O&P
0670	1″ diam. pipe size	Q-17	50	.320	L.F.	102	18.40	1.12	121.52	141
0680	1-1/4″ diam. pipe size		47	.340		114	19.60	1.19	134.79	156
0690	1-1/2″ diam. pipe size		45	.356		124	20.50	1.25	145.75	168
0700	2″ diam. pipe size		42	.381		128	22	1.34	151.34	175
0710	2-1/2″ diam. pipe size		34	.471		130	27	1.65	158.65	185
0720	3″ diam. pipe size		28	.571		151	33	2.01	186.01	218
0730	4″ diam. pipe size		22	.727		190	42	2.55	234.55	275
0740	5″ diam. pipe size		18	.889		243	51	3.12	297.12	345
0750	6″ diam. pipe size	Q-18	23	1.043		282	62.50	2.44	346.94	405
0760	8″ diam. pipe size		19	1.263		415	75.50	2.96	493.46	570
0770	10″ diam. pipe size		16	1.500		525	89.50	3.51	618.01	715
0780	12″ diam. pipe size		13	1.846		650	110	4.32	764.32	885
0790	14″ diam. pipe size		11	2.182		725	130	5.10	860.10	1,000
0800	16″ diam. pipe size		10	2.400		835	143	5.60	983.60	1,125
0810	18″ diam. pipe size		8	3		960	179	7	1,146	1,325
0820	20″ diam. pipe size		7	3.429		1,075	205	8	1,288	1,500
0830	24″ diam. pipe size		6	4		1,300	239	9.35	1,548.35	1,800
0900	For 1″ thick insulation, add					10%				
0940	For 1-1/2″ thick insulation, add					13%				
0980	For 2″ thick insulation, add					20%				
1500	Gland seal for system, 3/4″ diam. pipe size	Q-17	32	.500	Ea.	1,075	29	1.76	1,105.76	1,225
1510	1″ diam. pipe size		32	.500		1,075	29	1.76	1,105.76	1,225
1540	1-1/4″ diam. pipe size		30	.533		1,150	30.50	1.87	1,182.37	1,300
1550	1-1/2″ diam. pipe size		30	.533		1,150	30.50	1.87	1,182.37	1,300
1560	2″ diam. pipe size		28	.571		1,350	33	2.01	1,385.01	1,525
1570	2-1/2″ diam. pipe size		26	.615		1,450	35.50	2.16	1,487.66	1,650
1580	3″ diam. pipe size		24	.667		1,575	38.50	2.34	1,615.84	1,775
1590	4″ diam. pipe size		22	.727		1,850	42	2.55	1,894.55	2,100
1600	5″ diam. pipe size		19	.842		2,275	48.50	2.96	2,326.46	2,575
1610	6″ diam. pipe size	Q-18	26	.923		2,425	55	2.16	2,482.16	2,750
1620	8″ diam. pipe size		25	.960		2,800	57.50	2.25	2,859.75	3,200
1630	10″ diam. pipe size		23	1.043		3,275	62.50	2.44	3,339.94	3,725
1640	12″ diam. pipe size		21	1.143		3,625	68	2.67	3,695.67	4,100
1650	14″ diam. pipe size		19	1.263		4,075	75.50	2.96	4,153.46	4,600
1660	16″ diam. pipe size		18	1.333		4,800	79.50	3.12	4,882.62	5,425
1670	18″ diam. pipe size		16	1.500		5,100	89.50	3.51	5,193.01	5,750
1680	20″ diam. pipe size		14	1.714		5,800	102	4.01	5,906.01	6,525
1690	24″ diam. pipe size		12	2		6,425	119	4.68	6,548.68	7,250
2000	Elbow, 45° for system									
2020	3/4″ diam. pipe size	Q-17	14	1.143	Ea.	700	66	4.01	770.01	875
2040	1″ diam. pipe size		13	1.231		720	71	4.32	795.32	905
2050	1-1/4″ diam. pipe size		11	1.455		820	83.50	5.10	908.60	1,025
2060	1-1/2″ diam. pipe size		9	1.778		825	102	6.25	933.25	1,075
2070	2″ diam. pipe size		6	2.667		885	153	9.35	1,047.35	1,200
2080	2-1/2″ diam. pipe size		4	4		970	230	14.05	1,214.05	1,425
2090	3″ diam. pipe size		3.50	4.571		1,125	263	16.05	1,404.05	1,650
2100	4″ diam. pipe size		3	5.333		1,300	305	18.70	1,623.70	1,900
2110	5″ diam. pipe size		2.80	5.714		1,675	330	20	2,025	2,375
2120	6″ diam. pipe size	Q-18	4	6		1,900	360	14.05	2,274.05	2,625
2130	8″ diam. pipe size		3	8		2,750	475	18.70	3,243.70	3,750
2140	10″ diam. pipe size		2.40	10		3,375	595	23.50	3,993.50	4,625
2150	12″ diam. pipe size		2	12		4,450	715	28	5,193	6,000
2160	14″ diam. pipe size		1.80	13.333		5,525	795	31	6,351	7,300

33 61 Hydronic Energy Distribution

33 61 13 – Underground Hydronic Energy Distribution

33 61 13.20 Pipe Conduit, Prefabricated/Preinsulated		Crew	Daily Output	Labor-Hours	Unit	Material	2019 Bare Costs Labor	Equipment	Total	Total Incl O&P
2170	16" diam. pipe size	Q-18	1.60	15	Ea.	6,550	895	35	7,480	8,625
2180	18" diam. pipe size		1.30	18.462		8,250	1,100	43	9,393	10,700
2190	20" diam. pipe size		1	24		10,400	1,425	56	11,881	13,600
2200	24" diam. pipe size		.70	34.286		13,000	2,050	80	15,130	17,500
2260	For elbow, 90°, add					25%				
2300	For tee, straight, add					85%	30%			
2340	For tee, reducing, add					170%	30%			
2380	For weldolet, straight, add					50%				
2400	Polyurethane insulation, 1"									
2410	FRP carrier and casing									
2420	4"	Q-5	18	.889	L.F.	54.50	51		105.50	137
2422	6"	Q-6	23	1.043		95	62.50		157.50	198
2424	8"		19	1.263		155	75.50		230.50	283
2426	10"		16	1.500		209	89.50		298.50	365
2428	12"		13	1.846		272	110		382	465
2430	FRP carrier and PVC casing									
2440	4"	Q-5	18	.889	L.F.	22.50	51		73.50	102
2444	8"	Q-6	19	1.263		46.50	75.50		122	165
2446	10"		16	1.500		68.50	89.50		158	209
2448	12"		13	1.846		92.50	110		202.50	267
2450	PVC carrier and casing									
2460	4"	Q-1	36	.444	L.F.	11.65	25.50		37.15	51
2462	6"	"	29	.552		16.60	31.50		48.10	65.50
2464	8"	Q-2	36	.667		22	39.50		61.50	83
2466	10"		32	.750		31	44		75	101
2468	12"		31	.774		34.50	45.50		80	107

33 63 Steam Energy Distribution

33 63 13 – Underground Steam and Condensate Distribution Piping

33 63 13.10 Calcium Silicate Insulated System

		Crew	Daily Output	Labor-Hours	Unit	Material	Labor	Equipment	Total	Total Incl O&P
0010	**CALCIUM SILICATE INSULATED SYSTEM**									
0011	High temp. (1200 degrees F)									
2840	Steel casing with protective exterior coating									
2850	6-5/8" diameter	Q-18	52	.462	L.F.	134	27.50	1.08	162.58	190
2860	8-5/8" diameter		50	.480		145	28.50	1.12	174.62	203
2870	10-3/4" diameter		47	.511		170	30.50	1.19	201.69	234
2880	12-3/4" diameter		44	.545		185	32.50	1.28	218.78	254
2890	14" diameter		41	.585		209	35	1.37	245.37	284
2900	16" diameter		39	.615		224	36.50	1.44	261.94	305
2910	18" diameter		36	.667		247	40	1.56	288.56	335
2920	20" diameter		34	.706		278	42	1.65	321.65	370
2930	22" diameter		32	.750		385	45	1.76	431.76	495
2940	24" diameter		29	.828		440	49.50	1.94	491.44	560
2950	26" diameter		26	.923		505	55	2.16	562.16	640
2960	28" diameter		23	1.043		625	62.50	2.44	689.94	780
2970	30" diameter		21	1.143		670	68	2.67	740.67	840
2980	32" diameter		19	1.263		750	75.50	2.96	828.46	940
2990	34" diameter		18	1.333		755	79.50	3.12	837.62	950
3000	36" diameter		16	1.500		810	89.50	3.51	903.01	1,025
3040	For multi-pipe casings, add					10%				
3060	For oversize casings, add					2%				

33 63 Steam Energy Distribution

33 63 13 – Underground Steam and Condensate Distribution Piping

33 63 13.10 Calcium Silicate Insulated System		Crew	Daily Output	Labor-Hours	Unit	2019 Bare Costs Material	Labor	Equipment	Total	Total Incl O&P
3400	Steel casing gland seal, single pipe									
3420	6-5/8" diameter	Q-18	25	.960	Ea.	2,050	57.50	2.25	2,109.75	2,350
3440	8-5/8" diameter		23	1.043		2,400	62.50	2.44	2,464.94	2,725
3450	10-3/4" diameter		21	1.143		2,675	68	2.67	2,745.67	3,025
3460	12-3/4" diameter		19	1.263		3,150	75.50	2.96	3,228.46	3,600
3470	14" diameter		17	1.412		3,500	84.50	3.30	3,587.80	3,975
3480	16" diameter		16	1.500		4,100	89.50	3.51	4,193.01	4,650
3490	18" diameter		15	1.600		4,525	95.50	3.74	4,624.24	5,150
3500	20" diameter		13	1.846		5,025	110	4.32	5,139.32	5,700
3510	22" diameter		12	2		5,650	119	4.68	5,773.68	6,375
3520	24" diameter		11	2.182		6,350	130	5.10	6,485.10	7,175
3530	26" diameter		10	2.400		7,250	143	5.60	7,398.60	8,200
3540	28" diameter		9.50	2.526		8,175	151	5.90	8,331.90	9,225
3550	30" diameter		9	2.667		8,325	159	6.25	8,490.25	9,400
3560	32" diameter		8.50	2.824		9,350	169	6.60	9,525.60	10,600
3570	34" diameter		8	3		10,200	179	7	10,386	11,500
3580	36" diameter		7	3.429		10,800	205	8	11,013	12,200
3620	For multi-pipe casings, add					5%				
4000	Steel casing anchors, single pipe									
4020	6-5/8" diameter	Q-18	8	3	Ea.	1,825	179	7	2,011	2,300
4040	8-5/8" diameter		7.50	3.200		1,900	191	7.50	2,098.50	2,400
4050	10-3/4" diameter		7	3.429		2,500	205	8	2,713	3,075
4060	12-3/4" diameter		6.50	3.692		2,675	220	8.65	2,903.65	3,275
4070	14" diameter		6	4		3,150	239	9.35	3,398.35	3,850
4080	16" diameter		5.50	4.364		3,625	260	10.20	3,895.20	4,400
4090	18" diameter		5	4.800		4,125	286	11.25	4,422.25	4,975
4100	20" diameter		4.50	5.333		4,525	320	12.50	4,857.50	5,500
4110	22" diameter		4	6		5,050	360	14.05	5,424.05	6,125
4120	24" diameter		3.50	6.857		5,500	410	16.05	5,926.05	6,675
4130	26" diameter		3	8		6,225	475	18.70	6,718.70	7,575
4140	28" diameter		2.50	9.600		6,850	575	22.50	7,447.50	8,425
4150	30" diameter		2	12		7,275	715	28	8,018	9,100
4160	32" diameter		1.50	16		8,700	955	37.50	9,692.50	11,000
4170	34" diameter		1	24		9,775	1,425	56	11,256	12,900
4180	36" diameter		1	24		10,600	1,425	56	12,081	13,900
4220	For multi-pipe, add					5%	20%			
4800	Steel casing elbow									
4820	6-5/8" diameter	Q-18	15	1.600	Ea.	2,600	95.50	3.74	2,699.24	3,025
4830	8-5/8" diameter		15	1.600		2,800	95.50	3.74	2,899.24	3,225
4850	10-3/4" diameter		14	1.714		3,375	102	4.01	3,481.01	3,850
4860	12-3/4" diameter		13	1.846		3,975	110	4.32	4,089.32	4,550
4870	14" diameter		12	2		4,275	119	4.68	4,398.68	4,875
4880	16" diameter		11	2.182		4,675	130	5.10	4,810.10	5,350
4890	18" diameter		10	2.400		5,325	143	5.60	5,473.60	6,075
4900	20" diameter		9	2.667		5,675	159	6.25	5,840.25	6,500
4910	22" diameter		8	3		6,125	179	7	6,311	7,000
4920	24" diameter		7	3.429		6,725	205	8	6,938	7,725
4930	26" diameter		6	4		7,375	239	9.35	7,623.35	8,500
4940	28" diameter		5	4.800		7,975	286	11.25	8,272.25	9,225
4950	30" diameter		4	6		8,000	360	14.05	8,374.05	9,350
4960	32" diameter		3	8		9,100	475	18.70	9,593.70	10,700
4970	34" diameter		2	12		10,000	715	28	10,743	12,100
4980	36" diameter		2	12		10,600	715	28	11,343	12,800

33 63 13.10 Calcium Silicate Insulated System		Crew	Daily Output	Labor-Hours	Unit	Material	2019 Bare Costs Labor	Equipment	Total	Total Incl O&P
5500	Black steel service pipe, std. wt., 1" thick insulation									
5510	3/4" diameter pipe size	Q-17	54	.296	L.F.	77.50	17.05	1.04	95.59	112
5540	1" diameter pipe size		50	.320		80	18.40	1.12	99.52	117
5550	1-1/4" diameter pipe size		47	.340		87.50	19.60	1.19	108.29	127
5560	1-1/2" diameter pipe size		45	.356		92	20.50	1.25	113.75	133
5570	2" diameter pipe size		42	.381		101	22	1.34	124.34	145
5580	2-1/2" diameter pipe size		34	.471		106	27	1.65	134.65	158
5590	3" diameter pipe size		28	.571		111	33	2.01	146.01	174
5600	4" diameter pipe size		22	.727		125	42	2.55	169.55	204
5610	5" diameter pipe size		18	.889		136	51	3.12	190.12	230
5620	6" diameter pipe size	Q-18	23	1.043		147	62.50	2.44	211.94	257
6000	Black steel service pipe, std. wt., 1-1/2" thick insul.									
6010	3/4" diameter pipe size	Q-17	54	.296	L.F.	78.50	17.05	1.04	96.59	113
6040	1" diameter pipe size		50	.320		83.50	18.40	1.12	103.02	121
6050	1-1/4" diameter pipe size		47	.340		91	19.60	1.19	111.79	131
6060	1-1/2" diameter pipe size		45	.356		90	20.50	1.25	111.75	131
6070	2" diameter pipe size		42	.381		97	22	1.34	120.34	140
6080	2-1/2" diameter pipe size		34	.471		107	27	1.65	135.65	159
6090	3" diameter pipe size		28	.571		118	33	2.01	153.01	182
6100	4" diameter pipe size		22	.727		137	42	2.55	181.55	216
6110	5" diameter pipe size		18	.889		145	51	3.12	199.12	239
6120	6" diameter pipe size	Q-18	23	1.043		151	62.50	2.44	215.94	262
6130	8" diameter pipe size		19	1.263		208	75.50	2.96	286.46	345
6140	10" diameter pipe size		16	1.500		278	89.50	3.51	371.01	445
6150	12" diameter pipe size		13	1.846		345	110	4.32	459.32	545
6190	For 2" thick insulation, add					15%				
6220	For 2-1/2" thick insulation, add					25%				
6260	For 3" thick insulation, add					30%				
6800	Black steel service pipe, ex. hvy. wt., 1" thick insul.									
6820	3/4" diameter pipe size	Q-17	50	.320	L.F.	80	18.40	1.12	99.52	117
6840	1" diameter pipe size		47	.340		85	19.60	1.19	105.79	124
6850	1-1/4" diameter pipe size		44	.364		95	21	1.28	117.28	137
6860	1-1/2" diameter pipe size		42	.381		97	22	1.34	120.34	141
6870	2" diameter pipe size		40	.400		106	23	1.40	130.40	153
6880	2-1/2" diameter pipe size		31	.516		106	29.50	1.81	137.31	162
6890	3" diameter pipe size		27	.593		122	34	2.08	158.08	187
6900	4" diameter pipe size		21	.762		118	44	2.67	164.67	199
6910	5" diameter pipe size		17	.941		140	54	3.30	197.30	239
6920	6" diameter pipe size	Q-18	22	1.091		150	65	2.55	217.55	265
7400	Black steel service pipe, ex. hvy. wt., 1-1/2" thick insul.									
7420	3/4" diameter pipe size	Q-17	50	.320	L.F.	68.50	18.40	1.12	88.02	104
7440	1" diameter pipe size		47	.340		74.50	19.60	1.19	95.29	113
7450	1-1/4" diameter pipe size		44	.364		83	21	1.28	105.28	124
7460	1-1/2" diameter pipe size		42	.381		82	22	1.34	105.34	124
7470	2" diameter pipe size		40	.400		92	23	1.40	116.40	137
7480	2-1/2" diameter pipe size		31	.516		98	29.50	1.81	129.31	154
7490	3" diameter pipe size		27	.593		109	34	2.08	145.08	173
7500	4" diameter pipe size		21	.762		139	44	2.67	185.67	221
7510	5" diameter pipe size		17	.941		191	54	3.30	248.30	296
7520	6" diameter pipe size	Q-18	22	1.091		205	65	2.55	272.55	325
7530	8" diameter pipe size		18	1.333		249	79.50	3.12	331.62	395
7540	10" diameter pipe size		15	1.600		315	95.50	3.74	414.24	490
7550	12" diameter pipe size		13	1.846		375	110	4.32	489.32	585

33 63 13 – Underground Steam and Condensate Distribution Piping

33 63 13.10 Calcium Silicate Insulated System		Crew	Daily Output	Labor-Hours	Unit	2019 Bare Costs Material	Labor	Equipment	Total	Total Incl O&P
7590	For 2" thick insulation, add				L.F.	13%				
7640	For 2-1/2" thick insulation, add					18%				
7680	For 3" thick insulation, add					24%				

33 63 13.20 Combined Steam Pipe With Condensate Return		Crew	Daily Output	Labor-Hours	Unit	Material	Labor	Equipment	Total	Total Incl O&P
0010	**COMBINED STEAM PIPE WITH CONDENSATE RETURN**									
9010	8" and 4" in 24" case	Q-18	100	.240	L.F.	375	14.30	.56	389.86	430
9020	6" and 3" in 20" case		104	.231		380	13.75	.54	394.29	435
9030	3" and 1-1/2" in 16" case		110	.218		285	13	.51	298.51	335
9040	2" and 1-1/4" in 12-3/4" case		114	.211		240	12.55	.49	253.04	283
9050	1-1/2" and 1-1/4" in 10-3/4" case		116	.207		237	12.35	.48	249.83	280
9100	Steam pipe only (no return)									
9110	6" in 18" case	Q-18	108	.222	L.F.	310	13.25	.52	323.77	360
9120	4" in 14" case		112	.214		240	12.80	.50	253.30	284
9130	3" in 14" case		114	.211		227	12.55	.49	240.04	269
9140	2-1/2" in 12-3/4" case		118	.203		211	12.15	.48	223.63	252
9150	2" in 12-3/4" case		122	.197		200	11.75	.46	212.21	238

Division Notes

I	CREW	DAILY OUTPUT	LABOR-HOURS	UNIT	BARE COSTS MAT.	LABOR	EQUIP.	TOTAL	TOTAL INCL O&P

Estimating Tips

This section involves equipment and construction costs for air, noise, and odor pollution control systems. These systems may be interrelated and care must be taken that the complete systems are estimated. For example, air pollution equipment may include dust and air-entrained particles that have to be collected. The vacuum systems could be noisy, requiring silencers to reduce noise pollution, and the collected solids have to be disposed of to prevent solid pollution.

Reference Numbers

Reference numbers are shown at the beginning of some major classifications. These numbers refer to related items in the Reference Section. The reference information may be an estimating procedure, an alternate pricing method, or technical information.

44 11 16 – Fugitive Dust Barrier Systems

44 11 16.10 Dust Collection Systems		Crew	Daily Output	Labor-Hours	Unit	Material	Labor	Equipment	Total	Total Incl O&P
						2019 Bare Costs				
0010	DUST COLLECTION SYSTEMS Commercial/Industrial									
0120	Central vacuum units									
0130	Includes stand, filters and motorized shaker									
0200	500 CFM, 10" inlet, 2 HP	Q-20	2.40	8.333	Ea.	4,825	465		5,290	6,000
0220	1,000 CFM, 10" inlet, 3 HP		2.20	9.091		5,075	510		5,585	6,350
0240	1,500 CFM, 10" inlet, 5 HP		2	10		5,325	560		5,885	6,725
0260	3,000 CFM, 13" inlet, 10 HP		1.50	13.333		15,600	745		16,345	18,300
0280	5,000 CFM, 16" inlet, 2 @ 10 HP		1	20		16,600	1,125		17,725	20,000
1000	Vacuum tubing, galvanized									
1100	2-1/8" OD, 16 ga.	Q-9	440	.036	L.F.	3.43	1.99		5.42	6.80
1110	2-1/2" OD, 16 ga.		420	.038		3.79	2.09		5.88	7.35
1120	3" OD, 16 ga.		400	.040		4.76	2.19		6.95	8.60
1130	3-1/2" OD, 16 ga.		380	.042		6.75	2.31		9.06	10.95
1140	4" OD, 16 ga.		360	.044		7.05	2.44		9.49	11.45
1150	5" OD, 14 ga.		320	.050		12.65	2.74		15.39	18.05
1160	6" OD, 14 ga.		280	.057		14.15	3.13		17.28	20.50
1170	8" OD, 14 ga.		200	.080		21	4.39		25.39	29.50
1180	10" OD, 12 ga.		160	.100		42.50	5.50		48	55
1190	12" OD, 12 ga.		120	.133		54	7.30		61.30	70.50
1200	14" OD, 12 ga.		80	.200		61.50	10.95		72.45	84.50
1940	Hose, flexible wire reinforced rubber									
1956	3" diam.	Q-9	400	.040	L.F.	6.75	2.19		8.94	10.75
1960	4" diam.		360	.044		7.80	2.44		10.24	12.25
1970	5" diam.		320	.050		10.10	2.74		12.84	15.25
1980	6" diam.		280	.057		10.60	3.13		13.73	16.40
2000	90° elbow, slip fit									
2110	2-1/8" diam.	Q-9	70	.229	Ea.	13.65	12.55		26.20	34
2120	2-1/2" diam.		65	.246		19.05	13.50		32.55	41.50
2130	3" diam.		60	.267		26	14.65		40.65	50.50
2140	3-1/2" diam.		55	.291		32	15.95		47.95	59
2150	4" diam.		50	.320		39.50	17.55		57.05	70
2160	5" diam.		45	.356		75	19.50		94.50	112
2170	6" diam.		40	.400		104	22		126	148
2180	8" diam.		30	.533		197	29.50		226.50	262
2400	45° elbow, slip fit									
2410	2-1/8" diam.	Q-9	70	.229	Ea.	11.95	12.55		24.50	32
2420	2-1/2" diam.		65	.246		17.60	13.50		31.10	40
2430	3" diam.		60	.267		21	14.65		35.65	45.50
2440	3-1/2" diam.		55	.291		26.50	15.95		42.45	53.50
2450	4" diam.		50	.320		34.50	17.55		52.05	64.50
2460	5" diam.		45	.356		58.50	19.50		78	93.50
2470	6" diam.		40	.400		79	22		101	121
2480	8" diam.		35	.457		155	25		180	208
2800	90° TY, slip fit thru 6" diam.									
2810	2-1/8" diam.	Q-9	42	.381	Ea.	18.15	21		39.15	51.50
2820	2-1/2" diam.		39	.410		27	22.50		49.50	63.50
2830	3" diam.		36	.444		37.50	24.50		62	78
2840	3-1/2" diam.		33	.485		39.50	26.50		66	84
2850	4" diam.		30	.533		42.50	29.50		72	91.50
2860	5" diam.		27	.593		72.50	32.50		105	129
2870	6" diam.		24	.667		104	36.50		140.50	170
2880	8" diam., butt end					160			160	176
2890	10" diam., butt end					485			485	535

44 11 Particulate Control Equipment

44 11 16 – Fugitive Dust Barrier Systems

44 11 16.10 Dust Collection Systems		Crew	Daily Output	Labor-Hours	Unit	Material	2019 Bare Costs Labor	Equipment	Total	Total Incl O&P
2900	12" diam., butt end				Ea.	485			485	535
2910	14" diam., butt end					875			875	965
2920	6" x 4" diam., butt end					133			133	146
2930	8" x 4" diam., butt end					252			252	277
2940	10" x 4" diam., butt end					330			330	365
2950	12" x 4" diam., butt end					375			375	415
3100	90° elbow, butt end, segmented									
3110	8" diam., butt end, segmented				Ea.	160			160	176
3120	10" diam., butt end, segmented					490			490	540
3130	12" diam., butt end, segmented					620			620	680
3140	14" diam., butt end, segmented					775			775	855
3200	45° elbow, butt end, segmented									
3210	8" diam., butt end, segmented				Ea.	155			155	170
3220	10" diam., butt end, segmented					355			355	390
3230	12" diam., butt end, segmented					360			360	400
3240	14" diam., butt end, segmented					590			590	650
3400	All butt end fittings require one coupling per joint.									
3410	Labor for fitting included with couplings.									
3460	Compression coupling, galvanized, neoprene gasket									
3470	2-1/8" diam.	Q-9	44	.364	Ea.	13.70	19.95		33.65	45.50
3480	2-1/2" diam.		44	.364		13.70	19.95		33.65	45.50
3490	3" diam.		38	.421		19.70	23		42.70	56.50
3500	3-1/2" diam.		35	.457		22	25		47	62.50
3510	4" diam.		33	.485		24	26.50		50.50	67
3520	5" diam.		29	.552		27.50	30.50		58	76
3530	6" diam.		26	.615		31	34		65	85.50
3540	8" diam.		22	.727		58	40		98	125
3550	10" diam.		20	.800		86	44		130	161
3560	12" diam.		18	.889		106	49		155	191
3570	14" diam.		16	1		128	55		183	224
3800	Air gate valves, galvanized									
3810	2-1/8" diam.	Q-9	30	.533	Ea.	120	29.50		149.50	177
3820	2-1/2" diam.		28	.571		131	31.50		162.50	192
3830	3" diam.		26	.615		139	34		173	205
3840	4" diam.		23	.696		161	38		199	235
3850	6" diam.		18	.889		224	49		273	320

44 11 96 – Dry Plate Electrostatic Precipitator Equipment

44 11 96.10 Electrostatic Precipitators		Crew	Daily Output	Labor-Hours	Unit	Material	Labor	Equipment	Total	Total Incl O&P
0010	ELECTROSTATIC PRECIPITATORS									
4100	Smoke pollution control									
4110	Exhaust air volume									
4120	1 to 2.4 MCFM	Q-10	2.15	11.163	MCFM	35,900	635		36,535	40,500
4130	2.4 to 4.8 MCFM		2.70	8.889		56,500	505		57,005	63,000
4140	4.8 to 7.2 MCFM		3.60	6.667		77,000	380		77,380	85,000
4150	7.2 to 12 MCFM		5.40	4.444		133,500	253		133,753	147,500

Division Notes

I		CREW	DAILY OUTPUT	LABOR-HOURS	UNIT	BARE COSTS MAT.	LABOR	EQUIP.	TOTAL	TOTAL INCL O&P

Estimating Tips

This division contains information about water and wastewater equipment and systems, which was formerly located in Division 44. The main areas of focus are total wastewater treatment plants and components of wastewater treatment plants. Also included in this section are oil/water separators for wastewater treatment.

Reference Numbers

Reference numbers are shown at the beginning of some major classifications. These numbers refer to related items in the Reference Section. The reference information may be an estimating procedure, an alternate pricing method, or technical information.

46 23 Grit Removal And Handling Equipment

46 23 23 – Vortex Grit Removal Equipment

46 23 23.10 Rainwater Filters		Crew	Daily Output	Labor-Hours	Unit	Material	Labor	Equipment	Total	Total Incl O&P
						2019 Bare Costs				
0010	RAINWATER FILTERS									
0100	42 gal./min	B-21	3.50	8	Ea.	39,600	380	37	40,017	44,100
0200	65 gal./min		3.50	8		39,600	380	37	40,017	44,200
0300	208 gal./min		3.50	8		39,700	380	37	40,117	44,200

46 25 Oil and Grease Separation and Removal Equipment

46 25 13 – Coalescing Oil-Water Separators

46 25 13.20 Oil/Water Separators		Crew	Daily Output	Labor-Hours	Unit	Material	Labor	Equipment	Total	Total Incl O&P
0010	OIL/WATER SEPARATORS									
0020	Underground, tank only									
0030	Excludes excavation, backfill & piping									
0100	200 GPM	B-21	3.50	8	Ea.	39,400	380	37	39,817	43,900
0110	400 GPM		3.25	8.615		80,000	410	40	80,450	88,500
0120	600 GPM		2.75	10.182		87,500	485	47	88,032	97,000
0130	800 GPM		2.50	11.200		100,500	530	52	101,082	111,500
0140	1,000 GPM		2	14		116,000	665	65	116,730	129,000
0150	1,200 GPM		1.50	18.667		125,500	885	86.50	126,471.50	139,500
0160	1,500 GPM		1	28		146,000	1,325	130	147,455	162,500

46 25 16 – API Oil-Water Separators

46 25 16.10 Oil/Water Separators		Crew	Daily Output	Labor-Hours	Unit	Material	Labor	Equipment	Total	Total Incl O&P
0010	OIL/WATER SEPARATORS									
0020	Complete system, not including excavation and backfill									
0100	Treated capacity 0.2 C.F./second	B-22	1	30	Ea.	7,950	1,450	195	9,595	11,100
0200	0.5 C.F./second	B-13	.75	74.667		10,000	3,350	775	14,125	16,900
0300	1.0 C.F./second		.60	93.333		14,900	4,175	970	20,045	23,800
0400	2.4 to 3.0 C.F./second		.30	187		21,700	8,375	1,950	32,025	38,500
0500	11 C.F./second		.17	329		28,100	14,800	3,425	46,325	57,000
0600	22 C.F./second	B-13	.10	560		46,800	25,100	5,825	77,725	96,000

Assemblies Section

Table of Contents

Table of Contents

RSMeans data: Assemblies—How They Work

Assemblies estimating provides a fast and reasonably accurate way to develop construction costs. An assembly is the grouping of individual work items—with appropriate quantities—to provide a cost for a major construction component in a convenient unit of measure.

An assemblies estimate is often used during early stages of design development to compare the cost impact of various design alternatives on total building cost.

Assemblies estimates are also used as an efficient tool to verify construction estimates.

Assemblies estimates do not require a completed design or detailed drawings. Instead, they are based on the general size of the structure and other known parameters of the project. The degree of accuracy of an assemblies estimate is generally within +/- 15%.

Most assemblies consist of three major elements: a graphic, the system components, and the cost data itself. The **Graphic** is a visual representation showing the typical appearance of the assembly

1 Unique 12-character Identifier

Our assemblies are identified by a **unique 12-character identifier**. The assemblies are numbered using UNIFORMAT II, ASTM Standard E1557. The first 5 characters represent this system to Level 3. The last 7 characters represent further breakdown in order to arrange items in understandable groups of similar tasks. Line numbers are consistent across all of our publications, so a line number in any assemblies data set will always refer to the same item.

2 Reference Box

Information is available in the Reference Section to assist the estimator with estimating procedures, alternate pricing methods, and additional technical information.

The **Reference Box** indicates the exact location of this information in the Reference Section. The "R" stands for "reference," and the remaining characters are the line numbers.

3 Narrative Descriptions

Our assemblies descriptions appear in two formats: narrative and table. **Narrative descriptions** are shown in a hierarchical structure to make them readable. In order to read a complete description, read up through the indents to the top of the section. Include everything that is above and to the left that is not contradicted by information below.

D30 HVAC

D3010 Energy Supply

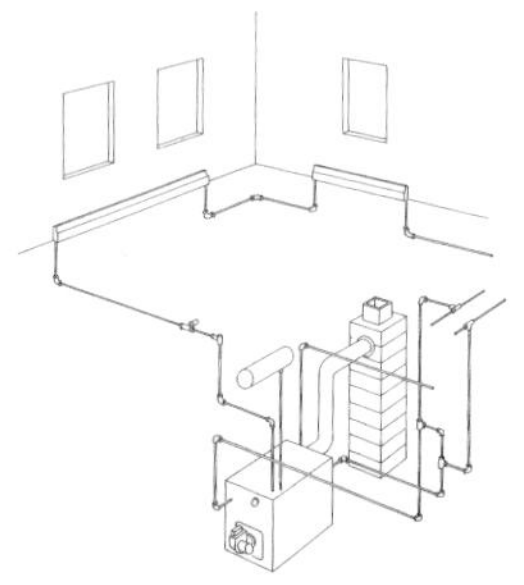

Basis for Heat Loss Estimate, Apartment Type Structures:

1. Masonry walls and flat roof are insulated. U factor is assumed at .08.
2. Window glass area taken as BOCA minimum, 1/10th of floor area. Double insulating glass with 1/4" air space, U = .65.
3. Infiltration = 0.3 C.F. per hour per S.F. of net wall.
4. Concrete floor loss is 2 BTUH per S.F.
5. Temperature difference taken as 70°F.
6. Ventilating or makeup air has not been included and must be added if desired. Air shafts are not used.

System Components	QUANTITY	UNIT	COST EACH MAT.	COST EACH INST.	COST EACH TOTAL
(1) SYSTEM D3010 510 1760					
HEATING SYSTEM, FIN TUBE RADIATION, FORCED HOT WATER					
1,000 S.F. AREA, 10,000 C.F. VOLUME					
Boiler, oil fired, CI, burner, ctrls/insul/breech/pipe/ftng/valves, 109 MBH	1.000	Ea.	4,287.50	4,243.75	8,531.25
Circulating pump, CI flange connection, 1/12 HP	1.000	Ea.	505	227	732
Expansion tank, painted steel, ASME 18 Gal capacity	1.000	Ea.	870	98.50	968.50
Storage tank, steel, above ground, 275 Gal capacity w/supports	1.000	Ea.	570	276	846
Copper tubing type L, solder joint, hanger 10' OC, 3/4" diam	100.000	L.F.	482	995	1,477
Radiation, 3/4" copper tube w/alum fin baseboard pkg, 7" high	30.000	L.F.	267	720	987
Pipe covering, calcium silicate w/cover, 1' wall, 3/4' diam	100.000	L.F.	463	750	1,213
TOTAL			7,444.50	7,310.25	14,754.75
COST PER S.F.			7.44	7.31	14.75

D3010 510	Apartment Building Heating - Fin Tube Radiation		COST PER S.F. MAT.	COST PER S.F. INST.	COST PER S.F. TOTAL
1740	Heating systems, fin tube radiation, forced hot water				
1760	1,000 S.F. area, 10,000 C.F. volume		7.45	7.30	14.75
1800	10,000 S.F. area, 100,000 C.F. volume	R235000-10	3.39	4.42	7.81
1840	(3) 20,000 S.F. area, 200,000 C.F. volume	(2)	3.70	4.96	8.66
1880	30,000 S.F. area, 300,000 C.F. volume	R235000-20	3.60	4.82	8.42
1890					

For supplemental customizable square foot estimating forms, visit: **RSMeans.com/2019books**

in question. It is frequently accompanied by additional explanatory technical information describing the class of items. The **System Components** is a listing of the individual tasks that make up the assembly, including the quantity and unit of measure for each item, along with the cost of material and installation. The **Assemblies data** below lists prices for other similar systems with dimensional and/or size variations.

All of our assemblies costs represent the cost for the installing contractor. An allowance for profit has been added to all material, labor, and equipment rental costs. A markup for labor burdens, including workers' compensation, fixed overhead, and business overhead, is included with installation costs.

The information in RSMeans cost data represents a "national average" cost. This data should be modified to the project location using the **City Cost Indexes** or **Location Factors** tables found in the Reference Section.

D30 HVAC

D3010 Energy Supply

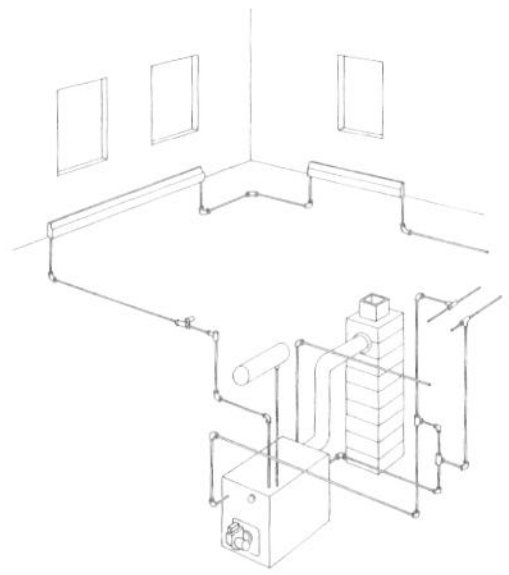

Basis for Heat Loss Estimate, Apartment Type Structures:

1. Masonry walls and flat roof are insulated. U factor is assumed at .08.
2. Window glass area taken as BOCA minimum, 1/10th of floor area. Double insulating glass with 1/4" air space, U = .65.
3. Infiltration = 0.3 C.F. per hour per S.F. of net wall.
4. Concrete floor loss is 2 BTUH per S.F.
5. Temperature difference taken as 70°F.
6. Ventilating or makeup air has not been included and must be added if desired. Air shafts are not used.

System Components (5)	QUANTITY	UNIT	COST EACH (4) MAT.	INST.	TOTAL
SYSTEM D3010 510 1760					
HEATING SYSTEM, FIN TUBE RADIATION, FORCED HOT WATER					
1,000 S.F. AREA, 10,000 C.F. VOLUME					
Boiler, oil fired, CI, burner, ctrls/insul/breech/pipe/ftng/valves, 109 MBH	1.000	Ea.	4,287.50	4,243.75	8,531.25
Circulating pump, CI flange connection, 1/12 HP	1.000	Ea.	505	227	732
Expansion tank, painted steel, ASME 18 Gal capacity	1.000	Ea.	870	98.50	968.50
Storage tank, steel, above ground, 275 Gal capacity w/supports	1.000	Ea.	570	276	846
Copper tubing type L, solder joint, hanger 10' OC, 3/4" diam	100.000	L.F.	482	995	1,477
Radiation, 3/4" copper tube w/alum fin baseboard pkg, 7" high	30.000	L.F.	267	720	987
Pipe covering, calcium silicate w/cover, 1' wall, 3/4' diam	100.000	L.F.	463	750	1,213
TOTAL			7,444.50	7,310.25	14,754.75
COST PER S.F.			7.44	7.31	14.75

D3010 510	Apartment Building Heating - Fin Tube Radiation		COST PER S.F. MAT.	INST.	TOTAL
1740	Heating systems, fin tube radiation, forced hot water				
1760	1,000 S.F. area, 10,000 C.F. volume		7.45	7.30	14.75
1800	10,000 S.F. area, 100,000 C.F. volume	R235000 -10	3.39	4.42	7.81
1840	20,000 S.F. area, 200,000 C.F. volume		3.70	4.96	8.66
1880	30,000 S.F. area, 300,000 C.F. volume	R235000 -20	3.60	4.82	8.42
1890					

4 Unit of Measure

All RSMeans data: Assemblies include a typical **Unit of Measure** used for estimating that item. For instance, for continuous footings or foundation walls the unit is linear feet (L.F.). For spread footings the unit is each (Ea.). The estimator needs to take special care that the unit in the data matches the unit in the takeoff. Abbreviations and unit conversions can be found in the Reference Section.

5 System Components

System components are listed separately to detail what is included in the development of the total system price.

RSMeans data: Assemblies— How They Work (Continued)

Sample Estimate

This sample demonstrates the elements of an estimate, including a tally of the RSMeans data lines. Published assemblies costs include all markups for labor burden and profit for the installing contractor. This estimate adds a summary of the markups applied by a general contractor on the installing contractor's work. These figures represent the total cost to the owner. The location factor with RSMeans data is applied at the bottom of the estimate to adjust the cost of the work to a specific location.

Project Name:	Interior Fit-out, ABC Office			
Location:	Anywhere, USA	Date: 1/1/2019		STD
Assembly Number	Description	Qty.	Unit	Subtotal
1 C1010 124 1200	Wood partition, 2 x 4 @ 16" OC w/5/8" FR gypsum board	560	S.F.	$2,889.60
C1020 114 1800	Metal door & frame, flush hollow core, 3'-0" x 7'-0"	2	Ea.	$2,560.00
C3010 230 0080	Painting, brushwork, primer & 2 coats	1,120	S.F.	$1,433.60
C3020 410 0140	Carpet, tufted, nylon, roll goods, 12' wide, 26 oz	240	S.F.	$885.60
C3030 210 6000	Acoustic ceilings, 24" x 48" tile, tee grid suspension	200	S.F.	$1,288.00
D5020 125 0560	Receptacles incl plate, box, conduit, wire, 20 A duplex	8	Ea.	$2,388.00
D5020 125 0720	Light switch incl plate, box, conduit, wire, 20 A single pole	2	Ea.	$583.00
D5020 210 0560	Fluorescent fixtures, recess mounted, 20 per 1000 SF	200	S.F.	$2,270.00
	Assembly Subtotal			**$14,297.80**
	Sales Tax @ 2	**5 %**		**$ 357.45**
	General Requirements @ 3	**7 %**		**$ 1,000.85**
	Subtotal A			**$15,656.09**
	GC Overhead @ 4	**5 %**		**$ 782.80**
	Subtotal B			**$16,438.90**
	GC Profit @ 5	**5 %**		**$ 821.94**
	Subtotal C			**$17,260.84**
	Adjusted by Location Factor 6	**113.9**		**$ 19,660.10**
	Architects Fee @ 7	**8 %**		**$ 1,572.81**
	Contingency @ 8	**15 %**		**$ 2,949.01**
	Project Total Cost			**$ 24,181.92**

This estimate is based on an interactive spreadsheet. You are free to download it and adjust it to your methodology. A copy of this spreadsheet is available at **RSMeans.com/2019books.**

❶ Work Performed

The body of the estimate shows the RSMeans data selected, including line numbers, a brief description of each item, its takeoff quantity and unit, and the total installed cost, including the installing contractor's overhead and profit.

❷ Sales Tax

If the work is subject to state or local sales taxes, the amount must be added to the estimate. In a conceptual estimate it can be assumed that one half of the total represents material costs. Therefore, apply the sales tax rate to 50% of the assembly subtotal.

❸ General Requirements

This item covers project-wide needs provided by the general contractor. These items vary by project but may include temporary facilities and utilities, security, testing, project cleanup, etc. In assemblies estimates a percentage is used—typically between 5% and 15% of project cost.

❹ General Contractor Overhead

This entry represents the general contractor's markup on all work to cover project administration costs.

❺ General Contractor Profit

This entry represents the GC's profit on all work performed. The value included here can vary widely by project and is influenced by the GC's perception of the project's financial risk and market conditions.

❻ Location Factor

RSMeans published data are based on national average costs. If necessary, adjust the total cost of the project using a location factor from the "Location Factor" table or the "City Cost Indexes" table found in the Reference Section. Use location factors if the work is general, covering the work of multiple trades. If the work is by a single trade (e.g., masonry) use the more specific data found in the City Cost Indexes.

To adjust costs by location factors, multiply the base cost by the factor and divide by 100.

❼ Architect's Fee

If appropriate, add the design cost to the project estimate. These fees vary based on project complexity and size. Typical design and engineering fees can be found in the Reference Section.

❽ Contingency

A factor for contingency may be added to any estimate to represent the cost of unknowns that may occur between the time that the estimate is performed and the time the project is constructed. The amount of the allowance will depend on the stage of design at which the estimate is done, as well as the contractor's assessment of the risk involved.

A Substructure

A2020 Basement Walls

A2020 220	Subdrainage Piping	COST PER L.F.		
		MAT.	INST.	TOTAL
2000	Piping, excavation & backfill excluded, PVC, perforated			
2110	3″ diameter	1.84	4.45	6.29
2130	4″ diameter	1.84	4.45	6.29
2140	5″ diameter	4.04	4.76	8.80
2150	6″ diameter	4.04	4.76	8.80
3000	Metal alum. or steel, perforated asphalt coated			
3150	6″ diameter	7.40	4.39	11.79
3160	8″ diameter	9.25	4.51	13.76
3170	10″ diameter	11.60	6	17.60
3180	12″ diameter	13	7.55	20.55
3220	18″ diameter	19.45	10.50	29.95
4000	Corrugated HDPE tubing			
4130	4″ diameter	.83	.83	1.66
4150	6″ diameter	2.10	1.10	3.20
4160	8″ diameter	4.79	1.42	6.21
4180	12″ diameter	7.60	4.90	12.50
4200	15″ diameter	10.95	5.55	16.50
4220	18″ diameter	13.70	7.80	21.50

D2020 Domestic Water Distribution

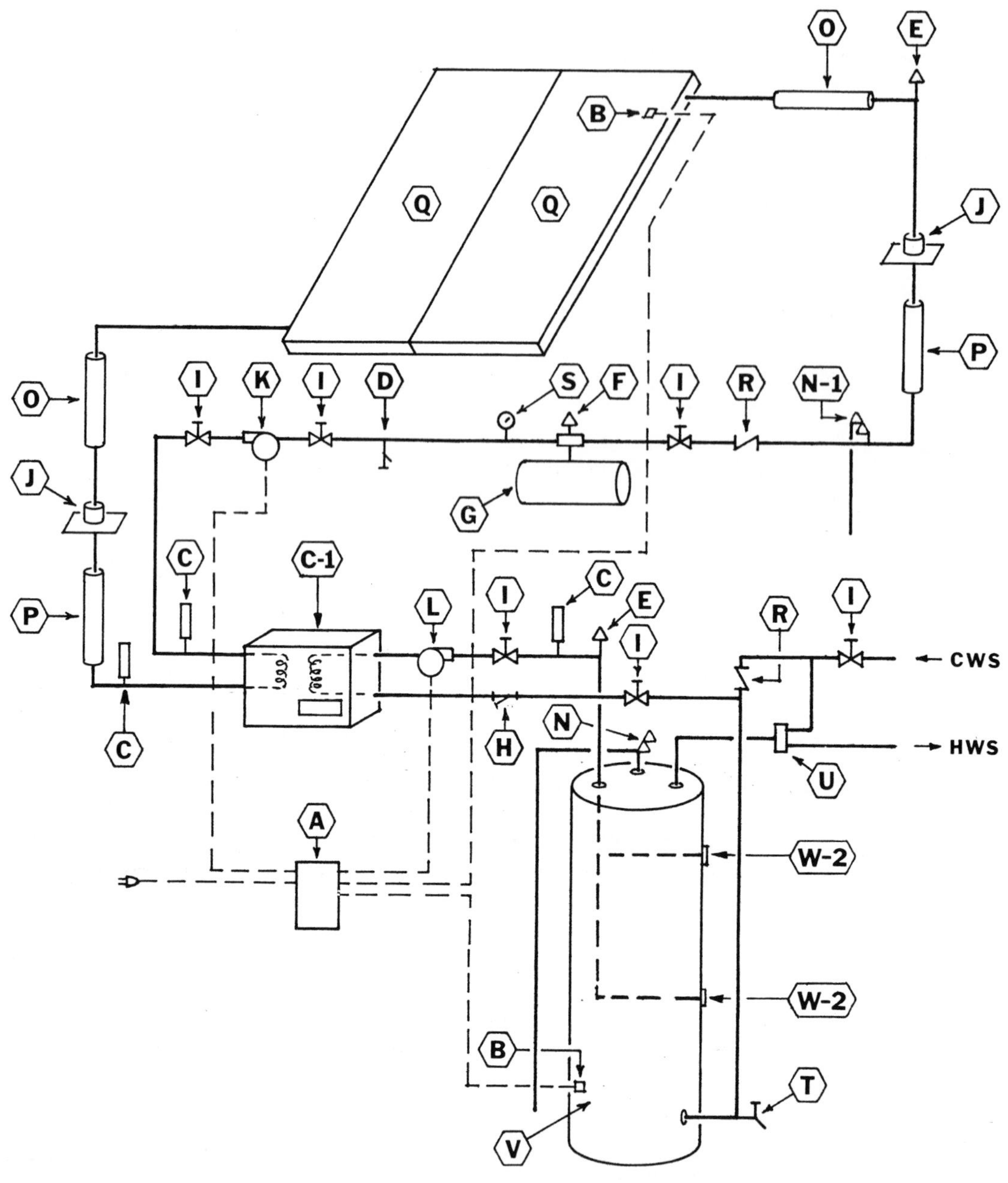

In this closed-loop indirect collection system, fluid with a low freezing temperature, such as propylene glycol, transports heat from the collectors to water storage. The transfer fluid is contained in a closed-loop consisting of collectors, supply and return piping, and a remote heat exchanger. The heat exchanger transfers heat energy from the fluid in the collector loop to potable water circulated in a storage loop. A typical two-or-three panel system contains 5 to 6 gallons of heat transfer fluid.

When the collectors become approximately 20°F warmer than the storage temperature, a controller activates the circulator on the collector and storage loops. The circulators will move the fluid and potable water through the heat exchanger until heat collection no longer occurs. At that point, the system shuts down. Since the heat transfer medium is a fluid with a very low freezing temperature, there is no need for it to be drained from the system between periods of collection.

D2020 Domestic Water Distribution

System Components

System Components	QUANTITY	UNIT	COST EACH MAT.	COST EACH INST.	COST EACH TOTAL
SYSTEM D2020 265 2760					
SOLAR, CLOSED LOOP, ADD-ON HOT WATER SYS., EXTERNAL HEAT EXCHANGER					
3/4" TUBING, TWO 3'X7' BLACK CHROME COLLECTORS					
A,B,G,L,K,M Heat exchanger fluid-fluid pkg incl 2 circulators, expansion tank, Check valve, relief valve, controller, hi temp cutoff, & 2 sensors	1.000	Ea.	900	545	1,445
C Thermometer, 2" dial	3.000	Ea.	73.50	142.50	216
D, T Fill & drain valve, brass, 3/4" connection	1.000	Ea.	11.65	31.50	43.15
E Air vent, manual, 1/8" fitting	2.000	Ea.	5.86	47	52.86
F Air purger	1.000	Ea.	52.50	63	115.50
H Strainer, Y type, bronze body, 3/4" IPS	1.000	Ea.	58.50	40.50	99
I Valve, gate, bronze, NRS, soldered 3/4" diam	6.000	Ea.	450	228	678
J Neoprene vent flashing	2.000	Ea.	21.90	76	97.90
N-1, N Relief valve temp & press, 150 psi 210°F self-closing 3/4" IPS	1.000	Ea.	26.50	25.50	52
O Pipe covering, urethane, ultraviolet cover, 1" wall 3/4" diam	20.000	L.F.	65.60	139	204.60
P Pipe covering, fiberglass, all service jacket, 1" wall, 3/4" diam	50.000	L.F.	53	277.50	330.50
Q Collector panel solar energy blk chrome on copper, 1/8" temp glass 3'x7'	2.000	Ea.	2,250	288	2,538
Roof clamps for solar energy collector panels	2.000	Set	6.36	39	45.36
R Valve, swing check, bronze, regrinding disc, 3/4" diam	2.000	Ea.	200	76	276
S Pressure gauge, 60 psi, 2" dial	1.000	Ea.	27	23.50	50.50
U Valve, water tempering, bronze, sweat connections, 3/4" diam	1.000	Ea.	159	38	197
W-2, V Tank water storage w/heating element, drain, relief valve, existing	1.000	Ea.			
Copper tubing type L, solder joint, hanger 10' OC 3/4" diam	20.000	L.F.	96.40	199	295.40
Copper tubing, type M, solder joint, hanger 10' OC 3/4" diam	70.000	L.F.	399	679	1,078
Sensor wire, #22-2 conductor multistranded	.500	C.L.F.	5.65	35.75	41.40
Solar energy heat transfer fluid, propylene glycol anti-freeze	6.000	Gal.	156	162	318
Wrought copper fittings & solder, 3/4" diam	76.000	Ea.	200.64	3,040	3,240.64
TOTAL			5,219.06	6,195.75	11,414.81

D2020 265 Solar, Closed Loop, Add-On Hot Water Systems

		COST EACH MAT.	COST EACH INST.	COST EACH TOTAL
2550	Solar, closed loop, add-on hot water system, external heat exchanger			
2570	3/8" tubing, 3 ea. 4' x 4'-4" vacuum tube collectors	5,900	5,800	11,700
2580	1/2" tubing, 4 ea. 4 x 4'-4" vacuum tube collectors (R235616 -60)	6,650	6,275	12,925
2600	2 ea. 3'x7' black chrome collectors	4,800	5,900	10,700
2620	3 ea. 3'x7' black chrome collectors	5,925	6,075	12,000
2640	2 ea. 3'x7' flat black collectors	4,375	5,925	10,300
2660	3 ea. 3'x7' flat black collectors	5,300	6,100	11,400
2700	3/4" tubing, 3 ea. 3'x7' black chrome collectors	6,350	6,350	12,700
2720	3 ea. 3'x7' flat black absorber plate collectors	5,725	6,375	12,100
2740	2 ea. 4'x9' flat black w/plastic glazing collectors	5,325	6,400	11,725
2760	2 ea. 3'x7' black chrome collectors	5,225	6,200	11,425
2780	1" tubing, 4 ea 2'x9' plastic absorber & glazing collectors	6,450	7,200	13,650
2800	4 ea. 3'x7' black chrome absorber collectors	8,300	7,225	15,525
2820	4 ea. 3'x7' flat black absorber collectors	7,500	7,275	14,775

D2020 Domestic Water Distribution

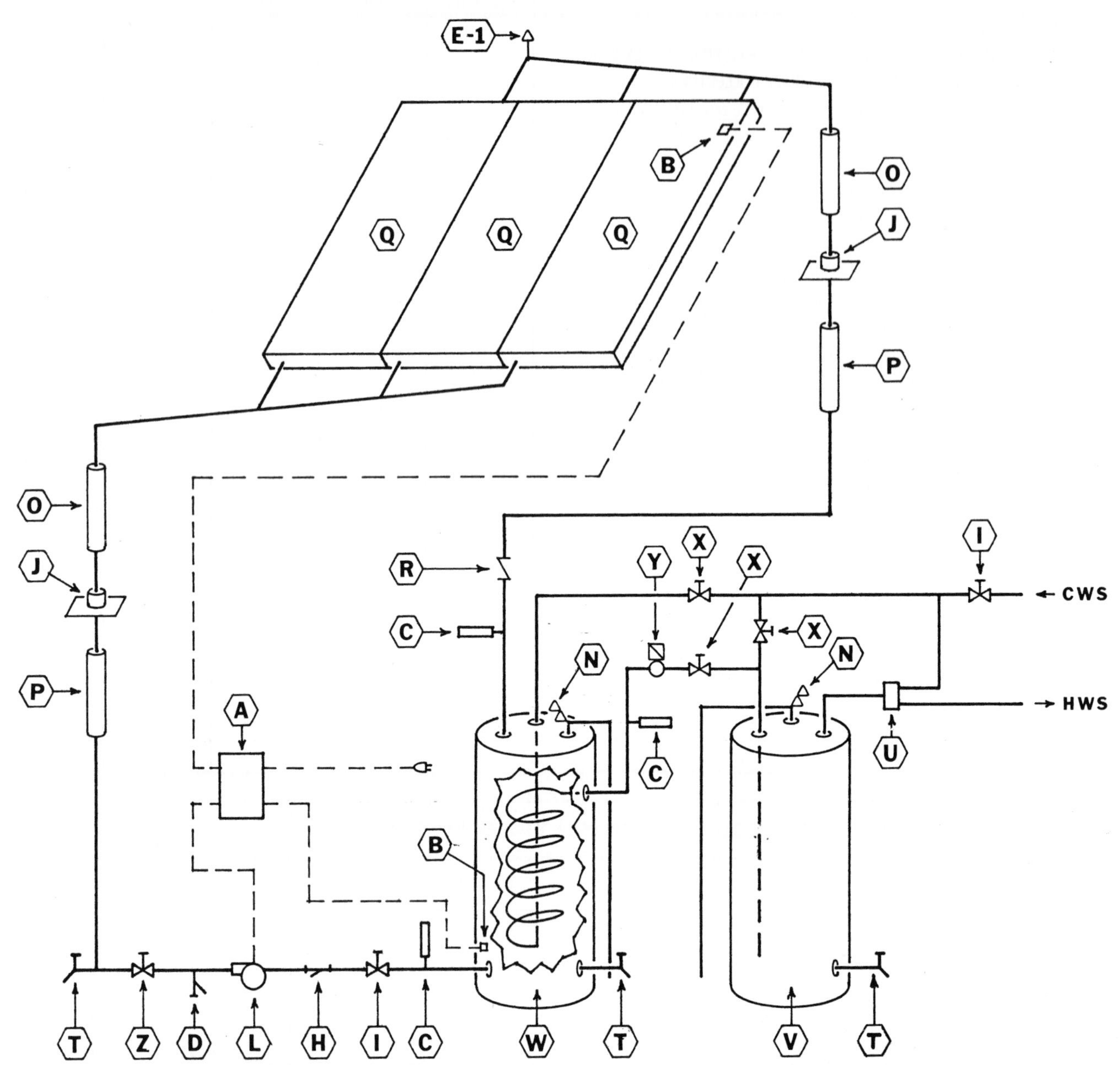

In the drainback indirect-collection system, the heat transfer fluid is distilled water contained in a loop consisting of collectors, supply and return piping, and an unpressurized holding tank. A large heat exchanger containing incoming potable water is immersed in the holding tank. When a controller activates solar collection, the distilled water is pumped through the collectors and heated and pumped back down to the holding tank. When the temperature differential between the water in the collectors and water in storage is such that collection no longer occurs, the pump turns off and gravity causes the distilled water in the collector loop to drain back to the holding tank. All the loop piping is pitched so that the water can drain out of the collectors and piping and not freeze there. As hot water is needed in the home, incoming water first flows through the holding tank with the immersed heat exchanger and is warmed and then flows through a conventional heater for any supplemental heating that is necessary.

D2020 Domestic Water Distribution

System Components	QUANTITY	UNIT	COST EACH MAT.	INST.	TOTAL
SYSTEM D2020 270 2760					
SOLAR, DRAINBACK, ADD ON, HOT WATER, IMMERSED HEAT EXCHANGER					
3/4" TUBING, THREE EA 3'X7' BLACK CHROME COLLECTOR					
A, B Differential controller 2 sensors, thermostat, solar energy system	1.000	Ea.	145	63	208
C Thermometer 2" dial	3.000	Ea.	73.50	142.50	216
D, T Fill & drain valve, brass, 3/4" connection	1.000	Ea.	11.65	31.50	43.15
E-1 Automatic air vent 1/8" fitting	1.000	Ea.	20.50	23.50	44
H Strainer, Y type, bronze body, 3/4" IPS	1.000	Ea.	58.50	40.50	99
I Valve, gate, bronze, NRS, soldered 3/4" diam	2.000	Ea.	150	76	226
J Neoprene vent flashing	2.000	Ea.	21.90	76	97.90
L Circulator, solar heated liquid, 1/20 HP	1.000	Ea.	295	114	409
N Relief valve temp. & press. 150 psi 210°F self-closing 3/4" IPS	1.000	Ea.	26.50	25.50	52
O Pipe covering, urethane, ultraviolet cover, 1" wall, 3/4" diam	20.000	L.F.	65.60	139	204.60
P Pipe covering, fiberglass, all service jacket, 1" wall, 3/4" diam	50.000	L.F.	53	277.50	330.50
Q Collector panel solar energy blk chrome on copper, 1/8" temp glas 3'x7'	3.000	Ea.	3,375	432	3,807
Roof clamps for solar energy collector panels	3.000	Set	9.54	58.50	68.04
R Valve, swing check, bronze, regrinding disc, 3/4" diam	1.000	Ea.	100	38	138
U Valve, water tempering, bronze sweat connections, 3/4" diam	1.000	Ea.	159	38	197
V Tank, water storage w/heating element, drain, relief valve, existing	1.000	Ea.			
W Tank, water storage immersed heat exchr elec 2"x1/2# insul 120 gal	1.000	Ea.	2,125	540	2,665
X Valve, globe, bronze, rising stem, 3/4" diam, soldered	3.000	Ea.	462	114	576
Y Flow control valve	1.000	Ea.	148	34.50	182.50
Z Valve, ball, bronze, solder 3/4" diam, solar loop flow control	1.000	Ea.	31.50	38	69.50
Copper tubing, type L, solder joint, hanger 10' OC 3/4" diam	20.000	L.F.	96.40	199	295.40
Copper tubing, type M, solder joint, hanger 10' OC 3/4" diam	70.000	L.F.	399	679	1,078
Sensor wire, #22-2 conductor, multistranded	.500	C.L.F.	5.65	35.75	41.40
Wrought copper fittings & solder, 3/4" diam	76.000	Ea.	200.64	3,040	3,240.64
TOTAL			8,032.88	6,255.75	14,288.63

D2020 270	Solar, Drainback, Hot Water Systems	COST EACH MAT.	INST.	TOTAL
2550	Solar, drainback, hot water, immersed heat exchanger (R235616-60)			
2560	3/8" tubing, 3 ea. 4' x 4'-4" vacuum tube collectors	7,550	5,625	13,175
2580	1/2" tubing, 4 ea 4' x 4'-4" vacuum tube collectors, 80 gal tank	8,325	6,100	14,425
2600	120 gal tank	8,250	6,175	14,425
2640	2 ea. 3'x7' blk chrome collectors, 80 gal tank	6,475	5,750	12,225
2660	3 ea. 3'x7' blk chrome collectors, 120 gal tank	7,525	5,975	13,500
2700	2 ea. 3'x7' flat blk collectors, 120 gal tank	6,000	5,825	11,825
2720	3 ea. 3'x7' flat blk collectors, 120 gal tank	6,925	6,000	12,925
2760	3/4" tubing, 3 ea 3'x7' black chrome collectors, 120 gal tank	8,025	6,250	14,275
2780	3 ea. 3'x7' flat black absorber collectors, 120 gal tank	7,425	6,275	13,700
2800	2 ea. 4'x9' flat blk w/plastic glazing collectors 120 gal tank	7,000	6,300	13,300
2840	1" tubing, 4 ea. 2'x9' plastic absorber & glazing collectors, 120 gal tank	8,250	7,100	15,350
2860	4 ea. 3'x7' black chrome absorber collectors, 120 gal tank	10,100	7,125	17,225
2880	4 ea. 3'x7' flat black absorber collectors, 120 gal tank	9,275	7,150	16,425

D2020 Domestic Water Distribution

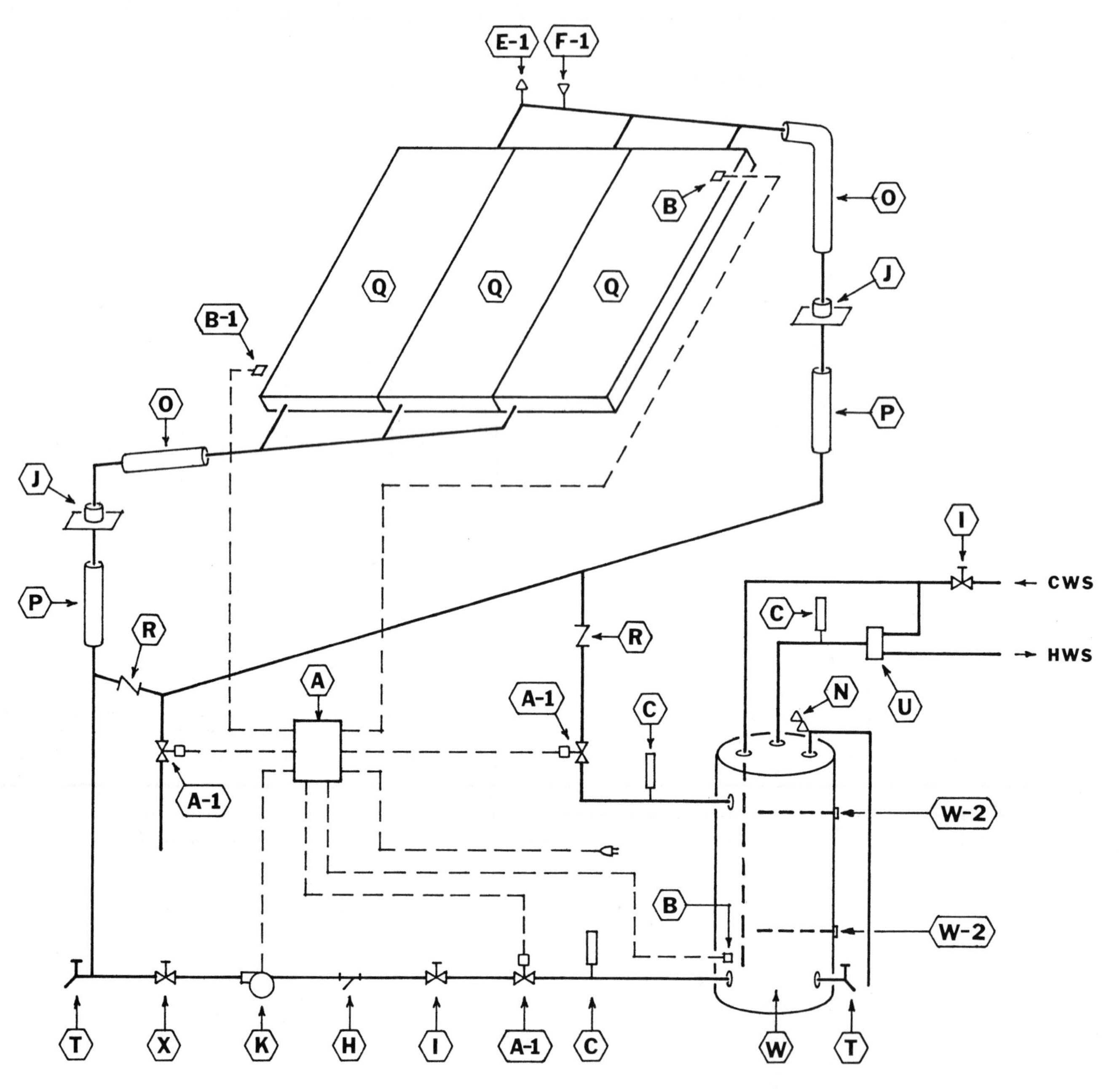

In the draindown direct-collection system, incoming domestic water is heated in the collectors. When the controller activates solar collection, domestic water is first heated as it flows through the collectors and is then pumped to storage. When conditions are no longer suitable for heat collection, the pump shuts off and the water in the loop drains down and out of the system by means of solenoid valves and properly pitched piping.

D2020 Domestic Water Distribution

System Components		QUANTITY	UNIT	COST EACH MAT.	COST EACH INST.	COST EACH TOTAL
	SYSTEM D2020 275 2760					
	SOLAR, DRAINDOWN, HOT WATER, DIRECT COLLECTION					
	3/4″ TUBING, THREE 3′X7′ BLACK CHROME COLLECTORS					
A, B	Differential controller, 2 sensors, thermostat, solar energy system	1.000	Ea.	145	63	208
A-1	Solenoid valve, solar heating loop, brass, 3/4″ diam, 24 volts	3.000	Ea.	459	252	711
B-1	Solar energy sensor, freeze prevention	1.000	Ea.	25.50	23.50	49
C	Thermometer, 2″ dial	3.000	Ea.	73.50	142.50	216
E-1	Vacuum relief valve, 3/4″ diam	1.000	Ea.	33	23.50	56.50
F-1	Air vent, automatic, 1/8″ fitting	1.000	Ea.	20.50	23.50	44
H	Strainer, Y type, bronze body, 3/4″ IPS	1.000	Ea.	58.50	40.50	99
I	Valve, gate, bronze, NRS, soldered, 3/4″ diam	2.000	Ea.	150	76	226
J	Vent flashing neoprene	2.000	Ea.	21.90	76	97.90
K	Circulator, solar heated liquid, 1/25 HP	1.000	Ea.	219	97.50	316.50
N	Relief valve temp & press 150 psi 210°F self-closing 3/4″ IPS	1.000	Ea.	26.50	25.50	52
O	Pipe covering, urethane, ultraviolet cover, 1″ wall, 3/4″ diam	20.000	L.F.	65.60	139	204.60
P	Pipe covering, fiberglass, all service jacket, 1″ wall, 3/4″ diam	50.000	L.F.	53	277.50	330.50
	Roof clamps for solar energy collector panels	3.000	Set	9.54	58.50	68.04
Q	Collector panel solar energy blk chrome on copper, 1/8″ temp glass 3′x7′	3.000	Ea.	3,375	432	3,807
R	Valve, swing check, bronze, regrinding disc, 3/4″ diam, soldered	2.000	Ea.	200	76	276
T	Drain valve, brass, 3/4″ connection	2.000	Ea.	23.30	63	86.30
U	Valve, water tempering, bronze, sweat connections, 3/4″ diam	1.000	Ea.	159	38	197
W-2, W	Tank, water storage elec elem 2″x1/2# insul 120 gal	1.000	Ea.	2,125	540	2,665
X	Valve, globe, bronze, rising stem, 3/4″ diam, soldered	1.000	Ea.	154	38	192
	Copper tubing, type L, solder joints, hangers 10′ OC 3/4″ diam	20.000	L.F.	96.40	199	295.40
	Copper tubing, type M, solder joints, hangers 10′ OC 3/4″ diam	70.000	L.F.	399	679	1,078
	Sensor wire, #22-2 conductor, multistranded	.500	C.L.F.	5.65	35.75	41.40
	Wrought copper fittings & solder, 3/4″ diam	76.000	Ea.	200.64	3,040	3,240.64
	TOTAL			8,098.53	6,459.25	14,557.78

D2020 275	Solar, Draindown, Hot Water Systems		COST EACH MAT.	COST EACH INST.	COST EACH TOTAL
2550	Solar, draindown, hot water				
2560	3/8″ tubing, 3 ea. 4′ x 4′-4″ vacuum tube collectors, 80 gal tank		7,450	6,150	13,600
2580	1/2″ tubing, 4 ea. 4′ x 4′-4″ vacuum tube collectors, 80 gal tank	R235616 -60	8,475	6,325	14,800
2600	120 gal tank		8,400	6,375	14,775
2640	2 ea. 3′x7′ black chrome collectors, 80 gal tank		6,625	5,950	12,575
2660	3 ea. 3′x7′ black chrome collectors, 120 gal tank		7,675	6,200	13,875
2700	2 ea. 3′x7′ flat black collectors, 120 gal tank		6,125	6,050	12,175
2720	3 ea. 3′x7′ flat black collectors, 120 gal tank		6,975	6,175	13,150
2760	3/4″ tubing, 3 ea. 3′x7′ black chrome collectors, 120 gal tank		8,100	6,450	14,550
2780	3 ea. 3′x7′ flat collectors, 120 gal tank		7,475	6,475	13,950
2800	2 ea. 4′x9′ flat black & plastic glazing collectors, 120 gal tank		7,075	6,500	13,575
2840	1″ tubing, 4 ea. 2′x9′ plastic absorber & glazing collectors, 120 gal tank		12,000	8,575	20,575
2860	4 ea. 3′x7′ black chrome absorber collectors, 120 gal tank		11,500	8,100	19,600
2880	4 ea. 3′x7′ flat black absorber collectors, 120 gal tank		9,550	7,350	16,900

D2020 Domestic Water Distribution

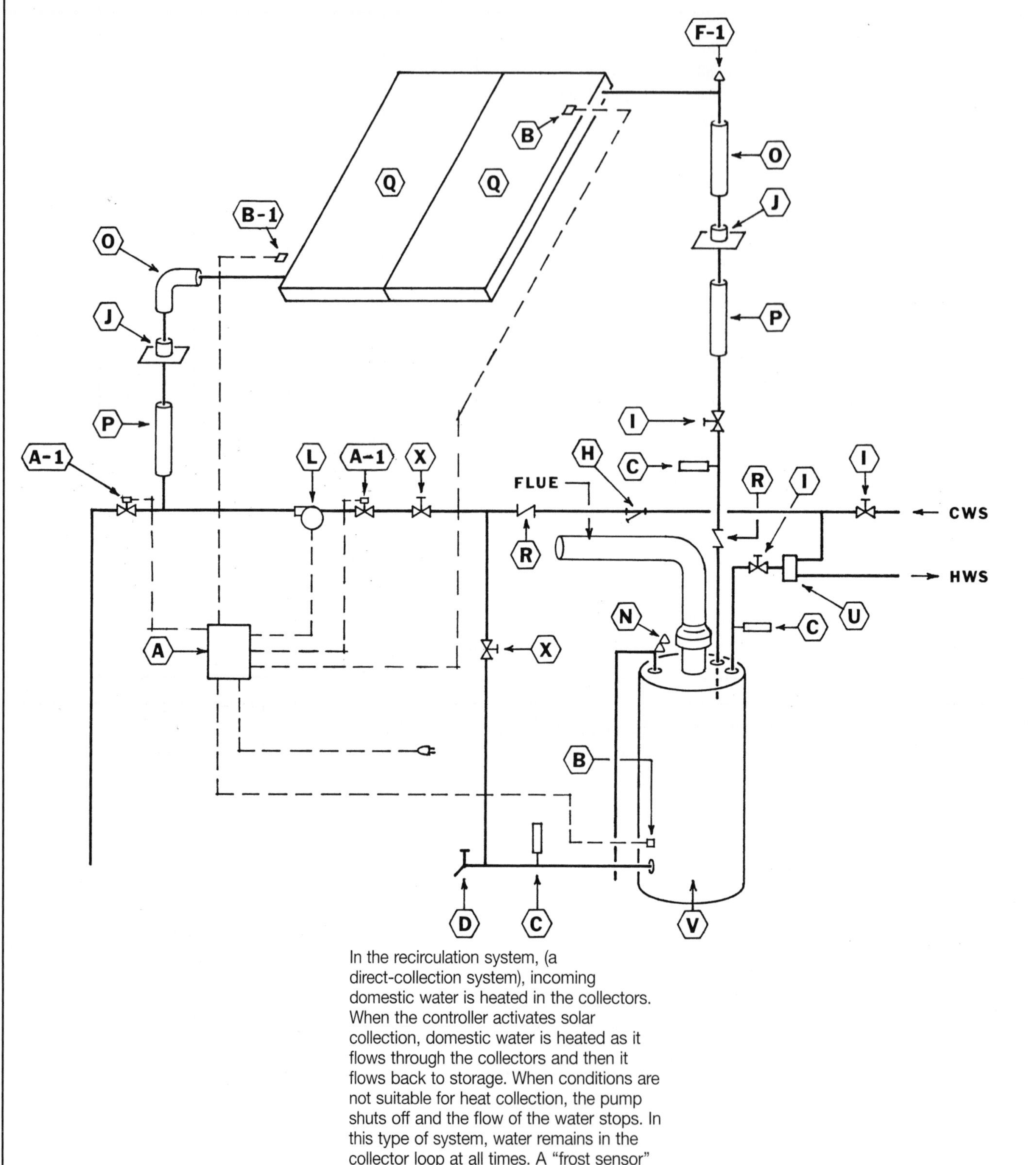

In the recirculation system, (a direct-collection system), incoming domestic water is heated in the collectors. When the controller activates solar collection, domestic water is heated as it flows through the collectors and then it flows back to storage. When conditions are not suitable for heat collection, the pump shuts off and the flow of the water stops. In this type of system, water remains in the collector loop at all times. A "frost sensor" at the collector activates the circulation of warm water from storage through the collectors when protection from freezing is required.

D2020 Domestic Water Distribution

System Components		QUANTITY	UNIT	COST EACH MAT.	COST EACH INST.	COST EACH TOTAL
	SYSTEM D2020 280 2820 SOLAR, RECIRCULATION, HOT WATER 3/4″ TUBING, TWO 3′X7′ BLACK CHROME COLLECTORS					
A, B	Differential controller 2 sensors, thermostat, for solar energy system	1.000	Ea.	145	63	208
A-1	Solenoid valve, solar heating loop, brass, 3/4″ IPS, 24 volts	2.000	Ea.	306	168	474
B-1	Solar energy sensor freeze prevention	1.000	Ea.	25.50	23.50	49
C	Thermometer, 2″ dial	3.000	Ea.	73.50	142.50	216
D	Drain valve, brass, 3/4″ connection	1.000	Ea.	11.65	31.50	43.15
F-1	Air vent, automatic, 1/8″ fitting	1.000	Ea.	20.50	23.50	44
H	Strainer, Y type, bronze body, 3/4″ IPS	1.000	Ea.	58.50	40.50	99
I	Valve, gate, bronze, 125 lb, soldered 3/4″ diam	3.000	Ea.	225	114	339
J	Vent flashing, neoprene	2.000	Ea.	21.90	76	97.90
L	Circulator, solar heated liquid, 1/20 HP	1.000	Ea.	295	114	409
N	Relief valve, temp & press 150 psi 210° F self-closing 3/4″ IPS	2.000	Ea.	53	51	104
O	Pipe covering, urethane, ultraviolet cover, 1″ wall, 3/4″ diam	20.000	L.F.	65.60	139	204.60
P	Pipe covering, fiberglass, all service jacket, 1″ wall 3/4″ diam	50.000	L.F.	53	277.50	330.50
Q	Collector panel solar energy blk chrome on copper, 1/8″ temp glass 3′x7′	2.000	Ea.	2,250	288	2,538
	Roof clamps for solar energy collector panels	2.000	Set	6.36	39	45.36
R	Valve, swing check, bronze, 125 lb, regrinding disc, soldered 3/4″ diam	2.000	Ea.	200	76	276
U	Valve, water tempering, bronze, sweat connections, 3/4″ diam	1.000	Ea.	159	38	197
V	Tank, water storage, w/heating element, drain, relief valve, existing	1.000	Ea.			
X	Valve, globe, bronze, 125 lb, 3/4″ diam	2.000	Ea.	308	76	384
	Copper tubing, type L, solder joints, hangers 10′ OC 3/4″ diam	20.000	L.F.	96.40	199	295.40
	Copper tubing, type M, solder joints, hangers 10′ OC 3/4″ diam	70.000	L.F.	399	679	1,078
	Wrought copper fittings & solder, 3/4″ diam	76.000	Ea.	200.64	3,040	3,240.64
	Sensor wire, #22-2 conductor, multistranded	.500	C.L.F.	5.65	35.75	41.40
	TOTAL			4,979.20	5,734.75	10,713.95

D2020 280	Solar, Recirculation, Domestic Hot Water Systems	COST EACH MAT.	COST EACH INST.	COST EACH TOTAL
2550	Solar, recirculation, hot water			
2560	3/8″ tubing, 3 ea. 4′ x 4′-4″ vacuum tube collectors	5,575	5,350	10,925
2580	1/2″ tubing, 4 ea. 4′ x 4′-4″ vacuum tube collectors R235616 -60	6,350	5,800	12,150
2640	2 ea. 3′x7′ black chrome collectors	4,500	5,450	9,950
2660	3 ea. 3′x7′ black chrome collectors	5,625	5,625	11,250
2700	2 ea. 3′x7′ flat black collectors	4,100	5,475	9,575
2720	3 ea. 3′x7′ flat black collectors	5,025	5,650	10,675
2760	3/4″ tubing, 3 ea. 3′x7′ black chrome collectors	6,100	5,900	12,000
2780	3 ea. 3′x7′ flat black absorber plate collectors	5,500	5,925	11,425
2800	2 ea. 4′x9′ flat black w/plastic glazing collectors	5,075	5,950	11,025
2820	2 ea. 3′x7′ black chrome collectors	4,975	5,725	10,700
2840	1″ tubing, 4 ea. 2′x9′ black plastic absorber & glazing collectors	6,500	6,750	13,250
2860	4 ea. 3′x7′ black chrome absorber collectors	8,350	6,775	15,125
2880	4 ea. 3′x7′ flat black absorber collectors	7,525	6,800	14,325

D2020 Domestic Water Distribution

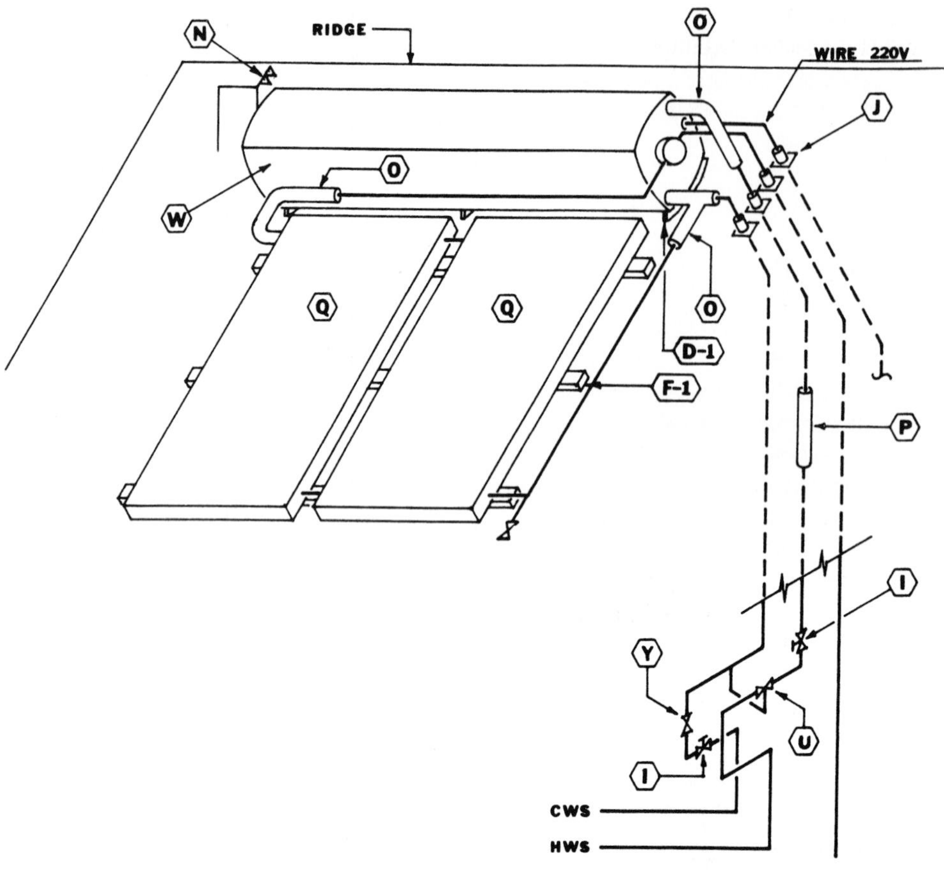

The thermosyphon domestic hot water system, a direct collection system, operates under city water pressure and does not require pumps for system operation. An insulated water storage tank is located above the collectors. As the sun heats the collectors, warm water in them rises by means of natural convection; the colder water in the storage tank flows into the collectors by means of gravity. As long as the sun is shining the water continues to flow through the collectors and to become warmer.

To prevent freezing, the system must be drained or the collectors covered with an insulated lid when the temperature drops below 32°F.

D2020 Domestic Water Distribution

System Components		QUANTITY	UNIT	COST EACH		
				MAT.	INST.	TOTAL
	SYSTEM D2020 285 0960					
	SOLAR, THERMOSYPHON, WATER HEATER					
	3/4" TUBING, TWO 3'X7' BLACK CHROME COLLECTORS					
D-1	Framing lumber, fir, 2" x 6" x 8', tank cradle	.008	M.B.F.	6.36	8	14.36
F-1	Framing lumber, fir, 2" x 4" x 24', sleepers	.016	M.B.F.	11.28	24	35.28
I	Valve, gate, bronze, 125 lb, soldered 1/2" diam	2.000	Ea.	131	63	194
J	Vent flashing, neoprene	4.000	Ea.	43.80	152	195.80
O	Pipe covering, urethane, ultraviolet cover, 1" wall, 1/2"diam	40.000	L.F.	86.80	272	358.80
P	Pipe covering fiberglass all service jacket 1" wall 1/2" diam	160.000	L.F.	155.20	848	1,003.20
Q	Collector panel solar, blk chrome on copper, 3/16" temp glass 3'-6" x 7.5'	2.000	Ea.	1,840	304	2,144
Y	Flow control valve, globe, bronze, 125#, soldered, 1/2" diam	1.000	Ea.	113	31.50	144.50
U	Valve, water tempering, bronze,sweat connections, 1/2" diam	1.000	Ea.	116	31.50	147.50
W	Tank, water storage, solar energy system, 80 Gal, 2" x 1/2 lb insul	1.000	Ea.	2,200	475	2,675
	Copper tubing type L, solder joints, hangers 10' OC 1/2" diam	150.000	L.F.	558	1,402.50	1,960.50
	Copper tubing type M, solder joints, hangers 10' OC 1/2" diam	50.000	L.F.	200.50	450	650.50
	Sensor wire, #22-2 gauge multistranded	.500	C.L.F.	5.65	35.75	41.40
	Wrought copper fittings & solder, 1/2" diam	75.000	Ea.	99	2,850	2,949
	TOTAL			5,566.59	6,947.25	12,513.84

D2020 285	Thermosyphon, Hot Water		COST EACH		
			MAT.	INST.	TOTAL
0960	Solar, thermosyphon, hot water, two collector system	R235616 -60	5,575	6,950	12,525
0970					

D2020 Domestic Water Distribution

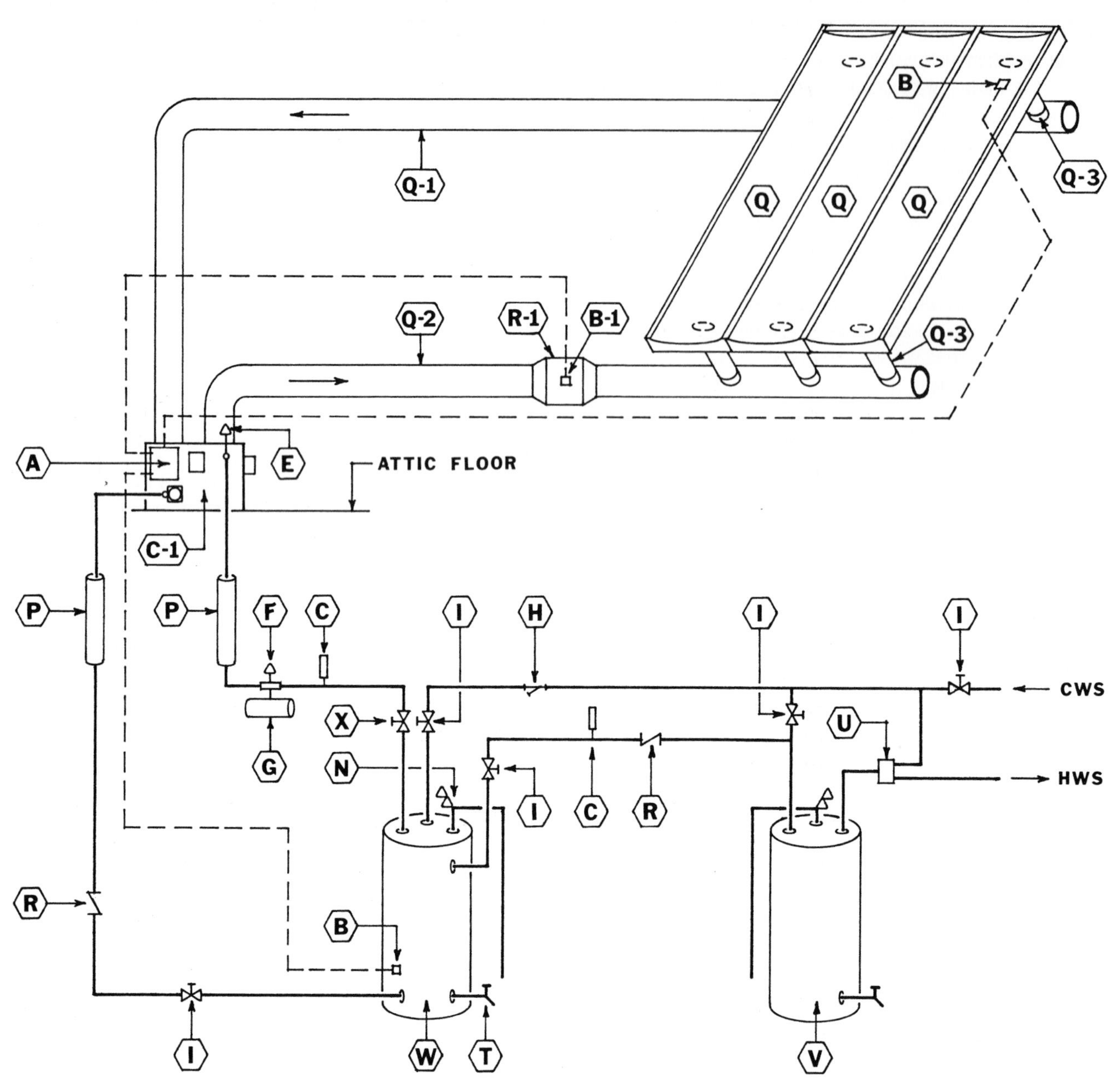

This domestic hot water pre-heat system includes heat exchanger with a circulating pump, blower, air-to-water, coil and controls, mounted in the upper collector manifold. Heat from the hot air coming out of the collectors is transferred through the heat exchanger. For each degree of DHW preheating gained, one degree less heating is needed from the fuel fired water heater. The system is simple, inexpensive to operate and can provide a substantial portion of DHW requirements for modest additional cost.

D2020 Domestic Water Distribution

System Components		QUANTITY	UNIT	COST EACH MAT.	COST EACH INST.	COST EACH TOTAL
	SYSTEM D2020 290 2560					
	SOLAR, HOT WATER, AIR TO WATER HEAT EXCHANGE					
	THREE COLLECTORS, OPTICAL BLACK ON ALUM., 7.5′ x 3.5′,80 GAL TANK					
A, B	Differential controller, 2 sensors, thermostat, solar energy system	1.000	Ea.	145	63	208
C	Thermometer, 2″ dial	2.000	Ea.	49	95	144
C-1	Heat exchanger, air to fluid, up flow 70 MBH	1.000	Ea.	400	390	790
E	Air vent, manual, for solar energy system 1/8″ fitting	1.000	Ea.	2.93	23.50	26.43
F	Air purger	1.000	Ea.	52.50	63	115.50
G	Expansion tank	1.000	Ea.	76.50	23.50	100
H	Strainer, Y type, bronze body, 1/2″ IPS	1.000	Ea.	47.50	38.50	86
I	Valve, gate, bronze, 125 lb, NRS, soldered 1/2″ diam	5.000	Ea.	327.50	157.50	485
N	Relief valve, temp & pressure solar 150 psi 210°F self-closing	1.000	Ea.	26.50	25.50	52
P	Pipe covering, fiberglass, all service jacket, 1″ wall, 1/2″ diam	60.000	L.F.	58.20	318	376.20
Q	Collector panel solar energy, air, black on alum plate, 7.5′ x 3.5′	3.000	Ea.	2,760	456	3,216
B-1, R-1	Shutter damper for solar heater circulator	1.000	Ea.	63	111	174
B-1, R-1	Shutter motor for solar heater circulator	1.000	Ea.	156	83.50	239.50
R	Backflow preventer, 1/2″ pipe size	2.000	Ea.	185	95	280
T	Drain valve, brass, 3/4″ connection	1.000	Ea.	11.65	31.50	43.15
U	Valve, water tempering, bronze, sweat connections, 1/2 ″ diam	1.000	Ea.	116	31.50	147.50
W	Tank, water storage, solar energy system, 80 Gal, 2″ x 2 lb insul	1.000	Ea.	2,200	475	2,675
V	Tank, water storage, w/heating element, drain, relief valve, existing	1.000	System			
X	Valve, globe, bronze, 125 lb, rising stem, 1/2″ diam	1.000	Ea.	113	31.50	144.50
	Copper tubing type M, solder joints, hangers 10′ OC, 1/2″ diam	50.000	L.F.	200.50	450	650.50
	Copper tubing type L, solder joints, hangers 10′ OC 1/2″ diam	10.000	L.F.	37.20	93.50	130.70
	Wrought copper fittings & solder, 1/2″ diam	10.000	Ea.	13.20	380	393.20
	Sensor wire, #22-2 conductor multistranded	.500	C.L.F.	5.65	35.75	41.40
Q-1, Q-2	Ductwork, fiberglass, aluminized jacket, 1-1/2″ thick, 8″ diam	32.000	S.F.	201.60	236.80	438.40
Q-3	Manifold for flush mount solar energy collector panels, air	6.000	L.F.	1,026	50.10	1,076.10
	TOTAL			8,274.43	3,758.65	12,033.08

D2020 290	Solar, Hot Water, Air To Water Heat Exchange		COST EACH MAT.	COST EACH INST.	COST EACH TOTAL
2550	Solar, hot water, air to water heat exchange				
2560	Three collectors, optical black on aluminum, 7.5′ x 3.5′, 80 Gal tank		8,275	3,750	12,025
2580	Four collectors, optical black on aluminum, 7.5′ x 3.5′, 80 Gal tank	R235616 -60	9,575	3,975	13,550
2600	Four collectors, optical black on aluminum, 7.5′ x 3.5′, 120 Gal tank		9,500	4,050	13,550

D2020 Domestic Water Distribution

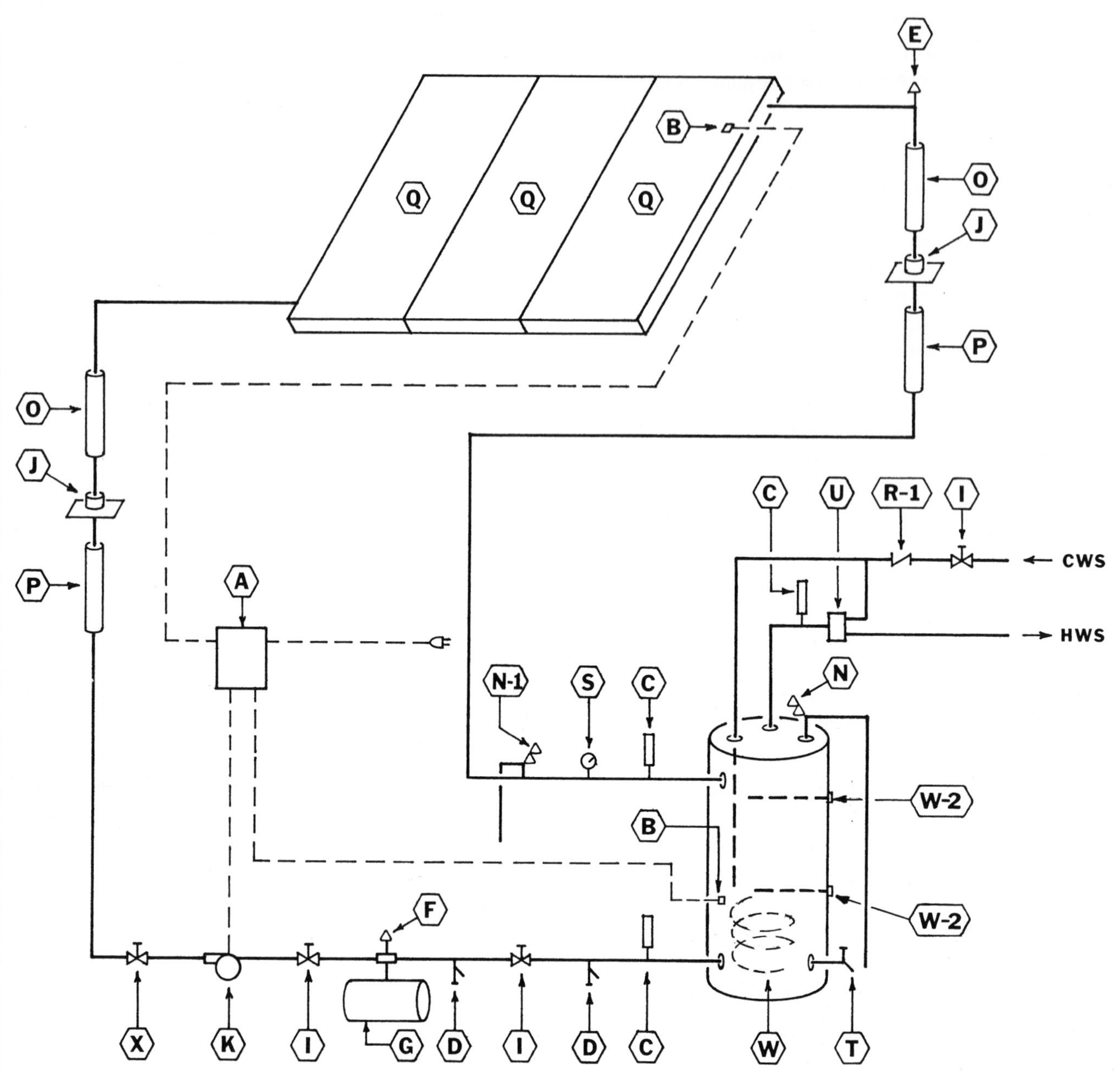

In this closed-loop indirect collection system, fluid with a low freezing temperature, such as propylene glycol, transports heat from the collectors to water storage. The transfer fluid is contained in a closed-loop consisting of collectors, supply and return piping, and a heat exchanger immersed in the storage tank. A typical two-or-three panel system contains 5 to 6 gallons of heat transfer fluid.

When the collectors become approximately 20°F warmer than the storage temperature, a controller activates the circulator. The circulator moves the fluid continuously through the collectors until the temperature difference between the collectors and storage is such that heat collection no longer occurs; at that point, the circulator shuts off. Since the heat transfer fluid has a very low freezing temperature, there is no need for it to be drained from the collectors between periods of collection.

D2020 Domestic Water Distribution

System Components

System Components	QUANTITY	UNIT	COST EACH MAT.	INST.	TOTAL
SYSTEM D2020 295 2760					
SOLAR, CLOSED LOOP, HOT WATER SYSTEM, IMMERSED HEAT EXCHANGER					
3/4" TUBING, THREE 3' X 7' BLACK CHROME COLLECTORS					
A, B Differential controller, 2 sensors, thermostat, solar energy system	1.000	Ea.	145	63	208
C Thermometer 2" dial	3.000	Ea.	73.50	142.50	216
D, T Fill & drain valves, brass, 3/4" connection	3.000	Ea.	34.95	94.50	129.45
E Air vent, manual, 1/8" fitting	1.000	Ea.	2.93	23.50	26.43
F Air purger	1.000	Ea.	52.50	63	115.50
G Expansion tank	1.000	Ea.	76.50	23.50	100
I Valve, gate, bronze, NRS, soldered 3/4" diam	3.000	Ea.	225	114	339
J Neoprene vent flashing	2.000	Ea.	21.90	76	97.90
K Circulator, solar heated liquid, 1/25 HP	1.000	Ea.	219	97.50	316.50
N-1, N Relief valve, temp & press 150 psi 210°F self-closing 3/4" IPS	2.000	Ea.	53	51	104
O Pipe covering, urethane ultraviolet cover, 1" wall, 3/4" diam	20.000	L.F.	65.60	139	204.60
P Pipe covering, fiberglass, all service jacket, 1" wall, 3/4" diam	50.000	L.F.	53	277.50	330.50
Roof clamps for solar energy collector panel	3.000	Set	9.54	58.50	68.04
Q Collector panel solar blk chrome on copper, 1/8" temp glass, 3'x7'	3.000	Ea.	3,375	432	3,807
R-1 Valve, swing check, bronze, regrinding disc, 3/4" diam, soldered	1.000	Ea.	100	38	138
S Pressure gauge, 60 psi, 2-1/2" dial	1.000	Ea.	27	23.50	50.50
U Valve, water tempering, bronze, sweat connections, 3/4" diam	1.000	Ea.	159	38	197
W-2, W Tank, water storage immersed heat exchr elec elem 2"x2# insul 120 Gal	1.000	Ea.	2,125	540	2,665
X Valve, globe, bronze, rising stem, 3/4" diam, soldered	1.000	Ea.	154	38	192
Copper tubing type L, solder joint, hanger 10' OC 3/4" diam	20.000	L.F.	96.40	199	295.40
Copper tubing, type M, solder joint, hanger 10' OC 3/4" diam	70.000	L.F.	399	679	1,078
Sensor wire, #22-2 conductor multistranded	.500	C.L.F.	5.65	35.75	41.40
Solar energy heat transfer fluid, propylene glycol, anti-freeze	6.000	Gal.	156	162	318
Wrought copper fittings & solder, 3/4" diam	76.000	Ea.	200.64	3,040	3,240.64
TOTAL			7,830.11	6,448.75	14,278.86

D2020 295	Solar, Closed Loop, Hot Water Systems	COST EACH MAT.	INST.	TOTAL
2550	Solar, closed loop, hot water system, immersed heat exchanger			
2560	3/8" tubing, 3 ea. 4' x 4'-4" vacuum tube collectors, 80 gal. tank	7,450	5,850	13,300
2580	1/2" tubing, 4 ea. 4' x 4'-4" vacuum tube collectors, 80 gal. tank	8,225	6,300	14,525
2600	120 gal. tank	8,150	6,375	14,525
2640	2 ea. 3'x7' black chrome collectors, 80 gal. tank (R235616-60)	6,425	6,025	12,450
2660	120 gal. tank	6,300	6,025	12,325
2700	2 ea. 3'x7' flat black collectors, 120 gal. tank	5,900	6,025	11,925
2720	3 ea. 3'x7' flat black collectors, 120 gal. tank	6,800	6,200	13,000
2760	3/4" tubing, 3 ea. 3'x7' black chrome collectors, 120 gal. tank	7,825	6,450	14,275
2780	3 ea. 3'x7' flat black collectors, 120 gal. tank	7,225	6,475	13,700
2800	2 ea. 4'x9' flat black w/plastic glazing collectors 120 gal. tank	6,800	6,500	13,300
2840	1" tubing, 4 ea. 2'x9' plastic absorber & glazing collectors 120 gal. tank	7,850	7,250	15,100
2860	4 ea. 3'x7' black chrome collectors, 120 gal. tank	9,700	7,300	17,000
2880	4 ea. 3'x7' flat black absorber collectors, 120 gal. tank	8,875	7,325	16,200

D3010 Energy Supply

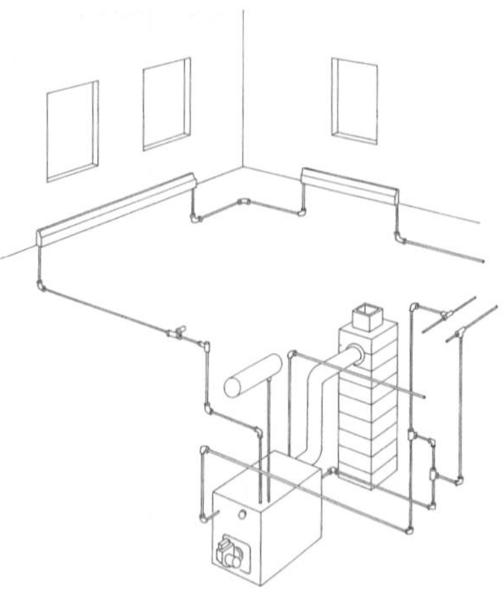

Basis for Heat Loss Estimate, Apartment Type Structures:

1. Masonry walls and flat roof are insulated. U factor is assumed at .08.
2. Window glass area taken as BOCA minimum, 1/10th of floor area. Double insulating glass with 1/4″ air space, U = .65.
3. Infiltration = 0.3 C.F. per hour per S.F. of net wall.
4. Concrete floor loss is 2 BTUH per S.F.
5. Temperature difference taken as 70°F.
6. Ventilating or makeup air has not been included and must be added if desired. Air shafts are not used.

System Components	QUANTITY	UNIT	COST EACH		
			MAT.	INST.	TOTAL
SYSTEM D3010 510 1760					
HEATING SYSTEM, FIN TUBE RADIATION, FORCED HOT WATER					
1,000 S.F. AREA, 10,000 C.F. VOLUME					
Boiler, oil fired, CI, burner, ctrls/insul/breech/pipe/ftng/valves, 109 MBH	1.000	Ea.	4,287.50	4,243.75	8,531.25
Circulating pump, CI flange connection, 1/12 HP	1.000	Ea.	505	227	732
Expansion tank, painted steel, ASME 18 Gal capacity	1.000	Ea.	870	98.50	968.50
Storage tank, steel, above ground, 275 Gal capacity w/supports	1.000	Ea.	570	276	846
Copper tubing type L, solder joint, hanger 10′ OC, 3/4″ diam	100.000	L.F.	482	995	1,477
Radiation, 3/4″ copper tube w/alum fin baseboard pkg, 7″ high	30.000	L.F.	267	720	987
Pipe covering, calcium silicate w/cover, 1′ wall, 3/4′ diam	100.000	L.F.	463	750	1,213
TOTAL			7,444.50	7,310.25	14,754.75
COST PER S.F.			7.44	7.31	14.75

D3010 510	Apartment Building Heating - Fin Tube Radiation		COST PER S.F.		
			MAT.	INST.	TOTAL
1740	Heating systems, fin tube radiation, forced hot water				
1760	1,000 S.F. area, 10,000 C.F. volume		7.45	7.30	14.75
1800	10,000 S.F. area, 100,000 C.F. volume	R235000 -10	3.39	4.42	7.81
1840	20,000 S.F. area, 200,000 C.F. volume		3.70	4.96	8.66
1880	30,000 S.F. area, 300,000 C.F. volume	R235000 -20	3.60	4.82	8.42
1890					

D3010 Energy Supply

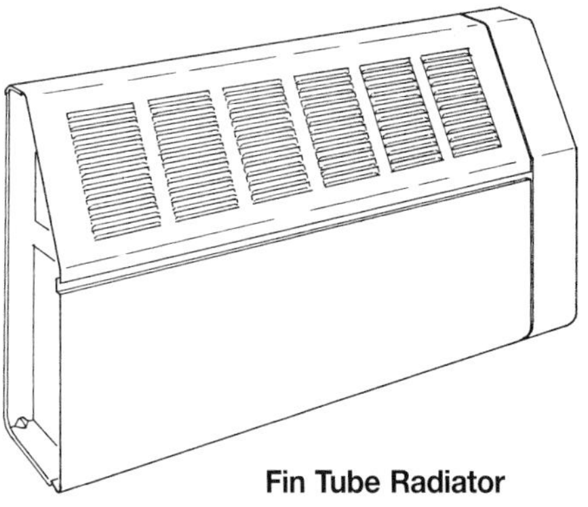
Fin Tube Radiator

Basis for Heat Loss Estimate, Factory or Commercial Type Structures:

1. Walls and flat roof are of lightly insulated concrete block or metal. U factor is assumed at .17.
2. Windows and doors are figured at 1-1/2 S.F. per linear foot of building perimeter. The U factor for flat single glass is 1.13.
3. Infiltration is approximately 5 C.F. per hour per S.F. of wall.
4. Concrete floor loss is 2 BTUH per S.F. of floor.
5. Temperature difference is assumed at 70°F.
6. Ventilation or makeup air has not been included and must be added if desired.

System Components	QUANTITY	UNIT	COST EACH		
			MAT.	INST.	TOTAL
SYSTEM D3010 520 1960					
HEATING SYSTEM, FIN TUBE RADIATION, FORCED HOT WATER					
1,000 S.F. BLDG., ONE FLOOR					
Boiler, oil fired, CI, burner/ctrls/insul/breech/pipe/ftngs/valves, 109 MBH	1.000	Ea.	4,287.50	4,243.75	8,531.25
Expansion tank, painted steel, ASME 18 Gal capacity	1.000	Ea.	3,075	115	3,190
Storage tank, steel, above ground, 550 Gal capacity w/supports	1.000	Ea.	4,875	510	5,385
Circulating pump, CI flanged, 1/8 HP	1.000	Ea.	900	227	1,127
Pipe, steel, black, sch. 40, threaded, cplg & hngr 10'OC, 1-1/2" diam.	260.000	L.F.	1,638	4,433	6,071
Pipe covering, calcium silicate w/cover , 1" wall, 1-1/2" diam	260.000	L.F.	1,219.40	2,015	3,234.40
Radiation, steel 1-1/4" tube & 4-1/4" fin w/cover & damper, wall hung	42.000	L.F.	1,953	1,617	3,570
Rough in, steel fin tube radiation w/supply & balance valves	4.000	Set	1,440	3,380	4,820
TOTAL			19,387.90	16,540.75	35,928.65
COST PER S.F.			19.39	16.54	35.93

D3010 520	Commercial Building Heating - Fin Tube Radiation		COST PER S.F.		
			MAT.	INST.	TOTAL
1940	Heating systems, fin tube radiation, forced hot water				
1960	1,000 S.F. bldg, one floor		19.40	16.50	35.90
2000	10,000 S.F., 100,000 C.F., total two floors	R235000 -10	4.81	6.30	11.11
2040	100,000 S.F., 1,000,000 C.F., total three floors		2.19	2.81	5
2080	1,000,000 S.F., 10,000,000 C.F., total five floors	R235000 -20	1.15	1.47	2.62
2090					

D3010 Energy Supply

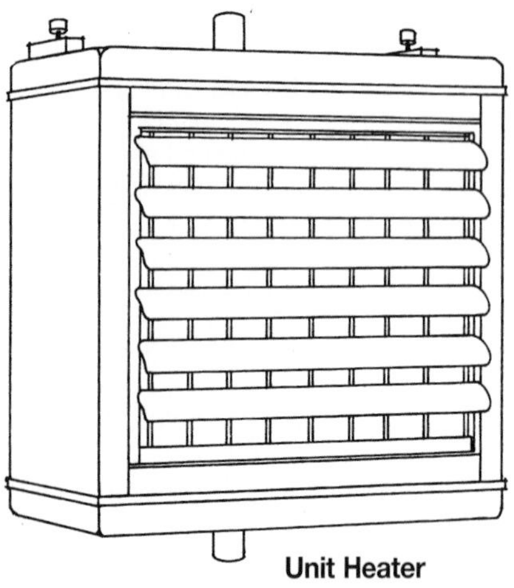
Unit Heater

Basis for Heat Loss Estimate, Factory or Commercial Type Structures:

1. Walls and flat roof are of lightly insulated concrete block or metal. U factor is assumed at .17.
2. Windows and doors are figured at 1-1/2 S.F. per linear foot of building perimeter. The U factor for flat single glass is 1.13.
3. Infiltration is approximately 5 C.F. per hour per S.F. of wall.
4. Concrete floor loss is 2 BTUH per S.F. of floor.
5. Temperature difference is assumed at 70°F.
6. Ventilation or makeup air has not been included and must be added if desired.

System Components	QUANTITY	UNIT	COST EACH		
			MAT.	INST.	TOTAL
SYSTEM D3010 530 1880					
HEATING SYSTEM, TERMINAL UNIT HEATERS, FORCED HOT WATER					
1,000 S.F. BLDG., ONE FLOOR					
Boiler oil fired, CI, burner/ctrls/insul/breech/pipe/ftngs/valves, 109 MBH	1.000	Ea.	4,287.50	4,243.75	8,531.25
Expansion tank, painted steel, ASME 18 Gal capacity	1.000	Ea.	3,075	115	3,190
Storage tank, steel, above ground, 550 Gal capacity w/supports	1.000	Ea.	4,875	510	5,385
Circulating pump, CI, flanged, 1/8 HP	1.000	Ea.	900	227	1,127
Pipe, steel, black, schedule 40, threaded, cplg & hngr 10′ OC 1-1/2″ diam	260.000	L.F.	1,638	4,433	6,071
Pipe covering, calcium silicate w/cover, 1″ wall, 1-1/2″ diam	260.000	L.F.	1,219.40	2,015	3,234.40
Unit heater, 1 speed propeller, horizontal, 200° EWT, 26.9 MBH	2.000	Ea.	1,330	346	1,676
Unit heater piping hookup with controls	2.000	Set	1,710	3,250	4,960
TOTAL			19,034.90	15,139.75	34,174.65
COST PER S.F.			19.03	15.14	34.17

D3010 530	Commercial Bldg. Heating - Terminal Unit Heaters		COST PER S.F.		
			MAT.	INST.	TOTAL
1860	Heating systems, terminal unit heaters, forced hot water				
1880	1,000 S.F. bldg., one floor		19	15.10	34.10
1920	10,000 S.F. bldg., 100,000 C.F. total two floors	R235000 -10	4.30	5.20	9.50
1960	100,000 S.F. bldg., 1,000,000 C.F. total three floors		2.23	2.57	4.80
2000	1,000,000 S.F. bldg., 10,000,000 C.F. total five floors	R235000 -20	1.47	1.65	3.12
2010					

D3010 Energy Supply

Many styles of active solar energy systems exist. Those shown on the following page represent the majority of systems now being installed in different regions of the country.

The five active domestic hot water (DHW) systems which follow are typically specified as two or three panel systems with additional variations being the type of glazing and size of the storage tanks. Various combinations have been costed for the user's evaluation and comparison. The basic specifications from which the following systems were developed satisfy the construction detail requirements specified in the HUD Intermediate Minimum Property Standards (IMPS). If these standards are not complied with, the renewable energy system's costs could be significantly lower than shown.

To develop the system's specifications and costs it was necessary to make a number of assumptions about the systems. Certain systems are more appropriate to one climatic region than another or the systems may require modifications to be usable in particular locations. Specific instances in which the systems are not appropriate throughout the country as specified, or in which modification will be needed include the following:

- The freeze protection mechanisms provided in the DHW systems vary greatly. In harsh climates, where freeze is a major concern, a closed-loop indirect collection system may be more appropriate than a direct collection system.
- The thermosyphon water heater system described cannot be used when temperatures drop below 32°F.
- In warm climates it may be necessary to modify the systems installed to prevent overheating.

For each renewable resource (solar) system a schematic diagram and descriptive summary of the system is provided along with a list of all the components priced as part of the system. The costs were developed based on these specifications.

Considerations affecting costs which may increase or decrease beyond the estimates presented here include the following:

- Special structural qualities (allowance for earthquake, future expansion, high winds, and unusual spans or shapes);
- Isolated building site or rough terrain that would affect the transportation of personnel, material, or equipment;
- Unusual climatic conditions during the construction process;
- Substitution of other materials or system components for those used in the system specifications.

D3010 Energy Supply

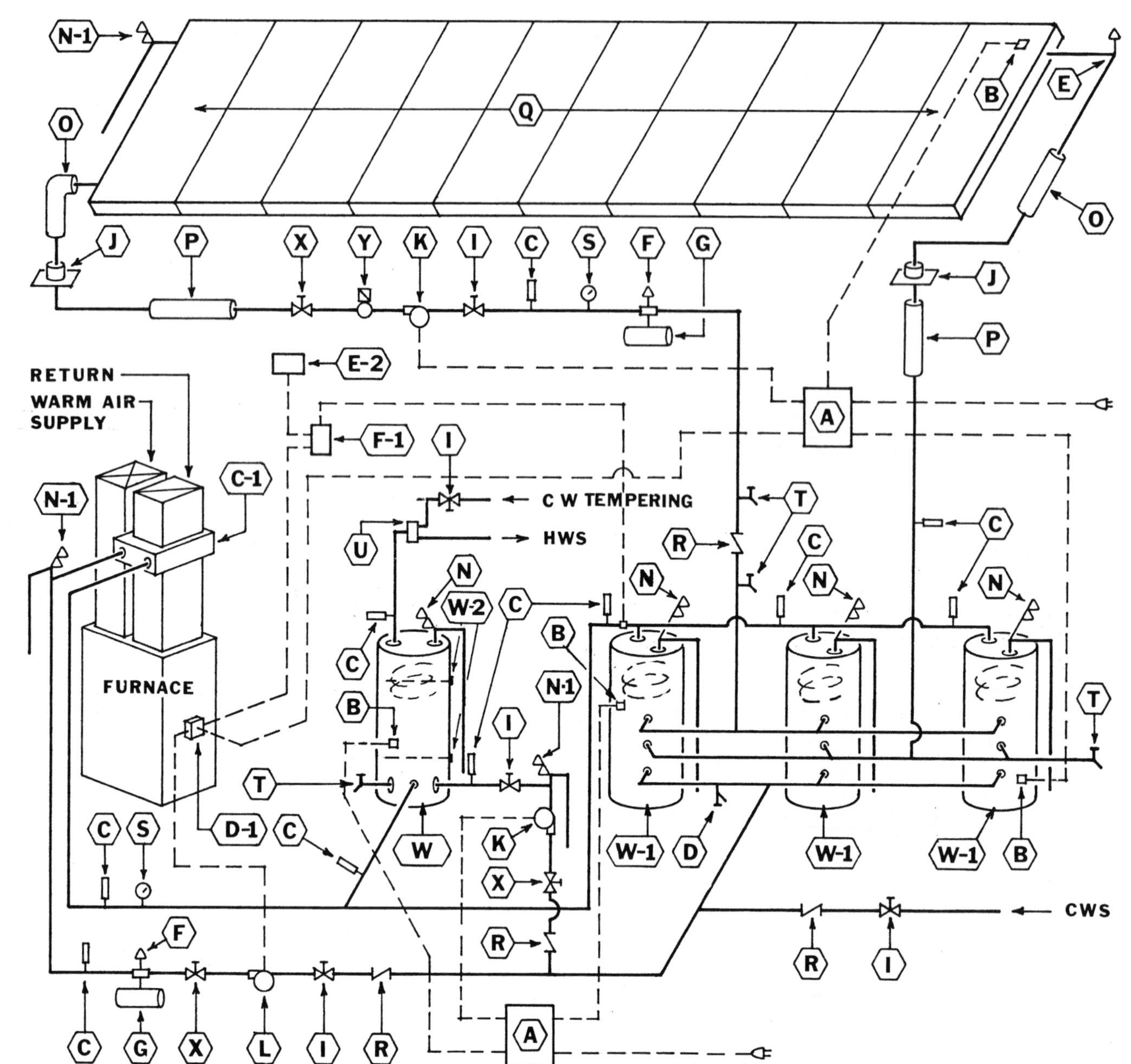

In this closed-loop indirect collection system, fluid with a low freezing temperature, propylene glycol, transports heat from the collectors to water storage. The transfer fluid is contained in a closed loop consisting of collectors, supply and return piping, and a heat exchanger immersed in the storage tank.

When the collectors become approximately 20°F warmer than the storage temperature, the controller activates the circulator. The circulator moves the fluid continuously until the temperature difference between fluid in the collectors and storage is such that the collection will no longer occur and then the circulator turns off. Since the heat transfer fluid has a very low freezing temperature, there is no need for it to be drained from the collectors between periods of collection.

D3010 Energy Supply

System Components	QUANTITY	UNIT	COST EACH		
			MAT.	INST.	TOTAL
SYSTEM D3010 650 2750					
SOLAR, CLOSED LOOP, SPACE/HOT WATER					
1″ TUBING, TEN 3′X7′ BLK CHROME ON COPPER ABSORBER COLLECTORS					
A, B Differential controller 2 sensors, thermostat, solar energy system	2.000	Ea.	290	126	416
C Thermometer, 2″ dial	10.000	Ea.	245	475	720
C-1 Heat exchanger, solar energy system, fluid to air, up flow, 80 MBH	1.000	Ea.	535	455	990
D, T Fill & drain valves, brass, 3/4″ connection	5.000	Ea.	58.25	157.50	215.75
D-1 Fan center	1.000	Ea.	127	159	286
E Air vent, manual, 1/8″ fitting	1.000	Ea.	2.93	23.50	26.43
E-2 Thermostat, 2 stage for sensing room temperature	1.000	Ea.	247	96	343
F Air purger	2.000	Ea.	105	126	231
F-1 Controller, liquid temperature, solar energy system	1.000	Ea.	126	152	278
G Expansion tank	2.000	Ea.	153	47	200
I Valve, gate, bronze, 125 lb, soldered, 1″ diam	5.000	Ea.	422.50	200	622.50
J Vent flashing, neoprene	2.000	Ea.	21.90	76	97.90
K Circulator, solar heated liquid, 1/25 HP	2.000	Ea.	438	195	633
L Circulator, solar heated liquid, 1/20 HP	1.000	Ea.	295	114	409
N Relief valve, temp & pressure 150 psi 210°F self-closing	4.000	Ea.	106	102	208
N-1 Relief valve, pressure poppet, bronze, 30 psi, 3/4″ IPS	3.000	Ea.	291	81	372
O Pipe covering, urethane, ultraviolet cover, 1″ wall, 1″ diam	50.000	L.F.	164	347.50	511.50
P Pipe covering, fiberglass, all service jacket, 1″ wall, 1″ diam	60.000	L.F.	63.60	333	396.60
Q Collector panel solar energy blk chrome on copper 1/8″ temp glass 3′x7′	10.000	Ea.	11,250	1,440	12,690
Roof clamps for solar energy collector panel	10.000	Set	31.80	195	226.80
R Valve, swing check, bronze, 125 lb, regrinding disc, 3/4″ & 1″ diam	4.000	Ea.	628	160	788
S Pressure gage, 0-60 psi, for solar energy system	2.000	Ea.	54	47	101
U Valve, water tempering, bronze, sweat connections, 3/4″ diam	1.000	Ea.	159	38	197
W-2, W-1, W Tank, water storage immersed heat xchr elec elem 2″x1/2# ins 120 gal	4.000	Ea.	8,500	2,160	10,660
X Valve, globe, bronze, 125 lb, rising stem, 1″ diam	3.000	Ea.	660	120	780
Y Valve, flow control	1.000	Ea.	148	34.50	182.50
Copper tubing, type M, solder joint, hanger 10′ OC 1″ diam	110.000	L.F.	1,017.50	1,193.50	2,211
Copper tubing, type L, solder joint, hanger 10′ OC 3/4″ diam	20.000	L.F.	96.40	199	295.40
Wrought copper fittings & solder, 3/4″ & 1″ diam	121.000	Ea.	871.20	5,747.50	6,618.70
Sensor, wire, #22-2 conductor, multistranded	.700	C.L.F.	7.91	50.05	57.96
Ductwork, galvanized steel, for heat exchanger	8.000	Lb.	5.04	70.40	75.44
Solar energy heat transfer fluid propylene glycol, anti-freeze	25.000	Gal.	650	675	1,325
TOTAL			27,770.03	15,395.45	43,165.48

D3010 650	Solar, Closed Loop, Space/Hot Water Systems	COST EACH		
		MAT.	INST.	TOTAL
2540	Solar, closed loop, space/hot water			
2550	1/2″ tubing, 12 ea. 4′x4′4″ vacuum tube collectors	26,700	14,300	41,000
2600	3/4″ tubing, 12 ea. 4′x4′4″ vacuum tube collectors R235616-60	27,900	14,800	42,700
2650	10 ea. 3′ x 7′ black chrome absorber collectors	26,800	14,400	41,200
2700	10 ea. 3′ x 7′ flat black absorber collectors	24,800	14,400	39,200
2750	1″ tubing, 10 ea. 3′ x 7′ black chrome absorber collectors	27,800	15,400	43,200
2800	10 ea. 3′ x 7′ flat black absorber collectors	25,800	15,500	41,300
2850	6 ea. 4′ x 9′ flat black w/plastic glazing collectors	23,700	15,400	39,100
2900	12 ea. 2′ x 9′ plastic absorber and glazing collectors	24,600	15,700	40,300

D3010 Energy Supply

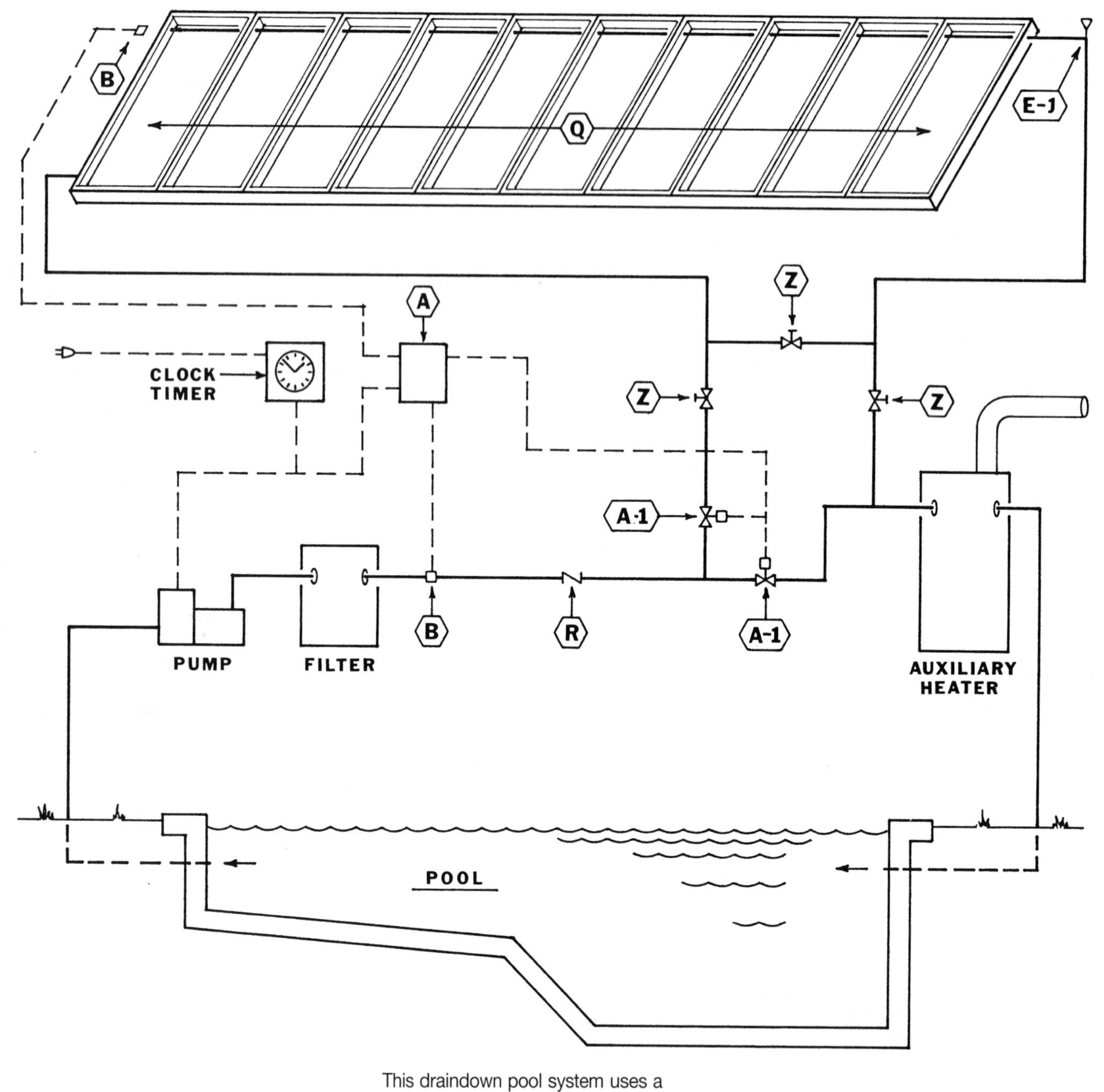

This draindown pool system uses a differential thermostat similar to those used in solar domestic hot water and space heating applications. To heat the pool, the pool water passes through the conventional pump-filter loop and then flows through the collectors. When collection is not possible, or when the pool temperature is reached, all water drains from the solar loop back to the pool through the existing piping. The modes are controlled by solenoid valves or other automatic valves in conjunction with a vacuum breaker relief valve, which facilitates draindown.

D3010 Energy Supply

System Components	QUANTITY	UNIT	COST EACH MAT.	COST EACH INST.	COST EACH TOTAL
SYSTEM D3010 660 2640					
SOLAR SWIMMING POOL HEATER, ROOF MOUNTED COLLECTORS					
TEN 4′ X 10′ FULLY WETTED UNGLAZED PLASTIC ABSORBERS					
A Differential thermostat/controller, 110V, adj pool pump system	1.000	Ea.	365	380	745
A-1 Solenoid valve, PVC, normally 1 open 1 closed (included)	2.000	Ea.			
B Sensor, thermistor type (included)	2.000	Ea.			
E-1 Valve, vacuum relief	1.000	Ea.	33	23.50	56.50
Q Collector panel, solar energy, plastic, liquid full wetted, 4′ x 10′	10.000	Ea.	3,650	2,730	6,380
R Valve, ball check, PVC, socket, 1-1/2″ diam	1.000	Ea.	116	38	154
Z Valve, ball, PVC, socket, 1-1/2″ diam	3.000	Ea.	279	114	393
Pipe, PVC, sch 40, 1-1/2″ diam	80.000	L.F.	864	1,680	2,544
Pipe fittings, PVC sch 40, socket joint, 1-1/2″ diam	10.000	Ea.	21.70	380	401.70
Sensor wire, #22-2 conductor, multistranded	.500	C.L.F.	5.65	35.75	41.40
Roof clamps for solar energy collector panels	10.000	Set	31.80	195	226.80
Roof strap, teflon for solar energy collector panels	26.000	L.F.	715	96.20	811.20
TOTAL			6,081.15	5,672.45	11,753.60

D3010 660	Solar Swimming Pool Heater Systems	COST EACH MAT.	COST EACH INST.	COST EACH TOTAL
2530	Solar swimming pool heater systems, roof mounted collectors			
2540	10 ea. 3′x7′ black chrome absorber, 1/8″ temp. glass	13,700	4,375	18,075
2560	10 ea. 4′x8′ black chrome absorber, 3/16″ temp. glass R235616-60	15,400	5,200	20,600
2580	10 ea. 3′8″x6′ flat black absorber, 3/16″ temp. glass	11,600	4,450	16,050
2600	10 ea. 4′x9′ flat black absorber, plastic glazing	14,200	5,425	19,625
2620	10 ea. 2′x9′ rubber absorber, plastic glazing	15,700	5,850	21,550
2640	10 ea. 4′x10′ fully wetted unglazed plastic absorber	6,075	5,675	11,750
2660	Ground mounted collectors			
2680	10 ea. 3′x7′ black chrome absorber, 1/8″ temp. glass	13,700	4,900	18,600
2700	10 ea. 4′x8′ black chrome absorber, 3/16″ temp. glass	15,400	5,725	21,125
2720	10 ea. 3′8″x6′ flat blk absorber, 3/16″ temp. glass	11,600	4,975	16,575
2740	10 ea. 4′x9′ flat blk absorber, plastic glazing	14,200	5,925	20,125
2760	10 ea. 2′x9′ rubber absorber, plastic glazing	15,600	6,175	21,775
2780	10 ea. 4′x10′ fully wetted unglazed plastic absorber	6,075	6,175	12,250

D3010 Energy Supply

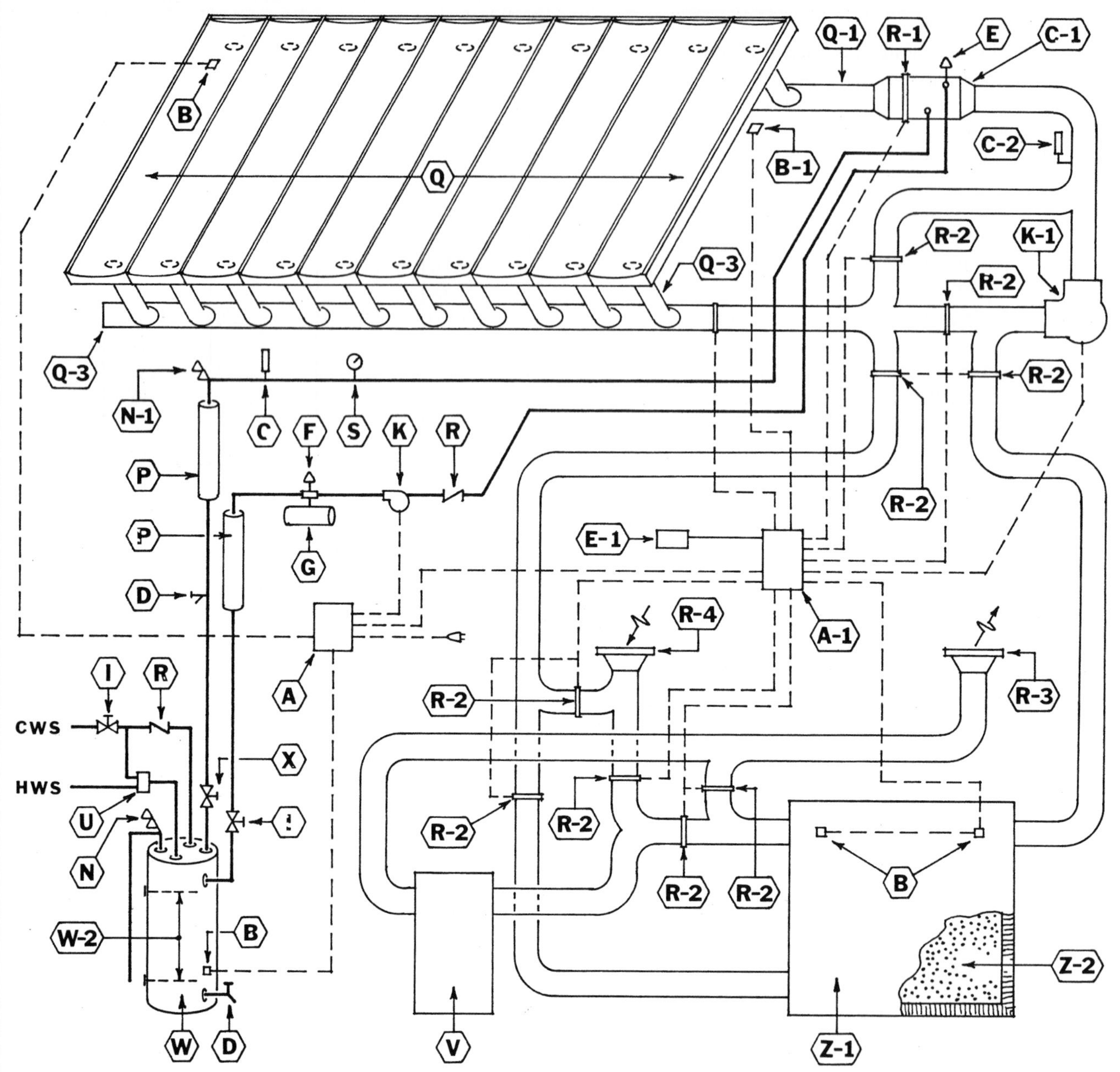

The complete Solar Air Heating System provides maximum savings of conventional fuel with both space heating and year-round domestic hot water heating. It allows for the home air conditioning to operate simultaneously and independently from the solar domestic water heating in summer. The system's modes of operation are:

Mode 1: The building is heated directly from the collectors with air circulated by the Solar Air Mover.

Mode 2: When heat is not needed in the building, the dampers change within the air mover to circulate the air from the collectors to the rock storage bin.

Mode 3: When heat is not available from the collector array and is available in rock storage, the air mover draws heated air from rock storage and directs it into the building. When heat is not available from the collectors or the rock storage bin, the auxiliary heating unit will provide heat for the building. The size of the collector array is typically 25% the size of the main floor area.

D3010 Energy Supply

System Components		QUANTITY	UNIT	COST EACH MAT.	COST EACH INST.	COST EACH TOTAL
	SYSTEM D3010 675 1210					
	SOLAR, SPACE/HOT WATER, AIR TO WATER HEAT EXCHANGE					
A, B	Differential controller 2 sensors thermos., solar energy sys liquid loop	1.000	Ea.	145	63	208
A-1, B	Differential controller 2 sensors 6 station solar energy sys air loop	1.000	Ea.	298	253	551
B-1	Solar energy sensor, freeze prevention	1.000	Ea.	25.50	23.50	49
C	Thermometer for solar energy system, 2″ dial	2.000	Ea.	49	95	144
C-1	Heat exchanger, solar energy system, air to fluid, up flow, 70 MBH	1.000	Ea.	400	390	790
D	Drain valve, brass, 3/4″ connection	2.000	Ea.	23.30	63	86.30
E	Air vent, manual, for solar energy system 1/8″ fitting	1.000	Ea.	2.93	23.50	26.43
E-1	Thermostat, 2 stage for sensing room temperature	1.000	Ea.	247	96	343
F	Air purger	1.000	Ea.	52.50	63	115.50
G	Expansion tank, for solar energy system	1.000	Ea.	76.50	23.50	100
I	Valve, gate, bronze, 125 lb, NRS, soldered 3/4″ diam	2.000	Ea.	150	76	226
K	Circulator, solar heated liquid, 1/25 HP	1.000	Ea.	219	97.50	316.50
N	Relief valve temp & press 150 psi 210°F self-closing, 3/4″ IPS	1.000	Ea.	26.50	25.50	52
N-1	Relief valve, pressure, poppet, bronze, 30 psi, 3/4″ IPS	1.000	Ea.	97	27	124
P	Pipe covering, fiberglass, all service jacket, 1″ wall, 3/4″ diam	60.000	L.F.	63.60	333	396.60
Q-3	Manifold for flush mount solar energy collector panels	20.000	L.F.	3,420	167	3,587
Q	Collector panel solar energy, air, black on alum. plate, 7.5′ x 3.5′	10.000	Ea.	9,200	1,520	10,720
R	Valve, swing check, bronze, 125 lb, regrinding disc, 3/4″ diam	2.000	Ea.	200	76	276
S	Pressure gage, 2″ dial, for solar energy system	1.000	Ea.	27	23.50	50.50
U	Valve, water tempering, bronze, sweat connections, 3/4″ diam	1.000	Ea.	159	38	197
W-2, W	Tank, water storage, solar, elec element 2″x1/2# insul, 80 Gal	1.000	Ea.	2,200	475	2,675
X	Valve, globe, bronze, 125 lb, soldered, 3/4″ diam	1.000	Ea.	154	38	192
	Copper tubing type L, solder joints, hangers 10′ OC 3/4″ diam	10.000	L.F.	48.20	99.50	147.70
	Copper tubing type M, solder joints, hangers 10′ OC 3/4″ diam	60.000	L.F.	342	582	924
	Wrought copper fittings & solder, 3/4″ diam	26.000	Ea.	68.64	1,040	1,108.64
	Sensor wire, #22-2 conductor multistranded	1.200	C.L.F.	13.56	85.80	99.36
Q-1	Duct work, rigid fiberglass, rectangular	400.000	S.F.	396	2,360	2,756
	Duct work, spiral preformed, steel, PVC coated both sides, 12″ x 10″	8.000	Ea.	204	444	648
R-2	Shutter/damper for solar heater circulator	9.000	Ea.	567	999	1,566
K-1	Shutter motor for solar heater circulator blower	9.000	Ea.	1,404	751.50	2,155.50
K-1	Fan, solar energy heated air circulator, space & DHW system	1.000	Ea.	1,750	2,675	4,425
Z-1	Tank, solar energy air storage, 6′-3″H 7′x7′ = 306 CF/2000 Gal	1.000	Ea.	16,000	1,725	17,725
Z-2	Crushed stone 1-1/2″	11.000	C.Y.	330	86.24	416.24
C-2	Themometer, remote probe, 2″ dial	1.000	Ea.	37	95	132
R-1	Solenoid valve	1.000	Ea.	153	84	237
V, R-4, R-3	Furnace, supply diffusers, return grilles, existing	1.000	Ea.	16,000	1,725	17,725
	TOTAL			54,549.23	16,742.04	71,291.27

D3010 675	Air To Water Heat Exchange	COST EACH MAT.	COST EACH INST.	COST EACH TOTAL
1210	Solar, air to water heat exchange, for space/hot water heating (R235616-60)	54,500	16,700	71,200
1220				

D3020 Heat Generating Systems

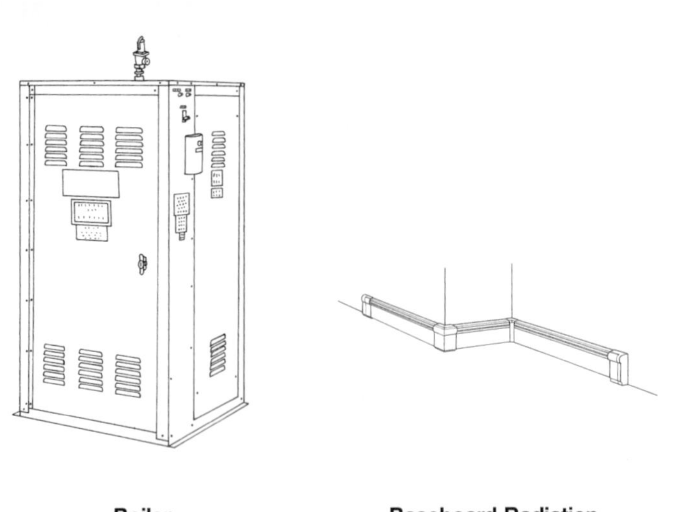

Boiler | Baseboard Radiation

Small Electric Boiler System Considerations:

1. Terminal units are fin tube baseboard radiation rated at 720 BTU/hr with 200° water temperature or 820 BTU/hr steam.
2. Primary use being for residential or smaller supplementary areas, the floor levels are based on 7-1/2′ ceiling heights.
3. All distribution piping is copper for boilers through 205 MBH. All piping for larger systems is steel pipe.

System Components	QUANTITY	UNIT	COST EACH MAT.	COST EACH INST.	COST EACH TOTAL
SYSTEM D3020 102 1120					
SMALL HEATING SYSTEM, HYDRONIC, ELECTRIC BOILER					
1,480 S.F., 61 MBH, STEAM, 1 FLOOR					
Boiler, electric steam, std cntrls, trim, ftngs and valves, 18 KW, 61.4 MBH	1.000	Ea.	5,335	1,925	7,260
Copper tubing type L, solder joint, hanger 10′OC, 1-1/4″ diam	160.000	L.F.	2,048	2,088	4,136
Radiation, 3/4″ copper tube w/alum fin baseboard pkg 7″ high	60.000	L.F.	534	1,440	1,974
Rough in baseboard panel or fin tube with valves & traps	10.000	Set	2,840	7,250	10,090
Pipe covering, calcium silicate w/cover, 1″ wall 1-1/4″ diam	160.000	L.F.	740.80	1,240	1,980.80
Low water cut-off, quick hookup, in gage glass tappings	1.000	Ea.	340	48	388
TOTAL			11,837.80	13,991	25,828.80
COST PER S.F.			8	9.45	17.45

D3020 102	Small Heating Systems, Hydronic, Electric Boilers		COST PER S.F. MAT.	COST PER S.F. INST.	COST PER S.F. TOTAL
1100	Small heating systems, hydronic, electric boilers				
1120	Steam, 1 floor, 1480 S.F., 61 M.B.H.		7.97	9.46	17.43
1160	3,000 S.F., 123 M.B.H.	R235000 -10	6.15	8.30	14.45
1200	5,000 S.F., 205 M.B.H.		5.05	7.65	12.70
1240	2 floors, 12,400 S.F., 512 M.B.H.	R235000 -20	4.10	7.60	11.70
1280	3 floors, 24,800 S.F., 1023 M.B.H.		4.69	7.50	12.19
1320	34,750 S.F., 1,433 M.B.H.	R235000 -30	4.31	7.30	11.61
1360	Hot water, 1 floor, 1,000 S.F., 41 M.B.H.		13.40	5.30	18.70
1400	2,500 S.F., 103 M.B.H.		9.20	9.45	18.65
1440	2 floors, 4,850 S.F., 205 M.B.H.		8.55	11.35	19.90
1480	3 floors, 9,700 S.F., 410 M.B.H.		9.05	11.75	20.80

D3020 Heat Generating Systems

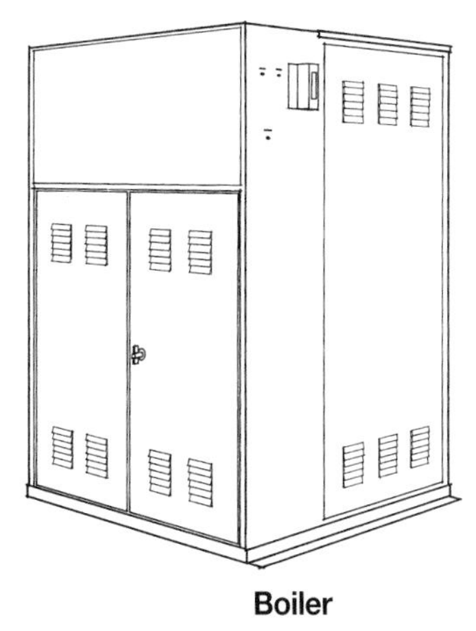
Boiler

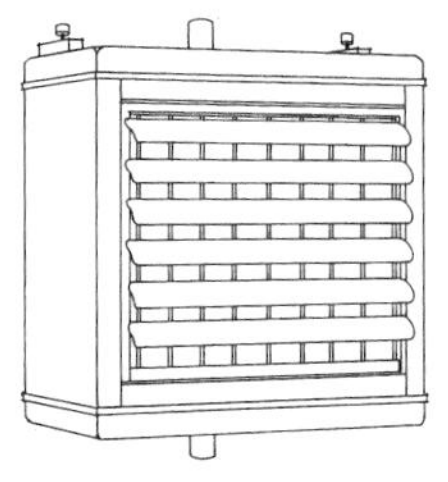
Unit Heater

Large Electric Boiler System Considerations:

1. Terminal units are all unit heaters of the same size. Quantities are varied to accommodate total requirements.
2. All air is circulated through the heaters a minimum of three times per hour.
3. As the capacities are adequate for commercial use, floor levels are based on 10′ ceiling heights.
4. All distribution piping is black steel pipe.

System Components	QUANTITY	UNIT	COST EACH MAT.	COST EACH INST.	COST EACH TOTAL
SYSTEM D3020 104 1240					
LARGE HEATING SYSTEM, HYDRONIC, ELECTRIC BOILER					
9,280 S.F., 135 KW, 461 MBH, 1 FLOOR					
Boiler, electric hot water, std ctrls, trim, ftngs, valves, 135 KW, 461 MBH	1.000	Ea.	12,487.50	4,200	16,687.50
Expansion tank, painted steel, 60 Gal capacity ASME	1.000	Ea.	5,225	230	5,455
Circulating pump, CI, close cpld, 50 GPM, 2 HP, 2″ pipe conn	1.000	Ea.	3,325	455	3,780
Unit heater, 1 speed propeller, horizontal, 200° EWT, 72.7 MBH	7.000	Ea.	6,755	1,757	8,512
Unit heater piping hookup with controls	7.000	Set	5,985	11,375	17,360
Pipe, steel, black, schedule 40, welded, 2-1/2″ diam	380.000	L.F.	6,270	11,517.80	17,787.80
Pipe covering, calcium silicate w/cover, 1″ wall, 2-1/2″ diam	380.000	L.F.	2,185	3,021	5,206
TOTAL			42,232.50	32,555.80	74,788.30
COST PER S.F.			4.55	3.51	8.06

D3020 104	Large Heating Systems, Hydronic, Electric Boilers		COST PER S.F. MAT.	COST PER S.F. INST.	COST PER S.F. TOTAL
1230	Large heating systems, hydronic, electric boilers				
1240	9,280 S.F., 135 K.W., 461 M.B.H., 1 floor		4.55	3.51	8.06
1280	14,900 S.F., 240 K.W., 820 M.B.H., 2 floors	R235000 -10	5.95	5.65	11.60
1320	18,600 S.F., 296 K.W., 1,010 M.B.H., 3 floors		5.90	6.15	12.05
1360	26,100 S.F., 420 K.W., 1,432 M.B.H., 4 floors	R235000 -20	5.80	6.05	11.85
1400	39,100 S.F., 666 K.W., 2,273 M.B.H., 4 floors		4.95	5.05	10
1440	57,700 S.F., 900 K.W., 3,071 M.B.H., 5 floors	R235000 -30	4.71	4.96	9.67
1480	111,700 S.F., 1,800 K.W., 6,148 M.B.H., 6 floors		4.15	4.26	8.41
1520	149,000 S.F., 2,400 K.W., 8,191 M.B.H., 8 floors		4.13	4.26	8.39
1560	223,300 S.F., 3,600 K.W., 12,283 M.B.H., 14 floors		4.39	4.83	9.22

D3020 Heat Generating Systems

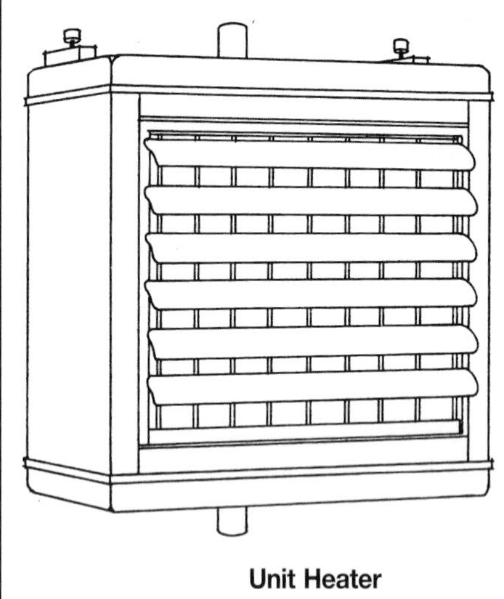

Unit Heater

Fossil Fuel Boiler System Considerations:

1. Terminal units are horizontal unit heaters. Quantities are varied to accommodate total heat loss per building.
2. Unit heater selection was determined by their capacity to circulate the building volume a minimum of three times per hour in addition to the BTU output.
3. Systems shown are forced hot water. Steam boilers cost slightly more than hot water boilers. However, this is compensated for by the smaller size or fewer terminal units required with steam.
4. Floor levels are based on 10′ story heights.
5. MBH requirements are gross boiler output.

System Components	QUANTITY	UNIT	COST EACH MAT.	COST EACH INST.	COST EACH TOTAL
SYSTEM D3020 108 1280					
HEATING SYSTEM, HYDRONIC, FOSSIL FUEL, TERMINAL UNIT HEATERS					
CAST IRON BOILER, GAS, 80 MBH, 1,070 S.F. BUILDING					
Boiler, gas, hot water, CI, burner, controls, insulation, breeching, 80 MBH	1.000	Ea.	2,231.25	2,100	4,331.25
Pipe, steel, black, schedule 40, threaded, cplg & hngr 10′OC, 2″ diam	200.000	L.F.	2,540	4,300	6,840
Unit heater, 1 speed propeller, horizontal, 200° EWT, 72.7 MBH	2.000	Ea.	1,930	502	2,432
Unit heater piping hookup with controls	2.000	Set	1,710	3,250	4,960
Expansion tank, painted steel, ASME, 18 Gal capacity	1.000	Ea.	3,075	115	3,190
Circulating pump, CI, flange connection, 1/12 HP	1.000	Ea.	505	227	732
Pipe covering, calcium silicate w/cover, 1″ wall, 2″ diam	200.000	L.F.	1,080	1,590	2,670
TOTAL			13,071.25	12,084	25,155.25
COST PER S.F.			12.22	11.29	23.51

D3020 108	Heating Systems, Unit Heaters		COST PER S.F. MAT.	COST PER S.F. INST.	COST PER S.F. TOTAL
1260	Heating systems, hydronic, fossil fuel, terminal unit heaters,				
1280	Cast iron boiler, gas, 80 M.B.H., 1,070 S.F. bldg.		12.24	11.31	23.55
1320	163 M.B.H., 2,140 S.F. bldg.	R235000-10	8.10	7.65	15.75
1360	544 M.B.H., 7,250 S.F. bldg.		5.85	5.40	11.25
1400	1,088 M.B.H., 14,500 S.F. bldg.	R235000-20	4.95	5.10	10.05
1440	3,264 M.B.H., 43,500 S.F. bldg.		3.98	3.85	7.83
1480	5,032 M.B.H., 67,100 S.F. bldg.	R235000-30	4.24	3.96	8.20
1520	Oil, 109 M.B.H., 1,420 S.F. bldg.		13.55	10.05	23.60
1560	235 M.B.H., 3,150 S.F. bldg.		7.85	7.10	14.95
1600	940 M.B.H., 12,500 S.F. bldg.		5.80	4.66	10.46
1640	1,600 M.B.H., 21,300 S.F. bldg.		5.10	4.47	9.57
1680	2,480 M.B.H., 33,100 S.F. bldg.		4.71	4.03	8.74
1720	3,350 M.B.H., 44,500 S.F. bldg.		4.56	4.10	8.66
1760	Coal, 148 M.B.H., 1,975 S.F. bldg.		96.50	6.55	103.05
1800	300 M.B.H., 4,000 S.F. bldg.		53	5.15	58.15
1840	2,360 M.B.H., 31,500 S.F. bldg.		12	4.08	16.08

D3020 Heat Generating Systems

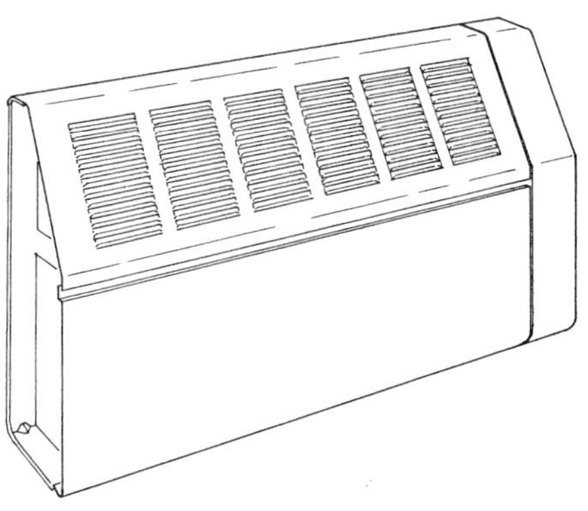

Fin Tube Radiator

Fossil Fuel Boiler System Considerations:

1. Terminal units are commercial steel fin tube radiation. Quantities are varied to accommodate total heat loss per building.
2. Systems shown are forced hot water. Steam boilers cost slightly more than hot water boilers. However, this is compensated for by the smaller size or fewer terminal units required with steam.
3. Floor levels are based on 10′ story heights.
4. MBH requirements are gross boiler output.

System Components	QUANTITY	UNIT	COST EACH MAT.	COST EACH INST.	COST EACH TOTAL
SYSTEM D3020 110 3240					
HEATING SYSTEM, HYDRONIC, FOSSIL FUEL, FIN TUBE RADIATION					
CAST IRON BOILER, GAS, 80 MBH, 1,070 S.F. BLDG.					
Boiler, gas, hot water, CI, burner, controls, insulation, breeching, 80 MBH	1.000	Ea.	2,231.25	2,100	4,331.25
Pipe, steel, black, schedule 40, threaded, 2″ diam	200.000	L.F.	2,540	4,300	6,840
Radiation,steel 1-1/4″ tube & 4-1/4″ fin w/cover & damper, wall hung	80.000	L.F.	3,720	3,080	6,800
Rough-in, wall hung steel fin radiation with supply & balance valves	8.000	Set	2,880	6,760	9,640
Expansion tank, painted steel, ASME, 18 Gal capacity	1.000	Ea.	3,075	115	3,190
Circulating pump, CI flanged, 1/12 HP	1.000	Ea.	505	227	732
Pipe covering, calcium silicate w/cover, 1″ wall, 2″ diam	200.000	L.F.	1,080	1,590	2,670
TOTAL			16,031.25	18,172	34,203.25
COST PER S.F.			14.98	16.98	31.96

D3020 110	Heating System, Fin Tube Radiation		COST PER S.F. MAT.	COST PER S.F. INST.	COST PER S.F. TOTAL
3230	Heating systems, hydronic, fossil fuel, fin tube radiation				
3240	Cast iron boiler, gas, 80 MBH, 1,070 S.F. bldg.		14.95	17.01	31.96
3280	169 M.B.H., 2,140 S.F. bldg.	R235000-10	9.35	10.75	20.10
3320	544 M.B.H., 7,250 S.F. bldg.		7.85	9.15	17
3360	1,088 M.B.H., 14,500 S.F. bldg.	R235000-20	7.05	9	16.05
3400	3,264 M.B.H., 43,500 S.F. bldg.		6.25	7.95	14.20
3440	5,032 M.B.H., 67,100 S.F. bldg.	R235000-30	6.55	8.05	14.60
3480	Oil, 109 M.B.H., 1,420 S.F. bldg.		18.40	18.55	36.95
3520	235 M.B.H., 3,150 S.F. bldg.		9.85	10.80	20.65
3560	940 M.B.H., 12,500 S.F. bldg.		7.95	8.60	16.55
3600	1,600 M.B.H., 21,300 S.F. bldg.		7.40	8.55	15.95
3640	2,480 M.B.H., 33,100 S.F. bldg.		7	8.10	15.10
3680	3,350 M.B.H., 44,500 S.F. bldg.		6.85	8.20	15.05
3720	Coal, 148 M.B.H., 1,975 S.F. bldg.		98.50	10.25	108.75
3760	300 M.B.H., 4,000 S.F. bldg.		55	8.80	63.80
3800	2,360 M.B.H., 31,500 S.F. bldg.		14.20	8.10	22.30
4080	Steel boiler, oil, 97 M.B.H., 1,300 S.F. bldg.		13	15.20	28.20
4120	315 M.B.H., 4,550 S.F. bldg.		6.95	7.75	14.70
4160	525 M.B.H., 7,000 S.F. bldg.		9.70	8.85	18.55
4200	1,050 M.B.H., 14,000 S.F. bldg.		8.90	8.85	17.75
4240	2,310 M.B.H., 30,800 S.F. bldg.		7.55	8.15	15.70
4280	3,150 M.B.H., 42,000 S.F. bldg.		7.40	8.25	15.65

D3020 Heat Generating Systems

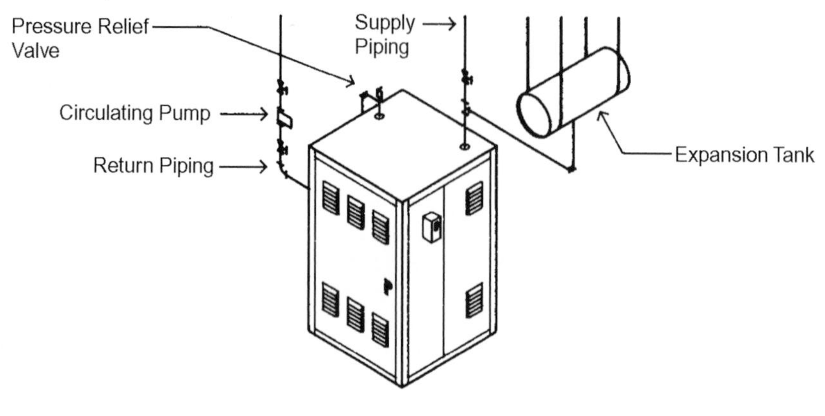

Electric Boiler, Hot Water

System Components	QUANTITY	UNIT	COST EACH MAT.	COST EACH INST.	COST EACH TOTAL
SYSTEM D3020 126 1010					
BOILER, ELECTRIC, HOT WATER, 15 kW, 52 MBH					
Boilers, electric, ASME, hot water, 15 KW, 52 MBH	1.000	Ea.	5,775	1,625	7,400
Pipe, black steel, Sch 40, threaded, W/coupling & hangers, 10′ OC, 3/4″ dia	10.000	L.F.	53.82	143.18	197
Pipe, black steel, Sch 40, threaded, W/coupling & hangers, 10′ OC, 1″ dia	20.000	L.F.	104.71	307.45	412.16
Elbow, 90°, black, straight, 3/4″ dia.	6.000	Ea.	29.46	324	353.46
Elbow, 90°, black, straight, 1″ dia.	6.000	Ea.	51.30	351	402.30
Tee, black, straight, 3/4″ dia.	1.000	Ea.	8.05	84	92.05
Tee, black, straight, 1″ dia.	2.000	Ea.	27.60	190	217.60
Tee, black, reducing, 1″ dia.	2.000	Ea.	44	190	234
Union, black with brass seat, 3/4″ dia.	1.000	Ea.	23	58.50	81.50
Union, black with brass seat, 1″ dia.	2.000	Ea.	59	126	185
Pipe nipples, 3/4″ diam.	3.000	Ea.	7.02	18.68	25.70
Pipe nipples, 1″ diam.	3.000	Ea.	7.31	21.45	28.76
Valves, bronze, gate, N.R.S., threaded, class 150, 3/4″ size	2.000	Ea.	196	76	272
Valves, bronze, gate, N.R.S., threaded, class 150, 1″ size	2.000	Ea.	246	80	326
Thermometer, stem type, 9″ case, 8″ stem, 3/4″ NPT	1.000	Ea.	392	55	447
Tank, steel, liquid expansion, ASME, painted, 15 gallon capacity	1.000	Ea.	870	81.50	951.50
Pump, circulating, bronze, flange connection, 3/4″ to 1-1/2″ size, 1/8 HP	1.000	Ea.	1,450	227	1,677
Insulation, fiberglass pipe covering, 1″ wall, 1″ IPS	20.000	L.F.	22.60	116	138.60
TOTAL			9,366.87	4,074.76	13,441.63

D3020 126	Electric Boiler, Hot Water	COST EACH MAT.	COST EACH INST.	COST EACH TOTAL
1010	Boiler, electric, hot water, 15 kW, 52 MBH	9,375	4,075	13,450
1020	30 kW, 103 MBH	10,100	4,275	14,375
1030	60 kW, 205 MBH	10,300	4,425	14,725
1040	120 kW, 410 MBH	12,100	5,300	17,400
1050	150 kW, 510 MBH	18,900	10,200	29,100
1060	210 kW, 716 MBH	21,200	11,200	32,400
1070	296 kW, 1010 MBH	29,800	11,500	41,300
1080	444 kW, 1515 MBH	34,700	15,500	50,200
1090	666 kW, 2273 MBH	41,800	16,700	58,500
1100	900 kW, 3071 MBH	55,000	19,600	74,600
1110	1320 kW, 4505 MBH	62,000	20,500	82,500
1120	2100 kW, 7167 MBH	92,500	24,900	117,400

D3020 Heat Generating Systems

D3020 126	Electric Boiler, Hot Water	COST EACH		
		MAT.	INST.	TOTAL
1130	2610 kW, 8905 MBH	111,000	30,000	141,000
1140	3600 kW, 12,283 MBH	135,000	36,000	171,000

D3020 128	Electric Boiler, Steam	COST EACH		
		MAT.	INST.	TOTAL
1010	Boiler, electric, steam, 18 kW, 61.4 MBH	6,200	3,900	10,100
1020	36 kW, 123 MBH	7,800	4,125	11,925
1030	60 kW, 205 MBH	19,000	8,400	27,400
1040	120 kW, 409 MBH	21,700	9,250	30,950
1050	150 kW, 512 MBH	25,500	13,900	39,400
1060	210 kW, 716 MBH	41,100	14,900	56,000
1070	300 kW, 1023 MBH	43,900	16,300	60,200
1080	510 kW, 1740 MBH	52,500	21,200	73,700
1090	720 kW, 2456 MBH	57,000	22,200	79,200
1100	1080 kW, 3685 MBH	64,000	26,500	90,500
1110	1260 kW, 4300 MBH	71,000	28,200	99,200
1120	1620 kW, 5527 MBH	75,000	31,900	106,900
1130	2070 kW, 7063 MBH	97,500	33,500	131,000
1140	2340 kW, 7984 MBH	125,000	38,000	163,000

D3020 Heat Generating Systems

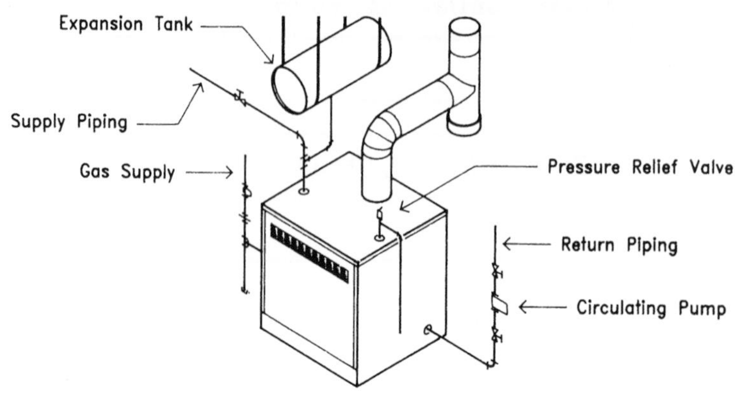

Cast Iron Boiler, Hot Water, Gas Fired

System Components	QUANTITY	UNIT	COST EACH MAT.	COST EACH INST.	COST EACH TOTAL
SYSTEM D3020 130 1010					
BOILER, CAST IRON, GAS, HOT WATER, 100 MBH					
Boilers, gas fired, std controls, CI, insulated, HW, gross output 100 MBH	1.000	Ea.	2,500	2,175	4,675
Pipe, black steel, Sch 40, threaded, W/coupling & hangers, 10′ OC, 3/4″ dia	20.000	L.F.	105.30	280.13	385.43
Pipe, black steel, Sch 40, threaded, W/coupling & hangers, 10′ OC, 1″ dia.	20.000	L.F.	104.71	307.45	412.16
Elbow, 90°, black, straight, 3/4″ dia.	9.000	Ea.	44.19	486	530.19
Elbow, 90°, black, straight, 1″ dia.	6.000	Ea.	51.30	351	402.30
Tee, black, straight, 3/4″ dia.	2.000	Ea.	16.10	168	184.10
Tee, black, straight, 1″ dia.	2.000	Ea.	27.60	190	217.60
Tee, black, reducing, 1″ dia.	2.000	Ea.	44	190	234
Pipe cap, black, 3/4″ dia.	1.000	Ea.	5.75	23.50	29.25
Union, black with brass seat, 3/4″ dia.	2.000	Ea.	46	117	163
Union, black with brass seat, 1″ dia.	2.000	Ea.	59	126	185
Pipe nipples, black, 3/4″ dia.	5.000	Ea.	11.70	31.13	42.83
Pipe nipples, black, 1″ dia.	3.000	Ea.	7.31	21.45	28.76
Valves, bronze, gate, N.R.S., threaded, class 150, 3/4″ size	2.000	Ea.	196	76	272
Valves, bronze, gate, N.R.S., threaded, class 150, 1″ size	2.000	Ea.	246	80	326
Gas cock, brass, 3/4″ size	1.000	Ea.	18.70	34.50	53.20
Thermometer, stem type, 9″ case, 8″ stem, 3/4″ NPT	2.000	Ea.	392	55	447
Tank, steel, liquid expansion, ASME, painted, 15 gallon capacity	1.000	Ea.	870	81.50	951.50
Pump, circulating, bronze, flange connection, 3/4″ to 1-1/2″ size, 1/8 HP	1.000	Ea.	1,450	227	1,677
Vent chimney, all fuel, pressure tight, double wall, SS, 6″ dia.	20.000	L.F.	1,220	440	1,660
Vent chimney, elbow, 90° fixed, 6″ dia.	2.000	Ea.	930	89	1,019
Vent chimney, Tee, 6″ dia.	2.000	Ea.	544	111	655
Vent chimney, ventilated roof thimble, 6″ dia.	1.000	Ea.	283	51.50	334.50
Vent chimney, flat roof flashing, 6″ dia.	1.000	Ea.	122	44.50	166.50
Vent chimney, stack cap, 6″ diameter	1.000	Ea.	335	29	364
Insulation, fiberglass pipe covering, 1″ wall, 1″ IPS	20.000	L.F.	22.60	116	138.60
TOTAL			9,652.26	5,901.66	15,553.92

D3020 130	Boiler, Cast Iron, Hot Water, Gas	COST EACH MAT.	COST EACH INST.	COST EACH TOTAL
1010	Boiler, cast iron, gas, hot water, 100 MBH	9,650	5,900	15,550
1020	200 MBH	11,500	6,750	18,250
1030	320 MBH	10,700	7,325	18,025
1040	440 MBH	14,900	9,475	24,375
1050	544 MBH	24,100	14,800	38,900
1060	765 MBH	28,700	15,900	44,600

D3020 Heat Generating Systems

D3020 130	Boiler, Cast Iron, Hot Water, Gas	COST EACH		
		MAT.	INST.	TOTAL
1070	1088 MBH	31,800	16,900	48,700
1080	1530 MBH	36,300	22,400	58,700
1090	2312 MBH	47,000	26,300	73,300
1100	2856 MBH	57,500	28,400	85,900
1110	3808 MBH	64,000	33,200	97,200
1120	4720 MBH	106,500	37,400	143,900
1130	6100 MBH	152,500	41,600	194,100
1140	6970 MBH	165,500	51,500	217,000

D3020 134	Boiler, Cast Iron, Steam, Gas	COST EACH		
		MAT.	INST.	TOTAL
1010	Boiler, cast iron, gas, steam, 100 MBH	7,875	5,700	13,575
1020	200 MBH	18,700	10,700	29,400
1030	320 MBH	20,500	12,100	32,600
1040	544 MBH	28,100	18,800	46,900
1050	765 MBH	31,100	19,600	50,700
1060	1275 MBH	36,200	21,600	57,800
1070	2675 MBH	53,000	31,400	84,400
1080	3570 MBH	64,500	35,200	99,700
1090	4720 MBH	98,000	37,600	135,600
1100	6100 MBH	121,500	43,700	165,200
1110	6970 MBH	133,000	51,500	184,500

D3020 136	Boiler, Cast Iron, Hot Water, Gas/Oil	COST EACH		
		MAT.	INST.	TOTAL
1010	Boiler, cast iron, gas & oil, hot water, 584 MBH	30,600	16,400	47,000
1020	876 MBH	38,700	17,600	56,300
1030	1168 MBH	48,300	19,900	68,200
1040	1460 MBH	57,500	23,900	81,400
1050	2044 MBH	63,500	24,800	88,300
1060	2628 MBH	75,000	29,900	104,900
1070	3210 MBH	79,000	32,200	111,200
1080	3796 MBH	85,500	33,200	118,700
1090	4672 MBH	99,500	34,800	134,300
1100	5256 MBH	123,000	42,200	165,200
1110	6000 MBH	146,000	43,000	189,000
1120	9800 MBH	182,500	73,500	256,000
1130	12,200 MBH	228,500	85,500	314,000
1140	13,500 MBH	248,500	91,000	339,500

D3020 138	Boiler, Cast Iron, Steam, Gas/Oil	COST EACH		
		MAT.	INST.	TOTAL
1010	Boiler, cast iron, gas & oil, steam, 810 MBH	35,000	21,300	56,300
1020	1360 MBH	40,400	22,700	63,100
1030	2040 MBH	49,500	29,200	78,700
1040	2700 MBH	52,500	32,400	84,900
1050	3270 MBH	61,000	36,200	97,200
1060	3770 MBH	97,500	36,900	134,400
1070	4650 MBH	108,500	42,000	150,500
1080	5520 MBH	124,500	44,300	168,800
1090	6100 MBH	130,000	46,600	176,600
1100	6970 MBH	137,500	57,000	194,500

D3020 Heat Generating Systems

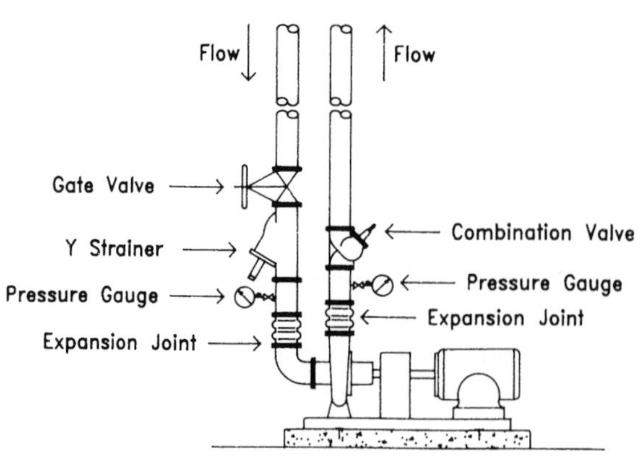

Base Mounted End-Suction Pump

System Components	QUANTITY	UNIT	COST EACH MAT.	COST EACH INST.	COST EACH TOTAL
SYSTEM D3020 330 1010					
PUMP, BASE MTD. WITH MOTOR, END-SUCTION, 2-1/2" SIZE, 3 HP, TO 150 GPM					
Pump, circulating, CI, base mounted, 2-1/2" size, 3 HP, to 150 GPM	1.000	Ea.	10,000	760	10,760
Pipe, black steel, Sch. 40, on yoke & roll hangers, 10' O.C., 2-1/2" dia	12.000	L.F.	198	363.72	561.72
Elbow, 90°, weld joint, steel, 2-1/2" pipe size	1.000	Ea.	38	178.70	216.70
Flange, weld neck, 150 LB, 2-1/2" pipe size	8.000	Ea.	324	714.88	1,038.88
Valve, iron body, gate, 125 lb., N.R.S., flanged, 2-1/2" size	1.000	Ea.	680	273	953
Strainer, Y type, iron body, flanged, 125 lb., 2-1/2" pipe size	1.000	Ea.	184	276	460
Multipurpose valve, CI body, 2" size	1.000	Ea.	570	96	666
Expansion joint, flanged spool, 6" F to F, 2-1/2" dia	2.000	Ea.	600	222	822
T-O-L, weld joint, socket, 1/4" pipe size, nozzle	2.000	Ea.	17.20	124.38	141.58
T-O-L, weld joint, socket, 1/2" pipe size, nozzle	2.000	Ea.	17.20	129.62	146.82
Control gauges, pressure or vacuum, 3-1/2" diameter dial	2.000	Ea.	41	48	89
Pressure/temperature relief plug, 316 SS, 3/4" OD, 7-1/2" insertion	2.000	Ea.	112	48	160
Insulation, fiberglass pipe covering, 1-1/2" wall, 2-1/2" IPS	12.000	L.F.	36	85.20	121.20
Water flow	1.000	Ea.	1,400	1,175	2,575
Pump balancing	1.000	Ea.		340	340
TOTAL			14,217.40	4,834.50	19,051.90

D3020 330	Circulating Pump Systems, End Suction	COST EACH MAT.	COST EACH INST.	COST EACH TOTAL
1010	Pump, base mtd with motor, end-suction, 2-1/2" size, 3 HP, to 150 GPM	14,200	4,850	19,050
1020	3" size, 5 HP, to 225 GPM	17,100	5,475	22,575
1030	4" size, 7-1/2 HP, to 350 GPM	20,300	6,475	26,775
1040	5" size, 15 HP, to 1000 GPM	29,000	9,575	38,575
1050	6" size, 25 HP, to 1550 GPM	39,200	11,500	50,700

D3020 340	Circulating Pump Systems, Double Suction	COST EACH MAT.	COST EACH INST.	COST EACH TOTAL
1010	Pump, base mtd w/motor, double suction, 6" size, 50 HP, to 1200 GPM	24,700	16,400	41,100
1020	8" size, 75 HP, to 2500 GPM	37,500	22,100	59,600
1030	8" size, 100 HP, to 3000 GPM	50,500	24,600	75,100
1040	10" size, 150 HP, to 4000 GPM	69,500	27,800	97,300

D3030 Cooling Generating Systems

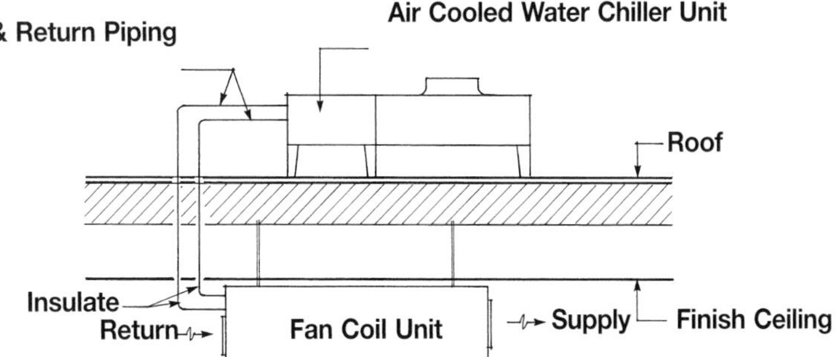

Design Assumptions: The chilled water, air cooled systems priced, utilize reciprocating hermetic compressors and propeller-type condenser fans. Piping with pumps and expansion tanks is included based on a two pipe system. No ducting is included and the fan-coil units are cooling only. Water treatment and balancing are not included. Chilled water piping is insulated. Area distribution is through the use of multiple fan coil units. Fewer but larger fan coil units with duct distribution would be approximately the same S.F. cost.

System Components	QUANTITY	UNIT	COST EACH MAT.	COST EACH INST.	COST EACH TOTAL
SYSTEM D3030 110 1200					
PACKAGED CHILLER, AIR COOLED, WITH FAN COIL UNIT					
APARTMENT CORRIDORS, 3,000 S.F., 5.50 TON					
Fan coil air conditioning unit, cabinet mounted & filters chilled water	1.000	Ea.	3,986.78	632.39	4,619.17
Water chiller, air conditioning unit, air cooled	1.000	Ea.	6,737.50	2,365	9,102.50
Chilled water unit coil connections	1.000	Ea.	1,550	1,725	3,275
Chilled water distribution piping	440.000	L.F.	11,000	23,320	34,320
TOTAL			23,274.28	28,042.39	51,316.67
COST PER S.F.			7.76	9.35	17.11
*Cooling requirements would lead to choosing a water cooled unit.					

D3030 110	Chilled Water, Air Cooled Condenser Systems	COST PER S.F. MAT.	COST PER S.F. INST.	COST PER S.F. TOTAL
1180	Packaged chiller, air cooled, with fan coil unit			
1200	Apartment corridors, 3,000 S.F., 5.50 ton	7.77	9.33	17.10
1240	6,000 S.F., 11.00 ton	6.20	7.45	13.65
1280	10,000 S.F., 18.33 ton	5.25	5.65	10.90
1320	20,000 S.F., 36.66 ton	4.03	4.05	8.08
1360	40,000 S.F., 73.33 ton	5	4.08	9.08
1440	Banks and libraries, 3,000 S.F., 12.50 ton	11.15	10.85	22
1480	6,000 S.F., 25.00 ton	10.05	8.80	18.85
1520	10,000 S.F., 41.66 ton	7.75	6.10	13.85
1560	20,000 S.F., 83.33 ton	8.35	5.50	13.85
1600	40,000 S.F., 167 ton*			
1680	Bars and taverns, 3,000 S.F., 33.25 ton	18.70	12.70	31.40
1720	6,000 S.F., 66.50 ton	16.40	9.60	26
1760	10,000 S.F., 110.83 ton	12.75	2.89	15.64
1800	20,000 S.F., 220 ton*			
1840	40,000 S.F., 440 ton*			
1920	Bowling alleys, 3,000 S.F., 17.00 ton	13.90	12.15	26.05
1960	6,000 S.F., 34.00 ton	10.50	8.50	19
2000	10,000 S.F., 56.66 ton	8.85	6.05	14.90
2040	20,000 S.F., 113.33 ton	9.30	5.55	14.85
2080	40,000 S.F., 227 ton*			
2160	Department stores, 3,000 S.F., 8.75 ton	10.70	10.40	21.10
2200	6,000 S.F., 17.50 ton	7.80	7.95	15.75
2240	10,000 S.F., 29.17 ton	6.30	5.70	12

D3030 Cooling Generating Systems

D3030 110	Chilled Water, Air Cooled Condenser Systems	COST PER S.F.		
		MAT.	INST.	TOTAL
2280	20,000 S.F., 58.33 ton	5.15	4.18	9.33
2320	40,000 S.F., 116.66 ton	6.10	4.21	10.31
2400	Drug stores, 3,000 S.F., 20.00 ton	15.65	12.55	28.20
2440	6000 S.F., 40.00 ton	12.25	9.15	21.40
2480	10,000 S.F., 66.66 ton	12.75	7.85	20.60
2520	20,000 S.F., 133.33 ton	11	5.85	16.85
2560	40,000 S.F., 267 ton*			
2640	Factories, 2,000 S.F., 10.00 ton	9.90	10.45	20.35
2680	6,000 S.F., 20.00 ton	8.80	8.40	17.20
2720	10,000 S.F., 33.33 ton	6.75	5.85	12.60
2760	20,000 S.F., 66.66 ton	7.35	5.25	12.60
2800	40,000 S.F., 133.33 ton	6.60	4.31	10.91
2880	Food supermarkets, 3,000 S.F., 8.50 ton	10.50	10.35	20.85
2920	6,000 S.F., 17.00 ton	7.70	7.95	15.65
2960	10,000 S.F., 28.33 ton	5.95	5.55	11.50
3000	20,000 S.F., 56.66 ton	4.93	4.10	9.03
3040	40,000 S.F., 113.33 ton	5.95	4.22	10.17
3120	Medical centers, 3,000 S.F., 7.00 ton	9.55	10.10	19.65
3160	6,000 S.F., 14.00 ton	6.95	7.70	14.65
3200	10,000 S.F., 23.33 ton	6.10	5.85	11.95
3240	20,000 S.F., 46.66 ton	4.78	4.28	9.06
3280	40,000 S.F., 93.33 ton	5.50	4.15	9.65
3360	Offices, 3,000 S.F., 9.50 ton	9.55	10.30	19.85
3400	6,000 S.F., 19.00 ton	8.60	8.30	16.90
3440	10,000 S.F., 31.66 ton	6.55	5.80	12.35
3480	20,000 S.F., 63.33 ton	7.40	5.35	12.75
3520	40,000 S.F., 126.66 ton	6.55	4.35	10.90
3600	Restaurants, 3,000 S.F., 15.00 ton	12.40	11.25	23.65
3640	6,000 S.F., 30.00 ton	10.10	8.55	18.65
3680	10,000 S.F., 50.00 ton	8.65	6.20	14.85
3720	20,000 S.F., 100.00 ton	9.65	5.85	15.50
3760	40,000 S.F., 200 ton*			
3840	Schools and colleges, 3,000 S.F., 11.50 ton	10.65	10.70	21.35
3880	6,000 S.F., 23.00 ton	9.55	8.65	18.20
3920	10,000 S.F., 38.33 ton	7.35	6	13.35
3960	20,000 S.F., 76.66 ton	8.25	5.55	13.80
4000	40,000 S.F., 153 ton*			

D3030 Cooling Generating Systems

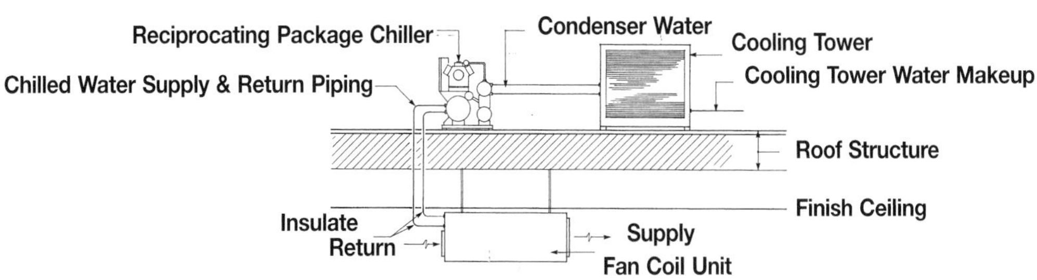

General: Water cooled chillers are available in the same sizes as air cooled units. They are also available in larger capacities.

Design Assumptions: The chilled water systems with water cooled condenser include reciprocating hermetic compressors, water cooling tower, pumps, piping and expansion tanks and are based on a two pipe system. Chilled water piping is insulated. No ducts are included and fan-coil units are cooling only. Area distribution is through use of multiple fan coil units. Fewer but larger fan coil units with duct distribution would be approximately the same S.F. cost. Water treatment and balancing are not included.

System Components	QUANTITY	UNIT	COST EACH MAT.	COST EACH INST.	COST EACH TOTAL
SYSTEM D3030 115 1320					
PACKAGED CHILLER, WATER COOLED, WITH FAN COIL UNIT					
APARTMENT CORRIDORS, 4,000 S.F., 7.33 TON					
Fan coil air conditioner unit, cabinet mounted & filters, chilled water	2.000	Ea.	5,313.53	842.84	6,156.37
Water chiller, water cooled, 1 compressor, hermetic scroll,	1.000	Ea.	5,908.20	4,122	10,030.20
Cooling tower, draw thru single flow, belt drive	1.000	Ea.	1,597.94	175.92	1,773.86
Cooling tower pumps & piping	1.000	System	798.97	414.15	1,213.12
Chilled water unit coil connections	2.000	Ea.	3,100	3,450	6,550
Chilled water distribution piping	520.000	L.F.	13,000	27,560	40,560
TOTAL			29,718.64	36,564.91	66,283.55
COST PER S.F.			7.43	9.14	16.57
*Cooling requirements would lead to choosing a water cooled unit.					

D3030 115	Chilled Water, Cooling Tower Systems	COST PER S.F. MAT.	COST PER S.F. INST.	COST PER S.F. TOTAL
1300	Packaged chiller, water cooled, with fan coil unit			
1320	Apartment corridors, 4,000 S.F., 7.33 ton	7.43	9.15	16.58
1360	6,000 S.F., 11.00 ton (R236000-20)	6.05	7.80	13.85
1400	10,000 S.F., 18.33 ton	6	5.85	11.85
1440	20,000 S.F., 26.66 ton	4.56	4.23	8.79
1480	40,000 S.F., 73.33 ton	5.55	4.23	9.78
1520	60,000 S.F., 110.00 ton	5.40	4.49	9.89
1600	Banks and libraries, 4,000 S.F., 16.66 ton	12.90	9.90	22.80
1640	6,000 S.F., 25.00 ton	10.55	8.70	19.25
1680	10,000 S.F., 41.66 ton	8.45	6.45	14.90
1720	20,000 S.F., 83.33 ton	9.45	5.80	15.25
1760	40,000 S.F., 166.66 ton	8.80	7.45	16.25
1800	60,000 S.F., 250.00 ton	9.05	7.90	16.95
1880	Bars and taverns, 4,000 S.F., 44.33 ton	19.70	12.30	32
1920	6,000 S.F., 66.50 ton	21	12.85	33.85
1960	10,000 S.F., 110.83 ton	19	9.80	28.80
2000	20,000 S.F., 221.66 ton	20	10.15	30.15
2040	40,000 S.F., 440 ton*			
2080	60,000 S.F., 660 ton*			
2160	Bowling alleys, 4,000 S.F., 22.66 ton	14.75	11.10	25.85
2200	6,000 S.F., 34.00 ton	12.40	9.35	21.75
2240	10,000 S.F., 56.66 ton	10.30	6.95	17.25
2280	20,000 S.F., 113.33 ton	10.70	6.35	17.05
2320	40,000 S.F., 226.66 ton	11.50	7.40	18.90

D3030 Cooling Generating Systems

D3030 115	Chilled Water, Cooling Tower Systems	COST PER S.F.		
		MAT.	INST.	TOTAL
2360	60,000 S.F., 340 ton			
2440	Department stores, 4,000 S.F., 11.66 ton	8.10	9.80	17.90
2480	6,000 S.F., 17.50 ton	9.55	8.10	17.65
2520	10,000 S.F., 29.17 ton	6.85	5.90	12.75
2560	20,000 S.F., 58.33 ton	5.40	4.46	9.86
2600	40,000 S.F., 116.66 ton	6.50	4.54	11.04
2640	60,000 S.F., 175.00 ton	7.85	7.25	15.10
2720	Drug stores, 4,000 S.F., 26.66 ton	15.50	11.15	26.65
2760	6,000 S.F., 40.00 ton	13.35	9.50	22.85
2800	10,000 S.F., 66.66 ton	13.65	8.70	22.35
2840	20,000 S.F., 133.33 ton	11.35	6.75	18.10
2880	40,000 S.F., 266.67 ton	12.05	8.40	20.45
2920	60,000 S.F., 400 ton*			
3000	Factories, 4,000 S.F., 13.33 ton	11.20	9.45	20.65
3040	6,000 S.F., 20.00 ton	9.45	8.30	17.75
3080	10,000 S.F., 33.33 ton	7.40	6.05	13.45
3120	20,000 S.F., 66.66 ton	7.65	5.60	13.25
3160	40,000 S.F., 133.33 ton	6.75	4.77	11.52
3200	60,000 S.F., 200.00 ton	8.20	7.60	15.80
3280	Food supermarkets, 4,000 S.F., 11.33 ton	8	9.75	17.75
3320	6,000 S.F., 17.00 ton	8.45	8.10	16.55
3360	10,000 S.F., 28.33 ton	6.70	5.85	12.55
3400	20,000 S.F., 56.66 ton	5.50	4.48	9.98
3440	40,000 S.F., 113.33 ton	6.45	4.57	11.02
3480	60,000 S.F., 170.00 ton	7.80	7.25	15.05
3560	Medical centers, 4.000 S.F., 9.33 ton	6.95	8.95	15.90
3600	6,000 S.F., 14.00 ton	8.35	7.80	16.15
3640	10,000 S.F., 23.33 ton	6.40	5.80	12.20
3680	20,000 S.F., 46.66 ton	5	4.38	9.38
3720	40,000 S.F., 93.33 ton	6.05	4.45	10.50
3760	60,000 S.F., 140.00 ton	6.80	7.35	14.15
3840	Offices, 4,000 S.F., 12.66 ton	10.85	9.40	20.25
3880	6,000 S.F., 19.00 ton	9.45	8.50	17.95
3920	10,000 S.F., 31.66 ton	7.40	6.15	13.55
3960	20,000 S.F., 63.33 ton	7.60	5.60	13.20
4000	40,000 S.F., 126.66 ton	7.75	7.05	14.80
4040	60,000 S.F., 190.00 ton	7.95	7.55	15.50
4120	Restaurants, 4,000 S.F., 20.00 ton	13.20	10.30	23.50
4160	6,000 S.F., 30.00 ton	11.20	8.75	19.95
4200	10,000 S.F., 50.00 ton	9.50	6.70	16.20
4240	20,000 S.F., 100.00 ton	10.55	6.25	16.80
4280	40,000 S.F., 200.00 ton	9.20	7.15	16.35
4320	60,000 S.F., 300.00 ton	10.05	8.20	18.25
4400	Schools and colleges, 4,000 S.F., 15.33 ton	12.25	9.75	22
4440	6,000 S.F., 23.00 ton	10.05	8.55	18.60
4480	10,000 S.F., 38.33 ton	8	6.30	14.30
4520	20,000 S.F., 76.66 ton	9.15	5.75	14.90
4560	40,000 S.F., 153.33 ton	8.35	7.30	15.65
4600	60,000 S.F., 230.00 ton	8.35	7.65	16

D3030 Cooling Generating Systems

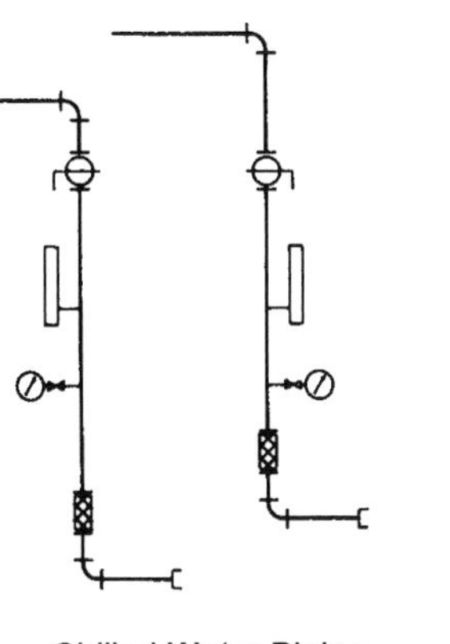
Chilled Water Piping

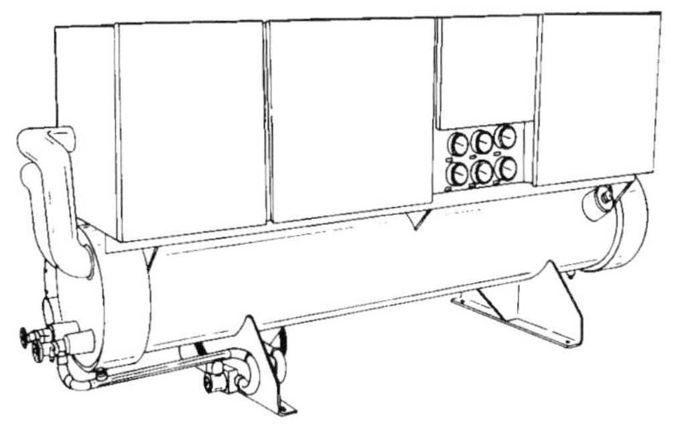
Reciprocating Chiller, Water Cooled

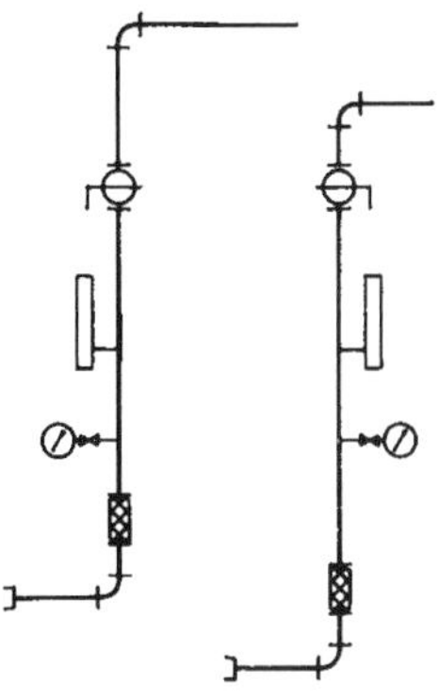
Condenser Water Piping

System Components	QUANTITY	UNIT	COST EACH MAT.	COST EACH INST.	COST EACH TOTAL
SYSTEM D3030 130 1010					
CHILLER, RECIPROCATING, WATER COOLED, STD. CONTROLS, 60 TON					
Water chiller, water cooled, 60 ton cooling	1.000	Ea.	36,500	11,700	48,200
Pipe, black steel, welded, Sch. 40, on yoke & roll hgrs, 10′ O.C., 4″ dia	40.000	L.F.	748	1,546.80	2,294.80
Elbow, 90°, weld joint, butt, 4″ pipe size	8.000	Ea.	492	2,282.80	2,774.80
Flange, welding neck, 150 lb., 4″ pipe size	16.000	Ea.	736	2,275.20	3,011.20
Valves, iron body, butterfly, lug type, lever actuator, 4″ size	4.000	Ea.	1,044	1,092	2,136
Expansion joints, bellows type, flange spool, 6″ F to F, 4″ dia	4.000	Ea.	1,460	656	2,116
T-O-L socket, weld joint, 1/4″ pipe size, nozzle	4.000	Ea.	34.40	248.76	283.16
T-O-L socket, weld joint, 3/4″ pipe size, nozzle	4.000	Ea.	39.60	271.76	311.36
Thermometers, stem type, 9″ case, 8″ stem, 3/4″ NPT	4.000	Ea.	784	110	894
Gauges, pressure or vacuum, 3-1/2″ diameter dial	4.000	Ea.	82	96	178
Insulation, fiberglass pipe covering, 1-1/2″ wall, 4″ IPS	20.000	L.F.	78.80	182	260.80
Chiller balancing	1.000	Ea.		655	655
TOTAL			41,998.80	21,116.32	63,115.12

D3030 130	Reciprocating Chiller, Water Cooled	COST EACH MAT.	COST EACH INST.	COST EACH TOTAL
1010	Chiller, reciprocating, water cooled, std. controls, 60 Ton	42,000	21,100	63,100
1020	100 Ton	75,000	26,800	101,800
1030	150 Ton	85,500	34,700	120,200
1040	200 Ton	115,000	36,600	151,600

D3030 135	Reciprocating Chiller, Air Cooled	COST EACH MAT.	COST EACH INST.	COST EACH TOTAL
1010	Chiller, reciprocating, air cooled, std. controls, 20 Ton	26,800	12,300	39,100
1020	30 Ton	33,200	13,000	46,200
1030	40 Ton	38,900	13,900	52,800

D3030 Cooling Generating Systems

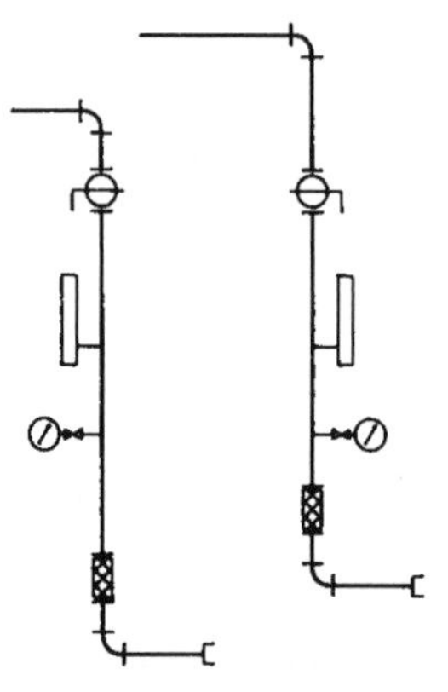
Chilled Water Piping

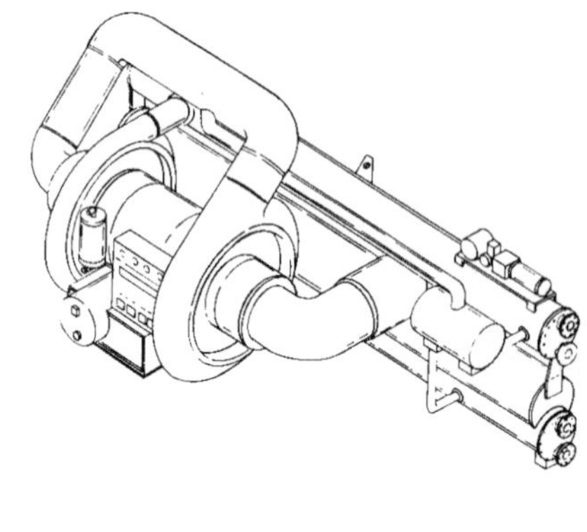
Centrifugal Chiller, Water Cooled, Hermetic

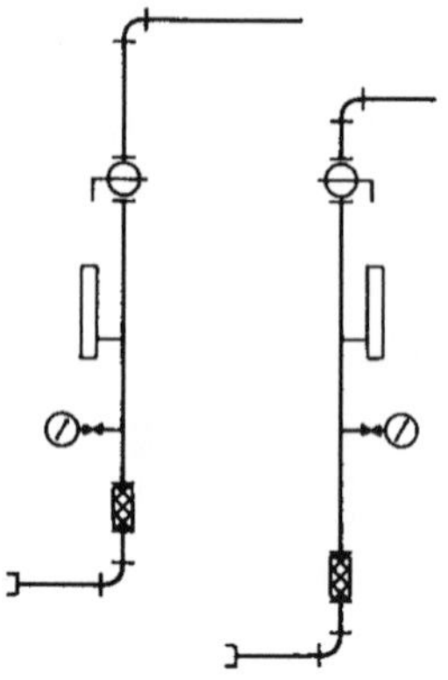
Condenser Water Piping

System Components	QUANTITY	UNIT	COST EACH MAT.	COST EACH INST.	COST EACH TOTAL
SYSTEM D3030 140 1010					
CHILLER, CENTRIF., WATER COOLED, PKGD. HERMETIC, STD. CONTROLS, 200 TON					
Water chiller, cntfgl, water cooled, not incl. tower, 200 ton	1.000	Ea.	145,500	15,700	161,200
Pipe, black steel, welded, Sch. 40, on yoke & roll hgrs, 10′ OC, 6″ dia	40.000	L.F.	1,880	2,428.40	4,308.40
Elbow, 90°, weld joint, 6″ pipe size	8.000	Ea.	1,088	3,498.80	4,586.80
Flange, welding neck, 150 lb., 6″ pipe size	16.000	Ea.	1,200	3,490.40	4,690.40
Valves, iron body, butterfly, lug type, lever actuator, 6″ size	4.000	Ea.	1,720	1,700	3,420
Expansion joints, bellows type, flange spool, 6″ F to F, 6″ dia	4.000	Ea.	1,840	812	2,652
T-O-L socket, weld jiont, 1/4″ pipe size, nozzle	4.000	Ea.	34.40	248.76	283.16
T-O-L socket, weld joint, 3/4″ pipe size, nozzle	4.000	Ea.	39.60	271.76	311.36
Thermometers, stem type, 9″ case, 8″ stem, 3/4″ NPT	4.000	Ea.	784	110	894
Gauges, pressure or vacuum, 3-1/2″ diameter dial	4.000	Ea.	82	96	178
Insulation, fiberglass pipe covering, 1-1/2″ wall, 6″ IPS	40.000	L.F.	168.80	464	632.80
Chiller balancing	1.000	Ea.		655	655
TOTAL			154,336.80	29,475.12	183,811.92

D3030 140	Centrifugal Chiller, Water Cooled	COST EACH MAT.	COST EACH INST.	COST EACH TOTAL
1010	Chiller, centrif., water cooled, pkgd. hermetic, std controls, 200 Ton	154,500	29,500	184,000
1020	400 Ton	187,000	44,600	231,600
1030	1,000 Ton	453,000	65,500	518,500
1040	1,500 Ton	609,500	81,500	691,000

D3030 Cooling Generating Systems

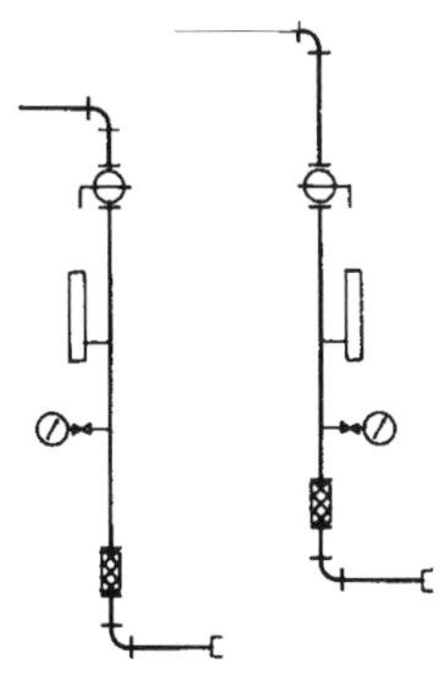

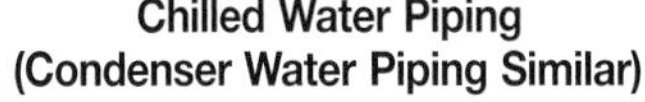

Chilled Water Piping
(Condenser Water Piping Similar)

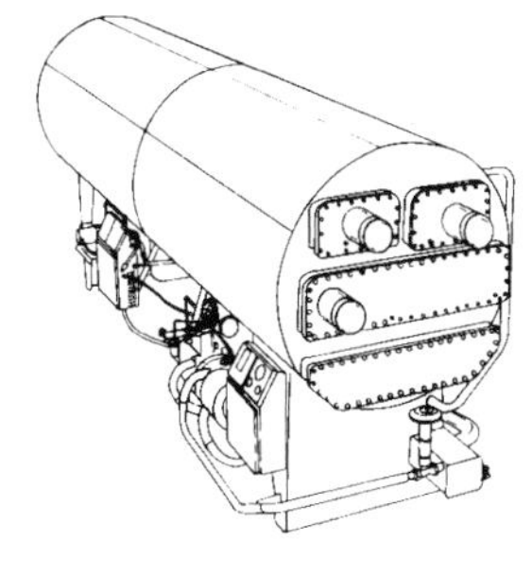

Gas Absorption Chiller, Water Cooled

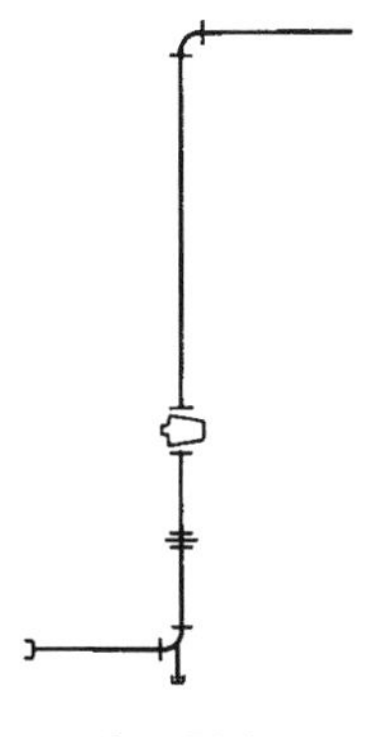

Gas Piping

System Components	QUANTITY	UNIT	COST EACH MAT.	COST EACH INST.	COST EACH TOTAL
SYSTEM D3030 145 1010					
CHILLER, GAS FIRED ABSORPTION, WATER COOLED, STD. CONTROLS, 800 TON					
Absorption water chiller, gas fired, water cooled, 800 ton	1.000	Ea.	661,500	34,800	696,300
Pipe, black steel, weld, Sch. 40, on yoke & roll hangers, 10′ O.C., 10″ dia	20.000	L.F.	1,930	1,821.40	3,751.40
Pipe, black steel, welded, Sch. 40, 14″ dia., (two rod roll type hanger)	44.000	L.F.	4,796	6,384.84	11,180.84
Pipe, black steel, weld, Sch. 40, on yoke & roll hangers, 10′ O.C., 4″ dia	20.000	L.F.	374	773.40	1,147.40
Elbow, 90°, weld joint, 10″ pipe size	4.000	Ea.	2,040	2,902	4,942
Elbow, 90°, weld joint, 14″ pipe size	6.000	Ea.	6,600	6,486	13,086
Elbow, 90°, weld joint, 4″ pipe size	3.000	Ea.	184.50	856.05	1,040.55
Flange, welding neck, 150 lb., 10″ pipe size	8.000	Ea.	1,640	2,922.40	4,562.40
Flange, welding neck, 150 lb., 14″ pipe size	14.000	Ea.	6,650	6,771.80	13,421.80
Flange, welding neck, 150 lb., 4″ pipe size	3.000	Ea.	138	426.60	564.60
Valves, iron body, butterfly, lug type, gear operated, 10″ size	2.000	Ea.	2,450	1,060	3,510
Valves, iron body, butterfly, lug type, gear operated, 14″ size	5.000	Ea.	14,125	4,625	18,750
Valves, semi-steel, lubricated plug valve, flanged, 4″ pipe size	1.000	Ea.	625	455	1,080
Expansion joints, bellows type, flange spool, 6″ F to F, 10″ dia	2.000	Ea.	1,470	552	2,022
Expansion joints, bellows type, flange spool, 10″ F to F, 14″ dia	2.000	Ea.	2,550	730	3,280
T-O-L socket, weld joint, 1/4″ pipe size, nozzle	4.000	Ea.	34.40	248.76	283.16
T-O-L socket, weld joint, 3/4″ pipe size, nozzle	4.000	Ea.	39.60	271.76	311.36
Thermometers, stem type, 9″ case, 8″ stem, 3/4″ NPT	4.000	Ea.	784	110	894
Gauges, pressure or vacuum, 3-1/2″ diameter dial	4.000	Ea.	82	96	178
Insulation, fiberglass pipe covering, 2″ wall, 10″ IPS	20.000	L.F.	188	364	552
Vent chimney, all fuel, pressure tight, double wall, SS, 12″ dia	20.000	L.F.	1,760	610	2,370
Chiller balancing	1.000	Ea.		655	655
TOTAL			709,960.50	73,922.01	783,882.51

D3030 145	Gas Absorption Chiller, Water Cooled	COST EACH MAT.	COST EACH INST.	COST EACH TOTAL
1010	Chiller, gas fired absorption, water cooled, std. ctrls, 800 Ton	710,000	74,000	784,000
1020	1000 Ton	929,500	95,500	1,025,000

D3030 150	Steam Absorption Chiller, Water Cooled	COST EACH MAT.	COST EACH INST.	COST EACH TOTAL
1010	Chiller, steam absorption, water cooled, std. ctrls, 750 Ton	790,500	87,000	877,500
1020	955 Ton	921,000	105,000	1,026,000
1030	1465 Ton	1,334,000	120,000	1,454,000
1040	1660 Ton	1,604,500	136,000	1,740,500

D3030 Cooling Generating Systems

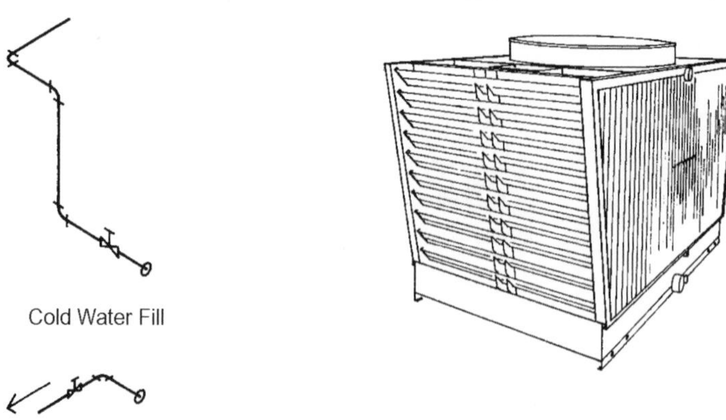

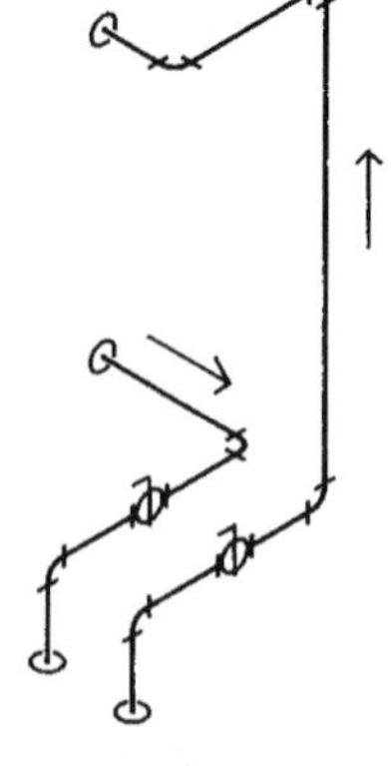

System Components	QUANTITY	UNIT	COST EACH MAT.	COST EACH INST.	COST EACH TOTAL
SYSTEM D3030 310 1010					
COOLING TOWER, GALVANIZED STEEL, PACKAGED UNIT, DRAW THRU, 60 TON					
Cooling towers, draw thru, single flow, belt drive, 60 tons	1.000	Ea.	13,080	1,440	14,520
Pipe, black steel, welded, Sch. 40, on yoke & roll hangers, 10′ OC, 4″ dia.	40.000	L.F.	748	1,546.80	2,294.80
Tubing, copper, Type L, couplings & hangers 10′ OC, 1-1/2″ dia.	20.000	L.F.	230	292	522
Elbow, 90°, copper, cu x cu, 1-1/2″ dia.	3.000	Ea.	50.70	175.50	226.20
Elbow, 90°, weld joint, 4″ pipe size	6.000	Ea.	369	1,712.10	2,081.10
Flange, welding neck, 150 lb., 4″ pipe size	6.000	Ea.	276	853.20	1,129.20
Valves, bronze, gate, N.R.S., soldered, 125 psi, 1-1/2″ size	1.000	Ea.	179	58.50	237.50
Valves, iron body, butterfly, lug type, lever actuator, 4″ size	2.000	Ea.	522	546	1,068
Electric heat trace system, 400 degree, 115v, 5 watts/L.F.	160.000	L.F.	1,992	216	2,208
Insulation, fiberglass pipe covering, , 1-1/2″ wall, 4″ IPS	40.000	L.F.	157.60	364	521.60
Insulation, fiberglass pipe covering, 1″ wall, 1-1/2″ IPS	20.000	L.F.	26.40	121	147.40
Cooling tower condenser control system	1.000	Ea.	6,525	3,125	9,650
Cooling tower balancing	1.000	Ea.		500	500
TOTAL			24,155.70	10,950.10	35,105.80

D3030 310	Galvanized Draw Through Cooling Tower	COST EACH MAT.	COST EACH INST.	COST EACH TOTAL
1010	Cooling tower, galvanized steel, packaged unit, draw thru, 60 Ton	24,200	11,000	35,200
1020	110 Ton	32,700	12,900	45,600
1030	300 Ton	47,900	19,400	67,300
1040	600 Ton	59,000	29,100	88,100
1050	1000 Ton	118,500	42,900	161,400

D3030 320	Fiberglass Draw Through Cooling Tower	COST EACH MAT.	COST EACH INST.	COST EACH TOTAL
1010	Cooling tower, fiberglass, packaged unit, draw thru, 100 Ton	26,900	11,100	38,000
1020	120 Ton	31,000	12,500	43,500
1030	140 Ton	36,000	16,600	52,600
1040	160 Ton	45,500	22,000	67,500
1050	180 Ton	63,500	31,800	95,300

D3030 330	Stainless Steel Draw Through Cooling Tower	COST EACH MAT.	COST EACH INST.	COST EACH TOTAL
1010	Cooling tower, stainless steel, packaged unit, draw thru, 60 Ton	44,700	11,000	55,700
1020	110 Ton	54,000	12,900	66,900
1030	300 Ton	87,500	19,400	106,900
1040	600 Ton	138,500	28,700	167,200
1050	1000 Ton	236,500	42,800	279,300

D3040 Distribution Systems

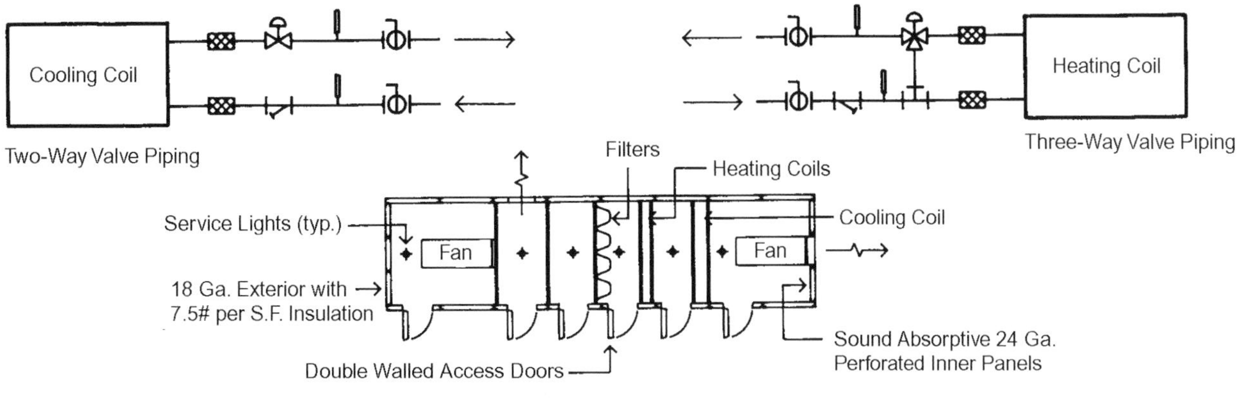

Built-Up Air Handling Unit

System Components	QUANTITY	UNIT	COST EACH MAT.	COST EACH INST.	COST EACH TOTAL
SYSTEM D3040 106 1010					
AHU, FIELD FAB., BLT. UP, COOL/HEAT COILS, FLTRS., CONST. VOL., 40,000 CFM					
AHU, Built-up, cool/heat coils, filters, mix box, constant volume	1.000	Ea.	110,000	7,175	117,175
Pipe, black steel, welded, Sch. 40, on yoke & roll hgrs, 10′ OC, 3″ dia.	26.000	L.F.	473.20	856.44	1,329.64
Pipe, black steel, welded, Sch. 40, on yoke & roll hgrs, 10′ OC, 4″ dia	20.000	L.F.	374	773.40	1,147.40
Elbow, 90°, weld joint, 3″ pipe size	5.000	Ea.	165	1,019	1,184
Elbow, 90°, weld joint, 4″ pipe size	4.000	Ea.	246	1,141.40	1,387.40
Tee, welded, reducing on outlet, 3″ by 2-1/2″ pipe	3.000	Ea.	489	1,066.35	1,555.35
Tee, welded, reducing on outlet, 4″ by 3″ pipe	2.000	Ea.	264	951	1,215
Reducer, eccentric, weld joint, 3″ pipe size	4.000	Ea.	320	714.80	1,034.80
Reducer, eccentric, weld joint, 4″ pipe size	4.000	Ea.	644	949.20	1,593.20
Flange, weld neck, 150 LB, 2-1/2″ pipe size	3.000	Ea.	121.50	268.08	389.58
Flange, weld neck, 150 lb., 3″ pipe size	14.000	Ea.	623	1,426.74	2,049.74
Flange, weld neck, 150 lb., 4″ pipe size	10.000	Ea.	460	1,422	1,882
Valves, iron body, butterfly, lug type, lever actuator, 3″ size	2.000	Ea.	466	342	808
Valves, iron body, butterfly, lug type, lever actuator, 4″ size	2.000	Ea.	522	546	1,068
Strainer, Y type, iron body, flanged, 125 lb., 3″ pipe size	1.000	Ea.	200	305	505
Strainer, Y type, iron body, flanged, 125 lb., 4″ pipe size	1.000	Ea.	345	460	805
Valve, electric motor actuated, iron body, 3 way, flanged, 2-1/2″ pipe	1.000	Ea.	1,275	370	1,645
Valve, electric motor actuated, iron body, two way, flanged, 3″ pipe size	1.000	Ea.	1,300	370	1,670
Expansion joints, bellows type, flange spool, 6″ F to F, 3″ dia	2.000	Ea.	680	242	922
Expansion joints, bellows type, flange spool, 6″ F to F, 4″ dia	2.000	Ea.	730	328	1,058
T-O-L socket, weld joint, 1/4″ pipe size, nozzle	4.000	Ea.	34.40	248.76	283.16
T-O-L socket, weld joint, 1/2″ pipe size, nozzle	12.000	Ea.	103.20	777.72	880.92
Thermometers, stem type, 9″ case, 8″ stem, 3/4″ NPT	4.000	Ea.	784	110	894
Pressure/temperature relief plug, 316 SS, 3/4″OD, 7-1/2″ insertion	4.000	Ea.	224	96	320
Insulation, fiberglass pipe covering, 1-1/2″ wall, 3″ IPS	26.000	L.F.	81.64	195	276.64
Insulation, fiberglass pipe covering, 1-1/2″ wall, 4″ IPS	20.000	L.F.	78.80	182	260.80
Fan coil control system	1.000	Ea.	1,450	700	2,150
Re-heat coil balancing	2.000	Ea.		284	284
Air conditioner balancing	1.000	Ea.		890	890
TOTAL			122,453.74	24,209.89	146,663.63

D3040 106	Field Fabricated Air Handling Unit	COST EACH MAT.	COST EACH INST.	COST EACH TOTAL
1010	AHU, Field fab, blt up, cool/heat coils, fltrs, const vol, 40,000 CFM	122,500	24,200	146,700
1020	60,000 CFM	129,000	37,400	166,400
1030	75,000 CFM	130,500	39,500	170,000

D3040 Distribution Systems

D3040 108	Field Fabricated VAV Air Handling Unit	COST EACH		
		MAT.	INST.	TOTAL
1010	AHU, Field fab, blt up, cool/heat coils, fltrs, VAV, 75,000 CFM	132,500	45,400	177,900
1020	100,000 CFM	210,500	60,000	270,500
1030	150,000 CFM	342,500	96,500	439,000

D3040 Distribution Systems

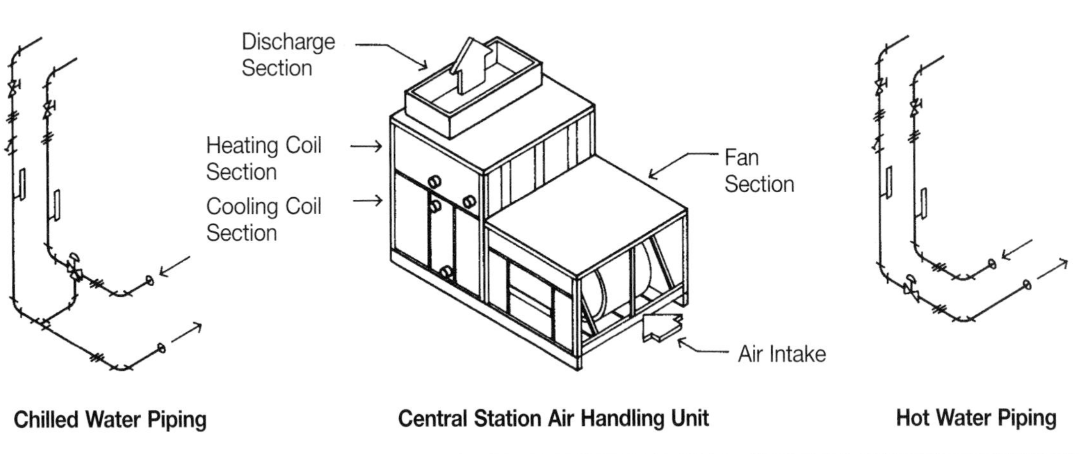

Chilled Water Piping | Central Station Air Handling Unit | Hot Water Piping

System Components	QUANTITY	UNIT	COST EACH MAT.	COST EACH INST.	COST EACH TOTAL
SYSTEM D3040 110 1010					
AHU, CENTRAL STATION, COOL/HEAT COILS, CONST. VOL., FLTRS., 2,000 CFM					
Central station air handling unit, chilled water, 2,000 CFM	1.000	Ea.	12,000	1,250	13,250
Pipe, black steel, Sch 40, threaded, W/cplgs & hangers, 10′ OC, 1″ dia.	26.000	L.F.	158.28	464.75	623.03
Pipe, black steel, Sch 40, threaded, W/cplg & hangers, 10′ OC, 1-1/4″ dia.	20.000	L.F.	148.20	399.10	547.30
Elbow, 90°, black, straight, 1″ dia.	4.000	Ea.	34.20	234	268.20
Elbow, 90°, black, straight, 1-1/4″ dia.	4.000	Ea.	56.80	248	304.80
Tee, black, reducing, 1″ dia.	8.000	Ea.	176	760	936
Tee, black, reducing, 1-1/4″ dia.	8.000	Ea.	312	780	1,092
Reducer, concentric, black, 1″ dia.	4.000	Ea.	63.80	202	265.80
Reducer, concentric, black, 1-1/4″ dia.	4.000	Ea.	71.60	210	281.60
Union, black with brass seat, 1″ dia.	4.000	Ea.	118	252	370
Union, black with brass seat, 1-1/4″ dia.	2.000	Ea.	85	130	215
Pipe nipple, black, 1″ dia	13.000	Ea.	31.66	92.95	124.61
Pipe nipple, black, 1-1/4″ dia	12.000	Ea.	34.20	92.10	126.30
Valves, bronze, gate, N.R.S., threaded, class 150, 1″ size	2.000	Ea.	246	80	326
Valves, bronze, gate, N.R.S., threaded, class 150, 1-1/4″ size	2.000	Ea.	332	101	433
Strainer, Y type, bronze body, screwed, 150 lb., 1″ pipe size	1.000	Ea.	68	45	113
Strainer, Y type, bronze body, screwed, 150 lb., 1-1/4″ pipe size	1.000	Ea.	138	51	189
Valve, electric motor actuated, brass, 3 way, screwed, 3/4″ pipe	1.000	Ea.	385	41.50	426.50
Valve, electric motor actuated, brass, 2 way, screwed, 1″ pipe size	1.000	Ea.	805	40	845
Thermometers, stem type, 9″ case, 8″ stem, 3/4″ NPT	4.000	Ea.	784	110	894
Pressure/temperature relief plug, 316 SS, 3/4″OD, 7-1/2″ insertion	4.000	Ea.	224	96	320
Insulation, fiberglass pipe covering, 1″ wall, 1″ IPS	26.000	L.F.	29.38	150.80	180.18
Insulation, fiberglass pipe covering, 1″ wall, 1-1/4″ IPS	20.000	L.F.	24.40	121	145.40
Fan coil control system	1.000	Ea.	1,450	700	2,150
Re-heat coil balancing	2.000	Ea.		284	284
Multizone A/C balancing	1.000	Ea.		665	665
TOTAL			17,775.52	7,600.20	25,375.72

D3040 110	Central Station Air Handling Unit	COST EACH MAT.	COST EACH INST.	COST EACH TOTAL
1010	AHU, Central station, cool/heat coils, constant vol, fltrs, 2,000 CFM	17,800	7,600	25,400
1020	5,000 CFM	26,900	9,775	36,675
1030	10,000 CFM	39,000	13,700	52,700
1040	15,000 CFM	71,500	18,400	89,900
1050	20,000 CFM	87,000	21,500	108,500

D3040 Distribution Systems

D3040 112	Central Station Air Handling Unit, VAV	COST EACH		
		MAT.	INST.	TOTAL
1010	AHU, Central station, cool/heat coils, VAV, fltrs, 5,000 CFM	31,100	10,200	41,300
1020	10,000 CFM	56,500	14,000	70,500
1030	15,000 CFM	80,000	18,200	98,200
1040	20,000 CFM	99,000	21,900	120,900
1050	30,000 CFM	116,500	25,200	141,700

D3040 Distribution Systems

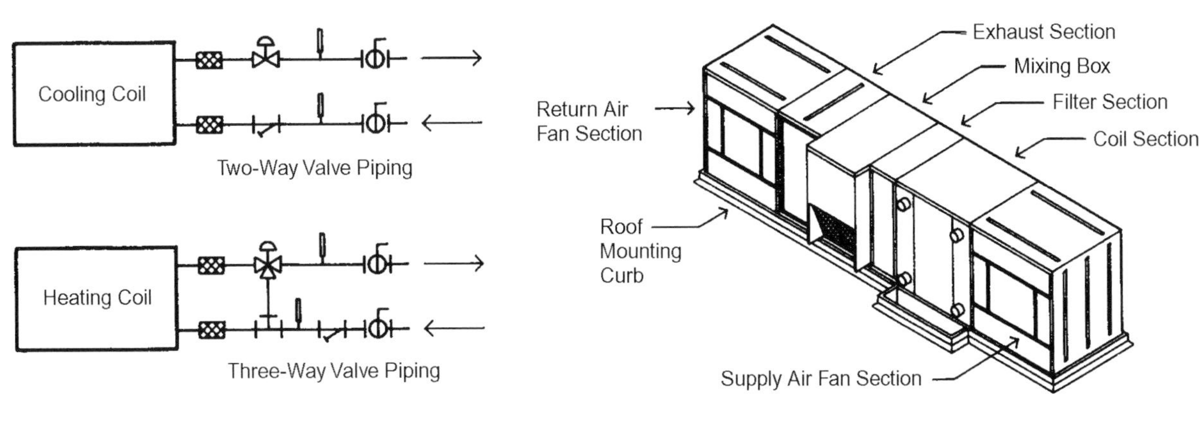

Rooftop Air Handling Unit

System Components	QUANTITY	UNIT	COST EACH MAT.	COST EACH INST.	COST EACH TOTAL
SYSTEM D3040 114 1010					
AHU, ROOFTOP, COOL/HEAT COILS, FLTRS., CONST. VOL., 2,000 CFM					
AHU, Built-up, cool/heat coils, filter, mix box, rooftop, constant volume	1.000	Ea.	26,200	2,150	28,350
Pipe, black steel, Sch 40, threaded, W/cplgs & hangers, 10′ OC, 1″ dia.	26.000	L.F.	158.28	464.75	623.03
Pipe, black steel, Sch 40, threaded, W/cplg & hangers, 10′ OC, 1-1/4″ dia.	20.000	L.F.	148.20	399.10	547.30
Elbow, 90°, black, straight, 1″ dia.	4.000	Ea.	34.20	234	268.20
Elbow, 90°, black, straight, 1-1/4″ dia.	4.000	Ea.	56.80	248	304.80
Tee, black, reducing, 1″ dia.	8.000	Ea.	176	760	936
Tee, black, reducing, 1-1/4″ dia.	8.000	Ea.	312	780	1,092
Reducer, black, concentric, 1″ dia.	4.000	Ea.	63.80	202	265.80
Reducer, black, concentric, 1-1/4″ dia.	4.000	Ea.	71.60	210	281.60
Union, black with brass seat, 1″ dia.	4.000	Ea.	118	252	370
Union, black with brass seat, 1-1/4″ dia.	2.000	Ea.	85	130	215
Pipe nipples, 1″ dia	13.000	Ea.	31.66	92.95	124.61
Pipe nipples, 1-1/4″ dia	12.000	Ea.	34.20	92.10	126.30
Valves, bronze, gate, N.R.S., threaded, class 150, 1″ size	2.000	Ea.	246	80	326
Valves, bronze, gate, N.R.S., threaded, class 150, 1-1/4″ size	2.000	Ea.	332	101	433
Strainer, Y type, bronze body, screwed, 150 lb., 1″ pipe size	1.000	Ea.	68	45	113
Strainer, Y type, bronze body, screwed, 150 lb., 1-1/4″ pipe size	1.000	Ea.	138	51	189
Valve, electric motor actuated, brass, 3 way, screwed, 3/4″ pipe	1.000	Ea.	385	41.50	426.50
Valve, electric motor actuated, brass, 2 way, screwed, 1″ pipe size	1.000	Ea.	805	40	845
Thermometers, stem type, 9″ case, 8″ stem, 3/4″ NPT	4.000	Ea.	784	110	894
Pressure/temperature relief plug, 316 SS, 3/4″OD, 7-1/2″ insertion	4.000	Ea.	224	96	320
Insulation, fiberglass pipe covering, 1″ wall, 1″ IPS	26.000	L.F.	29.38	150.80	180.18
Insulation, fiberglass pipe covering, 1″ wall, 1-1/4″ IPS	20.000	L.F.	24.40	121	145.40
Insulation pipe covering .016″ aluminum jacket finish, add	45.000	S.F.	52.65	288	340.65
Fan coil control system	1.000	Ea.	1,450	700	2,150
Re-heat coil balancing	2.000	Ea.		284	284
Rooftop unit heat/cool balancing	1.000	Ea.		520	520
TOTAL			32,028.17	8,643.20	40,671.37

D3040 114	Rooftop Air Handling Unit	COST EACH MAT.	COST EACH INST.	COST EACH TOTAL
1010	AHU, Rooftop, cool/heat coils, constant vol., fltrs, 2,000 CFM	32,000	8,650	40,650
1020	5,000 CFM	38,100	10,000	48,100

D3040 Distribution Systems

D3040 114	Rooftop Air Handling Unit	COST EACH		
		MAT.	INST.	TOTAL
1030	10,000 CFM	74,000	14,000	88,000
1040	15,000 CFM	106,000	18,400	124,400
1050	20,000 CFM	140,500	22,200	162,700

D3040 116	Rooftop Air Handling Unit, VAV	COST EACH		
		MAT.	INST.	TOTAL
1010	AHU, Rooftop, cool/heat coils, VAV, fltrs, 5,000 CFM	41,100	10,300	51,400
1020	10,000 CFM	81,500	14,200	95,700
1030	15,000 CFM	122,500	18,500	141,000
1040	20,000 CFM	155,500	22,400	177,900
1050	30,000 CFM	180,500	25,700	206,200

D3040 Distribution Systems

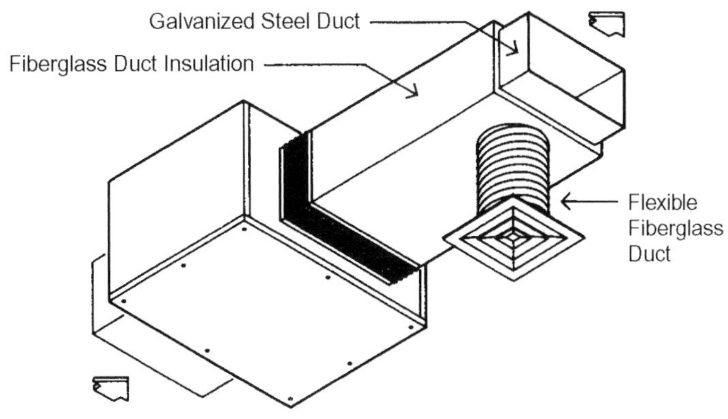

Horizontal Fan Coil Air Conditioning System

System Components	QUANTITY	UNIT	COST EACH MAT.	COST EACH INST.	COST EACH TOTAL
SYSTEM D3040 124 1010					
FAN COIL A/C SYSTEM, HORIZONTAL, W/HOUSING, CONTROLS, 2 PIPE, 1/2 TON					
Fan coil A/C, horizontal housing, filters, chilled water, 1/2 ton cooling	1.000	Ea.	1,525	173	1,698
Pipe, black steel, Sch 40, threaded, W/cplgs & hangers, 10′ OC, 3/4″ dia.	20.000	L.F.	100.62	267.68	368.30
Elbow, 90°, black, straight, 3/4″ dia.	6.000	Ea.	29.46	324	353.46
Tee, black, straight, 3/4″ dia.	2.000	Ea.	16.10	168	184.10
Union, black with brass seat, 3/4″ dia.	2.000	Ea.	46	117	163
Pipe nipples, 3/4″ diam	3.000	Ea.	7.02	18.68	25.70
Valves, bronze, gate, N.R.S., threaded, class 150, 3/4″ size	2.000	Ea.	196	76	272
Circuit setter, balance valve, bronze body, threaded, 3/4″ pipe size	1.000	Ea.	98.50	38.50	137
Insulation, fiberglass pipe covering, 1″ wall, 3/4″ IPS	20.000	L.F.	21.20	111	132.20
Ductwork, 12″ x 8″ fabricated, galvanized steel, 12 LF	55.000	Lb.	34.65	484	518.65
Insulation, ductwork, blanket type, fiberglass, 1″ thk, 1-1/2 LB density	40.000	S.F.	204	608	812
Diffusers, aluminum, OB damper, ceiling, perf, 24″x24″ panel size, 6″x6″	2.000	Ea.	362	93	455
Ductwork, flexible, fiberglass fabric, insulated, 1″thk, PE jacket, 6″ dia	16.000	L.F.	55.20	82.40	137.60
Round volume control damper 6″ dia.	2.000	Ea.	75	67	142
Fan coil unit balancing	1.000	Ea.		96	96
Re-heat coil balancing	1.000	Ea.		142	142
Diffuser/register, high, balancing	2.000	Ea.		266	266
TOTAL			2,770.75	3,132.26	5,903.01

D3040 118	Fan Coil A/C Unit, Two Pipe	COST EACH MAT.	COST EACH INST.	COST EACH TOTAL
1010	Fan coil A/C system, cabinet mounted, controls, 2 pipe, 1/2 Ton	1,150	1,425	2,575
1020	1 Ton	1,575	1,600	3,175
1030	1-1/2 Ton	1,625	1,625	3,250
1040	2 Ton	2,700	1,975	4,675
1050	3 Ton	3,525	2,050	5,575

D3040 120	Fan Coil A/C Unit, Two Pipe, Electric Heat	COST EACH MAT.	COST EACH INST.	COST EACH TOTAL
1010	Fan coil A/C system, cabinet mntd, elect. ht, controls, 2 pipe, 1/2 Ton	2,100	1,425	3,525
1020	1 Ton	2,500	1,600	4,100
1030	1-1/2 Ton	2,925	1,625	4,550
1040	2 Ton	4,375	2,000	6,375
1050	3 Ton	7,225	2,050	9,275

D3040 Distribution Systems

D3040 122	Fan Coil A/C Unit, Four Pipe	COST EACH		
		MAT.	INST.	TOTAL
1010	Fan coil A/C system, cabinet mounted, controls, 4 pipe, 1/2 Ton	1,950	2,775	4,725
1020	1 Ton	2,475	2,950	5,425
1030	1-1/2 Ton	2,550	2,975	5,525
1040	2 Ton	3,300	3,075	6,375
1050	3 Ton	4,575	3,325	7,900

D3040 124	Fan Coil A/C, Horizontal, Duct Mount, 2 Pipe	COST EACH		
		MAT.	INST.	TOTAL
1010	Fan coil A/C system, horizontal w/housing, controls, 2 pipe, 1/2 Ton	2,775	3,125	5,900
1020	1 Ton	3,700	5,000	8,700
1030	1-1/2 Ton	4,600	6,700	11,300
1040	2 Ton	6,025	8,325	14,350
1050	3 Ton	7,725	11,600	19,325
1060	3-1/2 Ton	8,025	11,800	19,825
1070	4 Ton	8,450	13,800	22,250
1080	5 Ton	10,300	17,800	28,100
1090	6 Ton	10,500	19,000	29,500
1100	7 Ton	10,600	19,400	30,000
1110	8 Ton	10,600	19,700	30,300
1120	10 Ton	11,300	20,200	31,500

D3040 126	Fan Coil A/C, Horiz., Duct Mount, 2 Pipe, Elec. Ht.	COST EACH		
		MAT.	INST.	TOTAL
1010	Fan coil A/C system, horiz. hsng, elect. ht, ctrls, 2 pipe, 1/2 Ton	3,025	3,125	6,150
1020	1 Ton	4,100	5,000	9,100
1030	1-1/2 Ton	4,975	6,700	11,675
1040	2 Ton	6,625	8,325	14,950
1050	3 Ton	10,400	11,800	22,200
1060	3-1/2 Ton	11,500	11,800	23,300
1070	4 Ton	13,000	13,800	26,800
1080	5 Ton	15,900	17,800	33,700
1090	6 Ton	16,100	19,000	35,100
1100	7 Ton	16,600	19,700	36,300
1110	8 Ton	16,900	20,300	37,200
1120	10 Ton	18,200	19,400	37,600

D3040 128	Fan Coil A/C, Horizontal, Duct Mount, 4 Pipe	COST EACH		
		MAT.	INST.	TOTAL
1010	Fan coil A/C system, horiz. w/cabinet, controls, 4 pipe, 1/2 Ton	3,900	4,475	8,375
1020	1 Ton	4,925	6,475	11,400
1030	1-1/2 Ton	5,925	8,050	13,975
1040	2 Ton	7,050	9,450	16,500
1050	3 Ton	9,075	13,100	22,175
1060	3.5 Ton	10,700	16,500	27,200
1070	4 Ton	10,700	16,700	27,400
1080	5 Ton	13,300	23,500	36,800
1090	6 Ton	14,900	28,400	43,300
1100	7 Ton	17,000	31,900	48,900
1110	8 Ton	17,500	34,200	51,700
1120	10 Ton	21,000	41,300	62,300

D3040 Distribution Systems

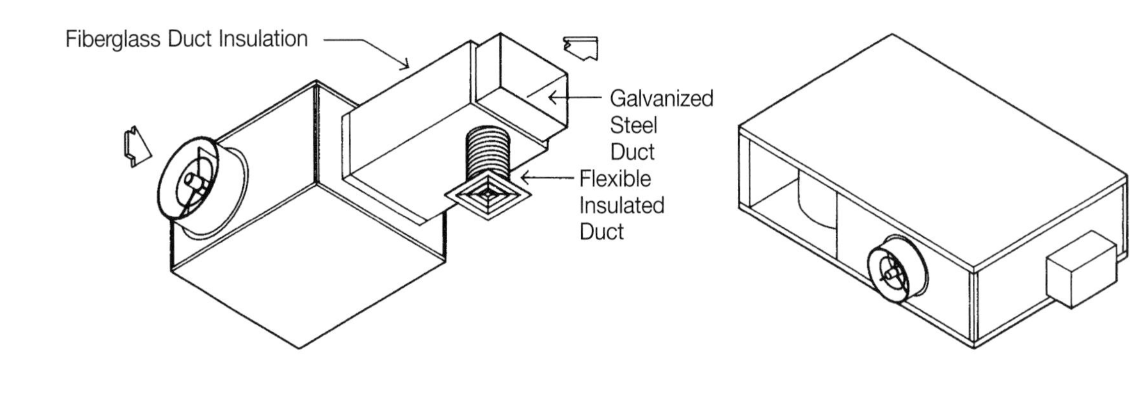

VAV Terminal

Fan Powered VAV Terminal

System Components	QUANTITY	UNIT	COST EACH MAT.	COST EACH INST.	COST EACH TOTAL
SYSTEM D3040 132 1010					
VAV TERMINAL, COOLING ONLY, WITH ACTUATOR/CONTROLS, 200 CFM					
Mixing box variable volume 300 to 600 CFM, cool only	1.000	Ea.	984.50	133.10	1,117.60
Ductwork, 12″ x 8″ fabricated, galvanized steel, 12 LF	55.000	Lb.	34.65	484	518.65
Insulation, ductwork, blanket type, fiberglass, 1″ thk, 1-1/2 LB density	40.000	S.F.	204	608	812
Diffusers, aluminum, OB damper, ceiling, perf, 24″x24″ panel size, 6″x6″	2.000	Ea.	362	93	455
Ductwork, flexible, fiberglass fabric, insulated, 1″thk, PE jacket, 6″ dia	16.000	L.F.	55.20	82.40	137.60
Round volume control damper 6″ dia.	2.000	Ea.	75	67	142
VAV box balancing	1.000	Ea.		89	89
Diffuser/register, high, balancing	2.000	Ea.		266	266
TOTAL			1,715.35	1,822.50	3,537.85

D3040 132	VAV Terminal, Cooling Only	COST EACH MAT.	COST EACH INST.	COST EACH TOTAL
1010	VAV Terminal, cooling only, with actuator/controls, 200 CFM	1,725	1,825	3,550
1020	400 CFM	2,250	3,500	5,750
1030	600 CFM	2,950	5,225	8,175
1040	800 CFM	3,250	6,600	9,850
1050	1000 CFM	3,725	7,600	11,325
1060	1250 CFM	4,475	9,925	14,400
1070	1500 CFM	4,950	11,900	16,850
1080	2000 CFM	6,500	16,800	23,300

D3040 134	VAV Terminal, Hot Water Reheat	COST EACH MAT.	COST EACH INST.	COST EACH TOTAL
1010	VAV Terminal, cooling, HW reheat, with actuator/controls, 200 CFM	3,300	3,000	6,300
1020	400 CFM	3,850	4,575	8,425
1030	600 CFM	4,525	6,400	10,925
1040	800 CFM	4,875	7,800	12,675
1050	1000 CFM	5,300	8,800	14,100
1060	1250 CFM	6,200	11,200	17,400
1070	1500 CFM	6,975	13,400	20,375
1080	2000 CFM	8,575	18,400	26,975

D3040 136	Fan Powered VAV Terminal, Cooling Only	COST EACH MAT.	COST EACH INST.	COST EACH TOTAL
1010	VAV Terminal, cooling, fan powrd, with actuator/controls, 200 CFM	2,525	1,825	4,350
1020	400 CFM	3,250	3,350	6,600

D3040 Distribution Systems

D3040 136	Fan Powered VAV Terminal, Cooling Only	COST EACH		
		MAT.	INST.	TOTAL
1030	600 CFM	3,900	5,225	9,125
1040	800 CFM	4,350	6,525	10,875
1050	1000 CFM	4,775	7,500	12,275
1060	1250 CFM	5,700	10,000	15,700
1070	1500 CFM	6,150	12,000	18,150
1080	2000 CFM	7,850	17,100	24,950

D3040 138	Fan Powered VAV Terminal, Hot Water Reheat	COST EACH		
		MAT.	INST.	TOTAL
1010	VAV Terminal, cool, HW reht, fan powrd, with actuator/ctrls, 200 CFM	4,250	3,000	7,250
1020	400 CFM	4,875	4,600	9,475
1030	600 CFM	5,550	6,400	11,950
1040	800 CFM	6,000	7,725	13,725
1050	1000 CFM	6,425	8,725	15,150
1060	1250 CFM	7,550	11,200	18,750
1070	1500 CFM	8,350	13,500	21,850
1080	2000 CFM	10,000	18,700	28,700

D3040 Distribution Systems

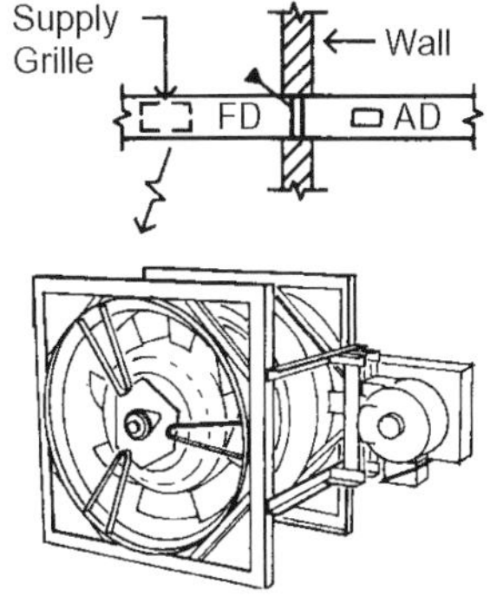

Axial Flow Centrifugal Fan Belt Drive

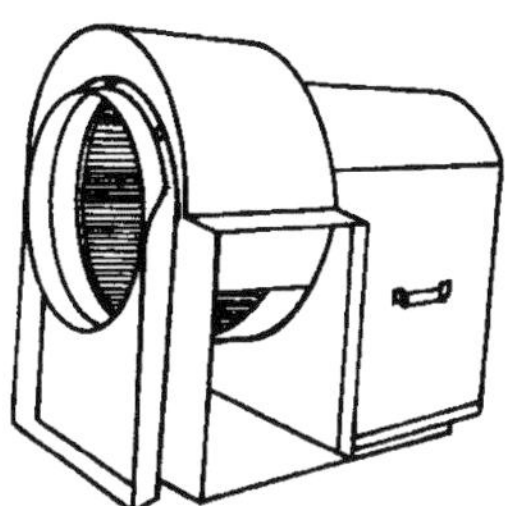

Belt Drive Utility Set

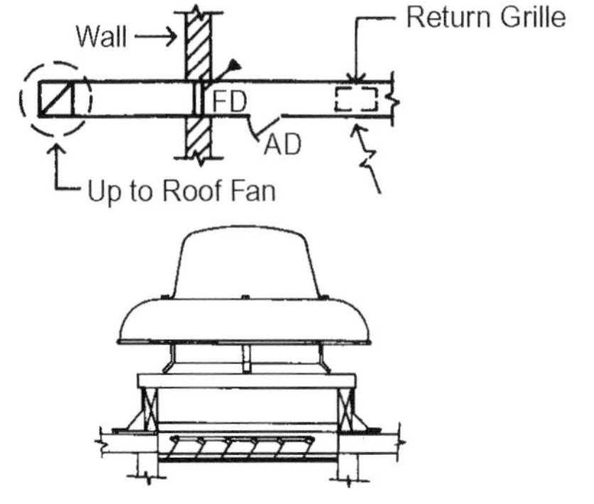

Centrifugal Roof Exhaust Fan Direct Drive

System Components	QUANTITY	UNIT	COST EACH MAT.	COST EACH INST.	COST EACH TOTAL
SYSTEM D3040 220 1010					
FAN SYSTEM, IN-LINE CENTRIFUGAL, 500 CFM					
Fans, in-line centrifugal, supply/exhaust, 500 CFM, 10" dia conn	1.000	Ea.	1,775	565	2,340
Louver, alum., w/screen, damper, mill fin, fxd blade, cont line, stormproof	4.050	S.F.	196.43	107.33	303.76
Motor for louver damper, electric or pneumatic	1.000	Ea.	560	53	613
Ductwork, 12" x 8" fabricated, galvanized steel, 60 LF	200.000	Lb.	166.40	2,197	2,363.40
Grilles, aluminum, air supply, single deflection, adjustable, 12" x 6"	1.000	Ea.	35.85	96	131.85
Fire damper, curtain type, vertical, 12" x 8"	1.000	Ea.	30.50	33.50	64
Duct access door, insulated, 10" x 10"	1.000	Ea.	24	67.50	91.50
Duct access door, insulated, 16" x 12"	1.000	Ea.	46	82.50	128.50
Pressure controller/switch, air flow controller	1.000	Ea.	199	64	263
Fan control switch and mounting base	1.000	Ea.	121	48	169
Fan balancing	1.000	Ea.		445	445
Diffuser/register balancing	3.000	Ea.		267	267
TOTAL			3,154.18	4,025.83	7,180.01

D3040 220	Centrifugal In Line Fan Systems	COST EACH MAT.	COST EACH INST.	COST EACH TOTAL
1010	Fan system, in-line centrifugal, 500 CFM	3,150	4,025	7,175
1020	1300 CFM	3,650	6,975	10,625
1030	1500 CFM	3,900	8,075	11,975
1040	2500 CFM	5,850	24,600	30,450
1050	3500 CFM	7,300	33,200	40,500
1060	5000 CFM	9,300	49,000	58,300
1070	7500 CFM	11,000	50,000	61,000
1080	10,000 CFM	13,100	58,500	71,600

D3040 230	Utility Set Fan Systems	COST EACH MAT.	COST EACH INST.	COST EACH TOTAL
1010	Utility fan set system, belt drive, 2000 CFM	2,825	13,000	15,825
1020	3500 CFM	4,675	23,100	27,775
1030	5000 CFM	5,825	32,000	37,825
1040	7500 CFM	8,275	37,200	45,475
1050	10,000 CFM	11,200	48,600	59,800
1060	15,000 CFM	11,600	49,600	61,200
1070	20,000 CFM	13,800	56,000	69,800

D3040 Distribution Systems

D3040 240	Roof Exhaust Fan Systems	COST EACH		
		MAT.	INST.	TOTAL
1010	Roof vent. system, centrifugal, alum., galv curb, BDD, 500 CFM	1,425	2,625	4,050
1020	800 CFM	1,825	5,100	6,925
1030	1500 CFM	2,600	7,575	10,175
1040	2750 CFM	3,725	17,000	20,725
1050	3500 CFM	4,500	22,000	26,500
1060	5000 CFM	6,175	36,400	42,575
1070	8500 CFM	8,200	46,500	54,700
1080	13,800 CFM	11,600	67,500	79,100

D3040 Distribution Systems

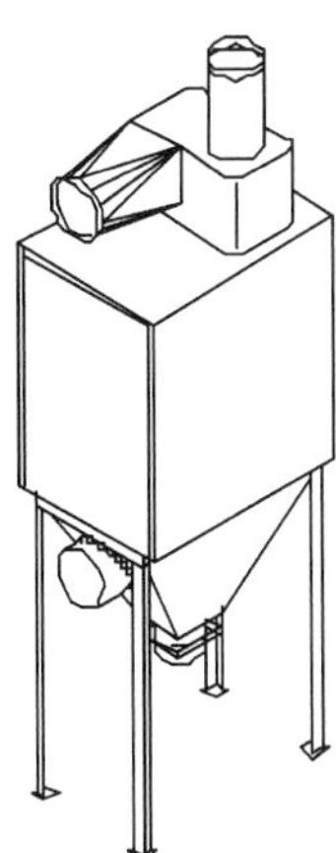

Commercial/Industrial Dust Collection System

System Components	QUANTITY	UNIT	COST EACH MAT.	COST EACH INST.	COST EACH TOTAL
SYSTEM D3040 250 1010					
COMMERCIAL/INDUSTRIAL VACUUM DUST COLLECTION SYSTEM, 500 CFM					
Dust collection central vac unit inc. stand, filter & shaker, 500 CFM 2 HP	1.000	Ea.	5,300	705	6,005
Galvanized tubing, 16 ga., 4" OD	80.000	L.F.	620	296	916
Galvanized tubing, 90° ell, 4" dia.	3.000	Ea.	130.50	79.50	210
Galvanized tubing, 45° ell, 4" dia.	4.000	Ea.	152	106	258
Galvanized tubing, TY slip fit, 4" dia.	4.000	Ea.	188	178	366
Flexible rubber hose, 4" dia.	32.000	L.F.	273.60	118.40	392
Galvanized air gate valve, 4" dia.	4.000	Ea.	708	232	940
Galvanized compression coupling, 4" dia.	36.000	Ea.	954	1,458	2,412
Pipe hangers, steel, 4" pipe size	16.000	Ea.	424	198.40	622.40
TOTAL			8,750.10	3,371.30	12,121.40

D3040 250	Commercial/Industrial Vacuum Dust Collection	COST EACH MAT.	COST EACH INST.	COST EACH TOTAL
1010	Commercial/industrial, vacuum dust collection system, 500 CFM	8,750	3,375	12,125
1020	1000 CFM	12,300	3,850	16,150
1030	1500 CFM	15,200	4,250	19,450
1040	3000 CFM	34,900	5,950	40,850
1050	5000 CFM	46,100	8,925	55,025

D3040 Distribution Systems

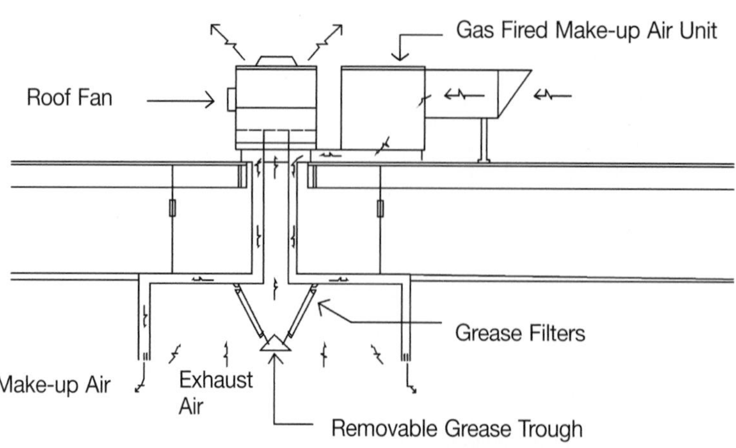

Commercial Kitchen Exhaust/Make-up Air Rooftop System

System Components	QUANTITY	UNIT	COST EACH		
			MAT.	INST.	TOTAL
SYSTEM D3040 260 1010					
COMMERCIAL KITCHEN EXHAUST/MAKE-UP AIR SYSTEM, ROOFTOP, GAS, 2000 CFM					
Make-up air rooftop unit, gas, 750 MBH	1.000	Ea.	5,908.20	163.74	6,071.94
Pipe, black steel, Sch 40, thrded, nipples, W/cplngs/hngrs, 10′ OC, 1″ dia.	20.000	L.F.	107.14	314.60	421.74
Elbow, 90°, black steel, straight, 3/4″ dia.	2.000	Ea.	9.82	108	117.82
Elbow, 90°, black steel, straight, 1″ dia.	3.000	Ea.	25.65	175.50	201.15
Tee, black steel, reducing, 1″ dia.	1.000	Ea.	22	95	117
Union, black with brass seat, 1″ dia.	1.000	Ea.	29.50	63	92.50
Pipe nipples, black, 3/4″ dia	2.000	Ea.	9.36	24.90	34.26
Pipe cap, malleable iron, black, 1″ dia.	1.000	Ea.	6.95	25.50	32.45
Gas cock, brass, 1″ size	1.000	Ea.	36.50	40	76.50
Ductwork, 20″ x 14″ fabricated, galv. steel, 30 LF	220.000	Lb.	136.40	1,793	1,929.40
Ductwork, 18″ x 10″ fabricated, galv. steel, 12 LF	72.000	Lb.	44.64	586.80	631.44
Ductwork, 30″ x 10″ fabricated, galv. steel, 30 LF	260.000	Lb.	161.20	2,119	2,280.20
Ductwork, 16″ x 10″ fabricated, galv. steel, 12 LF	70.000	Lb.	43.40	570.50	613.90
Kitchen ventilation, island style, water wash, Stainless Steel	5.000	L.F.	13,875	1,460	15,335
Control system, Make-up Air Unit/Exhauster	1.000	Ea.	3,250	2,700	5,950
Rooftop unit heat/cool balancing	1.000	Ea.		520	520
Roof fan balancing	1.000	Ea.		296	296
TOTAL			23,665.76	11,055.54	34,721.30

D3040 260	Kitchen Exhaust/Make-Up Air	COST EACH		
		MAT.	INST.	TOTAL
1010	Commercial kitchen exhaust/make-up air system, rooftop, gas, 2000 CFM	23,700	11,100	34,800
1020	3000 CFM	35,600	15,400	51,000
1030	5000 CFM	41,600	18,200	59,800
1040	8000 CFM	60,500	26,900	87,400
1050	12,000 CFM	79,000	31,600	110,600
1060	16,000 CFM	97,500	48,500	146,000

D3040 Distribution Systems

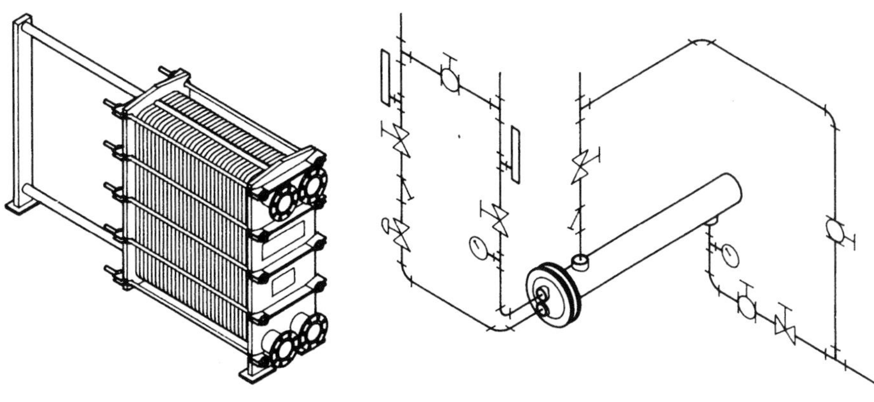

Plate Heat Exchanger   Shell and Tube Heat Exchanger

System Components	QUANTITY	UNIT	COST EACH		
			MAT.	INST.	TOTAL
SYSTEM D3040 620 1010					
SHELL & TUBE HEAT EXCHANGER, 40 GPM					
Heat exchanger, 4 pass, 3/4" O.D. copper tubes, by steam at 10 psi, 40 GPM	1.000	Ea.	6,450	345	6,795
Pipe, black steel, Sch 40, threaded, w/couplings & hangers, 10' OC, 2" dia.	40.000	L.F.	641.35	1,085.75	1,727.10
Elbow, 90°, straight, 2" dia.	16.000	Ea.	520	1,216	1,736
Reducer, black steel, concentric, 2" dia.	2.000	Ea.	74	130	204
Valves, bronze, gate, rising stem, threaded, class 150, 2" size	6.000	Ea.	1,614	414	2,028
Valves, bronze, globe, class 150, rising stem, threaded, 2" size	2.000	Ea.	1,450	138	1,588
Strainers, Y type, bronze body, screwed, 125 lb, 2" pipe size	2.000	Ea.	388	118	506
Valve, electric motor actuated, brass, 2 way, screwed, 1-1/2" pipe size	1.000	Ea.	905	59	964
Union, black with brass seat, 2" dia.	8.000	Ea.	496	644	1,140
Tee, black, straight, 2" dia.	9.000	Ea.	373.50	1,116	1,489.50
Tee, black, reducing run and outlet, 2" dia.	4.000	Ea.	240	496	736
Pipe nipple, black, 2" dia.	21.000	Ea.	6.35	10.75	17.10
Thermometers, stem type, 9" case, 8" stem, 3/4" NPT	2.000	Ea.	392	55	447
Gauges, pressure or vacuum, 3-1/2" diameter dial	2.000	Ea.	41	48	89
Insulation, fiberglass pipe covering, 1" wall, 2" IPS	40.000	L.F.	70.80	256	326.80
Heat exchanger control system	1.000	Ea.	3,175	2,450	5,625
Coil balancing	1.000	Ea.		142	142
TOTAL			16,837	8,723.50	25,560.50

D3040 610	Heat Exchanger, Plate Type	COST EACH		
		MAT.	INST.	TOTAL
1010	Plate heat exchanger, 400 GPM	65,000	17,500	82,500
1020	800 GPM	105,000	22,300	127,300
1030	1200 GPM	157,500	30,200	187,700
1040	1800 GPM	212,000	37,700	249,700

D3040 620	Heat Exchanger, Shell & Tube	COST EACH		
		MAT.	INST.	TOTAL
1010	Shell & tube heat exchanger, 40 GPM	16,800	8,725	25,525
1020	96 GPM	27,900	15,400	43,300
1030	240 GPM	52,000	20,800	72,800
1040	600 GPM	99,500	29,400	128,900

D3050 Terminal & Package Units

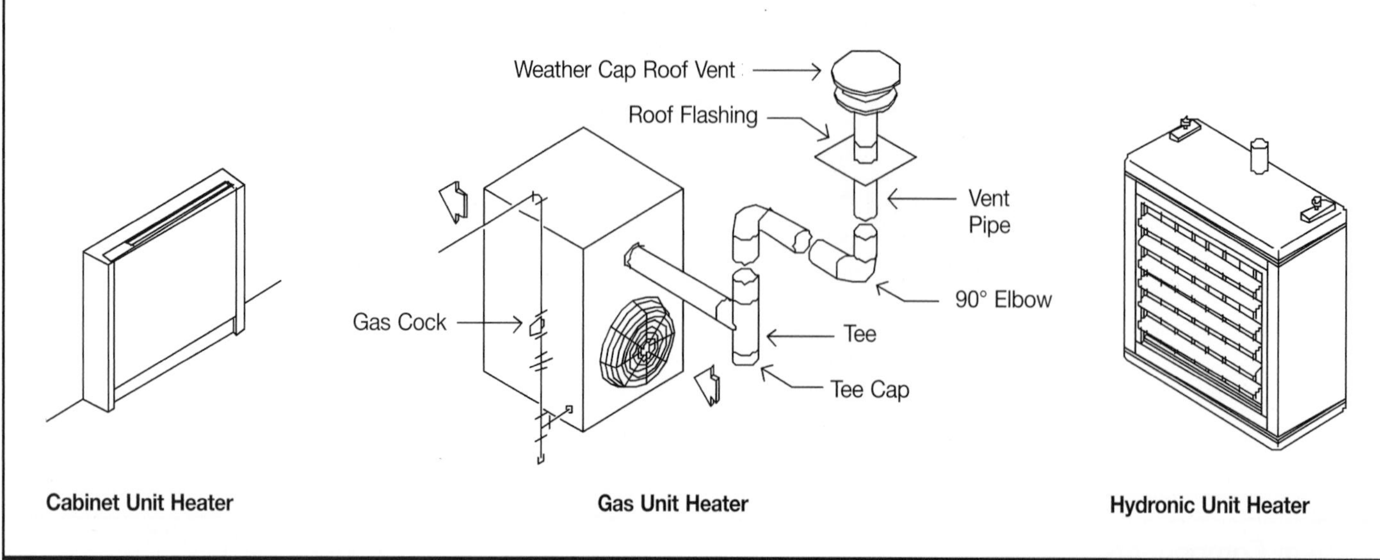

System Components	QUANTITY	UNIT	COST EACH MAT.	COST EACH INST.	COST EACH TOTAL
SYSTEM D3050 120 1010					
SPACE HEATER, SUSPENDED, GAS FIRED, PROPELLER FAN, 20 MBH					
Space heater, propeller fan, 20 MBH output	1.000	Ea.	1,875	163	2,038
Pipe, black steel, Sch 40, threaded, w/coupling & hangers, 10′ OC, 3/4″ dia	20.000	L.F.	98.28	261.45	359.73
Elbow, 90°, black steel, straight, 1/2″ dia.	3.000	Ea.	12.18	151.50	163.68
Elbow, 90°, black steel, straight, 3/4″ dia.	3.000	Ea.	14.73	162	176.73
Tee, black steel, reducing, 3/4″ dia.	1.000	Ea.	15.70	84	99.70
Union, black with brass seat, 3/4″ dia.	1.000	Ea.	23	58.50	81.50
Pipe nipples, black, 1/2″ dia	2.000	Ea.	8.60	24.10	32.70
Pipe nipples, black, 3/4″ dia	2.000	Ea.	4.68	12.45	17.13
Pipe cap, black, 3/4″ dia.	1.000	Ea.	5.75	23.50	29.25
Gas cock, brass, 3/4″ size	1.000	Ea.	18.70	34.50	53.20
Thermostat, 1 set back, electric, timed	1.000	Ea.	83	96	179
Wiring, thermostat hook-up,25′ of #18-3	1.000	Ea.	6.80	30	36.80
Vent chimney, prefab metal, U.L. listed, gas, double wall, galv. st, 4″ dia	12.000	L.F.	109.80	235.20	345
Vent chimney, gas, double wall, galv. steel, elbow 90°, 4″ dia.	2.000	Ea.	53	78	131
Vent chimney, gas, double wall, galv. steel, Tee, 4″ dia.	1.000	Ea.	37.50	51.50	89
Vent chimney, gas, double wall, galv. steel, T cap, 4″ dia.	1.000	Ea.	2.56	31.50	34.06
Vent chimney, gas, double wall, galv. steel, roof flashing, 4″ dia.	1.000	Ea.	9	39	48
Vent chimney, gas, double wall, galv. steel, top, 4″ dia.	1.000	Ea.	18.05	30.50	48.55
TOTAL			2,396.33	1,566.70	3,963.03

D3050 120	Unit Heaters, Gas	COST EACH MAT.	COST EACH INST.	COST EACH TOTAL
1010	Space heater, suspended, gas fired, propeller fan, 20 MBH	2,400	1,575	3,975
1020	60 MBH	2,575	1,625	4,200
1030	100 MBH	2,950	1,725	4,675
1040	160 MBH	3,475	1,900	5,375
1050	200 MBH	4,350	2,100	6,450
1060	280 MBH	5,175	2,250	7,425
1070	320 MBH	6,650	2,550	9,200

D3050 130	Unit Heaters, Hydronic	COST EACH MAT.	COST EACH INST.	COST EACH TOTAL
1010	Space heater, suspended, horiz. mount, HW, prop. fan, 20 MBH	2,375	1,675	4,050
1020	60 MBH	3,175	1,875	5,050

D3050 Terminal & Package Units

D3050 130	Unit Heaters, Hydronic	COST EACH		
		MAT.	INST.	TOTAL
1030	100 MBH	3,825	2,050	5,875
1040	150 MBH	4,300	2,275	6,575
1050	200 MBH	4,850	2,550	7,400
1060	300 MBH	5,875	3,000	8,875

D3050 140	Cabinet Unit Heaters, Hydronic	COST EACH		
		MAT.	INST.	TOTAL
1010	Unit heater, cabinet type, horizontal blower, hot water, 20 MBH	2,925	1,650	4,575
1020	60 MBH	4,250	1,800	6,050
1030	100 MBH	4,825	2,000	6,825
1040	120 MBH	4,825	2,050	6,875

D3050 Terminal & Package Units

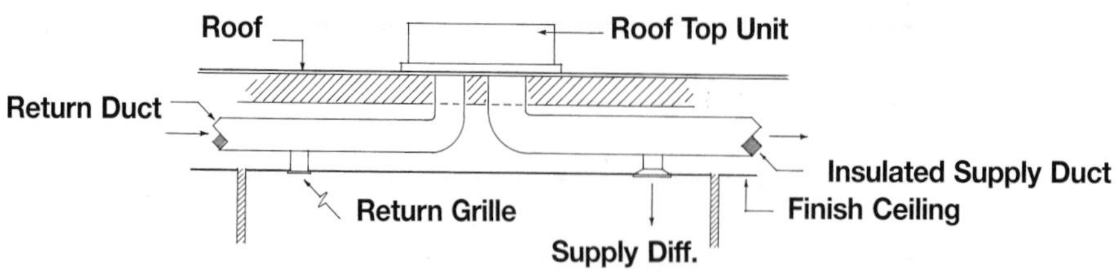

System Description: Rooftop single zone units are electric cooling and gas heat. Duct systems are low velocity, galvanized steel supply and return. Price variations between sizes are due to several factors. Jumps in the cost of the rooftop unit occur when the manufacturer shifts from the largest capacity unit on a small frame to the smallest capacity on the next larger frame, or changes from one compressor to two. As the unit capacity increases for larger areas the duct distribution grows in proportion. For most applications there is a tradeoff point where it is less expensive and more efficient to utilize smaller units with short simple distribution systems. Larger units also require larger initial supply and return ducts which can create a space problem. Supplemental heat may be desired in colder locations. The table below is based on one unit supplying the area listed. The 10,000 S.F. unit for bars and taverns is not listed because a nominal 110 ton unit would be required and this is above the normal single zone rooftop capacity.

System Components	QUANTITY	UNIT	COST EACH		
			MAT.	INST.	TOTAL
SYSTEM D3050 150 1280					
ROOFTOP, SINGLE ZONE, AIR CONDITIONER					
APARTMENT CORRIDORS, 500 S.F., .92 TON					
Rooftop air conditioner, 1 zone, electric cool, standard controls, curb	1.000	Ea.	1,529.50	908.50	2,438
Ductwork package for rooftop single zone units	1.000	System	268.64	1,288	1,556.64
TOTAL			1,798.14	2,196.50	3,994.64
COST PER S.F.			3.60	4.39	7.99
*Size would suggest multiple units.					

D3050 150	Rooftop Single Zone Unit Systems	COST PER S.F.		
		MAT.	INST.	TOTAL
1260	Rooftop, single zone, air conditioner			
1280	Apartment corridors, 500 S.F., .92 ton	3.60	4.40	8
1320	1,000 S.F., 1.83 ton (R236000-20)	3.57	4.37	7.94
1360	1500 S.F., 2.75 ton	2.33	3.60	5.93
1400	3,000 S.F., 5.50 ton	2.32	3.48	5.80
1440	5,000 S.F., 9.17 ton	2.39	3.16	5.55
1480	10,000 S.F., 18.33 ton	3.52	2.97	6.49
1560	Banks or libraries, 500 S.F., 2.08 ton	8.15	9.95	18.10
1600	1,000 S.F., 4.17 ton	5.30	8.20	13.50
1640	1,500 S.F., 6.25 ton	5.25	7.90	13.15
1680	3,000 S.F., 12.50 ton	5.45	7.15	12.60
1720	5,000 S.F., 20.80 ton	7.95	6.75	14.70
1760	10,000 S.F., 41.67 ton	6.25	6.70	12.95
1840	Bars and taverns, 500 S.F. 5.54 ton	12.35	13.25	25.60
1880	1,000 S.F., 11.08 ton	12.80	11.35	24.15
1920	1,500 S.F., 16.62 ton	12.25	10.55	22.80
1960	3,000 S.F., 33.25 ton	16.15	10.05	26.20
2000	5,000 S.F., 55.42 ton	13.80	10.05	23.85
2040	10,000 S.F., 110.83 ton*			
2080	Bowling alleys, 500 S.F., 2.83 ton	7.20	11.15	18.35
2120	1,000 S.F., 5.67 ton	7.15	10.75	17.90
2160	1,500 S.F., 8.50 ton	7.35	9.75	17.10
2200	3,000 S.F., 17.00 ton	7.10	9.35	16.45
2240	5,000 S.F., 28.33 ton	9.05	9.10	18.15
2280	10,000 S.F., 56.67 ton	7.90	9.10	17
2360	Department stores, 500 S.F., 1.46 ton	5.70	6.95	12.65

D3050 Terminal & Package Units

D3050 150	Rooftop Single Zone Unit Systems	COST PER S.F.		
		MAT.	INST.	TOTAL
2400	1,000 S.F., 2.92 ton	3.71	5.75	9.46
2440	1,500 S.F., 4.37 ton	3.68	5.50	9.18
2480	3,000 S.F., 8.75 ton	3.80	5	8.80
2520	5,000 S.F., 14.58 ton	3.65	4.81	8.46
2560	10,000 S.F., 29.17 ton	4.67	4.69	9.36
2640	Drug stores, 500 S.F., 3.33 ton	8.50	13.10	21.60
2680	1,000 S.F., 6.67 ton	8.40	12.65	21.05
2720	1,500 S.F., 10.00 ton	8.70	11.50	20.20
2760	3,000 S.F., 20.00 ton	12.80	10.80	23.60
2800	5,000 S.F., 33.33 ton	10.70	10.70	21.40
2840	10,000 S.F., 66.67 ton	9.30	10.70	20
2920	Factories, 500 S.F., 1.67 ton	6.50	7.95	14.45
2960	1,000 S.F., 3.33 ton	4.24	6.55	10.79
3000	1,500 S.F., 5.00 ton	4.20	6.30	10.50
3040	3,000 S.F., 10.00 ton	4.34	5.75	10.09
3080	5,000 S.F., 16.67 ton	4.17	5.50	9.67
3120	10,000 S.F., 33.33 ton	5.35	5.35	10.70
3200	Food supermarkets, 500 S.F., 1.42 ton	5.55	6.80	12.35
3240	1,000 S.F., 2.83 ton	3.58	5.55	9.13
3280	1,500 S.F., 4.25 ton	3.58	5.35	8.93
3320	3,000 S.F., 8.50 ton	3.69	4.88	8.57
3360	5,000 S.F., 14.17 ton	3.55	4.68	8.23
3400	10,000 S.F., 28.33 ton	4.54	4.56	9.10
3480	Medical centers, 500 S.F., 1.17 ton	4.56	5.60	10.16
3520	1,000 S.F., 2.33 ton	4.56	5.55	10.11
3560	1,500 S.F., 3.50 ton	2.97	4.60	7.57
3600	3,000 S.F., 7.00 ton	2.94	4.43	7.37
3640	5,000 S.F., 11.67 ton	3.04	4.02	7.06
3680	10,000 S.F., 23.33 ton	4.47	3.78	8.25
3760	Offices, 500 S.F., 1.58 ton	6.20	7.55	13.75
3800	1,000 S.F., 3.17 ton	4.04	6.25	10.29
3840	1,500 S.F., 4.75 ton	3.99	6	9.99
3880	3,000 S.F., 9.50 ton	4.12	5.45	9.57
3920	5,000 S.F., 15.83 ton	3.96	5.25	9.21
3960	10,000 S.F., 31.67 ton	5.05	5.10	10.15
4000	Restaurants, 500 S.F., 2.50 ton	9.75	11.95	21.70
4040	1,000 S.F., 5.00 ton	6.30	9.50	15.80
4080	1,500 S.F., 7.50 ton	6.50	8.60	15.10
4120	3,000 S.F., 15.00 ton	6.25	8.25	14.50
4160	5,000 S.F., 25.00 ton	9.60	8.10	17.70
4200	10,000 S.F., 50.00 ton	6.95	8.05	15
4240	Schools and colleges, 500 S.F., 1.92 ton	7.50	9.15	16.65
4280	1,000 S.F., 3.83 ton	4.88	7.55	12.43
4320	1,500 S.F., 5.75 ton	4.84	7.25	12.09
4360	3,000 S.F., 11.50 ton	4.99	6.60	11.59
4400	5,000 S.F., 19.17 ton	7.35	6.20	13.55
4440	10,000 S.F., 38.33 ton	5.75	6.15	11.90
5000				
9000	Components of ductwork packages for above systems, per ton of cooling:			
9010	Ductwork; galvanized steel, 120 pounds			
9020	Insulation; fiberglass 2" thick, FRK faced, 52 S.F.			
9030	Diffusers; aluminum, 24" x 12", one			
9040	Registers; aluminum, return with dampers, one			

D3050 Terminal & Package Units

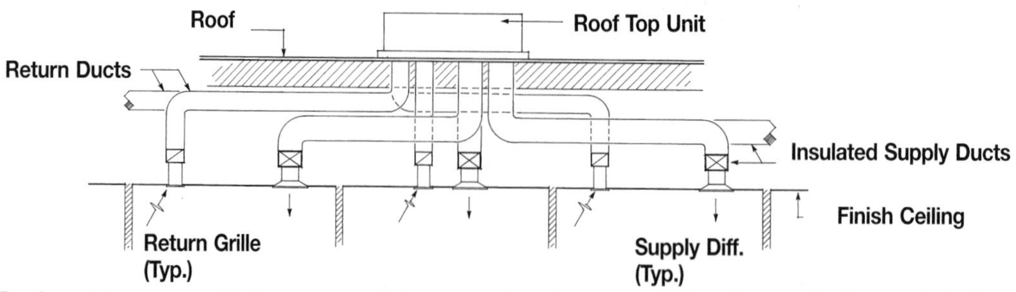

System Description: Rooftop units are multizone with up to 12 zones, and include electric cooling, gas heat, thermostats, filters, supply and return fans complete. Duct systems are low velocity, galvanized steel supply and return with insulated supplies.

Multizone units cost more per ton of cooling than single zone. However, they offer flexibility where load conditions are varied due to heat generating areas or exposure to radiational heating. For example, perimeter offices on the "sunny side" may require cooling at the same time "shady side" or central offices may require heating. It is possible to accomplish similar results using duct heaters in branches of the single zone unit. However, heater location could be a problem and total system operating energy efficiency could be lower.

System Components	QUANTITY	UNIT	COST EACH MAT.	COST EACH INST.	COST EACH TOTAL
SYSTEM D3050 155 1280					
ROOFTOP, MULTIZONE, AIR CONDITIONER					
APARTMENT CORRIDORS, 3,000 S.F., 5.50 TON					
Rooftop multizone unit, standard controls, curb	1.000	Ea.	31,460	2,112	33,572
Ductwork package for rooftop multizone units	1.000	System	2,117.50	14,712.50	16,830
TOTAL			33,577.50	16,824.50	50,402
COST PER S.F.			11.19	5.61	16.80
Note A: Small single zone unit recommended.					
Note B: A combination of multizone units recommended.					

D3050 155	Rooftop Multizone Unit Systems	COST PER S.F. MAT.	COST PER S.F. INST.	COST PER S.F. TOTAL
1240	Rooftop, multizone, air conditioner			
1260	Apartment corridors, 1,500 S.F., 2.75 ton. See Note A.			
1280	3,000 S.F., 5.50 ton (R236000-20)	11.20	5.60	16.80
1320	10,000 S.F., 18.30 ton	7.70	5.40	13.10
1360	15,000 S.F., 27.50 ton	8.15	5.35	13.50
1400	20,000 S.F., 36.70 ton	8	5.40	13.40
1440	25,000 S.F., 45.80 ton	7.45	5.40	12.85
1520	Banks or libraries, 1,500 S.F., 6.25 ton	25.50	12.75	38.25
1560	3,000 S.F., 12.50 ton	21.50	12.50	34
1600	10,000 S.F., 41.67 ton	16.90	12.30	29.20
1640	15,000 S.F., 62.50 ton	12.40	12.25	24.65
1680	20,000 S.F., 83.33 ton	12.40	12.25	24.65
1720	25,000 S.F., 104.00 ton	10.95	12.20	23.15
1800	Bars and taverns, 1,500 S.F., 16.62 ton	55	18.35	73.35
1840	3,000 S.F., 33.24 ton	46	17.70	63.70
1880	10,000 S.F., 110.83 ton	27	17.65	44.65
1920	15,000 S.F., 165 ton, See Note B.			
1960	20,000 S.F., 220 ton, See Note B.			
2000	25,000 S.F., 275 ton, See Note B.			
2080	Bowling alleys, 1,500 S.F., 8.50 ton	34.50	17.35	51.85
2120	3,000 S.F., 17.00 ton	29	16.95	45.95
2160	10,000 S.F., 56.70 ton	23	16.75	39.75
2200	15,000 S.F., 85.00 ton	16.90	16.65	33.55
2240	20,000 S.F., 113.00 ton	14.90	16.55	31.45
2280	25,000 S.F., 140.00 ton see Note B.			
2360	Department stores, 1,500 S.F., 4.37 ton, See Note A.			

D3050 Terminal & Package Units

D3050 155	Rooftop Multizone Unit Systems	COST PER S.F.		
		MAT.	INST.	TOTAL
2400	3,000 S.F., 8.75 ton	17.80	8.90	26.70
2440	10,000 S.F., 29.17 ton	12.70	8.55	21.25
2480	15,000 S.F., 43.75 ton	11.85	8.60	20.45
2520	20,000 S.F., 58.33 ton	8.70	8.55	17.25
2560	25,000 S.F., 72.92 ton	8.70	8.55	17.25
2640	Drug stores, 1,500 S.F., 10.00 ton	40.50	20.50	61
2680	3,000 S.F., 20.00 ton	28	19.65	47.65
2720	10,000 S.F., 66.66 ton	19.85	19.55	39.40
2760	15,000 S.F., 100.00 ton	17.55	19.50	37.05
2800	20,000 S.F., 135 ton, See Note B.			
2840	25,000 S.F., 165 ton, See Note B.			
2920	Factories, 1,500 S.F., 5 ton, See Note A.			
2960	3,000 S.F., 10.00 ton	20.50	10.20	30.70
3000	10,000 S.F., 33.33 ton	14.50	9.80	24.30
3040	15,000 S.F., 50.00 ton	13.55	9.85	23.40
3080	20,000 S.F., 66.66 ton	9.90	9.80	19.70
3120	25,000 S.F., 83.33 ton	9.95	9.80	19.75
3200	Food supermarkets, 1,500 S.F., 4.25 ton, See Note A.			
3240	3,000 S.F., 8.50 ton	17.30	8.65	25.95
3280	10,000 S.F., 28.33 ton	12.65	8.30	20.95
3320	15,000 S.F., 42.50 ton	11.50	8.40	19.90
3360	20,000 S.F., 56.67 ton	8.45	8.30	16.75
3400	25,000 S.F., 70.83 ton	8.45	8.30	16.75
3480	Medical centers, 1,500 S.F., 3.5 ton, See Note A.			
3520	3,000 S.F., 7.00 ton	14.25	7.15	21.40
3560	10,000 S.F., 23.33 ton	9.65	6.85	16.50
3600	15,000 S.F., 35.00 ton	10.15	6.85	17
3640	20,000 S.F., 46.66 ton	9.45	6.90	16.35
3680	25,000 S.F., 58.33 ton	6.95	6.85	13.80
3760	Offices, 1,500 S.F., 4.75 ton, See Note A.			
3800	3,000 S.F., 9.50 ton	19.35	9.70	29.05
3840	10,000 S.F., 31.66 ton	13.80	9.30	23.10
3880	15,000 S.F., 47.50 ton	12.85	9.35	22.20
3920	20,000 S.F., 63.33 ton	9.45	9.30	18.75
3960	25,000 S.F., 79.16 ton	9.45	9.30	18.75
4000	Restaurants, 1,500 S.F., 7.50 ton	30.50	15.30	45.80
4040	3,000 S.F., 15.00 ton	26	15	41
4080	10,000 S.F., 50.00 ton	20.50	14.80	35.30
4120	15,000 S.F., 75.00 ton	14.90	14.70	29.60
4160	20,000 S.F., 100.00 ton	13.15	14.65	27.80
4200	25,000 S.F., 125 ton, See Note B.			
4240	Schools and colleges, 1,500 S.F., 5.75 ton	23.50	11.70	35.20
4280	3,000 S.F., 11.50 ton	19.75	11.50	31.25
4320	10,000 S.F., 38.33 ton	15.55	11.35	26.90
4360	15,000 S.F., 57.50 ton	11.40	11.25	22.65
4400	20,000 S.F.,76.66 ton	11.40	11.25	22.65
4440	25,000 S.F., 95.83 ton	10.55	11.25	21.80
4450				
9000	Components of ductwork packages for above systems, per ton of cooling:			
9010	Ductwork; galvanized steel, 240 pounds			
9020	Insulation; fiberglass, 2" thick, FRK faced, 104 S.F.			
9030	Diffusers; aluminum, 24" x 12", two			
9040	Registers; aluminum, return with dampers, one			

D3050 Terminal & Package Units

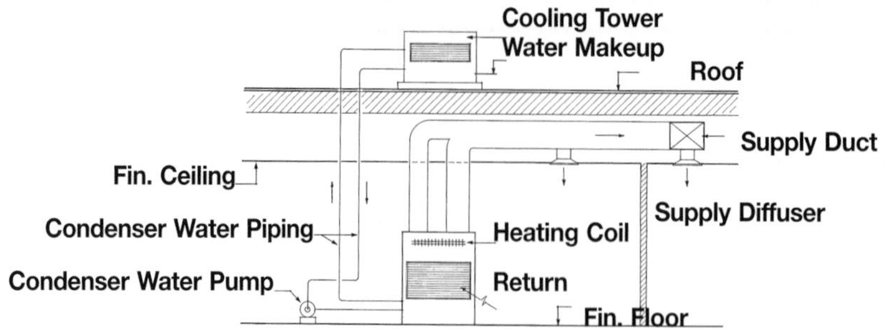

System Description: Self-contained, single package water cooled units include cooling tower, pump, piping allowance. Systems for 1000 S.F. and up include duct and diffusers to provide for even distribution of air. Smaller units distribute air through a supply air plenum, which is integral with the unit.

Returns are not ducted and supplies are not insulated.

Hot water or steam heating coils are included but piping to boiler and the boiler itself is not included.

Where local codes or conditions permit single pass cooling for the smaller units, deduct 10%.

System Components	QUANTITY	UNIT	COST EACH MAT.	COST EACH INST.	COST EACH TOTAL
SYSTEM D3050 160 1300					
SELF-CONTAINED, WATER COOLED UNIT					
APARTMENT CORRIDORS, 500 S.F., .92 TON					
Self-contained, water cooled, single package air conditioner unit	1.000	Ea.	1,394.40	722.40	2,116.80
Ductwork package for water or air cooled packaged units	1.000	System	63.48	966	1,029.48
Cooling tower, draw thru single flow, belt drive	1.000	Ea.	200.56	22.08	222.64
Cooling tower pumps & piping	1.000	System	100.28	51.98	152.26
TOTAL			1,758.72	1,762.46	3,521.18
COST PER S.F.			3.52	3.52	7.04

D3050 160	Self-contained, Water Cooled Unit Systems	COST PER S.F. MAT.	COST PER S.F. INST.	COST PER S.F. TOTAL
1280	Self-contained, water cooled unit	3.52	3.51	7.03
1300	Apartment corridors, 500 S.F., .92 ton	3.50	3.50	7
1320	1,000 S.F., 1.83 ton (R236000-20)	3.51	3.50	7.01
1360	3,000 S.F., 5.50 ton	2.77	2.94	5.71
1400	5,000 S.F., 9.17 ton	2.67	2.72	5.39
1440	10,000 S.F., 18.33 ton	3.85	2.43	6.28
1520	Banks or libraries, 500 S.F., 2.08 ton	7.70	3.62	11.32
1560	1,000 S.F., 4.17 ton	6.30	6.70	13
1600	3,000 S.F., 12.50 ton	6.05	6.20	12.25
1640	5,000 S.F., 20.80 ton	8.70	5.55	14.25
1680	10,000 S.F., 41.66 ton	7.05	5.55	12.60
1760	Bars and taverns, 500 S.F., 5.54 ton	15.85	6.10	21.95
1800	1,000 S.F., 11.08 ton	15.65	10.65	26.30
1840	3,000 S.F., 33.25 ton	20.50	8.40	28.90
1880	5,000 S.F., 55.42 ton	17.20	8.75	25.95
1920	10,000 S.F., 110.00 ton	16.70	8.65	25.35
2000	Bowling alleys, 500 S.F., 2.83 ton	10.50	4.92	15.42
2040	1,000 S.F., 5.66 ton	8.55	9.10	17.65
2080	3,000 S.F., 17.00 ton	11.90	7.55	19.45
2120	5,000 S.F., 28.33 ton	10.75	7.30	18.05
2160	10,000 S.F., 56.66 ton	9	7.40	16.40
2200	Department stores, 500 S.F., 1.46 ton	5.40	2.53	7.93
2240	1,000 S.F., 2.92 ton	4.40	4.67	9.07
2280	3,000 S.F., 8.75 ton	4.24	4.33	8.57
2320	5,000 S.F., 14.58 ton	6.10	3.88	9.98
2360	10,000 S.F., 29.17 ton	5.55	3.74	9.29

D3050 Terminal & Package Units

D3050 160	Self-contained, Water Cooled Unit Systems	COST PER S.F.		
		MAT.	INST.	TOTAL
2440	Drug stores, 500 S.F., 3.33 ton	12.30	5.80	18.10
2480	1,000 S.F., 6.66 ton	10.05	10.70	20.75
2520	3,000 S.F., 20.00 ton	13.95	8.90	22.85
2560	5,000 S.F., 33.33 ton	14.65	8.75	23.40
2600	10,000 S.F., 66.66 ton	10.60	8.80	19.40
2680	Factories, 500 S.F., 1.66 ton	6.15	2.89	9.04
2720	1,000 S.F. 3.37 ton	5.05	5.40	10.45
2760	3,000 S.F., 10.00 ton	4.84	4.96	9.80
2800	5,000 S.F., 16.66 ton	7	4.44	11.44
2840	10,000 S.F., 33.33 ton	6.30	4.28	10.58
2920	Food supermarkets, 500 S.F., 1.42 ton	5.25	2.46	7.71
2960	1,000 S.F., 2.83 ton	5.45	5.45	10.90
3000	3,000 S.F., 8.50 ton	4.28	4.55	8.83
3040	5,000 S.F., 14.17 ton	4.12	4.21	8.33
3080	10,000 S.F., 28.33 ton	5.40	3.63	9.03
3160	Medical centers, 500 S.F., 1.17 ton	4.30	2.03	6.33
3200	1,000 S.F., 2.33 ton	4.47	4.48	8.95
3240	3,000 S.F., 7.00 ton	3.51	3.74	7.25
3280	5,000 S.F., 11.66 ton	3.38	3.47	6.85
3320	10,000 S.F., 23.33 ton	4.88	3.11	7.99
3400	Offices, 500 S.F., 1.58 ton	5.85	2.75	8.60
3440	1,000 S.F., 3.17 ton	6.05	6.05	12.10
3480	3,000 S.F., 9.50 ton	4.60	4.72	9.32
3520	5,000 S.F., 15.83 ton	6.65	4.22	10.87
3560	10,000 S.F., 31.67 ton	6	4.06	10.06
3640	Restaurants, 500 S.F., 2.50 ton	9.25	4.34	13.59
3680	1,000 S.F., 5.00 ton	7.55	8	15.55
3720	3,000 S.F., 15.00 ton	7.25	7.45	14.70
3760	5,000 S.F., 25.00 ton	10.50	6.65	17.15
3800	10,000 S.F., 50.00 ton	6.50	6.10	12.60
3880	Schools and colleges, 500 S.F., 1.92 ton	7.10	3.33	10.43
3920	1,000 S.F., 3.83 ton	5.80	6.15	11.95
3960	3,000 S.F., 11.50 ton	5.55	5.70	11.25
4000	5,000 S.F., 19.17 ton	8.05	5.10	13.15
4040	10,000 S.F., 38.33 ton	6.45	5.10	11.55
4050				
9000	Components of ductwork packages for above systems, per ton of cooling:			
9010	Ductwork; galvanized steel, 108 pounds			
9020	Diffusers, aluminum, 24" x 12", two			

D3050 Terminal & Package Units

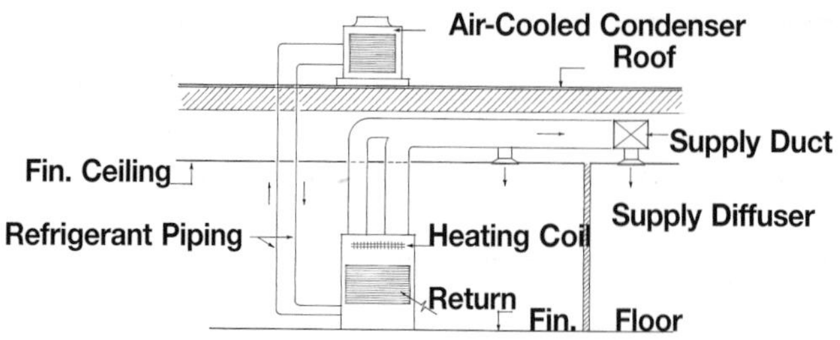

System Description: Self-contained air cooled units with remote air cooled condenser and interconnecting tubing. Systems for 1000 S.F. and up include duct and diffusers. Smaller units distribute air directly.

Returns are not ducted and supplies are not insulated.

Potential savings may be realized by using a single zone rooftop system or through-the-wall unit, especially in the smaller capacities, if the application permits.

Hot water or steam heating coils are included but piping to boiler and the boiler itself is not included.

Condenserless models are available for 15% less where remote refrigerant source is available.

System Components	QUANTITY	UNIT	COST EACH MAT.	COST EACH INST.	COST EACH TOTAL
SYSTEM D3050 165 1320					
SELF-CONTAINED, AIR COOLED UNIT					
APARTMENT CORRIDORS, 500 S.F., .92 TON					
Air cooled, package unit	1.000	Ea.	1,428	467.50	1,895.50
Ductwork package for water or air cooled packaged units	1.000	System	63.48	966	1,029.48
Refrigerant piping	1.000	System	363.40	634.80	998.20
Air cooled condenser, direct drive, propeller fan	1.000	Ea.	852.50	178.25	1,030.75
TOTAL			2,707.38	2,246.55	4,953.93
COST PER S.F.			5.41	4.49	9.90

D3050 165	Self-contained, Air Cooled Unit Systems	COST PER S.F. MAT.	COST PER S.F. INST.	COST PER S.F. TOTAL
1300	Self-contained, air cooled unit			
1320	Apartment corridors, 500 S.F., .92 ton	5.40	4.50	9.90
1360	1,000 S.F., 1.83 ton R236000-20	5.35	4.45	9.80
1400	3,000 S.F., 5.50 ton	4.92	4.18	9.10
1440	5,000 S.F., 9.17 ton	4.01	3.95	7.96
1480	10,000 S.F., 18.33 ton	3.40	3.63	7.03
1560	Banks or libraries, 500 S.F., 2.08 ton	11.75	5.75	17.50
1600	1,000 S.F., 4.17 ton	11.10	9.45	20.55
1640	3,000 S.F., 12.50 ton	9.10	9	18.10
1680	5,000 S.F., 20.80 ton	7.85	8.25	16.10
1720	10,000 S.F., 41.66 ton	8.65	8.10	16.75
1800	Bars and taverns, 500 S.F., 5.54 ton	27.50	12.55	40.05
1840	1,000 S.F., 11.08 ton	24	18.15	42.15
1880	3,000 S.F., 33.25 ton	23	15.70	38.70
1920	5,000 S.F., 55.42 ton	23	15.70	38.70
1960	10,000 S.F., 110.00 ton	23.50	15.60	39.10
2040	Bowling alleys, 500 S.F., 2.83 ton	16.15	7.85	24
2080	1,000 S.F., 5.66 ton	15.20	12.90	28.10
2120	3,000 S.F., 17.00 ton	10.55	11.25	21.80
2160	5,000 S.F., 28.33 ton	11.90	11	22.90
2200	10,000 S.F., 56.66 ton	12.05	11	23.05
2240	Department stores, 500 S.F., 1.46 ton	8.30	4.04	12.34
2280	1,000 S.F., 2.92 ton	7.80	6.65	14.45
2320	3,000 S.F., 8.75 ton	6.35	6.30	12.65

D3050 Terminal & Package Units

D3050 165	Self-contained, Air Cooled Unit Systems	COST PER S.F.		
		MAT.	INST.	TOTAL
2360	5,000 S.F., 14.58 ton	6.35	6.30	12.65
2400	10,000 S.F., 29.17 ton	6.15	5.65	11.80
2480	Drug stores, 500 S.F., 3.33 ton	19	9.25	28.25
2520	1,000 S.F., 6.66 ton	17.85	15.15	33
2560	3,000 S.F., 20.00 ton	12.55	13.25	25.80
2600	5,000 S.F., 33.33 ton	14	12.95	26.95
2640	10,000 S.F., 66.66 ton	14.30	12.95	27.25
2720	Factories, 500 S.F., 1.66 ton	9.65	4.66	14.31
2760	1,000 S.F., 3.33 ton	9	7.60	16.60
2800	3,000 S.F., 10.00 ton	7.30	7.20	14.50
2840	5,000 S.F., 16.66 ton	6.15	6.60	12.75
2880	10,000 S.F., 33.33 ton	7	6.45	13.45
2960	Food supermarkets, 500 S.F., 1.42 ton	8.10	3.92	12.02
3000	1,000 S.F., 2.83 ton	8.25	6.90	15.15
3040	3,000 S.F., 8.50 ton	7.60	6.45	14.05
3080	5,000 S.F., 14.17 ton	6.20	6.10	12.30
3120	10,000 S.F., 28.33 ton	5.95	5.50	11.45
3200	Medical centers, 500 S.F., 1.17 ton	6.70	3.24	9.94
3240	1,000 S.F., 2.33 ton	6.85	5.70	12.55
3280	3,000 S.F., 7.00 ton	6.25	5.30	11.55
3320	5,000 S.F., 16.66 ton	5.10	5.05	10.15
3360	10,000 S.F., 23.33 ton	4.34	4.63	8.97
3440	Offices, 500 S.F., 1.58 ton	9.05	4.39	13.44
3480	1,000 S.F., 3.16 ton	9.30	7.70	17
3520	3,000 S.F., 9.50 ton	6.95	6.85	13.80
3560	5,000 S.F., 15.83 ton	5.95	6.30	12.25
3600	10,000 S.F., 31.66 ton	6.65	6.15	12.80
3680	Restaurants, 500 S.F., 2.50 ton	14.20	6.90	21.10
3720	1,000 S.F., 5.00 ton	13.40	11.35	24.75
3760	3,000 S.F., 15.00 ton	10.95	10.80	21.75
3800	5,000 S.F., 25.00 ton	10.45	9.70	20.15
3840	10,000 S.F., 50.00 ton	11.15	9.65	20.80
3920	Schools and colleges, 500 S.F., 1.92 ton	10.90	5.30	16.20
3960	1,000 S.F., 3.83 ton	10.30	8.70	19
4000	3,000 S.F., 11.50 ton	8.35	8.30	16.65
4040	5,000 S.F., 19.17 ton	7.10	7.60	14.70
4080	10,000 S.F., 38.33 ton	7.95	7.45	15.40
4090				
9000	Components of ductwork packages for above systems, per ton of cooling:			
9010	Ductwork; galvanized steel, 102 pounds			
9020	Diffusers; aluminum, 24" x 12", two			

D3050 Terminal & Package Units

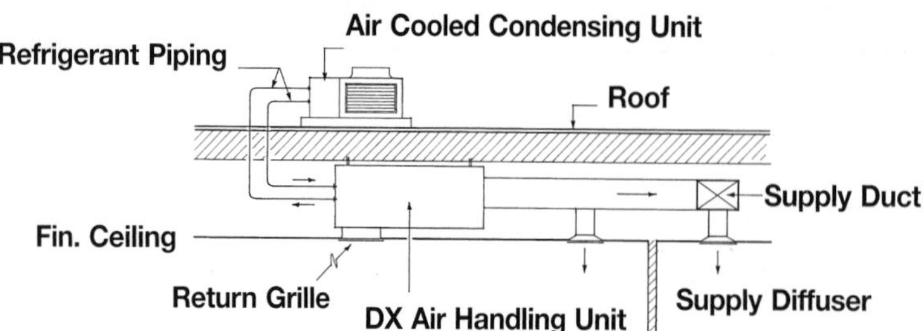

General: Split systems offer several important advantages which should be evaluated when a selection is to be made. They provide a greater degree of flexibility in component selection which permits an accurate match-up of the proper equipment size and type with the particular needs of the building. This allows for maximum use of modern energy saving concepts in heating and cooling. Outdoor installation of the air cooled condensing unit allows space savings in the building and also isolates the equipment operating sounds from building occupants.

Design Assumptions: The systems below are comprised of a direct expansion air handling unit and air cooled condensing unit with interconnecting copper tubing. Ducts and diffusers are also included for distribution of air. Systems are priced for cooling only. Heat can be added as desired either by putting hot water/steam coils into the air unit or into the duct supplying the particular area of need. Gas fired duct furnaces are also available. Refrigerant liquid line is insulated.

System Components	QUANTITY	UNIT	COST EACH MAT.	COST EACH INST.	COST EACH TOTAL
SYSTEM D3050 170 1280					
SPLIT SYSTEM, AIR COOLED CONDENSING UNIT					
APARTMENT CORRIDORS, 1,000 S.F., 1.80 TON					
Fan coil AC unit, cabinet mntd & filters direct expansion air cool	1.000	Ea.	457.50	168.36	625.86
Ductwork package, for split system, remote condensing unit	1.000	System	59.48	915	974.48
Refrigeration piping	1.000	System	384.30	1,052.25	1,436.55
Condensing unit, air cooled, incls compressor & standard controls	1.000	Ea.	1,433.50	677.10	2,110.60
TOTAL			2,334.78	2,812.71	5,147.49
COST PER S.F.			2.33	2.81	5.14
*Cooling requirements would lead to choosing a water cooled unit.					

D3050 170	Split Systems With Air Cooled Condensing Units	COST PER S.F. MAT.	COST PER S.F. INST.	COST PER S.F. TOTAL
1260	Split system, air cooled condensing unit			
1280	Apartment corridors, 1,000 S.F., 1.83 ton	2.33	2.83	5.16
1320	2,000 S.F., 3.66 ton (R236000 -20)	1.79	2.84	4.63
1360	5,000 S.F., 9.17 ton	1.66	3.54	5.20
1400	10,000 S.F., 18.33 ton	1.82	3.86	5.68
1440	20,000 S.F., 36.66 ton	2.17	3.93	6.10
1520	Banks and libraries, 1,000 S.F., 4.17 ton	4.08	6.45	10.53
1560	2,000 S.F., 8.33 ton	3.78	8.05	11.83
1600	5,000 S.F., 20.80 ton	4.13	8.80	12.93
1640	10,000 S.F., 41.66 ton	4.93	8.95	13.88
1680	20,000 S.F., 83.32 ton	5.95	9.30	15.25
1760	Bars and taverns, 1,000 S.F., 11.08 ton	8.55	12.70	21.25
1800	2,000 S.F., 22.16 ton	11.15	15.20	26.35
1840	5,000 S.F., 55.42 ton	10.85	14.25	25.10
1880	10,000 S.F., 110.84 ton	13.65	14.85	28.50
1920	20,000 S.F., 220 ton*			
2000	Bowling alleys, 1,000 S.F., 5.66 ton	5.65	12.05	17.70
2040	2,000 S.F., 11.33 ton	5.15	10.95	16.10
2080	5,000 S.F., 28.33 ton	5.60	11.95	17.55
2120	10,000 S.F., 56.66 ton	6.70	12.15	18.85
2160	20,000 S.F., 113.32 ton	9.20	13.20	22.40
2320	Department stores, 1,000 S.F., 2.92 ton	2.84	4.45	7.29
2360	2,000 S.F., 5.83 ton	2.90	6.20	9.10
2400	5,000 S.F., 14.58 ton	2.65	5.65	8.30

D3050 Terminal & Package Units

D3050 170	Split Systems With Air Cooled Condensing Units	COST PER S.F.		
		MAT.	INST.	TOTAL
2440	10,000 S.F., 29.17 ton	2.89	6.15	9.04
2480	20,000 S.F., 58.33 ton	3.45	6.25	9.70
2560	Drug stores, 1,000 S.F., 6.66 ton	6.65	14.15	20.80
2600	2,000 S.F., 13.32 ton	6.05	12.90	18.95
2640	5,000 S.F., 33.33 ton	7.90	14.30	22.20
2680	10,000 S.F., 66.66 ton	8.15	15	23.15
2720	20,000 S.F., 133.32 ton*			
2800	Factories, 1,000 S.F., 3.33 ton	3.25	5.05	8.30
2840	2,000 S.F., 6.66 ton	3.32	7.10	10.42
2880	5,000 S.F., 16.66 ton	3.31	7.05	10.36
2920	10,000 S.F., 33.33 ton	3.95	7.15	11.10
2960	20,000 S.F., 66.66 ton	4.08	7.50	11.58
3040	Food supermarkets, 1,000 S.F., 2.83 ton	2.76	4.32	7.08
3080	2,000 S.F., 5.66 ton	2.81	6	8.81
3120	5,000 S.F., 14.66 ton	2.56	5.50	8.06
3160	10,000 S.F., 28.33 ton	2.80	6	8.80
3200	20,000 S.F., 56.66 ton	3.35	6.10	9.45
3280	Medical centers, 1,000 S.F., 2.33 ton	2.43	3.48	5.91
3320	2,000 S.F., 4.66 ton	2.32	4.95	7.27
3360	5,000 S.F., 11.66 ton	2.11	4.51	6.62
3400	10,000 S.F., 23.33 ton	2.31	4.93	7.24
3440	20,000 S.F., 46.66 ton	2.75	5	7.75
3520	Offices, 1,000 S.F., 3.17 ton	3.08	4.82	7.90
3560	2,000 S.F., 6.33 ton	3.15	6.75	9.90
3600	5,000 S.F., 15.83 ton	2.88	6.15	9.03
3640	10,000 S.F., 31.66 ton	3.14	6.70	9.84
3680	20,000 S.F., 63.32 ton	3.87	7.10	10.97
3760	Restaurants, 1,000 S.F., 5.00 ton	4.98	10.65	15.63
3800	2,000 S.F., 10.00 ton	4.54	9.70	14.24
3840	5,000 S.F., 25.00 ton	4.96	10.55	15.51
3880	10,000 S.F., 50.00 ton	5.95	10.75	16.70
3920	20,000 S.F., 100.00 ton	8.15	11.65	19.80
4000	Schools and colleges, 1,000 S.F., 3.83 ton	3.75	5.95	9.70
4040	2,000 S.F., 7.66 ton	3.47	7.40	10.87
4080	5,000 S.F., 19.17 ton	3.81	8.10	11.91
4120	10,000 S.F., 38.33 ton	4.54	8.25	12.79
4160	20,000 S.F., 76.66 ton	4.69	8.60	13.29
5000				
9000	Components of ductwork packages for above systems, per ton of cooling:			
9010	Ductwork; galvanized steel, 102 pounds			
9020	Diffusers; aluminum, 24″ x 12″, two			

D3050 Terminal & Package Units

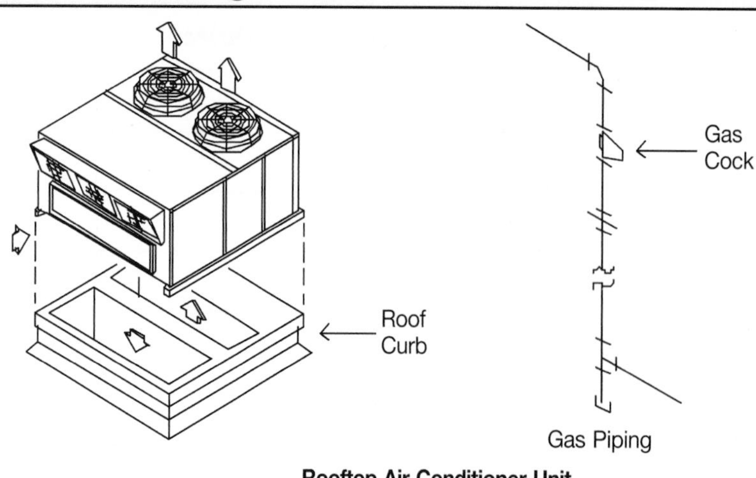

Rooftop Air Conditioner Unit

System Components	QUANTITY	UNIT	COST EACH MAT.	COST EACH INST.	COST EACH TOTAL
SYSTEM D3050 175 1010					
A/C, ROOFTOP, DX COOL, GAS HEAT, CURB, ECONOMIZER, FILTERS., 5 TON					
Roof top A/C, curb, economizer, sgl zone, elec cool, gas ht, 5 ton, 112 MBH	1.000	Ea.	4,850	2,475	7,325
Pipe, black steel, Sch 40, threaded, w/cplgs & hangers, 10′ OC, 1″ dia	20.000	L.F.	102.27	300.30	402.57
Elbow, 90°, black, straight, 3/4″ dia.	3.000	Ea.	14.73	162	176.73
Elbow, 90°, black, straight, 1″ dia.	3.000	Ea.	25.65	175.50	201.15
Tee, black, reducing, 1″ dia.	1.000	Ea.	22	95	117
Union, black with brass seat, 1″ dia.	1.000	Ea.	29.50	63	92.50
Pipe nipple, black, 3/4″ dia	2.000	Ea.	9.36	24.90	34.26
Pipe nipple, black, 1″ dia	2.000	Ea.	4.87	14.30	19.17
Cap, black, 1″ dia.	1.000	Ea.	6.95	25.50	32.45
Gas cock, brass, 1″ size	1.000	Ea.	36.50	40	76.50
Control system, pneumatic, Rooftop A/C unit	1.000	Ea.	3,250	2,700	5,950
Rooftop unit heat/cool balancing	1.000	Ea.		520	520
TOTAL			8,351.83	6,595.50	14,947.33

D3050 175	Rooftop Air Conditioner, Const. Volume	COST EACH MAT.	COST EACH INST.	COST EACH TOTAL
1010	A/C, Rooftop, DX cool, gas heat, curb, economizer, fltrs, 5 Ton	8,350	6,600	14,950
1020	7-1/2 Ton	9,925	6,850	16,775
1030	12-1/2 Ton	15,200	7,525	22,725
1040	18 Ton	19,000	8,225	27,225
1050	25 Ton	39,600	9,375	48,975
1060	40 Ton	52,000	12,500	64,500

D3050 180	Rooftop Air Conditioner, Variable Air Volume	COST EACH MAT.	COST EACH INST.	COST EACH TOTAL
1010	A/C, Rooftop, DX cool, gas heat, curb, ecmizr, fltrs, VAV, 12-1/2 Ton	18,200	8,050	26,250
1020	18 Ton	23,000	8,850	31,850
1030	25 Ton	44,500	10,200	54,700
1040	40 Ton	60,000	13,800	73,800
1050	60 Ton	84,500	18,800	103,300
1060	80 Ton	108,000	23,500	131,500

D3050 Terminal & Package Units

Computer rooms impose special requirements on air conditioning systems. A prime requirement is reliability, due to the potential monetary loss that could be incurred by a system failure. A second basic requirement is the tolerance of control with which temperature and humidity are regulated, and dust eliminated. As the air conditioning system reliability is so vital, the additional cost of reserve capacity and redundant components is often justified.

System Descriptions: Computer areas may be environmentally controlled by one of three methods as follows:

1. **Self-contained Units**
   These are units built to higher standards of performance and reliability. They usually contain alarms and controls to indicate component operation failure, filter change, etc. It should be remembered that these units in the room will occupy space that is relatively expensive to build and that all alterations and service of the equipment will also have to be accomplished within the computer area.
2. **Decentralized Air Handling Units** In operation these are similar to the self-contained units except that their cooling capability comes from remotely located refrigeration equipment as refrigerant or chilled water. As no compressors or refrigerating equipment are required in the air units, they are smaller and require less service than self-contained units. An added plus for this type of system occurs if some of the computer components themselves also require chilled water for cooling.
3. **Central System Supply** Cooling is obtained from a central source which, since it is not located within the computer room, may have excess capacity and permit greater flexibility without interfering with the computer components. System performance criteria must still be met.

Note: The costs shown below do not include an allowance for ductwork or piping.

D3050 185	Computer Room Cooling Units	COST EACH		
		MAT.	INST.	TOTAL
0560	Computer room unit, air cooled, includes remote condenser			
0580	3 ton	29,300	2,775	32,075
0600	5 ton	31,300	3,075	34,375
0620	8 ton	58,500	5,125	63,625
0640	10 ton	61,000	5,525	66,525
0660	15 ton	67,000	6,275	73,275
0680	20 ton	81,500	8,950	90,450
0700	23 ton	100,500	10,200	110,700
0800	Chilled water, for connection to existing chiller system			
0820	5 ton	21,200	1,875	23,075
0840	8 ton	21,800	2,775	24,575
0860	10 ton	21,900	2,825	24,725
0880	15 ton	24,000	2,875	26,875
0900	20 ton	25,500	3,000	28,500
0920	23 ton	27,100	3,400	30,500
1000	Glycol system, complete except for interconnecting tubing			
1020	3 ton	36,700	3,450	40,150
1040	5 ton	40,200	3,625	43,825
1060	8 ton	64,500	6,000	70,500
1080	10 ton	68,500	6,575	75,075
1100	15 ton	85,000	8,275	93,275
1120	20 ton	93,500	8,950	102,450
1140	23 ton	98,500	9,350	107,850
1240	Water cooled, not including condenser water supply or cooling tower			
1260	3 ton	29,800	2,225	32,025
1280	5 ton	32,400	2,550	34,950
1300	8 ton	53,000	4,175	57,175
1320	15 ton	64,500	5,125	69,625
1340	20 ton	69,000	5,650	74,650
1360	23 ton	72,500	6,150	78,650

D3050 201	Packaged A/C, Elec. Ht., Const. Volume	COST EACH		
		MAT.	INST.	TOTAL
1010	A/C Packaged, DX, air cooled, electric heat, constant vol., 5 Ton	9,875	6,425	16,300
1020	10 Ton	16,200	8,650	24,850

D3050 Terminal & Package Units

D3050 201	Packaged A/C, Elec. Ht., Const. Volume	COST EACH		
		MAT.	INST.	TOTAL
1030	20 Ton	25,000	11,500	36,500
1040	30 Ton	45,400	12,800	58,200
1050	40 Ton	59,500	14,400	73,900
1060	50 Ton	80,000	16,000	96,000

D3050 202	Packaged Air Conditioner, Elect. Heat, VAV	COST EACH		
		MAT.	INST.	TOTAL
1010	A/C Packaged, DX, air cooled, electric heat, VAV, 10 Ton	18,600	8,100	26,700
1020	20 Ton	29,200	10,700	39,900
1030	30 Ton	52,000	11,600	63,600
1040	40 Ton	66,500	13,400	79,900
1050	50 Ton	90,000	14,900	104,900
1060	60 Ton	101,000	17,400	118,400

D3050 203	Packaged A/C, Hot Wtr. Heat, Const. Volume	COST EACH		
		MAT.	INST.	TOTAL
1010	A/C Packaged, DX, air cooled, H/W heat, constant vol., 5 Ton	8,325	8,675	17,000
1020	10 Ton	12,800	11,300	24,100
1030	20 Ton	20,500	14,400	34,900
1040	30 Ton	37,600	18,500	56,100
1050	40 Ton	52,000	20,600	72,600
1060	50 Ton	68,500	22,300	90,800

D3050 204	Packaged A/C, Hot Wtr. Heat, VAV	COST EACH		
		MAT.	INST.	TOTAL
1010	A/C Packaged, DX, air cooled, H/W heat, VAV, 10 Ton	15,200	10,600	25,800
1020	20 Ton	25,000	13,900	38,900
1030	30 Ton	44,100	17,200	61,300
1040	40 Ton	59,500	18,600	78,100
1050	50 Ton	79,000	21,500	100,500
1060	60 Ton	92,000	24,200	116,200

D3050 Terminal & Package Units

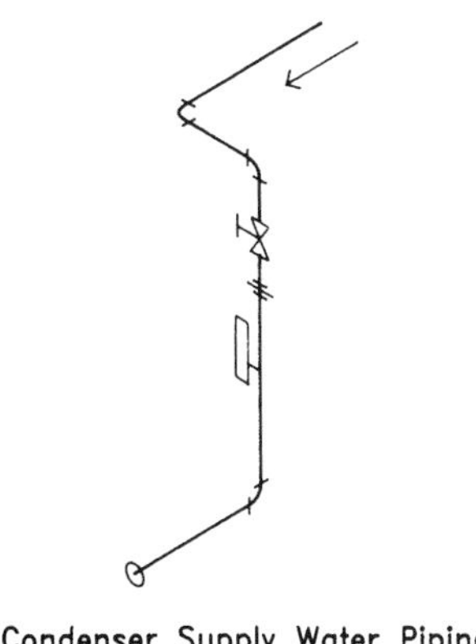

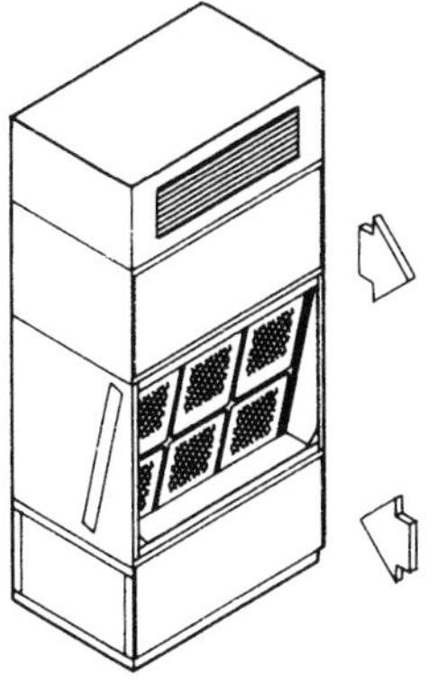

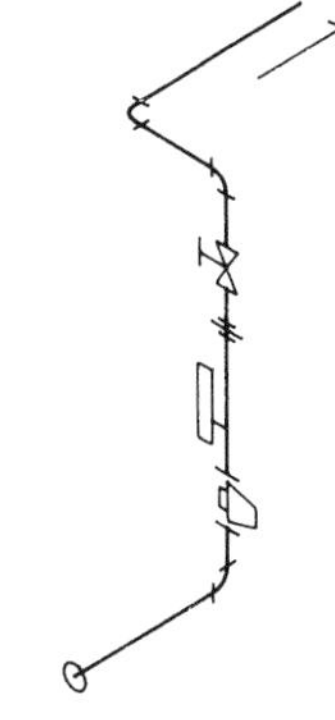

Self–Contained Air Conditioner Unit, Water Cooled

System Components	QUANTITY	UNIT	COST EACH MAT.	INST.	TOTAL
SYSTEM D3050 210 1010					
A/C, SELF CONTAINED, SINGLE PKG., WATER COOLED, ELECT. HEAT, 5 TON					
Self-contained, water cooled, elect. heat, not inc tower, 5 ton, const vol	1.000	Ea.	9,200	1,800	11,000
Pipe, black steel, Sch 40, threaded, w/cplg & hangers, 10′ OC, 1-1/4″ dia.	20.000	L.F.	122.55	330.03	452.58
Elbow, 90°, black, straight, 1-1/4″ dia.	6.000	Ea.	85.20	372	457.20
Tee, black, straight, 1-1/4″ dia.	2.000	Ea.	45	195	240
Tee, black, reducing, 1-1/4″ dia.	2.000	Ea.	78	195	273
Thermometers, stem type, 9″ case, 8″ stem, 3/4″ NPT	2.000	Ea.	392	55	447
Union, black with brass seat, 1-1/4″ dia.	2.000	Ea.	85	130	215
Pipe nipple, black, 1-1/4″ dia	3.000	Ea.	8.55	23.03	31.58
Valves, bronze, gate, N.R.S., threaded, class 150, 1-1/4″ size	2.000	Ea.	332	101	433
Circuit setter, bal valve, bronze body, threaded, 1-1/4″ pipe size	1.000	Ea.	187	51	238
Insulation, fiberglass pipe covering, 1″ wall, 1-1/4″ IPS	20.000	L.F.	24.40	121	145.40
Control system, pneumatic, A/C Unit with heat	1.000	Ea.	3,250	2,700	5,950
Re-heat coil balancing	1.000	Ea.		142	142
Rooftop unit heat/cool balancing	1.000	Ea.		520	520
TOTAL			13,809.70	6,735.06	20,544.76

D3050 210	AC Unit, Package, Elec. Ht., Water Cooled	COST EACH MAT.	INST.	TOTAL
1010	A/C, Self contained, single pkg., water cooled, elect. heat, 5 Ton	13,800	6,725	20,525
1020	10 Ton	21,300	8,300	29,600
1030	20 Ton	28,600	10,500	39,100
1040	30 Ton	49,500	11,500	61,000
1050	40 Ton	63,500	12,700	76,200
1060	50 Ton	84,500	15,200	99,700

D3050 215	AC Unit, Package, Elec. Ht., Water Cooled, VAV	COST EACH MAT.	INST.	TOTAL
1010	A/C, Self contained, single pkg., water cooled, elect. ht, VAV, 10 Ton	24,200	8,300	32,500
1020	20 Ton	33,500	10,500	44,000
1030	30 Ton	57,000	11,500	68,500
1040	40 Ton	71,500	12,700	84,200
1050	50 Ton	95,500	15,200	110,700
1060	60 Ton	107,500	18,300	125,800

D3050 Terminal & Package Units

D3050 220	AC Unit, Package, Hot Water Coil, Water Cooled	COST EACH		
		MAT.	INST.	TOTAL
1010	A/C, Self contn'd, single pkg., water cool, H/W ht, const. vol, 5 Ton	12,000	9,325	21,325
1020	10 Ton	17,500	11,100	28,600
1030	20 Ton	43,200	13,900	57,100
1040	30 Ton	56,000	17,500	73,500
1050	40 Ton	63,000	21,300	84,300
1060	50 Ton	74,500	26,200	100,700

D3050 225	AC Unit, Package, HW Coil, Water Cooled, VAV	COST EACH		
		MAT.	INST.	TOTAL
1010	A/C, Self contn'd, single pkg., water cool, H/W ht, VAV, 10 Ton	20,400	10,500	30,900
1020	20 Ton	29,400	13,500	42,900
1030	30 Ton	49,500	16,800	66,300
1040	40 Ton	65,000	18,300	83,300
1050	50 Ton	85,000	21,600	106,600
1060	60 Ton	98,000	23,400	121,400

D3050 Terminal & Package Units

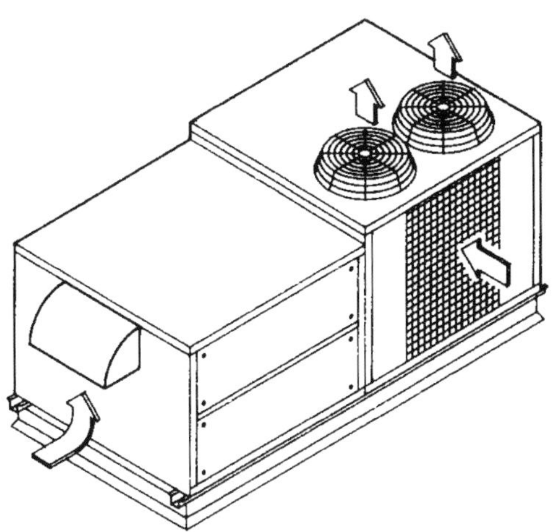

Heat Pump, Air/Air, Rooftop w/Economizer

Note: Water source units do not include an allowance for any type of water source or piping to it.

System Components	QUANTITY	UNIT	COST EACH MAT.	COST EACH INST.	COST EACH TOTAL
SYSTEM D3050 245 1010					
HEAT PUMP, ROOFTOP, AIR/AIR, CURB., ECONOMIZER., SUP. ELECT. HEAT, 10 TON					
Heat pump, air to air, rooftop, curb, economizer, elect ht, 10 ton cooling	1.000	Ea.	72,500	5,375	77,875
Ductwork, 32" x 12" fabricated, galvanized steel, 16 LF	176.000	Lb.	110.88	1,548.80	1,659.68
Insulation, ductwork, blanket type, fiberglass, 1" thk, 1-1/2 LB density	117.000	S.F.	596.70	1,778.40	2,375.10
Rooftop unit heat/cool balancing	1.000	Ea.		520	520
TOTAL			73,207.58	9,222.20	82,429.78

D3050 230	Heat Pump, Water Source, Central Station	COST EACH MAT.	COST EACH INST.	COST EACH TOTAL
1010	Heat pump, central station, water source, constant vol., 5 Ton	5,050	3,750	8,800
1020	10 Ton	10,500	5,275	15,775
1030	20 Ton	20,100	9,800	29,900
1040	30 Ton	29,100	14,300	43,400
1050	40 Ton	40,000	15,600	55,600
1060	50 Ton	45,200	20,900	66,100

D3050 235	Heat Pump, Water Source, Console	COST EACH MAT.	COST EACH INST.	COST EACH TOTAL
1020	Heat pump, console, water source, 1 Ton	2,750	2,075	4,825
1030	1-1/2 Ton	2,750	2,150	4,900
1040	2 Ton	3,850	2,525	6,375
1050	3 Ton	4,125	2,700	6,825
1060	3-1/2 Ton	4,375	2,800	7,175

D3050 240	Heat Pump, Horizontal, Duct Mounted	COST EACH MAT.	COST EACH INST.	COST EACH TOTAL
1020	Heat pump, horizontal, ducted, water source, 1 Ton	3,250	4,400	7,650
1030	1-1/2 Ton	3,450	5,525	8,975
1040	2 Ton	4,625	6,625	11,250
1050	3 Ton	5,300	8,675	13,975
1060	3-1/2 Ton	5,550	8,975	14,525

D3050 Terminal & Package Units

D3050 245	Heat Pump, Rooftop, Air/Air	COST EACH		
		MAT.	INST.	TOTAL
1010	Heat pump, rooftop, air/air, curb, economizer, sup. elect heat, 10 Ton	73,000	9,225	82,225
1020	20 Ton	108,500	14,900	123,400
1030	30 Ton	141,000	20,700	161,700
1040	40 Ton	157,000	24,000	181,000
1050	50 Ton	183,000	26,900	209,900
1060	60 Ton	193,000	30,400	223,400

D3050 Terminal & Package Units

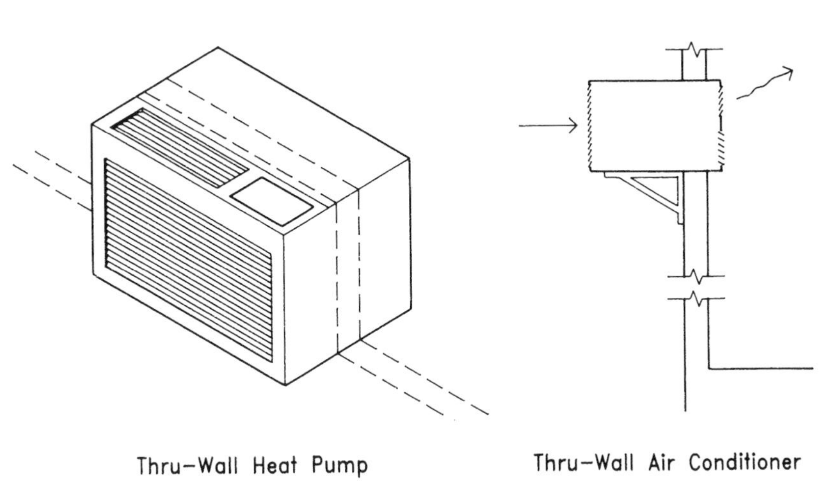

System Components	QUANTITY	UNIT	COST EACH MAT.	COST EACH INST.	COST EACH TOTAL
SYSTEM D3050 255 1010 A/C UNIT, THRU-THE-WALL, SUPPLEMENTAL ELECT. HEAT, CABINET, LOUVER, 1/2 TON					
A/C unit, thru-wall, electric heat., cabinet, louver, 1/2 ton	1.000	Ea.	775	230	1,005
TOTAL			775	230	1,005

D3050 255	Thru-Wall A/C Unit	COST EACH MAT.	COST EACH INST.	COST EACH TOTAL
1010	A/C Unit, thru-the-wall, supplemental elect. heat, cabinet, louver, 1/2 Ton	775	230	1,005
1020	3/4 Ton	1,075	276	1,351
1030	1 Ton	1,625	345	1,970
1040	1-1/2 Ton	2,675	575	3,250
1050	2 Ton	2,725	725	3,450

D3050 260	Thru-Wall Heat Pump	COST EACH MAT.	COST EACH INST.	COST EACH TOTAL
1010	Heat pump, thru-the-wall, cabinet, louver, 1/2 Ton	3,725	173	3,898
1020	3/4 Ton	4,200	230	4,430
1030	1 Ton	3,600	345	3,945
1040	Supplemental. elect. heat, 1-1/2 Ton	3,650	890	4,540
1050	2 Ton	3,625	920	4,545

D3090 Other HVAC Systems/Equip

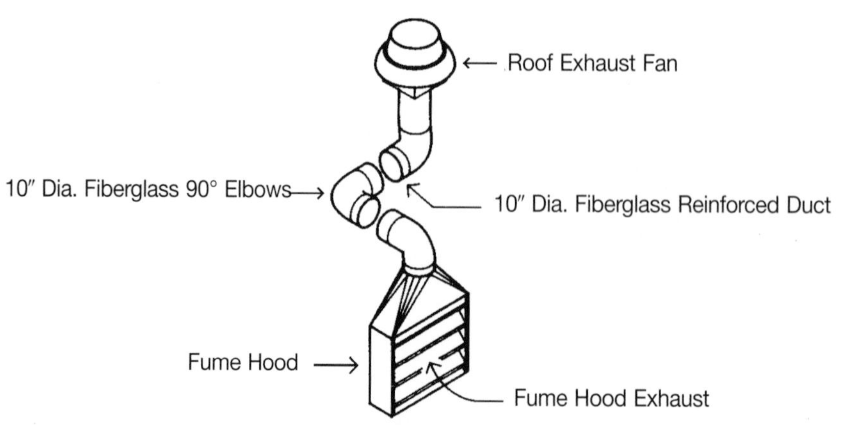

System Components	QUANTITY	UNIT	COST EACH MAT.	COST EACH INST.	COST EACH TOTAL
SYSTEM D3090 310 1010					
FUME HOOD EXHAUST SYSTEM, 3′ LONG, 1000 CFM					
Fume hood	3.000	L.F.	3,600	1,560	5,160
Ductwork, plastic, FM approved for acid fume & smoke, straight, 10″ dia	60.000	L.F.	1,320	942	2,262
Ductwork, plastic, FM approved for acid fume & smoke, elbow 90°, 10″ dia	3.000	Ea.	339	513	852
Fan, corrosive fume resistant plastic, 1,630 CFM, 1/2 HP	1.000	Ea.	7,200	425	7,625
Fan control switch & mounting base	1.000	Ea.	121	48	169
Roof fan balancing	1.000	Ea.		296	296
TOTAL			12,580	3,784	16,364

D3090 310	Fume Hood Exhaust Systems	COST EACH MAT.	COST EACH INST.	COST EACH TOTAL
1010	Fume hood exhaust system, 3′ long, 1000 CFM	12,600	3,775	16,375
1020	4′ long, 2000 CFM	14,100	4,775	18,875
1030	6′ long, 3500 CFM	18,800	7,150	25,950
1040	6′ long, 5000 CFM	28,300	8,375	36,675
1050	10′ long, 8000 CFM	35,900	11,500	47,400

D3090 Other HVAC Systems/Equip

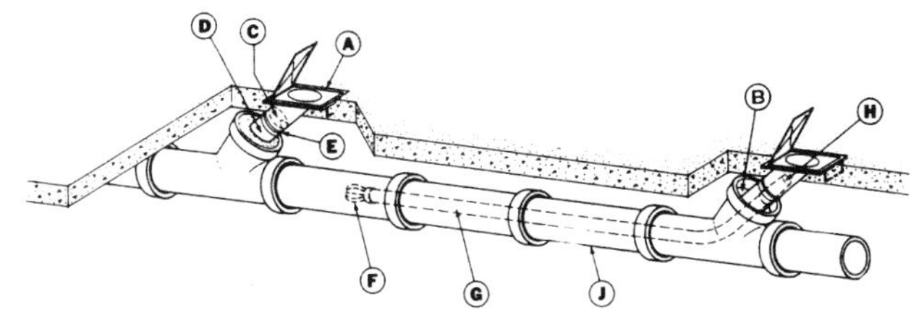

Cast Iron Garage Exhaust System

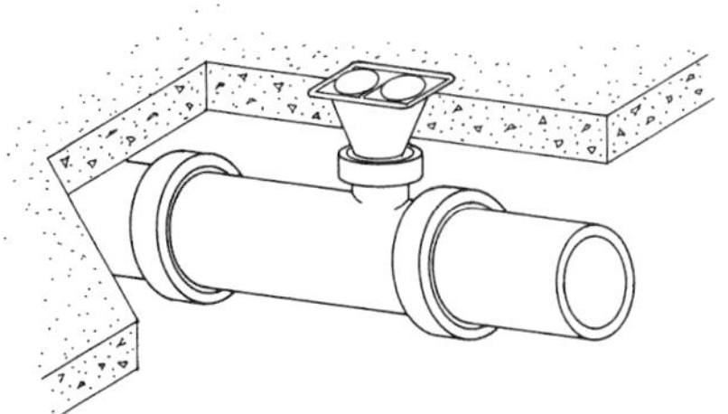
Dual Exhaust System

System Components	QUANTITY	UNIT	COST EACH MAT.	COST EACH INST.	COST EACH TOTAL
SYSTEM D3090 320 1040					
GARAGE, EXHAUST, SINGLE 3" EXHAUST OUTLET, CARS & LIGHT TRUCKS					
A Outlet top assy, for engine exhaust system with adapters and ftngs, 3" diam	1.000	Ea.	243	46.50	289.50
F Bullnose (guide) for engine exhaust system, 3" diam	1.000	Ea.	28		28
G Galvanized flexible tubing for engine exhaust system, 3" diam	8.000	L.F.	92.80		92.80
H Adapter for metal tubing end of engine exhaust system, 3" tail pipe	1.000	Ea.	49		49
J Pipe, sewer, cast iron, push-on joint, 8" diam.	18.000	L.F.	1,395	837	2,232
Excavating utility trench, chain trencher, 8" wide, 24" deep	18.000	L.F.		35.46	35.46
Backfill utility trench by hand, incl. compaction, 8" wide 24" deep	18.000	L.F.		49.86	49.86
Stand for blower, concrete over polystyrene core, 6" high	1.000	Ea.	5.55	23	28.55
AC&V duct spiral reducer 10"x8"	1.000	Ea.	23.50	41.50	65
AC&V duct spiral reducer 12"x10"	1.000	Ea.	23	55.50	78.50
AC&V duct spiral preformed 45° elbow, 8" diam	1.000	Ea.	8.85	47.50	56.35
AC&V utility fan, belt drive, 3 phase, 2000 CFM, 1 HP	2.000	Ea.	2,800	740	3,540
Safety switch, heavy duty fused, 240V, 3 pole, 30 amp	1.000	Ea.	96.50	224	320.50
TOTAL			4,765.20	2,100.32	6,865.52

D3090 320	Garage Exhaust Systems	COST PER BAY MAT.	COST PER BAY INST.	COST PER BAY TOTAL
1040	Garage, single 3" exhaust outlet, cars & light trucks, one bay	4,775	2,100	6,875
1060	Additional bays up to seven bays	1,200	560	1,760
1500	4" outlet, trucks, one bay	4,800	2,100	6,900
1520	Additional bays up to six bays	1,225	560	1,785
1600	5" outlet, diesel trucks, one bay	5,075	2,100	7,175
1650	Additional single bays up to six bays	1,650	660	2,310
1700	Two adjoining bays	5,075	2,100	7,175
2000	Dual exhaust, 3" outlets, pair of adjoining bays	6,025	2,625	8,650
2100	Additional pairs of adjoining bays	1,825	660	2,485

D3090 Other HVAC Systems/Equip

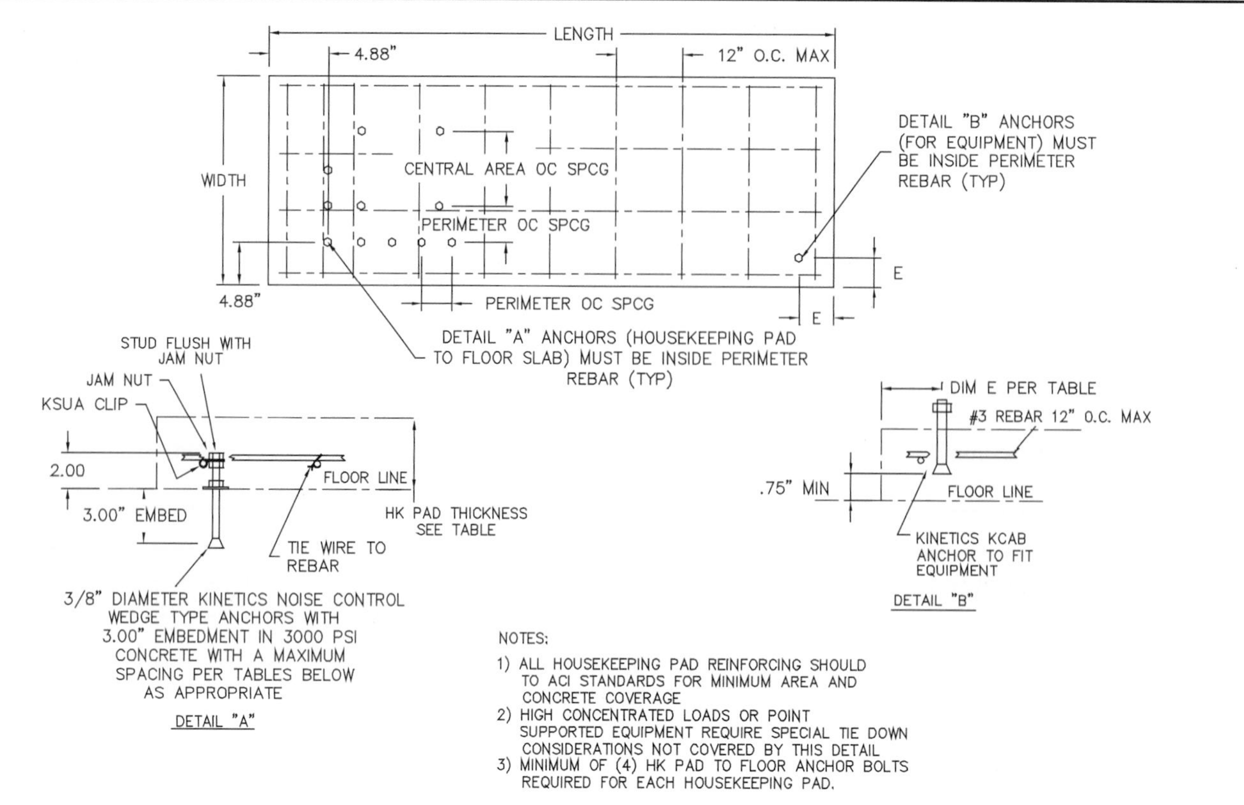

System Components	QUANTITY	UNIT	COST EACH MAT.	COST EACH INST.	COST EACH TOTAL
SYSTEM D3090 750 1000					
SEISMIC PAD FOR 20-38 SELF-CONTAINED UNIT					
Form equipment pad	28.000	L.F.	34.72	130.20	164.92
Reinforcing in hold down pad, #3 to #7	60.000	Lb.	34.20	37.80	72
Concrete 3000 PSI	.866	C.Y.	117.78		117.78
Place concrete pad	.866	C.Y.		34.73	34.73
Finish concrete equipment pad	47.000	S.F.		41.83	41.83
Drill for 3/8 inch slab anchors	10.000	Ea.	.60	99	99.60
Exisiting slab anchors 3/8 inch	10.000	Ea.	10.50	44.70	55.20
Equipment anchors 3/8 inch for plate/foot	8.000	Ea.	14.40	38.16	52.56
TOTAL			212.20	426.42	638.62

D3090 750	Seismic Concrete Equipment Pads	COST PER EACH MAT.	COST PER EACH INST.	COST PER EACH TOTAL
1000	Seismic pad for 20-38 tons Self-contained AC unit without economizer .25g	212	425	637
1010	0.5 g seismic acceleration	212	425	637
1020	0.75 g seismic acceleration	212	425	637
1030	1.0 g seismic acceleration	212	425	637
1040	1.5 g seismic acceleration	217	480	697
1050	2.0 g seismic acceleration	234	665	899
1060	2.5 g seismic acceleration	247	835	1,082
1100	Seismic pad for 40-80 tons Self contained AC unit without economizer .25g	335	620	955
1110	0.5 g seismic acceleration	335	620	955
1120	0.75 g seismic acceleration	335	620	955
1130	1.0 g seismic acceleration	335	620	955
1140	1.5 g seismic acceleration	365	880	1,245
1150	2.0 g seismic acceleration	425	1,375	1,800
1160	2.5 g seismic acceleration	495	1,950	2,445

D3090 Other HVAC Systems/Equip

D3090 750	Seismic Concrete Equipment Pads	COST PER EACH		
		MAT.	INST.	TOTAL
1200	Seismic pad for 80-110 tons Self contained AC unit without economizer .25g	365	670	1,035
1210	0.5 g seismic acceleration	365	670	1,035
1220	0.75 g seismic acceleration	365	670	1,035
1230	1.0 g seismic acceleration	365	670	1,035
1240	1.5 g seismic acceleration	400	950	1,350
1250	2.0 g seismic acceleration	470	1,500	1,970
1260	2.5 g seismic acceleration	545	2,125	2,670
1990	Seismic pad for self-contained AC units for isolator/snubber bolts			
2000	Seismic pad for 20-38 tons Self contained AC unit without economizer .25g	227	465	692
2010	0.5 g seismic acceleration	227	465	692
2020	0.75 g seismic acceleration	227	465	692
2030	1.0 g seismic acceleration	227	465	692
2040	1.5 g seismic acceleration	231	520	751
2050	2.0 g seismic acceleration	252	710	962
2060	2.5 g seismic acceleration	265	885	1,150
2100	Seismic pad for 40-80 tons Self contained AC unit without economizer .25g	350	665	1,015
2110	0.5 g seismic acceleration	350	665	1,015
2120	0.75 g seismic acceleration	350	665	1,015
2130	1.0 g seismic acceleration	350	665	1,015
2140	1.5 g seismic acceleration	380	930	1,310
2150	2.0 g seismic acceleration	450	1,425	1,875
2160	2.5 g seismic acceleration	525	2,025	2,550
2200	Seismic pad for 80-110 tons Self contained AC unit without economizer .25g	385	720	1,105
2210	0.5 g seismic acceleration	385	720	1,105
2220	0.75 g seismic acceleration	385	720	1,105
2230	1.0 g seismic acceleration	385	720	1,105
2240	1.5 g seismic acceleration	420	1,000	1,420
2250	2.0 g seismic acceleration	495	1,575	2,070
2260	2.5 g seismic acceleration	570	2,200	2,770
2990	Seismic pad for self-contained AC economizer units for plate/foot attachment			
3000	Seismic pad for 20-38 tons Self contained AC unit economizer .25 g-2.0 g	141	247	388
3010	2.5 g seismic acceleration	150	330	480
3100	Seismic pad for 40-80 tons Self contained AC unit economizer .25-1.5 g	189	330	519
3110	2.0 g seismic acceleration	195	380	575
3120	2.5 g seismic acceleration	203	450	653
3190	Seismic pad for self-contained AC unit economizer for isolator/snubber bolts			
3200	Seismic pad for 20-38 tons Self contained AC unit economizer .25 g-2.0 g	146	261	407
3210	2.5 g seismic acceleration	159	355	514
3300	Seismic pad for 40-80 tons Self contained AC unit economizer .25-1.5 g	199	355	554
3310	2.0 g seismic acceleration	205	410	615
3320	2.5 g seismic acceleration	214	480	694

G1030 Site Earthwork

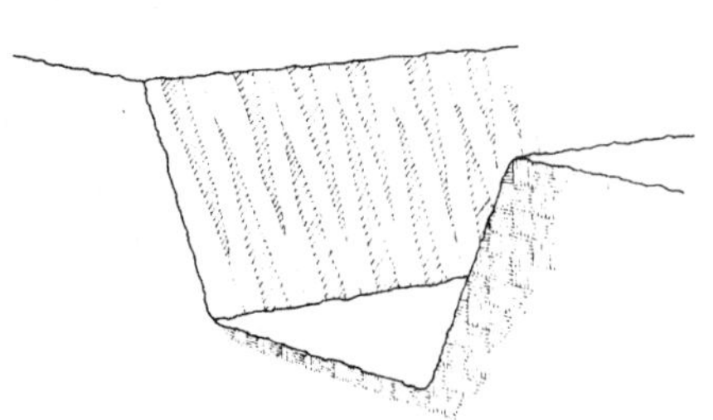

Trenching Systems are shown on a cost per linear foot basis. The systems include: excavation; backfill and removal of spoil; and compaction for various depths and trench bottom widths. The backfill has been reduced to accommodate a pipe of suitable diameter and bedding.

The slope for trench sides varies from none to 1:1.

The Expanded System Listing shows Trenching Systems that range from 2′ to 12′ in width. Depths range from 2′ to 25′.

System Components	QUANTITY	UNIT	COST PER L.F. EQUIP.	LABOR	TOTAL
SYSTEM G1030 805 1310					
TRENCHING, COMMON EARTH, NO SLOPE, 2′ WIDE, 2′ DP, 3/8 C.Y. BUCKET					
Excavation, trench, hyd. backhoe, track mtd., 3/8 C.Y. bucket	.148	B.C.Y.	.35	1.14	1.49
Backfill and load spoil, from stockpile	.153	L.C.Y.	.13	.35	.48
Compaction by vibrating plate, 6″ lifts, 4 passes	.118	E.C.Y.	.03	.42	.45
Remove excess spoil, 8 C.Y. dump truck, 2 mile roundtrip	.040	L.C.Y.	.12	.19	.31
TOTAL			.63	2.10	2.73

G1030 805	Trenching Common Earth	COST PER L.F. EQUIP.	LABOR	TOTAL
1310	Trenching, common earth, no slope, 2′ wide, 2′ deep, 3/8 C.Y. bucket	.63	2.10	2.73
1320	3′ deep, 3/8 C.Y. bucket	.90	3.15	4.05
1330	4′ deep, 3/8 C.Y. bucket	1.16	4.20	5.36
1340	6′ deep, 3/8 C.Y. bucket	1.61	5.45	7.06
1350	8′ deep, 1/2 C.Y. bucket	2.14	7.25	9.39
1360	10′ deep, 1 C.Y. bucket	3.51	8.60	12.11
1400	4′ wide, 2′ deep, 3/8 C.Y. bucket	1.40	4.16	5.56
1410	3′ deep, 3/8 C.Y. bucket	1.94	6.25	8.19
1420	4′ deep, 1/2 C.Y. bucket	2.40	7	9.40
1430	6′ deep, 1/2 C.Y. bucket	3.97	11.25	15.22
1440	8′ deep, 1/2 C.Y. bucket	6.65	14.55	21.20
1450	10′ deep, 1 C.Y. bucket	8.10	18.05	26.15
1460	12′ deep, 1 C.Y. bucket	10.45	23	33.45
1470	15′ deep, 1-1/2 C.Y. bucket	8.80	20.50	29.30
1480	18′ deep, 2-1/2 C.Y. bucket	12.40	29	41.40
1520	6′ wide, 6′ deep, 5/8 C.Y. bucket w/trench box	8.45	16.85	25.30
1530	8′ deep, 3/4 C.Y. bucket	11.25	22	33.25
1540	10′ deep, 1 C.Y. bucket	10.85	23	33.85
1550	12′ deep, 1-1/2 C.Y. bucket	11.50	24.50	36
1560	16′ deep, 2-1/2 C.Y. bucket	15.85	30.50	46.35
1570	20′ deep, 3-1/2 C.Y. bucket	21	36.50	57.50
1580	24′ deep, 3-1/2 C.Y. bucket	25	44	69
1640	8′ wide, 12′ deep, 1-1/2 C.Y. bucket w/trench box	15.95	31	46.95
1650	15′ deep, 1-1/2 C.Y. bucket	21	41	62
1660	18′ deep, 2-1/2 C.Y. bucket	22.50	41	63.50
1680	24′ deep, 3-1/2 C.Y. bucket	33.50	57	90.50
1730	10′ wide, 20′ deep, 3-1/2 C.Y. bucket w/trench box	27.50	54	81.50
1740	24′ deep, 3-1/2 C.Y. bucket	40	65	105
1780	12′ wide, 20′ deep, 3-1/2 C.Y. bucket w/trench box	42.50	68.50	111
1790	25′ deep, bucket	52.50	87.50	140
1800	1/2 to 1 slope, 2′ wide, 2′ deep, 3/8 C.Y. bucket	.90	3.15	4.05
1810	3′ deep, 3/8 C.Y. bucket	1.50	5.50	7

G1030 Site Earthwork

G1030 805	Trenching Common Earth	COST PER L.F.		
		EQUIP.	LABOR	TOTAL
1820	4' deep, 3/8 C.Y. bucket	2.25	8.40	10.65
1840	6' deep, 3/8 C.Y. bucket	3.90	13.65	17.55
1860	8' deep, 1/2 C.Y. bucket	6.20	22	28.20
1880	10' deep, 1 C.Y. bucket	12.10	30.50	42.60
2300	4' wide, 2' deep, 3/8 C.Y. bucket	1.67	5.20	6.87
2310	3' deep, 3/8 C.Y. bucket	2.56	8.65	11.21
2320	4' deep, 1/2 C.Y. bucket	3.43	10.65	14.08
2340	6' deep, 1/2 C.Y. bucket	6.75	20	26.75
2360	8' deep, 1/2 C.Y. bucket	12.95	29.50	42.45
2380	10' deep, 1 C.Y. bucket	18	41.50	59.50
2400	12' deep, 1 C.Y. bucket	23	56	79
2430	15' deep, 1-1/2 C.Y. bucket	25	60	85
2460	18' deep, 2-1/2 C.Y. bucket	43.50	93.50	137
2840	6' wide, 6' deep, 5/8 C.Y. bucket w/trench box	12.55	25	37.55
2860	8' deep, 3/4 C.Y. bucket	18.30	37.50	55.80
2880	10' deep, 1 C.Y. bucket	17.35	38	55.35
2900	12' deep, 1-1/2 C.Y. bucket	22	50	72
2940	16' deep, 2-1/2 C.Y. bucket	36.50	73	109.50
2980	20' deep, 3-1/2 C.Y. bucket	52.50	98	150.50
3020	24' deep, 3-1/2 C.Y. bucket	74	134	208
3100	8' wide, 12' deep, 1-1/2 C.Y. bucket w/trench box	27	57	84
3120	15' deep, 1-1/2 C.Y. bucket	39.50	82	121.50
3140	18' deep, 2-1/2 C.Y. bucket	50	97.50	147.50
3180	24' deep, 3-1/2 C.Y. bucket	83	147	230
3270	10' wide, 20' deep, 3-1/2 C.Y. bucket w/trench box	53.50	113	166.50
3280	24' deep, 3-1/2 C.Y. bucket	91.50	161	252.50
3370	12' wide, 20' deep, 3-1/2 C.Y. bucket w/trench box	76.50	131	207.50
3380	25' deep, 3-1/2 C.Y. bucket	107	188	295
3500	1 to 1 slope, 2' wide, 2' deep, 3/8 C.Y. bucket	1.16	4.21	5.37
3520	3' deep, 3/8 C.Y. bucket	3.24	9.60	12.84
3540	4' deep, 3/8 C.Y. bucket	3.34	12.65	15.99
3560	6' deep, 1/2 C.Y. bucket	3.89	13.65	17.54
3580	8' deep, 1/2 C.Y. bucket	7.75	27.50	35.25
3600	10' deep, 1 C.Y. bucket	20.50	52	72.50
3800	4' wide, 2' deep, 3/8 C.Y. bucket	1.94	6.25	8.19
3820	3' deep, 3/8 C.Y. bucket	3.17	11	14.17
3840	4' deep, 1/2 C.Y. bucket	4.44	14.30	18.74
3860	6' deep, 1/2 C.Y. bucket	9.50	28.50	38
3880	8' deep, 1/2 C.Y. bucket	19.30	44	63.30
3900	10' deep, 1 C.Y. bucket	28	64.50	92.50
3920	12' deep, 1 C.Y. bucket	41	93.50	134.50
3940	15' deep, 1-1/2 C.Y. bucket	41	99.50	140.50
3960	18' deep, 2-1/2 C.Y. bucket	60	129	189
4030	6' wide, 6' deep, 5/8 C.Y. bucket w/trench box	16.55	33.50	50.05
4040	8' deep, 3/4 C.Y. bucket	24	46	70
4050	10' deep, 1 C.Y. bucket	25	56	81
4060	12' deep, 1-1/2 C.Y. bucket	33.50	76	109.50
4070	16' deep, 2-1/2 C.Y. bucket	57	115	172
4080	20' deep, 3-1/2 C.Y. bucket	84.50	160	244.50
4090	24' deep, 3-1/2 C.Y. bucket	123	225	348
4500	8' wide, 12' deep, 1-1/2 C.Y. bucket w/trench box	38	82.50	120.50
4550	15' deep, 1-1/2 C.Y. bucket	57.50	124	181.50
4600	18' deep, 2-1/2 C.Y. bucket	76	151	227
4650	24' deep, 3-1/2 C.Y. bucket	132	238	370
4800	10' wide, 20' deep, 3-1/2 C.Y. bucket w/trench box	79.50	172	251.50
4850	24' deep, 3-1/2 C.Y. bucket	141	251	392
4950	12' wide, 20' deep, 3-1/2 C.Y. bucket w/trench box	111	194	305
4980	25' deep, 3-1/2 C.Y. bucket	160	284	444

G1030 Site Earthwork

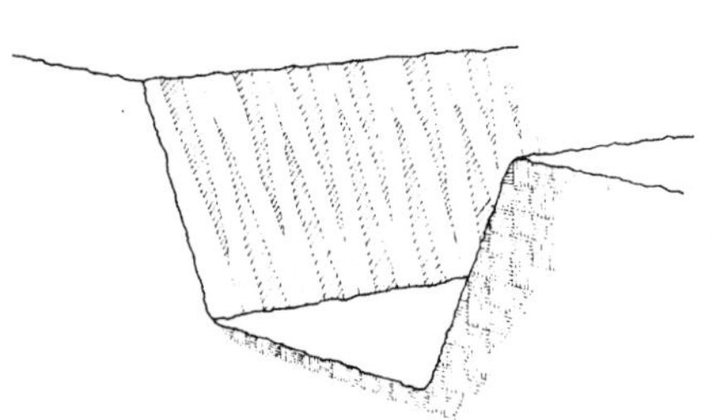

Trenching Systems are shown on a cost per linear foot basis. The systems include: excavation; backfill and removal of spoil; and compaction for various depths and trench bottom widths. The backfill has been reduced to accommodate a pipe of suitable diameter and bedding.

The slope for trench sides varies from none to 1:1.

The Expanded System Listing shows Trenching Systems that range from 2′ to 12′ in width. Depths range from 2′ to 25′.

System Components	QUANTITY	UNIT	COST PER L.F. EQUIP.	COST PER L.F. LABOR	COST PER L.F. TOTAL
SYSTEM G1030 806 1310					
TRENCHING, LOAM & SANDY CLAY, NO SLOPE, 2′ WIDE, 2′ DP, 3/8 C.Y. BUCKET					
Excavation, trench, hyd. backhoe, track mtd., 3/8 C.Y. bucket	.148	B.C.Y.	.32	1.06	1.38
Backfill and load spoil, from stockpile	.165	L.C.Y.	.14	.38	.52
Compaction by vibrating plate 18″ wide, 6″ lifts, 4 passes	.118	E.C.Y.	.03	.42	.45
Remove excess spoil, 8 C.Y. dump truck, 2 mile roundtrip	.042	L.C.Y.	.13	.20	.33
TOTAL			.62	2.06	2.68

G1030 806	Trenching Loam & Sandy Clay	COST PER L.F. EQUIP.	COST PER L.F. LABOR	COST PER L.F. TOTAL
1310	Trenching, loam & sandy clay, no slope, 2′ wide, 2′ deep, 3/8 C.Y. bucket	.62	2.06	2.68
1320	3′ deep, 3/8 C.Y. bucket	.97	3.37	4.34
1330	4′ deep, 3/8 C.Y. bucket	1.14	4.11	5.25
1340	6′ deep, 3/8 C.Y. bucket	1.75	4.90	6.65
1350	8′ deep, 1/2 C.Y. bucket	2.33	6.45	8.80
1360	10′ deep, 1 C.Y. bucket	2.53	6.90	9.45
1400	4′ wide, 2′ deep, 3/8 C.Y. bucket	1.39	4.09	5.50
1410	3′ deep, 3/8 C.Y. bucket	1.92	6.15	8.05
1420	4′ deep, 1/2 C.Y. bucket	2.39	6.90	9.30
1430	6′ deep, 1/2 C.Y. bucket	4.26	10.15	14.40
1440	8′ deep, 1/2 C.Y. bucket	6.50	14.35	21
1450	10′ deep, 1 C.Y. bucket	6.15	14.70	21
1460	12′ deep, 1 C.Y. bucket	7.70	18.30	26
1470	15′ deep, 1-1/2 C.Y. bucket	9.15	21.50	30.50
1480	18′ deep, 2-1/2 C.Y. bucket	10.85	23.50	34.50
1520	6′ wide, 6′ deep, 5/8 C.Y. bucket w/trench box	8.15	16.55	24.50
1530	8′ deep, 3/4 C.Y. bucket	10.85	21.50	32.50
1540	10′ deep, 1 C.Y. bucket	10	22	32
1550	12′ deep, 1-1/2 C.Y. bucket	11.25	24.50	36
1560	16′ deep, 2-1/2 C.Y. bucket	15.50	31	46.50
1570	20′ deep, 3-1/2 C.Y. bucket	19.30	36.50	56
1580	24′ deep, 3-1/2 C.Y. bucket	24	44.50	68.50
1640	8′ wide, 12′ deep, 1-1/4 C.Y. bucket w/trench box	15.75	31	47
1650	15′ deep, 1-1/2 C.Y. bucket	19.30	39.50	59
1660	18′ deep, 2-1/2 C.Y. bucket	23.50	44.50	68
1680	24′ deep, 3-1/2 C.Y. bucket	32.50	57.50	90
1730	10′ wide, 20′ deep, 3-1/2 C.Y. bucket w/trench box	33	58	91
1740	24′ deep, 3-1/2 C.Y. bucket	41	71.50	113
1780	12′ wide, 20′ deep, 3-1/2 C.Y. bucket w/trench box	39.50	69	109
1790	25′ deep, 3-1/2 C.Y. bucket	51.50	88.50	140
1800	1/2:1 slope, 2′ wide, 2′ deep, 3/8 C.Y. bucket	.88	3.08	3.96
1810	3′ deep, 3/8 C.Y. bucket	1.47	5.40	6.85
1820	4′ deep, 3/8 C.Y. bucket	2.21	8.25	10.45
1840	6′ deep, 3/8 C.Y. bucket	4.23	12.25	16.50

G1030 Site Earthwork

G1030 806	Trenching Loam & Sandy Clay	COST PER L.F.		
		EQUIP.	LABOR	TOTAL
1860	8′ deep, 1/2 C.Y. bucket	6.75	19.55	26.50
1880	10′ deep, 1 C.Y. bucket	8.65	24.50	33
2300	4′ wide, 2′ deep, 3/8 C.Y. bucket	1.65	5.10	6.75
2310	3′ deep, 3/8 C.Y. bucket	2.52	8.45	10.95
2320	4′ deep, 1/2 C.Y. bucket	3.39	10.45	13.85
2340	6′ deep, 1/2 C.Y. bucket	7.20	18	25
2360	8′ deep, 1/2 C.Y. bucket	12.60	29	41.50
2380	10′ deep, 1 C.Y. bucket	13.55	33.50	47
2400	12′ deep, 1 C.Y. bucket	22.50	56	78.50
2430	15′ deep, 1-1/2 C.Y. bucket	26	62	88
2460	18′ deep, 2-1/2 C.Y. bucket	43	95	138
2840	6′ wide, 6′ deep, 5/8 C.Y. bucket w/trench box	12	25	37
2860	8′ deep, 3/4 C.Y. bucket	17.60	36.50	54
2880	10′ deep, 1 C.Y. bucket	18	41	59
2900	12′ deep, 1-1/2 C.Y. bucket	22	50.50	72.50
2940	16′ deep, 2-1/2 C.Y. bucket	35.50	74	110
2980	20′ deep, 3-1/2 C.Y. bucket	51	99.50	151
3020	24′ deep, 3-1/2 C.Y. bucket	72	136	208
3100	8′ wide, 12′ deep, 1-1/2 C.Y. bucket w/trench box	26.50	57	83.50
3120	15′ deep, 1-1/2 C.Y. bucket	36	80	116
3140	18′ deep, 2-1/2 C.Y. bucket	49	99	148
3180	24′ deep, 3-1/2 C.Y. bucket	80.50	150	231
3270	10′ wide, 20′ deep, 3-1/2 C.Y. bucket w/trench box	64.50	121	186
3280	24′ deep, 3-1/2 C.Y. bucket	89	163	252
3320	12′ wide, 20′ deep, 3-1/2 C.Y. bucket w/trench box	71.50	132	204
3380	25′ deep, 3-1/2 C.Y. bucket w/trench box	97	177	274
3500	1:1 slope, 2′ wide, 2′ deep, 3/8 C.Y. bucket	1.14	4.12	5.25
3520	3′ deep, 3/8 C.Y. bucket	2.07	7.70	9.75
3540	4′ deep, 3/8 C.Y. bucket	3.26	12.35	15.60
3560	6′ deep, 1/2 C.Y. bucket	4.23	12.25	16.50
3580	8′ deep, 1/2 C.Y. bucket	11.15	32.50	43.50
3600	10′ deep, 1 C.Y. bucket	14.75	42	57
3800	4′ wide, 2′ deep, 3/8 C.Y. bucket	1.92	6.15	8.05
3820	3′ deep, 1/2 C.Y. bucket	3.11	10.75	13.85
3840	4′ deep, 1/2 C.Y. bucket	4.38	14.05	18.45
3860	6′ deep, 1/2 C.Y. bucket	10.20	26	36
3880	8′ deep, 1/2 C.Y. bucket	18.75	43.50	62.50
3900	10′ deep, 1 C.Y. bucket	21	52.50	73.50
3920	12′ deep, 1 C.Y. bucket	30	74.50	105
3940	15′ deep, 1-1/2 C.Y. bucket	42.50	103	146
3960	18′ deep, 2-1/2 C.Y. bucket	59	131	190
4030	6′ wide, 6′ deep, 5/8 C.Y. bucket w/trench box	15.80	33.50	49.50
4040	8′ deep, 3/4 C.Y. bucket	24.50	51.50	76
4050	10′ deep, 1 C.Y. bucket	26	60.50	86.50
4060	12′ deep, 1-1/2 C.Y. bucket	33	76	109
4070	16′ deep, 2-1/2 C.Y. bucket	55.50	117	173
4080	20′ deep, 3-1/2 C.Y. bucket	82.50	163	246
4090	24′ deep, 3-1/2 C.Y. bucket	120	228	350
4500	8′ wide, 12′ deep, 1-1/4 C.Y. bucket w/trench box	37.50	82.50	120
4550	15′ deep, 1-1/2 C.Y. bucket	53	120	173
4600	18′ deep, 2-1/2 C.Y. bucket	74.50	153	228
4650	24′ deep, 3-1/2 C.Y. bucket	128	241	370
4800	10′ wide, 20′ deep, 3-1/2 C.Y. bucket w/trench box	96.50	184	281
4850	24′ deep, 3-1/2 C.Y. bucket	137	255	390
4950	12′ wide, 20′ deep, 3-1/2 C.Y. bucket w/trench box	103	195	298
4980	25′ deep, 3-1/2 C.Y. bucket	155	288	445

G1030 Site Earthwork

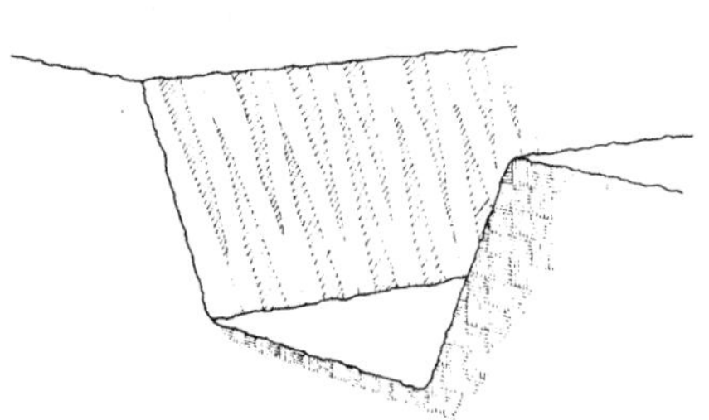

Trenching Systems are shown on a cost per linear foot basis. The systems include: excavation; backfill and removal of spoil; and compaction for various depths and trench bottom widths. The backfill has been reduced to accommodate a pipe of suitable diameter and bedding.

The slope for trench sides varies from none to 1:1.

The Expanded System Listing shows Trenching Systems that range from 2′ to 12′ in width. Depths range from 2′ to 25′.

System Components	QUANTITY	UNIT	COST PER L.F. EQUIP.	LABOR	TOTAL
SYSTEM G1030 807 1310					
TRENCHING, SAND & GRAVEL, NO SLOPE, 2′ WIDE, 2′ DEEP, 3/8 C.Y. BUCKET					
Excavation, trench, hyd. backhoe, track mtd., 3/8 C.Y. bucket	.148	B.C.Y.	.32	1.04	1.36
Backfill and load spoil, from stockpile	.140	L.C.Y.	.11	.32	.43
Compaction by vibrating plate 18″ wide, 6″ lifts, 4 passes	.118	E.C.Y.	.03	.42	.45
Remove excess spoil, 8 C.Y. dump truck, 2 mile roundtrip	.035	L.C.Y.	.11	.17	.28
TOTAL			.57	1.95	2.52

G1030 807	Trenching Sand & Gravel	COST PER L.F. EQUIP.	LABOR	TOTAL
1310	Trenching, sand & gravel, no slope, 2′ wide, 2′ deep, 3/8 C.Y. bucket	.57	1.95	2.52
1320	3′ deep, 3/8 C.Y. bucket	.91	3.23	4.14
1330	4′ deep, 3/8 C.Y. bucket	1.06	3.89	4.95
1340	6′ deep, 3/8 C.Y. bucket	1.65	4.67	6.30
1350	8′ deep, 1/2 C.Y. bucket	2.19	6.20	8.40
1360	10′ deep, 1 C.Y. bucket	2.35	6.50	8.85
1400	4′ wide, 2′ deep, 3/8 C.Y. bucket	1.27	3.84	5.10
1410	3′ deep, 3/8 C.Y. bucket	1.78	5.80	7.60
1420	4′ deep, 1/2 C.Y. bucket	2.20	6.50	8.70
1430	6′ deep, 1/2 C.Y. bucket	4.05	9.70	13.75
1440	8′ deep, 1/2 C.Y. bucket	6.10	13.65	19.75
1450	10′ deep, 1 C.Y. bucket	5.75	13.90	19.65
1460	12′ deep, 1 C.Y. bucket	7.25	17.30	24.50
1470	15′ deep, 1-1/2 C.Y. bucket	8.60	20	28.50
1480	18′ deep, 2-1/2 C.Y. bucket	10.25	22	32.50
1520	6′ wide, 6′ deep, 5/8 C.Y. bucket w/trench box	7.70	15.65	23.50
1530	8′ deep, 3/4 C.Y. bucket	10.30	20.50	31
1540	10′ deep, 1 C.Y. bucket	9.45	20.50	30
1550	12′ deep, 1-1/2 C.Y. bucket	10.60	23	33.50
1560	16′ deep, 2 C.Y. bucket	14.80	29	44
1570	20′ deep, 3-1/2 C.Y. bucket	18.25	34.50	53
1580	24′ deep, 3-1/2 C.Y. bucket	23	42	65
1640	8′ wide, 12′ deep, 1-1/2 C.Y. bucket w/trench box	14.80	29.50	44.50
1650	15′ deep, 1-1/2 C.Y. bucket	18	37.50	55.50
1660	18′ deep, 2-1/2 C.Y. bucket	22	42	64
1680	24′ deep, 3-1/2 C.Y. bucket	31	54	85
1730	10′ wide, 20′ deep, 3-1/2 C.Y. bucket w/trench box	31	54.50	85.50
1740	24′ deep, 3-1/2 C.Y. bucket	38.50	67	106
1780	12′ wide, 20′ deep, 3-1/2 C.Y. bucket w/trench box	37.50	64.50	102
1790	25′ deep, 3-1/2 C.Y. bucket	48.50	83	132
1800	1/2:1 slope, 2′ wide, 2′ deep, 3/8 C.Y. bucket	.82	2.92	3.74
1810	3′ deep, 3/8 C.Y. bucket	1.38	5.10	6.50
1820	4′ deep, 3/8 C.Y. bucket	2.05	7.85	9.90
1840	6′ deep, 3/8 C.Y. bucket	4.03	11.70	15.75

G1030 Site Earthwork

G1030 807	Trenching Sand & Gravel	COST PER L.F.		
		EQUIP.	LABOR	TOTAL
1860	8' deep, 1/2 C.Y. bucket	6.40	18.65	25
1880	10' deep, 1 C.Y. bucket	8.05	23	31
2300	4' wide, 2' deep, 3/8 C.Y. bucket	1.53	4.82	6.35
2310	3' deep, 3/8 C.Y. bucket	2.33	8	10.35
2320	4' deep, 1/2 C.Y. bucket	3.14	9.90	13.05
2340	6' deep, 1/2 C.Y. bucket	6.90	17.25	24
2360	8' deep, 1/2 C.Y. bucket	11.95	27.50	39.50
2380	10' deep, 1 C.Y. bucket	12.80	32	45
2400	12' deep, 1 C.Y. bucket	21.50	53	74.50
2430	15' deep, 1-1/2 C.Y. bucket	24.50	58.50	83
2460	18' deep, 2-1/2 C.Y. bucket	40.50	89.50	130
2840	6' wide, 6' deep, 5/8 C.Y. bucket w/trench box	11.35	24	35.50
2860	8' deep, 3/4 C.Y. bucket	16.75	35	52
2880	10' deep, 1 C.Y. bucket	17.05	39	56
2900	12' deep, 1-1/2 C.Y. bucket	21	47.50	68.50
2940	16' deep, 2 C.Y. bucket	34	69.50	104
2980	20' deep, 3-1/2 C.Y. bucket	48.50	93.50	142
3020	24' deep, 3-1/2 C.Y. bucket	68.50	128	197
3100	8' wide, 12' deep, 1-1/4 C.Y. bucket w/trench box	25	54	79
3120	15' deep, 1-1/2 C.Y. bucket	34	75.50	110
3140	18' deep, 2-1/2 C.Y. bucket	46.50	93	140
3180	24' deep, 3-1/2 C.Y. bucket	76.50	140	217
3270	10' wide, 20' deep, 3-1/2 C.Y. bucket w/trench box	61	113	174
3280	24' deep, 3-1/2 C.Y. bucket	84	153	237
3370	12' wide, 20' deep, 3-1/2 C.Y. bucket w/trench box	68	126	194
3380	25' deep, 3-1/2 C.Y. bucket	98	177	275
3500	1:1 slope, 2' wide, 2' deep, 3/8 C.Y. bucket	1.76	4.89	6.65
3520	3' deep, 3/8 C.Y. bucket	1.93	7.35	9.30
3540	4' deep, 3/8 C.Y. bucket	3.04	11.75	14.80
3560	6' deep, 3/8 C.Y. bucket	4.03	11.70	15.75
3580	8' deep, 1/2 C.Y. bucket	10.60	31	41.50
3600	10' deep, 1 C.Y. bucket	13.75	39.50	53.50
3800	4' wide, 2' deep, 3/8 C.Y. bucket	1.78	5.80	7.60
3820	3' deep, 3/8 C.Y. bucket	2.89	10.20	13.10
3840	4' deep, 1/2 C.Y. bucket	4.07	13.30	17.35
3860	6' deep, 1/2 C.Y. bucket	9.75	25	35
3880	8' deep, 1/2 C.Y. bucket	17.80	41.50	59.50
3900	10' deep, 1 C.Y. bucket	19.80	50	70
3920	12' deep, 1 C.Y. bucket	28.50	70.50	99
3940	15' deep, 1-1/2 C.Y. bucket	40.50	97	138
3960	18' deep, 2-1/2 C.Y. bucket	56	123	179
4030	6' wide, 6' deep, 5/8 C.Y. bucket w/trench box	15	32	47
4040	8' deep, 3/4 C.Y. bucket	23	49	72
4050	10' deep, 1 C.Y. bucket	24.50	57	81.50
4060	12' deep, 1-1/2 C.Y. bucket	31	71.50	103
4070	16' deep, 2 C.Y. bucket	53.50	110	164
4080	20' deep, 3-1/2 C.Y. bucket	78.50	153	232
4090	24' deep, 3-1/2 C.Y. bucket	114	214	330
4500	8' wide, 12' deep, 1-1/2 C.Y. bucket w/trench box	35.50	78	114
4550	15' deep, 1-1/2 C.Y. bucket	50	113	163
4600	18' deep, 2-1/2 C.Y. bucket	70.50	144	215
4650	24' deep, 3-1/2 C.Y. bucket	122	227	350
4800	10' wide, 20' deep, 3-1/2 C.Y. bucket w/trench box	91	173	264
4850	24' deep, 3-1/2 C.Y. bucket	130	239	370
4950	12' wide, 20' deep, 3-1/2 C.Y. bucket w/trench box	97.50	183	281
4980	25' deep, 3-1/2 C.Y. bucket	147	270	415

G1030 Site Earthwork

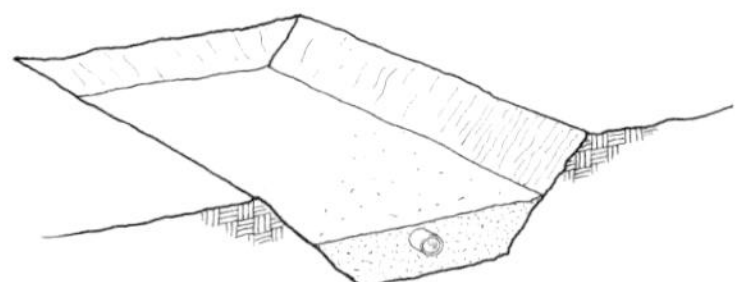

The Pipe Bedding System is shown for various pipe diameters. Compacted bank sand is used for pipe bedding and to fill 12″ over the pipe. No backfill is included. Various side slopes are shown to accommodate different soil conditions. Pipe sizes vary from 6″ to 84″ diameter.

System Components	QUANTITY	UNIT	COST PER L.F. MAT.	COST PER L.F. INST.	COST PER L.F. TOTAL
SYSTEM G1030 815 1440					
PIPE BEDDING, SIDE SLOPE 0 TO 1, 1′ WIDE, PIPE SIZE 6″ DIAMETER					
Borrow, bank sand, 2 mile haul, machine spread	.086	C.Y.	1.70	.67	2.37
Compaction, vibrating plate	.086	C.Y.		.22	.22
TOTAL			1.70	.89	2.59

G1030 815	Pipe Bedding	COST PER L.F. MAT.	COST PER L.F. INST.	COST PER L.F. TOTAL
1440	Pipe bedding, side slope 0 to 1, 1′ wide, pipe size 6″ diameter	1.70	.89	2.59
1460	2′ wide, pipe size 8″ diameter	3.68	1.94	5.62
1480	Pipe size 10″ diameter	3.76	1.98	5.74
1500	Pipe size 12″ diameter	3.84	2.03	5.87
1520	3′ wide, pipe size 14″ diameter	6.25	3.28	9.53
1540	Pipe size 15″ diameter	6.30	3.30	9.60
1560	Pipe size 16″ diameter	6.35	3.34	9.69
1580	Pipe size 18″ diameter	6.45	3.42	9.87
1600	4′ wide, pipe size 20″ diameter	9.20	4.85	14.05
1620	Pipe size 21″ diameter	9.30	4.89	14.19
1640	Pipe size 24″ diameter	9.45	4.99	14.44
1660	Pipe size 30″ diameter	9.65	5.10	14.75
1680	6′ wide, pipe size 32″ diameter	16.50	8.70	25.20
1700	Pipe size 36″ diameter	16.90	8.90	25.80
1720	7′ wide, pipe size 48″ diameter	27	14.15	41.15
1740	8′ wide, pipe size 60″ diameter	32.50	17.25	49.75
1760	10′ wide, pipe size 72″ diameter	45.50	24	69.50
1780	12′ wide, pipe size 84″ diameter	60	31.50	91.50
2140	Side slope 1/2 to 1, 1′ wide, pipe size 6″ diameter	3.17	1.68	4.85
2160	2′ wide, pipe size 8″ diameter	5.40	2.84	8.24
2180	Pipe size 10″ diameter	5.80	3.05	8.85
2200	Pipe size 12″ diameter	6.15	3.23	9.38
2220	3′ wide, pipe size 14″ diameter	8.85	4.67	13.52
2240	Pipe size 15″ diameter	9.05	4.78	13.83
2260	Pipe size 16″ diameter	9.30	4.90	14.20
2280	Pipe size 18″ diameter	9.75	5.15	14.90
2300	4′ wide, pipe size 20″ diameter	12.90	6.80	19.70
2320	Pipe size 21″ diameter	13.20	6.95	20.15
2340	Pipe size 24″ diameter	14	7.40	21.40
2360	Pipe size 30″ diameter	15.55	8.20	23.75
2380	6′ wide, pipe size 32″ diameter	23	12.10	35.10
2400	Pipe size 36″ diameter	24.50	12.85	37.35
2420	7′ wide, pipe size 48″ diameter	38	20	58
2440	8′ wide, pipe size 60″ diameter	48.50	25.50	74
2460	10′ wide, pipe size 72″ diameter	66	35	101
2480	12′ wide, pipe size 84″ diameter	86.50	45.50	132
2620	Side slope 1 to 1, 1′ wide, pipe size 6″ diameter	4.65	2.45	7.10
2640	2′ wide, pipe size 8″ diameter	7.15	3.77	10.92

G1030 Site Earthwork

G1030 815	Pipe Bedding	COST PER L.F.		
		MAT.	INST.	TOTAL
2660	Pipe size 10" diameter	7.80	4.11	11.91
2680	Pipe size 12" diameter	8.50	4.47	12.97
2700	3' wide, pipe size 14" diameter	11.50	6.05	17.55
2720	Pipe size 15" diameter	11.85	6.25	18.10
2740	Pipe size 16" diameter	12.25	6.45	18.70
2760	Pipe size 18" diameter	13.05	6.90	19.95
2780	4' wide, pipe size 20" diameter	16.60	8.75	25.35
2800	Pipe size 21" diameter	17.05	9	26.05
2820	Pipe size 24" diameter	18.50	9.75	28.25
2840	Pipe size 30" diameter	21.50	11.25	32.75
2860	6' wide, pipe size 32" diameter	29.50	15.50	45
2880	Pipe size 36" diameter	32	16.75	48.75
2900	7' wide, pipe size 48" diameter	49	26	75
2920	8' wide, pipe size 60" diameter	64	33.50	97.50
2940	10' wide, pipe size 72" diameter	87	45.50	132.50
2960	12' wide, pipe size 84" diameter	113	60	173

G3010 Water Supply

G3010 110	Water Distribution Piping	COST PER L.F.		
		MAT.	INST.	TOTAL
2000	Piping, excav. & backfill excl., ductile iron class 250, mech. joint			
2130	4" diameter	39.50	17.40	56.90
2150	6" diameter	47.50	22	69.50
2160	8" diameter	58.50	26	84.50
2170	10" diameter	78	30.50	108.50
2180	12" diameter	99	33	132
2210	16" diameter	114	47.50	161.50
2220	18" diameter	140	50.50	190.50
3000	Tyton joint			
3130	4" diameter	22.50	8.70	31.20
3150	6" diameter	25	10.45	35.45
3160	8" diameter	34.50	17.40	51.90
3170	10" diameter	51.50	19.15	70.65
3180	12" diameter	54	22	76
3210	16" diameter	63	30.50	93.50
3220	18" diameter	70	34.50	104.50
3230	20" diameter	73	39	112
3250	24" diameter	100	45.50	145.50
4000	Copper tubing, type K			
4050	3/4" diameter	7.70	3.41	11.11
4060	1" diameter	10.70	4.26	14.96
4080	1-1/2" diameter	16.30	5.15	21.45
4090	2" diameter	25	5.95	30.95
4110	3" diameter	53	10.20	63.20
4130	4" diameter	88.50	14.35	102.85
4150	6" diameter	144	26.50	170.50
5000	Polyvinyl chloride class 160, S.D.R. 26			
5130	1-1/2" diameter	.57	1.27	1.84
5150	2" diameter	.97	1.39	2.36
5160	4" diameter	3.08	6.25	9.33
5170	6" diameter	5.55	7.55	13.10
5180	8" diameter	9.35	9	18.35
6000	Polyethylene 160 psi, S.D.R. 7			
6050	3/4" diameter	.45	1.82	2.27
6060	1" diameter	.61	1.97	2.58
6080	1-1/2" diameter	.98	2.12	3.10
6090	2" diameter	1.90	2.61	4.51

G3020 Sanitary Sewer

G3020 110	Drainage & Sewage Piping	COST PER L.F.		
		MAT.	INST.	TOTAL
2000	Piping, excavation & backfill excluded, PVC, plain			
2130	4″ diameter	1.84	4.45	6.29
2150	6″ diameter	4.04	4.76	8.80
2160	8″ diameter	7.25	4.98	12.23
2900	Box culvert, precast, 8′ long			
3000	6′ x 3′	305	35	340
3020	6′ x 7′	365	39.50	404.50
3040	8′ x 3′	400	43.50	443.50
3060	8′ x 8′	475	49.50	524.50
3080	10′ x 3′	460	45	505
3100	10′ x 8′	700	61.50	761.50
3120	12′ x 3′	865	49.50	914.50
3140	12′ x 8′	1,050	73.50	1,123.50
4000	Concrete, nonreinforced			
4150	6″ diameter	8.45	13.15	21.60
4160	8″ diameter	9.25	15.50	24.75
4170	10″ diameter	10.30	16.15	26.45
4180	12″ diameter	12	17.40	29.40
4200	15″ diameter	16.85	19.35	36.20
4220	18″ diameter	20.50	24	44.50
4250	24″ diameter	24.50	35	59.50
4400	Reinforced, no gasket			
4580	12″ diameter	15.60	23.50	39.10
4600	15″ diameter	21	23.50	44.50
4620	18″ diameter	25.50	26	51.50
4650	24″ diameter	33.50	35	68.50
4670	30″ diameter	57.50	50.50	108
4680	36″ diameter	84.50	61.50	146
4690	42″ diameter	108	67.50	175.50
4700	48″ diameter	103	76	179
4720	60″ diameter	192	102	294
4730	72″ diameter	310	122	432
4740	84″ diameter	390	152	542
4800	With gasket			
4980	12″ diameter	17.15	12.85	30
5000	15″ diameter	23	13.50	36.50
5020	18″ diameter	28	14.20	42.20
5050	24″ diameter	41	15.85	56.85
5070	30″ diameter	66.50	50.50	117
5080	36″ diameter	95	61.50	156.50
5090	42″ diameter	93.50	60.50	154
5100	48″ diameter	117	76	193
5120	60″ diameter	248	97	345
5130	72″ diameter	335	122	457
5140	84″ diameter	470	203	673
5700	Corrugated metal, alum. or galv. bit. coated			
5760	8″ diameter	9.10	10.55	19.65
5770	10″ diameter	9.45	13.40	22.85
5780	12″ diameter	11.75	16.60	28.35
5800	15″ diameter	15.90	17.40	33.30
5820	18″ diameter	21	18.30	39.30
5850	24″ diameter	25	22	47
5870	30″ diameter	30.50	37	67.50
5880	36″ diameter	40.50	37	77.50
5900	48″ diameter	53.50	44.50	98
5920	60″ diameter	82	65	147
5930	72″ diameter	97.50	108	205.50
6000	Plain			

G3020 Sanitary Sewer

G3020 110	Drainage & Sewage Piping	COST PER L.F.		
		MAT.	INST.	TOTAL
6060	8″ diameter	8.35	9.80	18.15
6070	10″ diameter	9.40	12.40	21.80
6080	12″ diameter	10.55	15.80	26.35
6100	15″ diameter	12.20	15.80	28
6120	18″ diameter	15.20	16.95	32.15
6140	24″ diameter	23.50	19.90	43.40
6170	30″ diameter	30	34	64
6180	36″ diameter	42	34	76
6200	48″ diameter	61.50	40.50	102
6220	60″ diameter	96	62.50	158.50
6230	72″ diameter	169	162	331
6300	Steel or alum. oval arch, coated & paved invert			
6400	15″ equivalent diameter	14.70	17.40	32.10
6420	18″ equivalent diameter	19.20	23.50	42.70
6450	24″ equivalent diameter	25	28	53
6470	30″ equivalent diameter	30.50	35	65.50
6480	36″ equivalent diameter	36	44.50	80.50
6490	42″ equivalent diameter	44	49	93
6500	48″ equivalent diameter	56.50	59	115.50
6600	Plain			
6700	15″ equivalent diameter	14.25	15.45	29.70
6720	18″ equivalent diameter	17.20	19.90	37.10
6750	24″ equivalent diameter	25.50	23.50	49
6770	30″ equivalent diameter	31.50	41	72.50
6780	36″ equivalent diameter	42.50	41	83.50
6790	42″ equivalent diameter	55.50	48	103.50
6800	48″ equivalent diameter	68	59	127
8000	Polyvinyl chloride SDR 35			
8130	4″ diameter	1.84	4.45	6.29
8150	6″ diameter	4.04	4.76	8.80
8160	8″ diameter	7.25	4.98	12.23
8170	10″ diameter	12.55	6.55	19.10
8180	12″ diameter	14.30	6.75	21.05
8200	15″ diameter	16.65	9	25.65

G3030 Storm Sewer

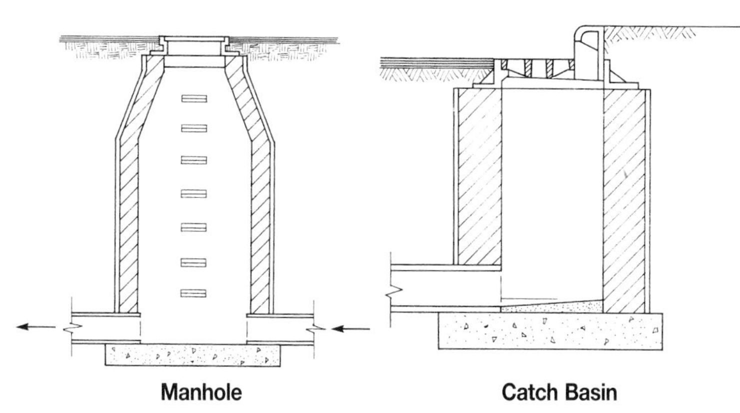

The Manhole and Catch Basin System includes: excavation with a backhoe; a formed concrete footing; frame and cover; cast iron steps and compacted backfill.

The Expanded System Listing shows manholes that have a 4′, 5′ and 6′ inside diameter riser. Depths range from 4′ to 14′. Construction material shown is either concrete, concrete block, precast concrete, or brick.

System Components	QUANTITY	UNIT	COST PER EACH MAT.	INST.	TOTAL
SYSTEM G3030 210 1920					
MANHOLE/CATCH BASIN, BRICK, 4′ I.D. RISER, 4′ DEEP					
Excavation, hydraulic backhoe, 3/8 C.Y. bucket	14.815	B.C.Y.		127.11	127.11
Trim sides and bottom of excavation	64.000	S.F.		66.56	66.56
Forms in place, manhole base, 4 uses	20.000	SFCA	16.60	115	131.60
Reinforcing in place footings, #4 to #7	.019	Ton	21.38	24.23	45.61
Concrete, 3000 psi	.925	C.Y.	125.80		125.80
Place and vibrate concrete, footing, direct chute	.925	C.Y.		52.78	52.78
Catch basin or MH, brick, 4′ ID, 4′ deep	1.000	Ea.	650	1,125	1,775
Catch basin or MH steps; heavy galvanized cast iron	1.000	Ea.	20.50	15.60	36.10
Catch basin or MH frame and cover	1.000	Ea.	300	252	552
Fill, granular	12.954	L.C.Y.	284.99		284.99
Backfill, spread with wheeled front end loader	12.954	L.C.Y.		33.55	33.55
Backfill compaction, 12″ lifts, air tamp	12.954	E.C.Y.		125.79	125.79
TOTAL			1,419.27	1,937.62	3,356.89

G3030 210	Manholes & Catch Basins	COST PER EACH MAT.	INST.	TOTAL
1920	Manhole/catch basin, brick, 4′ I.D. riser, 4′ deep	1,425	1,925	3,350
1940	6′ deep	1,975	2,675	4,650
1960	8′ deep	2,600	3,650	6,250
1980	10′ deep	3,025	4,525	7,550
3000	12′ deep	3,750	4,950	8,700
3020	14′ deep	4,550	6,900	11,450
3200	Block, 4′ I.D. riser, 4′ deep	1,225	1,550	2,775
3220	6′ deep	1,650	2,200	3,850
3240	8′ deep	2,175	3,025	5,200
3260	10′ deep	2,575	3,750	6,325
3280	12′ deep	3,250	4,775	8,025
3300	14′ deep	4,025	5,825	9,850
4620	Concrete, cast-in-place, 4′ I.D. riser, 4′ deep	1,375	2,650	4,025
4640	6′ deep	1,925	3,525	5,450
4660	8′ deep	2,700	5,125	7,825
4680	10′ deep	3,225	6,350	9,575
4700	12′ deep	4,000	7,875	11,875
4720	14′ deep	4,925	9,425	14,350
5820	Concrete, precast, 4′ I.D. riser, 4′ deep	1,725	1,400	3,125
5840	6′ deep	2,225	1,900	4,125

G3030 Storm Sewer

G3030 210	Manholes & Catch Basins	COST PER EACH		
		MAT.	INST.	TOTAL
5860	8′ deep	2,750	2,650	5,400
5880	10′ deep	3,275	3,250	6,525
5900	12′ deep	4,075	4,025	8,100
5920	14′ deep	4,975	5,100	10,075
6000	5′ I.D. riser, 4′ deep	2,775	1,575	4,350
6020	6′ deep	3,425	2,200	5,625
6040	8′ deep	4,375	2,900	7,275
6060	10′ deep	5,500	3,700	9,200
6080	12′ deep	6,750	4,725	11,475
6100	14′ deep	8,100	5,775	13,875
6200	6′ I.D. riser, 4′ deep	3,725	2,075	5,800
6220	6′ deep	4,675	2,750	7,425
6240	8′ deep	5,900	3,825	9,725
6260	10′ deep	7,350	4,850	12,200
6280	12′ deep	8,900	6,100	15,000
6300	14′ deep	10,600	7,425	18,025

G3060 Fuel Distribution

G3060 110	Gas Service Piping	COST PER L.F.		
		MAT.	INST.	TOTAL
2000	Piping, excavation & backfill excluded, polyethylene			
2070	1-1/4" diam, SDR 10	1.80	4.38	6.18
2090	2" diam, SDR 11	2.79	4.88	7.67
2110	3" diam, SDR 11.5	6.15	5.85	12
2130	4" diam,SDR 11	12.85	9.90	22.75
2150	6" diam, SDR 21	34.50	10.60	45.10
2160	8" diam, SDR 21	51.50	12.80	64.30
3000	Steel, schedule 40, plain end, tarred & wrapped			
3060	1" diameter	5.70	9.80	15.50
3090	2" diameter	8.95	10.50	19.45
3110	3" diameter	14.80	11.35	26.15
3130	4" diameter	19.25	17.75	37
3140	5" diameter	28	20.50	48.50
3150	6" diameter	34	25	59
3160	8" diameter	54	32.50	86.50

G3060 Fuel Distribution

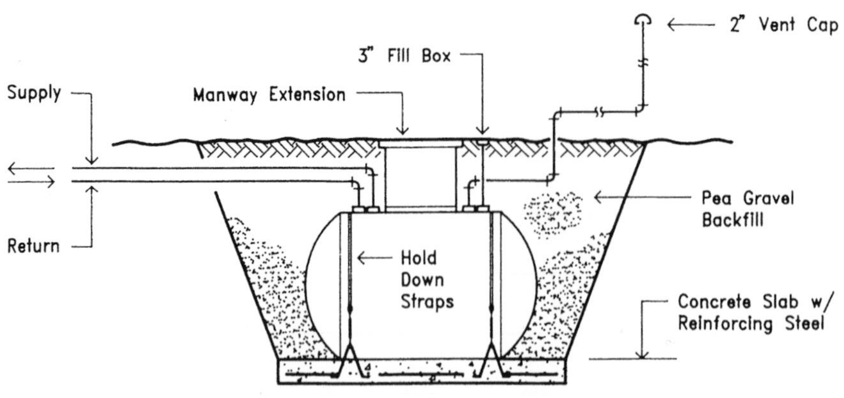

Fiberglass Underground Fuel Storage Tank, Double Wall

System Components	QUANTITY	UNIT	COST EACH MAT.	COST EACH INST.	COST EACH TOTAL
SYSTEM G3060 310 1010					
STORAGE TANK, FUEL, UNDERGROUND, DOUBLE WALL FIBERGLASS, 600 GAL.					
Excavating trench, no sheeting or dewatering, 6'-10 ' D, 3/4 CY hyd backhoe	18.000	Ea.		154.44	154.44
Corrosion resistance wrap & coat, small diam. pipe, 1" diam., add	120.000	Ea.	122.40		122.40
Fiberglass-reinforced elbows 90 & 45 degree, 2 inch	7.000	Ea.	322	626.50	948.50
Fiberglass-reinforced threaded adapters, 2 inch	2.000	Ea.	38.10	204	242.10
Remote tank gauging system, 30', 5" pointer travel	1.000	Ea.	6,325	385	6,710
Corrosion resistance wrap & coat, small diam. pipe. 4 " diam. add	30.000	Ea.	81		81
Pea gravel	9.000	Ea.	1,134	486	1,620
Reinforcing in hold down pad, #3 to #7	120.000	Ea.	68.40	75.60	144
Tank leak detector system, 8 channel, external monitoring	1.000	Ea.	1,250		1,250
Tubbing copper, type L, 3/4" diam.	120.000	Ea.	578.40	1,194	1,772.40
Elbow 90 degree ,wrought cu x cu 3/4 " diam.	4.000	Ea.	10.56	160	170.56
Pipe, black steel welded, sch 40, 3" diam.	30.000	Ea.	546	988.20	1,534.20
Elbow 90 degree, steel welded butt joint, 3 " diam.	3.000	Ea.	99	611.40	710.40
Fiberglass-reinforced plastic pipe, 2 inch	156.000	Ea.	1,123.20	787.80	1,911
Fiberglass-reinforced plastic pipe, 3 inch	36.000	Ea.	1,026	207	1,233
Fiberglass-reinforced elbows 90 & 45 degree, 3 inch	3.000	Ea.	447	357	804
Tank leak detection system, probes, well monitoring, liquid phase detection	2.000	Ea.	1,390		1,390
Double wall fiberglass annular space	1.000	Ea.	460		460
Flange, weld neck, 150 lb, 3 " pipe size	1.000	Ea.	44.50	101.91	146.41
Vent protector/breather, 2" diam.	1.000	Ea.	53	24	77
Corrosion resistance wrap & coat, small diam. pipe, 2" diam., add	36.000	Ea.	50.76		50.76
Concrete hold down pad, 6' x 5' x 8" thick	30.000	Ea.	108	49.80	157.80
600 gallon capacity, underground storage tank, double wall fiberglass	1.000	Ea.	9,750	570	10,320
Tank, for hold-downs 500-4,000 gal., add	1.000	Ea.	555	183	738
Foot valve, single poppet, 3/4" diam.	1.000	Ea.	105	42.50	147.50
Fuel fill box, locking inner cover, 3" diam.	1.000	Ea.	130	154	284
Fuel oil specialties, valve,ball chk, globe type fusible, 3/4" diam.	2.000	Ea.	185	77	262
TOTAL			26,002.32	7,439.15	33,441.47

G3060 310	Fiberglass Fuel Tank, Double Wall	COST EACH MAT.	COST EACH INST.	COST EACH TOTAL
1010	Storage tank, fuel, underground, double wall fiberglass, 600 Gal.	26,000	7,425	33,425
1020	2500 Gal.	39,900	10,200	50,100

G3060 Fuel Distribution

G3060 310	Fiberglass Fuel Tank, Double Wall	COST EACH		
		MAT.	INST.	TOTAL
1030	4000 Gal.	47,400	12,700	60,100
1040	6000 Gal.	56,000	14,200	70,200
1050	8000 Gal.	69,500	17,200	86,700
1060	10,000 Gal.	77,500	19,800	97,300
1070	15,000 Gal.	112,000	22,700	134,700
1080	20,000 Gal.	135,000	27,500	162,500
1090	25,000 Gal.	162,000	32,000	194,000
1100	30,000 Gal.	184,500	33,900	218,400

G3060 Fuel Distribution

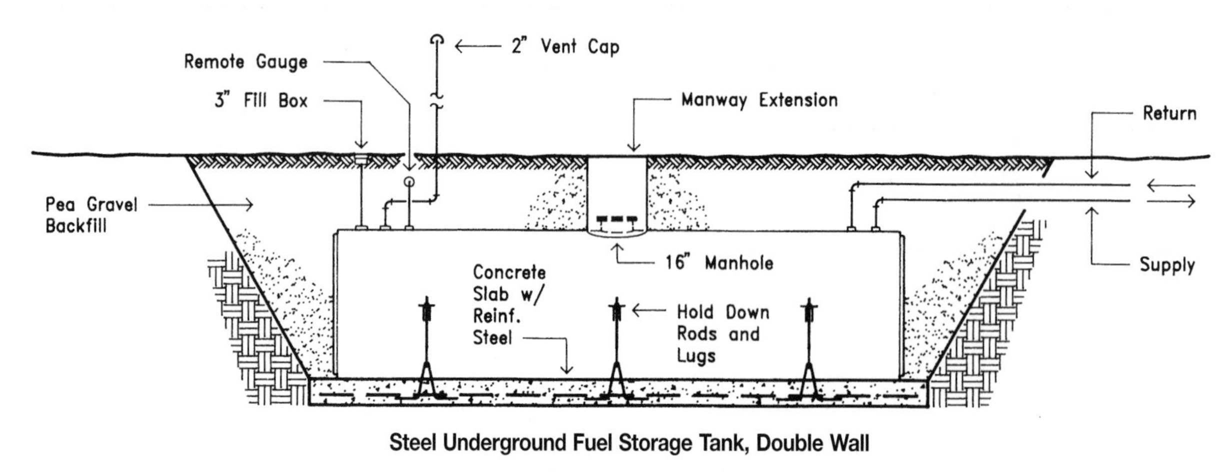

Steel Underground Fuel Storage Tank, Double Wall

System Components	QUANTITY	UNIT	COST EACH MAT.	INST.	TOTAL
SYSTEM G3060 320 1010					
STORAGE TANK, FUEL, UNDERGROUND, DOUBLE WALL STEEL, 500 GAL.					
Fiberglass-reinforced plastic pipe, 2 inch	156.000	Ea.	1,123.20	787.80	1,911
Fiberglass-reinforced plastic pipe, 3 inch	36.000	Ea.	1,026	207	1,233
Fiberglass-reinforced elbows 90 & 45 degree, 3 inch	3.000	Ea.	447	357	804
Tank leak detection system, probes, well monitoring, liquid phase detection	2.000	Ea.	1,390		1,390
Flange, weld neck, 150 lb, 3" pipe size	1.000	Ea.	44.50	101.91	146.41
Vent protector/breather, 2" diam.	1.000	Ea.	53	24	77
Concrete hold down pad, 6' x 5' x 8" thick	30.000	Ea.	108	49.80	157.80
Tanks, for hold-downs, 500-2,000 gal, add	1.000	Ea.	241	183	424
Double wall steel tank annular space	1.000	Ea.	325		325
Foot valve, single poppet, 3/4" diam.	1.000	Ea.	105	42.50	147.50
Fuel fill box, locking inner cover, 3" diam.	1.000	Ea.	130	154	284
Fuel oil specialties, valve, ball chk, globe type fusible, 3/4" diam.	2.000	Ea.	185	77	262
Pea gravel	9.000	Ea.	1,134	486	1,620
Reinforcing in hold down pad, #3 to #7	120.000	Ea.	68.40	75.60	144
Tank, for manways, add	1.000	Ea.	1,100		1,100
Tank leak detector system, 8 channel, external monitoring	1.000	Ea.	1,250		1,250
Tubbing copper, type L, 3/4" diam.	120.000	Ea.	578.40	1,194	1,772.40
Elbow 90 degree, wrought cu x cu 3/4" diam.	8.000	Ea.	21.12	320	341.12
Pipe, black steel welded, sch 40, 3" diam.	30.000	Ea.	546	988.20	1,534.20
Elbow 90 degree, steel welded butt joint, 3" diam.	3.000	Ea.	99	611.40	710.40
Excavating trench, no sheeting or dewatering, 6'-10' D, 3/4 CY hyd backhoe	18.000	Ea.		154.44	154.44
Fiberglass-reinforced elbows 90 & 45 degree, 2 inch	7.000	Ea.	322	626.50	948.50
Fiberglass-reinforced threaded adapters, 2 inch	2.000	Ea.	38.10	204	242.10
500 gallon capacity, underground storage tank, double wall steel	1.000	Ea.	3,300	575	3,875
Remote tank gauging system, 30', 5" pointer travel	1.000	Ea.	6,325	385	6,710
TOTAL			19,959.72	7,604.15	27,563.87

G3060 320	Steel Fuel Tank, Double Wall	COST EACH MAT.	INST.	TOTAL
1010	Storage tank, fuel, underground, double wall steel, 500 Gal.	20,000	7,600	27,600
1020	2000 Gal.	26,100	9,925	36,025
1030	4000 Gal.	35,500	12,800	48,300
1040	6000 Gal.	41,600	14,100	55,700
1050	8000 Gal.	46,500	17,500	64,000
1060	10,000 Gal.	53,000	20,000	73,000

G3060 Fuel Distribution

G3060 320	Steel Fuel Tank, Double Wall	COST EACH		
		MAT.	INST.	TOTAL
1070	15,000 Gal.	64,500	23,300	87,800
1080	20,000 Gal.	89,500	27,900	117,400
1090	25,000 Gal.	98,000	31,900	129,900
1100	30,000 Gal.	102,000	34,000	136,000
1110	40,000 Gal.	200,500	42,800	243,300
1120	50,000 Gal.	242,000	51,000	293,000

G3060 325	Steel Tank, Above Ground, Single Wall	COST EACH		
		MAT.	INST.	TOTAL
1010	Storage tank, fuel, above ground, single wall steel, 550 Gal.	15,600	5,325	20,925
1020	2000 Gal.	24,900	5,450	30,350
1030	5000 Gal.	35,800	8,350	44,150
1040	10,000 Gal.	57,500	8,875	66,375
1050	15,000 Gal.	72,000	9,150	81,150
1060	20,000 Gal.	90,000	9,450	99,450
1070	25,000 Gal.	103,000	9,675	112,675
1080	30,000 Gal.	121,000	10,100	131,100

G3060 330	Steel Tank, Above Ground, Double Wall	COST EACH		
		MAT.	INST.	TOTAL
1010	Storage tank, fuel, above ground, double wall steel, 500 Gal.	14,300	5,400	19,700
1020	2000 Gal.	19,000	5,525	24,525
1030	4000 Gal.	27,500	5,625	33,125
1040	6000 Gal.	31,000	8,650	39,650
1050	8000 Gal.	34,300	8,875	43,175
1060	10,000 Gal.	47,700	9,050	56,750
1070	15,000 Gal.	58,500	9,375	67,875
1080	20,000 Gal.	65,500	9,675	75,175
1090	25,000 Gal.	78,000	9,975	87,975
1100	30,000 Gal.	88,000	10,300	98,300

Reference Section

All the reference information is in one section, making it easy to find what you need to know and easy to use the data set on a daily basis. This section is visually identified by a vertical black bar on the page edges.

In this Reference Section, we've included Equipment Rental Costs, a listing of rental and operating costs; Crew Listings, a full listing of all crews and equipment and their costs; Historical Cost Indexes for cost comparisons over time; City Cost Indexes and Location Factors for adjusting costs to the region you are in; Reference Tables, where you will find explanations, estimating information and procedures, and technical data; Change Orders, information on pricing changes to contract documents; and an explanation of all the Abbreviations in the data set.

Table of Contents

Estimating Tips

- This section contains the average costs to rent and operate hundreds of pieces of construction equipment. This is useful information when one is estimating the time and material requirements of any particular operation in order to establish a unit or total cost. Bare equipment costs shown on a unit cost line include, not only rental, but also operating costs for equipment under normal use.

Rental Costs

- Equipment rental rates are obtained from the following industry sources throughout North America: contractors, suppliers, dealers, manufacturers, and distributors.
- Rental rates vary throughout the country, with larger cities generally having lower rates. Lease plans for new equipment are available for periods in excess of six months, with a percentage of payments applying toward purchase.
- Monthly rental rates vary from 2% to 5% of the purchase price of the equipment depending on the anticipated life of the equipment and its wearing parts.
- Weekly rental rates are about 1/3 of the monthly rates, and daily rental rates are about 1/3 of the weekly rate.
- Rental rates can also be treated as reimbursement costs for contractor-owned equipment. Owned equipment costs include depreciation, loan payments, interest, taxes, insurance, storage, and major repairs.

Operating Costs

- The operating costs include parts and labor for routine servicing, such as the repair and replacement of pumps, filters, and worn lines. Normal operating expendables, such as fuel, lubricants, tires, and electricity (where applicable), are also included.
- Extraordinary operating expendables with highly variable wear patterns, such as diamond bits and blades, are excluded. These costs can be found as material costs in the Unit Price section.
- The hourly operating costs listed do not include the operator's wages.

Equipment Cost/Day

- Any power equipment required by a crew is shown in the Crew Listings with a daily cost.
- This daily cost of equipment needed by a crew includes both the rental cost and the operating cost and is based on dividing the weekly rental rate by 5 (the number of working days in the week), then adding the hourly operating cost multiplied by 8 (the number of hours in a day). This "Equipment Cost/Day" is shown in the far right column of the Equipment Rental section.
- If equipment is needed for only one or two days, it is best to develop your own cost by including components for daily rent and hourly operating costs. This is important when the listed Crew for a task does not contain the equipment needed, such as a crane for lifting mechanical heating/cooling equipment up onto a roof.
- If the quantity of work is less than the crew's Daily Output shown for a Unit Price line item that includes a bare unit equipment cost, the recommendation is to estimate one day's rental cost and operating cost for equipment shown in the Crew Listing for that line item.
- Please note, in some cases the equipment description in the crew is followed by a time period in parenthesis. For example: (daily) or (monthly). In these cases the equipment cost/day is calculated by adding the rental cost per time period to the hourly operating cost multiplied by 8.

Mobilization, Demobilization Costs

- The cost to move construction equipment from an equipment yard or rental company to the job site and back again is not included in equipment rental costs listed in the Reference Section. It is also not included in the bare equipment cost of any Unit Price line item or in any equipment costs shown in the Crew Listings.
- Mobilization (to the site) and demobilization (from the site) costs can be found in the Unit Price section.
- If a piece of equipment is already at the job site, it is not appropriate to utilize mobilization or demobilization costs again in an estimate. ■

54 33 | Equipment Rental

Description		UNIT	HOURLY OPER. COST	RENT PER DAY	RENT PER WEEK	RENT PER MONTH	EQUIPMENT COST/DAY	
CONCRETE EQUIPMENT RENTAL without operators	R015433 -10							10
Bucket, concrete lightweight, 1/2 C.Y.		Ea.	.88	24.50	74	222	21.80	
1 C.Y.			.98	28.50	85.50	257	24.95	
1-1/2 C.Y.			1.24	38.50	115	345	32.90	
2 C.Y.			1.34	47	141	425	38.90	
8 C.Y.			6.49	265	795	2,375	210.90	
Cart, concrete, self-propelled, operator walking, 10 C.F.			2.86	58.50	176	530	58.10	
Operator riding, 18 C.F.			4.82	98.50	296	890	97.75	
Conveyer for concrete, portable, gas, 16″ wide, 26′ long			10.64	130	390	1,175	163.10	
46′ long			11.02	155	465	1,400	181.15	
56′ long			11.18	162	485	1,450	186.45	
Core drill, electric, 2-1/2 H.P., 1″ to 8″ bit diameter			1.57	56.50	170	510	46.55	
11 H.P., 8″ to 18″ cores			5.40	115	345	1,025	112.20	
Finisher, concrete floor, gas, riding trowel, 96″ wide			9.66	148	445	1,325	166.30	
Gas, walk-behind, 3 blade, 36″ trowel			2.04	24.50	73.50	221	31	
4 blade, 48″ trowel			3.07	28	84.50	254	41.50	
Float, hand-operated (Bull float), 48″ wide			.08	13.85	41.50	125	8.95	
Curb builder, 14 H.P., gas, single screw			14.04	285	855	2,575	283.30	
Double screw			15.04	340	1,025	3,075	325.35	
Floor grinder, concrete and terrazzo, electric, 22″ path			3.04	190	570	1,700	138.30	
Edger, concrete, electric, 7″ path			1.18	56.50	170	510	43.45	
Vacuum pick-up system for floor grinders, wet/dry			1.62	98.50	296	890	72.15	
Mixer, powered, mortar and concrete, gas, 6 C.F., 18 H.P.			7.42	127	380	1,150	135.35	
10 C.F., 25 H.P.			9.00	150	450	1,350	162	
16 C.F.			9.36	173	520	1,550	178.85	
Concrete, stationary, tilt drum, 2 C.Y.			7.23	243	730	2,200	203.85	
Pump, concrete, truck mounted, 4″ line, 80′ boom			29.88	1,000	3,025	9,075	844.05	
5″ line, 110′ boom			37.46	1,375	4,150	12,500	1,130	
Mud jack, 50 C.F. per hr.			6.45	123	370	1,100	125.60	
225 C.F. per hr.			8.55	145	435	1,300	155.40	
Shotcrete pump rig, 12 C.Y./hr.			13.97	218	655	1,975	242.75	
35 C.Y./hr.			15.80	242	725	2,175	271.40	
Saw, concrete, manual, gas, 18 H.P.			5.54	47.50	143	430	72.90	
Self-propelled, gas, 30 H.P.			7.90	78.50	235	705	110.15	
V-groove crack chaser, manual, gas, 6 H.P.			1.64	19	57	171	24.50	
Vibrators, concrete, electric, 60 cycle, 2 H.P.			.47	9.15	27.50	82.50	9.25	
3 H.P.			.56	11.65	35	105	11.50	
Gas engine, 5 H.P.			1.54	16.35	49	147	22.15	
8 H.P.			2.08	16.50	49.50	149	26.55	
Vibrating screed, gas engine, 8 H.P.			2.81	95	285	855	79.45	
Concrete transit mixer, 6 x 4, 250 H.P., 8 C.Y., rear discharge			50.72	615	1,850	5,550	775.80	
Front discharge			58.88	740	2,225	6,675	916.05	
6 x 6, 285 H.P., 12 C.Y., rear discharge			58.15	710	2,125	6,375	890.20	
Front discharge			60.59	750	2,250	6,750	934.70	
EARTHWORK EQUIPMENT RENTAL without operators	R015433 -10							20
Aggregate spreader, push type, 8′ to 12′ wide		Ea.	2.60	27.50	82.50	248	37.30	
Tailgate type, 8′ wide			2.55	33	99	297	40.15	
Earth auger, truck mounted, for fence & sign posts, utility poles			13.85	460	1,375	4,125	385.85	
For borings and monitoring wells			42.65	690	2,075	6,225	756.20	
Portable, trailer mounted			2.30	33	99	297	38.20	
Truck mounted, for caissons, water wells			85.39	2,900	8,725	26,200	2,428	
Horizontal boring machine, 12″ to 36″ diameter, 45 H.P.			22.77	188	565	1,700	295.15	
12″ to 48″ diameter, 65 H.P.			31.25	330	985	2,950	447	
Auger, for fence posts, gas engine, hand held			.45	6.35	19	57	7.40	
Excavator, diesel hydraulic, crawler mounted, 1/2 C.Y. cap.			21.72	450	1,350	4,050	443.75	
5/8 C.Y. capacity			29.04	590	1,775	5,325	587.30	
3/4 C.Y. capacity			32.66	700	2,100	6,300	681.25	
1 C.Y. capacity			41.21	700	2,100	6,300	749.70	

01 54 33 | Equipment Rental

			UNIT	HOURLY OPER. COST	RENT PER DAY	RENT PER WEEK	RENT PER MONTH	EQUIPMENT COST/DAY	
20	0200	1-1/2 C.Y. capacity	Ea.	48.59	865	2,600	7,800	908.70	20
	0300	2 C.Y. capacity		56.58	1,050	3,125	9,375	1,078	
	0320	2-1/2 C.Y. capacity		82.64	1,300	3,900	11,700	1,441	
	0325	3-1/2 C.Y. capacity		120.12	2,150	6,475	19,400	2,256	
	0330	4-1/2 C.Y. capacity		151.63	2,750	8,275	24,800	2,868	
	0335	6 C.Y. capacity		192.39	3,400	10,200	30,600	3,579	
	0340	7 C.Y. capacity		175.20	3,225	9,650	29,000	3,332	
	0342	Excavator attachments, bucket thumbs		3.40	250	750	2,250	177.20	
	0345	Grapples		3.14	215	645	1,925	154.10	
	0346	Hydraulic hammer for boom mounting, 4,000 ft. lb.		13.48	375	1,125	3,375	332.85	
	0347	5,000 ft. lb.		15.95	465	1,400	4,200	407.60	
	0348	8,000 ft. lb.		23.54	675	2,025	6,075	593.35	
	0349	12,000 ft. lb.		25.72	800	2,400	7,200	685.75	
	0350	Gradall type, truck mounted, 3 ton @ 15′ radius, 5/8 C.Y.		43.44	860	2,575	7,725	862.55	
	0370	1 C.Y. capacity		59.40	1,175	3,550	10,700	1,185	
	0400	Backhoe-loader, 40 to 45 H.P., 5/8 C.Y. capacity		11.90	207	620	1,850	219.20	
	0450	45 H.P. to 60 H.P., 3/4 C.Y. capacity		18.02	293	880	2,650	320.20	
	0460	80 H.P., 1-1/4 C.Y. capacity		20.36	375	1,125	3,375	387.85	
	0470	112 H.P., 1-1/2 C.Y. capacity		32.99	590	1,775	5,325	618.90	
	0482	Backhoe-loader attachment, compactor, 20,000 lb.		6.44	150	450	1,350	141.50	
	0485	Hydraulic hammer, 750 ft. lb.		3.68	103	310	930	91.45	
	0486	Hydraulic hammer, 1,200 ft. lb.		6.55	198	595	1,775	171.40	
	0500	Brush chipper, gas engine, 6″ cutter head, 35 H.P.		9.17	110	330	990	139.35	
	0550	Diesel engine, 12″ cutter head, 130 H.P.		23.67	335	1,000	3,000	389.40	
	0600	15″ cutter head, 165 H.P.		26.59	400	1,200	3,600	452.70	
	0750	Bucket, clamshell, general purpose, 3/8 C.Y.		1.40	41	123	370	35.80	
	0800	1/2 C.Y.		1.52	50.50	152	455	42.55	
	0850	3/4 C.Y.		1.64	57.50	173	520	47.75	
	0900	1 C.Y.		1.70	62.50	188	565	51.20	
	0950	1-1/2 C.Y.		2.79	87	261	785	74.50	
	1000	2 C.Y.		2.92	95	285	855	80.40	
	1010	Bucket, dragline, medium duty, 1/2 C.Y.		.82	24.50	74	222	21.35	
	1020	3/4 C.Y.		.78	26	78	234	21.85	
	1030	1 C.Y.		.80	27	81.50	245	22.70	
	1040	1-1/2 C.Y.		1.26	41.50	125	375	35.10	
	1050	2 C.Y.		1.29	45	135	405	37.35	
	1070	3 C.Y.		2.08	66	198	595	56.25	
	1200	Compactor, manually guided 2-drum vibratory smooth roller, 7.5 H.P.		7.22	207	620	1,850	181.75	
	1250	Rammer/tamper, gas, 8″		2.21	46.50	140	420	45.65	
	1260	15″		2.63	53.50	160	480	53.05	
	1300	Vibratory plate, gas, 18″ plate, 3,000 lb. blow		2.13	23.50	70.50	212	31.10	
	1350	21″ plate, 5,000 lb. blow		2.62	32.50	98	294	40.55	
	1370	Curb builder/extruder, 14 H.P., gas, single screw		14.03	283	850	2,550	282.25	
	1390	Double screw		15.04	340	1,025	3,075	325.30	
	1500	Disc harrow attachment, for tractor		.47	79.50	239	715	51.60	
	1810	Feller buncher, shearing & accumulating trees, 100 H.P.		39.20	815	2,450	7,350	803.55	
	1860	Grader, self-propelled, 25,000 lb.		33.35	760	2,275	6,825	721.80	
	1910	30,000 lb.		32.85	660	1,975	5,925	657.85	
	1920	40,000 lb.		51.89	1,275	3,825	11,500	1,180	
	1930	55,000 lb.		66.93	1,650	4,925	14,800	1,520	
	1950	Hammer, pavement breaker, self-propelled, diesel, 1,000 to 1,250 lb.		28.40	475	1,425	4,275	512.20	
	2000	1,300 to 1,500 lb.		42.80	975	2,925	8,775	927.40	
	2050	Pile driving hammer, steam or air, 4,150 ft. lb. @ 225 bpm		12.15	565	1,700	5,100	437.20	
	2100	8,750 ft. lb. @ 145 bpm		14.35	810	2,425	7,275	599.75	
	2150	15,000 ft. lb. @ 60 bpm		14.68	840	2,525	7,575	622.40	
	2200	24,450 ft. lb. @ 111 bpm		15.69	935	2,800	8,400	685.50	
	2250	Leads, 60′ high for pile driving hammers up to 20,000 ft. lb.		3.67	85.50	256	770	80.55	
	2300	90′ high for hammers over 20,000 ft. lb.	↓	5.45	152	455	1,375	134.60	

01 54 33 | Equipment Rental

			UNIT	HOURLY OPER. COST	RENT PER DAY	RENT PER WEEK	RENT PER MONTH	EQUIPMENT COST/DAY	
20	2350	Diesel type hammer, 22,400 ft. lb.	Ea.	17.82	475	1,425	4,275	427.55	20
	2400	41,300 ft. lb.		25.68	600	1,800	5,400	565.45	
	2450	141,000 ft. lb.		41.33	950	2,850	8,550	900.65	
	2500	Vib. elec. hammer/extractor, 200 kW diesel generator, 34 H.P.		41.37	690	2,075	6,225	745.95	
	2550	80 H.P.		73.03	1,000	3,000	9,000	1,184	
	2600	150 H.P.		135.14	1,925	5,775	17,300	2,236	
	2800	Log chipper, up to 22" diameter, 600 H.P.		46.26	665	2,000	6,000	770.10	
	2850	Logger, for skidding & stacking logs, 150 H.P.		43.53	835	2,500	7,500	848.25	
	2860	Mulcher, diesel powered, trailer mounted		18.04	220	660	1,975	276.35	
	2900	Rake, spring tooth, with tractor		14.72	360	1,075	3,225	332.75	
	3000	Roller, vibratory, tandem, smooth drum, 20 H.P.		7.81	150	450	1,350	152.45	
	3050	35 H.P.		10.13	252	755	2,275	232.05	
	3100	Towed type vibratory compactor, smooth drum, 50 H.P.		25.27	365	1,100	3,300	422.15	
	3150	Sheepsfoot, 50 H.P.		25.64	375	1,125	3,375	430.15	
	3170	Landfill compactor, 220 H.P.		70.01	1,575	4,750	14,300	1,510	
	3200	Pneumatic tire roller, 80 H.P.		12.92	390	1,175	3,525	338.35	
	3250	120 H.P.		19.39	640	1,925	5,775	540.10	
	3300	Sheepsfoot vibratory roller, 240 H.P.		62.21	1,375	4,125	12,400	1,323	
	3320	340 H.P.		83.82	2,075	6,250	18,800	1,921	
	3350	Smooth drum vibratory roller, 75 H.P.		23.34	635	1,900	5,700	566.70	
	3400	125 H.P.		27.61	715	2,150	6,450	650.90	
	3410	Rotary mower, brush, 60", with tractor		18.79	350	1,050	3,150	360.30	
	3420	Rototiller, walk-behind, gas, 5 H.P.		2.14	89	267	800	70.50	
	3422	8 H.P.		2.81	103	310	930	84.50	
	3440	Scrapers, towed type, 7 C.Y. capacity		6.44	123	370	1,100	125.50	
	3450	10 C.Y. capacity		7.20	165	495	1,475	156.65	
	3500	15 C.Y. capacity		7.40	190	570	1,700	173.20	
	3525	Self-propelled, single engine, 14 C.Y. capacity		133.29	2,575	7,725	23,200	2,611	
	3550	Dual engine, 21 C.Y. capacity		141.37	2,350	7,050	21,200	2,541	
	3600	31 C.Y. capacity		187.85	3,500	10,500	31,500	3,603	
	3640	44 C.Y. capacity		232.68	4,500	13,500	40,500	4,561	
	3650	Elevating type, single engine, 11 C.Y. capacity		61.87	1,125	3,350	10,100	1,165	
	3700	22 C.Y. capacity		114.62	2,325	6,950	20,900	2,307	
	3710	Screening plant, 110 H.P. w/5' x 10' screen		21.13	385	1,150	3,450	399.05	
	3720	5' x 16' screen		26.68	490	1,475	4,425	508.40	
	3850	Shovel, crawler-mounted, front-loading, 7 C.Y. capacity		218.65	3,800	11,400	34,200	4,029	
	3855	12 C.Y. capacity		336.90	5,275	15,800	47,400	5,855	
	3860	Shovel/backhoe bucket, 1/2 C.Y.		2.69	70.50	212	635	63.95	
	3870	3/4 C.Y.		2.67	79.50	238	715	68.90	
	3880	1 C.Y.		2.76	88	264	790	74.85	
	3890	1-1/2 C.Y.		2.95	103	310	930	85.65	
	3910	3 C.Y.		3.44	140	420	1,250	111.50	
	3950	Stump chipper, 18" deep, 30 H.P.		6.90	212	635	1,900	182.20	
	4110	Dozer, crawler, torque converter, diesel 80 H.P.		25.25	450	1,350	4,050	472	
	4150	105 H.P.		34.34	560	1,675	5,025	609.70	
	4200	140 H.P.		41.27	840	2,525	7,575	835.20	
	4260	200 H.P.		63.16	1,300	3,925	11,800	1,290	
	4310	300 H.P.		80.73	1,975	5,925	17,800	1,831	
	4360	410 H.P.		106.74	2,350	7,075	21,200	2,269	
	4370	500 H.P.		133.38	2,900	8,725	26,200	2,812	
	4380	700 H.P.		230.17	5,300	15,900	47,700	5,021	
	4400	Loader, crawler, torque conv., diesel, 1-1/2 C.Y., 80 H.P.		29.54	590	1,775	5,325	591.30	
	4450	1-1/2 to 1-3/4 C.Y., 95 H.P.		30.27	675	2,025	6,075	647.15	
	4510	1-3/4 to 2-1/4 C.Y., 130 H.P.		47.76	1,025	3,100	9,300	1,002	
	4530	2-1/2 to 3-1/4 C.Y., 190 H.P.		57.79	1,225	3,700	11,100	1,202	
	4560	3-1/2 to 5 C.Y., 275 H.P.		71.41	1,475	4,450	13,400	1,461	
	4610	Front end loader, 4WD, articulated frame, diesel, 1 to 1-1/4 C.Y., 70 H.P.		16.62	273	820	2,450	297	
	4620	1-1/2 to 1-3/4 C.Y., 95 H.P.	↓	20.00	320	965	2,900	352.95	

01 54 33 | Equipment Rental

			UNIT	HOURLY OPER. COST	RENT PER DAY	RENT PER WEEK	RENT PER MONTH	EQUIPMENT COST/DAY	
20	4650	1-3/4 to 2 C.Y., 130 H.P.	Ea.	21.07	385	1,150	3,450	398.55	20
	4710	2-1/2 to 3-1/2 C.Y., 145 H.P.		29.53	490	1,475	4,425	531.20	
	4730	3 to 4-1/2 C.Y., 185 H.P.		32.09	515	1,550	4,650	566.70	
	4760	5-1/4 to 5-3/4 C.Y., 270 H.P.		53.19	915	2,750	8,250	975.55	
	4810	7 to 9 C.Y., 475 H.P.		91.17	1,750	5,275	15,800	1,784	
	4870	9 to 11 C.Y., 620 H.P.		131.92	2,625	7,850	23,600	2,625	
	4880	Skid-steer loader, wheeled, 10 C.F., 30 H.P. gas		9.57	163	490	1,475	174.55	
	4890	1 C.Y., 78 H.P., diesel		18.44	410	1,225	3,675	392.50	
	4892	Skid-steer attachment, auger		.75	137	410	1,225	87.95	
	4893	Backhoe		.74	118	355	1,075	76.90	
	4894	Broom		.71	123	370	1,100	79.65	
	4895	Forks		.15	26.50	80	240	17.25	
	4896	Grapple		.72	107	320	960	69.80	
	4897	Concrete hammer		1.05	177	530	1,600	114.40	
	4898	Tree spade		.60	100	300	900	64.80	
	4899	Trencher		.65	108	325	975	70.20	
	4900	Trencher, chain, boom type, gas, operator walking, 12 H.P.		4.18	50	150	450	63.40	
	4910	Operator riding, 40 H.P.		16.69	360	1,075	3,225	348.50	
	5000	Wheel type, diesel, 4′ deep, 12″ wide		68.71	935	2,800	8,400	1,110	
	5100	6′ deep, 20″ wide		87.58	2,150	6,475	19,400	1,996	
	5150	Chain type, diesel, 5′ deep, 8″ wide		16.30	350	1,050	3,150	340.40	
	5200	Diesel, 8′ deep, 16″ wide		89.66	1,875	5,600	16,800	1,837	
	5202	Rock trencher, wheel type, 6″ wide x 18″ deep		47.12	700	2,100	6,300	797	
	5206	Chain type, 18″ wide x 7′ deep		104.70	3,075	9,200	27,600	2,678	
	5210	Tree spade, self-propelled		13.69	390	1,175	3,525	344.55	
	5250	Truck, dump, 2-axle, 12 ton, 8 C.Y. payload, 220 H.P.		23.95	247	740	2,225	339.60	
	5300	Three axle dump, 16 ton, 12 C.Y. payload, 400 H.P.		44.63	350	1,050	3,150	567.05	
	5310	Four axle dump, 25 ton, 18 C.Y. payload, 450 H.P.		50.00	510	1,525	4,575	705	
	5350	Dump trailer only, rear dump, 16-1/2 C.Y.		5.74	147	440	1,325	133.95	
	5400	20 C.Y.		6.20	165	495	1,475	148.60	
	5450	Flatbed, single axle, 1-1/2 ton rating		19.05	71.50	214	640	195.20	
	5500	3 ton rating		23.12	102	305	915	245.95	
	5550	Off highway rear dump, 25 ton capacity		62.86	1,425	4,250	12,800	1,353	
	5600	35 ton capacity		67.10	1,550	4,675	14,000	1,472	
	5610	50 ton capacity		84.12	1,775	5,325	16,000	1,738	
	5620	65 ton capacity		89.83	1,925	5,800	17,400	1,879	
	5630	100 ton capacity		121.61	2,850	8,550	25,700	2,683	
	6000	Vibratory plow, 25 H.P., walking	↓	6.79	60.50	181	545	90.55	
40	0010	**GENERAL EQUIPMENT RENTAL** without operators R015433 -10							40
	0020	Aerial lift, scissor type, to 20′ high, 1200 lb. capacity, electric	Ea.	3.49	50.50	151	455	58.15	
	0030	To 30′ high, 1,200 lb. capacity		3.78	67.50	202	605	70.70	
	0040	Over 30′ high, 1,500 lb. capacity		5.15	122	365	1,100	114.20	
	0070	Articulating boom, to 45′ high, 500 lb. capacity, diesel R015433 -15		9.95	275	825	2,475	244.60	
	0075	To 60′ high, 500 lb. capacity		13.70	510	1,525	4,575	414.60	
	0080	To 80′ high, 500 lb. capacity		16.10	560	1,675	5,025	463.80	
	0085	To 125′ high, 500 lb. capacity		18.40	775	2,325	6,975	612.20	
	0100	Telescoping boom to 40′ high, 500 lb. capacity, diesel		11.27	320	965	2,900	283.15	
	0105	To 45′ high, 500 lb. capacity		12.54	315	945	2,825	289.35	
	0110	To 60′ high, 500 lb. capacity		16.41	535	1,600	4,800	451.25	
	0115	To 80′ high, 500 lb. capacity		21.33	835	2,500	7,500	670.65	
	0120	To 100′ high, 500 lb. capacity		28.80	860	2,575	7,725	745.40	
	0125	To 120′ high, 500 lb. capacity		29.25	840	2,525	7,575	739	
	0195	Air compressor, portable, 6.5 CFM, electric		.91	12.65	38	114	14.85	
	0196	Gasoline		.65	18	54	162	16.05	
	0200	Towed type, gas engine, 60 CFM		9.46	51.50	155	465	106.70	
	0300	160 CFM		10.51	53	159	475	115.85	
	0400	Diesel engine, rotary screw, 250 CFM		12.12	118	355	1,075	167.95	
	0500	365 CFM	↓	16.05	142	425	1,275	213.35	

01 54 33 | Equipment Rental

		01 54 33 Equipment Rental	UNIT	HOURLY OPER. COST	RENT PER DAY	RENT PER WEEK	RENT PER MONTH	EQUIPMENT COST/DAY	
40	0550	450 CFM	Ea.	20.01	177	530	1,600	266.05	40
	0600	600 CFM		34.20	240	720	2,150	417.60	
	0700	750 CFM		34.72	252	755	2,275	428.80	
	0930	Air tools, breaker, pavement, 60 lb.		.57	10.35	31	93	10.75	
	0940	80 lb.		.57	10.65	32	96	10.90	
	0950	Drills, hand (jackhammer), 65 lb.		.67	17.35	52	156	15.80	
	0960	Track or wagon, swing boom, 4″ drifter		54.83	940	2,825	8,475	1,004	
	0970	5″ drifter		63.49	1,100	3,325	9,975	1,173	
	0975	Track mounted quarry drill, 6″ diameter drill		102.22	1,625	4,875	14,600	1,793	
	0980	Dust control per drill		1.04	25	75.50	227	23.45	
	0990	Hammer, chipping, 12 lb.		.60	27	81	243	21	
	1000	Hose, air with couplings, 50′ long, 3/4″ diameter		.07	10	30	90	6.55	
	1100	1″ diameter		.08	13	39	117	8.40	
	1200	1-1/2″ diameter		.22	35	105	315	22.75	
	1300	2″ diameter		.24	40	120	360	25.90	
	1400	2-1/2″ diameter		.36	56.50	170	510	36.85	
	1410	3″ diameter		.42	37.50	112	335	25.75	
	1450	Drill, steel, 7/8″ x 2′		.08	12.50	37.50	113	8.20	
	1460	7/8″ x 6′		.12	19	57	171	12.30	
	1520	Moil points		.03	4.73	14.20	42.50	3.05	
	1525	Pneumatic nailer w/accessories		.48	31	92.50	278	22.35	
	1530	Sheeting driver for 60 lb. breaker		.04	7.50	22.50	67.50	4.85	
	1540	For 90 lb. breaker		.13	10.15	30.50	91.50	7.15	
	1550	Spade, 25 lb.		.50	7.15	21.50	64.50	8.30	
	1560	Tamper, single, 35 lb.		.59	39.50	119	355	28.55	
	1570	Triple, 140 lb.		.89	59.50	179	535	42.90	
	1580	Wrenches, impact, air powered, up to 3/4″ bolt		.43	12.85	38.50	116	11.10	
	1590	Up to 1-1/4″ bolt		.58	23	69.50	209	18.50	
	1600	Barricades, barrels, reflectorized, 1 to 99 barrels		.03	5.50	16.55	49.50	3.55	
	1610	100 to 200 barrels		.02	4.27	12.80	38.50	2.75	
	1620	Barrels with flashers, 1 to 99 barrels		.03	6.20	18.55	55.50	4	
	1630	100 to 200 barrels		.03	4.95	14.85	44.50	3.20	
	1640	Barrels with steady burn type C lights		.05	8.15	24.50	73.50	5.30	
	1650	Illuminated board, trailer mounted, with generator		3.29	135	405	1,225	107.30	
	1670	Portable barricade, stock, with flashers, 1 to 6 units		.03	6.15	18.50	55.50	4	
	1680	25 to 50 units		.03	5.75	17.25	52	3.70	
	1685	Butt fusion machine, wheeled, 1.5 HP electric, 2″ - 8″ diameter pipe		2.64	203	610	1,825	143.10	
	1690	Tracked, 20 HP diesel, 4″ - 12″ diameter pipe		11.27	525	1,575	4,725	405.15	
	1695	83 HP diesel, 8″ - 24″ diameter pipe		51.47	2,525	7,575	22,700	1,927	
	1700	Carts, brick, gas engine, 1,000 lb. capacity		2.95	69.50	208	625	65.20	
	1800	1,500 lb., 7-1/2′ lift		2.92	65	195	585	62.40	
	1822	Dehumidifier, medium, 6 lb./hr., 150 CFM		1.19	74	222	665	53.95	
	1824	Large, 18 lb./hr., 600 CFM		2.20	140	420	1,250	101.60	
	1830	Distributor, asphalt, trailer mounted, 2,000 gal., 38 H.P. diesel		11.03	340	1,025	3,075	293.20	
	1840	3,000 gal., 38 H.P. diesel		12.90	365	1,100	3,300	323.25	
	1850	Drill, rotary hammer, electric		1.12	27	80.50	242	25.05	
	1860	Carbide bit, 1-1/2″ diameter, add to electric rotary hammer		.03	5.10	15.30	46	3.30	
	1865	Rotary, crawler, 250 H.P.		136.18	2,225	6,650	20,000	2,419	
	1870	Emulsion sprayer, 65 gal., 5 H.P. gas engine		2.78	103	310	930	84.20	
	1880	200 gal., 5 H.P. engine		7.25	173	520	1,550	161.95	
	1900	Floor auto-scrubbing machine, walk-behind, 28″ path		5.64	365	1,100	3,300	265.15	
	1930	Floodlight, mercury vapor, or quartz, on tripod, 1,000 watt		.46	22	66.50	200	16.95	
	1940	2,000 watt		.59	27.50	82	246	21.15	
	1950	Floodlights, trailer mounted with generator, one - 300 watt light		3.56	76	228	685	74.05	
	1960	two 1000 watt lights		4.50	84.50	254	760	86.80	
	2000	four 300 watt lights		4.26	96.50	290	870	92.05	
	2005	Foam spray rig, incl. box trailer, compressor, generator, proportioner		25.54	515	1,550	4,650	514.30	
	2015	Forklift, pneumatic tire, rough terr., straight mast, 5,000 lb., 12′ lift, gas	↓	18.65	212	635	1,900	276.20	

01 54 33 | Equipment Rental

			UNIT	HOURLY OPER. COST	RENT PER DAY	RENT PER WEEK	RENT PER MONTH	EQUIPMENT COST/DAY	
40	2025	8,000 lb, 12′ lift	Ea.	22.75	350	1,050	3,150	392	40
	2030	5,000 lb., 12′ lift, diesel		15.45	237	710	2,125	265.60	
	2035	8,000 lb, 12′ lift, diesel		16.75	268	805	2,425	295	
	2045	All terrain, telescoping boom, diesel, 5,000 lb., 10′ reach, 19′ lift		17.25	410	1,225	3,675	383	
	2055	6,600 lb., 29′ reach, 42′ lift		21.10	400	1,200	3,600	408.80	
	2065	10,000 lb., 31′ reach, 45′ lift		23.10	465	1,400	4,200	464.75	
	2070	Cushion tire, smooth floor, gas, 5,000 lb. capacity		8.25	76.50	230	690	112	
	2075	8,000 lb. capacity		11.37	89	267	800	144.30	
	2085	Diesel, 5,000 lb. capacity		7.75	83.50	250	750	112	
	2090	12,000 lb. capacity		12.05	130	390	1,175	174.40	
	2095	20,000 lb. capacity		17.25	150	450	1,350	228.05	
	2100	Generator, electric, gas engine, 1.5 kW to 3 kW		2.58	10.15	30.50	91.50	26.70	
	2200	5 kW		3.22	12.65	38	114	33.35	
	2300	10 kW		5.93	33.50	101	305	67.65	
	2400	25 kW		7.41	84.50	254	760	110.05	
	2500	Diesel engine, 20 kW		9.21	74	222	665	118.05	
	2600	50 kW		15.95	96.50	289	865	185.35	
	2700	100 kW		28.60	133	400	1,200	308.80	
	2800	250 kW		54.35	255	765	2,300	587.85	
	2850	Hammer, hydraulic, for mounting on boom, to 500 ft. lb.		2.90	90.50	271	815	77.40	
	2860	1,000 ft. lb.		4.61	135	405	1,225	117.85	
	2900	Heaters, space, oil or electric, 50 MBH		1.46	8	24	72	16.50	
	3000	100 MBH		2.72	11.85	35.50	107	28.85	
	3100	300 MBH		7.92	40	120	360	87.40	
	3150	500 MBH		13.16	45	135	405	132.25	
	3200	Hose, water, suction with coupling, 20′ long, 2″ diameter		.02	3.13	9.40	28	2.05	
	3210	3″ diameter		.03	4.10	12.30	37	2.70	
	3220	4″ diameter		.03	4.80	14.40	43	3.10	
	3230	6″ diameter		.11	17	51	153	11.10	
	3240	8″ diameter		.27	45	135	405	29.15	
	3250	Discharge hose with coupling, 50′ long, 2″ diameter		.01	1.33	4	12	.90	
	3260	3″ diameter		.01	2.22	6.65	19.95	1.40	
	3270	4″ diameter		.02	3.47	10.40	31	2.25	
	3280	6″ diameter		.06	8.65	26	78	5.70	
	3290	8″ diameter		.24	40	120	360	25.90	
	3295	Insulation blower		.83	6	18.05	54	10.25	
	3300	Ladders, extension type, 16′ to 36′ long		.18	31.50	95	285	20.45	
	3400	40′ to 60′ long		.64	112	335	1,000	72.10	
	3405	Lance for cutting concrete		2.21	64	192	575	56.05	
	3407	Lawn mower, rotary, 22″, 5 H.P.		1.06	25	75	225	23.45	
	3408	48″ self-propelled		2.90	127	380	1,150	99.20	
	3410	Level, electronic, automatic, with tripod and leveling rod		1.05	61.50	185	555	45.40	
	3430	Laser type, for pipe and sewer line and grade		2.18	153	460	1,375	109.40	
	3440	Rotating beam for interior control		.90	62	186	560	44.40	
	3460	Builder's optical transit, with tripod and rod		.10	17	51	153	11	
	3500	Light towers, towable, with diesel generator, 2,000 watt		4.27	98	294	880	92.95	
	3600	4,000 watt		4.51	102	305	915	97.10	
	3700	Mixer, powered, plaster and mortar, 6 C.F., 7 H.P.		2.06	21	62.50	188	28.95	
	3800	10 C.F., 9 H.P.		2.24	33.50	101	305	38.15	
	3850	Nailer, pneumatic		.48	32.50	97	291	23.25	
	3900	Paint sprayers complete, 8 CFM		.85	59.50	179	535	42.60	
	4000	17 CFM		1.60	107	320	960	76.80	
	4020	Pavers, bituminous, rubber tires, 8′ wide, 50 H.P., diesel		32.02	550	1,650	4,950	586.20	
	4030	10′ wide, 150 H.P.		95.91	1,875	5,650	17,000	1,897	
	4050	Crawler, 8′ wide, 100 H.P., diesel		87.85	2,000	5,975	17,900	1,898	
	4060	10′ wide, 150 H.P.		104.28	2,275	6,825	20,500	2,199	
	4070	Concrete paver, 12′ to 24′ wide, 250 H.P.		87.88	1,625	4,900	14,700	1,683	
	4080	Placer-spreader-trimmer, 24′ wide, 300 H.P.	↓	117.87	2,475	7,425	22,300	2,428	

	01 54 33 | Equipment Rental		Unit	Hourly Oper. Cost	Rent per Day	Rent per Week	Rent per Month	Equipment Cost/Day	
40	4100	Pump, centrifugal gas pump, 1-1/2" diam., 65 GPM	Ea.	3.93	52.50	157	470	62.90	40
	4200	2" diameter, 130 GPM		4.99	64.50	194	580	78.75	
	4300	3" diameter, 250 GPM		5.13	63.50	191	575	79.25	
	4400	6" diameter, 1,500 GPM		22.31	197	590	1,775	296.45	
	4500	Submersible electric pump, 1-1/4" diameter, 55 GPM		.40	17.65	53	159	13.80	
	4600	1-1/2" diameter, 83 GPM		.45	20.50	61	183	15.75	
	4700	2" diameter, 120 GPM		1.65	25	75.50	227	28.30	
	4800	3" diameter, 300 GPM		3.04	45	135	405	51.35	
	4900	4" diameter, 560 GPM		14.80	163	490	1,475	216.35	
	5000	6" diameter, 1,590 GPM		22.14	200	600	1,800	297.15	
	5100	Diaphragm pump, gas, single, 1-1/2" diameter		1.13	57.50	172	515	43.45	
	5200	2" diameter		3.99	70.50	212	635	74.30	
	5300	3" diameter		4.06	74.50	224	670	77.30	
	5400	Double, 4" diameter		6.05	155	465	1,400	141.40	
	5450	Pressure washer, 5 GPM, 3,000 psi		3.88	53.50	160	480	63.05	
	5460	7 GPM, 3,000 psi		4.95	63	189	565	77.40	
	5500	Trash pump, self-priming, gas, 2" diameter		3.83	23.50	71	213	44.80	
	5600	Diesel, 4" diameter		6.70	94.50	284	850	110.40	
	5650	Diesel, 6" diameter		16.90	165	495	1,475	234.20	
	5655	Grout pump		18.75	272	815	2,450	313	
	5700	Salamanders, L.P. gas fired, 100,000 BTU		2.89	14.15	42.50	128	31.65	
	5705	50,000 BTU		1.67	11.65	35	105	20.35	
	5720	Sandblaster, portable, open top, 3 C.F. capacity		.60	26.50	79.50	239	20.70	
	5730	6 C.F. capacity		1.01	39.50	118	355	31.65	
	5740	Accessories for above		.14	23	69	207	14.90	
	5750	Sander, floor		.77	19.35	58	174	17.75	
	5760	Edger		.52	15.85	47.50	143	13.65	
	5800	Saw, chain, gas engine, 18" long		1.76	22.50	67.50	203	27.55	
	5900	Hydraulic powered, 36" long		.78	67.50	202	605	46.65	
	5950	60" long		.78	69.50	208	625	47.85	
	6000	Masonry, table mounted, 14" diameter, 5 H.P.		1.32	81	243	730	59.20	
	6050	Portable cut-off, 8 H.P.		1.82	32.50	97.50	293	34	
	6100	Circular, hand held, electric, 7-1/4" diameter		.23	4.93	14.80	44.50	4.80	
	6200	12" diameter		.24	7.85	23.50	70.50	6.60	
	6250	Wall saw, w/hydraulic power, 10 H.P.		3.30	33.50	101	305	46.60	
	6275	Shot blaster, walk-behind, 20" wide		4.75	272	815	2,450	201	
	6280	Sidewalk broom, walk-behind		2.25	80	240	720	65.95	
	6300	Steam cleaner, 100 gallons per hour		3.35	80	240	720	74.80	
	6310	200 gallons per hour		4.35	96.50	290	870	92.75	
	6340	Tar kettle/pot, 400 gallons		16.53	76.50	230	690	178.20	
	6350	Torch, cutting, acetylene-oxygen, 150' hose, excludes gases		.45	14.85	44.50	134	12.50	
	6360	Hourly operating cost includes tips and gas		20.98				167.85	
	6410	Toilet, portable chemical		.13	22.50	67	201	14.45	
	6420	Recycle flush type		.16	27.50	83	249	17.90	
	6430	Toilet, fresh water flush, garden hose,		.19	33	99	297	21.35	
	6440	Hoisted, non-flush, for high rise		.16	27	81	243	17.45	
	6465	Tractor, farm with attachment		17.42	300	900	2,700	319.35	
	6480	Trailers, platform, flush deck, 2 axle, 3 ton capacity		1.69	21.50	65	195	26.55	
	6500	25 ton capacity		6.25	138	415	1,250	133	
	6600	40 ton capacity		8.06	197	590	1,775	182.50	
	6700	3 axle, 50 ton capacity		8.75	218	655	1,975	200.95	
	6800	75 ton capacity		11.11	290	870	2,600	262.90	
	6810	Trailer mounted cable reel for high voltage line work		5.90	282	845	2,525	216.20	
	6820	Trailer mounted cable tensioning rig		11.71	560	1,675	5,025	428.65	
	6830	Cable pulling rig		73.99	3,125	9,375	28,100	2,467	
	6850	Portable cable/wire puller, 8,000 lb. max. pulling capacity		3.71	205	615	1,850	152.70	
	6900	Water tank trailer, engine driven discharge, 5,000 gallons		7.18	153	460	1,375	149.45	
	6925	10,000 gallons	↓	9.78	208	625	1,875	203.25	

01 54 33 | Equipment Rental

			UNIT	HOURLY OPER. COST	RENT PER DAY	RENT PER WEEK	RENT PER MONTH	EQUIPMENT COST/DAY
40	6950	Water truck, off highway, 6,000 gallons	Ea.	71.97	810	2,425	7,275	1,061
	7010	Tram car for high voltage line work, powered, 2 conductor		6.90	152	455	1,375	146.20
	7020	Transit (builder's level) with tripod		.10	17	51	153	11
	7030	Trench box, 3,000 lb., 6′ x 8′		.56	93.50	281	845	60.70
	7040	7,200 lb., 6′ x 20′		.72	120	360	1,075	77.75
	7050	8,000 lb., 8′ x 16′		1.08	180	540	1,625	116.65
	7060	9,500 lb., 8′ x 20′		1.21	225	675	2,025	144.65
	7065	11,000 lb., 8′ x 24′		1.27	212	635	1,900	137.15
	7070	12,000 lb., 10′ x 20′		1.49	255	765	2,300	164.95
	7100	Truck, pickup, 3/4 ton, 2 wheel drive		9.26	59.50	179	535	109.90
	7200	4 wheel drive		9.51	76	228	685	121.70
	7250	Crew carrier, 9 passenger		12.70	88	264	790	154.35
	7290	Flat bed truck, 20,000 lb. GVW		15.31	128	385	1,150	199.45
	7300	Tractor, 4 x 2, 220 H.P.		22.32	208	625	1,875	303.55
	7410	330 H.P.		32.43	285	855	2,575	430.40
	7500	6 x 4, 380 H.P.		36.20	330	990	2,975	487.60
	7600	450 H.P.		44.36	400	1,200	3,600	594.90
	7610	Tractor, with A frame, boom and winch, 225 H.P.		24.81	283	850	2,550	368.50
	7620	Vacuum truck, hazardous material, 2,500 gallons		12.83	300	900	2,700	282.65
	7625	5,000 gallons		13.06	425	1,275	3,825	359.50
	7650	Vacuum, HEPA, 16 gallon, wet/dry		.85	17.15	51.50	155	17.10
	7655	55 gallon, wet/dry		.78	26.50	80	240	22.25
	7660	Water tank, portable		.73	155	465	1,400	98.85
	7690	Sewer/catch basin vacuum, 14 C.Y., 1,500 gallons		17.37	640	1,925	5,775	523.95
	7700	Welder, electric, 200 amp		3.83	16	48	144	40.20
	7800	300 amp		5.56	19.35	58	174	56.10
	7900	Gas engine, 200 amp		8.97	23	69	207	85.60
	8000	300 amp		10.16	24	72	216	95.65
	8100	Wheelbarrow, any size		.06	10.35	31	93	6.70
	8200	Wrecking ball, 4,000 lb.	↓	2.51	72	216	650	63.25
50	0010	**HIGHWAY EQUIPMENT RENTAL** without operators R015433 -10						
	0050	Asphalt batch plant, portable drum mixer, 100 ton/hr.	Ea.	88.67	1,500	4,475	13,400	1,604
	0060	200 ton/hr.		102.29	1,600	4,775	14,300	1,773
	0070	300 ton/hr.		120.22	1,875	5,600	16,800	2,082
	0100	Backhoe attachment, long stick, up to 185 H.P., 10-1/2′ long		.37	24.50	74	222	17.75
	0140	Up to 250 H.P., 12′ long		.41	27.50	83	249	19.90
	0180	Over 250 H.P., 15′ long		.57	37.50	113	340	27.15
	0200	Special dipper arm, up to 100 H.P., 32′ long		1.16	77	231	695	55.50
	0240	Over 100 H.P., 33′ long		1.45	96.50	290	870	69.60
	0280	Catch basin/sewer cleaning truck, 3 ton, 9 C.Y., 1,000 gal.		35.50	410	1,225	3,675	528.95
	0300	Concrete batch plant, portable, electric, 200 C.Y./hr.		24.26	540	1,625	4,875	519.05
	0520	Grader/dozer attachment, ripper/scarifier, rear mounted, up to 135 H.P.		3.16	61.50	184	550	62.10
	0540	Up to 180 H.P.		4.15	92.50	278	835	88.80
	0580	Up to 250 H.P.		5.87	148	445	1,325	135.95
	0700	Pvmt. removal bucket, for hyd. excavator, up to 90 H.P.		2.17	56.50	169	505	51.10
	0740	Up to 200 H.P.		2.31	72	216	650	61.70
	0780	Over 200 H.P.		2.53	88.50	265	795	73.20
	0900	Aggregate spreader, self-propelled, 187 H.P.		50.75	715	2,150	6,450	836
	1000	Chemical spreader, 3 C.Y.		3.18	46.50	139	415	53.20
	1900	Hammermill, traveling, 250 H.P.		67.55	2,250	6,750	20,300	1,890
	2000	Horizontal borer, 3″ diameter, 13 H.P. gas driven		5.43	57.50	172	515	77.85
	2150	Horizontal directional drill, 20,000 lb. thrust, 78 H.P. diesel		27.66	685	2,050	6,150	631.30
	2160	30,000 lb. thrust, 115 H.P.		34.00	1,050	3,125	9,375	897
	2170	50,000 lb. thrust, 170 H.P.		48.74	1,325	3,975	11,900	1,185
	2190	Mud trailer for HDD, 1,500 gallons, 175 H.P., gas		25.58	157	470	1,400	298.65
	2200	Hydromulcher, diesel, 3,000 gallon, for truck mounting		17.48	255	765	2,300	292.90
	2300	Gas, 600 gallon		7.52	107	320	960	124.15
	2400	Joint & crack cleaner, walk behind, 25 H.P.	↓	3.17	52.50	158	475	56.95

01 54 33 | Equipment Rental

			UNIT	HOURLY OPER. COST	RENT PER DAY	RENT PER WEEK	RENT PER MONTH	EQUIPMENT COST/DAY	
50	2500	Filler, trailer mounted, 400 gallons, 20 H.P.	Ea.	8.37	220	660	1,975	198.95	50
	3000	Paint striper, self-propelled, 40 gallon, 22 H.P.		6.78	163	490	1,475	152.25	
	3100	120 gallon, 120 H.P.		19.28	410	1,225	3,675	399.25	
	3200	Post drivers, 6" I-Beam frame, for truck mounting		12.45	390	1,175	3,525	334.60	
	3400	Road sweeper, self-propelled, 8' wide, 90 H.P.		36.01	690	2,075	6,225	703.10	
	3450	Road sweeper, vacuum assisted, 4 C.Y., 220 gallons		58.45	650	1,950	5,850	857.60	
	4000	Road mixer, self-propelled, 130 H.P.		46.37	800	2,400	7,200	851	
	4100	310 H.P.		75.23	2,100	6,275	18,800	1,857	
	4220	Cold mix paver, incl. pug mill and bitumen tank, 165 H.P.		95.25	2,250	6,725	20,200	2,107	
	4240	Pavement brush, towed		3.44	96.50	290	870	85.50	
	4250	Paver, asphalt, wheel or crawler, 130 H.P., diesel		94.52	2,200	6,600	19,800	2,076	
	4300	Paver, road widener, gas, 1' to 6', 67 H.P.		46.80	940	2,825	8,475	939.40	
	4400	Diesel, 2' to 14', 88 H.P.		56.55	1,125	3,350	10,100	1,122	
	4600	Slipform pavers, curb and gutter, 2 track, 75 H.P.		58.01	1,225	3,650	11,000	1,194	
	4700	4 track, 165 H.P.		35.79	815	2,450	7,350	776.35	
	4800	Median barrier, 215 H.P.		58.61	1,300	3,900	11,700	1,249	
	4901	Trailer, low bed, 75 ton capacity		10.74	273	820	2,450	249.90	
	5000	Road planer, walk behind, 10" cutting width, 10 H.P.		2.46	33.50	100	300	39.70	
	5100	Self-propelled, 12" cutting width, 64 H.P.		8.28	117	350	1,050	136.25	
	5120	Traffic line remover, metal ball blaster, truck mounted, 115 H.P.		46.70	785	2,350	7,050	843.60	
	5140	Grinder, truck mounted, 115 H.P.		51.04	810	2,425	7,275	893.35	
	5160	Walk-behind, 11 H.P.		3.57	54.50	164	490	61.35	
	5200	Pavement profiler, 4' to 6' wide, 450 H.P.		217.23	3,425	10,300	30,900	3,798	
	5300	8' to 10' wide, 750 H.P.		332.58	4,525	13,600	40,800	5,381	
	5400	Roadway plate, steel, 1" x 8' x 20'		.09	14.65	44	132	9.50	
	5600	Stabilizer, self-propelled, 150 H.P.		41.26	640	1,925	5,775	715.05	
	5700	310 H.P.		76.41	1,700	5,075	15,200	1,626	
	5800	Striper, truck mounted, 120 gallon paint, 460 H.P.		48.89	490	1,475	4,425	686.10	
	5900	Thermal paint heating kettle, 115 gallons		7.73	26.50	80	240	77.85	
	6000	Tar kettle, 330 gallon, trailer mounted		12.31	60	180	540	134.50	
	7000	Tunnel locomotive, diesel, 8 to 12 ton		29.85	600	1,800	5,400	598.75	
	7005	Electric, 10 ton		29.33	685	2,050	6,150	644.70	
	7010	Muck cars, 1/2 C.Y. capacity		2.30	26	77.50	233	33.95	
	7020	1 C.Y. capacity		2.52	33.50	101	305	40.35	
	7030	2 C.Y. capacity		2.66	37.50	113	340	43.90	
	7040	Side dump, 2 C.Y. capacity		2.88	46.50	140	420	51.05	
	7050	3 C.Y. capacity		3.86	51.50	154	460	61.70	
	7060	5 C.Y. capacity		5.64	66.50	199	595	84.90	
	7100	Ventilating blower for tunnel, 7-1/2 H.P.		2.14	51	153	460	47.75	
	7110	10 H.P.		2.43	53.50	160	480	51.40	
	7120	20 H.P.		3.55	69.50	208	625	70	
	7140	40 H.P.		6.16	91.50	275	825	104.25	
	7160	60 H.P.		8.71	98.50	295	885	128.70	
	7175	75 H.P.		10.40	153	460	1,375	175.20	
	7180	200 H.P.		20.85	300	905	2,725	347.75	
	7800	Windrow loader, elevating		54.10	1,350	4,025	12,100	1,238	
60	0010	**LIFTING AND HOISTING EQUIPMENT RENTAL** without operators R015433-10							60
	0150	Crane, flatbed mounted, 3 ton capacity	Ea.	14.45	195	585	1,750	232.65	
	0200	Crane, climbing, 106' jib, 6,000 lb. capacity, 410 fpm R312316-45		39.84	1,775	5,350	16,100	1,389	
	0300	101' jib, 10,250 lb. capacity, 270 fpm		46.57	2,275	6,800	20,400	1,733	
	0500	Tower, static, 130' high, 106' jib, 6,200 lb. capacity at 400 fpm		45.29	2,075	6,200	18,600	1,602	
	0520	Mini crawler spider crane, up to 24" wide, 1,990 lb. lifting capacity		12.54	535	1,600	4,800	420.30	
	0525	Up to 30" wide, 6,450 lb. lifting capacity		14.57	635	1,900	5,700	496.55	
	0530	Up to 52" wide, 6,680 lb. lifting capacity		23.17	775	2,325	6,975	650.40	
	0535	Up to 55" wide, 8,920 lb. lifting capacity		25.87	860	2,575	7,725	722	
	0540	Up to 66" wide, 13,350 lb. lifting capacity		35.03	1,325	4,000	12,000	1,080	
	0600	Crawler mounted, lattice boom, 1/2 C.Y., 15 tons at 12' radius		37.07	950	2,850	8,550	866.60	
	0700	3/4 C.Y., 20 tons at 12' radius		50.57	1,200	3,625	10,900	1,130	

01 54 33 | Equipment Rental

		01 54 33 Equipment Rental	UNIT	HOURLY OPER. COST	RENT PER DAY	RENT PER WEEK	RENT PER MONTH	EQUIPMENT COST/DAY	
60	0800	1 C.Y., 25 tons at 12′ radius	Ea.	67.62	1,400	4,225	12,700	1,386	60
	0900	1-1/2 C.Y., 40 tons at 12′ radius		66.51	1,425	4,300	12,900	1,392	
	1000	2 C.Y., 50 tons at 12′ radius		89.03	2,200	6,600	19,800	2,032	
	1100	3 C.Y., 75 tons at 12′ radius		75.49	1,875	5,650	17,000	1,734	
	1200	100 ton capacity, 60′ boom		86.17	1,975	5,950	17,900	1,879	
	1300	165 ton capacity, 60′ boom		106.42	2,275	6,850	20,600	2,221	
	1400	200 ton capacity, 70′ boom		138.63	3,175	9,550	28,700	3,019	
	1500	350 ton capacity, 80′ boom		182.75	4,075	12,200	36,600	3,902	
	1600	Truck mounted, lattice boom, 6 x 4, 20 tons at 10′ radius		39.88	915	2,750	8,250	869	
	1700	25 tons at 10′ radius		42.86	1,400	4,175	12,500	1,178	
	1800	8 x 4, 30 tons at 10′ radius		45.67	1,475	4,425	13,300	1,250	
	1900	40 tons at 12′ radius		48.70	1,550	4,625	13,900	1,315	
	2000	60 tons at 15′ radius		53.85	1,650	4,950	14,900	1,421	
	2050	82 tons at 15′ radius		59.60	1,775	5,350	16,100	1,547	
	2100	90 tons at 15′ radius		66.59	1,950	5,825	17,500	1,698	
	2200	115 tons at 15′ radius		75.12	2,175	6,525	19,600	1,906	
	2300	150 tons at 18′ radius		81.33	2,275	6,850	20,600	2,021	
	2350	165 tons at 18′ radius		87.30	2,425	7,275	21,800	2,153	
	2400	Truck mounted, hydraulic, 12 ton capacity		29.59	390	1,175	3,525	471.75	
	2500	25 ton capacity		36.46	485	1,450	4,350	581.70	
	2550	33 ton capacity		50.82	900	2,700	8,100	946.60	
	2560	40 ton capacity		49.62	900	2,700	8,100	937	
	2600	55 ton capacity		53.94	915	2,750	8,250	981.50	
	2700	80 ton capacity		75.94	1,475	4,400	13,200	1,487	
	2720	100 ton capacity		75.18	1,550	4,675	14,000	1,536	
	2740	120 ton capacity		103.12	1,825	5,500	16,500	1,925	
	2760	150 ton capacity		110.25	2,050	6,125	18,400	2,107	
	2800	Self-propelled, 4 x 4, with telescoping boom, 5 ton		15.18	230	690	2,075	259.45	
	2900	12-1/2 ton capacity		21.48	335	1,000	3,000	371.85	
	3000	15 ton capacity		34.52	535	1,600	4,800	596.20	
	3050	20 ton capacity		24.09	660	1,975	5,925	587.75	
	3100	25 ton capacity		36.80	615	1,850	5,550	664.40	
	3150	40 ton capacity		45.03	650	1,950	5,850	750.25	
	3200	Derricks, guy, 20 ton capacity, 60′ boom, 75′ mast		22.81	435	1,300	3,900	442.50	
	3300	100′ boom, 115′ mast		36.15	750	2,250	6,750	739.20	
	3400	Stiffleg, 20 ton capacity, 70′ boom, 37′ mast		25.48	565	1,700	5,100	543.85	
	3500	100′ boom, 47′ mast		39.43	910	2,725	8,175	860.50	
	3550	Helicopter, small, lift to 1,250 lb. maximum, w/pilot		99.44	3,525	10,600	31,800	2,916	
	3600	Hoists, chain type, overhead, manual, 3/4 ton		.15	.33	1	3	1.35	
	3900	10 ton		.79	6	18	54	9.90	
	4000	Hoist and tower, 5,000 lb. cap., portable electric, 40′ high		5.14	252	755	2,275	192.10	
	4100	For each added 10′ section, add		.12	19.65	59	177	12.75	
	4200	Hoist and single tubular tower, 5,000 lb. electric, 100′ high		6.98	350	1,050	3,150	265.80	
	4300	For each added 6′-6″ section, add		.21	34.50	103	310	22.25	
	4400	Hoist and double tubular tower, 5,000 lb., 100′ high		7.59	385	1,150	3,450	290.75	
	4500	For each added 6′-6″ section, add		.23	37.50	113	340	24.40	
	4550	Hoist and tower, mast type, 6,000 lb., 100′ high		8.26	400	1,200	3,600	306.10	
	4570	For each added 10′ section, add		.13	23	69	207	14.85	
	4600	Hoist and tower, personnel, electric, 2,000 lb., 100′ @ 125 fpm		17.55	1,075	3,200	9,600	780.40	
	4700	3,000 lb., 100′ @ 200 fpm		20.08	1,200	3,625	10,900	885.65	
	4800	3,000 lb., 150′ @ 300 fpm		22.28	1,350	4,075	12,200	993.30	
	4900	4,000 lb., 100′ @ 300 fpm		23.05	1,375	4,150	12,500	1,014	
	5000	6,000 lb., 100′ @ 275 fpm	▼	24.77	1,450	4,350	13,100	1,068	
	5100	For added heights up to 500′, add	L.F.	.01	1.67	5	15	1.10	
	5200	Jacks, hydraulic, 20 ton	Ea.	.05	2	6	18	1.60	
	5500	100 ton		.40	12	36	108	10.40	
	6100	Jacks, hydraulic, climbing w/50′ jackrods, control console, 30 ton cap.		2.17	145	435	1,300	104.40	
	6150	For each added 10′ jackrod section, add	▼	.05	3.33	10	30	2.40	

01 54 33 | Equipment Rental

		01 54 33 Equipment Rental	UNIT	HOURLY OPER. COST	RENT PER DAY	RENT PER WEEK	RENT PER MONTH	EQUIPMENT COST/DAY	
60	6300	50 ton capacity	Ea.	3.49	232	695	2,075	166.95	60
	6350	For each added 10′ jackrod section, add		.06	4	12	36	2.90	
	6500	125 ton capacity		9.13	610	1,825	5,475	438.05	
	6550	For each added 10′ jackrod section, add		.62	41	123	370	29.55	
	6600	Cable jack, 10 ton capacity with 200′ cable		1.82	122	365	1,100	87.60	
	6650	For each added 50′ of cable, add	↓	.22	14.65	44	132	10.55	
70	0010	**WELLPOINT EQUIPMENT RENTAL** without operators R015433-10							70
	0020	Based on 2 months rental							
	0100	Combination jetting & wellpoint pump, 60 H.P. diesel	Ea.	15.72	360	1,075	3,225	340.75	
	0200	High pressure gas jet pump, 200 H.P., 300 psi	"	33.93	305	920	2,750	455.45	
	0300	Discharge pipe, 8″ diameter	L.F.	.01	.58	1.74	5.20	.40	
	0350	12″ diameter		.01	.83	2.50	7.50	.60	
	0400	Header pipe, flows up to 150 GPM, 4″ diameter		.01	.53	1.59	4.77	.40	
	0500	400 GPM, 6″ diameter		.01	.62	1.86	5.60	.45	
	0600	800 GPM, 8″ diameter		.01	.83	2.50	7.50	.60	
	0700	1,500 GPM, 10″ diameter		.01	.88	2.64	7.90	.65	
	0800	2,500 GPM, 12″ diameter		.03	1.70	5.10	15.30	1.20	
	0900	4,500 GPM, 16″ diameter		.03	2.18	6.55	19.65	1.55	
	0950	For quick coupling aluminum and plastic pipe, add	↓	.03	2.27	6.80	20.50	1.65	
	1100	Wellpoint, 25′ long, with fittings & riser pipe, 1-1/2″ or 2″ diameter	Ea.	.07	4.52	13.55	40.50	3.25	
	1200	Wellpoint pump, diesel powered, 4″ suction, 20 H.P.		7.02	207	620	1,850	180.20	
	1300	6″ suction, 30 H.P.		9.42	257	770	2,300	229.35	
	1400	8″ suction, 40 H.P.		12.76	350	1,050	3,150	312.10	
	1500	10″ suction, 75 H.P.		18.83	410	1,225	3,675	395.65	
	1600	12″ suction, 100 H.P.		27.32	660	1,975	5,925	613.55	
	1700	12″ suction, 175 H.P.	↓	39.09	725	2,175	6,525	747.75	
80	0010	**MARINE EQUIPMENT RENTAL** without operators R015433-10							80
	0200	Barge, 400 ton, 30′ wide x 90′ long	Ea.	17.69	1,150	3,475	10,400	836.50	
	0240	800 ton, 45′ wide x 90′ long		22.21	1,425	4,275	12,800	1,033	
	2000	Tugboat, diesel, 100 H.P.		29.66	230	690	2,075	375.25	
	2040	250 H.P.		57.58	415	1,250	3,750	710.70	
	2080	380 H.P.		125.36	1,250	3,750	11,300	1,753	
	3000	Small work boat, gas, 16-foot, 50 H.P.		11.38	46.50	139	415	118.85	
	4000	Large, diesel, 48-foot, 200 H.P.	↓	74.90	1,325	3,975	11,900	1,394	

Crew No.	Bare Costs		Incl. Subs O&P		Cost Per Labor-Hour	

Crew A-1	Hr.	Daily	Hr.	Daily	Bare Costs	Incl. O&P
1 Building Laborer	$41.05	$328.40	$62.10	$496.80	$41.05	$62.10
1 Concrete Saw, Gas Manual		72.90		80.19	9.11	10.02
8 L.H., Daily Totals		$401.30		$576.99	$50.16	$72.12

Crew A-1A	Hr.	Daily	Hr.	Daily	Bare Costs	Incl. O&P
1 Skilled Worker	$53.40	$427.20	$81.25	$650.00	$53.40	$81.25
1 Shot Blaster, 20"		201.00		221.10	25.13	27.64
8 L.H., Daily Totals		$628.20		$871.10	$78.53	$108.89

Crew A-1B	Hr.	Daily	Hr.	Daily	Bare Costs	Incl. O&P
1 Building Laborer	$41.05	$328.40	$62.10	$496.80	$41.05	$62.10
1 Concrete Saw		110.15		121.17	13.77	15.15
8 L.H., Daily Totals		$438.55		$617.97	$54.82	$77.25

Crew A-1C	Hr.	Daily	Hr.	Daily	Bare Costs	Incl. O&P
1 Building Laborer	$41.05	$328.40	$62.10	$496.80	$41.05	$62.10
1 Chain Saw, Gas, 18"		27.55		30.31	3.44	3.79
8 L.H., Daily Totals		$355.95		$527.11	$44.49	$65.89

Crew A-1D	Hr.	Daily	Hr.	Daily	Bare Costs	Incl. O&P
1 Building Laborer	$41.05	$328.40	$62.10	$496.80	$41.05	$62.10
1 Vibrating Plate, Gas, 18"		31.10		34.21	3.89	4.28
8 L.H., Daily Totals		$359.50		$531.01	$44.94	$66.38

Crew A-1E	Hr.	Daily	Hr.	Daily	Bare Costs	Incl. O&P
1 Building Laborer	$41.05	$328.40	$62.10	$496.80	$41.05	$62.10
1 Vibrating Plate, Gas, 21"		40.55		44.60	5.07	5.58
8 L.H., Daily Totals		$368.95		$541.40	$46.12	$67.68

Crew A-1F	Hr.	Daily	Hr.	Daily	Bare Costs	Incl. O&P
1 Building Laborer	$41.05	$328.40	$62.10	$496.80	$41.05	$62.10
1 Rammer/Tamper, Gas, 8"		45.65		50.22	5.71	6.28
8 L.H., Daily Totals		$374.05		$547.01	$46.76	$68.38

Crew A-1G	Hr.	Daily	Hr.	Daily	Bare Costs	Incl. O&P
1 Building Laborer	$41.05	$328.40	$62.10	$496.80	$41.05	$62.10
1 Rammer/Tamper, Gas, 15"		53.05		58.35	6.63	7.29
8 L.H., Daily Totals		$381.45		$555.15	$47.68	$69.39

Crew A-1H	Hr.	Daily	Hr.	Daily	Bare Costs	Incl. O&P
1 Building Laborer	$41.05	$328.40	$62.10	$496.80	$41.05	$62.10
1 Exterior Steam Cleaner		74.80		82.28	9.35	10.29
8 L.H., Daily Totals		$403.20		$579.08	$50.40	$72.39

Crew A-1J	Hr.	Daily	Hr.	Daily	Bare Costs	Incl. O&P
1 Building Laborer	$41.05	$328.40	$62.10	$496.80	$41.05	$62.10
1 Cultivator, Walk-Behind, 5 H.P.		70.50		77.55	8.81	9.69
8 L.H., Daily Totals		$398.90		$574.35	$49.86	$71.79

Crew A-1K	Hr.	Daily	Hr.	Daily	Bare Costs	Incl. O&P
1 Building Laborer	$41.05	$328.40	$62.10	$496.80	$41.05	$62.10
1 Cultivator, Walk-Behind, 8 H.P.		84.50		92.95	10.56	11.62
8 L.H., Daily Totals		$412.90		$589.75	$51.61	$73.72

Crew A-1M	Hr.	Daily	Hr.	Daily	Bare Costs	Incl. O&P
1 Building Laborer	$41.05	$328.40	$62.10	$496.80	$41.05	$62.10
1 Snow Blower, Walk-Behind		65.95		72.55	8.24	9.07
8 L.H., Daily Totals		$394.35		$569.35	$49.29	$71.17

Crew No.	Bare Costs		Incl. Subs O&P		Cost Per Labor-Hour	

Crew A-2	Hr.	Daily	Hr.	Daily	Bare Costs	Incl. O&P
2 Laborers	$41.05	$656.80	$62.10	$993.60	$42.53	$64.23
1 Truck Driver (light)	45.50	364.00	68.50	548.00		
1 Flatbed Truck, Gas, 1.5 Ton		195.20		214.72	8.13	8.95
24 L.H., Daily Totals		$1216.00		$1756.32	$50.67	$73.18

Crew A-2A	Hr.	Daily	Hr.	Daily	Bare Costs	Incl. O&P
2 Laborers	$41.05	$656.80	$62.10	$993.60	$42.53	$64.23
1 Truck Driver (light)	45.50	364.00	68.50	548.00		
1 Flatbed Truck, Gas, 1.5 Ton		195.20		214.72		
1 Concrete Saw		110.15		121.17	12.72	14.00
24 L.H., Daily Totals		$1326.15		$1877.48	$55.26	$78.23

Crew A-2B	Hr.	Daily	Hr.	Daily	Bare Costs	Incl. O&P
1 Truck Driver (light)	$45.50	$364.00	$68.50	$548.00	$45.50	$68.50
1 Flatbed Truck, Gas, 1.5 Ton		195.20		214.72	24.40	26.84
8 L.H., Daily Totals		$559.20		$762.72	$69.90	$95.34

Crew A-3A	Hr.	Daily	Hr.	Daily	Bare Costs	Incl. O&P
1 Equip. Oper. (light)	$51.65	$413.20	$77.55	$620.40	$51.65	$77.55
1 Pickup Truck, 4x4, 3/4 Ton		121.70		133.87	15.21	16.73
8 L.H., Daily Totals		$534.90		$754.27	$66.86	$94.28

Crew A-3B	Hr.	Daily	Hr.	Daily	Bare Costs	Incl. O&P
1 Equip. Oper. (medium)	$55.10	$440.80	$82.70	$661.60	$51.05	$76.72
1 Truck Driver (heavy)	47.00	376.00	70.75	566.00		
1 Dump Truck, 12 C.Y., 400 H.P.		567.05		623.76		
1 F.E. Loader, W.M., 2.5 C.Y.		531.20		584.32	68.64	75.50
16 L.H., Daily Totals		$1915.05		$2435.68	$119.69	$152.23

Crew A-3C	Hr.	Daily	Hr.	Daily	Bare Costs	Incl. O&P
1 Equip. Oper. (light)	$51.65	$413.20	$77.55	$620.40	$51.65	$77.55
1 Loader, Skid Steer, 78 H.P.		392.50		431.75	49.06	53.97
8 L.H., Daily Totals		$805.70		$1052.15	$100.71	$131.52

Crew A-3D	Hr.	Daily	Hr.	Daily	Bare Costs	Incl. O&P
1 Truck Driver (light)	$45.50	$364.00	$68.50	$548.00	$45.50	$68.50
1 Pickup Truck, 4x4, 3/4 Ton		121.70		133.87		
1 Flatbed Trailer, 25 Ton		133.00		146.30	31.84	35.02
8 L.H., Daily Totals		$618.70		$828.17	$77.34	$103.52

Crew A-3E	Hr.	Daily	Hr.	Daily	Bare Costs	Incl. O&P
1 Equip. Oper. (crane)	$57.45	$459.60	$86.25	$690.00	$52.23	$78.50
1 Truck Driver (heavy)	47.00	376.00	70.75	566.00		
1 Pickup Truck, 4x4, 3/4 Ton		121.70		133.87	7.61	8.37
16 L.H., Daily Totals		$957.30		$1389.87	$59.83	$86.87

Crew A-3F	Hr.	Daily	Hr.	Daily	Bare Costs	Incl. O&P
1 Equip. Oper. (crane)	$57.45	$459.60	$86.25	$690.00	$52.23	$78.50
1 Truck Driver (heavy)	47.00	376.00	70.75	566.00		
1 Pickup Truck, 4x4, 3/4 Ton		121.70		133.87		
1 Truck Tractor, 6x4, 380 H.P.		487.60		536.36		
1 Lowbed Trailer, 75 Ton		249.90		274.89	53.70	59.07
16 L.H., Daily Totals		$1694.80		$2201.12	$105.93	$137.57

Crews - Standard

Crew No.	Bare Costs		Incl. Subs O&P		Cost Per Labor-Hour	
Crew A-3G	**Hr.**	**Daily**	**Hr.**	**Daily**	**Bare Costs**	**Incl. O&P**
1 Equip. Oper. (crane)	$57.45	$459.60	$86.25	$690.00	$52.23	$78.50
1 Truck Driver (heavy)	47.00	376.00	70.75	566.00		
1 Pickup Truck, 4x4, 3/4 Ton		121.70		133.87		
1 Truck Tractor, 6x4, 450 H.P.		594.90		654.39		
1 Lowbed Trailer, 75 Ton		249.90		274.89	60.41	66.45
16 L.H., Daily Totals		$1802.10		$2319.15	$112.63	$144.95

Crew A-3H	Hr.	Daily	Hr.	Daily	Bare Costs	Incl. O&P
1 Equip. Oper. (crane)	$57.45	$459.60	$86.25	$690.00	$57.45	$86.25
1 Hyd. Crane, 12 Ton (Daily)		708.90		779.79	88.61	97.47
8 L.H., Daily Totals		$1168.50		$1469.79	$146.06	$183.72

Crew A-3I	Hr.	Daily	Hr.	Daily	Bare Costs	Incl. O&P
1 Equip. Oper. (crane)	$57.45	$459.60	$86.25	$690.00	$57.45	$86.25
1 Hyd. Crane, 25 Ton (Daily)		785.85		864.43	98.23	108.05
8 L.H., Daily Totals		$1245.45		$1554.43	$155.68	$194.30

Crew A-3J	Hr.	Daily	Hr.	Daily	Bare Costs	Incl. O&P
1 Equip. Oper. (crane)	$57.45	$459.60	$86.25	$690.00	$57.45	$86.25
1 Hyd. Crane, 40 Ton (Daily)		1280.00		1408.00	160.00	176.00
8 L.H., Daily Totals		$1739.60		$2098.00	$217.45	$262.25

Crew A-3K	Hr.	Daily	Hr.	Daily	Bare Costs	Incl. O&P
1 Equip. Oper. (crane)	$57.45	$459.60	$86.25	$690.00	$53.33	$80.05
1 Equip. Oper. (oiler)	49.20	393.60	73.85	590.80		
1 Hyd. Crane, 55 Ton (Daily)		1299.00		1428.90		
1 P/U Truck, 3/4 Ton (Daily)		140.45		154.50	89.97	98.96
16 L.H., Daily Totals		$2292.65		$2864.20	$143.29	$179.01

Crew A-3L	Hr.	Daily	Hr.	Daily	Bare Costs	Incl. O&P
1 Equip. Oper. (crane)	$57.45	$459.60	$86.25	$690.00	$53.33	$80.05
1 Equip. Oper. (oiler)	49.20	393.60	73.85	590.80		
1 Hyd. Crane, 80 Ton (Daily)		2056.00		2261.60		
1 P/U Truck, 3/4 Ton (Daily)		140.45		154.50	137.28	151.01
16 L.H., Daily Totals		$3049.65		$3696.90	$190.60	$231.06

Crew A-3M	Hr.	Daily	Hr.	Daily	Bare Costs	Incl. O&P
1 Equip. Oper. (crane)	$57.45	$459.60	$86.25	$690.00	$53.33	$80.05
1 Equip. Oper. (oiler)	49.20	393.60	73.85	590.80		
1 Hyd. Crane, 100 Ton (Daily)		2179.00		2396.90		
1 P/U Truck, 3/4 Ton (Daily)		140.45		154.50	144.97	159.46
16 L.H., Daily Totals		$3172.65		$3832.20	$198.29	$239.51

Crew A-3N	Hr.	Daily	Hr.	Daily	Bare Costs	Incl. O&P
1 Equip. Oper. (crane)	$57.45	$459.60	$86.25	$690.00	$57.45	$86.25
1 Tower Crane (monthly)		1181.00		1299.10	147.63	162.39
8 L.H., Daily Totals		$1640.60		$1989.10	$205.07	$248.64

Crew A-3P	Hr.	Daily	Hr.	Daily	Bare Costs	Incl. O&P
1 Equip. Oper. (light)	$51.65	$413.20	$77.55	$620.40	$51.65	$77.55
1 A.T. Forklift, 31' reach, 45' lift		464.75		511.23	58.09	63.90
8 L.H., Daily Totals		$877.95		$1131.63	$109.74	$141.45

Crew A-3Q	Hr.	Daily	Hr.	Daily	Bare Costs	Incl. O&P
1 Equip. Oper. (light)	$51.65	$413.20	$77.55	$620.40	$51.65	$77.55
1 Pickup Truck, 4x4, 3/4 Ton		121.70		133.87		
1 Flatbed Trailer, 3 Ton		26.55		29.20	18.53	20.38
8 L.H., Daily Totals		$561.45		$783.48	$70.18	$97.93

Crew No.	Bare Costs		Incl. Subs O&P		Cost Per Labor-Hour	
Crew A-3R	**Hr.**	**Daily**	**Hr.**	**Daily**	**Bare Costs**	**Incl. O&P**
1 Equip. Oper. (light)	$51.65	$413.20	$77.55	$620.40	$51.65	$77.55
1 Forklift, Smooth Floor, 8,000 Lb.		144.30		158.73	18.04	19.84
8 L.H., Daily Totals		$557.50		$779.13	$69.69	$97.39

Crew A-4	Hr.	Daily	Hr.	Daily	Bare Costs	Incl. O&P
2 Carpenters	$51.65	$826.40	$78.15	$1250.40	$48.85	$73.68
1 Painter, Ordinary	43.25	346.00	64.75	518.00		
24 L.H., Daily Totals		$1172.40		$1768.40	$48.85	$73.68

Crew A-5	Hr.	Daily	Hr.	Daily	Bare Costs	Incl. O&P
2 Laborers	$41.05	$656.80	$62.10	$993.60	$41.54	$62.81
.25 Truck Driver (light)	45.50	91.00	68.50	137.00		
.25 Flatbed Truck, Gas, 1.5 Ton		48.80		53.68	2.71	2.98
18 L.H., Daily Totals		$796.60		$1184.28	$44.26	$65.79

Crew A-6	Hr.	Daily	Hr.	Daily	Bare Costs	Incl. O&P
1 Instrument Man	$53.40	$427.20	$81.25	$650.00	$51.52	$78.00
1 Rodman/Chainman	49.65	397.20	74.75	598.00		
1 Level, Electronic		45.40		49.94	2.84	3.12
16 L.H., Daily Totals		$869.80		$1297.94	$54.36	$81.12

Crew A-7	Hr.	Daily	Hr.	Daily	Bare Costs	Incl. O&P
1 Chief of Party	$65.10	$520.80	$98.10	$784.80	$56.05	$84.70
1 Instrument Man	53.40	427.20	81.25	650.00		
1 Rodman/Chainman	49.65	397.20	74.75	598.00		
1 Level, Electronic		45.40		49.94	1.89	2.08
24 L.H., Daily Totals		$1390.60		$2082.74	$57.94	$86.78

Crew A-8	Hr.	Daily	Hr.	Daily	Bare Costs	Incl. O&P
1 Chief of Party	$65.10	$520.80	$98.10	$784.80	$54.45	$82.21
1 Instrument Man	53.40	427.20	81.25	650.00		
2 Rodmen/Chainmen	49.65	794.40	74.75	1196.00		
1 Level, Electronic		45.40		49.94	1.42	1.56
32 L.H., Daily Totals		$1787.80		$2680.74	$55.87	$83.77

Crew A-9	Hr.	Daily	Hr.	Daily	Bare Costs	Incl. O&P
1 Asbestos Foreman	$57.85	$462.80	$89.30	$714.40	$57.41	$88.64
7 Asbestos Workers	57.35	3211.60	88.55	4958.80		
64 L.H., Daily Totals		$3674.40		$5673.20	$57.41	$88.64

Crew A-10A	Hr.	Daily	Hr.	Daily	Bare Costs	Incl. O&P
1 Asbestos Foreman	$57.85	$462.80	$89.30	$714.40	$57.52	$88.80
2 Asbestos Workers	57.35	917.60	88.55	1416.80		
24 L.H., Daily Totals		$1380.40		$2131.20	$57.52	$88.80

Crew A-10B	Hr.	Daily	Hr.	Daily	Bare Costs	Incl. O&P
1 Asbestos Foreman	$57.85	$462.80	$89.30	$714.40	$57.48	$88.74
3 Asbestos Workers	57.35	1376.40	88.55	2125.20		
32 L.H., Daily Totals		$1839.20		$2839.60	$57.48	$88.74

Crew A-10C	Hr.	Daily	Hr.	Daily	Bare Costs	Incl. O&P
3 Asbestos Workers	$57.35	$1376.40	$88.55	$2125.20	$57.35	$88.55
1 Flatbed Truck, Gas, 1.5 Ton		195.20		214.72	8.13	8.95
24 L.H., Daily Totals		$1571.60		$2339.92	$65.48	$97.50

Crew A-10D	Bare Costs Hr.	Bare Costs Daily	Incl. Subs O&P Hr.	Incl. Subs O&P Daily	Cost Per Labor-Hour Bare Costs	Cost Per Labor-Hour Incl. O&P
2 Asbestos Workers	$57.35	$917.60	$88.55	$1416.80	$55.34	$84.30
1 Equip. Oper. (crane)	57.45	459.60	86.25	690.00		
1 Equip. Oper. (oiler)	49.20	393.60	73.85	590.80		
1 Hydraulic Crane, 33 Ton		946.60		1041.26	29.58	32.54
32 L.H., Daily Totals		$2717.40		$3738.86	$84.92	$116.84

Crew A-11	Hr.	Daily	Hr.	Daily	Bare Costs	Incl. O&P
1 Asbestos Foreman	$57.85	$462.80	$89.30	$714.40	$57.41	$88.64
7 Asbestos Workers	57.35	3211.60	88.55	4958.80		
2 Chip. Hammers, 12 Lb., Elec.		42.00		46.20	.66	.72
64 L.H., Daily Totals		$3716.40		$5719.40	$58.07	$89.37

Crew A-12	Hr.	Daily	Hr.	Daily	Bare Costs	Incl. O&P
1 Asbestos Foreman	$57.85	$462.80	$89.30	$714.40	$57.41	$88.64
7 Asbestos Workers	57.35	3211.60	88.55	4958.80		
1 Trk-Mtd Vac, 14 CY, 1500 Gal.		523.95		576.35		
1 Flatbed Truck, 20,000 GVW		199.45		219.40	11.30	12.43
64 L.H., Daily Totals		$4397.80		$6468.94	$68.72	$101.08

Crew A-13	Hr.	Daily	Hr.	Daily	Bare Costs	Incl. O&P
1 Equip. Oper. (light)	$51.65	$413.20	$77.55	$620.40	$51.65	$77.55
1 Trk-Mtd Vac, 14 CY, 1500 Gal.		523.95		576.35		
1 Flatbed Truck, 20,000 GVW		199.45		219.40	90.42	99.47
8 L.H., Daily Totals		$1136.60		$1416.14	$142.07	$177.02

Crew B-1	Hr.	Daily	Hr.	Daily	Bare Costs	Incl. O&P
1 Labor Foreman (outside)	$43.05	$344.40	$65.10	$520.80	$41.72	$63.10
2 Laborers	41.05	656.80	62.10	993.60		
24 L.H., Daily Totals		$1001.20		$1514.40	$41.72	$63.10

Crew B-1A	Hr.	Daily	Hr.	Daily	Bare Costs	Incl. O&P
1 Labor Foreman (outside)	$43.05	$344.40	$65.10	$520.80	$41.72	$63.10
2 Laborers	41.05	656.80	62.10	993.60		
2 Cutting Torches		25.00		27.50		
2 Sets of Gases		335.70		369.27	15.03	16.53
24 L.H., Daily Totals		$1361.90		$1911.17	$56.75	$79.63

Crew B-1B	Hr.	Daily	Hr.	Daily	Bare Costs	Incl. O&P
1 Labor Foreman (outside)	$43.05	$344.40	$65.10	$520.80	$45.65	$68.89
2 Laborers	41.05	656.80	62.10	993.60		
1 Equip. Oper. (crane)	57.45	459.60	86.25	690.00		
2 Cutting Torches		25.00		27.50		
2 Sets of Gases		335.70		369.27		
1 Hyd. Crane, 12 Ton		471.75		518.92	26.01	28.62
32 L.H., Daily Totals		$2293.25		$3120.09	$71.66	$97.50

Crew B-1C	Hr.	Daily	Hr.	Daily	Bare Costs	Incl. O&P
1 Labor Foreman (outside)	$43.05	$344.40	$65.10	$520.80	$41.72	$63.10
2 Laborers	41.05	656.80	62.10	993.60		
1 Telescoping Boom Lift, to 60'		451.25		496.38	18.80	20.68
24 L.H., Daily Totals		$1452.45		$2010.78	$60.52	$83.78

Crew B-1D	Hr.	Daily	Hr.	Daily	Bare Costs	Incl. O&P
2 Laborers	$41.05	$656.80	$62.10	$993.60	$41.05	$62.10
1 Small Work Boat, Gas, 50 H.P.		118.85		130.74		
1 Pressure Washer, 7 GPM		77.40		85.14	12.27	13.49
16 L.H., Daily Totals		$853.05		$1209.47	$53.32	$75.59

Crew B-1E	Hr.	Daily	Hr.	Daily	Bare Costs	Incl. O&P
1 Labor Foreman (outside)	$43.05	$344.40	$65.10	$520.80	$41.55	$62.85
3 Laborers	41.05	985.20	62.10	1490.40		
1 Work Boat, Diesel, 200 H.P.		1394.00		1533.40		
2 Pressure Washers, 7 GPM		154.80		170.28	48.40	53.24
32 L.H., Daily Totals		$2878.40		$3714.88	$89.95	$116.09

Crew B-1F	Hr.	Daily	Hr.	Daily	Bare Costs	Incl. O&P
2 Skilled Workers	$53.40	$854.40	$81.25	$1300.00	$49.28	$74.87
1 Laborer	41.05	328.40	62.10	496.80		
1 Small Work Boat, Gas, 50 H.P.		118.85		130.74		
1 Pressure Washer, 7 GPM		77.40		85.14	8.18	8.99
24 L.H., Daily Totals		$1379.05		$2012.68	$57.46	$83.86

Crew B-1G	Hr.	Daily	Hr.	Daily	Bare Costs	Incl. O&P
2 Laborers	$41.05	$656.80	$62.10	$993.60	$41.05	$62.10
1 Small Work Boat, Gas, 50 H.P.		118.85		130.74	7.43	8.17
16 L.H., Daily Totals		$775.65		$1124.34	$48.48	$70.27

Crew B-1H	Hr.	Daily	Hr.	Daily	Bare Costs	Incl. O&P
2 Skilled Workers	$53.40	$854.40	$81.25	$1300.00	$49.28	$74.87
1 Laborer	41.05	328.40	62.10	496.80		
1 Small Work Boat, Gas, 50 H.P.		118.85		130.74	4.95	5.45
24 L.H., Daily Totals		$1301.65		$1927.54	$54.24	$80.31

Crew B-1J	Hr.	Daily	Hr.	Daily	Bare Costs	Incl. O&P
1 Labor Foreman (inside)	$41.55	$332.40	$62.85	$502.80	$41.30	$62.48
1 Laborer	41.05	328.40	62.10	496.80		
16 L.H., Daily Totals		$660.80		$999.60	$41.30	$62.48

Crew B-1K	Hr.	Daily	Hr.	Daily	Bare Costs	Incl. O&P
1 Carpenter Foreman (inside)	$52.15	$417.20	$78.90	$631.20	$51.90	$78.53
1 Carpenter	51.65	413.20	78.15	625.20		
16 L.H., Daily Totals		$830.40		$1256.40	$51.90	$78.53

Crew B-2	Hr.	Daily	Hr.	Daily	Bare Costs	Incl. O&P
1 Labor Foreman (outside)	$43.05	$344.40	$65.10	$520.80	$41.45	$62.70
4 Laborers	41.05	1313.60	62.10	1987.20		
40 L.H., Daily Totals		$1658.00		$2508.00	$41.45	$62.70

Crew B-2A	Hr.	Daily	Hr.	Daily	Bare Costs	Incl. O&P
1 Labor Foreman (outside)	$43.05	$344.40	$65.10	$520.80	$41.72	$63.10
2 Laborers	41.05	656.80	62.10	993.60		
1 Telescoping Boom Lift, to 60'		451.25		496.38	18.80	20.68
24 L.H., Daily Totals		$1452.45		$2010.78	$60.52	$83.78

Crew B-3	Hr.	Daily	Hr.	Daily	Bare Costs	Incl. O&P
1 Labor Foreman (outside)	$43.05	$344.40	$65.10	$520.80	$45.71	$68.92
2 Laborers	41.05	656.80	62.10	993.60		
1 Equip. Oper. (medium)	55.10	440.80	82.70	661.60		
2 Truck Drivers (heavy)	47.00	752.00	70.75	1132.00		
1 Crawler Loader, 3 C.Y.		1202.00		1322.20		
2 Dump Trucks, 12 C.Y., 400 H.P.		1134.10		1247.51	48.67	53.54
48 L.H., Daily Totals		$4530.10		$5877.71	$94.38	$122.45

Crew B-3A	Hr.	Daily	Hr.	Daily	Bare Costs	Incl. O&P
4 Laborers	$41.05	$1313.60	$62.10	$1987.20	$43.86	$66.22
1 Equip. Oper. (medium)	55.10	440.80	82.70	661.60		
1 Hyd. Excavator, 1.5 C.Y.		908.70		999.57	22.72	24.99
40 L.H., Daily Totals		$2663.10		$3648.37	$66.58	$91.21

Crew No.	Bare Costs		Incl. Subs O&P		Cost Per Labor-Hour	
Crew B-3B	**Hr.**	**Daily**	**Hr.**	**Daily**	**Bare Costs**	**Incl. O&P**
2 Laborers	$41.05	$656.80	$62.10	$993.60	$46.05	$69.41
1 Equip. Oper. (medium)	55.10	440.80	82.70	661.60		
1 Truck Driver (heavy)	47.00	376.00	70.75	566.00		
1 Backhoe Loader, 80 H.P.		387.85		426.63		
1 Dump Truck, 12 C.Y., 400 H.P.		567.05		623.76	29.84	32.82
32 L.H., Daily Totals		$2428.50		$3271.59	$75.89	$102.24
Crew B-3C	**Hr.**	**Daily**	**Hr.**	**Daily**	**Bare Costs**	**Incl. O&P**
3 Laborers	$41.05	$985.20	$62.10	$1490.40	$44.56	$67.25
1 Equip. Oper. (medium)	55.10	440.80	82.70	661.60		
1 Crawler Loader, 4 C.Y.		1461.00		1607.10	45.66	50.22
32 L.H., Daily Totals		$2887.00		$3759.10	$90.22	$117.47
Crew B-4	**Hr.**	**Daily**	**Hr.**	**Daily**	**Bare Costs**	**Incl. O&P**
1 Labor Foreman (outside)	$43.05	$344.40	$65.10	$520.80	$42.38	$64.04
4 Laborers	41.05	1313.60	62.10	1987.20		
1 Truck Driver (heavy)	47.00	376.00	70.75	566.00		
1 Truck Tractor, 220 H.P.		303.55		333.90		
1 Flatbed Trailer, 40 Ton		182.50		200.75	10.13	11.14
48 L.H., Daily Totals		$2520.05		$3608.66	$52.50	$75.18
Crew B-5	**Hr.**	**Daily**	**Hr.**	**Daily**	**Bare Costs**	**Incl. O&P**
1 Labor Foreman (outside)	$43.05	$344.40	$65.10	$520.80	$45.35	$68.41
4 Laborers	41.05	1313.60	62.10	1987.20		
2 Equip. Oper. (medium)	55.10	881.60	82.70	1323.20		
1 Air Compressor, 250 cfm		167.95		184.75		
2 Breakers, Pavement, 60 lb.		21.50		23.65		
2 -50' Air Hoses, 1.5"		45.50		50.05		
1 Crawler Loader, 3 C.Y.		1202.00		1322.20	25.66	28.23
56 L.H., Daily Totals		$3976.55		$5411.85	$71.01	$96.64
Crew B-5A	**Hr.**	**Daily**	**Hr.**	**Daily**	**Bare Costs**	**Incl. O&P**
1 Labor Foreman (outside)	$43.05	$344.40	$65.10	$520.80	$45.43	$68.51
6 Laborers	41.05	1970.40	62.10	2980.80		
2 Equip. Oper. (medium)	55.10	881.60	82.70	1323.20		
1 Equip. Oper. (light)	51.65	413.20	77.55	620.40		
2 Truck Drivers (heavy)	47.00	752.00	70.75	1132.00		
1 Air Compressor, 365 cfm		213.35		234.69		
2 Breakers, Pavement, 60 lb.		21.50		23.65		
8 -50' Air Hoses, 1"		67.20		73.92		
2 Dump Trucks, 8 C.Y., 220 H.P.		679.20		747.12	10.22	11.24
96 L.H., Daily Totals		$5342.85		$7656.57	$55.65	$79.76
Crew B-5B	**Hr.**	**Daily**	**Hr.**	**Daily**	**Bare Costs**	**Incl. O&P**
1 Powderman	$53.40	$427.20	$81.25	$650.00	$50.77	$76.48
2 Equip. Oper. (medium)	55.10	881.60	82.70	1323.20		
3 Truck Drivers (heavy)	47.00	1128.00	70.75	1698.00		
1 F.E. Loader, W.M., 2.5 C.Y.		531.20		584.32		
3 Dump Trucks, 12 C.Y., 400 H.P.		1701.15		1871.27		
1 Air Compressor, 365 cfm		213.35		234.69	50.95	56.05
48 L.H., Daily Totals		$4882.50		$6361.47	$101.72	$132.53

Crew No.	Bare Costs		Incl. Subs O&P		Cost Per Labor-Hour	
Crew B-5C	**Hr.**	**Daily**	**Hr.**	**Daily**	**Bare Costs**	**Incl. O&P**
3 Laborers	$41.05	$985.20	$62.10	$1490.40	$47.36	$71.33
1 Equip. Oper. (medium)	55.10	440.80	82.70	661.60		
2 Truck Drivers (heavy)	47.00	752.00	70.75	1132.00		
1 Equip. Oper. (crane)	57.45	459.60	86.25	690.00		
1 Equip. Oper. (oiler)	49.20	393.60	73.85	590.80		
2 Dump Trucks, 12 C.Y., 400 H.P.		1134.10		1247.51		
1 Crawler Loader, 4 C.Y.		1461.00		1607.10		
1 S.P. Crane, 4x4, 25 Ton		664.40		730.84	50.93	56.02
64 L.H., Daily Totals		$6290.70		$8150.25	$98.29	$127.35
Crew B-5D	**Hr.**	**Daily**	**Hr.**	**Daily**	**Bare Costs**	**Incl. O&P**
1 Labor Foreman (outside)	$43.05	$344.40	$65.10	$520.80	$45.56	$68.71
4 Laborers	41.05	1313.60	62.10	1987.20		
2 Equip. Oper. (medium)	55.10	881.60	82.70	1323.20		
1 Truck Driver (heavy)	47.00	376.00	70.75	566.00		
1 Air Compressor, 250 cfm		167.95		184.75		
2 Breakers, Pavement, 60 lb.		21.50		23.65		
2 -50' Air Hoses, 1.5"		45.50		50.05		
1 Crawler Loader, 3 C.Y.		1202.00		1322.20		
1 Dump Truck, 12 C.Y., 400 H.P.		567.05		623.76	31.31	34.44
64 L.H., Daily Totals		$4919.60		$6601.60	$76.87	$103.15
Crew B-6	**Hr.**	**Daily**	**Hr.**	**Daily**	**Bare Costs**	**Incl. O&P**
2 Laborers	$41.05	$656.80	$62.10	$993.60	$44.58	$67.25
1 Equip. Oper. (light)	51.65	413.20	77.55	620.40		
1 Backhoe Loader, 48 H.P.		320.20		352.22	13.34	14.68
24 L.H., Daily Totals		$1390.20		$1966.22	$57.92	$81.93
Crew B-6A	**Hr.**	**Daily**	**Hr.**	**Daily**	**Bare Costs**	**Incl. O&P**
.5 Labor Foreman (outside)	$43.05	$172.20	$65.10	$260.40	$47.07	$70.94
1 Laborer	41.05	328.40	62.10	496.80		
1 Equip. Oper. (medium)	55.10	440.80	82.70	661.60		
1 Vacuum Truck, 5000 Gal.		359.50		395.45	17.98	19.77
20 L.H., Daily Totals		$1300.90		$1814.25	$65.05	$90.71
Crew B-6B	**Hr.**	**Daily**	**Hr.**	**Daily**	**Bare Costs**	**Incl. O&P**
2 Labor Foremen (outside)	$43.05	$688.80	$65.10	$1041.60	$41.72	$63.10
4 Laborers	41.05	1313.60	62.10	1987.20		
1 S.P. Crane, 4x4, 5 Ton		259.45		285.39		
1 Flatbed Truck, Gas, 1.5 Ton		195.20		214.72		
1 Butt Fusion Mach., 4"-12" diam.		405.15		445.67	17.91	19.70
48 L.H., Daily Totals		$2862.20		$3974.58	$59.63	$82.80
Crew B-6C	**Hr.**	**Daily**	**Hr.**	**Daily**	**Bare Costs**	**Incl. O&P**
2 Labor Foremen (outside)	$43.05	$688.80	$65.10	$1041.60	$41.72	$63.10
4 Laborers	41.05	1313.60	62.10	1987.20		
1 S.P. Crane, 4x4, 12 Ton		371.85		409.04		
1 Flatbed Truck, Gas, 3 Ton		245.95		270.55		
1 Butt Fusion Mach., 8"-24" diam.		1927.00		2119.70	53.02	58.32
48 L.H., Daily Totals		$4547.20		$5828.08	$94.73	$121.42
Crew B-7	**Hr.**	**Daily**	**Hr.**	**Daily**	**Bare Costs**	**Incl. O&P**
1 Labor Foreman (outside)	$43.05	$344.40	$65.10	$520.80	$43.73	$66.03
4 Laborers	41.05	1313.60	62.10	1987.20		
1 Equip. Oper. (medium)	55.10	440.80	82.70	661.60		
1 Brush Chipper, 12", 130 H.P.		389.40		428.34		
1 Crawler Loader, 3 C.Y.		1202.00		1322.20		
2 Chain Saws, Gas, 36" Long		93.30		102.63	35.10	38.61
48 L.H., Daily Totals		$3783.50		$5022.77	$78.82	$104.64

Crew B-7A	Hr.	Daily	Hr.	Daily	Bare Costs	Incl. O&P
2 Laborers	$41.05	$656.80	$62.10	$993.60	$44.58	$67.25
1 Equip. Oper. (light)	51.65	413.20	77.55	620.40		
1 Rake w/Tractor		332.75		366.02		
2 Chain Saws, Gas, 18"		55.10		60.61	16.16	17.78
24 L.H., Daily Totals		$1457.85		$2040.64	$60.74	$85.03

Crew B-7B	Hr.	Daily	Hr.	Daily	Bare Costs	Incl. O&P
1 Labor Foreman (outside)	$43.05	$344.40	$65.10	$520.80	$44.19	$66.71
4 Laborers	41.05	1313.60	62.10	1987.20		
1 Equip. Oper. (medium)	55.10	440.80	82.70	661.60		
1 Truck Driver (heavy)	47.00	376.00	70.75	566.00		
1 Brush Chipper, 12", 130 H.P.		389.40		428.34		
1 Crawler Loader, 3 C.Y.		1202.00		1322.20		
2 Chain Saws, Gas, 36" Long		93.30		102.63		
1 Dump Truck, 8 C.Y., 220 H.P.		339.60		373.56	36.15	39.76
56 L.H., Daily Totals		$4499.10		$5962.33	$80.34	$106.47

Crew B-7C	Hr.	Daily	Hr.	Daily	Bare Costs	Incl. O&P
1 Labor Foreman (outside)	$43.05	$344.40	$65.10	$520.80	$44.19	$66.71
4 Laborers	41.05	1313.60	62.10	1987.20		
1 Equip. Oper. (medium)	55.10	440.80	82.70	661.60		
1 Truck Driver (heavy)	47.00	376.00	70.75	566.00		
1 Brush Chipper, 12", 130 H.P.		389.40		428.34		
1 Crawler Loader, 3 C.Y.		1202.00		1322.20		
2 Chain Saws, Gas, 36" Long		93.30		102.63		
1 Dump Truck, 12 C.Y., 400 H.P.		567.05		623.76	40.21	44.23
56 L.H., Daily Totals		$4726.55		$6212.52	$84.40	$110.94

Crew B-8	Hr.	Daily	Hr.	Daily	Bare Costs	Incl. O&P
1 Labor Foreman (outside)	$43.05	$344.40	$65.10	$520.80	$47.32	$71.26
2 Laborers	41.05	656.80	62.10	993.60		
2 Equip. Oper. (medium)	55.10	881.60	82.70	1323.20		
1 Equip. Oper. (oiler)	49.20	393.60	73.85	590.80		
2 Truck Drivers (heavy)	47.00	752.00	70.75	1132.00		
1 Hyd. Crane, 25 Ton		581.70		639.87		
1 Crawler Loader, 3 C.Y.		1202.00		1322.20		
2 Dump Trucks, 12 C.Y., 400 H.P.		1134.10		1247.51	45.59	50.15
64 L.H., Daily Totals		$5946.20		$7769.98	$92.91	$121.41

Crew B-9	Hr.	Daily	Hr.	Daily	Bare Costs	Incl. O&P
1 Labor Foreman (outside)	$43.05	$344.40	$65.10	$520.80	$41.45	$62.70
4 Laborers	41.05	1313.60	62.10	1987.20		
1 Air Compressor, 250 cfm		167.95		184.75		
2 Breakers, Pavement, 60 lb.		21.50		23.65		
2 -50' Air Hoses, 1.5"		45.50		50.05	5.87	6.46
40 L.H., Daily Totals		$1892.95		$2766.45	$47.32	$69.16

Crew B-9A	Hr.	Daily	Hr.	Daily	Bare Costs	Incl. O&P
2 Laborers	$41.05	$656.80	$62.10	$993.60	$43.03	$64.98
1 Truck Driver (heavy)	47.00	376.00	70.75	566.00		
1 Water Tank Trailer, 5000 Gal.		149.45		164.40		
1 Truck Tractor, 220 H.P.		303.55		333.90		
2 -50' Discharge Hoses, 3"		2.80		3.08	18.99	20.89
24 L.H., Daily Totals		$1488.60		$2060.98	$62.02	$85.87

Crew B-9B	Hr.	Daily	Hr.	Daily	Bare Costs	Incl. O&P
2 Laborers	$41.05	$656.80	$62.10	$993.60	$43.03	$64.98
1 Truck Driver (heavy)	47.00	376.00	70.75	566.00		
2 -50' Discharge Hoses, 3"		2.80		3.08		
1 Water Tank Trailer, 5000 Gal.		149.45		164.40		
1 Truck Tractor, 220 H.P.		303.55		333.90		
1 Pressure Washer		63.05		69.36	21.62	23.78
24 L.H., Daily Totals		$1551.65		$2130.34	$64.65	$88.76

Crew B-9D	Hr.	Daily	Hr.	Daily	Bare Costs	Incl. O&P
1 Labor Foreman (outside)	$43.05	$344.40	$65.10	$520.80	$41.45	$62.70
4 Common Laborers	41.05	1313.60	62.10	1987.20		
1 Air Compressor, 250 cfm		167.95		184.75		
2 -50' Air Hoses, 1.5"		45.50		50.05		
2 Air Powered Tampers		57.10		62.81	6.76	7.44
40 L.H., Daily Totals		$1928.55		$2805.61	$48.21	$70.14

Crew B-9E	Hr.	Daily	Hr.	Daily	Bare Costs	Incl. O&P
1 Cement Finisher	$48.90	$391.20	$72.20	$577.60	$44.98	$67.15
1 Laborer	41.05	328.40	62.10	496.80		
1 Chip. Hammers, 12 Lb., Elec.		21.00		23.10	1.31	1.44
16 L.H., Daily Totals		$740.60		$1097.50	$46.29	$68.59

Crew B-10	Hr.	Daily	Hr.	Daily	Bare Costs	Incl. O&P
1 Equip. Oper. (medium)	$55.10	$440.80	$82.70	$661.60	$50.42	$75.83
.5 Laborer	41.05	164.20	62.10	248.40		
12 L.H., Daily Totals		$605.00		$910.00	$50.42	$75.83

Crew B-10A	Hr.	Daily	Hr.	Daily	Bare Costs	Incl. O&P
1 Equip. Oper. (medium)	$55.10	$440.80	$82.70	$661.60	$50.42	$75.83
.5 Laborer	41.05	164.20	62.10	248.40		
1 Roller, 2-Drum, W.B., 7.5 H.P.		181.75		199.93	15.15	16.66
12 L.H., Daily Totals		$786.75		$1109.93	$65.56	$92.49

Crew B-10B	Hr.	Daily	Hr.	Daily	Bare Costs	Incl. O&P
1 Equip. Oper. (medium)	$55.10	$440.80	$82.70	$661.60	$50.42	$75.83
.5 Laborer	41.05	164.20	62.10	248.40		
1 Dozer, 200 H.P.		1290.00		1419.00	107.50	118.25
12 L.H., Daily Totals		$1895.00		$2329.00	$157.92	$194.08

Crew B-10C	Hr.	Daily	Hr.	Daily	Bare Costs	Incl. O&P
1 Equip. Oper. (medium)	$55.10	$440.80	$82.70	$661.60	$50.42	$75.83
.5 Laborer	41.05	164.20	62.10	248.40		
1 Dozer, 200 H.P.		1290.00		1419.00		
1 Vibratory Roller, Towed, 23 Ton		422.15		464.37	142.68	156.95
12 L.H., Daily Totals		$2317.15		$2793.36	$193.10	$232.78

Crew B-10D	Hr.	Daily	Hr.	Daily	Bare Costs	Incl. O&P
1 Equip. Oper. (medium)	$55.10	$440.80	$82.70	$661.60	$50.42	$75.83
.5 Laborer	41.05	164.20	62.10	248.40		
1 Dozer, 200 H.P.		1290.00		1419.00		
1 Sheepsft. Roller, Towed		430.15		473.17	143.35	157.68
12 L.H., Daily Totals		$2325.15		$2802.17	$193.76	$233.51

Crew B-10E	Hr.	Daily	Hr.	Daily	Bare Costs	Incl. O&P
1 Equip. Oper. (medium)	$55.10	$440.80	$82.70	$661.60	$50.42	$75.83
.5 Laborer	41.05	164.20	62.10	248.40		
1 Tandem Roller, 5 Ton		152.45		167.69	12.70	13.97
12 L.H., Daily Totals		$757.45		$1077.69	$63.12	$89.81

Crew No.	Bare Costs		Incl. Subs O&P		Cost Per Labor-Hour	
Crew B-10F	**Hr.**	**Daily**	**Hr.**	**Daily**	**Bare Costs**	**Incl. O&P**
1 Equip. Oper. (medium)	$55.10	$440.80	$82.70	$661.60	$50.42	$75.83
.5 Laborer	41.05	164.20	62.10	248.40		
1 Tandem Roller, 10 Ton		232.05		255.26	19.34	21.27
12 L.H., Daily Totals		$837.05		$1165.26	$69.75	$97.10

Crew B-10G	Hr.	Daily	Hr.	Daily	Bare Costs	Incl. O&P
1 Equip. Oper. (medium)	$55.10	$440.80	$82.70	$661.60	$50.42	$75.83
.5 Laborer	41.05	164.20	62.10	248.40		
1 Sheepsfoot Roller, 240 H.P.		1323.00		1455.30	110.25	121.28
12 L.H., Daily Totals		$1928.00		$2365.30	$160.67	$197.11

Crew B-10H	Hr.	Daily	Hr.	Daily	Bare Costs	Incl. O&P
1 Equip. Oper. (medium)	$55.10	$440.80	$82.70	$661.60	$50.42	$75.83
.5 Laborer	41.05	164.20	62.10	248.40		
1 Diaphragm Water Pump, 2"		74.30		81.73		
1 -20' Suction Hose, 2"		2.05		2.25		
2 -50' Discharge Hoses, 2"		1.80		1.98	6.51	7.16
12 L.H., Daily Totals		$683.15		$995.97	$56.93	$83.00

Crew B-10I	Hr.	Daily	Hr.	Daily	Bare Costs	Incl. O&P
1 Equip. Oper. (medium)	$55.10	$440.80	$82.70	$661.60	$50.42	$75.83
.5 Laborer	41.05	164.20	62.10	248.40		
1 Diaphragm Water Pump, 4"		141.40		155.54		
1 -20' Suction Hose, 4"		3.10		3.41		
2 -50' Discharge Hoses, 4"		4.50		4.95	12.42	13.66
12 L.H., Daily Totals		$754.00		$1073.90	$62.83	$89.49

Crew B-10J	Hr.	Daily	Hr.	Daily	Bare Costs	Incl. O&P
1 Equip. Oper. (medium)	$55.10	$440.80	$82.70	$661.60	$50.42	$75.83
.5 Laborer	41.05	164.20	62.10	248.40		
1 Centrifugal Water Pump, 3"		79.25		87.17		
1 -20' Suction Hose, 3"		2.70		2.97		
2 -50' Discharge Hoses, 3"		2.80		3.08	7.06	7.77
12 L.H., Daily Totals		$689.75		$1003.23	$57.48	$83.60

Crew B-10K	Hr.	Daily	Hr.	Daily	Bare Costs	Incl. O&P
1 Equip. Oper. (medium)	$55.10	$440.80	$82.70	$661.60	$50.42	$75.83
.5 Laborer	41.05	164.20	62.10	248.40		
1 Centr. Water Pump, 6"		296.45		326.10		
1 -20' Suction Hose, 6"		11.10		12.21		
2 -50' Discharge Hoses, 6"		11.40		12.54	26.58	29.24
12 L.H., Daily Totals		$923.95		$1260.85	$77.00	$105.07

Crew B-10L	Hr.	Daily	Hr.	Daily	Bare Costs	Incl. O&P
1 Equip. Oper. (medium)	$55.10	$440.80	$82.70	$661.60	$50.42	$75.83
.5 Laborer	41.05	164.20	62.10	248.40		
1 Dozer, 80 H.P.		472.00		519.20	39.33	43.27
12 L.H., Daily Totals		$1077.00		$1429.20	$89.75	$119.10

Crew B-10M	Hr.	Daily	Hr.	Daily	Bare Costs	Incl. O&P
1 Equip. Oper. (medium)	$55.10	$440.80	$82.70	$661.60	$50.42	$75.83
.5 Laborer	41.05	164.20	62.10	248.40		
1 Dozer, 300 H.P.		1831.00		2014.10	152.58	167.84
12 L.H., Daily Totals		$2436.00		$2924.10	$203.00	$243.68

Crew B-10N	Hr.	Daily	Hr.	Daily	Bare Costs	Incl. O&P
1 Equip. Oper. (medium)	$55.10	$440.80	$82.70	$661.60	$50.42	$75.83
.5 Laborer	41.05	164.20	62.10	248.40		
1 F.E. Loader, T.M., 1.5 C.Y.		591.30		650.43	49.27	54.20
12 L.H., Daily Totals		$1196.30		$1560.43	$99.69	$130.04

Crew B-10O	Hr.	Daily	Hr.	Daily	Bare Costs	Incl. O&P
1 Equip. Oper. (medium)	$55.10	$440.80	$82.70	$661.60	$50.42	$75.83
.5 Laborer	41.05	164.20	62.10	248.40		
1 F.E. Loader, T.M., 2.25 C.Y.		1002.00		1102.20	83.50	91.85
12 L.H., Daily Totals		$1607.00		$2012.20	$133.92	$167.68

Crew B-10P	Hr.	Daily	Hr.	Daily	Bare Costs	Incl. O&P
1 Equip. Oper. (medium)	$55.10	$440.80	$82.70	$661.60	$50.42	$75.83
.5 Laborer	41.05	164.20	62.10	248.40		
1 Crawler Loader, 3 C.Y.		1202.00		1322.20	100.17	110.18
12 L.H., Daily Totals		$1807.00		$2232.20	$150.58	$186.02

Crew B-10Q	Hr.	Daily	Hr.	Daily	Bare Costs	Incl. O&P
1 Equip. Oper. (medium)	$55.10	$440.80	$82.70	$661.60	$50.42	$75.83
.5 Laborer	41.05	164.20	62.10	248.40		
1 Crawler Loader, 4 C.Y.		1461.00		1607.10	121.75	133.93
12 L.H., Daily Totals		$2066.00		$2517.10	$172.17	$209.76

Crew B-10R	Hr.	Daily	Hr.	Daily	Bare Costs	Incl. O&P
1 Equip. Oper. (medium)	$55.10	$440.80	$82.70	$661.60	$50.42	$75.83
.5 Laborer	41.05	164.20	62.10	248.40		
1 F.E. Loader, W.M., 1 C.Y.		297.00		326.70	24.75	27.23
12 L.H., Daily Totals		$902.00		$1236.70	$75.17	$103.06

Crew B-10S	Hr.	Daily	Hr.	Daily	Bare Costs	Incl. O&P
1 Equip. Oper. (medium)	$55.10	$440.80	$82.70	$661.60	$50.42	$75.83
.5 Laborer	41.05	164.20	62.10	248.40		
1 F.E. Loader, W.M., 1.5 C.Y.		352.95		388.25	29.41	32.35
12 L.H., Daily Totals		$957.95		$1298.24	$79.83	$108.19

Crew B-10T	Hr.	Daily	Hr.	Daily	Bare Costs	Incl. O&P
1 Equip. Oper. (medium)	$55.10	$440.80	$82.70	$661.60	$50.42	$75.83
.5 Laborer	41.05	164.20	62.10	248.40		
1 F.E. Loader, W.M., 2.5 C.Y.		531.20		584.32	44.27	48.69
12 L.H., Daily Totals		$1136.20		$1494.32	$94.68	$124.53

Crew B-10U	Hr.	Daily	Hr.	Daily	Bare Costs	Incl. O&P
1 Equip. Oper. (medium)	$55.10	$440.80	$82.70	$661.60	$50.42	$75.83
.5 Laborer	41.05	164.20	62.10	248.40		
1 F.E. Loader, W.M., 5.5 C.Y.		975.55		1073.11	81.30	89.43
12 L.H., Daily Totals		$1580.55		$1983.11	$131.71	$165.26

Crew B-10V	Hr.	Daily	Hr.	Daily	Bare Costs	Incl. O&P
1 Equip. Oper. (medium)	$55.10	$440.80	$82.70	$661.60	$50.42	$75.83
.5 Laborer	41.05	164.20	62.10	248.40		
1 Dozer, 700 H.P.		5021.00		5523.10	418.42	460.26
12 L.H., Daily Totals		$5626.00		$6433.10	$468.83	$536.09

Crew B-10W	Hr.	Daily	Hr.	Daily	Bare Costs	Incl. O&P
1 Equip. Oper. (medium)	$55.10	$440.80	$82.70	$661.60	$50.42	$75.83
.5 Laborer	41.05	164.20	62.10	248.40		
1 Dozer, 105 H.P.		609.70		670.67	50.81	55.89
12 L.H., Daily Totals		$1214.70		$1580.67	$101.22	$131.72

Crew B-10X	Hr.	Daily	Hr.	Daily	Bare Costs	Incl. O&P
1 Equip. Oper. (medium)	$55.10	$440.80	$82.70	$661.60	$50.42	$75.83
.5 Laborer	41.05	164.20	62.10	248.40		
1 Dozer, 410 H.P.		2269.00		2495.90	189.08	207.99
12 L.H., Daily Totals		$2874.00		$3405.90	$239.50	$283.82

Crew No.	Bare Costs		Incl. Subs O&P		Cost Per Labor-Hour	
Crew B-10Y	**Hr.**	**Daily**	**Hr.**	**Daily**	**Bare Costs**	**Incl. O&P**
1 Equip. Oper. (medium)	$55.10	$440.80	$82.70	$661.60	$50.42	$75.83
.5 Laborer	41.05	164.20	62.10	248.40		
1 Vibr. Roller, Towed, 12 Ton		566.70		623.37	47.23	51.95
12 L.H., Daily Totals		$1171.70		$1533.37	$97.64	$127.78

Crew B-11A	Hr.	Daily	Hr.	Daily	Bare Costs	Incl. O&P
1 Equipment Oper. (med.)	$55.10	$440.80	$82.70	$661.60	$48.08	$72.40
1 Laborer	41.05	328.40	62.10	496.80		
1 Dozer, 200 H.P.		1290.00		1419.00	80.63	88.69
16 L.H., Daily Totals		$2059.20		$2577.40	$128.70	$161.09

Crew B-11B	Hr.	Daily	Hr.	Daily	Bare Costs	Incl. O&P
1 Equipment Oper. (light)	$51.65	$413.20	$77.55	$620.40	$46.35	$69.83
1 Laborer	41.05	328.40	62.10	496.80		
1 Air Powered Tamper		28.55		31.41		
1 Air Compressor, 365 cfm		213.35		234.69		
2 -50' Air Hoses, 1.5"		45.50		50.05	17.96	19.76
16 L.H., Daily Totals		$1029.00		$1433.34	$64.31	$89.58

Crew B-11C	Hr.	Daily	Hr.	Daily	Bare Costs	Incl. O&P
1 Equipment Oper. (med.)	$55.10	$440.80	$82.70	$661.60	$48.08	$72.40
1 Laborer	41.05	328.40	62.10	496.80		
1 Backhoe Loader, 48 H.P.		320.20		352.22	20.01	22.01
16 L.H., Daily Totals		$1089.40		$1510.62	$68.09	$94.41

Crew B-11J	Hr.	Daily	Hr.	Daily	Bare Costs	Incl. O&P
1 Equipment Oper. (med.)	$55.10	$440.80	$82.70	$661.60	$48.08	$72.40
1 Laborer	41.05	328.40	62.10	496.80		
1 Grader, 30,000 Lbs.		657.85		723.63		
1 Ripper, Beam & 1 Shank		88.80		97.68	46.67	51.33
16 L.H., Daily Totals		$1515.85		$1979.71	$94.74	$123.73

Crew B-11K	Hr.	Daily	Hr.	Daily	Bare Costs	Incl. O&P
1 Equipment Oper. (med.)	$55.10	$440.80	$82.70	$661.60	$48.08	$72.40
1 Laborer	41.05	328.40	62.10	496.80		
1 Trencher, Chain Type, 8' D		1837.00		2020.70	114.81	126.29
16 L.H., Daily Totals		$2606.20		$3179.10	$162.89	$198.69

Crew B-11L	Hr.	Daily	Hr.	Daily	Bare Costs	Incl. O&P
1 Equipment Oper. (med.)	$55.10	$440.80	$82.70	$661.60	$48.08	$72.40
1 Laborer	41.05	328.40	62.10	496.80		
1 Grader, 30,000 Lbs.		657.85		723.63	41.12	45.23
16 L.H., Daily Totals		$1427.05		$1882.04	$89.19	$117.63

Crew B-11M	Hr.	Daily	Hr.	Daily	Bare Costs	Incl. O&P
1 Equipment Oper. (med.)	$55.10	$440.80	$82.70	$661.60	$48.08	$72.40
1 Laborer	41.05	328.40	62.10	496.80		
1 Backhoe Loader, 80 H.P.		387.85		426.63	24.24	26.66
16 L.H., Daily Totals		$1157.05		$1585.04	$72.32	$99.06

Crew B-11N	Hr.	Daily	Hr.	Daily	Bare Costs	Incl. O&P
1 Labor Foreman (outside)	$43.05	$344.40	$65.10	$520.80	$48.36	$72.78
2 Equipment Operators (med.)	55.10	881.60	82.70	1323.20		
6 Truck Drivers (heavy)	47.00	2256.00	70.75	3396.00		
1 F.E. Loader, W.M., 5.5 C.Y.		975.55		1073.11		
1 Dozer, 410 H.P.		2269.00		2495.90		
6 Dump Trucks, Off Hwy., 50 Ton		10428.00		11470.80	189.90	208.89
72 L.H., Daily Totals		$17154.55		$20279.81	$238.26	$281.66

Crew No.	Bare Costs		Incl. Subs O&P		Cost Per Labor-Hour	
Crew B-11Q	**Hr.**	**Daily**	**Hr.**	**Daily**	**Bare Costs**	**Incl. O&P**
1 Equipment Operator (med.)	$55.10	$440.80	$82.70	$661.60	$50.42	$75.83
.5 Laborer	41.05	164.20	62.10	248.40		
1 Dozer, 140 H.P.		835.20		918.72	69.60	76.56
12 L.H., Daily Totals		$1440.20		$1828.72	$120.02	$152.39

Crew B-11R	Hr.	Daily	Hr.	Daily	Bare Costs	Incl. O&P
1 Equipment Operator (med.)	$55.10	$440.80	$82.70	$661.60	$50.42	$75.83
.5 Laborer	41.05	164.20	62.10	248.40		
1 Dozer, 200 H.P.		1290.00		1419.00	107.50	118.25
12 L.H., Daily Totals		$1895.00		$2329.00	$157.92	$194.08

Crew B-11S	Hr.	Daily	Hr.	Daily	Bare Costs	Incl. O&P
1 Equipment Operator (med.)	$55.10	$440.80	$82.70	$661.60	$50.42	$75.83
.5 Laborer	41.05	164.20	62.10	248.40		
1 Dozer, 300 H.P.		1831.00		2014.10		
1 Ripper, Beam & 1 Shank		88.80		97.68	159.98	175.98
12 L.H., Daily Totals		$2524.80		$3021.78	$210.40	$251.82

Crew B-11T	Hr.	Daily	Hr.	Daily	Bare Costs	Incl. O&P
1 Equipment Operator (med.)	$55.10	$440.80	$82.70	$661.60	$50.42	$75.83
.5 Laborer	41.05	164.20	62.10	248.40		
1 Dozer, 410 H.P.		2269.00		2495.90		
1 Ripper, Beam & 2 Shanks		135.95		149.54	200.41	220.45
12 L.H., Daily Totals		$3009.95		$3555.45	$250.83	$296.29

Crew B-11U	Hr.	Daily	Hr.	Daily	Bare Costs	Incl. O&P
1 Equipment Operator (med.)	$55.10	$440.80	$82.70	$661.60	$50.42	$75.83
.5 Laborer	41.05	164.20	62.10	248.40		
1 Dozer, 520 H.P.		2812.00		3093.20	234.33	257.77
12 L.H., Daily Totals		$3417.00		$4003.20	$284.75	$333.60

Crew B-11V	Hr.	Daily	Hr.	Daily	Bare Costs	Incl. O&P
3 Laborers	$41.05	$985.20	$62.10	$1490.40	$41.05	$62.10
1 Roller, 2-Drum, W.B., 7.5 H.P.		181.75		199.93	7.57	8.33
24 L.H., Daily Totals		$1166.95		$1690.33	$48.62	$70.43

Crew B-11W	Hr.	Daily	Hr.	Daily	Bare Costs	Incl. O&P
1 Equipment Operator (med.)	$55.10	$440.80	$82.70	$661.60	$47.18	$71.03
1 Common Laborer	41.05	328.40	62.10	496.80		
10 Truck Drivers (heavy)	47.00	3760.00	70.75	5660.00		
1 Dozer, 200 H.P.		1290.00		1419.00		
1 Vibratory Roller, Towed, 23 Ton		422.15		464.37		
10 Dump Trucks, 8 C.Y., 220 H.P.		3396.00		3735.60	53.21	58.53
96 L.H., Daily Totals		$9637.35		$12437.37	$100.39	$129.56

Crew B-11Y	Hr.	Daily	Hr.	Daily	Bare Costs	Incl. O&P
1 Labor Foreman (outside)	$43.05	$344.40	$65.10	$520.80	$45.96	$69.30
5 Common Laborers	41.05	1642.00	62.10	2484.00		
3 Equipment Operators (med.)	55.10	1322.40	82.70	1984.80		
1 Dozer, 80 H.P.		472.00		519.20		
2 Rollers, 2-Drum, W.B., 7.5 H.P.		363.50		399.85		
4 Vibrating Plates, Gas, 21"		162.20		178.42	13.86	15.24
72 L.H., Daily Totals		$4306.50		$6087.07	$59.81	$84.54

Crew B-12A	Hr.	Daily	Hr.	Daily	Bare Costs	Incl. O&P
1 Equip. Oper. (crane)	$57.45	$459.60	$86.25	$690.00	$49.25	$74.17
1 Laborer	41.05	328.40	62.10	496.80		
1 Hyd. Excavator, 1 C.Y.		749.70		824.67	46.86	51.54
16 L.H., Daily Totals		$1537.70		$2011.47	$96.11	$125.72

Crew No.	Bare Costs		Incl. Subs O&P		Cost Per Labor-Hour	
Crew B-12B	**Hr.**	**Daily**	**Hr.**	**Daily**	**Bare Costs**	**Incl. O&P**
1 Equip. Oper. (crane)	$57.45	$459.60	$86.25	$690.00	$49.25	$74.17
1 Laborer	41.05	328.40	62.10	496.80		
1 Hyd. Excavator, 1.5 C.Y.		908.70		999.57	56.79	62.47
16 L.H., Daily Totals		$1696.70		$2186.37	$106.04	$136.65
Crew B-12C	**Hr.**	**Daily**	**Hr.**	**Daily**	**Bare Costs**	**Incl. O&P**
1 Equip. Oper. (crane)	$57.45	$459.60	$86.25	$690.00	$49.25	$74.17
1 Laborer	41.05	328.40	62.10	496.80		
1 Hyd. Excavator, 2 C.Y.		1078.00		1185.80	67.38	74.11
16 L.H., Daily Totals		$1866.00		$2372.60	$116.63	$148.29
Crew B-12D	**Hr.**	**Daily**	**Hr.**	**Daily**	**Bare Costs**	**Incl. O&P**
1 Equip. Oper. (crane)	$57.45	$459.60	$86.25	$690.00	$49.25	$74.17
1 Laborer	41.05	328.40	62.10	496.80		
1 Hyd. Excavator, 3.5 C.Y.		2256.00		2481.60	141.00	155.10
16 L.H., Daily Totals		$3044.00		$3668.40	$190.25	$229.28
Crew B-12E	**Hr.**	**Daily**	**Hr.**	**Daily**	**Bare Costs**	**Incl. O&P**
1 Equip. Oper. (crane)	$57.45	$459.60	$86.25	$690.00	$49.25	$74.17
1 Laborer	41.05	328.40	62.10	496.80		
1 Hyd. Excavator, .5 C.Y.		443.75		488.13	27.73	30.51
16 L.H., Daily Totals		$1231.75		$1674.93	$76.98	$104.68
Crew B-12F	**Hr.**	**Daily**	**Hr.**	**Daily**	**Bare Costs**	**Incl. O&P**
1 Equip. Oper. (crane)	$57.45	$459.60	$86.25	$690.00	$49.25	$74.17
1 Laborer	41.05	328.40	62.10	496.80		
1 Hyd. Excavator, .75 C.Y.		681.25		749.38	42.58	46.84
16 L.H., Daily Totals		$1469.25		$1936.18	$91.83	$121.01
Crew B-12G	**Hr.**	**Daily**	**Hr.**	**Daily**	**Bare Costs**	**Incl. O&P**
1 Equip. Oper. (crane)	$57.45	$459.60	$86.25	$690.00	$49.25	$74.17
1 Laborer	41.05	328.40	62.10	496.80		
1 Crawler Crane, 15 Ton		866.60		953.26		
1 Clamshell Bucket, .5 C.Y.		42.55		46.81	56.82	62.50
16 L.H., Daily Totals		$1697.15		$2186.86	$106.07	$136.68
Crew B-12H	**Hr.**	**Daily**	**Hr.**	**Daily**	**Bare Costs**	**Incl. O&P**
1 Equip. Oper. (crane)	$57.45	$459.60	$86.25	$690.00	$49.25	$74.17
1 Laborer	41.05	328.40	62.10	496.80		
1 Crawler Crane, 25 Ton		1386.00		1524.60		
1 Clamshell Bucket, 1 C.Y.		51.20		56.32	89.83	98.81
16 L.H., Daily Totals		$2225.20		$2767.72	$139.07	$172.98
Crew B-12I	**Hr.**	**Daily**	**Hr.**	**Daily**	**Bare Costs**	**Incl. O&P**
1 Equip. Oper. (crane)	$57.45	$459.60	$86.25	$690.00	$49.25	$74.17
1 Laborer	41.05	328.40	62.10	496.80		
1 Crawler Crane, 20 Ton		1130.00		1243.00		
1 Dragline Bucket, .75 C.Y.		21.85		24.04	71.99	79.19
16 L.H., Daily Totals		$1939.85		$2453.84	$121.24	$153.36
Crew B-12J	**Hr.**	**Daily**	**Hr.**	**Daily**	**Bare Costs**	**Incl. O&P**
1 Equip. Oper. (crane)	$57.45	$459.60	$86.25	$690.00	$49.25	$74.17
1 Laborer	41.05	328.40	62.10	496.80		
1 Gradall, 5/8 C.Y.		862.55		948.80	53.91	59.30
16 L.H., Daily Totals		$1650.55		$2135.61	$103.16	$133.48

Crew No.	Bare Costs		Incl. Subs O&P		Cost Per Labor-Hour	
Crew B-12K	**Hr.**	**Daily**	**Hr.**	**Daily**	**Bare Costs**	**Incl. O&P**
1 Equip. Oper. (crane)	$57.45	$459.60	$86.25	$690.00	$49.25	$74.17
1 Laborer	41.05	328.40	62.10	496.80		
1 Gradall, 3 Ton, 1 C.Y.		1185.00		1303.50	74.06	81.47
16 L.H., Daily Totals		$1973.00		$2490.30	$123.31	$155.64
Crew B-12L	**Hr.**	**Daily**	**Hr.**	**Daily**	**Bare Costs**	**Incl. O&P**
1 Equip. Oper. (crane)	$57.45	$459.60	$86.25	$690.00	$49.25	$74.17
1 Laborer	41.05	328.40	62.10	496.80		
1 Crawler Crane, 15 Ton		866.60		953.26		
1 F.E. Attachment, .5 C.Y.		63.95		70.34	58.16	63.98
16 L.H., Daily Totals		$1718.55		$2210.41	$107.41	$138.15
Crew B-12M	**Hr.**	**Daily**	**Hr.**	**Daily**	**Bare Costs**	**Incl. O&P**
1 Equip. Oper. (crane)	$57.45	$459.60	$86.25	$690.00	$49.25	$74.17
1 Laborer	41.05	328.40	62.10	496.80		
1 Crawler Crane, 20 Ton		1130.00		1243.00		
1 F.E. Attachment, .75 C.Y.		68.90		75.79	74.93	82.42
16 L.H., Daily Totals		$1986.90		$2505.59	$124.18	$156.60
Crew B-12N	**Hr.**	**Daily**	**Hr.**	**Daily**	**Bare Costs**	**Incl. O&P**
1 Equip. Oper. (crane)	$57.45	$459.60	$86.25	$690.00	$49.25	$74.17
1 Laborer	41.05	328.40	62.10	496.80		
1 Crawler Crane, 25 Ton		1386.00		1524.60		
1 F.E. Attachment, 1 C.Y.		74.85		82.33	91.30	100.43
16 L.H., Daily Totals		$2248.85		$2793.74	$140.55	$174.61
Crew B-12O	**Hr.**	**Daily**	**Hr.**	**Daily**	**Bare Costs**	**Incl. O&P**
1 Equip. Oper. (crane)	$57.45	$459.60	$86.25	$690.00	$49.25	$74.17
1 Laborer	41.05	328.40	62.10	496.80		
1 Crawler Crane, 40 Ton		1392.00		1531.20		
1 F.E. Attachment, 1.5 C.Y.		85.65		94.22	92.35	101.59
16 L.H., Daily Totals		$2265.65		$2812.22	$141.60	$175.76
Crew B-12P	**Hr.**	**Daily**	**Hr.**	**Daily**	**Bare Costs**	**Incl. O&P**
1 Equip. Oper. (crane)	$57.45	$459.60	$86.25	$690.00	$49.25	$74.17
1 Laborer	41.05	328.40	62.10	496.80		
1 Crawler Crane, 40 Ton		1392.00		1531.20		
1 Dragline Bucket, 1.5 C.Y.		35.10		38.61	89.19	98.11
16 L.H., Daily Totals		$2215.10		$2756.61	$138.44	$172.29
Crew B-12Q	**Hr.**	**Daily**	**Hr.**	**Daily**	**Bare Costs**	**Incl. O&P**
1 Equip. Oper. (crane)	$57.45	$459.60	$86.25	$690.00	$49.25	$74.17
1 Laborer	41.05	328.40	62.10	496.80		
1 Hyd. Excavator, 5/8 C.Y.		587.30		646.03	36.71	40.38
16 L.H., Daily Totals		$1375.30		$1832.83	$85.96	$114.55
Crew B-12S	**Hr.**	**Daily**	**Hr.**	**Daily**	**Bare Costs**	**Incl. O&P**
1 Equip. Oper. (crane)	$57.45	$459.60	$86.25	$690.00	$49.25	$74.17
1 Laborer	41.05	328.40	62.10	496.80		
1 Hyd. Excavator, 2.5 C.Y.		1441.00		1585.10	90.06	99.07
16 L.H., Daily Totals		$2229.00		$2771.90	$139.31	$173.24
Crew B-12T	**Hr.**	**Daily**	**Hr.**	**Daily**	**Bare Costs**	**Incl. O&P**
1 Equip. Oper. (crane)	$57.45	$459.60	$86.25	$690.00	$49.25	$74.17
1 Laborer	41.05	328.40	62.10	496.80		
1 Crawler Crane, 75 Ton		1734.00		1907.40		
1 F.E. Attachment, 3 C.Y.		111.50		122.65	115.34	126.88
16 L.H., Daily Totals		$2633.50		$3216.85	$164.59	$201.05

Crew No.	Bare Costs		Incl. Subs O&P		Cost Per Labor-Hour	
Crew B-12V	**Hr.**	**Daily**	**Hr.**	**Daily**	**Bare Costs**	**Incl. O&P**
1 Equip. Oper. (crane)	$57.45	$459.60	$86.25	$690.00	$49.25	$74.17
1 Laborer	41.05	328.40	62.10	496.80		
1 Crawler Crane, 75 Ton		1734.00		1907.40		
1 Dragline Bucket, 3 C.Y.		56.25		61.88	111.89	123.08
16 L.H., Daily Totals		$2578.25		$3156.07	$161.14	$197.25

Crew B-12Y	Hr.	Daily	Hr.	Daily	Bare Costs	Incl. O&P
1 Equip. Oper. (crane)	$57.45	$459.60	$86.25	$690.00	$46.52	$70.15
2 Laborers	41.05	656.80	62.10	993.60		
1 Hyd. Excavator, 3.5 C.Y.		2256.00		2481.60	94.00	103.40
24 L.H., Daily Totals		$3372.40		$4165.20	$140.52	$173.55

Crew B-12Z	Hr.	Daily	Hr.	Daily	Bare Costs	Incl. O&P
1 Equip. Oper. (crane)	$57.45	$459.60	$86.25	$690.00	$46.52	$70.15
2 Laborers	41.05	656.80	62.10	993.60		
1 Hyd. Excavator, 2.5 C.Y.		1441.00		1585.10	60.04	66.05
24 L.H., Daily Totals		$2557.40		$3268.70	$106.56	$136.20

Crew B-13	Hr.	Daily	Hr.	Daily	Bare Costs	Incl. O&P
1 Labor Foreman (outside)	$43.05	$344.40	$65.10	$520.80	$44.84	$67.66
4 Laborers	41.05	1313.60	62.10	1987.20		
1 Equip. Oper. (crane)	57.45	459.60	86.25	690.00		
1 Equip. Oper. (oiler)	49.20	393.60	73.85	590.80		
1 Hyd. Crane, 25 Ton		581.70		639.87	10.39	11.43
56 L.H., Daily Totals		$3092.90		$4428.67	$55.23	$79.08

Crew B-13A	Hr.	Daily	Hr.	Daily	Bare Costs	Incl. O&P
1 Labor Foreman (outside)	$43.05	$344.40	$65.10	$520.80	$47.05	$70.89
2 Laborers	41.05	656.80	62.10	993.60		
2 Equipment Operators (med.)	55.10	881.60	82.70	1323.20		
2 Truck Drivers (heavy)	47.00	752.00	70.75	1132.00		
1 Crawler Crane, 75 Ton		1734.00		1907.40		
1 Crawler Loader, 4 C.Y.		1461.00		1607.10		
2 Dump Trucks, 8 C.Y., 220 H.P.		679.20		747.12	69.18	76.10
56 L.H., Daily Totals		$6509.00		$8231.22	$116.23	$146.99

Crew B-13B	Hr.	Daily	Hr.	Daily	Bare Costs	Incl. O&P
1 Labor Foreman (outside)	$43.05	$344.40	$65.10	$520.80	$44.84	$67.66
4 Laborers	41.05	1313.60	62.10	1987.20		
1 Equip. Oper. (crane)	57.45	459.60	86.25	690.00		
1 Equip. Oper. (oiler)	49.20	393.60	73.85	590.80		
1 Hyd. Crane, 55 Ton		981.50		1079.65	17.53	19.28
56 L.H., Daily Totals		$3492.70		$4868.45	$62.37	$86.94

Crew B-13C	Hr.	Daily	Hr.	Daily	Bare Costs	Incl. O&P
1 Labor Foreman (outside)	$43.05	$344.40	$65.10	$520.80	$44.84	$67.66
4 Laborers	41.05	1313.60	62.10	1987.20		
1 Equip. Oper. (crane)	57.45	459.60	86.25	690.00		
1 Equip. Oper. (oiler)	49.20	393.60	73.85	590.80		
1 Crawler Crane, 100 Ton		1879.00		2066.90	33.55	36.91
56 L.H., Daily Totals		$4390.20		$5855.70	$78.40	$104.57

Crew B-13D	Hr.	Daily	Hr.	Daily	Bare Costs	Incl. O&P
1 Laborer	$41.05	$328.40	$62.10	$496.80	$49.25	$74.17
1 Equip. Oper. (crane)	57.45	459.60	86.25	690.00		
1 Hyd. Excavator, 1 C.Y.		749.70		824.67		
1 Trench Box		77.75		85.53	51.72	56.89
16 L.H., Daily Totals		$1615.45		$2096.99	$100.97	$131.06

Crew B-13E	Hr.	Daily	Hr.	Daily	Bare Costs	Incl. O&P
1 Laborer	$41.05	$328.40	$62.10	$496.80	$49.25	$74.17
1 Equip. Oper. (crane)	57.45	459.60	86.25	690.00		
1 Hyd. Excavator, 1.5 C.Y.		908.70		999.57		
1 Trench Box		77.75		85.53	61.65	67.82
16 L.H., Daily Totals		$1774.45		$2271.90	$110.90	$141.99

Crew B-13F	Hr.	Daily	Hr.	Daily	Bare Costs	Incl. O&P
1 Laborer	$41.05	$328.40	$62.10	$496.80	$49.25	$74.17
1 Equip. Oper. (crane)	57.45	459.60	86.25	690.00		
1 Hyd. Excavator, 3.5 C.Y.		2256.00		2481.60		
1 Trench Box		77.75		85.53	145.86	160.45
16 L.H., Daily Totals		$3121.75		$3753.93	$195.11	$234.62

Crew B-13G	Hr.	Daily	Hr.	Daily	Bare Costs	Incl. O&P
1 Laborer	$41.05	$328.40	$62.10	$496.80	$49.25	$74.17
1 Equip. Oper. (crane)	57.45	459.60	86.25	690.00		
1 Hyd. Excavator, .75 C.Y.		681.25		749.38		
1 Trench Box		77.75		85.53	47.44	52.18
16 L.H., Daily Totals		$1547.00		$2021.70	$96.69	$126.36

Crew B-13H	Hr.	Daily	Hr.	Daily	Bare Costs	Incl. O&P
1 Laborer	$41.05	$328.40	$62.10	$496.80	$49.25	$74.17
1 Equip. Oper. (crane)	57.45	459.60	86.25	690.00		
1 Gradall, 5/8 C.Y.		862.55		948.80		
1 Trench Box		77.75		85.53	58.77	64.65
16 L.H., Daily Totals		$1728.30		$2221.13	$108.02	$138.82

Crew B-13I	Hr.	Daily	Hr.	Daily	Bare Costs	Incl. O&P
1 Laborer	$41.05	$328.40	$62.10	$496.80	$49.25	$74.17
1 Equip. Oper. (crane)	57.45	459.60	86.25	690.00		
1 Gradall, 3 Ton, 1 C.Y.		1185.00		1303.50		
1 Trench Box		77.75		85.53	78.92	86.81
16 L.H., Daily Totals		$2050.75		$2575.82	$128.17	$160.99

Crew B-13J	Hr.	Daily	Hr.	Daily	Bare Costs	Incl. O&P
1 Laborer	$41.05	$328.40	$62.10	$496.80	$49.25	$74.17
1 Equip. Oper. (crane)	57.45	459.60	86.25	690.00		
1 Hyd. Excavator, 2.5 C.Y.		1441.00		1585.10		
1 Trench Box		77.75		85.53	94.92	104.41
16 L.H., Daily Totals		$2306.75		$2857.43	$144.17	$178.59

Crew B-13K	Hr.	Daily	Hr.	Daily	Bare Costs	Incl. O&P
2 Equip. Opers. (crane)	$57.45	$919.20	$86.25	$1380.00	$57.45	$86.25
1 Hyd. Excavator, .75 C.Y.		681.25		749.38		
1 Hyd. Hammer, 4000 ft-lb		332.85		366.13		
1 Hyd. Excavator, .75 C.Y.		681.25		749.38	105.96	116.56
16 L.H., Daily Totals		$2614.55		$3244.89	$163.41	$202.81

Crew B-13L	Hr.	Daily	Hr.	Daily	Bare Costs	Incl. O&P
2 Equip. Opers. (crane)	$57.45	$919.20	$86.25	$1380.00	$57.45	$86.25
1 Hyd. Excavator, 1.5 C.Y.		908.70		999.57		
1 Hyd. Hammer, 5000 ft-lb		407.60		448.36		
1 Hyd. Excavator, .75 C.Y.		681.25		749.38	124.85	137.33
16 L.H., Daily Totals		$2916.75		$3577.30	$182.30	$223.58

Crew No.	Bare Costs		Incl. Subs O&P		Cost Per Labor-Hour	
Crew B-13M	**Hr.**	**Daily**	**Hr.**	**Daily**	**Bare Costs**	**Incl. O&P**
2 Equip. Opers. (crane)	$57.45	$919.20	$86.25	$1380.00	$57.45	$86.25
1 Hyd. Excavator, 2.5 C.Y.		1441.00		1585.10		
1 Hyd. Hammer, 8000 ft-lb		593.35		652.68		
1 Hyd. Excavator, 1.5 C.Y.		908.70		999.57	183.94	202.33
16 L.H., Daily Totals		$3862.25		$4617.35	$241.39	$288.58
Crew B-13N	**Hr.**	**Daily**	**Hr.**	**Daily**	**Bare Costs**	**Incl. O&P**
2 Equip. Opers. (crane)	$57.45	$919.20	$86.25	$1380.00	$57.45	$86.25
1 Hyd. Excavator, 3.5 C.Y.		2256.00		2481.60		
1 Hyd. Hammer, 12,000 ft-lb		685.75		754.33		
1 Hyd. Excavator, 1.5 C.Y.		908.70		999.57	240.65	264.72
16 L.H., Daily Totals		$4769.65		$5615.49	$298.10	$350.97
Crew B-14	**Hr.**	**Daily**	**Hr.**	**Daily**	**Bare Costs**	**Incl. O&P**
1 Labor Foreman (outside)	$43.05	$344.40	$65.10	$520.80	$43.15	$65.17
4 Laborers	41.05	1313.60	62.10	1987.20		
1 Equip. Oper. (light)	51.65	413.20	77.55	620.40		
1 Backhoe Loader, 48 H.P.		320.20		352.22	6.67	7.34
48 L.H., Daily Totals		$2391.40		$3480.62	$49.82	$72.51
Crew B-14A	**Hr.**	**Daily**	**Hr.**	**Daily**	**Bare Costs**	**Incl. O&P**
1 Equip. Oper. (crane)	$57.45	$459.60	$86.25	$690.00	$51.98	$78.20
.5 Laborer	41.05	164.20	62.10	248.40		
1 Hyd. Excavator, 4.5 C.Y.		2868.00		3154.80	239.00	262.90
12 L.H., Daily Totals		$3491.80		$4093.20	$290.98	$341.10
Crew B-14B	**Hr.**	**Daily**	**Hr.**	**Daily**	**Bare Costs**	**Incl. O&P**
1 Equip. Oper. (crane)	$57.45	$459.60	$86.25	$690.00	$51.98	$78.20
.5 Laborer	41.05	164.20	62.10	248.40		
1 Hyd. Excavator, 6 C.Y.		3579.00		3936.90	298.25	328.07
12 L.H., Daily Totals		$4202.80		$4875.30	$350.23	$406.27
Crew B-14C	**Hr.**	**Daily**	**Hr.**	**Daily**	**Bare Costs**	**Incl. O&P**
1 Equip. Oper. (crane)	$57.45	$459.60	$86.25	$690.00	$51.98	$78.20
.5 Laborer	41.05	164.20	62.10	248.40		
1 Hyd. Excavator, 7 C.Y.		3332.00		3665.20	277.67	305.43
12 L.H., Daily Totals		$3955.80		$4603.60	$329.65	$383.63
Crew B-14F	**Hr.**	**Daily**	**Hr.**	**Daily**	**Bare Costs**	**Incl. O&P**
1 Equip. Oper. (crane)	$57.45	$459.60	$86.25	$690.00	$51.98	$78.20
.5 Laborer	41.05	164.20	62.10	248.40		
1 Hyd. Shovel, 7 C.Y.		4029.00		4431.90	335.75	369.32
12 L.H., Daily Totals		$4652.80		$5370.30	$387.73	$447.52
Crew B-14G	**Hr.**	**Daily**	**Hr.**	**Daily**	**Bare Costs**	**Incl. O&P**
1 Equip. Oper. (crane)	$57.45	$459.60	$86.25	$690.00	$51.98	$78.20
.5 Laborer	41.05	164.20	62.10	248.40		
1 Hyd. Shovel, 12 C.Y.		5855.00		6440.50	487.92	536.71
12 L.H., Daily Totals		$6478.80		$7378.90	$539.90	$614.91
Crew B-14J	**Hr.**	**Daily**	**Hr.**	**Daily**	**Bare Costs**	**Incl. O&P**
1 Equip. Oper. (medium)	$55.10	$440.80	$82.70	$661.60	$50.42	$75.83
.5 Laborer	41.05	164.20	62.10	248.40		
1 F.E. Loader, 8 C.Y.		1784.00		1962.40	148.67	163.53
12 L.H., Daily Totals		$2389.00		$2872.40	$199.08	$239.37

Crew No.	Bare Costs		Incl. Subs O&P		Cost Per Labor-Hour	
Crew B-14K	**Hr.**	**Daily**	**Hr.**	**Daily**	**Bare Costs**	**Incl. O&P**
1 Equip. Oper. (medium)	$55.10	$440.80	$82.70	$661.60	$50.42	$75.83
.5 Laborer	41.05	164.20	62.10	248.40		
1 F.E. Loader, 10 C.Y.		2625.00		2887.50	218.75	240.63
12 L.H., Daily Totals		$3230.00		$3797.50	$269.17	$316.46
Crew B-15	**Hr.**	**Daily**	**Hr.**	**Daily**	**Bare Costs**	**Incl. O&P**
1 Equipment Oper. (med.)	$55.10	$440.80	$82.70	$661.60	$48.46	$72.93
.5 Laborer	41.05	164.20	62.10	248.40		
2 Truck Drivers (heavy)	47.00	752.00	70.75	1132.00		
2 Dump Trucks, 12 C.Y., 400 H.P.		1134.10		1247.51		
1 Dozer, 200 H.P.		1290.00		1419.00	86.58	95.23
28 L.H., Daily Totals		$3781.10		$4708.51	$135.04	$168.16
Crew B-16	**Hr.**	**Daily**	**Hr.**	**Daily**	**Bare Costs**	**Incl. O&P**
1 Labor Foreman (outside)	$43.05	$344.40	$65.10	$520.80	$43.04	$65.01
2 Laborers	41.05	656.80	62.10	993.60		
1 Truck Driver (heavy)	47.00	376.00	70.75	566.00		
1 Dump Truck, 12 C.Y., 400 H.P.		567.05		623.76	17.72	19.49
32 L.H., Daily Totals		$1944.25		$2704.16	$60.76	$84.50
Crew B-17	**Hr.**	**Daily**	**Hr.**	**Daily**	**Bare Costs**	**Incl. O&P**
2 Laborers	$41.05	$656.80	$62.10	$993.60	$45.19	$68.13
1 Equip. Oper. (light)	51.65	413.20	77.55	620.40		
1 Truck Driver (heavy)	47.00	376.00	70.75	566.00		
1 Backhoe Loader, 48 H.P.		320.20		352.22		
1 Dump Truck, 8 C.Y., 220 H.P.		339.60		373.56	20.62	22.68
32 L.H., Daily Totals		$2105.80		$2905.78	$65.81	$90.81
Crew B-17A	**Hr.**	**Daily**	**Hr.**	**Daily**	**Bare Costs**	**Incl. O&P**
2 Labor Foremen (outside)	$43.05	$688.80	$65.10	$1041.60	$44.12	$66.83
6 Laborers	41.05	1970.40	62.10	2980.80		
1 Skilled Worker Foreman (out)	55.40	443.20	84.25	674.00		
1 Skilled Worker	53.40	427.20	81.25	650.00		
80 L.H., Daily Totals		$3529.60		$5346.40	$44.12	$66.83
Crew B-17B	**Hr.**	**Daily**	**Hr.**	**Daily**	**Bare Costs**	**Incl. O&P**
2 Laborers	$41.05	$656.80	$62.10	$993.60	$45.19	$68.13
1 Equip. Oper. (light)	51.65	413.20	77.55	620.40		
1 Truck Driver (heavy)	47.00	376.00	70.75	566.00		
1 Backhoe Loader, 48 H.P.		320.20		352.22		
1 Dump Truck, 12 C.Y., 400 H.P.		567.05		623.76	27.73	30.50
32 L.H., Daily Totals		$2333.25		$3155.97	$72.91	$98.62
Crew B-18	**Hr.**	**Daily**	**Hr.**	**Daily**	**Bare Costs**	**Incl. O&P**
1 Labor Foreman (outside)	$43.05	$344.40	$65.10	$520.80	$41.72	$63.10
2 Laborers	41.05	656.80	62.10	993.60		
1 Vibrating Plate, Gas, 21"		40.55		44.60	1.69	1.86
24 L.H., Daily Totals		$1041.75		$1559.01	$43.41	$64.96
Crew B-19	**Hr.**	**Daily**	**Hr.**	**Daily**	**Bare Costs**	**Incl. O&P**
1 Pile Driver Foreman (outside)	$54.15	$433.20	$84.65	$677.20	$53.36	$82.15
4 Pile Drivers	52.15	1668.80	81.55	2609.60		
2 Equip. Oper. (crane)	57.45	919.20	86.25	1380.00		
1 Equip. Oper. (oiler)	49.20	393.60	73.85	590.80		
1 Crawler Crane, 40 Ton		1392.00		1531.20		
1 Lead, 90' High		134.60		148.06		
1 Hammer, Diesel, 22k ft-lb		427.55		470.31	30.53	33.59
64 L.H., Daily Totals		$5368.95		$7407.17	$83.89	$115.74

Crew B-19A	Hr.	Daily	Hr.	Daily	Bare Costs	Incl. O&P
1 Pile Driver Foreman (outside)	$54.15	$433.20	$84.65	$677.20	$53.36	$82.15
4 Pile Drivers	52.15	1668.80	81.55	2609.60		
2 Equip. Oper. (crane)	57.45	919.20	86.25	1380.00		
1 Equip. Oper. (oiler)	49.20	393.60	73.85	590.80		
1 Crawler Crane, 75 Ton		1734.00		1907.40		
1 Lead, 90' High		134.60		148.06		
1 Hammer, Diesel, 41k ft-lb		565.45		622.00	38.03	41.84
64 L.H., Daily Totals		$5848.85		$7935.06	$91.39	$123.99

Crew B-19B	Hr.	Daily	Hr.	Daily	Bare Costs	Incl. O&P
1 Pile Driver Foreman (outside)	$54.15	$433.20	$84.65	$677.20	$53.36	$82.15
4 Pile Drivers	52.15	1668.80	81.55	2609.60		
2 Equip. Oper. (crane)	57.45	919.20	86.25	1380.00		
1 Equip. Oper. (oiler)	49.20	393.60	73.85	590.80		
1 Crawler Crane, 40 Ton		1392.00		1531.20		
1 Lead, 90' High		134.60		148.06		
1 Hammer, Diesel, 22k ft-lb		427.55		470.31		
1 Barge, 400 Ton		836.50		920.15	43.60	47.96
64 L.H., Daily Totals		$6205.45		$8327.32	$96.96	$130.11

Crew B-19C	Hr.	Daily	Hr.	Daily	Bare Costs	Incl. O&P
1 Pile Driver Foreman (outside)	$54.15	$433.20	$84.65	$677.20	$53.36	$82.15
4 Pile Drivers	52.15	1668.80	81.55	2609.60		
2 Equip. Oper. (crane)	57.45	919.20	86.25	1380.00		
1 Equip. Oper. (oiler)	49.20	393.60	73.85	590.80		
1 Crawler Crane, 75 Ton		1734.00		1907.40		
1 Lead, 90' High		134.60		148.06		
1 Hammer, Diesel, 41k ft-lb		565.45		622.00		
1 Barge, 400 Ton		836.50		920.15	51.10	56.21
64 L.H., Daily Totals		$6685.35		$8855.20	$104.46	$138.36

Crew B-20	Hr.	Daily	Hr.	Daily	Bare Costs	Incl. O&P
1 Labor Foreman (outside)	$43.05	$344.40	$65.10	$520.80	$45.83	$69.48
1 Skilled Worker	53.40	427.20	81.25	650.00		
1 Laborer	41.05	328.40	62.10	496.80		
24 L.H., Daily Totals		$1100.00		$1667.60	$45.83	$69.48

Crew B-20A	Hr.	Daily	Hr.	Daily	Bare Costs	Incl. O&P
1 Labor Foreman (outside)	$43.05	$344.40	$65.10	$520.80	$49.44	$74.44
1 Laborer	41.05	328.40	62.10	496.80		
1 Plumber	63.15	505.20	94.75	758.00		
1 Plumber Apprentice	50.50	404.00	75.80	606.40		
32 L.H., Daily Totals		$1582.00		$2382.00	$49.44	$74.44

Crew B-21	Hr.	Daily	Hr.	Daily	Bare Costs	Incl. O&P
1 Labor Foreman (outside)	$43.05	$344.40	$65.10	$520.80	$47.49	$71.88
1 Skilled Worker	53.40	427.20	81.25	650.00		
1 Laborer	41.05	328.40	62.10	496.80		
.5 Equip. Oper. (crane)	57.45	229.80	86.25	345.00		
.5 S.P. Crane, 4x4, 5 Ton		129.72		142.70	4.63	5.10
28 L.H., Daily Totals		$1459.53		$2155.30	$52.13	$76.97

Crew B-21A	Hr.	Daily	Hr.	Daily	Bare Costs	Incl. O&P
1 Labor Foreman (outside)	$43.05	$344.40	$65.10	$520.80	$51.04	$76.80
1 Laborer	41.05	328.40	62.10	496.80		
1 Plumber	63.15	505.20	94.75	758.00		
1 Plumber Apprentice	50.50	404.00	75.80	606.40		
1 Equip. Oper. (crane)	57.45	459.60	86.25	690.00		
1 S.P. Crane, 4x4, 12 Ton		371.85		409.04	9.30	10.23
40 L.H., Daily Totals		$2413.45		$3481.03	$60.34	$87.03

Crew B-21B	Hr.	Daily	Hr.	Daily	Bare Costs	Incl. O&P
1 Labor Foreman (outside)	$43.05	$344.40	$65.10	$520.80	$44.73	$67.53
3 Laborers	41.05	985.20	62.10	1490.40		
1 Equip. Oper. (crane)	57.45	459.60	86.25	690.00		
1 Hyd. Crane, 12 Ton		471.75		518.92	11.79	12.97
40 L.H., Daily Totals		$2260.95		$3220.13	$56.52	$80.50

Crew B-21C	Hr.	Daily	Hr.	Daily	Bare Costs	Incl. O&P
1 Labor Foreman (outside)	$43.05	$344.40	$65.10	$520.80	$44.84	$67.66
4 Laborers	41.05	1313.60	62.10	1987.20		
1 Equip. Oper. (crane)	57.45	459.60	86.25	690.00		
1 Equip. Oper. (oiler)	49.20	393.60	73.85	590.80		
2 Cutting Torches		25.00		27.50		
2 Sets of Gases		335.70		369.27		
1 Lattice Boom Crane, 90 Ton		1698.00		1867.80	36.76	40.44
56 L.H., Daily Totals		$4569.90		$6053.37	$81.61	$108.10

Crew B-22	Hr.	Daily	Hr.	Daily	Bare Costs	Incl. O&P
1 Labor Foreman (outside)	$43.05	$344.40	$65.10	$520.80	$48.16	$72.84
1 Skilled Worker	53.40	427.20	81.25	650.00		
1 Laborer	41.05	328.40	62.10	496.80		
.75 Equip. Oper. (crane)	57.45	344.70	86.25	517.50		
.75 S.P. Crane, 4x4, 5 Ton		194.59		214.05	6.49	7.13
30 L.H., Daily Totals		$1639.29		$2399.15	$54.64	$79.97

Crew B-22A	Hr.	Daily	Hr.	Daily	Bare Costs	Incl. O&P
1 Labor Foreman (outside)	$43.05	$344.40	$65.10	$520.80	$47.20	$71.36
1 Skilled Worker	53.40	427.20	81.25	650.00		
2 Laborers	41.05	656.80	62.10	993.60		
1 Equipment Operator, Crane	57.45	459.60	86.25	690.00		
1 S.P. Crane, 4x4, 5 Ton		259.45		285.39		
1 Butt Fusion Mach., 4"-12" diam.		405.15		445.67	16.61	18.28
40 L.H., Daily Totals		$2552.60		$3585.46	$63.81	$89.64

Crew B-22B	Hr.	Daily	Hr.	Daily	Bare Costs	Incl. O&P
1 Labor Foreman (outside)	$43.05	$344.40	$65.10	$520.80	$47.20	$71.36
1 Skilled Worker	53.40	427.20	81.25	650.00		
2 Laborers	41.05	656.80	62.10	993.60		
1 Equip. Oper. (crane)	57.45	459.60	86.25	690.00		
1 S.P. Crane, 4x4, 5 Ton		259.45		285.39		
1 Butt Fusion Mach., 8"-24" diam.		1927.00		2119.70	54.66	60.13
40 L.H., Daily Totals		$4074.45		$5259.49	$101.86	$131.49

Crew B-22C	Hr.	Daily	Hr.	Daily	Bare Costs	Incl. O&P
1 Skilled Worker	$53.40	$427.20	$81.25	$650.00	$47.23	$71.67
1 Laborer	41.05	328.40	62.10	496.80		
1 Butt Fusion Mach., 2"-8" diam.		143.10		157.41	8.94	9.84
16 L.H., Daily Totals		$898.70		$1304.21	$56.17	$81.51

Crew B-23	Hr.	Daily	Hr.	Daily	Bare Costs	Incl. O&P
1 Labor Foreman (outside)	$43.05	$344.40	$65.10	$520.80	$41.45	$62.70
4 Laborers	41.05	1313.60	62.10	1987.20		
1 Drill Rig, Truck-Mounted		2428.00		2670.80		
1 Flatbed Truck, Gas, 3 Ton		245.95		270.55	66.85	73.53
40 L.H., Daily Totals		$4331.95		$5449.35	$108.30	$136.23

Crew No.	Bare Costs		Incl. Subs O&P		Cost Per Labor-Hour	
Crew B-23A	**Hr.**	**Daily**	**Hr.**	**Daily**	**Bare Costs**	**Incl. O&P**
1 Labor Foreman (outside)	$43.05	$344.40	$65.10	$520.80	$46.40	$69.97
1 Laborer	41.05	328.40	62.10	496.80		
1 Equip. Oper. (medium)	55.10	440.80	82.70	661.60		
1 Drill Rig, Truck-Mounted		2428.00		2670.80		
1 Pickup Truck, 3/4 Ton		109.90		120.89	105.75	116.32
24 L.H., Daily Totals		$3651.50		$4470.89	$152.15	$186.29
Crew B-23B	**Hr.**	**Daily**	**Hr.**	**Daily**	**Bare Costs**	**Incl. O&P**
1 Labor Foreman (outside)	$43.05	$344.40	$65.10	$520.80	$46.40	$69.97
1 Laborer	41.05	328.40	62.10	496.80		
1 Equip. Oper. (medium)	55.10	440.80	82.70	661.60		
1 Drill Rig, Truck-Mounted		2428.00		2670.80		
1 Pickup Truck, 3/4 Ton		109.90		120.89		
1 Centr. Water Pump, 6"		296.45		326.10	118.10	129.91
24 L.H., Daily Totals		$3947.95		$4796.98	$164.50	$199.87
Crew B-24	**Hr.**	**Daily**	**Hr.**	**Daily**	**Bare Costs**	**Incl. O&P**
1 Cement Finisher	$48.90	$391.20	$72.20	$577.60	$47.20	$70.82
1 Laborer	41.05	328.40	62.10	496.80		
1 Carpenter	51.65	413.20	78.15	625.20		
24 L.H., Daily Totals		$1132.80		$1699.60	$47.20	$70.82
Crew B-25	**Hr.**	**Daily**	**Hr.**	**Daily**	**Bare Costs**	**Incl. O&P**
1 Labor Foreman (outside)	$43.05	$344.40	$65.10	$520.80	$45.06	$67.99
7 Laborers	41.05	2298.80	62.10	3477.60		
3 Equip. Oper. (medium)	55.10	1322.40	82.70	1984.80		
1 Asphalt Paver, 130 H.P.		2076.00		2283.60		
1 Tandem Roller, 10 Ton		232.05		255.26		
1 Roller, Pneum. Whl., 12 Ton		338.35		372.19	30.07	33.08
88 L.H., Daily Totals		$6612.00		$8894.24	$75.14	$101.07
Crew B-25B	**Hr.**	**Daily**	**Hr.**	**Daily**	**Bare Costs**	**Incl. O&P**
1 Labor Foreman (outside)	$43.05	$344.40	$65.10	$520.80	$45.90	$69.22
7 Laborers	41.05	2298.80	62.10	3477.60		
4 Equip. Oper. (medium)	55.10	1763.20	82.70	2646.40		
1 Asphalt Paver, 130 H.P.		2076.00		2283.60		
2 Tandem Rollers, 10 Ton		464.10		510.51		
1 Roller, Pneum. Whl., 12 Ton		338.35		372.19	29.98	32.98
96 L.H., Daily Totals		$7284.85		$9811.09	$75.88	$102.20
Crew B-25C	**Hr.**	**Daily**	**Hr.**	**Daily**	**Bare Costs**	**Incl. O&P**
1 Labor Foreman (outside)	$43.05	$344.40	$65.10	$520.80	$46.07	$69.47
3 Laborers	41.05	985.20	62.10	1490.40		
2 Equip. Oper. (medium)	55.10	881.60	82.70	1323.20		
1 Asphalt Paver, 130 H.P.		2076.00		2283.60		
1 Tandem Roller, 10 Ton		232.05		255.26	48.08	52.89
48 L.H., Daily Totals		$4519.25		$5873.26	$94.15	$122.36
Crew B-25D	**Hr.**	**Daily**	**Hr.**	**Daily**	**Bare Costs**	**Incl. O&P**
1 Labor Foreman (outside)	$43.05	$344.40	$65.10	$520.80	$46.27	$69.76
3 Laborers	41.05	985.20	62.10	1490.40		
2.125 Equip. Oper. (medium)	55.10	936.70	82.70	1405.90		
.125 Truck Driver (heavy)	47.00	47.00	70.75	70.75		
.125 Truck Tractor, 6x4, 380 H.P.		60.95		67.05		
.125 Dist. Tanker, 3000 Gallon		40.41		44.45		
1 Asphalt Paver, 130 H.P.		2076.00		2283.60		
1 Tandem Roller, 10 Ton		232.05		255.26	48.19	53.01
50 L.H., Daily Totals		$4722.71		$6138.20	$94.45	$122.76

Crew No.	Bare Costs		Incl. Subs O&P		Cost Per Labor-Hour	
Crew B-25E	**Hr.**	**Daily**	**Hr.**	**Daily**	**Bare Costs**	**Incl. O&P**
1 Labor Foreman (outside)	$43.05	$344.40	$65.10	$520.80	$46.45	$70.03
3 Laborers	41.05	985.20	62.10	1490.40		
2.250 Equip. Oper. (medium)	55.10	991.80	82.70	1488.60		
.25 Truck Driver (heavy)	47.00	94.00	70.75	141.50		
.25 Truck Tractor, 6x4, 380 H.P.		121.90		134.09		
.25 Dist. Tanker, 3000 Gallon		80.81		88.89		
1 Asphalt Paver, 130 H.P.		2076.00		2283.60		
1 Tandem Roller, 10 Ton		232.05		255.26	48.28	53.11
52 L.H., Daily Totals		$4926.16		$6403.14	$94.73	$123.14
Crew B-26	**Hr.**	**Daily**	**Hr.**	**Daily**	**Bare Costs**	**Incl. O&P**
1 Labor Foreman (outside)	$43.05	$344.40	$65.10	$520.80	$45.75	$68.94
6 Laborers	41.05	1970.40	62.10	2980.80		
2 Equip. Oper. (medium)	55.10	881.60	82.70	1323.20		
1 Rodman (reinf.)	54.85	438.80	83.00	664.00		
1 Cement Finisher	48.90	391.20	72.20	577.60		
1 Grader, 30,000 Lbs.		657.85		723.63		
1 Paving Mach. & Equip.		2428.00		2670.80	35.07	38.57
88 L.H., Daily Totals		$7112.25		$9460.83	$80.82	$107.51
Crew B-26A	**Hr.**	**Daily**	**Hr.**	**Daily**	**Bare Costs**	**Incl. O&P**
1 Labor Foreman (outside)	$43.05	$344.40	$65.10	$520.80	$45.75	$68.94
6 Laborers	41.05	1970.40	62.10	2980.80		
2 Equip. Oper. (medium)	55.10	881.60	82.70	1323.20		
1 Rodman (reinf.)	54.85	438.80	83.00	664.00		
1 Cement Finisher	48.90	391.20	72.20	577.60		
1 Grader, 30,000 Lbs.		657.85		723.63		
1 Paving Mach. & Equip.		2428.00		2670.80		
1 Concrete Saw		110.15		121.17	36.32	39.95
88 L.H., Daily Totals		$7222.40		$9582.00	$82.07	$108.89
Crew B-26B	**Hr.**	**Daily**	**Hr.**	**Daily**	**Bare Costs**	**Incl. O&P**
1 Labor Foreman (outside)	$43.05	$344.40	$65.10	$520.80	$46.53	$70.08
6 Laborers	41.05	1970.40	62.10	2980.80		
3 Equip. Oper. (medium)	55.10	1322.40	82.70	1984.80		
1 Rodman (reinf.)	54.85	438.80	83.00	664.00		
1 Cement Finisher	48.90	391.20	72.20	577.60		
1 Grader, 30,000 Lbs.		657.85		723.63		
1 Paving Mach. & Equip.		2428.00		2670.80		
1 Concrete Pump, 110' Boom		1130.00		1243.00	43.92	48.31
96 L.H., Daily Totals		$8683.05		$11365.43	$90.45	$118.39
Crew B-26C	**Hr.**	**Daily**	**Hr.**	**Daily**	**Bare Costs**	**Incl. O&P**
1 Labor Foreman (outside)	$43.05	$344.40	$65.10	$520.80	$44.82	$67.56
6 Laborers	41.05	1970.40	62.10	2980.80		
1 Equip. Oper. (medium)	55.10	440.80	82.70	661.60		
1 Rodman (reinf.)	54.85	438.80	83.00	664.00		
1 Cement Finisher	48.90	391.20	72.20	577.60		
1 Paving Mach. & Equip.		2428.00		2670.80		
1 Concrete Saw		110.15		121.17	31.73	34.90
80 L.H., Daily Totals		$6123.75		$8196.76	$76.55	$102.46
Crew B-27	**Hr.**	**Daily**	**Hr.**	**Daily**	**Bare Costs**	**Incl. O&P**
1 Labor Foreman (outside)	$43.05	$344.40	$65.10	$520.80	$41.55	$62.85
3 Laborers	41.05	985.20	62.10	1490.40		
1 Berm Machine		325.30		357.83	10.17	11.18
32 L.H., Daily Totals		$1654.90		$2369.03	$51.72	$74.03

Crew B-28	Hr.	Daily	Hr.	Daily	Bare Costs	Incl. O&P
2 Carpenters	$51.65	$826.40	$78.15	$1250.40	$48.12	$72.80
1 Laborer	41.05	328.40	62.10	496.80		
24 L.H., Daily Totals		$1154.80		$1747.20	$48.12	$72.80

Crew B-29	Hr.	Daily	Hr.	Daily	Bare Costs	Incl. O&P
1 Labor Foreman (outside)	$43.05	$344.40	$65.10	$520.80	$44.84	$67.66
4 Laborers	41.05	1313.60	62.10	1987.20		
1 Equip. Oper. (crane)	57.45	459.60	86.25	690.00		
1 Equip. Oper. (oiler)	49.20	393.60	73.85	590.80		
1 Gradall, 5/8 C.Y.		862.55		948.80	15.40	16.94
56 L.H., Daily Totals		$3373.75		$4737.60	$60.25	$84.60

Crew B-30	Hr.	Daily	Hr.	Daily	Bare Costs	Incl. O&P
1 Equip. Oper. (medium)	$55.10	$440.80	$82.70	$661.60	$49.70	$74.73
2 Truck Drivers (heavy)	47.00	752.00	70.75	1132.00		
1 Hyd. Excavator, 1.5 C.Y.		908.70		999.57		
2 Dump Trucks, 12 C.Y., 400 H.P.		1134.10		1247.51	85.12	93.63
24 L.H., Daily Totals		$3235.60		$4040.68	$134.82	$168.36

Crew B-31	Hr.	Daily	Hr.	Daily	Bare Costs	Incl. O&P
1 Labor Foreman (outside)	$43.05	$344.40	$65.10	$520.80	$43.57	$65.91
3 Laborers	41.05	985.20	62.10	1490.40		
1 Carpenter	51.65	413.20	78.15	625.20		
1 Air Compressor, 250 cfm		167.95		184.75		
1 Sheeting Driver		7.15		7.87		
2 -50' Air Hoses, 1.5"		45.50		50.05	5.51	6.07
40 L.H., Daily Totals		$1963.40		$2879.06	$49.09	$71.98

Crew B-32	Hr.	Daily	Hr.	Daily	Bare Costs	Incl. O&P
1 Laborer	$41.05	$328.40	$62.10	$496.80	$51.59	$77.55
3 Equip. Oper. (medium)	55.10	1322.40	82.70	1984.80		
1 Grader, 30,000 Lbs.		657.85		723.63		
1 Tandem Roller, 10 Ton		232.05		255.26		
1 Dozer, 200 H.P.		1290.00		1419.00	68.12	74.93
32 L.H., Daily Totals		$3830.70		$4879.49	$119.71	$152.48

Crew B-32A	Hr.	Daily	Hr.	Daily	Bare Costs	Incl. O&P
1 Laborer	$41.05	$328.40	$62.10	$496.80	$50.42	$75.83
2 Equip. Oper. (medium)	55.10	881.60	82.70	1323.20		
1 Grader, 30,000 Lbs.		657.85		723.63		
1 Roller, Vibratory, 25 Ton		650.90		715.99	54.53	59.98
24 L.H., Daily Totals		$2518.75		$3259.63	$104.95	$135.82

Crew B-32B	Hr.	Daily	Hr.	Daily	Bare Costs	Incl. O&P
1 Laborer	$41.05	$328.40	$62.10	$496.80	$50.42	$75.83
2 Equip. Oper. (medium)	55.10	881.60	82.70	1323.20		
1 Dozer, 200 H.P.		1290.00		1419.00		
1 Roller, Vibratory, 25 Ton		650.90		715.99	80.87	88.96
24 L.H., Daily Totals		$3150.90		$3954.99	$131.29	$164.79

Crew B-32C	Hr.	Daily	Hr.	Daily	Bare Costs	Incl. O&P
1 Labor Foreman (outside)	$43.05	$344.40	$65.10	$520.80	$48.41	$72.90
2 Laborers	41.05	656.80	62.10	993.60		
3 Equip. Oper. (medium)	55.10	1322.40	82.70	1984.80		
1 Grader, 30,000 Lbs.		657.85		723.63		
1 Tandem Roller, 10 Ton		232.05		255.26		
1 Dozer, 200 H.P.		1290.00		1419.00	45.41	49.96
48 L.H., Daily Totals		$4503.50		$5897.09	$93.82	$122.86

Crew B-33A	Hr.	Daily	Hr.	Daily	Bare Costs	Incl. O&P
1 Equip. Oper. (medium)	$55.10	$440.80	$82.70	$661.60	$51.09	$76.81
.5 Laborer	41.05	164.20	62.10	248.40		
.25 Equip. Oper. (medium)	55.10	110.20	82.70	165.40		
1 Scraper, Towed, 7 C.Y.		125.50		138.05		
1.25 Dozers, 300 H.P.		2288.75		2517.63	172.45	189.69
14 L.H., Daily Totals		$3129.45		$3731.07	$223.53	$266.51

Crew B-33B	Hr.	Daily	Hr.	Daily	Bare Costs	Incl. O&P
1 Equip. Oper. (medium)	$55.10	$440.80	$82.70	$661.60	$51.09	$76.81
.5 Laborer	41.05	164.20	62.10	248.40		
.25 Equip. Oper. (medium)	55.10	110.20	82.70	165.40		
1 Scraper, Towed, 10 C.Y.		156.65		172.32		
1.25 Dozers, 300 H.P.		2288.75		2517.63	174.67	192.14
14 L.H., Daily Totals		$3160.60		$3765.34	$225.76	$268.95

Crew B-33C	Hr.	Daily	Hr.	Daily	Bare Costs	Incl. O&P
1 Equip. Oper. (medium)	$55.10	$440.80	$82.70	$661.60	$51.09	$76.81
.5 Laborer	41.05	164.20	62.10	248.40		
.25 Equip. Oper. (medium)	55.10	110.20	82.70	165.40		
1 Scraper, Towed, 15 C.Y.		173.20		190.52		
1.25 Dozers, 300 H.P.		2288.75		2517.63	175.85	193.44
14 L.H., Daily Totals		$3177.15		$3783.55	$226.94	$270.25

Crew B-33D	Hr.	Daily	Hr.	Daily	Bare Costs	Incl. O&P
1 Equip. Oper. (medium)	$55.10	$440.80	$82.70	$661.60	$51.09	$76.81
.5 Laborer	41.05	164.20	62.10	248.40		
.25 Equip. Oper. (medium)	55.10	110.20	82.70	165.40		
1 S.P. Scraper, 14 C.Y.		2611.00		2872.10		
.25 Dozer, 300 H.P.		457.75		503.52	219.20	241.12
14 L.H., Daily Totals		$3783.95		$4451.02	$270.28	$317.93

Crew B-33E	Hr.	Daily	Hr.	Daily	Bare Costs	Incl. O&P
1 Equip. Oper. (medium)	$55.10	$440.80	$82.70	$661.60	$51.09	$76.81
.5 Laborer	41.05	164.20	62.10	248.40		
.25 Equip. Oper. (medium)	55.10	110.20	82.70	165.40		
1 S.P. Scraper, 21 C.Y.		2541.00		2795.10		
.25 Dozer, 300 H.P.		457.75		503.52	214.20	235.62
14 L.H., Daily Totals		$3713.95		$4374.02	$265.28	$312.43

Crew B-33F	Hr.	Daily	Hr.	Daily	Bare Costs	Incl. O&P
1 Equip. Oper. (medium)	$55.10	$440.80	$82.70	$661.60	$51.09	$76.81
.5 Laborer	41.05	164.20	62.10	248.40		
.25 Equip. Oper. (medium)	55.10	110.20	82.70	165.40		
1 Elev. Scraper, 11 C.Y.		1165.00		1281.50		
.25 Dozer, 300 H.P.		457.75		503.52	115.91	127.50
14 L.H., Daily Totals		$2337.95		$2860.43	$167.00	$204.32

Crew B-33G	Hr.	Daily	Hr.	Daily	Bare Costs	Incl. O&P
1 Equip. Oper. (medium)	$55.10	$440.80	$82.70	$661.60	$51.09	$76.81
.5 Laborer	41.05	164.20	62.10	248.40		
.25 Equip. Oper. (medium)	55.10	110.20	82.70	165.40		
1 Elev. Scraper, 22 C.Y.		2307.00		2537.70		
.25 Dozer, 300 H.P.		457.75		503.52	197.48	217.23
14 L.H., Daily Totals		$3479.95		$4116.63	$248.57	$294.04

Column groups: Crew No. | Bare Costs (Hr., Daily) | Incl. Subs O&P (Hr., Daily) | Cost Per Labor-Hour (Bare Costs, Incl. O&P)

Crew No.	Bare Costs		Incl. Subs O&P		Cost Per Labor-Hour	
Crew B-33H	**Hr.**	**Daily**	**Hr.**	**Daily**	**Bare Costs**	**Incl. O&P**
.5 Laborer	$41.05	$164.20	$62.10	$248.40	$51.09	$76.81
1 Equipment Operator (med.)	55.10	440.80	82.70	661.60		
.25 Equipment Operator (med.)	55.10	110.20	82.70	165.40		
1 S.P. Scraper, 44 C.Y.		4561.00		5017.10		
.25 Dozer, 410 H.P.		567.25		623.98	366.30	402.93
14 L.H., Daily Totals		$5843.45		$6716.48	$417.39	$479.75
Crew B-33J	**Hr.**	**Daily**	**Hr.**	**Daily**	**Bare Costs**	**Incl. O&P**
1 Equipment Operator (med.)	$55.10	$440.80	$82.70	$661.60	$55.10	$82.70
1 S.P. Scraper, 14 C.Y.		2611.00		2872.10	326.38	359.01
8 L.H., Daily Totals		$3051.80		$3533.70	$381.48	$441.71
Crew B-33K	**Hr.**	**Daily**	**Hr.**	**Daily**	**Bare Costs**	**Incl. O&P**
1 Equipment Operator (med.)	$55.10	$440.80	$82.70	$661.60	$51.09	$76.81
.25 Equipment Operator (med.)	55.10	110.20	82.70	165.40		
.5 Laborer	41.05	164.20	62.10	248.40		
1 S.P. Scraper, 31 C.Y.		3603.00		3963.30		
.25 Dozer, 410 H.P.		567.25		623.98	297.88	327.66
14 L.H., Daily Totals		$4885.45		$5662.68	$348.96	$404.48
Crew B-34A	**Hr.**	**Daily**	**Hr.**	**Daily**	**Bare Costs**	**Incl. O&P**
1 Truck Driver (heavy)	$47.00	$376.00	$70.75	$566.00	$47.00	$70.75
1 Dump Truck, 8 C.Y., 220 H.P.		339.60		373.56	42.45	46.70
8 L.H., Daily Totals		$715.60		$939.56	$89.45	$117.44
Crew B-34B	**Hr.**	**Daily**	**Hr.**	**Daily**	**Bare Costs**	**Incl. O&P**
1 Truck Driver (heavy)	$47.00	$376.00	$70.75	$566.00	$47.00	$70.75
1 Dump Truck, 12 C.Y., 400 H.P.		567.05		623.76	70.88	77.97
8 L.H., Daily Totals		$943.05		$1189.76	$117.88	$148.72
Crew B-34C	**Hr.**	**Daily**	**Hr.**	**Daily**	**Bare Costs**	**Incl. O&P**
1 Truck Driver (heavy)	$47.00	$376.00	$70.75	$566.00	$47.00	$70.75
1 Truck Tractor, 6x4, 380 H.P.		487.60		536.36		
1 Dump Trailer, 16.5 C.Y.		133.95		147.35	77.69	85.46
8 L.H., Daily Totals		$997.55		$1249.70	$124.69	$156.21
Crew B-34D	**Hr.**	**Daily**	**Hr.**	**Daily**	**Bare Costs**	**Incl. O&P**
1 Truck Driver (heavy)	$47.00	$376.00	$70.75	$566.00	$47.00	$70.75
1 Truck Tractor, 6x4, 380 H.P.		487.60		536.36		
1 Dump Trailer, 20 C.Y.		148.60		163.46	79.53	87.48
8 L.H., Daily Totals		$1012.20		$1265.82	$126.53	$158.23
Crew B-34E	**Hr.**	**Daily**	**Hr.**	**Daily**	**Bare Costs**	**Incl. O&P**
1 Truck Driver (heavy)	$47.00	$376.00	$70.75	$566.00	$47.00	$70.75
1 Dump Truck, Off Hwy., 25 Ton		1353.00		1488.30	169.13	186.04
8 L.H., Daily Totals		$1729.00		$2054.30	$216.13	$256.79
Crew B-34F	**Hr.**	**Daily**	**Hr.**	**Daily**	**Bare Costs**	**Incl. O&P**
1 Truck Driver (heavy)	$47.00	$376.00	$70.75	$566.00	$47.00	$70.75
1 Dump Truck, Off Hwy., 35 Ton		1472.00		1619.20	184.00	202.40
8 L.H., Daily Totals		$1848.00		$2185.20	$231.00	$273.15
Crew B-34G	**Hr.**	**Daily**	**Hr.**	**Daily**	**Bare Costs**	**Incl. O&P**
1 Truck Driver (heavy)	$47.00	$376.00	$70.75	$566.00	$47.00	$70.75
1 Dump Truck, Off Hwy., 50 Ton		1738.00		1911.80	217.25	238.97
8 L.H., Daily Totals		$2114.00		$2477.80	$264.25	$309.73

Crew No.	Bare Costs		Incl. Subs O&P		Cost Per Labor-Hour	
Crew B-34H	**Hr.**	**Daily**	**Hr.**	**Daily**	**Bare Costs**	**Incl. O&P**
1 Truck Driver (heavy)	$47.00	$376.00	$70.75	$566.00	$47.00	$70.75
1 Dump Truck, Off Hwy., 65 Ton		1879.00		2066.90	234.88	258.36
8 L.H., Daily Totals		$2255.00		$2632.90	$281.88	$329.11
Crew B-34I	**Hr.**	**Daily**	**Hr.**	**Daily**	**Bare Costs**	**Incl. O&P**
1 Truck Driver (heavy)	$47.00	$376.00	$70.75	$566.00	$47.00	$70.75
1 Dump Truck, 18 C.Y., 450 H.P.		705.00		775.50	88.13	96.94
8 L.H., Daily Totals		$1081.00		$1341.50	$135.13	$167.69
Crew B-34J	**Hr.**	**Daily**	**Hr.**	**Daily**	**Bare Costs**	**Incl. O&P**
1 Truck Driver (heavy)	$47.00	$376.00	$70.75	$566.00	$47.00	$70.75
1 Dump Truck, Off Hwy., 100 Ton		2683.00		2951.30	335.38	368.91
8 L.H., Daily Totals		$3059.00		$3517.30	$382.38	$439.66
Crew B-34K	**Hr.**	**Daily**	**Hr.**	**Daily**	**Bare Costs**	**Incl. O&P**
1 Truck Driver (heavy)	$47.00	$376.00	$70.75	$566.00	$47.00	$70.75
1 Truck Tractor, 6x4, 450 H.P.		594.90		654.39		
1 Lowbed Trailer, 75 Ton		249.90		274.89	105.60	116.16
8 L.H., Daily Totals		$1220.80		$1495.28	$152.60	$186.91
Crew B-34L	**Hr.**	**Daily**	**Hr.**	**Daily**	**Bare Costs**	**Incl. O&P**
1 Equip. Oper. (light)	$51.65	$413.20	$77.55	$620.40	$51.65	$77.55
1 Flatbed Truck, Gas, 1.5 Ton		195.20		214.72	24.40	26.84
8 L.H., Daily Totals		$608.40		$835.12	$76.05	$104.39
Crew B-34M	**Hr.**	**Daily**	**Hr.**	**Daily**	**Bare Costs**	**Incl. O&P**
1 Equip. Oper. (light)	$51.65	$413.20	$77.55	$620.40	$51.65	$77.55
1 Flatbed Truck, Gas, 3 Ton		245.95		270.55	30.74	33.82
8 L.H., Daily Totals		$659.15		$890.95	$82.39	$111.37
Crew B-34N	**Hr.**	**Daily**	**Hr.**	**Daily**	**Bare Costs**	**Incl. O&P**
1 Truck Driver (heavy)	$47.00	$376.00	$70.75	$566.00	$51.05	$76.72
1 Equip. Oper. (medium)	55.10	440.80	82.70	661.60		
1 Truck Tractor, 6x4, 380 H.P.		487.60		536.36		
1 Flatbed Trailer, 40 Ton		182.50		200.75	41.88	46.07
16 L.H., Daily Totals		$1486.90		$1964.71	$92.93	$122.79
Crew B-34P	**Hr.**	**Daily**	**Hr.**	**Daily**	**Bare Costs**	**Incl. O&P**
1 Pipe Fitter	$63.95	$511.60	$95.95	$767.60	$54.85	$82.38
1 Truck Driver (light)	45.50	364.00	68.50	548.00		
1 Equip. Oper. (medium)	55.10	440.80	82.70	661.60		
1 Flatbed Truck, Gas, 3 Ton		245.95		270.55		
1 Backhoe Loader, 48 H.P.		320.20		352.22	23.59	25.95
24 L.H., Daily Totals		$1882.55		$2599.97	$78.44	$108.33
Crew B-34Q	**Hr.**	**Daily**	**Hr.**	**Daily**	**Bare Costs**	**Incl. O&P**
1 Pipe Fitter	$63.95	$511.60	$95.95	$767.60	$55.63	$83.57
1 Truck Driver (light)	45.50	364.00	68.50	548.00		
1 Equip. Oper. (crane)	57.45	459.60	86.25	690.00		
1 Flatbed Trailer, 25 Ton		133.00		146.30		
1 Dump Truck, 8 C.Y., 220 H.P.		339.60		373.56		
1 Hyd. Crane, 25 Ton		581.70		639.87	43.93	48.32
24 L.H., Daily Totals		$2389.50		$3165.33	$99.56	$131.89

Crew No.	Bare Costs		Incl. Subs O&P		Cost Per Labor-Hour	
Crew B-34R	**Hr.**	**Daily**	**Hr.**	**Daily**	**Bare Costs**	**Incl. O&P**
1 Pipe Fitter	$63.95	$511.60	$95.95	$767.60	$55.63	$83.57
1 Truck Driver (light)	45.50	364.00	68.50	548.00		
1 Equip. Oper. (crane)	57.45	459.60	86.25	690.00		
1 Flatbed Trailer, 25 Ton		133.00		146.30		
1 Dump Truck, 8 C.Y., 220 H.P.		339.60		373.56		
1 Hyd. Crane, 25 Ton		581.70		639.87		
1 Hyd. Excavator, 1 C.Y.		749.70		824.67	75.17	82.68
24 L.H., Daily Totals		$3139.20		$3990.00	$130.80	$166.25
Crew B-34S	**Hr.**	**Daily**	**Hr.**	**Daily**	**Bare Costs**	**Incl. O&P**
2 Pipe Fitters	$63.95	$1023.20	$95.95	$1535.20	$58.09	$87.22
1 Truck Driver (heavy)	47.00	376.00	70.75	566.00		
1 Equip. Oper. (crane)	57.45	459.60	86.25	690.00		
1 Flatbed Trailer, 40 Ton		182.50		200.75		
1 Truck Tractor, 6x4, 380 H.P.		487.60		536.36		
1 Hyd. Crane, 80 Ton		1487.00		1635.70		
1 Hyd. Excavator, 2 C.Y.		1078.00		1185.80	101.10	111.21
32 L.H., Daily Totals		$5093.90		$6349.81	$159.18	$198.43
Crew B-34T	**Hr.**	**Daily**	**Hr.**	**Daily**	**Bare Costs**	**Incl. O&P**
2 Pipe Fitters	$63.95	$1023.20	$95.95	$1535.20	$58.09	$87.22
1 Truck Driver (heavy)	47.00	376.00	70.75	566.00		
1 Equip. Oper. (crane)	57.45	459.60	86.25	690.00		
1 Flatbed Trailer, 40 Ton		182.50		200.75		
1 Truck Tractor, 6x4, 380 H.P.		487.60		536.36		
1 Hyd. Crane, 80 Ton		1487.00		1635.70	67.41	74.15
32 L.H., Daily Totals		$4015.90		$5164.01	$125.50	$161.38
Crew B-34U	**Hr.**	**Daily**	**Hr.**	**Daily**	**Bare Costs**	**Incl. O&P**
1 Truck Driver (heavy)	$47.00	$376.00	$70.75	$566.00	$49.33	$74.15
1 Equip. Oper. (light)	51.65	413.20	77.55	620.40		
1 Truck Tractor, 220 H.P.		303.55		333.90		
1 Flatbed Trailer, 25 Ton		133.00		146.30	27.28	30.01
16 L.H., Daily Totals		$1225.75		$1666.61	$76.61	$104.16
Crew B-34V	**Hr.**	**Daily**	**Hr.**	**Daily**	**Bare Costs**	**Incl. O&P**
1 Truck Driver (heavy)	$47.00	$376.00	$70.75	$566.00	$52.03	$78.18
1 Equip. Oper. (crane)	57.45	459.60	86.25	690.00		
1 Equip. Oper. (light)	51.65	413.20	77.55	620.40		
1 Truck Tractor, 6x4, 450 H.P.		594.90		654.39		
1 Equipment Trailer, 50 Ton		200.95		221.04		
1 Pickup Truck, 4x4, 3/4 Ton		121.70		133.87	38.23	42.05
24 L.H., Daily Totals		$2166.35		$2885.70	$90.26	$120.24
Crew B-34W	**Hr.**	**Daily**	**Hr.**	**Daily**	**Bare Costs**	**Incl. O&P**
5 Truck Drivers (heavy)	$47.00	$1880.00	$70.75	$2830.00	$49.83	$74.97
2 Equip. Opers. (crane)	57.45	919.20	86.25	1380.00		
1 Equip. Oper. (mechanic)	57.55	460.40	86.40	691.20		
1 Laborer	41.05	328.40	62.10	496.80		
4 Truck Tractors, 6x4, 380 H.P.		1950.40		2145.44		
2 Equipment Trailers, 50 Ton		401.90		442.09		
2 Flatbed Trailers, 40 Ton		365.00		401.50		
1 Pickup Truck, 4x4, 3/4 Ton		121.70		133.87		
1 S.P. Crane, 4x4, 20 Ton		587.75		646.52	47.59	52.35
72 L.H., Daily Totals		$7014.75		$9167.42	$97.43	$127.33

Crew No.	Bare Costs		Incl. Subs O&P		Cost Per Labor-Hour	
Crew B-35	**Hr.**	**Daily**	**Hr.**	**Daily**	**Bare Costs**	**Incl. O&P**
1 Labor Foreman (outside)	$43.05	$344.40	$65.10	$520.80	$51.22	$77.22
1 Skilled Worker	53.40	427.20	81.25	650.00		
1 Welder (plumber)	63.15	505.20	94.75	758.00		
1 Laborer	41.05	328.40	62.10	496.80		
1 Equip. Oper. (crane)	57.45	459.60	86.25	690.00		
1 Equip. Oper. (oiler)	49.20	393.60	73.85	590.80		
1 Welder, Electric, 300 amp		56.10		61.71		
1 Hyd. Excavator, .75 C.Y.		681.25		749.38	15.36	16.90
48 L.H., Daily Totals		$3195.75		$4517.48	$66.58	$94.11
Crew B-35A	**Hr.**	**Daily**	**Hr.**	**Daily**	**Bare Costs**	**Incl. O&P**
1 Labor Foreman (outside)	$43.05	$344.40	$65.10	$520.80	$49.76	$75.06
2 Laborers	41.05	656.80	62.10	993.60		
1 Skilled Worker	53.40	427.20	81.25	650.00		
1 Welder (plumber)	63.15	505.20	94.75	758.00		
1 Equip. Oper. (crane)	57.45	459.60	86.25	690.00		
1 Equip. Oper. (oiler)	49.20	393.60	73.85	590.80		
1 Welder, Gas Engine, 300 amp		95.65		105.22		
1 Crawler Crane, 75 Ton		1734.00		1907.40	32.67	35.94
56 L.H., Daily Totals		$4616.45		$6215.81	$82.44	$111.00
Crew B-36	**Hr.**	**Daily**	**Hr.**	**Daily**	**Bare Costs**	**Incl. O&P**
1 Labor Foreman (outside)	$43.05	$344.40	$65.10	$520.80	$47.07	$70.94
2 Laborers	41.05	656.80	62.10	993.60		
2 Equip. Oper. (medium)	55.10	881.60	82.70	1323.20		
1 Dozer, 200 H.P.		1290.00		1419.00		
1 Aggregate Spreader		37.30		41.03		
1 Tandem Roller, 10 Ton		232.05		255.26	38.98	42.88
40 L.H., Daily Totals		$3442.15		$4552.89	$86.05	$113.82
Crew B-36A	**Hr.**	**Daily**	**Hr.**	**Daily**	**Bare Costs**	**Incl. O&P**
1 Labor Foreman (outside)	$43.05	$344.40	$65.10	$520.80	$49.36	$74.30
2 Laborers	41.05	656.80	62.10	993.60		
4 Equip. Oper. (medium)	55.10	1763.20	82.70	2646.40		
1 Dozer, 200 H.P.		1290.00		1419.00		
1 Aggregate Spreader		37.30		41.03		
1 Tandem Roller, 10 Ton		232.05		255.26		
1 Roller, Pneum. Whl., 12 Ton		338.35		372.19	33.89	37.28
56 L.H., Daily Totals		$4662.10		$6248.27	$83.25	$111.58
Crew B-36B	**Hr.**	**Daily**	**Hr.**	**Daily**	**Bare Costs**	**Incl. O&P**
1 Labor Foreman (outside)	$43.05	$344.40	$65.10	$520.80	$49.07	$73.86
2 Laborers	41.05	656.80	62.10	993.60		
4 Equip. Oper. (medium)	55.10	1763.20	82.70	2646.40		
1 Truck Driver (heavy)	47.00	376.00	70.75	566.00		
1 Grader, 30,000 Lbs.		657.85		723.63		
1 F.E. Loader, Crl, 1.5 C.Y.		647.15		711.87		
1 Dozer, 300 H.P.		1831.00		2014.10		
1 Roller, Vibratory, 25 Ton		650.90		715.99		
1 Truck Tractor, 6x4, 450 H.P.		594.90		654.39		
1 Water Tank Trailer, 5000 Gal.		149.45		164.40	70.80	77.88
64 L.H., Daily Totals		$7671.65		$9711.17	$119.87	$151.74

Crew B-36C	Bare Costs Hr.	Bare Costs Daily	Incl. Subs O&P Hr.	Incl. Subs O&P Daily	Cost Per Labor-Hour Bare Costs	Cost Per Labor-Hour Incl. O&P
1 Labor Foreman (outside)	$43.05	$344.40	$65.10	$520.80	$51.07	$76.79
3 Equip. Oper. (medium)	55.10	1322.40	82.70	1984.80		
1 Truck Driver (heavy)	47.00	376.00	70.75	566.00		
1 Grader, 30,000 Lbs.		657.85		723.63		
1 Dozer, 300 H.P.		1831.00		2014.10		
1 Roller, Vibratory, 25 Ton		650.90		715.99		
1 Truck Tractor, 6x4, 450 H.P.		594.90		654.39		
1 Water Tank Trailer, 5000 Gal.		149.45		164.40	97.10	106.81
40 L.H., Daily Totals		$5926.90		$7344.11	$148.17	$183.60

Crew B-36D	Hr.	Daily	Hr.	Daily	Bare Costs	Incl. O&P
1 Labor Foreman (outside)	$43.05	$344.40	$65.10	$520.80	$52.09	$78.30
3 Equip. Oper. (medium)	55.10	1322.40	82.70	1984.80		
1 Grader, 30,000 Lbs.		657.85		723.63		
1 Dozer, 300 H.P.		1831.00		2014.10		
1 Roller, Vibratory, 25 Ton		650.90		715.99	98.12	107.93
32 L.H., Daily Totals		$4806.55		$5959.32	$150.20	$186.23

Crew B-37	Hr.	Daily	Hr.	Daily	Bare Costs	Incl. O&P
1 Labor Foreman (outside)	$43.05	$344.40	$65.10	$520.80	$43.15	$65.17
4 Laborers	41.05	1313.60	62.10	1987.20		
1 Equip. Oper. (light)	51.65	413.20	77.55	620.40		
1 Tandem Roller, 5 Ton		152.45		167.69	3.18	3.49
48 L.H., Daily Totals		$2223.65		$3296.09	$46.33	$68.67

Crew B-37A	Hr.	Daily	Hr.	Daily	Bare Costs	Incl. O&P
2 Laborers	$41.05	$656.80	$62.10	$993.60	$42.53	$64.23
1 Truck Driver (light)	45.50	364.00	68.50	548.00		
1 Flatbed Truck, Gas, 1.5 Ton		195.20		214.72		
1 Tar Kettle, T.M.		134.50		147.95	13.74	15.11
24 L.H., Daily Totals		$1350.50		$1904.27	$56.27	$79.34

Crew B-37B	Hr.	Daily	Hr.	Daily	Bare Costs	Incl. O&P
3 Laborers	$41.05	$985.20	$62.10	$1490.40	$42.16	$63.70
1 Truck Driver (light)	45.50	364.00	68.50	548.00		
1 Flatbed Truck, Gas, 1.5 Ton		195.20		214.72		
1 Tar Kettle, T.M.		134.50		147.95	10.30	11.33
32 L.H., Daily Totals		$1678.90		$2401.07	$52.47	$75.03

Crew B-37C	Hr.	Daily	Hr.	Daily	Bare Costs	Incl. O&P
2 Laborers	$41.05	$656.80	$62.10	$993.60	$43.27	$65.30
2 Truck Drivers (light)	45.50	728.00	68.50	1096.00		
2 Flatbed Trucks, Gas, 1.5 Ton		390.40		429.44		
1 Tar Kettle, T.M.		134.50		147.95	16.40	18.04
32 L.H., Daily Totals		$1909.70		$2666.99	$59.68	$83.34

Crew B-37D	Hr.	Daily	Hr.	Daily	Bare Costs	Incl. O&P
1 Laborer	$41.05	$328.40	$62.10	$496.80	$43.27	$65.30
1 Truck Driver (light)	45.50	364.00	68.50	548.00		
1 Pickup Truck, 3/4 Ton		109.90		120.89	6.87	7.56
16 L.H., Daily Totals		$802.30		$1165.69	$50.14	$72.86

Crew B-37E	Hr.	Daily	Hr.	Daily	Bare Costs	Incl. O&P
3 Laborers	$41.05	$985.20	$62.10	$1490.40	$45.84	$69.08
1 Equip. Oper. (light)	51.65	413.20	77.55	620.40		
1 Equip. Oper. (medium)	55.10	440.80	82.70	661.60		
2 Truck Drivers (light)	45.50	728.00	68.50	1096.00		
4 Barrels w/ Flasher		16.00		17.60		
1 Concrete Saw		110.15		121.17		
1 Rotary Hammer Drill		25.05		27.56		
1 Hammer Drill Bit		3.30		3.63		
1 Loader, Skid Steer, 30 H.P.		174.55		192.01		
1 Conc. Hammer Attach.		114.40		125.84		
1 Vibrating Plate, Gas, 18"		31.10		34.21		
2 Flatbed Trucks, Gas, 1.5 Ton		390.40		429.44	15.45	16.99
56 L.H., Daily Totals		$3432.15		$4819.85	$61.29	$86.07

Crew B-37F	Hr.	Daily	Hr.	Daily	Bare Costs	Incl. O&P
3 Laborers	$41.05	$985.20	$62.10	$1490.40	$42.16	$63.70
1 Truck Driver (light)	45.50	364.00	68.50	548.00		
4 Barrels w/ Flasher		16.00		17.60		
1 Concrete Mixer, 10 C.F.		162.00		178.20		
1 Air Compressor, 60 cfm		106.70		117.37		
1 -50' Air Hose, 3/4"		6.55		7.21		
1 Spade (Chipper)		8.30		9.13		
1 Flatbed Truck, Gas, 1.5 Ton		195.20		214.72	15.46	17.01
32 L.H., Daily Totals		$1843.95		$2582.63	$57.62	$80.71

Crew B-37G	Hr.	Daily	Hr.	Daily	Bare Costs	Incl. O&P
1 Labor Foreman (outside)	$43.05	$344.40	$65.10	$520.80	$43.15	$65.17
4 Laborers	41.05	1313.60	62.10	1987.20		
1 Equip. Oper. (light)	51.65	413.20	77.55	620.40		
1 Berm Machine		325.30		357.83		
1 Tandem Roller, 5 Ton		152.45		167.69	9.95	10.95
48 L.H., Daily Totals		$2548.95		$3653.93	$53.10	$76.12

Crew B-37H	Hr.	Daily	Hr.	Daily	Bare Costs	Incl. O&P
1 Labor Foreman (outside)	$43.05	$344.40	$65.10	$520.80	$43.15	$65.17
4 Laborers	41.05	1313.60	62.10	1987.20		
1 Equip. Oper. (light)	51.65	413.20	77.55	620.40		
1 Tandem Roller, 5 Ton		152.45		167.69		
1 Flatbed Truck, Gas, 1.5 Ton		195.20		214.72		
1 Tar Kettle, T.M.		134.50		147.95	10.04	11.05
48 L.H., Daily Totals		$2553.35		$3658.76	$53.19	$76.22

Crew B-37I	Hr.	Daily	Hr.	Daily	Bare Costs	Incl. O&P
3 Laborers	$41.05	$985.20	$62.10	$1490.40	$45.84	$69.08
1 Equip. Oper. (light)	51.65	413.20	77.55	620.40		
1 Equip. Oper. (medium)	55.10	440.80	82.70	661.60		
2 Truck Drivers (light)	45.50	728.00	68.50	1096.00		
4 Barrels w/ Flasher		16.00		17.60		
1 Concrete Saw		110.15		121.17		
1 Rotary Hammer Drill		25.05		27.56		
1 Hammer Drill Bit		3.30		3.63		
1 Air Compressor, 60 cfm		106.70		117.37		
1 -50' Air Hose, 3/4"		6.55		7.21		
1 Spade (Chipper)		8.30		9.13		
1 Loader, Skid Steer, 30 H.P.		174.55		192.01		
1 Conc. Hammer Attach.		114.40		125.84		
1 Concrete Mixer, 10 C.F.		162.00		178.20		
1 Vibrating Plate, Gas, 18"		31.10		34.21		
2 Flatbed Trucks, Gas, 1.5 Ton		390.40		429.44	20.51	22.56
56 L.H., Daily Totals		$3715.70		$5131.75	$66.35	$91.64

Crew No.	Bare Costs		Incl. Subs O&P		Cost Per Labor-Hour	
Crew B-37J	**Hr.**	**Daily**	**Hr.**	**Daily**	**Bare Costs**	**Incl. O&P**
1 Labor Foreman (outside)	$43.05	$344.40	$65.10	$520.80	$43.15	$65.17
4 Laborers	41.05	1313.60	62.10	1987.20		
1 Equip. Oper. (light)	51.65	413.20	77.55	620.40		
1 Air Compressor, 60 cfm		106.70		117.37		
1 -50' Air Hose, 3/4"		6.55		7.21		
2 Concrete Mixers, 10 C.F.		324.00		356.40		
2 Flatbed Trucks, Gas, 1.5 Ton		390.40		429.44		
1 Shot Blaster, 20"		201.00		221.10	21.43	23.57
48 L.H., Daily Totals		$3099.85		$4259.92	$64.58	$88.75
Crew B-37K	**Hr.**	**Daily**	**Hr.**	**Daily**	**Bare Costs**	**Incl. O&P**
1 Labor Foreman (outside)	$43.05	$344.40	$65.10	$520.80	$43.15	$65.17
4 Laborers	41.05	1313.60	62.10	1987.20		
1 Equip. Oper. (light)	51.65	413.20	77.55	620.40		
1 Air Compressor, 60 cfm		106.70		117.37		
1 -50' Air Hose, 3/4"		6.55		7.21		
2 Flatbed Trucks, Gas, 1.5 Ton		390.40		429.44		
1 Shot Blaster, 20"		201.00		221.10	14.68	16.15
48 L.H., Daily Totals		$2775.85		$3903.51	$57.83	$81.32
Crew B-38	**Hr.**	**Daily**	**Hr.**	**Daily**	**Bare Costs**	**Incl. O&P**
1 Labor Foreman (outside)	$43.05	$344.40	$65.10	$520.80	$46.38	$69.91
2 Laborers	41.05	656.80	62.10	993.60		
1 Equip. Oper. (light)	51.65	413.20	77.55	620.40		
1 Equip. Oper. (medium)	55.10	440.80	82.70	661.60		
1 Backhoe Loader, 48 H.P.		320.20		352.22		
1 Hyd. Hammer (1200 lb.)		171.40		188.54		
1 F.E. Loader, W.M., 4 C.Y.		566.70		623.37		
1 Pvmt. Rem. Bucket		61.70		67.87	28.00	30.80
40 L.H., Daily Totals		$2975.20		$4028.40	$74.38	$100.71
Crew B-39	**Hr.**	**Daily**	**Hr.**	**Daily**	**Bare Costs**	**Incl. O&P**
1 Labor Foreman (outside)	$43.05	$344.40	$65.10	$520.80	$43.15	$65.17
4 Laborers	41.05	1313.60	62.10	1987.20		
1 Equip. Oper. (light)	51.65	413.20	77.55	620.40		
1 Air Compressor, 250 cfm		167.95		184.75		
2 Breakers, Pavement, 60 lb.		21.50		23.65		
2 -50' Air Hoses, 1.5"		45.50		50.05	4.89	5.38
48 L.H., Daily Totals		$2306.15		$3386.84	$48.04	$70.56
Crew B-40	**Hr.**	**Daily**	**Hr.**	**Daily**	**Bare Costs**	**Incl. O&P**
1 Pile Driver Foreman (outside)	$54.15	$433.20	$84.65	$677.20	$53.36	$82.15
4 Pile Drivers	52.15	1668.80	81.55	2609.60		
2 Equip. Oper. (crane)	57.45	919.20	86.25	1380.00		
1 Equip. Oper. (oiler)	49.20	393.60	73.85	590.80		
1 Crawler Crane, 40 Ton		1392.00		1531.20		
1 Vibratory Hammer & Gen.		2236.00		2459.60	56.69	62.36
64 L.H., Daily Totals		$7042.80		$9248.40	$110.04	$144.51
Crew B-40B	**Hr.**	**Daily**	**Hr.**	**Daily**	**Bare Costs**	**Incl. O&P**
1 Labor Foreman (outside)	$43.05	$344.40	$65.10	$520.80	$45.48	$68.58
3 Laborers	41.05	985.20	62.10	1490.40		
1 Equip. Oper. (crane)	57.45	459.60	86.25	690.00		
1 Equip. Oper. (oiler)	49.20	393.60	73.85	590.80		
1 Lattice Boom Crane, 40 Ton		1315.00		1446.50	27.40	30.14
48 L.H., Daily Totals		$3497.80		$4738.50	$72.87	$98.72

Crew No.	Bare Costs		Incl. Subs O&P		Cost Per Labor-Hour	
Crew B-41	**Hr.**	**Daily**	**Hr.**	**Daily**	**Bare Costs**	**Incl. O&P**
1 Labor Foreman (outside)	$43.05	$344.40	$65.10	$520.80	$42.53	$64.28
4 Laborers	41.05	1313.60	62.10	1987.20		
.25 Equip. Oper. (crane)	57.45	114.90	86.25	172.50		
.25 Equip. Oper. (oiler)	49.20	98.40	73.85	147.70		
.25 Crawler Crane, 40 Ton		348.00		382.80	7.91	8.70
44 L.H., Daily Totals		$2219.30		$3211.00	$50.44	$72.98
Crew B-42	**Hr.**	**Daily**	**Hr.**	**Daily**	**Bare Costs**	**Incl. O&P**
1 Labor Foreman (outside)	$43.05	$344.40	$65.10	$520.80	$46.24	$70.37
4 Laborers	41.05	1313.60	62.10	1987.20		
1 Equip. Oper. (crane)	57.45	459.60	86.25	690.00		
1 Equip. Oper. (oiler)	49.20	393.60	73.85	590.80		
1 Welder	56.00	448.00	89.35	714.80		
1 Hyd. Crane, 25 Ton		581.70		639.87		
1 Welder, Gas Engine, 300 amp		95.65		105.22		
1 Horz. Boring Csg. Mch.		447.00		491.70	17.57	19.32
64 L.H., Daily Totals		$4083.55		$5740.39	$63.81	$89.69
Crew B-43	**Hr.**	**Daily**	**Hr.**	**Daily**	**Bare Costs**	**Incl. O&P**
1 Labor Foreman (outside)	$43.05	$344.40	$65.10	$520.80	$45.48	$68.58
3 Laborers	41.05	985.20	62.10	1490.40		
1 Equip. Oper. (crane)	57.45	459.60	86.25	690.00		
1 Equip. Oper. (oiler)	49.20	393.60	73.85	590.80		
1 Drill Rig, Truck-Mounted		2428.00		2670.80	50.58	55.64
48 L.H., Daily Totals		$4610.80		$5962.80	$96.06	$124.22
Crew B-44	**Hr.**	**Daily**	**Hr.**	**Daily**	**Bare Costs**	**Incl. O&P**
1 Pile Driver Foreman (outside)	$54.15	$433.20	$84.65	$677.20	$52.34	$80.68
4 Pile Drivers	52.15	1668.80	81.55	2609.60		
2 Equip. Oper. (crane)	57.45	919.20	86.25	1380.00		
1 Laborer	41.05	328.40	62.10	496.80		
1 Crawler Crane, 40 Ton		1392.00		1531.20		
1 Lead, 60' High		80.55		88.61		
1 Hammer, Diesel, 15K ft.-lbs.		622.40		684.64	32.73	36.01
64 L.H., Daily Totals		$5444.55		$7468.05	$85.07	$116.69
Crew B-45	**Hr.**	**Daily**	**Hr.**	**Daily**	**Bare Costs**	**Incl. O&P**
1 Equip. Oper. (medium)	$55.10	$440.80	$82.70	$661.60	$51.05	$76.72
1 Truck Driver (heavy)	47.00	376.00	70.75	566.00		
1 Dist. Tanker, 3000 Gallon		323.25		355.57		
1 Truck Tractor, 6x4, 380 H.P.		487.60		536.36	50.68	55.75
16 L.H., Daily Totals		$1627.65		$2119.53	$101.73	$132.47
Crew B-46	**Hr.**	**Daily**	**Hr.**	**Daily**	**Bare Costs**	**Incl. O&P**
1 Pile Driver Foreman (outside)	$54.15	$433.20	$84.65	$677.20	$46.93	$72.34
2 Pile Drivers	52.15	834.40	81.55	1304.80		
3 Laborers	41.05	985.20	62.10	1490.40		
1 Chain Saw, Gas, 36" Long		46.65		51.31	.97	1.07
48 L.H., Daily Totals		$2299.45		$3523.72	$47.91	$73.41
Crew B-47	**Hr.**	**Daily**	**Hr.**	**Daily**	**Bare Costs**	**Incl. O&P**
1 Blast Foreman (outside)	$43.05	$344.40	$65.10	$520.80	$45.25	$68.25
1 Driller	41.05	328.40	62.10	496.80		
1 Equip. Oper. (light)	51.65	413.20	77.55	620.40		
1 Air Track Drill, 4"		1004.00		1104.40		
1 Air Compressor, 600 cfm		417.60		459.36		
2 -50' Air Hoses, 3"		51.50		56.65	61.38	67.52
24 L.H., Daily Totals		$2559.10		$3258.41	$106.63	$135.77

Crew B-47A	Bare Costs Hr.	Bare Costs Daily	Incl. Subs O&P Hr.	Incl. Subs O&P Daily	Cost Per Labor-Hour Bare Costs	Cost Per Labor-Hour Incl. O&P
1 Drilling Foreman (outside)	$43.05	$344.40	$65.10	$520.80	$49.90	$75.07
1 Equip. Oper. (heavy)	57.45	459.60	86.25	690.00		
1 Equip. Oper. (oiler)	49.20	393.60	73.85	590.80		
1 Air Track Drill, 5"		1173.00		1290.30	48.88	53.76
24 L.H., Daily Totals		$2370.60		$3091.90	$98.78	$128.83

Crew B-47C	Hr.	Daily	Hr.	Daily	Bare Costs	Incl. O&P
1 Laborer	$41.05	$328.40	$62.10	$496.80	$46.35	$69.83
1 Equip. Oper. (light)	51.65	413.20	77.55	620.40		
1 Air Compressor, 750 cfm		428.80		471.68		
2 -50' Air Hoses, 3"		51.50		56.65		
1 Air Track Drill, 4"		1004.00		1104.40	92.77	102.05
16 L.H., Daily Totals		$2225.90		$2749.93	$139.12	$171.87

Crew B-47E	Hr.	Daily	Hr.	Daily	Bare Costs	Incl. O&P
1 Labor Foreman (outside)	$43.05	$344.40	$65.10	$520.80	$41.55	$62.85
3 Laborers	41.05	985.20	62.10	1490.40		
1 Flatbed Truck, Gas, 3 Ton		245.95		270.55	7.69	8.45
32 L.H., Daily Totals		$1575.55		$2281.74	$49.24	$71.30

Crew B-47G	Hr.	Daily	Hr.	Daily	Bare Costs	Incl. O&P
1 Labor Foreman (outside)	$43.05	$344.40	$65.10	$520.80	$44.20	$66.71
2 Laborers	41.05	656.80	62.10	993.60		
1 Equip. Oper. (light)	51.65	413.20	77.55	620.40		
1 Air Track Drill, 4"		1004.00		1104.40		
1 Air Compressor, 600 cfm		417.60		459.36		
2 -50' Air Hoses, 3"		51.50		56.65		
1 Gunite Pump Rig		313.00		344.30	55.82	61.40
32 L.H., Daily Totals		$3200.50		$4099.51	$100.02	$128.11

Crew B-47H	Hr.	Daily	Hr.	Daily	Bare Costs	Incl. O&P
1 Skilled Worker Foreman (out)	$55.40	$443.20	$84.25	$674.00	$53.90	$82.00
3 Skilled Workers	53.40	1281.60	81.25	1950.00		
1 Flatbed Truck, Gas, 3 Ton		245.95		270.55	7.69	8.45
32 L.H., Daily Totals		$1970.75		$2894.55	$61.59	$90.45

Crew B-48	Hr.	Daily	Hr.	Daily	Bare Costs	Incl. O&P
1 Labor Foreman (outside)	$43.05	$344.40	$65.10	$520.80	$46.36	$69.86
3 Laborers	41.05	985.20	62.10	1490.40		
1 Equip. Oper. (crane)	57.45	459.60	86.25	690.00		
1 Equip. Oper. (oiler)	49.20	393.60	73.85	590.80		
1 Equip. Oper. (light)	51.65	413.20	77.55	620.40		
1 Centr. Water Pump, 6"		296.45		326.10		
1 -20' Suction Hose, 6"		11.10		12.21		
1 -50' Discharge Hose, 6"		5.70		6.27		
1 Drill Rig, Truck-Mounted		2428.00		2670.80	48.95	53.85
56 L.H., Daily Totals		$5337.25		$6927.77	$95.31	$123.71

Crew B-49	Hr.	Daily	Hr.	Daily	Bare Costs	Incl. O&P
1 Labor Foreman (outside)	$43.05	$344.40	$65.10	$520.80	$48.68	$73.84
3 Laborers	41.05	985.20	62.10	1490.40		
2 Equip. Oper. (crane)	57.45	919.20	86.25	1380.00		
2 Equip. Oper. (oilers)	49.20	787.20	73.85	1181.60		
1 Equip. Oper. (light)	51.65	413.20	77.55	620.40		
2 Pile Drivers	52.15	834.40	81.55	1304.80		
1 Hyd. Crane, 25 Ton		581.70		639.87		
1 Centr. Water Pump, 6"		296.45		326.10		
1 -20' Suction Hose, 6"		11.10		12.21		
1 -50' Discharge Hose, 6"		5.70		6.27		
1 Drill Rig, Truck-Mounted		2428.00		2670.80	37.76	41.54
88 L.H., Daily Totals		$7606.55		$10153.25	$86.44	$115.38

Crew B-50	Hr.	Daily	Hr.	Daily	Bare Costs	Incl. O&P
2 Pile Driver Foremen (outside)	$54.15	$866.40	$84.65	$1354.40	$50.60	$77.95
6 Pile Drivers	52.15	2503.20	81.55	3914.40		
2 Equip. Oper. (crane)	57.45	919.20	86.25	1380.00		
1 Equip. Oper. (oiler)	49.20	393.60	73.85	590.80		
3 Laborers	41.05	985.20	62.10	1490.40		
1 Crawler Crane, 40 Ton		1392.00		1531.20		
1 Lead, 60' High		80.55		88.61		
1 Hammer, Diesel, 15K ft.-lbs.		622.40		684.64		
1 Air Compressor, 600 cfm		417.60		459.36		
2 -50' Air Hoses, 3"		51.50		56.65		
1 Chain Saw, Gas, 36" Long		46.65		51.31	23.31	25.64
112 L.H., Daily Totals		$8278.30		$11601.77	$73.91	$103.59

Crew B-51	Hr.	Daily	Hr.	Daily	Bare Costs	Incl. O&P
1 Labor Foreman (outside)	$43.05	$344.40	$65.10	$520.80	$42.13	$63.67
4 Laborers	41.05	1313.60	62.10	1987.20		
1 Truck Driver (light)	45.50	364.00	68.50	548.00		
1 Flatbed Truck, Gas, 1.5 Ton		195.20		214.72	4.07	4.47
48 L.H., Daily Totals		$2217.20		$3270.72	$46.19	$68.14

Crew B-52	Hr.	Daily	Hr.	Daily	Bare Costs	Incl. O&P
1 Carpenter Foreman (outside)	$53.65	$429.20	$81.15	$649.20	$47.48	$71.52
1 Carpenter	51.65	413.20	78.15	625.20		
3 Laborers	41.05	985.20	62.10	1490.40		
1 Cement Finisher	48.90	391.20	72.20	577.60		
.5 Rodman (reinf.)	54.85	219.40	83.00	332.00		
.5 Equip. Oper. (medium)	55.10	220.40	82.70	330.80		
.5 Crawler Loader, 3 C.Y.		601.00		661.10	10.73	11.81
56 L.H., Daily Totals		$3259.60		$4666.30	$58.21	$83.33

Crew B-53	Hr.	Daily	Hr.	Daily	Bare Costs	Incl. O&P
1 Equip. Oper. (light)	$51.65	$413.20	$77.55	$620.40	$51.65	$77.55
1 Trencher, Chain, 12 H.P.		63.40		69.74	7.92	8.72
8 L.H., Daily Totals		$476.60		$690.14	$59.58	$86.27

Crew B-54	Hr.	Daily	Hr.	Daily	Bare Costs	Incl. O&P
1 Equip. Oper. (light)	$51.65	$413.20	$77.55	$620.40	$51.65	$77.55
1 Trencher, Chain, 40 H.P.		348.50		383.35	43.56	47.92
8 L.H., Daily Totals		$761.70		$1003.75	$95.21	$125.47

Crew B-54A	Hr.	Daily	Hr.	Daily	Bare Costs	Incl. O&P
.17 Labor Foreman (outside)	$43.05	$58.55	$65.10	$88.54	$53.35	$80.14
1 Equipment Operator (med.)	55.10	440.80	82.70	661.60		
1 Wheel Trencher, 67 H.P.		1110.00		1221.00	118.59	130.45
9.36 L.H., Daily Totals		$1609.35		$1971.14	$171.94	$210.59

Crew B-54B	Hr.	Daily	Hr.	Daily	Bare Costs	Incl. O&P
.25 Labor Foreman (outside)	$43.05	$86.10	$65.10	$130.20	$52.69	$79.18
1 Equipment Operator (med.)	55.10	440.80	82.70	661.60		
1 Wheel Trencher, 150 H.P.		1996.00		2195.60	199.60	219.56
10 L.H., Daily Totals		$2522.90		$2987.40	$252.29	$298.74

Crew B-54C	Hr.	Daily	Hr.	Daily	Bare Costs	Incl. O&P
1 Laborer	$41.05	$328.40	$62.10	$496.80	$48.08	$72.40
1 Equipment Operator (med.)	55.10	440.80	82.70	661.60		
1 Wheel Trencher, 67 H.P.		1110.00		1221.00	69.38	76.31
16 L.H., Daily Totals		$1879.20		$2379.40	$117.45	$148.71

Crew No.	Bare Costs		Incl. Subs O&P		Cost Per Labor-Hour	
Crew B-54D	**Hr.**	**Daily**	**Hr.**	**Daily**	**Bare Costs**	**Incl. O&P**
1 Laborer	$41.05	$328.40	$62.10	$496.80	$48.08	$72.40
1 Equipment Operator (med.)	55.10	440.80	82.70	661.60		
1 Rock Trencher, 6" Width		797.00		876.70	49.81	54.79
16 L.H., Daily Totals		$1566.20		$2035.10	$97.89	$127.19

Crew B-54E	**Hr.**	**Daily**	**Hr.**	**Daily**	**Bare Costs**	**Incl. O&P**
1 Laborer	$41.05	$328.40	$62.10	$496.80	$48.08	$72.40
1 Equipment Operator (med.)	55.10	440.80	82.70	661.60		
1 Rock Trencher, 18" Width		2678.00		2945.80	167.38	184.11
16 L.H., Daily Totals		$3447.20		$4104.20	$215.45	$256.51

Crew B-55	**Hr.**	**Daily**	**Hr.**	**Daily**	**Bare Costs**	**Incl. O&P**
2 Laborers	$41.05	$656.80	$62.10	$993.60	$42.53	$64.23
1 Truck Driver (light)	45.50	364.00	68.50	548.00		
1 Truck-Mounted Earth Auger		756.20		831.82		
1 Flatbed Truck, Gas, 3 Ton		245.95		270.55	41.76	45.93
24 L.H., Daily Totals		$2022.95		$2643.97	$84.29	$110.17

Crew B-56	**Hr.**	**Daily**	**Hr.**	**Daily**	**Bare Costs**	**Incl. O&P**
1 Laborer	$41.05	$328.40	$62.10	$496.80	$46.35	$69.83
1 Equip. Oper. (light)	51.65	413.20	77.55	620.40		
1 Air Track Drill, 4"		1004.00		1104.40		
1 Air Compressor, 600 cfm		417.60		459.36		
1 -50' Air Hose, 3"		25.75		28.32	90.46	99.51
16 L.H., Daily Totals		$2188.95		$2709.28	$136.81	$169.33

Crew B-57	**Hr.**	**Daily**	**Hr.**	**Daily**	**Bare Costs**	**Incl. O&P**
1 Labor Foreman (outside)	$43.05	$344.40	$65.10	$520.80	$47.24	$71.16
2 Laborers	41.05	656.80	62.10	993.60		
1 Equip. Oper. (crane)	57.45	459.60	86.25	690.00		
1 Equip. Oper. (light)	51.65	413.20	77.55	620.40		
1 Equip. Oper. (oiler)	49.20	393.60	73.85	590.80		
1 Crawler Crane, 25 Ton		1386.00		1524.60		
1 Clamshell Bucket, 1 C.Y.		51.20		56.32		
1 Centr. Water Pump, 6"		296.45		326.10		
1 -20' Suction Hose, 6"		11.10		12.21		
20 -50' Discharge Hoses, 6"		114.00		125.40	38.72	42.60
48 L.H., Daily Totals		$4126.35		$5460.23	$85.97	$113.75

Crew B-58	**Hr.**	**Daily**	**Hr.**	**Daily**	**Bare Costs**	**Incl. O&P**
2 Laborers	$41.05	$656.80	$62.10	$993.60	$44.58	$67.25
1 Equip. Oper. (light)	51.65	413.20	77.55	620.40		
1 Backhoe Loader, 48 H.P.		320.20		352.22		
1 Small Helicopter, w/ Pilot		2916.00		3207.60	134.84	148.33
24 L.H., Daily Totals		$4306.20		$5173.82	$179.43	$215.58

Crew B-59	**Hr.**	**Daily**	**Hr.**	**Daily**	**Bare Costs**	**Incl. O&P**
1 Truck Driver (heavy)	$47.00	$376.00	$70.75	$566.00	$47.00	$70.75
1 Truck Tractor, 220 H.P.		303.55		333.90		
1 Water Tank Trailer, 5000 Gal.		149.45		164.40	56.63	62.29
8 L.H., Daily Totals		$829.00		$1064.30	$103.63	$133.04

Crew B-59A	**Hr.**	**Daily**	**Hr.**	**Daily**	**Bare Costs**	**Incl. O&P**
2 Laborers	$41.05	$656.80	$62.10	$993.60	$43.03	$64.98
1 Truck Driver (heavy)	47.00	376.00	70.75	566.00		
1 Water Tank Trailer, 5000 Gal.		149.45		164.40		
1 Truck Tractor, 220 H.P.		303.55		333.90	18.88	20.76
24 L.H., Daily Totals		$1485.80		$2057.90	$61.91	$85.75

Crew No.	Bare Costs		Incl. Subs O&P		Cost Per Labor-Hour	
Crew B-60	**Hr.**	**Daily**	**Hr.**	**Daily**	**Bare Costs**	**Incl. O&P**
1 Labor Foreman (outside)	$43.05	$344.40	$65.10	$520.80	$47.87	$72.07
2 Laborers	41.05	656.80	62.10	993.60		
1 Equip. Oper. (crane)	57.45	459.60	86.25	690.00		
2 Equip. Oper. (light)	51.65	826.40	77.55	1240.80		
1 Equip. Oper. (oiler)	49.20	393.60	73.85	590.80		
1 Crawler Crane, 40 Ton		1392.00		1531.20		
1 Lead, 60' High		80.55		88.61		
1 Hammer, Diesel, 15K ft.-lbs.		622.40		684.64		
1 Backhoe Loader, 48 H.P.		320.20		352.22	43.13	47.44
56 L.H., Daily Totals		$5095.95		$6692.67	$91.00	$119.51

Crew B-61	**Hr.**	**Daily**	**Hr.**	**Daily**	**Bare Costs**	**Incl. O&P**
1 Labor Foreman (outside)	$43.05	$344.40	$65.10	$520.80	$43.57	$65.79
3 Laborers	41.05	985.20	62.10	1490.40		
1 Equip. Oper. (light)	51.65	413.20	77.55	620.40		
1 Cement Mixer, 2 C.Y.		203.85		224.24		
1 Air Compressor, 160 cfm		115.85		127.44	7.99	8.79
40 L.H., Daily Totals		$2062.50		$2983.27	$51.56	$74.58

Crew B-62	**Hr.**	**Daily**	**Hr.**	**Daily**	**Bare Costs**	**Incl. O&P**
2 Laborers	$41.05	$656.80	$62.10	$993.60	$44.58	$67.25
1 Equip. Oper. (light)	51.65	413.20	77.55	620.40		
1 Loader, Skid Steer, 30 H.P.		174.55		192.01	7.27	8.00
24 L.H., Daily Totals		$1244.55		$1806.01	$51.86	$75.25

Crew B-62A	**Hr.**	**Daily**	**Hr.**	**Daily**	**Bare Costs**	**Incl. O&P**
2 Laborers	$41.05	$656.80	$62.10	$993.60	$44.58	$67.25
1 Equip. Oper. (light)	51.65	413.20	77.55	620.40		
1 Loader, Skid Steer, 30 H.P.		174.55		192.01		
1 Trencher Attachment		70.20		77.22	10.20	11.22
24 L.H., Daily Totals		$1314.75		$1883.22	$54.78	$78.47

Crew B-63	**Hr.**	**Daily**	**Hr.**	**Daily**	**Bare Costs**	**Incl. O&P**
4 Laborers	$41.05	$1313.60	$62.10	$1987.20	$43.17	$65.19
1 Equip. Oper. (light)	51.65	413.20	77.55	620.40		
1 Loader, Skid Steer, 30 H.P.		174.55		192.01	4.36	4.80
40 L.H., Daily Totals		$1901.35		$2799.61	$47.53	$69.99

Crew B-63B	**Hr.**	**Daily**	**Hr.**	**Daily**	**Bare Costs**	**Incl. O&P**
1 Labor Foreman (inside)	$41.55	$332.40	$62.85	$502.80	$43.83	$66.15
2 Laborers	41.05	656.80	62.10	993.60		
1 Equip. Oper. (light)	51.65	413.20	77.55	620.40		
1 Loader, Skid Steer, 78 H.P.		392.50		431.75	12.27	13.49
32 L.H., Daily Totals		$1794.90		$2548.55	$56.09	$79.64

Crew B-64	**Hr.**	**Daily**	**Hr.**	**Daily**	**Bare Costs**	**Incl. O&P**
1 Laborer	$41.05	$328.40	$62.10	$496.80	$43.27	$65.30
1 Truck Driver (light)	45.50	364.00	68.50	548.00		
1 Power Mulcher (small)		139.35		153.29		
1 Flatbed Truck, Gas, 1.5 Ton		195.20		214.72	20.91	23.00
16 L.H., Daily Totals		$1026.95		$1412.81	$64.18	$88.30

Crew B-65	**Hr.**	**Daily**	**Hr.**	**Daily**	**Bare Costs**	**Incl. O&P**
1 Laborer	$41.05	$328.40	$62.10	$496.80	$43.27	$65.30
1 Truck Driver (light)	45.50	364.00	68.50	548.00		
1 Power Mulcher (Large)		276.35		303.99		
1 Flatbed Truck, Gas, 1.5 Ton		195.20		214.72	29.47	32.42
16 L.H., Daily Totals		$1163.95		$1563.51	$72.75	$97.72

Crews - Standard

Crew No.	Bare Costs		Incl. Subs O&P		Cost Per Labor-Hour	

Crew B-66	Hr.	Daily	Hr.	Daily	Bare Costs	Incl. O&P
1 Equip. Oper. (light)	$51.65	$413.20	$77.55	$620.40	$51.65	$77.55
1 Loader-Backhoe, 40 H.P.		219.20		241.12	27.40	30.14
8 L.H., Daily Totals		$632.40		$861.52	$79.05	$107.69

Crew B-67	Hr.	Daily	Hr.	Daily	Bare Costs	Incl. O&P
1 Millwright	$54.15	$433.20	$78.70	$629.60	$52.90	$78.13
1 Equip. Oper. (light)	51.65	413.20	77.55	620.40		
1 R.T. Forklift, 5,000 Lb., diesel		265.60		292.16	16.60	18.26
16 L.H., Daily Totals		$1112.00		$1542.16	$69.50	$96.39

Crew B-67B	Hr.	Daily	Hr.	Daily	Bare Costs	Incl. O&P
1 Millwright Foreman (inside)	$54.65	$437.20	$79.40	$635.20	$54.40	$79.05
1 Millwright	54.15	433.20	78.70	629.60		
16 L.H., Daily Totals		$870.40		$1264.80	$54.40	$79.05

Crew B-68	Hr.	Daily	Hr.	Daily	Bare Costs	Incl. O&P
2 Millwrights	$54.15	$866.40	$78.70	$1259.20	$53.32	$78.32
1 Equip. Oper. (light)	51.65	413.20	77.55	620.40		
1 R.T. Forklift, 5,000 Lb., diesel		265.60		292.16	11.07	12.17
24 L.H., Daily Totals		$1545.20		$2171.76	$64.38	$90.49

Crew B-68A	Hr.	Daily	Hr.	Daily	Bare Costs	Incl. O&P
1 Millwright Foreman (inside)	$54.65	$437.20	$79.40	$635.20	$54.32	$78.93
2 Millwrights	54.15	866.40	78.70	1259.20		
1 Forklift, Smooth Floor, 8,000 Lb.		144.30		158.73	6.01	6.61
24 L.H., Daily Totals		$1447.90		$2053.13	$60.33	$85.55

Crew B-68B	Hr.	Daily	Hr.	Daily	Bare Costs	Incl. O&P
1 Millwright Foreman (inside)	$54.65	$437.20	$79.40	$635.20	$58.48	$86.50
2 Millwrights	54.15	866.40	78.70	1259.20		
2 Electricians	60.05	960.80	89.60	1433.60		
2 Plumbers	63.15	1010.40	94.75	1516.00		
1 R.T. Forklift, 5,000 Lb., gas		276.20		303.82	4.93	5.43
56 L.H., Daily Totals		$3551.00		$5147.82	$63.41	$91.93

Crew B-68C	Hr.	Daily	Hr.	Daily	Bare Costs	Incl. O&P
1 Millwright Foreman (inside)	$54.65	$437.20	$79.40	$635.20	$58.00	$85.61
1 Millwright	54.15	433.20	78.70	629.60		
1 Electrician	60.05	480.40	89.60	716.80		
1 Plumber	63.15	505.20	94.75	758.00		
1 R.T. Forklift, 5,000 Lb., gas		276.20		303.82	8.63	9.49
32 L.H., Daily Totals		$2132.20		$3043.42	$66.63	$95.11

Crew B-68D	Hr.	Daily	Hr.	Daily	Bare Costs	Incl. O&P
1 Labor Foreman (inside)	$41.55	$332.40	$62.85	$502.80	$44.75	$67.50
1 Laborer	41.05	328.40	62.10	496.80		
1 Equip. Oper. (light)	51.65	413.20	77.55	620.40		
1 R.T. Forklift, 5,000 Lb., gas		276.20		303.82	11.51	12.66
24 L.H., Daily Totals		$1350.20		$1923.82	$56.26	$80.16

Crew B-68E	Hr.	Daily	Hr.	Daily	Bare Costs	Incl. O&P
1 Struc. Steel Foreman (inside)	$56.50	$452.00	$90.10	$720.80	$56.10	$89.50
3 Struc. Steel Workers	56.00	1344.00	89.35	2144.40		
1 Welder	56.00	448.00	89.35	714.80		
1 Forklift, Smooth Floor, 8,000 Lb.		144.30		158.73	3.61	3.97
40 L.H., Daily Totals		$2388.30		$3738.73	$59.71	$93.47

Crew B-68F	Hr.	Daily	Hr.	Daily	Bare Costs	Incl. O&P
1 Skilled Worker Foreman (out)	$55.40	$443.20	$84.25	$674.00	$54.07	$82.25
2 Skilled Workers	53.40	854.40	81.25	1300.00		
1 R.T. Forklift, 5,000 Lb., gas		276.20		303.82	11.51	12.66
24 L.H., Daily Totals		$1573.80		$2277.82	$65.58	$94.91

Crew B-68G	Hr.	Daily	Hr.	Daily	Bare Costs	Incl. O&P
2 Structural Steel Workers	$56.00	$896.00	$89.35	$1429.60	$56.00	$89.35
1 R.T. Forklift, 5,000 Lb., gas		276.20		303.82	17.26	18.99
16 L.H., Daily Totals		$1172.20		$1733.42	$73.26	$108.34

Crew B-69	Hr.	Daily	Hr.	Daily	Bare Costs	Incl. O&P
1 Labor Foreman (outside)	$43.05	$344.40	$65.10	$520.80	$45.48	$68.58
3 Laborers	41.05	985.20	62.10	1490.40		
1 Equip. Oper. (crane)	57.45	459.60	86.25	690.00		
1 Equip. Oper. (oiler)	49.20	393.60	73.85	590.80		
1 Hyd. Crane, 80 Ton		1487.00		1635.70	30.98	34.08
48 L.H., Daily Totals		$3669.80		$4927.70	$76.45	$102.66

Crew B-69A	Hr.	Daily	Hr.	Daily	Bare Costs	Incl. O&P
1 Labor Foreman (outside)	$43.05	$344.40	$65.10	$520.80	$45.03	$67.72
3 Laborers	41.05	985.20	62.10	1490.40		
1 Equip. Oper. (medium)	55.10	440.80	82.70	661.60		
1 Concrete Finisher	48.90	391.20	72.20	577.60		
1 Curb/Gutter Paver, 2-Track		1194.00		1313.40	24.88	27.36
48 L.H., Daily Totals		$3355.60		$4563.80	$69.91	$95.08

Crew B-69B	Hr.	Daily	Hr.	Daily	Bare Costs	Incl. O&P
1 Labor Foreman (outside)	$43.05	$344.40	$65.10	$520.80	$45.03	$67.72
3 Laborers	41.05	985.20	62.10	1490.40		
1 Equip. Oper. (medium)	55.10	440.80	82.70	661.60		
1 Cement Finisher	48.90	391.20	72.20	577.60		
1 Curb/Gutter Paver, 4-Track		776.35		853.99	16.17	17.79
48 L.H., Daily Totals		$2937.95		$4104.39	$61.21	$85.51

Crew B-70	Hr.	Daily	Hr.	Daily	Bare Costs	Incl. O&P
1 Labor Foreman (outside)	$43.05	$344.40	$65.10	$520.80	$47.36	$71.36
3 Laborers	41.05	985.20	62.10	1490.40		
3 Equip. Oper. (medium)	55.10	1322.40	82.70	1984.80		
1 Grader, 30,000 Lbs.		657.85		723.63		
1 Ripper, Beam & 1 Shank		88.80		97.68		
1 Road Sweeper, S.P., 8' wide		703.10		773.41		
1 F.E. Loader, W.M., 1.5 C.Y.		352.95		388.25	32.19	35.41
56 L.H., Daily Totals		$4454.70		$5978.97	$79.55	$106.77

Crew B-70A	Hr.	Daily	Hr.	Daily	Bare Costs	Incl. O&P
1 Laborer	$41.05	$328.40	$62.10	$496.80	$52.29	$78.58
4 Equip. Oper. (medium)	55.10	1763.20	82.70	2646.40		
1 Grader, 40,000 Lbs.		1180.00		1298.00		
1 F.E. Loader, W.M., 2.5 C.Y.		531.20		584.32		
1 Dozer, 80 H.P.		472.00		519.20		
1 Roller, Pneum. Whl., 12 Ton		338.35		372.19	63.04	69.34
40 L.H., Daily Totals		$4613.15		$5916.90	$115.33	$147.92

Crew B-71	Hr.	Daily	Hr.	Daily	Bare Costs	Incl. O&P
1 Labor Foreman (outside)	$43.05	$344.40	$65.10	$520.80	$47.36	$71.36
3 Laborers	41.05	985.20	62.10	1490.40		
3 Equip. Oper. (medium)	55.10	1322.40	82.70	1984.80		
1 Pvmt. Profiler, 750 H.P.		5381.00		5919.10		
1 Road Sweeper, S.P., 8' wide		703.10		773.41		
1 F.E. Loader, W.M., 1.5 C.Y.		352.95		388.25	114.95	126.44
56 L.H., Daily Totals		$9089.05		$11076.75	$162.30	$197.80

Crew B-72	Hr.	Daily	Hr.	Daily	Bare Costs	Incl. O&P
1 Labor Foreman (outside)	$43.05	$344.40	$65.10	$520.80	$48.33	$72.78
3 Laborers	41.05	985.20	62.10	1490.40		
4 Equip. Oper. (medium)	55.10	1763.20	82.70	2646.40		
1 Pvmt. Profiler, 750 H.P.		5381.00		5919.10		
1 Hammermill, 250 H.P.		1890.00		2079.00		
1 Windrow Loader		1238.00		1361.80		
1 Mix Paver, 165 H.P.		2107.00		2317.70		
1 Roller, Pneum. Whl., 12 Ton		338.35		372.19	171.16	188.28
64 L.H., Daily Totals		$14047.15		$16707.38	$219.49	$261.05

Crew B-73	Hr.	Daily	Hr.	Daily	Bare Costs	Incl. O&P
1 Labor Foreman (outside)	$43.05	$344.40	$65.10	$520.80	$50.08	$75.35
2 Laborers	41.05	656.80	62.10	993.60		
5 Equip. Oper. (medium)	55.10	2204.00	82.70	3308.00		
1 Road Mixer, 310 H.P.		1857.00		2042.70		
1 Tandem Roller, 10 Ton		232.05		255.26		
1 Hammermill, 250 H.P.		1890.00		2079.00		
1 Grader, 30,000 Lbs.		657.85		723.63		
.5 F.E. Loader, W.M., 1.5 C.Y.		176.47		194.12		
.5 Truck Tractor, 220 H.P.		151.78		166.95		
.5 Water Tank Trailer, 5000 Gal.		74.72		82.20	78.75	86.62
64 L.H., Daily Totals		$8245.08		$10366.26	$128.83	$161.97

Crew B-74	Hr.	Daily	Hr.	Daily	Bare Costs	Incl. O&P
1 Labor Foreman (outside)	$43.05	$344.40	$65.10	$520.80	$49.81	$74.94
1 Laborer	41.05	328.40	62.10	496.80		
4 Equip. Oper. (medium)	55.10	1763.20	82.70	2646.40		
2 Truck Drivers (heavy)	47.00	752.00	70.75	1132.00		
1 Grader, 30,000 Lbs.		657.85		723.63		
1 Ripper, Beam & 1 Shank		88.80		97.68		
2 Stabilizers, 310 H.P.		3252.00		3577.20		
1 Flatbed Truck, Gas, 3 Ton		245.95		270.55		
1 Chem. Spreader, Towed		53.20		58.52		
1 Roller, Vibratory, 25 Ton		650.90		715.99		
1 Water Tank Trailer, 5000 Gal.		149.45		164.40		
1 Truck Tractor, 220 H.P.		303.55		333.90	84.40	92.84
64 L.H., Daily Totals		$8589.70		$10737.87	$134.21	$167.78

Crew B-75	Hr.	Daily	Hr.	Daily	Bare Costs	Incl. O&P
1 Labor Foreman (outside)	$43.05	$344.40	$65.10	$520.80	$50.21	$75.54
1 Laborer	41.05	328.40	62.10	496.80		
4 Equip. Oper. (medium)	55.10	1763.20	82.70	2646.40		
1 Truck Driver (heavy)	47.00	376.00	70.75	566.00		
1 Grader, 30,000 Lbs.		657.85		723.63		
1 Ripper, Beam & 1 Shank		88.80		97.68		
2 Stabilizers, 310 H.P.		3252.00		3577.20		
1 Dist. Tanker, 3000 Gallon		323.25		355.57		
1 Truck Tractor, 6x4, 380 H.P.		487.60		536.36		
1 Roller, Vibratory, 25 Ton		650.90		715.99	97.51	107.26
56 L.H., Daily Totals		$8272.40		$10236.44	$147.72	$182.79

Crew B-76	Hr.	Daily	Hr.	Daily	Bare Costs	Incl. O&P
1 Dock Builder Foreman (outside)	$54.15	$433.20	$84.65	$677.20	$53.22	$82.08
5 Dock Builders	52.15	2086.00	81.55	3262.00		
2 Equip. Oper. (crane)	57.45	919.20	86.25	1380.00		
1 Equip. Oper. (oiler)	49.20	393.60	73.85	590.80		
1 Crawler Crane, 50 Ton		2032.00		2235.20		
1 Barge, 400 Ton		836.50		920.15		
1 Hammer, Diesel, 15K ft.-lbs.		622.40		684.64		
1 Lead, 60' High		80.55		88.61		
1 Air Compressor, 600 cfm		417.60		459.36		
2 -50' Air Hoses, 3"		51.50		56.65	56.12	61.73
72 L.H., Daily Totals		$7872.55		$10354.61	$109.34	$143.81

Crew B-76A	Hr.	Daily	Hr.	Daily	Bare Costs	Incl. O&P
1 Labor Foreman (outside)	$43.05	$344.40	$65.10	$520.80	$44.37	$66.96
5 Laborers	41.05	1642.00	62.10	2484.00		
1 Equip. Oper. (crane)	57.45	459.60	86.25	690.00		
1 Equip. Oper. (oiler)	49.20	393.60	73.85	590.80		
1 Crawler Crane, 50 Ton		2032.00		2235.20		
1 Barge, 400 Ton		836.50		920.15	44.82	49.30
64 L.H., Daily Totals		$5708.10		$7440.95	$89.19	$116.26

Crew B-77	Hr.	Daily	Hr.	Daily	Bare Costs	Incl. O&P
1 Labor Foreman (outside)	$43.05	$344.40	$65.10	$520.80	$42.34	$63.98
3 Laborers	41.05	985.20	62.10	1490.40		
1 Truck Driver (light)	45.50	364.00	68.50	548.00		
1 Crack Cleaner, 25 H.P.		56.95		62.65		
1 Crack Filler, Trailer Mtd.		198.95		218.85		
1 Flatbed Truck, Gas, 3 Ton		245.95		270.55	12.55	13.80
40 L.H., Daily Totals		$2195.45		$3111.24	$54.89	$77.78

Crew B-78	Hr.	Daily	Hr.	Daily	Bare Costs	Incl. O&P
1 Labor Foreman (outside)	$43.05	$344.40	$65.10	$520.80	$42.13	$63.67
4 Laborers	41.05	1313.60	62.10	1987.20		
1 Truck Driver (light)	45.50	364.00	68.50	548.00		
1 Paint Striper, S.P., 40 Gallon		152.25		167.47		
1 Flatbed Truck, Gas, 3 Ton		245.95		270.55		
1 Pickup Truck, 3/4 Ton		109.90		120.89	10.59	11.64
48 L.H., Daily Totals		$2530.10		$3614.91	$52.71	$75.31

Crew B-78A	Hr.	Daily	Hr.	Daily	Bare Costs	Incl. O&P
1 Equip. Oper. (light)	$51.65	$413.20	$77.55	$620.40	$51.65	$77.55
1 Line Rem. (Metal Balls) 115 H.P.		843.60		927.96	105.45	116.00
8 L.H., Daily Totals		$1256.80		$1548.36	$157.10	$193.54

Crew B-78B	Hr.	Daily	Hr.	Daily	Bare Costs	Incl. O&P
2 Laborers	$41.05	$656.80	$62.10	$993.60	$42.23	$63.82
.25 Equip. Oper. (light)	51.65	103.30	77.55	155.10		
1 Pickup Truck, 3/4 Ton		109.90		120.89		
1 Line Rem.,11 H.P.,Walk Behind		61.35		67.48		
.25 Road Sweeper, S.P., 8' wide		175.78		193.35	19.28	21.21
18 L.H., Daily Totals		$1107.13		$1530.43	$61.51	$85.02

Crew B-78C	Hr.	Daily	Hr.	Daily	Bare Costs	Incl. O&P
1 Labor Foreman (outside)	$43.05	$344.40	$65.10	$520.80	$42.13	$63.67
4 Laborers	41.05	1313.60	62.10	1987.20		
1 Truck Driver (light)	45.50	364.00	68.50	548.00		
1 Paint Striper, T.M., 120 Gal.		686.10		754.71		
1 Flatbed Truck, Gas, 3 Ton		245.95		270.55		
1 Pickup Truck, 3/4 Ton		109.90		120.89	21.71	23.88
48 L.H., Daily Totals		$3063.95		$4202.15	$63.83	$87.54

Crew No.	Bare Costs		Incl. Subs O&P		Cost Per Labor-Hour	
Crew B-78D	**Hr.**	**Daily**	**Hr.**	**Daily**	**Bare Costs**	**Incl. O&P**
2 Labor Foremen (outside)	$43.05	$688.80	$65.10	$1041.60	$41.90	$63.34
7 Laborers	41.05	2298.80	62.10	3477.60		
1 Truck Driver (light)	45.50	364.00	68.50	548.00		
1 Paint Striper, T.M., 120 Gal.		686.10		754.71		
1 Flatbed Truck, Gas, 3 Ton		245.95		270.55		
3 Pickup Trucks, 3/4 Ton		329.70		362.67		
1 Air Compressor, 60 cfm		106.70		117.37		
1 -50' Air Hose, 3/4"		6.55		7.21		
1 Breaker, Pavement, 60 lb.		10.75		11.82	17.32	19.05
80 L.H., Daily Totals		$4737.35		$6591.52	$59.22	$82.39

Crew B-78E	Hr.	Daily	Hr.	Daily	Bare Costs	Incl. O&P
2 Labor Foremen (outside)	$43.05	$688.80	$65.10	$1041.60	$41.75	$63.13
9 Laborers	41.05	2955.60	62.10	4471.20		
1 Truck Driver (light)	45.50	364.00	68.50	548.00		
1 Paint Striper, T.M., 120 Gal.		686.10		754.71		
1 Flatbed Truck, Gas, 3 Ton		245.95		270.55		
4 Pickup Trucks, 3/4 Ton		439.60		483.56		
2 Air Compressors, 60 cfm		213.40		234.74		
2 -50' Air Hoses, 3/4"		13.10		14.41		
2 Breakers, Pavement, 60 lb.		21.50		23.65	16.87	18.56
96 L.H., Daily Totals		$5628.05		$7842.42	$58.63	$81.69

Crew B-78F	Hr.	Daily	Hr.	Daily	Bare Costs	Incl. O&P
2 Labor Foremen (outside)	$43.05	$688.80	$65.10	$1041.60	$41.65	$62.99
11 Laborers	41.05	3612.40	62.10	5464.80		
1 Truck Driver (light)	45.50	364.00	68.50	548.00		
1 Paint Striper, T.M., 120 Gal.		686.10		754.71		
1 Flatbed Truck, Gas, 3 Ton		245.95		270.55		
7 Pickup Trucks, 3/4 Ton		769.30		846.23		
3 Air Compressors, 60 cfm		320.10		352.11		
3 -50' Air Hoses, 3/4"		19.65		21.61		
3 Breakers, Pavement, 60 lb.		32.25		35.48	18.51	20.36
112 L.H., Daily Totals		$6738.55		$9335.08	$60.17	$83.35

Crew B-79	Hr.	Daily	Hr.	Daily	Bare Costs	Incl. O&P
1 Labor Foreman (outside)	$43.05	$344.40	$65.10	$520.80	$42.34	$63.98
3 Laborers	41.05	985.20	62.10	1490.40		
1 Truck Driver (light)	45.50	364.00	68.50	548.00		
1 Paint Striper, T.M., 120 Gal.		686.10		754.71		
1 Heating Kettle, 115 Gallon		77.85		85.64		
1 Flatbed Truck, Gas, 3 Ton		245.95		270.55		
2 Pickup Trucks, 3/4 Ton		219.80		241.78	30.74	33.82
40 L.H., Daily Totals		$2923.30		$3911.87	$73.08	$97.80

Crew B-79A	Hr.	Daily	Hr.	Daily	Bare Costs	Incl. O&P
1.5 Equip. Oper. (light)	$51.65	$619.80	$77.55	$930.60	$51.65	$77.55
.5 Line Remov. (Grinder) 115 H.P.		446.68		491.34		
1 Line Rem. (Metal Balls) 115 H.P.		843.60		927.96	107.52	118.28
12 L.H., Daily Totals		$1910.08		$2349.90	$159.17	$195.83

Crew B-79B	Hr.	Daily	Hr.	Daily	Bare Costs	Incl. O&P
1 Laborer	$41.05	$328.40	$62.10	$496.80	$41.05	$62.10
1 Set of Gases		167.85		184.63	20.98	23.08
8 L.H., Daily Totals		$496.25		$681.43	$62.03	$85.18

Crew B-79C	Hr.	Daily	Hr.	Daily	Bare Costs	Incl. O&P
1 Labor Foreman (outside)	$43.05	$344.40	$65.10	$520.80	$41.97	$63.44
5 Laborers	41.05	1642.00	62.10	2484.00		
1 Truck Driver (light)	45.50	364.00	68.50	548.00		
1 Paint Striper, T.M., 120 Gal.		686.10		754.71		
1 Heating Kettle, 115 Gallon		77.85		85.64		
1 Flatbed Truck, Gas, 3 Ton		245.95		270.55		
3 Pickup Trucks, 3/4 Ton		329.70		362.67		
1 Air Compressor, 60 cfm		106.70		117.37		
1 -50' Air Hose, 3/4"		6.55		7.21		
1 Breaker, Pavement, 60 lb.		10.75		11.82	26.14	28.75
56 L.H., Daily Totals		$3814.00		$5162.76	$68.11	$92.19

Crew B-79D	Hr.	Daily	Hr.	Daily	Bare Costs	Incl. O&P
2 Labor Foremen (outside)	$43.05	$688.80	$65.10	$1041.60	$42.11	$63.65
5 Laborers	41.05	1642.00	62.10	2484.00		
1 Truck Driver (light)	45.50	364.00	68.50	548.00		
1 Paint Striper, T.M., 120 Gal.		686.10		754.71		
1 Heating Kettle, 115 Gallon		77.85		85.64		
1 Flatbed Truck, Gas, 3 Ton		245.95		270.55		
4 Pickup Trucks, 3/4 Ton		439.60		483.56		
1 Air Compressor, 60 cfm		106.70		117.37		
1 -50' Air Hose, 3/4"		6.55		7.21		
1 Breaker, Pavement, 60 lb.		10.75		11.82	24.59	27.04
64 L.H., Daily Totals		$4268.30		$5804.45	$66.69	$90.69

Crew B-79E	Hr.	Daily	Hr.	Daily	Bare Costs	Incl. O&P
2 Labor Foremen (outside)	$43.05	$688.80	$65.10	$1041.60	$41.90	$63.34
7 Laborers	41.05	2298.80	62.10	3477.60		
1 Truck Driver (light)	45.50	364.00	68.50	548.00		
1 Paint Striper, T.M., 120 Gal.		686.10		754.71		
1 Heating Kettle, 115 Gallon		77.85		85.64		
1 Flatbed Truck, Gas, 3 Ton		245.95		270.55		
5 Pickup Trucks, 3/4 Ton		549.50		604.45		
2 Air Compressors, 60 cfm		213.40		234.74		
2 -50' Air Hoses, 3/4"		13.10		14.41		
2 Breakers, Pavement, 60 lb.		21.50		23.65	22.59	24.85
80 L.H., Daily Totals		$5159.00		$7055.34	$64.49	$88.19

Crew B-80	Hr.	Daily	Hr.	Daily	Bare Costs	Incl. O&P
1 Labor Foreman (outside)	$43.05	$344.40	$65.10	$520.80	$45.31	$68.31
1 Laborer	41.05	328.40	62.10	496.80		
1 Truck Driver (light)	45.50	364.00	68.50	548.00		
1 Equip. Oper. (light)	51.65	413.20	77.55	620.40		
1 Flatbed Truck, Gas, 3 Ton		245.95		270.55		
1 Earth Auger, Truck-Mtd.		385.85		424.44	19.74	21.72
32 L.H., Daily Totals		$2081.80		$2880.98	$65.06	$90.03

Crew B-80A	Hr.	Daily	Hr.	Daily	Bare Costs	Incl. O&P
3 Laborers	$41.05	$985.20	$62.10	$1490.40	$41.05	$62.10
1 Flatbed Truck, Gas, 3 Ton		245.95		270.55	10.25	11.27
24 L.H., Daily Totals		$1231.15		$1760.94	$51.30	$73.37

Crew B-80B	Hr.	Daily	Hr.	Daily	Bare Costs	Incl. O&P
3 Laborers	$41.05	$985.20	$62.10	$1490.40	$43.70	$65.96
1 Equip. Oper. (light)	51.65	413.20	77.55	620.40		
1 Crane, Flatbed Mounted, 3 Ton		232.65		255.91	7.27	8.00
32 L.H., Daily Totals		$1631.05		$2366.72	$50.97	$73.96

Crew No.	Bare Costs		Incl. Subs O&P		Cost Per Labor-Hour	
Crew B-80C	**Hr.**	**Daily**	**Hr.**	**Daily**	**Bare Costs**	**Incl. O&P**
2 Laborers	$41.05	$656.80	$62.10	$993.60	$42.53	$64.23
1 Truck Driver (light)	45.50	364.00	68.50	548.00		
1 Flatbed Truck, Gas, 1.5 Ton		195.20		214.72		
1 Manual Fence Post Auger, Gas		7.40		8.14	8.44	9.29
24 L.H., Daily Totals		$1223.40		$1764.46	$50.98	$73.52
Crew B-81	**Hr.**	**Daily**	**Hr.**	**Daily**	**Bare Costs**	**Incl. O&P**
1 Laborer	$41.05	$328.40	$62.10	$496.80	$47.72	$71.85
1 Equip. Oper. (medium)	55.10	440.80	82.70	661.60		
1 Truck Driver (heavy)	47.00	376.00	70.75	566.00		
1 Hydromulcher, T.M., 3000 Gal.		292.90		322.19		
1 Truck Tractor, 220 H.P.		303.55		333.90	24.85	27.34
24 L.H., Daily Totals		$1741.65		$2380.49	$72.57	$99.19
Crew B-81A	**Hr.**	**Daily**	**Hr.**	**Daily**	**Bare Costs**	**Incl. O&P**
1 Laborer	$41.05	$328.40	$62.10	$496.80	$43.27	$65.30
1 Truck Driver (light)	45.50	364.00	68.50	548.00		
1 Hydromulcher, T.M., 600 Gal.		124.15		136.57		
1 Flatbed Truck, Gas, 3 Ton		245.95		270.55	23.13	25.44
16 L.H., Daily Totals		$1062.50		$1451.91	$66.41	$90.74
Crew B-82	**Hr.**	**Daily**	**Hr.**	**Daily**	**Bare Costs**	**Incl. O&P**
1 Laborer	$41.05	$328.40	$62.10	$496.80	$46.35	$69.83
1 Equip. Oper. (light)	51.65	413.20	77.55	620.40		
1 Horiz. Borer, 6 H.P.		77.85		85.64	4.87	5.35
16 L.H., Daily Totals		$819.45		$1202.84	$51.22	$75.18
Crew B-82A	**Hr.**	**Daily**	**Hr.**	**Daily**	**Bare Costs**	**Incl. O&P**
2 Laborers	$41.05	$656.80	$62.10	$993.60	$46.35	$69.83
2 Equip. Opers. (light)	51.65	826.40	77.55	1240.80		
2 Dump Trucks, 8 C.Y., 220 H.P.		679.20		747.12		
1 Flatbed Trailer, 25 Ton		133.00		146.30		
1 Horiz. Dir. Drill, 20k lb. Thrust		631.30		694.43		
1 Mud Trailer for HDD, 1500 Gal.		298.65		328.51		
1 Pickup Truck, 4x4, 3/4 Ton		121.70		133.87		
1 Flatbed Trailer, 3 Ton		26.55		29.20		
1 Loader, Skid Steer, 78 H.P.		392.50		431.75	71.34	78.47
32 L.H., Daily Totals		$3766.10		$4745.59	$117.69	$148.30
Crew B-82B	**Hr.**	**Daily**	**Hr.**	**Daily**	**Bare Costs**	**Incl. O&P**
2 Laborers	$41.05	$656.80	$62.10	$993.60	$46.35	$69.83
2 Equip. Opers. (light)	51.65	826.40	77.55	1240.80		
2 Dump Trucks, 8 C.Y., 220 H.P.		679.20		747.12		
1 Flatbed Trailer, 25 Ton		133.00		146.30		
1 Horiz. Dir. Drill, 30k lb. Thrust		897.00		986.70		
1 Mud Trailer for HDD, 1500 Gal.		298.65		328.51		
1 Pickup Truck, 4x4, 3/4 Ton		121.70		133.87		
1 Flatbed Trailer, 3 Ton		26.55		29.20		
1 Loader, Skid Steer, 78 H.P.		392.50		431.75	79.64	87.61
32 L.H., Daily Totals		$4031.80		$5037.86	$125.99	$157.43

Crew No.	Bare Costs		Incl. Subs O&P		Cost Per Labor-Hour	
Crew B-82C	**Hr.**	**Daily**	**Hr.**	**Daily**	**Bare Costs**	**Incl. O&P**
2 Laborers	$41.05	$656.80	$62.10	$993.60	$46.35	$69.83
2 Equip. Opers. (light)	51.65	826.40	77.55	1240.80		
2 Dump Trucks, 8 C.Y., 220 H.P.		679.20		747.12		
1 Flatbed Trailer, 25 Ton		133.00		146.30		
1 Horiz. Dir. Drill, 50k lb. Thrust		1185.00		1303.50		
1 Mud Trailer for HDD, 1500 Gal.		298.65		328.51		
1 Pickup Truck, 4x4, 3/4 Ton		121.70		133.87		
1 Flatbed Trailer, 3 Ton		26.55		29.20		
1 Loader, Skid Steer, 78 H.P.		392.50		431.75	88.64	97.51
32 L.H., Daily Totals		$4319.80		$5354.66	$134.99	$167.33
Crew B-82D	**Hr.**	**Daily**	**Hr.**	**Daily**	**Bare Costs**	**Incl. O&P**
1 Equip. Oper. (light)	$51.65	$413.20	$77.55	$620.40	$51.65	$77.55
1 Mud Trailer for HDD, 1500 Gal.		298.65		328.51	37.33	41.06
8 L.H., Daily Totals		$711.85		$948.91	$88.98	$118.61
Crew B-83	**Hr.**	**Daily**	**Hr.**	**Daily**	**Bare Costs**	**Incl. O&P**
1 Tugboat Captain	$55.10	$440.80	$82.70	$661.60	$48.08	$72.40
1 Tugboat Hand	41.05	328.40	62.10	496.80		
1 Tugboat, 250 H.P.		710.70		781.77	44.42	48.86
16 L.H., Daily Totals		$1479.90		$1940.17	$92.49	$121.26
Crew B-84	**Hr.**	**Daily**	**Hr.**	**Daily**	**Bare Costs**	**Incl. O&P**
1 Equip. Oper. (medium)	$55.10	$440.80	$82.70	$661.60	$55.10	$82.70
1 Rotary Mower/Tractor		360.30		396.33	45.04	49.54
8 L.H., Daily Totals		$801.10		$1057.93	$100.14	$132.24
Crew B-85	**Hr.**	**Daily**	**Hr.**	**Daily**	**Bare Costs**	**Incl. O&P**
3 Laborers	$41.05	$985.20	$62.10	$1490.40	$45.05	$67.95
1 Equip. Oper. (medium)	55.10	440.80	82.70	661.60		
1 Truck Driver (heavy)	47.00	376.00	70.75	566.00		
1 Telescoping Boom Lift, to 80'		670.65		737.72		
1 Brush Chipper, 12", 130 H.P.		389.40		428.34		
1 Pruning Saw, Rotary		6.60		7.26	26.67	29.33
40 L.H., Daily Totals		$2868.65		$3891.32	$71.72	$97.28
Crew B-86	**Hr.**	**Daily**	**Hr.**	**Daily**	**Bare Costs**	**Incl. O&P**
1 Equip. Oper. (medium)	$55.10	$440.80	$82.70	$661.60	$55.10	$82.70
1 Stump Chipper, S.P.		182.20		200.42	22.77	25.05
8 L.H., Daily Totals		$623.00		$862.02	$77.88	$107.75
Crew B-86A	**Hr.**	**Daily**	**Hr.**	**Daily**	**Bare Costs**	**Incl. O&P**
1 Equip. Oper. (medium)	$55.10	$440.80	$82.70	$661.60	$55.10	$82.70
1 Grader, 30,000 Lbs.		657.85		723.63	82.23	90.45
8 L.H., Daily Totals		$1098.65		$1385.23	$137.33	$173.15
Crew B-86B	**Hr.**	**Daily**	**Hr.**	**Daily**	**Bare Costs**	**Incl. O&P**
1 Equip. Oper. (medium)	$55.10	$440.80	$82.70	$661.60	$55.10	$82.70
1 Dozer, 200 H.P.		1290.00		1419.00	161.25	177.38
8 L.H., Daily Totals		$1730.80		$2080.60	$216.35	$260.07
Crew B-87	**Hr.**	**Daily**	**Hr.**	**Daily**	**Bare Costs**	**Incl. O&P**
1 Laborer	$41.05	$328.40	$62.10	$496.80	$52.29	$78.58
4 Equip. Oper. (medium)	55.10	1763.20	82.70	2646.40		
2 Feller Bunchers, 100 H.P.		1607.10		1767.81		
1 Log Chipper, 22" Tree		770.10		847.11		
1 Dozer, 105 H.P.		609.70		670.67		
1 Chain Saw, Gas, 36" Long		46.65		51.31	75.84	83.42
40 L.H., Daily Totals		$5125.15		$6480.10	$128.13	$162.00

Crew No.	Bare Costs		Incl. Subs O&P		Cost Per Labor-Hour	
Crew B-88	**Hr.**	**Daily**	**Hr.**	**Daily**	**Bare Costs**	**Incl. O&P**
1 Laborer	$41.05	$328.40	$62.10	$496.80	$53.09	$79.76
6 Equip. Oper. (medium)	55.10	2644.80	82.70	3969.60		
2 Feller Bunchers, 100 H.P.		1607.10		1767.81		
1 Log Chipper, 22" Tree		770.10		847.11		
2 Log Skidders, 50 H.P.		1696.50		1866.15		
1 Dozer, 105 H.P.		609.70		670.67		
1 Chain Saw, Gas, 36" Long		46.65		51.31	84.47	92.91
56 L.H., Daily Totals		$7703.25		$9669.45	$137.56	$172.67
Crew B-89	**Hr.**	**Daily**	**Hr.**	**Daily**	**Bare Costs**	**Incl. O&P**
1 Equip. Oper. (light)	$51.65	$413.20	$77.55	$620.40	$48.58	$73.03
1 Truck Driver (light)	45.50	364.00	68.50	548.00		
1 Flatbed Truck, Gas, 3 Ton		245.95		270.55		
1 Concrete Saw		110.15		121.17		
1 Water Tank, 65 Gal.		98.85		108.74	28.43	31.28
16 L.H., Daily Totals		$1232.15		$1668.85	$77.01	$104.30
Crew B-89A	**Hr.**	**Daily**	**Hr.**	**Daily**	**Bare Costs**	**Incl. O&P**
1 Skilled Worker	$53.40	$427.20	$81.25	$650.00	$47.23	$71.67
1 Laborer	41.05	328.40	62.10	496.80		
1 Core Drill (Large)		112.20		123.42	7.01	7.71
16 L.H., Daily Totals		$867.80		$1270.22	$54.24	$79.39
Crew B-89B	**Hr.**	**Daily**	**Hr.**	**Daily**	**Bare Costs**	**Incl. O&P**
1 Equip. Oper. (light)	$51.65	$413.20	$77.55	$620.40	$48.58	$73.03
1 Truck Driver (light)	45.50	364.00	68.50	548.00		
1 Wall Saw, Hydraulic, 10 H.P.		46.60		51.26		
1 Generator, Diesel, 100 kW		308.80		339.68		
1 Water Tank, 65 Gal.		98.85		108.74		
1 Flatbed Truck, Gas, 3 Ton		245.95		270.55	43.76	48.14
16 L.H., Daily Totals		$1477.40		$1938.62	$92.34	$121.16
Crew B-90	**Hr.**	**Daily**	**Hr.**	**Daily**	**Bare Costs**	**Incl. O&P**
1 Labor Foreman (outside)	$43.05	$344.40	$65.10	$520.80	$45.44	$68.50
3 Laborers	41.05	985.20	62.10	1490.40		
2 Equip. Oper. (light)	51.65	826.40	77.55	1240.80		
2 Truck Drivers (heavy)	47.00	752.00	70.75	1132.00		
1 Road Mixer, 310 H.P.		1857.00		2042.70		
1 Dist. Truck, 2000 Gal.		293.20		322.52	33.60	36.96
64 L.H., Daily Totals		$5058.20		$6749.22	$79.03	$105.46
Crew B-90A	**Hr.**	**Daily**	**Hr.**	**Daily**	**Bare Costs**	**Incl. O&P**
1 Labor Foreman (outside)	$43.05	$344.40	$65.10	$520.80	$49.36	$74.30
2 Laborers	41.05	656.80	62.10	993.60		
4 Equip. Oper. (medium)	55.10	1763.20	82.70	2646.40		
2 Graders, 30,000 Lbs.		1315.70		1447.27		
1 Tandem Roller, 10 Ton		232.05		255.26		
1 Roller, Pneum. Whl., 12 Ton		338.35		372.19	33.68	37.05
56 L.H., Daily Totals		$4650.50		$6235.51	$83.04	$111.35
Crew B-90B	**Hr.**	**Daily**	**Hr.**	**Daily**	**Bare Costs**	**Incl. O&P**
1 Labor Foreman (outside)	$43.05	$344.40	$65.10	$520.80	$48.41	$72.90
2 Laborers	41.05	656.80	62.10	993.60		
3 Equip. Oper. (medium)	55.10	1322.40	82.70	1984.80		
1 Roller, Pneum. Whl., 12 Ton		338.35		372.19		
1 Road Mixer, 310 H.P.		1857.00		2042.70	45.74	50.31
48 L.H., Daily Totals		$4518.95		$5914.09	$94.14	$123.21

Crew No.	Bare Costs		Incl. Subs O&P		Cost Per Labor-Hour	
Crew B-90C	**Hr.**	**Daily**	**Hr.**	**Daily**	**Bare Costs**	**Incl. O&P**
1 Labor Foreman (outside)	$43.05	$344.40	$65.10	$520.80	$46.69	$70.35
4 Laborers	41.05	1313.60	62.10	1987.20		
3 Equip. Oper. (medium)	55.10	1322.40	82.70	1984.80		
3 Truck Drivers (heavy)	47.00	1128.00	70.75	1698.00		
3 Road Mixers, 310 H.P.		5571.00		6128.10	63.31	69.64
88 L.H., Daily Totals		$9679.40		$12318.90	$109.99	$139.99
Crew B-90D	**Hr.**	**Daily**	**Hr.**	**Daily**	**Bare Costs**	**Incl. O&P**
1 Labor Foreman (outside)	$43.05	$344.40	$65.10	$520.80	$45.82	$69.08
6 Laborers	41.05	1970.40	62.10	2980.80		
3 Equip. Oper. (medium)	55.10	1322.40	82.70	1984.80		
3 Truck Drivers (heavy)	47.00	1128.00	70.75	1698.00		
3 Road Mixers, 310 H.P.		5571.00		6128.10	53.57	58.92
104 L.H., Daily Totals		$10336.20		$13312.50	$99.39	$128.00
Crew B-90E	**Hr.**	**Daily**	**Hr.**	**Daily**	**Bare Costs**	**Incl. O&P**
1 Labor Foreman (outside)	$43.05	$344.40	$65.10	$520.80	$46.62	$70.26
4 Laborers	41.05	1313.60	62.10	1987.20		
3 Equip. Oper. (medium)	55.10	1322.40	82.70	1984.80		
1 Truck Driver (heavy)	47.00	376.00	70.75	566.00		
1 Road Mixer, 310 H.P.		1857.00		2042.70	25.79	28.37
72 L.H., Daily Totals		$5213.40		$7101.50	$72.41	$98.63
Crew B-91	**Hr.**	**Daily**	**Hr.**	**Daily**	**Bare Costs**	**Incl. O&P**
1 Labor Foreman (outside)	$43.05	$344.40	$65.10	$520.80	$49.07	$73.86
2 Laborers	41.05	656.80	62.10	993.60		
4 Equip. Oper. (medium)	55.10	1763.20	82.70	2646.40		
1 Truck Driver (heavy)	47.00	376.00	70.75	566.00		
1 Dist. Tanker, 3000 Gallon		323.25		355.57		
1 Truck Tractor, 6x4, 380 H.P.		487.60		536.36		
1 Aggreg. Spreader, S.P.		836.00		919.60		
1 Roller, Pneum. Whl., 12 Ton		338.35		372.19		
1 Tandem Roller, 10 Ton		232.05		255.26	34.64	38.11
64 L.H., Daily Totals		$5357.65		$7165.77	$83.71	$111.97
Crew B-91B	**Hr.**	**Daily**	**Hr.**	**Daily**	**Bare Costs**	**Incl. O&P**
1 Laborer	$41.05	$328.40	$62.10	$496.80	$48.08	$72.40
1 Equipment Oper. (med.)	55.10	440.80	82.70	661.60		
1 Road Sweeper, Vac. Assist.		857.60		943.36	53.60	58.96
16 L.H., Daily Totals		$1626.80		$2101.76	$101.68	$131.36
Crew B-91C	**Hr.**	**Daily**	**Hr.**	**Daily**	**Bare Costs**	**Incl. O&P**
1 Laborer	$41.05	$328.40	$62.10	$496.80	$43.27	$65.30
1 Truck Driver (light)	45.50	364.00	68.50	548.00		
1 Catch Basin Cleaning Truck		528.95		581.85	33.06	36.37
16 L.H., Daily Totals		$1221.35		$1626.65	$76.33	$101.67
Crew B-91D	**Hr.**	**Daily**	**Hr.**	**Daily**	**Bare Costs**	**Incl. O&P**
1 Labor Foreman (outside)	$43.05	$344.40	$65.10	$520.80	$47.52	$71.58
5 Laborers	41.05	1642.00	62.10	2484.00		
5 Equip. Oper. (medium)	55.10	2204.00	82.70	3308.00		
2 Truck Drivers (heavy)	47.00	752.00	70.75	1132.00		
1 Aggreg. Spreader, S.P.		836.00		919.60		
2 Truck Tractors, 6x4, 380 H.P.		975.20		1072.72		
2 Dist. Tankers, 3000 Gallon		646.50		711.15		
2 Pavement Brushes, Towed		171.00		188.10		
2 Rollers Pneum. Whl., 12 Ton		676.70		744.37	31.78	34.96
104 L.H., Daily Totals		$8247.80		$11080.74	$79.31	$106.55

Crew B-92	Hr.	Daily	Hr.	Daily	Bare Costs	Incl. O&P
1 Labor Foreman (outside)	$43.05	$344.40	$65.10	$520.80	$41.55	$62.85
3 Laborers	41.05	985.20	62.10	1490.40		
1 Crack Cleaner, 25 H.P.		56.95		62.65		
1 Air Compressor, 60 cfm		106.70		117.37		
1 Tar Kettle, T.M.		134.50		147.95		
1 Flatbed Truck, Gas, 3 Ton		245.95		270.55	17.00	18.70
32 L.H., Daily Totals		$1873.70		$2609.71	$58.55	$81.55

Crew B-93	Hr.	Daily	Hr.	Daily	Bare Costs	Incl. O&P
1 Equip. Oper. (medium)	$55.10	$440.80	$82.70	$661.60	$55.10	$82.70
1 Feller Buncher, 100 H.P.		803.55		883.90	100.44	110.49
8 L.H., Daily Totals		$1244.35		$1545.51	$155.54	$193.19

Crew B-94A	Hr.	Daily	Hr.	Daily	Bare Costs	Incl. O&P
1 Laborer	$41.05	$328.40	$62.10	$496.80	$41.05	$62.10
1 Diaphragm Water Pump, 2"		74.30		81.73		
1 -20' Suction Hose, 2"		2.05		2.25		
2 -50' Discharge Hoses, 2"		1.80		1.98	9.77	10.75
8 L.H., Daily Totals		$406.55		$582.76	$50.82	$72.85

Crew B-94B	Hr.	Daily	Hr.	Daily	Bare Costs	Incl. O&P
1 Laborer	$41.05	$328.40	$62.10	$496.80	$41.05	$62.10
1 Diaphragm Water Pump, 4"		141.40		155.54		
1 -20' Suction Hose, 4"		3.10		3.41		
2 -50' Discharge Hoses, 4"		4.50		4.95	18.63	20.49
8 L.H., Daily Totals		$477.40		$660.70	$59.67	$82.59

Crew B-94C	Hr.	Daily	Hr.	Daily	Bare Costs	Incl. O&P
1 Laborer	$41.05	$328.40	$62.10	$496.80	$41.05	$62.10
1 Centrifugal Water Pump, 3"		79.25		87.17		
1 -20' Suction Hose, 3"		2.70		2.97		
2 -50' Discharge Hoses, 3"		2.80		3.08	10.59	11.65
8 L.H., Daily Totals		$413.15		$590.02	$51.64	$73.75

Crew B-94D	Hr.	Daily	Hr.	Daily	Bare Costs	Incl. O&P
1 Laborer	$41.05	$328.40	$62.10	$496.80	$41.05	$62.10
1 Centr. Water Pump, 6"		296.45		326.10		
1 -20' Suction Hose, 6"		11.10		12.21		
2 -50' Discharge Hoses, 6"		11.40		12.54	39.87	43.86
8 L.H., Daily Totals		$647.35		$847.64	$80.92	$105.96

Crew C-1	Hr.	Daily	Hr.	Daily	Bare Costs	Incl. O&P
3 Carpenters	$51.65	$1239.60	$78.15	$1875.60	$49.00	$74.14
1 Laborer	41.05	328.40	62.10	496.80		
32 L.H., Daily Totals		$1568.00		$2372.40	$49.00	$74.14

Crew C-2	Hr.	Daily	Hr.	Daily	Bare Costs	Incl. O&P
1 Carpenter Foreman (outside)	$53.65	$429.20	$81.15	$649.20	$50.22	$75.97
4 Carpenters	51.65	1652.80	78.15	2500.80		
1 Laborer	41.05	328.40	62.10	496.80		
48 L.H., Daily Totals		$2410.40		$3646.80	$50.22	$75.97

Crew C-2A	Hr.	Daily	Hr.	Daily	Bare Costs	Incl. O&P
1 Carpenter Foreman (outside)	$53.65	$429.20	$81.15	$649.20	$49.76	$74.98
3 Carpenters	51.65	1239.60	78.15	1875.60		
1 Cement Finisher	48.90	391.20	72.20	577.60		
1 Laborer	41.05	328.40	62.10	496.80		
48 L.H., Daily Totals		$2388.40		$3599.20	$49.76	$74.98

Crew C-3	Hr.	Daily	Hr.	Daily	Bare Costs	Incl. O&P
1 Rodman Foreman (outside)	$56.85	$454.80	$86.05	$688.40	$51.25	$77.47
4 Rodmen (reinf.)	54.85	1755.20	83.00	2656.00		
1 Equip. Oper. (light)	51.65	413.20	77.55	620.40		
2 Laborers	41.05	656.80	62.10	993.60		
3 Stressing Equipment		31.20		34.32		
.5 Grouting Equipment		77.70		85.47	1.70	1.87
64 L.H., Daily Totals		$3388.90		$5078.19	$52.95	$79.35

Crew C-4	Hr.	Daily	Hr.	Daily	Bare Costs	Incl. O&P
1 Rodman Foreman (outside)	$56.85	$454.80	$86.05	$688.40	$55.35	$83.76
3 Rodmen (reinf.)	54.85	1316.40	83.00	1992.00		
3 Stressing Equipment		31.20		34.32	.97	1.07
32 L.H., Daily Totals		$1802.40		$2714.72	$56.33	$84.83

Crew C-4A	Hr.	Daily	Hr.	Daily	Bare Costs	Incl. O&P
2 Rodmen (reinf.)	$54.85	$877.60	$83.00	$1328.00	$54.85	$83.00
4 Stressing Equipment		41.60		45.76	2.60	2.86
16 L.H., Daily Totals		$919.20		$1373.76	$57.45	$85.86

Crew C-5	Hr.	Daily	Hr.	Daily	Bare Costs	Incl. O&P
1 Rodman Foreman (outside)	$56.85	$454.80	$86.05	$688.40	$54.70	$82.59
4 Rodmen (reinf.)	54.85	1755.20	83.00	2656.00		
1 Equip. Oper. (crane)	57.45	459.60	86.25	690.00		
1 Equip. Oper. (oiler)	49.20	393.60	73.85	590.80		
1 Hyd. Crane, 25 Ton		581.70		639.87	10.39	11.43
56 L.H., Daily Totals		$3644.90		$5265.07	$65.09	$94.02

Crew C-6	Hr.	Daily	Hr.	Daily	Bare Costs	Incl. O&P
1 Labor Foreman (outside)	$43.05	$344.40	$65.10	$520.80	$42.69	$64.28
4 Laborers	41.05	1313.60	62.10	1987.20		
1 Cement Finisher	48.90	391.20	72.20	577.60		
2 Gas Engine Vibrators		53.10		58.41	1.11	1.22
48 L.H., Daily Totals		$2102.30		$3144.01	$43.80	$65.50

Crew C-6A	Hr.	Daily	Hr.	Daily	Bare Costs	Incl. O&P
2 Cement Finishers	$48.90	$782.40	$72.20	$1155.20	$48.90	$72.20
1 Concrete Vibrator, Elec, 2 HP		9.25		10.18	.58	.64
16 L.H., Daily Totals		$791.65		$1165.38	$49.48	$72.84

Crew B-89C	Hr.	Daily	Hr.	Daily	Bare Costs	Incl. O&P
1 Cement Finisher	$48.90	$391.20	$72.20	$577.60	$48.90	$72.20
1 Masonry cut-off saw, gas		34.00		37.40	4.25	4.67
8 L.H., Daily Totals		$425.20		$615.00	$53.15	$76.88

Crew C-7	Hr.	Daily	Hr.	Daily	Bare Costs	Incl. O&P
1 Labor Foreman (outside)	$43.05	$344.40	$65.10	$520.80	$44.61	$67.15
5 Laborers	41.05	1642.00	62.10	2484.00		
1 Cement Finisher	48.90	391.20	72.20	577.60		
1 Equip. Oper. (medium)	55.10	440.80	82.70	661.60		
1 Equip. Oper. (oiler)	49.20	393.60	73.85	590.80		
2 Gas Engine Vibrators		53.10		58.41		
1 Concrete Bucket, 1 C.Y.		24.95		27.45		
1 Hyd. Crane, 55 Ton		981.50		1079.65	14.72	16.19
72 L.H., Daily Totals		$4271.55		$6000.31	$59.33	$83.34

Crew No.	Bare Costs		Incl. Subs O&P		Cost Per Labor-Hour	
Crew C-7A	**Hr.**	**Daily**	**Hr.**	**Daily**	**Bare Costs**	**Incl. O&P**
1 Labor Foreman (outside)	$43.05	$344.40	$65.10	$520.80	$42.79	$64.64
5 Laborers	41.05	1642.00	62.10	2484.00		
2 Truck Drivers (heavy)	47.00	752.00	70.75	1132.00		
2 Conc. Transit Mixers		1869.40		2056.34	29.21	32.13
64 L.H., Daily Totals		$4607.80		$6193.14	$72.00	$96.77
Crew C-7B	**Hr.**	**Daily**	**Hr.**	**Daily**	**Bare Costs**	**Incl. O&P**
1 Labor Foreman (outside)	$43.05	$344.40	$65.10	$520.80	$44.37	$66.96
5 Laborers	41.05	1642.00	62.10	2484.00		
1 Equipment Operator, Crane	57.45	459.60	86.25	690.00		
1 Equipment Oiler	49.20	393.60	73.85	590.80		
1 Conc. Bucket, 2 C.Y.		38.90		42.79		
1 Lattice Boom Crane, 165 Ton		2153.00		2368.30	34.25	37.67
64 L.H., Daily Totals		$5031.50		$6696.69	$78.62	$104.64
Crew C-7C	**Hr.**	**Daily**	**Hr.**	**Daily**	**Bare Costs**	**Incl. O&P**
1 Labor Foreman (outside)	$43.05	$344.40	$65.10	$520.80	$44.81	$67.63
5 Laborers	41.05	1642.00	62.10	2484.00		
2 Equipment Operators (med.)	55.10	881.60	82.70	1323.20		
2 F.E. Loaders, W.M., 4 C.Y.		1133.40		1246.74	17.71	19.48
64 L.H., Daily Totals		$4001.40		$5574.74	$62.52	$87.11
Crew C-7D	**Hr.**	**Daily**	**Hr.**	**Daily**	**Bare Costs**	**Incl. O&P**
1 Labor Foreman (outside)	$43.05	$344.40	$65.10	$520.80	$43.34	$65.47
5 Laborers	41.05	1642.00	62.10	2484.00		
1 Equip. Oper. (medium)	55.10	440.80	82.70	661.60		
1 Concrete Conveyer		186.45		205.10	3.33	3.66
56 L.H., Daily Totals		$2613.65		$3871.49	$46.67	$69.13
Crew C-8	**Hr.**	**Daily**	**Hr.**	**Daily**	**Bare Costs**	**Incl. O&P**
1 Labor Foreman (outside)	$43.05	$344.40	$65.10	$520.80	$45.59	$68.36
3 Laborers	41.05	985.20	62.10	1490.40		
2 Cement Finishers	48.90	782.40	72.20	1155.20		
1 Equip. Oper. (medium)	55.10	440.80	82.70	661.60		
1 Concrete Pump (Small)		844.05		928.46	15.07	16.58
56 L.H., Daily Totals		$3396.85		$4756.45	$60.66	$84.94
Crew C-8A	**Hr.**	**Daily**	**Hr.**	**Daily**	**Bare Costs**	**Incl. O&P**
1 Labor Foreman (outside)	$43.05	$344.40	$65.10	$520.80	$44.00	$65.97
3 Laborers	41.05	985.20	62.10	1490.40		
2 Cement Finishers	48.90	782.40	72.20	1155.20		
48 L.H., Daily Totals		$2112.00		$3166.40	$44.00	$65.97
Crew C-8B	**Hr.**	**Daily**	**Hr.**	**Daily**	**Bare Costs**	**Incl. O&P**
1 Labor Foreman (outside)	$43.05	$344.40	$65.10	$520.80	$44.26	$66.82
3 Laborers	41.05	985.20	62.10	1490.40		
1 Equip. Oper. (medium)	55.10	440.80	82.70	661.60		
1 Vibrating Power Screed		79.45		87.39		
1 Roller, Vibratory, 25 Ton		650.90		715.99		
1 Dozer, 200 H.P.		1290.00		1419.00	50.51	55.56
40 L.H., Daily Totals		$3790.75		$4895.19	$94.77	$122.38

Crew No.	Bare Costs		Incl. Subs O&P		Cost Per Labor-Hour	
Crew C-8C	**Hr.**	**Daily**	**Hr.**	**Daily**	**Bare Costs**	**Incl. O&P**
1 Labor Foreman (outside)	$43.05	$344.40	$65.10	$520.80	$45.03	$67.72
3 Laborers	41.05	985.20	62.10	1490.40		
1 Cement Finisher	48.90	391.20	72.20	577.60		
1 Equip. Oper. (medium)	55.10	440.80	82.70	661.60		
1 Shotcrete Rig, 12 C.Y./hr		242.75		267.02		
1 Air Compressor, 160 cfm		115.85		127.44		
4 -50' Air Hoses, 1"		33.60		36.96		
4 -50' Air Hoses, 2"		103.60		113.96	10.33	11.36
48 L.H., Daily Totals		$2657.40		$3795.78	$55.36	$79.08
Crew C-8D	**Hr.**	**Daily**	**Hr.**	**Daily**	**Bare Costs**	**Incl. O&P**
1 Labor Foreman (outside)	$43.05	$344.40	$65.10	$520.80	$46.16	$69.24
1 Laborer	41.05	328.40	62.10	496.80		
1 Cement Finisher	48.90	391.20	72.20	577.60		
1 Equipment Oper. (light)	51.65	413.20	77.55	620.40		
1 Air Compressor, 250 cfm		167.95		184.75		
2 -50' Air Hoses, 1"		16.80		18.48	5.77	6.35
32 L.H., Daily Totals		$1661.95		$2418.82	$51.94	$75.59
Crew C-8E	**Hr.**	**Daily**	**Hr.**	**Daily**	**Bare Costs**	**Incl. O&P**
1 Labor Foreman (outside)	$43.05	$344.40	$65.10	$520.80	$44.46	$66.86
3 Laborers	41.05	985.20	62.10	1490.40		
1 Cement Finisher	48.90	391.20	72.20	577.60		
1 Equipment Oper. (light)	51.65	413.20	77.55	620.40		
1 Shotcrete Rig, 35 C.Y./hr		271.40		298.54		
1 Air Compressor, 250 cfm		167.95		184.75		
4 -50' Air Hoses, 1"		33.60		36.96		
4 -50' Air Hoses, 2"		103.60		113.96	12.01	13.21
48 L.H., Daily Totals		$2710.55		$3843.41	$56.47	$80.07
Crew C-9	**Hr.**	**Daily**	**Hr.**	**Daily**	**Bare Costs**	**Incl. O&P**
1 Cement Finisher	$48.90	$391.20	$72.20	$577.60	$45.66	$68.49
2 Laborers	41.05	656.80	62.10	993.60		
1 Equipment Oper. (light)	51.65	413.20	77.55	620.40		
1 Grout Pump, 50 C.F./hr.		125.60		138.16		
1 Air Compressor, 160 cfm		115.85		127.44		
2 -50' Air Hoses, 1"		16.80		18.48		
2 -50' Air Hoses, 2"		51.80		56.98	9.69	10.66
32 L.H., Daily Totals		$1771.25		$2532.66	$55.35	$79.15
Crew C-10	**Hr.**	**Daily**	**Hr.**	**Daily**	**Bare Costs**	**Incl. O&P**
1 Laborer	$41.05	$328.40	$62.10	$496.80	$46.28	$68.83
2 Cement Finishers	48.90	782.40	72.20	1155.20		
24 L.H., Daily Totals		$1110.80		$1652.00	$46.28	$68.83
Crew C-10B	**Hr.**	**Daily**	**Hr.**	**Daily**	**Bare Costs**	**Incl. O&P**
3 Laborers	$41.05	$985.20	$62.10	$1490.40	$44.19	$66.14
2 Cement Finishers	48.90	782.40	72.20	1155.20		
1 Concrete Mixer, 10 C.F.		162.00		178.20		
2 Trowels, 48" Walk-Behind		83.00		91.30	6.13	6.74
40 L.H., Daily Totals		$2012.60		$2915.10	$50.31	$72.88
Crew C-10C	**Hr.**	**Daily**	**Hr.**	**Daily**	**Bare Costs**	**Incl. O&P**
1 Laborer	$41.05	$328.40	$62.10	$496.80	$46.28	$68.83
2 Cement Finishers	48.90	782.40	72.20	1155.20		
1 Trowel, 48" Walk-Behind		41.50		45.65	1.73	1.90
24 L.H., Daily Totals		$1152.30		$1697.65	$48.01	$70.74

Crew C-10D	Hr.	Daily	Hr.	Daily	Bare Costs	Incl. O&P
1 Laborer	$41.05	$328.40	$62.10	$496.80	$46.28	$68.83
2 Cement Finishers	48.90	782.40	72.20	1155.20		
1 Vibrating Power Screed		79.45		87.39		
1 Trowel, 48" Walk-Behind		41.50		45.65	5.04	5.54
24 L.H., Daily Totals		$1231.75		$1785.05	$51.32	$74.38

Crew C-10E	Hr.	Daily	Hr.	Daily	Bare Costs	Incl. O&P
1 Laborer	$41.05	$328.40	$62.10	$496.80	$46.28	$68.83
2 Cement Finishers	48.90	782.40	72.20	1155.20		
1 Vibrating Power Screed		79.45		87.39		
1 Cement Trowel, 96" Ride-On		166.30		182.93	10.24	11.26
24 L.H., Daily Totals		$1356.55		$1922.33	$56.52	$80.10

Crew C-10F	Hr.	Daily	Hr.	Daily	Bare Costs	Incl. O&P
1 Laborer	$41.05	$328.40	$62.10	$496.80	$46.28	$68.83
2 Cement Finishers	48.90	782.40	72.20	1155.20		
1 Telescoping Boom Lift, to 60'		451.25		496.38	18.80	20.68
24 L.H., Daily Totals		$1562.05		$2148.38	$65.09	$89.52

Crew C-11	Hr.	Daily	Hr.	Daily	Bare Costs	Incl. O&P
1 Struc. Steel Foreman (outside)	$58.00	$464.00	$92.50	$740.00	$55.63	$87.63
6 Struc. Steel Workers	56.00	2688.00	89.35	4288.80		
1 Equip. Oper. (crane)	57.45	459.60	86.25	690.00		
1 Equip. Oper. (oiler)	49.20	393.60	73.85	590.80		
1 Lattice Boom Crane, 150 Ton		2021.00		2223.10	28.07	30.88
72 L.H., Daily Totals		$6026.20		$8532.70	$83.70	$118.51

Crew C-12	Hr.	Daily	Hr.	Daily	Bare Costs	Incl. O&P
1 Carpenter Foreman (outside)	$53.65	$429.20	$81.15	$649.20	$51.18	$77.33
3 Carpenters	51.65	1239.60	78.15	1875.60		
1 Laborer	41.05	328.40	62.10	496.80		
1 Equip. Oper. (crane)	57.45	459.60	86.25	690.00		
1 Hyd. Crane, 12 Ton		471.75		518.92	9.83	10.81
48 L.H., Daily Totals		$2928.55		$4230.52	$61.01	$88.14

Crew C-13	Hr.	Daily	Hr.	Daily	Bare Costs	Incl. O&P
1 Struc. Steel Worker	$56.00	$448.00	$89.35	$714.80	$54.55	$85.62
1 Welder	56.00	448.00	89.35	714.80		
1 Carpenter	51.65	413.20	78.15	625.20		
1 Welder, Gas Engine, 300 amp		95.65		105.22	3.99	4.38
24 L.H., Daily Totals		$1404.85		$2160.01	$58.54	$90.00

Crew C-14	Hr.	Daily	Hr.	Daily	Bare Costs	Incl. O&P
1 Carpenter Foreman (outside)	$53.65	$429.20	$81.15	$649.20	$50.00	$75.38
5 Carpenters	51.65	2066.00	78.15	3126.00		
4 Laborers	41.05	1313.60	62.10	1987.20		
4 Rodmen (reinf.)	54.85	1755.20	83.00	2656.00		
2 Cement Finishers	48.90	782.40	72.20	1155.20		
1 Equip. Oper. (crane)	57.45	459.60	86.25	690.00		
1 Equip. Oper. (oiler)	49.20	393.60	73.85	590.80		
1 Hyd. Crane, 80 Ton		1487.00		1635.70	10.33	11.36
144 L.H., Daily Totals		$8686.60		$12490.10	$60.32	$86.74

Crew C-14A	Hr.	Daily	Hr.	Daily	Bare Costs	Incl. O&P
1 Carpenter Foreman (outside)	$53.65	$429.20	$81.15	$649.20	$51.42	$77.71
16 Carpenters	51.65	6611.20	78.15	10003.20		
4 Rodmen (reinf.)	54.85	1755.20	83.00	2656.00		
2 Laborers	41.05	656.80	62.10	993.60		
1 Cement Finisher	48.90	391.20	72.20	577.60		
1 Equip. Oper. (medium)	55.10	440.80	82.70	661.60		
1 Gas Engine Vibrator		26.55		29.20		
1 Concrete Pump (Small)		844.05		928.46	4.35	4.79
200 L.H., Daily Totals		$11155.00		$16498.86	$55.77	$82.49

Crew C-14B	Hr.	Daily	Hr.	Daily	Bare Costs	Incl. O&P
1 Carpenter Foreman (outside)	$53.65	$429.20	$81.15	$649.20	$51.33	$77.49
16 Carpenters	51.65	6611.20	78.15	10003.20		
4 Rodmen (reinf.)	54.85	1755.20	83.00	2656.00		
2 Laborers	41.05	656.80	62.10	993.60		
2 Cement Finishers	48.90	782.40	72.20	1155.20		
1 Equip. Oper. (medium)	55.10	440.80	82.70	661.60		
1 Gas Engine Vibrator		26.55		29.20		
1 Concrete Pump (Small)		844.05		928.46	4.19	4.60
208 L.H., Daily Totals		$11546.20		$17076.46	$55.51	$82.10

Crew C-14C	Hr.	Daily	Hr.	Daily	Bare Costs	Incl. O&P
1 Carpenter Foreman (outside)	$53.65	$429.20	$81.15	$649.20	$49.02	$74.05
6 Carpenters	51.65	2479.20	78.15	3751.20		
2 Rodmen (reinf.)	54.85	877.60	83.00	1328.00		
4 Laborers	41.05	1313.60	62.10	1987.20		
1 Cement Finisher	48.90	391.20	72.20	577.60		
1 Gas Engine Vibrator		26.55		29.20	.24	.26
112 L.H., Daily Totals		$5517.35		$8322.41	$49.26	$74.31

Crew C-14D	Hr.	Daily	Hr.	Daily	Bare Costs	Incl. O&P
1 Carpenter Foreman (outside)	$53.65	$429.20	$81.15	$649.20	$51.17	$77.32
18 Carpenters	51.65	7437.60	78.15	11253.60		
2 Rodmen (reinf.)	54.85	877.60	83.00	1328.00		
2 Laborers	41.05	656.80	62.10	993.60		
1 Cement Finisher	48.90	391.20	72.20	577.60		
1 Equip. Oper. (medium)	55.10	440.80	82.70	661.60		
1 Gas Engine Vibrator		26.55		29.20		
1 Concrete Pump (Small)		844.05		928.46	4.35	4.79
200 L.H., Daily Totals		$11103.80		$16421.26	$55.52	$82.11

Crew C-14E	Hr.	Daily	Hr.	Daily	Bare Costs	Incl. O&P
1 Carpenter Foreman (outside)	$53.65	$429.20	$81.15	$649.20	$49.85	$75.27
2 Carpenters	51.65	826.40	78.15	1250.40		
4 Rodmen (reinf.)	54.85	1755.20	83.00	2656.00		
3 Laborers	41.05	985.20	62.10	1490.40		
1 Cement Finisher	48.90	391.20	72.20	577.60		
1 Gas Engine Vibrator		26.55		29.20	.30	.33
88 L.H., Daily Totals		$4413.75		$6652.81	$50.16	$75.60

Crew C-14F	Hr.	Daily	Hr.	Daily	Bare Costs	Incl. O&P
1 Labor Foreman (outside)	$43.05	$344.40	$65.10	$520.80	$46.51	$69.17
2 Laborers	41.05	656.80	62.10	993.60		
6 Cement Finishers	48.90	2347.20	72.20	3465.60		
1 Gas Engine Vibrator		26.55		29.20	.37	.41
72 L.H., Daily Totals		$3374.95		$5009.20	$46.87	$69.57

Crews - Standard

Crew No.	Bare Costs		Incl. Subs O&P		Cost Per Labor-Hour	
Crew C-14G	**Hr.**	**Daily**	**Hr.**	**Daily**	**Bare Costs**	**Incl. O&P**
1 Labor Foreman (outside)	$43.05	$344.40	$65.10	$520.80	$45.82	$68.30
2 Laborers	41.05	656.80	62.10	993.60		
4 Cement Finishers	48.90	1564.80	72.20	2310.40		
1 Gas Engine Vibrator		26.55		29.20	.47	.52
56 L.H., Daily Totals		$2592.55		$3854.01	$46.30	$68.82
Crew C-14H	**Hr.**	**Daily**	**Hr.**	**Daily**	**Bare Costs**	**Incl. O&P**
1 Carpenter Foreman (outside)	$53.65	$429.20	$81.15	$649.20	$50.29	$75.79
2 Carpenters	51.65	826.40	78.15	1250.40		
1 Rodman (reinf.)	54.85	438.80	83.00	664.00		
1 Laborer	41.05	328.40	62.10	496.80		
1 Cement Finisher	48.90	391.20	72.20	577.60		
1 Gas Engine Vibrator		26.55		29.20	.55	.61
48 L.H., Daily Totals		$2440.55		$3667.20	$50.84	$76.40
Crew C-14L	**Hr.**	**Daily**	**Hr.**	**Daily**	**Bare Costs**	**Incl. O&P**
1 Carpenter Foreman (outside)	$53.65	$429.20	$81.15	$649.20	$48.05	$72.55
6 Carpenters	51.65	2479.20	78.15	3751.20		
4 Laborers	41.05	1313.60	62.10	1987.20		
1 Cement Finisher	48.90	391.20	72.20	577.60		
1 Gas Engine Vibrator		26.55		29.20	.28	.30
96 L.H., Daily Totals		$4639.75		$6994.40	$48.33	$72.86
Crew C-14M	**Hr.**	**Daily**	**Hr.**	**Daily**	**Bare Costs**	**Incl. O&P**
1 Carpenter Foreman (outside)	$53.65	$429.20	$81.15	$649.20	$49.74	$74.94
2 Carpenters	51.65	826.40	78.15	1250.40		
1 Rodman (reinf.)	54.85	438.80	83.00	664.00		
2 Laborers	41.05	656.80	62.10	993.60		
1 Cement Finisher	48.90	391.20	72.20	577.60		
1 Equip. Oper. (medium)	55.10	440.80	82.70	661.60		
1 Gas Engine Vibrator		26.55		29.20		
1 Concrete Pump (Small)		844.05		928.46	13.60	14.96
64 L.H., Daily Totals		$4053.80		$5754.06	$63.34	$89.91
Crew C-15	**Hr.**	**Daily**	**Hr.**	**Daily**	**Bare Costs**	**Incl. O&P**
1 Carpenter Foreman (outside)	$53.65	$429.20	$81.15	$649.20	$48.08	$72.35
2 Carpenters	51.65	826.40	78.15	1250.40		
3 Laborers	41.05	985.20	62.10	1490.40		
2 Cement Finishers	48.90	782.40	72.20	1155.20		
1 Rodman (reinf.)	54.85	438.80	83.00	664.00		
72 L.H., Daily Totals		$3462.00		$5209.20	$48.08	$72.35
Crew C-16	**Hr.**	**Daily**	**Hr.**	**Daily**	**Bare Costs**	**Incl. O&P**
1 Labor Foreman (outside)	$43.05	$344.40	$65.10	$520.80	$45.59	$68.36
3 Laborers	41.05	985.20	62.10	1490.40		
2 Cement Finishers	48.90	782.40	72.20	1155.20		
1 Equip. Oper. (medium)	55.10	440.80	82.70	661.60		
1 Gunite Pump Rig		313.00		344.30		
2 -50' Air Hoses, 3/4"		13.10		14.41		
2 -50' Air Hoses, 2"		51.80		56.98	6.75	7.42
56 L.H., Daily Totals		$2930.70		$4243.69	$52.33	$75.78
Crew C-16A	**Hr.**	**Daily**	**Hr.**	**Daily**	**Bare Costs**	**Incl. O&P**
1 Laborer	$41.05	$328.40	$62.10	$496.80	$48.49	$72.30
2 Cement Finishers	48.90	782.40	72.20	1155.20		
1 Equip. Oper. (medium)	55.10	440.80	82.70	661.60		
1 Gunite Pump Rig		313.00		344.30		
2 -50' Air Hoses, 3/4"		13.10		14.41		
2 -50' Air Hoses, 2"		51.80		56.98		
1 Telescoping Boom Lift, to 60'		451.25		496.38	25.91	28.50
32 L.H., Daily Totals		$2380.75		$3225.67	$74.40	$100.80

Crew No.	Bare Costs		Incl. Subs O&P		Cost Per Labor-Hour	
Crew C-17	**Hr.**	**Daily**	**Hr.**	**Daily**	**Bare Costs**	**Incl. O&P**
2 Skilled Worker Foremen (out)	$55.40	$886.40	$84.25	$1348.00	$53.80	$81.85
8 Skilled Workers	53.40	3417.60	81.25	5200.00		
80 L.H., Daily Totals		$4304.00		$6548.00	$53.80	$81.85
Crew C-17A	**Hr.**	**Daily**	**Hr.**	**Daily**	**Bare Costs**	**Incl. O&P**
2 Skilled Worker Foremen (out)	$55.40	$886.40	$84.25	$1348.00	$53.85	$81.90
8 Skilled Workers	53.40	3417.60	81.25	5200.00		
.125 Equip. Oper. (crane)	57.45	57.45	86.25	86.25		
.125 Hyd. Crane, 80 Ton		185.88		204.46	2.29	2.52
81 L.H., Daily Totals		$4547.32		$6838.71	$56.14	$84.43
Crew C-17B	**Hr.**	**Daily**	**Hr.**	**Daily**	**Bare Costs**	**Incl. O&P**
2 Skilled Worker Foremen (out)	$55.40	$886.40	$84.25	$1348.00	$53.89	$81.96
8 Skilled Workers	53.40	3417.60	81.25	5200.00		
.25 Equip. Oper. (crane)	57.45	114.90	86.25	172.50		
.25 Hyd. Crane, 80 Ton		371.75		408.93		
.25 Trowel, 48" Walk-Behind		10.38		11.41	4.66	5.13
82 L.H., Daily Totals		$4801.02		$7140.84	$58.55	$87.08
Crew C-17C	**Hr.**	**Daily**	**Hr.**	**Daily**	**Bare Costs**	**Incl. O&P**
2 Skilled Worker Foremen (out)	$55.40	$886.40	$84.25	$1348.00	$53.93	$82.01
8 Skilled Workers	53.40	3417.60	81.25	5200.00		
.375 Equip. Oper. (crane)	57.45	172.35	86.25	258.75		
.375 Hyd. Crane, 80 Ton		557.63		613.39	6.72	7.39
83 L.H., Daily Totals		$5033.98		$7420.14	$60.65	$89.40
Crew C-17D	**Hr.**	**Daily**	**Hr.**	**Daily**	**Bare Costs**	**Incl. O&P**
2 Skilled Worker Foremen (out)	$55.40	$886.40	$84.25	$1348.00	$53.97	$82.06
8 Skilled Workers	53.40	3417.60	81.25	5200.00		
.5 Equip. Oper. (crane)	57.45	229.80	86.25	345.00		
.5 Hyd. Crane, 80 Ton		743.50		817.85	8.85	9.74
84 L.H., Daily Totals		$5277.30		$7710.85	$62.83	$91.80
Crew C-17E	**Hr.**	**Daily**	**Hr.**	**Daily**	**Bare Costs**	**Incl. O&P**
2 Skilled Worker Foremen (out)	$55.40	$886.40	$84.25	$1348.00	$53.80	$81.85
8 Skilled Workers	53.40	3417.60	81.25	5200.00		
1 Hyd. Jack with Rods		104.40		114.84	1.30	1.44
80 L.H., Daily Totals		$4408.40		$6662.84	$55.10	$83.29
Crew C-18	**Hr.**	**Daily**	**Hr.**	**Daily**	**Bare Costs**	**Incl. O&P**
.125 Labor Foreman (outside)	$43.05	$43.05	$65.10	$65.10	$41.27	$62.43
1 Laborer	41.05	328.40	62.10	496.80		
1 Concrete Cart, 10 C.F.		58.10		63.91	6.46	7.10
9 L.H., Daily Totals		$429.55		$625.81	$47.73	$69.53
Crew C-19	**Hr.**	**Daily**	**Hr.**	**Daily**	**Bare Costs**	**Incl. O&P**
.125 Labor Foreman (outside)	$43.05	$43.05	$65.10	$65.10	$41.27	$62.43
1 Laborer	41.05	328.40	62.10	496.80		
1 Concrete Cart, 18 C.F.		97.75		107.53	10.86	11.95
9 L.H., Daily Totals		$469.20		$669.42	$52.13	$74.38
Crew C-20	**Hr.**	**Daily**	**Hr.**	**Daily**	**Bare Costs**	**Incl. O&P**
1 Labor Foreman (outside)	$43.05	$344.40	$65.10	$520.80	$44.04	$66.31
5 Laborers	41.05	1642.00	62.10	2484.00		
1 Cement Finisher	48.90	391.20	72.20	577.60		
1 Equip. Oper. (medium)	55.10	440.80	82.70	661.60		
2 Gas Engine Vibrators		53.10		58.41		
1 Concrete Pump (Small)		844.05		928.46	14.02	15.42
64 L.H., Daily Totals		$3715.55		$5230.86	$58.06	$81.73

Crew No.	Bare Costs		Incl. Subs O&P		Cost Per Labor-Hour	
Crew C-21	**Hr.**	**Daily**	**Hr.**	**Daily**	**Bare Costs**	**Incl. O&P**
1 Labor Foreman (outside)	$43.05	$344.40	$65.10	$520.80	$44.04	$66.31
5 Laborers	41.05	1642.00	62.10	2484.00		
1 Cement Finisher	48.90	391.20	72.20	577.60		
1 Equip. Oper. (medium)	55.10	440.80	82.70	661.60		
2 Gas Engine Vibrators		53.10		58.41		
1 Concrete Conveyer		186.45		205.10	3.74	4.12
64 L.H., Daily Totals		$3057.95		$4507.51	$47.78	$70.43
Crew C-22	**Hr.**	**Daily**	**Hr.**	**Daily**	**Bare Costs**	**Incl. O&P**
1 Rodman Foreman (outside)	$56.85	$454.80	$86.05	$688.40	$55.16	$83.44
4 Rodmen (reinf.)	54.85	1755.20	83.00	2656.00		
.125 Equip. Oper. (crane)	57.45	57.45	86.25	86.25		
.125 Equip. Oper. (oiler)	49.20	49.20	73.85	73.85		
.125 Hyd. Crane, 25 Ton		72.71		79.98	1.73	1.90
42 L.H., Daily Totals		$2389.36		$3584.48	$56.89	$85.34
Crew C-23	**Hr.**	**Daily**	**Hr.**	**Daily**	**Bare Costs**	**Incl. O&P**
2 Skilled Worker Foremen (out)	$55.40	$886.40	$84.25	$1348.00	$53.78	$81.61
6 Skilled Workers	53.40	2563.20	81.25	3900.00		
1 Equip. Oper. (crane)	57.45	459.60	86.25	690.00		
1 Equip. Oper. (oiler)	49.20	393.60	73.85	590.80		
1 Lattice Boom Crane, 90 Ton		1698.00		1867.80	21.23	23.35
80 L.H., Daily Totals		$6000.80		$8396.60	$75.01	$104.96
Crew C-23A	**Hr.**	**Daily**	**Hr.**	**Daily**	**Bare Costs**	**Incl. O&P**
1 Labor Foreman (outside)	$43.05	$344.40	$65.10	$520.80	$46.36	$69.88
2 Laborers	41.05	656.80	62.10	993.60		
1 Equip. Oper. (crane)	57.45	459.60	86.25	690.00		
1 Equip. Oper. (oiler)	49.20	393.60	73.85	590.80		
1 Crawler Crane, 100 Ton		1879.00		2066.90		
3 Conc. Buckets, 8 C.Y.		632.70		695.97	62.79	69.07
40 L.H., Daily Totals		$4366.10		$5558.07	$109.15	$138.95
Crew C-24	**Hr.**	**Daily**	**Hr.**	**Daily**	**Bare Costs**	**Incl. O&P**
2 Skilled Worker Foremen (out)	$55.40	$886.40	$84.25	$1348.00	$53.78	$81.61
6 Skilled Workers	53.40	2563.20	81.25	3900.00		
1 Equip. Oper. (crane)	57.45	459.60	86.25	690.00		
1 Equip. Oper. (oiler)	49.20	393.60	73.85	590.80		
1 Lattice Boom Crane, 150 Ton		2021.00		2223.10	25.26	27.79
80 L.H., Daily Totals		$6323.80		$8751.90	$79.05	$109.40
Crew C-25	**Hr.**	**Daily**	**Hr.**	**Daily**	**Bare Costs**	**Incl. O&P**
2 Rodmen (reinf.)	$54.85	$877.60	$83.00	$1328.00	$44.35	$69.70
2 Rodmen Helpers	33.85	541.60	56.40	902.40		
32 L.H., Daily Totals		$1419.20		$2230.40	$44.35	$69.70
Crew C-27	**Hr.**	**Daily**	**Hr.**	**Daily**	**Bare Costs**	**Incl. O&P**
2 Cement Finishers	$48.90	$782.40	$72.20	$1155.20	$48.90	$72.20
1 Concrete Saw		110.15		121.17	6.88	7.57
16 L.H., Daily Totals		$892.55		$1276.37	$55.78	$79.77
Crew C-28	**Hr.**	**Daily**	**Hr.**	**Daily**	**Bare Costs**	**Incl. O&P**
1 Cement Finisher	$48.90	$391.20	$72.20	$577.60	$48.90	$72.20
1 Portable Air Compressor, Gas		16.05		17.66	2.01	2.21
8 L.H., Daily Totals		$407.25		$595.26	$50.91	$74.41
Crew C-29	**Hr.**	**Daily**	**Hr.**	**Daily**	**Bare Costs**	**Incl. O&P**
1 Laborer	$41.05	$328.40	$62.10	$496.80	$41.05	$62.10
1 Pressure Washer		63.05		69.36	7.88	8.67
8 L.H., Daily Totals		$391.45		$566.15	$48.93	$70.77

Crew No.	Bare Costs		Incl. Subs O&P		Cost Per Labor-Hour	
Crew C-30	**Hr.**	**Daily**	**Hr.**	**Daily**	**Bare Costs**	**Incl. O&P**
1 Laborer	$41.05	$328.40	$62.10	$496.80	$41.05	$62.10
1 Concrete Mixer, 10 C.F.		162.00		178.20	20.25	22.27
8 L.H., Daily Totals		$490.40		$675.00	$61.30	$84.38
Crew C-31	**Hr.**	**Daily**	**Hr.**	**Daily**	**Bare Costs**	**Incl. O&P**
1 Cement Finisher	$48.90	$391.20	$72.20	$577.60	$48.90	$72.20
1 Grout Pump		313.00		344.30	39.13	43.04
8 L.H., Daily Totals		$704.20		$921.90	$88.03	$115.24
Crew C-32	**Hr.**	**Daily**	**Hr.**	**Daily**	**Bare Costs**	**Incl. O&P**
1 Cement Finisher	$48.90	$391.20	$72.20	$577.60	$44.98	$67.15
1 Laborer	41.05	328.40	62.10	496.80		
1 Crack Chaser Saw, Gas, 6 H.P.		24.50		26.95		
1 Vacuum Pick-Up System		72.15		79.36	6.04	6.64
16 L.H., Daily Totals		$816.25		$1180.71	$51.02	$73.79
Crew D-1	**Hr.**	**Daily**	**Hr.**	**Daily**	**Bare Costs**	**Incl. O&P**
1 Bricklayer	$51.05	$408.40	$77.95	$623.60	$45.73	$69.83
1 Bricklayer Helper	40.40	323.20	61.70	493.60		
16 L.H., Daily Totals		$731.60		$1117.20	$45.73	$69.83
Crew D-2	**Hr.**	**Daily**	**Hr.**	**Daily**	**Bare Costs**	**Incl. O&P**
3 Bricklayers	$51.05	$1225.20	$77.95	$1870.80	$47.23	$72.06
2 Bricklayer Helpers	40.40	646.40	61.70	987.20		
.5 Carpenter	51.65	206.60	78.15	312.60		
44 L.H., Daily Totals		$2078.20		$3170.60	$47.23	$72.06
Crew D-3	**Hr.**	**Daily**	**Hr.**	**Daily**	**Bare Costs**	**Incl. O&P**
3 Bricklayers	$51.05	$1225.20	$77.95	$1870.80	$47.02	$71.77
2 Bricklayer Helpers	40.40	646.40	61.70	987.20		
.25 Carpenter	51.65	103.30	78.15	156.30		
42 L.H., Daily Totals		$1974.90		$3014.30	$47.02	$71.77
Crew D-4	**Hr.**	**Daily**	**Hr.**	**Daily**	**Bare Costs**	**Incl. O&P**
1 Bricklayer	$51.05	$408.40	$77.95	$623.60	$45.88	$69.72
2 Bricklayer Helpers	40.40	646.40	61.70	987.20		
1 Equip. Oper. (light)	51.65	413.20	77.55	620.40		
1 Grout Pump, 50 C.F./hr.		125.60		138.16	3.92	4.32
32 L.H., Daily Totals		$1593.60		$2369.36	$49.80	$74.04
Crew D-5	**Hr.**	**Daily**	**Hr.**	**Daily**	**Bare Costs**	**Incl. O&P**
1 Bricklayer	51.05	408.40	77.95	623.60	51.05	77.95
8 L.H., Daily Totals		$408.40		$623.60	$51.05	$77.95
Crew D-6	**Hr.**	**Daily**	**Hr.**	**Daily**	**Bare Costs**	**Incl. O&P**
3 Bricklayers	$51.05	$1225.20	$77.95	$1870.80	$45.96	$70.16
3 Bricklayer Helpers	40.40	969.60	61.70	1480.80		
.25 Carpenter	51.65	103.30	78.15	156.30		
50 L.H., Daily Totals		$2298.10		$3507.90	$45.96	$70.16
Crew D-7	**Hr.**	**Daily**	**Hr.**	**Daily**	**Bare Costs**	**Incl. O&P**
1 Tile Layer	$47.85	$382.80	$70.55	$564.40	$42.80	$63.10
1 Tile Layer Helper	37.75	302.00	55.65	445.20		
16 L.H., Daily Totals		$684.80		$1009.60	$42.80	$63.10
Crew D-8	**Hr.**	**Daily**	**Hr.**	**Daily**	**Bare Costs**	**Incl. O&P**
3 Bricklayers	$51.05	$1225.20	$77.95	$1870.80	$46.79	$71.45
2 Bricklayer Helpers	40.40	646.40	61.70	987.20		
40 L.H., Daily Totals		$1871.60		$2858.00	$46.79	$71.45

Crews - Standard

Crew D-9	Hr.	Daily	Hr.	Daily	Bare Costs	Incl. O&P
3 Bricklayers	$51.05	$1225.20	$77.95	$1870.80	$45.73	$69.83
3 Bricklayer Helpers	40.40	969.60	61.70	1480.80		
48 L.H., Daily Totals		$2194.80		$3351.60	$45.73	$69.83

Crew D-10	Hr.	Daily	Hr.	Daily	Bare Costs	Incl. O&P
1 Bricklayer Foreman (outside)	$53.05	$424.40	$81.00	$648.00	$50.49	$76.72
1 Bricklayer	51.05	408.40	77.95	623.60		
1 Bricklayer Helper	40.40	323.20	61.70	493.60		
1 Equip. Oper. (crane)	57.45	459.60	86.25	690.00		
1 S.P. Crane, 4x4, 12 Ton		371.85		409.04	11.62	12.78
32 L.H., Daily Totals		$1987.45		$2864.24	$62.11	$89.51

Crew D-11	Hr.	Daily	Hr.	Daily	Bare Costs	Incl. O&P
1 Bricklayer Foreman (outside)	$53.05	$424.40	$81.00	$648.00	$48.17	$73.55
1 Bricklayer	51.05	408.40	77.95	623.60		
1 Bricklayer Helper	40.40	323.20	61.70	493.60		
24 L.H., Daily Totals		$1156.00		$1765.20	$48.17	$73.55

Crew D-12	Hr.	Daily	Hr.	Daily	Bare Costs	Incl. O&P
1 Bricklayer Foreman (outside)	$53.05	$424.40	$81.00	$648.00	$46.23	$70.59
1 Bricklayer	51.05	408.40	77.95	623.60		
2 Bricklayer Helpers	40.40	646.40	61.70	987.20		
32 L.H., Daily Totals		$1479.20		$2258.80	$46.23	$70.59

Crew D-13	Hr.	Daily	Hr.	Daily	Bare Costs	Incl. O&P
1 Bricklayer Foreman (outside)	$53.05	$424.40	$81.00	$648.00	$49.00	$74.46
1 Bricklayer	51.05	408.40	77.95	623.60		
2 Bricklayer Helpers	40.40	646.40	61.70	987.20		
1 Carpenter	51.65	413.20	78.15	625.20		
1 Equip. Oper. (crane)	57.45	459.60	86.25	690.00		
1 S.P. Crane, 4x4, 12 Ton		371.85		409.04	7.75	8.52
48 L.H., Daily Totals		$2723.85		$3983.03	$56.75	$82.98

Crew D-14	Hr.	Daily	Hr.	Daily	Bare Costs	Incl. O&P
3 Bricklayers	$51.05	$1225.20	$77.95	$1870.80	$48.39	$73.89
1 Bricklayer Helper	40.40	323.20	61.70	493.60		
32 L.H., Daily Totals		$1548.40		$2364.40	$48.39	$73.89

Crew E-1	Hr.	Daily	Hr.	Daily	Bare Costs	Incl. O&P
1 Welder Foreman (outside)	$58.00	$464.00	$92.50	$740.00	$55.22	$86.47
1 Welder	56.00	448.00	89.35	714.80		
1 Equip. Oper. (light)	51.65	413.20	77.55	620.40		
1 Welder, Gas Engine, 300 amp		95.65		105.22	3.99	4.38
24 L.H., Daily Totals		$1420.85		$2180.42	$59.20	$90.85

Crew E-2	Hr.	Daily	Hr.	Daily	Bare Costs	Incl. O&P
1 Struc. Steel Foreman (outside)	$58.00	$464.00	$92.50	$740.00	$55.52	$87.14
4 Struc. Steel Workers	56.00	1792.00	89.35	2859.20		
1 Equip. Oper. (crane)	57.45	459.60	86.25	690.00		
1 Equip. Oper. (oiler)	49.20	393.60	73.85	590.80		
1 Lattice Boom Crane, 90 Ton		1698.00		1867.80	30.32	33.35
56 L.H., Daily Totals		$4807.20		$6747.80	$85.84	$120.50

Crew E-3	Hr.	Daily	Hr.	Daily	Bare Costs	Incl. O&P
1 Struc. Steel Foreman (outside)	$58.00	$464.00	$92.50	$740.00	$56.67	$90.40
1 Struc. Steel Worker	56.00	448.00	89.35	714.80		
1 Welder	56.00	448.00	89.35	714.80		
1 Welder, Gas Engine, 300 amp		95.65		105.22	3.99	4.38
24 L.H., Daily Totals		$1455.65		$2274.82	$60.65	$94.78

Crew E-3A	Hr.	Daily	Hr.	Daily	Bare Costs	Incl. O&P
1 Struc. Steel Foreman (outside)	$58.00	$464.00	$92.50	$740.00	$56.67	$90.40
1 Struc. Steel Worker	56.00	448.00	89.35	714.80		
1 Welder	56.00	448.00	89.35	714.80		
1 Welder, Gas Engine, 300 amp		95.65		105.22		
1 Telescoping Boom Lift, to 40'		283.15		311.46	15.78	17.36
24 L.H., Daily Totals		$1738.80		$2586.28	$72.45	$107.76

Crew E-4	Hr.	Daily	Hr.	Daily	Bare Costs	Incl. O&P
1 Struc. Steel Foreman (outside)	$58.00	$464.00	$92.50	$740.00	$56.50	$90.14
3 Struc. Steel Workers	56.00	1344.00	89.35	2144.40		
1 Welder, Gas Engine, 300 amp		95.65		105.22	2.99	3.29
32 L.H., Daily Totals		$1903.65		$2989.61	$59.49	$93.43

Crew E-5	Hr.	Daily	Hr.	Daily	Bare Costs	Incl. O&P
2 Struc. Steel Foremen (outside)	$58.00	$928.00	$92.50	$1480.00	$55.87	$88.12
5 Struc. Steel Workers	56.00	2240.00	89.35	3574.00		
1 Equip. Oper. (crane)	57.45	459.60	86.25	690.00		
1 Welder	56.00	448.00	89.35	714.80		
1 Equip. Oper. (oiler)	49.20	393.60	73.85	590.80		
1 Lattice Boom Crane, 90 Ton		1698.00		1867.80		
1 Welder, Gas Engine, 300 amp		95.65		105.22	22.42	24.66
80 L.H., Daily Totals		$6262.85		$9022.61	$78.29	$112.78

Crew E-6	Hr.	Daily	Hr.	Daily	Bare Costs	Incl. O&P
3 Struc. Steel Foremen (outside)	$58.00	$1392.00	$92.50	$2220.00	$55.77	$88.04
9 Struc. Steel Workers	56.00	4032.00	89.35	6433.20		
1 Equip. Oper. (crane)	57.45	459.60	86.25	690.00		
1 Welder	56.00	448.00	89.35	714.80		
1 Equip. Oper. (oiler)	49.20	393.60	73.85	590.80		
1 Equip. Oper. (light)	51.65	413.20	77.55	620.40		
1 Lattice Boom Crane, 90 Ton		1698.00		1867.80		
1 Welder, Gas Engine, 300 amp		95.65		105.22		
1 Air Compressor, 160 cfm		115.85		127.44		
2 Impact Wrenches		37.00		40.70	15.21	16.73
128 L.H., Daily Totals		$9084.90		$13410.35	$70.98	$104.77

Crew E-7	Hr.	Daily	Hr.	Daily	Bare Costs	Incl. O&P
1 Struc. Steel Foreman (outside)	$58.00	$464.00	$92.50	$740.00	$55.87	$88.12
4 Struc. Steel Workers	56.00	1792.00	89.35	2859.20		
1 Equip. Oper. (crane)	57.45	459.60	86.25	690.00		
1 Equip. Oper. (oiler)	49.20	393.60	73.85	590.80		
1 Welder Foreman (outside)	58.00	464.00	92.50	740.00		
2 Welders	56.00	896.00	89.35	1429.60		
1 Lattice Boom Crane, 90 Ton		1698.00		1867.80		
2 Welders, Gas Engine, 300 amp		191.30		210.43	23.62	25.98
80 L.H., Daily Totals		$6358.50		$9127.83	$79.48	$114.10

Crew E-8	Hr.	Daily	Hr.	Daily	Bare Costs	Incl. O&P
1 Struc. Steel Foreman (outside)	$58.00	$464.00	$92.50	$740.00	$55.56	$87.50
4 Struc. Steel Workers	56.00	1792.00	89.35	2859.20		
1 Welder Foreman (outside)	58.00	464.00	92.50	740.00		
4 Welders	56.00	1792.00	89.35	2859.20		
1 Equip. Oper. (crane)	57.45	459.60	86.25	690.00		
1 Equip. Oper. (oiler)	49.20	393.60	73.85	590.80		
1 Equip. Oper. (light)	51.65	413.20	77.55	620.40		
1 Lattice Boom Crane, 90 Ton		1698.00		1867.80		
4 Welders, Gas Engine, 300 amp		382.60		420.86	20.01	22.01
104 L.H., Daily Totals		$7859.00		$11388.26	$75.57	$109.50

Crew E-9	Bare Costs Hr.	Bare Costs Daily	Incl. Subs O&P Hr.	Incl. Subs O&P Daily	Cost Per Labor-Hour Bare Costs	Cost Per Labor-Hour Incl. O&P
2 Struc. Steel Foremen (outside)	$58.00	$928.00	$92.50	$1480.00	$55.77	$88.04
5 Struc. Steel Workers	56.00	2240.00	89.35	3574.00		
1 Welder Foreman (outside)	58.00	464.00	92.50	740.00		
5 Welders	56.00	2240.00	89.35	3574.00		
1 Equip. Oper. (crane)	57.45	459.60	86.25	690.00		
1 Equip. Oper. (oiler)	49.20	393.60	73.85	590.80		
1 Equip. Oper. (light)	51.65	413.20	77.55	620.40		
1 Lattice Boom Crane, 90 Ton		1698.00		1867.80		
5 Welders, Gas Engine, 300 amp		478.25		526.08	17.00	18.70
128 L.H., Daily Totals		$9314.65		$13663.08	$72.77	$106.74

Crew E-10	Hr.	Daily	Hr.	Daily	Bare Costs	Incl. O&P
1 Welder Foreman (outside)	$58.00	$464.00	$92.50	$740.00	$57.00	$90.92
1 Welder	56.00	448.00	89.35	714.80		
1 Welder, Gas Engine, 300 amp		95.65		105.22		
1 Flatbed Truck, Gas, 3 Ton		245.95		270.55	21.35	23.48
16 L.H., Daily Totals		$1253.60		$1830.56	$78.35	$114.41

Crew E-11	Hr.	Daily	Hr.	Daily	Bare Costs	Incl. O&P
2 Painters, Struc. Steel	$44.65	$714.40	$73.00	$1168.00	$45.50	$71.41
1 Building Laborer	41.05	328.40	62.10	496.80		
1 Equip. Oper. (light)	51.65	413.20	77.55	620.40		
1 Air Compressor, 250 cfm		167.95		184.75		
1 Sandblaster, Portable, 3 C.F.		20.70		22.77		
1 Set Sand Blasting Accessories		14.90		16.39	6.36	7.00
32 L.H., Daily Totals		$1659.55		$2509.11	$51.86	$78.41

Crew E-11A	Hr.	Daily	Hr.	Daily	Bare Costs	Incl. O&P
2 Painters, Struc. Steel	$44.65	$714.40	$73.00	$1168.00	$45.50	$71.41
1 Building Laborer	41.05	328.40	62.10	496.80		
1 Equip. Oper. (light)	51.65	413.20	77.55	620.40		
1 Air Compressor, 250 cfm		167.95		184.75		
1 Sandblaster, Portable, 3 C.F.		20.70		22.77		
1 Set Sand Blasting Accessories		14.90		16.39		
1 Telescoping Boom Lift, to 60'		451.25		496.38	20.46	22.51
32 L.H., Daily Totals		$2110.80		$3005.48	$65.96	$93.92

Crew E-11B	Hr.	Daily	Hr.	Daily	Bare Costs	Incl. O&P
2 Painters, Struc. Steel	$44.65	$714.40	$73.00	$1168.00	$43.45	$69.37
1 Building Laborer	41.05	328.40	62.10	496.80		
2 Paint Sprayer, 8 C.F.M.		85.20		93.72		
1 Telescoping Boom Lift, to 60'		451.25		496.38	22.35	24.59
24 L.H., Daily Totals		$1579.25		$2254.90	$65.80	$93.95

Crew E-12	Hr.	Daily	Hr.	Daily	Bare Costs	Incl. O&P
1 Welder Foreman (outside)	$58.00	$464.00	$92.50	$740.00	$54.83	$85.03
1 Equip. Oper. (light)	51.65	413.20	77.55	620.40		
1 Welder, Gas Engine, 300 amp		95.65		105.22	5.98	6.58
16 L.H., Daily Totals		$972.85		$1465.62	$60.80	$91.60

Crew E-13	Hr.	Daily	Hr.	Daily	Bare Costs	Incl. O&P
1 Welder Foreman (outside)	$58.00	$464.00	$92.50	$740.00	$55.88	$87.52
.5 Equip. Oper. (light)	51.65	206.60	77.55	310.20		
1 Welder, Gas Engine, 300 amp		95.65		105.22	7.97	8.77
12 L.H., Daily Totals		$766.25		$1155.42	$63.85	$96.28

Crew E-14	Hr.	Daily	Hr.	Daily	Bare Costs	Incl. O&P
1 Welder Foreman (outside)	$58.00	$464.00	$92.50	$740.00	$58.00	$92.50
1 Welder, Gas Engine, 300 amp		95.65		105.22	11.96	13.15
8 L.H., Daily Totals		$559.65		$845.22	$69.96	$105.65

Crew E-16	Hr.	Daily	Hr.	Daily	Bare Costs	Incl. O&P
1 Welder Foreman (outside)	$58.00	$464.00	$92.50	$740.00	$57.00	$90.92
1 Welder	56.00	448.00	89.35	714.80		
1 Welder, Gas Engine, 300 amp		95.65		105.22	5.98	6.58
16 L.H., Daily Totals		$1007.65		$1560.02	$62.98	$97.50

Crew E-17	Hr.	Daily	Hr.	Daily	Bare Costs	Incl. O&P
1 Struc. Steel Foreman (outside)	$58.00	$464.00	$92.50	$740.00	$57.00	$90.92
1 Structural Steel Worker	56.00	448.00	89.35	714.80		
16 L.H., Daily Totals		$912.00		$1454.80	$57.00	$90.92

Crew E-18	Hr.	Daily	Hr.	Daily	Bare Costs	Incl. O&P
1 Struc. Steel Foreman (outside)	$58.00	$464.00	$92.50	$740.00	$56.22	$88.65
3 Structural Steel Workers	56.00	1344.00	89.35	2144.40		
1 Equipment Operator (med.)	55.10	440.80	82.70	661.60		
1 Lattice Boom Crane, 20 Ton		869.00		955.90	21.73	23.90
40 L.H., Daily Totals		$3117.80		$4501.90	$77.94	$112.55

Crew E-19	Hr.	Daily	Hr.	Daily	Bare Costs	Incl. O&P
1 Struc. Steel Foreman (outside)	$58.00	$464.00	$92.50	$740.00	$55.22	$86.47
1 Structural Steel Worker	56.00	448.00	89.35	714.80		
1 Equip. Oper. (light)	51.65	413.20	77.55	620.40		
1 Lattice Boom Crane, 20 Ton		869.00		955.90	36.21	39.83
24 L.H., Daily Totals		$2194.20		$3031.10	$91.42	$126.30

Crew E-20	Hr.	Daily	Hr.	Daily	Bare Costs	Incl. O&P
1 Struc. Steel Foreman (outside)	$58.00	$464.00	$92.50	$740.00	$55.58	$87.42
5 Structural Steel Workers	56.00	2240.00	89.35	3574.00		
1 Equip. Oper. (crane)	57.45	459.60	86.25	690.00		
1 Equip. Oper. (oiler)	49.20	393.60	73.85	590.80		
1 Lattice Boom Crane, 40 Ton		1315.00		1446.50	20.55	22.60
64 L.H., Daily Totals		$4872.20		$7041.30	$76.13	$110.02

Crew E-22	Hr.	Daily	Hr.	Daily	Bare Costs	Incl. O&P
1 Skilled Worker Foreman (out)	$55.40	$443.20	$84.25	$674.00	$54.07	$82.25
2 Skilled Workers	53.40	854.40	81.25	1300.00		
24 L.H., Daily Totals		$1297.60		$1974.00	$54.07	$82.25

Crew E-24	Hr.	Daily	Hr.	Daily	Bare Costs	Incl. O&P
3 Structural Steel Workers	$56.00	$1344.00	$89.35	$2144.40	$55.77	$87.69
1 Equipment Operator (med.)	55.10	440.80	82.70	661.60		
1 Hyd. Crane, 25 Ton		581.70		639.87	18.18	20.00
32 L.H., Daily Totals		$2366.50		$3445.87	$73.95	$107.68

Crew E-25	Hr.	Daily	Hr.	Daily	Bare Costs	Incl. O&P
1 Welder Foreman (outside)	$58.00	$464.00	$92.50	$740.00	$58.00	$92.50
1 Cutting Torch		12.50		13.75	1.56	1.72
8 L.H., Daily Totals		$476.50		$753.75	$59.56	$94.22

Crew E-26	Hr.	Daily	Hr.	Daily	Bare Costs	Incl. O&P
1 Struc. Steel Foreman (outside)	$58.00	$464.00	$92.50	$740.00	$57.37	$90.65
1 Struc. Steel Worker	56.00	448.00	89.35	714.80		
1 Welder	56.00	448.00	89.35	714.80		
.25 Electrician	60.05	120.10	89.60	179.20		
.25 Plumber	63.15	126.30	94.75	189.50		
1 Welder, Gas Engine, 300 amp		95.65		105.22	3.42	3.76
28 L.H., Daily Totals		$1702.05		$2643.51	$60.79	$94.41

Crew No.	Bare Costs		Incl. Subs O&P		Cost Per Labor-Hour	
Crew E-27	**Hr.**	**Daily**	**Hr.**	**Daily**	**Bare Costs**	**Incl. O&P**
1 Struc. Steel Foreman (outside)	$58.00	$464.00	$92.50	$740.00	$55.58	$87.42
5 Struc. Steel Workers	56.00	2240.00	89.35	3574.00		
1 Equip. Oper. (crane)	57.45	459.60	86.25	690.00		
1 Equip. Oper. (oiler)	49.20	393.60	73.85	590.80		
1 Hyd. Crane, 12 Ton		471.75		518.92		
1 Hyd. Crane, 80 Ton		1487.00		1635.70	30.61	33.67
64 L.H., Daily Totals		$5515.95		$7749.43	$86.19	$121.08

Crew F-3	Hr.	Daily	Hr.	Daily	Bare Costs	Incl. O&P
4 Carpenters	$51.65	$1652.80	$78.15	$2500.80	$52.81	$79.77
1 Equip. Oper. (crane)	57.45	459.60	86.25	690.00		
1 Hyd. Crane, 12 Ton		471.75		518.92	11.79	12.97
40 L.H., Daily Totals		$2584.15		$3709.72	$64.60	$92.74

Crew F-4	Hr.	Daily	Hr.	Daily	Bare Costs	Incl. O&P
4 Carpenters	$51.65	$1652.80	$78.15	$2500.80	$52.21	$78.78
1 Equip. Oper. (crane)	57.45	459.60	86.25	690.00		
1 Equip. Oper. (oiler)	49.20	393.60	73.85	590.80		
1 Hyd. Crane, 55 Ton		981.50		1079.65	20.45	22.49
48 L.H., Daily Totals		$3487.50		$4861.25	$72.66	$101.28

Crew F-5	Hr.	Daily	Hr.	Daily	Bare Costs	Incl. O&P
1 Carpenter Foreman (outside)	$53.65	$429.20	$81.15	$649.20	$52.15	$78.90
3 Carpenters	51.65	1239.60	78.15	1875.60		
32 L.H., Daily Totals		$1668.80		$2524.80	$52.15	$78.90

Crew F-6	Hr.	Daily	Hr.	Daily	Bare Costs	Incl. O&P
2 Carpenters	$51.65	$826.40	$78.15	$1250.40	$48.57	$73.35
2 Building Laborers	41.05	656.80	62.10	993.60		
1 Equip. Oper. (crane)	57.45	459.60	86.25	690.00		
1 Hyd. Crane, 12 Ton		471.75		518.92	11.79	12.97
40 L.H., Daily Totals		$2414.55		$3452.93	$60.36	$86.32

Crew F-7	Hr.	Daily	Hr.	Daily	Bare Costs	Incl. O&P
2 Carpenters	$51.65	$826.40	$78.15	$1250.40	$46.35	$70.13
2 Building Laborers	41.05	656.80	62.10	993.60		
32 L.H., Daily Totals		$1483.20		$2244.00	$46.35	$70.13

Crew G-1	Hr.	Daily	Hr.	Daily	Bare Costs	Incl. O&P
1 Roofer Foreman (outside)	$47.05	$376.40	$78.40	$627.20	$42.14	$70.20
4 Roofers Composition	45.05	1441.60	75.05	2401.60		
2 Roofer Helpers	33.85	541.60	56.40	902.40		
1 Application Equipment		181.15		199.26		
1 Tar Kettle/Pot		178.20		196.02		
1 Crew Truck		154.35		169.79	9.17	10.09
56 L.H., Daily Totals		$2873.30		$4496.27	$51.31	$80.29

Crew G-2	Hr.	Daily	Hr.	Daily	Bare Costs	Incl. O&P
1 Plasterer	$47.75	$382.00	$71.40	$571.20	$43.32	$65.02
1 Plasterer Helper	41.15	329.20	61.55	492.40		
1 Building Laborer	41.05	328.40	62.10	496.80		
1 Grout Pump, 50 C.F./hr.		125.60		138.16	5.23	5.76
24 L.H., Daily Totals		$1165.20		$1698.56	$48.55	$70.77

Crew No.	Bare Costs		Incl. Subs O&P		Cost Per Labor-Hour	
Crew G-2A	**Hr.**	**Daily**	**Hr.**	**Daily**	**Bare Costs**	**Incl. O&P**
1 Roofer Composition	$45.05	$360.40	$75.05	$600.40	$39.98	$64.52
1 Roofer Helper	33.85	270.80	56.40	451.20		
1 Building Laborer	41.05	328.40	62.10	496.80		
1 Foam Spray Rig, Trailer-Mtd.		514.30		565.73		
1 Pickup Truck, 3/4 Ton		109.90		120.89	26.01	28.61
24 L.H., Daily Totals		$1583.80		$2235.02	$65.99	$93.13

Crew G-3	Hr.	Daily	Hr.	Daily	Bare Costs	Incl. O&P
2 Sheet Metal Workers	$60.95	$975.20	$92.55	$1480.80	$51.00	$77.33
2 Building Laborers	41.05	656.80	62.10	993.60		
32 L.H., Daily Totals		$1632.00		$2474.40	$51.00	$77.33

Crew G-4	Hr.	Daily	Hr.	Daily	Bare Costs	Incl. O&P
1 Labor Foreman (outside)	$43.05	$344.40	$65.10	$520.80	$41.72	$63.10
2 Building Laborers	41.05	656.80	62.10	993.60		
1 Flatbed Truck, Gas, 1.5 Ton		195.20		214.72		
1 Air Compressor, 160 cfm		115.85		127.44	12.96	14.26
24 L.H., Daily Totals		$1312.25		$1856.56	$54.68	$77.36

Crew G-5	Hr.	Daily	Hr.	Daily	Bare Costs	Incl. O&P
1 Roofer Foreman (outside)	$47.05	$376.40	$78.40	$627.20	$40.97	$68.26
2 Roofers Composition	45.05	720.80	75.05	1200.80		
2 Roofer Helpers	33.85	541.60	56.40	902.40		
1 Application Equipment		181.15		199.26	4.53	4.98
40 L.H., Daily Totals		$1819.95		$2929.67	$45.50	$73.24

Crew G-6A	Hr.	Daily	Hr.	Daily	Bare Costs	Incl. O&P
2 Roofers Composition	$45.05	$720.80	$75.05	$1200.80	$45.05	$75.05
1 Small Compressor, Electric		14.85		16.34		
2 Pneumatic Nailers		44.70		49.17	3.72	4.09
16 L.H., Daily Totals		$780.35		$1266.31	$48.77	$79.14

Crew G-7	Hr.	Daily	Hr.	Daily	Bare Costs	Incl. O&P
1 Carpenter	$51.65	$413.20	$78.15	$625.20	$51.65	$78.15
1 Small Compressor, Electric		14.85		16.34		
1 Pneumatic Nailer		22.35		24.59	4.65	5.12
8 L.H., Daily Totals		$450.40		$666.12	$56.30	$83.27

Crew H-1	Hr.	Daily	Hr.	Daily	Bare Costs	Incl. O&P
2 Glaziers	$49.65	$794.40	$74.75	$1196.00	$52.83	$82.05
2 Struc. Steel Workers	56.00	896.00	89.35	1429.60		
32 L.H., Daily Totals		$1690.40		$2625.60	$52.83	$82.05

Crew H-2	Hr.	Daily	Hr.	Daily	Bare Costs	Incl. O&P
2 Glaziers	$49.65	$794.40	$74.75	$1196.00	$46.78	$70.53
1 Building Laborer	41.05	328.40	62.10	496.80		
24 L.H., Daily Totals		$1122.80		$1692.80	$46.78	$70.53

Crew H-3	Hr.	Daily	Hr.	Daily	Bare Costs	Incl. O&P
1 Glazier	$49.65	$397.20	$74.75	$598.00	$44.25	$67.20
1 Helper	38.85	310.80	59.65	477.20		
16 L.H., Daily Totals		$708.00		$1075.20	$44.25	$67.20

Crew H-4	Hr.	Daily	Hr.	Daily	Bare Costs	Incl. O&P
1 Carpenter	$51.65	$413.20	$78.15	$625.20	$48.21	$73.04
1 Carpenter Helper	38.85	310.80	59.65	477.20		
.5 Electrician	60.05	240.20	89.60	358.40		
20 L.H., Daily Totals		$964.20		$1460.80	$48.21	$73.04

Crews - Standard

Crew No.	Bare Costs		Incl. Subs O&P		Cost Per Labor-Hour	
Crew J-1	**Hr.**	**Daily**	**Hr.**	**Daily**	**Bare Costs**	**Incl. O&P**
3 Plasterers	$47.75	$1146.00	$71.40	$1713.60	$45.11	$67.46
2 Plasterer Helpers	41.15	658.40	61.55	984.80		
1 Mixing Machine, 6 C.F.		135.35		148.88	3.38	3.72
40 L.H., Daily Totals		$1939.75		$2847.28	$48.49	$71.18

Crew J-2	Hr.	Daily	Hr.	Daily	Bare Costs	Incl. O&P
3 Plasterers	$47.75	$1146.00	$71.40	$1713.60	$45.97	$68.52
2 Plasterer Helpers	41.15	658.40	61.55	984.80		
1 Lather	50.25	402.00	73.80	590.40		
1 Mixing Machine, 6 C.F.		135.35		148.88	2.82	3.10
48 L.H., Daily Totals		$2341.75		$3437.68	$48.79	$71.62

Crew J-3	Hr.	Daily	Hr.	Daily	Bare Costs	Incl. O&P
1 Terrazzo Worker	$47.85	$382.80	$70.55	$564.40	$44.10	$65.03
1 Terrazzo Helper	40.35	322.80	59.50	476.00		
1 Floor Grinder, 22" Path		138.30		152.13		
1 Terrazzo Mixer		178.85		196.74	19.82	21.80
16 L.H., Daily Totals		$1022.75		$1389.27	$63.92	$86.83

Crew J-4	Hr.	Daily	Hr.	Daily	Bare Costs	Incl. O&P
2 Cement Finishers	$48.90	$782.40	$72.20	$1155.20	$46.28	$68.83
1 Laborer	41.05	328.40	62.10	496.80		
1 Floor Grinder, 22" Path		138.30		152.13		
1 Floor Edger, 7" Path		43.45		47.80		
1 Vacuum Pick-Up System		72.15		79.36	10.58	11.64
24 L.H., Daily Totals		$1364.70		$1931.29	$56.86	$80.47

Crew J-4A	Hr.	Daily	Hr.	Daily	Bare Costs	Incl. O&P
2 Cement Finishers	$48.90	$782.40	$72.20	$1155.20	$44.98	$67.15
2 Laborers	41.05	656.80	62.10	993.60		
1 Floor Grinder, 22" Path		138.30		152.13		
1 Floor Edger, 7" Path		43.45		47.80		
1 Vacuum Pick-Up System		72.15		79.36		
1 Floor Auto Scrubber		265.15		291.67	16.22	17.84
32 L.H., Daily Totals		$1958.25		$2719.76	$61.20	$84.99

Crew J-4B	Hr.	Daily	Hr.	Daily	Bare Costs	Incl. O&P
1 Laborer	$41.05	$328.40	$62.10	$496.80	$41.05	$62.10
1 Floor Auto Scrubber		265.15		291.67	33.14	36.46
8 L.H., Daily Totals		$593.55		$788.47	$74.19	$98.56

Crew J-6	Hr.	Daily	Hr.	Daily	Bare Costs	Incl. O&P
2 Painters	$43.25	$692.00	$64.75	$1036.00	$44.80	$67.29
1 Building Laborer	41.05	328.40	62.10	496.80		
1 Equip. Oper. (light)	51.65	413.20	77.55	620.40		
1 Air Compressor, 250 cfm		167.95		184.75		
1 Sandblaster, Portable, 3 C.F.		20.70		22.77		
1 Set Sand Blasting Accessories		14.90		16.39	6.36	7.00
32 L.H., Daily Totals		$1637.15		$2377.11	$51.16	$74.28

Crew J-7	Hr.	Daily	Hr.	Daily	Bare Costs	Incl. O&P
2 Painters	$43.25	$692.00	$64.75	$1036.00	$43.25	$64.75
1 Floor Belt Sander		17.75		19.52		
1 Floor Sanding Edger		13.65		15.02	1.96	2.16
16 L.H., Daily Totals		$723.40		$1070.54	$45.21	$66.91

Crew No.	Bare Costs		Incl. Subs O&P		Cost Per Labor-Hour	
Crew K-1	**Hr.**	**Daily**	**Hr.**	**Daily**	**Bare Costs**	**Incl. O&P**
1 Carpenter	$51.65	$413.20	$78.15	$625.20	$48.58	$73.33
1 Truck Driver (light)	45.50	364.00	68.50	548.00		
1 Flatbed Truck, Gas, 3 Ton		245.95		270.55	15.37	16.91
16 L.H., Daily Totals		$1023.15		$1443.74	$63.95	$90.23

Crew K-2	Hr.	Daily	Hr.	Daily	Bare Costs	Incl. O&P
1 Struc. Steel Foreman (outside)	$58.00	$464.00	$92.50	$740.00	$53.17	$83.45
1 Struc. Steel Worker	56.00	448.00	89.35	714.80		
1 Truck Driver (light)	45.50	364.00	68.50	548.00		
1 Flatbed Truck, Gas, 3 Ton		245.95		270.55	10.25	11.27
24 L.H., Daily Totals		$1521.95		$2273.34	$63.41	$94.72

Crew L-1	Hr.	Daily	Hr.	Daily	Bare Costs	Incl. O&P
1 Electrician	$60.05	$480.40	$89.60	$716.80	$61.60	$92.17
1 Plumber	63.15	505.20	94.75	758.00		
16 L.H., Daily Totals		$985.60		$1474.80	$61.60	$92.17

Crew L-2	Hr.	Daily	Hr.	Daily	Bare Costs	Incl. O&P
1 Carpenter	$51.65	$413.20	$78.15	$625.20	$45.25	$68.90
1 Carpenter Helper	38.85	310.80	59.65	477.20		
16 L.H., Daily Totals		$724.00		$1102.40	$45.25	$68.90

Crew L-3	Hr.	Daily	Hr.	Daily	Bare Costs	Incl. O&P
1 Carpenter	$51.65	$413.20	$78.15	$625.20	$56.08	$84.61
.5 Electrician	60.05	240.20	89.60	358.40		
.5 Sheet Metal Worker	60.95	243.80	92.55	370.20		
16 L.H., Daily Totals		$897.20		$1353.80	$56.08	$84.61

Crew L-3A	Hr.	Daily	Hr.	Daily	Bare Costs	Incl. O&P
1 Carpenter Foreman (outside)	$53.65	$429.20	$81.15	$649.20	$56.08	$84.95
.5 Sheet Metal Worker	60.95	243.80	92.55	370.20		
12 L.H., Daily Totals		$673.00		$1019.40	$56.08	$84.95

Crew L-4	Hr.	Daily	Hr.	Daily	Bare Costs	Incl. O&P
2 Skilled Workers	$53.40	$854.40	$81.25	$1300.00	$48.55	$74.05
1 Helper	38.85	310.80	59.65	477.20		
24 L.H., Daily Totals		$1165.20		$1777.20	$48.55	$74.05

Crew L-5	Hr.	Daily	Hr.	Daily	Bare Costs	Incl. O&P
1 Struc. Steel Foreman (outside)	$58.00	$464.00	$92.50	$740.00	$56.49	$89.36
5 Struc. Steel Workers	56.00	2240.00	89.35	3574.00		
1 Equip. Oper. (crane)	57.45	459.60	86.25	690.00		
1 Hyd. Crane, 25 Ton		581.70		639.87	10.39	11.43
56 L.H., Daily Totals		$3745.30		$5643.87	$66.88	$100.78

Crew L-5A	Hr.	Daily	Hr.	Daily	Bare Costs	Incl. O&P
1 Struc. Steel Foreman (outside)	$58.00	$464.00	$92.50	$740.00	$56.86	$89.36
2 Structural Steel Workers	56.00	896.00	89.35	1429.60		
1 Equip. Oper. (crane)	57.45	459.60	86.25	690.00		
1 S.P. Crane, 4x4, 25 Ton		664.40		730.84	20.76	22.84
32 L.H., Daily Totals		$2484.00		$3590.44	$77.63	$112.20

Crew No.	Bare Costs		Incl. Subs O&P		Cost Per Labor-Hour	
Crew L-5B	**Hr.**	**Daily**	**Hr.**	**Daily**	**Bare Costs**	**Incl. O&P**
1 Struc. Steel Foreman (outside)	$58.00	$464.00	$92.50	$740.00	$58.29	$89.16
2 Structural Steel Workers	56.00	896.00	89.35	1429.60		
2 Electricians	60.05	960.80	89.60	1433.60		
2 Steamfitters/Pipefitters	63.95	1023.20	95.95	1535.20		
1 Equip. Oper. (crane)	57.45	459.60	86.25	690.00		
1 Equip. Oper. (oiler)	49.20	393.60	73.85	590.80		
1 Hyd. Crane, 80 Ton		1487.00		1635.70	20.65	22.72
72 L.H., Daily Totals		$5684.20		$8054.90	$78.95	$111.87
Crew L-6	**Hr.**	**Daily**	**Hr.**	**Daily**	**Bare Costs**	**Incl. O&P**
1 Plumber	$63.15	$505.20	$94.75	$758.00	$62.12	$93.03
.5 Electrician	60.05	240.20	89.60	358.40		
12 L.H., Daily Totals		$745.40		$1116.40	$62.12	$93.03
Crew L-7	**Hr.**	**Daily**	**Hr.**	**Daily**	**Bare Costs**	**Incl. O&P**
2 Carpenters	$51.65	$826.40	$78.15	$1250.40	$49.82	$75.20
1 Building Laborer	41.05	328.40	62.10	496.80		
.5 Electrician	60.05	240.20	89.60	358.40		
28 L.H., Daily Totals		$1395.00		$2105.60	$49.82	$75.20
Crew L-8	**Hr.**	**Daily**	**Hr.**	**Daily**	**Bare Costs**	**Incl. O&P**
2 Carpenters	$51.65	$826.40	$78.15	$1250.40	$53.95	$81.47
.5 Plumber	63.15	252.60	94.75	379.00		
20 L.H., Daily Totals		$1079.00		$1629.40	$53.95	$81.47
Crew L-9	**Hr.**	**Daily**	**Hr.**	**Daily**	**Bare Costs**	**Incl. O&P**
1 Labor Foreman (inside)	$41.55	$332.40	$62.85	$502.80	$46.59	$71.38
2 Building Laborers	41.05	656.80	62.10	993.60		
1 Struc. Steel Worker	56.00	448.00	89.35	714.80		
.5 Electrician	60.05	240.20	89.60	358.40		
36 L.H., Daily Totals		$1677.40		$2569.60	$46.59	$71.38
Crew L-10	**Hr.**	**Daily**	**Hr.**	**Daily**	**Bare Costs**	**Incl. O&P**
1 Struc. Steel Foreman (outside)	$58.00	$464.00	$92.50	$740.00	$57.15	$89.37
1 Structural Steel Worker	56.00	448.00	89.35	714.80		
1 Equip. Oper. (crane)	57.45	459.60	86.25	690.00		
1 Hyd. Crane, 12 Ton		471.75		518.92	19.66	21.62
24 L.H., Daily Totals		$1843.35		$2663.72	$76.81	$110.99
Crew L-11	**Hr.**	**Daily**	**Hr.**	**Daily**	**Bare Costs**	**Incl. O&P**
2 Wreckers	$41.05	$656.80	$63.55	$1016.80	$47.80	$72.72
1 Equip. Oper. (crane)	57.45	459.60	86.25	690.00		
1 Equip. Oper. (light)	51.65	413.20	77.55	620.40		
1 Hyd. Excavator, 2.5 C.Y.		1441.00		1585.10		
1 Loader, Skid Steer, 78 H.P.		392.50		431.75	57.30	63.03
32 L.H., Daily Totals		$3363.10		$4344.05	$105.10	$135.75
Crew M-1	**Hr.**	**Daily**	**Hr.**	**Daily**	**Bare Costs**	**Incl. O&P**
3 Elevator Constructors	$84.00	$2016.00	$124.90	$2997.60	$79.80	$118.65
1 Elevator Apprentice	67.20	537.60	99.90	799.20		
5 Hand Tools		49.50		54.45	1.55	1.70
32 L.H., Daily Totals		$2603.10		$3851.25	$81.35	$120.35

Crew No.	Bare Costs		Incl. Subs O&P		Cost Per Labor-Hour	
Crew M-3	**Hr.**	**Daily**	**Hr.**	**Daily**	**Bare Costs**	**Incl. O&P**
1 Electrician Foreman (outside)	$62.05	$496.40	$92.60	$740.80	$63.08	$94.16
1 Common Laborer	41.05	328.40	62.10	496.80		
.25 Equipment Operator (med.)	55.10	110.20	82.70	165.40		
1 Elevator Constructor	84.00	672.00	124.90	999.20		
1 Elevator Apprentice	67.20	537.60	99.90	799.20		
.25 S.P. Crane, 4x4, 20 Ton		146.94		161.63	4.32	4.75
34 L.H., Daily Totals		$2291.54		$3363.03	$67.40	$98.91
Crew M-4	**Hr.**	**Daily**	**Hr.**	**Daily**	**Bare Costs**	**Incl. O&P**
1 Electrician Foreman (outside)	$62.05	$496.40	$92.60	$740.80	$62.44	$93.23
1 Common Laborer	41.05	328.40	62.10	496.80		
.25 Equipment Operator, Crane	57.45	114.90	86.25	172.50		
.25 Equip. Oper. (oiler)	49.20	98.40	73.85	147.70		
1 Elevator Constructor	84.00	672.00	124.90	999.20		
1 Elevator Apprentice	67.20	537.60	99.90	799.20		
.25 S.P. Crane, 4x4, 40 Ton		187.56		206.32	5.21	5.73
36 L.H., Daily Totals		$2435.26		$3562.52	$67.65	$98.96
Crew Q-1	**Hr.**	**Daily**	**Hr.**	**Daily**	**Bare Costs**	**Incl. O&P**
1 Plumber	$63.15	$505.20	$94.75	$758.00	$56.83	$85.28
1 Plumber Apprentice	50.50	404.00	75.80	606.40		
16 L.H., Daily Totals		$909.20		$1364.40	$56.83	$85.28
Crew Q-1A	**Hr.**	**Daily**	**Hr.**	**Daily**	**Bare Costs**	**Incl. O&P**
.25 Plumber Foreman (outside)	$65.15	$130.30	$97.75	$195.50	$63.55	$95.35
1 Plumber	63.15	505.20	94.75	758.00		
10 L.H., Daily Totals		$635.50		$953.50	$63.55	$95.35
Crew Q-1C	**Hr.**	**Daily**	**Hr.**	**Daily**	**Bare Costs**	**Incl. O&P**
1 Plumber	$63.15	$505.20	$94.75	$758.00	$56.25	$84.42
1 Plumber Apprentice	50.50	404.00	75.80	606.40		
1 Equip. Oper. (medium)	55.10	440.80	82.70	661.60		
1 Trencher, Chain Type, 8' D		1837.00		2020.70	76.54	84.20
24 L.H., Daily Totals		$3187.00		$4046.70	$132.79	$168.61
Crew Q-2	**Hr.**	**Daily**	**Hr.**	**Daily**	**Bare Costs**	**Incl. O&P**
2 Plumbers	$63.15	$1010.40	$94.75	$1516.00	$58.93	$88.43
1 Plumber Apprentice	50.50	404.00	75.80	606.40		
24 L.H., Daily Totals		$1414.40		$2122.40	$58.93	$88.43
Crew Q-3	**Hr.**	**Daily**	**Hr.**	**Daily**	**Bare Costs**	**Incl. O&P**
1 Plumber Foreman (inside)	$63.65	$509.20	$95.50	$764.00	$60.11	$90.20
2 Plumbers	63.15	1010.40	94.75	1516.00		
1 Plumber Apprentice	50.50	404.00	75.80	606.40		
32 L.H., Daily Totals		$1923.60		$2886.40	$60.11	$90.20
Crew Q-4	**Hr.**	**Daily**	**Hr.**	**Daily**	**Bare Costs**	**Incl. O&P**
1 Plumber Foreman (inside)	$63.65	$509.20	$95.50	$764.00	$60.11	$90.20
1 Plumber	63.15	505.20	94.75	758.00		
1 Welder (plumber)	63.15	505.20	94.75	758.00		
1 Plumber Apprentice	50.50	404.00	75.80	606.40		
1 Welder, Electric, 300 amp		56.10		61.71	1.75	1.93
32 L.H., Daily Totals		$1979.70		$2948.11	$61.87	$92.13
Crew Q-5	**Hr.**	**Daily**	**Hr.**	**Daily**	**Bare Costs**	**Incl. O&P**
1 Steamfitter	$63.95	$511.60	$95.95	$767.60	$57.55	$86.35
1 Steamfitter Apprentice	51.15	409.20	76.75	614.00		
16 L.H., Daily Totals		$920.80		$1381.60	$57.55	$86.35

Crew No.	Bare Costs		Incl. Subs O&P		Cost Per Labor-Hour	
Crew Q-6	**Hr.**	**Daily**	**Hr.**	**Daily**	**Bare Costs**	**Incl. O&P**
2 Steamfitters	$63.95	$1023.20	$95.95	$1535.20	$59.68	$89.55
1 Steamfitter Apprentice	51.15	409.20	76.75	614.00		
24 L.H., Daily Totals		$1432.40		$2149.20	$59.68	$89.55

Crew No.	Bare Costs		Incl. Subs O&P		Cost Per Labor-Hour	
Crew Q-7	**Hr.**	**Daily**	**Hr.**	**Daily**	**Bare Costs**	**Incl. O&P**
1 Steamfitter Foreman (inside)	$64.45	$515.60	$96.70	$773.60	$60.88	$91.34
2 Steamfitters	63.95	1023.20	95.95	1535.20		
1 Steamfitter Apprentice	51.15	409.20	76.75	614.00		
32 L.H., Daily Totals		$1948.00		$2922.80	$60.88	$91.34

Crew No.	Bare Costs		Incl. Subs O&P		Cost Per Labor-Hour	
Crew Q-8	**Hr.**	**Daily**	**Hr.**	**Daily**	**Bare Costs**	**Incl. O&P**
1 Steamfitter Foreman (inside)	$64.45	$515.60	$96.70	$773.60	$60.88	$91.34
1 Steamfitter	63.95	511.60	95.95	767.60		
1 Welder (steamfitter)	63.95	511.60	95.95	767.60		
1 Steamfitter Apprentice	51.15	409.20	76.75	614.00		
1 Welder, Electric, 300 amp		56.10		61.71	1.75	1.93
32 L.H., Daily Totals		$2004.10		$2984.51	$62.63	$93.27

Crew No.	Bare Costs		Incl. Subs O&P		Cost Per Labor-Hour	
Crew Q-9	**Hr.**	**Daily**	**Hr.**	**Daily**	**Bare Costs**	**Incl. O&P**
1 Sheet Metal Worker	$60.95	$487.60	$92.55	$740.40	$54.85	$83.30
1 Sheet Metal Apprentice	48.75	390.00	74.05	592.40		
16 L.H., Daily Totals		$877.60		$1332.80	$54.85	$83.30

Crew No.	Bare Costs		Incl. Subs O&P		Cost Per Labor-Hour	
Crew Q-10	**Hr.**	**Daily**	**Hr.**	**Daily**	**Bare Costs**	**Incl. O&P**
2 Sheet Metal Workers	$60.95	$975.20	$92.55	$1480.80	$56.88	$86.38
1 Sheet Metal Apprentice	48.75	390.00	74.05	592.40		
24 L.H., Daily Totals		$1365.20		$2073.20	$56.88	$86.38

Crew No.	Bare Costs		Incl. Subs O&P		Cost Per Labor-Hour	
Crew Q-11	**Hr.**	**Daily**	**Hr.**	**Daily**	**Bare Costs**	**Incl. O&P**
1 Sheet Metal Foreman (inside)	$61.45	$491.60	$93.30	$746.40	$58.02	$88.11
2 Sheet Metal Workers	60.95	975.20	92.55	1480.80		
1 Sheet Metal Apprentice	48.75	390.00	74.05	592.40		
32 L.H., Daily Totals		$1856.80		$2819.60	$58.02	$88.11

Crew No.	Bare Costs		Incl. Subs O&P		Cost Per Labor-Hour	
Crew Q-12	**Hr.**	**Daily**	**Hr.**	**Daily**	**Bare Costs**	**Incl. O&P**
1 Sprinkler Installer	$61.50	$492.00	$92.50	$740.00	$55.35	$83.25
1 Sprinkler Apprentice	49.20	393.60	74.00	592.00		
16 L.H., Daily Totals		$885.60		$1332.00	$55.35	$83.25

Crew No.	Bare Costs		Incl. Subs O&P		Cost Per Labor-Hour	
Crew Q-13	**Hr.**	**Daily**	**Hr.**	**Daily**	**Bare Costs**	**Incl. O&P**
1 Sprinkler Foreman (inside)	$62.00	$496.00	$93.25	$746.00	$58.55	$88.06
2 Sprinkler Installers	61.50	984.00	92.50	1480.00		
1 Sprinkler Apprentice	49.20	393.60	74.00	592.00		
32 L.H., Daily Totals		$1873.60		$2818.00	$58.55	$88.06

Crew No.	Bare Costs		Incl. Subs O&P		Cost Per Labor-Hour	
Crew Q-14	**Hr.**	**Daily**	**Hr.**	**Daily**	**Bare Costs**	**Incl. O&P**
1 Asbestos Worker	$57.35	$458.80	$88.55	$708.40	$51.63	$79.70
1 Asbestos Apprentice	45.90	367.20	70.85	566.80		
16 L.H., Daily Totals		$826.00		$1275.20	$51.63	$79.70

Crew No.	Bare Costs		Incl. Subs O&P		Cost Per Labor-Hour	
Crew Q-15	**Hr.**	**Daily**	**Hr.**	**Daily**	**Bare Costs**	**Incl. O&P**
1 Plumber	$63.15	$505.20	$94.75	$758.00	$56.83	$85.28
1 Plumber Apprentice	50.50	404.00	75.80	606.40		
1 Welder, Electric, 300 amp		56.10		61.71	3.51	3.86
16 L.H., Daily Totals		$965.30		$1426.11	$60.33	$89.13

Crew No.	Bare Costs		Incl. Subs O&P		Cost Per Labor-Hour	
Crew Q-16	**Hr.**	**Daily**	**Hr.**	**Daily**	**Bare Costs**	**Incl. O&P**
2 Plumbers	$63.15	$1010.40	$94.75	$1516.00	$58.93	$88.43
1 Plumber Apprentice	50.50	404.00	75.80	606.40		
1 Welder, Electric, 300 amp		56.10		61.71	2.34	2.57
24 L.H., Daily Totals		$1470.50		$2184.11	$61.27	$91.00

Crew No.	Bare Costs		Incl. Subs O&P		Cost Per Labor-Hour	
Crew Q-17	**Hr.**	**Daily**	**Hr.**	**Daily**	**Bare Costs**	**Incl. O&P**
1 Steamfitter	$63.95	$511.60	$95.95	$767.60	$57.55	$86.35
1 Steamfitter Apprentice	51.15	409.20	76.75	614.00		
1 Welder, Electric, 300 amp		56.10		61.71	3.51	3.86
16 L.H., Daily Totals		$976.90		$1443.31	$61.06	$90.21

Crew No.	Bare Costs		Incl. Subs O&P		Cost Per Labor-Hour	
Crew Q-17A	**Hr.**	**Daily**	**Hr.**	**Daily**	**Bare Costs**	**Incl. O&P**
1 Steamfitter	$63.95	$511.60	$95.95	$767.60	$57.52	$86.32
1 Steamfitter Apprentice	51.15	409.20	76.75	614.00		
1 Equip. Oper. (crane)	57.45	459.60	86.25	690.00		
1 Hyd. Crane, 12 Ton		471.75		518.92		
1 Welder, Electric, 300 amp		56.10		61.71	21.99	24.19
24 L.H., Daily Totals		$1908.25		$2652.24	$79.51	$110.51

Crew No.	Bare Costs		Incl. Subs O&P		Cost Per Labor-Hour	
Crew Q-18	**Hr.**	**Daily**	**Hr.**	**Daily**	**Bare Costs**	**Incl. O&P**
2 Steamfitters	$63.95	$1023.20	$95.95	$1535.20	$59.68	$89.55
1 Steamfitter Apprentice	51.15	409.20	76.75	614.00		
1 Welder, Electric, 300 amp		56.10		61.71	2.34	2.57
24 L.H., Daily Totals		$1488.50		$2210.91	$62.02	$92.12

Crew No.	Bare Costs		Incl. Subs O&P		Cost Per Labor-Hour	
Crew Q-19	**Hr.**	**Daily**	**Hr.**	**Daily**	**Bare Costs**	**Incl. O&P**
1 Steamfitter	$63.95	$511.60	$95.95	$767.60	$58.38	$87.43
1 Steamfitter Apprentice	51.15	409.20	76.75	614.00		
1 Electrician	60.05	480.40	89.60	716.80		
24 L.H., Daily Totals		$1401.20		$2098.40	$58.38	$87.43

Crew No.	Bare Costs		Incl. Subs O&P		Cost Per Labor-Hour	
Crew Q-20	**Hr.**	**Daily**	**Hr.**	**Daily**	**Bare Costs**	**Incl. O&P**
1 Sheet Metal Worker	$60.95	$487.60	$92.55	$740.40	$55.89	$84.56
1 Sheet Metal Apprentice	48.75	390.00	74.05	592.40		
.5 Electrician	60.05	240.20	89.60	358.40		
20 L.H., Daily Totals		$1117.80		$1691.20	$55.89	$84.56

Crew No.	Bare Costs		Incl. Subs O&P		Cost Per Labor-Hour	
Crew Q-21	**Hr.**	**Daily**	**Hr.**	**Daily**	**Bare Costs**	**Incl. O&P**
2 Steamfitters	$63.95	$1023.20	$95.95	$1535.20	$59.77	$89.56
1 Steamfitter Apprentice	51.15	409.20	76.75	614.00		
1 Electrician	60.05	480.40	89.60	716.80		
32 L.H., Daily Totals		$1912.80		$2866.00	$59.77	$89.56

Crew No.	Bare Costs		Incl. Subs O&P		Cost Per Labor-Hour	
Crew Q-22	**Hr.**	**Daily**	**Hr.**	**Daily**	**Bare Costs**	**Incl. O&P**
1 Plumber	$63.15	$505.20	$94.75	$758.00	$56.83	$85.28
1 Plumber Apprentice	50.50	404.00	75.80	606.40		
1 Hyd. Crane, 12 Ton		471.75		518.92	29.48	32.43
16 L.H., Daily Totals		$1380.95		$1883.33	$86.31	$117.71

Crew No.	Bare Costs		Incl. Subs O&P		Cost Per Labor-Hour	
Crew Q-22A	**Hr.**	**Daily**	**Hr.**	**Daily**	**Bare Costs**	**Incl. O&P**
1 Plumber	$63.15	$505.20	$94.75	$758.00	$53.04	$79.72
1 Plumber Apprentice	50.50	404.00	75.80	606.40		
1 Laborer	41.05	328.40	62.10	496.80		
1 Equip. Oper. (crane)	57.45	459.60	86.25	690.00		
1 Hyd. Crane, 12 Ton		471.75		518.92	14.74	16.22
32 L.H., Daily Totals		$2168.95		$3070.13	$67.78	$95.94

Crews - Standard

Crew Q-23	Bare Costs Hr.	Bare Costs Daily	Incl. Subs O&P Hr.	Incl. Subs O&P Daily	Cost Per Labor-Hour Bare Costs	Cost Per Labor-Hour Incl. O&P
1 Plumber Foreman (outside)	$65.15	$521.20	$97.75	$782.00	$61.13	$91.73
1 Plumber	63.15	505.20	94.75	758.00		
1 Equip. Oper. (medium)	55.10	440.80	82.70	661.60		
1 Lattice Boom Crane, 20 Ton		869.00		955.90	36.21	39.83
24 L.H., Daily Totals		$2336.20		$3157.50	$97.34	$131.56

Crew R-1	Hr.	Daily	Hr.	Daily	Bare Costs	Incl. O&P
1 Electrician Foreman	$60.55	$484.40	$90.35	$722.80	$56.13	$83.76
3 Electricians	60.05	1441.20	89.60	2150.40		
2 Electrician Apprentices	48.05	768.80	71.70	1147.20		
48 L.H., Daily Totals		$2694.40		$4020.40	$56.13	$83.76

Crew R-1A	Hr.	Daily	Hr.	Daily	Bare Costs	Incl. O&P
1 Electrician	$60.05	$480.40	$89.60	$716.80	$54.05	$80.65
1 Electrician Apprentice	48.05	384.40	71.70	573.60		
16 L.H., Daily Totals		$864.80		$1290.40	$54.05	$80.65

Crew R-1B	Hr.	Daily	Hr.	Daily	Bare Costs	Incl. O&P
1 Electrician	$60.05	$480.40	$89.60	$716.80	$52.05	$77.67
2 Electrician Apprentices	48.05	768.80	71.70	1147.20		
24 L.H., Daily Totals		$1249.20		$1864.00	$52.05	$77.67

Crew R-1C	Hr.	Daily	Hr.	Daily	Bare Costs	Incl. O&P
2 Electricians	$60.05	$960.80	$89.60	$1433.60	$54.05	$80.65
2 Electrician Apprentices	48.05	768.80	71.70	1147.20		
1 Portable Cable Puller, 8000 lb.		152.70		167.97	4.77	5.25
32 L.H., Daily Totals		$1882.30		$2748.77	$58.82	$85.90

Crew R-2	Hr.	Daily	Hr.	Daily	Bare Costs	Incl. O&P
1 Electrician Foreman	$60.55	$484.40	$90.35	$722.80	$56.32	$84.11
3 Electricians	60.05	1441.20	89.60	2150.40		
2 Electrician Apprentices	48.05	768.80	71.70	1147.20		
1 Equip. Oper. (crane)	57.45	459.60	86.25	690.00		
1 S.P. Crane, 4x4, 5 Ton		259.45		285.39	4.63	5.10
56 L.H., Daily Totals		$3413.45		$4995.80	$60.95	$89.21

Crew R-3	Hr.	Daily	Hr.	Daily	Bare Costs	Incl. O&P
1 Electrician Foreman	$60.55	$484.40	$90.35	$722.80	$59.73	$89.23
1 Electrician	60.05	480.40	89.60	716.80		
.5 Equip. Oper. (crane)	57.45	229.80	86.25	345.00		
.5 S.P. Crane, 4x4, 5 Ton		129.72		142.70	6.49	7.13
20 L.H., Daily Totals		$1324.33		$1927.30	$66.22	$96.36

Crew R-4	Hr.	Daily	Hr.	Daily	Bare Costs	Incl. O&P
1 Struc. Steel Foreman (outside)	$58.00	$464.00	$92.50	$740.00	$57.21	$90.03
3 Struc. Steel Workers	56.00	1344.00	89.35	2144.40		
1 Electrician	60.05	480.40	89.60	716.80		
1 Welder, Gas Engine, 300 amp		95.65		105.22	2.39	2.63
40 L.H., Daily Totals		$2384.05		$3706.42	$59.60	$92.66

Crew R-5	Hr.	Daily	Hr.	Daily	Bare Costs	Incl. O&P
1 Electrician Foreman	$60.55	$484.40	$90.35	$722.80	$52.39	$78.78
4 Electrician Linemen	60.05	1921.60	89.60	2867.20		
2 Electrician Operators	60.05	960.80	89.60	1433.60		
4 Electrician Groundmen	38.85	1243.20	59.65	1908.80		
1 Crew Truck		154.35		169.79		
1 Flatbed Truck, 20,000 GVW		199.45		219.40		
1 Pickup Truck, 3/4 Ton		109.90		120.89		
.2 Hyd. Crane, 55 Ton		196.30		215.93		
.2 Hyd. Crane, 12 Ton		94.35		103.79		
.2 Earth Auger, Truck-Mtd.		77.17		84.89		
1 Tractor w/Winch		368.50		405.35	13.64	15.00
88 L.H., Daily Totals		$5810.02		$8252.42	$66.02	$93.78

Crew R-6	Hr.	Daily	Hr.	Daily	Bare Costs	Incl. O&P
1 Electrician Foreman	$60.55	$484.40	$90.35	$722.80	$52.39	$78.78
4 Electrician Linemen	60.05	1921.60	89.60	2867.20		
2 Electrician Operators	60.05	960.80	89.60	1433.60		
4 Electrician Groundmen	38.85	1243.20	59.65	1908.80		
1 Crew Truck		154.35		169.79		
1 Flatbed Truck, 20,000 GVW		199.45		219.40		
1 Pickup Truck, 3/4 Ton		109.90		120.89		
.2 Hyd. Crane, 55 Ton		196.30		215.93		
.2 Hyd. Crane, 12 Ton		94.35		103.79		
.2 Earth Auger, Truck-Mtd.		77.17		84.89		
1 Tractor w/Winch		368.50		405.35		
3 Cable Trailers		648.60		713.46		
.5 Tensioning Rig		214.32		235.76		
.5 Cable Pulling Rig		1233.50		1356.85	37.46	41.21
88 L.H., Daily Totals		$7906.44		$10558.49	$89.85	$119.98

Crew R-7	Hr.	Daily	Hr.	Daily	Bare Costs	Incl. O&P
1 Electrician Foreman	$60.55	$484.40	$90.35	$722.80	$42.47	$64.77
5 Electrician Groundmen	38.85	1554.00	59.65	2386.00		
1 Crew Truck		154.35		169.79	3.22	3.54
48 L.H., Daily Totals		$2192.75		$3278.59	$45.68	$68.30

Crew R-8	Hr.	Daily	Hr.	Daily	Bare Costs	Incl. O&P
1 Electrician Foreman	$60.55	$484.40	$90.35	$722.80	$53.07	$79.74
3 Electrician Linemen	60.05	1441.20	89.60	2150.40		
2 Electrician Groundmen	38.85	621.60	59.65	954.40		
1 Pickup Truck, 3/4 Ton		109.90		120.89		
1 Crew Truck		154.35		169.79	5.51	6.06
48 L.H., Daily Totals		$2811.45		$4118.27	$58.57	$85.80

Crew R-9	Hr.	Daily	Hr.	Daily	Bare Costs	Incl. O&P
1 Electrician Foreman	$60.55	$484.40	$90.35	$722.80	$49.51	$74.72
1 Electrician Lineman	60.05	480.40	89.60	716.80		
2 Electrician Operators	60.05	960.80	89.60	1433.60		
4 Electrician Groundmen	38.85	1243.20	59.65	1908.80		
1 Pickup Truck, 3/4 Ton		109.90		120.89		
1 Crew Truck		154.35		169.79	4.13	4.54
64 L.H., Daily Totals		$3433.05		$5072.68	$53.64	$79.26

Crew R-10	Hr.	Daily	Hr.	Daily	Bare Costs	Incl. O&P
1 Electrician Foreman	$60.55	$484.40	$90.35	$722.80	$56.60	$84.73
4 Electrician Linemen	60.05	1921.60	89.60	2867.20		
1 Electrician Groundman	38.85	310.80	59.65	477.20		
1 Crew Truck		154.35		169.79		
3 Tram Cars		438.60		482.46	12.35	13.59
48 L.H., Daily Totals		$3309.75		$4719.44	$68.95	$98.32

Crew R-11	Bare Costs Hr.	Bare Costs Daily	Incl. Subs O&P Hr.	Incl. Subs O&P Daily	Cost Per Labor-Hour Bare Costs	Cost Per Labor-Hour Incl. O&P
1 Electrician Foreman	$60.55	$484.40	$90.35	$722.80	$57.04	$85.30
4 Electricians	60.05	1921.60	89.60	2867.20		
1 Equip. Oper. (crane)	57.45	459.60	86.25	690.00		
1 Common Laborer	41.05	328.40	62.10	496.80		
1 Crew Truck		154.35		169.79		
1 Hyd. Crane, 12 Ton		471.75		518.92	11.18	12.30
56 L.H., Daily Totals		$3820.10		$5465.51	$68.22	$97.60

Crew R-12	Hr.	Daily	Hr.	Daily	Bare Costs	Incl. O&P
1 Carpenter Foreman (inside)	$52.15	$417.20	$78.90	$631.20	$48.55	$73.81
4 Carpenters	51.65	1652.80	78.15	2500.80		
4 Common Laborers	41.05	1313.60	62.10	1987.20		
1 Equip. Oper. (medium)	55.10	440.80	82.70	661.60		
1 Steel Worker	56.00	448.00	89.35	714.80		
1 Dozer, 200 H.P.		1290.00		1419.00		
1 Pickup Truck, 3/4 Ton		109.90		120.89	15.91	17.50
88 L.H., Daily Totals		$5672.30		$8035.49	$64.46	$91.31

Crew R-13	Hr.	Daily	Hr.	Daily	Bare Costs	Incl. O&P
1 Electrician Foreman	$60.55	$484.40	$90.35	$722.80	$57.95	$86.58
3 Electricians	60.05	1441.20	89.60	2150.40		
.25 Equip. Oper. (crane)	57.45	114.90	86.25	172.50		
1 Equipment Oiler	49.20	393.60	73.85	590.80		
.25 Hydraulic Crane, 33 Ton		236.65		260.32	5.63	6.20
42 L.H., Daily Totals		$2670.75		$3896.82	$63.59	$92.78

Crew R-15	Hr.	Daily	Hr.	Daily	Bare Costs	Incl. O&P
1 Electrician Foreman	$60.55	$484.40	$90.35	$722.80	$58.73	$87.72
4 Electricians	60.05	1921.60	89.60	2867.20		
1 Equipment Oper. (light)	51.65	413.20	77.55	620.40		
1 Telescoping Boom Lift, to 40'		283.15		311.46	5.90	6.49
48 L.H., Daily Totals		$3102.35		$4521.86	$64.63	$94.21

Crew R-15A	Hr.	Daily	Hr.	Daily	Bare Costs	Incl. O&P
1 Electrician Foreman	$60.55	$484.40	$90.35	$722.80	$52.40	$78.55
2 Electricians	60.05	960.80	89.60	1433.60		
2 Common Laborers	41.05	656.80	62.10	993.60		
1 Equip. Oper. (light)	51.65	413.20	77.55	620.40		
1 Telescoping Boom Lift, to 40'		283.15		311.46	5.90	6.49
48 L.H., Daily Totals		$2798.35		$4081.86	$58.30	$85.04

Crew R-18	Hr.	Daily	Hr.	Daily	Bare Costs	Incl. O&P
.25 Electrician Foreman	$60.55	$121.10	$90.35	$180.70	$52.70	$78.64
1 Electrician	60.05	480.40	89.60	716.80		
2 Electrician Apprentices	48.05	768.80	71.70	1147.20		
26 L.H., Daily Totals		$1370.30		$2044.70	$52.70	$78.64

Crew R-19	Hr.	Daily	Hr.	Daily	Bare Costs	Incl. O&P
.5 Electrician Foreman	$60.55	$242.20	$90.35	$361.40	$60.15	$89.75
2 Electricians	60.05	960.80	89.60	1433.60		
20 L.H., Daily Totals		$1203.00		$1795.00	$60.15	$89.75

Crew R-21	Hr.	Daily	Hr.	Daily	Bare Costs	Incl. O&P
1 Electrician Foreman	$60.55	$484.40	$90.35	$722.80	$60.05	$89.61
3 Electricians	60.05	1441.20	89.60	2150.40		
.1 Equip. Oper. (medium)	55.10	44.08	82.70	66.16		
.1 S.P. Crane, 4x4, 25 Ton		66.44		73.08	2.03	2.23
32.8 L.H., Daily Totals		$2036.12		$3012.44	$62.08	$91.84

Crew R-22	Hr.	Daily	Hr.	Daily	Bare Costs	Incl. O&P
.66 Electrician Foreman	$60.55	$319.70	$90.35	$477.05	$54.97	$82.02
2 Electricians	60.05	960.80	89.60	1433.60		
2 Electrician Apprentices	48.05	768.80	71.70	1147.20		
37.28 L.H., Daily Totals		$2049.30		$3057.85	$54.97	$82.02

Crew R-30	Hr.	Daily	Hr.	Daily	Bare Costs	Incl. O&P
.25 Electrician Foreman (outside)	$62.05	$124.10	$92.60	$185.20	$48.51	$72.91
1 Electrician	60.05	480.40	89.60	716.80		
2 Laborers (Semi-Skilled)	41.05	656.80	62.10	993.60		
26 L.H., Daily Totals		$1261.30		$1895.60	$48.51	$72.91

Crew R-31	Hr.	Daily	Hr.	Daily	Bare Costs	Incl. O&P
1 Electrician	$60.05	$480.40	$89.60	$716.80	$60.05	$89.60
1 Core Drill, Electric, 2.5 H.P.		46.55		51.20	5.82	6.40
8 L.H., Daily Totals		$526.95		$768.01	$65.87	$96.00

Crew W-41E	Hr.	Daily	Hr.	Daily	Bare Costs	Incl. O&P
.5 Plumber Foreman (outside)	$65.15	$260.60	$97.75	$391.00	$54.71	$82.29
1 Plumber	63.15	505.20	94.75	758.00		
1 Laborer	41.05	328.40	62.10	496.80		
20 L.H., Daily Totals		$1094.20		$1645.80	$54.71	$82.29

Historical Cost Indexes

The table below lists both the RSMeans® historical cost index based on Jan. 1, 1993 = 100 as well as the computed value of an index based on Jan. 1, 2019 costs. Since the Jan. 1, 2019 figure is estimated, space is left to write in the actual index figures as they become available through the quarterly *RSMeans Construction Cost Indexes*.

To compute the actual index based on Jan. 1, 2019 = 100, divide the historical cost index for a particular year by the actual Jan. 1, 2019 construction cost index. Space has been left to advance the index figures as the year progresses.

Year	Historical Cost Index Jan. 1, 1993 = 100		Current Index Based on Jan. 1, 2019 = 100		Year	Historical Cost Index Jan. 1, 1993 = 100	Current Index Based on Jan. 1, 2019 = 100		Year	Historical Cost Index Jan. 1, 1993 = 100	Current Index Based on Jan. 1, 2019 = 100	
	Est.	Actual	Est.	Actual		Actual	Est.	Actual		Actual	Est.	Actual
Oct 2019*					July 2004	143.7	63.2		July 1986	84.2	37.1	
July 2019*					2003	132.0	58.1		1985	82.6	36.3	
April 2019*					2002	128.7	56.6		1984	82.0	36.1	
Jan 2019*	227.3		100.0	100.0	2001	125.1	55.0		1983	80.2	35.3	
July 2018		222.9	98.1		2000	120.9	53.2		1982	76.1	33.5	
2017		213.6	94.0		1999	117.6	51.7		1981	70.0	30.8	
2016		207.3	91.2		1998	115.1	50.6		1980	62.9	27.7	
2015		206.2	90.7		1997	112.8	49.6		1979	57.8	25.4	
2014		204.9	90.1		1996	110.2	48.5		1978	53.5	23.5	
2013		201.2	88.5		1995	107.6	47.3		1977	49.5	21.8	
2012		194.6	85.6		1994	104.4	45.9		1976	46.9	20.6	
2011		191.2	84.1		1993	101.7	44.7		1975	44.8	19.7	
2010		183.5	80.7		1992	99.4	43.7		1974	41.4	18.2	
2009		180.1	79.2		1991	96.8	42.6		1973	37.7	16.6	
2008		180.4	79.4		1990	94.3	41.5		1972	34.8	15.3	
2007		169.4	74.5		1989	92.1	40.5		1971	32.1	14.1	
2006		162.0	71.3		1988	89.9	39.5		1970	28.7	12.6	
2005		151.6	66.7		1987	87.7	38.6		1969	26.9	11.8	

Adjustments to Costs

The "Historical Cost Index" can be used to convert national average building costs at a particular time to the approximate building costs for some other time.

Time Adjustment Using the Historical Cost Indexes:

$$\frac{\text{Index for Year A}}{\text{Index for Year B}} \times \text{Cost in Year B} = \text{Cost in Year A}$$

Example:

Estimate and compare construction costs for different years in the same city.

To estimate the national average construction cost of a building in 1970, knowing that it cost $900,000 in 2019:

INDEX in 1970 = 28.7

INDEX in 2019 = 227.3

$$\frac{\text{INDEX 1970}}{\text{INDEX 2019}} \times \text{Cost 2019} = \text{Cost 1970}$$

$$\frac{28.7}{227.3} \times \$900{,}000 = .126 \times \$900{,}000 = \$113{,}400$$

The construction cost of the building in 1970 was $113,400.

Note: The city cost indexes for Canada can be used to convert U.S. national averages to local costs in Canadian dollars.

Example:

To estimate and compare the cost of a building in Toronto, ON in 2019 with the known cost of $600,000 (US$) in New York, NY in 2019:

INDEX Toronto = 110.1

INDEX New York = 132.1

$$\frac{\text{INDEX Toronto}}{\text{INDEX New York}} \times \text{Cost New York} = \text{Cost Toronto}$$

$$\frac{110.1}{132.1} \times \$600{,}000 = .834 \times \$600{,}000 = \$500{,}076$$

The construction cost of the building in Toronto is $500,076 (CN$).

*Historical Cost Index updates and other resources are provided on the following website:
http://info.thegordiangroup.com/RSMeans.html

How to Use the City Cost Indexes

What you should know before you begin

RSMeans City Cost Indexes (CCI) are an extremely useful tool for when you want to compare costs from city to city and region to region.

This publication contains average construction cost indexes for 731 U.S. and Canadian cities covering over 930 three-digit zip code locations, as listed directly under each city.

Keep in mind that a City Cost Index number is a percentage ratio of a specific city's cost to the national average cost of the same item at a stated time period.

In other words, these index figures represent relative construction factors (or, if you prefer, multipliers) for material and installation costs, as well as the weighted average for Total In Place costs for each CSI MasterFormat division. Installation costs include both labor and equipment rental costs. When estimating equipment rental rates only for a specific location, use 01 54 33 EQUIPMENT RENTAL COSTS in the Reference Section.

The 30 City Average Index is the average of 30 major U.S. cities and serves as a national average.

Index figures for both material and installation are based on the 30 major city average of 100 and represent the cost relationship as of July 1, 2018. The index for each division is computed from representative material and labor quantities for that division. The weighted average for each city is a weighted total of the components listed above it. It does not include relative productivity between trades or cities.

As changes occur in local material prices, labor rates, and equipment rental rates (including fuel costs), the impact of these changes should be accurately measured by the change in the City Cost Index for each particular city (as compared to the 30 city average).

Therefore, if you know (or have estimated) building costs in one city today, you can easily convert those costs to expected building costs in another city.

In addition, by using the Historical Cost Index, you can easily convert national average building costs at a particular time to the approximate building costs for some other time. The City Cost Indexes can then be applied to calculate the costs for a particular city.

Quick calculations

Location Adjustment Using the City Cost Indexes:

$$\frac{\text{Index for City A}}{\text{Index for City B}} \times \text{Cost in City B} = \text{Cost in City A}$$

Time Adjustment for the National Average Using the Historical Cost Index:

$$\frac{\text{Index for Year A}}{\text{Index for Year B}} \times \text{Cost in Year B} = \text{Cost in Year A}$$

Adjustment from the National Average:

$$\frac{\text{Index for City A}}{100} \times \text{National Average Cost} = \text{Cost in City A}$$

Since each of the other RSMeans data sets contains many different items, any *one* item multiplied by the particular city index may give incorrect results. However, the larger the number of items compiled, the closer the results should be to actual costs for that particular city.

The City Cost Indexes for Canadian cities are calculated using Canadian material and equipment prices and labor rates in Canadian dollars. Therefore, indexes for Canadian cities can be used to convert U.S. national average prices to local costs in Canadian dollars.

How to use this section

1. Compare costs from city to city.

In using the RSMeans Indexes, remember that an index number is not a fixed number but a ratio: It's a percentage ratio of a building component's cost at any stated time to the national average cost of that same component at the same time period. Put in the form of an equation:

$$\frac{\text{Specific City Cost}}{\text{National Average Cost}} \times 100 = \text{City Index Number}$$

Therefore, when making cost comparisons between cities, do not subtract one city's index number from the index number of another city and read the result as a percentage difference. Instead, divide one city's index number by that of the other city. The resulting number may then be used as a multiplier to calculate cost differences from city to city.

The formula used to find cost differences between cities for the purpose of comparison is as follows:

$$\frac{\text{City A Index}}{\text{City B Index}} \times \text{City B Cost (Known)} = \text{City A Cost (Unknown)}$$

In addition, you can use RSMeans CCI to calculate and compare costs division by division between cities using the same basic formula. (Just be sure that you're comparing similar divisions.)

2. Compare a specific city's construction costs with the national average.

When you're studying construction location feasibility, it's advisable to compare a prospective project's cost index with an index of the national average cost.

For example, divide the weighted average index of construction costs of a specific city by that of the 30 City Average, which = 100.

$$\frac{\text{City Index}}{100} = \text{\% of National Average}$$

As a result, you get a ratio that indicates the relative cost of construction in that city in comparison with the national average.

3. Convert U.S. national average to actual costs in Canadian City.

$$\frac{\text{Index for Canadian City}}{100} \times \text{National Average Cost} = \text{Cost in Canadian City in \$ CAN}$$

4. **Adjust construction cost data based on a national average.**

When you use a source of construction cost data which is based on a national average (such as RSMeans cost data), it is necessary to adjust those costs to a specific location.

$$\frac{\text{City Index}}{100} \times \frac{\text{Cost Based on}}{\text{National Average Costs}} = \frac{\text{City Cost}}{\text{(Unknown)}}$$

5. **When applying the City Cost Indexes to demolition projects, use the appropriate division installation index.** For example, for removal of existing doors and windows, use the Division 8 (Openings) index.

What you might like to know about how we developed the Indexes

The information presented in the CCI is organized according to the Construction Specifications Institute (CSI) MasterFormat 2016 classification system.

To create a reliable index, RSMeans researched the building type most often constructed in the United States and Canada. Because it was concluded that no one type of building completely represented the building construction industry, nine different types of buildings were combined to create a composite model.

The exact material, labor, and equipment quantities are based on detailed analyses of these nine building types, and then each quantity is weighted in proportion to expected usage. These various material items, labor hours, and equipment rental rates are thus combined to form a composite building representing as closely as possible the actual usage of materials, labor, and equipment in the North American building construction industry.

The following structures were chosen to make up that composite model:

1. Factory, 1 story
2. Office, 2–4 stories
3. Store, Retail
4. Town Hall, 2–3 stories
5. High School, 2–3 stories
6. Hospital, 4–8 stories
7. Garage, Parking
8. Apartment, 1–3 stories
9. Hotel/Motel, 2–3 stories

For the purposes of ensuring the timeliness of the data, the components of the index for the composite model have been streamlined. They currently consist of:

- specific quantities of 66 commonly used construction materials;
- specific labor-hours for 21 building construction trades; and
- specific days of equipment rental for 6 types of construction equipment (normally used to install the 66 material items by the 21 trades.) Fuel costs and routine maintenance costs are included in the equipment cost.

Material and equipment price quotations are gathered quarterly from cities in the United States and Canada. These prices and the latest negotiated labor wage rates for 21 different building trades are used to compile the quarterly update of the City Cost Index.

The 30 major U.S. cities used to calculate the national average are:

Atlanta, GA	Memphis, TN
Baltimore, MD	Milwaukee, WI
Boston, MA	Minneapolis, MN
Buffalo, NY	Nashville, TN
Chicago, IL	New Orleans, LA
Cincinnati, OH	New York, NY
Cleveland, OH	Philadelphia, PA
Columbus, OH	Phoenix, AZ
Dallas, TX	Pittsburgh, PA
Denver, CO	St. Louis, MO
Detroit, MI	San Antonio, TX
Houston, TX	San Diego, CA
Indianapolis, IN	San Francisco, CA
Kansas City, MO	Seattle, WA
Los Angeles, CA	Washington, DC

What the CCI does not indicate

The weighted average for each city is a total of the divisional components weighted to reflect typical usage. It does not include the productivity variations between trades or cities.

In addition, the CCI does not take into consideration factors such as the following:

- managerial efficiency
- competitive conditions
- automation
- restrictive union practices
- unique local requirements
- regional variations due to specific building codes

	DIVISION	UNITED STATES			ALABAMA															
		30 CITY			ANNISTON			BIRMINGHAM			BUTLER			DECATUR			DOTHAN			
		AVERAGE			362			350 - 352			369			356			363			
		MAT.	INST.	TOTAL	MAT.	INST.	TOTAL	MAT.	INST.	TOTAL	MAT.	INST.	TOTAL	MAT.	INST.	TOTAL	MAT.	INST.	TOTAL	
015433	CONTRACTOR EQUIPMENT		100.0	100.0		104.5	104.5		104.7	104.7		102.1	102.1		104.5	104.5		102.1	102.1	
0241, 31 - 34	SITE & INFRASTRUCTURE, DEMOLITION	100.0	100.0	100.0	90.9	92.7	92.1	92.1	93.1	92.8	105.4	88.1	93.3	86.0	92.0	90.2	103.0	88.7	93.1	
0310	Concrete Forming & Accessories	100.0	100.0	100.0	87.3	65.3	68.6	90.7	66.5	70.1	83.8	65.9	68.6	91.5	61.4	65.9	92.2	68.2	71.8	
0320	Concrete Reinforcing	100.0	100.0	100.0	86.9	72.3	79.5	96.0	72.3	83.9	91.9	72.9	82.2	89.8	68.7	79.1	91.9	72.3	82.0	
0330	Cast-in-Place Concrete	100.0	100.0	100.0	87.8	66.3	79.8	102.6	68.4	89.8	85.6	66.6	78.5	100.7	66.1	87.8	85.6	68.0	79.0	
03	CONCRETE	100.0	100.0	100.0	88.4	68.6	79.6	90.6	69.9	81.4	89.3	69.1	80.3	90.4	66.1	79.6	88.6	70.6	80.6	
04	MASONRY	100.0	100.0	100.0	85.1	69.1	75.2	86.0	67.7	74.7	89.3	69.1	76.8	84.1	68.6	74.5	90.6	60.2	71.8	
05	METALS	100.0	100.0	100.0	99.8	94.5	98.2	99.6	95.0	98.2	98.7	95.5	97.7	101.9	93.2	99.3	98.7	95.4	97.7	
06	WOOD, PLASTICS & COMPOSITES	100.0	100.0	100.0	84.9	65.9	74.6	94.0	65.9	78.7	79.4	65.9	72.1	98.1	60.4	77.6	91.4	68.7	79.1	
07	THERMAL & MOISTURE PROTECTION	100.0	100.0	100.0	94.3	63.9	81.1	95.9	67.3	83.4	94.4	67.3	82.6	95.1	66.4	82.6	94.4	65.0	81.6	
08	OPENINGS	100.0	100.0	100.0	96.8	67.7	90.0	103.3	67.7	95.1	96.8	67.8	90.1	109.4	63.7	98.8	96.9	69.3	90.5	
0920	Plaster & Gypsum Board	100.0	100.0	100.0	90.5	65.5	74.0	92.0	65.5	74.5	87.1	65.5	72.9	100.4	59.8	73.6	97.8	68.3	78.4	
0950, 0980	Ceilings & Acoustic Treatment	100.0	100.0	100.0	80.8	65.5	70.4	82.1	65.5	70.8	80.8	65.5	70.4	82.4	59.8	67.1	80.8	68.3	72.3	
0960	Flooring	100.0	100.0	100.0	83.8	80.6	82.9	94.8	76.3	89.4	88.6	80.6	86.3	85.6	80.6	84.2	94.0	82.2	90.6	
0970, 0990	Wall Finishes & Painting/Coating	100.0	100.0	100.0	90.3	58.9	71.7	88.5	60.1	71.7	90.3	48.0	65.3	80.4	61.4	69.2	90.3	80.9	84.8	
09	FINISHES	100.0	100.0	100.0	80.0	67.8	73.4	86.5	67.4	76.1	83.0	66.7	74.1	82.3	64.7	72.7	85.9	72.3	78.4	
COVERS	DIVS. 10 - 14, 25, 28, 41, 43, 44, 46	100.0	100.0	100.0	100.0	70.2	93.4	100.0	83.4	96.3	100.0	77.0	94.9	100.0	69.5	93.3	100.0	72.0	93.8	
21, 22, 23	FIRE SUPPRESSION, PLUMBING & HVAC	100.0	100.0	100.0	100.7	50.4	80.4	100.0	66.4	86.4	98.1	65.7	85.0	100.1	64.7	85.8	98.1	65.4	84.9	
26, 27, 3370	ELECTRICAL, COMMUNICATIONS & UTIL.	100.0	100.0	100.0	96.6	58.4	77.2	95.3	61.2	77.9	98.7	65.4	81.8	91.2	64.0	77.4	97.5	75.3	86.2	
MF2016	WEIGHTED AVERAGE	100.0	100.0	100.0	95.2	67.4	83.3	96.5	71.7	85.9	95.5	71.6	85.3	96.6	70.1	85.2	95.6	72.9	85.9	

	DIVISION	ALABAMA																	
		EVERGREEN			GADSDEN			HUNTSVILLE			JASPER			MOBILE			MONTGOMERY		
		364			359			357 - 358			355			365 - 366			360 - 361		
		MAT.	INST.	TOTAL	MAT.	INST.	TOTAL	MAT.	INST.	TOTAL	MAT.	INST.	TOTAL	MAT.	INST.	TOTAL	MAT.	INST.	TOTAL
015433	CONTRACTOR EQUIPMENT		102.1	102.1		104.5	104.5		104.5	104.5		104.5	104.5		102.1	102.1		102.1	102.1
0241, 31 - 34	SITE & INFRASTRUCTURE, DEMOLITION	105.9	88.0	93.4	91.3	92.7	92.3	85.7	92.6	90.5	91.4	92.8	92.4	98.1	89.1	91.8	96.0	88.5	90.8
0310	Concrete Forming & Accessories	80.7	66.7	68.8	84.0	65.8	68.5	91.5	65.2	69.1	88.8	66.1	69.5	91.3	66.7	70.4	94.5	65.9	70.1
0320	Concrete Reinforcing	92.0	72.8	82.2	95.4	72.3	83.6	89.8	73.7	81.6	89.8	72.4	81.0	89.6	72.2	80.8	97.6	71.8	84.5
0330	Cast-in-Place Concrete	85.6	66.3	78.4	100.7	67.3	88.3	98.0	67.0	86.5	111.6	67.1	95.1	89.7	66.3	81.0	86.8	67.6	79.7
03	CONCRETE	89.6	69.3	80.6	94.6	69.2	83.3	89.2	69.0	80.3	98.3	69.2	85.4	85.3	69.2	78.2	83.7	69.2	77.3
04	MASONRY	89.3	68.7	76.5	82.6	65.8	72.3	85.2	65.4	73.0	80.4	70.3	74.1	87.9	58.8	69.9	85.2	59.2	69.2
05	METALS	98.7	95.4	97.7	99.8	95.0	98.4	101.9	95.3	99.9	99.8	94.9	98.3	100.7	94.7	98.9	99.6	94.4	98.0
06	WOOD, PLASTICS & COMPOSITES	76.0	67.4	71.3	88.5	65.9	76.3	98.1	65.3	80.3	95.2	65.9	79.3	89.9	67.4	77.7	93.3	65.9	78.5
07	THERMAL & MOISTURE PROTECTION	94.4	66.5	82.2	95.3	66.8	82.9	95.0	66.6	82.6	95.3	62.4	81.0	93.9	64.9	81.3	92.5	65.1	80.5
08	OPENINGS	96.8	68.6	90.3	105.9	67.7	97.0	109.1	67.7	99.5	105.9	67.7	97.0	99.5	68.6	92.3	98.8	67.7	91.6
0920	Plaster & Gypsum Board	86.5	67.0	73.7	92.6	65.5	74.7	100.4	64.8	77.0	97.2	65.5	76.3	94.2	67.0	76.3	89.9	65.5	73.8
0950, 0980	Ceilings & Acoustic Treatment	80.8	67.0	71.4	79.0	65.5	69.8	84.3	64.8	71.1	79.0	65.5	69.8	86.1	67.0	73.1	84.8	65.5	71.7
0960	Flooring	86.4	80.6	84.7	82.2	77.3	80.8	85.6	77.3	83.2	84.1	80.6	83.1	93.5	63.7	84.8	94.0	63.7	85.2
0970, 0990	Wall Finishes & Painting/Coating	90.3	48.0	65.3	80.4	58.9	67.7	80.4	67.3	72.6	80.4	58.9	67.7	93.6	49.9	67.7	91.3	58.9	72.1
09	FINISHES	82.3	67.4	74.2	80.0	67.1	72.9	82.7	67.5	74.4	81.1	68.1	74.0	85.8	64.2	74.0	86.1	64.4	74.2
COVERS	DIVS. 10 - 14, 25, 28, 41, 43, 44, 46	100.0	70.1	93.4	100.0	82.8	96.2	100.0	82.6	96.2	100.0	71.4	93.7	100.0	83.7	96.4	100.0	83.0	96.2
21, 22, 23	FIRE SUPPRESSION, PLUMBING & HVAC	98.1	65.5	84.9	102.0	62.8	86.2	100.1	63.5	85.3	102.0	66.2	87.5	100.0	62.8	84.9	100.0	65.0	85.8
26, 27, 3370	ELECTRICAL, COMMUNICATIONS & UTIL.	95.9	65.4	80.4	91.2	61.2	75.9	92.1	65.7	78.7	90.9	58.4	74.4	99.4	58.1	78.4	99.9	73.9	86.7
MF2016	WEIGHTED AVERAGE	95.2	71.4	85.0	96.6	70.6	85.5	96.5	71.4	85.7	97.1	71.0	85.9	96.1	68.8	84.4	95.6	71.4	85.2

	DIVISION	ALABAMA									ALASKA								
		PHENIX CITY			SELMA			TUSCALOOSA			ANCHORAGE			FAIRBANKS			JUNEAU		
		368			367			354			995 - 996			997			998		
		MAT.	INST.	TOTAL	MAT.	INST.	TOTAL	MAT.	INST.	TOTAL	MAT.	INST.	TOTAL	MAT.	INST.	TOTAL	MAT.	INST.	TOTAL
015433	CONTRACTOR EQUIPMENT		102.1	102.1		102.1	102.1		104.5	104.5		115.9	115.9		115.9	115.9		115.9	115.9
0241, 31 - 34	SITE & INFRASTRUCTURE, DEMOLITION	109.8	88.5	95.0	102.9	88.5	92.9	86.3	92.7	90.8	121.2	130.7	127.8	120.6	130.7	127.7	139.5	130.7	133.3
0310	Concrete Forming & Accessories	87.2	66.1	69.2	85.0	65.9	68.8	91.4	66.0	69.8	116.4	117.7	117.5	120.5	117.9	118.3	123.6	117.7	118.5
0320	Concrete Reinforcing	91.9	69.2	80.4	91.9	72.4	82.0	89.8	72.3	80.9	152.4	121.3	136.6	141.8	121.3	131.4	127.6	121.3	124.4
0330	Cast-in-Place Concrete	85.6	67.7	78.9	85.6	66.6	78.5	102.1	67.6	89.3	97.6	117.2	104.9	108.1	117.6	111.7	120.4	117.2	119.2
03	CONCRETE	92.3	68.9	81.9	88.1	69.0	79.6	91.1	69.4	81.5	108.3	117.3	112.3	101.3	117.6	108.5	115.2	117.3	116.1
04	MASONRY	89.2	69.1	76.8	92.5	69.1	78.1	84.3	66.3	73.2	183.5	119.5	144.0	185.0	119.8	144.7	157.6	119.5	134.1
05	METALS	98.6	94.2	97.3	98.6	95.1	97.6	101.1	95.0	99.3	119.7	104.8	115.2	120.8	104.9	116.0	117.1	104.8	113.4
06	WOOD, PLASTICS & COMPOSITES	84.6	65.8	74.4	81.6	65.9	73.1	98.1	65.9	80.6	114.0	115.8	115.0	125.8	115.8	120.4	123.4	115.8	119.3
07	THERMAL & MOISTURE PROTECTION	94.8	67.8	83.0	94.2	67.3	82.5	95.1	67.0	82.9	168.7	117.6	146.4	179.9	118.7	153.2	184.7	117.6	155.5
08	OPENINGS	96.8	66.8	89.8	96.8	67.7	90.0	109.1	67.7	99.5	130.6	116.5	127.3	135.8	116.5	131.4	130.4	116.5	127.2
0920	Plaster & Gypsum Board	91.7	65.3	74.3	89.9	65.5	73.8	100.4	65.5	77.4	142.8	116.1	125.2	172.8	116.1	135.4	155.4	116.1	129.5
0950, 0980	Ceilings & Acoustic Treatment	80.8	65.3	70.3	80.8	65.5	70.4	84.3	65.5	71.5	132.6	116.1	121.4	127.9	116.1	119.9	138.9	116.1	123.4
0960	Flooring	90.6	80.6	87.7	89.1	80.6	86.6	85.6	77.3	83.2	125.1	122.6	124.4	122.1	122.6	122.2	129.4	122.6	127.4
0970, 0990	Wall Finishes & Painting/Coating	90.3	81.0	84.8	90.3	58.9	71.7	80.4	58.9	67.7	108.7	113.2	111.3	113.7	117.7	116.1	110.8	113.2	112.2
09	FINISHES	84.7	70.2	76.8	83.2	67.8	74.8	82.6	67.2	74.2	125.7	118.3	121.7	128.6	118.9	123.3	129.3	118.3	123.3
COVERS	DIVS. 10 - 14, 25, 28, 41, 43, 44, 46	100.0	82.9	96.2	100.0	70.7	93.6	100.0	82.9	96.2	100.0	109.3	102.0	100.0	109.4	102.1	100.0	109.3	102.0
21, 22, 23	FIRE SUPPRESSION, PLUMBING & HVAC	98.1	65.0	84.7	98.1	64.3	84.5	100.1	66.5	86.5	100.6	106.1	102.8	100.4	110.3	104.4	100.6	106.1	102.8
26, 27, 3370	ELECTRICAL, COMMUNICATIONS & UTIL.	98.1	65.7	81.6	97.1	73.9	85.3	91.7	61.2	76.2	113.8	108.8	111.3	116.7	108.8	112.7	100.4	108.8	104.7
MF2016	WEIGHTED AVERAGE	96.1	72.0	85.8	95.3	72.4	85.5	96.6	71.5	85.8	117.5	113.6	115.8	118.2	114.7	116.7	116.5	113.6	115.3

City Cost Indexes

DIVISION		ALASKA			ARIZONA															
		KETCHIKAN			CHAMBERS			FLAGSTAFF			GLOBE			KINGMAN			MESA/TEMPE		
		999			865			860			855			864			852		
		MAT.	INST.	TOTAL	MAT.	INST.	TOTAL	MAT.	INST.	TOTAL	MAT.	INST.	TOTAL	MAT.	INST.	TOTAL	MAT.	INST.	TOTAL
015433	CONTRACTOR EQUIPMENT		115.9	115.9		91.1	91.1		91.1	91.1		92.2	92.2		91.1	91.1		92.2	92.2
0241, 31 - 34	SITE & INFRASTRUCTURE, DEMOLITION	175.1	130.7	144.1	71.9	95.0	88.0	90.9	95.0	93.8	102.3	95.9	97.8	71.8	95.0	88.0	92.4	95.9	94.9
0310	Concrete Forming & Accessories	112.3	117.7	116.9	97.1	68.0	72.3	102.5	69.2	74.1	93.4	68.0	71.8	95.3	67.9	72.0	96.5	68.1	72.3
0320	Concrete Reinforcing	108.4	121.3	114.9	103.2	79.1	91.0	103.0	79.0	90.8	109.0	79.1	93.8	103.4	79.1	91.0	109.7	79.1	94.2
0330	Cast-in-Place Concrete	218.8	117.2	181.0	89.7	69.4	82.2	89.8	69.4	82.2	84.1	69.0	78.5	89.5	69.4	82.0	84.8	69.1	79.0
03	CONCRETE	173.5	117.3	148.6	89.8	70.3	81.2	108.8	70.9	92.0	104.4	70.3	89.3	89.5	70.3	81.0	94.9	70.4	84.0
04	MASONRY	192.7	119.5	147.5	95.2	60.7	73.9	95.3	59.0	72.9	100.4	60.6	75.8	95.2	60.7	73.9	100.6	60.6	75.9
05	METALS	121.0	104.8	116.1	98.1	72.4	90.3	98.6	72.3	90.7	101.8	73.2	93.1	98.8	72.3	90.8	102.1	73.3	93.4
06	WOOD, PLASTICS & COMPOSITES	116.0	115.8	115.9	102.8	69.3	84.6	109.3	70.9	88.4	94.2	69.4	80.8	97.5	69.3	82.2	98.1	69.4	82.5
07	THERMAL & MOISTURE PROTECTION	184.9	117.6	155.6	96.8	70.1	85.1	98.8	69.6	86.1	98.7	68.7	85.7	96.7	69.9	85.0	98.5	68.4	85.4
08	OPENINGS	131.5	116.5	128.0	108.2	66.4	98.5	108.3	72.2	99.9	97.2	66.7	90.2	108.4	68.7	99.2	97.3	68.8	90.7
0920	Plaster & Gypsum Board	158.3	116.1	130.5	92.6	68.6	76.8	96.2	70.2	79.1	88.0	68.6	75.2	84.6	68.6	74.1	90.8	68.6	76.2
0950, 0980	Ceilings & Acoustic Treatment	121.3	116.1	117.8	107.2	68.6	81.1	108.1	70.2	82.4	93.5	68.6	76.6	108.1	68.6	81.3	93.5	68.6	76.6
0960	Flooring	122.1	122.6	122.2	89.9	58.8	80.9	92.1	58.8	82.4	104.0	58.8	90.9	88.7	58.8	80.0	105.6	58.8	92.0
0970, 0990	Wall Finishes & Painting/Coating	113.7	113.2	113.4	88.5	68.3	76.6	88.5	68.3	76.6	92.8	68.3	78.3	88.5	68.3	76.6	92.8	68.3	78.3
09	FINISHES	129.1	118.3	123.2	90.1	66.1	77.0	93.2	67.0	78.9	96.5	66.2	80.0	88.7	66.1	76.4	96.3	66.2	79.9
COVERS	DIVS. 10 - 14, 25, 28, 41, 43, 44, 46	100.0	109.3	102.0	100.0	83.0	96.3	100.0	83.1	96.3	100.0	83.2	96.3	100.0	82.9	96.2	100.0	83.2	96.3
21, 22, 23	FIRE SUPPRESSION, PLUMBING & HVAC	98.6	106.1	101.6	97.7	77.0	89.3	100.2	77.2	90.9	96.6	77.0	88.7	97.7	77.0	89.3	100.0	77.2	90.8
26, 27, 3370	ELECTRICAL, COMMUNICATIONS & UTIL.	116.6	108.8	112.7	102.7	83.6	93.0	101.7	62.2	81.6	100.5	64.4	82.1	102.7	64.4	83.2	97.2	64.4	80.5
MF2016	WEIGHTED AVERAGE	128.4	113.6	122.0	97.2	74.4	87.4	101.0	71.7	88.4	99.5	71.8	87.6	97.2	71.8	86.3	98.6	72.0	87.2

DIVISION		ARIZONA												ARKANSAS					
		PHOENIX			PRESCOTT			SHOW LOW			TUCSON			BATESVILLE			CAMDEN		
		850, 853			863			859			856 - 857			725			717		
		MAT.	INST.	TOTAL	MAT.	INST.	TOTAL	MAT.	INST.	TOTAL	MAT.	INST.	TOTAL	MAT.	INST.	TOTAL	MAT.	INST.	TOTAL
015433	CONTRACTOR EQUIPMENT		93.2	93.2		91.1	91.1		92.2	92.2		92.2	92.2		91.9	91.9		91.9	91.9
0241, 31 - 34	SITE & INFRASTRUCTURE, DEMOLITION	92.8	94.5	94.0	78.6	95.0	90.0	104.5	95.9	98.5	87.9	95.9	93.5	76.7	89.7	85.8	80.5	89.7	86.9
0310	Concrete Forming & Accessories	101.2	71.6	76.0	98.7	68.0	72.5	100.6	68.1	72.9	96.9	69.3	73.4	80.1	56.4	59.9	79.6	56.3	59.8
0320	Concrete Reinforcing	107.8	79.2	93.2	103.0	79.1	90.9	109.7	79.1	94.2	91.0	79.0	84.9	93.3	65.3	79.1	98.6	65.0	81.5
0330	Cast-in-Place Concrete	84.5	69.3	78.8	89.8	69.4	82.2	84.2	69.1	78.5	87.3	69.1	80.5	74.8	76.4	75.4	78.1	76.4	77.5
03	CONCRETE	97.6	72.1	86.3	95.0	70.3	84.1	106.8	70.3	90.7	93.7	70.9	83.6	74.0	65.7	70.3	76.7	65.7	71.8
04	MASONRY	93.5	62.7	74.5	95.3	60.7	73.9	100.4	60.6	75.8	87.5	58.9	69.8	92.1	42.3	61.4	104.9	36.9	62.9
05	METALS	103.5	75.2	94.9	98.7	72.3	90.7	101.6	73.2	93.0	102.8	73.2	93.8	98.6	72.8	90.8	103.4	72.4	94.0
06	WOOD, PLASTICS & COMPOSITES	101.1	72.9	85.7	104.4	69.5	85.4	102.7	69.4	84.6	98.3	71.0	83.5	92.9	57.2	73.5	91.0	57.2	72.6
07	THERMAL & MOISTURE PROTECTION	99.8	69.2	86.5	97.4	70.2	85.5	99.0	69.0	85.9	99.5	68.1	85.9	100.3	57.4	81.6	97.0	55.8	79.1
08	OPENINGS	98.6	72.8	92.6	108.3	66.7	98.7	96.5	68.8	90.1	93.7	72.3	88.7	102.0	53.1	90.7	105.6	54.9	93.8
0920	Plaster & Gypsum Board	98.0	71.9	80.8	93.0	68.8	77.1	92.6	68.6	76.8	95.3	70.2	78.8	78.6	56.4	64.0	87.9	56.4	67.1
0950, 0980	Ceilings & Acoustic Treatment	104.1	71.9	82.3	106.4	68.8	80.9	93.5	68.6	76.6	94.3	70.2	78.0	85.7	56.4	65.8	80.0	56.4	64.0
0960	Flooring	107.7	58.8	93.5	90.7	58.8	81.5	107.3	58.8	93.2	96.1	58.8	85.3	86.3	53.0	76.6	92.4	35.5	75.9
0970, 0990	Wall Finishes & Painting/Coating	97.5	65.0	78.3	88.5	68.3	76.6	92.8	68.3	78.3	93.3	68.3	78.5	90.7	53.5	68.7	94.5	53.5	70.2
09	FINISHES	100.9	68.5	83.2	90.7	66.2	77.3	98.3	66.2	80.8	94.3	67.2	79.5	76.9	54.9	64.9	80.4	51.7	64.7
COVERS	DIVS. 10 - 14, 25, 28, 41, 43, 44, 46	100.0	84.7	96.6	100.0	82.9	96.2	100.0	83.2	96.3	100.0	83.4	96.3	100.0	65.9	92.5	100.0	65.9	92.5
21, 22, 23	FIRE SUPPRESSION, PLUMBING & HVAC	99.9	78.6	91.2	100.2	77.0	90.8	96.6	77.0	88.7	100.0	77.1	90.7	96.6	52.2	78.6	96.5	56.9	80.5
26, 27, 3370	ELECTRICAL, COMMUNICATIONS & UTIL.	103.0	62.2	82.2	101.4	64.4	82.6	97.5	64.4	80.7	99.5	59.5	79.2	95.8	60.7	78.0	93.9	57.2	75.3
MF2016	WEIGHTED AVERAGE	100.0	73.0	88.4	98.7	71.7	87.1	99.7	71.9	87.8	97.6	71.4	86.4	92.6	60.0	78.7	94.8	59.6	79.7

DIVISION		ARKANSAS																	
		FAYETTEVILLE			FORT SMITH			HARRISON			HOT SPRINGS			JONESBORO			LITTLE ROCK		
		727			729			726			719			724			720 - 722		
		MAT.	INST.	TOTAL	MAT.	INST.	TOTAL	MAT.	INST.	TOTAL	MAT.	INST.	TOTAL	MAT.	INST.	TOTAL	MAT.	INST.	TOTAL
015433	CONTRACTOR EQUIPMENT		91.9	91.9		91.9	91.9		91.9	91.9		91.9	91.9		115.7	115.7		91.9	91.9
0241, 31 - 34	SITE & INFRASTRUCTURE, DEMOLITION	76.1	89.7	85.6	81.3	89.8	87.2	81.5	89.7	87.2	83.8	89.7	87.9	100.6	107.5	105.4	89.2	90.1	89.8
0310	Concrete Forming & Accessories	75.9	56.5	59.4	94.2	61.7	66.5	84.3	56.3	60.5	77.3	56.5	59.6	83.2	56.5	60.5	90.9	62.0	66.3
0320	Concrete Reinforcing	93.4	62.6	77.7	94.5	65.9	80.0	92.9	65.4	78.9	96.7	65.3	80.8	90.1	65.5	77.6	100.0	67.9	83.7
0330	Cast-in-Place Concrete	74.8	76.5	75.4	85.6	76.7	82.2	83.0	76.4	80.5	79.9	76.5	78.6	81.5	77.7	80.1	82.9	76.8	80.6
03	CONCRETE	73.7	65.3	70.0	81.1	68.3	75.4	80.6	65.7	74.0	79.9	65.8	73.6	78.2	67.3	73.4	81.9	68.9	76.1
04	MASONRY	83.5	41.8	57.8	90.1	62.1	72.8	92.6	43.3	62.1	78.8	42.0	56.1	84.8	39.9	57.1	86.0	62.1	71.2
05	METALS	98.6	72.0	90.5	100.8	74.6	92.8	99.7	72.8	91.6	103.3	72.9	94.1	95.0	87.8	92.8	101.2	75.6	93.4
06	WOOD, PLASTICS & COMPOSITES	89.2	57.2	71.8	110.0	63.6	84.8	98.9	57.2	76.2	88.4	57.2	71.4	97.1	57.5	75.6	96.5	63.6	78.6
07	THERMAL & MOISTURE PROTECTION	101.1	57.2	82.0	101.5	63.6	85.0	100.6	57.7	81.9	97.3	57.3	79.9	105.8	56.6	84.4	95.9	63.9	81.9
08	OPENINGS	102.0	54.9	91.1	104.1	60.1	93.8	102.8	52.0	91.0	105.5	54.8	93.8	107.8	56.0	95.8	97.2	59.7	88.5
0920	Plaster & Gypsum Board	78.0	56.4	63.8	84.8	63.0	70.4	83.9	56.4	65.8	86.6	56.4	66.7	89.9	56.4	67.8	94.5	63.0	73.8
0950, 0980	Ceilings & Acoustic Treatment	85.7	56.4	65.8	87.4	63.0	70.9	87.4	56.4	66.4	80.0	56.4	64.0	90.7	56.4	67.4	89.5	63.0	71.5
0960	Flooring	83.4	51.9	74.3	92.9	73.3	87.2	88.6	53.0	78.3	91.3	51.9	79.8	62.6	47.0	58.1	94.0	82.1	90.5
0970, 0990	Wall Finishes & Painting/Coating	90.7	53.5	68.7	90.7	53.5	68.7	90.7	53.5	68.7	94.5	53.5	70.2	80.6	53.5	64.6	91.2	53.5	68.9
09	FINISHES	76.1	54.5	64.3	80.3	63.1	70.9	79.0	54.7	65.8	80.2	54.7	66.3	74.9	53.8	63.4	86.3	64.9	74.6
COVERS	DIVS. 10 - 14, 25, 28, 41, 43, 44, 46	100.0	65.9	92.5	100.0	78.4	95.2	100.0	67.0	92.7	100.0	65.9	92.5	100.0	64.9	92.3	100.0	78.4	95.3
21, 22, 23	FIRE SUPPRESSION, PLUMBING & HVAC	96.6	61.2	82.3	100.1	49.1	79.5	96.6	48.5	77.1	96.5	52.5	78.7	100.4	52.2	80.9	99.8	52.6	80.7
26, 27, 3370	ELECTRICAL, COMMUNICATIONS & UTIL.	90.1	53.1	71.3	93.3	58.4	75.6	94.5	58.9	76.4	95.8	66.8	81.1	99.4	60.9	79.9	100.4	66.8	83.3
MF2016	WEIGHTED AVERAGE	91.5	60.7	78.3	95.1	63.6	81.6	94.0	59.1	79.0	94.2	61.0	79.9	94.7	62.7	81.0	95.4	65.9	82.8

	DIVISION	ARKANSAS												CALIFORNIA					
		PINE BLUFF			RUSSELLVILLE			TEXARKANA			WEST MEMPHIS			ALHAMBRA			ANAHEIM		
		716			728			718			723			917 - 918			928		
		MAT.	INST.	TOTAL	MAT.	INST.	TOTAL	MAT.	INST.	TOTAL	MAT.	INST.	TOTAL	MAT.	INST.	TOTAL	MAT.	INST.	TOTAL
015433	CONTRACTOR EQUIPMENT		91.9	91.9		91.9	91.9		93.2	93.2		115.7	115.7		97.7	97.7		101.8	101.8
0241, 31 - 34	SITE & INFRASTRUCTURE, DEMOLITION	85.8	89.8	88.6	77.9	89.7	86.1	96.6	91.9	93.3	107.4	107.7	107.6	99.9	110.3	107.1	100.5	110.6	107.5
0310	Concrete Forming & Accessories	77.0	61.9	64.1	80.8	56.3	59.9	82.6	56.4	60.3	88.5	56.8	61.5	115.3	140.0	136.3	101.2	140.1	134.3
0320	Concrete Reinforcing	98.4	67.9	82.9	94.0	65.2	79.4	98.0	65.3	81.4	90.1	63.7	76.7	102.5	130.1	116.5	95.4	129.9	112.9
0330	Cast-in-Place Concrete	79.9	76.7	78.7	78.4	76.4	77.6	87.1	76.4	83.2	85.5	77.8	82.6	83.4	129.8	100.7	85.2	133.2	103.1
03	CONCRETE	80.6	68.8	75.3	76.9	65.6	71.9	79.0	65.7	73.1	84.7	67.1	76.9	94.0	133.3	111.4	94.2	134.6	112.1
04	MASONRY	110.4	62.1	80.5	89.4	40.9	59.4	92.9	41.3	61.0	73.6	39.8	52.7	111.7	141.9	130.4	78.5	140.2	116.6
05	METALS	104.2	75.4	95.5	98.6	72.6	90.7	96.0	72.8	89.0	94.1	87.4	92.0	82.3	110.5	90.8	106.1	111.7	107.8
06	WOOD, PLASTICS & COMPOSITES	88.0	63.6	74.8	94.2	57.2	74.1	95.6	57.2	74.7	103.0	57.5	78.3	102.8	137.1	121.4	97.9	137.4	119.3
07	THERMAL & MOISTURE PROTECTION	97.4	64.0	82.9	101.4	57.0	82.0	98.0	57.1	80.2	106.3	56.6	84.7	97.2	133.2	112.9	106.8	136.3	119.7
08	OPENINGS	106.8	60.5	96.0	102.0	54.1	90.9	111.8	54.1	98.4	105.3	54.8	93.5	89.8	135.7	100.4	106.2	135.9	113.1
0920	Plaster & Gypsum Board	86.3	63.0	71.0	78.6	56.4	64.0	89.2	56.4	67.6	92.1	56.4	68.6	96.3	138.3	124.0	111.2	138.3	129.1
0950, 0980	Ceilings & Acoustic Treatment	80.0	63.0	68.5	85.7	56.4	65.8	83.4	56.4	65.1	88.8	56.4	66.8	112.6	138.3	130.0	112.6	138.3	130.0
0960	Flooring	91.1	73.3	85.9	85.9	53.0	76.3	93.2	44.8	79.1	64.9	47.0	59.7	100.0	120.4	105.9	97.3	120.4	104.0
0970, 0990	Wall Finishes & Painting/Coating	94.5	53.5	70.2	90.7	53.5	68.7	94.5	53.5	70.2	80.6	53.5	64.6	105.4	121.1	114.7	90.7	121.1	108.7
09	FINISHES	80.2	63.1	70.9	77.1	54.8	64.9	82.4	53.4	66.6	76.2	54.0	64.1	100.8	134.3	119.1	98.1	134.5	118.0
COVERS	DIVS. 10 - 14, 25, 28, 41, 43, 44, 46	100.0	78.4	95.2	100.0	65.9	92.5	100.0	65.9	92.5	100.0	66.7	92.7	100.0	119.6	104.3	100.0	120.1	104.4
21, 22, 23	FIRE SUPPRESSION, PLUMBING & HVAC	100.0	52.6	80.8	96.6	51.9	78.5	100.0	55.6	82.1	96.9	64.2	83.7	96.5	131.7	110.8	100.0	131.8	112.8
26, 27, 3370	ELECTRICAL, COMMUNICATIONS & UTIL.	94.0	58.9	76.2	93.3	58.9	75.8	95.7	58.9	77.0	100.9	62.7	81.5	117.4	130.5	124.1	90.2	111.0	100.8
MF2016	WEIGHTED AVERAGE	96.7	64.6	82.9	92.7	59.6	78.5	95.6	60.4	80.5	94.3	65.5	81.9	96.8	129.4	110.8	98.9	126.9	110.9

	DIVISION	CALIFORNIA																	
		BAKERSFIELD			BERKELEY			EUREKA			FRESNO			INGLEWOOD			LONG BEACH		
		932 - 933			947			955			936 - 938			903 - 905			906 - 908		
		MAT.	INST.	TOTAL	MAT.	INST.	TOTAL	MAT.	INST.	TOTAL	MAT.	INST.	TOTAL	MAT.	INST.	TOTAL	MAT.	INST.	TOTAL
015433	CONTRACTOR EQUIPMENT		99.8	99.8		102.8	102.8		99.6	99.6		99.8	99.8		99.0	99.0		99.0	99.0
0241, 31 - 34	SITE & INFRASTRUCTURE, DEMOLITION	97.6	108.4	105.1	111.6	110.7	110.9	111.8	107.4	108.8	101.6	107.8	106.0	90.5	106.7	101.8	97.6	106.7	104.0
0310	Concrete Forming & Accessories	101.0	138.9	133.3	112.6	167.2	159.1	109.4	152.9	146.4	98.3	152.1	144.1	104.1	140.3	134.9	99.2	140.3	134.2
0320	Concrete Reinforcing	103.5	129.9	116.9	89.1	131.8	110.8	103.9	131.3	117.8	86.5	130.8	109.0	103.0	130.1	116.8	102.1	130.1	116.4
0330	Cast-in-Place Concrete	89.3	132.3	105.3	111.5	133.6	119.7	92.5	129.8	106.3	96.8	129.4	108.9	79.5	132.3	99.2	90.5	132.3	106.1
03	CONCRETE	89.6	133.7	109.1	102.8	147.1	122.4	104.6	139.3	120.0	93.5	138.7	113.6	89.2	134.4	109.2	98.7	134.4	114.5
04	MASONRY	92.5	139.6	121.6	124.9	156.3	144.3	103.7	152.4	133.8	97.2	146.3	127.5	75.6	142.1	116.6	84.7	142.1	120.1
05	METALS	98.8	111.1	102.5	105.7	114.0	108.2	105.8	113.6	108.2	99.1	112.1	103.0	91.7	112.1	97.9	91.6	112.1	97.8
06	WOOD, PLASTICS & COMPOSITES	92.8	136.2	116.4	106.7	172.6	142.5	111.8	156.0	135.8	98.8	156.0	129.9	96.9	137.5	119.0	90.7	137.5	116.1
07	THERMAL & MOISTURE PROTECTION	103.9	128.2	114.5	107.4	157.2	129.1	110.6	150.6	128.0	95.0	134.4	112.2	100.6	134.9	115.5	100.9	134.9	115.7
08	OPENINGS	96.1	133.1	104.7	92.7	159.5	108.2	105.5	141.6	113.9	97.0	144.0	108.0	89.4	135.9	100.2	89.4	135.9	100.2
0920	Plaster & Gypsum Board	97.4	137.0	123.5	113.1	174.1	153.3	116.2	157.4	143.3	94.3	157.4	135.8	105.2	138.3	127.0	101.3	138.3	125.7
0950, 0980	Ceilings & Acoustic Treatment	104.7	137.0	126.6	98.5	174.1	149.8	116.0	157.4	144.0	103.9	157.4	140.1	110.8	138.3	129.5	110.8	138.3	129.5
0960	Flooring	102.9	120.4	108.0	115.9	142.0	123.5	101.1	142.0	113.0	102.1	118.0	106.7	101.8	120.4	107.2	98.9	120.4	105.1
0970, 0990	Wall Finishes & Painting/Coating	92.1	109.3	102.3	113.3	162.7	142.5	92.1	137.6	119.0	104.1	123.9	115.8	105.2	121.1	114.6	105.2	121.1	114.6
09	FINISHES	94.0	132.5	115.0	107.6	163.7	138.3	103.1	150.9	129.2	93.5	145.3	121.8	104.5	134.6	120.9	103.5	134.6	120.4
COVERS	DIVS. 10 - 14, 25, 28, 41, 43, 44, 46	100.0	120.2	104.4	100.0	129.9	106.6	100.0	126.7	105.9	100.0	126.7	105.9	100.0	120.4	104.5	100.0	120.4	104.5
21, 22, 23	FIRE SUPPRESSION, PLUMBING & HVAC	100.1	129.0	111.8	96.7	168.0	125.6	96.5	129.1	109.7	100.2	130.3	112.3	96.1	131.8	110.5	96.1	131.8	110.5
26, 27, 3370	ELECTRICAL, COMMUNICATIONS & UTIL.	107.1	110.7	109.0	102.3	161.0	132.1	96.8	124.3	110.8	96.8	108.0	102.5	100.3	130.5	115.7	100.0	130.5	115.5
MF2016	WEIGHTED AVERAGE	98.0	125.3	109.7	102.3	151.2	123.3	102.1	133.2	115.4	97.7	129.3	111.2	94.2	129.6	109.4	95.9	129.6	110.3

	DIVISION	CALIFORNIA																	
		LOS ANGELES			MARYSVILLE			MODESTO			MOJAVE			OAKLAND			OXNARD		
		900 - 902			959			953			935			946			930		
		MAT.	INST.	TOTAL	MAT.	INST.	TOTAL	MAT.	INST.	TOTAL	MAT.	INST.	TOTAL	MAT.	INST.	TOTAL	MAT.	INST.	TOTAL
015433	CONTRACTOR EQUIPMENT		105.2	105.2		99.6	99.6		99.6	99.6		99.8	99.8		102.8	102.8		98.7	98.7
0241, 31 - 34	SITE & INFRASTRUCTURE, DEMOLITION	96.1	112.4	107.4	108.3	107.3	107.6	103.0	107.3	106.0	94.7	108.4	104.3	116.5	110.7	112.4	101.9	106.5	105.1
0310	Concrete Forming & Accessories	102.2	140.6	134.9	99.8	152.5	144.6	96.2	152.3	143.9	108.3	138.8	134.3	101.9	167.2	157.5	100.5	140.2	134.3
0320	Concrete Reinforcing	103.8	130.2	117.2	103.9	130.6	117.5	107.6	130.5	119.3	104.1	129.9	117.2	91.1	131.8	111.8	102.3	129.8	116.3
0330	Cast-in-Place Concrete	82.1	133.4	101.2	103.3	129.5	113.1	92.5	129.4	106.2	83.7	132.3	101.8	105.8	133.6	116.2	97.5	132.5	110.6
03	CONCRETE	96.5	135.0	113.6	105.0	138.9	120.0	96.1	138.8	115.1	85.4	133.6	106.8	103.5	147.1	122.8	93.5	134.3	111.6
04	MASONRY	91.0	142.1	122.5	104.7	141.7	127.5	102.1	141.7	126.6	95.4	139.6	122.7	132.8	156.3	147.3	98.3	138.0	122.8
05	METALS	97.8	114.5	102.9	105.3	112.3	107.4	102.2	112.1	105.2	96.7	111.0	101.0	101.8	114.0	105.5	94.6	111.3	99.6
06	WOOD, PLASTICS & COMPOSITES	99.9	137.7	120.4	98.8	156.0	129.9	94.3	156.0	127.8	98.4	136.2	118.9	94.6	172.6	137.0	93.3	137.4	117.3
07	THERMAL & MOISTURE PROTECTION	96.9	134.8	113.4	110.0	139.1	122.7	109.5	135.5	120.8	100.8	125.6	111.6	105.9	157.2	128.3	104.2	134.3	117.3
08	OPENINGS	98.9	136.1	107.5	104.7	144.9	114.1	103.5	144.9	113.1	92.0	133.1	101.6	92.8	159.5	108.3	94.6	135.9	104.2
0920	Plaster & Gypsum Board	100.5	138.3	125.4	108.7	157.4	140.8	111.2	157.4	141.6	103.8	137.0	125.7	107.2	174.1	151.3	97.8	138.3	124.5
0950, 0980	Ceilings & Acoustic Treatment	117.7	138.3	131.7	115.1	157.4	143.8	112.6	157.4	143.0	106.1	137.0	127.1	101.3	174.1	150.6	107.7	138.3	128.4
0960	Flooring	100.0	120.4	105.9	96.8	123.5	104.6	97.2	129.9	106.7	102.3	120.4	107.5	109.6	142.0	119.0	94.9	120.4	102.3
0970, 0990	Wall Finishes & Painting/Coating	101.9	121.1	113.2	92.1	137.6	119.0	92.1	137.6	119.0	88.9	109.3	100.9	113.3	162.7	142.5	88.9	115.6	104.7
09	FINISHES	105.5	134.8	121.5	100.1	147.7	126.1	99.6	148.8	126.5	93.8	132.5	114.9	105.9	163.7	137.5	91.3	133.9	114.6
COVERS	DIVS. 10 - 14, 25, 28, 41, 43, 44, 46	100.0	121.1	104.6	100.0	126.7	105.9	100.0	126.7	105.9	100.0	120.2	104.4	100.0	129.9	106.6	100.0	120.4	104.5
21, 22, 23	FIRE SUPPRESSION, PLUMBING & HVAC	99.9	131.9	112.8	96.5	130.3	110.2	100.0	130.3	112.2	96.6	129.0	109.7	100.2	168.0	127.6	100.1	131.8	112.9
26, 27, 3370	ELECTRICAL, COMMUNICATIONS & UTIL.	98.6	130.5	114.8	93.6	122.4	108.3	96.0	108.4	102.3	95.3	110.7	103.1	101.5	161.0	131.8	101.0	117.5	109.4
MF2016	WEIGHTED AVERAGE	98.7	130.4	112.3	101.2	131.3	114.1	100.2	129.4	112.7	94.7	125.2	107.7	102.7	151.2	123.5	97.2	127.1	110.0

City Cost Indexes

	DIVISION	CALIFORNIA																	
		PALM SPRINGS			PALO ALTO			PASADENA			REDDING			RICHMOND			RIVERSIDE		
		922			943			910 - 912			960			948			925		
		MAT.	INST.	TOTAL	MAT.	INST.	TOTAL	MAT.	INST.	TOTAL	MAT.	INST.	TOTAL	MAT.	INST.	TOTAL	MAT.	INST.	TOTAL
015433	CONTRACTOR EQUIPMENT		100.7	100.7		102.8	102.8		97.7	97.7		99.6	99.6		102.8	102.8		100.7	100.7
0241, 31 - 34	SITE & INFRASTRUCTURE, DEMOLITION	92.1	108.6	103.6	107.8	110.7	109.8	96.5	110.3	106.1	129.0	107.3	113.9	115.7	110.6	112.2	99.0	108.6	105.7
0310	Concrete Forming & Accessories	98.0	140.1	133.8	100.0	167.2	157.2	103.7	140.0	134.6	102.8	152.7	145.3	115.4	166.8	159.2	101.8	140.1	134.4
0320	Concrete Reinforcing	109.3	129.9	119.8	89.1	131.1	110.5	103.4	130.1	116.9	132.1	130.6	131.3	89.1	131.3	110.5	106.2	129.9	118.2
0330	Cast-in-Place Concrete	81.2	133.2	100.6	94.4	133.6	109.0	79.1	129.8	98.0	101.6	129.5	112.0	108.5	133.5	117.8	88.4	133.2	105.0
03	CONCRETE	88.5	134.6	108.9	92.8	147.0	116.9	89.7	133.3	109.0	108.0	139.0	121.8	105.4	146.8	123.8	94.4	134.6	112.2
04	MASONRY	76.7	139.8	115.7	105.9	146.3	130.8	97.9	141.9	125.1	132.2	141.7	138.1	124.6	150.4	140.6	77.7	139.8	116.1
05	METALS	106.7	111.7	108.2	99.4	113.3	103.6	82.3	110.5	90.9	112.5	112.3	112.5	99.4	113.2	103.6	106.2	111.7	107.8
06	WOOD, PLASTICS & COMPOSITES	92.6	137.4	116.9	92.2	172.6	135.9	88.1	137.1	114.7	110.1	156.3	135.2	110.0	172.6	144.0	97.9	137.4	119.3
07	THERMAL & MOISTURE PROTECTION	106.1	136.2	119.2	105.1	154.4	126.6	96.8	133.2	112.6	131.9	139.2	135.1	105.8	154.5	127.0	107.0	136.2	119.7
08	OPENINGS	102.1	135.9	109.9	92.8	159.5	108.3	89.8	135.7	100.4	118.1	143.8	124.0	92.8	159.5	108.3	104.8	135.9	112.0
0920	Plaster & Gypsum Board	106.2	138.3	127.4	105.7	174.1	150.7	90.3	138.3	121.9	108.9	157.8	141.1	114.2	174.1	153.7	110.5	138.3	128.8
0950, 0980	Ceilings & Acoustic Treatment	109.2	138.3	128.9	99.4	174.1	150.0	112.6	138.3	130.0	148.5	157.8	154.8	99.4	174.1	150.0	118.7	138.3	132.0
0960	Flooring	99.4	120.4	105.5	108.2	142.0	118.0	94.2	120.4	101.8	93.5	132.9	105.0	118.2	142.0	125.1	100.9	120.4	106.6
0970, 0990	Wall Finishes & Painting/Coating	89.1	121.1	108.1	113.3	162.7	142.5	105.4	121.1	114.7	100.0	137.6	122.3	113.3	162.7	142.5	89.1	121.1	108.1
09	FINISHES	96.6	134.5	117.3	104.3	163.7	136.8	98.0	134.3	117.8	106.4	149.4	129.8	109.0	163.7	138.9	99.8	134.5	118.7
COVERS	DIVS. 10 - 14, 25, 28, 41, 43, 44, 46	100.0	120.2	104.4	100.0	129.9	106.6	100.0	119.6	104.3	100.0	126.8	105.9	100.0	129.8	106.6	100.0	120.2	104.4
21, 22, 23	FIRE SUPPRESSION, PLUMBING & HVAC	96.5	131.8	110.8	96.7	168.7	125.8	96.5	131.7	110.8	100.4	130.3	112.5	96.7	159.7	122.2	100.0	131.8	112.8
26, 27, 3370	ELECTRICAL, COMMUNICATIONS & UTIL.	93.2	113.5	103.5	101.4	169.9	136.2	114.1	130.5	122.4	103.9	122.4	113.3	102.0	136.5	119.5	90.0	113.5	101.9
MF2016	WEIGHTED AVERAGE	96.9	127.1	109.8	98.5	151.4	121.2	94.8	129.4	109.7	109.1	131.5	118.7	101.8	145.2	120.4	98.8	127.1	110.9

	DIVISION	CALIFORNIA																	
		SACRAMENTO			SALINAS			SAN BERNARDINO			SAN DIEGO			SAN FRANCISCO			SAN JOSE		
		942, 956 - 958			939			923 - 924			919 - 921			940 - 941			951		
		MAT.	INST.	TOTAL	MAT.	INST.	TOTAL	MAT.	INST.	TOTAL	MAT.	INST.	TOTAL	MAT.	INST.	TOTAL	MAT.	INST.	TOTAL
015433	CONTRACTOR EQUIPMENT		101.6	101.6		99.8	99.8		100.7	100.7		102.5	102.5		110.9	110.9		101.3	101.3
0241, 31 - 34	SITE & INFRASTRUCTURE, DEMOLITION	97.8	116.3	110.7	115.4	107.9	110.2	78.4	108.6	99.5	104.7	107.8	106.8	117.5	114.2	115.2	134.2	102.3	112.0
0310	Concrete Forming & Accessories	100.2	155.0	146.9	103.8	155.3	147.7	105.4	140.1	134.9	104.4	128.0	124.5	105.7	167.7	158.5	102.0	166.9	157.3
0320	Concrete Reinforcing	84.6	130.9	108.2	102.8	131.0	117.2	106.2	129.9	118.2	100.1	129.8	115.2	103.7	132.6	118.4	94.9	131.4	113.5
0330	Cast-in-Place Concrete	89.5	130.5	104.7	96.3	129.8	108.8	61.0	133.2	87.9	90.6	123.5	102.9	118.1	133.8	123.9	107.5	132.7	116.9
03	CONCRETE	95.1	140.1	115.0	103.0	140.4	119.6	69.5	134.6	98.3	100.1	125.8	111.5	114.7	148.1	129.5	102.0	147.0	121.9
04	MASONRY	106.8	147.2	131.8	95.4	147.3	127.5	84.0	139.8	118.5	89.8	132.3	116.0	142.9	153.6	149.5	132.3	150.4	143.5
05	METALS	97.6	107.5	100.6	99.2	113.2	103.4	106.1	111.7	107.8	97.8	113.7	102.6	106.7	122.6	111.5	100.3	119.0	106.0
06	WOOD, PLASTICS & COMPOSITES	88.9	159.3	127.1	98.1	159.0	131.2	101.6	137.4	121.0	99.4	124.9	113.2	97.0	172.6	138.1	106.5	172.4	142.3
07	THERMAL & MOISTURE PROTECTION	115.1	141.0	126.4	101.5	145.2	120.5	105.1	136.2	118.7	101.2	120.3	109.5	111.4	155.7	130.7	105.9	155.6	127.6
08	OPENINGS	105.8	146.7	115.3	95.7	152.0	108.8	102.1	135.9	110.0	98.3	126.1	104.7	98.5	159.8	112.7	94.6	159.4	109.7
0920	Plaster & Gypsum Board	103.2	160.5	140.9	98.9	160.5	139.5	111.8	138.3	129.3	96.3	125.3	115.4	105.3	174.1	150.6	107.3	174.1	151.3
0950, 0980	Ceilings & Acoustic Treatment	99.4	160.5	140.8	106.1	160.5	143.0	112.6	138.3	130.0	123.1	125.3	124.6	108.6	174.1	153.0	113.6	174.1	154.6
0960	Flooring	108.2	127.6	113.9	96.9	140.6	109.6	102.8	120.8	108.0	103.6	120.4	108.4	110.0	142.0	119.3	90.8	142.0	105.7
0970, 0990	Wall Finishes & Painting/Coating	109.9	137.6	126.3	89.7	162.7	132.9	89.1	121.1	108.1	103.0	121.1	113.7	111.2	179.4	151.6	92.4	162.7	134.0
09	FINISHES	103.2	150.4	128.9	93.4	155.2	127.1	98.0	134.5	117.9	105.5	125.5	116.4	107.5	165.6	139.2	98.9	163.6	134.2
COVERS	DIVS. 10 - 14, 25, 28, 41, 43, 44, 46	100.0	127.6	106.1	100.0	127.1	106.0	100.0	120.2	104.4	100.0	117.0	103.7	100.0	130.0	106.6	100.0	129.4	106.5
21, 22, 23	FIRE SUPPRESSION, PLUMBING & HVAC	100.1	129.9	112.2	96.6	136.3	112.6	96.5	131.8	110.8	99.9	128.4	111.4	100.1	185.2	134.5	100.0	168.6	127.7
26, 27, 3370	ELECTRICAL, COMMUNICATIONS & UTIL.	97.0	122.4	109.9	96.3	127.6	112.2	93.2	111.8	102.7	100.6	102.7	101.7	101.6	176.3	139.6	98.9	169.9	135.0
MF2016	WEIGHTED AVERAGE	100.1	132.7	114.1	98.3	135.7	114.3	94.5	126.8	108.4	99.6	120.8	108.7	106.3	158.1	128.5	102.1	151.7	123.4

	DIVISION	CALIFORNIA																	
		SAN LUIS OBISPO			SAN MATEO			SAN RAFAEL			SANTA ANA			SANTA BARBARA			SANTA CRUZ		
		934			944			949			926 - 927			931			950		
		MAT.	INST.	TOTAL	MAT.	INST.	TOTAL	MAT.	INST.	TOTAL	MAT.	INST.	TOTAL	MAT.	INST.	TOTAL	MAT.	INST.	TOTAL
015433	CONTRACTOR EQUIPMENT		99.8	99.8		102.8	102.8		102.0	102.0		100.7	100.7		99.8	99.8		101.3	101.3
0241, 31 - 34	SITE & INFRASTRUCTURE, DEMOLITION	107.5	108.4	108.1	113.6	110.7	111.5	110.5	116.2	114.5	90.5	108.6	103.1	101.9	108.4	106.4	133.8	102.2	111.7
0310	Concrete Forming & Accessories	110.0	140.2	135.7	105.9	167.1	158.0	111.1	167.0	158.7	105.7	140.1	135.0	101.1	140.1	134.3	102.0	155.6	147.6
0320	Concrete Reinforcing	104.1	129.8	117.2	89.1	131.6	110.7	89.7	131.6	111.0	109.8	129.9	120.1	102.3	129.8	116.3	117.4	131.1	124.3
0330	Cast-in-Place Concrete	103.6	132.4	114.3	105.0	133.6	115.7	122.2	132.6	126.1	77.9	133.2	98.5	97.2	132.4	110.3	106.7	131.6	116.0
03	CONCRETE	101.0	134.3	115.8	102.0	147.0	122.0	124.7	146.4	134.3	86.1	134.6	107.6	93.4	134.3	111.5	104.3	141.4	120.7
04	MASONRY	97.1	137.3	121.9	124.3	153.6	142.4	101.5	153.6	133.7	73.6	140.2	114.7	95.7	137.3	121.4	136.1	147.5	143.1
05	METALS	97.4	111.2	101.6	99.3	113.7	103.7	100.4	110.0	103.3	106.2	111.7	107.9	95.1	111.2	100.0	107.8	117.6	110.8
06	WOOD, PLASTICS & COMPOSITES	100.5	137.4	120.6	99.7	172.6	139.3	97.4	172.4	138.1	103.5	137.4	121.9	93.3	137.4	117.3	106.5	159.2	135.1
07	THERMAL & MOISTURE PROTECTION	101.6	133.7	115.6	105.5	155.9	127.5	109.9	156.2	130.0	106.5	136.3	119.5	101.1	133.7	115.3	105.5	147.5	123.8
08	OPENINGS	93.9	133.8	103.2	92.8	159.5	108.3	103.4	159.4	116.4	101.4	135.9	109.4	95.5	135.9	104.9	95.9	152.1	108.9
0920	Plaster & Gypsum Board	104.7	138.3	126.8	110.8	174.1	152.5	112.7	174.1	153.2	113.3	138.3	129.8	97.8	138.3	124.5	115.3	160.5	145.1
0950, 0980	Ceilings & Acoustic Treatment	106.1	138.3	127.9	99.4	174.1	150.0	106.2	174.1	152.2	112.6	138.3	130.0	107.7	138.3	128.4	113.4	160.5	145.3
0960	Flooring	103.1	120.4	108.1	111.8	142.0	120.6	123.4	140.6	128.4	103.4	120.4	108.3	96.0	120.4	103.1	94.8	140.6	108.1
0970, 0990	Wall Finishes & Painting/Coating	88.9	115.6	104.7	113.3	162.7	142.5	109.3	169.9	145.2	89.1	121.1	108.1	88.9	115.6	104.7	92.5	162.7	134.0
09	FINISHES	95.1	133.9	116.3	106.5	163.7	137.8	109.1	164.1	139.2	99.5	134.5	118.6	91.8	133.9	114.8	101.2	155.2	130.7
COVERS	DIVS. 10 - 14, 25, 28, 41, 43, 44, 46	100.0	126.4	105.8	100.0	129.9	106.6	100.0	129.2	106.4	100.0	120.2	104.4	100.0	119.5	104.3	100.0	127.4	106.0
21, 22, 23	FIRE SUPPRESSION, PLUMBING & HVAC	96.6	131.8	110.8	96.7	163.6	123.8	96.7	185.1	132.5	96.5	131.8	110.8	100.1	131.8	112.9	100.0	136.7	114.8
26, 27, 3370	ELECTRICAL, COMMUNICATIONS & UTIL.	95.3	113.1	104.3	101.4	163.0	132.7	98.3	124.3	111.5	93.2	112.8	103.2	94.2	114.2	104.3	98.1	127.6	113.1
MF2016	WEIGHTED AVERAGE	97.5	126.6	110.0	100.9	150.2	122.1	103.9	149.3	123.4	96.5	127.0	109.6	96.5	126.7	109.4	104.0	136.0	117.7

	DIVISION	CALIFORNIA															COLORADO		
		SANTA ROSA			STOCKTON			SUSANVILLE			VALLEJO			VAN NUYS			ALAMOSA		
		954			952			961			945			913 - 916			811		
		MAT.	INST.	TOTAL	MAT.	INST.	TOTAL	MAT.	INST.	TOTAL	MAT.	INST.	TOTAL	MAT.	INST.	TOTAL	MAT.	INST.	TOTAL
015433	CONTRACTOR EQUIPMENT		100.1	100.1		99.6	99.6		99.6	99.6		102.0	102.0		97.7	97.7		93.1	93.1
0241, 31 - 34	SITE & INFRASTRUCTURE, DEMOLITION	104.5	107.3	106.5	102.8	107.3	105.9	137.1	107.3	116.4	99.1	116.1	111.0	114.9	110.3	111.7	141.1	88.5	104.5
0310	Concrete Forming & Accessories	99.0	166.2	156.2	100.1	154.6	146.5	104.0	152.5	145.3	101.7	165.9	156.4	110.6	140.0	135.6	101.3	62.7	68.4
0320	Concrete Reinforcing	104.8	131.5	118.4	107.6	130.5	119.3	132.1	130.4	131.2	90.8	131.4	111.5	103.4	130.1	116.9	111.5	69.8	90.3
0330	Cast-in-Place Concrete	101.5	131.1	112.5	90.1	129.5	104.8	92.5	129.5	106.3	97.4	131.7	110.2	83.5	129.8	100.7	100.1	73.5	90.2
03	CONCRETE	105.0	145.9	123.1	95.3	139.9	115.0	110.9	138.9	123.3	99.6	145.6	120.0	104.4	133.3	117.2	107.8	68.4	90.4
04	MASONRY	102.3	152.5	133.3	102.0	141.7	126.5	130.7	139.2	136.0	77.4	152.5	123.8	111.7	141.9	130.4	129.7	60.6	87.0
05	METALS	106.5	115.1	109.1	102.4	112.2	105.4	111.4	112.1	111.6	100.3	109.3	103.1	81.5	110.5	90.3	100.3	78.4	93.7
06	WOOD, PLASTICS & COMPOSITES	94.1	172.2	136.5	99.9	159.0	132.0	111.9	156.0	135.9	86.6	172.4	133.2	97.2	137.1	118.9	99.6	63.0	79.7
07	THERMAL & MOISTURE PROTECTION	106.9	152.9	126.9	110.0	138.1	122.2	133.8	138.4	135.8	107.9	153.6	127.8	97.9	133.2	113.3	107.7	68.2	90.5
08	OPENINGS	103.0	159.3	116.1	103.5	146.6	113.5	119.0	144.9	125.0	105.1	159.4	117.7	89.6	135.7	100.3	98.5	66.4	91.0
0920	Plaster & Gypsum Board	108.2	174.1	151.6	111.2	160.5	143.7	109.6	157.4	141.1	106.9	174.1	151.2	94.4	138.3	123.3	81.5	61.9	68.6
0950, 0980	Ceilings & Acoustic Treatment	112.6	174.1	154.3	122.1	160.5	148.1	140.0	157.4	151.8	108.1	174.1	152.8	109.2	138.3	128.9	103.8	61.9	75.4
0960	Flooring	99.9	133.5	109.7	97.2	129.9	106.7	94.0	132.9	105.3	117.5	142.0	124.6	97.3	120.4	104.0	108.4	71.3	97.6
0970, 0990	Wall Finishes & Painting/Coating	89.1	169.9	136.9	92.1	137.6	119.0	100.0	137.6	122.3	110.3	169.9	145.6	105.4	121.1	114.7	102.7	60.8	77.9
09	FINISHES	98.6	162.5	133.5	101.5	150.6	128.3	106.0	149.3	129.6	105.8	164.1	137.6	100.1	134.3	118.8	98.4	63.5	79.4
COVERS	DIVS. 10 - 14, 25, 28, 41, 43, 44, 46	100.0	128.3	106.2	100.0	127.0	106.0	100.0	126.7	105.9	100.0	128.9	106.4	100.0	119.6	104.3	100.0	84.8	96.6
21, 22, 23	FIRE SUPPRESSION, PLUMBING & HVAC	96.5	184.5	132.1	100.0	130.3	112.2	96.9	130.3	110.4	100.2	147.9	119.5	96.5	131.7	110.8	96.5	66.0	84.2
26, 27, 3370	ELECTRICAL, COMMUNICATIONS & UTIL.	93.5	124.3	109.2	96.0	114.3	105.3	104.3	122.4	113.5	94.5	127.9	111.4	114.1	130.5	122.4	97.0	64.8	80.6
MF2016	WEIGHTED AVERAGE	100.8	148.5	121.2	100.3	130.8	113.4	108.8	131.2	118.4	99.5	141.5	117.5	97.9	129.4	111.4	102.3	68.8	87.9

	DIVISION	COLORADO																	
		BOULDER			COLORADO SPRINGS			DENVER			DURANGO			FORT COLLINS			FORT MORGAN		
		803			808 - 809			800 - 802			813			805			807		
		MAT.	INST.	TOTAL	MAT.	INST.	TOTAL	MAT.	INST.	TOTAL	MAT.	INST.	TOTAL	MAT.	INST.	TOTAL	MAT.	INST.	TOTAL
015433	CONTRACTOR EQUIPMENT		95.2	95.2		93.1	93.1		99.4	99.4		93.1	93.1		95.2	95.2		95.2	95.2
0241, 31 - 34	SITE & INFRASTRUCTURE, DEMOLITION	99.4	95.8	96.9	101.3	91.3	94.3	106.3	102.3	103.5	134.6	88.5	102.5	112.4	95.7	100.7	101.9	95.5	97.4
0310	Concrete Forming & Accessories	102.1	73.6	77.8	92.8	65.1	69.2	101.4	64.8	70.3	107.4	62.5	69.2	99.8	68.6	73.3	102.6	62.4	68.4
0320	Concrete Reinforcing	98.3	69.8	83.8	97.5	69.9	83.5	97.5	69.9	83.5	111.5	69.7	90.3	98.3	69.7	83.8	98.5	69.7	83.9
0330	Cast-in-Place Concrete	112.5	72.8	97.7	115.5	73.9	100.0	120.5	71.3	102.2	115.1	73.4	99.6	127.0	71.4	106.3	110.3	72.7	96.3
03	CONCRETE	101.1	73.0	88.7	104.5	69.6	89.1	107.2	68.6	90.1	110.1	68.2	91.5	112.0	70.3	93.5	99.6	67.9	85.6
04	MASONRY	101.3	63.3	77.8	103.2	63.6	78.7	104.4	63.8	79.3	116.7	60.6	82.1	119.2	60.9	83.2	116.7	60.9	82.2
05	METALS	90.8	77.4	86.7	93.8	78.1	89.0	96.1	78.6	90.8	100.3	78.2	93.6	92.0	77.3	87.6	90.5	77.3	86.5
06	WOOD, PLASTICS & COMPOSITES	106.7	76.3	90.2	96.1	65.5	79.5	104.3	65.3	83.1	109.5	63.0	84.2	104.2	70.9	86.1	106.7	62.5	82.7
07	THERMAL & MOISTURE PROTECTION	106.0	72.7	91.5	106.7	71.4	91.4	104.4	70.9	89.8	107.7	68.2	90.5	106.3	70.4	90.7	105.9	69.6	90.1
08	OPENINGS	97.5	73.4	91.9	101.7	67.8	93.8	104.5	67.7	96.0	105.4	66.4	96.4	97.4	70.4	91.2	97.4	66.1	90.1
0920	Plaster & Gypsum Board	116.5	76.0	89.8	100.0	64.8	76.8	111.7	64.8	80.8	95.3	61.9	73.3	110.7	70.5	84.2	116.5	61.9	80.5
0950, 0980	Ceilings & Acoustic Treatment	86.2	76.0	79.3	93.0	64.8	73.9	97.5	64.8	75.3	103.8	61.9	75.4	86.2	70.5	75.5	86.2	61.9	69.7
0960	Flooring	108.3	71.3	97.6	99.7	71.3	91.5	105.9	71.3	95.9	113.4	71.3	101.2	104.6	71.3	95.0	108.7	71.3	97.9
0970, 0990	Wall Finishes & Painting/Coating	99.4	63.0	77.8	99.1	63.0	77.7	106.2	63.0	80.6	102.7	60.8	77.9	99.4	60.8	76.5	99.4	60.8	76.5
09	FINISHES	100.5	72.2	85.0	97.0	65.5	79.8	102.3	65.3	82.1	101.0	63.5	80.6	99.1	68.2	82.2	100.6	63.2	80.2
COVERS	DIVS. 10 - 14, 25, 28, 41, 43, 44, 46	100.0	86.1	96.9	100.0	84.7	96.6	100.0	84.2	96.5	100.0	84.7	96.6	100.0	84.7	96.6	100.0	83.8	96.4
21, 22, 23	FIRE SUPPRESSION, PLUMBING & HVAC	96.6	74.6	87.7	100.2	73.8	89.5	100.0	73.6	89.4	96.5	57.2	80.6	100.1	73.3	89.2	96.6	73.3	87.2
26, 27, 3370	ELECTRICAL, COMMUNICATIONS & UTIL.	98.0	80.0	88.9	101.4	74.9	87.9	103.2	80.2	91.5	96.4	51.5	73.6	98.0	77.2	87.4	98.3	80.0	89.0
MF2016	WEIGHTED AVERAGE	97.7	75.9	88.3	100.0	73.0	88.4	101.8	74.4	90.0	102.7	65.0	86.6	101.1	73.7	89.4	98.3	72.9	87.4

	DIVISION	COLORADO																	
		GLENWOOD SPRINGS			GOLDEN			GRAND JUNCTION			GREELEY			MONTROSE			PUEBLO		
		816			804			815			806			814			810		
		MAT.	INST.	TOTAL	MAT.	INST.	TOTAL	MAT.	INST.	TOTAL	MAT.	INST.	TOTAL	MAT.	INST.	TOTAL	MAT.	INST.	TOTAL
015433	CONTRACTOR EQUIPMENT		96.3	96.3		95.2	95.2		96.3	96.3		95.2	95.2		94.7	94.7		93.1	93.1
0241, 31 - 34	SITE & INFRASTRUCTURE, DEMOLITION	150.4	96.4	112.8	112.9	95.8	101.0	134.0	96.6	107.9	98.0	95.7	96.4	143.8	92.5	108.0	125.5	88.9	100.0
0310	Concrete Forming & Accessories	98.2	52.4	59.2	95.2	63.4	68.1	106.3	73.0	78.0	97.7	72.6	76.3	97.7	62.7	67.9	103.6	65.4	71.1
0320	Concrete Reinforcing	110.3	69.7	89.7	98.5	69.7	83.9	110.7	69.8	89.9	98.3	69.7	83.8	110.2	69.8	89.6	106.6	69.8	87.9
0330	Cast-in-Place Concrete	100.1	72.3	89.7	110.4	74.1	96.9	110.8	73.7	97.0	106.1	72.7	93.7	100.1	72.9	90.0	99.4	74.2	90.0
03	CONCRETE	112.8	63.3	90.8	109.5	68.9	91.5	106.7	73.2	91.8	96.3	72.6	85.8	104.1	68.2	88.2	97.1	69.9	85.0
04	MASONRY	101.6	60.8	76.4	119.7	63.3	84.9	136.8	64.0	91.8	112.5	60.9	80.7	109.5	60.6	79.3	97.8	61.6	75.4
05	METALS	99.9	77.9	93.3	90.7	77.3	86.7	101.5	78.2	94.5	92.0	77.4	87.6	99.1	78.3	92.8	103.2	79.0	95.9
06	WOOD, PLASTICS & COMPOSITES	94.6	49.2	69.9	98.5	62.5	79.0	107.2	75.7	90.1	101.3	76.2	87.7	95.6	62.8	77.8	102.3	65.8	82.5
07	THERMAL & MOISTURE PROTECTION	107.6	66.7	89.8	107.1	71.3	91.6	106.6	71.6	91.3	105.5	71.0	90.5	107.7	68.2	90.5	106.0	69.5	90.1
08	OPENINGS	104.4	58.8	93.8	97.5	66.1	90.2	105.1	73.1	97.7	97.4	73.7	91.9	105.6	66.3	96.5	100.3	67.9	92.8
0920	Plaster & Gypsum Board	122.5	48.0	73.4	108.0	61.9	77.6	135.9	75.2	96.0	109.2	76.0	87.3	80.6	61.9	68.3	84.9	64.8	71.7
0950, 0980	Ceilings & Acoustic Treatment	103.0	48.0	65.7	86.2	61.9	69.7	103.0	75.2	84.2	86.2	76.0	79.3	103.8	61.9	75.4	109.8	64.8	79.3
0960	Flooring	107.5	71.3	97.0	102.3	71.3	93.3	112.8	71.3	100.7	103.5	71.3	94.2	110.8	71.3	99.4	109.5	75.5	99.7
0970, 0990	Wall Finishes & Painting/Coating	102.7	60.8	77.9	99.4	63.0	77.8	102.7	63.0	79.2	99.4	60.8	76.5	102.7	60.8	77.9	102.7	60.8	77.9
09	FINISHES	104.3	55.4	77.6	98.9	64.1	79.9	106.0	71.8	87.3	97.8	71.3	83.3	99.0	63.4	79.6	98.6	66.3	80.9
COVERS	DIVS. 10 - 14, 25, 28, 41, 43, 44, 46	100.0	82.6	96.2	100.0	84.6	96.6	100.0	86.3	97.0	100.0	85.3	96.8	100.0	84.4	96.6	100.0	85.4	96.8
21, 22, 23	FIRE SUPPRESSION, PLUMBING & HVAC	96.5	67.0	84.6	96.6	74.6	87.7	99.9	75.8	90.1	100.1	73.3	89.2	96.5	75.8	88.1	99.9	73.8	89.4
26, 27, 3370	ELECTRICAL, COMMUNICATIONS & UTIL.	93.9	51.5	72.3	98.4	80.0	89.0	96.1	51.5	73.4	98.0	77.2	87.4	96.1	51.5	73.4	97.0	64.8	80.6
MF2016	WEIGHTED AVERAGE	102.5	65.4	86.6	99.8	73.7	88.6	104.5	72.2	90.7	98.3	74.7	88.2	101.4	69.3	87.6	100.5	71.4	88.0

	DIVISION	COLORADO			CONNECTICUT															
		SALIDA			BRIDGEPORT			BRISTOL			HARTFORD			MERIDEN			NEW BRITAIN		
		812			066			060			061			064			060		
		MAT.	INST.	TOTAL	MAT.	INST.	TOTAL	MAT.	INST.	TOTAL	MAT.	INST.	TOTAL	MAT.	INST.	TOTAL	MAT.	INST.	TOTAL
015433	**CONTRACTOR EQUIPMENT**		94.7	94.7		97.9	97.9		97.9	97.9		97.9	97.9		98.3	98.3		97.9	97.9
0241, 31 - 34	**SITE & INFRASTRUCTURE, DEMOLITION**	134.3	92.4	105.1	103.3	102.3	102.6	102.4	102.3	102.3	97.8	102.3	100.9	100.2	103.1	102.2	102.6	102.3	102.4
0310	Concrete Forming & Accessories	106.2	52.5	60.5	98.1	119.5	116.3	98.1	119.3	116.2	99.2	119.3	116.3	97.9	119.3	116.1	98.6	119.3	116.2
0320	Concrete Reinforcing	109.9	69.7	89.5	125.5	145.1	135.4	125.5	145.1	135.4	120.3	145.1	132.9	125.5	145.1	135.4	125.5	145.1	135.4
0330	Cast-in-Place Concrete	114.7	72.8	99.1	102.1	128.8	112.1	95.7	128.7	108.0	100.2	128.7	110.8	92.1	128.7	105.7	97.2	128.7	108.9
03	**CONCRETE**	105.3	63.5	86.8	100.1	126.2	111.7	97.1	126.1	109.9	97.3	126.1	110.1	95.4	126.1	109.0	97.9	126.1	110.4
04	**MASONRY**	138.1	60.6	90.2	116.6	130.9	125.4	107.4	130.9	121.9	106.8	130.9	121.7	107.0	130.9	121.8	109.7	130.9	122.8
05	**METALS**	98.7	78.1	92.5	87.9	115.7	96.4	87.9	115.6	96.3	92.3	115.6	99.4	85.5	115.6	94.6	84.6	115.6	94.0
06	**WOOD, PLASTICS & COMPOSITES**	103.6	49.3	74.1	100.8	117.5	109.9	100.8	117.5	109.9	91.0	117.5	105.4	100.8	117.5	109.9	100.8	117.5	109.9
07	**THERMAL & MOISTURE PROTECTION**	106.5	66.7	89.2	97.8	125.8	110.0	97.9	121.9	108.4	102.6	121.9	111.0	97.9	121.9	108.4	97.9	122.2	108.5
08	**OPENINGS**	98.6	58.8	89.3	98.4	122.7	104.0	98.4	122.7	104.0	96.5	122.7	102.5	100.6	122.7	105.7	98.4	122.7	104.0
0920	Plaster & Gypsum Board	80.9	48.0	59.2	116.7	117.7	117.4	116.7	117.7	117.4	104.7	117.7	113.3	117.8	117.7	117.7	116.7	117.7	117.4
0950, 0980	Ceilings & Acoustic Treatment	103.8	48.0	66.0	97.9	117.7	111.3	97.9	117.7	111.3	94.1	117.7	110.1	101.1	117.7	112.4	97.9	117.7	111.3
0960	Flooring	116.1	71.3	103.1	90.6	131.2	102.4	90.6	128.7	101.6	94.9	131.2	105.4	90.6	128.7	101.6	90.6	128.7	101.6
0970, 0990	Wall Finishes & Painting/Coating	102.7	60.8	77.9	89.8	128.8	112.9	89.8	128.8	112.9	94.3	128.8	114.8	89.8	128.8	112.9	89.8	128.8	112.9
09	**FINISHES**	99.5	55.5	75.5	90.9	122.2	108.0	90.9	121.8	107.8	92.0	122.2	108.5	91.7	121.8	108.1	90.9	121.8	107.8
COVERS	**DIVS. 10 - 14, 25, 28, 41, 43, 44, 46**	100.0	83.0	96.3	100.0	113.4	102.9	100.0	113.4	102.9	100.0	113.4	102.9	100.0	113.4	102.9	100.0	113.4	102.9
21, 22, 23	**FIRE SUPPRESSION, PLUMBING & HVAC**	96.5	65.0	83.8	100.1	118.5	107.5	100.1	118.5	107.5	100.0	118.5	107.5	96.6	118.5	105.5	100.1	118.5	107.5
26, 27, 3370	**ELECTRICAL, COMMUNICATIONS & UTIL.**	96.3	64.8	80.3	96.0	107.6	101.9	96.0	106.2	101.2	95.7	109.3	102.6	96.0	108.4	102.3	96.1	106.2	101.3
MF2016	**WEIGHTED AVERAGE**	102.0	66.6	86.8	97.6	118.4	106.5	96.8	118.1	105.9	97.2	118.5	106.4	95.5	118.4	105.4	96.4	118.1	105.7

	DIVISION	CONNECTICUT																	
		NEW HAVEN			NEW LONDON			NORWALK			STAMFORD			WATERBURY			WILLIMANTIC		
		065			063			068			069			067			062		
		MAT.	INST.	TOTAL	MAT.	INST.	TOTAL	MAT.	INST.	TOTAL	MAT.	INST.	TOTAL	MAT.	INST.	TOTAL	MAT.	INST.	TOTAL
015433	**CONTRACTOR EQUIPMENT**		98.3	98.3		98.3	98.3		97.9	97.9		97.9	97.9		97.9	97.9		97.9	97.9
0241, 31 - 34	**SITE & INFRASTRUCTURE, DEMOLITION**	102.5	103.1	102.9	94.9	102.8	100.4	103.1	102.3	102.5	103.7	102.3	102.8	103.1	102.3	102.6	103.0	102.1	102.3
0310	Concrete Forming & Accessories	97.9	119.4	116.2	97.9	119.3	116.1	98.1	119.5	116.3	98.1	119.8	116.6	98.1	119.4	116.2	98.1	119.2	116.0
0320	Concrete Reinforcing	125.5	145.1	135.4	98.3	145.1	122.1	125.5	145.2	135.5	125.5	145.3	135.5	125.5	145.1	135.4	125.5	145.1	135.4
0330	Cast-in-Place Concrete	98.9	128.7	110.0	84.3	128.7	100.8	100.5	130.0	111.5	102.1	130.1	112.6	102.1	128.7	112.0	95.4	127.0	107.1
03	**CONCRETE**	111.9	126.1	118.2	85.8	126.0	103.7	99.4	126.6	111.4	100.1	126.8	112.0	100.1	126.1	111.6	97.0	125.4	109.6
04	**MASONRY**	107.7	130.9	122.0	106.0	130.9	121.4	107.1	132.2	122.6	108.0	132.2	122.9	108.0	130.9	122.1	107.3	130.9	121.9
05	**METALS**	84.8	115.6	94.1	84.5	115.6	93.9	87.9	115.9	96.4	87.9	116.2	96.5	87.9	115.6	96.3	87.7	115.4	96.1
06	**WOOD, PLASTICS & COMPOSITES**	100.8	117.5	109.9	100.8	117.5	109.9	100.8	117.5	109.9	100.8	117.5	109.9	100.8	117.5	109.9	100.8	117.5	109.9
07	**THERMAL & MOISTURE PROTECTION**	98.0	122.5	108.7	97.8	121.9	108.3	98.0	126.4	110.3	97.9	126.4	110.3	97.9	122.5	108.6	98.1	121.3	108.2
08	**OPENINGS**	98.4	122.7	104.0	100.8	122.7	105.9	98.4	122.7	104.0	98.4	122.7	104.0	98.4	122.7	104.0	100.8	122.7	105.9
0920	Plaster & Gypsum Board	116.7	117.7	117.4	116.7	117.7	117.4	116.7	117.7	117.4	116.7	117.7	117.4	116.7	117.7	117.4	116.7	117.7	117.4
0950, 0980	Ceilings & Acoustic Treatment	97.9	117.7	111.3	96.0	117.7	110.7	97.9	117.7	111.3	97.9	117.7	111.3	97.9	117.7	111.3	96.0	117.7	110.7
0960	Flooring	90.6	131.2	102.4	90.6	128.7	101.6	90.6	128.7	101.6	90.6	131.2	102.4	90.6	131.2	102.4	90.6	125.2	100.6
0970, 0990	Wall Finishes & Painting/Coating	89.8	128.8	112.9	89.8	128.8	112.9	89.8	128.8	112.9	89.8	128.8	112.9	89.8	128.8	112.9	89.8	128.8	112.9
09	**FINISHES**	91.0	122.2	108.0	90.1	121.8	107.4	90.9	121.8	107.8	91.0	122.2	108.0	90.8	122.2	107.9	90.6	121.2	107.3
COVERS	**DIVS. 10 - 14, 25, 28, 41, 43, 44, 46**	100.0	113.4	102.9	100.0	113.4	102.9	100.0	113.4	102.9	100.0	113.6	103.0	100.0	113.4	103.0	100.0	113.4	102.9
21, 22, 23	**FIRE SUPPRESSION, PLUMBING & HVAC**	100.1	118.5	107.5	96.6	118.5	105.5	100.1	118.5	107.5	100.1	118.6	107.5	100.1	118.5	107.5	100.1	118.4	107.5
26, 27, 3370	**ELECTRICAL, COMMUNICATIONS & UTIL.**	96.0	108.6	102.4	93.3	108.4	101.0	96.0	107.4	101.8	96.0	153.8	125.4	95.6	109.7	102.8	96.0	109.3	102.8
MF2016	**WEIGHTED AVERAGE**	98.1	118.5	106.9	93.6	118.4	104.2	97.0	118.6	106.3	97.2	125.2	109.2	97.1	118.6	106.3	97.0	118.2	106.1

	DIVISION	D.C.			DELAWARE									FLORIDA					
		WASHINGTON			DOVER			NEWARK			WILMINGTON			DAYTONA BEACH			FORT LAUDERDALE		
		200 - 205			199			197			198			321			333		
		MAT.	INST.	TOTAL	MAT.	INST.	TOTAL	MAT.	INST.	TOTAL	MAT.	INST.	TOTAL	MAT.	INST.	TOTAL	MAT.	INST.	TOTAL
015433	**CONTRACTOR EQUIPMENT**		105.6	105.6		121.3	121.3		121.3	121.3		121.5	121.5		102.1	102.1		95.2	95.2
0241, 31 - 34	**SITE & INFRASTRUCTURE, DEMOLITION**	103.8	95.9	98.3	100.7	113.6	109.7	102.3	113.6	110.2	104.9	114.0	111.2	113.4	88.9	96.3	92.5	76.8	81.5
0310	Concrete Forming & Accessories	96.7	74.0	77.3	97.3	101.9	101.2	96.7	101.9	101.2	95.3	101.9	101.0	98.4	62.7	68.0	92.4	63.3	67.6
0320	Concrete Reinforcing	100.9	88.7	94.7	109.1	118.4	113.8	103.3	118.4	111.0	107.4	118.4	113.0	93.3	62.6	77.7	94.5	61.4	77.7
0330	Cast-in-Place Concrete	112.6	80.6	100.7	100.6	107.5	103.1	90.0	107.5	96.5	102.3	107.5	104.3	88.1	65.9	79.8	90.1	62.0	79.6
03	**CONCRETE**	106.8	79.8	94.8	99.0	107.7	102.9	93.4	107.7	99.8	99.5	107.7	103.2	83.7	65.6	75.6	83.9	64.2	75.2
04	**MASONRY**	100.4	80.9	88.3	97.5	100.1	99.1	101.3	100.1	100.5	97.1	100.1	98.9	88.4	59.5	70.6	93.3	62.3	74.1
05	**METALS**	101.1	95.8	99.5	103.0	123.7	109.2	103.9	123.7	109.9	103.0	123.7	109.2	102.1	89.0	98.1	98.9	89.2	96.0
06	**WOOD, PLASTICS & COMPOSITES**	95.0	72.7	82.9	88.7	100.2	95.0	91.1	100.2	96.0	81.7	100.2	91.8	96.4	61.8	77.6	80.2	65.2	72.1
07	**THERMAL & MOISTURE PROTECTION**	102.7	84.9	95.0	104.7	112.7	108.2	108.2	112.7	110.2	103.5	112.7	107.5	100.8	65.7	85.5	101.1	62.5	84.3
08	**OPENINGS**	99.3	75.9	93.9	90.4	110.4	95.1	93.4	110.4	97.4	88.8	110.4	93.9	95.1	60.9	87.2	96.4	62.4	88.5
0920	Plaster & Gypsum Board	105.7	71.9	83.5	97.8	100.2	99.4	95.9	100.2	98.7	96.8	100.2	99.1	95.0	61.2	72.8	109.2	64.7	79.9
0950, 0980	Ceilings & Acoustic Treatment	105.6	71.9	82.8	99.1	100.2	99.9	96.6	100.2	99.1	90.3	100.2	97.0	83.7	61.2	68.5	86.1	64.7	71.6
0960	Flooring	96.4	78.1	91.1	90.4	109.0	95.8	87.7	109.0	93.9	91.5	109.0	96.6	101.0	56.7	88.2	96.8	76.7	91.0
0970, 0990	Wall Finishes & Painting/Coating	99.1	75.0	84.8	88.6	122.2	108.5	84.6	122.2	106.8	86.8	122.2	107.7	97.1	63.1	77.0	88.4	58.6	70.8
09	**FINISHES**	97.9	74.0	84.9	91.5	104.5	98.6	87.7	104.5	96.9	90.1	104.5	97.9	90.4	61.0	74.4	90.5	65.1	76.7
COVERS	**DIVS. 10 - 14, 25, 28, 41, 43, 44, 46**	100.0	96.3	99.2	100.0	105.4	101.2	100.0	105.4	101.2	100.0	105.4	101.2	100.0	82.7	96.2	100.0	82.7	96.2
21, 22, 23	**FIRE SUPPRESSION, PLUMBING & HVAC**	100.1	90.3	96.1	100.0	122.0	108.9	100.2	122.0	109.0	100.1	122.0	108.9	99.9	76.7	90.6	100.0	68.2	87.1
26, 27, 3370	**ELECTRICAL, COMMUNICATIONS & UTIL.**	100.6	102.4	101.5	96.0	112.4	104.3	98.5	112.4	105.6	97.3	112.4	105.0	95.5	60.3	77.6	92.7	69.1	80.7
MF2016	**WEIGHTED AVERAGE**	101.0	87.7	95.3	98.2	112.3	104.2	98.3	112.3	104.3	98.2	112.3	104.2	96.3	70.3	85.2	95.3	69.5	84.2

	DIVISION	FLORIDA																	
		FORT MYERS			GAINESVILLE			JACKSONVILLE			LAKELAND			MELBOURNE			MIAMI		
		339, 341			326, 344			320, 322			338			329			330 - 332, 340		
		MAT.	INST.	TOTAL	MAT.	INST.	TOTAL	MAT.	INST.	TOTAL	MAT.	INST.	TOTAL	MAT.	INST.	TOTAL	MAT.	INST.	TOTAL
015433	CONTRACTOR EQUIPMENT		102.1	102.1		102.1	102.1		102.1	102.1		102.1	102.1		102.1	102.1		95.2	95.2
0241, 31 - 34	SITE & INFRASTRUCTURE, DEMOLITION	104.3	89.0	93.7	122.6	88.7	99.0	113.3	88.9	96.3	106.2	88.9	94.2	121.2	89.1	98.8	93.9	77.1	82.2
0310	Concrete Forming & Accessories	88.5	62.1	66.0	93.5	56.9	62.4	98.2	60.1	65.8	85.3	61.5	65.0	94.7	62.9	67.6	97.7	63.3	68.4
0320	Concrete Reinforcing	95.5	78.9	87.1	99.0	60.7	79.5	93.3	60.8	76.8	97.9	78.3	87.9	94.4	67.4	80.7	101.4	60.9	80.8
0330	Cast-in-Place Concrete	93.9	63.4	82.6	101.2	63.0	87.0	89.0	64.0	79.7	96.1	65.2	84.6	106.2	66.0	91.3	87.0	63.2	78.2
03	CONCRETE	84.6	67.1	76.9	93.9	61.6	79.6	84.1	63.4	74.9	86.2	67.4	77.8	94.5	66.5	82.1	82.5	64.6	74.5
04	MASONRY	87.1	64.8	73.3	100.7	55.9	73.0	88.1	56.5	68.6	101.9	58.3	74.9	85.0	59.6	69.3	92.6	54.7	69.2
05	METALS	101.3	95.2	99.4	101.0	87.9	97.0	100.5	88.1	96.8	101.2	94.2	99.1	111.1	90.8	104.9	99.1	88.3	95.9
06	WOOD, PLASTICS & COMPOSITES	77.1	63.2	69.6	90.0	56.2	71.6	96.4	60.3	76.8	72.4	61.0	66.2	91.8	61.8	75.5	90.3	65.2	76.7
07	THERMAL & MOISTURE PROTECTION	100.8	64.4	85.0	101.2	60.7	83.6	101.1	61.5	83.9	100.8	63.0	84.3	101.3	64.3	85.2	101.0	61.3	83.7
08	OPENINGS	97.7	65.2	90.2	94.8	57.4	86.1	95.1	59.7	86.9	97.7	64.0	89.8	94.4	62.0	86.9	99.9	62.4	91.2
0920	Plaster & Gypsum Board	105.3	62.7	77.2	91.8	55.4	67.8	95.0	59.7	71.8	101.6	60.4	74.5	91.8	61.2	71.7	95.5	64.7	75.2
0950, 0980	Ceilings & Acoustic Treatment	81.6	62.7	68.8	78.1	55.4	62.7	83.7	59.7	67.4	81.6	60.4	67.3	82.9	61.2	68.2	84.9	64.7	71.2
0960	Flooring	93.7	79.8	89.7	98.4	56.7	86.3	101.0	58.5	88.7	91.6	58.4	82.0	98.6	57.8	86.8	98.8	58.4	87.1
0970, 0990	Wall Finishes & Painting/Coating	92.6	63.6	75.4	97.1	63.6	77.3	97.1	63.6	77.3	92.6	63.6	75.4	97.1	83.0	88.8	88.9	58.6	70.9
09	FINISHES	90.3	65.3	76.6	89.1	56.9	71.6	90.5	59.7	73.7	89.2	60.6	73.6	89.8	63.4	75.4	89.1	61.6	74.1
COVERS	DIVS. 10 - 14, 25, 28, 41, 43, 44, 46	100.0	79.4	95.5	100.0	81.1	95.8	100.0	78.2	95.2	100.0	80.4	95.7	100.0	82.8	96.2	100.0	83.0	96.2
21, 22, 23	FIRE SUPPRESSION, PLUMBING & HVAC	98.1	57.5	81.7	98.8	62.3	84.0	99.9	62.3	84.7	98.1	59.4	82.5	99.9	75.3	90.0	100.0	62.5	84.8
26, 27, 3370	ELECTRICAL, COMMUNICATIONS & UTIL.	94.7	61.5	77.8	95.8	57.2	76.2	95.2	63.4	79.1	93.0	60.0	76.2	96.6	64.3	80.1	96.5	72.3	84.2
MF2016	WEIGHTED AVERAGE	95.6	68.3	83.9	97.8	64.9	83.7	96.1	66.5	83.4	96.2	67.1	83.8	99.2	71.2	87.2	95.8	67.4	83.7

	DIVISION	FLORIDA																	
		ORLANDO			PANAMA CITY			PENSACOLA			SARASOTA			ST. PETERSBURG			TALLAHASSEE		
		327 - 328, 347			324			325			342			337			323		
		MAT.	INST.	TOTAL	MAT.	INST.	TOTAL	MAT.	INST.	TOTAL	MAT.	INST.	TOTAL	MAT.	INST.	TOTAL	MAT.	INST.	TOTAL
015433	CONTRACTOR EQUIPMENT		102.1	102.1		102.1	102.1		102.1	102.1		102.1	102.1		102.1	102.1		102.1	102.1
0241, 31 - 34	SITE & INFRASTRUCTURE, DEMOLITION	113.8	89.2	96.6	126.6	88.2	99.8	126.6	88.6	100.1	117.2	89.0	97.5	107.8	88.5	94.4	106.8	88.7	94.2
0310	Concrete Forming & Accessories	102.4	62.5	68.4	97.5	60.3	65.8	95.4	63.6	68.3	94.3	61.6	66.5	91.8	59.3	64.1	100.6	56.9	63.4
0320	Concrete Reinforcing	98.5	67.4	82.7	97.5	69.8	83.4	100.0	69.2	84.3	93.7	78.3	85.8	97.9	78.2	87.9	99.1	60.7	79.6
0330	Cast-in-Place Concrete	102.6	65.8	88.9	93.5	58.8	80.6	115.6	63.9	96.4	103.7	65.4	89.5	97.1	61.0	83.7	93.1	62.8	81.9
03	CONCRETE	89.7	66.2	79.3	92.0	63.2	79.2	101.5	66.4	85.9	93.1	67.5	81.8	87.6	64.9	77.5	87.2	61.5	75.8
04	MASONRY	92.3	59.5	72.1	92.4	57.4	70.8	113.1	57.4	78.7	90.1	58.8	70.7	138.6	52.1	85.2	90.1	55.4	68.7
05	METALS	99.1	90.4	96.5	101.8	91.4	98.6	102.8	90.8	99.2	103.0	93.6	100.2	102.1	93.9	99.6	99.6	88.0	96.1
06	WOOD, PLASTICS & COMPOSITES	93.5	61.8	76.3	95.2	64.2	78.4	93.1	64.2	77.4	94.4	61.0	76.3	81.5	61.0	70.4	95.5	56.2	74.1
07	THERMAL & MOISTURE PROTECTION	105.8	65.7	88.3	101.5	58.8	82.9	101.4	61.8	84.1	98.7	63.2	83.3	101.0	58.8	82.6	98.7	61.2	82.4
08	OPENINGS	99.4	62.0	90.7	93.2	63.7	86.4	93.2	63.7	86.4	100.0	63.6	91.5	96.5	64.0	88.9	100.3	57.4	90.3
0920	Plaster & Gypsum Board	96.0	61.2	73.1	94.3	63.7	74.1	102.8	63.7	77.0	100.9	60.4	74.2	107.7	60.4	76.6	99.3	55.4	70.4
0950, 0980	Ceilings & Acoustic Treatment	90.0	61.2	70.5	82.9	63.7	69.9	82.9	63.7	69.9	88.7	60.4	69.6	83.5	60.4	67.9	88.3	55.4	66.0
0960	Flooring	97.2	58.4	85.9	100.6	75.0	93.2	96.4	56.7	84.9	103.5	56.7	89.9	95.7	56.7	84.4	99.0	56.7	86.7
0970, 0990	Wall Finishes & Painting/Coating	93.5	61.9	74.8	97.1	63.6	77.3	97.1	63.6	77.3	98.2	63.6	77.7	92.6	63.6	75.4	92.3	63.6	75.3
09	FINISHES	92.1	61.2	75.2	91.5	63.5	76.2	91.1	62.1	75.3	95.8	60.4	76.5	91.7	58.9	73.8	91.2	56.8	72.4
COVERS	DIVS. 10 - 14, 25, 28, 41, 43, 44, 46	100.0	82.7	96.2	100.0	76.5	94.8	100.0	81.0	95.8	100.0	80.6	95.7	100.0	77.1	95.0	100.0	78.7	95.3
21, 22, 23	FIRE SUPPRESSION, PLUMBING & HVAC	100.1	57.1	82.7	99.9	51.5	80.4	99.9	63.2	85.1	99.9	58.8	83.3	100.0	55.7	82.1	100.0	66.6	86.5
26, 27, 3370	ELECTRICAL, COMMUNICATIONS & UTIL.	99.8	64.7	81.9	94.3	57.2	75.4	97.9	51.9	74.5	95.8	60.0	77.6	93.0	62.1	77.3	99.9	57.2	78.2
MF2016	WEIGHTED AVERAGE	97.9	67.1	84.7	97.6	64.2	83.3	100.3	66.4	85.8	98.7	67.0	85.1	99.0	65.2	84.5	97.3	65.7	83.7

	DIVISION	FLORIDA						GEORGIA											
		TAMPA			WEST PALM BEACH			ALBANY			ATHENS			ATLANTA			AUGUSTA		
		335 - 336, 346			334, 349			317, 398			306			300 - 303, 399			308 - 309		
		MAT.	INST.	TOTAL	MAT.	INST.	TOTAL	MAT.	INST.	TOTAL	MAT.	INST.	TOTAL	MAT.	INST.	TOTAL	MAT.	INST.	TOTAL
015433	CONTRACTOR EQUIPMENT		102.1	102.1		95.2	95.2		96.1	96.1		94.6	94.6		96.5	96.5		94.6	94.6
0241, 31 - 34	SITE & INFRASTRUCTURE, DEMOLITION	108.3	89.0	94.8	89.3	76.8	80.6	102.3	79.4	86.3	101.7	94.9	97.0	98.7	95.3	96.3	95.0	95.3	95.2
0310	Concrete Forming & Accessories	94.5	61.9	66.7	95.5	62.9	67.7	88.5	67.7	70.8	91.7	44.3	51.4	96.3	71.9	75.5	92.7	72.7	75.7
0320	Concrete Reinforcing	94.5	78.3	86.2	97.1	59.2	77.8	90.4	73.3	81.7	98.9	65.4	81.9	98.2	73.4	85.6	99.3	68.0	83.4
0330	Cast-in-Place Concrete	94.9	65.8	84.1	85.6	61.8	76.8	85.7	67.7	79.0	109.6	68.5	94.3	113.1	71.6	97.6	103.5	70.1	91.1
03	CONCRETE	86.3	67.8	78.1	81.0	63.6	73.3	81.2	70.4	76.4	102.1	57.7	82.4	104.3	72.7	90.3	94.6	71.6	84.4
04	MASONRY	94.2	59.3	72.6	92.8	53.3	68.4	91.9	67.7	77.0	76.6	77.8	77.3	89.9	69.7	77.4	90.1	69.2	77.2
05	METALS	101.1	94.3	99.1	98.0	87.7	94.9	102.9	97.5	101.2	97.6	79.5	92.1	98.4	84.6	94.2	97.2	80.5	92.2
06	WOOD, PLASTICS & COMPOSITES	85.3	61.0	72.1	85.0	65.2	74.3	79.5	68.4	73.5	95.0	36.8	63.4	97.8	73.1	84.4	96.5	75.4	85.0
07	THERMAL & MOISTURE PROTECTION	101.3	63.5	84.8	100.8	61.2	83.5	98.6	68.7	85.6	100.2	69.5	86.9	101.7	72.4	89.0	100.0	71.3	87.5
08	OPENINGS	97.7	64.0	89.8	96.0	62.0	88.1	87.6	70.5	83.6	93.0	50.9	83.3	97.6	73.4	92.0	93.1	73.3	88.5
0920	Plaster & Gypsum Board	110.1	60.4	77.4	113.8	64.7	81.5	103.8	68.0	80.2	97.4	35.5	56.6	99.6	72.6	81.8	98.7	75.1	83.1
0950, 0980	Ceilings & Acoustic Treatment	86.1	60.4	68.7	81.6	64.7	70.2	81.9	68.0	72.5	101.2	35.5	56.7	94.0	72.6	79.5	102.2	75.1	83.8
0960	Flooring	96.8	58.4	85.6	98.6	56.7	86.5	102.3	65.9	91.8	94.0	84.7	91.3	97.0	68.0	88.6	94.1	65.9	85.9
0970, 0990	Wall Finishes & Painting/Coating	92.6	63.6	75.4	88.4	58.6	70.8	94.5	96.5	95.6	101.8	96.5	98.7	106.2	96.5	100.5	101.8	80.6	89.2
09	FINISHES	93.0	60.8	75.4	90.6	61.1	74.5	92.4	70.0	80.2	97.7	55.2	74.5	97.8	73.5	84.5	97.4	72.5	83.8
COVERS	DIVS. 10 - 14, 25, 28, 41, 43, 44, 46	100.0	80.8	95.8	100.0	82.7	96.2	100.0	83.3	96.3	100.0	79.6	95.5	100.0	86.4	97.0	100.0	83.9	96.5
21, 22, 23	FIRE SUPPRESSION, PLUMBING & HVAC	100.0	60.0	83.8	98.1	59.9	82.7	99.9	69.9	87.8	96.6	67.3	84.8	100.0	72.0	88.7	100.1	63.6	85.4
26, 27, 3370	ELECTRICAL, COMMUNICATIONS & UTIL.	92.7	64.7	78.5	93.8	69.1	81.2	95.3	63.4	79.1	97.0	65.4	81.0	96.6	72.6	84.4	97.6	69.8	83.4
MF2016	WEIGHTED AVERAGE	96.8	68.1	84.5	94.3	66.0	82.2	95.2	72.6	85.5	96.8	67.9	84.4	99.0	75.6	89.0	97.1	72.7	86.6

DIVISION		GEORGIA																	
		COLUMBUS			DALTON			GAINESVILLE			MACON			SAVANNAH			STATESBORO		
		318 - 319			307			305			310 - 312			313 - 314			304		
		MAT.	INST.	TOTAL	MAT.	INST.	TOTAL	MAT.	INST.	TOTAL	MAT.	INST.	TOTAL	MAT.	INST.	TOTAL	MAT.	INST.	TOTAL
015433	CONTRACTOR EQUIPMENT		96.1	96.1		110.1	110.1		94.6	94.6		105.9	105.9		97.0	97.0		97.7	97.7
0241, 31 - 34	SITE & INFRASTRUCTURE, DEMOLITION	102.2	79.5	86.4	101.5	100.3	100.7	101.7	94.9	97.0	103.9	94.6	97.4	102.1	80.9	87.3	102.6	81.4	87.8
0310	Concrete Forming & Accessories	88.5	67.8	70.8	84.6	65.5	68.3	95.2	41.8	49.7	88.3	67.8	70.9	92.0	70.4	73.6	79.5	52.1	56.2
0320	Concrete Reinforcing	90.3	73.3	81.7	98.4	62.5	80.2	98.7	65.3	81.7	91.5	73.3	82.3	97.3	68.0	82.4	98.0	36.6	66.8
0330	Cast-in-Place Concrete	85.4	67.8	78.8	106.5	67.2	91.8	115.2	68.3	97.7	84.2	69.7	78.8	89.0	67.6	81.0	109.4	67.2	93.7
03	CONCRETE	81.0	70.5	76.4	101.4	67.4	86.3	103.9	56.4	82.9	80.6	71.2	76.5	82.6	70.7	77.3	101.7	57.0	81.9
04	MASONRY	91.9	67.8	77.0	77.4	77.1	77.2	85.4	77.8	80.7	104.2	67.9	81.8	86.8	67.7	75.0	79.4	77.8	78.4
05	METALS	102.5	97.9	101.1	98.9	93.8	97.3	96.9	78.9	91.4	97.6	98.0	97.8	98.6	95.3	97.6	102.5	84.0	96.9
06	WOOD, PLASTICS & COMPOSITES	79.5	68.4	73.5	78.2	67.1	72.1	98.9	34.1	63.7	85.0	68.4	76.0	88.2	72.3	79.6	71.5	48.5	59.0
07	THERMAL & MOISTURE PROTECTION	98.5	69.4	85.8	102.4	71.7	89.0	100.2	69.2	86.7	97.0	71.5	85.9	96.4	68.9	84.5	100.8	68.1	86.6
08	OPENINGS	87.6	70.5	83.6	94.2	66.9	87.9	93.0	49.4	82.9	87.4	70.5	83.5	95.1	71.3	89.6	95.1	49.5	84.5
0920	Plaster & Gypsum Board	103.8	68.0	80.2	85.0	66.6	72.9	99.9	32.7	55.6	108.4	68.0	81.8	102.7	72.0	82.5	85.8	47.5	60.6
0950, 0980	Ceilings & Acoustic Treatment	81.9	68.0	72.5	117.4	66.6	83.0	101.2	32.7	54.8	77.1	68.0	70.9	88.3	72.0	77.3	111.8	47.5	68.2
0960	Flooring	102.3	65.9	91.8	94.5	84.7	91.6	95.5	84.7	92.3	81.0	65.9	76.6	98.3	65.9	88.9	111.2	84.7	103.5
0970, 0990	Wall Finishes & Painting/Coating	94.5	96.5	95.6	91.6	71.9	80.0	101.8	96.5	98.7	96.5	96.5	96.5	91.5	86.3	88.4	100.0	71.9	83.4
09	FINISHES	92.3	70.0	80.1	107.5	70.2	87.2	98.3	53.6	73.9	82.3	70.0	75.6	92.3	71.2	80.8	110.2	59.4	82.5
COVERS	DIVS. 10 - 14, 25, 28, 41, 43, 44, 46	100.0	83.3	96.3	100.0	28.4	84.2	100.0	36.2	85.9	100.0	83.4	96.3	100.0	83.3	96.3	100.0	42.2	87.3
21, 22, 23	FIRE SUPPRESSION, PLUMBING & HVAC	100.0	66.2	86.3	96.7	62.0	82.7	96.6	67.1	84.7	100.0	69.3	87.6	100.1	67.1	86.7	97.1	67.8	85.3
26, 27, 3370	ELECTRICAL, COMMUNICATIONS & UTIL.	95.5	69.4	82.2	106.2	63.7	84.6	97.0	69.3	82.9	94.0	64.2	78.9	99.1	70.2	84.4	97.3	58.6	77.6
MF2016	WEIGHTED AVERAGE	95.1	72.7	85.5	98.7	71.0	86.8	97.4	66.5	84.2	93.9	74.0	85.4	95.6	73.1	86.0	98.9	65.8	84.7

DIVISION		GEORGIA						HAWAII									IDAHO		
		VALDOSTA			WAYCROSS			HILO			HONOLULU			STATES & POSS., GUAM			BOISE		
		316			315			967			968			969			836 - 837		
		MAT.	INST.	TOTAL	MAT.	INST.	TOTAL	MAT.	INST.	TOTAL	MAT.	INST.	TOTAL	MAT.	INST.	TOTAL	MAT.	INST.	TOTAL
015433	CONTRACTOR EQUIPMENT		96.1	96.1		96.1	96.1		100.4	100.4		100.4	100.4		165.0	165.0		97.6	97.6
0241, 31 - 34	SITE & INFRASTRUCTURE, DEMOLITION	111.5	79.4	89.1	108.5	80.6	89.0	149.7	107.5	120.3	158.4	107.5	122.9	194.0	103.5	130.9	87.5	97.0	94.1
0310	Concrete Forming & Accessories	79.7	42.0	47.6	81.5	64.7	67.2	105.8	123.7	121.0	118.7	123.7	122.9	108.3	53.0	61.2	98.0	79.8	82.5
0320	Concrete Reinforcing	92.4	61.2	76.5	92.4	61.4	76.7	137.5	126.6	131.9	159.9	126.6	142.9	237.2	28.2	131.0	105.1	78.8	91.8
0330	Cast-in-Place Concrete	83.9	67.5	77.8	94.8	67.4	84.6	170.9	125.6	154.1	137.5	125.6	133.1	148.2	98.5	129.7	90.6	83.1	87.8
03	CONCRETE	85.4	56.7	72.7	88.4	66.7	78.8	137.6	123.7	131.5	130.8	123.7	127.7	138.8	65.7	106.4	93.8	81.0	88.1
04	MASONRY	97.2	77.8	85.2	98.0	77.8	85.5	148.6	120.0	130.9	135.5	120.0	125.9	205.8	35.5	100.6	125.2	87.2	101.7
05	METALS	102.1	93.1	99.3	101.1	88.8	97.4	119.9	106.6	115.9	134.3	106.6	125.9	155.0	76.6	131.3	104.5	82.1	97.7
06	WOOD, PLASTICS & COMPOSITES	68.1	33.9	49.5	69.6	65.2	67.2	113.9	124.5	119.6	131.3	124.5	127.6	126.6	55.4	87.9	94.6	78.5	85.9
07	THERMAL & MOISTURE PROTECTION	98.8	67.5	85.2	98.5	70.3	86.2	132.2	120.2	127.0	150.1	120.2	137.1	155.5	59.8	113.9	97.5	85.5	92.3
08	OPENINGS	84.2	48.2	75.9	84.6	62.9	79.5	116.3	123.0	117.9	130.9	123.0	129.1	120.7	45.0	103.1	99.2	73.1	93.1
0920	Plaster & Gypsum Board	95.8	32.5	54.1	95.8	64.7	75.4	112.1	124.9	120.6	155.7	124.9	135.5	228.9	43.7	106.9	94.2	77.9	83.5
0950, 0980	Ceilings & Acoustic Treatment	80.2	32.5	47.9	78.3	64.7	69.1	137.5	124.9	129.0	144.3	124.9	131.2	254.9	43.7	111.8	101.2	77.9	85.4
0960	Flooring	96.1	86.3	93.2	97.3	84.7	93.6	110.8	140.7	119.5	130.2	140.7	133.3	130.1	40.8	104.2	93.2	79.6	89.3
0970, 0990	Wall Finishes & Painting/Coating	94.5	96.5	95.6	94.5	71.9	81.1	94.7	140.6	121.9	105.1	140.6	126.1	100.0	32.4	60.0	93.4	40.5	62.1
09	FINISHES	90.0	53.8	70.2	89.6	69.3	78.5	110.5	129.0	120.6	125.7	129.0	127.5	183.1	49.4	110.1	91.4	75.7	82.8
COVERS	DIVS. 10 - 14, 25, 28, 41, 43, 44, 46	100.0	79.1	95.4	100.0	49.8	89.0	100.0	112.6	102.8	100.0	112.6	102.8	100.0	67.3	92.8	100.0	88.6	97.5
21, 22, 23	FIRE SUPPRESSION, PLUMBING & HVAC	100.0	69.3	87.6	97.8	64.2	84.2	100.4	112.0	105.1	100.4	112.0	105.1	103.1	34.3	75.3	100.0	74.8	89.8
26, 27, 3370	ELECTRICAL, COMMUNICATIONS & UTIL.	93.8	52.9	73.0	97.5	54.0	75.4	111.0	124.3	117.8	112.8	124.3	118.7	162.9	36.8	98.8	97.3	69.8	83.3
MF2016	WEIGHTED AVERAGE	95.3	66.2	82.8	95.4	68.4	83.8	116.3	118.3	117.1	120.9	118.3	119.8	138.8	52.6	101.9	99.7	79.4	91.0

DIVISION		IDAHO															ILLINOIS		
		COEUR D'ALENE			IDAHO FALLS			LEWISTON			POCATELLO			TWIN FALLS			BLOOMINGTON		
		838			834			835			832			833			617		
		MAT.	INST.	TOTAL	MAT.	INST.	TOTAL	MAT.	INST.	TOTAL	MAT.	INST.	TOTAL	MAT.	INST.	TOTAL	MAT.	INST.	TOTAL
015433	CONTRACTOR EQUIPMENT		92.5	92.5		97.6	97.6		92.5	92.5		97.6	97.6		97.6	97.6		104.7	104.7
0241, 31 - 34	SITE & INFRASTRUCTURE, DEMOLITION	86.5	91.0	89.6	85.4	96.8	93.4	93.3	91.9	92.3	88.7	97.0	94.5	95.6	96.7	96.4	97.0	100.8	99.6
0310	Concrete Forming & Accessories	105.8	82.0	85.5	91.8	78.6	80.6	110.9	83.4	87.5	98.1	79.5	82.2	99.2	78.5	81.5	84.1	119.2	114.0
0320	Concrete Reinforcing	113.2	99.0	106.0	107.2	77.4	92.0	113.2	99.0	106.0	105.6	78.8	92.0	107.6	77.3	92.2	96.1	105.3	100.8
0330	Cast-in-Place Concrete	98.0	84.1	92.8	86.2	83.3	85.2	101.8	84.5	95.4	93.1	83.0	89.4	95.6	83.3	91.0	98.1	117.0	105.2
03	CONCRETE	99.8	85.7	93.6	86.1	80.2	83.5	103.3	86.5	95.8	93.0	80.8	87.6	100.4	80.2	91.4	91.9	116.7	102.9
04	MASONRY	126.8	85.0	100.9	120.2	84.6	98.2	127.2	86.5	102.0	122.7	87.2	100.8	125.4	84.6	100.2	110.9	124.2	119.2
05	METALS	98.2	89.0	95.4	112.6	81.4	103.1	97.6	89.2	95.0	112.7	81.7	103.3	112.7	81.2	103.2	94.9	123.8	103.6
06	WOOD, PLASTICS & COMPOSITES	97.6	82.1	89.2	88.1	78.5	82.9	103.7	82.1	92.0	94.6	78.5	85.9	95.7	78.5	86.3	83.4	116.3	101.3
07	THERMAL & MOISTURE PROTECTION	151.8	84.1	122.3	96.6	75.4	87.4	152.2	85.5	123.2	97.2	76.4	88.2	98.1	82.7	91.4	98.1	115.2	105.5
08	OPENINGS	116.4	78.1	107.5	102.1	72.8	95.3	110.1	82.0	103.6	99.9	68.1	92.5	102.9	63.6	93.8	93.3	122.1	100.0
0920	Plaster & Gypsum Board	170.4	81.8	112.0	80.8	77.9	78.9	172.2	81.8	112.7	82.2	77.9	79.4	84.4	77.9	80.1	88.8	116.9	107.3
0950, 0980	Ceilings & Acoustic Treatment	136.3	81.8	99.4	104.7	77.9	86.6	136.3	81.8	99.4	109.8	77.9	88.2	107.2	77.9	87.4	89.7	116.9	108.1
0960	Flooring	131.9	79.6	116.7	93.1	79.6	89.2	135.1	79.6	119.0	96.5	79.6	91.6	97.7	79.6	92.5	87.3	124.9	98.2
0970, 0990	Wall Finishes & Painting/Coating	112.5	74.5	90.0	93.4	40.5	62.1	112.5	73.2	89.3	93.3	40.5	62.1	93.4	40.5	62.1	88.7	138.3	118.0
09	FINISHES	155.0	80.7	114.4	89.7	75.0	81.7	156.3	81.2	115.3	92.3	75.7	83.2	93.1	75.0	83.2	86.9	122.5	106.3
COVERS	DIVS. 10 - 14, 25, 28, 41, 43, 44, 46	100.0	93.6	98.6	100.0	87.7	97.3	100.0	94.6	98.8	100.0	88.6	97.5	100.0	87.7	97.3	100.0	106.7	101.5
21, 22, 23	FIRE SUPPRESSION, PLUMBING & HVAC	99.5	81.7	92.3	100.9	73.4	89.7	100.7	86.9	95.1	99.9	74.8	89.7	99.9	69.6	87.6	96.5	107.7	101.0
26, 27, 3370	ELECTRICAL, COMMUNICATIONS & UTIL.	89.3	83.5	86.3	88.9	69.4	79.0	87.4	80.1	83.7	94.7	67.6	80.9	90.3	69.4	79.7	94.2	92.5	93.3
MF2016	WEIGHTED AVERAGE	106.8	84.4	97.2	99.2	78.2	90.2	106.9	85.7	97.8	100.7	78.6	91.2	101.9	77.2	91.3	95.2	112.2	102.5

DIVISION		ILLINOIS																	
		CARBONDALE			CENTRALIA			CHAMPAIGN			CHICAGO			DECATUR			EAST ST. LOUIS		
		629			628			618 - 619			606 - 608			625			620 - 622		
		MAT.	INST.	TOTAL	MAT.	INST.	TOTAL	MAT.	INST.	TOTAL	MAT.	INST.	TOTAL	MAT.	INST.	TOTAL	MAT.	INST.	TOTAL
015433	CONTRACTOR EQUIPMENT		113.3	113.3		113.3	113.3		105.6	105.6		98.9	98.9		105.6	105.6		113.3	113.3
0241, 31 - 34	SITE & INFRASTRUCTURE, DEMOLITION	98.5	101.5	100.6	98.8	103.4	102.0	106.2	101.6	103.0	103.6	103.7	103.7	93.5	101.2	98.9	100.9	102.0	101.7
0310	Concrete Forming & Accessories	89.6	108.2	105.4	91.1	114.7	111.2	90.6	116.7	112.8	99.5	162.0	152.7	91.8	115.9	112.3	87.0	111.5	107.9
0320	Concrete Reinforcing	90.2	106.3	98.4	90.2	107.0	98.8	96.1	100.6	98.4	100.2	154.7	128.0	88.5	100.9	94.8	90.1	107.0	98.7
0330	Cast-in-Place Concrete	88.9	102.3	93.8	89.3	121.6	101.3	113.8	111.9	113.1	126.4	154.6	136.9	97.0	113.7	103.2	90.8	121.5	102.3
03	CONCRETE	79.1	107.3	91.6	79.5	117.1	96.2	104.2	112.8	108.0	109.4	157.3	130.6	89.3	113.1	99.9	80.5	115.7	96.1
04	MASONRY	76.0	112.4	98.5	76.0	125.5	106.6	133.9	121.8	126.4	104.5	166.2	142.6	71.8	120.2	101.7	76.3	125.5	106.7
05	METALS	103.2	130.7	111.6	103.3	133.8	112.5	94.9	119.7	102.4	102.1	146.6	115.6	107.2	119.2	110.9	104.4	133.5	113.2
06	WOOD, PLASTICS & COMPOSITES	88.7	105.4	97.8	91.2	110.9	101.9	90.4	114.9	103.7	101.1	161.7	134.0	89.8	114.9	103.4	85.8	106.9	97.3
07	THERMAL & MOISTURE PROTECTION	95.0	102.3	98.2	95.1	113.2	102.9	98.9	114.7	105.8	93.8	152.0	119.1	100.7	113.0	106.1	95.1	112.5	102.6
08	OPENINGS	89.2	116.9	95.6	89.2	120.0	96.3	93.9	116.8	99.2	104.4	170.9	119.9	100.3	116.9	104.2	89.3	117.7	95.9
0920	Plaster & Gypsum Board	95.3	105.7	102.2	96.6	111.3	106.3	90.6	115.4	107.0	94.8	163.5	140.1	97.8	115.4	109.4	93.8	107.2	102.6
0950, 0980	Ceilings & Acoustic Treatment	84.6	105.7	98.9	84.6	111.3	102.7	89.7	115.4	107.1	99.8	163.5	143.0	90.5	115.4	107.4	84.6	107.2	99.9
0960	Flooring	111.2	113.8	112.0	112.0	127.7	116.5	90.4	114.7	97.4	98.7	164.8	117.9	99.5	115.2	104.1	110.2	127.7	115.3
0970, 0990	Wall Finishes & Painting/Coating	109.1	107.7	108.3	109.1	114.6	112.3	88.7	115.4	104.5	104.9	162.3	138.9	100.8	114.4	108.9	109.1	114.6	112.3
09	FINISHES	92.8	108.1	101.1	93.2	117.1	106.3	88.8	116.6	104.0	98.4	163.5	133.9	92.8	116.2	105.6	92.4	114.4	104.4
COVERS	DIVS. 10 - 14, 25, 28, 41, 43, 44, 46	100.0	102.3	100.5	100.0	104.3	100.9	100.0	105.4	101.2	100.0	128.1	106.2	100.0	104.8	101.1	100.0	103.8	100.8
21, 22, 23	FIRE SUPPRESSION, PLUMBING & HVAC	96.4	106.5	100.5	96.4	98.8	97.4	96.5	106.3	100.4	99.8	138.0	115.3	99.9	96.4	98.5	99.9	99.3	99.7
26, 27, 3370	ELECTRICAL, COMMUNICATIONS & UTIL.	91.2	108.8	100.2	92.5	108.9	100.8	97.4	92.6	94.9	97.4	134.1	116.1	94.9	89.8	92.3	92.1	102.1	97.2
MF2016	WEIGHTED AVERAGE	93.1	109.8	100.2	93.3	113.0	101.8	98.7	109.8	103.5	101.7	146.1	120.7	97.2	107.0	101.4	94.4	111.3	101.6

DIVISION		ILLINOIS																	
		EFFINGHAM			GALESBURG			JOLIET			KANKAKEE			LA SALLE			NORTH SUBURBAN		
		624			614			604			609			613			600 - 603		
		MAT.	INST.	TOTAL	MAT.	INST.	TOTAL	MAT.	INST.	TOTAL	MAT.	INST.	TOTAL	MAT.	INST.	TOTAL	MAT.	INST.	TOTAL
015433	CONTRACTOR EQUIPMENT		105.6	105.6		104.7	104.7		95.4	95.4		95.4	95.4		104.7	104.7		95.4	95.4
0241, 31 - 34	SITE & INFRASTRUCTURE, DEMOLITION	97.8	100.4	99.6	99.5	99.8	99.7	99.3	101.7	101.0	93.1	101.3	98.8	98.9	101.3	100.6	98.6	102.0	101.0
0310	Concrete Forming & Accessories	96.3	116.6	113.6	90.5	119.5	115.2	98.8	161.1	151.8	92.1	146.0	138.0	104.5	126.0	122.8	98.1	158.8	149.8
0320	Concrete Reinforcing	91.4	96.2	93.9	95.5	108.6	102.1	100.2	140.6	120.7	101.0	138.9	120.3	95.7	137.5	116.9	100.2	153.3	127.2
0330	Cast-in-Place Concrete	96.7	109.2	101.3	101.0	111.5	104.9	115.0	148.1	127.3	107.1	138.1	118.6	100.9	125.0	109.9	115.0	154.5	129.7
03	CONCRETE	90.0	111.0	99.3	94.7	115.4	103.9	102.2	151.9	124.3	96.0	141.3	116.1	95.6	128.2	110.0	102.2	155.4	125.8
04	MASONRY	80.2	114.1	101.2	111.1	123.7	118.9	103.7	160.0	138.5	100.0	148.4	129.9	111.1	145.6	132.4	100.7	162.1	138.6
05	METALS	104.2	114.5	107.3	94.9	124.1	103.7	100.0	135.6	110.8	100.0	133.8	110.2	95.0	142.8	109.4	101.1	142.0	113.5
06	WOOD, PLASTICS & COMPOSITES	92.1	117.3	105.8	90.3	118.7	105.7	99.2	162.3	133.5	92.1	144.6	120.6	106.5	123.3	115.6	98.1	157.7	130.5
07	THERMAL & MOISTURE PROTECTION	100.1	109.8	104.4	98.3	112.1	104.3	98.1	148.2	119.9	97.1	141.1	116.2	98.5	128.9	111.7	98.4	148.9	120.3
08	OPENINGS	95.3	116.2	100.2	93.3	118.6	99.2	101.9	166.5	116.9	94.8	156.8	109.2	93.3	135.2	103.0	102.0	168.4	117.4
0920	Plaster & Gypsum Board	97.8	117.9	111.0	90.6	119.3	109.5	87.3	164.1	137.9	85.5	146.0	125.3	98.3	124.1	115.2	90.9	159.5	136.1
0950, 0980	Ceilings & Acoustic Treatment	84.6	117.9	107.1	89.7	119.3	109.8	99.4	164.1	143.3	99.4	146.0	131.0	89.7	124.1	113.0	99.4	159.5	140.1
0960	Flooring	100.5	116.9	105.3	90.2	124.9	100.3	96.7	154.8	113.5	93.7	155.7	111.7	96.7	122.9	104.3	97.2	164.8	116.8
0970, 0990	Wall Finishes & Painting/Coating	100.8	112.2	107.6	88.7	101.1	96.0	97.0	167.3	138.6	97.0	138.3	121.4	88.7	136.5	117.0	99.0	168.2	139.9
09	FINISHES	91.9	117.4	105.8	88.1	119.4	105.2	93.7	162.0	131.0	92.2	147.5	122.4	91.1	125.3	109.8	94.5	161.8	131.2
COVERS	DIVS. 10 - 14, 25, 28, 41, 43, 44, 46	100.0	77.2	95.0	100.0	101.4	100.3	100.0	125.0	105.5	100.0	117.8	103.9	100.0	102.7	100.6	100.0	121.3	104.7
21, 22, 23	FIRE SUPPRESSION, PLUMBING & HVAC	96.5	104.7	99.8	96.5	105.7	100.2	99.9	133.0	113.3	96.4	132.4	111.0	96.5	127.0	108.8	99.8	136.4	114.6
26, 27, 3370	ELECTRICAL, COMMUNICATIONS & UTIL.	92.9	108.8	101.0	95.1	83.8	89.3	96.5	136.5	116.9	91.8	133.8	113.2	92.2	133.9	113.4	96.4	135.7	116.4
MF2016	WEIGHTED AVERAGE	95.7	109.2	101.5	95.9	109.5	101.8	99.7	142.2	117.9	96.4	136.1	113.4	96.1	128.9	110.2	99.8	144.1	118.8

DIVISION		ILLINOIS																	
		PEORIA			QUINCY			ROCK ISLAND			ROCKFORD			SOUTH SUBURBAN			SPRINGFIELD		
		615 - 616			623			612			610 - 611			605			626 - 627		
		MAT.	INST.	TOTAL	MAT.	INST.	TOTAL	MAT.	INST.	TOTAL	MAT.	INST.	TOTAL	MAT.	INST.	TOTAL	MAT.	INST.	TOTAL
015433	CONTRACTOR EQUIPMENT		104.7	104.7		105.6	105.6		104.7	104.7		104.7	104.7		95.4	95.4		105.6	105.6
0241, 31 - 34	SITE & INFRASTRUCTURE, DEMOLITION	99.9	100.7	100.5	96.7	101.2	99.9	97.6	99.6	99.0	99.3	101.8	101.0	98.6	101.7	100.7	98.4	101.6	100.6
0310	Concrete Forming & Accessories	93.6	119.2	115.4	94.1	116.3	113.0	92.1	101.9	100.4	97.9	131.3	126.4	98.1	158.8	149.8	93.0	116.6	113.0
0320	Concrete Reinforcing	93.1	107.7	100.5	91.0	88.5	89.7	95.5	101.8	98.7	88.1	138.0	113.5	100.2	153.3	127.2	91.3	100.9	96.2
0330	Cast-in-Place Concrete	98.0	119.0	105.8	96.8	109.6	101.6	98.8	101.3	99.8	100.3	130.5	111.5	115.0	154.4	129.6	92.1	111.1	99.2
03	CONCRETE	92.1	117.7	103.4	89.6	109.7	98.5	92.7	102.7	97.2	92.8	132.6	110.5	102.2	155.3	125.7	87.5	112.5	98.6
04	MASONRY	110.8	124.0	119.0	100.2	114.1	108.8	110.9	102.4	105.6	86.3	145.3	122.8	100.7	162.1	138.6	81.7	121.6	106.4
05	METALS	97.5	125.0	105.8	104.3	113.0	106.9	94.9	119.6	102.4	97.5	142.8	111.2	101.1	142.0	113.5	104.6	119.3	109.1
06	WOOD, PLASTICS & COMPOSITES	98.8	116.5	108.4	89.6	117.3	104.6	92.0	100.4	96.5	98.8	127.2	114.2	98.1	157.7	130.5	91.0	114.9	104.0
07	THERMAL & MOISTURE PROTECTION	99.0	115.6	106.2	100.1	108.5	103.8	98.2	101.2	99.5	101.4	132.6	115.0	98.4	148.9	120.3	103.0	114.6	108.1
08	OPENINGS	98.8	123.0	104.4	96.0	114.9	100.4	93.3	106.3	96.3	98.8	140.5	108.5	102.0	168.4	117.4	100.8	117.4	104.6
0920	Plaster & Gypsum Board	95.1	117.0	109.5	96.6	117.9	110.6	90.6	100.5	97.1	95.1	128.0	116.8	90.9	159.5	136.1	99.5	115.4	110.0
0950, 0980	Ceilings & Acoustic Treatment	94.8	117.0	109.8	84.6	117.9	107.1	89.7	100.5	97.0	94.8	128.0	117.3	99.4	159.5	140.1	91.0	115.4	107.5
0960	Flooring	93.7	123.5	102.3	99.5	111.5	103.0	91.3	97.8	93.2	93.7	129.1	103.9	97.2	164.8	116.8	102.0	115.2	105.8
0970, 0990	Wall Finishes & Painting/Coating	88.7	138.3	118.0	100.8	114.4	108.9	88.7	98.0	94.2	88.7	146.5	122.9	99.0	168.2	139.9	102.9	113.1	108.9
09	FINISHES	90.8	122.2	107.9	91.4	116.5	105.1	88.4	100.4	94.9	90.8	132.2	113.4	94.5	161.8	131.2	97.3	116.4	107.7
COVERS	DIVS. 10 - 14, 25, 28, 41, 43, 44, 46	100.0	106.6	101.4	100.0	79.8	95.5	100.0	99.5	99.9	100.0	116.5	103.6	100.0	121.3	104.7	100.0	105.3	101.2
21, 22, 23	FIRE SUPPRESSION, PLUMBING & HVAC	99.9	102.4	100.9	96.5	103.5	99.3	96.5	99.7	97.8	100.0	118.7	107.6	99.8	136.4	114.6	99.9	102.4	101.0
26, 27, 3370	ELECTRICAL, COMMUNICATIONS & UTIL.	96.1	93.7	94.9	90.6	84.2	87.3	87.4	95.4	91.5	96.4	128.7	112.8	96.4	135.7	116.4	97.4	84.2	90.7
MF2016	WEIGHTED AVERAGE	97.8	111.5	103.7	96.3	105.1	100.1	94.9	102.1	98.0	96.8	128.7	110.5	99.8	144.0	118.7	97.9	107.7	102.1

City Cost Indexes

DIVISION		INDIANA																	
		ANDERSON			BLOOMINGTON			COLUMBUS			EVANSVILLE			FORT WAYNE			GARY		
		460			474			472			476 - 477			467 - 468			463 - 464		
		MAT.	INST.	TOTAL	MAT.	INST.	TOTAL	MAT.	INST.	TOTAL	MAT.	INST.	TOTAL	MAT.	INST.	TOTAL	MAT.	INST.	TOTAL
015433	**CONTRACTOR EQUIPMENT**		97.4	97.4		83.9	83.9		83.9	83.9		113.7	113.7		97.4	97.4		97.4	97.4
0241, 31 - 34	**SITE & INFRASTRUCTURE, DEMOLITION**	99.4	93.7	95.5	87.3	92.1	90.6	84.0	91.9	89.5	92.7	121.2	112.6	100.5	93.6	95.7	100.1	97.3	98.2
0310	Concrete Forming & Accessories	96.5	79.6	82.1	101.2	80.1	83.2	95.2	79.9	82.2	94.4	80.8	82.8	94.4	75.7	78.5	96.6	112.6	110.2
0320	Concrete Reinforcing	95.6	85.4	90.4	89.4	85.1	87.2	89.8	87.2	88.5	97.9	83.4	90.5	95.6	80.2	87.8	95.6	112.7	104.3
0330	Cast-in-Place Concrete	103.7	77.1	93.8	99.3	76.2	90.7	98.8	74.1	89.6	94.9	85.7	91.5	110.2	76.2	97.6	108.4	113.7	110.4
03	**CONCRETE**	91.8	80.3	86.7	97.1	79.2	89.2	96.4	78.7	88.6	97.3	83.1	91.0	94.7	77.4	87.0	94.0	112.7	102.3
04	**MASONRY**	87.3	75.6	80.1	88.1	72.0	78.1	87.9	73.1	78.7	83.7	78.9	80.7	90.4	72.9	79.6	88.7	109.2	101.3
05	**METALS**	98.8	89.1	95.9	99.5	74.2	91.9	99.6	74.4	91.9	92.5	84.0	90.0	98.8	87.2	95.3	98.8	105.1	100.7
06	**WOOD, PLASTICS & COMPOSITES**	97.5	79.8	87.9	112.3	80.6	95.1	106.6	80.4	92.4	93.2	79.9	86.0	97.2	76.0	85.6	94.7	111.0	103.5
07	**THERMAL & MOISTURE PROTECTION**	109.6	77.2	95.5	96.7	77.7	88.5	96.1	78.1	88.2	100.9	84.0	93.6	109.4	78.9	96.1	108.1	106.1	107.2
08	**OPENINGS**	95.3	78.0	91.3	99.7	78.6	94.8	95.9	79.0	92.0	93.8	78.2	90.2	95.3	73.2	90.1	95.3	114.2	99.7
0920	Plaster & Gypsum Board	104.6	79.5	88.1	98.6	80.7	86.8	95.5	80.6	85.7	94.0	79.0	84.1	103.7	75.6	85.2	97.3	111.6	106.7
0950, 0980	Ceilings & Acoustic Treatment	91.9	79.5	83.5	79.6	80.7	80.4	79.6	80.6	80.3	83.4	79.0	80.4	91.9	75.6	80.8	91.9	111.6	105.2
0960	Flooring	95.1	77.5	90.0	99.1	83.1	94.4	94.0	83.1	90.9	94.0	73.8	88.1	95.1	73.5	88.9	95.1	114.3	100.7
0970, 0990	Wall Finishes & Painting/Coating	94.0	67.6	78.4	85.4	79.5	81.9	85.4	79.5	81.9	91.0	86.6	88.4	94.0	72.1	81.0	94.0	121.4	110.2
09	**FINISHES**	91.8	77.9	84.2	90.1	80.7	85.0	88.2	80.6	84.0	88.8	79.9	84.0	91.5	75.1	82.5	90.6	113.9	103.3
COVERS	**DIVS. 10 - 14, 25, 28, 41, 43, 44, 46**	100.0	89.2	97.6	100.0	85.8	96.9	100.0	85.8	96.9	100.0	93.1	98.5	100.0	89.1	97.6	100.0	105.1	101.1
21, 22, 23	**FIRE SUPPRESSION, PLUMBING & HVAC**	99.9	78.7	91.4	99.8	78.9	91.3	96.3	78.6	89.2	100.0	80.4	92.0	99.9	73.8	89.4	99.9	105.1	102.0
26, 27, 3370	**ELECTRICAL, COMMUNICATIONS & UTIL.**	86.5	85.2	85.8	99.2	86.6	92.8	98.4	87.0	92.6	95.2	83.4	89.2	87.2	76.7	81.8	98.1	110.7	104.5
MF2016	**WEIGHTED AVERAGE**	95.9	81.8	89.8	97.7	80.3	90.3	96.1	80.4	89.3	95.4	84.8	90.9	96.4	78.2	88.6	97.2	108.3	102.0

DIVISION		INDIANA																	
		INDIANAPOLIS			KOKOMO			LAFAYETTE			LAWRENCEBURG			MUNCIE			NEW ALBANY		
		461 - 462			469			479			470			473			471		
		MAT.	INST.	TOTAL	MAT.	INST.	TOTAL	MAT.	INST.	TOTAL	MAT.	INST.	TOTAL	MAT.	INST.	TOTAL	MAT.	INST.	TOTAL
015433	**CONTRACTOR EQUIPMENT**		85.3	85.3		97.4	97.4		83.9	83.9		103.6	103.6		95.5	95.5		93.3	93.3
0241, 31 - 34	**SITE & INFRASTRUCTURE, DEMOLITION**	100.2	90.9	93.7	95.8	93.6	94.3	84.8	92.0	89.9	82.3	107.3	99.8	87.3	92.6	91.0	79.5	94.5	89.9
0310	Concrete Forming & Accessories	101.2	85.1	87.5	99.7	78.0	81.2	92.6	80.9	82.6	91.6	76.6	78.9	92.3	79.1	81.0	90.4	77.4	79.3
0320	Concrete Reinforcing	95.6	87.6	91.5	86.4	87.3	86.9	89.4	85.3	87.3	88.7	77.2	82.9	98.9	85.3	92.0	90.0	83.0	86.4
0330	Cast-in-Place Concrete	100.6	84.7	94.7	102.7	80.9	94.6	99.4	80.0	92.2	92.9	74.4	86.0	104.3	76.2	93.9	95.9	74.3	87.9
03	**CONCRETE**	97.5	84.7	91.8	88.9	81.2	85.5	96.7	80.9	89.7	89.8	76.6	84.0	95.3	79.7	88.4	95.2	77.5	87.4
04	**MASONRY**	88.0	79.1	82.5	87.0	73.6	78.7	93.3	75.5	82.3	73.2	71.8	72.3	89.8	75.7	81.1	79.8	68.1	72.6
05	**METALS**	96.1	75.7	89.9	95.2	89.6	93.5	97.9	74.4	90.8	94.5	84.4	91.4	101.3	88.9	97.6	96.5	82.0	92.1
06	**WOOD, PLASTICS & COMPOSITES**	98.6	86.0	91.8	100.8	77.4	88.1	103.6	81.5	91.6	91.3	76.5	83.3	104.9	79.4	91.0	93.7	78.6	85.5
07	**THERMAL & MOISTURE PROTECTION**	99.8	81.4	91.8	108.5	76.5	94.6	96.1	79.4	88.8	101.7	76.2	90.6	99.2	78.5	90.2	88.2	72.2	81.3
08	**OPENINGS**	104.1	82.2	99.0	90.3	77.3	87.3	94.4	78.9	90.8	95.8	73.8	90.7	93.0	77.8	89.5	93.4	77.5	89.7
0920	Plaster & Gypsum Board	96.4	85.8	89.4	109.5	77.1	88.1	93.3	81.7	85.7	71.9	76.5	74.9	94.0	79.5	84.5	91.9	78.5	83.0
0950, 0980	Ceilings & Acoustic Treatment	94.9	85.8	88.7	91.9	77.1	81.9	76.2	81.7	80.0	86.0	76.5	79.5	79.6	79.5	79.6	83.4	78.5	80.1
0960	Flooring	96.9	81.8	92.5	99.0	91.3	96.8	93.0	81.8	89.8	69.0	82.9	73.1	93.4	77.5	88.8	91.4	55.8	81.1
0970, 0990	Wall Finishes & Painting/Coating	99.3	79.5	87.6	94.0	69.7	79.6	85.4	83.5	84.3	86.2	74.5	79.3	85.4	67.6	74.8	91.0	68.0	77.4
09	**FINISHES**	96.0	84.2	89.6	93.4	79.6	85.8	86.9	81.5	83.9	78.6	77.7	78.1	87.4	77.6	82.0	88.0	72.3	79.5
COVERS	**DIVS. 10 - 14, 25, 28, 41, 43, 44, 46**	100.0	90.6	97.9	100.0	86.4	97.0	100.0	88.4	97.5	100.0	84.8	96.6	100.0	88.1	97.4	100.0	83.5	96.4
21, 22, 23	**FIRE SUPPRESSION, PLUMBING & HVAC**	99.9	79.7	91.7	96.5	78.8	89.3	96.3	77.9	88.9	97.2	74.5	88.0	99.8	78.6	91.2	96.5	76.0	88.2
26, 27, 3370	**ELECTRICAL, COMMUNICATIONS & UTIL.**	101.9	87.0	94.3	90.9	79.1	84.9	97.8	80.5	89.0	93.1	72.9	82.9	91.0	75.9	83.3	93.7	76.4	84.9
MF2016	**WEIGHTED AVERAGE**	98.7	83.0	92.0	94.1	81.0	88.4	95.8	80.1	89.1	92.6	78.5	86.5	96.1	80.3	89.3	93.9	77.2	86.7

DIVISION		INDIANA									IOWA								
		SOUTH BEND			TERRE HAUTE			WASHINGTON			BURLINGTON			CARROLL			CEDAR RAPIDS		
		465 - 466			478			475			526			514			522 - 524		
		MAT.	INST.	TOTAL	MAT.	INST.	TOTAL	MAT.	INST.	TOTAL	MAT.	INST.	TOTAL	MAT.	INST.	TOTAL	MAT.	INST.	TOTAL
015433	**CONTRACTOR EQUIPMENT**		108.9	108.9		113.7	113.7		113.7	113.7		101.4	101.4		101.4	101.4		98.2	98.2
0241, 31 - 34	**SITE & INFRASTRUCTURE, DEMOLITION**	99.3	94.3	95.8	94.4	121.6	113.4	94.4	121.8	113.5	98.5	96.5	97.1	87.8	97.3	94.4	100.1	96.0	97.3
0310	Concrete Forming & Accessories	95.5	78.1	80.7	95.2	79.2	81.6	96.2	82.1	84.2	94.4	96.8	96.4	81.9	79.8	80.1	100.5	86.3	88.4
0320	Concrete Reinforcing	95.8	86.0	90.8	97.9	85.4	91.6	90.5	84.9	87.7	94.4	100.6	97.6	95.1	88.3	91.6	95.1	82.5	88.7
0330	Cast-in-Place Concrete	103.6	79.9	94.8	91.8	80.6	87.7	99.9	86.0	94.7	107.9	56.4	88.7	107.9	84.4	99.2	108.2	85.9	99.9
03	**CONCRETE**	91.0	81.7	86.9	100.2	81.0	91.7	106.0	84.1	96.3	93.6	83.7	89.2	92.4	83.7	88.5	93.7	86.1	90.3
04	**MASONRY**	92.9	75.6	82.2	91.0	74.8	81.0	83.9	78.6	80.7	99.7	73.4	83.5	101.3	74.0	84.4	105.3	81.7	90.7
05	**METALS**	102.3	104.8	103.0	93.2	85.2	90.8	87.7	85.2	87.0	85.5	99.9	89.8	85.5	95.8	88.6	87.9	93.9	89.7
06	**WOOD, PLASTICS & COMPOSITES**	95.8	77.2	85.7	95.5	78.9	86.5	95.9	81.5	88.1	93.0	101.0	97.4	78.7	82.9	81.0	100.4	86.1	92.6
07	**THERMAL & MOISTURE PROTECTION**	102.4	80.4	92.8	101.0	81.4	92.5	100.9	84.2	93.6	103.9	79.2	93.1	104.2	76.3	92.1	105.0	82.3	95.1
08	**OPENINGS**	95.8	77.7	91.6	94.3	77.5	90.4	91.2	79.4	88.4	96.5	98.6	97.0	101.0	82.2	96.6	101.5	83.6	97.3
0920	Plaster & Gypsum Board	94.5	76.9	82.9	94.0	78.0	83.4	94.3	80.6	85.3	105.0	101.2	102.5	100.5	82.6	88.7	110.1	86.0	94.3
0950, 0980	Ceilings & Acoustic Treatment	90.3	76.9	81.2	83.4	78.0	79.7	79.2	80.6	80.1	95.0	101.2	99.2	95.0	82.6	86.6	97.5	86.0	89.7
0960	Flooring	93.6	89.1	92.3	94.0	78.2	89.4	94.9	79.3	90.3	95.3	71.4	88.4	89.6	82.2	87.5	109.6	87.1	103.0
0970, 0990	Wall Finishes & Painting/Coating	88.9	84.7	86.4	91.0	81.9	85.6	91.0	86.6	88.4	92.3	87.5	89.5	92.3	85.8	88.4	93.9	72.6	81.3
09	**FINISHES**	90.0	80.5	84.9	88.8	79.2	83.6	88.6	81.9	85.0	92.4	92.4	92.4	88.7	81.2	84.6	97.8	85.1	90.9
COVERS	**DIVS. 10 - 14, 25, 28, 41, 43, 44, 46**	100.0	90.3	97.9	100.0	90.5	97.9	100.0	93.3	98.5	100.0	92.1	98.3	100.0	85.8	96.9	100.0	94.0	98.7
21, 22, 23	**FIRE SUPPRESSION, PLUMBING & HVAC**	99.9	76.7	90.5	100.0	78.2	91.2	96.5	80.4	90.0	96.7	85.7	92.2	96.7	72.6	86.9	100.1	83.0	93.2
26, 27, 3370	**ELECTRICAL, COMMUNICATIONS & UTIL.**	97.7	86.0	91.8	93.5	85.5	89.4	94.0	83.5	88.6	100.3	74.0	86.9	101.0	78.8	89.7	97.9	82.0	89.8
MF2016	**WEIGHTED AVERAGE**	97.4	83.6	91.5	96.2	83.8	90.9	94.6	85.4	90.7	95.1	86.3	91.3	94.9	81.3	89.1	97.4	85.8	92.4

City Cost Indexes

DIVISION		IOWA																	
		COUNCIL BLUFFS			CRESTON			DAVENPORT			DECORAH			DES MOINES			DUBUQUE		
		515			508			527 - 528			521			500 - 503, 509			520		
		MAT.	INST.	TOTAL	MAT.	INST.	TOTAL	MAT.	INST.	TOTAL	MAT.	INST.	TOTAL	MAT.	INST.	TOTAL	MAT.	INST.	TOTAL
015433	CONTRACTOR EQUIPMENT		97.5	97.5		101.4	101.4		101.4	101.4		101.4	101.4		103.2	103.2		97.0	97.0
0241, 31 - 34	SITE & INFRASTRUCTURE, DEMOLITION	103.9	93.0	96.3	95.1	98.4	97.4	98.7	99.4	99.2	97.1	96.3	96.5	101.1	99.9	100.3	98.0	93.2	94.7
0310	Concrete Forming & Accessories	81.2	75.6	76.4	76.5	82.6	81.7	100.0	98.6	98.8	91.9	73.7	76.4	93.4	86.1	87.2	82.7	83.8	83.6
0320	Concrete Reinforcing	97.0	82.1	89.4	94.5	88.3	91.3	95.1	101.9	98.6	94.4	88.0	91.1	99.0	85.7	92.3	93.7	82.3	87.9
0330	Cast-in-Place Concrete	112.5	81.4	100.9	114.7	62.5	95.3	104.2	98.0	101.9	105.0	78.2	95.0	97.6	90.5	95.0	105.9	85.4	98.3
03	CONCRETE	95.8	79.7	88.6	94.0	77.3	86.6	91.8	99.4	95.2	91.6	78.7	85.9	87.3	88.3	87.7	90.6	84.8	88.0
04	MASONRY	106.7	77.8	88.8	99.3	81.6	88.3	102.1	94.1	97.2	120.9	76.2	93.3	87.3	85.5	86.2	106.3	70.3	84.1
05	METALS	92.6	93.5	92.9	87.0	96.3	89.8	87.9	106.4	93.5	85.6	95.4	88.6	92.6	97.2	94.0	86.5	93.1	88.5
06	WOOD, PLASTICS & COMPOSITES	77.5	74.2	75.7	70.6	82.9	77.3	100.4	98.1	99.1	89.8	72.9	80.7	87.7	85.3	86.4	79.3	84.6	82.2
07	THERMAL & MOISTURE PROTECTION	104.3	76.9	92.4	105.0	81.5	94.8	104.4	94.2	99.9	104.1	73.4	90.8	96.6	85.9	92.0	104.6	78.6	93.3
08	OPENINGS	100.5	77.6	95.2	111.0	84.0	104.7	101.5	99.5	101.0	99.5	80.2	95.0	98.9	88.2	96.4	100.5	84.8	96.9
0920	Plaster & Gypsum Board	100.5	73.9	82.9	96.1	82.6	87.2	110.1	98.2	102.3	103.8	72.3	83.1	93.5	85.0	87.9	100.5	84.5	90.0
0950, 0980	Ceilings & Acoustic Treatment	95.0	73.9	80.7	88.5	82.6	84.5	97.5	98.2	98.0	95.0	72.3	79.6	91.9	85.0	87.3	95.0	84.5	87.9
0960	Flooring	88.5	87.1	88.1	83.9	71.4	80.3	97.7	92.4	96.2	94.9	71.4	88.1	97.9	94.8	97.0	100.8	71.4	92.3
0970, 0990	Wall Finishes & Painting/Coating	89.1	72.6	79.3	83.2	76.2	79.1	92.3	92.9	92.6	92.3	85.8	88.4	88.8	85.8	87.0	93.1	80.5	85.7
09	FINISHES	89.4	77.1	82.7	82.5	79.9	81.1	94.3	96.8	95.7	92.0	74.4	82.4	90.2	87.7	88.8	93.1	81.2	86.6
COVERS	DIVS. 10 - 14, 25, 28, 41, 43, 44, 46	100.0	89.4	97.7	100.0	88.4	97.4	100.0	97.6	99.5	100.0	86.2	97.0	100.0	94.4	98.8	100.0	92.9	98.4
21, 22, 23	FIRE SUPPRESSION, PLUMBING & HVAC	100.1	74.4	89.7	96.6	80.2	90.0	100.1	96.0	98.5	96.7	75.3	88.0	99.8	84.7	93.7	100.1	77.3	90.9
26, 27, 3370	ELECTRICAL, COMMUNICATIONS & UTIL.	103.2	82.3	92.6	93.8	78.8	86.2	95.9	89.7	92.8	97.9	48.4	72.7	105.4	84.4	94.7	101.7	78.8	90.1
MF2016	WEIGHTED AVERAGE	98.2	80.8	90.7	95.2	83.1	90.0	96.5	96.9	96.7	95.9	75.9	87.3	96.0	88.4	92.8	96.5	81.9	90.3

DIVISION		IOWA																	
		FORT DODGE			MASON CITY			OTTUMWA			SHENANDOAH			SIBLEY			SIOUX CITY		
		505			504			525			516			512			510 - 511		
		MAT.	INST.	TOTAL	MAT.	INST.	TOTAL	MAT.	INST.	TOTAL	MAT.	INST.	TOTAL	MAT.	INST.	TOTAL	MAT.	INST.	TOTAL
015433	CONTRACTOR EQUIPMENT		101.4	101.4		101.4	101.4		97.0	97.0		97.5	97.5		101.4	101.4		101.4	101.4
0241, 31 - 34	SITE & INFRASTRUCTURE, DEMOLITION	103.5	95.1	97.6	103.6	96.2	98.4	98.4	91.0	93.3	102.2	93.1	95.9	108.2	96.2	99.8	109.9	97.3	101.1
0310	Concrete Forming & Accessories	77.1	74.1	74.5	81.0	73.3	74.4	89.8	73.7	76.1	82.9	79.2	79.8	83.5	38.7	45.4	100.5	74.4	78.3
0320	Concrete Reinforcing	94.5	88.0	91.2	94.3	88.0	91.1	94.4	100.6	97.5	97.0	88.3	92.6	97.0	87.9	92.4	95.1	79.3	87.1
0330	Cast-in-Place Concrete	107.6	43.0	83.6	107.6	73.3	94.9	108.6	65.6	92.6	108.8	85.0	99.9	106.6	57.1	88.1	107.2	71.9	94.1
03	CONCRETE	89.5	66.7	79.4	89.7	76.8	84.0	93.1	76.6	85.8	93.4	83.6	89.1	92.4	55.5	76.0	93.1	75.3	85.2
04	MASONRY	98.1	58.6	73.7	110.9	75.0	88.7	102.6	54.0	72.6	106.2	76.0	87.6	124.9	58.8	84.1	99.4	68.4	80.2
05	METALS	87.1	95.0	89.5	87.1	95.4	89.6	85.5	100.1	89.9	91.6	96.3	93.0	85.7	94.3	88.3	87.9	91.4	89.0
06	WOOD, PLASTICS & COMPOSITES	71.1	82.9	77.5	75.0	72.9	73.9	86.9	79.7	83.0	79.3	80.8	80.1	80.1	35.1	55.7	100.4	74.3	86.2
07	THERMAL & MOISTURE PROTECTION	104.3	67.9	88.4	103.7	74.0	90.8	104.8	68.7	89.1	103.5	74.4	90.8	103.8	55.8	82.9	104.4	72.6	90.6
08	OPENINGS	104.8	73.7	97.6	96.8	80.2	93.0	101.0	83.9	97.0	92.1	78.7	89.0	97.7	47.3	86.0	101.5	75.5	95.4
0920	Plaster & Gypsum Board	96.1	82.6	87.2	96.1	72.3	80.5	101.4	79.5	87.0	100.5	80.6	87.4	100.5	33.5	56.3	110.1	73.8	86.2
0950, 0980	Ceilings & Acoustic Treatment	88.5	82.6	84.5	88.5	72.3	77.6	95.0	79.5	84.5	95.0	80.6	85.2	95.0	33.5	53.3	97.5	73.8	81.4
0960	Flooring	85.2	71.4	81.2	87.2	71.4	82.7	103.8	71.4	94.4	89.2	76.4	85.5	90.5	71.4	85.0	97.7	74.0	90.9
0970, 0990	Wall Finishes & Painting/Coating	83.2	83.4	83.3	83.2	85.8	84.7	93.1	85.8	88.8	89.1	85.8	87.1	92.3	85.8	88.4	92.3	68.2	78.0
09	FINISHES	84.3	75.5	79.5	84.9	74.1	79.0	94.3	74.8	83.6	89.5	79.3	83.9	91.7	47.7	67.7	95.7	73.5	83.6
COVERS	DIVS. 10 - 14, 25, 28, 41, 43, 44, 46	100.0	82.4	96.1	100.0	85.8	96.9	100.0	82.8	96.2	100.0	85.8	96.9	100.0	76.3	94.8	100.0	91.1	98.0
21, 22, 23	FIRE SUPPRESSION, PLUMBING & HVAC	96.6	69.5	85.6	96.6	78.6	89.3	96.7	72.6	86.9	96.7	85.7	92.3	96.7	67.7	84.9	100.1	78.1	91.2
26, 27, 3370	ELECTRICAL, COMMUNICATIONS & UTIL.	100.2	71.1	85.4	99.3	48.4	73.4	100.1	72.7	86.2	97.9	80.4	89.0	97.9	48.4	72.7	97.9	75.8	86.7
MF2016	WEIGHTED AVERAGE	94.9	74.1	86.0	94.7	76.2	86.8	95.7	76.3	87.4	95.4	83.8	90.4	96.1	63.3	82.1	97.1	78.7	89.2

DIVISION		IOWA						KANSAS											
		SPENCER			WATERLOO			BELLEVILLE			COLBY			DODGE CITY			EMPORIA		
		513			506 - 507			669			677			678			668		
		MAT.	INST.	TOTAL	MAT.	INST.	TOTAL	MAT.	INST.	TOTAL	MAT.	INST.	TOTAL	MAT.	INST.	TOTAL	MAT.	INST.	TOTAL
015433	CONTRACTOR EQUIPMENT		101.4	101.4		101.4	101.4		106.5	106.5		106.5	106.5		106.5	106.5		104.7	104.7
0241, 31 - 34	SITE & INFRASTRUCTURE, DEMOLITION	108.2	94.9	99.0	108.8	97.4	100.9	109.9	96.7	100.7	107.2	97.4	100.3	109.6	96.4	100.4	102.3	94.2	96.7
0310	Concrete Forming & Accessories	89.7	38.5	46.1	92.1	70.6	73.8	96.9	55.7	61.8	99.3	62.9	68.3	92.6	62.7	67.2	87.5	67.4	70.4
0320	Concrete Reinforcing	97.0	87.9	92.3	95.1	82.6	88.7	105.6	103.6	104.6	105.2	103.6	104.4	102.5	103.6	103.1	104.1	103.8	103.9
0330	Cast-in-Place Concrete	106.6	67.4	92.0	115.3	86.5	104.6	119.6	86.0	107.1	116.6	89.8	106.6	118.7	89.5	107.8	115.7	89.9	106.1
03	CONCRETE	92.8	59.0	77.8	95.3	79.3	88.2	110.0	75.9	94.9	107.3	80.6	95.5	108.7	80.3	96.1	102.6	82.6	93.7
04	MASONRY	124.9	58.8	84.1	98.9	76.7	85.2	90.4	59.7	71.5	99.5	66.0	78.8	108.0	65.9	82.0	96.0	67.4	78.3
05	METALS	85.7	94.1	88.2	89.4	93.8	90.7	88.9	99.3	92.0	89.2	100.2	92.5	90.7	99.1	93.3	88.6	100.3	92.1
06	WOOD, PLASTICS & COMPOSITES	86.8	35.1	58.7	88.7	66.2	76.5	97.0	52.6	72.9	101.1	59.0	78.2	92.5	59.0	74.3	87.3	64.8	75.1
07	THERMAL & MOISTURE PROTECTION	104.8	58.9	84.8	104.0	79.3	93.2	93.3	63.9	80.5	99.6	67.2	85.5	99.6	69.9	86.6	91.4	76.8	85.0
08	OPENINGS	109.0	47.3	94.7	97.3	75.1	92.1	96.2	63.7	88.7	100.2	71.0	93.4	100.1	67.2	92.5	94.2	72.8	89.2
0920	Plaster & Gypsum Board	101.4	33.5	56.7	104.9	65.4	78.9	89.2	51.4	64.3	99.3	58.0	72.1	92.9	58.0	69.9	86.2	63.9	71.5
0950, 0980	Ceilings & Acoustic Treatment	95.0	33.5	53.3	91.1	65.4	73.7	85.4	51.4	62.4	82.9	58.0	66.0	82.9	58.0	66.0	85.4	63.9	70.8
0960	Flooring	93.3	71.4	86.9	92.4	81.6	89.3	90.5	69.3	84.4	86.2	69.3	81.3	82.4	69.3	78.6	86.1	69.3	81.2
0970, 0990	Wall Finishes & Painting/Coating	92.3	58.7	72.4	83.2	83.4	83.3	92.0	56.5	71.0	92.8	56.5	71.3	92.8	56.5	71.3	92.0	56.5	71.0
09	FINISHES	92.6	43.5	65.8	88.4	73.0	80.0	85.9	56.7	70.0	86.0	62.0	72.9	84.1	62.0	72.0	83.2	65.4	73.5
COVERS	DIVS. 10 - 14, 25, 28, 41, 43, 44, 46	100.0	76.2	94.8	100.0	91.5	98.1	100.0	83.6	96.4	100.0	86.4	97.0	100.0	86.4	97.0	100.0	87.0	97.1
21, 22, 23	FIRE SUPPRESSION, PLUMBING & HVAC	96.7	69.7	85.7	100.0	81.9	92.7	96.4	71.3	86.3	96.6	72.1	86.7	100.1	72.1	88.8	96.4	74.9	87.7
26, 27, 3370	ELECTRICAL, COMMUNICATIONS & UTIL.	99.6	48.4	73.6	95.7	63.1	79.1	99.2	64.2	81.4	96.7	69.5	82.9	93.8	69.5	81.5	96.4	68.5	82.2
MF2016	WEIGHTED AVERAGE	97.7	63.6	83.1	96.2	79.3	89.0	96.6	72.1	86.1	97.1	75.7	87.9	98.3	75.4	88.5	94.9	77.2	87.3

DIVISION		KANSAS																	
		FORT SCOTT			HAYS			HUTCHINSON			INDEPENDENCE			KANSAS CITY			LIBERAL		
		667			676			675			673			660 - 662			679		
		MAT.	INST.	TOTAL	MAT.	INST.	TOTAL	MAT.	INST.	TOTAL	MAT.	INST.	TOTAL	MAT.	INST.	TOTAL	MAT.	INST.	TOTAL
015433	CONTRACTOR EQUIPMENT		105.6	105.6		106.5	106.5		106.5	106.5		106.5	106.5		103.1	103.1		106.5	106.5
0241, 31 - 34	SITE & INFRASTRUCTURE, DEMOLITION	99.0	94.2	95.7	111.5	96.9	101.3	91.1	97.4	95.5	110.5	97.4	101.4	93.8	94.5	94.3	111.4	97.0	101.3
0310	Concrete Forming & Accessories	105.5	82.1	85.6	96.8	60.5	65.9	87.3	58.1	62.4	107.9	69.6	75.3	101.7	97.2	97.8	93.0	60.4	65.3
0320	Concrete Reinforcing	103.4	103.7	103.6	102.5	103.6	103.1	102.5	103.6	103.1	101.9	103.7	102.8	100.3	105.3	102.9	104.0	103.6	103.8
0330	Cast-in-Place Concrete	107.2	84.8	98.9	92.0	86.2	89.8	85.2	89.4	86.8	119.2	89.6	108.2	91.8	98.6	94.4	92.0	86.0	89.7
03	CONCRETE	98.1	87.5	93.4	98.9	78.2	89.7	82.3	78.2	80.5	109.8	83.5	98.2	90.3	99.5	94.4	100.7	78.1	90.7
04	MASONRY	97.0	57.3	72.5	107.5	59.8	78.0	99.5	65.9	78.7	96.7	65.9	77.7	98.1	98.0	98.1	105.6	59.7	77.3
05	METALS	88.6	99.7	91.9	88.8	99.9	92.2	88.6	99.0	91.8	88.6	99.6	91.9	95.6	105.2	98.5	89.1	99.2	92.1
06	WOOD, PLASTICS & COMPOSITES	108.2	88.9	97.7	97.8	59.0	76.7	87.2	52.9	68.5	112.3	67.8	88.1	103.8	96.8	100.0	93.2	59.0	74.6
07	THERMAL & MOISTURE PROTECTION	92.5	74.4	84.6	100.0	64.6	84.6	98.3	66.2	84.3	99.7	77.0	89.8	92.0	99.8	95.4	100.1	64.0	84.4
08	OPENINGS	94.2	86.1	92.3	100.1	67.2	92.4	100.1	63.8	91.6	98.1	71.9	92.0	95.4	96.4	95.6	100.1	67.2	92.5
0920	Plaster & Gypsum Board	91.6	88.7	89.7	96.5	58.0	71.1	91.6	51.6	65.3	107.8	67.0	80.9	85.5	96.8	92.9	93.8	58.0	70.2
0950, 0980	Ceilings & Acoustic Treatment	85.4	88.7	87.7	82.9	58.0	66.0	82.9	51.6	61.7	82.9	67.0	72.1	85.4	96.8	93.1	82.9	58.0	66.0
0960	Flooring	99.7	69.1	90.9	85.0	69.3	80.4	79.7	69.3	76.6	90.4	69.1	84.2	80.9	97.5	85.7	82.7	68.1	78.5
0970, 0990	Wall Finishes & Painting/Coating	93.6	79.5	85.3	92.8	56.5	71.3	92.8	56.5	71.3	92.8	56.5	71.3	99.6	100.9	100.3	92.8	56.5	71.3
09	FINISHES	88.2	80.5	84.0	85.8	60.5	72.0	81.6	58.4	68.9	88.6	67.7	77.2	83.7	97.5	91.3	85.0	60.2	71.5
COVERS	DIVS. 10 - 14, 25, 28, 41, 43, 44, 46	100.0	86.7	97.1	100.0	84.3	96.5	100.0	85.7	96.8	100.0	87.4	97.2	100.0	94.5	98.8	100.0	84.3	96.5
21, 22, 23	FIRE SUPPRESSION, PLUMBING & HVAC	96.4	68.3	85.1	96.6	68.9	85.4	96.6	70.7	86.2	96.6	72.0	86.7	99.8	98.8	99.4	96.6	70.3	86.0
26, 27, 3370	ELECTRICAL, COMMUNICATIONS & UTIL.	95.6	69.5	82.4	95.8	69.5	82.4	91.1	62.4	76.5	93.1	74.6	83.7	101.4	97.9	99.6	93.8	69.5	81.5
MF2016	WEIGHTED AVERAGE	94.8	78.2	87.7	96.3	73.5	86.5	92.4	73.1	84.1	97.0	77.9	88.8	96.0	98.6	97.1	96.2	73.7	86.5

DIVISION		KANSAS									KENTUCKY								
		SALINA			TOPEKA			WICHITA			ASHLAND			BOWLING GREEN			CAMPTON		
		674			664 - 666			670 - 672			411 - 412			421 - 422			413 - 414		
		MAT.	INST.	TOTAL	MAT.	INST.	TOTAL	MAT.	INST.	TOTAL	MAT.	INST.	TOTAL	MAT.	INST.	TOTAL	MAT.	INST.	TOTAL
015433	CONTRACTOR EQUIPMENT		106.5	106.5		104.7	104.7		106.5	106.5		100.4	100.4		93.3	93.3		99.7	99.7
0241, 31 - 34	SITE & INFRASTRUCTURE, DEMOLITION	100.1	96.9	97.9	96.9	94.0	94.9	96.2	98.1	97.5	115.0	83.0	92.7	79.5	94.2	89.7	88.5	95.5	93.4
0310	Concrete Forming & Accessories	89.1	55.8	60.7	98.8	67.8	72.4	94.8	55.0	60.9	87.0	92.2	91.4	86.6	77.8	79.1	89.2	83.2	84.1
0320	Concrete Reinforcing	101.9	103.4	102.7	99.9	102.7	101.3	100.0	101.0	100.5	91.4	96.9	94.2	88.7	79.0	83.8	89.6	95.7	92.7
0330	Cast-in-Place Concrete	103.0	86.2	96.8	96.5	87.9	93.3	96.3	74.6	88.2	87.3	97.5	91.1	86.6	71.5	81.0	96.6	69.7	86.6
03	CONCRETE	95.9	76.0	87.1	92.0	82.0	87.5	91.6	71.2	82.6	91.0	95.9	93.2	89.5	76.2	83.6	93.0	80.9	87.6
04	MASONRY	122.3	56.0	81.4	91.8	65.6	75.6	96.6	50.5	68.1	89.6	94.0	92.3	91.6	70.6	78.6	89.0	55.9	68.5
05	METALS	90.5	99.6	93.3	92.5	100.4	94.9	92.5	97.8	94.1	95.6	108.8	99.6	97.2	83.8	93.1	96.5	90.7	94.7
06	WOOD, PLASTICS & COMPOSITES	88.7	52.6	69.1	102.0	66.9	82.9	98.6	52.9	73.8	75.3	90.2	83.4	87.9	78.4	82.7	86.3	90.9	88.8
07	THERMAL & MOISTURE PROTECTION	99.0	62.8	83.2	96.3	76.1	87.6	97.8	59.4	81.1	92.1	90.9	91.5	88.2	79.6	84.5	101.1	69.7	87.4
08	OPENINGS	100.1	63.7	91.6	103.6	75.2	97.0	104.5	63.3	94.9	92.6	91.6	92.4	93.4	80.5	90.4	94.6	89.9	93.5
0920	Plaster & Gypsum Board	91.6	51.4	65.1	97.9	66.0	76.9	94.8	51.6	66.4	60.0	90.2	79.9	87.3	78.2	81.3	87.3	90.2	89.2
0950, 0980	Ceilings & Acoustic Treatment	82.9	51.4	61.6	91.7	66.0	74.3	86.7	51.6	63.0	78.7	90.2	86.5	83.4	78.2	79.9	83.4	90.2	88.0
0960	Flooring	81.0	69.3	77.6	95.7	69.3	88.0	92.8	69.3	86.0	74.3	84.0	77.1	89.4	64.0	82.0	91.5	66.5	84.3
0970, 0990	Wall Finishes & Painting/Coating	92.8	56.5	71.3	96.1	69.5	80.4	91.5	56.5	70.8	92.9	92.8	92.8	91.0	90.4	90.7	91.0	55.3	69.9
09	FINISHES	82.8	56.7	68.6	92.5	67.2	78.7	89.6	56.3	71.5	75.9	90.7	84.0	86.7	76.3	81.0	87.4	78.1	82.4
COVERS	DIVS. 10 - 14, 25, 28, 41, 43, 44, 46	100.0	84.3	96.5	100.0	81.6	95.9	100.0	83.7	96.4	100.0	90.0	97.8	100.0	85.6	96.8	100.0	48.8	88.7
21, 22, 23	FIRE SUPPRESSION, PLUMBING & HVAC	100.1	70.9	88.3	100.0	74.7	89.8	99.8	68.5	87.2	96.3	86.9	92.5	100.0	78.8	91.4	96.6	78.1	89.1
26, 27, 3370	ELECTRICAL, COMMUNICATIONS & UTIL.	93.6	71.9	82.5	100.1	73.6	86.6	97.5	71.8	84.5	91.4	90.4	90.9	94.0	76.5	85.1	91.4	90.3	90.9
MF2016	WEIGHTED AVERAGE	96.9	72.8	86.6	97.0	77.7	88.7	96.7	70.8	85.6	93.2	92.0	92.7	94.5	78.9	87.8	94.4	79.9	88.2

DIVISION		KENTUCKY																	
		CORBIN			COVINGTON			ELIZABETHTOWN			FRANKFORT			HAZARD			HENDERSON		
		407 - 409			410			427			406			417 - 418			424		
		MAT.	INST.	TOTAL	MAT.	INST.	TOTAL	MAT.	INST.	TOTAL	MAT.	INST.	TOTAL	MAT.	INST.	TOTAL	MAT.	INST.	TOTAL
015433	CONTRACTOR EQUIPMENT		99.7	99.7		103.6	103.6		93.3	93.3		99.7	99.7		99.7	99.7		113.7	113.7
0241, 31 - 34	SITE & INFRASTRUCTURE, DEMOLITION	91.3	96.0	94.6	83.8	107.3	100.2	73.9	93.9	87.8	89.4	96.5	94.4	86.3	96.6	93.5	82.5	120.8	109.2
0310	Concrete Forming & Accessories	85.5	78.1	79.2	84.7	72.7	74.5	81.3	74.0	75.1	102.4	79.3	82.8	85.6	84.1	84.4	92.3	79.1	81.1
0320	Concrete Reinforcing	92.1	92.3	92.2	88.3	80.9	84.5	89.2	84.3	86.7	96.9	82.2	89.5	90.0	92.5	91.3	88.8	81.1	84.9
0330	Cast-in-Place Concrete	92.8	74.3	85.9	92.4	82.1	88.6	78.3	69.0	74.8	94.1	76.4	87.6	92.9	71.7	85.0	76.5	87.0	80.4
03	CONCRETE	84.1	79.6	82.1	91.4	78.2	85.6	81.6	74.5	78.5	87.5	79.2	83.8	89.8	81.5	86.1	86.7	82.5	84.9
04	MASONRY	81.0	61.3	68.8	103.0	73.5	84.8	76.5	63.0	68.2	79.5	71.8	74.7	87.5	58.0	69.3	94.8	80.4	85.9
05	METALS	92.2	89.3	91.3	94.4	88.9	92.8	96.4	84.9	92.9	95.0	86.0	92.3	96.5	89.7	94.4	87.5	85.8	87.0
06	WOOD, PLASTICS & COMPOSITES	75.0	79.8	77.6	83.7	70.2	76.4	82.5	76.1	79.0	102.1	79.8	90.0	82.9	90.9	87.2	90.8	78.0	83.8
07	THERMAL & MOISTURE PROTECTION	103.6	71.5	89.6	101.9	73.7	89.7	87.6	70.5	80.2	102.5	75.7	90.8	101.0	71.5	88.1	100.2	85.1	93.7
08	OPENINGS	89.2	71.9	85.2	96.6	73.9	91.4	93.4	74.7	89.0	96.7	79.6	92.7	95.0	89.1	93.6	91.6	79.4	88.7
0920	Plaster & Gypsum Board	93.1	78.9	83.7	68.6	70.0	69.5	86.1	75.8	79.3	95.2	78.9	84.4	86.1	90.2	88.8	91.0	77.0	81.8
0950, 0980	Ceilings & Acoustic Treatment	79.8	78.9	79.2	86.0	70.0	75.2	83.4	75.8	78.3	90.1	78.9	82.5	83.4	90.2	88.0	79.2	77.0	77.7
0960	Flooring	86.0	66.5	80.3	66.6	86.3	72.3	86.8	76.3	83.8	95.1	66.5	86.8	89.8	66.5	83.0	93.0	78.6	88.9
0970, 0990	Wall Finishes & Painting/Coating	90.7	63.6	74.7	86.2	73.0	78.4	91.0	69.9	78.6	97.5	94.1	95.5	91.0	55.3	69.9	91.0	89.0	89.9
09	FINISHES	81.3	74.6	77.6	77.6	74.8	76.1	85.3	73.7	79.0	89.1	78.9	83.5	86.6	78.7	82.3	86.8	80.1	83.2
COVERS	DIVS. 10 - 14, 25, 28, 41, 43, 44, 46	100.0	91.7	98.2	100.0	87.7	97.3	100.0	71.5	93.7	100.0	58.6	90.9	100.0	49.6	88.9	100.0	56.9	90.5
21, 22, 23	FIRE SUPPRESSION, PLUMBING & HVAC	96.7	74.9	87.9	97.2	77.7	89.3	96.7	77.8	89.0	100.0	78.7	91.4	96.6	79.2	89.6	96.7	78.1	89.2
26, 27, 3370	ELECTRICAL, COMMUNICATIONS & UTIL.	90.3	90.3	90.3	95.2	74.2	84.5	91.3	78.2	84.6	102.2	73.2	87.5	91.4	90.4	90.9	93.5	75.5	84.3
MF2016	WEIGHTED AVERAGE	91.1	79.6	86.2	94.4	79.8	88.1	91.3	76.7	85.1	95.5	78.7	88.3	93.8	80.6	88.1	92.2	82.4	88.0

DIVISION		KENTUCKY																	
		LEXINGTON			LOUISVILLE			OWENSBORO			PADUCAH			PIKEVILLE			SOMERSET		
		403 - 405			400 - 402			423			420			415 - 416			425 - 426		
		MAT.	INST.	TOTAL	MAT.	INST.	TOTAL	MAT.	INST.	TOTAL	MAT.	INST.	TOTAL	MAT.	INST.	TOTAL	MAT.	INST.	TOTAL
015433	CONTRACTOR EQUIPMENT		99.7	99.7		93.3	93.3		113.7	113.7		113.7	113.7		100.4	100.4		99.7	99.7
0241, 31 - 34	SITE & INFRASTRUCTURE, DEMOLITION	93.3	97.8	96.4	88.6	94.1	92.4	92.6	121.4	112.7	85.3	120.9	110.1	126.4	82.1	95.5	78.8	96.0	90.8
0310	Concrete Forming & Accessories	97.8	75.6	78.9	94.4	76.9	79.5	90.7	79.0	80.7	88.6	80.8	82.0	96.2	88.0	89.3	86.7	73.4	75.4
0320	Concrete Reinforcing	101.2	82.8	91.8	96.9	82.6	89.6	88.8	79.5	84.1	89.4	82.9	86.1	91.9	96.8	94.4	89.2	62.1	75.4
0330	Cast-in-Place Concrete	95.0	86.9	92.0	89.9	70.6	82.7	89.3	86.4	88.2	81.6	83.1	82.1	96.0	92.1	94.6	76.5	90.1	81.6
03	CONCRETE	86.9	81.2	84.4	85.1	76.1	81.1	98.5	82.0	91.2	91.2	82.3	87.2	104.5	92.2	99.0	76.8	77.8	77.2
04	MASONRY	79.6	72.6	75.3	75.7	68.0	70.9	87.8	80.3	83.1	90.6	80.1	84.1	87.9	85.0	86.1	82.4	64.0	71.1
05	METALS	94.6	86.1	92.0	95.9	84.6	92.5	89.0	85.6	87.9	86.0	87.6	86.5	95.5	108.6	99.5	96.4	78.3	90.9
06	WOOD, PLASTICS & COMPOSITES	91.3	73.6	81.7	87.9	78.4	82.7	88.5	78.0	82.8	86.0	80.7	83.1	85.7	90.2	88.2	83.4	73.6	78.1
07	THERMAL & MOISTURE PROTECTION	103.9	75.5	91.6	100.0	72.8	88.2	101.0	78.8	91.3	100.3	84.3	93.3	92.9	82.8	88.5	100.2	71.8	87.8
08	OPENINGS	89.4	76.0	86.3	87.4	75.5	84.6	91.6	79.0	88.7	90.9	81.3	88.7	93.2	88.2	92.1	94.1	68.0	88.0
0920	Plaster & Gypsum Board	102.7	72.5	82.8	95.9	78.2	84.3	89.5	77.0	81.2	88.5	79.8	82.8	63.7	90.2	81.2	86.1	72.5	77.1
0950, 0980	Ceilings & Acoustic Treatment	83.2	72.5	76.0	88.3	78.2	81.5	79.2	77.0	77.7	79.2	79.8	79.6	78.7	90.2	86.5	83.4	72.5	76.0
0960	Flooring	90.8	66.5	83.7	93.5	64.0	84.9	92.5	64.0	84.2	91.4	78.6	87.7	78.4	66.5	74.9	90.1	66.5	83.2
0970, 0990	Wall Finishes & Painting/Coating	90.7	81.8	85.4	96.9	69.9	80.9	91.0	89.0	89.9	91.0	75.1	81.6	92.9	94.1	93.6	91.0	69.9	78.6
09	FINISHES	84.7	74.1	78.9	88.2	73.8	80.4	87.0	76.9	81.5	86.2	79.7	82.7	78.4	85.0	82.0	86.0	71.6	78.1
COVERS	DIVS. 10 - 14, 25, 28, 41, 43, 44, 46	100.0	90.3	97.9	100.0	87.5	97.2	100.0	94.9	98.9	100.0	86.3	97.0	100.0	49.2	88.8	100.0	88.9	97.6
21, 22, 23	FIRE SUPPRESSION, PLUMBING & HVAC	100.1	78.0	91.2	100.0	77.2	90.8	100.0	78.8	91.4	96.7	78.4	89.3	96.3	84.1	91.4	96.7	74.5	87.7
26, 27, 3370	ELECTRICAL, COMMUNICATIONS & UTIL.	92.8	76.5	84.5	95.5	76.5	85.9	93.5	75.5	84.4	95.8	77.5	86.5	94.4	90.4	92.4	91.9	90.3	91.1
MF2016	WEIGHTED AVERAGE	93.4	79.6	87.5	93.2	77.7	86.6	94.6	83.1	89.7	92.5	83.8	88.8	95.8	87.6	92.3	91.7	77.8	85.7

DIVISION		LOUISIANA																	
		ALEXANDRIA			BATON ROUGE			HAMMOND			LAFAYETTE			LAKE CHARLES			MONROE		
		713 - 714			707 - 708			704			705			706			712		
		MAT.	INST.	TOTAL	MAT.	INST.	TOTAL	MAT.	INST.	TOTAL	MAT.	INST.	TOTAL	MAT.	INST.	TOTAL	MAT.	INST.	TOTAL
015433	CONTRACTOR EQUIPMENT		93.2	93.2		90.4	90.4		90.9	90.9		90.9	90.9		90.4	90.4		93.2	93.2
0241, 31 - 34	SITE & INFRASTRUCTURE, DEMOLITION	102.9	92.1	95.4	103.4	90.4	94.4	100.1	89.6	92.8	101.1	91.8	94.7	101.9	91.0	94.3	102.9	92.0	95.3
0310	Concrete Forming & Accessories	78.8	61.6	64.1	94.4	73.4	76.5	77.8	56.2	59.4	94.8	69.5	73.3	95.5	69.6	73.4	78.4	60.9	63.5
0320	Concrete Reinforcing	99.9	55.6	77.4	90.5	55.6	72.8	91.7	55.3	73.2	93.0	55.6	74.0	93.0	55.9	74.1	98.8	55.6	76.8
0330	Cast-in-Place Concrete	91.1	68.1	82.5	97.0	75.6	89.1	92.8	64.2	82.1	92.3	67.8	83.2	97.2	67.9	86.3	91.1	65.9	81.7
03	CONCRETE	83.9	63.5	74.9	88.1	71.7	80.8	87.7	59.6	75.2	88.7	66.9	79.0	91.0	67.0	80.4	83.8	62.5	74.3
04	MASONRY	109.4	61.6	79.9	91.6	64.0	74.5	94.4	53.6	69.2	94.4	64.0	75.6	93.9	68.2	78.0	104.1	60.3	77.1
05	METALS	94.7	71.0	87.5	98.1	74.3	90.9	89.7	68.7	83.3	89.0	69.3	83.0	89.0	69.7	83.1	94.6	70.8	87.4
06	WOOD, PLASTICS & COMPOSITES	91.0	60.5	74.5	99.7	75.1	86.3	83.4	55.8	68.4	104.2	70.0	85.6	102.4	70.0	84.8	90.4	60.5	74.2
07	THERMAL & MOISTURE PROTECTION	98.5	66.3	84.5	97.0	68.7	84.6	96.2	62.4	81.5	96.8	68.1	84.3	96.6	69.1	84.6	98.5	65.5	84.2
08	OPENINGS	113.4	60.3	101.1	101.2	72.3	94.5	98.2	56.5	88.5	101.9	63.9	93.1	101.9	63.9	93.1	113.4	58.7	100.7
0920	Plaster & Gypsum Board	86.8	59.8	69.0	97.1	74.9	82.4	97.9	55.0	69.6	106.1	69.6	82.0	106.1	69.6	82.0	86.5	59.8	68.9
0950, 0980	Ceilings & Acoustic Treatment	80.9	59.8	66.6	95.3	74.9	81.5	98.8	55.0	69.1	97.1	69.6	78.4	98.0	69.6	78.7	80.9	59.8	66.6
0960	Flooring	91.3	63.5	83.2	92.1	59.0	82.5	88.7	57.1	79.6	97.2	62.9	87.3	97.2	65.9	88.1	90.9	56.9	81.0
0970, 0990	Wall Finishes & Painting/Coating	94.5	61.3	74.9	91.5	61.3	73.7	95.5	61.4	75.3	95.5	61.3	75.3	95.5	61.3	75.3	94.5	61.3	74.9
09	FINISHES	81.5	61.6	70.6	87.8	69.9	78.0	89.0	56.4	71.2	92.3	67.6	78.8	92.5	68.2	79.2	81.4	59.9	69.7
COVERS	DIVS. 10 - 14, 25, 28, 41, 43, 44, 46	100.0	80.7	95.8	100.0	85.5	96.8	100.0	80.1	95.6	100.0	85.0	96.7	100.0	85.0	96.7	100.0	80.6	95.7
21, 22, 23	FIRE SUPPRESSION, PLUMBING & HVAC	100.0	64.7	85.7	100.0	65.5	86.1	96.7	62.3	82.8	100.1	65.6	86.2	100.1	65.9	86.3	100.0	63.3	85.2
26, 27, 3370	ELECTRICAL, COMMUNICATIONS & UTIL.	92.6	63.0	77.5	96.5	59.5	77.7	93.9	71.3	82.4	95.0	65.3	79.9	94.6	67.6	80.8	94.3	58.7	76.2
MF2016	WEIGHTED AVERAGE	96.7	66.6	83.8	96.6	69.7	85.1	93.8	64.5	81.2	95.6	68.9	84.1	95.8	69.8	84.6	96.6	65.1	83.1

DIVISION		LOUISIANA									MAINE								
		NEW ORLEANS			SHREVEPORT			THIBODAUX			AUGUSTA			BANGOR			BATH		
		700 - 701			710 - 711			703			043			044			045		
		MAT.	INST.	TOTAL	MAT.	INST.	TOTAL	MAT.	INST.	TOTAL	MAT.	INST.	TOTAL	MAT.	INST.	TOTAL	MAT.	INST.	TOTAL
015433	CONTRACTOR EQUIPMENT		86.7	86.7		93.2	93.2		90.9	90.9		97.9	97.9		97.9	97.9		97.9	97.9
0241, 31 - 34	SITE & INFRASTRUCTURE, DEMOLITION	101.2	92.9	95.4	105.1	92.0	96.0	102.4	91.6	94.9	87.6	98.5	95.2	91.2	99.8	97.2	89.1	98.5	95.7
0310	Concrete Forming & Accessories	97.9	70.6	74.6	94.0	61.6	66.5	89.0	65.8	69.3	98.4	78.7	81.6	92.9	78.0	80.2	88.7	78.7	80.2
0320	Concrete Reinforcing	94.7	56.0	75.0	100.1	55.4	77.4	91.7	55.1	73.1	105.3	82.9	93.9	94.5	83.1	88.7	93.5	82.6	88.0
0330	Cast-in-Place Concrete	93.1	70.7	84.7	94.0	66.1	83.6	99.7	65.8	87.1	90.5	110.3	97.8	73.2	111.1	87.3	73.2	110.2	87.0
03	CONCRETE	95.7	68.0	83.4	87.3	62.9	76.5	92.6	64.5	80.2	91.8	90.9	91.4	85.8	90.9	88.0	85.8	90.8	88.0
04	MASONRY	98.4	65.6	78.2	94.7	57.7	71.8	118.4	52.8	77.9	102.9	90.3	95.2	115.5	93.5	101.9	122.6	85.9	99.9
05	METALS	100.4	63.1	89.1	98.9	70.7	90.3	89.7	68.4	83.2	92.3	91.1	91.9	83.1	92.2	85.9	81.6	90.8	84.4
06	WOOD, PLASTICS & COMPOSITES	99.4	73.0	85.0	98.0	61.3	78.0	90.9	67.4	78.1	95.3	77.6	85.7	94.4	75.1	83.9	88.4	77.6	82.6
07	THERMAL & MOISTURE PROTECTION	95.2	69.7	84.1	96.7	64.9	82.9	96.4	63.8	82.2	106.2	76.0	93.1	102.9	82.9	94.2	102.8	75.4	90.9
08	OPENINGS	100.8	66.0	92.7	107.2	58.2	95.8	102.8	58.6	92.5	100.3	80.7	95.8	98.8	79.3	94.3	98.8	80.7	94.6
0920	Plaster & Gypsum Board	96.9	72.6	80.9	95.4	60.6	72.5	99.1	66.9	77.9	114.0	76.6	89.4	117.1	74.0	88.7	112.3	76.6	88.8
0950, 0980	Ceilings & Acoustic Treatment	102.1	72.6	82.1	87.6	60.6	69.3	98.8	66.9	77.2	105.6	76.6	86.0	90.9	74.0	79.5	90.0	76.6	80.9
0960	Flooring	102.3	65.7	91.7	97.6	56.9	85.8	94.6	39.4	78.6	88.2	47.3	76.3	80.9	102.1	87.0	79.4	47.3	70.0
0970, 0990	Wall Finishes & Painting/Coating	100.2	62.2	77.7	89.8	61.3	72.9	96.7	61.4	75.8	95.0	91.3	92.8	90.0	91.3	90.7	90.0	91.3	90.7
09	FINISHES	98.7	69.0	82.5	88.2	60.4	73.1	91.1	60.4	74.4	92.1	72.9	81.6	86.4	83.1	84.6	85.1	72.9	78.4
COVERS	DIVS. 10 - 14, 25, 28, 41, 43, 44, 46	100.0	85.8	96.9	100.0	84.0	96.5	100.0	83.4	96.4	100.0	99.3	99.9	100.0	101.8	100.4	100.0	99.3	99.9
21, 22, 23	FIRE SUPPRESSION, PLUMBING & HVAC	100.1	63.8	85.4	99.9	64.1	85.4	96.7	63.8	83.4	100.0	74.1	89.5	100.1	71.6	88.6	96.7	74.1	87.5
26, 27, 3370	ELECTRICAL, COMMUNICATIONS & UTIL.	100.6	71.3	85.7	99.9	67.1	83.2	92.7	71.3	81.8	96.7	78.6	87.5	94.5	71.1	82.6	92.7	78.6	85.6
MF2016	WEIGHTED AVERAGE	99.4	69.5	86.6	97.8	66.4	84.3	96.2	66.4	83.4	96.7	83.1	90.9	94.3	83.5	89.7	93.2	82.6	88.6

	DIVISION	MAINE																	
		HOULTON			KITTERY			LEWISTON			MACHIAS			PORTLAND			ROCKLAND		
		047			039			042			046			040 - 041			048		
		MAT.	INST.	TOTAL	MAT.	INST.	TOTAL	MAT.	INST.	TOTAL	MAT.	INST.	TOTAL	MAT.	INST.	TOTAL	MAT.	INST.	TOTAL
015433	CONTRACTOR EQUIPMENT		97.9	97.9		97.9	97.9		97.9	97.9		97.9	97.9		97.9	97.9		97.9	97.9
0241, 31 - 34	SITE & INFRASTRUCTURE, DEMOLITION	90.9	98.5	96.2	81.7	98.5	93.4	88.9	99.8	96.5	90.3	98.5	96.0	89.7	99.8	96.7	86.9	98.5	95.0
0310	Concrete Forming & Accessories	96.8	78.5	81.2	87.9	79.0	80.3	98.6	78.0	81.1	93.8	78.4	80.7	99.9	78.0	81.3	94.9	78.7	81.1
0320	Concrete Reinforcing	94.5	81.1	87.7	89.3	82.6	85.9	116.1	83.1	99.3	94.5	81.1	87.7	104.7	83.1	93.7	94.5	82.9	88.6
0330	Cast-in-Place Concrete	73.2	110.2	86.9	71.9	110.4	86.2	74.7	111.1	88.2	73.2	110.1	86.9	90.8	111.1	98.4	74.7	110.3	87.9
03	CONCRETE	86.8	90.4	88.4	80.8	91.0	85.3	85.6	90.9	88.0	86.2	90.4	88.1	91.2	90.9	91.0	83.6	90.8	86.8
04	MASONRY	98.2	61.9	75.8	113.2	85.9	96.4	98.8	93.5	95.5	98.2	61.9	75.8	102.5	93.5	96.9	92.3	90.3	91.1
05	METALS	81.8	88.8	83.9	78.5	91.1	82.3	86.2	92.2	88.0	81.8	88.8	83.9	92.3	92.2	92.3	81.7	91.1	84.5
06	WOOD, PLASTICS & COMPOSITES	98.6	77.6	87.2	91.1	77.6	83.8	100.5	75.1	86.7	95.3	77.6	85.7	97.5	75.1	85.3	96.2	77.6	86.1
07	THERMAL & MOISTURE PROTECTION	103.0	69.1	88.3	103.7	74.5	91.0	102.7	82.9	94.1	102.9	68.4	87.9	106.0	82.9	95.9	102.6	76.4	91.2
08	OPENINGS	98.8	80.7	94.6	98.4	84.3	95.2	101.8	79.3	96.6	98.8	80.7	94.6	100.9	79.3	95.9	98.8	80.7	94.6
0920	Plaster & Gypsum Board	119.6	76.6	91.3	108.0	76.6	87.4	122.2	74.0	90.5	117.8	76.6	90.7	111.0	74.0	86.6	117.8	76.6	90.7
0950, 0980	Ceilings & Acoustic Treatment	90.0	76.6	80.9	101.6	76.6	84.7	100.3	74.0	82.5	90.0	76.6	80.9	104.7	74.0	83.9	90.0	76.6	80.9
0960	Flooring	82.1	44.2	71.1	85.2	49.3	74.8	83.5	102.1	88.9	81.3	44.2	70.6	87.7	102.1	91.9	81.7	44.2	70.8
0970, 0990	Wall Finishes & Painting/Coating	90.0	91.3	90.7	83.3	104.8	96.1	90.0	91.3	90.7	90.0	91.3	90.7	93.6	91.3	92.2	90.0	91.3	90.7
09	FINISHES	87.0	72.4	79.0	87.9	74.7	80.7	89.4	83.1	85.9	86.4	72.4	78.8	91.8	83.1	87.0	86.2	72.4	78.7
COVERS	DIVS. 10 - 14, 25, 28, 41, 43, 44, 46	100.0	99.3	99.8	100.0	99.3	99.9	100.0	101.8	100.4	100.0	99.3	99.9	100.0	101.8	100.4	100.0	99.3	99.9
21, 22, 23	FIRE SUPPRESSION, PLUMBING & HVAC	96.7	74.0	87.5	96.6	81.6	90.5	100.1	71.6	88.6	96.7	74.0	87.5	100.0	71.6	88.5	96.7	74.1	87.5
26, 27, 3370	ELECTRICAL, COMMUNICATIONS & UTIL.	96.2	78.6	87.3	90.1	78.6	84.2	96.2	75.3	85.6	96.2	78.6	87.3	97.9	75.3	86.4	96.1	78.6	87.2
MF2016	WEIGHTED AVERAGE	92.8	79.7	87.2	91.3	84.6	88.5	94.7	84.1	90.1	92.6	79.7	87.1	96.9	84.1	91.4	91.9	83.1	88.1

	DIVISION	MAINE			MARYLAND														
		WATERVILLE			ANNAPOLIS			BALTIMORE			COLLEGE PARK			CUMBERLAND			EASTON		
		049			214			210 - 212			207 - 208			215			216		
		MAT.	INST.	TOTAL	MAT.	INST.	TOTAL	MAT.	INST.	TOTAL	MAT.	INST.	TOTAL	MAT.	INST.	TOTAL	MAT.	INST.	TOTAL
015433	CONTRACTOR EQUIPMENT		97.9	97.9		104.4	104.4		104.4	104.4		110.1	110.1		104.4	104.4		104.4	104.4
0241, 31 - 34	SITE & INFRASTRUCTURE, DEMOLITION	90.8	98.5	96.2	103.0	93.4	96.3	100.3	97.5	98.4	101.3	98.0	99.0	93.6	94.3	94.1	100.8	91.5	94.3
0310	Concrete Forming & Accessories	88.2	78.7	80.1	98.8	74.3	77.9	99.7	78.4	81.5	83.4	77.1	78.0	91.7	84.2	85.3	89.8	74.9	77.1
0320	Concrete Reinforcing	94.5	82.9	88.6	113.0	89.7	101.2	100.7	89.8	95.2	100.9	96.7	98.7	95.4	90.0	92.6	94.6	89.6	92.0
0330	Cast-in-Place Concrete	73.2	110.3	87.0	124.7	73.7	105.7	114.1	79.0	101.0	119.0	77.4	103.5	102.7	85.2	96.2	114.1	66.5	96.4
03	CONCRETE	87.3	90.8	88.9	107.9	78.2	94.7	106.6	81.5	95.5	106.4	82.1	95.6	93.5	86.7	90.5	102.0	75.8	90.4
04	MASONRY	109.3	90.3	97.6	96.6	69.1	79.6	99.4	74.6	84.1	113.9	75.1	89.9	97.9	87.3	91.3	111.7	59.8	79.7
05	METALS	81.8	91.1	84.6	103.1	103.1	103.1	101.3	98.3	100.4	87.8	108.8	94.1	98.3	104.7	100.3	98.6	99.9	99.0
06	WOOD, PLASTICS & COMPOSITES	87.8	77.6	82.3	96.2	75.9	85.2	102.9	79.0	89.9	78.1	76.3	77.2	88.4	83.3	85.6	86.2	82.3	84.1
07	THERMAL & MOISTURE PROTECTION	103.0	76.0	91.2	104.6	78.5	93.3	103.1	82.0	93.9	103.9	83.0	94.8	101.7	84.1	94.1	101.9	75.0	90.2
08	OPENINGS	98.8	80.7	94.6	100.1	81.4	95.7	98.8	83.1	95.1	94.7	83.4	92.1	99.7	87.1	96.8	98.1	84.9	95.0
0920	Plaster & Gypsum Board	112.3	76.6	88.8	103.0	75.6	84.9	104.8	78.4	87.4	96.7	75.8	83.0	105.0	83.1	90.6	105.0	82.2	89.9
0950, 0980	Ceilings & Acoustic Treatment	90.0	76.6	80.9	87.9	75.6	79.6	105.5	78.4	87.2	109.0	75.8	86.5	103.9	83.1	89.9	103.9	82.2	89.2
0960	Flooring	79.0	47.3	69.8	89.3	78.2	86.1	101.2	78.2	94.5	88.6	79.1	85.8	89.1	96.2	91.2	88.2	78.2	85.3
0970, 0990	Wall Finishes & Painting/Coating	90.0	91.3	90.7	91.7	76.6	82.7	104.1	77.3	88.2	100.0	74.4	84.9	96.9	86.9	91.0	96.9	74.4	83.6
09	FINISHES	85.1	72.9	78.5	86.5	74.8	80.1	101.7	77.9	88.7	94.6	76.5	84.8	94.9	86.7	90.4	95.1	76.1	84.7
COVERS	DIVS. 10 - 14, 25, 28, 41, 43, 44, 46	100.0	99.2	99.8	100.0	86.1	96.9	100.0	88.6	97.5	100.0	89.0	97.6	100.0	90.5	97.9	100.0	69.8	93.3
21, 22, 23	FIRE SUPPRESSION, PLUMBING & HVAC	96.7	74.1	87.5	100.0	82.7	93.0	100.0	83.8	93.5	96.7	88.1	93.2	96.4	73.6	87.2	96.4	74.4	87.5
26, 27, 3370	ELECTRICAL, COMMUNICATIONS & UTIL.	96.2	78.6	87.2	100.2	88.2	94.1	100.4	89.0	94.6	100.1	101.0	100.5	97.9	81.0	89.3	97.4	63.8	80.3
MF2016	WEIGHTED AVERAGE	93.1	83.1	88.8	100.5	83.1	93.0	101.2	85.0	94.2	97.7	88.6	93.8	97.1	85.5	92.1	98.8	76.0	89.0

	DIVISION	MARYLAND															MASSACHUSETTS		
		ELKTON			HAGERSTOWN			SALISBURY			SILVER SPRING			WALDORF			BOSTON		
		219			217			218			209			206			020 - 022, 024		
		MAT.	INST.	TOTAL	MAT.	INST.	TOTAL	MAT.	INST.	TOTAL	MAT.	INST.	TOTAL	MAT.	INST.	TOTAL	MAT.	INST.	TOTAL
015433	CONTRACTOR EQUIPMENT		104.4	104.4		104.4	104.4		104.4	104.4		101.8	101.8		101.8	101.8		102.0	102.0
0241, 31 - 34	SITE & INFRASTRUCTURE, DEMOLITION	87.6	93.0	91.3	91.8	94.4	93.6	100.7	91.5	94.3	89.4	89.6	89.5	95.9	89.6	91.5	95.1	101.3	99.4
0310	Concrete Forming & Accessories	95.7	94.0	94.2	90.8	80.9	82.4	104.1	51.6	59.4	91.4	76.3	78.6	98.3	76.3	79.6	103.0	137.0	131.9
0320	Concrete Reinforcing	94.6	115.3	105.1	95.4	90.0	92.6	94.6	67.0	80.6	99.7	96.5	98.1	100.3	96.5	98.4	105.8	152.0	129.3
0330	Cast-in-Place Concrete	92.3	73.1	85.2	97.8	85.2	93.1	114.1	64.6	95.7	121.8	77.8	105.4	136.5	77.8	114.6	91.1	141.0	109.7
03	CONCRETE	85.9	91.2	88.2	89.7	85.2	87.7	102.9	60.7	84.2	104.4	81.6	94.3	115.3	81.7	100.4	97.1	140.3	116.2
04	MASONRY	96.6	68.4	79.2	103.4	87.3	93.4	111.4	56.3	77.4	113.4	75.4	89.9	97.4	75.4	83.8	108.3	148.1	132.9
05	METALS	98.6	112.4	102.8	98.5	104.7	100.4	98.6	91.3	96.4	92.2	104.6	96.0	92.2	104.6	96.0	94.3	133.0	106.0
06	WOOD, PLASTICS & COMPOSITES	94.1	103.0	98.9	87.5	78.9	82.8	105.2	52.8	76.8	84.9	75.6	79.9	92.1	75.6	83.2	103.2	136.5	121.3
07	THERMAL & MOISTURE PROTECTION	101.4	81.1	92.5	102.0	85.2	94.7	102.2	70.1	88.2	106.8	88.0	98.6	107.3	88.0	98.9	113.5	138.6	124.4
08	OPENINGS	98.1	103.0	99.2	98.0	83.1	94.6	98.3	62.5	90.0	86.7	83.0	85.9	87.4	83.0	86.4	97.8	143.1	108.4
0920	Plaster & Gypsum Board	108.9	103.5	105.3	105.0	78.6	87.6	116.2	51.8	73.8	103.1	75.8	85.1	106.4	75.8	86.3	107.2	137.4	127.1
0950, 0980	Ceilings & Acoustic Treatment	103.9	103.5	103.6	104.9	78.6	87.1	103.9	51.8	68.6	117.5	75.8	89.3	117.5	75.8	89.3	106.3	137.4	127.4
0960	Flooring	90.5	78.2	87.0	88.7	96.2	90.9	94.1	78.2	89.5	94.6	79.1	90.1	98.2	79.1	92.6	93.4	169.1	115.4
0970, 0990	Wall Finishes & Painting/Coating	96.9	74.4	83.6	96.9	74.4	83.6	96.9	74.4	83.6	107.2	74.4	87.8	107.2	74.4	87.8	97.4	153.2	130.4
09	FINISHES	95.5	90.4	92.7	94.8	82.7	88.2	98.5	57.8	76.2	95.1	75.9	84.6	96.9	76.0	85.5	101.6	145.2	125.4
COVERS	DIVS. 10 - 14, 25, 28, 41, 43, 44, 46	100.0	57.7	90.7	100.0	90.0	97.8	100.0	78.0	95.2	100.0	84.6	96.6	100.0	84.6	96.6	100.0	117.3	103.8
21, 22, 23	FIRE SUPPRESSION, PLUMBING & HVAC	96.4	80.4	90.0	99.9	84.9	93.8	96.4	72.3	86.7	96.7	87.9	93.1	96.7	87.9	93.1	100.1	127.5	111.2
26, 27, 3370	ELECTRICAL, COMMUNICATIONS & UTIL.	99.1	89.0	94.0	97.7	81.0	89.2	96.2	61.6	78.6	97.2	101.0	99.1	94.8	101.0	97.9	100.8	131.1	116.2
MF2016	WEIGHTED AVERAGE	96.0	87.6	92.4	97.5	87.0	93.0	99.3	68.5	86.1	96.9	87.4	92.8	97.7	87.4	93.3	99.4	133.3	113.9

DIVISION		MASSACHUSETTS																	
		BROCKTON			BUZZARDS BAY			FALL RIVER			FITCHBURG			FRAMINGHAM			GREENFIELD		
		023			025			027			014			017			013		
		MAT.	INST.	TOTAL	MAT.	INST.	TOTAL	MAT.	INST.	TOTAL	MAT.	INST.	TOTAL	MAT.	INST.	TOTAL	MAT.	INST.	TOTAL
015433	CONTRACTOR EQUIPMENT		100.3	100.3		100.3	100.3		101.2	101.2		97.9	97.9		99.7	99.7		97.9	97.9
0241, 31 - 34	SITE & INFRASTRUCTURE, DEMOLITION	91.9	102.7	99.4	82.5	102.9	96.7	90.9	102.8	99.2	84.5	102.6	97.1	81.4	102.5	96.1	88.1	101.9	97.7
0310	Concrete Forming & Accessories	99.6	124.3	120.6	97.3	123.9	119.9	99.6	124.0	120.4	93.0	121.6	117.3	100.5	124.5	120.9	91.3	120.8	116.4
0320	Concrete Reinforcing	104.9	146.8	126.2	84.1	125.8	105.3	104.9	125.8	115.5	88.3	146.6	118.0	88.3	146.9	118.1	92.0	123.8	108.1
0330	Cast-in-Place Concrete	84.6	133.5	102.8	70.3	133.3	93.7	81.8	133.7	101.2	78.7	133.3	99.0	78.7	135.3	99.7	80.9	119.4	95.2
03	CONCRETE	90.2	130.8	108.2	76.5	126.9	98.9	88.9	127.1	105.8	78.3	129.2	100.9	81.1	131.5	103.4	81.7	120.1	98.7
04	MASONRY	101.9	138.0	124.2	94.8	138.0	121.5	102.5	137.9	124.4	100.3	134.7	121.6	106.8	139.0	126.7	105.0	121.6	115.2
05	METALS	89.5	126.4	100.7	84.7	117.7	94.7	89.5	118.0	98.1	87.6	122.8	98.2	87.6	126.5	99.4	89.9	111.6	96.4
06	WOOD, PLASTICS & COMPOSITES	99.2	123.1	112.2	95.9	123.1	110.7	99.2	123.4	112.3	96.4	119.9	109.2	103.2	122.9	113.9	94.4	123.1	110.0
07	THERMAL & MOISTURE PROTECTION	104.2	129.8	115.3	103.0	127.0	113.5	104.0	126.2	113.7	101.3	122.7	110.6	101.4	129.9	113.8	101.4	112.1	106.1
08	OPENINGS	99.5	132.1	107.0	95.4	122.4	101.7	99.5	121.8	104.6	101.5	130.3	108.2	91.8	132.0	101.1	101.7	122.1	106.4
0920	Plaster & Gypsum Board	90.2	123.4	112.1	86.0	123.4	110.7	90.2	123.4	112.1	110.9	120.1	117.0	114.0	123.4	120.2	111.8	123.4	119.5
0950, 0980	Ceilings & Acoustic Treatment	106.5	123.4	118.0	90.2	123.4	112.7	106.5	123.4	118.0	94.5	120.1	111.9	94.5	123.4	114.1	103.0	123.4	116.9
0960	Flooring	88.9	167.4	111.6	86.7	167.4	110.1	87.9	167.4	110.9	85.8	167.4	109.5	87.2	167.4	110.5	85.0	139.6	100.9
0970, 0990	Wall Finishes & Painting/Coating	91.0	142.9	121.7	91.0	142.9	121.7	91.0	142.9	121.7	87.2	142.9	120.1	88.1	142.9	120.5	87.2	112.2	102.0
09	FINISHES	90.8	134.6	114.7	85.8	134.6	112.4	90.5	134.8	114.7	86.5	132.7	111.7	87.2	134.4	113.0	88.4	124.4	108.0
COVERS	DIVS. 10 - 14, 25, 28, 41, 43, 44, 46	100.0	111.9	102.6	100.0	112.0	102.6	100.0	112.5	102.7	100.0	106.4	101.4	100.0	111.6	102.5	100.0	105.0	101.1
21, 22, 23	FIRE SUPPRESSION, PLUMBING & HVAC	100.3	106.1	102.6	96.8	105.8	100.4	100.3	105.8	102.5	97.0	106.8	101.0	97.0	123.7	107.8	97.0	100.6	98.5
26, 27, 3370	ELECTRICAL, COMMUNICATIONS & UTIL.	98.0	100.1	99.1	95.3	102.6	99.0	97.9	100.1	99.0	97.7	106.6	102.3	94.6	127.4	111.3	97.7	98.5	98.1
MF2016	WEIGHTED AVERAGE	96.1	119.2	106.0	91.1	117.7	102.5	95.9	117.4	105.1	93.0	118.7	104.0	92.4	127.0	107.2	94.3	110.9	101.4

DIVISION		MASSACHUSETTS																	
		HYANNIS			LAWRENCE			LOWELL			NEW BEDFORD			PITTSFIELD			SPRINGFIELD		
		026			019			018			027			012			010 - 011		
		MAT.	INST.	TOTAL	MAT.	INST.	TOTAL	MAT.	INST.	TOTAL	MAT.	INST.	TOTAL	MAT.	INST.	TOTAL	MAT.	INST.	TOTAL
015433	CONTRACTOR EQUIPMENT		100.3	100.3		100.3	100.3		97.9	97.9		101.2	101.2		97.9	97.9		97.9	97.9
0241, 31 - 34	SITE & INFRASTRUCTURE, DEMOLITION	88.4	102.9	98.5	93.3	102.7	99.8	92.2	102.8	99.6	89.4	102.8	98.7	93.3	101.3	98.9	92.7	101.7	98.9
0310	Concrete Forming & Accessories	91.4	123.9	119.0	101.7	124.6	121.1	98.2	125.7	121.6	99.6	124.1	120.4	98.2	105.7	104.6	98.4	120.7	117.4
0320	Concrete Reinforcing	84.1	125.8	105.3	109.3	138.5	124.2	110.2	138.3	124.5	104.9	125.9	115.6	91.3	117.4	104.6	110.2	123.8	117.1
0330	Cast-in-Place Concrete	77.2	133.3	98.1	91.0	133.6	106.9	82.8	135.3	102.3	72.1	133.8	95.1	90.3	116.6	100.1	86.2	119.4	98.5
03	CONCRETE	82.3	126.9	102.1	94.6	129.5	110.1	86.5	130.4	106.0	84.4	127.2	103.3	87.8	111.2	98.2	88.1	120.1	102.3
04	MASONRY	100.9	138.0	123.8	112.5	138.8	128.7	99.3	137.8	123.1	100.8	138.6	124.1	99.9	113.8	108.5	99.6	121.6	113.2
05	METALS	86.2	117.7	95.7	90.2	123.4	100.3	90.2	120.0	99.2	89.5	118.1	98.2	90.0	108.6	95.7	92.7	111.5	98.4
06	WOOD, PLASTICS & COMPOSITES	89.0	123.1	107.5	103.7	123.1	114.2	102.9	123.1	113.9	99.2	123.4	112.3	102.9	105.5	104.3	102.9	123.1	113.9
07	THERMAL & MOISTURE PROTECTION	103.5	127.0	113.8	102.1	130.0	114.2	101.8	130.3	114.2	104.0	126.4	113.7	101.9	107.2	104.2	101.8	112.1	106.3
08	OPENINGS	96.0	122.4	102.1	95.5	129.9	103.5	102.8	129.9	109.0	99.5	126.6	105.8	102.8	110.7	104.6	102.8	122.1	107.2
0920	Plaster & Gypsum Board	81.4	123.4	109.1	116.4	123.4	121.0	116.4	123.4	121.0	90.2	123.4	112.1	116.4	105.3	109.1	116.4	123.4	121.0
0950, 0980	Ceilings & Acoustic Treatment	99.7	123.4	115.8	104.9	123.4	117.5	104.9	123.4	117.5	106.5	123.4	118.0	104.9	105.3	105.2	104.9	123.4	117.5
0960	Flooring	84.2	167.4	108.3	87.8	167.4	110.9	87.8	167.4	110.9	87.9	167.4	110.9	88.2	134.2	101.5	87.2	139.6	102.4
0970, 0990	Wall Finishes & Painting/Coating	91.0	142.9	121.7	87.3	142.9	120.2	87.2	142.9	120.1	91.0	142.9	121.7	87.2	112.2	102.0	88.0	112.2	102.3
09	FINISHES	86.6	134.6	112.8	90.4	134.6	114.6	90.4	135.4	115.0	90.4	134.8	114.6	90.5	111.7	102.1	90.3	124.4	108.9
COVERS	DIVS. 10 - 14, 25, 28, 41, 43, 44, 46	100.0	112.0	102.6	100.0	112.1	102.7	100.0	113.1	102.9	100.0	112.5	102.7	100.0	101.5	100.3	100.0	105.0	101.1
21, 22, 23	FIRE SUPPRESSION, PLUMBING & HVAC	100.3	106.5	102.8	100.1	121.9	108.9	100.1	123.4	109.5	100.3	105.7	102.5	100.1	96.3	98.5	100.1	100.6	100.3
26, 27, 3370	ELECTRICAL, COMMUNICATIONS & UTIL.	95.7	100.1	97.9	96.9	125.9	111.6	97.3	122.2	110.0	98.8	100.1	99.4	97.3	98.5	97.9	97.3	95.2	96.3
MF2016	WEIGHTED AVERAGE	93.4	117.5	103.7	96.7	125.8	109.2	95.8	125.4	108.5	95.3	117.6	104.9	96.0	105.1	99.9	96.4	110.4	102.4

DIVISION		MASSACHUSETTS			MICHIGAN														
		WORCESTER			ANN ARBOR			BATTLE CREEK			BAY CITY			DEARBORN			DETROIT		
		015 - 016			481			490			487			481			482		
		MAT.	INST.	TOTAL	MAT.	INST.	TOTAL	MAT.	INST.	TOTAL	MAT.	INST.	TOTAL	MAT.	INST.	TOTAL	MAT.	INST.	TOTAL
015433	CONTRACTOR EQUIPMENT		97.9	97.9		111.9	111.9		99.0	99.0		111.9	111.9		111.9	111.9		96.4	96.4
0241, 31 - 34	SITE & INFRASTRUCTURE, DEMOLITION	92.6	102.6	99.6	79.5	96.7	91.5	93.7	86.0	88.3	71.2	96.2	88.7	79.3	96.8	91.5	92.9	100.7	98.4
0310	Concrete Forming & Accessories	98.7	121.5	118.1	97.3	104.3	103.2	96.4	79.5	82.0	97.3	83.7	85.8	97.1	104.9	103.7	101.7	108.2	107.2
0320	Concrete Reinforcing	110.2	146.6	128.7	102.3	106.5	104.5	94.8	83.3	88.9	102.3	105.7	104.0	102.3	106.6	104.5	103.1	108.3	105.7
0330	Cast-in-Place Concrete	85.7	133.3	103.4	81.0	99.2	87.8	82.0	93.8	86.4	77.5	85.9	80.6	79.2	100.0	87.0	86.1	101.5	91.8
03	CONCRETE	87.9	129.2	106.2	85.1	103.7	93.3	82.0	84.9	83.3	83.5	89.7	86.2	84.2	104.3	93.1	93.6	105.1	98.7
04	MASONRY	99.1	134.7	121.1	99.9	98.4	99.0	97.3	80.1	86.7	99.5	79.3	87.0	99.7	100.3	100.1	102.0	101.9	102.0
05	METALS	92.8	122.7	101.8	104.3	116.6	108.1	105.4	84.5	99.1	105.0	114.9	108.0	104.4	116.7	108.1	106.9	95.7	103.5
06	WOOD, PLASTICS & COMPOSITES	103.4	119.9	112.3	92.4	106.0	99.8	92.1	77.8	84.3	92.4	84.3	88.0	92.4	106.0	99.8	99.2	109.8	105.0
07	THERMAL & MOISTURE PROTECTION	101.8	122.7	110.9	109.7	100.6	105.7	101.5	81.3	92.7	106.9	83.2	96.6	107.9	103.1	105.8	106.1	106.2	106.1
08	OPENINGS	102.8	130.3	109.2	96.3	101.6	97.5	88.4	76.3	85.6	96.3	85.9	93.9	96.3	101.9	97.6	98.7	104.1	99.9
0920	Plaster & Gypsum Board	116.4	120.1	118.8	100.4	105.8	104.0	90.2	73.8	79.4	100.4	83.5	89.3	100.4	105.8	104.0	97.3	109.9	105.6
0950, 0980	Ceilings & Acoustic Treatment	104.9	120.1	115.2	85.1	105.8	99.1	77.8	73.8	75.1	86.1	83.5	84.3	85.1	105.8	99.1	95.9	109.9	105.4
0960	Flooring	87.8	160.1	108.8	89.4	110.8	95.7	89.7	72.8	84.8	89.4	79.5	86.6	89.0	105.0	93.6	93.7	106.0	97.3
0970, 0990	Wall Finishes & Painting/Coating	87.2	142.9	120.1	83.5	98.0	92.1	86.8	78.3	81.7	83.5	81.2	82.1	83.5	96.3	91.1	91.4	97.9	95.3
09	FINISHES	90.4	131.2	112.7	89.4	105.1	98.0	83.3	77.4	80.1	89.2	82.2	85.4	89.2	104.3	97.5	96.3	107.1	102.2
COVERS	DIVS. 10 - 14, 25, 28, 41, 43, 44, 46	100.0	106.4	101.4	100.0	95.2	98.9	100.0	95.7	99.0	100.0	89.8	97.8	100.0	95.6	99.0	100.0	102.9	100.6
21, 22, 23	FIRE SUPPRESSION, PLUMBING & HVAC	100.1	106.8	102.8	100.1	93.4	97.4	100.1	84.5	93.8	100.1	80.4	92.1	100.1	102.3	101.0	100.0	103.7	101.5
26, 27, 3370	ELECTRICAL, COMMUNICATIONS & UTIL.	97.3	106.6	102.1	98.5	105.5	102.1	94.9	79.6	87.1	97.6	84.5	90.9	98.5	98.4	98.5	102.5	101.3	101.9
MF2016	WEIGHTED AVERAGE	96.4	118.5	105.9	97.2	101.7	99.1	95.3	82.5	89.8	96.6	87.5	92.7	97.0	102.9	99.5	100.2	103.0	101.4

	DIVISION	MICHIGAN																	
		FLINT			GAYLORD			GRAND RAPIDS			IRON MOUNTAIN			JACKSON			KALAMAZOO		
		484 - 485			497			493, 495			498 - 499			492			491		
		MAT.	INST.	TOTAL	MAT.	INST.	TOTAL	MAT.	INST.	TOTAL	MAT.	INST.	TOTAL	MAT.	INST.	TOTAL	MAT.	INST.	TOTAL
015433	CONTRACTOR EQUIPMENT		111.9	111.9		106.8	106.8		99.0	99.0		93.8	93.8		106.8	106.8		99.0	99.0
0241, 31 - 34	SITE & INFRASTRUCTURE, DEMOLITION	69.3	96.3	88.1	88.3	82.8	84.5	92.5	85.9	87.9	96.9	93.2	94.3	111.2	85.5	93.3	94.0	86.0	88.4
0310	Concrete Forming & Accessories	100.2	84.1	86.5	95.1	74.3	77.4	95.3	78.0	80.5	87.5	79.7	80.9	91.8	84.6	85.7	96.4	79.3	81.9
0320	Concrete Reinforcing	102.3	106.1	104.2	88.4	89.7	89.0	99.5	83.0	91.1	88.1	87.3	87.7	85.8	106.1	96.1	94.8	78.4	86.5
0330	Cast-in-Place Concrete	81.6	88.3	84.1	81.8	80.6	81.3	86.1	92.0	88.3	96.8	69.7	86.7	81.6	92.5	85.7	83.6	93.7	87.4
03	CONCRETE	85.5	90.8	87.8	79.0	80.9	79.8	85.7	83.6	84.7	87.0	78.2	83.1	74.4	92.3	82.3	85.0	84.0	84.6
04	MASONRY	99.9	87.7	92.4	107.9	74.2	87.1	90.6	75.7	81.4	93.7	79.4	84.8	87.9	88.8	88.4	95.9	80.1	86.1
05	METALS	104.4	115.5	107.8	106.9	111.6	108.3	102.3	84.0	96.7	106.3	91.8	101.9	107.1	113.9	109.2	105.4	82.7	98.5
06	WOOD, PLASTICS & COMPOSITES	96.2	83.0	89.1	85.5	73.8	79.1	95.8	76.3	85.2	81.7	79.8	80.7	84.2	82.1	83.1	92.1	77.8	84.3
07	THERMAL & MOISTURE PROTECTION	107.2	86.9	98.4	100.0	70.6	87.2	103.4	72.7	90.0	104.0	76.2	91.9	99.3	90.8	95.6	101.5	81.3	92.7
08	OPENINGS	96.3	85.0	93.7	87.9	80.1	86.1	99.0	76.0	93.6	94.6	71.1	89.2	87.1	88.0	87.3	88.4	75.6	85.4
0920	Plaster & Gypsum Board	102.6	82.2	89.1	90.0	72.1	78.2	97.1	72.2	80.7	48.2	79.8	69.0	88.5	80.6	83.3	90.2	73.8	79.4
0950, 0980	Ceilings & Acoustic Treatment	85.1	82.2	83.1	77.0	72.1	73.6	89.6	72.2	77.8	74.0	79.8	77.9	77.0	80.6	79.4	77.8	73.8	75.1
0960	Flooring	89.4	88.9	89.3	83.0	88.5	84.6	93.6	78.7	89.3	101.9	90.3	98.5	81.7	81.3	81.6	89.7	72.8	84.8
0970, 0990	Wall Finishes & Painting/Coating	83.5	80.4	81.7	83.1	79.5	81.0	93.6	78.9	84.9	100.5	71.5	83.3	83.1	96.3	90.9	86.8	78.3	81.7
09	FINISHES	88.9	84.0	86.2	82.7	76.9	79.6	89.5	77.6	83.0	82.7	80.7	81.6	83.8	84.3	84.1	83.3	77.4	80.1
COVERS	DIVS. 10 - 14, 25, 28, 41, 43, 44, 46	100.0	90.6	97.9	100.0	79.2	95.4	100.0	95.4	99.0	100.0	87.5	97.3	100.0	93.5	98.6	100.0	95.7	99.0
21, 22, 23	FIRE SUPPRESSION, PLUMBING & HVAC	100.1	85.1	94.1	96.9	80.6	90.3	100.1	81.4	92.5	96.8	84.8	91.9	96.9	87.6	93.1	100.1	80.0	92.0
26, 27, 3370	ELECTRICAL, COMMUNICATIONS & UTIL.	98.5	91.3	94.8	92.7	74.3	83.4	100.3	85.0	92.5	99.2	82.6	90.8	96.7	105.5	101.2	94.7	75.9	85.2
MF2016	WEIGHTED AVERAGE	96.9	90.8	94.3	94.3	81.3	88.8	97.1	81.6	90.5	96.2	83.0	90.6	93.8	93.0	93.5	95.6	80.7	89.2

	DIVISION	MICHIGAN															MINNESOTA		
		LANSING			MUSKEGON			ROYAL OAK			SAGINAW			TRAVERSE CITY			BEMIDJI		
		488 - 489			494			480, 483			486			496			566		
		MAT.	INST.	TOTAL	MAT.	INST.	TOTAL	MAT.	INST.	TOTAL	MAT.	INST.	TOTAL	MAT.	INST.	TOTAL	MAT.	INST.	TOTAL
015433	CONTRACTOR EQUIPMENT		111.9	111.9		99.0	99.0		91.9	91.9		111.9	111.9		93.8	93.8		100.3	100.3
0241, 31 - 34	SITE & INFRASTRUCTURE, DEMOLITION	91.9	96.5	95.1	91.7	85.9	87.7	83.2	97.2	92.9	72.2	96.2	88.9	82.2	91.1	88.4	95.5	99.8	98.5
0310	Concrete Forming & Accessories	94.2	77.0	79.6	96.8	79.2	81.8	92.8	105.1	103.3	97.3	81.9	84.2	87.5	72.3	74.6	85.8	84.3	84.5
0320	Concrete Reinforcing	105.3	105.9	105.6	95.4	83.0	89.1	92.7	89.8	91.2	102.3	105.7	104.0	89.5	82.5	85.9	97.1	101.4	99.3
0330	Cast-in-Place Concrete	99.5	87.2	95.0	81.7	91.9	85.5	70.9	100.3	81.8	80.0	85.8	82.2	75.5	76.9	76.0	102.4	105.7	103.6
03	CONCRETE	90.5	87.2	89.0	80.5	84.1	82.1	73.1	100.2	85.1	84.6	88.8	86.5	71.5	76.5	73.7	88.8	96.0	92.0
04	MASONRY	94.7	86.3	89.5	94.5	75.7	82.9	93.1	99.4	97.0	101.4	79.3	87.7	91.9	75.3	81.6	99.1	102.4	101.2
05	METALS	102.3	115.0	106.1	102.9	83.8	97.1	108.0	91.0	102.9	104.4	114.7	107.5	106.3	89.4	101.2	88.7	116.8	97.2
06	WOOD, PLASTICS & COMPOSITES	90.1	73.3	81.0	89.2	78.1	83.2	88.1	106.0	97.8	89.1	82.0	85.2	81.7	72.2	76.6	69.9	79.2	75.0
07	THERMAL & MOISTURE PROTECTION	106.3	85.5	97.3	100.4	73.2	88.5	105.4	102.7	104.2	108.2	83.1	97.3	102.9	67.9	87.7	105.8	92.2	99.9
08	OPENINGS	100.5	79.4	95.6	87.7	77.6	85.3	96.1	101.3	97.3	94.4	84.6	92.1	94.6	66.4	88.0	101.9	102.3	102.0
0920	Plaster & Gypsum Board	96.3	72.2	80.4	72.6	74.1	73.6	97.9	105.8	103.1	100.4	81.1	87.7	48.2	72.1	63.9	104.3	79.0	87.7
0950, 0980	Ceilings & Acoustic Treatment	84.9	72.2	76.3	77.8	74.1	75.3	84.4	105.8	98.9	85.1	81.1	82.4	74.0	72.1	72.7	121.3	79.0	92.6
0960	Flooring	96.2	81.3	91.9	88.4	74.7	84.4	86.9	105.0	92.1	89.4	79.5	86.6	101.9	88.5	98.0	92.3	88.4	91.2
0970, 0990	Wall Finishes & Painting/Coating	95.0	78.9	85.5	85.0	78.9	81.4	85.1	96.3	91.7	83.5	81.2	82.1	100.5	40.5	65.0	89.1	106.5	99.4
09	FINISHES	88.8	76.8	82.2	79.8	77.2	78.4	88.3	104.2	97.0	89.1	80.9	84.6	81.6	71.8	76.2	96.0	86.9	91.1
COVERS	DIVS. 10 - 14, 25, 28, 41, 43, 44, 46	100.0	90.7	97.9	100.0	95.5	99.0	100.0	100.6	100.1	100.0	89.5	97.7	100.0	85.6	96.8	100.0	93.9	98.6
21, 22, 23	FIRE SUPPRESSION, PLUMBING & HVAC	100.0	85.5	94.1	99.9	81.4	92.4	96.9	103.6	99.6	100.1	80.0	92.0	96.8	80.0	90.0	96.8	82.0	90.8
26, 27, 3370	ELECTRICAL, COMMUNICATIONS & UTIL.	99.1	89.8	94.4	95.2	74.7	84.8	100.7	101.3	101.0	96.6	86.6	91.5	94.4	75.8	84.9	104.1	101.1	102.6
MF2016	WEIGHTED AVERAGE	97.9	88.7	93.9	94.1	80.3	88.2	95.2	100.6	97.5	96.5	87.3	92.6	93.2	78.3	86.8	96.0	95.4	95.8

	DIVISION	MINNESOTA																	
		BRAINERD			DETROIT LAKES			DULUTH			MANKATO			MINNEAPOLIS			ROCHESTER		
		564			565			556 - 558			560			553 - 555			559		
		MAT.	INST.	TOTAL	MAT.	INST.	TOTAL	MAT.	INST.	TOTAL	MAT.	INST.	TOTAL	MAT.	INST.	TOTAL	MAT.	INST.	TOTAL
015433	CONTRACTOR EQUIPMENT		102.9	102.9		100.3	100.3		103.2	103.2		102.9	102.9		107.6	107.6		103.2	103.2
0241, 31 - 34	SITE & INFRASTRUCTURE, DEMOLITION	97.2	104.8	102.5	93.7	100.0	98.1	98.5	102.9	101.6	93.9	104.4	101.2	94.9	108.4	104.3	96.5	102.5	100.7
0310	Concrete Forming & Accessories	86.3	85.0	85.2	82.6	84.2	83.9	96.9	97.0	97.0	94.6	93.7	93.8	99.4	117.1	114.4	98.3	97.2	97.4
0320	Concrete Reinforcing	95.9	101.8	98.9	97.1	101.4	99.3	96.3	101.6	99.0	95.8	109.0	102.5	97.5	112.5	105.1	94.1	108.7	101.5
0330	Cast-in-Place Concrete	111.4	110.3	111.0	99.4	108.6	102.9	99.3	99.6	99.4	102.5	99.3	101.3	103.3	117.7	108.7	99.1	98.0	98.7
03	CONCRETE	92.7	98.0	95.0	86.5	96.9	91.1	89.6	99.8	94.1	88.4	99.4	93.3	98.7	117.1	106.8	89.6	100.6	94.5
04	MASONRY	123.4	115.3	118.4	123.3	108.0	113.8	101.6	107.9	105.5	111.9	109.3	110.3	112.2	122.0	118.3	103.3	105.7	104.8
05	METALS	89.7	117.1	98.0	88.6	116.2	97.0	91.6	119.0	99.9	89.5	120.6	98.9	91.6	126.4	102.2	91.4	122.2	100.7
06	WOOD, PLASTICS & COMPOSITES	86.0	76.4	80.8	66.8	76.6	72.1	93.8	94.4	94.1	95.9	91.6	93.6	96.7	113.9	106.0	97.8	95.6	96.6
07	THERMAL & MOISTURE PROTECTION	103.9	103.9	103.9	105.6	101.2	103.7	105.3	104.4	104.9	104.4	93.6	99.7	103.0	118.5	109.8	109.1	92.5	101.9
08	OPENINGS	88.3	100.7	91.2	101.8	100.8	101.6	106.3	106.5	106.3	92.9	110.8	97.1	101.6	125.0	107.0	100.6	113.8	103.6
0920	Plaster & Gypsum Board	91.4	76.3	81.5	103.4	76.3	85.5	94.4	94.8	94.6	96.0	92.0	93.3	97.4	114.4	108.6	103.5	95.9	98.5
0950, 0980	Ceilings & Acoustic Treatment	56.8	76.3	70.0	121.3	76.3	90.8	93.2	94.8	94.2	56.8	92.0	80.6	98.2	114.4	109.2	93.3	95.9	95.1
0960	Flooring	91.1	88.4	90.3	91.0	88.4	90.3	96.5	119.7	103.3	92.9	84.6	90.5	99.7	118.5	105.2	93.2	84.6	90.7
0970, 0990	Wall Finishes & Painting/Coating	83.6	106.5	97.1	89.1	85.6	87.0	89.5	107.2	100.0	94.4	102.5	99.2	97.4	127.1	115.0	85.3	100.8	94.5
09	FINISHES	80.8	87.0	84.2	95.4	84.5	89.4	91.5	101.6	97.0	82.5	93.0	88.2	100.5	118.2	110.2	91.2	95.8	93.7
COVERS	DIVS. 10 - 14, 25, 28, 41, 43, 44, 46	100.0	95.5	99.0	100.0	95.3	99.0	100.0	96.2	99.2	100.0	95.1	98.9	100.0	107.3	101.6	100.0	100.3	100.1
21, 22, 23	FIRE SUPPRESSION, PLUMBING & HVAC	96.1	85.7	91.9	96.8	84.7	91.9	99.8	96.5	98.4	96.1	84.8	91.5	99.9	113.0	105.2	99.9	93.5	97.3
26, 27, 3370	ELECTRICAL, COMMUNICATIONS & UTIL.	101.7	101.8	101.7	103.8	70.4	86.8	97.4	101.8	99.6	108.3	92.4	100.2	105.4	113.0	109.3	101.3	92.4	96.7
MF2016	WEIGHTED AVERAGE	95.0	98.6	96.5	96.7	92.2	94.8	97.1	102.7	99.5	95.1	98.0	96.4	99.7	116.5	106.9	97.1	100.3	98.4

DIVISION		MINNESOTA															MISSISSIPPI		
		SAINT PAUL			ST. CLOUD			THIEF RIVER FALLS			WILLMAR			WINDOM			BILOXI		
		550 - 551			563			567			562			561			395		
		MAT.	INST.	TOTAL	MAT.	INST.	TOTAL	MAT.	INST.	TOTAL	MAT.	INST.	TOTAL	MAT.	INST.	TOTAL	MAT.	INST.	TOTAL
015433	CONTRACTOR EQUIPMENT		103.2	103.2		102.9	102.9		100.3	100.3		102.9	102.9		102.9	102.9		102.7	102.7
0241, 31 - 34	SITE & INFRASTRUCTURE, DEMOLITION	93.3	103.8	100.7	92.5	105.8	101.8	94.5	99.6	98.0	91.9	104.7	100.8	85.8	103.8	98.4	107.9	89.9	95.4
0310	Concrete Forming & Accessories	98.5	116.8	114.1	83.7	112.9	108.6	86.5	83.4	83.8	83.5	89.1	88.2	87.8	84.3	84.8	94.1	65.7	69.9
0320	Concrete Reinforcing	95.5	109.2	102.5	95.9	109.2	102.7	97.5	101.2	99.4	95.5	109.0	102.4	95.5	107.9	101.8	86.0	50.0	67.7
0330	Cast-in-Place Concrete	100.5	116.4	106.4	98.2	115.1	104.5	101.5	82.7	94.5	99.8	81.7	93.0	86.3	85.6	86.0	120.0	66.8	100.2
03	CONCRETE	93.3	115.9	103.4	84.4	113.6	97.4	87.6	87.5	87.6	84.4	91.2	87.4	75.8	90.2	82.1	98.9	65.1	83.9
04	MASONRY	102.5	122.0	114.6	107.8	121.0	115.9	99.1	102.4	101.1	112.2	115.3	114.1	122.7	88.3	101.4	97.8	60.2	74.6
05	METALS	91.6	124.1	101.4	90.4	122.0	99.9	88.8	115.5	96.9	89.5	120.4	98.9	89.4	118.4	98.2	91.5	84.2	89.3
06	WOOD, PLASTICS & COMPOSITES	95.4	113.6	105.3	83.1	109.7	97.5	71.0	79.2	75.5	82.8	82.2	82.5	87.4	82.2	84.6	95.3	66.4	79.6
07	THERMAL & MOISTURE PROTECTION	103.8	119.3	110.5	104.1	113.1	108.0	106.7	90.3	99.6	103.9	103.4	103.7	103.9	84.5	95.4	97.1	63.2	82.3
08	OPENINGS	98.8	123.7	104.6	93.4	121.6	99.9	101.9	102.3	102.0	90.3	104.7	93.6	93.9	104.7	96.4	98.6	56.6	88.9
0920	Plaster & Gypsum Board	99.0	114.4	109.2	91.4	110.5	104.0	104.0	79.0	87.6	91.4	82.3	85.4	91.4	82.3	85.4	105.2	65.9	79.3
0950, 0980	Ceilings & Acoustic Treatment	96.2	114.4	108.5	56.8	110.5	93.2	121.3	79.0	92.6	56.8	82.3	74.0	56.8	82.3	74.0	94.2	65.9	75.0
0960	Flooring	93.1	124.6	102.2	87.9	118.5	96.8	91.9	88.4	90.9	89.3	88.4	89.1	91.8	88.4	90.8	93.5	56.2	82.7
0970, 0990	Wall Finishes & Painting/Coating	91.3	127.0	112.5	94.4	127.0	113.7	89.1	85.6	87.0	89.1	85.6	87.0	89.1	102.5	97.0	83.0	48.1	62.4
09	FINISHES	92.0	119.3	106.9	80.3	115.5	99.5	95.9	84.6	89.7	80.3	87.7	84.4	80.5	86.8	83.9	88.0	62.4	74.0
COVERS	DIVS. 10 - 14, 25, 28, 41, 43, 44, 46	100.0	106.6	101.5	100.0	101.1	100.2	100.0	93.7	98.6	100.0	96.1	99.1	100.0	92.4	98.3	100.0	73.4	94.1
21, 22, 23	FIRE SUPPRESSION, PLUMBING & HVAC	99.9	112.7	105.1	99.5	110.9	104.1	96.8	81.7	90.7	96.1	103.7	99.2	96.1	82.0	90.4	100.0	56.5	82.4
26, 27, 3370	ELECTRICAL, COMMUNICATIONS & UTIL.	102.1	113.0	107.6	101.7	113.0	107.4	101.1	70.4	85.5	101.7	85.8	93.6	108.3	92.3	100.2	99.7	55.0	77.0
MF2016	WEIGHTED AVERAGE	97.2	115.8	105.2	94.4	114.0	102.8	95.6	89.4	92.9	93.3	99.9	96.1	93.7	92.3	93.1	97.3	64.6	83.3

DIVISION		MISSISSIPPI																	
		CLARKSDALE			COLUMBUS			GREENVILLE			GREENWOOD			JACKSON			LAUREL		
		386			397			387			389			390 - 392			394		
		MAT.	INST.	TOTAL	MAT.	INST.	TOTAL	MAT.	INST.	TOTAL	MAT.	INST.	TOTAL	MAT.	INST.	TOTAL	MAT.	INST.	TOTAL
015433	CONTRACTOR EQUIPMENT		102.7	102.7		102.7	102.7		102.7	102.7		102.7	102.7		102.7	102.7		102.7	102.7
0241, 31 - 34	SITE & INFRASTRUCTURE, DEMOLITION	100.0	88.5	92.0	106.8	89.4	94.7	105.8	89.9	94.7	103.0	88.2	92.7	102.7	89.9	93.8	112.2	88.6	95.7
0310	Concrete Forming & Accessories	83.9	44.8	50.6	82.6	47.2	52.5	80.6	63.8	66.3	92.9	45.0	52.1	93.2	66.1	70.2	82.8	60.7	64.0
0320	Concrete Reinforcing	101.0	66.0	83.2	92.1	66.4	79.0	101.6	67.3	84.2	101.0	66.1	83.3	95.6	51.9	73.4	92.7	32.8	62.3
0330	Cast-in-Place Concrete	102.8	50.0	83.1	122.4	55.3	97.4	105.9	56.6	87.5	110.5	49.7	87.9	104.5	65.9	90.2	119.7	51.2	94.2
03	CONCRETE	91.6	52.4	74.2	100.2	55.5	80.4	96.8	63.6	82.1	97.8	52.4	77.7	93.0	65.3	80.7	102.6	54.4	81.2
04	MASONRY	90.6	40.2	59.4	123.7	46.2	75.8	134.1	68.2	93.4	91.3	40.0	59.6	103.4	59.1	76.1	119.8	42.9	72.3
05	METALS	92.9	86.8	91.1	88.6	89.6	88.9	93.9	90.8	93.0	92.9	86.5	91.0	98.1	85.0	94.2	88.7	75.5	84.7
06	WOOD, PLASTICS & COMPOSITES	80.8	45.7	61.7	80.6	46.5	62.1	77.4	64.6	70.5	93.8	45.7	67.7	94.9	67.8	80.2	81.8	66.4	73.4
07	THERMAL & MOISTURE PROTECTION	96.3	48.6	75.5	97.0	53.5	78.1	96.8	63.4	82.3	96.7	50.8	76.8	95.2	62.6	81.0	97.3	54.4	78.6
08	OPENINGS	97.2	49.2	86.0	98.2	49.6	87.0	96.9	60.0	88.3	97.2	49.2	86.0	99.9	58.2	90.2	95.4	53.3	85.6
0920	Plaster & Gypsum Board	93.4	44.6	61.3	96.6	45.5	63.0	93.1	64.1	74.0	104.6	44.6	65.1	90.0	67.3	75.1	96.6	65.9	76.4
0950, 0980	Ceilings & Acoustic Treatment	86.2	44.6	58.0	88.1	45.5	59.2	90.0	64.1	72.5	86.2	44.6	58.0	94.8	67.3	76.2	88.1	65.9	73.1
0960	Flooring	97.7	46.5	82.9	87.1	52.2	76.9	96.1	46.5	81.7	103.5	46.5	87.0	91.7	57.9	81.9	85.7	46.5	74.3
0970, 0990	Wall Finishes & Painting/Coating	93.8	48.1	66.8	83.0	48.1	62.4	93.8	49.1	67.3	93.8	48.1	66.8	85.0	49.1	63.8	83.0	48.1	62.4
09	FINISHES	89.9	45.2	65.5	83.8	47.5	63.9	90.7	59.0	73.4	93.5	45.2	67.1	87.0	63.2	74.0	83.9	58.1	69.8
COVERS	DIVS. 10 - 14, 25, 28, 41, 43, 44, 46	100.0	48.9	88.8	100.0	50.0	89.0	100.0	72.6	94.0	100.0	48.9	88.8	100.0	72.9	94.0	100.0	34.9	85.7
21, 22, 23	FIRE SUPPRESSION, PLUMBING & HVAC	98.5	52.4	79.8	98.1	54.2	80.4	100.0	58.2	83.1	98.5	52.8	80.0	100.0	59.0	83.4	98.2	48.0	77.9
26, 27, 3370	ELECTRICAL, COMMUNICATIONS & UTIL.	95.7	42.6	68.7	97.2	55.0	75.7	95.7	56.1	75.6	95.7	39.9	67.4	101.3	56.1	78.3	98.7	57.7	77.8
MF2016	WEIGHTED AVERAGE	95.2	54.4	77.7	97.0	58.4	80.5	98.6	66.0	84.6	96.5	54.2	78.3	98.0	65.4	84.0	97.2	56.9	79.9

DIVISION		MISSISSIPPI									MISSOURI								
		MCCOMB			MERIDIAN			TUPELO			BOWLING GREEN			CAPE GIRARDEAU			CHILLICOTHE		
		396			393			388			633			637			646		
		MAT.	INST.	TOTAL	MAT.	INST.	TOTAL	MAT.	INST.	TOTAL	MAT.	INST.	TOTAL	MAT.	INST.	TOTAL	MAT.	INST.	TOTAL
015433	CONTRACTOR EQUIPMENT		102.7	102.7		102.7	102.7		102.7	102.7		110.5	110.5		110.5	110.5		104.5	104.5
0241, 31 - 34	SITE & INFRASTRUCTURE, DEMOLITION	99.1	88.4	91.6	103.5	90.0	94.1	97.5	88.4	91.2	90.6	94.7	93.5	92.1	94.6	93.9	104.0	92.8	96.2
0310	Concrete Forming & Accessories	82.6	46.3	51.7	79.9	64.2	66.5	81.2	46.9	52.0	97.3	90.2	91.3	89.6	80.7	82.0	85.0	96.1	94.5
0320	Concrete Reinforcing	93.3	34.5	63.4	92.1	51.9	71.7	98.9	66.1	82.2	96.5	99.3	98.0	97.7	78.2	87.8	100.7	105.6	103.2
0330	Cast-in-Place Concrete	106.2	49.5	85.1	113.7	66.5	96.1	102.8	69.7	90.5	90.0	97.0	92.6	89.1	86.8	88.2	97.2	87.8	93.7
03	CONCRETE	89.5	47.5	70.9	94.1	64.6	81.0	91.2	60.2	77.4	92.1	95.7	93.7	91.2	84.3	88.1	99.4	95.6	97.7
04	MASONRY	125.1	39.6	72.3	97.3	59.6	74.0	123.0	42.7	73.3	110.9	97.4	102.5	107.5	80.2	90.6	99.2	92.9	95.3
05	METALS	88.8	74.8	84.6	89.7	84.9	88.3	92.8	86.8	91.0	97.3	118.9	103.8	98.4	109.0	101.6	84.8	111.2	92.8
06	WOOD, PLASTICS & COMPOSITES	80.6	48.1	62.9	77.8	64.6	70.6	78.0	46.5	60.9	101.3	89.8	95.1	93.1	78.7	85.3	93.9	97.5	95.9
07	THERMAL & MOISTURE PROTECTION	96.5	50.9	76.6	96.7	62.6	81.8	96.3	52.1	77.1	99.7	99.5	99.6	99.1	85.3	93.1	93.4	94.2	93.8
08	OPENINGS	98.3	42.5	85.3	98.0	56.1	88.3	97.2	52.5	86.8	100.7	97.0	99.8	100.7	75.4	94.8	87.9	97.7	90.2
0920	Plaster & Gypsum Board	96.6	47.1	64.0	96.6	64.1	75.2	93.1	45.5	61.7	97.7	89.9	92.6	97.1	78.4	84.8	98.8	97.3	97.8
0950, 0980	Ceilings & Acoustic Treatment	88.1	47.1	60.3	90.0	64.1	72.5	86.2	45.5	58.6	91.9	89.9	90.5	91.9	78.4	82.8	88.9	97.3	94.6
0960	Flooring	87.1	46.5	75.3	85.7	56.2	77.1	96.3	46.5	81.9	96.9	100.1	97.9	93.6	87.4	91.8	99.1	103.0	100.2
0970, 0990	Wall Finishes & Painting/Coating	83.0	48.1	62.4	83.0	48.1	62.4	93.8	46.4	65.8	97.0	109.1	104.1	97.0	68.8	80.3	93.8	105.9	101.0
09	FINISHES	83.1	46.5	63.1	83.4	61.2	71.3	89.4	46.1	65.8	97.4	93.5	95.3	96.3	79.5	87.1	98.1	98.8	98.5
COVERS	DIVS. 10 - 14, 25, 28, 41, 43, 44, 46	100.0	51.9	89.4	100.0	73.0	94.0	100.0	50.0	89.0	100.0	81.4	95.9	100.0	91.8	98.2	100.0	82.4	96.1
21, 22, 23	FIRE SUPPRESSION, PLUMBING & HVAC	98.1	51.8	79.4	100.0	58.7	83.3	98.6	53.8	80.5	96.6	100.4	98.1	100.0	99.8	99.9	96.7	101.5	98.6
26, 27, 3370	ELECTRICAL, COMMUNICATIONS & UTIL.	95.8	56.7	75.9	98.7	57.0	77.5	95.5	54.9	74.9	96.5	78.7	87.5	96.5	100.9	98.7	93.9	77.7	85.7
MF2016	WEIGHTED AVERAGE	95.4	54.5	77.9	95.6	65.1	82.5	96.5	58.2	80.1	97.5	96.0	96.9	98.1	91.8	95.4	94.5	95.4	94.9

		MISSOURI																	
	DIVISION	COLUMBIA			FLAT RIVER			HANNIBAL			HARRISONVILLE			JEFFERSON CITY			JOPLIN		
		652			636			634			647			650 - 651			648		
		MAT.	INST.	TOTAL	MAT.	INST.	TOTAL	MAT.	INST.	TOTAL	MAT.	INST.	TOTAL	MAT.	INST.	TOTAL	MAT.	INST.	TOTAL
015433	CONTRACTOR EQUIPMENT		113.3	113.3		110.5	110.5		110.5	110.5		104.5	104.5		113.3	113.3		108.0	108.0
0241, 31 - 34	SITE & INFRASTRUCTURE, DEMOLITION	103.1	98.4	99.8	93.3	94.5	94.1	88.3	94.6	92.7	95.5	94.1	94.5	102.5	98.4	99.6	105.0	97.6	99.8
0310	Concrete Forming & Accessories	83.1	80.1	80.5	104.0	89.2	91.4	95.4	84.8	86.4	82.3	101.7	98.8	95.1	80.0	82.2	96.3	76.3	79.3
0320	Concrete Reinforcing	90.7	96.0	93.4	97.7	105.9	101.9	96.0	99.3	97.7	100.3	111.1	105.8	98.0	96.0	97.0	103.9	86.8	95.2
0330	Cast-in-Place Concrete	87.0	84.6	86.1	93.0	93.3	93.1	85.2	96.3	89.4	99.6	102.4	100.6	92.2	84.6	89.4	105.3	77.8	95.0
03	CONCRETE	81.6	86.3	83.7	95.1	95.1	95.1	88.4	93.1	90.5	95.3	104.2	99.2	88.5	86.2	87.5	98.2	79.8	90.0
04	MASONRY	139.0	86.5	106.6	107.9	78.0	89.4	103.0	96.6	99.0	93.9	101.5	98.6	97.7	86.5	90.8	93.0	81.9	86.1
05	METALS	94.0	116.3	100.8	97.2	120.4	104.3	97.3	118.5	103.7	85.2	114.7	94.1	93.6	116.2	100.4	87.8	99.8	91.4
06	WOOD, PLASTICS & COMPOSITES	80.9	77.6	79.1	110.7	90.0	99.4	99.3	83.5	90.7	90.4	101.6	96.5	95.8	77.6	85.9	106.0	75.7	89.5
07	THERMAL & MOISTURE PROTECTION	96.6	86.1	92.0	99.9	93.3	97.1	99.5	96.1	98.0	92.6	103.2	97.2	104.0	86.1	96.2	92.6	83.5	88.6
08	OPENINGS	96.5	82.3	93.2	100.7	99.1	100.3	100.7	86.2	97.3	87.6	104.8	91.6	96.9	82.3	93.5	88.9	78.5	86.5
0920	Plaster & Gypsum Board	84.6	77.1	79.7	104.1	90.0	94.8	97.4	83.4	88.2	94.8	101.5	99.2	93.9	77.1	82.8	105.7	74.9	85.4
0950, 0980	Ceilings & Acoustic Treatment	97.8	77.1	83.8	91.9	90.0	90.6	91.9	83.4	86.1	88.9	101.5	97.5	97.3	77.1	83.6	89.8	74.9	79.7
0960	Flooring	90.6	96.3	92.2	100.2	87.4	96.5	96.3	100.1	97.4	94.3	104.2	97.2	98.0	73.9	91.0	125.7	74.6	110.8
0970, 0990	Wall Finishes & Painting/Coating	91.6	80.2	84.9	97.0	73.6	83.2	97.0	96.6	96.7	98.1	105.6	102.5	88.4	80.2	83.6	93.4	77.9	84.2
09	FINISHES	84.9	82.2	83.4	99.4	86.3	92.3	96.9	87.8	92.0	95.7	102.6	99.4	90.8	78.3	84.0	105.1	76.6	89.6
COVERS	DIVS. 10 - 14, 25, 28, 41, 43, 44, 46	100.0	94.6	98.8	100.0	92.3	98.3	100.0	80.5	95.7	100.0	84.5	96.6	100.0	94.6	98.8	100.0	81.2	95.9
21, 22, 23	FIRE SUPPRESSION, PLUMBING & HVAC	99.9	98.1	99.2	96.6	99.0	97.6	96.6	99.9	97.9	96.6	103.0	99.2	99.9	98.1	99.2	100.1	72.7	89.0
26, 27, 3370	ELECTRICAL, COMMUNICATIONS & UTIL.	93.7	81.3	87.4	100.8	100.9	100.8	95.4	78.7	86.9	100.7	101.0	100.9	98.9	81.3	90.0	91.9	67.7	79.6
MF2016	WEIGHTED AVERAGE	96.2	91.3	94.1	98.5	96.2	97.5	96.5	94.0	95.4	94.0	102.6	97.7	96.3	90.8	94.0	95.9	79.7	88.9

		MISSOURI																	
	DIVISION	KANSAS CITY			KIRKSVILLE			POPLAR BLUFF			ROLLA			SEDALIA			SIKESTON		
		640 - 641			635			639			654 - 655			653			638		
		MAT.	INST.	TOTAL	MAT.	INST.	TOTAL	MAT.	INST.	TOTAL	MAT.	INST.	TOTAL	MAT.	INST.	TOTAL	MAT.	INST.	TOTAL
015433	CONTRACTOR EQUIPMENT		105.5	105.5		100.4	100.4		102.7	102.7		113.3	113.3		103.1	103.1		102.7	102.7
0241, 31 - 34	SITE & INFRASTRUCTURE, DEMOLITION	96.8	98.2	97.8	93.4	89.5	90.6	79.9	93.3	89.2	102.1	98.9	99.9	100.5	93.6	95.7	83.5	93.9	90.8
0310	Concrete Forming & Accessories	96.0	102.2	101.2	87.6	80.6	81.6	87.8	78.8	80.1	90.4	96.6	95.7	88.4	80.4	81.6	88.8	78.9	80.4
0320	Concrete Reinforcing	98.7	111.2	105.0	96.7	85.7	91.1	99.7	78.2	88.7	91.2	96.1	93.7	89.9	110.5	100.3	99.0	78.2	88.4
0330	Cast-in-Place Concrete	103.2	103.9	103.5	92.9	84.8	89.9	71.5	86.0	76.9	89.1	97.5	92.2	93.0	83.5	89.5	76.4	86.0	80.0
03	CONCRETE	98.2	104.8	101.1	108.1	84.1	97.5	82.2	82.4	82.3	83.3	98.2	89.9	97.2	87.8	93.0	85.9	82.5	84.4
04	MASONRY	99.7	103.7	102.2	114.4	86.6	97.3	105.8	75.7	87.2	112.7	87.2	96.9	119.0	83.3	96.9	105.6	75.7	87.1
05	METALS	94.1	113.4	99.9	96.9	102.1	98.5	97.5	98.7	97.9	93.4	116.9	100.5	92.4	112.2	98.4	97.9	98.8	98.2
06	WOOD, PLASTICS & COMPOSITES	101.3	101.5	101.4	85.8	78.8	82.0	84.8	78.7	81.5	88.9	98.4	94.0	82.3	79.0	80.5	86.5	78.7	82.3
07	THERMAL & MOISTURE PROTECTION	92.7	104.9	98.0	106.0	92.9	100.3	104.2	84.1	95.4	96.8	94.5	95.8	102.9	89.7	97.1	104.3	84.1	95.5
08	OPENINGS	94.4	104.8	96.8	106.2	81.1	100.4	107.2	75.4	99.8	96.5	93.7	95.9	101.7	88.8	98.7	107.2	75.4	99.8
0920	Plaster & Gypsum Board	101.3	101.5	101.5	92.2	78.5	83.2	92.6	78.4	83.3	87.0	98.4	94.5	80.6	78.5	79.2	94.4	78.4	83.9
0950, 0980	Ceilings & Acoustic Treatment	92.1	101.5	98.5	90.2	78.5	82.3	91.9	78.4	82.8	97.8	98.4	98.2	97.8	78.5	84.7	91.9	78.4	82.8
0960	Flooring	100.3	104.2	101.4	75.2	99.7	82.3	88.5	87.4	88.1	94.1	99.7	95.7	72.9	99.7	80.7	89.0	87.4	88.6
0970, 0990	Wall Finishes & Painting/Coating	96.8	105.9	102.2	92.8	80.4	85.4	92.0	68.8	78.3	91.6	92.6	92.2	91.6	105.9	100.1	92.0	68.8	78.3
09	FINISHES	99.0	102.8	101.1	96.4	83.4	89.3	95.9	78.6	86.4	86.4	96.7	92.0	83.6	85.5	84.7	96.6	78.9	86.9
COVERS	DIVS. 10 - 14, 25, 28, 41, 43, 44, 46	100.0	99.0	99.8	100.0	80.0	95.6	100.0	90.1	97.8	100.0	98.7	99.7	100.0	89.7	97.7	100.0	90.3	97.9
21, 22, 23	FIRE SUPPRESSION, PLUMBING & HVAC	100.0	103.0	101.2	96.6	99.3	97.7	96.6	97.0	96.8	96.5	100.8	98.2	96.4	97.0	96.7	96.6	97.0	96.8
26, 27, 3370	ELECTRICAL, COMMUNICATIONS & UTIL.	102.3	101.0	101.7	95.5	78.6	86.9	95.7	100.8	98.3	92.3	81.3	86.7	93.3	101.0	97.2	94.9	100.8	97.9
MF2016	WEIGHTED AVERAGE	98.1	103.6	100.5	100.2	88.8	95.3	96.3	89.2	93.3	94.3	96.6	95.3	96.6	93.7	95.4	96.9	89.4	93.7

		MISSOURI									MONTANA								
	DIVISION	SPRINGFIELD			ST. JOSEPH			ST. LOUIS			BILLINGS			BUTTE			GREAT FALLS		
		656 - 658			644 - 645			630 - 631			590 - 591			597			594		
		MAT.	INST.	TOTAL	MAT.	INST.	TOTAL	MAT.	INST.	TOTAL	MAT.	INST.	TOTAL	MAT.	INST.	TOTAL	MAT.	INST.	TOTAL
015433	CONTRACTOR EQUIPMENT		105.6	105.6		104.5	104.5		111.0	111.0		100.5	100.5		100.3	100.3		100.3	100.3
0241, 31 - 34	SITE & INFRASTRUCTURE, DEMOLITION	102.7	96.1	98.1	99.0	91.9	94.1	97.7	99.2	98.7	95.2	97.5	96.8	102.1	96.3	98.0	105.8	97.2	99.8
0310	Concrete Forming & Accessories	96.4	77.5	80.3	95.1	91.5	92.1	100.8	102.6	102.3	99.1	66.4	71.3	85.8	66.5	69.4	99.1	65.7	70.7
0320	Concrete Reinforcing	87.4	95.6	91.6	97.5	110.6	104.1	89.3	107.2	98.4	90.7	80.6	85.6	98.4	81.1	89.6	90.8	80.6	85.6
0330	Cast-in-Place Concrete	94.6	75.9	87.6	97.8	98.2	97.9	100.5	103.5	101.6	115.6	69.8	98.6	127.2	69.8	105.9	134.5	69.7	110.4
03	CONCRETE	93.4	81.2	88.0	94.4	98.0	96.0	98.0	104.8	101.1	92.9	71.0	83.2	96.5	71.1	85.3	101.5	70.6	87.8
04	MASONRY	90.6	81.6	85.0	95.7	90.5	92.5	92.8	109.7	103.2	121.7	73.8	92.1	117.6	82.2	95.7	121.8	73.8	92.1
05	METALS	98.5	102.5	99.8	90.7	113.4	97.6	103.2	121.0	108.5	109.1	87.1	102.4	103.2	87.8	98.6	106.4	87.1	100.5
06	WOOD, PLASTICS & COMPOSITES	90.3	77.2	83.2	106.0	91.2	98.0	102.7	100.8	101.6	92.6	63.3	76.7	78.5	63.3	70.3	93.9	62.5	76.8
07	THERMAL & MOISTURE PROTECTION	101.0	77.7	90.8	93.0	90.5	91.9	97.5	106.0	101.2	105.3	69.7	89.8	105.0	73.3	91.2	105.7	71.0	90.6
08	OPENINGS	104.0	86.7	100.0	91.9	98.8	93.5	99.3	105.8	100.8	100.8	65.0	92.5	99.0	65.0	91.1	101.9	64.6	93.3
0920	Plaster & Gypsum Board	88.0	76.6	80.5	107.1	90.9	96.4	102.7	101.1	101.6	114.5	62.7	80.4	112.1	62.7	79.6	122.5	61.8	82.5
0950, 0980	Ceilings & Acoustic Treatment	97.8	76.6	83.5	96.6	90.9	92.7	93.6	101.1	98.7	90.6	62.7	71.7	96.5	62.7	73.6	98.2	61.8	73.5
0960	Flooring	93.0	73.9	87.5	104.0	102.4	103.5	99.3	100.1	99.5	92.9	78.0	88.6	90.8	94.0	91.7	97.5	78.0	91.8
0970, 0990	Wall Finishes & Painting/Coating	86.1	102.0	95.5	93.8	105.6	100.8	98.0	109.1	104.5	88.8	69.5	77.4	88.0	69.5	77.1	88.0	69.5	77.1
09	FINISHES	88.5	79.9	83.8	101.4	95.0	97.9	99.9	102.4	101.3	90.5	68.3	78.4	90.4	71.5	80.1	94.5	67.7	79.9
COVERS	DIVS. 10 - 14, 25, 28, 41, 43, 44, 46	100.0	92.4	98.3	100.0	96.0	99.1	100.0	102.3	100.5	100.0	90.3	97.9	100.0	90.3	97.9	100.0	90.2	97.9
21, 22, 23	FIRE SUPPRESSION, PLUMBING & HVAC	100.0	71.0	88.2	100.1	89.4	95.8	100.0	106.1	102.5	100.0	73.2	89.2	100.1	71.5	88.5	100.1	69.6	87.8
26, 27, 3370	ELECTRICAL, COMMUNICATIONS & UTIL.	97.5	70.5	83.7	100.8	77.7	89.0	98.9	100.9	99.9	98.5	69.9	84.0	105.6	70.9	88.0	97.9	69.1	83.2
MF2016	WEIGHTED AVERAGE	97.7	81.0	90.5	96.8	92.8	95.1	99.6	105.8	102.3	100.8	75.0	89.7	100.7	76.2	90.2	102.1	74.0	90.1

	DIVISION	MONTANA																	
		HAVRE			HELENA			KALISPELL			MILES CITY			MISSOULA			WOLF POINT		
		595			596			599			593			598			592		
		MAT.	INST.	TOTAL	MAT.	INST.	TOTAL	MAT.	INST.	TOTAL	MAT.	INST.	TOTAL	MAT.	INST.	TOTAL	MAT.	INST.	TOTAL
015433	CONTRACTOR EQUIPMENT		100.3	100.3		100.3	100.3		100.3	100.3		100.3	100.3		100.3	100.3		100.3	100.3
0241, 31 - 34	SITE & INFRASTRUCTURE, DEMOLITION	109.5	96.4	100.3	95.4	96.3	96.0	92.0	96.2	94.9	98.5	96.3	96.9	84.5	96.0	92.5	116.0	96.4	102.3
0310	Concrete Forming & Accessories	78.6	66.4	68.2	100.3	66.5	71.5	89.1	67.6	70.8	97.4	66.6	71.2	89.1	66.5	69.8	89.9	65.9	69.5
0320	Concrete Reinforcing	99.2	81.1	90.0	104.7	79.9	92.1	101.0	85.9	93.3	98.8	81.2	89.8	100.1	85.9	92.9	100.2	80.4	90.2
0330	Cast-in-Place Concrete	137.1	69.7	112.0	99.5	69.8	88.4	110.5	69.8	95.3	120.9	69.8	101.9	93.7	69.7	84.8	135.5	69.8	111.1
03	CONCRETE	103.9	71.0	89.3	89.0	70.9	81.0	86.7	72.4	80.4	93.6	71.1	83.6	76.1	71.9	74.2	107.0	70.7	90.9
04	MASONRY	118.6	82.2	96.1	110.3	82.2	92.9	116.4	82.2	95.2	123.5	82.2	98.0	140.6	82.2	104.5	124.7	82.2	98.4
05	METALS	99.2	87.7	95.7	105.2	87.4	99.8	99.1	89.4	96.1	98.3	87.9	95.2	99.6	89.3	96.5	98.4	87.6	95.2
06	WOOD, PLASTICS & COMPOSITES	69.7	63.3	66.2	93.3	63.3	77.0	82.1	64.7	72.6	90.8	63.3	75.9	82.1	63.3	71.9	81.5	62.5	71.2
07	THERMAL & MOISTURE PROTECTION	105.4	67.3	88.8	100.3	73.3	88.5	104.6	74.0	91.3	105.0	68.8	89.2	104.1	72.8	90.5	106.0	68.7	89.8
08	OPENINGS	99.5	65.0	91.5	99.8	64.8	91.7	99.5	66.9	91.9	99.0	65.0	91.1	99.0	66.1	91.4	99.0	64.4	91.0
0920	Plaster & Gypsum Board	107.5	62.7	78.0	108.5	62.7	78.3	112.1	64.0	80.5	121.8	62.7	82.9	112.1	62.7	79.6	115.7	61.8	80.2
0950, 0980	Ceilings & Acoustic Treatment	96.5	62.7	73.6	100.6	62.7	74.9	96.5	64.0	74.5	94.8	62.7	73.0	96.5	62.7	73.6	94.8	61.8	72.4
0960	Flooring	88.4	94.0	90.0	101.6	94.0	99.4	92.5	94.0	92.9	97.4	94.0	96.4	92.5	94.0	92.9	93.9	94.0	93.9
0970, 0990	Wall Finishes & Painting/Coating	88.0	69.5	77.1	93.5	69.5	79.3	88.0	69.5	77.1	88.0	69.5	77.1	88.0	55.0	68.5	88.0	69.5	77.1
09	FINISHES	89.7	71.5	79.8	97.5	71.4	83.2	90.3	72.3	80.5	93.2	71.5	81.4	89.8	69.9	78.9	92.7	71.0	80.8
COVERS	DIVS. 10 - 14, 25, 28, 41, 43, 44, 46	100.0	90.3	97.9	100.0	90.3	97.9	100.0	90.5	97.9	100.0	90.3	97.9	100.0	90.3	97.9	100.0	90.3	97.9
21, 22, 23	FIRE SUPPRESSION, PLUMBING & HVAC	96.6	70.4	86.0	100.1	71.5	88.5	96.6	69.7	85.8	96.6	76.9	88.6	100.1	69.7	87.8	96.6	76.9	88.6
26, 27, 3370	ELECTRICAL, COMMUNICATIONS & UTIL.	97.9	70.9	84.2	104.7	70.9	87.5	102.3	69.8	85.7	97.9	76.3	86.9	103.4	69.7	86.3	97.9	76.3	86.9
MF2016	WEIGHTED AVERAGE	99.5	75.7	89.3	100.1	76.1	89.8	97.4	76.2	88.3	98.4	77.9	89.6	97.9	75.7	88.4	100.5	77.8	90.8

	DIVISION	NEBRASKA																	
		ALLIANCE			COLUMBUS			GRAND ISLAND			HASTINGS			LINCOLN			MCCOOK		
		693			686			688			689			683 - 685			690		
		MAT.	INST.	TOTAL	MAT.	INST.	TOTAL	MAT.	INST.	TOTAL	MAT.	INST.	TOTAL	MAT.	INST.	TOTAL	MAT.	INST.	TOTAL
015433	CONTRACTOR EQUIPMENT		97.8	97.8		104.7	104.7		104.7	104.7		104.7	104.7		104.7	104.7		104.7	104.7
0241, 31 - 34	SITE & INFRASTRUCTURE, DEMOLITION	101.0	100.8	100.9	105.7	95.0	98.2	110.6	95.2	99.9	109.4	95.0	99.3	98.8	95.2	96.3	103.9	94.9	97.7
0310	Concrete Forming & Accessories	87.4	56.6	61.1	96.6	77.0	79.9	96.2	70.7	74.5	99.5	74.5	78.2	92.5	77.4	79.6	92.8	57.3	62.6
0320	Concrete Reinforcing	110.9	89.9	100.2	100.7	88.4	94.5	100.1	78.3	89.0	100.1	78.3	89.1	101.2	78.6	89.7	103.3	78.5	90.7
0330	Cast-in-Place Concrete	108.2	83.9	99.1	109.4	84.0	99.9	115.8	79.7	102.4	115.8	75.9	101.0	89.3	84.3	87.4	117.1	75.9	101.7
03	CONCRETE	114.1	72.5	95.7	97.3	82.5	90.7	101.9	76.4	90.6	102.1	76.8	90.9	85.1	81.1	83.3	103.7	69.1	88.4
04	MASONRY	107.1	79.0	89.7	112.2	78.9	91.7	105.0	75.3	86.7	113.4	75.4	89.9	92.5	79.8	84.6	102.9	78.9	88.1
05	METALS	98.8	85.1	94.6	89.8	98.2	92.4	91.5	94.1	92.3	92.2	93.9	92.7	93.4	94.6	93.8	93.6	94.0	93.7
06	WOOD, PLASTICS & COMPOSITES	84.6	50.1	65.9	96.1	77.0	85.7	95.2	68.2	80.5	99.1	74.2	85.6	96.3	77.0	85.8	92.8	51.1	70.1
07	THERMAL & MOISTURE PROTECTION	103.8	67.7	88.0	104.5	81.7	94.6	104.6	79.6	93.7	104.7	79.0	93.5	100.5	81.8	92.4	98.7	77.4	89.4
08	OPENINGS	92.8	59.8	85.1	93.2	75.9	89.2	93.2	68.7	87.5	93.2	71.8	88.2	102.5	70.8	95.1	93.3	57.4	84.9
0920	Plaster & Gypsum Board	80.1	48.9	59.5	94.5	76.5	82.6	93.6	67.4	76.3	95.4	73.5	81.0	103.7	76.5	85.8	90.6	49.8	63.8
0950, 0980	Ceilings & Acoustic Treatment	93.7	48.9	63.3	88.3	76.5	80.3	88.3	67.4	74.1	88.3	73.5	78.3	92.2	76.5	81.6	89.7	49.8	62.7
0960	Flooring	92.6	90.3	92.0	85.8	90.3	87.1	85.5	81.5	84.4	86.8	84.0	86.0	99.5	89.5	96.6	91.1	90.3	90.9
0970, 0990	Wall Finishes & Painting/Coating	157.3	53.0	95.6	75.6	61.7	67.3	75.6	64.1	68.8	75.6	61.7	67.3	97.7	81.4	88.1	88.7	46.2	63.5
09	FINISHES	91.7	60.9	74.9	84.8	77.8	81.0	84.9	71.4	77.5	85.5	74.8	79.7	94.5	80.1	86.6	89.1	60.8	73.6
COVERS	DIVS. 10 - 14, 25, 28, 41, 43, 44, 46	100.0	85.6	96.8	100.0	85.2	96.7	100.0	87.8	97.3	100.0	84.9	96.7	100.0	88.7	97.5	100.0	85.8	96.9
21, 22, 23	FIRE SUPPRESSION, PLUMBING & HVAC	96.6	76.7	88.5	96.6	77.0	88.7	100.0	81.3	92.5	96.6	76.2	88.3	99.9	81.3	92.4	96.5	76.9	88.6
26, 27, 3370	ELECTRICAL, COMMUNICATIONS & UTIL.	92.1	67.2	79.5	95.6	83.4	89.4	94.2	67.3	80.5	93.5	81.4	87.3	106.4	67.3	86.5	94.2	67.3	80.5
MF2016	WEIGHTED AVERAGE	98.9	74.7	88.5	95.6	82.7	90.1	97.0	78.5	89.1	96.7	80.0	89.5	97.1	81.1	90.2	96.6	74.8	87.3

	DIVISION	NEBRASKA												NEVADA					
		NORFOLK			NORTH PLATTE			OMAHA			VALENTINE			CARSON CITY			ELKO		
		687			691			680 - 681			692			897			898		
		MAT.	INST.	TOTAL	MAT.	INST.	TOTAL	MAT.	INST.	TOTAL	MAT.	INST.	TOTAL	MAT.	INST.	TOTAL	MAT.	INST.	TOTAL
015433	CONTRACTOR EQUIPMENT		94.1	94.1		104.7	104.7		94.1	94.1		97.6	97.6		97.6	97.6		97.6	97.6
0241, 31 - 34	SITE & INFRASTRUCTURE, DEMOLITION	86.4	94.1	91.7	105.2	94.7	97.9	90.4	94.3	93.1	88.0	99.7	96.2	88.2	97.7	94.8	71.3	96.3	88.7
0310	Concrete Forming & Accessories	83.3	76.1	77.2	95.3	77.2	79.9	95.8	75.5	78.5	83.8	54.4	58.8	104.0	78.3	82.1	108.8	99.1	100.5
0320	Concrete Reinforcing	100.8	68.4	84.3	102.8	78.4	90.4	97.7	78.5	88.0	103.3	68.2	85.5	108.1	119.5	113.9	116.1	117.9	117.0
0330	Cast-in-Place Concrete	109.9	73.1	96.2	117.1	62.0	96.6	92.6	79.8	87.9	103.2	55.3	85.4	97.0	83.6	92.0	93.9	74.7	86.7
03	CONCRETE	95.9	74.2	86.3	103.8	73.2	90.2	86.5	78.0	82.7	101.1	58.2	82.1	95.1	87.3	91.7	92.8	93.4	93.0
04	MASONRY	117.8	78.9	93.8	91.1	78.4	83.2	95.6	80.2	86.1	102.5	78.3	87.6	118.0	70.2	88.5	125.0	69.1	90.5
05	METALS	93.1	80.4	89.2	92.8	93.2	92.9	93.4	84.6	90.8	104.7	79.2	97.0	104.9	95.7	102.2	108.2	94.7	104.1
06	WOOD, PLASTICS & COMPOSITES	80.0	76.6	78.1	94.8	79.1	86.3	94.5	74.5	83.6	79.4	48.8	62.8	92.3	76.4	83.7	105.4	105.3	105.3
07	THERMAL & MOISTURE PROTECTION	104.2	79.7	93.5	98.6	79.5	90.3	101.3	80.5	92.2	99.3	74.9	88.7	111.0	79.7	97.4	107.2	75.7	93.5
08	OPENINGS	94.6	70.7	89.1	92.6	74.2	88.3	100.6	75.3	94.7	94.8	55.0	85.6	99.4	79.7	94.8	104.5	95.3	102.4
0920	Plaster & Gypsum Board	94.3	76.5	82.6	90.6	78.6	82.7	101.6	74.3	83.6	91.9	47.9	62.9	103.2	75.8	85.2	109.5	105.5	106.9
0950, 0980	Ceilings & Acoustic Treatment	103.1	76.5	85.1	89.7	78.6	82.2	104.1	74.3	83.9	106.1	47.9	66.7	102.3	75.8	84.4	102.7	105.5	104.6
0960	Flooring	106.9	90.3	102.1	92.2	81.5	89.1	101.8	90.3	98.4	118.5	90.3	110.3	97.8	69.4	89.6	101.1	69.4	91.9
0970, 0990	Wall Finishes & Painting/Coating	132.0	61.7	90.4	88.7	60.4	72.0	101.1	74.0	85.1	157.4	62.6	101.3	96.7	81.4	87.7	93.9	81.4	86.5
09	FINISHES	101.4	77.4	88.3	89.4	76.9	82.6	98.0	77.8	87.0	109.7	61.0	83.1	95.2	76.1	84.8	93.2	93.1	93.1
COVERS	DIVS. 10 - 14, 25, 28, 41, 43, 44, 46	100.0	84.8	96.7	100.0	62.4	91.7	100.0	87.5	97.2	100.0	57.2	90.6	100.0	102.0	100.4	100.0	92.9	98.4
21, 22, 23	FIRE SUPPRESSION, PLUMBING & HVAC	96.3	76.0	88.1	99.9	75.3	90.0	99.9	76.8	90.6	96.1	75.2	87.6	100.1	78.7	91.4	98.3	78.7	90.4
26, 27, 3370	ELECTRICAL, COMMUNICATIONS & UTIL.	94.5	83.4	88.8	92.4	67.3	79.6	102.7	83.4	92.9	89.7	67.3	78.3	98.9	90.4	94.6	96.2	90.4	93.3
MF2016	WEIGHTED AVERAGE	96.9	79.2	89.3	96.6	77.3	88.3	96.9	80.8	90.0	98.8	70.8	86.8	100.4	84.2	93.5	100.3	87.4	94.8

DIVISION		NEVADA									NEW HAMPSHIRE								
		ELY			LAS VEGAS			RENO			CHARLESTON			CLAREMONT			CONCORD		
		893			889 - 891			894 - 895			036			037			032 - 033		
		MAT.	INST.	TOTAL	MAT.	INST.	TOTAL	MAT.	INST.	TOTAL	MAT.	INST.	TOTAL	MAT.	INST.	TOTAL	MAT.	INST.	TOTAL
015433	CONTRACTOR EQUIPMENT		97.6	97.6		97.6	97.6		97.6	97.6		97.9	97.9		97.9	97.9		97.9	97.9
0241, 31 - 34	SITE & INFRASTRUCTURE, DEMOLITION	77.4	97.5	91.4	80.7	100.1	94.2	77.1	97.7	91.5	83.7	100.4	95.3	77.8	100.4	93.5	91.1	100.6	97.7
0310	Concrete Forming & Accessories	101.8	104.5	104.1	102.9	107.0	106.4	98.3	78.3	81.3	85.6	85.0	85.0	91.4	85.0	85.9	95.3	94.5	94.6
0320	Concrete Reinforcing	114.8	118.1	116.5	105.8	122.0	114.0	108.5	120.9	114.8	89.3	86.2	87.7	89.3	86.2	87.7	104.0	86.6	95.2
0330	Cast-in-Place Concrete	100.8	98.1	99.8	97.6	107.9	101.4	106.6	83.6	98.0	86.1	113.0	96.1	79.1	113.0	91.7	105.0	113.8	108.3
03	CONCRETE	101.0	104.1	102.4	96.4	109.3	102.1	100.6	87.6	94.8	89.2	95.0	91.8	81.8	95.0	87.6	97.4	99.7	98.4
04	MASONRY	130.4	76.4	97.0	117.2	97.1	104.8	124.1	70.2	90.8	93.8	94.8	94.4	94.2	94.8	94.6	101.1	101.3	101.2
05	METALS	108.2	97.6	105.0	116.3	101.3	111.8	109.7	96.2	105.6	84.8	89.5	86.2	84.8	89.5	86.2	90.8	90.4	90.7
06	WOOD, PLASTICS & COMPOSITES	95.2	107.5	101.9	93.0	105.3	99.7	88.9	76.4	82.1	89.4	81.7	85.2	95.8	81.7	88.1	93.8	93.6	93.7
07	THERMAL & MOISTURE PROTECTION	107.7	94.8	102.1	121.5	98.9	111.7	107.2	79.7	95.2	103.2	104.5	103.7	102.9	104.5	103.6	109.6	109.4	109.5
08	OPENINGS	104.5	96.5	102.6	103.4	110.0	105.0	102.2	80.0	97.1	98.4	81.8	94.5	99.5	81.8	95.4	94.4	88.4	93.0
0920	Plaster & Gypsum Board	104.1	107.7	106.5	97.1	105.5	102.6	92.6	75.8	81.5	108.0	80.8	90.1	109.2	80.8	90.5	107.7	93.1	98.1
0950, 0980	Ceilings & Acoustic Treatment	102.7	107.7	106.1	108.6	105.5	106.5	106.1	75.8	85.6	101.6	80.8	87.5	101.6	80.8	87.5	101.7	93.1	95.8
0960	Flooring	98.8	69.4	90.2	90.5	99.9	93.2	95.5	69.4	88.0	83.7	104.3	89.7	86.0	104.3	91.3	93.2	104.3	96.5
0970, 0990	Wall Finishes & Painting/Coating	93.9	115.0	106.4	96.4	115.0	107.4	93.9	81.4	86.5	83.3	92.4	88.7	83.3	92.4	88.7	91.1	92.4	91.8
09	FINISHES	92.2	99.8	96.4	90.1	106.5	99.1	90.1	76.1	82.5	86.6	88.5	87.6	87.0	88.5	87.8	90.1	95.8	93.2
COVERS	DIVS. 10 - 14, 25, 28, 41, 43, 44, 46	100.0	72.3	93.9	100.0	101.8	100.4	100.0	102.0	100.4	100.0	90.3	97.9	100.0	90.3	97.9	100.0	104.6	101.0
21, 22, 23	FIRE SUPPRESSION, PLUMBING & HVAC	98.3	98.3	98.3	100.1	101.3	100.6	100.0	78.7	91.4	96.6	83.9	91.5	96.6	83.9	91.5	100.0	86.3	94.4
26, 27, 3370	ELECTRICAL, COMMUNICATIONS & UTIL.	96.4	95.2	95.8	100.6	110.4	105.6	96.8	90.4	93.6	91.2	77.1	84.0	91.2	77.1	84.0	92.5	77.1	84.7
MF2016	WEIGHTED AVERAGE	101.6	95.6	99.0	102.7	104.1	103.3	101.5	84.2	94.1	92.5	88.6	90.8	91.7	88.6	90.3	96.1	92.4	94.5

DIVISION		NEW HAMPSHIRE															NEW JERSEY		
		KEENE			LITTLETON			MANCHESTER			NASHUA			PORTSMOUTH			ATLANTIC CITY		
		034			035			031			030			038			082, 084		
		MAT.	INST.	TOTAL	MAT.	INST.	TOTAL	MAT.	INST.	TOTAL	MAT.	INST.	TOTAL	MAT.	INST.	TOTAL	MAT.	INST.	TOTAL
015433	CONTRACTOR EQUIPMENT		97.9	97.9		97.9	97.9		97.9	97.9		97.9	97.9		97.9	97.9		95.4	95.4
0241, 31 - 34	SITE & INFRASTRUCTURE, DEMOLITION	90.7	100.4	97.5	77.9	99.3	92.9	88.8	100.6	97.0	92.2	100.5	98.0	85.9	101.0	96.4	88.6	104.0	99.3
0310	Concrete Forming & Accessories	90.0	85.2	85.9	101.7	79.0	82.4	95.8	94.7	94.9	98.1	94.7	95.2	87.0	94.2	93.1	112.3	141.1	136.8
0320	Concrete Reinforcing	89.3	86.2	87.7	90.0	86.2	88.1	104.4	86.6	95.4	111.5	86.6	98.9	89.3	86.6	87.9	80.2	141.7	111.4
0330	Cast-in-Place Concrete	86.5	113.1	96.4	77.6	104.0	87.4	98.0	116.0	104.7	81.8	115.9	94.5	77.6	115.8	91.8	81.4	139.4	103.0
03	CONCRETE	88.8	95.1	91.6	81.2	89.2	84.7	94.2	100.5	97.0	88.7	100.4	93.9	81.4	100.3	89.8	87.3	139.0	110.2
04	MASONRY	96.4	94.8	95.4	106.4	79.0	89.5	99.9	101.3	100.8	98.9	101.3	100.4	94.4	100.5	98.1	105.2	137.8	125.3
05	METALS	85.5	89.7	86.8	85.5	89.5	86.7	92.3	90.7	91.8	90.7	90.5	90.7	87.0	91.4	88.3	97.7	114.5	102.8
06	WOOD, PLASTICS & COMPOSITES	94.0	81.7	87.3	105.9	81.7	92.8	93.9	93.6	93.7	104.1	93.6	98.4	90.7	93.6	92.2	122.6	141.9	133.1
07	THERMAL & MOISTURE PROTECTION	103.8	104.5	104.1	103.1	97.3	100.5	105.2	109.7	107.1	104.2	109.7	106.6	103.7	108.5	105.8	98.8	134.5	114.3
08	OPENINGS	97.0	85.4	94.3	100.4	81.8	96.1	99.3	92.0	97.6	101.4	91.3	99.1	102.0	81.7	97.3	98.7	137.3	107.7
0920	Plaster & Gypsum Board	108.3	80.8	90.2	123.2	80.8	95.3	105.3	93.1	97.2	117.7	93.1	101.5	108.0	93.1	98.2	117.5	142.7	134.1
0950, 0980	Ceilings & Acoustic Treatment	101.6	80.8	87.5	101.6	80.8	87.5	102.6	93.1	96.1	113.6	93.1	99.7	102.6	93.1	96.1	90.9	142.7	126.0
0960	Flooring	85.7	104.3	91.1	96.0	103.2	98.1	94.1	108.6	98.3	89.3	104.3	93.7	83.8	104.3	89.8	92.5	164.3	113.4
0970, 0990	Wall Finishes & Painting/Coating	83.3	106.1	96.8	83.3	92.4	88.7	94.3	106.1	101.3	83.3	104.9	96.1	83.3	104.9	96.1	81.1	145.5	119.2
09	FINISHES	88.4	90.0	89.3	91.9	84.3	87.8	91.9	98.2	95.3	93.1	97.1	95.3	87.5	96.9	92.6	87.7	147.1	120.1
COVERS	DIVS. 10 - 14, 25, 28, 41, 43, 44, 46	100.0	96.6	99.3	100.0	97.5	99.5	100.0	104.6	101.0	100.0	104.6	101.0	100.0	104.3	100.9	100.0	116.6	103.7
21, 22, 23	FIRE SUPPRESSION, PLUMBING & HVAC	96.6	84.0	91.5	96.6	75.7	88.2	100.0	86.3	94.4	100.1	86.3	94.5	100.1	85.7	94.3	99.7	129.2	111.6
26, 27, 3370	ELECTRICAL, COMMUNICATIONS & UTIL.	91.2	77.1	84.0	92.1	50.3	70.8	91.3	77.1	84.1	93.1	77.1	85.0	91.6	77.1	84.2	94.5	143.6	119.4
MF2016	WEIGHTED AVERAGE	93.0	89.2	91.3	93.0	80.1	87.4	96.2	93.0	94.8	95.9	92.8	94.6	93.3	92.2	92.8	96.4	132.6	111.9

DIVISION		NEW JERSEY																	
		CAMDEN			DOVER			ELIZABETH			HACKENSACK			JERSEY CITY			LONG BRANCH		
		081			078			072			076			073			077		
		MAT.	INST.	TOTAL	MAT.	INST.	TOTAL	MAT.	INST.	TOTAL	MAT.	INST.	TOTAL	MAT.	INST.	TOTAL	MAT.	INST.	TOTAL
015433	CONTRACTOR EQUIPMENT		95.4	95.4		97.9	97.9		97.9	97.9		97.9	97.9		95.4	95.4		95.1	95.1
0241, 31 - 34	SITE & INFRASTRUCTURE, DEMOLITION	89.8	104.0	99.7	106.0	105.9	105.9	110.9	105.9	107.4	107.0	106.1	106.4	97.4	106.0	103.4	102.2	105.2	104.3
0310	Concrete Forming & Accessories	102.7	141.0	135.3	97.9	150.2	142.4	110.3	150.2	144.3	97.9	150.1	142.3	102.2	150.2	143.1	102.8	142.7	136.8
0320	Concrete Reinforcing	105.2	132.2	118.9	81.2	151.8	117.1	81.2	151.8	117.1	81.2	151.8	117.1	105.3	151.8	129.0	81.2	151.4	116.9
0330	Cast-in-Place Concrete	78.9	137.5	100.7	82.4	134.8	101.9	70.8	138.0	95.8	80.6	137.9	101.9	64.3	134.9	90.6	71.5	137.5	96.1
03	CONCRETE	87.4	136.6	109.2	86.0	143.5	111.4	83.0	144.6	110.3	84.4	144.5	111.0	80.6	143.3	108.4	84.7	140.7	109.6
04	MASONRY	96.5	137.8	122.0	90.5	148.2	126.1	107.5	148.2	132.6	94.4	148.2	127.6	86.7	148.2	124.7	100.0	138.3	123.7
05	METALS	103.2	111.1	105.6	95.4	123.3	103.8	96.9	123.3	104.9	95.4	123.2	103.8	101.0	120.3	106.9	95.5	119.2	102.7
06	WOOD, PLASTICS & COMPOSITES	109.9	141.9	127.2	98.6	150.2	126.6	114.4	150.2	133.9	98.6	150.2	126.6	99.5	150.2	127.1	100.6	143.3	123.8
07	THERMAL & MOISTURE PROTECTION	98.6	135.5	114.7	109.0	145.5	124.9	109.2	146.0	125.2	108.7	138.4	121.6	108.4	145.5	124.6	108.6	132.4	119.0
08	OPENINGS	100.8	135.1	108.8	104.1	145.7	113.8	102.5	145.7	112.5	101.9	145.7	112.1	100.5	145.7	111.0	96.5	141.2	106.9
0920	Plaster & Gypsum Board	112.5	142.7	132.4	108.5	151.3	136.7	116.1	151.3	139.3	108.5	151.3	136.7	111.3	151.3	137.6	110.3	144.3	132.7
0950, 0980	Ceilings & Acoustic Treatment	99.5	142.7	128.8	87.3	151.3	130.7	89.2	151.3	131.3	87.3	151.3	130.7	96.9	151.3	133.8	87.3	144.3	125.9
0960	Flooring	88.9	164.3	110.8	81.5	190.7	113.2	86.3	190.7	116.6	81.5	190.7	113.2	82.5	190.7	113.9	82.7	182.9	111.8
0970, 0990	Wall Finishes & Painting/Coating	81.1	145.5	119.2	83.0	145.5	120.0	83.0	145.5	120.0	83.0	145.5	120.0	83.1	145.5	120.1	83.1	145.5	120.1
09	FINISHES	87.4	147.1	120.0	84.4	157.8	124.4	87.8	157.8	126.0	84.2	157.8	124.4	86.7	158.1	125.7	85.3	151.3	121.3
COVERS	DIVS. 10 - 14, 25, 28, 41, 43, 44, 46	100.0	116.6	103.7	100.0	133.7	107.4	100.0	133.7	107.4	100.0	133.8	107.4	100.0	133.8	107.4	100.0	116.8	103.7
21, 22, 23	FIRE SUPPRESSION, PLUMBING & HVAC	99.9	129.1	111.7	99.8	138.3	115.4	100.1	137.9	115.4	99.8	138.3	115.4	100.1	138.3	115.5	99.8	134.0	113.6
26, 27, 3370	ELECTRICAL, COMMUNICATIONS & UTIL.	99.1	135.3	117.5	95.6	141.9	119.1	96.2	141.9	119.4	95.6	142.5	119.4	100.2	142.5	121.7	95.3	133.1	114.5
MF2016	WEIGHTED AVERAGE	97.5	130.7	111.8	96.1	139.6	114.7	97.3	139.7	115.4	95.9	139.6	114.6	96.3	139.4	114.8	95.6	133.6	111.9

	DIVISION	NEW JERSEY																	
		NEW BRUNSWICK			NEWARK			PATERSON			POINT PLEASANT			SUMMIT			TRENTON		
		088 - 089			070 - 071			074 - 075			087			079			085 - 086		
		MAT.	INST.	TOTAL	MAT.	INST.	TOTAL	MAT.	INST.	TOTAL	MAT.	INST.	TOTAL	MAT.	INST.	TOTAL	MAT.	INST.	TOTAL
015433	CONTRACTOR EQUIPMENT		95.1	95.1		97.9	97.9		97.9	97.9		95.1	95.1		97.9	97.9		95.1	95.1
0241, 31 - 34	SITE & INFRASTRUCTURE, DEMOLITION	101.2	105.6	104.3	110.5	105.9	107.3	108.9	106.1	106.9	102.7	105.2	104.5	108.3	105.9	106.6	88.1	105.2	100.0
0310	Concrete Forming & Accessories	106.0	149.9	143.4	101.8	150.2	143.0	99.9	150.1	142.6	100.1	142.7	136.3	100.7	150.2	142.8	101.3	142.5	136.3
0320	Concrete Reinforcing	81.1	151.5	116.9	103.6	151.8	128.1	105.3	151.8	129.0	81.1	151.4	116.8	81.2	151.8	117.1	106.5	119.6	113.1
0330	Cast-in-Place Concrete	100.7	140.4	115.4	91.0	138.0	108.5	82.0	137.9	102.8	100.7	139.6	115.1	68.5	138.0	94.3	99.1	137.4	113.3
03	CONCRETE	103.9	145.0	122.1	93.0	144.6	115.8	88.7	144.5	113.4	103.5	141.4	120.3	80.3	144.6	108.8	97.0	135.2	113.9
04	MASONRY	104.4	143.1	128.3	93.1	148.2	127.2	90.9	148.2	126.3	93.9	138.3	121.3	93.5	148.2	127.3	97.1	138.3	122.6
05	METALS	97.7	119.6	104.4	103.0	123.3	109.1	96.2	123.2	104.4	97.7	119.2	104.2	95.4	123.3	103.8	102.9	108.1	104.5
06	WOOD, PLASTICS & COMPOSITES	115.3	150.2	134.3	99.4	150.2	127.0	101.3	150.2	127.9	107.0	143.3	126.7	102.5	150.2	128.4	99.0	143.3	123.1
07	THERMAL & MOISTURE PROTECTION	99.1	141.6	117.6	110.6	146.0	126.0	109.1	138.4	121.8	99.1	134.8	114.6	109.4	146.0	125.3	99.6	136.5	115.7
08	OPENINGS	93.8	145.7	105.9	100.3	145.7	110.8	107.2	145.7	116.1	95.6	141.2	106.2	108.3	145.7	117.0	100.0	133.8	107.8
0920	Plaster & Gypsum Board	114.8	151.3	138.8	104.9	151.3	135.4	111.3	151.3	137.6	109.6	144.3	132.4	110.3	151.3	137.3	103.6	144.3	130.4
0950, 0980	Ceilings & Acoustic Treatment	90.9	151.3	131.8	95.8	151.3	133.4	96.9	151.3	133.8	90.9	144.3	127.1	87.3	151.3	130.7	94.9	144.3	128.4
0960	Flooring	90.2	190.7	119.4	94.0	190.7	122.1	82.5	190.7	113.9	87.8	164.3	110.0	82.7	190.7	114.0	94.9	182.9	120.4
0970, 0990	Wall Finishes & Painting/Coating	81.1	145.5	119.2	88.0	145.5	122.0	83.0	145.5	120.0	81.1	145.5	119.2	83.0	145.5	120.0	88.5	145.5	122.2
09	FINISHES	87.7	157.7	125.9	90.4	157.8	127.2	86.8	157.8	125.5	86.2	148.1	120.0	85.3	157.8	124.9	89.7	151.3	123.3
COVERS	DIVS. 10 - 14, 25, 28, 41, 43, 44, 46	100.0	133.6	107.4	100.0	133.8	107.4	100.0	133.8	107.4	100.0	113.9	103.1	100.0	133.7	107.4	100.0	116.8	103.7
21, 22, 23	FIRE SUPPRESSION, PLUMBING & HVAC	99.7	136.5	114.6	100.1	137.9	115.4	100.1	138.3	115.5	99.7	134.0	113.5	99.8	137.9	115.2	100.1	133.7	113.7
26, 27, 3370	ELECTRICAL, COMMUNICATIONS & UTIL.	95.1	137.0	116.4	104.5	142.5	123.8	100.2	141.9	121.4	94.5	133.1	114.1	96.2	141.9	119.4	103.0	131.8	117.7
MF2016	WEIGHTED AVERAGE	98.3	137.7	115.2	99.6	139.7	116.8	97.7	139.5	115.6	97.7	133.3	113.0	96.2	139.7	114.8	99.2	131.4	113.0

	DIVISION	NEW JERSEY			NEW MEXICO														
		VINELAND			ALBUQUERQUE			CARRIZOZO			CLOVIS			FARMINGTON			GALLUP		
		080, 083			870 - 872			883			881			874			873		
		MAT.	INST.	TOTAL	MAT.	INST.	TOTAL	MAT.	INST.	TOTAL	MAT.	INST.	TOTAL	MAT.	INST.	TOTAL	MAT.	INST.	TOTAL
015433	CONTRACTOR EQUIPMENT		95.4	95.4		110.9	110.9		110.9	110.9		110.9	110.9		110.9	110.9		110.9	110.9
0241, 31 - 34	SITE & INFRASTRUCTURE, DEMOLITION	92.9	104.0	100.6	91.1	102.3	98.9	110.0	102.3	104.6	97.3	102.3	100.8	97.5	102.3	100.9	106.6	102.3	103.6
0310	Concrete Forming & Accessories	97.1	141.2	134.7	99.2	66.9	71.7	96.5	66.9	71.3	96.5	66.8	71.2	99.3	66.9	71.7	99.3	66.9	71.7
0320	Concrete Reinforcing	80.2	135.0	108.0	108.9	71.9	90.1	113.5	71.9	92.3	114.7	71.9	92.9	119.1	71.9	95.1	113.9	71.9	92.6
0330	Cast-in-Place Concrete	87.8	137.6	106.3	91.4	69.1	83.1	94.8	69.1	85.2	94.7	69.1	85.2	92.3	69.1	83.7	86.8	69.1	80.2
03	CONCRETE	91.8	137.3	112.0	87.5	69.8	79.6	111.4	69.8	93.0	100.1	69.8	86.6	90.6	69.8	81.4	95.5	69.8	84.1
04	MASONRY	93.5	138.3	121.2	105.7	58.7	76.7	106.9	58.7	77.1	106.9	58.7	77.1	114.3	58.7	80.0	100.9	58.7	74.9
05	METALS	97.6	112.7	102.2	102.5	90.6	98.9	101.6	90.6	98.3	101.3	90.5	98.0	100.3	90.6	97.4	99.5	90.6	96.8
06	WOOD, PLASTICS & COMPOSITES	103.5	141.9	124.3	104.0	68.8	84.9	94.6	68.8	80.6	94.6	68.8	80.6	104.1	68.8	84.9	104.1	68.8	84.9
07	THERMAL & MOISTURE PROTECTION	98.5	134.7	114.3	101.8	70.0	88.0	102.8	70.0	88.5	101.5	70.0	87.8	102.1	70.0	88.1	103.4	70.0	88.8
08	OPENINGS	95.2	136.1	104.7	100.9	67.7	93.2	98.7	67.7	91.5	98.8	67.7	91.6	103.3	67.7	95.1	103.4	67.7	95.1
0920	Plaster & Gypsum Board	108.1	142.7	130.9	107.8	67.7	81.4	80.9	67.7	72.2	80.9	67.7	72.2	98.8	67.7	78.3	98.8	67.7	78.3
0950, 0980	Ceilings & Acoustic Treatment	90.9	142.7	126.0	103.0	67.7	79.1	103.8	67.7	79.4	103.8	67.7	79.4	102.6	67.7	79.0	102.6	67.7	79.0
0960	Flooring	86.9	164.3	109.4	87.8	68.0	82.0	97.5	68.0	88.9	97.5	68.0	88.9	89.2	68.0	83.1	89.2	68.0	83.1
0970, 0990	Wall Finishes & Painting/Coating	81.1	145.5	119.2	96.1	51.9	69.9	93.4	51.9	68.8	93.4	51.9	68.8	90.8	51.9	67.7	90.8	51.9	67.7
09	FINISHES	85.0	147.2	119.0	89.8	65.5	76.5	92.9	65.5	78.0	91.7	65.5	77.4	88.8	65.5	76.1	90.0	65.5	76.6
COVERS	DIVS. 10 - 14, 25, 28, 41, 43, 44, 46	100.0	116.8	103.7	100.0	85.1	96.7	100.0	85.1	96.7	100.0	85.1	96.7	100.0	85.1	96.7	100.0	85.1	96.7
21, 22, 23	FIRE SUPPRESSION, PLUMBING & HVAC	99.7	129.4	111.7	100.1	68.7	87.4	97.9	68.7	86.1	97.9	68.4	85.9	99.9	68.7	87.3	97.8	68.7	86.1
26, 27, 3370	ELECTRICAL, COMMUNICATIONS & UTIL.	94.5	143.6	119.4	88.9	85.8	87.3	90.8	85.8	88.3	88.5	85.8	87.1	86.9	85.8	86.4	86.2	85.8	86.0
MF2016	WEIGHTED AVERAGE	95.8	132.3	111.4	97.1	75.0	87.6	100.2	75.0	89.4	98.0	74.9	88.1	97.7	75.0	88.0	97.3	75.0	87.7

	DIVISION	NEW MEXICO																	
		LAS CRUCES			LAS VEGAS			ROSWELL			SANTA FE			SOCORRO			TRUTH/CONSEQUENCES		
		880			877			882			875			878			879		
		MAT.	INST.	TOTAL	MAT.	INST.	TOTAL	MAT.	INST.	TOTAL	MAT.	INST.	TOTAL	MAT.	INST.	TOTAL	MAT.	INST.	TOTAL
015433	CONTRACTOR EQUIPMENT		86.2	86.2		110.9	110.9		110.9	110.9		110.9	110.9		110.9	110.9		86.2	86.2
0241, 31 - 34	SITE & INFRASTRUCTURE, DEMOLITION	98.0	81.1	86.3	96.8	102.3	100.6	99.5	102.3	101.5	100.8	102.3	101.8	93.1	102.3	99.5	112.8	81.2	90.8
0310	Concrete Forming & Accessories	93.2	65.8	69.9	99.3	66.9	71.7	96.5	66.9	71.3	97.8	66.9	71.5	99.3	66.9	71.7	97.2	65.8	70.5
0320	Concrete Reinforcing	110.5	71.7	90.8	115.9	71.9	93.5	114.7	71.9	92.9	108.3	71.9	89.8	118.2	71.9	94.6	111.1	71.7	91.1
0330	Cast-in-Place Concrete	89.5	61.9	79.2	89.7	69.1	82.1	94.7	69.1	85.2	95.3	69.1	85.5	87.9	69.1	80.9	96.4	61.9	83.5
03	CONCRETE	79.5	66.4	73.7	88.4	69.8	80.2	100.9	69.8	87.1	87.9	69.8	79.9	87.4	69.8	79.6	82.6	66.4	75.4
04	MASONRY	102.3	58.4	75.2	101.2	58.7	75.0	118.1	58.7	81.4	94.3	58.7	72.3	101.1	58.7	74.9	98.2	58.4	73.6
05	METALS	100.1	82.8	94.9	99.2	90.6	96.6	102.5	90.6	98.9	96.3	90.6	94.6	99.5	90.6	96.8	99.0	82.8	94.1
06	WOOD, PLASTICS & COMPOSITES	83.9	67.8	75.1	104.1	68.8	84.9	94.6	68.8	80.6	99.1	68.8	82.7	104.1	68.8	84.9	95.4	67.8	80.4
07	THERMAL & MOISTURE PROTECTION	88.5	65.3	78.4	101.7	70.0	87.9	101.7	70.0	87.9	104.5	70.0	89.5	101.6	70.0	87.8	89.5	65.3	78.9
08	OPENINGS	94.1	67.1	87.9	99.8	67.7	92.3	98.6	67.7	91.5	100.2	67.7	92.6	99.6	67.7	92.2	92.9	67.1	86.9
0920	Plaster & Gypsum Board	79.0	67.7	71.6	98.8	67.7	78.3	80.9	67.7	72.2	111.2	67.7	82.6	98.8	67.7	78.3	99.7	67.7	78.7
0950, 0980	Ceilings & Acoustic Treatment	87.0	67.7	73.9	102.6	67.7	79.0	103.8	67.7	79.4	101.4	67.7	78.6	102.6	67.7	79.0	86.9	67.7	73.9
0960	Flooring	127.8	68.0	110.4	89.2	68.0	83.1	97.5	68.0	88.9	99.4	68.0	90.3	89.2	68.0	83.1	117.2	68.0	103.0
0970, 0990	Wall Finishes & Painting/Coating	82.6	51.9	64.4	90.8	51.9	67.7	93.4	51.9	68.8	98.2	51.9	70.8	90.8	51.9	67.7	83.3	51.9	64.7
09	FINISHES	100.5	64.7	81.0	88.6	65.5	76.0	91.8	65.5	77.4	96.1	65.5	79.4	88.6	65.5	76.0	98.7	64.7	80.2
COVERS	DIVS. 10 - 14, 25, 28, 41, 43, 44, 46	100.0	82.6	96.2	100.0	85.1	96.7	100.0	85.1	96.7	100.0	85.1	96.7	100.0	85.1	96.7	100.0	82.6	96.2
21, 22, 23	FIRE SUPPRESSION, PLUMBING & HVAC	100.3	68.5	87.4	97.8	68.7	86.1	99.9	68.7	87.3	100.0	68.7	87.4	97.8	68.7	86.1	97.7	68.5	85.9
26, 27, 3370	ELECTRICAL, COMMUNICATIONS & UTIL.	90.5	85.8	88.1	88.4	85.8	87.1	89.8	85.8	87.8	101.6	85.8	93.6	86.7	85.8	86.2	90.5	85.8	88.1
MF2016	WEIGHTED AVERAGE	95.6	71.7	85.4	95.8	75.0	86.9	99.5	75.0	89.0	97.6	75.0	87.9	95.5	75.0	86.7	95.2	71.7	85.2

	DIVISION	NEW MEXICO			NEW YORK														
		TUCUMCARI			ALBANY			BINGHAMTON			BRONX			BROOKLYN			BUFFALO		
		884			120 - 122			137 - 139			104			112			140 - 142		
		MAT.	INST.	TOTAL	MAT.	INST.	TOTAL	MAT.	INST.	TOTAL	MAT.	INST.	TOTAL	MAT.	INST.	TOTAL	MAT.	INST.	TOTAL
015433	CONTRACTOR EQUIPMENT		110.9	110.9		115.2	115.2		120.3	120.3		106.8	106.8		111.8	111.8		99.5	99.5
0241, 31 - 34	SITE & INFRASTRUCTURE, DEMOLITION	96.9	102.3	100.6	82.2	104.6	97.8	95.4	93.4	94.0	96.2	114.2	108.8	119.7	125.6	123.8	105.2	101.8	102.8
0310	Concrete Forming & Accessories	96.5	66.8	71.2	98.5	107.8	106.4	100.0	94.0	94.9	99.4	193.1	179.2	106.0	193.1	180.1	101.1	118.9	116.3
0320	Concrete Reinforcing	112.5	71.9	91.8	106.2	116.1	111.3	102.8	109.4	106.1	99.4	179.6	140.2	104.1	246.0	176.2	109.1	115.0	112.1
0330	Cast-in-Place Concrete	94.7	69.1	85.2	80.0	116.1	93.4	104.1	108.3	105.7	77.3	175.4	113.8	106.1	173.8	131.3	115.0	124.3	118.5
03	CONCRETE	99.4	69.8	86.3	86.2	113.1	98.1	93.6	103.8	98.1	83.4	183.7	127.8	105.5	193.3	144.4	110.6	119.5	114.6
04	MASONRY	117.9	58.7	81.3	87.4	116.9	105.6	106.7	106.2	106.4	87.5	189.3	150.4	117.1	189.2	161.7	123.5	123.6	123.5
05	METALS	101.3	90.5	98.0	92.3	125.9	102.5	93.6	134.4	105.9	84.7	173.7	111.6	101.5	174.0	123.4	100.6	107.1	102.6
06	WOOD, PLASTICS & COMPOSITES	94.6	68.8	80.6	95.6	104.7	100.5	103.7	90.7	96.6	103.5	194.1	152.7	105.9	193.8	153.6	99.6	118.2	109.7
07	THERMAL & MOISTURE PROTECTION	101.4	70.0	87.8	107.0	111.5	108.9	108.9	96.2	103.4	108.8	171.4	136.0	109.1	170.6	135.9	106.0	113.6	109.3
08	OPENINGS	98.6	67.7	91.4	97.2	105.0	99.0	92.4	95.3	93.1	94.0	202.1	119.1	89.8	201.9	115.8	99.1	112.0	102.1
0920	Plaster & Gypsum Board	80.9	67.7	72.2	103.7	104.7	104.4	105.9	90.2	95.5	95.9	196.6	162.2	102.3	196.6	164.4	103.6	118.6	113.5
0950, 0980	Ceilings & Acoustic Treatment	103.8	67.7	79.4	99.5	104.7	103.1	96.5	90.2	92.2	94.3	196.6	163.6	92.3	196.6	162.9	93.3	118.6	110.4
0960	Flooring	97.5	68.0	88.9	93.5	111.3	98.7	101.8	103.9	102.4	96.1	188.7	123.0	108.8	188.7	132.0	98.3	120.6	104.8
0970, 0990	Wall Finishes & Painting/Coating	93.4	51.9	68.8	93.4	103.9	99.6	87.8	106.9	99.1	104.8	166.6	141.3	114.1	166.6	145.1	98.9	118.6	110.6
09	FINISHES	91.6	65.5	77.4	90.4	107.6	99.8	91.7	96.5	94.3	97.5	190.5	148.2	104.7	190.2	151.4	98.7	119.9	110.3
COVERS	DIVS. 10 - 14, 25, 28, 41, 43, 44, 46	100.0	85.1	96.7	100.0	102.5	100.6	100.0	98.0	99.6	100.0	138.0	108.4	100.0	137.2	108.2	100.0	106.4	101.4
21, 22, 23	FIRE SUPPRESSION, PLUMBING & HVAC	97.9	68.4	85.9	100.1	111.8	104.8	100.6	98.7	99.8	100.2	176.4	131.0	99.8	176.3	130.7	100.0	99.7	99.9
26, 27, 3370	ELECTRICAL, COMMUNICATIONS & UTIL.	90.8	85.8	88.3	99.6	107.6	103.6	100.3	100.9	100.6	91.4	186.7	139.9	100.2	186.7	144.2	103.7	102.8	103.3
MF2016	WEIGHTED AVERAGE	98.7	74.9	88.5	95.0	111.5	102.1	97.3	102.8	99.7	93.3	176.8	129.1	101.8	178.9	134.9	103.0	110.0	106.0

	DIVISION	NEW YORK																	
		ELMIRA			FAR ROCKAWAY			FLUSHING			GLENS FALLS			HICKSVILLE			JAMAICA		
		148 - 149			116			113			128			115, 117, 118			114		
		MAT.	INST.	TOTAL	MAT.	INST.	TOTAL	MAT.	INST.	TOTAL	MAT.	INST.	TOTAL	MAT.	INST.	TOTAL	MAT.	INST.	TOTAL
015433	CONTRACTOR EQUIPMENT		123.0	123.0		111.8	111.8		111.8	111.8		115.2	115.2		111.8	111.8		111.8	111.8
0241, 31 - 34	SITE & INFRASTRUCTURE, DEMOLITION	104.3	93.8	97.0	123.0	125.6	124.8	123.1	125.6	124.8	73.0	104.2	94.8	112.7	124.4	120.9	117.3	125.6	123.1
0310	Concrete Forming & Accessories	85.0	97.4	95.5	92.4	193.1	178.1	96.3	193.1	178.7	83.0	100.7	98.0	88.9	161.5	150.7	96.3	193.1	178.7
0320	Concrete Reinforcing	108.5	108.2	108.4	104.1	246.0	176.2	105.9	246.0	177.1	101.8	116.0	109.0	104.1	179.4	142.4	104.1	246.0	176.2
0330	Cast-in-Place Concrete	103.2	107.8	104.9	114.8	173.8	136.7	114.8	173.8	136.7	77.1	114.0	90.9	97.4	166.2	123.0	106.1	173.8	131.3
03	CONCRETE	93.0	105.2	98.4	111.8	193.3	147.9	112.2	193.3	148.1	79.9	109.1	92.9	97.3	166.2	127.9	104.9	193.3	144.1
04	MASONRY	110.3	106.8	108.2	121.8	189.2	163.4	115.8	189.2	161.2	93.3	113.6	105.8	111.6	176.9	152.0	119.9	189.2	162.7
05	METALS	96.8	136.7	108.9	101.5	174.0	123.5	101.5	174.0	123.5	86.4	125.3	98.2	103.0	173.6	124.4	101.5	174.0	123.5
06	WOOD, PLASTICS & COMPOSITES	84.8	95.6	90.7	88.8	193.8	145.8	93.6	193.8	148.0	85.0	97.0	91.5	85.3	158.4	125.0	93.6	193.8	148.0
07	THERMAL & MOISTURE PROTECTION	105.1	96.4	101.3	109.0	170.6	135.8	109.0	170.6	135.9	99.9	109.2	103.9	108.6	161.9	131.8	108.8	170.6	135.7
08	OPENINGS	98.6	97.7	98.4	88.5	201.9	114.8	88.5	201.9	114.8	92.0	100.9	94.0	88.9	183.2	110.8	88.5	201.9	114.8
0920	Plaster & Gypsum Board	98.0	95.4	96.3	91.1	196.6	160.6	93.6	196.6	161.4	94.8	96.8	96.1	90.8	160.2	136.5	93.6	196.6	161.4
0950, 0980	Ceilings & Acoustic Treatment	99.9	95.4	96.9	81.3	196.6	159.4	81.3	196.6	159.4	86.5	96.8	93.5	80.3	160.2	134.5	81.3	196.6	159.4
0960	Flooring	88.8	103.9	93.2	103.7	188.7	128.3	105.2	188.7	129.4	84.9	111.3	92.6	102.6	185.7	126.7	105.2	188.7	129.4
0970, 0990	Wall Finishes & Painting/Coating	98.0	94.0	95.6	114.1	166.6	145.1	114.1	166.6	145.1	86.9	102.0	95.9	114.1	166.6	145.1	114.1	166.6	145.1
09	FINISHES	91.7	98.0	95.1	99.9	190.2	149.2	100.7	190.2	149.6	83.0	102.0	93.4	98.5	165.8	135.2	100.3	190.2	149.4
COVERS	DIVS. 10 - 14, 25, 28, 41, 43, 44, 46	100.0	99.1	99.8	100.0	137.2	108.2	100.0	137.2	108.2	100.0	98.1	99.6	100.0	134.8	107.7	100.0	137.2	108.2
21, 22, 23	FIRE SUPPRESSION, PLUMBING & HVAC	96.7	97.2	96.9	96.3	176.3	128.6	96.3	176.3	128.6	96.7	108.9	101.6	99.8	163.2	125.4	96.3	176.3	128.6
26, 27, 3370	ELECTRICAL, COMMUNICATIONS & UTIL.	99.7	104.3	102.0	107.2	186.7	147.6	107.2	186.7	147.6	94.1	107.6	100.9	99.6	144.4	122.4	98.6	186.7	143.4
MF2016	WEIGHTED AVERAGE	97.6	103.9	100.3	102.2	178.9	135.1	102.0	178.9	135.0	90.6	108.7	98.4	99.8	160.5	125.8	100.2	178.9	134.0

	DIVISION	NEW YORK																	
		JAMESTOWN			KINGSTON			LONG ISLAND CITY			MONTICELLO			MOUNT VERNON			NEW ROCHELLE		
		147			124			111			127			105			108		
		MAT.	INST.	TOTAL	MAT.	INST.	TOTAL	MAT.	INST.	TOTAL	MAT.	INST.	TOTAL	MAT.	INST.	TOTAL	MAT.	INST.	TOTAL
015433	CONTRACTOR EQUIPMENT		92.7	92.7		111.8	111.8		111.8	111.8		111.8	111.8		106.8	106.8		106.8	106.8
0241, 31 - 34	SITE & INFRASTRUCTURE, DEMOLITION	105.8	94.2	97.7	143.9	121.2	128.1	120.9	125.6	124.2	139.1	121.1	126.6	101.1	110.1	107.3	100.8	110.0	107.2
0310	Concrete Forming & Accessories	85.0	92.2	91.1	84.3	133.6	126.2	100.7	193.1	179.4	91.3	133.6	127.3	89.4	141.8	134.0	104.8	138.2	133.2
0320	Concrete Reinforcing	108.7	114.4	111.6	102.3	164.9	134.1	104.1	246.0	176.2	101.4	164.9	133.7	98.1	182.1	140.8	98.3	181.9	140.8
0330	Cast-in-Place Concrete	107.0	106.6	106.8	104.1	147.5	120.2	109.5	173.8	133.4	97.5	147.5	116.1	86.3	151.4	110.5	86.3	151.1	110.4
03	CONCRETE	96.1	101.0	98.3	100.7	143.2	119.5	108.0	193.3	145.8	95.8	143.2	116.8	91.9	152.2	118.6	91.3	150.5	117.5
04	MASONRY	120.0	104.7	110.6	107.8	154.7	136.8	113.9	189.2	160.5	100.1	154.7	133.8	92.8	156.2	132.0	92.8	156.2	132.0
05	METALS	94.2	102.5	96.7	93.8	136.5	106.7	101.5	174.0	123.4	93.7	136.7	106.7	84.5	165.6	109.0	84.7	164.9	109.0
06	WOOD, PLASTICS & COMPOSITES	83.4	88.7	86.3	86.2	127.6	108.7	99.9	193.8	150.9	93.1	127.6	111.8	92.4	136.3	116.2	111.5	132.4	122.9
07	THERMAL & MOISTURE PROTECTION	104.5	96.7	101.1	123.5	146.8	133.6	109.0	170.6	135.8	123.1	146.8	133.4	109.8	148.9	126.8	109.9	146.3	125.7
08	OPENINGS	98.4	95.7	97.8	93.7	146.1	105.8	88.5	201.9	114.8	89.5	146.0	102.6	94.0	169.4	111.5	94.1	167.2	111.1
0920	Plaster & Gypsum Board	85.7	88.3	87.4	96.9	128.5	117.7	98.5	196.6	163.1	97.5	128.5	117.9	90.9	137.1	121.3	104.3	133.1	123.3
0950, 0980	Ceilings & Acoustic Treatment	96.5	88.3	90.9	77.1	128.5	112.0	81.3	196.6	159.4	77.1	128.5	111.9	92.6	137.1	122.7	92.6	133.1	120.0
0960	Flooring	91.4	103.9	95.0	102.2	166.0	120.7	106.8	188.7	130.6	104.5	166.0	122.4	87.9	185.7	116.3	95.3	169.5	116.9
0970, 0990	Wall Finishes & Painting/Coating	99.6	101.0	100.4	113.6	130.5	123.6	114.1	166.6	145.1	113.6	126.6	121.3	103.2	166.6	140.7	103.2	166.6	140.7
09	FINISHES	90.2	94.3	92.5	96.5	138.5	119.4	101.6	190.2	150.0	96.9	138.0	119.3	94.5	151.4	125.5	98.4	145.8	124.3
COVERS	DIVS. 10 - 14, 25, 28, 41, 43, 44, 46	100.0	98.5	99.7	100.0	119.7	104.3	100.0	137.2	108.2	100.0	120.6	104.5	100.0	131.4	106.9	100.0	114.8	103.3
21, 22, 23	FIRE SUPPRESSION, PLUMBING & HVAC	96.6	95.3	96.0	96.6	136.7	112.8	99.8	176.3	130.7	96.6	141.0	114.6	96.8	144.9	116.2	96.8	144.6	116.1
26, 27, 3370	ELECTRICAL, COMMUNICATIONS & UTIL.	98.6	97.2	97.9	95.5	119.2	107.5	99.1	186.7	143.7	95.5	119.2	107.5	89.6	170.1	130.5	89.6	145.2	117.8
MF2016	WEIGHTED AVERAGE	97.8	97.8	97.8	98.9	136.0	114.8	101.5	178.9	134.7	97.4	136.9	114.3	93.5	151.3	118.3	93.9	146.0	116.2

DIVISION		NEW YORK																	
		NEW YORK			NIAGARA FALLS			PLATTSBURGH			POUGHKEEPSIE			QUEENS			RIVERHEAD		
		100 - 102			143			129			125 - 126			110			119		
		MAT.	INST.	TOTAL	MAT.	INST.	TOTAL	MAT.	INST.	TOTAL	MAT.	INST.	TOTAL	MAT.	INST.	TOTAL	MAT.	INST.	TOTAL
015433	CONTRACTOR EQUIPMENT		104.8	104.8		92.7	92.7		96.4	96.4		111.8	111.8		111.8	111.8		111.8	111.8
0241, 31 - 34	SITE & INFRASTRUCTURE, DEMOLITION	105.1	111.5	109.6	108.2	95.2	99.1	112.4	101.3	104.7	140.1	120.3	126.3	116.0	125.6	122.7	113.8	123.9	120.8
0310	Concrete Forming & Accessories	108.9	192.3	179.9	85.0	116.3	111.6	88.4	93.9	93.1	84.3	172.3	159.2	89.1	193.1	177.6	93.5	160.5	150.5
0320	Concrete Reinforcing	105.2	183.4	144.9	107.3	115.2	111.3	106.3	115.3	110.8	102.3	164.7	134.0	105.9	246.0	177.1	106.1	221.8	164.9
0330	Cast-in-Place Concrete	89.3	174.6	121.0	110.7	127.7	117.0	94.3	103.5	97.7	100.8	139.3	115.2	100.8	173.8	128.0	99.1	164.4	123.4
03	CONCRETE	96.2	183.5	134.9	98.5	119.3	107.7	92.8	100.6	96.3	98.0	157.8	124.5	100.6	193.3	141.7	97.9	171.0	130.3
04	MASONRY	101.2	187.1	154.2	128.3	127.8	128.0	87.7	99.4	94.9	100.1	142.5	126.3	108.1	189.2	158.2	117.3	175.8	153.5
05	METALS	95.6	173.1	119.1	96.9	103.1	98.7	90.2	100.0	93.2	93.8	136.3	106.7	101.5	174.0	123.4	103.4	161.6	121.1
06	WOOD, PLASTICS & COMPOSITES	109.2	194.2	155.4	83.4	111.4	98.6	91.7	90.7	91.2	86.2	184.5	139.6	85.4	193.8	144.3	90.5	158.4	127.4
07	THERMAL & MOISTURE PROTECTION	111.6	170.7	137.3	104.6	113.1	108.3	117.8	99.9	110.0	123.4	147.6	133.9	108.6	170.6	135.6	109.6	158.8	131.0
08	OPENINGS	97.1	203.0	121.7	98.5	107.7	100.6	99.4	97.4	98.9	93.7	175.5	112.7	88.5	201.9	114.8	88.9	176.6	109.3
0920	Plaster & Gypsum Board	101.2	196.6	164.0	85.7	111.6	102.8	114.7	90.0	98.4	96.9	187.0	156.2	90.8	196.6	160.5	91.9	160.2	136.9
0950, 0980	Ceilings & Acoustic Treatment	109.6	196.6	168.5	96.5	111.6	106.7	106.5	90.0	95.3	77.1	187.0	151.6	81.3	196.6	159.4	81.2	160.2	134.7
0960	Flooring	97.3	188.7	123.8	91.4	115.8	98.5	107.4	108.6	107.8	102.2	164.0	120.1	102.6	188.7	127.6	103.7	172.5	123.6
0970, 0990	Wall Finishes & Painting/Coating	101.4	166.6	140.0	99.6	114.4	108.4	108.3	97.6	102.0	113.6	126.6	121.3	114.1	166.6	145.1	114.1	166.6	145.1
09	FINISHES	102.3	190.0	150.2	90.4	116.2	104.5	94.9	96.3	95.7	96.3	169.0	136.0	98.9	190.2	148.7	99.1	162.7	133.8
COVERS	DIVS. 10 - 14, 25, 28, 41, 43, 44, 46	100.0	143.1	109.5	100.0	104.8	101.1	100.0	94.6	98.8	100.0	118.8	104.1	100.0	137.2	108.2	100.0	122.6	105.0
21, 22, 23	FIRE SUPPRESSION, PLUMBING & HVAC	100.1	175.3	130.5	96.6	107.5	101.0	96.6	102.6	99.0	96.6	123.0	107.3	99.8	176.3	130.7	99.9	157.9	123.4
26, 27, 3370	ELECTRICAL, COMMUNICATIONS & UTIL.	97.7	186.7	142.9	97.2	101.3	99.3	92.1	94.0	93.1	95.5	124.2	110.1	99.6	186.7	143.9	101.2	138.2	120.0
MF2016	WEIGHTED AVERAGE	99.0	176.2	132.1	98.8	110.2	103.7	95.6	99.0	97.0	98.1	140.2	116.1	99.8	178.9	133.8	100.5	156.8	124.6

DIVISION		NEW YORK																	
		ROCHESTER			SCHENECTADY			STATEN ISLAND			SUFFERN			SYRACUSE			UTICA		
		144 - 146			123			103			109			130 - 132			133 - 135		
		MAT.	INST.	TOTAL	MAT.	INST.	TOTAL	MAT.	INST.	TOTAL	MAT.	INST.	TOTAL	MAT.	INST.	TOTAL	MAT.	INST.	TOTAL
015433	CONTRACTOR EQUIPMENT		119.9	119.9		115.2	115.2		106.8	106.8		106.8	106.8		115.2	115.2		115.2	115.2
0241, 31 - 34	SITE & INFRASTRUCTURE, DEMOLITION	92.2	108.7	103.7	83.0	104.6	98.0	105.1	114.2	111.5	98.0	108.8	105.5	94.0	103.2	100.4	72.5	102.7	93.5
0310	Concrete Forming & Accessories	107.6	101.9	102.8	98.9	107.7	106.4	88.8	193.3	177.8	98.1	144.7	137.8	99.1	91.6	92.7	100.1	88.7	90.4
0320	Concrete Reinforcing	110.9	108.0	109.4	100.8	116.1	108.6	99.4	222.2	161.8	98.3	156.4	127.8	103.8	109.2	106.5	103.8	103.2	103.5
0330	Cast-in-Place Concrete	101.1	105.3	102.7	93.1	116.0	101.7	86.3	175.6	119.5	83.7	143.0	105.8	96.6	105.3	99.9	88.4	103.9	94.2
03	CONCRETE	97.9	105.4	101.2	93.5	113.0	102.1	93.7	190.3	136.5	88.8	145.1	113.8	96.8	100.8	98.6	95.0	98.0	96.3
04	MASONRY	101.4	105.6	104.0	90.2	116.9	106.7	99.5	189.3	155.0	92.7	145.2	125.1	99.4	103.2	101.7	90.5	101.5	97.3
05	METALS	103.4	121.5	108.9	90.3	125.8	101.1	82.9	173.9	110.5	82.9	132.9	98.0	97.0	120.6	104.1	95.1	118.3	102.1
06	WOOD, PLASTICS & COMPOSITES	108.4	101.4	104.6	103.5	104.7	104.1	90.9	194.1	147.0	103.0	146.6	126.7	100.3	88.5	93.9	100.3	85.1	92.0
07	THERMAL & MOISTURE PROTECTION	105.3	101.8	103.8	101.5	111.5	105.8	109.2	171.4	136.3	109.7	144.3	124.8	103.6	96.8	100.6	92.1	96.7	94.1
08	OPENINGS	102.1	101.3	101.9	97.4	105.0	99.1	94.0	202.1	119.1	94.1	154.5	108.1	94.5	92.1	94.0	97.4	88.8	95.4
0920	Plaster & Gypsum Board	103.7	101.6	102.3	104.1	104.7	104.5	91.0	196.6	160.5	95.2	147.7	129.8	95.9	88.1	90.8	95.9	84.6	88.5
0950, 0980	Ceilings & Acoustic Treatment	102.4	101.6	101.8	93.2	104.7	101.0	94.3	196.6	163.6	92.6	147.7	129.9	96.5	88.1	90.8	96.5	84.6	88.4
0960	Flooring	90.0	108.5	95.4	92.1	111.3	97.7	91.5	188.7	119.7	91.4	177.3	116.3	90.7	95.4	92.0	88.5	95.5	90.6
0970, 0990	Wall Finishes & Painting/Coating	97.1	101.0	99.4	86.9	103.9	97.0	104.8	166.6	141.3	103.2	130.3	119.2	90.4	101.3	96.9	84.2	101.3	94.3
09	FINISHES	94.1	103.4	99.2	88.4	107.6	98.8	96.2	190.5	147.7	95.8	150.1	125.4	90.5	92.6	91.6	88.9	90.4	89.7
COVERS	DIVS. 10 - 14, 25, 28, 41, 43, 44, 46	100.0	100.4	100.1	100.0	102.5	100.6	100.0	138.0	108.4	100.0	117.0	103.8	100.0	97.0	99.3	100.0	91.6	98.1
21, 22, 23	FIRE SUPPRESSION, PLUMBING & HVAC	100.0	90.1	96.0	100.1	107.7	103.2	100.2	176.4	131.0	96.8	126.5	108.8	100.2	96.6	98.8	100.2	94.6	98.0
26, 27, 3370	ELECTRICAL, COMMUNICATIONS & UTIL.	102.0	92.1	97.0	98.6	107.6	103.2	91.4	186.7	139.9	95.8	117.4	106.8	100.3	103.5	101.9	98.2	103.5	100.9
MF2016	WEIGHTED AVERAGE	100.3	101.3	100.8	95.5	110.6	101.9	94.9	177.8	130.5	93.6	133.5	110.7	97.8	100.8	99.1	95.9	98.9	97.2

DIVISION		NEW YORK									NORTH CAROLINA								
		WATERTOWN			WHITE PLAINS			YONKERS			ASHEVILLE			CHARLOTTE			DURHAM		
		136			106			107			287 - 288			281 - 282			277		
		MAT.	INST.	TOTAL	MAT.	INST.	TOTAL	MAT.	INST.	TOTAL	MAT.	INST.	TOTAL	MAT.	INST.	TOTAL	MAT.	INST.	TOTAL
015433	CONTRACTOR EQUIPMENT		115.2	115.2		106.8	106.8		106.8	106.8		100.9	100.9		100.9	100.9		106.2	106.2
0241, 31 - 34	SITE & INFRASTRUCTURE, DEMOLITION	80.1	103.2	96.2	96.0	110.1	105.8	102.7	110.1	107.8	95.8	81.1	85.5	98.0	81.3	86.4	101.0	90.1	93.4
0310	Concrete Forming & Accessories	85.2	95.6	94.1	103.3	152.6	145.3	103.5	149.2	142.4	88.6	62.1	66.1	95.5	61.7	66.7	94.8	62.1	67.0
0320	Concrete Reinforcing	104.5	109.2	106.9	98.3	182.1	140.9	102.2	182.0	142.8	92.3	67.8	79.9	96.5	68.6	82.3	95.8	68.6	82.0
0330	Cast-in-Place Concrete	103.0	107.7	104.7	76.7	151.4	104.5	85.8	151.4	110.2	104.8	69.7	91.7	106.8	69.0	92.7	104.2	69.7	91.3
03	CONCRETE	107.8	103.4	105.9	83.1	157.1	115.9	91.4	155.6	119.9	91.6	67.4	80.9	91.5	67.1	80.7	92.5	67.5	81.4
04	MASONRY	91.7	107.0	101.2	92.2	156.2	131.7	96.3	156.2	133.3	86.5	62.4	71.6	90.2	61.0	72.2	85.6	62.4	71.3
05	METALS	95.2	120.6	102.9	84.4	165.6	109.0	92.3	165.5	114.4	99.0	90.1	96.3	99.7	90.5	96.9	115.1	90.4	107.6
06	WOOD, PLASTICS & COMPOSITES	81.7	92.5	87.6	109.1	151.0	131.9	109.0	146.6	129.4	87.4	60.7	72.9	92.3	60.7	75.1	90.3	60.7	74.2
07	THERMAL & MOISTURE PROTECTION	92.4	99.9	95.7	109.5	150.4	127.3	109.9	150.0	127.4	100.3	63.9	84.5	94.5	63.3	80.9	105.4	63.9	87.3
08	OPENINGS	97.4	96.5	97.2	94.1	177.5	113.5	97.8	173.0	115.2	90.9	61.6	84.1	97.8	61.7	89.4	97.4	61.7	89.1
0920	Plaster & Gypsum Board	86.7	92.2	90.4	98.2	152.2	133.7	101.4	147.7	131.9	107.3	59.5	75.8	104.4	59.5	74.8	94.1	59.5	71.3
0950, 0980	Ceilings & Acoustic Treatment	96.5	92.2	93.6	92.6	152.2	132.9	108.7	147.7	135.1	81.6	59.5	66.6	87.3	59.5	68.5	84.3	59.5	67.5
0960	Flooring	81.8	95.5	85.8	93.7	185.7	120.4	93.2	188.7	120.9	94.7	62.1	85.2	95.2	62.1	85.6	98.3	62.1	87.8
0970, 0990	Wall Finishes & Painting/Coating	84.2	95.4	90.9	103.2	166.6	140.7	103.2	166.6	140.7	103.7	58.6	77.0	97.4	58.6	74.4	98.9	58.6	75.1
09	FINISHES	86.2	95.1	91.1	96.6	160.0	131.2	100.5	158.0	131.9	88.6	61.5	73.8	88.7	61.1	73.6	87.0	61.5	73.1
COVERS	DIVS. 10 - 14, 25, 28, 41, 43, 44, 46	100.0	98.4	99.6	100.0	129.6	106.5	100.0	126.8	105.9	100.0	83.4	96.3	100.0	83.0	96.3	100.0	83.4	96.3
21, 22, 23	FIRE SUPPRESSION, PLUMBING & HVAC	100.2	89.8	96.0	100.3	144.9	118.3	100.3	144.8	118.3	100.5	61.5	84.7	99.9	61.9	84.6	100.6	61.5	84.8
26, 27, 3370	ELECTRICAL, COMMUNICATIONS & UTIL.	100.2	92.4	96.2	89.6	170.1	130.5	95.8	159.8	128.4	101.0	59.3	79.8	100.1	61.8	80.6	95.2	58.8	76.7
MF2016	WEIGHTED AVERAGE	97.7	99.2	98.3	93.3	153.5	119.1	97.3	151.2	120.5	96.3	67.0	83.8	97.0	67.3	84.3	99.3	67.7	85.7

	DIVISION	NORTH CAROLINA																	
		ELIZABETH CITY			FAYETTEVILLE			GASTONIA			GREENSBORO			HICKORY			KINSTON		
		279			283			280			270, 272 - 274			286			285		
		MAT.	INST.	TOTAL	MAT.	INST.	TOTAL	MAT.	INST.	TOTAL	MAT.	INST.	TOTAL	MAT.	INST.	TOTAL	MAT.	INST.	TOTAL
015433	CONTRACTOR EQUIPMENT		110.3	110.3		106.2	106.2		100.9	100.9		106.2	106.2		106.2	106.2		106.2	106.2
0241, 31 - 34	SITE & INFRASTRUCTURE, DEMOLITION	105.5	91.8	96.0	95.1	90.0	91.5	95.6	81.4	85.7	100.9	90.1	93.4	94.7	88.8	90.6	93.6	88.7	90.2
0310	Concrete Forming & Accessories	81.0	62.5	65.2	88.2	61.6	65.6	94.6	62.4	67.2	94.6	62.1	66.9	85.2	62.1	65.6	81.8	61.6	64.6
0320	Concrete Reinforcing	93.9	71.7	82.6	95.9	68.6	82.0	92.7	68.6	80.5	94.7	68.6	81.4	92.3	68.6	80.3	91.9	68.6	80.0
0330	Cast-in-Place Concrete	104.4	71.4	92.1	109.9	68.9	94.6	102.4	69.8	90.3	103.4	69.7	90.8	104.7	69.7	91.7	101.1	68.9	89.1
03	CONCRETE	92.6	68.8	82.1	93.5	67.0	81.7	90.3	67.7	80.3	92.0	67.5	81.1	91.4	67.5	80.8	88.5	67.0	79.0
04	MASONRY	96.8	61.0	74.7	89.6	61.0	72.0	90.5	62.4	73.1	82.9	62.4	70.2	75.9	62.4	67.6	82.6	61.0	69.3
05	METALS	101.0	92.6	98.5	119.6	90.4	110.8	99.5	90.6	96.8	107.4	90.3	102.3	99.0	90.4	96.4	97.7	90.3	95.5
06	WOOD, PLASTICS & COMPOSITES	75.0	61.4	67.6	86.6	60.7	72.5	95.3	60.7	76.5	90.1	60.7	74.1	82.3	60.7	70.6	79.3	60.7	69.2
07	THERMAL & MOISTURE PROTECTION	104.6	63.2	86.5	99.8	63.3	83.9	100.5	63.9	84.6	105.2	63.9	87.2	100.7	63.9	84.7	100.5	63.3	84.3
08	OPENINGS	94.7	62.7	87.3	91.0	61.7	84.2	94.3	61.7	86.7	97.4	61.7	89.1	90.9	61.7	84.2	91.0	61.7	84.2
0920	Plaster & Gypsum Board	88.8	59.5	69.5	111.2	59.5	77.2	113.2	59.5	77.9	95.5	59.5	71.8	107.3	59.5	75.8	107.1	59.5	75.8
0950, 0980	Ceilings & Acoustic Treatment	84.3	59.5	67.5	83.3	59.5	67.2	85.0	59.5	67.7	84.3	59.5	67.5	81.6	59.5	66.6	85.0	59.5	67.7
0960	Flooring	90.4	79.8	87.3	94.8	62.1	85.3	98.0	79.8	92.7	98.3	62.1	87.8	94.6	79.8	90.3	91.7	79.8	88.3
0970, 0990	Wall Finishes & Painting/Coating	98.9	58.6	75.1	103.7	58.6	77.0	103.7	58.6	77.0	98.9	58.6	75.1	103.7	58.6	77.0	103.7	58.6	77.0
09	FINISHES	84.3	65.2	73.9	89.5	61.1	74.0	91.0	65.1	76.8	87.2	61.5	73.2	88.8	65.1	75.8	88.4	64.7	75.5
COVERS	DIVS. 10 - 14, 25, 28, 41, 43, 44, 46	100.0	81.7	96.0	100.0	83.0	96.2	100.0	83.4	96.4	100.0	80.7	95.7	100.0	83.4	96.3	100.0	82.9	96.2
21, 22, 23	FIRE SUPPRESSION, PLUMBING & HVAC	97.0	59.9	82.0	100.3	60.8	84.3	100.5	60.4	84.3	100.5	61.5	84.7	97.0	60.4	82.2	97.0	59.7	81.9
26, 27, 3370	ELECTRICAL, COMMUNICATIONS & UTIL.	94.9	66.2	80.3	100.6	58.8	79.3	100.5	61.9	80.9	94.3	59.3	76.5	98.7	61.9	80.0	98.5	57.2	77.5
MF2016	WEIGHTED AVERAGE	96.1	69.2	84.6	100.0	67.3	86.0	97.0	67.7	84.4	97.7	67.7	84.8	94.7	68.3	83.4	94.3	67.1	82.7

	DIVISION	NORTH CAROLINA															NORTH DAKOTA		
		MURPHY			RALEIGH			ROCKY MOUNT			WILMINGTON			WINSTON-SALEM			BISMARCK		
		289			275 - 276			278			284			271			585		
		MAT.	INST.	TOTAL	MAT.	INST.	TOTAL	MAT.	INST.	TOTAL	MAT.	INST.	TOTAL	MAT.	INST.	TOTAL	MAT.	INST.	TOTAL
015433	CONTRACTOR EQUIPMENT		100.9	100.9		106.2	106.2		106.2	106.2		100.9	100.9		106.2	106.2		100.3	100.3
0241, 31 - 34	SITE & INFRASTRUCTURE, DEMOLITION	97.0	79.6	84.9	100.9	90.1	93.3	103.5	90.1	94.1	96.9	81.0	85.8	101.2	90.1	93.5	100.9	98.4	99.2
0310	Concrete Forming & Accessories	95.2	61.5	66.5	96.0	61.6	66.7	87.0	61.8	65.5	89.9	61.6	65.8	96.5	62.1	67.3	103.9	73.4	77.9
0320	Concrete Reinforcing	91.9	66.1	78.8	96.8	68.6	82.5	93.9	68.6	81.1	93.1	68.6	80.6	94.7	68.6	81.4	103.8	94.5	99.1
0330	Cast-in-Place Concrete	108.4	68.9	93.7	107.1	68.9	92.9	102.1	69.0	89.8	104.3	68.9	91.2	106.0	69.7	92.5	102.2	83.9	95.4
03	CONCRETE	94.4	66.5	82.0	91.2	67.0	80.5	93.2	67.1	81.6	91.5	67.0	80.7	93.3	67.5	81.9	90.6	81.4	86.5
04	MASONRY	78.9	61.0	67.9	81.6	61.0	68.9	75.3	61.0	66.5	76.5	61.0	66.9	83.2	62.4	70.3	98.4	82.4	88.5
05	METALS	96.8	89.5	94.6	99.7	90.4	96.9	100.2	90.5	97.3	98.5	90.4	96.0	104.5	90.4	100.2	95.2	92.6	94.4
06	WOOD, PLASTICS & COMPOSITES	96.0	60.6	76.7	90.4	60.7	74.3	81.8	60.7	70.4	89.2	60.7	73.7	90.1	60.7	74.1	102.1	69.5	84.4
07	THERMAL & MOISTURE PROTECTION	100.5	63.3	84.3	99.4	63.3	83.7	105.1	63.3	86.9	100.3	63.3	84.2	105.2	63.9	87.2	105.1	83.2	95.6
08	OPENINGS	90.9	61.1	84.0	98.7	61.7	90.1	93.9	61.7	86.5	91.0	61.7	84.2	97.4	61.7	89.1	106.3	79.3	100.0
0920	Plaster & Gypsum Board	112.5	59.3	77.5	91.1	59.5	70.3	90.3	59.5	70.0	109.2	59.5	76.5	95.5	59.5	71.8	100.0	69.0	79.6
0950, 0980	Ceilings & Acoustic Treatment	81.6	59.3	66.5	82.2	59.5	66.8	82.6	59.5	67.0	83.3	59.5	67.2	84.3	59.5	67.5	109.7	69.0	82.1
0960	Flooring	98.2	79.8	92.9	93.3	62.1	84.3	93.9	79.8	89.8	95.4	62.1	85.7	98.3	62.1	87.8	91.3	51.4	79.8
0970, 0990	Wall Finishes & Painting/Coating	103.7	58.6	77.0	92.1	58.6	72.3	98.9	58.6	75.1	103.7	58.6	77.0	98.9	58.6	75.1	93.7	59.0	73.2
09	FINISHES	90.6	64.6	76.4	85.6	61.1	72.3	85.3	64.7	74.1	89.4	61.1	74.0	87.2	61.5	73.2	95.6	68.3	80.7
COVERS	DIVS. 10 - 14, 25, 28, 41, 43, 44, 46	100.0	82.9	96.2	100.0	83.0	96.2	100.0	83.0	96.2	100.0	83.0	96.2	100.0	83.4	96.3	100.0	89.1	97.6
21, 22, 23	FIRE SUPPRESSION, PLUMBING & HVAC	97.0	59.6	81.9	100.0	60.8	84.1	97.0	59.7	81.9	100.5	60.8	84.5	100.5	61.5	84.7	99.9	75.5	90.1
26, 27, 3370	ELECTRICAL, COMMUNICATIONS & UTIL.	101.9	57.2	79.2	97.6	57.7	77.3	96.8	58.8	77.5	101.3	57.2	78.9	94.3	59.3	76.5	97.5	74.4	85.7
MF2016	WEIGHTED AVERAGE	95.5	66.2	82.9	96.4	67.1	83.8	95.3	67.5	83.4	95.9	66.3	83.2	97.4	67.8	84.7	98.2	80.0	90.4

	DIVISION	NORTH DAKOTA																	
		DEVILS LAKE			DICKINSON			FARGO			GRAND FORKS			JAMESTOWN			MINOT		
		583			586			580 - 581			582			584			587		
		MAT.	INST.	TOTAL	MAT.	INST.	TOTAL	MAT.	INST.	TOTAL	MAT.	INST.	TOTAL	MAT.	INST.	TOTAL	MAT.	INST.	TOTAL
015433	CONTRACTOR EQUIPMENT		100.3	100.3		100.3	100.3		100.3	100.3		100.3	100.3		100.3	100.3		100.3	100.3
0241, 31 - 34	SITE & INFRASTRUCTURE, DEMOLITION	106.3	98.5	100.8	113.8	98.4	103.1	97.6	98.4	98.2	109.8	98.4	101.9	105.3	98.5	100.5	107.4	98.4	101.1
0310	Concrete Forming & Accessories	102.4	73.5	77.8	91.5	73.4	76.1	97.7	73.5	77.1	95.6	73.4	76.7	93.2	73.4	76.4	91.1	73.4	76.0
0320	Concrete Reinforcing	100.2	94.4	97.3	101.1	94.5	97.8	94.6	94.8	94.7	98.7	94.4	96.5	100.7	94.8	97.7	102.1	94.5	98.2
0330	Cast-in-Place Concrete	122.1	84.0	107.9	110.4	83.9	100.5	99.9	83.9	93.9	110.4	83.9	100.5	120.5	84.0	106.9	110.4	83.9	100.5
03	CONCRETE	100.2	81.4	91.9	99.1	81.4	91.2	92.9	81.5	87.8	96.7	81.4	89.9	98.9	81.5	91.2	95.3	81.4	89.1
04	MASONRY	114.6	80.9	93.8	116.6	82.4	95.5	103.2	87.1	93.3	108.9	80.9	91.6	127.0	87.2	102.4	107.2	82.4	91.9
05	METALS	91.5	92.7	91.9	91.4	92.5	91.7	94.0	93.0	93.7	91.4	92.4	91.7	91.4	93.1	92.0	91.7	92.5	92.0
06	WOOD, PLASTICS & COMPOSITES	99.9	69.5	83.4	86.2	69.5	77.1	94.9	69.5	81.1	91.0	69.5	79.3	88.3	69.5	78.0	85.9	69.5	77.0
07	THERMAL & MOISTURE PROTECTION	106.3	82.8	96.1	106.9	83.2	96.6	106.1	85.0	96.9	106.5	82.8	96.2	106.1	85.0	96.9	106.2	83.2	96.2
08	OPENINGS	103.1	79.3	97.6	103.1	79.3	97.6	100.2	79.3	95.4	101.7	79.3	96.5	103.1	79.3	97.6	101.9	79.3	96.6
0920	Plaster & Gypsum Board	114.9	69.0	84.7	105.2	69.0	81.3	98.2	69.0	78.9	106.4	69.0	81.7	106.1	69.0	81.6	105.2	69.0	81.3
0950, 0980	Ceilings & Acoustic Treatment	101.8	69.0	79.6	101.8	69.0	79.6	95.2	69.0	77.4	101.8	69.0	79.6	101.8	69.0	79.6	101.8	69.0	79.6
0960	Flooring	95.3	51.4	82.6	88.8	51.4	77.9	100.0	51.4	85.9	90.6	51.4	79.3	89.5	51.4	78.4	88.5	51.4	77.8
0970, 0990	Wall Finishes & Painting/Coating	88.1	57.8	70.2	88.1	57.8	70.2	92.8	69.1	78.8	88.1	67.8	76.1	88.1	57.8	70.2	88.1	59.0	70.9
09	FINISHES	93.8	68.0	79.7	91.4	68.0	78.6	94.9	69.2	80.9	91.6	69.1	79.3	90.8	68.0	78.4	90.6	68.1	78.3
COVERS	DIVS. 10 - 14, 25, 28, 41, 43, 44, 46	100.0	89.1	97.6	100.0	89.1	97.6	100.0	89.1	97.6	100.0	89.1	97.6	100.0	89.1	97.6	100.0	89.1	97.6
21, 22, 23	FIRE SUPPRESSION, PLUMBING & HVAC	96.7	80.0	90.0	96.7	74.9	87.9	99.9	75.6	90.1	100.2	73.8	89.5	96.7	73.8	87.5	100.2	73.6	89.4
26, 27, 3370	ELECTRICAL, COMMUNICATIONS & UTIL.	95.5	69.9	82.5	103.8	69.5	86.4	102.1	71.1	86.3	98.9	69.9	84.2	95.5	69.9	82.5	101.9	74.4	87.9
MF2016	WEIGHTED AVERAGE	98.3	80.2	90.5	99.0	79.1	90.5	98.2	80.3	90.5	98.4	79.0	90.1	98.3	79.6	90.3	98.3	79.6	90.3

	DIVISION	NORTH DAKOTA			OHIO														
		WILLISTON			AKRON			ATHENS			CANTON			CHILLICOTHE			CINCINNATI		
		588			442 - 443			457			446 - 447			456			451 - 452		
		MAT.	INST.	TOTAL	MAT.	INST.	TOTAL	MAT.	INST.	TOTAL	MAT.	INST.	TOTAL	MAT.	INST.	TOTAL	MAT.	INST.	TOTAL
015433	CONTRACTOR EQUIPMENT		100.3	100.3		91.4	91.4		87.5	87.5		91.4	91.4		98.4	98.4		96.0	96.0
0241, 31 - 34	SITE & INFRASTRUCTURE, DEMOLITION	107.8	96.2	99.7	99.8	99.1	99.3	115.7	89.8	97.6	99.9	98.8	99.2	102.0	100.0	100.6	97.5	99.0	98.6
0310	Concrete Forming & Accessories	97.4	73.1	76.7	101.2	85.2	87.6	94.3	79.2	81.4	101.2	75.7	79.5	97.1	83.9	85.8	100.6	82.1	84.9
0320	Concrete Reinforcing	103.1	94.5	98.7	95.4	93.7	94.6	90.7	91.8	91.2	95.4	77.2	86.2	88.0	91.4	89.7	93.3	80.0	86.6
0330	Cast-in-Place Concrete	110.4	83.8	100.5	107.0	91.7	101.3	111.5	98.3	106.6	108.0	89.0	100.9	101.2	95.2	99.0	96.6	77.5	89.5
03	CONCRETE	96.6	81.2	89.8	100.7	88.4	95.2	102.5	87.7	95.9	101.2	80.3	91.9	97.1	89.4	93.7	96.1	80.3	89.1
04	MASONRY	102.1	82.4	89.9	94.2	91.3	92.4	72.3	98.5	88.5	94.9	80.9	86.2	78.7	92.4	87.2	81.3	81.0	81.1
05	METALS	91.6	92.2	91.8	93.6	81.1	89.8	100.1	81.4	94.4	93.6	74.4	87.8	92.3	90.8	91.8	94.5	82.4	90.8
06	WOOD, PLASTICS & COMPOSITES	92.7	69.5	80.1	108.4	83.5	94.8	87.7	73.9	80.2	108.7	73.9	89.8	100.7	80.3	89.6	100.7	82.3	90.7
07	THERMAL & MOISTURE PROTECTION	106.5	83.2	96.3	103.0	93.7	99.0	98.9	93.5	96.5	104.1	89.0	97.5	101.0	90.9	96.6	98.8	82.0	91.5
08	OPENINGS	103.2	79.3	97.6	109.5	85.1	103.8	99.1	76.8	93.9	103.1	71.3	95.8	90.3	80.3	88.0	98.4	78.1	93.7
0920	Plaster & Gypsum Board	106.4	69.0	81.7	101.8	82.9	89.3	96.9	73.0	81.2	102.7	73.0	83.2	100.3	80.1	87.0	99.9	82.2	88.2
0950, 0980	Ceilings & Acoustic Treatment	101.8	69.0	79.6	94.9	82.9	86.8	110.6	73.0	85.1	94.9	73.0	80.1	103.2	80.1	87.6	97.0	82.2	87.0
0960	Flooring	91.6	51.4	79.9	98.1	86.1	94.6	121.9	77.8	109.1	98.3	76.6	92.0	99.5	77.8	93.2	100.8	81.2	95.1
0970, 0990	Wall Finishes & Painting/Coating	88.1	57.8	70.2	95.1	93.3	94.0	103.6	91.5	96.4	95.1	70.9	80.8	100.8	91.5	95.3	100.4	73.2	84.3
09	FINISHES	91.8	68.0	78.8	99.0	86.3	92.1	101.9	79.4	89.7	99.2	74.6	85.7	99.3	82.7	90.3	98.5	81.2	89.1
COVERS	DIVS. 10 - 14, 25, 28, 41, 43, 44, 46	100.0	89.0	97.6	100.0	92.2	98.3	100.0	89.2	97.6	100.0	89.8	97.7	100.0	88.7	97.5	100.0	89.8	97.8
21, 22, 23	FIRE SUPPRESSION, PLUMBING & HVAC	96.7	74.8	87.9	100.0	89.8	95.9	96.6	82.6	91.0	100.0	79.6	91.8	97.2	94.5	96.1	100.0	77.2	90.8
26, 27, 3370	ELECTRICAL, COMMUNICATIONS & UTIL.	99.2	70.4	84.6	96.5	83.9	90.1	97.0	91.9	94.4	95.8	87.1	91.4	96.5	91.9	94.2	95.4	72.9	84.0
MF2016	WEIGHTED AVERAGE	97.4	79.0	89.6	99.5	88.3	94.7	98.2	86.4	93.2	98.9	81.4	91.4	95.4	90.8	93.4	96.9	80.7	89.9

	DIVISION	OHIO																	
		CLEVELAND			COLUMBUS			DAYTON			HAMILTON			LIMA			LORAIN		
		441			430 - 432			453 - 454			450			458			440		
		MAT.	INST.	TOTAL	MAT.	INST.	TOTAL	MAT.	INST.	TOTAL	MAT.	INST.	TOTAL	MAT.	INST.	TOTAL	MAT.	INST.	TOTAL
015433	CONTRACTOR EQUIPMENT		92.5	92.5		92.0	92.0		91.6	91.6		98.4	98.4		90.9	90.9		91.4	91.4
0241, 31 - 34	SITE & INFRASTRUCTURE, DEMOLITION	98.4	97.4	97.7	101.0	92.5	95.1	97.6	99.3	98.7	97.5	99.5	98.9	109.2	89.5	95.4	99.0	99.2	99.2
0310	Concrete Forming & Accessories	100.6	89.2	90.9	98.6	77.9	81.0	98.9	79.4	82.3	99.0	79.6	82.5	94.2	78.5	80.8	101.3	76.6	80.3
0320	Concrete Reinforcing	95.9	91.8	93.8	99.2	81.0	90.0	93.3	81.4	87.3	93.3	79.5	86.3	90.7	81.6	86.1	95.4	94.0	94.7
0330	Cast-in-Place Concrete	103.7	98.7	101.8	100.6	82.4	93.8	86.7	83.0	85.3	92.9	83.4	89.4	102.4	92.2	98.6	101.8	94.2	99.0
03	CONCRETE	100.9	92.4	97.1	98.6	80.0	90.3	88.3	80.9	85.0	91.2	81.4	86.8	95.3	83.8	90.2	98.3	85.4	92.6
04	MASONRY	100.7	97.9	99.0	84.3	86.9	85.9	77.6	78.0	77.9	78.0	82.9	81.0	97.4	79.8	86.5	90.8	95.0	93.4
05	METALS	95.2	83.7	91.7	94.5	80.8	90.3	93.8	77.6	88.9	93.8	86.5	91.6	100.1	81.0	94.3	94.2	81.8	90.5
06	WOOD, PLASTICS & COMPOSITES	102.3	86.7	93.8	97.8	76.7	86.3	104.6	79.4	90.9	103.4	79.4	90.3	87.6	77.3	82.0	108.4	72.1	88.7
07	THERMAL & MOISTURE PROTECTION	100.1	97.8	99.1	93.6	84.9	89.9	105.1	81.7	94.9	101.1	82.2	92.9	98.5	86.0	93.0	104.0	93.6	99.5
08	OPENINGS	97.7	86.3	95.0	99.3	74.4	93.5	97.5	76.8	92.7	95.2	77.0	91.0	99.1	75.9	93.7	103.1	78.9	97.5
0920	Plaster & Gypsum Board	101.4	86.1	91.3	97.3	76.1	83.4	102.0	79.2	87.0	102.0	79.2	87.0	96.9	76.5	83.5	101.8	71.2	81.7
0950, 0980	Ceilings & Acoustic Treatment	90.4	86.1	87.5	91.4	76.1	81.1	105.0	79.2	87.5	104.1	79.2	87.2	109.7	76.5	87.2	94.9	71.2	78.9
0960	Flooring	99.5	90.5	96.9	94.2	77.8	89.5	103.2	74.8	95.0	100.6	81.2	95.0	120.8	79.8	108.9	98.3	92.8	96.7
0970, 0990	Wall Finishes & Painting/Coating	97.7	93.3	95.1	95.8	81.7	87.5	100.8	72.6	84.1	100.8	73.2	84.5	103.7	78.5	88.8	95.1	93.3	94.0
09	FINISHES	99.3	89.6	94.0	94.7	77.9	85.5	100.7	77.6	88.1	99.7	79.2	88.5	101.0	78.2	88.5	99.0	80.3	88.8
COVERS	DIVS. 10 - 14, 25, 28, 41, 43, 44, 46	100.0	98.0	99.6	100.0	89.5	97.7	100.0	86.9	97.1	100.0	87.1	97.2	100.0	86.2	97.0	100.0	92.8	98.4
21, 22, 23	FIRE SUPPRESSION, PLUMBING & HVAC	100.0	92.1	96.8	100.0	85.9	94.3	100.8	83.0	93.6	100.6	76.6	90.9	96.6	93.8	95.5	100.0	90.4	96.1
26, 27, 3370	ELECTRICAL, COMMUNICATIONS & UTIL.	96.1	93.0	94.5	98.4	82.2	90.1	94.1	77.5	85.7	94.4	77.3	85.7	97.3	77.5	87.2	95.9	80.1	87.9
MF2016	WEIGHTED AVERAGE	98.7	92.2	95.9	97.4	83.2	91.3	95.9	81.3	89.6	95.9	81.5	89.7	98.3	83.8	92.1	98.4	86.8	93.4

	DIVISION	OHIO																	
		MANSFIELD			MARION			SPRINGFIELD			STEUBENVILLE			TOLEDO			YOUNGSTOWN		
		448 - 449			433			455			439			434 - 436			444 - 445		
		MAT.	INST.	TOTAL	MAT.	INST.	TOTAL	MAT.	INST.	TOTAL	MAT.	INST.	TOTAL	MAT.	INST.	TOTAL	MAT.	INST.	TOTAL
015433	CONTRACTOR EQUIPMENT		91.4	91.4		91.8	91.8		91.6	91.6		95.6	95.6		95.6	95.6		91.4	91.4
0241, 31 - 34	SITE & INFRASTRUCTURE, DEMOLITION	95.4	99.0	97.9	95.3	95.8	95.6	97.9	99.2	98.8	140.7	103.8	114.9	99.0	95.9	96.8	99.6	98.9	99.1
0310	Concrete Forming & Accessories	90.3	75.6	77.8	96.5	79.4	81.9	98.9	79.3	82.2	97.5	81.2	83.6	100.2	86.7	88.7	101.2	78.8	82.1
0320	Concrete Reinforcing	86.9	81.3	84.1	91.4	81.3	86.3	93.3	81.4	87.3	89.2	97.6	93.4	99.2	85.1	92.0	95.4	87.7	91.5
0330	Cast-in-Place Concrete	99.0	90.4	95.8	87.2	90.9	88.6	89.0	82.8	86.7	94.7	92.2	93.7	95.5	92.3	94.3	105.9	89.9	100.0
03	CONCRETE	92.9	81.5	87.8	85.3	83.6	84.5	89.4	80.7	85.5	89.4	87.3	88.5	93.1	88.4	91.0	100.2	83.8	92.9
04	MASONRY	93.3	90.8	91.7	86.5	91.8	89.7	77.8	77.5	77.6	76.8	91.4	85.8	91.7	90.2	90.8	94.5	88.5	90.8
05	METALS	94.5	76.7	89.1	93.6	79.8	89.4	93.8	77.6	88.9	90.2	82.7	87.9	94.3	86.2	91.8	93.7	78.6	89.1
06	WOOD, PLASTICS & COMPOSITES	94.8	72.1	82.5	95.0	76.8	85.1	106.0	79.4	91.5	90.2	78.6	83.9	99.3	86.4	92.3	108.4	76.8	91.2
07	THERMAL & MOISTURE PROTECTION	102.2	90.9	97.3	90.5	91.0	90.7	104.9	81.5	94.7	101.9	89.5	96.5	90.9	92.2	91.5	104.2	90.7	98.4
08	OPENINGS	104.2	73.1	97.0	93.1	75.6	89.0	95.6	76.8	91.3	93.3	80.7	90.4	95.7	83.7	92.9	103.1	77.4	97.2
0920	Plaster & Gypsum Board	94.6	71.2	79.2	98.5	76.1	83.8	102.0	79.2	87.0	96.7	77.6	84.1	100.3	86.0	90.9	101.8	76.1	84.8
0950, 0980	Ceilings & Acoustic Treatment	95.7	71.2	79.1	97.6	76.1	83.1	105.0	79.2	87.5	94.7	77.6	83.1	97.6	86.0	89.8	94.9	76.1	82.1
0960	Flooring	93.3	95.8	94.0	93.6	95.8	94.3	103.2	74.8	95.0	121.5	93.3	113.3	94.0	93.8	93.9	98.3	93.0	96.8
0970, 0990	Wall Finishes & Painting/Coating	95.1	79.9	86.1	100.2	79.9	88.2	100.8	72.6	84.1	113.6	89.4	99.3	100.2	89.3	93.8	95.1	82.1	87.4
09	FINISHES	96.4	78.8	86.8	96.1	81.8	88.3	100.7	77.5	88.0	113.9	83.5	97.3	96.8	88.0	92.0	99.1	81.1	89.3
COVERS	DIVS. 10 - 14, 25, 28, 41, 43, 44, 46	100.0	90.2	97.8	100.0	87.4	97.2	100.0	86.7	97.1	100.0	88.0	97.4	100.0	92.9	98.4	100.0	90.4	97.9
21, 22, 23	FIRE SUPPRESSION, PLUMBING & HVAC	96.6	89.4	93.7	96.6	95.2	96.0	100.8	82.8	93.5	97.0	93.4	95.5	100.1	95.1	98.0	100.0	85.8	94.3
26, 27, 3370	ELECTRICAL, COMMUNICATIONS & UTIL.	93.6	93.0	93.3	92.5	93.0	92.7	94.1	82.2	88.1	87.4	109.1	98.4	98.6	106.0	102.3	95.9	76.2	85.9
MF2016	WEIGHTED AVERAGE	96.5	86.4	92.1	93.4	88.5	91.3	95.9	81.8	89.8	95.3	92.1	94.0	96.7	92.8	95.0	98.7	84.0	92.4

	DIVISION	OHIO			OKLAHOMA														
		ZANESVILLE			ARDMORE			CLINTON			DURANT			ENID			GUYMON		
		437 - 438			734			736			747			737			739		
		MAT.	INST.	TOTAL	MAT.	INST.	TOTAL	MAT.	INST.	TOTAL	MAT.	INST.	TOTAL	MAT.	INST.	TOTAL	MAT.	INST.	TOTAL
015433	CONTRACTOR EQUIPMENT		91.8	91.8		82.7	82.7		81.8	81.8		81.8	81.8		81.8	81.8		81.8	81.8
0241, 31 - 34	SITE & INFRASTRUCTURE, DEMOLITION	98.3	95.5	96.4	99.4	96.0	97.0	100.8	94.5	96.4	94.3	93.0	93.4	102.5	94.5	96.9	105.2	95.0	98.1
0310	Concrete Forming & Accessories	93.6	78.0	80.3	86.2	56.6	61.0	84.9	56.6	60.8	82.6	56.4	60.3	88.2	56.8	61.5	91.1	56.7	61.8
0320	Concrete Reinforcing	90.9	95.2	93.1	87.7	67.2	77.3	88.2	67.2	77.5	93.2	65.5	79.1	87.6	67.3	77.3	88.2	65.1	76.4
0330	Cast-in-Place Concrete	91.9	88.8	90.7	87.0	72.4	81.6	84.1	72.4	79.7	90.2	72.3	83.5	84.1	72.4	79.8	84.1	72.1	79.6
03	CONCRETE	89.1	84.6	87.1	80.8	63.8	73.2	80.3	63.8	73.0	84.2	63.4	75.0	80.8	63.9	73.3	83.3	63.3	74.4
04	MASONRY	83.5	86.2	85.2	94.6	57.5	71.7	118.1	57.5	80.6	87.3	63.4	72.5	100.2	57.5	73.8	97.1	56.6	72.1
05	METALS	94.9	85.1	92.0	92.9	61.2	83.3	92.9	61.2	83.3	93.6	60.7	83.7	94.4	61.3	84.4	93.4	59.7	83.2
06	WOOD, PLASTICS & COMPOSITES	90.3	76.8	82.9	99.6	56.0	75.9	98.7	56.0	75.5	89.3	56.0	71.2	102.0	56.0	77.0	105.4	56.0	78.5
07	THERMAL & MOISTURE PROTECTION	90.7	85.8	88.6	101.5	64.5	85.4	101.6	64.5	85.4	97.1	66.1	83.6	101.7	64.5	85.5	102.1	64.0	85.5
08	OPENINGS	93.1	79.7	90.0	105.9	56.0	94.3	105.9	56.0	94.3	98.1	55.5	88.2	107.1	56.7	95.4	106.0	55.5	94.3
0920	Plaster & Gypsum Board	94.5	76.1	82.4	93.7	55.3	68.4	93.4	55.3	68.3	79.6	55.3	63.6	94.3	55.3	68.6	94.3	55.3	68.6
0950, 0980	Ceilings & Acoustic Treatment	97.6	76.1	83.1	83.2	55.3	64.3	83.2	55.3	64.3	79.4	55.3	63.1	83.2	55.3	64.3	83.2	55.3	64.3
0960	Flooring	91.8	77.8	87.8	87.3	59.2	79.1	86.2	59.2	78.3	91.4	53.4	80.4	87.8	73.1	83.6	89.3	59.2	80.6
0970, 0990	Wall Finishes & Painting/Coating	100.2	91.5	95.0	84.6	45.1	61.2	84.6	45.1	61.2	92.9	45.1	64.6	84.6	45.1	61.2	84.6	43.4	60.2
09	FINISHES	95.2	78.5	86.1	81.0	54.9	66.8	80.8	54.9	66.7	81.4	53.8	66.3	81.5	57.8	68.6	82.4	56.1	68.0
COVERS	DIVS. 10 - 14, 25, 28, 41, 43, 44, 46	100.0	84.5	96.6	100.0	79.0	95.4	100.0	79.0	95.4	100.0	79.0	95.4	100.0	79.0	95.4	100.0	79.0	95.4
21, 22, 23	FIRE SUPPRESSION, PLUMBING & HVAC	96.6	91.5	94.5	96.7	67.8	85.0	96.7	67.8	85.0	96.8	67.7	85.0	100.2	67.8	87.1	96.7	65.0	83.9
26, 27, 3370	ELECTRICAL, COMMUNICATIONS & UTIL.	92.6	91.9	92.2	93.3	70.9	81.9	94.4	70.9	82.4	95.5	70.9	83.0	94.4	70.9	82.4	96.0	67.2	81.4
MF2016	WEIGHTED AVERAGE	94.0	87.1	91.0	93.8	66.2	82.0	95.0	66.1	82.6	93.2	66.3	81.7	95.5	66.5	83.1	94.9	64.8	82.0

	DIVISION	OKLAHOMA																	
		LAWTON			MCALESTER			MIAMI			MUSKOGEE			OKLAHOMA CITY			PONCA CITY		
		735			745			743			744			730 - 731			746		
		MAT.	INST.	TOTAL	MAT.	INST.	TOTAL	MAT.	INST.	TOTAL	MAT.	INST.	TOTAL	MAT.	INST.	TOTAL	MAT.	INST.	TOTAL
015433	CONTRACTOR EQUIPMENT		82.7	82.7		81.8	81.8		93.2	93.2		93.2	93.2		83.0	83.0		81.8	81.8
0241, 31 - 34	SITE & INFRASTRUCTURE, DEMOLITION	98.6	96.0	96.8	87.6	95.1	92.8	88.9	92.4	91.3	89.3	92.2	91.4	96.7	96.5	96.6	94.9	95.4	95.2
0310	Concrete Forming & Accessories	90.9	56.8	61.9	80.7	42.5	48.2	93.5	56.6	62.1	97.9	56.3	62.5	91.1	63.5	67.6	89.1	56.8	61.6
0320	Concrete Reinforcing	87.8	67.3	77.4	92.9	65.1	78.7	91.4	67.3	79.1	92.3	64.7	78.2	97.1	67.3	81.9	92.3	67.2	79.5
0330	Cast-in-Place Concrete	81.3	72.4	78.0	79.1	72.0	76.4	82.9	73.4	79.3	83.8	73.2	79.9	81.3	74.3	78.7	92.6	72.4	85.1
03	CONCRETE	77.5	63.9	71.5	75.3	56.9	67.2	79.8	65.2	73.3	81.3	64.5	73.9	80.5	67.6	74.8	86.3	63.9	76.4
04	MASONRY	97.0	57.5	72.5	104.4	57.4	75.3	90.1	57.5	70.0	106.1	50.2	71.6	99.6	57.5	73.5	82.9	57.5	67.2
05	METALS	97.9	61.3	86.8	93.6	59.6	83.3	93.6	76.4	88.4	95.1	74.4	88.8	91.5	61.4	82.4	93.6	61.3	83.8
06	WOOD, PLASTICS & COMPOSITES	104.5	56.0	78.1	86.8	37.2	59.9	102.1	56.1	77.1	106.6	56.1	79.2	92.9	65.0	77.7	97.5	56.0	75.0
07	THERMAL & MOISTURE PROTECTION	101.4	64.5	85.4	96.7	62.4	81.8	97.2	65.4	83.3	97.5	62.8	82.4	93.3	65.8	81.3	97.3	64.7	83.1
08	OPENINGS	108.9	56.7	96.8	98.1	45.2	85.8	98.1	56.1	88.3	99.3	55.6	89.1	103.4	61.6	93.7	98.1	56.0	88.3
0920	Plaster & Gypsum Board	95.8	55.3	69.1	78.7	36.0	50.6	85.7	55.3	65.7	87.5	55.3	66.3	100.2	64.5	76.7	84.5	55.3	65.3
0950, 0980	Ceilings & Acoustic Treatment	90.0	55.3	66.5	79.4	36.0	50.0	79.4	55.3	63.1	87.9	55.3	65.8	91.5	64.5	73.2	79.4	55.3	63.1
0960	Flooring	89.6	73.1	84.8	90.3	59.2	81.3	97.4	53.4	84.6	99.9	34.3	80.8	91.1	73.1	85.9	94.5	59.2	84.2
0970, 0990	Wall Finishes & Painting/Coating	84.6	45.1	61.2	92.9	43.4	63.6	92.9	43.4	63.6	92.9	43.4	63.6	88.6	45.1	62.9	92.9	45.1	64.6
09	FINISHES	83.1	57.8	69.3	80.4	43.7	60.3	83.3	53.7	67.1	86.0	50.3	66.5	86.5	63.1	73.7	83.1	55.9	68.2
COVERS	DIVS. 10 - 14, 25, 28, 41, 43, 44, 46	100.0	79.0	95.4	100.0	77.0	94.9	100.0	77.9	95.1	100.0	77.9	95.1	100.0	80.0	95.6	100.0	79.0	95.4
21, 22, 23	FIRE SUPPRESSION, PLUMBING & HVAC	100.2	67.8	87.1	96.8	65.0	83.9	96.8	65.0	83.9	100.2	62.7	85.1	100.1	67.8	87.0	96.8	65.0	83.9
26, 27, 3370	ELECTRICAL, COMMUNICATIONS & UTIL.	96.0	70.9	83.3	94.0	68.6	81.1	95.4	68.7	81.8	93.6	68.8	81.0	101.5	70.9	86.0	93.6	67.2	80.2
MF2016	WEIGHTED AVERAGE	95.9	66.6	83.3	92.4	61.9	79.3	92.8	66.5	81.5	95.1	64.4	81.9	95.2	68.2	83.6	93.2	65.2	81.2

	DIVISION	OKLAHOMA												OREGON					
		POTEAU			SHAWNEE			TULSA			WOODWARD			BEND			EUGENE		
		749			748			740 - 741			738			977			974		
		MAT.	INST.	TOTAL	MAT.	INST.	TOTAL	MAT.	INST.	TOTAL	MAT.	INST.	TOTAL	MAT.	INST.	TOTAL	MAT.	INST.	TOTAL
015433	CONTRACTOR EQUIPMENT		91.9	91.9		81.8	81.8		93.2	93.2		81.8	81.8		99.8	99.8		99.8	99.8
0241, 31 - 34	SITE & INFRASTRUCTURE, DEMOLITION	76.1	87.9	84.3	97.9	95.4	96.1	95.5	91.4	92.7	101.2	94.5	96.5	114.4	103.2	106.6	103.9	103.2	103.4
0310	Concrete Forming & Accessories	87.1	56.4	60.9	82.5	56.7	60.6	97.9	57.5	63.5	85.0	44.6	50.6	106.4	96.8	98.2	102.9	96.8	97.7
0320	Concrete Reinforcing	93.3	67.2	80.1	92.3	67.2	79.5	92.5	67.3	79.7	87.6	67.2	77.3	100.4	111.7	106.1	104.8	111.7	108.3
0330	Cast-in-Place Concrete	82.9	73.3	79.3	95.6	72.4	87.0	91.3	75.3	85.4	84.1	72.3	79.7	115.1	99.9	109.4	111.4	99.9	107.1
03	CONCRETE	81.9	65.0	74.4	87.9	63.9	77.2	86.4	66.3	77.5	80.6	58.3	70.7	106.0	99.9	103.3	97.7	99.9	98.7
04	MASONRY	90.4	57.5	70.1	105.8	57.5	75.9	91.0	57.5	70.3	90.0	57.5	69.9	101.6	101.6	101.6	98.5	101.6	100.4
05	METALS	93.6	76.1	88.3	93.5	61.3	83.7	98.1	76.5	91.5	93.0	61.1	83.3	101.0	94.9	99.2	101.7	94.9	99.7
06	WOOD, PLASTICS & COMPOSITES	94.1	56.1	73.5	89.1	56.0	71.1	105.8	57.2	79.4	98.8	39.8	66.8	101.1	96.1	98.4	96.9	96.1	96.5
07	THERMAL & MOISTURE PROTECTION	97.3	65.4	83.4	97.3	64.0	82.8	97.5	65.6	83.6	101.6	62.9	84.8	116.5	98.7	108.8	115.6	101.7	109.5
08	OPENINGS	98.1	56.1	88.3	98.1	56.0	88.3	101.0	57.3	90.9	105.9	47.1	92.2	99.2	99.3	99.2	99.5	99.3	99.5
0920	Plaster & Gypsum Board	82.6	55.3	64.6	79.6	55.3	63.6	87.5	56.4	67.0	93.4	38.7	57.3	119.9	95.8	104.0	118.1	95.8	103.4
0950, 0980	Ceilings & Acoustic Treatment	79.4	55.3	63.1	79.4	55.3	63.1	87.9	56.4	66.6	83.2	38.7	53.0	95.9	95.8	95.9	96.9	95.8	96.2
0960	Flooring	93.9	53.4	82.1	91.4	59.2	82.0	98.7	63.7	88.5	86.2	56.5	77.5	97.8	104.2	99.7	96.3	104.2	98.6
0970, 0990	Wall Finishes & Painting/Coating	92.9	45.1	64.6	92.9	43.4	63.6	92.9	52.6	69.1	84.6	45.1	61.2	96.1	71.4	81.5	96.1	71.4	81.5
09	FINISHES	81.1	53.9	66.3	81.6	54.7	66.9	85.9	57.4	70.3	80.9	44.8	61.2	96.5	95.2	95.8	94.9	95.2	95.1
COVERS	DIVS. 10 - 14, 25, 28, 41, 43, 44, 46	100.0	79.3	95.4	100.0	79.0	95.4	100.0	78.0	95.2	100.0	77.3	95.0	100.0	98.3	99.6	100.0	98.3	99.6
21, 22, 23	FIRE SUPPRESSION, PLUMBING & HVAC	96.8	65.0	83.9	96.8	67.8	85.1	100.2	65.1	86.0	96.7	67.8	85.0	96.6	107.5	101.0	100.0	100.8	100.3
26, 27, 3370	ELECTRICAL, COMMUNICATIONS & UTIL.	93.7	68.7	81.0	95.6	70.9	83.0	95.5	68.7	81.9	95.9	70.9	83.2	100.1	93.7	96.8	98.7	93.7	96.2
MF2016	WEIGHTED AVERAGE	92.4	66.1	81.1	94.6	66.1	82.4	96.0	67.1	83.6	93.9	63.4	80.8	100.7	99.8	100.3	99.8	98.5	99.3

	DIVISION	OREGON																	
		KLAMATH FALLS			MEDFORD			PENDLETON			PORTLAND			SALEM			VALE		
		976			975			978			970 - 972			973			979		
		MAT.	INST.	TOTAL	MAT.	INST.	TOTAL	MAT.	INST.	TOTAL	MAT.	INST.	TOTAL	MAT.	INST.	TOTAL	MAT.	INST.	TOTAL
015433	CONTRACTOR EQUIPMENT		99.8	99.8		99.8	99.8		97.3	97.3		99.8	99.8		99.8	99.8		97.3	97.3
0241, 31 - 34	SITE & INFRASTRUCTURE, DEMOLITION	118.7	103.2	107.9	112.0	103.2	105.9	112.1	96.3	101.1	106.9	103.2	104.3	98.0	103.2	101.6	98.7	96.2	97.0
0310	Concrete Forming & Accessories	99.4	96.6	97.0	98.4	96.6	96.9	100.1	97.0	97.5	104.1	96.8	97.9	105.0	96.8	98.0	106.6	95.7	97.3
0320	Concrete Reinforcing	100.4	111.6	106.1	102.1	111.6	106.9	99.5	111.7	105.7	105.5	111.7	108.7	113.5	111.7	112.6	97.1	111.5	104.4
0330	Cast-in-Place Concrete	115.1	99.8	109.4	115.0	99.8	109.4	115.9	100.3	110.1	114.5	99.9	109.0	105.2	99.9	103.2	91.2	100.6	94.7
03	CONCRETE	108.6	99.8	104.7	103.5	99.8	101.9	91.1	100.2	95.1	99.3	99.9	99.6	95.2	99.9	97.3	75.8	99.6	86.4
04	MASONRY	114.5	101.6	106.5	95.4	101.6	99.2	105.6	101.7	103.2	100.0	101.6	101.0	104.3	101.6	102.6	103.4	101.7	102.3
05	METALS	101.0	94.6	99.1	101.3	94.6	99.3	108.4	95.5	104.5	102.9	94.9	100.5	109.2	94.9	104.9	108.3	94.2	104.0
06	WOOD, PLASTICS & COMPOSITES	91.9	96.1	94.2	90.8	96.1	93.7	94.0	96.2	95.2	98.1	96.1	97.0	93.1	96.1	94.7	102.6	96.2	99.1
07	THERMAL & MOISTURE PROTECTION	116.8	95.3	107.4	116.4	95.3	107.2	109.0	97.0	103.8	115.6	98.8	108.3	112.0	99.1	106.4	108.4	92.8	101.6
08	OPENINGS	99.2	99.3	99.3	102.0	99.3	101.4	95.5	99.4	96.4	97.4	99.3	97.9	101.2	99.3	100.7	95.4	89.2	94.0
0920	Plaster & Gypsum Board	114.8	95.8	102.3	114.2	95.8	102.1	100.3	95.8	97.4	118.0	95.8	103.4	113.0	95.8	101.7	106.4	95.8	99.4
0950, 0980	Ceilings & Acoustic Treatment	103.6	95.8	98.3	110.2	95.8	100.5	68.4	95.8	87.0	98.8	95.8	96.8	103.9	95.8	98.4	68.4	95.8	87.0
0960	Flooring	95.0	104.2	97.6	94.4	104.2	97.2	65.5	104.2	76.7	94.2	104.2	97.1	99.1	104.2	100.6	67.5	104.2	78.1
0970, 0990	Wall Finishes & Painting/Coating	96.1	67.0	78.9	96.1	67.0	78.9	87.6	76.2	80.9	95.9	76.2	84.3	93.4	75.1	82.6	87.6	71.4	78.0
09	FINISHES	96.9	94.7	95.7	97.1	94.7	95.8	68.8	95.8	83.5	94.7	95.7	95.2	95.1	95.6	95.4	69.3	95.3	83.5
COVERS	DIVS. 10 - 14, 25, 28, 41, 43, 44, 46	100.0	98.2	99.6	100.0	98.2	99.6	100.0	98.2	99.6	100.0	98.3	99.6	100.0	98.3	99.6	100.0	98.5	99.7
21, 22, 23	FIRE SUPPRESSION, PLUMBING & HVAC	96.6	107.5	101.0	100.0	107.5	103.0	98.3	114.5	104.9	100.0	104.7	101.9	100.0	107.5	103.1	98.3	72.6	87.9
26, 27, 3370	ELECTRICAL, COMMUNICATIONS & UTIL.	98.8	80.5	89.5	102.2	80.5	91.2	91.1	99.2	95.2	98.9	101.5	100.2	106.6	93.7	100.0	91.1	69.7	80.2
MF2016	WEIGHTED AVERAGE	101.5	97.8	99.9	101.3	97.8	99.8	96.8	101.7	98.9	100.2	100.4	100.3	101.7	99.9	100.9	94.5	87.9	91.7

	DIVISION	PENNSYLVANIA																	
		ALLENTOWN			ALTOONA			BEDFORD			BRADFORD			BUTLER			CHAMBERSBURG		
		181			166			155			167			160			172		
		MAT.	INST.	TOTAL	MAT.	INST.	TOTAL	MAT.	INST.	TOTAL	MAT.	INST.	TOTAL	MAT.	INST.	TOTAL	MAT.	INST.	TOTAL
015433	CONTRACTOR EQUIPMENT		115.2	115.2		115.2	115.2		113.2	113.2		115.2	115.2		115.2	115.2		114.4	114.4
0241, 31 - 34	SITE & INFRASTRUCTURE, DEMOLITION	92.7	101.8	99.0	95.7	101.6	99.8	100.6	99.3	99.7	91.7	100.5	97.9	87.0	102.4	97.7	88.0	99.8	96.2
0310	Concrete Forming & Accessories	98.5	111.2	109.3	83.5	84.9	84.7	82.3	83.5	83.3	85.8	99.2	97.2	85.0	96.7	95.0	85.0	78.9	79.8
0320	Concrete Reinforcing	103.8	116.6	110.3	100.6	108.8	104.8	99.5	108.8	104.2	102.8	108.8	105.9	101.3	122.8	112.2	103.4	115.5	109.5
0330	Cast-in-Place Concrete	87.6	102.8	93.2	97.6	89.3	94.5	94.0	88.3	91.9	93.2	93.2	93.2	86.2	98.9	90.9	95.1	96.5	95.6
03	CONCRETE	91.6	109.9	99.7	86.8	91.9	89.1	94.9	90.9	93.1	93.4	99.6	96.2	78.9	103.0	89.6	99.8	92.9	96.8
04	MASONRY	94.2	95.1	94.8	97.9	85.1	90.0	109.3	81.9	92.4	95.3	83.7	88.1	99.9	94.9	96.8	95.1	80.8	86.3
05	METALS	97.3	121.3	104.6	91.4	115.6	98.7	100.7	114.1	104.8	95.2	114.0	100.9	91.1	122.5	100.6	96.4	119.6	103.4
06	WOOD, PLASTICS & COMPOSITES	99.6	114.8	107.8	78.0	84.1	81.3	80.6	84.1	82.5	84.2	104.2	95.0	79.5	96.8	88.9	85.2	77.8	81.1
07	THERMAL & MOISTURE PROTECTION	103.6	110.8	106.7	102.4	92.5	98.1	100.6	89.8	95.9	103.4	91.0	98.0	101.9	97.1	99.8	93.0	85.4	89.7
08	OPENINGS	94.5	110.4	98.2	88.2	88.1	88.2	95.5	88.1	93.7	94.5	98.0	95.3	88.2	102.2	91.4	91.2	83.8	89.5
0920	Plaster & Gypsum Board	94.2	115.1	108.0	85.6	83.6	84.3	100.1	83.6	89.2	86.7	104.2	98.2	85.6	96.6	92.9	102.2	77.1	85.7
0950, 0980	Ceilings & Acoustic Treatment	88.8	115.1	106.6	91.4	83.6	86.1	101.3	83.6	89.3	91.5	104.2	100.1	92.3	96.6	95.2	92.1	77.1	81.9
0960	Flooring	90.7	98.3	92.9	84.2	102.3	89.4	96.7	107.2	99.7	85.0	107.2	91.5	85.1	108.9	92.0	88.1	80.6	85.9
0970, 0990	Wall Finishes & Painting/Coating	90.4	104.5	98.8	86.0	109.0	99.6	94.8	108.5	102.9	90.4	108.5	101.1	86.0	109.0	99.6	91.5	103.4	98.5
09	FINISHES	88.8	108.9	99.8	86.4	90.0	88.4	99.0	89.9	94.1	86.6	102.2	95.2	86.4	100.0	93.8	85.9	81.1	83.3
COVERS	DIVS. 10 - 14, 25, 28, 41, 43, 44, 46	100.0	101.6	100.4	100.0	96.4	99.2	100.0	95.1	98.9	100.0	98.1	99.6	100.0	99.3	99.8	100.0	94.6	98.8
21, 22, 23	FIRE SUPPRESSION, PLUMBING & HVAC	100.2	117.1	107.1	99.8	85.5	94.0	96.5	86.3	92.4	96.8	92.6	95.1	96.3	97.6	96.8	96.7	91.7	94.7
26, 27, 3370	ELECTRICAL, COMMUNICATIONS & UTIL.	99.6	98.1	98.8	90.0	110.3	100.3	97.8	110.3	104.2	93.6	110.3	102.1	90.6	110.4	100.6	91.0	87.9	89.4
MF2016	WEIGHTED AVERAGE	96.7	108.4	101.7	93.2	95.1	94.0	98.2	94.3	96.5	94.9	99.5	96.8	91.3	103.1	96.4	94.9	91.6	93.4

	DIVISION	PENNSYLVANIA																	
		DOYLESTOWN			DUBOIS			ERIE			GREENSBURG			HARRISBURG			HAZLETON		
		189			158			164 - 165			156			170 - 171			182		
		MAT.	INST.	TOTAL	MAT.	INST.	TOTAL	MAT.	INST.	TOTAL	MAT.	INST.	TOTAL	MAT.	INST.	TOTAL	MAT.	INST.	TOTAL
015433	CONTRACTOR EQUIPMENT		93.4	93.4		113.2	113.2		115.2	115.2		113.2	113.2		114.4	114.4		115.2	115.2
0241, 31 - 34	SITE & INFRASTRUCTURE, DEMOLITION	106.7	89.8	94.9	105.4	99.7	101.4	92.6	101.9	99.1	96.8	101.2	99.9	88.2	99.9	96.3	86.1	101.6	96.9
0310	Concrete Forming & Accessories	82.7	130.2	123.2	81.8	86.0	85.4	97.8	88.2	89.6	88.4	91.0	90.6	96.3	87.9	89.1	80.4	91.3	89.7
0320	Concrete Reinforcing	100.4	154.7	128.0	98.8	122.8	111.0	102.8	110.4	106.7	98.8	122.5	110.9	113.1	114.2	113.7	100.8	116.3	108.7
0330	Cast-in-Place Concrete	82.7	133.4	101.6	90.6	96.8	92.9	95.9	91.2	94.2	87.2	98.4	91.4	93.7	98.7	95.5	82.7	98.3	88.5
03	CONCRETE	87.3	134.6	108.2	97.1	97.3	97.2	85.7	94.4	89.6	90.4	100.2	94.7	94.4	97.5	95.8	84.3	99.3	90.9
04	MASONRY	97.6	128.4	116.6	109.8	95.4	100.9	86.9	88.7	88.0	120.0	90.9	102.1	89.9	85.6	87.3	107.2	91.4	97.4
05	METALS	94.9	123.9	103.7	100.7	119.6	106.4	91.6	116.3	99.1	100.6	120.5	106.6	103.0	119.3	107.9	97.0	120.5	104.1
06	WOOD, PLASTICS & COMPOSITES	79.8	129.4	106.7	79.5	84.1	82.0	96.0	86.8	91.0	86.8	89.6	88.3	98.0	87.6	92.4	78.3	89.5	84.4
07	THERMAL & MOISTURE PROTECTION	100.8	132.7	114.7	100.9	95.5	98.6	102.9	91.0	97.7	100.5	94.8	98.0	95.8	102.6	98.8	102.9	103.5	103.2
08	OPENINGS	96.9	137.0	106.2	95.5	91.2	94.5	88.3	90.5	88.9	95.4	98.3	96.1	98.0	88.9	95.9	95.0	96.0	95.3
0920	Plaster & Gypsum Board	84.7	130.2	114.7	98.9	83.6	88.8	94.2	86.4	89.1	101.1	89.3	93.4	108.1	87.2	94.3	85.1	89.2	87.8
0950, 0980	Ceilings & Acoustic Treatment	88.1	130.2	116.6	101.3	83.6	89.3	88.8	86.4	87.2	100.5	89.3	92.9	99.7	87.2	91.2	89.8	89.2	89.4
0960	Flooring	75.2	144.0	95.2	96.4	107.2	99.5	91.1	102.3	94.3	100.3	80.6	94.6	92.0	93.5	92.4	82.3	92.3	85.2
0970, 0990	Wall Finishes & Painting/Coating	90.0	147.5	124.0	94.8	109.0	103.2	96.5	94.8	95.5	94.8	109.0	103.2	96.6	87.0	90.9	90.4	106.8	100.1
09	FINISHES	80.5	134.3	109.9	99.2	91.4	95.0	89.8	91.0	90.5	99.7	90.6	94.8	91.1	88.2	89.5	84.9	92.8	89.2
COVERS	DIVS. 10 - 14, 25, 28, 41, 43, 44, 46	100.0	115.3	103.4	100.0	97.1	99.4	100.0	97.9	99.5	100.0	98.4	99.6	100.0	96.9	99.3	100.0	97.4	99.4
21, 22, 23	FIRE SUPPRESSION, PLUMBING & HVAC	96.3	135.6	112.2	96.5	89.5	93.7	99.8	95.7	98.1	96.5	89.3	93.6	100.1	93.7	97.5	96.8	97.7	97.2
26, 27, 3370	ELECTRICAL, COMMUNICATIONS & UTIL.	93.0	135.6	114.7	98.4	110.3	104.5	91.7	95.4	93.6	98.4	110.3	104.5	98.2	87.9	93.0	94.4	92.2	93.3
MF2016	WEIGHTED AVERAGE	94.0	129.2	109.1	98.7	98.3	98.5	93.1	96.2	94.5	98.2	98.7	98.4	97.8	94.9	96.5	94.4	98.3	96.1

	DIVISION	PENNSYLVANIA																	
		INDIANA			JOHNSTOWN			KITTANNING			LANCASTER			LEHIGH VALLEY			MONTROSE		
		157			159			162			175 - 176			180			188		
		MAT.	INST.	TOTAL	MAT.	INST.	TOTAL	MAT.	INST.	TOTAL	MAT.	INST.	TOTAL	MAT.	INST.	TOTAL	MAT.	INST.	TOTAL
015433	CONTRACTOR EQUIPMENT		113.2	113.2		113.2	113.2		115.2	115.2		114.4	114.4		115.2	115.2		115.2	115.2
0241, 31 - 34	SITE & INFRASTRUCTURE, DEMOLITION	95.0	100.3	98.7	101.1	100.5	100.7	89.5	102.0	98.2	80.0	100.1	94.0	90.0	102.0	98.4	88.7	101.7	97.8
0310	Concrete Forming & Accessories	82.8	95.5	93.6	81.8	84.8	84.4	85.0	92.2	91.1	87.0	88.8	88.5	92.0	111.6	108.7	81.3	91.8	90.2
0320	Concrete Reinforcing	98.1	122.9	110.7	99.5	122.3	111.1	101.3	122.6	112.2	103.0	114.2	108.7	100.8	111.1	106.0	105.6	119.6	112.7
0330	Cast-in-Place Concrete	85.5	97.1	89.8	94.8	89.1	92.7	89.6	97.4	92.5	80.8	99.5	87.8	89.6	102.0	94.2	87.8	95.7	90.7
03	CONCRETE	88.0	101.7	94.1	95.8	94.0	95.0	81.4	100.4	89.8	87.5	98.2	92.3	90.6	108.9	98.7	89.3	99.2	93.7
04	MASONRY	105.6	95.5	99.3	106.4	85.5	93.5	102.5	90.7	95.2	100.9	88.6	93.3	94.3	98.5	96.9	94.2	93.5	93.7
05	METALS	100.7	120.7	106.8	100.7	118.7	106.2	91.2	121.4	100.3	96.4	119.5	103.4	97.0	119.7	103.8	95.2	121.2	103.1
06	WOOD, PLASTICS & COMPOSITES	81.3	96.7	89.7	79.5	84.1	82.0	79.5	91.9	86.2	88.1	87.6	87.8	90.9	113.5	103.2	79.1	89.5	84.8
07	THERMAL & MOISTURE PROTECTION	100.4	96.9	98.8	100.6	91.0	96.4	102.0	95.0	99.0	92.4	103.8	97.4	103.3	112.8	107.4	102.9	93.5	98.8
08	OPENINGS	95.5	98.2	96.1	95.5	91.2	94.5	88.2	99.5	90.8	91.2	88.9	90.7	95.0	108.5	98.1	91.8	94.5	92.4
0920	Plaster & Gypsum Board	100.4	96.6	97.9	98.7	83.6	88.8	85.6	91.6	89.6	104.6	87.2	93.1	87.5	113.8	104.8	85.5	89.2	87.9
0950, 0980	Ceilings & Acoustic Treatment	101.3	96.6	98.1	100.5	83.6	89.1	92.3	91.6	91.8	92.1	87.2	88.8	89.8	113.8	106.1	91.5	89.2	89.9
0960	Flooring	97.2	107.2	100.1	96.4	80.6	91.8	85.1	107.2	91.5	89.1	98.9	91.9	87.7	97.3	90.5	83.0	107.2	90.0
0970, 0990	Wall Finishes & Painting/Coating	94.8	109.0	103.2	94.8	109.0	103.2	86.0	109.0	99.6	91.5	88.9	90.0	90.4	103.6	98.2	90.4	106.8	100.1
09	FINISHES	98.8	98.9	98.9	98.6	86.3	91.9	86.5	96.1	91.7	86.0	89.9	88.1	87.0	108.3	98.6	85.6	95.4	91.0
COVERS	DIVS. 10 - 14, 25, 28, 41, 43, 44, 46	100.0	98.4	99.7	100.0	96.3	99.2	100.0	98.1	99.6	100.0	97.6	99.5	100.0	104.3	101.0	100.0	98.1	99.6
21, 22, 23	FIRE SUPPRESSION, PLUMBING & HVAC	96.5	87.3	92.8	96.5	80.7	90.1	96.3	92.7	94.9	96.7	94.8	95.9	96.8	118.6	105.6	96.8	96.9	96.9
26, 27, 3370	ELECTRICAL, COMMUNICATIONS & UTIL.	98.4	110.3	104.5	98.4	110.3	104.5	90.0	110.3	100.3	92.3	95.6	94.0	94.4	132.5	113.8	93.6	97.6	95.6
MF2016	WEIGHTED AVERAGE	97.1	100.1	98.4	98.2	94.2	96.5	91.8	100.4	95.5	93.5	96.8	94.9	95.0	113.6	103.0	93.8	99.2	96.1

	DIVISION	PENNSYLVANIA																	
		NEW CASTLE			NORRISTOWN			OIL CITY			PHILADELPHIA			PITTSBURGH			POTTSVILLE		
		161			194			163			190 - 191			150 - 152			179		
		MAT.	INST.	TOTAL	MAT.	INST.	TOTAL	MAT.	INST.	TOTAL	MAT.	INST.	TOTAL	MAT.	INST.	TOTAL	MAT.	INST.	TOTAL
015433	CONTRACTOR EQUIPMENT		115.2	115.2		99.0	99.0		115.2	115.2		99.3	99.3		102.0	102.0		114.4	114.4
0241, 31 - 34	SITE & INFRASTRUCTURE, DEMOLITION	87.4	102.3	97.8	95.3	99.3	98.1	85.9	100.8	96.3	97.3	100.0	99.2	100.8	99.9	100.2	82.9	100.3	95.1
0310	Concrete Forming & Accessories	85.0	96.2	94.5	83.0	128.7	121.9	85.0	95.3	93.8	100.2	140.0	134.1	98.4	97.2	97.3	78.8	90.1	88.4
0320	Concrete Reinforcing	100.1	103.1	101.6	101.9	154.7	128.7	101.3	97.7	99.5	105.2	126.5	116.0	100.0	123.1	111.8	102.2	117.4	109.9
0330	Cast-in-Place Concrete	87.0	96.9	90.7	85.0	131.0	102.1	84.5	97.1	89.2	93.3	135.3	108.9	93.5	99.2	95.6	86.1	99.1	91.0
03	CONCRETE	79.2	98.7	87.9	87.9	133.0	107.9	77.7	97.5	86.5	98.8	134.8	114.8	96.5	102.3	99.1	91.2	99.3	94.8
04	MASONRY	99.6	94.6	96.5	108.7	124.6	118.5	98.8	90.6	93.7	98.9	131.3	118.9	101.3	99.6	100.2	94.6	89.4	91.4
05	METALS	91.2	115.0	98.4	99.6	123.8	106.9	91.2	113.9	98.0	101.3	113.5	105.0	102.2	107.4	103.8	96.7	121.2	104.1
06	WOOD, PLASTICS & COMPOSITES	79.5	96.8	88.9	76.1	129.3	105.0	79.5	96.8	88.9	96.8	141.8	121.2	99.7	97.2	98.3	77.8	87.6	83.1
07	THERMAL & MOISTURE PROTECTION	101.9	94.6	98.7	107.7	131.1	117.9	101.8	93.3	98.1	103.3	136.4	117.7	100.8	98.4	99.8	92.5	102.4	96.8
08	OPENINGS	88.2	94.1	89.5	88.7	137.0	99.9	88.2	97.7	90.4	101.3	137.3	109.6	97.8	102.5	98.9	91.3	95.7	92.3
0920	Plaster & Gypsum Board	85.6	96.6	92.9	83.8	130.2	114.3	85.6	96.6	92.9	97.8	142.8	127.4	98.7	96.9	97.5	99.1	87.2	91.3
0950, 0980	Ceilings & Acoustic Treatment	92.3	96.6	95.2	91.5	130.2	117.7	92.3	96.6	95.2	101.6	142.8	129.5	95.1	96.9	96.3	92.1	87.2	88.8
0960	Flooring	85.1	108.9	92.0	83.2	144.0	100.8	85.1	107.2	91.5	97.5	145.7	111.5	104.9	107.5	105.7	85.2	107.2	91.6
0970, 0990	Wall Finishes & Painting/Coating	86.0	109.0	99.6	85.8	147.5	122.3	86.0	109.0	99.6	91.8	152.1	127.5	97.5	111.4	105.7	91.5	106.8	100.6
09	FINISHES	86.4	99.8	93.7	83.3	133.3	110.6	86.3	98.9	93.2	98.4	142.8	122.7	100.4	100.2	100.3	84.3	93.9	89.6
COVERS	DIVS. 10 - 14, 25, 28, 41, 43, 44, 46	100.0	99.3	99.8	100.0	111.3	102.5	100.0	98.6	99.7	100.0	120.0	104.4	100.0	101.5	100.3	100.0	99.5	99.9
21, 22, 23	FIRE SUPPRESSION, PLUMBING & HVAC	96.3	99.0	97.4	96.6	133.6	111.6	96.3	96.3	96.3	100.2	140.0	116.3	99.9	98.3	99.2	96.7	97.9	97.2
26, 27, 3370	ELECTRICAL, COMMUNICATIONS & UTIL.	90.6	100.6	95.7	93.4	145.5	119.9	92.3	110.3	101.5	100.4	160.5	131.0	100.8	110.8	105.9	90.6	93.5	92.1
MF2016	WEIGHTED AVERAGE	91.4	100.3	95.2	94.8	130.0	109.9	91.3	100.3	95.1	100.1	135.3	115.2	99.9	102.2	100.9	93.4	98.4	95.6

	DIVISION	PENNSYLVANIA																	
		READING			SCRANTON			STATE COLLEGE			STROUDSBURG			SUNBURY			UNIONTOWN		
		195 - 196			184 - 185			168			183			178			154		
		MAT.	INST.	TOTAL	MAT.	INST.	TOTAL	MAT.	INST.	TOTAL	MAT.	INST.	TOTAL	MAT.	INST.	TOTAL	MAT.	INST.	TOTAL
015433	CONTRACTOR EQUIPMENT		121.3	121.3		115.2	115.2		114.4	114.4		115.2	115.2		115.2	115.2		113.2	113.2
0241, 31 - 34	SITE & INFRASTRUCTURE, DEMOLITION	99.6	112.5	108.6	93.2	101.8	99.2	83.8	100.2	95.3	87.7	101.9	97.6	94.7	101.2	99.3	95.5	101.1	99.4
0310	Concrete Forming & Accessories	98.5	88.5	90.0	98.6	89.0	90.5	84.5	85.1	85.0	86.6	93.1	92.1	89.5	87.1	87.5	76.4	96.5	93.5
0320	Concrete Reinforcing	103.3	152.9	128.5	103.8	119.7	111.9	102.0	108.9	105.5	104.1	119.8	112.1	105.0	115.5	110.4	98.8	122.9	111.1
0330	Cast-in-Place Concrete	76.2	99.7	84.9	91.4	95.8	93.1	88.2	89.4	88.6	86.2	97.2	90.3	94.2	97.6	95.4	85.5	98.4	90.3
03	CONCRETE	87.1	104.7	94.9	93.4	98.0	95.4	93.8	92.1	93.0	88.0	100.3	93.5	94.6	97.1	95.7	87.7	102.7	94.3
04	MASONRY	97.7	93.4	95.0	94.6	95.8	95.4	100.2	85.7	91.2	92.4	98.3	96.1	94.9	82.7	87.4	122.0	97.4	106.8
05	METALS	99.9	133.7	110.1	99.3	121.3	105.9	95.0	115.8	101.3	97.0	121.7	104.5	96.4	120.2	103.6	100.4	121.0	106.7
06	WOOD, PLASTICS & COMPOSITES	93.7	84.8	88.8	99.6	85.7	92.1	86.1	84.1	85.0	85.1	89.5	87.5	86.0	87.6	86.9	73.7	96.7	86.2
07	THERMAL & MOISTURE PROTECTION	108.1	106.2	107.3	103.5	93.9	99.3	102.5	100.4	101.6	103.1	93.9	99.1	93.5	95.3	94.3	100.2	97.8	99.2
08	OPENINGS	93.2	102.3	95.3	94.5	92.4	94.0	91.6	88.1	90.8	95.1	97.3	95.6	91.3	92.5	91.6	95.4	102.2	97.0
0920	Plaster & Gypsum Board	93.2	84.3	87.4	95.9	85.3	88.9	87.4	83.6	84.9	86.3	89.2	88.2	98.4	87.2	91.0	96.9	96.6	96.7
0950, 0980	Ceilings & Acoustic Treatment	82.9	84.3	83.9	96.5	85.3	88.9	88.9	83.6	85.3	88.1	89.2	88.8	88.7	87.2	87.7	100.5	96.6	97.9
0960	Flooring	87.7	101.2	91.6	90.7	96.0	92.2	88.1	101.2	91.9	85.5	97.2	88.9	86.0	107.2	92.1	93.5	107.2	97.5
0970, 0990	Wall Finishes & Painting/Coating	84.6	104.5	96.4	90.4	111.1	102.7	90.4	109.0	101.4	90.4	106.8	100.1	91.5	106.8	100.6	94.8	112.9	105.5
09	FINISHES	84.5	91.0	88.0	90.5	91.9	91.3	86.2	89.8	88.2	85.8	94.5	90.5	85.1	92.2	89.0	97.1	99.8	98.6
COVERS	DIVS. 10 - 14, 25, 28, 41, 43, 44, 46	100.0	97.6	99.5	100.0	96.5	99.2	100.0	96.5	99.2	100.0	99.2	99.8	100.0	94.9	98.9	100.0	99.1	99.8
21, 22, 23	FIRE SUPPRESSION, PLUMBING & HVAC	100.2	109.2	103.8	100.2	98.2	99.4	96.8	93.4	95.4	96.8	100.1	98.2	96.7	89.3	93.7	96.5	93.7	95.4
26, 27, 3370	ELECTRICAL, COMMUNICATIONS & UTIL.	99.5	93.6	96.5	99.6	97.6	98.6	92.7	110.3	101.7	94.4	145.2	120.2	91.0	94.1	92.6	95.4	110.3	103.0
MF2016	WEIGHTED AVERAGE	96.5	104.0	99.7	97.4	98.9	98.1	94.5	97.0	95.6	94.3	107.3	99.9	94.3	95.0	94.6	97.2	102.2	99.4

DIVISION		PENNSYLVANIA																	
		WASHINGTON			WELLSBORO			WESTCHESTER			WILKES-BARRE			WILLIAMSPORT			YORK		
		153			169			193			186 - 187			177			173 - 174		
		MAT.	INST.	TOTAL	MAT.	INST.	TOTAL	MAT.	INST.	TOTAL	MAT.	INST.	TOTAL	MAT.	INST.	TOTAL	MAT.	INST.	TOTAL
015433	CONTRACTOR EQUIPMENT		113.2	113.2		115.2	115.2		99.0	99.0		115.2	115.2		115.2	115.2		114.4	114.4
0241, 31 - 34	SITE & INFRASTRUCTURE, DEMOLITION	95.6	101.4	99.6	95.1	101.2	99.3	101.4	100.3	100.7	85.7	101.8	96.9	85.7	101.4	96.6	83.8	100.1	95.1
0310	Concrete Forming & Accessories	82.9	96.6	94.6	85.2	86.5	86.3	89.5	130.0	124.0	89.3	87.6	87.9	86.3	88.0	87.7	82.0	88.7	87.7
0320	Concrete Reinforcing	98.8	123.0	111.1	102.0	119.5	110.9	101.0	154.6	128.3	102.8	119.7	111.4	104.2	114.1	109.3	105.0	114.2	109.7
0330	Cast-in-Place Concrete	85.5	98.6	90.4	92.4	90.7	91.8	94.1	133.1	108.7	82.7	95.6	87.5	79.7	92.5	84.5	86.5	99.5	91.3
03	CONCRETE	88.1	102.8	94.6	96.0	95.0	95.6	95.8	134.4	112.9	85.1	97.3	90.5	82.3	95.4	88.1	92.3	98.2	94.9
04	MASONRY	105.3	97.5	100.5	101.4	83.1	90.1	103.2	128.4	118.8	107.5	95.5	100.1	86.9	88.0	87.6	95.9	88.6	91.3
05	METALS	100.4	121.4	106.8	95.1	120.5	102.8	99.6	123.9	106.9	95.2	121.2	103.1	96.4	118.9	103.2	98.1	119.4	104.5
06	WOOD, PLASTICS & COMPOSITES	81.4	96.7	89.7	83.6	86.8	85.4	82.7	129.3	108.0	87.6	84.1	85.7	82.5	87.6	85.3	81.3	87.6	84.7
07	THERMAL & MOISTURE PROTECTION	100.4	97.8	99.2	103.7	88.4	97.0	108.2	128.0	116.8	102.9	93.4	98.8	92.8	90.7	91.9	92.6	103.8	97.5
08	OPENINGS	95.4	102.2	96.9	94.5	92.6	94.0	88.7	137.0	99.9	91.8	91.5	91.7	91.3	92.2	91.5	91.2	88.9	90.7
0920	Plaster & Gypsum Board	100.2	96.6	97.8	86.1	86.4	86.3	84.4	130.2	114.5	87.0	83.6	84.7	99.1	87.2	91.3	99.5	87.2	91.4
0950, 0980	Ceilings & Acoustic Treatment	100.5	96.6	97.9	88.9	86.4	87.2	91.5	130.2	117.7	91.5	83.6	86.1	92.1	87.2	88.8	91.1	87.2	88.5
0960	Flooring	97.3	107.2	100.2	84.7	107.2	91.2	86.0	144.0	102.8	86.4	96.0	89.2	84.9	89.7	86.3	86.5	98.9	90.1
0970, 0990	Wall Finishes & Painting/Coating	94.8	112.9	105.5	90.4	106.8	100.1	85.8	147.5	122.3	90.4	104.5	98.8	91.5	104.5	99.2	91.5	87.0	88.8
09	FINISHES	98.7	99.8	99.3	86.3	92.2	89.5	84.6	134.3	111.7	86.6	90.1	88.5	84.8	89.2	87.2	84.6	89.6	87.3
COVERS	DIVS. 10 - 14, 25, 28, 41, 43, 44, 46	100.0	99.1	99.8	100.0	98.2	99.6	100.0	117.6	103.9	100.0	96.2	99.2	100.0	97.4	99.4	100.0	97.6	99.5
21, 22, 23	FIRE SUPPRESSION, PLUMBING & HVAC	96.5	95.3	96.0	96.8	89.3	93.8	96.6	135.5	112.4	96.8	98.8	97.6	96.7	94.3	95.7	100.1	94.8	98.0
26, 27, 3370	ELECTRICAL, COMMUNICATIONS & UTIL.	97.8	110.4	104.2	93.6	97.6	95.6	93.2	135.6	114.8	94.4	92.2	93.3	91.4	85.4	88.4	92.3	84.9	88.5
MF2016	WEIGHTED AVERAGE	96.9	102.7	99.4	95.5	95.2	95.4	95.8	129.9	110.4	94.1	97.9	95.7	92.1	94.6	93.2	94.9	95.3	95.1

DIVISION		PUERTO RICO			RHODE ISLAND						SOUTH CAROLINA								
		SAN JUAN			NEWPORT			PROVIDENCE			AIKEN			BEAUFORT			CHARLESTON		
		009			028			029			298			299			294		
		MAT.	INST.	TOTAL	MAT.	INST.	TOTAL	MAT.	INST.	TOTAL	MAT.	INST.	TOTAL	MAT.	INST.	TOTAL	MAT.	INST.	TOTAL
015433	CONTRACTOR EQUIPMENT		86.5	86.5		100.1	100.1		100.1	100.1		105.8	105.8		105.8	105.8		105.8	105.8
0241, 31 - 34	SITE & INFRASTRUCTURE, DEMOLITION	129.4	89.2	101.4	86.8	102.2	97.5	86.7	102.2	97.5	126.6	89.0	100.4	121.8	88.4	98.5	107.3	88.7	94.3
0310	Concrete Forming & Accessories	90.1	21.3	31.6	99.5	121.2	118.0	99.7	121.2	118.0	90.6	63.7	67.7	89.6	39.1	46.6	89.0	63.7	67.5
0320	Concrete Reinforcing	198.3	18.3	106.8	104.9	118.9	112.0	105.8	118.9	112.5	92.7	62.4	77.3	91.8	26.5	58.6	91.6	67.7	79.4
0330	Cast-in-Place Concrete	94.9	32.2	71.6	68.9	120.7	88.1	88.1	120.7	100.2	91.9	66.1	82.3	91.8	65.7	82.1	107.8	66.1	92.3
03	CONCRETE	96.6	25.6	65.1	82.8	120.1	99.3	90.0	120.1	103.4	99.2	66.0	84.5	96.7	48.8	75.5	93.3	66.9	81.6
04	MASONRY	87.7	20.4	46.1	96.0	124.0	113.3	100.4	124.0	115.0	78.2	65.3	70.3	92.4	65.3	75.7	93.8	65.3	76.2
05	METALS	108.8	38.4	87.5	89.5	112.3	96.4	92.3	112.3	98.3	100.5	89.4	97.1	100.5	79.7	94.2	102.4	91.1	98.9
06	WOOD, PLASTICS & COMPOSITES	97.6	20.7	55.8	99.2	120.3	110.7	97.7	120.3	110.0	87.9	65.6	75.8	86.9	33.3	57.8	85.7	65.6	74.8
07	THERMAL & MOISTURE PROTECTION	131.4	25.4	85.3	103.6	119.5	110.5	103.7	119.5	110.6	97.1	64.9	83.1	96.7	56.9	79.4	95.7	64.5	82.1
08	OPENINGS	155.2	19.0	123.6	99.5	120.2	104.3	100.8	120.2	105.3	95.8	62.6	88.1	95.8	38.8	82.6	99.4	63.8	91.1
0920	Plaster & Gypsum Board	170.9	18.6	70.6	89.5	120.6	109.9	99.7	120.6	113.4	92.0	64.6	73.9	95.6	31.4	53.3	96.7	64.6	75.5
0950, 0980	Ceilings & Acoustic Treatment	254.2	18.6	94.5	93.6	120.6	111.9	98.6	120.6	113.5	82.6	64.6	70.4	85.1	31.4	48.7	85.1	64.6	71.2
0960	Flooring	197.7	20.8	146.4	87.9	130.5	100.2	91.6	130.5	102.9	98.2	92.8	96.6	99.7	78.4	93.5	99.3	79.8	93.7
0970, 0990	Wall Finishes & Painting/Coating	186.1	22.2	89.1	91.0	118.7	107.4	95.3	118.7	109.1	97.0	67.8	79.7	97.0	58.4	74.1	97.0	67.8	79.7
09	FINISHES	191.9	21.4	98.8	87.9	123.1	107.1	89.2	123.1	107.7	89.6	69.7	78.7	90.6	47.1	66.9	88.7	67.5	77.1
COVERS	DIVS. 10 - 14, 25, 28, 41, 43, 44, 46	100.0	19.5	82.3	100.0	106.1	101.3	100.0	106.1	101.3	100.0	70.6	93.5	100.0	81.4	95.9	100.0	70.6	93.5
21, 22, 23	FIRE SUPPRESSION, PLUMBING & HVAC	103.5	17.9	68.9	100.3	113.2	105.5	100.1	113.2	105.4	97.0	56.0	80.5	97.0	56.0	80.4	100.5	58.3	83.4
26, 27, 3370	ELECTRICAL, COMMUNICATIONS & UTIL.	120.6	16.9	67.9	98.8	97.9	98.3	97.4	97.9	97.6	94.0	63.4	78.4	97.4	31.6	64.0	95.9	58.3	76.8
MF2016	WEIGHTED AVERAGE	117.8	27.2	79.0	94.6	113.7	102.8	96.2	113.7	103.7	96.9	67.9	84.4	97.5	56.0	79.7	98.0	67.6	84.9

DIVISION		SOUTH CAROLINA															SOUTH DAKOTA		
		COLUMBIA			FLORENCE			GREENVILLE			ROCK HILL			SPARTANBURG			ABERDEEN		
		290 - 292			295			296			297			293			574		
		MAT.	INST.	TOTAL	MAT.	INST.	TOTAL	MAT.	INST.	TOTAL	MAT.	INST.	TOTAL	MAT.	INST.	TOTAL	MAT.	INST.	TOTAL
015433	CONTRACTOR EQUIPMENT		105.8	105.8		105.8	105.8		105.8	105.8		105.8	105.8		105.8	105.8		100.3	100.3
0241, 31 - 34	SITE & INFRASTRUCTURE, DEMOLITION	106.5	88.7	94.1	116.1	88.5	96.8	111.2	88.8	95.6	109.0	88.0	94.4	111.0	88.8	95.5	100.5	98.2	98.9
0310	Concrete Forming & Accessories	90.9	63.7	67.7	78.4	63.6	65.8	88.5	63.7	67.4	86.8	63.4	66.9	91.0	63.7	67.8	95.5	77.8	80.4
0320	Concrete Reinforcing	96.7	67.7	81.9	91.2	67.7	79.3	91.1	66.0	78.4	91.9	67.7	79.6	91.1	67.7	79.2	105.3	72.4	88.5
0330	Cast-in-Place Concrete	109.0	66.1	93.1	91.8	65.9	82.2	91.8	66.1	82.2	91.8	66.0	82.2	91.8	66.1	82.2	112.8	81.2	101.1
03	CONCRETE	91.7	66.9	80.7	91.7	66.7	80.6	90.4	66.6	79.9	88.5	66.7	78.8	90.6	66.9	80.1	98.1	78.6	89.5
04	MASONRY	87.6	65.3	73.8	78.3	65.3	70.3	76.0	65.3	69.4	99.5	65.3	78.4	78.3	65.3	70.3	114.9	74.9	90.2
05	METALS	99.4	91.1	96.9	101.3	90.5	98.0	101.3	91.1	98.2	100.5	90.8	97.6	101.3	91.1	98.2	91.4	82.4	88.7
06	WOOD, PLASTICS & COMPOSITES	90.0	65.6	76.7	73.6	65.6	69.3	85.3	65.6	74.6	84.0	65.6	74.0	89.0	65.6	76.3	97.6	76.4	86.1
07	THERMAL & MOISTURE PROTECTION	91.8	64.5	79.9	96.1	64.5	82.4	96.0	64.5	82.3	95.8	60.2	80.3	96.0	64.5	82.3	102.9	79.4	92.7
08	OPENINGS	98.1	63.8	90.1	95.9	63.8	88.4	95.8	63.8	88.4	95.8	63.8	88.4	95.8	63.8	88.4	99.5	63.8	91.2
0920	Plaster & Gypsum Board	97.3	64.6	75.7	86.4	64.6	72.0	91.1	64.6	73.6	90.8	64.6	73.5	93.9	64.6	74.6	108.3	76.1	87.1
0950, 0980	Ceilings & Acoustic Treatment	87.6	64.6	72.0	83.4	64.6	70.6	82.6	64.6	70.4	82.6	64.6	70.4	82.6	64.6	70.4	93.9	76.1	81.8
0960	Flooring	91.4	78.5	87.6	90.7	78.4	87.1	97.0	78.4	91.6	95.9	78.4	90.8	98.4	78.4	92.6	91.9	45.3	78.4
0970, 0990	Wall Finishes & Painting/Coating	94.9	67.8	78.9	97.0	67.8	79.7	97.0	67.8	79.7	97.0	67.8	79.7	97.0	67.8	79.7	87.8	34.3	56.1
09	FINISHES	86.7	67.2	76.1	85.6	67.2	75.6	87.5	67.2	76.4	86.9	67.2	76.1	88.3	67.2	76.8	89.1	67.9	77.5
COVERS	DIVS. 10 - 14, 25, 28, 41, 43, 44, 46	100.0	70.6	93.5	100.0	70.6	93.5	100.0	70.6	93.5	100.0	70.6	93.5	100.0	70.6	93.5	100.0	86.5	97.0
21, 22, 23	FIRE SUPPRESSION, PLUMBING & HVAC	100.0	57.7	82.9	100.5	57.7	83.2	100.5	57.7	83.2	97.0	56.1	80.5	100.5	57.7	83.2	100.1	55.4	82.0
26, 27, 3370	ELECTRICAL, COMMUNICATIONS & UTIL.	96.2	61.1	78.4	94.0	61.1	77.3	96.0	61.3	78.3	96.0	61.4	78.4	96.0	61.3	78.3	100.2	65.0	82.3
MF2016	WEIGHTED AVERAGE	96.5	67.9	84.2	96.2	67.8	84.0	96.2	67.9	84.1	96.0	67.3	83.7	96.4	67.9	84.2	98.3	71.5	86.8

	DIVISION	SOUTH DAKOTA																	
		MITCHELL			MOBRIDGE			PIERRE			RAPID CITY			SIOUX FALLS			WATERTOWN		
		573			576			575			577			570 - 571			572		
		MAT.	INST.	TOTAL	MAT.	INST.	TOTAL	MAT.	INST.	TOTAL	MAT.	INST.	TOTAL	MAT.	INST.	TOTAL	MAT.	INST.	TOTAL
015433	**CONTRACTOR EQUIPMENT**		100.3	100.3		100.3	100.3		100.3	100.3		100.3	100.3		101.2	101.2		100.3	100.3
0241, 31 - 34	**SITE & INFRASTRUCTURE, DEMOLITION**	97.2	97.9	97.7	97.2	98.0	97.7	100.3	97.4	98.2	98.8	97.4	97.8	90.9	99.7	97.1	97.1	98.0	97.7
0310	Concrete Forming & Accessories	94.6	45.4	52.7	85.1	45.8	51.7	95.7	46.6	53.9	103.2	58.1	64.8	100.0	78.4	81.6	81.6	76.6	77.3
0320	Concrete Reinforcing	104.7	72.6	88.4	107.4	72.5	89.7	107.6	96.7	102.0	98.4	96.9	97.6	105.2	97.1	101.1	101.8	72.5	86.9
0330	Cast-in-Place Concrete	109.7	53.8	88.9	109.7	79.5	98.4	110.2	79.1	98.6	108.8	79.4	97.9	87.1	80.3	84.6	109.7	79.4	98.4
03	**CONCRETE**	95.8	54.5	77.5	95.5	63.6	81.4	94.8	67.8	82.9	95.2	73.1	85.4	84.1	82.8	83.5	94.7	77.5	87.1
04	**MASONRY**	102.5	77.1	86.8	111.2	74.4	88.4	113.0	74.1	89.0	110.9	75.6	89.1	93.2	77.1	83.2	138.5	75.3	99.4
05	**METALS**	90.4	82.2	87.9	90.5	82.7	88.1	92.7	88.9	91.5	93.1	89.8	92.1	94.2	92.0	93.5	90.4	82.6	88.0
06	**WOOD, PLASTICS & COMPOSITES**	96.5	34.6	62.9	84.5	34.4	57.3	102.2	34.9	65.6	102.1	49.3	73.4	98.8	76.4	86.6	80.4	76.4	78.2
07	**THERMAL & MOISTURE PROTECTION**	102.7	74.9	90.6	102.7	77.2	91.6	104.6	74.2	91.3	103.4	80.1	93.3	101.6	84.2	94.0	102.5	78.8	92.2
08	**OPENINGS**	98.6	40.8	85.2	101.2	40.1	87.0	100.7	58.1	90.8	103.4	66.1	94.7	102.3	81.0	97.3	98.6	63.3	90.4
0920	Plaster & Gypsum Board	106.8	33.2	58.3	100.6	32.9	56.0	102.8	33.4	57.1	107.5	48.3	68.5	98.5	76.1	83.7	98.5	76.1	83.7
0950, 0980	Ceilings & Acoustic Treatment	89.7	33.2	51.4	93.9	32.9	52.6	91.4	33.4	52.1	94.8	48.3	63.3	92.8	76.1	81.5	89.7	76.1	80.5
0960	Flooring	91.6	45.3	78.1	87.3	48.0	75.9	97.5	32.3	78.6	91.3	75.1	86.6	97.0	77.5	91.4	86.0	45.3	74.2
0970, 0990	Wall Finishes & Painting/Coating	87.8	37.5	58.0	87.8	38.5	58.6	95.5	96.2	95.9	87.8	96.2	92.8	99.9	96.2	97.7	87.8	34.3	56.1
09	**FINISHES**	87.8	42.3	63.0	86.6	42.8	62.7	92.3	46.2	67.2	89.0	63.4	75.0	93.6	79.8	86.1	85.0	66.6	74.9
COVERS	**DIVS. 10 - 14, 25, 28, 41, 43, 44, 46**	100.0	83.7	96.4	100.0	83.6	96.4	100.0	83.7	96.4	100.0	85.3	96.8	100.0	88.3	97.4	100.0	87.6	97.3
21, 22, 23	**FIRE SUPPRESSION, PLUMBING & HVAC**	96.6	50.1	77.8	96.6	70.6	86.1	100.0	78.4	91.3	100.1	78.5	91.4	100.0	71.6	88.5	96.6	53.2	79.1
26, 27, 3370	**ELECTRICAL, COMMUNICATIONS & UTIL.**	98.5	64.9	81.5	100.2	40.7	70.0	103.6	48.9	75.8	96.7	48.9	72.4	100.4	65.0	82.4	97.7	65.0	81.0
MF2016	**WEIGHTED AVERAGE**	96.0	62.4	81.6	96.6	64.4	82.8	98.7	69.4	86.2	98.0	73.3	87.4	96.3	79.2	89.0	97.1	70.8	85.8

	DIVISION	TENNESSEE																	
		CHATTANOOGA			COLUMBIA			COOKEVILLE			JACKSON			JOHNSON CITY			KNOXVILLE		
		373 - 374			384			385			383			376			377 - 379		
		MAT.	INST.	TOTAL	MAT.	INST.	TOTAL	MAT.	INST.	TOTAL	MAT.	INST.	TOTAL	MAT.	INST.	TOTAL	MAT.	INST.	TOTAL
015433	**CONTRACTOR EQUIPMENT**		107.6	107.6		102.1	102.1		102.1	102.1		108.3	108.3		101.4	101.4		101.4	101.4
0241, 31 - 34	**SITE & INFRASTRUCTURE, DEMOLITION**	104.2	98.6	100.3	90.1	88.8	89.2	95.8	85.6	88.7	99.3	98.7	98.9	110.3	85.2	92.8	90.8	88.0	88.8
0310	Concrete Forming & Accessories	95.7	58.9	64.3	80.8	64.0	66.5	81.0	32.5	39.7	87.6	41.9	48.7	82.1	59.0	62.4	94.1	63.4	67.9
0320	Concrete Reinforcing	95.2	67.0	80.9	88.7	66.8	77.6	88.7	66.4	77.4	88.7	67.3	77.8	95.9	63.6	79.5	95.2	63.6	79.1
0330	Cast-in-Place Concrete	98.5	63.3	85.4	91.6	62.0	80.6	103.9	40.5	80.3	101.4	67.8	88.9	79.2	58.7	71.6	92.5	64.8	82.2
03	**CONCRETE**	94.2	63.6	80.7	90.1	65.5	79.2	100.0	43.7	75.0	91.0	57.6	76.2	104.7	61.4	85.5	91.8	65.5	80.1
04	**MASONRY**	100.1	57.5	73.8	112.7	56.7	78.1	107.9	40.3	66.1	113.1	42.5	69.5	112.9	44.1	70.4	77.9	51.7	61.7
05	**METALS**	95.3	88.8	93.4	94.0	89.5	92.6	94.1	87.9	92.2	96.4	89.2	94.2	92.6	86.9	90.9	95.8	87.1	93.2
06	**WOOD, PLASTICS & COMPOSITES**	106.5	57.9	80.1	69.1	65.2	67.0	69.3	30.3	48.1	84.4	40.8	60.7	78.8	64.4	71.0	93.3	64.4	77.6
07	**THERMAL & MOISTURE PROTECTION**	95.7	62.1	81.1	94.3	62.9	80.6	94.8	50.4	75.5	96.7	55.3	78.7	91.9	54.6	75.6	89.5	61.2	77.2
08	**OPENINGS**	104.1	59.2	93.6	93.1	54.9	84.2	93.1	36.5	79.9	99.8	45.2	87.2	100.3	61.3	91.3	97.6	56.8	88.1
0920	Plaster & Gypsum Board	83.1	57.2	66.1	91.2	64.7	73.8	91.2	28.8	50.1	93.6	39.6	58.1	101.7	63.9	76.8	109.5	63.9	79.5
0950, 0980	Ceilings & Acoustic Treatment	107.2	57.2	73.3	80.2	64.7	69.7	80.2	28.8	45.4	89.6	39.6	55.8	102.6	63.9	76.4	103.4	63.9	76.7
0960	Flooring	98.0	57.1	86.1	82.0	55.8	74.4	82.0	51.7	73.2	81.3	56.1	74.0	93.3	45.4	79.4	98.3	51.3	84.7
0970, 0990	Wall Finishes & Painting/Coating	97.7	61.1	76.0	84.9	57.9	69.0	84.9	57.9	69.0	86.6	57.9	69.6	94.4	60.3	74.2	94.4	59.1	73.5
09	**FINISHES**	96.6	58.2	75.6	85.9	61.9	72.8	86.4	37.2	59.6	86.2	44.6	63.5	100.5	56.9	76.7	93.5	60.7	75.6
COVERS	**DIVS. 10 - 14, 25, 28, 41, 43, 44, 46**	100.0	69.2	93.2	100.0	68.7	93.1	100.0	36.6	86.0	100.0	62.5	91.7	100.0	77.3	95.0	100.0	81.0	95.8
21, 22, 23	**FIRE SUPPRESSION, PLUMBING & HVAC**	100.1	60.6	84.2	98.0	75.3	88.8	98.0	68.5	86.1	100.1	60.0	83.9	99.9	56.8	82.5	99.9	61.8	84.5
26, 27, 3370	**ELECTRICAL, COMMUNICATIONS & UTIL.**	101.2	84.9	92.9	93.0	51.8	72.0	94.5	62.8	78.4	99.4	52.8	75.7	91.8	41.8	66.4	97.4	56.3	76.5
MF2016	**WEIGHTED AVERAGE**	98.9	69.6	86.3	94.7	67.9	83.2	96.1	57.3	79.5	97.5	59.7	81.3	99.1	59.9	82.3	95.6	65.2	82.6

	DIVISION	TENNESSEE									TEXAS								
		MCKENZIE			MEMPHIS			NASHVILLE			ABILENE			AMARILLO			AUSTIN		
		382			375, 380 - 381			370 - 372			795 - 796			790 - 791			786 - 787		
		MAT.	INST.	TOTAL	MAT.	INST.	TOTAL	MAT.	INST.	TOTAL	MAT.	INST.	TOTAL	MAT.	INST.	TOTAL	MAT.	INST.	TOTAL
015433	**CONTRACTOR EQUIPMENT**		102.1	102.1		101.5	101.5		105.2	105.2		93.2	93.2		93.2	93.2		92.7	92.7
0241, 31 - 34	**SITE & INFRASTRUCTURE, DEMOLITION**	95.4	85.8	88.7	87.8	94.9	92.8	101.9	97.5	98.8	94.1	91.8	92.5	93.1	91.1	91.7	96.9	90.8	92.7
0310	Concrete Forming & Accessories	88.6	34.8	42.8	94.5	66.3	70.5	95.6	64.9	69.4	95.2	60.8	65.9	97.2	53.0	59.6	95.1	54.4	60.5
0320	Concrete Reinforcing	88.8	67.4	77.9	102.8	67.9	85.1	101.3	67.0	83.9	95.6	52.7	73.8	106.3	51.0	78.2	90.0	48.5	68.9
0330	Cast-in-Place Concrete	101.6	53.2	83.6	95.5	78.0	89.0	91.1	65.6	81.6	84.8	66.8	78.1	81.6	66.7	76.1	94.6	66.9	84.3
03	**CONCRETE**	98.6	49.3	76.7	96.7	71.8	85.6	94.9	66.8	82.5	79.5	62.3	71.9	81.3	58.4	71.1	87.3	58.7	74.6
04	**MASONRY**	111.8	46.1	71.2	95.5	57.5	72.0	87.7	57.0	68.8	94.3	61.9	74.3	93.1	61.4	73.5	92.9	61.9	73.8
05	**METALS**	94.1	88.5	92.4	89.9	84.1	88.2	103.5	85.8	98.1	102.7	70.7	93.0	94.3	69.8	86.9	105.3	68.8	94.2
06	**WOOD, PLASTICS & COMPOSITES**	78.2	32.6	53.4	97.0	67.1	80.7	97.6	65.6	80.2	102.6	62.5	80.8	100.8	52.3	74.5	93.2	54.0	71.9
07	**THERMAL & MOISTURE PROTECTION**	94.8	52.6	76.4	88.6	68.6	79.9	94.4	63.9	81.1	99.3	65.2	84.5	97.9	63.4	82.9	93.7	65.0	81.2
08	**OPENINGS**	93.1	38.2	80.3	100.5	63.0	91.8	98.9	64.4	90.9	102.9	58.2	92.5	103.7	52.2	91.8	104.0	50.9	91.7
0920	Plaster & Gypsum Board	94.5	31.2	52.8	94.1	66.4	75.8	93.8	64.7	74.7	88.6	61.9	71.0	95.9	51.4	66.6	90.4	53.1	65.8
0950, 0980	Ceilings & Acoustic Treatment	80.2	31.2	47.0	98.9	66.4	76.9	98.7	64.7	75.7	90.4	61.9	71.1	97.5	51.4	66.3	89.4	53.1	64.8
0960	Flooring	84.9	55.8	76.4	97.2	57.1	85.6	97.4	57.1	85.7	92.5	69.9	85.9	96.5	65.9	87.6	92.3	65.9	84.6
0970, 0990	Wall Finishes & Painting/Coating	84.9	45.3	61.5	94.0	60.9	74.4	99.5	69.1	81.5	94.4	53.0	69.9	89.4	53.0	67.9	94.7	45.4	65.5
09	**FINISHES**	87.6	38.6	60.8	98.1	63.6	79.3	98.3	63.5	79.3	82.7	61.8	71.3	89.7	55.0	70.8	87.8	55.1	70.0
COVERS	**DIVS. 10 - 14, 25, 28, 41, 43, 44, 46**	100.0	24.5	83.4	100.0	81.9	96.0	100.0	81.7	96.0	100.0	80.5	95.7	100.0	73.8	94.2	100.0	77.8	95.1
21, 22, 23	**FIRE SUPPRESSION, PLUMBING & HVAC**	98.0	59.9	82.5	100.0	71.9	88.6	100.0	75.5	90.1	100.3	53.3	81.3	100.0	52.2	80.7	100.0	60.0	83.8
26, 27, 3370	**ELECTRICAL, COMMUNICATIONS & UTIL.**	94.3	56.7	75.2	103.2	65.1	83.9	94.1	62.8	78.2	96.1	55.3	75.4	99.8	60.3	79.7	93.7	58.7	75.9
MF2016	**WEIGHTED AVERAGE**	96.2	56.0	79.0	97.3	71.1	86.1	98.4	71.1	86.7	96.0	62.8	81.8	95.7	61.1	80.9	97.4	62.7	82.5

DIVISION		TEXAS																	
		BEAUMONT			BROWNWOOD			BRYAN			CHILDRESS			CORPUS CHRISTI			DALLAS		
		776 - 777			768			778			792			783 - 784			752 - 753		
		MAT.	INST.	TOTAL	MAT.	INST.	TOTAL	MAT.	INST.	TOTAL	MAT.	INST.	TOTAL	MAT.	INST.	TOTAL	MAT.	INST.	TOTAL
015433	CONTRACTOR EQUIPMENT		97.3	97.3		93.2	93.2		97.3	97.3		93.2	93.2		103.7	103.7		106.9	106.9
0241, 31 - 34	SITE & INFRASTRUCTURE, DEMOLITION	87.9	96.0	93.6	102.7	91.7	95.0	79.1	96.2	91.0	104.7	90.1	94.5	141.6	87.3	103.7	106.8	96.8	99.9
0310	Concrete Forming & Accessories	100.2	56.1	62.7	99.8	60.4	66.2	80.2	59.8	62.8	93.7	60.2	65.2	99.2	53.6	60.4	96.5	62.9	67.9
0320	Concrete Reinforcing	90.3	63.8	76.8	86.5	52.6	69.3	92.5	52.6	72.2	95.8	52.5	73.8	82.4	52.4	67.1	92.0	53.8	72.6
0330	Cast-in-Place Concrete	97.8	67.0	86.3	93.0	60.3	80.8	78.3	61.2	71.9	87.0	60.2	77.0	119.7	69.1	100.9	93.0	72.5	85.3
03	CONCRETE	97.4	62.2	81.8	87.9	59.8	75.4	80.9	60.0	71.6	86.9	59.7	74.9	97.9	60.8	81.5	96.4	66.2	83.0
04	MASONRY	94.3	63.0	75.0	124.4	61.9	85.8	123.8	61.9	85.6	98.3	61.4	75.5	84.5	62.0	70.6	98.2	60.7	75.0
05	METALS	103.7	76.5	95.4	103.6	70.2	93.5	103.8	72.3	94.3	100.2	70.1	91.1	101.6	85.3	96.7	103.1	84.0	97.3
06	WOOD, PLASTICS & COMPOSITES	111.1	55.6	81.0	109.6	62.5	84.0	76.8	61.0	68.2	101.9	62.5	80.5	118.0	53.0	82.7	98.8	64.7	80.2
07	THERMAL & MOISTURE PROTECTION	93.7	66.1	81.7	93.0	63.4	80.1	86.2	65.6	77.2	100.0	61.9	83.4	100.2	63.9	84.4	89.1	67.1	79.5
08	OPENINGS	97.1	56.6	87.7	102.4	58.2	92.1	97.9	57.6	88.5	98.2	58.2	88.9	103.7	51.5	91.6	97.5	60.4	88.9
0920	Plaster & Gypsum Board	104.4	54.8	71.7	86.6	61.9	70.3	91.4	60.4	71.0	88.2	61.9	70.9	99.4	51.9	68.1	94.2	63.7	74.1
0950, 0980	Ceilings & Acoustic Treatment	99.7	54.8	69.3	79.3	61.9	67.5	92.7	60.4	70.8	88.7	61.9	70.5	92.4	51.9	65.0	89.6	63.7	72.1
0960	Flooring	116.6	77.2	105.2	83.6	56.4	75.7	87.4	69.2	82.1	90.9	56.4	80.9	104.8	70.0	94.7	98.1	70.4	90.0
0970, 0990	Wall Finishes & Painting/Coating	92.7	55.3	70.6	91.7	53.0	68.8	90.0	58.9	71.6	94.4	53.0	69.9	112.2	45.4	72.7	101.4	55.2	74.0
09	FINISHES	94.3	59.4	75.3	78.1	59.1	67.7	82.4	61.1	70.7	83.1	59.1	70.0	97.7	55.4	74.6	95.8	63.5	78.2
COVERS	DIVS. 10 - 14, 25, 28, 41, 43, 44, 46	100.0	79.3	95.4	100.0	74.9	94.5	100.0	78.5	95.3	100.0	74.9	94.5	100.0	79.4	95.5	100.0	81.5	95.9
21, 22, 23	FIRE SUPPRESSION, PLUMBING & HVAC	100.1	64.4	85.7	96.6	50.6	78.0	96.7	63.8	83.4	96.8	52.9	79.0	100.1	49.8	79.8	100.0	60.7	84.1
26, 27, 3370	ELECTRICAL, COMMUNICATIONS & UTIL.	98.9	64.9	81.6	95.3	47.3	70.9	96.9	64.9	80.6	96.1	57.1	76.3	90.5	65.0	77.5	98.5	61.3	79.6
MF2016	WEIGHTED AVERAGE	98.8	67.1	85.2	97.4	60.2	81.4	95.5	66.4	83.1	95.8	61.8	81.2	99.7	63.0	84.0	99.1	67.7	85.6

DIVISION		TEXAS																	
		DEL RIO			DENTON			EASTLAND			EL PASO			FORT WORTH			GALVESTON		
		788			762			764			798 - 799, 885			760 - 761			775		
		MAT.	INST.	TOTAL	MAT.	INST.	TOTAL	MAT.	INST.	TOTAL	MAT.	INST.	TOTAL	MAT.	INST.	TOTAL	MAT.	INST.	TOTAL
015433	CONTRACTOR EQUIPMENT		92.7	92.7		102.8	102.8		93.2	93.2		93.2	93.2		93.2	93.2		110.3	110.3
0241, 31 - 34	SITE & INFRASTRUCTURE, DEMOLITION	121.8	90.1	99.7	102.1	83.7	89.3	105.6	89.8	94.6	92.7	92.5	92.6	98.4	92.5	94.3	103.9	94.8	97.6
0310	Concrete Forming & Accessories	95.7	52.5	58.9	107.9	60.5	67.5	100.6	60.2	66.2	95.3	58.2	63.7	99.5	61.0	66.7	89.3	57.3	62.1
0320	Concrete Reinforcing	82.9	48.9	65.6	87.9	52.5	69.9	86.7	52.5	69.4	98.3	52.5	75.1	93.1	52.3	72.4	92.0	61.4	76.5
0330	Cast-in-Place Concrete	128.9	60.2	103.3	71.9	61.4	68.0	98.3	60.2	84.1	75.9	69.0	73.3	85.7	66.9	78.7	104.0	62.2	88.4
03	CONCRETE	118.6	55.5	90.7	69.0	61.3	65.6	92.2	59.7	77.8	77.5	61.8	70.6	83.8	62.3	74.3	98.7	61.8	82.4
04	MASONRY	99.5	61.8	76.2	133.1	60.6	88.3	94.2	61.9	74.2	85.5	64.0	72.2	93.8	60.5	73.2	89.5	62.0	72.5
05	METALS	101.6	68.0	91.4	103.2	85.5	97.8	103.4	70.0	93.3	101.3	70.5	92.0	105.3	70.7	94.8	105.6	91.8	101.4
06	WOOD, PLASTICS & COMPOSITES	98.2	51.9	73.0	122.3	62.7	89.9	117.0	62.5	87.4	90.5	58.6	73.2	101.2	62.5	80.2	93.5	57.5	73.9
07	THERMAL & MOISTURE PROTECTION	97.0	63.8	82.5	90.9	63.9	79.2	93.4	63.4	80.3	95.1	65.6	82.2	90.0	65.5	79.4	85.5	66.2	77.1
08	OPENINGS	98.9	49.2	87.3	120.9	57.9	106.2	71.7	58.2	68.6	97.8	54.3	87.7	102.0	58.1	91.8	101.9	57.7	91.6
0920	Plaster & Gypsum Board	96.0	50.9	66.3	91.3	61.9	71.9	86.6	61.9	70.3	97.2	57.8	71.2	93.6	61.9	72.7	98.4	56.6	70.9
0950, 0980	Ceilings & Acoustic Treatment	88.8	50.9	63.2	84.4	61.9	69.1	79.3	61.9	67.5	91.1	57.8	68.5	89.3	61.9	70.7	97.0	56.6	69.6
0960	Flooring	89.9	56.4	80.2	79.2	56.4	72.6	106.4	56.4	91.9	93.1	72.1	87.0	104.6	66.6	93.6	102.7	69.2	93.0
0970, 0990	Wall Finishes & Painting/Coating	99.5	43.7	66.5	101.9	50.2	71.3	93.1	53.0	69.4	99.7	50.9	70.8	97.2	53.2	71.1	101.6	55.1	74.1
09	FINISHES	90.3	51.8	69.3	77.0	58.9	67.1	85.1	59.1	70.9	88.4	59.7	72.7	90.7	61.2	74.6	91.9	58.6	73.7
COVERS	DIVS. 10 - 14, 25, 28, 41, 43, 44, 46	100.0	75.2	94.5	100.0	80.8	95.8	100.0	73.2	94.1	100.0	77.9	95.1	100.0	80.5	95.7	100.0	79.8	95.6
21, 22, 23	FIRE SUPPRESSION, PLUMBING & HVAC	96.6	60.7	82.1	96.6	57.1	80.6	96.6	50.6	78.0	99.9	65.9	86.2	99.9	56.6	82.4	96.6	63.9	83.4
26, 27, 3370	ELECTRICAL, COMMUNICATIONS & UTIL.	92.2	61.3	76.5	97.5	60.1	78.5	95.2	60.0	77.3	94.8	51.7	72.9	97.0	60.0	78.2	98.6	65.0	81.5
MF2016	WEIGHTED AVERAGE	100.6	62.0	84.0	97.5	64.4	83.3	94.0	61.8	80.2	94.6	64.7	81.8	97.4	64.1	83.1	98.5	68.1	85.5

DIVISION		TEXAS																	
		GIDDINGS			GREENVILLE			HOUSTON			HUNTSVILLE			LAREDO			LONGVIEW		
		789			754			770 - 772			773			780			756		
		MAT.	INST.	TOTAL	MAT.	INST.	TOTAL	MAT.	INST.	TOTAL	MAT.	INST.	TOTAL	MAT.	INST.	TOTAL	MAT.	INST.	TOTAL
015433	CONTRACTOR EQUIPMENT		92.7	92.7		103.6	103.6		102.2	102.2		97.3	97.3		92.7	92.7		94.6	94.6
0241, 31 - 34	SITE & INFRASTRUCTURE, DEMOLITION	106.7	90.4	95.3	97.7	86.8	90.1	102.8	94.6	97.0	94.4	95.9	95.4	100.0	90.9	93.6	96.1	93.3	94.2
0310	Concrete Forming & Accessories	93.2	52.6	58.6	88.0	58.7	63.1	94.5	58.6	64.0	87.2	55.6	60.3	95.8	53.6	59.9	83.9	60.4	63.9
0320	Concrete Reinforcing	83.4	49.8	66.4	92.4	52.6	72.1	91.8	52.8	72.0	92.7	52.6	72.3	82.9	52.4	67.4	91.5	52.2	71.5
0330	Cast-in-Place Concrete	109.1	60.5	91.0	92.5	61.4	80.9	96.5	63.3	84.2	107.7	61.1	90.4	92.1	67.0	82.7	107.6	60.3	90.0
03	CONCRETE	95.0	55.9	77.6	89.3	60.4	76.5	97.9	60.6	81.3	105.8	58.0	84.6	89.2	59.0	75.8	106.3	59.7	85.6
04	MASONRY	106.7	61.9	79.1	154.1	60.5	96.3	88.7	62.9	72.8	124.6	61.9	85.9	92.5	61.9	73.6	150.1	60.5	94.7
05	METALS	101.1	68.9	91.3	100.6	84.1	95.6	108.4	77.8	99.1	103.7	72.1	94.1	104.1	70.1	93.8	93.2	69.1	85.9
06	WOOD, PLASTICS & COMPOSITES	97.4	51.9	72.7	87.7	60.2	72.7	98.9	59.0	77.2	86.1	55.6	69.5	98.2	52.9	73.6	81.6	62.6	71.3
07	THERMAL & MOISTURE PROTECTION	97.4	64.5	83.1	89.1	63.0	77.7	87.2	67.3	78.6	87.3	65.0	77.6	96.2	64.8	82.6	90.6	62.5	78.3
08	OPENINGS	98.0	50.1	86.9	94.1	56.9	85.4	105.1	56.5	93.8	97.8	54.0	87.7	99.0	51.2	87.9	85.1	57.8	78.7
0920	Plaster & Gypsum Board	95.1	50.9	66.0	84.9	59.2	68.0	97.7	58.0	71.5	96.0	54.8	68.9	96.7	51.9	67.2	84.0	61.9	69.4
0950, 0980	Ceilings & Acoustic Treatment	88.8	50.9	63.2	85.0	59.2	67.5	93.9	58.0	69.6	92.7	54.8	67.0	92.2	51.9	64.9	80.8	61.9	68.0
0960	Flooring	89.5	56.4	79.9	91.8	56.4	81.5	103.8	72.1	94.6	91.7	56.4	81.5	89.8	65.9	82.8	96.9	56.4	85.1
0970, 0990	Wall Finishes & Painting/Coating	99.5	45.4	67.5	95.8	53.0	70.5	99.4	58.9	75.5	90.0	58.9	71.6	99.5	45.4	67.5	86.0	50.2	64.8
09	FINISHES	89.0	52.0	68.8	90.3	57.7	72.5	97.9	60.9	77.7	85.1	55.6	69.0	89.4	54.4	70.3	95.5	58.8	75.5
COVERS	DIVS. 10 - 14, 25, 28, 41, 43, 44, 46	100.0	71.8	93.8	100.0	78.1	95.2	100.0	81.9	96.0	100.0	71.2	93.7	100.0	77.4	95.0	100.0	80.7	95.7
21, 22, 23	FIRE SUPPRESSION, PLUMBING & HVAC	96.7	62.8	83.0	96.7	59.3	81.6	100.1	65.5	86.1	96.7	63.8	83.4	100.1	60.2	84.0	96.6	58.8	81.3
26, 27, 3370	ELECTRICAL, COMMUNICATIONS & UTIL.	89.6	57.0	73.0	95.2	60.1	77.4	100.7	68.1	84.1	96.9	60.6	78.4	92.3	58.5	75.1	95.6	52.1	73.5
MF2016	WEIGHTED AVERAGE	97.0	61.9	82.0	98.2	64.5	83.8	100.7	67.9	86.6	99.4	64.3	84.4	97.1	62.7	82.4	98.4	62.6	83.0

DIVISION		TEXAS																	
		LUBBOCK			LUFKIN			MCALLEN			MCKINNEY			MIDLAND			ODESSA		
		793 - 794			759			785			750			797			797		
		MAT.	INST.	TOTAL	MAT.	INST.	TOTAL	MAT.	INST.	TOTAL	MAT.	INST.	TOTAL	MAT.	INST.	TOTAL	MAT.	INST.	TOTAL
015433	**CONTRACTOR EQUIPMENT**		105.6	105.6		94.6	94.6		103.8	103.8		103.6	103.6		105.6	105.6		93.2	93.2
0241, 31 - 34	**SITE & INFRASTRUCTURE, DEMOLITION**	118.6	91.0	99.4	91.0	94.8	93.7	146.4	87.3	105.2	94.1	86.8	89.0	121.6	90.0	99.6	94.3	92.5	93.0
0310	Concrete Forming & Accessories	94.4	54.5	60.4	87.1	55.7	60.3	99.7	52.6	59.6	87.2	58.7	63.0	98.2	60.7	66.3	95.1	60.8	65.9
0320	Concrete Reinforcing	96.9	52.8	74.4	93.1	68.7	80.7	82.5	52.3	67.2	92.4	52.6	72.2	97.9	52.7	74.9	95.6	52.6	73.7
0330	Cast-in-Place Concrete	85.0	70.2	79.5	96.2	60.1	82.8	130.0	61.6	104.6	86.7	61.4	77.3	90.3	67.8	81.9	84.8	66.8	78.1
03	**CONCRETE**	78.4	61.6	71.0	98.3	60.2	81.4	105.8	57.7	84.5	84.4	60.4	73.8	82.3	63.6	74.0	79.5	62.3	71.9
04	**MASONRY**	93.7	62.3	74.3	115.0	61.8	82.1	99.2	62.0	76.2	165.6	60.5	100.7	110.1	61.5	80.1	94.3	61.4	74.0
05	**METALS**	106.4	86.3	100.3	100.8	74.0	92.7	101.4	85.1	96.5	100.6	84.2	95.6	104.7	85.8	99.0	102.1	70.6	92.5
06	**WOOD, PLASTICS & COMPOSITES**	102.2	53.6	75.8	89.0	55.7	70.9	116.6	52.0	81.5	86.6	60.2	72.2	106.9	62.7	82.9	102.6	62.5	80.8
07	**THERMAL & MOISTURE PROTECTION**	88.6	65.3	78.5	90.3	61.6	77.8	100.1	64.7	84.7	88.9	63.0	77.6	88.9	65.1	78.5	99.3	64.2	84.1
08	**OPENINGS**	110.9	53.8	97.6	64.5	58.2	63.1	102.7	50.0	90.5	94.1	56.9	85.4	112.4	58.3	99.8	102.9	58.2	92.5
0920	Plaster & Gypsum Board	88.8	52.6	64.9	83.1	54.8	64.5	100.4	50.9	67.8	84.9	59.2	68.0	90.6	61.9	71.7	88.6	61.9	71.0
0950, 0980	Ceilings & Acoustic Treatment	91.3	52.6	65.0	76.5	54.8	61.8	92.2	50.9	64.3	85.0	59.2	67.5	89.6	61.9	70.8	90.4	61.9	71.1
0960	Flooring	86.4	71.3	82.0	129.5	56.4	108.3	104.4	75.2	95.9	91.4	56.4	81.2	87.6	65.9	81.3	92.5	65.9	84.8
0970, 0990	Wall Finishes & Painting/Coating	104.9	55.2	75.5	86.0	53.0	66.5	112.2	43.7	71.7	95.8	53.0	70.5	104.9	53.0	74.2	94.4	53.0	69.9
09	**FINISHES**	85.1	57.1	69.8	103.7	55.5	77.4	98.1	55.7	74.9	89.9	57.7	72.3	85.8	61.1	72.3	82.7	61.0	70.9
COVERS	**DIVS. 10 - 14, 25, 28, 41, 43, 44, 46**	100.0	79.9	95.6	100.0	71.5	93.7	100.0	77.5	95.0	100.0	80.8	95.8	100.0	75.3	94.6	100.0	75.0	94.5
21, 22, 23	**FIRE SUPPRESSION, PLUMBING & HVAC**	99.7	54.3	81.3	96.6	59.4	81.6	96.7	49.8	77.7	96.7	61.6	82.5	96.3	48.5	76.9	100.3	53.7	81.4
26, 27, 3370	**ELECTRICAL, COMMUNICATIONS & UTIL.**	94.9	61.6	77.9	96.9	62.9	79.6	90.3	32.4	60.8	95.3	60.1	77.4	94.9	60.4	77.3	96.2	60.3	78.0
MF2016	**WEIGHTED AVERAGE**	97.5	64.4	83.3	95.5	64.2	82.1	100.6	57.9	82.3	98.0	65.1	83.9	98.1	63.7	83.3	96.0	63.3	82.0

DIVISION		TEXAS																	
		PALESTINE			SAN ANGELO			SAN ANTONIO			TEMPLE			TEXARKANA			TYLER		
		758			769			781 - 782			765			755			757		
		MAT.	INST.	TOTAL	MAT.	INST.	TOTAL	MAT.	INST.	TOTAL	MAT.	INST.	TOTAL	MAT.	INST.	TOTAL	MAT.	INST.	TOTAL
015433	**CONTRACTOR EQUIPMENT**		94.6	94.6		93.2	93.2		93.5	93.5		93.2	93.2		94.6	94.6		94.6	94.6
0241, 31 - 34	**SITE & INFRASTRUCTURE, DEMOLITION**	96.5	93.6	94.5	98.9	92.1	94.2	98.8	93.5	95.1	87.3	92.0	90.6	85.5	96.0	92.9	95.4	93.4	94.0
0310	Concrete Forming & Accessories	78.2	60.4	63.1	100.2	52.9	59.9	96.4	53.9	60.2	103.8	52.6	60.2	94.4	61.0	66.0	88.8	60.5	64.8
0320	Concrete Reinforcing	90.8	52.5	71.3	86.4	52.6	69.2	92.1	49.1	70.2	86.6	52.3	69.2	90.6	52.3	71.2	91.5	52.2	71.5
0330	Cast-in-Place Concrete	87.9	60.3	77.7	87.8	66.8	80.0	93.7	69.4	84.7	71.9	60.3	67.6	88.6	66.9	80.5	105.7	60.3	88.8
03	**CONCRETE**	100.7	59.8	82.6	83.9	58.7	72.7	93.9	59.3	78.6	71.6	56.2	64.8	90.8	62.3	78.2	105.8	59.8	85.4
04	**MASONRY**	109.4	60.5	79.2	121.0	61.9	84.5	95.5	62.0	74.8	132.2	61.9	88.8	169.9	60.5	102.3	159.7	60.5	98.4
05	**METALS**	100.5	69.2	91.1	103.8	70.6	93.7	106.5	66.7	94.4	103.5	69.7	93.2	93.1	69.7	86.0	100.4	69.2	91.0
06	**WOOD, PLASTICS & COMPOSITES**	79.5	62.6	70.3	109.9	51.9	78.4	99.5	53.0	74.3	119.9	51.9	83.0	94.1	62.6	77.0	90.8	62.6	75.5
07	**THERMAL & MOISTURE PROTECTION**	90.9	63.5	79.0	92.8	64.4	80.4	92.3	66.7	81.2	92.3	63.4	79.7	90.0	64.8	79.0	90.7	63.5	78.9
08	**OPENINGS**	64.5	58.3	63.0	102.4	52.3	90.8	101.7	50.8	89.9	68.3	52.3	64.5	85.0	58.2	78.8	64.4	58.2	63.0
0920	Plaster & Gypsum Board	81.3	61.9	68.5	86.6	50.9	63.1	98.5	51.9	67.9	86.6	50.9	63.1	88.6	61.9	71.0	83.1	61.9	69.1
0950, 0980	Ceilings & Acoustic Treatment	76.5	61.9	66.6	79.3	50.9	60.1	91.3	51.9	64.6	79.3	50.9	60.1	80.8	61.9	68.0	76.5	61.9	66.6
0960	Flooring	120.4	56.4	101.8	83.7	56.4	75.8	98.5	70.0	90.2	108.0	56.4	93.0	105.3	65.9	93.8	131.8	56.4	110.0
0970, 0990	Wall Finishes & Painting/Coating	86.0	53.0	66.5	91.7	53.0	68.8	99.4	45.4	67.5	93.1	45.4	64.9	86.0	53.0	66.5	86.0	53.0	66.5
09	**FINISHES**	101.4	59.2	78.3	77.8	52.8	64.2	99.0	55.4	75.2	84.3	52.0	66.6	98.0	61.1	77.8	104.7	59.2	79.8
COVERS	**DIVS. 10 - 14, 25, 28, 41, 43, 44, 46**	100.0	75.1	94.5	100.0	77.6	95.1	100.0	79.3	95.4	100.0	73.8	94.2	100.0	80.7	95.8	100.0	80.7	95.7
21, 22, 23	**FIRE SUPPRESSION, PLUMBING & HVAC**	96.6	58.0	81.0	96.6	53.4	79.1	100.1	61.1	84.3	96.6	54.4	79.5	96.6	56.8	80.5	96.6	59.5	81.6
26, 27, 3370	**ELECTRICAL, COMMUNICATIONS & UTIL.**	93.0	49.6	70.9	99.7	55.9	77.4	93.7	61.2	77.2	96.5	53.3	74.5	96.8	55.9	76.0	95.6	55.0	75.0
MF2016	**WEIGHTED AVERAGE**	95.1	62.0	80.9	97.1	60.8	81.6	99.3	63.5	83.9	92.5	60.0	78.5	97.5	63.6	83.0	98.6	63.2	83.5

DIVISION		TEXAS															UTAH		
		VICTORIA			WACO			WAXAHACKIE			WHARTON			WICHITA FALLS			LOGAN		
		779			766 - 767			751			774			763			843		
		MAT.	INST.	TOTAL	MAT.	INST.	TOTAL	MAT.	INST.	TOTAL	MAT.	INST.	TOTAL	MAT.	INST.	TOTAL	MAT.	INST.	TOTAL
015433	**CONTRACTOR EQUIPMENT**		109.0	109.0		93.2	93.2		103.6	103.6		110.3	110.3		93.2	93.2		96.9	96.9
0241, 31 - 34	**SITE & INFRASTRUCTURE, DEMOLITION**	108.9	92.0	97.1	95.8	92.4	93.4	95.9	87.0	89.7	113.9	94.2	100.2	96.6	92.5	93.7	99.7	94.1	95.8
0310	Concrete Forming & Accessories	88.5	52.7	58.1	102.2	60.9	67.0	87.2	60.7	64.6	84.0	54.5	58.9	102.2	60.9	67.0	101.1	67.5	72.5
0320	Concrete Reinforcing	88.4	52.4	70.1	86.3	48.6	67.1	92.4	52.6	72.2	91.9	52.4	71.8	86.3	51.1	68.4	105.6	86.4	95.8
0330	Cast-in-Place Concrete	116.8	62.1	96.5	77.8	66.8	73.7	91.5	61.6	80.4	120.0	61.1	98.1	83.1	66.8	77.1	86.4	75.2	82.2
03	**CONCRETE**	107.0	58.2	85.3	77.6	61.6	70.5	88.2	61.4	76.3	111.4	58.6	88.0	80.1	62.1	72.1	100.7	74.0	88.8
04	**MASONRY**	104.0	62.0	78.0	92.5	61.9	73.6	154.7	60.6	96.5	90.9	61.9	73.0	92.9	61.4	73.5	108.1	67.8	83.2
05	**METALS**	103.8	88.2	99.1	105.8	68.8	94.6	100.6	84.4	95.7	105.6	87.8	100.2	105.8	70.7	95.2	104.3	83.6	98.0
06	**WOOD, PLASTICS & COMPOSITES**	96.5	52.0	72.3	118.0	62.5	87.8	86.6	62.8	73.6	86.6	54.2	69.0	118.0	62.5	87.8	87.0	67.8	76.6
07	**THERMAL & MOISTURE PROTECTION**	88.9	62.8	77.6	93.2	65.5	81.2	89.0	63.2	77.8	85.8	65.5	77.0	93.2	64.5	80.7	99.7	71.7	87.5
08	**OPENINGS**	101.7	52.5	90.3	79.7	57.2	74.5	94.1	58.3	85.8	101.9	53.8	90.7	79.7	58.2	74.7	94.4	66.8	88.0
0920	Plaster & Gypsum Board	94.6	50.9	65.9	87.0	61.9	70.4	85.3	61.9	69.9	93.5	53.3	67.0	87.0	61.9	70.4	81.1	66.9	71.8
0950, 0980	Ceilings & Acoustic Treatment	97.8	50.9	66.1	81.0	61.9	68.0	86.7	61.9	69.9	97.0	53.3	67.4	81.0	61.9	68.0	104.7	66.9	79.1
0960	Flooring	101.4	56.4	88.3	107.3	65.9	95.3	91.4	56.4	81.2	100.0	68.8	90.9	108.0	74.3	98.3	97.5	59.7	86.5
0970, 0990	Wall Finishes & Painting/Coating	101.9	58.9	76.5	93.1	53.0	69.4	95.8	53.0	70.5	101.6	57.2	75.3	95.1	53.0	70.2	93.4	55.7	71.1
09	**FINISHES**	89.4	53.5	69.8	84.9	61.0	71.9	90.5	59.3	73.4	91.0	56.8	72.3	85.3	62.7	73.0	92.4	64.4	77.1
COVERS	**DIVS. 10 - 14, 25, 28, 41, 43, 44, 46**	100.0	71.1	93.6	100.0	80.5	95.7	100.0	81.0	95.8	100.0	71.3	93.7	100.0	74.9	94.5	100.0	83.7	96.4
21, 22, 23	**FIRE SUPPRESSION, PLUMBING & HVAC**	96.7	63.7	83.3	100.1	60.1	83.9	96.7	59.3	81.6	96.6	63.3	83.2	100.1	53.3	81.2	99.9	69.9	87.8
26, 27, 3370	**ELECTRICAL, COMMUNICATIONS & UTIL.**	103.7	55.3	79.1	99.8	55.4	77.2	95.2	60.1	77.4	102.6	60.6	81.2	101.5	55.3	78.0	95.7	70.5	82.9
MF2016	**WEIGHTED AVERAGE**	100.5	64.4	85.0	94.3	63.9	81.3	98.1	65.1	83.9	100.8	65.8	85.8	94.9	62.7	81.1	99.4	73.1	88.1

DIVISION		UTAH												VERMONT					
		OGDEN			PRICE			PROVO			SALT LAKE CITY			BELLOWS FALLS			BENNINGTON		
		842, 844			845			846 - 847			840 - 841			051			052		
		MAT.	INST.	TOTAL	MAT.	INST.	TOTAL	MAT.	INST.	TOTAL	MAT.	INST.	TOTAL	MAT.	INST.	TOTAL	MAT.	INST.	TOTAL
015433	CONTRACTOR EQUIPMENT		96.9	96.9		95.9	95.9		95.9	95.9		96.8	96.8		97.9	97.9		97.9	97.9
0241, 31 - 34	SITE & INFRASTRUCTURE, DEMOLITION	87.6	94.1	92.1	97.1	92.2	93.7	95.9	92.5	93.5	87.4	94.0	92.0	90.1	100.7	97.5	89.5	100.7	97.3
0310	Concrete Forming & Accessories	101.1	67.5	72.5	103.5	66.0	71.6	102.8	67.5	72.8	103.4	67.5	72.9	96.5	104.2	103.1	93.9	104.1	102.6
0320	Concrete Reinforcing	105.2	86.4	95.7	113.2	86.4	99.6	114.2	86.4	100.1	107.5	86.4	96.8	85.4	84.3	84.9	85.4	84.3	84.8
0330	Cast-in-Place Concrete	87.7	75.2	83.1	86.5	73.8	81.8	86.5	75.2	82.3	95.8	75.2	88.2	91.3	112.8	99.3	91.3	112.7	99.3
03	CONCRETE	90.6	74.0	83.2	101.8	72.8	88.9	100.2	73.9	88.6	109.2	74.0	93.6	88.1	103.4	94.9	87.9	103.3	94.7
04	MASONRY	103.0	67.8	81.3	113.9	66.5	84.6	114.0	67.8	85.5	116.7	67.8	86.5	101.6	88.0	93.2	112.4	88.0	97.3
05	METALS	104.8	83.6	98.4	101.5	83.5	96.0	102.5	83.6	96.8	108.2	83.6	100.8	85.1	89.6	86.4	85.0	89.4	86.3
06	WOOD, PLASTICS & COMPOSITES	87.0	67.8	76.6	90.3	65.9	77.1	88.6	67.8	77.3	89.0	67.8	77.5	106.9	109.8	108.5	103.8	109.8	107.1
07	THERMAL & MOISTURE PROTECTION	98.5	71.7	86.8	101.6	70.2	87.9	101.6	71.7	88.6	106.2	71.7	91.2	98.7	90.7	95.2	98.7	90.7	95.2
08	OPENINGS	94.4	66.8	88.0	98.3	78.5	93.7	98.3	66.8	91.0	96.2	66.8	89.4	102.4	99.3	101.7	102.4	99.3	101.7
0920	Plaster & Gypsum Board	81.1	66.9	71.8	83.9	65.0	71.4	81.7	66.9	72.0	93.1	66.9	75.9	104.9	109.8	108.1	103.4	109.8	107.6
0950, 0980	Ceilings & Acoustic Treatment	104.7	66.9	79.1	104.7	65.0	77.8	104.7	66.9	79.1	96.1	66.9	76.3	97.0	109.8	105.7	97.0	109.8	105.7
0960	Flooring	95.4	59.7	85.0	98.6	58.0	86.8	98.4	59.7	87.1	99.3	59.7	87.8	91.2	102.2	94.4	90.4	102.2	93.8
0970, 0990	Wall Finishes & Painting/Coating	93.4	55.7	71.1	93.4	55.7	71.1	93.4	55.7	71.1	96.5	56.8	73.0	87.1	104.2	97.2	87.1	104.2	97.2
09	FINISHES	90.6	64.4	76.3	93.6	62.8	76.8	93.1	64.4	77.4	92.5	64.5	77.2	86.3	104.8	96.4	85.9	104.8	96.2
COVERS	DIVS. 10 - 14, 25, 28, 41, 43, 44, 46	100.0	83.7	96.4	100.0	83.4	96.3	100.0	83.7	96.4	100.0	83.7	96.4	100.0	93.7	98.6	100.0	93.6	98.6
21, 22, 23	FIRE SUPPRESSION, PLUMBING & HVAC	99.9	69.9	87.8	98.1	64.6	84.5	99.9	69.9	87.8	100.1	69.9	87.8	96.7	89.2	93.7	96.7	89.2	93.7
26, 27, 3370	ELECTRICAL, COMMUNICATIONS & UTIL.	96.0	70.5	83.0	101.0	70.5	85.5	96.3	70.3	83.1	98.6	70.5	84.3	98.9	81.5	90.0	98.9	55.8	77.0
MF2016	WEIGHTED AVERAGE	97.5	73.1	87.0	100.0	71.7	87.9	99.8	72.9	88.3	102.0	73.1	89.6	94.2	93.6	94.0	94.6	90.0	92.6

DIVISION		VERMONT																	
		BRATTLEBORO			BURLINGTON			GUILDHALL			MONTPELIER			RUTLAND			ST. JOHNSBURY		
		053			054			059			056			057			058		
		MAT.	INST.	TOTAL	MAT.	INST.	TOTAL	MAT.	INST.	TOTAL	MAT.	INST.	TOTAL	MAT.	INST.	TOTAL	MAT.	INST.	TOTAL
015433	CONTRACTOR EQUIPMENT		97.9	97.9		97.9	97.9		97.9	97.9		97.9	97.9		97.9	97.9		97.9	97.9
0241, 31 - 34	SITE & INFRASTRUCTURE, DEMOLITION	91.1	100.7	97.8	93.0	100.9	98.5	89.1	96.2	94.0	95.2	100.6	99.0	93.5	100.6	98.5	89.2	99.6	96.5
0310	Concrete Forming & Accessories	96.7	104.2	103.1	98.0	81.9	84.3	94.1	97.5	97.0	97.6	103.6	102.7	97.0	81.9	84.2	92.4	97.9	97.1
0320	Concrete Reinforcing	84.5	84.3	84.4	104.8	84.2	94.3	86.2	84.1	85.1	104.9	84.2	94.4	106.8	84.2	95.3	84.5	84.2	84.3
0330	Cast-in-Place Concrete	94.2	112.8	101.1	110.1	112.5	111.0	88.5	103.8	94.2	107.8	112.6	109.6	89.5	112.5	98.0	88.5	104.0	94.3
03	CONCRETE	90.2	103.4	96.0	100.0	93.2	97.0	85.6	97.1	90.7	99.0	103.0	100.7	90.6	93.1	91.7	85.3	97.4	90.7
04	MASONRY	112.0	88.0	97.2	113.8	88.1	97.9	112.3	73.0	88.0	104.3	88.1	94.3	92.5	88.1	89.8	140.6	73.0	98.8
05	METALS	85.0	89.6	86.4	92.3	88.8	91.2	85.1	88.5	86.1	90.2	88.9	89.8	90.5	88.7	90.0	85.1	88.9	86.2
06	WOOD, PLASTICS & COMPOSITES	107.2	109.8	108.6	100.8	80.6	89.8	102.8	109.8	106.6	96.5	109.8	103.8	107.4	80.6	92.8	97.5	109.8	104.2
07	THERMAL & MOISTURE PROTECTION	98.8	90.7	95.3	105.9	87.6	97.9	98.5	83.8	92.1	105.7	90.8	99.2	98.9	87.6	94.0	98.4	83.8	92.1
08	OPENINGS	102.4	99.6	101.8	104.1	79.6	98.4	102.4	95.7	100.9	106.0	95.7	103.6	105.6	79.6	99.5	102.4	95.7	100.9
0920	Plaster & Gypsum Board	104.9	109.8	108.1	106.4	79.7	88.8	110.7	109.8	110.1	106.7	109.8	108.7	105.3	79.7	88.4	112.2	109.8	110.6
0950, 0980	Ceilings & Acoustic Treatment	97.0	109.8	105.7	101.0	79.7	86.6	97.0	109.8	105.7	106.9	109.8	108.9	101.5	79.7	86.7	97.0	109.8	105.7
0960	Flooring	91.3	102.2	94.5	98.1	102.2	99.3	94.1	102.2	96.5	100.4	102.2	100.9	91.2	102.2	94.4	97.4	102.2	98.8
0970, 0990	Wall Finishes & Painting/Coating	87.1	104.2	97.2	96.4	90.7	93.0	87.1	90.7	89.2	95.3	90.7	92.5	87.1	90.7	89.2	87.1	90.7	89.2
09	FINISHES	86.5	104.8	96.5	92.5	86.1	89.0	87.9	99.5	94.3	94.2	103.4	99.2	87.4	86.1	86.7	89.1	99.5	94.8
COVERS	DIVS. 10 - 14, 25, 28, 41, 43, 44, 46	100.0	93.7	98.6	100.0	90.4	97.9	100.0	88.5	97.5	100.0	93.6	98.6	100.0	90.4	97.9	100.0	88.5	97.5
21, 22, 23	FIRE SUPPRESSION, PLUMBING & HVAC	96.7	89.2	93.7	100.1	69.4	87.7	96.7	61.5	82.5	96.6	69.4	85.6	100.2	69.4	87.8	96.7	61.5	82.5
26, 27, 3370	ELECTRICAL, COMMUNICATIONS & UTIL.	98.9	81.5	90.0	98.7	55.0	76.5	98.9	55.8	77.0	98.1	55.0	76.2	98.9	55.0	76.6	98.9	55.8	77.0
MF2016	WEIGHTED AVERAGE	95.0	93.6	94.4	99.1	80.6	91.2	94.5	80.2	88.3	97.7	85.3	92.4	96.3	80.5	89.5	95.8	80.5	89.3

DIVISION		VERMONT			VIRGINIA														
		WHITE RIVER JCT.			ALEXANDRIA			ARLINGTON			BRISTOL			CHARLOTTESVILLE			CULPEPER		
		050			223			222			242			229			227		
		MAT.	INST.	TOTAL	MAT.	INST.	TOTAL	MAT.	INST.	TOTAL	MAT.	INST.	TOTAL	MAT.	INST.	TOTAL	MAT.	INST.	TOTAL
015433	CONTRACTOR EQUIPMENT		97.9	97.9		107.5	107.5		106.2	106.2		106.2	106.2		110.4	110.4		106.2	106.2
0241, 31 - 34	SITE & INFRASTRUCTURE, DEMOLITION	93.5	99.7	97.8	114.3	92.2	98.9	124.3	89.9	100.4	108.9	89.2	95.2	113.4	91.7	98.3	112.0	89.9	96.6
0310	Concrete Forming & Accessories	91.3	98.4	97.4	90.2	70.2	73.1	89.0	70.1	73.0	85.1	65.6	68.5	83.5	47.6	52.9	80.8	72.9	74.0
0320	Concrete Reinforcing	85.4	84.2	84.8	83.9	87.4	85.7	94.7	87.4	91.0	94.7	69.8	82.0	94.1	70.1	81.9	94.7	87.4	91.0
0330	Cast-in-Place Concrete	94.2	104.8	98.2	110.1	76.5	97.6	107.1	76.5	95.7	106.7	45.4	83.9	111.0	76.9	98.3	109.7	76.4	97.3
03	CONCRETE	91.9	97.9	94.6	99.6	76.8	89.5	103.8	76.8	91.9	100.2	61.0	82.8	100.8	63.8	84.4	97.8	78.0	89.0
04	MASONRY	126.0	74.4	94.1	89.4	72.3	78.9	103.1	72.3	84.1	92.7	46.3	64.1	116.9	55.7	79.1	104.6	72.3	84.7
05	METALS	85.1	88.9	86.2	103.5	100.4	102.6	102.2	100.4	101.6	101.0	92.2	98.3	101.3	94.1	99.1	101.3	100.0	100.9
06	WOOD, PLASTICS & COMPOSITES	100.6	109.8	105.6	91.8	68.4	79.1	88.3	68.4	77.5	80.9	71.8	76.0	79.4	40.8	58.5	78.1	72.1	74.8
07	THERMAL & MOISTURE PROTECTION	98.9	84.5	92.6	102.7	79.8	92.7	104.7	79.8	93.9	104.0	59.5	84.6	103.6	68.0	88.1	103.8	79.7	93.3
08	OPENINGS	102.4	95.7	100.9	97.4	73.6	91.9	95.6	73.6	90.5	98.4	65.7	90.8	96.7	54.6	86.9	97.0	75.9	92.1
0920	Plaster & Gypsum Board	102.2	109.8	107.2	101.7	67.4	79.1	98.3	67.4	77.9	94.8	70.9	79.1	94.8	38.4	57.7	94.9	71.2	79.3
0950, 0980	Ceilings & Acoustic Treatment	97.0	109.8	105.7	93.4	67.4	75.8	91.7	67.4	75.2	90.8	70.9	77.4	90.8	38.4	55.3	91.7	71.2	77.8
0960	Flooring	89.3	102.2	93.0	97.0	76.9	91.2	95.5	76.9	90.1	92.1	57.6	82.1	90.6	57.6	81.0	90.6	76.9	86.7
0970, 0990	Wall Finishes & Painting/Coating	87.1	90.7	89.2	115.2	73.6	90.6	115.2	75.1	91.4	101.5	55.6	74.4	101.5	58.9	76.3	115.2	75.1	91.4
09	FINISHES	85.7	99.9	93.5	93.1	70.9	81.0	93.1	71.1	81.1	89.8	64.0	75.7	89.4	48.6	67.1	90.1	73.3	80.9
COVERS	DIVS. 10 - 14, 25, 28, 41, 43, 44, 46	100.0	89.0	97.6	100.0	87.3	97.2	100.0	84.9	96.7	100.0	76.0	94.7	100.0	80.4	95.7	100.0	85.3	96.8
21, 22, 23	FIRE SUPPRESSION, PLUMBING & HVAC	96.7	62.2	82.8	100.4	87.2	95.1	100.4	87.2	95.1	97.0	46.8	76.7	97.0	69.1	85.7	97.0	87.2	93.0
26, 27, 3370	ELECTRICAL, COMMUNICATIONS & UTIL.	98.9	55.7	76.9	96.2	96.7	96.4	93.9	99.3	96.6	95.8	37.2	66.0	95.8	70.0	82.7	98.2	99.3	98.7
MF2016	WEIGHTED AVERAGE	95.8	81.0	89.4	99.3	84.2	92.8	100.1	84.3	93.3	97.9	59.4	81.4	99.1	68.0	85.8	98.4	84.9	92.6

City Cost Indexes

	DIVISION	VIRGINIA																	
		FAIRFAX			FARMVILLE			FREDERICKSBURG			GRUNDY			HARRISONBURG			LYNCHBURG		
		220 - 221			239			224 - 225			246			228			245		
		MAT.	INST.	TOTAL	MAT.	INST.	TOTAL	MAT.	INST.	TOTAL	MAT.	INST.	TOTAL	MAT.	INST.	TOTAL	MAT.	INST.	TOTAL
015433	CONTRACTOR EQUIPMENT		106.2	106.2		110.4	110.4		106.2	106.2		106.2	106.2		106.2	106.2		106.2	106.2
0241, 31 - 34	SITE & INFRASTRUCTURE, DEMOLITION	123.0	89.9	100.0	109.8	90.7	96.5	111.5	89.7	96.3	106.7	88.5	94.0	120.0	88.3	97.9	107.5	90.0	95.3
0310	Concrete Forming & Accessories	83.7	70.1	72.1	97.8	45.2	53.0	83.7	67.1	69.6	88.1	38.1	45.6	79.8	38.7	44.8	85.1	71.5	73.5
0320	Concrete Reinforcing	94.7	87.4	91.0	90.0	69.7	79.7	95.4	82.5	88.8	93.4	45.3	69.0	94.7	58.9	76.5	94.1	70.6	82.1
0330	Cast-in-Place Concrete	107.1	76.5	95.7	106.2	86.0	98.7	108.7	76.2	96.6	106.7	51.0	86.0	107.1	53.8	87.3	106.7	75.7	95.2
03	CONCRETE	103.5	76.7	91.6	99.3	65.7	84.4	97.6	74.5	87.3	98.8	46.0	75.4	101.3	49.8	78.5	98.7	74.3	87.9
04	MASONRY	103.0	72.3	84.0	101.3	51.0	70.2	103.7	68.2	81.7	94.2	54.6	69.7	101.5	59.1	75.3	109.0	64.4	81.5
05	METALS	101.4	100.3	101.1	98.9	90.6	96.4	101.4	98.4	100.5	101.0	75.1	93.2	101.3	87.2	97.0	101.2	94.2	99.1
06	WOOD, PLASTICS & COMPOSITES	80.9	68.4	74.1	94.1	39.5	64.4	80.9	66.5	73.1	83.8	31.4	55.3	77.1	34.9	54.2	80.9	73.2	76.7
07	THERMAL & MOISTURE PROTECTION	104.5	75.6	91.9	103.1	54.9	82.1	103.8	77.1	92.2	104.0	47.4	79.3	104.2	62.9	86.2	103.8	71.1	89.6
08	OPENINGS	95.6	73.9	90.6	96.2	44.4	84.2	96.7	70.5	90.6	98.4	30.9	82.7	97.0	44.3	84.8	97.0	66.4	89.9
0920	Plaster & Gypsum Board	94.9	67.4	76.8	103.8	37.0	59.8	94.9	65.4	75.5	94.8	29.4	51.7	94.8	33.0	54.1	94.8	72.3	80.0
0950, 0980	Ceilings & Acoustic Treatment	91.7	67.4	75.2	86.3	37.0	52.9	91.7	65.4	73.9	90.8	29.4	49.2	90.8	33.0	51.6	90.8	72.3	78.3
0960	Flooring	92.4	77.7	88.1	92.5	52.2	80.8	92.4	74.8	87.3	93.4	29.2	74.8	90.5	76.9	86.5	92.1	67.6	85.0
0970, 0990	Wall Finishes & Painting/Coating	115.2	78.0	93.2	101.8	58.7	76.3	115.2	58.5	81.6	101.5	32.1	60.4	115.2	49.6	76.4	101.5	55.6	74.4
09	FINISHES	91.7	71.5	80.7	88.7	46.3	65.6	90.6	66.9	77.7	90.0	34.4	59.6	90.6	45.3	65.8	89.6	69.2	78.5
COVERS	DIVS. 10 - 14, 25, 28, 41, 43, 44, 46	100.0	84.9	96.7	100.0	76.2	94.8	100.0	79.8	95.6	100.0	74.3	94.4	100.0	61.2	91.5	100.0	79.5	95.5
21, 22, 23	FIRE SUPPRESSION, PLUMBING & HVAC	97.0	87.2	93.0	97.0	53.1	79.3	97.0	81.7	90.8	97.0	66.1	84.5	97.0	66.8	84.8	97.0	69.4	85.8
26, 27, 3370	ELECTRICAL, COMMUNICATIONS & UTIL.	96.9	99.3	98.1	89.9	70.0	79.8	94.1	94.0	94.1	95.8	70.0	82.7	96.0	88.5	92.2	96.9	69.4	82.9
MF2016	WEIGHTED AVERAGE	99.2	84.2	92.8	97.1	62.8	82.4	97.9	80.6	90.5	97.8	59.2	81.2	98.7	66.0	84.7	98.4	73.7	87.8

	DIVISION	VIRGINIA																	
		NEWPORT NEWS			NORFOLK			PETERSBURG			PORTSMOUTH			PULASKI			RICHMOND		
		236			233 - 235			238			237			243			230 - 232		
		MAT.	INST.	TOTAL	MAT.	INST.	TOTAL	MAT.	INST.	TOTAL	MAT.	INST.	TOTAL	MAT.	INST.	TOTAL	MAT.	INST.	TOTAL
015433	CONTRACTOR EQUIPMENT		110.4	110.4		111.0	111.0		110.4	110.4		110.3	110.3		106.2	106.2		110.4	110.4
0241, 31 - 34	SITE & INFRASTRUCTURE, DEMOLITION	108.7	91.7	96.9	105.4	92.7	96.6	112.4	91.7	97.9	107.2	91.0	95.9	106.0	89.0	94.2	106.3	91.7	96.1
0310	Concrete Forming & Accessories	97.1	61.3	66.6	103.1	61.4	67.6	90.5	62.5	66.6	86.5	48.0	53.7	88.1	41.2	48.2	94.5	80.7	82.7
0320	Concrete Reinforcing	89.8	67.3	78.3	98.0	67.3	82.4	89.4	70.7	79.9	89.4	66.6	77.8	93.4	86.2	89.7	98.8	70.7	84.5
0330	Cast-in-Place Concrete	103.2	76.2	93.1	108.0	77.2	96.6	109.6	77.4	97.6	102.2	61.2	86.9	106.7	84.6	98.5	97.1	77.5	89.8
03	CONCRETE	96.4	69.3	84.4	98.6	69.7	85.8	101.9	70.8	88.1	95.3	58.0	78.7	98.8	66.2	84.3	93.1	79.1	86.9
04	MASONRY	96.2	63.2	75.8	98.4	63.2	76.6	109.6	63.2	80.9	101.9	48.4	68.9	88.1	55.4	67.9	90.5	63.2	73.6
05	METALS	101.1	93.5	98.8	103.0	93.6	100.1	98.9	95.1	97.8	100.1	91.9	97.6	101.1	96.2	99.6	103.0	95.2	100.6
06	WOOD, PLASTICS & COMPOSITES	92.8	60.3	75.1	101.3	60.3	79.0	84.0	61.2	71.6	80.2	46.1	61.7	83.8	32.9	56.2	93.1	85.7	89.1
07	THERMAL & MOISTURE PROTECTION	103.1	70.9	89.1	100.6	71.1	87.8	103.0	72.1	89.6	103.0	60.8	84.7	104.0	57.9	83.9	100.6	74.8	89.4
08	OPENINGS	96.6	59.0	87.9	98.3	60.3	89.5	96.0	65.8	89.0	96.7	48.8	85.6	98.4	44.1	85.8	100.1	79.4	95.3
0920	Plaster & Gypsum Board	104.7	58.4	74.2	99.9	58.4	72.5	98.2	59.4	72.6	98.6	43.8	62.5	94.8	30.9	52.7	101.7	84.6	90.4
0950, 0980	Ceilings & Acoustic Treatment	90.5	58.4	68.8	94.8	58.4	70.1	87.1	59.4	68.3	90.5	43.8	58.9	90.8	30.9	50.2	92.7	84.6	87.2
0960	Flooring	92.5	67.6	85.3	90.6	67.6	83.9	88.7	72.7	84.0	85.8	57.6	77.6	93.4	57.6	83.0	93.6	72.7	87.5
0970, 0990	Wall Finishes & Painting/Coating	101.8	58.9	76.4	101.5	58.9	76.3	101.8	58.9	76.4	101.8	58.9	76.4	101.5	49.6	70.8	99.6	58.9	75.5
09	FINISHES	89.5	61.7	74.3	89.0	61.7	74.1	87.1	63.3	74.1	86.7	49.5	66.4	90.0	42.0	63.8	91.6	77.8	84.1
COVERS	DIVS. 10 - 14, 25, 28, 41, 43, 44, 46	100.0	79.5	95.5	100.0	79.5	95.5	100.0	82.2	96.1	100.0	65.2	92.3	100.0	74.6	94.4	100.0	84.8	96.7
21, 22, 23	FIRE SUPPRESSION, PLUMBING & HVAC	100.5	63.5	85.5	100.1	66.0	86.3	97.0	68.5	85.5	100.5	63.2	85.4	97.0	67.5	85.1	100.0	68.5	87.2
26, 27, 3370	ELECTRICAL, COMMUNICATIONS & UTIL.	92.4	64.7	78.3	94.8	61.9	78.1	92.5	70.0	81.1	90.8	61.9	76.1	95.8	85.4	90.5	96.9	70.0	83.2
MF2016	WEIGHTED AVERAGE	97.9	69.7	85.8	98.8	70.0	86.5	97.9	72.4	87.0	97.4	63.1	82.7	97.5	68.3	85.0	98.3	76.4	88.9

	DIVISION	VIRGINIA									WASHINGTON								
		ROANOKE			STAUNTON			WINCHESTER			CLARKSTON			EVERETT			OLYMPIA		
		240 - 241			244			226			994			982			985		
		MAT.	INST.	TOTAL	MAT.	INST.	TOTAL	MAT.	INST.	TOTAL	MAT.	INST.	TOTAL	MAT.	INST.	TOTAL	MAT.	INST.	TOTAL
015433	CONTRACTOR EQUIPMENT		106.2	106.2		110.4	110.4		106.2	106.2		92.5	92.5		102.5	102.5		102.5	102.5
0241, 31 - 34	SITE & INFRASTRUCTURE, DEMOLITION	106.8	89.9	95.1	110.0	90.6	96.5	118.7	89.8	98.6	101.5	90.4	93.7	93.2	110.8	105.5	93.0	110.8	105.4
0310	Concrete Forming & Accessories	94.3	71.6	75.0	87.8	49.4	55.1	82.1	68.1	70.2	103.1	66.0	71.5	110.9	101.2	102.7	100.2	101.1	100.9
0320	Concrete Reinforcing	94.5	70.6	82.3	94.1	86.1	90.0	94.1	82.5	88.2	104.8	98.7	101.7	111.1	111.6	111.4	118.0	111.6	114.8
0330	Cast-in-Place Concrete	121.2	85.7	108.0	111.0	86.5	101.9	107.1	65.6	91.7	81.1	85.3	82.7	99.4	108.9	103.0	87.2	108.8	95.2
03	CONCRETE	103.2	77.8	92.0	100.2	70.5	87.0	100.8	71.3	87.7	88.5	78.7	84.2	92.1	105.1	97.9	84.9	105.0	93.8
04	MASONRY	95.6	64.4	76.3	104.5	54.4	73.6	99.0	71.2	81.8	99.6	93.6	95.9	117.9	103.8	109.2	110.3	103.8	106.3
05	METALS	103.3	94.3	100.6	101.2	95.9	99.6	101.4	98.3	100.4	91.0	88.0	90.1	110.0	96.0	105.8	108.5	95.6	104.6
06	WOOD, PLASTICS & COMPOSITES	92.7	73.2	82.1	83.8	44.7	62.5	79.4	66.7	72.5	103.6	60.2	80.0	107.5	99.8	103.3	93.6	99.8	97.0
07	THERMAL & MOISTURE PROTECTION	103.8	74.5	91.0	103.6	56.3	83.0	104.4	76.7	92.3	156.1	85.2	125.2	113.6	106.8	110.6	113.1	104.3	109.3
08	OPENINGS	97.4	66.4	90.2	97.0	50.9	86.3	98.5	70.3	91.9	119.0	66.9	106.9	108.2	103.6	107.2	110.7	102.1	108.7
0920	Plaster & Gypsum Board	101.7	72.3	82.4	94.8	42.4	60.3	94.9	65.6	75.6	151.9	58.9	90.7	109.4	99.9	103.1	103.2	99.9	101.0
0950, 0980	Ceilings & Acoustic Treatment	93.4	72.3	79.1	90.8	42.4	58.0	91.7	65.6	74.0	106.8	58.9	74.4	109.1	99.9	102.8	103.6	99.9	101.1
0960	Flooring	97.0	67.6	88.5	93.0	34.2	75.9	91.9	76.9	87.5	88.3	80.1	85.9	105.2	99.3	103.5	94.0	99.3	95.6
0970, 0990	Wall Finishes & Painting/Coating	101.5	55.6	74.4	101.5	31.5	60.1	115.2	79.6	94.1	83.8	74.0	78.0	91.1	95.6	93.8	90.2	95.6	93.4
09	FINISHES	92.1	69.2	79.6	89.8	43.7	64.6	91.1	70.3	79.7	107.3	67.7	85.7	103.8	99.7	101.5	92.3	99.7	96.3
COVERS	DIVS. 10 - 14, 25, 28, 41, 43, 44, 46	100.0	79.4	95.5	100.0	77.2	95.0	100.0	84.3	96.6	100.0	91.2	98.1	100.0	100.4	100.1	100.0	100.8	100.2
21, 22, 23	FIRE SUPPRESSION, PLUMBING & HVAC	100.4	65.2	86.2	97.0	59.2	81.7	97.0	83.2	91.4	97.0	82.4	91.1	100.1	102.8	101.2	100.0	102.8	101.1
26, 27, 3370	ELECTRICAL, COMMUNICATIONS & UTIL.	95.8	56.1	75.6	94.8	70.1	82.2	94.5	94.0	94.3	84.8	97.9	91.5	104.5	101.5	102.9	102.8	100.1	101.4
MF2016	WEIGHTED AVERAGE	99.7	71.6	87.7	98.3	65.6	84.3	98.6	81.3	91.2	98.9	83.9	92.5	103.4	102.6	103.1	100.9	102.3	101.5

DIVISION		WASHINGTON																	
		RICHLAND			SEATTLE			SPOKANE			TACOMA			VANCOUVER			WENATCHEE		
		993			980 - 981, 987			990 - 992			983 - 984			986			988		
		MAT.	INST.	TOTAL	MAT.	INST.	TOTAL	MAT.	INST.	TOTAL	MAT.	INST.	TOTAL	MAT.	INST.	TOTAL	MAT.	INST.	TOTAL
015433	CONTRACTOR EQUIPMENT		92.5	92.5		104.0	104.0		92.5	92.5		102.5	102.5		98.8	98.8		102.5	102.5
0241, 31 - 34	SITE & INFRASTRUCTURE, DEMOLITION	103.5	90.6	94.5	100.0	110.0	107.0	103.0	90.5	94.3	96.4	110.8	106.4	106.7	98.1	100.7	105.8	108.5	107.7
0310	Concrete Forming & Accessories	103.2	82.2	85.3	107.0	103.6	104.1	108.2	81.7	85.6	101.4	101.1	101.1	101.7	93.2	94.5	103.2	78.6	82.3
0320	Concrete Reinforcing	100.5	98.1	99.3	114.5	111.7	113.0	101.2	98.0	99.6	109.7	111.5	110.6	110.7	110.8	110.7	110.6	98.7	104.6
0330	Cast-in-Place Concrete	81.3	86.9	83.4	107.8	108.7	108.1	84.5	86.7	85.3	102.2	108.8	104.7	114.2	99.0	108.6	104.4	93.6	100.4
03	CONCRETE	88.2	86.5	87.4	105.0	106.3	105.6	90.1	86.1	88.4	93.9	105.0	98.8	103.8	98.1	101.2	101.8	87.3	95.4
04	MASONRY	100.6	85.5	91.3	125.2	103.8	112.0	101.2	85.5	91.5	116.6	102.4	107.8	116.6	90.6	100.6	120.0	96.1	105.2
05	METALS	91.3	87.8	90.3	111.7	99.1	107.9	93.5	87.3	91.6	111.8	95.5	106.9	109.2	96.0	105.2	109.2	88.2	102.9
06	WOOD, PLASTICS & COMPOSITES	103.8	80.2	91.0	101.3	102.7	102.1	112.7	80.2	95.0	97.0	99.8	98.5	89.5	92.2	91.0	98.8	75.5	86.1
07	THERMAL & MOISTURE PROTECTION	157.2	86.7	126.5	109.0	106.0	107.7	153.7	87.0	124.7	113.3	104.1	109.3	113.3	96.9	106.1	112.8	90.8	103.2
08	OPENINGS	117.0	76.9	107.7	108.4	105.2	107.7	117.6	76.8	108.1	109.0	102.1	107.4	105.0	96.6	103.0	108.5	75.1	100.7
0920	Plaster & Gypsum Board	151.9	79.5	104.2	107.8	102.9	104.5	144.5	79.5	101.7	109.2	99.9	103.0	107.4	92.4	97.5	112.5	74.9	87.7
0950, 0980	Ceilings & Acoustic Treatment	113.5	79.5	90.4	120.3	102.9	108.5	108.7	79.5	88.9	112.5	99.9	103.9	109.6	92.4	97.9	104.7	74.9	84.5
0960	Flooring	88.6	80.1	86.1	105.3	99.3	103.5	87.7	80.1	85.5	98.2	99.3	98.5	103.9	81.7	97.5	101.0	80.1	94.9
0970, 0990	Wall Finishes & Painting/Coating	83.8	77.0	79.8	101.9	95.6	98.2	83.9	78.4	80.7	91.1	95.6	93.8	93.8	73.8	82.0	91.1	72.8	80.3
09	FINISHES	108.9	80.4	93.3	109.6	101.4	105.1	106.8	80.5	92.4	102.5	99.7	101.0	100.5	88.5	93.9	103.2	77.1	89.0
COVERS	DIVS. 10 - 14, 25, 28, 41, 43, 44, 46	100.0	94.1	98.7	100.0	101.2	100.3	100.0	94.1	98.7	100.0	100.8	100.2	100.0	96.3	99.2	100.0	93.1	98.5
21, 22, 23	FIRE SUPPRESSION, PLUMBING & HVAC	100.6	110.9	104.8	100.1	112.5	105.1	100.5	85.5	94.4	100.1	102.8	101.2	100.2	100.0	100.1	96.7	94.1	95.6
26, 27, 3370	ELECTRICAL, COMMUNICATIONS & UTIL.	82.6	95.1	89.0	103.7	115.4	109.6	81.2	78.9	80.0	104.3	100.1	102.2	109.5	98.5	103.9	105.2	96.8	100.9
MF2016	WEIGHTED AVERAGE	99.6	92.2	96.4	106.0	107.4	106.6	99.9	84.5	93.3	103.8	102.1	103.1	104.8	96.2	101.1	104.1	91.0	98.5

DIVISION		WASHINGTON			WEST VIRGINIA														
		YAKIMA			BECKLEY			BLUEFIELD			BUCKHANNON			CHARLESTON			CLARKSBURG		
		989			258 - 259			247 - 248			262			250 - 253			263 - 264		
		MAT.	INST.	TOTAL	MAT.	INST.	TOTAL	MAT.	INST.	TOTAL	MAT.	INST.	TOTAL	MAT.	INST.	TOTAL	MAT.	INST.	TOTAL
015433	CONTRACTOR EQUIPMENT		102.5	102.5		106.2	106.2		106.2	106.2		106.2	106.2		106.2	106.2		106.2	106.2
0241, 31 - 34	SITE & INFRASTRUCTURE, DEMOLITION	98.9	109.3	106.2	99.1	91.3	93.6	100.6	91.3	94.1	106.8	91.2	95.9	98.6	92.3	94.2	107.6	91.2	96.2
0310	Concrete Forming & Accessories	102.0	96.1	97.0	83.6	89.7	88.8	85.2	89.5	88.8	84.6	85.9	85.7	94.9	89.6	90.4	82.1	86.0	85.4
0320	Concrete Reinforcing	110.2	98.2	104.1	88.8	89.4	89.1	93.0	84.9	88.9	93.7	84.7	89.1	96.0	89.4	92.7	93.7	98.0	95.9
0330	Cast-in-Place Concrete	109.2	85.6	100.4	103.1	95.1	100.1	104.4	95.0	100.9	104.0	94.4	100.5	99.6	96.1	98.3	114.0	90.0	105.1
03	CONCRETE	98.5	92.4	95.8	90.8	92.4	91.5	94.7	91.6	93.3	97.7	89.7	94.1	89.5	92.8	91.0	101.7	90.5	96.8
04	MASONRY	109.2	82.8	92.9	88.7	90.9	90.1	89.7	90.9	90.5	101.4	86.5	92.2	85.8	92.3	89.8	105.0	86.5	93.6
05	METALS	109.9	88.2	103.3	96.3	103.3	98.5	101.2	101.7	101.4	101.4	101.6	101.5	95.1	103.5	97.7	101.4	106.1	102.8
06	WOOD, PLASTICS & COMPOSITES	97.4	99.8	98.7	81.7	89.9	86.2	82.8	89.9	86.7	82.1	85.4	83.9	93.6	88.8	91.0	78.5	85.4	82.2
07	THERMAL & MOISTURE PROTECTION	113.5	87.7	102.2	106.1	88.4	98.4	103.7	88.4	97.1	104.1	88.6	97.4	102.3	88.9	96.5	104.0	88.3	97.2
08	OPENINGS	108.5	99.1	106.3	96.5	84.9	93.8	98.9	83.9	95.4	98.9	81.5	94.9	96.9	84.3	93.9	98.9	84.4	95.6
0920	Plaster & Gypsum Board	108.8	99.9	102.9	97.3	89.5	92.2	94.1	89.5	91.1	94.5	84.9	88.2	101.7	88.4	92.9	91.7	84.9	87.2
0950, 0980	Ceilings & Acoustic Treatment	106.6	99.9	102.0	81.9	89.5	87.1	89.1	89.5	89.4	90.8	84.9	86.8	93.8	88.4	90.1	90.8	84.9	86.8
0960	Flooring	99.0	80.1	93.5	91.3	101.6	94.3	89.9	101.6	93.3	89.6	96.4	91.6	97.3	101.6	98.5	88.6	96.4	90.9
0970, 0990	Wall Finishes & Painting/Coating	91.1	78.4	83.6	92.3	94.6	93.7	101.5	92.2	96.0	101.5	90.3	94.9	90.7	94.6	93.0	101.5	90.3	94.9
09	FINISHES	101.7	91.3	96.0	86.2	92.7	89.7	88.1	92.4	90.5	89.0	88.2	88.5	92.4	92.4	92.4	88.2	88.2	88.2
COVERS	DIVS. 10 - 14, 25, 28, 41, 43, 44, 46	100.0	97.3	99.4	100.0	92.1	98.3	100.0	92.1	98.3	100.0	90.8	98.0	100.0	92.1	98.3	100.0	90.8	98.0
21, 22, 23	FIRE SUPPRESSION, PLUMBING & HVAC	100.1	109.8	104.0	97.2	93.2	95.6	97.0	85.9	92.5	97.0	91.7	94.8	100.1	94.3	97.8	97.0	91.8	94.9
26, 27, 3370	ELECTRICAL, COMMUNICATIONS & UTIL.	107.4	95.1	101.2	93.1	86.2	89.6	94.8	86.2	90.4	96.2	91.5	93.8	98.1	86.2	92.0	96.2	91.5	93.8
MF2016	WEIGHTED AVERAGE	104.0	96.6	100.8	94.9	92.1	93.7	96.7	90.2	93.9	98.1	90.7	94.9	96.1	92.5	94.6	98.7	91.4	95.5

DIVISION		WEST VIRGINIA																	
		GASSAWAY			HUNTINGTON			LEWISBURG			MARTINSBURG			MORGANTOWN			PARKERSBURG		
		266			255 - 257			249			254			265			261		
		MAT.	INST.	TOTAL	MAT.	INST.	TOTAL	MAT.	INST.	TOTAL	MAT.	INST.	TOTAL	MAT.	INST.	TOTAL	MAT.	INST.	TOTAL
015433	CONTRACTOR EQUIPMENT		106.2	106.2		106.2	106.2		106.2	106.2		106.2	106.2		106.2	106.2		106.2	106.2
0241, 31 - 34	SITE & INFRASTRUCTURE, DEMOLITION	104.3	91.2	95.2	103.7	92.3	95.8	116.2	91.2	98.8	102.9	92.0	95.3	101.6	92.0	94.9	109.8	92.2	97.5
0310	Concrete Forming & Accessories	84.0	89.2	88.4	94.7	91.7	92.2	82.4	89.2	88.1	83.6	81.7	82.0	82.4	86.1	85.6	86.5	88.3	88.0
0320	Concrete Reinforcing	93.7	94.2	94.0	90.1	95.8	93.0	93.7	84.9	89.2	88.8	97.9	93.4	93.7	98.0	95.9	93.0	84.7	88.8
0330	Cast-in-Place Concrete	108.9	94.4	103.5	112.4	96.5	106.5	104.4	94.9	100.9	108.0	90.8	101.6	104.0	94.5	100.5	106.3	91.4	100.8
03	CONCRETE	98.1	92.8	95.8	96.0	95.0	95.5	104.4	91.4	98.6	94.3	88.8	91.9	94.4	92.1	93.4	99.7	89.8	95.3
04	MASONRY	105.9	89.7	95.9	89.0	95.4	92.9	93.6	90.9	91.9	90.5	85.7	87.5	122.8	86.5	100.4	80.0	87.0	84.3
05	METALS	101.4	104.9	102.4	98.8	105.9	100.9	101.3	101.4	101.3	96.7	105.0	99.2	101.5	106.2	102.9	102.1	101.6	101.9
06	WOOD, PLASTICS & COMPOSITES	81.0	89.9	85.8	92.8	91.2	91.9	78.8	89.9	84.8	81.7	80.9	81.3	78.8	85.4	82.4	82.2	88.1	85.4
07	THERMAL & MOISTURE PROTECTION	103.7	90.0	97.8	106.3	90.4	99.4	104.7	88.4	97.6	106.4	83.0	96.2	103.8	88.6	97.2	103.8	89.1	97.4
08	OPENINGS	97.2	86.0	94.6	95.7	87.0	93.7	98.9	83.9	95.4	98.2	76.0	93.1	100.1	84.4	96.5	97.8	82.7	94.3
0920	Plaster & Gypsum Board	93.2	89.5	90.8	104.1	90.8	95.4	91.7	89.5	90.3	97.4	80.2	86.1	91.7	84.9	87.2	94.8	87.6	90.1
0950, 0980	Ceilings & Acoustic Treatment	90.8	89.5	89.9	82.8	90.8	88.2	90.8	89.5	89.9	82.8	80.2	81.0	90.8	84.9	86.8	90.8	87.6	88.7
0960	Flooring	89.4	101.6	93.0	98.6	106.7	100.9	88.7	101.6	92.4	91.3	102.3	94.5	88.7	96.4	90.9	92.5	95.0	93.2
0970, 0990	Wall Finishes & Painting/Coating	101.5	94.6	97.4	92.3	92.2	92.3	101.5	66.2	80.6	92.3	85.9	88.6	101.5	90.3	94.9	101.5	85.9	92.3
09	FINISHES	88.4	92.4	90.6	89.4	94.7	92.3	89.2	89.6	89.4	86.6	85.4	86.0	87.8	88.2	88.0	89.9	89.3	89.6
COVERS	DIVS. 10 - 14, 25, 28, 41, 43, 44, 46	100.0	91.3	98.1	100.0	92.5	98.3	100.0	66.7	92.7	100.0	83.0	96.3	100.0	85.2	96.7	100.0	91.5	98.1
21, 22, 23	FIRE SUPPRESSION, PLUMBING & HVAC	97.0	85.3	92.2	100.7	92.6	97.4	97.0	93.1	95.4	97.2	82.7	91.4	97.0	91.8	94.9	100.4	90.3	96.3
26, 27, 3370	ELECTRICAL, COMMUNICATIONS & UTIL.	96.2	86.2	91.1	96.8	91.8	94.3	92.4	86.2	89.2	98.9	78.7	88.7	96.3	91.5	93.9	96.3	90.4	93.3
MF2016	WEIGHTED AVERAGE	98.0	90.5	94.8	97.5	94.3	96.1	98.4	90.5	95.0	96.4	86.1	92.0	98.5	91.5	95.5	98.2	90.7	95.0

DIVISION		WEST VIRGINIA									WISCONSIN								
		PETERSBURG			ROMNEY			WHEELING			BELOIT			EAU CLAIRE			GREEN BAY		
		268			267			260			535			547			541 - 543		
		MAT.	INST.	TOTAL	MAT.	INST.	TOTAL	MAT.	INST.	TOTAL	MAT.	INST.	TOTAL	MAT.	INST.	TOTAL	MAT.	INST.	TOTAL
015433	CONTRACTOR EQUIPMENT		106.2	106.2		106.2	106.2		106.2	106.2		101.8	101.8		102.6	102.6		100.3	100.3
0241, 31 - 34	SITE & INFRASTRUCTURE, DEMOLITION	100.9	92.2	94.8	103.8	92.2	95.7	110.5	92.0	97.6	95.8	107.0	103.6	96.1	104.0	101.7	99.7	100.2	100.0
0310	Concrete Forming & Accessories	85.5	88.8	88.3	81.7	89.1	88.0	88.2	86.1	86.4	97.6	98.9	98.7	96.3	97.4	97.3	106.4	100.3	101.2
0320	Concrete Reinforcing	93.0	94.0	93.5	93.7	98.2	96.0	92.5	98.0	95.3	96.4	141.1	119.2	93.9	117.5	105.9	91.9	112.6	102.4
0330	Cast-in-Place Concrete	104.0	94.3	100.4	108.9	85.7	100.3	106.3	94.5	101.9	103.9	103.2	103.6	100.5	97.6	99.4	103.9	102.9	103.5
03	CONCRETE	94.5	92.5	93.6	98.0	90.4	94.6	99.8	92.1	96.4	95.8	107.6	101.0	90.8	101.1	95.3	93.8	103.4	98.1
04	MASONRY	96.7	86.5	90.4	94.0	89.7	91.4	104.2	86.4	93.2	98.5	102.4	100.9	91.3	98.1	95.5	122.4	99.9	108.5
05	METALS	101.5	104.2	102.3	101.6	106.2	102.9	102.2	106.2	103.4	101.1	113.0	104.7	91.2	106.7	95.9	93.7	105.1	97.1
06	WOOD, PLASTICS & COMPOSITES	83.1	89.9	86.8	77.9	89.9	84.4	83.8	85.4	84.7	98.6	96.9	97.6	102.0	97.0	99.3	109.6	100.0	104.4
07	THERMAL & MOISTURE PROTECTION	103.8	84.5	95.4	103.9	86.1	96.2	104.3	88.8	97.5	99.8	96.1	98.2	103.5	97.2	100.8	105.7	100.2	103.3
08	OPENINGS	100.1	86.0	96.8	100.0	86.9	97.0	98.6	84.4	95.3	100.4	111.4	103.0	105.4	102.2	104.7	101.3	104.2	102.0
0920	Plaster & Gypsum Board	94.5	89.5	91.2	91.4	89.5	90.2	94.8	84.9	88.3	93.7	97.2	96.0	105.7	97.2	100.1	104.2	100.4	101.7
0950, 0980	Ceilings & Acoustic Treatment	90.8	89.5	89.9	90.8	89.5	89.9	90.8	84.9	86.8	82.5	97.2	92.5	87.3	97.2	94.0	80.7	100.4	94.0
0960	Flooring	90.5	96.4	92.2	88.5	102.3	92.5	93.4	96.4	94.3	94.1	122.4	102.3	81.9	110.1	90.1	97.6	118.7	103.7
0970, 0990	Wall Finishes & Painting/Coating	101.5	90.3	94.9	101.5	90.3	94.9	101.5	90.3	94.9	94.3	106.3	101.4	83.0	78.4	80.3	91.5	82.1	85.9
09	FINISHES	88.7	90.9	89.9	87.9	92.1	90.2	90.1	88.2	89.1	90.0	103.7	97.5	86.0	97.7	92.4	90.2	102.0	96.6
COVERS	DIVS. 10 - 14, 25, 28, 41, 43, 44, 46	100.0	56.0	90.3	100.0	91.2	98.1	100.0	86.0	96.9	100.0	100.7	100.2	100.0	96.1	99.1	100.0	97.7	99.5
21, 22, 23	FIRE SUPPRESSION, PLUMBING & HVAC	97.0	85.6	92.4	97.0	85.4	92.3	100.4	91.8	96.9	100.1	97.6	99.1	100.1	90.2	96.1	100.3	85.3	94.2
26, 27, 3370	ELECTRICAL, COMMUNICATIONS & UTIL.	99.4	78.8	88.9	98.6	78.8	88.5	93.6	91.5	92.5	99.7	86.3	92.9	103.6	86.5	94.9	98.0	83.2	90.5
MF2016	WEIGHTED AVERAGE	97.7	87.7	93.4	97.9	89.2	94.2	99.3	91.5	96.0	98.7	101.4	99.9	96.8	96.5	96.7	98.7	95.8	97.5

DIVISION		WISCONSIN																	
		KENOSHA			LA CROSSE			LANCASTER			MADISON			MILWAUKEE			NEW RICHMOND		
		531			546			538			537			530, 532			540		
		MAT.	INST.	TOTAL	MAT.	INST.	TOTAL	MAT.	INST.	TOTAL	MAT.	INST.	TOTAL	MAT.	INST.	TOTAL	MAT.	INST.	TOTAL
015433	CONTRACTOR EQUIPMENT		99.8	99.8		102.6	102.6		101.8	101.8		101.8	101.8		89.6	89.6		102.9	102.9
0241, 31 - 34	SITE & INFRASTRUCTURE, DEMOLITION	101.7	103.6	103.0	89.9	104.0	99.8	94.8	106.3	102.9	93.7	106.9	102.9	93.7	95.2	94.8	94.6	104.7	101.6
0310	Concrete Forming & Accessories	107.6	109.1	108.9	83.0	97.2	95.1	96.9	97.8	97.7	101.9	98.3	98.9	100.1	110.5	109.0	91.3	95.1	94.5
0320	Concrete Reinforcing	96.2	113.7	105.1	93.5	109.3	101.5	97.8	109.3	103.7	100.6	109.5	105.1	99.9	113.9	107.0	91.0	117.5	104.5
0330	Cast-in-Place Concrete	113.0	105.8	110.4	90.3	99.4	93.7	103.3	102.7	103.1	101.3	102.5	101.7	90.5	110.4	97.9	104.6	105.8	105.0
03	CONCRETE	100.6	108.5	104.1	82.4	100.2	90.3	95.5	101.5	98.1	93.7	101.7	97.3	95.4	110.1	101.9	88.8	102.9	95.0
04	MASONRY	96.2	110.3	104.9	90.4	98.1	95.2	98.6	102.3	100.9	97.1	101.2	99.6	100.6	113.0	108.3	117.5	100.5	107.0
05	METALS	102.1	104.2	102.7	91.1	103.7	95.0	98.5	101.2	99.3	104.0	102.1	103.5	101.5	96.5	100.0	91.5	106.3	96.0
06	WOOD, PLASTICS & COMPOSITES	106.0	108.8	107.5	86.2	97.0	92.1	97.7	96.9	97.3	98.4	96.9	97.5	101.3	109.2	105.6	90.8	92.8	91.9
07	THERMAL & MOISTURE PROTECTION	100.2	106.9	103.1	102.8	96.1	99.9	99.5	96.3	98.1	96.8	102.8	99.4	102.3	109.0	105.2	104.1	103.1	103.7
08	OPENINGS	94.7	108.5	97.9	105.4	94.7	102.9	96.2	94.6	95.8	103.5	101.7	103.0	101.4	109.5	103.3	90.6	93.0	91.1
0920	Plaster & Gypsum Board	82.7	109.5	100.4	100.7	97.2	98.4	93.0	97.2	95.8	105.1	97.2	99.9	98.0	109.5	105.6	92.1	93.2	92.8
0950, 0980	Ceilings & Acoustic Treatment	82.5	109.5	100.8	86.5	97.2	93.8	79.1	97.2	91.4	89.5	97.2	94.7	89.7	109.5	103.1	55.9	93.2	81.2
0960	Flooring	111.8	118.7	113.8	76.0	115.4	87.4	93.7	114.7	99.8	93.4	114.7	99.6	100.9	114.7	104.9	91.2	118.4	99.1
0970, 0990	Wall Finishes & Painting/Coating	104.4	121.9	114.8	83.0	83.9	83.5	94.3	100.3	97.8	93.6	106.3	101.1	106.1	122.7	115.9	94.4	83.9	88.1
09	FINISHES	94.4	112.6	104.3	83.0	99.3	91.9	89.1	101.3	95.7	90.2	102.1	96.7	99.4	112.8	106.7	81.4	98.0	90.5
COVERS	DIVS. 10 - 14, 25, 28, 41, 43, 44, 46	100.0	99.4	99.9	100.0	94.7	98.8	100.0	86.0	96.9	100.0	98.6	99.7	100.0	102.7	100.6	100.0	97.1	99.4
21, 22, 23	FIRE SUPPRESSION, PLUMBING & HVAC	100.3	98.1	99.4	100.1	90.1	96.0	96.6	91.2	94.4	100.0	97.4	98.9	100.0	105.6	102.3	96.1	91.0	94.0
26, 27, 3370	ELECTRICAL, COMMUNICATIONS & UTIL.	100.1	99.5	99.8	103.9	86.5	95.1	99.4	86.9	93.1	100.7	96.3	98.5	99.3	101.6	100.5	101.6	86.9	94.2
MF2016	WEIGHTED AVERAGE	99.4	104.6	101.6	95.2	95.9	95.5	96.9	96.7	96.8	99.1	100.3	99.6	99.6	105.9	102.3	94.7	97.0	95.7

DIVISION		WISCONSIN																	
		OSHKOSH			PORTAGE			RACINE			RHINELANDER			SUPERIOR			WAUSAU		
		549			539			534			545			548			544		
		MAT.	INST.	TOTAL	MAT.	INST.	TOTAL	MAT.	INST.	TOTAL	MAT.	INST.	TOTAL	MAT.	INST.	TOTAL	MAT.	INST.	TOTAL
015433	CONTRACTOR EQUIPMENT		100.3	100.3		101.8	101.8		101.8	101.8		100.3	100.3		102.9	102.9		100.3	100.3
0241, 31 - 34	SITE & INFRASTRUCTURE, DEMOLITION	91.5	100.1	97.5	85.8	106.2	100.0	95.6	107.4	103.8	103.5	100.1	101.1	91.4	104.4	100.4	87.3	100.7	96.6
0310	Concrete Forming & Accessories	88.8	97.4	96.1	88.9	98.4	97.0	97.9	109.1	107.4	86.4	97.3	95.7	89.3	91.0	90.8	88.3	100.1	98.3
0320	Concrete Reinforcing	92.1	112.5	102.4	97.9	109.5	103.8	96.4	113.7	105.2	92.3	109.5	101.0	91.0	109.9	100.6	92.3	109.3	100.9
0330	Cast-in-Place Concrete	96.3	100.9	98.0	88.7	101.3	93.4	101.9	105.6	103.3	109.3	100.9	106.1	98.5	101.2	99.5	89.8	95.2	91.8
03	CONCRETE	84.2	101.4	91.8	83.3	101.5	91.3	94.9	108.4	100.8	95.6	100.8	97.9	83.6	98.2	90.1	79.4	100.1	88.6
04	MASONRY	104.9	99.8	101.8	97.4	102.3	100.4	98.5	110.3	105.8	122.2	99.7	108.3	116.7	102.5	107.9	104.3	99.7	101.5
05	METALS	91.7	104.4	95.6	99.2	105.3	101.0	102.8	104.2	103.3	91.6	103.3	95.1	92.4	104.2	96.0	91.4	103.8	95.2
06	WOOD, PLASTICS & COMPOSITES	87.9	97.0	92.8	87.1	96.9	92.4	98.9	108.8	104.3	85.3	97.0	91.6	89.0	88.2	88.6	87.3	100.0	94.2
07	THERMAL & MOISTURE PROTECTION	104.6	85.6	96.4	98.9	103.0	100.7	99.8	105.6	102.3	105.6	85.4	96.8	103.8	98.6	101.5	104.4	97.8	101.5
08	OPENINGS	97.7	99.0	98.0	96.4	102.6	97.8	100.4	108.5	102.3	97.7	94.6	97.0	90.0	93.8	90.9	97.9	99.9	98.4
0920	Plaster & Gypsum Board	90.8	97.2	95.0	85.6	97.2	93.3	93.7	109.5	104.1	90.8	97.2	95.0	92.0	88.5	89.7	90.8	100.4	97.1
0950, 0980	Ceilings & Acoustic Treatment	80.7	97.2	91.9	81.6	97.2	92.2	82.5	109.5	100.8	80.7	97.2	91.9	56.8	88.5	78.3	80.7	100.4	94.0
0960	Flooring	89.3	118.7	97.8	90.0	118.4	98.3	94.1	118.7	101.2	88.7	118.4	97.3	92.3	127.2	102.4	89.2	118.4	97.7
0970, 0990	Wall Finishes & Painting/Coating	89.1	107.0	99.7	94.3	100.3	97.8	94.3	119.0	108.9	89.1	82.1	84.9	83.6	108.6	98.3	89.1	85.4	86.9
09	FINISHES	85.1	102.9	94.8	86.9	102.3	95.3	90.0	112.3	102.1	86.0	100.1	93.7	80.8	99.2	90.8	84.8	102.3	94.4
COVERS	DIVS. 10 - 14, 25, 28, 41, 43, 44, 46	100.0	87.3	97.2	100.0	87.4	97.2	100.0	99.4	99.9	100.0	87.6	97.3	100.0	95.8	99.1	100.0	97.7	99.5
21, 22, 23	FIRE SUPPRESSION, PLUMBING & HVAC	96.8	84.9	92.0	96.6	98.0	97.2	100.1	98.2	99.3	96.8	90.6	94.3	96.1	93.0	94.8	96.8	90.8	94.4
26, 27, 3370	ELECTRICAL, COMMUNICATIONS & UTIL.	102.2	81.1	91.5	103.2	96.3	99.7	99.5	100.4	100.0	101.6	80.0	90.6	106.6	101.0	103.7	103.4	80.0	91.5
MF2016	WEIGHTED AVERAGE	94.8	94.2	94.5	95.3	100.5	97.5	98.8	105.0	101.5	97.3	94.6	96.1	94.5	98.7	96.3	94.1	95.8	94.8

City Cost Indexes

	DIVISION	WYOMING																	
		CASPER			CHEYENNE			NEWCASTLE			RAWLINS			RIVERTON			ROCK SPRINGS		
		826			820			827			823			825			829 - 831		
		MAT.	INST.	TOTAL	MAT.	INST.	TOTAL	MAT.	INST.	TOTAL	MAT.	INST.	TOTAL	MAT.	INST.	TOTAL	MAT.	INST.	TOTAL
015433	CONTRACTOR EQUIPMENT		97.6	97.6		97.6	97.6		97.6	97.6		97.6	97.6		97.6	97.6		97.6	97.6
0241, 31 - 34	SITE & INFRASTRUCTURE, DEMOLITION	96.3	94.7	95.2	91.6	94.7	93.8	84.8	94.2	91.4	98.1	94.2	95.4	91.8	94.1	93.4	88.8	94.2	92.6
0310	Concrete Forming & Accessories	98.5	54.3	60.9	99.9	67.2	72.0	90.9	72.5	75.2	94.9	72.7	76.0	89.8	61.9	66.1	96.9	62.8	67.9
0320	Concrete Reinforcing	109.6	84.9	97.0	105.0	84.9	94.8	112.9	84.7	98.6	112.6	84.8	98.4	113.6	84.7	98.9	113.6	83.0	98.0
0330	Cast-in-Place Concrete	104.4	76.0	93.8	98.6	76.0	90.2	99.6	75.0	90.4	99.7	75.1	90.5	99.6	75.0	90.5	99.6	75.0	90.4
03	CONCRETE	97.7	67.9	84.5	95.3	73.7	85.7	95.5	75.7	86.7	108.2	75.9	93.9	103.1	71.0	88.9	95.9	71.1	84.9
04	MASONRY	99.3	64.7	77.9	102.2	63.8	78.5	98.9	60.5	75.2	98.9	60.5	75.2	98.9	60.5	75.2	159.3	56.3	95.7
05	METALS	98.1	80.9	92.9	100.4	80.9	94.5	96.7	80.5	91.8	96.7	80.7	91.9	96.8	80.5	91.9	97.5	79.7	92.1
06	WOOD, PLASTICS & COMPOSITES	94.5	49.0	69.8	93.4	66.5	78.8	83.7	74.8	78.9	87.7	74.8	80.7	82.6	60.4	70.6	92.9	61.7	75.9
07	THERMAL & MOISTURE PROTECTION	103.3	66.2	87.2	98.1	67.9	85.0	99.8	65.5	84.8	101.3	76.8	90.6	100.7	69.4	87.0	99.9	68.3	86.1
08	OPENINGS	106.0	60.0	95.3	109.6	69.7	100.3	114.0	74.3	104.7	113.6	74.3	104.5	113.8	66.3	102.8	114.4	66.7	103.3
0920	Plaster & Gypsum Board	103.3	47.6	66.6	92.9	65.6	74.9	89.9	74.1	79.5	90.2	74.1	79.6	89.9	59.4	69.8	102.4	60.7	74.9
0950, 0980	Ceilings & Acoustic Treatment	105.4	47.6	66.2	99.3	65.6	76.5	101.2	74.1	82.9	101.2	74.1	82.9	101.2	59.4	72.9	101.2	60.7	73.7
0960	Flooring	104.2	72.0	94.8	102.2	72.0	93.5	96.1	46.0	81.6	98.9	46.0	83.6	95.6	62.3	85.9	101.2	46.0	85.2
0970, 0990	Wall Finishes & Painting/Coating	94.7	59.9	74.1	98.9	59.9	75.8	95.5	62.1	75.7	95.5	62.1	75.7	95.5	62.1	75.7	95.5	62.1	75.7
09	FINISHES	96.7	56.6	74.8	93.7	67.0	79.1	89.2	66.5	76.8	91.3	66.5	77.8	89.9	61.3	74.3	92.5	58.8	74.1
COVERS	DIVS. 10 - 14, 25, 28, 41, 43, 44, 46	100.0	93.7	98.6	100.0	89.0	97.6	100.0	93.8	98.6	100.0	93.8	98.6	100.0	79.0	95.4	100.0	84.3	96.5
21, 22, 23	FIRE SUPPRESSION, PLUMBING & HVAC	100.0	71.3	88.4	99.9	71.3	88.4	98.1	70.5	86.9	98.1	70.5	86.9	98.1	70.5	86.9	99.9	70.5	88.0
26, 27, 3370	ELECTRICAL, COMMUNICATIONS & UTIL.	96.7	61.8	78.9	98.1	62.7	80.1	96.9	63.9	80.1	96.9	63.9	80.1	96.9	60.6	78.4	95.2	78.5	86.7
MF2016	WEIGHTED AVERAGE	99.3	69.5	86.5	99.6	72.1	87.8	98.1	72.3	87.1	100.3	72.7	88.5	99.4	69.7	86.6	101.9	71.6	88.9

	DIVISION	WYOMING												CANADA					
		SHERIDAN			WHEATLAND			WORLAND			YELLOWSTONE NAT'L PA			BARRIE, ONTARIO			BATHURST, NEW BRUNSWICK		
		828			822			824			821								
		MAT.	INST.	TOTAL	MAT.	INST.	TOTAL	MAT.	INST.	TOTAL	MAT.	INST.	TOTAL	MAT.	INST.	TOTAL	MAT.	INST.	TOTAL
015433	CONTRACTOR EQUIPMENT		97.6	97.6		97.6	97.6		97.6	97.6		97.6	97.6		103.0	103.0		102.7	102.7
0241, 31 - 34	SITE & INFRASTRUCTURE, DEMOLITION	92.1	94.7	93.9	89.4	94.2	92.7	86.8	94.2	91.9	86.9	94.2	92.0	118.2	99.7	105.3	106.1	95.5	98.7
0310	Concrete Forming & Accessories	98.1	62.7	68.0	92.6	49.7	56.1	92.7	62.7	67.2	92.7	62.8	67.2	124.2	84.6	90.5	103.9	59.4	66.1
0320	Concrete Reinforcing	113.6	84.8	98.9	112.9	84.2	98.3	113.6	84.7	98.9	115.6	84.7	99.9	177.7	89.4	132.8	144.5	59.2	101.1
0330	Cast-in-Place Concrete	103.0	76.0	92.9	104.0	74.9	93.2	99.6	75.0	90.4	99.6	75.0	90.4	156.6	84.6	129.8	111.1	57.9	91.3
03	CONCRETE	103.4	71.6	89.3	100.3	65.3	84.8	95.7	71.3	84.9	95.9	71.3	85.0	136.7	85.8	114.1	107.5	59.9	86.4
04	MASONRY	99.2	62.0	76.2	99.2	52.2	70.2	98.9	60.5	75.2	98.9	60.5	75.2	165.9	91.2	119.7	160.0	58.7	97.4
05	METALS	100.2	80.7	94.3	96.6	79.6	91.5	96.8	80.4	91.8	97.4	79.9	92.1	110.2	92.4	104.8	118.4	74.3	105.0
06	WOOD, PLASTICS & COMPOSITES	95.0	60.4	76.2	85.5	44.2	63.1	85.6	61.7	72.6	85.6	61.7	72.6	118.5	83.2	99.3	99.7	59.6	77.9
07	THERMAL & MOISTURE PROTECTION	101.0	66.7	86.1	100.0	59.5	82.4	99.8	66.4	85.3	99.2	66.4	84.9	114.6	88.5	103.3	110.9	60.1	88.8
08	OPENINGS	114.5	66.3	103.3	112.5	57.4	99.7	114.2	67.0	103.2	106.8	66.7	97.5	92.6	83.2	90.4	85.8	52.7	78.1
0920	Plaster & Gypsum Board	112.7	59.4	77.6	89.9	42.7	58.8	89.9	60.7	70.7	90.1	60.7	70.7	151.9	82.7	106.3	121.5	58.4	79.9
0950, 0980	Ceilings & Acoustic Treatment	104.0	59.4	73.8	101.2	42.7	61.6	101.2	60.7	73.7	102.0	60.7	74.0	95.1	82.7	86.7	115.9	58.4	76.9
0960	Flooring	99.5	62.3	88.7	97.6	45.2	82.4	97.6	46.0	82.6	97.6	46.0	82.6	118.9	90.2	110.6	100.1	42.9	83.5
0970, 0990	Wall Finishes & Painting/Coating	97.7	59.9	75.3	95.5	59.9	74.5	95.5	62.1	75.7	95.5	62.1	75.7	105.0	86.0	93.7	110.7	49.2	74.3
09	FINISHES	96.5	61.7	77.5	89.9	48.1	67.1	89.7	58.8	72.8	89.8	58.9	72.9	109.7	85.9	96.7	105.7	55.4	78.2
COVERS	DIVS. 10 - 14, 25, 28, 41, 43, 44, 46	100.0	94.9	98.9	100.0	83.2	96.3	100.0	77.8	95.1	100.0	77.8	95.1	139.2	66.4	123.2	131.1	59.1	115.3
21, 22, 23	FIRE SUPPRESSION, PLUMBING & HVAC	98.1	71.3	87.3	98.1	70.5	86.9	98.1	70.5	86.9	98.1	70.5	86.9	103.2	95.6	100.1	103.1	66.2	88.2
26, 27, 3370	ELECTRICAL, COMMUNICATIONS & UTIL.	99.2	60.6	79.6	96.9	78.5	87.5	96.9	78.5	87.5	95.9	88.2	92.0	115.0	84.9	99.7	108.3	57.2	82.3
MF2016	WEIGHTED AVERAGE	100.9	70.6	87.9	98.8	68.1	85.6	98.3	71.8	87.0	97.6	73.2	87.1	116.0	89.3	104.6	110.4	63.9	90.5

	DIVISION	CANADA																	
		BRANDON, MANITOBA			BRANTFORD, ONTARIO			BRIDGEWATER, NOVA SCOTIA			CALGARY, ALBERTA			CAP-DE-LA-MADELEINE, QUEBEC			CHARLESBOURG, QUEBEC		
		MAT.	INST.	TOTAL	MAT.	INST.	TOTAL	MAT.	INST.	TOTAL	MAT.	INST.	TOTAL	MAT.	INST.	TOTAL	MAT.	INST.	TOTAL
015433	CONTRACTOR EQUIPMENT		104.7	104.7		103.1	103.1		102.7	102.7		132.6	132.6		103.5	103.5		103.5	103.5
0241, 31 - 34	SITE & INFRASTRUCTURE, DEMOLITION	128.6	98.0	107.3	117.9	100.1	105.5	102.3	97.4	98.9	127.2	125.7	126.2	98.1	98.9	98.7	98.1	98.9	98.7
0310	Concrete Forming & Accessories	142.7	68.0	79.1	124.4	91.7	96.5	97.0	69.8	73.9	119.4	97.4	100.6	130.4	81.4	88.7	130.4	81.4	88.7
0320	Concrete Reinforcing	184.3	56.1	119.1	171.2	88.1	128.9	145.8	49.0	96.6	136.9	84.2	110.1	145.8	74.6	109.6	145.8	74.6	109.6
0330	Cast-in-Place Concrete	104.8	72.4	92.7	127.3	105.0	119.0	130.8	69.0	107.8	137.0	107.0	125.8	102.7	90.2	98.1	102.7	90.2	98.1
03	CONCRETE	115.2	68.4	94.5	119.1	95.9	108.8	117.4	66.9	95.0	121.8	99.3	111.8	105.1	83.8	95.6	105.1	83.8	95.6
04	MASONRY	213.4	62.0	119.8	166.5	95.4	122.6	162.1	67.0	103.3	204.0	88.8	132.8	162.8	77.9	110.4	162.8	77.9	110.4
05	METALS	133.1	79.1	116.8	116.8	93.1	109.6	116.1	76.9	104.2	134.9	104.4	125.7	115.2	86.0	106.4	115.2	86.0	106.4
06	WOOD, PLASTICS & COMPOSITES	154.7	68.8	108.0	121.8	90.3	104.7	90.4	69.4	79.0	99.9	97.2	98.4	133.5	81.3	105.2	133.5	81.3	105.2
07	THERMAL & MOISTURE PROTECTION	128.6	70.6	103.4	121.0	94.0	109.3	115.4	69.6	95.5	128.7	96.5	114.6	114.2	86.5	102.2	114.2	86.5	102.2
08	OPENINGS	102.4	61.6	92.9	89.7	89.2	89.6	84.2	63.1	79.3	82.9	86.0	83.7	91.1	74.4	87.2	91.1	74.4	87.2
0920	Plaster & Gypsum Board	113.7	67.5	83.3	113.1	90.0	97.9	118.4	68.4	85.5	125.1	96.0	106.0	142.5	80.7	101.8	142.5	80.7	101.8
0950, 0980	Ceilings & Acoustic Treatment	122.4	67.5	85.2	104.7	90.0	94.7	104.7	68.4	80.1	150.3	96.0	113.5	104.7	80.7	88.4	104.7	80.7	88.4
0960	Flooring	131.3	64.0	111.8	113.6	90.2	106.8	96.1	60.9	85.9	121.3	86.7	111.2	113.6	88.8	106.5	113.6	88.8	106.5
0970, 0990	Wall Finishes & Painting/Coating	117.6	55.5	80.8	109.8	94.4	100.7	109.8	60.9	80.8	115.0	109.2	111.6	109.8	85.8	95.6	109.8	85.8	95.6
09	FINISHES	119.9	66.2	90.6	106.6	92.0	98.6	101.7	67.5	83.0	124.6	97.2	109.7	110.2	83.6	95.7	110.2	83.6	95.7
COVERS	DIVS. 10 - 14, 25, 28, 41, 43, 44, 46	131.1	61.4	115.8	131.1	68.4	117.3	131.1	62.1	115.9	131.1	94.0	123.0	131.1	77.2	119.3	131.1	77.2	119.3
21, 22, 23	FIRE SUPPRESSION, PLUMBING & HVAC	103.3	80.7	94.1	103.1	98.5	101.3	103.1	81.3	94.3	104.9	91.6	99.5	103.5	86.1	96.5	103.5	86.1	96.5
26, 27, 3370	ELECTRICAL, COMMUNICATIONS & UTIL.	110.5	64.7	87.2	107.3	84.5	95.7	111.7	60.1	85.5	102.0	96.2	99.1	106.3	66.8	86.2	106.3	66.8	86.2
MF2016	WEIGHTED AVERAGE	120.9	72.4	100.1	113.1	93.1	104.5	111.2	72.2	94.5	119.0	97.6	109.8	110.7	82.1	98.4	110.7	82.1	98.4

	DIVISION	CANADA																	
		CHARLOTTETOWN, PRINCE EDWARD ISLAND			CHICOUTIMI, QUEBEC			CORNER BROOK, NEWFOUNDLAND			CORNWALL, ONTARIO			DALHOUSIE, NEW BRUNSWICK			DARTMOUTH, NOVA SCOTIA		
		MAT.	INST.	TOTAL	MAT.	INST.	TOTAL	MAT.	INST.	TOTAL	MAT.	INST.	TOTAL	MAT.	INST.	TOTAL	MAT.	INST.	TOTAL
015433	CONTRACTOR EQUIPMENT		120.9	120.9		103.3	103.3		103.2	103.2		103.1	103.1		103.0	103.0		102.0	102.0
0241, 31 - 34	SITE & INFRASTRUCTURE, DEMOLITION	133.4	108.0	115.7	103.1	98.4	99.8	132.7	95.4	106.7	116.1	99.6	104.6	101.4	95.7	97.4	119.9	97.0	103.9
0310	Concrete Forming & Accessories	117.5	54.4	63.8	133.0	89.2	95.7	119.3	77.7	83.9	122.2	84.7	90.3	103.1	59.7	66.1	108.9	69.8	75.6
0320	Concrete Reinforcing	164.4	47.9	105.2	109.0	93.9	101.3	169.0	50.2	108.6	171.2	87.8	128.8	146.3	59.2	102.1	176.7	49.0	111.8
0330	Cast-in-Place Concrete	147.6	58.3	114.4	104.7	94.9	101.0	122.0	65.5	101.0	114.5	95.7	107.5	106.9	58.1	88.7	117.7	68.8	99.5
03	CONCRETE	130.6	56.5	97.8	99.3	92.1	96.1	145.1	69.4	111.6	113.0	89.5	102.6	109.7	60.0	87.7	128.4	66.8	101.1
04	MASONRY	176.6	56.0	102.1	162.3	88.4	116.7	209.3	75.9	126.9	165.5	87.4	117.2	163.7	58.7	98.8	223.5	67.0	126.8
05	METALS	141.8	81.5	123.6	118.1	91.2	110.0	133.6	76.0	116.1	116.6	91.8	109.1	109.7	74.6	99.1	134.2	76.6	116.7
06	WOOD, PLASTICS & COMPOSITES	103.7	53.8	76.6	135.4	89.8	110.6	131.2	83.8	105.5	120.2	84.0	100.5	97.6	59.6	76.9	118.4	69.3	91.7
07	THERMAL & MOISTURE PROTECTION	129.6	58.4	98.6	111.4	96.0	104.7	133.2	67.8	104.8	120.8	88.6	106.8	120.7	60.1	94.3	130.2	69.5	103.8
08	OPENINGS	85.3	47.0	76.4	89.7	76.1	86.6	108.6	68.5	99.3	91.1	82.9	89.2	87.3	52.7	79.3	93.2	63.0	86.2
0920	Plaster & Gypsum Board	118.6	52.0	74.8	141.4	89.3	107.1	145.6	83.2	104.6	168.5	83.5	112.5	126.5	58.4	81.6	140.7	68.4	93.1
0950, 0980	Ceilings & Acoustic Treatment	128.3	52.0	76.6	115.1	89.3	97.6	123.2	83.2	96.1	107.2	83.5	91.2	107.0	58.4	74.1	130.0	68.4	88.3
0960	Flooring	116.7	57.1	99.4	115.8	88.8	108.0	113.2	51.2	95.2	113.6	88.9	106.5	102.3	65.6	91.7	108.1	60.9	94.4
0970, 0990	Wall Finishes & Painting/Coating	116.4	40.9	71.7	110.7	103.2	106.3	117.4	57.9	82.2	109.8	87.9	96.8	113.4	49.2	75.4	117.4	60.9	84.0
09	FINISHES	118.8	53.8	83.3	112.6	91.6	101.1	119.8	72.3	93.8	115.1	86.0	99.2	106.5	60.0	81.1	117.6	67.5	90.2
COVERS	DIVS. 10 - 14, 25, 28, 41, 43, 44, 46	131.1	59.3	115.3	131.1	80.9	120.1	131.1	61.9	115.9	131.1	66.0	116.8	131.1	59.1	115.3	131.1	62.1	115.9
21, 22, 23	FIRE SUPPRESSION, PLUMBING & HVAC	103.4	60.3	86.0	103.1	80.0	93.7	103.3	67.4	88.8	103.5	96.4	100.7	103.2	66.2	88.2	103.3	81.3	94.4
26, 27, 3370	ELECTRICAL, COMMUNICATIONS & UTIL.	107.2	48.6	77.4	105.0	83.1	93.9	107.9	52.8	79.9	107.6	85.4	96.3	108.5	54.1	80.8	112.1	60.1	85.7
MF2016	WEIGHTED AVERAGE	120.1	61.8	95.1	110.3	87.3	100.4	125.0	70.2	101.5	113.1	89.6	103.0	109.8	64.1	90.2	121.8	72.1	100.5

	DIVISION	CANADA																	
		EDMONTON, ALBERTA			FORT MCMURRAY, ALBERTA			FREDERICTON, NEW BRUNSWICK			GATINEAU, QUEBEC			GRANBY, QUEBEC			HALIFAX, NOVA SCOTIA		
		MAT.	INST.	TOTAL	MAT.	INST.	TOTAL	MAT.	INST.	TOTAL	MAT.	INST.	TOTAL	MAT.	INST.	TOTAL	MAT.	INST.	TOTAL
015433	CONTRACTOR EQUIPMENT		132.2	132.2		105.3	105.3		120.6	120.6		103.5	103.5		103.5	103.5		119.0	119.0
0241, 31 - 34	SITE & INFRASTRUCTURE, DEMOLITION	124.4	125.5	125.1	122.9	100.7	107.4	112.2	109.3	110.2	97.9	98.9	98.6	98.4	98.9	98.8	104.1	109.4	107.8
0310	Concrete Forming & Accessories	121.1	97.4	100.9	122.1	89.8	94.6	122.4	60.2	69.5	130.4	81.3	88.6	130.4	81.3	88.6	117.8	83.3	88.5
0320	Concrete Reinforcing	137.3	84.2	110.3	158.5	84.1	120.6	161.2	59.5	109.5	154.2	74.5	113.7	154.2	74.5	113.7	155.1	72.1	112.9
0330	Cast-in-Place Concrete	131.0	106.9	122.0	169.0	101.1	143.7	108.4	58.7	89.9	101.3	90.2	97.2	104.7	90.2	99.3	102.5	84.6	95.8
03	CONCRETE	119.2	99.2	110.3	136.8	93.0	117.4	112.3	61.3	89.7	105.4	83.7	95.8	107.0	83.7	96.6	107.9	83.1	96.9
04	MASONRY	199.0	88.8	130.9	206.9	85.5	131.9	183.6	60.0	107.2	162.7	77.9	110.3	163.0	77.9	110.4	181.0	85.3	121.9
05	METALS	142.1	104.2	130.6	145.7	91.6	129.3	142.1	86.5	125.3	115.2	85.9	106.3	115.4	85.8	106.5	142.0	97.6	128.6
06	WOOD, PLASTICS & COMPOSITES	98.8	97.2	97.9	116.0	89.3	101.5	109.0	59.8	82.3	133.5	81.3	105.2	133.5	81.3	105.2	104.1	82.5	92.4
07	THERMAL & MOISTURE PROTECTION	140.5	96.5	121.3	129.5	92.0	113.2	131.1	60.5	100.4	114.2	86.5	102.2	114.2	85.0	101.5	134.2	84.9	112.7
08	OPENINGS	82.0	86.0	82.9	91.1	81.6	88.9	86.7	51.7	78.6	91.1	69.9	86.2	91.1	69.9	86.2	87.1	75.2	84.4
0920	Plaster & Gypsum Board	122.4	96.0	105.0	113.6	88.6	97.1	121.7	58.4	80.0	112.4	80.7	91.5	112.4	80.7	91.5	111.5	81.7	91.9
0950, 0980	Ceilings & Acoustic Treatment	145.4	96.0	112.0	112.3	88.6	96.3	132.9	58.4	82.4	104.7	80.7	88.4	104.7	80.7	88.4	142.1	81.7	101.2
0960	Flooring	122.0	86.7	111.7	113.6	86.7	105.8	118.0	68.5	103.6	113.6	88.8	106.5	113.6	88.8	106.5	111.2	81.9	102.7
0970, 0990	Wall Finishes & Painting/Coating	114.8	109.2	111.5	109.9	90.6	98.5	116.0	62.7	84.4	109.8	85.8	95.6	109.8	85.8	95.6	119.2	90.0	101.9
09	FINISHES	119.6	97.2	107.4	109.5	89.6	98.6	120.1	62.1	88.5	105.8	83.6	93.7	105.8	83.6	93.7	115.1	84.2	98.2
COVERS	DIVS. 10 - 14, 25, 28, 41, 43, 44, 46	131.1	94.0	123.0	131.1	90.4	122.2	131.1	59.7	115.4	131.1	77.2	119.3	131.1	77.2	119.3	131.1	67.4	117.1
21, 22, 23	FIRE SUPPRESSION, PLUMBING & HVAC	104.8	91.6	99.5	103.6	95.3	100.2	103.5	75.3	92.1	103.5	86.1	96.5	103.1	86.1	96.2	104.5	80.9	95.0
26, 27, 3370	ELECTRICAL, COMMUNICATIONS & UTIL.	105.7	96.2	100.9	101.7	79.4	90.4	106.5	71.4	88.6	106.3	66.8	86.2	106.9	66.8	86.5	108.2	84.4	96.1
MF2016	WEIGHTED AVERAGE	119.7	97.5	110.2	122.1	90.2	108.4	117.9	71.2	97.9	110.4	81.9	98.2	110.6	81.8	98.3	117.1	85.8	103.7

	DIVISION	CANADA																	
		HAMILTON, ONTARIO			HULL, QUEBEC			JOLIETTE, QUEBEC			KAMLOOPS, BRITISH COLUMBIA			KINGSTON, ONTARIO			KITCHENER, ONTARIO		
		MAT.	INST.	TOTAL	MAT.	INST.	TOTAL	MAT.	INST.	TOTAL	MAT.	INST.	TOTAL	MAT.	INST.	TOTAL	MAT.	INST.	TOTAL
015433	CONTRACTOR EQUIPMENT		120.9	120.9		103.5	103.5		103.5	103.5		106.7	106.7		105.3	105.3		105.5	105.5
0241, 31 - 34	SITE & INFRASTRUCTURE, DEMOLITION	109.6	114.9	113.3	97.9	98.9	98.6	98.5	98.9	98.8	120.7	102.6	108.1	116.1	103.4	107.3	96.2	104.9	102.3
0310	Concrete Forming & Accessories	122.2	94.0	98.2	130.4	81.3	88.6	130.4	81.4	88.7	121.0	84.6	90.0	122.3	84.8	90.4	113.8	86.8	90.9
0320	Concrete Reinforcing	151.4	98.6	124.5	154.2	74.5	113.7	145.8	74.6	109.6	114.3	78.0	95.9	171.2	87.8	128.8	105.9	98.4	102.1
0330	Cast-in-Place Concrete	113.9	102.2	109.5	101.3	90.2	97.2	105.5	90.2	99.9	91.4	94.7	92.6	114.5	95.7	107.5	111.0	97.3	105.9
03	CONCRETE	110.0	98.3	104.8	105.4	83.7	95.8	106.4	83.8	96.4	114.2	87.4	102.3	114.8	89.5	103.6	96.6	92.7	94.9
04	MASONRY	167.0	99.7	125.4	162.7	77.9	110.3	163.1	77.9	110.5	170.0	85.7	117.9	172.4	87.5	119.9	143.0	96.9	114.5
05	METALS	141.7	105.7	130.8	115.4	85.9	106.5	115.4	86.0	106.5	117.3	88.2	108.5	118.0	91.8	110.1	130.3	96.3	120.0
06	WOOD, PLASTICS & COMPOSITES	102.6	93.1	97.4	133.5	81.3	105.2	133.5	81.3	105.2	103.4	83.0	92.3	120.2	84.1	100.6	108.5	84.8	95.6
07	THERMAL & MOISTURE PROTECTION	124.7	100.0	114.0	114.2	86.5	102.1	114.2	86.5	102.1	130.7	84.1	110.4	120.8	89.8	107.3	115.3	96.8	107.2
08	OPENINGS	83.5	91.9	85.4	91.1	69.9	86.2	91.1	74.4	87.2	87.9	80.8	86.3	91.1	82.6	89.1	81.8	85.6	82.7
0920	Plaster & Gypsum Board	118.6	92.3	101.3	112.4	80.7	91.5	142.5	80.7	101.8	98.6	82.1	87.8	171.0	83.6	113.5	102.6	84.4	90.6
0950, 0980	Ceilings & Acoustic Treatment	134.8	92.3	106.0	104.7	80.7	88.4	104.7	80.7	88.4	104.7	82.1	89.4	119.1	83.6	95.1	107.6	84.4	91.8
0960	Flooring	116.4	95.6	110.4	113.6	88.8	106.5	113.6	88.8	106.5	112.8	50.8	94.8	113.6	88.9	106.5	101.4	95.6	99.7
0970, 0990	Wall Finishes & Painting/Coating	110.4	103.5	106.3	109.8	85.8	95.6	109.8	85.8	95.6	109.8	78.4	91.2	109.8	81.3	92.9	103.8	93.3	97.6
09	FINISHES	115.6	95.3	104.5	105.8	83.6	93.7	110.2	83.6	95.7	106.1	78.0	90.8	117.8	85.4	100.1	99.6	88.7	93.6
COVERS	DIVS. 10 - 14, 25, 28, 41, 43, 44, 46	131.1	91.8	122.5	131.1	77.2	119.3	131.1	77.2	119.3	131.1	86.2	121.2	131.1	66.0	116.8	131.1	89.0	121.9
21, 22, 23	FIRE SUPPRESSION, PLUMBING & HVAC	104.9	90.5	99.1	103.1	86.1	96.2	103.1	86.1	96.3	103.1	89.5	97.6	103.5	96.6	100.7	104.0	89.2	98.1
26, 27, 3370	ELECTRICAL, COMMUNICATIONS & UTIL.	106.3	99.6	102.9	108.2	66.8	87.2	106.9	66.8	86.5	110.2	76.0	92.8	107.6	84.2	95.7	107.7	97.1	102.3
MF2016	WEIGHTED AVERAGE	116.1	98.1	108.4	110.5	81.9	98.2	110.9	82.1	98.5	113.0	85.7	101.3	114.1	89.7	103.6	109.4	93.4	102.5

	DIVISION	CANADA																	
		LAVAL, QUEBEC			LETHBRIDGE, ALBERTA			LLOYDMINSTER, ALBERTA			LONDON, ONTARIO			MEDICINE HAT, ALBERTA			MONCTON, NEW BRUNSWICK		
		MAT.	INST.	TOTAL	MAT.	INST.	TOTAL	MAT.	INST.	TOTAL	MAT.	INST.	TOTAL	MAT.	INST.	TOTAL	MAT.	INST.	TOTAL
015433	CONTRACTOR EQUIPMENT		103.5	103.5		105.3	105.3		105.3	105.3		122.4	122.4		105.3	105.3		102.7	102.7
0241, 31 - 34	SITE & INFRASTRUCTURE, DEMOLITION	98.4	98.9	98.8	115.5	101.3	105.6	115.4	100.7	105.2	104.1	114.7	111.5	114.2	100.8	104.8	105.5	95.8	98.8
0310	Concrete Forming & Accessories	130.6	81.3	88.6	123.6	89.9	94.9	121.7	80.3	86.5	122.4	88.4	93.4	123.6	80.3	86.7	103.9	62.0	68.2
0320	Concrete Reinforcing	154.2	74.5	113.7	158.5	84.1	120.6	158.5	84.1	120.6	151.4	97.3	123.9	158.5	84.0	120.6	144.5	70.8	107.0
0330	Cast-in-Place Concrete	104.7	90.2	99.3	126.8	101.1	117.2	117.6	97.5	110.1	120.4	100.1	112.8	117.6	97.5	110.1	107.0	68.2	92.6
03	CONCRETE	107.0	83.7	96.7	117.3	93.1	106.6	112.9	87.5	101.6	113.0	94.9	105.0	113.0	87.5	101.7	105.6	66.9	88.4
04	MASONRY	162.9	77.9	110.4	180.7	85.5	121.9	162.0	79.2	110.9	182.7	97.9	130.3	162.0	79.2	110.9	159.7	60.1	98.2
05	METALS	115.3	85.9	106.4	140.2	91.6	125.5	117.3	91.5	109.5	141.7	107.0	131.2	117.5	91.4	109.6	118.4	84.9	108.2
06	WOOD, PLASTICS & COMPOSITES	133.6	81.3	105.2	119.5	89.3	103.1	116.0	79.7	96.3	109.2	86.3	96.7	119.5	79.7	97.9	99.7	61.5	78.9
07	THERMAL & MOISTURE PROTECTION	114.7	86.5	102.4	126.7	92.0	111.6	123.0	87.5	107.6	124.5	97.6	112.8	130.0	87.5	111.5	115.6	64.2	93.2
08	OPENINGS	91.1	69.9	86.2	91.1	81.6	88.9	91.1	76.4	87.7	80.2	87.0	81.8	91.1	76.4	87.7	85.8	60.2	79.8
0920	Plaster & Gypsum Board	112.7	80.7	91.6	103.7	88.6	93.7	99.9	78.8	86.0	123.9	85.3	98.5	102.1	78.8	86.7	121.5	60.3	81.2
0950, 0980	Ceilings & Acoustic Treatment	104.7	80.7	88.4	112.3	88.6	96.3	104.7	78.8	87.1	142.7	85.3	103.8	104.7	78.8	87.1	115.9	60.3	78.2
0960	Flooring	113.6	88.8	106.5	113.6	86.7	105.8	113.6	86.7	105.8	112.8	95.6	107.8	113.6	86.7	105.8	100.1	67.2	90.5
0970, 0990	Wall Finishes & Painting/Coating	109.8	85.8	95.6	109.7	99.0	103.4	109.9	77.1	90.5	113.8	100.1	105.7	109.7	77.1	90.4	110.7	50.4	75.1
09	FINISHES	105.9	83.6	93.7	107.2	90.5	98.1	105.2	80.9	92.0	118.5	90.6	103.3	105.4	80.9	92.0	105.7	61.7	81.6
COVERS	DIVS. 10 - 14, 25, 28, 41, 43, 44, 46	131.1	77.2	119.3	131.1	90.4	122.2	131.1	87.3	121.5	131.1	90.7	122.2	131.1	87.3	121.5	131.1	60.7	115.6
21, 22, 23	FIRE SUPPRESSION, PLUMBING & HVAC	104.1	86.1	96.8	103.4	92.1	98.8	103.5	92.0	98.9	104.9	87.8	98.0	103.1	88.8	97.3	103.1	68.6	89.1
26, 27, 3370	ELECTRICAL, COMMUNICATIONS & UTIL.	105.4	66.8	85.8	103.1	79.4	91.1	100.9	79.4	90.0	103.5	97.1	100.3	100.9	79.4	90.0	112.0	59.0	85.1
MF2016	WEIGHTED AVERAGE	110.7	81.9	98.3	117.2	89.7	105.4	111.6	86.5	100.8	116.8	95.6	107.7	111.7	85.8	100.6	110.6	68.1	92.4

	DIVISION	CANADA																	
		MONTREAL, QUEBEC			MOOSE JAW, SASKATCHEWAN			NEW GLASGOW, NOVA SCOTIA			NEWCASTLE, NEW BRUNSWICK			NORTH BAY, ONTARIO			OSHAWA, ONTARIO		
		MAT.	INST.	TOTAL	MAT.	INST.	TOTAL	MAT.	INST.	TOTAL	MAT.	INST.	TOTAL	MAT.	INST.	TOTAL	MAT.	INST.	TOTAL
015433	CONTRACTOR EQUIPMENT		121.9	121.9		101.7	101.7		102.0	102.0		102.7	102.7		102.4	102.4		105.5	105.5
0241, 31 - 34	SITE & INFRASTRUCTURE, DEMOLITION	113.8	112.2	112.7	115.3	95.0	101.2	114.1	97.0	102.2	106.1	95.5	98.7	129.0	98.7	107.9	107.0	103.9	104.8
0310	Concrete Forming & Accessories	126.2	89.7	95.1	106.8	56.1	63.7	108.8	69.8	75.6	103.9	59.7	66.2	143.6	82.3	91.4	119.2	87.0	91.8
0320	Concrete Reinforcing	134.8	94.0	114.0	111.8	63.2	87.1	169.0	49.0	108.0	144.5	59.2	101.2	199.7	87.3	142.6	167.8	93.3	129.9
0330	Cast-in-Place Concrete	123.6	96.6	113.5	114.1	65.7	96.1	117.7	68.8	99.5	111.1	58.0	91.4	113.4	82.6	101.9	128.3	85.4	112.3
03	CONCRETE	112.8	93.6	104.3	99.2	61.7	82.6	127.5	66.8	100.6	107.5	60.0	86.5	129.4	83.7	109.2	117.7	87.9	104.5
04	MASONRY	168.0	88.4	118.9	160.7	57.6	97.0	208.9	67.0	121.2	160.0	58.7	97.4	215.0	83.6	133.8	146.0	89.9	111.4
05	METALS	149.0	102.6	134.9	114.2	75.9	102.6	131.6	76.6	114.9	118.4	74.5	105.1	132.4	91.2	119.9	119.9	95.0	112.4
06	WOOD, PLASTICS & COMPOSITES	108.8	90.2	98.7	100.7	54.7	75.7	118.4	69.3	91.7	99.7	59.6	77.9	157.8	82.6	117.0	115.5	85.9	99.4
07	THERMAL & MOISTURE PROTECTION	118.8	96.2	109.0	113.5	61.8	91.0	130.2	69.5	103.8	115.6	60.1	91.4	136.7	85.0	114.2	116.2	89.3	104.5
08	OPENINGS	85.7	78.2	83.9	87.1	52.9	79.2	93.2	63.0	86.2	85.8	52.7	78.1	100.8	80.5	96.1	86.9	87.1	86.9
0920	Plaster & Gypsum Board	117.7	89.3	99.0	96.1	53.4	68.0	139.1	68.4	92.6	121.5	58.4	79.9	136.4	82.1	100.6	105.7	85.4	92.4
0950, 0980	Ceilings & Acoustic Treatment	140.2	89.3	105.7	104.7	53.4	70.0	122.4	68.4	85.8	115.9	58.4	76.9	122.4	82.1	95.1	103.3	85.4	91.2
0960	Flooring	114.0	91.0	107.3	103.9	55.5	89.8	108.1	60.9	94.4	100.1	65.6	90.0	131.3	88.9	119.0	104.3	98.1	102.5
0970, 0990	Wall Finishes & Painting/Coating	112.2	103.2	106.9	109.8	62.6	81.9	117.4	60.9	84.0	110.7	49.2	74.3	117.4	87.3	99.6	103.8	107.5	106.0
09	FINISHES	116.5	92.2	103.2	101.7	56.4	76.9	115.9	67.5	89.4	105.7	59.9	80.7	122.9	84.3	101.9	100.7	91.0	95.4
COVERS	DIVS. 10 - 14, 25, 28, 41, 43, 44, 46	131.1	81.8	120.3	131.1	58.7	115.2	131.1	62.1	115.9	131.1	59.1	115.3	131.1	64.7	116.5	131.1	88.6	121.8
21, 22, 23	FIRE SUPPRESSION, PLUMBING & HVAC	105.1	80.1	95.0	103.5	71.9	90.7	103.3	81.3	94.4	103.1	66.2	88.2	103.3	94.5	99.7	104.0	100.0	102.4
26, 27, 3370	ELECTRICAL, COMMUNICATIONS & UTIL.	107.2	83.1	95.0	109.2	57.7	83.0	108.2	60.1	83.8	107.7	57.2	82.0	108.9	85.3	96.9	108.8	87.6	98.0
MF2016	WEIGHTED AVERAGE	118.1	89.9	106.0	109.0	65.6	90.4	119.9	72.1	99.4	110.5	64.5	90.8	122.8	87.4	107.7	111.6	92.9	103.6

	DIVISION	CANADA																	
		OTTAWA, ONTARIO			OWEN SOUND, ONTARIO			PETERBOROUGH, ONTARIO			PORTAGE LA PRAIRIE, MANITOBA			PRINCE ALBERT, SASKATCHEWAN			PRINCE GEORGE, BRITISH COLUMBIA		
		MAT.	INST.	TOTAL	MAT.	INST.	TOTAL	MAT.	INST.	TOTAL	MAT.	INST.	TOTAL	MAT.	INST.	TOTAL	MAT.	INST.	TOTAL
015433	CONTRACTOR EQUIPMENT		123.1	123.1		103.0	103.0		103.1	103.1		105.3	105.3		101.7	101.7		106.7	106.7
0241, 31 - 34	SITE & INFRASTRUCTURE, DEMOLITION	108.6	114.7	112.8	118.2	99.6	105.2	117.9	99.5	105.1	116.6	98.4	103.9	111.0	95.2	100.0	124.1	102.6	109.1
0310	Concrete Forming & Accessories	123.5	88.4	93.6	124.2	80.9	87.3	124.4	83.3	89.4	123.8	67.6	75.9	106.8	55.9	63.5	112.2	79.7	84.5
0320	Concrete Reinforcing	135.1	97.3	115.9	177.7	89.4	132.8	171.2	87.8	128.8	158.5	56.1	106.4	116.8	63.1	89.5	114.3	78.0	95.9
0330	Cast-in-Place Concrete	121.6	102.9	114.6	156.6	78.9	127.7	127.3	84.3	111.3	117.6	71.9	100.6	103.4	65.6	89.3	114.5	94.7	107.1
03	CONCRETE	111.8	96.0	104.8	136.7	82.2	112.5	119.1	84.9	103.9	107.1	68.1	89.8	94.8	61.6	80.1	124.4	85.2	107.0
04	MASONRY	164.0	100.4	124.7	165.9	88.9	118.3	166.5	89.9	119.2	165.2	61.0	100.8	159.9	57.6	96.7	172.0	85.7	118.7
05	METALS	142.5	108.4	132.2	110.2	92.2	104.7	116.8	91.9	109.3	117.5	79.3	105.9	114.2	75.7	102.6	117.3	88.2	108.5
06	WOOD, PLASTICS & COMPOSITES	107.8	85.2	95.5	118.5	79.5	97.3	121.8	81.4	99.8	119.5	68.8	91.9	100.7	54.7	75.7	103.4	76.3	88.7
07	THERMAL & MOISTURE PROTECTION	131.6	98.8	117.4	114.6	85.8	102.1	121.0	90.6	107.8	114.0	70.2	94.9	113.4	60.7	90.4	124.2	83.3	106.4
08	OPENINGS	88.5	86.3	88.0	92.6	79.9	89.6	89.7	82.3	88.0	91.1	61.7	84.2	86.1	52.9	78.4	87.9	77.1	85.4
0920	Plaster & Gypsum Board	120.1	84.1	96.4	151.9	78.9	103.8	113.1	80.8	91.8	101.7	67.5	79.2	96.1	53.4	68.0	98.6	75.2	83.2
0950, 0980	Ceilings & Acoustic Treatment	136.4	84.1	101.0	95.1	78.9	84.1	104.7	80.8	88.5	104.7	67.5	79.5	104.7	53.4	70.0	104.7	75.2	84.7
0960	Flooring	110.7	91.1	105.0	118.9	90.2	110.6	113.6	88.9	106.5	113.6	64.0	99.2	103.9	55.5	89.8	109.2	69.7	97.7
0970, 0990	Wall Finishes & Painting/Coating	110.6	94.9	101.3	105.0	86.0	93.7	109.8	89.4	97.7	109.9	55.5	77.7	109.8	53.4	76.4	109.8	78.4	91.2
09	FINISHES	117.0	89.1	101.8	109.7	83.2	95.2	106.6	85.0	94.8	105.3	66.0	83.8	101.7	55.4	76.4	105.0	77.3	89.9
COVERS	DIVS. 10 - 14, 25, 28, 41, 43, 44, 46	131.1	89.3	121.9	139.2	65.2	122.9	131.1	66.2	116.8	131.1	61.1	115.7	131.1	58.7	115.2	131.1	85.5	121.1
21, 22, 23	FIRE SUPPRESSION, PLUMBING & HVAC	104.9	89.5	98.7	103.2	94.4	99.6	103.1	97.9	101.0	103.1	80.1	93.8	103.5	64.8	87.9	103.1	89.5	97.6
26, 27, 3370	ELECTRICAL, COMMUNICATIONS & UTIL.	103.4	97.9	100.6	116.4	84.2	100.0	107.3	84.9	95.9	108.9	55.8	81.9	109.2	57.7	83.0	107.4	76.0	91.4
MF2016	WEIGHTED AVERAGE	116.9	96.4	108.1	116.1	87.6	103.9	113.1	89.4	102.9	111.6	70.9	94.1	108.2	63.9	89.2	113.9	85.0	101.5

	DIVISION	CANADA																	
		QUEBEC CITY, QUEBEC			RED DEER, ALBERTA			REGINA, SASKATCHEWAN			RIMOUSKI, QUEBEC			ROUYN-NORANDA, QUEBEC			SAINT HYACINTHE, QUEBEC		
		MAT.	INST.	TOTAL	MAT.	INST.	TOTAL	MAT.	INST.	TOTAL	MAT.	INST.	TOTAL	MAT.	INST.	TOTAL	MAT.	INST.	TOTAL
015433	**CONTRACTOR EQUIPMENT**		122.8	122.8		105.3	105.3		127.7	127.7		103.5	103.5		103.5	103.5		103.5	103.5
0241, 31 - 34	**SITE & INFRASTRUCTURE, DEMOLITION**	116.2	112.1	113.3	114.2	100.8	104.8	127.0	119.3	121.6	98.2	98.5	98.4	97.9	98.9	98.6	98.4	98.9	98.8
0310	Concrete Forming & Accessories	123.5	89.8	94.8	136.5	80.3	88.6	125.4	93.8	98.5	130.4	89.3	95.4	130.4	81.3	88.6	130.4	81.3	88.6
0320	Concrete Reinforcing	135.0	94.0	114.2	158.5	84.0	120.6	137.5	91.4	114.1	110.0	93.9	101.8	154.2	74.5	113.7	154.2	74.5	113.7
0330	Cast-in-Place Concrete	118.2	96.3	110.1	117.6	97.5	110.1	138.1	99.2	123.6	106.6	94.9	102.2	101.3	90.2	97.2	104.7	90.2	99.3
03	**CONCRETE**	110.2	93.7	102.9	113.8	87.5	102.2	121.2	96.1	110.1	102.7	92.1	98.0	105.4	83.7	95.8	107.0	83.7	96.7
04	**MASONRY**	162.6	88.4	116.8	162.0	79.2	110.9	192.6	89.2	128.7	162.5	88.4	116.7	162.7	77.9	110.3	162.9	77.9	110.4
05	**METALS**	142.3	104.4	130.8	117.5	91.4	109.6	148.8	103.1	134.9	115.0	91.3	107.8	115.4	85.9	106.5	115.4	85.9	106.5
06	**WOOD, PLASTICS & COMPOSITES**	113.7	90.1	100.9	119.5	79.7	97.9	110.3	94.6	101.8	133.5	89.8	109.7	133.5	81.3	105.2	133.5	81.3	105.2
07	**THERMAL & MOISTURE PROTECTION**	119.0	96.2	109.1	140.7	87.5	117.5	139.2	88.5	117.1	114.2	96.0	106.3	114.2	86.5	102.1	114.4	86.5	102.3
08	**OPENINGS**	87.4	85.6	87.0	91.1	76.4	87.7	86.3	81.8	85.3	90.7	76.1	87.3	91.1	69.9	86.2	91.1	69.9	86.2
0920	Plaster & Gypsum Board	121.7	89.3	100.4	102.1	78.8	86.7	131.2	93.8	106.6	142.3	89.3	107.4	112.2	80.7	91.4	112.2	80.7	91.4
0950, 0980	Ceilings & Acoustic Treatment	139.5	89.3	105.5	104.7	78.8	87.1	147.1	93.8	111.0	103.8	89.3	94.0	103.8	80.7	88.1	103.8	80.7	88.1
0960	Flooring	112.3	91.0	106.1	116.3	86.7	107.7	123.9	95.6	115.7	115.0	88.8	107.4	113.6	88.8	106.5	113.6	88.8	106.5
0970, 0990	Wall Finishes & Painting/Coating	122.0	103.2	110.9	109.7	77.1	90.4	115.3	91.4	101.2	113.0	103.2	107.2	109.8	85.8	95.6	109.8	85.8	95.6
09	**FINISHES**	116.5	92.2	103.2	106.1	80.9	92.4	128.0	94.9	109.9	110.7	91.6	100.2	105.7	83.6	93.6	105.7	83.6	93.6
COVERS	**DIVS. 10 - 14, 25, 28, 41, 43, 44, 46**	131.1	81.6	120.2	131.1	87.3	121.5	131.1	70.8	117.8	131.1	80.9	120.1	131.1	77.2	119.3	131.1	77.2	119.3
21, 22, 23	**FIRE SUPPRESSION, PLUMBING & HVAC**	104.8	80.1	94.8	103.1	88.8	97.3	104.4	88.8	98.1	103.1	80.0	93.8	103.1	86.1	96.2	99.7	86.1	94.2
26, 27, 3370	**ELECTRICAL, COMMUNICATIONS & UTIL.**	104.5	83.1	93.6	100.9	79.4	90.0	107.6	93.7	100.6	106.9	83.1	94.8	106.9	66.8	86.5	107.5	66.8	86.8
MF2016	**WEIGHTED AVERAGE**	116.4	90.3	105.2	112.2	85.8	100.8	122.1	94.2	110.1	110.3	87.3	100.4	110.4	81.9	98.1	109.8	81.9	97.8

	DIVISION	CANADA																	
		SAINT JOHN, NEW BRUNSWICK			SARNIA, ONTARIO			SASKATOON, SASKATCHEWAN			SAULT STE MARIE, ONTARIO			SHERBROOKE, QUEBEC			SOREL, QUEBEC		
		MAT.	INST.	TOTAL	MAT.	INST.	TOTAL	MAT.	INST.	TOTAL	MAT.	INST.	TOTAL	MAT.	INST.	TOTAL	MAT.	INST.	TOTAL
015433	**CONTRACTOR EQUIPMENT**		102.7	102.7		103.1	103.1		101.7	101.7		103.1	103.1		103.5	103.5		103.5	103.5
0241, 31 - 34	**SITE & INFRASTRUCTURE, DEMOLITION**	106.2	97.2	99.9	116.5	99.6	104.7	112.7	97.9	102.4	106.5	99.2	101.4	98.4	98.9	98.8	98.5	98.9	98.8
0310	Concrete Forming & Accessories	124.2	65.0	73.8	123.1	90.1	95.0	106.8	93.2	95.2	112.3	86.2	90.1	130.4	81.3	88.6	130.4	81.4	88.7
0320	Concrete Reinforcing	144.5	71.4	107.3	122.2	89.2	105.4	117.1	91.3	104.0	109.8	87.9	98.7	154.2	74.5	113.7	145.8	74.6	109.6
0330	Cast-in-Place Concrete	109.5	70.0	94.8	117.4	97.2	109.9	109.7	96.0	104.6	105.5	83.0	97.1	104.7	90.2	99.3	105.5	90.2	99.9
03	**CONCRETE**	108.2	68.9	90.8	108.7	92.6	101.6	98.5	93.8	96.4	95.3	85.8	91.1	107.0	83.7	96.7	106.4	83.8	96.4
04	**MASONRY**	181.3	70.0	112.6	178.2	92.5	125.2	172.6	89.2	121.1	163.4	91.9	119.2	163.0	77.9	110.4	163.1	77.9	110.5
05	**METALS**	118.3	86.2	108.6	116.8	92.3	109.4	110.6	89.6	104.3	116.0	94.0	109.3	115.2	85.9	106.3	115.4	86.0	106.5
06	**WOOD, PLASTICS & COMPOSITES**	123.8	63.2	90.9	121.0	89.3	103.8	98.0	94.0	95.8	107.8	87.1	96.6	133.5	81.3	105.2	133.5	81.3	105.2
07	**THERMAL & MOISTURE PROTECTION**	116.0	71.0	96.4	121.1	94.1	109.3	116.8	87.4	104.0	119.8	89.4	106.6	114.2	86.5	102.2	114.2	86.5	102.1
08	**OPENINGS**	85.7	60.0	79.7	92.4	86.0	90.9	87.1	81.4	85.8	84.2	85.3	84.4	91.1	69.9	86.2	91.1	74.4	87.2
0920	Plaster & Gypsum Board	135.3	62.1	87.1	137.6	89.0	105.6	113.0	93.8	100.4	104.4	86.8	92.8	112.2	80.7	91.4	142.3	80.7	101.7
0950, 0980	Ceilings & Acoustic Treatment	121.0	62.1	81.1	108.9	89.0	95.4	124.0	93.8	103.6	104.7	86.8	92.5	103.8	80.7	88.1	103.8	80.7	88.1
0960	Flooring	111.6	67.2	98.7	113.6	97.2	108.9	106.8	95.6	103.6	106.7	94.2	103.1	113.6	88.8	106.5	113.6	88.8	106.5
0970, 0990	Wall Finishes & Painting/Coating	110.7	82.7	94.1	109.8	101.1	104.7	113.4	91.4	100.4	109.8	93.7	100.3	109.8	85.8	95.6	109.8	85.8	95.6
09	**FINISHES**	112.1	66.8	87.4	111.0	92.9	101.1	109.9	94.4	101.4	102.7	88.7	95.0	105.7	83.6	93.6	110.0	83.6	95.6
COVERS	**DIVS. 10 - 14, 25, 28, 41, 43, 44, 46**	131.1	61.8	115.9	131.1	67.5	117.1	131.1	69.6	117.6	131.1	87.7	121.6	131.1	77.2	119.3	131.1	77.2	119.3
21, 22, 23	**FIRE SUPPRESSION, PLUMBING & HVAC**	103.1	76.5	92.3	103.1	103.8	103.4	104.3	88.7	98.0	103.1	92.3	98.7	103.5	86.1	96.5	103.1	86.1	96.3
26, 27, 3370	**ELECTRICAL, COMMUNICATIONS & UTIL.**	114.5	83.7	98.8	110.2	87.1	98.4	107.4	93.7	100.4	108.9	85.3	96.9	106.9	66.8	86.5	106.9	66.8	86.5
MF2016	**WEIGHTED AVERAGE**	113.0	75.6	96.9	113.2	93.7	104.8	109.6	90.8	101.5	108.6	90.0	100.6	110.6	81.9	98.3	110.9	82.1	98.5

	DIVISION	CANADA																	
		ST CATHARINES, ONTARIO			ST JEROME, QUEBEC			ST. JOHN'S, NEWFOUNDLAND			SUDBURY, ONTARIO			SUMMERSIDE, PRINCE EDWARD ISLAND			SYDNEY, NOVA SCOTIA		
		MAT.	INST.	TOTAL	MAT.	INST.	TOTAL	MAT.	INST.	TOTAL	MAT.	INST.	TOTAL	MAT.	INST.	TOTAL	MAT.	INST.	TOTAL
015433	**CONTRACTOR EQUIPMENT**		103.4	103.4		103.5	103.5		122.7	122.7		103.4	103.4		102.0	102.0		102.0	102.0
0241, 31 - 34	**SITE & INFRASTRUCTURE, DEMOLITION**	96.3	101.4	99.9	97.9	98.9	98.6	119.1	111.9	114.0	96.5	101.0	99.6	123.9	94.3	103.3	110.2	97.0	101.0
0310	Concrete Forming & Accessories	111.8	93.0	95.8	130.4	81.3	88.6	121.0	86.4	91.5	107.8	88.1	91.1	109.1	54.0	62.2	108.8	69.8	75.6
0320	Concrete Reinforcing	106.8	98.4	102.5	154.2	74.5	113.7	168.2	84.7	125.8	107.7	102.2	104.9	166.8	47.8	106.3	169.0	49.0	108.0
0330	Cast-in-Place Concrete	106.0	99.1	103.4	101.3	90.2	97.2	132.0	99.6	119.9	106.9	95.8	102.8	111.8	56.4	91.2	90.9	68.8	82.7
03	**CONCRETE**	94.2	96.2	95.1	105.4	83.7	95.8	124.6	91.7	110.0	94.5	93.4	94.0	135.3	54.9	99.7	115.0	66.8	93.7
04	**MASONRY**	142.6	99.2	115.8	162.7	77.9	110.3	187.8	91.7	128.5	142.7	95.6	113.6	208.2	56.0	114.2	206.6	67.0	120.3
05	**METALS**	119.8	96.3	112.7	115.4	85.9	106.5	144.7	100.2	131.2	119.2	95.8	112.2	131.6	70.0	112.9	131.6	76.6	114.9
06	**WOOD, PLASTICS & COMPOSITES**	106.0	92.5	98.7	133.5	81.3	105.2	105.1	84.3	93.8	101.9	87.2	93.9	119.0	53.4	83.3	118.4	69.3	91.7
07	**THERMAL & MOISTURE PROTECTION**	115.2	100.9	109.0	114.2	86.5	102.1	141.3	96.5	121.8	114.7	95.6	106.4	129.6	58.9	98.8	130.2	69.5	103.8
08	**OPENINGS**	81.3	90.2	83.4	91.1	69.9	86.2	85.0	75.4	82.8	82.0	85.7	82.8	104.7	46.8	91.2	93.2	63.0	86.2
0920	Plaster & Gypsum Board	94.2	92.3	92.9	112.2	80.7	91.4	138.3	83.2	102.0	95.6	86.8	89.8	140.0	52.0	82.1	139.1	68.4	92.6
0950, 0980	Ceilings & Acoustic Treatment	103.3	92.3	95.8	103.8	80.7	88.1	136.9	83.2	100.6	98.2	86.8	90.5	122.4	52.0	74.7	122.4	68.4	85.8
0960	Flooring	100.2	92.4	98.0	113.6	88.8	106.5	117.5	53.0	98.8	98.4	94.2	97.2	108.1	57.1	93.3	108.1	60.9	94.4
0970, 0990	Wall Finishes & Painting/Coating	103.8	103.5	103.6	109.8	85.8	95.6	118.0	101.2	108.0	103.8	94.4	98.2	117.4	40.9	72.1	117.4	60.9	84.0
09	**FINISHES**	97.2	94.1	95.5	105.7	83.6	93.6	123.2	81.7	100.5	95.9	89.7	92.5	117.0	53.4	82.3	115.9	67.5	89.4
COVERS	**DIVS. 10 - 14, 25, 28, 41, 43, 44, 46**	131.1	69.2	117.5	131.1	77.2	119.3	131.1	68.6	117.4	131.1	89.0	121.9	131.1	58.2	115.1	131.1	62.1	115.9
21, 22, 23	**FIRE SUPPRESSION, PLUMBING & HVAC**	104.0	89.1	98.0	103.1	86.1	96.2	104.5	84.8	96.5	103.0	88.6	97.2	103.3	60.2	85.9	103.3	81.3	94.4
26, 27, 3370	**ELECTRICAL, COMMUNICATIONS & UTIL.**	109.3	97.8	103.5	107.6	66.8	86.8	105.8	83.4	94.4	109.2	98.3	103.6	107.3	48.6	77.4	108.2	60.1	83.8
MF2016	**WEIGHTED AVERAGE**	107.3	94.3	101.7	110.4	81.9	98.2	120.8	88.8	107.0	106.9	93.1	101.0	122.3	59.4	95.3	118.1	72.1	98.4

	DIVISION	CANADA																	
		THUNDER BAY, ONTARIO			TIMMINS, ONTARIO			TORONTO, ONTARIO			TROIS RIVIERES, QUEBEC			TRURO, NOVA SCOTIA			VANCOUVER, BRITISH COLUMBIA		
		MAT.	INST.	TOTAL	MAT.	INST.	TOTAL	MAT.	INST.	TOTAL	MAT.	INST.	TOTAL	MAT.	INST.	TOTAL	MAT.	INST.	TOTAL
015433	**CONTRACTOR EQUIPMENT**		103.4	103.4		103.1	103.1		122.5	122.5		102.9	102.9		102.7	102.7		142.3	142.3
0241, 31 - 34	**SITE & INFRASTRUCTURE, DEMOLITION**	100.9	101.3	101.2	117.9	99.1	104.8	116.0	114.5	114.9	110.5	98.4	102.1	102.5	97.4	99.0	114.8	130.8	125.9
0310	Concrete Forming & Accessories	119.2	91.3	95.5	124.4	82.3	88.6	121.0	99.1	102.4	151.5	81.4	91.8	97.0	69.8	73.9	122.2	92.8	97.1
0320	Concrete Reinforcing	95.5	97.9	96.7	171.2	87.3	128.5	151.4	100.3	125.4	169.0	74.5	121.0	145.8	49.0	96.6	122.5	93.4	107.7
0330	Cast-in-Place Concrete	116.7	98.4	109.9	127.3	82.8	110.7	113.9	111.9	113.1	94.1	90.1	92.6	132.2	69.0	108.7	110.9	92.3	104.0
03	**CONCRETE**	101.3	95.0	98.5	119.1	83.8	103.4	109.9	104.4	107.5	117.3	83.7	102.4	118.0	66.9	95.4	107.7	94.0	101.6
04	**MASONRY**	143.2	99.4	116.1	166.5	83.6	115.3	166.8	106.1	129.3	210.6	77.9	128.6	162.2	67.0	103.3	161.7	86.3	115.1
05	**METALS**	119.8	95.3	112.4	116.8	91.5	109.1	142.3	108.9	132.2	130.9	85.6	117.2	116.1	76.9	104.2	136.4	111.8	129.0
06	**WOOD, PLASTICS & COMPOSITES**	115.5	89.9	101.6	121.8	82.6	100.5	105.7	96.9	100.9	173.4	81.3	123.4	90.4	69.4	79.0	103.3	92.8	97.6
07	**THERMAL & MOISTURE PROTECTION**	115.5	98.2	108.0	121.0	85.0	105.4	132.3	106.4	121.0	128.9	86.5	110.4	115.4	69.6	95.5	137.7	88.0	116.1
08	**OPENINGS**	80.6	88.1	82.4	89.7	80.5	87.6	82.1	96.0	85.4	102.4	74.4	95.9	84.2	63.1	79.3	85.0	89.7	86.1
0920	Plaster & Gypsum Board	120.6	89.6	100.2	113.1	82.1	92.7	116.8	96.2	103.2	169.8	80.7	111.1	118.4	68.4	85.5	118.0	91.5	100.5
0950, 0980	Ceilings & Acoustic Treatment	98.2	89.6	92.4	104.7	82.1	89.4	133.9	96.2	108.4	120.7	80.7	93.6	104.7	68.4	80.1	143.7	91.5	108.3
0960	Flooring	104.3	99.0	102.8	113.6	88.9	106.5	109.1	101.4	106.9	131.3	88.8	119.0	96.1	60.9	85.9	127.9	90.7	117.1
0970, 0990	Wall Finishes & Painting/Coating	103.8	95.3	98.8	109.8	87.3	96.5	107.0	107.5	107.3	117.4	85.8	98.7	109.8	60.9	80.8	115.0	87.6	98.8
09	**FINISHES**	101.6	93.2	97.0	106.6	84.4	94.5	112.2	100.4	105.7	126.7	83.5	103.1	101.7	67.5	83.0	122.8	92.2	106.1
COVERS	**DIVS. 10 - 14, 25, 28, 41, 43, 44, 46**	131.1	69.3	117.5	131.1	64.8	116.5	131.1	93.9	122.9	131.1	77.2	119.3	131.1	62.1	115.9	131.1	91.7	122.5
21, 22, 23	**FIRE SUPPRESSION, PLUMBING & HVAC**	104.0	89.3	98.1	103.1	94.5	99.6	104.8	97.2	101.7	103.3	86.1	96.3	103.1	81.3	94.3	104.8	77.1	93.6
26, 27, 3370	**ELECTRICAL, COMMUNICATIONS & UTIL.**	107.7	96.5	102.0	108.9	85.3	96.9	103.2	100.0	101.6	108.5	66.8	87.3	106.6	60.1	83.0	101.7	80.5	90.9
MF2016	**WEIGHTED AVERAGE**	108.5	93.7	102.1	113.2	87.5	102.2	115.8	102.4	110.1	120.7	82.0	104.1	110.8	72.2	94.2	115.4	91.6	105.2

	DIVISION	CANADA																	
		VICTORIA, BRITISH COLUMBIA			WHITEHORSE, YUKON			WINDSOR, ONTARIO			WINNIPEG, MANITOBA			YARMOUTH, NOVA SCOTIA			YELLOWKNIFE, NWT		
		MAT.	INST.	TOTAL	MAT.	INST.	TOTAL	MAT.	INST.	TOTAL	MAT.	INST.	TOTAL	MAT.	INST.	TOTAL	MAT.	INST.	TOTAL
015433	**CONTRACTOR EQUIPMENT**		109.9	109.9		132.7	132.7		103.4	103.4		126.9	126.9		102.0	102.0		131.8	131.8
0241, 31 - 34	**SITE & INFRASTRUCTURE, DEMOLITION**	124.0	106.8	112.0	139.5	120.4	126.2	92.6	101.3	98.6	116.0	115.2	115.5	113.9	97.0	102.1	148.2	125.6	132.4
0310	Concrete Forming & Accessories	111.9	85.9	89.8	130.4	58.1	68.8	119.2	89.6	94.0	126.4	65.6	74.6	108.8	69.8	75.6	132.6	77.7	85.9
0320	Concrete Reinforcing	114.5	93.2	103.7	153.0	65.1	108.3	104.7	97.2	100.9	140.0	61.1	99.9	169.0	49.0	108.0	150.2	67.7	108.2
0330	Cast-in-Place Concrete	113.7	88.9	104.5	159.0	71.9	126.6	108.6	99.9	105.3	138.6	72.6	114.0	116.5	68.8	98.8	157.1	89.1	131.8
03	**CONCRETE**	125.6	88.5	109.1	139.7	66.0	107.0	95.6	94.7	95.2	121.8	68.8	98.3	126.9	66.8	100.3	138.6	81.3	113.2
04	**MASONRY**	175.6	86.2	120.4	240.3	59.2	128.4	142.7	98.4	115.3	188.9	64.9	112.3	208.8	67.0	121.2	235.0	70.3	133.2
05	**METALS**	111.7	91.4	105.6	149.2	91.8	131.8	119.8	95.8	112.5	148.8	89.3	130.8	131.6	76.6	114.9	147.3	94.7	131.4
06	**WOOD, PLASTICS & COMPOSITES**	101.8	84.6	92.5	123.9	56.7	87.4	115.5	88.2	100.7	110.1	66.2	86.3	118.4	69.3	91.7	132.9	79.2	103.7
07	**THERMAL & MOISTURE PROTECTION**	127.0	87.6	109.8	143.3	64.2	108.9	115.3	97.6	107.6	132.2	69.9	105.1	130.2	69.5	103.8	141.8	80.3	115.0
08	**OPENINGS**	88.6	80.9	86.8	102.8	54.4	91.5	80.4	87.6	82.1	85.6	59.8	79.6	93.2	63.0	86.2	97.1	67.4	90.2
0920	Plaster & Gypsum Board	106.4	83.8	91.5	172.9	54.6	95.0	106.6	87.8	94.2	124.1	64.2	84.6	139.1	68.4	92.6	179.7	77.7	112.5
0950, 0980	Ceilings & Acoustic Treatment	106.2	83.8	91.0	168.6	54.6	91.4	98.2	87.8	91.2	147.1	64.2	90.9	122.4	68.4	85.8	164.1	77.7	105.5
0960	Flooring	112.2	69.7	99.9	130.8	57.0	109.4	104.3	96.4	102.0	118.6	68.8	104.2	108.1	60.9	94.4	124.1	85.7	112.9
0970, 0990	Wall Finishes & Painting/Coating	113.4	87.6	98.1	121.7	54.5	82.0	103.8	96.3	99.4	111.5	53.4	77.1	117.4	60.9	84.0	124.4	77.9	96.9
09	**FINISHES**	108.4	83.7	94.9	145.3	57.2	97.2	99.2	91.5	95.0	123.1	65.2	91.5	115.9	67.5	89.4	142.1	78.7	107.5
COVERS	**DIVS. 10 - 14, 25, 28, 41, 43, 44, 46**	131.1	66.1	116.8	131.1	61.3	115.8	131.1	68.8	117.4	131.1	63.8	116.3	131.1	62.1	115.9	131.1	64.8	116.5
21, 22, 23	**FIRE SUPPRESSION, PLUMBING & HVAC**	103.2	75.2	91.9	103.6	73.2	91.3	104.0	89.2	98.0	104.8	64.3	88.5	103.3	81.3	94.4	104.1	90.4	98.6
26, 27, 3370	**ELECTRICAL, COMMUNICATIONS & UTIL.**	107.4	80.0	93.4	124.4	58.5	90.9	112.5	97.9	105.1	107.0	63.2	84.7	108.2	60.1	83.8	121.7	79.3	100.2
MF2016	**WEIGHTED AVERAGE**	113.7	84.2	101.0	131.9	70.5	105.6	107.8	93.5	101.7	121.1	71.2	99.7	119.8	72.1	99.4	130.4	85.0	110.9

Location Factors - Commercial

Costs shown in RSMeans cost data publications are based on national averages for materials and installation. To adjust these costs to a specific location, simply multiply the base cost by the factor and divide by 100 for that city. The data is arranged alphabetically by state and postal zip code numbers. For a city not listed, use the factor for a nearby city with similar economic characteristics.

STATE/ZIP	CITY	MAT.	INST.	TOTAL
ALABAMA				
350-352	Birmingham	96.5	71.7	85.9
354	Tuscaloosa	96.6	71.5	85.8
355	Jasper	97.1	71.0	85.9
356	Decatur	96.6	70.1	85.2
357-358	Huntsville	96.5	71.4	85.7
359	Gadsden	96.6	70.6	85.5
360-361	Montgomery	95.6	71.4	85.2
362	Anniston	95.2	67.4	83.3
363	Dothan	95.6	72.9	85.9
364	Evergreen	95.2	71.4	85.0
365-366	Mobile	96.1	68.8	84.4
367	Selma	95.3	72.4	85.5
368	Phenix City	96.1	72.0	85.8
369	Butler	95.5	71.6	85.3
ALASKA				
995-996	Anchorage	117.5	113.6	115.8
997	Fairbanks	118.2	114.7	116.7
998	Juneau	116.5	113.6	115.3
999	Ketchikan	128.4	113.6	122.0
ARIZONA				
850,853	Phoenix	100.0	73.0	88.4
851,852	Mesa/Tempe	98.6	72.0	87.2
855	Globe	99.5	71.8	87.6
856-857	Tucson	97.6	71.4	86.4
859	Show Low	99.7	71.9	87.8
860	Flagstaff	101.0	71.7	88.4
863	Prescott	98.7	71.7	87.1
864	Kingman	97.2	71.8	86.3
865	Chambers	97.2	74.4	87.4
ARKANSAS				
716	Pine Bluff	96.7	64.6	82.9
717	Camden	94.8	59.6	79.7
718	Texarkana	95.6	60.4	80.5
719	Hot Springs	94.2	61.0	79.9
720-722	Little Rock	95.4	65.9	82.8
723	West Memphis	94.3	65.5	81.9
724	Jonesboro	94.7	62.7	81.0
725	Batesville	92.6	60.0	78.7
726	Harrison	94.0	59.1	79.0
727	Fayetteville	91.5	60.7	78.3
728	Russellville	92.7	59.6	78.5
729	Fort Smith	95.1	63.6	81.6
CALIFORNIA				
900-902	Los Angeles	98.7	130.4	112.3
903-905	Inglewood	94.2	129.6	109.4
906-908	Long Beach	95.9	129.6	110.3
910-912	Pasadena	94.8	129.4	109.7
913-916	Van Nuys	97.9	129.4	111.4
917-918	Alhambra	96.8	129.4	110.8
919-921	San Diego	99.6	120.8	108.7
922	Palm Springs	96.9	127.1	109.8
923-924	San Bernardino	94.5	126.8	108.4
925	Riverside	98.8	127.1	110.9
926-927	Santa Ana	96.5	127.0	109.6
928	Anaheim	98.9	126.9	110.9
930	Oxnard	97.2	127.1	110.0
931	Santa Barbara	96.5	126.7	109.4
932-933	Bakersfield	98.0	125.3	109.7
934	San Luis Obispo	97.5	126.6	110.0
935	Mojave	94.7	125.2	107.7
936-938	Fresno	97.7	129.3	111.2
939	Salinas	98.3	135.7	114.3
940-941	San Francisco	106.3	158.1	128.5
942,956-958	Sacramento	100.1	132.7	114.1
943	Palo Alto	98.5	151.4	121.2
944	San Mateo	100.9	150.2	122.1
945	Vallejo	99.5	141.5	117.5
946	Oakland	102.7	151.2	123.5
947	Berkeley	102.3	151.2	123.3
948	Richmond	101.8	145.2	120.4
949	San Rafael	103.9	149.3	123.4
950	Santa Cruz	104.0	136.0	117.7

STATE/ZIP	CITY	MAT.	INST.	TOTAL
CALIFORNIA (CONT'D)				
951	San Jose	102.1	151.7	123.4
952	Stockton	100.3	130.8	113.4
953	Modesto	100.2	129.4	112.7
954	Santa Rosa	100.8	148.5	121.2
955	Eureka	102.1	133.2	115.4
959	Marysville	101.2	131.3	114.1
960	Redding	109.1	131.5	118.7
961	Susanville	108.8	131.2	118.4
COLORADO				
800-802	Denver	101.8	74.4	90.0
803	Boulder	97.7	75.9	88.3
804	Golden	99.8	73.7	88.6
805	Fort Collins	101.1	73.7	89.4
806	Greeley	98.3	74.7	88.2
807	Fort Morgan	98.3	72.9	87.4
808-809	Colorado Springs	100.0	73.0	88.4
810	Pueblo	100.5	71.4	88.0
811	Alamosa	102.3	68.8	87.9
812	Salida	102.0	66.6	86.8
813	Durango	102.7	65.0	86.6
814	Montrose	101.4	69.3	87.6
815	Grand Junction	104.5	72.2	90.7
816	Glenwood Springs	102.5	65.4	86.6
CONNECTICUT				
060	New Britain	96.4	118.1	105.7
061	Hartford	97.2	118.5	106.4
062	Willimantic	97.0	118.2	106.1
063	New London	93.6	118.4	104.2
064	Meriden	95.5	118.4	105.4
065	New Haven	98.1	118.5	106.9
066	Bridgeport	97.6	118.4	106.5
067	Waterbury	97.1	118.6	106.3
068	Norwalk	97.0	118.6	106.3
069	Stamford	97.2	125.2	109.2
D.C.				
200-205	Washington	101.0	87.7	95.3
DELAWARE				
197	Newark	98.3	112.3	104.3
198	Wilmington	98.2	112.3	104.2
199	Dover	98.2	112.3	104.2
FLORIDA				
320,322	Jacksonville	96.1	66.5	83.4
321	Daytona Beach	96.3	70.3	85.2
323	Tallahassee	97.3	65.7	83.7
324	Panama City	97.6	64.2	83.3
325	Pensacola	100.3	66.4	85.8
326,344	Gainesville	97.8	64.9	83.7
327-328,347	Orlando	97.9	67.1	84.7
329	Melbourne	99.2	71.2	87.2
330-332,340	Miami	95.8	67.4	83.7
333	Fort Lauderdale	95.3	69.5	84.2
334,349	West Palm Beach	94.3	66.0	82.2
335-336,346	Tampa	96.8	68.1	84.5
337	St. Petersburg	99.0	65.2	84.5
338	Lakeland	96.2	67.1	83.8
339,341	Fort Myers	95.6	68.3	83.9
342	Sarasota	98.7	67.0	85.1
GEORGIA				
300-303,399	Atlanta	99.0	75.6	89.0
304	Statesboro	98.9	65.8	84.7
305	Gainesville	97.4	66.5	84.2
306	Athens	96.8	67.9	84.4
307	Dalton	98.7	71.0	86.8
308-309	Augusta	97.1	72.7	86.6
310-312	Macon	93.9	74.0	85.4
313-314	Savannah	95.6	73.1	86.0
315	Waycross	95.4	68.4	83.8
316	Valdosta	95.3	66.2	82.8
317,398	Albany	95.2	72.6	85.5
318-319	Columbus	95.1	72.7	85.5

STATE/ZIP	CITY	MAT.	INST.	TOTAL
HAWAII				
967	Hilo	116.3	118.3	117.1
968	Honolulu	120.9	118.3	119.8
STATES & POSS.				
969	Guam	138.8	52.6	101.9
IDAHO				
832	Pocatello	100.7	78.6	91.2
833	Twin Falls	101.9	77.2	91.3
834	Idaho Falls	99.2	78.2	90.2
835	Lewiston	106.9	85.7	97.8
836-837	Boise	99.7	79.4	91.0
838	Coeur d'Alene	106.8	84.4	97.2
ILLINOIS				
600-603	North Suburban	99.8	144.1	118.8
604	Joliet	99.7	142.2	117.9
605	South Suburban	99.8	144.0	118.7
606-608	Chicago	101.7	146.1	120.7
609	Kankakee	96.4	136.1	113.4
610-611	Rockford	96.8	128.7	110.5
612	Rock Island	94.9	102.1	98.0
613	La Salle	96.1	128.9	110.2
614	Galesburg	95.9	109.5	101.8
615-616	Peoria	97.8	111.5	103.7
617	Bloomington	95.2	112.2	102.5
618-619	Champaign	98.7	109.8	103.5
620-622	East St. Louis	94.4	111.3	101.6
623	Quincy	96.3	105.1	100.1
624	Effingham	95.7	109.2	101.5
625	Decatur	97.2	107.0	101.4
626-627	Springfield	97.9	107.7	102.1
628	Centralia	93.3	113.0	101.8
629	Carbondale	93.1	109.8	100.2
INDIANA				
460	Anderson	95.9	81.8	89.8
461-462	Indianapolis	98.7	83.0	92.0
463-464	Gary	97.2	108.3	102.0
465-466	South Bend	97.4	83.6	91.5
467-468	Fort Wayne	96.4	78.2	88.6
469	Kokomo	94.1	81.0	88.4
470	Lawrenceburg	92.6	78.5	86.5
471	New Albany	93.9	77.2	86.7
472	Columbus	96.1	80.4	89.3
473	Muncie	96.1	80.3	89.3
474	Bloomington	97.7	80.3	90.3
475	Washington	94.6	85.4	90.7
476-477	Evansville	95.4	84.8	90.9
478	Terre Haute	96.2	83.8	90.9
479	Lafayette	95.8	80.1	89.1
IOWA				
500-503,509	Des Moines	96.0	88.4	92.8
504	Mason City	94.7	76.2	86.8
505	Fort Dodge	94.9	74.1	86.0
506-507	Waterloo	96.2	79.3	89.0
508	Creston	95.2	83.1	90.0
510-511	Sioux City	97.1	78.7	89.2
512	Sibley	96.1	63.3	82.1
513	Spencer	97.7	63.6	83.1
514	Carroll	94.9	81.3	89.1
515	Council Bluffs	98.2	80.8	90.7
516	Shenandoah	95.4	83.8	90.4
520	Dubuque	96.5	81.9	90.3
521	Decorah	95.9	75.9	87.3
522-524	Cedar Rapids	97.4	85.8	92.4
525	Ottumwa	95.7	76.3	87.4
526	Burlington	95.1	86.3	91.3
527-528	Davenport	96.5	96.9	96.7
KANSAS				
660-662	Kansas City	96.0	98.6	97.1
664-666	Topeka	97.0	77.7	88.7
667	Fort Scott	94.8	78.2	87.7
668	Emporia	94.9	77.2	87.3
669	Belleville	96.6	72.1	86.1
670-672	Wichita	96.7	70.8	85.6
673	Independence	97.0	77.9	88.8
674	Salina	96.9	72.8	86.6
675	Hutchinson	92.4	73.1	84.1
676	Hays	96.3	73.5	86.5
677	Colby	97.1	75.7	87.9

STATE/ZIP	CITY	MAT.	INST.	TOTAL
KANSAS (CONT'D)				
678	Dodge City	98.3	75.4	88.5
679	Liberal	96.2	73.7	86.5
KENTUCKY				
400-402	Louisville	93.2	77.7	86.6
403-405	Lexington	93.4	79.6	87.5
406	Frankfort	95.5	78.7	88.3
407-409	Corbin	91.1	79.6	86.2
410	Covington	94.4	79.8	88.1
411-412	Ashland	93.2	92.0	92.7
413-414	Campton	94.4	79.9	88.2
415-416	Pikeville	95.8	87.6	92.3
417-418	Hazard	93.8	80.6	88.1
420	Paducah	92.5	83.8	88.8
421-422	Bowling Green	94.5	78.9	87.8
423	Owensboro	94.6	83.1	89.7
424	Henderson	92.2	82.4	88.0
425-426	Somerset	91.7	77.8	85.7
427	Elizabethtown	91.3	76.7	85.1
LOUISIANA				
700-701	New Orleans	99.4	69.5	86.6
703	Thibodaux	96.2	66.4	83.4
704	Hammond	93.8	64.5	81.2
705	Lafayette	95.6	68.9	84.1
706	Lake Charles	95.8	69.8	84.6
707-708	Baton Rouge	96.6	69.7	85.1
710-711	Shreveport	97.8	66.4	84.3
712	Monroe	96.6	65.1	83.1
713-714	Alexandria	96.7	66.6	83.8
MAINE				
039	Kittery	91.3	84.6	88.5
040-041	Portland	96.9	84.1	91.4
042	Lewiston	94.7	84.1	90.1
043	Augusta	96.7	83.1	90.9
044	Bangor	94.3	83.5	89.7
045	Bath	93.2	82.6	88.6
046	Machias	92.6	79.7	87.1
047	Houlton	92.8	79.7	87.2
048	Rockland	91.9	83.1	88.1
049	Waterville	93.1	83.1	88.8
MARYLAND				
206	Waldorf	97.7	87.4	93.3
207-208	College Park	97.7	88.6	93.8
209	Silver Spring	96.9	87.4	92.8
210-212	Baltimore	101.2	85.0	94.2
214	Annapolis	100.5	83.1	93.0
215	Cumberland	97.1	85.5	92.1
216	Easton	98.8	76.0	89.0
217	Hagerstown	97.5	87.0	93.0
218	Salisbury	99.3	68.5	86.1
219	Elkton	96.0	87.6	92.4
MASSACHUSETTS				
010-011	Springfield	96.4	110.4	102.4
012	Pittsfield	96.0	105.1	99.9
013	Greenfield	94.3	110.9	101.4
014	Fitchburg	93.0	118.7	104.0
015-016	Worcester	96.4	118.5	105.9
017	Framingham	92.4	127.0	107.2
018	Lowell	95.8	125.4	108.5
019	Lawrence	96.7	125.8	109.2
020-022, 024	Boston	99.4	133.3	113.9
023	Brockton	96.1	119.2	106.0
025	Buzzards Bay	91.1	117.7	102.5
026	Hyannis	93.4	117.5	103.7
027	New Bedford	95.3	117.6	104.9
MICHIGAN				
480,483	Royal Oak	95.2	100.6	97.5
481	Ann Arbor	97.2	101.7	99.1
482	Detroit	100.2	103.0	101.4
484-485	Flint	96.9	90.8	94.3
486	Saginaw	96.5	87.3	92.6
487	Bay City	96.6	87.5	92.7
488-489	Lansing	97.9	88.7	93.9
490	Battle Creek	95.3	82.5	89.8
491	Kalamazoo	95.6	80.7	89.2
492	Jackson	93.8	93.0	93.5
493,495	Grand Rapids	97.1	81.6	90.5
494	Muskegon	94.1	80.3	88.2

Location Factors - Commercial

STATE/ZIP	CITY	MAT.	INST.	TOTAL
MICHIGAN (CONT'D)				
496	Traverse City	93.2	78.3	86.8
497	Gaylord	94.3	81.3	88.8
498-499	Iron Mountain	96.2	83.0	90.6
MINNESOTA				
550-551	Saint Paul	97.2	115.8	105.2
553-555	Minneapolis	99.7	116.5	106.9
556-558	Duluth	97.1	102.7	99.5
559	Rochester	97.1	100.3	98.4
560	Mankato	95.1	98.0	96.4
561	Windom	93.7	92.3	93.1
562	Willmar	93.3	99.9	96.1
563	St. Cloud	94.4	114.0	102.8
564	Brainerd	95.0	98.6	96.5
565	Detroit Lakes	96.7	92.2	94.8
566	Bemidji	96.0	95.4	95.8
567	Thief River Falls	95.6	89.4	92.9
MISSISSIPPI				
386	Clarksdale	95.2	54.4	77.7
387	Greenville	98.6	66.0	84.6
388	Tupelo	96.5	58.2	80.1
389	Greenwood	96.5	54.2	78.3
390-392	Jackson	98.0	65.4	84.0
393	Meridian	95.6	65.1	82.5
394	Laurel	97.2	56.9	79.9
395	Biloxi	97.3	64.6	83.3
396	McComb	95.4	54.5	77.9
397	Columbus	97.0	58.4	80.5
MISSOURI				
630-631	St. Louis	99.6	105.8	102.3
633	Bowling Green	97.5	96.0	96.9
634	Hannibal	96.5	94.0	95.4
635	Kirksville	100.2	88.8	95.3
636	Flat River	98.5	96.2	97.5
637	Cape Girardeau	98.1	91.8	95.4
638	Sikeston	96.9	89.4	93.7
639	Poplar Bluff	96.3	89.2	93.3
640-641	Kansas City	98.1	103.6	100.5
644-645	St. Joseph	96.8	92.8	95.1
646	Chillicothe	94.5	95.4	94.9
647	Harrisonville	94.0	102.6	97.7
648	Joplin	95.9	79.7	88.9
650-651	Jefferson City	96.3	90.8	94.0
652	Columbia	96.2	91.3	94.1
653	Sedalia	96.6	93.7	95.4
654-655	Rolla	94.3	96.6	95.3
656-658	Springfield	97.7	81.0	90.5
MONTANA				
590-591	Billings	100.8	75.0	89.7
592	Wolf Point	100.5	77.8	90.8
593	Miles City	98.4	77.9	89.6
594	Great Falls	102.1	74.0	90.1
595	Havre	99.5	75.7	89.3
596	Helena	100.1	76.1	89.8
597	Butte	100.7	76.2	90.2
598	Missoula	97.9	75.7	88.4
599	Kalispell	97.4	76.2	88.3
NEBRASKA				
680-681	Omaha	96.9	80.8	90.0
683-685	Lincoln	97.1	81.1	90.2
686	Columbus	95.6	82.7	90.1
687	Norfolk	96.9	79.2	89.3
688	Grand Island	97.0	78.5	89.1
689	Hastings	96.7	80.0	89.5
690	McCook	96.6	74.8	87.3
691	North Platte	96.6	77.3	88.3
692	Valentine	98.8	70.8	86.8
693	Alliance	98.9	74.7	88.5
NEVADA				
889-891	Las Vegas	102.7	104.1	103.3
893	Ely	101.6	95.6	99.0
894-895	Reno	101.5	84.2	94.1
897	Carson City	100.4	84.2	93.5
898	Elko	100.3	87.4	94.8
NEW HAMPSHIRE				
030	Nashua	95.9	92.8	94.6
031	Manchester	96.2	93.0	94.8

STATE/ZIP	CITY	MAT.	INST.	TOTAL
NEW HAMPSHIRE (CONT'D)				
032-033	Concord	96.1	92.4	94.5
034	Keene	93.0	89.2	91.3
035	Littleton	93.0	80.1	87.4
036	Charleston	92.5	88.6	90.8
037	Claremont	91.7	88.6	90.3
038	Portsmouth	93.3	92.2	92.8
NEW JERSEY				
070-071	Newark	99.6	139.7	116.8
072	Elizabeth	97.3	139.7	115.4
073	Jersey City	96.3	139.4	114.8
074-075	Paterson	97.7	139.5	115.6
076	Hackensack	95.9	139.6	114.6
077	Long Branch	95.6	133.6	111.9
078	Dover	96.1	139.6	114.7
079	Summit	96.2	139.7	114.8
080,083	Vineland	95.8	132.3	111.4
081	Camden	97.5	130.7	111.8
082,084	Atlantic City	96.4	132.6	111.9
085-086	Trenton	99.2	131.4	113.0
087	Point Pleasant	97.7	133.3	113.0
088-089	New Brunswick	98.3	137.7	115.2
NEW MEXICO				
870-872	Albuquerque	97.1	75.0	87.6
873	Gallup	97.3	75.0	87.7
874	Farmington	97.7	75.0	88.0
875	Santa Fe	97.6	75.0	87.9
877	Las Vegas	95.8	75.0	86.9
878	Socorro	95.5	75.0	86.7
879	Truth/Consequences	95.2	71.7	85.2
880	Las Cruces	95.6	71.7	85.4
881	Clovis	98.0	74.9	88.1
882	Roswell	99.5	75.0	89.0
883	Carrizozo	100.2	75.0	89.4
884	Tucumcari	98.7	74.9	88.5
NEW YORK				
100-102	New York	99.0	176.2	132.1
103	Staten Island	94.9	177.8	130.5
104	Bronx	93.3	176.8	129.1
105	Mount Vernon	93.5	151.3	118.3
106	White Plains	93.3	153.5	119.1
107	Yonkers	97.3	151.2	120.5
108	New Rochelle	93.9	146.0	116.2
109	Suffern	93.6	133.5	110.7
110	Queens	99.8	178.9	133.8
111	Long Island City	101.5	178.9	134.7
112	Brooklyn	101.8	178.9	134.9
113	Flushing	102.0	178.9	135.0
114	Jamaica	100.2	178.9	134.0
115,117,118	Hicksville	99.8	160.5	125.8
116	Far Rockaway	102.2	178.9	135.1
119	Riverhead	100.5	156.8	124.6
120-122	Albany	95.0	111.5	102.1
123	Schenectady	95.5	110.6	101.9
124	Kingston	98.9	136.0	114.8
125-126	Poughkeepsie	98.1	140.2	116.1
127	Monticello	97.4	136.9	114.3
128	Glens Falls	90.6	108.7	98.4
129	Plattsburgh	95.6	99.0	97.0
130-132	Syracuse	97.8	100.8	99.1
133-135	Utica	95.9	98.9	97.2
136	Watertown	97.7	99.2	98.3
137-139	Binghamton	97.3	102.8	99.7
140-142	Buffalo	103.0	110.0	106.0
143	Niagara Falls	98.8	110.2	103.7
144-146	Rochester	100.3	101.3	100.8
147	Jamestown	97.8	97.8	97.8
148-149	Elmira	97.6	103.9	100.3
NORTH CAROLINA				
270,272-274	Greensboro	97.7	67.7	84.8
271	Winston-Salem	97.4	67.8	84.7
275-276	Raleigh	96.4	67.1	83.8
277	Durham	99.3	67.7	85.7
278	Rocky Mount	95.3	67.5	83.4
279	Elizabeth City	96.1	69.2	84.6
280	Gastonia	97.0	67.7	84.4
281-282	Charlotte	97.0	67.3	84.3
283	Fayetteville	100.0	67.3	86.0
284	Wilmington	95.9	66.3	83.2
285	Kinston	94.3	67.1	82.7

Location Factors - Commercial

STATE/ZIP	CITY	MAT.	INST.	TOTAL
NORTH CAROLINA (CONT'D)				
286	Hickory	94.7	68.3	83.4
287-288	Asheville	96.3	67.0	83.8
289	Murphy	95.5	66.2	82.9
NORTH DAKOTA				
580-581	Fargo	98.2	80.3	90.5
582	Grand Forks	98.4	79.0	90.1
583	Devils Lake	98.3	80.2	90.5
584	Jamestown	98.3	79.6	90.3
585	Bismarck	98.2	80.0	90.4
586	Dickinson	99.0	79.1	90.5
587	Minot	98.3	79.6	90.3
588	Williston	97.4	79.0	89.6
OHIO				
430-432	Columbus	97.4	83.2	91.3
433	Marion	93.4	88.5	91.3
434-436	Toledo	96.7	92.8	95.0
437-438	Zanesville	94.0	87.1	91.0
439	Steubenville	95.3	92.1	94.0
440	Lorain	98.4	86.8	93.4
441	Cleveland	98.7	92.2	95.9
442-443	Akron	99.5	88.3	94.7
444-445	Youngstown	98.7	84.0	92.4
446-447	Canton	98.9	81.4	91.4
448-449	Mansfield	96.5	86.4	92.1
450	Hamilton	95.9	81.5	89.7
451-452	Cincinnati	96.9	80.7	89.9
453-454	Dayton	95.9	81.3	89.6
455	Springfield	95.9	81.8	89.8
456	Chillicothe	95.4	90.8	93.4
457	Athens	98.2	86.4	93.2
458	Lima	98.3	83.8	92.1
OKLAHOMA				
730-731	Oklahoma City	95.2	68.2	83.6
734	Ardmore	93.8	66.2	82.0
735	Lawton	95.9	66.6	83.3
736	Clinton	95.0	66.1	82.6
737	Enid	95.5	66.5	83.1
738	Woodward	93.9	63.4	80.8
739	Guymon	94.9	64.8	82.0
740-741	Tulsa	96.0	67.1	83.6
743	Miami	92.8	66.5	81.5
744	Muskogee	95.1	64.4	81.9
745	McAlester	92.4	61.9	79.3
746	Ponca City	93.2	65.2	81.2
747	Durant	93.2	66.3	81.7
748	Shawnee	94.6	66.1	82.4
749	Poteau	92.4	66.1	81.1
OREGON				
970-972	Portland	100.2	100.4	100.3
973	Salem	101.7	99.9	100.9
974	Eugene	99.8	98.5	99.3
975	Medford	101.3	97.8	99.8
976	Klamath Falls	101.5	97.8	99.9
977	Bend	100.7	99.8	100.3
978	Pendleton	96.8	101.7	98.9
979	Vale	94.5	87.9	91.7
PENNSYLVANIA				
150-152	Pittsburgh	99.9	102.2	100.9
153	Washington	96.9	102.7	99.4
154	Uniontown	97.2	102.2	99.4
155	Bedford	98.2	94.3	96.5
156	Greensburg	98.2	98.7	98.4
157	Indiana	97.1	100.1	98.4
158	Dubois	98.7	98.3	98.5
159	Johnstown	98.2	94.2	96.5
160	Butler	91.3	103.1	96.4
161	New Castle	91.4	100.3	95.2
162	Kittanning	91.8	100.4	95.5
163	Oil City	91.3	100.3	95.1
164-165	Erie	93.1	96.2	94.5
166	Altoona	93.2	95.1	94.0
167	Bradford	94.9	99.5	96.8
168	State College	94.5	97.0	95.6
169	Wellsboro	95.5	95.2	95.4
170-171	Harrisburg	97.8	94.9	96.5
172	Chambersburg	94.9	91.6	93.4
173-174	York	94.9	95.3	95.1
175-176	Lancaster	93.5	96.8	94.9

STATE/ZIP	CITY	MAT.	INST.	TOTAL
PENNSYLVANIA (CONT'D)				
177	Williamsport	92.1	94.6	93.2
178	Sunbury	94.3	95.0	94.6
179	Pottsville	93.4	98.4	95.6
180	Lehigh Valley	95.0	113.6	103.0
181	Allentown	96.7	108.4	101.7
182	Hazleton	94.4	98.3	96.1
183	Stroudsburg	94.3	107.3	99.9
184-185	Scranton	97.4	98.9	98.1
186-187	Wilkes-Barre	94.1	97.9	95.7
188	Montrose	93.8	99.2	96.1
189	Doylestown	94.0	129.2	109.1
190-191	Philadelphia	100.1	135.3	115.2
193	Westchester	95.8	129.9	110.4
194	Norristown	94.8	130.0	109.9
195-196	Reading	96.5	104.0	99.7
PUERTO RICO				
009	San Juan	117.8	27.2	79.0
RHODE ISLAND				
028	Newport	94.6	113.7	102.8
029	Providence	96.2	113.7	103.7
SOUTH CAROLINA				
290-292	Columbia	96.5	67.9	84.2
293	Spartanburg	96.4	67.9	84.2
294	Charleston	98.0	67.6	84.9
295	Florence	96.2	67.8	84.0
296	Greenville	96.2	67.9	84.1
297	Rock Hill	96.0	67.3	83.7
298	Aiken	96.9	67.9	84.4
299	Beaufort	97.5	56.0	79.7
SOUTH DAKOTA				
570-571	Sioux Falls	96.3	79.2	89.0
572	Watertown	97.1	70.8	85.8
573	Mitchell	96.0	62.4	81.6
574	Aberdeen	98.3	71.5	86.8
575	Pierre	98.7	69.4	86.2
576	Mobridge	96.6	64.4	82.8
577	Rapid City	98.0	73.3	87.4
TENNESSEE				
370-372	Nashville	98.4	71.1	86.7
373-374	Chattanooga	98.9	69.6	86.3
375,380-381	Memphis	97.3	71.1	86.1
376	Johnson City	99.1	59.9	82.3
377-379	Knoxville	95.6	65.2	82.6
382	McKenzie	96.2	56.0	79.0
383	Jackson	97.5	59.7	81.3
384	Columbia	94.7	67.9	83.2
385	Cookeville	96.1	57.3	79.5
TEXAS				
750	McKinney	98.0	65.1	83.9
751	Waxahachie	98.1	65.1	83.9
752-753	Dallas	99.1	67.7	85.6
754	Greenville	98.2	64.5	83.8
755	Texarkana	97.5	63.6	83.0
756	Longview	98.4	62.6	83.0
757	Tyler	98.6	63.2	83.5
758	Palestine	95.1	62.0	80.9
759	Lufkin	95.5	64.2	82.1
760-761	Fort Worth	97.4	64.1	83.1
762	Denton	97.5	64.4	83.3
763	Wichita Falls	94.9	62.7	81.1
764	Eastland	94.0	61.8	80.2
765	Temple	92.5	60.0	78.5
766-767	Waco	94.3	63.9	81.3
768	Brownwood	97.4	60.2	81.4
769	San Angelo	97.1	60.8	81.6
770-772	Houston	100.7	67.9	86.6
773	Huntsville	99.4	64.3	84.4
774	Wharton	100.8	65.8	85.8
775	Galveston	98.5	68.1	85.5
776-777	Beaumont	98.8	67.1	85.2
778	Bryan	95.5	66.4	83.1
779	Victoria	100.5	64.4	85.0
780	Laredo	97.1	62.7	82.4
781-782	San Antonio	99.3	63.5	83.9
783-784	Corpus Christi	99.7	63.0	84.0
785	McAllen	100.6	57.9	82.3
786-787	Austin	97.4	62.7	82.5

Location Factors - Commercial

STATE/ZIP	CITY	MAT.	INST.	TOTAL
TEXAS (CONT'D)				
788	Del Rio	100.6	62.0	84.0
789	Giddings	97.0	61.9	82.0
790-791	Amarillo	95.7	61.1	80.9
792	Childress	95.8	61.8	81.2
793-794	Lubbock	97.5	64.4	83.3
795-796	Abilene	96.0	62.8	81.8
797	Midland	98.1	63.7	83.3
798-799,885	El Paso	94.6	64.7	81.8
UTAH				
840-841	Salt Lake City	102.0	73.1	89.6
842,844	Ogden	97.5	73.1	87.0
843	Logan	99.4	73.1	88.1
845	Price	100.0	71.7	87.9
846-847	Provo	99.8	72.9	88.3
VERMONT				
050	White River Jct.	95.8	81.0	89.4
051	Bellows Falls	94.2	93.6	94.0
052	Bennington	94.6	90.0	92.6
053	Brattleboro	95.0	93.6	94.4
054	Burlington	99.1	80.6	91.2
056	Montpelier	97.7	85.3	92.4
057	Rutland	96.3	80.5	89.5
058	St. Johnsbury	95.8	80.5	89.3
059	Guildhall	94.5	80.2	88.3
VIRGINIA				
220-221	Fairfax	99.2	84.2	92.8
222	Arlington	100.1	84.3	93.3
223	Alexandria	99.3	84.2	92.8
224-225	Fredericksburg	97.9	80.6	90.5
226	Winchester	98.6	81.3	91.2
227	Culpeper	98.4	84.9	92.6
228	Harrisonburg	98.7	66.0	84.7
229	Charlottesville	99.1	68.0	85.8
230-232	Richmond	98.3	76.4	88.9
233-235	Norfolk	98.8	70.0	86.5
236	Newport News	97.9	69.7	85.8
237	Portsmouth	97.4	63.1	82.7
238	Petersburg	97.9	72.4	87.0
239	Farmville	97.1	62.8	82.4
240-241	Roanoke	99.7	71.6	87.7
242	Bristol	97.9	59.4	81.4
243	Pulaski	97.5	68.3	85.0
244	Staunton	98.3	65.6	84.3
245	Lynchburg	98.4	73.7	87.8
246	Grundy	97.8	59.2	81.2
WASHINGTON				
980-981,987	Seattle	106.0	107.4	106.6
982	Everett	103.4	102.6	103.1
983-984	Tacoma	103.8	102.1	103.1
985	Olympia	100.9	102.3	101.5
986	Vancouver	104.8	96.2	101.1
988	Wenatchee	104.1	91.0	98.5
989	Yakima	104.0	96.6	100.8
990-992	Spokane	99.9	84.5	93.3
993	Richland	99.6	92.2	96.4
994	Clarkston	98.9	83.9	92.5
WEST VIRGINIA				
247-248	Bluefield	96.7	90.2	93.9
249	Lewisburg	98.4	90.5	95.0
250-253	Charleston	96.1	92.5	94.6
254	Martinsburg	96.4	86.1	92.0
255-257	Huntington	97.5	94.3	96.1
258-259	Beckley	94.9	92.1	93.7
260	Wheeling	99.3	91.5	96.0
261	Parkersburg	98.2	90.7	95.0
262	Buckhannon	98.1	90.7	94.9
263-264	Clarksburg	98.7	91.4	95.5
265	Morgantown	98.5	91.5	95.5
266	Gassaway	98.0	90.5	94.8
267	Romney	97.9	89.2	94.2
268	Petersburg	97.7	87.7	93.4
WISCONSIN				
530,532	Milwaukee	99.6	105.9	102.3
531	Kenosha	99.4	104.6	101.6
534	Racine	98.8	105.0	101.5
535	Beloit	98.7	101.4	99.9
537	Madison	99.1	100.3	99.6

STATE/ZIP	CITY	MAT.	INST.	TOTAL
WISCONSIN (CONT'D)				
538	Lancaster	96.9	96.7	96.8
539	Portage	95.3	100.5	97.5
540	New Richmond	94.7	97.0	95.7
541-543	Green Bay	98.7	95.8	97.5
544	Wausau	94.1	95.8	94.8
545	Rhinelander	97.3	94.6	96.1
546	La Crosse	95.2	95.9	95.5
547	Eau Claire	96.8	96.5	96.7
548	Superior	94.5	98.7	96.3
549	Oshkosh	94.8	94.2	94.5
WYOMING				
820	Cheyenne	99.6	72.1	87.8
821	Yellowstone Nat'l Park	97.6	73.2	87.1
822	Wheatland	98.8	68.1	85.6
823	Rawlins	100.3	72.7	88.5
824	Worland	98.3	71.8	87.0
825	Riverton	99.4	69.7	86.6
826	Casper	99.3	69.5	86.5
827	Newcastle	98.1	72.3	87.1
828	Sheridan	100.9	70.6	87.9
829-831	Rock Springs	101.9	71.6	88.9
CANADIAN FACTORS (reflect Canadian currency)				
ALBERTA				
	Calgary	119.0	97.6	109.8
	Edmonton	119.7	97.5	110.2
	Fort McMurray	122.1	90.2	108.4
	Lethbridge	117.2	89.7	105.4
	Lloydminster	111.6	86.5	100.8
	Medicine Hat	111.7	85.8	100.6
	Red Deer	112.2	85.8	100.8
BRITISH COLUMBIA				
	Kamloops	113.0	85.7	101.3
	Prince George	113.9	85.0	101.5
	Vancouver	115.4	91.6	105.2
	Victoria	113.7	84.2	101.0
MANITOBA				
	Brandon	120.9	72.4	100.1
	Portage la Prairie	111.6	70.9	94.1
	Winnipeg	121.1	71.2	99.7
NEW BRUNSWICK				
	Bathurst	110.4	63.9	90.5
	Dalhousie	109.8	64.1	90.2
	Fredericton	117.9	71.2	97.9
	Moncton	110.6	68.1	92.4
	Newcastle	110.5	64.5	90.8
	Saint John	113.0	75.6	96.9
NEWFOUNDLAND				
	Corner Brook	125.0	70.2	101.5
	St. John's	120.8	88.8	107.0
NORTHWEST TERRITORIES				
	Yellowknife	130.4	85.0	110.9
NOVA SCOTIA				
	Bridgewater	111.2	72.2	94.5
	Dartmouth	121.8	72.1	100.5
	Halifax	117.1	85.8	103.7
	New Glasgow	119.9	72.1	99.4
	Sydney	118.1	72.1	98.4
	Truro	110.8	72.2	94.2
	Yarmouth	119.8	72.1	99.4
ONTARIO				
	Barrie	116.0	89.3	104.6
	Brantford	113.1	93.1	104.5
	Cornwall	113.1	89.6	103.0
	Hamilton	116.1	98.1	108.4
	Kingston	114.1	89.7	103.6
	Kitchener	109.4	93.4	102.5
	London	116.8	95.6	107.7
	North Bay	122.8	87.4	107.7
	Oshawa	111.6	92.9	103.6
	Ottawa	116.9	96.4	108.1
	Owen Sound	116.1	87.6	103.9
	Peterborough	113.1	89.4	102.9
	Sarnia	113.2	93.7	104.8

STATE/ZIP	CITY	MAT.	INST.	TOTAL
ONTARIO (CONT'D)				
	Sault Ste. Marie	108.6	90.0	100.6
	St. Catharines	107.3	94.3	101.7
	Sudbury	106.9	93.1	101.0
	Thunder Bay	108.5	93.7	102.1
	Timmins	113.2	87.5	102.2
	Toronto	115.8	102.4	110.1
	Windsor	107.8	93.5	101.7
PRINCE EDWARD ISLAND				
	Charlottetown	120.1	61.8	95.1
	Summerside	122.3	59.4	95.3
QUEBEC				
	Cap-de-la-Madeleine	110.7	82.1	98.4
	Charlesbourg	110.7	82.1	98.4
	Chicoutimi	110.3	87.3	100.4
	Gatineau	110.4	81.9	98.2
	Granby	110.6	81.8	98.3
	Hull	110.5	81.9	98.2
	Joliette	110.9	82.1	98.5
	Laval	110.7	81.9	98.3
	Montreal	118.1	89.9	106.0
	Quebec City	116.4	90.3	105.2
	Rimouski	110.3	87.3	100.4
	Rouyn-Noranda	110.4	81.9	98.1
	Saint-Hyacinthe	109.8	81.9	97.8
	Sherbrooke	110.6	81.9	98.3
	Sorel	110.9	82.1	98.5
	Saint-Jerome	110.4	81.9	98.2
	Trois-Rivieres	120.7	82.0	104.1
SASKATCHEWAN				
	Moose Jaw	109.0	65.6	90.4
	Prince Albert	108.2	63.9	89.2
	Regina	122.1	94.2	110.1
	Saskatoon	109.6	90.8	101.5
YUKON				
	Whitehorse	131.9	70.5	105.6

R011105-05 Tips for Accurate Estimating

1. Use pre-printed or columnar forms for orderly sequence of dimensions and locations and for recording telephone quotations.
2. Use only the front side of each paper or form except for certain pre-printed summary forms.
3. Be consistent in listing dimensions: For example, length x width x height. This helps in rechecking to ensure that, the total length of partitions is appropriate for the building area.
4. Use printed (rather than measured) dimensions where given.
5. Add up multiple printed dimensions for a single entry where possible.
6. Measure all other dimensions carefully.
7. Use each set of dimensions to calculate multiple related quantities.
8. Convert foot and inch measurements to decimal feet when listing. Memorize decimal equivalents to .01 parts of a foot (1/8″ equals approximately .01′).
9. Do not "round off" quantities until the final summary.
10. Mark drawings with different colors as items are taken off.
11. Keep similar items together, different items separate.
12. Identify location and drawing numbers to aid in future checking for completeness.
13. Measure or list everything on the drawings or mentioned in the specifications.
14. It may be necessary to list items not called for to make the job complete.
15. Be alert for: Notes on plans such as N.T.S. (not to scale); changes in scale throughout the drawings; reduced size drawings; discrepancies between the specifications and the drawings.
16. Develop a consistent pattern of performing an estimate. For example:
    a. Start the quantity takeoff at the lower floor and move to the next higher floor.
    b. Proceed from the main section of the building to the wings.
    c. Proceed from south to north or vice versa, clockwise or counterclockwise.
    d. Take off floor plan quantities first, elevations next, then detail drawings.
17. List all gross dimensions that can be either used again for different quantities, or used as a rough check of other quantities for verification (exterior perimeter, gross floor area, individual floor areas, etc.).
18. Utilize design symmetry or repetition (repetitive floors, repetitive wings, symmetrical design around a center line, similar room layouts, etc.). Note: Extreme caution is needed here so as not to omit or duplicate an area.
19. Do not convert units until the final total is obtained. For instance, when estimating concrete work, keep all units to the nearest cubic foot, then summarize and convert to cubic yards.
20. When figuring alternatives, it is best to total all items involved in the basic system, then total all items involved in the alternates. Therefore you work with positive numbers in all cases. When adds and deducts are used, it is often confusing whether to add or subtract a portion of an item; especially on a complicated or involved alternate.

R011105-10 Unit Gross Area Requirements

The figures in the table below indicate typical ranges in square feet as a function of the "occupant" unit. This table is best used in the preliminary design stages to help determine the probable size requirement for the total project.

Building Type	Unit	Gross Area in S.F.		
		1/4	Median	3/4
Apartments	Unit	660	860	1,100
Auditorium & Play Theaters	Seat	18	25	38
Bowling Alleys	Lane		940	
Churches & Synagogues	Seat	20	28	39
Dormitories	Bed	200	230	275
Fraternity & Sorority Houses	Bed	220	315	370
Garages, Parking	Car	325	355	385
Hospitals	Bed	685	850	1,075
Hotels	Rental Unit	475	600	710
Housing for the elderly	Unit	515	635	755
Housing, Public	Unit	700	875	1,030
Ice Skating Rinks	Total	27,000	30,000	36,000
Motels	Rental Unit	360	465	620
Nursing Homes	Bed	290	350	450
Restaurants	Seat	23	29	39
Schools, Elementary	Pupil	65	77	90
Junior High & Middle	↓	85	110	129
Senior High	↓	102	130	145
Vocational	↓	110	135	195
Shooting Ranges	Point		450	
Theaters & Movies	Seat		15	

R011105-20 Floor Area Ratios

Table below lists commonly used gross to net area and net to gross area ratios expressed in % for various building types.

Building Type	Gross to Net Ratio	Net to Gross Ratio	Building Type	Gross to Net Ratio	Net to Gross Ratio
Apartment	156	64	School Buildings (campus type)		
Bank	140	72	Administrative	150	67
Church	142	70	Auditorium	142	70
Courthouse	162	61	Biology	161	62
Department Store	123	81	Chemistry	170	59
Garage	118	85	Classroom	152	66
Hospital	183	55	Dining Hall	138	72
Hotel	158	63	Dormitory	154	65
Laboratory	171	58	Engineering	164	61
Library	132	76	Fraternity	160	63
Office	135	75	Gymnasium	142	70
Restaurant	141	70	Science	167	60
Warehouse	108	93	Service	120	83
			Student Union	172	59

The gross area of a building is the total floor area based on outside dimensions.

The net area of a building is the usable floor area for the function intended and excludes such items as stairways, corridors and mechanical rooms. In the case of a commercial building, it might be considered as the "leasable area."

R011105-30 Occupancy Determinations

Function of Space	SF/Person Required
Accessory storage areas, mechanical equipment rooms	300
Agriculture Building	300
Aircraft Hangars	500
Airport Terminal	
Baggage claim	20
Baggage handling	300
Concourse	100
Waiting areas	15
Assembly	
Gaming floors (keno, slots, etc.)	11
Exhibit Gallery and Museum	30
Assembly w/ fixed seats	load determined by seat number
Assembly w/o fixed seats	
Concentrated (chairs only-not fixed)	7
Standing space	5
Unconcentrated (tables and chairs)	15
Bowling centers, allow 5 persons for each lane including 15 feet of runway, and for additional areas	7
Business areas	100
Courtrooms-other than fixed seating areas	40
Day care	35
Dormitories	50
Educational	
Classroom areas	20
Shops and other vocational room areas	50
Exercise rooms	50
Fabrication and Manufacturing areas where hazardous materials are used	200
Industrial areas	100
Institutional areas	
Inpatient treatment areas	240
Outpatient areas	100
Sleeping areas	120
Kitchens commercial	200
Library	
Reading rooms	50
Stack area	100
Mercantile	
Areas on other floors	60
Basement and grade floor areas	30
Storage, stock, shipping areas	300
Parking garages	200
Residential	200
Skating rinks, swimming pools	
Rink and pool	50
Decks	15
Stages and platforms	15
Warehouses	500

R011105-50 Metric Conversion Factors

Description: This table is primarily for converting customary U.S. units in the left hand column to SI metric units in the right hand column. In addition, conversion factors for some commonly encountered Canadian and non-SI metric units are included.

	If You Know		Multiply By		To Find
Length	Inches	x	25.4[a]	=	Millimeters
	Feet	x	0.3048[a]	=	Meters
	Yards	x	0.9144[a]	=	Meters
	Miles (statute)	x	1.609	=	Kilometers
Area	Square inches	x	645.2	=	Square millimeters
	Square feet	x	0.0929	=	Square meters
	Square yards	x	0.8361	=	Square meters
Volume (Capacity)	Cubic inches	x	16,387	=	Cubic millimeters
	Cubic feet	x	0.02832	=	Cubic meters
	Cubic yards	x	0.7646	=	Cubic meters
	Gallons (U.S. liquids)[b]	x	0.003785	=	Cubic meters[c]
	Gallons (Canadian liquid)[b]	x	0.004546	=	Cubic meters[c]
	Ounces (U.S. liquid)[b]	x	29.57	=	Milliliters[c, d]
	Quarts (U.S. liquid)[b]	x	0.9464	=	Liters[c, d]
	Gallons (U.S. liquid)[b]	x	3.785	=	Liters[c, d]
Force	Kilograms force[d]	x	9.807	=	Newtons
	Pounds force	x	4.448	=	Newtons
	Pounds force	x	0.4536	=	Kilograms force[d]
	Kips	x	4448	=	Newtons
	Kips	x	453.6	=	Kilograms force[d]
Pressure, Stress, Strength (Force per unit area)	Kilograms force per square centimeter[d]	x	0.09807	=	Megapascals
	Pounds force per square inch (psi)	x	0.006895	=	Megapascals
	Kips per square inch	x	6.895	=	Megapascals
	Pounds force per square inch (psi)	x	0.07031	=	Kilograms force per square centimeter[d]
	Pounds force per square foot	x	47.88	=	Pascals
	Pounds force per square foot	x	4.882	=	Kilograms force per square meter[d]
Flow	Cubic feet per minute	x	0.4719	=	Liters per second
	Gallons per minute	x	0.0631	=	Liters per second
	Gallons per hour	x	1.05	=	Milliliters per second
Bending Moment Or Torque	Inch-pounds force	x	0.01152	=	Meter-kilograms force[d]
	Inch-pounds force	x	0.1130	=	Newton-meters
	Foot-pounds force	x	0.1383	=	Meter-kilograms force[d]
	Foot-pounds force	x	1.356	=	Newton-meters
	Meter-kilograms force[d]	x	9.807	=	Newton-meters
Mass	Ounces (avoirdupois)	x	28.35	=	Grams
	Pounds (avoirdupois)	x	0.4536	=	Kilograms
	Tons (metric)	x	1000	=	Kilograms
	Tons, short (2000 pounds)	x	907.2	=	Kilograms
	Tons, short (2000 pounds)	x	0.9072	=	Megagrams[e]
Mass per Unit Volume	Pounds mass per cubic foot	x	16.02	=	Kilograms per cubic meter
	Pounds mass per cubic yard	x	0.5933	=	Kilograms per cubic meter
	Pounds mass per gallon (U.S. liquid)[b]	x	119.8	=	Kilograms per cubic meter
	Pounds mass per gallon (Canadian liquid)[b]	x	99.78	=	Kilograms per cubic meter
Temperature	Degrees Fahrenheit		(F-32)/1.8	=	Degrees Celsius
	Degrees Fahrenheit		(F+459.67)/1.8	=	Degrees Kelvin
	Degrees Celsius		C+273.15	=	Degrees Kelvin

[a]The factor given is exact
[b]One U.S. gallon = 0.8327 Canadian gallon
[c]1 liter = 1000 milliliters = 1000 cubic centimeters
1 cubic decimeter = 0.001 cubic meter
[d]Metric but not SI unit
[e]Called "tonne" in England and "metric ton" in other metric countries

R011105-60 Weights and Measures

Measures of Length
1 Mile = 1760 Yards = 5280 Feet
1 Yard = 3 Feet = 36 inches
1 Foot = 12 Inches
1 Mil = 0.001 Inch
1 Fathom = 2 Yards = 6 Feet
1 Rod = 5.5 Yards = 16.5 Feet
1 Hand = 4 Inches
1 Span = 9 Inches
1 Micro-inch = One Millionth Inch or 0.000001 Inch
1 Micron = One Millionth Meter + 0.00003937 Inch

Surveyor's Measure
1 Mile = 8 Furlongs = 80 Chains
1 Furlong = 10 Chains = 220 Yards
1 Chain = 4 Rods = 22 Yards = 66 Feet = 100 Links
1 Link = 7.92 Inches

Square Measure
1 Square Mile = 640 Acres = 6400 Square Chains
1 Acre = 10 Square Chains = 4840 Square Yards = 43,560 Sq. Ft.
1 Square Chain = 16 Square Rods = 484 Square Yards = 4356 Sq. Ft.
1 Square Rod = 30.25 Square Yards = 272.25 Square Feet = 625 Square Lines
1 Square Yard = 9 Square Feet
1 Square Foot = 144 Square Inches
An Acre equals a Square 208.7 Feet per Side

Cubic Measure
1 Cubic Yard = 27 Cubic Feet
1 Cubic Foot = 1728 Cubic Inches
1 Cord of Wood = 4 x 4 x 8 Feet = 128 Cubic Feet
1 Perch of Masonry = 16½ x 1½ x 1 Foot = 24.75 Cubic Feet

Avoirdupois or Commercial Weight
1 Gross or Long Ton = 2240 Pounds
1 Net or Short Ton = 2000 Pounds
1 Pound = 16 Ounces = 7000 Grains
1 Ounce = 16 Drachms = 437.5 Grains
1 Stone = 14 Pounds

Power
1 British Thermal Unit per Hour = 0.2931 Watts
1 Ton (Refrigeration) = 3.517 Kilowatts
1 Horsepower (Boiler) = 9.81 Kilowatts
1 Horsepower (550 ft-lb/s) = 0.746 Kilowatts

Shipping Measure
For Measuring Internal Capacity of a Vessel:
1 Register Ton = 100 Cubic Feet

For Measurement of Cargo:
Approximately 40 Cubic Feet of Merchandise is considered a Shipping Ton, unless that bulk would weigh more than 2000 Pounds, in which case Freight Charge may be based upon weight.

40 Cubic Feet = 32.143 U.S. Bushels = 31.16 Imp. Bushels

Liquid Measure
1 Imperial Gallon = 1.2009 U.S. Gallon = 277.42 Cu. In.
1 Cubic Foot = 7.48 U.S. Gallons

R011110-10 Architectural Fees

Tabulated below are typical percentage fees by project size, for good professional architectural service. Fees may vary from those listed depending upon degree of design difficulty and economic conditions in any particular area.

Rates can be interpolated horizontally and vertically. Various portions of the same project requiring different rates should be adjusted proportionately. For alterations, add 50% to the fee for the first $500,000 of project cost and add 25% to the fee for project cost over $500,000.

Architectural fees tabulated below include Structural, Mechanical and Electrical Engineering Fees. They do not include the fees for special consultants such as kitchen planning, security, acoustical, interior design, etc

Civil Engineering fees are included in the Architectural fee for project sites requiring minimal design such as city sites. However, separate Civil Engineering fees must be added when utility connections require design, drainage calculations are needed, stepped foundations are required, or provisions are required to protect adjacent wetlands.

	Total Project Size in Thousands of Dollars						
Building Types	100	250	500	1,000	5,000	10,000	50,000
Factories, garages, warehouses, repetitive housing	9.0%	8.0%	7.0%	6.2%	5.3%	4.9%	4.5%
Apartments, banks, schools, libraries, offices, municipal buildings	12.2	12.3	9.2	8.0	7.0	6.6	6.2
Churches, hospitals, homes, laboratories, museums, research	15.0	13.6	12.7	11.9	9.5	8.8	8.0
Memorials, monumental work, decorative furnishings	—	16.0	14.5	13.1	10.0	9.0	8.3

R011110-30 Engineering Fees

Typical **Structural Engineering Fees** based on type of construction and total project size. These fees are included in Architectural Fees.

	Total Project Size (in thousands of dollars)			
Type of Construction	$500	$500-$1,000	$1,000-$5,000	Over $5000
Industrial buildings, factories & warehouses	Technical payroll times 2.0 to 2.5	1.60%	1.25%	1.00%
Hotels, apartments, offices, dormitories, hospitals, public buildings, food stores	↓	2.00%	1.70%	1.20%
Museums, banks, churches and cathedrals	↓	2.00%	1.75%	1.25%
Thin shells, prestressed concrete, earthquake resistive	↓	2.00%	1.75%	1.50%
Parking ramps, auditoriums, stadiums, convention halls, hangars & boiler houses	↓	2.50%	2.00%	1.75%
Special buildings, major alterations, underpinning & future expansion	↓	Add to above 0.5%	Add to above 0.5%	Add to above 0.5%

For complex reinforced concrete or unusually complicated structures, add 20% to 50%.

Typical **Mechanical and Electrical Engineering Fees** are based on the size of the subcontract. The fee structure for both is shown below. These fees are included in Architectural Fees.

	Subcontract Size							
Type of Construction	$25,000	$50,000	$100,000	$225,000	$350,000	$500,000	$750,000	$1,000,000
Simple structures	6.4%	5.7%	4.8%	4.5%	4.4%	4.3%	4.2%	4.1%
Intermediate structures	8.0	7.3	6.5	5.6	5.1	5.0	4.9	4.8
Complex structures	10.1	9.0	9.0	8.0	7.5	7.5	7.0	7.0

For renovations, add 15% to 25% to applicable fee.

R012153-10 Repair and Remodeling

Cost figures are based on new construction utilizing the most cost-effective combination of labor, equipment and material with the work scheduled in proper sequence to allow the various trades to accomplish their work in an efficient manner.

The costs for repair and remodeling work must be modified due to the following factors that may be present in any given repair and remodeling project.

1. Equipment usage curtailment due to the physical limitations of the project, with only hand-operated equipment being used.
2. Increased requirement for shoring and bracing to hold up the building while structural changes are being made and to allow for temporary storage of construction materials on above-grade floors.
3. Material handling becomes more costly due to having to move within the confines of an enclosed building. For multi-story construction, low capacity elevators and stairwells may be the only access to the upper floors.
4. Large amount of cutting and patching and attempting to match the existing construction is required. It is often more economical to remove entire walls rather than create many new door and window openings. This sort of trade-off has to be carefully analyzed.
5. Cost of protection of completed work is increased since the usual sequence of construction usually cannot be accomplished.
6. Economies of scale usually associated with new construction may not be present. If small quantities of components must be custom fabricated due to job requirements, unit costs will naturally increase. Also, if only small work areas are available at a given time, job scheduling between trades becomes difficult and subcontractor quotations may reflect the excessive start-up and shut-down phases of the job.
7. Work may have to be done on other than normal shifts and may have to be done around an existing production facility which has to stay in production during the course of the repair and remodeling.
8. Dust and noise protection of adjoining non-construction areas can involve substantial special protection and alter usual construction methods.
9. Job may be delayed due to unexpected conditions discovered during demolition or removal. These delays ultimately increase construction costs.
10. Piping and ductwork runs may not be as simple as for new construction. Wiring may have to be snaked through walls and floors.
11. Matching "existing construction" may be impossible because materials may no longer be manufactured. Substitutions may be expensive.
12. Weather protection of existing structure requires additional temporary structures to protect building at openings.
13. On small projects, because of local conditions, it may be necessary to pay a tradesman for a minimum of four hours for a task that is completed in one hour.

All of the above areas can contribute to increased costs for a repair and remodeling project. Each of the above factors should be considered in the planning, bidding and construction stage in order to minimize the increased costs associated with repair and remodeling jobs.

R012153-60 Security Factors

Contractors entering, working in, and exiting secure facilities often lose productive time during a normal workday. The recommended allowances in this section are intended to provide for the loss of productivity by increasing labor costs. Note that different costs are associated with searches upon entry only and searches upon entry and exit. Time spent in a queue is unpredictable and not part of these allowances. Contractors should plan ahead for this situation.

Security checkpoints are designed to reflect the level of security required to gain access or egress. An extreme example is when contractors, along with any materials, tools, equipment, and vehicles, must be physically searched and have all materials, tools, equipment, and vehicles inventoried and documented prior to both entry and exit.

Physical searches without going through the documentation process represent the next level and take up less time.

Electronic searches—passing through a detector or x-ray machine with no documentation of materials, tools, equipment, and vehicles—take less time than physical searches.

Visual searches of materials, tools, equipment, and vehicles represent the next level of security.

Finally, access by means of an ID card or displayed sticker takes the least amount of time.

Another consideration is if the searches described above are performed each and every day, or if they are performed only on the first day with access granted by ID card or displayed sticker for the remainder of the project. The figures for this situation have been calculated to represent the initial check-in as described and subsequent entry by ID card or displayed sticker for up to 20 days on site. For the situation described above, where the time period is beyond 20 days, the impact on labor cost is negligible.

There are situations where tradespeople must be accompanied by an escort and observed during the work day. The loss of freedom of movement will slow down productivity for the tradesperson. Costs for the observer have not been included. Those costs are normally born by the owner.

R012909-80 Sales Tax by State

State sales tax on materials is tabulated below (5 states have no sales tax). Many states allow local jurisdictions, such as a county or city, to levy additional sales tax.

Some projects may be sales tax exempt, particularly those constructed with public funds.

State	Tax (%)	State	Tax (%)	State	Tax (%)	State	Tax (%)
Alabama	4	Illinois	6.25	Montana	0	Rhode Island	7
Alaska	0	Indiana	7	Nebraska	5.5	South Carolina	6
Arizona	5.6	Iowa	6	Nevada	6.85	South Dakota	4.5
Arkansas	6.5	Kansas	6.5	New Hampshire	0	Tennessee	7
California	7.25	Kentucky	6	New Jersey	6.625	Texas	6.25
Colorado	2.9	Louisiana	4.45	New Mexico	5.125	Utah	5.95
Connecticut	6.35	Maine	5.5	New York	4	Vermont	6
Delaware	0	Maryland	6	North Carolina	4.75	Virginia	5.3
District of Columbia	5.75	Massachusetts	6.25	North Dakota	5	Washington	6.5
Florida	6	Michigan	6	Ohio	5.75	West Virginia	6
Georgia	4	Minnesota	6.875	Oklahoma	4.5	Wisconsin	5
Hawaii	4	Mississippi	7	Oregon	0	Wyoming	4
Idaho	6	Missouri	4.225	Pennsylvania	6	Average	5.10%

Sales Tax by Province (Canada)

GST - a value-added tax, which the government imposes on most goods and services provided in or imported into Canada. PST - a retail sales tax, which five of the provinces impose on the prices of most goods and some services. QST - a value-added tax, similar to the federal GST, which Quebec imposes. HST - Three provinces have combined their retail sales taxes with the federal GST into one harmonized tax.

Province	PST (%)	QST (%)	GST(%)	HST(%)
Alberta	0	0	5	0
British Columbia	7	0	5	0
Manitoba	8	0	5	0
New Brunswick	0	0	0	15
Newfoundland	0	0	0	15
Northwest Territories	0	0	5	0
Nova Scotia	0	0	0	15
Ontario	0	0	0	13
Prince Edward Island	0	0	0	15
Quebec	0	9.975	5	0
Saskatchewan	6	0	5	0
Yukon	0	0	5	0

R012909-85 Unemployment Taxes and Social Security Taxes

State unemployment tax rates vary not only from state to state, but also with the experience rating of the contractor. The federal unemployment tax rate is 6.0% of the first $7,000 of wages. This is reduced by a credit of up to 5.4% for timely payment to the state. The minimum federal unemployment tax is 0.6% after all credits.

Social security (FICA) for 2019 is estimated at time of publication to be 7.65% of wages up to $128,400.

R012909-86 Unemployment Tax by State

Information is from the U.S. Department of Labor, state unemployment tax rates.

State	Tax (%)	State	Tax (%)	State	Tax (%)	State	Tax (%)
Alabama	6.74	Illinois	7.75	Montana	6.12	Rhode Island	9.79
Alaska	5.4	Indiana	7.474	Nebraska	5.4	South Carolina	5.46
Arizona	8.91	Iowa	8	Nevada	5.4	South Dakota	9.5
Arkansas	6.0	Kansas	7.6	New Hampshire	7.5	Tennessee	10.0
California	6.2	Kentucky	10.0	New Jersey	5.8	Texas	7.5
Colorado	8.9	Louisiana	6.2	New Mexico	5.4	Utah	7.2
Connecticut	6.8	Maine	5.4	New York	8.5	Vermont	8.4
Delaware	8.0	Maryland	7.50	North Carolina	5.76	Virginia	6.27
District of Columbia	7	Massachusetts	11.13	North Dakota	10.72	Washington	5.7
Florida	5.4	Michigan	10.3	Ohio	8.7	West Virginia	7.5
Georgia	5.4	Minnesota	9.0	Oklahoma	5.5	Wisconsin	12.0
Hawaii	5.6	Mississippi	5.4	Oregon	5.4	Wyoming	8.8
Idaho	5.4	Missouri	9.75	Pennsylvania	10.89	Median	7.47 %

R012909-90 Overtime

One way to improve the completion date of a project or eliminate negative float from a schedule is to compress activity duration times. This can be achieved by increasing the crew size or working overtime with the proposed crew.

To determine the costs of working overtime to compress activity duration times, consider the following examples. Below is an overtime efficiency and cost chart based on a five, six, or seven day week with an eight through twelve hour day. Payroll percentage increases for time and one half and double times are shown for the various working days.

Days per Week	Hours per Day	Production Efficiency					Payroll Cost Factors	
		1st Week	2nd Week	3rd Week	4th Week	Average 4 Weeks	@ 1-1/2 Times	@ 2 Times
5	8	100%	100%	100%	100%	100%	1.000	1.000
	9	100	100	95	90	96	1.056	1.111
	10	100	95	90	85	93	1.100	1.200
	11	95	90	75	65	81	1.136	1.273
	12	90	85	70	60	76	1.167	1.333
6	8	100	100	95	90	96	1.083	1.167
	9	100	95	90	85	93	1.130	1.259
	10	95	90	85	80	88	1.167	1.333
	11	95	85	70	65	79	1.197	1.394
	12	90	80	65	60	74	1.222	1.444
7	8	100	95	85	75	89	1.143	1.286
	9	95	90	80	70	84	1.183	1.365
	10	90	85	75	65	79	1.214	1.429
	11	85	80	65	60	73	1.240	1.481
	12	85	75	60	55	69	1.262	1.524

R013113-40 Builder's Risk Insurance

Builder's risk insurance is insurance on a building during construction. Premiums are paid by the owner or the contractor. Blasting, collapse and underground insurance would raise total insurance costs.

R013113-50 General Contractor's Overhead

There are two distinct types of overhead on a construction project: Project overhead and main office overhead. Project overhead includes those costs at a construction site not directly associated with the installation of construction materials. Examples of project overhead costs include the following:

1. Superintendent
2. Construction office and storage trailers
3. Temporary sanitary facilities
4. Temporary utilities
5. Security fencing
6. Photographs
7. Cleanup
8. Performance and payment bonds

The above project overhead items are also referred to as general requirements and therefore are estimated in Division 1. Division 1 is the first division listed in the CSI MasterFormat but it is usually the last division estimated. The sum of the costs in Divisions 1 through 49 is referred to as the sum of the direct costs.

All construction projects also include indirect costs. The primary components of indirect costs are the contractor's main office overhead and profit. The amount of the main office overhead expense varies depending on the following:

1. Owner's compensation
2. Project managers' and estimators' wages
3. Clerical support wages
4. Office rent and utilities
5. Corporate legal and accounting costs
6. Advertising
7. Automobile expenses
8. Association dues
9. Travel and entertainment expenses

These costs are usually calculated as a percentage of annual sales volume. This percentage can range from 35% for a small contractor doing less than $500,000 to 5% for a large contractor with sales in excess of $100 million.

R013113-55 Installing Contractor's Overhead

Installing contractors (subcontractors) also incur costs for general requirements and main office overhead.

Included within the total incl. overhead and profit costs is a percent mark-up for overhead that includes:

1. Compensation and benefits for office staff and project managers
2. Office rent, utilities, business equipment, and maintenance
3. Corporate legal and accounting costs
4. Advertising
5. Vehicle expenses (for office staff and project managers)
6. Association dues
7. Travel, entertainment
8. Insurance
9. Small tools and equipment

R013113-60 Workers' Compensation Insurance Rates by Trade

The table below tabulates the national averages for workers' compensation insurance rates by trade and type of building. The average "Insurance Rate" is multiplied by the "% of Building Cost" for each trade. This produces the "Workers' Compensation" cost by % of total labor cost, to be added for each trade by building type to determine the weighted average workers' compensation rate for the building types analyzed.

Trade	Insurance Rate (% Labor Cost)				% of Building Cost			Workers' Compensation		
	Range			Average	Office Bldgs.	Schools & Apts.	Mfg.	Office Bldgs.	Schools & Apts.	Mfg.
Excavation, Grading, etc.	2.3 %	to	19.2%	8.5%	4.8%	4.9%	4.5%	0.41%	0.42%	0.38%
Piles & Foundations	3.9	to	27.3	13.4	7.1	5.2	8.7	0.95	0.70	1.17
Concrete	3.5	to	25.6	11.8	5.0	14.8	3.7	0.59	1.75	0.44
Masonry	3.6	to	50.4	13.8	6.9	7.5	1.9	0.95	1.04	0.26
Structural Steel	4.9	to	43.8	21.2	10.7	3.9	17.6	2.27	0.83	3.73
Miscellaneous & Ornamental Metals	2.8	to	27.0	10.6	2.8	4.0	3.6	0.30	0.42	0.38
Carpentry & Millwork	3.7	to	31.9	13.0	3.7	4.0	0.5	0.48	0.52	0.07
Metal or Composition Siding	5.1	to	113.7	19.0	2.3	0.3	4.3	0.44	0.06	0.82
Roofing	5.7	to	103.7	29.0	2.3	2.6	3.1	0.67	0.75	0.90
Doors & Hardware	3.1	to	31.9	11.0	0.9	1.4	0.4	0.10	0.15	0.04
Sash & Glazing	3.9	to	24.3	12.1	3.5	4.0	1.0	0.42	0.48	0.12
Lath & Plaster	2.7	to	34.4	10.7	3.3	6.9	0.8	0.35	0.74	0.09
Tile, Marble & Floors	2.2	to	21.5	8.7	2.6	3.0	0.5	0.23	0.26	0.04
Acoustical Ceilings	1.9	to	29.7	8.5	2.4	0.2	0.3	0.20	0.02	0.03
Painting	3.7	to	36.7	11.2	1.5	1.6	1.6	0.17	0.18	0.18
Interior Partitions	3.7	to	31.9	13.0	3.9	4.3	4.4	0.51	0.56	0.57
Miscellaneous Items	2.1	to	97.7	11.2	5.2	3.7	9.7	0.58	0.42	1.09
Elevators	1.4	to	11.4	4.7	2.1	1.1	2.2	0.10	0.05	0.10
Sprinklers	2.0	to	16.4	6.7	0.5	—	2.0	0.03	—	0.13
Plumbing	1.5	to	14.6	6.3	4.9	7.2	5.2	0.31	0.45	0.33
Heat., Vent., Air Conditioning	3.0	to	17.0	8.3	13.5	11.0	12.9	1.12	0.91	1.07
Electrical	1.9	to	11.3	5.2	10.1	8.4	11.1	0.53	0.44	0.58
Total	1.4 %	to	113.7%	—	100.0%	100.0%	100.0%	11.71%	11.15%	12.52%
Overall Weighted Average 11.79%										

Workers' Compensation Insurance Rates by States

The table below lists the weighted average Workers' Compensation base rate for each state with a factor comparing this with the national average of 11.8%.

State	Weighted Average	Factor	State	Weighted Average	Factor	State	Weighted Average	Factor
Alabama	13.9%	129	Kentucky	11.3%	105	North Dakota	7.4%	69
Alaska	11.0	102	Louisiana	20.0	185	Ohio	6.2	57
Arizona	9.1	84	Maine	9.8	91	Oklahoma	8.7	81
Arkansas	5.7	53	Maryland	11.2	104	Oregon	9.0	83
California	23.6	219	Massachusetts	9.1	84	Pennsylvania	24.9	231
Colorado	8.1	75	Michigan	8.0	74	Rhode Island	10.6	98
Connecticut	17.0	157	Minnesota	16.2	150	South Carolina	20.3	188
Delaware	10.7	99	Mississippi	11.4	106	South Dakota	10.3	95
District of Columbia	9.1	84	Missouri	14.0	130	Tennessee	7.9	73
Florida	11.3	105	Montana	7.8	72	Texas	6.2	57
Georgia	32.9	305	Nebraska	13.4	124	Utah	7.4	69
Hawaii	9.2	85	Nevada	8.9	82	Vermont	10.8	100
Idaho	9.2	85	New Hampshire	11.3	105	Virginia	7.4	69
Illinois	20.1	186	New Jersey	15.4	143	Washington	8.9	82
Indiana	3.7	34	New Mexico	13.7	127	West Virginia	4.7	44
Iowa	12.1	112	New York	19.1	177	Wisconsin	13.3	123
Kansas	6.6	61	North Carolina	17.1	158	Wyoming	6.3	58
Weighted Average for U.S. is 11.8% of payroll = 100%								

The weighted average skilled worker rate for 35 trades is 11.8%. For bidding purposes, apply the full value of Workers' Compensation directly to total labor costs, or if labor is 38%, materials 42% and overhead and profit 20% of total cost, carry 38/80 x 11.8% = 6.0% of cost (before overhead and profit) into overhead. Rates vary not only from state to state but also with the experience rating of the contractor.

Rates are the most current available at the time of publication.

R013113-80 Performance Bond

This table shows the cost of a Performance Bond for a construction job scheduled to be completed in 12 months. Add 1% of the premium cost per month for jobs requiring more than 12 months to complete. The rates are "standard" rates offered to contractors that the bonding company considers financially sound and capable of doing the work. Preferred rates are offered by some bonding companies based upon financial strength of the contractor. Actual rates vary from contractor to contractor and from bonding company to bonding company. Contractors should prequalify through a bonding agency before submitting a bid on a contract that requires a bond.

Contract Amount	Building Construction Class B Projects	Highways & Bridges Class A New Construction	Highways & Bridges Class A-1 Highway Resurfacing
First $ 100,000 bid	$25.00 per M	$15.00 per M	$9.40 per M
Next 400,000 bid	$ 2,500 plus $15.00 per M	$ 1,500 plus $10.00 per M	$ 940 plus $7.20 per M
Next 2,000,000 bid	8,500 plus 10.00 per M	5,500 plus 7.00 per M	3,820 plus 5.00 per M
Next 2,500,000 bid	28,500 plus 7.50 per M	19,500 plus 5.50 per M	15,820 plus 4.50 per M
Next 2,500,000 bid	47,250 plus 7.00 per M	33,250 plus 5.00 per M	28,320 plus 4.50 per M
Over 7,500,000 bid	64,750 plus 6.00 per M	45,750 plus 4.50 per M	39,570 plus 4.00 per M

General Requirements R0154 Construction Aids

R015423-10 Steel Tubular Scaffolding

On new construction, tubular scaffolding is efficient up to 60′ high or five stories. Above this it is usually better to use a hung scaffolding if construction permits. Swing scaffolding operations may interfere with tenants. In this case, the tubular is more practical at all heights.

In repairing or cleaning the front of an existing building the cost of tubular scaffolding per S.F. of building front increases as the height increases above the first tier. The first tier cost is relatively high due to leveling and alignment.

The minimum efficient crew for erecting and dismantling is three workers. They can set up and remove 18 frame sections per day up to 5 stories high. For 6 to 12 stories high, a crew of four is most efficient. Use two or more on top and two on the bottom for handing up or hoisting. They can also set up and remove 18 frame sections per day. At 7' horizontal spacing, this will run about 800 S.F. per day of erecting and dismantling. Time for placing and removing planks must be added to the above. A crew of three can place and remove 72 planks per day up to 5 stories. For over 5 stories, a crew of four can place and remove 80 planks per day.

The table below shows the number of pieces required to erect tubular steel scaffolding for 1000 S.F. of building frontage. This area is made up of a scaffolding system that is 12 frames (11 bays) long by 2 frames high.

For jobs under twenty-five frames, add 50% to rental cost. Rental rates will be lower for jobs over three months duration. Large quantities for long periods can reduce rental rates by 20%.

Description of Component	Number of Pieces for 1000 S.F. of Building Front	Unit
5′ Wide Standard Frame, 6′-4″ High	24	Ea.
Leveling Jack & Plate	24	↓
Cross Brace	44	↓
Side Arm Bracket, 21″	12	↓
Guardrail Post	12	↓
Guardrail, 7′ section	22	↓
Stairway Section	2	↓
Stairway Starter Bar	1	↓
Stairway Inside Handrail	2	↓
Stairway Outside Handrail	2	↓
Walk-Thru Frame Guardrail	2	↓

Scaffolding is often used as falsework over 15′ high during construction of cast-in-place concrete beams and slabs. Two foot wide scaffolding is generally used for heavy beam construction. The span between frames depends upon the load to be carried with a maximum span of 5′.

Heavy duty shoring frames with a capacity of 10,000#/leg can be spaced up to 10′ O.C. depending upon form support design and loading.

Scaffolding used as horizontal shoring requires less than half the material required with conventional shoring.

On new construction, erection is done by carpenters.

Rolling towers supporting horizontal shores can reduce labor and speed the job. For maintenance work, catwalks with spans up to 70′ can be supported by the rolling towers.

R015433-10 Contractor Equipment

Rental Rates shown elsewhere in the data set pertain to late model high quality machines in excellent working condition, rented from equipment dealers. Rental rates from contractors may be substantially lower than the rental rates from equipment dealers depending upon economic conditions; for older, less productive machines, reduce rates by a maximum of 15%. Any overtime must be added to the base rates. For shift work, rates are lower. Usual rule of thumb is 150% of one shift rate for two shifts; 200% for three shifts.

For periods of less than one week, operated equipment is usually more economical to rent than renting bare equipment and hiring an operator.

Costs to move equipment to a job site (mobilization) or from a job site (demobilization) are not included in rental rates, nor in any Equipment costs on any Unit Price line items or crew listings. These costs can be found elsewhere. If a piece of equipment is already at a job site, it is not appropriate to utilize mob/demob costs in an estimate again.

Rental rates vary throughout the country with larger cities generally having lower rates. Lease plans for new equipment are available for periods in excess of six months with a percentage of payments applying toward purchase.

Rental rates can also be treated as reimbursement costs for contractor-owned equipment. Owned equipment costs include depreciation, loan payments, interest, taxes, insurance, storage, and major repairs.

Monthly rental rates vary from 2% to 5% of the cost of the equipment depending on the anticipated life of the equipment and its wearing parts. Weekly rates are about 1/3 the monthly rates and daily rental rates are about 1/3 the weekly rates.

The hourly operating costs for each piece of equipment include costs to the user such as fuel, oil, lubrication, normal expendables for the equipment, and a percentage of the mechanic's wages chargeable to maintenance. The hourly operating costs listed do not include the operator's wages.

The daily cost for equipment used in the standard crews is figured by dividing the weekly rate by five, then adding eight times the hourly operating cost to give the total daily equipment cost, not including the operator. This figure is in the right hand column of the Equipment listings under Equipment Cost/Day.

Pile Driving rates shown for the pile hammer and extractor do not include leads, cranes, boilers or compressors. Vibratory pile driving requires an added field specialist during set-up and pile driving operation for the electric model. The hydraulic model requires a field specialist for set-up only. Up to 125 reuses of sheet piling are possible using vibratory drivers. For normal conditions, crane capacity for hammer type and size is as follows.

Crane Capacity	Hammer Type and Size		
	Air or Steam	Diesel	Vibratory
25 ton	to 8,750 ft.-lb.		70 H.P.
40 ton	15,000 ft.-lb.	to 32,000 ft.-lb.	170 H.P.
60 ton	25,000 ft.-lb.		300 H.P.
100 ton		112,000 ft.-lb.	

Cranes should be specified for the job by size, building and site characteristics, availability, performance characteristics, and duration of time required.

Backhoes & Shovels rent for about the same as equivalent size cranes but maintenance and operating expenses are higher. The crane operator's rate must be adjusted for high boom heights. Average adjustments: for 150′ boom add 2% per hour; over 185′, add 4% per hour; over 210′, add 6% per hour; over 250′, add 8% per hour and over 295′, add 12% per hour.

Tower Cranes of the climbing or static type have jibs from 50′ to 200′ and capacities at maximum reach range from 4,000 to 14,000 pounds. Lifting capacities increase up to maximum load as the hook radius decreases.

Typical rental rates, based on purchase price, are about 2% to 3% per month.

Erection and dismantling run between 500 and 2000 labor hours. Climbing operation takes 10 labor hours per 20′ climb. Crane dead time is about 5 hours per 40′ climb. If crane is bolted to side of the building add cost of ties and extra mast sections. Climbing cranes have from 80′ to 180′ of mast while static cranes have 80′ to 800′ of mast.

Truck Cranes can be converted to tower cranes by using tower attachments. Mast heights over 400′ have been used.

A single 100′ high material **Hoist and Tower** can be erected and dismantled in about 400 labor hours; a double 100′ high hoist and tower in about 600 labor hours. Erection times for additional heights are 3 and 4 labor hours per vertical foot respectively up to 150′, and 4 to 5 labor hours per vertical foot over 150′ high. A 40′ high portable Buck hoist takes about 160 labor hours to erect and dismantle. Additional heights take 2 labor hours per vertical foot to 80′ and 3 labor hours per vertical foot for the next 100′. Most material hoists do not meet local code requirements for carrying personnel.

A 150′ high **Personnel Hoist** requires about 500 to 800 labor hours to erect and dismantle. Budget erection time at 5 labor hours per vertical foot for all trades. Local code requirements or labor scarcity requiring overtime can add up to 50% to any of the above erection costs.

Earthmoving Equipment: The selection of earthmoving equipment depends upon the type and quantity of material, moisture content, haul distance, haul road, time available, and equipment available. Short haul cut and fill operations may require dozers only, while another operation may require excavators, a fleet of trucks, and spreading and compaction equipment. Stockpiled material and granular material are easily excavated with front end loaders. Scrapers are most economically used with hauls between 300′ and 1-1/2 miles if adequate haul roads can be maintained. Shovels are often used for blasted rock and any material where a vertical face of 8′ or more can be excavated. Special conditions may dictate the use of draglines, clamshells, or backhoes. Spreading and compaction equipment must be matched to the soil characteristics, the compaction required and the rate the fill is being supplied.

R015433-15 Heavy Lifting

Hydraulic Climbing Jacks

The use of hydraulic heavy lift systems is an alternative to conventional type crane equipment. The lifting, lowering, pushing, or pulling mechanism is a hydraulic climbing jack moving on a square steel jackrod from 1-5/8″ to 4″ square, or a steel cable. The jackrod or cable can be vertical or horizontal, stationary or movable, depending on the individual application. When the jackrod is stationary, the climbing jack will climb the rod and push or pull the load along with itself. When the climbing jack is stationary, the jackrod is movable with the load attached to the end and the climbing jack will lift or lower the jackrod with the attached load. The heavy lift system is normally operated by a single control lever located at the hydraulic pump.

The system is flexible in that one or more climbing jacks can be applied wherever a load support point is required, and the rate of lift synchronized.

Economic benefits have been demonstrated on projects such as: erection of ground assembled roofs and floors, complete bridge spans, girders and trusses, towers, chimney liners and steel vessels, storage tanks, and heavy machinery. Other uses are raising and lowering offshore work platforms, caissons, tunnel sections and pipelines.

R015436-50 Mobilization

Costs to move rented construction equipment to a job site from an equipment dealer's or contractor's yard (mobilization) or off the job site (demobilization) are not included in the rental or operating rates, nor in the equipment cost on a unit price line or in a crew listing. These costs can be found consolidated in the Mobilization section of the data and elsewhere in particular site work sections. If a piece of equipment is already on the job site, it is not appropriate to include mob/demob costs in a new estimate that requires use of that equipment. The following table identifies approximate sizes of rented construction equipment that would be hauled on a towed trailer. Because this listing is not all-encompassing, the user can infer as to what size trailer might be required for a piece of equipment not listed.

3-ton Trailer	20-ton Trailer	40-ton Trailer	50-ton Trailer
20 H.P. Excavator	110 H.P. Excavator	200 H.P. Excavator	270 H.P. Excavator
50 H.P. Skid Steer	165 H.P. Dozer	300 H.P. Dozer	Small Crawler Crane
35 H.P. Roller	150 H.P. Roller	400 H.P. Scraper	500 H.P. Scraper
40 H.P. Trencher	Backhoe	450 H.P. Art. Dump Truck	500 H.P. Art. Dump Truck

Existing Conditions — R0241 Demolition

R024119-10 Demolition Defined

Whole Building Demolition - Demolition of the whole building with no concern for any particular building element, component, or material type being demolished. This type of demolition is accomplished with large pieces of construction equipment that break up the structure, load it into trucks and haul it to a disposal site, but disposal or dump fees are not included. Demolition of below-grade foundation elements, such as footings, foundation walls, grade beams, slabs on grade, etc., is not included. Certain mechanical equipment containing flammable liquids or ozone-depleting refrigerants, electric lighting elements, communication equipment components, and other building elements may contain hazardous waste, and must be removed, either selectively or carefully, as hazardous waste before the building can be demolished.

Foundation Demolition - Demolition of below-grade foundation footings, foundation walls, grade beams, and slabs on grade. This type of demolition is accomplished by hand or pneumatic hand tools, and does not include saw cutting, or handling, loading, hauling, or disposal of the debris.

Gutting - Removal of building interior finishes and electrical/mechanical systems down to the load-bearing and sub-floor elements of the rough building frame, with no concern for any particular building element, component, or material type being demolished. This type of demolition is accomplished by hand or pneumatic hand tools, and includes loading into trucks, but not hauling, disposal or dump fees, scaffolding, or shoring. Certain mechanical equipment containing flammable liquids or ozone-depleting refrigerants, electric lighting elements, communication equipment components, and other building elements may contain hazardous waste, and must be removed, either selectively or carefully, as hazardous waste, before the building is gutted.

Selective Demolition - Demolition of a selected building element, component, or finish, with some concern for surrounding or adjacent elements, components, or finishes (see the first Subdivision (s) at the beginning of appropriate Divisions). This type of demolition is accomplished by hand or pneumatic hand tools, and does not include handling, loading, storing, hauling, or disposal of the debris, scaffolding, or shoring. "Gutting" methods may be used in order to save time, but damage that is caused to surrounding or adjacent elements, components, or finishes may have to be repaired at a later time.

Careful Removal - Removal of a piece of service equipment, building element or component, or material type, with great concern for both the removed item and surrounding or adjacent elements, components or finishes. The purpose of careful removal may be to protect the removed item for later re-use, preserve a higher salvage value of the removed item, or replace an item while taking care to protect surrounding or adjacent elements, components, connections, or finishes from cosmetic and/or structural damage. An approximation of the time required to perform this type of removal is 1/3 to 1/2 the time it would take to install a new item of like kind (see Reference Number R220105-10). This type of removal is accomplished by hand or pneumatic hand tools, and does not include loading, hauling, or storing the removed item, scaffolding, shoring, or lifting equipment.

Cutout Demolition - Demolition of a small quantity of floor, wall, roof, or other assembly, with concern for the appearance and structural integrity of the surrounding materials. This type of demolition is accomplished by hand or pneumatic hand tools, and does not include saw cutting, handling, loading, hauling, or disposal of debris, scaffolding, or shoring.

Rubbish Handling - Work activities that involve handling, loading or hauling of debris. Generally, the cost of rubbish handling must be added to the cost of all types of demolition, with the exception of whole building demolition.

Minor Site Demolition - Demolition of site elements outside the footprint of a building. This type of demolition is accomplished by hand or pneumatic hand tools, or with larger pieces of construction equipment, and may include loading a removed item onto a truck (check the Crew for equipment used). It does not include saw cutting, hauling or disposal of debris, and, sometimes, handling or loading.

R024119-20 Dumpsters

Dumpster rental costs on construction sites are presented in two ways.

The cost per week rental includes the delivery of the dumpster; its pulling or emptying once per week, and its final removal. The assumption is made that the dumpster contractor could choose to empty a dumpster by simply bringing in an empty unit and removing the full one. These costs also include the disposal of the materials in the dumpster.

The Alternate Pricing can be used when actual planned conditions are not approximated by the weekly numbers. For example, these lines can be used when a dumpster is needed for 4 weeks and will need to be emptied 2 or 3 times per week. Conversely the Alternate Pricing lines can be used when a dumpster will be rented for several weeks or months but needs to be emptied only a few times over this period.

Existing Conditions — R0265 Underground Storage Tank Removal

R026510-20 Underground Storage Tank Removal

Underground Storage Tank Removal can be divided into two categories: Non-Leaking and Leaking. Prior to removing an underground storage tank, tests should be made, with the proper authorities present, to determine whether a tank has been leaking or the surrounding soil has been contaminated.

To safely remove Liquid Underground Storage Tanks:

1. Excavate to the top of the tank.
2. Disconnect all piping.
3. Open all tank vents and access ports.
4. Remove all liquids and/or sludge.
5. Purge the tank with an inert gas.
6. Provide access to the inside of the tank and clean out the interior using proper personal protective equipment (PPE).
7. Excavate soil surrounding the tank using proper PPE for on-site personnel.
8. Pull and properly dispose of the tank.
9. Clean up the site of all contaminated material.
10. Install new tanks or close the excavation.

R028213-20 Asbestos Removal Process

Asbestos removal is accomplished by a specialty contractor who understands the federal and state regulations regarding the handling and disposal of the material. The process of asbestos removal is divided into many individual steps. An accurate estimate can be calculated only after all the steps have been priced.

The steps are generally as follows:

1. Obtain an asbestos abatement plan from an industrial hygienist.
2. Monitor the air quality in and around the removal area and along the path of travel between the removal area and transport area. This establishes the background contamination.
3. Construct a two part decontamination chamber at entrance to removal area.
4. Install a HEPA filter to create a negative pressure in the removal area.
5. Install wall, floor and ceiling protection as required by the plan, usually 2 layers of fireproof 6 mil polyethylene.
6. Industrial hygienist visually inspects work area to verify compliance with plan.
7. Provide temporary supports for conduit and piping affected by the removal process.
8. Proceed with asbestos removal and bagging process. Monitor air quality as described in Step #2. Discontinue operations when contaminate levels exceed applicable standards.
9. Document the legal disposal of materials in accordance with EPA standards.
10. Thoroughly clean removal area including all ledges, crevices and surfaces.
11. Post abatement inspection by industrial hygienist to verify plan compliance.
12. Provide a certificate from a licensed industrial hygienist attesting that contaminate levels are within acceptable standards before returning area to regular use.

R078413-30 Firestopping

Firestopping is the sealing of structural, mechanical, electrical and other penetrations through fire-rated assemblies. The basic components of firestop systems are safing insulation and firestop sealant on both sides of wall penetrations and the top side of floor penetrations.

Pipe penetrations are assumed to be through concrete, grout, or joint compound and can be sleeved or unsleeved. Costs for the penetrations and sleeves are not included. An annular space of 1″ is assumed. Escutcheons are not included.

A metallic pipe is assumed to be copper, aluminum, cast iron or similar metallic material. An insulated metallic pipe is assumed to be covered with a thermal insulating jacket of varying thickness and materials.

A non-metallic pipe is assumed to be PVC, CPVC, FR Polypropylene or similar plastic piping material. Intumescent firestop sealants or wrap strips are included. Collars on both sides of wall penetrations and a sheet metal plate on the underside of floor penetrations are included.

Ductwork is assumed to be sheet metal, stainless steel or similar metallic material. Duct penetrations are assumed to be through concrete, grout or joint compound. Costs for penetrations and sleeves are not included. An annular space of 1/2″ is assumed.

Multi-trade openings include costs for sheet metal forms, firestop mortar, wrap strips, collars and sealants as necessary.

Structural penetrations joints are assumed to be 1/2″ or less. CMU walls are assumed to be within 1-1/2″ of the metal deck. Drywall walls are assumed to be tight to the underside of metal decking.

Metal panel, glass or curtain wall systems include a spandrel area of 5′ filled with mineral wool foil-faced insulation. Fasteners and stiffeners are included.

R099100-10 Painting Estimating Techniques

Proper estimating methodology is needed to obtain an accurate painting estimate. There is no known reliable shortcut or square foot method. The following steps should be followed:

- List all surfaces to be painted, with an accurate quantity (area) of each. Items having similar surface condition, finish, application method and accessibility may be grouped together.
- List all the tasks required for each surface to be painted, including surface preparation, masking, and protection of adjacent surfaces. Surface preparation may include minor repairs, washing, sanding and puttying.
- Select the proper Means line for each task. Review and consider all adjustments to labor and materials for type of paint and location of work. Apply the height adjustment carefully. For instance, when applying the adjustment for work over 8' high to a wall that is 12' high, apply the adjustment only to the area between 8' and 12' high, and not to the entire wall.

When applying more than one percent (%) adjustment, apply each to the base cost of the data, rather than applying one percentage adjustment on top of the other.

When estimating the cost of painting walls and ceilings remember to add the brushwork for all cut-ins at inside corners and around windows and doors as a LF measure. One linear foot of cut-in with a brush equals one square foot of painting.

All items for spray painting include the labor for roll-back.

Deduct for openings greater than 100 SF or openings that extend from floor to ceiling and are greater than 5' wide. Do not deduct small openings.

The cost of brushes, rollers, ladders and spray equipment are considered part of a painting contractor's overhead, and should not be added to the estimate. The cost of rented equipment such as scaffolding and swing staging should be added to the estimate.

Plumbing R2201 Operation & Maintenance of Plumbing

R220102-20 Labor Adjustment Factors

Labor Adjustment Factors are provided for Divisions 21, 22, and 23 to assist the mechanical estimator account for the various complexities and special conditions of any particular project. While a single percentage has been entered on each line of Division 22 01 02.20, it should be understood that these are just suggested midpoints of ranges of values commonly used by mechanical estimators. They may be increased or decreased depending on the severity of the special conditions.

The group for "existing occupied buildings" has been the subject of requests for explanation. Actually there are two stages to this group: buildings that are existing and "finished" but unoccupied, and those that also are occupied. Buildings that are "finished" may result in higher labor costs due to the workers having to be more careful not to damage finished walls, ceilings, floors, etc. and may necessitate special protective coverings and barriers. Also corridor bends and doorways may not accommodate long pieces of pipe or larger pieces of equipment. Work above an already hung ceiling can be very time consuming. The addition of occupants may force the work to be done on premium time (nights and/or weekends), eliminate the possible use of some preferred tools such as pneumatic drivers, powder charged drivers, etc. The estimator should evaluate the access to the work area and just how the work is going to be accomplished to arrive at an increase in labor costs over "normal" new construction productivity.

R220105-10 Demolition (Selective vs. Removal for Replacement)

Demolition can be divided into two basic categories.

One type of demolition involves the removal of material with no concern for its replacement. The labor-hours to estimate this work are found under "Selective Demolition" in the Fire Protection, Plumbing and HVAC Divisions. It is selective in that individual items or all the material installed as a system or trade grouping such as plumbing or heating systems are removed. This may be accomplished by the easiest way possible, such as sawing, torch cutting, or sledge hammering as well as simple unbolting.

The second type of demolition is the removal of some items for repair or replacement. This removal may involve careful draining, opening of unions, disconnecting and tagging electrical connections, capping pipes/ducts to prevent entry of debris or leakage of the material contained as well as transporting the item away from its in-place location to a truck/dumpster. An approximation of the time required to accomplish this type of demolition is to use half of the time indicated as necessary to install a new unit. For example: installation of a new pump might be listed as requiring 6 labor-hours so if we had to estimate the removal of the old pump we would allow an additional 3 hours for a total of 9 hours. That is, the complete replacement of a defective pump with a new pump would be estimated to take 9 labor-hours.

R220523-80 Valve Materials

VALVE MATERIALS

Bronze:
Bronze is one of the oldest materials used to make valves. It is most commonly used in hot and cold water systems and other non-corrosive services. It is often used as a seating surface in larger iron body valves to ensure tight closure.

Carbon Steel:
Carbon steel is a high strength material. Therefore, valves made from this metal are used in higher pressure services, such as steam lines up to 600 psi at 850° F. Many steel valves are available with butt-weld ends for economy and are generally used in high pressure steam service as well as other higher pressure non-corrosive services.

Forged Steel:
Valves from tough carbon steel are used in service up to 2000 psi and temperatures up to 1000° F in Gate, Globe and Check valves.

Iron:
Valves are normally used in medium to large pipe lines to control non-corrosive fluid and gases, where pressures do not exceed 250 psi at 450° F or 500 psi cold water, oil or gas.

Stainless Steel:
Developed steel alloys can be used in over 90% corrosive services.

Plastic PVC:
This is used in a great variety of valves generally in high corrosive service with lower temperatures and pressures.

VALVE SERVICE PRESSURES

Pressure ratings on valves provide an indication of the safe operating pressure for a valve at some elevated temperature. This temperature is dependent upon the materials used and the fabrication of the valve. When specific data is not available, a good "rule-of-thumb" to follow is the temperature of saturated steam on the primary rating indicated on the valve body. Example: The valve has the number 150S printed on the side indicating 150 psi and hence, a maximum operating temperature of 367° F (temperature of saturated steam and 150 psi).

DEFINITIONS

1. "WOG" – Water, oil, gas (cold working pressures).
2. "SWP" – Steam working pressure.
3. 100% area (full port) – means the area through the valve is equal to or greater than the area of standard pipe.
4. "Standard Opening" – means that the area through the valve is less than the area of standard pipe and therefore these valves should be used only where restriction of flow is unimportant.
5. "Round Port" – means the valve has a full round opening through the plug and body, of the same size and area as standard pipe.
6. "Rectangular Port" – valves have rectangular shaped ports through the plug body. The area of the port is either equal to 100% of the area of standard pipe, or restricted (standard opening). In either case it is clearly marked.
7. "ANSI" – American National Standards Institute.

R220523-90 Valve Selection Considerations

INTRODUCTION: In any piping application, valve performance is critical. Valves should be selected to give the best performance at the lowest cost.

The following is a list of performance characteristics generally expected of valves.

1. Stopping flow or starting it.
2. Throttling flow (Modulation).
3. Flow direction changing.
4. Checking backflow (Permitting flow in only one direction).
5. Relieving or regulating pressure.

In order to properly select the right valve, some facts must be determined.

A. What liquid or gas will flow through the valve?

B. Does the fluid contain suspended particles?

C. Does the fluid remain in liquid form at all times?

D. Which metals does fluid corrode?

E. What are the pressure and temperature limits? (As temperature and pressure rise, so will the price of the valve.)

F. Is there constant line pressure?

G. Is the valve merely an on-off valve?

H. Will checking of backflow be required?

I. Will the valve operate frequently or infrequently?

Valves are classified by design type into such classifications as Gate, Globe, Angle, Check, Ball, Butterfly and Plug. They are also classified by end connection, stem, pressure restrictions and material such as bronze, cast iron, etc. Each valve has a specific use. A quality valve used correctly will provide a lifetime of trouble- free service, but a high quality valve installed in the wrong service may require frequent attention.

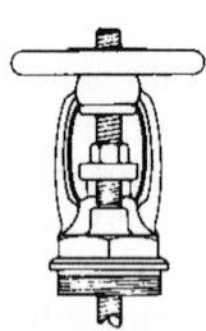

STEM TYPES
(OS & Y)—Rising Stem-Outside Screw and Yoke

Offers a visual indication of whether the valve is open or closed. Recommended where high temperatures, corrosives, and solids in the line might cause damage to inside-valve stem threads. The stem threads are engaged by the yoke bushing so the stem rises through the hand wheel as it is turned.

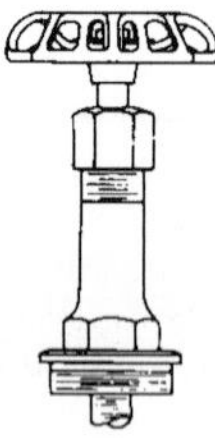

(R.S.)—Rising Stem-Inside Screw

Adequate clearance for operation must be provided because both the hand wheel and the stem rise.
The valve wedge position is indicated by the position of the stem and hand wheel.

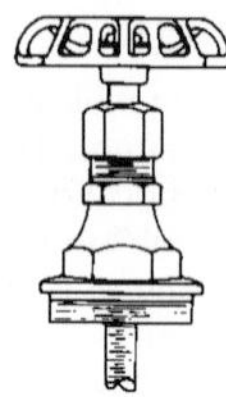

(N.R.S.)—Non-Rising Stem-Inside Screw

A minimum clearance is required for operating this type of valve. Excessive wear or damage to stem threads inside the valve may be caused by heat, corrosion, and solids. Because the hand wheel and stem do not rise, wedge position cannot be visually determined.

VALVE TYPES
Gate Valves

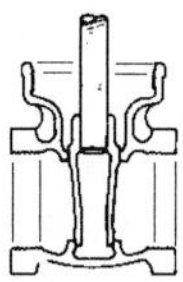

Provide full flow, minute pressure drop, minimum turbulence and minimum fluid trapped in the line.
They are normally used where operation is infrequent.

Globe Valves

Globe valves are designed for throttling and/or frequent operation with positive shut-off. Particular attention must be paid to the several types of seating materials available to avoid unnecessary wear. The seats must be compatible with the fluid in service and may be composition or metal. The configuration of the Globe valve opening causes turbulence which results in increased resistance. Most bronze Globe valves are rising stem-inside screw, but they are also available on O.S. & Y.

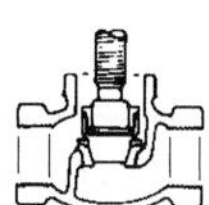

Angle Valves

The fundamental difference between the Angle valve and the Globe valve is the fluid flow through the Angle valve. It makes a 90° turn and offers less resistance to flow than the Globe valve while replacing an elbow. An Angle valve thus reduces the number of joints and installation time.

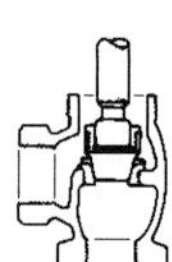

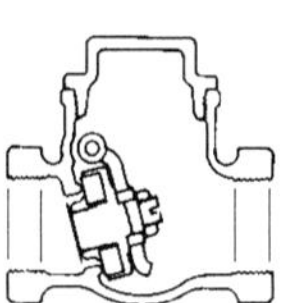

Check Valves

Check valves are designed to prevent backflow by automatically seating when the direction of fluid is reversed.
Swing Check valves are generally installed with Gate-valves, as they provide comparable full flow. Usually recommended for lines where flow velocities are low and should not be used on lines with pulsating flow. Recommended for horizontal installation, or in vertical lines only where flow is upward.

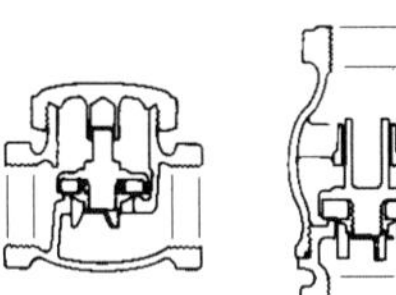

Lift Check Valves

These are commonly used with Globe and Angle valves since they have similar diaphragm seating arrangements and are recommended for preventing backflow of steam, air, gas and water, and on vapor lines with high flow velocities. For horizontal lines, horizontal lift checks should be used and vertical lift checks for vertical lines.

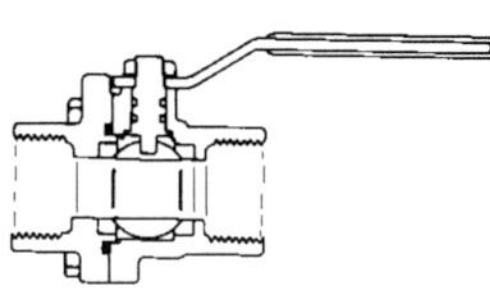

Ball Valves

Ball valves are light and easily installed, yet because of modern elastomeric seats, provide tight closure. Flow is controlled by rotating up to 90° a drilled ball which fits tightly against resilient seals. This ball seats with flow in either direction, and valve handle indicates the degree of opening. Recommended for frequent operation readily adaptable to automation, ideal for installation where space is limited.

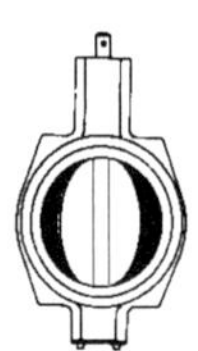

Butterfly Valves

Butterfly valves provide bubble-tight closure with excellent throttling characteristics. They can be used for full-open, closed and for throttling applications.

The Butterfly valve consists of a disc within the valve body which is controlled by a shaft. In its closed position, the valve disc seals against a resilient seat. The disc position throughout the full 90° rotation is visually indicated by the position of the operator.

A Butterfly valve is only a fraction of the weight of a Gate valve and requires no gaskets between flanges in most cases. Recommended for frequent operation and adaptable to automation where space is limited.

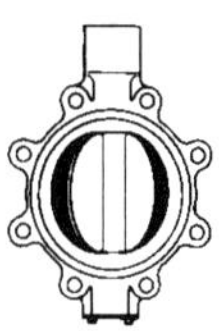

Wafer and Lug type bodies when installed between two pipe flanges, can be easily removed from the line. The pressure of the bolted flanges holds the valve in place.
Locating lugs makes installation easier.

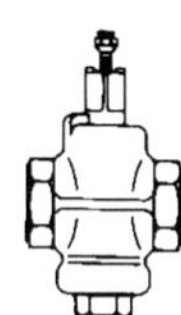

Plug Valves

Lubricated plug valves, because of the wide range of service to which they are adapted, may be classified as all purpose valves. They can be safely used at all pressure and vacuums, and at all temperatures up to the limits of available lubricants. They are the most satisfactory valves for the handling of gritty suspensions and many other destructive, erosive, corrosive and chemical solutions.

R221113-50 Pipe Material Considerations

1. Malleable fittings should be used for gas service.
2. Malleable fittings are used where there are stresses/strains due to expansion and vibration.
3. Cast fittings may be broken as an aid to disassembling heating lines frozen by long use, temperature and minerals.
4. A cast iron pipe is extensively used for underground and submerged service.
5. Type M (light wall) copper tubing is available in hard temper only and is used for nonpressure and less severe applications than K and L.
6. Type L (medium wall) copper tubing, available hard or soft for interior service.
7. Type K (heavy wall) copper tubing, available in hard or soft temper for use where conditions are severe. For underground and interior service.
8. Hard drawn tubing requires fewer hangers or supports but should not be bent. Silver brazed fittings are recommended, but soft solder is normally used.
9. Type DMV (very light wall) copper tubing designed for drainage, waste and vent plus other non-critical pressure services.

Domestic/Imported Pipe and Fittings Costs

The prices shown in this publication for steel/cast iron pipe and steel, cast iron, and malleable iron fittings are based on domestic production sold at the normal trade discounts. The above listed items of foreign manufacture may be available at prices 1/3 to 1/2 of those shown. Some imported items after minor machining or finishing operations are being sold as domestic to further complicate the system.

Caution: Most pipe prices in this data set also include a coupling and pipe hangers which for the larger sizes can add significantly to the per foot cost and should be taken into account when comparing "book cost" with the quoted supplier's cost.

R221113-70 Piping to 10′ High

When taking off pipe, it is important to identify the different material types and joining procedures, as well as distances between supports and components required for proper support.

During the takeoff, measure through all fittings. Do not subtract the lengths of the fittings, valves, or strainers, etc. This added length plus the final rounding of the totals will compensate for nipples and waste.

When rounding off totals always increase the actual amount to correspond with manufacturer's shipping lengths.

A. Both red brass and yellow brass pipe are normally furnished in 12′ lengths, plain end. The Unit Price section includes in the linear foot costs two field threads and one coupling per 10′ length. A carbon steel clevis type hanger assembly every 10′ is also prorated into the linear foot costs, including both material and labor.

B. Cast iron soil pipe is furnished in either 5′ or 10′ lengths. For pricing purposes, the Unit Price section features 10′ lengths with a joint and a carbon steel clevis hanger assembly every 5′ prorated into the per foot costs of both material and labor.

Three methods of joining are considered: lead and oakum poured joints or push-on gasket type joints for the bell and spigot pipe and a joint clamp for the no-hub soil pipe. The labor and material costs for each of these individual joining procedures are also prorated into the linear costs per foot.

C. Copper tubing covers types K, L, M, and DWV which are furnished in 20′ lengths. Means pricing data is based on a tubing cut each length and a coupling and two soft soldered joints every 10′. A carbon steel, clevis type hanger assembly every 10′ is also prorated into the per foot costs. The prices for refrigeration tubing are for materials only. Labor for full lengths may be based on the type L labor but short cut measures in tight areas can increase the installation labor-hours from 20 to 40%.

D. Corrosion-resistant piping does not lend itself to one particular standard of hanging or support assembly due to its diversity of application and placement. The several varieties of corrosion-resistant piping do not include any material or labor costs for hanger assemblies (See the Unit Price section for appropriate selection).

E. Glass pipe is furnished in standard lengths either 5′ or 10′ long, beaded on one end. Special orders for diverse lengths beaded on both ends are also available. For pricing purposes, R.S. Means features 10′ lengths with a coupling and a carbon steel band hanger assembly every 10′ prorated into the per foot linear costs.

Glass pipe is also available with conical ends and standard lengths ranging from 6″ through 3′ in 6″ increments, then up to 10′ in 12″ increments. Special lengths can be customized for particular installation requirements.

For pricing purposes, Means has based the labor and material pricing on 10′ lengths. Included in these costs per linear foot are the prorated costs for a flanged assembly every 10′ consisting of two flanges, a gasket, two insertable seals, and the required number of bolts and nuts. A carbon steel band hanger assembly based on 10′ center lines has also been prorated into the costs per foot for labor and materials.

F. Plastic pipe of several compositions and joining methods are considered. Fiberglass reinforced pipe (FRP) is priced, based on 10′ lengths (20′ lengths are also available), with coupling and epoxy joints every 10′. FRP is furnished in both "General Service" and "High Strength." A carbon steel clevis hanger assembly, 3 for every 10′, is built into the prorated labor and material costs on a per foot basis.

The PVC and CPVC pipe schedules 40, 80 and 120, plus SDR ratings are all based on 20′ lengths with a coupling installed every 10′, as well as a carbon steel clevis hanger assembly every 3′. The PVC and ABS type DWV piping is based on 10′ lengths with solvent weld couplings every 10′, and with carbon steel clevis hanger assemblies, 3 for every 10′. The rest of the plastic piping in this section is based on flexible 100′ coils and does not include any coupling or supports.

This section ends with PVC drain and sewer piping based on 10′ lengths with bell and spigot ends and 0-ring type, push-on joints.

G. Stainless steel piping includes both weld end and threaded piping, both in the type 304 and 316 specification and in the following schedules, 5, 10, 40, 80, and 160. Although this piping is usually furnished in 20′ lengths, this cost grouping has a joint (either heli-arc butt-welded or threads and coupling) every 10′. A carbon steel clevis type hanger assembly is also included at 10′ intervals and prorated into the linear foot costs.

H. Carbon steel pipe includes both black and galvanized. This section encompasses schedules 40 (standard) and 80 (extra heavy).

Several common methods of joining steel pipe — such as thread and coupled, butt welded, and flanged (150 lb. weld neck flanges) are also included.

For estimating purposes, it is assumed that the piping is purchased in 20′ lengths and that a compatible joint is made up every 10′. These joints are prorated into the labor and material costs per linear foot. The following hanger and support assemblies every 10′ are also included: carbon steel clevis for the T & C pipe, and single rod roll type for both the welded and flanged piping. All of these hangers are oversized to accommodate pipe insulation 3/4″ thick through 5″ pipe size and 1-1/2″ thick from 6″ through 12″ pipe size.

I. Grooved joint steel pipe is priced both black and galvanized, in schedules 10, 40, and 80, furnished in 20′ lengths. This section describes two joining methods: cut groove and roll groove. The schedule 10 piping is roll-grooved, while the heavier schedules are cut-grooved. The labor and material costs are prorated into per linear foot prices, including a coupled joint every 10′, as well as a carbon steel clevis hanger assembly.

Notes:

The pipe hanger assemblies mentioned in the preceding paragraphs include the described hanger; appropriately sized steel, box-type insert and nut; plus 18″ of threaded hanger rod.

C clamps are used when the pipe is to be supported from steel shapes rather than anchored in the slab. C clamps are slightly less costly than inserts. However, to save time in estimating, it is advisable to use the given line number cost, rather than substituting a C clamp for the insert.

Add to piping labor for elevated installation:

10′ to 14.5′ high	10%	30′ to 34.5′ high	40%
15′ to 19.5′ high	20%	35′ to 39.5′ high	50%
20′ to 24.5′ high	25%	Over 40′ and higher	55%
25′ to 29.5′ high	35%		

When using the percentage adds for elevated piping installations as shown above, bear in mind that the given heights are for the pipe supports, even though the insert, anchor, or clamp may be several feet higher than the pipe itself.

An allowance has been included in the piping installation time for testing and minor tightening of leaking joints, fittings, stuffing boxes, packing glands, etc. For extraordinary test requirements such as x-rays, prolonged pressure or demonstration tests, a percentage of the piping labor, based on the estimator's experience, must be added to the labor total. A testing service specializing in weld x-rays should be consulted for pricing if it is an estimate requirement. Equipment installation time includes start-up with associated adjustments.

R230500-10 Subcontractors

On the unit cost section of the R.S. Means Cost Data, the last column is entitled "Total Incl. O&P". This is normally the cost of the installing contractor. In the HVAC Division, this is the cost of the mechanical contractor. If the particular work being estimated is to be performed by a sub to the mechanical contractor, the mechanical's profit and handling charge (usually 10%) is added to the total of the last column.

R233100-30 Duct Fabrication/Installation

The labor cost for fabricated sheet metal duct includes both the cost of fabrication and installation of the duct. The split is approximately 40% for fabrication, 60% for installation. It is for this reason that the percentage add for elevated installation of fabricated duct is less than the percentage add for prefabricated/preformed duct.

Example: assume a piece of duct cost $100 installed (labor only)

Sheet Metal Fabrication = 40% = $40
Installation = 60% = $60

The add for elevated installation is:

Based on total labor (fabrication & installation) = $100 x 6% = $6.00

Based on installation cost only (Material purchased prefabricated) = $60 x 10% = $6.00

The $6.00 markup (10′ to 15′ high) is the same.

R233100-40 Steel Sheet Metal Calculator (Weight in Lb./Ft. of Length)

Gauge	26	24	22	20	18	16
Wt.-Lb./S.F.	.906	1.156	1.406	1.656	2.156	2.656
SMACNA Max. Dimension – Long Side	*	30″	54″	84″	85″ Up	*
Sum-2 sides						
2	.3	.40	.50	.60	.80	.90
3	.5	.65	.80	.90	1.1	1.4
4	.7	.85	1.0	1.2	1.5	1.8
5	.8	1.1	1.3	1.5	1.9	2.3
6	1.0	1.3	1.5	1.7	2.3	2.7
7	1.2	1.5	1.8	2.0	2.7	3.2
8	1.3	1.7	2.0	2.3	3.0	3.6
9	1.5	1.9	2.3	2.6	3.4	4.1
10	1.7	2.2	2.5	2.9	3.8	4.5
11	1.8	2.4	2.8	3.2	4.2	5.0
12	2.0	2.6	3.0	3.5	4.6	5.4
13	2.2	2.8	3.3	3.8	4.9	5.9
14	2.3	3.0	3.5	4.1	5.3	6.3
15	2.5	3.2	3.8	4.4	5.7	6.8
16	2.7	3.4	4.0	4.6	6.1	7.2
17	2.8	3.7	4.3	4.9	6.5	7.7
18	3.0	3.9	4.5	5.2	6.8	8.1
19	3.2	4.1	4.8	5.5	7.2	8.6
20	3.3	4.3	5.0	5.8	7.6	9.0
21	3.5	4.5	5.3	6.1	8.0	9.5
22	3.7	4.7	5.5	6.4	8.4	9.9
23	3.8	5.0	5.8	6.7	8.7	10.4
24	4.0	5.2	6.0	7.0	9.1	10.8
25	4.2	5.4	6.3	7.3	9.5	11.3
26	4.3	5.6	6.5	7.5	9.9	11.7
27	4.5	5.8	6.8	7.8	10.3	12.2
28	4.7	6.0	7.0	8.1	10.6	12.6
29	4.8	6.2	7.3	8.4	11.0	13.1
30	5.0	6.5	7.5	8.7	11.4	13.5
31	5.2	6.7	7.8	9.0	11.8	14.0
32	5.3	6.9	8.0	9.3	12.2	14.4
33	5.5	7.1	8.3	9.6	12.5	14.9
34	5.7	7.3	8.5	9.9	12.9	15.3
35	5.8	7.5	8.8	10.2	13.3	15.8
36	6.0	7.8	9.0	10.4	13.7	16.2
37	6.2	8.0	9.3	10.7	14.1	16.7
38	6.3	8.2	9.5	11.0	14.4	17.1
39	6.5	8.4	9.8	11.3	14.8	17.6
40	6.7	8.6	10.0	11.6	15.2	18.0
41	6.8	8.8	10.3	11.9	15.6	18.5
42	7.0	9.0	10.5	12.2	16.0	18.9
43	7.2	9.2	10.8	12.5	16.3	19.4
44	7.3	9.5	11.0	12.8	16.7	19.8
45	7.5	9.7	11.3	13.1	17.1	20.3
46	7.7	9.9	11.5	13.3	17.5	20.7
47	7.8	10.1	11.8	13.6	17.9	21.2
48	8.0	10.3	12.0	13.9	18.2	21.6
49	8.2	10.5	12.3	14.2	18.6	22.1
50	8.3	10.7	12.5	14.5	19.0	22.5
51	8.5	11.0	12.8	14.8	19.4	23.0
52	8.7	11.2	13.0	15.1	19.8	23.4
53	8.8	11.4	13.3	15.4	20.1	23.9
54	9.0	11.6	13.5	15.7	20.5	24.3
55	9.2	11.8	13.8	16.0	20.9	24.8

Gauge	26	24	22	20	18	16
Wt.-Lb./S.F.	.906	1.156	1.406	1.656	2.156	2.656
SMACNA Max. Dimension – Long Side	*	30″	54″	84″	85″ Up	*
Sum-2 Sides						
56	9.3	12.0	14.0	16.2	21.3	25.2
57	9.5	12.3	14.3	16.5	21.7	25.7
58	9.7	12.5	14.5	16.8	22.0	26.1
59	9.8	12.7	14.8	17.1	22.4	26.6
60	10.0	12.9	15.0	17.4	22.8	27.0
61	10.2	13.1	15.3	17.7	23.2	27.5
62	10.3	13.3	15.5	18.0	23.6	27.9
63	10.5	13.5	15.8	18.3	24.0	28.4
64	10.7	13.7	16.0	18.6	24.3	28.8
65	10.8	13.9	16.3	18.9	24.7	29.3
66	11.0	14.1	16.5	19.1	25.1	29.7
67	11.2	14.3	16.8	19.4	25.5	30.2
68	11.3	14.6	17.0	19.7	25.8	30.6
69	11.5	14.8	17.3	20.0	26.2	31.1
70	11.7	15.0	17.5	20.3	26.6	31.5
71	11.8	15.2	17.8	20.6	27.0	32.0
72	12.0	15.4	18.0	20.9	27.4	32.4
73	12.2	15.6	18.3	21.2	27.7	32.9
74	12.3	15.8	18.5	21.5	28.1	33.3
75	12.5	16.1	18.8	21.8	28.5	33.8
76	12.7	16.3	19.0	22.0	28.9	34.2
77	12.8	16.5	19.3	22.3	29.3	34.7
78	13.0	16.7	19.5	22.6	29.6	35.1
79	13.2	16.9	19.8	22.9	30.0	35.6
80	13.3	17.1	20.0	23.2	30.4	36.0
81	13.5	17.3	20.3	23.5	30.8	36.5
82	13.7	17.5	20.5	23.8	31.2	36.9
83	13.8	17.8	20.8	24.1	31.5	37.4
84	14.0	18.0	21.0	24.4	31.9	37.8
85	14.2	18.2	21.3	24.7	32.3	38.3
86	14.3	18.4	21.5	24.9	32.7	38.7
87	14.5	18.6	21.8	25.2	33.1	39.2
88	14.7	18.8	22.0	25.5	33.4	39.6
89	14.8	19.0	22.3	25.8	33.8	40.1
90	15.0	19.3	22.5	26.1	34.2	40.5
91	15.2	19.5	22.8	26.4	34.6	41.0
92	15.3	19.7	23.0	26.7	35.0	41.4
93	15.5	19.9	23.3	27.0	35.3	41.9
94	15.7	20.1	23.5	27.3	35.7	42.3
95	15.8	20.3	23.8	27.6	36.1	42.8
96	16.0	20.5	24.0	27.8	36.5	43.2
97	16.2	20.8	24.3	28.1	36.9	43.7
98	16.3	21.0	24.5	28.4	37.2	44.1
99	16.5	21.2	24.8	28.7	37.6	44.6
100	16.7	21.4	25.0	29.0	38.0	45.0
101	16.8	21.6	25.3	29.3	38.4	45.5
102	17.0	21.8	25.5	29.6	38.8	45.9
103	17.2	22.0	25.8	29.9	39.1	46.4
104	17.3	22.3	26.0	30.2	39.5	46.8
105	17.5	22.5	26.3	30.5	39.9	47.3
106	17.7	22.7	26.5	30.7	40.3	47.7
107	17.8	22.9	26.8	31.0	40.7	48.2
108	18.0	23.1	27.0	31.3	41.0	48.6
109	18.2	23.3	27.3	31.6	41.4	49.1
110	18.3	23.5	27.5	31.9	41.8	49.5

Example: If duct is 34″ x 20″ x 15′ long, 34″ is greater than 30″ maximum, for 24 ga. so must be 22 ga. 34″ + 20″ = 54″ going across from 54″ find 13.5 lb. per foot. 13.5 x 15′ = 202.5 lbs. For S.F. of surface area 202.5 ÷ 1.406 = 144 S.F.

Note: Figures include an allowance for scrap.
***Do Not** use unless engineer specified. 26GA is very light and sometimes used for toilet exhaust. 16 GA is heavy plate and mostly specified for plenums and hoods.

R233100-50 Ductwork Packages (per Ton of Cooling)

System	Sheet Metal	Insulation	Diffusers	Return Register
Roof Top Unit Single Zone	120 Lbs.	52 S.F.	1	1
Roof Top Unit Multizone	240 Lbs.	104 S.F.	2	1
Self-contained Air or Water Cooled	108 Lbs.	—	2	—
Split System Air Cooled	102 Lbs.	—	2	—

Systems reflect most common usage.
Refer to system graphics for duct layout.

Heating, Ventilating & A.C. R2337 Air Outlets & Inlets

R233700-60 Diffuser Evaluation

CFM = V × An × K where V = Outlet velocity in feet per minute. An = Neck area in square feet and K = Diffuser delivery factor. An undersized diffuser for a desired CFM will produce a high velocity and noise level. When air moves past people at a velocity in excess of 25 FPM, an annoying draft is felt. An oversized diffuser will result in low velocity with poor mixing. Consideration must be given to avoid vertical stratification or horizontal areas of stagnation.

R235000-10 Heating Systems

Heating Systems

The basic function of a heating system is to bring an enclosed volume up to a desired temperature and then maintain that temperature within a reasonable range. To accomplish this, the selected system must have sufficient capacity to offset transmission losses resulting from the temperature difference on the interior and exterior of the enclosing walls in addition to losses due to cold air infiltration through cracks, crevices and around doors and windows. The amount of heat to be furnished is dependent upon the building size, construction, temperature difference, air leakage, use, shape, orientation and exposure. Air circulation is also an important consideration. Circulation will prevent stratification which could result in heat losses through uneven temperatures at various levels. For example, the most efficient use of unit heaters can usually be achieved by circulating the space volume through the total number of units once every 20 minutes or 3 times an hour. This general rule must, of course, be adapted for special cases such as large buildings with low ratios of heat transmitting surface to cubical volume. The type of occupancy of a building will have considerable bearing on the number of heat transmitting units and the location selected. It is axiomatic, however, that the basis of any successful heating system is to provide the maximum amount of heat at the points of maximum heat loss such as exposed walls, windows, and doors. Large roof areas, wind direction, and wide doorways create problems of excessive heat loss and require special consideration and treatment.

Heat Transmission

Heat transfer is an important parameter to be considered during selection of the exterior wall style, material and window area. A high rate of transfer will permit greater heat loss during the wintertime with the resultant increase in heating energy costs and a greater rate of heat gain in the summer with proportionally greater cooling cost. Several terms are used to describe various aspects of heat transfer. However, for general estimating purposes this data set lists U values for systems of construction materials. U is the "overall heat transfer coefficient." It is defined as the heat flow per hour through one square foot when the temperature difference in the air on either side of the structure wall, roof, ceiling or floor is one degree Fahrenheit. The structural segment may be a single homogeneous material or a composite.

Total heat transfer is found using the following equation:

$$Q = AU(T_2 - T_1) \text{ where}$$

- **Q** = Heat flow, BTU per hour
- **A** = Area, square feet
- **U** = Overall heat transfer coefficient
- $(T_2 - T_1)$ = Difference in temperature of air on each side of the construction component. (Also abbreviated TD)

Note that heat can flow through all surfaces of any building and this flow is in addition to heat gain or loss due to ventilation, infiltration and generation (appliances, machinery, people).

R235000-20 Heating Approximations for Quick Estimating

Oil Piping & Boiler Room Piping

- Small System . . . 20 to 30% of Boiler
- Complex System with Pumps, Headers, Etc. . . . 80 to 110% of Boiler

Breeching With Insulation:

- Small . . . 10 to 15% of Boiler
- Large . . . 15 to 25% of Boiler

Coils: . . . 15 to 30% of Containing Unit

Balancing (Independent) . . . 1/2% of H.V.A.C. Estimate

Quality/Complexity Adjustment: For all heating installations add these adjustments to the estimate to more closely allow for the equipment and conditions of the particular job under consideration.

- Economy installation, add . . . 0 to 5% of System
- Good quality, medium complexity, add . . . 5 to 15% of System
- Above average quality and complexity, add . . . 15 to 25% of System

R235000-30 The Basics of a Heating System

The function of a heating system is to achieve and maintain a desired temperature in a room or building by replacing the amount of heat being dissipated. There are four kinds of heating systems: hot-water, steam, warm-air and electric resistance. Each has certain essential and similar elements with the exception of electric resistance heating.

The basic elements of a heating system are:

A. A **combustion chamber** in which fuel is burned and heat transferred to a conveying medium.

B. The **"fluid"** used for conveying the heat (water, steam or air).

C. **Conductors** or pipes for transporting the fluid to specific desired locations.

D. A means of disseminating the heat, sometimes called **terminal units.**

A. The **combustion chamber** in a furnace heats air which is then distributed. This is called a warm-air system.

The combustion chamber in a boiler heats water which is either distributed as hot water or steam and this is termed a hydronic system.

The maximum allowable working pressures are limited by ASME "Code for Heating Boilers" to 15 PSI for steam and 160 PSI for hot water heating boilers, with a maximum temperature limitation of 250° F. Hot water boilers are generally rated for a working pressure of 30 PSI. High pressure boilers are governed by the ASME "Code for Power Boilers" which is used almost universally for boilers operating over 15 PSIG. High pressure boilers used for a combination of heating/process loads are usually designed for 150 PSIG.

Boiler ratings are usually indicated as either Gross or Net Output. The Gross Load is equal to the Net Load plus a piping and pickup allowance. When this allowance cannot be determined, divide the gross output rating by 1.25 for a value equal to or greater than the net heat loss requirement of the building.

B. Of the three **fluids** used, steam carries the greatest amount of heat per unit volume. This is due to the fact that it gives up its latent heat of vaporization at a temperature considerably above room temperature. Another advantage is that the pressure to produce a positive circulation is readily available. Piping conducts the steam to terminal units and returns condensate to the boiler.

The **steam system** is well adapted to large buildings because of its positive circulation, its comparatively economical installation and its ability to deliver large quantities of heat. Nearly all large office buildings, stores, hotels, and industrial buildings are so heated, in addition to many residences.

Hot water, when used as the heat carrying fluid, gives up a portion of its sensible heat and then returns to the boiler or heating apparatus for reheating. As the heat conveyed by each pound of water is about one-fiftieth of the heat conveyed by a pound of steam, it is necessary to circulate about fifty times as much water as steam by weight (although only one-thirtieth as much by volume). The hot water system is usually, although not necessarily, designed to operate at temperatures below that of the ordinary steam system and so the amount of heat transfer surface must be correspondingly greater. A temperature of 190° F to 200° F is normally the maximum. Circulation in small buildings may depend on the difference in density between hot water and the cool water returning to the boiler; circulating pumps are normally used to maintain a desired rate of flow. Pumps permit a greater degree of flexibility and better control.

In **warm-air** furnace systems, cool air is taken from one or more points in the building, passed over the combustion chamber and flue gas passages and then distributed through a duct system. A disadvantage of this system is that the ducts take up much more building volume than steam or hot water pipes. Advantages of this system are the relative ease with which humidification can be accomplished by the evaporation of water as the air circulates through the heater, and the lack of need for expensive disseminating units as the warm air simply becomes part of the interior atmosphere of the building.

C. Conductors (pipes and ducts) have been lightly treated in the discussion of conveying fluids. For more detailed information such as sizing and distribution methods, the reader is referred to technical publications such as the American Society of Heating, Refrigerating and Air-Conditioning Engineers "Handbook of Fundamentals."

D. Terminal units come in an almost infinite variety of sizes and styles, but the basic principles of operation are very limited. As previously mentioned, warm-air systems require only a simple register or diffuser to mix heated air with that present in the room. Special application items such as radiant coils and infrared heaters are available to meet particular conditions but are not usually considered for general heating needs. Most heating is accomplished by having air flow over coils or pipes containing th heat transporting medium (steam, hot-water, electricity). These units, while varied, may be separated into two general types, (1) radiator/convectors and (2) unit heaters.

Radiator/convectors may be cast, fin-tube or pipe assemblies. They may be direct, indirect, exposed, concealed or mounted within a cabinet enclosure, upright or baseboard style. These units are often collectively referred to as "radiators" or "radiation" although none gives off heat either entirely by radiation or by convection but rather a combination of both. The air flows over the units as a gravity "current." It is necessary to have one or more heat-emitting units in each room. The most efficient placement is low along an outside wall or under a window to counteract the cold coming into the room and achieve an even distribution.

In contrast to radiator/convectors which operate most effectively against the walls of smaller rooms, **unit heaters** utilize a fan to move air over heating coils and are very effective in locations of relatively large volume. Unit heaters, while usually suspended overhead, may be floor mounted. They also may take in fresh outside air for ventilation. The heat distributed by unit heaters may be from a remote source and conveyed by a fluid or it may be from the combustion of fuel in each individual heater. In the latter case the only piping required would be for fuel, however, a vent for the product of combustion would be necessary.

The following list gives may of the advantages of unit heaters for applications other than office or residential:

a. Large capacity so smaller number of units are required, **b.** Piping system simplified, **c.** Space saved where they are located overhead out of the way, **d.** Rapid heating directed where needed with effective wide distribution, **e.** Difference between floor and ceiling temperature reduced, **f.** Circulation of air obtained, and ventilation with introduction of fresh air possible, **g.** Heat output flexible and easily controlled.

R235616-60 Solar Heating (Space and Hot Water)

Collectors should face as close to due South as possible, but variations of up to 20 degrees on either side of true South are acceptable. Local climate and collector type may influence the choice between east or west deviations. Obviously they should be located so they are not shaded from the sun's rays. Incline collectors at a slope of latitude minus 5 degrees for domestic hot water and latitude plus 15 degrees for space heating.

Flat plate collectors consist of a number of components as follows: Insulation to reduce heat loss through the bottom and sides of the collector. The enclosure which contains all the components in this assembly is usually weatherproof and prevents dust, wind and water from coming in contact with the absorber plate. The cover plate usually consists of one or more layers of a variety of glass or plastic and reduces the reradiation by creating an air space which traps the heat between the cover and the absorber plates.

The absorber plate must have a good thermal bond with the fluid passages. The absorber plate is usually metallic and treated with a surface coating which improves absorptivity. Black or dark paints or selective coatings are used for this purpose, and the design of this passage and plate combination helps determine a solar system's effectiveness.

Heat transfer fluid passage tubes are attached above and below or integral with an absorber plate for the purpose of transferring thermal energy from the absorber plate to a heat transfer medium. The heat exchanger is a device for transferring thermal energy from one fluid to another.

Piping and storage tanks should be well insulated to minimize heat losses.

Size domestic water heating storage tanks to hold 20 gallons of water per user, minimum, plus 10 gallons per dishwasher or washing machine. For domestic water heating an optimum collector size is approximately 3/4 square foot of area per gallon of water storage. For space heating of residences and small commercial applications the collector is commonly sized between 30% and 50% of the internal floor area. For space heating of large commercial applications, collector areas less than 30% of the internal floor area can still provide significant heat reductions.

A supplementary heat source is recommended for Northern states for December through February.

The solar energy transmission per square foot of collector surface varies greatly with the material used. Initial cost, heat transmittance and useful life are obviously interrelated.

R236000-10 Air Conditioning

General: The purpose of air conditioning is to control the environment of a space so that comfort is provided for the occupants and/or conditions are suitable for the processes or equipment contained therein. The several items which should be evaluated to define system objectives are:

- Temperature Control
- Humidity Control
- Cleanliness
- Odor, smoke and fumes
- Ventilation

Efforts to control the above parameters must also include consideration of the degree or tolerance of variation, the noise level introduced, the velocity of air motion and the energy requirements to accomplish the desired results.

The variation in **temperature** and **humidity** is a function of the sensor and the controller. The controller reacts to a signal from the sensor and produces the appropriate suitable response in either the terminal unit, the conductor of the transporting medium (air, steam, chilled water, etc.), or the source (boiler, evaporating coils, etc.).

The **noise level** is a by-product of the energy supplied to moving components of the system. Those items which usually contribute the most noise are pumps, blowers, fans, compressors and diffusers. The level of noise can be partially controlled through use of vibration pads, isolators, proper sizing, shields, baffles and sound absorbing liners.

Some **air motion** is necessary to prevent stagnation and stratification. The maximum acceptable velocity varies with the degree of heating or cooling which is taking place. Most people feel air moving past them at velocities in excess of 25 FPM as an annoying draft. However, velocities up to 45 FPM may be acceptable in certain cases. Ventilation, expressed as air changes per hour and percentage of fresh air, is usually an item regulated by local codes.

Selection of the system to be used for a particular application is usually a trade-off. In some cases the building size, style, or room available for mechanical use limits the range of possibilities. Prime factors influencing the decision are first cost and total life (operating, maintenance and replacement costs). The accuracy with which each parameter is determined will be an important measure of the reliability of the decision and subsequent satisfactory operation of the installed system.

Heat delivery may be desired from an air conditioning system. Heating capability usually is added as follows: A gas fired burner or hot water/steam/electric coils may be added to the air handling unit directly and heat all air equally. For limited or localized heat requirements the water/steam/electric coils may be inserted into the duct branch supplying the cold areas. Gas fired duct furnaces are also available.

Note: When water or steam coils are used the cost of the piping and boiler must also be added. For a rough estimate use the cost per square foot of the appropriate sized hydronic system with unit heaters. This will provide a cost for the boiler and piping, and the unit heaters of the system would equate to the approximate cost of the heating coils.

R236000-20 Air Conditioning Requirements

BTUs per hour per S.F. of floor area and S.F. per ton of air conditioning.

Type of Building	BTU/Hr per S.F.	S.F. per Ton	Type of Building	BTU/Hr per S.F.	S.F. per Ton	Type of Building	BTU/Hr per S.F.	S.F. per Ton
Apartments, Individual	26	450	Dormitory, Rooms	40	300	Libraries	50	240
Corridors	22	550	Corridors	30	400	Low Rise Office, Exterior	38	320
Auditoriums & Theaters	40	300/18*	Dress Shops	43	280	Interior	33	360
Banks	50	240	Drug Stores	80	150	Medical Centers	28	425
Barber Shops	48	250	Factories	40	300	Motels	28	425
Bars & Taverns	133	90	High Rise Office—Ext. Rms.	46	263	Office (small suite)	43	280
Beauty Parlors	66	180	Interior Rooms	37	325	Post Office, Individual Office	42	285
Bowling Alleys	68	175	Hospitals, Core	43	280	Central Area	46	260
Churches	36	330/20*	Perimeter	46	260	Residences	20	600
Cocktail Lounges	68	175	Hotel, Guest Rooms	44	275	Restaurants	60	200
Computer Rooms	141	85	Corridors	30	400	Schools & Colleges	46	260
Dental Offices	52	230	Public Spaces	55	220	Shoe Stores	55	220
Dept. Stores, Basement	34	350	Industrial Plants, Offices	38	320	Shop'g. Ctrs., Supermarkets	34	350
Main Floor	40	300	General Offices	34	350	Retail Stores	48	250
Upper Floor	30	400	Plant Areas	40	300	Specialty	60	200

*Persons per ton
12,000 BTU = 1 ton of air conditioning

R236000-30 Psychrometric Table

Dewpoint or Saturation Temperature (F)

Relative humidity (%)	32	35	40	45	50	55	60	65	70	75	80	85	90	95	100
100	32	35	40	45	50	55	60	65	70	75	80	85	90	95	100
90	30	33	37	42	47	52	57	62	67	72	77	82	87	92	97
80	27	30	34	39	44	49	54	58	64	68	73	78	83	88	93
70	24	27	31	36	40	45	50	55	60	64	69	74	79	84	88
60	20	24	28	32	36	41	46	51	55	60	65	69	74	79	83
50	16	20	24	28	33	36	41	46	50	55	60	64	69	73	78
40	12	15	18	23	27	31	35	40	45	49	53	58	62	67	71
30	8	10	14	18	21	25	29	33	37	42	46	50	54	59	62
20	6	7	8	9	13	16	20	24	28	31	35	40	43	48	52
10	4	4	5	5	6	8	9	10	13	17	20	24	27	30	34

Dry bulb temperature (F) (columns: 32 35 40 45 50 55 60 65 70 75 80 85 90 95 100)

This table shows the relationship between RELATIVE HUMIDITY, DRY BULB TEMPERATURE AND DEWPOINT.

As an example, assume that the thermometer in a room reads 75° F, and we know that the relative humidity is 50%. The chart shows the dewpoint temperature to be 55° F. That is, any surface colder than 55° F will "sweat" or collect condensing moisture. This surface could be the outside of an uninsulated chilled water pipe in the summertime, or the inside surface of a wall or deck in the wintertime. After determining the extreme ambient parameters, the table at the left is useful in determining which surfaces need insulation or vapor barrier protection.

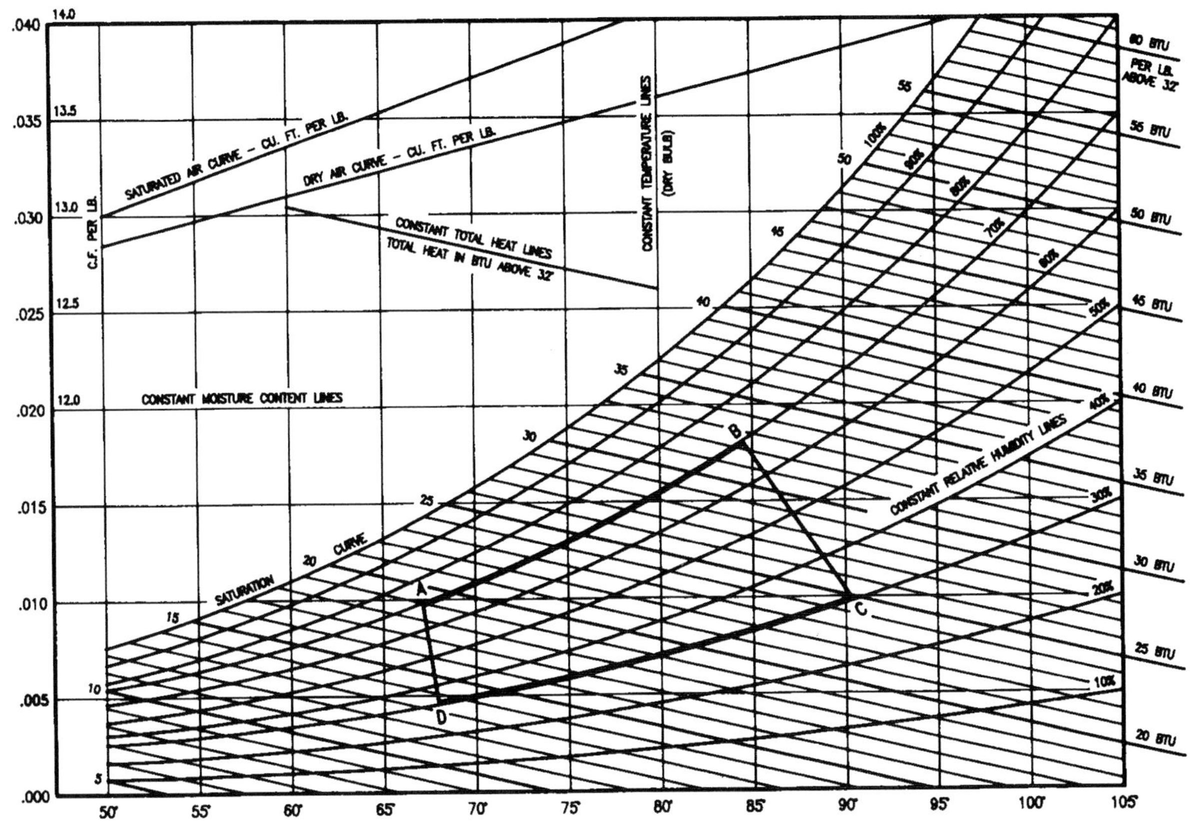

TEMPERATURE DEGREES FAHRENHEIT
TOTAL PRESSURE = 14.696 LB. PER SQ. IN. ABS.

Psychrometric chart showing different variables based on one pound of dry air. Space marked A B C D is temperature-humidity range which is most comfortable for majority of people.

R236000-90 Quality/Complexity Adjustment for Air Conditioning Systems

Economy installation, add	0 to 5%
Good quality, medium complexity, add	5 to 15%
Above average quality and complexity, add	15 to 25%

Add the above adjustments to the estimate to more closely allow for the equipment and conditions of the particular job under consideration.

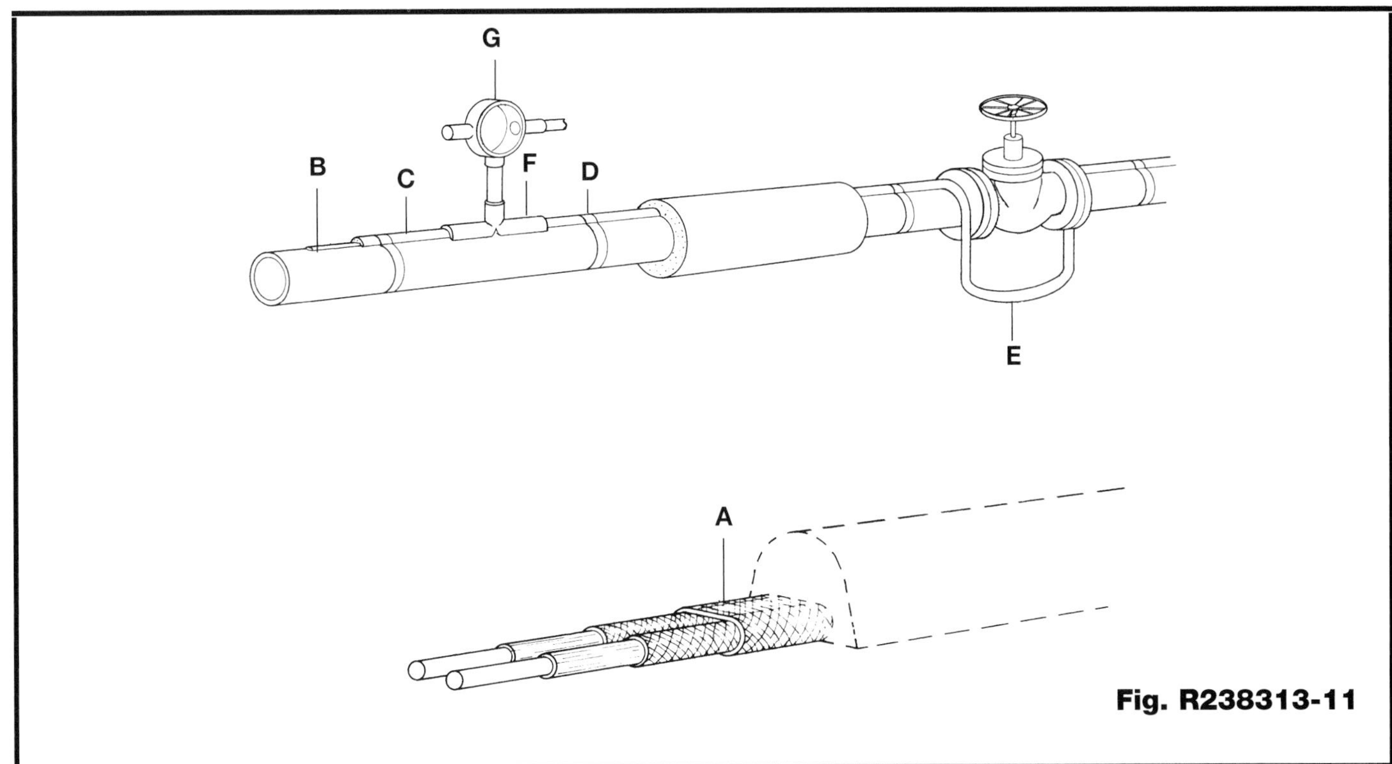

Fig. R238313-11

R238313-10 Heat Trace Systems

Before you can determine the cost of a HEAT TRACE installation, the method of attachment must be established. There are (4) common methods:

1. Cable is simply attached to the pipe with polyester tape every 12′.
2. Cable is attached with a continuous cover of 2″ wide aluminum tape.
3. Cable is attached with factory extruded heat transfer cement and covered with metallic raceway with clips every 10′.
4. Cable is attached between layers of pipe insulation using either clips or polyester tape.

In all of the above methods each component of the system must be priced individually.

Example: Components for method 3 must include:

A. Heat trace cable by voltage and watts per linear foot.
B. Heat transfer cement, 1 gallon per 60 linear feet of cover.
C. Metallic raceway by size and type.
D. Raceway clips by size of pipe.

When taking off linear foot lengths of cable add the following for each valve in the system. (E)

SCREWED OR WELDED VALVE:

Size		Length
1/2″	=	6″
3/4″	=	9″
1″	=	1′ -0″
1-1/2″	=	1′ -6″
2″	=	2′
2-1/2″	=	2′ -6″
3″	=	2′ -6″
4″	=	4′ -0″
6″	=	7′ -0″
8″	=	9′ -6″
10″	=	12′ -6″
12″	=	15′ -0″
14″	=	18′ -0″
16″	=	21′ -6″
18″	=	25′ -6″
20″	=	28′ -6″
24″	=	34′ -0″
30″	=	40′ -0″

FLANGED VALVE:

Size		Length
1/2″	=	1′ -0″
3/4″	=	1′ -6″
1″	=	2′ -0″
1-1/2″	=	2′ -6″
2″	=	2′ -6″
2-1/2″	=	3′ -0″
3″	=	3′ -6″
4″	=	4′ -0″
6″	=	8′ -0″
8″	=	11′ -0″
10″	=	14′ -0″
12″	=	16′ -6″
14″	=	19′ -6″
16″	=	23′ -0″
18″	=	27′ -0″
20″	=	30′ -0″
24″	=	36′ -0″
30″	=	42′ -0″

BUTTERFLY VALVES:

Size		Length
1/2″	=	0′
3/4″	=	0′
1″	=	1′ -0″
1-1/2″	=	1′ -6″
2″	=	2′ -0″
2-1/2″	=	2′ -6″
3″	=	2′ -6″
4″	=	3′ -0″
6″	=	3′ -6″
8″	=	4′ -0″
10″	=	4′ -0″
12″	=	5′ -0″
14″	=	5′ -6″
16″	=	6′ -0″
18″	=	6′ -6″
20″	=	7′ -0″
24″	=	8′ -0″
30″	=	10′ -0″

R238313-10 Heat Trace Systems (cont.)

Add the following quantities of heat transfer cement to linear foot totals for each valve:

Nominal Valve Size	Gallons of Cement per Valve
1/2″	0.14
3/4″	0.21
1″	0.29
1-1/2″	0.36
2″	0.43
2-1/2″	0.70
3″	0.71
4″	1.00
6″	1.43
8″	1.48
10″	1.50
12″	1.60
14″	1.75
16″	2.00
18″	2.25
20″	2.50
24″	3.00
30″	3.75

The following must be added to the list of components to accurately price HEAT TRACE systems:

1. Expediter fitting and clamp fasteners (F)
2. Junction box and nipple connected to expediter fitting (G)
3. Field installed terminal blocks within junction box
4. Ground lugs
5. Piping from power source to expediter fitting
6. Controls
7. Thermostats
8. Branch wiring
9. Cable splices
10. End of cable terminations
11. Branch piping fittings and boxes

Deduct the following percentages from labor if cable lengths in the same area exceed:

150′ to 250′	10%	351′ to 500′	20%
251′ to 350′	15%	Over 500′	25%

Add the following percentages to labor for elevated installations:

15′ to 20′ high	10%	31′ to 35′ high	40%
21′ to 25′ high	20%	36′ to 40′ high	50%
26′ to 30′ high	30%	Over 40′ high	60%

R238313-20 Spiral-Wrapped Heat Trace Cable (Pitch Table)

In order to increase the amount of heat, occasionally heat trace cable is wrapped in a spiral fashion around a pipe; increasing the number of feet of heater cable per linear foot of pipe.

Engineers first determine the heat loss per foot of pipe (based on the insulating material, its thickness, and the temperature differential across it). A ratio is then calculated by the formula:

$$\text{Feet of Heat Trace per Foot of Pipe} = \frac{\text{Watts/Foot of Heat Loss}}{\text{Watts/Foot of the Cable}}$$

The linear distance between wraps (pitch) is then taken from a chart or table. Generally, the pitch is listed on a drawing leaving the estimator to calculate the total length of heat tape required. An approximation may be taken from this table.

	Feet of Heat Trace Per Foot of Pipe															
	Nominal Pipe Size in Inches															
Pitch In Inches	1	1¼	1½	2	2½	3	4	6	8	10	12	14	16	18	20	24
3.5	1.80															
4	1.65															
5	1.46	1.60	1.80													
6	1.34	1.45	1.55	1.75												
7	1.25	1.35	1.43	1.57	1.75											
8	1.20	1.28	1.34	1.45	1.60	1.80										
9	1.16	1.23	1.28	1.37	1.51	1.68										
10	1.13	1.19	1.24	1.32	1.44	1.57	1.82									
15	1.06	1.08	1.10	1.15	1.21	1.29	1.42	1.78								
20	1.04	1.05	1.06	1.08	1.13	1.17	1.25	1.49	1.73							
25		1.04	1.04	1.06	1.08	1.11	1.17	1.33	1.51	1.72						
30				1.04	1.05	1.07	1.12	1.24	1.37	1.54	1.70	1.80				
35					1.06	1.06	1.09	1.17	1.28	1.42	1.54	1.64	1.78			
40						1.05	1.07	1.14	1.22	1.33	1.44	1.52	1.64	1.75		
50							1.05	1.09	1.15	1.22	1.29	1.35	1.44	1.53	1.64	1.83
60								1.06	1.11	1.16	1.21	1.25	1.31	1.39	1.46	1.62
70								1.05	1.08	1.12	1.17	1.19	1.24	1.30	1.35	1.47
80									1.06	1.09	1.13	1.15	1.19	1.24	1.28	1.38
90									1.04	1.06	1.10	1.13	1.16	1.19	1.23	1.32
100										1.05	1.08	1.10	1.13	1.15	1.19	1.23

Note: Common practice would normally limit the lower end of the table to 5% of additional heat and above 80% an engineer would likely opt for two (2) parallel cables.

R260519-90 Wire

Wire quantities are taken off by either measuring each cable run or by extending the conduit and raceway quantities times the number of conductors in the raceway. Ten percent should be added for waste and tie-ins. Keep in mind that the unit of measure of wire is C.L.F. not L.F. as in raceways so the formula would read:

$$\frac{(\text{L.F. Raceway x No. of Conductors}) \times 1.10}{100} = \text{C.L.F.}$$

Price per C.L.F. of wire includes:
1. Setting up wire coils or spools on racks
2. Attaching wire to pull in means
3. Measuring and cutting wire
4. Pulling wire into a raceway
5. Identifying and tagging

Price does not include:
1. Connections to breakers, panelboards, or equipment
2. Splices

Job Conditions: Productivity is based on new construction to a height of 15′ using rolling staging in an unobstructed area. Material staging is assumed to be within 100′ of work being performed.

Economy of Scale: If more than three wires at a time are being pulled, deduct the following percentages from the labor of that grouping:

4-5 wires	25%
6-10 wires	30%
11-15 wires	35%
over 15	40%

If a wire pull is less than 100′ in length and is interrupted several times by boxes, lighting outlets, etc., it may be necessary to add the following lengths to each wire being pulled:

Junction box to junction box	2 L.F.
Lighting panel to junction box	6 L.F.
Distribution panel to sub panel	8 L.F.
Switchboard to distribution panel	12 L.F.
Switchboard to motor control center	20 L.F.
Switchboard to cable tray	40 L.F.

Measure of Drops and Riser: It is important when taking off wire quantities to include the wire for drops to electrical equipment. If heights of electrical equipment are not clearly stated, use the following guide:

	Bottom A.F.F.	Top A.F.F.	Inside Cabinet
Safety switch to 100A	5′	6′	2′
Safety switch 400 to 600A	4′	6′	3′
100A panel 12 to 30 circuit	4′	6′	3′
42 circuit panel	3′	6′	4′
Switch box	3′	3′6″	1′
Switchgear	0′	8′	8′
Motor control centers	0′	8′	8′
Transformers - wall mount	4′	8′	2′
Transformers - floor mount	0′	12′	4′

R312319-90 Wellpoints

A single stage wellpoint system is usually limited to dewatering an average 15′ depth below normal ground water level. Multi-stage systems are employed for greater depth with the pumping equipment installed only at the lowest header level. Ejectors with unlimited lift capacity can be economical when two or more stages of wellpoints can be replaced or when horizontal clearance is restricted, such as in deep trenches or tunneling projects, and where low water flows are expected. Wellpoints are usually spaced on 2-1/2′ to 10′ centers along a header pipe. Wellpoint spacing, header size, and pump size are all determined by the expected flow as dictated by soil conditions.

In almost all soils encountered in wellpoint dewatering, the wellpoints may be jetted into place. Cemented soils and stiff clays may require sand wicks about 12″ in diameter around each wellpoint to increase efficiency and eliminate weeping into the excavation. These sand wicks require 1/2 to 3 C.Y. of washed filter sand and are installed by using a 12″ diameter steel casing and hole puncher jetted into the ground 2′ deeper than the wellpoint. Rock may require predrilled holes.

Labor required for the complete installation and removal of a single stage wellpoint system is in the range of 3/4 to 2 labor-hours per linear foot of header, depending upon jetting conditions, wellpoint spacing, etc.

Continuous pumping is necessary except in some free draining soil where temporary flooding is permissible (as in trenches which are backfilled after each day's work). Good practice requires provision of a stand-by pump during the continuous pumping operation.

Systems for continuous trenching below the water table should be installed three to four times the length of expected daily progress to ensure uninterrupted digging, and header pipe size should not be changed during the job.

For pervious free draining soils, deep wells in place of wellpoints may be economical because of lower installation and maintenance costs. Daily production ranges between two to three wells per day, for 25′ to 40′ depths, to one well per day for depths over 50′.

Detailed analysis and estimating for any dewatering problem is available at no cost from wellpoint manufacturers. Major firms will quote "sufficient equipment" quotes or their affiliates will offer lump sum proposals to cover complete dewatering responsibility.

Description for 200′ System with 8″ Header		Quantities
Equipment & Material	Wellpoints 25′ long, 2″ diameter @ 5′ O.C.	40 Each
	Header pipe, 8″ diameter	200 L.F.
	Discharge pipe, 8″ diameter	100 L.F.
	8″ valves	3 Each
	Combination jetting & wellpoint pump (standby)	1 Each
	Wellpoint pump, 8″ diameter	1 Each
	Transportation to and from site	1 Day
	Fuel for 30 days x 60 gal./day	1800 Gallons
	Lubricants for 30 days x 16 lbs./day	480 Lbs.
	Sand for points	40 C.Y.
Labor	Technician to supervise installation	1 Week
	Labor for installation and removal of system	300 Labor-hours
	4 Operators straight time 40 hrs./wk. for 4.33 wks.	693 Hrs.
	4 Operators overtime 2 hrs./wk. for 4.33 wks.	35 Hrs.

R314116-40 Wood Sheet Piling

Wood sheet piling may be used for depths to 20′ where there is no ground water. If moderate ground water is encountered Tongue & Groove sheeting will help to keep it out. When considerable ground water is present, steel sheeting must be used.

For estimating purposes on trench excavation, sizes are as follows:

Depth	Sheeting	Wales	Braces	B.F. per S.F.
To 8′	3 x 12′s	6 x 8′s, 2 line	6 x 8′s, @ 10′	4.0 @ 8′
8′ x 12′	3 x 12′s	10 x 10′s, 2 line	10 x 10′s, @ 9′	5.0 average
12′ to 20′	3 x 12′s	12 x 12′s, 3 line	12 x 12′s, @ 8′	7.0 average

Sheeting to be toed in at least 2′ depending upon soil conditions. A five person crew with an air compressor and sheeting driver can drive and brace 440 SF/day at 8′ deep, 360 SF/day at 12′ deep, and 320 SF/day at 16′ deep. For normal soils, piling can be pulled in 1/3 the time to install. Pulling difficulty increases with the time in the ground. Production can be increased by high pressure jetting.

R314116-45 Steel Sheet Piling

Limiting weights are 22 to 38#/S.F. of wall surface with 27#/S.F. average for usual types and sizes. (Weights of piles themselves are from 30.7#/L.F. to 57#/L.F. but they are 15″ to 21″ wide.) Lightweight sections 12″ to 28″ wide from 3 ga. to 12 ga. thick are also available for shallow excavations. Piles may be driven two at a time with an impact or vibratory hammer (use vibratory to pull) hung from a crane without leads. A reasonable estimate of the life of steel sheet piling is 10 uses with up to 125 uses possible if a vibratory hammer is used. Used piling costs from 50% to 80% of new piling depending on location and market conditions. Sheet piling and H piles can be rented for about 30% of the delivered mill price for the first month and 5% per month thereafter. Allow 1 labor-hour per pile for cleaning and trimming after driving. These costs increase with depth and hydrostatic head. Vibratory drivers are faster in wet granular soils and are excellent for pile extraction. Pulling difficulty increases with the time in the ground and may cost more than driving. It is often economical to abandon the sheet piling, especially if it can be used as the outer wall form. Allow about 1/3 additional length or more for toeing into ground. Add bracing, waler and strut costs. Waler costs can equal the cost per ton of sheeting.

Utilities R3311 Water Utility Distribution Piping

R331113-80 Piping Designations

There are several systems currently in use to describe pipe and fittings. The following paragraphs will help to identify and clarify classifications of piping systems used for water distribution.

Piping may be classified by schedule. Piping schedules include 5S, 10S, 10, 20, 30, Standard, 40, 60, Extra Strong, 80, 100, 120, 140, 160 and Double Extra Strong. These schedules are dependent upon the pipe wall thickness. The wall thickness of a particular schedule may vary with pipe size.

Ductile iron pipe for water distribution is classified by Pressure Classes such as Class 150, 200, 250, 300 and 350. These classes are actually the rated water working pressure of the pipe in pounds per square inch (psi). The pipe in these pressure classes is designed to withstand the rated water working pressure plus a surge allowance of 100 psi.

The American Water Works Association (AWWA) provides standards for various types of **plastic pipe.** C-900 is the specification for polyvinyl chloride (PVC) piping used for water distribution in sizes ranging from 4″ through 12″. C-901 is the specification for polyethylene (PE) pressure pipe, tubing and fittings used for water distribution in sizes ranging from 1/2″ through 3″. C-905 is the specification for PVC piping sizes 14″ and greater.

PVC pressure-rated pipe is identified using the standard dimensional ratio (SDR) method. This method is defined by the American Society for Testing and Materials (ASTM) Standard D 2241. This pipe is available in SDR numbers 64, 41, 32.5, 26, 21, 17, and 13.5. A pipe with an SDR of 64 will have the thinnest wall while a pipe with an SDR of 13.5 will have the thickest wall. When the pressure rating (PR) of a pipe is given in psi, it is based on a line supplying water at 73 degrees F.

The National Sanitation Foundation (NSF) seal of approval is applied to products that can be used with potable water. These products have been tested to ANSI/NSF Standard 14.

Valves and strainers are classified by American National Standards Institute (ANSI) Classes. These Classes are 125, 150, 200, 250, 300, 400, 600, 900, 1500 and 2500. Within each class there is an operating pressure range dependent upon temperature. Design parameters should be compared to the appropriate material dependent, pressure-temperature rating chart for accurate valve selection.

Change Orders

Change Order Considerations

A change order is a written document usually prepared by the design professional and signed by the owner, the architect/engineer, and the contractor. A change order states the agreement of the parties to: an addition, deletion, or revision in the work; an adjustment in the contract sum, if any; or an adjustment in the contract time, if any. Change orders, or "extras", in the construction process occur after execution of the construction contract and impact architects/ engineers, contractors, and owners.

Change orders that are properly recognized and managed can ensure orderly, professional, and profitable progress for everyone involved in the project. There are many causes for change orders and change order requests. In all cases, change orders or change order requests should be addressed promptly and in a precise and prescribed manner. The following paragraphs include information regarding change order pricing and procedures.

The Causes of Change Orders

Reasons for issuing change orders include:

- Unforeseen field conditions that require a change in the work
- Correction of design discrepancies, errors, or omissions in the contract documents
- Owner-requested changes, either by design criteria, scope of work, or project objectives
- Completion date changes for reasons unrelated to the construction process
- Changes in building code interpretations, or other public authority requirements that require a change in the work
- Changes in availability of existing or new materials and products

Procedures

Properly written contract documents must include the correct change order procedures for all parties—owners, design professionals, and contractors—to follow in order to avoid costly delays and litigation.

Being "in the right" is not always a sufficient or acceptable defense. The contract provisions requiring notification and documentation must be adhered to within a defined or reasonable time frame.

The appropriate method of handling change orders is by a written proposal and acceptance by all parties involved. Prior to starting work on a project, all parties should identify their authorized agents who may sign and accept change orders, as well as any limits placed on their authority.

Time may be a critical factor when the need for a change arises. For such cases, the contractor might be directed to proceed on a "time and materials" basis, rather than wait for all paperwork to be processed—a delay that could impede progress. In this situation, the contractor must still follow the prescribed change order procedures including, but not limited to, notification and documentation.

Lack of documentation can be very costly, especially if legal judgments are to be made, and if certain field personnel are no longer available. For time and material change orders, the contractor should keep accurate daily records of all labor and material allocated to the change.

Owners or awarding authorities who do considerable and continual building construction (such as the federal government) realize the inevitability of change orders for numerous reasons, both predictable and unpredictable. As a result, the federal government, the American Institute of Architects (AIA), the Engineers Joint Contract Documents Committee (EJCDC), and other contractor, legal, and technical organizations have developed standards and procedures to be followed by all parties to achieve contract continuance and timely completion, while being financially fair to all concerned.

Pricing Change Orders

When pricing change orders, regardless of their cause, the most significant factor is when the change occurs. The need for a change may be perceived in the field or requested by the architect/engineer *before* any of the actual installation has begun, or may evolve or appear *during* construction when the item of work in question is partially installed. In the latter cases, the original sequence of construction is disrupted, along with all contiguous and supporting systems. Change orders cause the greatest impact when they occur *after* the installation has been completed and must be uncovered, or even replaced. Post-completion changes may be caused by necessary design changes, product failure, or changes in the owner's requirements that are not discovered until the building or the systems begin to function.

Specified procedures of notification and record keeping must be adhered to and enforced regardless of the stage of construction: *before, during,* or *after* installation. Some bidding documents anticipate change orders by requiring that unit prices including overhead and profit percentages—for additional as well as deductible changes—be listed. Generally these unit prices do not fully take into account the ripple effect, or impact on other trades, and should be used for general guidance only.

When pricing change orders, it is important to classify the time frame in which the change occurs. There are two basic time frames for change orders: *pre-installation change orders,* which occur before the start of construction, and *post-installation change orders,* which involve reworking after the original installation. Change orders that occur between these stages may be priced according to the extent of work completed using a combination of techniques developed for pricing *pre-* and *post-installation* changes.

Factors To Consider When Pricing Change Orders

As an estimator begins to prepare a change order, the following questions should be reviewed to determine their impact on the final price.

General

- *Is the change order work* pre-installation *or* post-installation?

  Change order work costs vary according to how much of the installation has been completed. Once workers have the project scoped in their minds, even though they have not started, it can be difficult to refocus. Consequently they may spend more than the normal amount of time understanding the change. Also, modifications to work in place, such as trimming or refitting, usually take more time than was initially estimated. The greater the amount of work in place, the more reluctant workers are to change it. Psychologically they may resent the change and as a result the rework takes longer than normal. Post-installation change order estimates must include demolition of existing work as required to accomplish the change. If the work is performed at a later time, additional obstacles, such as building finishes, may be present which must be protected. Regardless of whether the change occurs pre-installation or post-installation, attempt to isolate the identifiable factors and price them separately. For example, add shipping costs that may be required pre-installation or any demolition required post-installation. Then analyze the potential impact on productivity of psychological and/or learning curve factors and adjust the output rates accordingly. One approach is to break down the typical workday into segments and quantify the impact on each segment.

Change Order Installation Efficiency

The labor-hours expressed (for new construction) are based on average installation time, using an efficiency level. For change order situations, adjustments to this efficiency level should reflect the daily labor-hour allocation for that particular occurrence.

- *Will the change substantially delay the original completion date?*

  A significant change in the project may cause the original completion date to be extended. The extended schedule may subject the contractor to new wage rates dictated by relevant labor contracts. Project supervision and other project overhead must also be extended beyond the original completion date. The schedule extension may also put installation into a new weather season. For example, underground piping scheduled for October installation was delayed until January. As a result, frost penetrated the trench area, thereby changing the degree of difficulty of the task. Changes and delays may have a ripple effect throughout the project. This effect must be analyzed and negotiated with the owner.

- *What is the net effect of a deduct change order?*

  In most cases, change orders resulting in a deduction or credit reflect only bare costs. The contractor may retain the overhead and profit based on the original bid.

Materials

- *Will you have to pay more or less for the new material, required by the change order, than you paid for the original purchase?*

  The same material prices or discounts will usually apply to materials purchased for change orders as new construction. In some instances, however, the contractor may forfeit the advantages of competitive pricing for change orders. Consider the following example:

  A contractor purchased over $20,000 worth of fan coil units for an installation and obtained the maximum discount. Some time later it was determined the project required an additional matching unit. The contractor has to purchase this unit from the original supplier to ensure a match. The supplier at this time may not discount the unit because of the small quantity, and he is no longer in a competitive situation. The impact of quantity on purchase can add between 0% and 25% to material prices and/or subcontractor quotes.

- *If materials have been ordered or delivered to the job site, will they be subject to a cancellation charge or restocking fee?*

  Check with the supplier to determine if ordered materials are subject to a cancellation charge. Delivered materials not used as a result of a change order may be subject to a restocking fee if returned to the supplier. Common restocking charges run between 20% and 40%. Also, delivery charges to return the goods to the supplier must be added.

Labor

- *How efficient is the existing crew at the actual installation?*

  Is the same crew that performed the initial work going to do the change order? Possibly the change consists of the installation of a unit identical to one already installed; therefore, the change should take less time. Be sure to consider this potential productivity increase and modify the productivity rates accordingly.

- *If the crew size is increased, what impact will that have on supervision requirements?*

  Under most bargaining agreements or management practices, there is a point at which a working foreman is replaced by a nonworking foreman. This replacement increases project overhead by adding a nonproductive worker. If additional workers are added to accelerate the project or to perform changes while maintaining the schedule, be sure to add additional supervision time if warranted. Calculate the hours involved and the additional cost directly if possible.

- *What are the other impacts of increased crew size?*

  The larger the crew, the greater the potential for productivity to decrease. Some of the factors that cause this productivity loss are: overcrowding (producing restrictive conditions in the working space) and possibly a shortage of any special tools and equipment required. Such factors affect not only the crew working on the elements directly involved in the change order, but other crews whose movements may also be hampered. As the crew increases, check its basic composition for changes by the addition or deletion of apprentices or nonworking foreman, and quantify the potential effects of equipment shortages or other logistical factors.

- *As new crews, unfamiliar with the project, are brought onto the site, how long will it take them to become oriented to the project requirements?*

  The orientation time for a new crew to become 100% effective varies with the site and type of project. Orientation is easiest at a new construction site and most difficult at existing, very restrictive renovation sites. The type of work also affects orientation time. When all elements of the work are exposed, such as concrete or masonry work, orientation is decreased. When the work is concealed or less visible, such as existing electrical systems, orientation takes longer. Usually orientation can be accomplished in one day or less. Costs for added orientation should be itemized and added to the total estimated cost.

- *How much actual production can be gained by working overtime?*

  Short term overtime can be used effectively to accomplish more work in a day. However, as overtime is scheduled to run beyond several weeks, studies have shown marked decreases in output. The following chart shows the effect of long term overtime on worker efficiency. If the anticipated change requires extended overtime to keep the job on schedule, these factors can be used as a guide to predict the impact on time and cost. Add project overhead, particularly supervision, that may also be incurred.

Days per Week	Hours per Day	Production Efficiency					Payroll Cost Factors	
		1st Week	2nd Week	3rd Week	4th Week	Average 4 Weeks	@ 1-1/2 Times	@ 2 Times
5	8	100%	100%	100%	100%	100%	100%	100%
	9	100	100	95	90	96.25	105.6	111.1
	10	100	95	90	85	92.50	110.0	120.0
	11	95	90	75	65	81.25	113.6	127.3
	12	90	85	70	60	76.25	116.7	133.3
6	8	100	100	95	90	96.25	108.3	116.7
	9	100	95	90	85	92.50	113.0	125.9
	10	95	90	85	80	87.50	116.7	133.3
	11	95	85	70	65	78.75	119.7	139.4
	12	90	80	65	60	73.75	122.2	144.4
7	8	100	95	85	75	88.75	114.3	128.6
	9	95	90	80	70	83.75	118.3	136.5
	10	90	85	75	65	78.75	121.4	142.9
	11	85	80	65	60	72.50	124.0	148.1
	12	85	75	60	55	68.75	126.2	152.4

Effects of Overtime

Caution: Under many labor agreements, Sundays and holidays are paid at a higher premium than the normal overtime rate.

The use of long-term overtime is counterproductive on almost any construction job; that is, the longer the period of overtime, the lower the actual production rate. Numerous studies have been conducted, and while they have resulted in slightly different numbers, all reach the same conclusion. The figure above tabulates the effects of overtime work on efficiency.

As illustrated, there can be a difference between the *actual* payroll cost per hour and the *effective* cost per hour for overtime work. This is due to the reduced production efficiency with the increase in weekly hours beyond 40. This difference between actual and effective cost results from overtime work over a prolonged period. Short-term overtime work does not result in as great a reduction in efficiency and, in such cases, effective cost may not vary significantly from the actual payroll cost. As the total hours per week are increased on a regular basis, more time is lost due to fatigue, lowered morale, and an increased accident rate.

As an example, assume a project where workers are working 6 days a week, 10 hours per day. From the figure above (based on productivity studies), the average effective productive hours over a 4-week period are:

$$0.875 \times 60 = 52.5$$

Depending upon the locale and day of week, overtime hours may be paid at time and a half or double time. For time and a half, the overall (average) *actual* payroll cost (including regular and overtime hours) is determined as follows:

$$\frac{40 \text{ reg. hrs.} + (20 \text{ overtime hrs.} \times 1.5)}{60 \text{ hrs.}} = 1.167$$

Based on 60 hours, the payroll cost per hour will be 116.7% of the normal rate at 40 hours per week. However, because the effective production (efficiency) for 60 hours is reduced to the equivalent of 52.5 hours, the effective cost of overtime is calculated as follows:

For time and a half:

$$\frac{40 \text{ reg. hrs.} + (20 \text{ overtime hrs.} \times 1.5)}{52.5 \text{ hrs.}} = 1.33$$

The installed cost will be 133% of the normal rate (for labor).

Thus, when figuring overtime, the actual cost per unit of work will be higher than the apparent overtime payroll dollar increase, due to the reduced productivity of the longer work week. These efficiency calculations are true only for those cost factors determined by hours worked. Costs that are applied weekly or monthly, such as equipment rentals, will not be similarly affected.

Equipment

- *What equipment is required to complete the change order?*

  Change orders may require extending the rental period of equipment already on the job site, or the addition of special equipment brought in to accomplish the change work. In either case, the additional rental charges and operator labor charges must be added.

Summary

The preceding considerations and others you deem appropriate should be analyzed and applied to a change order estimate. The impact of each should be quantified and listed on the estimate to form an audit trail.

Change orders that are properly identified, documented, and managed help to ensure the orderly, professional, and profitable progress of the work. They also minimize potential claims or disputes at the end of the project.

Back by customer demand!

You asked and we listened. For customer convenience and estimating ease, we have made the 2019 Project Costs available for download at **RSMeans.com/2019books**. You will also find sample estimates, an RSMeans data overview video, and a book registration form to receive quarterly data updates throughout 2019.

Estimating Tips

- The cost figures available in the download were derived from hundreds of projects contained in the RSMeans database of completed construction projects. They include the contractor's overhead and profit. The figures have been adjusted to January of the current year.
- These projects were located throughout the U.S. and reflect a tremendous variation in square foot (S.F.) costs. This is due to differences, not only in labor and material costs, but also in individual owners' requirements. For instance, a bank in a large city would have different features than one in a rural area. This is true of all the different types of buildings analyzed. Therefore, caution should be exercised when using these Project Costs. For example, for courthouses, costs in the database are local courthouse costs and will not apply to the larger, more elaborate federal courthouses.
- None of the figures "go with" any others. All individual cost items were computed and tabulated separately. Thus, the sum of the median figures for plumbing, HVAC, and electrical will not normally total up to the total mechanical and electrical costs arrived at by separate analysis and tabulation of the projects.
- Each building was analyzed as to total and component costs and percentages. The figures were arranged in ascending order with the results tabulated as shown. The 1/4 column shows that 25% of the projects had lower costs and 75% had higher. The 3/4 column shows that 75% of the projects had lower costs and 25% had higher. The median column shows that 50% of the projects had lower costs and 50% had higher.
- Project Costs are useful in the conceptual stage when no details are available. As soon as details become available in the project design, the square foot approach should be discontinued and the project should be priced as to its particular components. When more precision is required, or for estimating the replacement cost of specific buildings, the current edition of *Square Foot Costs with RSMeans data* should be used.
- In using the figures in this section, it is recommended that the median column be used for preliminary figures if no additional information is available. The median figures, when multiplied by the total city construction cost index figures (see City Cost Indexes) and then multiplied by the project size modifier at the end of this section, should present a fairly accurate base figure, which would then have to be adjusted in view of the estimator's experience, local economic conditions, code requirements, and the owner's particular requirements. There is no need to factor in the percentage figures, as these should remain constant from city to city.
- The editors of this data would greatly appreciate receiving cost figures on one or more of your recent projects, which would then be included in the averages for next year. All cost figures received will be kept confidential, except that they will be averaged with other similar projects to arrive at square foot cost figures for next year.

See the website above for details and the discount available for submitting one or more of your projects.

50 17 00 | Project Costs

				UNIT COSTS			% OF TOTAL			
			UNIT	1/4	MEDIAN	3/4	1/4	MEDIAN	3/4	
01	0000	**Auto Sales with Repair**	S.F.							01
	0100	Architectural		104	116	126	59.8%	59.2%	63.7%	
	0200	Plumbing		8.70	9.10	12.15	5%	5%	5%	
	0300	Mechanical		11.65	15.60	17.20	6.7%	6.7%	8.6%	
	0400	Electrical		17.90	22	27.50	10.3%	10.3%	12.1%	
	0500	Total Project Costs		174	182	187				
02	0000	**Banking Institutions**	S.F.							02
	0100	Architectural		157	192	234	60.4%	60.4%	65.5%	
	0200	Plumbing		6.30	8.80	12.25	2.4%	2.4%	3%	
	0300	Mechanical		12.55	17.35	20.50	4.8%	4.8%	5.9%	
	0400	Electrical		30.50	37	57	11.7%	11.7%	12.6%	
	0500	Total Project Costs		260	293	360				
03	0000	**Court House**	S.F.							03
	0100	Architectural		82.50	162	162	54.6%	54.5%	58.3%	
	0200	Plumbing		3.12	3.12	3.12	2.1%	2.1%	1.1%	
	0300	Mechanical		19.50	19.50	19.50	12.9%	12.9%	7%	
	0400	Electrical		25	25	25	16.6%	16.6%	9%	
	0500	Total Project Costs		151	278	278				
04	0000	**Data Centers**	S.F.							04
	0100	Architectural		187	187	187	68%	67.9%	68%	
	0200	Plumbing		10.20	10.20	10.20	3.7%	3.7%	3.7%	
	0300	Mechanical		26	26	26	9.5%	9.4%	9.5%	
	0400	Electrical		24.50	24.50	24.50	8.9%	9%	8.9%	
	0500	Total Project Costs		275	275	275				
05	0000	**Detention Centers**	S.F.							05
	0100	Architectural		173	183	194	59.2%	59.2%	59%	
	0200	Plumbing		18.25	22	27	6.3%	6.2%	7.1%	
	0300	Mechanical		23	33	39.50	7.9%	7.9%	10.6%	
	0400	Electrical		38	45	58.50	13%	13%	14.5%	
	0500	Total Project Costs		292	310	365				
06	0000	**Fire Stations**	S.F.							06
	0100	Architectural		95	121	171	47%	50%	52.4%	
	0200	Plumbing		9.95	13.50	15.65	4.9%	4.9%	5.8%	
	0300	Mechanical		13.45	18.40	25.50	6.7%	6.6%	8%	
	0400	Electrical		22.50	28.50	32.50	11.1%	11%	12.3%	
	0500	Total Project Costs		202	231	300				
07	0000	**Gymnasium**	S.F.							07
	0100	Architectural		86.50	114	114	64.6%	64.4%	57.3%	
	0200	Plumbing		2.12	6.95	6.95	1.6%	1.6%	3.5%	
	0300	Mechanical		3.25	29	29	2.4%	2.4%	14.6%	
	0400	Electrical		10.65	20.50	20.50	7.9%	7.9%	10.3%	
	0500	Total Project Costs		134	199	199				
08	0000	**Hospitals**	S.F.							08
	0100	Architectural		105	172	187	42.9%	43%	47.1%	
	0200	Plumbing		7.70	14.70	32	3.1%	3.1%	4%	
	0300	Mechanical		51	57.50	74.50	20.8%	20.8%	15.8%	
	0400	Electrical		23	46.50	60.50	9.4%	9.5%	12.7%	
	0500	Total Project Costs		245	365	395				
09	0000	**Industrial Buildings**	S.F.							09
	0100	Architectural		44	70	227	56.4%	56.3%	68.6%	
	0200	Plumbing		1.70	6.40	12.95	2.2%	2.2%	6.3%	
	0300	Mechanical		4.71	8.95	42.50	6%	6%	8.8%	
	0400	Electrical		7.20	8.20	68.50	9.2%	9.2%	8%	
	0500	Total Project Costs		78	102	425				
10	0000	**Medical Clinics & Offices**	S.F.							10
	0100	Architectural		87.50	120	158	53.7%	51%	56.3%	
	0200	Plumbing		8.50	12.85	20.50	5.2%	5.2%	6%	
	0300	Mechanical		14.20	22	45	8.7%	8.8%	10.3%	
	0400	Electrical		19.55	26	36.50	12%	11.9%	12.2%	
	0500	Total Project Costs		163	213	286				

50 17 00 | Project Costs

			UNIT	UNIT COSTS			% OF TOTAL			
				1/4	MEDIAN	3/4	1/4	MEDIAN	3/4	
11	0000	**Mixed Use**	S.F.							11
	0100	Architectural		86.50	126	207	47%	47.5%	59.4%	
	0200	Plumbing		6	9.15	11.55	3.3%	3.2%	4.3%	
	0300	Mechanical		14.80	24	46.50	8%	7.8%	11.3%	
	0400	Electrical		15.25	24	40.50	8.3%	8.3%	11.3%	
	0500	Total Project Costs	↓	184	212	335				
12	0000	**Multi-Family Housing**	S.F.							12
	0100	Architectural		75	125	167	60.5%	67.6%	56.6%	
	0200	Plumbing		6.40	12.75	15.10	5.2%	5.6%	5.8%	
	0300	Mechanical		6.95	11.70	37.50	5.6%	6.3%	5.3%	
	0400	Electrical		10.10	18	22.50	8.1%	8.8%	8.1%	
	0500	Total Project Costs	↓	124	221	282				
13	0000	**Nursing Home & Assisted Living**	S.F.							13
	0100	Architectural		70	92	116	58.3%	58.4%	59%	
	0200	Plumbing		7.55	11.35	12.50	6.3%	5.9%	7.3%	
	0300	Mechanical		6.20	9.15	17.95	5.2%	5.2%	5.9%	
	0400	Electrical		10.25	16.20	22.50	8.5%	8.6%	10.4%	
	0500	Total Project Costs	↓	120	156	188				
14	0000	**Office Buildings**	S.F.							14
	0100	Architectural		92.50	126	177	60.1%	60%	64.6%	
	0200	Plumbing		4.98	7.85	15.40	3.2%	3.1%	4%	
	0300	Mechanical		10.75	17.65	25.50	7%	6.8%	9.1%	
	0400	Electrical		12.35	21	34	8%	7.9%	10.8%	
	0500	Total Project Costs	↓	154	195	285				
15	0000	**Parking Garage**	S.F.							15
	0100	Architectural		31	38	39.50	82.7%	82.1%	82.6%	
	0200	Plumbing		1.02	1.07	2	2.7%	2.7%	2.3%	
	0300	Mechanical		.79	1.22	4.62	2.1%	2.1%	2.7%	
	0400	Electrical		2.72	2.98	6.25	7.3%	7.1%	6.5%	
	0500	Total Project Costs	↓	37.50	46	49.50				
16	0000	**Parking Garage/Mixed Use**	S.F.							16
	0100	Architectural		100	110	112	61%	61.2%	64.3%	
	0200	Plumbing		3.22	4.22	6.45	2%	2%	2.5%	
	0300	Mechanical		13.80	15.50	22.50	8.4%	8.4%	9.1%	
	0400	Electrical		14.45	21	21.50	8.8%	8.8%	12.3%	
	0500	Total Project Costs	↓	164	171	177				
17	0000	**Police Stations**	S.F.							17
	0100	Architectural		113	127	160	53.3%	54%	48.5%	
	0200	Plumbing		15	18	18.10	7.1%	7%	6.9%	
	0300	Mechanical		34	47.50	49	16%	16.1%	18.1%	
	0400	Electrical		25.50	28	29.50	12%	12.1%	10.7%	
	0500	Total Project Costs	↓	212	262	297				
18	0000	**Police/Fire**	S.F.							18
	0100	Architectural		110	110	340	67.9%	68.2%	65.9%	
	0200	Plumbing		8.65	9.15	34	5.3%	5.5%	5.5%	
	0300	Mechanical		13.55	21.50	77.50	8.4%	8.4%	12.9%	
	0400	Electrical		15.40	19.70	88.50	9.5%	9.6%	11.8%	
	0500	Total Project Costs	↓	162	167	610				
19	0000	**Public Assembly Buildings**	S.F.							19
	0100	Architectural		115	156	218	62.5%	63%	61.7%	
	0200	Plumbing		5.95	8.75	12.90	3.2%	3%	3.5%	
	0300	Mechanical		13.60	22.50	34.50	7.4%	8%	8.9%	
	0400	Electrical		18.60	25.50	40.50	10.1%	10.5%	10.1%	
	0500	Total Project Costs	↓	184	253	360				
20	0000	**Recreational**	S.F.							20
	0100	Architectural		108	170	231	56.3%	55.7%	59.2%	
	0200	Plumbing		8.35	15.35	24.50	4.3%	4.6%	5.3%	
	0300	Mechanical		12.90	19.60	31	6.7%	6.9%	6.8%	
	0400	Electrical		15.80	28	39	8.2%	7.7%	9.8%	
	0500	Total Project Costs	↓	192	287	435				

50 17 00 | Project Costs

	50 17 00 | Project Costs		Unit	Unit Costs 1/4	Unit Costs Median	Unit Costs 3/4	% of Total 1/4	% of Total Median	% of Total 3/4	
21	0000	**Restaurants**	S.F.							21
	0100	Architectural		123	198	245	60.6%	77.9%	59.1%	
	0200	Plumbing		9.95	31	39	4.9%	14.6%	9.3%	
	0300	Mechanical		14.55	19.30	47	7.2%	11.2%	5.8%	
	0400	Electrical		14.45	30.50	51	7.1%	17.9%	9.1%	
	0500	Total Project Costs		203	335	420				
22	0000	**Retail**	S.F.							22
	0100	Architectural		54	60.50	109	73%	59.9%	64.4%	
	0200	Plumbing		5.65	7.85	9.95	7.6%	6.2%	8.4%	
	0300	Mechanical		5.15	7.40	9.05	7%	5.6%	7.9%	
	0400	Electrical		7.20	11.55	18.50	9.7%	7.9%	12.3%	
	0500	Total Project Costs		74	94	148				
23	0000	**Schools**	S.F.							23
	0100	Architectural		94	120	160	58%	58.8%	55.6%	
	0200	Plumbing		7.50	10.40	15.15	4.6%	4.7%	4.8%	
	0300	Mechanical		17.85	24.50	36.50	11%	11.1%	11.3%	
	0400	Electrical		17.45	24	30.50	10.8%	11%	11.1%	
	0500	Total Project Costs		162	216	286				
24	0000	**University, College & Private School Classroom & Admin Buildings**	S.F.							24
	0100	Architectural		121	150	188	60.2%	61%	54%	
	0200	Plumbing		6.90	10.70	15.10	3.4%	3.4%	3.8%	
	0300	Mechanical		26	37.50	45	12.9%	12.9%	13.5%	
	0400	Electrical		19.50	27.50	33.50	9.7%	9.8%	9.9%	
	0500	Total Project Costs		201	278	370				
25	0000	**University, College & Private School Dormitories**	S.F.							25
	0100	Architectural		79	139	147	67.5%	67.1%	62.6%	
	0200	Plumbing		10.45	14.80	22	8.9%	8.9%	6.7%	
	0300	Mechanical		4.69	19.95	31.50	4%	4%	9%	
	0400	Electrical		5.55	19.35	29.50	4.7%	4.8%	8.7%	
	0500	Total Project Costs		117	222	263				
26	0000	**University, College & Private School Science, Eng. & Lab Buildings**	S.F.							26
	0100	Architectural		136	144	188	48.7%	49.1%	50.5%	
	0200	Plumbing		9.35	14.20	26	3.4%	3.4%	5%	
	0300	Mechanical		42.50	67	68.50	15.2%	15.3%	23.5%	
	0400	Electrical		27.50	32	37.50	9.9%	9.8%	11.2%	
	0500	Total Project Costs		279	285	320				
27	0000	**University, College & Private School Student Union Buildings**	S.F.							27
	0100	Architectural		108	283	283	50.9%	60%	54.4%	
	0200	Plumbing		16.25	16.25	24	7.7%	4.3%	3.1%	
	0300	Mechanical		31	50	50	14.6%	9.7%	9.6%	
	0400	Electrical		27	47	47	12.7%	13.3%	9%	
	0500	Total Project Costs		212	520	520				
28	0000	**Warehouses**	S.F.							28
	0100	Architectural		46	72	171	66.7%	67.4%	58.5%	
	0200	Plumbing		2.40	5.15	9.90	3.5%	3.5%	4.2%	
	0300	Mechanical		2.84	16.20	25.50	4.1%	4.1%	13.2%	
	0400	Electrical	S.F.	5.15	19.40	32.50	7.5%	7.5%	15.8%	
	0500	Total Project Costs	S.F.	69	123	238				

Square Foot Project Size Modifier

One factor that affects the S.F. cost of a particular building is the size. In general, for buildings built to the same specifications in the same locality, the larger building will have the lower S.F. cost. This is due mainly to the decreasing contribution of the exterior walls plus the economy of scale usually achievable in larger buildings. The Area Conversion Scale shown below will give a factor to convert costs for the typical size building to an adjusted cost for the particular project.

The Square Foot Base Size lists the median costs, most typical project size in our accumulated data, and the range in size of the projects.

The Size Factor for your project is determined by dividing your project area in S.F. by the typical project size for the particular Building Type. With this factor, enter the Area Conversion Scale at the appropriate Size Factor and determine the appropriate cost multiplier for your building size.

Example: Determine the cost per S.F. for a 152,600 S.F. Multi-family housing.

$$\frac{\text{Proposed building area} = 152{,}600\text{ S.F.}}{\text{Typical size from below} = 76{,}300\text{ S.F.}} = 2.00$$

Enter Area Conversion scale at 2.0, intersect curve, read horizontally the appropriate cost multiplier of .94. Size adjusted cost becomes .94 x \$194.00 = \$182.36 based on national average costs.

Note: For Size Factors less than .50, the Cost Multiplier is 1.1
For Size Factors greater than 3.5, the Cost Multiplier is .90

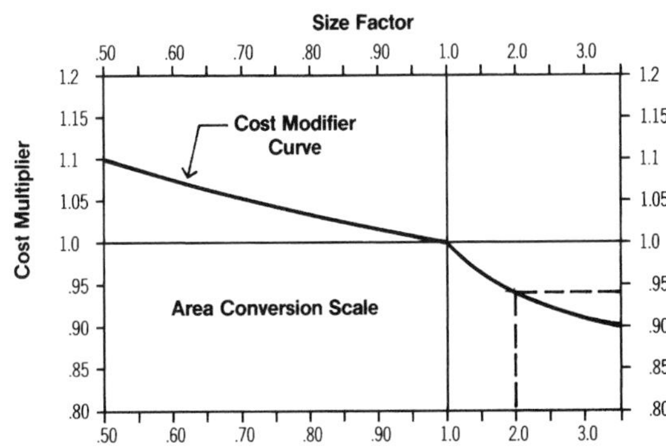

System	Median Cost (Total Project Costs)	Typical Size Gross S.F. (Median of Projects)	Typical Range (Low – High) (Projects)
Auto Sales with Repair	\$182.00	24,900	4,700 – 29,300
Banking Institutions	293.00	9,300	3,300 – 38,100
Detention Centers	310.00	37,800	12,300 – 183,300
Fire Stations	231.00	12,300	6,300 – 29,600
Hospitals	365.00	87,100	22,400 – 410,300
Industrial Buildings	\$102.00	22,100	5,100 – 200,600
Medical Clinics & Offices	213.00	22,500	2,300 – 327,000
Mixed Use	212.00	27,200	7,200 – 109,800
Multi-Family Housing	221.00	54,700	2,500 – 1,161,500
Nursing Home & Assisted Living	156.00	38,200	1,500 – 242,600
Office Buildings	195.00	20,600	1,100 – 930,000
Parking Garage	46.00	151,800	99,900 – 287,000
Parking Garage/Mixed Use	171.00	254,200	5,300 – 318,000
Police Stations	262.00	28,500	15,400 – 88,600
Public Assembly Buildings	253.00	22,600	2,200 – 235,300
Recreational	287.00	19,900	1,000 – 223,800
Restaurants	335.00	6,100	5,500 – 42,000
Retail	94.00	28,700	5,200 – 84,300
Schools	216.00	73,500	1,300 – 410,800
University, College & Private School Classroom & Admin Buildings	278.00	48,300	9,400 – 196,200
University, College & Private School Dormitories	222.00	28,900	1,500 – 126,900
University, College & Private School Science, Eng. & Lab Buildings	285.00	73,400	25,700 – 117,600
Warehouses	123.00	10,400	600 – 303,800

Abbreviations

A	Area Square Feet; Ampere
AAFES	Army and Air Force Exchange Service
ABS	Acrylonitrile Butadiene Stryrene; Asbestos Bonded Steel
A.C., AC	Alternating Current; Air-Conditioning; Asbestos Cement; Plywood Grade A & C
ACI	American Concrete Institute
ACR	Air Conditioning Refrigeration
ADA	Americans with Disabilities Act
AD	Plywood, Grade A & D
Addit.	Additional
Adh.	Adhesive
Adj.	Adjustable
af	Audio-frequency
AFFF	Aqueous Film Forming Foam
AFUE	Annual Fuel Utilization Efficiency
AGA	American Gas Association
Agg.	Aggregate
A.H., Ah	Ampere Hours
A hr.	Ampere-hour
A.H.U., AHU	Air Handling Unit
A.I.A.	American Institute of Architects
AIC	Ampere Interrupting Capacity
Allow.	Allowance
alt., alt	Alternate
Alum.	Aluminum
a.m.	Ante Meridiem
Amp.	Ampere
Anod.	Anodized
ANSI	American National Standards Institute
APA	American Plywood Association
Approx.	Approximate
Apt.	Apartment
Asb.	Asbestos
A.S.B.C.	American Standard Building Code
Asbe.	Asbestos Worker
ASCE	American Society of Civil Engineers
A.S.H.R.A.E.	American Society of Heating, Refrig. & AC Engineers
ASME	American Society of Mechanical Engineers
ASTM	American Society for Testing and Materials
Attchmt.	Attachment
Avg., Ave.	Average
AWG	American Wire Gauge
AWWA	American Water Works Assoc.
Bbl.	Barrel
B&B, BB	Grade B and Better; Balled & Burlapped
B&S	Bell and Spigot
B.&W.	Black and White
b.c.c.	Body-centered Cubic
B.C.Y.	Bank Cubic Yards
BE	Bevel End
B.F.	Board Feet
Bg. cem.	Bag of Cement
BHP	Boiler Horsepower; Brake Horsepower
B.I.	Black Iron
bidir.	bidirectional
Bit., Bitum.	Bituminous
Bit., Conc.	Bituminous Concrete
Bk.	Backed
Bkrs.	Breakers
Bldg., bldg	Building
Blk.	Block
Bm.	Beam
Boil.	Boilermaker
bpm	Blows per Minute
BR	Bedroom
Brg., brng.	Bearing
Brhe.	Bricklayer Helper
Bric.	Bricklayer
Brk., brk	Brick
brkt	Bracket
Brs.	Brass
Brz.	Bronze
Bsn.	Basin
Btr.	Better
BTU	British Thermal Unit
BTUH	BTU per Hour
Bu.	Bushels
BUR	Built-up Roofing
BX	Interlocked Armored Cable
°C	Degree Centigrade
c	Conductivity, Copper Sweat
C	Hundred; Centigrade
C/C	Center to Center, Cedar on Cedar
C-C	Center to Center
Cab	Cabinet
Cair.	Air Tool Laborer
Cal.	Caliper
Calc	Calculated
Cap.	Capacity
Carp.	Carpenter
C.B.	Circuit Breaker
C.C.A.	Chromate Copper Arsenate
C.C.F.	Hundred Cubic Feet
cd	Candela
cd/sf	Candela per Square Foot
CD	Grade of Plywood Face & Back
CDX	Plywood, Grade C & D, exterior glue
Cefi.	Cement Finisher
Cem.	Cement
CF	Hundred Feet
C.F.	Cubic Feet
CFM	Cubic Feet per Minute
CFRP	Carbon Fiber Reinforced Plastic
c.g.	Center of Gravity
CHW	Chilled Water; Commercial Hot Water
C.I., CI	Cast Iron
C.I.P., CIP	Cast in Place
Circ.	Circuit
C.L.	Carload Lot
CL	Chain Link
Clab.	Common Laborer
Clam	Common Maintenance Laborer
C.L.F.	Hundred Linear Feet
CLF	Current Limiting Fuse
CLP	Cross Linked Polyethylene
cm	Centimeter
CMP	Corr. Metal Pipe
CMU	Concrete Masonry Unit
CN	Change Notice
Col.	Column
CO_2	Carbon Dioxide
Comb.	Combination
comm.	Commercial, Communication
Compr.	Compressor
Conc.	Concrete
Cont., cont	Continuous; Continued, Container
Corkbd.	Cork Board
Corr.	Corrugated
Cos	Cosine
Cot	Cotangent
Cov.	Cover
C/P	Cedar on Paneling
CPA	Control Point Adjustment
Cplg.	Coupling
CPM	Critical Path Method
CPVC	Chlorinated Polyvinyl Chloride
C.Pr.	Hundred Pair
CRC	Cold Rolled Channel
Creos.	Creosote
Crpt.	Carpet & Linoleum Layer
CRT	Cathode-ray Tube
CS	Carbon Steel, Constant Shear Bar Joist
Csc	Cosecant
C.S.F.	Hundred Square Feet
CSI	Construction Specifications Institute
CT	Current Transformer
CTS	Copper Tube Size
Cu	Copper, Cubic
Cu. Ft.	Cubic Foot
cw	Continuous Wave
C.W.	Cool White; Cold Water
Cwt.	100 Pounds
C.W.X.	Cool White Deluxe
C.Y.	Cubic Yard (27 cubic feet)
C.Y./Hr.	Cubic Yard per Hour
Cyl.	Cylinder
d	Penny (nail size)
D	Deep; Depth; Discharge
Dis., Disch.	Discharge
Db	Decibel
Dbl.	Double
DC	Direct Current
DDC	Direct Digital Control
Demob.	Demobilization
d.f.t.	Dry Film Thickness
d.f.u.	Drainage Fixture Units
D.H.	Double Hung
DHW	Domestic Hot Water
DI	Ductile Iron
Diag.	Diagonal
Diam., Dia	Diameter
Distrib.	Distribution
Div.	Division
Dk.	Deck
D.L.	Dead Load; Diesel
DLH	Deep Long Span Bar Joist
dlx	Deluxe
Do.	Ditto
DOP	Dioctyl Phthalate Penetration Test (Air Filters)
Dp., dp	Depth
D.P.S.T.	Double Pole, Single Throw
Dr.	Drive
DR	Dimension Ratio
Drink.	Drinking
D.S.	Double Strength
D.S.A.	Double Strength A Grade
D.S.B.	Double Strength B Grade
Dty.	Duty
DWV	Drain Waste Vent
DX	Deluxe White, Direct Expansion
dyn	Dyne
e	Eccentricity
E	Equipment Only; East; Emissivity
Ea.	Each
EB	Encased Burial
Econ.	Economy
E.C.Y	Embankment Cubic Yards
EDP	Electronic Data Processing
EIFS	Exterior Insulation Finish System
E.D.R.	Equiv. Direct Radiation
Eq.	Equation
EL	Elevation
Elec.	Electrician; Electrical
Elev.	Elevator; Elevating
EMT	Electrical Metallic Conduit; Thin Wall Conduit
Eng.	Engine, Engineered
EPDM	Ethylene Propylene Diene Monomer
EPS	Expanded Polystyrene
Eqhv.	Equip. Oper., Heavy
Eqlt.	Equip. Oper., Light
Eqmd.	Equip. Oper., Medium
Eqmm.	Equip. Oper., Master Mechanic
Eqol.	Equip. Oper., Oilers
Equip.	Equipment
ERW	Electric Resistance Welded

Abbreviations

E.S.	Energy Saver
Est.	Estimated
esu	Electrostatic Units
E.W.	Each Way
EWT	Entering Water Temperature
Excav.	Excavation
excl	Excluding
Exp., exp	Expansion, Exposure
Ext., ext	Exterior; Extension
Extru.	Extrusion
f.	Fiber Stress
F	Fahrenheit; Female; Fill
Fab., fab	Fabricated; Fabric
FBGS	Fiberglass
F.C.	Footcandles
f.c.c.	Face-centered Cubic
f'c.	Compressive Stress in Concrete; Extreme Compressive Stress
F.E.	Front End
FEP	Fluorinated Ethylene Propylene (Teflon)
F.G.	Flat Grain
F.H.A.	Federal Housing Administration
Fig.	Figure
Fin.	Finished
FIPS	Female Iron Pipe Size
Fixt.	Fixture
FJP	Finger jointed and primed
Fl. Oz.	Fluid Ounces
Flr.	Floor
Flrs.	Floors
FM	Frequency Modulation; Factory Mutual
Fmg.	Framing
FM/UL	Factory Mutual/Underwriters Labs
Fdn.	Foundation
FNPT	Female National Pipe Thread
Fori.	Foreman, Inside
Foro.	Foreman, Outside
Fount.	Fountain
fpm	Feet per Minute
FPT	Female Pipe Thread
Fr	Frame
F.R.	Fire Rating
FRK	Foil Reinforced Kraft
FSK	Foil/Scrim/Kraft
FRP	Fiberglass Reinforced Plastic
FS	Forged Steel
FSC	Cast Body; Cast Switch Box
Ft., ft	Foot; Feet
Ftng.	Fitting
Ftg.	Footing
Ft lb.	Foot Pound
Furn.	Furniture
FVNR	Full Voltage Non-Reversing
FVR	Full Voltage Reversing
FXM	Female by Male
Fy.	Minimum Yield Stress of Steel
g	Gram
G	Gauss
Ga.	Gauge
Gal., gal.	Gallon
Galv., galv	Galvanized
GC/MS	Gas Chromatograph/Mass Spectrometer
Gen.	General
GFI	Ground Fault Interrupter
GFRC	Glass Fiber Reinforced Concrete
Glaz.	Glazier
GPD	Gallons per Day
gpf	Gallon per Flush
GPH	Gallons per Hour
gpm, GPM	Gallons per Minute
GR	Grade
Gran.	Granular
Grnd.	Ground
GVW	Gross Vehicle Weight
GWB	Gypsum Wall Board
H	High Henry
HC	High Capacity
H.D., HD	Heavy Duty; High Density
H.D.O.	High Density Overlaid
HDPE	High Density Polyethylene Plastic
Hdr.	Header
Hdwe.	Hardware
H.I.D., HID	High Intensity Discharge
Help.	Helper Average
HEPA	High Efficiency Particulate Air Filter
Hg	Mercury
HIC	High Interrupting Capacity
HM	Hollow Metal
HMWPE	High Molecular Weight Polyethylene
HO	High Output
Horiz.	Horizontal
H.P., HP	Horsepower; High Pressure
H.P.F.	High Power Factor
Hr.	Hour
Hrs./Day	Hours per Day
HSC	High Short Circuit
Ht.	Height
Htg.	Heating
Htrs.	Heaters
HVAC	Heating, Ventilation & Air-Conditioning
Hvy.	Heavy
HW	Hot Water
Hyd.; Hydr.	Hydraulic
Hz	Hertz (cycles)
I.	Moment of Inertia
IBC	International Building Code
I.C.	Interrupting Capacity
ID	Inside Diameter
I.D.	Inside Dimension; Identification
I.F.	Inside Frosted
I.M.C.	Intermediate Metal Conduit
In.	Inch
Incan.	Incandescent
Incl.	Included; Including
Int.	Interior
Inst.	Installation
Insul., insul	Insulation/Insulated
I.P.	Iron Pipe
I.P.S., IPS	Iron Pipe Size
IPT	Iron Pipe Threaded
I.W.	Indirect Waste
J	Joule
J.I.C.	Joint Industrial Council
K	Thousand; Thousand Pounds; Heavy Wall Copper Tubing, Kelvin
K.A.H.	Thousand Amp. Hours
kcmil	Thousand Circular Mils
KD	Knock Down
K.D.A.T.	Kiln Dried After Treatment
kg	Kilogram
kG	Kilogauss
kgf	Kilogram Force
kHz	Kilohertz
Kip	1000 Pounds
KJ	Kilojoule
K.L.	Effective Length Factor
K.L.F.	Kips per Linear Foot
Km	Kilometer
KO	Knock Out
K.S.F.	Kips per Square Foot
K.S.I.	Kips per Square Inch
kV	Kilovolt
kVA	Kilovolt Ampere
kVAR	Kilovar (Reactance)
KW	Kilowatt
KWh	Kilowatt-hour
L	Labor Only; Length; Long; Medium Wall Copper Tubing
Lab.	Labor
lat	Latitude
Lath.	Lather
Lav.	Lavatory
lb.; #	Pound
L.B., LB	Load Bearing; L Conduit Body
L. & E.	Labor & Equipment
lb./hr.	Pounds per Hour
lb./L.F.	Pounds per Linear Foot
lbf/sq.in.	Pound-force per Square Inch
L.C.L.	Less than Carload Lot
L.C.Y.	Loose Cubic Yard
Ld.	Load
LE	Lead Equivalent
LED	Light Emitting Diode
L.F.	Linear Foot
L.F. Hdr	Linear Feet of Header
L.F. Nose	Linear Foot of Stair Nosing
L.F. Rsr	Linear Foot of Stair Riser
Lg.	Long; Length; Large
L & H	Light and Heat
LH	Long Span Bar Joist
L.H.	Labor Hours
L.L., LL	Live Load
L.L.D.	Lamp Lumen Depreciation
lm	Lumen
lm/sf	Lumen per Square Foot
lm/W	Lumen per Watt
LOA	Length Over All
log	Logarithm
L-O-L	Lateralolet
long.	Longitude
L.P., LP	Liquefied Petroleum; Low Pressure
L.P.F.	Low Power Factor
LR	Long Radius
L.S.	Lump Sum
Lt.	Light
Lt. Ga.	Light Gauge
L.T.L.	Less than Truckload Lot
Lt. Wt.	Lightweight
L.V.	Low Voltage
M	Thousand; Material; Male; Light Wall Copper Tubing
M^2CA	Meters Squared Contact Area
m/hr.; M.H.	Man-hour
mA	Milliampere
Mach.	Machine
Mag. Str.	Magnetic Starter
Maint.	Maintenance
Marb.	Marble Setter
Mat; Mat'l.	Material
Max.	Maximum
MBF	Thousand Board Feet
MBH	Thousand BTU's per hr.
MC	Metal Clad Cable
MCC	Motor Control Center
M.C.F.	Thousand Cubic Feet
MCFM	Thousand Cubic Feet per Minute
M.C.M.	Thousand Circular Mils
MCP	Motor Circuit Protector
MD	Medium Duty
MDF	Medium-density fibreboard
M.D.O.	Medium Density Overlaid
Med.	Medium
MF	Thousand Feet
M.F.B.M.	Thousand Feet Board Measure
Mfg.	Manufacturing
Mfrs.	Manufacturers
mg	Milligram
MGD	Million Gallons per Day
MGPH	Million Gallons per Hour
MH, M.H.	Manhole; Metal Halide; Man-Hour
MHz	Megahertz
Mi.	Mile
MI	Malleable Iron; Mineral Insulated
MIPS	Male Iron Pipe Size
mj	Mechanical Joint
m	Meter
mm	Millimeter
Mill.	Millwright
Min., min.	Minimum, Minute

Misc.	Miscellaneous
ml	Milliliter, Mainline
M.L.F.	Thousand Linear Feet
Mo.	Month
Mobil.	Mobilization
Mog.	Mogul Base
MPH	Miles per Hour
MPT	Male Pipe Thread
MRGWB	Moisture Resistant Gypsum Wallboard
MRT	Mile Round Trip
ms	Millisecond
M.S.F.	Thousand Square Feet
Mstz.	Mosaic & Terrazzo Worker
M.S.Y.	Thousand Square Yards
Mtd., mtd., mtd	Mounted
Mthe.	Mosaic & Terrazzo Helper
Mtng.	Mounting
Mult.	Multi; Multiply
MUTCD	Manual on Uniform Traffic Control Devices
M.V.A.	Million Volt Amperes
M.V.A.R.	Million Volt Amperes Reactance
MV	Megavolt
MW	Megawatt
MXM	Male by Male
MYD	Thousand Yards
N	Natural; North
nA	Nanoampere
NA	Not Available; Not Applicable
N.B.C.	National Building Code
NC	Normally Closed
NEMA	National Electrical Manufacturers Assoc.
NEHB	Bolted Circuit Breaker to 600V.
NFPA	National Fire Protection Association
NLB	Non-Load-Bearing
NM	Non-Metallic Cable
nm	Nanometer
No.	Number
NO	Normally Open
N.O.C.	Not Otherwise Classified
Nose.	Nosing
NPT	National Pipe Thread
NQOD	Combination Plug-on/Bolt on Circuit Breaker to 240V.
N.R.C., NRC	Noise Reduction Coefficient/ Nuclear Regulator Commission
N.R.S.	Non Rising Stem
ns	Nanosecond
NTP	Notice to Proceed
nW	Nanowatt
OB	Opposing Blade
OC	On Center
OD	Outside Diameter
O.D.	Outside Dimension
ODS	Overhead Distribution System
O.G.	Ogee
O.H.	Overhead
O&P	Overhead and Profit
Oper.	Operator
Opng.	Opening
Orna.	Ornamental
OSB	Oriented Strand Board
OS&Y	Outside Screw and Yoke
OSHA	Occupational Safety and Health Act
Ovhd.	Overhead
OWG	Oil, Water or Gas
Oz.	Ounce
P.	Pole; Applied Load; Projection
p.	Page
Pape.	Paperhanger
P.A.P.R.	Powered Air Purifying Respirator
PAR	Parabolic Reflector
P.B., PB	Push Button
Pc., Pcs.	Piece, Pieces
P.C.	Portland Cement; Power Connector
P.C.F.	Pounds per Cubic Foot
PCM	Phase Contrast Microscopy
PDCA	Painting and Decorating Contractors of America
P.E., PE	Professional Engineer; Porcelain Enamel; Polyethylene; Plain End
P.E.C.I.	Porcelain Enamel on Cast Iron
Perf.	Perforated
PEX	Cross Linked Polyethylene
Ph.	Phase
P.I.	Pressure Injected
Pile.	Pile Driver
Pkg.	Package
Pl.	Plate
Plah.	Plasterer Helper
Plas.	Plasterer
plf	Pounds Per Linear Foot
Pluh.	Plumber Helper
Plum.	Plumber
Ply.	Plywood
p.m.	Post Meridiem
Pntd.	Painted
Pord.	Painter, Ordinary
pp	Pages
PP, PPL	Polypropylene
P.P.M.	Parts per Million
Pr.	Pair
P.E.S.B.	Pre-engineered Steel Building
Prefab.	Prefabricated
Prefin.	Prefinished
Prop.	Propelled
PSF, psf	Pounds per Square Foot
PSI, psi	Pounds per Square Inch
PSIG	Pounds per Square Inch Gauge
PSP	Plastic Sewer Pipe
Pspr.	Painter, Spray
Psst.	Painter, Structural Steel
P.T.	Potential Transformer
P. & T.	Pressure & Temperature
Ptd.	Painted
Ptns.	Partitions
Pu	Ultimate Load
PVC	Polyvinyl Chloride
Pvmt.	Pavement
PRV	Pressure Relief Valve
Pwr.	Power
Q	Quantity Heat Flow
Qt.	Quart
Quan., Qty.	Quantity
Q.C.	Quick Coupling
r	Radius of Gyration
R	Resistance
R.C.P.	Reinforced Concrete Pipe
Rect.	Rectangle
recpt.	Receptacle
Reg.	Regular
Reinf.	Reinforced
Req'd.	Required
Res.	Resistant
Resi.	Residential
RF	Radio Frequency
RFID	Radio-frequency Identification
Rgh.	Rough
RGS	Rigid Galvanized Steel
RHW	Rubber, Heat & Water Resistant; Residential Hot Water
rms	Root Mean Square
Rnd.	Round
Rodm.	Rodman
Rofc.	Roofer, Composition
Rofp.	Roofer, Precast
Rohe.	Roofer Helpers (Composition)
Rots.	Roofer, Tile & Slate
R.O.W.	Right of Way
RPM	Revolutions per Minute
R.S.	Rapid Start
Rsr	Riser
RT	Round Trip
S.	Suction; Single Entrance; South
SBS	Styrene Butadiere Styrene
SC	Screw Cover
SCFM	Standard Cubic Feet per Minute
Scaf.	Scaffold
Sch., Sched.	Schedule
S.C.R.	Modular Brick
S.D.	Sound Deadening
SDR	Standard Dimension Ratio
S.E.	Surfaced Edge
Sel.	Select
SER, SEU	Service Entrance Cable
S.F.	Square Foot
S.F.C.A.	Square Foot Contact Area
S.F. Flr.	Square Foot of Floor
S.F.G.	Square Foot of Ground
S.F. Hor.	Square Foot Horizontal
SFR	Square Feet of Radiation
S.F. Shlf.	Square Foot of Shelf
S4S	Surface 4 Sides
Shee.	Sheet Metal Worker
Sin.	Sine
Skwk.	Skilled Worker
SL	Saran Lined
S.L.	Slimline
Sldr.	Solder
SLH	Super Long Span Bar Joist
S.N.	Solid Neutral
SO	Stranded with oil resistant inside insulation
S-O-L	Socketolet
sp	Standpipe
S.P.	Static Pressure; Single Pole; Self-Propelled
Spri.	Sprinkler Installer
spwg	Static Pressure Water Gauge
S.P.D.T.	Single Pole, Double Throw
SPF	Spruce Pine Fir; Sprayed Polyurethane Foam
S.P.S.T.	Single Pole, Single Throw
SPT	Standard Pipe Thread
Sq.	Square; 100 Square Feet
Sq. Hd.	Square Head
Sq. In.	Square Inch
S.S.	Single Strength; Stainless Steel
S.S.B.	Single Strength B Grade
sst, ss	Stainless Steel
Sswk.	Structural Steel Worker
Sswl.	Structural Steel Welder
St.; Stl.	Steel
STC	Sound Transmission Coefficient
Std.	Standard
Stg.	Staging
STK	Select Tight Knot
STP	Standard Temperature & Pressure
Stpi.	Steamfitter, Pipefitter
Str.	Strength; Starter; Straight
Strd.	Stranded
Struct.	Structural
Sty.	Story
Subj.	Subject
Subs.	Subcontractors
Surf.	Surface
Sw.	Switch
Swbd.	Switchboard
S.Y.	Square Yard
Syn.	Synthetic
S.Y.P.	Southern Yellow Pine
Sys.	System
t.	Thickness
T	Temperature; Ton
Tan	Tangent
T.C.	Terra Cotta
T & C	Threaded and Coupled
T.D.	Temperature Difference
TDD	Telecommunications Device for the Deaf
T.E.M.	Transmission Electron Microscopy
temp	Temperature, Tempered, Temporary
TFFN	Nylon Jacketed Wire

Abbreviations

TFE	Tetrafluoroethylene (Teflon)
T. & G.	Tongue & Groove; Tar & Gravel
Th.,Thk.	Thick
Thn.	Thin
Thrded	Threaded
Tilf.	Tile Layer, Floor
Tilh.	Tile Layer, Helper
THHN	Nylon Jacketed Wire
THW.	Insulated Strand Wire
THWN	Nylon Jacketed Wire
T.L.,TL	Truckload
T.M.	Track Mounted
Tot.	Total
T-O-L	Threadolet
tmpd	Tempered
TPO	Thermoplastic Polyolefin
T.S.	Trigger Start
Tr.	Trade
Transf.	Transformer
Trhv.	Truck Driver, Heavy
Trlr	Trailer
Trlt.	Truck Driver, Light
TTY	Teletypewriter
TV	Television
T.W.	Thermoplastic Water Resistant Wire
UCI	Uniform Construction Index
UF	Underground Feeder
UGND	Underground Feeder
UHF	Ultra High Frequency
U.I.	United Inch
U.L., UL	Underwriters Laboratory
Uld.	Unloading
Unfin.	Unfinished
UPS	Uninterruptible Power Supply
URD	Underground Residential Distribution
US	United States
USGBC	U.S. Green Building Council
USP	United States Primed
UTMCD	Uniform Traffic Manual For Control Devices
UTP	Unshielded Twisted Pair
V	Volt
VA	Volt Amperes
VAT	Vinyl Asbestos Tile
V.C.T.	Vinyl Composition Tile
VAV	Variable Air Volume
VC	Veneer Core
VDC	Volts Direct Current
Vent.	Ventilation
Vert.	Vertical
V.F.	Vinyl Faced
V.G.	Vertical Grain
VHF	Very High Frequency
VHO	Very High Output
Vib.	Vibrating
VLF	Vertical Linear Foot
VOC	Volatile Organic Compound
Vol.	Volume
VRP	Vinyl Reinforced Polyester
W	Wire; Watt; Wide; West
w/	With
W.C., WC	Water Column; Water Closet
W.F.	Wide Flange
W.G.	Water Gauge
Wldg.	Welding
W. Mile	Wire Mile
W-O-L	Weldolet
W.R.	Water Resistant
Wrck.	Wrecker
WSFU	Water Supply Fixture Unit
W.S.P.	Water, Steam, Petroleum
WT., Wt.	Weight
WWF	Welded Wire Fabric
XFER	Transfer
XFMR	Transformer
XHD	Extra Heavy Duty
XHHW	Cross-Linked Polyethylene Wire
XLPE	Insulation
XLP	Cross-linked Polyethylene
Xport	Transport
Y	Wye
yd	Yard
yr	Year
Δ	Delta
%	Percent
~	Approximately
Ø	Phase; diameter
@	At
#	Pound; Number
<	Less Than
>	Greater Than
Z	Zone

A

B

C

D

E

F

G

Q

R

S

V

W

X

Y

Z

Notes

Notes

Notes

Notes

Notes

Division Notes

	CREW	DAILY OUTPUT	LABOR-HOURS	UNIT	BARE COSTS MAT.	BARE COSTS LABOR	BARE COSTS EQUIP.	BARE COSTS TOTAL	TOTAL INCL O&P

Division Notes

		CREW	DAILY OUTPUT	LABOR-HOURS	UNIT	BARE COSTS MAT.	LABOR	EQUIP.	TOTAL	TOTAL INCL O&P

Division Notes

	CREW	DAILY OUTPUT	LABOR-HOURS	UNIT	BARE COSTS MAT.	LABOR	EQUIP.	TOTAL	TOTAL INCL O&P

Division Notes

	CREW	DAILY OUTPUT	LABOR-HOURS	UNIT	BARE COSTS MAT.	BARE COSTS LABOR	BARE COSTS EQUIP.	BARE COSTS TOTAL	TOTAL INCL O&P

Division Notes

		CREW	DAILY OUTPUT	LABOR-HOURS	UNIT	BARE COSTS MAT.	BARE COSTS LABOR	BARE COSTS EQUIP.	BARE COSTS TOTAL	TOTAL INCL O&P

Cost Data Selection Guide

ne following table provides definitive information on the content of each cost data publication. The number of lines of data provided in each unit price or ssemblies division, as well as the number of crews, is listed for each data set. The presence of other elements such as reference tables, square foot models, quipment rental costs, historical cost indexes, and city cost indexes, is also indicated. You can use the table to help select the RSMeans data set that has e quantity and type of information you most need in your work.

Unit Cost Divisions	Building Construction	Mechanical	Electrical	Commercial Renovation	Square Foot	Site Work Landsc.	Green Building	Interior	Concrete Masonry	Open Shop	Heavy Construction	Light Commercial	Facilities Construction	Plumbing	Residential
1	584	406	427	531	0	516	200	326	467	583	521	273	1056	416	178
2	779	278	86	735	0	995	207	397	218	778	737	479	1222	285	274
3	1744	340	230	1265	0	1536	1041	354	2273	1744	1929	537	2027	316	444
4	961	21	0	921	0	726	180	615	1159	929	616	534	1176	0	448
5	1889	158	155	1093	0	852	1787	1106	729	1889	1025	979	1906	204	746
6	2453	18	18	2111	0	110	589	1528	281	2449	123	2141	2125	22	2661
7	1596	215	128	1634	0	580	763	532	523	1593	26	1329	1697	227	1049
8	2140	80	3	2733	0	255	1140	1813	105	2142	0	2328	2966	0	1552
9	2107	86	45	1931	0	309	455	2193	412	2048	15	1756	2356	54	1521
10	1089	17	10	685	0	232	32	899	136	1089	34	589	1180	237	224
11	1097	201	166	541	0	135	56	925	29	1064	0	231	1117	164	110
12	548	0	2	298	0	219	147	1551	14	515	0	273	1574	23	217
13	744	149	158	253	0	366	125	254	78	720	267	109	760	115	104
14	273	36	0	223	0	0	0	257	0	273	0	12	293	16	6
21	130	0	41	37	0	0	0	296	0	130	0	121	668	688	259
22	1165	7559	160	1226	0	1573	1063	849	20	1154	1682	875	7506	9416	719
23	1198	7001	581	940	0	157	901	789	38	1181	110	890	5240	1918	485
25	0	0	14	14	0	0	0	0	0	0	0	0	0	0	0
26	1512	491	10455	1293	0	811	644	1159	55	1438	600	1360	10236	399	636
27	94	0	447	101	0	0	0	71	0	94	39	67	388	0	56
28	143	79	223	124	0	0	28	97	0	127	0	70	209	57	41
31	1511	733	610	807	0	3266	289	7	1218	1456	3282	605	1570	660	614
32	838	49	8	905	0	4475	355	406	315	809	1891	440	1752	142	487
33	1248	1080	534	252	0	3040	38	0	239	525	3090	128	1707	2089	154
34	107	0	47	4	0	190	0	0	31	62	221	0	136	0	0
35	18	0	0	0	0	327	0	0	0	18	442	0	84	0	0
41	62	0	0	33	0	8	0	22	0	61	31	0	68	14	0
44	75	79	0	0	0	0	0	0	0	0	0	0	75	75	0
46	23	16	0	0	0	274	261	0	0	23	264	0	33	33	0
48	8	0	36	2	0	0	21	0	0	8	15	8	21	0	8
Totals	26136	19092	14584	20692	0	20952	10322	16446	8340	24902	16960	16134	51148	17570	12993

Assem Div	Building Construction	Mechanical	Elec-trical	Commercial Renovation	Square Foot	Site Work Landscape	Assemblies	Green Building	Interior	Concrete Masonry	Heavy Construction	Light Commercial	Facilities Construction	Plumbing	Asm Div	Residential
A		15	0	188	164	577	598	0	0	536	571	154	24	0	1	378
B		0	0	848	2554	0	5661	56	329	1976	368	2094	174	0	2	211
C		0	0	647	954	0	1334	0	1641	146	0	844	251	0	3	588
D		1057	941	712	1858	72	2538	330	824	0	0	1345	1104	1088	4	851
E		0	0	86	261	0	301	0	5	0	0	258	5	0	5	391
F		0	0	0	114	0	143	0	0	0	0	114	0	0	6	357
G		527	447	318	312	3378	792	0	0	535	1349	205	293	677	7	307
															8	760
															9	80
															10	0
															11	0
															12	0
Totals		1599	1388	2799	6217	4027	11367	386	2799	3193	2288	5014	1851	1765		3923

Reference Section	Building Construction Costs	Mechanical	Electrical	Commercial Renovation	Square Foot	Site Work Landscape	Assem.	Green Building	Interior	Concrete Masonry	Open Shop	Heavy Construction	Light Commercial	Facilities Construction	Plumbing	Resi.
Reference Tables	yes	yes	yes	yes	no	yes	yes	yes	yes	yes	yes	yes	yes	yes	yes	yes
Models					111			25					50			28
Crews	582	582	582	561		582		582	582	582	560	582	560	561	582	560
Equipment Rental Costs	yes	yes	yes	yes		yes		yes	yes	yes	yes	yes	yes	yes	yes	yes
Historical Cost Indexes	yes	yes	yes	yes	yes	yes	yes	yes	yes	yes	yes	yes	yes	yes	yes	no
City Cost Indexes	yes	yes	yes	yes	yes	yes	yes	yes	yes	yes	yes	yes	yes	yes	yes	yes

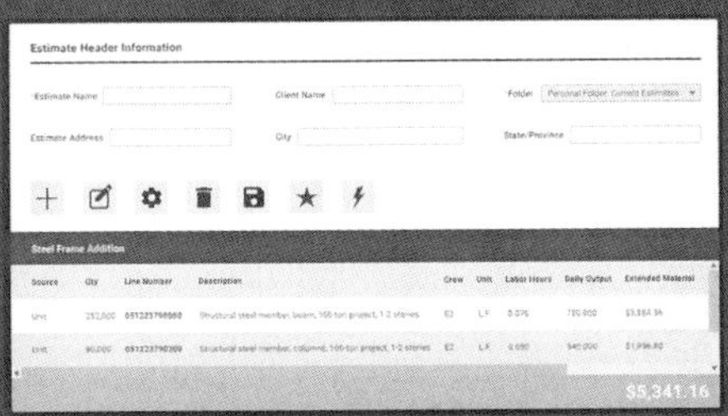

rsmeans.com/core

RSMeans Data Online **Core**

The Core tier of RSMeans Data Online provides reliable construction cost data along with the tools necessary to quickly access costs at the material or task level. Users can create unit line estimates with RSMeans data from Gordian and share them with ease from the web-based application.

Key Features:

- Search
- Estimate
- Share

Data Available:

16 datasets and packages curated by project type

RSMeans Data Online **Complete**

The Complete tier of RSMeans Data Online is designed to provide the latest in construction costs with comprehensive tools for projects of varying scopes. Unit, assembly or square foot price data is available 24/7 from the web-based application. Harness the power of RSMeans data from Gordian with expanded features and powerful tools like the square foot model estimator and cost trends analysis.

Key Features:

- Alerts
- Square Foot Estimator
- Trends

Data Available:

20 datasets and packages curated by project type

rsmeans.com/complete

RSMeans Data Online **Complete Plus**

The Complete Plus tier of RSMeans Data Online was developed to take project planning and estimating to the next level with predictive cost data. Users gain access to an exclusive set of tools and features and the most comprehensive database in the industry. Leverage everything RSMeans Data Online has to offer with the all-inclusive Complete Plus tier.

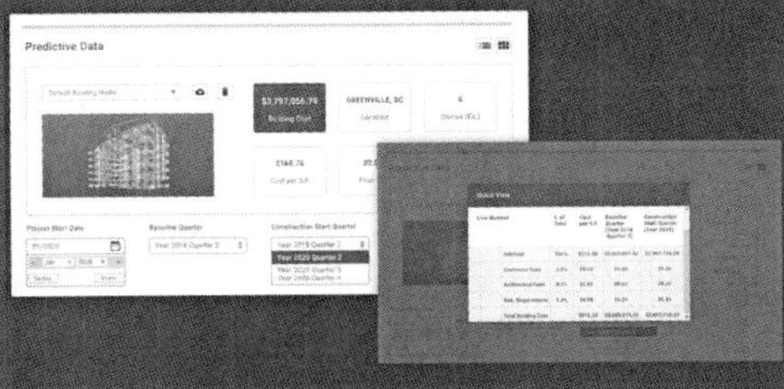

rsmeans.com/completeplus

Key Features:

- Predictive Cost Data
- Life Cycle Costing
- Full History

Data Available:

Full library and renovation models

2019 Seminar Schedule

877-620-6245

Note: call for exact dates, locations, and details as some cities are subject to change.

Location	Dates
Seattle, WA	January and August
Dallas/Ft. Worth, TX	January
Austin, TX	February
Jacksonville, FL	February
Anchorage, AK	March and September
Las Vegas, NV	March
Washington, DC	April and September
Charleston, SC	April
Toronto	May
Denver, CO	May
San Francisco, CA	June
Bethesda, MD	June
Dallas, TX	September
Raleigh, NC	October
Baltimore, MD	November
Orlando, FL	November
San Diego, CA	December
San Antonio, TX	December

Gordian also offers a suite of online RSMeans data self-paced offerings.
Check our website RSMeans.com/products/training.aspx more information.

Facilities Construction Estimating

In this two-day course, professionals working in facilities management can get help with their daily challenges to establish budgets for all phases of a project.

Some of what you'll learn:

- Determining the full scope of a project
- Identifying the scope of risks and opportunities
- Creative solutions to estimating issues
- Organizing estimates for presentation and discussion
- Special techniques for repair/remodel and maintenance projects
- Negotiating project change orders

Who should attend: facility managers, engineers, contractors, facility tradespeople, planners, and project managers.

Mechanical & Electrical Estimating

This two-day course teaches attendees how to prepare more accurate and complete mechanical/electrical estimates, avoid the pitfalls of omission and double-counting, and understand the composition and rationale within the RSMeans mechanical/electrical database.

Some of what you'll learn:

- The unique way mechanical and electrical systems are interrelated
- M&E estimates—conceptual, planning, budgeting, and bidding stages
- Order of magnitude, square foot, assemblies, and unit price estimating
- Comparative cost analysis of equipment and design alternatives

Who should attend: architects, engineers, facilities managers, mechanical and electrical contractors, and others who need a highly reliable method for developing, understanding, and evaluating mechanical and electrical contracts.

Construction Cost Estimating: Concepts and Practice

This one or two day introductory course to improve estimating skills and effectiveness starts with the details of interpreting bid documents and ends with the summary of the estimate and bid submission.

Some of what you'll learn:

- Using the plans and specifications to create estimates
- The takeoff process—deriving all tasks with correct quantities
- Developing pricing using various sources; how subcontractor pricing fits in
- Summarizing the estimate to arrive at the final number
- Formulas for area and cubic measure, adding waste and adjusting productivity to specific projects
- Evaluating subcontractors' proposals and prices
- Adding insurance and bonds
- Understanding how labor costs are calculated
- Submitting bids and proposals

Who should attend: project managers, architects, engineers, owners' representatives, contractors, and anyone who's responsible for budgeting or estimating construction projects.

Assessing Scope of Work for Facilities Construction Estimating

This two-day practical training program addresses the vital importance of understanding the scope of projects in order to produce accurate cost estimates for facility repair and remodeling.

Some of what you'll learn:

- Discussions of site visits, plans/specs, record drawings of facilities, and site-specific lists
- Review of CSI divisions, including means, methods, materials, and the challenges of scoping each topic
- Exercises in scope identification and scope writing for accurate estimating of projects
- Hands-on exercises that require scope, take-off, and pricing

Who should attend: corporate and government estimators, planners, facility managers, and others who need to produce accurate project estimates.

Maintenance & Repair Estimating for Facilities

This two-day course teaches attendees how to plan, budget, and estimate the cost of ongoing and preventive maintenance and repair for existing buildings and grounds.

Some of what you'll learn:

- The most financially favorable maintenance, repair, and replacement scheduling and estimating
- Auditing and value engineering facilities
- Preventive planning and facilities upgrading
- Determining both in-house and contract-out service costs
- Annual, asset-protecting M&R plan

Who should attend: facility managers, maintenance supervisors, buildings and grounds superintendents, plant managers, planners, estimators, and others involved in facilities planning and budgeting.

Practical Project Management for Construction Professionals

In this two-day course, acquire the essential knowledge and develop the skills to effectively and efficiently execute the day-to-day responsibilities of the construction project manager.

Some of what you'll learn:

- General conditions of the construction contract
- Contract modifications: change orders and construction change directives
- Negotiations with subcontractors and vendors
- Effective writing: notification and communications
- Dispute resolution: claims and liens

Who should attend: architects, engineers, owners' representatives, and project managers.

Life Cycle Cost Estimating for Facility Asset Managers

Life Cycle Cost Estimating will take the attendee through choosing the correct RSMeans database to use and then correctly applying RSMeans data to their specific life cycle application. Conceptual estimating through RSMeans new building models, conceptual estimating of major existing building projects through RSMeans renovation models, pricing specific renovation elements, estimating repair, replacement and preventive maintenance costs today and forward up to 30 years will be covered.

Some of what you'll learn:

- Cost implications of managing assets
- Planning projects and initial & life cycle costs
- How to use RSMeans data online

Who should attend: facilities owners and managers and anyone involved in the financial side of the decision making process in the planning, design, procurement, and operation of facility real assets.

Please bring a laptop with ability to access the internet.

Building Systems and the Construction Process

This one-day course was written to assist novices and those outside the industry in obtaining a solid understanding of the construction process - from both a building systems and construction administration approach.

Some of what you'll learn:

- Various systems used and how components come together to create a building
- Start with foundation and end with the physical systems of the structure such as HVAC and Electrical
- Focus on the process from start of design through project closeout

This training session requires you to bring a laptop computer to class.

Who should attend: building professionals or novices to help make the crossover to the construction industry; suited for anyone responsible for providing high level oversight on construction projects.

Training for our Online Estimating Solution

Construction estimating is vital to the decision-making process at each state of every project. Our online solution works the way you do. It's systematic, flexible and intuitive. In this one-day class you will see how you can estimate any phase of any project faster and better.

Some of what you'll learn:

- Customizing our online estimating solution
- Making the most of RSMeans "Circle Reference" numbers
- How to integrate your cost data
- Generating reports, exporting estimates to MS Excel, sharing, collaborating and more

Also offered as a self-paced or on-site training program!

Training for our CD Estimating Solution

This one-day course helps users become more familiar with the functionality of the CD. Each menu, icon, screen, and function found in the program is explained in depth. Time is devoted to hands-on estimating exercises.

Some of what you'll learn:

- Searching the database using all navigation methods
- Exporting RSMeans data to your preferred spreadsheet format
- Viewing crews, assembly components, and much more
- Automatically regionalizing the database

This training session requires you to bring a laptop computer to class.

When you register for this course you will receive an outline for your laptop requirements.

Also offered as a self-paced or on-site training program!

Site Work Estimating with RSMeans data

This one-day program focuses directly on site work costs. Accurately scoping, quantifying, and pricing site preparation, underground utility work, and improvements to exterior site elements are often the most difficult estimating tasks on any project. Some of what you'll learn:

- Evaluation of site work and understanding site scope including: site clearing, grading, excavation, disposal and trucking of materials, backfill and compaction, underground utilities, paving, sidewalks, and seeding & planting.
- Unit price site work estimates—Correct use of RSMeans site work cost data to develop a cost estimate.
- Using and modifying assemblies—Save valuable time when estimating site work activities using custom assemblies.

Who should attend: Engineers, contractors, estimators, project managers, owner's representatives, and others who are concerned with the proper preparation and/or evaluation of site work estimates.

Please bring a laptop with ability to access the internet.

Facilities Estimating Using the CD

This two-day class combines hands-on skill-building with best estimating practices and real-life problems. You will learn key concepts, tips, pointers, and guidelines to save time and avoid cost oversights and errors.

Some of what you'll learn:

- Estimating process concepts
- Customizing and adapting RSMeans cost data
- Establishing scope of work to account for all known variables
- Budget estimating: when, why, and how
- Site visits: what to look for and what you can't afford to overlook
- How to estimate repair and remodeling variables

This training session requires you to bring a laptop computer to class.

Who should attend: facility managers, architects, engineers, contractors, facility tradespeople, planners, project managers, and anyone involved with JOC, SABRE, or IDIQ.

Registration Information

Register early to save up to $100!!!

Register 45+ days before date of a class and save $50 off each class. This savings cannot be combined with any other promotional or discounting of the regular price of classes!

How to register

By Phone
Register by phone at 877-620-6245

Online
Register online at
RSMeans.com/products/seminars.aspx

Note: Purchase Orders or Credits Cards are required to register.

Two-day seminar registration fee - $1,200.

One-Day Construction Cost Estimating or Building Systems and the Construction Process - $765.

Government pricing

All federal government employees save off the regular seminar price. Other promotional discounts cannot be combined with the government discount. Call 781-422-5115 for government pricing.

CANCELLATION POLICY:

If you are unable to attend a seminar, substitutions may be made at any time before the session starts by notifying the seminar registrar at 1-781-422-5115 or your sales representative.

If you cancel twenty-one (21) days or more prior to the seminar, there will be no penalty and your registration fees will be refunded. These cancellations must be received by the seminar registrar or your sales representative and will be confirmed to be eligible for cancellation.

If you cancel fewer than twenty-one (21) days prior to the seminar, you will forfeit the registration fee.

In the unfortunate event of an RSMeans cancellation, RSMeans will work with you to reschedule your attendance in the same seminar at a later date or will fully refund your registration fee. RSMeans cannot be responsible for any non-refundable travel expenses incurred by you or another as a result of your registration, attendance at, or cancellation of an RSMeans seminar.

Any on-demand training modules are not eligible for cancellation, substitution, transfer, return or refund.

AACE approved courses

Many seminars described and offered here have been approved for 14 hours (1.4 recertification credits) of credit by the AACE International Certification Board toward meeting the continuing education requirements for recertification as a Certified Cost Engineer/Certified Cost Consultant.

AIA Continuing Education

We are registered with the AIA Continuing Education System (AIA/CES) and are committed to developing quality learning activities in accordance with the CES criteria. Many seminars meet the AIA/CES criteria for Quality Level 2. AIA members may receive 14 learning units (LUs) for each two-day RSMeans course.

Daily course schedule

The first day of each seminar session begins at 8:30 a.m. and ends at 4:30 p.m. The second day begins at 8:00 a.m. and ends at 4:00 p.m. Participants are urged to bring a hand-held calculator since many actual problems will be worked out in each session.

Continental breakfast

Your registration includes the cost of a continental breakfast and a morning and afternoon refreshment break. These informal segments allow you to discuss topics of mutual interest with other seminar attendees. (You are free to make your own lunch and dinner arrangements.)

Hotel/transportation arrangements

We arrange to hold a block of rooms at most host hotels. To take advantage of special group rates when making your reservation, be sure to mention that you are attending the RSMeans Institute data seminar. You are, of course, free to stay at the lodging place of your choice. (Hotel reservations and transportation arrangements should be made directly by seminar attendees.)

Important

Class sizes are limited, so please register as soon as possible.

Note: Pricing subject to change.